Complete Solutions Guide for

PRECALCULUS
Seventh Edition

PRECALCULUS WITH LIMITS

Larson / Hostetler

Dianna L. Zook

Indiana University – Purdue University
Fort Wayne, Indiana

Houghton Mifflin Company   Boston   New York

CONTENTS

Houghton Mifflin Company hereby grants you permission to reproduce the Houghton Mifflin material contained in this work in classroom quantities, solely for use with the accompanying Houghton Mifflin textbook. All reproductions must include the Houghton Mifflin copyright notice, and no fee may be collected except to cover the cost of duplication. If you wish to make any other use of this material, including reproducing or transmitting the material or portions thereof in any form or by any electronic or mechanical means, including any information storage or retrieval system, you must obtain prior written permission from Houghton Mifflin Company unless such use is expressly permitted by federal copyright law. If you wish to reproduce material acknowledging a rights holder other than Houghton Mifflin Company, you must obtain permission from the rights holder. Address inquiries to College Permissions, Houghton Mifflin Company, 222 Berkeley Street, Boston, MA 02116-3764.

Printed in the United States of America

ISBN-13: 978-0-618-64350-9
ISBN-10: 0-618-64350-8

3456789-CS-10 09 08 07

PART I

CHAPTER 1
Functions and Their Graphs

C H A P T E R   1
Functions and Their Graphs

Section 1.1   Rectangular Coordinates

■ You should be able to use the point-plotting method of graphing.

■ You should be able to find x- and y-intercepts.

    (a) To find the x-intercepts, let $y = 0$ and solve for x.

    (b) To find the y-intercepts, let $x = 0$ and solve for y.

■ You should be able to test for symmetry.

    (a) To test for x-axis symmetry, replace y with $-y$.

    (b) To test for y-axis symmetry, replace x with $-x$.

    (c) To test for origin symmetry, replace x with $-x$ and y with $-y$.

■ You should know the standard equation of a circle with center (h, k) and radius r:

$$(x - h)^2 + (y - k)^2 = r^2$$

Vocabulary Check

1. (a) **v** horizontal real number line   (b) **vi** vertical real number line

    (c) **i** point of intersection of vertical axis and horizontal axis   (d) **iv** four regions of the coordinate plane

    (e) **iii** directed distance from the y-axis   (f) **ii** directed distance from the x-axis

2. Cartesian   **3.** Distance Formula   **4.** Midpoint Formula

---

1. A: $(2, 6)$, B: $(-6, -2)$, C: $(4, -4)$, D: $(-3, 2)$ **2.** A: $\left(\frac{3}{2}, -4\right)$; B: $(0, -2)$; C: $\left(-3, \frac{5}{2}\right)$, D: $(-6, 0)$

3.

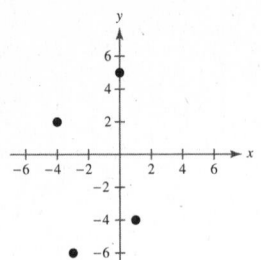

4.

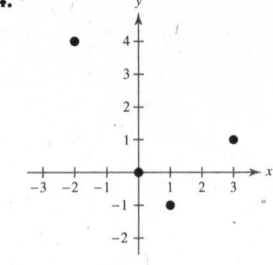

5.

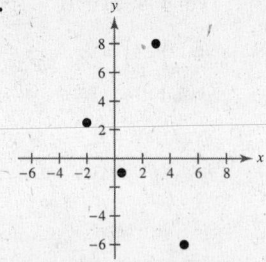

6.

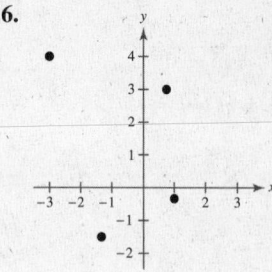

7. $(-3, 4)$ **8.** $(4, -8)$ **9.** $(-5, -5)$ **10.** $(-12, 0)$

11. $x > 0$ and $y < 0$ in Quadrant IV.   **12.** $x < 0$ and $y < 0$ in Quadrant III.   **13.** $x = -4$ and $y > 0$ in Quadrant II.

14. $x > 2$ and $y = 3$ in Quadrant I.   **15.** $y < -5$ in Quadrants III and IV.   **16.** $x > 4$ in Quadrants I and IV.

17. $(x, -y)$ is in the second Quadrant means that (x, y) is in Quadrant III.

18. If $(-x, y)$ is in Quadrant IV, then (x, y) must be in Quadrant III.

19. (x, y), $xy > 0$ means x and y have the same signs. This occurs in Quadrants I and III.

20. If $xy < 0$, then x and y have opposite signs. This happens in Quadrants II and IV.

21.

Year, x	Number of stores, y
1996	3054
1997	3406
1998	3599
1999	3985
2000	4189
2001	4414
2002	4688
2003	4906

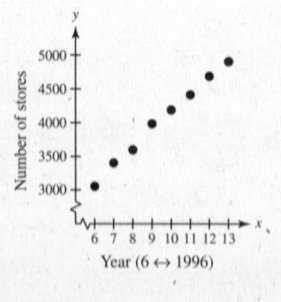

22.

Month, x	Temperature, y
1	-39
2	-39
3	-29
4	-5
5	17
6	27
7	35
8	32
9	22
10	8
11	-23
12	-34

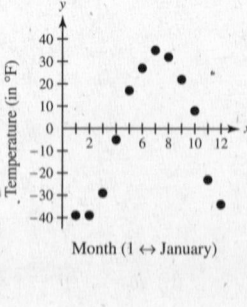

Month (1 ↔ January)

23. $d = |5 - (-3)| = 8$ **24.** $d = |1 - 8| = |-7| = 7$ **25.** $d = |2 - (-3)| = 5$ **26.** $d = |-4 - 6|$
$$= |-10| = 10$$

27. (a) The distance between $(0, 2)$ and $(4, 2)$ is 4.

The distance between $(4, 2)$ and $(4, 5)$ is 3.

The distance between $(0, 2)$ and $(4, 5)$ is

$$\sqrt{(4 - 0)^2 + (5 - 2)^2} = \sqrt{16 + 9} = \sqrt{25} = 5.$$

(b) $4^2 + 3^2 = 16 + 9 = 25 = 5^2$

28. (a) $(1, 0), (13, 5)$

Distance $= \sqrt{(13 - 1)^2 + (5 - 0)^2}$

$= \sqrt{12^2 + 5^2} = \sqrt{169} = 13$

$(13, 5), (13, 0)$

Distance $= |5 - 0| = |5| = 5$

$(1, 0), (13, 0)$

Distance $= |1 - 13| = |-12| = 12$

(b) $5^2 + 12^2 = 25 + 144 = 169 = 13^2$

29. (a) The distance between $(-1, 1)$ and $(9, 1)$ is 10.

The distance between $(9, 1)$ and $(9, 4)$ is 3.

The distance between $(-1, 1)$ and $(9, 4)$ is

$\sqrt{(9 - (-1))^2 + (4 - 1)^2} = \sqrt{100 + 9} = \sqrt{109}.$

(b) $10^2 + 3^2 = 109 = \left(\sqrt{109}\right)^2$

30. (a) $(1, 5), (5, -2)$

Distance $= \sqrt{(1 - 5)^2 + (5 - (-2))^2}$

$= \sqrt{(-4)^2 + (7)^2} = \sqrt{16 + 49} = \sqrt{65}$

$(1, 5), (1, -2)$

Distance $= |5 - (-2)| = |5 + 2| = |7| = 7$

$(1, -2), (5, -2)$

Distance $= |1 - 5| = |-4| = 4$

(b) $4^2 + 7^2 = 16 + 49 = 65 = \left(\sqrt{65}\right)^2$

31. (a)

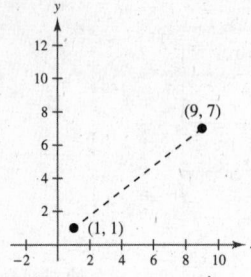

(b) $d = \sqrt{(9 - 1)^2 + (7 - 1)^2}$

$= \sqrt{64 + 36} = 10$

(c) $\left(\dfrac{9 + 1}{2}, \dfrac{7 + 1}{2}\right) = (5, 4)$

32. (a)

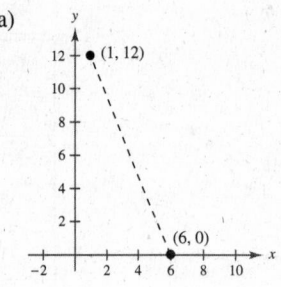

(b) $d = \sqrt{(1 - 6)^2 + (12 - 0)^2}$

$= \sqrt{25 + 144} = 13$

(c) $\left(\dfrac{1 + 6}{2}, \dfrac{12 + 0}{2}\right) = \left(\dfrac{7}{2}, 6\right)$

33. (a)

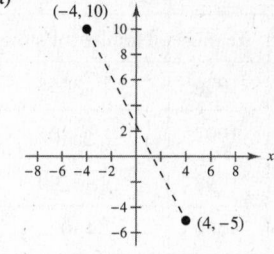

(b) $d = \sqrt{(4 + 4)^2 + (-5 - 10)^2}$

$= \sqrt{64 + 225} = 17$

(c) $\left(\dfrac{4 - 4}{2}, \dfrac{-5 + 10}{2}\right) = \left(0, \dfrac{5}{2}\right)$

34. (a)

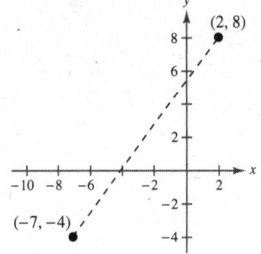

(b) $d = \sqrt{(-7 - 2)^2 + (-4 - 8)^2}$

$= \sqrt{81 + 144} = 15$

(c) $\left(\dfrac{-7 + 2}{2}, \dfrac{-4 + 8}{2}\right) = \left(-\dfrac{5}{2}, 2\right)$

35. (a)

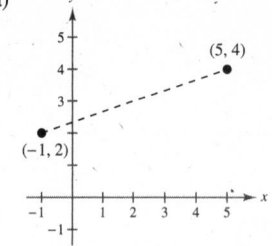

(b) $d = \sqrt{(5 + 1)^2 + (4 - 2)^2}$

$= \sqrt{36 + 4} = 2\sqrt{10}$

(c) $\left(\dfrac{-1 + 5}{2}, \dfrac{2 + 4}{2}\right) = (2, 3)$

36. (a)

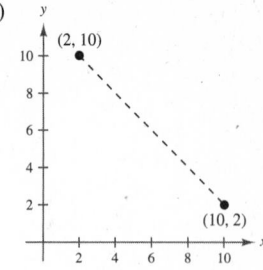

(b) $d = \sqrt{(2 - 10)^2 + (10 - 2)^2}$

$= \sqrt{64 + 64} = 8\sqrt{2}$

(c) $\left(\dfrac{2 + 10}{2}, \dfrac{10 + 2}{2}\right) = (6, 6)$

37. (a)

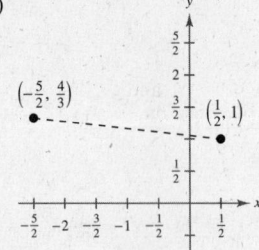

(b) $d = \sqrt{\left(\dfrac{1}{2} + \dfrac{5}{2}\right)^2 + \left(1 - \dfrac{4}{3}\right)^2}$

$\quad = \sqrt{9 + \dfrac{1}{9}} = \dfrac{\sqrt{82}}{3}$

(c) $\left(\dfrac{-(5/2) + (1/2)}{2}, \dfrac{(4/3) + 1}{2}\right) = \left(-1, \dfrac{7}{6}\right)$

38. (a)

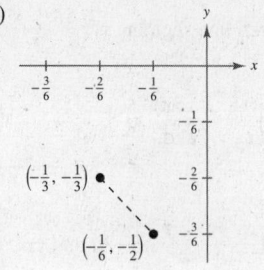

(b) $d = \sqrt{\left(-\dfrac{1}{3} + \dfrac{1}{6}\right)^2 + \left(-\dfrac{1}{3} + \dfrac{1}{2}\right)^2}$

$\quad = \sqrt{\dfrac{1}{36} + \dfrac{1}{36}} = \dfrac{\sqrt{2}}{6}$

(c) $\left(\dfrac{-\dfrac{1}{3} + \left(-\dfrac{1}{6}\right)}{2}, \dfrac{-\dfrac{1}{3} + \left(-\dfrac{1}{2}\right)}{2}\right) = \left(-\dfrac{1}{4}, -\dfrac{5}{12}\right)$

39. (a)

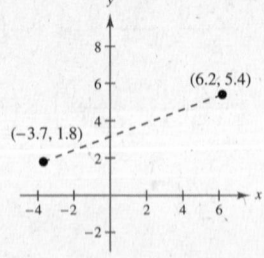

(b) $d = \sqrt{(6.2 + 3.7)^2 + (5.4 - 1.8)^2}$

$\quad = \sqrt{98.01 + 12.96} = \sqrt{110.97}$

(c) $\left(\dfrac{6.2 - 3.7}{2}, \dfrac{5.4 + 1.8}{2}\right) = (1.25, 3.6)$

40. (a)

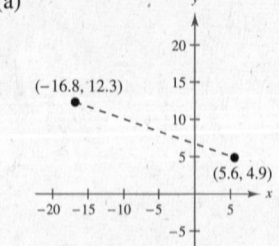

(b) $d = \sqrt{(-16.8 - 5.6)^2 + (12.3 - 4.9)^2}$

$\quad = \sqrt{501.76 + 54.76} = \sqrt{556.52}$

(c) $\left(\dfrac{-16.8 + 5.6}{2}, \dfrac{12.3 + 4.9}{2}\right) = (-5.6, 8.6)$

41. $d_1 = \sqrt{(4 - 2)^2 + (0 - 1)^2} = \sqrt{5}$

$d_2 = \sqrt{(4 + 1)^2 + (0 + 5)^2} = \sqrt{50}$

$d_3 = \sqrt{(2 + 1)^2 + (1 + 5)^2} = \sqrt{45}$

$\left(\sqrt{5}\right)^2 + \left(\sqrt{45}\right)^2 = \left(\sqrt{50}\right)^2$

42. $d_1 = \sqrt{(1 - 3)^2 + (-3 - 2)^2} = \sqrt{4 + 25} = \sqrt{29}$

$d_2 = \sqrt{(3 + 2)^2 + (2 - 4)^2} = \sqrt{25 + 4} = \sqrt{29}$

$d_3 = \sqrt{(1 + 2)^2 + (-3 - 4)^2} = \sqrt{9 + 49} = \sqrt{58}$

$d_1 = d_2$

43. Since $x_m = \dfrac{x_1 + x_2}{2}$ and $y_m = \dfrac{y_1 + y_2}{2}$ we have:

$$2x_m = x_1 + x_2 \qquad 2y_m = y_1 + y_2$$

$$2x_m - x_1 = x_2 \qquad 2y_m - y_1 = y_2$$

Thus, $(x_2, y_2) = (2x_m - x_1, 2y_m - y_1)$.

44. (a) $(x_2, y_2) = (2x_m - x_1, 2y_m - y_1)$

$\quad = (2 \cdot 4 - 1, 2(-.1) - (-2)) = (7, 0)$

(b) $(x_2, y_2) = (2x_m - x_1, 2y_m - y_1)$

$\quad = (2 \cdot 2 - (-5), 2 \cdot 4 - 11) = (9, -3)$

45. The midpoint of the given line segment is $\left(\dfrac{x_1 + x_2}{2}, \dfrac{y_1 + y_2}{2}\right)$.

The midpoint between (x_1, y_1) and $\left(\dfrac{x_1 + x_2}{2}, \dfrac{y_1 + y_2}{2}\right)$ is $\left(\dfrac{x_1 + \dfrac{x_1 + x_2}{2}}{2}, \dfrac{y_1 + \dfrac{y_1 + y_2}{2}}{2}\right) = \left(\dfrac{3x_1 + x_2}{4}, \dfrac{3y_1 + y_2}{4}\right)$.

The midpoint between $\left(\dfrac{x_1 + x_2}{2}, \dfrac{y_1 + y_2}{2}\right)$ and (x_2, y_2) is $\left(\dfrac{\dfrac{x_1 + x_2}{2} + x_2}{2}, \dfrac{\dfrac{y_1 + y_2}{2} + y_2}{2}\right) = \left(\dfrac{x_1 + 3x_2}{4}, \dfrac{y_1 + 3y_2}{4}\right)$.

Thus, the three points are

$$\left(\dfrac{3x_1 + x_2}{4}, \dfrac{3y_1 + y_2}{4}\right), \left(\dfrac{x_1 + x_2}{2}, \dfrac{y_1 + y_2}{2}\right), \text{ and } \left(\dfrac{x_1 + 3x_2}{4}, \dfrac{y_1 + 3y_2}{4}\right).$$

46. (a) $\left(\dfrac{3x_1 + x_2}{4}, \dfrac{3y_1 + y_2}{4}\right) = \left(\dfrac{3 \cdot 1 + 4}{4}, \dfrac{3(-2) - 1}{4}\right)$

$= \left(\dfrac{7}{4}, -\dfrac{7}{4}\right)$

$\left(\dfrac{x_1 + x_2}{2}, \dfrac{y_1 + y_2}{2}\right) = \left(\dfrac{1 + 4}{2}, \dfrac{-2 - 1}{2}\right) = \left(\dfrac{5}{2}, -\dfrac{3}{2}\right)$

$\left(\dfrac{x_1 + 3x_2}{4}, \dfrac{y_1 + 3y_2}{4}\right) = \left(\dfrac{1 + 3 \cdot 4}{4}, \dfrac{-2 + 3(-1)}{4}\right)$

$= \left(\dfrac{13}{4}, -\dfrac{5}{4}\right)$

(b) $\left(\dfrac{3x_1 + x_2}{4}, \dfrac{3y_1 + y_2}{4}\right) = \left(\dfrac{3(-2) + 0}{4}, \dfrac{3(-3) + 0}{4}\right)$

$= \left(-\dfrac{3}{2}, -\dfrac{9}{4}\right)$

$\left(\dfrac{x_1 + x_2}{2}, \dfrac{y_1 + y_2}{2}\right) = \left(\dfrac{-2 + 0}{2}, \dfrac{-3 + 0}{2}\right) = \left(-1, -\dfrac{3}{2}\right)$

$\left(\dfrac{x_1 + 3x_2}{4}, \dfrac{y_1 + 3y_2}{4}\right) = \left(\dfrac{-2 + 0}{4}, \dfrac{-3 + 0}{4}\right) = \left(-\dfrac{1}{2}, -\dfrac{3}{4}\right)$

47. $d = \sqrt{(42 - 18)^2 + (50 - 12)^2}$

$= \sqrt{24^2 + 38^2}$

$= \sqrt{2020}$

$= 2\sqrt{505}$

≈ 45 yards

48. Distance $= \sqrt{120^2 + 150^2}$

$= \sqrt{36{,}900}$

$= 30\sqrt{41}$

≈ 192.09 kilometers

The plane flies about 192 kilometers.

49. $\left(\dfrac{2001 + 2003}{2}, \dfrac{3433 + 4174}{2}\right) = (2002, 3803.5)$

In 2002, the sales for Big Lots was approximately $3803.5 million.

50. $\dfrac{\$1987 + \$2800}{2} = \dfrac{\$4787}{2}$

$\approx \$2393.50$ million

51. $(-2 + 2, -4 + 5) = (0, 1)$

$(2 + 2, -3 + 5) = (4, 2)$

$(-1 + 2, -1 + 5) = (1, 4)$

52. $(-3 + 6, 6 - 3) = (3, 3)$

$(-5 + 6, 3 - 3) = (1, 0)$

$(-3 + 6, 0 - 3) = (3, -3)$

$(-1 + 6, 3 - 3) = (5, 0)$

53. $(-7 + 4, -2 + 8) = (-3, 6)$

$(-2 + 4, 2 + 8) = (2, 10)$

$(-2 + 4, -4 + 8) = (2, 4)$

$(-7 + 4, -4 + 8) = (-3, 4)$

54. $(5 - 10, 8 - 6) = (-5, 2)$

$(3 - 10, 6 - 6) = (-7, 0)$

$(7 - 10, 6 - 6) = (-3, 0)$

$(5 - 10, 2 - 6) = (-5, -4)$

55. The highest price of butter is approximately $3.31 per pound. This occurred in 2001.

56. Price of butter in 1995 $\approx$ \$1.75

Highest price of butter = \$3.31 in 2001

Percent change $= \dfrac{3.31 - 1.75}{1.75} \approx 89.1\%$

57. $\left[\dfrac{2400 - 700}{700}\right](100) \approx 242.9\%$ increase

58. (a) Cost during Super Bowl XXVII (1993) $\approx$ \$850,000

Cost during Super Bowl XXIII (1989) $\approx$ \$700,000

Increase = \$850,000 $-$ \$700,000 $\approx$ \$150,000

Percent increase $= \dfrac{\$150,000}{\$700,000} \approx 0.214$, or 21.4\%

(b) Cost during Super Bowl XXXVII (2003) $\approx$ \$2,100,000

Increase = \$2,100,000 $-$ \$850,000 = \$1,250,000

Percent increase $= \dfrac{\$1,250,000}{\$850,000} \approx 1.47$, or 147\%

59. (a) The number of artists elected each year seems to be nearly steady except for the first few years. Between 6 and 8 artists will be elected in 2008.

(b) Elections for inclusion in the Rock and Roll Hall of Fame began in 1986.

60. (a) The minimum wage had the greatest increase in the 1990s.

(b) Minimum wage in 1990: \$3.80

Minimum wage in 1995: \$4.25

Percent increase: $\left(\dfrac{\$4.25 - \$3.80}{\$3.80}\right)(100) \approx 11.8\%$

Minimum wage in 1995: \$4.25

Minimum wage in 2000: \$5.15

Percent increase: $\left(\dfrac{\$5.15 - \$4.25}{\$4.25}\right)(100) \approx 21.2\%$

(c) \$5.15 + 0.212(\$5.15) $\approx$ \$6.24

(d) The political nature of the minimum wage makes it difficult to predict, but this does seem like a reasonable value.

61. (1996, 18,546), (2004, 21,900)

By Exercise 45 we have the following:

$\left(\dfrac{3(1996) + 2004}{4}, \dfrac{3(18,546) + 21,900}{4}\right) = (1998, 19,384.5)$

$\left(\dfrac{1996 + 2004}{2}, \dfrac{18,546 + 21,900}{2}\right) = (2000, 20,223)$

$\left(\dfrac{1996 + 3(2004)}{4}, \dfrac{18,546 + 3(21,900)}{4}\right) = (2002, 21,061.5)$

Year	Sales for Coca-Cola Company
1998	\$19,384.5 million
2000	\$20,223 million
2002	\$21,061.5 million

62. (a)

x	y
22	53
29	74
35	57
40	66
44	79
48	90
53	76
58	93
65	83
76	99

(b) The point $(65, 83)$ represents an entrance exam score of 65.

(c) No. There are many variables that will affect the final exam score.

63. $V = \dfrac{4}{3}\pi r^3$

$5.96 = \dfrac{4}{3}\pi r^3$

$17.88 = 4\pi r^3$

$\dfrac{17.88}{4\pi} = r^3$

$r = \sqrt[3]{\dfrac{4.47}{\pi}} \approx 1.12$ inches

64. $V = \pi r^2 h$

$h = \dfrac{V}{\pi r^2} = \dfrac{603.2}{\pi(2)^2} \approx 48$ feet

65.

$3S = 129$

$S = 43$ centimeters

$h^2 + \left(\dfrac{S}{2}\right)^2 = S^2$

$h^2 = \dfrac{3S^2}{4}$

$h = \dfrac{\sqrt{3}\,S}{2}$

$A = \dfrac{1}{2}bh = \dfrac{1}{2}S\left(\dfrac{\sqrt{3}\,S}{2}\right) = \dfrac{\sqrt{3}\,S^2}{4}$

When $S = 43$ centimeters,

$A = \dfrac{\sqrt{3}\,(43)^2}{4} \approx 800.64$ square centimeters.

66. $S = \pi R \sqrt{R^2 + h^2}$

$1617 = (\pi)(14)\sqrt{14^2 + h^2}$

$\dfrac{1617}{14\pi} = \sqrt{196 + h^2}$

$\left(\dfrac{1617}{14\pi}\right)^2 = 196 + h^2$

$\left(\dfrac{1617}{14\pi}\right)^2 - 196 = h^2$

$\sqrt{\left(\dfrac{1617}{14\pi}\right)^2 - 196} = h$

$h \approx 33.995 \approx 34$ centimeters

67. (a)

(b) $l = 1.5w$

$P = 2l + 2w$

$= 2(1.5w) + 2w$

$= 5w$

(c) $25 = 5w$

$5 = w$

Width: $w = 5$ meters

Length: $l = 1.5w = 7.5$ meters

Dimensions: 7.5 meters $\times$ 5 meters

68. (a)

(b) $w = 1.25h = \dfrac{5}{4}h$

$V = l \cdot w \cdot h = (16)\left(\dfrac{5}{4}h\right)(h)$

$V = 20h^2$

(c) $V = 2000 = 20h^2$

$100 = h^2 \implies h = 10$ in.

$w = \left(\dfrac{5}{4}\right)(10) = \dfrac{25}{2} = 12.5$ in.

$l = 16$ in.

Dimensions: 16 inches $\times$ 12.5 inches $\times$ 10 inches

69. (a)

Year, x	Pieces of mail, y (in billions)
1996	183
1997	191
1998	197
1999	202
2000	208
2001	207
2002	203
2003	202

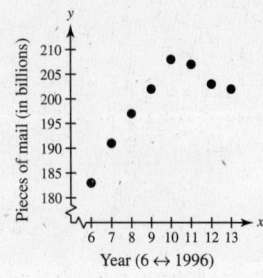

(b) The greatest decrease occurred in 2002.

(c) Answers will vary. Technology now enables us to transport information in ways other than by mail. The internet is one example.

70. (a)

Year, x	Men's teams, M	Women's teams, W
1994	858	859
1995	868	864
1996	866	874
1997	865	879
1998	895	911
1999	926	940
2000	932	956
2001	937	958
2002	936	975
2003	967	1009

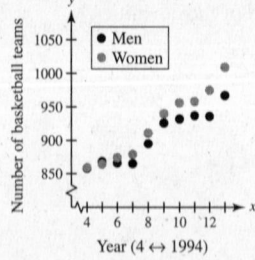

(b) In 1994, the number of men's and women's teams were nearly equal.

(c) In 2003, the difference between the number of teams was greatest: $1009 - 967 = 42$ teams.

71.

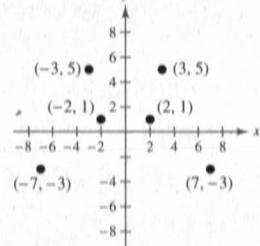

(a) The point is reflected through the y-axis.

(b) The point is reflected through the x-axis.

(c) The point is reflected through the origin.

72. (a)

First Set

$$d(A, B) = \sqrt{(2-2)^2 + (3-6)^2} = \sqrt{9} = 3$$

$$d(B, C) = \sqrt{(2-6)^2 + (6-3)^2} = \sqrt{16+9} = 5$$

$$d(A, C) = \sqrt{(2-6)^2 + (3-3)^2} = \sqrt{16} = 4$$

Since $3^2 + 4^2 = 5^2$, A, B, and C are the vertices of a right triangle.

Second Set

$$d(A, B) = \sqrt{(8-5)^2 + (3-2)^2} = \sqrt{10}$$

$$d(B, C) = \sqrt{(5-2)^2 + (2-1)^2} = \sqrt{10}$$

$$d(A, C) = \sqrt{(8-2)^2 + (3-1)^2} = \sqrt{40}$$

A, B, and C are the vertices of an isosceles triangle or are collinear: $\sqrt{10} + \sqrt{10} = 2\sqrt{10} = \sqrt{40}$.

(b)

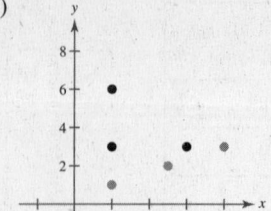

First set: Not collinear

Second set: The points are collinear.

(c) If A, B, and C are collinear, then two of the distances will add up to the third distance.

73. False, you would have to use the Midpoint Formula 15 times.

74. True. Two sides of the triangle have lengths $\sqrt{149}$ and the third side has a length of $\sqrt{18}$.

75. No. It depends on the magnitude of the quantities measured.

76. Use the Midpoint Formula to prove the diagonals of the parallelogram bisect each other.

$$\left(\frac{b+a}{2}, \frac{c+0}{2}\right) = \left(\frac{a+b}{2}, \frac{c}{2}\right)$$

$$\left(\frac{a+b+0}{2}, \frac{c+0}{2}\right) = \left(\frac{a+b}{2}, \frac{c}{2}\right)$$

77. Since (x_0, y_0) lies in Quadrant II, $(x_0, -y_0)$ must lie in Quadrant III. Matches (b).

78. Since (x_0, y_0) lies in Quadrant II, $(-2x_0, y_0)$ must lie in Quadrant I. Matches (c).

79. Since (x_0, y_0) lies in Quadrant II, $\left(x_0, \frac{1}{2}y_0\right)$ must lie in Quadrant II. Matches (d).

80. Since (x_0, y_0) lies in Quadrant II, $(-x_0, -y_0)$ must lie in Quadrant IV. Matches (a).

81. $2x + 1 = 7x - 4$

$-5x = -5$

$x = 1$

82. $\frac{1}{3}x + 2 = 5 - \frac{1}{6}x$

$\frac{1}{3}x + \frac{1}{6}x = 5 - 2$

$\frac{1}{2}x = 3$

$x = 6$

83. $x^2 - 4x - 7 = 0$

$x^2 - 4x = 7$

$x^2 - 4x + 4 = 7 + 4$

$(x - 2)^2 = 11$

$x - 2 = \pm\sqrt{11}$

$x = 2 \pm \sqrt{11}$

84. $2x^2 + 3x - 8 = 0$

$x = \dfrac{-3 \pm \sqrt{(3)^2 - (4)(2)(-8)}}{(2)(2)}$

$x = \dfrac{-3 \pm \sqrt{9 + 64}}{4}$

$x = \dfrac{-3 \pm \sqrt{73}}{4}$

85. $3x + 1 < 2(2 - x)$

$3x + 1 < 4 - 2x$

$5x < 3$

$x < \frac{3}{5}$

86. $3x - 8 \geq \frac{1}{2}(10x + 7)$

$2(3x - 8) \geq 10x + 7$

$6x - 16 \geq 10x + 7$

$-4x \geq 23$

$x \leq -\frac{23}{4}$

87. $|x - 18| < 4$

$-4 < x - 18 < 4$

$14 < x < 22$

88. $|2x + 15| \geq 11$

$2x + 15 \geq 11$ or $2x + 15 \leq -11$

$2x \geq 11 - 15$ $2x \leq -11 - 15$

$2x \geq -4$ $2x \leq -26$

$x \geq -2$ $x \leq -13$

Section 1.2   Graphs of Equations

You should know the following important facts about lines.

- The graph of $y = mx + b$ is a straight line. It is called a linear equation in two variables.

  (a) The slope (steepness) is m.

  (b) The y-intercept is $(0, b)$.

- The slope of the line through (x_1, y_1) and (x_2, y_2) is

  $$m = \frac{y_2 - y_1}{x_2 - x_1} = \frac{\text{change in } y}{\text{change in } x} = \frac{\text{rise}}{\text{run}}.$$

- (a) If $m > 0$, the line rises from left to right.

  (b) If $m = 0$, the line is horizontal.

  (c) If $m < 0$, the line falls from left to right.

  (d) If m is undefined, the line is vertical.

- Equations of Lines

  (a) Slope-Intercept Form: $y = mx + b$

  (b) Point-Slope Form: $y - y_1 = m(x - x_1)$

  (c) Two-Point Form: $y - y_1 = \dfrac{y_2 - y_1}{x_2 - x_1}(x - x_1)$

  (d) General Form: $Ax + By + C = 0$

  (e) Vertical Line: $x = a$

  (f) Horizontal Line: $y = b$

- Given two distinct nonvertical lines

  $L_1: y = m_1 x + b_1$ and $L_2: y = m_2 x + b_2$

  (a) L_1 is parallel to L_2 if and only if $m_1 = m_2$ and $b_1 \neq b_2$.

  (b) L_1 is perpendicular to L_2 if and only if $m_1 = -1/m_2$.

Vocabulary Check

1. solution or solution point

2. graph

3. intercepts

4. y-axis

5. circle; (h, k); r

6. numerical

1. $y = \sqrt{x + 4}$

    (a) $(0, 2)$: $2 \overset{?}{=} \sqrt{0 + 4}$

              $2 = 2$

    Yes, the point *is* on the graph.

    (b) $(5, 3)$: $3 \overset{?}{=} \sqrt{5 + 4}$

              $3 = \sqrt{9}$

    Yes, the point *is* on the graph.

2. $y = x^2 - 3x + 2$

    (a) $(2, 0)$: $(2)^2 - 3(2) + 2 \overset{?}{=} 0$

               $4 - 6 + 2 \overset{?}{=} 0$

                     $0 = 0$

    Yes, the point *is* on the graph.

    (b) $(-2, 8)$: $(-2)^2 - 3(-2) + 2 \overset{?}{=} 8$

                 $4 + 6 + 2 \overset{?}{=} 8$

                    $12 \neq 8$

    No, the point *is not* on the graph.

3. $y = 4 - |x - 2|$

    (a) $(1, 5)$: $5 \overset{?}{=} 4 - |1 - 2|$

              $5 \neq 4 - 1$

    No, the point *is not* on the graph.

    (b) $(6, 0)$: $0 \overset{?}{=} 4 - |6 - 2|$

             $0 = 4 - 4$

    Yes, the point *is* on the graph.

4. $y = \frac{1}{3}x^3 - 2x^2$

    (a) $\left(2, -\frac{16}{3}\right)$: $\frac{1}{3}(2)^3 - 2(2)^2 \overset{?}{=} -\frac{16}{3}$

                 $\frac{1}{3} \cdot 8 - 2 \cdot 4 \overset{?}{=} -\frac{16}{3}$

                    $\frac{8}{3} - 8 \overset{?}{=} -\frac{16}{3}$

                  $\frac{8}{3} - \frac{24}{3} \overset{?}{=} -\frac{16}{3}$

                      $-\frac{16}{3} = -\frac{16}{3}$

    Yes, the point *is* on the graph.

    (b) $(-3, 9)$: $\frac{1}{3}(-3)^3 - 2(-3)^2 \overset{?}{=} 9$

               $\frac{1}{3}(-27) - 2(9) \overset{?}{=} 9$

                  $-9 - 18 \overset{?}{=} 9$

                   $-27 \neq 9$

    No, the point *is not* on the graph.

5. $y = -2x + 5$

x	-1	0	1	2	$\frac{5}{2}$
y	7	5	3	1	0
(x, y)	$(-1, 7)$	$(0, 5)$	$(1, 3)$	$(2, 1)$	$\left(\frac{5}{2}, 0\right)$

6. $y = \frac{3}{4}x - 1$

x	-2	0	1	$\frac{4}{3}$	2
y	$-\frac{5}{2}$	-1	$-\frac{1}{4}$	0	$\frac{1}{2}$
x, y	$\left(-2, -\frac{5}{2}\right)$	$(0, -1)$	$\left(1, -\frac{1}{4}\right)$	$\left(\frac{4}{3}, 0\right)$	$\left(2, \frac{1}{2}\right)$

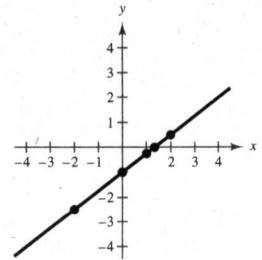

7. $y = x^2 - 3x$

x	-1	0	1	2	3
y	4	0	-2	-2	0
(x, y)	$(-1, 4)$	$(0, 0)$	$(1, -2)$	$(2, -2)$	$(3, 0)$

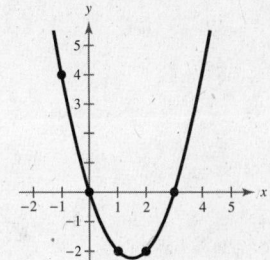

8. $5 - x^2$

x	-2	-1	0	1	2
y	1	4	5	4	1
x, y	$(-2, 1)$	$(-1, 4)$	$(0, 5)$	$(1, 4)$	$(2, 1)$

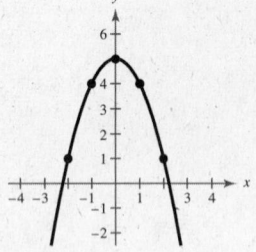

9. $y = 16 - 4x^2$

x-intercepts: $0 = 16 - 4x^2$

$\qquad\qquad 4x^2 = 16$

$\qquad\qquad x^2 = 4$

$\qquad\qquad x = \pm 2$

$\qquad\qquad (-2, 0), (2, 0)$

y-intercept: $y = 16 - 4(0)^2 = 16$

$\qquad\qquad (0, 16)$

10. $y = (x + 3)^2$

x-intercept: $0 = (x + 3)^2$

$\qquad\qquad 0 = x + 3$

$\qquad\qquad x = -3$

$\qquad\qquad (-3, 0)$

y-intercept: $y = (0 + 3)^2$

$\qquad\qquad y = 3^2$

$\qquad\qquad y = 9$

$\qquad\qquad (0, 9)$

11. $y = 5x - 6$

x-intercept: $0 = 5x - 6$

$\qquad\qquad 6 = 5x$

$\qquad\qquad \frac{6}{5} = x$

$\qquad\qquad \left(\frac{6}{5}, 0\right)$

y-intercept: $y = 5(0) - 6 = -6$

$\qquad\qquad (0, -6)$

12. $y = 8 - 3x$

x-intercept: $0 = 8 - 3x$

$\qquad\qquad 3x = 8$

$\qquad\qquad x = \frac{8}{3}$

$\qquad\qquad \left(\frac{8}{3}, 0\right)$

y-intercept: $y = 8 - 3(0) = 8$

$\qquad\qquad (0, 8)$

13. $y = \sqrt{x + 4}$

x-intercept: $0 = \sqrt{x + 4}$

$\qquad\qquad 0 = x + 4$

$\qquad\qquad -4 = x$

$\qquad\qquad (-4, 0)$

y-intercept: $y = \sqrt{0 + 4} = 2$

$\qquad\qquad (0, 2)$

14. $y = \sqrt{2x - 1}$

x-intercept: $\qquad 0 = \sqrt{2x - 1}$

$\qquad\qquad 2x - 1 = 0$

$\qquad\qquad x = \frac{1}{2}$

$\qquad\qquad \left(\frac{1}{2}, 0\right)$

y-intercept: $y = \sqrt{2(0) - 1}$

$\qquad\qquad = \sqrt{-1}$ There is no real solution.

There is no y-intercept.

15. $y = |3x - 7|$

    x-intercept: $0 = |3x - 7|$

                   $0 = 3x - 7$

                   $\frac{7}{3} = 0$

                   $\left(\frac{7}{3}, 0\right)$

    y-intercept: $y = |3(0) - 7| = 7$

                   $(0, 7)$

16. $y = -|x + 10|$

    x-intercept: $0 = -|x + 10|$

                   $x + 10 = 0$

                     $x = -10$

                     $(-10, 0)$

    y-intercept: $y = -|0 + 10|$

                     $= -|10| = -10$

                     $(0, -10)$

17. $y = 2x^3 - 4x^2$

    x-intercepts: $0 = 2x^3 - 4x^2$

                    $0 = 2x^2(x - 2)$

                    $x = 0$ or $x = 2$

                    $(0, 0), (2, 0)$

    y-intercept: $y = 2(0)^3 - 4(0)^2$

                    $y = 0$

                    $(0, 0)$

18. $y = x^4 - 25$

    x-intercept: $0 = x^4 - 25$

                   $x^4 = 25$

                   $x = \pm\sqrt[4]{5^2} = \pm\sqrt{5}$

                   $\left(\pm\sqrt{5}, 0\right)$

    y-intercept: $y = (0)^4 - 25 = -25$

                   $(0, -25)$

19. $y^2 = 6 - x$

    x-intercept: $0 = 6 - x$

                 $x = 6$

                 $(6, 0)$

    y-intercepts: $y^2 = 6 - 0$

                 $y = \pm\sqrt{6}$

                 $\left(0, \sqrt{6}\right), \left(0, -\sqrt{6}\right)$

20. $y^2 = x + 1$

    x-intercept: $0 = x + 1$

                 $x = -1$

                 $(-1, 0)$

    y-intercepts: $y^2 = 0 + 1$

                 $y = \pm 1$

                 $(0, 1), (0, -1)$

21. y-axis symmetry

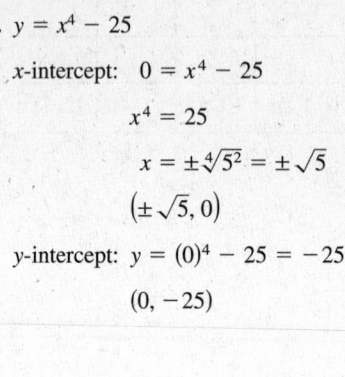

22.

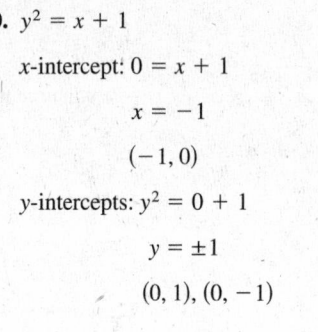

23. Origin symmetry

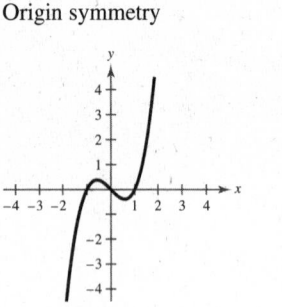

24.

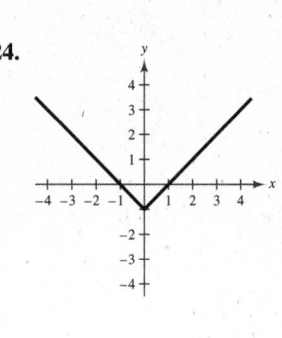

25. $x^2 - y = 0$

    $(-x)^2 - y = 0 \implies x^2 - y = 0 \implies$ y-axis symmetry

    $x^2 - (-y) = 0 \implies x^2 + y = 0 \implies$ No x-axis symmetry

    $(-x)^2 - (-y) = 0 \implies x^2 + y = 0 \implies$ No origin symmetry

26. $x - y^2 = 0$

    $x - (-y)^2 = 0$

    $x - y^2 = 0$

    x-axis symmetry

27. $y = x^3$

$\quad y = (-x)^3 \implies y = -x^3 \implies$ No y-axis symmetry

$\quad -y = x^3 \implies y = -x^3 \implies$ No x-axis symmetry

$\quad -y = (-x)^3 \implies -y = -x^3 \implies y = x^3 \implies$ Origin symmetry

28. $y = x^4 - x^2 + 3$

$\quad y = (-x)^4 - (-x)^2 + 3 \implies y = x^4 - x^2 + 3 \implies y$-axis symmetry

$\quad -y = x^4 - x^2 + 3 \implies y = -x^4 + x^2 - 3 \implies$ No x-axis symmetry

$\quad -y = (-x)^4 - (-x)^2 + 3 \implies y = -x^4 + x^2 - 3 \implies$ No origin symmetry

29. $y = \dfrac{x}{x^2 + 1}$

$\quad y = \dfrac{-x}{(-x)^2 + 1} \implies y = \dfrac{-x}{x^2 + 1} \implies$ No y-axis symmetry

$\quad -y = \dfrac{x}{x^2 + 1} \implies y = \dfrac{-x}{x^2 + 1} \implies$ No x-axis symmetry

$\quad -y = \dfrac{-x}{(-x)^2 + 1} \implies -y = \dfrac{-x}{x^2 + 1} \implies y = \dfrac{x}{x^2 + 1} \implies$ Origin symmetry

30. $y = \dfrac{1}{1 + x^2}$

$\quad y = \dfrac{1}{1 + (-x)^2} \implies y = \dfrac{1}{1 + x^2} \implies y$-axis symmetry

$\quad -y = \dfrac{1}{1 + x^2} \implies y = \dfrac{-1}{1 + x^2} \implies$ No x-axis symmetry

$\quad -y = \dfrac{1}{1 + (-x)^2} \implies y = \dfrac{-1}{1 + x^2} \implies$ No origin symmetry

31. $xy^2 + 10 = 0$

$\quad (-x)y^2 + 10 = 0 \implies -xy^2 + 10 = 0 \implies$ No y-axis symmetry

$\quad x(-y)^2 + 10 = 0 \implies xy^2 + 10 = 0 \implies x$-axis symmetry

$\quad (-x)(-y)^2 + 10 = 0 \implies -xy^2 + 10 = 0 \implies$ No origin symmetry

32. $xy = 4$

$\quad (-x)y = 4 \implies xy = -4 \implies$ No y-axis symmetry

$\quad x(-y) = 4 \implies xy = -4 \implies$ No x-axis symmetry

$\quad (-x)(-y) = 4 \implies xy = 4 \implies$ Origin symmetry

33. $y = -3x + 1$

x-intercept: $\left(\frac{1}{3}, 0\right)$

y-intercept: $(0, 1)$

No axis or origin symmetry

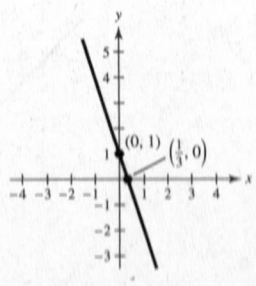

34. $y = 2x - 3$

x-intercept: $\left(\frac{3}{2}, 0\right)$

y-intercept: $(0, -3)$

No symmetry

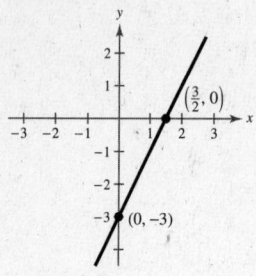

35. $y = x^2 - 2x$

Intercepts: $(0, 0)$, $(2, 0)$

No axis or origin symmetry

x	-1	0	1	2	3
y	3	0	-1	0	3

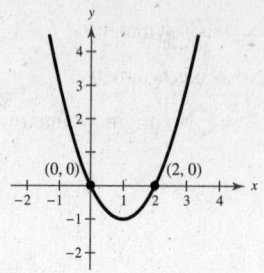

36. $y = -x^2 - 2x$

x-intercept: $(-2, 0)$, $(0, 0)$

y-intercept: $(0, 0)$

No symmetry

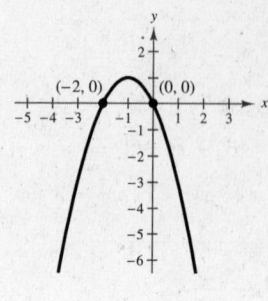

37. $y = x^3 + 3$

Intercepts: $(0, 3)$, $\left(\sqrt[3]{-3}, 0\right)$

No axis or origin symmetry

x	-2	-1	0	1	2
y	-5	2	3	4	11

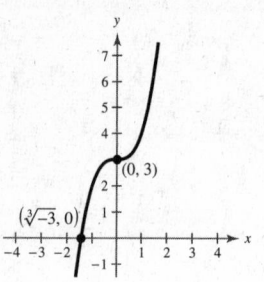

38. $y = x^3 - 1$

x-intercept: $(1, 0)$

y-intercept: $(0, -1)$

No symmetry

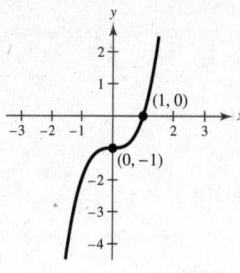

39. $y = \sqrt{x - 3}$

Domain: $[3, \infty)$

Intercept: $(3, 0)$

No axis or origin symmetry

x	3	4	7	12
y	0	1	2	3

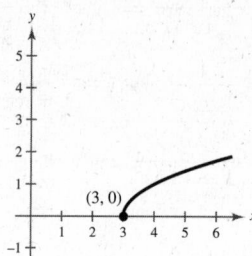

40. $y = \sqrt{1 - x}$

Domain: $(-\infty, 1]$

x-intercept: $(1, 0)$

y-intercept: $(0, 1)$

No symmetry

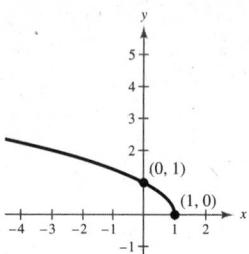

41. $y = |x - 6|$

Intercepts: $(0, 6)$, $(6, 0)$

No axis or origin symmetry

x	-2	0	2	4	6	8	10
y	8	6	4	2	0	2	4

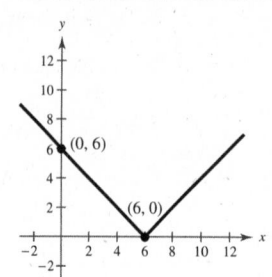

42. $y = 1 - |x|$

x-intercepts: $(\pm 1, 0)$

y-intercept: $(0, 1)$

y-axis symmetry

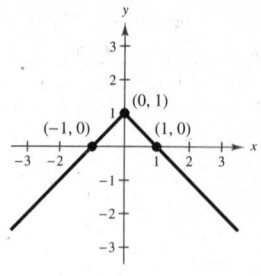

43. $x = y^2 - 1$

Intercepts: $(0, -1), (0, 1), (-1, 0)$

x-axis symmetry

x	-1	0	3
y	0	± 1	± 2

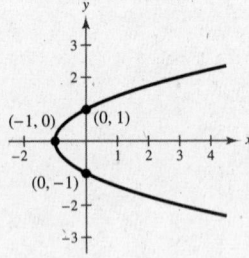

44. $x = y^2 - 5$

x-intercept: $(-5, 0)$

y-intercept: $\left(0, \pm\sqrt{5}\right)$

x-axis symmetry

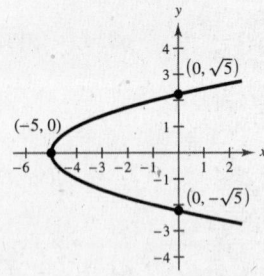

45. $y = 3 - \frac{1}{2}x$

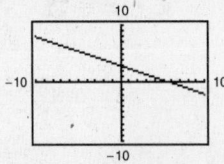

Intercepts: $(6, 0), (0, 3)$

46. $y = \frac{2}{3}x - 1$

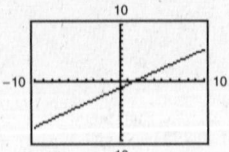

Intercepts: $(0, -1), \left(\frac{3}{2}, 0\right)$

47. $y = x^2 - 4x + 3$

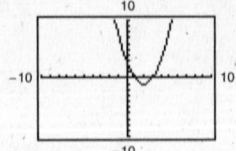

Intercepts: $(3, 0), (1, 0), (0, 3)$

48. $y = x^2 + x - 2$

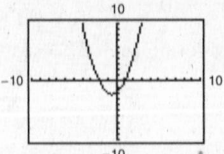

Intercepts: $(-2, 0), (1, 0), (0, -2)$

49. $y = \dfrac{2x}{x - 1}$

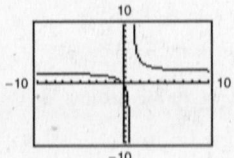

Intercept: $(0, 0)$

50. $y = \dfrac{4}{x^2 + 1}$

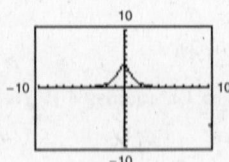

Intercept: $(0, 4)$

51. $y = \sqrt[3]{x}$

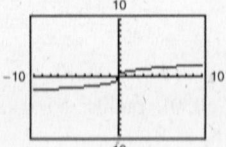

Intercept: $(0, 0)$

52. $y = \sqrt[3]{x + 1}$

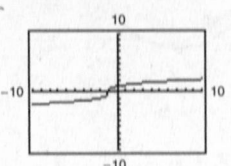

Intercepts: $(-1, 0), (0, 1)$

53. $y = x\sqrt{x + 6}$

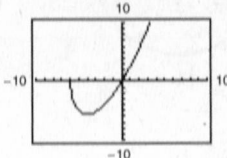

Intercepts: $(0, 0), (-6, 0)$

54. $y = (6 - x)\sqrt{x}$

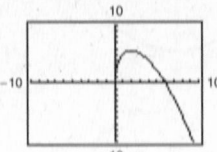

Intercepts: $(0, 0), (6, 0)$

55. $y = |x + 3|$

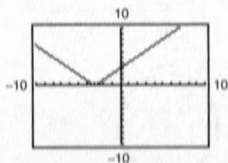

Intercepts: $(-3, 0), (0, 3)$

56. $y = 2 - |x|$

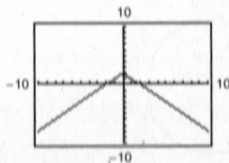

Intercepts: $(\pm 2, 0), (0, 2)$

57. Center: $(0, 0)$; radius: 4

Standard form:

$$(x - 0)^2 + (y - 0)^2 = 4^2$$
$$x^2 + y^2 = 16$$

58. $(x - 0)^2 + (y - 0)^2 = 5^2$

$x^2 + y^2 = 25$

59. Center: $(2, -1)$; radius: 4

Standard form:

$(x - 2)^2 + (y - (-1))^2 = 4^2$

$(x - 2)^2 + (y + 1)^2 = 16$

60. $(x - (-7))^2 + (y - (-4))^2 = 7^2$

$(x + 7)^2 + (y + 4)^2 = 49$

61. Center: $(-1, 2)$; solution point: $(0, 0)$

$(x - (-1))^2 + (y - 2)^2 = r^2$

$(0 + 1)^2 + (0 - 2)^2 = r^2 \Rightarrow 5 = r^2$

Standard form: $(x + 1)^2 + (y - 2)^2 = 5$

62. $r = \sqrt{(3 - (-1))^2 + (-2 - 1)^2}$

$= \sqrt{4^2 + (-3)^2} = \sqrt{25} = 5$

$(x - 3)^2 + (y - (-2))^2 = 5^2$

$(x - 3)^2 + (y + 2)^2 = 25$

63. Endpoints of a diameter: $(0, 0), (6, 8)$

Center: $\left(\dfrac{0 + 6}{2}, \dfrac{0 + 8}{2}\right) = (3, 4)$

$(x - 3)^2 + (y - 4)^2 = r^2$

$(0 - 3)^2 + (0 - 4)^2 = r^2 \Rightarrow 25 = r^2$

Standard form: $(x - 3)^2 + (y - 4)^2 = 25$

64. $r = \dfrac{1}{2}\sqrt{(-4 - 4)^2 + (-1 - 1)^2}$

$= \dfrac{1}{2}\sqrt{(-8)^2 + (-2)^2}$

$= \dfrac{1}{2}\sqrt{64 + 4}$

$= \dfrac{1}{2}\sqrt{68} = \left(\dfrac{1}{2}\right)(2)\sqrt{17} = \sqrt{17}$

Midpoint of diameter (center of circle):

$\left(\dfrac{-4 + 4}{2}, \dfrac{-1 + 1}{2}\right) = (0, 0)$

$(x - 0)^2 + (y - 0)^2 = \left(\sqrt{17}\right)^2$

$x^2 + y^2 = 17$

65. $x^2 + y^2 = 25$

Center: $(0, 0)$, radius: 5

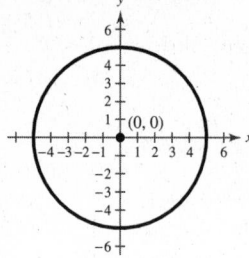

66. $x^2 + y^2 = 16$

Center: $(0, 0)$, radius: 4

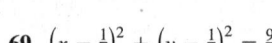

67. $(x - 1)^2 + (y + 3)^2 = 9$

Center: $(1, -3)$, radius: 3

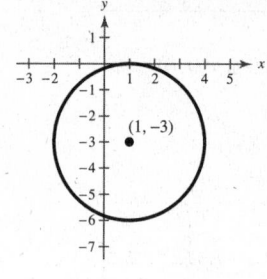

68. $x^2 + (y - 1)^2 = 1$

Center: $(0, 1)$, radius: 1

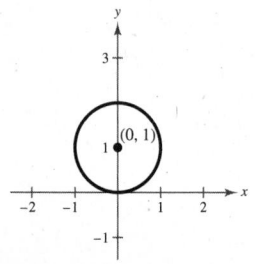

69. $\left(x - \dfrac{1}{2}\right)^2 + \left(y - \dfrac{1}{2}\right)^2 = \dfrac{9}{4}$

Center: $\left(\dfrac{1}{2}, \dfrac{1}{2}\right)$, radius: $\dfrac{3}{2}$

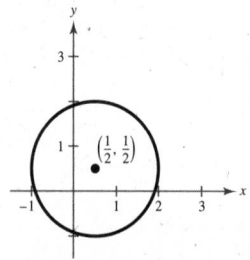

70. $(x - 2)^2 + (y + 3)^2 = \dfrac{16}{9}$

Center: $(2, -3)$, radius: $\dfrac{4}{3}$

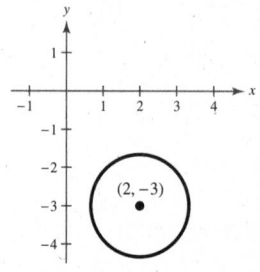

71. $y = 225{,}000 - 20{,}000t,\ 0 \le t \le 8$

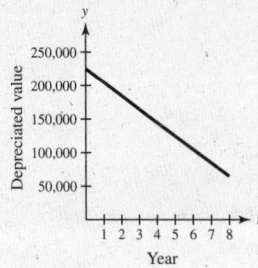

72. $y = 8100 - 929t,\ 0 \le t \le 6$

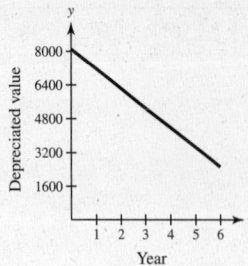

73. (a)

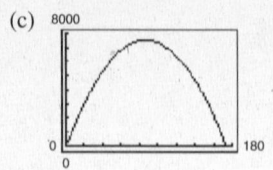

(b) $2x + 2y = \dfrac{1040}{3}$

$2y = \dfrac{1040}{3} - 2x$

$y = \dfrac{520}{3} - x$

$A = xy = x\left(\dfrac{520}{3} - x\right)$

(c)

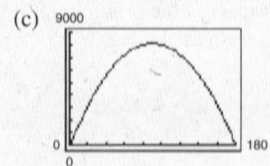

(d) When $x = y = 86\frac{2}{3}$ yards, the area is a maximum of $7511\frac{1}{9}$ square yards.

(e) A regulation NFL playing field is 120 yards long and $53\frac{1}{3}$ yards wide. The actual area is 6400 square yards.

74. (a)

(c)

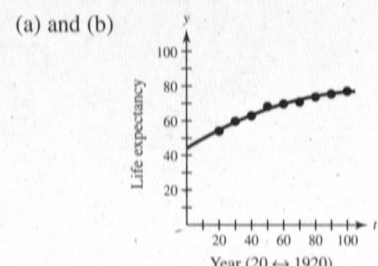

(b) $P = 360$ meters so:

$2x + 2y = 360$

$w = y = 180 - x$

$A = lw = x(180 - x)$

(d) $x = 90$ and $y = 90$

A square will give the maximum area of 8100 square meters.

(e) The dimensions of a Major League Soccer field can vary between 110 and 120 yards in length and between 70 and 80 yards in width.

75. $y = -0.0025t^2 + 0.574t + 44.25,\ 20 \le t \le 100$

(a) and (b)

(c) For the year 1948, let $t = 48$: $y \approx 66.0$ years.

(d) For the year 2005, let $t = 105$: $y \approx 77.0$ years.

For the year 2010, let $t = 110$: $y \approx 77.1$ years.

(e) No. The graph reaches a maximum of $y \approx 77.2$ years when $t \approx 114.8$, or during the year 2014. After this time, the model has life expectancy decreasing, which is not realistic.

76. (a)

x	5	10	20	30	40	50	60	70	80	90	100
y	430.43	107.33	26.56	11.60	6.36	3.94	2.62	1.83	1.31	0.96	0.71

(b)

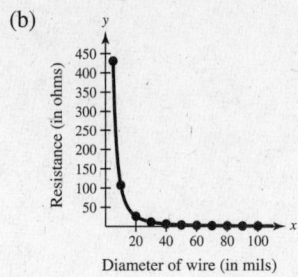

(c) When $x = 85.5$,

$$y = \frac{10{,}770}{85.5^2} - 0.37 = 1.10327.$$

(d) As the diameter of the wire increases, the resistance decreases.

77. False. A graph is symmetric with respect to the x-axis if, whenever (x, y) is on the graph, $(x, -y)$ is also on the graph.

78. True. The graph can have no intercepts, one, two or many. For example, a circle centered at the origin has two y-intercepts. A circle of radius 1, centered at $(7, 7)$, has no y-intercepts.

79. The viewing window is incorrect. Change the viewing window. Examples will vary. For example, $y = x^2 + 20$ will not appear in the standard window setting.

80. $y = ax^2 + bx^3$

(a) $y = a(-x)^2 + b(-x)^3$

   $= ax^2 - bx^3$

   To be symmetric with respect to the y-axis; a can be any non-zero real number, b must be zero.

(b) $-y = a(-x)^2 + b(-x)^3$

   $-y = ax^2 - bx^3$

   $y = -ax^2 + bx^3$

   To be symmetric with respect to the origin; a must be zero, b can be any non-zero real number.

81. $9x^5 + 4x^3 - 7$

Terms: $9x^5, 4x^3, -7$

82. $-(7 \times 7 \times 7 \times 7) = -(7)^4 = -7^4$

83. $\sqrt{18x} - \sqrt{2x} = 3\sqrt{2x} - \sqrt{2x} = 2\sqrt{2x}$

84. $\sqrt[4]{x^5} = \sqrt[4]{x \cdot x^4} = |x| \sqrt[4]{x}$

85. $\dfrac{70}{\sqrt{7x}} = \dfrac{70}{\sqrt{7x}} \cdot \dfrac{\sqrt{7x}}{\sqrt{7x}} = \dfrac{70\sqrt{7x}}{7x} = \dfrac{10\sqrt{7x}}{x}$

86. $\dfrac{55}{\sqrt{20} - 3} = \dfrac{55}{\sqrt{20} - 3} \cdot \dfrac{\sqrt{20} + 3}{\sqrt{20} + 3}$

$= \dfrac{55(\sqrt{20} + 3)}{20 - 9} = \dfrac{55(\sqrt{20} + 3)}{11}$

$= 5(\sqrt{20} + 3) = 5(2\sqrt{5} + 3)$

87. $\sqrt[6]{t^2} = t^{2/6} = |t|^{1/3} = \sqrt[3]{|t|}$

88. $\sqrt[3]{\sqrt{y}} = (y^{1/2})^{1/3} = y^{1/6} = \sqrt[6]{y}$

Section 1.3   Linear Equations in Two Variables

You should know the following important facts about lines.

- ■   The graph of $y = mx + b$ is a straight line. It is called a linear equation in two variables.

  (a) The slope (steepness) is m.

  (b) The y-intercept is $(0, b)$.

- ■   The slope of the line through (x_1, y_1) and (x_2, y_2) is

  $$m = \frac{y_2 - y_1}{x_2 - x_1} = \frac{\text{change in } y}{\text{change in } x} = \frac{\text{rise}}{\text{run}}.$$

- ■   (a) If $m > 0$, the line rises from left to right.

  (b) If $m = 0$, the line is horizontal.

  (c) If $m < 0$, the line falls from left to right.

  (d) If m is undefined, the line is vertical.

- ■   Equations of Lines

  (a) Slope-Intercept Form: $y = mx + b$

  (b) Point-Slope Form: $y - y_1 = m(x - x_1)$

  (c) Two-Point Form: $y - y_1 = \dfrac{y_2 - y_1}{x_2 - x_1}(x - x_1)$

  (d) General Form: $Ax + By + C = 0$

  (e) Vertical Line: $x = a$

  (f) Horizontal Line: $y = b$

- ■   Given two distinct nonvertical lines

  $L_1 : y = m_1 x + b_1$ and $L_2 : y = m_2 x + b_2$

  (a) L_1 is parallel to L_2 if and only if $m_1 = m_2$ and $b_1 \neq b_2$.

  (b) L_1 is perpendicular to L_2 if and only if $m_1 = -1/m_2$.

Vocabulary Check

1. linear

2. slope

3. parallel

4. perpendicular

5. rate or rate of change

6. linear extrapolation

7. (a) $Ax + By + C = 0$ (iii) general form

   (b) $x = a$ (i) vertical line

   (c) $y = b$ (v) horizontal line

   (d) $y = mx + b$ (ii) slope-intercept form

   (e) $y - y_1 = m(x - x_1)$ (iv) point-slope form

---

1. (a) $m = \frac{2}{3}$. Since the slope is positive, the line rises. Matches L_2.

   (b) m is undefined. The line is vertical. Matches L_3.

   (c) $m = -2$. The line falls. Matches L_1.

2. (a) $m = 0$. The line is horizontal. Matches L_2.

   (b) $m = -\frac{3}{4}$. Because the slope is negative, the line falls. Matches L_1.

   (c) $m = 1$. Because the slope is positive, the line rises. Matches L_3.

3.

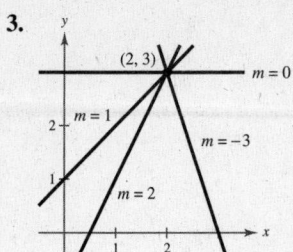

4.

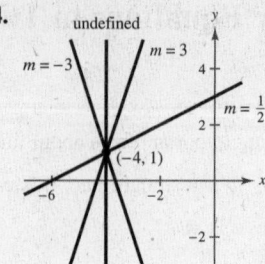

5. Two points on the line: $(0, 0)$ and $(4, 6)$

$$\text{Slope} = \frac{\text{rise}}{\text{run}} = \frac{6}{4} = \frac{3}{2}$$

6. The line appears to go through $(1, 0)$ and $(3, 5)$.

$$\text{Slope} = \frac{y_2 - y_1}{x_2 - x_1} = \frac{5 - 0}{3 - 1} = \frac{5}{2}$$

7. Two points on the line: $(0, 8)$ and $(2, 0)$

$$\text{Slope} = \frac{\text{rise}}{\text{run}} = \frac{-8}{2} = -4$$

8. The line appears to go through $(0, 7)$ and $(7, 0)$.

$$\text{Slope} = \frac{y_2 - y_1}{x_2 - x_1} = \frac{0 - 7}{7 - 0} = -1$$

9. $y = 5x + 3$

Slope: $m = 5$

y-intercept: $(0, 3)$

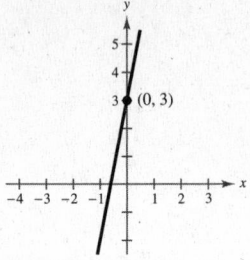

10. $y = x - 10$

Slope: $m = 1$

y-intercept: $(0, -10)$

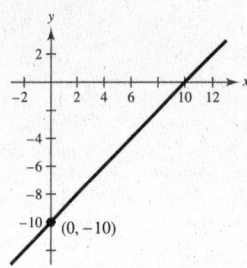

11. $y = -\frac{1}{2}x + 4$

Slope: $m = -\frac{1}{2}$

y-intercept: $(0, 4)$

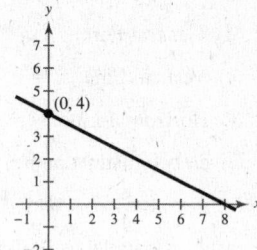

12. $y = -\frac{3}{2}x + 6$

Slope: $m = -\frac{3}{2}$

y-intercept: $(0, 6)$

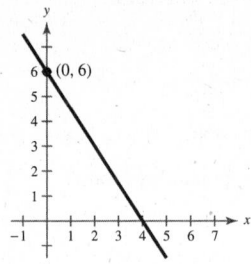

13. $5x - 2 = 0$

$x = \frac{2}{5}$, vertical line

Slope: undefined

No y-intercept

14. $3y + 5 = 0$

$$3y = -5$$

$$y = -\frac{5}{3}$$

Slope: $m = 0$

y-intercept: $\left(0, -\frac{5}{3}\right)$

15. $7x + 6y = 30$

$\quad\quad y = -\frac{7}{6}x + 5$

Slope: $m = -\frac{7}{6}$

y-intercept: $(0, 5)$

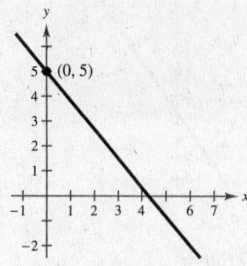

16. $2x + 3y = 9$

$\quad\quad 3y = -2x + 9$

$\quad\quad\quad y = -\frac{2}{3}x + 3$

Slope: $m = -\frac{2}{3}$

y-intercept: $(0, 3)$

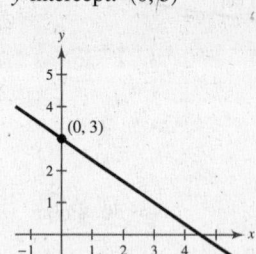

17. $y - 3 = 0$

$\quad\quad y = 3$, horizontal line

Slope: $m = 0$

y-intercept: $(0, 3)$

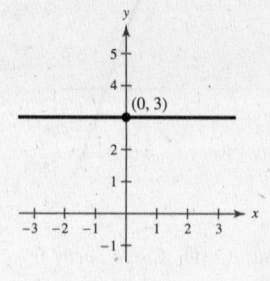

18. $y + 4 = 0$

$\quad\quad y = -4$

Slope: $m = 0$

y-intercept: $(0, -4)$

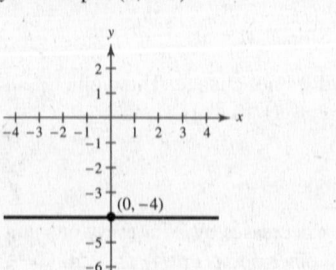

19. $x + 5 = 0$

$\quad\quad x = -5$

Slope: undefined (vertical line)

No y-intercept

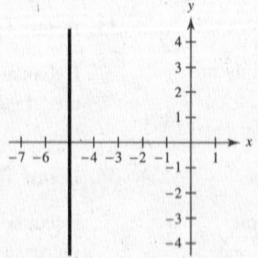

20. $x - 2 = 0$

$\quad\quad x = 2$

Slope: undefined (vertical line)

y-intercept: none

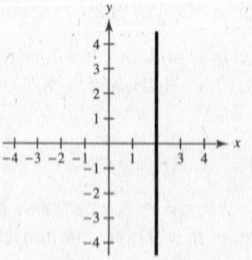

21. $m = \dfrac{6 - (-2)}{1 - (-3)} = \dfrac{8}{4} = 2$

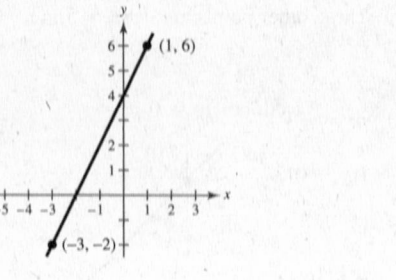

22. Slope $= \dfrac{-4 - 4}{4 - 2} = -4$

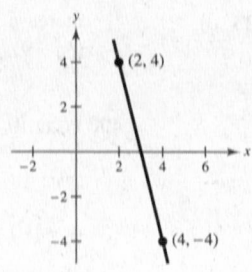

23. $m = \dfrac{4 - (-1)}{-6 - (-6)} = \dfrac{5}{0}$

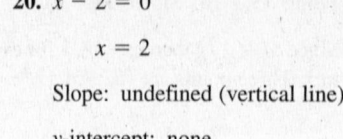

m is undefined.

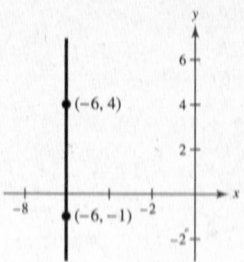

24. Slope $= \dfrac{0 - (-10)}{-4 - 0} = -\dfrac{5}{2}$

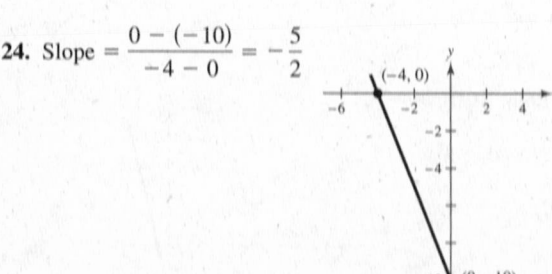

25. $m = \dfrac{-\frac{1}{3} - \left(-\frac{4}{3}\right)}{-\frac{3}{2} - \frac{11}{2}} = -\dfrac{1}{7}$

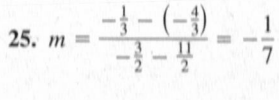

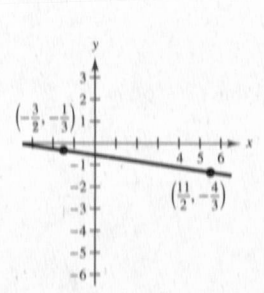

26. Slope $= \dfrac{-\frac{1}{4} - \frac{3}{4}}{\frac{5}{4} - \frac{7}{8}} = \dfrac{-1}{\frac{3}{8}} = -\dfrac{8}{3}$

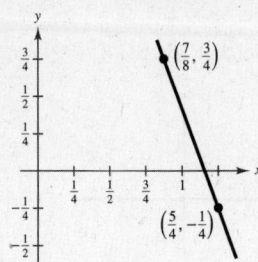

27. $m = \dfrac{1.6 - 3.1}{-5.2 - 4.8} = \dfrac{-1.5}{-10} = 0.15$

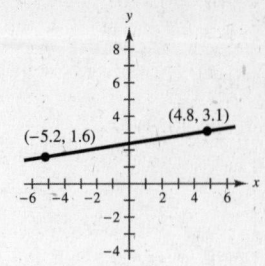

28. Slope $= \dfrac{-2.6 - (-8.3)}{2.25 - (-1.75)} = 1.425$

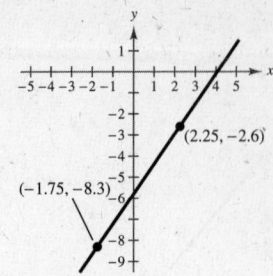

29. Point: $(2, 1)$, Slope: $m = 0$

Since $m = 0$, y does not change. Three points are $(0, 1)$, $(3, 1)$, and $(-1, 1)$.

30. Point: $(-4, 1)$, Slope is undefined.

Because m is undefined, x does not change. Three other points are: $(-4, 0), (-4, 3), (-4, 5)$.

31. Point: $(5, -6)$, Slope: $m = 1$

Since $m = 1$, y increases by 1 for every one unit increase in x. Three points are $(6, -5), (7, -4),$ and $(8, -3)$.

32. Point: $(10, -6)$, Slope: $m = -1$

Because $m = -1$, y decreases by 1 for every one unit increase in x. Three other points are: $(0, 4), (9, -5),$ $(11, -7)$.

33. Point: $(-8, 1)$, Slope is undefined.

Since m is undefined, x does not change. Three points are $(-8, 0), (-8, 2),$ and $(-8, 3)$.

34. Point: $(-3, -1)$, Slope: $m = 0$

Because $m = 0$, y does not change. Three other points are: $(-4, -1), (-2, -1), (0, -1)$.

35. Point: $(-5, 4)$, Slope: $m = 2$

Since $m = 2 = \frac{2}{1}$, y increases by 2 for every one unit increase in x. Three additional points are $(-4, 6), (-3, 8),$ and $(-2, 10)$.

36. Point: $(0, -9)$, Slope: $m = -2$

Because $m = -2$, y decreases by 2 for every one unit increase in x. Three other points are: $(-2, -5), (1, -11),$ $(3, -15)$.

37. Point: $(7, -2)$, Slope: $m = \frac{1}{2}$

Since $m = \frac{1}{2}$, y increases by 1 unit for every two unit increase in x. Three additional points are $(9, -1), (11, 0),$ and $(13, 1)$.

38. Point: $(-1, -6)$, Slope: $m = -\frac{1}{2}$

Because $m = -\frac{1}{2}$, y decreases by 1 for every 2 unit increase in x. Three other points are: $(-3, -5), (1, -7),$ $(5, -9)$.

39. Point $(0, -2)$; $m = 3$

$y + 2 = 3(x - 0)$

$y = 3x - 2$

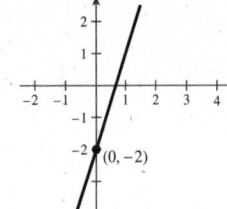

40. Point $(0, 10)$; $m = -1$

$y - 10 = -1(x - 0)$

$y - 10 = -x$

$y = -x + 10$

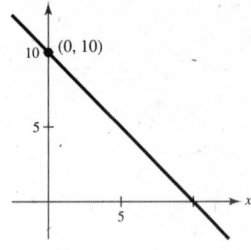

41. Point $(-3, 6)$; $m = -2$

$y - 6 = -2(x + 3)$

$y = -2x$

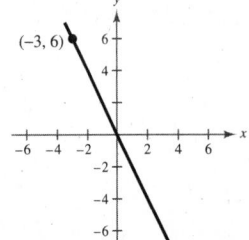

42. Point $(0, 0)$; $m = 4$

$y - 0 = 4(x - 0)$

$y = 4x$

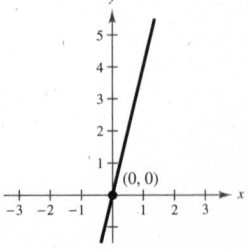

43. Point $(4, 0)$; $m = -\frac{1}{3}$

$$y - 0 = -\frac{1}{3}(x - 4)$$

$$y = -\frac{1}{3}x + \frac{4}{3}$$

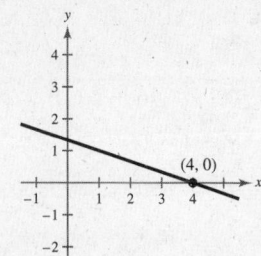

44. Point $(-2, -5)$; $m = \frac{3}{4}$

$$y + 5 = \frac{3}{4}(x + 2)$$

$$4y + 20 = 3x + 6$$

$$4y = 3x - 14$$

$$y = \frac{3}{4}x - \frac{7}{2}$$

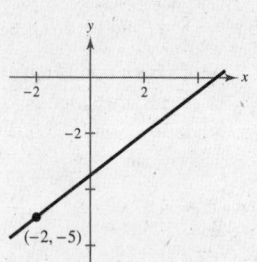

45. Point $(6, -1)$; m is undefined.

The line is vertical.

$$x = 6$$

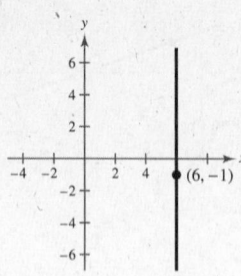

46. Point $(-10, 4)$; m is undefined.

Because the slope is undefined, the line is a vertical line passing through $x = -10$, which is the equation.

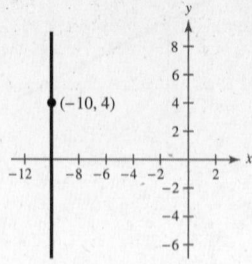

47. Point $\left(4, \frac{5}{2}\right)$; $m = 0$

The line is horizontal.

$$y = \frac{5}{2}$$

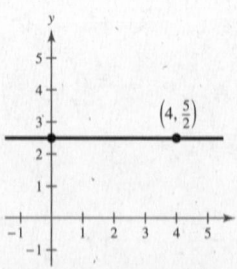

48. Point $\left(-\frac{1}{2}, \frac{3}{2}\right)$; $m = 0$

$$y - \frac{3}{2} = 0\left(x + \frac{1}{2}\right)$$

$$y - \frac{3}{2} = 0$$

$$y = \frac{3}{2}$$

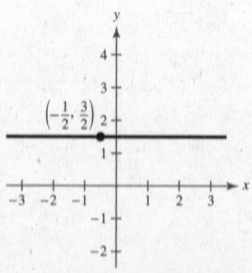

49. Point $(-5.1, 1.8)$; $m = 5$

$$y - 1.8 = 5(x - (-5.1))$$

$$y = 5x + 27.3$$

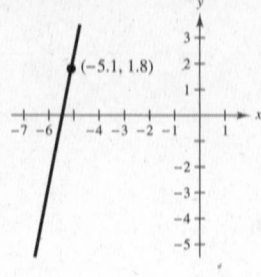

50. Point $(2.3, -8.5)$; $m = -\frac{5}{2}$

$$y - (-8.5) = -\frac{5}{2}(x - 2.3)$$

$$y + 8.5 = -2.5x + 5.75$$

$$y = -2.5x - 2.75$$

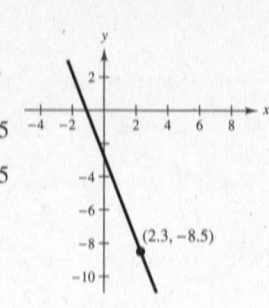

51. $(5, -1)$ and $(-5, 5)$

$$y + 1 = \frac{5 + 1}{-5 - 5}(x - 5)$$

$$y = -\frac{3}{5}(x - 5) - 1$$

$$y = -\frac{3}{5}x + 2$$

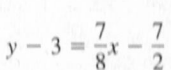
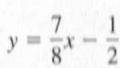

52. $(4, 3), (-4, -4)$

$$y - 3 = \frac{-4 - 3}{-4 - 4}(x - 4)$$

$$y - 3 = \frac{7}{8}(x - 4)$$

$$y - 3 = \frac{7}{8}x - \frac{7}{2}$$

$$y = \frac{7}{8}x - \frac{1}{2}$$

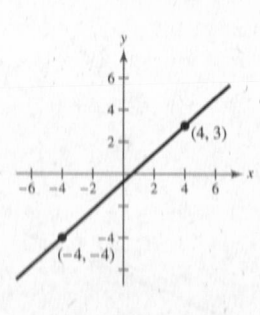

53. $(-8, 1)$ and $(-8, 7)$

Since both points have $x = -8$, the slope is undefined, and the line is vertical.

$x = -8$

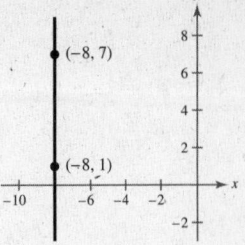

54. $(-1, 4), (6, 4)$

$y - 4 = \dfrac{4 - 4}{6 - (-1)}(x + 1)$

$y - 4 = 0(x + 1)$

$y - 4 = 0$

$y = 4$

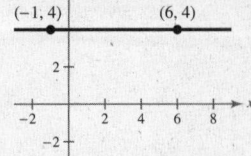

55. $\left(2, \dfrac{1}{2}\right)$ and $\left(\dfrac{1}{2}, \dfrac{5}{4}\right)$

$y - \dfrac{1}{2} = \dfrac{\frac{5}{4} - \frac{1}{2}}{\frac{1}{2} - 2}(x - 2)$

$y = -\dfrac{1}{2}(x - 2) + \dfrac{1}{2}$

$y = -\dfrac{1}{2}x + \dfrac{3}{2}$

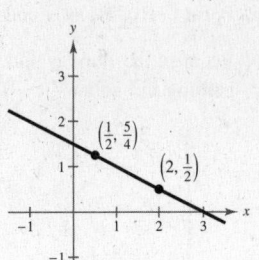

56. $(1, 1), \left(6, -\dfrac{2}{3}\right)$

$y - 1 = \dfrac{-\frac{2}{3} - 1}{6 - 1}(x - 1)$

$y - 1 = -\dfrac{1}{3}(x - 1)$

$y - 1 = -\dfrac{1}{3}x + \dfrac{1}{3}$

$y = -\dfrac{1}{3}x + \dfrac{4}{3}$

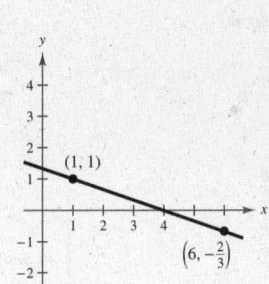

57. $\left(-\dfrac{1}{10}, -\dfrac{3}{5}\right)$ and $\left(\dfrac{9}{10}, -\dfrac{9}{5}\right)$

$y - \left(-\dfrac{3}{5}\right) = \dfrac{-\frac{9}{5} - \left(-\frac{3}{5}\right)}{\frac{9}{10} - \left(-\frac{1}{10}\right)}\left(x - \left(-\dfrac{1}{10}\right)\right)$

$y = -\dfrac{6}{5}\left(x + \dfrac{1}{10}\right) - \dfrac{3}{5}$

$y = -\dfrac{6}{5}x - \dfrac{18}{25}$

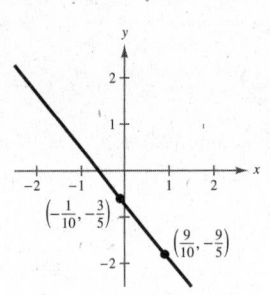

58. $\left(\dfrac{3}{4}, \dfrac{3}{2}\right), \left(-\dfrac{4}{3}, \dfrac{7}{4}\right)$

$y - \dfrac{3}{2} = \dfrac{\frac{7}{4} - \frac{3}{2}}{-\frac{4}{3} - \frac{3}{4}}\left(x - \dfrac{3}{4}\right)$

$y - \dfrac{3}{2} = \dfrac{\frac{1}{4}}{-\frac{25}{12}}\left(x - \dfrac{3}{4}\right)$

$y - \dfrac{3}{2} = -\dfrac{3}{25}\left(x - \dfrac{3}{4}\right)$

$y - \dfrac{3}{2} = -\dfrac{3}{25}x + \dfrac{9}{100}$

$y = -\dfrac{3}{25}x + \dfrac{159}{100}$

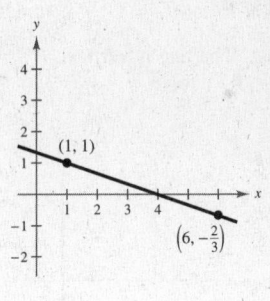

59. $(1, 0.6)$ and $(-2, -0.6)$

$y - 0.6 = \dfrac{-0.6 - 0.6}{-2 - 1}(x - 1)$

$y = 0.4(x - 1) + 0.6$

$y = 0.4x + 0.2$

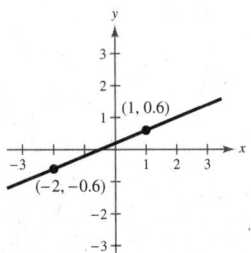

60. $(-8, 0.6), (2, -2.4)$

$y - 0.6 = \dfrac{-2.4 - 0.6}{2 - (-8)}(x + 8)$

$y - 0.6 = -\dfrac{3}{10}(x + 8)$

$10y - 6 = -3(x + 8)$

$10y - 6 = -3x - 24$

$10y = -3x - 18$

$y = -\dfrac{3}{10}x - \dfrac{9}{5}$ or $y = -0.3x - 1.8$

61. $(2, -1)$ and $\left(\dfrac{1}{3}, -1\right)$

$y + 1 = \dfrac{-1 - (-1)}{\frac{1}{3} - 2}(x - 2)$

$y + 1 = 0$

$\qquad y = -1$

The line is horizontal.

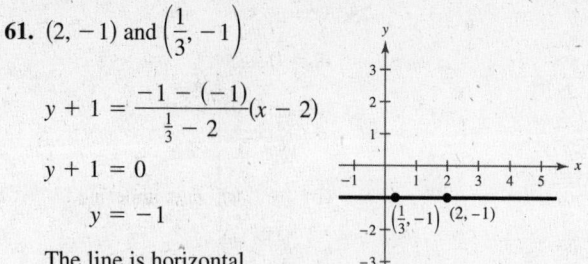

62. $\left(\dfrac{1}{5}, -2\right), (-6, -2)$

$y + 2 = \dfrac{-2 - (-2)}{-6 - \frac{1}{5}}(x + 6)$

$y + 2 = \dfrac{0}{-6 - \frac{1}{5}}(x + 6)$

$y + 2 = 0$

$\qquad y = -2$

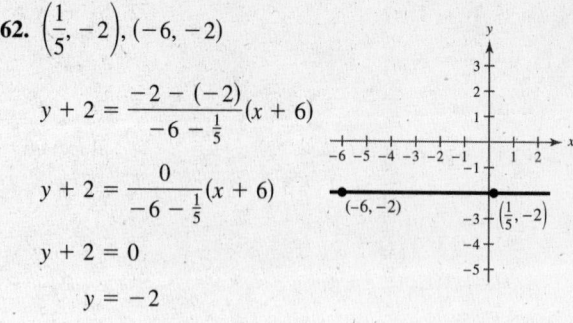

63. $\left(\dfrac{7}{3}, -8\right)$ and $\left(\dfrac{7}{3}, 1\right)$

$m = \dfrac{1 - (-8)}{\frac{7}{3} - \frac{7}{3}} = \dfrac{9}{0}$ and is undefined.

$x = \dfrac{7}{3}$

The line is vertical.

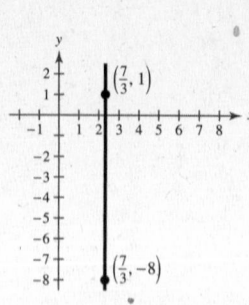

64. $(1.5, -2), (1.5, 0.2)$

$y + 2 = \dfrac{-2 - 0.2}{1.5 - 1.5}(x - 1.5)$

$y + 2 = \dfrac{-2 - 0.2}{0}(x - 1.5)$

The slope is undefined. The line is vertical.

$x = 1.5$

65. L_1: $(0, -1), (5, 9)$

Slope of L_1: $m = \dfrac{9 + 1}{5 - 0} = 2$

L_2: $(0, 3), (4, 1)$

Slope of L_2: $m = \dfrac{1 - 3}{4 - 0} = -\dfrac{1}{2}$

L_1 and L_2 are perpendicular.

66. L_1: $(-2, -1), (1, 5)$

$m_1 = \dfrac{5 - (-1)}{1 - (-2)} = \dfrac{6}{3} = 2$

L_2: $(1, 3), (5, -5)$

$m_2 = \dfrac{-5 - 3}{5 - 1} = \dfrac{-8}{4} = -2$

The lines are neither parallel nor perpendicular.

67. L_1: $(3, 6), (-6, 0)$

Slope of L_1: $m = \dfrac{0 - 6}{-6 - 3} = \dfrac{2}{3}$

L_2: $(0, -1), \left(5, \dfrac{7}{3}\right)$

Slope of L_2: $m = \dfrac{\frac{7}{3} + 1}{5 - 0} = \dfrac{2}{3}$

L_1 and L_2 are parallel.

68. L_1: $(4, 8), (-4, 2)$

$m_1 = \dfrac{2 - 8}{-4 - 4} = \dfrac{-6}{-8} = \dfrac{3}{4}$

L_2: $(3, -5), \left(-1, \dfrac{1}{3}\right)$

$m_2 = \dfrac{\frac{1}{3} - (-5)}{-1 - 3} = \dfrac{\frac{16}{3}}{-4} = -\dfrac{4}{3}$

The lines are perpendicular.

69. $4x - 2y = 3$

$\qquad y = 2x - \dfrac{3}{2}$

Slope: $m = 2$

(a) $(2, 1), m = 2$

$y - 1 = 2(x - 2)$

$\qquad y = 2x - 3$

(b) $(2, 1), m = -\dfrac{1}{2}$

$y - 1 = -\dfrac{1}{2}(x - 2)$

$\qquad y = -\dfrac{1}{2}x + 2$

70. $x + y = 7$

$\qquad y = -x + 7$

Slope: $m = -1$

(a) $m = -1, (-3, 2)$

$y - 2 = -1(x + 3)$

$y - 2 = -x - 3$

$\qquad y = -x - 1$

(b) $m = 1, (-3, 2)$

$y - 2 = 1(x + 3)$

$\qquad y = x + 5$

71. $3x + 4y = 7$

$$y = -\tfrac{3}{4}x + \tfrac{7}{4}$$

Slope: $m = -\tfrac{3}{4}$

(a) $\left(-\tfrac{2}{3}, \tfrac{7}{8}\right), m = -\tfrac{3}{4}$

$$y - \tfrac{7}{8} = -\tfrac{3}{4}\left(x - \left(-\tfrac{2}{3}\right)\right)$$

$$y = -\tfrac{3}{4}x + \tfrac{3}{8}$$

(b) $\left(-\tfrac{2}{3}, \tfrac{7}{8}\right), m = \tfrac{4}{3}$

$$y - \tfrac{7}{8} = \tfrac{4}{3}\left(x - \left(-\tfrac{2}{3}\right)\right)$$

$$y = \tfrac{4}{3}x + \tfrac{127}{72}$$

72. $5x + 3y = 0$

$$3y = -5x$$

$$y = -\tfrac{5}{3}x$$

Slope: $m = -\tfrac{5}{3}$

(a) $m = -\tfrac{5}{3}, \left(\tfrac{7}{8}, \tfrac{3}{4}\right)$

$$y - \tfrac{3}{4} = -\tfrac{5}{3}\left(x - \tfrac{7}{8}\right)$$

$$24y - 18 = -40\left(x - \tfrac{7}{8}\right)$$

$$24y - 18 = -40x + 35$$

$$24y = -40x + 53$$

$$y = -\tfrac{5}{3}x + \tfrac{53}{24}$$

(b) $m = \tfrac{3}{5}, \left(\tfrac{7}{8}, \tfrac{3}{4}\right)$

$$y - \tfrac{3}{4} = \tfrac{3}{5}\left(x - \tfrac{7}{8}\right)$$

$$40y - 30 = 24\left(x - \tfrac{7}{8}\right)$$

$$40y - 30 = 24x - 21$$

$$40y = 24x + 9$$

$$y = \tfrac{3}{5}x + \tfrac{9}{40}$$

73. $y = -3$

$m = 0$

(a) $(-1, 0)$ and $m = 0$

$$y = 0$$

(b) $(-1, 0)$, m is undefined.

$$x = -1$$

74. $y = 1$

Slope: $m = 0$

(a) $m = 0, (4, -2)$

$$y + 2 = 0(x - 4)$$

$$y + 2 = 0$$

$$y = -2$$

(b) The reciprocal of 0 is undefined. The line is vertical, passing through $(4, -2)$.

$$x = 4$$

75. $x = 4$

m is undefined.

(a) $(2, 5)$, m is undefined. The line is vertical, passing through $(2, 5)$.

$$x = 2$$

(b) $(2, 5), m = 0$

$$y = 5$$

76. $x = -2$

Slope: undefined

(a) The original line is the vertical line through $x = -2$. The line parallel to this line containing $(-5, 1)$ is the vertical line $x = -5$.

(b) A perpendicular to a vertical line is a horizontal line, whose slope is 0. The horizontal line containing $(-5, 1)$ is the line $y = 1$.

77. $x - y = 4$

$$y = x - 4$$

Slope: $m = 1$

(a) $(2.5, 6.8), m = 1$

$$y - 6.8 = 1(x - 2.5)$$

$$y = x + 4.3$$

(b) $(2.5, 6.8), m = -1$

$$y - 6.8 = (-1)(x - 2.5)$$

$$y = -x + 9.3$$

78. $6x + 2y = 9$

$$2y = -6x + 9$$

$$y = -3x + \tfrac{9}{2}$$

Slope: $m = -3$

(a) $(-3.9, -1.4), m = -3$

$$y - (-1.4) = -3(x - (-3.9))$$

$$y + 1.4 = -3x - 11.7$$

$$y = -3x - 13.1$$

(b) $(-3.9, -1.4), m = \tfrac{1}{3}$

$$y - (-1.4) = \tfrac{1}{3}(x - (-3.9))$$

$$y + 1.4 = \tfrac{1}{3}x + 1.3$$

$$y = \tfrac{1}{3}x - 0.1$$

79. $\dfrac{x}{2} + \dfrac{y}{3} = 1$

$3x + 2y - 6 = 0$

80. $(-3, 0), (0, 4)$

$\dfrac{x}{-3} + \dfrac{y}{4} = 1$

$(-12)\dfrac{x}{-3} + (-12)\dfrac{y}{4} = (-12) \cdot 1$

$4x - 3y + 12 = 0$

81. $\dfrac{x}{-1/6} + \dfrac{y}{-2/3} = 1$

$6x + \dfrac{3}{2}y = -1$

$12x + 3y + 2 = 0$

82. $\left(\dfrac{2}{3}, 0\right), (0, -2)$

$\dfrac{x}{2/3} + \dfrac{y}{-2} = 1$

$\dfrac{3x}{2} - \dfrac{y}{2} = 1$

$3x - y - 2 = 0$

83. $\dfrac{x}{c} + \dfrac{y}{c} = 1,\ c \neq 0$

$x + y = c$

$1 + 2 = c$

$3 = c$

$x + y = 3$

$x + y - 3 = 0$

84. $(d, 0), (0, d), (-3, 4)$

$\dfrac{x}{d} + \dfrac{y}{d} = 1$

$x + y = d$

$-3 + 4 = d$

$1 = d$

$x + y = 1$

$x + y - 1 = 0$

85. (a) $y = 2x$

    (b) $y = -2x$

    (c) $y = \dfrac{1}{2}x$

    (b) and (c) are perpendicular.

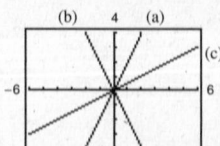

86. (a) $y = \dfrac{2}{3}x$

    (b) $y = -\dfrac{3}{2}x$

    (c) $y = \dfrac{2}{3}x + 2$

    (a) is parallel to (c). (b) is perpendicular to (a) and (c).

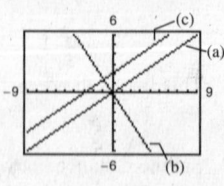

87. (a) $y = -\dfrac{1}{2}x$

    (b) $y = -\dfrac{1}{2}x + 3$

    (c) $y = 2x - 4$

    (a) and (b) are parallel. (c) is perpendicular to (a) and (b).

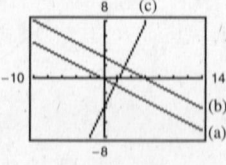

88. (a) $y = x - 8$

    (b) $y = x + 1$

    (c) $y = -x + 3$

    (a) is parallel to (b). (c) is perpendicular to (a) and (b).

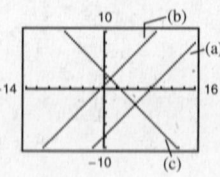

89. Set the distance between $(4, -1)$ and (x, y) equal to the distance between $(-2, 3)$ and (x, y).

$$\sqrt{(x - 4)^2 + [y - (-1)]^2} = \sqrt{[x - (-2)]^2 + (y - 3)^2}$$

$$(x - 4)^2 + (y + 1)^2 = (x + 2)^2 + (y - 3)^2$$

$$x^2 - 8x + 16 + y^2 + 2y + 1 = x^2 + 4x + 4 + y^2 - 6y + 9$$

$$-8x + 2y + 17 = 4x - 6y + 13$$

$$0 = 12x - 8y - 4$$

$$0 = 4(3x - 2y - 1)$$

$$0 = 3x - 2y - 1$$

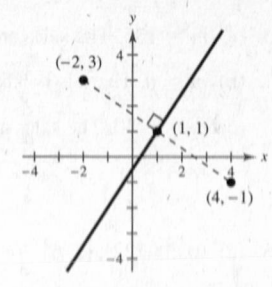

This line is the perpendicular bisector of the line segment connecting $(4, -1)$ and $(-2, 3)$.

90. Set the distance between $(6, 5)$ and (x, y) equal to the distance between $(1, -8)$ and (x, y).

$$\sqrt{(x - 6)^2 + (y - 5)^2} = \sqrt{(x - 1)^2 + (y - (-8))^2}$$

$$(x - 6)^2 + (y - 5)^2 = (x - 1)^2 + (y + 8)^2$$

$$x^2 - 12x + 36 + y^2 - 10y + 25 = x^2 - 2x + 1 + y^2 + 16y + 64$$

$$x^2 + y^2 - 12x - 10y + 61 = x^2 + y^2 - 2x + 16y + 65$$

$$-12x - 10y + 61 = -2x + 16y + 65$$

$$-10x - 26y - 4 = 0$$

$$-2(5x + 13y + 2) = 0$$

$$5x + 13y + 2 = 0$$

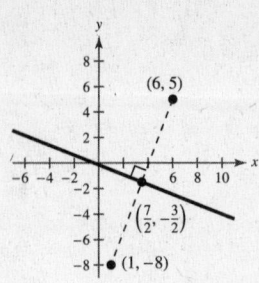

91. Set the distance between $\left(3, \frac{5}{2}\right)$ and (x, y) equal to the distance between $(-7, 1)$ and (x, y).

$$\sqrt{(x - 3)^2 + \left(y - \frac{5}{2}\right)^2} = \sqrt{[x - (-7)]^2 + (y - 1)^2}$$

$$(x - 3)^2 + \left(y - \frac{5}{2}\right)^2 = (x + 7)^2 + (y - 1)^2$$

$$x^2 - 6x + 9 + y^2 - 5y + \frac{25}{4} = x^2 + 14x + 49 + y^2 - 2y + 1$$

$$-6x - 5y + \frac{61}{4} = 14x - 2y + 50$$

$$-24x - 20y + 61 = 56x - 8y + 200$$

$$80x + 12y + 139 = 0$$

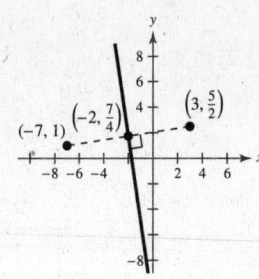

This line is the perpendicular bisector of the line segment connecting $\left(3, \frac{5}{2}\right)$ and $(-7, 1)$.

92. Set the distance between $\left(-\frac{1}{2}, -4\right)$ and (x, y) equal to the distance between $\left(\frac{7}{2}, \frac{5}{4}\right)$ and (x, y).

$$\sqrt{\left(x - \left(-\frac{1}{2}\right)\right)^2 + (y - (-4))^2} = \sqrt{\left(x - \frac{7}{2}\right)^2 + \left(y - \frac{5}{4}\right)^2}$$

$$\left(x + \frac{1}{2}\right)^2 + (y + 4)^2 = \left(x - \frac{7}{2}\right)^2 + \left(y - \frac{5}{4}\right)^2$$

$$x^2 + x + \frac{1}{4} + y^2 + 8y + 16 = x^2 - 7x + \frac{49}{4} + y^2 - \frac{5}{2}y + \frac{25}{16}$$

$$x^2 + y^2 + x + 8y + \frac{65}{4} = x^2 + y^2 - 7x - \frac{5}{2}y + \frac{221}{16}$$

$$x + 8y + \frac{65}{4} = -7x - \frac{5}{2}y + \frac{221}{16}$$

$$8x + \frac{21}{2}y + \frac{39}{16} = 0$$

$$128x + 168y + 39 = 0$$

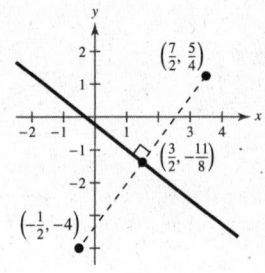

93. (a) $m = 135$. The sales are increasing 135 units per year.

(b) $m = 0$. There is no change in sales during the year.

(c) $m = -40$. The sales are decreasing 40 units per year.

94. (a) $m = 400$. The revenues are increasing 400 units per day.

(b) $m = 100$. The revenues are increasing 100 units per day.

(c) $m = 0$. There is no change in revenue during the day. (Revenue remains constant.)

95. (a) $(0, 55{,}722), (2, 61{,}768)$: $m = \dfrac{61{,}768 - 55{,}722}{2 - 0} = 3023$

$(6, 69{,}277), (8, 74{,}380)$: $m = \dfrac{74{,}380 - 69{,}277}{8 - 6} = 2551.5$

$(2, 61{,}768), (4, 64{,}993)$: $m = \dfrac{64{,}993 - 61{,}768}{4 - 2} = 1612.5$

$(8, 74{,}380), (10, 79{,}839)$: $m = \dfrac{79{,}839 - 74{,}380}{10 - 8} = 2729.5$

$(4, 64{,}993), (6, 69{,}277)$: $m = \dfrac{69{,}277 - 64{,}993}{6 - 4} = 2142$

$(10, 79{,}839), (12, 83{,}944)$: $m = \dfrac{83{,}944 - 79{,}839}{12 - 10} = 2052.5$

The average salary increased the most from 1990 to 1992 and the least from 1992 to 1994.

—CONTINUED—

95. **—CONTINUED—**

(b) $(0, 55{,}722)$, $(12, 83{,}944)$: $m = \dfrac{83{,}944 - 55{,}722}{12 - 0} \approx \$2351.83$

(c) The average salary for senior high school principals increased by $2351.83 per year over the 12 years between 1990 and 2002.

96. (a) The greatest increase of $16.2 million is between 2002 and 2003. The least increase of $5.4 million is between 2000 and 2001.

(b) Slope $= \dfrac{99.2 - 16.6}{13 - 4} = 9.18$

(c) Each year the net profit increases by $9.18 million.

97. $y = \dfrac{6}{100}x$

$y = \dfrac{6}{100}(200) = 12$ feet

98. (a) and (b)

x	300	600	900	1200	1500	1800	2100
y	-25	-50	-75	-100	-125	-150	-175

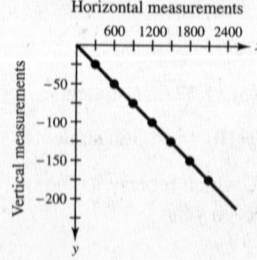

(c) $m = \dfrac{-50 - (-25)}{600 - 300} = \dfrac{-25}{300} = -\dfrac{1}{12}$

$y - (-50) = -\dfrac{1}{12}(x - 600)$

$y + 50 = -\dfrac{1}{12}x + 50$

$y = -\dfrac{1}{12}x$

(d) Since $m = -\dfrac{1}{12}$, for every change in the horizontal measurement of 12 units, the vertical measurement decreases by 1.

(e) $\dfrac{1}{12} \approx 0.083 = 8.3\%$ grade

99. $(5, 2540)$, $m = -125$

$V - 2540 = -125(t - 5)$

$V - 2540 = -125t + 625$

$V = -125t + 3165,\ 5 \le t \le 10$

100. $(5, 156)$, $m = 4.50$

$V - 156 = 4.50(t - 5)$

$V - 156 = 4.50t - 22.5$

$V = 4.5t + 133.5,\ 5 \le t \le 10$

101. Matches graph (b).

The slope is -20, which represents the decrease in the amount of the loan each week. The y-intercept is $(0, 200)$, which represents the original amount of the loan.

102. Matches graph (c).

The slope is 2, which represents the increase in the hourly wage for each unit produced. The y-intercept is $(0, 8.5)$, which represents the hourly rate if the employee produces no units.

103. Matches graph (a).

The slope is 0.32, which represents the increase in travel cost for each mile driven. The y-intercept is $(0, 30)$, which represents the fixed cost of $30 per day for meals. This amount does not depend on the number of miles driven.

104. Matches graph (d).

The slope is -100, which represents the amount by which the computer depreciates each year. The y-intercept is $(0, 750)$, which represents the original purchase price.

105. $(5, 0.18)$, $(13, 4.04)$: $m = \dfrac{4.04 - 0.18}{13 - 5} = 0.4825$

$$y - 0.18 = 0.4825(t - 5)$$

$$y = 0.4825t - 2.2325$$

For 2008, use $t = 18$: $y(18) \approx \$6.45$

For 2010, use $t = 20$: $y(20) \approx \$7.42$

106. $t = 9$ represents 1999, $(9, 4076)$.

$t = 13$ represents 2003, $(13, 1078)$.

$$m = \dfrac{4076 - 1078}{9 - 13} = \dfrac{-2998}{4} = -749.5$$

$$N = -749.5t + 10,821.5$$

$t = 18$ represents 2008:

$N = -749.5(18) + 10,821.5 = -2669.5$ stores

$t = 20$ represents 2010:

$N = -749.5(20) + 10,821.5 = -4168.5$ stores

These answers are not reasonable because they are negative.

107. Using the points $(0, 875)$ and $(5, 0)$, where the first coordinate represents the year t and the second coordinate represents the value V, we have

$$m = \dfrac{0 - 875}{5 - 0} = -175$$

$$V = -175t + 875, \quad 0 \le t \le 5.$$

108. $(0, 25{,}000)$ and $(10, 2000)$

$$m = \dfrac{2000 - 25000}{10 - 0} = -2300$$

$$V = -2300t + 25{,}000, \quad 0 \le t \le 10$$

109. (a) $(0, 40{,}571)$, $(4, 41{,}289)$:

$$m = \dfrac{41{,}289 - 40{,}571}{4 - 0} = 179.5$$

$$y = 179.5t + 40{,}571$$

(b) For 2008, use $t = 8$: $y(8) = 42{,}007$ students.

For 2010, use $t = 10$: $y(10) = 42{,}366$ students.

(c) The slope is $m = 179.5$, which represents the increase in the number of students each year.

110. (a) Average annual salary change from 1990 to 2003:

$$\dfrac{48{,}673 - 36{,}531}{13 - 0} = \dfrac{12{,}142}{13} = 934 \text{ students per year}$$

(c) $m = 934$, $b = 36{,}531$, so $N(t) = 934t + 36{,}531$.

The slope, 934, represents the average annual change in enrollment.

(b) Using (a) to estimate the enrollment in:

1994: $36{,}531 + 4(934) = 40{,}267$ students

1998: $36{,}531 + 8(934) = 44{,}003$ students

2002: $36{,}531 + 12(934) = 47{,}739$ students

(d) Answers will vary.

111. Sale price = List price $-$ 15% of the list price

$$S = L - 0.15L$$

$$S = 0.85L$$

112. $W = 0.75x + 11.50$

113. (a) $C = 36{,}500 + 5.25t + 11.50t$

$$= 16.75t + 36{,}500$$

(c) $P = R - C$

$$= 27t - (16.75t + 36{,}500)$$

$$= 10.25t - 36{,}500$$

(b) $R = 27t$

(d) $\quad 0 = 10.25t - 36{,}500$

$$36{,}500 = 10.25t$$

$$t \approx 3561 \text{ hours}$$

114. (580, 50) and (625, 47)

(a) $m = \dfrac{47 - 50}{625 - 580} = \dfrac{-3}{45} = -\dfrac{1}{15}$

$x - 50 = -\dfrac{1}{15}(p - 580)$

$x - 50 = -\dfrac{1}{15}p + \dfrac{116}{3}$

$x = -\dfrac{1}{15}p + \dfrac{266}{3}$

(b) $x = -\dfrac{1}{15}(655) + \dfrac{266}{3} = 45$ units

(c) $x = -\dfrac{1}{15}(595) + \dfrac{266}{3} = 49$ units

115. (a)

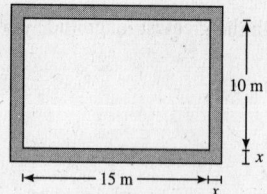

(b) $y = 2(15 + 2x) + 2(10 + 2x) = 8x + 50$

(c)

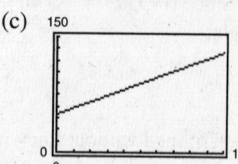

(d) Since $m = 8$, each 1-meter increase in x will increase y by 8 meters.

116. $W = 0.07S + 2500$

117. $C = 0.38x + 120$

118.

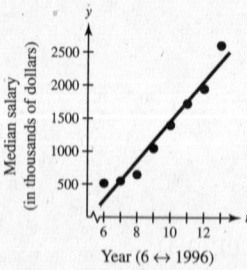

Using a calculator, the linear regression line is $y = 300.3t - 1547.4$. Choosing the points $(7, 550)$ and $(10, 1400)$:

$$m = \dfrac{1400 - 550}{10 - 7} = \dfrac{850}{3} = 283.3$$

$y - 550 = 283.3(t - 7)$

$y = 283.3t - 1433.1$

The answer varies depending on the points chosen to estimate the line.

119. (a) and (b)

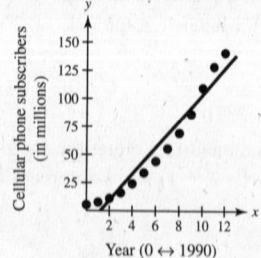

(c) Answers will vary. Find two points on your line and then find the equation of the line through your points. Sample answer: $y \approx 11.72x - 14.08$

(d) Answers will vary. Sample answer: The y-intercept should represent the number of initial subscribers. In this case, since b is negative, it cannot be interpreted as such. The slope of 11.72 represents the increase in the number of subscribers per year (in millions).

(e) The model is a fairly good fit to the data.

(f) Answers will vary. Sample answer:

$y(18) \approx 11.72(18) - 14.08$

$= 196.88$ million subscribers in 2008

120. (a) and (b)

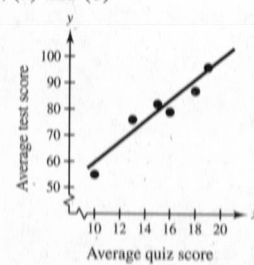

(c) Two approximate points on the line are $(10, 55)$ and $(19, 96)$.

$$m = \dfrac{96 - 55}{19 - 10} = \dfrac{41}{9}$$

$y - 55 = \dfrac{41}{9}(x - 10)$

$y = \dfrac{41}{9}x + \dfrac{85}{9}$

(d) $y = \dfrac{41}{9}(17) + \dfrac{85}{9} \approx 87$

(e) Each point will shift four units upward, so the best-fitting line will move four units upward. The slope remains the same, as the new line is parallel to the old, but the y-intercept becomes

$\left(0, \dfrac{85}{9} + 4\right) = \left(0, \dfrac{121}{9}\right)$

so the new equation is $y = \dfrac{41}{9}x + \dfrac{121}{9}$.

121. False. The slope with the greatest magnitude corresponds to the steepest line.

122. $(-8, 2)$ and $(-1, 4) : m_1 = \dfrac{4 - 2}{-1 - (-8)} = \dfrac{2}{7}$

$(0, -4)$ and $(-7, 7) : m_2 = \dfrac{7 - (-4)}{-7 - 0} = \dfrac{11}{-7}$

False, the lines are not parallel.

123. Using the Distance Formula, we have $AB = 6$, $BC = \sqrt{40}$, and $AC = 2$. Since $6^2 + 2^2 = \left(\sqrt{40}\right)^2$, the triangle is a right triangle.

124. On a vertical line, all the points have the same x-value, so when you evaluate $m = \dfrac{y_2 - y_1}{x_2 - x_1}$, you would have a zero in the denominator, and division by zero is undefined.

125. No. The slope cannot be determined without knowing the scale on the y-axis. The slopes will be the same if the scale on the y-axis of (a) is $2\frac{1}{2}$ and the scale on the y-axis of (b) is 1. Then the slope of both is $\frac{5}{4}$.

126. Since $|-4| > \left|\frac{5}{2}\right|$, the steeper line is the one with a slope of -4. The slope with the greatest magnitude corresponds to the steepest line.

127. The V-intercept measures the initial cost and the slope measures annual depreciation.

128. No, the slopes of two perpendicular lines have opposite signs. (Assume that neither line is vertical or horizontal.)

129. $y = 8 - 3x$ is a linear equation with slope $m = -3$ and y-intercept $(0, 8)$. Matches graph (d).

130. $y = 8 - \sqrt{x}$

Intercepts: $(64, 0), (0, 8)$

Matches graph (c).

131. $y = \frac{1}{2}x^2 + 2x + 1$ is a quadratic equation. Its graph is a parabola with vertex $(-2, -1)$ and y-intercept $(0, 1)$. Matches graph (a).

132. $y = |x + 2| - 1$

Intercepts: $(-1, 0), (-3, 0), (0, 1)$

Matches graph (b).

133. $-7(3 - x) = 14(x - 1)$

$-21 + 7x = 14x - 14$

$-7x = 7$

$x = -1$

134. $\dfrac{8}{2x - 7} = \dfrac{4}{9 - 4x}$

$8(9 - 4x) = 4(2x - 7)$

$72 - 32x = 8x - 28$

$-40x = -100$

$x = \dfrac{5}{2}$

135. $2x^2 - 21x + 49 = 0$

$(2x - 7)(x - 7) = 0$

$2x - 7 = 0$ or $x - 7 = 0$

$x = \dfrac{7}{2}$ or $x = 7$

136. $x^2 - 8x + 3 = 0$

$x = \dfrac{-b \pm \sqrt{b^2 - 4ac}}{2a}$

$= \dfrac{-(-8) \pm \sqrt{(-8)^2 - 4(1)(3)}}{2(1)}$

$= \dfrac{8 \pm \sqrt{52}}{2} = 4 \pm \sqrt{13}$

137. $\sqrt{x - 9} + 15 = 0$

$\sqrt{x - 9} = -15$

No real solution

The square root of $x - 9$ cannot be negative.

138. $3x - 16\sqrt{x} + 5 = 0$

$\left(3\sqrt{x} - 1\right)\left(\sqrt{x} - 5\right) = 0$

$3\sqrt{x} - 1 = 0 \Rightarrow x = \frac{1}{9}$

$\sqrt{x} - 5 = 0 \Rightarrow x = 25$

139. Answers will vary.

Section 1.4   Functions

- ■ Given a set or an equation, you should be able to determine if it represents a function.

- ■ Know that functions can be represented in four ways: verbally, numerically, graphically, and algebraically.

- ■ Given a function, you should be able to do the following.

  (a) Find the domain and range.

  (b) Evaluate it at specific values.

- ■ You should be able to use function notation.

Vocabulary Check

1. domain; range; function

2. verbally; numerically; graphically; algebraically

3. independent; dependent

4. piecewise-defined

5. implied domain

6. difference quotient

---

1. Yes, the relationship is a function. Each domain value is matched with only one range value.

2. No, it is not a function. The domain value of -1 is matched with two output values.

3. No, the relationship is not a function. The domain values are each matched with three range values.

4. Yes, it is a function. Each domain value is matched with only one range value.

5. Yes, it does represent a function. Each input value is matched with only one output value.

6. No, the table does not represent a function. The input values of 0 and 1 are each matched with two different output values.

7. No, it does not represent a function. The input values of 10 and 7 are each matched with two output values.

8. Yes, the table does represent a function. Each input value is matched with only one output value.

9. (a) Each element of A is matched with exactly one element of B, so it does represent a function.

   (b) The element 1 in A is matched with two elements, -2 and 1 of B, so it does not represent a function.

   (c) Each element of A is matched with exactly one element of B, so it does represent a function.

   (d) The element 2 in A is not matched with an element of B, so the relation does not represent a function.

10. (a) The element c in A is matched with two elements, 2 and 3 of B, so it is not a function.

   (b) Each element of A is matched with exactly one element of B, so it does represent a function.

   (c) This is not a function from A to B (it represents a function from B to A instead).

   (d) Each element of A is matched with exactly one element of B, so it does represent a function.

11. Each is a function. For each year there corresponds one and only one circulation.

12. Reading from the graph, $f(1998)$ is approximately 11 million.

13. $x^2 + y^2 = 4 \implies y = \pm\sqrt{4 - x^2}$

   No, y *is not* a function of x.

14. $x = y^2 \implies y = \pm\sqrt{x}$

   Thus, y *is not* a function of x.

15. $x^2 + y = 4 \implies y = 4 - x^2$

Yes, *y is* a function of *x*.

16. $x + y^2 = 4 \implies y = \pm\sqrt{4 - x}$

Thus, *y is not* a function of *x*.

17. $2x + 3y = 4 \implies y = \frac{1}{3}(4 - 2x)$

Yes, *y is* a function of *x*.

18. $(x - 2)^2 + y^2 = 4 \implies y = \pm\sqrt{4 - (x - 2)^2}$

Thus, *y is not* a function of *x*.

19. $y^2 = x^2 - 1 \implies y = \pm\sqrt{x^2 - 1}$

Thus, *y is not* a function of *x*.

20. $y = \sqrt{x + 5}$

Yes, *y is* a function of *x*.

21. $y = |4 - x|$

Yes, *y is* a function of *x*.

22. $|y| = 4 - x \implies y = 4 - x$ or $y = -(4 - x)$

Thus, *y is not* a function of *x*.

23. $x = 14$

Thus, this *is not* a function of *x*.

24. $y = -75$ or $y = -75 + 0x$

y is a function of *x*.

25. $f(x) = 2x - 3$

(a) $f(1) = 2(1) - 3 = -1$

(b) $f(-3) = 2(-3) - 3 = -9$

(c) $f(x - 1) = 2(x - 1) - 3 = 2x - 5$

26. $g(y) = 7 - 3y$

(a) $g(0) = 7 - 3(0) = 7$

(b) $g\left(\frac{7}{3}\right) = 7 - 3\left(\frac{7}{3}\right) = 0$

(c) $g(s + 2) = 7 - 3(s + 2)$

$\qquad = 7 - 3s - 6 = 1 - 3s$

27. $V(r) = \frac{4}{3}\pi r^3$

(a) $V(3) = \frac{4}{3}\pi(3)^3 = \frac{4}{3}\pi(27) = 36\pi$

(b) $V\left(\frac{3}{2}\right) = \frac{4}{3}\pi\left(\frac{3}{2}\right)^3 = \frac{4}{3}\pi\left(\frac{27}{8}\right) = \frac{9}{2}\pi$

(c) $V(2r) = \frac{4}{3}\pi(2r)^3 = \frac{4}{3}\pi(8r^3) = \frac{32}{3}\pi r^3$

28. $h(t) = t^2 - 2t$

(a) $h(2) = 2^2 - 2(2) = 0$

(b) $h(1.5) = (1.5)^2 - 2(1.5) = -0.75$

(c) $h(x + 2) = (x + 2)^2 - 2(x + 2) = x^2 + 2x$

29. $f(y) = 3 - \sqrt{y}$

(a) $f(4) = 3 - \sqrt{4} = 1$

(b) $f(0.25) = 3 - \sqrt{0.25} = 2.5$

(c) $f(4x^2) = 3 - \sqrt{4x^2} = 3 - 2|x|$

30. $f(x) = \sqrt{x + 8} + 2$

(a) $f(-8) = \sqrt{(-8) + 8} + 2 = 2$

(b) $f(1) = \sqrt{(1) + 8} + 2 = 5$

(c) $f(x - 8) = \sqrt{(x - 8) + 8} + 2 = \sqrt{x} + 2$

31. $q(x) = \dfrac{1}{x^2 - 9}$

(a) $q(0) = \dfrac{1}{0^2 - 9} = -\dfrac{1}{9}$

(b) $q(3) = \dfrac{1}{3^2 - 9}$ is undefined.

(c) $q(y + 3) = \dfrac{1}{(y + 3)^2 - 9} = \dfrac{1}{y^2 + 6y}$

32. $q(t) = \dfrac{2t^2 + 3}{t^2}$

(a) $q(2) = \dfrac{2(2)^2 + 3}{(2)^2} = \dfrac{8 + 3}{4} = \dfrac{11}{4}$

(b) $q(0) = \dfrac{2(0)^2 + 3}{(0)^2}$

Division by zero is undefined.

(c) $q(-x) = \dfrac{2(-x)^2 + 3}{(-x)^2} = \dfrac{2x^2 + 3}{x^2}$

33. $f(x) = \dfrac{|x|}{x}$

   (a) $f(2) = \dfrac{|2|}{2} = 1$

   (b) $f(-2) = \dfrac{|-2|}{-2} = -1$

   (c) $f(x - 1) = \dfrac{|x - 1|}{x - 1} = \begin{cases} -1 & \text{if } x < 1 \\ 1 & \text{if } x > 1 \end{cases}$

34. $f(x) = |x| + 4$

   (a) $f(2) = |2| + 4 = 6$

   (b) $f(-2) = |-2| + 4 = 6$

   (c) $f(x^2) = |x^2| + 4 = x^2 + 4$

35. $f(x) = \begin{cases} 2x + 1, & x < 0 \\ 2x + 2, & x \geq 0 \end{cases}$

   (a) $f(-1) = 2(-1) + 1 = -1$

   (b) $f(0) = 2(0) + 2 = 2$

   (c) $f(2) = 2(2) + 2 = 6$

36. $f(x) = \begin{cases} x^2 + 2, & x \leq 1 \\ 2x^2 + 2, & x > 1 \end{cases}$

   (a) $f(-2) = (-2)^2 + 2 = 6$

   (b) $f(1) = (1)^2 + 2 = 3$

   (c) $f(2) = 2(2)^2 + 2 = 10$

37. $f(x) = \begin{cases} 3x - 1, & x < -1 \\ 4, & -1 \leq x \leq 1 \\ x^2, & x > 1 \end{cases}$

   (a) $f(-2) = 3(-2) - 1 = -7$

   (b) $f\left(-\dfrac{1}{2}\right) = 4$

   (c) $f(3) = 3^2 = 9$

38. $f(x) = \begin{cases} 4 - 5x, & x \leq -2 \\ 0, & -2 < x \leq 2 \\ x^2 + 1 & x > 2 \end{cases}$

   (a) $f(-3) = 4 - 5(-3) = 19$

   (b) $f(4) = (4)^2 + 1 = 17$

   (c) $f(-1) = 0$

39. $f(x) = x^2 - 3$

$f(-2) = (-2)^2 - 3 = 1$

$f(-1) = (-1)^2 - 3 = -2$

$f(0) = (0)^2 - 3 = -3$

$f(1) = (1)^2 - 3 = -2$

$f(2) = (2)^2 - 3 = 1$

x	-2	-1	0	1	2
$f(x)$	1	-2	-3	-2	1

40. $g(x) = \sqrt{x - 3}$

$g(3) = \sqrt{3 - 3} = 0$

$g(4) = \sqrt{4 - 3} = 1$

$g(5) = \sqrt{5 - 3} = \sqrt{2}$

$g(6) = \sqrt{6 - 3} = \sqrt{3}$

$g(7) = \sqrt{7 - 3} = 2$

x	3	4	5	6	7
$g(x)$	0	1	$\sqrt{2}$	$\sqrt{3}$	2

41. $h(t) = \dfrac{1}{2}|t + 3|$

$h(-5) = \dfrac{1}{2}|-5 + 3| = 1$

$h(-4) = \dfrac{1}{2}|-4 + 3| = \dfrac{1}{2}$

$h(-3) = \dfrac{1}{2}|-3 + 3| = 0$

$h(-2) = \dfrac{1}{2}|-2 + 3| = \dfrac{1}{2}$

$h(-1) = \dfrac{1}{2}|-1 + 3| = 1$

t	-5	-4	-3	-2	-1
$h(t)$	1	$\dfrac{1}{2}$	0	$\dfrac{1}{2}$	1

42. $f(s) = \dfrac{|s - 2|}{s - 2}$

$f(0) = \dfrac{|0 - 2|}{0 - 2} = \dfrac{2}{-2} = -1$

$f(1) = \dfrac{|1 - 2|}{1 - 2} = \dfrac{1}{-1} = -1$

$f\left(\dfrac{3}{2}\right) = \dfrac{|(3/2) - 2|}{(3/2) - 2} = \dfrac{1/2}{-1/2} = -1$

$f\left(\dfrac{5}{2}\right) = \dfrac{|(5/2) - 2|}{(5/2) - 2} = \dfrac{1/2}{1/2} = 1$

$f(4) = \dfrac{|4 - 2|}{4 - 2} = \dfrac{2}{2} = 1$

s	0	1	$\frac{3}{2}$	$\frac{5}{2}$	4
$f(s)$	-1	-1	-1	1	1

43. $f(x) = \begin{cases} -\frac{1}{2}x + 4, & x \leq 0 \\ (x - 2)^2, & x > 0 \end{cases}$

x	-2	-1	0	1	2
$f(x)$	5	$\frac{9}{2}$	4	1	0

$f(-2) = -\dfrac{1}{2}(-2) + 4 = 5$

$f(-1) = -\dfrac{1}{2}(-1) + 4 = 4\frac{1}{2} = \dfrac{9}{2}$

$f(0) = -\dfrac{1}{2}(0) + 4 = 4$

$f(1) = (1 - 2)^2 = 1$

$f(2) = (2 - 2)^2 = 0$

44. $f(x) = \begin{cases} 9 - x^2, & x < 3 \\ x - 3, & x \ge 3 \end{cases}$

$f(1) = 9 - (1)^2 = 8$

$f(2) = 9 - (2)^2 = 5$

$f(3) = (3) - 3 = 0$

$f(4) = (4) - 3 = 1$

$f(5) = (5) - 3 = 2$

x	1	2	3	4	5
$f(x)$	8	5	0	1	2

45. $15 - 3x = 0$

$3x = 15$

$x = 5$

46. $f(x) = 5x + 1$

$5x + 1 = 0$

$x = -\dfrac{1}{5}$

47. $\dfrac{3x - 4}{5} = 0$

$3x - 4 = 0$

$x = \dfrac{4}{3}$

48. $f(x) = \dfrac{12 - x^2}{5}$

$\dfrac{12 - x^2}{5} = 0$

$x^2 = 12$

$x = \pm\sqrt{12} = \pm 2\sqrt{3}$

49. $x^2 - 9 = 0$

$x^2 = 9$

$x = \pm 3$

50. $f(x) = x^2 - 8x + 15$

$x^2 - 8x + 15 = 0$

$(x - 5)(x - 3) = 0$

$x - 5 = 0 \Rightarrow x = 5$

$x - 3 = 0 \Rightarrow x = 3$

51. $x^3 - x = 0$

$x(x^2 - 1) = 0$

$x(x + 1)(x - 1) = 0$

$x = 0, \; x = -1, \text{ or } x = 1$

52. $f(x) = x^3 - x^2 - 4x + 4$

$x^3 - x^2 - 4x + 4 = 0$

$x^2(x - 1) - 4(x - 1) = 0$

$(x - 1)(x^2 - 4) = 0$

$x - 1 = 0 \Rightarrow x = 1$

$x^2 - 4 = 0 \Rightarrow x = \pm 2$

53. $f(x) = g(x)$

$x^2 + 2x + 1 = 3x + 3$

$x^2 - x - 2 = 0$

$(x + 1)(x - 2) = 0$

$x = -1 \text{ or } x = 2$

54. $f(x) = g(x)$

$x^4 - 2x^2 = 2x^2$

$x^4 - 4x^2 = 0$

$x^2(x^2 - 4) = 0$

$x^2(x + 2)(x - 2) = 0$

$x^2 = 0 \Rightarrow x = 0$

$x + 2 = 0 \Rightarrow x = -2$

$x - 2 = 0 \Rightarrow x = 2$

55. $f(x) = g(x)$

$\sqrt{3x} + 1 = x + 1$

$\sqrt{3x} = x$

$3x = x^2$

$0 = x^2 - 3x$

$0 = x(x - 3)$

$x = 0 \text{ or } x = 3$

56. $f(x) = g(x)$

$\sqrt{x} - 4 = 2 - x$

$x + \sqrt{x} - 6 = 0$

$(\sqrt{x} + 3)(\sqrt{x} - 2) = 0$

$\sqrt{x} + 3 = 0 \Rightarrow \sqrt{x} = -3,$ which is a contradiction, since $\sqrt{x}$ represents the principal square root.

$\sqrt{x} - 2 = 0 \Rightarrow \sqrt{x} = 2 \Rightarrow x = 4$

57. $f(x) = 5x^2 + 2x - 1$

Since $f(x)$ is a polynomial, the domain is all real numbers x.

58. $f(x) = 1 - 2x^2$

Because $f(x)$ is a polynomial, the domain is all real numbers x.

59. $h(t) = \dfrac{4}{t}$

Domain: All real numbers t except $t = 0$

60. $s(y) = \dfrac{3y}{y + 5}$

$y + 5 \neq 0$

$y \neq -5$

The domain is all real numbers y except $y = -5$.

61. $g(y) = \sqrt{y - 10}$

Domain: $y - 10 \geq 0$

$y \geq 10$

62. $f(t) = \sqrt[3]{t + 4}$

Because $f(t)$ is a cube root, the domain is all real numbers t.

63. $f(x) = \sqrt[4]{1 - x^2}$

Domain: $1 - x^2 \geq 0$

By solving this inequality, we conclude that $-1 \leq x \leq 1$ or $[-1, 1]$.

64. $f(x) = \sqrt[4]{x^2 + 3x}$

$x^2 + 3x \geq 0$

$x(x + 3) \geq 0$

By solving this inequality, we conclude that $x \leq -3$ or $x \geq 0$ or $(-\infty, -3] \cup [0, \infty)$.

65. $g(x) = \dfrac{1}{x} - \dfrac{3}{x + 2}$

Domain: All real numbers x except $x = 0$, $x = -2$

66. $h(x) = \dfrac{10}{x^2 - 2x}$

$x^2 - 2x \neq 0$

$x(x - 2) \neq 0$

The domain is all real numbers x except $x = 0$, $x = 2$

67. $f(s) = \dfrac{\sqrt{s - 1}}{s - 4}$

Domain: $s - 1 \geq 0 \Rightarrow s \geq 1$ and $s \neq 4$

The domain consists of all real numbers s, such that $s \geq 1$ and $s \neq 4$.

68. $f(x) = \dfrac{\sqrt{x + 6}}{6 + x}$

Domain: $x + 6 \geq 0 \Rightarrow x \geq -6$ and $x \neq -6$

The domain is all real numbers x such that $x > -6$ or $(-6, \infty)$.

69. $f(x) = \dfrac{x - 4}{\sqrt{x}}$

The domain is all real numbers such that $x > 0$ or $(0, \infty)$.

70. $f(x) = \dfrac{x - 5}{\sqrt{x^2 - 9}}$

$x^2 - 9 > 0$

$(x + 3)(x - 3) > 0$

Test intervals: $(-\infty, -3), (-3, 3), (3, \infty)$

The domain is all real numbers $x < -3$ or $x > 3$ or $(-\infty, -3) \cup (3, \infty)$.

71. $f(x) = x^2$

$\{(-2, 4), (-1, 1), (0, 0), (1, 1), (2, 4)\}$

72. $f(x) = x^2 - 3$

$f(-2) = (-2)^2 - 3 = 1$

$f(-1) = (-1)^2 - 3 = -2$

$f(0) = (0)^2 - 3 = -3$

$f(1) = (1)^2 - 3 = -2$

$f(2) = (2)^2 - 3 = 1$

$\{(-2, 1), (-1, -2), (0, -3), (1, -2), (2, 1)\}$

73. $f(x) = |x| + 2$

$\{(-2, 4), (-1, 3), (0, 2), (1, 3), (2, 4)\}$

74. $f(x) = |x + 1|$

$\{(-2, 1), (-1, 0), (0, 1), (1, 2), (2, 3)\}$

75. By plotting the points, we have a parabola, so $g(x) = cx^2$. Since $(-4, -32)$ is on the graph, we have $-32 = c(-4)^2 \implies c = -2$. Thus, $g(x) = -2x^2$.

76. By plotting the data, you can see that they represent a line, or $f(x) = cx$. Because $(0, 0)$ and $\left(1, \frac{1}{4}\right)$ are on the line, the slope is $\frac{1}{4}$. Thus, $f(x) = \frac{1}{4}x$.

77. Since the function is undefined at 0, we have $r(x) = c/x$. Since $(-4, -8)$ is on the graph, we have $-8 = c/-4 \implies c = 32$. Thus, $r(x) = 32/x$.

78. By plotting the data, you can see that they represent $h(x) = c\sqrt{|x|}$. Because $\sqrt{|-4|} = 2$ and $\sqrt{|-1|} = 1$, and the corresponding y-values are 6 and 3, $c = 3$ and $h(x) = 3\sqrt{|x|}$.

79.
$$f(x) = x^2 - x + 1$$
$$f(2 + h) = (2 + h)^2 - (2 + h) + 1$$
$$= 4 + 4h + h^2 - 2 - h + 1$$
$$= h^2 + 3h + 3$$
$$f(2) = (2)^2 - 2 + 1 = 3$$
$$f(2 + h) - f(2) = h^2 + 3h$$
$$\frac{f(2 + h) - f(2)}{h} = \frac{h^2 + 3h}{h} = h + 3, \ h \neq 0$$

80.
$$f(x) = 5x - x^2$$
$$f(5 + h) = 5(5 + h) - (5 + h)^2$$
$$= 25 + 5h - (25 + 10h + h^2)$$
$$= 25 + 5h - 25 - 10h - h^2$$
$$= -h^2 - 5h$$
$$f(5) = 5(5) - (5)^2$$
$$= 25 - 25 = 0$$
$$\frac{f(5 + h) - f(5)}{h} = \frac{-h^2 - 5h}{h}$$
$$= \frac{-h(h + 5)}{h} = -(h + 5), h \neq 0$$

81.
$$f(x) = x^3 + 3x$$
$$f(x + h) = (x + h)^3 + 3(x + h)$$
$$= x^3 + 3x^2h + 3xh^2 + h^3 + 3x + 3h$$
$$\frac{f(x + h) - f(x)}{h} = \frac{(x^3 + 3x^2h + 3xh^2 + h^3 + 3x + 3h) - (x^3 + 3x)}{h}$$
$$= \frac{h(3x^2 + 3xh + h^2 + 3)}{h}$$
$$= 3x^2 + 3xh + h^2 + 3, \ h \neq 0$$

82.
$$f(x) = 4x^2 - 2x$$
$$f(x + h) = 4(x + h)^2 - 2(x + h)$$
$$= 4(x^2 + 2xh + h^2) - 2x - 2h$$
$$= 4x^2 + 8xh + 4h^2 - 2x - 2h$$
$$\frac{f(x + h) - f(x)}{h} = \frac{4x^2 + 8xh + 4h^2 - 2x - 2h - 4x^2 + 2x}{h}$$
$$= \frac{8xh + 4h^2 - 2h}{h}$$
$$= \frac{h(8x + 4h - 2)}{h}$$
$$= 8x + 4h - 2, \ h \neq 0$$

83. $g(x) = \dfrac{1}{x^2}$

$$\frac{g(x) - g(3)}{x - 3} = \frac{\dfrac{1}{x^2} - \dfrac{1}{9}}{x - 3}$$
$$= \frac{9 - x^2}{9x^2(x - 3)}$$
$$= \frac{-(x + 3)(x - 3)}{9x^2(x - 3)}$$
$$= -\frac{x + 3}{9x^2}, \ x \neq 3$$

84.
$$f(t) = \frac{1}{t-2}$$

$$f(1) = \frac{1}{1-2} = -1$$

$$\frac{f(t)-f(1)}{t-1} = \frac{\frac{1}{t-2}-(-1)}{t-1} = \frac{1+(t-2)}{(t-2)(t-1)} = \frac{(t-1)}{(t-2)(t-1)} = \frac{1}{t-2}, \quad t \neq 1$$

85. $f(x) = \sqrt{5x}$

$$\frac{f(x)-f(5)}{x-5} = \frac{\sqrt{5x}-5}{x-5}$$

86.
$$f(x) = x^{2/3} + 1$$

$$f(8) = 8^{2/3} + 1 = 5$$

$$\frac{f(x)-f(8)}{x-8} = \frac{x^{2/3}+1-5}{x-8} = \frac{x^{2/3}-4}{x-8}$$

87. $A = s^2$ and $P = 4s \Rightarrow \dfrac{P}{4} = s$

$$A = \left(\frac{P}{4}\right)^2 = \frac{P^2}{16}$$

88. $A = \pi r^2, \ C = 2\pi r$

$$r = \frac{C}{2\pi}$$

$$A = \pi\left(\frac{C}{2\pi}\right)^2 = \frac{C^2}{4\pi}$$

89. (a)

Height, x	Volume, V
1	484
2	800
3	972
4	1024
5	980
6	864

The volume is maximum when $x = 4$ and $V = 1024$ cubic centimeters.

(b)

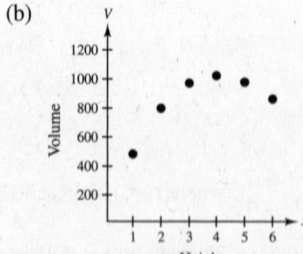

V is a function of x.

(c) $V = x(24 - 2x)^2$

Domain: $0 < x < 12$

90. (a) The maximum profit is $3375.

(b)

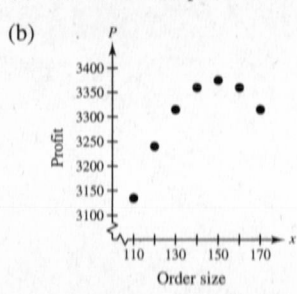

Yes, P is a function of x.

(c) Profit = Revenue − Cost

= (price per unit)(number of units) − (cost)(number of units)

= $[90 - (x - 100)(0.15)]x - 60x, \ x > 100$

= $(90 - 0.15x + 15)x - 60x$

= $(105 - 0.15x)x - 60x$

= $105x - 0.15x^2 - 60x$

= $45x - 0.15x^2, \ x > 100$

91. $A = \frac{1}{2}bh = \frac{1}{2}xy$

Since $(0, y)$, $(2, 1)$, and $(x, 0)$ all lie on the same line, the slopes between any pair are equal.

$$\frac{1-y}{2-0} = \frac{0-1}{x-2}$$

$$\frac{1-y}{2} = \frac{-1}{x-2}$$

$$y = \frac{2}{x-2} + 1$$

$$y = \frac{x}{x-2}$$

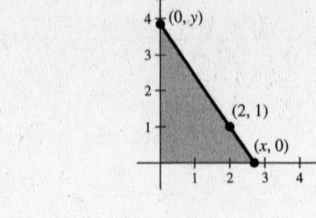

Therefore,

$$A = \frac{1}{2}x\left(\frac{x}{x-2}\right) = \frac{x^2}{2(x-2)}.$$

The domain of A includes x-values such that $x^2/[2(x-2)] > 0$. By solving this inequality, we find that the domain is $x > 2$.

92. $A = l \cdot w = (2x)y = 2xy$

But $y = \sqrt{36 - x^2}$, so $A = 2x\sqrt{36 - x^2}$, $0 < x < 6$.

93. $y = -\frac{1}{10}x^2 + 3x + 6$

$y(30) = -\frac{1}{10}(30)^2 + 3(30) + 6 = 6$ feet

If the child holds a glove at a height of 5 feet, then the ball *will* be over the child's head since it will be at a height of 6 feet.

94. $d(t) = \begin{cases} 5.0t + 37, & 0 \le t \le 7 \\ 18.7t - 64, & 0 \le t \le 12 \end{cases}$ where $t = 1$ represents 1991.

1991: $t = 1$ and $d(1) = 5.0(1) + 37 = 42$ billion dollars $= \$42,000,000,000$

1992: $t = 2$ and $d(2) = 5.0(2) + 37 = 47$ billion dollars $= \$47,000,000,000$

1993: $t = 3$ and $d(3) = 5.0(3) + 37 = 52$ billion dollars $= \$52,000,000,000$

1994: $t = 4$ and $d(4) = 5.0(4) + 37 = 57$ billion dollars $= \$57,000,000,000$

1995: $t = 5$ and $d(5) = 5.0(5) + 37 = 62$ billion dollars $= \$62,000,000,000$

1996: $t = 6$ and $d(6) = 5.0(6) + 37 = 67$ billion dollars $= \$67,000,000,000$

1997: $t = 7$ and $d(7) = 5.0(7) + 37 = 72$ billion dollars $= \$72,000,000,000$

1998: $t = 8$ and $d(8) = 18.7(8) - 64 = 85.6$ billion dollars $= \$85,600,000,000$

1999: $t = 9$ and $d(9) = 18.7(9) - 64 = 104.3$ billion dollars $= \$104,300,000,000$

2000: $t = 10$ and $d(10) = 18.7(10) - 64 = 123$ billion dollars $= \$123,000,000,000$

2001: $t = 11$ and $d(11) = 18.7(11) - 64 = 141.7$ billion dollars $= \$141,700,000,000$

2002: $t = 12$ and $d(12) = 18.7(12) - 64 = 160.4$ billion dollars $= \$160,400,000,000$

95. $p(t) = \begin{cases} 0.182t^2 + 0.57t + 27.3, & 0 \le t \le 7 \\ 2.50t + 21.3, & 8 \le t \le 12 \end{cases}$

Year	Function Value	Price
1990	$p(0) = 27.3$	$27,300
1991	$p(1) = 28.052$	$28,052
1992	$p(2) = 29.168$	$29,168
1993	$p(3) = 30.648$	$30,648
1994	$p(4) = 32.492$	$32,492
1995	$p(5) = 34.7$	$34,700
1996	$p(6) = 37.272$	$37,272
1997	$p(7) = 40.208$	$40,208
1998	$p(8) = 41.3$	$41,300
1999	$p(9) = 43.8$	$43,800
2000	$p(10) = 46.3$	$46,300
2001	$p(11) = 48.8$	$48,800
2002	$p(12) = 51.3$	$51,300

96. (a) $V = l \cdot w \cdot h = x \cdot y \cdot x = x^2 y$ where $4x + y = 108$.
Thus, $y = 108 - 4x$ and

$$V = x^2(108 - 4x) = 108x^2 - 4x^3.$$

Domain: $0 < x < 27$

(b)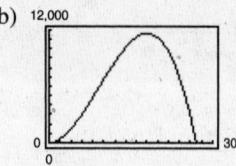

(c) The dimensions that will maximize the volume of the package are $18 \times 18 \times 36$. From the graph, the maximum volume occurs when $x = 18$. To find the dimension for y, use the equation $y = 108 - 4x$.

$$y = 108 - 4x = 108 - 4(18) = 108 - 72 = 36$$

97. (a) Cost = variable costs + fixed costs

$$C = 12.30x + 98,000$$

(b) Revenue = price per unit $\times$ number of units

$$R = 17.98x$$

(c) Profit = Revenue $-$ Cost

$$P = 17.98x - (12.30x + 98,000)$$

$$P = 5.68x - 98,000$$

98. (a) *Model:* (Total cost) = (Fixed costs) + (Variable costs)

Labels: Total cost = C

Fixed cost = 6000

Variable costs = $0.95x$

Equation: $C = 6000 + 0.95x$

(b) $\overline{C} = \dfrac{C}{x} = \dfrac{6000 + 0.95x}{x}$

99. (a) $R = n(\text{rate}) = n[8.00 - 0.05(n - 80)], \ n \ge 80$

$$R = 12.00n - 0.05n^2 = 12n - \frac{n^2}{20} = \frac{240n - n^2}{20}, \ n \ge 80$$

(b)

n	90	100	110	120	130	140	150
$R(n)$	$675	$700	$715	$720	$715	$700	$675

The revenue is maximum when 120 people take the trip.

100. $F(y) = 149.76\sqrt{10}\, y^{5/2}$

(a)

y	5	10	20	30	40
$F(y)$	26,474.08	149,760.00	847,170.49	2,334,527.36	4,792,320

The force, in tons, of the water against the dam increases with the depth of the water.

(b) It appears that approximately 21 feet of water would produce 1,000,000 tons of force.

(c) $1,000,000 = 149.76\sqrt{10}\, y^{5/2}$

$$\frac{1,000,000}{149.76\sqrt{10}} = y^{5/2}$$

$$2111.56 \approx y^{5/2}$$

$$21.37 \text{ feet} \approx y$$

101. (a)

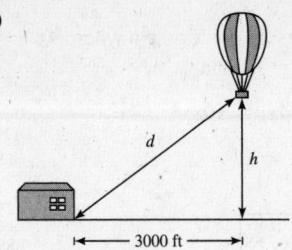

- 3000 ft -

(b) $(3000)^2 + h^2 = d^2$

$$h = \sqrt{d^2 - (3000)^2}$$

Domain: $d \geq 3000$ (since both $d \geq 0$ and $d^2 - (3000)^2 \geq 0$)

102. (a) $\dfrac{f(2003) - f(1996)}{2003 - 1996} = \dfrac{126 - 116}{7} = \dfrac{10}{7} = 1.428$

The number of threatened and endangered fish species increased, on average, by 1.428 per year from 1996 to 2003.

(b) $y = \begin{cases} 2x + 104, & 6 \leq x \leq 7 \\ 2x + 103, & 8 \leq x \leq 11 \\ 126, & 12 \leq x \leq 13 \end{cases}$

(d) The algebraic model is an excellent fit to the actual data.

(e) The calculator model is

$y \approx 1.55x + 107$.

It also gives a good fit, but not as good as the algebraic model.

(c)

Year $6 \leftrightarrow 1996$	Actual Number of Fish Species	Number from the Algebraic Model	Number from the Calculator Model
6	116	116	116
7	118	118	118
8	119	119	119
9	121	121	121
10	123	123	122
11	125	125	124
12	126	126	126
13	126	126	127

103. False. The range is $[-1, \infty)$.

104. True. The set represents a function. Each x-value is mapped to exactly one y-value.

105. The domain is the set of inputs of the function, and the range is the set of outputs.

106. Since $f(x)$ is a function of an even root, the radicand cannot be negative. $g(x)$ is an odd root, therefore the radicand can be any real number. Therefore, the domain of f is all real numbers x and the domain of g is all real numbers x such that $x \geq 2$.

107. (a) Yes. The amount that you pay in sales tax will increase as the price of the item purchased increases.

(b) No. The length of time that you study the night before an exam does not necessarily determine your score on the exam.

108. (a) No. During the course of a year, for example, your salary may remain constant while your savings account balance may vary. That is, there may be two or more outputs (savings account balances) for one input (salary).

(b) Yes. The greater the height from which the ball is dropped, the greater the speed with which the ball will strike the ground.

109. $\dfrac{t}{3} + \dfrac{t}{5} = 1$

$15\left(\dfrac{t}{3} + \dfrac{t}{5}\right) = 15(1)$

$5t + 3t = 15$

$8t = 15$

$t = \dfrac{15}{8}$

110. $\dfrac{3}{t} + \dfrac{5}{t} = 1$

$\dfrac{8}{t} = 1$

$8 = t$

111.
$$\frac{3}{x(x + 1)} - \frac{4}{x} = \frac{1}{x + 1}$$

$$x(x + 1)\left[\frac{3}{x(x + 1)} - \frac{4}{x}\right] = x(x + 1)\left(\frac{1}{x + 1}\right)$$

$$3 - 4(x + 1) = x$$

$$3 - 4x - 4 = x$$

$$-1 = 5x$$

$$-\frac{1}{5} = x$$

112. $\dfrac{12}{x} - 3 = \dfrac{4}{x} + 9$

$$\frac{12}{x} - \frac{4}{x} = 9 + 3$$

$$\frac{8}{x} = 12$$

$$\frac{8}{12} = x$$

$$x = \frac{2}{3}$$

113. $(-2, -5)$ and $(4, -1)$

$$m = \frac{-1 - (-5)}{4 - (-2)} = \frac{4}{6} = \frac{2}{3}$$

$$y - (-5) = \frac{2}{3}(x - (-2))$$

$$y + 5 = \frac{2}{3}x + \frac{4}{3}$$

$$3y + 15 = 2x + 4$$

$$2x - 3y - 11 = 0$$

114.   Slope $= \dfrac{9 - 0}{1 - 10} = \dfrac{9}{-9} = -1$

$$m = -1$$

$$y - 0 = (-1)(x - 10)$$

$$y = -x + 10$$

$$x + y - 10 = 0$$

115. $(-6, 5)$ and $(3, -5)$

$$m = \frac{-5 - 5}{3 - (-6)} = -\frac{10}{9}$$

$$y - 5 = -\frac{10}{9}(x - (-6))$$

$$9y - 45 = -10x - 60$$

$$10x + 9y + 15 = 0$$

116.   Slope $= \dfrac{-(1/3) - 3}{(11/2) - (-1/2)}$

$$= \frac{-10/3}{12/2} = -\frac{10}{3} \cdot \frac{1}{6} = -\frac{5}{9}$$

$$m = -\frac{5}{9}$$

$$y - 3 = -\frac{5}{9}\left(x - \left(-\frac{1}{2}\right)\right)$$

$$y - 3 = -\frac{5}{9}x - \frac{5}{18}$$

$$18y - 54 = -10x - 5$$

$$10x + 18y - 49 = 0$$

Section 1.5   Analyzing Graphs of Functions

- ■ You should be able to determine the domain and range of a function from its graph.

- ■ You should be able to use the vertical line test for functions.

- ■ You should be able to find the zeros of a function.

- ■ You should be able to determine when a function is constant, increasing, or decreasing.

- ■ You should be able to approximate relative minimums and relative maximums from the graph of a function.

- ■ You should know that f is

    (a) odd if $f(-x) = -f(x)$.   (b) even if $f(-x) = f(x)$.

Vocabulary Check

1. ordered pairs

2. vertical line test

3. zeros

4. decreasing

5. maximum

6. average rate of change; secant

7. odd

8. even

---

1. Domain: $(-\infty, -1] \cup [1, \infty)$

Range: $[0, \infty)$

2. Domain: $(-\infty, \infty)$

Range: $[0, \infty)$

3. Domain: $[-4, 4]$

Range: $[0, 4]$

4. Domain: $(-\infty, 1), (1, \infty)$

Range: $-1, 1$

5. (a) $f(-2) = 0$ (b) $f(-1) = -1$

(c) $f\left(\frac{1}{2}\right) = 0$ (d) $f(1) = -3$

6. (a) $f(-1) = 4$ (b) $f(2) = 4$

(c) $f(0) = 2$ (d) $f(1) = 0$

7. (a) $f(-2) = -3$ (b) $f(1) = 0$

(c) $f(0) = 1$ (d) $f(2) = -3$

8. (a) $f(2) = 0$ (b) $f(1) = 1$

(c) $f(3) = 2$ (d) $f(-1) = 3$

9. $y = \frac{1}{2}x^2$

A vertical line intersects the graph just once, so *y is* a function of *x*.

10. $y = \frac{1}{4}x^3$

A vertical line intersects the graph no more than once, so *y is* a function of *x*.

11. $x - y^2 = 1 \implies y = \pm\sqrt{x - 1}$

y is not a function of *x*. Some vertical lines cross the graph twice.

12. $x^2 + y^2 = 25$

A vertical line intersects the graph more than once, so *y is not* a function of *x*.

13. $x^2 = 2xy - 1$

A vertical line intersects the graph just once, so *y is* a function of *x*.

14. $x = |y + 2|$

A vertical line intersects the graph more than once, so *y is not* a function of *x*.

15. $2x^2 - 7x - 30 = 0$

$(2x + 5)(x - 6) = 0$

$2x + 5 = 0$ or $x - 6 = 0$

$x = -\frac{5}{2}$ or $x = 6$

16. $f(x) = 3x^2 + 22x - 16$

$0 = (3x - 2)(x + 8)$

$3x - 2 = 0 \implies x = \frac{2}{3}$

$x + 8 = 0 \implies x = -8$

17. $\dfrac{x}{9x^2 - 4} = 0$

$x = 0$

18. $f(x) = \dfrac{x^2 - 9x + 14}{4x}$

$0 = \dfrac{x^2 - 9x + 14}{4x}$

$0 = (x - 7)(x - 2)$

$x - 7 = 0 \implies x = 7$

$x - 2 = 0 \implies x = 2$

19. $\dfrac{1}{2}x^3 - x = 0$

$x^3 - 2x = 2(0)$

$x(x^2 - 2) = 0$

$x = 0$ or $x^2 - 2 = 0$

$x^2 = 2$

$x = \pm\sqrt{2}$

20. $f(x) = x^3 - 4x^2 - 9x + 36$

$0 = x^3 - 4x^2 - 9x + 36$

$0 = x^2(x - 4) - 9(x - 4)$

$0 = (x - 4)(x^2 - 9)$

$x - 4 = 0 \implies x = 4$

$x^2 - 9 = 0 \implies x = \pm 3$

21. $4x^3 - 24x^2 - x + 6 = 0$

$4x^2(x - 6) - 1(x - 6) = 0$

$(x - 6)(4x^2 - 1) = 0$

$(x - 6)(2x + 1)(2x - 1) = 0$

$x - 6 = 0, \quad 2x + 1 = 0, \quad 2x - 1 = 0$

$x = 6, \qquad x = -\frac{1}{2}, \qquad x = \frac{1}{2}$

22. $f(x) = 9x^4 - 25x^2$

$0 = 9x^4 - 25x^2$

$0 = x^2(9x^2 - 25)$

$x^2 = 0 \implies x = 0$

$9x^2 - 25 = 0 \implies x = \pm\frac{5}{3}$

23. $\sqrt{2x} - 1 = 0$

$\sqrt{2x} = 1$

$2x = 1$

$x = \frac{1}{2}$

24. $f(x) = \sqrt{3x + 2}$

$0 = \sqrt{3x + 2}$

$0 = 3x + 2$

$-\frac{2}{3} = x$

25. (a)

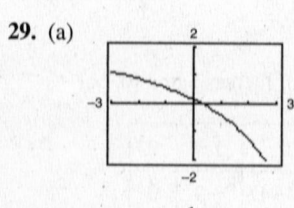

Zero: $x = -\frac{5}{3}$

(b) $3 + \frac{5}{x} = 0$

$3x + 5 = 0$

$x = -\frac{5}{3}$

26. (a)

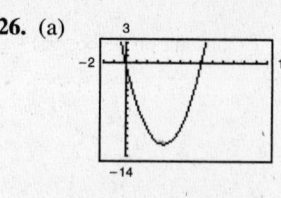

Zeros: $x = 0, x = 7$

(b) $f(x) = x(x - 7)$

$0 = x(x - 7)$

$x = 0$

$x - 7 = 0 \Rightarrow x = 7$

27. (a)

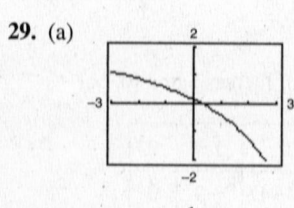

Zero: $x = -\frac{11}{2}$

(b) $\sqrt{2x + 11} = 0$

$2x + 11 = 0$

$x = -\frac{11}{2}$

28. (a)

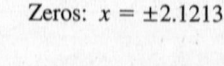

Zero: $x = 26$

(b) $f(x) = \sqrt{3x - 14} - 8$

$0 = \sqrt{3x - 14} - 8$

$8 = \sqrt{3x - 14}$

$64 = 3x - 14$

$x = 26$

29. (a)

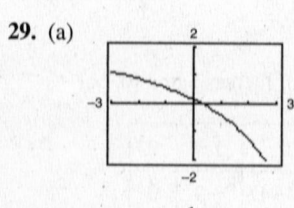

Zero: $x = \frac{1}{3}$

(b) $\frac{3x - 1}{x - 6} = 0$

$3x - 1 = 0$

$x = \frac{1}{3}$

30. (a)

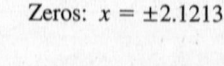

Zeros: $x = \pm 2.1213$

(b) $f(x) = \frac{2x^2 - 9}{3 - x}$

$0 = \frac{2x^2 - 9}{3 - x}$

$2x^2 - 9 = 0 \Rightarrow x = \pm\frac{3\sqrt{2}}{2} = \pm 2.1213$

31. $f(x) = \frac{3}{2}x$

f is increasing on $(-\infty, \infty)$.

32. $f(x) = x^2 - 4x$

The graph is decreasing on $(-\infty, 2)$ and increasing on $(2, \infty)$.

33. $f(x) = x^3 - 3x^2 + 2$

f is increasing on $(-\infty, 0)$ and $(2, \infty)$.

f is decreasing on $(0, 2)$.

34. $f(x) = \sqrt{x^2 - 1}$

The graph is decreasing on $(-\infty, -1)$ and increasing on $(1, \infty)$.

35. $f(x) = \begin{cases} x + 3, & x \le 0 \\ 3, & 0 < x \le 2 \\ 2x + 1, & x > 2 \end{cases}$

f is increasing on $(-\infty, 0)$ and $(2, \infty)$.

f is constant on $(0, 2)$.

36. $f(x) = \begin{cases} 2x + 1, & x \le -1 \\ x^2 - 2, & x > -1 \end{cases}$

The graph is decreasing on $(-1, 0)$ and increasing on $(-\infty, -1)$ and $(0, \infty)$.

37. $f(x) = |x + 1| + |x - 1|$

  f is increasing on $(1, \infty)$.

  f is constant on $(-1, 1)$.

  f is decreasing on $(-\infty, -1)$.

38. The graph is decreasing on $(-2, -1)$ and $(-1, 0)$ and increasing on $(-\infty, -2)$ and $(0, \infty)$.

39. $f(x) = 3$

(a) 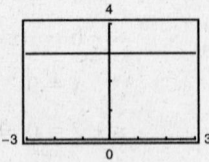   Constant on $(-\infty, \infty)$

(b)

x	-2	-1	0	1	2
$f(x)$	3	3	3	3	3

40. $g(x) = x$

(a) 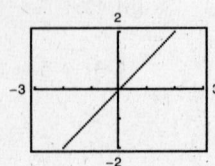   Increasing on $(-\infty, \infty)$

(b)

x	-2	-1	0	1	2
$g(x)$	-2	-1	0	1	2

41. $g(s) = \dfrac{s^2}{4}$

(a)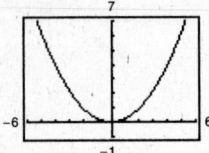

Decreasing on $(-\infty, 0)$; Increasing on $(0, \infty)$

(b)

s	-4	-2	0	2	4
$g(s)$	4	1	0	1	4

42. $h(x) = x^2 - 4$

(a)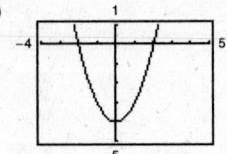

Decreasing on $(-\infty, 0)$; Increasing on $(0, \infty)$

(b)

x	-2	-1	0	1	2
$h(x)$	0	-3	-4	-3	0

43. $f(t) = -t^4$

(a)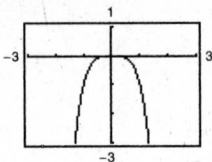

Increasing on $(-\infty, 0)$; Decreasing on $(0, \infty)$

(b)

t	-2	-1	0	1	2
$f(t)$	-16	-1	0	-1	-16

44. $f(x) = 3x^4 - 6x^2$

(a)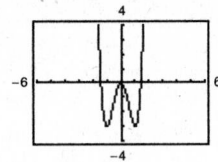

Increasing on $(-1, 0)$, $(1, \infty)$; Decreasing on $(-\infty, -1)$, $(0, 1)$

(b)

x	-2	-1	0	1	2
$f(x)$	24	-3	0	-3	24

45. $f(x) = \sqrt{1 - x}$

(a) 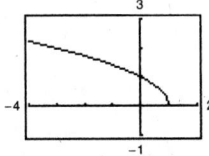   Decreasing on $(-\infty, 1)$

(b)

x	-3	-2	-1	0	1
$f(x)$	2	$\sqrt{3}$	$\sqrt{2}$	1	0

46. $f(x) = x\sqrt{x + 3}$

(a)

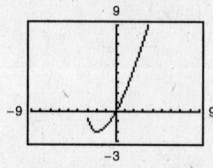

Increasing on $(-2, \infty)$; Decreasing on $(-3, -2)$

(b)

x	-3	-2	-1	0	1
$f(x)$	0	-2	-1.414	0	2

47. $f(x) = x^{3/2}$

(a)

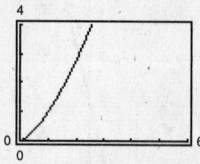

Increasing on $(0, \infty)$

(b)

x	0	1	2	3	4
$f(x)$	0	1	2.8	5.2	8

48. $f(x) = x^{2/3}$

(a)

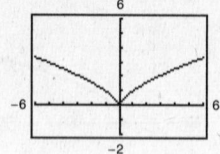

(b)

x	-2	-1	0	1	2
$f(x)$	1.59	1	0	1	1.59

Decreasing on $(-\infty, 0)$; Increasing on $(0, \infty)$

49. $f(x) = (x - 4)(x + 2)$

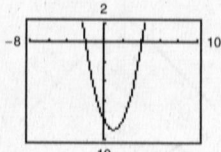

Relative minimum: $(1, -9)$

50. $f(x) = 3x^2 - 2x - 5$

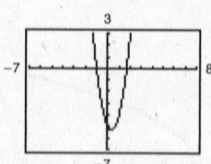

Relative minimum: $\left(\frac{1}{3}, -\frac{16}{3}\right)$ or $(0.33, -5.33)$

51. $f(x) = -x^2 + 3x - 2$

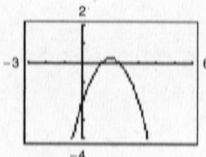

Relative maximum: $(1.5, 0.25)$

52. $f(x) = -2x^2 + 9x$

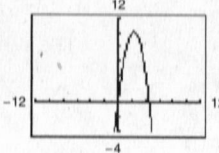

Relative maximum: $(2.25, 10.125)$

53. $f(x) = x(x - 2)(x + 3)$

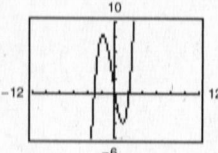

Relative minimum: $(1.12, -4.06)$

Relative maximum: $(-1.79, 8.21)$

54. $f(x) = x^3 - 3x^2 - x + 1$

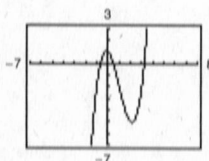

Relative maximum: $(-0.15, 1.08)$

Relative minimum: $(2.15, -5.08)$

55. $f(x) = 4 - x$

$f(x) \geq 0$ on $(-\infty, 4]$.

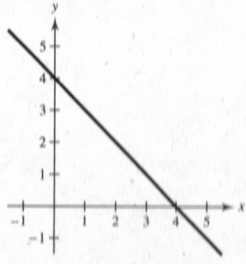

56. $f(x) = 4x + 2$

$f(x) \geq 0$

$4x + 2 \geq 0$

$4x \geq -2$

$x \geq -\frac{1}{2}$

$\left[-\frac{1}{2}, \infty\right)$

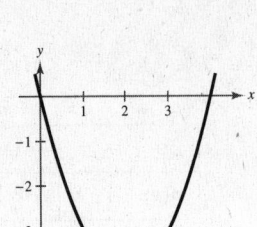

57. $f(x) = x^2 + x$

$f(x) \geq 0$ on $(-\infty, -1]$ and $[0, \infty)$.

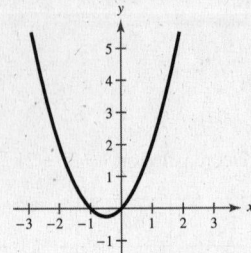

58. $f(x) = x^2 - 4x$

$f(x) \geq 0$

$x^2 - 4x \geq 0$

$x(x - 4) \geq 0$

$(-\infty, 0], [4, \infty)$

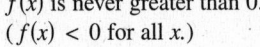

59. $f(x) = \sqrt{x - 1}$

$f(x) \geq 0$ on $[1, \infty)$.

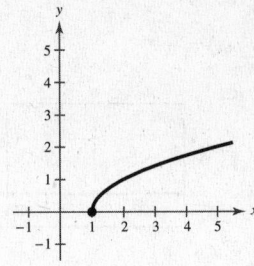

60. $f(x) = \sqrt{x + 2}$

$f(x) \geq 0$

$\sqrt{x + 2} \geq 0$

$x + 2 \geq 0$

$x \geq -2$

$[-2, \infty)$

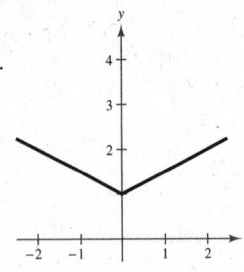

61. $f(x) = -(1 + |x|)$

$f(x)$ is never greater than 0.
($f(x) < 0$ for all x.)

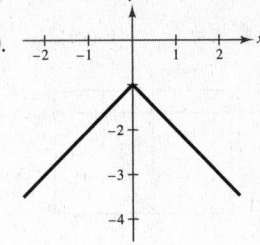

62. $f(x) = \frac{1}{2}(2 + |x|)$

$f(x)$ is always greater than 0.
$(-\infty, \infty)$

63. $f(x) = -2x + 15$

$\dfrac{f(3) - f(0)}{3 - 0} = \dfrac{9 - 15}{3} = -2$

The average rate of change from $x_1 = 0$ to $x_2 = 3$ is -2.

64. $f(x) = 3x + 8$

$\dfrac{f(3) - f(0)}{3 - 0} = \dfrac{17 - 8}{3} = \dfrac{9}{3} = 3$

The average rate of change from $x_1 = 0$ to $x_2 = 3$ is 3.

65. $f(x) = x^2 + 12x - 4$

$\dfrac{f(5) - f(1)}{5 - 1} = \dfrac{81 - 9}{4} = 18$

The average rate of change from $x_1 = 1$ to $x_2 = 5$ is 18.

66. $f(x) = x^2 - 2x + 8$

$\dfrac{f(5) - f(1)}{5 - 1} = \dfrac{23 - 7}{4} = \dfrac{16}{4} = 4$

The average rate of change from $x_1 = 1$ to $x_2 = 5$ is 4.

67. $f(x) = x^3 - 3x^2 - x$

$\dfrac{f(3) - f(1)}{3 - 1} = \dfrac{-3 - (-3)}{2} = 0$

The average rate of change from $x_1 = 1$ to $x_2 = 3$ is 0.

68. $f(x) = -x^3 + 6x^2 + x$

$$\frac{f(6) - f(1)}{6 - 1} = \frac{6 - 6}{5} = \frac{0}{5} = 0$$

The average rate of change from $x_1 = 1$ to $x_2 = 6$ is 0.

69. $f(x) = -\sqrt{x - 2} + 5$

$$\frac{f(11) - f(3)}{11 - 3} = \frac{2 - 4}{8} = -\frac{1}{4}$$

The average rate of change from $x_1 = 3$ to $x_2 = 11$ is $-\frac{1}{4}$.

70. $f(x) = -\sqrt{x + 1} + 3$

$$\frac{f(8) - f(3)}{8 - 3} = \frac{0 - 1}{5} = -\frac{1}{5}$$

The average rate of change from $x_1 = 3$ to $x_2 = 8$ is $-\frac{1}{5}$.

71. $f(x) = x^6 - 2x^2 + 3$

$f(-x) = (-x)^6 - 2(-x)^2 + 3$

$\quad\quad = x^6 - 2x^2 + 3$

$\quad\quad = f(x)$

The function is even.
y-axis symmetry

72. $h(x) = x^3 - 5$

$h(-x) = (-x)^3 - 5$

$\quad\quad = -x^3 - 5$

$\quad\quad \neq h(x)$

$\quad\quad \neq -h(x)$

The function is neither odd nor even. No symmetry

73. $g(x) = x^3 - 5x$

$g(-x) = (-x)^3 - 5(-x)$

$\quad\quad = -x^3 + 5x$

$\quad\quad = -g(x)$

The function is odd.
Origin symmetry

74. $f(x) = x\sqrt{1 - x^2}$

$f(-x) = -x\sqrt{1 - (-x)^2}$

$\quad\quad = -x\sqrt{1 - x^2}$

$\quad\quad = -f(x)$

The function is odd.
Origin symmetry

75. $f(t) = t^2 + 2t - 3$

$f(-t) = (-t)^2 + 2(-t) - 3$

$\quad\quad = t^2 - 2t - 3$

$\quad\quad \neq f(t), \neq -f(t)$

The function is neither even nor odd. No symmetry

76. $g(s) = 4s^{2/3}$

$g(-s) = 4(-s)^{2/3}$

$\quad\quad = 4s^{2/3}$

$\quad\quad = g(s)$

The function is even.
y-axis symmetry

77. $h = \text{top} - \text{bottom}$

$\quad = (-x^2 + 4x - 1) - 2$

$\quad = -x^2 + 4x - 3$

78. $h = \text{top} - \text{bottom}$

$\quad = 3 - (4x - x^2)$

$\quad = 3 - 4x + x^2$

79. $h = \text{top} - \text{bottom}$

$\quad = (4x - x^2) - 2x$

$\quad = 2x - x^2$

80. $h = \text{top} - \text{bottom}$

$\quad = 2 - \sqrt[3]{x}$

81. $L = \text{right} - \text{left}$

$\quad = \frac{1}{2}y^2 - 0 = \frac{1}{2}y^2$

82. $L = \text{right} - \text{left}$

$\quad = 2 - \sqrt[3]{2y}$

83. $L = \text{right} - \text{left}$

$\quad = 4 - y^2$

84. $L = \text{right} - \text{left}$

$\quad = \frac{2}{y} - 0$

$\quad = \frac{2}{y}$

85. $L = -0.294x^2 + 97.744x - 664.875,\ 20 \le x \le 90$

(a)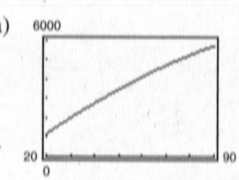

(b) $L = 2000$ when $x \approx 29.9645 \approx 30$ watts.

86. (a)

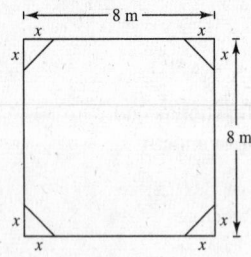

(b) The model is an excellent fit.

(c) The temperature is increasing from 6 A.M. until noon ($x = 0$ to $x = 6$). Then it decreases until 2 A.M. ($x = 6$ to $x = 20$). Then the temperature increases until 6 A.M. ($x = 20$ to $x = 24$).

(d) The maximum temperature according to the model is about 63.93°F. According to the data, it is 64°F. The minimum temperature according to the model is about 33.98°F. According to the data, it is 34°F.

(e) Answers may vary. Temperatures will depend upon the weather patterns, which usually change from day to day.

87. (a) For the average salaries of college professors, a scale of $10,000 would be appropriate.

(b) For the population of the United States, use a scale of 10,000,000.

(c) For the percent of the civilian workforce that is unemployed, use a scale of 1%.

88.

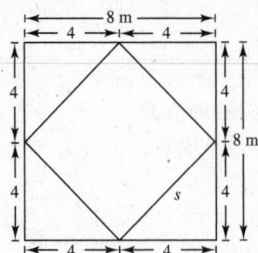

(b)

Range: $32 \leq A \leq 64$

(a) $A = (8)(8) - 4\left(\frac{1}{2}\right)(x)(x)$

$\quad = 64 - 2x^2$

Domain: $0 \leq x \leq 4$

(c) When $x = 4$, the resulting figure is a square.

By the Pythagorean Theorem,

$4^2 + 4^2 = s^2 \implies s = \sqrt{32} = 4\sqrt{2}$ meters.

89. $r = 15.639t^3 - 104.75t^2 + 303.5t - 301,\ 2 \leq t \leq 7$

(a)

(b) $\dfrac{r(7) - r(2)}{7 - 2} = \dfrac{2054.927 - 12.112}{5} = 408.563$

The average rate of change from 2002 to 2007 is $408.563 billion per year. The estimated revenue is increasing each year at a rapid pace.

90. (a)

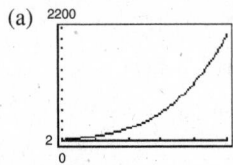

(b) The average rate of change from 1992 to 2002:

$\dfrac{F(12) - F(2)}{12 - 2} = \dfrac{580.78 - 433.5}{12 - 2}$

$\qquad = \dfrac{147.28}{10} = 14.728$

The number of foreign students increased at a steady rate of 14.728 thousand students each year.

(c) The five-year period of least average rate of change was 1992 to 1997.

$\dfrac{F(7) - F(2)}{7 - 2} = \dfrac{463.74 - 433.5}{7 - 2} = \dfrac{30.24}{5} = 6.05$

The five-year period of greatest increase was 1997 to 2002.

$\dfrac{F(12) - F(7)}{12 - 7} = \dfrac{580.78 - 463.74}{12 - 7} = \dfrac{117.04}{5} = 23.4$

The least rate of change was about 6.05 thousand students from 1992 to 1997.

The greatest rate of change was about 23.4 thousand students from 1997 to 2002.

91. $s_0 = 6, v_0 = 64$

(a) $s = -16t^2 + 64t + 6$

(b)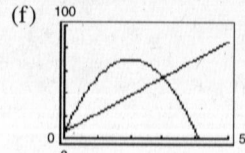

(c) $\dfrac{s(3) - s(0)}{3 - 0} = \dfrac{54 - 6}{3} = 16$

(d) The average rate of change of the height of the object with respect to time over the interval $t_1 = 0$ to $t_2 = 3$ is 16 feet per second.

(e) $s(0) = 6, m = 16$

Secant line: $y - 6 = 16(t - 0)$

$$y = 16t + 6$$

(f)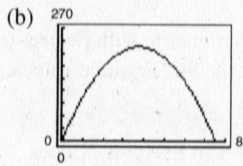

93. $v_0 = 120, s_0 = 0$

(a) $s = -16t^2 + 120t$

(b)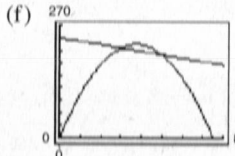

(c) $\dfrac{s(5) - s(3)}{5 - 3} = \dfrac{200 - 216}{2} = -8$

(d) The average decrease in the height of the object over the interval $t_1 = 3$ to $t_2 = 5$ is 8 feet per second.

(e) $s(5) = 200, m = -8$

Secant line: $y - 200 = -8(t - 5)$

$$y = -8t + 240$$

(f)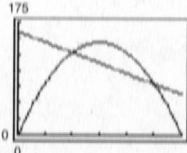

92. (a) $s = -16t^2 + 72t + 6.5$

(b)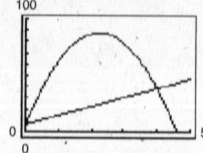

(c) The average rate of change from $t = 0$ to $t = 4$:

$$\frac{s(4) - s(0)}{4 - 0} = \frac{38.5 - 6.5}{4} = \frac{32}{4} = 8 \text{ feet per second}$$

(d) The slope of the secant line through $(0, s(0))$ and $(4, s(4))$ is positive. The average rate of change of the position of the object from $t = 0$ to $t = 4$ is 8 feet per second.

(e) The equation of the secant line:

$$m = 8, \quad y = 8t + 6.5$$

(f) The graph is shown in (b).

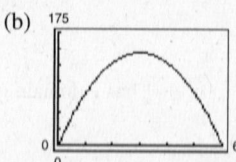

94. (a) $s = -16t^2 + 96t$

(b)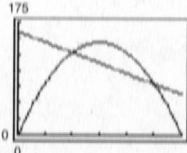

(c) The average rate of change from $t = 2$ to $t = 5$:

$$\frac{s(5) - s(2)}{5 - 2} = \frac{80 - 128}{3}$$

$$= -\frac{48}{3} = -16 \text{ feet per second}$$

(d) The slope of the secant line through $(2, s(2))$ and $(5, s(5))$ is negative. The average rate of change of the position of the object from $t = 2$ to $t = 5$ is -16 feet per second.

(e) The equation of the secant line: $m = -16$

Using $(2, s(2)) = (2, 128)$ we have

$$y - 128 = -16(t - 2)$$

$$y = -16t + 160.$$

(f) The graph is shown in (b).

95. $v_0 = 0, s_0 = 120$

(a) $s = -16t^2 + 120$

(b)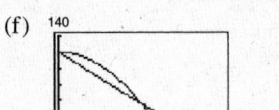

(c) $\dfrac{s(2) - s(0)}{2 - 0} = \dfrac{56 - 120}{2} = -32$

(d) On the interval $t_1 = 0$ to $t_2 = 2$, the height of the object is decreasing at a rate of 32 feet per second.

(e) $s(0) = 120, m = -32$

Secant line: $y - 120 = -32(t - 0)$

$$y = -32t + 120$$

(f)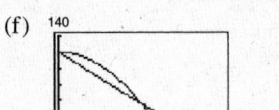

96. (a) $s = -16t^2 + 80$

(b)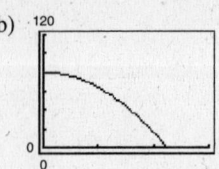

(c) The average rate of change from $t = 1$ to $t = 2$:

$$\frac{s(2) - s(1)}{2 - 1} = \frac{16 - 64}{1} = -\frac{48}{1} = -48 \text{ feet per second}$$

(d) The slope of the secant line through $(1, s(1))$ and $(2, s(2))$ is negative. The average rate of change of the position of the object from $t = 1$ to $t = 2$ is -48 feet per second.

(e) The equation of the tangent line: $m = -48$

Using $(1, s(1)) = (1, 64)$ we have

$$y - 64 = -48(t - 1)$$

$$y = -48t + 112.$$

(f) The graph is shown in (b).

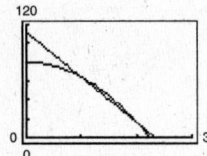

97. False. The function $f(x) = \sqrt{x^2 + 1}$ has a domain of all real numbers.

98. False. An odd function is symmetric with respect to the origin, so its domain must include negative values.

99. (a) Even. The graph is a reflection in the x-axis.

(b) Even. The graph is a reflection in the y-axis.

(c) Even. The graph is a vertical translation of f.

(d) Neither. The graph is a horizontal translation of f.

100. Yes, the graph of $x = y^2 + 1$ in Exercise 11 does represent x as a function of y. Each y-value corresponds to only one x-value.

101. $\left(-\frac{3}{2}, 4\right)$

(a) If f is even, another point is $\left(\frac{3}{2}, 4\right)$.

(b) If f is odd, another point is $\left(\frac{3}{2}, -4\right)$.

102. $\left(-\frac{5}{3}, -7\right)$

(a) If f is even, another point is $\left(\frac{5}{3}, -7\right)$.

(b) If f is odd, another point is $\left(\frac{5}{3}, 7\right)$.

103. $(4, 9)$

(a) If f is even, another point is $(-4, 9)$.

(b) If f is odd, another point is $(-4, -9)$.

104. $(5, -1)$

(a) If f is even, another point is $(-5, -1)$.

(b) If f is odd, another point is $(-5, 1)$.

105. (a) $y = x$ 　　　　　　 (b) $y = x^2$ 　　　　　　 (c) $y = x^3$

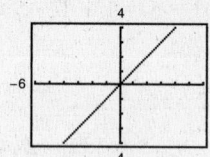 　　 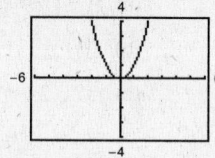 　　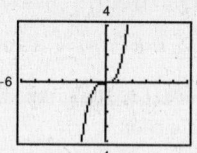

(d) $y = x^4$ 　　　　　　 (e) $y = x^5$ 　　　　　　 (f) $y = x^6$

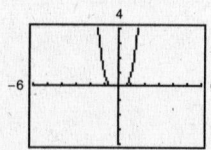 　　 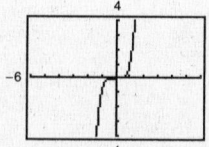 　　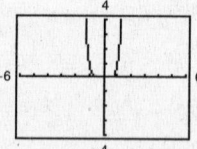

All the graphs pass through the origin. The graphs of the odd powers of x are symmetric with respect to the origin and the graphs of the even powers are symmetric with respect to the y-axis. As the powers increase, the graphs become flatter in the interval $-1 < x < 1$.

106. The graph of $y = x^7$ will pass through the origin and will be symmetric with the origin.
The graph of $y = x^8$ will pass through the origin and will be symmetric with respect to the y-axis.

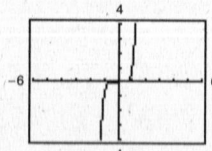 　　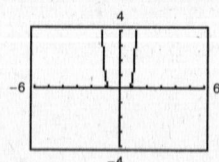

107. $x^2 - 10x = 0$

$x(x - 10) = 0$

$x = 0$ or $x = 10$

108. $100 - (x - 5)^2 = 0$

$(x - 5)^2 = 100$

$x - 5 = \pm 10$

$x - 5 = -10 \implies x = -5$

$x - 5 = 10 \implies x = 15$

109. $x^3 - x = 0$

$x(x^2 - 1) = 0$

$x = 0$ or $x^2 - 1 = 0$

$x^2 = 1$

$x = \pm 1$

110. $16x^2 - 40x + 25 = 0$

$(4x - 5)(4x - 5) = 0$

$4x - 5 = 0 \implies x = \frac{5}{4}$

111. $f(x) = 5x - 8$

(a) $f(9) = 5(9) - 8 = 37$

(b) $f(-4) = 5(-4) - 8 = -28$

(c) $f(x - 7) = 5(x - 7) - 8 = 5x - 35 - 8 = 5x - 43$

112. $f(x) = x^2 - 10x$

(a) $f(4) = (4)^2 - 10(4) = 16 - 40 = -24$

(b) $f(-8) = (-8)^2 - 10(-8) = 64 + 80 = 144$

(c) $f(x - 4) = (x - 4)^2 - 10(x - 4)$

$= x^2 - 8x + 16 - 10x + 40 = x^2 - 18x + 56$

113. $f(x) = \sqrt{x - 12} - 9$

    (a) $f(12) = \sqrt{12 - 12} - 9 = 0 - 9 = -9$

    (b) $f(40) = \sqrt{40 - 12} - 9 = \sqrt{28} - 9 = 2\sqrt{7} - 9$

    (c) $f\left(-\sqrt{36}\right)$ does not exist. The given value is not in the domain of the function.

114. $f(x) = x^4 - x - 5$

    (a) $f(-1) = (-1)^4 - (-1) - 5 = 1 + 1 - 5 = -3$

    (b) $f\left(\frac{1}{2}\right) = \left(\frac{1}{2}\right)^4 - \frac{1}{2} - 5 = -\frac{87}{16}$

    (c) $f\left(2\sqrt{3}\right) = \left(2\sqrt{3}\right)^4 - 2\sqrt{3} - 5$

$$= 16(9) - 2\sqrt{3} - 5 = 139 - 2\sqrt{3}$$

115.
$$f(x) = x^2 - 2x + 9$$
$$f(3 + h) = (3 + h)^2 - 2(3 + h) + 9$$
$$= 9 + 6h + h^2 - 6 - 2h + 9$$
$$= h^2 + 4h + 12$$
$$f(3) = 3^2 - 2(3) + 9 = 12$$
$$\frac{f(3 + h) - f(3)}{h} = \frac{(h^2 + 4h + 12) - (12)}{h} = \frac{h^2 + 4h}{h} = \frac{h(h + 4)}{h} = h + 4, \quad h \neq 0$$

116. $f(x) = 5 + 6x - x^2$, $\dfrac{f(6 + h) - f(6)}{h}$, $h \neq 0$

$$\frac{f(6 + h) - f(6)}{h} = \frac{5 + 6(6 + h) - (6 + h)^2 - 5 - 6(6) + 6^2}{h}$$

$$= \frac{5 + 36 + 6h - 36 - 12h - h^2 - 5 - 36 + 36}{h}$$

$$= \frac{-h^2 - 6h}{h} = -h - 6, \quad h \neq 0$$

Section 1.6   A Library of Parent Functions

■   You should be able to identify and graph the following types of functions:

    (a)   Linear functions like $f(x) = ax + b$

    (b)   Squaring functions like $f(x) = x^2$

    (c)   Cubic functions like $f(x) = x^3$

    (d)   Square root functions like $f(x) = \sqrt{x}$

    (e)   Reciprocal functions like $f(x) = \dfrac{1}{x}$

    (f)   Constant functions like $f(x) = c$

    (g)   Absolute value functions like $f(x) = |x|$

    (h)   Step and piecewise-defined functions like $f(x) = [\![x]\!]$

■   You should be able to determine the following about these parent functions:

    (a)   Domain and range

    (b) x-intercept(s) and y-intercept

    (c)   Symmetries

    (d)   Where it is increasing, decreasing, or constant

    (e)   If it is odd, even or neither

    (f)   Relative maximums and relative minimums

Vocabulary Check

1. $f(x) = [\![x]\!]$

(g) greatest integer function

2. $f(x) = x$

(i) identity function

3. $f(x) = \dfrac{1}{x}$

(h) reciprocal function

4. $f(x) = x^2$

(a) squaring function

5. $f(x) = \sqrt{x}$

(b) square root function

6. $f(x) = c$

(e) constant function

7. $f(x) = |x|$

(f) absolute value function

8. $f(x) = x^3$

(c) cubic function

9. $f(x) = ax + b$

(d) linear function

---

1. (a) $f(1) = 4,\ f(0) = 6$

$(1, 4)$ and $(0, 6)$

$m = \dfrac{6 - 4}{0 - 1} = -2$

$y - 6 = -2(x - 0)$

$y = -2x + 6$

$f(x) = -2x + 6$

(b)

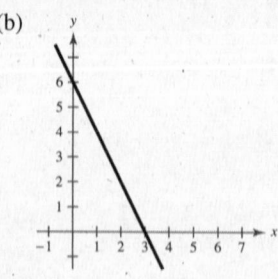

2. (a) $f(-3) = -8,\ f(1) = 2$

$(-3, -8),\ (1, 2)$

$m = \dfrac{2 - (-8)}{1 - (-3)} = \dfrac{10}{4} = \dfrac{5}{2}$

$f(x) - 2 = \dfrac{5}{2}(x - 1)$

$f(x) = \dfrac{5}{2}x - \dfrac{1}{2}$

(b)

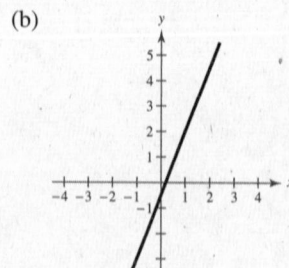

3. (a) $f(5) = -4,\ f(-2) = 17$

$(5, -4)$ and $(-2, 17)$

$m = \dfrac{17 - (-4)}{-2 - 5} = \dfrac{21}{-7} = -3$

$y - (-4) = -3(x - 5)$

$y + 4 = -3x + 15$

$y = -3x + 11$

$f(x) = -3x + 11$

(b)

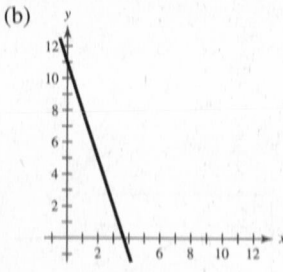

4. (a) $f(3) = 9,\ f(-1) = -11$

$(3, 9),\ (-1, -11)$

$m = \dfrac{-11 - 9}{-1 - 3} = \dfrac{-20}{-4} = 5$

$f(x) - 9 = 5(x - 3)$

$f(x) = 5x - 6$

(b)

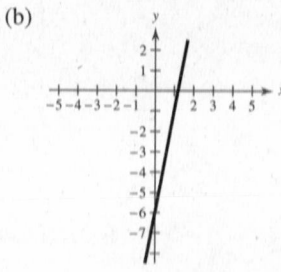

5. (a) $f(-5) = -1, f(5) = -1$

$(-5, -1)$ and $(5, -1)$

$m = \dfrac{-1 - (-1)}{5 - (-5)} = \dfrac{0}{10} = 0$

$y - (-1) = 0(x - (-5))$

$y + 1 = 0$

$y = -1$

$f(x) = -1$

(b)

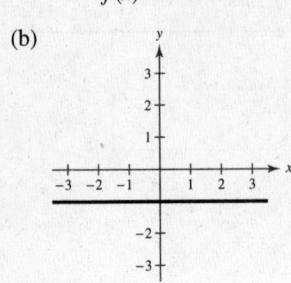

6. (a) $f(-10) = 12, f(16) = -1$

$(-10, 12), (16, -1)$

$m = \dfrac{-1 - 12}{16 - (-10)} = \dfrac{-13}{26} = -\dfrac{1}{2}$

$f(x) - (-1) = -\dfrac{1}{2}(x - 16)$

$f(x) = -\dfrac{1}{2}x + 7$

(b)

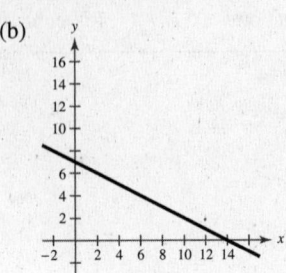

7. (a) $f\left(\dfrac{1}{2}\right) = -6, f(4) = -3$

$\left(\dfrac{1}{2}, -6\right)$ and $(4, -3)$

$m = \dfrac{-3 - (-6)}{4 - (1/2)} = \dfrac{3}{7/2} = \dfrac{6}{7}$

$y - (-3) = \dfrac{6}{7}(x - 4)$

$y + 3 = \dfrac{6}{7}x - \dfrac{24}{7}$

$y = \dfrac{6}{7}x - \dfrac{45}{7}$

$f(x) = \dfrac{6}{7}x - \dfrac{45}{7}$

(b)

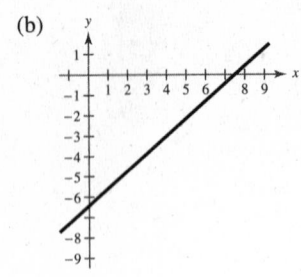

8. (a) $f\left(\dfrac{2}{3}\right) = -\dfrac{15}{2}, f(-4) = -11$

$\left(\dfrac{2}{3}, -\dfrac{15}{2}\right), (-4, -11)$

$m = \dfrac{-11 - (-15/2)}{-4 - (2/3)}$

$= \dfrac{-7/2}{-14/3} = \left(-\dfrac{7}{2}\right) \cdot \left(-\dfrac{3}{14}\right) = \dfrac{3}{4}$

$f(x) - (-11) = \dfrac{3}{4}(x - (-4))$

$f(x) = \dfrac{3}{4}x - 8$

(b)

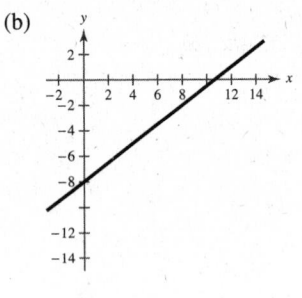

9. $f(x) = -x - \dfrac{3}{4}$

10. $f(x) = 3x - \dfrac{5}{2}$

11. $f(x) = -\dfrac{1}{6}x - \dfrac{5}{2}$

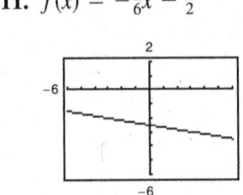

12. $f(x) = \frac{5}{6} - \frac{2}{3}x$

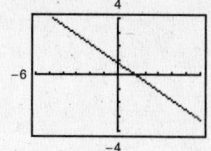

13. $f(x) = x^2 - 2x$

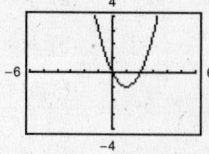

14. $f(x) = -x^2 + 8x$

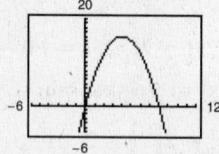

15. $h(x) = -x^2 + 4x + 12$

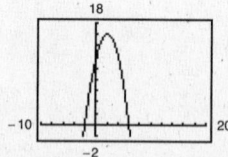

16. $g(x) = x^2 - 6x - 16$

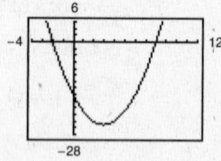

17. $f(x) = x^3 - 1$

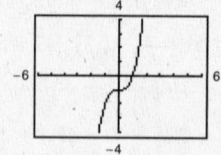

18. $f(x) = 8 - x^3$

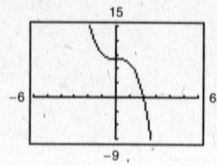

19. $f(x) = (x - 1)^3 + 2$

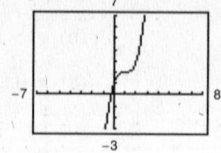

20. $g(x) = 2(x + 3)^3 + 1$

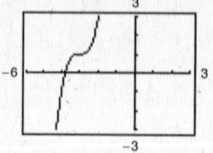

21. $f(x) = 4\sqrt{x}$

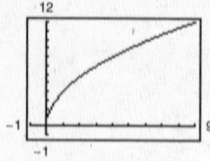

22. $f(x) = 4 - 2\sqrt{x}$

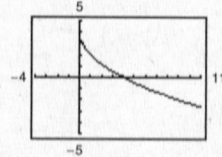

23. $g(x) = 2 - \sqrt{x + 4}$

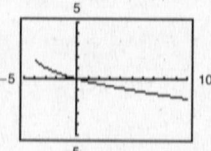

24. $h(x) = \sqrt{x + 2} + 3$

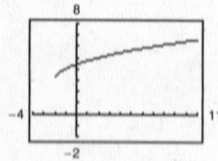

25. $f(x) = -\dfrac{1}{x}$

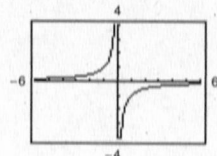

26. $f(x) = 4 + \dfrac{1}{x}$

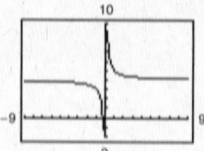

27. $h(x) = \dfrac{1}{x + 2}$

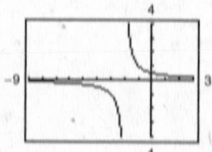

28. $k(x) = \dfrac{1}{x - 3}$

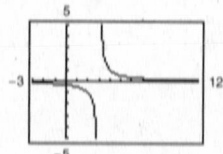

29. $f(x) = [\![x]\!]$

(a) $f(2.1) = 2$

(b) $f(2.9) = 2$

(c) $f(-3.1) = -4$

(d) $f\left(\dfrac{7}{2}\right) = 3$

30. $g(x) = 2[\![x]\!]$

   (a) $g(-3) = 2[\![-3]\!] = 2(-3) = -6$

   (b) $g(0.25) = 2[\![0.25]\!] = 2(0) = 0$

   (c) $g(9.5) = 2[\![9.5]\!] = 2(9) = 18$

   (d) $g\left(\frac{11}{3}\right) = 2\left[\!\left[\frac{11}{3}\right]\!\right] = 2(3) = 6$

31. $h(x) = [\![x + 3]\!]$

   (a) $h(-2) = [\![1]\!] = 1$

   (b) $h\left(\frac{1}{2}\right) = [\![3.5]\!] = 3$

   (c) $h(4.2) = [\![7.2]\!] = 7$

   (d) $h(-21.6) = [\![-18.6]\!] = -19$

32. $f(x) = 4[\![x]\!] + 7$

   (a) $f(0) = 4[\![0]\!] + 7 = 4(0) + 7 = 7$

   (b) $f(-1.5) = 4[\![-1.5]\!] + 7 = 4(-2) + 7 = -1$

   (c) $f(6) = 4[\![6]\!] + 7 = 4(6) + 7 = 31$

   (d) $f\left(\frac{5}{3}\right) = 4\left[\!\left[\frac{5}{3}\right]\!\right] + 7 = 4(1) + 7 = 11$

33. $h(x) = [\![3x - 1]\!]$

   (a) $h(2.5) = [\![6.5]\!] = 6$

   (b) $h(-3.2) = [\![-10.6]\!] = -11$

   (c) $h\left(\frac{7}{3}\right) = [\![6]\!] = 6$

   (d) $h\left(-\frac{21}{3}\right) = [\![-22]\!] = -22$

34. $k(x) = \left[\!\left[\frac{1}{2}x + 6\right]\!\right]$

   (a) $k(5) = \left[\!\left[\frac{1}{2}(5) + 6\right]\!\right] = [\![8.5]\!] = 8$

   (b) $k(-6.1) = \left[\!\left[\frac{1}{2}(-6.1) + 6\right]\!\right] = [\![2.95]\!] = 2$

   (c) $k(0.1) = \left[\!\left[\frac{1}{2}(0.1) + 6\right]\!\right] = [\![6.05]\!] = 6$

   (d) $k(15) = \left[\!\left[\frac{1}{2}(15) + 6\right]\!\right] = [\![13.5]\!] = 13$

35. $g(x) = 3[\![x - 2]\!] + 5$

   (a) $g(-2.7) = 3[\![-4.7]\!] + 5 = 3(-5) + 5 = -10$

   (b) $g(-1) = 3[\![-3]\!] + 5 = 3(-3) + 5 = -4$

   (c) $g(0.8) = 3[\![-1.2]\!] + 5 = 3(-2) + 5 = -1$

   (d) $g(14.5) = 3[\![12.5]\!] + 5 = 3(12) + 5 = 41$

36. $g(x) = -7[\![x + 4]\!] + 6$

   (a) $g\left(\frac{1}{8}\right) = -7\left[\!\left[\frac{1}{8} + 4\right]\!\right] + 6$

        $= -7\left[\!\left[4\frac{1}{8}\right]\!\right] + 6 = -7(4) + 6 = -22$

   (b) $g(9) = -7[\![9 + 4]\!] + 6$

        $= -7[\![13]\!] + 6 = -7(13) + 6 = -85$

   (c) $g(-4) = -7[\![-4 + 4]\!] + 6$

        $= -7[\![0]\!] + 6 = -7(0) + 6 = 6$

   (d) $g\left(\frac{3}{2}\right) = -7\left[\!\left[\frac{3}{2} + 4\right]\!\right] + 6$

        $= -7\left[\!\left[5\frac{1}{2}\right]\!\right] + 6 = -7(5) + 6 = -29$

37. $g(x) = -[\![x]\!]$

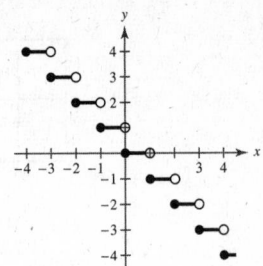

38. $g(x) = 4[\![x]\!]$

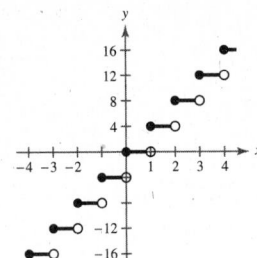

39. $g(x) = [\![x]\!] - 2$

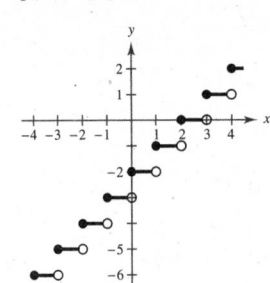

40. $g(x) = [\![x]\!] - 1$

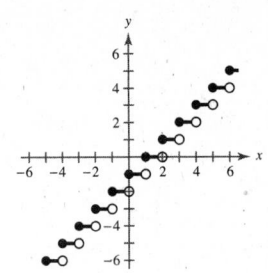

41. $g(x) = [\![x + 1]\!]$

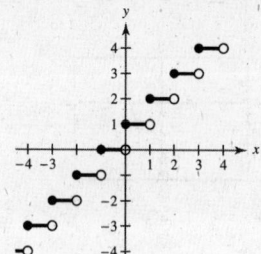

42. $g(x) = [\![x - 3]\!]$

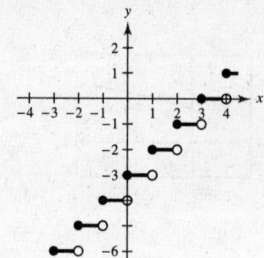

43. $f(x) = \begin{cases} 2x + 3, & x < 0 \\ 3 - x, & x \geq 0 \end{cases}$

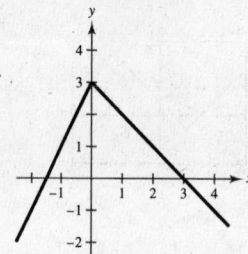

44. $g(x) = \begin{cases} x + 6, & x \leq -4 \\ \frac{1}{2}x - 4, & x > -4 \end{cases}$

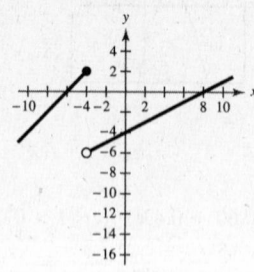

45. $f(x) = \begin{cases} \sqrt{4 + x}, & x < 0 \\ \sqrt{4 - x}, & x \geq 0 \end{cases}$

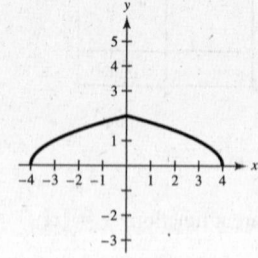

46. $f(x) = \begin{cases} 1 - (x - 1)^2, & x \leq 2 \\ \sqrt{x - 2}, & x > 2 \end{cases}$

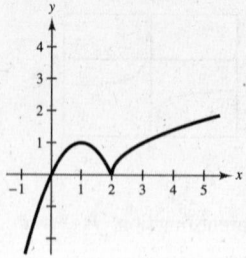

47. $f(x) = \begin{cases} x^2 + 5, & x \leq 1 \\ -x^2 + 4x + 3, & x > 1 \end{cases}$

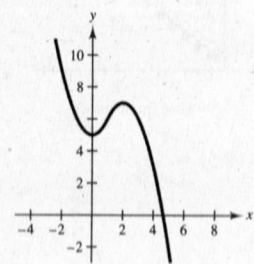

48. $h(x) = \begin{cases} 3 - x^2, & x < 0 \\ x^2 + 2, & x \geq 0 \end{cases}$

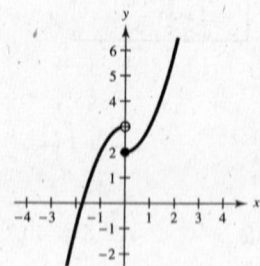

49. $h(x) = \begin{cases} 4 - x^2, & x < -2 \\ 3 + x, & -2 \leq x < 0 \\ x^2 + 1, & x \geq 0 \end{cases}$

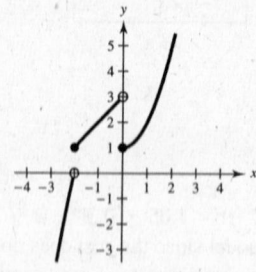

50. $k(x) = \begin{cases} 2x + 1, & x \leq -1 \\ 2x^2 - 1, & -1 < x \leq 1 \\ 1 - x^2, & x 1 \end{cases}$

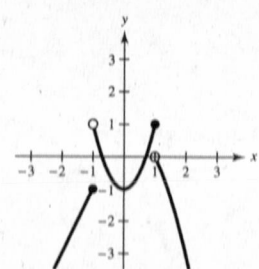

51. $s(x) = 2\left(\frac{1}{4}x - [\![\frac{1}{4}x]\!]\right)$

(a)

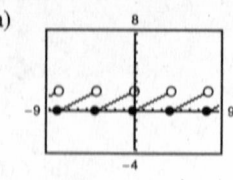

(b) Domain: $(-\infty, \infty)$

Range: $[0, 2)$

(c) Sawtooth pattern

52. (a) $g(x) = 2\left(\frac{1}{4}x - [\![\frac{1}{4}x]\!]\right)^2$

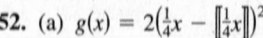
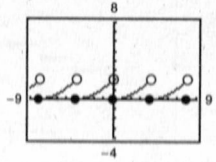

(b) Domain: $(-\infty, \infty)$

Range: $[0, 2)$

(c) Sawtooth pattern

53. (a) Parent function: $f(x) = |x|$

(b) $g(x) = |x + 2| - 1$

(c)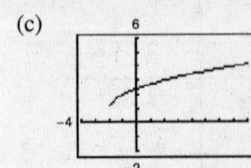

54. (a) Parent function: $y = \sqrt{x}$

(b) $y = 1 + \sqrt{x + 2}$

(c)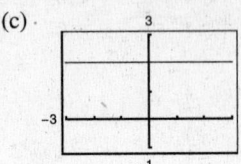

55. (a) Parent function: $f(x) = x^3$

(b) $g(x) = (x - 1)^3 - 2$

(c)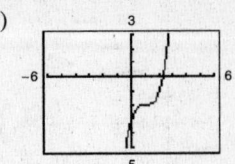

56. (a) Parent function: $y = \dfrac{1}{x}$

(b) $y = \dfrac{1}{x} - 2$

(c)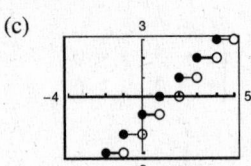

57. (a) Parent function: $f(x) = c$

(b) $g(x) = 2$

(c)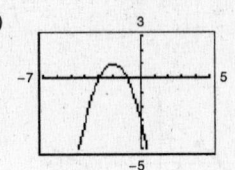

58. (a) Parent function: $y = x^2$

(b) $y = 1 - (x + 2)^2$

(c)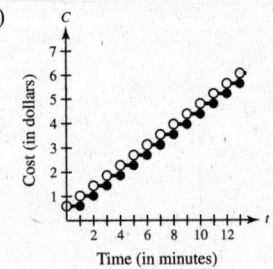

59. (a) Parent function: $f(x) = x$

(b) $g(x) = x - 2$

(c)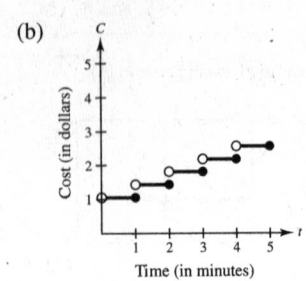

60. (a) Parent function: $y = [\![x]\!]$

(b) $y = [\![x - 1]\!]$

(c)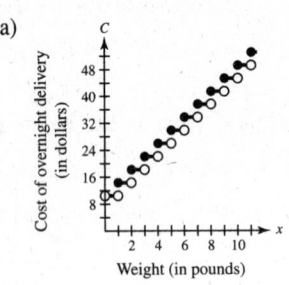

61. $C = 0.60 - 0.42[\![1 - t]\!], t > 0$

(a)

Cost (in dollars)

Time (in minutes)

(b) $C(12.5) = \$5.64$

62. (a) $C_2(t) = 1.05 - 0.38[\![-(t - 1)]\!]$ is the appropriate model since the cost does not increase until after the next minute of conversation has started.

(b)

Cost (in dollars)

Time (in minutes)

$C = 1.05 - 0.38[\![-17.75]\!] = \$7.89$

63. $C = 10.75 + 3.95[\![x]\!], \quad x > 0$

(a)

Cost of overnight delivery (in dollars)

Weight (in pounds)

(b) $C(10.33) = 10.75 + 3.95(10) = \$50.25$

64. (a) Model: (Total cost) = (Flat rate) + (Rate per pound)

Labels: Total cost = C

Flat rate = 9.80

Rate per pound = $2.50[\![x]\!]$, $x > 0$

Equation: $C = 9.80 + 2.50[\![x]\!]$, $x > 0$

(b)

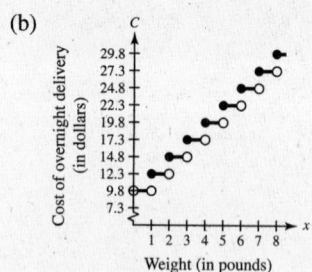

Weight (in pounds)

65. $W(h) = \begin{cases} 12h, & 0 < h \le 40 \\ 18(h - 40) + 480, & h > 40 \end{cases}$

(a) $W(30) = 12(30) = \$360$

$W(40) = 12(40) = \$480$

$W(45) = 18(5) + 480 = \$570$

$W(50) = 18(10) + 480 = \$660$

(b) $W(h) = \begin{cases} 12h, & 0 < h \le 45 \\ 18(h - 45) + 540, & h > 45 \end{cases}$

66. For the first two hours the slope is 1. For the next six hours, the slope is 2. For the final hour, the slope is $\frac{1}{2}$.

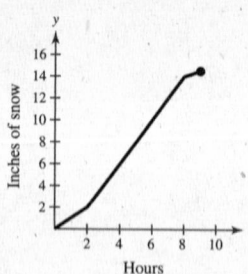

Hours

$f(t) = \begin{cases} t, & 0 \le t \le 2 \\ 2t - 2, & 2 < t \le 8 \\ \frac{1}{2}t + 10, & 8 < t \le 9 \end{cases}$

To find $f(t) = 2t - 2$, use $m = 2$ and $(2, 2)$.

$y - 2 = 2(t - 2) \Rightarrow y = 2t - 2$

To find $f(t) = \frac{1}{2}t + 10$, use $m = \frac{1}{2}$ and $(8, 14)$.

$y - 14 = \frac{1}{2}(t - 8) \Rightarrow y = \frac{1}{2}t + 10$

Total accumulation = 14.5 inches

67. (a) The domain of $f(x) = -1.97x + 26.3$ is $6 < x \le 12$. One way to see this is to notice that this is the equation of a line with negative slope, so the function values are decreasing as x increases, which matches the data for the corresponding part of the table. The domain of $f(x) = 0.505x^2 - 1.47x + 6.3$ is then $1 \le x \le 6$.

(b)

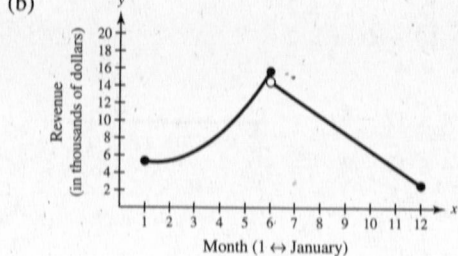

Month (1 ↔ January)

(c) $f(5) = 0.505(5)^2 - 1.47(5) + 6.3$

$= 0.505(25) - 7.35 + 6.3 = 11.575$

$f(11) = -1.97(11) + 26.3 = 4.63$

These values represent the income in thousands of dollars for the months of May and November, respectively.

(d) The model values are very close to the actual values.

Month, x	1	2	3	4	5	6	7	8	9	10	11	12
Revenue, y	5.2	5.6	6.6	8.3	11.5	15.8	12.8	10.1	8.6	6.9	4.5	2.7
Model, $f(x)$	5.3	5.4	6.4	8.5	11.6	15.7	12.5	10.5	8.6	6.6	4.6	2.7

68.

Interval	Intake Pipe	Drainpipe 1	Drainpipe 2
$[0, 5]$	Open	Closed	Closed
$[5, 10]$	Open	Open	Closed
$[10, 20]$	Closed	Closed	Closed
$[20, 30]$	Closed	Closed	Open
$[30, 40]$	Open	Open	Open
$[40, 45]$	Open	Closed	Open
$[45, 50]$	Open	Open	Open
$[50, 60]$	Open	Open	Closed

69. False. A piecewise-defined function is a function that is defined by two or more equations over a specified domain. That domain may or may not include x- and y-intercepts.

70. True. $f(x) = 2[\![x]\!]$, $1 \le x < 4$ is equivalent to the given piecewise function.

71. For the line through $(0, 6)$ and $(3, 2)$: $m = \dfrac{6 - 2}{0 - 3} = -\dfrac{4}{3}$

$$y - 6 = -\frac{4}{3}(x - 0) \implies y = -\frac{4}{3}x + 6$$

For the line through $(3, 2)$ and $(8, 0)$: $m = \dfrac{2 - 0}{3 - 8} = -\dfrac{2}{5}$

$$y - 0 = -\frac{2}{5}(x - 8) \implies y = -\frac{2}{5}x + \frac{16}{5}$$

$$f(x) = \begin{cases} -\frac{4}{3}x + 6, & 0 \le x \le 3 \\ -\frac{2}{5}x + \frac{16}{5}, & 3 < x \le 8 \end{cases}$$

Note that the respective domains can also be $0 \le x < 3$ and $3 \le x \le 8$.

72. $f(x) = \begin{cases} x^2, & x \le 2 \\ 7 - x, & x > 2 \end{cases}$

73. $3x + 4 \le 12 - 5x$

$8x + 4 \le 12$

$8x \le 8$

$x \le 1$

74. $2x + 1 > 6x - 9$

$10 > 4x$

$\frac{5}{2} > x$ or $x < \frac{5}{2}$

75. L_1: $(-2, -2)$ and $(2, 10)$

$$m_1 = \frac{10 - (-2)}{2 - (-2)} = \frac{12}{4} = 3$$

L_2: $(-1, 3)$ and $(3, 9)$

$$m_2 = \frac{9 - 3}{3 - (-1)} = \frac{6}{4} = \frac{3}{2}$$

The lines are neither parallel nor perpendicular.

76. L_1: $(-1, -7), (4, 3)$

$$m_1 = \frac{3 - (-7)}{4 - (-1)} = \frac{10}{5} = 2$$

L_2: $(1, 5), (-2, -7)$

$$m_2 = \frac{5 - (-7)}{1 - (-2)} = \frac{12}{3} = 4$$

Because the slopes are neither the same nor negative reciprocals, the lines L_1 and L_2 are neither parallel nor perpendicular.

Section 1.7   Transformations of Functions

■ You should know the basic types of transformations.

Let $y = f(x)$ and let c be a positive real number.

1.	$h(x) = f(x) + c$	Vertical shift c units upward
2.	$h(x) = f(x) - c$	Vertical shift c units downward
3.	$h(x) = f(x - c)$	Horizontal shift c units to the right
4.	$h(x) = f(x + c)$	Horizontal shift c units to the left
5.	$h(x) = -f(x)$	Reflection in the x-axis
6.	$h(x) = f(-x)$	Reflection in the y-axis
7.	$h(x) = cf(x), c > 1$	Vertical stretch
8.	$h(x) = cf(x), 0 < c < 1$	Vertical shrink
9.	$h(x) = f(cx), c > 1$	Horizontal shrink
10.	$h(x) = f(cx), 0 < c < 1$	Horizontal stretch

Vocabulary Check

1. rigid

2. $-f(x);\ f(-x)$

3. nonrigid

4. horizontal shrink; horizontal stretch

5. vertical stretch; vertical shrink

6. (a) iv   (b) ii   (c) iii   (d) i

---

1. (a) $f(x) = |x| + c$ Vertical shifts

  $c = -1 : f(x) = |x| - 1$ 1 unit down

  $c = 1 : f(x) = |x| + 1$ 1 unit up

  $c = 3 : f(x) = |x| + 3$ 3 units up

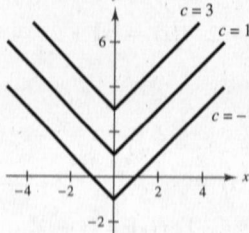

  (b) $f(x) = |x - c|$ Horizontal shifts

  $c = -1 : f(x) = |x + 1|$ 1 unit left

  $c = 1 : f(x) = |x - 1|$ 1 unit right

  $c = 3 : f(x) = |x - 3|$ 3 units right

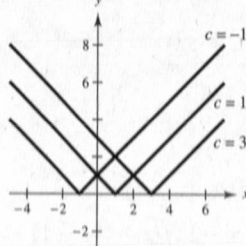

  (c) $f(x) = |x + 4| + c$ Horizontal shift four units left and a vertical shift

  $c = -1 : f(x) = |x + 4| - 1$ 1 unit down

  $c = 1 : f(x) = |x + 4| + 1$ 1 unit up

  $c = 3 : f(x) = |x + 4| + 3$ 3 units up

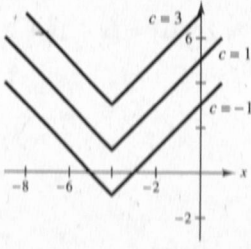

2. (a) $f(x) = \sqrt{x} + c$ Vertical shifts

$c = -3 : f(x) = \sqrt{x} - 3$ 3 units down

$c = -1 : f(x) = \sqrt{x} - 1$ 1 unit down

$c = 1 : f(x) = \sqrt{x} + 1$ 1 unit up

$c = 3 : f(x) = \sqrt{x} + 3$ 3 units up

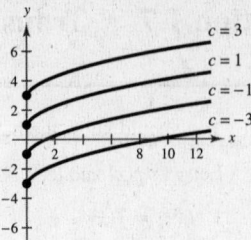

(b) $f(x) = \sqrt{x - c}$ Horizontal shifts

$c = -3 : f(x) = \sqrt{x + 3}$ 3 units left

$c = -1 : f(x) = \sqrt{x + 1}$ 1 unit left

$c = 1 : f(x) = \sqrt{x - 1}$ 1 unit right

$c = 3 : f(x) = \sqrt{x - 3}$ 3 units right

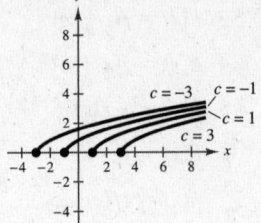

(c) $f(x) = \sqrt{x - 3} + c$ Horizontal shift 3 units right and a vertical shift

$c = -3 : f(x) = \sqrt{x - 3} - 3$ 3 units down

$c = -1 : f(x) = \sqrt{x - 3} - 1$ 1 unit down

$c = 1 : f(x) = \sqrt{x - 3} + 1$ 1 unit up

$c = 3 : f(x) = \sqrt{x - 3} + 3$ 3 units up

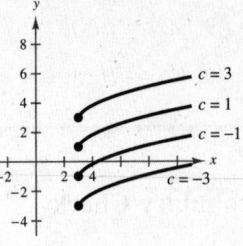

3. (a) $f(x) = [\![x]\!] + c$ Vertical shifts

$c = -2 : f(x) = [\![x]\!] - 2$ 2 units down

$c = 0 : f(x) = [\![x]\!]$ Parent function

$c = 2 : f(x) = [\![x]\!] + 2$ 2 units up

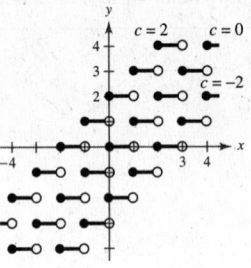

(b) $f(x) = [\![x + c]\!]$ Horizontal shifts

$c = -2 : f(x) = [\![x - 2]\!]$ 2 units right

$c = 0 : f(x) = [\![x]\!]$ Parent function

$c = 2 : f(x) = [\![x + 2]\!]$ 2 units left

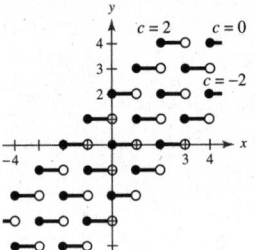

(c) $f(x) = [\![x - 1]\!] + c$ Horizontal shift 1 unit right and a vertical shift

$c = -2 : f(x) = [\![x - 1]\!] - 2$ 2 units down

$c = 0 : f(x) = [\![x - 1]\!]$

$c = 2 : f(x) = [\![x - 1]\!] + 2$ 2 units up

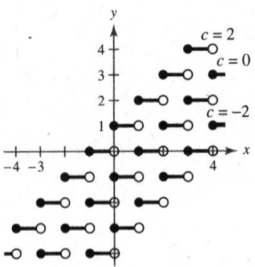

4. (a) $f(x) = \begin{cases} x^2 + c, & x < 0 \\ -x^2 + c, & x \geq 0 \end{cases}$

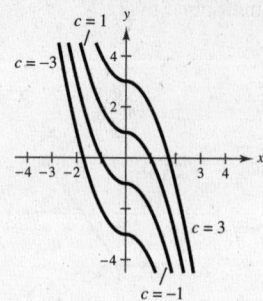

(b) $f(x) = \begin{cases} (x + c)^2, & x < 0 \\ -(x + c)^2, & x \geq 0 \end{cases}$

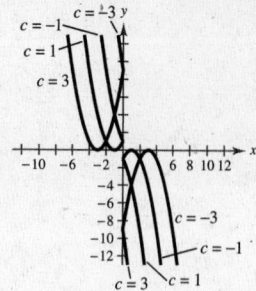

5. (a) $y = f(x) + 2$

Vertical shift 2 units upward

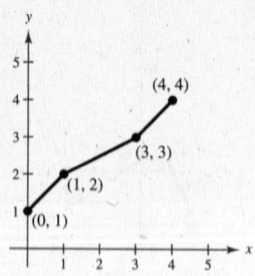

(b) $y = f(x - 2)$

Horizontal shift 2 units to the right

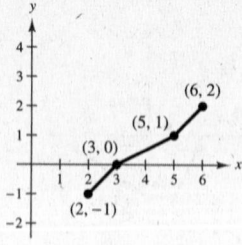

(c) $y = 2f(x)$

Vertical stretch (each y-value is multiplied by 2)

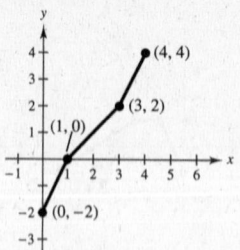

(d) $y = -f(x)$

Reflection in the x-axis

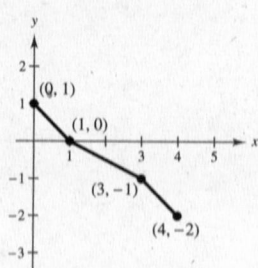

(e) $y = f(x + 3)$

Horizontal shift 3 units to the left

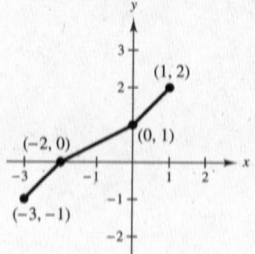

(f) $y = f(-x)$

Reflection in the y-axis

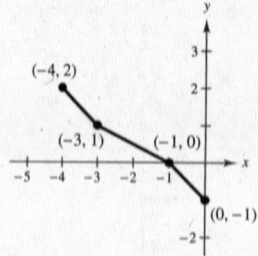

(g) $y = f\left(\frac{1}{2}x\right)$

Horizontal stretch (each x-value is multiplied by 2)

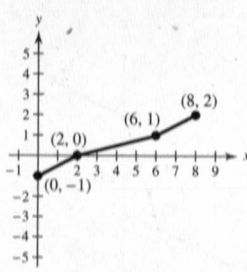

6. (a) $y = f(-x)$

Reflection in the y-axis

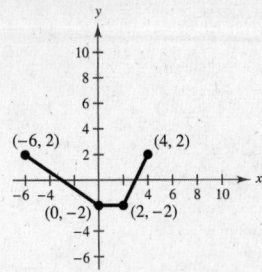

(b) $y = f(x) + 4$

Vertical shift 4 units upward

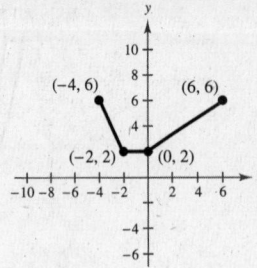

(c) $y = 2f(x)$

Vertical stretch (each y-value is multiplied by 2)

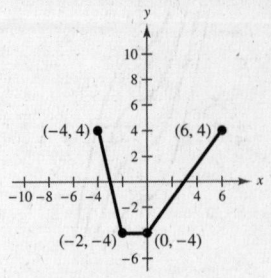

(d) $y = -f(x - 4)$

Reflection in the x-axis and a horizontal shift 4 units to the right

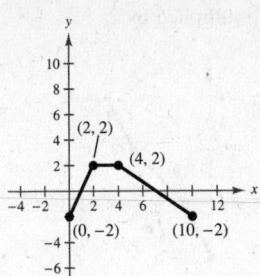

(e) $y = f(x) - 3$

Vertical shift 3 units downward

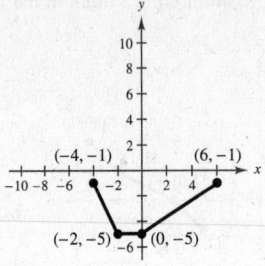

(f) $y = -f(x) - 1$

Reflection in the x-axis and a vertical shift 1 unit downward

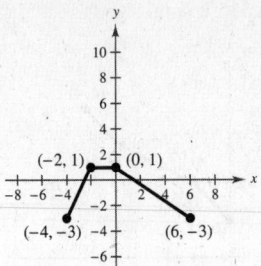

(g) $y = f(2x)$

Horizontal shrink (each x-value is divided by 2)

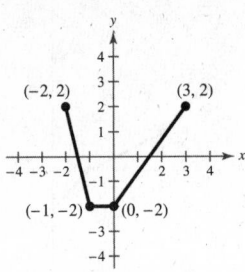

7. (a) $y = f(x) - 1$

Vertical shift 1 unit downward

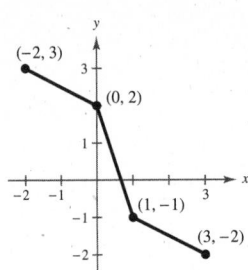

(b) $y = f(x - 1)$

Horizontal shift 1 unit to the right

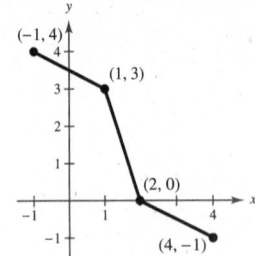

(c) $y = f(-x)$

Reflection about the y-axis

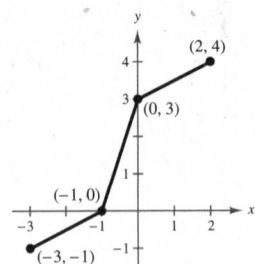

—CONTINUED—

7. —CONTINUED—

(d) $y = f(x + 1)$

Horizontal shift 1 unit to the left

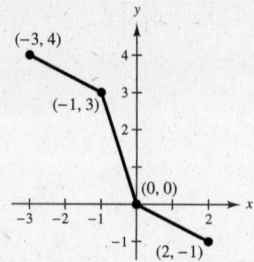

(e) $y = -f(x - 2)$

Reflection about the x-axis and a horizontal shift 2 units to the right

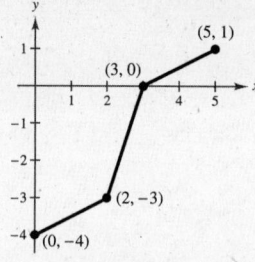

(f) $y = \frac{1}{2} f(x)$

Vertical shrink $\left(\text{each } y\text{-value is multiplied by } \frac{1}{2}\right)$

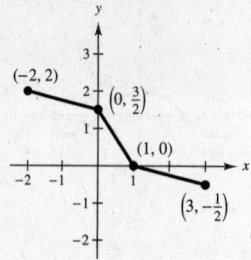

(g) $y = f(2x)$

Horizontal shrink $\left(\text{each } x\text{-value is multiplied by } \frac{1}{2}\right)$

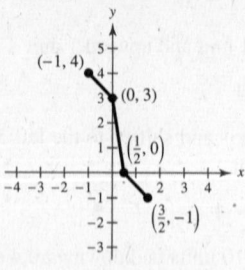

8. (a) $y = f(x - 5)$

Horizontal shift 5 units to the right

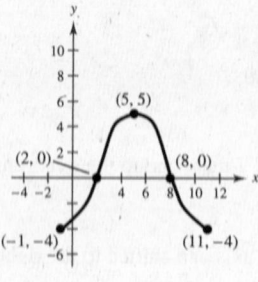

(b) $y = -f(x) + 3$

Reflection in the x-axis and a vertical shift 3 units upward

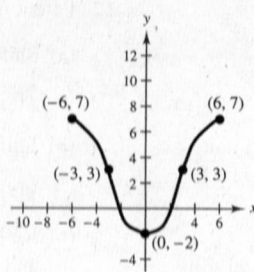

(c) $y = \frac{1}{3} f(x)$

Vertical shrink $\left(\text{each } y\text{-value is multiplied by } \frac{1}{3}\right)$

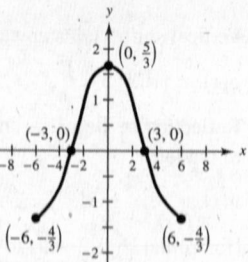

(d) $y = -f(x + 1)$

Reflection in the x-axis and a horizontal shift 1 unit to the left

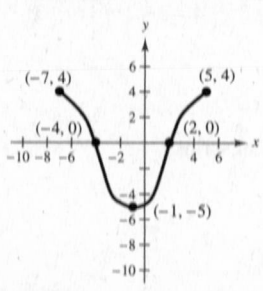

(e) $y = f(-x)$

Reflection in the y-axis

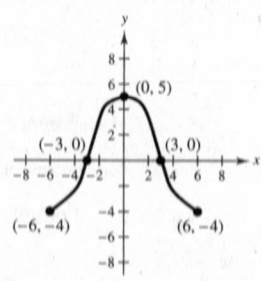

(f) $y = f(x) - 10$

Vertical shift 10 units downward

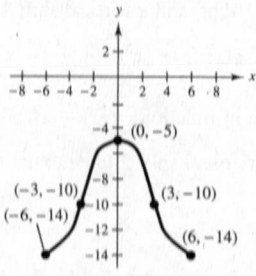

—CONTINUED—

8. **—CONTINUED—**

(g) $y = f\left(\frac{1}{3}x\right)$

Horizontal stretch (each x-value is multiplied by 3)

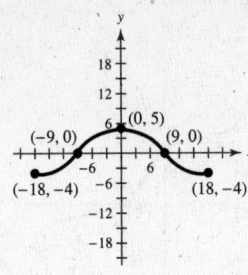

9. Parent function: $f(x) = x^2$

(a) Vertical shift 1 unit downward

$g(x) = x^2 - 1$

(b) Reflection about the x-axis, horizontal shift 1 unit to the left, and a vertical shift 1 unit upward

$g(x) = -(x + 1)^2 + 1$

(c) Reflection about the x-axis, horizontal shift 2 units to the right, and a vertical shift 6 units upward

$g(x) = -(x - 2)^2 + 6$

(d) Horizontal shift 5 units to the right and a vertical shift 3 units downward

$g(x) = (x - 5)^2 - 3$

10. Parent function: $f(x) = x^3$

(a) Reflected in the x-axis and shifted upward 1 unit

$g(x) = -x^3 + 1 = 1 - x^3$

(b) Shifted to the right 1 unit and upward 1 unit

$g(x) = (x - 1)^3 + 1$

(c) Reflected in the x-axis and shifted to the left 3 units and downward 1 unit

$g(x) = -(x + 3)^3 - 1$

(d) Shifted to the right 10 units and downward 4 units

$g(x) = (x - 10)^3 - 4$

11. Parent function: $f(x) = |x|$

(a) Vertical shift 5 units upward

$g(x) = |x| + 5$

(b) Reflection in the x-axis and a horizontal shift 3 units to the left

$g(x) = -|x + 3|$

(c) Horizontal shift 2 units to the right and a vertical shift 4 units downward

$g(x) = |x - 2| - 4$

(d) Reflection in the x-axis, horizontal shift 6 units to the right, and a vertical shift 1 unit downward

$g(x) = -|x - 6| - 1$

12. Parent function: $f(x) = \sqrt{x}$

(a) Shifted down 3 units

$g(x) = \sqrt{x} - 3$

(b) Shifted downward 7 units and to the left 1 unit

$g(x) = \sqrt{x + 1} - 7$

(c) Reflected in the x-axis and shifted to the right 5 units and upward 5 units

$g(x) = -\sqrt{x - 5} + 5$

(d) Reflected about the x- and y-axis and shifted to the right 3 units and downward 4 units

$g(x) = -\sqrt{-x + 3} - 4 = -\sqrt{-(x - 3)} - 4$

13. Parent function: $f(x) = x^3$

Horizontal shift 2 units to the right: $y = (x - 2)^3$

14. Parent function: $y = x$

Transformation: vertical shrink

Formula: $y = \frac{1}{2}x$

15. Parent function: $f(x) = x^2$

Reflection in the x-axis: $y = -x^2$

16. Parent function: $y = [\![x]\!]$

Transformation: vertical shift

Formula: $y = [\![x]\!] + 4$

17. Parent function: $f(x) = \sqrt{x}$

Reflection in the x-axis and a vertical shift 1 unit upward:
$y = -\sqrt{x} + 1$

18. Parent function: $y = |x|$

Transformation: horizontal shift

Formula: $y = |x + 2|$

19. $g(x) = 12 - x^2$

(a) Parent function: $f(x) = x^2$

(b) Reflection in the x-axis and a vertical shift 12 units upward

(c)

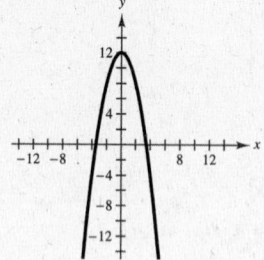

(d) $g(x) = 12 - f(x)$

20. $g(x) = (x - 8)^2$

(a) Parent function: $f(x) = y = x^2$

(b) Horizontal shift of 8 units to the right

(c)

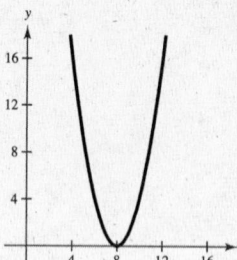

(d) $g(x) = f(x - 8)$

21. $g(x) = x^3 + 7$

(a) Parent function: $f(x) = x^3$

(b) Vertical shift 7 units upward

(c)

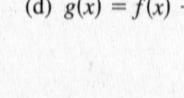

(d) $g(x) = f(x) + 7$

22. $g(x) = -x^3 - 1$

(a) Parent function: $f(x) = x^3$

(b) Reflection in the x-axis; vertical shift of 1 unit downward

(c)

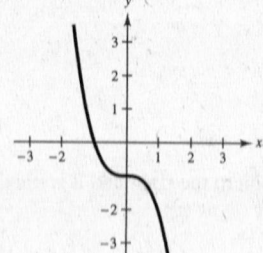

(d) $g(x) = -f(x) - 1$

23. $g(x) = \frac{2}{3}x^2 + 4$

(a) Parent function: $f(x) = x^2$

(b) Vertical shrink of two-thirds, and a vertical shift 4 units upward

(c)

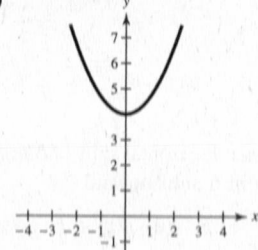

(d) $g(x) = \frac{2}{3}f(x) + 4$

24. $g(x) = 2(x - 7)^2$

(a) Parent function: $f(x) = x^2$

(b) Vertical stretch of 2 and a horizontal shift 7 units to the right of $f(x) = x^2$

(c)

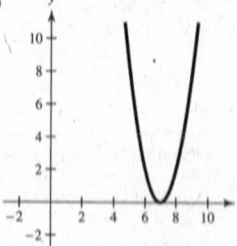

(d) $g(x) = 2f(x - 7)$

25. $g(x) = 2 - (x + 5)^2$

    (a) Parent function: $f(x) = x^2$

    (b) Reflection in the *x*-axis, horizontal shift 5 units to the left, and a vertical shift 2 units upward

    (c) 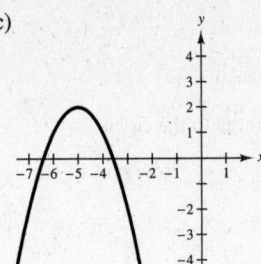   (d) $g(x) = 2 - f(x + 5)$

26. $g(x) = -(x + 10)^2 + 5$

    (a) Parent function: $f(x) = x^2$

    (b) Reflection in the *x*-axis; horizontal shift of 10 units to the left; vertical shift of 5 units upward

    (c) 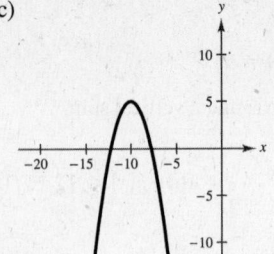   (d) $g(x) = -f(x + 10) + 5$

27. $g(x) = \sqrt{3x}$

    (a) Parent function: $f(x) = \sqrt{x}$

    (b) Horizontal shrink by $\frac{1}{3}$

    (c) 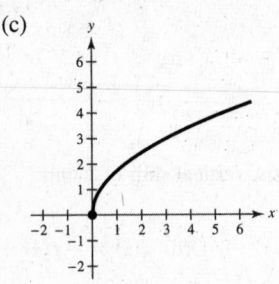   (d) $g(x) = f(3x)$

28. $g(x) = \sqrt{\frac{1}{4}x}$

    (a) Parent function: $f(x) = \sqrt{x}$

    (b) Horizontal stretch of 4, $f(x) = \sqrt{x}$

    (c) 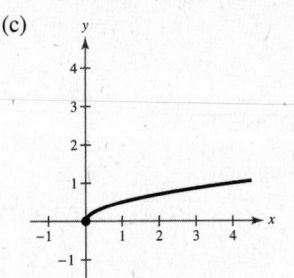   (d) $g(x) = f\left(\frac{1}{4}x\right)$

29. $g(x) = (x - 1)^3 + 2$

    (a) Parent function: $f(x) = x^3$

    (b) Horizontal shift 1 unit to the right and a vertical shift 2 units upward

    (c) 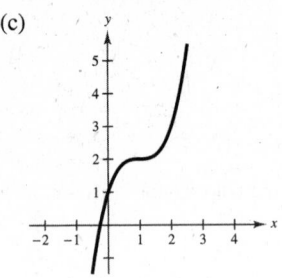   (d) $g(x) = f(x - 1) + 2$

30. $g(x) = (x + 3)^3 - 10$

    (a) Parent function: $f(x) = x^3$

    (b) Horizontal shift of 3 units to the left; vertical shift of 10 units downward

    (c) 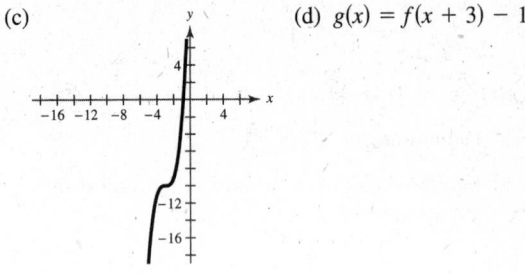   (d) $g(x) = f(x + 3) - 10$

31. $g(x) = -|x| - 2$

    (a) Parent function: $f(x) = |x|$

    (b) Reflection in the *x*-axis; vertical shift 2 units downward

    (c) 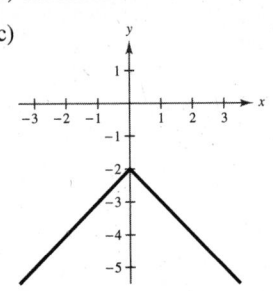   (d) $g(x) = -f(x) - 2$

32. $g(x) = 6 - |x + 5|$

    (a) Parent function: $f(x) = |x|$

    (b) Reflection in the *x*-axis; horizontal shift of 5 units to the left; vertical shift of 6 units upward

    (c) 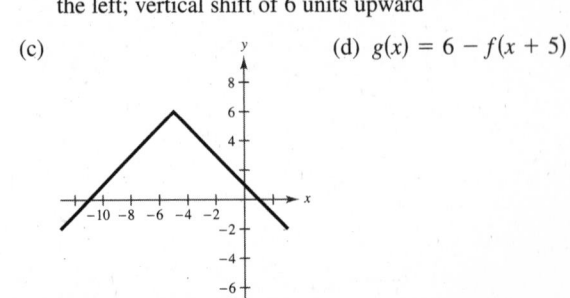   (d) $g(x) = 6 - f(x + 5)$

33. $g(x) = -|x + 4| + 8$

(a) Parent function: $f(x) = |x|$

(b) Reflection in the x-axis, horizontal shift 4 units to the left, and a vertical shift 8 units upward

(c)

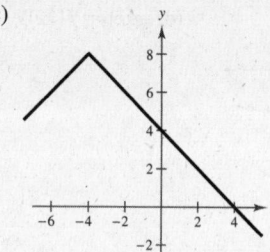

(d) $g(x) = -f(x + 4) + 8$

34. $g(x) = |-x + 3| + 9$

(a) Parent function: $f(x) = |x|$

(b) Reflection in the y-axis; horizontal shift of 3 units to the right; vertical shift of 9 units upward

(c)

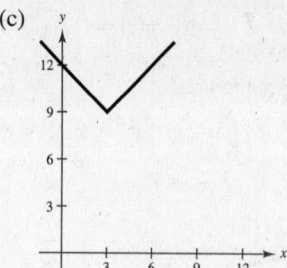

(d) $g(x) = f(-(x - 3)) + 9$

35. $g(x) = 3 - [\![x]\!]$

(a) Parent function: $f(x) = [\![x]\!]$

(b) Reflection in the x-axis and a vertical shift 3 units up

(c)

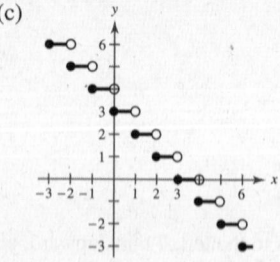

(d) $g(x) = 3 - f(x)$

36. $g(x) = 2[\![x + 5]\!]$

(a) Parent function: $f(x) = [\![x]\!]$

(b) Horizontal shift of 5 units to the left; vertical stretch (each y-value is multiplied by 2)

(c)

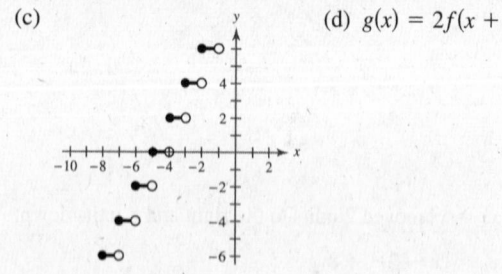

(d) $g(x) = 2f(x + 5)$

37. $g(x) = \sqrt{x - 9}$

(a) Parent function: $f(x) = \sqrt{x}$

(b) Horizontal shift 9 units to the right

(c)

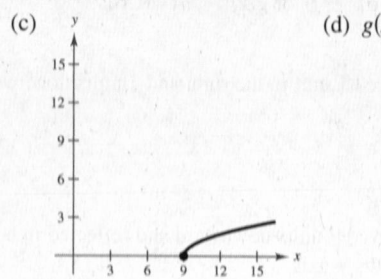

(d) $g(x) = f(x - 9)$

38. $g(x) = \sqrt{x + 4} + 8$

(a) Parent function: $f(x) = \sqrt{x}$

(b) Horizontal shift of 4 units to the left; vertical shift of 8 units upward

(c)

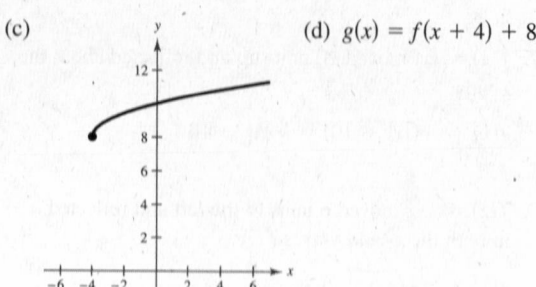

(d) $g(x) = f(x + 4) + 8$

39. $g(x) = \sqrt{7 - x} - 2$ or $g(x) = \sqrt{-(x - 7)} - 2$

(a) Parent function: $f(x) = \sqrt{x}$

(b) Reflection in the y-axis, horizontal shift 7 units to the right, and a vertical shift 2 units downward

(c)

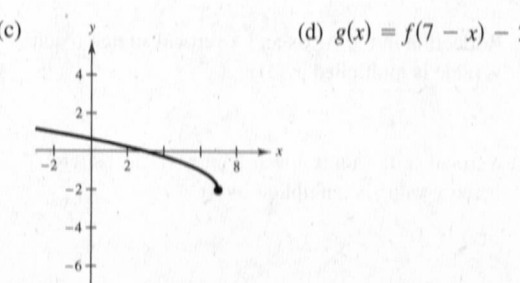

(d) $g(x) = f(7 - x) - 2$

40. $g(x) = -\sqrt{x+1} - 6$

(a) Parent function: $f(x) = \sqrt{x}$

(b) Reflection in the x-axis; horizontal shift of 1 unit to the left; vertical shift of 6 units downward

(c)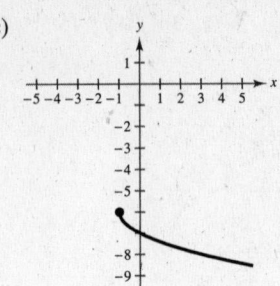

(d) $g(x) = -f(x+1) - 6$

41. $g(x) = \sqrt{\frac{1}{2}x} - 4$

(a) Parent function: $f(x) = \sqrt{x}$

(b) Horizontal stretch (each x-value is multiplied by 2) and a vertical shift 4 units down

(c)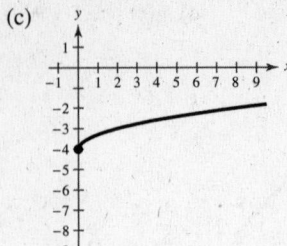

(d) $g(x) = f\left(\frac{1}{2}x\right) - 4$

42. $g(x) = \sqrt{3x} + 1$

(a) Parent function: $f(x) = \sqrt{x}$

(b) Horizontal shrink $\left(\text{each } x\text{-value is multiplied by } \frac{1}{3}\right)$; vertical shift of 1 unit upward

(c)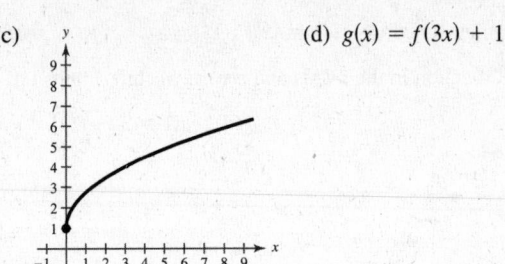

(d) $g(x) = f(3x) + 1$

43. $f(x) = x^2$ moved 2 units to the right and 8 units down.

$g(x) = (x-2)^2 - 8$

44. $f(x) = x^2$ moved 3 units to the left, 7 units upward, and reflected in the x-axis (in that order)

$g(x) = -(x+3)^2 - 7$

45. $f(x) = x^3$ moved 13 units to the right.

$g(x) = (x-13)^3$

46. $f(x) = x^3$ moved 6 units to the left, 6 units downward, and reflected in the y-axis (in that order)

$g(x) = (-x+6)^3 - 6$ or $g(x) = -(x-6)^3 - 6$

47. $f(x) = |x|$ moved 10 units up and reflected about the x-axis.

$g(x) = -(|x| + 10) = -|x| - 10$

48. $f(x) = |x|$ moved 1 unit to the right and 7 units downward

$g(x) = |x-1| - 7$

49. $f(x) = \sqrt{x}$ moved 6 units to the left and reflected in both the x- and y-axes.

$g(x) = -\sqrt{-x+6}$

50. $f(x) = \sqrt{x}$ moved 9 units downward and reflected in both the x-axis and the y-axis

$g(x) = -\left(\sqrt{-x} - 9\right)$

51. $f(x) = x^2$

(a) Reflection in the x-axis and a vertical stretch (each y-value is multiplied by 3)

$g(x) = -3x^2$

(b) Vertical shift 3 units upward and a vertical stretch (each y-value is multiplied by 4)

$g(x) = 4x^2 + 3$

52. $f(x) = x^3$

(a) Vertical shrink $\left(\text{each } y\text{-value is multiplied by } \frac{1}{4}\right)$

$g(x) = \frac{1}{4}x^3$

(b) Reflection in the x-axis and a vertical stretch (each y-value is multiplied by 2)

$g(x) = -2x^3$

53. $f(x) = |x|$

   (a) Reflection in the x-axis and a vertical shrink $\left(\text{each } y\text{-value is multiplied by } \frac{1}{2}\right)$

      $g(x) = -\frac{1}{2}|x|$

   (b) Vertical stretch (each y-value is multiplied by 3) and a vertical shift 3 units downward

      $g(x) = 3|x| - 3$

54. $f(x) = \sqrt{x}$

   (a) Vertical stretch (each y-value is multiplied by 8)

      $g(x) = 8\sqrt{x}$

   (b) Reflection in the x-axis and a vertical shrink $\left(\text{each } y\text{-value is multiplied by } \frac{1}{4}\right)$

      $g(x) = -\frac{1}{4}\sqrt{x}$

55. Parent function: $f(x) = x^3$

Vertical stretch (each y-value is multiplied by 2)

$g(x) = 2x^3$

56. Parent function: $f(x) = |x|$

Vertical stretch (each y-value is multiplied by 6)

$g(x) = 6|x|$

57. Parent function: $f(x) = x^2$

Reflection in the x-axis; vertical shrink $\left(\text{each } y\text{-value is multiplied by } \frac{1}{2}\right)$

$g(x) = -\frac{1}{2}x^2$

58. Parent function: $y = [\![x]\!]$

Horizontal stretch (each x-value is multiplied by 2)

$g(x) = [\![\frac{1}{2}x]\!]$

59. Parent function: $f(x) = \sqrt{x}$

Reflection in the y-axis; vertical shrink $\left(\text{each } y\text{-value is multiplied by } \frac{1}{2}\right)$

$g(x) = \frac{1}{2}\sqrt{-x}$

60. Parent function: $f(x) = |x|$

Reflection in the x-axis; vertical shift of 2 units downward; vertical stretch (each y-value is multiplied by 2)

$g(x) = -2|x| - 2$

61. Parent function: $f(x) = x^3$

Reflection in the x-axis, horizontal shift 2 units to the right and a vertical shift 2 units upward

$g(x) = -(x - 2)^3 + 2$

62. Parent function: $f(x) = |x|$

Horizontal shift of 4 units to the left and a vertical shift of 2 units downward

$g(x) = |x + 4| - 2$

63. Parent function: $f(x) = \sqrt{x}$

Reflection in the x-axis and a vertical shift 3 units downward

$g(x) = -\sqrt{x} - 3$

64. Parent function: $f(x) = x^2$

Horizontal shift of 2 units to the right and a vertical shift of 4 units upward.

$g(x) = (x - 2)^2 + 4$

65. (a) $g(x) = f(x) + 2$

Vertical shift 2 units upward

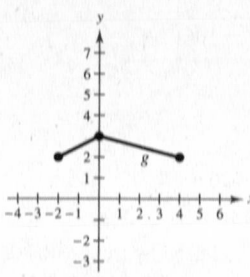

    (b) $g(x) = f(x) - 1$

Vertical shift 1 unit downward

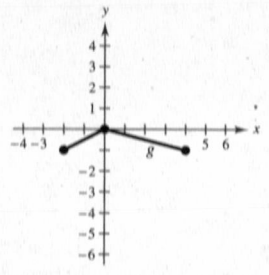

    (c) $g(x) = f(-x)$

Reflection in the y-axis

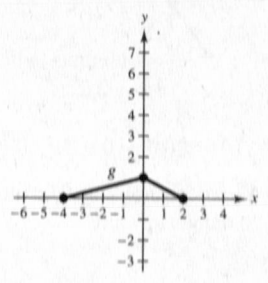

—CONTINUED—

65. **—CONTINUED—**

(d) $g(x) = -2f(x)$

Reflection in the *x*-axis and a vertical stretch (each *y*-value is multiplied by 2)

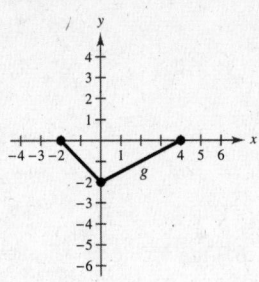

(e) $g(x) = f(4x)$

Horizontal shrink (each *x*-value is multiplied by $\frac{1}{4}$)

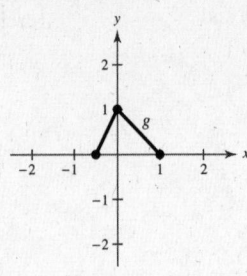

(f) $g(x) = f\left(\frac{1}{2}x\right)$

Horizontal stretch (each *x*-value is multiplied by 2)

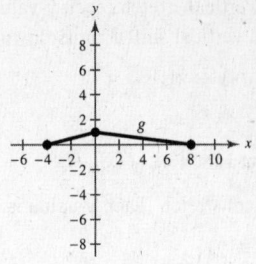

66. (a) $g(x) = f(x) - 5$

Vertical shift 5 units downward

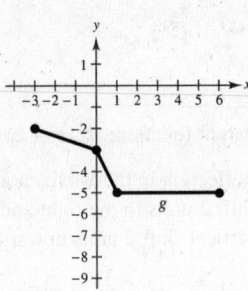

(b) $g(x) = f(x) + \frac{1}{2}$

Vertical shift $\frac{1}{2}$ unit upward

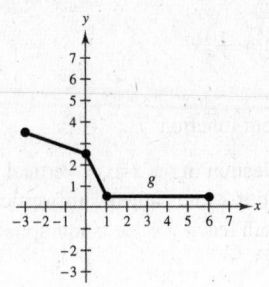

(c) $g(x) = f(-x)$

Reflection in the *y*-axis

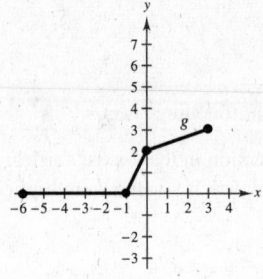

(d) $g(x) = -4f(x)$

Reflection in the *x*-axis and a vertical stretch (each *y*-value is multiplied by 4)

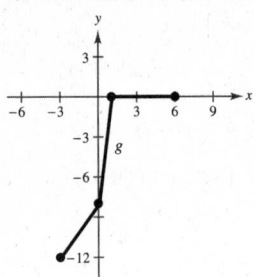

(e) $g(x) = f(2x) + 1$

Horizontal shrink (each *x*-value is multiplied by $\frac{1}{2}$) and a vertical shift 1 unit upward

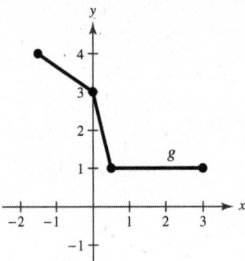

(f) $g(x) = f\left(\frac{1}{4}x\right) - 2$

Horizontal stretch (each *x*-value is multiplied by 4) and a vertical shift 2 units downward

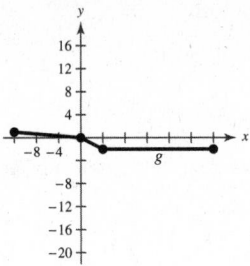

67. $F = f(t) = 20.6 + 0.035t^2, \ 0 \le t \le 22$

(a) A vertical shrink by 0.035 and a vertical shift of 20.6 units upward

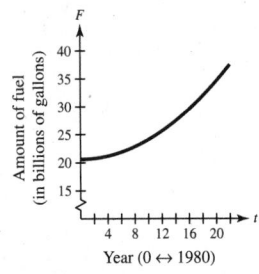

Year ($0 \leftrightarrow 1980$)

(b) $\dfrac{f(22) - f(0)}{22 - 0} = \dfrac{37.54 - 20.6}{22} = 0.77$

The average increase in fuel used by trucks was 0.77 billion gallons per year between 1980 and 2002.

(c) $g(t) = 20.6 + 0.035(t + 10)^2 = f(t + 10)$

This represents a horizontal shift 10 units to the left.

(d) $g(20) = 52.1$ billion gallons

Yes. There are many factors involved here. The number of trucks on the road continues to increase but are more fuel efficient. The availability and the cost of overseas and domestic fuel also plays a role in usage.

68. (a) The graph is a horizontal shift 20.396 units to the left of the graph of the common function $f(x) = x^2$ and a vertical shrink by a factor of 0.0054.

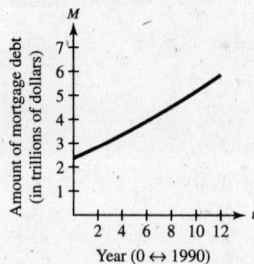

(b) $f(t) = 0.0054(t + 30.396)^2$

By shifting the graph 10 units to the left, you obtain $t = 0$ represents 1990.

69. True, since $|x| = |-x|$, the graphs of $f(x) = |x| + 6$ and $f(x) = |-x| + 6$ are identical.

70. False. The point $(-2, -67)$ lies on the transformation.

71. (a) The profits were only $\frac{3}{4}$ as large as expected:

$$g(t) = \tfrac{3}{4}f(t)$$

(b) The profits were $10,000 greater than predicted:

$$g(t) = f(t) + 10,000$$

(c) There was a two-year delay: $g(t) = f(t - 2)$

72. If you consider the x-axis to be a mirror, each of the y-values of the graph of $y = -f(x)$ is the mirror image of each of the y-values of the graph of $y = f(x)$.

73. $y = f(x + 2) - 1$

Horizontal shift 2 units to the left and a vertical shift 1 unit downward

$$(0, 1) \rightarrow (0 - 2, 1 - 1) = (-2, 0)$$

$$(1, 2) \rightarrow (1 - 2, 2 - 1) = (-1, 1)$$

$$(2, 3) \rightarrow (2 - 2, 3 - 1) = (0, 2)$$

74. Answers will vary.

(a) is probably simpler to graph by plotting points and (b) is probably simpler to graph by translating the graph of $y = x^2$.

75. $\dfrac{4}{x} + \dfrac{4}{1 - x} = \dfrac{4(1 - x) + 4x}{x(1 - x)} = \dfrac{4 - 4x + 4x}{x(1 - x)} = \dfrac{4}{x(1 - x)}$

76. $\dfrac{2}{x + 5} - \dfrac{2}{x - 5}$

$$\dfrac{2}{x + 5} - \dfrac{2}{x - 5} = \dfrac{2(x - 5) - 2(x + 5)}{(x + 5)(x - 5)}$$

$$= \dfrac{2x - 10 - 2x - 10}{(x + 5)(x - 5)} = \dfrac{-20}{(x + 5)(x - 5)}$$

77. $\dfrac{3}{x - 1} - \dfrac{2}{x(x - 1)} = \dfrac{3x - 2}{x(x - 1)}$

78. $\dfrac{x}{x - 5} + \dfrac{1}{2}$

$$\dfrac{x}{x - 5} + \dfrac{1}{2} = \dfrac{2x + x - 5}{2(x - 5)} = \dfrac{3x - 5}{2(x - 5)}$$

79. $(x - 4)\left(\dfrac{1}{\sqrt{x^2 - 4}}\right) = \dfrac{x - 4}{\sqrt{x^2 - 4}} = \dfrac{(x - 4)\sqrt{x^2 - 4}}{x^2 - 4}$

80. $\left(\dfrac{x}{x^2 - 4}\right)\left(\dfrac{x^2 - x - 2}{x^2}\right)$

$$\left(\dfrac{x}{x^2 - 4}\right)\left(\dfrac{x^2 - x - 2}{x^2}\right) = \dfrac{x(x - 2)(x + 1)}{x^2(x - 2)(x + 2)}$$

$$= \dfrac{x + 1}{x(x + 2)}, \quad x \neq 2$$

81. $(x^2 - 9) \div \left(\dfrac{x+3}{5}\right) = \dfrac{(x+3)(x-3)}{1} \cdot \dfrac{5}{x+3} = 5(x-3), \ x \neq -3$

82. $\left(\dfrac{x}{x^2 - 3x - 28}\right) \div \left(\dfrac{x^2 + 3x}{x^2 + 5x + 4}\right)$

$\left(\dfrac{x}{x^2 - 3x - 28}\right) \div \left(\dfrac{x^2 + 3x}{x^2 + 5x + 4}\right) = \left(\dfrac{x}{x^2 - 3x - 28}\right) \cdot \left(\dfrac{x^2 + 5x + 4}{x^2 + 3x}\right)$

$$= \dfrac{x(x+4)(x+1)}{(x-7)(x+4)x(x+3)} = \dfrac{x+1}{(x-7)(x+3)}, \ x \neq -4, -1, 0$$

83. $f(x) = x^2 - 6x + 11$

(a) $f(-3) = (-3)^2 - 6(-3) + 11 = 38$

(b) $f\left(-\frac{1}{2}\right) = \left(-\frac{1}{2}\right)^2 - 6\left(-\frac{1}{2}\right) + 11 = \frac{1}{4} + 3 + 11 = \frac{57}{4}$

(c) $f(x - 3) = (x - 3)^2 - 6(x - 3) + 11 = x^2 - 6x + 9 - 6x + 18 + 11 = x^2 - 12x + 38$

84. $f(x) = \sqrt{x + 10} - 3$

(a) $f(-10) = \sqrt{-10 + 10} - 3$ (b) $f(26) = \sqrt{26 + 10} - 3$ (c) $f(x - 10) = \sqrt{x - 10 + 10} - 3$

           $= -3$ $= \sqrt{36} - 3 = 3$ $= \sqrt{x} - 3$

85. $f(x) = \dfrac{2}{11 - x}$

Domain: All real numbers except $x = 11$

86. $f(x) = \dfrac{\sqrt{x - 3}}{x - 8}$

Domain: $x \geq 3, \ x \neq 8$ or $[3, 8) \cup (8, \infty)$

87. $f(x) = \sqrt{81 - x^2}$

       $81 - x^2 \geq 0$

$(9 + x)(9 - x) \geq 0$

Critical numbers: $x = \pm 9$

Test intervals: $(-\infty, -9), (-9, 9), (9, \infty)$

Test: Is $81 - x^2 \geq 0$?

Solution: $[-9, 9]$

Domain of $f(x)$: $-9 \leq x \leq 9$

88. $f(x) = \sqrt[3]{4 - x^2}$

Domain: All real numbers

Section 1.8   Combinations of Functions: Composite Functions

- ■   Given two functions, f and g, you should be able to form the following functions (if defined):

  1. Sum: $(f + g)(x) = f(x) + g(x)$

  2. Difference: $(f - g)(x) = f(x) - g(x)$

  3. Product: $(fg)(x) = f(x)g(x)$

  4. Quotient: $(f/g)(x) = f(x)/g(x), \ g(x) \neq 0$

  5. Composition of f with g: $(f \circ g)(x) = f(g(x))$

  6. Composition of g with f: $(g \circ f)(x) = g(f(x))$

Vocabulary Check

1. addition, subtraction, multiplication, division

2. composition

3. $g(x)$

4. inner; outer

1.

x	0	1	2	3
f	2	3	1	2
g	-1	0	$\frac{1}{2}$	0
$f + g$	1	3	$\frac{3}{2}$	2

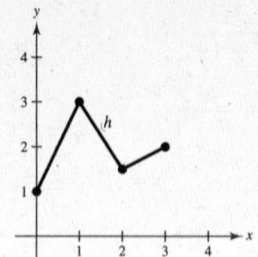

2.

x	-2	-1	0	1	2
$f(x)$	-2	0	-1	-1	1
$g(x)$	1	1	0	2	2
$h(x) = (f + g)(x)$	-1	1	-1	1	3

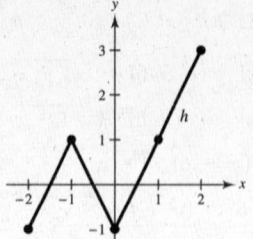

3.

x	-2	0	1	2	4
f	2	0	1	2	4
g	4	2	1	0	2
$f + g$	6	2	2	2	6

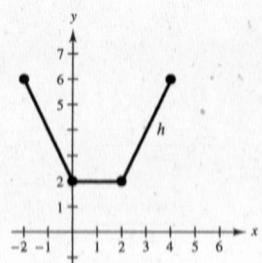

4. The domain common to both functions is $[-1, 1]$, which is the domain of the sum.

x	-1	0	1
$f(x)$	0	1.5	3
$g(x)$	-1	-2	1
$h(x) = f(x) + g(x)$	-1	-0.5	4

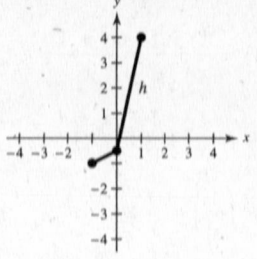

5. $f(x) = x + 2, g(x) = x - 2$

(a) $(f + g)(x) = f(x) + g(x) = (x + 2) + (x - 2) = 2x$

(b) $(f - g)(x) = f(x) - g(x) = (x + 2) - (x - 2) = 4$

(c) $(fg)(x) = f(x) \cdot g(x) = (x + 2)(x - 2) = x^2 - 4$

(d) $\left(\dfrac{f}{g}\right)(x) = \dfrac{f(x)}{g(x)} = \dfrac{x + 2}{x - 2}$

Domain: all real numbers x except $x = 2$

6. $f(x) = 2x - 5, g(x) = 2 - x$

(a) $(f + g)(x) = 2x - 5 + 2 - x = x - 3$

(b) $(f - g)(x) = 2x - 5 - (2 - x)$

$\qquad = 2x - 5 - 2 + x = 3x - 7$

(c) $(fg)(x) = (2x - 5)(2 - x)$

$\qquad = 4x - 2x^2 - 10 + 5x = -2x^2 + 9x - 10$

(d) $\left(\dfrac{f}{g}\right)(x) = \dfrac{2x - 5}{2 - x}$

Domain: all real numbers x except $x = 2$

7. $f(x) = x^2, g(x) = 4x - 5$

    (a) $(f + g)(x) = f(x) + g(x)$

$$= x^2 + (4x - 5) = x^2 + 4x - 5$$

    (b) $(f - g)(x) = f(x) - g(x)$

$$= x^2 - (4x - 5) = x^2 - 4x + 5$$

    (c) $(fg)(x) = f(x) \cdot g(x) = x^2(4x - 5) = 4x^3 - 5x^2$

    (d) $\left(\dfrac{f}{g}\right)(x) = \dfrac{f(x)}{g(x)} = \dfrac{x^2}{4x - 5}$

        Domain: all real numbers x except $x = \dfrac{5}{4}$

8. $f(x) = 2x - 5, g(x) = 4$

    (a) $(f + g)(x) = 2x - 5 + 4 = 2x - 1$

    (b) $(f - g)(x) = 2x - 5 - 4 = 2x - 9$

    (c) $(fg)(x) = (2x - 5)(4) = 8x - 20$

    (d) $\left(\dfrac{f}{g}\right)(x) = \dfrac{2x - 5}{4} = \dfrac{1}{2}x - \dfrac{5}{4}$

        Domain: all real numbers x

9. $f(x) = x^2 + 6, g(x) = \sqrt{1 - x}$

    (a) $(f + g)(x) = f(x) + g(x) = (x^2 + 6) + \sqrt{1 - x}$

    (b) $(f - g)(x) = f(x) - g(x) = (x^2 + 6) - \sqrt{1 - x}$

    (c) $(fg)(x) = f(x) \cdot g(x) = (x^2 + 6)\sqrt{1 - x}$

    (d) $\left(\dfrac{f}{g}\right)(x) = \dfrac{f(x)}{g(x)} = \dfrac{x^2 + 6}{\sqrt{1 - x}} = \dfrac{(x^2 + 6)\sqrt{1 - x}}{1 - x},$

        Domain: $x < 1$

10. $f(x) = \sqrt{x^2 - 4}, \; g(x) = \dfrac{x^2}{x^2 + 1}$

    (a) $(f + g)(x) = \sqrt{x^2 - 4} + \dfrac{x^2}{x^2 + 1}$

    (b) $(f - g)(x) = \sqrt{x^2 - 4} - \dfrac{x^2}{x^2 + 1}$

    (c) $(fg)(x) = \sqrt{x^2 - 4}\left(\dfrac{x^2}{x^2 + 1}\right) = \dfrac{x^2\sqrt{x^2 - 4}}{x^2 + 1}$

    (d) $\left(\dfrac{f}{g}\right)(x) = \sqrt{x^2 - 4} \div \dfrac{x^2}{x^2 + 1} = \dfrac{(x^2 + 1)\sqrt{x^2 - 4}}{x^2}$

        Domain: $x^2 - 4 \geq 0$

$$x^2 \geq 4 \implies x \geq 2 \text{ or } x \leq -2$$

        Domain: $|x| \geq 2$

11. $f(x) = \dfrac{1}{x}, g(x) = \dfrac{1}{x^2}$

    (a) $(f + g)(x) = f(x) + g(x) = \dfrac{1}{x} + \dfrac{1}{x^2} = \dfrac{x + 1}{x^2}$

    (b) $(f - g)(x) = f(x) - g(x) = \dfrac{1}{x} - \dfrac{1}{x^2} = \dfrac{x - 1}{x^2}$

    (c) $(fg)(x) = f(x) \cdot g(x) = \dfrac{1}{x}\left(\dfrac{1}{x^2}\right) = \dfrac{1}{x^3}$

    (d) $\left(\dfrac{f}{g}\right)(x) = \dfrac{f(x)}{g(x)} = \dfrac{1/x}{1/x^2} = \dfrac{x^2}{x} = x$

        Domain: all real numbers x except $x = 0$

12. $f(x) = \dfrac{x}{x + 1}, g(x) = x^3$

    (a) $(f + g)(x) = \dfrac{x}{x + 1} + x^3 = \dfrac{x + x^4 + x^3}{x + 1}$

    (b) $(f - g)(x) = \dfrac{x}{x + 1} - x^3 = \dfrac{x - x^4 - x^3}{x + 1}$

    (c) $(fg)(x) = \dfrac{x}{x + 1} \cdot x^3 = \dfrac{x^4}{x + 1}$

    (d) $\left(\dfrac{f}{g}\right)(x) = \dfrac{x}{x + 1} \div x^3 = \dfrac{x}{x + 1} \cdot \dfrac{1}{x^3} = \dfrac{1}{x^2(x + 1)}$

        Domain: all real numbers x except $x = 0$ and $x = -1$

For Exercises 13–24, $f(x) = x^2 + 1$ and $g(x) = x - 4$.

13. $(f + g)(2) = f(2) + g(2) = (2^2 + 1) + (2 - 4) = 3$

14. $(f - g)(-1) = f(-1) - g(-1)$

$$= (-1)^2 + 1 - (-1 - 4)$$

$$= 1 + 1 - (-5)$$

$$= 7$$

15. $(f - g)(0) = f(0) - g(0) = (0^2 + 1) - (0 - 4) = 5$

16. $(f + g)(1) = f(1) + g(1)$
$$= (1)^2 + 1 + (1) - 4$$
$$= -1$$

17. $(f - g)(3t) = f(3t) - g(3t) = [(3t)^2 + 1] - (3t - 4)$
$$= 9t^2 - 3t + 5$$

18. $(f + g)(t - 2) = f(t - 2) + g(t - 2)$
$$= (t - 2)^2 + 1 + (t - 2) - 4$$
$$= t^2 - 4t + 4 + 1 + t - 2 - 4$$
$$= t^2 - 3t - 1$$

19. $(fg)(6) = f(6)g(6) = (6^2 + 1)(6 - 4) = 74$

20. $(fg)(-6) = f(-6) \cdot g(-6)$
$$= [(-6)^2 + 1][(-6) - 4]$$
$$= (37)(-10)$$
$$= -370$$

21. $\left(\dfrac{f}{g}\right)(5) = \dfrac{f(5)}{g(5)} = \dfrac{5^2 + 1}{5 - 4} = 26$

22. $\left(\dfrac{f}{g}\right)(0) = \dfrac{f(0)}{g(0)} = \dfrac{0^2 + 1}{0 - 4} = -\dfrac{1}{4}$

23. $\left(\dfrac{f}{g}\right)(-1) - g(3) = \dfrac{f(-1)}{g(-1)} - g(3)$
$$= \dfrac{(-1)^2 + 1}{-1 - 4} - (3 - 4)$$
$$= -\dfrac{2}{5} + 1 = \dfrac{3}{5}$$

24. $(fg)(5) + f(4) = f(5)g(5) + f(4)$
$$= (5^2 + 1)(5 - 4) + (4^2 + 1)$$
$$= 26 \cdot 1 + 17$$
$$= 43$$

25. $f(x) = \tfrac{1}{2}x, g(x) = x - 1, (f + g)(x) = \tfrac{3}{2}x - 1$

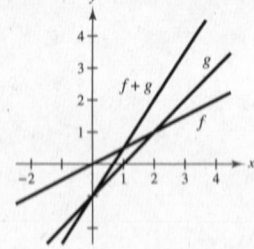

26. $f(x) = \tfrac{1}{3}x, \ g(x) = -x + 4$
$(f + g)(x) = \tfrac{1}{3}x - x + 4 = -\tfrac{2}{3}x + 4$

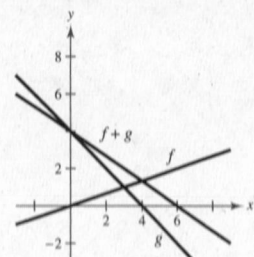

27. $f(x) = x^2, g(x) = -2x, (f + g)(x) = x^2 - 2x$

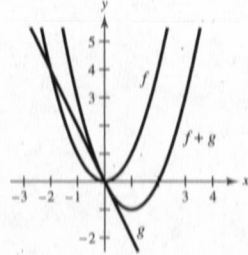

28. $f(x) = 4 - x^2, g(x) = x$
$(f + g)(x) = 4 - x^2 + x = 4 + x - x^2$

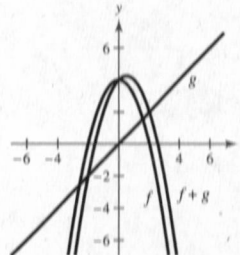

29. $f(x) = 3x$, $g(x) = -\dfrac{x^3}{10}$, $(f+g)(x) = 3x - \dfrac{x^3}{10}$

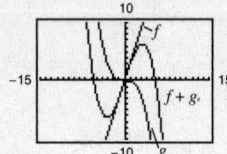

For $0 \le x \le 2$, $f(x)$ contributes most to the magnitude.

For $x > 6$, $g(x)$ contributes most to the magnitude.

30. $f(x) = \dfrac{x}{2}$, $g(x) = \sqrt{x}$, $(f+g)(x) = \dfrac{x}{2} + \sqrt{x}$

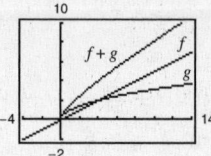

$g(x)$ contributes most to the magnitude of the sum for $0 \le x \le 2$. $f(x)$ contributes most to the magnitude of the sum for $x > 6$.

31. $f(x) = x^2$, $g(x) = x - 1$

(a) $(f \circ g)(x) = f(g(x)) = f(x-1) = (x-1)^2$

(b) $(g \circ f)(x) = g(f(x)) = g(x^2) = x^2 - 1$

(c) $(f \circ f)(x) = f(f(x)) = f(x^2) = (x^2)^2 = x^4$

32. $f(x) = 3x + 5$, $g(x) = 5 - x$

(a) $(f \circ g)(x) = f(g(x))$

$\qquad = f(5 - x) = 3(5 - x) + 5 = 20 - 3x$

(b) $(g \circ f)(x) = g(f(x))$

$\qquad = g(3x + 5) = 5 - (3x + 5) = -3x$

(c) $(f \circ f)(x) = f(f(x))$

$\qquad = f(3x + 5) = 3(3x + 5) + 5 = 9x + 20$

33. $f(x) = \sqrt[3]{x - 1}$, $g(x) = x^3 + 1$

(a) $(f \circ g)(x) = f(g(x))$

$\qquad = f(x^3 + 1)$

$\qquad = \sqrt[3]{(x^3 + 1) - 1}$

$\qquad = \sqrt[3]{x^3} = x$

(b) $(g \circ f)(x) = g(f(x))$

$\qquad = g\left(\sqrt[3]{x - 1}\right)$

$\qquad = \left(\sqrt[3]{x - 1}\right)^3 + 1$

$\qquad = (x - 1) + 1 = x$

(c) $(f \circ f)(x) = f(f(x))$

$\qquad = f\left(\sqrt[3]{x - 1}\right)$

$\qquad = \sqrt[3]{\sqrt[3]{x - 1} - 1}$

34. $f(x) = x^3$, $g(x) = \dfrac{1}{x}$

(a) $(f \circ g)(x) = f(g(x)) = f\left(\dfrac{1}{x}\right) = \left(\dfrac{1}{x}\right)^3 = \dfrac{1}{x^3}$

(b) $(g \circ f)(x) = g(f(x)) = g(x^3) = \dfrac{1}{x^3}$

(c) $(f \circ f)(x) = f(f(x)) = f(x^3) = (x^3)^3 = x^9$

35. $f(x) = \sqrt{x + 4}$ Domain: $x \ge -4$

$g(x) = x^2$ Domain: all real numbers x

(a) $(f \circ g)(x) = f(g(x)) = f(x^2) = \sqrt{x^2 + 4}$

Domain: all real numbers x

(b) $(g \circ f)(x) = g(f(x))$

$\qquad = g\left(\sqrt{x + 4}\right) = \left(\sqrt{x + 4}\right)^2 = x + 4$

Domain: $x \ge -4$

36. $f(x) = \sqrt[3]{x - 5}$ Domain: all real numbers x

$g(x) = x^3 + 1$ all real numbers x

(a) $(f \circ g)(x) = f(g(x))$

$\qquad = f(x^3 + 1) = \sqrt[3]{x^3 + 1 - 5} = \sqrt[3]{x^3 - 4}$

Domain: all real numbers x

(b) $(g \circ f)(x) = g(f(x))$

$\qquad = g\left(\sqrt[3]{x - 5}\right)$

$\qquad = \left(\sqrt[3]{x - 5}\right)^3 + 1$

$\qquad = x - 5 + 1 = x - 4$

Domain: all real numbers x

37. $f(x) = x^2 + 1$ Domain: all real numbers x

$g(x) = \sqrt{x}$ Domain: $x \geq 0$

(a) $(f \circ g)(x) = f(g(x)) = f(\sqrt{x}) = (\sqrt{x})^2 + 1 = x + 1$

Domain: $x \geq 0$

(b) $(g \circ f)(x) = g(f(x)) = g(x^2 + 1) = \sqrt{x^2 + 1}$

Domain: all real numbers x

38. $f(x) = x^{2/3}$ Domain: all real numbers x

$g(x) = x^6$ Domain: all real numbers x

(a) $(f \circ g)(x) = f(g(x)) = f(x^6) = (x^6)^{2/3} = x^4$

Domain: all real numbers x

(b) $(g \circ f)(x) = g(f(x)) = g(x^{2/3}) = (x^{2/3})^6 = x^4$

Domain: all real numbers x

39. $f(x) = |x|$ Domain: all real numbers x

$g(x) = x + 6$ Domain: all real numbers x

(a) $(f \circ g)(x) = f(g(x)) = f(x + 6) = |x + 6|$

Domain: all real numbers x

(b) $(g \circ f)(x) = g(f(x)) = g(|x|) = |x| + 6$

Domain: all real numbers x

40. $f(x) = |x - 4|$ Domain: all real numbers x

$g(x) = 3 - x$ Domain: all real numbers x

(a) $(f \circ g)(x) = f(g(x))$

$\qquad = f(3 - x) = |(3 - x) - 4| = |-x - 1|$

Domain: all real numbers x

(b) $(g \circ f)(x) = g(f(x))$

$\qquad = g(|x - 4|) = 3 - (|x - 4|) = 3 - |x - 4|$

Domain: all real numbers x

41. $f(x) = \dfrac{1}{x}$ Domain: all real numbers x except $x = 0$

$g(x) = x + 3$ Domain: all real numbers x

(a) $(f \circ g)(x) = f(g(x)) = f(x + 3) = \dfrac{1}{x + 3}$

Domain: all real numbers x except $x = -3$

(b) $(g \circ f)(x) = g(f(x)) = g\left(\dfrac{1}{x}\right) = \dfrac{1}{x} + 3$

Domain: all real numbers x except $x = 0$

42. $f(x) = \dfrac{3}{x^2 - 1}$ Domain: all real numbers x except $x = \pm 1$

$g(x) = x + 1$ Domain: all real numbers x

(a) $(f \circ g)(x) = f(g(x))$

$\qquad = f(x + 1)$

$\qquad = \dfrac{3}{(x + 1)^2 - 1}$

$\qquad = \dfrac{3}{x^2 + 2x + 1 - 1}$

$\qquad = \dfrac{3}{x^2 + 2x}$

Domain: all real numbers x except $x = 0$ and $x = -2$

(b) $(g \circ f)(x) = g(f(x))$

$\qquad = g\left(\dfrac{3}{x^2 - 1}\right)$

$\qquad = \dfrac{3}{x^2 - 1} + 1$

$\qquad = \dfrac{3 + x^2 - 1}{x^2 - 1}$

$\qquad = \dfrac{x^2 + 2}{x^2 - 1}$

Domain: all real numbers x except $x = \pm 1$

43. (a) $(f + g)(3) = f(3) + g(3) = 2 + 1 = 3$

(b) $\left(\dfrac{f}{g}\right)(2) = \dfrac{f(2)}{g(2)} = \dfrac{0}{2} = 0$

44. (a) $(f - g)(1) = f(1) - g(1) = 2 - 3 = -1$

(b) $(fg)(4) = f(4) \cdot g(4) = 4 \cdot 0 = 0$

45. (a) $(f \circ g)(2) = f(g(2)) = f(2) = 0$

(b) $(g \circ f)(2) = g(f(2)) = g(0) = 4$

46. (a) $(f \circ g)(1) = f(g(1)) = f(3) = 2$

(b) $(g \circ f)(3) = g(f(3)) = g(2) = 2$

47. $h(x) = (2x^2 + 1)^2$

One possibility: Let $f(x) = x^2$ and $g(x) = 2x + 1$, then $(f \circ g)(x) = h(x)$.

48. $h(x) = (1 - x)^3$

One possibility: Let $g(x) = 1 - x$ and $f(x) = x^3$, then $(f \circ g)(x) = h(x)$.

49. $h(x) = \sqrt[3]{x^2 - 4}$

One possibility: Let $f(x) = \sqrt[3]{x}$ and $g(x) = x^2 - 4$, then $(f \circ g)(x) = h(x)$.

50. $h(x) = \sqrt{9 - x}$

One possibility: Let $g(x) = 9 - x$ and $f(x) = \sqrt{x}$, then $(f \circ g)(x) = h(x)$.

51. $h(x) = \dfrac{1}{x + 2}$

One possibility: Let $f(x) = 1/x$ and $g(x) = x + 2$, then $(f \circ g)(x) = h(x)$.

52. $h(x) = \dfrac{4}{(5x + 2)^2}$

One possibility: Let $g(x) = 5x + 2$ and $f(x) = \dfrac{4}{x^2}$, then $(f \circ g)(x) = h(x)$.

53. $h(x) = \dfrac{-x^2 + 3}{4 - x^2}$

One possibility: Let $f(x) = \dfrac{x + 3}{4 + x}$ and $g(x) = -x^2$, then $(f \circ g)(x) = h(x)$.

54. $h(x) = \dfrac{27x^3 + 6x}{10 - 27x^3}$

One possibility: Let $g(x) = x^3$ and $f(x) = \dfrac{27x + 6\sqrt[3]{x}}{10 - 27x}$, then $(f \circ g)(x) = h(x)$.

55. $T(x) = R(x) + B(x) = \frac{3}{4}x + \frac{1}{15}x^2$

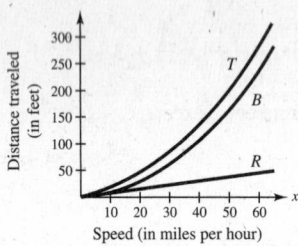

56. (a) Total sales $= R_1 + R_2$

$$= 480 - 8t - 0.8t^2 + 254 + 0.78t$$

$$= 734 - 7.22t - 0.8t^2$$

(b)

57. (a) $c(t) = \dfrac{p(t) + b(t) - d(t)}{p(t)} \times 100$

(b) $c(5)$ represents the percent change in the population in the year 2005.

58. (a) $p(t) = d(t) + c(t)$

(b) $p(5)$ represents the number of dogs and cats in 2005.

(c) $h(t) = \dfrac{p(t)}{n(t)} = \dfrac{d(t) + c(t)}{n(t)}$

$h(t)$ represents the number of dogs and cats at time t compared to the population at time t or the number of dogs and cats per capita.

59. $A(t) = 3.36t^2 - 59.8t + 735$, $N(t) = 1.95t^2 - 42.2t + 603$

(a) $(A + N)(t) = A(t) + N(t) = 5.31t^2 - 102.0t + 1338$

This represents the combined Army and Navy personnel (in thousands) from 1990 to 2002, where $t = 0$ corresponds to 1990.

$(A + N)(4) = 1014.96$ thousand

$(A + N)(8) = 861.84$ thousand

$(A + N)(12) = 878.64$ thousand

(b) $(A - N)(t) = A(t) - N(t) = 1.41t^2 - 17.6t + 132$

This represents the number of Army personnel (in thousands) more than the number of Navy personnel from 1990 to 2002, where $t = 0$ corresponds to 1990.

$(A - N)(4) = 84.16$ thousand

$(A - N)(8) = 81.44$ thousand

$(A - N)(12) = 123.84$ thousand

60. (a) $h(t) = \dfrac{E(t)}{P(t)} = \dfrac{25.95t^2 - 231.2t + 3356}{3.02t + 252.0}$

$h(t)$ represents the millions of dollars spent on exercise equipment compared to the millions of people in the U.S., or the amount spent per capita.

(b) $h(7) = 11.0169$ dollars spent per person in 1997

$h(10) = 12.895$ dollars spent per person in 2000

$h(12) = 14.982$ dollars spent per person in 2002

61.

Year	y_1	y_2	y_3
1995	146.2	329.1	44.8
1996	152.0	344.1	48.1
1997	162.2	359.9	52.1
1998	175.2	382.0	55.6
1999	184.4	412.1	57.8
2000	194.7	449.0	57.4
2001	205.5	496.1	57.8

(a) $y_1 \approx 10.20t + 92.7$

$y_2 \approx 3.357t^2 - 26.46t + 379.5$

$y_3 \approx -0.465t^2 + 9.71t + 7.4$

(b) $y_1 + y_2 + y_3 \approx 2.892t^2 - 6.55t + 479.6$

This sum represents the total spent on health services and supplies for the years 1995 through 2001. It includes out-of-pocket payments, insurance premiums, and other types of payments.

(c)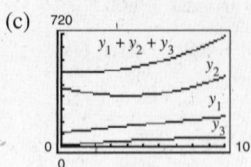

(d) For 2008 use $t = 18$:

$(y_1 + y_2 + y_3)(18) \approx \$1298.708$ billion

For 2010 use $t = 20$:

$(y_1 + y_2 + y_3)(20) \approx \$1505.4$ billion

62. (a) T is a function of t since for each time t there corresponds one and only one temperature T.

(b) $T(4) \approx 60°$; $T(15) \approx 72°$

(c) $H(t) = T(t - 1)$; All the temperature changes would be one hour later.

(d) $H(t) = T(t) - 1$; The temperature would be decreased by one degree.

(e) The points at the endpoints of the individual functions that form each "piece" appear to be $(0, 60)$, $(6, 60)$, $(7, 72)$, $(20, 72)$, $(21, 60)$, and $(24, 60)$. Note that the value $t = 24$ is chosen for the last ordered pair because that is when the day ends and the cycle starts over.

From $t = 0$ to $t = 6$: This is the constant function $T(t) = 60$.

From $t = 6$ to $t = 7$: Use the points $(6, 60)$ and $(7, 72)$.

$$m = \frac{72 - 60}{7 - 6} = 12$$

$y - 60 = 12(x - 6) \implies y = 12x - 12$, or $T(t) = 12t - 12$

From $t = 7$ to $t = 20$: This is the constant function $T(t) = 72$.

From $t = 20$ to $t = 21$: Use the points $(20, 72)$ and $(21, 60)$.

$$m = \frac{72 - 60}{20 - 21} = -12$$

$y - 60 = -12(x - 21) \implies y = -12x + 312$, or $T(t) = -12t + 312$

From $t = 21$ to $t = 24$: This is the constant function $T(t) = 60$.

A piecewise-defined function is $T(t) = \begin{cases} 60, & 0 \leq t \leq 6 \\ 12t - 12, & 6 < t < 7 \\ 72, & 7 \leq t \leq 20 \\ -12t + 312, & 20 < t < 21 \\ 60, & 21 \leq t \leq 24 \end{cases}$

Note that the endpoints of each domain interval can be ascribed to the function on either side of it.

63. (a) $r(x) = \dfrac{x}{2}$

(b) $A(r) = \pi r^2$

(c) $(A \circ r)(x) = A(r(x)) = A\left(\dfrac{x}{2}\right) = \pi\left(\dfrac{x}{2}\right)^2$

$(A \circ r)(x)$ represents the area of the circular base of the tank on the square foundation with side length x.

64. $(A \circ r)(t) = A(r(t)) = A(0.6t) = \pi(0.6t)^2 = 0.36\pi t^2$

$A \circ r$ represents the area of the circle at time t.

65. (a) $N(T(t)) = N(3t + 2)$

$= 10(3t + 2)^2 - 20(3t + 2) + 600$

$= 10(9t^2 + 12t + 4) - 60t - 40 + 600$

$= 90t^2 + 60t + 600$

$= 30(3t^2 + 2t + 20), \ 0 \le t \le 6$

This represents the number of bacteria in the food as a function of time.

(b) $30(3t^2 + 2t + 20) = 1500$

$3t^2 + 2t + 20 = 50$

$3t^2 + 2t - 30 = 0$

By the Quadratic Formula, $t \approx -3.513$ or 2.846. Choosing the positive value for t, we have $t \approx 2.846$ hours.

66. $C(x) = 60x + 750, \ x(t) = 50t$

(a) $(C \circ x)(t) = C(x(t))$

$= C(50t)$

$= 60(50t) + 750$

$= 3000t + 750$

$(C \circ x)(t)$ represents the cost of production as a function of time.

(b) Find t when $(C \circ x)(t) = 15,000$.

$15,000 = 3000t + 750$

$t = 4.75$ hours

The cost of production for 4 hours 45 minutes is $15,000.

67. (a) $f(g(x)) = f(0.03x) = 0.03x - 500,000$

(b) $g(f(x)) = g(x - 500,000) = 0.03(x - 500,000)$

$g(f(x))$ represents your bonus of 3% of an amount over $500,000.

68. (a) $R(p) = p - 2000$ the cost of the car after the factory rebate

(c) $(R \circ S)(p) = R(0.9p) = 0.9p - 2000$

$(S \circ R)(p) = S(p - 2000)$

$= 0.9(p - 2000) = 0.9p - 1800$

$(R \circ S)(p)$ represents the factory rebate *after* the dealership discount.

$(S \circ R)(p)$ represents the dealership discount after the factory rebate.

(b) $S(p) = 0.9p$ the cost of the car with the dealership discount

(d) $(R \circ S)(p) = (R \circ S)(20,500)$

$= 0.9(20,500) - 2000 = \$16,450$

$(S \circ R)(p) = (S \circ R)(20,500)$

$= 0.9(20,500) - 1800 = \$16,650$

$(S \circ R)(p)$ will always be larger. Observe the formulas in (c).

69. False. $(f \circ g)(x) = 6x + 1$ and $(g \circ f)(x) = 6x + 6$.

70. True. The range of g must be a subset of the domain of f for $(f \circ g)(x)$ to be defined.

71. Let $f(x)$ and $g(x)$ be two odd functions and define $h(x) = f(x)g(x)$. Then

$$h(-x) = f(-x)g(-x)$$

$$= [-f(x)][-g(x)] \quad \text{since } f \text{ and } g \text{ are odd}$$

$$= f(x)g(x)$$

$$= h(x).$$

Thus, $h(x)$ is even.

Let $f(x)$ and $g(x)$ be two even functions and define $h(x) = f(x)g(x)$. Then

$$h(-x) = f(-x)g(-x)$$

$$= f(x)g(x) \quad \text{since } f \text{ and } g \text{ are even}$$

$$= h(x).$$

Thus, $h(x)$ is even.

72. Let $f(x)$ be an odd function, $g(x)$ be an even function, and define $h(x) = f(x)g(x)$. Then

$$h(-x) = f(-x)g(-x)$$

$$= [-f(x)]g(x) \quad \text{since } f \text{ is odd and } g \text{ is even}$$

$$= -f(x)g(x)$$

$$= -h(x).$$

Thus, h is odd and the product of an odd function and an even function is odd.

73. $f(x) = 3x - 4$

$$\frac{f(x + h) - f(x)}{h} = \frac{[3(x + h) - 4] - (3x - 4)}{h}$$

$$= \frac{3x + 3h - 4 - 3x + 4}{h}$$

$$= \frac{3h}{h}$$

$$= 3, \; h \neq 0$$

74. $f(x) = 1 - x^2$

$$f(x + h) = 1 - (x + h)^2$$

$$= 1 - (x^2 + 2hx + h^2)$$

$$= 1 - x^2 - 2hx - h^2$$

$$\frac{f(x + h) - f(x)}{h} = \frac{1 - x^2 - 2hx - h^2 - (1 - x^2)}{h}$$

$$= \frac{-2hx - h^2}{h} = -2x - h, \; h \neq 0$$

75. $f(x) = \dfrac{4}{x}$

$$\frac{f(x + h) - f(x)}{h} = \frac{\dfrac{4}{x + h} - \dfrac{4}{x}}{h} = \frac{\dfrac{4x - 4(x + h)}{x(x + h)}}{\dfrac{h}{1}}$$

$$= \frac{4x - 4x - 4h}{x(x + h)} \cdot \frac{1}{h} = \frac{-4h}{x(x + h)} \cdot \frac{1}{h} = \frac{-4}{x(x + h)}, \; h \neq 0$$

76. $f(x) = \sqrt{2x + 1}$

$$f(x + h) = \sqrt{2(x + h) + 1}$$

$$\frac{f(x + h) - f(x)}{h} = \frac{\sqrt{2(x + h) + 1} - \sqrt{2x + 1}}{h}$$

$$= \frac{\sqrt{2(x + h) + 1} - \sqrt{2x + 1}}{h} \cdot \frac{\sqrt{2(x + h) + 1} + \sqrt{2x + 1}}{\sqrt{2(x + h) + 1} + \sqrt{2x + 1}}$$

$$= \frac{[2(x + h) + 1] - (2x + 1)}{h\left(\sqrt{2(x + h) + 1} + \sqrt{2x + 1}\right)}$$

$$= \frac{2x + 2h + 1 - 2x - 1}{h\left(\sqrt{2(x + h) + 1} + \sqrt{2x + 1}\right)}$$

$$= \frac{2}{\sqrt{2(x + h) + 1} + \sqrt{2x + 1}}, \; h \neq 0$$

77. Point: $(2, -4)$

Slope: $m = 3$

$$y - (-4) = 3(x - 2)$$

$$y + 4 = 3x - 6$$

$$3x - y - 10 = 0$$

78. $(-6, 3), m = -1$

$$y - 3 = (-1)(x - (-6))$$

$$y - 3 = -x - 6$$

$$x + y + 3 = 0$$

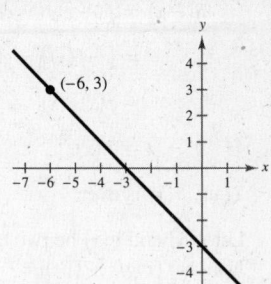

79. Point: $(8, -1)$

Slope: $m = -\frac{3}{2}$

$$y - (-1) = -\frac{3}{2}(x - 8)$$

$$y + 1 = -\frac{3}{2}x + 12$$

$$2y + 2 = -3x + 24$$

$$3x + 2y - 22 = 0$$

80. $(7, 0), m = \frac{5}{7}$

$$y - 0 = \frac{5}{7}(x - 7)$$

$$7y = 5x - 35$$

$$5x - 7y - 35 = 0$$

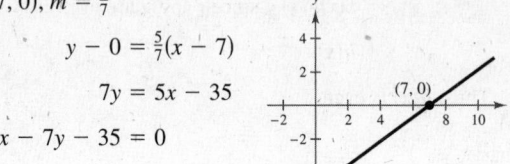

Section 1.9   Inverse Functions

- Two functions f and g are inverses of each other if $f(g(x)) = x$ for every x in the domain of g and $g(f(x)) = x$ for every x in the domain of f.

- A function f has an inverse function if and only if no **horizontal** line crosses the graph of f at more than one point.

- The graph of f^{-1} is a reflection of the graph of f about the line $y = x$.

- Be able to find the inverse of a function, if it exists.

  1. Use the Horizontal Line Test to see if f^{-1} exists.

  2. Replace $f(x)$ with y.

  3. Interchange x and y and solve for y.

  4. Replace y with $f^{-1}(x)$.

Vocabulary Check

1. inverse; f-inverse

2. range; domain

3. $y = x$

4. one-to-one

5. Horizontal

1. $f(x) = 6x$

$$f^{-1}(x) = \frac{x}{6} = \frac{1}{6}x$$

$$f(f^{-1}(x)) = f\left(\frac{x}{6}\right) = 6\left(\frac{x}{6}\right) = x$$

$$f^{-1}(f(x)) = f^{-1}(6x) = \frac{6x}{6} = x$$

2. $f(x) = \frac{1}{3}x$

$$f^{-1}(x) = 3x$$

$$f(f^{-1}(x)) = f(3x) = \frac{1}{3}(3x) = x$$

$$f^{-1}(f(x)) = f^{-1}\left(\frac{1}{3}x\right) = 3\left(\frac{1}{3}x\right) = x$$

3. $f(x) = x + 9$

    $f^{-1}(x) = x - 9$

    $f(f^{-1}(x)) = f(x - 9) = (x - 9) + 9 = x$

    $f^{-1}(f(x)) = f^{-1}(x + 9) = (x + 9) - 9 = x$

4. $f(x) = x - 4$

    $f^{-1}(x) = x + 4$

    $f(f^{-1}(x)) = f(x + 4) = x + 4 - 4 = x$

    $f^{-1}(f(x)) = f^{-1}(x - 4) = x - 4 + 4 = x$

5. $f(x) = 3x + 1$

    $f^{-1}(x) = \dfrac{x - 1}{3}$

    $f(f^{-1}(x)) = f\left(\dfrac{x - 1}{3}\right) = 3\left(\dfrac{x - 1}{3}\right) + 1 = x$

    $f^{-1}(f(x)) = f^{-1}(3x + 1) = \dfrac{(3x + 1) - 1}{3} = x$

6. $f(x) = \dfrac{x - 1}{5}$

    $f^{-1}(x) = 5x + 1$

    $f(f^{-1}(x)) = f(5x + 1) = \dfrac{5x + 1 - 1}{5} = \dfrac{5x}{5} = x$

    $f^{-1}(f(x)) = f^{-1}\left(\dfrac{x - 1}{5}\right) = 5\left(\dfrac{x - 1}{5}\right) + 1$

             $= x - 1 + 1 = x$

7. $f(x) = \sqrt[3]{x}$

    $f^{-1}(x) = x^3$

    $f(f^{-1}(x)) = f(x^3) = \sqrt[3]{x^3} = x$

    $f^{-1}(f(x)) = f^{-1}(\sqrt[3]{x}) = (\sqrt[3]{x})^3 = x$

8. $f(x) = x^5$

    $f^{-1}(x) = \sqrt[5]{x}$

    $f(f^{-1}(x)) = f(\sqrt[5]{x}) = (\sqrt[5]{x})^5 = x$

    $f^{-1}(f(x)) = f^{-1}(x^5) = \sqrt[5]{x^5} = x$

9. The inverse is a line through $(-1, 0)$.
    Matches graph (c).

10. The inverse is a line through $(0, 6)$ and $(6, 0)$.
    Matches graph (b).

11. The inverse is half a parabola starting at $(1, 0)$.
    Matches graph (a).

12. The inverse is a third-degree equation through $(0, 0)$.
    Matches graph (d).

13. $f(x) = 2x,\ g(x) = \dfrac{x}{2}$

    (a) $f(g(x)) = f\left(\dfrac{x}{2}\right) = 2\left(\dfrac{x}{2}\right) = x$

        $g(f(x)) = g(2x) = \dfrac{2x}{2} = x$

    (b)

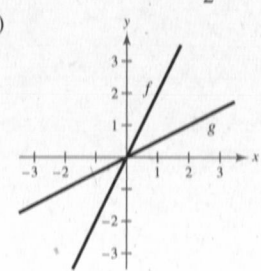

14. $f(x) = x - 5,\ g(x) = x + 5$

    (a) $f(g(x)) = f(x + 5) = (x + 5) - 5 = x$

        $g(f(x)) = g(x - 5) = (x - 5) + 5 = x$

    (b)

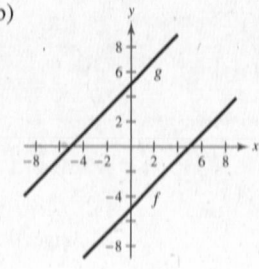

15. $f(x) = 7x + 1,\ g(x) = \dfrac{x - 1}{7}$

    (a) $f(g(x)) = f\left(\dfrac{x - 1}{7}\right) = 7\left(\dfrac{x - 1}{7}\right) + 1 = x$

        $g(f(x)) = g(7x + 1) = \dfrac{(7x + 1) - 1}{7} = x$

    (b)

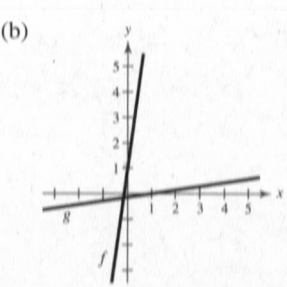

16. $f(x) = 3 - 4x$, $g(x) = \dfrac{3 - x}{4}$

(a) $f(g(x)) = f\left(\dfrac{3 - x}{4}\right) = 3 - 4\left(\dfrac{3 - x}{4}\right)$

$\qquad = 3 - (3 - x) = x$

$g(f(x)) = g(3 - 4x) = \dfrac{3 - (3 - 4x)}{4} = \dfrac{4x}{4} = x$

(b)

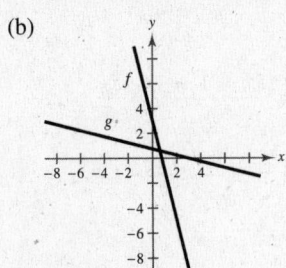

17. $f(x) = \dfrac{x^3}{8}$, $g(x) = \sqrt[3]{8x}$

(a) $f(g(x)) = f\left(\sqrt[3]{8x}\right) = \dfrac{\left(\sqrt[3]{8x}\right)^3}{8} = \dfrac{8x}{8} = x$

$g(f(x)) = g\left(\dfrac{x^3}{8}\right) = \sqrt[3]{8\left(\dfrac{x^3}{8}\right)} = \sqrt[3]{x^3} = x$

(b)

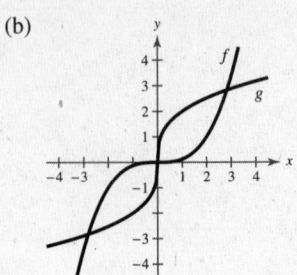

18. $f(x) = \dfrac{1}{x}$, $g(x) = \dfrac{1}{x}$

(a) $f(g(x)) = f\left(\dfrac{1}{x}\right) = \dfrac{1}{1/x} = 1 \div \dfrac{1}{x} = 1 \cdot \dfrac{x}{1} = x$

$g(f(x)) = g\left(\dfrac{1}{x}\right) = \dfrac{1}{1/x} = 1 \div \dfrac{1}{x} = 1 \cdot \dfrac{x}{1} = x$

(b)

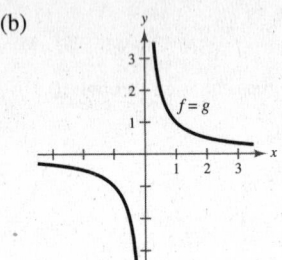

19. $f(x) = \sqrt{x - 4}$, $g(x) = x^2 + 4, x \geq 0$

(a) $f(g(x)) = f(x^2 + 4)$, $x \geq 0$

$\qquad = \sqrt{(x^2 + 4) - 4} = x$

$g(f(x)) = g\left(\sqrt{x - 4}\right)$

$\qquad = \left(\sqrt{x - 4}\right)^2 + 4 = x$

(b)

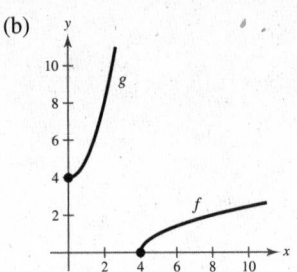

20. $f(x) = 1 - x^3$, $g(x) = \sqrt[3]{1 - x}$

(a) $f(g(x)) = f\left(\sqrt[3]{1 - x}\right) = 1 - \left(\sqrt[3]{1 - x}\right)^3$

$\qquad = 1 - (1 - x) = x$

$g(f(x)) = g(1 - x^3) = \sqrt[3]{1 - (1 - x^3)} = \sqrt[3]{x^3} = x$

(b)

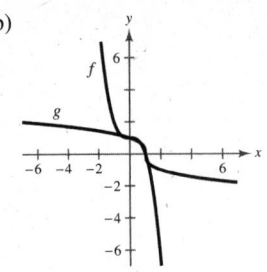

21. $f(x) = 9 - x^2, x \geq 0$; $g(x) = \sqrt{9 - x}, x \leq 9$

(a) $f(g(x)) = f\left(\sqrt{9 - x}\right)$, $x \leq 9$

$\qquad = 9 - \left(\sqrt{9 - x}\right)^2 = x$

$g(f(x)) = g(9 - x^2)$, $x \geq 0$

$\qquad = \sqrt{9 - (9 - x^2)} = x$

(b)

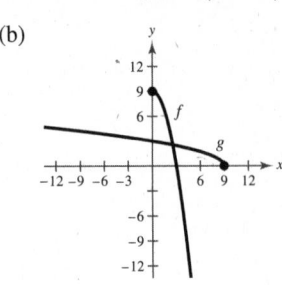

22. $f(x) = \dfrac{1}{1 + x},\ x \geq 0;\ g(x) = \dfrac{1 - x}{x},\ 0 < x \leq 1$

   (a) $f(g(x)) = f\left(\dfrac{1 - x}{x}\right) = \dfrac{1}{1 + \left(\dfrac{1 - x}{x}\right)} = \dfrac{1}{\dfrac{x}{x} + \dfrac{1 - x}{x}} = \dfrac{1}{\dfrac{1}{x}} = x$

        $g(f(x)) = g\left(\dfrac{1}{1 + x}\right) = \dfrac{1 - \left(\dfrac{1}{1 + x}\right)}{\left(\dfrac{1}{1 + x}\right)} = \dfrac{\dfrac{1 + x}{1 + x} - \dfrac{1}{1 + x}}{\dfrac{1}{1 + x}} = \dfrac{\dfrac{x}{1 + x}}{\dfrac{1}{1 + x}} = \dfrac{x}{1 + x} \cdot \dfrac{x + 1}{1} = x$

(b)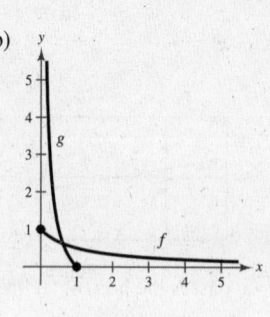

23. $f(x) = \dfrac{x - 1}{x + 5},\ g(x) = -\dfrac{5x + 1}{x - 1}$

   (a) $f(g(x)) = f\left(-\dfrac{5x + 1}{x - 1}\right)$

        $= \dfrac{\left(-\dfrac{5x + 1}{x - 1} - 1\right)}{\left(-\dfrac{5x + 1}{x - 1} + 5\right)} \cdot \dfrac{x - 1}{x - 1} = \dfrac{-(5x + 1) - (x - 1)}{-(5x + 1) + 5(x - 1)} = \dfrac{-6x}{-6} = x$

        $g(f(x)) = g\left(\dfrac{x - 1}{x + 5}\right)$

        $= -\dfrac{\left[5\left(\dfrac{x - 1}{x + 5}\right) + 1\right]}{\left[\dfrac{x - 1}{x + 5} - 1\right]} \cdot \dfrac{x + 5}{x + 5} = -\dfrac{5(x - 1) + (x + 5)}{(x - 1) - (x + 5)} = -\dfrac{6x}{-6} = x$

(b)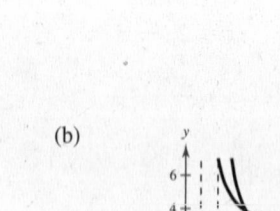

24. $f(x) = \dfrac{x + 3}{x - 2},\ g(x) = \dfrac{2x + 3}{x - 1}$

   (a) $f(g(x)) = f\left(\dfrac{2x + 3}{x - 1}\right) = \dfrac{\dfrac{2x + 3}{x - 1} + 3}{\dfrac{2x + 3}{x - 1} - 2} = \dfrac{\dfrac{2x + 3 + 3x - 3}{x - 1}}{\dfrac{2x + 3 - 2x + 2}{x - 1}} = \dfrac{5x}{5} = x$

        $g(f(x)) = g\left(\dfrac{x + 3}{x - 2}\right) = \dfrac{2\left(\dfrac{x + 3}{x - 2}\right) + 3}{\dfrac{x + 3}{x - 2} - 1} = \dfrac{\dfrac{2x + 6 + 3x - 6}{x - 2}}{\dfrac{x + 3 - x + 2}{x - 2}} = \dfrac{5x}{5} = x$

(b)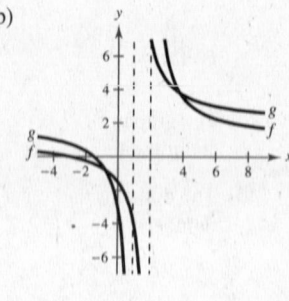

25. No, $\{(-2, -1), (1, 0), (2, 1), (1, 2), (-2, 3), (-6, 4)\}$ does not represent a function. -2 and 1 are paired with two different values.

26. Yes, $\{(10, -3), (6, -2), (4, -1), (1, 0), (-3, 2), (10, 2)\}$ does represent a function.

27.

x	-2	0	2	4	6	8
$f^{-1}(x)$	-2	-1	0	1	2	3

28.

x	-10	-7	-4	-1	2	5
$f^{-1}(x)$	-3	-2	-1	0	1	2

29. Yes, since no horizontal line crosses the graph of f at more than one point, f *has* an inverse.

30. No, because some horizontal lines intersect the graph twice, f *does not* have an inverse.

31. No, since some horizontal lines cross the graph of f twice, f *does not* have an inverse.

32. Yes, because no horizontal lines intersect the graph at more than one point, f *has* an inverse.

33. $g(x) = \dfrac{4 - x}{6}$

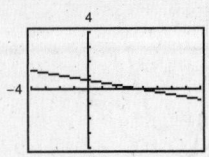

g passes the horizontal line test, so g *has* an inverse.

34. $f(x) = 10$

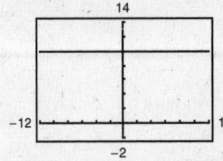

f does not pass the horizontal line test, so f *does not* have an inverse.

35. $h(x) = |x + 4| - |x - 4|$

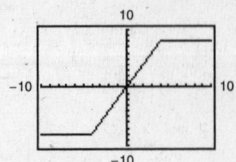

h does not pass the horizontal line test, so h *does not* have an inverse.

36. $g(x) = (x + 5)^3$

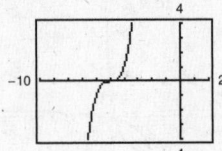

g passes the horizontal line test, so g *has* an inverse.

37. $f(x) = -2x\sqrt{16 - x^2}$

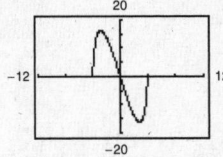

f does not pass the horizontal line test, so f *does not* have an inverse.

38. $f(x) = \frac{1}{8}(x + 2)^2 - 1$

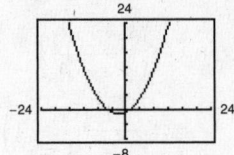

f does not pass the horizontal line test, so f *does not* have an inverse.

39. (a) $f(x) = 2x - 3$

$y = 2x - 3$

$x = 2y - 3$

$y = \dfrac{x + 3}{2}$

$f^{-1}(x) = \dfrac{x + 3}{2}$

(b)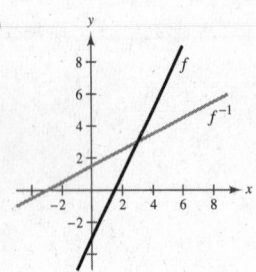

(c) The graph of f^{-1} is the reflection of the graph of f about the line $y = x$.

(d) The domains and ranges of f and f^{-1} are all real numbers.

40. (a) $f(x) = 3x + 1$

$y = 3x + 1$

$x = 3y + 1$

$\dfrac{x - 1}{3} = y$

$f^{-1}(x) = \dfrac{x - 1}{3}$

(b)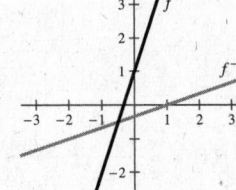

(c) The graph of f^{-1} is the reflection of f in the line $y = x$.

(d) The domains and ranges of f and f^{-1} are all real numbers.

41. (a) $f(x) = x^5 - 2$

$y = x^5 - 2$

$x = y^5 - 2$

$y = \sqrt[5]{x + 2}$

$f^{-1}(x) = \sqrt[5]{x + 2}$

(b)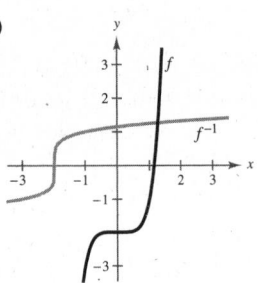

(c) The graph of f^{-1} is the reflection of the graph of f about the line $y = x$.

(d) The domains and ranges of f and f^{-1} are all real numbers.

42. (a) $f(x) = x^3 + 1$

$y = x^3 + 1$

$x = y^3 + 1$

$x - 1 = y^3$

$\sqrt[3]{x - 1} = y$

$f^{-1}(x) = \sqrt[3]{x - 1}$

(b)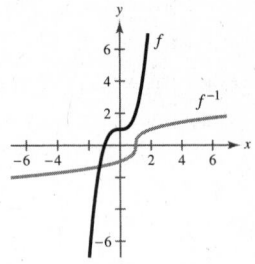

(c) The graph of f^{-1} is the reflection of f in the line $y = x$.

(d) The domains and ranges of f and f^{-1} are all real numbers.

43. (a) $f(x) = \sqrt{x}$ (b)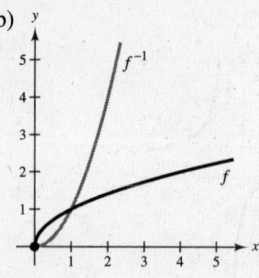

$$y = \sqrt{x}$$

$$x = \sqrt{y}$$

$$y = x^2$$

$$f^{-1}(x) = x^2, \ x \geq 0$$

(c) The graph of f^{-1} is the reflection of the graph of f about the line $y = x$.

(d) The domains and ranges of f and f^{-1} are $[0, \infty)$.

44. (a) $f(x) = x^2, \ x \geq 0$ (b)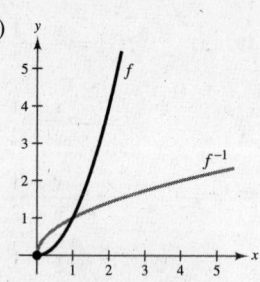

$$y = x^2$$

$$x = y^2$$

$$\sqrt{x} = y$$

$$f^{-1}(x) = \sqrt{x}$$

(c) The graph of f^{-1} is the reflection of f in the line $y = x$.

(d) The domains and ranges of f and f^{-1} are $[0, \infty)$.

45. (a) $f(x) = \sqrt{4 - x^2}, \ 0 \leq x \leq 2$

$$y = \sqrt{4 - x^2}$$

$$x = \sqrt{4 - y^2}$$

$$x^2 = 4 - y^2$$

$$y^2 = 4 - x^2$$

$$y = \sqrt{4 - x^2}$$

$$f^{-1}(x) = \sqrt{4 - x^2}, \ 0 \leq x \leq 2$$

(b)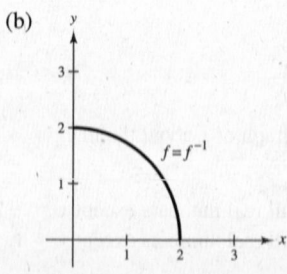

(c) The graph of f^{-1} is the same as the graph of f.

(d) The domains and ranges of f and f^{-1} are $[0, 2]$.

46. (a) $f(x) = x^2 - 2, \ x \leq 0$

$$y = x^2 - 2$$

$$x = y^2 - 2$$

$$\pm\sqrt{x + 2} = y$$

$$f^{-1}(x) = -\sqrt{x + 2}$$

(b)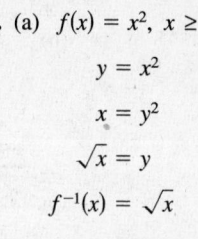

(c) The graph of f^{-1} is the reflection of f in the line $y = x$.

(d) $[-2, \infty)$ is the range of f and domain of f^{-1}. $(-\infty, 0]$ is the domain of f and the range of f^{-1}.

47. (a) $f(x) = \dfrac{4}{x}$ (b)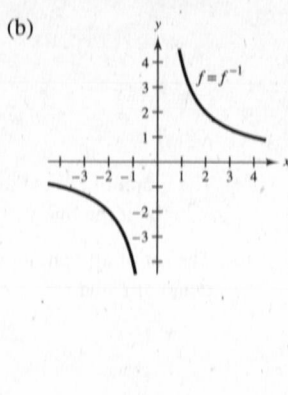

$$y = \frac{4}{x}$$

$$x = \frac{4}{y}$$

$$xy = 4$$

$$y = \frac{4}{x}$$

$$f^{-1}(x) = \frac{4}{x}$$

(c) The graph of f^{-1} is the same as the graph of f.

(d) The domains and ranges of f and f^{-1} are all real numbers except for 0.

48. (a) $f(x) = -\dfrac{2}{x}$ (b)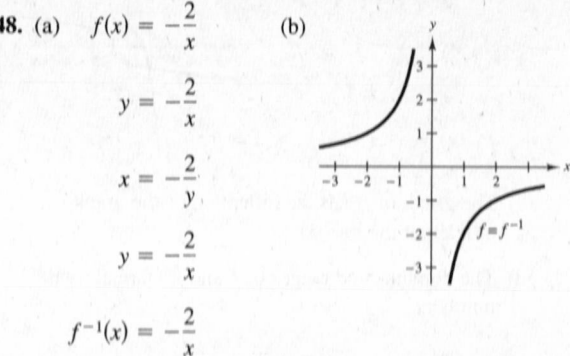

$$y = -\frac{2}{x}$$

$$x = -\frac{2}{y}$$

$$y = -\frac{2}{x}$$

$$f^{-1}(x) = -\frac{2}{x}$$

(c) The graphs are the same.

(d) $(-\infty, 0) \cup (0, \infty)$ is the domain and range of f and f^{-1}.

49. (a) $f(x) = \dfrac{x+1}{x-2}$

$$y = \frac{x+1}{x-2}$$

$$x = \frac{y+1}{y-2}$$

$$x(y-2) = y+1$$

$$xy - 2x = y + 1$$

$$xy - y = 2x + 1$$

$$y(x-1) = 2x + 1$$

$$y = \frac{2x+1}{x-1}$$

$$f^{-1}(x) = \frac{2x+1}{x-1}$$

(b)

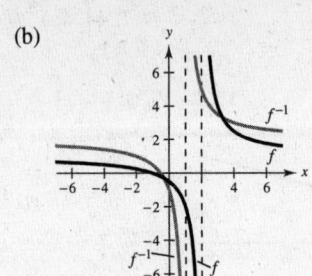

(c) The graph of f^{-1} is the reflection of the graph of f about the line $y = x$.

(d) The domain of f and the range of f^{-1} is all real numbers except 2. The range of f and the domain of f^{-1} is all real numbers except 1.

50. (a) $f(x) = \dfrac{x-3}{x+2}$

$$y = \frac{x-3}{x+2}$$

$$x = \frac{y-3}{y+2}$$

$$xy + 2x - y + 3 = 0$$

$$y(x-1) = -2x - 3$$

$$y = \frac{-2x-3}{x-1}$$

$$f^{-1}(x) = \frac{-2x-3}{x-1}$$

(b)

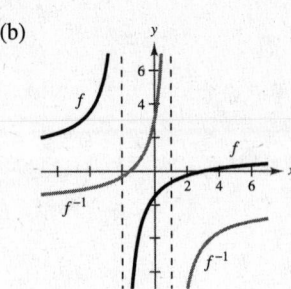

(c) The graph of f^{-1} is the reflection of the graph of f about the line $y = x$.

(d) The domain of f and the range of f^{-1} is all real numbers except $x = -2$. The range of f and the domain of f^{-1} is all real numbers except $x = 1$.

51. (a) $f(x) = \sqrt[3]{x-1}$

$$y = \sqrt[3]{x-1}$$

$$x = \sqrt[3]{y-1}$$

$$x^3 = y - 1$$

$$y = x^3 + 1$$

$$f^{-1}(x) = x^3 + 1$$

(b)

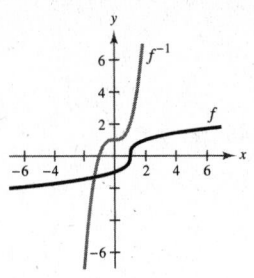

(c) The graph of f^{-1} is the reflection of the graph of f about the line $y = x$.

(d) The domains and ranges of f and f^{-1} are all real numbers.

52. (a) $f(x) = x^{3/5}$

$$y = x^{3/5}$$

$$x = y^{3/5}$$

$$x^{5/3} = \left(y^{3/5}\right)^{5/3}$$

$$x^{5/3} = y$$

$$f^{-1}(x) = x^{5/3}$$

(b)

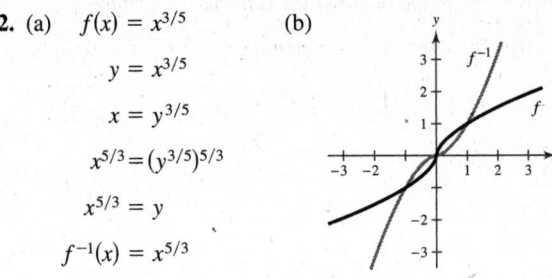

(c) The graph of f^{-1} is the reflection of the graph of f about the line $y = x$.

(d) The set of all real numbers is the domain and range of f and f^{-1}.

53. (a)
$$f(x) = \frac{6x + 4}{4x + 5}$$

$$y = \frac{6x + 4}{4x + 5}$$

$$x = \frac{6y + 4}{4y + 5}$$

$$x(4y + 5) = 6y + 4$$

$$4xy + 5x = 6y + 4$$

$$4xy - 6y = -5x + 4$$

$$y(4x - 6) = -5x + 4$$

$$y = \frac{-5x + 4}{4x - 6}$$

$$f^{-1}(x) = \frac{-5x + 4}{4x - 6} = \frac{5x - 4}{6 - 4x}$$

(b)

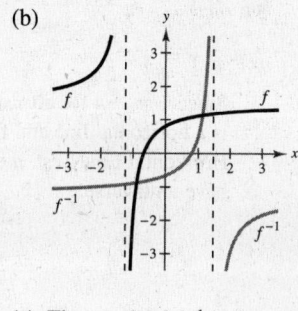

(c) The graph of f^{-1} is the graph of f reflected about the line $y = x$.

(d) The domain of f and the range of f^{-1} is all real numbers except $-\frac{5}{4}$.
The range of f and the domain of f^{-1} is all real numbers except $\frac{3}{2}$.

54. (a)
$$f(x) = \frac{8x - 4}{2x + 6}$$

$$y = \frac{8x - 4}{2x + 6}$$

$$x = \frac{8y - 4}{2y + 6}$$

$$2xy + 6x = 8y - 4$$

$$y(2x - 8) = -6x - 4$$

$$y = \frac{-6x - 4}{2x - 8}$$

(b)

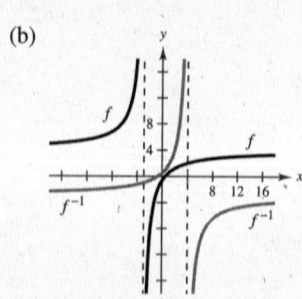

(c) The graph of f^{-1} is the graph of f reflected about the line $y = x$.

(d) The domain of f and the range of f^{-1} is the set of all real numbers x except $x = -3$.
The domain of f^{-1} and the range of f is the set of all real numbers x except $x = 4$.

55. $f(x) = x^4$

$$y = x^4$$

$$x = y^4$$

$$y = \pm \sqrt[4]{x}$$

This does not represent y as a function of x. f does not have an inverse.

56. $f(x) = \frac{1}{x^2}$

$$y = \frac{1}{x^2}$$

$$x = \frac{1}{y^2}$$

$$y^2 = \frac{1}{x}$$

$$y = \pm \sqrt{\frac{1}{x}}$$

This does not represent y as a function of x. f does not have an inverse.

57. $g(x) = \frac{x}{8}$

$$y = \frac{x}{8}$$

$$x = \frac{y}{8}$$

$$y = 8x$$

This is a function of x, so g has an inverse.

$$g^{-1}(x) = 8x$$

58. $f(x) = 3x + 5$

$y = 3x + 5$

$x = 3y + 5$

$x - 5 = 3y$

$\dfrac{x - 5}{3} = y$

This is a function of x, so f has an inverse.

$f^{-1}(x) = \dfrac{x - 5}{3}$

59. $p(x) = -4$

$y = -4$

Since $y = -4$ for all x, the graph is a horizontal line and fails the Horizontal Line Test. p does not have an inverse.

60. $f(x) = \dfrac{3x + 4}{5}$

$y = \dfrac{3x + 4}{5}$

$x = \dfrac{3y + 4}{5}$

$5x = 3y + 4$

$5x - 4 = 3y$

$\dfrac{5x - 4}{3} = y$

This is a function of x, so f has an inverse.

$f^{-1}(x) = \dfrac{5x - 4}{3}$

61. $f(x) = (x + 3)^2, \ x \geq -3 \Rightarrow y \geq 0$

$y = (x + 3)^2, \ x \geq -3, \ y \geq 0$

$x = (y + 3)^2, \ y \geq -3, \ x \geq 0$

$\sqrt{x} = y + 3, \ y \geq -3, \ x \geq 0$

$y = \sqrt{x} - 3, \ x \geq 0, \ y \geq -3$

This is a function of x, so f has an inverse.

$f^{-1}(x) = \sqrt{x} - 3, \ x \geq 0$

62. $q(x) = (x - 5)^2$

$y = (x - 5)^2$

$x = (y - 5)^2$

$\pm\sqrt{x} = y - 5$

$5 \pm \sqrt{x} = y$

This does not represent y as a function of x, so q does not have an inverse.

63. $f(x) = \begin{cases} x + 3, & x < 0 \\ 6 - x, & x \geq 0 \end{cases}$

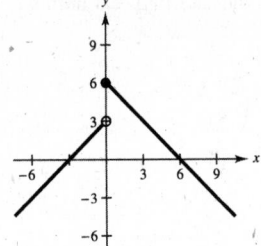

The graph fails the Horizontal Line Test, so $f(x)$ does not have an inverse.

64. $f(x) = \begin{cases} -x, & x \leq 0 \\ x^2 - 3x, & x > 0 \end{cases}$

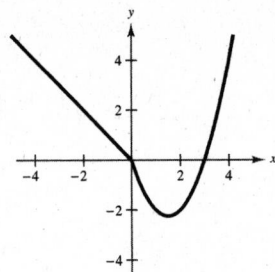

The graph fails the Horizontal Line Test, so f does not have an inverse.

65. $h(x) = -\dfrac{4}{x^2}$

The graph fails the Horizontal Line Test so h does not have an inverse.

66. $f(x) = |x - 2|, \ x \leq 2 \Rightarrow y \geq 0$

$y = |x - 2|, \ x \leq 2, \ y \geq 0$

$x = |y - 2|, \ y \leq 2, \ x \geq 0$

$x = y - 2 \quad \text{or} \quad -x = y - 2$

$2 + x = y \quad \quad \text{or} \ 2 - x = y$

The portion that satisfies the conditions $y \leq 2$ and $x \geq 0$ is $2 - x = y$. This is a function of x, so f has an inverse.

$f^{-1}(x) = 2 - x, \ x \geq 0$

67. $f(x) = \sqrt{2x + 3} \Rightarrow x \geq -\dfrac{3}{2},\ y \geq 0$

$y = \sqrt{2x + 3},\ x \geq -\dfrac{3}{2},\ y \geq 0$

$x = \sqrt{2y + 3},\ y \geq -\dfrac{3}{2},\ x \geq 0$

$x^2 = 2y + 3,\ x \geq 0,\ y \geq -\dfrac{3}{2}$

$y = \dfrac{x^2 - 3}{2},\ x \geq 0,\ y \geq -\dfrac{3}{2}$

This is a function of x, so f has an inverse.

$f^{-1}(x) = \dfrac{x^2 - 3}{2},\ x \geq 0$

68. $f(x) = \sqrt{x - 2} \Rightarrow x \geq 2,\ y \geq 0$

$y = \sqrt{x - 2},\ x \geq 2,\ y \geq 0$

$x = \sqrt{y - 2},\ y \geq 2,\ x \geq 0$

$x^2 = y - 2,\ x \geq 0,\ y \geq 2$

$x^2 + 2 = y,\ x \geq 0,\ y \geq 2$

This is a function of x, so f has an inverse.

$f^{-1}(x) = x^2 + 2,\ x \geq 0$

In Exercises 69–74, $f(x) = \frac{1}{8}x - 3$, $f^{-1}(x) = 8(x + 3)$, $g(x) = x^3$, $g^{-1}(x) = \sqrt[3]{x}$.

69. $(f^{-1} \circ g^{-1})(1) = f^{-1}(g^{-1}(1))$

$= f^{-1}\left(\sqrt[3]{1}\right)$

$= 8\left(\sqrt[3]{1} + 3\right) = 32$

70. $(g^{-1} \circ f^{-1})(-3) = g^{-1}(f^{-1}(-3))$

$= g^{-1}(8(-3 + 3))$

$= g^{-1}(0) = \sqrt[3]{0} = 0$

71. $(f^{-1} \circ f^{-1})(6) = f^{-1}(f^{-1}(6))$

$= f^{-1}(8[6 + 3])$

$= 8[8(6 + 3) + 3] = 600$

72. $(g^{-1} \circ g^{-1})(-4) = g^{-1}(g^{-1}(-4))$

$= g^{-1}\left(\sqrt[3]{-4}\right)$

$= \sqrt[3]{\sqrt[3]{-4}} = \sqrt[9]{-4}$

73. $(f \circ g)(x) = f(g(x)) = f(x^3) = \frac{1}{8}x^3 - 3$

$y = \frac{1}{8}x^3 - 3$

$x = \frac{1}{8}y^3 - 3$

$x + 3 = \frac{1}{8}y^3$

$8(x + 3) = y^3$

$\sqrt[3]{8(x + 3)} = y$

$(f \circ g)^{-1}(x) = 2\sqrt[3]{x + 3}$

74. $g^{-1} \circ f^{-1} = g^{-1}(f^{-1}(x))$

$= g^{-1}(8(x + 3))$

$= \sqrt[3]{8(x + 3)}$

$= 2\sqrt[3]{x + 3}$

In Exercises 75–78, $f(x) = x + 4$, $f^{-1}(x) = x - 4$, $g(x) = 2x - 5$, $g^{-1}(x) = \dfrac{x + 5}{2}$.

75. $(g^{-1} \circ f^{-1})(x) = g^{-1}(f^{-1}(x))$

$= g^{-1}(x - 4)$

$= \dfrac{(x - 4) + 5}{2}$

$= \dfrac{x + 1}{2}$

76. $(f^{-1} \circ g^{-1})(x) = f^{-1}(g^{-1}(x))$

$= f^{-1}\left(\dfrac{x + 5}{2}\right)$

$= \dfrac{x + 5}{2} - 4$

$= \dfrac{x + 5 - 8}{2}$

$= \dfrac{x - 3}{2}$

77. $(f \circ g)(x) = f(g(x))$

$= f(2x - 5)$

$= (2x - 5) + 4$

$= 2x - 1$

$(f \circ g)^{-1}(x) = \dfrac{x + 1}{2}$

Note: Comparing Exercises 75 and 77, we see that $(f \circ g)^{-1}(x) = (g^{-1} \circ f^{-1})(x)$.

78. $(g \circ f)(x) = g(f(x))$

$$= g(x + 4)$$

$$= 2(x + 4) - 5$$

$$= 2x + 8 - 5$$

$$= 2x + 3$$

$$y = 2x + 3$$

$$x = 2y + 3$$

$$x - 3 = 2y$$

$$\frac{x - 3}{2} = y$$

$$(g \circ f)^{-1}(x) = \frac{x - 3}{2}$$

79. (a) $f^{-1}(108,209) = 11$

(b) f^{-1} represents the year for a given number of house-holds in the United States.

(c) $y \approx 1578.68t + 90,183.63$

(d)
$$y = 1578.68t + 90,183.63$$

$$t = 1578.68y + 90,183.63$$

$$\frac{t - 90,183.63}{1578.68} = y$$

$$f^{-1}(t) = \frac{t - 90,183.63}{1578.68}$$

(e) $f^{-1}(117,022) \approx 17$

(f) $f^{-1}(108,209) \approx 11.418$

This is close to the value of 11 in the table.

80. (a) Yes, f^{-1} exists.

(b) f^{-1} represents the time in years for a given total sales.

(c) $f^{-1}(1825) = 10$

(d) No. f^{-1} would not exist since $f(12) = 2794$ and $f(14) = 2794$. The function would fail the Horizontal Line Test.

81. (a) Yes. Since the values of f increase each year, no two f-values are paired with the same t-value so f does have an inverse.

(b) f^{-1} would represent the year that a given number of miles was traveled by motor vehicles.

(c) Since $f(8) = 2632, f^{-1}(2632) = 8$.

(d) No. Since the new value is the same as the value given for 2000, f would not pass the Horizontal Line Test and would not have an inverse.

82. (a)
$$y = 8 + 0.75x$$

$$x = 8 + 0.75y$$

$$x - 8 = 0.75y$$

$$\frac{x - 8}{0.75} = y$$

$$f^{-1}(x) = \frac{x - 8}{0.75}$$

(b) $x =$ hourly wage, $y =$ number of units produced

(c) $y = \dfrac{22.25 - 8}{0.75} = 19$ units

83. (a)
$$y = 0.03x^2 + 245.50, \ 0 < x < 100$$

$$\Rightarrow 245.50 < y < 545.50$$

$$x = 0.03y^2 + 245.50$$

$$x - 245.50 = 0.03y^2$$

$$\frac{x - 245.50}{0.03} = y^2$$

$$\sqrt{\frac{x - 245.50}{0.03}} = y, \ 245.50 < x < 545.50$$

$$f^{-1}(x) = \sqrt{\frac{x - 245.50}{0.03}}$$

$x =$ temperature in degrees Fahrenheit

$y =$ percent load for a diesel engine

(b)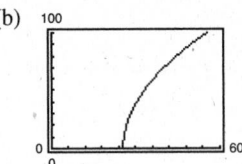

(c)
$$0.03x^2 + 245.50 \leq 500$$

$$0.03x^2 \leq 254.50$$

$$x^2 \leq 8483.33$$

$$x \leq 92.10$$

Thus, $0 < x \leq 92.10$.

84. (a)

$$x = 1.25y + 1.60(50 - y)$$

$$x = 1.25y + 80 - 1.60y$$

$$x - 80 = -0.35y$$

$$\frac{x - 80}{-0.35} = y$$

$$y = \frac{80 - x}{0.35}$$

$x = $ total cost

$y = $ number of pounds of less
expensive ground beef

(b)

$$0 \le y \le 50$$

$$0 \le \frac{80 - x}{0.35} \le 50$$

$$0 \le 80 - x \le 17.5$$

$$-80 \le -x \le -62.5$$

$$62.5 \le x \le 80$$

(c) $\dfrac{80 - 73}{0.35} = y = 20$ pounds of the less expensive ground beef

85. False. $f(x) = x^2$ is even and does not have an inverse.

86. True. If $f(x)$ has an inverse and it has a y-intercept at $(0, b)$, then the point $(b, 0)$ must be a point on the graph of $f^{-1}(x)$.

87. Let $(f \circ g)(x) = y$. Then $x = (f \circ g)^{-1}(y)$. Also,

$$(f \circ g)(x) = y \implies f(g(x)) = y$$

$$g(x) = f^{-1}(y)$$

$$x = g^{-1}(f^{-1}(y))$$

$$x = (g^{-1} \circ f^{-1})(y).$$

Since f and g are both one-to-one functions,

$$(f \circ g)^{-1} = g^{-1} \circ f^{-1}.$$

88. Let $f(x)$ be a one-to-one odd function. Then $f^{-1}(x)$ exists and $f(-x) = -f(x)$. Letting (x, y) be any point on the graph of $f(x) \implies (-x, -y)$ is also on the graph of $f(x)$ and $f^{-1}(-y) = -x = -f^{-1}(y)$. Therefore, $f^{-1}(x)$ is also an odd function.

89.

x	1	3	4	6
f	1	2	6	7

x	1	2	6	7
$f^{-1}(x)$	1	3	4	6

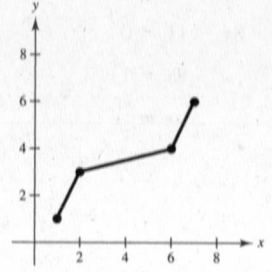

90.

x	-2	-1	1	3
$f(x)$	-5	-2	2	3

x	-5	-2	2	3
$f^{-1}(x)$	-2	-1	1	3

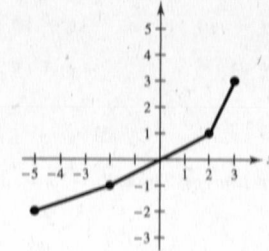

91.

x	-2	-1	3	4
f	6	0	-2	-3

x	-3	-2	0	6
$f^{-1}(x)$	4	3	-1	-2

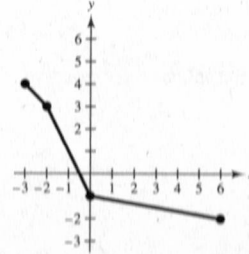

92.

x	-4	-2	0	3
$f(x)$	3	4	0	-1

The graph does not pass the Horizontal Line Test, so $f^{-1}(x)$ does not exist.

93. If $f(x) = k(2 - x - x^3)$ has an inverse and $f^{-1}(3) = -2$, then $f(-2) = 3$. Thus,

$$f(-2) = k(2 - (-2) - (-2)^3) = 3$$

$$k(2 + 2 + 8) = 3$$

$$12k = 3$$

$$k = \frac{3}{12} = \frac{1}{4}$$

So, $k = \frac{1}{4}$.

94. $f(x) = k(x^3 + 3x - 4)$

$$y = k(x^3 + 3x - 4)$$

$$x = k(y^3 + 3y - 4)$$

$$-5 = k[(2)^3 + 3(2) - 4]$$

$$-5 = 10k$$

$$-\frac{1}{2} = k$$

95. $x^2 = 64$

$$x = \pm\sqrt{64} = \pm 8$$

96. $(x - 5)^2 = 8$

$$x - 5 = \pm\sqrt{8}$$

$$x = 5 \pm 2\sqrt{2}$$

97. $4x^2 - 12x + 9 = 0$

$$(2x - 3)^2 = 0$$

$$2x - 3 = 0$$

$$x = \frac{3}{2}$$

98. $9x^2 + 12x + 3 = 0$

$$(9x + 3)(x + 1) = 0$$

$$9x + 3 = 0 \implies x = -\frac{1}{3}$$

$$x + 1 = 0 \implies x = -1$$

99. $x^2 - 6x + 4 = 0$ Complete the square.

$$x^2 - 6x = -4$$

$$x^2 - 6x + 9 = -4 + 9$$

$$(x - 3)^2 = 5$$

$$x - 3 = \pm\sqrt{5}$$

$$x = 3 \pm \sqrt{5}$$

100. $2x^2 - 4x - 6 = 0$

$$2(x^2 - 2x - 3) = 0$$

$$2(x + 1)(x - 3) = 0$$

$$x + 1 = 0 \implies x = -1$$

$$x - 3 = 0 \implies x = 3$$

101. $50 + 5x = 3x^2$

$$0 = 3x^2 - 5x - 50$$

$$0 = (3x + 10)(x - 5)$$

$$3x + 10 = 0 \implies x = -\frac{10}{3}$$

$$x - 5 = 0 \implies x = 5$$

102. $2x^2 + 4x - 9 = 2(x - 1)^2$

$$2x^2 + 4x - 9 = 2(x^2 - 2x + 1)$$

$$2x^2 + 4x - 9 = 2x^2 - 4x + 2$$

$$8x - 11 = 0$$

$$8x = 11$$

$$x = \frac{11}{8}$$

103. Let $2n = $ first positive even integer. Then $2n + 2 = $ next positive even integer.

$$2n(2n + 2) = 288$$

$$4n^2 + 4n - 288 = 0$$

$$4(n^2 + n - 72) = 0$$

$$4(n + 9)(n - 8) = 0$$

$$n + 9 = 0 \implies n = -9$$ Not a solution since the integers are positive.

$$n - 8 = 0 \implies n = 8$$

So, $2n = 16$ and $2n + 2 = 18$.

104. Given $h = 2b$ and $A = 10$

$$A = \tfrac{1}{2}bh$$

$$10 = \tfrac{1}{2}b(2b)$$

$$10 = b^2$$

$$\sqrt{10} = b \text{ and } h = 2b = 2\sqrt{10}$$

The base is $\sqrt{10}$ feet and the height is $2\sqrt{10}$ feet.

Section 1.10   Mathematical Modeling and Variation

You should know the following the following terms and formulas.
- ■ Direct variation (varies directly, directly proportional)
    (a) $y = kx$ (b) $y = kx^n$ (as nth power)
- ■ Inverse variation (varies inversely, inversely proportional)
    (a) $y = k/x$ (b) $y = k/(x^n)$ (as nth power)
- ■ Joint variation (varies jointly, jointly proportional)
    (a) $z = kxy$ (b) $z = kx^n y^m$ (as nth power of x and mth power of y)
- ■ k is called the constant of proportionality.
- ■ Least Squares Regression Line $y = ax + b$. Use your calculator or computer to enter
    the data points and to find the "best-fitting" linear model.

Vocabulary Check

1. variation; regression   2. sum of square differences   3. correlation coefficient

4. directly proportional   5. constant of variation   6. directly proportional

7. inverse   8. combined   9. jointly proportional

1. $y = 1767.0t + 123,916$

Year	Actual Number (in thousands)	Model (in thousands)
1992	128,105	127,450
1993	129,200	129,217
1994	131,056	130,984
1995	132,304	132,751
1996	133,943	134,518
1997	136,297	136,285
1998	137,673	138,052
1999	139,368	139,819
2000	142,583	141,586
2001	143,734	143,353
2002	144,863	145,120

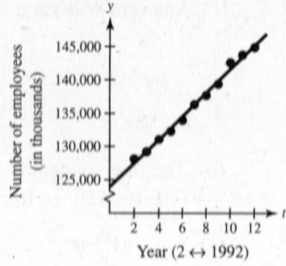

Year (2 ↔ 1992)

The model is a good fit for the actual data.

2.

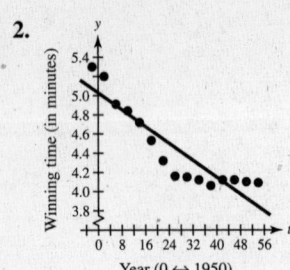

The model is not a "good fit" for the actual data. It appears that another type of model may be a better fit.

3.

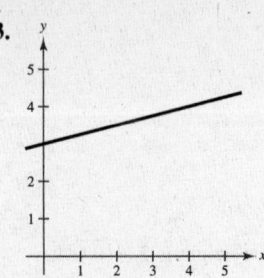

Using the points (0, 3) and (4, 4), we have $y = \frac{1}{4}x + 3$.

4.

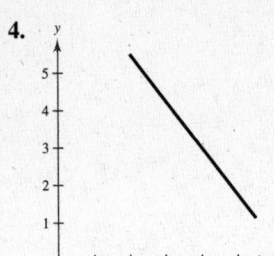

The line appears to pass through (2, 5.5) and (6, 0.5), so its equation is $y = -\frac{5}{4}x + 8$.

5.

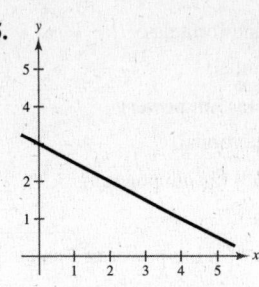

Using the points (2, 2) and (4, 1), we have $y = -\frac{1}{2}x + 3$.

6.

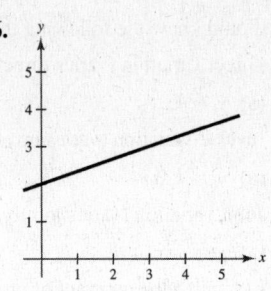

The line appears to pass through (0, 2) and (3, 3) so its equation is $y = \frac{1}{3}x + 2$.

7. (a) and (b)

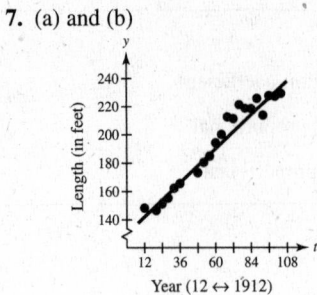

$y \approx t + 130$

(c) $y \approx 1.03t + 130.27$

(d) The models are similar.

(e) When $t = 108$, we have:

Model in part (b): 238 feet

Model in part (c): 241.51 feet

(f) Answers will vary.

8. (a) and (b)

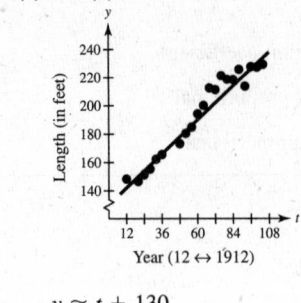

(b) The line appears to pass through (7, 1151.6) and (10, 1906.0), so the equation is about $y = 251.5x - 609$.

(c) $y = 251.56x - 608.79$

(d) Answers will vary.

(e) Using the model in (b), $y = 251.5(15) - 609 = \$3164.6$ million.

Using the model in (c), $y = 251.56(15) - 608.79 = \$3165.2$ million.

(f) Answers will vary.

9. (a) and (c)

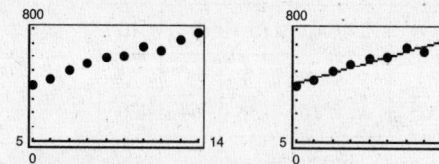

The model is a good fit to the actual data. ($r \approx 0.98$)

(b) $S \approx 38.4t + 224$

(d) For 2005, use $t = 15$: $S \approx \$800.4$ million

For 2007, use $t = 17$: $S \approx \$877.3$ million

(e) Each year the annual gross ticket sales for Broadway shows in New York City increase by approximately $38.4 million.

10. (a) $y = 0.4306x + 67.708$

(b)

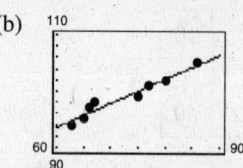

The model is a good fit to the data. ($r \approx 0.97$)

(c) $y = 0.4306(90) + 67.708 = 106.5$ million

(d) For every increase of one million households with cable TV, there is a 0.43 million increase in the number of households with color TV.

11. The graph appears to represent $y = 4/x$, so y varies inversely as x.

12. The graph appears to represent $y = \frac{3}{2}x$ which is a direct variation.

13. $k = 1$

x	2	4	6	8	10
$y = kx^2$	4	16	36	64	100

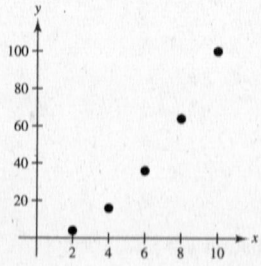

14. $k = 2$

x	2	4	6	8	10
$y = kx^2$	8	32	72	128	200

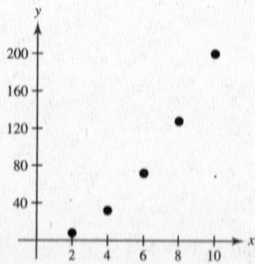

15. $k = \frac{1}{2}$

x	2	4	6	8	10
$y = kx^2$	2	8	18	32	50

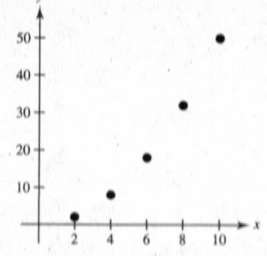

16. $k = \frac{1}{4}$

x	2	4	6	8	10
$y = kx^2$	1	4	9	16	25

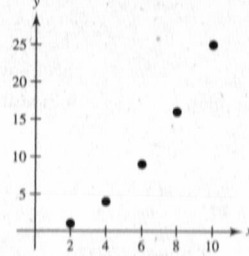

17. $k = 2$

x	2	4	6	8	10
$y = \dfrac{k}{x^2}$	$\dfrac{1}{2}$	$\dfrac{1}{8}$	$\dfrac{1}{18}$	$\dfrac{1}{32}$	$\dfrac{1}{50}$

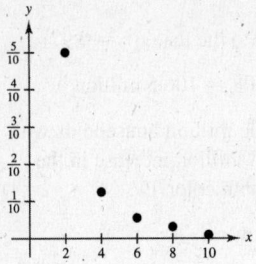

18. $k = 5$

x	2	4	6	8	10
$y = \dfrac{k}{x^2}$	$\dfrac{5}{4}$	$\dfrac{5}{16}$	$\dfrac{5}{36}$	$\dfrac{5}{64}$	$\dfrac{1}{20}$

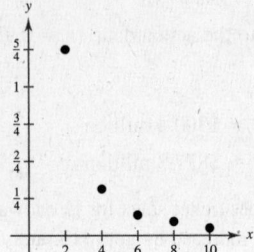

19. $k = 10$

x	2	4	6	8	10
$y = \dfrac{k}{x^2}$	$\dfrac{5}{2}$	$\dfrac{5}{8}$	$\dfrac{5}{18}$	$\dfrac{5}{32}$	$\dfrac{1}{10}$

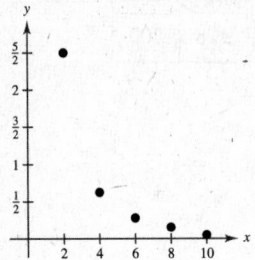

20. $k = 20$

x	2	4	6	8	10
$y = \dfrac{k}{x^2}$	5	$\dfrac{5}{4}$	$\dfrac{5}{9}$	$\dfrac{5}{16}$	$\dfrac{1}{5}$

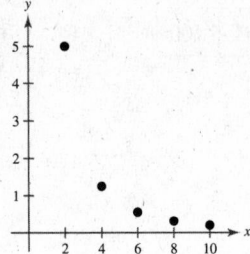

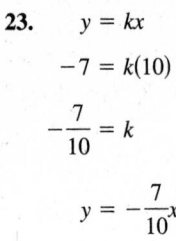

21. The table represents the equation $y = 5/x$.

22. The table represents the equation $y = \frac{2}{5}x$.

23.
$$y = kx$$
$$-7 = k(10)$$
$$-\frac{7}{10} = k$$
$$y = -\frac{7}{10}x$$

This equation checks with the other points given in the table.

24.
$$y = \frac{k}{x}$$
$$24 = \frac{k}{5}$$
$$120 = k$$

Thus, $y = 120/x$. This equation checks with the other points given in the table.

25.
$$y = kx$$
$$12 = k(5)$$
$$\frac{12}{5} = k$$
$$y = \frac{12}{5}x$$

26.
$$y = kx$$
$$14 = k(2)$$
$$7 = k$$
$$y = 7x$$

27.
$$y = kx$$
$$2050 = k(10)$$
$$205 = k$$
$$y = 205x$$

28.
$$y = kx$$
$$580 = k(6)$$
$$\frac{290}{3} = k$$
$$y = \frac{290}{3}x$$

29.
$$I = kP$$
$$87.50 = k(2500)$$
$$0.035 = k$$
$$I = 0.035P$$

30. $I = kP$

 $187.50 = k(5000)$

 $0.0375 = k$

  $I = 0.0375P$

31. $y = kx$

 $33 = k(13)$

 $\frac{33}{13} = k$

 $y = \frac{33}{13}x$

When $x = 10$ inches,
$y \approx 25.4$ centimeters.

When $x = 20$ inches,
$y \approx 50.8$ centimeters.

32. $y = kx$

 $53 = k(14)$

 $\frac{53}{14} = k$

 $y = \frac{53}{14}x$

5 gallons: $y = \frac{53}{14}(5) \approx 18.9$ liters

25 gallons: $y = \frac{53}{14}(25) \approx 94.6$ liters

33. $y = kx$

 $5520 = k(150{,}000)$

 $0.0368 = k$

 $y = 0.0368x$

 $y = 0.0368(200{,}000)$

  $= \$7360$

The property tax is \$7360.

34. $y = kx$

 $10.22 = k(145.99)$

 $0.07 \approx k$

 $y = 0.07x$

 $y = 0.07(540.50)$

 $y \approx 37.84$

The sales tax is \$37.84.

35. $d = kF$

 $0.15 = k(265)$

 $\frac{3}{5300} = k$

 $d = \frac{3}{5300}F$

(a) $d = \frac{3}{5300}(90) \approx 0.05$ meter

(b) $0.1 = \frac{3}{5300}F$

 $\frac{530}{3} = F$

 $F = 176\frac{2}{3}$ newtons

36. $d = kF$

 $0.12 = k(220)$

 $\frac{3}{5500} = k$

 $d = \frac{3}{5500}F$

 $0.16 = \frac{3}{5500}F$

 $\frac{880}{3} = F$

The required force is $293\frac{1}{3}$ newtons.

37. $d = kF$

 $1.9 = k(25) \implies k = 0.076$

 $d = 0.076F$

When the distance compressed is
3 inches, we have

 $3 = 0.076F$

 $F \approx 39.47.$

No child over 39.47 pounds should
use the toy.

38. $d = kF$

 $1 = k(15)$

 $k = \frac{1}{15}$

 $d = \frac{1}{15}F$

 $\frac{8}{2} = \frac{1}{15}F$

 $F = 60$ lb per spring

Combined lifting force $= 2F$

     $= 120$ lbs

39. $A = kr^2$

40. $V = ke^3$

41. $y = \dfrac{k}{x^2}$

42. $h = \dfrac{k}{\sqrt{s}}$

43. $F = \dfrac{kg}{r^2}$

44. $z = kx^2y^3$

45. $P = \dfrac{k}{V}$

46. $R = k(T - T_e)$

47. $F = \dfrac{km_1m_2}{r^2}$

48. $R = kS(S - L)$

49. $A = \frac{1}{2}bh$

The area of a triangle is jointly
proportional to its base and height.

50. $S = 4\pi r^2$

The surface area of a sphere
varies directly as the square of
the radius r.

51. $V = \frac{4}{3}\pi r^3$

The volume of a sphere varies
directly as the cube of its radius.

52. $V = \pi r^2 h$

The volume of a right circular
cylinder is jointly proportional to
the height and the square of the
radius.

53. $r = \dfrac{d}{t}$

Average speed is directly
proportional to the distance and
inversely proportional to the time.

54. $\omega = \sqrt{\dfrac{kg}{W}}$

ω varies directly as the square root
of g and inversely as the square
root of W. (**Note:** The constant of
proportionality is $\sqrt{k}$.)

55. $A = kr^2$

$9\pi = k(3)^2$

$\pi = k$

$A = \pi r^2$

56. $y = \dfrac{k}{x}$

$3 = \dfrac{k}{25}$

$75 = k$

$y = \dfrac{75}{x}$

57. $y = \dfrac{k}{x}$

$7 = \dfrac{k}{4}$

$28 = k$

$y = \dfrac{28}{x}$

58. $z = kxy$

$64 = k(4)(8)$

$2 = k$

$z = 2xy$

59. $F = krs^3$

$4158 = k(11)(3)^3$

$k = 14$

$F = 14rs^3$

60. $P = \dfrac{kx}{y^2}$

$\dfrac{28}{3} = \dfrac{k(42)}{9^2}$

$\dfrac{28}{3} \cdot \dfrac{81}{42} = k$

$\dfrac{2 \cdot 27}{3} = k$

$18 = k$

$P = \dfrac{18x}{y^2}$

61. $z = \dfrac{kx^2}{y}$

$6 = \dfrac{k(6)^2}{4}$

$\dfrac{24}{36} = k$

$\dfrac{2}{3} = k$

$z = \dfrac{2/3 x^2}{y} = \dfrac{2x^2}{3y}$

62. $v = \dfrac{kpq}{s^2}$

$1.5 = \dfrac{k(4.1)(6.3)}{(1.2)^2}$

$\dfrac{(1.5)(1.44)}{(4.1)(6.3)} = k$

$\dfrac{2.16}{25.83} = k$

$k = \dfrac{24}{287}$

$v = \dfrac{24pq}{287s^2}$

63. $d = kv^2$

$0.02 = k\left(\dfrac{1}{4}\right)^2$

$k = 0.32$

$d = 0.32v^2$

$0.12 = 0.32v^2$

$v^2 = \dfrac{0.12}{0.32} = \dfrac{3}{8}$

$v = \dfrac{\sqrt{3}}{2\sqrt{2}} = \dfrac{\sqrt{6}}{4} \approx 0.61$ mi/hr

64. $d = kv^2$

If the velocity is doubled:

$d = k(2v)^2$

$d = k \cdot 4v^2$

$\dfrac{4kv^2}{kv^2} = 4$

d increases by a factor of 4 when velocity is doubled.

65. $r = \dfrac{kl}{A}, \; A = \pi r^2 = \dfrac{\pi d^2}{4}$

$r = \dfrac{4kl}{\pi d^2}$

$66.17 = \dfrac{4(1000)k}{\pi\left(\dfrac{0.0126}{12}\right)^2}$

$k \approx 5.73 \times 10^{-8}$

$r = \dfrac{4(5.73 \times 10^{-8})l}{\pi\left(\dfrac{0.0126}{12}\right)^2}$

$33.5 = \dfrac{4(5.73 \times 10^{-8})l}{\pi\left(\dfrac{0.0126}{12}\right)^2}$

$\dfrac{33.5\pi\left(\dfrac{0.0126}{12}\right)^2}{4(5.73 \times 10^{-8})} = l$

$l \approx 506$ feet

66. From Exercise 65:

$k \approx 5.73 \times 10^{-8}$.

$r = \dfrac{4(5.73 \times 10^{-8})l}{\pi d^2}$

$d = \sqrt{\dfrac{4(5.73 \times 10^{-8})l}{\pi r}}$

$d = \sqrt{\dfrac{4(5.73 \times 10^{-8})(14)}{\pi(0.05)}}$

$d \approx 0.0045$ feet $= 0.054$ inch

67. $W = kmh$

$2116.8 = k(120)(1.8)$

$k = \dfrac{2116.8}{(120)(1.8)} = 9.8$

$W = 9.8mh$

When $m = 100$ kilograms and $h = 1.5$ meters, we have
$W = 9.8(100)(1.5) = 1470$ joules.

68. $P = kA = k(\pi r^2) = k\pi\left(\dfrac{d}{2}\right)^2$

$8.78 = k\pi\left(\dfrac{9}{2}\right)^2$

$\dfrac{4(8.78)}{81\pi} = k$

$k \approx 0.138$

However, we do not obtain $11.78 when $d = 12$ inches.

$P = 0.138\pi\left(\dfrac{12}{2}\right)^2 \approx \$15.61$

Instead, $k = \dfrac{11.78}{36\pi} \approx 0.104$.

For the 15-inch pizza, we have $k = \dfrac{4(14.18)}{225\pi} \approx 0.080$.

The price is not directly proportional to the surface area.
The best buy is the 15-inch pizza.

69. $v = \dfrac{k}{A}$

$v = \dfrac{k}{0.75A} = \dfrac{4}{3}\left(\dfrac{k}{A}\right)$

The velocity is increased by one-third.

70. Load $= \dfrac{kwd^2}{l}$

(a) load $= \dfrac{k(2w)d^2}{2l} = \dfrac{kwd^2}{l}$

The safe load is unchanged.

(b) load $= \dfrac{k(2w)(2d)^2}{l} = \dfrac{8kwd^2}{l}$

The safe load is eight times as great.

(c) load $= \dfrac{k(2w)(2d)^2}{2l} = \dfrac{4kwd^2}{l}$

The safe load is four times as great.

(d) load $= \dfrac{kw(d/2)^2}{l} = \dfrac{(1/4)kwd^2}{l}$

The safe load is one-fourth as great.

71. (a)

(b) Yes, the data appears to be modeled (approximately) by the inverse proportion model.

$4.2 = \dfrac{k_1}{1000}$	$1.9 = \dfrac{k_2}{2000}$	$1.4 = \dfrac{k_3}{3000}$	$1.2 = \dfrac{k_4}{4000}$	$0.9 = \dfrac{k_5}{5000}$
$4200 = k_1$	$3800 = k_2$	$4200 = k_3$	$4800 = k_4$	$4500 = k_5$

—CONTINUED—

71. —CONTINUED—

(c) Mean: $k = \dfrac{4200 + 3800 + 4200 + 4800 + 4500}{5} = 4300$, Model: $C = \dfrac{4300}{d}$

(d)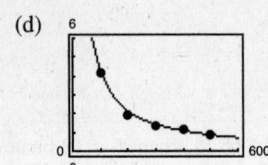

(e) $3 = \dfrac{4300}{d}$

$d = \dfrac{4300}{3} = 1433\dfrac{1}{3}$ meters

---

72. (a)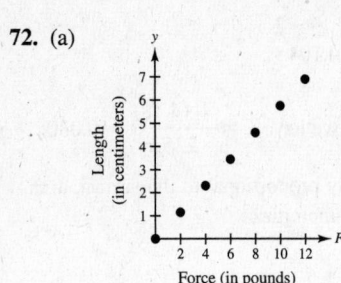

(b) It appears to fit Hooke's Law.

$k \approx \dfrac{6.9}{12} = 0.575$

(c) $y = kF$

$9 = 0.575F$

$F \approx 15.7$ pounds

73. $y = \dfrac{262.76}{x^{2.12}}$

(a)

(b) $y = \dfrac{262.76}{(25)^{2.12}}$

≈ 0.2857 microwatts per sq. cm.

74. $I = \dfrac{k}{d^2}$

When the distance is doubled:

$I = \dfrac{k}{(2d)^2} = \dfrac{k}{4d^2}.$

The illumination is one-fourth as great. The model given in Exercise 73 is very close to $I = k/d^2$. The difference is probably due to measurement error.

---

75. False. y will increase if k is positive and y will decrease if k is negative.

76. False. E is jointly proportional (not "directly proportional") to the mass of an object and the square of its velocity.

77. False. The closer the value of $|r|$ is to 1, the better the fit.

---

78. (a) The data shown could be represented by a linear model which would be a good approximation.

(b) The points do not follow a linear pattern. A linear model would be a poor approximation. A quadratic model would be better.

(c) The points do not follow a linear pattern. A linear model would be a poor approximation.

(d) The data shown could be represented by a linear model which would be a good approximation.

---

79. The accuracy of the model in predicting prize winnings is questionable because the model is based on limited data.

80. Answers will vary.

---

81. $3x + 2 > 17$

$3x > 15$

$x > 5$

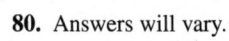

82. $-7x + 10 \le -1 + x$

$-8x \le -11$

$x \ge \dfrac{11}{8}$

83. $|2x - 1| < 9$

$-9 < 2x - 1 < 9$

$-8 < 2x < 10$

$-4 < x < 5$

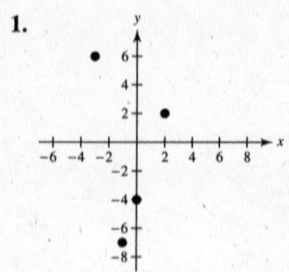

84. $|4 - 3x| + 7 \geq 12$

$|4 - 3x| \geq 5$

$4 - 3x \leq -5 \quad \text{or} \quad 4 - 3x \geq 5$

$-3x \leq -9 \quad \text{or} \quad -3x \geq 1$

$x \geq 3 \quad \text{or} \quad x \leq -\frac{1}{3}$

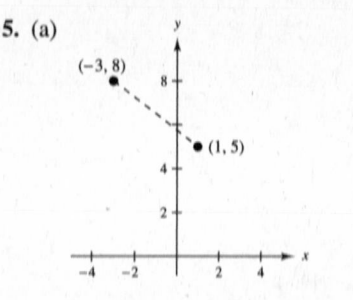

85. $f(x) = \dfrac{x^2 + 5}{x - 3}$

   (a) $f(0) = \dfrac{0^2 + 5}{0 - 3} = -\dfrac{5}{3}$

   (b) $f(-3) = \dfrac{(-3)^2 + 5}{-3 - 3} = \dfrac{14}{-6} = -\dfrac{7}{3}$

   (c) $f(4) = \dfrac{4^2 + 5}{4 - 3} = 21$

86. $f(x) = \begin{cases} -x^2 + 10, & x \geq -2 \\ 6x^2 - 1, & x < -2 \end{cases}$

   (a) $f(-2) = -(-2)^2 + 10 = -4 + 10 = 6$

   (b) $f(1) = -(1)^2 + 10 = -1 + 10 = 9$

   (c) $f(-8) = 6(-8)^2 - 1 = 384 - 1 = 383$

87. Answers will vary.

Review Exercises for Chapter 1

1.

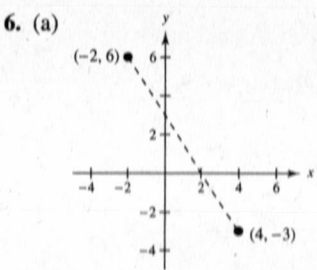

2.

3. $x > 0$ and $y = -2$ in Quadrant IV.

4. $y > 0$ in Quadrants I and II.

5. (a)

   (b) $d = \sqrt{(-3 - 1)^2 + (8 - 5)^2} = \sqrt{16 + 9} = 5$

   (c) Midpoint: $\left(\dfrac{-3 + 1}{2}, \dfrac{8 + 5}{2} \right) = \left(-1, \dfrac{13}{2} \right)$

6. (a)

   (b) $d = \sqrt{(-2 - 4)^2 + (6 + 3)^2}$

$= \sqrt{36 + 81} = \sqrt{117} = 3\sqrt{13}$

   (c) Midpoint: $\left(\dfrac{-2 + 4}{2}, \dfrac{6 - 3}{2} \right) = \left(1, \dfrac{3}{2} \right)$

7. (a)

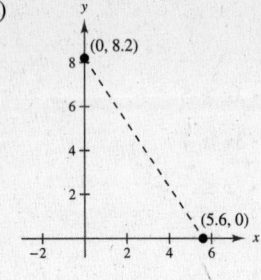

(b) $d = \sqrt{(5.6 - 0)^2 + (0 - 8.2)^2}$

$\quad = \sqrt{31.36 + 67.24} = \sqrt{98.6} \approx 9.9$

(c) Midpoint: $\left(\dfrac{0 + 5.6}{2}, \dfrac{8.2 + 0}{2}\right) = (2.8, 4.1)$

8. (a)

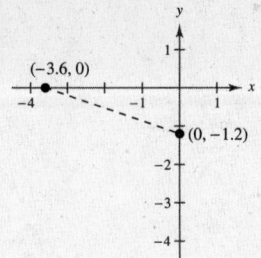

(b) $d = \sqrt{(0 + 3.6)^2 + (-1.2 - 0)^2}$

$\quad = \sqrt{14.4} \approx 3.8$

(c) $\left(\dfrac{0 - 3.6}{2}, \dfrac{-1.2 + 0}{2}\right) = (-1.8, -0.6)$

9. $(4 - 2, 8 - 3) = (2, 5)$

$(6 - 2, 8 - 3) = (4, 5)$

$(4 - 2, 3 - 3) = (2, 0)$

$(6 - 2, 3 - 3) = (4, 0)$

10. Original: $(0, 1), (3, 3), (0, 5), (-3, 3)$

New: $(0 - 4, 1 + 5), (3 - 4, 3 + 5), (0 - 4, 5 + 5), (-3 - 4, 3 + 5) = (-4, 6), (-1, 8), (-4, 10), (-7, 8)$

11. $(2001, 539.1), (2003, 773.8)$

$\left(\dfrac{2001 + 2003}{2}, \dfrac{539.1 + 773.8}{2}\right) = (2002, 656.45)$

In 2002, the sales were approximately \$656.45 million.

12. (a)

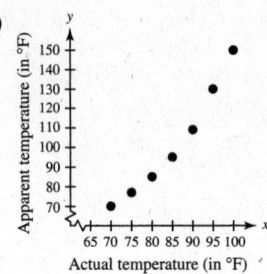

(b) Change in apparent temperature $= 150°F - 70°F$

$= 80°F$

13. $\dfrac{4}{3}\pi r^3 = 47{,}712.94$

$r = \sqrt[3]{\dfrac{47{,}712.94(3)}{4\pi}}$

$r \approx 22.5$ centimeters

14. (a)

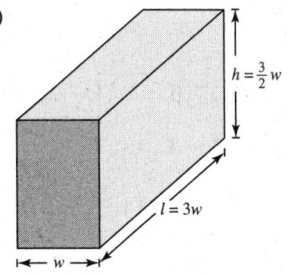

(b) $V = l \cdot w \cdot h$

$2304 = (3w) \cdot w \cdot \left(\tfrac{3}{2}w\right)$

$2304 = \tfrac{9}{2}w^3$

$512 = w^3$

$8 = w \implies w = 8$ inches

$l = 3(8) = 24$ inches

$h = \tfrac{3}{2}(8) = 12$ inches

15. $y = 3x - 5$

x	-2	-1	0	1	2
y	-11	-8	-5	-2	1

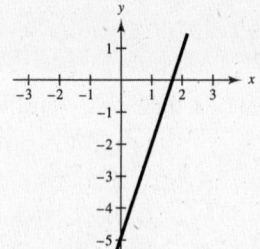

16. $y = -\frac{1}{2}x + 2$

x	-4	-2	0	2	4
y	4	3	2	1	0

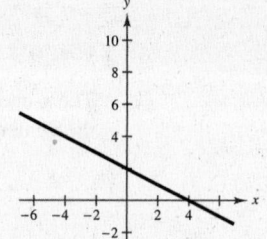

17. $y = x^2 - 3x$

x	-1	0	1	2	3	4
y	4	0	-2	-2	0	4

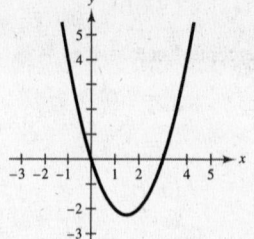

18. $y = 2x^2 - x - 9$

x	-2	-1	0	1	2	3
y	1	-6	-9	-8	-3	6

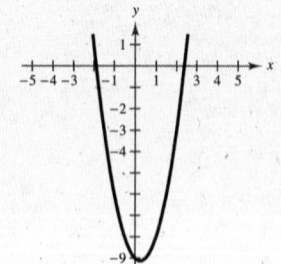

19. $y - 2x - 3 = 0$

$y = 2x + 3$

Line with x-intercept $\left(-\frac{3}{2}, 0\right)$ and y-intercept $(0, 3)$

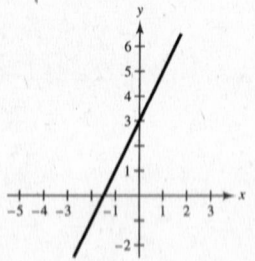

20. $3x + 2y + 6 = 0$

$2y = -3x - 6$

$y = -\frac{3}{2}x - 3$

Line with x-intercept $(-2, 0)$ and y-intercept $(0, -3)$

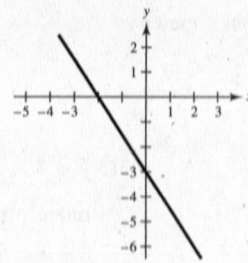

21. $y = \sqrt{5 - x}$

Domain: $(-\infty, 5]$

x	5	4	1	-4
y	0	1	2	3

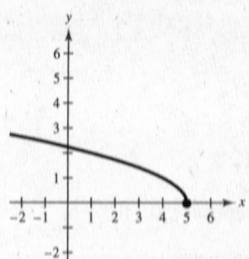

22. $y = \sqrt{x + 2}$, domain: $[-2, \infty)$

x	-2	0	2	7
y	0	$\sqrt{2}$	2	3

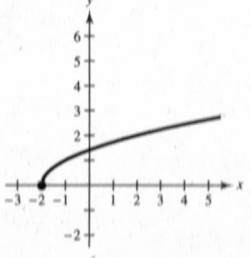

23. $y + 2x^2 = 0$

$y = -2x^2$ is a parabola.

x	0	±1	±2
y	0	-2	-8

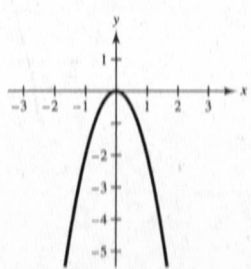

24. $y = x^2 - 4x$ is a parabola.

x	-1	0	1	2	3	4
y	5	0	-3	-4	-3	0

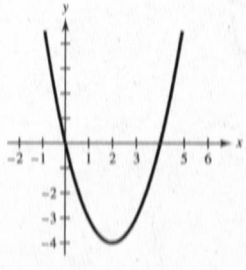

25. $y = 2x + 7$

x-intercept: Let $y = 0$.

$$0 = 2x + 7$$

$$x = -\frac{7}{2}$$

$$\left(-\frac{7}{2}, 0\right)$$

y-intercept: Let $x = 0$.

$$y = 2(0) + 7$$

$$y = 7$$

$$(0, 7)$$

26. $y = |x + 1| - 3$

$$0 = |x + 1| - 3$$

For $x + 1 > 0$, $0 = x + 1 - 3$, or $2 = x$.

For $x + 1 < 0$, $0 = -(x + 1) - 3$, or $-4 = x$.

$$y = |x + 1| - 3$$

$$y = |0 + 1| - 3 \text{ or } y = -2$$

The *x*-intercepts are $(2, 0)$ and $(-4, 0)$; the *y*-intercept is $(0, -2)$.

27. $y = (x - 3)^2 - 4$

x-intercepts: $0 = (x - 3)^2 - 4 \implies (x - 3)^2 = 4$

$$\implies x - 3 = \pm 2$$

$$\implies x = 3 \pm 2$$

$$\implies x = 5 \text{ or } x = 1$$

y-intercept: $y = (0 - 3)^2 - 4 \implies y = 9 - 4 \implies y = 5$

The *x*-intercepts are $(1, 0)$ and $(5, 0)$.
The *y*-intercept is $(0, 5)$.

28. $y = x\sqrt{4 - x^2}$

x-intercepts: $0 = x\sqrt{4 - x^2}$

$$x = 0 \qquad \sqrt{4 - x^2} = 0$$

$$4 - x^2 = 0$$

$$x = \pm 2$$

$$(0, 0), (-2, 0), (2, 0)$$

y-intercept: $y = 0 \cdot \sqrt{4 - 0} = 0$

$$(0, 0)$$

29. $y = -4x + 1$

Intercepts: $\left(\frac{1}{4}, 0\right), (0, 1)$

$y = -4(-x) + 1 \implies y = 4x + 1 \implies$ No *y*-axis symmetry

$-y = -4x + 1 \implies y = 4x - 1 \implies$ No *x*-axis symmetry

$-y = -4(-x) + 1 \implies y = -4x - 1 \implies$ No origin symmetry

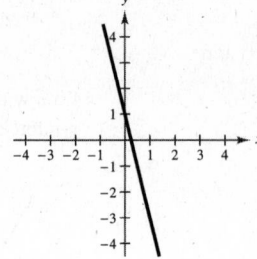

30. $y = 5x - 6$

Intercepts: $\left(\frac{6}{5}, 0\right), (0, -6)$

No symmetry

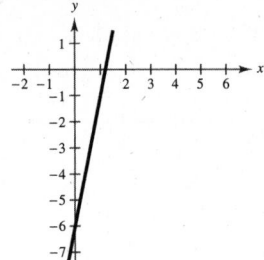

31. $y = 5 - x^2$

Intercepts: $\left(\pm\sqrt{5}, 0\right), (0, 5)$

$y = 5 - (-x)^2 \implies y = 5 - x^2 \implies$ *y*-axis symmetry

$-y = 5 - x^2 \implies y = -5 + x^2 \implies$ No *x*-axis symmetry

$-y = 5 - (-x)^2 \implies y = -5 + x^2 \implies$ No origin symmetry

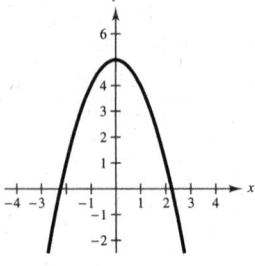

32. $y = x^2 - 10$

Intercepts: $\left(\pm\sqrt{10}, 0\right), (0, -10)$

y-axis symmetry

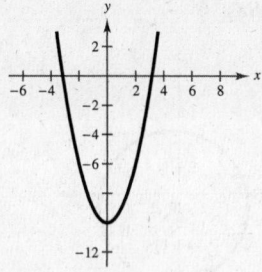

33. $y = x^3 + 3$

Intercepts: $\left(-\sqrt[3]{3}, 0\right), (0, 3)$

$y = (-x)^3 + 3 \implies y = -x^3 + 3 \implies$ No y-axis symmetry

$-y = x^3 + 3 \implies y = -x^3 - 3 \implies$ No x-axis symmetry

$-y = (-x)^3 + 3 \implies y = x^3 - 3 \implies$ No origin symmetry

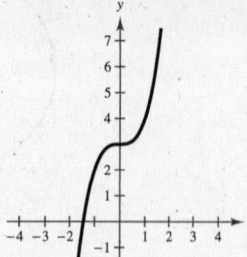

34. $y = -6 - x^3$

Intercepts: $\left(\sqrt[3]{-6}, 0\right), (0, -6)$

No symmetry

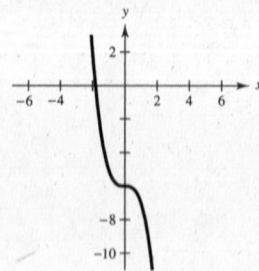

35. $y = \sqrt{x + 5}$

Domain: $[-5, \infty)$

Intercepts: $(-5, 0), \left(0, \sqrt{5}\right)$

$y = \sqrt{-x + 5} \implies$ No y-axis symmetry

$-y = \sqrt{x + 5} \implies y = -\sqrt{x + 5} \implies$ No x-axis symmetry

$-y = \sqrt{-x + 5} \implies y = -\sqrt{-x + 5} \implies$ No origin symmetry

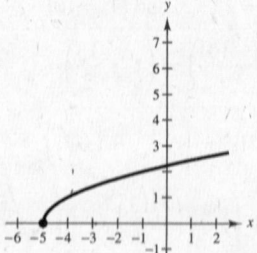

36. $y = |x| + 9$

Intercepts: $(0, 9)$

y-axis symmetry

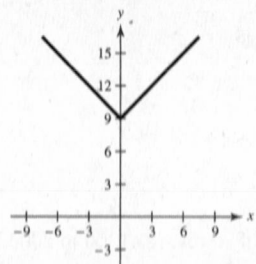

37. $x^2 + y^2 = 9$

Center: $(0, 0)$

Radius: 3

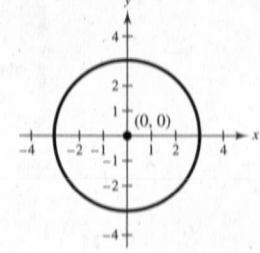

38. $x^2 + y^2 = 4$

Center: $(0, 0)$

Radius: 2

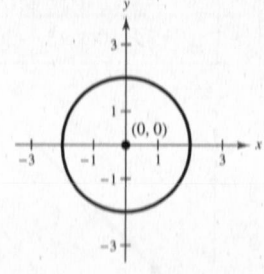

39.
$$(x + 2)^2 + y^2 = 16$$
$$(x - (-2))^2 + (y - 0)^2 = 4^2$$
Center: $(-2, 0)$
Radius: 4

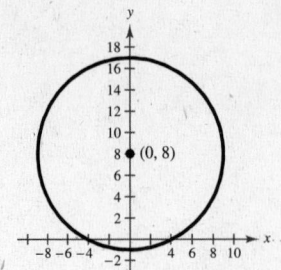

40. $x^2 + (y - 8)^2 = 81$
Center: $(0, 8)$
Radius: 9

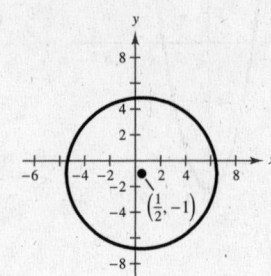

41. $\left(x - \frac{1}{2}\right)^2 + (y + 1)^2 = 36$
$$\left(x - \frac{1}{2}\right)^2 + (y - (-1))^2 = 6^2$$
Center: $\left(\frac{1}{2}, -1\right)$
Radius: 6

42.
$$(x + 4)^2 + \left(y - \frac{3}{2}\right)^2 = 100$$
$$(x - (-4))^2 + \left(y - \frac{3}{2}\right)^2 = 100$$
Center: $\left(-4, \frac{3}{2}\right)$
Radius: 10

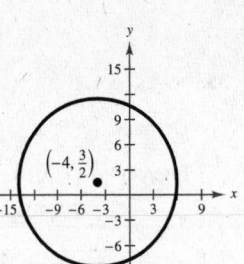

43. Endpoints of a diameter: $(0, 0)$ and $(4, -6)$

Center: $\left(\dfrac{0 + 4}{2}, \dfrac{0 + (-6)}{2}\right) = (2, -3)$

Radius: $r = \sqrt{(2 - 0)^2 + (-3 - 0)^2} = \sqrt{4 + 9} = \sqrt{13}$

Standard form: $(x - 2)^2 + (y - (-3))^2 = \left(\sqrt{13}\right)^2$
$$(x - 2)^2 + (y + 3)^2 = 13$$

44. Endpoints of a diameter: $(-2, -3)$ and $(4, -10)$

Center: $\left(\dfrac{-2 + 4}{2}, \dfrac{-3 + (-10)}{2}\right) = \left(1, -\dfrac{13}{2}\right)$

Radius: $r = \sqrt{(1 - (-2))^2 + \left(-\dfrac{13}{2} - (-3)\right)^2} = \sqrt{9 + \dfrac{49}{4}} = \sqrt{\dfrac{85}{4}}$

Standard form: $(x - 1)^2 + \left(y - \left(-\dfrac{13}{2}\right)\right)^2 = \left(\sqrt{\dfrac{85}{4}}\right)^2$
$$(x - 1)^2 + \left(y + \dfrac{13}{2}\right)^2 = \dfrac{85}{4}$$

45. $F = \frac{5}{4}x$, $0 \le x \le 20$

(a)

x	0	4	8	12	16	20
F	0	5	10	15	20	25

(b)

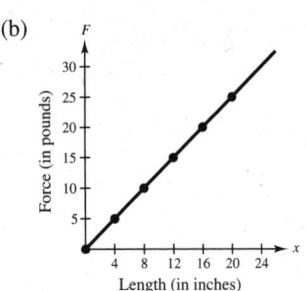

(c) When $x = 10$, $F = \frac{50}{4} = 12.5$ pounds.

46. (a)

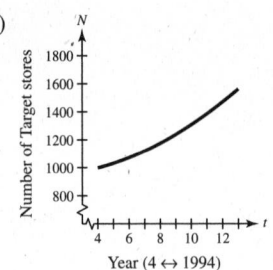

(b) $z = 9.94$; The number of stores was 1300 in 2003.

47. $y = 6$

Horizontal line, $m = 0$

y-intercept: $(0, 6)$

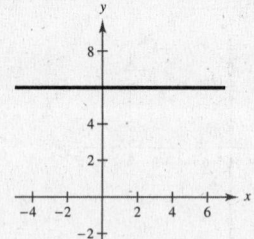

48. $x = -3$

Slope: m is undefined.

y-intercept: none

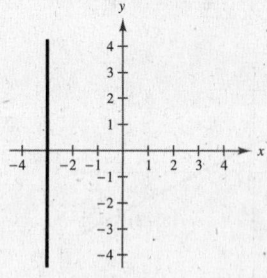

49. $y = 3x + 13$

Slope: $m = 3 = \frac{3}{1}$

y-intercept: $(0, 13)$

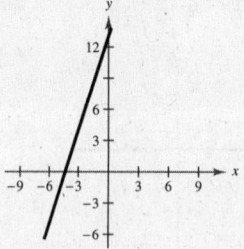

50. $y = -10x + 9$

Slope: $m = -10$

y-intercept: $(0, 9)$

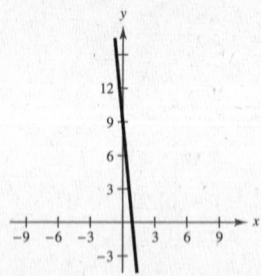

51. $(3, -4), (-7, 1)$

$$m = \frac{1 - (-4)}{-7 - 3} = \frac{5}{-10} = -\frac{1}{2}$$

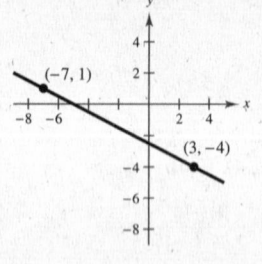

52. $(-1, 8), (6, 5)$

$$m = \frac{5 - 8}{6 - (-1)} = -\frac{3}{7}$$

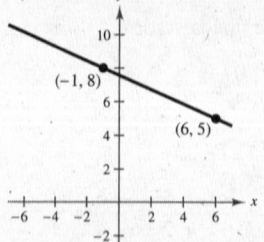

53. $(-4.5, 6), (2.1, 3)$

$$m = \frac{3 - 6}{2.1 - (-4.5)}$$

$$= \frac{-3}{6.6} = -\frac{30}{66} = -\frac{5}{11}$$

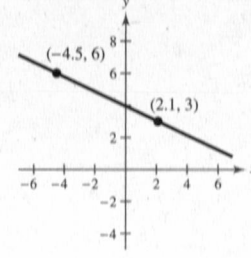

54. $(-3, 2), (8, 2)$

$$m = \frac{2 - 2}{-3 - 8} = \frac{0}{-11} = 0$$

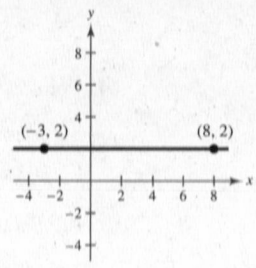

55. $(0, -5), m = \frac{3}{2}$

$$y - (-5) = \frac{3}{2}(x - 0)$$

$$y + 5 = \frac{3}{2}x$$

$$y = \frac{3}{2}x - 5$$

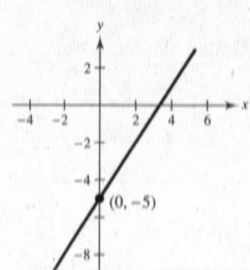

56. $(-2, 6)$, $m = 0$

$$y - 6 = 0(x - (-2))$$

$$y - 6 = 0$$

$$y = 6$$

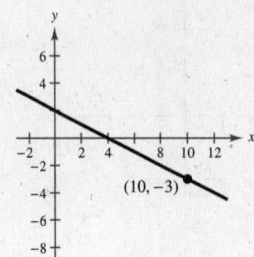

57. $(10, -3)$, $m = -\frac{1}{2}$

$$y - (-3) = -\frac{1}{2}(x - 10)$$

$$y + 3 = -\frac{1}{2}x + 5$$

$$y = -\frac{1}{2}x + 2$$

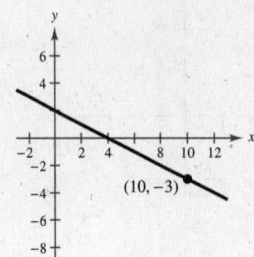

58. $(-8, 5)$, m is undefined.

The line is vertical.

$$x = -8$$

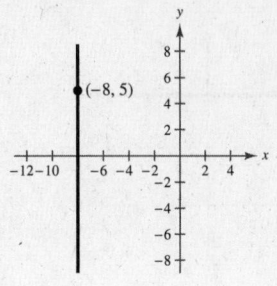

59. $(0, 0)$, $(0, 10)$

$$m = \frac{10 - 0}{0 - 0} = \frac{10}{0}, \quad \text{undefined}$$

The line is vertical.

$$x = 0$$

60. $(2, 5)$, $(-2, -1)$

$$m = \frac{-1 - 5}{-2 - 2} = \frac{-6}{-4} = \frac{3}{2}$$

$$y - 5 = \frac{3}{2}(x - 2)$$

$$2y - 10 = 3x - 6$$

$$2y = 3x + 4$$

$$y = \frac{3}{2}x + 2$$

61. $(-1, 4)$, $(2, 0)$

$$m = \frac{0 - 4}{2 - (-1)} = -\frac{4}{3}$$

$$y - 4 = -\frac{4}{3}(x - (-1))$$

$$y - 4 = -\frac{4}{3}x - \frac{4}{3}$$

$$y = -\frac{4}{3}x + \frac{8}{3}$$

62. $(11, -2)$, $(6, -1)$

$$m = \frac{-1 - (-2)}{6 - 11} = -\frac{1}{5}$$

$$y - (-2) = -\frac{1}{5}(x - 11)$$

$$5y + 10 = -x + 11$$

$$5y = -x + 1$$

$$y = -\frac{1}{5}x + \frac{1}{5}$$

63. Point: $(3, -2)$

$$5x - 4y = 8 \implies y = \frac{5}{4}x - 2 \text{ and } m = \frac{5}{4}$$

(a) Parallel slope: $m = \frac{5}{4}$

$$y - (-2) = \frac{5}{4}(x - 3)$$

$$y + 2 = \frac{5}{4}x - \frac{15}{4}$$

$$y = \frac{5}{4}x - \frac{23}{4}$$

(b) Perpendicular slope: $m = -\frac{4}{5}$

$$y - (-2) = -\frac{4}{5}(x - 3)$$

$$y + 2 = -\frac{4}{5}x + \frac{12}{5}$$

$$y = -\frac{4}{5}x + \frac{2}{5}$$

64. Point: $(-8, 3)$, $2x + 3y = 5$

$$3y = 5 - 2x$$

$$y = \frac{5}{3} - \frac{2}{3}x$$

(a) Parallel slope: $m = -\frac{2}{3}$

$$y - 3 = -\frac{2}{3}(x + 8)$$

$$3y - 9 = -2x - 16$$

$$3y = -2x - 7$$

$$y = -\frac{2}{3}x - \frac{7}{3}$$

(b) Perpendicular slope: $m = \frac{3}{2}$

$$y - 3 = \frac{3}{2}(x + 8)$$

$$2y - 6 = 3x + 24$$

$$2y = 3x + 30$$

$$y = \frac{3}{2}x + 15$$

65. $(6, 12{,}500)$ $m = 850$

$$y - 12{,}500 = 850(t - 6)$$

$$y - 12{,}500 = 850t - 5100$$

$$y = 850t + 7400, \ '6 \le t \le 11$$

66. $(6, 72.95)$, $m = 5.15$

$$V - 72.95 = 5.15(t - 6)$$

$$V - 72.95 = 5.15t - 30.9$$

$$V = 5.15t + 42.05, \ 6 \le t \le 11$$

67. $16x - y^4 = 0$

$$y^4 = 16x$$

$$y = \pm 2\sqrt[4]{x}$$

No, y is not a function of x. Some x-values correspond to two y-values.

68. $2x - y - 3 = 0$

$$2x - 3 = y$$

Yes, the equation represents y as a function of x.

69. $y = \sqrt{1 - x}$

Yes. Each x-value, $x \le 1$, corresponds to only one y-value so y is a function of x.

70. $|y| = x + 2$ corresponds to $y = x + 2$ or $-y = x + 2$. No, y is not a function of x. Some x-values correspond to two y-values.

71. $f(x) = x^2 + 1$

(a) $f(2) = (2)^2 + 1 = 5$

(b) $f(-4) = (-4)^2 + 1 = 17$

(c) $f(t^2) = (t^2)^2 + 1 = t^4 + 1$

(d) $f(t + 1) = (t + 1)^2 + 1$

$$= t^2 + 2t + 2$$

72. $h(x) = \begin{cases} 2x + 1, & x \le -1 \\ x^2 + 2, & x > -1 \end{cases}$

(a) $h(-2) = 2(-2) + 1 = -3$

(b) $h(-1) = 2(-1) + 1 = -1$

(c) $h(0) = 0^2 + 2 = 2$

(d) $h(2) = 2^2 + 2 = 6$

73. $f(x) = \sqrt{25 - x^2}$

Domain: $25 - x^2 \ge 0$

$(5 + x)(5 - x) \ge 0$

Critical numbers: $x = \pm 5$

Test intervals: $(-\infty, -5), (-5, 5), (5, \infty)$

Test: Is $25 - x^2 \ge 0$?

Solution set: $-5 \le x \le 5$

Thus, the domain is all real numbers x such that $-5 \le x \le 5$, or $[-5, 5]$.

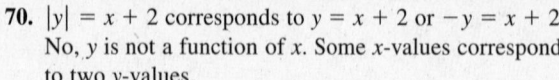

74. $f(x) = 3x + 4$

Domain: all real numbers

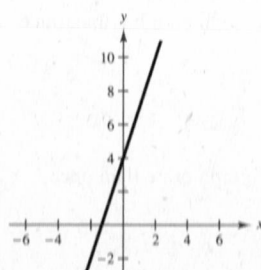

75. $h(x) = \dfrac{x}{x^2 - x - 6}$

$$= \dfrac{x}{(x + 2)(x - 3)}$$

Domain: All real numbers x except $x = -2, 3$

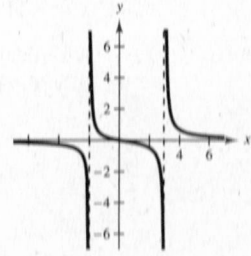

76. $h(t) = |t + 1|$

Domain: all real numbers

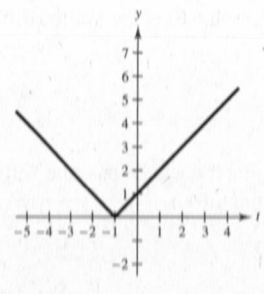

77. $v(t) = -32t + 48$

    (a) $v(1) = 16$ feet per second

    (b) $0 = -32t + 48$

        $t = \frac{48}{32} = 1.5$ seconds

    (c) $v(2) = -16$ feet per second

78. (a) Model: (40% of $(50 - x)$) + (100% of x) = (amount of acid in final mixture)

        Amount of acid in final mixture $= f(x)$

        $f(x) = 0.4(50 - x) + 1.0x = 20 + 0.6x$

    (b) Domain: $0 \le x \le 50$ (c) $20 + 0.6x = 50\%(50)$

        Range: $20 \le y \le 50$ $20 + 0.6x = 25$

                                                      $0.6x = 5$

                                                      $x = 8\frac{1}{3}$ liters

79. $f(x) = 2x^2 + 3x - 1$

$$\frac{f(x + h) - f(x)}{h} = \frac{[2(x + h)^2 + 3(x + h) - 1] - (2x^2 + 3x - 1)}{h}$$

$$= \frac{2x^2 + 4xh + 2h^2 + 3x + 3h - 1 - 2x^2 - 3x + 1}{h}$$

$$= \frac{h(4x + 2h + 3)}{h}$$

$$= 4x + 2h + 3, \ \ h \ne 0$$

80. $$f(x) = x^3 - 5x^2 + x$$

$$f(x + h) = (x + h)^3 - 5(x + h)^2 + (x + h)$$

$$= x^3 + 3x^2h + 3xh^2 + h^3 - 5x^2 - 10xh - 5h^2 + x + h$$

$$\frac{f(x + h) - f(x)}{h} = \frac{x^3 + 3x^2h + 3xh^2 + h^3 - 5x^2 - 10xh - 5h^2 + x + h - x^3 + 5x^2 - x}{h}$$

$$= \frac{3x^2h + 3xh^2 + h^3 - 10xh - 5h^2 + h}{h}$$

$$= \frac{h(3x^2 + 3xh + h^2 - 10x - 5h + 1)}{h}$$

$$= 3x^2 + 3xh + h^2 - 10x - 5h + 1, \ \ h \ne 0$$

81. $y = (x - 3)^2$

The graph passes the Vertical Line Test. y *is* a function of x.

82. $y = -\frac{3}{5}x^3 - 2x + 1$

A vertical line intersects the graph no more than once, so y *is* a function of x.

83. $x - 4 = y^2$

The graph does not pass the Vertical Line Test. y *is not* a function of x.

84. $x = -|4 - y|$

A vertical line intersects the graph more than once, so y *is not* a function of x.

85. $f(x) = 3x^2 - 16x + 21$

$3x^2 - 16x + 21 = 0$

$(3x - 7)(x - 3) = 0$

$3x - 7 = 0$ or $x - 3 = 0$

$x = \dfrac{7}{3}$ or $x = 3$

86. $f(x) = 5x^2 + 4x - 1$

$5x^2 + 4x - 1 = 0$

$(5x - 1)(x + 1) = 0$

$5x - 1 = 0 \Rightarrow x = \dfrac{1}{5}$

$x + 1 = 0 \Rightarrow x = -1$

87. $f(x) = \dfrac{8x + 3}{11 - x}$

$\dfrac{8x + 3}{11 - x} = 0$

$8x + 3 = 0$

$x = -\dfrac{3}{8}$

88. $f(x) = x^3 - x^2 - 25x + 25$

$x^3 - x^2 - 25x + 25 = 0$

$x^2(x - 1) - 25(x - 1) = 0$

$(x - 1)(x^2 - 25) = 0$

$x - 1 = 0 \Rightarrow x = 1$

$x^2 - 25 = 0 \Rightarrow x = \pm 5$

89. $f(x) = |x| + |x + 1|$

f is increasing on $(0, \infty)$.

f is decreasing on $(-\infty, -1)$.

f is constant on $(-1, 0)$.

90. Increasing on $(-2, 0)$ and $(2, \infty)$

Decreasing on $(-\infty, -2)$ and $(0, 2)$

91. $f(x) = -x^2 + 2x + 1$

Relative maximum: $(1, 2)$

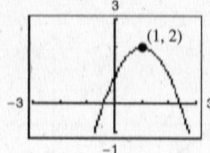

92. $f(x) = x^4 - 4x^2 - 2$

Relative minimum: $(-1.41, -6), (1.41, -6)$

Relative maximum: $(0, -2)$

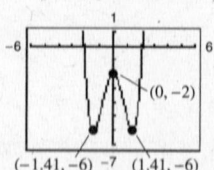

93. $f(x) = x^3 - 6x^4$

Relative maximum: $(0.125, 0.000488) \approx (0.13, 0.00)$

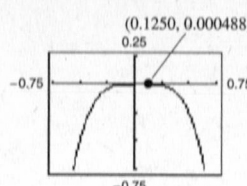

94. $f(x) = x^3 - 4x^2 + x - 1$

Relative minimum: $(2.54, -7.88)$

Relative maximum: $(0.13, -0.94)$

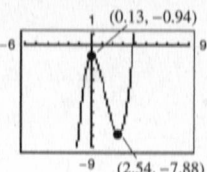

95. $f(x) = -x^2 + 8x - 4$

$\dfrac{f(4) - f(0)}{4 - 0} = \dfrac{12 - (-4)}{4} = 4$

The average rate of change of f from $x_1 = 0$ to $x_2 = 4$ is 4.

96. $f(x) = x^3 + 12x - 2, \; x_1 = 0, \; x_2 = 4$

$\dfrac{f(x_2) - f(x_1)}{x_2 - x_1} = \dfrac{f(4) - f(0)}{4 - 0}$

$= \dfrac{110 - (-2)}{4} = \dfrac{112}{4} = 28$

The average rate of change from $x = 0$ to $x = 4$ is 28.

97. $f(x) = 2 - \sqrt{x + 1}$

$$\frac{f(7) - f(3)}{7 - 3} = \frac{\left(2 - \sqrt{8}\right) - (2 - 2)}{4}$$

$$= \frac{2 - 2\sqrt{2}}{4} = \frac{1 - \sqrt{2}}{2}$$

The average rate of change of f from $x_1 = 3$ to $x_2 = 7$ is $\left(1 - \sqrt{2}\right)/2$.

98. $f(x) = 1 - \sqrt{x + 3},\ x_1 = 1,\ x_2 = 6$

$$\frac{f(x_2) - f(x_1)}{x_2 - x_1} = \frac{f(6) - f(1)}{6 - 1}$$

$$= \frac{-2 - (-1)}{5} = \frac{-2 + 1}{5} = -\frac{1}{5} = -0.2$$

The average rate of change from $x = 1$ to $x = 6$ is -0.2.

99. $f(x) = x^5 + 4x - 7$

$f(-x) = (-x)^5 + 4(-x) - 7$

$\quad = -x^5 - 4x - 7$

$\quad \neq f(x)$

$\quad \neq -f(x)$

Neither even nor odd

100. $f(x) = x^4 - 20x^2$

$f(-x) = (-x)^4 - 20(-x)^2 = x^4 - 20x^2 = f(x)$

The function is even.

101. $f(x) = 2x\sqrt{x^2 + 3}$

$f(-x) = 2(-x)\sqrt{(-x)^2 + 3}$

$\quad = -2x\sqrt{x^2 + 3}$

$\quad = -f(x)$

f is odd.

102. $f(x) = \sqrt[5]{6x^2}$

$f(-x) = \sqrt[5]{6(-x)^2} = \sqrt[5]{6x^2} = f(x)$

The function is even.

103. $f(2) = -6,\ f(-1) = 3$

Points: $(2, -6),\ (-1, 3)$

$m = \dfrac{3 - (-6)}{-1 - 2} = \dfrac{9}{-3} = -3$

$y - (-6) = -3(x - 2)$

$\quad y + 6 = -3x + 6$

$\qquad\quad y = -3x$

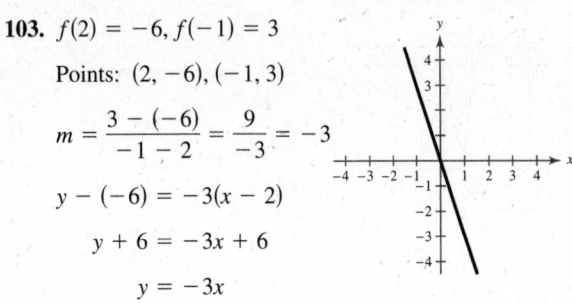

104. $f(0) = -5,\ f(4) = -8$

$(0, -5),\ (4, -8)$

$m = \dfrac{-8 - (-5)}{4 - 0} = -\dfrac{3}{4}$

$y - (-5) = -\dfrac{3}{4}(x - 0)$

$\qquad\quad y = -\dfrac{3}{4}x - 5$

$\quad f(x) = -\dfrac{3}{4}x - 5$

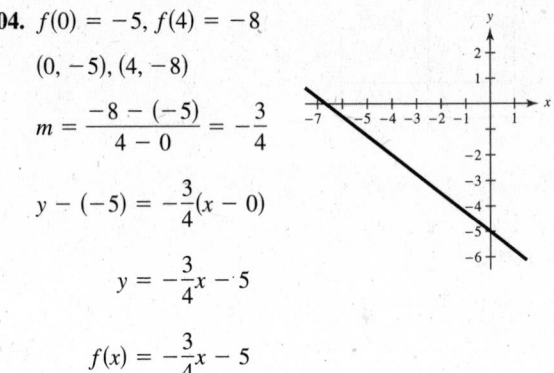

105. $f(x) = 3 - x^2$

Intercepts: $(0, 3),\ \left(\pm\sqrt{3}, 0\right)$

y-axis symmetry

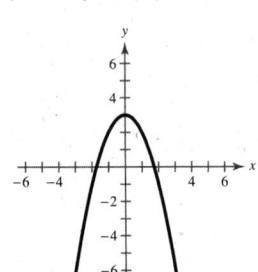

106. $h(x) = x^3 - 2$

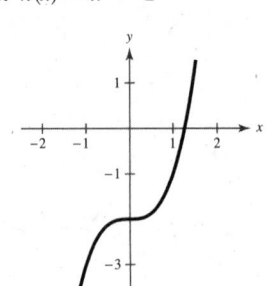

107. $f(x) = -\sqrt{x}$

Domain: $x \geq 0$

Intercepts: $(0, 0)$

x	0	1	4	9
y	0	-1	-2	-3

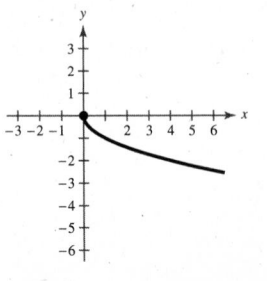

108. $f(x) = \sqrt{x+1}$

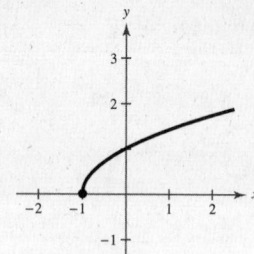

109. $g(x) = \dfrac{3}{x}$

No intercepts

Origin symmetry

x	-3	-1	1	3
y	-1	-3	3	1

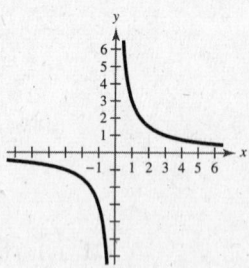

110. $g(x) = \dfrac{1}{x+5}$

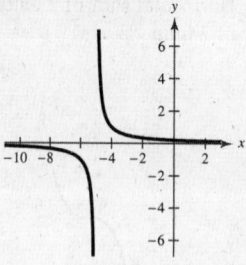

111. $f(x) = [\![x]\!] - 2$

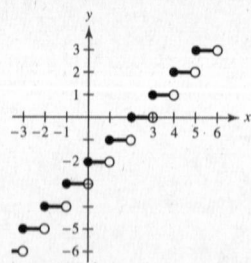

112. $g(x) = [\![x+4]\!]$

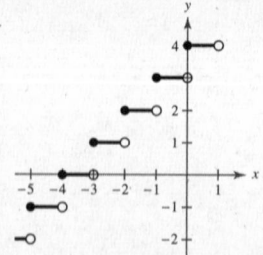

113. $f(x) = \begin{cases} 5x - 3, & x \geq -1 \\ -4x + 5, & x < -1 \end{cases}$

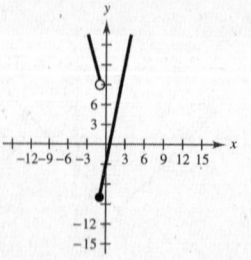

114. $f(x) = \begin{cases} x^2 - 2, & x < -2 \\ 5, & -2 \leq x \leq 0 \\ 8x - 5, & x > 0 \end{cases}$

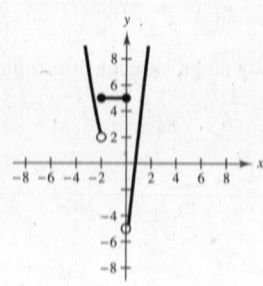

115. Common function: $f(x) = x^3$

Horizontal shift 4 units to the left
and a vertical shift 4 units upward

116. The graph of $y = \sqrt{x}$ was shifted
upward 4 units.

117. (a) $f(x) = x^2$

(b) $h(x) = x^2 - 9$

Vertical shift 9 units downward

(c)

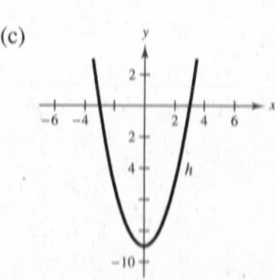

(d) $h(x) = f(x) - 9$

118. (a) $f(x) = x^3$

(b) $h(x) = (x - 2)^3 + 2$
Horizontal shift of 2 units to the right; vertical shift of 2 units upward

(c)

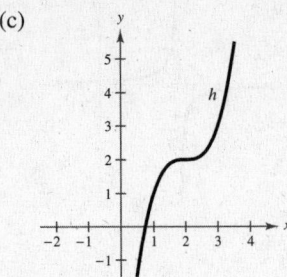

(d) $h(x) = f(x - 2) + 2$

119. (a) $f(x) = \sqrt{x}$

(b) $h(x) = \sqrt{x - 7}$
Horizontal shift 7 units to the right

(c)

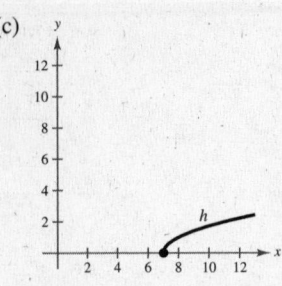

(d) $h(x) = f(x - 7)$

120. (a) $f(x) = |x|$

(b) $h(x) = |x + 3| - 5$
Horizontal shift of 3 units to the left; vertical shift of 5 units downward

(c)

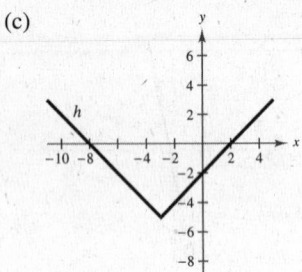

(d) $h(x) = f(x + 3) - 5$

121. (a) $f(x) = x^2$

(b) $h(x) = -(x + 3)^2 + 1$
Reflection in the x-axis, a horizontal shift 3 units to the left, and a vertical shift 1 unit upward

(c)

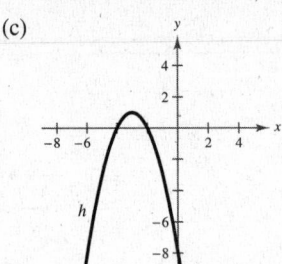

(d) $h(x) = -f(x + 3) + 1$

122. (a) $f(x) = x^3$

(b) $h(x) = -(x - 5)^3 - 5$
Reflection in the x-axis; horizontal shift of 5 units to the right; vertical shift of 5 units downward

(c)

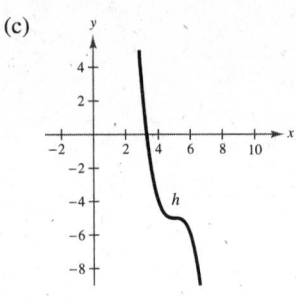

(d) $h(x) = -f(x - 5) - 5$

123. (a) $f(x) = [\![x]\!]$

(b) $h(x) = -[\![x]\!] + 6$
Reflection in the x-axis and a vertical shift 6 units upward

(c)

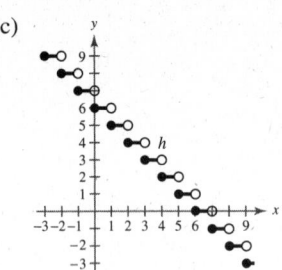

(d) $h(x) = -f(x) + 6$

124. (a) $f(x) = \sqrt{x}$

(b) $h(x) = -\sqrt{x+1} + 9$

Reflection in the x-axis, a horizontal shift 1 unit to the left, and a vertical shift 9 units upward

(c)

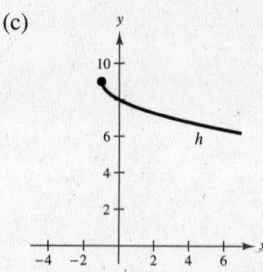

(d) $h(x) = -f(x+1) + 9$

125. (a) $f(x) = |x|$

(b) $h(x) = -|-x+4| + 6$

Reflection in both the x- and y-axes; horizontal shift of 4 units to the right; vertical shift of 6 units upward

(c)

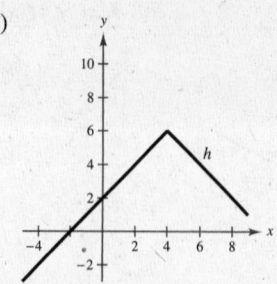

(d) $h(x) = -f(-(x-4)) + 6 = -f(-x+4) + 6$

126. (a) $f(x) = x^2$

(b) $h(x) = -(x+1)^2 - 3$

Reflection in the x-axis; horizontal shift of 1 unit to the left; vertical shift of 3 units downward

(c)

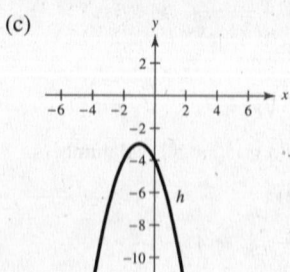

(d) $h(x) = -f(x+1) - 3$

127. (a) $f(x) = [\![x]\!]$

(b) $h(x) = 5[\![x-9]\!]$

Horizontal shift 9 units to the right and a vertical stretch (each y-value is multiplied by 5)

(c)

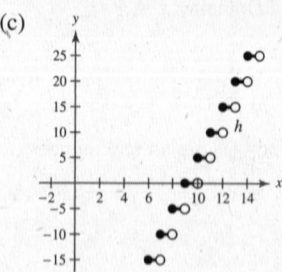

(d) $h(x) = 5f(x-9)$

128. (a) $f(x) = x^3$

(b) $h(x) = -\frac{1}{3}x^3$

Reflection in the x-axis; vertical shrink (each y-value is multiplied by $\frac{1}{3}$)

(c)

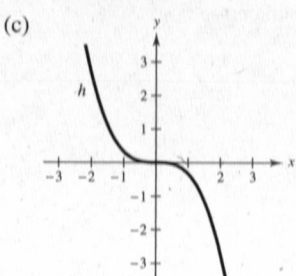

(d) $h(x) = -\frac{1}{3}f(x)$

129. (a) $f(x) = \sqrt{x}$

(b) $h(x) = -2\sqrt{x-4}$

Reflection in the x-axis, a vertical stretch (each y-value is multiplied by 2), and a horizontal shift 4 units to the right

(c)

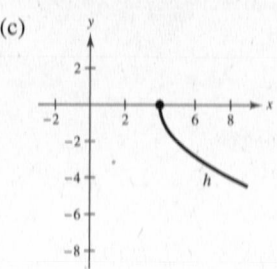

(d) $h(x) = -2f(x-4)$

130. (a) $f(x) = |x|$

(b) $h(x) = \frac{1}{2}|x| - 1$

Vertical shrink $\left(\text{each } y\text{-value is multiplied by } \frac{1}{2}\right)$; vertical shift of 1 unit downward

(c)

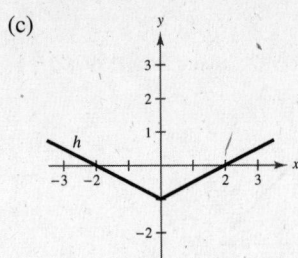

(d) $h(x) = \frac{1}{2}f(x) - 1$

131. $f(x) = x^2 + 3,\ g(x) = 2x - 1$

(a) $(f + g)(x) = (x^2 + 3) + (2x - 1) = x^2 + 2x + 2$

(b) $(f - g)(x) = (x^2 + 3) - (2x - 1) = x^2 - 2x + 4$

(c) $(fg)(x) = (x^2 + 3)(2x - 1) = 2x^3 - x^2 + 6x - 3$

(d) $\left(\dfrac{f}{g}\right)(x) = \dfrac{x^2 + 3}{2x - 1}$,   Domain: $x \neq \dfrac{1}{2}$

132. $f(x) = x^2 - 4,\ g(x) = \sqrt{3 - x}$

(a) $(f + g)(x) = f(x) + g(x) = x^2 - 4 + \sqrt{3 - x}$

(b) $(f - g)(x) = f(x) - g(x) = x^2 - 4 - \sqrt{3 - x}$

(c) $(fg)(x) = f(x)g(x) = (x^2 - 4)\left(\sqrt{3 - x}\right)$

(d) $(f/g)(x) = \dfrac{f(x)}{g(x)} = \dfrac{x^2 - 4}{\sqrt{3 - x}}$, $x < 3$

133. $f(x) = \frac{1}{3}x - 3,\ g(x) = 3x + 1$

The domains of $f(x)$ and $g(x)$ are all real numbers.

(a) $(f \circ g)(x) = f(g(x))$

$= f(3x + 1)$

$= \frac{1}{3}(3x + 1) - 3$

$= x + \frac{1}{3} - 3$

$= x - \frac{8}{3}$

Domain: all real numbers

(b) $(g \circ f)(x) = g(f(x))$

$= g\left(\frac{1}{3}x - 3\right)$

$= 3\left(\frac{1}{3}x - 3\right) + 1$

$= x - 9 + 1$

$= x - 8$

Domain: all real numbers

134. $f(x) = x^3 - 4,\ g(x) = \sqrt[3]{x + 7}$

The domains of $f(x)$ and $g(x)$ are all real numbers.

(a) $(f \circ g)(x) = f(g(x))$

$= \left(\sqrt[3]{x + 7}\right)^3 - 4$

$= x + 7 - 4$

$= x + 3$

Domain: all real numbers

(b) $(g \circ f)(x) = g(f(x))$

$= \sqrt[3]{(x^3 - 4) + 7}$

$= \sqrt[3]{x^3 + 3}$

Domain: all real numbers

135. $h(x) = (6x - 5)^3$

Answer is not unique.

One possibility: Let $f(x) = x^3$ and $g(x) = 6x - 5$.

$f(g(x)) = f(6x - 5) = (6x - 5)^3 = h(x)$

136. $h(x) = \sqrt[3]{x + 2}$

Answer is not unique.

One possibility: Let $g(x) = x + 2$ and $f(x) = \sqrt[3]{x}$.

$f(g(x)) = f(x + 2) = \sqrt[3]{x + 2} = h(x)$

137. $v(t) = -31.86t^2 + 233.6t + 2594$

$d(t) = -4.18t^2 + 571.0t - 3706$

(a) $(v + d)(t) = v(t) + d(t)$

$$= -36.04t^2 + 804.6t - 1112$$

$(v + d)(t)$ represents the combined factory sales (in millions of dollars) for VCRs and DVD players from 1997 to 2003.

(b)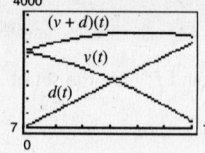

(c) $(v + d)(10) = \$3330$ million

138. (a) $N(T(t)) = 25(2t + 1)^2 - 50(2t + 1) + 300, \quad 2 \le t \le 20$

$$= 25(4t^2 + 4t + 1) - 100t - 50 + 300$$

$$= 100t^2 + 100t + 25 - 100t + 250$$

$$= 100t^2 + 275$$

The composition $N(T(t))$ represents the number of bacteria in the food as a function of time.

(b) When $N = 750$,

$$750 = 100t^2 + 275$$

$$100t^2 = 475$$

$$t^2 = 4.75$$

$$t = 2.18 \text{ hours.}$$

After about 2.18 hours, the bacterial count will reach 750.

139. $f(x) = x - 7$

$f^{-1}(x) = x + 7$

$f(f^{-1}(x)) = f(x + 7) = (x + 7) - 7 = x$

$f^{-1}(f(x)) = f^{-1}(x - 7) = (x - 7) + 7 = x$

140. $f(x) = x + 5$

$y = x + 5$

$x = y + 5$

$y = x - 5$

$f^{-1}(x) = x - 5$

$f(f^{-1}(x)) = f(x - 5) = x - 5 + 5 = x$

$f^{-1}(f(x)) = f^{-1}(x + 5) = x + 5 - 5 = x$

141. The graph passes the Horizontal Line Test. The function has an inverse.

142. No, the function does not have an inverse because some horizontal lines intersect the graph twice.

143. $f(x) = 4 - \frac{1}{3}x$

The graph passes the Horizontal Line Test. The function has an inverse.

144. No, the function does not have an inverse because some horizontal lines intersect the graph twice.

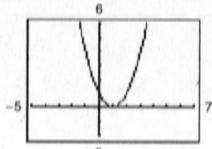

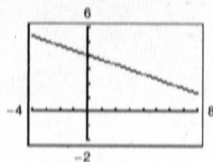

145. $h(t) = \dfrac{2}{t - 3}$

The graph passes the Horizontal Line Test. The function has an inverse.

146. Yes, the function has an inverse because no horizontal lines intersect the graph at more than one point.

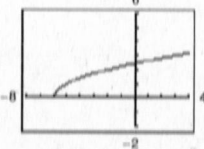

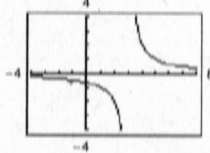

147. (a) $f(x) = \frac{1}{2}x - 3$

$y = \frac{1}{2}x - 3$

$x = \frac{1}{2}y - 3$

$x + 3 = \frac{1}{2}y$

$2(x + 3) = y$

$f^{-1}(x) = 2x + 6$

(c) The graph of f^{-1} is the reflection of the graph of f about the line $y = x$.

(b)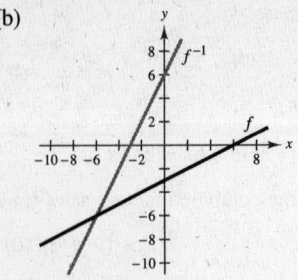

(d) The domains and ranges of f and f^{-1} are the set of all real numbers.

148. $f(x) = 5x - 7$

(a) $y = 5x - 7$

$x = 5y - 7$

$x + 7 = 5y$

$\dfrac{x + 7}{5} = y$

$f^{-1}(x) = \dfrac{x + 7}{5}$

(c) The graph of f^{-1} is the reflection of the graph of f across the line $y = x$.

(b)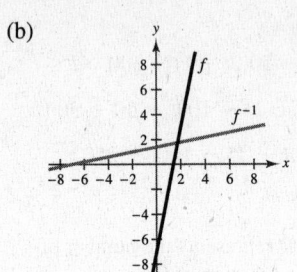

(d) The domains and ranges of f and f^{-1} are the set of all real numbers.

149. (a) $f(x) = \sqrt{x + 1}$

$y = \sqrt{x + 1}$

$x = \sqrt{y + 1}$

$x^2 = y + 1$

$x^2 - 1 = y$

$f^{-1}(x) = x^2 - 1, \; x \geq 0$

Note: The inverse must have a restricted domain.

(c) The graph of f^{-1} is the reflection of the graph of f about the line $y = x$.

(b)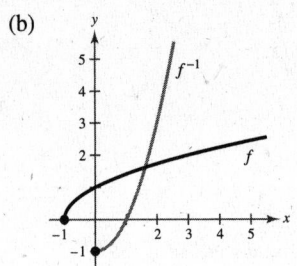

(d) The domain of f and the range of f^{-1} is $[-1, \infty)$. The range of f and the domain of f^{-1} is $[0, \infty)$.

150. $f(x) = x^3 + 2$

(a) $y = x^3 + 2$

$x = y^3 + 2$

$x - 2 = y^3$

$\sqrt[3]{x - 2} = y$

$f^{-1}(x) = \sqrt[3]{x - 2}$

(c) The graph of f^{-1} is the reflection of the graph of f across the line $y = x$.

(b)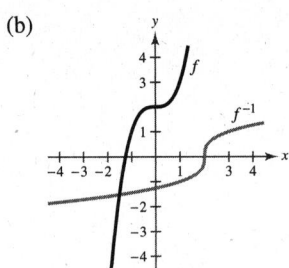

(d) The domains and ranges of f and f^{-1} are the set of all real numbers.

151. $f(x) = 2(x - 4)^2$ is increasing on $[4, \infty)$.

Let $f(x) = 2(x - 4)^2$, $x \geq 4$ and $y \geq 0$.

$$y = 2(x - 4)^2$$

$$x = 2(y - 4)^2, \ x \geq 0, \ y \geq 4$$

$$\frac{x}{2} = (y - 4)^2$$

$$\sqrt{\frac{x}{2}} = y - 4$$

$$\sqrt{\frac{x}{2}} + 4 = y$$

$$f^{-1}(x) = \sqrt{\frac{x}{2}} + 4, \ x \geq 0$$

152. $f(x) = |x - 2|$

Increasing on $[2, \infty)$

Let $f(x) = x - 2$, $x \geq 2$, $y \geq 0$.

$$y = x - 2$$

$$x = y - 2, \ x \geq 0, \ y \geq 2$$

$$x + 2 = y, \ x \geq 0, \ y \geq 2$$

$$f^{-1}(x) = x + 2, \ x \geq 0$$

153. $I = 2.09t + 37.2$

(a)

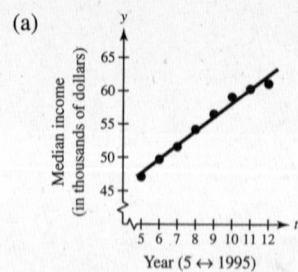

(b) The model is a good fit to the actual data.

154. (a)

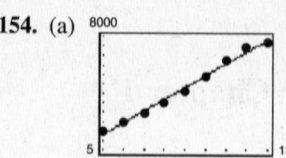

(b) $S = 627t - 346$

The model is a good fit to the actual data.

(c) $S = 627.02(18) - 346 = \$10,940.36$ million

(d) The factory sales of electronic gaming software in the U.S. increases by $627.02 million each year.

155. $D = km$

$4 = 2.5k$

$1.6 = k$

In 2 miles:

$D = 1.6(2) = 3.2$ kilometers

In 10 miles:

$D = 1.6(10) = 16$ kilometers

156. $P = kS^3$

$750 = k(27)^3$

$k = 0.03810395$

$P = 0.03810395(40)^3$

$= 2438.7$ kilowatts

157. $F = ks^2$

If speed is doubled,

$F = k(2s)^2$

$F = 4ks^2$.

Thus, the force will be changed by a factor of 4.

158. $x = \dfrac{k}{p}$

$800 = \dfrac{k}{5}$

$k = 4000$

$x = \dfrac{4000}{6} \approx 667$ boxes

159. $T = \dfrac{k}{r}$

$3 = \dfrac{k}{65}$

$k = 3(65) = 195$

$T = \dfrac{195}{r}$

When $r = 80$ mph,

$T = \dfrac{195}{80} = 2.4375$ hours

≈ 2 hours, 26 minutes.

160. $C = khw^2$

$28.80 = k(16)(6)^2$

$k = 0.05$

$C = (0.05)(14)(8)^2$

$= \$44.80$

161. False. The graph is reflected in the x-axis, shifted 9 units to the left, then shifted 13 units down.

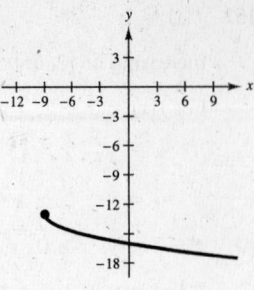

162. True. If $f(x) = x^3$ and $g(x) = \sqrt[3]{x}$, then the domain of g is all real numbers, which is equal to the range of f and vice versa.

163. True. If $y = kx$, then

$$x = \frac{1}{k}y.$$

164. The Vertical Line Test is used to determine if a graph of y is a function of x. The Horizontal Line Test is used to determine if a function has an inverse function.

165. A function from a Set A to a Set B is a relation that assigns to each element x in the Set A exactly one element y in the Set B.

Problem Solving for Chapter 1

1. (a) $W_1 = 0.07x + 2000$

(b) $W_2 = 0.05x + 2300$

(d) If you think you can sell \$20,000 per month, keep your current job with the higher commission rate. For sales over \$15,000 it pays more than the other job.

(c)

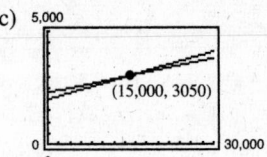

Point of intersection: (15,000, 3050)

Both jobs pay the same, \$3050, if you sell \$15,000 per month.

2. Mapping numbers onto letters is *not* a function. Each number between 2 and 9 is mapped to more than one letter.

$\{(2, A), (2, B), (2, C), (3, D), (3, E), (3, F), (4, G), (4, H), (4, I), (5, J), (5, K), (5, L),$

$(6, M), (6, N), (6, O), (7, P), (7, Q), (7, R), (7, S), (8, T), (8, U), (8, V), (9, W), (9, X), (9, Y), (9, Z)\}$

Mapping letters onto numbers *is* a function. Each letter is only mapped to one number.

$\{(A, 2), (B, 2), (C, 2), (D, 3), (E, 3), (F, 3), (G, 4), (H, 4), (I, 4), (J, 5), (K, 5), (L, 5),$

$(M, 6), (N, 6), (O, 6), (P, 7), (Q, 7), (R, 7), (S, 7), (T, 8), (U, 8), (V, 8), (W, 9), (X, 9), (Y, 9), (Z, 9)\}$

3. (a) Let $f(x)$ and $g(x)$ be two even functions. Then define $h(x) = f(x) \pm g(x)$.

$h(-x) = f(-x) \pm g(-x)$

$= f(x) \pm g(x)$ since f and g are even

$= h(x)$

So, $h(x)$ is also even.

(b) Let $f(x)$ and $g(x)$ be two odd functions. Then define $h(x) = f(x) \pm g(x)$.

$h(-x) = f(-x) \pm g(-x)$

$= -f(x) \mp g(x)$ since f and g are odd

$= -h(x)$

So, $h(x)$ is also odd. (If $f(x) \neq g(x)$)

(c) Let $f(x)$ be odd and $g(x)$ be even. Then define $h(x) = f(x) \pm g(x)$.

$h(-x) = f(-x) \pm g(-x)$

$= -f(x) \pm g(x)$ since f is odd and g is even

$\neq h(x)$

$\neq -h(x)$

So, $h(x)$ is neither odd nor even.

4. $f(x) = x$ $g(x) = x$

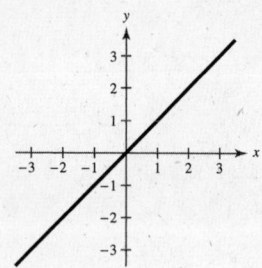   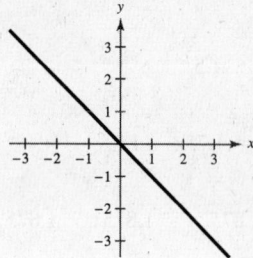

$(f \circ f)(x) = x$ and $(g \circ g)(x) = x$

These are the only two linear functions that are their own inverse functions since m has to equal $1/m$ for this to be true.

5. $f(x) = a_{2n}x^{2n} + a_{2n-2}x^{2n-2} + \cdots + a_2 x^2 + a_0$

$f(-x) = a_{2n}(-x)^{2n} + a_{2n-2}(-x)^{2n-2} + \cdots + a_2(-x)^2 + a_0$

$\qquad = a_{2n}x^{2n} + a_{2n-2}x^{2n-2} + \cdots + a_2 x^2 + a_0$

$\qquad = f(x)$

Therefore, $f(x)$ is even.

6. It appears, from the drawing, that the triangles are equal; thus $(x, y) = (6, 8)$. The line between $(2.5, 2)$ and $(6, 8)$ is $y = \frac{12}{7}x - \frac{16}{7}$. The line between $(9.5, 2)$ and $(6, 8)$ is $y = -\frac{12}{7}x + \frac{128}{7}$. The path of the ball is:

$$f(x) = \begin{cases} \frac{12}{7}x - \frac{16}{7}, & 2.5 \le x \le 6 \\ -\frac{12}{7}x + \frac{128}{7}, & 6 < x \le 9.5 \end{cases}$$

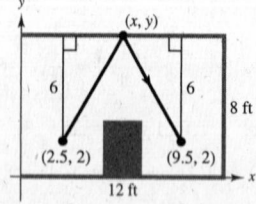

7. (a) April 11: 10 hours

April 12: 24 hours

April 13: 24 hours

April 14: $23\frac{2}{3}$ hours

Total: $81\frac{2}{3}$ hours

(c) $D = -\dfrac{180}{7}t + 3400$

Domain: $0 \le t \le \dfrac{1190}{9}$

Range: $0 \le D \le 3400$

(b) Speed $= \dfrac{\text{distance}}{\text{time}} = \dfrac{2100}{81\frac{2}{3}} = \dfrac{180}{7} = 25\frac{5}{7}$ mph

(d)

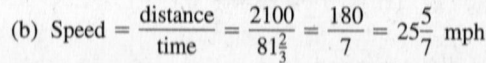
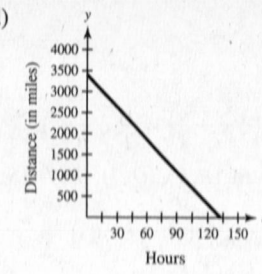

8. (a) $\dfrac{f(x_2) - f(x_1)}{x_2 - x_1} = \dfrac{f(2) - f(1)}{2 - 1} = \dfrac{1 - 0}{1} = 1$

(b) $\dfrac{f(x_2) - f(x_1)}{x_2 - x_1} = \dfrac{f(1.5) - f(1)}{1.5 - 1} = \dfrac{0.75 - 0}{0.5} = 1.5$

(c) $\dfrac{f(x_2) - f(x_1)}{x_2 - x_1} = \dfrac{f(1.25) - f(1)}{1.25 - 1} = \dfrac{0.4375 - 0}{0.25} = 1.75$

(d) $\dfrac{f(x_2) - f(x_1)}{x_2 - x_1} = \dfrac{f(1.125) - f(1)}{1.125 - 1} = \dfrac{0.234375 - 0}{0.125} = 1.875$

(e) $\dfrac{f(x_2) - f(x_1)}{x_2 - x_1} = \dfrac{f(1.0625) - f(1)}{1.0625 - 1} = \dfrac{0.12109375 - 0}{0.625} = 1.9375$

(f) Yes, the average rate of change appears to be approaching 2.

—CONTINUED—

8. —CONTINUED—

(g) a. $(1, 0), (2, 1), m = 1, y = x - 1$

b. $(1, 0), (1.5, 0.75), m = \dfrac{0.75}{0.5} = 1.5, y = 1.5x - 1.5$

c. $(1, 0), (1.25, 0.4375), m = \dfrac{0.4375}{0.25} = 1.75, y = 1.75x - 1.75$

d. $(1, 0), (1.125, 0.234375), m = \dfrac{0.234375}{0.125} = 1.875, y = 1.875x - 1.875$

e. $(1, 0), (1.0625, 0.12109375), m = \dfrac{0.12109375}{0.0625} = 1.9375, y = 1.9375x - 1.9375$

(h) $(1, f(1)) = (1, 0), m \to 2, y = 2(x - 1), y = 2x - 2$

9. (a)–(d) Use $f(x) = 4x$ and $g(x) = x + 6$.

(a) $(fg)(x) = f(x + 6) = 4(x + 6) = 4x + 24$

(b) $(f \circ g)^{-1}(x) = \dfrac{x - 24}{4} = \dfrac{1}{4}x - 6$

(c) $f^{-1}(x) = \dfrac{1}{4}x$

$g^{-1}(x) = x - 6$

(d) $(g^{-1} \circ f^{-1})(x) = g^{-1}\left(\dfrac{1}{4}x\right) = \dfrac{1}{4}x - 6$

(e) $f(x) = x^3 + 1$ and $g(x) = 2x$

$(f \circ g)(x) = f(2x) = (2x)^3 + 1 = 8x^3 + 1$

$(f \circ g)^{-1}(x) = \sqrt[3]{\dfrac{x - 1}{8}} = \dfrac{1}{2}\sqrt[3]{x - 1}$

$f^{-1}(x) = \sqrt[3]{x - 1}$

$g^{-1}(x) = \dfrac{1}{2}x$

$(g^{-1} \circ f^{-1})(x) = g^{-1}\left(\sqrt[3]{x - 1}\right) = \dfrac{1}{2}\sqrt[3]{x - 1}$

(f) Answers will vary.

(g) Conjecture: $(f \circ g)^{-1}(x) = (g^{-1} \circ f^{-1})(x)$

10. (a) The length of the trip in the water is $\sqrt{2^2 + x^2}$, and the length of the trip over land is $\sqrt{1 + (3 - x)^2}$. Hence, the total time is

$$T(x) = \dfrac{\sqrt{4 + x^2}}{2} + \dfrac{\sqrt{1 + (3 - x)^2}}{4} \text{ hours.}$$

(b) Domain of $T(x)$: $0 \le x \le 3$

(c)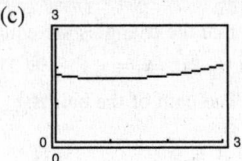

(d) $T(x)$ is a minimum when $x = 1$.

(e) To reach point Q in the shortest amount of time, you should row to a point one mile down the coast, and then walk the rest of the way.

11. $H(x) = \begin{cases} 1, & x \ge 0 \\ 0, & x < 0 \end{cases}$

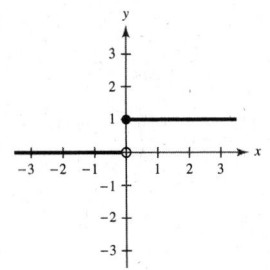

—CONTINUED—

11. —CONTINUED—

(a) $H(x) - 2$

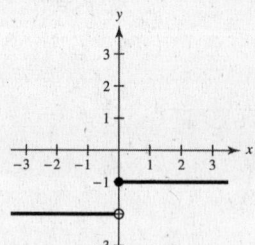

(b) $H(x - 2)$

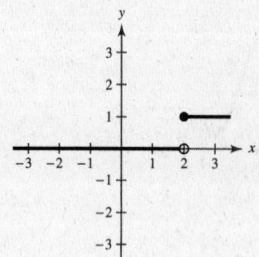

(c) $-H(x)$

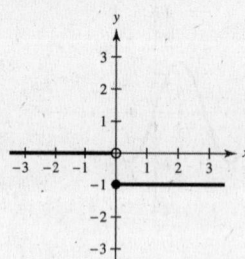

(d) $H(-x)$

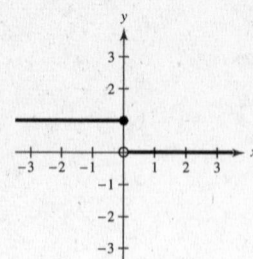

(e) $\frac{1}{2}H(x)$

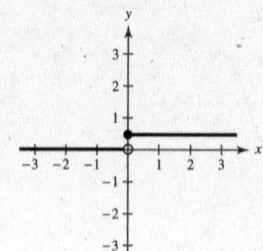

(f) $-H(x - 2) + 2$

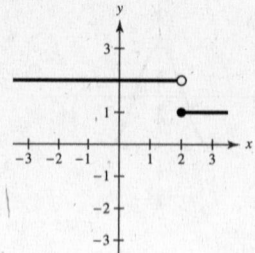

12. $f(x) = y = \dfrac{1}{1 - x}$

(a) Domain: all $x \neq 1$

   Range: all $y \neq 0$

(b) $f(f(x)) = f\left(\dfrac{1}{1 - x}\right)$

$$= \dfrac{1}{1 - \left(\dfrac{1}{1 - x}\right)} = \dfrac{1}{\dfrac{1 - x - 1}{1 - x}}$$

$$= \dfrac{1 - x}{-x} = \dfrac{x - 1}{x}$$

Domain: all $x \neq 0, 1$

(c) $f(f(f(x))) = f\left(\dfrac{x - 1}{x}\right) = \dfrac{1}{1 - \left(\dfrac{x - 1}{x}\right)} = \dfrac{1}{\dfrac{1}{x}} = x$

Domain: all $x \neq 0, 1$

The graph is not a line. It has holes at $(0, 0)$ and $(1, 1)$.

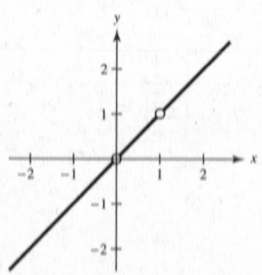

13. $(f \circ (g \circ h))(x) = f((g \circ h)(x))$

$$= f(g(h(x)))$$

$$= (f \circ g \circ h)(x)$$

$$((f \circ g) \circ h)(x) = (f \circ g)(h(x))$$

$$= f(g(h(x)))$$

$$= (f \circ g \circ h)(x)$$

14. (a) $f(x + 1)$

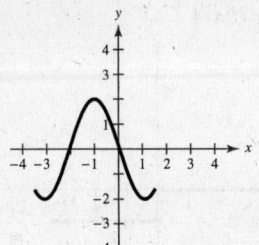

(b) $f(x) + 1$

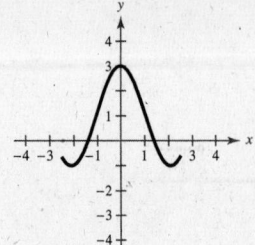

(c) $2f(x)$

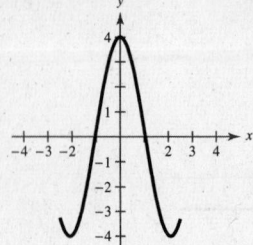

(d) $f(-x)$

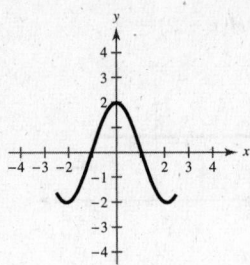

(e) $-f(x)$

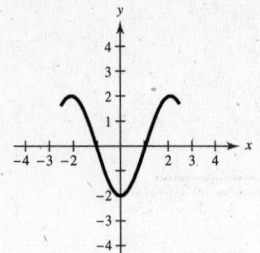

(f) $|f(x)|$

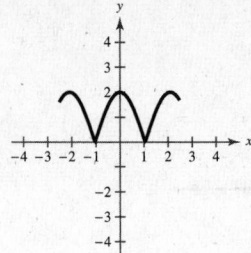

(g) $f(|x|)$

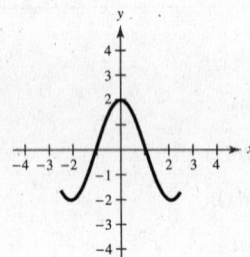

15.

x	$f(x)$	$f^{-1}(x)$
-4	—	2
-3	4	1
-2	1	0
-1	0	—
0	-2	-1
1	-3	-2
2	-4	—
3	—	—
4	—	-3

(a)

x	$f(f^{-1}(x))$
-4	$f(f^{-1}(-4)) = f(2) = -4$
-2	$f(f^{-1}(-2)) = f(0) = -2$
0	$f(f^{-1}(0)) = f(-1) = 0$
4	$f(f^{-1}(4)) = f(-3) = 4$

(b)

x	$(f + f^{-1})(x)$
-3	$f(-3) + f^{-1}(-3) = 4 + 1 = 5$
-2	$f(-2) + f^{-1}(-2) = 1 + 0 = 1$
0	$f(0) + f^{-1}(0) = -2 + (-1) = -3$
1	$f(1) + f^{-1}(1) = -3 + (-2) = -5$

(c)

x	$(f \cdot f^{-1})(x)$
-3	$f(-3)f^{-1}(-3) = (4)(1) = 4$
-2	$f(-2)f^{-1}(-2) = (1)(0) = 0$
0	$f(0)f^{-1}(0) = (-2)(-1) = 2$
1	$f(1)f^{-1}(1) = (-3)(-2) = 6$

(d)

x	$|f^{-1}(x)|$
-4	$|f^{-1}(-4)| = |2| = 2$
-3	$|f^{-1}(-3)| = |1| = 1$
0	$|f^{-1}(0)| = |-1| = 1$
4	$|f^{-1}(4)| = |-3| = 3$

Chapter 1   Practice Test

1. Given the points $(-3, 4)$ and $(5, -6)$, find (a) the midpoint of the line segment joining the points, and (b) the distance between the points.

2. Graph $y = \sqrt{7 - x}$.

3. Write the standard equation of the circle with center $(-3, 5)$ and radius 6.

4. Find the equation of the line through $(2, 4)$ and $(3, -1)$.

5. Find the equation of the line with slope $m = 4/3$ and y-intercept $b = -3$.

6. Find the equation of the line through $(4, 1)$ perpendicular to the line $2x + 3y = 0$.

7. If it costs a company \$32 to produce 5 units of a product and \$44 to produce 9 units, how much does it cost to produce 20 units? (Assume that the cost function is linear.)

8. Given $f(x) = x^2 - 2x + 1$, find $f(x - 3)$.

9. Given $f(x) = 4x - 11$, find $\dfrac{f(x) - f(3)}{x - 3}$

10. Find the domain and range of $f(x) = \sqrt{36 - x^2}$.

11. Which equations determine y as a function of x?

    (a) $6x - 5y + 4 = 0$

    (b) $x^2 + y^2 = 9$

    (c) $y^3 = x^2 + 6$

12. Sketch the graph of $f(x) = x^2 - 5$.

13. Sketch the graph of $f(x) = |x + 3|$.

14. Sketch the graph of $f(x) = \begin{cases} 2x + 1, & \text{if } x \geq 0, \\ x^2 - x, & \text{if } x < 0. \end{cases}$

15. Use the graph of $f(x) = |x|$ to graph the following:

    (a) $f(x + 2)$

    (b) $-f(x) + 2$

16. Given $f(x) = 3x + 7$ and $g(x) = 2x^2 - 5$, find the following:

   (a) $(g - f)(x)$

   (b) $(fg)(x)$

17. Given $f(x) = x^2 - 2x + 16$ and $g(x) = 2x + 3$, find $f(g(x))$.

18. Given $f(x) = x^3 + 7$, find $f^{-1}(x)$.

19. Which of the following functions have inverses?

   (a) $f(x) = |x - 6|$

   (b) $f(x) = ax + b, \ a \neq 0$

   (c) $f(x) = x^3 - 19$

20. Given $f(x) = \sqrt{\dfrac{3 - x}{x}}, \ 0 < x \leq 3$, find $f^{-1}(x)$.

Exercises 21–23, true or false?

21. $y = 3x + 7$ and $y = \frac{1}{3}x - 4$ are perpendicular.

22. $(f \circ g)^{-1} = g^{-1} \circ f^{-1}$

23. If a function has an inverse, then it must pass both the Vertical Line Test and the Horizontal Line Test.

24. If z varies directly as the cube of x and inversely as the square root of y, and $z = -1$ when $x = -1$ and $y = 25$, find z in terms of x and y.

25. Use your calculator to find the least square regression line for the data.

x	-2	-1	0	1	2	3
y	1	2.4	3	3.1	4	4.7

C H A P T E R 2
Polynomial and Rational Functions

CHAPTER 2
Polynomial and Rational Functions

Section 2.1 Quadratic Functions and Models

You should know the following facts about parabolas.

- $f(x) = ax^2 + bx + c$, $a \neq 0$, is a quadratic function, and its graph is a parabola.

- If $a > 0$, the parabola opens upward and the vertex is the point with the minimum y-value.
  If $a < 0$, the parabola opens downward and the vertex is the point with the maximum y-value.

- The vertex is $(-b/2a, f(-b/2a))$.

- To find the x-intercepts (if any), solve
  $$ax^2 + bx + c = 0.$$

- The standard form of the equation of a parabola is
  $$f(x) = a(x - h)^2 + k$$
  where $a \neq 0$.

  (a) The vertex is (h, k).

  (b) The axis is the vertical line $x = h$.

Vocabulary Check

1. nonnegative integer; real

2. quadratic; parabola

3. axis or axis of symmetry

4. positive; minimum

5. negative; maximum

---

1. $f(x) = (x - 2)^2$ opens upward and has vertex $(2, 0)$. Matches graph (g).

2. $f(x) = (x + 4)^2$ opens upward and has vertex $(-4, 0)$. Matches graph (c).

3. $f(x) = x^2 - 2$ opens upward and has vertex $(0, -2)$. Matches graph (b).

4. $f(x) = 3 - x^2$ opens downward and has vertex $(0, 3)$. Matches graph (h).

5. $f(x) = 4 - (x - 2)^2 = -(x - 2)^2 + 4$ opens downward and has vertex $(2, 4)$. Matches graph (f).

6. $f(x) = (x + 1)^2 - 2$ opens upward and has vertex $(-1, -2)$. Matches graph (a).

7. $f(x) = -(x - 3)^2 - 2$ opens downward and has vertex $(3, -2)$. Matches graph (e).

8. $f(x) = -(x - 4)^2$ opens downward and has vertex $(4, 0)$. Matches graph (d).

9. (a) $y = \frac{1}{2}x^2$

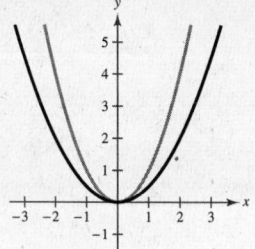

Vertical shrink

(b) $y = -\frac{1}{8}x^2$

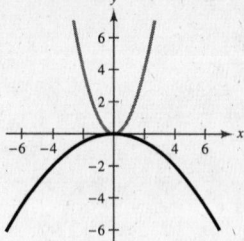

Vertical shrink and reflection in the x-axis

(c) $y = \frac{3}{2}x^2$

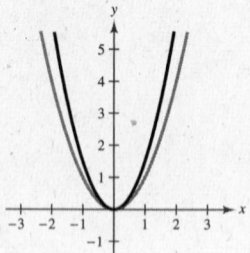

Vertical stretch

(d) $y = -3x^2$

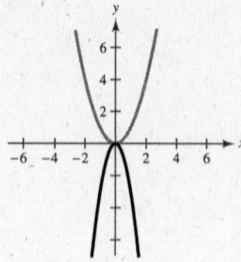

Vertical stretch and reflection in the x-axis

10. (a) $y = x^2 + 1$

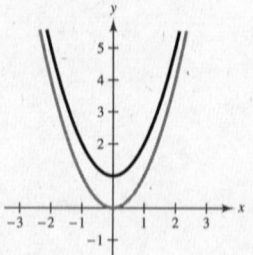

Vertical translation one unit upward

(b) $y = x^2 - 1$

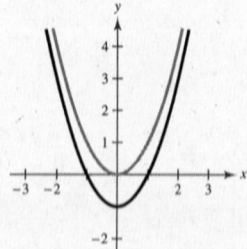

Vertical translation one unit downward

(c) $y = x^2 + 3$

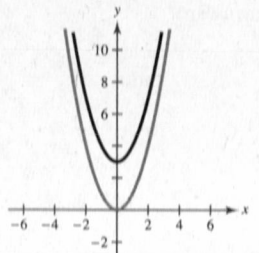

Vertical translation three units upward

(d) $y = x^2 - 3$

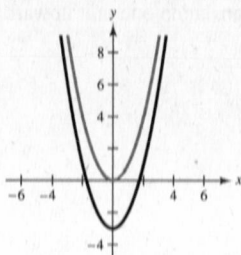

Vertical translation three units downward

11. (a) $y = (x - 1)^2$

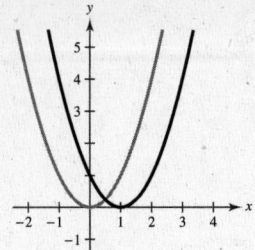

Horizontal translation one unit to the right

(b) $y = (3x)^2 + 1$

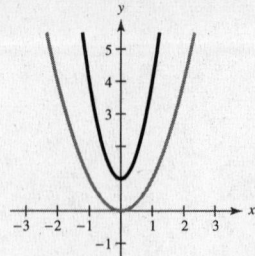

Horizontal shrink and a vertical translation one unit upward

(c) $y = \left(\frac{1}{3}x\right)^2 - 3$

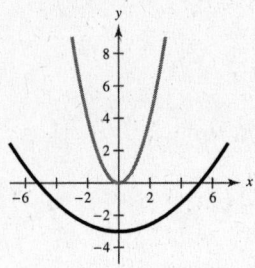

Horizontal stretch and a vertical translation three units downward

(d) $y = (x + 3)^2$

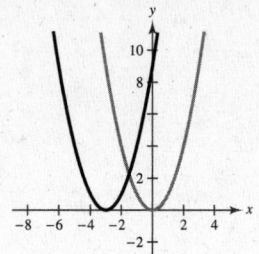

Horizontal translation three units to the left

12. (a) $y = -\frac{1}{2}(x - 2)^2 + 1$

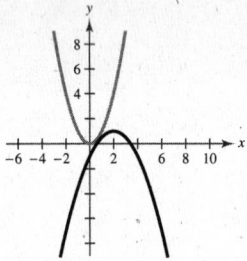

Horizontal translation two units to the right, vertical shrink $\left(\text{each } y\text{-value is multiplied by } \frac{1}{2}\right)$, reflection in the x-axis, and vertical translation one unit upward

(b) $y = \left[\frac{1}{2}(x - 1)\right]^2 - 3$

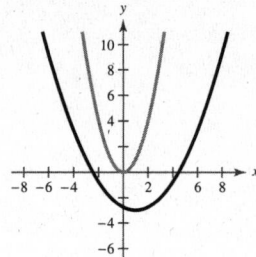

Horizontal translation one unit to the right, horizontal stretch (each x-value is multiplied by 2), and vertical translation three units downward

(c) $y = -\frac{1}{2}(x + 2)^2 - 1$

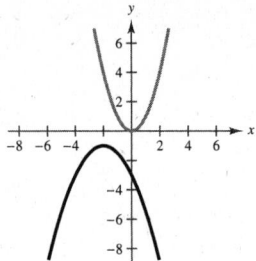

Horizontal translation two units to the left, vertical shrink $\left(\text{each } y\text{-value is multiplied by } \frac{1}{2}\right)$, reflection in x-axis, and vertical translation one unit downward

(d) $y = [2(x + 1)]^2 + 4$

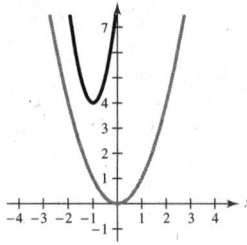

Horizontal translation one unit to the left, horizontal shrink $\left(\text{each } x\text{-value is multiplied by } \frac{1}{2}\right)$, and vertical translation four units upward

13. $f(x) = x^2 - 5$

Vertex: $(0, -5)$

Axis of symmetry: $x = 0$ or the y-axis

Find x-intercepts:

$x^2 - 5 = 0$

$x^2 = 5$

$x = \pm\sqrt{5}$

x-intercepts:

$\left(-\sqrt{5}, 0\right), \left(\sqrt{5}, 0\right)$

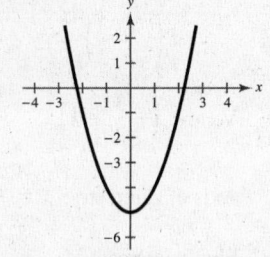

14. $h(x) = 25 - x^2$

Vertex: $(0, 25)$

Axis of symmetry: $x = 0$

Find x-intercepts:

$25 - x^2 = 0$

$x^2 = 25$

$x = \pm 5$

x-intercepts: $(\pm 5, 0)$

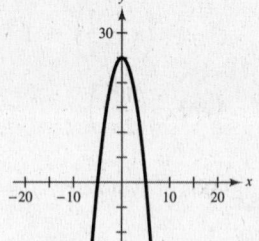

15. $f(x) = \frac{1}{2}x^2 - 4 = \frac{1}{2}(x - 0)^2 - 4$

Vertex: $(0, -4)$

Axis of symmetry: $x = 0$ or the y-axis

Find x-intercepts:

$\frac{1}{2}x^2 - 4 = 0$

$x^2 = 8$

$x = \pm\sqrt{8} = \pm 2\sqrt{2}$

x-intercepts:

$\left(-2\sqrt{2}, 0\right), \left(2\sqrt{2}, 0\right)$

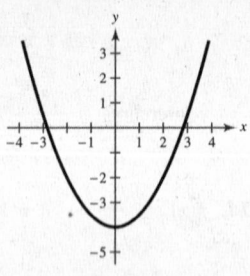

16. $f(x) = 16 = \frac{1}{4}x^2 = -\frac{1}{4}x^2 + 16$

Vertex: $(0, 16)$

Axis of symmetry: $x = 0$

Find x-intercepts:

$16 - \frac{1}{4}x^2 = 0$

$x^2 = 64$

$x = \pm 8$

x-intercepts: $(\pm 8, 0)$

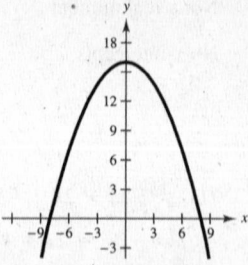

17. $f(x) = (x + 5)^2 - 6$

Vertex: $(-5, -6)$

Axis of symmetry: $x = -5$

Find x-intercepts:

$(x + 5)^2 - 6 = 0$

$(x + 5)^2 = 6$

$x + 5 = \pm\sqrt{6}$

$x = -5 \pm \sqrt{6}$

x-intercepts: $\left(-5 - \sqrt{6}, 0\right), \left(-5 + \sqrt{6}, 0\right)$

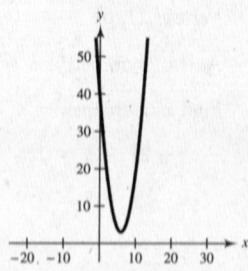

18. $f(x) = (x - 6)^2 + 3$

Vertex: $(6, 3)$

Axis of symmetry: $x = 6$

Find x-intercepts:

$(x - 6)^2 + 3 = 0$

$(x - 6)^2 = -3$

Not possible for real x

No x-intercepts

19. $h(x) = x^2 - 8x + 16 = (x - 4)^2$

Vertex: $(4, 0)$

Axis of symmetry: $x = 4$

x-intercept: $(4, 0)$

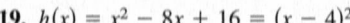

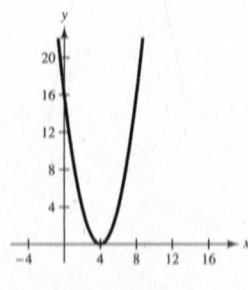

20. $g(x) = x^2 + 2x + 1 = (x + 1)^2$

Vertex: $(-1, 0)$

Axis of symmetry: $x = -1$

x-intercept: $(-1, 0)$

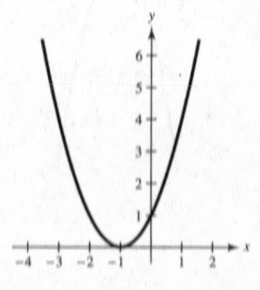

21. $f(x) = x^2 - x + \dfrac{5}{4}$

$\qquad = \left(x^2 - x + \dfrac{1}{4}\right) - \dfrac{1}{4} + \dfrac{5}{4}$

$\qquad = \left(x - \dfrac{1}{2}\right)^2 + 1$

Vertex: $\left(\dfrac{1}{2}, 1\right)$

Axis of symmetry: $x = \dfrac{1}{2}$

Find x-intercepts:

$x^2 - x + \dfrac{5}{4} = 0$

$\qquad x = \dfrac{1 \pm \sqrt{1 - 5}}{2}$

Not a real number

No x-intercepts

22. $f(x) = x^2 + 3x + \dfrac{1}{4}$

$\qquad = \left(x^2 + 3x + \dfrac{9}{4}\right) - \dfrac{9}{4} + \dfrac{1}{4}$

$\qquad = \left(x + \dfrac{3}{2}\right)^2 - 2$

Vertex: $\left(-\dfrac{3}{2}, -2\right)$

Axis of symmetry: $x = -\dfrac{3}{2}$

Find x-intercepts:

$x^2 + 3x + \dfrac{1}{4} = 0$

$\qquad x = \dfrac{-3 \pm \sqrt{9 - 1}}{2}$

$\qquad = -\dfrac{3}{2} \pm \sqrt{2}$

x-intercepts: $\left(-\dfrac{3}{2} \pm \sqrt{2}, 0\right)$

23. $f(x) = -x^2 + 2x + 5$

$\qquad = -(x^2 - 2x + 1) - (-1) + 5$

$\qquad = -(x - 1)^2 + 6$

Vertex: $(1, 6)$

Axis of symmetry: $x = 1$

Find x-intercepts:

$-x^2 + 2x + 5 = 0$

$x^2 - 2x - 5 = 0$

$\qquad x = \dfrac{2 \pm \sqrt{4 + 20}}{2}$

$\qquad = 1 \pm \sqrt{6}$

x-intercepts: $\left(1 - \sqrt{6}, 0\right), \left(1 + \sqrt{6}, 0\right)$

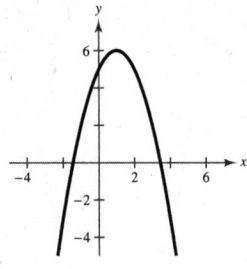

24. $f(x) = -x^2 - 4x + 1 = -(x^2 + 4x) + 1$

$\qquad\qquad\qquad = -(x^2 + 4x + 4) - (-4) + 1$

$\qquad\qquad\qquad = -(x + 2)^2 + 5$

Vertex: $(-2, 5)$

Axis of symmetry: $x = -2$

Find x-intercepts: $-x^2 - 4x + 1 = 0$

$\qquad\qquad\qquad\qquad x^2 + 4x - 1 = 0$

$\qquad\qquad x = \dfrac{-4 \pm \sqrt{16 + 4}}{2}$

$\qquad\qquad\quad = -2 \pm \sqrt{5}$

x-intercepts: $\left(-2 \pm \sqrt{5}, 0\right)$

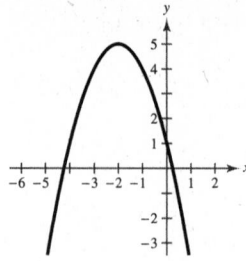

25. $h(x) = 4x^2 - 4x + 21$

$$= 4\left(x^2 - x + \frac{1}{4}\right) - 4\left(\frac{1}{4}\right) + 21$$

$$= 4\left(x - \frac{1}{2}\right)^2 + 20$$

Vertex: $\left(\frac{1}{2}, 20\right)$

Axis of symmetry: $x = \frac{1}{2}$

Find x-intercepts:

$4x^2 - 4x + 21 = 0$

$$x = \frac{4 \pm \sqrt{16 - 336}}{2(4)}$$

Not a real number $\Rightarrow$ No x-intercepts

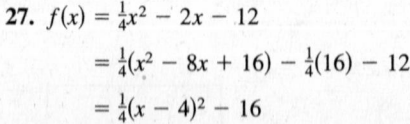

26. $f(x) = 2x^2 - x + 1$

$$= 2\left(x^2 - \frac{1}{2}x\right) + 1$$

$$= 2\left(x - \frac{1}{4}\right)^2 - 2\left(\frac{1}{16}\right) + 1$$

$$= 2\left(x - \frac{1}{4}\right)^2 + \frac{7}{8}$$

Vertex: $\left(\frac{1}{4}, \frac{7}{8}\right)$

Axis of symmetry: $x = \frac{1}{4}$

Find x-intercepts:

$2x^2 - x + 1 = 0$

$$x = \frac{1 \pm \sqrt{1 - 8}}{2(2)}$$

Not a real number

No x-intercepts

27. $f(x) = \frac{1}{4}x^2 - 2x - 12$

$$= \frac{1}{4}(x^2 - 8x + 16) - \frac{1}{4}(16) - 12$$

$$= \frac{1}{4}(x - 4)^2 - 16$$

Vertex: $(4, -16)$

Axis of symmetry: $x = 4$

Find x-intercepts:

$\frac{1}{4}x^2 - 2x - 12 = 0$

$x^2 - 8x - 48 = 0$

$(x + 4)(x - 12) = 0$

$x = -4$ or $x = 12$

x-intercepts: $(-4, 0), (12, 0)$

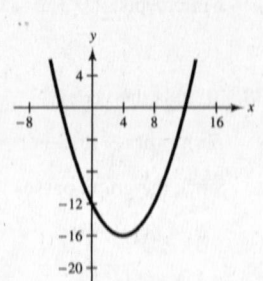

28. $f(x) = -\frac{1}{3}x^2 + 3x - 6$

$$= -\frac{1}{3}(x^2 - 9x) - 6$$

$$= -\frac{1}{3}\left(x^2 - 9x + \frac{81}{4}\right) + \frac{1}{3}\left(\frac{81}{4}\right) - 6$$

$$= -\frac{1}{3}\left(x - \frac{9}{2}\right)^2 + \frac{3}{4}$$

Vertex: $\left(\frac{9}{2}, \frac{3}{4}\right)$

Axis of symmetry: $x = \frac{9}{2}$

Find x-intercepts:

$-\frac{1}{3}x^2 + 3x - 6 = 0$

$x^2 - 9x + 18 = 0$

$(x - 3)(x - 6) = 0$

x-intercepts: $(3, 0), (6, 0)$

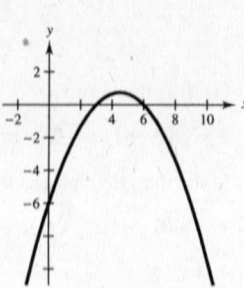

29. $f(x) = -(x^2 + 2x - 3) = -(x + 1)^2 + 4$

Vertex: $(-1, 4)$

Axis of symmetry: $x = -1$

x-intercepts: $(-3, 0), (1, 0)$

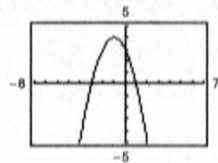

30. $f(x) = -(x^2 + x - 30)$

$$= -(x^2 + x) + 30$$

$$= -\left(x^2 + x + \frac{1}{4}\right) + \frac{1}{4} + 30$$

$$= -\left(x + \frac{1}{2}\right)^2 + \frac{121}{4}$$

Vertex: $\left(-\frac{1}{2}, \frac{121}{4}\right)$

Axis of symmetry: $x = -\frac{1}{2}$

x-intercepts: $(-6, 0), (5, 0)$

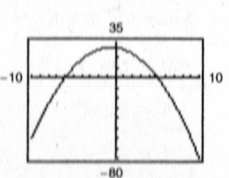

31. $g(x) = x^2 + 8x + 11 = (x + 4)^2 - 5$

Vertex: $(-4, -5)$

Axis of symmetry: $x = -4$

x-intercepts: $(-4 \pm \sqrt{5}, 0)$

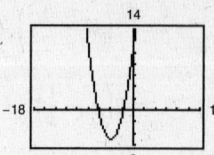

32. $f(x) = x^2 + 10x + 14$

$= (x^2 + 10x + 25) - 25 + 14$

$= (x + 5)^2 - 11$

Vertex: $(-5, -11)$

Axis of symmetry: $x = -5$

x-intercepts: $(-5 \pm \sqrt{11}, 0)$

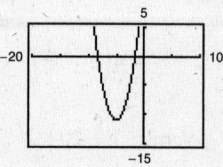

33. $f(x) = 2x^2 - 16x + 31$

$= 2(x - 4)^2 - 1$

Vertex: $(4, -1)$

Axis of symmetry: $x = 4$

x-intercepts: $\left(4 \pm \frac{1}{2}\sqrt{2}, 0\right)$

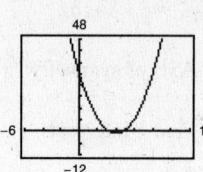

34. $f(x) = -4x^2 + 24x - 41$

$= -4(x^2 - 6x) - 41$

$= -4(x^2 - 6x + 9) + 36 - 41$

$= -4(x - 3)^2 - 5$

Vertex: $(3, -5)$

Axis of symmetry: $x = 3$

No x-intercepts

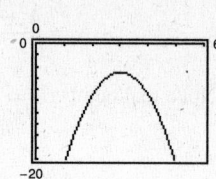

35. $g(x) = \frac{1}{2}(x^2 + 4x - 2) = \frac{1}{2}(x + 2)^2 - 3$

Vertex: $(-2, -3)$

Axis of symmetry: $x = -2$

x-intercepts: $(-2 \pm \sqrt{6}, 0)$

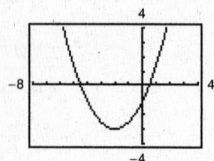

36. $f(x) = \frac{3}{5}(x^2 + 6x - 5)$

$= \frac{3}{5}(x^2 + 6x + 9) - \frac{27}{5} - 3$

$= \frac{3}{5}(x + 3)^2 - \frac{42}{5}$

Vertex: $\left(-3, -\frac{42}{5}\right)$

Axis of symmetry: $x = -3$

x-intercepts: $(-3 \pm \sqrt{14}, 0)$

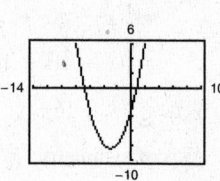

37. $(1, 0)$ is the vertex.

$y = a(x - 1)^2 + 0 = a(x - 1)^2$

Since the graph passes through the point $(0, 1)$, we have:

$1 = a(0 - 1)^2$

$1 = a$

$y = 1(x - 1)^2 = (x - 1)^2$

38. $(0, 1)$ is the vertex.

$f(x) = a(x - 0)^2 + 1 = ax^2 + 1$

Since the graph passes through $(1, 0)$,

$0 = a(1)^2 + 1$

$-1 = a.$

So, $y = -x^2 + 1.$

39. $(-1, 4)$ is the vertex.

$y = a(x + 1)^2 + 4$

Since the graph passes through the point $(1, 0)$, we have:

$0 = a(1 + 1)^2 + 4$

$-4 = 4a$

$-1 = a$

$y = -1(x + 1)^2 + 4 = -(x + 1)^2 + 4$

40. $(-2, -1)$ is the vertex.

$f(x) = a(x + 2)^2 - 1$

Since the graph passes through $(0, 3)$,

$3 = a(0 + 2)^2 - 1$

$3 = 4a - 1$

$4 = 4a$

$1 = a.$

So, $y = (x + 2)^2 - 1.$

41. $(-2, 2)$ is the vertex.

$y = a(x + 2)^2 + 2$

Since the graph passes through the point $(-1, 0)$, we have:

$0 = a(-1 + 2)^2 + 2$

$-2 = a$

$y = -2(x + 2)^2 + 2$

42. $(2, 0)$ is the vertex.

$f(x) = a(x - 2)^2 + 0 = a(x - 2)^2$

Since the graph passes through $(3, 2)$,

$2 = a(3 - 2)^2$

$2 = a.$

So, $y = 2(x - 2)^2.$

43. $(-2, 5)$ is the vertex.

$f(x) = a(x + 2)^2 + 5$

Since the graph passes through the point $(0, 9)$, we have:

$9 = a(0 + 2)^2 + 5$

$4 = 4a$

$1 = a$

$f(x) = 1(x + 2)^2 + 5 = (x + 2)^2 + 5$

44. $(4, -1)$ is the vertex.

$f(x) = a(x - 4)^2 - 1$

Since the graph passes through $(2, 3)$,

$3 = a(2 - 4)^2 - 1$

$3 = 4a - 1$

$4 = 4a$

$1 = a.$

So, $f(x) = (x - 4)^2 - 1.$

45. $(3, 4)$ is the vertex.

$f(x) = a(x - 3)^2 + 4$

Since the graph passes through the point $(1, 2)$, we have:

$2 = a(1 - 3)^2 + 4$

$-2 = 4a$

$-\frac{1}{2} = a$

$f(x) = -\frac{1}{2}(x - 3)^2 + 4$

46. $(2, 3)$ is the vertex.

$f(x) = a(x - 2)^2 + 3$

Since the graph passes through $(0, 2)$,

$2 = a(0 - 2)^2 + 3$

$2 = 4a + 3$

$-1 = 4a$

$-\frac{1}{4} = a.$

So, $f(x) = -\frac{1}{4}(x - 2)^2 + 3.$

47. $(5, 12)$ is the vertex.

$f(x) = a(x - 5)^2 + 12$

Since the graph passes through the point $(7, 15)$, we have:

$15 = a(7 - 5)^2 + 12$

$3 = 4a \implies a = \frac{3}{4}$

$f(x) = \frac{3}{4}(x - 5)^2 + 12$

48. $(-2, -2)$ is the vertex.

$f(x) = a(x + 2)^2 - 2$

Since the graph passes through $(-1, 0)$,

$0 = a(-1 + 2)^2 - 2$

$0 = a - 2$

$2 = a.$

So, $f(x) = 2(x + 2)^2 - 2.$

49. $\left(-\frac{1}{4}, \frac{3}{2}\right)$ is the vertex.

$f(x) = a\left(x + \frac{1}{4}\right)^2 + \frac{3}{2}$

Since the graph passes through the point $(-2, 0)$, we have:

$0 = a\left(-2 + \frac{1}{4}\right)^2 + \frac{3}{2}$

$-\frac{3}{2} = \frac{49}{16}a \implies a = -\frac{24}{49}$

$f(x) = -\frac{24}{49}\left(x + \frac{1}{4}\right)^2 + \frac{3}{2}$

50. $\left(\frac{5}{2}, -\frac{3}{4}\right)$ is the vertex.

$f(x) = a\left(x - \frac{5}{2}\right)^2 - \frac{3}{4}$

Since the graph passes through $(-2, 4)$,

$4 = a\left(-2 - \frac{5}{2}\right)^2 - \frac{3}{4}$

$4 = \frac{81}{4}a - \frac{3}{4}$

$\frac{19}{4} = \frac{81}{4}a$

$\frac{19}{81} = a.$

So, $f(x) = \frac{19}{81}\left(x - \frac{5}{2}\right)^2 - \frac{3}{4}.$

51. $\left(-\frac{5}{2}, 0\right)$ is the vertex.

$f(x) = a\left(x + \frac{5}{2}\right)^2$

Since the graph passes through the point $\left(-\frac{7}{2}, -\frac{16}{3}\right)$, we have:

$-\frac{16}{3} = a\left(-\frac{7}{2} + \frac{5}{2}\right)^2$

$-\frac{16}{3} = a$

$f(x) = -\frac{16}{3}\left(x + \frac{5}{2}\right)^2$

52. $(6, 6)$ is the vertex.

$f(x) = a(x - 6)^2 + 6$

Since the graph passes through $\left(\frac{61}{10}, \frac{3}{2}\right)$,

$\frac{3}{2} = a\left(\frac{61}{10} - 6\right)^2 + 6$

$\frac{3}{2} = \frac{1}{100}a + 6$

$-\frac{9}{2} = \frac{1}{100}a$

$-450 = a.$

So, $f(x) = -450(x - 6)^2 + 6.$

53. $y = x^2 - 16$

x-intercepts: $(\pm 4, 0)$

$0 = x^2 - 16$

$x^2 = 16$

$x = \pm 4$

54. $y = x^2 - 6x + 9$

x-intercept: $(3, 0)$

$0 = x^2 - 6x + 9$

$0 = (x - 3)^2$

$x - 3 = 0 \implies x = 3$

55. $y = x^2 - 4x - 5$

x-intercepts: $(5, 0), (-1, 0)$

$0 = x^2 - 4x - 5$

$0 = (x - 5)(x + 1)$

$x = 5$ or $x = -1$

56. $y = 2x^2 + 5x - 3$

x-intercepts: $\left(\frac{1}{2}, 0\right), (-3, 0)$

$0 = 2x^2 + 5x - 3$

$0 = (2x - 1)(x + 3)$

$2x - 1 = 0 \implies x = \frac{1}{2}$

$x + 3 = 0 \implies x = -3$

57. $f(x) = x^2 - 4x$

x-intercepts: $(0, 0), (4, 0)$

$0 = x^2 - 4x$

$0 = x(x - 4)$

$x = 0$ or $x = 4$

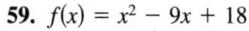

The x-intercepts and the solutions of $f(x) = 0$ are the same.

58. $f(x) = -2x^2 + 10x$

x-intercepts: $(0, 0), (5, 0)$

$0 = -2x^2 + 10x$

$0 = -2x(x - 5)$

$-2x = 0 \implies x = 0$

$x - 5 = 0 \implies x = 5$

The x-intercepts and the solutions of $f(x) = 0$ are the same.

59. $f(x) = x^2 - 9x + 18$

x-intercepts: $(3, 0), (6, 0)$

$0 = x^2 - 9x + 18$

$0 = (x - 3)(x - 6)$

$x = 3$ or $x = 6$

The x-intercepts and the solutions of $f(x) = 0$ are the same.

60. $f(x) = x^2 - 8x - 20$

x-intercepts: $(-2, 0), (10, 0)$

$0 = x^2 - 8x - 20$

$0 = (x + 2)(x - 10)$

$x + 2 = 0 \implies x = -2$

$x - 10 = 0 \implies x = 10$

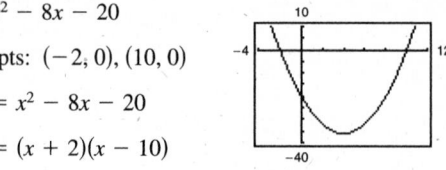

The x-intercepts and the solutions of $f(x) = 0$ are the same.

61. $f(x) = 2x^2 - 7x - 30$

x-intercepts: $\left(-\frac{5}{2}, 0\right), (6, 0)$

$0 = 2x^2 - 7x - 30$

$0 = (2x + 5)(x - 6)$

$x = -\frac{5}{2}$ or $x = 6$

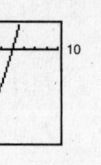

The x-intercepts and the solutions of $f(x) = 0$ are the same.

62. $f(x) = 4x^2 + 25x - 21$

x-intercepts: $(-7, 0), \left(\frac{3}{4}, 0\right)$

$0 = 4x^2 + 25x - 21$

$0 = (x + 7)(4x - 3)$

$x + 7 = 0 \implies x = -7$

$4x - 3 = 0 \implies x = \frac{3}{4}$

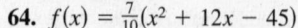

The x-intercepts and the solutions of $f(x) = 0$ are the same.

63. $f(x) = -\frac{1}{2}(x^2 - 6x - 7)$

x-intercepts: $(-1, 0), (7, 0)$

$0 = -\frac{1}{2}(x^2 - 6x - 7)$

$0 = x^2 - 6x - 7$

$0 = (x + 1)(x - 7)$

$x = -1$ or $x = 7$

The x-intercepts and the solutions of $f(x) = 0$ are the same.

64. $f(x) = \frac{7}{10}(x^2 + 12x - 45)$

x-intercepts: $(-15, 0), (3, 0)$

$0 = \frac{7}{10}(x^2 + 12x - 45)$

$0 = (x + 15)(x - 3)$

$x + 15 = 0 \implies x = -15$

$x - 3 = 0 \implies x = 3$

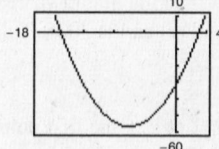

The x-intercepts and the solutions of $f(x) = 0$ are the same.

65. $f(x) = [x - (-1)](x - 3)$ ⠀⠀⠀ opens upward

$\quad = (x + 1)(x - 3)$

$\quad = x^2 - 2x - 3$

$g(x) = -[x - (-1)](x - 3)$ ⠀⠀⠀ opens downward

$\quad = -(x + 1)(x - 3)$

$\quad = -(x^2 - 2x - 3)$

$\quad = -x^2 + 2x + 3$

Note: $f(x) = a(x + 1)(x - 3)$ has x-intercepts $(-1, 0)$ and $(3, 0)$ for all real numbers $a \neq 0$.

66. $f(x) = [x - (-5)](x - 5)$

$\quad = (x + 5)(x - 5)$

$\quad = x^2 - 25$, opens upward

$g(x) = -f(x)$, opens downward

$g(x) = -x^2 + 25$

Note: $f(x) = a(x^2 - 25)$ has x-intercepts $(-5, 0)$ and $(5, 0)$ for all real numbers $a \neq 0$.

67. $f(x) = (x - 0)(x - 10)$ ⠀⠀⠀ opens upward

$\quad = x^2 - 10x$

$g(x) = -(x - 0)(x - 10)$ ⠀⠀⠀ opens downward

$\quad = -x^2 + 10x$

Note: $f(x) = a(x - 0)(x - 10) = ax(x - 10)$ has x-intercepts $(0, 0)$ and $(10, 0)$ for all real numbers $a \neq 0$.

68. $f(x) = (x - 4)(x - 8)$

$\quad = x^2 - 12x + 32$, opens upward

$g(x) = -f(x)$, opens downward

$g(x) = -x^2 + 12x - 32$

Note: $f(x) = a(x - 4)(x - 8)$ has x-intercepts $(4, 0)$ and $(8, 0)$ for all real numbers $a \neq 0$.

69. $f(x) = [x - (-3)]\left[x - \left(-\frac{1}{2}\right)\right](2)$ ⠀⠀opens upward

$\quad = (x + 3)\left(x + \frac{1}{2}\right)(2)$

$\quad = (x + 3)(2x + 1)$

$\quad = 2x^2 + 7x + 3$

$g(x) = -(2x^2 + 7x + 3)$ ⠀⠀⠀ opens downward

$\quad = -2x^2 - 7x - 3$

Note: $f(x) = a(x + 3)(2x + 1)$ has x-intercepts $(-3, 0)$ and $\left(-\frac{1}{2}, 0\right)$ for all real numbers $a \neq 0$.

70. $f(x) = 2\left[x - \left(-\frac{5}{2}\right)\right](x - 2)$

$\quad = 2\left(x + \frac{5}{2}\right)(x - 2)$

$\quad = 2\left(x^2 + \frac{1}{2}x - 5\right)$

$\quad = 2x^2 + x - 10$, opens upward

$g(x) = -f(x)$, opens downward

$g(x) = -2x^2 - x + 10$

Note: $f(x) = a\left(x + \frac{5}{2}\right)(x - 2)$ has x-intercepts $\left(-\frac{5}{2}, 0\right)$ and $(2, 0)$ for all real numbers $a \neq 0$.

71. Let x = the first number and y = the second number. Then the sum is

$$x + y = 110 \implies y = 110 - x.$$

The product is $P(x) = xy = x(110 - x) = 110x - x^2$.

$$P(x) = -x^2 + 110x$$
$$= -(x^2 - 110x + 3025 - 3025)$$
$$= -[(x - 55)^2 - 3025]$$
$$= -(x - 55)^2 + 3025$$

The maximum value of the product occurs at the vertex of $P(x)$ and is 3025. This happens when $x = y = 55$.

72. Let x = first number and y = second number. Then, $x + y = S$, $y = S - x$. The product is

$$P(x) = xy = x(S - x).$$
$$P(x) = Sx - x^2$$
$$= -x^2 + Sx$$
$$= -\left(x^2 - Sx + \frac{S^2}{4} - \frac{S^2}{4}\right)$$
$$= -\left(x - \frac{S}{2}\right)^2 + \frac{S^2}{4}$$

The maximum value of the product occurs at the vertex of $P(x)$ and is $S^2/4$. This happens when $x = y = S/2$.

73. Let x = the first number and y = the second number. Then the sum is

$$x + 2y = 24 \implies y = \frac{24 - x}{2}.$$

The product is $P(x) = xy = x\left(\dfrac{24 - x}{2}\right)$.

$$P(x) = \frac{1}{2}(-x^2 + 24x)$$
$$= -\frac{1}{2}(x^2 - 24x + 144 - 144)$$
$$= -\frac{1}{2}[(x - 12)^2 - 144] = -\frac{1}{2}(x - 12)^2 + 72$$

The maximum value of the product occurs at the vertex of $P(x)$ and is 72. This happens when $x = 12$ and $y = (24 - 12)/2 = 6$. Thus, the numbers are 12 and 6.

74. Let x = the first number and y = the second number.

Then the sum is $x + 3y = 42 \implies y = \dfrac{42 - x}{3}$.

The product is $P(x) = xy = x\left(\dfrac{42 - x}{3}\right)$.

$$P(x) = \frac{1}{3}(-x^2 + 42x)$$
$$= -\frac{1}{3}(x^2 - 42x + 441 - 441)$$
$$= -\frac{1}{3}[(x - 21)^2 - 441] = -\frac{1}{3}(x - 21)^2 + 147$$

The maximum value of the product occurs at the vertex of $P(x)$ and is 147. This happens when $x = 21$ and $y = \dfrac{42 - 21}{3} = 7$. Thus, the numbers are 21 and 7.

75. (a)

$$4x + 3y = 200 \implies y = \frac{1}{3}(200 - 4x) = \frac{4}{3}(50 - x)$$

$$A = 2xy = 2x\left[\frac{4}{3}(50 - x)\right] = \frac{8}{3}x(50 - x) = \frac{8x(50 - x)}{3}$$

(c)

This area is maximum when $x = 25$ feet and $y = \dfrac{100}{3} = 33\frac{1}{3}$ feet.

(b)

x	A
5	600
10	$1066\frac{2}{3}$
15	1400
20	1600
25	$1666\frac{2}{3}$
30	1600

This area is maximum when $x = 25$ feet and $y = \dfrac{100}{3} = 33\frac{1}{3}$ feet.

—CONTINUED—

75. —CONTINUED—

(d) $A = \dfrac{8}{3}x(50 - x)$

$\qquad = -\dfrac{8}{3}(x^2 - 50x)$

$\qquad = -\dfrac{8}{3}(x^2 - 50x + 625 - 625)$

$\qquad = -\dfrac{8}{3}[(x - 25)^2 - 625]$

$\qquad = -\dfrac{8}{3}(x - 25)^2 + \dfrac{5000}{3}$

The maximum area occurs at the vertex and is 5000/3 square feet. This happens when $x = 25$ feet and $y = (200 - 4(25))/3 = 100/3$ feet. The dimensions are $2x = 50$ feet by $33\frac{1}{3}$ feet.

(e) They are all identical.

$\qquad x = 25$ feet and $y = 33\frac{1}{3}$ feet

76. (a) Radius of semicircular ends of track: $r = \dfrac{1}{2}y$

Distance around two semicircular parts of track:

$\qquad d = 2\pi r = 2\pi\left(\dfrac{1}{2}y\right) = \pi y$

(b) Distance traveled around track in one lap:

$\qquad d = \pi y + 2x = 200$

$\qquad\qquad \pi y = 200 - 2x$

$\qquad\qquad\quad y = \dfrac{200 - 2x}{\pi}$

(c) Area of rectangular region:

$\qquad A = xy = x\left(\dfrac{200 - 2x}{\pi}\right)$

$\qquad\quad = \dfrac{1}{\pi}(200x - 2x^2)$

$\qquad\quad = -\dfrac{2}{\pi}(x^2 - 100x)$

$\qquad\quad = -\dfrac{2}{\pi}(x^2 - 100x + 2500 - 2500)$

$\qquad\quad = -\dfrac{2}{\pi}(x - 50)^2 + \dfrac{5000}{\pi}$

The area is maximum when $x = 50$ and

$\qquad y = \dfrac{200 - 2(50)}{\pi} = \dfrac{100}{\pi}$.

77. $y = -\dfrac{4}{9}x^2 + \dfrac{24}{9}x + 12$

The vertex occurs at $-\dfrac{b}{2a} = \dfrac{-24/9}{2(-4/9)} = 3$. The maximum height is $y(3) = -\dfrac{4}{9}(3)^2 + \dfrac{24}{9}(3) + 12 = 16$ feet.

78. $y = -\dfrac{16}{2025}x^2 + \dfrac{9}{5}x + 1.5$

(a) The ball height when it is punted is the y-intercept.

$\qquad y = -\dfrac{16}{2025}(0)^2 + \dfrac{9}{5}(0) + 1.5 = 1.5$ feet

(b) The vertex occurs at $x = -\dfrac{b}{2a} = -\dfrac{9/5}{2(-16/2025)} = \dfrac{3645}{32}$.

The maximum height is $f\left(\dfrac{3645}{32}\right) = -\dfrac{16}{2025}\left(\dfrac{3645}{32}\right)^2 + \dfrac{9}{5}\left(\dfrac{3645}{32}\right) + 1.5$

$\qquad\qquad = -\dfrac{6561}{64} + \dfrac{6561}{32} + 1.5 = -\dfrac{6561}{64} + \dfrac{13{,}122}{64} + \dfrac{96}{64} = \dfrac{6657}{64}$ feet ≈ 104.02 feet.

—CONTINUED—

78. —CONTINUED—

(c) The length of the punt is the positive x-intercept.

$$0 = -\frac{16}{2025}x^2 + \frac{9}{5}x + 1.5$$

$$x = \frac{-(9/5) \pm \sqrt{(9/5)^2 - (4)(1.5)(-16/2025)}}{-32/2025} \approx \frac{1.8 \pm 1.81312}{-0.01580247}$$

$x \approx -0.83031$ or $x \approx 228.64$

The punt is approximately 228.64 ft.

79. $C = 800 - 10x + 0.25x^2 = 0.25x^2 - 10x + 800$

The vertex occurs at $x = -\dfrac{b}{2a} = -\dfrac{-10}{2(0.25)} = 20.$

The cost is minimum when $x = 20$ fixtures.

80. $C = 100,000 - 110x + 0.045x^2$

The vertex occurs at $x = -\dfrac{-110}{2(0.045)} \approx 1222.$

The cost is minimum when $x \approx 1222$ units.

81. $P = -0.0002x^2 + 140x - 250,000$

The vertex occurs at $x = -\dfrac{b}{2a} = -\dfrac{140}{2(-0.0002)} = 350,000.$

The profit is maximum when $x = 350,000$ units.

82. $P = 230 + 20x - 0.5x^2$

The vertex occurs at $x = -\dfrac{b}{2a} = -\dfrac{20}{2(-0.5)} = 20.$

Because x is in hundreds of dollars, $20 \times 100 = 2000$ dollars is the amount spent on advertising that gives maximum profit.

83. $R(p) = -25p^2 + 1200p$

(a) $R(20) = \$14,000$ thousand

$R(25) = \$14,375$ thousand

$R(30) = \$13,500$ thousand

(b) The revenue is a maximum at the vertex.

$$-\frac{b}{2a} = \frac{-1200}{2(-25)} = 24$$

$R(24) = 14,400$

The unit price that will yield a maximum revenue of $14,400 thousand is $24.

84. $R(p) = -12p^2 + 150p$

(a) $R(\$4) = -12(\$4)^2 + 150(\$4) = \$408$

$R(\$6) = -12(\$6)^2 + 150(\$6) = \$468$

$R(\$8) = -12(\$8)^2 + 150(\$8) = \$432$

(b) The vertex occurs at

$$p = -\frac{b}{2a} = -\frac{150}{2(-12)} = \$6.25$$

Revenue is maximum when price $= \$6.25$ per pet.

The maximum revenue is

$$f(\$6.25) = -12(\$6.25)^2 + 150(\$6.25) = \$468.75.$$

85. $C = 4299 - 1.8t - 1.36t^2, \ 0 \le t \le 43$

(a)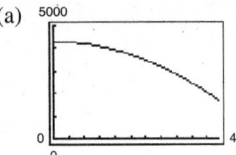

(c) $C(40) = 2051$

Annually: $\dfrac{209,128,094(2051)}{48,308,590} \approx 8879$ cigarettes

Daily: $\dfrac{8879}{366} \approx 24$ cigarettes

(b) Vertex $\approx (0, 4299)$

The vertex occurs when $y \approx 4299$ which is the maximum average annual consumption. The warnings may not have had an immediate effect, but over time they and other findings about the health risks and the increased cost of cigarettes have had an effect.

86. (a) and (c)

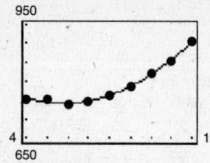

(b) $y = 4.303x^2 - 49.948x + 886.28$

(d) 1996

(e) Vertex occurs at

$$x = -\frac{b}{2a} = \frac{49.948}{2(4.303)} = 5.8$$

Minimum occurs at year ≈ 1996.

(f) $x = 18$

$y = 4.303(18)^2 - 49.948(18) + 886.28 = 1381.388$

There will be approximately 1,381,000 hairdressers and cosmetologists in 2008.

87. (a)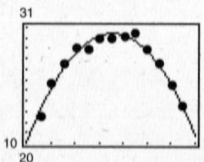

(b) $0.002s^2 + 0.005s - 0.029 = 10$

$$2s^2 + 5s - 29 = 10,000$$

$$2s^2 + 5s - 10,029 = 0$$

$a = 2, b = 5, c = -10,029$

$$s = \frac{-5 \pm \sqrt{5^2 - 4(2)(-10,029)}}{2(2)}$$

$$s = \frac{-5 \pm \sqrt{80,257}}{4}$$

$s \approx -72.1, 69.6$

The maximum speed if power is not to exceed 10 horsepower is 69.6 miles per hour.

88. (a) and (c)

(b) $y = -0.0082x^2 + 0.746x + 13.47$

(d) The maximum of the graph is at $x \approx 45.5$, or about 45.5 mi/h. Algebraically, the maximum occurs at

$$x = -\frac{b}{2a} = \frac{-0.746}{2(-0.0082)} \approx 45.5 \text{ mi/h}.$$

89. True. The equation $-12x^2 - 1 = 0$ has no real solution, so the graph has no x-intercepts.

90. True. The vertex of $f(x)$ is $\left(-\frac{5}{4}, \frac{53}{4}\right)$ and the vertex of $g(x)$ is $\left(-\frac{5}{4}, -\frac{71}{4}\right)$.

91. $f(x) = ax^2 + bx + c$

$$= a\left(x^2 + \frac{b}{a}x\right) + c$$

$$= a\left(x^2 + \frac{b}{a}x + \frac{b^2}{4a^2} - \frac{b^2}{4a^2}\right) + c$$

$$= a\left(x + \frac{b}{2a}\right)^2 - \frac{b^2}{4a} + c$$

$$= a\left(x - \left(-\frac{b}{2a}\right)\right)^2 + \frac{4ac - b^2}{4a}$$

$$f\left(-\frac{b}{2a}\right) = a\left(\frac{b^2}{4a^2}\right) + b\left(-\frac{b}{2a}\right) + c$$

$$= \frac{b^2}{4a} - \frac{b^2}{2a} + c$$

$$= \frac{b^2 - 2b^2 + 4ac}{4a} = \frac{4ac - b^2}{4a}$$

So, the vertex occurs at $\left(-\frac{b}{2a}, \frac{4ac - b^2}{4a}\right) = \left(-\frac{b}{2a}, f\left(-\frac{b}{2a}\right)\right)$.

92. Conditions (a) and (d) are preferable because profits would be increasing.

93. Yes. A graph of a quadratic equation whose vertex is $(0, 0)$ has only one x-intercept.

94. If $f(x) = ax^2 + bx + c$ has two real zeros, then by the Quadratic Formula they are

$$x = \frac{-b \pm \sqrt{b^2 - 4ac}}{2a}.$$

The average of the zeros of f is

$$\frac{\dfrac{-b - \sqrt{b^2 - 4ac}}{2a} + \dfrac{-b + \sqrt{b^2 - 4ac}}{2a}}{2} = \frac{\dfrac{-2b}{2a}}{2} = -\frac{b}{2a}.$$

This is the x-coordinate of the vertex of the graph.

95. $(-4, 3)$ and $(2, 1)$

$$m = \frac{1 - 3}{2 - (-4)} = \frac{-2}{6} = -\frac{1}{3}$$

$$y - 1 = -\frac{1}{3}(x - 2)$$

$$y - 1 = -\frac{1}{3}x + \frac{2}{3}$$

$$y = -\frac{1}{3}x + \frac{5}{3}$$

96. $\left(\frac{7}{2}, 2\right), m = \frac{3}{2}$

$$y - 2 = \frac{3}{2}\left(x - \frac{7}{2}\right)$$

$$y - 2 = \frac{3}{2}x - \frac{21}{4}$$

$$y = \frac{3}{2}x - \frac{13}{4}$$

97. $4x + 5y = 10 \implies y = -\frac{4}{5}x + 2$ and $m = -\frac{4}{5}$

The slope of the perpendicular line through $(0, 3)$ is $m = \frac{5}{4}$ and the y-intercept is $b = 3$.

$$y = \frac{5}{4}x + 3$$

98. $y = -3x + 2$

$m = -3$

For a parallel line, $m = -3$. So, for $(-8, 4)$, the line is

$$y - 4 = -3(x - (-8))$$

$$y - 4 = -3x - 24$$

$$y = -3x - 20.$$

For Exercises 99–104, let $f(x) = 14x - 3$, and $g(x) = 8x^2$.

99. $(f + g)(-3) = f(-3) + g(-3)$

$$= [14(-3) - 3] + 8(-3)^2 = 27$$

100. $(g - f)(2) = 8(2)^2 - 14(2) + 3 = 32 - 28 + 3 = 7$

101. $(fg)\left(-\frac{4}{7}\right) = f\left(-\frac{4}{7}\right)g\left(-\frac{4}{7}\right)$

$$= \left[14\left(-\frac{4}{7}\right) - 3\right]\left[8\left(-\frac{4}{7}\right)^2\right]$$

$$= (-11)\left(\frac{128}{49}\right) = -\frac{1408}{49}$$

102. $\left(\frac{f}{g}\right)(-1.5) = \frac{14(-1.5) - 3}{8(-1.5)^2} = \frac{-24}{18} = -\frac{4}{3}$

103. $(f \circ g)(-1) = f(g(-1)) = f(8) = 14(8) - 3 = 109$

104. $(g \circ f)(0) = g(f(0)) = g(14(0) - 3) = g(-3)$

$$= 8(-3)^2 = 72$$

105. Answers will vary.

Section 2.2   Polynomial Functions of Higher Degree

You should know the following basic principles about polynomials.

- $f(x) = a_n x^n + a_{n-1} x^{n-1} + \cdots + a_2 x^2 + a_1 x + a_0, \; a_n \neq 0$, is a polynomial function of degree n.
- If f is of odd degree and

  (a) $a_n > 0$, then

  1. $f(x) \to \infty$ as $x \to \infty$.
  2. $f(x) \to -\infty$ as $x \to -\infty$.

  (b) $a_n < 0$, then

  1. $f(x) \to -\infty$ as $x \to \infty$.
  2. $f(x) \to \infty$ as $x \to -\infty$.

- If f is of even degree and

  (a) $a_n > 0$, then

  1. $f(x) \to \infty$ as $x \to \infty$.
  2. $f(x) \to \infty$ as $x \to -\infty$.

  (b) $a_n < 0$, then

  1. $f(x) \to -\infty$ as $x \to \infty$.
  2. $f(x) \to -\infty$ as $x \to -\infty$.

- The following are equivalent for a polynomial function.

  (a) $x = a$ is a zero of a function.

  (b) $x = a$ is a solution of the polynomial equation $f(x) = 0$.

  (c) $(x - a)$ is a factor of the polynomial.

  (d) $(a, 0)$ is an x-intercept of the graph of f.

- A polynomial of degree n has at most n distinct zeros and at most $n - 1$ turning points.
- A factor $(x - a)^k, \; k > 1$, yields a repeated zero of $x = a$ of multiplicity k.

  (a) If k is odd, the graph crosses the x-axis at $x = a$.

  (b) If k is even, the graph just touches the x-axis at $x = a$.

- If f is a polynomial function such that $a < b$ and $f(a) \neq f(b)$, then f takes on every value between $f(a)$ and $f(b)$ in the interval $[a, b]$.
- If you can find a value where a polynomial is positive and another value where it is negative, then there is at least one real zero between the values.

Vocabulary Check

1. continuous

2. Leading Coefficient Test

3. $n; n - 1$

4. solution; $(x - a)$; x-intercept

5. touches; crosses

6. standard

7. Intermediate Value

---

1. $f(x) = -2x + 3$ is a line with y-intercept $(0, 3)$. Matches graph (c).

2. $f(x) = x^2 - 4x$ is a parabola with intercepts $(0, 0)$ and $(4, 0)$ and opens upward. Matches graph (g).

3. $f(x) = -2x^2 - 5x$ is a parabola with x-intercepts $(0, 0)$ and $\left(-\frac{5}{2}, 0\right)$ and opens downward. Matches graph (h).

4. $f(x) = 2x^3 - 3x + 1$ has intercepts $(0, 1), (1, 0)$, $\left(-\frac{1}{2} - \frac{1}{2}\sqrt{3}, 0\right)$ and $\left(-\frac{1}{2} + \frac{1}{2}\sqrt{3}, 0\right)$. Matches graph (f).

5. $f(x) = -\frac{1}{4}x^4 + 3x^2$ has intercepts $(0, 0)$ and $(\pm 2\sqrt{3}, 0)$. Matches graph (a).

6. $f(x) = -\frac{1}{3}x^3 + x^2 - \frac{4}{3}$ has y-intercept $\left(0, -\frac{4}{3}\right)$. Matches graph (e).

7. $f(x) = x^4 + 2x^3$ has intercepts $(0, 0)$ and $(-2, 0)$. Matches graph (d).

8. $f(x) = \frac{1}{5}x^5 - 2x^3 + \frac{9}{5}x$ has intercepts $(0, 0), (1, 0)$, $(-1, 0), (3, 0), (-3, 0)$. Matches graph (b).

9. $y = x^3$

(a) $f(x) = (x - 2)^3$

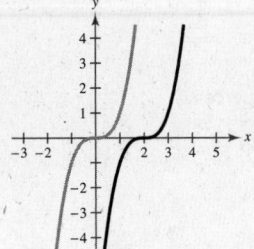

Horizontal shift two units to the right

(b) $f(x) = x^3 - 2$

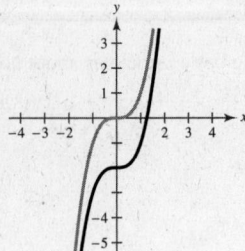

Vertical shift two units downward

(c) $f(x) = -\frac{1}{2}x^3$

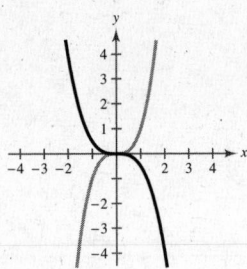

Reflection in the *x*-axis and a vertical shrink

(d) $f(x) = (x - 2)^3 - 2$

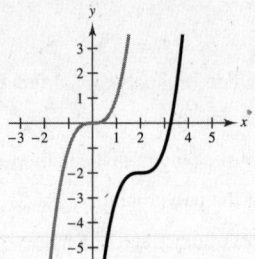

Horizontal shift two units to the right and
a vertical shift two units downward

10. $y = x^5$

(a) $f(x) = (x + 1)^5$

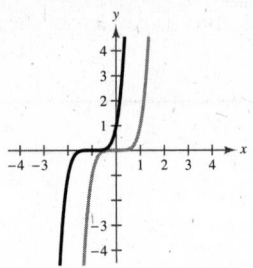

Horizontal shift one unit to the left

(b) $f(x) = x^5 + 1$

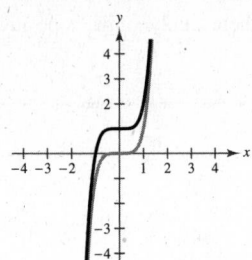

Vertical shift one unit upward

(c) $f(x) = 1 - \frac{1}{2}x^5$

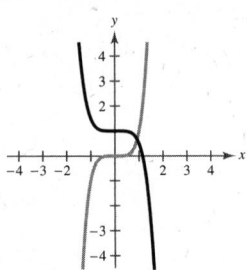

Reflection in the *x*-axis, vertical shrink (each *y*-value
is multiplied by $\frac{1}{2}$), and vertical shift one unit upward

(d) $f(x) = -\frac{1}{2}(x + 1)^5$

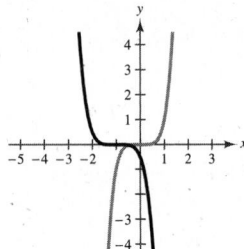

Reflection in the *x*-axis, vertical shrink (each *y*-value is
multiplied by $\frac{1}{2}$), and horizontal shift one unit to the left

11. $y = x^4$

(a) $f(x) = (x + 3)^4$

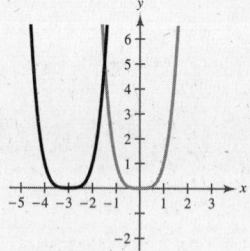

Horizontal shift three units to the left

(b) $f(x) = x^4 - 3$

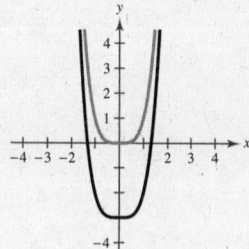

Vertical shift three units downward

(c) $f(x) = 4 - x^4$

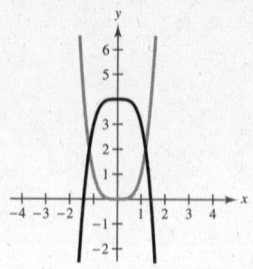

Reflection in the x-axis and then a vertical shift four units upward

(d) $f(x) = \frac{1}{2}(x - 1)^4$

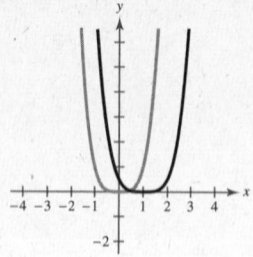

Horizontal shift one unit to the right and a vertical shrink (each y-value is multiplied by $\frac{1}{2}$)

(e) $f(x) = (2x)^4 + 1$

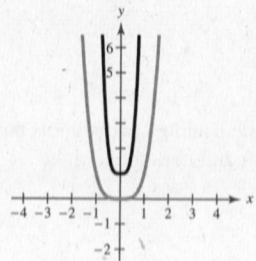

Vertical shift one unit upward and a horizontal shrink (each y-value is multiplied by $\frac{1}{2}$)

(f) $f(x) = \left(\frac{1}{2}x\right)^4 - 2$

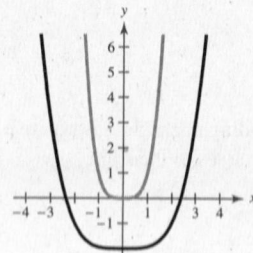

Vertical shift two units downward and a horizontal stretch (each y-value is multiplied by $\frac{1}{2}$)

12. $y = x^6$

(a) $f(x) = -\frac{1}{8}x^6$

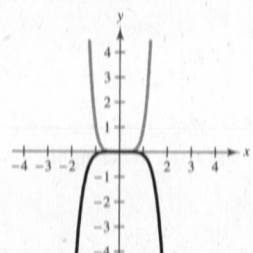

Vertical shrink (each y-value is multiplied by $\frac{1}{8}$) and reflection in the x-axis

(b) $f(x) = (x + 2)^6 - 4$

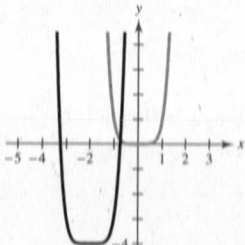

Horizontal shift two units to the left and vertical shift four units downward

—CONTINUED—

12. —CONTINUED—

(c) $f(x) = x^6 - 4$

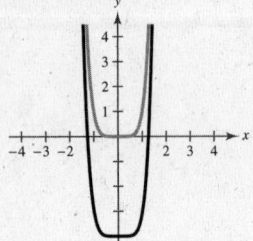

Vertical shift four units downward

(e) $f(x) = \left(\frac{1}{4}x\right)^6 - 2$

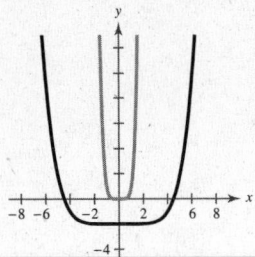

Horizontal stretch (each x-value is multiplied by 4), and vertical shift two units downward

(d) $f(x) = -\frac{1}{4}x^6 + 1$

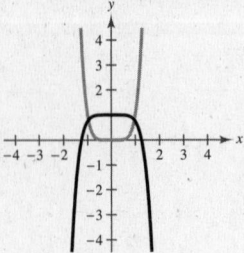

Reflection in the x-axis, vertical shrink $\left(\text{each } y\text{-value is multiplied by } \frac{1}{4}\right)$, and vertical shift one unit upward

(f) $f(x) = (2x)^6 - 1$

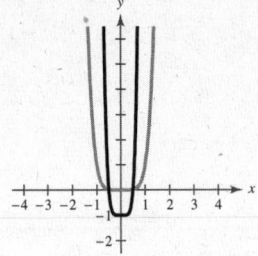

Horizontal shrink $\left(\text{each } x\text{-value is multiplied by } \frac{1}{2}\right)$, and vertical shift one unit downward

13. $f(x) = \frac{1}{3}x^3 + 5x$

Degree: 3

Leading coefficient: $\frac{1}{3}$

The degree is odd and the leading coefficient is positive. The graph falls to the left and rises to the right.

14. $f(x) = 2x^2 - 3x + 1$

Degree: 2

Leading coefficient: 2

The degree is even and the leading coefficient is positive. The graph rises to the left and rises to the right.

15. $g(x) = 5 - \frac{7}{2}x - 3x^2$

Degree: 2

Leading coefficient: -3

The degree is even and the leading coefficient is negative. The graph falls to the left and falls to the right.

16. $h(x) = 1 - x^6$

Degree: 6

Leading coefficient: -1

The degree is even and the leading coefficient is negative. The graph falls to the left and falls to the right.

17. $f(x) = -2.1x^5 + 4x^3 - 2$

Degree: 5

Leading coefficient: -2.1

The degree is odd and the leading coefficient is negative. The graph rises to the left and falls to the right.

18. $f(x) = 2x^5 - 5x + 7.5$

Degree: 5

Leading coefficient: 2

The degree is odd and the leading coefficient is positive. The graph falls to the left and rises to the right.

19. $f(x) = 6 - 2x + 4x^2 - 5x^3$

Degree: 3

Leading coefficient: -5

The degree is odd and the leading coefficient is negative. The graph rises to the left and falls to the right.

20. $f(x) = \dfrac{3x^4 - 2x + 5}{4}$

Degree: 4

Leading coefficient: $\frac{3}{4}$

The degree is even and the leading coefficient is positive. The graph rises to the left and rises to the right.

21. $h(t) = -\frac{2}{3}(t^2 - 5t + 3)$

Degree: 2

Leading coefficient: $-\frac{2}{3}$

The degree is even and the leading coefficient is negative. The graph falls to the left and falls to the right.

22. $f(s) = -\frac{7}{8}(s^3 + 5s^2 - 7s + 1)$

Degree: 3

Leading coefficient: $-\frac{7}{8}$

The degree is odd and the leading coefficient is negative. The graph rises to the left and falls to the right.

23. $f(x) = 3x^3 - 9x + 1;\ g(x) = 3x^3$

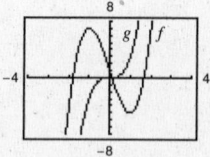

24. $f(x) = -\frac{1}{3}(x^3 - 3x + 2),\ g(x) = -\frac{1}{3}x^3$

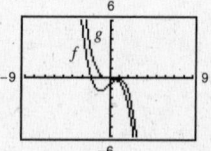

25. $f(x) = -(x^4 - 4x^3 + 16x);\ g(x) = -x^4$

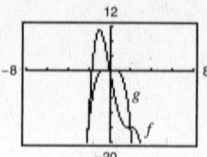

26. $f(x) = 3x^4 - 6x^2,\ g(x) = 3x^4$

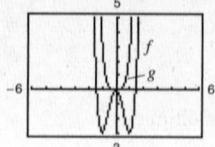

27. $f(x) = x^2 - 25$

(a) $0 = x^2 - 25 = (x + 5)(x - 5)$

Zeros: $x = \pm 5$

(b) Each zero has a multiplicity of 1 (odd multiplicity).

Turning point: 1 (the vertex of the parabola)

(c)

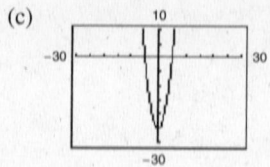

28. (a) $f(x) = 49 - x^2$

$0 = (7 - x)(7 + x)$

$x = \pm 7$, both with multiplicity 1

(b) Multiplicity of $x = 7$ is 1.

Multiplicity of $x = -7$ is 1.

There is one turning point.

(c)

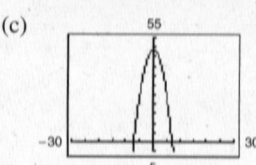

29. $h(t) = t^2 - 6t + 9$

(a) $0 = t^2 - 6t + 9 = (t - 3)^2$

Zero: $t = 3$

(b) $t = 3$ has a multiplicity of 2 (even multiplicity).

Turning point: 1 (the vertex of the parabola)

(c)

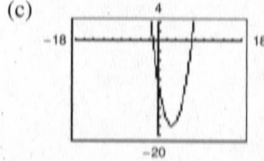

30. (a) $f(x) = x^2 + 10x + 25$

$0 = (x + 5)^2$

$x = -5$, with multiplicity 2

(b) The multiplicity of $x = -5$ is 2.

There is one turning point.

(c)

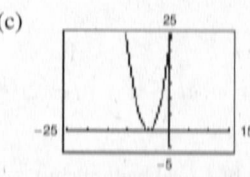

31. $f(x) = \frac{1}{3}x^2 + \frac{1}{3}x - \frac{2}{3}$

    (a) $0 = \frac{1}{3}x^2 + \frac{1}{3}x - \frac{2}{3}$

          $= \frac{1}{3}(x^2 + x - 2)$

          $= \frac{1}{3}(x + 2)(x - 1)$

      Zeros: $x = -2, x = 1$

    (b) Each zero has a multiplicity of 1 (odd multiplicity).

      Turning point: 1 (the vertex of the parabola)

(c)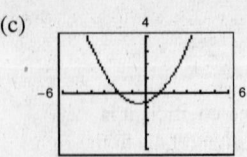

32. (a) $f(x) = \frac{1}{2}x^2 + \frac{5}{2}x - \frac{3}{2}$

      $a = \frac{1}{2}, b = \frac{5}{2}, c = -\frac{3}{2}$

      $x = \dfrac{-\frac{5}{2} \pm \sqrt{\left(\frac{5}{2}\right)^2 - 4\left(\frac{1}{2}\right)\left(-\frac{3}{2}\right)}}{1}$

        $= -\frac{5}{2} \pm \sqrt{\dfrac{37}{4}}$

        $= \dfrac{-5 \pm \sqrt{37}}{2}$, both with multiplicity 1

(b) The multiplicity of $\dfrac{-5 + \sqrt{37}}{2}$ is 1.

The multiplicity of $\dfrac{-5 + - \sqrt{37}}{2}$ is 1.

There is one turning point.

(c)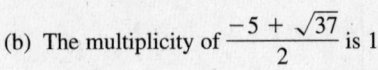

33. $f(x) = 3x^3 - 12x^2 + 3x$

    (a) $0 = 3x^3 - 12x^2 + 3x = 3x(x^2 - 4x + 1)$

      Zeros: $x = 0, x = 2 \pm \sqrt{3}$ (by the Quadratic Formula)

    (b) Each zero has a multiplicity of 1 (odd multiplicity).

      Turning points: 2

(c)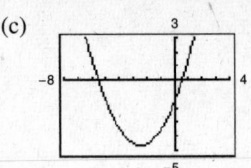

34. (a) $g(x) = 5x(x^2 - 2x - 1)$

      $0 = 5x(x^2 - 2x - 1)$

      $0 = x(x^2 - 2x - 1)$

      For $x^2 - 2x - 1$, $a = 1, b = -2, c = -1$.

      $x = \dfrac{-(-2) \pm \sqrt{(-2)^2 - 4(1)(-1)}}{2(1)}$

        $= \dfrac{2 \pm \sqrt{8}}{2}$

        $= 1 \pm \sqrt{2}$

      The zeros are $0, 1 + \sqrt{2}$, and $1 - \sqrt{2}$, all with multiplicity 1.

(b) The multiplicity of $x = 0$ is 1.

The multiplicity of $x = 1 + \sqrt{2}$ is 1.

The multiplicity of $x = 1 - \sqrt{2}$ is 1.

There are two turning points.

(c)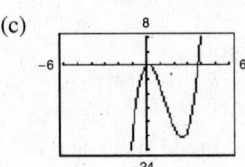

35. $f(t) = t^3 - 4t^2 + 4t$

    (a) $0 = t^3 - 4t^2 + 4t = t(t^2 - 4t + 4) = t(t - 2)^2$

      Zeros: $t = 0, t = 2$

    (b) $t = 0$ has a multiplicity of 1 (odd multiplicity).

      $t = 2$ has a multiplicity of 2 (even multiplicity).

      Turning points: 2

(c)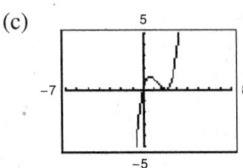

36. (a) $f(x) = x^4 - x^3 - 20x^2$

$0 = x^2(x^2 - x - 20)$

$0 = x^2(x + 4)(x - 5)$

$x = 0, -4, 5$

0 with multiplicity 2, -4 and 5 with multiplicity 1.

(b) The multiplicity of $x = 0$ is 2.

The multiplicity of $x = 5$ is 1.

The multiplicity of $x = -4$ is 1.

There are three turning points.

(c)

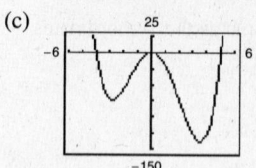

37. $g(t) = t^5 - 6t^3 + 9t$

(a) $0 = t^5 - 6t^3 + 9t = t(t^4 - 6t^2 + 9) = t(t^2 - 3)^2$

$= t(t + \sqrt{3})^2(t - \sqrt{3})^2$

Zeros: $t = 0, t = \pm\sqrt{3}$

(b) $t = 0$ has a multiplicity of 1 (odd multiplicity).

$t = \pm\sqrt{3}$ each have a multiplicity of 2 (even multiplicity).

Turning points: 4

(c)

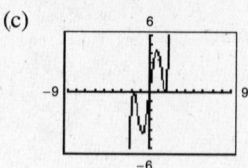

38. (a) $f(x) = x^5 + x^3 - 6x$

$0 = x(x^4 + x^2 - 6)$

$0 = x(x^2 + 3)(x^2 - 2)$

$x = 0, \pm\sqrt{2}$, all with multiplicity 1

(b) The multiplicity of $x = 0$ is 1.

The multiplicity of $x = \sqrt{2}$ is 1.

The multiplicity of $x = -\sqrt{2}$ is 1.

There are two turning points.

(c)

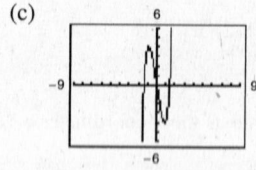

39. $f(x) = 5x^4 + 15x^2 + 10$

(a) $0 = 5x^4 + 15x^2 + 10$

$= 5(x^4 + 3x^2 + 2)$

$= 5(x^2 + 1)(x^2 + 2)$

No real zeros

(b) Turning point: 1

(c)

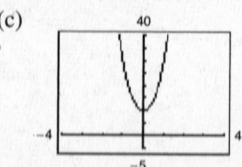

40. (a) $f(x) = 2x^4 - 2x^2 - 40$

$0 = 2x^4 - 2x^2 - 40$

$0 = 2(x^2 + 4)(x + \sqrt{5})(x - \sqrt{5})$

$x = \pm\sqrt{5}$, both with multiplicity 1

(b) The multiplicity of $x = \sqrt{5}$ is 1.

The multiplicity of $x = -\sqrt{5}$ is 1.

There is one turning point.

(c)

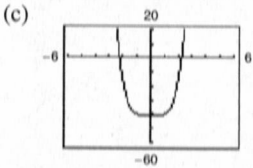

41. $g(x) = x^3 + 3x^2 - 4x - 12$

(a) $0 = x^3 + 3x^2 - 4x - 12 = x^2(x + 3) - 4(x + 3)$

$= (x^2 - 4)(x + 3) = (x - 2)(x + 2)(x + 3)$

Zeros: $x = \pm 2, x = -3$

(b) Each zero has a multiplicity of 1 (odd multiplicity).

Turning points: 2

(c)

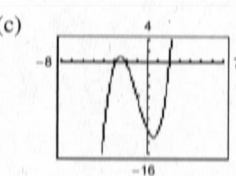

42. (a) $f(x) = x^3 - 4x^2 - 25x + 100$

$$0 = x^2(x - 4) - 25(x - 4)$$

$$0 = (x^2 - 25)(x - 4)$$

$$0 = (x + 5)(x - 5)(x - 4)$$

$x = \pm 5, 4$, all with multiplicity 1

(b) The multiplicity of $x = 5$ is 1.

The multiplicity of $x = -5$ is 1.

The multiplicity of $x = 4$ is 1.

There are two turning points.

(c)

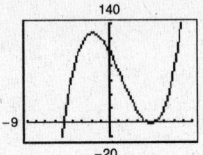

44. $y = 4x^3 + 4x^2 - 8x - 8$

(a)

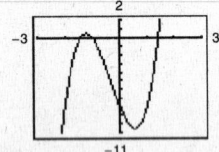

(b) $(-1, 0), (-1.414214, 0), (1.414214, 0)$

(c) $0 = 4x^3 + 4x^2 - 8x - 8$

$$0 = 4x^2(x + 1) - 8(x + 1)$$

$$0 = (4x^2 - 8)(x + 1)$$

$$0 = 4(x^2 - 2)(x + 1)$$

$$x = \pm\sqrt{2}, -1$$

(d) The intercepts match part (b).

46. $y = \frac{1}{4}x^3(x^2 - 9)$

(a)

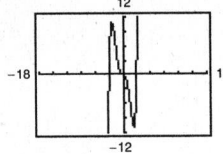

(b) $(0, 0), (3, 0), (-3, 0)$

(c) $0 = \frac{1}{4}x^3(x^2 - 9)$

$$x = 0, \pm 3$$

x-intercepts: $(0, 0), (\pm 3, 0)$

(d) The intercepts match part (b).

43. $y = 4x^3 - 20x^2 + 25x$

(a)

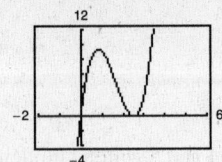

(b) x-intercepts: $(0, 0), \left(\frac{5}{2}, 0\right)$

(c) $0 = 4x^3 - 20x^2 + 25x$

$$0 = x(2x - 5)^2$$

$$x = 0 \text{ or } x = \frac{5}{2}$$

(d) The solutions are the same as the x-coordinates of the x-intercepts.

45. $y = x^5 - 5x^3 + 4x$

(a)

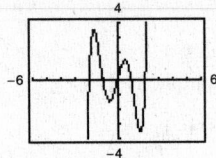

(b) x-intercepts: $(0, 0), (\pm 1, 0), (\pm 2, 0)$

(c) $0 = x^5 - 5x^3 + 4x$

$$0 = x(x^2 - 1)(x^2 - 4)$$

$$0 = x(x + 1)(x - 1)(x + 2)(x - 2)$$

$$x = 0, \pm 1, \pm 2$$

(d) The solutions are the same as the x-coordinates of the x-intercepts.

47. $f(x) = (x - 0)(x - 10)$

$$f(x) = x^2 - 10x$$

Note: $f(x) = a(x - 0)(x - 10) = ax(x - 10)$ has zeros 0 and 10 for all real numbers $a \neq 0$.

48. $f(x) = (x - 0)(x - (-3))$

$$= x(x + 3)$$

$$= x^2 + 3x$$

Note: $f(x) = ax(x + 3)$ has zeros 0 and -3 for all real numbers a.

49. $f(x) = (x - 2)(x - (-6))$

$= (x - 2)(x + 6)$

$= x^2 + 4x - 12$

Note: $f(x) = a(x - 2)(x + 6)$ has zeros 2 and -6 for all real numbers $a \neq 0$.

50. $f(x) = (x - (-4))(x - 5)$

$= (x + 4)(x - 5)$

$= x^2 - x - 20$

Note: $f(x) = a(x + 4)(x - 5)$ has zeros -4 and 5 for all real numbers a.

51. $f(x) = (x - 0)(x - (-2))(x - (-3))$

$= x(x + 2)(x + 3)$

$= x^3 + 5x^2 + 6x$

Note: $f(x) = ax(x + 2)(x + 3)$ has zeros $0, -2, -3$ for all real numbers $a \neq 0$.

52. $f(x) = (x - 0)(x - 2)(x - 5)$

$= x(x - 2)(x - 5)$

$= x(x^2 - 7x + 10)$

$= x^3 - 7x^2 + 10x$

Note: $f(x) = ax(x - 2)(x - 5)$ has zeros $0, 2, 5$ for all real numbers a.

53. $f(x) = (x - 4)(x + 3)(x - 3)(x - 0)$

$= (x - 4)(x^2 - 9)x$

$= x^4 - 4x^3 - 9x^2 + 36x$

Note: $f(x) = a(x^4 - 4x^3 - 9x^2 + 36x)$ has these zeros for all real numbers $a \neq 0$.

54. $f(x) = (x - (-2))(x - (-1))(x - 0)(x - 1)(x - 2)$

$= x(x + 2)(x + 1)(x - 1)(x - 2)$

$= x(x^2 - 4)(x^2 - 1)$

$= x(x^4 - 5x^2 + 4)$

$= x^5 - 5x^3 + 4x$

Note: $f(x) = ax(x + 2)(x + 1)(x - 1)(x - 2)$ has zeros $-2, -1, 0, 1, 2$ for all real numbers a.

55. $f(x) = \left[x - \left(1 + \sqrt{3}\right)\right]\left[x - \left(1 - \sqrt{3}\right)\right]$

$= \left[(x - 1) - \sqrt{3}\right]\left[(x - 1) + \sqrt{3}\right]$

$= (x - 1)^2 - \left(\sqrt{3}\right)^2$

$= x^2 - 2x + 1 - 3$

$= x^2 - 2x - 2$

Note: $f(x) = a(x^2 - 2x - 2)$ has these zeros for all real numbers $a \neq 0$.

56. $f(x) = (x - 2)\left[x - \left(4 + \sqrt{5}\right)\right]\left[x - \left(4 - \sqrt{5}\right)\right]$

$= (x - 2)\left[(x - 4) - \sqrt{5}\right]\left[(x - 4) + \sqrt{5}\right]$

$= (x - 2)\left[(x - 4)^2 - 5\right]$

$= x(x - 4)^2 - 5x - 2(x - 4)^2 + 10$

$= x^3 - 8x^2 + 16x - 5x - 2x^2 + 16x - 32 + 10$

$= x^3 - 10x^2 + 27x - 22$

Note: $f(x) = a(x^3 - 10x^2 + 27x - 22)$ has these zeros for all real numbers a.

57. $f(x) = (x - (-2))(x - (-2))$

$= (x + 2)^2 = x^2 + 4x + 4$

Note: $f(x) = a(x^2 + 4x + 4)$, $a \neq 0$, has degree 2 and zero $x = -2$.

58. $f(x) = [x - (-8)][x - (-4)]$

$= (x + 8)(x + 4) = x^2 + 12x + 32$

Note: $f(x) = a(x^2 + 12x + 32)$, $a \neq 0$, has degree 2 and zeros $x = -8$ and -4.

59. $f(x) = (x - (-3))(x - 0)(x - 1)$

$= x(x + 3)(x - 1) = x^3 + 2x^2 - 3x$

Note: $f(x) = a(x^3 + 2x^2 - 3x)$, $a \neq 0$, has degree 3 and zeros $x = -3, 0, 1$.

60. $f(x) = (x + 2)(x - 4)(x - 7)$

$= (x + 2)(x^2 - 11x + 28) = x^3 - 9x^2 + 6x + 56$

Note: $f(x) = a(x^3 - 9x^2 + 6x + 56)$, $a \neq 0$, has degree 3 and zeros $x = -2, 4,$ and 7.

61. $f(x) = (x - 0)\left(x - \sqrt{3}\right)\left(x - \left(-\sqrt{3}\right)\right)$

$= x\left(x - \sqrt{3}\right)\left(x + \sqrt{3}\right) = x^3 - 3x$

Note: $f(x) = a(x^3 - 3x)$, $a \neq 0$, has degree 3 and zeros $x = 0, \sqrt{3}, -\sqrt{3}$.

62. $f(x) = (x - 9)^3 = x^3 - 27x^2 + 243x - 729$

Note: $f(x) = a(x^3 - 27x^2 + 243x - 729)$, $a \neq 0$, has degree 3 and zero $x = 9$.

63. $f(x) = (x - (-5))^2(x - 1)(x - 2) = x^4 + 7x^3 - 3x^2 - 55x + 50$

or $f(x) = (x - (-5))(x - 1)^2(x - 2) = x^4 + x^3 - 15x^2 + 23x - 10$

or $f(x) = (x - (-5))(x - 1)(x - 2)^2 = x^4 - 17x^2 + 36x - 20$

Note: Any nonzero scalar multiple of these functions would also have degree 4 and zeros $x = -5, 1, 2$.

64. $f(x) = (x + 4)(x + 1)(x - 3)(x - 6) = x^4 - 4x^3 - 23x^2 + 54x + 72$

Note: $f(x) = a(x^4 - 4x^3 - 23x^2 + 54x + 72), a \neq 0$, has degree 4 and zeros $x = -4, -1, 3,$ and 6.

65. $f(x) = x^4(x + 4) = x^5 + 4x^4$

or $f(x) = x^3(x + 4)^2 = x^5 + 8x^4 + 16x^3$

or $f(x) = x^2(x + 4)^3 = x^5 + 12x^4 + 48x^3 + 64x^2$

or $f(x) = x(x + 4)^4 = x^5 + 16x^4 + 96x^3 + 256x^2 + 256x$

Note: Any nonzero scalar multiple of these functions would also have degree 5 and zeros $x = 0$ and -4.

66. $f(x) = (x + 3)^2(x - 1)(x - 5)(x - 6) = x^5 - 6x^4 - 22x^3 + 108x^2 + 189x - 270$

or $f(x) = (x + 3)(x - 1)^2(x - 5)(x - 6) = x^5 - 10x^4 + 14x^3 + 88x^2 - 183x + 90$

or $f(x) = (x + 3)(x - 1)(x - 5)^2(x - 6) = x^5 - 14x^4 + 50x^3 + 68x^2 - 555x + 450$

or $f(x) = (x + 3)(x - 1)(x - 5)(x - 6)^2 = x^5 - 15x^4 + 59x^3 + 63x^2 - 648x + 540$

Note: Any nonzero multiple of these functions would also have degree 5 and zeros $x = -3, 1, 5,$ and 6.

67. $f(x) = x^3 - 9x = x(x^2 - 9) = x(x + 3)(x - 3)$

(a) Falls to the left; rises to the right

(b) Zeros: $0, -3, 3$

(c)

x	-3	-2	-1	0	1	2	3
$f(x)$	0	10	8	0	-8	-10	0

(d)

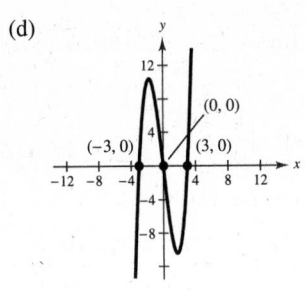

68. $g(x) = x^4 - 4x^2 = x^2(x + 2)(x - 2)$

(a) Rises to the left; rises to the right

(b) Zeros: $-2, 0, 2$

(c)

x	± 0.5	± 1	± 1.5	± 2.5
$g(x)$	-0.94	-3	-3.94	14.1

(d)

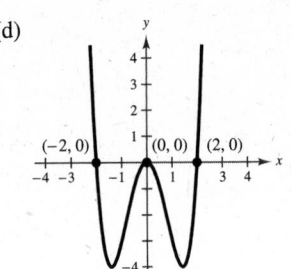

69. $f(t) = \frac{1}{4}(t^2 - 2t + 15) = \frac{1}{4}(t - 1)^2 + \frac{7}{2}$

(a) Rises to the left; rises to the right

(b) No real zero (no x-intercepts)

(c)

t	-1	0	1	2	3
$f(t)$	4.5	3.75	3.5	3.75	4.5

(d) The graph is a parabola with vertex $\left(1, \frac{7}{2}\right)$.

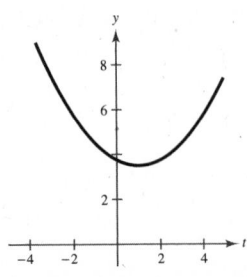

70. $g(x) = -x^2 + 10x - 16 = -(x - 2)(x - 8)$

(a) Falls to the left; falls to the right

(b) Zeros: 2, 8

(c)

x	1	3	5	7	9
$g(x)$	-7	5	9	5	-7

(d)

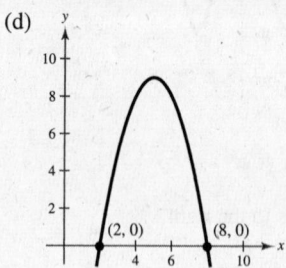

71. $f(x) = x^3 - 3x^2 = x^2(x - 3)$

(a) Falls to the left; rises to the right

(b) Zeros: 0, 3

(c)

x	-1	0	1	2	3
$f(x)$	-4	0	-2	-4	0

(d)

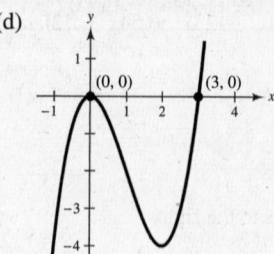

72. $f(x) = 1 - \frac{1}{8}x^3$

(a) Rises to the left; falls to the right

(b) Zero: 1

(c)

x	-2	-1	0	1	2
$f(x)$	9	2	1	0	-7

(d)

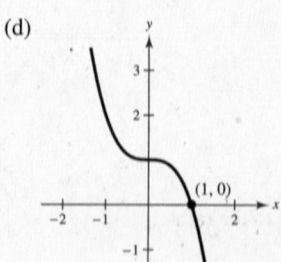

73. $f(x) = 3x^3 - 15x^2 + 18x = 3x(x - 2)(x - 3)$

(a) Falls to the left; rises to the right

(b) Zeros: 0, 2, 3

(c)

x	0	1	2	2.5	3	3.5
$f(x)$	0	6	0	-1.875	0	7.875

(d)

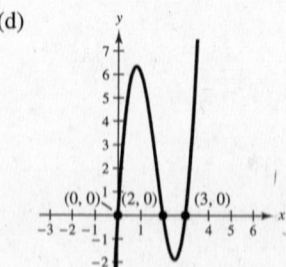

74. $f(x) = -4x^3 + 4x^2 + 15x$

$= -x(4x^2 - 4x - 15)$

$= -x(2x - 5)(2x + 3)$

(a) Rises to the left; falls to the right

(b) Zeros: $-\frac{3}{2}, 0, \frac{5}{2}$

(c)

x	-3	-2	-1	0	1	2	3
$f(x)$	99	18	-7	0	15	14	-27

(d)

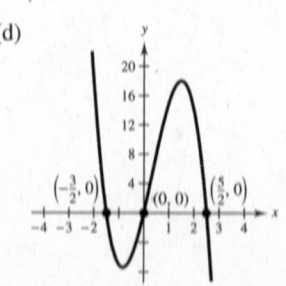

75. $f(x) = -5x^2 - x^3 = -x^2(5 + x)$

(a) Rises to the left; falls to the right

(b) Zeros: 0, -5

(c)

x	-5	-4	-3	-2	-1	0	1
$f(x)$	0	-16	-18	-12	-4	0	-6

(d)

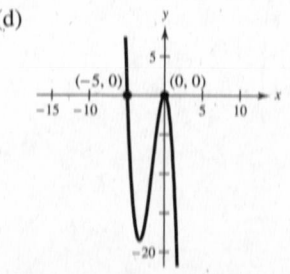

76. $f(x) = -48x^2 + 3x^4$

$\quad = 3x^2(x^2 - 16)$

(a) Rises to the left; rises to the right

(b) Zeros: $0, \pm 4$

(c)

x	-5	-4	-3	-2	-1	0	1	2	3	4	5
$f(x)$	675	0	-189	-144	-45	0	-45	-144	-189	0	675

(d)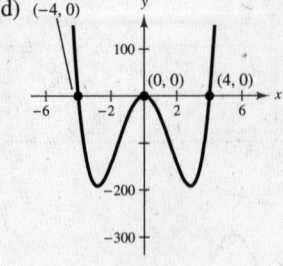

77. $f(x) = x^2(x - 4)$

(a) Falls to the left; rises to the right

(b) Zeros: $0, 4$

(c)

x	-1	0	1	2	3	4	5
$f(x)$	-5	0	-3	-8	-9	0	25

(d)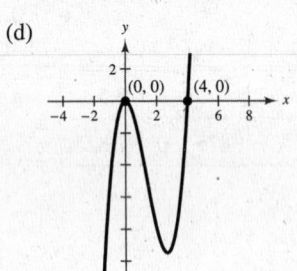

78. $h(x) = \frac{1}{3}x^3(x - 4)^2$

(a) Falls to the left; rises to the right

(b) Zeros: $0, 4$

(c)

x	-1	0	1	2	3	4	5
$h(x)$	$-\frac{25}{3}$	0	3	$\frac{32}{3}$	9	0	$\frac{125}{3}$

(d)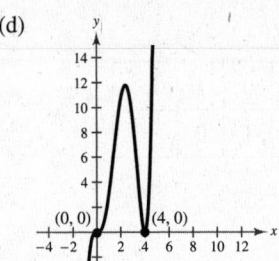

79. $g(t) = -\frac{1}{4}(t - 2)^2(t + 2)^2$

(a) Falls to the left; falls to the right

(b) Zeros: $2, -2$

(c)

t	-3	-2	-1	0	1	2	3
$g(t)$	$-\frac{25}{4}$	0	$-\frac{9}{4}$	-4	$-\frac{9}{4}$	0	$-\frac{25}{4}$

(d)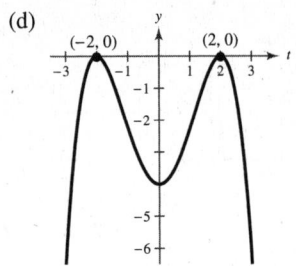

80. $g(x) = \frac{1}{10}(x + 1)^2(x - 3)^3$

(a) Falls to the left; rises to the right

(b) Zeros: $-1, 3$

(c)

x	-2	-1	0	1	2	4
$g(x)$	-12.5	0	-2.7	-3.2	-0.9	2.5

(d)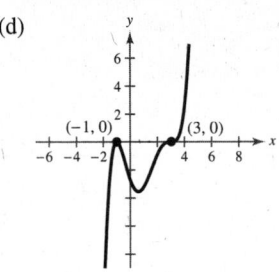

81. $f(x) = x^3 - 4x = x(x + 2)(x - 2)$

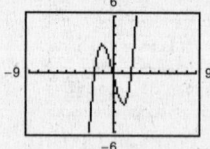

Zeros: 0, −2, 2 all of multiplicity 1

82. $f(x) = \frac{1}{4}x^4 - 2x^2$

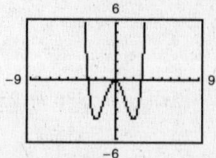

Zeros: −2.828 and 2.828 with multiplicity 1; 0, with multiplicity 2

83. $g(x) = \frac{1}{5}(x + 1)^2(x - 3)(2x - 9)$

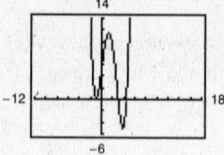

Zeros: −1 of multiplicity 2; 3 of multiplicity 1; $\frac{9}{2}$ of multiplicity 1

84. $h(x) = \frac{1}{5}(x + 2)^2(3x - 5)^2$

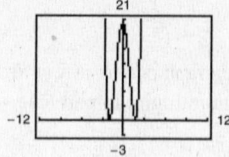

Zeros: $-2, \frac{5}{3}$, both with multiplicity 2

85. $f(x) = x^3 - 3x^2 + 3$

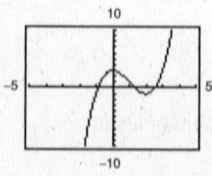

The function has three zeros. They are in the intervals $(-1, 0)$, $(1, 2)$ and $(2, 3)$. They are $x \approx -0.879, 1.347, 2.532$.

x	y_1
−3	−51
−2	−17
−1	−1
0	3
1	1
2	−1
3	3
4	19

86. $f(x) = 0.11x^3 - 2.07x^2 + 9.81x - 6.88$

The function has three zeros. They are in the intervals $(0, 1)$, $(6, 7)$, and $(11, 12)$. They are approximately 0.845, 6.385, and 11.588.

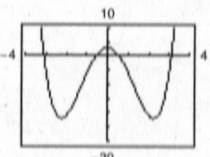

x	y	x	y
0	−6.88	7	−1.91
1	0.97	8	−4.56
2	5.34	9	−6.07
3	6.89	10	−5.78
4	6.28	11	−3.03
5	4.17	12	2.84
6	1.12		

87. $g(x) = 3x^4 + 4x^3 - 3$

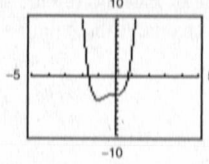

The function has two zeros. They are in the intervals $(-2, -1)$ and $(0, 1)$. They are $x \approx -1.585, 0.779$.

x	y_1
−4	509
−3	132
−2	13
−1	−4
0	−3
1	4
2	77
3	348

88. $h(x) = x^4 - 10x^2 + 3$

The function has four zeros. They are in the intervals $(-4, -3)$, $(-1, 0)$, $(0, 1)$, and $(3, 4)$. They are approximately ± 3.113 and ± 0.556.

x	y
−4	99
−3	−6
−2	−21
−1	−6
0	3
1	−6
2	−21
3	−6
4	99

89. (a) Volume $= l \cdot w \cdot h$

height $= x$

length $=$ width $= 36 - 2x$

Thus, $V(x) = (36 - 2x)(36 - 2x)(x) = x(36 - 2x)^2$.

(b) Domain: $0 < x < 18$

The length and width must be positive.

(d)

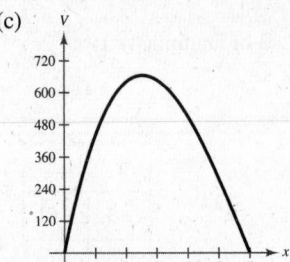

The maximum point on the graph occurs at $x = 6$.
This agrees with the maximum found in part (c).

(c)

Box Height	Box Width	Box Volume, V
1	$36 - 2(1)$	$1[36 - 2(1)]^2 = 1156$
2	$36 - 2(2)$	$2[36 - 2(2)]^2 = 2048$
3	$36 - 2(3)$	$3[36 - 2(3)]^2 = 2700$
4	$36 - 2(4)$	$4[36 - 2(4)]^2 = 3136$
5	$36 - 2(5)$	$5[36 - 2(5)]^2 = 3380$
6	$36 - 2(6)$	$6[36 - 2(6)]^2 = 3456$
7	$36 - 2(7)$	$7[36 - 2(7)]^2 = 3388$

The volume is a maximum of 3456 cubic inches when the height is 6 inches and the length and width are each 24 inches. So the dimensions are $6 \times 24 \times 24$ inches.

90. (a) Volume $= l \cdot w \cdot h = (24 - 2x)(24 - 4x)x$

$= 2(12 - x) \cdot 4(6 - x)x$

$= 8x(12 - x)(6 - x)$

(b) $x > 0, \quad 12 - x > 0, \quad 6 - x > 0$

$x < 12 \qquad x < 6$

Domain: $0 < x < 6$

(c)

$x \approx 2.6$ corresponds to a maximum of about 665 cubic inches.

91. (a) $A = l \cdot w = (12 - 2x)(x) = -2x^2 + 12x$ square inches

(b) 16 feet $= 192$ inches

$V = l \cdot w \cdot h$

$= (12 - 2x)(x)(192)$

$= -384x^2 + 2304x$ cubic inches

(c) Since x and $12 - 2x$ cannot be negative, we have $0 < x < 6$ inches for the domain.

(d)

x	V
0	0
1	1920
2	3072
3	3456
4	3072
5	1920
6	0

When $x = 3$, the volume is a maximum with $V = 3456$ in.³. The dimensions of the gutter cross-section are 3 inches $\times$ 6 inches $\times$ 3 inches.

(e)

Maximum: $(3, 3456)$

The maximum value is the same.

(f) No. The volume is a product of the constant length and the cross-sectional area. The value of x would remain the same; only the value of V would change if the length was changed.

92. (a) $V = \frac{4}{3}\pi r^3 + \pi r^2(4r)$

  $V = \frac{4}{3}\pi r^3 + 4\pi r^3$

  $= \frac{16}{3}\pi r^3$

(b) $r \geq 0$

(d) $V = 120 \text{ ft}^3 = \frac{16}{3}\pi r^3$

  $r = 1.93 \text{ ft}$

  length $= 4r = 7.72 \text{ ft}$

(c)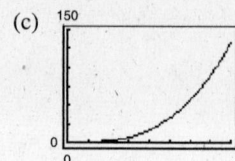

93. $y_1 = 0.139t^3 - 4.42t^2 + 51.1t - 39$

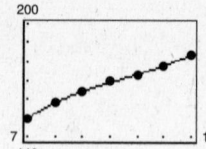

The model is a good fit to the actual data.

94. $y = 0.056t^3 - 1.73t^2 + 23.8t + 29$

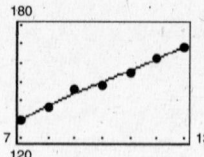

The data fit the model closely.

95. Midwest: $y_1(18) = \$259.368$ thousand $= \$259,368$

South: $y_2(18) = \$223.472$ thousand $= \$223,472$

Since the models are both cubic functions with positive leading coefficients, both will increase without bound as t increases, thus should only be used for short term projections.

96. Answers will vary.

Example: The median price of homes in the South are all lower than those in the Midwest. The curves do not intersect.

97. $G = -0.003t^3 + 0.137t^2 + 0.458t - 0.839, \ 2 \leq t \leq 34$

(a)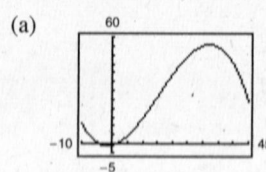

(b) The tree is growing most rapidly at $t \approx 15$.

(c) $y = -0.009t^2 + 0.274t + 0.458$

  $-\dfrac{b}{2a} = \dfrac{-0.274}{2(-0.009)} \approx 15.222$

  $y(15.222) \approx 2.543$

  Vertex $\approx (15.22, 2.54)$

(d) The x-value of the vertex in part (c) is approximately equal to the value found in part (b).

98. $R = \frac{1}{100,000}(-x^3 + 600x^2)$

The point of diminishing returns (where the graph changes from curving upward to curving downward) occurs when $x = 200$. The point is $(200, 160)$ which corresponds to spending $2,000,000 on advertising to obtain a revenue of $160 million.

99. False. A fifth degree polynomial can have at most four turning points.

100. True. $f(x) = (x - 1)^6$ has one repeated solution.

101. True. A polynomial of degree 7 with a negative leading coefficient rises to the left and falls to the right.

102. (a) Degree: 3
  Leading coefficient: Positive

(b) Degree: 2
  Leading coefficient: Positive

(c) Degree: 4
  Leading coefficient: Positive

(d) Degree: 5
  Leading coefficient: Positive

103. $f(x) = x^4$; $f(x)$ is even.

(a) $g(x) = f(x) + 2$

Vertical shift two units upward

$g(-x) = f(-x) + 2$

$\qquad = f(x) + 2$

$\qquad = g(x)$

Even

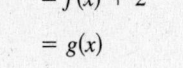

(b) $g(x) = f(x + 2)$

Horizontal shift two units to the left

Neither odd nor even

(c) $g(x) = f(-x) = (-x)^4 = x^4$

Reflection in the y-axis. The graph looks the same.

Even

(d) $g(x) = -f(x) = -x^4$

Reflection in the x-axis

Even

(e) $g(x) = f\left(\frac{1}{2}x\right) = \frac{1}{16}x^4$

Horizontal stretch

Even

(f) $g(x) = \frac{1}{2}f(x) = \frac{1}{2}x^4$

Vertical shrink

Even

(g) $g(x) = f(x^{3/4}) = (x^{3/4})^4 = x^3, x \geq 0$

Neither odd nor even

(h) $g(x) = (f \circ f)(x) = f(f(x)) = f(x^4) = (x^4)^4 = x^{16}$

Even

104. (a) $y_1 = -\frac{1}{3}(x - 2)^5 + 1$ is decreasing.

$y_2 = \frac{3}{5}(x + 2)^5 - 3$ is increasing.

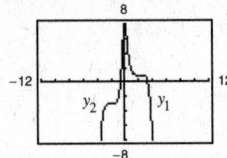

(b) The graph is either always increasing or always decreasing. The behavior is determined by a. If $a > 0$, $g(x)$ will always be increasing. If $a < 0$, $g(x)$ will always be decreasing.

(c) $H(x) = x^5 - 3x^3 + 2x + 1$

Since $H(x)$ is not always increasing or always decreasing, $H(x)$ cannot be written in the form $a(x - h)^5 + k$.

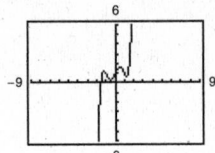

105. $5x^2 + 7x - 24 = (5x - 8)(x + 3)$

106. $6x^3 - 61x^2 + 10x = x(6x^2 - 61x + 10)$

$\qquad\qquad\qquad\qquad = x(6x - 1)(x - 10)$

107. $4x^4 - 7x^3 - 15x^2 = x^2(4x^2 - 7x - 15)$

$\qquad\qquad\qquad\qquad = x^2(4x + 5)(x - 3)$

108. $y^3 + 216 = y^3 + 6^3$

$\qquad\qquad = (y + 6)(y^2 - 6y + 36)$

109. $\quad 2x^2 - x - 28 = 0$

$(2x + 7)(x - 4) = 0$

$\qquad 2x + 7 = 0 \Rightarrow x = -\frac{7}{2}$

$\qquad\quad x - 4 = 0 \Rightarrow x = 4$

110. $\quad 3x^2 - 22x - 16 = 0$

$(3x + 2)(x - 8) = 0$

$\qquad 3x + 2 = 0 \quad$ or $\quad x - 8 = 0$

$\qquad\quad x = -\frac{2}{3} \quad$ or $\qquad\quad x = 8$

111. $12x^2 + 11x - 5 = 0$

$\quad (3x - 1)(4x + 5) = 0$

$\qquad 3x - 1 = 0 \implies x = \frac{1}{3}$

$\qquad 4x + 5 = 0 \implies x = -\frac{5}{4}$

112. $x^2 + 24x + 144 = 0$

$\quad (x + 12)^2 = 0$

$\qquad x + 12 = 0$

$\qquad\quad x = -12$

113. $\qquad\qquad x^2 - 2x - 21 = 0$

$\quad (x^2 - 2x + (-1)^2) - 21 - 1 = 0$

$\qquad\qquad (x - 1)^2 - 22 = 0$

$\qquad\qquad\quad (x - 1)^2 = 22$

$\qquad\qquad\quad x - 1 = \pm\sqrt{22}$

$\qquad\qquad\qquad x = 1 \pm \sqrt{22}$

114. $\quad x^2 - 8x + 2 = 0$

$\qquad x^2 - 8x = -2$

$\quad x^2 - 8x + 16 = -2 + 16$

$\qquad (x - 4)^2 = 14$

$\qquad x - 4 = \pm\sqrt{14}$

$\qquad\quad x = 4 \pm \sqrt{14}$

115. $\qquad\qquad 2x^2 + 5x - 20 = 0$

$\qquad\qquad 2\left(x^2 + \frac{5}{2}x\right) - 20 = 0$

$\quad 2\left(x^2 + \frac{5}{2}x + \left(\frac{5}{4}\right)^2\right) - 20 - \frac{25}{8} = 0$

$\qquad\qquad 2\left(x + \frac{5}{4}\right)^2 - \frac{185}{8} = 0$

$\qquad\qquad\qquad \left(x + \frac{5}{4}\right)^2 = \frac{185}{16}$

$\qquad\qquad\qquad x + \frac{5}{4} = \pm\frac{\sqrt{185}}{4}$

$\qquad\qquad\qquad x = \frac{-5 \pm \sqrt{185}}{4}$

116. $3x^2 + 4x - 9 = 0$

$\quad x^2 + \frac{4}{3}x - 3 = 0$

$\quad x^2 + \frac{4}{3}x = 3$

$\quad x^2 + \frac{4}{3}x + \frac{4}{9} = 3 + \frac{4}{9}$

$\qquad \left(x + \frac{2}{3}\right)^2 = \frac{31}{9}$

$\qquad x + \frac{2}{3} = \pm\sqrt{\frac{31}{9}}$

$\qquad x = -\frac{2}{3} \pm \frac{\sqrt{31}}{3}$

$\qquad x = \frac{-2 \pm \sqrt{31}}{3}$

117. $f(x) = (x + 4)^2$

Common function: $y = x^2$

Transformation: Horizontal shift four units to the left

118. $f(x) = 3 - x^2$

Reflection in the x-axis and vertical shift of three units upward of $y = x^2$

119. $f(x) = \sqrt{x + 1} - 5$

Common function: $y = \sqrt{x}$

Transformation: Horizontal shift one unit to the left and a vertical shift five units downward

120. $f(x) = 7 - \sqrt{x - 6}$

Horizontal shift of six units to the right, reflection in the x-axis, and vertical shift of seven units upward of $y = \sqrt{x}$

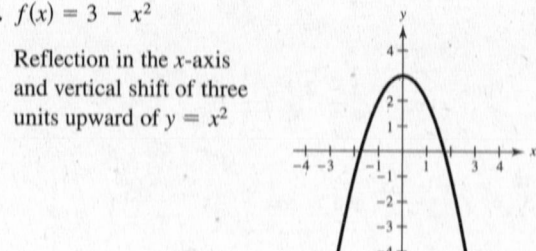

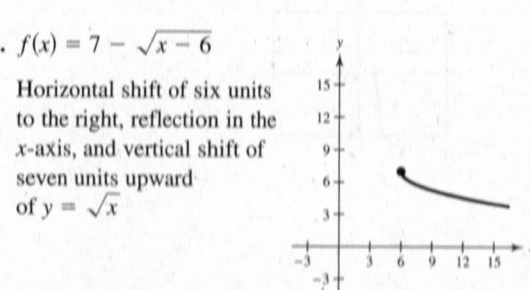

121. $f(x) = 2[\![x]\!] + 9$

Common function: $y = [\![x]\!]$

Transformation: Vertical stretch (each y-value is multiplied by 2), then a vertical shift nine units upward

122. $f(x) = 10 - \frac{1}{3}[\![x + 3]\!]$

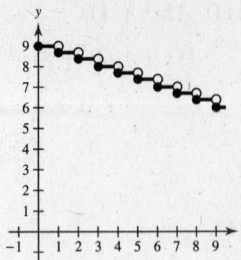

Horizontal shift of three units to the left, vertical shrink $\left(\text{each } y\text{-value is multiplied by } \frac{1}{3}\right)$, reflection in the x-axis and vertical shift of ten units upward of $y = [\![x]\!]$

Section 2.3   Polynomial and Synthetic Division

You should know the following basic techniques and principles of polynomial division.

- ■ The Division Algorithm (Long Division of Polynomials)
- ■ Synthetic Division
- ■ $f(k)$ is equal to the remainder of $f(x)$ divided by $(x - k)$ (the Remainder Theorem).
- ■ $f(k) = 0$ if and only if $(x - k)$ is a factor of $f(x)$.

Vocabulary Check

1. $f(x)$ is the dividend; $d(x)$ is the divisor; $g(x)$ is the quotient; $r(x)$ is the remainder

2. improper; proper   **3.** synthetic division   **4.** factor   **5.** remainder

1. $y_1 = \dfrac{x^2}{x + 2}$ and $y_2 = x - 2 + \dfrac{4}{x + 2}$

$$
\begin{array}{r}
x - 2 \\
x + 2 \overline{\smash{)}\, x^2 + 0x + 0} \\
\underline{x^2 + 2x} \\
-2x + 0 \\
\underline{-2x - 4} \\
4
\end{array}
$$

Thus, $\dfrac{x^2}{x + 2} = x - 2 + \dfrac{4}{x + 2}$ and $y_1 = y_2$.

2. $y_1 = \dfrac{x^4 - 3x^2 - 1}{x^2 + 5}$ and $y_2 = x^2 - 8 + \dfrac{39}{x^2 + 5}$

$$
\begin{array}{r}
x^2 - 8 \\
x^2 + 5 \overline{\smash{)}\, x^4 - 3x^2 - 1} \\
\underline{x^4 + 5x^2} \\
-8x^2 - 1 \\
\underline{-8x^2 - 40} \\
39
\end{array}
$$

Thus, $\dfrac{x^4 - 3x^2 - 1}{x^2 + 5} = x^2 - 8 + \dfrac{39}{x^2 + 5}$ and $y_1 = y_2$.

3. $y_1 = \dfrac{x^5 - 3x^3}{x^2 + 1}$ and $y_2 = x^3 - 4x + \dfrac{4x}{x^2 + 1}$

(a) and (b)

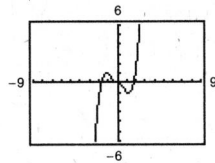

(c)
$$
\begin{array}{r}
x^3 - 4x \\
x^2 + 0x + 1 \overline{\smash{)}\, x^5 + 0x^4 - 3x^3 + 0x^2 + 0x + 0} \\
\underline{x^5 + 0x^4 + x^3} \\
-4x^3 + 0x^2 + 0x \\
\underline{-4x^3 + 0x^2 - 4x} \\
4x + 0
\end{array}
$$

Thus, $\dfrac{x^5 - 3x^3}{x^2 + 1} = x^3 - 4x + \dfrac{4x}{x^2 + 1}$ and $y_1 = y_2$.

4. $y_1 = \dfrac{x^3 - 2x^2 + 5}{x^2 + x + 1}$ and $y_2 = x - 3 + \dfrac{2(x + 4)}{x^2 + x + 1}$

(a) and (b)

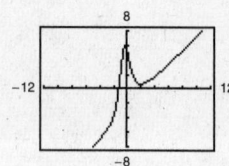

(c)
$$
\begin{array}{r}
x - 3 \\
x^2 + x + 1 \overline{\smash{\big)}\ x^3 - 2x^2 + 0x + 5} \\
\underline{x^3 + x^2 + x} \\
-3x^2 - x + 5 \\
\underline{-3x^2 - 3x - 3} \\
2x + 8
\end{array}
$$

Thus, $\dfrac{x^3 - 2x^2 + 5}{x^2 + x + 1} = x - 3 + \dfrac{2(x + 4)}{x^2 + x + 1}$ and $y_1 = y_2$.

5.
$$
\begin{array}{r}
2x + 4 \\
x + 3 \overline{\smash{\big)}\ 2x^2 + 10x + 12} \\
\underline{2x^2 + 6x} \\
4x + 12 \\
\underline{4x + 12} \\
0
\end{array}
$$

$\dfrac{2x^2 + 10x + 12}{x + 3} = 2x + 4$

6.
$$
\begin{array}{r}
5x + 3 \\
x - 4 \overline{\smash{\big)}\ 5x^2 - 17x - 12} \\
\underline{5x^2 - 20x} \\
3x - 12 \\
\underline{3x - 12} \\
0
\end{array}
$$

$\dfrac{5x^2 - 17x - 12}{x - 4} = 5x + 3$

7.
$$
\begin{array}{r}
x^2 - 3x + 1 \\
4x + 5 \overline{\smash{\big)}\ 4x^3 - 7x^2 - 11x + 5} \\
\underline{4x^3 + 5x^2} \\
-12x^2 - 11x \\
\underline{-12x^2 - 15x} \\
4x + 5 \\
\underline{4x + 5} \\
0
\end{array}
$$

$\dfrac{4x^3 - 7x^2 - 11x + 5}{4x + 5} = x^2 - 3x + 1$

8.
$$
\begin{array}{r}
2x^2 - 4x + 3 \\
3x - 2 \overline{\smash{\big)}\ 6x^3 - 16x^2 + 17x - 6} \\
\underline{6x^3 - 4x^2} \\
-12x^2 + 17x \\
\underline{-12x^2 + 8x} \\
9x - 6 \\
\underline{9x - 6} \\
0
\end{array}
$$

$\dfrac{6x^3 - 16x^2 + 17x - 6}{3x - 2} = 2x^2 - 4x + 3$

9.
$$
\begin{array}{r}
x^3 + 3x^2 - 1 \\
x + 2 \overline{\smash{\big)}\ x^4 + 5x^3 + 6x^2 - x - 2} \\
\underline{x^4 + 2x^3} \\
3x^3 + 6x^2 \\
\underline{3x^3 + 6x^2} \\
- x - 2 \\
\underline{- x - 2} \\
0
\end{array}
$$

$\dfrac{x^4 + 5x^3 + 6x^2 - x - 2}{x + 2} = x^3 + 3x^2 - 1$

10.
$$
\begin{array}{r}
x^2 + 7x + 18 \\
x - 3 \overline{\smash{\big)}\ x^3 + 4x^2 - 3x - 12} \\
\underline{x^3 - 3x^2} \\
7x^2 - 3x \\
\underline{7x^2 - 21x} \\
18x - 12 \\
\underline{18x - 54} \\
42
\end{array}
$$

$\dfrac{x^3 + 4x^2 - 3x - 12}{x - 3} = x^2 + 7x + 18 + \dfrac{42}{x - 3}$

11.

$$\begin{array}{r} 7 \\ x+2\overline{)\,7x+\ \ 3} \\ \underline{7x+14} \\ -11 \end{array}$$

$$\frac{7x+3}{x+2}=7-\frac{11}{x+2}$$

12.

$$\begin{array}{r} 4 \\ 2x+1\overline{)\,8x-5} \\ \underline{8x+4} \\ -9 \end{array}$$

$$\frac{8x-5}{2x+1}=4-\frac{9}{2x+1}$$

13.

$$\begin{array}{r} 3x+5 \\ 2x^2+0x+1\overline{)\,6x^3+10x^2+\ \ x+8} \\ \underline{6x^3+\ \ 0x^2+3x} \\ 10x^2-2x+8 \\ \underline{10x^2+0x+5} \\ -2x+3 \end{array}$$

$$\frac{6x^3+10x^2+x+8}{2x^2+1}=3x+5-\frac{2x-3}{2x^2+1}$$

14.

$$\begin{array}{r} x \\ x^2+0x+1\overline{)\,x^3+0x^2+0x-9} \\ \underline{x^3+0x^2+\ \ x} \\ -x-9 \end{array}$$

$$\frac{x^3-9}{x^2+1}=x-\frac{x+9}{x^2+1}$$

15.

$$\begin{array}{r} x^2+2x+\ \ 4 \\ x^2-2x+3\overline{)\,x^4+0x^3+3x^2+0x+\ \ 1} \\ \underline{x^4-2x^3+3x^2} \\ 2x^3+0x^2+0x \\ \underline{2x^3-4x^2+6x} \\ 4x^2-6x+\ \ 1 \\ \underline{4x^2-8x+12} \\ 2x-11 \end{array}$$
$\Rightarrow \dfrac{x^4+3x^2+1}{x^2-2x+3}=x^2+2x+4+\dfrac{2x-11}{x^2-2x+3}$

16.

$$\begin{array}{r} x^2 \\ x^3+0x^2+0x-1\overline{)\,x^5+0x^4+0x^3+0x^2+0x+7} \\ \underline{x^5+0x^4+0x^3-\ \ x^2} \\ x^2\quad\quad\ +7 \end{array}$$

$$\frac{x^5+7}{x^3-1}=x^2+\frac{x^2+7}{x^3-1}$$

17.

$$\begin{array}{r} x+3 \\ x^3-3x^2+3x-1\overline{)\,x^4+0x^3+0x^2+0x+0} \\ \underline{x^4-3x^3+3x^2-\ \ x} \\ 3x^3-3x^2+\ \ x+0 \\ \underline{3x^3-9x^2+9x-3} \\ 6x^2-8x+3 \end{array}$$

$$\frac{x^4}{(x-1)^3}=x+3+\frac{6x^2-8x+3}{(x-1)^3}$$

18.

$$\begin{array}{r} 2x \\ x^2-2x+1\overline{)\,2x^3-4x^2-15x+5} \\ \underline{2x^3-4x^2+\ \ 2x} \\ -17x+5 \end{array}$$

$$\frac{2x^3-4x^2-15x+5}{(x-1)^2}=2x-\frac{17x-5}{x^2-2x+1}$$

19. $\begin{array}{r|rrrr} 5 & 3 & -17 & 15 & -25 \\ & & 15 & -10 & 25 \\ \hline & 3 & -2 & 5 & 0 \end{array}$

$$\frac{3x^3-17x^2+15x-25}{x-5}=3x^2-2x+5$$

20.

$$
\begin{array}{r|rrrr}
-3 & 5 & 18 & 7 & -6 \\
   &   & -15 & -9 & 6 \\
\hline
   & 5 & 3 & -2 & 0
\end{array}
$$

$$\frac{5x^3 + 18x^2 + 7x - 6}{x + 3} = 5x^2 + 3x - 2$$

21.

$$
\begin{array}{r|rrrr}
-2 & 4 & 8 & -9 & -18 \\
   &   & -8 & 0 & 18 \\
\hline
   & 4 & 0 & -9 & 0
\end{array}
$$

$$\frac{4x^3 + 8x^2 - 9x - 18}{x + 2} = 4x^2 - 9$$

22.

$$
\begin{array}{r|rrrr}
2 & 9 & -18 & -16 & 32 \\
  &   & 18 & 0 & -32 \\
\hline
  & 9 & 0 & -16 & 0
\end{array}
$$

$$\frac{9x^3 - 18x^2 - 16x + 32}{x - 2} = 9x^2 - 16$$

23.

$$
\begin{array}{r|rrrr}
-10 & -1 & 0 & 75 & -250 \\
    &   & 10 & -100 & 250 \\
\hline
    & -1 & 10 & -25 & 0
\end{array}
$$

$$\frac{-x^3 + 75x - 250}{x + 10} = -x^2 + 10x - 25$$

24.

$$
\begin{array}{r|rrrr}
6 & 3 & -16 & 0 & -72 \\
  &   & 18 & 12 & 72 \\
\hline
  & 3 & 2 & 12 & 0
\end{array}
$$

$$\frac{3x^3 - 16x^2 - 72}{x - 6} = 3x^2 + 2x + 12$$

25.

$$
\begin{array}{r|rrrr}
4 & 5 & -6 & 0 & 8 \\
  &   & 20 & 56 & 224 \\
\hline
  & 5 & 14 & 56 & 232
\end{array}
$$

$$\frac{5x^3 - 6x^2 + 8}{x - 4} = 5x^2 + 14x + 56 + \frac{232}{x - 4}$$

26.

$$
\begin{array}{r|rrrr}
-2 & 5 & 0 & 6 & 8 \\
   &   & -10 & 20 & -52 \\
\hline
   & 5 & -10 & 26 & -44
\end{array}
$$

$$\frac{5x^3 + 6x + 8}{x + 2} = 5x^2 - 10x + 26 - \frac{44}{x + 2}$$

27.

$$
\begin{array}{r|rrrrr}
6 & 10 & -50 & 0 & 0 & -800 \\
  &   & 60 & 60 & 360 & 2160 \\
\hline
  & 10 & 10 & 60 & 360 & 1360
\end{array}
$$

$$\frac{10x^4 - 50x^3 - 800}{x - 6} = 10x^3 + 10x^2 + 60x + 360 + \frac{1360}{x - 6}$$

28.

$$
\begin{array}{r|rrrrrr}
-3 & 1 & -13 & 0 & 0 & -120 & 80 \\
   &   & -3 & 48 & -144 & 432 & -936 \\
\hline
   & 1 & -16 & 48 & -144 & 312 & -856
\end{array}
$$

$$\frac{x^5 - 13x^4 - 120x + 80}{x + 3} = x^4 - 16x^3 + 48x^2 - 144x + 312 - \frac{856}{x + 3}$$

29.

$$
\begin{array}{r|rrrr}
-8 & 1 & 0 & 0 & 512 \\
   &   & -8 & 64 & -512 \\
\hline
   & 1 & -8 & 64 & 0
\end{array}
$$

$$\frac{x^3 + 512}{x + 8} = x^2 - 8x + 64$$

30.

$$
\begin{array}{r|rrrr}
9 & 1 & 0 & 0 & -729 \\
  &   & 9 & 81 & 729 \\
\hline
  & 1 & 9 & 81 & 0
\end{array}
$$

$$\frac{x^3 - 729}{x - 9} = x^2 + 9x + 81$$

31.

$$
\begin{array}{r|rrrrr}
2 & -3 & 0 & 0 & 0 & 0 \\
  &   & -6 & -12 & -24 & -48 \\
\hline
  & -3 & -6 & -12 & -24 & -48
\end{array}
$$

$$\frac{-3x^4}{x - 2} = -3x^3 - 6x^2 - 12x - 24 - \frac{48}{x - 2}$$

32.

$$
\begin{array}{r|rrrrr}
-2 & -3 & 0 & 0 & 0 & 0 \\
   &   & 6 & -12 & 24 & -48 \\
\hline
   & -3 & 6 & -12 & 24 & -48
\end{array}
$$

$$\frac{-3x^4}{x + 2} = -3x^3 + 6x^2 - 12x + 24 - \frac{48}{x + 2}$$

33.

$$
\begin{array}{r|rrrrr}
6 & -1 & 0 & 0 & 180 & 0 \\
  &   & -6 & -36 & -216 & -216 \\
\hline
  & -1 & -6 & -36 & -36 & -216
\end{array}
$$

$$\frac{180x - x^4}{x - 6} = -x^3 - 6x^2 - 36x - 36 - \frac{216}{x - 6}$$

34.

$$
\begin{array}{r|rrrr}
-1 & -1 & 2 & -3 & 5 \\
   &   & 1 & -3 & 6 \\
\hline
   & -1 & 3 & -6 & 11
\end{array}
$$

$$\frac{5 - 3x + 2x^2 - x^3}{x + 1} = -x^2 + 3x - 6 + \frac{11}{x + 1}$$

35. $-\frac{1}{2}$ | 4   16 -23 -15

$$\begin{array}{r|rrrr} -\frac{1}{2} & 4 & 16 & -23 & -15 \\ & & -2 & -7 & 15 \\ \hline & 4 & 14 & -30 & 0 \end{array}$$

$$\frac{4x^3 + 16x^2 - 23x - 15}{x + (1/2)} = 4x^2 + 14x - 30$$

36.

$$\begin{array}{r|rrrr} \frac{3}{2} & 3 & -4 & 0 & 5 \\ & & \frac{9}{2} & \frac{3}{4} & \frac{9}{8} \\ \hline & 3 & \frac{1}{2} & \frac{3}{4} & \frac{49}{8} \end{array}$$

$$\frac{3x^3 - 4x^2 + 5}{x - (3/2)} = 3x^2 + \frac{1}{2}x + \frac{3}{4} + \frac{49}{8x - 12}$$

37. $f(x) = x^3 - x^2 - 14x + 11, \ k = 4$

$$\begin{array}{r|rrrr} 4 & 1 & -1 & -14 & 11 \\ & & 4 & 12 & -8 \\ \hline & 1 & 3 & -2 & 3 \end{array}$$

$$f(x) = (x - 4)(x^2 + 3x - 2) + 3$$
$$f(4) = 4^3 - 4^2 - 14(4) + 11 = 3$$

38. $f(x) = x^3 - 5x^2 - 11x + 8, \ k = -2$

$$\begin{array}{r|rrrr} -2 & 1 & -5 & -11 & 8 \\ & & -2 & 14 & -6 \\ \hline & 1 & -7 & 3 & 2 \end{array}$$

$$f(x) = (x + 2)(x^2 - 7x + 3) + 2$$
$$f(-2) = (-2)^3 - 5(-2)^2 - 11(-2) + 8$$
$$= -8 - 20 + 22 + 8 = 2$$

39. $f(x) = 15x^4 + 10x^3 - 6x^2 + 14, \ k = -\frac{2}{3}$

$$\begin{array}{r|rrrrr} -\frac{2}{3} & 15 & 10 & -6 & 0 & 14 \\ & & -10 & 0 & 4 & -\frac{8}{3} \\ \hline & 15 & 0 & -6 & 4 & \frac{34}{3} \end{array}$$

$$f(x) = \left(x + \frac{2}{3}\right)(15x^3 - 6x + 4) + \frac{34}{3}$$
$$f\left(-\frac{2}{3}\right) = 15\left(-\frac{2}{3}\right)^4 + 10\left(-\frac{2}{3}\right)^3 - 6\left(-\frac{2}{3}\right)^2 + 14 = \frac{34}{3}$$

40. $f(x) = 10x^3 - 22x^2 - 3x + 4, \ k = \frac{1}{5}$

$$\begin{array}{r|rrrr} \frac{1}{5} & 10 & -22 & -3 & 4 \\ & & 2 & -4 & -\frac{7}{5} \\ \hline & 10 & -20 & -7 & \frac{13}{5} \end{array}$$

$$f(x) = \left(x - \frac{1}{5}\right)(10x^2 - 20x - 7) + \frac{13}{5}$$
$$f\left(\frac{1}{5}\right) = 10\left(\frac{1}{5}\right)^3 - 22\left(\frac{1}{5}\right)^2 - 3\left(\frac{1}{5}\right) + 4$$
$$= \frac{2}{25} - \frac{22}{25} - \frac{3}{5} + 4 = \frac{65}{25} = \frac{13}{5}$$

41. $f(x) = x^3 + 3x^2 - 2x - 14, \ k = \sqrt{2}$

$$\begin{array}{r|rrrr} \sqrt{2} & 1 & 3 & -2 & -14 \\ & & \sqrt{2} & 2 + 3\sqrt{2} & 6 \\ \hline & 1 & 3 + \sqrt{2} & 3\sqrt{2} & -8 \end{array}$$

$$f(x) = (x - \sqrt{2})\left[x^2 + (3 + \sqrt{2})x + 3\sqrt{2}\right] - 8$$
$$f(\sqrt{2}) = (\sqrt{2})^3 + 3(\sqrt{2})^2 - 2\sqrt{2} - 14 = -8$$

42. $f(x) = x^3 + 2x^2 - 5x - 4, \ k = -\sqrt{5}$

$$\begin{array}{r|rrrr} -\sqrt{5} & 1 & 2 & -5 & -4 \\ & & -\sqrt{5} & -2\sqrt{5} + 5 & 10 \\ \hline & 1 & 2 - \sqrt{5} & -2\sqrt{5} & 6 \end{array}$$

$$f(x) = (x + \sqrt{5})\left[x^2 + (2 - \sqrt{5})x - 2\sqrt{5}\right] + 6$$
$$f(-\sqrt{5}) = (-\sqrt{5})^3 + 2(-\sqrt{5})^2 - 5(-\sqrt{5}) - 4$$
$$= -5\sqrt{5} + 10 + 5\sqrt{5} - 4 = 6$$

43. $f(x) = -4x^3 + 6x^2 + 12x + 4, \ k = 1 - \sqrt{3}$

$$\begin{array}{r|rrrr} 1 - \sqrt{3} & -4 & 6 & 12 & 4 \\ & & -4 + 4\sqrt{3} & -10 + 2\sqrt{3} & -4 \\ \hline & -4 & 2 + 4\sqrt{3} & 2 + 2\sqrt{3} & 0 \end{array}$$

$$f(x) = \left[x - (1 - \sqrt{3})\right]\left[-4x^2 + (2 + 4\sqrt{3})x + (2 + 2\sqrt{3})\right] + 0$$
$$f(1 - \sqrt{3}) = -4(1 - \sqrt{3})^3 + 6(1 - \sqrt{3})^2 + 12(1 - \sqrt{3}) + 4 = 0$$

44. $f(x) = -3x^3 + 8x^2 + 10x - 8, \ k = 2 + \sqrt{2}$

$$\begin{array}{r|rrrr} 2 + \sqrt{2} & -3 & 8 & 10 & -8 \\ & & -6 - 3\sqrt{2} & -2 - 4\sqrt{2} & 8 \\ \hline & -3 & 2 - 3\sqrt{2} & 8 - 4\sqrt{2} & 0 \end{array}$$

$$f(x) = (x - 2 - \sqrt{2})\left[-3x^2 + (2 - 3\sqrt{2})x + 8 - 4\sqrt{2}\right] + 0$$
$$f(2 + \sqrt{2}) = -3(2 + \sqrt{2})^3 + 8(2 + \sqrt{2})^2 + 10(2 + \sqrt{2}) - 8$$
$$= -3(20 + 14\sqrt{2}) + 8(6 + 4\sqrt{2}) + 10(2 + \sqrt{2}) - 8$$
$$= -60 - 42\sqrt{2} + 48 + 32\sqrt{2} + 20 + 10\sqrt{2} - 8$$
$$= 0$$

45. $f(x) = 4x^3 - 13x + 10$

(a)
$$
\begin{array}{r|rrrr}
1 & 4 & 0 & -13 & 10 \\
  &   & 4 & 4 & -9 \\
\hline
  & 4 & 4 & -9 & 1
\end{array}
$$

$f(1) = 1$

(b)
$$
\begin{array}{r|rrrr}
-2 & 4 & 0 & -13 & 10 \\
   &   & -8 & 16 & -6 \\
\hline
   & 4 & -8 & 3 & 4
\end{array}
$$

$f(-2) = 4$

(c)
$$
\begin{array}{r|rrrr}
\frac{1}{2} & 4 & 0 & -13 & 10 \\
   &   & 2 & 1 & -6 \\
\hline
   & 4 & 2 & -12 & 4
\end{array}
$$

$f\left(\frac{1}{2}\right) = 4$

(d)
$$
\begin{array}{r|rrrr}
8 & 4 & 0 & -13 & 10 \\
  &   & 32 & 256 & 1944 \\
\hline
  & 4 & 32 & 243 & 1954
\end{array}
$$

$f(8) = 1954$

46. $g(x) = x^6 - 4x^4 + 3x^2 + 2$

(a)
$$
\begin{array}{r|rrrrrrr}
2 & 1 & 0 & -4 & 0 & 3 & 0 & 2 \\
  &   & 2 & 4 & 0 & 0 & 6 & 12 \\
\hline
  & 1 & 2 & 0 & 0 & 3 & 6 & 14
\end{array}
$$

$g(2) = 14$

(b)
$$
\begin{array}{r|rrrrrrr}
-4 & 1 & 0 & -4 & 0 & 3 & 0 & 2 \\
   &   & -4 & 16 & -48 & 192 & -780 & 3120 \\
\hline
   & 1 & -4 & 12 & -48 & 195 & -780 & 3122
\end{array}
$$

$g(-4) = 3122$

(c)
$$
\begin{array}{r|rrrrrrr}
3 & 1 & 0 & -4 & 0 & 3 & 0 & 2 \\
  &   & 3 & 9 & 15 & 45 & 144 & 432 \\
\hline
  & 1 & 3 & 5 & 15 & 48 & 144 & 434
\end{array}
$$

$g(3) = 434$

(d)
$$
\begin{array}{r|rrrrrrr}
-1 & 1 & 0 & -4 & 0 & 3 & 0 & 2 \\
   &   & -1 & 1 & 3 & -3 & 0 & 0 \\
\hline
   & 1 & -1 & -3 & 3 & 0 & 0 & 2
\end{array}
$$

$g(-1) = 2$

47. $h(x) = 3x^3 + 5x^2 - 10x + 1$

(a)
$$
\begin{array}{r|rrrr}
3 & 3 & 5 & -10 & 1 \\
  &   & 9 & 42 & 96 \\
\hline
  & 3 & 14 & 32 & 97
\end{array}
$$

$h(3) = 97$

(b)
$$
\begin{array}{r|rrrr}
\frac{1}{3} & 3 & 5 & -10 & 1 \\
   &   & 1 & 2 & -\frac{8}{3} \\
\hline
   & 3 & 6 & -8 & -\frac{5}{3}
\end{array}
$$

$h\left(\frac{1}{3}\right) = -\frac{5}{3}$

(c)
$$
\begin{array}{r|rrrr}
-2 & 3 & 5 & -10 & 1 \\
   &   & -6 & 2 & 16 \\
\hline
   & 3 & -1 & -8 & 17
\end{array}
$$

$h(-2) = 17$

(d)
$$
\begin{array}{r|rrrr}
-5 & 3 & 5 & -10 & 1 \\
   &   & -15 & 50 & -200 \\
\hline
   & 3 & -10 & 40 & -199
\end{array}
$$

$h(-5) = -199$

48. $f(x) = 0.4x^4 - 1.6x^3 + 0.7x^2 - 2$

(a)
$$
\begin{array}{r|rrrrr}
1 & 0.4 & -1.6 & 0.7 & 0 & -2 \\
  &   & 0.4 & -1.2 & -0.5 & -0.5 \\
\hline
  & 0.4 & -1.2 & -0.5 & -0.5 & -2.5
\end{array}
$$

$f(1) = -2.5$

(b)
$$
\begin{array}{r|rrrrr}
-2 & 0.4 & -1.6 & 0.7 & 0 & -2 \\
   &   & -0.8 & 4.8 & -11 & 22 \\
\hline
   & 0.4 & -2.4 & 5.5 & -11 & 20
\end{array}
$$

$f(-2) = 20$

(c)
$$
\begin{array}{r|rrrrr}
5 & 0.4 & -1.6 & 0.7 & 0 & -2 \\
  &   & 2.0 & 2.0 & 13.5 & 67.5 \\
\hline
  & 0.4 & 0.4 & 2.7 & 13.5 & 65.5
\end{array}
$$

$f(5) = 65.5$

(d)
$$
\begin{array}{r|rrrrr}
-10 & 0.4 & -1.6 & 0.7 & 0 & -2 \\
    &   & -4.0 & 56.0 & -567 & 5670 \\
\hline
    & 0.4 & -5.6 & 56.7 & -567 & 5668
\end{array}
$$

$f(-10) = 5668$

49.
$$
\begin{array}{r|rrrr}
2 & 1 & 0 & -7 & 6 \\
  &   & 2 & 4 & -6 \\
\hline
  & 1 & 2 & -3 & 0
\end{array}
$$

$$
\begin{aligned}
x^3 - 7x + 6 &= (x - 2)(x^2 + 2x - 3) \\
&= (x - 2)(x + 3)(x - 1)
\end{aligned}
$$

Zeros: $2, -3, 1$

50.
$$
\begin{array}{r|rrrr}
-4 & 1 & 0 & -28 & -48 \\
   &   & -4 & 16 & 48 \\
\hline
   & 1 & -4 & -12 & 0
\end{array}
$$

$$
\begin{aligned}
x^3 - 28x - 48 &= (x + 4)(x^2 - 4x - 12) \\
&= (x + 4)(x - 6)(x + 2)
\end{aligned}
$$

Zeros: $-4, -2, 6$

51. $\frac{1}{2}$ | 2   −15   27   −10
                    1   −7   10
                 2 −14   20   0

$2x^3 - 15x^2 + 27x - 10$
$= \left(x - \frac{1}{2}\right)\left(2x^2 - 14x + 20\right)$
$= (2x - 1)(x - 2)(x - 5)$

Zeros: $\frac{1}{2}, 2, 5$

52. $\frac{2}{3}$ | 48 −80   41   −6
                    32 −32   6
                 48 −48   9   0

$48x^3 - 80x^2 + 41x - 6 = \left(x - \frac{2}{3}\right)\left(48x^2 - 48x + 9\right)$
$= \left(x - \frac{2}{3}\right)(4x - 3)(12x - 3)$
$= (3x - 2)(4x - 3)(4x - 1)$

Zeros: $\frac{2}{3}, \frac{3}{4}, \frac{1}{4}$

53. $\sqrt{3}$ | 1   2   −3.   −6
                     $\sqrt{3}$ $3 + 2\sqrt{3}$ 6
                  1 $2 + \sqrt{3}$ $2\sqrt{3}$ 0

$-\sqrt{3}$ | 1 $2 + \sqrt{3}$ $2\sqrt{3}$
                     $-\sqrt{3}$ $-2\sqrt{3}$
                  1   2   0

$x^3 + 2x^2 - 3x - 6 = \left(x - \sqrt{3}\right)\left(x + \sqrt{3}\right)(x + 2)$

Zeros: $\pm\sqrt{3}, -2$

54. $\sqrt{2}$ | 1   2   −2   −4
                     $\sqrt{2}$ $2\sqrt{2} + 2$ 4
                  1 $2 + \sqrt{2}$ $2\sqrt{2}$ 0

$-\sqrt{2}$ | 1 $2 + \sqrt{2}$ $2\sqrt{2}$
                     $-\sqrt{2}$ $-2\sqrt{2}$
                  1   2   0

$x^3 + 2x^2 - 2x - 4 = \left(x - \sqrt{2}\right)(x + 2)\left(x + \sqrt{2}\right)$

Zeros: $-2, -\sqrt{2}, \sqrt{2}$

55. $1 + \sqrt{3}$ | 1   −3   0   2
                          $1 + \sqrt{3}$ $1 - \sqrt{3}$ −2
                       1 $-2 + \sqrt{3}$ $1 - \sqrt{3}$ 0

$1 - \sqrt{3}$ | 1 $-2 + \sqrt{3}$ $1 - \sqrt{3}$
                          $1 - \sqrt{3}$ $-1 + \sqrt{3}$
                       1   −1   0

$x^3 - 3x^2 + 2 = \left[x - \left(1 + \sqrt{3}\right)\right]\left[x - \left(1 - \sqrt{3}\right)\right](x - 1)$
$= (x - 1)\left(x - 1 - \sqrt{3}\right)\left(x - 1 + \sqrt{3}\right)$

Zeros: $1, 1 \pm \sqrt{3}$

56. $2 - \sqrt{5}$ | 1   −1   −13   −3
                          $2 - \sqrt{5}$ $7 - 3\sqrt{5}$ 3
                       1 $1 - \sqrt{5}$ $-6 - 3\sqrt{5}$ 0

$2 + \sqrt{5}$ | 1 $1 - \sqrt{5}$ $-6 - 3\sqrt{5}$
                          $2 + \sqrt{5}$ $6 + 3\sqrt{5}$
                       1   3   0

$x^3 - x^2 - 13x - 3 = \left(x - 2 + \sqrt{5}\right)\left(x - 2 - \sqrt{5}\right)(x + 3)$

Zeros: $2 - \sqrt{5}, 2 + \sqrt{5}, -3$

57. $f(x) = 2x^3 + x^2 - 5x + 2$; Factors: $(x + 2), (x - 1)$

(a) -2 | 2   1   −5   2
                  −4   6   −2
               2   −3   1   0

    1 | 2   −3   1
              2   −1
           2 −1   0

Both are factors of $f(x)$ since the remainders are zero.

(b) The remaining factor of $f(x)$ is $(2x - 1)$.

(c) $f(x) = (2x - 1)(x + 2)(x - 1)$

(d) Zeros: $\frac{1}{2}, -2, 1$

(e)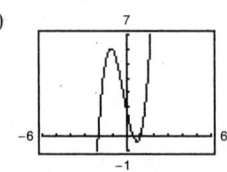

58. $f(x) = 3x^3 + 2x^2 - 19x + 6$; Factors: $(x + 3), (x - 2)$

(a)
$$\begin{array}{r|rrrr} -3 & 3 & 2 & -19 & 6 \\ & & -9 & 21 & -6 \\ \hline & 3 & -7 & 2 & 0 \end{array}$$

$$\begin{array}{r|rrr} 2 & 3 & -7 & 2 \\ & & 6 & -2 \\ \hline & 3 & -1 & 0 \end{array}$$

(b) The remaining factor is $(3x - 1)$.

(c) $f(x) = 3x^3 + 2x^2 - 19x + 6$

$= (3x - 1)(x + 3)(x - 2)$

(d) Zeros: $\frac{1}{3}, -3, 2$

(e)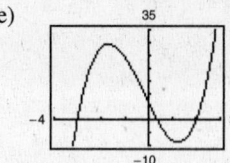

59. $f(x) = x^4 - 4x^3 - 15x^2 + 58x - 40$; Factors: $(x - 5), (x + 4)$

(a)
$$\begin{array}{r|rrrrr} 5 & 1 & -4 & -15 & 58 & -40 \\ & & 5 & 5 & -50 & 40 \\ \hline & 1 & 1 & -10 & 8 & 0 \end{array}$$

$$\begin{array}{r|rrrr} -4 & 1 & 1 & -10 & 8 \\ & & -4 & 12 & -8 \\ \hline & 1 & -3 & 2 & 0 \end{array}$$

Both are factors of $f(x)$ since the remainders are zero.

(b) $x^2 - 3x + 2 = (x - 1)(x - 2)$

The remaining factors are $(x - 1)$ and $(x - 2)$.

(c) $f(x) = (x - 1)(x - 2)(x - 5)(x + 4)$

(d) Zeros: $1, 2, 5, -4$

(e)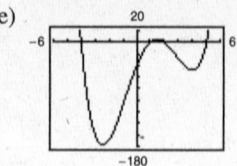

60. $f(x) = 8x^4 - 14x^3 - 71x^2 - 10x + 24$; Factors: $(x + 2), (x - 4)$

(a)
$$\begin{array}{r|rrrrr} -2 & 8 & -14 & -71 & -10 & 24 \\ & & -16 & 60 & 22 & -24 \\ \hline & 8 & -30 & -11 & 12 & 0 \end{array}$$

$$\begin{array}{r|rrrr} 4 & 8 & -30 & -11 & 12 \\ & & 32 & 8 & -12 \\ \hline & 8 & 2 & -3 & 0 \end{array}$$

(b) $8x^2 + 2x - 3 = (4x + 3)(2x - 1)$

The remaining factors are $(4x + 3)$ and $(2x - 1)$.

(c) $f(x) = (4x + 3)(2x - 1)(x + 2)(x - 4)$

(d) Zeros: $-\frac{3}{4}, \frac{1}{2}, -2, 4$

(e)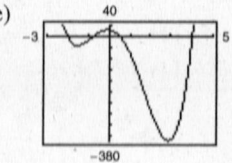

61. $f(x) = 6x^3 + 41x^2 - 9x - 14$; Factors: $(2x + 1), (3x - 2)$

(a)
$$\begin{array}{r|rrrr} -\frac{1}{2} & 6 & 41 & -9 & -14 \\ & & -3 & -19 & 14 \\ \hline & 6 & 38 & -28 & 0 \end{array}$$

$$\begin{array}{r|rrr} \frac{2}{3} & 6 & 38 & -28 \\ & & 4 & 28 \\ \hline & 6 & 42 & 0 \end{array}$$

Both are factors since the remainders are zero.

(c) $f(x) = (x + 7)(2x + 1)(3x - 2)$

(b) $6x + 42 = 6(x + 7)$

This shows that $\dfrac{f(x)}{\left(x + \frac{1}{2}\right)\left(x - \frac{2}{3}\right)} = 6(x + 7)$,

so $\dfrac{f(x)}{(2x + 1)(3x - 2)} = x + 7$.

The remaining factor is $(x + 7)$.

(d) Zeros: $-7, -\dfrac{1}{2}, \dfrac{2}{3}$

(e)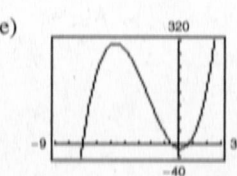

62. $f(x) = 10x^3 - 11x^2 - 72x + 45$;

Factors: $(2x + 5), (5x - 3)$

(a) $-\frac{5}{2}$ | $\begin{array}{rrrr} 10 & -11 & -72 & 45 \\ & -25 & 90 & -45 \\ \hline 10 & -36 & 18 & 0 \end{array}$

$\frac{3}{5}$ | $\begin{array}{rrrr} 10 & -36 & 18 & \\ & 6 & -18 & \\ \hline 10 & -30 & 0 \end{array}$

(b) $10x - 30 = 10(x - 3)$

This shows that $\dfrac{f(x)}{\left(x + \frac{5}{2}\right)\left(x - \frac{3}{5}\right)} = 10(x - 3)$,

so $\dfrac{f(x)}{(2x + 5)(5x - 3)} = x - 3$.

The remaining factor is $(x - 3)$.

(c) $f(x) = (x - 3)(2x + 5)(5x - 3)$

(d) Zeros: $3, -\dfrac{5}{2}, \dfrac{3}{5}$

(e)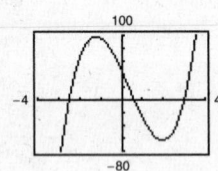

63. $f(x) = 2x^3 - x^2 - 10x + 5$;

Factors: $(2x - 1), \left(x + \sqrt{5}\right)$

(a) $\frac{1}{2}$ | $\begin{array}{rrrr} 2 & -1 & -10 & 5 \\ & 1 & 0 & -5 \\ \hline 2 & 0 & -10 & 0 \end{array}$

$-\sqrt{5}$ | $\begin{array}{rrr} 2 & 0 & -10 \\ & -2\sqrt{5} & 10 \\ \hline 2 & -2\sqrt{5} & 0 \end{array}$

Both are factors since the remainders are zero.

(b) $2x - 2\sqrt{5} = 2\left(x - \sqrt{5}\right)$

This shows that $\dfrac{f(x)}{\left(x - \frac{1}{2}\right)\left(x + \sqrt{5}\right)} = 2\left(x - \sqrt{5}\right)$,

so $\dfrac{f(x)}{(2x - 1)\left(x + \sqrt{5}\right)} = x - \sqrt{5}$.

The remaining factor is $\left(x - \sqrt{5}\right)$.

(c) $f(x) = \left(x + \sqrt{5}\right)\left(x - \sqrt{5}\right)(2x - 1)$

(d) Zeros: $-\sqrt{5}, \sqrt{5}, \dfrac{1}{2}$

(e)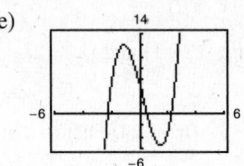

64. $f(x) = x^3 + 3x^2 - 48x - 144$; Factors: $\left(x + 4\sqrt{3}\right), (x + 3)$

(a) -3 | $\begin{array}{rrrr} 1 & 3 & -48 & -144 \\ & -3 & 0 & 144 \\ \hline 1 & 0 & -48 & 0 \end{array}$

$-4\sqrt{3}$ | $\begin{array}{rrr} 1 & 0 & -48 \\ & -4\sqrt{3} & 48 \\ \hline 1 & -4\sqrt{3} & 0 \end{array}$

(b) The remaining factor is $\left(x - 4\sqrt{3}\right)$.

(c) $f(x) = \left(x - 4\sqrt{3}\right)\left(x + 4\sqrt{3}\right)(x + 3)$

(d) Zeros: $\pm 4\sqrt{3}, -3$

(e)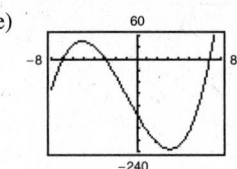

65. $f(x) = x^3 - 2x^2 - 5x + 10$

(a) The zeros of f are 2 and $\approx \pm 2.236$.

(b) An exact zero is $x = 2$.

(c) 2 | $\begin{array}{rrrr} 1 & -2 & -5 & 10 \\ & 2 & 0 & -10 \\ \hline 1 & 0 & -5 & 0 \end{array}$

$f(x) = (x - 2)(x^2 - 5)$

$= (x - 2)\left(x - \sqrt{5}\right)\left(x + \sqrt{5}\right)$

66. $g(x) = x^3 - 4x^2 - 2x + 8$

(a) The zeros of g are $x = 4, x \approx -1.414, x \approx 1.414$.

(b) $x = 4$ is an exact zero.

(c) 4 | $\begin{array}{rrrr} 1 & -4 & -2 & 8 \\ & 4 & 0 & -8 \\ \hline 1 & 0 & -2 & 0 \end{array}$

$f(x) = (x - 4)(x^2 - 2)$

$= (x - 4)\left(x - \sqrt{2}\right)\left(x + \sqrt{2}\right)$

67. $h(t) = t^3 - 2t^2 - 7t + 2$

   (a) The zeros of h are $t = -2, t \approx 3.732, t \approx 0.268$.

   (b) An exact zero is $t = -2$.

   (c) -2 | $\begin{array}{rrrr} 1 & -2 & -7 & 2 \\ & -2 & 8 & -2 \\ \hline 1 & -4 & 1 & 0 \end{array}$

   $h(t) = (t + 2)(t^2 - 4t + 1)$

   By the Quadratic Formula, the zeros of $t^2 - 4t + 1$ are $2 \pm \sqrt{3}$. Thus,

$$h(t) = (t + 2)\left[t - \left(2 + \sqrt{3}\right)\right]\left[t - \left(2 - \sqrt{3}\right)\right]$$
$$= (t + 2)\left(t - 2 - \sqrt{3}\right)\left(t - 2 + \sqrt{3}\right).$$

68. $f(s) = s^3 - 12s^2 + 40s - 24$

   (a) The zeros of f are $s = 6, s \approx 0.764, s \approx 5.236$

   (b) $s = 6$ is an exact zero.

   (c) 6 | $\begin{array}{rrrr} 1 & -12 & 40 & -24 \\ & 6 & -36 & 24 \\ \hline 1 & -6 & 4 & 0 \end{array}$

   $f(s) = (s - 6)(s^2 - 6s + 4)$
$$= (s - 6)\left[s - \left(3 + \sqrt{5}\right)\right]\left[s - \left(3 - \sqrt{5}\right)\right]$$

69. $\dfrac{4x^3 - 8x^2 + x + 3}{2x - 3}$

   $\frac{3}{2}$ | $\begin{array}{rrrr} 4 & -8 & 1 & 3 \\ & 6 & -3 & -3 \\ \hline 4 & -2 & -2 & 0 \end{array}$

$$\frac{4x^3 - 8x^2 + x + 3}{x - \frac{3}{2}} = 4x^2 - 2x - 2 = 2(2x^2 - x - 1)$$

Thus, $\dfrac{4x^3 - 8x^2 + x + 3}{2x - 3} = 2x^2 - x - 1, x \neq \dfrac{3}{2}$.

70. $\dfrac{x^3 + x^2 - 64x - 64}{x + 8}$

   -8 | $\begin{array}{rrrr} 1 & 1 & -64 & -64 \\ & -8 & 56 & 64 \\ \hline 1 & -7 & -8 & 0 \end{array}$

$$\frac{x^3 + x^2 - 64x - 64}{x + 8} = x^2 - 7x - 8, x \neq -8$$

71. $\dfrac{x^4 + 6x^3 + 11x^2 + 6x}{x^2 + 3x + 2} = \dfrac{x^4 + 6x^3 + 11x^2 + 6x}{(x + 1)(x + 2)}$

   -1 | $\begin{array}{rrrrr} 1 & 6 & 11 & 6 & 0 \\ & -1 & -5 & -6 & 0 \\ \hline 1 & 5 & 6 & 0 & 0 \end{array}$

   -2 | $\begin{array}{rrrr} 1 & 5 & 6 & 0 \\ & -2 & -6 & 0 \\ \hline 1 & 3 & 0 & 0 \end{array}$

$$\frac{x^4 + 6x^3 + 11x^2 + 6x}{(x + 1)(x + 2)} = x^2 + 3x, x \neq -2, -1$$

72. $\dfrac{x^4 + 9x^3 - 5x^2 - 36x + 4}{x^2 - 4} = \dfrac{x^4 + 9x^3 - 5x^2 - 36x + 4}{(x + 2)(x - 2)}$

   2 | $\begin{array}{rrrrr} 1 & 9 & -5 & -36 & 4 \\ & 2 & 22 & 34 & -4 \\ \hline 1 & 11 & 17 & -2 & 0 \end{array}$

   -2 | $\begin{array}{rrrr} 1 & 11 & 17 & -2 \\ & -2 & -18 & 2 \\ \hline 1 & 9 & -1 & 0 \end{array}$

$$\frac{x^4 + 9x^3 - 5x^2 - 36x + 4}{x^2 - 4} = x^2 + 9x - 1, x \neq \pm 2$$

73. (a) and (b)

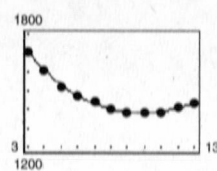

—CONTINUED—

73. —CONTINUED—

(c) $M \approx -0.242t^3 + 12.43t^2 - 173.4t + 2118$

Year, t	Military Personnel	M
3	1705	1703
4	1611	1608
5	1518	1532
6	1472	1473
7	1439	1430
8	1407	1402
9	1386	1388
10	1384	1385
11	1385	1393
12	1412	1409
13	1434	1433

The model is a good fit to the actual data.

(d)
$$
\begin{array}{r|rrrr}
18 & -0.242 & 12.43 & -173.4 & 2118 \\
   &   & -4.356 & 145.332 & -505.224 \\
\hline
   & -0.242 & 8.074 & -28.068 & 1612.776
\end{array}
$$

$M(18) \approx 1613$ thousand

No, this model should not be used to predict the number of military personnel in the future. It predicts an increase in military personnel until 2024 and then it decreases and will approach negative infinity quickly.

74. (a) and (b)

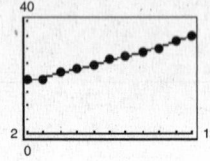

(b) $R = 0.0026t^3 - 0.0292t^2 + 1.558t + 15.632$

(c)
$$
\begin{array}{r|rrrr}
18 & 0.0026 & -0.0292 & 1.558 & 15.632 \\
   &   & 0.0468 & 0.3168 & 33.7464 \\
\hline
   & 0.0026 & 0.0176 & 1.8748 & 49.3784
\end{array}
$$

For the year 2008, the model predicts a monthly rate of about $49.38.

75. False. If $(7x + 4)$ is a factor of f, then $-\frac{4}{7}$ is a zero of f.

76. True.

$$
\begin{array}{r|rrrrrrr}
\frac{1}{2} & 6 & 1 & -92 & 45 & 184 & 4 & -48 \\
            &   & 3 & 2   & -45 & 0 & 92 & 48 \\
\hline
            & 6 & 4 & -90 & 0 & 184 & 96 & 0
\end{array}
$$

$f(x) = (2x - 1)(x + 1)(x - 2)(x - 3)(3x + 2)(x + 4)$

77. True. The degree of the numerator is greater than the degree of the denominator.

78. $f(x) = (x - k)q(x) + r$

(a) $k = 2$, $r = 5$, $q(x) =$ any quadratic $ax^2 + bx + c$ where $a > 0$. One example:

$f(x) = (x - 2)x^2 + 5 = x^3 - 2x^2 + 5$

(b) $k = -3$, $r = 1$, $q(x) =$ any quadratic $ax^2 + bx + c$ where $a < 0$. One example:

$f(x) = (x + 3)(-x^2) + 1 = -x^3 - 3x^2 + 1$

79.
$$
\begin{array}{r}
x^{2n} + 6x^n + 9 \\
x^n + 3 \overline{)\ x^{3n} + 9x^{2n} + 27x^n + 27} \\
\underline{x^{3n} + 3x^{2n}} \\
6x^{2n} + 27x^n \\
\underline{6x^{2n} + 18x^n} \\
9x^n + 27 \\
\underline{9x^n + 27} \\
0
\end{array}
$$

$$\frac{x^{3n} + 9x^{2n} + 27x^n + 27}{x^n + 3} = x^{2n} + 6x^n + 9$$

80.
$$
\begin{array}{r}
x^{2n} - x^n + 3 \\
x^n - 2 \overline{)\ x^{3n} - 3x^{2n} + 5x^n - 6} \\
\underline{x^{3n} - 2x^{2n}} \\
-x^{2n} + 5x^n \\
\underline{-x^{2n} + 2x^n} \\
3x^n - 6 \\
\underline{3x^n - 6} \\
0
\end{array}
$$

$$\frac{x^{3n} - 3x^{2n} + 5x^n - 6}{x^n - 2} = x^{2n} - x^n + 3$$

81. A divisor divides evenly into a dividend if the remainder is zero.

82. You can check polynomial division by multiplying the quotient by the divisor. This should yield the original dividend if the multiplication was performed correctly.

83.
$$
\begin{array}{r|rrrr}
5 & 1 & 4 & -3 & c \\
  &   & 5 & 45 & 210 \\
\hline
  & 1 & 9 & 42 & c + 210
\end{array}
$$

To divide evenly, $c + 210$ must equal zero. Thus, c must equal -210.

84.
$$
\begin{array}{r|rrrrrr}
-2 & 1 & 0 & 0 & -2 & 1 & c \\
   &   & -2 & 4 & -8 & 20 & -42 \\
\hline
   & 1 & -2 & 4 & -10 & 21 & c - 42
\end{array}
$$

To divide evenly, $c - 42$ must equal zero. Thus, c must equal 42.

85. $f(x) = (x + 3)^2(x - 3)(x + 1)^3$

The remainder when $k = -3$ is zero since $(x + 3)$ is a factor of $f(x)$.

86. In this case it is easier to evaluate $f(2)$ directly because $f(x)$ is in factored form. To evaluate using synthetic division you would have to expand each factor and then multiply it all out.

87. $9x^2 - 25 = 0$

$(3x - 5)(3x + 5) = 0$

$3x - 5 = 0 \implies x = \dfrac{5}{3}$

$3x + 5 = 0 \implies x = -\dfrac{5}{3}$

88. $16x^2 - 21 = 0$

$16x^2 = 21$

$x^2 = \dfrac{21}{16}$

$x = \pm\sqrt{\dfrac{21}{16}}$

$x = \pm\dfrac{\sqrt{21}}{4}$

89. $5x^2 - 3x - 14 = 0$

$(5x + 7)(x - 2) = 0$

$5x + 7 = 0 \implies x = -\dfrac{7}{5}$

$x - 2 = 0 \implies x = 2$

90. $8x^2 - 22x + 15 = 0$

$(4x - 5)(2x - 3) = 0$

$4x - 5 = 0$ or $2x - 3 = 0$

$x = \tfrac{5}{4}$ or $x = \tfrac{3}{2}$

91. $2x^2 + 6x + 3 = 0$

$$x = \frac{-b \pm \sqrt{b^2 - 4ac}}{2a} = \frac{-6 \pm \sqrt{6^2 - 4(2)(3)}}{2(2)} = \frac{-6 \pm \sqrt{12}}{4}$$

$$= \frac{-3 \pm \sqrt{3}}{2}$$

92. $x^2 + 3x - 3 = 0$

$$x = \frac{-3 \pm \sqrt{3^2 - 4(1)(-3)}}{2(1)} = \frac{-3 \pm \sqrt{21}}{2}$$

93. $f(x) = (x - 0)(x - 3)(x - 4)$

$\qquad = x(x - 3)(x - 4) = x(x^2 - 7x + 12)$

$\qquad = x^3 - 7x^2 + 12x$

Note: Any nonzero scalar multiple of $f(x)$ would also have these zeros.

94. $f(x) = (x - (-6))(x - 1)$

$\qquad = (x + 6)(x - 1)$

$\qquad = x^2 + 5x - 6$

Note: Any nonzero scalar multiple of $f(x)$ would also have these zeros.

95. $f(x) = [x - (-3)]\left[x - \left(1 + \sqrt{2}\right)\right]\left[x - \left(1 - \sqrt{2}\right)\right]$

$\qquad = (x + 3)\left[(x - 1) - \sqrt{2}\right]\left[(x - 1) + \sqrt{2}\right]$

$\qquad = (x + 3)\left[(x - 1)^2 - \left(\sqrt{2}\right)^2\right]$

$\qquad = (x + 3)(x^2 - 2x - 1)$

$\qquad = x^3 + x^2 - 7x - 3$

Note: Any nonzero scalar multiple of $f(x)$ would also have these zeros.

96. $f(x) = (x - 1)\left[x - (-2)\right]\left[x - \left(2 + \sqrt{3}\right)\right]\left[x - \left(2 - \sqrt{3}\right)\right]$

$\qquad = (x - 1)(x + 2)\left[(x - 2) - \sqrt{3}\right]\left[(x - 2) + \sqrt{3}\right]$

$\qquad = (x^2 + x - 2)\left[(x - 2)^2 - \left(\sqrt{3}\right)^2\right]$

$\qquad = (x^2 + x - 2)(x^2 - 4x + 1)$

$\qquad = x^4 - 3x^3 - 5x^2 + 9x - 2$

Note: Any nonzero scalar multiple of $f(x)$ would also have these zeros.

Section 2.4　Complex Numbers

■ Standard form: $a + bi$.

If $b = 0$, then $a + bi$ is a real number.

If $a = 0$ and $b \neq 0$, then $a + bi$ is a pure imaginary number.

■ Equality of Complex Numbers: $a + bi = c + di$ if and only if $a = c$ and $b = d$

■ Operations on complex numbers

(a) Addition: $(a + bi) + (c + di) = (a + c) + (b + d)i$

(b) Subtraction: $(a + bi) - (c + di) = (a - c) + (b - d)i$

(c) Multiplication: $(a + bi)(c + di) = (ac - bd) + (ad + bc)i$

(d) Division: $\dfrac{a + bi}{c + di} = \dfrac{a + bi}{c + di} \cdot \dfrac{c - di}{c - di} = \dfrac{ac + bd}{c^2 + d^2} + \dfrac{bc - ad}{c^2 + d^2}i$

■ The complex conjugate of $a + bi$ is $a - bi$:

$\qquad (a + bi)(a - bi) = a^2 + b^2$

■ The additive inverse of $a + bi$ is $-a - bi$.

■ $\sqrt{-a} = \sqrt{a}\,i$ for $a > 0$.

Vocabulary Check

1. (a) iii　　(b) i　　(c) ii

2. $\sqrt{-1}$; -1

3. principal square

4. complex conjugates

1. $a + bi = -10 + 6i$

$a = -10$

$b = 6$

2. $a + bi = 13 + 4i$

$a = 13$

$b = 4$

3. $(a - 1) + (b + 3)i = 5 + 8i$

$a - 1 = 5 \implies a = 6$

$b + 3 = 8 \implies b = 5$

4. $(a + 6) + 2bi = 6 - 5i$

$2b = -5$

$b = -\frac{5}{2}$

$a + 6 = 6$

$a = 0$

5. $4 + \sqrt{-9} = 4 + 3i$

6. $3 + \sqrt{-16} = 3 + 4i$

7. $2 - \sqrt{-27} = 2 - \sqrt{27}i$

$\quad\quad = 2 - 3\sqrt{3}i$

8. $1 + \sqrt{-8} = 1 + 2\sqrt{2}i$

9. $\sqrt{-75} = \sqrt{75}i = 5\sqrt{3}i$

10. $\sqrt{-4} = 2i$

11. $8 = 8 + 0i = 8$

12. 45

13. $-6i + i^2 = -6i - 1$

$\quad\quad = -1 - 6i$

14. $-4i^2 + 2i = -4(-1) + 2i$

$\quad\quad\quad = 4 + 2i$

15. $\sqrt{-0.09} = \sqrt{0.09}i$

$\quad\quad = 0.3i$

16. $\sqrt{-0.0004} = 0.02i$

17. $(5 + i) + (6 - 2i) = 11 - i$

18. $(13 - 2i) + (-5 + 6i) = 8 + 4i$

19. $(8 - i) - (4 - i) = 8 - i - 4 + i$

$\quad\quad = 4$

20. $(3 + 2i) - (6 + 13i) = 3 + 2i - 6 - 13i$

$\quad\quad\quad = -3 - 11i$

21. $\left(-2 + \sqrt{-8}\right) + \left(5 - \sqrt{-50}\right) = -2 + 2\sqrt{2}i + 5 - 5\sqrt{2}i$

$\quad\quad\quad = 3 - 3\sqrt{2}i$

22. $\left(8 + \sqrt{-18}\right) - \left(4 + 3\sqrt{2}i\right) = 8 + 3\sqrt{2}i - 4 - 3\sqrt{2}i$

$\quad\quad\quad\quad = 4$

23. $13i - (14 - 7i) = 13i - 14 + 7i$

$\quad\quad\quad = -14 + 20i$

24. $22 + (-5 + 8i) + 10i = 17 + 18i$

25. $-\left(\frac{3}{2} + \frac{5}{2}i\right) + \left(\frac{5}{3} + \frac{11}{3}i\right) = -\frac{3}{2} - \frac{5}{2}i + \frac{5}{3} + \frac{11}{3}i$

$\quad\quad\quad = -\frac{9}{6} - \frac{15}{6}i + \frac{10}{6} + \frac{22}{6}i$

$\quad\quad\quad = \frac{1}{6} + \frac{7}{6}i$

26. $(1.6 + 3.2i) + (-5.8 + 4.3i) = -4.2 + 7.5i$

27. $(1 + i)(3 - 2i) = 3 - 2i + 3i - 2i^2$

$\quad\quad = 3 + i + 2 = 5 + i$

28. $(6 - 2i)(2 - 3i) = 12 - 18i - 4i + 6i^2$

$\quad\quad\quad = 12 - 22i - 6 = 6 - 22i$

29. $6i(5 - 2i) = 30i - 12i^2 = 30i + 12$

$\quad\quad = 12 + 30i$

30. $-8i(9 + 4i) = -72i - 32i^2$

$\quad\quad = 32 - 72i$

31. $\left(\sqrt{14} + \sqrt{10}i\right)\left(\sqrt{14} - \sqrt{10}i\right) = 14 - 10i^2$

$\quad\quad\quad = 14 + 10 = 24$

32. $\left(\sqrt{3} + \sqrt{15}i\right)\left(\sqrt{3} - \sqrt{15}i\right) = 3 - 15i^2$
$$= 3 - 15(-1)$$
$$= 3 + 15 = 18$$

33. $(4 + 5i)^2 = 16 + 40i + 25i^2$
$$= 16 + 40i - 25$$
$$= -9 + 40i$$

34. $(2 - 3i)^2 = 4 - 12i + 9i^2$
$$= 4 - 9 - 12i$$
$$= -5 - 12i$$

35. $(2 + 3i)^2 + (2 - 3i)^2 = 4 + 12i + 9i^2 + 4 - 12i + 9i^2$
$$= 4 + 12i - 9 + 4 - 12i - 9$$
$$= -10$$

36. $(1 - 2i)^2 - (1 + 2i)^2 = 1 - 4i + 4i^2 - (1 + 4i + 4i^2)$
$$= 1 - 4i + 4i^2 - 1 - 4i - 4i^2$$
$$= -8i$$

37. The complex conjugate of $6 + 3i$ is $6 - 3i$.
$$(6 + 3i)(6 - 3i) = 36 - (3i)^2 = 36 + 9 = 45$$

38. The complex conjugate of $7 - 12i$ is $7 + 12i$.
$$(7 - 12i)(7 + 12i) = 49 - 144i^2$$
$$= 49 - (-144)$$
$$= 193$$

39. The complex conjugate of $-1 - \sqrt{5}i$ is $-1 + \sqrt{5}i$.
$$\left(-1 - \sqrt{5}i\right)\left(-1 + \sqrt{5}i\right) = (-1)^2 - \left(\sqrt{5}i\right)^2$$
$$= 1 + 5 = 6$$

40. The complex conjugate of $-3 + \sqrt{2}i$ is $-3 - \sqrt{2}i$.
$$\left(-3 + \sqrt{2}i\right)\left(-3 - \sqrt{2}i\right) = 9 - 2i^2$$
$$= 9 - (-2)$$
$$= 11$$

41. The complex conjugate of $\sqrt{-20} = 2\sqrt{5}i$ is $-2\sqrt{5}i$.
$$\left(2\sqrt{5}i\right)\left(-2\sqrt{5}i\right) = -20i^2 = 20$$

42. The complex conjugate of $\sqrt{-15} = \sqrt{15}i$ is $-\sqrt{15}i$.
$$\left(\sqrt{15}i\right)\left(-\sqrt{15}i\right) = -15i^2 = -(-15) = 15$$

43. The complex conjugate of $\sqrt{8}$ is $\sqrt{8}$.
$$\left(\sqrt{8}\right)\left(\sqrt{8}\right) = 8$$

44. The complex conjugate of $1 + \sqrt{8}$ is $1 + \sqrt{8}$.
$$\left(1 + \sqrt{8}\right)\left(1 + \sqrt{8}\right) = 1 + 2\sqrt{8} + 8$$
$$= 9 + 4\sqrt{2}$$

45. $\dfrac{5}{i} = \dfrac{5}{i} \cdot \dfrac{-i}{-i} = \dfrac{-5i}{1} = -5i$

46. $-\dfrac{14}{2i} \cdot \dfrac{-2i}{-2i} = \dfrac{28i}{-4i^2} = \dfrac{28i}{4} = 7i$

47. $\dfrac{2}{4 - 5i} = \dfrac{2}{4 - 5i} \cdot \dfrac{4 + 5i}{4 + 5i}$
$$= \dfrac{2(4 + 5i)}{16 + 25} = \dfrac{8 + 10i}{41} = \dfrac{8}{41} + \dfrac{10}{41}i$$

48. $\dfrac{5}{1 - i} \cdot \dfrac{1 + i}{1 + i} = \dfrac{5 + 5i}{1 - i^2} = \dfrac{5 + 5i}{2} = \dfrac{5}{2} + \dfrac{5}{2}i$

49. $\dfrac{3 + i}{3 - i} = \dfrac{3 + i}{3 - i} \cdot \dfrac{3 + i}{3 + i}$
$$= \dfrac{9 + 6i + i^2}{9 + 1} = \dfrac{8 + 6i}{10} = \dfrac{4}{5} + \dfrac{3}{5}i$$

50. $\dfrac{6 - 7i}{1 - 2i} \cdot \dfrac{1 + 2i}{1 + 2i} = \dfrac{6 + 12i - 7i - 14i^2}{1 - 4i^2}$
$$= \dfrac{20 + 5i}{5} = \dfrac{20}{5} + \dfrac{5}{5}i = 4 + i$$

51. $\dfrac{6 - 5i}{i} = \dfrac{6 - 5i}{i} \cdot \dfrac{-i}{-i}$
$$= \dfrac{-6i + 5i^2}{1} = -5 - 6i$$

52. $\dfrac{8 + 16i}{2i} \cdot \dfrac{-2i}{-2i} = \dfrac{-16i - 32i^2}{-4i^2} = 8 - 4i$

53. $\dfrac{3i}{(4 - 5i)^2} = \dfrac{3i}{16 - 40i + 25i^2} = \dfrac{3i}{-9 - 40i} \cdot \dfrac{-9 + 40i}{-9 + 40i}$

$\qquad = \dfrac{-27i + 120i^2}{81 + 1600} = \dfrac{-120 - 27i}{1681}$

$\qquad = -\dfrac{120}{1681} - \dfrac{27}{1681}i$

54. $\dfrac{5i}{(2 + 3i)^2} = \dfrac{5i}{4 + 12i + 9i^2}$

$\qquad = \dfrac{5i}{-5 + 12i} \cdot \dfrac{-5 - 12i}{-5 - 12i}$

$\qquad = \dfrac{-25i - 60i^2}{25 - 144i^2}$

$\qquad = \dfrac{60 - 25i}{169} = \dfrac{60}{169} - \dfrac{25}{169}i$

55. $\dfrac{2}{1 + i} - \dfrac{3}{1 - i} = \dfrac{2(1 - i) - 3(1 + i)}{(1 + i)(1 - i)}$

$\qquad = \dfrac{2 - 2i - 3 - 3i}{1 + 1}$

$\qquad = \dfrac{-1 - 5i}{2}$

$\qquad = -\dfrac{1}{2} - \dfrac{5}{2}i$

56. $\dfrac{2i}{2 + i} + \dfrac{5}{2 - i} = \dfrac{2i(2 - i)}{(2 + i)(2 - i)} + \dfrac{5(2 + i)}{(2 + i)(2 - i)}$

$\qquad = \dfrac{4i - 2i^2 + 10 + 5i}{4 - i^2}$

$\qquad = \dfrac{12 + 9i}{5}$

$\qquad = \dfrac{12}{5} + \dfrac{9}{5}i$

57. $\dfrac{i}{3 - 2i} + \dfrac{2i}{3 + 8i} = \dfrac{i(3 + 8i) + 2i(3 - 2i)}{(3 - 2i)(3 + 8i)}$

$\qquad = \dfrac{3i + 8i^2 + 6i - 4i^2}{9 + 24i - 6i - 16i^2}$

$\qquad = \dfrac{4i^2 + 9i}{9 + 18i + 16}$

$\qquad = \dfrac{-4 + 9i}{25 + 18i} \cdot \dfrac{25 - 18i}{25 - 18i}$

$\qquad = \dfrac{-100 + 72i + 225i - 162i^2}{625 + 324}$

$\qquad = \dfrac{-100 + 297i + 162}{949}$

$\qquad = \dfrac{62 + 297i}{949} = \dfrac{62}{949} + \dfrac{297}{949}i$

58. $\dfrac{1 + i}{i} - \dfrac{3}{4 - i} = \dfrac{(1 + i)(4 - i) - 3i}{i(4 - i)}$

$\qquad = \dfrac{4 - i + 4i - i^2 - 3i}{4i - i^2}$

$\qquad = \dfrac{5}{1 + 4i} \cdot \dfrac{1 - 4i}{1 - 4i}$

$\qquad = \dfrac{5 - 20i}{1 - 16i^2}$

$\qquad = \dfrac{5}{17} - \dfrac{20}{17}i$

59. $\sqrt{-6} \cdot \sqrt{-2} = \left(\sqrt{6}i\right)\left(\sqrt{2}i\right) = \sqrt{12}i^2 = \left(2\sqrt{3}\right)(-1)$

$\qquad = -2\sqrt{3}$

60. $\sqrt{-5} \cdot \sqrt{-10} = \left(\sqrt{5}i\right)\left(\sqrt{10}i\right)$

$\qquad = \sqrt{50}i^2 = 5\sqrt{2}(-1) = -5\sqrt{2}$

61. $\left(\sqrt{-10}\right)^2 = \left(\sqrt{10}i\right)^2 = 10i^2 = -10$

62. $\left(\sqrt{-75}\right)^2 = \left(\sqrt{75}i\right)^2 = 75i^2 = -75$

63. $\left(3 + \sqrt{-5}\right)\left(7 - \sqrt{-10}\right) = \left(3 + \sqrt{5}i\right)\left(7 - \sqrt{10}i\right)$

$\qquad = 21 - 3\sqrt{10}i + 7\sqrt{5}i - \sqrt{50}i^2$

$\qquad = \left(21 + \sqrt{50}\right) + \left(7\sqrt{5} - 3\sqrt{10}\right)i$

$\qquad = \left(21 + 5\sqrt{2}\right) + \left(7\sqrt{5} - 3\sqrt{10}\right)i$

64. $\left(2 - \sqrt{-6}\right)^2 = \left(2 - \sqrt{6}i\right)\left(2 - \sqrt{6}i\right)$

$\qquad\qquad = 4 - 2\sqrt{6}i - 2\sqrt{6}i + 6i^2$

$\qquad\qquad = 4 - 2\sqrt{6}i - 2\sqrt{6}i + 6(-1)$

$\qquad\qquad = 4 - 6 - 4\sqrt{6}i$

$\qquad\qquad = -2 - 4\sqrt{6}i$

65. $x^2 - 2x + 2 = 0; \; a = 1, \; b = -2, \; c = 2$

$x = \dfrac{-(-2) \pm \sqrt{(-2)^2 - 4(1)(2)}}{2(1)}$

$\quad = \dfrac{2 \pm \sqrt{-4}}{2}$

$\quad = \dfrac{2 \pm 2i}{2}$

$\quad = 1 \pm i$

66. $x^2 + 6x + 10 = 0; \; a = 1, b = 6, c = 10$

$x = \dfrac{-6 \pm \sqrt{6^2 - 4(1)(10)}}{2(1)}$

$\quad = \dfrac{-6 \pm \sqrt{-4}}{2}$

$\quad = -3 \pm i$

67. $4x^2 + 16x + 17 = 0; \; a = 4, \; b = 16, \; c = 17$

$x = \dfrac{-16 \pm \sqrt{(16)^2 - 4(4)(17)}}{2(4)}$

$\quad = \dfrac{-16 \pm \sqrt{-16}}{8}$

$\quad = \dfrac{-16 \pm 4i}{8} = -2 \pm \dfrac{1}{2}i$

68. $9x^2 - 6x + 37 = 0; \; a = 9, b = -6, c = 37$

$x = \dfrac{-(-6) \pm \sqrt{(-6)^2 - 4(9)(37)}}{2(9)}$

$\quad = \dfrac{6 \pm \sqrt{-1296}}{18}$

$\quad = \dfrac{1}{3} \pm \dfrac{36i}{18} = \dfrac{1}{3} \pm 2i$

69. $4x^2 + 16x + 15 = 0; \; a = 4, \; b = 16, \; c = 15$

$x = \dfrac{-16 \pm \sqrt{(16)^2 - 4(4)(15)}}{2(4)}$

$\quad = \dfrac{-16 \pm \sqrt{16}}{8} = \dfrac{-16 \pm 4}{8}$

$x = -\dfrac{12}{8} = -\dfrac{3}{2} \;$ or $\; x = -\dfrac{20}{8} = -\dfrac{5}{2}$

70. $16t^2 - 4t + 3 = 0; \; a = 16, \; b = -4, \; c = 3$

$t = \dfrac{-(-4) \pm \sqrt{(-4)^2 - 4(16)(3)}}{2(16)}$

$\quad = \dfrac{4 \pm \sqrt{-176}}{32}$

$\quad = \dfrac{4 \pm 4\sqrt{11}i}{32}$

$\quad = \dfrac{1}{8} \pm \dfrac{\sqrt{11}}{8}i$

71. $\dfrac{3}{2}x^2 - 6x + 9 = 0 \qquad$ Multiply both sides by 2.

$3x^2 - 12x + 18 = 0$

$x = \dfrac{-(-12) \pm \sqrt{(-12)^2 - 4(3)(18)}}{2(3)}$

$\quad = \dfrac{12 \pm \sqrt{-72}}{6}$

$\quad = \dfrac{12 \pm 6\sqrt{2}i}{6} = 2 \pm \sqrt{2}i$

72. $\dfrac{7}{8}x^2 - \dfrac{3}{4}x + \dfrac{5}{16} = 0$

$14x^2 - 12x + 5 = 0; \; a = 14, b = -12, c = 5$

$x = \dfrac{-(-12) \pm \sqrt{(-12)^2 - 4(14)(5)}}{2(14)}$

$\quad = \dfrac{12 \pm \sqrt{-136}}{28}$

$\quad = \dfrac{12 \pm 2i\sqrt{34}}{28}$

$\quad = \dfrac{3}{7} \pm \dfrac{\sqrt{34}}{14}i$

73. $1.4x^2 - 2x - 10 = 0 \qquad$ Multiply both sides by 5.

$7x^2 - 10x - 50 = 0$

$x = \dfrac{-(-10) \pm \sqrt{(-10)^2 - 4(7)(-50)}}{2(7)}$

$\quad = \dfrac{10 \pm \sqrt{1500}}{14} = \dfrac{10 \pm 10\sqrt{15}}{14}$

$\quad = \dfrac{5 \pm 5\sqrt{15}}{7} = \dfrac{5}{7} \pm \dfrac{5\sqrt{15}}{7}$

74. $4.5x^2 - 3x + 12 = 0; a = 4.5, b = -3, c = 12$

$$x = \frac{-(-3) \pm \sqrt{(-3)^2 - 4(4.5)(12)}}{2(4.5)}$$

$$= \frac{3 \pm \sqrt{-207}}{9} = \frac{3 \pm 3i\sqrt{23}}{9} = \frac{1}{3} \pm \frac{\sqrt{23}}{3}i$$

75. $-6i^3 + i^2 = -6i^2i + i^2$

$$= -6(-1)i + (-1)$$

$$= 6i - 1$$

$$= -1 + 6i$$

76. $4i^2 - 2i^3 = -4 + 2i$

77. $-5i^5 = -5i^2i^2i$

$$= -5(-1)(-1)i = -5i$$

78. $(-i)^3 = (-1)(i^3) = (-1)(-i) = i$

79. $\left(\sqrt{-75}\right)^3 = \left(5\sqrt{3}i\right)^3$

$$= 5^3\left(\sqrt{3}\right)^3 i^3$$

$$= 125\left(3\sqrt{3}\right)(-1)i$$

$$= -375\sqrt{3}i$$

80. $\left(\sqrt{-2}\right)^6 = \left(\sqrt{2}i\right)^6 = 8i^6 = 8i^4i^2 = -8$

81. $\dfrac{1}{i^3} = \dfrac{1}{-i} = \dfrac{1}{-i} \cdot \dfrac{i}{i} = \dfrac{i}{-i^2} = \dfrac{i}{1} = i$

82. $\dfrac{1}{(2i)^3} = \dfrac{1}{8i^3} = \dfrac{1}{-8i} \cdot \dfrac{8i}{8i} = \dfrac{8i}{-64i^2} = \dfrac{1}{8}i$

83. (a) $z_1 = 9 + 16i, z_2 = 20 - 10i$

(b) $\dfrac{1}{z} = \dfrac{1}{z_1} + \dfrac{1}{z_2} = \dfrac{1}{9 + 16i} + \dfrac{1}{20 - 10i} = \dfrac{20 - 10i + 9 + 16i}{(9 + 16i)(20 - 10i)} = \dfrac{29 + 6i}{340 + 230i}$

$$z = \left(\dfrac{340 + 230i}{29 + 6i}\right)\left(\dfrac{29 - 6i}{29 - 6i}\right) = \dfrac{11{,}240 + 4630i}{877} = \dfrac{11{,}240}{877} + \dfrac{4630}{877}i$$

84. (a) $(2)^3 = 8$

(b) $\left(-1 + \sqrt{3}i\right)^3 = (-1)^3 + 3(-1)^2\left(\sqrt{3}i\right) + 3(-1)\left(\sqrt{3}i\right)^2 + \left(\sqrt{3}i\right)^3$

$$= -1 + 3\sqrt{3}i - 9i^2 + 3\sqrt{3}i^3$$

$$= -1 + 3\sqrt{3}i + 9 - 3\sqrt{3}i$$

$$= 8$$

(c) $\left(-1 - \sqrt{3}i\right)^3 = (-1)^3 + 3(-1)^2\left(-\sqrt{3}i\right) + 3(-1)\left(-\sqrt{3}i\right)^2 + \left(-\sqrt{3}i\right)^3$

$$= -1 - 3\sqrt{3}i - 9i^2 - 3\sqrt{3}i^3$$

$$= -1 - 3\sqrt{3}i + 9 + 3\sqrt{3}i$$

$$= 8$$

85. (a) $2^4 = 16$

(b) $(-2)^4 = 16$

(c) $(2i)^4 = 2^4i^4 = 16(1) = 16$

(d) $(-2i)^4 = (-2)^4i^4 = 16(1) = 16$

86. (a) $i^{40} = (i^4)^{10} = (1)^{10} = 1$

(b) $i^{25} = (i^4)^6 \cdot i = (1)^6i = i$

(c) $i^{50} = (i^4)^{12}(i^2) = (1)(-1) = -1$

(d) $i^{67} = (i^4)^{16}(i^3) = (1)(-i) = -i$

87. False, if $b = 0$ then $a + bi = a - bi = a$.

That is, if the complex number is real, the number equals its conjugate.

88. True

$$x^4 - x^2 + 14 = 56$$

$$\left(-i\sqrt{6}\right)^4 - \left(-i\sqrt{6}\right)^2 + 14 \overset{?}{=} 56$$

$$36 + 6 + 14 \overset{?}{=} 56$$

$$56 = 56$$

89. False

$$i^{44} + i^{150} - i^{74} - i^{109} + i^{61} = (i^4)^{11} + (i^4)^{37}(i^2) - (i^4)^{18}(i^2) - (i^4)^{27}(i) + (i^4)^{15}(i)$$

$$= (1)^{11} + (1)^{37}(-1) - (1)^{18}(-1) - (1)^{27}(i) + (1)^{15}(i)$$

$$= 1 + (-1) + 1 - i + i = 1$$

90. $\sqrt{-6}\sqrt{-6} = \sqrt{6}i\sqrt{6}i = 6i^2 = -6$

91. $(a_1 + b_1 i)(a_2 + b_2 i) = a_1 a_2 + a_1 b_2 i + a_2 b_1 i + b_1 b_2 i^2$

$$= (a_1 a_2 - b_1 b_2) + (a_1 b_2 + a_2 b_1)i$$

The complex conjugate of this product is $(a_1 a_2 - b_1 b_2) - (a_1 b_2 + a_2 b_1)i$.

The product of the complex conjugates is:

$$(a_1 - b_1 i)(a_2 - b_2 i) = a_1 a_2 - a_1 b_2 i - a_2 b_1 i + b_1 b_2 i^2$$

$$= (a_1 a_2 - b_1 b_2) - (a_1 b_2 + a_2 b_1)i$$

Thus, the complex conjugate of the product of two complex numbers is the product of their complex conjugates.

92. $(a_1 + b_1 i) + (a_2 + b_2 i) = (a_1 + a_2) + (b_1 + b_2)i$

The complex conjugate of this sum is $(a_1 + a_2) - (b_1 + b_2)i$.

The sum of the complex conjugates is $(a_1 - b_1 i) + (a_2 - b_2 i) = (a_1 + a_2) - (b_1 + b_2)i$.

Thus, the complex conjugate of the sum of two complex numbers is the sum of their complex conjugates.

93. $(4 + 3x) + (8 - 6x - x^2) = -x^2 - 3x + 12$

94. $(x^3 - 3x^2) - (6 - 2x - 4x^2) = x^3 - 3x^2 - 6 + 2x + 4x^2$

$$= x^3 + x^2 + 2x - 6$$

95. $\left(3x - \frac{1}{2}\right)(x + 4) = 3x^2 + 12x - \frac{1}{2}x - 2$

$$= 3x^2 + \frac{23}{2}x - 2$$

96. $(2x - 5)^2 = (2x)^2 - 2(2x)(5) + (5)^2$

$$= 4x^2 - 20x + 25$$

97. $-x - 12 = 19$

$$-x = 31$$

$$x = -31$$

98. $8 - 3x = -34$

$$-3x = -42$$

$$x = 14$$

99. $4(5x - 6) - 3(6x + 1) = 0$

$$20x - 24 - 18x - 3 = 0$$

$$2x - 27 = 0$$

$$2x = 27$$

$$x = \frac{27}{2}$$

100. $5[x - (3x + 11)] = 20x - 15$

$$5x - 15x - 55 = 20x - 15$$

$$-30x = 40$$

$$x = \frac{40}{-30} = -\frac{4}{3}$$

101.

$$V = \frac{4}{3}\pi a^2 b$$

$$3V = 4\pi a^2 b$$

$$\frac{3V}{4\pi b} = a^2$$

$$\sqrt{\frac{3V}{4\pi b}} = a$$

$$a = \frac{1}{2}\sqrt{\frac{3V}{\pi b}} = \frac{\sqrt{3V\pi b}}{2\pi b}$$

102. $F = \alpha \dfrac{m_1 m_2}{r^2}$

$$r^2 = \alpha \frac{m_1 m_2}{F}$$

$$r = \sqrt{\frac{\alpha m_1 m_2}{F}} = \frac{\sqrt{\alpha m_1 m_2}}{\sqrt{F}} \cdot \frac{\sqrt{F}}{\sqrt{F}} = \frac{\sqrt{\alpha m_1 m_2 F}}{F}$$

103. Let x = # liters withdrawn and replaced.

$$0.50(5 - x) + 1.00x = 0.60(5)$$

$$2.50 - 0.50x + 1.00x = 3.00$$

$$0.50x = 0.50$$

$$x = 1 \text{ liter}$$

Section 2.5   Zeros of Polynomial Functions

- ■ You should know that if f is a polynomial of degree $n > 0$, then f has at least one zero in the complex number system.
- ■ You should know the Linear Factorization Theorem.
- ■ You should know the Rational Zero Test.
- ■ You should know shortcuts for the Rational Zero Test. Possible rational zeros $= \dfrac{\text{factors of constant term}}{\text{factors of leading coefficient}}$

  (a) Use a graphing or programmable calculator.

  (b) Sketch a graph.

  (c) After finding a root, use synthetic division to reduce the degree of the polynomial.
- ■ You should know that if $a + bi$ is a complex zero of a polynomial f, with real coefficients, then $a - bi$ is also a complex zero of f.
- ■ You should know the difference between a factor that is irreducible over the rationals (such as $x^2 - 7$) and a factor that is irreducible over the reals (such as $x^2 + 9$).
- ■ You should know Descartes's Rule of Signs. (For a polynomial with real coefficients and a non-zero constant term.)

  (a) The number of positive real zeros of f is either equal to the number of variations of sign of f or is less than that number by an even integer.

  (b) The number of negative real zeros of f is either equal to the number of variations in sign of $f(-x)$ or is less than that number by an even integer.

  (c) When there is only one variation in sign, there is exactly one positive (or negative) real zero.
- ■ You should be able to observe the last row obtained from synthetic division in order to determine upper or lower bounds.

  (a) If the test value is positive and all of the entries in the last row are positive or zero, then the test value is an upper bound.

  (b) If the test value is negative and the entries in the last row alternate from positive to negative, then the test value is a lower bound. (Zero entries count as positive or negative.)

Vocabulary Check

1. Fundamental Theorem of Algebra   **2.** Linear Factorization Theorem   **3.** Rational Zero

4. conjugate   **5.** irreducible; reals   **6.** Descarte's Rule of Signs

7. lower; upper

---

1. $f(x) = x(x - 6)^2$

The zeros are: $x = 0, x = 6$

2. $f(x) = x^2(x + 3)(x^2 - 1) = x^2(x + 3)(x + 1)(x - 1)$

The five zeros are: $0, 0, -3, \pm 1$

3. $g(x) = (x - 2)(x + 4)^3$

The zeros are: $x = 2, x = -4$

4. $f(x) = (x + 5)(x - 8)^2$

The three zeros are: $-5, 8, 8$

5. $f(x) = (x + 6)(x + i)(x - i)$

The three zeros are:

$$x = -6, x = -i, x = i$$

6. $h(t) = (t - 3)(t - 2)(t - 3i)(t + 3i)$

The four zeros are: $3, 2, \pm 3i$

7. $f(x) = x^3 + 3x^2 - x - 3$

Possible rational zeros: $\pm 1, \pm 3$

Zeros shown on graph: $-3, -1, 1$

8. $f(x) = x^3 - 4x^2 - 4x + 16$

Possible rational zeros: $\pm 1, \pm 2, \pm 4, \pm 8, \pm 16$

Zeros shown on graph: $-2, 2, 4$

9. $f(x) = 2x^4 - 17x^3 + 35x^2 + 9x - 45$

Possible rational zeros: $\pm 1, \pm 3, \pm 5, \pm 9, \pm 15, \pm 45,$
$\pm \frac{1}{2}, \pm \frac{3}{2}, \pm \frac{5}{2}, \pm \frac{9}{2}, \pm \frac{15}{2}, \pm \frac{45}{2}$

Zeros shown on graph: $-1, \frac{3}{2}, 3, 5$

10. $f(x) = 4x^5 - 8x^4 - 5x^3 + 10x^2 + x - 2$

Possible rational zeros: $\pm 1, \pm 2, \pm \frac{1}{2}, \pm \frac{1}{4}$

Zeros shown on graph: $-1, -\frac{1}{2}, \frac{1}{2}, 1, 2$

11. $f(x) = x^3 - 6x^2 + 11x - 6$

Possible rational zeros: $\pm 1, \pm 2, \pm 3, \pm 6$

$$
\begin{array}{r|rrr}
1 & 1 & -6 & 11 & -6 \\
  &   & 1 & -5 & 6 \\
\hline
  & 1 & -5 & 6 & 0
\end{array}
$$

$x^3 - 6x^2 + 11x - 6 = (x - 1)(x^2 - 5x + 6)$
$$= (x - 1)(x - 2)(x - 3)$$

Thus, the rational zeros are 1, 2, and 3.

12. $f(x) = x^3 - 7x - 6$

Possible rational zeros: $\pm 1, \pm 2, \pm 3, \pm 6$

$$
\begin{array}{r|rrr}
3 & 1 & 0 & -7 & -6 \\
  &   & 3 & 9 & 6 \\
\hline
  & 1 & 3 & 2 & 0
\end{array}
$$

$f(x) = (x - 3)(x^2 + 3x + 2) = (x - 3)(x + 2)(x + 1)$

Thus, the rational zeros are $-2, -1, 3$.

13. $g(x) = x^3 - 4x^2 - x + 4 = x^2(x - 4) - 1(x - 4)$
$$= (x - 4)(x^2 - 1)$$
$$= (x - 4)(x - 1)(x + 1)$$

Thus, the rational zeros of $g(x)$ are 4 and ± 1.

14. $h(x) = x^3 - 9x^2 + 20x - 12$

Possible rational zeros: $\pm 1, \pm 2, \pm 3, \pm 4, \pm 6, \pm 12$

$$
\begin{array}{r|rrr}
1 & 1 & -9 & 20 & -12 \\
  &   & 1 & -8 & 12 \\
\hline
  & 1 & -8 & 12 &   0
\end{array}
$$

$h(x) = (x - 1)(x^2 - 8x + 12)$
$$= (x - 1)(x - 2)(x - 6)$$

Thus, the rational zeros are 1, 2, 6.

15. $h(t) = t^3 + 12t^2 + 21t + 10$

Possible rational zeros: $\pm 1, \pm 2, \pm 5, \pm 10$

$$
\begin{array}{r|rrr}
-1 & 1 & 12 & 21 & 10 \\
   &   & -1 & -11 & -10 \\
\hline
   & 1 & 11 & 10 & 0
\end{array}
$$

$t^3 + 12t^2 + 21t + 10 = (t + 1)(t^2 + 11t + 10)$
$$= (t + 1)(t + 1)(t + 10)$$
$$= (t + 1)^2(t + 10)$$

Thus, the rational zeros are -1 and -10.

16. $p(x) = x^3 - 9x^2 + 27x - 27$

Possible rational zeros: $\pm 1, \pm 3, \pm 9, \pm 27$

$$
\begin{array}{r|rrr}
3 & 1 & -9 & 27 & -27 \\
  &   & 3 & -18 & 27 \\
\hline
  & 1 & -6 & 9 &   0
\end{array}
$$

$f(x) = (x - 3)(x^2 - 6x + 9)$
$$= (x - 3)(x - 3)(x - 3)$$

Thus, the rational zero is 3.

17. $C(x) = 2x^3 + 3x^2 - 1$

Possible rational zeros: $\pm 1, \pm \frac{1}{2}$

$$
\begin{array}{r|rrr}
-1 & 2 & 3 & 0 & -1 \\
   &   & -2 & -1 & 1 \\
\hline
   & 2 & 1 & -1 & 0
\end{array}
$$

$2x^3 + 3x^2 - 1 = (x + 1)(2x^2 + x - 1)$
$$= (x + 1)(x + 1)(2x - 1)$$
$$= (x + 1)^2(2x - 1)$$

Thus, the rational zeros are -1 and $\frac{1}{2}$.

18. $f(x) = 3x^3 - 19x^2 + 33x - 9$

Possible rational zeros: $\pm 1, \pm 3, \pm 9, \pm\frac{1}{3}$

$$
\begin{array}{r|rrrr}
3 & 3 & -19 & 33 & -9 \\
  &   & 9   & -30 & 9 \\
\hline
  & 3 & -10 & 3 & 0
\end{array}
$$

$f(x) = (x - 3)(3x^2 - 10x + 3) = (x - 3)(3x - 1)(x - 3)$

Thus, the rational zeros are $3, \frac{1}{3}$.

19. $f(x) = 9x^4 - 9x^3 - 58x^2 + 4x + 24$

Possible rational zeros: $\pm 1, \pm 2, \pm 3, \pm 4, \pm 6, \pm 8, \pm 12, \pm 24,$
$\pm\frac{1}{3}, \pm\frac{2}{3}, \pm\frac{4}{3}, \pm\frac{8}{3}, \pm\frac{1}{9}, \pm\frac{2}{9}, \pm\frac{4}{9}, \pm\frac{8}{9}$

$$
\begin{array}{r|rrrrr}
-2 & 9 & -9 & -58 & 4 & 24 \\
   &   & -18 & 54 & 8 & -24 \\
\hline
   & 9 & -27 & -4 & 12 & 0
\end{array}
$$

$$
\begin{array}{r|rrrr}
3 & 9 & -27 & -4 & 12 \\
  &   & 27 & 0 & -12 \\
\hline
  & 9 & 0 & -4 & 0
\end{array}
$$

$9x^4 - 9x^3 - 58x^2 + 4x - 24$

$\quad = (x + 2)(x - 3)(9x^2 - 4)$

$\quad = (x + 2)(x - 3)(3x - 2)(3x + 2)$

Thus, the rational zeros are $-2, 3,$ and $\pm\frac{2}{3}$.

20. $f(x) = 2x^4 - 15x^3 + 23x^2 + 15x - 25$

Possible rational zeros: $\pm 1, \pm 5, \pm 25, \pm\frac{1}{2}, \pm\frac{5}{2}, \pm\frac{25}{2}$

$$
\begin{array}{r|rrrrr}
5 & 2 & -15 & 23 & 15 & -25 \\
  &   & 10 & -25 & -10 & 25 \\
\hline
  & 2 & -5 & -2 & 5 & 0
\end{array}
$$

$$
\begin{array}{r|rrrr}
1 & 2 & -5 & -2 & 5 \\
  &   & 2 & -3 & -5 \\
\hline
  & 2 & -3 & -5 & 0
\end{array}
$$

$$
\begin{array}{r|rrr}
-1 & 2 & -3 & -5 \\
   &   & -2 & 5 \\
\hline
   & 2 & -5 & 0
\end{array}
$$

$f(x) = (x - 5)(x - 1)(x + 1)(2x - 5)$

Thus, the rational zeros are $5, 1, -1, \frac{5}{2}$.

21. $z^4 - z^3 - 2z - 4 = 0$

Possible rational zeros: $\pm 1, \pm 2, \pm 4$

$$
\begin{array}{r|rrrrr}
-1 & 1 & -1 & 0 & -2 & -4 \\
   &   & -1 & 2 & -2 & 4 \\
\hline
   & 1 & -2 & 2 & -4 & 0
\end{array}
$$

$$
\begin{array}{r|rrrr}
2 & 1 & -2 & 2 & -4 \\
  &   & 2 & 0 & 4 \\
\hline
  & 1 & 0 & 2 & 0
\end{array}
$$

$z^4 - z^3 - 2z - 4 = (z + 1)(z - 2)(z^2 + 2)$

The only real zeros are -1 and 2.

22. $x^4 - 13x^2 - 12x = 0$

$x(x^3 - 13x - 12) = 0$

Possible rational zeros of $x^3 - 13x - 12$:
$\pm 1, \pm 2, \pm 3, \pm 4, \pm 6, \pm 12$

$$
\begin{array}{r|rrrr}
-1 & 1 & 0 & -13 & -12 \\
   &   & -1 & 1 & 12 \\
\hline
   & 1 & -1 & -12 & 0
\end{array}
$$

$x(x + 1)(x^2 - x - 12) = 0$

$x(x + 1)(x - 4)(x + 3) = 0$

The real zeros are $0, -1, 4, -3$.

23. $2y^4 + 7y^3 - 26y^2 + 23y - 6 = 0$

Possible rational zeros: $\pm 1, \pm 2, \pm 3, \pm 6, \pm\frac{1}{2}, \pm\frac{3}{2}$

$$
\begin{array}{r|rrrrr}
1 & 2 & 7 & -26 & 23 & -6 \\
  &   & 2 & 9 & -17 & 6 \\
\hline
  & 2 & 9 & -17 & 6 & 0
\end{array}
$$

$$
\begin{array}{r|rrrr}
-6 & 2 & 9 & -17 & 6 \\
   &   & -12 & 18 & -6 \\
\hline
   & 2 & -3 & 1 & 0
\end{array}
$$

$2y^4 + 7y^3 - 26y^2 + 23y - 6 = (y - 1)(y + 6)(2y^2 - 3y + 1) = (y - 1)(y + 6)(2y - 1)(y - 1) = (y - 1)^2(y + 6)(2y - 1)$

The only real zeros are $1, -6,$ and $\frac{1}{2}$.

24. $x^5 - x^4 - 3x^3 + 5x^2 - 2x = 0$

$x(x^4 - x^3 - 3x^2 + 5x - 2) = 0$

Possible rational zeros of $x^4 - x^3 - 3x^2 + 5x - 2$:

$\pm 1, \pm 2$

$$
\begin{array}{r|rrrrr}
1 & 1 & -1 & -3 & 5 & -2 \\
  &   & 1 & 0 & -3 & 2 \\
\hline
  & 1 & 0 & -3 & 2 & 0 \\
\end{array}
$$

$$
\begin{array}{r|rrrr}
-2 & 1 & 0 & -3 & 2 \\
   &   & -2 & 4 & -2 \\
\hline
   & 1 & -2 & 1 & 0 \\
\end{array}
$$

$x(x - 1)(x + 2)(x^2 - 2x + 1) = 0$

$x(x - 1)(x + 2)(x - 1)(x - 1) = 0$

The real zeros are $-2, 0, 1$.

26. $f(x) = -3x^3 + 20x^2 - 36x + 16$

(a) Possible rational zeros: $\pm 1, \pm 2, \pm 4, \pm 8, \pm 16, \pm \frac{1}{3},$
$\pm \frac{2}{3}, \pm \frac{4}{3}, \pm \frac{8}{3}, \pm \frac{16}{3}$

(b)

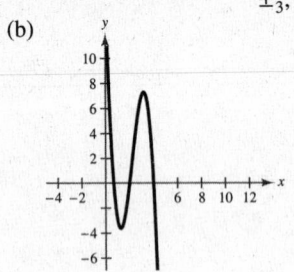

(c) Real zeros: $\frac{2}{3}, 2, 4$

28. $f(x) = 4x^3 - 12x^2 - x + 15$

(a) Possible rational zeros: $\pm 1, \pm 3, \pm 5, \pm 15, \pm \frac{1}{2}, \pm \frac{3}{2},$
$\pm \frac{5}{2}, \pm \frac{15}{2}, \pm \frac{1}{4}, \pm \frac{3}{4}, \pm \frac{5}{4}, \pm \frac{15}{4}$

(b)

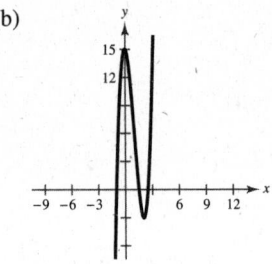

(c) Real zeros: $-1, \frac{3}{2}, \frac{5}{2}$

30. $f(x) = 4x^4 - 17x^2 + 4$

(a) Possible rational zeros: $\pm 1, \pm 2, \pm 4, \pm \frac{1}{2}, \pm \frac{1}{4}$

(b)

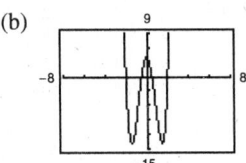

(c) Real zeros: $\pm 2, \pm \frac{1}{2}$

25. $f(x) = x^3 + x^2 - 4x - 4$

(a) Possible rational zeros: $\pm 1, \pm 2, \pm 4$

(b)

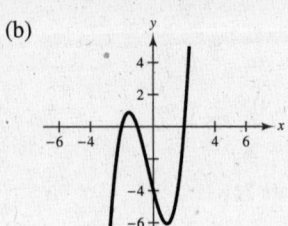

(c) The zeros are: $-2, -1, 2$

27. $f(x) = -4x^3 + 15x^2 - 8x - 3$

(a) Possible rational zeros: $\pm 1, \pm 3, \pm \frac{1}{2}, \pm \frac{3}{2}, \pm \frac{1}{4}, \pm \frac{3}{4}$

(b)

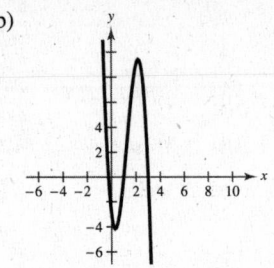

(c) The zeros are: $-\frac{1}{4}, 1, 3$

29. $f(x) = -2x^4 + 13x^3 - 21x^2 + 2x + 8$

(a) Possible rational zeros: $\pm 1, \pm 2, \pm 4, \pm 8, \pm \frac{1}{2}$

(b)

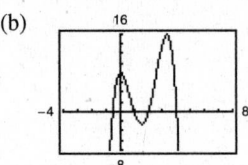

(c) The zeros are: $-\frac{1}{2}, 1, 2, 4$

31. $f(x) = 32x^3 - 52x^2 + 17x + 3$

(a) Possible rational zeros: $\pm 1, \pm 3, \pm \frac{1}{2}, \pm \frac{3}{2}, \pm \frac{1}{4}, \pm \frac{3}{4},$
$\pm \frac{1}{6}, \pm \frac{3}{8}, \pm \frac{1}{16}, \pm \frac{3}{16}, \pm \frac{1}{32}, \pm \frac{3}{32}$

(b)

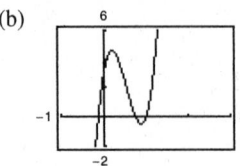

(c) The zeros are: $-\frac{1}{8}, \frac{3}{4}, 1$

32. $f(x) = 4x^3 + 7x^2 - 11x - 18$

(a) Possible rational zeros: $\pm 1, \pm 2, \pm 3, \pm 6, \pm 9, \pm 18,$
$\pm\frac{1}{2}, \pm\frac{3}{2}, \pm\frac{9}{2}, \pm\frac{1}{4}, \pm\frac{3}{4}, \pm\frac{9}{4}$

(b)

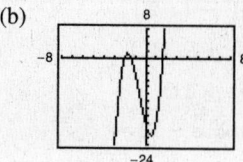

(c) Real zeros: $-2, \dfrac{1}{8} \pm \dfrac{\sqrt{145}}{8}$

34. $P(t) = t^4 - 7t^2 + 12$

(a) $t = \pm 2, \pm 1.732$

(b)

$$
\begin{array}{r|rrrrr}
2 & 1 & 0 & -7 & 0 & 12 \\
  &   & 2 & 4 & -6 & -12 \\
\hline
  & 1 & 2 & -3 & -6 & 0
\end{array}
$$

$$
\begin{array}{r|rrrr}
-2 & 1 & 2 & -3 & -6 \\
   &   & -2 & 0 & 6 \\
\hline
   & 1 & 0 & -3 & 0
\end{array}
$$

(c) $P(t) = (t - 2)(t + 2)(t^2 - 3)$

$= (t - 2)(t + 2)\left(t - \sqrt{3}\right)\left(t + \sqrt{3}\right)$

36. $g(x) = 6x^4 - 11x^3 - 51x^2 + 99x - 27$

(a) $x = \pm 3, 1.5, 0.333$

(b)

$$
\begin{array}{r|rrrrr}
3 & 6 & -11 & -51 & 99 & -27 \\
  &   & 18 & 21 & -90 & 27 \\
\hline
  & 6 & 7 & -30 & 9 & 0
\end{array}
$$

$$
\begin{array}{r|rrrr}
-3 & 6 & 7 & -30 & 9 \\
   &   & -18 & 33 & -9 \\
\hline
   & 6 & -11 & 3 & 0
\end{array}
$$

(c) $g(x) = (x - 3)(x + 3)(6x^2 - 11x + 3)$

$= (x - 3)(x + 3)(3x - 1)(2x - 3)$

38. $f(x) = (x - 4)(x - 3i)(x + 3i)$

$= (x - 4)(x^2 + 9)$

$= x^3 - 4x^2 + 9x - 36$

Note: $f(x) = a(x^3 - 4x^2 + 9x - 36)$, where a is any real number, has the zeros 4, $3i$ and $-3i$.

33. $f(x) = x^4 - 3x^2 + 2$

(a) From the calculator we have $x = \pm 1$ and $x \approx \pm 1.414$.

(b) An exact zero is $x = 1$.

$$
\begin{array}{r|rrrrr}
1 & 1 & 0 & -3 & 0 & 2 \\
  &   & 1 & 1 & -2 & -2 \\
\hline
  & 1 & 1 & -2 & -2 & 0
\end{array}
$$

(c)

$$
\begin{array}{r|rrrr}
-1 & 1 & 1 & -2 & -2 \\
   &   & -1 & 0 & 2 \\
\hline
   & 1 & 0 & -2 & 0
\end{array}
$$

$f(x) = (x - 1)(x + 1)(x^2 - 2)$

$= (x - 1)(x + 1)\left(x - \sqrt{2}\right)\left(x + \sqrt{2}\right)$

35. $h(x) = x^5 - 7x^4 + 10x^3 + 14x^2 - 24x$

(a) $h(x) = x(x^4 - 7x^3 + 10x^2 + 14x - 24)$

From the calculator we have $x = 0, 3, 4$ and $x \approx \pm 1.414$.

(b) An exact zero is $x = 3$.

$$
\begin{array}{r|rrrrr}
3 & 1 & -7 & 10 & 14 & -24 \\
  &   & 3 & -12 & -6 & 24 \\
\hline
  & 1 & -4 & -2 & 8 & 0
\end{array}
$$

(c)

$$
\begin{array}{r|rrrr}
4 & 1 & -4 & -2 & 8 \\
  &   & 4 & 0 & -8 \\
\hline
  & 1 & 0 & -2 & 0
\end{array}
$$

$h(x) = x(x - 3)(x - 4)(x^2 - 2)$

$= x(x - 3)(x - 4)\left(x - \sqrt{2}\right)\left(x + \sqrt{2}\right)$

37. $f(x) = (x - 1)(x - 5i)(x + 5i)$

$= (x - 1)(x^2 + 25)$

$= x^3 - x^2 + 25x - 25$

Note: $f(x) = a(x^3 - x^2 + 25x - 25)$, where a is any nonzero real number, has the zeros 1 and $\pm 5i$.

39. $f(x) = (x - 6)[x - (-5 + 2i)][x - (-5 - 2i)]$

$= (x - 6)[(x + 5) - 2i][(x + 5) + 2i]$

$= (x - 6)[(x + 5)^2 - (2i)^2]$

$= (x - 6)(x^2 + 10x + 25 + 4)$

$= (x - 6)(x^2 + 10x + 29)$

$= x^3 + 4x^2 - 31x - 174$

Note: $f(x) = a(x^3 + 4x^2 - 31x - 174)$, where a is any nonzero real number, has the zeros 6, and $-5 \pm 2i$.

40. $f(x) = (x - 2)(x - 4 - i)(x - 4 + i)$

$\quad = (x - 2)(x^2 - 8x + 17)$

$\quad = x^3 - 10x^2 + 33x - 34$

Note: $f(x) = a(x^3 - 10x^2 + 33x - 34)$ where a is any real number, has the zeros $2, 4 \pm i$.

41. If $3 + \sqrt{2}i$ is a zero, so is its conjugate, $3 - \sqrt{2}i$.

$f(x) = (3x - 2)(x + 1)\big[x - (3 + \sqrt{2}i)\big]\big[x - (3 - \sqrt{2}i)\big]$

$\quad = (3x - 2)(x + 1)\big[(x - 3) - \sqrt{2}i\big]\big[(x - 3) + \sqrt{2}i\big]$

$\quad = (3x^2 + x - 2)\big[(x - 3)^2 - (\sqrt{2}i)^2\big]$

$\quad = (3x^2 + x - 2)(x^2 - 6x + 9 + 2)$

$\quad = (3x^2 + x - 2)(x^2 - 6x + 11)$

$\quad = 3x^4 - 17x^3 + 25x^2 + 23x - 22$

Note: $f(x) = a(3x^4 - 17x^3 + 25x^2 + 23x - 22)$, where a is any nonzero real number, has the zeros $\frac{2}{3}, -1$, and $3 \pm \sqrt{2}i$.

42. If $1 + \sqrt{3}i$ is a zero, so is its conjugate, $1 - \sqrt{3}i$.

$f(x) = (x + 5)^2\big(x - 1 - \sqrt{3}i\big)\big(x - 1 + \sqrt{3}i\big)$

$\quad = (x^2 + 10x + 25)(x^2 - 2x + 4)$

$\quad = x^4 + 8x^3 + 9x^2 - 10x + 100$

Note: $f(x) = a(x^4 + 8x^3 + 9x^2 - 10x + 100)$, where a is any real number, has the zeros $-5, -5, 1 \pm \sqrt{3}i$.

43. $f(x) = x^4 + 6x^2 - 27$

(a) $f(x) = (x^2 + 9)(x^2 - 3)$

(b) $f(x) = (x^2 + 9)(x + \sqrt{3})(x - \sqrt{3})$

(c) $f(x) = (x + 3i)(x - 3i)(x + \sqrt{3})(x - \sqrt{3})$

44. $f(x) = x^4 - 2x^3 - 3x^2 + 12x - 18$

$$
\begin{array}{r}
x^2 - 2x + 3 \\
x - 6 \,\overline{)\, x^4 - 2x^3 - 3x^2 + 12x - 18} \\
\underline{x^4 \qquad\quad - 6x^2} \\
-2x^3 + 3x^2 + 12x \\
\underline{-2x^3 \qquad\quad + 12x} \\
3x^2 \qquad\quad - 18 \\
\underline{3x^2 \qquad\quad - 18} \\
0
\end{array}
$$

(a) $f(x) = (x^2 - 6)(x^2 - 2x + 3)$

(b) $f(x) = \big(x + \sqrt{6}\big)\big(x - \sqrt{6}\big)(x^2 - 2x + 3)$

(c) $f(x) = \big(x + \sqrt{6}\big)\big(x - \sqrt{6}\big)\big(x - 1 - \sqrt{2}i\big)\big(x - 1 + \sqrt{2}i\big)$

45. $f(x) = x^4 - 4x^3 + 5x^2 - 2x - 6$

$$
\begin{array}{r}
x^2 - 2x + 3 \\
x^2 - 2x - 2 \,\overline{)\, x^4 - 4x^3 + 5x^2 - 2x - 6} \\
\underline{x^4 - 2x^3 - 2x^2} \\
-2x^3 + 7x^2 - 2x \\
\underline{-2x^3 + 4x^2 + 4x} \\
3x^2 - 6x - 6 \\
\underline{3x^2 - 6x - 6} \\
0
\end{array}
$$

$f(x) = (x^2 - 2x - 2)(x^2 - 2x + 3)$

(a) $f(x) = (x^2 - 2x - 2)(x^2 - 2x + 3)$

(b) $f(x) = \big(x - 1 + \sqrt{3}\big)\big(x - 1 - \sqrt{3}\big)(x^2 - 2x + 3)$

(c) $f(x) = \big(x - 1 + \sqrt{3}\big)\big(x - 1 - \sqrt{3}\big)\big(x - 1 + \sqrt{2}\,i\big)\big(x - 1 - \sqrt{2}\,i\big)$

Note: Use the Quadratic Formula for (b) and (c).

46. $f(x) = x^4 - 3x^3 - x^2 - 12x - 20$

$$
\begin{array}{r}
x^2 - 3x - 5 \\
x^2 + 4 \,\overline{)\, x^4 - 3x^3 - x^2 - 12x - 20} \\
\underline{x^4 \qquad\quad + 4x^2} \\
-3x^3 - 5x^2 - 12x \\
\underline{-3x^3 \qquad\quad - 12x} \\
-5x^2 \qquad\quad - 20 \\
\underline{-5x^2 \qquad\quad - 20} \\
0
\end{array}
$$

(a) $f(x) = (x^2 + 4)(x^2 - 3x - 5)$

(b) $f(x) = (x^2 + 4)\left(x - \dfrac{3 + \sqrt{29}}{2}\right)\left(x - \dfrac{3 - \sqrt{29}}{2}\right)$

(c) $f(x) = (x + 2i)(x - 2i)\left(x - \dfrac{3 + \sqrt{29}}{2}\right)\left(x - \dfrac{3 - \sqrt{29}}{2}\right)$

47. $f(x) = 2x^3 + 3x^2 + 50x + 75$

Since $5i$ is a zero, so is $-5i$.

$$
\begin{array}{r|rrrr}
5i & 2 & 3 & 50 & 75 \\
   &   & 10i & -50 + 15i & -75 \\
\hline
   & 2 & 3 + 10i & 15i & 0
\end{array}
$$

$$
\begin{array}{r|rrr}
-5i & 2 & 3 + 10i & 15i \\
    &   & -10i & -15i \\
\hline
    & 2 & 3 & 0
\end{array}
$$

The zero of $2x + 3$ is $x = -\frac{3}{2}$. The zeros of $f(x)$ are $x = -\frac{3}{2}$ and $x = \pm 5i$.

Alternate Solution

Since $x = \pm 5i$ are zeros of $f(x)$, $(x + 5i)(x - 5i) = x^2 + 25$ is a factor of $f(x)$. By long division we have:

$$
\begin{array}{r}
2x + 3 \\
x^2 + 0x + 25 \overline{\smash{)}2x^3 + 3x^2 + 50x + 75} \\
\underline{2x^3 + 0x^2 + 50x} \\
3x^2 + 0x + 75 \\
\underline{3x^2 + 0x + 75} \\
0
\end{array}
$$

Thus, $f(x) = (x^2 + 25)(2x + 3)$ and the zeros of f are $x = \pm 5i$ and $x = -\frac{3}{2}$.

48. $f(x) = x^3 + x^2 + 9x + 9$

Since $3i$ is a zero, so is $-3i$.

$$
\begin{array}{r|rrrr}
3i & 1 & 1 & 9 & 9 \\
   &   & 3i & -9 + 3i & -9 \\
\hline
   & 1 & 1 + 3i & 3i & 0
\end{array}
$$

$$
\begin{array}{r|rrr}
-3i & 1 & 1 + 3i & 3i \\
    &   & -3i & -3i \\
\hline
    & 1 & 1 & 0
\end{array}
$$

The zero of $x + 1$ is $x = -1$. The zeros of f are $x = -1$ and $x = \pm 3i$.

49. $f(x) = 2x^4 - x^3 + 7x^2 - 4x - 4$

Since $2i$ is a zero, so is $-2i$.

$$
\begin{array}{r|rrrrr}
2i & 2 & -1 & 7 & -4 & -4 \\
   &   & 4i & -8 - 2i & 4 - 2i & 4 \\
\hline
   & 2 & -1 + 4i & -1 - 2i & -2i & 0
\end{array}
$$

$$
\begin{array}{r|rrrr}
-2i & 2 & -1 + 4i & -1 - 2i & -2i \\
    &   & -4i & 2i & 2i \\
\hline
    & 2 & -1 & -1 & 0
\end{array}
$$

The zeros of $2x^2 - x - 1 = (2x + 1)(x - 1)$ are $x = -\frac{1}{2}$ and $x = 1$. The zeros of $f(x)$ are $x = \pm 2i$, $x = -\frac{1}{2}$, and $x = 1$.

Alternate Solution

Since $x = \pm 2i$ are zeros of $f(x)$, $(x + 2i)(x - 2i) = x^2 + 4$ is a factor of $f(x)$. By long division we have:

$$
\begin{array}{r}
2x^2 - x - 1 \\
x^2 + 0x + 4 \overline{\smash{)}2x^4 - x^3 + 7x^2 - 4x - 4} \\
\underline{2x^4 + 0x^3 + 8x^2} \\
-x^3 - x^2 - 4x \\
\underline{-x^3 + 0x^2 - 4x} \\
-x^2 + 0x - 4 \\
\underline{-x^2 + 0x - 4} \\
0
\end{array}
$$

Thus, $f(x) = (x^2 + 4)(2x^2 - x - 1)$

$$= (x + 2i)(x - 2i)(2x + 1)(x - 1)$$

and the zeros of $f(x)$ are $x = \pm 2i$, $x = -\frac{1}{2}$, and $x = 1$.

50. $g(x) = x^3 - 7x^2 - x + 87$

Since $5 + 2i$ is a zero, so is $5 - 2i$.

$$
\begin{array}{r|rrrr}
5 + 2i & 1 & -7 & -1 & 87 \\
       &   & 5 + 2i & -14 + 6i & -87 \\
\hline
       & 1 & -2 + 2i & -15 + 6i & 0
\end{array}
$$

$$
\begin{array}{r|rrr}
5 - 2i & 1 & -2 + 2i & -15 + 6i \\
       &   & 5 - 2i & 15 - 6i \\
\hline
       & 1 & 3 & 0
\end{array}
$$

The zero of $x + 3$ is $x = -3$. The zeros of f are $x = -3, 5 \pm 2i$.

51. $g(x) = 4x^3 + 23x^2 + 34x - 10$

Since $-3 + i$ is a zero, so is $-3 - i$.

$$
\begin{array}{r|rrrr}
-3+i & 4 & 23 & 34 & -10 \\
 & & -12+4i & -37-i & 10 \\
\hline
 & 4 & 11+4i & -3-i & 0 \\
\end{array}
$$

$$
\begin{array}{r|rrr}
-3-i & 4 & 11+4i & -3-i \\
 & & -12-4i & 3+i \\
\hline
 & 4 & -1 & 0 \\
\end{array}
$$

The zero of $4x - 1$ is $x = \frac{1}{4}$. The zeros of $g(x)$ are $x = -3 \pm i$ and $x = \frac{1}{4}$.

Alternate Solution

Since $-3 \pm i$ are zeros of $g(x)$,

$$
\begin{aligned}
[x - (-3+i)][x - (-3-i)] &= [(x+3) - i][(x+3) + i] \\
&= (x+3)^2 - i^2 \\
&= x^2 + 6x + 10
\end{aligned}
$$

is a factor of $g(x)$. By long division we have:

$$
\begin{array}{r}
4x - 1 \\
x^2 + 6x + 10 \overline{\smash{)}\, 4x^3 + 23x^2 + 34x - 10} \\
\underline{4x^3 + 24x^2 + 40x} \\
-x^2 - 6x - 10 \\
\underline{-x^2 - 6x - 10} \\
0
\end{array}
$$

Thus, $g(x) = (x^2 + 6x + 10)(4x - 1)$ and the zeros of $g(x)$ are $x = -3 \pm i$ and $x = \frac{1}{4}$.

52. $h(x) = 3x^3 - 4x^2 + 8x + 8$

Since $1 - \sqrt{3}i$ is a zero, so is $1 + \sqrt{3}i$.

$$
\begin{array}{r|rrrr}
1-\sqrt{3}i & 3 & -4 & 8 & 8 \\
 & & 3-3\sqrt{3}i & -10-2\sqrt{3}i & -8 \\
\hline
 & 3 & -1-3\sqrt{3}i & -2-2\sqrt{3}i & 0 \\
\end{array}
$$

$$
\begin{array}{r|rrr}
1+\sqrt{3}i & 3 & -1-3\sqrt{3}i & -2-2\sqrt{3}i \\
 & & 3+3\sqrt{3}i & 2+2\sqrt{3}i \\
\hline
 & 3 & 2 & 0 \\
\end{array}
$$

The zero of $3x + 2$ is $x = -\frac{2}{3}$. The zeros of f are $x = -\frac{2}{3}, 1 \pm \sqrt{3}i$.

53. $f(x) = x^4 + 3x^3 - 5x^2 - 21x + 22$

Since $-3 + \sqrt{2}i$ is a zero, so is $-3 - \sqrt{2}i$, and

$$
\begin{aligned}
\left[x - \left(-3 + \sqrt{2}i\right)\right]\left[x - \left(-3 - \sqrt{2}i\right)\right] &= \left[(x+3) - \sqrt{2}i\right]\left[(x+3) + \sqrt{2}i\right] \\
&= (x+3)^2 - \left(\sqrt{2}i\right)^2 \\
&= x^2 + 6x + 11
\end{aligned}
$$

is a factor of $f(x)$. By long division, we have:

$$
\begin{array}{r}
x^2 - 3x + 2 \\
x^2 + 6x + 11 \overline{\smash{)}\, x^4 + 3x^3 - 5x^2 - 21x + 22} \\
\underline{x^4 + 6x^3 + 11x^2} \\
-3x^3 - 16x^2 - 21x \\
\underline{-3x^3 - 18x^2 - 33x} \\
2x^2 + 12x + 22 \\
\underline{2x^2 + 12x + 22} \\
0
\end{array}
$$

Thus,

$$
\begin{aligned}
f(x) &= (x^2 + 6x + 11)(x^2 - 3x + 2) \\
&= (x^2 + 6x + 11)(x - 1)(x - 2)
\end{aligned}
$$

and the zeros of f are $x = -3 \pm \sqrt{2}i, x = 1,$ and $x = 2$.

54. $f(x) = x^3 + 4x^2 + 14x + 20$

Since $-1 - 3i$ is a zero, so is $-1 + 3i$.

$$
\begin{array}{r|rrrr}
-1 - 3i & 1 & 4 & 14 & 20 \\
 & & -1 - 3i & -12 - 6i & -20 \\
\hline
 & 1 & 3 - 3i & 2 - 6i & 0 \\
\end{array}
$$

$$
\begin{array}{r|rrr}
-1 + 3i & 1 & 3 - 3i & 2 - 6i \\
 & & -1 + 3i & -2 + 6i \\
\hline
 & 1 & 2 & 0 \\
\end{array}
$$

The zero of $x + 2$ is $x = -2$.

The zeros of f are $x = -2, -1 \pm 3i$.

55. $f(x) = x^2 + 25$

$\qquad = (x + 5i)(x - 5i)$

The zeros of $f(x)$ are $x = \pm 5i$.

56. $f(x) = x^2 - x + 56$

By the Quadratic Formula, the zeros of $f(x)$ are

$$x = \frac{1 \pm \sqrt{1 - 224}}{2} = \frac{1 \pm \sqrt{223}i}{2}.$$

$$f(x) = \left(x - \frac{1 - \sqrt{223}i}{2}\right)\left(x - \frac{1 + \sqrt{223}i}{2}\right)$$

57. $h(x) = x^2 - 4x + 1$

By the Quadratic Formula, the zeros of $h(x)$ are

$$x = \frac{4 \pm \sqrt{16 - 4}}{2} = 2 \pm \sqrt{3}.$$

$$
\begin{aligned}
h(x) &= \left[x - \left(2 + \sqrt{3}\right)\right]\left[x - \left(2 - \sqrt{3}\right)\right] \\
&= \left(x - 2 - \sqrt{3}\right)\left(x - 2 + \sqrt{3}\right)
\end{aligned}
$$

58. $g(x) = x^2 + 10x + 23$

By the Quadratic Formula, the zeros of $f(x)$ are

$$x = \frac{-10 \pm \sqrt{100 - 92}}{2} = \frac{-10 \pm \sqrt{8}}{2} = -5 \pm \sqrt{2}.$$

$$g(x) = \left(x + 5 + \sqrt{2}\right)\left(x + 5 - \sqrt{2}\right)$$

59. $f(x) = x^4 - 81$

$\qquad = (x^2 - 9)(x^2 + 9)$

$\qquad = (x + 3)(x - 3)(x + 3i)(x - 3i)$

The zeros of $f(x)$ are $x = \pm 3$ and $x = \pm 3i$.

60. $f(y) = y^4 - 625$

$\qquad = (y^2 + 25)(y^2 - 25)$

Zeros: $y = \pm 5, \pm 5i$

$f(y) = (y + 5)(y - 5)(y + 5i)(y - 5i)$

61. $f(z) = z^2 - 2z + 2$

By the Quadratic Formula, the zeros of $f(z)$ are

$$z = \frac{2 \pm \sqrt{4 - 8}}{2} = 1 \pm i.$$

$$
\begin{aligned}
f(z) &= [z - (1 + i)][z - (1 - i)] \\
&= (z - 1 - i)(z - 1 + i)
\end{aligned}
$$

62. $h(x) = x^3 - 3x^2 + 4x - 2$

Possible rational zeros: $\pm 1, \pm 2$

$$
\begin{array}{r|rrrr}
1 & 1 & -3 & 4 & -2 \\
 & & 1 & -2 & 2 \\
\hline
 & 1 & -2 & 2 & 0 \\
\end{array}
$$

By the Quadratic Formula, the zeros of $x^2 - 2x + 2$

are $x = \dfrac{2 \pm \sqrt{4 - 8}}{2} = 1 \pm i.$

Zeros: $x = 1, 1 \pm i$

$h(x) = (x - 1)(x - 1 - i)(x - 1 + i)$

63. $g(x) = x^3 - 6x^2 + 13x - 10$

Possible rational zeros: $\pm 1, \pm 2, \pm 5, \pm 10$

$$
\begin{array}{r|rrrr}
2 & 1 & -6 & 13 & -10 \\
 & & 2 & -8 & 10 \\
\hline
 & 1 & -4 & 5 & 0 \\
\end{array}
$$

By the Quadratic Formula, the zeros of $x^2 - 4x + 5$ are

$$x = \frac{4 \pm \sqrt{16 - 20}}{2} = 2 \pm i.$$

The zeros of $g(x)$ are $x = 2$ and $x = 2 \pm i$.

$$
\begin{aligned}
g(x) &= (x - 2)[x - (2 + i)][x - (2 - i)] \\
&= (x - 2)(x - 2 - i)(x - 2 + i)
\end{aligned}
$$

64. $f(x) = x^3 - 2x^2 - 11x + 52$

Possible rational zeros: $\pm 1, \pm 2, \pm 4, \pm 13, \pm 26$

$$
\begin{array}{r|rrrr}
-4 & 1 & -2 & -11 & 52 \\
   &   & -4 & 24 & -52 \\
\hline
   & 1 & -6 & 13 & 0
\end{array}
$$

By the Quadratic Formula, the zeros of $x^2 - 6x + 13$ are $x = \dfrac{6 \pm \sqrt{36 - 52}}{2} = 3 \pm 2i.$

Zeros: $x = -4, 3 \pm 2i$

$f(x) = (x + 4)(x - 3 - 2i)(x - 3 + 2i)$

65. $h(x) = x^3 - x + 6$

Possible rational zeros: $\pm 1, \pm 2, \pm 3, \pm 6$

$$
\begin{array}{r|rrrr}
-2 & 1 & 0 & -1 & 6 \\
   &   & -2 & 4 & -6 \\
\hline
   & 1 & -2 & 3 & 0
\end{array}
$$

By the Quadratic Formula, the zeros of $x^2 - 2x + 3$ are

$x = \dfrac{2 \pm \sqrt{4 - 12}}{2} = 1 \pm \sqrt{2}\,i.$

The zeros of $h(x)$ are $x = -2$ and $x = 1 \pm \sqrt{2}\,i.$

$h(x) = [x - (-2)]\big[x - \big(1 + \sqrt{2}\,i\big)\big]\big[x - \big(1 - \sqrt{2}\,i\big)\big]$

$\quad = (x + 2)\big(x - 1 - \sqrt{2}\,i\big)\big(x - 1 + \sqrt{2}\,i\big)$

66. $h(x) = x^3 + 9x^2 + 27x + 35$

Possible rational zeros: $\pm 1, \pm 5, \pm 7, \pm 35$

$$
\begin{array}{r|rrrr}
-5 & 1 & 9 & 27 & 35 \\
   &   & -5 & -20 & -35 \\
\hline
   & 1 & 4 & 7 & 0
\end{array}
$$

By the Quadratic Formula, the zeros of $x^2 + 4x + 7$

are $x = \dfrac{-4 \pm \sqrt{16 - 28}}{2} = -2 \pm \sqrt{3}i.$

Zeros: $-5, -2 \pm \sqrt{3}i$

$h(x) = (x + 5)\big(x + 2 + \sqrt{3}i\big)\big(x + 2 - \sqrt{3}i\big)$

67. $f(x) = 5x^3 - 9x^2 + 28x + 6$

Possible rational zeros:
$\pm 1, \pm 2, \pm 3, \pm 6, \pm \frac{1}{5}, \pm \frac{2}{5}, \pm \frac{3}{5}, \pm \frac{6}{5}$

$$
\begin{array}{r|rrrr}
-\frac{1}{5} & 5 & -9 & 28 & 6 \\
   &   & -1 & 2 & -6 \\
\hline
   & 5 & -10 & 30 & 0
\end{array}
$$

By the Quadratic Formula, the zeros of
$5x^2 - 10x + 30 = 5(x^2 - 2x + 6)$ are

$x = \dfrac{2 \pm \sqrt{4 - 24}}{2} = 1 \pm \sqrt{5}\,i.$

The zeros of $f(x)$ are $x = -\frac{1}{5}$ and $x = 1 \pm \sqrt{5}\,i.$

$f(x) = \big[x - \big(-\tfrac{1}{5}\big)\big](5)\big[x - \big(1 + \sqrt{5}\,i\big)\big]\big[x - \big(1 - \sqrt{5}\,i\big)\big]$

$\quad = (5x + 1)\big(x - 1 - \sqrt{5}\,i\big)\big(x - 1 + \sqrt{5}\,i\big)$

68. $g(x) = 3x^3 - 4x^2 + 8x + 8$

Possible rational zeros:
$\pm 1, \pm 2, \pm 4, \pm 8, \pm \frac{1}{3}, \pm \frac{2}{3}, \pm \frac{4}{3}, \pm \frac{8}{3}$

$$
\begin{array}{r|rrrr}
-\frac{2}{3} & 3 & -4 & 8 & 8 \\
   &   & -2 & 4 & -8 \\
\hline
   & 3 & -6 & 12 & 0
\end{array}
$$

By the Quadratic Formula, the zeros of
$3x^2 - 6x + 12 = 3(x^2 - 2x + 4)$ are

$x = \dfrac{2 \pm \sqrt{4 - 16}}{2} = 1 \pm \sqrt{3}i.$

Zeros: $x = -\frac{2}{3}, 1 \pm \sqrt{3}i$

$g(x) = (3x + 2)\big(x - 1 + \sqrt{3}i\big)\big(x - 1 - \sqrt{3}i\big)$

69. $g(x) = x^4 - 4x^3 + 8x^2 - 16x + 16$

Possible rational zeros: $\pm 1, \pm 2, \pm 4, \pm 8, \pm 16$

$$
\begin{array}{r|rrrrr}
2 & 1 & -4 & 8 & -16 & 16 \\
  &   & 2 & -4 & 8 & -16 \\
\hline
  & 1 & -2 & 4 & -8 & 0 \\
\end{array}
$$

$$
\begin{array}{r|rrrr}
2 & 1 & -2 & 4 & -8 \\
  &   & 2 & 0 & 8 \\
\hline
  & 1 & 0 & 4 & 0 \\
\end{array}
$$

$g(x) = (x - 2)(x - 2)(x^2 + 4) = (x - 2)^2(x + 2i)(x - 2i)$

The zeros of $g(x)$ are 2 and $\pm 2i$.

70. $h(x) = x^4 + 6x^3 + 10x^2 + 6x + 9$

Possible rational zeros: $\pm 1, \pm 3, \pm 9$

$$
\begin{array}{r|rrrrr}
-3 & 1 & 6 & 10 & 6 & 9 \\
   &   & -3 & -9 & -3 & -9 \\
\hline
   & 1 & 3 & 1 & 3 & 0 \\
\end{array}
$$

$$
\begin{array}{r|rrrr}
-3 & 1 & 3 & 1 & 3 \\
   &   & -3 & 0 & -3 \\
\hline
   & 1 & 0 & 1 & 0 \\
\end{array}
$$

The zeros of $x^2 + 1$ are $x = \pm i$.

Zeros: $x = -3, \pm i$

$h(x) = (x + 3)^2(x + i)(x - i)$

71. $f(x) = x^4 + 10x^2 + 9$

$\qquad = (x^2 + 1)(x^2 + 9)$

$\qquad = (x + i)(x - i)(x + 3i)(x - 3i)$

The zeros of $f(x)$ are $x = \pm i$ and $x = \pm 3i$.

72. $f(x) = x^4 + 29x^2 + 100$

$\qquad = (x^2 + 25)(x^2 + 4)$

Zeros: $x = \pm 2i, \pm 5i$

$f(x) = (x + 2i)(x - 2i)(x + 5i)(x - 5i)$

73. $f(x) = x^3 + 24x^2 + 214x + 740$

Possible rational zeros: $\pm 1, \pm 2, \pm 4, \pm 5, \pm 10, \pm 20, \pm 37,$
$\qquad\qquad \pm 74, \pm 148, \pm 185, \pm 370, \pm 740$

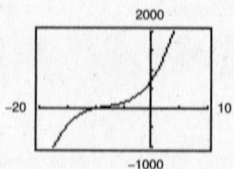

Based on the graph, try $x = -10$.

$$
\begin{array}{r|rrrr}
-10 & 1 & 24 & 214 & 740 \\
    &   & -10 & -140 & -740 \\
\hline
    & 1 & 14 & 74 & 0 \\
\end{array}
$$

By the Quadratic Formula, the zeros of $x^2 + 14x + 74$ are

$$x = \frac{-14 \pm \sqrt{196 - 296}}{2} = -7 \pm 5i.$$

The zeros of $f(x)$ are $x = -10$ and $x = -7 \pm 5i$.

74. $f(s) = 2s^3 - 5s^2 + 12s - 5$

Possible rational zeros: $\pm 1, \pm 5, \pm \frac{1}{2}, \pm \frac{5}{2}$

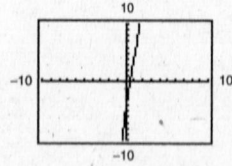

Based on the graph, try $s = \frac{1}{2}$.

$$
\begin{array}{r|rrrr}
\frac{1}{2} & 2 & -5 & 12 & -5 \\
            &   & 1 & -2 & 5 \\
\hline
            & 2 & -4 & 10 & 0 \\
\end{array}
$$

By the Quadratic Formula, the zeros of $2(s^2 - 2s + 5)$ are

$$s = \frac{2 \pm \sqrt{4 - 20}}{2} = 1 \pm 2i.$$

The zeros of $f(s)$ are $s = \frac{1}{2}$ and $s = 1 \pm 2i$.

75. $f(x) = 16x^3 - 20x^2 - 4x + 15$

Possible rational zeros:

$\pm 1, \pm 3, \pm 5, \pm 15, \pm \frac{1}{2}, \pm \frac{3}{2}, \pm \frac{5}{2}, \pm \frac{15}{2}, \pm \frac{1}{4}, \pm \frac{3}{4},$
$\pm \frac{5}{4}, \pm \frac{15}{4}, \pm \frac{1}{8}, \pm \frac{3}{8}, \pm \frac{5}{8}, \pm \frac{15}{8}, \pm \frac{1}{16}, \pm \frac{3}{16}, \pm \frac{5}{16}, \pm \frac{15}{16}$

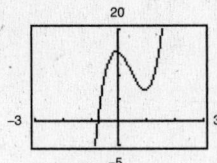

Based on the graph, try $x = -\frac{3}{4}$.

$$
\begin{array}{r|rrrr}
-\frac{3}{4} & 16 & -20 & -4 & 15 \\
 & & -12 & 24 & -15 \\
\hline
 & 16 & -32 & 20 & 0
\end{array}
$$

By the Quadratic Formula, the zeros of

$16x^2 - 32x + 20 = 4(4x^2 - 8x + 5)$ are

$$x = \frac{8 \pm \sqrt{64 - 80}}{8} = 1 \pm \frac{1}{2}i.$$

The zeros of $f(x)$ are $x = -\frac{3}{4}$ and $x = 1 \pm \frac{1}{2}i$.

76. $f(x) = 9x^3 - 15x^2 + 11x - 5$

Possible rational zeros: $\pm 1, \pm 5, \pm \frac{1}{3}, \pm \frac{5}{3}, \pm \frac{1}{9}, \pm \frac{5}{9}$

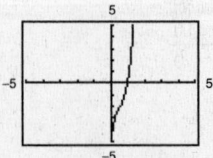

Based on the graph, try $x = 1$.

$$
\begin{array}{r|rrrr}
1 & 9 & -15 & 11 & -5 \\
 & & 9 & -6 & 5 \\
\hline
 & 9 & -6 & 5 & 0
\end{array}
$$

By the Quadratic Formula, the zeros of $9x^2 - 6x + 5$ are

$$x = \frac{6 \pm \sqrt{36 - 180}}{18} = \frac{1}{3} \pm \frac{2}{3}i.$$

The zeros of $f(x)$ are $x = 1$ and $x = \frac{1}{3} \pm \frac{2}{3}i$.

77. $f(x) = 2x^4 + 5x^3 + 4x^2 + 5x + 2$

Possible rational zeros: $\pm 1, \pm 2, \pm \frac{1}{2}$

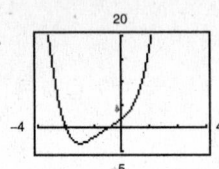

Based on the graph, try $x = -2$ and $x = -\frac{1}{2}$.

$$
\begin{array}{r|rrrrr}
-2 & 2 & 5 & 4 & 5 & 2 \\
 & & -4 & -2 & -4 & -2 \\
\hline
 & 2 & 1 & 2 & 1 & 0
\end{array}
$$

$$
\begin{array}{r|rrrr}
-\frac{1}{2} & 2 & 1 & 2 & 1 \\
 & & -1 & 0 & -1 \\
\hline
 & 2 & 0 & 2 & 0
\end{array}
$$

The zeros of $2x^2 + 2 = 2(x^2 + 1)$ are $x = \pm i$.

The zeros of $f(x)$ are $x = -2$, $x = -\frac{1}{2}$, and $x = \pm i$.

78. $g(x) = x^5 - 8x^4 + 28x^3 - 56x^2 + 64x - 32$

Possible rational zeros: $\pm 1, \pm 2, \pm 4, \pm 8, \pm 16, \pm 32$

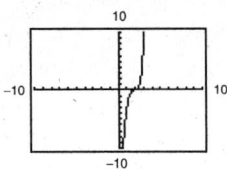

Based on the graph, try $x = 2$.

$$
\begin{array}{r|rrrrrr}
2 & 1 & -8 & 28 & -56 & 64 & -32 \\
 & & 2 & -12 & 32 & -48 & 32 \\
\hline
 & 1 & -6 & 16 & -24 & 16 & 0
\end{array}
$$

$$
\begin{array}{r|rrrrr}
2 & 1 & -6 & 16 & -24 & 16 \\
 & & 2 & -8 & 16 & -16 \\
\hline
 & 1 & -4 & 8 & -8 & 0
\end{array}
$$

$$
\begin{array}{r|rrrr}
2 & 1 & -4 & 8 & -8 \\
 & & 2 & -4 & 8 \\
\hline
 & 1 & -2 & 4 & 0
\end{array}
$$

By the Quadratic Formula, the zeros of $x^2 - 2x + 4$ are

$$x = \frac{2 \pm \sqrt{4 - 16}}{2} = 1 \pm \sqrt{3}i.$$

The zeros of $g(x)$ are $x = 2$ and $x = 1 \pm \sqrt{3}i$.

79. $g(x) = 5x^5 + 10x = 5x(x^4 + 2)$

Let $f(x) = x^4 + 2$.

Sign variations: 0, positive zeros: 0

$f(-x) = x^4 + 2$

Sign variations: 0, negative zeros: 0

80. $h(x) = 4x^2 - 8x + 3$

Sign variations: 2, positive zeros: 2 or 0

$h(-x) = 4x^2 + 8x + 3$

Sign variations: 0, negative zeros: 0

81. $h(x) = 3x^4 + 2x^2 + 1$

Sign variations: 0, positive zeros: 0

$h(-x) = 3x^4 + 2x^2 + 1$

Sign variations: 0, negative zeros: 0

82. $h(x) = 2x^4 - 3x + 2$

Sign variations: 2, positive zeros: 2 or 0

$h(-x) = 2x^4 + 3x + 2$

Sign variations: 0, negative zeros: 0

83. $g(x) = 2x^3 - 3x^2 - 3$

Sign variations: 1, positive zeros: 1

$g(-x) = -2x^3 - 3x^2 - 3$

Sign variations: 0, negative zeros: 0

84. $f(x) = 4x^3 - 3x^2 + 2x - 1$

Sign variations: 3, positive zeros: 3 or 1

$f(-x) = -4x^3 - 3x^2 - 2x - 1$

Sign variations: 0, negative zeros: 0

85. $f(x) = -5x^3 + x^2 - x + 5$

Sign variations: 3, positive zeros: 3 or 1

$f(-x) = 5x^3 + x^2 + x + 5$

Sign variations: 0, negative zeros: 0

86. $f(x) = 3x^3 + 2x^2 + x + 3$

Sign variations: 0, positive zeros: 0

$f(-x) = -3x^3 + 2x^2 - x + 3$

Sign variations: 3, negative zeros: 3 or 1

87. $f(x) = x^4 - 4x^3 + 15$

(a)
$$
\begin{array}{r|rrrrr}
4 & 1 & -4 & 0 & 0 & 15 \\
  &   & 4 & 0 & 0 & 0 \\
\hline
  & 1 & 0 & 0 & 0 & 15
\end{array}
$$

4 is an upper bound.

(b)
$$
\begin{array}{r|rrrrr}
-1 & 1 & -4 & 0 & 0 & 15 \\
   &   & -1 & 5 & -5 & 5 \\
\hline
   & 1 & -5 & 5 & -5 & 20
\end{array}
$$

-1 is a lower bound.

88. $f(x) = 2x^3 - 3x^2 - 12x + 8$

(a)
$$
\begin{array}{r|rrrr}
4 & 2 & -3 & -12 & 8 \\
  &   & 8 & 20 & 32 \\
\hline
  & 2 & 5 & 8   & 40
\end{array}
$$

4 is an upper bound.

(b)
$$
\begin{array}{r|rrrr}
-3 & 2 & -3 & -12 & 8 \\
   &   & -6 & 27 & -45 \\
\hline
   & 2 & -9 & 15 & -37
\end{array}
$$

-3 is a lower bound.

89. $f(x) = x^4 - 4x^3 + 16x - 16$

(a)
$$
\begin{array}{r|rrrrr}
5 & 1 & -4 & 0 & 16 & -16 \\
  &   & 5 & 5 & 25 & 205 \\
\hline
  & 1 & 1 & 5 & 41 & 189
\end{array}
$$

5 is an upper bound.

(b)
$$
\begin{array}{r|rrrrr}
-3 & 1 & -4 & 0 & 16 & -16 \\
   &   & -3 & 21 & -63 & 141 \\
\hline
   & 1 & -7 & 21 & -47 & 125
\end{array}
$$

-3 is a lower bound.

90. $f(x) = 2x^4 - 8x + 3$

(a)
$$
\begin{array}{r|rrrrr}
3 & 2 & 0 & 0 & -8 & 3 \\
  &   & 6 & 18 & 54 & 138 \\
\hline
  & 2 & 6 & 18 & 46 & 141
\end{array}
$$

3 is an upper bound.

(b)
$$
\begin{array}{r|rrrrr}
-4 & 2 & 0 & 0   & -8   & 3 \\
   &   & -8 & 32 & -128 & 544 \\
\hline
   & 2 & -8 & 32 & -136 & 547
\end{array}
$$

-3 is a lower bound.

91. $f(x) = 4x^3 - 3x - 1$

Possible rational zeros: $\pm 1, \pm\frac{1}{2}, \pm\frac{1}{4}$

$$
\begin{array}{r|rrrr}
1 & 4 & 0 & -3 & -1 \\
  &   & 4 & 4 & 1 \\
\hline
  & 4 & 4 & 1 & 0
\end{array}
$$

$4x^3 - 3x - 1 = (x - 1)(4x^2 + 4x + 1)$

$\qquad\qquad\qquad = (x - 1)(2x + 1)^2$

Thus, the zeros are 1 and $-\frac{1}{2}$.

92. $f(z) = 12z^3 - 4z^2 - 27z + 9$

Possible rational zeros: $\pm 1, \pm 3, \pm 9, \pm\frac{1}{2}, \pm\frac{3}{2}, \pm\frac{9}{2}, \pm\frac{1}{3},$
$\pm\frac{1}{4}, \pm\frac{3}{4}, \pm\frac{9}{4}, \pm\frac{1}{6}, \pm\frac{1}{12}$

$$
\begin{array}{r|rrrr}
\frac{3}{2} & 12 & -4 & -27 & 9 \\
            &   & 18 & 21 & -9 \\
\hline
            & 12 & 14 & -6 & 0
\end{array}
$$

$f(z) = 2\left(z - \frac{3}{2}\right)(6z^2 + 7z - 3)$

$\qquad = (2z - 3)(3z - 1)(2z + 3)$

Real zeros: $-\frac{3}{2}, \frac{1}{3}, \frac{3}{2}$

93. $f(y) = 4y^3 + 3y^2 + 8y + 6$

Possible rational zeros: $\pm 1, \pm 2, \pm 3, \pm 6, \pm\frac{1}{2}, \pm\frac{3}{2}, \pm\frac{1}{4}, \pm\frac{3}{4}$

$$
\begin{array}{r|rrrr}
-\frac{3}{4} & 4 & 3 & 8 & 6 \\
             &   & -3 & 0 & -6 \\
\hline
             & 4 & 0 & 8 & 0
\end{array}
$$

$4y^3 + 3y^2 + 8y + 6 = \left(y + \frac{3}{4}\right)(4y^2 + 8)$

$\qquad\qquad\qquad\qquad = \left(y + \frac{3}{4}\right)4(y^2 + 2)$

$\qquad\qquad\qquad\qquad = (4y + 3)(y^2 + 2)$

Thus, the only real zero is $-\frac{3}{4}$.

94. $g(x) = 3x^3 - 2x^2 + 15x - 10$

Possible rational zeros: $\pm 1, \pm 2, \pm 5, \pm 10, \pm\frac{1}{3}, \pm\frac{2}{3}, \pm\frac{5}{3}, \pm\frac{10}{3}$

$$
\begin{array}{r|rrrr}
\frac{2}{3} & 3 & -2 & 15 & -10 \\
            &   & 2 & 0 & 10 \\
\hline
            & 3 & 0 & 15 & 0
\end{array}
$$

$g(x) = \left(x - \frac{2}{3}\right)(3x^2 + 15) = (3x - 2)(x^2 + 5)$

Thus, the only real zero is $\frac{2}{3}$.

95. $P(x) = x^4 - \frac{25}{4}x^2 + 9$

$\qquad = \frac{1}{4}(4x^4 - 25x^2 + 36)$

$\qquad = \frac{1}{4}(4x^2 - 9)(x^2 - 4)$

$\qquad = \frac{1}{4}(2x + 3)(2x - 3)(x + 2)(x - 2)$

The rational zeros are $\pm\frac{3}{2}$ and ± 2.

96. $f(x) = \frac{1}{2}(2x^3 - 3x^2 - 23x + 12)$

Possible rational zeros: $\pm 1, \pm 2, \pm 3, \pm 4, \pm 6, \pm 12, \pm\frac{1}{2}, \pm\frac{3}{2}$

$$
\begin{array}{r|rrrr}
4 & 2 & -3 & -23 & 12 \\
  &   & 8 & 20 & -12 \\
\hline
  & 2 & 5 & -3 & 0
\end{array}
$$

$f(x) = \frac{1}{2}(x - 4)(2x^2 + 5x - 3) = \frac{1}{2}(x - 4)(2x - 1)(x + 3)$

The rational zeros are $-3, \frac{1}{2}$, and 4.

97. $f(x) = x^3 - \frac{1}{4}x^2 - x + \frac{1}{4}$

$\qquad = \frac{1}{4}(4x^3 - x^2 - 4x + 1)$

$\qquad = \frac{1}{4}[x^2(4x - 1) - 1(4x - 1)]$

$\qquad = \frac{1}{4}(4x - 1)(x^2 - 1)$

$\qquad = \frac{1}{4}(4x - 1)(x + 1)(x - 1)$

The rational zeros are $\frac{1}{4}$ and ± 1.

98. $f(z) = \frac{1}{6}(6z^3 + 11z^2 - 3z - 2)$

Possible rational zeros: $\pm 1, \pm 2, \pm\frac{1}{2}, \pm\frac{1}{3}, \pm\frac{2}{3}, \pm\frac{1}{6}$

$$
\begin{array}{r|rrrr}
-2 & 6 & 11 & -3 & -2 \\
   &   & -12 & 2 & 2 \\
\hline
   & 6 & -1 & -1 & 0
\end{array}
$$

$f(z) = \frac{1}{6}(z + 2)(6z^2 - z - 1)$

$\qquad = \frac{1}{6}(z + 2)(3z + 1)(2z - 1)$

Rational zeros: $-2, -\frac{1}{3}, \frac{1}{2}$

99. $f(x) = x^3 - 1 = (x - 1)(x^2 + x + 1)$

Rational zeros: 1 $(x = 1)$

Irrational zeros: 0

Matches (d).

100. $f(x) = x^3 - 2$

$\qquad = \left(x - \sqrt[3]{2}\right)\left(x^2 + \sqrt[3]{2}x + \sqrt[3]{4}\right)$

Rational zeros: 0

Irrational zeros: 1 $\left(x = \sqrt[3]{2}\right)$

Matches (a).

101. $f(x) = x^3 - x = x(x + 1)(x - 1)$

Rational zeros: 3 $(x = 0, \pm 1)$

Irrational zeros: 0

Matches (b).

102. $f(x) = x^3 - 2x$

$\qquad = x(x^2 - 2)$

$\qquad = x\left(x + \sqrt{2}\right)\left(x - \sqrt{2}\right)$

Rational zeros: 1 $(x = 0)$

Irrational zeros: 2 $\left(x = \pm\sqrt{2}\right)$

Matches (c).

103. (a)

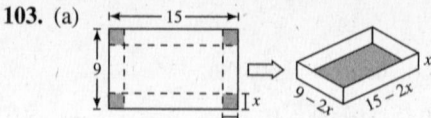

(c)

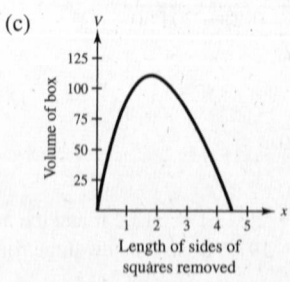

Length of sides of
squares removed

The volume is maximum when $x \approx 1.82$.

The dimensions are: length $\approx 15 - 2(1.82) = 11.36$

$\qquad\qquad\qquad\quad$ width $\approx 9 - 2(1.82) = 5.36$

$\qquad\qquad\qquad\quad$ height $= x \approx 1.82$

1.82 cm $\times$ 5.36 cm $\times$ 11.36 cm

(b) $V = l \cdot w \cdot h = (15 - 2x)(9 - 2x)x$

$\qquad = x(9 - 2x)(15 - 2x)$

Since length, width, and height must be positive, we have $0 < x < \frac{9}{2}$ for the domain.

(d) $56 = x(9 - 2x)(15 - 2x)$

$\quad 56 = 135x - 48x^2 + 4x^3$

$\quad\; 0 = 4x^3 - 48x^2 + 135x - 56$

The zeros of this polynomial are $\frac{1}{2}, \frac{7}{2}$, and 8.
x cannot equal 8 since it is not in the domain of V.
[The length cannot equal -1 and the width cannot equal -7. The product of $(8)(-1)(-7) = 56$ so it showed up as an extraneous solution.]

Thus, the volume is 56 cubic centimeters when $x = \frac{1}{2}$ centimeter or $x = \frac{7}{2}$ centimeters.

104. (a) Combined length and width:

$4x + y = 120 \implies y = 120 - 4x$

Volume $= l \cdot w \cdot h = x^2 y$

$\qquad\qquad\quad = x^2(120 - 4x)$

$\qquad\qquad\quad = 4x^2(30 - x)$

(b) 18,000

Dimensions with maximum volume:

20 in. $\times$ 20 in. $\times$ 40 in.

(c) $\qquad\qquad 13,500 = 4x^2(30 - x)$

$4x^3 - 120x^2 + 13,500 = 0$

$\qquad x^3 - 30x^2 + 3375 = 0$

$$
\begin{array}{r|rrrr}
15 & 1 & -30 & 0 & 3375 \\
   &   & 15 & -225 & -3375 \\
\hline
   & 1 & -15 & -225 & 0
\end{array}
$$

$(x - 15)(x^2 - 15x - 225) = 0$

Using the Quadratic Formula, $x = 15, \dfrac{15 \pm 15\sqrt{5}}{2}$.

The value of $\dfrac{15 - 15\sqrt{5}}{2}$ is not possible because it is negative.

105.
$$P = -76x^3 + 4830x^2 - 320,000, \ 0 \le x \le 60$$

$$2,500,000 = -76x^3 + 4830x^2 - 320,000$$

$$76x^3 - 4830x^2 + 2,820,000 = 0$$

The zeros of this equation are $x \approx 46.1$, $x \approx 38.4$, and $x \approx -21.0$. Since $0 \le x \le 60$, we disregard $x \approx -21.0$. The smaller remaining solution is $x \approx 38.4$. The advertising expense is $384,000.

106.
$$P = -45x^3 + 2500x^2 - 275,000$$

$$800,000 = -45x^3 + 2500x^2 - 275,000$$

$$0 = 45x^3 - 2500x^2 + 1,075,000$$

$$0 = 9x^3 - 500x^2 + 215,000$$

The zeros of this equation are $x \approx -18.0$, $x \approx 31.5$, and $x \approx 42.0$. Because $0 \le x \le 50$, disregard $x \approx -18.02$. The smaller remaining solution is $x \approx 31.5$, or an advertising expense of $315,000.

107. (a) Current bin: $V = 2 \times 3 \times 4 = 24$ cubic feet

New bin: $V = 5(24) = 120$ cubic feet

$$V = (2 + x)(3 + x)(4 + x) = 120$$

(b) $x^3 + 9x^2 + 26x + 24 = 120$

$$x^3 + 9x^2 + 26x - 96 = 0$$

The only real zero of this polynomial is $x = 2$. All the dimensions should be increased by 2 feet, so the new bin will have dimensions of 4 feet by 5 feet by 6 feet.

108. (a) $A = (250 + x)(160 + x) = (1.5)(160)(250)$

$$= 60,000$$

(b) $60,000 = x^2 + 410x + 40,000$

$$0 = x^2 + 410x - 20,000$$

$$x = \frac{-410 \pm \sqrt{410^2 - (4)(1)(-20,000)}}{2(1)}$$

$$= \frac{-410 \pm \sqrt{248,100}}{2}$$

x must be positive, so

$$x = \frac{-410 + \sqrt{248,100}}{2}$$

$$\approx 44.05.$$

The new length is $250 + 44.05 = 294.05$ ft and the new width is $160 + 44.05 = 204.05$ ft, so the new dimensions are 204.05 ft $\times$ 294.05 ft.

(c) $A = (250 + 2x)(160 + x) = 60,000$

$$2x^2 + 570x - 20,000 = 0$$

$$x = \frac{-570 \pm \sqrt{570^2 - (4)(2)(-20,000)}}{2(2)}$$

x must be positive, so

$$x = \frac{-570 + \sqrt{484,900}}{4} \approx 31.6.$$

The new length is $250 + 2(31.6) = 313.2$ ft and the new width is $160 + (31.6) = 191.6$ ft, so the new dimensions are 191.6 ft $\times$ 313.2 ft.

109. $C = 100\left(\dfrac{200}{x^2} + \dfrac{x}{x + 30}\right), \ x \ge 1$

C is minimum when $3x^3 - 40x^2 - 2400x - 36000 = 0$.

The only real zero is $x \approx 40$ or 4000 units.

110. $h(t) = -16t^2 + 48t + 6$

Let $h = 64$ and solve for t.

$$64 = -16t^2 + 48t + 6$$

$$16t^2 - 48t + 58 = 0$$

By the Quadratic Formula we have $t = \dfrac{48 \pm i\sqrt{1408}}{32}$.

Since the equation yields only imaginary zeros, it is *not* possible for the ball to have reached a height of 64 feet.

111.

$$P = R - C = xp - C$$

$$= x(140 - 0.0001x) - (80x + 150,000)$$

$$= -0.0001x^2 + 60x - 150,000$$

$$9,000,000 = -0.0001x^2 + 60x - 150,000$$

Thus, $0 = 0.0001x^2 - 60x + 9,150,000.$

$$x = \frac{60 \pm \sqrt{-60}}{0.0002} = 300,000 \pm 10,000\sqrt{15}\,i$$

Since the solutions are both complex, it is not possible to determine a price p that would yield a profit of 9 million dollars.

112. (a) $A \approx 0.0167t^3 - 0.508t^2 + 5.60t - 13.4$

(b) 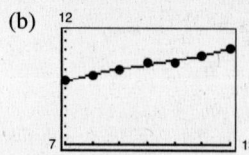   The model is a good fit to the actual data.

(c) $A = 8.5$ when $t \approx 10$ which corresponds to the year 2000.

(d) $A = 9$ when $t \approx 11$ which corresponds to the year 2001.

(e) Yes. The degree of A is odd and the leading coefficient is positive, so as x increases, A will increase. This implies that attendance will continue to grow.

113. False. The most nonreal complex zeros it can have is two and the Linear Factorization Theorem guarantees that there are 3 linear factors, so one zero must be real.

114. False. f does not have real coefficients.

115. $g(x) = -f(x)$. This function would have the same zeros as $f(x)$ so r_1, r_2, and r_3 are also zeros of $g(x)$.

116. $g(x) = 3f(x)$. This function has the same zeros as f because it is a vertical stretch of f. The zeros of g are r_1, r_2, and r_3.

117. $g(x) = f(x - 5)$. The graph of $g(x)$ is a horizontal shift of the graph of $f(x)$ five units to the right so the zeros of $g(x)$ are $5 + r_1$, $5 + r_2$, and $5 + r_3$.

118. $g(x) = f(2x)$. Note that x is a zero of g if and only if $2x$ is a zero of f. The zeros of g are $\frac{r_1}{2}$, $\frac{r_2}{2}$, and $\frac{r_3}{2}$.

119. $g(x) = 3 + f(x)$. Since $g(x)$ is a vertical shift of the graph of $f(x)$, the zeros of $g(x)$ cannot be determined.

120. $g(x) = f(-x)$. Note that x is a zero of g if and only if $-x$ is a zero of f. The zeros of g are $-r_1$, $-r_2$, and $-r_3$.

121. $f(x) = x^4 - 4x^2 + k$

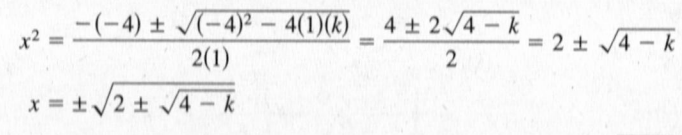

$$x^2 = \frac{-(-4) \pm \sqrt{(-4)^2 - 4(1)(k)}}{2(1)} = \frac{4 \pm 2\sqrt{4 - k}}{2} = 2 \pm \sqrt{4 - k}$$

$$x = \pm\sqrt{2 \pm \sqrt{4 - k}}$$

(a) For there to be four distinct real roots, both $4 - k$ and $2 \pm \sqrt{4 - k}$ must be positive. This occurs when $0 < k < 4$. Thus, some possible k-values are $k = 1$, $k = 2$, $k = 3$, $k = \frac{1}{2}$, $k = \sqrt{2}$, etc.

(b) For there to be two real roots, each of multiplicity 2, $4 - k$ must equal zero. Thus, $k = 4$.

(c) For there to be two real zeros and two complex zeros, $2 + \sqrt{4 - k}$ must be positive and $2 - \sqrt{4 - k}$ must be negative. This occurs when $k < 0$. Thus, some possible k-values are $k = -1$, $k = -2$, $k = -\frac{1}{2}$, etc.

(d) For there to be four complex zeros, $2 \pm \sqrt{4 - k}$ must be nonreal. This occurs when $k > 4$. Some possible k-values are $k = 5$, $k = 6$, $k = 7.4$, etc.

122. (a) $g(x) = f(x - 2)$

No. This function is a horizontal shift of $f(x)$. Note that x is a zero of g if and only if $x - 2$ is a zero of f; the number of real and complex zeros is not affected by a horizontal shift.

(b) $g(x) = f(2x)$

No. Since x is a zero of g if and only if $2x$ is a zero of f, the number of real and complex zeros of g is the same as the number of real and complex zeros of f.

123. Zeros: $-2, \frac{1}{2}, 3$

$$f(x) = -(x+2)(2x-1)(x-3)$$

$$= -2x^3 + 3x^2 + 11x - 6$$

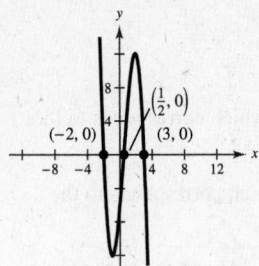

Any nonzero scalar multiple of f would have the same three zeros. Let $g(x) = af(x)$, $a > 0$. There are infinitely many possible functions for f.

124.

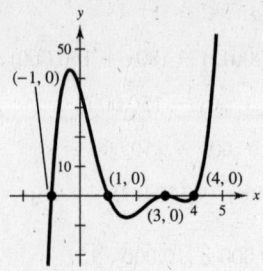

125. Answers will vary. Some of the factoring techniques are:

1. Factor out the greatest common factor.

2. Use special product formulas.

$$a^2 - b^2 = (a+b)(a-b)$$

$$a^2 + 2ab + b^2 = (a+b)^2$$

$$a^2 - 2ab + b^2 = (a-b)^2$$

$$a^3 + b^3 = (a+b)(a^2 - ab + b^2)$$

$$a^3 - b^3 = (a-b)(a^2 + ab + b^2)$$

3. Factor by grouping, if possible.

4. Factor general trinomials with binomial factors by "guess-and-test" or by the grouping method.

5. Use the Rational Zero Test together with synthetic division to factor a polynomial.

6. Use Descartes's Rule of Signs to determine the number of real zeros. Then find any zeros and use them to factor the polynomial.

7. Find any upper and lower bounds for the real zeros to eliminate some of the possible rational zeros. Then test the remaining candidates by synthetic division and use any zeros to factor the polynomial.

126. (a) Zeros of $f(x)$: $-2, 1, 4$

(b) The graph touches the x-axis at $x = 1$

(c) The least possible degree of the function is 4 because there are at least 4 real zeros (1 is repeated) and a function can have at most the number of real zeros equal to the degree of the function. The degree cannot be odd by the definition of multiplicity.

(d) The leading coefficient of f is positive. From the information in the table, you can conclude that the graph will eventually rise to the left and to the right.

(e) Answers may vary. One possibility is:

$$f(x) = (x-1)^2(x-(-2))(x-4)$$

$$= (x-1)^2(x+2)(x-4)$$

$$= (x^2 - 2x + 1)(x^2 - 2x - 8)$$

$$= x^4 - 4x^3 - 3x^2 + 14x - 8$$

(f)

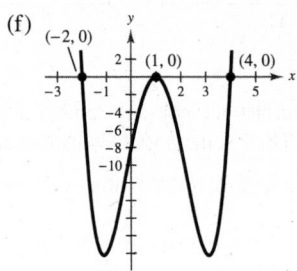

127. (a) $f(x) = \left(x - \sqrt{b}\,i\right)\left(x + \sqrt{b}\,i\right) = x^2 + b$

(b) $f(x) = [x - (a+bi)][x - (a-bi)]$

$$= [(x-a) - bi][(x-a) + bi]$$

$$= (x-a)^2 - (bi)^2$$

$$= x^2 - 2ax + a^2 + b^2$$

128. (a) $f(x)$ cannot have this graph since it also has a zero at $x = 0$.

(b) $g(x)$ cannot have this graph since it is a quadratic function. Its graph is a parabola.

(c) $h(x)$ is the correct function. It has two real zeros, $x = 2$ and $x = 3.5$, and it has a degree of four, needed to yield three turning points.

(d) $k(x)$ cannot have this graph since it also has a zero at $x = -1$. In addition, since it is only of degree three, it would have at most two turning points.

129. $(-3 + 6i) - (8 - 3i) = -3 + 6i - 8 + 3i = -11 + 9i$ **130.** $(12 - 5i) + 16i = 12 + 11i$

131. $(6 - 2i)(1 + 7i) = 6 + 42i - 2i - 14i^2 = 20 + 40i$ **132.** $(9 - 5i)(9 + 5i) = 81 - 25i^2 = 81 + 25 = 106$

133. $g(x) = f(x - 2)$

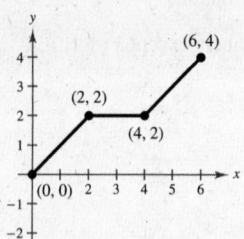

Horizontal shift two units
to the right

134. $g(x) = f(x) - 2$

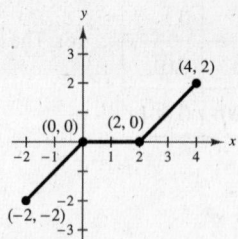

Vertical shift two units downward

135. $g(x) = 2f(x)$

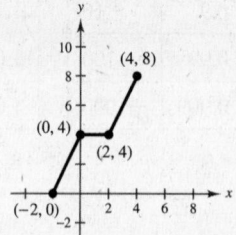

Vertical stretch (each y-value is
multiplied by 2)

136. $g(x) = f(-x)$

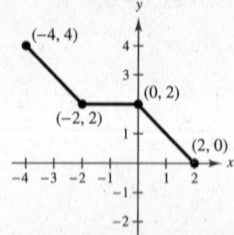

Reflection in the y-axis

137. $g(x) = f(2x)$

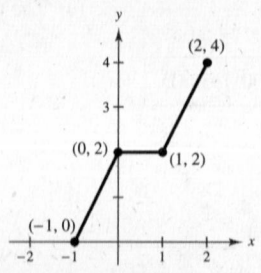

Horizontal shrink $\left(\text{each } x\text{-value is}\right.$
multiplied by $\left.\frac{1}{2}\right)$

138. $g(x) = f\left(\frac{1}{2}x\right)$

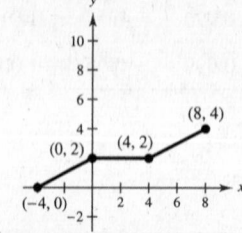

Horizontal stretch (each x-value
is multiplied by 2)

Section 2.6   Rational Functions

- ■   You should know the following basic facts about rational functions.

  (a) A function of the form $f(x) = N(x)/D(x)$, $D(x) \neq 0$, where $N(x)$ and $D(x)$ are polynomials, is called
  a rational function.

  (b) The domain of a rational function is the set of all real numbers except those which make the denominator zero.

  (c) If $f(x) = N(x)/D(x)$ is in reduced form, and a is a value such that $D(a) = 0$, then the line $x = a$ is a
  vertical asymptote of the graph of f. $f(x) \rightarrow \infty$ or $f(x) \rightarrow -\infty$ as $x \rightarrow a$.

  (d) The line $y = b$ is a horizontal asymptote of the graph of f if $f(x) \rightarrow b$ as $x \rightarrow \infty$ or $x \rightarrow -\infty$.

  (e) Let $f(x) = \dfrac{N(x)}{D(x)} = \dfrac{a_n x^n + a_{n-1}x^{n-1} + \cdots + a_1 x + a_0}{b_m x^m + b_{m-1}x^{m-1} + \cdots + b_1 x + b_0}$ where $N(x)$ and $D(x)$ have no common factors.

    1. If $n < m$, then the x-axis $(y = 0)$ is a horizontal asymptote.

    2. If $n = m$, then $y = \dfrac{a_n}{b_m}$ is a horizontal asymptote.

    3. If $n > m$, then there are no horizontal asymptotes.

Vocabulary Check

1. rational functions   2. vertical asymptote   3. horizontal asymptote   4. slant asymptote

1. $f(x) = \dfrac{1}{x - 1}$

(a)

x	$f(x)$	x	$f(x)$	x	$f(x)$
0.5	-2	1.5	2	5	0.25
0.9	-10	1.1	10	10	$0.\overline{1}$
0.99	-100	1.01	100	100	$0.\overline{01}$
0.999	-1000	1.001	1000	1000	$0.\overline{001}$

(b) The zero of the denominator is $x = 1$, so $x = 1$ is a vertical asymptote. The degree of the numerator is less than the degree of the denominator so the x-axis, or $y = 0$, is a horizontal asymptote.

(c) The domain is all real numbers except $x = 1$.

2. $f(x) = \dfrac{5x}{x - 1}$

(a)

x	$f(x)$	x	$f(x)$	x	$f(x)$
0.5	-5	1.5	15	5	6.25
0.9	-45	1.1	55	10	$5.\overline{55}$
0.99	-495	1.01	505	100	$5.\overline{05}$
0.999	-4995	1.001	5005	1000	$5.\overline{005}$

(b) The zero of the denominator is $x = 1$, so $x = 1$ is a vertical asymptote. The degree of the numerator is equal to the degree of the denominator, so the line $y = \frac{5}{1} = 5$ is a horizontal asymptote.

(c) The domain is all real numbers except $x = 1$.

3. $f(x) = \dfrac{3x^2}{x^2 - 1}$

(a)

x	$f(x)$	x	$f(x)$	x	$f(x)$
0.5	-1	1.5	5.4	5	3.125
0.9	-12.79	1.1	17.29	10	$3.\overline{03}$
0.99	-147.8	1.01	152.3	100	$3.\overline{0003}$
0.999	-1498	1.001	1502	1000	3

(b) The zeros of the denominator are $x = \pm 1$ so both $x = 1$ and $x = -1$ are vertical asymptotes. Since the degree of the numerator equals the degree of the denominator, $y = \frac{3}{1} = 3$ is a horizontal asymptote.

(c) The domain is all real numbers except $x = \pm 1$.

4. $f(x) = \dfrac{4x}{x^2 - 1}$

(a)

x	$f(x)$	x	$f(x)$	x	$f(x)$
0.5	$-2.\overline{66}$	1.5	4.8	5	$0.\overline{833}$
0.9	-18.95	1.1	20.95	10	$0.\overline{40}$
0.99	-199	1.01	201	100	0.04
0.999	-1999	1.001	2001	1000	0.004

(b) The zeros of the denominator are $x = \pm 1$ so both $x = 1$ and $x = -1$ are vertical asymptotes. Because the degree of the numerator is less than the degree of the denominator, the x-axis or $y = 0$ is a horizontal asymptote.

(c) The domain is all real numbers except $x = \pm 1$.

5. $f(x) = \dfrac{1}{x^2}$

Domain: all real numbers except $x = 0$

Vertical asymptote: $x = 0$

Horizontal asymptote: $y = 0$

[Degree of $N(x) < $ degree of $D(x)$]

6. $f(x) = \dfrac{4}{(x - 2)^3}$

Domain: all real numbers except $x = 2$

Vertical asymptote: $x = 2$

Horizontal asymptote: $y = 0$

[Degree of $N(x) < $ degree of $D(x)$]

7. $f(x) = \dfrac{2 + x}{2 - x} = \dfrac{x + 2}{-x + 2}$

Domain: all real numbers except $x = 2$

Vertical asymptote: $x = 2$

Horizontal asymptote: $y = -1$

[Degree of $N(x)$ = degree of $D(x)$]

8. $f(x) = \dfrac{1 - 5x}{1 + 2x} = \dfrac{-5x + 1}{2x + 1}$

Domain: all real numbers except $x = -\dfrac{1}{2}$

Vertical asymptote: $x = -\dfrac{1}{2}$

Horizontal asymptote: $y = -\dfrac{5}{2}$

[Degree of $N(x)$ = degree of $D(x)$]

9. $f(x) = \dfrac{x^3}{x^2 - 1}$

Domain: all real numbers except $x = \pm 1$

Vertical asymptotes: $x = \pm 1$

Horizontal asymptote: None

[Degree of $N(x)$ > degree of $D(x)$]

10. $f(x) = \dfrac{2x^2}{x + 1}$

Domain: all real numbers except $x = -1$

Vertical asymptote: $x = -1$

Horizontal asymptote: None

[Degree of $N(x)$ > degree of $D(x)$]

11. $f(x) = \dfrac{3x^2 + 1}{x^2 + x + 9}$

Domain: All real numbers. The denominator has no real zeros. [Try the Quadratic Formula on the denominator.]

Vertical asymptote: None

Horizontal asymptote: $y = 3$

[Degree of $N(x)$ = degree of $D(x)$]

12. $f(x) = \dfrac{3x^2 + x - 5}{x^2 + 1}$

Domain: All real numbers. The denominator has no real zeros. [Try the Quadratic Formula on the denominator.]

Vertical asymptote: None

Horizontal asymptote: $y = 3$

[Degree of $N(x)$ = degree of $D(x)$]

13. $f(x) = \dfrac{2}{x + 3}$

Vertical asymptote: $y = -3$

Horizontal asymptote: $y = 0$

Matches graph (d).

14. $f(x) = \dfrac{1}{x - 5}$

Vertical asymptote: $x = 5$

Horizontal asymptote: $y = 0$

Matches graph (a).

15. $f(x) = \dfrac{x - 1}{x - 4}$

Vertical asymptote: $x = 4$

Horizontal asymptote: $y = 1$

Matches graph (c).

16. $f(x) = -\dfrac{x + 2}{x + 4}$

Vertical asymptote: $x = -4$

Horizontal asymptote: $y = -1$

Matches graph (b).

17. $g(x) = \dfrac{x^2 - 1}{x + 1} = \dfrac{(x - 1)(x + 1)}{x + 1}$

The only zero of $g(x)$ is $x = 1$.

$x = -1$ makes $g(x)$ undefined.

18. $h(x) = 2 + \dfrac{5}{x^2 + 2}$

$$0 = 2 + \dfrac{5}{x^2 + 2}$$

$$-2 = \dfrac{5}{x^2 + 2}$$

$$-2(x^2 + 2) = 5$$

$$x^2 = -\dfrac{5}{2} - 2$$

No real solution, $h(x)$ has no real zeros.

19. $f(x) = 1 - \dfrac{3}{x - 3}$

$$1 - \dfrac{3}{x - 3} = 0$$

$$1 = \dfrac{3}{x - 3}$$

$$x - 3 = 3$$

$x = 6$ is a zero of $f(x)$.

20. $g(x) = \dfrac{x^3 - 8}{x^2 + 1}$

$$\dfrac{x^3 - 8}{x^2 + 1} = 0$$

$$x^3 - 8 = 0$$

$$x^3 = 8$$

$$x = 2$$

$x = 2$ is a real zero of $g(x)$.

21. $f(x) = \dfrac{x - 4}{x^2 - 16} = \dfrac{1}{x + 4}, \; x \neq 4$

Domain: all real numbers x except $x = \pm 4$

Horizontal asymptote: $y = 0$

[Degree of $N(x)$ < degree of $D(x)$]

Vertical asymptote: $x = -4$ (Since $x - 4$ is a common factor of $N(x)$ and $D(x)$, $x = 4$ is not a vertical asymptote of $f(x)$.)

22. $f(x) = \dfrac{x + 3}{x^2 - 9} = \dfrac{x + 3}{(x + 3)(x - 3)} = \dfrac{1}{x - 3}, \; x \neq -3$

Domain: all real numbers x except $x = \pm 3$

The degree of the numerator is less than the degree of the denominator, so the graph has the line $y = 0$ as a horizontal asymptote.

Vertical asymptote: $x = 3$ (Since $x + 3$ is a common factor of $N(x)$ and $D(x)$, $x = -3$ is not a vertical asymptote of $f(x)$.)

23. $f(x) = \dfrac{x^2 - 1}{x^2 - 2x - 3} = \dfrac{(x + 1)(x - 1)}{(x + 1)(x - 3)} = \dfrac{x - 1}{x - 3}, \; x \neq -1$

Domain: all real numbers x except $x = -1$ and $x = 3$

Horizontal asymptote: $y = 1$

[Degree of $N(x)$ = degree of $D(x)$]

Vertical asymptote: $x = 3$
(Since $x + 1$ is a common factor of $N(x)$ and $D(x)$, $x = -1$ is not a vertical asymptote of $f(x)$.)

24. $f(x) = \dfrac{x^2 - 4}{x^2 - 3x + 2}$

$$= \dfrac{(x - 2)(x + 2)}{(x - 2)(x - 1)} = \dfrac{x + 2}{x - 1}, \; x \neq 2$$

Domain: all real numbers x except $x = 1$ and $x = 2$

Horizontal asymptote: $y = 1$

[Degree of $N(x)$ = degree of $D(x)$]

Vertical asymptote: $x = 1$ (Since $x - 2$ is a common factor of $N(x)$ and $D(x)$, $x = 2$ is not a vertical asymptote of $f(x)$.)

25. $f(x) = \dfrac{x^2 - 3x - 4}{2x^2 + x - 1}$

$$= \dfrac{(x + 1)(x - 4)}{(2x - 1)(x + 1)} = \dfrac{x - 4}{2x - 1}, \; x \neq -1$$

Domain: all real numbers x except $x = \dfrac{1}{2}$ and $x = -1$

Horizontal asymptote: $y = \dfrac{1}{2}$

[Degree of $N(x)$ = degree of $D(x)$]

Vertical asymptote: $x = \frac{1}{2}$ (Since $x + 1$ is a common factor of $N(x)$ and $D(x)$, $x = -1$ is not a vertical asymptote of $f(x)$.)

26. $f(x) = \dfrac{6x^2 - 11x + 3}{6x^2 - 7x - 3}$

$$= \dfrac{(2x - 3)(3x - 1)}{(2x - 3)(3x + 1)} = \dfrac{3x - 1}{3x + 1}, \; x \neq \dfrac{3}{2}$$

Domain: all real numbers x except $x = \dfrac{3}{2}$ or $x = -\dfrac{1}{3}$

Horizontal asymptote: $y = 1$

[Degree of $N(x)$ = degree of $D(x)$]

Vertical asymptote: $x = -\frac{1}{3}$ (Since $2x - 3$ is a common factor of $N(x)$ and $D(x)$, $x = \frac{3}{2}$ is not a vertical asymptote of $f(x)$.)

27. $f(x) = \dfrac{1}{x+2}$

(a) Domain: all real numbers x except $x = -2$

(b) y-intercept: $\left(0, \dfrac{1}{2}\right)$

(c) Vertical asymptote: $x = -2$
Horizontal asymptote: $y = 0$

(d)

x	-4	-3	-1	0	1
y	$-\frac{1}{2}$	-1	1	$\frac{1}{2}$	$\frac{1}{3}$

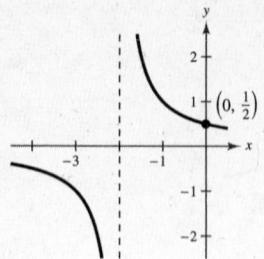

28. $f(x) = \dfrac{1}{x-3}$

(a) Domain: all real numbers x except $x = 3$

(b) y-intercept: $\left(0, -\dfrac{1}{3}\right)$

(c) Vertical asymptote: $x = 3$
Horizontal asymptote: $y = 0$

(d)

x	0	1	2	4	5	6
y	$-\frac{1}{3}$	$-\frac{1}{2}$	-1	1	$\frac{1}{2}$	$\frac{1}{3}$

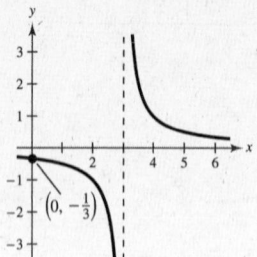

29. $h(x) = \dfrac{-1}{x+2}$

(a) Domain: all real numbers x except $x = -2$

(b) y-intercept: $\left(0, -\dfrac{1}{2}\right)$

(c) Vertical asymptote: $x = -2$
Horizontal asymptote: $y = 0$

(d)

x	-4	-3	-1	0
y	$\frac{1}{2}$	1	-1	$-\frac{1}{2}$

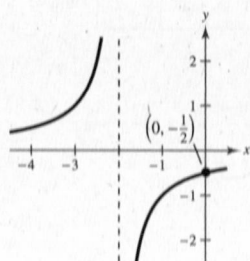

Note: This is the graph of $f(x) = \dfrac{1}{x+2}$
(Exercise 27) reflected about the x-axis.

30. $g(x) = \dfrac{1}{3-x} = -\dfrac{1}{x-3}$

(a) Domain: all real numbers x except $x = 3$

(b) y-intercept: $\left(0, \dfrac{1}{3}\right)$

(c) Vertical asymptote: $x = 3$
Horizontal asymptote: $y = 0$

(d)

x	0	1	2	4	5	6
y	$\frac{1}{3}$	$\frac{1}{2}$	1	-1	$-\frac{1}{2}$	$-\frac{1}{3}$

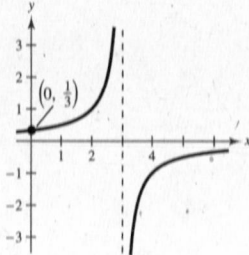

Note: This is the graph of $f(x) = \dfrac{1}{x-3}$
(Exercise 28) reflected about the x-axis.

31. $C(x) = \dfrac{5 + 2x}{1 + x} = \dfrac{2x + 5}{x + 1}$

(a) Domain: all real numbers x except $x = -1$

(b) x-intercept: $\left(-\dfrac{5}{2}, 0\right)$

y-intercept: $(0, 5)$

(c) Vertical asymptote: $x = -1$
Horizontal asymptote: $y = 2$

(d)

x	-4	-3	-2	0	1	2
$C(x)$	1	$\frac{1}{2}$	-1	5	$\frac{7}{2}$	3

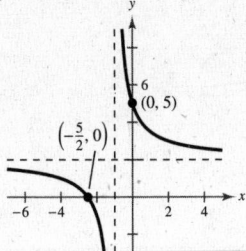

32. $P(x) = \dfrac{1 - 3x}{1 - x} = \dfrac{3x - 1}{x - 1}$

(a) Domain: all real numbers x except $x = 1$

(b) x-intercept: $\left(\dfrac{1}{3}, 0\right)$

y-intercept: $(0, 1)$

(c) Vertical asymptote: $x = 1$
Horizontal asymptote: $y = 3$

(d)

x	-1	0	2	3
y	2	1	5	4

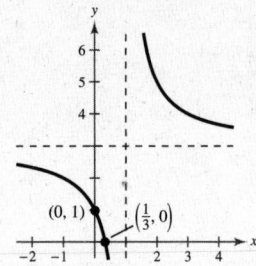

33. $f(x) = \dfrac{x^2}{x^2 + 9}$

(a) Domain: all real numbers x

(b) Intercept: $(0, 0)$

(c) Horizontal asymptote: $y = 1$

(d)

x	± 1	± 2	± 3
y	$\frac{1}{10}$	$\frac{4}{13}$	$\frac{1}{2}$

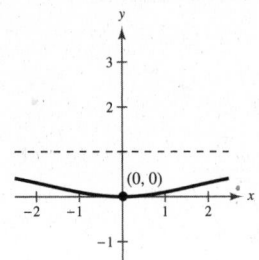

34. $f(t) = \dfrac{1 - 2t}{t} = -\dfrac{2t - 1}{t}$

(a) Domain: all real numbers t except $t = 0$

(b) t-intercept: $\left(\dfrac{1}{2}, 0\right)$

(c) Vertical asymptote: $t = 0$
Horizontal asymptote: $y = -2$

(d)

t	-2	-1	$\frac{1}{2}$	1	2
y	$-\frac{5}{2}$	-3	0	-1	$-\frac{3}{2}$

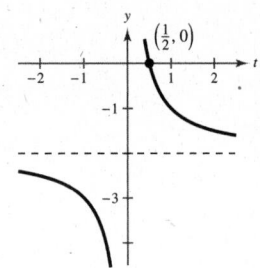

35. $g(s) = \dfrac{s}{s^2 + 1}$

    (a) Domain: all real numbers s

    (b) Intercept: $(0, 0)$

    (c) Horizontal asymptote: $y = 0$

    (d)

s	-2	-1	0	1	2
$g(s)$	$-\frac{2}{5}$	$-\frac{1}{2}$	0	$\frac{1}{2}$	$\frac{2}{5}$

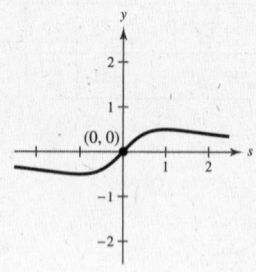

36. $f(x) = -\dfrac{1}{(x - 2)^2}$

    (a) Domain: all real numbers x except $x = 2$

    (b) y-intercept: $\left(0, -\dfrac{1}{4}\right)$

    (c) Vertical asymptote: $x = 2$
        Horizontal asymptote: $y = 0$

    (d)

x	0	$\frac{1}{2}$	1	$\frac{3}{2}$	$\frac{5}{2}$	3	$\frac{7}{2}$	4
y	$-\frac{1}{4}$	$-\frac{4}{9}$	-1	-4	-4	-1	$-\frac{4}{9}$	$-\frac{1}{4}$

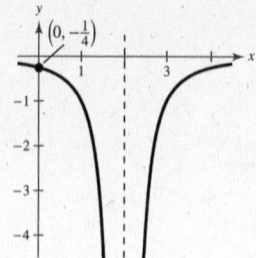

37. $h(x) = \dfrac{x^2 - 5x + 4}{x^2 - 4} = \dfrac{(x - 1)(x - 4)}{(x + 2)(x - 2)}$

    (a) Domain: all real numbers x except $x = \pm 2$

    (b) x-intercepts: $(1, 0)$, $(4, 0)$
        y-intercept: $(0, -1)$

    (c) Vertical asymptotes: $x = -2$, $x = 2$
        Horizontal asymptote: $y = 1$

    (d)

x	-4	-3	-1	0	1	3	4
y	$\frac{10}{3}$	$\frac{28}{5}$	$-\frac{10}{3}$	-1	0	$-\frac{2}{5}$	0

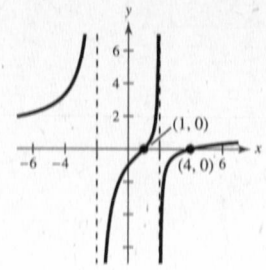

38. $g(x) = \dfrac{x^2 - 2x - 8}{x^2 - 9} = \dfrac{(x - 4)(x + 2)}{(x - 3)(x + 3)}$

    (a) Domain: all real numbers x except $x = \pm 3$

    (b) y-intercept: $\left(0, \dfrac{8}{9}\right)$
        x-intercepts: $(4, 0)$, $(-2, 0)$

    (c) Vertical asymptotes: $x = \pm 3$
        Horizontal asymptote: $y = 1$

    (d)

x	-5	-4	-2	0	2	4	5
y	$\frac{27}{16}$	$\frac{16}{7}$	0	$\frac{8}{9}$	$\frac{8}{5}$	0	$\frac{7}{16}$

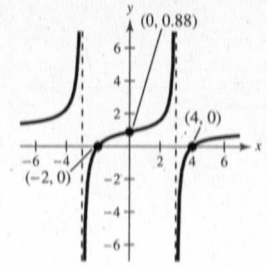

39. $f(x) = \dfrac{2x^2 - 5x - 3}{x^3 - 2x^2 - x + 2} = \dfrac{(2x + 1)(x - 3)}{(x - 2)(x + 1)(x - 1)}$

(a) Domain: all real numbers x except $x = 2$, $x = \pm 1$

(b) x-intercepts: $\left(-\dfrac{1}{2}, 0\right)$, $(3, 0)$

y-intercept: $\left(0, -\dfrac{3}{2}\right)$

(c) Vertical asymptotes: $x = 2$, $x = -1$ and $x = 1$
Horizontal asymptotes: $y = 0$

(d)

x	-3	-2	0	1.5	3	4
$f(x)$	$-\dfrac{3}{4}$	$-\dfrac{5}{4}$	$-\dfrac{3}{2}$	$\dfrac{48}{5}$	0	$\dfrac{3}{10}$

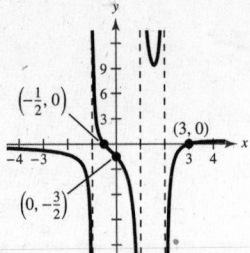

40. $f(x) = \dfrac{x^2 - x - 2}{x^3 - 2x^2 - 5x + 6} = \dfrac{(x + 1)(x - 2)}{(x - 1)(x + 2)(x - 3)}$

(a) Domain: all real numbers x except $x = 1$, $x = -2$, or $x = 3$

(b) x-intercepts: $(-1, 0)$, $(2, 0)$

y-intercept: $\left(0, -\dfrac{1}{3}\right)$

(c) Vertical asymptotes: $x = -2$, $x = 1$, $x = 3$
Horizontal asymptote: $y = 0$

(d)

x	-4	-3	-1	0	2	4
y	$-\dfrac{9}{35}$	$-\dfrac{5}{12}$	0	$-\dfrac{1}{3}$	0	$\dfrac{5}{9}$

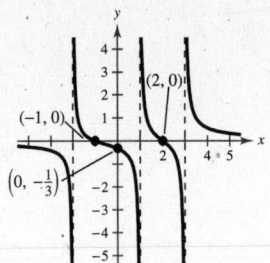

41. $f(x) = \dfrac{x^2 + 3x}{x^2 + x - 6} = \dfrac{x(x + 3)}{(x + 3)(x - 2)} = \dfrac{x}{x - 2}$, $x \neq -3$

(a) Domain: all real numbers x except $x = -3$ and $x = 2$

(b) Intercept: $(0, 0)$

(c) Vertical asymptote: $x = 2$
Horizontal asymptote: $y = 1$

(d)

x	-1	0	1	3	4
y	$\dfrac{1}{3}$	0	-1	3	2

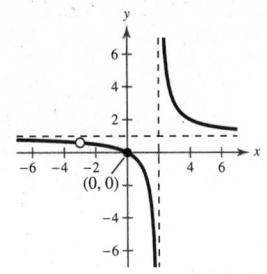

42. $f(x) = \dfrac{5(x + 4)}{x^2 + x - 12} = \dfrac{5(x + 4)}{(x + 4)(x - 3)} = \dfrac{5}{x - 3}$, $x \neq -4$

(a) Domain: all real numbers x except $x = -4$ or $x = 3$

(b) y-intercept: $\left(0, -\dfrac{5}{3}\right)$

x-intercept: none

(c) Vertical asymptote: $x = 3$
Horizontal asymptote: $y = 0$

(d)

x	-2	0	2	5	7
y	-1	$-\dfrac{5}{3}$	-5	$\dfrac{5}{2}$	$\dfrac{5}{4}$

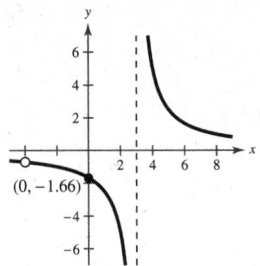

43. $f(x) = \dfrac{2x^2 - 5x + 2}{2x^2 - x - 6}$

$= \dfrac{(2x-1)(x-2)}{(2x+3)(x-2)} = \dfrac{2x-1}{2x+3}, \quad x \neq 2$

(a) Domain: all real numbers x except $x = 2$ and

$x = -\dfrac{3}{2}$

(b) x-intercept: $\left(\dfrac{1}{2}, 0\right)$

y-intercept: $\left(0, -\dfrac{1}{3}\right)$

(c) Vertical asymptote: $x = -\dfrac{3}{2}$

Horizontal asymptote: $y = 1$

(d)

x	-3	-2	-1	0	1
y	$\frac{7}{3}$	5	-3	$-\frac{1}{3}$	$\frac{1}{5}$

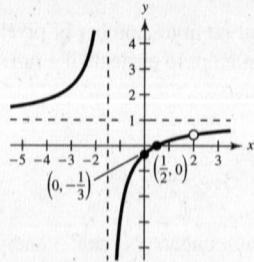

44. $f(x) = \dfrac{3x^2 - 8x + 4}{2x^2 - 3x - 2}$

$= \dfrac{(x-2)(3x-2)}{(x-2)(2x+1)} = \dfrac{3x-2}{2x+1}, \quad x \neq 2$

(a) Domain: all real numbers x except $x = 2$ or $x = -\dfrac{1}{2}$

(b) y-intercept: $(0, -2)$

x-intercept: $\left(\dfrac{2}{3}, 0\right)$

(c) Vertical asymptote: $x = -\dfrac{1}{2}$

Horizontal asymptote: $y = \dfrac{3}{2}$

(d)

x	-3	-1	0	$\frac{2}{3}$	3
y	$\frac{11}{5}$	5	-2	0	1

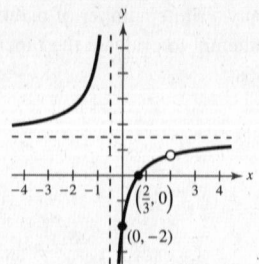

45. $f(t) = \dfrac{t^2 - 1}{t + 1} = \dfrac{(t+1)(t-1)}{t+1} = t - 1, \quad t \neq -1$

(a) Domain: all real numbers t except $t = -1$

(b) t-intercept: $(1, 0)$

y-intercept: $(0, -1)$

(c) Vertical asymptote: none

Horizontal asymptote: none

(d)

t	-3	-2	0	1
y	-4	-3	-1	0

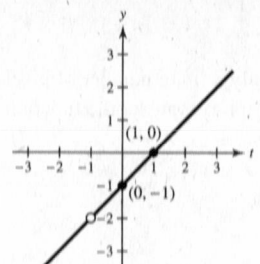

46. $f(x) = \dfrac{x^2 - 16}{x - 4} = \dfrac{(x-4)(x+4)}{x-4} = x + 4, \quad x \neq 4$

(a) Domain: all real numbers $x \neq 4$

(b) y-intercept: $(0, 4)$

x-intercept: $(-4, 0)$

(c) Vertical asymptote: none

Horizontal asymptote: none

(d)

x	-6	-4	0	5
y	-2	0	4	9

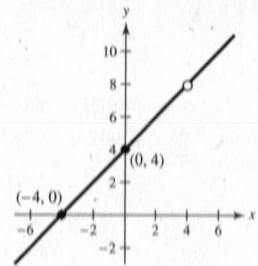

47. $f(x) = \dfrac{x^2 - 1}{x + 1}$, $g(x) = x - 1$

(a) Domain of f: all real numbers x except $x = -1$

Domain of g: all real numbers x

(b) Because $(x + 1)$ is a factor of both the numerator and the denominator of f, $x = -1$ is not a vertical asymptote. f has no vertical asymptotes.

(c)

x	-3	-2	-1.5	-1	-0.5	0	1
$f(x)$	-4	-3	-2.5	Undef.	-1.5	-1	0
$g(x)$	-4	-3	-2.5	-2	-1.5	-1	0

(d)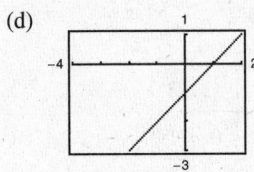

(e) Because there are only a finite number of pixels, the utility may not attempt to evaluate the function where it does not exist.

48. $f(x) = \dfrac{x^2(x - 2)}{x^2 - 2x}$, $g(x) = x$

(a) Domain of f: all real numbers x except 0 and 2

Domain of g: all real numbers x

(b) Since $x^2 - 2x$ is a factor of both the numerator and the denominator of f, neither $x = 0$ nor $x = 2$ is a vertical asymptote of f. Thus, f has no vertical asymptotes.

(c)

x	-1	0	1	1.5	2	2.5	3
$f(x)$	-1	Undef.	1	1.5	Undef.	2.5	3
$g(x)$	-1	0	1	1.5	2	2.5	3

(d)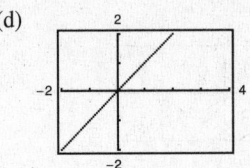

(e) Because there are only a finite number of pixels, the utility may not attempt to evaluate the function where it does not exist.

49. $f(x) = \dfrac{x - 2}{x^2 - 2x}$, $g(x) = \dfrac{1}{x}$

(a) Domain of f: all real numbers x except $x = 0$ and $x = 2$

Domain of g: all real numbers x except $x = 0$

(b) Because $(x - 2)$ is a factor of both the numerator and the denominator of f, $x = 2$ is not a vertical asymptote. The only vertical asymptote of f is $x = 0$.

(c)

x	-0.5	0	0.5	1	1.5	2	3
$f(x)$	-2	Undef.	2	1	$\frac{2}{3}$	Undef.	$\frac{1}{3}$
$g(x)$	-2	Undef.	2	1	$\frac{2}{3}$	$\frac{1}{2}$	$\frac{1}{3}$

(d)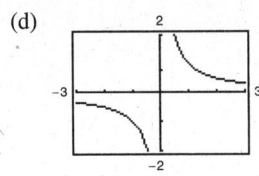

(e) Because there are only a finite number of pixels, the utility may not attempt to evaluate the function where it does not exist.

50. $f(x) = \dfrac{2x - 6}{x^2 - 7x + 12}$, $g(x) = \dfrac{2}{x - 4}$

(a) Domain of f: all real numbers x except 3 and 4

Domain of g: all real numbers x except 4

(b) Since $x - 3$ is a factor of both the numerator and the denominator of f, $x = 3$ is not a vertical asymptote of f. Thus, f has $x = 4$ as its only vertical asymptote.

(c)

x	0	1	2	3	4	5	6
$f(x)$	$-\frac{1}{2}$	$-\frac{2}{3}$	-1	Undef.	Undef.	2	1
$g(x)$	$-\frac{1}{2}$	$-\frac{2}{3}$	-1	-2	Undef.	2	1

(d)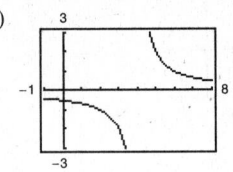

(e) Because there are only a finite number of pixels, the utility may not attempt to evaluate the function where it does not exist.

51. $h(x) = \dfrac{x^2 - 4}{x} = x - \dfrac{4}{x}$

(a) Domain: all real numbers x except $x = 0$

(b) Intercepts: $(2, 0), (-2, 0)$

(c) Vertical asymptote: $x = 0$
Slant asymptote: $y = x$

(d)

x	-3	-1	1	3
y	$-\frac{5}{3}$	3	-3	$\frac{5}{3}$

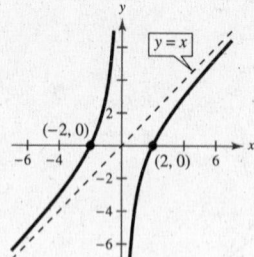

52. $g(x) = \dfrac{x^2 + 5}{x} = x + \dfrac{5}{x}$

(a) Domain: all real numbers x except $x = 0$

(b) No intercepts

(c) Vertical asymptote: $x = 0$
Slant asymptote: $y = x$

(d)

x	-3	-2	-1	1	2	3
y	$-\frac{14}{3}$	$-\frac{9}{2}$	-6	6	$\frac{9}{2}$	$\frac{14}{3}$

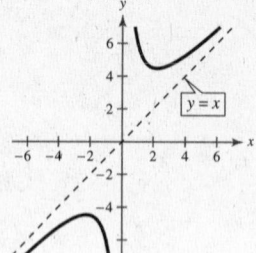

53. $f(x) = \dfrac{2x^2 + 1}{x} = 2x + \dfrac{1}{x}$

(a) Domain: all real numbers x except $x = 0$

(b) No intercepts

(c) Vertical asymptote: $x = 0$
Slant asymptote: $y = 2x$

(d)

x	-4	-2	2	4	6
$f(x)$	$-\frac{33}{4}$	$-\frac{9}{2}$	$\frac{9}{2}$	$\frac{33}{4}$	$\frac{73}{6}$

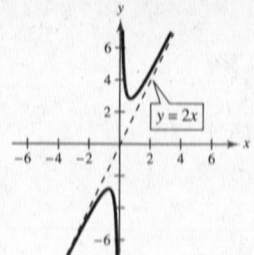

54. $f(x) = \dfrac{1 - x^2}{x} = -x + \dfrac{1}{x}$

(a) Domain: all real numbers x except $x = 0$

(b) x-intercepts: $(-1, 0), (1, 0)$

(c) Vertical asymptote: $x = 0$
Slant asymptote: $y = -x$

(d)

x	-6	-4	-2	2	4	6
$f(x)$	$\frac{35}{6}$	$\frac{15}{4}$	$\frac{3}{2}$	$-\frac{3}{2}$	$-\frac{15}{4}$	$-\frac{35}{6}$

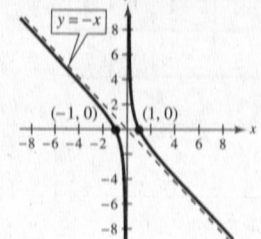

55. $g(x) = \dfrac{x^2 + 1}{x} = x + \dfrac{1}{x}$

(a) Domain: all real numbers x except $x = 0$

(b) No intercepts

(c) Vertical asymptote: $x = 0$
Slant asymptote: $y = x$

(d)

x	-4	-2	2	4	6
$g(x)$	$-\frac{17}{4}$	$-\frac{5}{2}$	$\frac{5}{2}$	$\frac{17}{4}$	$\frac{37}{6}$

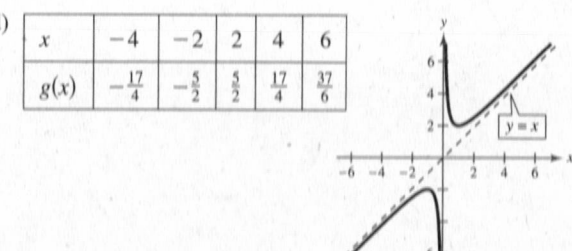

56. $h(x) = \dfrac{x^2}{x-1} = x + 1 + \dfrac{1}{x-1}$

(a) Domain: all real numbers x except $x = 1$

(b) Intercept: $(0, 0)$

(c) Vertical asymptote: $x = 1$
Slant asymptote: $y = x + 1$

(d)

x	-4	-2	2	4	6
$h(x)$	$-\frac{16}{5}$	$-\frac{4}{3}$	4	$\frac{16}{3}$	$\frac{36}{5}$

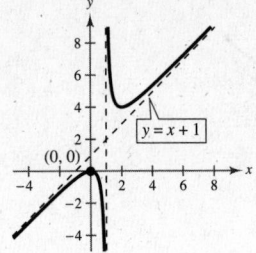

57. $f(t) = -\dfrac{t^2+1}{t+5} = -t + 5 - \dfrac{26}{t+5}$

(a) Domain: all real numbers t except $t = -5$

(b) Intercept: $\left(0, -\dfrac{1}{5}\right)$

(c) Vertical asymptote: $t = -5$
Slant asymptote: $y = -t + 5$

(d)

t	-7	-6	-4	-3	0
y	25	37	-17	-5	$-\frac{1}{5}$

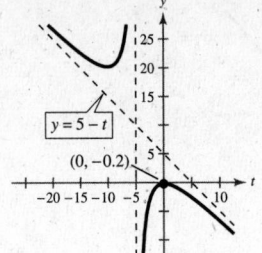

58. $f(x) = \dfrac{x^2}{3x+1} = \dfrac{1}{3}x - \dfrac{1}{9} + \dfrac{1}{9(3x+1)}$

(a) Domain: all real numbers x except $x = -\dfrac{1}{3}$

(b) Intercept: $(0, 0)$

(c) Vertical asymptote: $x = -\dfrac{1}{3}$

Slant asymptote: $y = \dfrac{1}{3}x - \dfrac{1}{9}$

(d)

x	-3	-2	-1	$-\frac{1}{2}$	0	2
y	$-\frac{9}{8}$	$-\frac{4}{5}$	$-\frac{1}{2}$	$-\frac{1}{2}$	0	$\frac{4}{7}$

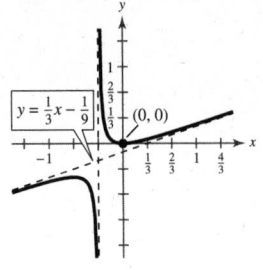

59. $f(x) = \dfrac{x^3}{x^2-1} = x + \dfrac{x}{x^2-1}$

(a) Domain: all real numbers x except $x = \pm 1$

(b) Intercept: $(0, 0)$

(c) Vertical asymptotes: $x = \pm 1$
Slant asymptote: $y = x$

(d)

x	-4	-2	0	2	4
$f(x)$	$-\frac{64}{15}$	$-\frac{8}{3}$	0	$\frac{8}{3}$	$\frac{64}{15}$

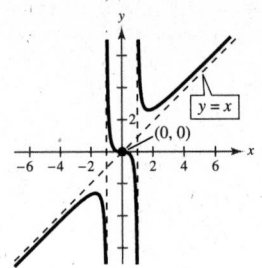

60. $g(x) = \dfrac{x^3}{2x^2 - 8} = \dfrac{1}{2}x + \dfrac{4x}{2x^2 - 8}$

(a) Domain: all real numbers x except $x = \pm 2$

(b) Intercept: $(0, 0)$

(c) Vertical asymptotes: $x = \pm 2$

Slant asymptote: $y = \dfrac{1}{2}x$

(d)

x	-6	-4	-1	1	4	6
$g(x)$	$-\frac{27}{8}$	$-\frac{8}{3}$	$\frac{1}{6}$	$-\frac{1}{6}$	$\frac{8}{3}$	$\frac{27}{8}$

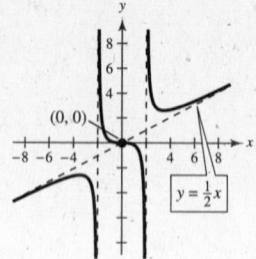

61. $f(x) = \dfrac{x^2 - x + 1}{x - 1} = x + \dfrac{1}{x - 1}$

(a) Domain: all real numbers x except $x = 1$

(b) y-intercept: $(0, -1)$

(c) Vertical asymptote: $x = 1$
Slant asymptote: $y = x$

(d)

x	-4	-2	0	2	4
$f(x)$	$-\frac{21}{5}$	$-\frac{7}{3}$	-1	3	$\frac{13}{3}$

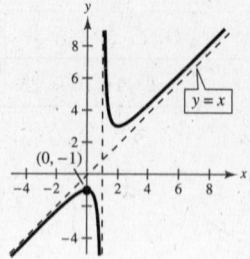

62. $f(x) = \dfrac{2x^2 - 5x + 5}{x - 2} = 2x - 1 + \dfrac{3}{x - 2}$

(a) Domain: all real numbers x except $x = 2$

(b) y-intercept: $\left(0, -\dfrac{5}{2}\right)$

(c) Vertical asymptote: $x = 2$
Slant asymptote: $y = 2x - 1$

(d)

x	-6	-3	1	3	6	7
y	$-\frac{107}{8}$	$-\frac{38}{5}$	-2	8	$\frac{47}{4}$	$\frac{68}{5}$

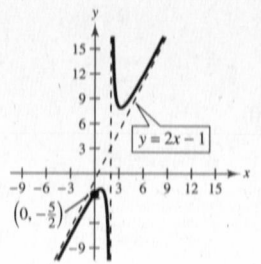

63. $f(x) = \dfrac{2x^3 - x^2 - 2x + 1}{x^2 + 3x + 2}$

$\quad = \dfrac{(2x - 1)(x + 1)(x - 1)}{(x + 1)(x + 2)}$

$\quad = \dfrac{(2x - 1)(x - 1)}{x + 2}, \quad x \neq -1$

$\quad = \dfrac{2x^2 - 3x + 1}{x + 2}$

$\quad = 2x - 7 + \dfrac{15}{x + 2}, \quad x \neq -1$

(a) Domain: all real numbers x except $x = -1$ and $x = -2$

(b) y-intercept: $\left(0, \dfrac{1}{2}\right)$

x-intercepts: $\left(\dfrac{1}{2}, 0\right), (1, 0)$

(c) Vertical asymptote: $x = -2$
Slant asymptote: $y = 2x - 7$

(d)

x	-4	-3	$-\frac{3}{2}$	0	1
y	$-\frac{45}{2}$	-28	20	$\frac{1}{2}$	0

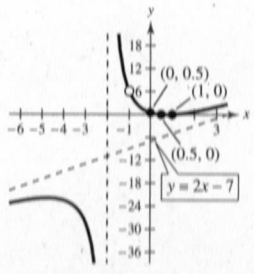

64. $f(x) = \dfrac{2x^3 + x^2 - 8x - 4}{x^2 - 3x + 2} = \dfrac{(x-2)(x+2)(2x+1)}{(x-2)(x-1)}$

$= 2x + 7 + \dfrac{9}{x-1}, x \neq 2$

(a) Domain: all real numbers x except $x = 1$ or $x = 2$

(b) y-intercept: $(0, -2)$

x-intercepts: $(-2, 0), \left(-\dfrac{1}{2}, 0\right)$

(c) Vertical asymptote: $x = 1$
Slant asymptote: $y = 2x + 7$

(d)

x	-3	-2	-1	0	$\frac{1}{2}$	$\frac{3}{2}$	3	4
y	$-\frac{5}{4}$	0	$\frac{1}{2}$	-2	-10	28	$\frac{35}{2}$	18

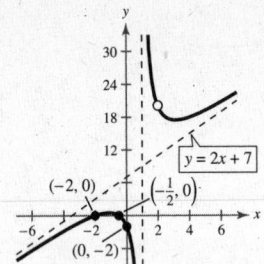

65. $f(x) = \dfrac{x^2 + 5x + 8}{x + 3} = x + 2 + \dfrac{2}{x+3}$

Domain: all real numbers x except $x = -3$

y-intercept: $\left(0, \dfrac{8}{3}\right)$

Vertical asymptote: $x = -3$

Slant asymptote: $y = x + 2$

Line: $y = x + 2$

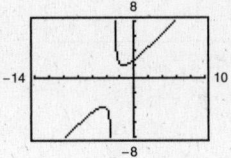

66. $f(x) = \dfrac{2x^2 + x}{x + 1} = 2x - 1 + \dfrac{1}{x+1}$

Domain: all real numbers x except $x = -1$

Vertical asymptote: $x = -1$

Slant asymptote: $y = 2x - 1$

Line: $y = 2x - 1$

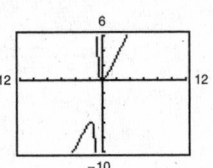

67. $g(x) = \dfrac{1 + 3x^2 - x^3}{x^2} = \dfrac{1}{x^2} + 3 - x = -x + 3 + \dfrac{1}{x^2}$

Domain: all real numbers x except $x = 0$

Vertical asymptote: $x = 0$

Slant asymptote: $y = -x + 3$

Line: $y = -x + 3$

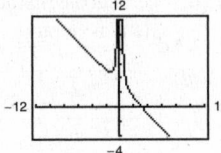

68. $h(x) = \dfrac{12 - 2x - x^2}{2(4 + x)} = -\dfrac{1}{2}x + 1 + \dfrac{2}{4+x}$

Domain: all real numbers x except $x = -4$

Vertical asymptote: $x = -4$

Slant asymptote: $y = -\dfrac{1}{2}x + 1$

Line: $y = -\dfrac{1}{2}x + 1$

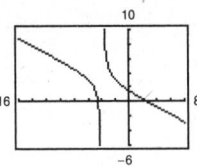

69. $y = \dfrac{x+1}{x-3}$

(a) x-intercept: $(-1, 0)$

(b) $\quad 0 = \dfrac{x+1}{x-3}$

$0 = x + 1$

$-1 = x$

70. (a) x-intercept: $(0, 0)$

(b) $0 = \dfrac{2x}{x-3}$

$0 = 2x$

$0 = x$

71. $y = \dfrac{1}{x} - x$

(a) x-intercepts: $(\pm 1, 0)$ (b) $0 = \dfrac{1}{x} - x$

$x = \dfrac{1}{x}$

$x^2 = 1$

$x = \pm 1$

72. (a) *x*-intercepts: $(1, 0), (2, 0)$

(b) $0 = x - 3 + \dfrac{2}{x}$

$0 = x^2 - 3x + 2$

$0 = (x - 1)(x - 2)$

$x = 1, x = 2$

73. $C = \dfrac{255p}{100 - p}, \; 0 \le p < 100$

(a)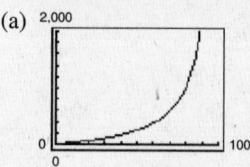

(b) $C(10) = \dfrac{255(10)}{100 - 10} \approx 28.33$ million dollars

$C(40) = \dfrac{255(40)}{100 - 40} = 170$ million dollars

$C(75) = \dfrac{255(75)}{100 - 75} = 765$ million dollars

(c) $C \to \infty$ as $x \to 100$. No, it would not be possible to remove 100% of the pollutants.

74. $C = \dfrac{25,000p}{100 - p}, \; 0 \le p < 100$

(a)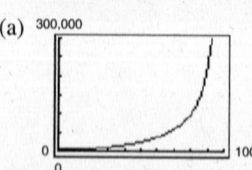

(b) $C = \dfrac{25,000(15)}{100 - 15} \approx 4411.76$

The cost would be $4411.76.

$C = \dfrac{25,000(50)}{100 - 50} = 25,000$

The cost would be $25,000.

$C = \dfrac{25,000(90)}{100 - 90} = 225,000$

The cost would be $225,000.

(c) $C \to \infty$ as $x \to 100$. No. The model is undefined for $p = 100$.

75. $N = \dfrac{20(5 + 3t)}{1 + 0.04t}, \; t \ge 0$

(a) $N(5) \approx 333$ deer

$N(10) = 500$ deer

$N(25) = 800$ deer

(b) The herd is limited by the horizontal asymptote:

$N = \dfrac{60}{0.04} = 1500$ deer

76. (a) $0.25(50) + 0.75(x) = C(50 + x)$

$C = \dfrac{12.50 + 0.75x}{50 + x} \cdot \dfrac{4}{4}$

$C = \dfrac{50 + 3x}{4(50 + x)} = \dfrac{3x + 50}{4(x + 50)}$

(b) Domain: $x \ge 0$ and $x \le 1000 - 50$

Thus, $0 \le x \le 950$. Using interval notation, the domain is $[0, 950]$.

(c)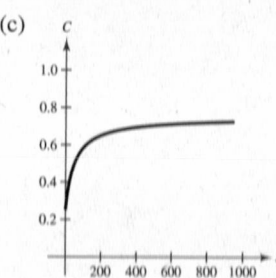

(d) As the tank is filled, the concentration increases more slowly. It approaches the horizontal asymptote of $C = \frac{3}{4} = 0.75$.

77. (a) $A = xy$ and

$$(x - 4)(y - 2) = 30$$

$$y - 2 = \frac{30}{x - 4}$$

$$y = 2 + \frac{30}{x - 4} = \frac{2x + 22}{x - 4}$$

Thus, $A = xy = x\left(\frac{2x + 22}{x - 4}\right) = \frac{2x(x + 11)}{x - 4}$.

(b) Domain: Since the margins on the left and right are each 2 inches, $x > 4$. In interval notation, the domain is $(4, \infty)$.

(c)

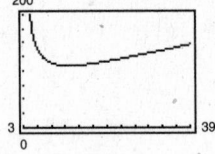

x	5	6	7	8	9	10	11	12	13	14	15
y_1 (Area)	160	102	84	76	72	70	69.143	69	69.333	70	70.909

The area is minimum when $x \approx 11.75$ inches and $y \approx 5.87$ inches.

The area is minimum when x is approximately 12.

78. $A = xy$ and

$$(x - 3)(y - 2) = 64$$

$$y - 2 = \frac{64}{x - 3}$$

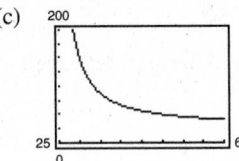

$$y = 2 + \frac{64}{x - 3} = \frac{2x + 58}{x - 3}$$

Thus, $A = xy = x\left(\frac{2x + 58}{x - 3}\right) = \frac{2x(x + 29)}{x - 3}$, $x > 3$.

By graphing the area function, we see that A is minimum when $x \approx 12.8$ inches and $y \approx 8.5$ inches.

79. (a) Let $t_1 =$ time from Akron to Columbus and $t_2 =$ time from Columbus back to Akron.

$$xt_1 = 100 \implies t_1 = \frac{100}{x}$$

$$yt_2 = 100 \implies t_2 = \frac{100}{y}$$

$$50(t_1 + t_2) = 200$$

$$t_1 + t_2 = 4$$

$$\frac{100}{x} + \frac{100}{y} = 4$$

$$100y + 100x = 4xy$$

$$25y + 25x = xy$$

$$25x = xy - 25y$$

$$25x = y(x - 25)$$

Thus, $y = \frac{25x}{x - 25}$.

(b) Vertical asymptote: $x = 25$

Horizontal asymptote: $y = 25$

(c)

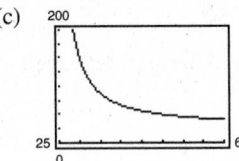

(d)

x	30	35	40	45	50	55	60
y	150	87.5	66.67	56.25	50	45.83	42.86

(e) Yes. You would expect the average speed for the round trip to be the average of the average speeds for the two parts of the trip.

(f) No. At 20 miles per hour you would use more time in one direction than is required for the round trip at an average speed of 50 miles per hour.

80. (a)

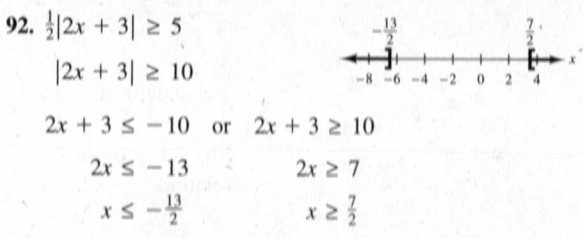

(b) $S = \dfrac{5.816(18)^2 - 130.68}{0.004(18)^2 + 1.00} = 763.81$

The sales in 2008 is estimated to be $763,810,000.

(c) Probably not. The graph has a horizontal asymptote at $S = \dfrac{5.816}{0.004} \approx 1454$ million dollars.

Future sales may exceed this limiting value.

81. False. Polynomial functions do not have vertical asymptotes.

82. False. The graph of $f(x) = \dfrac{x}{x^2 + 1}$ crosses $y = 0$, which is a horizontal asymptote.

83. Vertical asymptote: None $\implies$ The denominator is not zero for any value of x (unless the numerator is also zero there).

Horizontal asymptote: $y = 2 \implies$ The degree of the numerator equals the degree of the denominator.

$f(x) = \dfrac{2x^2}{x^2 + 1}$ is one possible function. There are many correct answers.

84. Vertical asymptotes: $x = -2,\ x = 1 \implies (x + 2)(x - 1)$ are factors of the denominator.

Horizontal asymptotes: None $\implies$ The degree of the numerator is greater than the degree of the denominator.

$f(x) = \dfrac{x^3}{(x + 2)(x - 1)}$ is one possible function. There are many correct answers.

85. $x^2 - 15x + 56 = (x - 8)(x - 7)$

86. $3x^2 + 23x - 36 = (3x - 4)(x + 9)$

87. $x^3 - 5x^2 + 4x - 20 = (x - 5)(x^2 + 4)$
$= (x - 5)(x + 2i)(x - 2i)$

88. $x^3 + 6x^2 - 2x - 12 = x^2(x + 6) - 2(x + 6)$
$= (x + 6)(x^2 - 2)$
$= (x + 6)(x + \sqrt{2})(x - \sqrt{2})$

89. $10 - 3x \le 0$
$3x \ge 10$
$x \ge \frac{10}{3}$

90. $5 - 2x > 5(x + 1)$
$5 - 2x > 5x + 5$
$-7x > 0$
$x < 0$

91. $|4(x - 2)| < 20$
$-20 < 4x - 8 < 20$
$-12 < 4x < 28$
$-3 < x < 7$

92. $\frac{1}{2}|2x + 3| \ge 5$
$|2x + 3| \ge 10$
$2x + 3 \le -10 \quad \text{or} \quad 2x + 3 \ge 10$
$2x \le -13 \qquad\qquad 2x \ge 7$
$x \le -\frac{13}{2} \qquad\qquad x \ge \frac{7}{2}$

93. Answers will vary.

Section 2.7   Nonlinear Inequalities

■ You should be able to solve inequalities.

  (a) Find the critical number.

    1. Values that make the expression zero

    2. Values that make the expression undefined

  (b) Test one value in each test interval on the real number line resulting from the critical numbers.

  (c) Determine the solution intervals.

Vocabulary Check

1. critical; test intervals   **2.** zeros; undefined values   **3.** $P = R - C$

---

1. $x^2 - 3 < 0$

  (a) $x = 3$

    $(3)^2 - 3 \overset{?}{<} 0$

        $6 \not< 0$

    No, $x = 3$ *is not* a solution.

  (b) $x = 0$

    $(0)^2 - 3 \overset{?}{<} 0$

        $-3 < 0$

    Yes, $x = 0$ *is* a solution.

  (c) $x = \frac{3}{2}$

    $\left(\frac{3}{2}\right)^2 - 3 \overset{?}{<} 0$

        $-\frac{3}{4} < 0$

    Yes, $x = \frac{3}{2}$ *is* a solution.

  (d) $x = -5$

    $(-5)^2 - 3 \overset{?}{<} 0$

        $22 \not< 0$

    No, $x = -5$ *is not* a solution.

2. $x^2 - x - 12 \geq 0$

  (a) $x = 5$

    $(5)^2 - (5) - 12 \overset{?}{\geq} 0$

            $8 \geq 0$

    Yes, $x = 5$ *is* a solution.

  (b) $x = 0$

    $(0)^2 - 0 - 12 \overset{?}{\geq} 0$

        $-12 \not\geq 0$

    No, $x = 0$ *is not* a solution.

  (c) $x = -4$

    $(-4)^2 - (-4) - 12 \overset{?}{\geq} 0$

        $16 + 4 - 12 \overset{?}{\geq} 0$

            $8 \geq 0$

    Yes, $x = -4$ *is* a solution.

  (d) $x = -3$

    $(-3)^2 - (-3) - 12 \overset{?}{\geq} 0$

        $9 + 3 - 12 \overset{?}{\geq} 0$

            $0 \geq 0$

    Yes, $x = -3$ *is* a solution.

3. $\dfrac{x + 2}{x - 4} \geq 3$

  (a) $x = 5$

    $\dfrac{5 + 2}{5 - 4} \overset{?}{\geq} 3$

        $7 \geq 3$

    Yes, $x = 5$ *is* a solution.

  (b) $x = 4$

    $\dfrac{4 + 2}{4 - 4} \overset{?}{\geq} 3$

    $\dfrac{6}{0}$ is undefined.

    No, $x = 4$ *is not* a solution.

  (c) $x = -\dfrac{9}{2}$

    $\dfrac{-\frac{9}{2} + 2}{-\frac{9}{2} - 4} \overset{?}{\geq} 3$

        $\dfrac{5}{17} \not\geq 3$

    No, $x = -\dfrac{9}{2}$ *is not* a solution.

  (d) $x = \dfrac{9}{2}$

    $\dfrac{\frac{9}{2} + 2}{\frac{9}{2} - 4} \overset{?}{\geq} 3$

        $13 \geq 3$

    Yes, $x = \dfrac{9}{2}$ *is* a solution.

4. $\dfrac{3x^2}{x^2 + 4} < 1$

(a) $x = -2$

$\dfrac{3(-2)^2}{(-2)^2 + 4} \overset{?}{<} 1$

$\dfrac{12}{8} \not< 1$

No, $x = -2$ *is not* a solution.

(b) $x = -1$

$\dfrac{3(-1)^2}{(-1)^2 + 4} \overset{?}{<} 1$

$\dfrac{3}{5} < 1$

Yes, $x = -1$ *is* a solution.

(c) $x = 0$

$\dfrac{3(0)^2}{(0)^2 + 4} \overset{?}{<} 1$

$0 < 1$

Yes, $x = 0$ *is* a solution.

(d) $x = 3$

$\dfrac{3(3)^2}{(3)^2 + 4} \overset{?}{<} 1$

$\dfrac{27}{13} \not< 1$

No, $x = 3$ *is not* a solution.

5. $2x^2 - x - 6 = (2x + 3)(x - 2)$

$2x + 3 = 0 \implies x = -\dfrac{3}{2}$

$x - 2 = 0 \implies x = 2$

Critical numbers: $x = -\dfrac{3}{2}, x = 2$

6. $9x^3 - 25x^2 = 0$

$x^2(9x - 25) = 0$

$x^2 = 0 \implies x = 0$

$9x - 25 = 0 \implies x = \dfrac{25}{9}$

The critical numbers are 0 and $\dfrac{25}{9}$.

7. $2 + \dfrac{3}{x - 5} = \dfrac{2(x - 5) + 3}{x - 5}$

$= \dfrac{2x - 7}{x - 5}$

$2x - 7 = 0 \implies x = \dfrac{7}{2}$

$x - 5 = 0 \implies x = 5$

Critical numbers: $x = \dfrac{7}{2}, x = 5$

8. $\dfrac{x}{x + 2} - \dfrac{2}{x - 1} = \dfrac{x(x - 1) - 2(x + 2)}{(x + 2)(x - 1)}$

$= \dfrac{x^2 - x - 2x - 4}{(x + 2)(x - 1)}$

$= \dfrac{(x - 4)(x + 1)}{(x + 2)(x - 1)}$

$(x - 4)(x + 1) = 0$

$x - 4 = 0 \implies x = 4$

$x + 1 = 0 \implies x = -1$

$(x + 2)(x - 1) = 0$

$x + 2 = 0 \implies x = -2$

$x - 1 = 0 \implies x = 1$

The critical numbers are $-2, -1, 1, 4$.

9. $x^2 \le 9$

$x^2 - 9 \le 0$

$(x + 3)(x - 3) \le 0$

Critical numbers: $x = \pm 3$

Test intervals: $(-\infty, -3), (-3, 3), (3, \infty)$

Test: Is $(x + 3)(x - 3) \le 0$?

Interval	x-Value	Value of $x^2 - 9$	Conclusion
$(-\infty, -3)$	$x = -4$	$16 - 9 = 7$	Positive
$(-3, 3)$	$x = 0$	$0 - 9 = -9$	Negative
$(3, \infty)$	$x = 4$	$16 - 9 = 7$	Positive

Solution set: $[-3, 3]$

10. $x^2 < 36$

$x^2 - 36 < 0$

$(x + 6)(x - 6) < 0$

Critical numbers: $x = -6, x = 6$

Test intervals: $(-\infty, -6) \implies (x + 6)(x - 6) > 0$

$(-6, 6) \implies (x + 6)(x - 6) < 0$

$(6, \infty) \implies (x + 6)(x - 6) > 0$

Solution interval: $(-6, 6)$

11. $(x + 2)^2 < 25$

   $x^2 + 4x + 4 < 25$

   $x^2 + 4x - 21 < 0$

   $(x + 7)(x - 3) < 0$

   Critical numbers: $x = -7, x = 3$

   Test intervals: $(-\infty, -7), (-7, 3), (3, \infty)$

   Test: Is $(x + 7)(x - 3) < 0$?

Interval	x-Value	Value of $(x + 7)(x - 3)$	Conclusion
$(-\infty, -7)$	$x = -10$	$(-3)(-13) = 39$	Positive
$(-7, 3)$	$x = 0$	$(7)(-3) = -21$	Negative
$(3, \infty)$	$x = 5$	$(12)(2) = 24$	Positive

   Solution set: $(-7, 3)$

12. $(x - 3)^2 \geq 1$

   $x^2 - 6x + 8 \geq 0$

   $(x - 2)(x - 4) \geq 0$

   Critical numbers: $x = 2, x = 4$

   Test intervals: $(-\infty, 2) \implies (x - 2)(x - 4) > 0$

   $(2, 4) \implies (x - 2)(x - 4) < 0$

   $(4, \infty) \implies (x - 2)(x - 4) > 0$

   Solution intervals: $(-\infty, 2] \cup [4, \infty)$

13. $x^2 + 4x + 4 \geq 9$

   $x^2 + 4x - 5 \geq 0$

   $(x + 5)(x - 1) \geq 0$

   Critical numbers: $x = -5, x = 1$

   Test intervals: $(-\infty, -5), (-5, 1), (1, \infty)$

   Test: Is $(x + 5)(x - 1) \geq 0$?

Interval	x-Value	Value of $(x + 5)(x - 1)$	Conclusion
$(-\infty, -5)$	$x = -6$	$(-1)(-7) = 7$	Positive
$(-5, 1)$	$x = 0$	$(5)(-1) = -5$	Negative
$(1, \infty)$	$x = 2$	$(7)(1) = 7$	Positive

   Solution set: $(-\infty, -5] \cup [1, \infty)$

14. $x^2 - 6x + 9 < 16$

   $x^2 - 6x - 7 < 0$

   $(x + 1)(x - 7) < 0$

   Critical numbers: $x = -1, x = 7$

   Test intervals: $(-\infty, -1) \implies (x + 1)(x - 7) > 0$

   $(-1, 7) \implies (x + 1)(x - 7) < 0$

   $(7, \infty) \implies (x + 1)(x - 7) > 0$

   Solution interval: $(-1, 7)$

15. $x^2 + x < 6$

   $x^2 + x - 6 < 0$

   $(x + 3)(x - 2) < 0$

   Critical numbers: $x = -3, x = 2$

   Test intervals: $(-\infty, -3), (-3, 2), (2, \infty)$

   Test: Is $(x + 3)(x - 2) < 0$?

Interval	x-Value	Value of $(x + 3)(x - 2)$	Conclusion
$(-\infty, -3)$	$x = -4$	$(-1)(-6) = 6$	Positive
$(-3, 2)$	$x = 0$	$(3)(-2) = -6$	Negative
$(2, \infty)$	$x = 3$	$(6)(1) = 6$	Positive

   Solution set: $(-3, 2)$

16. $x^2 + 2x > 3$

   $x^2 + 2x - 3 > 0$

   $(x + 3)(x - 1) > 0$

   Critical numbers: $x = -3, x = 1$

   Test intervals: $(-\infty, -3) \implies (x + 3)(x - 1) > 0$

   $(-3, 1) \implies (x + 3)(x - 1) < 0$

   $(1, \infty) \implies (x + 3)(x - 1) > 0$

   Solution intervals: $(-\infty, -3) \cup (1, \infty)$

17. $x^2 + 2x - 3 < 0$

$(x + 3)(x - 1) < 0$

Critical numbers: $x = -3, x = 1$

Test intervals: $(-\infty, -3), (-3, 1), (1, \infty)$

Test: Is $(x + 3)(x - 1) < 0$?

Interval	x-Value	Value of $(x + 3)(x - 1)$	Conclusion
$(-\infty, -3)$	$x = -4$	$(-1)(-5) = 5$	Positive
$(-3, 1)$	$x = 0$	$(3)(-1) = -3$	Negative
$(1, \infty)$	$x = 2$	$(5)(1) = 5$	Positive

Solution set: $(-3, 1)$

18. $x^2 - 4x - 1 > 0$

$x = \dfrac{4 \pm \sqrt{16 + 4}}{2} = 2 \pm \sqrt{5}$

Critical numbers: $x = 2 - \sqrt{5}, x = 2 + \sqrt{5}$

Test intervals: $(-\infty, 2 - \sqrt{5}) \Rightarrow x^2 - 4x - 1 > 0$

$\ (2 - \sqrt{5}, 2 + \sqrt{5}) \Rightarrow x^2 - 4x - 1 < 0$

$\ (2 + \sqrt{5}, \infty) \Rightarrow x^2 - 4x - 1 > 0$

Solution intervals: $(-\infty, 2 - \sqrt{5}) \cup (2 + \sqrt{5}, \infty)$

19. $x^2 + 8x - 5 \geq 0$

$x^2 + 8x - 5 = 0$ \qquad Complete the square.

$x^2 + 8x + 16 = 5 + 16$

$(x + 4)^2 = 21$

$x + 4 = \pm\sqrt{21}$

$x = -4 \pm \sqrt{21}$

Critical numbers: $x = -4 \pm \sqrt{21}$

Test intervals: $(-\infty, -4 - \sqrt{21}), (-4 - \sqrt{21}, -4 + \sqrt{21}), (-4 + \sqrt{21}, \infty)$

Test: Is $x^2 + 8x - 5 \geq 0$?

Interval	x-Value	Value of $x^2 + 8x - 5$	Conclusion
$(-\infty, -4 - \sqrt{21})$	$x = -10$	$100 - 80 - 5 = 15$	Positive
$(-4 - \sqrt{21}, -4 + \sqrt{21})$	$x = 0$	$0 + 0 - 5 = -5$	Negative
$(-4 + \sqrt{21}, \infty)$	$x = 2$	$4 + 16 - 5 = 15$	Positive

Solution set: $(-\infty < -4 - \sqrt{21}] \cup [-4 + \sqrt{21}, \infty)$

20. $-2x^2 + 6x + 15 \leq 0$

$2x^2 - 6x - 15 \geq 0$

$x = \dfrac{-(-6) \pm \sqrt{(-6)^2 - 4(2)(-15)}}{2(2)} = \dfrac{6 \pm \sqrt{156}}{4} = \dfrac{6 \pm 2\sqrt{39}}{4} = \dfrac{3}{2} \pm \dfrac{\sqrt{39}}{2}$

Critical numbers: $x = \dfrac{3}{2} - \dfrac{\sqrt{39}}{2}, x = \dfrac{3}{2} + \dfrac{\sqrt{39}}{2}$

Test intervals: $\left(-\infty, \dfrac{3}{2} - \dfrac{\sqrt{39}}{2}\right) \Rightarrow -2x^2 + 6x + 15 < 0$

$\ \left(\dfrac{3}{2} - \dfrac{\sqrt{39}}{2}, \dfrac{3}{2} + \dfrac{\sqrt{39}}{2}\right) \Rightarrow -2x^2 + 6x + 15 > 0$

$\ \left(\dfrac{3}{2} + \dfrac{\sqrt{39}}{2}, \infty\right) \Rightarrow -2x^2 + 6x + 15 < 0$

Solution interval: $\left(-\infty, \dfrac{3}{2} - \dfrac{\sqrt{39}}{2}\right] \cup \left[\dfrac{3}{2} + \dfrac{\sqrt{39}}{2}, \infty\right)$

21. $x^3 - 3x^2 - x + 3 > 0$

$x^2(x - 3) - 1(x - 3) > 0$

$(x^2 - 1)(x - 3) > 0$

$(x + 1)(x - 1)(x - 3) > 0$

Critical numbers: $x = \pm 1, x = 3$

Test intervals: $(-\infty, -1), (-1, 1), (1, 3), (3, \infty)$

Test: Is $(x + 1)(x - 1)(x - 3) > 0$?

Interval	x-Value	Value of $(x + 1)(x - 1)(x - 3)$	Conclusion
$(-\infty, -1)$	$x = -2$	$(-1)(-3)(-5) = -15$	Negative
$(-1, 1)$	$x = 0$	$(1)(-1)(-3) = 3$	Positive
$(1, 3)$	$x = 2$	$(3)(1)(-1) = -3$	Negative
$(3, \infty)$	$x = 4$	$(5)(3)(1) = 15$	Positive

Solution set: $(-1, 1) \cup (3, \infty)$

22. $x^3 + 2x^2 - 4x - 8 \le 0$

$x^2(x + 2) - 4(x + 2) \le 0$

$(x + 2)(x^2 - 4) \le 0$

Critical numbers: $x = -2, x = 2$

Test intervals: $(-\infty, -2) \implies x^3 + 2x^2 - 4x - 8 < 0$

$(-2, 2) \implies x^3 + 2x^2 - 4x - 8 < 0$

$(2, \infty) \implies x^3 + 2x^2 - 4x - 8 > 0$

Solution interval: $(-\infty, 2]$

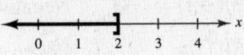

23. $x^3 - 2x^2 - 9x - 2 \ge -20$

$x^3 - 2x^2 - 9x + 18 \ge 0$

$x^2(x - 2) - 9(x - 2) \ge 0$

$(x - 2)(x^2 - 9) \ge 0$

$(x - 2)(x + 3)(x - 3) \ge 0$

Critical numbers: $x = 2, x = \pm 3$

Test intervals: $(-\infty, -3), (-3, 2), (2, 3), (3, \infty)$

Test: Is $(x - 2)(x + 3)(x - 3) \ge 0$?

Interval	x-Value	Value of $(x - 2)(x + 3)(x - 3)$	Conclusion
$(-\infty, -3)$	$x = -4$	$(-6)(-1)(-7) = -42$	Negative
$(-3, 2)$	$x = 0$	$(-2)(3)(-3) = 18$	Positive
$(2, 3)$	$x = 2.5$	$(0.5)(5.5)(-0.5) = -1.375$	Negative
$(3, \infty)$	$x = 4$	$(2)(7)(1) = 14$	Positive

Solution set: $[-3, 2] \cup [3, \infty)$

24. $2x^3 + 13x^2 - 8x - 46 \ge 6$

$2x^3 + 13x^2 - 8x - 52 \ge 0$

$x^2(2x + 13) - 4(2x + 13) \ge 0$

$(2x + 13)(x^2 - 4) \ge 0$

Critical numbers: $x = -\frac{13}{2}, x = -2, x = 2$

Test intervals: $\left(-\infty, -\frac{13}{2}\right) \implies 2x^3 + 13x^2 - 8x - 52 < 0$

$\left(-\frac{13}{2}, -2\right) \implies 2x^3 + 13x^2 - 8x - 52 > 0$

$(-2, 2) \implies 2x^3 + 13x^2 - 8x - 52 < 0$

$(2, \infty) \implies 2x^3 + 13x^2 - 8x - 52 > 0$

Solution interval: $\left[-\frac{13}{2}, -2\right], [2, \infty)$

25. $4x^2 - 4x + 1 \leq 0$

$(2x - 1)^2 \leq 0$

Critical number: $x = \dfrac{1}{2}$

Test intervals: $\left(-\infty, \dfrac{1}{2}\right), \left(\dfrac{1}{2}, \infty\right)$

Test: Is $(2x - 1)^2 \leq 0$?

Interval	x-Value	Value of $(2x - 1)^2$	Conclusion
$\left(-\infty, \dfrac{1}{2}\right)$	$x = 0$	$(-1)^2 = 1$	Positive
$\left(\dfrac{1}{2}, \infty\right)$	$x = 1$	$(1)^2 = 1$	Positive

Solution set: $x = \dfrac{1}{2}$

26. $x^2 + 3x + 8 > 0$

The critical numbers are imaginary:

$$-\dfrac{3}{2} \pm \dfrac{i\sqrt{23}}{2}$$

So the set of real numbers is the solution set.

27. $4x^3 - 6x^2 < 0$

$2x^2(2x - 3) < 0$

Critical numbers: $x = 0, x = \frac{3}{2}$

Test intervals: $(-\infty, 0), \left(0, \frac{3}{2}\right), \left(\frac{3}{2}, \infty\right)$

Test: Is $2x^2(2x - 3) < 0$?

By testing an x-value in each test interval in the inequality, we see that the solution set is: $(-\infty, 0) \cup \left(0, \frac{3}{2}\right)$

28. $4x^3 - 12x^2 > 0$

$4x^2(x - 3) > 0$

Critical numbers: $x = 0, x = 3$

Test intervals: $(-\infty, 0) \implies 4x^2(x - 3) < 0$

$\qquad (0, 3) \implies 4x^2(x - 3) < 0$

$\qquad (3, \infty) \implies 4x^2(x - 3) > 0$

Solution interval: $(3, \infty)$

29. $x^3 - 4x \geq 0$

$x(x + 2)(x - 2) \geq 0$

Critical numbers: $x = 0, x = \pm 2$

Test intervals: $(-\infty, -2), (-2, 0), (0, 2), (2, \infty)$

Test: Is $x(x + 2)(x - 2) \geq 0$?

By testing an x-value in each test interval in the inequality, we see that the solution set is: $[-2, 0] \cup [2, \infty)$

30. $2x^3 - x^4 \leq 0$

$x^3(2 - x) \leq 0$

Critical numbers: $x = 0, x = 2$

Test intervals: $(-\infty, 0) \implies x^3(2 - x) < 0$

$\qquad (0, 2) \implies x^3(2 - x) > 0$

$\qquad (2, \infty) \implies x^3(2 - x) < 0$

Solution intervals: $(-\infty, 0] \cup [2, \infty)$

31. $(x - 1)^2(x + 2)^3 \geq 0$

Critical numbers: $x = 1, x = -2$

Test intervals: $(-\infty, -2), (-2, 1), (1, \infty)$

Test: Is $(x - 1)^2(x + 3)^3 \geq 0$?

By testing an x-value in each test interval in the inequality, we see that the solution set is: $[-2, \infty)$

32. $x^4(x - 3) \leq 0$

Critical numbers: $x = 0, x = 3$

Test intervals: $(-\infty, 0) \implies x^4(x - 3) < 0$

$\qquad (0, 3) \implies x^4(x - 3) < 0$

$\qquad (3, \infty) \implies x^4(x - 3) > 0$

Solution intervals: $(-\infty, 0] \cup [0, 3]$ or $(-\infty, 3]$

33. $y = -x^2 + 2x + 3$

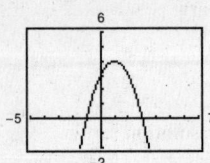

(a) $y \leq 0$ when $x \leq -1$ or $x \geq 3$.

(b) $y \geq 3$ when $0 \leq x \leq 2$.

34. $y = \dfrac{1}{2}x^2 - 2x + 1$

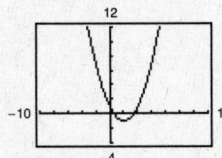

(a) $y \leq 0$

$\dfrac{1}{2}x^2 - 2x + 1 \leq 0$

$x^2 - 4x + 2 \leq 0$

$x = \dfrac{-(-4) \pm \sqrt{(-4)^2 - 4(1)(2)}}{2(1)}$

$= \dfrac{4 \pm \sqrt{8}}{2} = 2 \pm \sqrt{2}$

$y \leq 0$ when $2 - \sqrt{2} \leq x \leq 2 + \sqrt{2}$.

(b) $y \geq 7$

$\dfrac{1}{2}x^2 - 2x + 1 \geq 7$

$x^2 - 4x - 12 \geq 0$

$(x - 6)(x + 2) \geq 0$

$y \geq 7$ when $x \leq -2, x \geq 6$.

35. $y = \frac{1}{8}x^3 - \frac{1}{2}x$

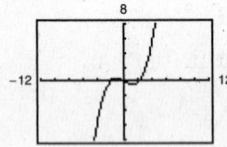

(a) $y \geq 0$ when $-2 \leq x \leq 0, 2 \leq x < \infty$.

(b) $y \leq 6$ when $x \leq 4$.

36. $y = x^3 - x^2 - 16x + 16$

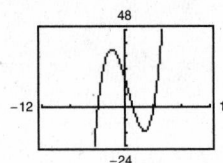

(a) $y \leq 0$

$x^3 - x^2 - 16x + 16 \leq 0$

$x^2(x - 1) - 16(x - 1) \leq 0$

$(x - 1)(x^2 - 16) \leq 0$

$y \leq 0$ when $-\infty < x \leq -4, 1 \leq x \leq 4$.

(b) $y \geq 36$

$x^3 - x^2 - 16x + 16 \geq 36$

$x^3 - x^2 - 16x - 20 \geq 0$

$(x + 2)(x - 5)(x + 2) \geq 0$

$y \geq 36$ when $x = -2, 5 \leq x < \infty$.

37. $\dfrac{1}{x} - x > 0$

$\dfrac{1 - x^2}{x} > 0$

Critical numbers: $x = 0, x = \pm 1$

Test intervals: $(-\infty, -1), (-1, 0), (0, 1), (1, \infty)$

Test: Is $\dfrac{1 - x^2}{x} > 0$?

By testing an x-value in each test interval in the inequality, we see that the solution set is: $(-\infty, -1) \cup (0, 1)$

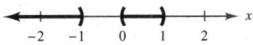

38. $\dfrac{1}{x} - 4 < 0$

$\dfrac{1 - 4x}{x} < 0$

Critical numbers: $x = 0, x = \dfrac{1}{4}$

Test intervals: $(-\infty, 0) \Longrightarrow \dfrac{1 - 4x}{x} < 0$

$\left(0, \dfrac{1}{4}\right) \Longrightarrow \dfrac{1 - 4x}{x} > 0$

$\left(\dfrac{1}{4}, \infty\right) \Longrightarrow \dfrac{1 - 4x}{x} < 0$

Solution interval: $(-\infty, 0) \cup \left(\dfrac{1}{4}, \infty\right)$

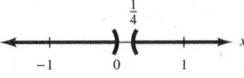

39. $\dfrac{x+6}{x+1} - 2 < 0$

$$\dfrac{x+6-2(x+1)}{x+1} < 0$$

$$\dfrac{4-x}{x+1} < 0$$

Critical numbers: $x = -1, x = 4$

Test intervals: $(-\infty, -1), (-1, 4), (4, \infty)$

Test: Is $\dfrac{4-x}{x+1} < 0$?

By testing an x-value in each test interval in the inequality, we see that the solution set is: $(-\infty, -1) \cup (4, \infty)$

40. $\dfrac{x+12}{x+2} - 3 \geq 0$

$$\dfrac{x+12-3(x+2)}{x+2} \geq 0$$

$$\dfrac{6-2x}{x+2} \geq 0$$

Critical numbers: $x = -2, x = 3$

Test intervals: $(-\infty, -2) \implies \dfrac{6-2x}{x+2} < 0$

$$(-2, 3) \implies \dfrac{6-2x}{x+2} > 0$$

$$(3, \infty) \implies \dfrac{6-2x}{x+2} < 0$$

Solution interval: $(-2, 3]$

41. $\dfrac{3x-5}{x-5} > 4$

$$\dfrac{3x-5}{x-5} - 4 > 0$$

$$\dfrac{3x-5-4(x-5)}{x-5} > 0$$

$$\dfrac{15-x}{x-5} > 0$$

Critical numbers: $x = 5, x = 15$

Test intervals: $(-\infty, 5), (5, 15), (15, \infty)$

Test: Is $\dfrac{15-x}{x-5} > 0$?

By testing an x-value in each test interval in the inequality, we see that the solution set is: $(5, 15)$

42. $\dfrac{5+7x}{1+2x} < 4$

$$\dfrac{5+7x-4(1+2x)}{1+2x} < 0$$

$$\dfrac{1-x}{1+2x} < 0$$

Critical numbers: $x = -\dfrac{1}{2}, x = 1$

Test intervals: $\left(-\infty, -\dfrac{1}{2}\right) \implies \dfrac{1-x}{1+2x} < 0$

$$\left(-\dfrac{1}{2}, 1\right) \implies \dfrac{1-x}{1+2x} > 0$$

$$(1, \infty) \implies \dfrac{1-x}{1+2x} < 0$$

Solution intervals: $\left(-\infty, -\dfrac{1}{2}\right) \cup (1, \infty)$

43.
$$\frac{4}{x + 5} > \frac{1}{2x + 3}$$

$$\frac{4}{x + 5} - \frac{1}{2x + 3} > 0$$

$$\frac{4(2x + 3) - (x + 5)}{(x + 5)(2x + 3)} > 0$$

$$\frac{7x + 7}{(x + 5)(2x + 3)} > 0$$

Critical numbers: $x = -1, x = -5, x = -\frac{3}{2}$

Test intervals: $(-\infty, -5), \left(-5, -\frac{3}{2}\right),$
$$\left(-\frac{3}{2}, -1\right), (-1, \infty)$$

Test: Is $\dfrac{7(x + 1)}{(x + 5)(2x + 3)} > 0?$

By testing an x-value in each test interval in the inequality, we see that the solution set is: $\left(-5, -\frac{3}{2}\right) \cup (-1, \infty)$

44.
$$\frac{5}{x - 6} > \frac{3}{x + 2}$$

$$\frac{5(x + 2) - 3(x - 6)}{(x - 6)(x + 2)} > 0$$

$$\frac{2x + 28}{(x - 6)(x + 2)} > 0$$

Critical numbers: $x = -14, x = -2, x = 6$

Test intervals: $(-\infty, -14) \Longrightarrow \dfrac{2x + 28}{(x - 6)(x + 2)} < 0$

$(-14, -2) \Longrightarrow \dfrac{2x + 28}{(x - 6)(x + 2)} > 0$

$(-2, 6) \Longrightarrow \dfrac{2x + 28}{(x - 6)(x + 2)} < 0$

$(6, \infty) \Longrightarrow \dfrac{2x + 28}{(x - 6)(x + 2)} > 0$

Solution intervals: $(-14, -2) \cup (6, \infty)$

45.
$$\frac{1}{x - 3} \le \frac{9}{4x + 3}$$

$$\frac{1}{x - 3} - \frac{9}{4x + 3} \le 0$$

$$\frac{4x + 3 - 9(x - 3)}{(x - 3)(4x + 3)} \le 0$$

$$\frac{30 - 5x}{(x - 3)(4x + 3)} \le 0$$

Critical numbers: $x = 3, x = -\frac{3}{4}, x = 6$

Test intervals: $\left(-\infty, -\frac{3}{4}\right), \left(-\frac{3}{4}, 3\right), (3, 6), (6, \infty)$

Test: Is $\dfrac{30 - 5x}{(x - 3)(4x + 3)} \le 0?$

By testing an x-value in each test interval in the inequality, we see that the solution set is: $\left(-\frac{3}{4}, 3\right) \cup [6, \infty)$

46.
$$\frac{1}{x} \ge \frac{1}{x + 3}$$

$$\frac{1(x + 3) - 1(x)}{x(x + 3)} \ge 0$$

$$\frac{3}{x(x + 3)} \ge 0$$

Critical numbers: $x = -3, x = 0$

Test intervals: $(-\infty, -3) \Longrightarrow \dfrac{3}{x(x + 3)} > 0$

$(-3, 0) \Longrightarrow \dfrac{3}{x(x + 3)} < 0$

$(0, \infty) \Longrightarrow \dfrac{3}{x(x + 3)} > 0$

Solution intervals: $(-\infty, -3) \cup (0, \infty)$

47. $\dfrac{x^2 + 2x}{x^2 - 9} \le 0$

$\dfrac{x(x + 2)}{(x + 3)(x - 3)} \le 0$

Critical numbers: $x = 0, x = -2, x = \pm 3$

Test intervals:
$(-\infty, -3), (-3, -2), (-2, 0), (0, 3), (3, \infty)$

Test: Is $\dfrac{x(x + 2)}{(x + 3)(x - 3)} \le 0$?

By testing an x-value in each test interval in the inequality, we see that the solution set is: $(-3, -2] \cup [0, 3)$

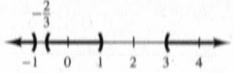

48. $\dfrac{x^2 + x - 6}{x} \ge 0$

$\dfrac{(x + 3)(x - 2)}{x} \ge 0$

Critical numbers: $x = -3, x = 0, x = 2$

Test intervals: $(-\infty, -3) \implies \dfrac{(x + 3)(x - 2)}{x} < 0$

$(-3, 0) \implies \dfrac{(x + 3)(x - 2)}{x} > 0$

$(0, 2) \implies \dfrac{(x + 3)(x - 2)}{x} < 0$

$(2, \infty) \implies \dfrac{(x + 3)(x - 2)}{x} > 0$

Solution intervals: $[-3, 0) \cup [2, \infty)$

49. $\dfrac{5}{x - 1} - \dfrac{2x}{x + 1} < 1$

$\dfrac{5}{x - 1} - \dfrac{2x}{x + 1} - 1 < 0$

$\dfrac{5(x + 1) - 2x(x - 1) - (x - 1)(x + 1)}{(x - 1)(x + 1)} < 0$

$\dfrac{5x + 5 - 2x^2 + 2x - x^2 + 1}{(x - 1)(x + 1)} < 0$

$\dfrac{-3x^2 + 7x + 6}{(x - 1)(x + 1)} < 0$

$\dfrac{-(3x + 2)(x - 3)}{(x - 1)(x + 1)} < 0$

Critical numbers: $x = -\dfrac{2}{3}, x = 3, x = \pm 1$

Test intervals: $(-\infty, -1), \left(-1, -\dfrac{2}{3}\right), \left(-\dfrac{2}{3}, 1\right), (1, 3), (3, \infty)$

Test: Is $\dfrac{-(3x + 2)(x - 3)}{(x - 1)(x + 1)} < 0$?

By testing an x-value in each test interval in the inequality, we see that the solution set is: $(-\infty, -1) \cup \left(-\dfrac{2}{3}, 1\right) \cup (3, \infty)$

50. $\dfrac{3x}{x - 1} \le \dfrac{x}{x + 4} + 3$

$\dfrac{3x(x + 4) - x(x - 1) - 3(x + 4)(x - 1)}{(x - 1)(x + 4)} \le 0$

$\dfrac{-x^2 + 4x + 12}{(x - 1)(x + 4)} \le 0$

$\dfrac{-(x - 6)(x + 2)}{(x - 1)(x + 4)} \le 0$

Critical numbers: $x = -4, x = -2, x = 1, x = 6$

Test intervals: $(-\infty, -4) \implies \dfrac{-(x - 6)(x + 2)}{(x - 1)(x + 4)} < 0$

$(-4, -2) \implies \dfrac{-(x - 6)(x + 2)}{(x - 1)(x + 4)} > 0$

$(-2, 1) \implies \dfrac{-(x - 6)(x + 2)}{(x - 1)(x + 4)} < 0$

$(1, 6) \implies \dfrac{-(x - 6)(x + 2)}{(x - 1)(x + 4)} > 0$

$(6, \infty) \implies \dfrac{-(x - 6)(x + 2)}{(x - 1)(x + 4)} < 0$

Solution intervals: $(-\infty, -4), [-2, 1), [6, \infty)$

51. $y = \dfrac{3x}{x - 2}$

(a) $y \le 0$ when $0 \le x < 2$.

(b) $y \ge 6$ when $2 < x \le 4$.

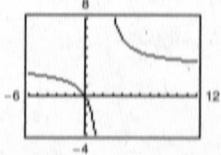

52. $y = \dfrac{2(x-2)}{x+1}$

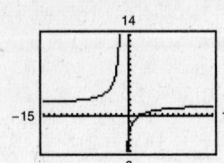

(a) $y \le 0$

$$\frac{2(x-2)}{x+1} \le 0$$

$y \le 0$ when $-1 < x \le 2$.

(b) $y \ge 8$

$$\frac{2(x-2)}{x+1} \ge 8$$

$$\frac{2(x-2) - 8(x+1)}{x+1} \ge 0$$

$$\frac{-6x - 12}{x+1} \ge 0$$

$$\frac{-6(x+2)}{x+1} \ge 0$$

$y \ge 8$ when $-2 \le x < -1$.

53. $y = \dfrac{2x^2}{x^2+4}$

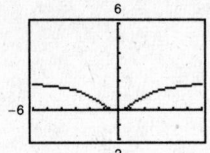

(a) $y \ge 1$ when $x \le -2$ or $x \ge 2$.

This can also be expressed as $|x| \ge 2$.

(b) $y \le 2$ for all real numbers x.

This can also be expressed as $-\infty < x < \infty$.

54. $y = \dfrac{5x}{x^2+4}$

(a) $y \ge 1$

$$\frac{5x}{x^2+4} \ge 1$$

$$\frac{5x - (x^2+4)}{x^2+4} \ge 0$$

$$\frac{-(x-4)(x-1)}{x^2+4} \ge 0$$

$y \ge 1$ when $1 \le x \le 4$.

(b) $y \le 0$

$$\frac{5x}{x^2+4} \le 0$$

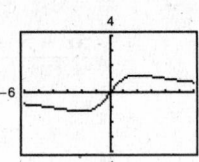

$y \le 0$ when $-\infty < x \le 0$.

55. $4 - x^2 \ge 0$

$(2 + x)(2 - x) \ge 0$

Critical numbers: $x = \pm 2$

Test intervals: $(-\infty, -2), (-2, 2), (2, \infty)$

Test: Is $4 - x^2 \ge 0$?

By testing an x-value in each test interval in the inequality, we see that the domain set is: $[-2, 2]$

56. $x^2 - 4 \ge 0$

$(x + 2)(x - 2) \ge 0$

Critical numbers: $x = -2, x = 2$

Test intervals: $(-\infty, -2) \implies (x + 2)(x - 2) > 0$

$\qquad\qquad (-2, 2) \implies (x + 2)(x - 2) < 0$

$\qquad\qquad (2, \infty) \implies (x + 2)(x - 2) > 0$

Domain: $(-\infty, -2] \cup [2, \infty)$

57. $x^2 - 7x + 12 \ge 0$

$(x - 3)(x - 4) \ge 0$

Critical numbers: $x = 3, x = 4$

Test intervals: $(-\infty, 3), (3, 4), (4, \infty)$

Test: Is $(x - 3)(x - 4) \ge 0$?

By testing an x-value in each test interval in the inequality, we see that the domain set is: $(-\infty, 3] \cup [4, \infty)$

58. $144 - 9x^2 \ge 0$

$9(4 - x)(4 + x) \ge 0$

Critical numbers: $x = -4, x = 4$

Test intervals: $(-\infty, -4) \implies 9(4 - x)(4 + x) < 0$

$\qquad\qquad (-4, 4) \implies 9(4 - x)(4 + x) > 0$

$\qquad\qquad (4, \infty) \implies 9(4 - x)(4 + x) < 0$

Domain: $[-4, 4]$

59. $\dfrac{x}{x^2 - 2x - 35} \geq 0$

$\dfrac{x}{(x + 5)(x - 7)} \geq 0$

Critical numbers: $x = 0, x = -5, x = 7$

Test intervals: $(-\infty, -5), (-5, 0), (0, 7), (7, \infty)$

Test: Is $\dfrac{x}{(x + 5)(x - 7)} \geq 0$?

By testing an x-value in each test interval in the inequality, we see that the domain set is: $(-5, 0] \cup (7, \infty)$

60. $\dfrac{x}{x^2 - 9} \geq 0$

$\dfrac{x}{(x + 3)(x - 3)} \geq 0$

Critical numbers: $x = -3, x = 0, x = 3$

Test intervals: $(-\infty, -3) \implies \dfrac{x}{(x + 3)(x - 3)} < 0$

$(-3, 0) \implies \dfrac{x}{(x + 3)(x - 3)} > 0$

$(0, 3) \implies \dfrac{x}{(x + 3)(x - 3)} < 0$

$(3, \infty) \implies \dfrac{x}{(x + 3)(x - 3)} > 0$

Domain: $(-3, 0] \cup (3, \infty)$

61. $0.4x^2 + 5.26 < 10.2$

$0.4x^2 - 4.94 < 0$

$0.4(x^2 - 12.35) < 0$

Critical numbers: $x \approx \pm 3.51$

Test intervals: $(-\infty, -3.51), (-3.51, 3.51), (3.51, \infty)$

By testing an x-value in each test interval in the inequality, we see that the solution set is: $(-3.51, 3.51)$

62. $-1.3x^2 + 3.78 > 2.12$

$-1.3x^2 + 1.66 > 0$

Critical numbers: ± 1.13

Test intervals: $(-\infty, -1.13), (-1.13, 1.13), (1.13, \infty)$

Solution set: $(-1.13, 1.13)$

63. $-0.5x^2 + 12.5x + 1.6 > 0$

The zeros are $x = \dfrac{-12.5 \pm \sqrt{(12.5)^2 - 4(-0.5)(1.6)}}{2(-0.5)}$.

Critical numbers: $x \approx -0.13, x \approx 25.13$

Test intervals: $(-\infty, -0.13), (-0.13, 25.13), (25.13, \infty)$

By testing an x-value in each test interval in the inequality, we see that the solution set is: $(-0.13, 25.13)$

64. $1.2x^2 + 4.8x + 3.1 < 5.3$

$1.2x^2 + 4.8x - 2.2 < 0$

Critical numbers: $-4.42, 0.42$

Test intervals: $(-\infty, -4.42), (-4.42, 0.42), (0.42, \infty)$

Solution set: $(-4.42, 0.42)$

65. $\dfrac{1}{2.3x - 5.2} > 3.4$

$\dfrac{1}{2.3x - 5.2} - 3.4 > 0$

$\dfrac{1 - 3.4(2.3x - 5.2)}{2.3x - 5.2} > 0$

$\dfrac{-7.82x + 18.68}{2.3x - 5.2} > 0$

Critical numbers: $x \approx 2.39, x \approx 2.26$

Test intervals: $(-\infty, 2.26), (2.26, 2.39), (2.39, \infty)$

By testing an x-value in each test interval in the inequality, we see that the solution set is: $(2.26, 2.39)$

66. $\dfrac{2}{3.1x - 3.7} > 5.8$

$\dfrac{2 - 5.8(3.1x - 3.7)}{3.1x - 3.7} > 0$

$\dfrac{23.46 - 17.98x}{3.1x - 3.7} > 0$

Critical numbers: $x \approx 1.19, x \approx 1.30$

Test intervals: $(-\infty, 1.19) \implies \dfrac{23.46 - 17.98x}{3.1x - 3.7} < 0$

$(1.19, 1.30) \implies \dfrac{23.46 - 17.98x}{3.1x - 3.7} > 0$

$(1.30, \infty) \implies \dfrac{23.46 - 17.98x}{3.1x - 3.7} < 0$

Solution interval: $(1.19, 1.30)$

67. $s = -16t^2 + v_0t + s_0 = -16t^2 + 160t$

(a) $-16t^2 + 160t = 0$

$-16t(t - 10) = 0$

$t = 0, t = 10$

It will be back on the ground in 10 seconds.

(b) $-16t^2 + 160t > 384$

$-16t^2 + 160t - 384 > 0$

$-16(t^2 - 10t + 24) > 0$

$t^2 - 10t + 24 < 0$

$(t - 4)(t - 6) < 0$

$4 < t < 6$ seconds

68. $s = -16t^2 + v_0t + s_0 = -16t^2 + 128t$

(a) $-16t^2 + 128t = 0$

$-16t(t - 8) = 0$

$-16t = 0 \implies t = 0$

$t - 8 = 0 \implies t = 8$

It will be back on the ground in 8 seconds.

(b) $-16t^2 + 128t < 128$

$-16t^2 + 128t - 128 < 0$

Critical numbers: $4 - 2\sqrt{2}, 4 + 2\sqrt{2}$

Test intervals:
$\left(-\infty, 4 - 2\sqrt{2}\right), \left(4 - 2\sqrt{2}, 4 + 2\sqrt{2}\right),$
$\left(4 + 2\sqrt{2}, \infty\right)$

Solution set: 0 seconds $\leq t < 4 - 2\sqrt{2}$ seconds
and $4 - 2\sqrt{2}$ seconds $< t \leq 8$ seconds

69. $2L + 2W = 100 \implies W = 50 - L$

$LW \geq 500$

$L(50 - L) \geq 500$

$-L^2 + 50L - 500 \geq 0$

By the Quadratic Formula we have:

Critical numbers: $L = 25 \pm 5\sqrt{5}$

Test: Is $-L^2 + 50L - 500 \geq 0$?

Solution set: $25 - 5\sqrt{5} \leq L \leq 25 + 5\sqrt{5}$

13.8 meters $\leq L \leq 36.2$ meters

70. $2L + 2W = 440 \implies W = 220 - L$

$LW \geq 8000$

$L(220 - L) \geq 8000$

$-L^2 + 220L - 8000 \geq 0$

By the Quadratic Formula we have:

Critical numbers: $L = 110 \pm 10\sqrt{41}$

Test: Is $-L^2 + 220L - 8000 \geq 0$?

Solution set: $110 - 10\sqrt{41} \leq L \leq 110 + 10\sqrt{41}$

45.97 feet $\leq L \leq 174.03$ feet

71. $R = x(75 - 0.0005x)$ and $C = 30x + 250,000$

$P = R - C$

$= (75x - 0.0005x^2) - (30x + 250,000)$

$= -0.0005x^2 + 45x - 250,000$

$P \geq 750,000$

$-0.0005x^2 + 45x - 250,000 \geq 750,000$

$-0.0005x^2 + 45x - 1,000,000 \geq 0$

Critical numbers: $x = 40,000, x = 50,000$ (These were obtained by using the Quadratic Formula.)

Test intervals: $(0, 40,000), (40,000, 50,000), (50,000, \infty)$

By testing x-values in each test interval in the inequality, we see that the solution set is $[40,000, 50,000]$ or $40,000 \leq x \leq 50,000$. The price per unit is

$p = \dfrac{R}{x} = 75 - 0.0005x.$

For $x = 40,000, p = \$55$. For $x = 50,000, p = \$50$.
Therefore, for $40,000 \leq x \leq 50,000, \$50.00 \leq p \leq \$55.00$.

72. What is the price per unit?

When $x = 90,000$:

$R = \$2,880,000 \implies \dfrac{2,880,000}{90,000} = \$32$ per unit

When $x = 100,000$:

$R = \$3,000,000 \implies \dfrac{3,000,000}{100,000} = \$30$ per unit

Solution interval: $\$30.00 \leq p \leq \$32.00$

73. $C = 0.0031t^3 - 0.216t^2 + 5.54t + 19.1, \ 0 \le t \le 23$

(a)

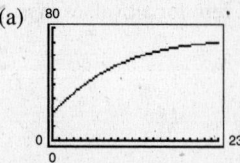

(b)

t	C
24	70.5
26	71.6
28	72.9
30	74.6
32	76.8
34	79.6

C will be greater than 75% when $t \approx 31$, which corresponds to 2011.

(c) $C = 75$ when $t \approx 30.41$.

(d)

t	C
36	83.2
37	85.4
38	87.8
39	90.5
40	93.5
41	96.8
42	100.4
43	104.4

C will be between 85% and 100% when t is between 37 and 42. These values correspond to the years 2017 to 2022.

(e) $85 \le C \le 100$ when $36.82 \le t \le 41.89$ or $37 \le t \le 42$.

(f) The model is a third-degree polynomial and as $t \to \infty, C \to \infty$.

74. (a)

d	4	6	8	10	12
Load	2223.9	5593.9	10,312	16,378	23,792

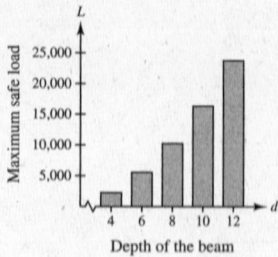

(b)
$$2000 \le 168.5d^2 - 472.1$$
$$2472.1 \le 168.5d^2$$
$$14.67 \le d^2$$
$$3.83 \le d$$

The minimum depth is 3.83 inches.

75.
$$\frac{1}{R} = \frac{1}{R_1} + \frac{1}{2}$$
$$2R_1 = 2R + RR_1$$
$$2R_1 = R(2 + R_1)$$
$$\frac{2R_1}{2 + R_1} = R$$

Since $R \ge 1$, we have
$$\frac{2R_1}{2 + R_1} \ge 1$$
$$\frac{2R_1}{2 + R_1} - 1 \ge 0$$
$$\frac{R_1 - 2}{2 + R_1} \ge 0.$$

Since $R_1 > 0$, the only critical number is $R_1 = 2$. The inequality is satisfied when $R_1 \ge 2$ ohms.

76. (a) $N = -0.03t^2 + 9.6t + 172$
$$= 220 \Rightarrow t = 5$$

So the number of master's degrees earned by women exceeded 220,000 in 1995.

(c) $N = -0.03t^2 + 9.6t + 172$
$$= 320 \Rightarrow t = 16.2$$

So the number of master's degrees earned by women will exceed 320,000 in 2006.

(b) and (d)

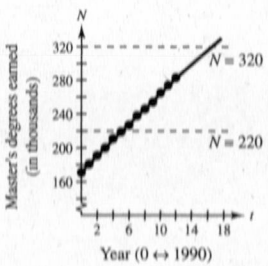

77. True

$x^3 - 2x^2 - 11x + 12 = (x + 3)(x - 1)(x - 4)$

The test intervals are $(-\infty, -3)$, $(-3, 1)$, $(1, 4)$, and $(4, \infty)$.

78. True

The y-values are greater than zero for all values of x.

79. $x^2 + bx + 4 = 0$

To have at least one real solution, $b^2 - 16 \geq 0$. This occurs when $b \leq -4$ or $b \geq 4$. This can be written as $(-\infty, -4] \cup [4, \infty)$.

80. $x^2 + bx - 4 = 0$

To have at least one real solution,

$b^2 - 4(1)(-4) \geq 0$

$b^2 + 16 \geq 0.$

This inequality is true for all real values of b. Thus, the interval for b such that the equation has at least one real solution is $(-\infty, \infty)$.

81. $3x^2 + bx + 10 = 0$

To have at least one real solution, $b^2 - 4(3)(10) \geq 0$.

$b^2 - 120 \geq 0$

$(b + \sqrt{120})(b - \sqrt{120}) \geq 0$

Critical numbers: $b = \pm\sqrt{120} = \pm 2\sqrt{30}$

Test intervals:
$(-\infty, -2\sqrt{30})$, $(-2\sqrt{30}, 2\sqrt{30})$, $(2\sqrt{30}, \infty)$

Test: Is $b^2 - 120 \geq 0$?

Solution set: $(-\infty, -2\sqrt{30}] \cup [2\sqrt{30}, \infty)$

82. $2x^2 + bx + 5 = 0$

To have at least one real solution,

$b^2 - 4(2)(5) \geq 0$

$b^2 - 40 \geq 0.$

This occurs when $b \leq -2\sqrt{10}$ or $b \geq 2\sqrt{10}$. Thus, the interval for b such that the equation has at least one real solution is $(-\infty, -2\sqrt{10}] \cup [2\sqrt{10}, \infty)$.

83. (a) If $a > 0$ and $c \leq 0$, then b can be any real number. If $a > 0$ and $c > 0$, then for $b^2 - 4ac$ to be greater than or equal to zero, b is restricted to $b < -2\sqrt{ac}$ or $b > 2\sqrt{ac}$.

(b) The center of the interval for b in Exercises 79–82 is 0.

84. (a) $x = a, x = b$

(b)

	−	+	+
	−	−	+
	+	−	+

$\xrightarrow{\quad\bullet\qquad\qquad\bullet\qquad} x$
$\quad\quad a \qquad\qquad b$

(c) The real zeros of the polynomial

85. $4x^2 + 20x + 25 = (2x + 5)^2$

86. $(x + 3)^2 - 16 = [(x + 3) + 4][(x + 3) - 4]$

$= (x + 7)(x - 1)$

87. $x^2(x + 3) - 4(x + 3) = (x^2 - 4)(x + 3)$

$= (x + 2)(x - 2)(x + 3)$

88. $2x^4 - 54x = 2x(x^3 - 27)$

$= 2x(x - 3)(x^2 + 3x + 9)$

89. Area $=$ (length)(width)

$= (2x + 1)(x)$

$= 2x^2 + x$

90. Area $= \frac{1}{2}$(base)(height)

$= \frac{1}{2}(b)(3b + 2)$

$= \frac{3}{2}b^2 + b$

Review Exercises for Chapter 2

1. (a) $y = 2x^2$

Vertical stretch

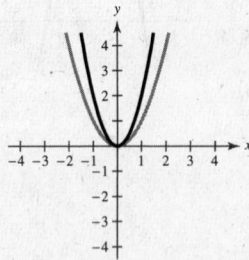

(b) $y = -2x^2$

Vertical stretch and a reflection in the x-axis

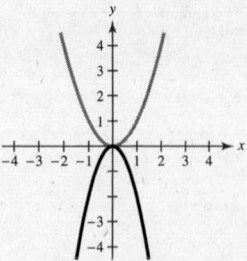

(c) $y = x^2 + 2$

Vertical shift two units upward

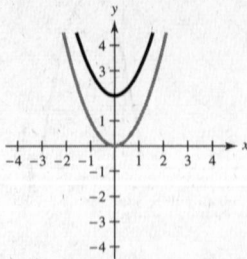

(d) $y = (x + 2)^2$

Horizontal shift two units to the left

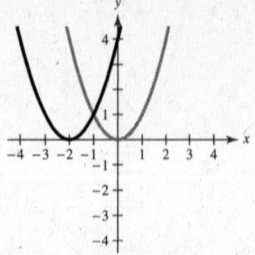

2. (a) $y = x^2 - 4$

Vertical shift four units downward

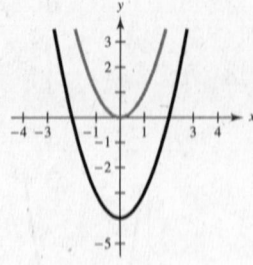

(b) $y = 4 - x^2$

Reflection in the x-axis and a vertical shift four units upward

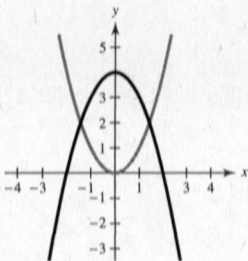

(c) $y = (x - 3)^2$

Horizontal shift three units to the right

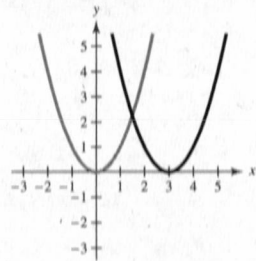

(d) $y = \frac{1}{2}x^2 - 1$

Vertical shrink (each y-value is multiplied by $\frac{1}{2}$), and a vertical shift one unit downward

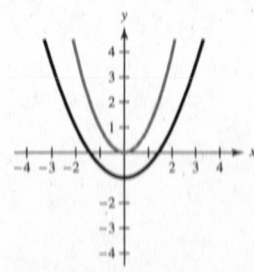

3. $g(x) = x^2 - 2x$

$\qquad = x^2 - 2x + 1 - 1$

$\qquad = (x - 1)^2 - 1$

Vertex: $(1, -1)$

Axis of symmetry: $x = 1$

$0 = x^2 - 2x = x(x - 2)$

x-intercepts: $(0, 0), (2, 0)$

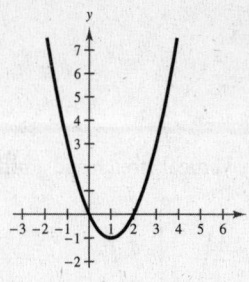

4. $f(x) = 6x - x^2$

$\qquad = -(x^2 - 6x + 9 - 9)$

$\qquad = -(x - 3)^2 + 9$

Vertex: $(3, 9)$

Axis of symmetry: $x = 3$

$0 = 6x - x^2 = x(6 - x)$

x-intercepts: $(0, 0), (6, 0)$

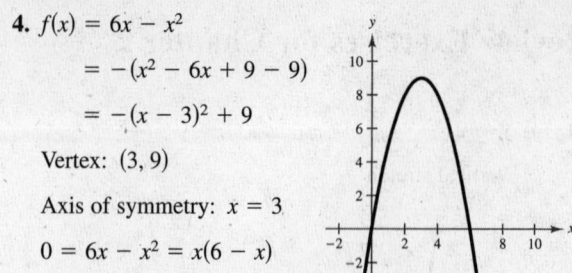

5. $f(x) = x^2 + 8x + 10$

$\qquad = x^2 + 8x + 16 - 16 + 10$

$\qquad = (x + 4)^2 - 6$

Vertex: $(-4, -6)$

Axis of symmetry: $x = -4$

$\qquad 0 = (x + 4)^2 - 6$

$(x + 4)^2 = 6$

$\quad x + 4 = \pm\sqrt{6}$

$\qquad x = -4 \pm \sqrt{6}$

x-intercepts: $\left(-4 \pm \sqrt{6}, 0\right)$

6. $h(x) = 3 + 4x - x^2$

$\qquad = -(x^2 - 4x - 3)$

$\qquad = -(x^2 - 4x + 4 - 4 - 3)$

$\qquad = -[(x - 2)^2 - 7]$

$\qquad = -(x - 2)^2 + 7$

Vertex: $(2, 7)$

Axis of symmetry: $x = 2$

$0 = 3 + 4x - x^2$

$0 = x^2 - 4x - 3$

$x = \dfrac{-(-4) \pm \sqrt{(-4)^2 - 4(1)(-3)}}{2(1)}$

$\quad = \dfrac{4 \pm \sqrt{28}}{2} = 2 \pm \sqrt{7}$

x-intercepts: $\left(2 \pm \sqrt{7}, 0\right)$

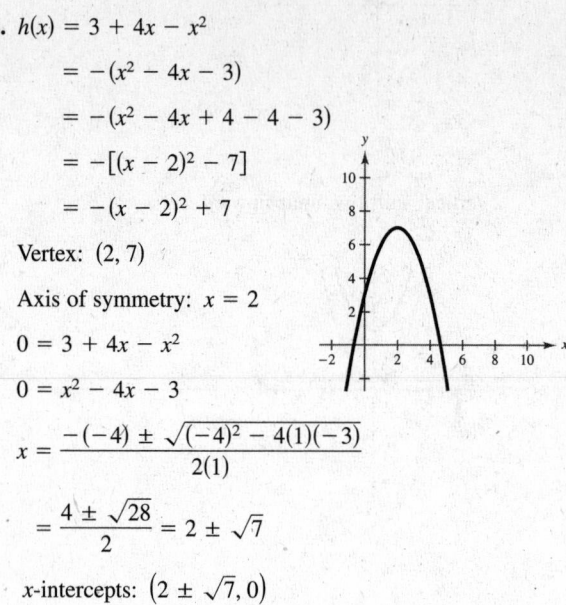

7. $f(t) = -2t^2 + 4t + 1$

$\qquad = -2(t^2 - 2t + 1 - 1) + 1$

$\qquad = -2[(t - 1)^2 - 1] + 1$

$\qquad = -2(t - 1)^2 + 3$

Vertex: $(1, 3)$

Axis of symmetry: $t = 1$

$\qquad 0 = -2(t - 1)^2 + 3$

$2(t - 1)^2 = 3$

$\quad t - 1 = \pm\sqrt{\dfrac{3}{2}}$

$\qquad t = 1 \pm \dfrac{\sqrt{6}}{2}$

t-intercepts: $\left(1 \pm \dfrac{\sqrt{6}}{2}, 0\right)$

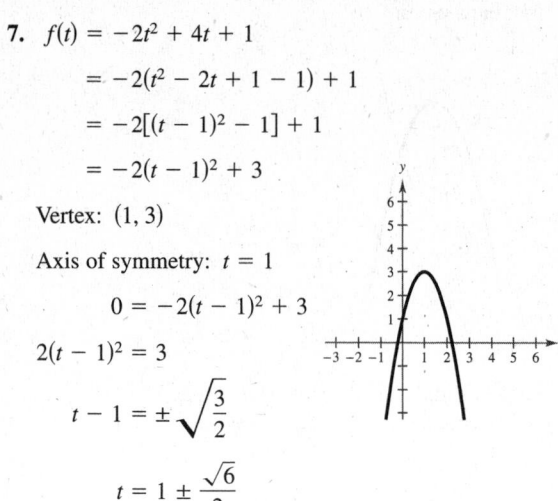

8. $f(x) = x^2 - 8x + 12$

$\qquad = x^2 - 8x + 16 - 16 + 12$

$\qquad = (x - 4)^2 - 4$

Vertex: $(4, -4)$

Axis of symmetry: $x = 4$

$0 = x^2 - 8x + 12$

$0 = (x - 2)(x - 6)$

x-intercepts: $(2, 0), (6, 0)$

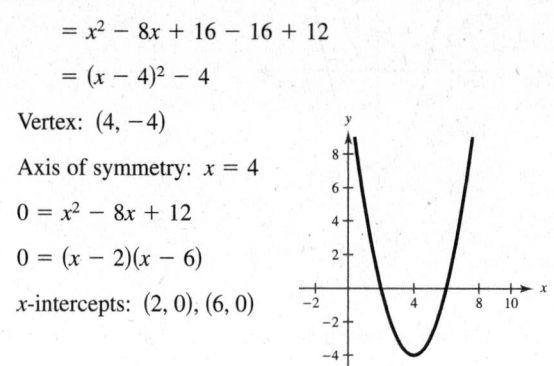

9. $h(x) = 4x^2 + 4x + 13$

$= 4(x^2 + x) + 13$

$= 4\left(x^2 + x + \dfrac{1}{4} - \dfrac{1}{4}\right) + 13$

$= 4\left(x^2 + x + \dfrac{1}{4}\right) - 1 + 13$

$= 4\left(x + \dfrac{1}{2}\right)^2 + 12$

Vertex: $\left(-\dfrac{1}{2}, 12\right)$

Axis of symmetry: $x = -\dfrac{1}{2}$

$0 = 4\left(x + \dfrac{1}{2}\right)^2 + 12$

$\left(x + \dfrac{1}{2}\right)^2 = -3$

No real zeros

x-intercepts: none

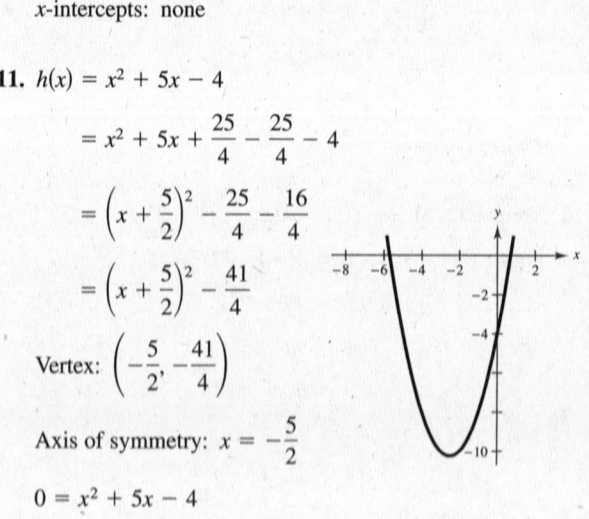

10. $f(x) = x^2 - 6x + 1$

$= x^2 - 6x + 9 - 9 + 1$

$= (x - 3)^2 - 8$

Vertex: $(3, -8)$

Axis of symmetry: $x = 3$

$0 = x^2 - 6x + 1$

$x = \dfrac{-(-6) \pm \sqrt{(-6)^2 - 4(1)(1)}}{2(1)}$

$= \dfrac{6 \pm \sqrt{32}}{2} = 3 \pm 2\sqrt{2}$

x-intercepts: $\left(3 \pm 2\sqrt{2}, 0\right)$

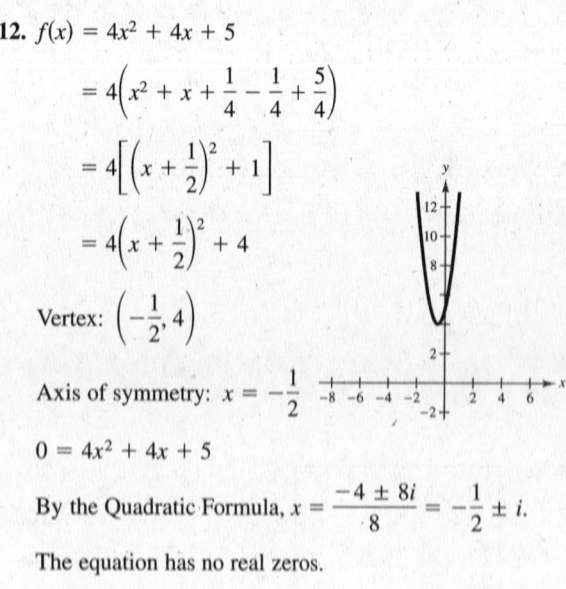

11. $h(x) = x^2 + 5x - 4$

$= x^2 + 5x + \dfrac{25}{4} - \dfrac{25}{4} - 4$

$= \left(x + \dfrac{5}{2}\right)^2 - \dfrac{25}{4} - \dfrac{16}{4}$

$= \left(x + \dfrac{5}{2}\right)^2 - \dfrac{41}{4}$

Vertex: $\left(-\dfrac{5}{2}, -\dfrac{41}{4}\right)$

Axis of symmetry: $x = -\dfrac{5}{2}$

$0 = x^2 + 5x - 4$

By the Quadratic Formula, $x = \dfrac{-5 \pm \sqrt{41}}{2}$.

x-intercepts: $\left(\dfrac{-5 \pm \sqrt{41}}{2}, 0\right)$

12. $f(x) = 4x^2 + 4x + 5$

$= 4\left(x^2 + x + \dfrac{1}{4} - \dfrac{1}{4} + \dfrac{5}{4}\right)$

$= 4\left[\left(x + \dfrac{1}{2}\right)^2 + 1\right]$

$= 4\left(x + \dfrac{1}{2}\right)^2 + 4$

Vertex: $\left(-\dfrac{1}{2}, 4\right)$

Axis of symmetry: $x = -\dfrac{1}{2}$

$0 = 4x^2 + 4x + 5$

By the Quadratic Formula, $x = \dfrac{-4 \pm 8i}{8} = -\dfrac{1}{2} \pm i$.

The equation has no real zeros.

x-intercepts: None

13. $f(x) = \dfrac{1}{3}(x^2 + 5x - 4)$

$= \dfrac{1}{3}\left(x^2 + 5x + \dfrac{25}{4} - \dfrac{25}{4} - 4\right)$

$= \dfrac{1}{3}\left[\left(x + \dfrac{5}{2}\right)^2 - \dfrac{41}{4}\right]$

$= \dfrac{1}{3}\left(x + \dfrac{5}{2}\right)^2 - \dfrac{41}{12}$

Vertex: $\left(-\dfrac{5}{2}, -\dfrac{41}{12}\right)$

Axis of symmetry: $x = -\dfrac{5}{2}$

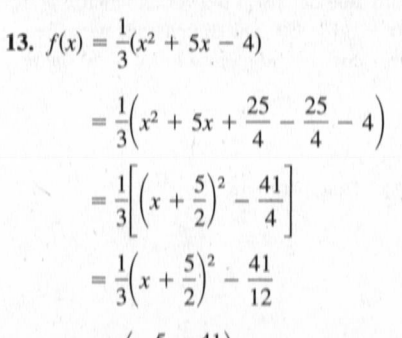

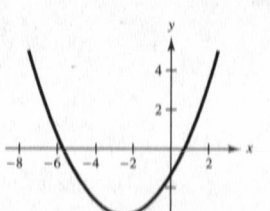

$0 = x^2 + 5x - 4$

By the Quadratic Formula, $x = \dfrac{-5 \pm \sqrt{41}}{2}$.

x-intercepts: $\left(\dfrac{-5 \pm \sqrt{41}}{2}, 0\right)$

14. $f(x) = \dfrac{1}{2}(6x^2 - 24x + 22)$

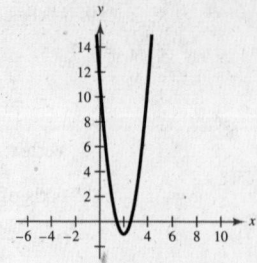

$= 3x^2 - 12x + 11$

$= 3(x^2 - 4x + 4 - 4) + 11$

$= 3(x - 2)^2 + 3(-4) + 11$

$= 3(x - 2)^2 - 1$

Vertex: $(2, -1)$

Axis of symmetry: $x = 2$

$0 = 3x^2 - 12x + 11$

$x = \dfrac{-(-12) \pm \sqrt{(-12)^2 - 4(3)(11)}}{2(3)} = \dfrac{12 \pm \sqrt{12}}{6} = 2 \pm \dfrac{\sqrt{3}}{3}$

x-intercepts: $\left(2 \pm \dfrac{\sqrt{3}}{3}, 0\right)$

15. Vertex: $(4, 1) \implies f(x) = a(x - 4)^2 + 1$

Point: $(2, -1) \implies -1 = a(2 - 4)^2 + 1$

$\qquad\qquad\qquad -2 = 4a$

$\qquad\qquad\qquad -\dfrac{1}{2} = a$

Thus, $f(x) = -\dfrac{1}{2}(x - 4)^2 + 1$.

16. Vertex: $(2, 2) \implies f(x) = a(x - 2)^2 + 2$

Point: $(0, 3) \implies 3 = a(0 - 2)^2 + 2$

$\qquad\qquad\qquad 3 = 4a + 2$

$\qquad\qquad\qquad 1 = 4a$

$\qquad\qquad\qquad \dfrac{1}{4} = a$

$\qquad f(x) = \dfrac{1}{4}(x - 2)^2 + 2$

17. Vertex: $(1, -4) \implies f(x) = a(x - 1)^2 - 4$

Point: $(2, -3) \implies -3 = a(2 - 1)^2 - 4$

$\qquad\qquad\qquad 1 = a$

Thus, $f(x) = (x - 1)^2 - 4$.

18. Vertex: $(2, 3) \implies f(x) = a(x - 2)^2 + 3$

Point: $(-1, 6) \implies 6 = a(-1 - 2)^2 + 3$

$\qquad\qquad\qquad 6 = 9a + 3$

$\qquad\qquad\qquad 3 = 9a$

$\qquad\qquad\qquad \dfrac{1}{3} = a$

$\qquad f(x) = \dfrac{1}{3}(x - 2)^2 + 3$

19. (a)

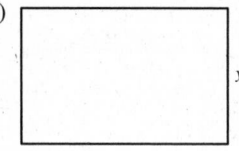

(b) $2x + 2y = 200$

$x + y = 100$

$y = 100 - x$

Area $= xy$

$= x(100 - x)$

$= 100x - x^2$

(c) Area $= 100x - x^2$

$= -(x^2 - 100x + 2500 - 2500)$

$= -[(x - 50)^2 - 2500]$

$= -(x - 50)^2 + 2500$

The maximum area occurs at the vertex when $x = 50$ and $y = 100 - 50 = 50$. The dimensions with the maximum area are $x = 50$ meters and $y = 50$ meters.

20. $R = -10p^2 + 800p$

(a) $R(20) = \$12{,}000$

$R(25) = \$13{,}750$

$R(30) = \$15{,}000$

(b) The maximum revenue occurs at the vertex of the parabola.

$-\dfrac{b}{2a} = \dfrac{-800}{2(-10)} = \$40$

$R(40) = \$16{,}000$

The revenue is maximum when the price is $40 per unit.
The maximum revenue is $16,000.

21. $C = 70,000 - 120x + 0.055x^2$

The minimum cost occurs at the vertex of the parabola.

Vertex: $-\dfrac{b}{2a} = -\dfrac{-120}{2(0.055)} \approx 1091$ units

Approximately 1091 units should be produced each day to yield a minimum cost.

22. $26 = -0.107x^2 + 5.68x - 48.5$

$0 = -0.107x^2 + 5.68x - 74.5$

$x = \dfrac{-5.68 \pm \sqrt{(5.68)^2 - 4(-0.107)(-74.5)}}{2(-0.107)}$

$x \approx 23.7, \ 29.4$

The age of the bride is approximately 24 years when the age of the groom is 26 years.

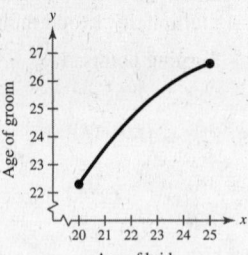

23. $y = x^3, \ f(x) = -(x - 4)^3$

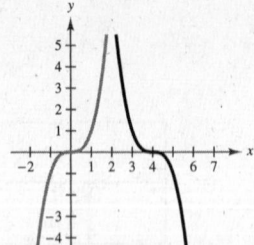

Transformation: Reflection in the x-axis and a horizontal shift four units to the right

24. $y = x^3, \ f(x) = -4x^3$

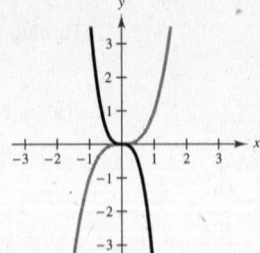

$f(x)$ is a reflection in the x-axis and a vertical stretch of the graph of $y = x^3$.

25. $y = x^4, \ f(x) = 2 - x^4$

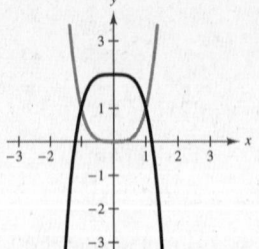

Transformation: Reflection in the x-axis and a vertical shift two units upward

26. $y = x^4, \ f(x) = 2(x - 2)^4$

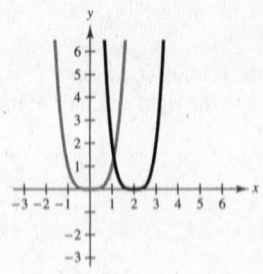

$f(x)$ is a shift to the right two units and a vertical stretch of the graph of $y = x^4$.

27. $y = x^5, \ f(x) = (x - 3)^5$

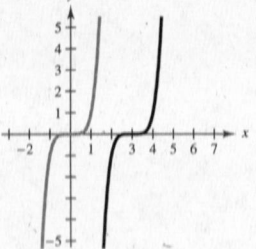

Transformation: Horizontal shift three units to the right

28. $y = x^5, \ f(x) = \frac{1}{2}x^5 + 3$

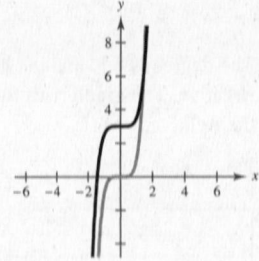

$f(x)$ is a vertical shrink and a vertical shift three units upward of the graph of $y = x^5$.

29. $f(x) = -x^2 + 6x + 9$

The degree is even and the leading coefficient is negative. The graph falls to the left and falls to the right.

30. $f(x) = \frac{1}{2}x^3 + 2x$

The degree is odd and the leading coefficient is positive. The graph falls to the left and rises to the right.

31. $g(x) = \frac{3}{4}(x^4 + 3x^2 + 2)$

The degree is even and the leading coefficient is positive. The graph rises to the left and rises to the right.

32. $h(x) = -x^5 - 7x^2 + 10x$

The degree is odd and the leading coefficient is negative. The graph rises to the left and falls to the right.

33. $f(x) = 2x^2 + 11x - 21$

$0 = 2x^2 + 11x - 21$

$= (2x - 3)(x + 7)$

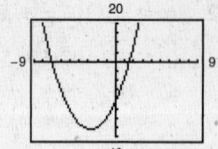

Zeros: $x = \frac{3}{2}, -7$, all of multiplicity 1 (odd multiplicity)

Turning points: 1

34. $f(x) = x(x + 3)^2$

$0 = x(x + 3)^2$

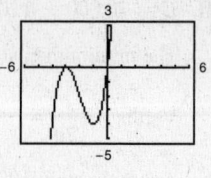

Zeros: $x = 0$ of multiplicity 1 (odd multiplicity)

$x = -3$ of multiplicity 2 (even multiplicity)

Turning points: 2

35. $f(t) = t^3 - 3t$

$0 = t^3 - 3t$

$0 = t(t^2 - 3)$

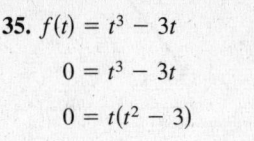

Zeros: $t = 0, \pm\sqrt{3}$ all of multiplicity 1 (odd multiplicity)

Turning points: 2

36. $f(x) = x^3 - 8x^2$

$0 = x^3 - 8x^2$

$0 = x^2(x - 8)$

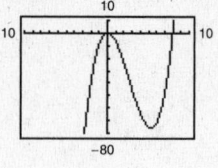

Zeros: $x = 0$ of multiplicity 2 (even multiplicity)

$x = 8$ of multiplicity 1 (odd multiplicity)

Turning points: 2

37. $f(x) = -12x^3 + 20x^2$

$0 = -12x^3 + 20x^2$

$0 = -4x^2(3x - 5)$

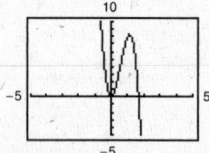

Zeros: $x = 0$ of multiplicity 2 (even multiplicity)

$x = \frac{5}{3}$ of multiplicity 1 (odd multiplicity)

Turning points: 2

38. $g(x) = x^4 - x^3 - 2x^2$

$0 = x^4 - x^3 - 2x^2$

$0 = x^2(x^2 - x - 2)$

$= x^2(x + 1)(x - 2)$

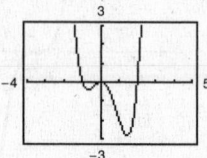

Zeros: $x = 0$ of multiplicity 2 (even multiplicity)

$x = -1$ of multiplicity 1 (odd multiplicity)

$x = 2$ of multiplicity 1 (odd multiplicity)

Turning points: 3

39. $f(x) = -x^3 + x^2 - 2$

(a) The degree is odd and the leading coefficient is negative. The graph rises to the left and falls to the right.

(b) Zero: $x = -1$

(c)

x	-3	-2	-1	0	1	2
$f(x)$	34	10	0	-2	-2	-6

(d)

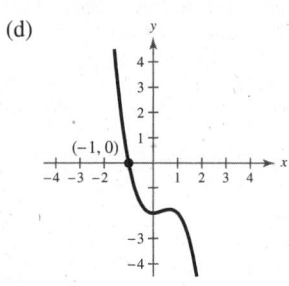

40. $g(x) = 2x^3 + 4x^2$

(a) The degree is odd and the leading coefficient, 2, is positive. The graph rises to the right and falls to the left.

(b) $g(x) = 2x^3 + 4x^2$

$0 = 2x^3 + 4x^2$

$0 = 2x^2(x + 2)$

$0 = x^2(x + 2)$

The zeros are 0 and -2.

(c)

x	-3	-2	-1	0	1
$g(x)$	-18	0	2	0	6

(d)

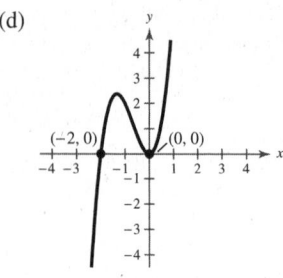

41. $f(x) = x(x^3 + x^2 - 5x + 3)$

  (a) The degree is even and the leading coefficient is positive. The graph rises to the left and rises to the right.

  (b) Zeros: $x = 0, 1, -3$

  (c)

x	-4	-3	-2	-1	0	1	2	3
$f(x)$	100	0	-18	-8	0	0	10	72

  (d)

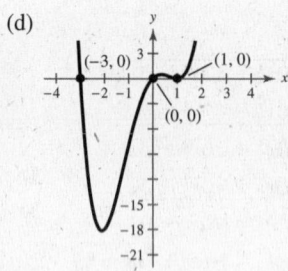

42. $h(x) = 3x^2 - x^4$

  (a) The degree is even and the leading coefficient, -1, is negative. The graph falls to the left and falls to the right.

  (b) $g(x) = 3x^2 - x^4$

$$0 = 3x^2 - x^4$$

$$0 = x^2(3 - x^2)$$

The zeros are 0, $-\sqrt{3}$, and $\sqrt{3}$.

  (c)

x	-2	-1	0	1	2
$h(x)$	-4	2	0	2	-4

  (d)

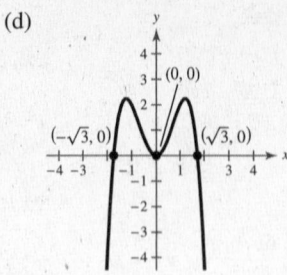

43. (a) $f(x) = 3x^3 - x^2 + 3$

x	-3	-2	-1	0	1	2	3
$f(x)$	-87	-25	-1	3	5	23	75

  (b) The zero is in the interval $[-1, 0]$.

    Zero: $x \approx -0.900$

44. (a) $f(x) = 0.25x^3 - 3.65x + 6.12$

x	-6	-5	-4	-3	-2
$f(x)$	-25.98	-6.88	4.72	10.32	11.42

x	-1	0	1	2	3	4
$f(x)$	9.52	6.12	2.72	0.82	1.92	7.52

  (b) The only zero is in the interval $(-5, -4)$. It is $x \approx -4.479$.

45. (a) $f(x) = x^4 - 5x - 1$

x	-3	-2	-1	0	1	2	3
$f(x)$	95	25	5	-1	-5	5	65

  (b) There are two zeros, one in the interval $[-1, 0]$ and one in the interval $[1, 2]$

    Zeros: $x \approx -0.200$, $x \approx 1.772$

46. (a) $f(x) = 7x^4 + 3x^3 - 8x^2 + 2$

x	-3	-2	-1	0	1	2
$f(x)$	416	58	-2	2	4	106

  (b) There are zeros in the intervals $(-2, -1)$ and $(-1, 0)$. They are $x \approx -1.211$ and $x \approx -0.509$.

47.

$$
\begin{array}{r}
8x + 5 \\
3x - 2 \overline{)\, 24x^2 - x - 8} \\
24x^2 - 16x \\
\hline
15x - 8 \\
15x - 10 \\
\hline
2
\end{array}
$$

Thus, $\dfrac{24x^2 - x - 8}{3x - 2} = 8x + 5 + \dfrac{2}{3x - 2}$.

48.

$$
\begin{array}{r}
\frac{4}{3} \\
3x - 2 \overline{)\, 4x + 7} \\
4x - \frac{8}{3} \\
\hline
\frac{29}{3}
\end{array}
$$

$$\frac{4x + 7}{3x - 2} = \frac{4}{3} + \frac{29}{3(3x - 2)}$$

49.

$$
\begin{array}{r}
5x + 2 \\
x^2 - 3x + 1 \overline{\smash{\big)}\ 5x^3 - 13x^2 - \ x + 2} \\
\underline{5x^3 - 15x^2 + 5x} \\
2x^2 - 6x + 2 \\
\underline{2x^2 - 6x + 2} \\
0
\end{array}
$$

Thus, $\dfrac{5x^3 - 13x^2 - x + 2}{x^2 - 3x + 1} = 5x + 2.$

50.

$$
\begin{array}{r}
3x^2 \qquad + 3 \\
x^2 - 1 \overline{\smash{\big)}\ 3x^4 + 0x^3 + 0x^2 + 0x + 0} \\
\underline{3x^4 \qquad - 3x^2} \\
3x^2 \qquad + 0 \\
\underline{3x^2 \qquad - 3} \\
3
\end{array}
$$

$\dfrac{3x^4}{x^2 - 1} = 3x^2 + 3 + \dfrac{3}{x^2 - 1}$

51.

$$
\begin{array}{r}
x^2 - 3x + 2 \\
x^2 + 0x + 2 \overline{\smash{\big)}\ x^4 - 3x^3 + 4x^2 - 6x + 3} \\
\underline{x^4 + 0x^3 + 2x^2} \\
-3x^3 + 2x^2 - 6x \\
\underline{-3x^3 + 0x^2 - 6x} \\
2x^2 + 0x + 3 \\
\underline{2x^2 + 0x + 4} \\
-1
\end{array}
$$

Thus, $\dfrac{x^4 - 3x^3 + 4x^2 - 6x + 3}{x^2 + 2} = x^2 - 3x + 2 - \dfrac{1}{x^2 + 2}.$

52.

$$
\begin{array}{r}
3x^2 + 5x + 8 \\
2x^2 + 0x - 1 \overline{\smash{\big)}\ 6x^4 + 10x^3 + 13x^2 - 5x + 2} \\
\underline{6x^4 + \ 0x^3 - \ 3x^2} \\
10x^3 + 16x^2 - 5x \\
\underline{10x^3 + \ 0x^2 - 5x} \\
16x^2 - 0x + 2 \\
\underline{16x^2 + 0x - 8} \\
10
\end{array}
$$

$\dfrac{6x^4 + 10x^3 + 13x^2 - 5x + 2}{2x^2 - 1} = 3x^2 + 5x + 8 + \dfrac{10}{2x^2 - 1}$

53.

$$
\begin{array}{r|rrrrr}
2 & 6 & -4 & -27 & 18 & 0 \\
  &   & 12 & 16 & -22 & -8 \\
\hline
  & 6 & 8 & -11 & -4 & -8
\end{array}
$$

Thus,

$\dfrac{6x^4 - 4x^3 - 27x^2 + 18x}{x - 2} = 6x^3 + 8x^2 - 11x - 4 - \dfrac{8}{x - 2}.$

54.

$$
\begin{array}{r|rrrr}
5 & 0.1 & 0.3 & 0 & -0.5 \\
  &   & 0.5 & 4 & 20 \\
\hline
  & 0.1 & 0.8 & 4 & 19.5
\end{array}
$$

$\dfrac{0.1x^3 + 0.3x^2 - 0.5}{x - 5} = 0.1x^2 + 0.8x + 4 + \dfrac{19.5}{x - 5}$

55.

$$
\begin{array}{r|rrrr}
4 & 2 & -19 & 38 & 24 \\
  &   & 8 & -44 & -24 \\
\hline
  & 2 & -11 & -6 & 0
\end{array}
$$

Thus, $\dfrac{2x^3 - 19x^2 + 38x + 24}{x - 4} = 2x^2 - 11x - 6.$

56.

$$
\begin{array}{r|rrrr}
-3 & 3 & 20 & 29 & -12 \\
   &   & -9 & -33 & 12 \\
\hline
   & 3 & 11 & -4 & 0
\end{array}
$$

$\dfrac{3x^3 + 20x^2 + 29x - 12}{x + 3} = 3x^2 + 11x - 4$

57. $f(x) = 20x^4 + 9x^3 - 14x^2 - 3x$

(a)
$$
\begin{array}{r|rrrrr}
-1 & 20 & 9 & -14 & -3 & 0 \\
   &   & -20 & 11 & 3 & 0 \\
\hline
   & 20 & -11 & -3 & 0 & 0
\end{array}
$$

Yes, $x = -1$ is a zero of f.

(b)
$$
\begin{array}{r|rrrrr}
\frac{3}{4} & 20 & 9 & -14 & -3 & 0 \\
   &   & 15 & 18 & 3 & 0 \\
\hline
   & 20 & 24 & 4 & 0 & 0
\end{array}
$$

Yes, $x = \frac{3}{4}$ is a zero of f.

(c)
$$
\begin{array}{r|rrrrr}
0 & 20 & 9 & -14 & -3 & 0 \\
  &   & 0 & 0 & 0 & 0 \\
\hline
  & 20 & 9 & -14 & -3 & 0
\end{array}
$$

Yes, $x = 0$ is a zero of f.

(d)
$$
\begin{array}{r|rrrrr}
1 & 20 & 9 & -14 & -3 & 0 \\
  &   & 20 & 29 & 15 & 12 \\
\hline
  & 20 & 29 & 15 & 12 & 12
\end{array}
$$

No, $x = 1$ is not a zero of f.

58. $f(x) = 3x^3 - 8x^2 - 20x + 16$

(a)

$$
\begin{array}{r|rrrr}
4 & 3 & -8 & -20 & 16 \\
  &   & 12 & 16 & -16 \\
\hline
  & 3 & 4 & -4 & 0
\end{array}
$$

Yes, $x = 4$ is a zero of f.

(c)

$$
\begin{array}{r|rrrr}
\frac{2}{3} & 3 & -8 & -20 & 16 \\
  &   & 2 & -4 & -16 \\
\hline
  & 3 & -6 & -24 & 0
\end{array}
$$

Yes, $x = \frac{2}{3}$ is a zero of f.

(b)

$$
\begin{array}{r|rrrr}
-4 & 3 & -8 & -20 & 16 \\
  &   & -12 & 80 & -240 \\
\hline
  & 3 & -20 & 60 & -224
\end{array}
$$

No, $x = -4$ is not a zero of f.

(d)

$$
\begin{array}{r|rrrr}
-1 & 3 & -8 & -20 & 16 \\
  &   & -3 & 11 & 9 \\
\hline
  & 3 & -11 & -9 & 25
\end{array}
$$

No, $x = -1$ is not a zero of f.

59. $f(x) = x^4 + 10x^3 - 24x^2 + 20x + 44$

(a)

$$
\begin{array}{r|rrrrr}
-3 & 1 & 10 & -24 & 20 & 44 \\
  &   & -3 & -21 & 135 & -465 \\
\hline
  & 1 & 7 & -45 & 155 & -421
\end{array}
$$

Thus, $f(-3) = -421$.

(b)

$$
\begin{array}{r|rrrrr}
-1 & 1 & 10 & -24 & 20 & 44 \\
  &   & -1 & -9 & 33 & -53 \\
\hline
  & 1 & 9 & -33 & 53 & -9
\end{array}
$$

$f(-1) = -9$

60. $g(t) = 2t^5 - 5t^4 - 8t + 20$

(a)

$$
\begin{array}{r|rrrrrr}
-4 & 2 & -5 & 0 & 0 & -8 & 20 \\
  &   & -8 & 52 & -208 & 832 & -3296 \\
\hline
  & 2 & -13 & 52 & -208 & 824 & -3276
\end{array}
$$

Thus, $g(-4) = -3276$.

(b)

$$
\begin{array}{r|rrrrrr}
\sqrt{2} & 2 & -5 & 0 & 0 & -8 & 20 \\
  &   & 2\sqrt{2} & -5\sqrt{2}+4 & -10+4\sqrt{2} & -10\sqrt{2}+8 & -20 \\
\hline
  & 2 & -5+2\sqrt{2} & -5\sqrt{2}+4 & -10+4\sqrt{2} & -10\sqrt{2} & 0
\end{array}
$$

Thus, $g\left(\sqrt{2}\right) = 0$.

61. $f(x) = x^3 + 4x^2 - 25x - 28$; Factor: $(x - 4)$

(a)

$$
\begin{array}{r|rrrr}
4 & 1 & 4 & -25 & -28 \\
  &   & 4 & 32 & 28 \\
\hline
  & 1 & 8 & 7 & 0
\end{array}
$$

Yes, $x - 4$ is a factor of $f(x)$.

(b) $x^2 + 8x + 7 = (x + 7)(x + 1)$

The remaining factors of f are $(x + 7)$ and $(x + 1)$.

(c) $f(x) = x^3 + 4x^2 - 25x - 28$

$= (x + 7)(x + 1)(x - 4)$

(d) Zeros: $-7, -1, 4$

(e)

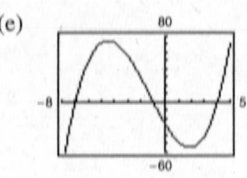

62. $f(x) = 2x^3 + 11x^2 - 21x - 90$

(a)

$$
\begin{array}{r|rrrr}
-6 & 2 & 11 & -21 & -90 \\
  &   & -12 & 6 & 90 \\
\hline
  & 2 & -1 & -15 & 0
\end{array}
$$

Yes, $(x + 6)$ is a factor of $f(x)$.

(b) $2x^2 - x - 15 = (2x + 5)(x - 3)$

The remaining factors are $(2x + 5)$ and $(x - 3)$.

(c) $f(x) = (2x + 5)(x - 3)(x + 6)$

(d) Zeros: $x = -\frac{5}{2}, 3, -6$

(e)

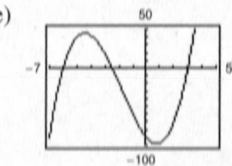

63. $f(x) = x^4 - 4x^3 - 7x^2 + 22x + 24$

Factors: $(x + 2), (x - 3)$

(a)
$$
\begin{array}{r|rrrrr}
-2 & 1 & -4 & -7 & 22 & 24 \\
   &   & -2 & 12 & -10 & -24 \\
\hline
   & 1 & -6 & 5 & 12 & 0
\end{array}
$$

$$
\begin{array}{r|rrrr}
3 & 1 & -6 & 5 & 12 \\
  &   & 3 & -9 & -12 \\
\hline
  & 1 & -3 & -4 & 0
\end{array}
$$

Both are factors since the remainders are zero.

(b) $x^2 - 3x - 4 = (x + 1)(x - 4)$

The remaining factors are $(x + 1)$ and $(x - 4)$.

(c) $f(x) = (x + 1)(x - 4)(x + 2)(x - 3)$

(d) Zeros: $-2, -1, 3, 4$

(e)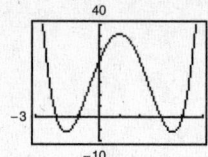

64. $f(x) = x^4 - 11x^3 + 41x^2 - 61x + 30$

(a)
$$
\begin{array}{r|rrrrr}
2 & 1 & -11 & 41 & -61 & 30 \\
  &   & 2 & -18 & 46 & -30 \\
\hline
  & 1 & -9 & 23 & -15 & 0
\end{array}
$$

$$
\begin{array}{r|rrrr}
5 & 1 & -9 & 23 & -15 \\
  &   & 5 & -20 & 15 \\
\hline
  & 1 & -4 & 3 & 0
\end{array}
$$

Yes, $(x - 2)$ and $(x - 5)$ are both factors of $f(x)$.

(b) $x^2 - 4x + 3 = (x - 1)(x - 3)$

The remaining factors are $(x - 1)$ and $(x - 3)$.

(c) $f(x) = (x - 1)(x - 3)(x - 2)(x - 5)$

(d) Zeros: $x = 1, 2, 3, 5$

(e)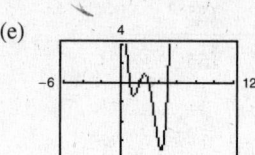

65. $6 + \sqrt{-4} = 6 + 2i$

66. $3 - \sqrt{-25} = 3 - 5i$

67. $i^2 + 3i = -1 + 3i$

68. $-5i + i^2 = -1 - 5i$

69. $(7 + 5i) + (-4 + 2i) = (7 - 4) + (5i + 2i) = 3 + 7i$

70. $\left(\dfrac{\sqrt{2}}{2} - \dfrac{\sqrt{2}}{2}i\right) - \left(\dfrac{\sqrt{2}}{2} + \dfrac{\sqrt{2}}{2}i\right) = \dfrac{\sqrt{2}}{2} - \dfrac{\sqrt{2}}{2}i - \dfrac{\sqrt{2}}{2} - \dfrac{\sqrt{2}}{2}i = -2\left(\dfrac{\sqrt{2}}{2}i\right) = -\sqrt{2}i$

71. $5i(13 - 8i) = 65i - 40i^2 = 40 + 65i$

72. $(1 + 6i)(5 - 2i) = 5 - 2i + 30i - 12i^2$

$$= 5 + 28i + 12$$

$$= 17 + 28i$$

73. $(10 - 8i)(2 - 3i) = 20 - 30i - 16i + 24i^2$

$$= -4 - 46i$$

74. $i(6 + i)(3 - 2i) = i(18 - 12i + 3i - 2i^2)$

$$= i(20 - 9i)$$

$$= 20i - 9i^2$$

$$= 9 + 20i$$

75. $\dfrac{6 + i}{4 - i} = \dfrac{6 + i}{4 - i} \cdot \dfrac{4 + i}{4 + i}$

$$= \dfrac{24 + 10i + i^2}{16 + 1}$$

$$= \dfrac{23 + 10i}{17}$$

$$= \dfrac{23}{17} + \dfrac{10}{17}i$$

76. $\dfrac{3 + 2i}{5 + i} = \dfrac{3 + 2i}{5 + i} \cdot \dfrac{5 - i}{5 - i}$

$$= \dfrac{15 - 3i + 10i - 2i^2}{25 - i^2}$$

$$= \dfrac{17 + 7i}{26}$$

$$= \dfrac{17}{26} + \dfrac{7i}{26}$$

77. $\dfrac{4}{2-3i} + \dfrac{2}{1+i} = \dfrac{4}{2-3i} \cdot \dfrac{2+3i}{2+3i} + \dfrac{2}{1+i} \cdot \dfrac{1-i}{1-i}$

$\qquad = \dfrac{8+12i}{4+9} + \dfrac{2-2i}{1+1}$

$\qquad = \dfrac{8}{13} + \dfrac{12}{13}i + 1 - i$

$\qquad = \left(\dfrac{8}{13} + 1\right) + \left(\dfrac{12}{13}i - i\right)$

$\qquad = \dfrac{21}{13} - \dfrac{1}{13}i$

78. $\dfrac{1}{2+i} - \dfrac{5}{1+4i} = \dfrac{(1+4i) - 5(2+i)}{(2+i)(1+4i)}$

$\qquad = \dfrac{1+4i-10-5i}{2+8i+i+4i^2}$

$\qquad = \dfrac{-9-i}{-2+9i} \cdot \dfrac{(-2-9i)}{(-2-9i)}$

$\qquad = \dfrac{18+81i+2i+9i^2}{4-81i^2}$

$\qquad = \dfrac{9+83i}{85} = \dfrac{9}{85} + \dfrac{83i}{85}$

79. $3x^2 + 1 = 0$

$\qquad 3x^2 = -1$

$\qquad x^2 = -\dfrac{1}{3}$

$\qquad x = \pm\sqrt{-\dfrac{1}{3}}$

$\qquad = \pm\sqrt{\dfrac{1}{3}}\,i = \pm\dfrac{\sqrt{3}}{3}i$

80. $2 + 8x^2 = 0$

$\qquad 8x^2 = -2$

$\qquad x^2 = -\dfrac{1}{4}$

$\qquad x = \pm\dfrac{1}{2}i$

81. $x^2 - 2x + 10 = 0$

$\qquad x^2 - 2x + 1 = -10 + 1$

$\qquad (x-1)^2 = -9$

$\qquad x - 1 = \pm\sqrt{-9}$

$\qquad x = 1 \pm 3i$

82. $6x^2 + 3x + 27 = 0$

$\qquad x = \dfrac{-b \pm \sqrt{b^2 - 4ac}}{2a}$

$\qquad = \dfrac{-3 \pm \sqrt{3^2 - 4(6)(27)}}{2(6)}$

$\qquad = \dfrac{-3 \pm \sqrt{-639}}{12}$

$\qquad = \dfrac{-3 \pm 3i\sqrt{71}}{12} = -\dfrac{1}{4} \pm \dfrac{\sqrt{71}}{4}i$

83. $f(x) = 3x(x-2)^2$

Zeros: $x = 0, x = 2$

84. $f(x) = (x-4)(x+9)^2$

Zeros: $x = -9, 4$

85. $f(x) = x^2 - 9x + 8$

$\qquad = (x-1)(x-8)$

Zeros: $x = 1, x = 8$

86. $f(x) = x^3 + 6x$

$\qquad = x(x^2 + 6)$

Zeros: $x = 0, \pm\sqrt{6}i$

87. $f(x) = (x+4)(x-6)(x-2i)(x+2i)$

Zeros: $x = -4, x = 6, x = 2i, x = -2i$

88. $f(x) = (x-8)(x-5)^2(x-3+i)(x-3-i)$

Zeros: $x = 5, 8, 3 \pm i$

89. $f(x) = -4x^3 + 8x^2 - 3x + 15$

Possible rational zeros: $\pm 1, \pm 3, \pm 5, \pm 15, \pm\frac{1}{2}, \pm\frac{3}{2}, \pm\frac{5}{2}, \pm\frac{15}{2},$
$\pm\frac{1}{4}, \pm\frac{3}{4}, \pm\frac{5}{4}, \pm\frac{15}{4}$

90. $f(x) = 3x^4 + 4x^3 - 5x^2 - 8$

Possible rational zeros: $\pm 1, \pm 2, \pm 4, \pm 8, \pm\frac{1}{3}, \pm\frac{2}{3}, \pm\frac{4}{3}, \pm\frac{8}{3}$

91. $f(x) = x^3 - 2x^2 - 21x - 18$

Possible rational zeros: $\pm 1, \pm 2, \pm 3, \pm 6, \pm 9, \pm 18$

$$
\begin{array}{r|rrr}
-1 & 1 & -2 & -21 & -18 \\
   &   & -1 & 3 & 18 \\
\hline
   & 1 & -3 & -18 & 0
\end{array}
$$

$$
\begin{aligned}
x^3 - 2x^2 - 21x - 18 &= (x + 1)(x^2 - 3x - 18) \\
&= (x + 1)(x - 6)(x + 3)
\end{aligned}
$$

The zeros of $f(x)$ are $x = -1$, $x = 6$, and $x = -3$.

92. $f(x) = 3x^3 - 20x^2 + 7x + 30$

Possible rational zeros: $\pm 1, \pm 2, \pm 3, \pm 5, \pm 6, \pm 10, \pm 15,$ $\pm 30, \pm\frac{1}{3}, \pm\frac{2}{3}, \pm\frac{5}{3}, \pm\frac{10}{3}$

$$
\begin{array}{r|rrr}
-1 & 3 & -20 & 7 & 30 \\
   &   & -3 & 23 & -30 \\
\hline
   & 3 & -23 & 30 & 0
\end{array}
$$

So, $f(x) = 3x^3 - 20x^2 + 7x + 30$

$$
\begin{aligned}
&= (x + 1)(3x^2 - 23x + 30) \\
&= (x + 1)(3x - 5)(x - 6) \\
0 &= (x + 1)(3x - 5)(x - 6).
\end{aligned}
$$

Zeros: $x = -1, \frac{5}{3}, 6$

93. $f(x) = x^3 - 10x^2 + 17x - 8$

Possible rational zeros: $\pm 1, \pm 2, \pm 4, \pm 8$

$$
\begin{array}{r|rrr}
1 & 1 & -10 & 17 & -8 \\
  &   & 1 & -9 & 8 \\
\hline
  & 1 & -9 & 8 & 0
\end{array}
$$

$$
\begin{aligned}
x^3 - 10x^2 + 17x - 8 &= (x - 1)(x^2 - 9x + 8) \\
&= (x - 1)(x - 1)(x - 8) \\
&= (x - 1)^2(x - 8)
\end{aligned}
$$

The zeros of $f(x)$ are $x = 1$ and $x = 8$.

94. $f(x) = x^3 + 9x^2 + 24x + 20$

Possible rational zeros: $\pm 1, \pm 2, \pm 4, \pm 5, \pm 10, \pm 20$

$$
\begin{array}{r|rrr}
-5 & 1 & 9 & 24 & 20 \\
   &   & -5 & -20 & -20 \\
\hline
   & 1 & 4 & 4 & 0
\end{array}
$$

So, $f(x) = x^3 + 9x^2 + 24x + 20$

$$
\begin{aligned}
&= (x + 5)(x^2 + 4x + 4) \\
&= (x + 5)(x + 2)^2.
\end{aligned}
$$

Zeros: $x = -5, -2$

95. $f(x) = x^4 + x^3 - 11x^2 + x - 12$

Possible rational zeros: $\pm 1, \pm 2, \pm 3, \pm 4, \pm 6, \pm 12$

$$
\begin{array}{r|rrrr}
3 & 1 & 1 & -11 & 1 & -12 \\
  &   & 3 & 12 & 3 & 12 \\
\hline
  & 1 & 4 & 1 & 4 & 0
\end{array}
$$

$$
\begin{array}{r|rrrr}
-4 & 1 & 4 & 1 & 4 \\
   &   & -4 & 0 & -4 \\
\hline
   & 1 & 0 & 1 & 0
\end{array}
$$

$$
x^4 + x^3 - 11x^2 + x - 12 = (x - 3)(x + 4)(x^2 + 1)
$$

The real zeros of $f(x)$ are $x = 3$, and $x = -4$.

96. $f(x) = 25x^4 + 25x^3 - 154x^2 - 4x + 24$

Possible rational zeros: $\pm 1, \pm 2, \pm 3, \pm 4, \pm 6, \pm 8, \pm 12, \pm 24, \pm\frac{1}{5}, \pm\frac{2}{5}, \pm\frac{3}{5}, \pm\frac{4}{5}, \pm\frac{6}{5}, \pm\frac{8}{5}, \pm\frac{12}{5},$ $\pm\frac{24}{5}, \pm\frac{1}{25}, \pm\frac{2}{25}, \pm\frac{3}{25}, \pm\frac{4}{25}, \pm\frac{6}{25}, \pm\frac{8}{25}, \pm\frac{12}{25}, \pm\frac{24}{25}$

$$
\begin{array}{r|rrrr}
-3 & 25 & 25 & -154 & -4 & 24 \\
   &   & -75 & 150 & 12 & -24 \\
\hline
   & 25 & -50 & -4 & 8 & 0
\end{array}
$$

$$
\begin{array}{r|rrrr}
2 & 25 & -50 & -4 & 8 \\
  &   & 50 & 0 & -8 \\
\hline
  & 25 & 0 & -4 & 0
\end{array}
$$

So, $f(x) = 25x^4 + 25x^3 - 154x^2 - 4x + 24$

$$
\begin{aligned}
&= (x + 3)(x - 2)(25x^2 - 4) \\
&= (x + 3)(x - 2)(5x + 2)(5x - 2).
\end{aligned}
$$

Zeros: $x = -3, 2, \pm\frac{2}{5}$

97. $f(x) = 3\left(x - \frac{2}{3}\right)(x - 4)\left(x - \sqrt{3}i\right)\left(x + \sqrt{3}i\right)$ Since $\sqrt{3}i$ is a zero, so is $-\sqrt{3}i$.

$\qquad = (3x - 2)(x - 4)(x^2 + 3)$ Multiply by 3 to clear the fraction.

$\qquad = (3x^2 - 14x + 8)(x^2 + 3)$

$\qquad = 3x^4 - 14x^3 + 17x^2 - 42x + 24$

Note: $f(x) = a(3x^4 - 14x^3 + 17x^2 - 42x + 24)$, where a is any real nonzero number, has zeros $\frac{2}{3}$, 4, and $\pm\sqrt{3}i$.

98. Since $1 - 2i$ is a zero and the coefficients are real, $1 + 2i$ must also be a zero.

$f(x) = (x - 2)(x + 3)(x - 1 + 2i)(x - 1 - 2i)$

$\qquad = (x^2 + x - 6)[(x - 1)^2 + 4]$

$\qquad = (x^2 + x - 6)(x^2 - 2x + 5)$

$\qquad = x^4 - x^3 - 3x^2 + 17x - 30$

99. $f(x) = x^3 - 4x^2 + x - 4$, Zero: i

Since i is a zero, so is $-i$.

i	1	-4	1	-4
		i	$-1 - 4i$	4
	1	$-4 + i$	$-4i$	0

$-i$	1	$-4 + i$	$-4i$
		$-i$	$4i$
	1	-4	0

$f(x) = (x - i)(x + i)(x - 4)$, Zeros: $x = \pm i, 4$

100. $h(x) = -x^3 + 2x^2 - 16x + 32$

Since $-4i$ is a zero, so is $4i$.

$-4i$	-1	2	-16	32
		$4i$	$16 - 8i$	-32
	-1	$2 + 4i$	$-8i$	0

$4i$	-1	$2 + 4i$	$-8i$
		$-4i$	$8i$
	-1	2	0

$h(x) = (x + 4i)(x - 4i)(-x + 2)$

Zeros: $x = \pm 4i, 2$

101. $g(x) = 2x^4 - 3x^3 - 13x^2 + 37x - 15$, Zero: $2 + i$

Since $2 + i$ is a zero, so is $2 - i$.

$2 + i$	2	-3	-13	37	-15
		$4 + 2i$	$5i$	$-31 - 3i$	15
	2	$1 + 2i$	$-13 + 5i$	$6 - 3i$	0

$2 - i$	2	$1 + 2i$	$-13 + 5i$	$6 - 3i$
		$4 - 2i$	$10 - 5i$	$-6 + 3i$
	2	5	-3	0

$g(x) = [x - (2 + i)][x - (2 - i)](2x^2 + 5x - 3)$

$\qquad = (x - 2 - i)(x - 2 + i)(2x - 1)(x + 3)$

Zeros: $x = 2 \pm i, \frac{1}{2}, -3$

102. $f(x) = 4x^4 - 11x^3 + 14x^2 - 6x$

$\qquad = x(4x^3 - 11x^2 + 14x - 6)$

One zero is $x = 0$. Since $1 - i$ is a zero, so is $1 + i$.

$1 - i$	4	-11	14	-6
		$4 - 4i$	$-11 + 3i$	6
	4	$-7 - 4i$	$3 + 3i$	0

$1 + i$	4	$-7 - 4i$	$3 + 3i$
		$4 + 4i$	$-3 - 3i$
	4	-3	0

$f(x) = x[x - (1 - i)][x - (1 + i)](4x - 3)$

$\qquad = x(x - 1 + i)(x - 1 - i)(4x - 3)$

Zeros: $0, \frac{3}{4}, 1 + i, 1 - i$

103. $f(x) = x^3 + 4x^2 - 5x$

$\qquad = x(x^2 + 4x - 5)$

$\qquad = x(x + 5)(x - 1)$

Zeros: $x = 0, -5, 1$

104. $g(x) = x^3 - 7x^2 + 36$

$$\begin{array}{r|rrrr} -2 & 1 & -7 & 0 & 36 \\ & & -2 & 18 & -36 \\ \hline & 1 & -9 & 18 & 0 \end{array}$$

The zeros of $x^2 - 9x + 18 = (x - 3)(x - 6)$ are $x = 3, 6$. The zeros of $g(x)$ are $-2, 3, 6$.

$g(x) = (x + 2)(x - 3)(x - 6)$

105. $g(x) = x^4 + 4x^3 - 3x^2 + 40x + 208$, Zero: $x = -4$

$$\begin{array}{r|rrrrr} -4 & 1 & 4 & -3 & 40 & 208 \\ & & -4 & 0 & 12 & -208 \\ \hline & 1 & 0 & -3 & 52 & 0 \end{array}$$

$$\begin{array}{r|rrrr} -4 & 1 & 0 & -3 & 52 \\ & & -4 & 16 & -52 \\ \hline & 1 & -4 & 13 & 0 \end{array}$$

$g(x) = (x + 4)^2(x^2 - 4x + 13)$

By the Quadratic Formula the zeros of $x^2 - 4x + 13$ are $x = 2 \pm 3i$. The zeros of $g(x)$ are $x = -4$ of multiplicity 2, and $x = 2 \pm 3i$.

$g(x) = (x + 4)^2[x - (2 + 3i)][x - (2 - 3i)]$

$\quad = (x + 4)^2(x - 2 - 3i)(x - 2 + 3i)$

106. $f(x) = x^4 + 8x^3 + 8x^2 - 72x - 153$

$$\begin{array}{r|rrrrr} 3 & 1 & 8 & 8 & -72 & -153 \\ & & 3 & 33 & 123 & 153 \\ \hline & 1 & 11 & 41 & 51 & 0 \end{array}$$

$$\begin{array}{r|rrrr} -3 & 1 & 11 & 41 & 51 \\ & & -3 & -24 & -51 \\ \hline & 1 & 8 & 17 & \end{array}$$

By the Quadratic Formula, the zeros of $x^2 + 8x + 17$ are

$$x = \frac{-8 \pm \sqrt{(8)^2 - 4(1)(17)}}{2(1)} = \frac{-8 \pm \sqrt{-4}}{2} = -4 \pm i.$$

The zeros of $f(x)$ are $-3, 3, -4 - i, -4 + i$.

$f(x) = (x + 3)(x - 3)(x + 4 - i)(x + 4 + i)$

107. $g(x) = 5x^3 + 3x^2 - 6x + 9$

$g(x)$ has two variations in sign, so g has either two or no positive real zeros.

$g(-x) = -5x^3 + 3x^2 + 6x + 9$

$g(-x)$ has one variation in sign, so g has one negative real zero.

108. $h(x) = -2x^5 + 4x^3 - 2x^2 + 5$

$h(x)$ has three variations in sign, so h has either three or one positive real zeros.

$h(-x) = -2(-x)^5 + 4(-x)^3 - 2(-x)^2 + 5$

$\quad = 2x^5 - 4x^3 - 2x^2 + 5$

$h(-x)$ has two variations in sign, so h has either two or no negative real zeros.

109. $f(x) = 4x^3 - 3x^2 + 4x - 3$

(a) $\begin{array}{r|rrrr} 1 & 4 & -3 & 4 & -3 \\ & & 4 & 1 & 5 \\ \hline & 4 & 1 & 5 & 2 \end{array}$

Since the last row has all positive entries, $x = 1$ is an upper bound.

(b) $\begin{array}{r|rrrr} -\frac{1}{4} & 4 & -3 & 4 & -3 \\ & & -1 & 1 & -\frac{5}{4} \\ \hline & 4 & -4 & 5 & -\frac{17}{4} \end{array}$

Since the last row entries alternate in sign, $x = -\frac{1}{4}$ is a lower bound.

110. $g(x) = 2x^3 - 5x^2 - 14x + 8$

(a) $\begin{array}{r|rrrr} 8 & 2 & -5 & -14 & 8 \\ & & 16 & 88 & 592 \\ \hline & 2 & 11 & 74 & 600 \end{array}$

Since the last row has all positive entries, $x = 8$ is an upper bound.

(b) $\begin{array}{r|rrrr} -4 & 2 & -5 & -14 & 8 \\ & & -8 & 52 & -152 \\ \hline & 2 & -13 & 38 & -144 \end{array}$

Since the last row entries alternate in sign, $x = 4$ is a lower bound.

111. $f(x) = \dfrac{5x}{x + 12}$

Domain: all real numbers x
except $x = -12$

112. $f(x) = \dfrac{3x^2}{1 + 3x}$

$1 + 3x = 0$

$3x = -1$

$x = -\dfrac{1}{3}$

Domain: all real numbers x
except $x = -\frac{1}{3}$

113. $f(x) = \dfrac{8}{x^2 - 10x + 24}$

$= \dfrac{8}{(x - 4)(x - 6)}$

Domain: all real numbers x
except $x = 4$ and $x = 6$

114. $f(x) = \dfrac{x^2 - x - 2}{x^2 + 4}$

Domain: all real numbers

115. $f(x) = \dfrac{4}{x + 3}$

Vertical asymptote: $x = -3$

Horizontal asymptote: $y = 0$

116. $f(x) = \dfrac{2x^2 + 5x - 3}{x^2 + 2}$

Vertical asymptote: none

Horizontal asymptote: $y = 2$

117. $h(x) = \dfrac{2x - 10}{x^2 - 2x - 15}$

$= \dfrac{2(x - 5)}{(x + 3)(x - 5)}$

$= \dfrac{2}{x + 3}, \; x \neq 5$

Vertical asymptote: $x = -3$

Horizontal asymptote: $y = 0$

118. $h(x) = \dfrac{x^3 - 4x^2}{x^2 + 3x + 2} = \dfrac{x^2(x - 4)}{(x + 2)(x + 1)}$

Vertical asymptotes: $x = -2, x = -1$

Horizontal asymptotes: none

119. $f(x) = \dfrac{-5}{x^2}$

(a) Domain: all real numbers x except $x = 0$

(b) No intercepts

(c) Vertical asymptote: $x = 0$
    Horizontal asymptote: $y = 0$

(d)

x	± 3	± 2	± 1
y	$-\frac{5}{9}$	$-\frac{5}{4}$	-5

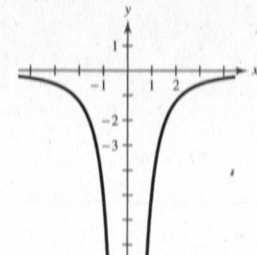

120. $f(x) = \dfrac{4}{x}$

(a) Domain: all real numbers x except $x = 0$

(b) No intercepts

(c) Vertical asymptote: $x = 0$
    Horizontal asymptote: $y = 0$

(d)

x	-3	-2	-1	1	2	3
y	$-\frac{4}{3}$	-2	-4	4	2	$\frac{4}{3}$

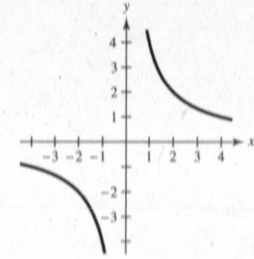

121. $g(x) = \dfrac{2 + x}{1 - x} = -\dfrac{x + 2}{x - 1}$

    (a) Domain: all real numbers x except $x = 1$

    (b) x-intercept: $(-2, 0)$
        y-intercept: $(0, 2)$

    (c) Vertical asymptote: $x = 1$
        Horizontal asymptote: $y = -1$

    (d)

x	-1	0	2	3
y	$\frac{1}{2}$	2	-4	$-\frac{5}{2}$

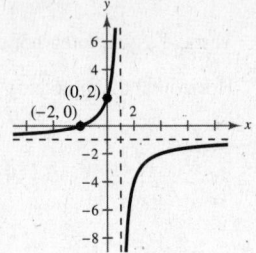

122. $h(x) = \dfrac{x - 3}{x - 2}$

    (a) Domain: all real numbers x except $x = 2$

    (b) x-intercept: $(3, 0)$
        y-intercept: $\left(0, \dfrac{3}{2}\right)$

    (c) Vertical asymptote: $x = 2$
        Horizontal asymptote: $y = 1$

    (d)

x	-1	0	1	3	4	5
y	$\frac{4}{3}$	$\frac{3}{2}$	2	0	$\frac{1}{2}$	$\frac{2}{3}$

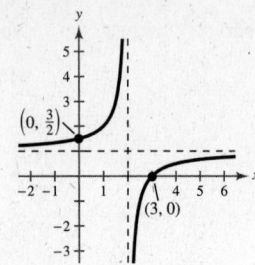

123. $p(x) = \dfrac{x^2}{x^2 + 1}$

    (a) Domain: all real numbers x

    (b) Intercept: $(0, 0)$

    (c) Horizontal asymptote: $y = 1$

    (d)

x	± 3	± 2	± 1	0
y	$\frac{9}{10}$	$\frac{4}{5}$	$\frac{1}{2}$	0

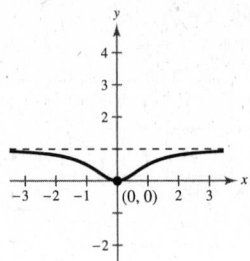

124. $f(x) = \dfrac{2x}{x^2 + 4}$

    (a) Domain: all real numbers x

    (b) Intercept: $(0, 0)$

    (c) Horizontal asymptote: $y = 0$

    (d)

x	-2	-1	0	1	2
y	$-\frac{1}{2}$	$-\frac{2}{5}$	0	$\frac{2}{5}$	$\frac{1}{2}$

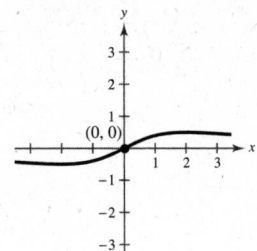

125. $f(x) = \dfrac{x}{x^2 + 1}$

    (a) Domain: all real numbers x

    (b) Intercept: $(0, 0)$

    (c) Horizontal asymptote: $y = 0$

    (d)

x	-2	-1	0	1	2
y	$-\frac{2}{5}$	$-\frac{1}{2}$	0	$\frac{1}{2}$	$\frac{2}{5}$

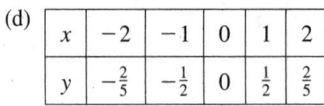
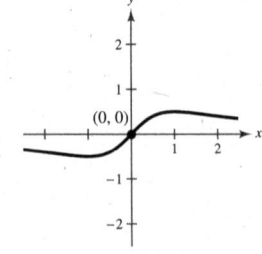

126. $h(x) = \dfrac{4}{(x-1)^2}$

(a) Domain: all real numbers x except $x = 1$

(b) y-intercept: $(0, 4)$

(c) Vertical asymptote: $x = 1$
Horizontal asymptote: $y = 0$

(d)

x	-2	-1	0	2	3	4
y	$\frac{4}{9}$	1	4	4	1	$\frac{4}{9}$

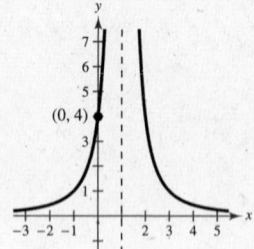

127. $f(x) = \dfrac{-6x^2}{x^2 + 1}$

(a) Domain: all real numbers x

(b) Intercept: $(0, 0)$

(c) Horizontal asymptote: $y = -6$

(d)

x	± 3	± 2	± 1	0
y	$-\frac{27}{5}$	$-\frac{24}{5}$	-3	0

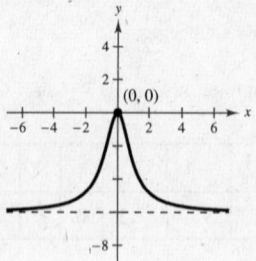

128. $y = \dfrac{2x^2}{x^2 - 4}$

(a) Domain: all real numbers x except $x = \pm 2$

(b) Intercept: $(0, 0)$

(c) Vertical asymptotes: $x = 2, x = -2$
Horizontal asymptote: $y = 2$

(d)

x	± 5	± 4	± 3	± 1	0
y	$\frac{50}{21}$	$\frac{8}{3}$	$\frac{18}{5}$	$-\frac{2}{3}$	0

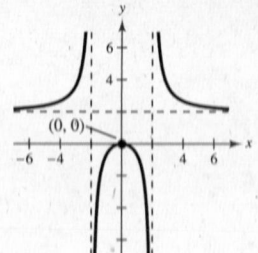

129. $f(x) = \dfrac{6x^2 - 11x + 3}{3x^2 - x}$

$\qquad = \dfrac{(3x - 1)(2x - 3)}{x(3x - 1)} = \dfrac{2x - 3}{x}, \; x \neq \dfrac{1}{3}$

(a) Domain: all real numbers x except $x = 0$ and $x = \dfrac{1}{3}$

(b) x-intercept: $\left(\dfrac{3}{2}, 0\right)$
y-intercept: none

(c) Vertical asymptote: $x = 0$
Horizontal asymptote: $y = 2$

(d)

x	-2	-1	1	2	3	4
y	$\frac{7}{2}$	5	-1	$\frac{1}{2}$	1	$\frac{5}{4}$

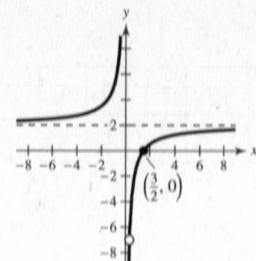

130. $f(x) = \dfrac{6x^2 - 7x + 2}{4x^2 - 1}$

$= \dfrac{(2x - 1)(3x - 2)}{(2x - 1)(2x + 1)} = \dfrac{3x - 2}{2x + 1}, \; x \neq \dfrac{1}{2}$

(a) Domain: all real numbers x except $x \neq \pm\dfrac{1}{2}$

(b) y-intercept: $(0, -2)$

x-intercept: $\left(\dfrac{2}{3}, 0\right)$

(c) Vertical asymptote: $x = -\dfrac{1}{2}$

Horizontal asymptote: $y = \dfrac{3}{2}$

(d)

x	-3	-2	-1	0	$\frac{2}{3}$	1	2
y	$\frac{11}{5}$	$\frac{8}{3}$	5	-2	0	$\frac{1}{3}$	$\frac{4}{5}$

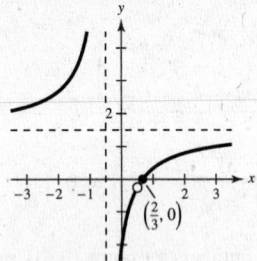

131. $f(x) = \dfrac{2x^3}{x^2 + 1} = 2x - \dfrac{2x}{x^2 + 1}$

(a) Domain: all real numbers x

(b) Intercept: $(0, 0)$

(c) Slant asymptote: $y = 2x$

(d)

x	-2	-1	0	1	2
y	$-\frac{16}{5}$	-1	0	1	$\frac{16}{5}$

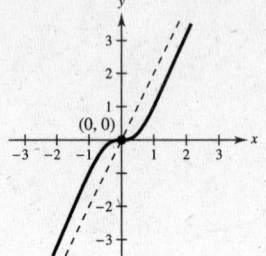

132. $f(x) = \dfrac{x^2 + 1}{x + 1}$

(a) Domain: all real numbers x except $x = -1$

(b) y-intercept: $(0, 1)$

(c) Vertical asymptote: $x = -1$

Using long division, $f(x) = \dfrac{x^2 + 1}{x + 1} = x - 1 + \dfrac{2}{x + 1}$.

Slant asymptote: $y = x - 1$

(d)

x	-6	-2	$-\frac{3}{2}$	$-\frac{1}{2}$	0	4
y	$-\frac{37}{5}$	-5	$-\frac{13}{2}$	$\frac{5}{2}$	1	$\frac{17}{5}$

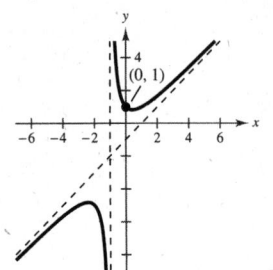

133. $f(x) = \dfrac{3x^3 - 2x^2 - 3x + 2}{3x^2 - x - 4}$

$= \dfrac{(3x - 2)(x + 1)(x - 1)}{(3x - 4)(x + 1)}$

$= \dfrac{(3x - 2)(x - 1)}{3x - 4}$

$= x - \dfrac{1}{3} + \dfrac{2/3}{3x - 4}, \quad x \neq -1$

(a) Domain: all real numbers x except $x = -1$, $x = \dfrac{4}{3}$

(b) x-intercepts: $(1, 0)$ and $\left(\dfrac{2}{3}, 0\right)$

y-intercept: $\left(0, -\dfrac{1}{2}\right)$

(c) Vertical asymptote: $x = \dfrac{4}{3}$

Slant asymptote: $y = x - \dfrac{1}{3}$

(d)

x	-3	-2	0	1	2	3
y	$-\dfrac{44}{13}$	$-\dfrac{12}{5}$	$-\dfrac{1}{2}$	0	2	$\dfrac{14}{5}$

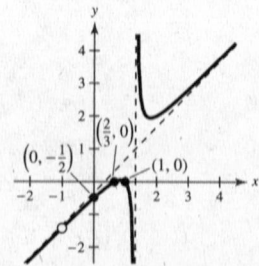

134. $f(x) = \dfrac{3x^3 - 4x^2 - 12x + 16}{3x^2 + 5x - 2}$

$= \dfrac{(x - 2)(x + 2)(3x - 4)}{(x + 2)(3x - 1)}$

$= \dfrac{(x - 2)(3x - 4)}{3x - 1}, x \neq -2$

(a) Domain: all real x except $x = -2$ or $x = \dfrac{1}{3}$

(b) y-intercept: $(0, -8)$

x-intercepts: $\left(\dfrac{4}{3}, 0\right)$, $(2, 0)$

(c) Vertical asymptote: $x = \dfrac{1}{3}$

Using long division,

$$f(x) = \dfrac{3x^2 - 10x + 8}{3x - 1} = x - 3 + \dfrac{5}{3x - 1}.$$

Slant asymptote: $y = x - 3$

(d)

x	-4	-1	0	1	2	4
y	$-\dfrac{96}{13}$	$-\dfrac{21}{4}$	-8	$\dfrac{1}{2}$	0	$\dfrac{16}{11}$

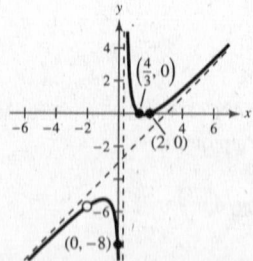

135. $\overline{C} = \dfrac{C}{x} = \dfrac{0.5x + 500}{x}, \quad 0 < x$

Horizontal asymptote: $\overline{C} = \dfrac{0.5}{1} = 0.5$

As x increases, the average cost per unit approaches the horizontal asymptote, $\overline{C} = 0.5 = \$0.50$.

136. $C = \dfrac{528p}{100 - p}, \quad 0 \leq p < 100$

(a)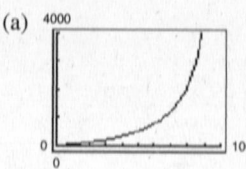

(b) When $p = 25$, $C = \dfrac{528(25)}{100 - 25} = \$176$ million.

When $p = 50$, $C = \dfrac{528(50)}{100 - 50} = \$528$ million.

When $p = 75$, $C = \dfrac{528(75)}{100 - 75} = \$1584$ million.

(c) As $p \to 100$, $C \to \infty$. No, it is not possible.

137. (a)

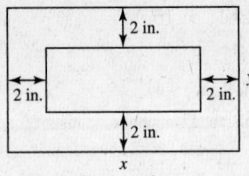

(b) The area of print is $(x - 4)(y - 4)$, which is 30 square inches.

$$(x - 4)(y - 4) = 30$$

$$y - 4 = \frac{30}{x - 4}$$

$$y = \frac{30}{x - 4} + 4$$

$$y = \frac{30 + 4(x - 4)}{x - 4}$$

$$y = \frac{4x + 14}{x - 4}$$

$$y = \frac{2(2x + 7)}{x - 4}$$

$$\text{Total area} = xy = x\left[\frac{2(2x + 7)}{x - 4}\right] = \frac{2x(2x + 7)}{x - 4}$$

(c) Because the horizontal margins total 4 inches, x must be greater than 4 inches. The domain is $x > 4$.

(d)

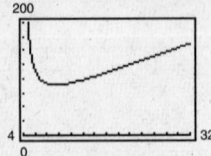

The minimum area occurs when $x \approx 9.477$ inches, so

$$y \approx \frac{2(2 \cdot 9.477 + 7)}{9.477 - 4} \approx 9.477 \text{ inches.}$$

The least amount of paper used is for a page size of about 9.48 inches by 9.48 inches.

138. $y = \dfrac{18.47x - 2.96}{0.23x + 1}, \ 0 < x$

The limiting amount of CO_2 uptake is determined by the horizontal asymptote,

$$y = \frac{18.47}{0.23} \approx 80.3 \text{ mg/dm}^2/\text{hr.}$$

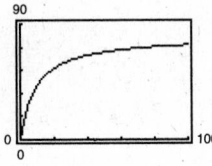

139.

$$6x^2 + 5x < 4$$

$$6x^2 + 5x - 4 < 0$$

$$(3x + 4)(2x - 1) < 0$$

Critical numbers: $x = -\frac{4}{3}, x = \frac{1}{2}$

Test intervals: $\left(-\infty, -\frac{4}{3},\right), \left(-\frac{4}{3}, \frac{1}{2}\right), \left(\frac{1}{2}, \infty\right)$

Test: Is $(3x + 4)(2x - 1) < 0$?

By testing an x-value in each test interval in the inequality, we see that the solution set is: $\left(-\frac{4}{3}, \frac{1}{2}\right)$

140.

$$2x^2 + x \geq 15$$

$$2x^2 + x - 15 \geq 0$$

$$(2x - 5)(x + 3) \geq 0$$

Critical numbers: $x = \frac{5}{2}, x = -3$

Test intervals: $(-\infty, -3) \implies (2x - 5)(x + 3) > 0$

$\left(-3, \frac{5}{2}\right) \implies (2x - 5)(x + 3) < 0$

$\left(\frac{5}{2}, \infty\right) \implies (2x - 5)(x + 3) > 0$

Solution interval: $\left(-\infty, -3\right] \cup \left[\frac{5}{2}, \infty\right)$

141.

$$x^3 - 16x \geq 0$$

$$x(x + 4)(x - 4) \geq 0$$

Critical numbers: $x = 0, x = \pm 4$

Test intervals: $(-\infty, -4), (-4, 0), (0, 4), (4, \infty)$

Test: Is $x(x + 4)(x - 4) \geq 0$?

By testing an x-value in each test interval in the inequality, we see that the solution set is: $[-4, 0] \cup [4, \infty)$.

142. $12x^3 - 20x^2 < 0$

$4x^2(3x - 5) < 0$

Critical numbers: $x = 0, x = \frac{5}{3}$

Test intervals: $(-\infty, 0) \Rightarrow 12x^3 - 20x^2 < 0$

$\left(0, \frac{5}{3}\right) \Rightarrow 12x^3 - 20x^2 < 0$

$\left(\frac{5}{3}, \infty\right) \Rightarrow 12x^3 - 20x^2 > 0$

Solution interval: $(-\infty, 0) \cup \left(0, \frac{5}{3}\right)$

143.
$$\frac{2}{x+1} \leq \frac{3}{x-1}$$

$$\frac{2(x-1) - 3(x+1)}{(x+1)(x-1)} \leq 0$$

$$\frac{2x - 2 - 3x - 3}{(x+1)(x-1)} \leq 0$$

$$\frac{-(x+5)}{(x+1)(x-1)} \leq 0$$

Critical numbers: $x = -5, x = \pm 1$

Test intervals: $(-\infty, -5), (-5, -1), (-1, 1), (1, \infty)$

Test: Is $\dfrac{-(x+5)}{(x+1)(x-1)} \leq 0$?

By testing an x-value in each test interval in the inequality, we see that the solution set is: $[-5, -1) \cup (1, \infty)$

144. $\dfrac{x-5}{3-x} < 0$

Critical numbers: $x = 5, x = 3$

Test intervals: $(-\infty, 3) \Rightarrow \dfrac{x-5}{3-x} < 0$

$(3, 5) \Rightarrow \dfrac{x-5}{3-x} > 0$

$(5, \infty) \Rightarrow \dfrac{x-5}{3-x} < 0$

Solution intervals: $(-\infty, 3) \cup (5, \infty)$

145.
$$\frac{x^2 + 7x + 12}{x} \geq 0$$

$$\frac{(x+4)(x+3)}{x} \geq 0$$

Critical numbers: $x = -4, x = -3, x = 0$

Test intervals: $(-\infty, -4), (-4, -3), (-3, 0), (0, \infty)$

Test: Is $\dfrac{(x+4)(x+3)}{x} \geq 0$?

By testing an x-value in each test interval in the inequality, we see that the solution set is: $[-4, -3] \cup (0, \infty)$

146.
$$\frac{1}{x-2} > \frac{1}{x}$$

$$\frac{1}{x-2} - \frac{1}{x} > 0$$

Critical numbers: $x = 2, x = 0$

Test intervals: $(-\infty, 0) \Rightarrow \dfrac{1}{x-2} - \dfrac{1}{x} > 0$

$(0, 2) \Rightarrow \dfrac{1}{x-2} - \dfrac{1}{x} < 0$

$(2, \infty) \Rightarrow \dfrac{1}{x-2} - \dfrac{1}{x} > 0$

Solution interval: $(-\infty, 0) \cup (2, \infty)$

147. $5000(1 + r)^2 > 5500$

$(1 + r)^2 > 1.1$

$1 + r > 1.0488$

$r > 0.0488$

$r > 4.9\%$

148.
$$P = \frac{1000(1 + 3t)}{5 + t}$$

$$2000 \leq \frac{1000(1 + 3t)}{5 + t}$$

$$2000(5 + t) \leq 1000(1 + 3t)$$

$$10,000 + 2000t \leq 1000 + 3000t$$

$$-1000t \leq -9000$$

$$t \geq 9 \text{ days}$$

149. False. A fourth-degree polynomial can have at most four zeros and complex zeros occur in conjugate pairs.

150. False. (See Exercise 123.)
The domain of

$$f(x) = \frac{1}{x^2 + 1}$$

is the set of all real numbers x.

151. The maximum (or minimum) value of a quadratic function is located at its graph's vertex. To find the vertex, either write the equation in standard form or use the formula

$$\left(-\frac{b}{2a}, f\left(-\frac{b}{2a}\right)\right).$$

If the leading coefficient is positive, the vertex is a minimum. If the leading coefficient is negative, the vertex is a maximum.

152. Answers will vary. Sample answer:

Polynomials of degree $n > 0$ with real coefficients can be written as the product of linear and quadratic factors with real coefficients, where the quadratic factors have no real zeros.

Setting the factors equal to zero and solving for the variable can find the zeros of a polynomial function.

To solve an equation is to find all the values of the variable for which the equation is true.

153. An asymptote of a graph is a line to which the graph becomes arbitrarily close as x increases or decreases without bound.

Problem Solving for Chapter 2

1. $f(x) = ax^3 + bx^2 + cx + d$

$$
\begin{array}{r}
ax^2 + (ak + b)x + (ak^2 + bk + c) \\
x - k \overline{\smash)ax^3 + bx^2 + cx + d} \\
\underline{ax^3 - akx^2} \\
(ak + b)x^2 + cx \\
\underline{(ak + b)x^2 - (ak^2 + bk)x} \\
(ak^2 + bk + c)x + d \\
\underline{(ak^2 + bk + c)x - (ak^3 + bk^2 + ck)} \\
(ak^3 + bk^2 + ck + d)
\end{array}
$$

Thus, $f(x) = ax^3 + bx^2 + cx + d = (x - k)[ax^2 + (ak + b)x + (ak^2 + bx + c)] + ak^3 + bk^2 + ck + d$ and

$f(k) = ak^3 + bk^2 + ck + d$. Since the remainder $r = ak^3 + bk^2 + ck + d$, $f(k) = r$.

2. (a)

y	$y^3 + y^2$
1	2
2	12
3	36
4	80
5	150
6	252
7	392
8	576
9	810
10	1100

(b) $x^3 + x^2 = 252 \implies x = 6$

(c) $x^3 + 2x^2 = 288; a = 1, b = 2 \implies \dfrac{a^2}{b^3} = \dfrac{1}{8}$

$$\frac{1}{8}x^3 + \frac{1}{8}(2x^2) = \frac{1}{8}(288)$$

$$\left(\frac{x}{2}\right)^3 + \left(\frac{x}{2}\right)^2 = 36 \implies \frac{x}{2} = 3 \implies x = 6$$

(d) $3x^3 + x^2 = 90; a = 3, b = 1 \implies \dfrac{a^2}{b^3} = 9$

$$9(3x^3) + 9x^2 = 9(90)$$

$$(3x)^3 + (3x)^2 = 810 \implies 3x = 9 \implies x = 3$$

(e) $2x^3 + 5x^2 = 2500; a = 2, b = 5 \implies \dfrac{a^2}{b^3} = \dfrac{4}{125}$

$$\frac{4}{125}(2x^3) + \frac{4}{125}(5x^2) = \frac{4}{125}(2500)$$

$$\left(\frac{2x}{5}\right)^3 + \left(\frac{2x}{5}\right)^2 = 80 \implies \frac{2x}{5} = 4 \implies x = 10$$

(f) $7x^3 + 6x^2 = 1728; a = 7, b = 6 \implies \dfrac{a^2}{b^3} = \dfrac{49}{216}$

$$\frac{49}{216}(7x^3) + \frac{49}{216}(6x^2) = \frac{49}{216}(1728)$$

$$\left(\frac{7x}{6}\right)^3 + \left(\frac{7x}{6}\right)^2 = 392 \implies \frac{7x}{6} = 7 \implies x = 6$$

(g) $10x^3 + 3x^2 = 297; a = 10, b = 3 \implies \dfrac{a^2}{b^3} = \dfrac{100}{27}$

$$\frac{100}{27}(10x^3) + \frac{100}{27}(3x^2) = \frac{100}{27}(297)$$

$$\left(\frac{10x}{3}\right)^3 + \left(\frac{10x}{3}\right)^2 = 1100 \implies \frac{10x}{3} = 10 \implies x = 3$$

3. $V = l \cdot w \cdot h = x^2(x + 3)$

$x^2(x + 3) = 20$

$x^3 + 3x^2 - 20 = 0$

Possible rational zeros: $\pm 1, \pm 2, \pm 4, \pm 5, \pm 10, \pm 20$

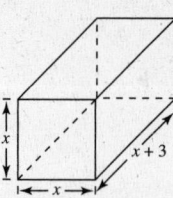

$$\begin{array}{r|rrrr} 2 & 1 & 3 & 0 & -20 \\ & & 2 & 10 & 20 \\ \hline & 1 & 5 & 10 & 0 \end{array}$$

$(x - 2)(x^2 + 5x + 10) = 0$

$x = 2$ or $x = \dfrac{-5 \pm \sqrt{15}i}{2}$

Choosing the real positive value for x we have: $x = 2$ and $x + 3 = 5$. The dimensions of the mold are 2 inches $\times$ 2 inches $\times$ 5 inches.

4. False. Since $f(x) = d(x)q(x) + r(x)$, we have $\dfrac{f(x)}{d(x)} = q(x) + \dfrac{r(x)}{d(x)}$.

The statement should be corrected to read $f(-1) = 2$ since $\dfrac{f(x)}{x + 1} = q(x) + \dfrac{f(-1)}{x + 1}$.

5. (a) $y = ax^2 + bx + c$

$(0, -4)$: $-4 = a(0)^2 + b(0) + c$

$-4 = c$

$(4, 0)$: $0 = a(4)^2 + b(4) - 4$

$0 = 16a + 4b - 4 = 4(4a + b - 1)$

$0 = 4a + b - 1$ or $b = 1 - 4a$

$(1, 0)$: $0 = a(1)^2 + b(1) - 4$

$4 = a + b$

$4 = a + (1 - 4a)$

$4 = 1 - 3a$

$-3 = -3a$

$a = -1$

$b = 1 - 4(-1) = 5$

$y = -x^2 + 5x - 4$

(b) Enter the data points $(0, -4)$, $(1, 0)$, $(2, 2)$, $(4, 0)$, $(6, -10)$ and use the regression feature to obtain $y = -x^2 + 5x - 4$.

6. (a) Slope $= \dfrac{9 - 4}{3 - 2} = 5$

Slope of tangent line is less than 5.

(b) Slope $= \dfrac{4 - 1}{2 - 1} = 3$

Slope of tangent line is greater than 3.

(c) Slope $= \dfrac{4.41 - 4}{2.1 - 2} = 4.1$

Slope of tangent line is less than 4.1.

(d) Slope $= \dfrac{f(2 + h) - f(2)}{(2 + h) - 2}$

$= \dfrac{(2 + h)^2 - 4}{h}$

$= \dfrac{4h + h^2}{h}$

$= 4 + h, h \neq 0$

(e) Slope $= 4 + h, \quad h \neq 0$

$4 + (-1) = 3$

$4 + 1 = 5$

$4 + 0.1 = 4.1$

The results are the same as in (a)–(c).

(f) Letting h get closer and closer to 0, the slope approaches 4. Hence, the slope at $(2, 4)$ is 4.

7. $f(x) = (x - k)q(x) + r$

(a) Cubic, passes through $(2, 5)$, rises to the right

One possibility:

$$f(x) = (x - 2)x^2 + 5$$
$$= x^3 - 2x^2 + 5$$

(b) Cubic, passes through $(-3, 1)$, falls to the right

One possibility:

$$f(x) = -(x + 3)x^2 + 1$$
$$= -x^3 - 3x^2 + 1$$

8. (a) $z_m = \dfrac{1}{z}$

$$= \frac{1}{1 + i} = \frac{1}{1 + i} \cdot \frac{1 - i}{1 - i}$$
$$= \frac{1 - i}{2} = \frac{1}{2} - \frac{1}{2}i$$

(b) $z_m = \dfrac{1}{z}$

$$= \frac{1}{3 - i} = \frac{1}{3 - i} \cdot \frac{3 + i}{3 + i}$$
$$= \frac{3 + i}{10} = \frac{3}{10} + \frac{1}{10}i$$

(c) $z_m = \dfrac{1}{z} = \dfrac{1}{-2 + 8i}$

$$= \frac{1}{-2 + 8i} \cdot \frac{-2 - 8i}{-2 - 8i}$$
$$= \frac{-2 - 8i}{68} = -\frac{1}{34} - \frac{2}{17}i$$

9. $(a + bi)(a - bi) = a^2 - abi + abi - b^2i^2 = a^2 + b^2$

Since a and b are real numbers, $a^2 + b^2$ is also a real number.

10. $f(x) = \dfrac{ax + b}{cx + d}$

Vertical asymptote: $x = -\dfrac{d}{c}$

Horizontal asymptote: $y = \dfrac{a}{c}$

(i) $a > 0, b < 0, c > 0, d < 0$

Both the vertical asymptote and the horizontal asymptote are positive. Matches graph (d).

(ii) $a > 0, b > 0, c < 0, d < 0$

Both the vertical asymptote and the horizontal asymptote are negative. Matches graph (b).

(iii) $a < 0, b > 0, c > 0, d < 0$

The vertical asymptote is positive and the horizontal asymptote is negative. Matches graph (a).

(iv) $a > 0, b < 0, c > 0, d > 0$

The vertical asymptote is negative and the horizontal asymptote is positive. Matches graph (c).

11. $f(x) = \dfrac{ax}{(x - b)^2}$

(a) $b \neq 0 \implies x = b$ is a vertical asymptote. a causes a vertical stretch if $|a| > 1$ and a vertical shrink if $0 < |a| < 1$. For $|a| > 1$, the graph becomes wider as $|a|$ increases. When a is negative the graph is reflected about the x-axis.

(b) $a \neq 0$. Varying the value of b varies the vertical asymptote of the graph of f. For $b > 0$, the graph is translated to the right. For $b < 0$, the graph is reflected in the x-axis and is translated to the left.

12. (a)

Age, x	Near point, y
16	3.0
32	4.7
44	9.8
50	19.7
60	39.4

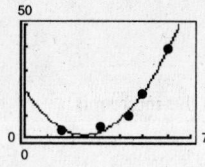

$$y \approx 0.0313x^2 - 1.586x + 21.02$$

(b) $\dfrac{1}{y} \approx -0.007x + 0.44$

$$y \approx \dfrac{1}{-0.007x + 0.44}$$

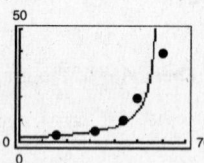

(c)

Age, x	Near point, y	Quadratic Model	Rational Model
16	3.0	3.66	3.05
32	4.7	2.32	4.63
44	9.8	11.83	7.58
50	19.7	19.97	11.11
60	39.4	38.54	50.00

The models are fairly good fits to the data. The quadratic model seems to be a better fit for older ages and the rational model a better fit for younger ages.

(d) For $x = 25$, the quadratic model yields $y \approx 0.9325$ inches and the rational model yields $y \approx 3.774$ inches.

(e) The reciprocal model cannot be used to predict the near point for a person who is 70 years old because it results in a negative value ($y \approx -20$). The quadratic model yields $y \approx 63.37$ inches.

Chapter 2   Practice Test

1. Sketch the graph of $f(x) = x^2 - 6x + 5$ and identify the vertex and the intercepts.

2. Find the number of units x that produce a minimum cost C if
   $C = 0.01x^2 - 90x + 15,000$.

3. Find the quadratic function that has a maximum at $(1, 7)$ and passes through the point $(2, 5)$.

4. Find two quadratic functions that have x-intercepts $(2, 0)$ and $\left(\frac{4}{3}, 0\right)$.

5. Use the leading coefficient test to determine the right and left end behavior of the graph of the polynomial function $f(x) = -3x^5 + 2x^3 - 17$.

6. Find all the real zeros of $f(x) = x^5 - 5x^3 + 4x$.

7. Find a polynomial function with 0, 3, and -2 as zeros.

8. Sketch $f(x) = x^3 - 12x$.

9. Divide $3x^4 - 7x^2 + 2x - 10$ by $x - 3$ using long division.

10. Divide $x^3 - 11$ by $x^2 + 2x - 1$.

11. Use synthetic division to divide $3x^5 + 13x^4 + 12x - 1$ by $x + 5$.

12. Use synthetic division to find $f(-6)$ given $f(x) = 7x^3 + 40x^2 - 12x + 15$.

13. Find the real zeros of $f(x) = x^3 - 19x - 30$.

14. Find the real zeros of $f(x) = x^4 + x^3 - 8x^2 - 9x - 9$.

15. List all possible rational zeros of the function $f(x) = 6x^3 - 5x^2 + 4x - 15$.

16. Find the rational zeros of the polynomial $f(x) = x^3 - \frac{20}{3}x^2 + 9x - \frac{10}{3}$.

17. Write $f(x) = x^4 + x^3 + 5x - 10$ as a product of linear factors.

18. Find a polynomial with real coefficients that has 2, $3 + i$, and $3 - 2i$ as zeros.

19. Use synthetic division to show that $3i$ is a zero of $f(x) = x^3 + 4x^2 + 9x + 36$.

20. Sketch the graph of $f(x) = \dfrac{x - 1}{2x}$ and label all intercepts and asymptotes.

21. Find all the asymptotes of $f(x) = \dfrac{8x^2 - 9}{x^2 + 1}$.

22. Find all the asymptotes of $f(x) = \dfrac{4x^2 - 2x + 7}{x - 1}$.

23. Given $z_1 = 4 - 3i$ and $z_2 = -2 + i$, find the following:

(a) $z_1 - z_2$

(b) $z_1 z_2$

(c) z_1/z_2

24. Solve the inequality: $x^2 - 49 \le 0$

25. Solve the inequality: $\dfrac{x + 3}{x - 7} \ge 0$

C H A P T E R 3
Exponential and Logarithmic Functions

C H A P T E R 3
Exponential and Logarithmic Functions

Section 3.1   Exponential Functions and Their Graphs

- ■ You should know that a function of the form $f(x) = a^x$, where $a > 0$, $a \neq 1$, is called an exponential function with base a.
- ■ You should be able to graph exponential functions.
- ■ You should know formulas for compound interest.

   (a) For n compoundings per year: $A = P\left(1 + \dfrac{r}{n}\right)^{nt}$.

   (b) For continuous compoundings: $A = Pe^{rt}$.

Vocabulary Check

1. algebraic

2. transcendental

3. natural exponential; natural

4. $A = P\left(1 + \dfrac{r}{n}\right)^{nt}$

5. $A = Pe^{rt}$

---

1. $f(5.6) = (3.4)^{5.6} \approx 946.852$

2. $f(x) = 2.3^x = 2.3^{3/2} \approx 3.488$

3. $f(-\pi) = 5^{-\pi} \approx 0.006$

4. $f(x) = \left(\frac{2}{3}\right)^{5x} = \left(\frac{2}{3}\right)^{5(0.3)} \approx 0.544$

5. $g(x) = 5000(2^x) = 5000(2^{-1.5})$
    ≈ 1767.767

6. $f(x) = 200(1.2)^{12x}$
    $= 200(1.2)^{12 \cdot 24}$
    $\approx 1.274 \times 10^{25}$

7. $f(x) = 2^x$

Increasing

Asymptote: $y = 0$

Intercept: $(0, 1)$

Matches graph (d).

8. $f(x) = 2^x + 1$ rises to the right.

Asymptote: $y = 1$

Intercept: $(0, 2)$

Matches graph (c).

9. $f(x) = 2^{-x}$

Decreasing

Asymptote: $y = 0$

Intercept: $(0, 1)$

Matches graph (a).

10. $f(x) = 2^{x-2}$ rises to the right.

Asymptote: $y = 0$

Intercept: $\left(0, \frac{1}{4}\right)$

Matches graph (b).

11. $f(x) = \left(\frac{1}{2}\right)^x$

x	-2	-1	0	1	2
$f(x)$	4	2	1	0.5	0.25

Asymptote: $y = 0$

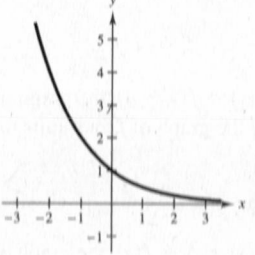

12. $f(x) = \left(\frac{1}{2}\right)^{-x} = 2^x$

x	-2	-1	0	1	2
f(x)	0.25	0.5	1	2	4

Asymptote: $y = 0$

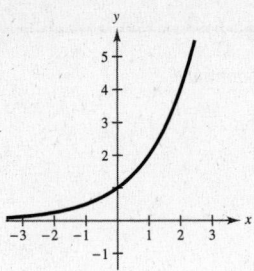

13. $f(x) = 6^{-x}$

x	-2	-1	0	1	2
f(x)	36	6	1	0.167	0.028

Asymptote: $y = 0$

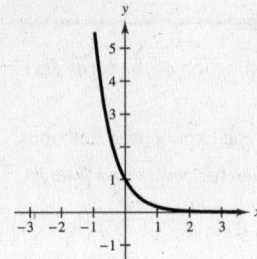

14. $f(x) = 6^x$

x	-2	-1	0	1	2
f(x)	0.028	0.167	1	6	36

Asymptote: $y = 0$

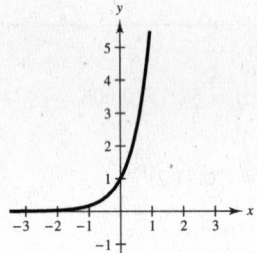

15. $f(x) = 2^{x-1}$

x	-2	-1	0	1	2
f(x)	0.125	0.25	0.5	1	2

Asymptote: $y = 0$

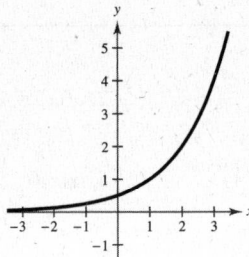

16. $f(x) = 4^{x-3} + 3$

x	-1	0	1	2	3
f(x)	3.004	3.016	3.063	3.25	4

Asymptote: $y = 3$

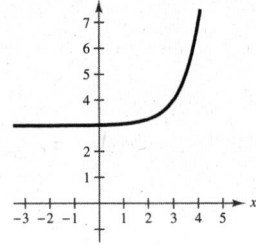

17. $f(x) = 3^x$, $g(x) = 3^{x-4}$

Because $g(x) = f(x - 4)$, the graph of g can be obtained by shifting the graph of f four units to the right.

18. $f(x) = 4^x$, $g(x) = 4^x + 1$

Because $g(x) = f(x) + 1$, the graph of g can be obtained by shifting the graph of f one unit upward.

19. $f(x) = -2^x$, $g(x) = 5 - 2^x$

Because $g(x) = 5 + f(x)$, the graph of g can be obtained by shifting the graph of f five units upward.

20. $f(x) = 10^x$, $g(x) = 10^{-x+3}$

Because $g(x) = f(-x + 3)$, the graph of g can be obtained by reflecting the graph of f in the y-axis and shifting f three units to the right. (**Note:** This is equivalent to shifting f three units to the left and then reflecting the graph in the y-axis.)

21. $f(x) = \left(\frac{7}{2}\right)^x$, $g(x) = -\left(\frac{7}{2}\right)^{-x+6}$

Because $g(x) = -f(-x + 6)$, the graph of g can be obtained by reflecting the graph of f in the x-axis and y-axis and shifting f six units to the right. (**Note:** This is equivalent to shifting f six units to the left and then reflecting the graph in the x-axis and y-axis.)

22. $f(x) = 0.3^x$, $g(x) = -0.3^x + 5$

$g(x) = -f(x) + 5$, hence the graph of g can be obtained by reflecting the graph of f in the x-axis and shifting the resulting graph five units upward.

23. $y = 2^{-x^2}$

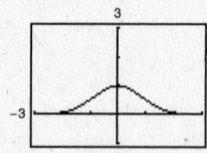

24. $y = 3^{-|x|}$

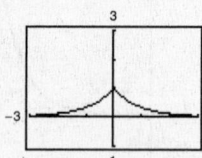

25. $f(x) = 3^{x-2} + 1$

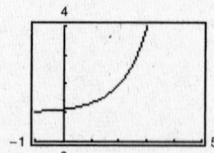

26. $y = 4^{x+1} - 2$

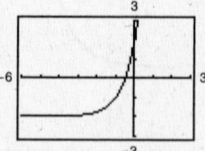

27. $f\left(\frac{3}{4}\right) = e^{-3/4} \approx 0.472$

28. $f(x) = e^x = e^{3.2} \approx 24.533$

29. $f(10) = 2e^{-5(10)} \approx 3.857 \times 10^{-22}$

30. $f(x) = 1.5e^{(1/2)x}$

$= 1.5e^{120} \approx 1.956 \times 10^{52}$

31. $f(6) = 5000e^{0.06(6)} \approx 7166.647$

32. $f(x) = 250e^{0.05x}$

$= 250e^{0.05(20)} \approx 679,570$

33. $f(x) = e^x$

x	-2	-1	0	1	2
$f(x)$	0.135	0.368	1	2.718	7.389

Asymptote: $y = 0$

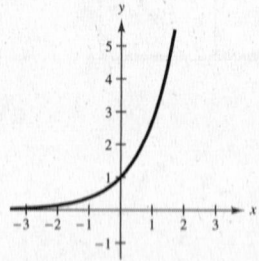

34. $f(x) = e^{-x}$

x	-2	-1	0	1	2
$f(x)$	7.389	2.718	1	0.368	0.135

Asymptote: $y = 0$

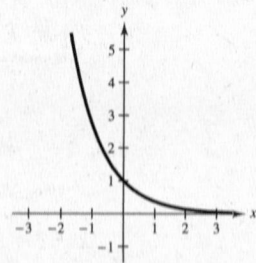

35. $f(x) = 3e^{x+4}$

x	-8	-7	-6	-5	-4
$f(x)$	0.055	0.149	0.406	1.104	3

Asymptote: $y = 0$

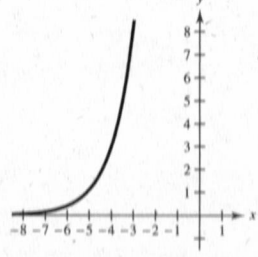

36. $f(x) = 2e^{-0.5x}$

x	-2	-1	0	1	2
$f(x)$	5.437	3.297	2	1.213	0.736

Asymptote: $y = 0$

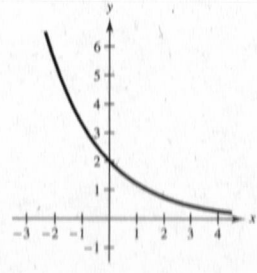

37. $f(x) = 2e^{x-2} + 4$

x	-2	-1	0	1	2
$f(x)$	4.037	4.100	4.271	4.736	6

Asymptote: $y = 4$

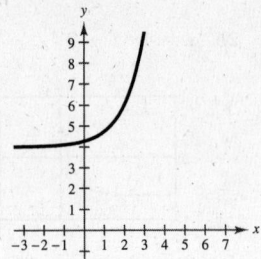

38. $f(x) = 2 + e^{x-5}$

x	0	2	4	5	6
$f(x)$	2.007	2.050	2.368	3	4.718

Asymptote: $y = 2$

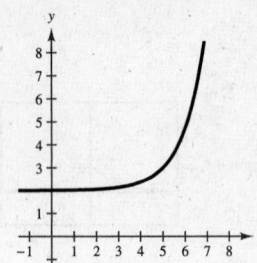

39. $y = 1.08^{-5x}$

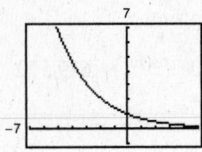

40. $y = 1.08^{5x}$

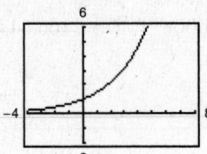

41. $s(t) = 2e^{0.12t}$

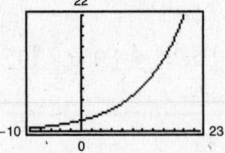

42. $s(t) = 3e^{-0.2t}$

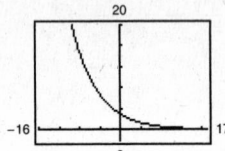

43. $g(x) = 1 + e^{-x}$

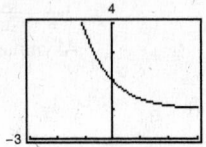

44. $h(x) = e^{x-2}$

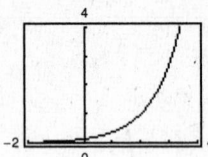

45. $3^{x+1} = 27$

$3^{x+1} = 3^3$

$x + 1 = 3$

$x = 2$

46. $2^{x-3} = 16$

$2^{x-3} = 2^4$

$x - 3 = 4$

$x = 7$

47. $2^{x-2} = \frac{1}{32}$

$2^{x-2} = 2^{-5}$

$x - 2 = -5$

$x = -3$

48. $\left(\frac{1}{5}\right)^{x+1} = 125$

$\left(\frac{1}{5}\right)^{x+1} = 5^3$

$\left(\frac{1}{5}\right)^{x+1} = \left(\frac{1}{5}\right)^{-3}$

$x + 1 = -3$

$x = -4$

49. $e^{3x+2} = e^3$

$3x + 2 = 3$

$3x = 1$

$x = \frac{1}{3}$

50. $e^{2x-1} = e^4$

$2x - 1 = 4$

$2x = 5$

$x = \frac{5}{2}$

51. $e^{x^2-3} = e^{2x}$

$x^2 - 3 = 2x$

$x^2 - 2x - 3 = 0$

$(x - 3)(x + 1) = 0$

$x = 3$ or $x = -1$

52. $e^{x^2+6} = e^{5x}$

$x^2 + 6 = 5x$

$x^2 - 5x + 6 = 0$

$(x - 3)(x - 2) = 0$

$x = 3$ or $x = 2$

53. $P = \$2500$, $r = 2.5\%$, $t = 10$ years

Compounded n times per year: $A = P\left(1 + \dfrac{r}{n}\right)^{nt} = 2500\left(1 + \dfrac{0.025}{n}\right)^{10n}$

Compounded continuously: $A = Pe^{rt} = 2500e^{0.025(10)}$

n	1	2	4	12	365	Continuous Compounding
A	\$3200.21	\$3205.09	\$3207.57	\$3209.23	\$3210.04	\$3210.06

54. $P = \$1000$, $r = 4\%$, $t = 10$ years

Compounded n times per year: $A = 1000\left(1 + \dfrac{0.04}{n}\right)^{10n}$

Compounded continuously: $A = 1000e^{0.04(10)}$

n	1	2	4	12	365	Continuous Compounding
A	\$1480.24	\$1485.95	\$1488.86	\$1490.83	\$1491.79	\$1491.82

55. $P = \$2500$, $r = 3\%$, $t = 20$ years

Compounded n times per year: $A = P\left(1 + \dfrac{r}{n}\right)^{nt} = 2500\left(1 + \dfrac{0.03}{n}\right)^{20n}$

Compounded continuously: $A = Pe^{rt} = 2500e^{0.03(20)}$

n	1	2	4	12	365	Continuous Compounding
A	\$4515.28	\$4535.05	\$4545.11	\$4551.89	\$4555.18	\$4555.30

56. $P = \$1000$, $r = 6\%$, $t = 40$ years

Compounded n times per year: $A = 1000\left(1 + \dfrac{0.06}{n}\right)^{40n}$

Compounded continuously: $A = 1000e^{0.06(40)}$

n	1	2	4	12	365	Continuous Compounding
A	\$10,285.72	\$10,640.89	\$10,828.46	\$10,957.45	\$11,021.00	\$11,023.18

57. $A = Pe^{rt} = 12{,}000e^{0.04t}$

t	10	20	30	40	50
A	\$17,901.90	\$26,706.49	\$39,841.40	\$59,436.39	\$88,668.67

58. $A = Pe^{rt} = 12{,}000e^{0.06t}$

t	10	20	30	40	50
A	\$21,865.43	\$39,841.40	\$72,595.77	\$132,278.12	\$241,026.44

59. $A = Pe^{rt} = 12{,}000e^{0.065t}$

t	10	20	30	40	50
A	\$22,986.49	\$44,031.56	\$84,344.25	\$161,564.86	\$309,484.08

60. $A = Pe^{rt} = 12{,}000e^{0.035t}$

t	10	20	30	40	50
A	\$17,028.81	\$24,165.03	\$34,291.81	\$48,662.40	\$69,055.23

61. $A = 25{,}000e^{(0.0875)(25)}$

    $\approx \$222{,}822.57$

62. $A = 5000e^{(0.075)(50)}$

    $\approx \$212{,}605.41$

63. $C(10) = 23.95(1.04)^{10} \approx \$35.45$

64. $p = 5000\left(1 - \dfrac{4}{4 + e^{-0.002x}}\right)$

(a)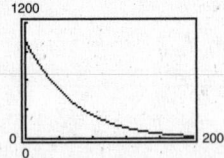

(b) When $x = 500$:

$$p = 5000\left(1 - \frac{4}{4 + e^{-0.002(500)}}\right) \approx \$421.12$$

(c) Since $(600, 350.13)$ is on the graph in part (a), it appears that the greatest price that will still yield a demand of at least 600 units is about \$350.

65. $V(t) = 100e^{4.6052t}$

(a) $V(1) \approx 10{,}000.298$ computers

(b) $V(1.5) \approx 10{,}004.472$ computers

(c) $V(2) \approx 1{,}000{,}059.63$ computers

66. (a) $P = 152.26e^{-0.0039t}$

Since the growth rate is negative, $-0.0039 = -0.39\%$, the population is decreasing.

(b) In 1998, $t = 8$ and the population is given by $P(8) = 152.26e^{-0.0039(8)} = 147.58$ million.

In 2000, $t = 10$ and the population is given by $P(10) = 152.26e^{-0.0039(10)} = 146.44$ million.

(c) In 2010, $t = 20$ and the population is given by $P(20) = 152.26e^{-0.0039(20)} = 140.84$ million.

67. $Q = 25\left(\frac{1}{2}\right)^{t/1599}$

(a) $Q(0) = 25$ grams

(b) $Q(1000) \approx 16.21$ grams

(c)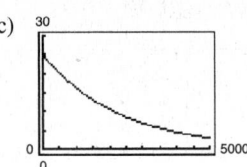

68. $Q = 10\left(\frac{1}{2}\right)^{t/5715}$

(a) When $t = 0$: $Q = 10\left(\frac{1}{2}\right)^{0/5715}$

                 $= 10(1) = 10$ grams

(b) When $t = 2000$: $Q = 10\left(\frac{1}{2}\right)^{2000/5715}$

                     ≈ 7.85 grams

(c)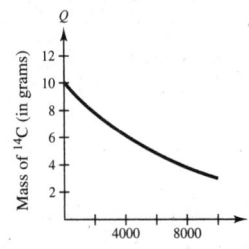

69. $y = \dfrac{100}{1 + 7e^{-0.069x}}$

(a)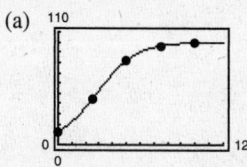

(b)

x	Sample Data	Model
0	12	12.5
25	44	44.5
50	81	81.82
75	96	96.19
100	99	99.3

(c) When $x = 36$:

$$y = \frac{100}{1 + 7e^{-0.069(36)}} \approx 63.14\%.$$

(d) $\dfrac{2}{3}(100) = \dfrac{100}{1 + 7e^{-0.069x}}$ when

$x \approx 38$ masses.

70. (a)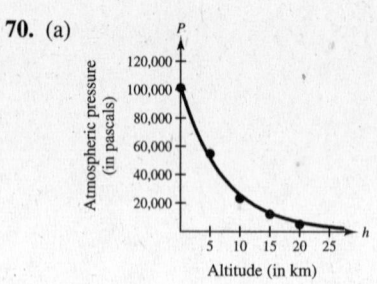

Altitude (in km)

(b) $p = 107,428e^{-0.150h}$

$\quad = 107,428e^{-0.150(8)}$

$\quad = 32,357$ pascals

71. True. The line $y = -2$ is a horizontal asymptote for the graph of $f(x) = 10^x - 2$.

72. False, $e \neq \frac{271,801}{99,990}$. e is an irrational number.

73. $f(x) = 3^{x-2}$

$\quad = 3^x 3^{-2}$

$\quad = 3^x \left(\dfrac{1}{3^2} \right)$

$\quad = \dfrac{1}{9}(3^x)$

$\quad = h(x)$

Thus, $f(x) \neq g(x)$, but $f(x) = h(x)$.

74. $g(x) = 2^{2x+6}$

$\quad = 2^{2x} \cdot 2^6$

$\quad = 64(2^{2x})$

$\quad = 64(2^2)^x$

$\quad = 64(4^x)$

$\quad = h(x)$

Thus, $g(x) = h(x)$ but $g(x) \neq f(x)$.

75. $f(x) = 16(4^{-x})$ and $f(x) = 16(4^{-x})$

$\quad = 4^2(4^{-x}) \qquad\qquad = 16(2^2)^{-x}$

$\quad = 4^{2-x} \qquad\qquad\quad = 16(2^{-2x})$

$\quad = \left(\dfrac{1}{4} \right)^{-(2-x)} \qquad = h(x)$

$\quad = \left(\dfrac{1}{4} \right)^{x-2}$

$\quad = g(x)$

Thus, $f(x) = g(x) = h(x)$.

76. $f(x) = 5^{-x} + 3$

$g(x) = 5^{3-x} = 5^3 \cdot 5^{-x}$

$h(x) = -5^{x-3} = -(5^x \cdot 5^{-3})$

Thus, none are equal.

77. $y = 3^x$ and $y = 4^x$

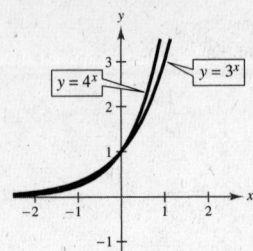

x	-2	-1	0	1	2
3^x	$\frac{1}{9}$	$\frac{1}{3}$	1	3	9
4^x	$\frac{1}{16}$	$\frac{1}{4}$	1	4	16

(a) $4^x < 3^x$ when $x < 0$.

(b) $4^x > 3^x$ when $x > 0$.

78. (a) $f(x) = x^2 e^{-x}$ (b) $g(x) = x2^{3-x}$

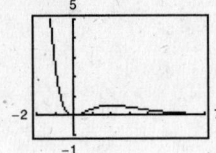   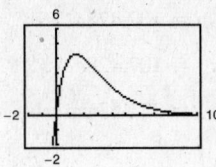

Decreasing: $(-\infty, 0),\ (2, \infty)$ Decreasing: $(1.44, \infty)$

Increasing: $(0, 2)$ Increasing: $(-\infty, 1.44)$

Relative maximum: $(2, 4e^{-2})$ Relative maximum: $(1.44, 4.25)$

Relative minimum: $(0, 0)$

79. $f(x) = \left(1 + \dfrac{0.5}{x}\right)^x$ and $g(x) = e^{0.5}$ (Horizontal line)   **80.** The functions (c) 3^x and (d) 2^{-x} are exponential.

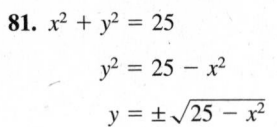

As $x \to \infty,\ f(x) \to g(x)$.

As $x \to -\infty,\ f(x) \to g(x)$.

81. $x^2 + y^2 = 25$

$\qquad y^2 = 25 - x^2$

$\qquad y = \pm\sqrt{25 - x^2}$

82. $x - |y| = 2$

$\qquad x - 2 = |y|$

$\qquad y = x - 2$ and $y = -(x - 2),\ x \geq 2$

83. $f(x) = \dfrac{2}{9 + x}$

Vertical asymptote: $x = -9$

Horizontal asymptote: $y = 0$

x	-11	-10	-8	-7
$f(x)$	-1	-2	2	1

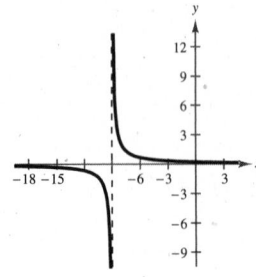

84. $f(x) = \sqrt{7 - x}$

Domain: $(-\infty, 7]$

x	-9	-2	3	6	7
y	4	3	2	1	0

85. Answers will vary.

Section 3.2   Logarithmic Functions and Their Graphs

- You should know that a function of the form $y = \log_a x$, where $a > 0$, $a \neq 1$, and $x > 0$, is called a logarithm of x to base a.

- You should be able to convert from logarithmic form to exponential form and vice versa.

  $$y = \log_a x \iff a^y = x$$

- You should know the following properties of logarithms.

  (a) $\log_a 1 = 0$ since $a^0 = 1$.

  (b) $\log_a a = 1$ since $a^1 = a$.

  (c) $\log_a a^x = x$ since $a^x = a^x$.

  (d) $a^{\log_a x} = x$ Inverse Property

  (e) If $\log_a x = \log_a y$, then $x = y$.

- You should know the definition of the natural logarithmic function.

  $$\log_e x = \ln x, \, x > 0$$

- You should know the properties of the natural logarithmic function.

  (a) $\ln 1 = 0$ since $e^0 = 1$.

  (b) $\ln e = 1$ since $e^1 = e$.

  (c) $\ln e^x = x$ since $e^x = e^x$.

  (d) $e^{\ln x} = x$ Inverse Property

  (e) If $\ln x = \ln y$, then $x = y$.

- You should be able to graph logarithmic functions.

Vocabulary Check

1. logarithmic

2. 10

3. natural; e

4. $a^{\log_a x} = x$

5. $x = y$

---

1. $\log_4 64 = 3 \implies 4^3 = 64$

2. $\log_3 81 = 4 \implies 3^4 = 81$

3. $\log_7 \frac{1}{49} = -2 \implies 7^{-2} = \frac{1}{49}$

4. $\log \frac{1}{1000} = -3 \implies 10^{-3} = \frac{1}{1000}$

5. $\log_{32} 4 = \frac{2}{5} \implies 32^{2/5} = 4$

6. $\log_{16} 8 = \frac{3}{4} \implies 16^{3/4} = 8$

7. $\log_{36} 6 = \frac{1}{2} \implies 36^{1/2} = 6$

8. $\log_8 4 = \frac{2}{3} \implies 8^{2/3} = 4$

9. $5^3 = 125 \implies \log_5 125 = 3$

10. $8^2 = 64 \implies \log_8 64 = 2$

11. $81^{1/4} = 3 \implies \log_{81} 3 = \frac{1}{4}$

12. $9^{3/2} = 27 \implies \log_9 27 = \frac{3}{2}$

13. $6^{-2} = \frac{1}{36} \implies \log_6 \frac{1}{36} = -2$

14. $4^{-3} = \frac{1}{64} \implies \log_4 \frac{1}{64} = -3$

15. $7^0 = 1 \implies \log_7 1 = 0$

16. $10^{-3} = 0.001 \implies \log_{10} 0.001 = -3$

17. $f(x) = \log_2 x$

$f(16) = \log_2 16 = 4$ since $2^4 = 16$

18. $f(x) = \log_{16} x$

$f(4) = \log_{16} 4 = \frac{1}{2}$ since $16^{1/2} = 4$

19. $f(x) = \log_7 x$

$f(1) = \log_7 1 = 0$ since $7^0 = 1$

20. $f(x) = \log x$

$f(10) = \log 10 = 1$ since $10^1 = 10$

21. $g(x) = \log_a x$

$g(a^2) = \log_a a^2$

$= 2$ by the Inverse Property

22. $g(x) = \log_b x$

$g(b^{-3}) = \log_b b^{-3} = -3$ since

$b^{-3} = b^{-3}$

23. $f(x) = \log x$

$f\left(\frac{4}{5}\right) = \log\left(\frac{4}{5}\right) \approx -0.097$

24. $f(x) = \log x$

$f\left(\frac{1}{500}\right) = \log \frac{1}{500} \approx -2.699$

25. $f(x) = \log x$

$f(12.5) \approx 1.097$

26. $f(x) = \log x$

$f(75.25) \approx 1.877$

27. $\log_3 3^4 = 4$ since $3^4 = 3^4$

28. $\log_{1.5} 1$

Since $1.5^0 = 1$, $\log_{1.5} 1 = 0$.

29. $\log_\pi \pi = 1$ since $\pi^1 = \pi$.

30. $9^{\log_9 15}$

Since $a^{\log_a x} = x$, $9^{\log_9 15} = 15$.

31. $f(x) = \log_4 x$

Domain: $x > 0 \implies$ The domain is $(0, \infty)$.

x-intercept: $(1, 0)$

Vertical asymptote: $x = 0$

$y = \log_4 x \implies 4^y = x$

x	$\frac{1}{4}$	1	4	2
$f(x)$	-1	0	1	$\frac{1}{2}$

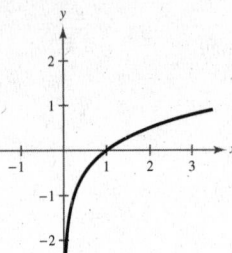

32. $g(x) = \log_6 x$

Domain: $(0, \infty)$

x-intercept: $(1, 0)$

Vertical asymptote: $x = 0$

$y = \log_6 x \implies 6^y = x$

x	$\frac{1}{6}$	1	$\sqrt{6}$	6
y	-1	0	$\frac{1}{2}$	1

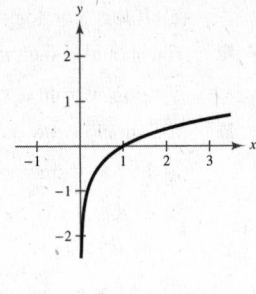

33. $y = -\log_3 x + 2$

Domain: $(0, \infty)$

x-intercept:

$-\log_3 x + 2 = 0$

$2 = \log_3 x$

$3^2 = x$

$9 = x$

The x-intercept is $(9, 0)$.

Vertical asymptote: $x = 0$

$y = -\log_3 x + 2$

$\log_3 x = 2 - y \implies 3^{2-y} = x$

x	27	9	3	1	$\frac{1}{3}$
y	-1	0	1	2	3

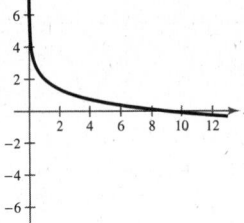

34. $h(x) = \log_4(x - 3)$

Domain: $x - 3 > 0 \implies x > 3$

The domain is $(3, \infty)$.

x-intercept:

$\log_4(x - 3) = 0$

$4^0 = x - 3$

$1 = x - 3$

$4 = x$

The x-intercept is $(4, 0)$.

Vertical asymptote: $x - 3 = 0 \implies x = 3$

$y = \log_4(x - 3) \implies 4^y + 3 = x$

x	$3\frac{1}{4}$	4	7	19
y	-1	0	1	2

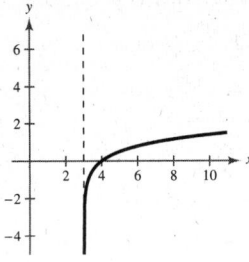

35. $f(x) = -\log_6(x + 2)$

Domain: $x + 2 > 0 \implies x > -2$

The domain is $(-2, \infty)$.

x-intercept:

$$0 = -\log_6(x + 2)$$

$$0 = \log_6(x + 2)$$

$$6^0 = x + 2$$

$$1 = x + 2$$

$$-1 = x$$

The x-intercept is $(-1, 0)$.

Vertical asymptote: $x + 2 = 0 \implies x = -2$

$$y = -\log_6(x + 2)$$

$$-y = \log_6(x + 2)$$

$$6^{-y} - 2 = x$$

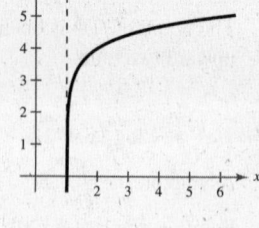

x	4	-1	$-1\frac{5}{6}$	$-1\frac{35}{36}$
$f(x)$	-1	0	1	2

36. $y = \log_5(x - 1) + 4$

Domain: $x - 1 > 0 \implies x > 1$

The domain is $(1, \infty)$.

x-intercept:

$$\log_5(x - 1) + 4 = 0$$

$$\log_5(x - 1) = -4$$

$$5^{-4} = x - 1$$

$$\frac{1}{625} = x - 1$$

$$\frac{626}{625} = x$$

The x-intercept is $\left(\frac{626}{625}, 0\right)$.

Vertical asymptote: $x - 1 = 0 \implies x = 1$

$$y = \log_5(x - 1) + 4 \implies 5^{y-4} + 1 = x$$

x	1.00032	1.0016	1.008	1.04	1.2
y	-1	0	1	2	3

37. $y = \log\left(\frac{x}{5}\right)$

Domain: $\frac{x}{5} > 0 \implies x > 0$

The domain is $(0, \infty)$.

x-intercept:

$$\log\left(\frac{x}{5}\right) = 0$$

$$\frac{x}{5} = 10^0$$

$$\frac{x}{5} = 1 \implies x = 5$$

The x-intercept is $(5, 0)$.

Vertical asymptote: $\frac{x}{5} = 0 \implies x = 0$

The vertical asymptote is the y-axis.

x	1	2	3	4	5	6	7
y	-0.70	-0.40	-0.22	-0.10	0	0.08	0.15

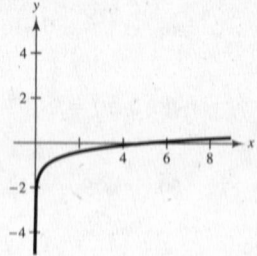

38. $y = \log(-x)$

Domain: $-x > 0 \implies x < 0$

The domain is $(-\infty, 0)$.

x-intercept: $\log(-x) = 0$

$$10^0 = -x$$

$$-1 = x$$

The x-intercept is $(-1, 0)$.

Vertical asymptote: $x = 0$

$$y = \log(-x) \implies -10^y = x$$

x	$-\frac{1}{100}$	$-\frac{1}{10}$	-1	-10
y	-2	-1	0	1

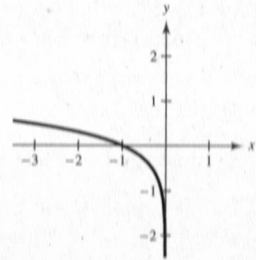

39. $f(x) = \log_3 x + 2$

Asymptote: $x = 0$

Point on graph: $(1, 2)$

Matches graph (c).

The graph of $f(x)$ is obtained by shifting the graph of $g(x)$ upward two units.

40. $f(x) = -\log_3 x$

Asymptote: $x = 0$

Point on graph: $(1, 0)$

Matches graph (f).

$f(x)$ reflects $g(x)$ in the x-axis.

41. $f(x) = -\log_3(x + 2)$

Asymptote: $x = -2$

Point on graph: $(-1, 0)$

Matches graph (d).

The graph of $f(x)$ is obtained by reflecting the graph of $g(x)$ about the x-axis and shifting the graph two units to the left.

42. $f(x) = \log_3(x - 1)$

Asymptote: $x = 1$

Point on graph: $(2, 0)$

Matches graph (e).

$f(x)$ shifts $g(x)$ one unit to the right.

43. $f(x) = \log_3(1 - x) = \log_3[-(x - 1)]$

Asymptote: $x = 1$

Point on graph: $(0, 0)$

Matches graph (b).

The graph of $f(x)$ is obtained by reflecting the graph of $g(x)$ about the y-axis and shifting the graph one unit to the right.

44. $f(x) = -\log_3(-x)$

Asymptote: $x = 0$

Point on graph: $(-1, 0)$

Matches graph (a).

$f(x)$ reflects $g(x)$ in the x-axis then reflects that graph in the y-axis.

45. $\ln \frac{1}{2} = -0.693 \ldots \implies e^{-0.693 \ldots} = \frac{1}{2}$

46. $\ln \frac{2}{5} = -0.916 \ldots \implies e^{-0.916 \ldots} = \frac{2}{5}$

47. $\ln 4 = 1.386 \ldots \implies e^{1.386 \ldots} = 4$

48. $\ln 10 = 2.302 \ldots \implies e^{2.302 \ldots} = 10$

49. $\ln 250 = 5.521 \ldots \implies e^{5.521 \ldots} = 250$

50. $\ln 679 = 6.520 \ldots \implies e^{6.520 \ldots} = 679$

51. $\ln 1 = 0 \implies e^0 = 1$

52. $\ln e = 1 \implies e^1 = e$

53. $e^3 = 20.0855 \ldots \implies \ln 20.0855 \ldots = 3$

54. $e^2 = 7.3890 \ldots \implies \ln 7.3890 \ldots = 2$

55. $e^{1/2} = 1.6487 \ldots \implies \ln 1.6487 \ldots = \frac{1}{2}$

56. $e^{1/3} = 1.3956 \ldots \implies \ln 1.3956 \ldots = \frac{1}{3}$

57. $e^{-0.5} = 0.6065 \ldots \implies \ln 0.6065 \ldots = -0.5$

58. $e^{-4.1} = 0.0165 \ldots \implies \ln 0.0165 \ldots = -4.1$

59. $e^x = 4 \implies \ln 4 = x$

60. $e^{2x} = 3 \implies \ln 3 = 2x$

61. $f(x) = \ln x$

$f(18.42) = \ln 18.42 \approx 2.913$

62. $f(x) = 3 \ln x$

$f(0.32) = 3 \ln 0.32 \approx -3.418$

63. $g(x) = 2 \ln x$

$g(0.75) = 2 \ln 0.75 \approx -0.575$

64. $g(x) = -\ln x$

$g\left(\frac{1}{2}\right) = -\ln \frac{1}{2} \approx 0.693$

65. $g(x) = \ln x$

$g(e^3) = \ln e^3 = 3$ by the Inverse Property

66. $g(x) = \ln x$

$g(e^{-2}) = \ln e^{-2} = -2$

67. $g(x) = \ln x$

$g(e^{-2/3}) = \ln e^{-2/3} = -\frac{2}{3}$ by the Inverse Property

68. $g(x) = \ln x$

$g(e^{-5/2}) = \ln e^{-5/2} = -\frac{5}{2}$

69. $f(x) = \ln(x - 1)$

Domain: $x - 1 > 0 \Rightarrow x > 1$

The domain is $(1, \infty)$.

x-intercept:

$0 = \ln(x - 1)$

$e^0 = x - 1$

$2 = x$

The x-intercept is $(2, 0)$.

Vertical asymptote: $x - 1 = 0 \Rightarrow x = 1$

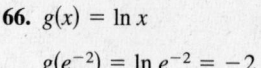

x	1.5	2	3	4
$f(x)$	-0.69	0	0.69	1.10

70. $h(x) = \ln(x + 1)$

Domain: $x + 1 > 0 \Rightarrow x > -1$

The domain is $(-1, \infty)$.

x-intercept:

$\ln(x + 1) = 0$

$e^0 = x + 1$

$1 = x + 1$

$0 = x$

The x-intercept is $(0, 0)$.

Vertical asymptote: $x + 1 = 0 \Rightarrow x = -1$

$y = \ln(x + 1) \Rightarrow e^y - 1 = x$

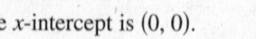

x	-0.39	0	1.72	6.39	19.09
y	$-\frac{1}{2}$	0	1	2	3

71. $g(x) = \ln(-x)$

Domain: $-x > 0 \Rightarrow x < 0$

The domain is $(-\infty, 0)$.

x-intercept:

$0 = \ln(-x)$

$e^0 = -x$

$-1 = x$

The x-intercept is $(-1, 0)$.

Vertical asymptote: $-x = 0 \Rightarrow x = 0$

x	-0.5	-1	-2	-3
$g(x)$	-0.69	0	0.69	1.10

72. $f(x) = \ln(3 - x)$

Domain: $3 - x > 0 \Rightarrow x < 3$

The domain is $(-\infty, 3)$.

x-intercept:

$\ln(3 - x) = 0$

$e^0 = 3 - x$

$1 = 3 - x$

$2 = x$

The x-intercept is $(2, 0)$.

Vertical asymptote: $3 - x = 0 \Rightarrow x = 3$

$y = \ln(3 - x) \Rightarrow 3 - e^y = x$

x	2.95	2.86	2.63	2	0.28
y	-3	-2	-1	0	1

73. $y_1 = \log(x + 1)$

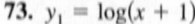

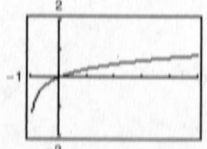

74. $f(x) = \log(x - 1)$

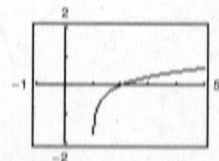

75. $y_1 = \ln(x - 1)$

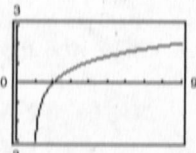

76. $f(x) = \ln(x + 2)$

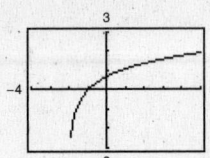

77. $y = \ln x + 2$

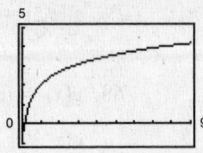

78. $f(x) = 3 \ln x - 1$

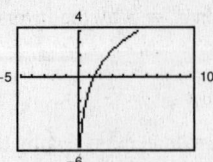

79. $\log_2(x + 1) = \log_2 4$

$x + 1 = 4$

$x = 3$

80. $\log_2(x - 3) = \log_2 9$

$x - 3 = 9$

$x = 12$

81. $\log(2x + 1) = \log 15$

$2x + 1 = 15$

$x = 7$

82. $\log(5x + 3) = \log 12$

$5x + 3 = 12$

$5x = 9$

$x = \frac{9}{5}$

83. $\ln(x + 2) = \ln 6$

$x + 2 = 6$

$x = 4$

84. $\ln(x - 4) = \ln 2$

$x - 4 = 2$

$x = 6$

85. $\ln(x^2 - 2) = \ln 23$

$x^2 - 2 = 23$

$x^2 = 25$

$x = \pm 5$

86. $\ln(x^2 - x) = \ln 6$

$x^2 - x = 6$

$x^2 - x - 6 = 0$

$(x - 3)(x + 2) = 0$

$x = -2 \text{ or } x = 3$

87. $t = 12.542 \ln\left(\dfrac{x}{x - 1000}\right), x > 1000$

(a) When $x = \$1100.65$:

$t = 12.542 \ln\left(\dfrac{1100.65}{1100.65 - 1000}\right) \approx 30 \text{ years}$

When $x = \$1254.68$:

$t = 12.542 \ln\left(\dfrac{1254.68}{1254.68 - 1000}\right) \approx 20 \text{ years}$

(b) Total amounts: $(1100.65)(12)(30) = \$396,234.00$

$(1254.68)(12)(20) = \$301,123.20$

(c) Interest charges: $396,234 - 150,000 = \$246,234$

$301,123.20 - 150,000 = \$151,123.20$

(d) The vertical asymptote is $x = 1000$. The closer the payment is to $1000 per month, the longer the length of the mortgage will be. Also, the monthly payment must be greater than $1000.

88. $t = \dfrac{\ln K}{0.095}$

(a)

K	1	2	4	6	8	10	12
t	0	7.3	14.6	18.9	21.9	24.2	26.2

The number of years required to multiply the original investment by K increases with K. However, the larger the value of K, the fewer the years required to increase the value of the investment by an additional multiple of the original investment.

(b)

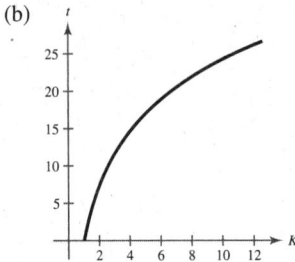

89. $f(t) = 80 - 17 \log(t + 1), 0 \le t \le 12$

(a)

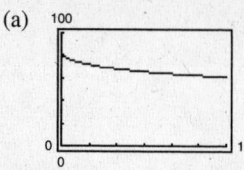

(b) $f(0) = 80 - 17 \log 1 = 80.0$

(c) $f(4) = 80 - 17 \log 5 \approx 68.1$

(d) $f(10) = 80 - 17 \log 11 \approx 62.3$

90. $\beta = 10 \log\left(\dfrac{I}{10^{-12}}\right)$

(a) $\beta = 10 \log\left(\dfrac{1}{10^{-12}}\right) = 10 \log(10^{12}) = 120$ decibels

(b) $\beta = 10 \log\left(\dfrac{10^{-2}}{10^{-12}}\right) = 10 \log(10^{10}) = 100$ decibels

(c) No, the difference is due to the logarithmic relationship between intensity and number of decibels.

91. False. Reflecting $g(x)$ about the line $y = x$ will determine the graph of $f(x)$.

92. True, $\log_3 27 = 3 \implies 3^3 = 27$.

93. $f(x) = 3^x$, $g(x) = \log_3 x$

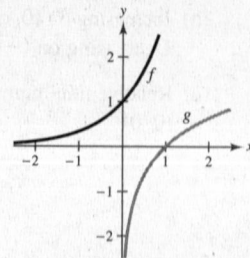

f and g are inverses. Their graphs are reflected about the line $y = x$.

94. $f(x) = 5^x$, $g(x) = \log_5 x$

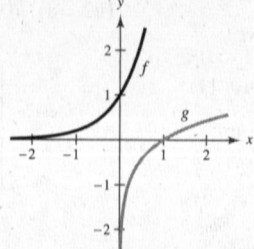

f and g are inverses. Their graphs are reflected about the line $y = x$.

95. $f(x) = e^x$, $g(x) = \ln x$

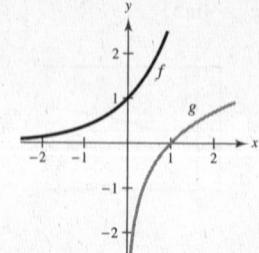

f and g are inverses. Their graphs are reflected about the line $y = x$.

96. $f(x) = 10^x$, $g(x) = \log_{10} x$

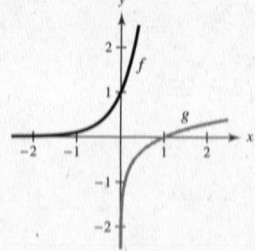

f and g are inverses. Their graphs are reflected about the line $y = x$.

97. (a) $f(x) = \ln x$, $g(x) = \sqrt{x}$

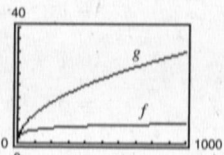

The natural log function grows at a slower rate than the square root function.

(b) $f(x) = \ln x$, $g(x) = \sqrt[4]{x}$

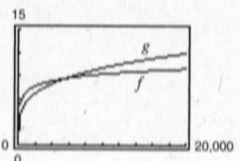

The natural log function grows at a slower rate than the fourth root function.

98. $f(x) = \dfrac{\ln x}{x}$

(a)

x	1	5	10	10^2	10^4	10^6
$f(x)$	0	0.322	0.230	0.046	0.00092	0.0000138

(b) As $x \to \infty$, $f(x) \to 0$.

(c)

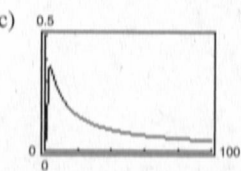

99. (a) False. If y were an exponential function of x, then $y = a^x$, but $a^1 = a$, not 0. Because one point is $(1, 0)$, y is not an exponential function of x.

(c) True. $x = a^y$

For $a = 2$, $x = 2^y$.

$y = 0, 2^0 = 1$

$y = 1, 2^1 = 2$

$y = 3, 2^3 = 8$

(b) True. $y = \log_a x$

For $a = 2$, $y = \log_2 x$.

$x = 1, \log_2 1 = 0$

$x = 2, \log_2 2 = 1$

$x = 8, \log_2 8 = 3$

(d) False. If y were a linear function of x, the slope between $(1, 0)$ and $(2, 1)$ and the slope between $(2, 1)$ and $(8, 3)$ would be the same. However,

$$m_1 = \frac{1 - 0}{2 - 1} = 1 \text{ and } m_2 = \frac{3 - 1}{8 - 2} = \frac{2}{6} = \frac{1}{3}.$$

Therefore, y is not a linear function of x.

100. $y = \log_a x \Rightarrow a^y = x$, so, for example, if $a = -2$, there is no value of y for which $(-2)^y = -4$. If $a = 1$, then every power of a is equal to 1, so x could only be 1. So, $\log_a x$ is defined only for $0 < a < 1$ and $a > 1$.

101. $f(x) = |\ln x|$

(a)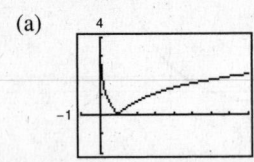

(b) Increasing on $(1, \infty)$
Decreasing on $(0, 1)$

(c) Relative minimum:
$(1, 0)$

102. (a) $h(x) = \ln(x^2 + 1)$

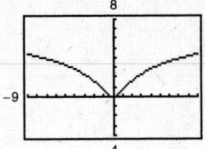

(b) Increasing on $(0, \infty)$
Decreasing on $(-\infty, 0)$

(c) Relative minimum:
$(0, 0)$

For Exercises 103–108, use $f(x) = 3x + 2$ and $g(x) = x^3 - 1$.

103. $(f + g)(2) = f(2) + g(2)$

$= [3(2) + 2] + [(2)^3 - 1]$

$= 8 + 7$

$= 15$

104. $f(x) - g(x) = 3x + 2 - (x^3 - 1)$

$= 3x + 2 - x^3 + 1$

$= 3x - x^3 + 3$

Therefore,

$(f - g)(-1) = 3(-1) - (-1)^3 + 3$

$= -3 + 1 + 3$

$= 1.$

105. $(fg)(6) = f(6)g(6)$

$= [3(6) + 2][(6)^3 - 1]$

$= (20)(215)$

$= 4300$

106. $\dfrac{f(x)}{g(x)} = \dfrac{3x + 2}{x^3 - 1}$

Therefore, $\left(\dfrac{f}{g}\right)(0) = \dfrac{3 \cdot 0 + 2}{0^3 - 1} = -2.$

107. $(f \circ g)(7) = f(g(7))$

$= f((7)^3 - 1)$

$= f(342)$

$= 3(342) + 2$

$= 1028$

108. $(g \circ f)(x) = g(f(x)) = g(3x + 2) = (3x + 2)^3 - 1$

Therefore,

$(g \circ f)(-3) = (3 \cdot (-3) + 2)^3 - 1$

$= -7^3 - 1 = -344.$

Section 3.3   Properties of Logarithms

- You should know the following properties of logarithms.

  (a) $\log_a x = \dfrac{\log_b x}{\log_b a}$ $\log_a x = \dfrac{\log_{10} x}{\log_{10} a}$ $\log_a x = \dfrac{\ln x}{\ln a}$

  (b) $\log_a(uv) = \log_a u + \log_a v$ $\ln(uv) = \ln u + \ln v$

  (c) $\log_a\left(\dfrac{u}{v}\right) = \log_a u - \log_a v$ $\ln\left(\dfrac{u}{v}\right) = \ln u - \ln v$

  (d) $\log_a u^n = n \log_a u$ $\ln u^n = n \ln u$

- You should be able to rewrite logarithmic expressions using these properties.

Vocabulary Check

1. change-of-base

2. $\dfrac{\log x}{\log a} = \dfrac{\ln x}{\ln a}$

3. $\log_a(uv) = \log_a u + \log_a v$
This is the Product Property. Matches (c).

4. $\ln u^n = n \ln u$
This is the Power Property. Matches (a).

5. $\log_a \dfrac{u}{v} = \log_a u - \log_a v$

This is the Quotient Property. Matches (b).

---

1. (a) $\log_5 x = \dfrac{\log x}{\log 5}$

(b) $\log_5 x = \dfrac{\ln x}{\ln 5}$

2. (a) $\log_3 x = \dfrac{\log x}{\log 3}$

(b) $\log_3 x = \dfrac{\ln x}{\ln 3}$

3. (a) $\log_{1/5} x = \dfrac{\log x}{\log(1/5)}$

(b) $\log_{1/5} x = \dfrac{\ln x}{\ln(1/5)}$

4. (a) $\log_{1/3} x = \dfrac{\log x}{\log(1/3)}$

(b) $\log_{1/3} x = \dfrac{\ln x}{\ln(1/3)}$

5. (a) $\log_x \dfrac{3}{10} = \dfrac{\log(3/10)}{\log x}$

(b) $\log_x \dfrac{3}{10} = \dfrac{\ln(3/10)}{\ln x}$

6. (a) $\log_x \dfrac{3}{4} = \dfrac{\log(3/4)}{\log x}$

(b) $\log_x \dfrac{3}{4} = \dfrac{\ln(3/4)}{\ln x}$

7. (a) $\log_{2.6} x = \dfrac{\log x}{\log 2.6}$

(b) $\log_{2.6} x = \dfrac{\ln x}{\ln 2.6}$

8. (a) $\log_{7.1} x = \dfrac{\log x}{\log 7.1}$

(b) $\log_{7.1} x = \dfrac{\ln x}{\ln 7.1}$

9. $\log_3 7 = \dfrac{\log 7}{\log 3} = \dfrac{\ln 7}{\ln 3} \approx 1.771$

10. $\log_7 4 = \dfrac{\log 4}{\log 7} = \dfrac{\ln 4}{\ln 7} \approx 0.712$

11. $\log_{1/2} 4 = \dfrac{\log 4}{\log(1/2)} = \dfrac{\ln 4}{\ln(1/2)} = -2.000$

12. $\log_{1/4} 5 = \dfrac{\log 5}{\log(1/4)} = \dfrac{\ln 5}{\ln(1/4)} \approx -1.161$

13. $\log_9(0.4) = \dfrac{\log 0.4}{\log 9} = \dfrac{\ln 0.4}{\ln 9} \approx -0.417$

14. $\log_{20} 0.125 = \dfrac{\log 0.125}{\log 20} = \dfrac{\ln 0.125}{\ln 20} \approx -0.694$

15. $\log_{15} 1250 = \dfrac{\log 1250}{\log 15} = \dfrac{\ln 1250}{\ln 15} \approx 2.633$

16. $\log_3 0.015 = \dfrac{\log 0.015}{\log 3} = \dfrac{\ln 0.015}{\ln 3} \approx -3.823$

17. $\log_4 8 = \dfrac{\log_2 8}{\log_2 4} = \dfrac{\log_2 2^3}{\log_2 2^2} = \dfrac{3}{2}$

18. $\log_2(4^2 \cdot 3^4) = \log_2 4^2 + \log_2 3^4$

$= 2 \log_2 4 + 4 \log_2 3$

$= 2 \log_2 2^2 + 4 \log_2 3$

$= 4 \log_2 2 + 4 \log_2 3$

$= 4 + 4 \log_2 3$

19. $\log_5 \frac{1}{250} = \log_5\left(\frac{1}{125} \cdot \frac{1}{2}\right)$

$= \log_5 \frac{1}{125} + \log_5 \frac{1}{2}$

$= \log_5 5^{-3} + \log_5 2^{-1}$

$= -3 - \log_5 2$

20. $\log \frac{9}{300} = \log \frac{3}{100}$

$= \log 3 - \log 100$

$= \log 3 - \log 10^2$

$= \log 3 - 2 \log 10$

$= \log 3 - 2$

21. $\ln(5e^6) = \ln 5 + \ln e^6$

$= \ln 5 + 6$

$= 6 + \ln 5$

22. $\ln \frac{6}{e^2} = \ln 6 - \ln e^2$

$= \ln 6 - 2 \ln e$

$= \ln 6 - 2$

23. $\log_3 9 = 2 \log_3 3 = 2$

24. $\log_5 \frac{1}{125} = \log_5 5^{-3} = -3 \log_5 5 = -3(1) = -3$

25. $\log_2 \sqrt[4]{8} = \frac{1}{4} \log_2 2^3 = \frac{3}{4} \log_2 2 = \frac{3}{4}(1) = \frac{3}{4}$

26. $\log_6 \sqrt[3]{6} = \log_6 6^{1/3} = \frac{1}{3} \log_6 6 = \frac{1}{3}(1) = \frac{1}{3}$

27. $\log_4 16^{1.2} = 1.2(\log_4 16) = 1.2 \log_4 4^2 = 1.2(2) = 2.4$

28. $\log_3 81^{-0.2} = -0.2 \log_3 81$

$= -0.2 \log_3 3^4$

$= -0.2(4) = -0.8$

29. $\log_3(-9)$ is undefined. -9 is not in the domain of $\log_3 x$.

30. $\log_2(-16)$ is undefined because -16 is not in the domain of $\log_2 x$.

31. $\ln e^{4.5} = 4.5$

32. $3 \ln e^4 = (3)(4) \ln e$

$= 12(1)$

$= 12$

33. $\ln \frac{1}{\sqrt{e}} = \ln 1 - \ln \sqrt{e}$

$= 0 - \frac{1}{2} \ln e$

$= 0 - \frac{1}{2}(1)$

$= -\frac{1}{2}$

34. $\ln \sqrt[4]{e^3} = \ln e^{3/4}$

$= \frac{3}{4} \ln e$

$= \frac{3}{4}(1)$

$= \frac{3}{4}$

35. $\ln e^2 + \ln e^5 = 2 + 5 = 7$

36. $2 \ln e^6 - \ln e^5 = \ln e^{12} - \ln e^5$

$= \ln \frac{e^{12}}{e^5}$

$= \ln e^7$

$= 7$

37. $\log_5 75 - \log_5 3 = \log_5 \frac{75}{3}$

$= \log_5 25$

$= \log_5 5^2$

$= 2 \log_5 5$

$= 2$

38. $\log_4 2 + \log_4 32 = \log_4 4^{1/2} + \log_4 4^{5/2}$

$= \frac{1}{2} \log_4 4 + \frac{5}{2} \log_4 4$

$= \frac{1}{2}(1) + \frac{5}{2}(1)$

$= 3$

39. $\log_4 5x = \log_4 5 + \log_4 x$

40. $\log_3 10z = \log_3 10 + \log_3 z$

41. $\log_8 x^4 = 4 \log_8 x$

42. $\log \dfrac{y}{2} = \log y - \log 2$

43. $\log_5 \dfrac{5}{x} = \log_5 5 - \log_5 x$

$\quad = 1 - \log_5 x$

44. $\log_6 z^{-3} = -3 \log_6 z$

45. $\ln \sqrt{z} = \ln z^{1/2} = \dfrac{1}{2} \ln z$

46. $\ln \sqrt[3]{t} = \ln t^{1/3} = \dfrac{1}{3} \ln t$

47. $\ln xyz^2 = \ln x + \ln y + \ln z^2$

$\quad = \ln x + \ln y + 2 \ln z$

48. $\log 4x^2 y = \log 4 + \log x^2 + \log y$

$\quad = \log 4 + 2 \log x + \log y$

49. $\ln z(z - 1)^2 = \ln z + \ln(z - 1)^2$

$\quad = \ln z + 2 \ln(z - 1), \ z > 1$

50. $\ln\left(\dfrac{x^2 - 1}{x^3}\right) = \ln(x^2 - 1) - \ln x^3$

$\quad = \ln[(x + 1)(x - 1)] - \ln x^3$

$\quad = \ln(x + 1) + \ln(x - 1) - 3 \ln x$

51. $\log_2 \dfrac{\sqrt{a - 1}}{9} = \log_2 \sqrt{a - 1} - \log_2 9$

$\quad = \dfrac{1}{2} \log_2(a - 1) - \log_2 3^2$

$\quad = \dfrac{1}{2} \log_2(a - 1) - 2 \log_2 3, \ a > 1$

52. $\ln \dfrac{6}{\sqrt{x^2 + 1}} = \ln 6 - \ln \sqrt{x^2 + 1}$

$\quad = \ln 6 - \ln(x^2 + 1)^{1/2}$

$\quad = \ln 6 - \dfrac{1}{2} \ln(x^2 + 1)$

53. $\ln \sqrt[3]{\dfrac{x}{y}} = \dfrac{1}{3} \ln \dfrac{x}{y}$

$\quad = \dfrac{1}{3}[\ln x - \ln y]$

$\quad = \dfrac{1}{3} \ln x - \dfrac{1}{3} \ln y$

54. $\ln \sqrt{\dfrac{x^2}{y^3}} = \ln\left(\dfrac{x^2}{y^3}\right)^{1/2} = \dfrac{1}{2} \ln\left(\dfrac{x^2}{y^3}\right)$

$\quad = \dfrac{1}{2}(\ln x^2 - \ln y^3)$

$\quad = \dfrac{1}{2}(2 \ln x - 3 \ln y)$

$\quad = \ln x - \dfrac{3}{2} \ln y$

55. $\ln\left(\dfrac{x^4 \sqrt{y}}{z^5}\right) = \ln x^4 \sqrt{y} - \ln z^5$

$\quad = \ln x^4 + \ln \sqrt{y} - \ln z^5$

$\quad = 4 \ln x + \dfrac{1}{2} \ln y - 5 \ln z$

56. $\log_2 \dfrac{\sqrt{x} y^4}{z^4} = \log_2 \sqrt{x} y^4 - \log_2 z^4$

$\quad = \log_2 \sqrt{x} + \log_2 y^4 - \log_2 z^4$

$\quad = \dfrac{1}{2} \log_2 x + 4 \log_2 y - 4 \log_2 z$

57. $\log_5\left(\dfrac{x^2}{y^2 z^3}\right) = \log_5 x^2 - \log_5 y^2 z^3$

$\quad = \log_5 x^2 - (\log_5 y^2 + \log_5 z^3)$

$\quad = 2 \log_5 x - 2 \log_5 y - 3 \log_5 z$

58. $\log \dfrac{xy^4}{z^5} = \log xy^4 - \log z^5$

$\quad = \log x + \log y^4 - \log z^5$

$\quad = \log x + 4 \log y - 5 \log z$

59. $\ln \sqrt[4]{x^3(x^2 + 3)} = \dfrac{1}{4} \ln x^3(x^2 + 3)$

$\quad = \dfrac{1}{4}[\ln x^3 + \ln(x^2 + 3)]$

$\quad = \dfrac{1}{4}[3 \ln x + \ln(x^2 + 3)]$

$\quad = \dfrac{3}{4} \ln x + \dfrac{1}{4} \ln(x^2 + 3)$

60. $\ln \sqrt{x^2(x + 2)} = \ln[x^2(x + 2)]^{1/2}$

$\quad = \ln[x(x + 2)^{1/2}]$

$\quad = \ln x + \ln(x + 2)^{1/2}$

$\quad = \ln x + \dfrac{1}{2} \ln(x + 2)$

61. $\ln x + \ln 3 = \ln 3x$

62. $\ln y + \ln t = \ln yt = \ln ty$

63. $\log_4 z - \log_4 y = \log_4 \dfrac{z}{y}$

64. $\log_5 8 - \log_5 t = \log_5 \dfrac{8}{t}$

65. $2 \log_2(x + 4) = \log_2(x + 4)^2$

66. $\dfrac{2}{3} \log_7(z - 2) = \log_7(z - 2)^{2/3}$

67. $\dfrac{1}{4} \log_3 5x = \log_3(5x)^{1/4} = \log_3 \sqrt[4]{5x}$

68. $-4 \log_6 2x = \log_6(2x)^{-4} = \log_6 \dfrac{1}{16x^4}$

69. $\ln x - 3 \ln(x + 1) = \ln x - \ln(x + 1)^3$

$$= \ln \dfrac{x}{(x + 1)^3}$$

70. $2 \ln 8 + 5 \ln(z - 4) = \ln 8^2 + \ln(z - 4)^5$

$$= \ln 64 + \ln(z - 4)^5$$

$$= \ln 64(z - 4)^5$$

71. $\log x - 2 \log y + 3 \log z = \log x - \log y^2 + \log z^3$

$$= \log \dfrac{x}{y^2} + \log z^3 = \log \dfrac{xz^3}{y^2}$$

72. $3 \log_3 x + 4 \log_3 y - 4 \log_3 z = \log_3 x^3 + \log_3 y^4 - \log_3 z^4$

$$= \log_3 x^3 y^4 - \log_3 z^4$$

$$= \log_3 \dfrac{x^3 y^4}{z^4}$$

73. $\ln x - 4[\ln(x + 2) + \ln(x - 2)] = \ln x - 4 \ln(x + 2)(x - 2)$

$$= \ln x - 4 \ln(x^2 - 4)$$

$$= \ln x - \ln(x^2 - 4)^4$$

$$= \ln \dfrac{x}{(x^2 - 4)^4}$$

74. $4[\ln z + \ln(z + 5)] - 2 \ln(z - 5) = 4[\ln z(z + 5)] - \ln(z - 5)^2$

$$= \ln[z(z + 5)]^4 - \ln(z - 5)^2$$

$$= \ln \dfrac{z^4(z + 5)^4}{(z - 5)^2}$$

75. $\dfrac{1}{3}[2 \ln(x + 3) + \ln x - \ln(x^2 - 1)] = \dfrac{1}{3}[\ln(x + 3)^2 + \ln x - \ln(x^2 - 1)]$

$$= \dfrac{1}{3}[\ln x(x + 3)^2 - \ln(x^2 - 1)]$$

$$= \dfrac{1}{3} \ln \dfrac{x(x + 3)^2}{x^2 - 1}$$

$$= \ln \sqrt[3]{\dfrac{x(x + 3)^2}{x^2 - 1}}$$

76. $2[3 \ln x - \ln(x + 1) - \ln(x - 1)] = 2[\ln x^3 - \ln(x + 1) - \ln(x - 1)]$

$$= 2[\ln x^3 - [\ln(x + 1) + \ln(x - 1)]]$$

$$= 2[\ln x^3 - [\ln(x + 1)(x - 1)]]$$

$$= 2 \ln \dfrac{x^3}{x^2 - 1}$$

$$= \ln\left(\dfrac{x^3}{x^2 - 1}\right)^2$$

77. $\frac{1}{3}\left[\log_8 y + 2\log_8(y+4)\right] - \log_8(y-1) = \frac{1}{3}\left[\log_8 y + \log_8(y+4)^2\right] - \log_8(y-1)$

$$= \frac{1}{3}\log_8 y(y+4)^2 - \log_8(y-1)$$

$$= \log_8 \sqrt[3]{y(y+4)^2} - \log_8(y-1)$$

$$= \log_8\left(\frac{\sqrt[3]{y(y+4)^2}}{y-1}\right)$$

78. $\frac{1}{2}\left[\log_4(x+1) + 2\log_4(x-1)\right] + 6\log_4 x = \frac{1}{2}\left[\log_4(x+1) + \log_4(x-1)^2\right] + \log_4 x^6$

$$= \frac{1}{2}\left[\log_4(x+1)(x-1)^2\right] + \log_4 x^6$$

$$= \log_4\left[\sqrt{x+1}\,(x-1)\right] + \log_4 x^6$$

$$= \log_4\left[x^6(x-1)\sqrt{x+1}\right]$$

79. $\log_2 \frac{32}{4} = \log_2 32 - \log_2 4 \neq \frac{\log_2 32}{\log_2 4}$

The second and third expressions are equal by Property 2.

80. $\log_7 \sqrt{70} = \frac{1}{2}\log_7 70 = \frac{1}{2}\left[\log_7 7 + \log_7 10\right]$

$$= \frac{1}{2}\left[1 + \log_7 10\right]$$

$$= \frac{1}{2} + \frac{1}{2}\log_7 10$$

$$= \frac{1}{2} + \log_7 \sqrt{10} \text{ by Property 1 and Property 3}$$

81. $\beta = 10\log\left(\frac{I}{10^{-12}}\right)$

$$= 10\left[\log I - \log 10^{-12}\right]$$

$$= 10\left[\log I + 12\right]$$

$$= 120 + 10\log I$$

When $I = 10^{-6}$:

$\beta = 120 + 10\log 10^{-6}$

$$= 120 + 10(-6)$$

$$= 60 \text{ decibels}$$

82. $\beta = 10\log\left(\frac{I}{10^{-12}}\right)$

$\text{Difference} = 10\log\left(\frac{3.16 \times 10^{-5}}{10^{-12}}\right) - 10\log\left(\frac{1.26 \times 10^{-7}}{10^{-12}}\right)$

$$= 10(\log(3.16 \times 10^7) - \log(1.26 \times 10^5))$$

$$= 10\left(\log\left(\frac{3.16 \times 10^7}{1.26 \times 10^5}\right)\right)$$

$$= 10(\log(2.5079 \times 10^2))$$

$$= 10(\log(250.79))$$

$$= 24 \text{ dB}$$

83. $\beta = 120 + 10\log(2I)$

$$= 120 + 10(\log 2 + \log I)$$

$$= (120 + 10\log I) + 10\log 2$$

With both stereos playing, the music is $10\log 2 \approx 3$ decibels louder.

84. $f(t) = 90 - 15 \log(t + 1),\ 0 \le t \le 12$

  (a) $f(t) = 90 - \log(t + 1)^{15}$

  (b) $f(0) = 90$

  (c) $f(4) = 90 - 15 \cdot \log(4 + 1) = 79.5$

  (d) $f(12) = 90 - 15 \cdot \log(12 + 1) = 73.3$

  (e)

  (f) The average score will be 75 when $t = 9$ months. See graph in (e).

  (g) $75 = 90 - 15 \log(t + 1)$

  $-15 = -15 \log(t + 1)$

  $1 = \log(t + 1)$

  $10^1 = t + 1$

  $t = 9$ months

85. By using the regression feature on a graphing calculator we obtain $y \approx 256.24 - 20.8 \ln x$.

86. (a)

  (b) $T - 21 = 54.4(0.964)^t$

  $T = 54.4(0.964)^t + 21$

  See graph in (a).

  (d) $\dfrac{1}{T - 21} = 0.0012t + 0.016$

  $T = \dfrac{1}{0.0012t + 0.016} + 21$

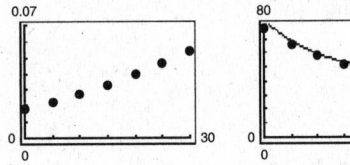

(c)

t (in minutes)	T (°C)	$T - 21$ (°C)	$\ln(T - 21)$	$1/(T - 21)$
0	78	57	4.043	0.0175
5	66	45	3.807	0.0222
10	57.5	36.5	3.597	0.0274
15	51.2	30.2	3.408	0.0331
20	46.3	25.3	3.231	0.0395
25	42.5	21.5	3.068	0.0465
30	39.6	18.6	2.923	0.0538

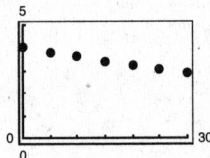

$\ln(T - 21) = -0.037t + 4$

$T = e^{-0.037t + 4} + 21$

This graph is identical to T in (b).

  (e) Since the scatter plot of the original data is so nicely exponential, there is no need to do the transformations unless one desires to deal with smaller numbers. The transformations did not make the problem simpler.

  Taking logs of temperatures led to a linear scatter plot because the log function increases very slowly as the x-values increase. Taking the reciprocals of the temperatures led to a linear scatter plot because of the asymptotic nature of the reciprocal function.

87. $f(x) = \ln x$

  False, $f(0) \ne 0$ since 0 is not in the domain of $f(x)$.

  $f(1) = \ln 1 = 0$

88. $f(ax) = f(a) + f(x),\ a > 0,\ x > 0$

  True, because $f(ax) = \ln ax = \ln a + \ln x = f(a) + f(x)$.

89. False. $f(x) - f(2) = \ln x - \ln 2 = \ln \dfrac{x}{2} \neq \ln(x - 2)$

90. $\sqrt{f(x)} = \dfrac{1}{2} f(x)$; false

$\sqrt{f(x)} = \sqrt{\ln x}$ can't be simplified further.

$f(\sqrt{x}) = \ln \sqrt{x} = \ln x^{1/2} = \dfrac{1}{2} \ln x = \dfrac{1}{2} f(x)$

91. False.

$f(u) = 2f(v) \implies \ln u = 2 \ln v \implies \ln u = \ln v^2 \implies u = v^2$

92. If $f(x) < 0$, then $0 < x < 1$.

True

93. Let $x = \log_b u$ and $y = \log_b v$, then $b^x = u$ and $b^y = v$.

$\dfrac{u}{v} = \dfrac{b^x}{b^y} = b^{x-y}$

Then $\log_b(u/v) = \log_b(b^{x-y}) = x - y = \log_b u - \log_b v$.

94. Let $x = \log_b u$, then $u = b^x$ and $u^n = b^{nx}$.

$\log_b u^n = \log_b b^{nx} = nx = n \log_b u$

95. $f(x) = \log_2 x = \dfrac{\log x}{\log 2} = \dfrac{\ln x}{\ln 2}$

96. $f(x) = \log_4 x = \dfrac{\log x}{\log 4} = \dfrac{\ln x}{\ln 4}$

97. $f(x) = \log_{1/2} x$

$= \dfrac{\log x}{\log(1/2)} = \dfrac{\ln x}{\ln(1/2)}$

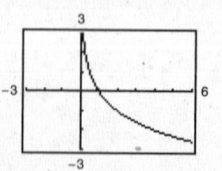

98. $f(x) = \log_{1/4} x$

$= \dfrac{\log x}{\log(1/4)} = \dfrac{\ln x}{\ln(1/4)}$

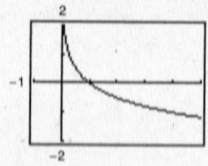

99. $f(x) = \log_{11.8} x$

$= \dfrac{\log x}{\log 11.8} = \dfrac{\ln x}{\ln 11.8}$

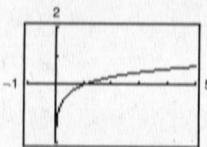

100. $f(x) = \log_{12.4} x$

$= \dfrac{\log x}{\log 12.4} = \dfrac{\ln x}{\ln 12.4}$

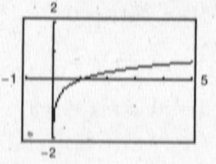

101. $f(x) = \ln \dfrac{x}{2}$, $g(x) = \dfrac{\ln x}{\ln 2}$, $h(x) = \ln x - \ln 2$

$f(x) = h(x)$ by Property 2

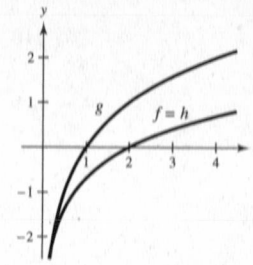

102. $\ln 2 \approx 0.6931, \ln 3 \approx 1.0986, \ln 5 \approx 1.6094$

$\ln 2 \approx 0.6931$

$\ln 3 \approx 1.0986$

$\ln 4 = \ln(2 \cdot 2) = \ln 2 + \ln 2 \approx 0.6931 + 0.6931 = 1.3862$

$\ln 5 \approx 1.6094$

$\ln 6 = \ln(2 \cdot 3) = \ln 2 + \ln 3 \approx 0.6931 + 1.0986 = 1.7917$

$\ln 8 = \ln 2^3 = 3 \ln 2 \approx 3(0.6931) = 2.0793$

$\ln 9 = \ln 3^2 = 2 \ln 3 \approx 2(1.0986) = 2.1972$

$\ln 10 = \ln(5 \cdot 2) = \ln 5 + \ln 2 \approx 1.6094 + 0.6931 = 2.3025$

$\ln 12 = \ln(2^2 \cdot 3) = \ln 2^2 + \ln 3 = 2 \ln 2 + \ln 3 \approx 2(0.6931) + 1.0986 = 2.4848$

$\ln 15 = \ln(5 \cdot 3) = \ln 5 + \ln 3 \approx 1.6094 + 1.0986 = 2.7080$

$\ln 16 = \ln 2^4 = 4 \ln 2 \approx 4(0.6931) = 2.7724$

$\ln 18 = \ln(3^2 \cdot 2) = \ln 3^2 + \ln 2 = 2 \ln 3 + \ln 2 \approx 2(1.0986) + 0.6931 = 2.8903$

$\ln 20 = \ln(5 \cdot 2^2) = \ln 5 + \ln 2^2 = \ln 5 + 2 \ln 2 \approx 1.6094 + 2(0.6931) = 2.9956$

103. $\dfrac{24xy^{-2}}{16x^{-3}y} = \dfrac{24xx^3}{16yy^2} = \dfrac{3x^4}{2y^3}, x \neq 0$

104. $\left(\dfrac{2x^2}{3y}\right)^{-3} = \left(\dfrac{3y}{2x^2}\right)^3 = \dfrac{(3y)^3}{(2x^2)^3} = \dfrac{27y^3}{8x^6}$

105. $(18x^3y^4)^{-3}(18x^3y^4)^3 = \dfrac{(18x^3y^4)^3}{(18x^3y^4)^3} = 1$ if $x \neq 0, y \neq 0$.

106. $xy(x^{-1} + y^{-1})^{-1} = \dfrac{xy}{x^{-1} + y^{-1}}$

$= \dfrac{xy}{(1/x) + (1/y)}$

$= \dfrac{xy}{(y + x)/xy} = \dfrac{(xy)^2}{x + y}$

107. $3x^2 + 2x - 1 = 0$

$(3x - 1)(x + 1) = 0$

$3x - 1 = 0 \Rightarrow x = \frac{1}{3}$

$x + 1 = 0 \Rightarrow x = -1$

108. $4x^2 - 5x + 1 = 0$

$(4x - 1)(x - 1) = 0$

$4x - 1 = 0 \Rightarrow x = \frac{1}{4}$

$x - 1 = 0 \Rightarrow x = 1$

The zeros are $x = \frac{1}{4}, 1$.

109. $\dfrac{2}{3x + 1} = \dfrac{x}{4}$

$(3x + 1)(x) = (2)(4)$

$3x^2 + x - 8 = 0$

$x = \dfrac{-1 \pm \sqrt{1^2 - 4(3)(-8)}}{2(3)}$

$= \dfrac{-1 \pm \sqrt{97}}{6}$

110. $\dfrac{5}{x - 1} = \dfrac{2x}{3}$

$5(3) = 2x(x - 1)$

$15 = 2x^2 - 2x$

$0 = 2x^2 - 2x - 15$

$\dfrac{-(-2) \pm \sqrt{(-2)^2 - 4(2)(-15)}}{2(2)} = x$

$\dfrac{2 \pm \sqrt{124}}{4} = x$

$\dfrac{1 \pm \sqrt{31}}{2} = x$

The zeros are $\dfrac{1 \pm \sqrt{31}}{2}$.

Section 3.4 Exponential and Logarithmic Equations

■ To solve an exponential equation, isolate the exponential expression, then take the logarithm of both sides. Then solve for the variable.

1. $\log_a a^x = x$ 2. $\ln e^x = x$

■ To solve a logarithmic equation, rewrite it in exponential form. Then solve for the variable.

1. $a^{\log_a x} = x$ 2. $e^{\ln x} = x$

■ If $a > 0$ and $a \neq 1$ we have the following:

1. $\log_a x = \log_a y \iff x = y$

2. $a^x = a^y \iff x = y$

■ Check for extraneous solutions.

Vocabulary Check

1. solve

2. (a) $x = y$ (b) $x = y$
      (c) x (d) x

3. extraneous

---

1. $4^{2x-7} = 64$

(a) $x = 5$

$4^{2(5)-7} = 4^3 = 64$

Yes, $x = 5$ *is* a solution.

(b) $x = 2$

$4^{2(2)-7} = 4^{-3} = \frac{1}{64} \neq 64$

No, $x = 2$ *is not* a solution.

2. $2^{3x+1} = 32$

(a) $x = -1$

$2^{3(-1)+1} = 2^{-2} = \frac{1}{4}$

No, $x = -1$ *is not* a solution.

(b) $x = 2$

$2^{3(2)+1} = 2^7 = 128$

No, $x = 2$ *is not* a solution.

3. $3e^{x+2} = 75$

(a) $x = -2 + e^{25}$

$3e^{(-2+e^{25})+2} = 3e^{e^{25}} \neq 75$

No, $x = -2 + e^{25}$ *is not* a solution.

(b) $x = -2 + \ln 25$

$3e^{(-2+\ln 25)+2} = 3e^{\ln 25} = 3(25) = 75$

Yes, $x = -2 + \ln 25$ *is* a solution.

(c) $x \approx 1.219$

$3e^{1.219+2} = 3e^{3.219} \approx 75$

Yes, $x \approx 1.219$ *is* a solution.

4. $2e^{5x+2} = 12$

(a) $x = \dfrac{1}{5}(-2 + \ln 6)$

$2e^{5[(1/5)(-2+\ln 6)]+2} = 2e^{-2+\ln 6+2}$

$= 2e^{\ln 6} = 2 \cdot 6 = 12$

Yes, $x = \dfrac{1}{5}(-2 + \ln 6)$ *is* a solution.

(b) $x = \dfrac{\ln 6}{5 \ln 2}$

$2e^{5[\ln 6/(5 \ln 2)]+2} = 2e^{(\ln 6/\ln 2)+2}$

$\approx 2e^{2.585+2}$

$\approx 2 \cdot 97.9995 = 195.999$

No, $x = \dfrac{\ln 6}{5 \ln 2}$ *is not* a solution.

(c) $x = -0.0416$

$2e^{5(-0.0416)+2} = 2e^{1.792} \approx 2(6.00144) \approx 12$

Yes, $x = -0.0416$ *is* an approximate solution.

5. $\log_4(3x) = 3 \implies 3x = 4^3 \implies 3x = 64$

  (a) $x \approx 21.333$

  $3(21.333) \approx 64$

  Yes, 21.333 *is* an approximate solution.

  (b) $x = -4$

  $3(-4) = -12 \neq 64$

  No, $x = -4$ *is not* a solution.

  (c) $x = \frac{64}{3}$

  $3\left(\frac{64}{3}\right) = 64$

  Yes, $x = \frac{64}{3}$ *is* a solution.

6. $\log_2(x + 3) = 10$

  (a) $x = 1021$

  $\log_2(1021 + 3) = \log_2(1024)$

  Since $2^{10} = 1024$, $x = 1021$ *is* a solution.

  (b) $x = 17$

  $\log_2(17 + 3) = \log_2(20)$

  Since $2^{10} \neq 20$, $x = 17$ *is not* a solution.

  (c) $x = 10^2 - 3 = 97$

  $\log_2(97 + 3) = \log_2(100)$

  Since $2^{10} \neq 100$, $10^2 - 3$ *is not* a solution.

7. $\ln(2x + 3) = 5.8$

  (a) $x = \frac{1}{2}(-3 + \ln 5.8)$

  $\ln\left[2\left(\frac{1}{2}\right)(-3 + \ln 5.8) + 3\right] = \ln(\ln 5.8) \neq 5.8$

  No, $x = \frac{1}{2}(-3 + \ln 5.8)$ *is not* a solution.

  (b) $x = \frac{1}{2}(-3 + e^{5.8})$

  $\ln\left[2\left(\frac{1}{2}\right)(-3 + e^{5.8}) + 3\right] = \ln(e^{5.8}) = 5.8$

  Yes, $x = \frac{1}{2}(-3 + e^{5.8})$ *is* a solution.

  (c) $x \approx 163.650$

  $\ln[2(163.650) + 3] = \ln 330.3 \approx 5.8$

  Yes, $x \approx 163.650$ *is* an approximate solution.

8. $\ln(x - 1) = 3.8$

  (a) $x = 1 + e^{3.8}$

  $\ln(1 + e^{3.8} - 1) = \ln e^{3.8} = 3.8$

  Yes, $x = 1 + e^{3.8}$ *is* a solution.

  (b) $x \approx 45.701$

  $\ln(45.701 - 1) = \ln(44.701) \approx 3.8$

  Yes, $x \approx 45.701$ *is* an approximate solution.

  (c) $x = 1 + \ln 3.8$

  $\ln(1 + \ln 3.8 - 1) = \ln(\ln 3.8) \approx 0.289$

  No, $x = 1 + \ln 3.8$ *is not* a solution.

9. $4^x = 16$

  $4^x = 4^2$

  $x = 2$

10. $3^x = 243$

  $3^x = 3^5$

  $x = 5$

11. $\left(\frac{1}{2}\right)^x = 32$

  $2^{-x} = 2^5$

  $-x = 5$

  $x = -5$

12. $\left(\frac{1}{4}\right)^x = 64$

  $4^{-x} = 4^3$

  $-x = 3$

  $x = -3$

13. $\ln x - \ln 2 = 0$

  $\ln x = \ln 2$

  $x = 2$

14. $\ln x - \ln 5 = 0$

  $\ln x = \ln 5$

  $x = 5$

15. $e^x = 2$

  $\ln e^x = \ln 2$

  $x = \ln 2$

  $x \approx 0.693$

16. $e^x = 4$

  $\ln e^x = \ln 4$

  $x = \ln 4$

  $x \approx 1.386$

17. $\ln x = -1$

  $e^{\ln x} = e^{-1}$

  $x = e^{-1}$

  $x \approx 0.368$

18. $\ln x = -7$

  $e^{\ln x} = e^{-7}$

  $x = e^{-7}$

  $x \approx 0.000912$

19. $\log_4 x = 3$

  $4^{\log_4 x} = 4^3$

  $x = 4^3$

  $x = 64$

20. $\log_5 x = -3$

  $x = 5^{-3}$

  $x = \frac{1}{125}$ or 0.008

21. $f(x) = g(x)$

$2^x = 8$

$2^x = 2^3$

$x = 3$

Point of intersection:
$(3, 8)$

22. $f(x) = g(x)$

$27^x = 9$

$27^x = 27^{2/3}$

$x = \frac{2}{3}$

Point of intersection:
$\left(\frac{2}{3}, 9\right)$

23. $f(x) = g(x)$

$\log_3 x = 2$

$x = 3^2$

$x = 9$

Point of intersection:
$(9, 2)$

24. $f(x) = g(x)$

$\ln(x - 4) = 0$

$e^{\ln(x-4)} = e^0$

$x - 4 = 1$

$x = 5$

Point of intersection: $(5, 0)$

25. $e^x = e^{x^2-2}$

$x = x^2 - 2$

$0 = x^2 - x - 2$

$0 = (x + 1)(x - 2)$

$x = -1$ or $x = 2$

26. $e^{2x} = e^{x^2-8}$

$2x = x^2 - 8$

$x^2 - 2x - 8 = 0$

$(x - 4)(x + 2) = 0$

$x = -2, \ x = 4$

27. $e^{x^2-3} = e^{x-2}$

$x^2 - 3 = x - 2$

$x^2 - x - 1 = 0$

By the Quadratic Formula
$x \approx 1.618$ or $x \approx -0.618$.

28. $e^{-x^2} = e^{x^2-2x}$

$-x^2 = x^2 - 2x$

$2x^2 - 2x = 0$

$2x(x - 1) = 0$

$x = 0, \ x = 1$

29. $4(3^x) = 20$

$3^x = 5$

$\log_3 3^x = \log_3 5$

$x = \log_3 5 = \dfrac{\log 5}{\log 3}$ or $\dfrac{\ln 5}{\ln 3}$

$x \approx 1.465$

30. $2(5^x) = 32$

$5^x = 16$

$x = \log_5 16$

$x = \dfrac{\ln 16}{\ln 5}$

$x \approx 1.723$

31. $2e^x = 10$

$e^x = 5$

$\ln e^x = \ln 5$

$x = \ln 5 \approx 1.609$

32. $4e^x = 91$

$e^x = \frac{91}{4}$

$\ln e^x = \ln \frac{91}{4}$

$x = \ln \frac{91}{4} \approx 3.125$

33. $e^x - 9 = 19$

$e^x = 28$

$\ln e^x = \ln 28$

$x = \ln 28 \approx 3.332$

34. $6^x + 10 = 47$

$6^x = 37$

$x = \log_6 37$

$x = \dfrac{\ln 37}{\ln 6}$

$x \approx 2.015$

35. $3^{2x} = 80$

$\ln 3^{2x} = \ln 80$

$2x \ln 3 = \ln 80$

$x = \dfrac{\ln 80}{2 \ln 3} \approx 1.994$

36. $6^{5x} = 3000$

$\ln 6^{5x} = \ln 3000$

$(5x) \ln 6 = \ln 3000$

$5x = \dfrac{\ln 3000}{\ln 6}$

$x = \dfrac{\ln 3000}{5 \ln 6} \approx 0.894$

37. $5^{-t/2} = 0.20$

$5^{-t/2} = \dfrac{1}{5}$

$5^{-t/2} = 5^{-1}$

$-\dfrac{t}{2} = -1$

$t = 2$

38. $4^{-3t} = 0.10$

$\ln 4^{-3t} = \ln 0.10$

$(-3t) \ln 4 = \ln 0.10$

$-3t = \dfrac{\ln 0.10}{\ln 4}$

$t = -\dfrac{\ln 0.10}{3 \ln 4} \approx 0.554$

39. $3^{x-1} = 27$

$3^{x-1} = 3^3$

$x - 1 = 3$

$x = 4$

40. $2^{x-3} = 32$

$x - 3 = \log_2 32$

$x - 3 = 5$

$x = 8$

41.
$$2^{3-x} = 565$$
$$\ln 2^{3-x} = \ln 565$$
$$(3 - x) \ln 2 = \ln 565$$
$$3 \ln 2 - x \ln 2 = \ln 565$$
$$-x \ln 2 = \ln 565 - 3 \ln 2$$
$$x \ln 2 = 3 \ln 2 - \ln 565$$
$$x = \frac{3 \ln 2 - \ln 565}{\ln 2}$$
$$= 3 - \frac{\ln 565}{\ln 2} \approx -6.142$$

42.
$$8^{-2-x} = 431$$
$$\ln 8^{-2-x} = \ln 431$$
$$(-2 - x) \ln 8 = \ln 431$$
$$-2 \ln 8 - x \ln 8 = \ln 431$$
$$-x \ln 8 = \ln 431 + \ln 8^2$$
$$x \ln 8 = -\ln 431 - \ln 64$$
$$x = \frac{-\ln 431 - \ln 64}{\ln 8} \approx -4.917$$

43. $8(10^{3x}) = 12$
$$10^{3x} = \frac{12}{8}$$
$$\log 10^{3x} = \log\left(\frac{3}{2}\right)$$
$$3x = \log\left(\frac{3}{2}\right)$$
$$x = \frac{1}{3}\log\left(\frac{3}{2}\right)$$
$$\approx 0.059$$

44. $5(10^{x-6}) = 7$
$$10^{x-6} = \frac{7}{5}$$
$$\log 10^{x-6} = \log \frac{7}{5}$$
$$x - 6 = \log \frac{7}{5}$$
$$x = 6 + \log \frac{7}{5}$$
$$\approx 6.146$$

45. $3(5^{x-1}) = 21$
$$5^{x-1} = 7$$
$$\ln 5^{x-1} = \ln 7$$
$$(x - 1) \ln 5 = \ln 7$$
$$x - 1 = \frac{\ln 7}{\ln 5}$$
$$x = 1 + \frac{\ln 7}{\ln 5} \approx 2.209$$

46. $8(3^{6-x}) = 40$
$$3^{6-x} = 5$$
$$\ln 3^{6-x} = \ln 5$$
$$(6 - x) \ln 3 = \ln 5$$
$$6 - x = \frac{\ln 5}{\ln 3}$$
$$-x = \frac{\ln 5}{\ln 3} - 6$$
$$x = 6 - \frac{\ln 5}{\ln 3} \approx 4.535$$

47. $e^{3x} = 12$
$$3x = \ln 12$$
$$x = \frac{\ln 12}{3} \approx 0.828$$

48. $e^{2x} = 50$
$$\ln e^{2x} = \ln 50$$
$$2x = \ln 50$$
$$x = \frac{\ln 50}{2} \approx 1.956$$

49. $500e^{-x} = 300$
$$e^{-x} = \frac{3}{5}$$
$$-x = \ln \frac{3}{5}$$
$$x = -\ln \frac{3}{5}$$
$$= \ln \frac{5}{3} \approx 0.511$$

50. $1000e^{-4x} = 75$
$$e^{-4x} = \frac{3}{40}$$
$$\ln e^{-4x} = \ln \frac{3}{40}$$
$$-4x = \ln \frac{3}{40}$$
$$x = -\frac{1}{4} \ln \frac{3}{40}$$
$$\approx 0.648$$

51. $7 - 2e^x = 5$
$$-2e^x = -2$$
$$e^x = 1$$
$$x = \ln 1 = 0$$

52. $-14 + 3e^x = 11$
$$3e^x = 25$$
$$e^x = \frac{25}{3}$$
$$\ln e^x = \ln \frac{25}{3}$$
$$x = \ln \frac{25}{3}$$
$$\approx 2.120$$

53. $6(2^{3x-1}) - 7 = 9$

$6(2^{3x-1}) = 16$

$2^{3x-1} = \dfrac{8}{3}$

$\log_2 2^{3x-1} = \log_2\left(\dfrac{8}{3}\right)$

$3x - 1 = \log_2\left(\dfrac{8}{3}\right) = \dfrac{\log(8/3)}{\log 2}$ or $\dfrac{\ln(8/3)}{\ln 2}$

$x = \dfrac{1}{3}\left[\dfrac{\log(8/3)}{\log 2} + 1\right] \approx 0.805$

54. $8(4^{6-2x}) + 13 = 41$

$8(4^{6-2x}) = 28$

$4^{6-2x} = 3.5$

$6 - 2x = \log_4 3.5$

$6 - 2x = \dfrac{\ln 3.5}{\ln 4}$

$-2x = -6 + \dfrac{\ln 3.5}{\ln 4}$

$x = 3 - \dfrac{\ln 3.5}{2\ln 4} \approx 2.548$

55. $e^{2x} - 4e^x - 5 = 0$

$(e^x + 1)(e^x - 5) = 0$

$e^x = -1$ or $e^x = 5$

(No solution) $x = \ln 5 \approx 1.609$

56. $e^{2x} - 5e^x + 6 = 0$

$(e^x - 2)(e^x - 3) = 0$

$e^x = 2$ or $e^x = 3$

$x = \ln 2 \approx 0.693$ or $x = \ln 3 \approx 1.099$

57. $e^{2x} - 3e^x - 4 = 0$

$(e^x + 1)(e^x - 4) = 0$

$e^x + 1 = 0 \implies e^x = -1$

Not possible since $e^x > 0$ for all x.

$e^x - 4 = 0 \implies e^x = 4 \implies x = \ln 4 \approx 1.386$

58. $e^{2x} + 9e^x + 36 = 0$

$(e^x)^2 + 9e^x + 36 = 0$

Because the discriminant is $9^2 - 4(1)(36) = -63$, there is no solution.

59. $\dfrac{500}{100 - e^{x/2}} = 20$

$500 = 20(100 - e^{x/2})$

$25 = 100 - e^{x/2}$

$e^{x/2} = 75$

$\dfrac{x}{2} = \ln 75$

$x = 2\ln 75 \approx 8.635$

60. $\dfrac{400}{1 + e^{-x}} = 350$

$400 = 350(1 + e^{-x})$

$\dfrac{8}{7} = 1 + e^{-x}$

$\dfrac{8}{7} - 1 = e^{-x}$

$\dfrac{1}{7} = e^{-x}$

$\ln\dfrac{1}{7} = \ln e^{-x}$

$-x = \ln\dfrac{1}{7}$

$-x = \ln 7^{-1}$

$-x = -\ln 7$

$x = \ln 7 \approx 1.946$

61. $\dfrac{3000}{2 + e^{2x}} = 2$

$3000 = 2(2 + e^{2x})$

$1500 = 2 + e^{2x}$

$1498 = e^{2x}$

$\ln 1498 = 2x$

$x = \dfrac{\ln 1498}{2} \approx 3.656$

62. $\dfrac{119}{e^{6x} - 14} = 7$

$119 = 7(e^{6x} - 14)$

$17 = e^{6x} - 14$

$31 = e^{6x}$

$\ln 31 = \ln e^{6x}$

$\ln 31 = 6x$

$x = \dfrac{\ln 31}{6} \approx 0.572$

63. $\left(1 + \dfrac{0.065}{365}\right)^{365t} = 4$

$\ln\left(1 + \dfrac{0.065}{365}\right)^{365t} = \ln 4$

$365t \ln\left(1 + \dfrac{0.065}{365}\right) = \ln 4$

$t = \dfrac{\ln 4}{365 \ln\left(1 + \frac{0.065}{365}\right)} \approx 21.330$

64. $\left(4 - \dfrac{2.471}{40}\right)^{9t} = 21$

$3.938225^{9t} = 21$

$\ln 3.938225^{9t} = \ln 21$

$9t \ln 3.938225 = \ln 21$

$t = \dfrac{\ln 21}{9 \ln 3.938225} \approx 0.247$

65. $\left(1 + \dfrac{0.10}{12}\right)^{12t} = 2$

$\ln\left(1 + \dfrac{0.10}{12}\right)^{12t} = \ln 2$

$12t \ln\left(1 + \dfrac{0.10}{12}\right) = \ln 2$

$t = \dfrac{\ln 2}{12 \ln\left(1 + \frac{0.10}{12}\right)} \approx 6.960$

66. $\left(16 - \dfrac{0.878}{26}\right)^{3t} = 30$

$\ln\left(16 - \dfrac{0.878}{26}\right)^{3t} = \ln 30$

$3t \ln\left(16 - \dfrac{0.878}{26}\right) = \ln 30$

$t = \dfrac{\ln 30}{3 \ln\left(16 - \frac{0.878}{26}\right)} \approx 0.409$

67. $g(x) = 6e^{1-x} - 25$

Algebraically:

$6e^{1-x} = 25$

$e^{1-x} = \dfrac{25}{6}$

$1 - x = \ln\left(\dfrac{25}{6}\right)$

$x = 1 - \ln\left(\dfrac{25}{6}\right)$

$x \approx -0.427$

The zero is $x \approx -0.427$.

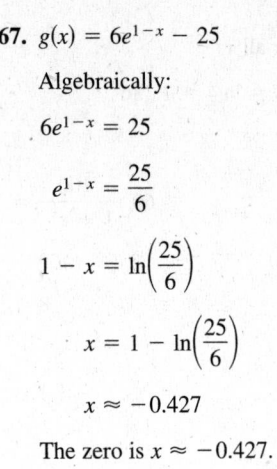

68. $f(x) = -4e^{-x-1} + 15$

$0 = -4e^{-x-1} + 15$

$-15 = -4e^{-x-1}$

$3.75 = e^{-x-1}$

$\ln 3.75 = -x - 1$

$1 + \ln 3.75 = -x$

$-1 - \ln 3.75 = x$

$-2.322 \approx x$

The zero is -2.322.

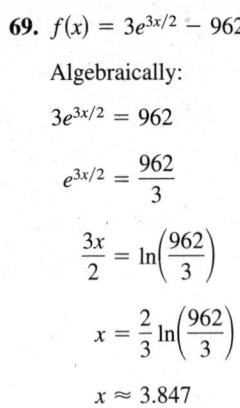

69. $f(x) = 3e^{3x/2} - 962$

Algebraically:

$3e^{3x/2} = 962$

$e^{3x/2} = \dfrac{962}{3}$

$\dfrac{3x}{2} = \ln\left(\dfrac{962}{3}\right)$

$x = \dfrac{2}{3} \ln\left(\dfrac{962}{3}\right)$

$x \approx 3.847$

The zero is $x \approx 3.847$.

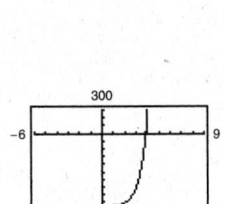

70. $g(x) = 8e^{-2x/3} - 11$

$8e^{-2x/3} = 11$

$e^{-2x/3} = 1.375$

$-\dfrac{2x}{3} = \ln 1.375$

$x = -1.5 \ln 1.375$

$x \approx -0.478$

The zero is -0.478.

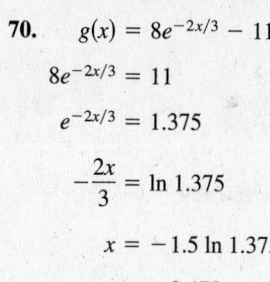

71. $g(t) = e^{0.09t} - 3$

Algebraically:

$e^{0.09t} = 3$

$0.09t = \ln 3$

$t = \dfrac{\ln 3}{0.09}$

$t \approx 12.207$

The zero is $t \approx 12.207$.

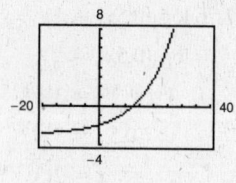

72. $f(x) = -e^{1.8x} + 7$

$-e^{1.8x} + 7 = 0$

$-e^{1.8x} = -7$

$e^{1.8x} = 7$

$1.8x = \ln 7$

$x = \dfrac{\ln 7}{1.8}$

$x \approx 1.081$

The zero is 1.081.

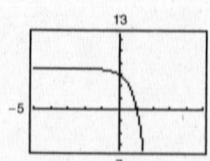

73. $h(t) = e^{0.125t} - 8$

Algebraically:

$e^{0.125t} - 8 = 0$

$e^{0.125t} = 8$

$0.125t = \ln 8$

$t = \dfrac{\ln 8}{0.125}$

$t \approx 16.636$

The zero is $t \approx 16.636$.

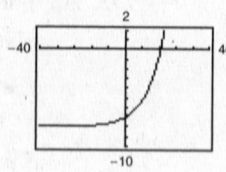

74. $f(x) = e^{2.724x} - 29$

$e^{2.724x} = 29$

$2.724x = \ln 29$

$x = \dfrac{\ln 29}{2.724}$

$x \approx 1.236$

The zero is 1.236.

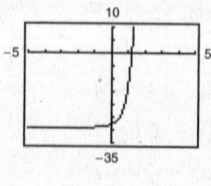

75. $\ln x = -3$

$x = e^{-3} \approx 0.050$

76. $\ln x = 2$

$e^{\ln x} = e^2$

$x = e^2 \approx 7.389$

77. $\ln 2x = 2.4$

$2x = e^{2.4}$

$x = \dfrac{e^{2.4}}{2} \approx 5.512$

78. $\ln 4x = 1$

$e^{\ln 4x} = e^1$

$4x = e$

$x = \dfrac{e}{4} \approx 0.680$

79. $\log x = 6$

$x = 10^6$

$= 1,000,000.000$

80. $\log 3z = 2$

$10^{\log 3z} = 10^2$

$3z = 100$

$z = \dfrac{100}{3} \approx 33.333$

81. $3 \ln 5x = 10$

$\ln 5x = \dfrac{10}{3}$

$5x = e^{10/3}$

$x = \dfrac{e^{10/3}}{5} \approx 5.606$

82. $2 \ln x = 7$

$\ln x = \dfrac{7}{2}$

$e^{\ln x} = e^{7/2}$

$x = e^{7/2} \approx 33.115$

83. $\ln \sqrt{x + 2} = 1$

$\sqrt{x + 2} = e^1$

$x + 2 = e^2$

$x = e^2 - 2$

≈ 5.389

84. $\ln \sqrt{x - 8} = 5$

$e^{\ln \sqrt{x-8}} = e^5$

$\sqrt{x - 8} = e^5$

$x - 8 = e^{10}$

$x = e^{10} + 8$

$\approx 22,034.466$

85. $7 + 3 \ln x = 5$

$3 \ln x = -2$

$\ln x = -\dfrac{2}{3}$

$x = e^{-2/3}$

≈ 0.513

86. $2 - 6 \ln x = 10$

$-6 \ln x = 8$

$\ln x = -\dfrac{4}{3}$

$e^{\ln x} = e^{-4/3}$

$x = e^{-4/3}$

≈ 0.264

87. $6 \log_3(0.5x) = 11$

$$\log_3(0.5x) = \frac{11}{6}$$

$$3^{\log_3(0.5x)} = 3^{11/6}$$

$$0.5x = 3^{11/6}$$

$$x = 2(3^{11/6}) \approx 14.988$$

88. $5 \log_{10}(x - 2) = 11$

$$\log_{10}(x - 2) = \frac{11}{5}$$

$$10^{\log_{10}(x-2)} = 10^{11/5}$$

$$x - 2 = 10^{11/5}$$

$$x = 10^{11/5} + 2 \approx 160.489$$

89. $\ln x - \ln(x + 1) = 2$

$$\ln\left(\frac{x}{x + 1}\right) = 2$$

$$\frac{x}{x + 1} = e^2$$

$$x = e^2(x + 1)$$

$$x = e^2 x + e^2$$

$$x - e^2 x = e^2$$

$$x(1 - e^2) = e^2$$

$$x = \frac{e^2}{1 - e^2} \approx -1.157$$

This negative value is extraneous. The equation has no solution.

90. $\ln x + \ln(x + 1) = 1$

$$\ln[x(x + 1)] = 1$$

$$e^{\ln[x(x+1)]} = e^1$$

$$x(x + 1) = e^1$$

$$x^2 + x - e = 0$$

$$x = \frac{-1 \pm \sqrt{1 + 4e}}{2}$$

The only solution is $x = \dfrac{-1 + \sqrt{1 + 4e}}{2} \approx 1.223$.

91. $\ln x + \ln(x - 2) = 1$

$$\ln[x(x - 2)] = 1$$

$$x(x - 2) = e^1$$

$$x^2 - 2x - e = 0$$

$$x = \frac{2 \pm \sqrt{4 + 4e}}{2}$$

$$= \frac{2 \pm 2\sqrt{1 + e}}{2} = 1 \pm \sqrt{1 + e}$$

The negative value is extraneous. The only solution is $x = 1 + \sqrt{1 + e} \approx 2.928$.

92. $\ln x + \ln(x + 3) = 1$

$$\ln[x(x + 3)] = 1$$

$$e^{\ln[x(x+3)]} = e^1$$

$$x(x + 3) = e^1$$

$$x^2 + 3x - e = 0$$

$$x = \frac{-3 \pm \sqrt{9 + 4e}}{2}$$

The only solution is $x = \dfrac{-3 + \sqrt{9 + 4e}}{2} \approx 0.729$.

93. $\ln(x + 5) = \ln(x - 1) - \ln(x + 1)$

$$\ln(x + 5) = \ln\left(\frac{x - 1}{x + 1}\right)$$

$$x + 5 = \frac{x - 1}{x + 1}$$

$$(x + 5)(x + 1) = x - 1$$

$$x^2 + 6x + 5 = x - 1$$

$$x^2 + 5x + 6 = 0$$

$$(x + 2)(x + 3) = 0$$

$$x = -2 \text{ or } x = -3$$

Both of these solutions are extraneous, so the equation has no solution.

94. $\ln(x + 1) - \ln(x - 2) = \ln x$

$$\ln\left(\frac{x + 1}{x - 2}\right) = \ln x$$

$$\frac{x + 1}{x - 2} = x$$

$$x + 1 = x^2 - 2x$$

$$0 = x^2 - 3x - 1$$

$$\frac{-(-3) \pm \sqrt{(-3)^2 - 4(1)(-1)}}{2(1)} = x$$

$$\frac{3 \pm \sqrt{13}}{2} = x$$

$$3.303 \approx x$$

(The negative apparent solution is extraneous.)

95. $\log_2(2x - 3) = \log_2(x + 4)$

$$2x - 3 = x + 4$$

$$x = 7$$

96. $\log(x - 6) = \log(2x + 1)$

$$x - 6 = 2x + 1$$
$$-7 = x$$

The apparent solution $x = -7$ is extraneous, because the domain of the logarithm function is positive numbers, and $-7 - 6$ and $2(-7) + 1$ are negative. There is no solution.

97. $\log(x + 4) - \log x = \log(x + 2)$

$$\log\left(\frac{x + 4}{x}\right) = \log(x + 2)$$

$$\frac{x + 4}{x} = x + 2$$

$$x + 4 = x^2 + 2x$$

$$0 = x^2 + x - 4$$

$$x = \frac{-1 \pm \sqrt{17}}{2} \quad \text{Quadratic Formula}$$

Choosing the positive value of x (the negative value is extraneous), we have

$$x = \frac{-1 + \sqrt{17}}{2} \approx 1.562.$$

98. $\log_2 x + \log_2(x + 2) = \log_2(x + 6)$

$$\log_2[x(x + 2)] = \log_2(x + 6)$$

$$x(x + 2) = x + 6$$

$$x^2 + x - 6 = 0$$

$$(x + 3)(x - 2) = 0$$

$$x = -3 \text{ or } x = 2$$

The value $x = -3$ is extraneous. The only solution is $x = 2$.

99. $\log_4 x - \log_4(x - 1) = \frac{1}{2}$

$$\log_4\left(\frac{x}{x - 1}\right) = \frac{1}{2}$$

$$4^{\log_4[x/(x-1)]} = 4^{1/2}$$

$$\frac{x}{x - 1} = 4^{1/2}$$

$$x = 2(x - 1)$$

$$x = 2x - 2$$

$$-x = -2$$

$$x = 2$$

100. $\log_3 x + \log_3(x - 8) = 2$

$$\log_3[x(x - 8)] = 2$$

$$3^{\log_3(x^2 - 8x)} = 3^2$$

$$x^2 - 8x = 9$$

$$x^2 - 8x - 9 = 0$$

$$(x - 9)(x + 1) = 0$$

$$x = 9 \text{ or } x = -1$$

The value $x = -1$ is extraneous. The only solution is $x = 9$.

101. $\log 8x - \log\left(1 + \sqrt{x}\right) = 2$

$$\log \frac{8x}{1 + \sqrt{x}} = 2$$

$$\frac{8x}{1 + \sqrt{x}} = 10^2$$

$$8x = 100\left(1 + \sqrt{x}\right)$$

$$2x = 25\left(1 + \sqrt{x}\right) = 25 + 25\sqrt{x}$$

$$2x - 25 = 25\sqrt{x}$$

$$(2x - 25)^2 = \left(25\sqrt{x}\right)^2$$

$$4x^2 - 100x + 625 = 625x$$

$$4x^2 - 725x + 625 = 0$$

$$x = \frac{725 \pm \sqrt{725^2 - 4(4)(625)}}{2(4)} = \frac{725 \pm \sqrt{515,625}}{8} = \frac{25\left(29 \pm 5\sqrt{33}\right)}{8}$$

$$x \approx 0.866 \text{ (extraneous) or } x \approx 180.384$$

The only solution is $x = \dfrac{25\left(29 + 5\sqrt{33}\right)}{8} \approx 180.384$.

102. $\log 4x - \log\left(12 + \sqrt{x}\right) = 2$

$$\log\left(\frac{4x}{12 + \sqrt{x}}\right) = 2$$

$$10^{\log(4x/(12 + \sqrt{x}))} = 10^2$$

$$\frac{4x}{12 + \sqrt{x}} = 100$$

$$4x = 100\left(12 + \sqrt{x}\right)$$

$$4x = 1200 + 100\sqrt{x}$$

$$4x - 1200 = 100\sqrt{x}$$

$$x - 300 = 25\sqrt{x}$$

$$(x - 300)^2 = \left(25\sqrt{x}\right)^2$$

$$x^2 - 600x + 90{,}000 = 625x$$

$$x^2 - 1225x + 90{,}000 = 0$$

$$x = \frac{1225 \pm \sqrt{(-1225)^2 - 4(1)(90{,}000)}}{2}$$

$$x = \frac{1225 \pm \sqrt{1{,}140{,}625}}{2}$$

$$x = \frac{1225 \pm 125\sqrt{73}}{2}$$

$$x \approx 78.500 \text{ (extraneous)} \quad \text{or} \quad x \approx 1146.500$$

The only solution is $x = \dfrac{1225 + 125\sqrt{73}}{2} \approx 1146.500$.

103. $y_1 = 7$

$y_2 = 2^x$

From the graph we have
$x \approx 2.807$ when $y = 7$.
Algebraically:

$$2^x = 7$$

$$\ln 2^x = \ln 7$$

$$x \ln 2 = \ln 7$$

$$x = \frac{\ln 7}{\ln 2} \approx 2.807$$

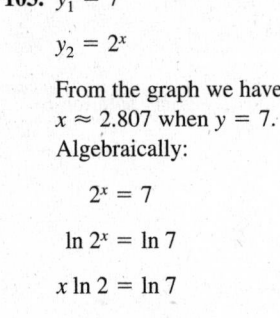

104. $500 = 1500e^{-x/2}$

$$\frac{1}{3} = e^{-x/2}$$

$$\ln\frac{1}{3} = -\frac{x}{2}$$

$$-2\ln\frac{1}{3} = x$$

$$2.197 \approx x$$

The solution is $x \approx 2.197$.

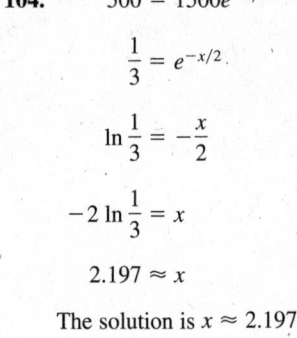

105. $y_1 = 3$

$y_2 = \ln x$

From the graph we have
$x \approx 20.086$ when $y = 3$.
Algebraically:

$$3 - \ln x = 0$$

$$\ln x = 3$$

$$x = e^3 \approx 20.086$$

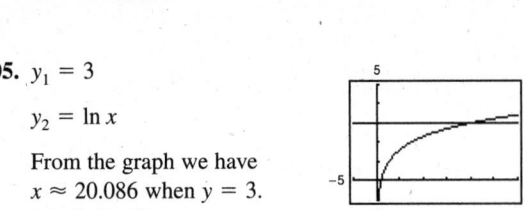
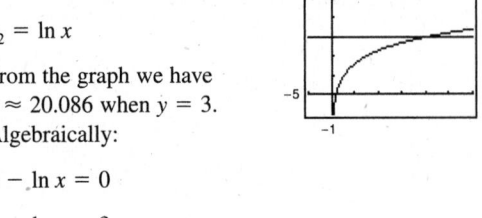

106. $10 - 4\ln(x - 2) = 0$

$$-4\ln(x - 2) = -10$$

$$\ln(x - 2) = 2.5$$

$$e^{\ln(x-2)} = e^{2.5}$$

$$x - 2 = e^{2.5}$$

$$x = e^{2.5} + 2$$

$$x \approx 14.182$$

The solution is $x \approx 14.182$.

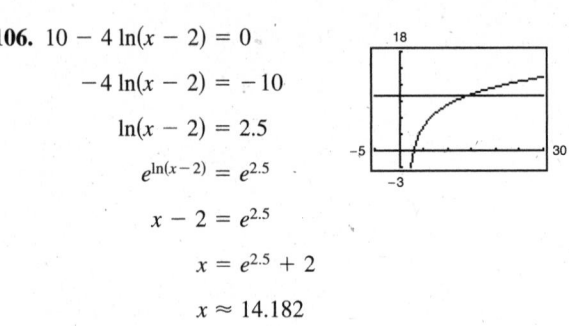

107. (a) $A = Pe^{rt}$

$5000 = 2500e^{0.085t}$

$2 = e^{0.085t}$

$\ln 2 = 0.085t$

$\dfrac{\ln 2}{0.085} = t$

$t \approx 8.2$ years

(b) $A = Pe^{rt}$

$7500 = 2500e^{0.085t}$

$3 = e^{0.085t}$

$\ln 3 = 0.085t$

$\dfrac{\ln 3}{0.085} = t$

$t \approx 12.9$ years

108. (a) $r = 0.12$

$A = Pe^{rt}$

$5000 = 2500e^{0.12t}$

$2 = e^{0.12t}$

$\ln 2 = \ln e^{0.12t}$

$\ln 2 = 0.12t$

$\dfrac{\ln 2}{0.12} = t$

$t \approx 5.8$ years

(b) $r = 0.12$

$A = Pe^{rt}$

$7500 = 2500e^{0.12t}$

$3 = e^{0.12t}$

$\ln 3 = \ln e^{0.12t}$

$\ln 3 = 0.12t$

$\dfrac{\ln 3}{0.12} = t$

$t = 9.2$ years

109. $p = 500 - 0.5(e^{0.004x})$

(a) $p = 350$

$350 = 500 - 0.5(e^{0.004x})$

$300 = e^{0.004x}$

$0.004x = \ln 300$

$x \approx 1426$ units

(b) $p = 300$

$300 = 500 - 0.5(e^{0.004x})$

$400 = e^{0.004x}$

$0.004x = \ln 400$

$x \approx 1498$ units

110. $p = 5000\left(1 - \dfrac{4}{4 + e^{-0.002x}}\right)$

(a) When $p = \$600$:

$600 = 5000\left(1 - \dfrac{4}{4 + e^{-0.002x}}\right)$

$0.12 = 1 - \dfrac{4}{4 + e^{-0.002x}}$

$\dfrac{4}{4 + e^{-0.002x}} = 0.88$

$4 = 3.52 + 0.88e^{-0.002x}$

$0.48 = 0.88e^{-0.002x}$

$\dfrac{6}{11} = e^{-0.002x}$

$\ln \dfrac{6}{11} = \ln e^{-0.002x}$

$\ln \dfrac{6}{11} = -0.002x$

$x = -\dfrac{\ln(6/11)}{0.002} \approx 303$ units

(b) When $p = \$400$:

$400 = 5000\left(1 - \dfrac{4}{4 + e^{-0.002x}}\right)$

$0.08 = 1 - \dfrac{4}{4 + e^{-0.002x}}$

$\dfrac{4}{4 + e^{-0.002x}} = 0.92$

$4 = 3.68 + 0.92e^{-0.002x}$

$0.32 = 0.92e^{-0.002x}$

$\dfrac{8}{23} = e^{-0.002x}$

$\ln \dfrac{8}{23} = \ln e^{-0.002x}$

$\ln \dfrac{8}{23} = -0.002x$

$x = -\dfrac{\ln(8/23)}{0.002} \approx 528$ units

111. $V = 6.7e^{-48.1/t}$, $t \geq 0$

(a)

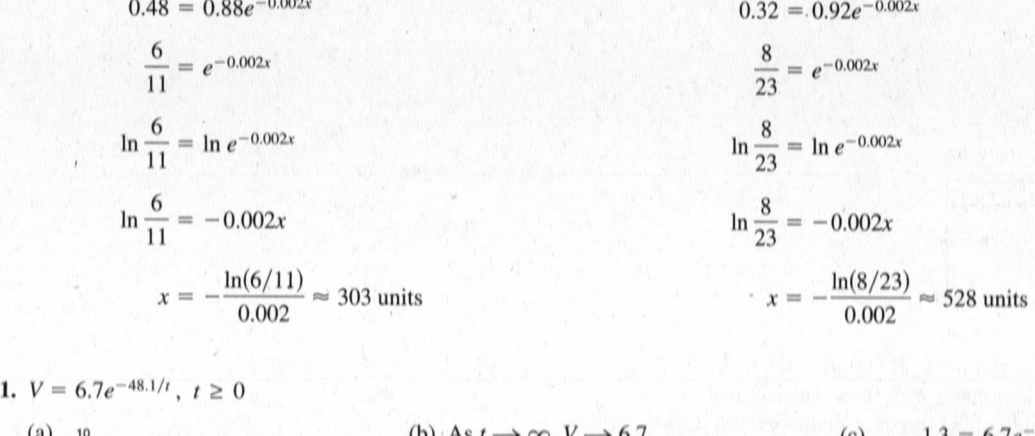

(b) As $t \to \infty$, $V \to 6.7$.

Horizontal asymptote: $V = 6.7$

The yield will approach
6.7 million cubic feet per acre.

(c) $1.3 = 6.7e^{-48.1/t}$

$\dfrac{1.3}{6.7} = e^{-48.1/t}$

$\ln\left(\dfrac{13}{67}\right) = \dfrac{-48.1}{t}$

$t = \dfrac{-48.1}{\ln(13/67)} \approx 29.3$ years

112. $N = 68(10^{-0.04x})$

When $N = 21$:

$$21 = 68(10^{-0.04x})$$

$$\frac{21}{68} = 10^{-0.04x}$$

$$\log_{10} \frac{21}{68} = -0.04x$$

$$x = -\frac{\log_{10}(21/68)}{0.04} \approx 12.76 \text{ inches}$$

113. $y = 7312 - 630.0 \ln t, \ 5 \le t \le 12$

$$7312 - 630.0 \ln t = 5800$$

$$-630.0 \ln t = -1512$$

$$\ln t = 2.4$$

$$t = e^{2.4} \approx 11$$

$t \approx 11$ corresponds to the year 2001.

114. $y = 4381 + 1883.6 \ln t, \ 5 \le t \le 13$

$$9000 = 4381 + 1883.6 \ln t$$

$$4619 = 1883.6 \ln t$$

$$\ln t = \frac{4619}{1883.6} = 2.45222$$

$$t = e^{2.45222} = 11.6$$

Since $t = 5$ represents 1995, $t = 11.6$ indicates that the number of daily fee golf facilities in the U.S. reached 9000 in 2001.

115. (a) From the graph shown in the textbook, we see horizontal asymptotes at $y = 0$ and $y = 100$. These represent the lower and upper percent bounds; the range falls between 0% and 100%.

(b) Males

$$50 = \frac{100}{1 + e^{-0.6114(x-69.71)}}$$

$$1 + e^{-0.6114(x-69.71)} = 2$$

$$e^{-0.6114(x-69.71)} = 1$$

$$-0.6114(x - 69.71) = \ln 1$$

$$-0.6114(x - 69.71) = 0$$

$$x = 69.71 \text{ inches}$$

Females

$$50 = \frac{100}{1 + e^{-0.66607(x-64.51)}}$$

$$1 + e^{-0.66607(x-64.51)} = 2$$

$$e^{-0.6667(x-64.51)} = 1$$

$$-0.66607(x - 64.51) = \ln 1$$

$$-0.66607(x - 64.51) = 0$$

$$x = 64.51 \text{ inches}$$

116. $P = \dfrac{0.83}{1 + e^{-0.2n}}$

(a)

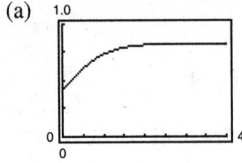

(b) Horizontal asymptotes: $P = 0, P = 0.83$
The upper asymptote, $P = 0.83$, indicates that the proportion of correct responses will approach 0.83 as the number of trials increases.

(c) When $P = 60\%$ or $P = 0.60$:

$$0.60 = \frac{0.83}{1 + e^{-0.2n}}$$

$$1 + e^{-0.2n} = \frac{0.83}{0.60}$$

$$e^{-0.2n} = \frac{0.83}{0.60} - 1$$

$$\ln e^{-0.2n} = \ln\left(\frac{0.83}{0.60} - 1\right)$$

$$-0.2n = \ln\left(\frac{0.83}{0.60} - 1\right)$$

$$n = -\frac{\ln\left(\dfrac{0.83}{0.60} - 1\right)}{0.2} \approx 5 \text{ trials}$$

117. $y = -3.00 + 11.88 \ln x + \dfrac{36.94}{x}$

(a)

x	0.2	0.4	0.6	0.8	1.0
y	162.6	78.5	52.5	40.5	33.9

(b)

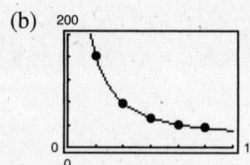

The model seems to fit the data well.

(c) When $y = 30$:

$$30 = -3.00 + 11.88 \ln x + \frac{36.94}{x},$$

Add the graph of $y = 30$ to the graph in part (a) and estimate the point of intersection of the two graphs. We find that $x \approx 1.20$ meters.

(d) No, it is probably not practical to lower the number of *gs* experienced during impact to less than 23 because the required distance traveled at $y = 23$ is $x \approx 2.27$ meters. It is probably not practical to design a car allowing a passenger to move forward 2.27 meters (or 7.45 feet) during an impact.

118. $T = 20[1 + 7(2^{-h})]$

(a) From the graph in the textbook we see a horizontal asymptote at $T = 20$. This represents the room temperature.

(b)
$$100 = 20[1 + 7(2^{-h})]$$
$$5 = 1 + 7(2^{-h})$$
$$4 = 7(2^{-h})$$
$$\frac{4}{7} = 2^{-h}$$
$$\ln\left(\frac{4}{7}\right) = \ln 2^{-h}$$
$$\ln\left(\frac{4}{7}\right) = -h \ln 2$$
$$\frac{\ln(4/7)}{-\ln 2} = h$$
$$h \approx 0.81 \text{ hour}$$

119. $\log_a(uv) = \log_a u + \log_a v$

True by Property 1 in Section 3.3.

120. $\log_a(u + v) = (\log_a u)(\log_a v)$

False.

$$2.04 \approx \log_{10}(10 + 100) \neq (\log_{10} 10)(\log_{10} 100) = 2$$

121. $\log_a(u - v) = \log_a u - \log_a v$

False.

$$1.95 \approx \log(100 - 10)$$
$$\neq \log 100 - \log 10 = 1$$

122. $\log_a\left(\dfrac{u}{v}\right) = \log_a u - \log_a v$

True by Property 2 in Section 3.3.

123. Yes, a logarithmic equation can have more than one extraneous solution. See Exercise 93.

124. $A = Pe^{rt}$

(a) $A = (2P)e^{rt} = 2(Pe^{rt})$ This doubles your money.

(b) $A = Pe^{(2r)t} = Pe^{rt}e^{rt} = e^{rt}(Pe^{rt})$

(c) $A = Pe^{r(2t)} = Pe^{rt}e^{rt} = e^{rt}(Pe^{rt})$

Doubling the interest rate yields the same result as doubling the number of years.

If $2 > e^{rt}$ (i.e., $rt < \ln 2$), then doubling your investment would yield the most money. If $rt > \ln 2$, then doubling either the interest rate or the number of years would yield more money.

125. Yes.

Time to Double	Time to Quadruple
$2P = Pe^{rt}$	$4P = Pe^{rt}$
$2 = e^{rt}$	$4 = e^{rt}$
$\ln 2 = rt$	$\ln 4 = rt$
$\dfrac{\ln 2}{r} = t$	$\dfrac{2\ln 2}{r} = t$

Thus, the time to quadruple is twice as long as the time to double.

126. (a) When solving an exponential equation, rewrite the original equation in a form that allows you to use the One-to-One Property $a^x = a^y$ if and only if $x = y$ or rewrite the original equation in logarithmic form and use the Inverse Property $\log_a a^x = x$.

(b) When solving a logarithmic equation, rewrite the original equation in a form that allows you to use the One-to-One Property $\log_a x = \log_a y$ if and only if $x = y$ or rewrite the original equation in exponential form and use the Inverse Property $a^{\log_a x} = x$.

127. $\sqrt{48x^2y^5} = \sqrt{16x^2y^4 3y}$
$\qquad = 4|x|y^2\sqrt{3y}$

128. $\sqrt{32} - 2\sqrt{25} = \sqrt{16 \cdot 2} - 2(5)$
$\qquad = 4\sqrt{2} - 10$

129. $\sqrt[3]{25}\sqrt[3]{15} = \sqrt[3]{375}$
$\qquad = \sqrt[3]{125 \cdot 3} = 5\sqrt[3]{3}$

130. $\dfrac{3}{\sqrt{10} - 2} = \dfrac{3}{\sqrt{10} - 2} \cdot \dfrac{\sqrt{10} + 2}{\sqrt{10} + 2}$

$\qquad = \dfrac{3(\sqrt{10} + 2)}{10 - 4}$

$\qquad = \dfrac{3(\sqrt{10} + 2)}{6}$

$\qquad = \dfrac{\sqrt{10} + 2}{2}$

$\qquad = \dfrac{1}{2}\sqrt{10} + 1$

131. $f(x) = |x| + 9$

Domain: all real numbers x

y-intercept: $(0, 9)$

y-axis symmetry

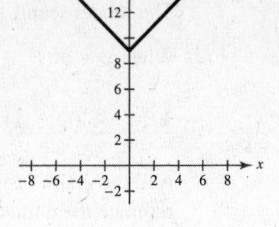

x	0	± 1	± 2	± 3
y	9	10	11	12

132.

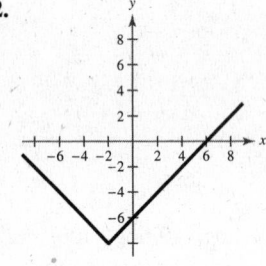

133. $g(x) = \begin{cases} 2x, & x < 0 \\ -x^2 + 4, & x \geq 0 \end{cases}$

Domain: all real numbers x

x-intercept: $(2, 0)$

y-intercept: $(0, 4)$

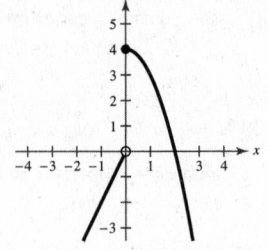

x	-3	-2	-1	-0.5	0	1	2	3
y	-6	-4	-2	-1	4	3	2	-5

134.

135. $\log_6 9 = \dfrac{\log_{10} 9}{\log_{10} 6} = \dfrac{\ln 9}{\ln 6} \approx 1.226$

136. $\log_3 4 = \dfrac{\log_{10} 4}{\log_{10} 3} = \dfrac{\ln 4}{\ln 3} \approx 1.262$

137. $\log_{3/4} 5 = \dfrac{\log_{10} 5}{\log_{10}(3/4)} = \dfrac{\ln 5}{\ln(3/4)} \approx -5.595$

138. $\log_8 22 = \dfrac{\log_{10} 22}{\log_{10} 8} = \dfrac{\ln 22}{\ln 8} \approx 1.486$

Section 3.5   Exponential and Logarithmic Models

- ■ You should be able to solve growth and decay problems.
  - (a) Exponential growth if $b > 0$ and $y = ae^{bx}$.
  - (b) Exponential decay if $b > 0$ and $y = ae^{-bx}$.
- ■ You should be able to use the Gaussian model
  $$y = ae^{-(x-b)^2/c}.$$
- ■ You should be able to use the logistic growth model
  $$y = \frac{a}{1 + be^{-rx}}.$$
- ■ You should be able to use the logarithmic models
  $$y = a + b \ln x, \quad y = a + b \log x.$$

Vocabulary Check

1. $y = ae^{bx}$; $y = ae^{-bx}$

2. $y = a + b \ln x$; $y = a + b \log x$

3. normally distributed

4. bell; average value

5. sigmoidal

1. $y = 2e^{x/4}$

This is an exponential growth model. Matches graph (c).

2. $y = 6e^{-x/4}$

This is an exponential decay model. Matches graph (e).

3. $y = 6 + \log(x + 2)$

This is a logarithmic function shifted up six units and left two units. Matches graph (b).

4. $y = 3e^{-(x-2)^2/5}$

This is a Gaussian model. Matches graph (a).

5. $y = \ln(x + 1)$

This is a logarithmic model shifted left one unit. Matches graph (d).

6. $y = \dfrac{4}{1 + e^{-2x}}$

This is a logistic growth model. Matches graph (f).

7. Since $A = 1000e^{0.035t}$, the time to double is given by $2000 = 1000e^{0.035t}$ and we have
$$2 = e^{0.035t}$$
$$\ln 2 = \ln e^{0.035t}$$
$$\ln 2 = 0.035t$$
$$t = \frac{\ln 2}{0.035} \approx 19.8 \text{ years.}$$

Amount after 10 years: $A = 1000e^{0.35} \approx \$1419.07$

8. Since $A = 750e^{0.105t}$, the time to double is given by $1500 = 750e^{0.105t}$, and we have
$$1500 = 750e^{0.105t}$$
$$2 = e^{0.105t}$$
$$\ln 2 = \ln e^{0.105t}$$
$$\ln 2 = 0.105t$$
$$t = \frac{\ln 2}{0.105} \approx 6.60 \text{ years.}$$

Amount after 10 years: $A = 750e^{0.105(10)} \approx \$2143.24$

9. Since $A = 750e^{rt}$ and $A = 1500$ when $t = 7.75$, we have the following.

$$1500 = 750e^{7.75r}$$

$$2 = e^{7.75r}$$

$$\ln 2 = \ln e^{7.75r}$$

$$\ln 2 = 7.75r$$

$$r = \frac{\ln 2}{7.75} \approx 0.089438 = 8.9438\%$$

Amount after 10 years: $A = 750e^{0.089438(10)} \approx \$1834.37$

10. Since $A = 10,000e^{rt}$ and $A = 20,000$ when $t = 12$, we have

$$20,000 = 10,000e^{12r}$$

$$2 = e^{12r}$$

$$\ln 2 = \ln e^{12r}$$

$$\ln 2 = 12r$$

$$r = \frac{\ln 2}{12} \approx 0.057762 = 5.7762\%.$$

Amount after 10 years:

$$A = 10,000e^{0.057762(10)} \approx \$17,817.97$$

11. Since $A = 500e^{rt}$ and $A = \$1505.00$ when $t = 10$, we have the following.

$$1505.00 = 500e^{10r}$$

$$r = \frac{\ln(1505.00/500)}{10} \approx 0.110 = 11.0\%$$

The time to double is given by

$$1000 = 500e^{0.110t}$$

$$t = \frac{\ln 2}{0.110} \approx 6.3 \text{ years.}$$

12. Since $A = 600e^{rt}$ and $A = 19,205$ when $t = 10$, we have

$$19,205 = 600e^{10r}$$

$$\frac{19,205}{600} = e^{10r}$$

$$\ln\left(\frac{19,205}{600}\right) = \ln e^{10r}$$

$$\ln\left(\frac{19,205}{600}\right) = 10r$$

$$r = \frac{\ln(19,205/600)}{10} \approx 0.3466 \text{ or } 34.66\%.$$

The time to double is given by

$$1200 = 600e^{0.3466t}$$

$$t = \frac{\ln 2}{0.3466} \approx 2 \text{ years.}$$

13. Since $A = Pe^{0.045t}$ and $A = 10,000.00$ when $t = 10$, we have the following.

$$10,000.00 = Pe^{0.045(10)}$$

$$\frac{10,000.00}{e^{0.045(10)}} = P \approx \$6376.28$$

The time to double is given by $t = \dfrac{\ln 2}{0.045} \approx 15.40$ years.

14. Since $A = Pe^{0.02t}$ and $A = 2000$ when $t = 10$, we have

$$2000 = Pe^{0.02(10)}$$

$$P = \frac{2000}{e^{0.02(10)}} = \$1637.46.$$

The time to double is given by $t = \dfrac{\ln 2}{0.02} = 34.7$ years.

15. $500,000 = P\left(1 + \dfrac{0.075}{12}\right)^{12(20)}$

$$P = \frac{500,000}{\left(1 + \dfrac{0.075}{12}\right)^{12(20)}}$$

$$= \frac{500,000}{1.00625^{240}} \approx \$112,087.09$$

16. $A = P\left(1 + \dfrac{r}{n}\right)^{nt}$

$$500,000 = P\left(1 + \dfrac{0.12}{12}\right)^{12(40)}$$

$$P = \$4214.16$$

17. $P = 1000, r = 11\%$

(a) $\qquad n = 1$

$$(1 + 0.11)^t = 2$$

$$t \ln 1.11 = \ln 2$$

$$t = \frac{\ln 2}{\ln 1.11} \approx 6.642 \text{ years}$$

(c) $\qquad n = 365$

$$\left(1 + \frac{0.11}{365}\right)^{365t} = 2$$

$$365t \ln\left(1 + \frac{0.11}{365}\right) = \ln 2$$

$$t = \frac{\ln 2}{365 \ln\left(1 + \frac{0.11}{365}\right)} \approx 6.302 \text{ years}$$

(b) $\qquad n = 12$

$$\left(1 + \frac{0.11}{12}\right)^{12t} = 2$$

$$12t \ln\left(1 + \frac{0.11}{12}\right) = \ln 2$$

$$t = \frac{\ln 2}{12 \ln\left(1 + \frac{0.11}{12}\right)} \approx 6.330 \text{ years}$$

(d) Compounded continuously

$$e^{0.11t} = 2$$

$$0.11t = \ln 2$$

$$t = \frac{\ln 2}{0.11} \approx 6.301 \text{ years}$$

18. $P = 1000, r = 10.5\% = 0.105$

(a) $n = 1$

$$t = \frac{\ln 2}{\ln(1 + 0.105)} \approx 6.94 \text{ years}$$

(c) $n = 365$

$$t = \frac{\ln 2}{365 \ln\left(1 + \frac{0.105}{365}\right)} \approx 6.602 \text{ years}$$

(b) $n = 12$

$$t = \frac{\ln 2}{12 \ln\left(1 + \frac{0.105}{12}\right)} \approx 6.63 \text{ years}$$

(d) Compounded continuously

$$t = \frac{\ln 2}{0.105} \approx 6.601 \text{ years}$$

19. $3P = Pe^{rt}$

$$3 = e^{rt}$$

$$\ln 3 = rt$$

$$\frac{\ln 3}{r} = t$$

r	2%	4%	6%	8%	10%	12%
$t = \dfrac{\ln 3}{r}$ (years)	54.93	27.47	18.31	13.73	10.99	9.16

20.

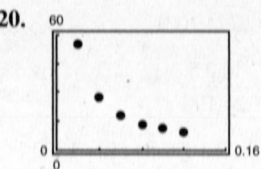

Using the power regression feature of a graphing utility, $t = 1.099r^{-1}$.

21.

$$3P = P(1 + r)^t$$

$$3 = (1 + r)^t$$

$$\ln 3 = \ln(1 + r)^t$$

$$\ln 3 = t \ln(1 + r)$$

$$\frac{\ln 3}{\ln(1 + r)} = t$$

r	2%	4%	6%	8%	10%	12%
$t = \dfrac{\ln 3}{\ln(1 + r)}$ (years)	55.48	28.01	18.85	14.27	11.53	9.69

22.

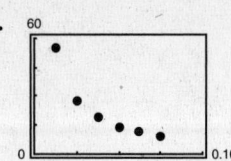

Using the power regression feature of a graphing utility, $t = 1.222r^{-1}$.

23. Continuous compounding results in faster growth.

$$A = 1 + 0.075[\![t]\!] \text{ and } A = e^{0.07t}$$

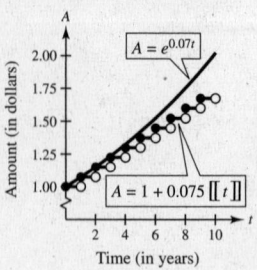

24.

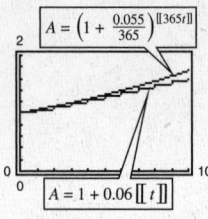

From the graph, $5\frac{1}{2}\%$ compounded daily grows faster than 6% simple interest.

25. $\frac{1}{2}C = Ce^{k(1599)}$

$0.5 = e^{k(1599)}$

$\ln 0.5 = \ln e^{k(1599)}$

$\ln 0.5 = k(1599)$

$k = \dfrac{\ln 0.5}{1599}$

Given $C = 10$ grams after 1000 years, we have

$$y = 10e^{[(\ln 0.5)/1599](1000)}$$

≈ 6.48 grams.

26. $\frac{1}{2}C = Ce^{k(1599)}$

$\dfrac{1}{2} = e^{k(1599)}$

$\ln \dfrac{1}{2} = \ln e^{k(1599)}$

$\ln \dfrac{1}{2} = k(1599)$

$k = \dfrac{\ln(1/2)}{1599}$

Given $y = 1.5$ grams after 1000 years, we have

$$1.5 = Ce^{[\ln(1/2)/1599](1000)}$$

$C \approx 2.31$ grams.

27. $\frac{1}{2}C = Ce^{k(5715)}$

$0.5 = e^{k(5715)}$

$\ln 0.5 = \ln e^{k(5715)}$

$\ln 0.5 = k(5715)$

$k = \dfrac{\ln 0.5}{5715}$

Given $y = 2$ grams after 1000 years, we have

$$2 = Ce^{[(\ln 0.5)/5715](1000)}$$

$C \approx 2.26$ grams.

28. $\frac{1}{2}C = Ce^{k(5715)}$

$\dfrac{1}{2} = e^{k(5715)}$

$\ln \dfrac{1}{2} = \ln e^{k(5715)}$

$\ln \dfrac{1}{2} = k(5715)$

$k = \dfrac{\ln(1/2)}{5715}$

Given $C = 3$ grams, after 1000 years we have

$$y = 3e^{[\ln(1/2)/5715](1000)}$$

$y \approx 2.66$ grams.

29. $\frac{1}{2}C = Ce^{k(24,100)}$

$0.5 = e^{k(24,100)}$

$\ln 0.5 = \ln e^{k(24,100)}$

$\ln 0.5 = k(24,100)$

$k = \dfrac{\ln 0.5}{24,100}$

Given $y = 2.1$ grams after 1000 years, we have

$$2.1 = Ce^{[(\ln 0.5)/24,100](1000)}$$

$C \approx 2.16$ grams.

30. $\frac{1}{2}C = Ce^{k(24,100)}$

$\frac{1}{2} = e^{k(24,100)}$

$\ln\frac{1}{2} = \ln e^{k(24,100)}$

$\ln\frac{1}{2} = k(24,100)$

$k = \dfrac{\ln(1/2)}{24,100}$

Given $y = 0.4$ grams after 1000 years, we have

$0.4 = Ce^{[\ln(1/2)/24,100](1000)}$

$C \approx 0.41$ grams.

31.

$y = ae^{bx}$

$1 = ae^{b(0)} \implies 1 = a$

$10 = e^{b(3)}$

$\ln 10 = 3b$

$\dfrac{\ln 10}{3} = b \implies b \approx 0.7675$

Thus, $y = e^{0.7675x}$.

32.

$y = ae^{bx}$

$\frac{1}{2} = ae^{b(0)} \implies a = \frac{1}{2}$

$5 = \frac{1}{2}e^{b(4)}$

$10 = e^{4b}$

$\ln 10 = \ln e^{4b}$

$\ln 10 = 4b$

$\dfrac{\ln 10}{4} = b \implies b \approx 0.5756$

Thus, $y = \frac{1}{2}e^{0.5756x}$.

33.

$y = ae^{bx}$

$5 = ae^{b(0)} \implies 5 = a$

$1 = 5e^{b(4)}$

$\frac{1}{5} = e^{4b}$

$\ln\left(\frac{1}{5}\right) = 4b$

$\dfrac{\ln(1/5)}{4} = b \implies b \approx -0.4024$

Thus, $y = 5e^{-0.4024x}$.

34.

$y = ae^{bx}$

$1 = ae^{b(0)} \implies 1 = a$

$\frac{1}{4} = e^{b(3)}$

$\ln\left(\frac{1}{4}\right) = \ln e^{3b}$

$\ln\left(\frac{1}{4}\right) = 3b$

$\dfrac{\ln(1/4)}{3} = b \implies b \approx -0.4621$

Thus, $y = e^{-0.4621x}$.

35. $P = 2430e^{-0.0029t}$

(a) Since the exponent is negative, this is an exponential decay model. The population is decreasing.

(b) For 2000, let $t = 0$: $P = 2430$ thousand people

For 2003, let $t = 3$: $P \approx 2408.95$ thousand people

(c) 2.3 million = 2300 thousand

$2300 = 2430e^{-0.0029t}$

$\dfrac{2300}{2430} = e^{-0.0029t}$

$\ln\left(\dfrac{2300}{2430}\right) = -0.0029t$

$t = \dfrac{\ln(2300/2430)}{-0.0029} \approx 18.96$

The population will reach 2.3 million (according to the model) during the later part of the year 2018.

36.

Country	2000	2010
Bulgaria	7.8	7.1
Canada	31.3	34.3
China	1268.9	1347.6
United Kingdom	59.5	61.2
United States	282.3	309.2

—CONTINUED—

36. **—CONTINUED—**

(a) Bulgaria:

$$a = 7.8$$

$$7.1 = 7.8e^{b(10)}$$

$$\ln \frac{7.1}{7.8} = 10b \implies b = -0.0094$$

For 2030, use $t = 30$.

$$y = 7.8e^{-0.0094(30)} = 5.88 \text{ million}$$

China:

$$a = 1268.9$$

$$1347.6 = 1268.9e^{b(10)}$$

$$\ln \frac{1347.6}{1268.9} = 10b \implies b \approx 0.00602$$

For 2030, use $t = 30$.

$$y = 1268.9e^{0.00602(30)} \approx 1520.06 \text{ million}$$

United Kingdom:

$$a = 59.5$$

$$61.2 = 59.5e^{b(10)}$$

$$\ln \frac{61.2}{59.5} = 10b \implies b \approx 0.00282$$

For 2030, use $t = 30$.

$$y = 59.5e^{0.00282(30)} \approx 64.7 \text{ million}$$

Canada:

$$a = 31.3$$

$$34.3 = 31.3e^{b(10)}$$

$$\ln \frac{34.3}{31.3} = 10b \implies b \approx 0.00915$$

For 2030, use $t = 30$.

$$y = 31.3e^{0.00915(30)} \approx 41.2 \text{ million}$$

United States:

$$a = 282.3$$

$$309.2 = 282.3e^{b(10)}$$

$$\ln \frac{309.2}{282.3} = 10b \implies b \approx 0.0091$$

For 2030, use $t = 30$.

$$y = 282.3e^{0.0091(30)} \approx 370.9 \text{ million}$$

(b) The constant b determines the growth rates. The greater the rate of growth, the greater the value of b.

(c) The constant b determines whether the population is increasing ($b > 0$) or decreasing ($b < 0$).

37. $y = 4080e^{kt}$

When $t = 3$, $y = 10,000$:

$$10,000 = 4080e^{k(3)}$$

$$\frac{10,000}{4080} = e^{3k}$$

$$\ln\left(\frac{10,000}{4080}\right) = 3k$$

$$k = \frac{\ln(10,000/4080)}{3} \approx 0.2988$$

When $t = 24$: $y = 4080e^{0.2988(24)} \approx 5,309,734$ hits

38. $y = 10e^{kt}$

$$65 = 10e^{k(14)}$$

$$\ln\left(\frac{65}{10}\right) = 14k \implies k = 0.1337$$

For 2010, $t = 20$:

$$y = 10e^{(0.1337)(20)} = \$144.98 \text{ million}$$

39. $N = 100e^{kt}$

$300 = 100e^{5k}$

$3 = e^{5k}$

$\ln 3 = \ln e^{5k}$

$\ln 3 = 5k$

$k = \dfrac{\ln 3}{5} \approx 0.2197$

$N = 100e^{0.2197t}$

$200 = 100e^{0.2197t}$

$t = \dfrac{\ln 2}{0.2197} \approx 3.15$ hours

40. $N = 250e^{kt}$

$280 = 250e^{k(10)}$

$1.12 = e^{10k}$

$k = \dfrac{\ln 1.12}{10}$

$N = 250e^{[(\ln 1.12)/10]t}$

$500 = 250e^{[(\ln 1.12)/10]t}$

$2 = e^{[(\ln 1.12)/10]t}$

$\ln 2 = \left(\dfrac{\ln 1.12}{10}\right)t$

$t = \dfrac{\ln 2}{(\ln 1.12)/10} \approx 61.16$ hours

41. $R = \dfrac{1}{10^{12}}e^{-t/8223}$

(a) $R = \dfrac{1}{8^{14}}$

$\dfrac{1}{10^{12}}e^{-t/8223} = \dfrac{1}{8^{14}}$

$e^{-t/8223} = \dfrac{10^{12}}{8^{14}}$

$-\dfrac{t}{8223} = \ln\left(\dfrac{10^{12}}{8^{14}}\right)$

$t = -8223 \ln\left(\dfrac{10^{12}}{8^{14}}\right) \approx 12{,}180$ years old

(b) $\dfrac{1}{10^{12}}e^{-t/8223} = \dfrac{1}{13^{11}}$

$e^{-t/8223} = \dfrac{10^{12}}{13^{11}}$

$-\dfrac{t}{8223} = \ln\left(\dfrac{10^{12}}{13^{11}}\right)$

$t = -8223 \ln\left(\dfrac{10^{12}}{13^{11}}\right) \approx 4797$ years old

42. $y = Ce^{kt}$

$\dfrac{1}{2}C = Ce^{5715k}$

$\ln\dfrac{1}{2} = 5715k$

$k = \dfrac{\ln(1/2)}{5715}$

The ancient charcoal has only 15% as much radioactive carbon.

$0.15C = Ce^{[(\ln 0.5)/5715]t}$

$\ln 0.15 = \dfrac{\ln 0.5}{5715}t$

$t = \dfrac{5715 \ln 0.15}{\ln 0.5} \approx 15{,}642$ years

43.   (0, 30,788), (2, 18,000)

(a) $m = \dfrac{18{,}000 - 30{,}788}{2 - 0} = -6394$

$b = 30{,}788$

Linear model:
$V = -6394t + 30{,}788$

(b) $a = 30{,}788$

$18{,}000 = 30{,}788e^{k(2)}$

$\dfrac{4500}{7697} = e^{2k}$

$\ln\left(\dfrac{4500}{7697}\right) = 2k$

$k = \dfrac{1}{2}\ln\left(\dfrac{4500}{7697}\right) \approx -0.268$

Exponential model: $V = 30{,}788e^{-0.268t}$

(c)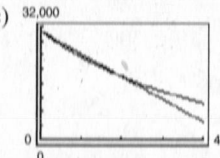

The exponential model depreciates faster in the first two years.

—CONTINUED—

43. **—CONTINUED—**

(d)

t	1	3
$V = -6394t + 30{,}788$	\$24,394	\$11,606
$V = 30{,}788e^{-0.268t}$	\$23,550	\$13,779

(e) The linear model gives a higher value for the car for the first two years, then the exponential model yields a higher value. If the car is less than two years old, the seller would most likely want to use the linear model and the buyer the exponential model. If it is more than two years old, the opposite is true.

44. $(0, 1150), (2, 550)$

(a) $m = \dfrac{550 - 1150}{2 - 0} = -300$

$V = -300t + 1150$

(c)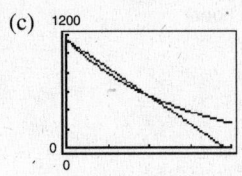

The exponential model depreciates faster in the first two years.

(e) The slope of the linear model means that the computer depreciates \$300 per year, then loses all value in the third year. The exponential model depreciates faster in the first two years but maintains value longer.

(b) $550 = 1150e^{k(2)}$

$\ln\left(\dfrac{550}{1150}\right) = 2k \implies k \approx -0.369$

$V = 1150e^{-0.369t}$

(d)

t	1	3
$V = -300t + 1100$	\$850	\$250
$V = 1150e^{-0.369t}$	\$795	\$380

45. $S(t) = 100(1 - e^{kt})$

(a) $15 = 100(1 - e^{k(1)})$

$-85 = -100e^k$

$\dfrac{85}{100} = e^k$

$0.85 = e^k$

$\ln 0.85 = \ln e^k$

$k = \ln 0.85$

$k \approx -0.1625$

$S(t) = 100(1 - e^{-0.1625t})$

(b)

(c) $S(5) = 100(1 - e^{-0.1625(5)}) \approx 55.625 = 55{,}625$ units

46. $N = 30(1 - e^{kt})$

(a) $N = 19, t = 20$

$19 = 30(1 - e^{20k})$

$30e^{20k} = 11$

$e^{20k} = \dfrac{11}{30}$

$\ln e^{20k} = \ln\left(\dfrac{11}{30}\right)$

$20k = \ln\left(\dfrac{11}{30}\right)$

$k = -0.050$

So, $N = 30(1 - e^{-0.050})$.

(b) $N = 25$

$25 = 30(1 - e^{-0.050t})$

$\dfrac{5}{30} = e^{-0.050t}$

$\ln\left(\dfrac{5}{30}\right) = \ln e^{-0.050t}$

$\ln\left(\dfrac{5}{30}\right) = -0.050t$

$t = \dfrac{\ln(5/30)}{-0.050} = 36$ days

47. $y = 0.0266e^{-(x-100)^2/450}$, $70 \le x \le 116$

(a)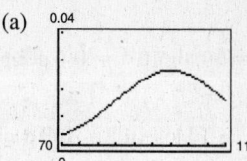

(b) The average IQ score of an adult student is 100.

48. (a)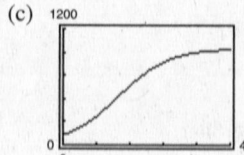

(b) The average number of hours per week a student uses the tutor center is 5.4.

49. $p(t) = \dfrac{1000}{1 + 9e^{-0.1656t}}$

(a) $p(5) = \dfrac{1000}{1 + 9e^{-0.1656(5)}} \approx 203$ animals

(b) $500 = \dfrac{1000}{1 + 9e^{-0.1656t}}$

$1 + 9e^{-0.1656t} = 2$

$9e^{-0.1656t} = 1$

$e^{-0.1656t} = \dfrac{1}{9}$

$t = -\dfrac{\ln(1/9)}{0.1656} \approx 13$ months

(c)

The horizontal asymptotes are $p = 0$ and $p = 1000$. The asymptote with the larger p-value, $p = 1000$, indicates that the population size will approach 1000 as time increases.

50. $S = \dfrac{500,000}{1 + 0.6e^{kt}}$

(a) $300,000 = \dfrac{500,000}{1 + 0.6e^{4k}}$

$1 + 0.6e^{4k} = \dfrac{5}{3}$

$0.6e^{4k} = \dfrac{2}{3}$

$e^{4k} = \dfrac{10}{9}$

$4k = \ln\left(\dfrac{10}{9}\right)$

$k = \dfrac{1}{4}\ln\left(\dfrac{10}{9}\right) \approx 0.0263$

So, $S = \dfrac{500,000}{1 + 0.6e^{0.0263t}}$.

(b) When $t = 8$:

$S = \dfrac{500,000}{1 + 0.6e^{0.0263(8)}} \approx 287,273$ units sold.

51. $R = \log\dfrac{I}{I_0} = \log I$ since $I_0 = 1$.

(a) $7.9 = \log I \Rightarrow I = 10^{7.9} \approx 79,432,823$

(b) $8.3 = \log I \Rightarrow I = 10^{8.3} \approx 199,526,231$

(c) $4.2 = \log I \Rightarrow I = 10^{4.2} \approx 15,849$

52. $R = \log\dfrac{I}{I_0} = \log I$ since $I_0 = 1$.

(a) $R = \log 80,500,000 = 7.91$

(b) $R = \log 48,275,000 = 7.68$

(c) $R = \log 251,200 = 5.40$

53. $\beta = 10\log\dfrac{I}{I_0}$ where $I_0 = 10^{-12}$ watt/m^2.

(a) $\beta = 10\log\dfrac{10^{-10}}{10^{-12}} = 10\log 10^2 = 20$ decibels

(c) $\beta = 10\log\dfrac{10^{-8}}{10^{-12}} = 10\log 10^4 = 40$ decibels

(b) $\beta = 10\log\dfrac{10^{-5}}{10^{-12}} = 10\log 10^7 = 70$ decibels

(d) $\beta = 10\log\dfrac{1}{10^{-12}} = 10\log 10^{12} = 120$ decibels

54. $\beta(I) = 10 \log \dfrac{I}{I_0}$ where $I_0 = 10^{-12}$ watt/m^2

(a) $\beta(10^{-11}) = 10 \log \dfrac{10^{-11}}{10^{-12}} = 10 \log 10^1 = 10$ decibels

(b) $\beta(10^2) = 10 \log \dfrac{10^2}{10^{-12}} = 10 \log 10^{14} = 140$ decibels

(c) $\beta(10^{-4}) = 10 \log \dfrac{10^{-4}}{10^{-12}} = 10 \log 10^8 = 80$ decibels

(d) $\beta(10^{-2}) = 10 \log \dfrac{10^{-2}}{10^{-12}} = 10 \log 10^{10} = 100$ decibels

55.
$$\beta = 10 \log \dfrac{I}{I_0}$$

$$\dfrac{\beta}{10} = \log \dfrac{I}{I_0}$$

$$10^{\beta/10} = 10^{\log I/I_0}$$

$$10^{\beta/10} = \dfrac{I}{I_0}$$

$$I = I_0 10^{\beta/10}$$

$$\% \text{ decrease} = \dfrac{I_0 10^{9.3} - I_0 10^{8.0}}{I_0 10^{9.3}} \times 100 \approx 95\%$$

56.
$$\beta = 10 \log_{10} \dfrac{I}{I_0}$$

$$10^{\beta/10} = \dfrac{I}{I_0}$$

$$I = I_0 10^{\beta/10}$$

$$\% \text{ decrease} = \dfrac{I_0 10^{8.8} - I_0 10^{7.2}}{I_0 10^{8.8}} \times 100 \approx 97\%$$

57. $\text{pH} = -\log[\text{H}^+]$

$-\log(2.3 \times 10^{-5}) \approx 4.64$

58. $\text{pH} = -\log[\text{H}^+]$

$-\log[11.3 \times 10^{-6}] \approx 4.95$

59.
$$5.8 = -\log[\text{H}^+]$$

$$-5.8 = \log[\text{H}^+]$$

$$10^{-5.8} = 10^{\log[\text{H}^+]}$$

$$10^{-5.8} = [\text{H}^+]$$

$$[\text{H}^+] \approx 1.58 \times 10^{-6} \text{ mole per liter}$$

60.
$$3.2 = -\log[\text{H}^+]$$

$$10^{-3.2} = [\text{H}^+]$$

$$[\text{H}^+] \approx 6.3 \times 10^{-4} \text{ mole per liter}$$

61.
$$2.9 = -\log[\text{H}^+]$$

$$-2.9 = \log[\text{H}^+]$$

$$[\text{H}^+] = 10^{-2.9} \text{ for the apple juice}$$

$$8.0 = -\log[\text{H}^+]$$

$$-8.0 = \log[\text{H}^+]$$

$$[\text{H}^+] = 10^{-8} \text{ for the drinking water}$$

$$\dfrac{10^{-2.9}}{10^{-8}} = 10^{5.1} \text{ times the hydrogen ion}$$
$$\text{concentration of drinking water}$$

62.
$$\text{pH} - 1 = -\log[\text{H}^+]$$

$$-(\text{pH} - 1) = \log[\text{H}^+]$$

$$10^{-(\text{pH} - 1)} = [\text{H}^+]$$

$$10^{-\text{pH} + 1} = [\text{H}^+]$$

$$10^{-\text{pH}} \cdot 10 = [\text{H}^+]$$

The hydrogen ion concentration is increased by a factor of 10.

63. $t = -10 \ln \dfrac{T - 70}{98.6 - 70}$

At 9:00 A.M. we have:

$t = -10 \ln \dfrac{85.7 - 70}{98.6 - 70} \approx 6$ hours

From this you can conclude that the person died at 3:00 A.M.

64. Interest: $u = M - \left(M - \dfrac{Pr}{12}\right)\left(1 + \dfrac{r}{12}\right)^{12t}$

Principal: $v = \left(M - \dfrac{Pr}{12}\right)\left(1 + \dfrac{r}{12}\right)^{12t}$

(a) $P = 120,000$, $t = 35$, $r = 0.075$, $M = 809.39$

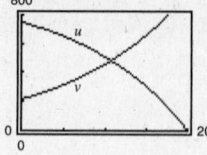

(b) In the early years of the mortgage, the majority of the monthly payment goes toward interest. The principal and interest are nearly equal when $t \approx 26$ years.

(c) $P = 120,000$, $t = 20$, $r = 0.075$, $M = 966.71$

The interest is still the majority of the monthly payment in the early years. Now the principal and interest are nearly equal when $t \approx 10.729 \approx 11$ years.

65. $u = 120,000\left[\dfrac{0.075t}{1 - \left(\dfrac{1}{1 + 0.075/12}\right)^{12t}} - 1\right]$

(a)

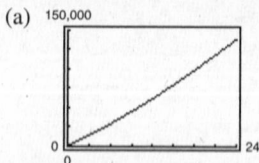

(b) From the graph, $u = \$120,000$ when $t \approx 21$ years. It would take approximately 37.6 years to pay $240,000 in interest. Yes, it is possible to pay twice as much in interest charges as the size of the mortgage. It is especially likely when the interest rates are higher.

66. $t_1 = 40.757 + 0.556s - 15.817 \ln s$

$t_2 = 1.2259 + 0.0023s^2$

(a) Linear model: $t_3 = 0.2729s - 6.0143$

Exponential model: $t_4 = 1.5385e^{0.02913s}$ or $t_4 = 1.5385(1.0296)^s$

(b)

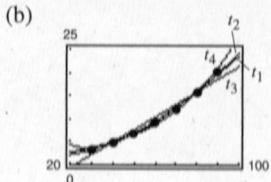

(c)

s	30	40	50	60	70	80	90
t_1	3.6	4.6	6.7	9.4	12.5	15.9	19.6
t_2	3.3	4.9	7.0	9.5	12.5	15.9	19.9
t_3	2.2	4.9	7.6	10.4	13.1	15.8	18.5
t_4	3.7	4.9	6.6	8.8	11.8	15.8	21.2

Note: Table values will vary slightly depending on the model used for t_4.

(d) Model t_1: $S_1 = |3.4 - 3.6| + |5 - 4.6| + |7 - 6.7| + |9.3 - 9.4| + |12 - 12.5| +$
$|15.8 - 15.9| + |20 - 19.6| = 2.0$

Model t_2: $S_2 = |3.4 - 3.3| + |5 - 4.9| + |7 - 7| + |9.3 - 9.5| + |12 - 12.5| +$
$|15.8 - 15.9| + |20 - 19.9| = 1.1$

Model t_3: $S_3 = |3.4 - 2.2| + |5 - 4.9| + |7 - 7.6| + |9.3 - 10.4| + |12 - 13.1| +$
$|15.8 - 15.8| + |20 - 18.5| = 5.6$

Model t_4: $S_4 = |3.4 - 3.7| + |5 - 4.9| + |7 - 6.6| + |9.3 - 8.9| + |12 - 11.9| +$
$|15.8 - 15.9| + |20 - 21.2| = 2.6$

The quadratic model, t_2, best fits the data.

67. False. The domain can be the set of real numbers for a logistic growth function.

68. False. A logistic growth function never has an x-intercept.

69. False. The graph of $f(x)$ is the graph of $g(x)$ shifted upward five units.

70. True. Powers of e are always positive, so if $a > 0$, a Gaussian model will always be greater than 0, and if $a < 0$, a Gaussian model will always be less than 0.

71. (a) Logarithmic

(b) Logistic

(c) Exponential (decay)

(d) Linear

(e) None of the above (appears to be a combination of a linear and a quadratic)

(f) Exponential (growth)

72. Answers will vary.

73. $(-1, 2), (0, 5)$

(a)

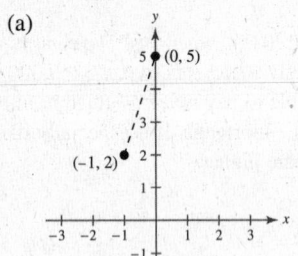

(b) $d = \sqrt{(0 - (-1))^2 + (5 - 2)^2} = \sqrt{1^2 + 3^2} = \sqrt{10}$

(c) Midpoint: $\left(\dfrac{-1 + 0}{2}, \dfrac{2 + 5}{2}\right) = \left(-\dfrac{1}{2}, \dfrac{7}{2}\right)$

(d) $m = \dfrac{5 - 2}{0 - (-1)} = \dfrac{3}{1} = 3$

74. $(4, -3), (-6, 1)$

(a)

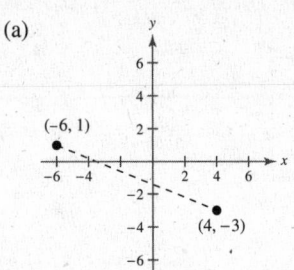

(b) $d = \sqrt{(-6 - 4)^2 + (1 - (-3))^2}$

$= \sqrt{100 + 16} = \sqrt{116} = 2\sqrt{29}$

(c) Midpoint: $\left(\dfrac{-6 + 4}{2}, \dfrac{-3 + 1}{2}\right) = (-1, -1)$

(d) $m = \dfrac{-3 - 1}{4 - (-6)} = \dfrac{-4}{10} = -\dfrac{2}{5}$

75. $(3, 3), (14, -2)$

(a)

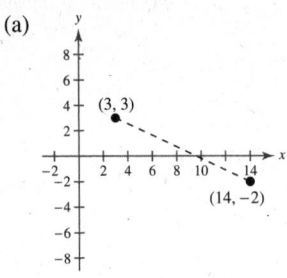

(b) $d = \sqrt{(14 - 3)^2 + (-2 - 3)^2}$

$= \sqrt{11^2 + (-5)^2} = \sqrt{146}$

(c) Midpoint: $\left(\dfrac{3 + 14}{2}, \dfrac{3 + (-2)}{2}\right) = \left(\dfrac{17}{2}, \dfrac{1}{2}\right)$

(d) $m = \dfrac{-2 - 3}{14 - 3} = -\dfrac{5}{11}$

76. $(10, 4), (7, 0)$

(a)

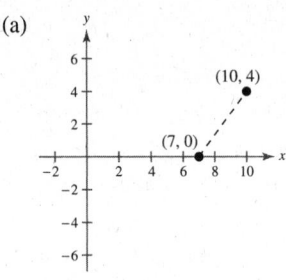

(b) $d = \sqrt{(10 - 7)^2 + (4 - 0)^2}$

$= \sqrt{9 + 16} = \sqrt{25} = 5$

(c) Midpoint: $\left(\dfrac{7 + 10}{2}, \dfrac{0 + 4}{2}\right) = \left(\dfrac{17}{2}, 2\right)$

(d) $m = \dfrac{4 - 0}{10 - 7} = \dfrac{4}{3}$

77. $\left(\frac{1}{2}, -\frac{1}{4}\right), \left(\frac{3}{4}, 0\right)$

(a)

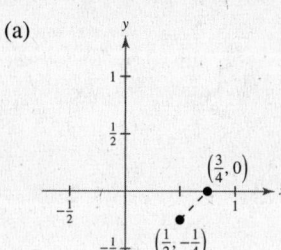

(b) $d = \sqrt{\left(\frac{3}{4} - \frac{1}{2}\right)^2 + \left(0 - \left(-\frac{1}{4}\right)\right)^2}$

$= \sqrt{\left(\frac{1}{4}\right)^2 + \left(\frac{1}{4}\right)^2} = \sqrt{\frac{1}{8}}$

(c) Midpoint: $\left(\frac{(1/2) + (3/4)}{2}, \frac{(-1/4) + 0}{2}\right) = \left(\frac{5}{8}, -\frac{1}{8}\right)$

(d) $m = \frac{0 - (-1/4)}{(3/4) - (1/2)} = \frac{1/4}{1/4} = 1$

78. $\left(\frac{7}{3}, \frac{1}{6}\right), \left(-\frac{2}{3}, -\frac{1}{3}\right)$

(a)

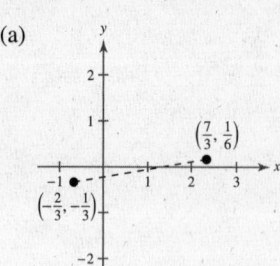

(b) $d = \sqrt{\left(-\frac{2}{3} - \frac{7}{3}\right)^2 + \left(-\frac{1}{3} - \frac{1}{6}\right)^2}$

$= \sqrt{(-3)^2 + \left(\frac{1}{2}\right)^2} = \sqrt{9.25}$

(c) Midpoint:

$\left(\frac{(-2/3) + (7/3)}{2}, \frac{(-1/3) + (1/6)}{2}\right) = \left(\frac{5}{6}, -\frac{1}{12}\right)$

(d) $m = \frac{(-1/3) - (1/6)}{(-2/3) - (7/3)} = \frac{-1/2}{-3} = \frac{1}{6}$

79. $y = 10 - 3x$

Line

Slope: $m = -3$

y-intercept: $(0, 10)$

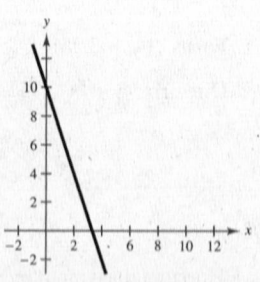

80. $y = -4x - 1$

Line

Slope: $m = -4$

y-intercept: $(0, -1)$

81. $y = -2x^2 - 3$

$y = -2(x - 0)^2 - 3$

Parabola

Vertex: $(0, -3)$

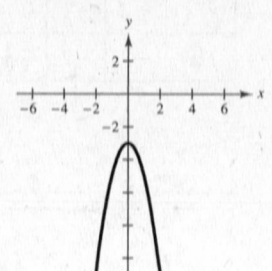

82. $y = 2x^2 - 7x - 30$

$= (2x + 5)(x - 6)$

$= 2\left(x - \frac{7}{4}\right)^2 - \frac{289}{8}$

Parabola

Vertex: $\left(\frac{7}{4}, -\frac{289}{8}\right)$

x-intercepts: $\left(-\frac{5}{2}, 0\right), (6, 0)$

83. $3x^2 - 4y = 0$

$3x^2 = 4y$

$x^2 = \frac{4}{3}y$

Parabola

Vertex: $(0, 0)$

Focus: $\left(0, \frac{1}{3}\right)$

Directrix: $y = -\frac{1}{3}$

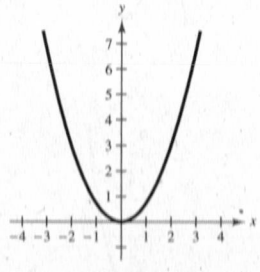

84. $-x^2 - 8y = 0$

$x^2 = -8y$

Parabola

Vertex: $(0, 0)$

Focus: $(0, -2)$

Directrix: $y = 2$

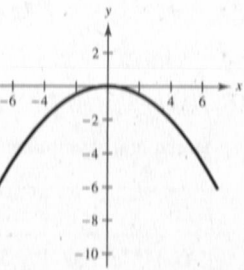

85. $y = \dfrac{4}{1 - 3x}$

Vertical asymptote: $x = \dfrac{1}{3}$

Horizontal asymptote: $y = 0$

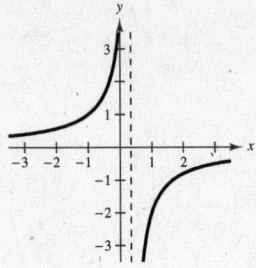

86. $y = \dfrac{x^2}{-x - 2} = -x + 2 + \dfrac{4}{-x - 2}$

Vertical asymptote: $x = -2$

Slant asymptote: $y = -x + 2$

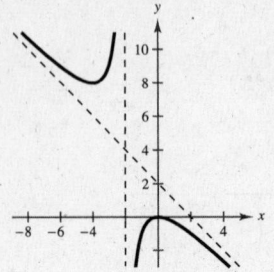

87. $x^2 + (y - 8)^2 = 25$

Circle

Center: $(0, 8)$

Radius: 5

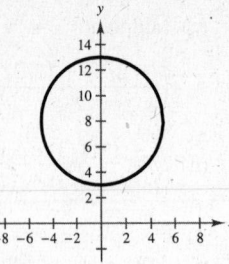

88. $(x - 4)^2 + (y + 7) = 4$

$$(x - 4)^2 = -y - 7 + 4$$

$$(x - 4)^2 = -(y + 3)$$

Parabola

Vertex: $(4, -3)$

$P = -\dfrac{1}{4}$

Focus: $(4, -3.25)$

Directrix: $y = -2.75$

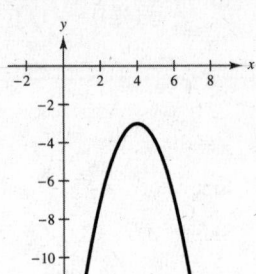

89. $f(x) = 2^{x-1} + 5$

Horizontal asymptote: $y = 5$

x	-5	-3	-1	0	1	3	5
$f(x)$	5.02	5.06	5.3	5.5	6	9	21

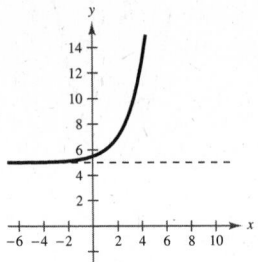

90. $f(x) = -2^{-x-1} - 1$

Horizontal asymptote: $y = -1$

x	-2	-1	0	1	2
$f(x)$	-3	-2	$-\frac{3}{2}$	$-\frac{5}{4}$	$-\frac{9}{8}$

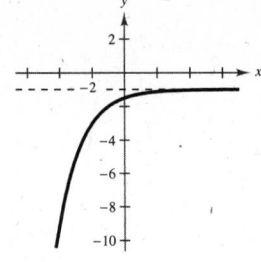

91. $f(x) = 3^x - 4$

Horizontal asymptote: $y = -4$

x	-4	-2	-1	0	1	2
$f(x)$	-3.99	-3.89	-3.67	-3	-1	5

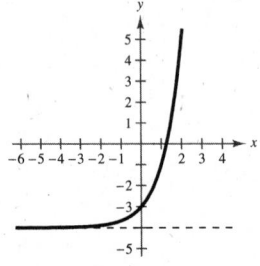

92. $f(x) = -3^x + 4$

Horizontal asymptote: $y = 4$

x	-2	-1	0	1	2
$f(x)$	$3\frac{8}{9}$	$3\frac{2}{3}$	3	1	-5

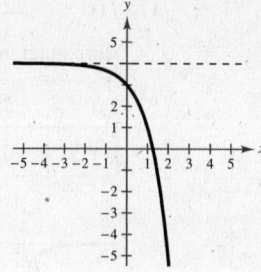

93. Answers will vary.

Review Exercises for Chapter 3

1. $f(x) = 6.1^x$

$f(2.4) = 6.1^{2.4} \approx 76.699$

2. $f(x) = 30^x$

$f(\sqrt{3}) = 30^{\sqrt{3}} = 361.784$

3. $f(x) = 2^{-0.5x}$

$f(\pi) = 2^{-0.5(\pi)} \approx 0.337$

4. $f(x) = 1278^{x/5}$

$f(1) = 1278^{1/5} = 4.181$

5. $f(x) = 7(0.2^x)$

$f(-\sqrt{11}) = 7(0.2^{-\sqrt{11}})$

≈ 1456.529

6. $f(x) = -14(5^x)$

$f(-0.8) = -14(5^{-0.8}) = -3.863$

7. $f(x) = 4^x$

Intercept: $(0, 1)$

Horizontal asymptote: x-axis

Increasing on: $(-\infty, \infty)$

Matches graph (c).

8. $f(x) = 4^{-x}$

Intercept: $(0,1)$

Horizontal asymptote: $y = 0$

Decreasing on: $(-\infty, \infty)$

Matches graph (d).

9. $f(x) = -4^x$

Intercept: $(0, -1)$

Horizontal asymptote: x-axis

Decreasing on: $(-\infty, \infty)$

Matches graph (a).

10. $f(x) = 4^x + 1$

Intercept: $(0, 2)$

Horizontal asymptote: $y = 1$

Increasing on: $(-\infty, \infty)$

Matches graph (b).

11. $f(x) = 5^x$

$g(x) = 5^{x-1}$

Since $g(x) = f(x - 1)$, the graph of g can be obtained by shifting the graph of f one unit to the right.

12. $f(x) = 4^x$, $g(x) = 4^x - 3$

Because $g(x) = f(x) - 3$, the graph of g can be obtained by shifting the graph of f three units downward.

13. $f(x) = \left(\frac{1}{2}\right)^x$

$g(x) = -\left(\frac{1}{2}\right)^{x+2}$

Since $g(x) = -f(x + 2)$, the graph of g can be obtained by reflecting the graph of f about the x-axis and shifting $-f$ two units to the left.

14. $f(x) = \left(\frac{2}{3}\right)^x$, $g(x) = 8 - \left(\frac{2}{3}\right)^x$

Because $g(x) = -f(x) + 8$, the graph of g can be obtained by reflecting the graph of f in the x-axis and shifting the graph of f eight units upward.

15. $f(x) = 4^{-x} + 4$

Horizontal asymptote: $y = 4$

x	-1	0	1	2	3
$f(x)$	8	5	4.25	4.063	4.016

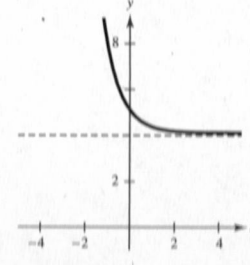

16. $f(x) = -4^x - 3$

Horizontal asymptote: $y = -3$

x	-2	-1	0	1	2
$f(x)$	-3.063	-3.25	-4	-7	-19

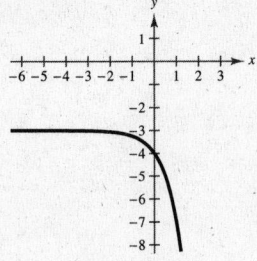

17. $f(x) = -2.65^{x+1}$

Horizontal asymptote: $y = 0$

x	-2	-1	0	1	2
$f(x)$	-0.377	-1	-2.65	-7.023	-18.61

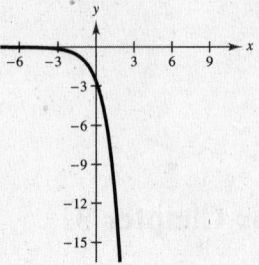

18. $f(x) = 2.65^{x-1}$

Horizontal asymptote: $y = 0$

x	-3	-1	0	1	3
$f(x)$	0.020	0.142	0.377	1	7.023

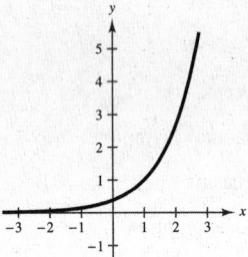

19. $f(x) = 5^{x-2} + 4$

Horizontal asymptote: $y = 4$

x	-1	0	1	2	3
$f(x)$	4.008	4.04	4.2	5	9

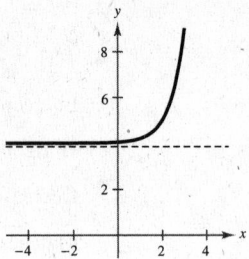

20. $f(x) = 2^{x-6} - 5$

Horizontal asymptote: $y = -5$

x	0	5	6	7	8	9
$f(x)$	-4.984	-4.5	-4	-3	-1	3

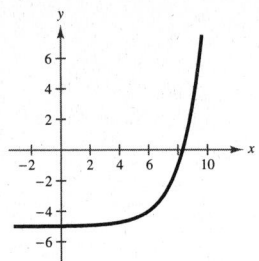

21. $f(x) = \left(\frac{1}{2}\right)^{-x} + 3 = 2^x + 3$

Horizontal asymptote: $y = 3$

x	-2	-1	0	1	2
$f(x)$	3.25	3.5	4	5	7

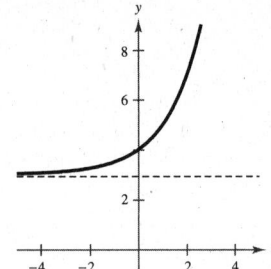

22. $f(x) = \left(\frac{1}{8}\right)^{x+2} - 5$

Horizontal asymptote: $y = -5$

x	-3	-2	-1	0	2
$f(x)$	3	-4	-4.875	-4.984	-5

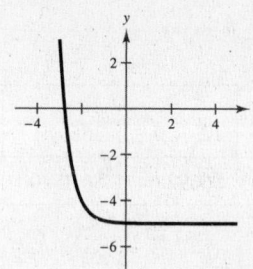

23. $3^{x+2} = \frac{1}{9}$

$3^{x+2} = 3^{-2}$

$x + 2 = -2$

$x = -4$

24. $\left(\frac{1}{3}\right)^{x-2} = 81$

$\left(\frac{1}{3}\right)^{x-2} = 3^4$

$\left(\frac{1}{3}\right)^{x-2} = \left(\frac{1}{3}\right)^{-4}$

$x - 2 = -4$

$x = -2$

25. $e^{5x-7} = e^{15}$

$5x - 7 = 15$

$5x = 22$

$x = \frac{22}{5}$

26. $e^{8-2x} = e^{-3}$

$8 - 2x = -3$

$-2x = -11$

$x = \frac{11}{2}$

27. $e^8 \approx 2980.958$

28. $e^{5/8} \approx 1.868$

29. $e^{-1.7} \approx 0.183$

30. $e^{0.278} = 1.320$

31. $h(x) = e^{-x/2}$

x	-2	-1	0	1	2
$h(x)$	2.72	1.65	1	0.61	0.37

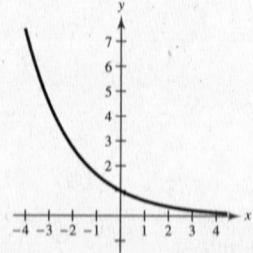

32. $h(x) = 2 - e^{-x/2}$

x	-2	-1	0	1	2
y	-0.72	0.35	1	1.39	1.63

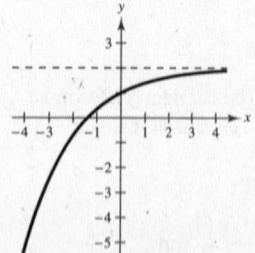

33. $f(x) = e^{x+2}$

x	-3	-2	-1	0	1
$f(x)$	0.37	1	2.72	7.39	20.09

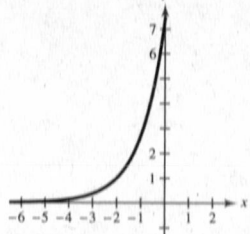

34. $s(t) = 4e^{-2/t}, t > 0$

t	$\frac{1}{2}$	1	2	3	4
y	0.07	0.54	1.47	2.05	2.43

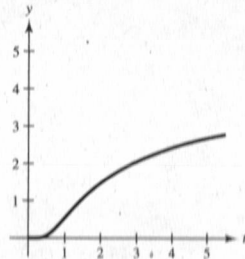

35. $A = 3500\left(1 + \dfrac{0.065}{n}\right)^{10n}$ or $A = 3500e^{(0.065)(10)}$

n	1	2	4	12	365	Continuous Compounding
A	$6569.98	$6635.43	$6669.46	$6692.64	$6704.00	$6704.39

36. $A = 2000\left(1 + \dfrac{0.05}{n}\right)^{30n}$ or $A = 2000e^{(0.05)(30)}$

n	1	2	4	12	365	Continuous
A	$8643.88	$8799.58	$8880.43	$8935.49	$8962.46	$8963.38

37. $F(t) = 1 - e^{-t/3}$

   (a) $F\left(\frac{1}{2}\right) \approx 0.154$ (b) $F(2) \approx 0.487$ (c) $F(5) \approx 0.811$

38. $V(t) = 14{,}000\left(\frac{3}{4}\right)^t$

   (a)

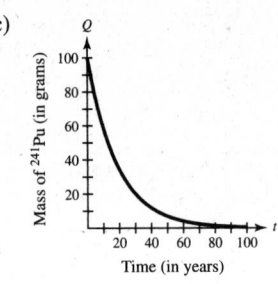

   (b) $V(2) = 14{,}000\left(\frac{3}{4}\right)^2 = \$7875$

   (c) According to the model, the car depreciates most
      rapidly at the beginning. Yes, this is realistic.

39. (a) $A = 50{,}000e^{(0.0875)(35)} \approx \$1{,}069{,}047.14$

   (b) The doubling time is $\dfrac{\ln 2}{0.0875} \approx 7.9$ years.

40. $Q = 100\left(\frac{1}{2}\right)^{t/14.4}$

   (a) For $t = 0$: $Q = 100\left(\frac{1}{2}\right)^{0/14.4} = 100$ grams   (b) For $t = 10$: $Q = 100\left(\frac{1}{2}\right)^{10/14.4} \approx 61.79$ grams

   (c)

Q vs *t* (Time (in years)); vertical axis: Mass of ^{241}Pu (in grams)

41. $4^3 = 64$

    $\log_4 64 = 3$

42. $25^{3/2} = 125$

    $\log_{25} 125 = \frac{3}{2}$

43. $e^{0.8} = 2.2255\ldots$

    $\ln 2.2255\ldots = 0.8$

44. $e^0 = 1$

    $\ln 1 = 0$

45. $f(x) = \log x$

    $f(1000) = \log 1000$

            $= \log 10^3 = 3$

46. $\log_9 3 = \log_9 9^{1/2} = \frac{1}{2}$

47. $g(x) = \log_2 x$

$g\left(\frac{1}{8}\right) = \log_2\left(\frac{1}{8}\right) = \log_2 2^{-3} = -3$

48. $f(x) = \log_4 x$

$f\left(\frac{1}{4}\right) = \log_4 \frac{1}{4} = -1$

49. $\log_4(x + 7) = \log_4 14$

$x + 7 = 14$

$x = 7$

50. $\log_8(3x - 10) = \log_8 5$

$3x - 10 = 5$

$3x = 15$

$x = 5$

51. $\ln(x + 9) = \ln 4$

$x + 9 = 4$

$x = -5$

52. $\ln(2x - 1) = \ln(11)$

$2x - 1 = 11$

$2x = 12$

$x = 6$

53. $g(x) = \log_7 x \implies x = 7^y$

Domain: $(0, \infty)$

x-intercept: $(1, 0)$

Vertical asymptote: $x = 0$

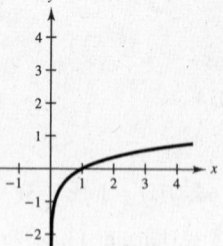

x	$\frac{1}{7}$	1	7	49
$g(x)$	-1	0	1	2

54. $g(x) = \log_5 x \implies 5^y = x$

Domain: $(0, \infty)$

$\log_5 x = 0$

$x = 5^0$

$x = 1$

x-intercept: $(1, 0)$

Vertical asymptote: $x = 0$

x	$\frac{1}{25}$	$\frac{1}{5}$	1	5	25
$g(x)$	-2	-1	0	1	2

55. $f(x) = \log\left(\frac{x}{3}\right) \implies \frac{x}{3} = 10^y \implies x = 3(10^y)$

Domain: $(0, \infty)$

x-intercept: $(3, 0)$

Vertical asymptote: $x = 0$

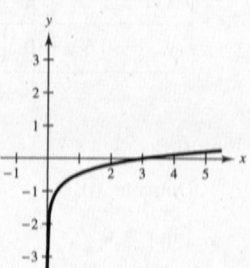

x	0.03	0.3	3	30
$f(x)$	-2	-1	0	1

56. $f(x) = 6 + \log x$

Domain: $(0, \infty)$

$6 + \log x = 0$

$\log x = -6$

$x = 10^{-6}$

$x = 0.000001$

x-intercept: $(0.000001, 0)$

Vertical asymptote: $x = 0$

x	1	2	4	6	8	10
$f(x)$	6	6.3	6.6	6.8	6.9	7

57. $f(x) = 4 - \log(x + 5)$

Domain: $(-5, \infty)$

x-intercept: $(9995, 0)$

Since $4 - \log(x + 5) = 0 \implies \log(x + 5) = 4$

$x + 5 = 10^4$

$x = 10^4 - 5 = 9995.$

Vertical asymptote: $x = -5$

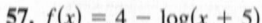

x	-4	-3	-2	-1	0	1
$f(x)$	4	3.70	3.52	3.40	3.30	3.22

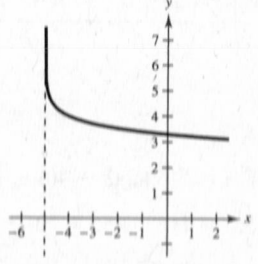

58. $f(x) = \log(x - 3) + 1$

Domain: $(3, \infty)$

$\log(x - 3) + 1 = 0$

$\log(x - 3) = -1$

$x - 3 = 10^{-1}$

$x = 3.1$

x-intercept: $(3.1, 0)$

Vertical asymptote: $x = 3$

x	4	5	6	7	8
$f(x)$	1	1.3	1.5	1.6	1.7

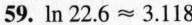

59. $\ln 22.6 \approx 3.118$

60. $\ln 0.98 \approx -0.020$

61. $\ln e^{-12} = -12$

62. $\ln e^7 = 7$

63. $\ln\left(\sqrt{7} + 5\right) \approx 2.034$

64. $\ln\left(\dfrac{\sqrt{3}}{8}\right) \approx -1.530$

65. $f(x) = \ln x + 3$

Domain: $(0, \infty)$

x-intercept: $\ln x + 3 = 0$

$\ln x = -3$

$x = e^{-3}$

$(e^{-3}, 0)$

Vertical asymptote: $x = 0$

x	1	2	3	$\frac{1}{2}$	$\frac{1}{4}$
$f(x)$	3	3.69	4.10	2.31	1.61

66. $f(x) = \ln(x - 3)$

Domain: $(3, \infty)$

$\ln(x - 3) = 0$

$x - 3 = e^0$

$x = 4$

x-intercept: $(4, 0)$

Vertical asymptote: $x = 3$

x	3.5	4	4.5	5	5.5
y	-0.69	0	0.41	0.69	0.92

67. $h(x) = \ln(x^2) = 2\ln|x|$

Domain: $(-\infty, 0) \cup (0, \infty)$

x-intercepts: $(\pm 1, 0)$

Vertical asymptote: $x = 0$

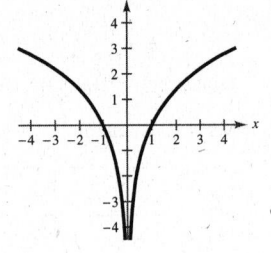

x	± 0.5	± 1	± 2	± 3	± 4
y	-1.39	0	1.39	2.20	2.77

68. $f(x) = \frac{1}{4}\ln x$

Domain: $(0, \infty)$

$\frac{1}{4}\ln x = 0$

$\ln x = 0$

$x = e^0$

$x = 1$

x-intercept: $(1, 0)$

Vertical asymptote: $x = 0$

x	$\frac{1}{2}$	1	$\frac{3}{2}$	2	$\frac{5}{2}$	3
y	-0.17	0	0.10	0.17	0.23	0.27

69. $h = 116\log(a + 40) - 176$

$h(55) = 116\log(55 + 40) - 176$

≈ 53.4 inches

70. $s = 25 - \dfrac{13\ln(10/12)}{\ln 3}$

≈ 27.16 miles

71. $\log_4 9 = \dfrac{\log 9}{\log 4} \approx 1.585$

$\log_4 9 = \dfrac{\ln 9}{\ln 4} \approx 1.585$

72. $\log_{12} 200 = \dfrac{\log 200}{\log 12} \approx 2.132$

$\log_{12} 200 = \dfrac{\ln 200}{\ln 12} \approx 2.132$

73. $\log_{1/2} 5 = \dfrac{\log 5}{\log(1/2)} \approx -2.322$

$\log_{1/2} 5 = \dfrac{\ln 5}{\ln(1/2)} \approx -2.322$

74. $\log_3 0.28 = \dfrac{\log 0.28}{\log 3} \approx -1.159$

$\log_3 0.28 = \dfrac{\ln 0.28}{\ln 3} \approx -1.159$

75. $\log 18 = \log(2 \cdot 3^2)$

$\quad = \log 2 + 2 \log 3$

$\quad \approx 1.255$

76. $\log_2 \dfrac{1}{12} = \log_2 1 - \log_2 12 = 0 - \log(2^2 \cdot 3)$

$\quad = -2 \log_2 2^2 - \log_2 3 = -2 - \dfrac{\log 3}{\log 2}$

$\quad \approx -3.585$

77. $\ln 20 = \ln(2^2 \cdot 5)$

$\quad = 2 \ln 2 + \ln 5 \approx 2.996$

78. $\ln 3e^{-4} = \ln 3 + \ln e^{-4}$

$\quad = \ln 3 - 4$

$\quad \approx -2.90$

79. $\log_5 5x^2 = \log_5 5 + \log_5 x^2$

$\quad = 1 + 2 \log_5 x$

80. $\log_{10} 7x^4 = \log 7 + \log x^4$

$\quad = \log 7 + 4 \log x$

81. $\log_3 \dfrac{6}{\sqrt[3]{x}} = \log_3 6 - \log_3 \sqrt[3]{x}$

$\quad = \log_3(3 \cdot 2) - \log_3 x^{1/3}$

$\quad = \log_3 3 + \log_3 2 - \dfrac{1}{3} \log_3 x$

$\quad = 1 + \log_3 2 - \dfrac{1}{3} \log_3 x$

82. $\log_7 \dfrac{\sqrt{x}}{4} = \log_7 \sqrt{x} - \log_7 4$

$\quad = \log_7 x^{1/2} - \log_7 4$

$\quad = \dfrac{1}{2} \log_7 x - \log_7 4$

83. $\ln x^2 y^2 z = \ln x^2 + \ln y^2 + \ln z$

$\quad = 2 \ln x + 2 \ln y + \ln z$

84. $\ln 3xy^2 = \ln 3 + \ln x + \ln y^2$

$\quad = \ln 3 + \ln x + 2 \ln y$

85. $\ln\left(\dfrac{x+3}{xy}\right) = \ln(x+3) - \ln xy$

$\quad = \ln(x+3) - [\ln x + \ln y]$

$\quad = \ln(x+3) - \ln x - \ln y$

86. $\ln\left(\dfrac{y-1}{4}\right)^2 = 2 \ln\left(\dfrac{y-1}{4}\right)$

$\quad = 2 \ln(y-1) - 2 \ln 4$

$\quad = 2 \ln(y-1) - \ln 16, \quad y > 1$

87. $\log_2 5 + \log_2 x = \log_2 5x$

88. $\log_6 y - 2 \log_6 z = \log_6 y - \log_6 z^2$

$\quad = \log_6 \dfrac{y}{z^2}$

89. $\ln x - \dfrac{1}{4} \ln y = \ln x - \ln \sqrt[4]{y} = \ln\left(\dfrac{x}{\sqrt[4]{y}}\right)$

90. $3 \ln x + 2 \ln(x+1) = \ln x^3 + \ln(x+1)^2$

$\quad = \ln x^3(x+1)^2$

91. $\dfrac{1}{3} \log_8(x+4) + 7 \log_8 y = \log_8 \sqrt[3]{x+4} + \log_8 y^7$

$\quad = \log_8\left(y^7 \sqrt[3]{x+4}\right)$

92. $-2 \log x - 5 \log(x+6) = \log x^{-2} - \log(x+6)^5$

$\quad = \log \dfrac{x^{-2}}{(x+6)^5}$

$\quad = \log \dfrac{1}{x^2(x+6)^5}$

57. $\mathbf{u} = \langle -3, 4 \rangle = -3\mathbf{i} + 4\mathbf{j}$

59. Initial point: $(3, 4)$

Terminal point: $(9, 8)$

$\mathbf{u} = (9 - 3)\mathbf{i} + (8 - 4)\mathbf{j} = 6\mathbf{i} + 4\mathbf{j}$

61. $\mathbf{v} = -10\mathbf{i} + 10\mathbf{j}$

$\|\mathbf{v}\| = \sqrt{(-10)^2 + (10)^2} = \sqrt{200} = 10\sqrt{2}$

$\tan \theta = \dfrac{10}{-10} = -1 \Rightarrow \theta = 135°$ since

$\mathbf{v}$ is in Quadrant II.

$\mathbf{v} = 10\sqrt{2}(\mathbf{i} \cos 135° + \mathbf{j} \sin 135°)$

63. $\mathbf{v} = 7(\cos 60° \, \mathbf{i} + \sin 60° \, \mathbf{j})$

$\|\mathbf{v}\| = 7$

$\theta = 60°$

65. $\mathbf{v} = 5\mathbf{i} + 4\mathbf{j}$

$\|\mathbf{v}\| = \sqrt{5^2 + 4^2} = \sqrt{41}$

$\tan \theta = \dfrac{4}{5} \Rightarrow \theta \approx 38.7°$

67. $\mathbf{v} = -3\mathbf{i} - 3\mathbf{j}$

$\|\mathbf{v}\| = \sqrt{(-3)^2 + (-3)^2} = 3\sqrt{2}$

$\tan \theta = \dfrac{-3}{-3} = 1 \Rightarrow \theta = 225°$

69. Magnitude of resultant:

$c = \sqrt{85^2 + 50^2 - 2(85)(50) \cos 165°}$

≈ 133.92 pounds

Let θ be the angle between the resultant and the 85-pound force.

$\cos \theta \approx \dfrac{(133.92)^2 + 85^2 - 50^2}{2(133.92)(85)}$

≈ 0.9953

$\Rightarrow \theta \approx 5.6°$

71. Airspeed: $\mathbf{u} = 430(\cos 45°\mathbf{i} - \sin 45°\mathbf{j}) = 215\sqrt{2}(\mathbf{i} - \mathbf{j})$

Wind: $\mathbf{w} = 35(\cos 60° + \sin 60° \, \mathbf{j}) = \dfrac{35}{2}(\mathbf{i} + \sqrt{3}\mathbf{j})$

Groundspeed: $\mathbf{u} + \mathbf{w} = \left(215\sqrt{2} + \dfrac{35}{2}\right)\mathbf{i} + \left(\dfrac{35\sqrt{3}}{2} - 215\sqrt{2}\right)\mathbf{j}$

$\|\mathbf{u} + \mathbf{w}\| = \sqrt{\left(215\sqrt{2} + \dfrac{35}{2}\right)^2 + \left(\dfrac{35\sqrt{3}}{2} - 215\sqrt{2}\right)^2}$

≈ 422.30 miles per hour

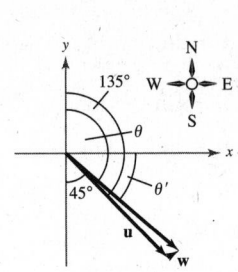

Bearing: $\tan \theta' = \dfrac{17.5\sqrt{3} - 215\sqrt{2}}{215\sqrt{2} + 17.5}$

$\theta' \approx -40.4°$

$\theta = 90° + |\theta'| = 130.4°$

73. $\mathbf{u} = \langle 6, 7 \rangle, \mathbf{v} = \langle -3, 9 \rangle$

$\mathbf{u} \cdot \mathbf{v} = 6(-3) + 7(9) = 45$

75. $\mathbf{u} = 3\mathbf{i} + 7\mathbf{j}, \mathbf{v} = 11\mathbf{i} - 5\mathbf{j}$

$\mathbf{u} \cdot \mathbf{v} = 3(11) + 7(-5) = -2$

77. $\mathbf{u} = \langle -3, 4 \rangle$

$2\mathbf{u} = \langle -6, 8 \rangle$

$2\mathbf{u} \cdot \mathbf{u} = (-6)(-3) + 8(4) = 50$

The result is a scalar.

111. $-4(5^x) = -68$

$5^x = 17$

$\ln 5^x = \ln 17$

$x \ln 5 = \ln 17$

$x = \dfrac{\ln 17}{\ln 5} \approx 1.760$

112. $2(12^x) = 190$

$12^x = 95$

$\ln 12^x = \ln 95$

$x \ln 12 = \ln 95$

$x = \dfrac{\ln 95}{\ln 12} \approx 1.833$

113. $e^{2x} - 7e^x + 10 = 0$

$(e^x - 2)(e^x - 5) = 0$

$e^x = 2 \qquad$ or $\qquad e^x = 5$

$\ln e^x = \ln 2 \qquad\qquad \ln e^x = \ln 5$

$x = \ln 2 \approx 0.693 \qquad x = \ln 5 \approx 1.609$

114. $e^{2x} - 6e^x + 8 = 0$

$(e^x - 2)(e^x - 4) = 0$

$e^x = 2 \quad$ or $\quad e^x = 4$

$x = \ln 2 \qquad x = \ln 4$

$x \approx 0.693 \qquad x \approx 1.386$

115. $2^{0.6x} - 3x = 0$

Graph $y_1 = 2^{0.6x} - 3x$.

The x-intercepts are at
$x \approx 0.392$ and at $x \approx 7.480$.

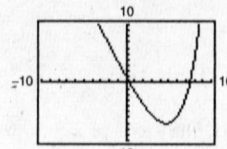

116. $4^{-0.2x} + x = 0$

Graph $y_1 = 4^{-0.2x} + x$.

The x-intercepts are at
$x \approx -7.038$ and at
$x \approx -1.527$.

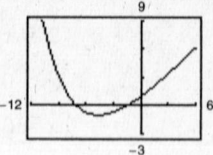

117. $25e^{-0.3x} = 12$

Graph $y_1 = 25e^{-0.3x}$ and
$y_2 = 12$.

The graphs intersect at
$x \approx 2.447$.

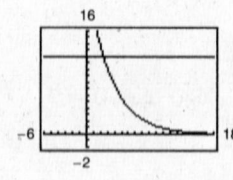

118. $4e^{1.2x} = 9$

Graph $y_1 = 4e^{1.2x}$ and $y_2 = 9$.

The graphs intersect at
$x \approx 0.676$.

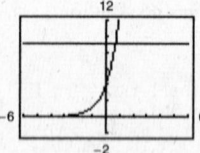

119. $\ln 3x = 8.2$

$e^{\ln 3x} = e^{8.2}$

$3x = e^{8.2}$

$x = \dfrac{e^{8.2}}{3} \approx 1213.650$

120. $\ln 5x = 7.2$

$5x = e^{7.2}$

$x = \dfrac{e^{7.2}}{5} \approx 267.886$

121. $2 \ln 4x = 15$

$\ln 4x = \dfrac{15}{2}$

$e^{\ln 4x} = e^{7.5}$

$4x = e^{7.5}$

$x = \dfrac{1}{4}e^{7.5} \approx 452.011$

122. $4 \ln 3x = 15$

$\ln 3x = \dfrac{15}{4}$

$3x = e^{15/4}$

$x = \dfrac{e^{15/4}}{3} \approx 14.174$

123. $\ln x - \ln 3 = 2$

$\ln \dfrac{x}{3} = 2$

$e^{\ln(x/3)} = e^2$

$\dfrac{x}{3} = e^2$

$x = 3e^2 \approx 22.167$

124. $\ln \sqrt{x + 8} = 3$

$\dfrac{1}{2} \ln(x + 8) = 3$

$\ln(x + 8) = 6$

$x + 8 = e^6$

$x = e^6 - 8 \approx 395.429$

125. $\ln\sqrt{x+1} = 2$

$$\frac{1}{2}\ln(x+1) = 2$$

$$\ln(x+1) = 4$$

$$e^{\ln(x+1)} = e^4$$

$$x + 1 = e^4$$

$$x = e^4 - 1 \approx 53.598$$

126. $\ln x - \ln 5 = 4$

$$\ln\frac{x}{5} = 4$$

$$\frac{x}{5} = e^4$$

$$x = 5e^4 \approx 272.991$$

127. $\log_8(x-1) = \log_8(x-2) - \log_8(x+2)$

$$\log_8(x-1) = \log_8\left(\frac{x-2}{x+2}\right)$$

$$x - 1 = \frac{x-2}{x+2}$$

$$(x-1)(x+2) = x - 2$$

$$x^2 + x - 2 = x - 2$$

$$x^2 = 0$$

$$x = 0$$

Since $x = 0$ is not in the domain of $\log_8(x-1)$ or of $\log_8(x-2)$, it is an extraneous solution. The equation has no solution.

128. $\log_6(x+2) - \log_6 x = \log_6(x+5)$

$$\log_6\left(\frac{x+2}{x}\right) = \log_6(x+5)$$

$$\frac{x+2}{x} = x + 5$$

$$x + 2 = x^2 + 5x$$

$$0 = x^2 + 4x - 2$$

$$x = -2 \pm \sqrt{6}, \text{ Quadratic Formula}$$

Only $x = -2 + \sqrt{6} \approx 0.449$ is a valid solution.

129. $\log(1 - x) = -1$

$$1 - x = 10^{-1}$$

$$1 - \frac{1}{10} = x$$

$$x = 0.900$$

130. $\log(-x - 4) = 2$

$$-x - 4 = 10^2$$

$$-x = 100 + 4$$

$$x = -104$$

131. $2\ln(x+3) + 3x = 8$

Graph $y_1 = 2\ln(x+3) + 3x$ and $y_2 = 8$.

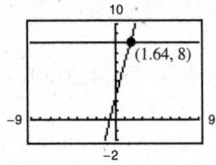

The graphs intersect at approximately $(1.643, 8)$. The solution of the equation is $x \approx 1.643$.

132. $6\log(x^2+1) - x = 0$

Graph $y_1 = 6\log(x^2+1) - x$.

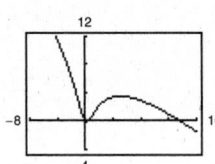

The x-intercepts are at $x = 0$, $x \approx 0.416$, and $x \approx 13.627$.

133. $4\ln(x+5) - x = 10$

Graph $y_1 = 4\ln(x+5) - x$ and $y_2 = 10$.

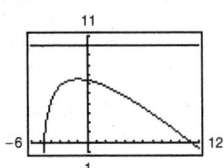

The graphs do not intersect. The equation has no solution.

113. Cube roots of $8 = 8(\cos 0 + i \sin 0)$, $k = 0, 1, 2$

(a) and (c)

$$\sqrt[3]{8}\left[\cos\left(\frac{0 + 2\pi k}{3}\right) + i \sin\left(\frac{0 + 2\pi k}{3}\right)\right]$$

$k = 0$: $2(\cos 0 + i \sin 0) = 2$

$k = 1$: $2\left(\cos\frac{2\pi}{3} + i \sin\frac{2\pi}{3}\right) = -1 + \sqrt{3}i$

$k = 2$: $2\left(\cos\frac{4\pi}{3} + i \sin\frac{4\pi}{3}\right) = -1 - \sqrt{3}i$

(b)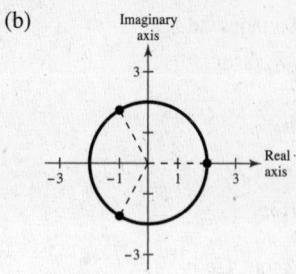

115. $x^4 + 81 = 0$

$x^4 = -81$ Solve by finding the fourth roots of -81.

$-81 = 81(\cos \pi + i \sin \pi)$

$$\sqrt[4]{-81} = \sqrt[4]{81}\left[\cos\left(\frac{\pi + 2\pi k}{4}\right) + i \sin\left(\frac{\pi + 2\pi k}{4}\right)\right], \ k = 0, 1, 2, 3$$

$k = 0$: $3\left(\cos\frac{\pi}{4} + i \sin\frac{\pi}{4}\right) = \frac{3\sqrt{2}}{2} + \frac{3\sqrt{2}}{2}i$

$k = 1$: $3\left(\cos\frac{3\pi}{4} + i \sin\frac{3\pi}{4}\right) = -\frac{3\sqrt{2}}{2} + \frac{3\sqrt{2}}{2}i$

$k = 2$: $3\left(\cos\frac{5\pi}{4} + i \sin\frac{5\pi}{4}\right) = -\frac{3\sqrt{2}}{2} - \frac{3\sqrt{2}}{2}i$

$k = 3$: $3\left(\cos\frac{7\pi}{4} + i \sin\frac{7\pi}{4}\right) = \frac{3\sqrt{2}}{2} - \frac{3\sqrt{2}}{2}i$

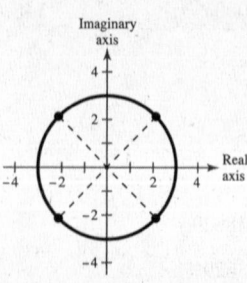

117. $x^3 + 8i = 0$

$x^3 = -8i$ Solve by finding the cube roots of $-8i$.

$-8i = 8\left(\cos\frac{3\pi}{2} + i \sin\frac{3\pi}{2}\right)$

$$\sqrt[3]{-8i} = \sqrt[3]{8}\left[\cos\left(\frac{\frac{3\pi}{2} + 2\pi k}{3}\right) + i \sin\left(\frac{\frac{3\pi}{2} + 2\pi k}{3}\right)\right], \ k = 0, 1, 2$$

$k = 0$: $2\left(\cos\frac{\pi}{2} + i \sin\frac{\pi}{2}\right) = 2i$

$k = 1$: $2\left(\cos\frac{7\pi}{6} + i \sin\frac{7\pi}{6}\right) = -\sqrt{3} - i$

$k = 2$: $2\left(\cos\frac{11\pi}{6} + i \sin\frac{11\pi}{6}\right) = \sqrt{3} - i$

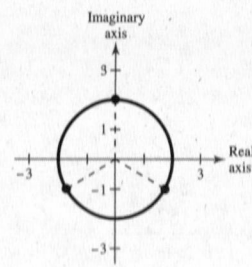

119. True. Sin 90° is defined in the Law of Sines.

121. True, by the definition of a unit vector.

$$\mathbf{u} = \frac{\mathbf{v}}{\|\mathbf{v}\|} \ \text{ so } \ \mathbf{v} = \|\mathbf{v}\|\mathbf{u}$$

123. False. $x = \sqrt{3} + i$ is a solution to $x^3 - 8i = 0$, not $x^2 - 8i = 0$.

Also, $(\sqrt{3} + i)^2 - 8i = 2 + (2\sqrt{3} - 8)i \neq 0$.

125. $a^2 = b^2 + c^2 - 2bc \cos A$

$b^2 = a^2 + c^2 - 2ac \cos B$

$c^2 = a^2 + b^2 - 2ab \cos C$

145.
$$P = 3499e^{0.0135t}$$

4.5 million = 4500 thousand

$$4500 = 3499e^{0.0135t}$$

$$\frac{4500}{3499} = e^{0.0135t}$$

$$\ln\left(\frac{4500}{3499}\right) = 0.0135t$$

$$t = \frac{\ln(4500/3499)}{0.0135} \approx 18.6 \text{ years}$$

According to this model, the population of South Carolina will reach 4.5 million during the year 2008.

146.
$$y = Ce^{kt}$$

$$\frac{1}{2}C = Ce^{(250,000)k}$$

$$\ln\frac{1}{2} = \ln e^{(250,000)k}$$

$$\ln\frac{1}{2} = 250,000k$$

$$k = \frac{\ln(1/2)}{250,000}$$

When $t = 5000$, we have

$$y = Ce^{[\ln(1/2)/250,000](5000)} \approx 0.986C = 98.6\%C.$$

After 5000 years, approximately 98.6% of the radioactive uranium II will remain.

147. (a)
$$20,000 = 10,000e^{r(5)}$$

$$2 = e^{5r}$$

$$\ln 2 = 5r$$

$$\frac{\ln 2}{5} = r$$

$$r \approx 0.138629$$

$$= 13.8629\%$$

(b) $A = 10,000e^{0.138629}$

$$\approx \$11,486.98$$

148. $N_0 = 2000$ and $N_3 = 1400$ so $N = 2000e^{kt}$ and:

$$1400 = 2000e^{3k}$$

$$\frac{7}{10} = e^{3k}$$

$$3k = \ln\left(\frac{7}{10}\right)$$

$$k = \frac{\ln(7/10)}{3} = -0.11889$$

The population one year ago:

$$N(4) = 2000e^{-0.11889(4)}$$

$$= 1243 \text{ bats}$$

149. $y = 0.0499e^{-(x-71)^2/128}$,

$$40 \le x \le 100$$

(a) Graph $y_1 = 0.0499e^{-(x-71)^2/128}$.

(b) The average test score is 71.

150. $N = \dfrac{157}{1 + 5.4e^{-0.12t}}$

(a) When $N = 50$:

$$50 = \frac{157}{1 + 5.4e^{-0.12t}}$$

$$1 + 5.4e^{-0.12t} = \frac{157}{50}$$

$$5.4e^{-0.12t} = \frac{107}{50}$$

$$e^{-0.12t} = \frac{107}{270}$$

$$-0.12t = \ln\frac{107}{270}$$

$$t = \frac{\ln(107/270)}{-0.12} \approx 7.7 \text{ weeks}$$

(b) When $N = 75$:

$$75 = \frac{157}{1 + 5.4e^{-0.12t}}$$

$$1 + 5.4e^{-0.12t} = \frac{157}{75}$$

$$5.4e^{-0.12t} = \frac{82}{75}$$

$$e^{-0.12t} = \frac{82}{405}$$

$$-0.12t = \ln\frac{82}{405}$$

$$t = \frac{\ln(82/405)}{-0.12} \approx 13.3 \text{ weeks}$$

151. $\beta = 10 \log\left(\dfrac{I}{10^{-16}}\right)$

  $125 = 10 \log\left(\dfrac{I}{10^{-16}}\right)$

  $12.5 = \log\left(\dfrac{I}{10^{-16}}\right)$

  $10^{12.5} = \dfrac{I}{10^{-16}}$

  $I = 10^{-3.5}$ watt/cm^2

152. $R = \log I$ since $I_0 = 1$.

  (a) $\log I = 8.4$

    $I = 10^{8.4} \approx 251{,}188{,}643$

  (b) $\log I = 6.85$

    $I = 10^{6.85} \approx 7{,}079{,}458$

  (c) $\log I = 9.1$

    $I = 10^{9.1} \approx 1{,}258{,}925{,}412$

153. True. By the inverse properties, $\log_b b^{2x} = 2x$.

154. False. $\ln x + \ln y = \ln(xy) \neq \ln(x + y)$

155. Since graphs (b) and (d) represent exponential decay, b and d are negative.

  Since graph (a) and (c) represent exponential growth, a and c are positive.

Problem Solving for Chapter 3

1. $y = a^x$

  $y_1 = 0.5^x$

  $y_2 = 1.2^x$

  $y_3 = 2.0^x$

  $y_4 = x$

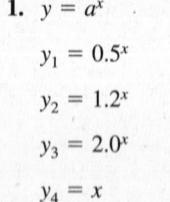 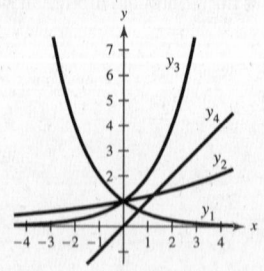 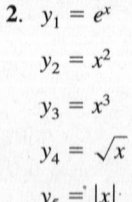

  The curves $y = 0.5^x$ and $y = 1.2^x$ cross the line $y = x$. From checking the graphs it appears that $y = x$ will cross $y = a^x$ for $0 \leq a \leq 1.44$.

2. $y_1 = e^x$

  $y_2 = x^2$

  $y_3 = x^3$

  $y_4 = \sqrt{x}$

  $y_5 = |x|$

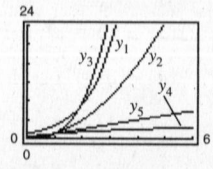

  The function that increases at the fastest rate for "large" values of x is $y_1 = e^x$. (*Note:* One of the intersection points of $y = e^x$ and $y = x^3$ is approximately $(4.536, 93)$ and past this point $e^x > x^3$. This is not shown on the graph above.)

3. The exponential function, $y = e^x$, increases at a faster rate than the polynomial function $y = x^n$.

4. It usually implies rapid growth.

5. (a) $f(u + v) = a^{u+v}$

    $= a^u \cdot a^v$

    $= f(u) \cdot f(v)$

  (b) $f(2x) = a^{2x}$

    $= (a^x)^2$

    $= [f(x)]^2$

6. $[f(x)]^2 - [g(x)]^2 = \left(\dfrac{e^x + e^{-x}}{2}\right)^2 - \left(\dfrac{e^x - e^{-x}}{2}\right)^2$

    $= \left(\dfrac{e^{2x} + 2 + e^{-2x}}{4}\right) - \left(\dfrac{e^{2x} - 2 + e^{-2x}}{4}\right)$

    $= \dfrac{4}{4}$

    $= 1$

7. (a)

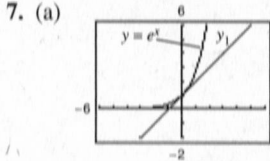

  (b)

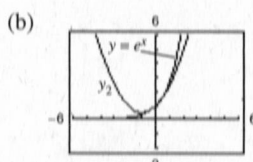

  (c)

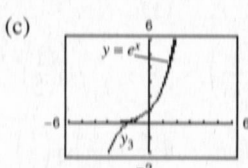

8. $y_4 = 1 + \dfrac{x}{1!} + \dfrac{x^2}{2!} + \dfrac{x^3}{3!} + \dfrac{x^4}{4!}$

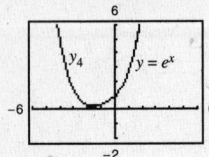

As more terms are added, the polynomial approaches e^x.

$$e^x = 1 + \frac{x}{1!} + \frac{x^2}{2!} + \frac{x^3}{3!} + \frac{x^4}{4!} + \frac{x^5}{5!} + \cdots$$

9.

$$f(x) = e^x - e^{-x}$$
$$y = e^x - e^{-x}$$
$$x = e^y - e^{-y}$$
$$x = \frac{e^{2y} - 1}{e^y}$$
$$xe^y = e^{2y} - 1$$
$$e^{2y} - xe^y - 1 = 0$$
$$e^y = \frac{x \pm \sqrt{x^2 + 4}}{2} \qquad \text{Quadratic Formula}$$

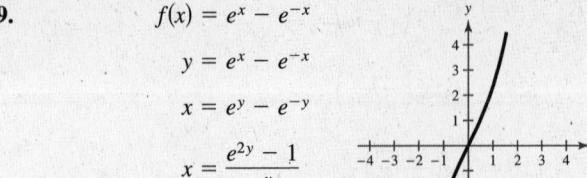

Choosing the positive quantity for e^y we have

$$y = \ln\left(\frac{x + \sqrt{x^2 + 4}}{2}\right). \text{ Thus, } f^{-1}(x) = \ln\left(\frac{x + \sqrt{x^2 + 4}}{2}\right).$$

10. $f(x) = \dfrac{a^x + 1}{a^x - 1}, a > 0, a \neq 1$

$$x = \frac{a^y + 1}{a^y - 1}$$
$$x(a^y - 1) = a^y + 1$$
$$xa^y - a^y = x + 1$$
$$a^y(x - 1) = x + 1$$
$$a^y = \frac{x + 1}{x - 1}$$
$$y = \log_a\left(\frac{x + 1}{x - 1}\right) = \frac{\ln\left(\dfrac{x + 1}{x - 1}\right)}{\ln a} = f^{-1}(x)$$

11. Answer (c). $y = 6(1 - e^{-x^2/2})$

The graph passes through $(0, 0)$ and neither (a) nor (b) pass through the origin. Also, the graph has y-axis symmetry and a horizontal asymptote at $y = 6$.

12. (a) The steeper curve represents the investment earning compound interest, because compound interest earns more than simple interest. With simple interest there is no compounding so the growth is linear.

(b) Compound interest formula: $A = 500\left(1 + \frac{0.07}{1}\right)^{(1)t} = 500(1.07)^t$

Simple interest formula: $A = Prt + P = 500(0.07)^t + 500$

(c) One should choose compound interest since the earnings would be higher.

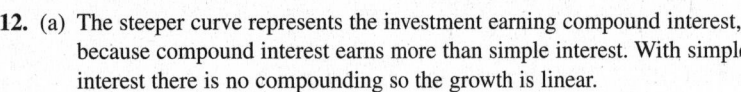

13. $y_1 = c_1\left(\dfrac{1}{2}\right)^{t/k_1}$ and $y_2 = c_2\left(\dfrac{1}{2}\right)^{t/k_2}$

$$c_1\left(\frac{1}{2}\right)^{t/k_1} = c_2\left(\frac{1}{2}\right)^{t/k_2}$$
$$\frac{c_1}{c_2} = \left(\frac{1}{2}\right)^{(t/k_2 - t/k_1)}$$
$$\ln\left(\frac{c_1}{c_2}\right) = \left(\frac{t}{k_2} - \frac{t}{k_1}\right)\ln\left(\frac{1}{2}\right)$$
$$\ln c_1 - \ln c_2 = t\left(\frac{1}{k_2} - \frac{1}{k_1}\right)\ln\left(\frac{1}{2}\right)$$
$$t = \frac{\ln c_1 - \ln c_2}{[(1/k_2) - (1/k_1)]\ln(1/2)}$$

14. $B = B_0 a^{kt}$ through $(0, 500)$ and $(2, 200)$

$$B_0 = 500$$
$$200 = 500a^{k(2)}$$
$$\frac{2}{5} = a^{2k}$$
$$\log_a\left(\frac{2}{5}\right) = 2k$$
$$\frac{1}{2}\log_a\left(\frac{2}{5}\right) = k$$

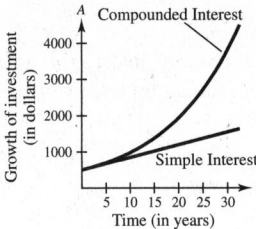

$$B = 500a^{[(1/2)\log_a(2/5)]t} = 500[a^{\log_a(2/5)}]^{t/2} = 500\left(\frac{2}{5}\right)^{t/2}$$

15. (a) $y \approx 252.606(1.0310)^t$

(b) $y \approx 400.88t^2 - 1464.6t + 291{,}782$

(c)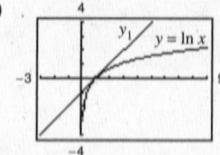

(d) Both models appear to be "good fits" for the data, but neither would be reliable to predict the population of the United States in 2010. The exponential model approaches infinity rapidly.

16. Let $\log_a x = m$ and $\log_{a/b} x = n$. Then $x = a^m$ and $x = (a/b)^n$.

$$a^m = \left(\frac{a}{b}\right)^n$$

$$a^{m/n} = \frac{a}{b}$$

$$a^{m/n - 1} = \frac{1}{b}$$

$$\log_a \frac{1}{b} = \frac{m}{n} - 1$$

$$1 + \log_a \frac{1}{b} = \frac{m}{n}$$

$$1 + \log_a \frac{1}{b} = \frac{\log_a x}{\log_{a/b} x}$$

17.
$$(\ln x)^2 = \ln x^2$$
$$(\ln x)^2 - 2\ln x = 0$$
$$\ln x(\ln x - 2) = 0$$
$$\ln x = 0 \quad \text{or} \quad \ln x = 2$$
$$x = 1 \quad \text{or} \quad x = e^2$$

18. $y = \ln x$

$$y_1 = x - 1$$
$$y_2 = (x - 1) - \tfrac{1}{2}(x - 1)^2$$
$$y_3 = (x - 1) - \tfrac{1}{2}(x - 1)^2 + \tfrac{1}{3}(x - 1)^3$$

(a) 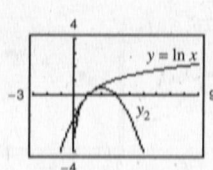   (b)   (c)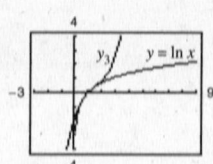

19. $y_4 = (x - 1) - \tfrac{1}{2}(x - 1)^2 + \tfrac{1}{3}(x - 1)^3 - \tfrac{1}{4}(x - 1)^4$

The pattern implies that

$$\ln x = (x - 1) - \tfrac{1}{2}(x - 1)^2 + \tfrac{1}{3}(x - 1)^3 - \tfrac{1}{4}(x - 1)^4 + \ldots$$

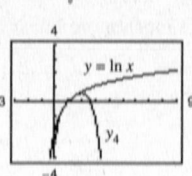

20.
$y = ab^x$	$y = ax^b$
$\ln y = \ln(ab^x)$	$\ln y = \ln(ax^b)$
$\ln y = \ln a + \ln b^x$	$\ln y = \ln a + \ln x^b$
$\ln y = \ln a + x \ln b$	$\ln y = \ln a + b \ln x$
$\ln y = (\ln b)x + \ln a$	$\ln y = b \ln x + \ln a$
Slope: $m = \ln b$	Slope: $m = b$
y-intercept: $(0, \ln a)$	y-intercept: $(0, \ln a)$

21. $y = 80.4 - 11 \ln x$

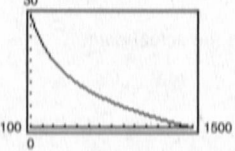

$$y(300) = 80.4 - 11 \ln 300 \approx 17.7 \ \text{ft}^3/\text{min}$$

22. (a) $\dfrac{450}{30} = 15$ cubic feet per minute

(b) $15 = 80.4 - 11 \ln x$

$11 \ln x = 65.4$

$\ln x = \dfrac{65.4}{11}$

$x = e^{65.4/11}$

$x \approx 382$ cubic feet of air space per child.

(c) Total air space required: $382(30) = 11,460$ cubic feet

Let x = floor space in square feet and h = 30 feet.

$V = xh$

$11,460 = x(30)$

$x = 382$

If the ceiling height is 30 feet, the minimum number of square feet of floor space required is 382 square feet.

23. (a)

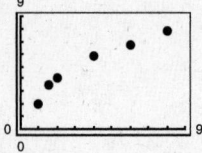

(b) The data could best be modeled by a logarithmic model.

(c) The shape of the curve looks much more logarithmic than linear or exponential.

(d) $y \approx 2.1518 + 2.7044 \ln x$

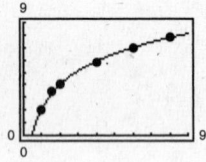

(e) The model is a good fit to the actual data.

24. (a)

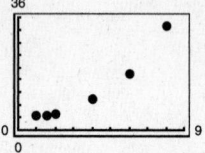

(b) The data could best be modeled by an exponential model.

(c) The data scatter plot looks exponential.

(d) $y \approx 3.114(1.341)^x$

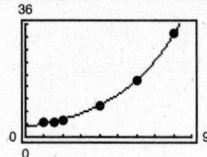

(e) The model graph hits every point of the scatter plot.

25. (a)

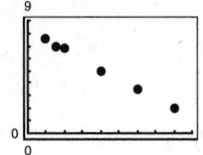

(b) The data could best be modeled by a linear model.

(c) The shape of the curve looks much more linear than exponential or logarithmic.

(d) $y \approx -0.7884x + 8.2566$

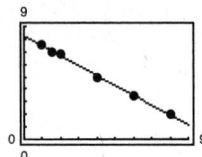

(e) The model is a good fit to the actual data.

26. (a)

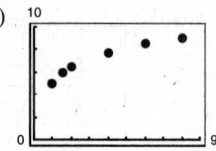

(b) The data could best be modeled by a logarithmic model.

(c) The data scatter plot looks logarithmic.

(d) $y \approx 5.099 + 1.92 \ln(x)$

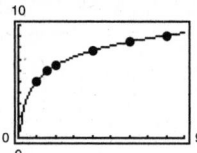

(e) The model graph hits every point of the scatter plot.

Chapter 3   Practice Test

1. Solve for x: $x^{3/5} = 8$.

2. Solve for x: $3^{x-1} = \frac{1}{81}$.

3. Graph $f(x) = 2^{-x}$.

4. Graph $g(x) = e^x + 1$.

5. If \$5000 is invested at 9% interest, find the amount after three years if the interest is compounded

   (a) monthly.   (b) quarterly.   (c) continuously.

6. Write the equation in logarithmic form: $7^{-2} = \frac{1}{49}$.

7. Solve for x: $x - 4 = \log_2 \frac{1}{64}$.

8. Given $\log_b 2 = 0.3562$ and $\log_b 5 = 0.8271$, evaluate $\log_b \sqrt[3]{8/25}$.

9. Write $5 \ln x - \frac{1}{2} \ln y + 6 \ln z$ as a single logarithm.

10. Using your calculator and the change of base formula, evaluate $\log_9 28$.

11. Use your calculator to solve for N: $\log_{10} N = 0.6646$

12. Graph $y = \log_4 x$.

13. Determine the domain of $f(x) = \log_3(x^2 - 9)$.

14. Graph $y = \ln(x - 2)$.

15. True or false: $\dfrac{\ln x}{\ln y} = \ln(x - y)$

16. Solve for x: $5^x = 41$

17. Solve for x: $x - x^2 = \log_5 \frac{1}{25}$

18. Solve for x: $\log_2 x + \log_2(x - 3) = 2$

19. Solve for x: $\dfrac{e^x + e^{-x}}{3} = 4$

20. Six thousand dollars is deposited into a fund at an annual interest rate of 13%. Find the time required for the investment to double if the interest is compounded continuously.

C H A P T E R   4
Trigonometry

C H A P T E R   4
Trigonometry

Section 4.1 · Radian and Degree Measure

You should know the following basic facts about angles, their measurement, and their applications.

■ Types of Angles:

   (a) Acute: Measure between 0° and 90°.

   (b) Right: Measure 90°.

   (c) Obtuse: Measure between 90° and 180°.

   (d) Straight: Measure 180°.

■ α and β are complementary if $\alpha + \beta = 90°$. They are supplementary if $\alpha + \beta = 180°$.

■ Two angles in standard position that have the same terminal side are called coterminal angles.

■ To convert degrees to radians, use $1° = \pi/180$ radians.

■ To convert radians to degrees, use 1 radian $= (180/\pi)°$.

■ $1' =$ one minute $= 1/60$ of $1°$.

■ $1'' =$ one second $= 1/60$ of $1' = 1/3600$ of $1°$.

■ The length of a circular arc is $s = r\theta$ where θ is measured in radians.

■ Linear speed $= \dfrac{\text{arc length}}{\text{time}} = \dfrac{s}{t}$

■ Angular speed $= \theta/t = s/rt$

Vocabulary Check

1. Trigonometry   **2.** angle

3. coterminal   **4.** radian

5. acute; obtuse   **6.** complementary; supplementary

7. degree   **8.** linear

9. angular   **10.** $A = \frac{1}{2}r^2\theta$

1.

The angle shown is approximately 2 radians.

2.

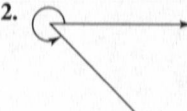

The angle shown is approximately 5.5 radians.

3.

The angle shown is approximately -3 radians.

4.

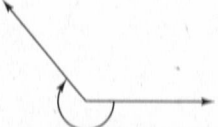

The angle shown is approximately -4 radians.

5.

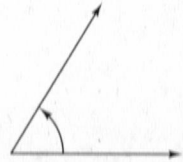

The angle shown is approximately 1 radian.

6.

The angle shown is approximately 6.5 radians.

7. (a) Since $0 < \frac{\pi}{5} < \frac{\pi}{2}$; $\frac{\pi}{5}$ lies in Quadrant I.

   (b) Since $\pi < \frac{7\pi}{5} < \frac{3\pi}{2}$; $\frac{7\pi}{5}$ lies in Quadrant III.

8. (a) Since $\pi < \frac{11\pi}{8} < \frac{3\pi}{2}$; $\frac{11\pi}{8}$ lies in Quadrant III.

   (b) Since $\pi < \frac{9\pi}{8} < \frac{3\pi}{2}$; $\frac{9\pi}{8}$ lies in Quadrant III.

9. (a) Since $-\frac{\pi}{2} < -\frac{\pi}{12} < 0$; $-\frac{\pi}{12}$ lies in Quadrant IV.

   (b) Since $-\pi < -2 < -\frac{\pi}{2}$; -2 lies in Quadrant III.

10. (a) Since $-\frac{\pi}{2} < -1 < 0$; -1 lies in Quadrant IV.

   (b) Since $-\frac{3\pi}{2} < -\frac{11\pi}{9} < -\pi$, $-\frac{11\pi}{9}$ lies in Quadrant II.

11. (a) Since $\pi < 3.5 < \frac{3\pi}{2}$; 3.5 lies in Quadrant III.

   (b) Since $\frac{\pi}{2} < 2.25 < \pi$; 2.25 lies in Quadrant II.

12. (a) Since $\frac{3\pi}{2} < 6.02 < 2\pi$; 6.02 lies in Quadrant IV.

   (b) Since $-\frac{3\pi}{2} < -4.25 < -\pi$; 4.25 lies in Quadrant II.

13. (a) $\frac{5\pi}{4}$

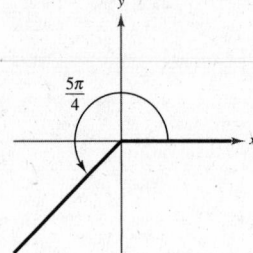

14. (a) $-\frac{7\pi}{4}$

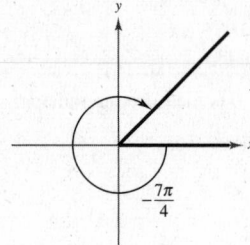

15. (a) $\frac{11\pi}{6}$

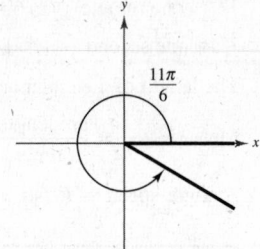

(b) $-\frac{2\pi}{3}$

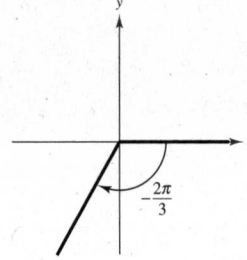

(b) $\frac{5\pi}{2}$

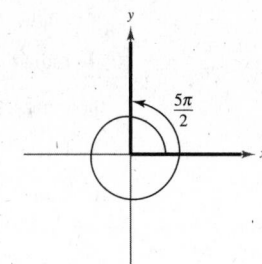

(b) -3

16. (a) 4

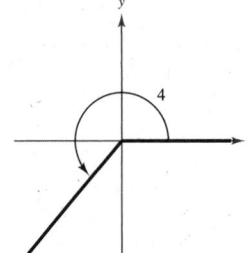

(b) 7π

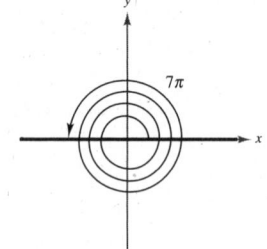

17. (a) Coterminal angles for $\dfrac{\pi}{6}$

$$\dfrac{\pi}{6} + 2\pi = \dfrac{13\pi}{6}$$

$$\dfrac{\pi}{6} - 2\pi = -\dfrac{11\pi}{6}$$

(b) Coterminal angles for $\dfrac{5\pi}{6}$

$$\dfrac{5\pi}{6} + 2\pi = \dfrac{17\pi}{6}$$

$$\dfrac{5\pi}{6} - 2\pi = -\dfrac{7\pi}{6}$$

18. (a) $\dfrac{7\pi}{6} + 2\pi = \dfrac{19\pi}{6}$

$$\dfrac{7\pi}{6} - 2\pi = -\dfrac{5\pi}{6}$$

(b) $-\dfrac{11\pi}{6} + 2\pi = \dfrac{\pi}{6}$

$$-\dfrac{11\pi}{6} - 2\pi = -\dfrac{23\pi}{6}$$

19. (a) Coterminal angles for $\dfrac{2\pi}{3}$

$$\dfrac{2\pi}{3} + 2\pi = \dfrac{8\pi}{3}$$

$$\dfrac{2\pi}{3} - 2\pi = -\dfrac{4\pi}{3}$$

(b) Coterminal angles for $\dfrac{\pi}{12}$

$$\dfrac{\pi}{12} + 2\pi = \dfrac{25\pi}{12}$$

$$\dfrac{\pi}{12} - 2\pi = -\dfrac{23\pi}{12}$$

20. (a) $-\dfrac{9\pi}{4} + 2\pi = -\dfrac{\pi}{4}$

$$-\dfrac{9\pi}{4} + 4\pi = \dfrac{7\pi}{4}$$

(b) $-\dfrac{2\pi}{15} + 2\pi = \dfrac{28\pi}{15}$

$$-\dfrac{2\pi}{15} - 2\pi = -\dfrac{32\pi}{15}$$

21. (a) Complement: $\dfrac{\pi}{2} - \dfrac{\pi}{3} = \dfrac{\pi}{6}$

Supplement: $\pi - \dfrac{\pi}{3} = \dfrac{2\pi}{3}$

(b) Complement: Not possible, $\dfrac{3\pi}{4}$ is greater than $\dfrac{\pi}{2}$.

Supplement: $\pi - \dfrac{3\pi}{4} = \dfrac{\pi}{4}$

22. (a) Complement: $\dfrac{\pi}{2} - \dfrac{\pi}{12} = \dfrac{5\pi}{12}$

Supplement: $\pi - \dfrac{\pi}{12} = \dfrac{11\pi}{12}$

(b) Complement: Not possible, $\dfrac{11\pi}{12}$ is greater than $\dfrac{\pi}{2}$.

Supplement: $\pi - \dfrac{11\pi}{12} = \dfrac{\pi}{12}$

23. (a) Complement: $\dfrac{\pi}{2} - 1 \approx 0.57$

Supplement: $\pi - 1 \approx 2.14$

(b) Complement: Not possible, 2 is greater than $\dfrac{\pi}{2}$.

Supplement: $\pi - 2 \approx 1.14$

24. (a) Complement: Not possible, 3 is greater than $\dfrac{\pi}{2}$.

Supplement: $\pi - 3 \approx 0.14$

(b) Complement: $\dfrac{\pi}{2} - 1.5 \approx 0.07$

Supplement: $\pi - 1.5 \approx 1.64$

25.

The angle shown is approximately 210°.

26.

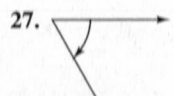

The angle shown is approximately 120°.

27.

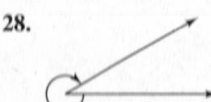

The angle shown is approximately −60°.

28.

The angle shown is approximately −330°.

29.

The angle shown is approximately 165°.

30.

The angle shown is approximately 10°.

31. (a) Since 90° < 130° < 180°, 130° lies in Quadrant II.

(b) Since 270° < 285° < 360°, 285° lies in Quadrant IV.

32. (a) Since 0° < 8.3° < 90°, 8.3° lies in Quadrant I.

(b) Since 180° < 257° 30′ < 270°, 257° 30′ lies in Quadrant III.

33. (a) Since −180° < −132° 50′ < −90°, −132° 50′ lies in Quadrant III.

(b) Since −360° < −336° < −270°, −336° lies in Quadrant I.

34. (a) Since −270° < −260° < −180°, −260° lies in Quadrant II.

(b) Since −90° < −3.4° < 0°, −3.4° lies in Quadrant IV.

35. (a) 30°

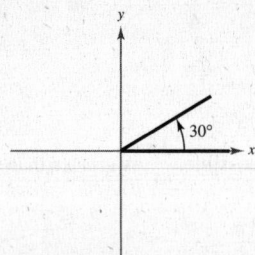

(b) 150°

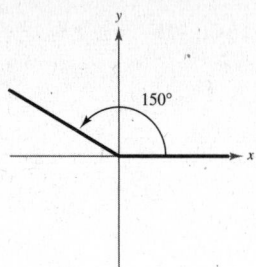

36. (a) −270°

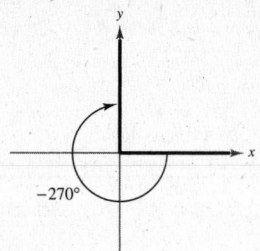

(b) −120°

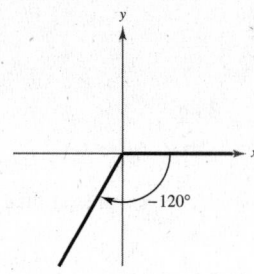

37. (a) 405°

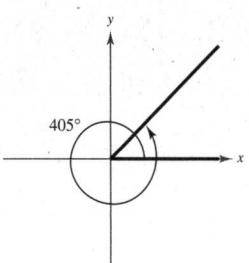

(b) 480°

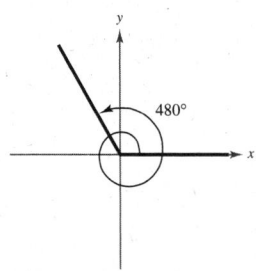

38. (a) −750°

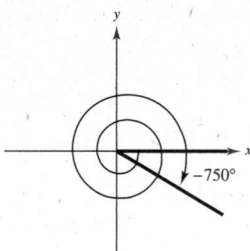

(b) −600°

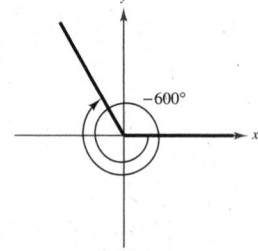

39. (a) Coterminal angles for 45°

$45° + 360° = 405°$

$45° - 360° = -315°$

(b) Coterminal angles for $-36°$

$-36° + 360° = 324°$

$-36° - 360° = -396°$

40. (a) $120° + 360° = 480°$

$120° - 360° = -240°$

(b) $-420° + 720° = 300°$

$-420° + 360° = -60°$

41. (a) Coterminal angles for 240°

$240° + 360° = 600°$

$240° - 360° = -120°$

(b) Coterminal angles for $-180°$

$-180° + 360° = 180°$

$-180° - 360° = -540°$

42. (a) $-420° + 720° = 300°$

$-420° + 360° = -60°$

(b) $230° + 360° = 590°$

$230° - 360° = -130°$

43. (a) Complement: $90° - 18° = 72°$

Supplement: $180° - 18° = 162°$

(b) Complement: Not possible, 115° is greater than 90°.

Supplement: $1180° - 115° = 65°$

44. (a) Complement: $90° - 3° = 87°$

Supplement: $180° - 3° = 177°$

(b) Complement: $90° - 64° = 26°$

Supplement: $180° - 64° = 116°$

45. (a) Complement: $90° - 79° = 11°$

Supplement: $180° - 79° = 101°$

(b) Complement: Not possible, 150° is greater than 90°.

Supplement: $180° - 150° = 30°$

46. (a) Complement: Not possible, 130° is greater than 90°.

Supplement: $180° - 130° = 50°$

(b) Complement: Not possible, 170° is greater than 90°.

Supplement: $180° - 170° = 10°$

47. (a) $30° = 30\left(\dfrac{\pi}{180}\right) = \dfrac{\pi}{6}$

(b) $150° = 150\left(\dfrac{\pi}{180}\right) = \dfrac{5\pi}{6}$

48. (a) $315° = 315°\left(\dfrac{\pi}{180°}\right) = \dfrac{7\pi}{4}$

(b) $120° = 120°\left(\dfrac{\pi}{180°}\right) = \dfrac{2\pi}{3}$

49. (a) $-20° = -20\left(\dfrac{\pi}{180}\right) = -\dfrac{\pi}{9}$

(b) $-240° = -240\left(\dfrac{\pi}{180}\right) = -\dfrac{4\pi}{3}$

50. (a) $-270° = -270°\left(\dfrac{\pi}{180°}\right) = -\dfrac{3\pi}{2}$

(b) $144° = 144°\left(\dfrac{\pi}{180°}\right) = \dfrac{4\pi}{5}$

51. (a) $\dfrac{3\pi}{2} = \dfrac{3\pi}{2}\left(\dfrac{180}{\pi}\right)° = 270°$

(b) $\dfrac{7\pi}{6} = \dfrac{7\pi}{6}\left(\dfrac{180}{\pi}\right)° = 210°$

52. (a) $-\dfrac{7\pi}{12} = -\dfrac{7\pi}{12}\left(\dfrac{180°}{\pi}\right) = -105°$

(b) $\dfrac{\pi}{9} = \dfrac{\pi}{9}\left(\dfrac{180°}{\pi}\right) = 20°$

53. (a) $\dfrac{7\pi}{3} = \dfrac{7\pi}{3}\left(\dfrac{180}{\pi}\right)° = 420°$

(b) $-\dfrac{11\pi}{30} = -\dfrac{11\pi}{30}\left(\dfrac{180}{\pi}\right)° = -66°$

54. (a) $\dfrac{11\pi}{6} = \dfrac{11\pi}{6}\left(\dfrac{180°}{\pi}\right) = 330°$

(b) $\dfrac{34\pi}{15} = \dfrac{34\pi}{15}\left(\dfrac{180°}{\pi}\right) = 408°$

55. $115° = 115\left(\dfrac{\pi}{180}\right)$

≈ 2.007 radians

56. $87.4° = 87.4°\left(\dfrac{\pi}{180°}\right)$

≈ 1.525 radians

57. $-216.35° = -216.35\left(\dfrac{\pi}{180}\right) \approx -3.776$ radians

58. $-48.27° = -48.27°\left(\dfrac{\pi}{180°}\right) \approx -0.842$ radians

59. $532° = 532\left(\dfrac{\pi}{180}\right) \approx 9.285$ radians

60. $345° = 345°\left(\dfrac{\pi}{180°}\right) \approx 6.021$ radians

61. $-0.83° = -0.83\left(\dfrac{\pi}{180}\right) \approx -0.014$ radian

62. $0.54° = 0.54°\left(\dfrac{\pi}{180°}\right) \approx 0.009$ radians

63. $\dfrac{\pi}{7} = \dfrac{\pi}{7}\left(\dfrac{180°}{\pi}\right)° \approx 25.714°$

64. $\dfrac{5\pi}{11} = \dfrac{5\pi}{11}\left(\dfrac{180°}{\pi}\right)° \approx 81.818°$

65. $\dfrac{15\pi}{8} = \dfrac{15\pi}{8}\left(\dfrac{180°}{\pi}\right)° = 337.500°$

66. $\dfrac{13\pi}{2} = \dfrac{13\pi}{2}\left(\dfrac{180°}{\pi}\right)° = 1170.000°$

67. $-4.2\pi = -4.2\pi\left(\dfrac{180°}{\pi}\right)°$
$= -756.000°$

68. $4.8\pi = 4.8\pi\left(\dfrac{180°}{\pi}\right)° = 864.000°$

69. $-2 = -2\left(\dfrac{180°}{\pi}\right)° \approx -114.592°$

70. $-0.57 = -0.57\left(\dfrac{180°}{\pi}\right)° \approx -32.659°$

71. (a) $54° \, 45' = 54° + \left(\dfrac{45}{60}\right)° = 54.75°$

(b) $-128° \, 30' = -128° - \left(\dfrac{30}{60}\right)° = -128.5°$

72. (a) $245°10' = 245° + \left(\dfrac{10}{60}\right)° \approx 245° + 0.167° = 245.167°$

(b) $2°12' = 2° + \left(\dfrac{12}{60}\right)° = 2° + 0.2° = 2.2°$

73. (a) $85° \, 18' \, 30'' = \left(85 + \dfrac{18}{60} + \dfrac{30}{3600}\right)° \approx 85.308°$

(b) $330° \, 25'' = \left(330 + \dfrac{25}{3600}\right)° \approx 330.007°$

74. (a) $-135° \, 36'' = -135° - \left(\dfrac{36}{3600}\right)°$
$= -135° - 0.01° = -135.01°$

(b) $-408° \, 16' \, 20'' = -\left(408° + \left(\dfrac{16}{60}\right)° + \left(\dfrac{20}{3600}\right)°\right)$
$\approx -(408° + 0.2667° + 0.0056°)$
$\approx -408.272°$

75. (a) $240.6° = 240° + 0.6(60)' = 240° \, 36'$

(b) $-145.8° = -[145° + 0.8(60')] = -145° \, 48'$

76. (a) $-345.12° = -(345° + (0.12)(60'))$
$= -(345° + 7' + 0.2(60''))$
$= -345° \, 7' \, 12''$

(b) $0.45° = 0° + (0.45)(60') = 0° + 27' = 0° \, 27'$

77. (a) $2.5° = 2° \, 30'$

(b) $-3.58° = -(3° + (0.58)(60'))$
$= -(3° + 34' + 0.8(60''))$
$= -3° \, 34' \, 48''$

78. (a) $-0.355° = -(0° + (0.355)(60'))$
$= -(0° + 21' + (0.3)(60''))$
$= -(0° + 21' + 18'') = -0° \, 21' \, 18''$

(b) $0.7865 = 0° + (0.7865)(60')$
$= 0° + 47' + (0.19)(60'')$
$= 0° + 47' + 11.4'' = 0° \, 47' \, 11.4''$

79. $s = r\theta$
$6 = 5\theta$
$\theta = \dfrac{6}{5}$ radians

80. $s = r\theta$
$29 = 10\,\theta$
$\theta = \dfrac{29}{10}$ radians

81. $s = r\theta$
$32 = 7\theta$
$\theta = \dfrac{32}{7} = 4\dfrac{4}{7}$ radians

82. $s = r\theta$
$60 = 75\theta$
$\theta = \dfrac{60}{75} = \dfrac{4}{5}$ radians

Because the angle represented is clockwise, this angle is $-\dfrac{4}{5}$ radians.

83. $s = r\theta$
$6 = 27\theta$
$\theta = \dfrac{6}{27} = \dfrac{2}{9}$ radians

84. $r = 14$ feet, $s = 8$ feet
$\theta = \dfrac{s}{r} = \dfrac{8}{14} = \dfrac{4}{7}$ radians

85. $s = r\theta$

$25 = 14.5\theta$

$\theta = \dfrac{25}{14.5} = \dfrac{50}{29}$ radians

86. $r = 80$ kilometers,
$s = 160$ kilometers

$\theta = \dfrac{s}{r} = \dfrac{160}{80} = 2$ radians

87. $s = r\theta$, θ in radians

$s = 15(180)\left(\dfrac{\pi}{180}\right) = 15\pi$ inches

≈ 47.12 inches

88. $r = 9$ feet, $\theta = 60° = \dfrac{\pi}{3}$

$s = r\theta = 9\left(\dfrac{\pi}{3}\right) = 3\pi$ feet

≈ 9.42 feet

89. $s = r\theta$, θ in radians

$s = 3(1) = 3$ meters

90. $r = 20$ centimeters, $\theta = \dfrac{\pi}{4}$

$s = r\theta = 20\left(\dfrac{\pi}{4}\right) = 5\pi$ centimeters

≈ 15.71 centimeters

91. $A = \dfrac{1}{2}r^2\theta$

$A = \dfrac{1}{2}(4)^2\left(\dfrac{\pi}{3}\right) = \dfrac{8\pi}{3}$ square inches

≈ 8.38 square inches

92. $r = 12$ mm, $\theta = \dfrac{\pi}{4}$

$A = \dfrac{1}{2}r^2\theta = \dfrac{1}{2}(12)^2\left(\dfrac{\pi}{4}\right)$

$= 18\pi$ mm^2

≈ 56.55 mm^2

93. $A = \dfrac{1}{2}r^2\theta$

$A = \dfrac{1}{2}(2.5)^2(225)\left(\dfrac{\pi}{180}\right)$

≈ 12.27 square feet

94. $r = 1.4$ miles, $\theta = 330°$

$A = \dfrac{1}{2}(1.4)^2\left(\dfrac{330°}{180°}\right)\pi = \dfrac{21.56}{12}\pi \approx 5.64$ square miles

95. $\theta = 41° \, 15' \, 50'' - 32° \, 47' \, 39''$

$\approx 8.46972° \approx 0.14782$ radian

$s = r\theta \approx 4000(0.14782) \approx 591.3$ miles

96. $r = 4000$ miles

$\theta = 47° \, 37' \, 18'' - 37° \, 47' \, 36'' = 9° \, 49' \, 42''$

≈ 0.1715 radian

$s = r\theta \approx (4000)(0.1715) = 686.2$ miles

97. $\theta = \dfrac{s}{r} = \dfrac{450}{6378} \approx 0.071$ radian $\approx 4.04°$

98. $r = 3189$ kilometers

$s = r\theta$

$400 = 6378\theta$

$\dfrac{400}{6378} = \theta$

$0.062716 \approx \theta$

The difference in latitude is about 0.062716 radians $\approx 3.59°$.

99. $\theta = \dfrac{s}{r} = \dfrac{2.5}{6} = \dfrac{25}{60} = \dfrac{5}{12}$ radian

100. $\theta = \dfrac{s}{r} = \dfrac{24}{5} = 4.8$ radians $= 4.8\left(\dfrac{180°}{\pi}\right) \approx 275°$

101. (a) 65 miles per hour $= \dfrac{65(5280)}{60} = 5720$ feet per minute

The circumference of the tire is $C = 2.5\pi$ feet.

The number of revolutions per minute is

$r = \dfrac{5720}{2.5\pi} \approx 728.3$ revolutions per minute

(b) The angular speed is $\dfrac{\theta}{t}$.

$\theta = \dfrac{5720}{2.5\pi}(2\pi) = 4576$ radians

Angular speed $= \dfrac{4576 \text{ radians}}{1 \text{ minute}} = 4576$ radians per minute

102. Linear velocity for either pulley: $1700(2\pi) = 3400\pi$ inches per minute

(a) Angular speed of motor pulley: $\omega = \dfrac{v}{r} = \dfrac{3400\pi}{1} = 3400\pi$ radians per minute

Angular speed of the saw arbor: $\omega = \dfrac{v}{r} = \dfrac{3400\pi}{2} = 1700\pi$ radians per minute

(b) Revolutions per minute of the saw arbor: $\dfrac{1700\pi}{2\pi} = 850$ revolutions per minute

103. (a) Angular speed $= \dfrac{(5200)(2\pi) \text{ radians}}{1 \text{ minute}}$

$= 10{,}400\pi$ radians per minute

$\approx 32{,}672.56$ radians per minute

(b) Linear speed $= \dfrac{\left(\dfrac{7.25}{2} \text{ in.}\right)\left(\dfrac{1 \text{ ft}}{12 \text{ in.}}\right)(5200)(2\pi)}{1 \text{ minute}}$

$= 3141\dfrac{2}{3}\pi$ feet per minute

≈ 9869.84 feet per minute

104. (a) 4 rpm $= 4(2\pi)$ radians per minute

$= 8\pi$ radians per minute

≈ 25.13 radians per minute

(b) $r = 25$ ft

$\dfrac{r\theta}{t} = 200\pi$ feet per minute

Linear speed $\approx 25(25.13274)$ feet per minute

≈ 628.32 feet per minute

105. (a) $(200)(2\pi) \le$ Angular speed $\le (500)(2\pi)$ radians per minute

Interval: $[400\pi, 1000\pi]$ radians per minute

(b) $(6)(200)(2\pi) \le$ Linear speed $\le (6)(500)(2\pi)$ centimeters per minute

Interval: $[2400\pi, 6000\pi]$ centimeters per minute

106. $A = \dfrac{1}{2}\theta(R^2 - r^2)$

$R = 25$

$r = 25 - 14 = 11$

$A = \dfrac{1}{2}\left(\dfrac{125}{180}\right)\pi \cdot (25^2 - 11^2) = 175\pi \approx 549.8$ square inches

107. $A = \dfrac{1}{2}r^2\theta$

$= \dfrac{1}{2}(35)^2(140°)\left(\dfrac{\pi}{180°}\right)$

$\approx 476.39\pi$ square meters

≈ 1496.62 square meters

108. (a) Arc length of larger sprocket in feet: $s = r\theta$

$$s = \dfrac{1}{3}(2\pi) = \dfrac{2\pi}{3} \text{ feet}$$

Therefore, the chain moves $2\pi/3$ feet, as does the smaller rear sprocket. Thus, the angle θ of the smaller sprocket is $(r = 2 \text{ inches} = 2/12 \text{ feet})$.

$$\theta = \dfrac{s}{r} = \dfrac{(2\pi)/3 \text{ feet}}{2/12 \text{ feet}} = 4\pi \text{ and the arc length of the tire in feet is:}$$

$$s = \theta r$$

$$s = (4\pi)\left(\dfrac{14}{12}\right) = \dfrac{14\pi}{3} \text{ feet}$$

$$\text{Speed} = \dfrac{s}{t} = \dfrac{(14\pi)/3}{1 \text{ second}} = \dfrac{14\pi}{3} \text{ feet per second}$$

$$\dfrac{14\pi \text{ feet}}{3 \text{ seconds}} \times \dfrac{3600 \text{ seconds}}{1 \text{ hour}} \times \dfrac{1 \text{ mile}}{5280 \text{ feet}} \approx 10 \text{ miles per hour}$$

—CONTINUED—

108. **—CONTINUED—**

(b) Since the arc length of the tire is $(14\pi)/3$ feet and the cyclist is pedaling at a rate of one revolution per second, we have:

$$\text{Distance} = \left(\frac{14\pi}{3} \frac{\text{feet}}{\text{revolutions}}\right)\left(\frac{1 \text{ mile}}{5280 \text{ feet}}\right)(n \text{ revolutions}) = \frac{7\pi}{7920}n \text{ miles}$$

(c) Distance = Rate · Time

$$= \left(\frac{14\pi}{3} \text{ feet per second}\right)\left(\frac{1 \text{ mile}}{5280 \text{ feet}}\right)(t \text{ seconds}) = \frac{7\pi}{7920}t \text{ miles}$$

(d) The functions are both linear.

109. False. An angle measure of 4π radians corresponds to two complete revolutions from the initial to the terminal side of an angle.

110. True. If α and β are coterminal angles, then $\alpha = \beta + n(360°)$ where n is an integer. The difference between α and β is $\alpha - \beta = n(360°) = 2\pi n$.

111. False. The terminal side of $-1260°$ lies on the negative x-axis.

112. (a) An angle is in standard position if its vertex is at the origin and its initial side is on the positive x-axis.

(b) A negative angle is generated by a clockwise rotation of the terminal side.

(c) Two angles in standard position with the same terminal sides are coterminal.

(d) An obtuse angle measures between $90°$ and $180°$.

113. Increases, since the linear speed is proportional to the radius.

114. $1 \text{ radian} = \left(\frac{180}{\pi}\right)^{\circ} \approx 57.3°,$
so one radian is much larger than one degree.

115. The arc length is increasing. In order for the angle θ to remain constant as the radius r increases, the arc length s must increase in proportion to r, as can be seen from the formula $s = r\theta$.

116. The area of a circle is $A = \pi r^2 \Rightarrow \pi = \dfrac{A}{r^2}$. The circumference of a circle is $C = 2\pi r$.

$$C = 2\left(\frac{A}{r^2}\right)r$$

$$C = \frac{2A}{r}$$

$$\frac{Cr}{2} = A$$

For a sector, $C = s = r\theta$. Thus, $A = \dfrac{(r\theta)r}{2} = \dfrac{1}{2}\theta r^2$ for a sector.

117. $\dfrac{4}{4\sqrt{2}} = \dfrac{4}{4\sqrt{2}} \cdot \dfrac{\sqrt{2}}{\sqrt{2}} = \dfrac{4\sqrt{2}}{8} = \dfrac{\sqrt{2}}{2}$

118. $\dfrac{5\sqrt{5}}{2\sqrt{10}} = \dfrac{5}{2}\sqrt{\dfrac{5}{10}} = \dfrac{5}{2}\sqrt{\dfrac{1}{2}} = \dfrac{5}{2\sqrt{2}} \cdot \dfrac{\sqrt{2}}{\sqrt{2}} = \dfrac{5\sqrt{2}}{4}$

119. $\sqrt{2^2 + 6^2} = \sqrt{4 + 36} = \sqrt{40} = \sqrt{4 \cdot 10} = 2\sqrt{10}$

120. $\sqrt{17^2 - 9^2} = \sqrt{289 - 81}$
$= \sqrt{208} = \sqrt{16 \cdot 13} = 4\sqrt{13}$

121. $f(x) = (x - 2)^5$

Graph of $y = x^5$ shifted
to the right by two units

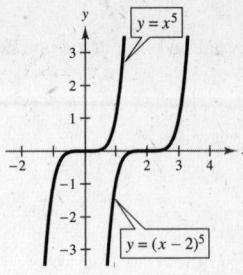

122. $f(x) = x^5 - 4$

Vertical shift four units
downward

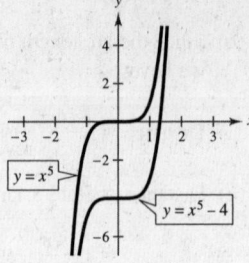

123. $f(x) = 2 - x^5$

Graph of $y = x^5$ reflected
in x-axis and shifted
upward by two units

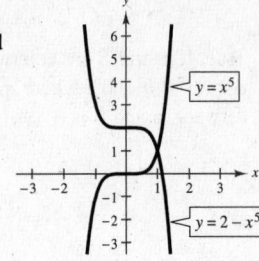

124. $f(x) = -(x + 3)^5$

Reflection in the x-axis
and a horizontal shift
three units to the left

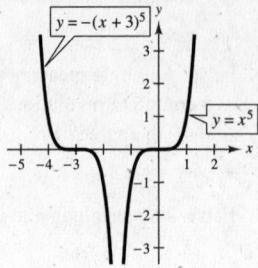

Section 4.2   Trigonometric Functions: The Unit Circle

■ You should know the definition of the trigonometric functions in terms of the unit circle. Let t be a real number and (x, y) the point on the unit circle corresponding to t.

$$\sin t = y \qquad\qquad \csc t = \frac{1}{y}, \quad y \neq 0$$

$$\cos t = x \qquad\qquad \sec t = \frac{1}{x}, \quad x \neq 0$$

$$\tan t = \frac{y}{x}, \quad x \neq 0 \qquad\qquad \cot t = \frac{x}{y}, \quad y \neq 0$$

■ The cosine and secant functions are even.

$$\cos(-t) = \cos t \qquad\qquad \sec(-t) = \sec t$$

■ The other four trigonometric functions are odd.

$$\sin(-t) = -\sin t \qquad\qquad \csc(-t) = -\csc t$$

$$\tan(-t) = -\tan t \qquad\qquad \cot(-t) = -\cot t$$

■ Be able to evaluate the trigonometric functions with a calculator.

Vocabulary Check

1. unit circle

2. periodic

3. period

4. odd; even

1. $x = -\dfrac{8}{17}, y = \dfrac{15}{17}$

$\sin\theta = y = \dfrac{15}{17}$ $\csc\theta = \dfrac{1}{y} = \dfrac{17}{15}$

$\cos\theta = x = -\dfrac{8}{17}$ $\sec\theta = \dfrac{1}{x} = -\dfrac{17}{8}$

$\tan\theta = \dfrac{y}{x} = -\dfrac{15}{8}$ $\cot\theta = \dfrac{x}{y} = -\dfrac{8}{15}$

2. $x = \dfrac{12}{13}, y = \dfrac{5}{13}$

$\sin\theta = y = \dfrac{5}{13}$ $\csc\theta = \dfrac{1}{y} = \dfrac{13}{5}$

$\cos\theta = x = \dfrac{12}{13}$ $\sec\theta = \dfrac{1}{x} = \dfrac{13}{12}$

$\tan\theta = \dfrac{y}{x} = \dfrac{5}{12}$ $\cot\theta = \dfrac{x}{y} = \dfrac{12}{5}$

3. $x = \dfrac{12}{13}, y = -\dfrac{5}{13}$

$\sin\theta = y = -\dfrac{5}{13}$ $\csc\theta = \dfrac{1}{y} = -\dfrac{13}{5}$

$\cos\theta = x = \dfrac{12}{13}$ $\sec\theta = \dfrac{1}{x} = \dfrac{13}{12}$

$\tan\theta = \dfrac{y}{x} = -\dfrac{5}{12}$ $\cot\theta = \dfrac{x}{y} = -\dfrac{12}{5}$

4. $x = -\dfrac{4}{5}, y = -\dfrac{3}{5}$

$\sin\theta = y = -\dfrac{3}{5}$ $\csc\theta = \dfrac{1}{y} = -\dfrac{5}{3}$

$\cos\theta = x = -\dfrac{4}{5}$ $\sec\theta = \dfrac{1}{x} = -\dfrac{5}{4}$

$\tan\theta = \dfrac{y}{x} = \dfrac{3}{4}$ $\cot\theta = \dfrac{x}{y} = \dfrac{4}{3}$

5. $t = \dfrac{\pi}{4}$ corresponds to $\left(\dfrac{\sqrt{2}}{2}, \dfrac{\sqrt{2}}{2}\right)$.

6. $t = \dfrac{\pi}{3}, (x\,y) = \left(\dfrac{1}{2}, \dfrac{\sqrt{3}}{2}\right)$

7. $t = \dfrac{7\pi}{6}$ corresponds to $\left(-\dfrac{\sqrt{3}}{2}, -\dfrac{1}{2}\right)$.

8. $t = \dfrac{5\pi}{4}, (x, y) = \left(-\dfrac{\sqrt{2}}{2}, -\dfrac{\sqrt{2}}{2}\right)$

9. $t = \dfrac{4\pi}{3}$ corresponds to $\left(-\dfrac{1}{2}, -\dfrac{\sqrt{3}}{2}\right)$.

10. $t = \dfrac{5\pi}{3}, (x, y) = \left(\dfrac{1}{2}, -\dfrac{\sqrt{3}}{2}\right)$

11. $t = \dfrac{3\pi}{2}$ corresponds to $(0, -1)$.

12. $t = \pi, (x, y) = (\neg 1, 0)$

13. $t = \dfrac{\pi}{4}$ corresponds to $\left(\dfrac{\sqrt{2}}{2}, \dfrac{\sqrt{2}}{2}\right)$.

$\sin t = y = \dfrac{\sqrt{2}}{2}$

$\cos t = x = \dfrac{\sqrt{2}}{2}$

$\tan t = \dfrac{y}{x} = 1$

14. $t = \dfrac{\pi}{3}, (x, y) = \left(\dfrac{1}{2}, \dfrac{\sqrt{3}}{2}\right)$

$\sin\dfrac{\pi}{3} = \dfrac{\sqrt{3}}{2}$

$\cos\dfrac{\pi}{3} = \dfrac{1}{2}$

$\tan\dfrac{\pi}{3} = \dfrac{\sqrt{3}/2}{1/2} = \sqrt{3}$

15. $t = -\dfrac{\pi}{6}$ corresponds to $\left(\dfrac{\sqrt{3}}{2}, -\dfrac{1}{2}\right)$.

$\sin t = y = -\dfrac{1}{2}$

$\cos t = x = \dfrac{\sqrt{3}}{2}$

$\tan t = \dfrac{y}{x} = -\dfrac{1}{\sqrt{3}} = -\dfrac{\sqrt{3}}{3}$

16. $t = -\dfrac{\pi}{4}, (x, y) = \left(\dfrac{\sqrt{2}}{2}, -\dfrac{\sqrt{2}}{2}\right)$

$\sin\left(-\dfrac{\pi}{4}\right) = -\dfrac{\sqrt{2}}{2}$

$\cos\left(-\dfrac{\pi}{4}\right) = \dfrac{\sqrt{2}}{2}$

$\tan\left(-\dfrac{\pi}{4}\right) = \dfrac{-\sqrt{2}/2}{\sqrt{2}/2} = -1$

17. $t = -\dfrac{7\pi}{4}$ corresponds to $\left(\dfrac{\sqrt{2}}{2}, \dfrac{\sqrt{2}}{2}\right)$.

$$\sin t = y = \frac{\sqrt{2}}{2}$$

$$\cos t = x = \frac{\sqrt{2}}{2}$$

$$\tan t = \frac{y}{x} = 1$$

18. $t = -\dfrac{4\pi}{3}, (x, y) = \left(-\dfrac{1}{2}, \dfrac{\sqrt{3}}{2}\right)$

$$\sin\left(-\frac{4\pi}{3}\right) = \frac{\sqrt{3}}{2}$$

$$\cos\left(-\frac{4\pi}{3}\right) = -\frac{1}{2}$$

$$\tan\left(-\frac{4\pi}{3}\right) = \frac{\sqrt{3}/2}{-1/2} = -\sqrt{3}$$

19. $t = \dfrac{11\pi}{6}$ corresponds to $\left(\dfrac{\sqrt{3}}{2}, -\dfrac{1}{2}\right)$.

$$\sin t = y = -\frac{1}{2}$$

$$\cos t = x = \frac{\sqrt{3}}{2}$$

$$\tan t = \frac{y}{x} = -\frac{1}{\sqrt{3}} = -\frac{\sqrt{3}}{3}$$

20. $t = \dfrac{5\pi}{3}, (x, y) = \left(\dfrac{1}{2}, -\dfrac{\sqrt{3}}{2}\right)$

$$\sin \frac{5\pi}{3} = -\frac{\sqrt{3}}{2}$$

$$\cos \frac{5\pi}{3} = \frac{1}{2}$$

$$\tan \frac{5\pi}{3} = \frac{-\sqrt{3}/2}{1/2} = -\sqrt{3}$$

21. $t = -\dfrac{3\pi}{2}$ corresponds to $(0, 1)$.

$$\sin t = y = 1$$

$$\cos t = x = 0$$

$$\tan t = \frac{y}{x} \text{ is undefined.}$$

22. $t = -2\pi, (x, y) = (1, 0)$

$$\sin(-2\pi) = 0$$

$$\cos(-2\pi) = 1$$

$$\tan(-2\pi) = \frac{0}{1} = 0$$

23. $t = \dfrac{3\pi}{4}$ corresponds to $\left(-\dfrac{\sqrt{2}}{2}, \dfrac{\sqrt{2}}{2}\right)$.

$$\sin t = y = \frac{\sqrt{2}}{2} \qquad \csc t = \frac{1}{y} = \sqrt{2}$$

$$\cos t = x = -\frac{\sqrt{2}}{2} \qquad \sec t = \frac{1}{x} = -\sqrt{2}$$

$$\tan t = \frac{y}{x} = -1 \qquad \cot t = \frac{x}{y} = -1$$

24. $t = \dfrac{5\pi}{6}, (x, y) = \left(-\dfrac{\sqrt{3}}{2}, \dfrac{1}{2}\right)$

$$\sin \frac{5\pi}{6} = \frac{1}{2} \qquad \csc \frac{5\pi}{6} = \frac{1}{\sin t} = 2$$

$$\cos \frac{5\pi}{6} = -\frac{\sqrt{3}}{2} \qquad \sec \frac{5\pi}{6} = \frac{1}{\cos t} = -\frac{2\sqrt{3}}{3}$$

$$\tan \frac{5\pi}{6} = \frac{1/2}{-\sqrt{3}/2} = -\frac{\sqrt{3}}{3} \qquad \cot \frac{5\pi}{6} = \frac{1}{\tan t} = -\sqrt{3}$$

25. $t = -\dfrac{\pi}{2}$ corresponds to $(0, -1)$.

$$\sin t = y = -1 \qquad \csc t = \frac{1}{y} = -1$$

$$\cos t = x = 0 \qquad \sec t = \frac{1}{x} \text{ is undefined.}$$

$$\tan t = \frac{y}{x} \text{ is undefined.} \qquad \cot t = \frac{x}{y} = 0$$

26. $t = \dfrac{3\pi}{2}, (x, y) = (0, -1)$

$$\sin \frac{3\pi}{2} = -1 \qquad \csc \frac{3\pi}{2} = \frac{1}{\sin t} = -1$$

$$\cos \frac{3\pi}{2} = 0 \qquad \sec \frac{3\pi}{2} \text{ is undefined.}$$

$$\tan \frac{3\pi}{2} \text{ is undefined.} \qquad \cot \frac{3\pi}{2} = \frac{0}{-1} = 0$$

27. $t = \dfrac{4\pi}{3}$ corresponds to $\left(-\dfrac{1}{2}, -\dfrac{\sqrt{3}}{2}\right)$.

$$\sin t = y = -\dfrac{\sqrt{3}}{2} \qquad \csc t = \dfrac{1}{y} = -\dfrac{2\sqrt{3}}{3}$$

$$\cos t = x = -\dfrac{1}{2} \qquad \sec t = \dfrac{1}{x} = -2$$

$$\tan t = \dfrac{y}{x} = \sqrt{3} \qquad \cot t = \dfrac{x}{y} = \dfrac{\sqrt{3}}{3}$$

28. $t = \dfrac{7\pi}{4}, (x, y) = \left(\dfrac{\sqrt{2}}{2}, -\dfrac{\sqrt{2}}{2}\right)$

$$\sin \dfrac{7\pi}{4} = -\dfrac{\sqrt{2}}{2} \qquad \csc \dfrac{7\pi}{4} = \dfrac{1}{\sin t} = -\sqrt{2}$$

$$\cos \dfrac{7\pi}{4} = \dfrac{\sqrt{2}}{2} \qquad \sec \dfrac{7\pi}{4} = \dfrac{1}{\cos t} = \sqrt{2}$$

$$\tan \dfrac{7\pi}{4} = \dfrac{-\sqrt{2}/2}{\sqrt{2}/2} = -1 \qquad \cot \dfrac{7\pi}{4} = \dfrac{1}{\tan t} = -1$$

29. $\sin 5\pi = \sin \pi = 0$

30. $\cos 5\pi = \cos \pi = -1$

31. $\cos \dfrac{8\pi}{3} = \cos \dfrac{2\pi}{3} = -\dfrac{1}{2}$

32. $\sin \dfrac{9\pi}{4} = \sin \dfrac{\pi}{4} = \dfrac{\sqrt{2}}{2}$

33. $\cos\left(-\dfrac{15\pi}{2}\right) = \cos \dfrac{\pi}{2} = 0$

34. $\sin \dfrac{19\pi}{6} = \sin \dfrac{7\pi}{6} = -\dfrac{1}{2}$

35. $\sin\left(-\dfrac{9\pi}{4}\right) = \sin\left(\dfrac{7\pi}{4}\right) = -\dfrac{\sqrt{2}}{2}$

36. $\cos\left(-\dfrac{8\pi}{3}\right) = \cos \dfrac{4\pi}{3} = -\dfrac{1}{2}$

37. $\sin t = \dfrac{1}{3}$

(a) $\sin(-t) = -\sin t = -\dfrac{1}{3}$

(b) $\csc(-t) = -\csc t = -3$

38. $\sin(-t) = \dfrac{3}{8}$

(a) $\sin t = -\sin(-t) = -\dfrac{3}{8}$

(b) $\csc t = \dfrac{1}{\sin t} = -\dfrac{8}{3}$

39. $\cos(-t) = -\dfrac{1}{5}$

(a) $\cos t = \cos(-t) = -\dfrac{1}{5}$

(b) $\sec(-t) = \dfrac{1}{\cos(-t)} = -5$

40. $\cos t = -\dfrac{3}{4}$

(a) $\cos(-t) = \cos t = -\dfrac{3}{4}$

(b) $\sec(-t) = \sec t = \dfrac{1}{\cos t} = -\dfrac{4}{3}$

41. $\sin t = \dfrac{4}{5}$

(a) $\sin(\pi - t) = \sin t = \dfrac{4}{5}$

(b) $\sin(t + \pi) = -\sin t = -\dfrac{4}{5}$

42. $\cos t = \dfrac{4}{5}$

(a) $\cos(\pi - t) = -\cos t = -\dfrac{4}{5}$

(b) $\cos(t + \pi) = -\cos t = -\dfrac{4}{5}$

43. $\sin \dfrac{\pi}{4} \approx 0.7071$

44. $\tan \dfrac{\pi}{3} \approx 1.7321$

45. $\csc 1.3 = \dfrac{1}{\sin 1.3} \approx 1.0378$

46. $\cot 1 = \dfrac{1}{\tan 1} \approx 0.6421$

47. $\cos(-1.7) \approx -0.1288$

48. $\cos(-2.5) \approx -0.8011$

49. $\csc 0.8 = \dfrac{1}{\sin 0.8} \approx 1.3940$

50. $\sec 1.8 = \dfrac{1}{0051.8} \approx -4.4014$

51. $\sec 22.8 = \dfrac{1}{\cos 22.8} \approx -1.4486$

52. $\sin(-0.9) \approx -0.7833$

53. (a) $\sin 5 = y \approx -1$

(b) $\cos 2 = x \approx -0.4$

54. (a) $\sin 0.75 = y \approx 0.7$

(b) $\cos 2.5 = x \approx -0.8$

55. (a) $\sin t = 0.25$

$\qquad t \approx 0.25$ or 2.89

(b) $\cos t = -0.25$

$\qquad t \approx 1.82$ or 4.46

56. (a) $\sin t = -0.75$

$\qquad t \approx 4.0$ or $t \approx 5.4$

(b) $\cos t = 0.75$

$\qquad t \approx 0.7$ or $t \approx 5.6$

57. $y(t) = \frac{1}{4}e^{-t}\cos 6t$

(a)

t	0	$\frac{1}{4}$	$\frac{1}{2}$	$\frac{3}{4}$	1
y	0.25	0.0138	-0.1501	-0.0249	0.0883

(b) From the table feature of a graphing utility we see that $y \approx 0$ when $t \approx 5$ seconds.

(c) As t increases, the displacement oscillates but decreases in amplitude.

58. $y(t) = \frac{1}{4}\cos 6t$

(a) $y(0) = \frac{1}{4}\cos 0 = 0.2500$ feet

(b) $y\left(\frac{1}{4}\right) = \frac{1}{4}\cos\frac{3}{2} \approx 0.0177$ feet

(c) $y\left(\frac{1}{2}\right) = \frac{1}{4}\cos 3 \approx -0.2475$ feet

59. False. $\sin(-t) = -\sin t$ means the function is odd, not that the sine of a negative angle is a negative number.

For example: $\sin\left(-\frac{3\pi}{2}\right) = -\sin\left(\frac{3\pi}{2}\right) = -(-1) = 1.$

Even though the angle is negative, the sine value is positive.

60. True. $\tan a = \tan(a - 6\pi)$ since the period of tan is π.

61. (a) The points have y-axis symmetry.

(b) $\sin t_1 = \sin(\pi - t_1)$ since they have the same y-value.

(c) $\cos(\pi - t_1) = -\cos t_1$ since the x-values have the opposite signs.

62. $\cos\theta = x = \cos(-\theta)$

$\sec\theta = \frac{1}{x} = \sec(-\theta)$

So $\sec\theta$ and $\cos\theta$ are even.

$\sin\theta = y$

$\sin(-\theta) = -y = -\sin\theta$

$\csc\theta = \frac{1}{y}$

$\csc(-\theta) = -\frac{1}{y} = -\csc\theta$

So $\sin\theta$ and $\csc\theta$ are odd.

$\tan\theta = \frac{y}{x}$

$\tan(-\theta) = \frac{-y}{x} = -\tan\theta$

$\cot\theta = \frac{x}{y}$

$\cot(-\theta) = \frac{x}{-y} = -\cot\theta$

So $\tan\theta$ and $\cot\theta$ are odd.

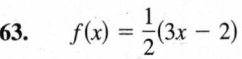

63. $f(x) = \frac{1}{2}(3x - 2)$

$y = \frac{1}{2}(3x - 2)$

$x = \frac{1}{2}(3y - 2)$

$2x = 3y - 2$

$\frac{2}{3}x + \frac{2}{3} = y$

$f^{-1}(x) = \frac{2}{3}x + \frac{2}{3} = \frac{2}{3}(x + 1)$

64. $f(x) = \frac{1}{4}x^3 + 1$

$$y = \frac{1}{4}x^3 + 1$$

$$x = \frac{1}{4}y^3 + 1$$

$$x - 1 = \frac{1}{4}y^3$$

$$4(x - 1) = y^3$$

$$y = \sqrt[3]{4(x - 1)}$$

$$f^{-1}(x) = \sqrt[3]{4(x - 1)}$$

65. $f(x) = \sqrt{x^2 - 4},\ x \geq 2$

$$y = \sqrt{x^2 - 4}$$

$$x = \sqrt{y^2 - 4}$$

$$x^2 = y^2 - 4$$

$$\pm\sqrt{x^2 + 4} = y$$

$$f^{-1}(x) = \sqrt{x^2 + 4},\ x \geq 0$$

66. $f(x) = \frac{x + 2}{x - 4}$

$$y = \frac{x + 2}{x - 4}$$

$$x = \frac{y + 2}{y - 4}$$

$$x(y - 4) = y + 2$$

$$xy - 4x = y + 2$$

$$xy - y = 4x + 2$$

$$y(x - 1) = 4x + 2$$

$$y = \frac{2(2x + 1)}{x - 1}$$

$$f^{-1}(x) = \frac{2(2x + 1)}{x - 1}$$

67. $f(x) = \frac{2x}{x - 3}$

Intercept: $(0, 0)$

Vertical asymptote: $x = 3$

Horizontal asymptote: $y = 2$

x	-1	0	1	2	4	5	6
y	$\frac{1}{2}$	0	-1	-4	8	5	4

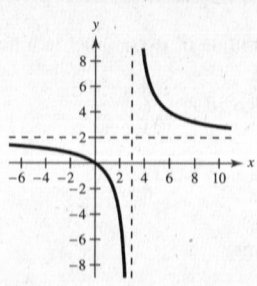

68. $f(x) = \frac{5x}{x^2 + x - 6} = \frac{5x}{(x + 3)(x - 2)},\ x \neq -3, 2$

Horizontal asymptote: $x = 0$

Vertical asymptote: $x = -3,\ x = 2$

Intercept: $(0, 0)$

x	-6	-4	-2	0	1	3	5
y	$-\frac{5}{4}$	$-\frac{10}{3}$	$\frac{5}{2}$	0	$-\frac{5}{4}$	$\frac{5}{2}$	$\frac{25}{24}$

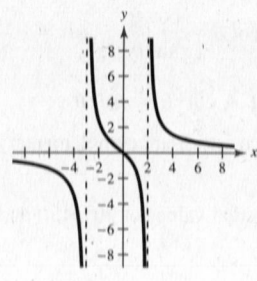

69. $f(x) = \frac{x^2 + 3x - 10}{2x^2 - 8} = \frac{(x + 5)(x - 2)}{2(x + 2)(x - 2)} = \frac{x + 5}{2(x + 2)},\ x \neq 2$

Intercepts: $(-5, 0),\ \left(0, \frac{5}{4}\right)$

Vertical asymptote: $x = -2$

Horizontal asymptote: $y = \frac{1}{2}$

Hole in the graph at $\left(2, \frac{7}{8}\right)$

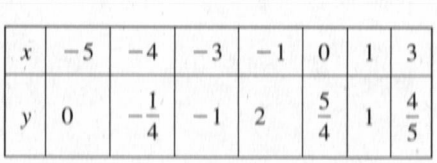

x	-5	-4	-3	-1	0	1	3
y	0	$-\frac{1}{4}$	-1	2	$\frac{5}{4}$	1	$\frac{4}{5}$

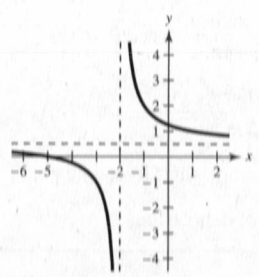

70. $f(x) = \dfrac{x^3 - 6x^2 + x - 1}{2x^2 - 5x - 8} = \dfrac{1}{2}x - \dfrac{7}{4} - \dfrac{15(x+4)}{4(2x^2 - 5x - 8)}$

x	-4	-3	$-\frac{3}{2}$	-1	0	2	3	4	7
y	$-\frac{15}{4}$	$-\frac{17}{5}$	$-\frac{155}{32}$	9	$\frac{1}{8}$	$\frac{3}{2}$	5	$-\frac{29}{4}$	1

Vertical asymptote: $2x^2 - 5x - 8 = 0$

$$x = \frac{5 \pm \sqrt{(-5)^2 - (4)(2)(-8)}}{2(2)}$$

$$x = \frac{5 \pm \sqrt{89}}{4}; x \approx -1.11, x \approx 3.61$$

Slant asymptote: $y = \dfrac{1}{2}x - \dfrac{7}{4}$

y-intercept: $\left(0, \dfrac{1}{8}\right)$

x-intercept: $(\approx 5.86, 0)$

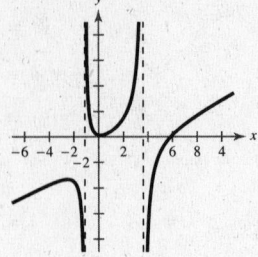

Section 4.3 Right Triangle Trigonometry

■ You should know the right triangle definition of trigonometric functions.

(a) $\sin \theta = \dfrac{\text{opp}}{\text{hyp}}$ (b) $\cos \theta = \dfrac{\text{adj}}{\text{hyp}}$ (c) $\tan \theta = \dfrac{\text{opp}}{\text{adj}}$

(d) $\csc \theta = \dfrac{\text{hyp}}{\text{opp}}$ (e) $\sec \theta = \dfrac{\text{hyp}}{\text{adj}}$ (f) $\cot \theta = \dfrac{\text{adj}}{\text{opp}}$

■ You should know the following identities.

(a) $\sin \theta = \dfrac{1}{\csc \theta}$ (b) $\csc \theta = \dfrac{1}{\sin \theta}$ (c) $\cos \theta = \dfrac{1}{\sec \theta}$

(d) $\sec \theta = \dfrac{1}{\cos \theta}$ (e) $\tan \theta = \dfrac{1}{\cot \theta}$ (f) $\cot \theta = \dfrac{1}{\tan \theta}$

(g) $\tan \theta = \dfrac{\sin \theta}{\cos \theta}$ (h) $\cot \theta = \dfrac{\cos \theta}{\sin \theta}$ (i) $\sin^2 \theta + \cos^2 \theta = 1$

(j) $1 + \tan^2 \theta = \sec^2 \theta$ (k) $1 + \cot^2 \theta = \csc^2 \theta$

■ You should know that two acute angles α and β are complementary if $\alpha + \beta = 90°$, and that cofunctions of complementary angles are equal.

■ You should know the trigonometric function values of $30°$, $45°$, and $60°$, or be able to construct triangles from which you can determine them.

Vocabulary Check

1. (i) $\dfrac{\text{hypotenuse}}{\text{adjacent}} = \sec \theta$ (v)   (ii) $\dfrac{\text{adjacent}}{\text{opposite}} = \cot \theta$ (iv)   (iii) $\dfrac{\text{hypotenuse}}{\text{opposite}} = \csc \theta$ (vi)

(iv) $\dfrac{\text{adjacent}}{\text{hypotenuse}} = \cos \theta$ (iii)   (v) $\dfrac{\text{opposite}}{\text{hypotenuse}} = \sin \theta$ (i)   (vi) $\dfrac{\text{opposite}}{\text{adjacent}} = \tan \theta$ (ii)

2. opposite; adjacent; hypotenuse

3. elevation; depression

1. hyp $= \sqrt{6^2 + 8^2} = \sqrt{36 + 64} = \sqrt{100} = 10$

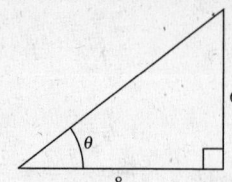

$\sin \theta = \dfrac{\text{opp}}{\text{hyp}} = \dfrac{6}{10} = \dfrac{3}{5}$ $\csc \theta = \dfrac{\text{hyp}}{\text{opp}} = \dfrac{10}{6} = \dfrac{5}{3}$

$\cos \theta = \dfrac{\text{adj}}{\text{hyp}} = \dfrac{8}{10} = \dfrac{4}{5}$ $\sec \theta = \dfrac{\text{hyp}}{\text{adj}} = \dfrac{10}{8} = \dfrac{5}{4}$

$\tan \theta = \dfrac{\text{opp}}{\text{adj}} = \dfrac{6}{8} = \dfrac{3}{4}$ $\cot \theta = \dfrac{\text{adj}}{\text{opp}} = \dfrac{8}{6} = \dfrac{4}{3}$

2.

adj $= \sqrt{13^2 - 5^2} = \sqrt{169 - 25} = 12$

$\sin \theta = \dfrac{\text{opp}}{\text{hyp}} = \dfrac{5}{13}$ $\csc \theta = \dfrac{\text{hyp}}{\text{opp}} = \dfrac{13}{5}$

$\cos \theta = \dfrac{\text{adj}}{\text{hyp}} = \dfrac{12}{13}$ $\sec \theta = \dfrac{\text{hyp}}{\text{adj}} = \dfrac{13}{12}$

$\tan \theta = \dfrac{\text{opp}}{\text{adj}} = \dfrac{5}{12}$ $\cot \theta = \dfrac{\text{adj}}{\text{opp}} = \dfrac{12}{5}$

3. adj $= \sqrt{41^2 - 9^2} = \sqrt{1681 - 81} = \sqrt{1600} = 40$

$\sin \theta = \dfrac{\text{opp}}{\text{hyp}} = \dfrac{9}{41}$ $\csc \theta = \dfrac{\text{hyp}}{\text{opp}} = \dfrac{41}{9}$

$\cos \theta = \dfrac{\text{adj}}{\text{hyp}} = \dfrac{40}{41}$ $\sec \theta = \dfrac{\text{hyp}}{\text{adj}} = \dfrac{41}{40}$

$\tan \theta = \dfrac{\text{opp}}{\text{adj}} = \dfrac{9}{40}$ $\cot \theta = \dfrac{\text{adj}}{\text{opp}} = \dfrac{40}{9}$

4.

hyp $= \sqrt{4^2 + 4^2} = \sqrt{32} = 4\sqrt{2}$

$\sin \theta = \dfrac{\text{opp}}{\text{hyp}} = \dfrac{4}{4\sqrt{2}} = \dfrac{1}{\sqrt{2}} = \dfrac{\sqrt{2}}{2}$ $\csc \theta = \dfrac{\text{hyp}}{\text{opp}} = \dfrac{4\sqrt{2}}{4} = \sqrt{2}$

$\cos \theta = \dfrac{\text{adj}}{\text{hyp}} = \dfrac{4}{4\sqrt{2}} = \dfrac{1}{\sqrt{2}} = \dfrac{\sqrt{2}}{2}$ $\sec \theta = \dfrac{\text{hyp}}{\text{adj}} = \dfrac{4\sqrt{2}}{4} = \sqrt{2}$

$\tan \theta = \dfrac{\text{opp}}{\text{adj}} = \dfrac{4}{4} = 1$ $\cot \theta = \dfrac{\text{adj}}{\text{opp}} = \dfrac{4}{4} = 1$

5. adj $= \sqrt{3^2 - 1^2} = \sqrt{8} = 2\sqrt{2}$

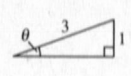

$\sin \theta = \dfrac{\text{opp}}{\text{hyp}} = \dfrac{1}{3}$ $\csc \theta = \dfrac{\text{hyp}}{\text{opp}} = 3$

$\cos \theta = \dfrac{\text{adj}}{\text{hyp}} = \dfrac{2\sqrt{2}}{3}$ $\sec \theta = \dfrac{\text{hyp}}{\text{adj}} = \dfrac{3}{2\sqrt{2}} = \dfrac{3\sqrt{2}}{4}$

$\tan \theta = \dfrac{\text{opp}}{\text{adj}} = \dfrac{1}{2\sqrt{2}} = \dfrac{\sqrt{2}}{4}$ $\cot \theta = \dfrac{\text{adj}}{\text{opp}} = 2\sqrt{2}$

adj $= \sqrt{6^2 - 2^2} = \sqrt{32} = 4\sqrt{2}$

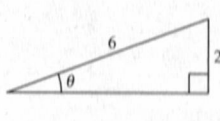

$\sin \theta = \dfrac{\text{opp}}{\text{hyp}} = \dfrac{2}{6} = \dfrac{1}{3}$ $\csc \theta = \dfrac{\text{hyp}}{\text{opp}} = \dfrac{6}{2} = 3$

$\cos \theta = \dfrac{\text{adj}}{\text{hyp}} = \dfrac{4\sqrt{2}}{6} = \dfrac{2\sqrt{2}}{3}$ $\sec \theta = \dfrac{\text{hyp}}{\text{adj}} = \dfrac{6}{4\sqrt{2}} = \dfrac{3}{2\sqrt{2}} = \dfrac{3\sqrt{2}}{4}$

$\tan \theta = \dfrac{\text{opp}}{\text{adj}} = \dfrac{2}{4\sqrt{2}} = \dfrac{1}{2\sqrt{2}} = \dfrac{\sqrt{2}}{4}$ $\cot \theta = \dfrac{\text{adj}}{\text{opp}} = \dfrac{4\sqrt{2}}{2} = 2\sqrt{2}$

The function values are the same since the triangles are similar and the corresponding sides are proportional.

6.

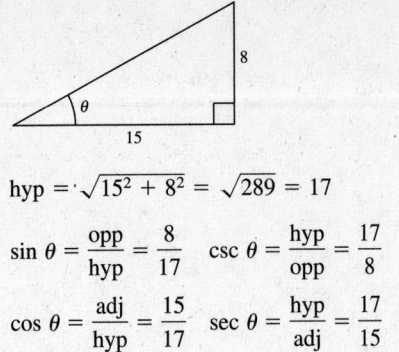

$$\text{hyp} = \sqrt{15^2 + 8^2} = \sqrt{289} = 17$$

$$\sin \theta = \frac{\text{opp}}{\text{hyp}} = \frac{8}{17} \qquad \csc \theta = \frac{\text{hyp}}{\text{opp}} = \frac{17}{8}$$

$$\cos \theta = \frac{\text{adj}}{\text{hyp}} = \frac{15}{17} \qquad \sec \theta = \frac{\text{hyp}}{\text{adj}} = \frac{17}{15}$$

$$\tan \theta = \frac{\text{opp}}{\text{adj}} = \frac{8}{15} \qquad \cot \theta = \frac{\text{adj}}{\text{opp}} = \frac{15}{8}$$

$$\text{hyp} = \sqrt{7.5^2 + 4^2} = \frac{17}{2}$$

$$\sin \theta = \frac{\text{opp}}{\text{hyp}} = \frac{4}{17/2} = \frac{8}{17} \qquad \csc \theta = \frac{\text{hyp}}{\text{opp}} = \frac{17/2}{4} = \frac{17}{8}$$

$$\cos \theta = \frac{\text{adj}}{\text{hyp}} = \frac{7.5}{17/2} = \frac{15}{17} \qquad \sec \theta = \frac{\text{hyp}}{\text{adj}} = \frac{17/2}{7.5} = \frac{17}{15}$$

$$\tan \theta = \frac{\text{opp}}{\text{adj}} = \frac{4}{7.5} = \frac{8}{15} \qquad \cot \theta = \frac{\text{adj}}{\text{opp}} = \frac{7.5}{4} = \frac{15}{8}$$

The function values are the same because the triangles are similar, and corresponding sides are proportional.

7. $\text{opp} = \sqrt{5^2 - 4^2} = 3$

$$\sin \theta = \frac{\text{opp}}{\text{hyp}} = \frac{3}{5} \qquad \csc \theta = \frac{\text{hyp}}{\text{opp}} = \frac{5}{3}$$

$$\cos \theta = \frac{\text{adj}}{\text{hyp}} = \frac{4}{5} \qquad \sec \theta = \frac{\text{hyp}}{\text{adj}} = \frac{5}{4}$$

$$\tan \theta = \frac{\text{opp}}{\text{adj}} = \frac{3}{4} \qquad \cot \theta = \frac{\text{adj}}{\text{opp}} = \frac{4}{3}$$

$$\text{opp} = \sqrt{1.25^2 - 1^2} = 0.75$$

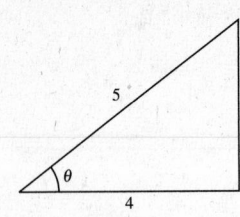

$$\sin \theta = \frac{\text{opp}}{\text{hyp}} = \frac{0.75}{1.25} = \frac{3}{5} \qquad \csc \theta = \frac{\text{hyp}}{\text{opp}} = \frac{1.25}{0.75} = \frac{5}{3}$$

$$\cos \theta = \frac{\text{adj}}{\text{hyp}} = \frac{1}{1.25} = \frac{4}{5} \qquad \sec \theta = \frac{\text{hyp}}{\text{adj}} = \frac{1.25}{1} = \frac{5}{4}$$

$$\tan \theta = \frac{\text{opp}}{\text{adj}} = \frac{0.75}{1} = \frac{3}{4} \qquad \cot \theta = \frac{\text{adj}}{\text{opp}} = \frac{1}{0.75} = \frac{4}{3}$$

The function values are the same since the triangles are similar and the corresponding sides are proportional.

8.

$$\text{hyp} = \sqrt{1^2 + 2^2} = \sqrt{5}$$

$$\sin \theta = \frac{\text{opp}}{\text{hyp}} = \frac{1}{\sqrt{5}} = \frac{\sqrt{5}}{5} \qquad \csc \theta = \frac{\text{hyp}}{\text{opp}} = \frac{\sqrt{5}}{1} = \sqrt{5}$$

$$\cos \theta = \frac{\text{adj}}{\text{hyp}} = \frac{2}{\sqrt{5}} = \frac{2\sqrt{5}}{5} \qquad \sec \theta = \frac{\text{hyp}}{\text{adj}} = \frac{\sqrt{5}}{2}$$

$$\tan \theta = \frac{\text{opp}}{\text{adj}} = \frac{1}{2} \qquad \cot \theta = \frac{\text{adj}}{\text{opp}} = \frac{2}{1} = 2$$

$$\text{hyp} = \sqrt{3^2 + 6^2} = 3\sqrt{5}$$

$$\sin \theta = \frac{3}{3\sqrt{5}} = \frac{1}{\sqrt{5}} = \frac{\sqrt{5}}{5} \qquad \csc \theta = \frac{3\sqrt{5}}{3} = \sqrt{5}$$

$$\cos \theta = \frac{6}{3\sqrt{5}} = \frac{2}{\sqrt{5}} = \frac{2\sqrt{5}}{5} \qquad \sec \theta = \frac{3\sqrt{5}}{6} = \frac{\sqrt{5}}{2}$$

$$\tan \theta = \frac{3}{6} = \frac{1}{2} \qquad \cot \theta = \frac{6}{3} = 2$$

The function values are the same because the triangles are similar, and corresponding sides are proportional.

9. Given: $\sin \theta = \dfrac{3}{4} = \dfrac{\text{opp}}{\text{hyp}}$

$3^2 + (\text{adj})^2 = 4^2$

$\text{adj} = \sqrt{7}$

$\cos \theta = \dfrac{\text{adj}}{\text{hyp}} = \dfrac{\sqrt{7}}{4}$

$\tan \theta = \dfrac{\text{opp}}{\text{adj}} = \dfrac{3\sqrt{7}}{7}$

$\csc \theta = \dfrac{\text{hyp}}{\text{opp}} = \dfrac{4}{3}$

$\sec \theta = \dfrac{\text{hyp}}{\text{adj}} = \dfrac{4\sqrt{7}}{7}$

$\cot \theta = \dfrac{\text{adj}}{\text{opp}} = \dfrac{\sqrt{7}}{3}$

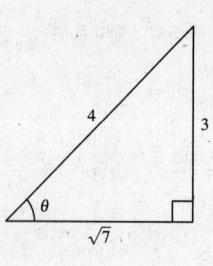

10. Given: $\cos \theta = \dfrac{5}{7} = \dfrac{\text{adj}}{\text{hyp}}$

$\text{opp} = \sqrt{7^2 - 5^2} = \sqrt{24} = 2\sqrt{6}$

$\sin \theta = \dfrac{\text{opp}}{\text{hyp}} = \dfrac{2\sqrt{6}}{7}$

$\tan \theta = \dfrac{\text{opp}}{\text{adj}} = \dfrac{2\sqrt{6}}{5}$

$\csc \theta = \dfrac{\text{hyp}}{\text{opp}} = \dfrac{7}{2\sqrt{6}} = \dfrac{7\sqrt{6}}{12}$

$\sec \theta = \dfrac{\text{hyp}}{\text{adj}} = \dfrac{7}{5}$

$\cot \theta = \dfrac{\text{adj}}{\text{opp}} = \dfrac{5}{2\sqrt{6}} = \dfrac{5\sqrt{6}}{12}$

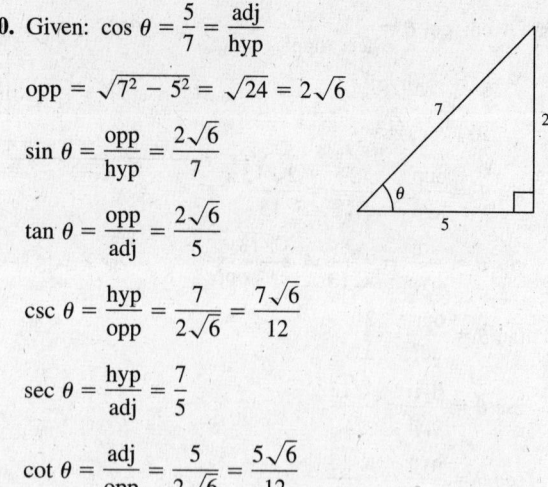

11. Given: $\sec \theta = 2 = \dfrac{2}{1} = \dfrac{\text{hyp}}{\text{adj}}$

$(\text{opp})^2 + 1^2 = 2^2$

$\text{opp} = \sqrt{3}$

$\sin \theta = \dfrac{\text{opp}}{\text{hyp}} = \dfrac{\sqrt{3}}{2}$

$\cos \theta = \dfrac{\text{adj}}{\text{hyp}} = \dfrac{1}{2}$

$\tan \theta = \dfrac{\text{opp}}{\text{adj}} = \sqrt{3}$

$\csc \theta = \dfrac{\text{hyp}}{\text{opp}} = \dfrac{2\sqrt{3}}{3}$

$\cot \theta = \dfrac{\text{adj}}{\text{opp}} = \dfrac{\sqrt{3}}{3}$

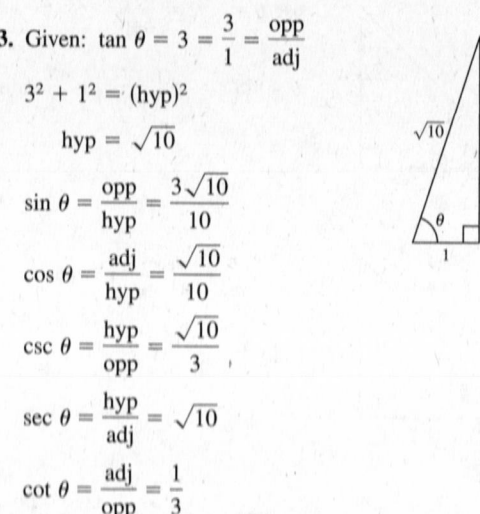

12. Given: $\cot \theta = \dfrac{5}{1} = \dfrac{\text{adj}}{\text{opp}}$

$\text{hyp} = \sqrt{5^2 + 1^2} = \sqrt{26}$

$\sin \theta = \dfrac{\text{opp}}{\text{hyp}} = \dfrac{1}{\sqrt{26}} = \dfrac{\sqrt{26}}{26}$

$\cos \theta = \dfrac{\text{adj}}{\text{hyp}} = \dfrac{5}{\sqrt{26}} = \dfrac{5\sqrt{26}}{26}$

$\tan \theta = \dfrac{\text{opp}}{\text{adj}} = \dfrac{1}{5}$

$\csc \theta = \dfrac{\text{hyp}}{\text{opp}} = \dfrac{\sqrt{26}}{1} = \sqrt{26}$

$\sec \theta = \dfrac{\text{hyp}}{\text{adj}} = \dfrac{\sqrt{26}}{5}$

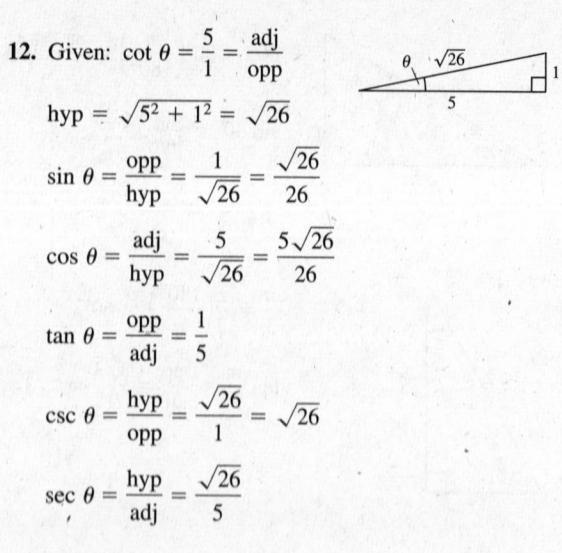

13. Given: $\tan \theta = 3 = \dfrac{3}{1} = \dfrac{\text{opp}}{\text{adj}}$

$3^2 + 1^2 = (\text{hyp})^2$

$\text{hyp} = \sqrt{10}$

$\sin \theta = \dfrac{\text{opp}}{\text{hyp}} = \dfrac{3\sqrt{10}}{10}$

$\cos \theta = \dfrac{\text{adj}}{\text{hyp}} = \dfrac{\sqrt{10}}{10}$

$\csc \theta = \dfrac{\text{hyp}}{\text{opp}} = \dfrac{\sqrt{10}}{3}$

$\sec \theta = \dfrac{\text{hyp}}{\text{adj}} = \sqrt{10}$

$\cot \theta = \dfrac{\text{adj}}{\text{opp}} = \dfrac{1}{3}$

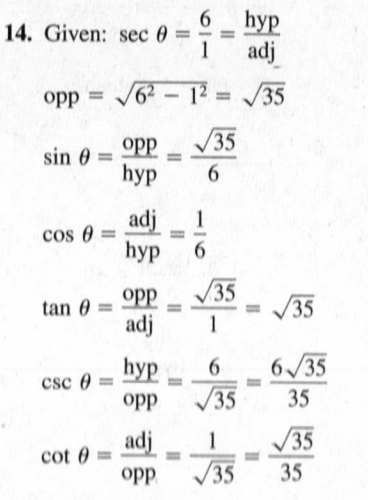

14. Given: $\sec \theta = \dfrac{6}{1} = \dfrac{\text{hyp}}{\text{adj}}$

$\text{opp} = \sqrt{6^2 - 1^2} = \sqrt{35}$

$\sin \theta = \dfrac{\text{opp}}{\text{hyp}} = \dfrac{\sqrt{35}}{6}$

$\cos \theta = \dfrac{\text{adj}}{\text{hyp}} = \dfrac{1}{6}$

$\tan \theta = \dfrac{\text{opp}}{\text{adj}} = \dfrac{\sqrt{35}}{1} = \sqrt{35}$

$\csc \theta = \dfrac{\text{hyp}}{\text{opp}} = \dfrac{6}{\sqrt{35}} = \dfrac{6\sqrt{35}}{35}$

$\cot \theta = \dfrac{\text{adj}}{\text{opp}} = \dfrac{1}{\sqrt{35}} = \dfrac{\sqrt{35}}{35}$

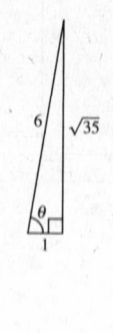

15. Given: $\cot \theta = \dfrac{3}{2} = \dfrac{\text{adj}}{\text{opp}}$

$2^2 + 3^2 = (\text{hyp})^2$

$\text{hyp} = \sqrt{13}$

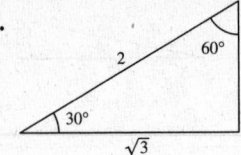

$\sin \theta = \dfrac{\text{opp}}{\text{hyp}} = \dfrac{2}{\sqrt{13}} = \dfrac{2\sqrt{13}}{13}$

$\cos \theta = \dfrac{\text{adj}}{\text{hyp}} = \dfrac{3}{\sqrt{13}} = \dfrac{3\sqrt{13}}{13}$

$\tan \theta = \dfrac{\text{opp}}{\text{adj}} = \dfrac{2}{3}$

$\csc \theta = \dfrac{\text{hyp}}{\text{opp}} = \dfrac{\sqrt{13}}{2}$

$\sec \theta = \dfrac{\text{hyp}}{\text{adj}} = \dfrac{\sqrt{13}}{3}$

16. Given: $\csc \theta = \dfrac{17}{4} = \dfrac{\text{hyp}}{\text{opp}}$

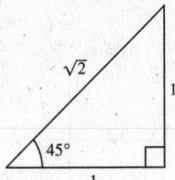

$\text{adj} = \sqrt{17^2 - 4^2} = \sqrt{273}$

$\sin \theta = \dfrac{\text{opp}}{\text{hyp}} = \dfrac{4}{17}$

$\cos \theta = \dfrac{\text{adj}}{\text{hyp}} = \dfrac{\sqrt{273}}{17}$

$\tan \theta = \dfrac{\text{opp}}{\text{adj}} = \dfrac{4}{\sqrt{273}} = \dfrac{4\sqrt{273}}{273}$

$\sec \theta = \dfrac{\text{hyp}}{\text{adj}} = \dfrac{17}{\sqrt{273}} = \dfrac{17\sqrt{273}}{273}$

$\cot \theta = \dfrac{\text{adj}}{\text{opp}} = \dfrac{\sqrt{273}}{4}$

17.
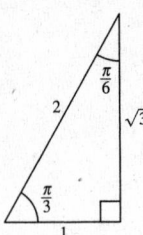

$30° = 30°\left(\dfrac{\pi}{180°}\right) = \dfrac{\pi}{6}$ radian

$\sin 30° = \dfrac{\text{opp}}{\text{hyp}} = \dfrac{1}{2}$

18.
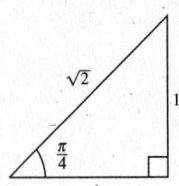

$45° = 45°\left(\dfrac{\pi}{180°}\right) = \dfrac{\pi}{4}$ radian

$\cos 45° = \dfrac{\text{adj}}{\text{hyp}} = \dfrac{1}{\sqrt{2}} = \dfrac{\sqrt{2}}{2}$

19.
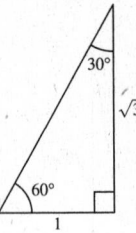

$\dfrac{\pi}{3} = \dfrac{\pi}{3}\left(\dfrac{180°}{\pi}\right) = 60°$

$\tan \dfrac{\pi}{3} = \dfrac{\text{opp}}{\text{adj}} = \dfrac{\sqrt{3}}{1} = \sqrt{3}$

20.
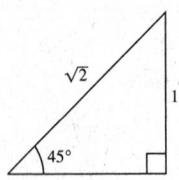

$\dfrac{\pi}{4} = \dfrac{\pi}{4}\left(\dfrac{180°}{\pi}\right) = 45°$

$\sec \dfrac{\pi}{4} = \dfrac{\text{hyp}}{\text{adj}} = \dfrac{\sqrt{2}}{1} = \sqrt{2}$

21.
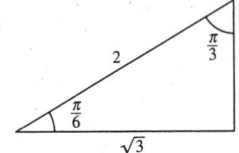

$\cot \theta = \dfrac{\sqrt{3}}{3} = \dfrac{1}{\sqrt{3}} = \dfrac{\text{adj}}{\text{opp}}$

$\theta = 60° = \dfrac{\pi}{3}$ radian

22.
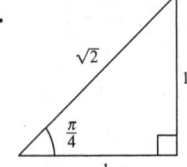

$\csc \theta = \sqrt{2} = \dfrac{\text{hyp}}{\text{opp}}$

$\theta = 45° = 45°\left(\dfrac{\pi}{180°}\right) = \dfrac{\pi}{4}$ radian

23.
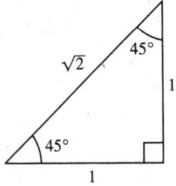

$\dfrac{\pi}{6} = \dfrac{\pi}{6}\left(\dfrac{180°}{\pi}\right) = 30°$

$\cos \dfrac{\pi}{6} = \dfrac{\text{adj}}{\text{hyp}} = \dfrac{\sqrt{3}}{2}$

24.
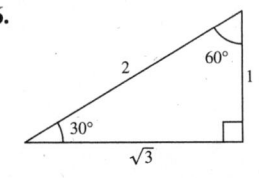

$\theta = \dfrac{\pi}{4} = \dfrac{\pi}{4}\left(\dfrac{180°}{\pi}\right) = 45°$

$\sin \dfrac{\pi}{4} = \dfrac{\text{opp}}{\text{hyp}} = \dfrac{1}{\sqrt{2}} = \dfrac{\sqrt{2}}{2}$

25.

$\cot \theta = 1 = \dfrac{1}{1} = \dfrac{\text{adj}}{\text{opp}}$

$\theta = 45° = 45°\left(\dfrac{\pi}{180°}\right)$

$= \dfrac{\pi}{4}$ radian

26.

$\tan \theta = \dfrac{\sqrt{3}}{3} = \dfrac{1}{\sqrt{3}} = \dfrac{\text{opp}}{\text{adj}}$

$\theta = 30° = 30°\left(\dfrac{\pi}{180°}\right)$

$= \dfrac{\pi}{6}$ radian

27. $\sin 60° = \dfrac{\sqrt{3}}{2}$, $\cos 60° = \dfrac{1}{2}$

(a) $\tan 60° = \dfrac{\sin 60°}{\cos 60°} = \sqrt{3}$

(b) $\sin 30° = \cos 60° = \dfrac{1}{2}$

(c) $\cos 30° = \sin 60° = \dfrac{\sqrt{3}}{2}$

(d) $\cot 60° = \dfrac{\cos 60°}{\sin 60°} = \dfrac{1}{\sqrt{3}} = \dfrac{\sqrt{3}}{3}$

28. $\sin 30° = \dfrac{1}{2}$, $\tan 30° = \dfrac{\sqrt{3}}{3}$

(a) $\csc 30° = \dfrac{1}{\sin 30°} = 2$

(b) $\cot 60° = \tan(90° - 60°) = \tan 30° = \dfrac{\sqrt{3}}{3}$

(c) $\cos 30° = \dfrac{\sin 30°}{\tan 30°} = \dfrac{\frac{1}{2}}{\frac{\sqrt{3}}{3}} = \dfrac{3}{2\sqrt{3}} = \dfrac{\sqrt{3}}{2}$

(d) $\cot 30° = \dfrac{1}{\tan 30°} = \dfrac{3}{\sqrt{3}} = \dfrac{3\sqrt{3}}{3} = \sqrt{3}$

29. $\csc \theta = \dfrac{\sqrt{13}}{2}$, $\sec \theta = \dfrac{\sqrt{13}}{3}$

(a) $\sin \theta = \dfrac{1}{\csc \theta} = \dfrac{2}{\sqrt{13}} = \dfrac{2\sqrt{13}}{13}$

(b) $\cos \theta = \dfrac{1}{\sec \theta} = \dfrac{3}{\sqrt{13}} = \dfrac{3\sqrt{13}}{13}$

(c) $\tan \theta = \dfrac{\sin \theta}{\cos \theta} = \dfrac{\frac{2\sqrt{13}}{13}}{\frac{3\sqrt{13}}{13}} = \dfrac{2}{3}$

(d) $\sec(90° - \theta) = \csc \theta = \dfrac{\sqrt{13}}{2}$

30. $\sec \theta = 5$, $\tan \theta = 2\sqrt{6}$

(a) $\cos \theta = \dfrac{1}{\sec \theta} = \dfrac{1}{5}$

(b) $\cot \theta = \dfrac{1}{\tan \theta} = \dfrac{1}{2\sqrt{6}} = \dfrac{\sqrt{6}}{12}$

(c) $\cot(90° - \theta) = \tan \theta = 2\sqrt{6}$

(d) $\sin \theta = \tan \theta \cos \theta = \left(2\sqrt{6}\right)\left(\dfrac{1}{5}\right) = \dfrac{2\sqrt{6}}{5}$

31. $\cos \alpha = \dfrac{1}{3}$

(a) $\sec \alpha = \dfrac{1}{\cos \alpha} = 3$

(b) $\sin^2\alpha + \cos^2\alpha = 1$

$\sin^2\alpha + \left(\dfrac{1}{3}\right)^2 = 1$

$\sin^2 \alpha = \dfrac{8}{9}$

$\sin \alpha = \dfrac{2\sqrt{2}}{3}$

(c) $\cot \alpha = \dfrac{\cos \alpha}{\sin \alpha} = \dfrac{\frac{1}{3}}{\frac{2\sqrt{2}}{3}} = \dfrac{1}{2\sqrt{2}} = \dfrac{\sqrt{2}}{4}$

(d) $\sin(90° - \alpha) = \cos \alpha = \dfrac{1}{3}$

32. $\tan \beta = 5$

(a) $\cot \beta = \dfrac{1}{\tan \beta} = \dfrac{1}{5}$

(b) $\cos \beta = \dfrac{1}{\sec \beta} = \dfrac{1}{\sqrt{1 + \tan^2 \beta}}$

$= \dfrac{1}{\sqrt{1 + 5^2}} = \dfrac{1}{\sqrt{26}} = \dfrac{\sqrt{26}}{26}$

(c) $\tan(90° - \beta) = \cot \beta = \dfrac{1}{\tan \beta} = \dfrac{1}{5}$

(d) $\csc \beta = \sqrt{1 + \cot^2 \beta}$

$= \sqrt{1 + \left(\dfrac{1}{5}\right)^2}$

$= \sqrt{1 + \dfrac{1}{25}} = \sqrt{\dfrac{26}{25}} = \dfrac{\sqrt{26}}{5}$

33. $\tan \theta \cot \theta = \tan \theta \left(\dfrac{1}{\tan \theta}\right) = 1$

34. $\cos \theta \sec \theta = \cos \theta \left(\dfrac{1}{\cos \theta}\right) = 1$

35. $\tan \alpha \cos \alpha = \left(\dfrac{\sin \alpha}{\cos \alpha}\right) \cos \alpha = \sin \alpha$

36. $\cot \alpha \sin \alpha = \dfrac{\cos \alpha}{\sin \alpha} \sin \alpha = \cos \alpha$

37. $(1 + \cos \theta)(1 - \cos \theta) = 1 - \cos^2 \theta$
$$= (\sin^2 \theta + \cos^2 \theta) - \cos^2 \theta$$
$$= \sin^2 \theta$$

38. $(1 + \sin \theta)(1 - \sin \theta) = 1 - \sin^2 \theta = \cos^2 \theta$

39. $(\sec \theta + \tan \theta)(\sec \theta - \tan \theta) = \sec^2 \theta - \tan^2 \theta$
$$= (1 + \tan^2 \theta) - \tan^2 \theta$$
$$= 1$$

40. $\sin^2 \theta - \cos^2 \theta = \sin^2 \theta - (1 - \sin^2 \theta)$
$$= \sin^2 \theta - 1 + \sin^2 \theta$$
$$= 2 \sin^2 \theta - 1$$

41. $\dfrac{\sin \theta}{\cos \theta} + \dfrac{\cos \theta}{\sin \theta} = \dfrac{\sin^2 \theta + \cos^2 \theta}{\sin \theta \cos \theta}$
$$= \dfrac{1}{\sin \theta \cos \theta}$$
$$= \dfrac{1}{\sin \theta} \cdot \dfrac{1}{\cos \theta}$$
$$= \csc \theta \sec \theta$$

42. $\dfrac{\tan \beta + \cot \beta}{\tan \beta} = \dfrac{\tan \beta}{\tan \beta} + \dfrac{\cot \beta}{\tan \beta}$
$$= 1 + \dfrac{\cot \beta}{\left(\dfrac{1}{\cot \beta}\right)}$$
$$= 1 + \cot^2 \beta = \csc^2 \beta$$

43. (a) $\sin 10° \approx 0.1736$

(b) $\cos 80° \approx 0.1736$

Note: $\cos 80° = \sin(90° - 80°) = \sin 10°.$

44. (a) $\tan 23.5° \approx 0.4348$

(b) $\cot 66.5° = \dfrac{1}{\tan 66.5°} \approx 0.4348$

45. (a) $\sin 16.35° \approx 0.2815$

(b) $\csc 16.35° = \dfrac{1}{\sin 16.35°} \approx 3.5523$

46. (a) $\cos 16° \, 18' = \cos\left(16 + \dfrac{18}{60}\right)° \approx 0.9598$

(b) $\sin 73° \, 56' = \sin\left(73 + \dfrac{56}{60}\right)° \approx 0.9609$

47. (a) $\sec 42° \, 12' = \sec 42.2° = \dfrac{1}{\cos 42.2°} \approx 1.3499$

(b) $\csc 48° \, 7' = \dfrac{1}{\sin\left(48 + \frac{7}{60}\right)°} \approx 1.3432$

48. (a) $\cos 4° \, 50' \, 15'' = \cos\left(4 + \dfrac{50}{60} + \dfrac{15}{3600}\right)°$
$$\approx 0.9964$$

(b) $\sec 4° \, 50' \, 15'' = \dfrac{1}{\cos 4° \, 50' \, 15''}$
$$\approx 1.0036$$

49. (a) $\cot 11° \, 15' = \dfrac{1}{\tan 11.25°} \approx 5.0273$

(b) $\tan 11° \, 15' = \tan 11.25° \approx 0.1989$

50. (a) $\sec 56° \, 8' \, 10'' = \sec\left(56 + \dfrac{8}{60} = \dfrac{10}{3600}\right)° \approx 1.7946$

(b) $\cos 56° \, 8' \, 10'' = \cos\left(56 + \dfrac{8}{60} + \dfrac{10}{3600}\right)° \approx 0.5572$

51. (a) $\csc 32° \, 40' \, 3'' = \dfrac{1}{\sin 32.6675°} \approx 1.8527$

(b) $\tan 44° \, 28' \, 16'' \approx \tan 44.4711° \approx 0.9817$

52. (a) $\sec\left(\dfrac{9}{5} \cdot 20 + 32\right)° \approx 2.6695$

(b) $\cot\left(\dfrac{9}{5} \cdot 30 + 32\right)° \approx 0.0699$

53. (a) $\sin \theta = \dfrac{1}{2} \implies \theta = 30° = \dfrac{\pi}{6}$

(b) $\csc \theta = 2 \implies \theta = 30° = \dfrac{\pi}{6}$

54. (a) $\cos \theta = \dfrac{\sqrt{2}}{2} \implies \theta = 45° = \dfrac{\pi}{4}$

(b) $\tan \theta = 1 \implies \theta = 45° = \dfrac{\pi}{4}$

55. (a) $\sec \theta = 2 \implies \theta = 60° = \dfrac{\pi}{3}$

(b) $\cot \theta = 1 \implies \theta = 45° = \dfrac{\pi}{4}$

56. (a) $\tan \theta = \sqrt{3} \implies \theta = 60° = \dfrac{\pi}{3}$

(b) $\cos \theta = \dfrac{1}{2} \implies \theta = 60° = \dfrac{\pi}{3}$

57. (a)

$\csc \theta = \dfrac{2\sqrt{3}}{3} \implies \theta = 60° = \dfrac{\pi}{3}$

(b) $\sin \theta = \dfrac{\sqrt{2}}{2} \implies \theta = 45° = \dfrac{\pi}{4}$

58. (a) $\cot \theta = \dfrac{\sqrt{3}}{3}$

$\tan \theta = \dfrac{3}{\sqrt{3}} = \sqrt{3} \implies \theta = 60° = \dfrac{\pi}{3}$

(b) $\sec \theta = \sqrt{2}$

$\cos \theta = \dfrac{1}{\sqrt{2}} = \dfrac{\sqrt{2}}{2} \implies \theta = 45° = \dfrac{\pi}{4}$

59.

$\tan 30° = \dfrac{30}{x}$

$\dfrac{1}{\sqrt{3}} = \dfrac{30}{x}$

$x = 30\sqrt{3}$

60. $\sin 60° = \dfrac{y}{18}$

$y = 18 \sin 60° = 18\left(\dfrac{\sqrt{3}}{2}\right) = 9\sqrt{3}$

61.

$\tan 60° = \dfrac{32}{x}$

$\sqrt{3} = \dfrac{32}{x}$

$\sqrt{3}x = 32$

$x = \dfrac{32}{\sqrt{3}} = \dfrac{32\sqrt{3}}{3}$

62. $\sin 45° = \dfrac{20}{r}$

$r = \dfrac{20}{\sin 45°} = \dfrac{20}{\sqrt{2}/2} = 20\sqrt{2}$

63. $\tan 82° = \dfrac{x}{45}$

$x = 45 \tan 82°$

Height of the building:

$123 + 45 \tan 82° \approx 443.2 \text{ meters}$

Distance between friends:

$\cos 82° = \dfrac{45}{y} \implies y = \dfrac{45}{\cos 82°}$

$\approx 323.34 \text{ meters}$

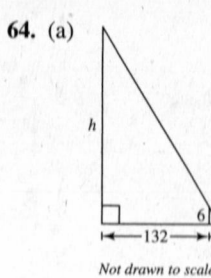

64. (a)

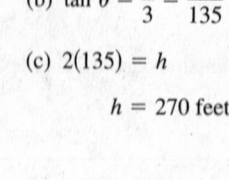

Not drawn to scale

(b) $\tan \theta = \dfrac{6}{3} = \dfrac{h}{135}$

(c) $2(135) = h$

$h = 270 \text{ feet}$

65.

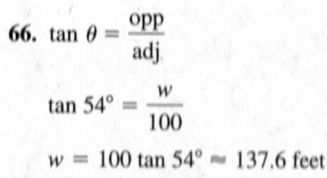

$\sin \theta = \dfrac{1500}{3000} = \dfrac{1}{2}$

$\theta = 0° = \dfrac{\pi}{6}$

66. $\tan \theta = \dfrac{\text{opp}}{\text{adj}}$

$\tan 54° = \dfrac{w}{100}$

$w = 100 \tan 54° \approx 137.6 \text{ feet}$

67.

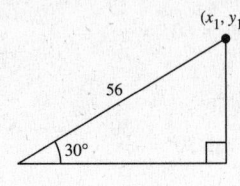

(a) $\sin 23° = \dfrac{145}{x}$

$x = \dfrac{145}{\sin 23°} \approx 371.1 \text{ feet}$

(b) $\tan 23° = \dfrac{145}{y}$

$y = \dfrac{145}{\tan 23°} \approx 341.6 \text{ feet}$

(c) Moving down the line:

$\dfrac{145/\sin 23°}{6} \approx 61.85 \text{ feet per second}$

Dropping vertically:

$\dfrac{145}{6} \approx 24.17 \text{ feet per second}$

68. Let $h =$ the height of the mountain.

Let $x =$ the horizontal distance from where the 9° angle of elevation is sighted to the point at that level directly below the mountain peak.

Then $\tan 3.5° = \dfrac{h}{x + 13}$ and $\tan 9° = \dfrac{h}{x}$.

$\tan 9° = \dfrac{h}{x} \Longrightarrow x = \dfrac{h}{\tan 9°}$

Substitute $x = \dfrac{h}{\tan 9°}$ into the expression for $\tan 3.5°$.

$\tan 3.5° = \dfrac{h}{\dfrac{h}{\tan 9°} + 13}$

$\tan 3.5° = \dfrac{h \tan 9°}{h + 13 \tan 9°}$

$h \tan 3.5° + 13 \tan 9° \tan 3.5° = h \tan 9°$

$13 \tan 9° \tan 3.5° = h(\tan 9° - \tan 3.5°)$

$\dfrac{13 \tan 9° \tan 3.5°}{\tan 9° - \tan 3.5°} = h$

$1.2953 \approx h$

The mountain is about 1.3 miles high.

69.

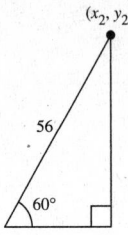

$\sin 30° = \dfrac{y_1}{56}$

$y_1 = (\sin 30°)(56) = \left(\dfrac{1}{2}\right)(56) = 28$

$\cos 30° = \dfrac{x_1}{56}$

$x_1 = \cos 30°(56) = \dfrac{\sqrt{3}}{2}(56) = 28\sqrt{3}$

$(x_1, y_1) = \left(28\sqrt{3}, 28\right)$

$\sin 60° = \dfrac{y_2}{56}$

$y_2 = \sin 60°(56) = \left(\dfrac{\sqrt{3}}{2}\right)(56) = 28\sqrt{3}$

$\cos 60° = \dfrac{x_2}{56}$

$x_2 = (\cos 60°)(56) = \left(\dfrac{1}{2}\right)(56) = 28$

$(x_2, y_2) = \left(28, 28\sqrt{3}\right)$

70. $\tan 3° = \dfrac{x}{15}$

$x = 15 \tan 3°$

$d = 5 + 2x = 5 + 2(15 \tan 3°) \approx 6.57 \text{ centimeters}$

71. (a)

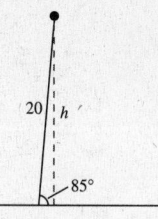

(b) $\sin 85° = \dfrac{h}{20}$

(c) $h = 20 \sin 85° \approx 19.9$ meters

(d) The side of the triangle labeled h will become shorter.

(e)

Angle, θ	Height (in meters)
80°	19.7
70°	18.8
60°	17.3
50°	15.3
40°	12.9
30°	10.0
20°	6.8
10°	3.5

(f) The height of the balloon decreases as θ decreases.

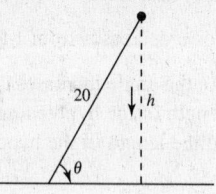

72. $x \approx 9.4, \; y \approx 3.4$

$\sin 20° = \dfrac{y}{10} \approx 0.34$ $\cot 20° = \dfrac{x}{y} \approx 2.75$

$\cos 20° = \dfrac{x}{10} \approx 0.94$ $\sec 20° = \dfrac{10}{x} \approx 1.06$

$\tan 20° = \dfrac{y}{x} \approx 0.36$ $\csc 20° = \dfrac{10}{y} \approx 2.92$

73. True,

$$\csc x = \dfrac{1}{\sin x} \implies \sin 60° \csc 60° = \sin 60°\left(\dfrac{1}{\sin 60°}\right) = 1$$

74. True, $\sec 30° = \csc 60°$ because $\sec(90° - \theta) = \csc \theta$.

75. False, $\dfrac{\sqrt{2}}{2} + \dfrac{\sqrt{2}}{2} = \sqrt{2} \neq 1$

76. True, $\cot^2 10° - \csc^2 10° = -1$ because

$1 + \cot^2 \theta = \csc^2 \theta$

$\cot^2 \theta = \csc^2 \theta - 1$

$\cot^2 \theta - \csc^2 \theta = -1.$

77. False, $\dfrac{\sin 60°}{\sin 30°} = \dfrac{\cos 30°}{\sin 30°} = \cot 30° \approx 1.7321;$

$\sin 2° \approx 0.0349$

78. False, $\tan[(5°)^2] \neq \tan^2(5°)$.

$\tan[(5°)^2] = \tan 25° \approx 0.4663$

$\tan^2(5°) = (\tan 5°)(\tan 5°) \approx 0.0077$

79. This is true because the corresponding sides of similar triangles are proportional.

80. Yes. Given $\tan \theta$, $\sec \theta$ can be found from the identity $1 + \tan^2 \theta = \sec^2 \theta$.

81. (a)

θ	0.1	0.2	0.3	0.4	0.5
$\sin \theta$	0.0998	0.1987	0.2955	0.3894	0.4794

(b) In the interval $(0, 0.5]$, $\theta > \sin \theta$.

(c) As θ approaches 0, $\sin \theta$ approaches θ.

82. (a)

θ	0°	18°	36°	54°	72°	90°
$\sin \theta$	0	0.3090	0.5878	0.8090	0.9511	1
$\cos \theta$	1	0.9511	0.8090	0.5878	0.3090	0

—CONTINUED—

82. —CONTINUED—

(b) $\sin \theta$ increases from 0 to 1 as θ increases from $0°$ to $90°$.

(c) $\cos \theta$ decreases from 1 to 0 as θ increases from $0°$ to $90°$.

(d) As the angle increases the length of the side opposite the angle increases relative to the length of the hypotenuse and the length of the side adjacent to the angle decreases relative to the length of the hypotenuse. Thus the sine increases and the cosine decreases.

83. $\dfrac{x^2 - 6x}{x^2 + 4x - 12} \cdot \dfrac{x^2 + 12x + 36}{x^2 - 36} = \dfrac{x(x - 6)}{(x + 6)(x - 2)} \cdot \dfrac{(x + 6)(x + 6)}{(x + 6)(x - 6)}$

$$= \dfrac{x}{x - 2}, x \neq \pm 6$$

84. $\dfrac{2t^2 + 5t - 12}{9 - 4t^2} \div \dfrac{t^2 - 16}{4t^2 + 12t + 9} = \dfrac{2t^2 + 5t - 12}{9 - 4t^2} \cdot \dfrac{4t^2 + 12t + 9}{t^2 - 16}$

$$= \dfrac{(2t - 3)(t + 4)}{(3 + 2t)(3 - 2t)} \cdot \dfrac{(2t + 3)(2t + 3)}{(t + 4)(t - 4)} = -\dfrac{(2t + 3)}{(t - 4)} = \dfrac{2t + 3}{4 - t}, t \neq \pm\dfrac{3}{2}, -4$$

85. $\dfrac{3}{x + 2} - \dfrac{2}{x - 2} + \dfrac{x}{x^2 + 4x + 4} = \dfrac{3(x + 2)(x - 2) - 2(x + 2)^2 + x(x - 2)}{(x - 2)(x + 2)^2}$

$$= \dfrac{3(x^2 - 4) - 2(x^2 + 4x + 4) + x^2 - 2x}{(x - 2)(x + 2)^2}$$

$$= \dfrac{2x^2 - 10x - 20}{(x - 2)(x + 2)^2} = \dfrac{2(x^2 - 5x - 10)}{(x - 2)(x + 2)^2}$$

86. $\dfrac{\left(\dfrac{3}{x} - \dfrac{1}{4}\right)}{\left(\dfrac{12}{x} - 1\right)} = \dfrac{\dfrac{12 - x}{4x}}{\dfrac{12 - x}{x}} = \dfrac{12 - x}{4x} \cdot \dfrac{x}{12 - x} = \dfrac{1}{4}, x \neq 0, 12$

Section 4.4   Trigonometric Functions of Any Angle

■ Know the Definitions of Trigonometric Functions of Any Angle.

If θ is in standard position, (x, y) a point on the terminal side and $r = \sqrt{x^2 + y^2} \neq 0$, then:

$$\sin \theta = \dfrac{y}{r} \qquad\qquad \csc \theta = \dfrac{r}{y}, y \neq 0$$

$$\cos \theta = \dfrac{x}{r} \qquad\qquad \sec \theta = \dfrac{r}{x}, x \neq 0$$

$$\tan \theta = \dfrac{y}{x}, x \neq 0 \qquad \cot \theta = \dfrac{x}{y}, y \neq 0$$

■ You should know the signs of the trigonometric functions in each quadrant.

■ You should know the trigonometric function values of the quadrant angles 0, $\dfrac{\pi}{2}$, π, and $\dfrac{3\pi}{2}$.

■ You should be able to find reference angles.

■ You should be able to evaluate trigonometric functions of any angle. (Use reference angles.)

■ You should know that the period of sine and cosine is 2π.

Vocabulary Check

1. $\sin \theta = \dfrac{y}{r}$

2. $\csc \theta$

3. $\tan \theta = \dfrac{y}{x}$

4. $\dfrac{r}{x}$

5. $\dfrac{x}{r} = \cos \theta$

6. $\cot \theta$

7. reference

1. (a) $(x, y) = (4, 3)$

$r = \sqrt{16 + 9} = 5$

$\sin \theta = \dfrac{y}{r} = \dfrac{3}{5}$ $\csc \theta = \dfrac{r}{y} = \dfrac{5}{3}$

$\cos \theta = \dfrac{x}{r} = \dfrac{4}{5}$ $\sec \theta = \dfrac{r}{x} = \dfrac{5}{4}$

$\tan \theta = \dfrac{y}{x} = \dfrac{3}{4}$ $\cot \theta = \dfrac{x}{y} = \dfrac{4}{3}$

(b) $(x, y) = (8, -15)$

$r = \sqrt{64 + 225} = 17$

$\sin \theta = \dfrac{y}{r} = -\dfrac{15}{17}$ $\csc \theta = \dfrac{r}{y} = -\dfrac{17}{15}$

$\cos \theta = \dfrac{x}{r} = \dfrac{8}{17}$ $\sec \theta = \dfrac{r}{x} = \dfrac{17}{8}$

$\tan \theta = \dfrac{y}{x} = -\dfrac{15}{8}$ $\cot \theta = \dfrac{x}{y} = -\dfrac{8}{15}$

2. (a) $x = -12, y = -5$

$r = \sqrt{(-12)^2 + (-5)^2} = 13$

$\sin \theta = \dfrac{y}{r} = -\dfrac{5}{13}$

$\cos \theta = \dfrac{x}{r} = -\dfrac{12}{13}$

$\tan \theta = \dfrac{y}{x} = \dfrac{-5}{-12} = \dfrac{5}{12}$

$\csc \theta = \dfrac{r}{y} = \dfrac{13}{-5} = \dfrac{13}{-5}$

$\sec \theta = \dfrac{r}{x} = \dfrac{13}{-12} = -\dfrac{13}{12}$

$\cot \theta = \dfrac{x}{y} = \dfrac{-12}{-5} = \dfrac{12}{5}$

(b) $x = -1, y = 1$

$r = \sqrt{(-1)^2 + 1^2} = \sqrt{2}$

$\sin \theta = \dfrac{y}{r} = \dfrac{1}{\sqrt{2}} = \dfrac{\sqrt{2}}{2}$

$\cos \theta = \dfrac{x}{r} = \dfrac{-1}{\sqrt{2}} = -\dfrac{\sqrt{2}}{2}$

$\tan \theta = \dfrac{y}{x} = \dfrac{1}{-1} = -1$

$\csc \theta = \dfrac{r}{y} = \dfrac{\sqrt{2}}{1} = \sqrt{2}$

$\sec \theta = \dfrac{r}{x} = \dfrac{\sqrt{2}}{-1} = -\sqrt{2}$

$\cot \theta = \dfrac{x}{y} = \dfrac{-1}{1} = -1$

3. (a) $(x, y) = \left(-\sqrt{3}, -1\right)$

$r = \sqrt{3 + 1} = 2$

$\sin \theta = \dfrac{y}{r} = -\dfrac{1}{2}$ $\csc \theta = \dfrac{r}{y} = -2$

$\cos \theta = \dfrac{x}{r} = -\dfrac{\sqrt{3}}{2}$ $\sec \theta = \dfrac{r}{x} = -\dfrac{2\sqrt{3}}{3}$

$\tan \theta = \dfrac{y}{x} = \dfrac{\sqrt{3}}{3}$ $\cot \theta = \dfrac{x}{y} = \sqrt{3}$

(b) $(x, y) = (-4, 1)$

$r = \sqrt{16 + 1} = \sqrt{17}$

$\sin \theta = \dfrac{y}{r} = \dfrac{\sqrt{17}}{17}$ $\csc \theta = \dfrac{r}{y} = \sqrt{17}$

$\cos \theta = \dfrac{x}{r} = -\dfrac{4\sqrt{17}}{17}$ $\sec \theta = \dfrac{r}{x} = -\dfrac{\sqrt{17}}{4}$

$\tan \theta = \dfrac{y}{x} = -\dfrac{1}{4}$ $\cot \theta = \dfrac{x}{y} = -4$

4. (a) $x = 3, y = 1$

$r = \sqrt{3^2 + 1^2} = \sqrt{10}$

$\sin \theta = \dfrac{y}{r} = \dfrac{1}{\sqrt{10}} = \dfrac{\sqrt{10}}{10}$

$\cos \theta = \dfrac{x}{r} = \dfrac{3}{\sqrt{10}} = \dfrac{3\sqrt{10}}{10}$

$\tan \theta = \dfrac{y}{x} = \dfrac{1}{3}$

$\csc \theta = \dfrac{r}{y} = \dfrac{\sqrt{10}}{1} = \sqrt{10}$

$\sec \theta = \dfrac{r}{x} = \dfrac{\sqrt{10}}{3}$

$\cot \theta = \dfrac{x}{y} = \dfrac{3}{1} = 3$

(b) $x = 4, y = -4$

$r = \sqrt{4^2 + (-4)^2} = 4\sqrt{2}$

$\sin \theta = \dfrac{y}{r} = \dfrac{-4}{4\sqrt{2}} = -\dfrac{\sqrt{2}}{2}$

$\cos \theta = \dfrac{x}{r} = \dfrac{4}{4\sqrt{2}} = \dfrac{\sqrt{2}}{2}$

$\tan \theta = \dfrac{y}{x} = \dfrac{-4}{4} = -1$

$\csc \theta = \dfrac{r}{y} = \dfrac{4\sqrt{2}}{-4} = -\sqrt{2}$

$\sec \theta = \dfrac{r}{x} = \dfrac{4\sqrt{2}}{4} = \sqrt{2}$

$\cot \theta = \dfrac{x}{y} = \dfrac{4}{-4} = -1$

5. $(x, y) = (7, 24)$

$r = \sqrt{49 + 576} = 25$

$\sin \theta = \dfrac{y}{r} = \dfrac{24}{25}$

$\cos \theta = \dfrac{x}{r} = \dfrac{7}{25}$

$\tan \theta = \dfrac{y}{x} = \dfrac{24}{7}$

$\csc \theta = \dfrac{r}{y} = \dfrac{25}{24}$

$\sec \theta = \dfrac{r}{x} = \dfrac{25}{7}$

$\cot \theta = \dfrac{x}{y} = \dfrac{7}{24}$

6. $x = 8, y = 15$

$r = \sqrt{8^2 + 15^2} = 17$

$\sin \theta = \dfrac{y}{r} = \dfrac{15}{17}$

$\cos \theta = \dfrac{x}{r} = \dfrac{8}{17}$

$\tan \theta = \dfrac{y}{x} = \dfrac{15}{8}$

$\csc \theta = \dfrac{r}{y} = \dfrac{17}{15}$

$\sec \theta = \dfrac{r}{x} = \dfrac{17}{8}$

$\cot \theta = \dfrac{x}{y} = \dfrac{8}{15}$

7. $(x, y) = (-4, 10)$

$r = \sqrt{16 + 100} = 2\sqrt{29}$

$\sin \theta = \dfrac{y}{r} = \dfrac{5\sqrt{29}}{29}$

$\cos \theta = \dfrac{x}{r} = -\dfrac{2\sqrt{29}}{29}$

$\tan \theta = \dfrac{y}{x} = -\dfrac{5}{2}$

$\csc \theta = \dfrac{r}{y} = \dfrac{\sqrt{29}}{5}$

$\sec \theta = \dfrac{r}{x} = -\dfrac{\sqrt{29}}{2}$

$\cot \theta = \dfrac{x}{y} = -\dfrac{2}{5}$

8. $x = -5, y = -2$

$r = \sqrt{(-5)^2 + (-2)^2} = \sqrt{29}$

$\sin \theta = \dfrac{y}{r} = \dfrac{-2}{\sqrt{29}} = -\dfrac{2\sqrt{29}}{29}$

$\cos \theta = \dfrac{x}{r} = \dfrac{-5}{\sqrt{29}} = -\dfrac{5\sqrt{29}}{29}$

$\tan \theta = \dfrac{y}{x} = \dfrac{-2}{-5} = \dfrac{2}{5}$

$\csc \theta = \dfrac{r}{y} = \dfrac{\sqrt{29}}{-2} = -\dfrac{\sqrt{29}}{2}$

$\sec \theta = \dfrac{r}{x} = \dfrac{\sqrt{29}}{-5} = -\dfrac{\sqrt{29}}{5}$

$\cot \theta = \dfrac{x}{y} = \dfrac{-5}{-2} = \dfrac{5}{2}$

9. $(x, y) = (-3.5, 6.8)$

$r = \sqrt{12.25 + 46.24} = \dfrac{\sqrt{5849}}{10}$

$\sin \theta = \dfrac{y}{r} = \dfrac{68\sqrt{5849}}{5849} \approx 0.9$

$\cos \theta = \dfrac{x}{r} = -\dfrac{35\sqrt{5849}}{5849} \approx -0.5$

$\tan \theta = \dfrac{y}{x} = -\dfrac{68}{35} \approx -1.9$

$\csc \theta = \dfrac{r}{y} = \dfrac{\sqrt{5849}}{68} \approx 1.1$

$\sec \theta = \dfrac{r}{x} = -\dfrac{\sqrt{5849}}{35} \approx -2.2$

$\cot \theta = \dfrac{x}{y} = -\dfrac{35}{68} \approx -0.5$

10. $x = 3\frac{1}{2} = \frac{7}{2}, y = -7\frac{3}{4} = -\frac{31}{4}$

$$r = \sqrt{\left(\frac{7}{2}\right)^2 + \left(-\frac{31}{4}\right)^2} = \frac{\sqrt{1157}}{4}$$

$\sin\theta = \dfrac{y}{r} = \dfrac{-31/4}{\sqrt{1157}/4} = -\dfrac{31\sqrt{1157}}{1157} \approx -0.9$ $\csc\theta = \dfrac{r}{y} = \dfrac{\sqrt{1157}/4}{-31/4} = -\dfrac{\sqrt{1157}}{31} \approx -1.1$

$\cos\theta = \dfrac{x}{r} = \dfrac{7/2}{\sqrt{1157}/4} = \dfrac{14\sqrt{1157}}{1157} \approx 0.4$ $\sec\theta = \dfrac{r}{x} = \dfrac{\sqrt{1157}/4}{7/2} = \dfrac{\sqrt{1157}}{14} \approx 2.4$

$\tan\theta = \dfrac{y}{x} = \dfrac{-31/4}{7/2} = -\dfrac{31}{14} \approx -2.2$ $\cot\theta = \dfrac{x}{y} = \dfrac{7/2}{-31/4} = -\dfrac{14}{31} \approx -0.5$

11. $\sin\theta < 0 \implies \theta$ lies in Quadrant III or in Quadrant IV.

$\cos\theta < 0 \implies \theta$ lies in Quadrant II or in Quadrant III.

$\sin\theta < 0$ *and* $\cos\theta < 0 \implies \theta$ lies in Quadrant III.

12. $\sin\theta > 0$ and $\cos\theta > 0$

$\dfrac{y}{r} > 0$ and $\dfrac{x}{r} > 0$

Quadrant I

13. $\sin\theta > 0 \implies \theta$ lies in Quadrant I or in Quadrant II.

$\tan\theta < 0 \implies \theta$ lies in Quadrant II or in Quadrant IV.

$\sin\theta > 0$ *and* $\tan\theta < 0 \implies \theta$ lies in Quadrant II.

14. $\sec\theta > 0$ and $\cot\theta < 0$

$\dfrac{r}{x} > 0$ and $\dfrac{x}{y} < 0$

Quadrant IV

15. $\sin\theta = \dfrac{y}{r} = \dfrac{3}{5} \implies x^2 = 25 - 9 = 16$

θ in Quadrant II $\implies x = -4$

$\sin\theta = \dfrac{y}{r} = \dfrac{3}{5}$ $\csc\theta = \dfrac{r}{y} = \dfrac{5}{3}$

$\cos\theta = \dfrac{x}{r} = -\dfrac{4}{5}$ $\sec\theta = \dfrac{r}{x} = -\dfrac{5}{4}$

$\tan\theta = \dfrac{y}{x} = -\dfrac{3}{4}$ $\cot\theta = \dfrac{x}{y} = -\dfrac{4}{3}$

16. $\cos\theta = \dfrac{x}{r} = \dfrac{-4}{5} \implies y^2 = 25 - 16 = 9$

θ in Quadrant III $\implies y = -3$

$\sin\theta = \dfrac{y}{r} = -\dfrac{3}{5}$ $\csc\theta = -\dfrac{5}{3}$

$\cos\theta = \dfrac{x}{r} = -\dfrac{4}{5}$ $\sec\theta = -\dfrac{5}{4}$

$\tan\theta = \dfrac{y}{x} = \dfrac{3}{4}$ $\cot\theta = \dfrac{4}{3}$

17. $\tan\theta = \dfrac{y}{x} = \dfrac{-15}{8}$

$\sin\theta < 0$ and $\tan\theta < 0 \implies \theta$ is in Quadrant IV $\implies$ $y < 0$ and $x > 0$.

$x = 8, y = -15, r = 17$

$\sin\theta = \dfrac{y}{r} = -\dfrac{15}{17}$ $\csc\theta = \dfrac{r}{y} = -\dfrac{17}{15}$

$\cos\theta = \dfrac{x}{r} = \dfrac{8}{17}$ $\sec\theta = \dfrac{r}{x} = \dfrac{17}{8}$

$\tan\theta = \dfrac{y}{x} = -\dfrac{15}{8}$ $\cot\theta = \dfrac{x}{y} = -\dfrac{8}{15}$

18. $\cos\theta = \dfrac{x}{r} = \dfrac{8}{17} \implies y = |15|$

$\tan\theta < 0 \implies y = -15$

$\sin\theta = \dfrac{y}{r} = \dfrac{-15}{17} = -\dfrac{15}{17}$ $\csc\theta = -\dfrac{17}{15}$

$\cos\theta = \dfrac{x}{r} = \dfrac{8}{17}$ $\sec\theta = \dfrac{17}{8}$

$\tan\theta = \dfrac{y}{x} = \dfrac{-15}{8} = -\dfrac{15}{8}$ $\cot\theta = -\dfrac{8}{15}$

19. $\cot \theta = \dfrac{x}{y} = -\dfrac{3}{1} = \dfrac{3}{-1}$

$\cos \theta > 0 \implies \theta$ is in Quadrant IV $\implies x$ is positive;
$x = 3, y = -1, r = \sqrt{10}$

$\sin \theta = \dfrac{y}{r} = -\dfrac{\sqrt{10}}{10}$ $\qquad$ $\csc \theta = \dfrac{r}{y} = -\sqrt{10}$

$\cos \theta = \dfrac{x}{r} = \dfrac{3\sqrt{10}}{10}$ $\qquad$ $\sec \theta = \dfrac{r}{x} = \dfrac{\sqrt{10}}{3}$

$\tan \theta = \dfrac{y}{x} = -\dfrac{1}{3}$ $\qquad$ $\cot \theta = \dfrac{x}{y} = -3$

20. $\csc \theta = \dfrac{r}{y} = \dfrac{4}{1} \implies x = \left| \sqrt{15} \right|$

$\cot \theta < 0 \implies x = -\sqrt{15}$

$\sin \theta = \dfrac{y}{r} = \dfrac{1}{4}$ $\qquad$ $\csc \theta = 4$

$\cos \theta = \dfrac{x}{r} = -\dfrac{\sqrt{15}}{4}$ $\qquad$ $\sec \theta = -\dfrac{4\sqrt{15}}{15}$

$\tan \theta = \dfrac{y}{x} = -\dfrac{\sqrt{15}}{15}$ $\qquad$ $\cot \theta = -\sqrt{15}$

21. $\sec \theta = \dfrac{r}{x} = \dfrac{2}{-1} \implies y^2 = 4 - 1 = 3$

$\sin \theta > 0 \implies \theta$ is in Quadrant II $\implies y = \sqrt{3}$

$\sin \theta = \dfrac{y}{r} = \dfrac{\sqrt{3}}{2}$ $\qquad$ $\csc \theta = \dfrac{r}{y} = \dfrac{2\sqrt{3}}{3}$

$\cos \theta = \dfrac{x}{r} = -\dfrac{1}{2}$ $\qquad$ $\sec \theta = \dfrac{r}{x} = -2$

$\tan \theta = \dfrac{y}{x} = -\sqrt{3}$ $\qquad$ $\cot \theta = \dfrac{x}{y} = -\dfrac{\sqrt{3}}{3}$

22. $\sin \theta = 0 \implies \theta = 0 + \pi n$

$\sec \theta = -1 \implies \theta = \pi + 2\pi m$

$y = 0, x = -r$

$\sin \theta = 0$ $\qquad$ $\csc \theta = \dfrac{r}{y}$ is undefined

$\cos \theta = \dfrac{x}{r} = \dfrac{-r}{r} = -1$ $\qquad$ $\sec \theta = \dfrac{r}{x} = -1$

$\tan \theta = \dfrac{y}{x} = 0$ $\qquad$ $\cot \theta = \dfrac{x}{y}$ is undefined

23. $\cot \theta$ is undefined, $\dfrac{\pi}{2} \le \theta \le \dfrac{3\pi}{2} \implies y = 0 \implies \theta = \pi$

$\sin \pi = 0$ $\qquad$ $\csc \pi$ is undefined

$\cos \pi = -1$ $\qquad$ $\sec \pi = -1$

$\tan \pi = 0$ $\qquad$ $\cot \pi$ is undefined

24. $\tan \theta$ is undefined $\implies \theta = n\pi + \dfrac{\pi}{2}$

$\pi \le \theta \le 2\pi \implies \theta = \dfrac{3\pi}{2}, x = 0, y = -r$

$\sin \theta = \dfrac{y}{r} = \dfrac{-r}{r} = -1$ $\qquad$ $\csc \theta = \dfrac{r}{y} = -1$

$\cos \theta = \dfrac{x}{r} = \dfrac{0}{r} = 0$ $\qquad$ $\sec \theta = \dfrac{r}{x}$ is undefined.

$\tan \theta = \dfrac{y}{x}$ is undefined. $\qquad$ $\cot \theta = \dfrac{x}{y} = \dfrac{0}{y} = 0$

25. To find a point on the terminal side of θ use any point on the line $y = -x$ that lies in Quadrant II. $(-1, 1)$ is one such point.

$x = -1, y = 1, r = \sqrt{2}$

$\sin \theta = \dfrac{1}{\sqrt{2}} = \dfrac{\sqrt{2}}{2}$

$\cos \theta = -\dfrac{1}{\sqrt{2}} = -\dfrac{\sqrt{2}}{2}$

$\tan \theta = -1$

$\csc \theta = \sqrt{2}$

$\sec \theta = -\sqrt{2}$

$\cot \theta = -1$

26. Let $x > 0$.

$\left(-x, -\dfrac{1}{3}x \right)$, Quadrant III

$r = \sqrt{x^2 + \dfrac{1}{9}x^2} = \dfrac{\sqrt{10}x}{3}$

$\sin \theta = \dfrac{y}{r} = \dfrac{(-1/3)x}{(\sqrt{10}x)/3} = -\dfrac{\sqrt{10}}{10}$

$\cos \theta = \dfrac{x}{r} = \dfrac{-x}{(\sqrt{10}x)/3} = -\dfrac{3\sqrt{10}}{10}$

$\tan \theta = \dfrac{y}{x} = \dfrac{(-1/3)x}{-x} = \dfrac{1}{3}$

$\csc \theta = \dfrac{r}{y} = \dfrac{(\sqrt{10}x)/3}{(-1/3)x} = -\sqrt{10}$

$\sec \theta = \dfrac{r}{x} = \dfrac{(\sqrt{10}x)/3}{-x} = -\dfrac{\sqrt{10}}{3}$

$\cot \theta = \dfrac{x}{y} = \dfrac{-x}{(-1/3)x} = 3$

27. To find a point on the terminal side of θ, use any point on the line $y = 2x$ that lies in Quadrant III. $(-1, -2)$ is one such point.

$x = -1, y = -2, r = \sqrt{5}$

$\sin \theta = -\dfrac{2}{\sqrt{5}} = -\dfrac{2\sqrt{5}}{5}$

$\cos \theta = -\dfrac{1}{\sqrt{5}} = -\dfrac{\sqrt{5}}{5}$

$\tan \theta = \dfrac{-2}{-1} = 2$

$\csc \theta = \dfrac{\sqrt{5}}{-2} = -\dfrac{\sqrt{5}}{2}$

$\sec \theta = \dfrac{\sqrt{5}}{-1} = -\sqrt{5}$

$\cot \theta = \dfrac{-1}{-2} = \dfrac{1}{2}$

28. Let $x > 0$.

$4x + 3y = 0 \Longrightarrow y = -\dfrac{4}{3}x$

$\left(x, -\dfrac{4}{3}x\right)$, Quadrant IV

$r = \sqrt{x^2 + \dfrac{16}{9}x^2} = \dfrac{5}{3}x$

$\sin \theta = \dfrac{y}{r} = \dfrac{(-4/3)x}{(5/3)x} = -\dfrac{4}{5}$ $\qquad \csc \theta = -\dfrac{5}{4}$

$\cos \theta = \dfrac{x}{r} = \dfrac{x}{(5/3)x} = \dfrac{3}{5}$ $\qquad \sec \theta = \dfrac{5}{3}$

$\tan \theta = \dfrac{y}{x} = \dfrac{(-4/3)x}{x} = -\dfrac{4}{3}$ $\qquad \tan \theta = -\dfrac{3}{4}$

29. $(x, y) = (-1, 0), r = 1$

$\sin \pi = \dfrac{y}{r} = 0$

30. $\csc \dfrac{3\pi}{2} = \dfrac{r}{y} = \dfrac{1}{-1} = -1$

since $\dfrac{3\pi}{2}$ corresponds to $(0, -1)$.

31. $(x, y) = (0, -1), r = 1$

$\sec \dfrac{3\pi}{2} = \dfrac{r}{x} = \dfrac{1}{0} \Longrightarrow$ undefined

32. $\sec \pi = \dfrac{r}{x} = \dfrac{1}{-1} = -1$

since $\dfrac{3\pi}{2}$ corresponds to $(-1, 0)$.

33. $(x, y) = (0, 1), r = 1$

$\sin \dfrac{\pi}{2} = \dfrac{y}{r} = 1$

34. $\cot \pi = \dfrac{x}{y} = \dfrac{-1}{0}$ (undefined)

since π corresponds to $(-1, 0)$.

35. $(x, y) = (-1, 0), r = 1$

$\csc \pi = \dfrac{r}{y} = \dfrac{1}{0} \Longrightarrow$ undefined

36. $\cot \dfrac{\pi}{2} = \dfrac{x}{y} = \dfrac{0}{1} = 0$

since $\dfrac{\pi}{2}$ corresponds to $(0, 1)$.

37. $\theta = 203°$

$\theta' = 203° - 180° = 23°$

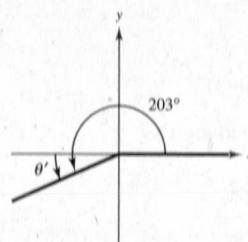

38. $\theta = 309°$

$\theta' = 360° - 309° = 51°$

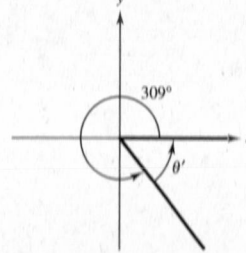

39.
$$\theta = -245°$$
$$360° - 245° = 115° \text{ (coterminal angle)}$$
$$\theta' = 180° - 115° = 65°$$

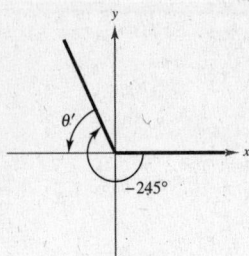

40. $\theta = -145°$ is coterminal with 215°.
$$\theta' = 215° - 180° = 35°$$

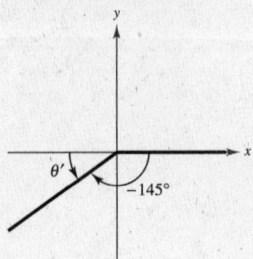

41. $\theta = \dfrac{2\pi}{3}$

$$\theta' = \pi - \dfrac{2\pi}{3} = \dfrac{\pi}{3}$$

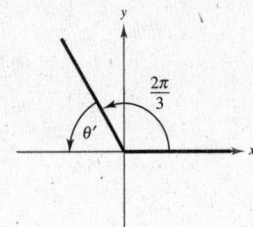

42. $\theta = \dfrac{7\pi}{4}$

$$\theta' = 2\pi - \dfrac{7\pi}{4} = \dfrac{\pi}{4}$$

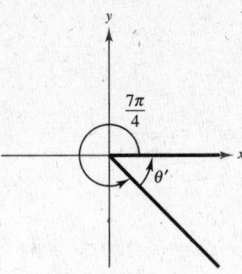

43. $\theta = 3.5$

$$\theta' = 3.5 - \pi$$

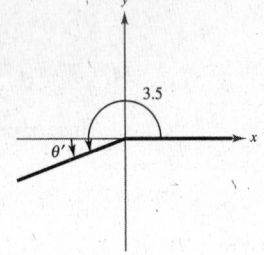

44. $\theta = \dfrac{11\pi}{3}$ is coterminal

with $\dfrac{5\pi}{3}$.

$$\theta' = 2\pi - \dfrac{5\pi}{3} = \dfrac{\pi}{3}$$

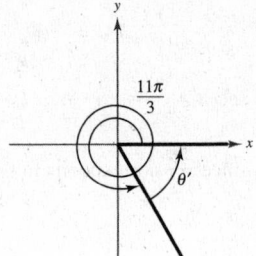

45. $\theta = 225°$, $\theta' = 360° - 225° = 45°$, Quadrant III

$$\sin 225° = -\sin 45° = -\dfrac{\sqrt{2}}{2}$$

$$\cos 225° = -\cos 45° = -\dfrac{\sqrt{2}}{2}$$

$$\tan 225° = \tan 45° = 1$$

46. $\theta = 300°$, $\theta' = 360° - 300° = 60°$, Quadrant IV

$$\sin 300° = -\sin 60° = -\dfrac{\sqrt{3}}{2}$$

$$\cos 300° = \cos 60° = \dfrac{1}{2}$$

$$\tan 300° = -\tan 60° = -\sqrt{3}$$

47. $\theta = 750°$ is coterminal with 30°.

$$\theta' = 30°, \text{ Quadrant I}$$

$$\sin 750° = \sin 30° = \dfrac{1}{2}$$

$$\cos 750° = \cos 30° = \dfrac{\sqrt{3}}{2}$$

$$\tan 750° = \tan 30° = \dfrac{\sqrt{3}}{3}$$

48. $\theta = -405°$ is coterminal with 315°.

$$\theta' = 360° - 315° = 45°, \text{ Quadrant IV}$$

$$\sin(-405°) = -\sin 45° = -\dfrac{\sqrt{2}}{2}$$

$$\cos(-405°) = \cos 45° = \dfrac{\sqrt{2}}{2}$$

$$\tan(-405°) = -\tan 45° = -1$$

49. $\theta = -150°$ is coterminal with $210°$.

$\theta' = 210° - 180° = 30°$, Quadrant III

$\sin(-150°) = -\sin 30° = -\dfrac{1}{2}$

$\cos(-150°) = -\cos 30° = -\dfrac{\sqrt{3}}{2}$

$\tan(-150°) = \tan 30° = \dfrac{\sqrt{3}}{3}$

50. $\theta = -840°$ is coterminal with $240°$.

$\theta' = 240° - 180° = 60°$, Quadrant III

$\sin(-840°) = -\sin 60° = -\dfrac{\sqrt{3}}{2}$

$\cos(-840°) = -\cos 60° = -\dfrac{1}{2}$

$\tan(-840°) = \tan 60° = \sqrt{3}$

51. $\theta = \dfrac{4\pi}{3}$, $\theta' = \dfrac{\pi}{3}$, Quadrant III

$\sin\dfrac{4\pi}{3} = -\sin\dfrac{\pi}{3} = -\dfrac{\sqrt{3}}{2}$

$\cos\dfrac{4\pi}{3} = -\cos\dfrac{\pi}{3} = -\dfrac{1}{2}$

$\tan\dfrac{4\pi}{3} = \tan\dfrac{\pi}{3} = \sqrt{3}$

52. $\theta = \dfrac{\pi}{4}$, $\theta' = \dfrac{\pi}{4}$, Quadrant I

$\sin\dfrac{\pi}{4} = \dfrac{\sqrt{2}}{2}$

$\cos\dfrac{\pi}{4} = \dfrac{\sqrt{2}}{2}$

$\tan\dfrac{\pi}{4} = 1$

53. $\theta = -\dfrac{\pi}{6}$, $\theta' = \dfrac{\pi}{6}$, Quadrant IV

$\sin\left(-\dfrac{\pi}{6}\right) = -\sin\dfrac{\pi}{6} = -\dfrac{1}{2}$

$\cos\left(-\dfrac{\pi}{6}\right) = \cos\dfrac{\pi}{6} = \dfrac{\sqrt{3}}{2}$

$\tan\left(-\dfrac{\pi}{6}\right) = -\tan\dfrac{\pi}{6} = -\dfrac{\sqrt{3}}{3}$

54. $\theta = -\dfrac{\pi}{2}$ is coterminal with $\dfrac{3\pi}{2}$.

$\sin\left(-\dfrac{\pi}{2}\right) = \sin\dfrac{3\pi}{2} = -1$

$\cos\left(-\dfrac{\pi}{2}\right) = \cos\dfrac{3\pi}{2} = 0$

$\tan\left(-\dfrac{\pi}{2}\right) = \tan\dfrac{3\pi}{2}$ is undefined.

55. $\theta = \dfrac{11\pi}{4}$ is coterminal with $\dfrac{3}{4}\pi$.

$\theta' = \pi - \dfrac{3}{4}\pi = \dfrac{\pi}{4}$, Quadrant II

$\sin\dfrac{11\pi}{4} = \sin\dfrac{\pi}{4} = \dfrac{\sqrt{2}}{2}$

$\cos\dfrac{11\pi}{4} = -\cos\dfrac{\pi}{4} = -\dfrac{\sqrt{2}}{2}$

$\tan\dfrac{11\pi}{4} = -\tan\dfrac{\pi}{4} = -1$

56. $\theta = \dfrac{10\pi}{3}$ is coterminal with $\dfrac{4\pi}{3}$.

$\theta' = \dfrac{4\pi}{3} - \pi = \dfrac{\pi}{3}$, Quadrant III

$\sin\dfrac{10\pi}{3} = -\sin\dfrac{\pi}{3} = -\dfrac{\sqrt{3}}{2}$

$\cos\dfrac{10\pi}{3} = -\cos\dfrac{\pi}{3} = -\dfrac{1}{2}$

$\tan\dfrac{10\pi}{3} = \tan\dfrac{\pi}{3} = \sqrt{3}$

57. $\theta = -\dfrac{3\pi}{2}$ is coterminal with $\dfrac{\pi}{2}$, $\theta' = \dfrac{\pi}{2}$.

$\sin\left(-\dfrac{3\pi}{2}\right) = \sin\dfrac{\pi}{2} = 1$

$\cos\left(-\dfrac{3\pi}{2}\right) = \cos\dfrac{\pi}{2} = 0$

$\tan\left(-\dfrac{3\pi}{2}\right) = \tan\dfrac{\pi}{2}$ which is undefined.

58. $\theta = -\dfrac{25\pi}{4}$ is coterminal with $\dfrac{7\pi}{4}$.

$\theta' = 2\pi - \dfrac{7\pi}{4} = \dfrac{\pi}{4}$ in Quadrant IV.

$\sin\left(-\dfrac{25\pi}{4}\right) = -\sin\left(\dfrac{\pi}{4}\right) = -\dfrac{\sqrt{2}}{2}$

$\cos\left(-\dfrac{25\pi}{4}\right) = \cos\left(\dfrac{\pi}{4}\right) = \dfrac{\sqrt{2}}{2}$

$\tan\left(-\dfrac{25\pi}{4}\right) = -\tan\left(\dfrac{\pi}{4}\right) = -1$

59. $\sin \theta = -\dfrac{3}{5}$

$\sin^2 \theta + \cos^2 \theta = 1$

$\cos^2 \theta = 1 - \sin^2 \theta$

$\cos^2 \theta = 1 - \left(-\dfrac{3}{5}\right)^2$

$\cos^2 \theta = 1 - \dfrac{9}{25}$

$\cos^2 \theta = \dfrac{16}{25}$

$\cos \theta > 0$ in Quadrant IV.

$\cos \theta = \dfrac{4}{5}$

60. $\cot \theta = -3$

$1 + \cot^2 \theta = \csc^2 \theta$

$1 + (-3)^2 = \csc^2 \theta$

$10 = \csc^2 \theta$

$\csc \theta > 0$ in Quadrant II.

$\sqrt{10} = \csc \theta$

$\csc \theta = \dfrac{1}{\sin \theta}$

$\sin \theta = \dfrac{1}{\csc \theta} = \dfrac{1}{\sqrt{10}} = \dfrac{\sqrt{10}}{10}$

61. $\tan \theta = \dfrac{3}{2}$

$\sec^2 \theta = 1 + \tan^2 \theta$

$\sec^2 \theta = 1 + \left(\dfrac{3}{2}\right)^2$

$\sec^2 \theta = 1 + \dfrac{9}{4}$

$\sec^2 \theta = \dfrac{13}{4}$

$\sec \theta < 0$ in Quadrant III.

$\sec \theta = -\dfrac{\sqrt{13}}{2}$

62. $\csc \theta = -2$

$1 + \cot^2 \theta = \csc^2 \theta$

$\cot^2 \theta = \csc^2 \theta - 1$

$\cot^2 \theta = (-2)^2 - 1$

$\cot^2 \theta = 3$

$\cot \theta < 0$ in Quadrant IV.

$\cot \theta = -\sqrt{3}$

63. $\cos \theta = \dfrac{5}{8}$

$\cos \theta = \dfrac{1}{\sec \theta} \implies \sec \theta = \dfrac{1}{\cos \theta}$

$\sec \theta = \dfrac{1}{5/8} = \dfrac{8}{5}$

64. $\sec \theta = -\dfrac{9}{4}$

$1 + \tan^2 \theta = \sec^2 \theta$

$\tan^2 \theta = \sec^2 \theta - 1$

$\tan^2 \theta = \left(-\dfrac{9}{4}\right)^2 - 1$

$\tan^2 \theta = \dfrac{65}{16}$

$\tan \theta > 0$ in Quadrant III.

$\tan \theta = \dfrac{\sqrt{65}}{4}$

65. $\sin 10° \approx 0.1736$

66. $\sec 225° = \dfrac{1}{\cos 225°} \approx -1.4142$

67. $\cos(-110°) \approx -0.3420$

68. $\csc(-330°) = \dfrac{1}{\sin(-330°)} = 2.0000$

69. $\tan 304° \approx -1.4826$

70. $\cot 178° = \dfrac{1}{\tan 178°} = -28.6363$

71. $\sec 72° = \dfrac{1}{\cos 72°} \approx 3.2361$

72. $\tan(-188°) = -0.1405$

73. $\tan 4.5 \approx 4.6373$

74. $\cot 1.35 = \dfrac{1}{\tan 1.35} \approx 0.2245$

75. $\tan \dfrac{\pi}{9} \approx 0.3640$

76. $\tan\left(-\dfrac{\pi}{9}\right) \approx -0.3640$

77. $\sin(-0.65) \approx -0.6052$

78. $\sec 0.29 = \dfrac{1}{\cos 0.29} \approx 1.0436$

79. $\cot\left(-\dfrac{11\pi}{8}\right) = \dfrac{1}{\tan\left(-\dfrac{11\pi}{8}\right)} \approx -0.4142$

80. $\csc\left(-\dfrac{15\pi}{14}\right) = \dfrac{1}{\sin\left(-\dfrac{15\pi}{14}\right)} \approx 4.4940$

81. (a) $\sin \theta = \dfrac{1}{2} \implies$ reference angle is 30° or $\dfrac{\pi}{6}$ and θ is in Quadrant I or Quadrant II.

Values in degrees: 30°, 150°

Values in radians: $\dfrac{\pi}{6}, \dfrac{5\pi}{6}$

(b) $\sin \theta = -\dfrac{1}{2} \implies$ reference angle is 30° or $\dfrac{\pi}{6}$ and θ is in Quadrant III or Quadrant IV.

Values in degrees: 210°, 330°

Values in radians: $\dfrac{7\pi}{6}, \dfrac{11\pi}{6}$

82. (a) $\cos \theta = \dfrac{\sqrt{2}}{2} \implies$ reference angle is 45° or $\dfrac{\pi}{4}$ and θ is in Quadrant I or IV.

Values in degrees: 45°, 315°

Values in radians: $\dfrac{\pi}{4}, \dfrac{7\pi}{4}$

(b) $\cos \theta = -\dfrac{\sqrt{2}}{2} \implies$ reference angle is 45° or $\dfrac{\pi}{4}$ and θ is in Quadrant II or III.

Values in degrees: 135°, 225°

Values in radians: $\dfrac{3\pi}{4}, \dfrac{5\pi}{4}$

83. (a) $\csc \theta = \dfrac{2\sqrt{3}}{3} \implies$ reference angle is 60° or $\dfrac{\pi}{3}$ and θ is in Quadrant I or Quadrant II.

Values in degrees: 60°, 120°

Values in radians: $\dfrac{\pi}{3}, \dfrac{2\pi}{3}$

(b) $\cot \theta = -1 \implies$ reference angle is 45° or $\dfrac{\pi}{4}$ and θ is in Quadrant II or Quadrant IV.

Values in degrees: 135°, 315°

Values in radians: $\dfrac{3\pi}{4}, \dfrac{7\pi}{4}$

84. (a) $\sec \theta = 2 \implies$ reference angle is 60° or $\dfrac{\pi}{3}$ and θ is in Quadrant I or IV.

Values in degrees: 60°, 300°

Values in radians: $\dfrac{\pi}{3}, \dfrac{5\pi}{3}$

(b) $\sec \theta = -2 \implies$ reference angle is 60° or $\dfrac{\pi}{3}$ and θ is in Quadrant II or III.

Values in degrees: 120°, 240°

Values in radians: $\dfrac{2\pi}{3}, \dfrac{4\pi}{3}$

85. (a) $\tan \theta = 1 \implies$ reference angle is 45° or $\dfrac{\pi}{4}$ and θ is in Quadrant I or Quadrant III.

Values in degrees: 45°, 225°

Values in radians: $\dfrac{\pi}{4}, \dfrac{5\pi}{4}$

(b) $\cot \theta = -\sqrt{3} \implies$ reference angle is 30° or $\dfrac{\pi}{6}$ and θ is in Quadrant II or Quadrant IV.

Values in degrees: 150°, 330°

Values in radians: $\dfrac{5\pi}{6}, \dfrac{11\pi}{6}$

86. (a) $\sin \theta = \dfrac{\sqrt{3}}{2} \implies$ reference angle is 60° or $\dfrac{\pi}{3}$ and θ is in Quadrant I or II.

Values in degrees: 60°, 120°

Values in radians: $\dfrac{\pi}{3}, \dfrac{2\pi}{3}$

(b) $\sin \theta = -\dfrac{\sqrt{3}}{2} \implies$ reference angle is 60° or $\dfrac{\pi}{3}$ and θ is in Quadrant III or IV.

Values in degrees: 240°, 300°

Values in radians: $\dfrac{4\pi}{3}, \dfrac{5\pi}{3}$

87. (a) New York City:

$N \approx 22.099 \sin(0.522t - 2.219) + 55.008$

Fairbanks:

$F \approx 36.641 \sin(0.502t - 1.831) + 25.610$

(b)

Month	New York City	Fairbanks
February	34.6°	−1.4°
March	41.6°	13.9°
May	63.4°	48.6°
June	72.5°	59.5°
August	75.5°	55.6°
September	68.6°	41.7°
November	46.8°	6.5°

(c) The periods are about the same for both models, approximately 12 months.

88. $S = 23.1 + 0.442t + 4.3 \cos \dfrac{\pi t}{6}$

(a) For February 2006, $t = 2$.

$S = 23.1 + 0.442(2) + 4.3 \cos \dfrac{\pi(2)}{6} \approx 26{,}134$ units

(b) For February 2007, $t = 14$.

$S = 23.1 + 0.442(14) + 4.3 \cos \dfrac{\pi(14)}{6} \approx 31{,}438$ units

(c) For June 2006, $t = 6$.

$S = 23.1 + 0.442(6) + 4.3 \cos \dfrac{\pi(6)}{6} \approx 21{,}452$ units

(d) For June 2007, $t = 18$.

$S = 23.1 + 0.442(18) + 4.3 \cos \dfrac{\pi(18)}{6} \approx 26{,}756$ units

89. $y(t) = 2 \cos 6t$

(a) $y(0) = 2 \cos 0 = 2$ centimeters

(b) $y\left(\dfrac{1}{4}\right) = 2 \cos\left(\dfrac{3}{2}\right) \approx 0.14$ centimeter

(c) $y\left(\dfrac{1}{2}\right) = 2 \cos 3 \approx -1.98$ centimeters

90. $y(t) = 2e^{-t} \cos 6t$

(a) $t = 0$

$y(0) = 2e^{-0} \cos 0 = 2$ centimeters

(b) $t = \dfrac{1}{4}$

$y\left(\tfrac{1}{4}\right) = 2e^{-1/4} \cos\left(6 \cdot \tfrac{1}{4}\right) \approx 0.11$ centimeters

(c) $t = \dfrac{1}{2}$

$y\left(\tfrac{1}{2}\right) = 2e^{-1/2} \cos\left(6 \cdot \tfrac{1}{2}\right) \approx -1.2$ centimeters

91. $I = 5e^{-2t} \sin t$

$I(0.7) = 5e^{-1.4} \sin 0.7 \approx 0.79$ ampere

92. $\sin \theta = \dfrac{6}{d} \Rightarrow d = \dfrac{6}{\sin \theta}$

(a) $\theta = 30°$

$$d = \dfrac{6}{\sin 30°} = \dfrac{6}{1/2} = 12 \text{ miles}$$

(b) $\theta = 90°$

$$d = \dfrac{6}{\sin 90°} = \dfrac{6}{1} = 6 \text{ miles}$$

(c) $\theta = 120°$

$$d = \dfrac{6}{\sin 120°} \approx 6.9 \text{ miles}$$

93. False. In each of the four quadrants, the sign of the secant function and the cosine function will be the same since they are reciprocals of each other.

94. False. For example, if $n = 1$ and $\theta = 225°$, $0 \le 135 \le 360$, but $360°n - \theta = 135°$ is not the reference angle. The reference angle would be $45°$. For θ in Quadrant II, $\theta' = 180° - \theta$. For θ in Quadrant III, $\theta' = \theta - 180°$. For θ in Quadrant IV, $\theta' = 360° - \theta$.

95. As θ increases from $0°$ to $90°$, x decreases from 12 cm to 0 cm and y increases from 0 cm to 12 cm.

Therefore, $\sin \theta = \dfrac{y}{12}$ increases from 0 to 1 and $\cos \theta = \dfrac{x}{12}$ decreases from 1 to 0. Thus,

$\tan \theta = \dfrac{y}{x}$ increases without bound, and when $\theta = 90°$ the tangent is undefined.

96. Determine the trigonometric function of the reference angle and prefix the appropriate sign.

97. $y = x^2 + 3x - 4 = (x + 4)(x - 1)$

x-intercepts: $(-4, 0), (1, 0)$

y-intercept: $(0, -4)$

No asymptotes

Domain: All real numbers x

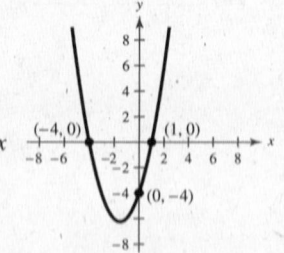

98. $y = 2x^2 - 5x = x(2x - 5)$

x-intercepts: $(0, 0), \left(\frac{5}{2}, 0\right)$

y-intercepts: $(0, 0)$

No asymptotes

Domain: All real numbers x

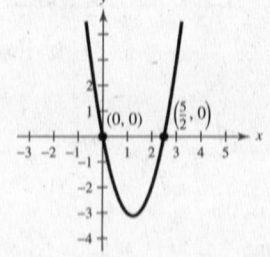

99. $f(x) = x^3 + 8$

x-intercept: $(-2, 0)$

y-intercept: $(0, 8)$

No asymptotes

Domain: All real numbers x

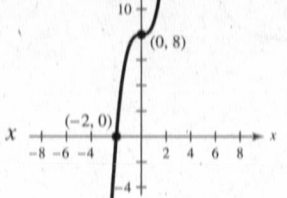

100. $g(x) = x^4 + 2x^2 - 3 = (x^2 + 3)(x^2 - 1)$

$\qquad\qquad\quad = (x^2 + 3)(x + 1)(x - 1)$

x-intercepts: $(-1, 0), (1, 0)$

y-intercepts: $(0, -3)$

No asymptotes

Domain: All real numbers x

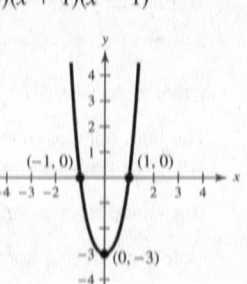

101. $f(x) = \dfrac{x - 7}{x^2 + 4x + 4} = \dfrac{x - 7}{(x + 2)^2}$

x-intercept: $(7, 0)$

y-intercept: $\left(0, -\dfrac{7}{4}\right)$

Vertical asymptote: $x = -2$

Horizontal asymptote: $y = 0$

Domain: All real numbers except $x = -2$

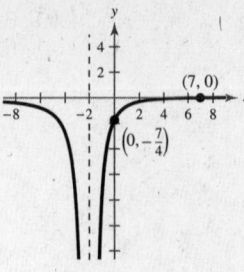

102. $h(x) = \dfrac{x^2 - 1}{x + 5} = \dfrac{(x + 1)(x - 1)}{x + 5}$

x-intercepts: $(-1, 0), (1, 0),$

To find the y-intercept, let $x = 0$: $\dfrac{0^2 - 1}{0 + 5} = -\dfrac{1}{5}$

y-intercept: $\left(0, -\dfrac{1}{5}\right)$

Vertical asymptote: $x = -5$

To find the slant asymptote, use long division:

$$\frac{x^2 - 1}{x + 5} = x - 5 + \frac{24}{x + 5}$$

Slant asymptote: $y = x - 5$

Domain: All real numbers except $x = -5$

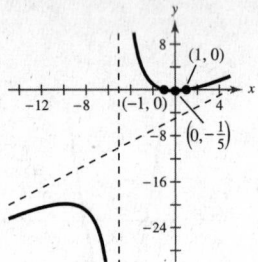

103. $y = 2^{x-1}$

y-intercept: $\left(0, \dfrac{1}{2}\right)$

Horizontal asymptote: $y = 0$

Domain: All real numbers x

x	-1	0	1	2	3
y	$\frac{1}{4}$	$\frac{1}{2}$	1	2	4

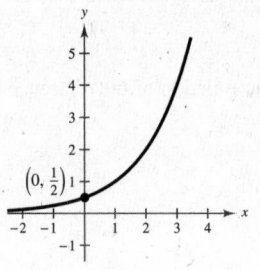

104. $y = 3^{x+1} + 2$

This is an exponential function (always positive) translated two units upward. There are no x-intercepts.

To find the y-intercept, let $x = 0$:

$y = 3^{0+1} + 2 = 3 + 2 = 5$

y-intercepts: $(0, 5)$

The horizontal asymptote is the horizontal asymptote of $y = 3^{x+1}$ translated two units upward.

Horizontal asymptote: $y = 2$

Domain: All real numbers x

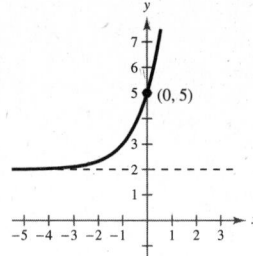

105. $y = \ln x^4$

Domain: All real numbers except $x = 0$

x-intercepts: $(\pm 1, 0)$

Vertical asymptote: $x = 0$

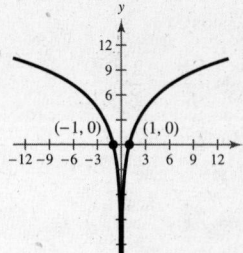

106. $y = \log_{10}(x + 2)$

To find the x-intercept, let $y = 0$:

$0 = \log_{10}(x + 2) \implies 10^0 = x + 2 \implies x = -1$

x-intercepts: $(-1, 0)$

To find the y-intercept, let $x = 0$:

$y = \log_{10}(x + 2) = \log_{10} 2 \approx 0.301$

y-intercepts: $(0, 0.301)$

The vertical asymptote is the horizontal asymptote of $y = \log_{10} x$ translated two units to the left.

Vertical asymptote: $x = -2$

Domain: All real numbers x such that $x > -2$

Section 4.5 Graphs of Sine and Cosine Functions

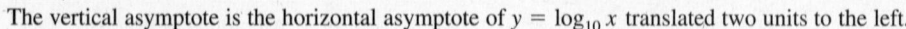

- ■ You should be able to graph $y = a \sin(bx - c)$ and $y = a \cos(bx - c)$. (Assume $b > 0$.)

- ■ Amplitude: $|a|$

- ■ Period: $\dfrac{2\pi}{b}$

- ■ Shift: Solve $bx - c = 0$ and $bx - c = 2\pi$.

- ■ Key increments: $\dfrac{1}{4}$ (period)

Vocabulary Check

1. cycle

2. amplitude

3. $\dfrac{2\pi}{b}$

4. phase shift

5. vertical shift

1. $y = 3 \sin 2x$

Period: $\dfrac{2\pi}{2} = \pi$

Amplitude: $|3| = 3$

2. $y = 2 \cos 3x$

Period: $\dfrac{2\pi}{b} = \dfrac{2\pi}{3}$

Amplitude: $|a| = 2$

3. $y = \dfrac{5}{2} \cos \dfrac{x}{2}$

Period: $\dfrac{2\pi}{1/2} = 4\pi$

Amplitude: $\left|\dfrac{5}{2}\right| = \dfrac{5}{2}$

4. $y = -3 \sin \dfrac{x}{3}$

Period: $\dfrac{2\pi}{b} = \dfrac{2\pi}{1/3} = 6\pi$

Amplitude: $|a| = |-3| = 3$

5. $y = \dfrac{1}{2} \sin \dfrac{\pi x}{3}$

Period: $\dfrac{2\pi}{\pi/3} = 6$

Amplitude: $\left|\dfrac{1}{2}\right| = \dfrac{1}{2}$

6. $y = \dfrac{3}{2} \cos \dfrac{\pi x}{2}$

Period: $\dfrac{2\pi}{b} = \dfrac{2\pi}{\pi/2} = 4$

Amplitude: $|a| = \dfrac{3}{2}$

7. $y = -2 \sin x$

Period: $\dfrac{2\pi}{1} = 2\pi$

Amplitude: $|-2| = 2$

8. $y = -\cos \dfrac{2x}{3}$

Period: $\dfrac{2\pi}{b} = \dfrac{2\pi}{2/3} = 3\pi$

Amplitude: $|a| = |-1| = 1$

9. $y = 3 \sin 10x$

Period: $\dfrac{2\pi}{10} = \dfrac{\pi}{5}$

Amplitude: $|3| = 3$

10. $y = \dfrac{1}{3} \sin 8x$

Period: $\dfrac{2\pi}{b} = \dfrac{2\pi}{8} = \dfrac{\pi}{4}$

Amplitude: $|a| = \dfrac{1}{3}$

11. $y = \dfrac{1}{2} \cos \dfrac{2x}{3}$

Period: $\dfrac{2\pi}{2/3} = 3\pi$

Amplitude: $\left|\dfrac{1}{2}\right| = \dfrac{1}{2}$

12. $y = \dfrac{5}{2} \cos \dfrac{x}{4}$

Period: $\dfrac{2\pi}{b} = \dfrac{2\pi}{1/4} = 8\pi$

Amplitude: $|a| = \dfrac{5}{2}$

13. $y = \dfrac{1}{4} \sin 2\pi x$

Period: $\dfrac{2\pi}{2\pi} = 1$

Amplitude: $\left|\dfrac{1}{4}\right| = \dfrac{1}{4}$

14. $y = \dfrac{2}{3} \cos \dfrac{\pi x}{10}$

Period: $\dfrac{2\pi}{b} = \dfrac{2\pi}{\pi/10} = 20$

Amplitude: $|a| = \dfrac{2}{3}$

15. $f(x) = \sin x$

$g(x) = \sin(x - \pi)$

The graph of g is a horizontal shift to the right π units of the graph of f (a phase shift).

16. $f(x) = \cos x$, $g(x) = \cos(x + \pi)$

g is a horizontal shift of f π units to the left.

17. $f(x) = \cos 2x$

$g(x) = -\cos 2x$

The graph of g is a reflection in the x-axis of the graph of f.

18. $f(x) = \sin 3x$, $g(x) = \sin(-3x)$

g is a reflection of f about the y-axis.

19. $f(x) = \cos x$

$g(x) = \cos 2x$

The period of f is twice that of g.

20. $f(x) = \sin x$, $g(x) = \sin 3x$

The period of g is one-third the period of f.

21. $f(x) = \sin 2x$

$f(x) = 3 + \sin 2x$

The graph of g is a vertical shift three units upward of the graph of f.

22. $f(x) = \cos 4x$, $g(x) = -2 + \cos 4x$
g is a vertical shift of f two units downward.

23. The graph of g has twice the amplitude as the graph of f. The period is the same.

24. The period of g is one-third the period of f.

25. The graph of g is a horizontal shift π units to the right of the graph of f.

26. Shift the graph of f two units upward to obtain the graph of g.

27. $f(x) = -2 \sin x$

Period: $\dfrac{2\pi}{b} = \dfrac{2\pi}{1} = 2\pi$

Amplitude: 2

Symmetry: origin

Key points:

Intercept	Minimum	Intercept	Maximum	Intercept
$(0, 0)$	$\left(\dfrac{\pi}{2}, -2\right)$	$(\pi, 0)$	$\left(\dfrac{3\pi}{2}, 0\right)$	$(2\pi, 0)$

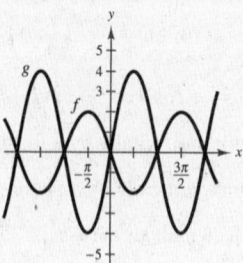

Since $g(x) = 4 \sin x = (-2)f(x)$, generate key points for the graph of $g(x)$ by multiplying the y-coordinate of each key point of $f(x)$ by -2.

28. $f(x) = \sin x$

Period: $\dfrac{2\pi}{b} = \dfrac{2\pi}{1} = 2\pi$

Amplitude: 1

Symmetry: origin

Key points:

Intercept	Maximum	Intercept	Minimum	Intercept
$(0, 0)$	$\left(\dfrac{\pi}{2}, 1\right)$	$(\pi, 0)$	$\left(\dfrac{3\pi}{2}, -1\right)$	$(2\pi, 0)$

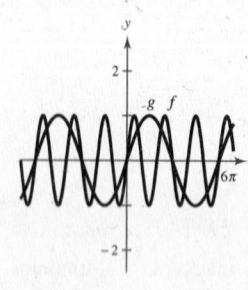

Since $g(x) = \sin\left(\dfrac{x}{3}\right) = f\left(\dfrac{x}{3}\right)$, the graph of $g(x)$ is the graph of $f(x)$, but stretched horizontally by a factor of 3.
Generate key points for the graph of $g(x)$ by multiplying the x-coordinate of each key point of $f(x)$ by 3.

29. $f(x) = \cos x$

Period: $\dfrac{2\pi}{b} = \dfrac{2\pi}{1} = 2\pi$

Amplitude: 1

Symmetry: y-axis

Key points:

Maximum	Intercept	Minimum	Intercept	Maximum
$(0, 1)$	$\left(\dfrac{\pi}{2}, 0\right)$	$(\pi, -1)$	$\left(\dfrac{3\pi}{2}, 0\right)$	$(2\pi, 1)$

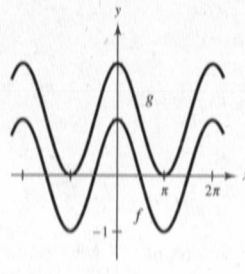

Since $g(x) = 1 + \cos(x) = f(x) + 1$, the graph of $g(x)$ is the graph of $f(x)$, but translated upward by one unit.
Generate key points for the graph of $g(x)$ by adding 1 to the y-coordinate of each key point of $f(x)$.

30. $f(x) = 2 \cos 2x$

Period: $\dfrac{2\pi}{b} = \dfrac{2\pi}{2} = \pi$

Amplitude: 2

Symmetry: y-axis

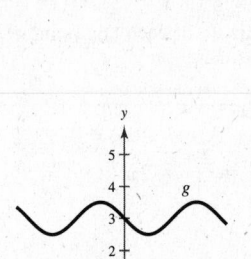

Key points:

Maximum	Intercept	Minimum	Intercept	Maximum
$(0, 2)$	$\left(\dfrac{\pi}{4}, 0\right)$	$\left(\dfrac{\pi}{2}, -2\right)$	$\left(\dfrac{3\pi}{4}, 0\right)$	$(\pi, 2)$

Since $g(x) = -\cos 4x = -\frac{1}{2}f(2x)$, the graph of $g(x)$ is the graph of $f(x)$, but

   i) shrunk horizontally by a factor of 2,

   ii) shrunk vertically by a factor of $\frac{1}{2}$, and

   iii) reflected about the x-axis.

Generate key points for the graph of $g(x)$ by

   i) dividing the x-coordinate of each key point of $f(x)$ by 2, and

   ii) dividing the y-coordinate of each key point of $f(x)$ by -2.

31. $f(x) = -\dfrac{1}{2}\sin\dfrac{x}{2}$

Period: $\dfrac{2\pi}{b} = \dfrac{2\pi}{1/2} = 4\pi$

Amplitude: $\dfrac{1}{2}$

Symmetry: origin

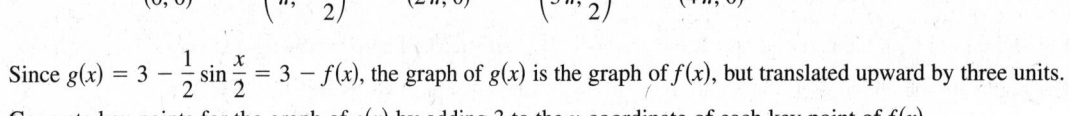

Key points:

Intercept	Minimum	Intercept	Maximum	Intercept
$(0, 0)$	$\left(\pi, -\dfrac{1}{2}\right)$	$(2\pi, 0)$	$\left(3\pi, \dfrac{1}{2}\right)$	$(4\pi, 0)$

Since $g(x) = 3 - \dfrac{1}{2}\sin\dfrac{x}{2} = 3 - f(x)$, the graph of $g(x)$ is the graph of $f(x)$, but translated upward by three units.

Generate key points for the graph of $g(x)$ by adding 3 to the y-coordinate of each key point of $f(x)$.

32. $f(x) = 4 \sin \pi x$

Period: $\dfrac{2\pi}{b} = \dfrac{2\pi}{\pi} = 2$

Amplitude: 4

Symmetry: origin

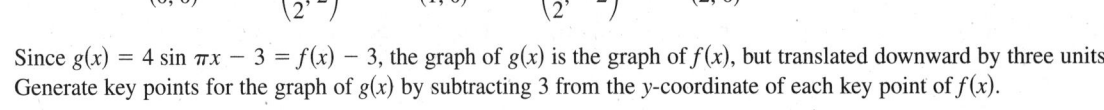

Key points:

Intercept	Maximum	Intercept	Minimum	Intercept
$(0, 0)$	$\left(\dfrac{1}{2}, 2\right)$	$(1, 0)$	$\left(\dfrac{3}{2}, -2\right)$	$(2, 0)$

Since $g(x) = 4 \sin \pi x - 3 = f(x) - 3$, the graph of $g(x)$ is the graph of $f(x)$, but translated downward by three units.

Generate key points for the graph of $g(x)$ by subtracting 3 from the y-coordinate of each key point of $f(x)$.

33. $f(x) = 2 \cos x$

Period: $\dfrac{2\pi}{b} = \dfrac{2\pi}{1} = 2\pi$

Amplitude: 2

Symmetry: y-axis

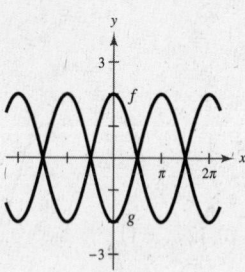

Key points: Maximum   Intercept   Minimum   Intercept   Maximum

$(0, 2)$ $\left(\dfrac{\pi}{2}, 0\right)$ $(\pi, -2)$ $\left(\dfrac{3\pi}{2}, 0\right)$ $(2\pi, 2)$

Since $g(x) = 2 \cos(x + \pi) = f(x + \pi)$, the graph of $g(x)$ is the graph of $f(x)$, but with a phase shift (horizontal translation) of $-\pi$. Generate key points for the graph of $g(x)$ by shifting each key point of $f(x)$ π units to the left.

34. $f(x) = -\cos x$

Period: $\dfrac{2\pi}{b} = \dfrac{2\pi}{1} = 2\pi$

Amplitude: 1

Symmetry: y-axis

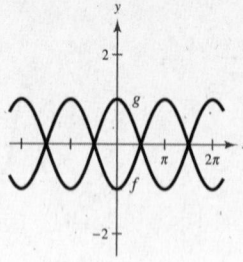

Key points: Minimum   Intercept   Maximum   Intercept   Minimum

$(0, -1)$ $\left(\dfrac{\pi}{2}, 0\right)$ $(\pi, 1)$ $\left(\dfrac{3\pi}{2}, 0\right)$ $(2\pi, -1)$

Since $g(x) = -\cos(x - \pi) = f(x - \pi)$, the graph of $g(x)$ is the graph of $f(x)$, but with a phase shift (horizontal translation) of π. Generate key points for the graph of $g(x)$ by shifting each key point of $f(x)$ π units to the right.

35. $y = 3 \sin x$

Period: 2π

Amplitude: 3

Key points:

$(0, 0), \left(\dfrac{\pi}{2}, 3\right), (\pi, 0),$

$\left(\dfrac{3\pi}{2}, -3\right), (2\pi, 0)$

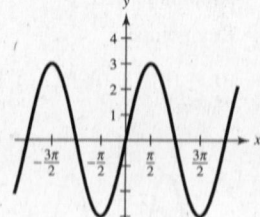

36. $y = \dfrac{1}{4} \sin x$

Period: 2π

Amplitude: $\dfrac{1}{4}$

Key points:

$(0, 0), \left(\dfrac{\pi}{2}, \dfrac{1}{4}\right), (\pi, 0),$

$\left(\dfrac{3\pi}{2}, -\dfrac{1}{4}\right), (2\pi, 0)$

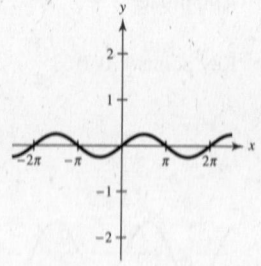

37. $y = \dfrac{1}{3} \cos x$

Period: 2π

Amplitude: $\dfrac{1}{3}$

Key points:

$\left(0, \dfrac{1}{3}\right), \left(\dfrac{\pi}{2}, 0\right), \left(\pi, -\dfrac{1}{3}\right),$

$\left(\dfrac{3\pi}{2}, 0\right), \left(2\pi, \dfrac{1}{3}\right)$

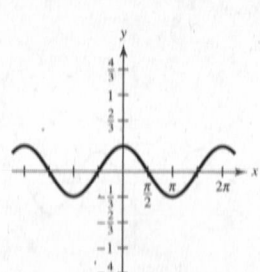

38. $y = 4 \cos x$

Period: 2π

Amplitude: 4

Key points:

$(0, 4), \left(\dfrac{\pi}{2}, 0\right), (\pi, -4),$

$\left(\dfrac{3\pi}{2}, 0\right), (2\pi, 4)$

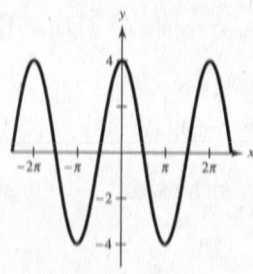

39. $y = \cos \dfrac{x}{2}$

Period: 4π

Amplitude: 1

Key points:

$(0, 1),\ (\pi, 0),\ (2\pi, -1),$

$(3\pi, 0),\ (4\pi, 1)$

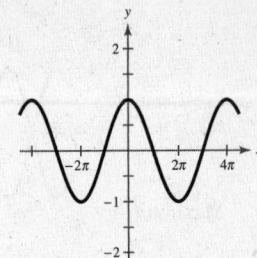

40. $y = \sin 4x$

Period: $\dfrac{\pi}{2}$

Amplitude: 1

Key points:

$(0, 0),\ \left(\dfrac{\pi}{8}, 1\right),\ \left(\dfrac{\pi}{4}, 0\right),$

$\left(\dfrac{3\pi}{8}, -1\right),\ \left(\dfrac{\pi}{2}, 0\right)$

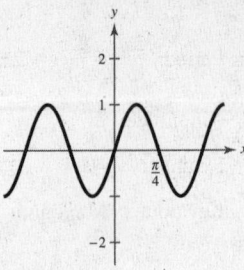

41. $y = \cos 2\pi x$

Period: $\dfrac{2\pi}{2\pi} = 1$

Amplitude: 1

Key points:

$(0, 1),\ \left(\dfrac{1}{4}, 0\right),\ \left(\dfrac{1}{2}, -1\right),\ \left(\dfrac{3}{4}, 0\right)$

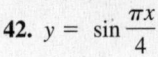

42. $y = \sin \dfrac{\pi x}{4}$

Period: $\dfrac{2\pi}{\pi/4} = 8$

Amplitude: 1

Key points:

$(0, 0),\ (2, 1),\ (4, 0),$

$(6, -1),\ (8, 0)$

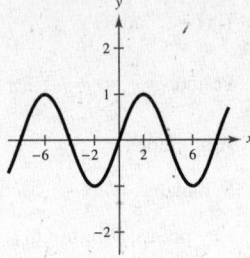

43. $y = -\sin \dfrac{2\pi x}{3}$; $a = -1,\ b = \dfrac{2\pi}{3},\ c = 0$

Period: $\dfrac{2\pi}{2\pi/3} = 3$

Amplitude: 1

Key points: $(0, 0),\ \left(\dfrac{3}{4}, -1\right),\ \left(\dfrac{3}{2}, 0\right),\ \left(\dfrac{9}{4}, 1\right),\ (3, 0)$

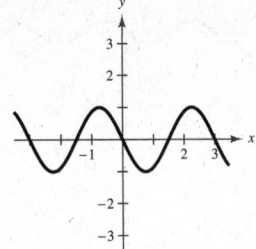

44. $y = -10 \cos \dfrac{\pi x}{6}$

Period: $\dfrac{2\pi}{\pi/6} = 12$

Amplitude: 10

Key points:

$(0, -10),\ (3, 0),\ (6, 10),\ (9, 0),\ (12, -10)$

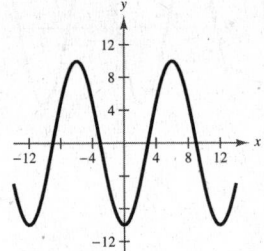

45. $y = \sin\left(x - \dfrac{\pi}{4}\right)$; $a = 1,\ b = 1,\ c = \dfrac{\pi}{4}$

Period: 2π

Amplitude: 1

Shift: Set $x - \dfrac{\pi}{4} = 0$ and $x - \dfrac{\pi}{4} = 2\pi$

$x = \dfrac{\pi}{4}$ \qquad $x = \dfrac{9\pi}{4}$

Key points: $\left(\dfrac{\pi}{4}, 0\right),\ \left(\dfrac{3\pi}{4}, 1\right),\ \left(\dfrac{5\pi}{4}, 0\right),\ \left(\dfrac{7\pi}{4}, -1\right),\ \left(\dfrac{9\pi}{4}, 0\right)$

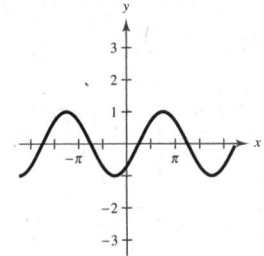

46. $y = \sin(x - \pi)$

Period: 2π

Amplitude: 1

Shift: Set $x - \pi = 0$ and $x - \pi = 2\pi$

$\qquad\qquad x = \pi$ $x = 3\pi$

Key points: $(\pi, 0), \left(\dfrac{3\pi}{2}, 1\right), (2\pi, 0), \left(\dfrac{5\pi}{2}, -1\right), (3\pi, 0)$

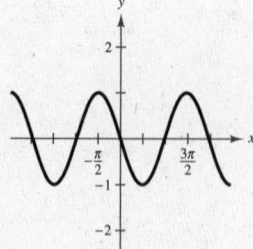

47. $y = 3\cos(x + \pi)$

Period: 2π

Amplitude: 3

Shift: Set $x + \pi = 0$ and $x + \pi = 2\pi$

$\qquad\qquad x = -\pi$ $x = \pi$

Key points: $(-\pi, 3), \left(-\dfrac{\pi}{2}, 0\right), (0, -3), \left(\dfrac{\pi}{2}, 0\right), (\pi, 3)$

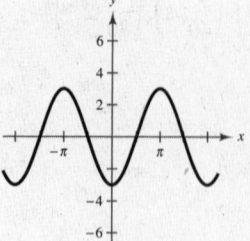

48. $y = 4\cos\left(x + \dfrac{\pi}{4}\right)$

Period: 2π

Amplitude: 4

Shift: Set $x + \dfrac{\pi}{4} = 0$ and $x + \dfrac{\pi}{4} = 2\pi$

$\qquad\qquad x = -\dfrac{\pi}{4}$ $x = \dfrac{7\pi}{4}$

Key points: $\left(-\dfrac{\pi}{4}, 4\right), \left(\dfrac{\pi}{4}, 0\right), \left(\dfrac{3\pi}{4}, -4\right), \left(\dfrac{5\pi}{4}, 0\right), \left(\dfrac{7\pi}{4}, 4\right)$

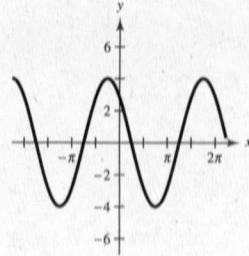

49. $y = 2 - \sin\dfrac{2\pi x}{3}$

Period: 3

Amplitude: 1

Key points: $(0, 2), \left(\dfrac{3}{4}, 1\right), \left(\dfrac{3}{2}, 2\right), \left(\dfrac{9}{4}, 3\right), (3, 2)$

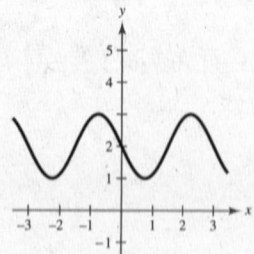

50. $y = -3 + 5\cos\dfrac{\pi t}{12}$

Period: $\dfrac{2\pi}{\pi/12} = 24$

Amplitude: 5

Key points: $(0, 2), (6, -3), (12, -8), (18, -3), (24, 2)$

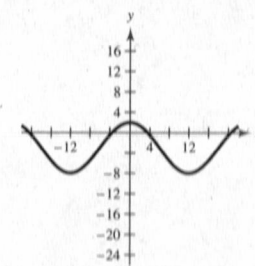

51. $y = 2 + \dfrac{1}{10} \cos 60\pi x$

Period: $\dfrac{2\pi}{60\pi} = \dfrac{1}{30}$

Amplitude: $\dfrac{1}{10}$

Vertical shift two units upward

Key points:

$(0, 2.1), \left(\dfrac{1}{120}, 2\right), \left(\dfrac{1}{60}, 1.9\right), \left(\dfrac{1}{40}, 2\right), \left(\dfrac{1}{30}, 2.1\right)$

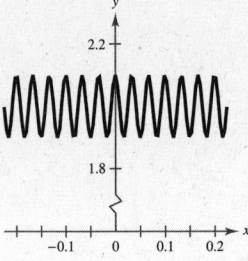

52. $y = 2 \cos x - 3$

Period: 2π

Amplitude: 2

Key points:

$(0, -1), \left(\dfrac{\pi}{2}, -3\right), (\pi, -5), \left(\dfrac{3\pi}{2}, -3\right), (2\pi, -1)$

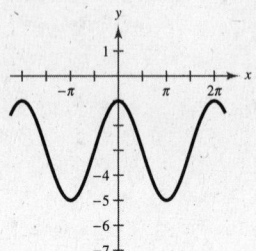

53. $y = 3 \cos(x + \pi) - 3$

Period: 2π

Amplitude: 3

Shift: Set $x + \pi = 0$ and $x + \pi = 2\pi$

$\qquad\qquad x = -\pi \qquad\qquad\quad x = \pi$

Key points: $(-\pi, 0), \left(-\dfrac{\pi}{2}, -3\right), (0, -6), \left(\dfrac{\pi}{2}, -3\right), (\pi, 0)$

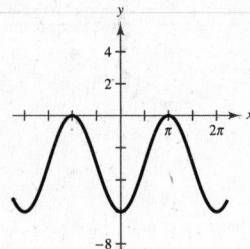

54. $y = 4 \cos\left(x + \dfrac{\pi}{4}\right) + 4$

Period: 2π

Amplitude: 4

Shift: Set $x + \dfrac{\pi}{4} = 0$ and $x + \dfrac{\pi}{4} = 2\pi$

$\qquad\qquad x = -\dfrac{\pi}{4} \qquad\qquad\quad x = \dfrac{7\pi}{4}$

Key points: $\left(-\dfrac{\pi}{4}, 8\right), \left(\dfrac{\pi}{4}, 4\right), \left(\dfrac{3\pi}{4}, 0\right), \left(\dfrac{5\pi}{4}, 4\right), \left(\dfrac{7\pi}{4}, 8\right)$

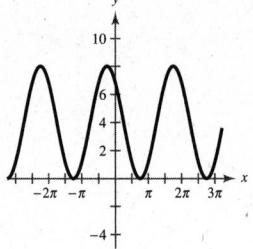

55. $y = \dfrac{2}{3} \cos\left(\dfrac{x}{2} - \dfrac{\pi}{4}\right);\ a = \dfrac{2}{3},\ b = \dfrac{1}{2}, c = \dfrac{\pi}{4}$

Period: 4π

Amplitude: $\dfrac{2}{3}$

Shift: $\dfrac{x}{2} - \dfrac{\pi}{4} = 0$ and $\dfrac{x}{2} - \dfrac{\pi}{4} = 2\pi$

$\qquad\quad x = \dfrac{\pi}{2} \qquad\qquad x = \dfrac{9\pi}{2}$

Key points: $\left(\dfrac{\pi}{2}, \dfrac{2}{3}\right), \left(\dfrac{3\pi}{2}, 0\right), \left(\dfrac{5\pi}{2}, \dfrac{-2}{3}\right), \left(\dfrac{7\pi}{2}, 0\right), \left(\dfrac{9\pi}{2}, \dfrac{2}{3}\right)$

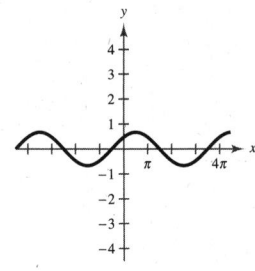

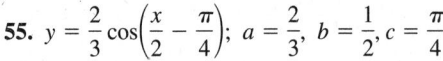

56. $y = -3 \cos(6x + \pi)$

Period: $\dfrac{2\pi}{6} = \dfrac{\pi}{3}$

Amplitude: 3

Shift: Set $6x + \pi = 0$ and $6x + \pi = 2\pi$

$x = -\dfrac{\pi}{6}$ $x = \dfrac{\pi}{6}$

Key points: $\left(-\dfrac{\pi}{6}, -3\right), \left(-\dfrac{\pi}{12}, 0\right), (0, 3), \left(\dfrac{\pi}{12}, 0\right), \left(\dfrac{\pi}{6}, -3\right)$

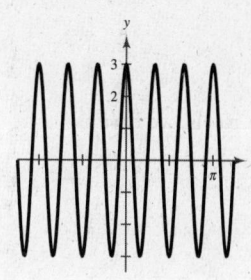

57. $y = -2 \sin(4x + \pi)$

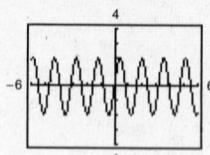

58. $y = -4 \sin\left(\dfrac{2}{3}x - \dfrac{\pi}{3}\right)$

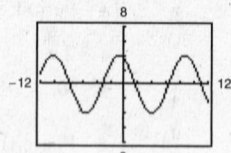

59. $y = \cos\left(2\pi x - \dfrac{\pi}{2}\right) + 1$

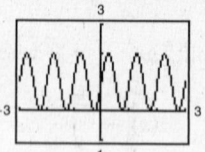

60. $y = 3 \cos\left(\dfrac{\pi x}{2} + \dfrac{\pi}{2}\right) - 2$

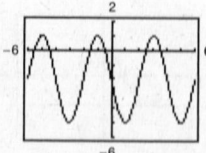

61. $y = -0.1 \sin\left(\dfrac{\pi x}{10} + \pi\right)$

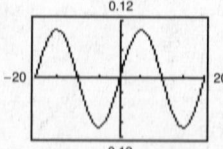

62. $y = \dfrac{1}{100} \sin 120\,\pi t$

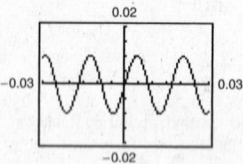

63. $f(x) = a \cos x + d$

Amplitude: $\dfrac{1}{2}[3 - (-1)] = 2 \implies a = 2$

Vertical shift one unit upward of

$g(x) = 2 \cos x \implies d = 1$

Thus, $f(x) = 2 \cos x + 1$.

64. $f(x) = a \cos x + d$

Amplitude: $\dfrac{1 - (-3)}{2} = 2$

$1 = 2 \cos 0 + d$

$d = 1 - 2 = -1$

$a = 2, d = -1$

65. $f(x) = a \cos x + d$

Amplitude: $\dfrac{1}{2}[8 - 0] = 4$

Since $f(x)$ is the graph of $g(x) = 4 \cos x$ reflected in the x-axis and shifted vertically four units upward, we have $a = -4$ and $d = 4$. Thus, $f(x) = -4 \cos x + 4$.

66. $f(x) = a \cos x + d$

Amplitude: $\dfrac{-2 - (-4)}{2} = 1$

Reflected in the x-axis: $a = -1$

$-4 = -1 \cos 0 + d$

$d = -3$

$a = -1, d = -3$

67. $y = a \sin(bx - c)$

Amplitude: $|a| = |3|$

Since the graph is reflected in the x-axis, we have $a = -3$.

Period: $\dfrac{2\pi}{b} = \pi \Rightarrow b = 2$

Phase shift: $c = 0$

Thus, $y = -3 \sin 2x$.

68. $y = a \sin(bx - c)$

Amplitude: $2 \Rightarrow a = 2$

Period: 4π

$\dfrac{2\pi}{b} = 4\pi \Rightarrow b = \dfrac{1}{2}$

Phase shift: $c = 0$

$a = 2,\ b = \dfrac{1}{2},\ c = 0$

69. $y = a \sin(bx - c)$

Amplitude: $a = 2$

Period: $2\pi \Rightarrow b = 1$

Phase shift: $bx - c = 0$ when $x = -\dfrac{\pi}{4}$

$$(1)\left(\dfrac{-\pi}{4}\right) - c = 0 \Rightarrow c = -\dfrac{\pi}{4}$$

Thus, $y = 2 \sin\left(x + \dfrac{\pi}{4}\right)$.

70. $y = a \sin(bx - c)$

Amplitude: $2 \Rightarrow a = 2$

Period: 2

$\dfrac{2\pi}{b} = 4 \Rightarrow b = \dfrac{\pi}{2}$

Phase shift: $\dfrac{c}{b} = -1 \Rightarrow c = -\dfrac{\pi}{2}$

$a = 2,\ b = \dfrac{\pi}{2},\ c = -\dfrac{\pi}{2}$

71. $y_1 = \sin x$

$y_2 = -\dfrac{1}{2}$

In the interval $[-2\pi, 2\pi]$,

$\sin x = -\dfrac{1}{2}$ when $x = -\dfrac{5\pi}{6}, -\dfrac{\pi}{6}, \dfrac{7\pi}{6}, \dfrac{11\pi}{6}$.

72. $y_1 = \cos x$

$y_2 = -1$

$y_1 = y_2$ when $x = \pi, -\pi$

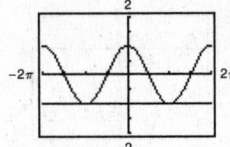

73. $y = 0.85 \sin \dfrac{\pi t}{3}$

(a) Time for one cycle $= \dfrac{2\pi}{\pi/3} = 6$ sec

(b) Cycles per min $= \dfrac{60}{6} = 10$ cycles per min

(c) Amplitude: 0.85; Period: 6

Key points: $(0, 0),\ \left(\dfrac{3}{2}, 0.85\right),\ (3, 0),\ \left(\dfrac{9}{2}, -0.85\right),\ (6, 0)$

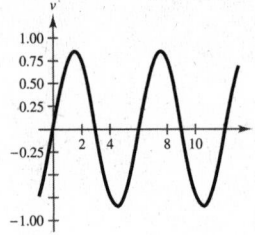

74. $v = 1.75 \sin \dfrac{\pi t}{2}$

(a) Period $= \dfrac{2\pi}{\pi/2} = 4$ seconds

(b) $\dfrac{1 \text{ cycle}}{4 \text{ seconds}} \cdot \dfrac{60 \text{ seconds}}{1 \text{ minute}} = 15$ cycles per minute

(c)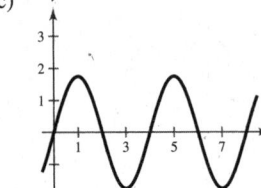

75. $y = 0.001 \sin 880\pi t$

(a) Period: $\dfrac{2\pi}{880\pi} = \dfrac{1}{440}$ seconds

(b) $f = \dfrac{1}{p} = 440$ cycles per second

77. (a) $a = \dfrac{1}{2}[\text{high} - \text{low}] = \dfrac{1}{2}[83.5 - 29.6] = 26.95$

$p = 2[\text{high time} - \text{low time}] = 2[7 - 1] = 12$

$b = \dfrac{2\pi}{p} = \dfrac{2\pi}{12} = \dfrac{\pi}{6}$

$\dfrac{c}{b} = 7 \Rightarrow c = 7\left(\dfrac{\pi}{6}\right) \approx 3.67$

$d = \dfrac{1}{2}[\text{high} + \text{low}] = \dfrac{1}{2}[83.5 + 29.6] = 56.55$

$C(t) = 56.55 + 26.95 \cos\left(\dfrac{\pi t}{6} - 3.67\right)$

(b)

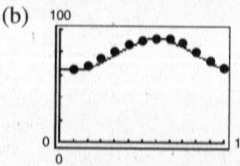

The model is a good fit.

78. (a) and (c)

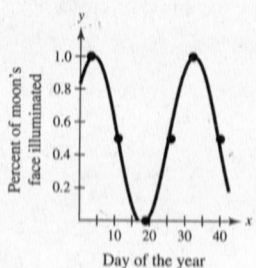

Reasonably good fit

(d) Period is 29.6 days.

(e) March 12 $\Rightarrow$ $x = 71$. $y = 0.44 = 44\%$

The Naval observatory says that 50% of the moon's face will be illuminated on March 12, 2007.

76. $P = 100 - 20 \cos \dfrac{5\pi t}{3}$

(a) Period: $\dfrac{2\pi}{(5\pi)/3} = \dfrac{6}{5}$ seconds

(b) $\dfrac{1 \text{ heartbeat}}{6/5 \text{ seconds}} \cdot \dfrac{60 \text{ seconds}}{1 \text{ minute}} = 50$ heartbeats per minute

(c)

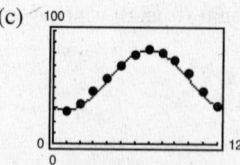

The model is a good fit.

(d) Tallahassee average maximum: $77.90°$

Chicago average maximum: $56.55°$

The constant term, d, gives the average maximum temperature.

(e) The period for both models is $\dfrac{2\pi}{\pi/6} = 12$ months.

This is as we expected since one full period is one year.

(f) Chicago has the greater variability in temperature throughout the year. The amplitude, a, determines this variability since it is $\dfrac{1}{2}[\text{high temp} - \text{low temp}]$.

(b) Vertical shift: $\dfrac{1}{2} \Rightarrow d = \dfrac{1}{2}$

Amplitude: $\dfrac{1}{2} \Rightarrow a = \dfrac{1}{2}$

Period: $\dfrac{8 + 8 + 7 + 6 + 8}{5} = 7.4$ (average length of interval in data)

$\dfrac{2\pi}{b} = 4(7.4) = 29.6$

$b = \dfrac{2\pi}{29.6} \approx 0.21$

Horizontal shift: $0.21(3 - 7.4) + C = 0$

$C = 0.92$

$y = \dfrac{1}{2} + \dfrac{1}{2} \sin(0.21x + 0.92)$

79. $C = 30.3 + 21.6 \sin\left(\dfrac{2\pi t}{365} + 10.9\right)$

(a) Period $= \dfrac{2\pi}{\dfrac{2\pi}{365}} = 365$

Yes, this is what is expected because there are 365 days in a year.

(b) The average daily fuel consumption is given by the amount of the vertical shift (from 0) which is given by the constant 30.3.

(c)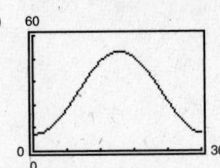

The consumption exceeds 40 gallons per day when $124 < x < 252$.

80. (a) Period $= \dfrac{2\pi}{\left(\dfrac{\pi}{6}\right)} = 12$ minutes

The wheel takes 12 minutes to revolve once.

(b) Amplitude: 50 feet

The radius of the wheel is 50 feet.

(c)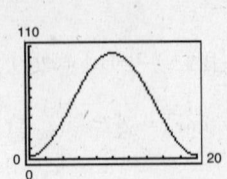

81. False. The graph of $\sin(x + 2\pi)$ is the graph of $\sin(x)$ translated to the *left* by one period, and the graphs are indeed identical.

82. False. $y = \frac{1}{2}\cos 2x$ has an amplitude that is **half** that of $y = \cos x$. For $y = a \cos bx$, the amplitude is $|a|$.

83. True.

Since $\cos x = \sin\left(x + \dfrac{\pi}{2}\right)$, $y = -\cos x = -\sin\left(x + \dfrac{\pi}{2}\right)$, and so is a reflection in the x-axis of $y = \sin\left(x + \dfrac{\pi}{2}\right)$.

84. Answers will vary.

85.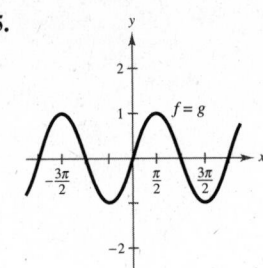

Since the graphs are the same, the conjecture is that

$\sin(x) = \cos\left(x - \dfrac{\pi}{2}\right)$.

86. $f(x) = \sin x,\ g(x) = -\cos\left(x + \dfrac{\pi}{2}\right)$

x	0	$\dfrac{\pi}{2}$	π	$\dfrac{3\pi}{2}$	2π
$\sin x$	0	1	0	-1	0
$-\cos\left(x + \dfrac{\pi}{2}\right)$	0	1	0	-1	0

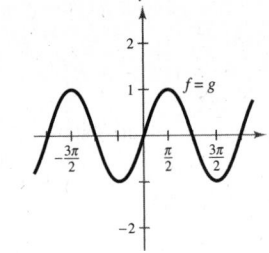

Conjecture: $\sin x = -\cos\left(x + \dfrac{\pi}{2}\right)$

87. (a)

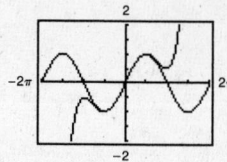

The graphs are nearly the same for $-\dfrac{\pi}{2} < x < \dfrac{\pi}{2}$.

(b)

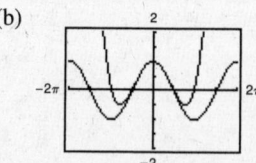

The graphs are nearly the same for $-\dfrac{\pi}{2} < x < \dfrac{\pi}{2}$.

(c) $\sin x \approx x - \dfrac{x^3}{3!} + \dfrac{x^5}{5!} - \dfrac{x^7}{7!}$

$\cos x \approx 1 - \dfrac{x^2}{2!} + \dfrac{x^4}{4!} - \dfrac{x^6}{6!}$

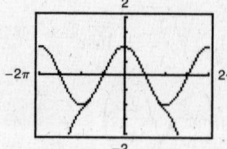 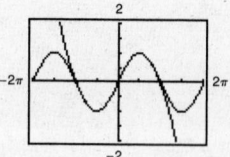

The graphs now agree over a wider range, $-\dfrac{3\pi}{4} < x < \dfrac{3\pi}{4}$.

88. (a) $\sin \dfrac{1}{2} \approx \dfrac{1}{2} - \dfrac{(1/2)^3}{3!} + \dfrac{(1/2)^5}{5!} \approx 0.4794$

$\sin \dfrac{1}{2} \approx 0.4794$ (by calculator)

(c) $\sin \dfrac{\pi}{6} \approx 1 - \dfrac{(\pi/6)^3}{3!} + \dfrac{(\pi/6)^5}{5!} \approx 0.5000$

$\sin \dfrac{\pi}{6} = 0.5$ (by calculator)

(e) $\cos 1 \approx 1 - \dfrac{1}{2!} + \dfrac{1}{4!} \approx 0.5417$

$\cos 1 \approx 0.5403$ (by calculator)

(b) $\sin 1 \approx 1 - \dfrac{1}{3!} + \dfrac{1}{5!} \approx 0.8417$

$\sin 1 \approx 0.8415$ (by calculator)

(d) $\cos(-0.5) \approx 1 - \dfrac{(-0.5)^2}{2!} + \dfrac{(-0.5)^4}{4!} \approx 0.8776$

$\cos(-0.5) \approx 0.8776$ (by calculator)

(f) $\cos \dfrac{\pi}{4} \approx 1 - \dfrac{(\pi/4)^2}{2!} + \dfrac{(\pi/4)^2}{4!} = 0.7074$

$\cos \dfrac{\pi}{4} \approx 0.7071$ (by calculator)

The error in the approximation is not the same in each case. The error appears to increase as x moves farther away from 0.

89. $\log_{10} \sqrt{x - 2} = \log_{10}(x - 2)^{1/2} = \tfrac{1}{2} \log_{10}(x - 2)$

90. $\log_2[x^2(x - 3)] = \log_2 x^2 + \log_2(x - 3)$
$= 2 \log_2 x + \log_2(x - 3)$

91. $\ln \dfrac{t^3}{t - 1} = \ln t^3 - \ln(t - 1) = 3 \ln t - \ln(t - 1)$

92. $\ln \sqrt{\dfrac{z}{z^2 + 1}} = \dfrac{1}{2} \ln \left(\dfrac{z}{z^2 + 1} \right) = \dfrac{1}{2}[\ln z - \ln(z^2 + 1)]$
$= \dfrac{1}{2} \ln z - \dfrac{1}{2} \ln(z^2 + 1)$

93. $\tfrac{1}{2}(\log_{10} x + \log_{10} y) = \tfrac{1}{2} \log_{10}(xy)$
$= \log_{10} \sqrt{xy}$

94. $2 \log_2 x + \log_2(xy) = \log_2 x^2 + \log_2(xy)$
$= \log_2 x^2(xy)$
$= \log_2 x^3 y$

95. $\ln 3x - 4 \ln y = \ln 3x - \ln y^4$
$= \ln \left(\dfrac{3x}{y^4} \right)$

96. $\frac{1}{2}(\ln 2x - 2 \ln x) + 3 \ln x = \frac{1}{2}(\ln 2x - \ln x^2) + \ln x^3$

$$= \frac{1}{2}\left(\ln \frac{2x}{x^2}\right) + \ln x^3$$

$$= \ln \sqrt{\frac{2x}{x^2}} + \ln x^3$$

$$= \ln\left(x^3 \sqrt{\frac{2x}{x^2}}\right)$$

$$= \ln\left(x^2 \sqrt{2x}\right)$$

97. Answers will vary.

Section 4.6 Graphs of Other Trigonometric Functions

■ You should be able to graph

$y = a \tan(bx - c)$ $\qquad$ $y = a \cot(bx - c)$

$y = a \sec(bx - c)$ $\qquad$ $y = a \csc(bx - c)$

■ When graphing $y = a \sec(bx - c)$ or $y = a \csc(bx - c)$ you should first graph $y = a \cos(bx - c)$ or $y = a \sin(bx - c)$ because

(a) The x-intercepts of sine and cosine are the vertical asymptotes of cosecant and secant.

(b) The maximums of sine and cosine are the local minimums of cosecant and secant.

(c) The minimums of sine and cosine are the local maximums of cosecant and secant.

■ You should be able to graph using a damping factor.

Vocabulary Check

1. vertical

2. reciprocal

3. damping

4. π

5. $x \neq n\pi$

6. $(-\infty, -1] \cup [1, \infty)$

7. 2π

---

1. $y = \sec 2x$

Period: $\dfrac{2\pi}{2} = \pi$

Matches graph (e).

2. $y = \tan \dfrac{x}{2}$

Period: $\dfrac{\pi}{b} = \dfrac{\pi}{1/2} = 2\pi$

Asymptotes: $x = -\pi, x = \pi$

Matches graph (c).

3. $y = \dfrac{1}{2} \cot \pi x$

Period: $\dfrac{\pi}{\pi} = 1$

Matches graph (a).

4. $y = -\csc x$

Period: 2π

Matches graph (d).

5. $y = \dfrac{1}{2} \sec \dfrac{\pi x}{2}$

Period: $\dfrac{2\pi}{b} = \dfrac{2\pi}{\pi/2} = 4$

Asymptotes: $x = -1, x = 1$

Matches graph (f).

6. $y = -2 \sec \dfrac{\pi x}{2}$

Period: $\dfrac{2\pi}{b} = \dfrac{2\pi}{\pi/2} = 4$

Asymptotes: $x = -1, x = 1$

Reflected in x-axis

Matches graph (b).

7. $y = \dfrac{1}{3}\tan x$

Period: π

Two consecutive asymptotes:

$x = -\dfrac{\pi}{2}$ and $x = \dfrac{\pi}{2}$

x	$-\dfrac{\pi}{4}$	0	$\dfrac{\pi}{4}$
y	$-\dfrac{1}{3}$	0	$\dfrac{1}{3}$

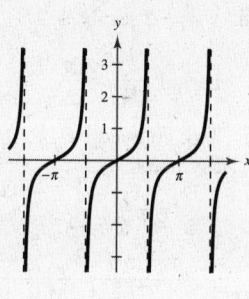

8. $y = \dfrac{1}{4}\tan x$

Period: π

Two consecutive asymptotes:

$x = -\dfrac{\pi}{2}, x = \dfrac{\pi}{2}$

x	$-\dfrac{\pi}{4}$	0	$\dfrac{\pi}{4}$
y	$-\dfrac{1}{4}$	0	$\dfrac{1}{4}$

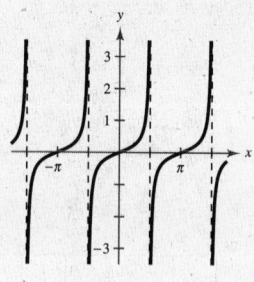

9. $y = \tan 3x$

Period: $\dfrac{\pi}{3}$

Two consecutive asymptotes:

$3x = -\dfrac{\pi}{2} \implies x = -\dfrac{\pi}{6}$

$3x - \dfrac{\pi}{2} \implies x = \dfrac{\pi}{6}$

x	$-\dfrac{\pi}{12}$	0	$\dfrac{\pi}{12}$
y	-1	0	1

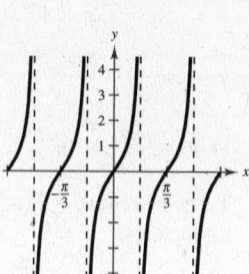

10. $y = -3\tan \pi x$

Period: $\dfrac{\pi}{\pi} = 1$

Two consecutive asymptotes:

$x = -\dfrac{1}{2}, x = \dfrac{1}{2}$

x	$-\dfrac{1}{4}$	0	$\dfrac{1}{4}$
y	3	0	-3

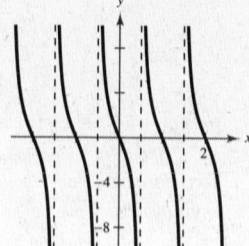

11. $y = -\dfrac{1}{2}\sec x$

Period: 2π

Two consecutive asymptotes:

$x = -\dfrac{\pi}{2}, x = \dfrac{\pi}{2}$

x	$-\dfrac{\pi}{3}$	0	$\dfrac{\pi}{3}$
y	-1	$-\dfrac{1}{2}$	-1

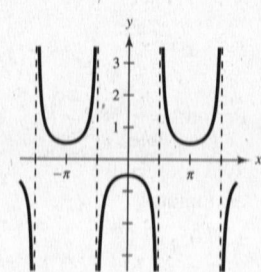

12. $y = \dfrac{1}{4}\sec x$

Period: 2π

Two consecutive asymptotes:

$x = -\dfrac{\pi}{2}, x = \dfrac{\pi}{2}$

x	$-\dfrac{\pi}{3}$	0	$\dfrac{\pi}{3}$
y	$\dfrac{1}{2}$	$\dfrac{1}{4}$	$\dfrac{1}{2}$

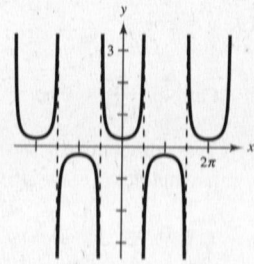

13. $y = \csc \pi x$

Period: $\dfrac{2\pi}{\pi} = 2$

Two consecutive asymptotes:

$x = 0, x = 1$

x	$\dfrac{1}{6}$	$\dfrac{1}{2}$	$\dfrac{5}{6}$
y	2	1	2

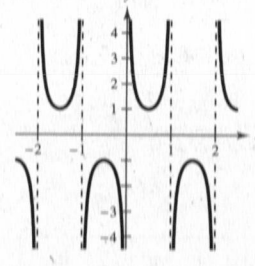

14. $y = 3\csc 4x$

Period: $\dfrac{2\pi}{4} = \dfrac{\pi}{2}$

Two consecutive asymptotes:

$x = 0, x = \dfrac{\pi}{4}$

x	$\dfrac{\pi}{24}$	$\dfrac{\pi}{8}$	$\dfrac{5\pi}{24}$
y	6	3	6

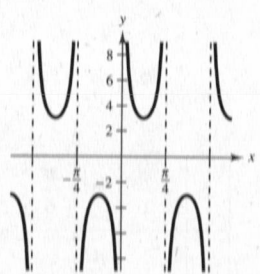

15. $y = \sec \pi x - 1$

Period: $\dfrac{2\pi}{\pi} = 2$

Two consecutive asymptotes:

$x = -\dfrac{1}{2}, x = \dfrac{1}{2}$

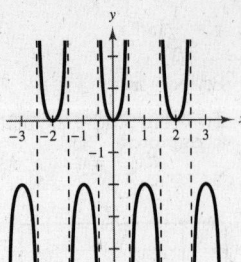

x	$-\dfrac{1}{3}$	0	$\dfrac{1}{3}$
y	1	0	1

16. $y = -2 \sec 4x + 2$

Period: $\dfrac{2\pi}{4} = \dfrac{\pi}{2}$

Two consecutive asymptotes:

$x = -\dfrac{\pi}{8}, x = \dfrac{\pi}{8}$

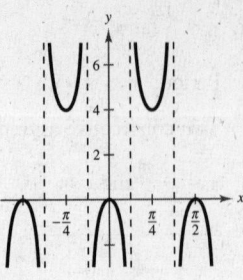

x	$-\dfrac{\pi}{12}$	0	$\dfrac{\pi}{12}$
y	-2	0	-2

17. $y = \csc \dfrac{x}{2}$

Period: $\dfrac{2\pi}{1/2} = 4\pi$

Two consecutive asymptotes:

$x = 0, x = 2\pi$

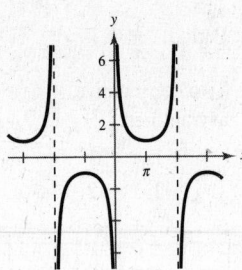

x	$\dfrac{\pi}{3}$	π	$\dfrac{5\pi}{3}$
y	2	1	2

18. $y = \csc \dfrac{x}{3}$

Period: $\dfrac{2\pi}{1/3} = 6\pi$

Two consecutive asymptotes:

$x = 0, x = 3\pi$

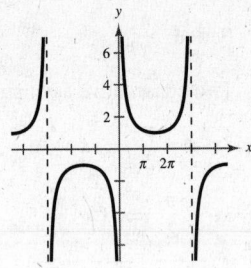

x	$\dfrac{\pi}{2}$	$\dfrac{3\pi}{2}$	$\dfrac{5\pi}{2}$
y	2	1	2

19. $y = \cot \dfrac{x}{2}$

Period: $\dfrac{\pi}{1/2} = 2\pi$

Two consecutive asymptotes:

$\dfrac{x}{2} = 0 \implies x = 0$

$\dfrac{x}{2} = \pi \implies x = 2\pi$

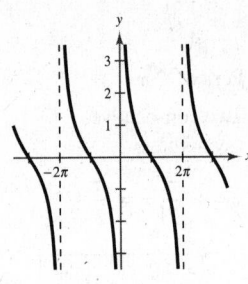

x	$\dfrac{\pi}{2}$	π	$\dfrac{3\pi}{2}$
y	1	0	-1

20. $y = 3 \cot \dfrac{\pi x}{2}$

Period: $\dfrac{\pi}{\pi/2} = 2$

Two consecutive asymptotes:

$x = 0, x = 2$

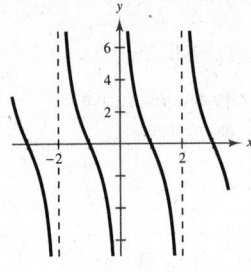

x	$\dfrac{1}{2}$	1	$\dfrac{3}{2}$
y	3	0	-3

21. $y = \dfrac{1}{2} \sec 2x$

Period: $\dfrac{2\pi}{2} = \pi$

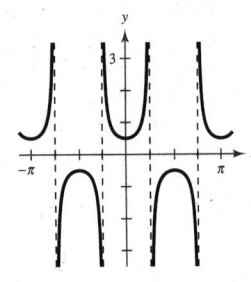

x	$-\dfrac{\pi}{6}$	0	$\dfrac{\pi}{6}$
y	1	$\dfrac{1}{2}$	1

22. $y = -\dfrac{1}{2} \tan x$

Period: π

Two consecutive asymptotes:

$x = -\dfrac{\pi}{2}, x = \dfrac{\pi}{2}$

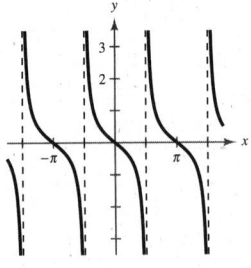

x	$-\dfrac{\pi}{4}$	0	$\dfrac{\pi}{4}$
y	$\dfrac{1}{2}$	0	$-\dfrac{1}{2}$

23. $y = \tan \dfrac{\pi x}{4}$

Period: $\dfrac{\pi}{\pi/4} = 4$

Two consecutive asymptotes:

$\dfrac{\pi x}{4} = -\dfrac{\pi}{2} \implies x = -2$

$\dfrac{\pi x}{4} = \dfrac{\pi}{2} \implies x = 2$

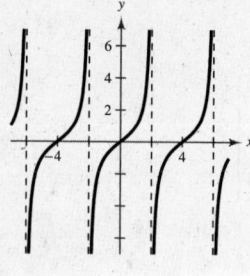

x	-1	0	1
y	-1	0	1

24. $y = \tan(x + \pi)$

Period: π

Two consecutive asymptotes:

$x = -\dfrac{\pi}{2},\ x = \dfrac{\pi}{2}$

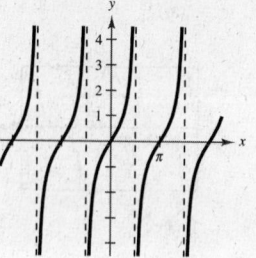

x	$-\dfrac{\pi}{4}$	0	$\dfrac{\pi}{4}$
y	-1	0	1

25. $y = \csc(\pi - x)$

Period: 2π

Two consecutive asymptotes:

$x = 0,\ x = \pi$

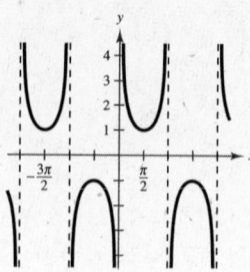

x	$\dfrac{\pi}{6}$	$\dfrac{\pi}{2}$	$\dfrac{5\pi}{6}$
y	2	1	2

26. $y = \csc(2x - \pi)$

Period: $\dfrac{2\pi}{2} = \pi$

Two consecutive asymptotes:

$x = 0,\ x = \dfrac{\pi}{2}$

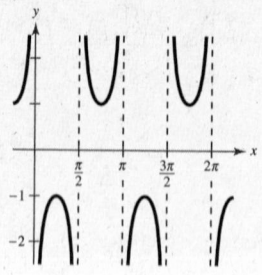

x	$\dfrac{\pi}{12}$	$\dfrac{\pi}{4}$	$\dfrac{5\pi}{12}$
y	-2	-1	-2

27. $y = 2\sec(x + \pi)$

Period: 2π

Two consecutive asymptotes:

$x = -\dfrac{\pi}{2},\ x = \dfrac{\pi}{2}$

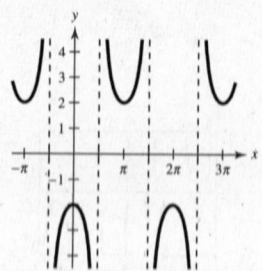

x	$-\dfrac{\pi}{3}$	0	$\dfrac{\pi}{3}$
y	-4	-2	-4

28. $y = -\sec \pi x + 1$

Period: $\dfrac{2\pi}{\pi} = 2$

Two consecutive asymptotes:

$x = -\dfrac{1}{2},\ x = \dfrac{1}{2}$

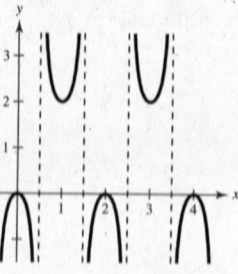

x	$-\dfrac{1}{3}$	0	$\dfrac{1}{3}$
y	-1	0	1

29. $y = \dfrac{1}{4}\csc\left(x + \dfrac{\pi}{4}\right)$

Period: 2π

Two consecutive asymptotes:

$x = -\dfrac{\pi}{4},\ x = \dfrac{3\pi}{4}$

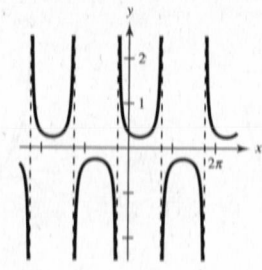

x	$-\dfrac{\pi}{12}$	$\dfrac{\pi}{4}$	$\dfrac{7\pi}{12}$
y	$\dfrac{1}{2}$	$\dfrac{1}{4}$	$\dfrac{1}{2}$

30. $y = 2\cot\left(x + \dfrac{\pi}{2}\right)$

Period: π

Two consecutive asymptotes:

$x = -\dfrac{\pi}{2},\ x = \dfrac{\pi}{2}$

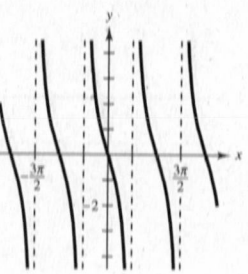

x	$-\dfrac{\pi}{4}$	0	$\dfrac{\pi}{4}$
y	2	0	-2

31. $y = \tan \dfrac{x}{3}$

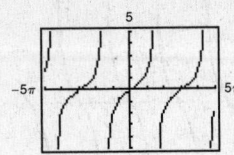

32. $y = -\tan 2x$

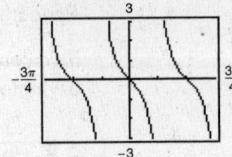

33. $y = -2 \sec 4x = \dfrac{-2}{\cos 4x}$

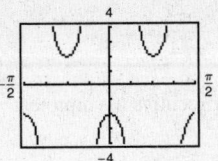

34. $y = \sec \pi x \implies y = \dfrac{1}{\cos(\pi x)}$

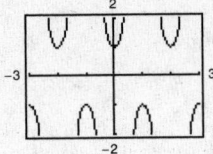

35. $y = \tan\left(x - \dfrac{\pi}{4}\right)$

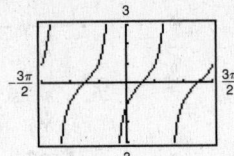

36. $y = \dfrac{1}{4} \cot\left(x - \dfrac{\pi}{2}\right)$

$= \dfrac{1}{4 \tan\left(x - \dfrac{\pi}{2}\right)}$

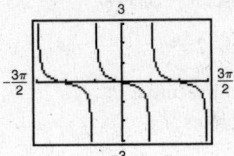

37. $y = -\csc(4x - \pi)$

$y = \dfrac{-1}{\sin(4x - \pi)}$

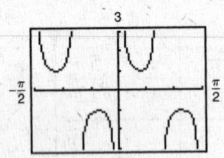

38. $y = 2 \sec(2x - \pi) \implies$

$y = \dfrac{2}{\cos(2x - \pi)}$

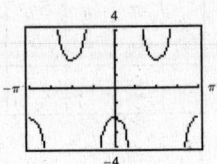

39. $y = 0.1 \tan\left(\dfrac{\pi x}{4} + \dfrac{\pi}{4}\right)$

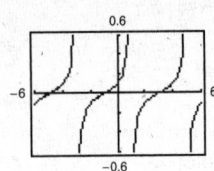

40. $y = \dfrac{1}{3} \sec\left(\dfrac{\pi x}{2} + \dfrac{\pi}{2}\right) \implies y = \dfrac{1}{3 \cos\left(\dfrac{\pi x}{2} + \dfrac{\pi}{2}\right)}$

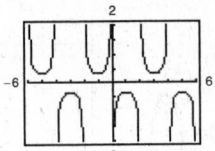

41. $\tan x = 1$

$x = -\dfrac{7\pi}{4},\ -\dfrac{3\pi}{4},\ \dfrac{\pi}{4},\ \dfrac{5\pi}{4}$

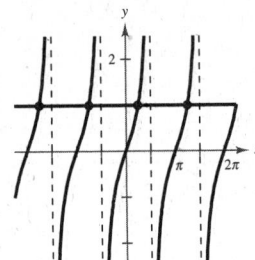

42. $\tan x = \sqrt{3}$

$x = -\dfrac{5\pi}{3},\ -\dfrac{2\pi}{3},\ \dfrac{\pi}{3},\ \dfrac{4\pi}{3}$

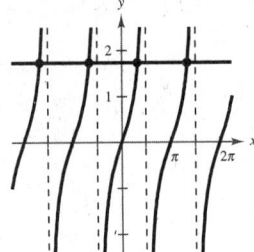

43. $\cot x = -\dfrac{\sqrt{3}}{3}$

$x = -\dfrac{4\pi}{3},\ -\dfrac{\pi}{3},\ \dfrac{2\pi}{3},\ \dfrac{5\pi}{3}$

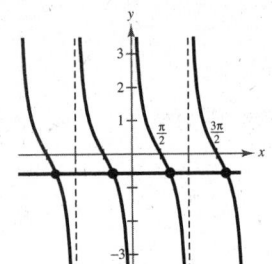

44. $\cot x = 1$

$$x = -\frac{7\pi}{4}, -\frac{3\pi}{4}, \frac{\pi}{4}, \frac{5\pi}{4}$$

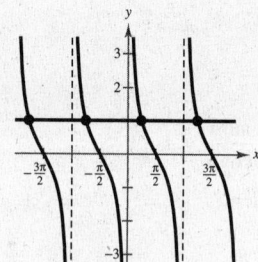

45. $\sec x = -2$

$$x = \pm\frac{2\pi}{3}, \pm\frac{4\pi}{3}$$

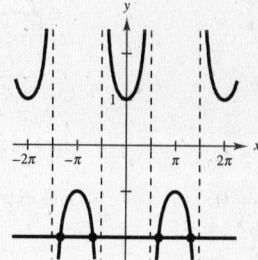

46. $\sec x = 2$

$$x = -\frac{5\pi}{3}, -\frac{\pi}{3}, \frac{\pi}{3}, \frac{5\pi}{3}$$

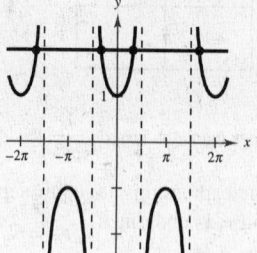

47. $\csc x = \sqrt{2}$

$$x = -\frac{7\pi}{4}, -\frac{5\pi}{4}, \frac{\pi}{4}, \frac{3\pi}{4}$$

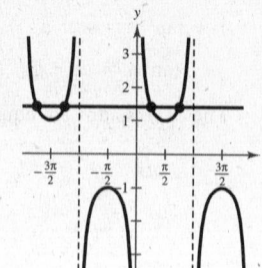

48. $\csc x = -\frac{2\sqrt{3}}{3}$

$$x = -\frac{2\pi}{3}, -\frac{\pi}{3}, \frac{4\pi}{3}, \frac{5\pi}{3}$$

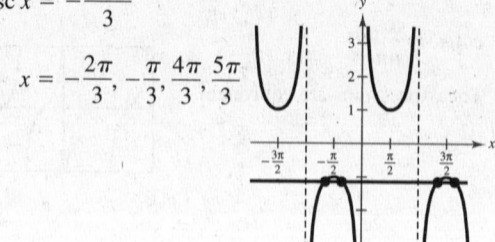

49. $f(x) = \sec x = \dfrac{1}{\cos x}$

$$f(-x) = \sec(-x)$$

$$= \frac{1}{\cos(-x)}$$

$$= \frac{1}{\cos x}$$

$$= f(x)$$

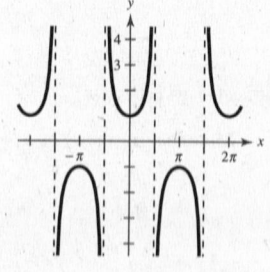

Thus, $f(x) = \sec x$ is an even function and the graph has y-axis symmetry.

50. $f(x) = \tan x$

$$\tan(-x) = -\tan x$$

Thus, the function is odd and the graph of $y = \tan x$ is symmetric about the origin.

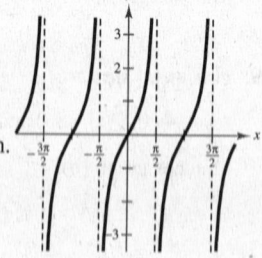

51. $f(x) = 2 \sin x$

$$g(x) = \frac{1}{2} \csc x$$

(a)

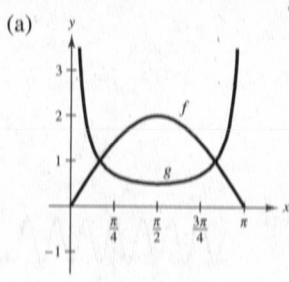

(b) $f > g$ on the interval, $\dfrac{\pi}{6} < x < \dfrac{5\pi}{6}$

(c) As $x \to \pi$, $f(x) = 2 \sin x \to 0$ and $g(x) = \frac{1}{2} \csc x \to \pm\infty$ since $g(x)$ is the reciprocal of $f(x)$.

52. $f(x) = \tan \dfrac{\pi x}{2}$, $g(x) = \dfrac{1}{2} \sec \dfrac{\pi x}{2}$

(a)

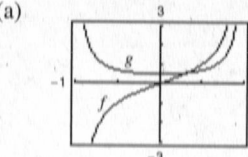

(b) The interval in which $f < g$ is $\left(-1, \frac{1}{3}\right)$.

(c) The interval in which $2f < 2g$ is $\left(-1, \frac{1}{3}\right)$, which is the same interval as part (b).

53. $y_1 = \sin x \csc x$ and $y_2 = 1$

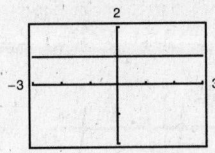

$\sin x \csc x = \sin x \left(\dfrac{1}{\sin x}\right) = 1,\ \sin x \neq 0$

The expressions are equivalent except when $\sin x = 0$ and y_1 is undefined.

54. $y_1 = \sin x \sec x,\ y_2 = \tan x$

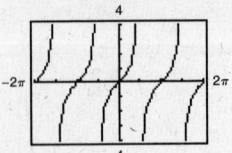

$\sin x \sec x = \sin x \dfrac{1}{\cos x} = \dfrac{\sin x}{\cos x} = \tan x$

The expressions are equivalent.

55. $y_1 = \dfrac{\cos x}{\sin x}$ and $y_2 = \cot x = \dfrac{1}{\tan x}$

$\cot x = \dfrac{\cos x}{\sin x}$

The expressions are equivalent.

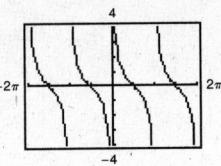

56. $y_1 = \sec^2 x - 1,\ y_2 = \tan^2 x$

$1 + \tan^2 x = \sec^2 x$

$\tan^2 x = \sec^2 x - 1$

The expressions are equivalent.

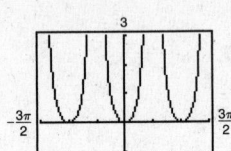

57. $f(x) = |x \cos x|$

As $x \to 0,\ f(x) \to 0$ and $f(x) > 0$.

Matches graph (d).

58. $f(x) = x \sin x$

Matches graph (a) as $x \to 0,\ f(x) \to 0$.

59. $g(x) = |x| \sin x$

As $x \to 0,\ g(x) \to 0$ and $g(x)$ is odd.

Matches graph (b).

60. $g(x) = |x| \cos x$

Matches graph (c) as $x \to 0,\ g(x) \to 0$.

61. $f(x) = \sin x + \cos\left(x + \dfrac{\pi}{2}\right)$

$g(x) = 0$

$f(x) = g(x)$

The graph is the line $y = 0$.

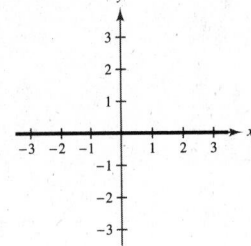

62. $f(x) = \sin x - \cos\left(x + \dfrac{\pi}{2}\right)$

$g(x) = 2 \sin x$

It appears that $f(x) = g(x)$. That is,

$\sin x - \cos\left(x + \dfrac{\pi}{2}\right) = 2 \sin x$.

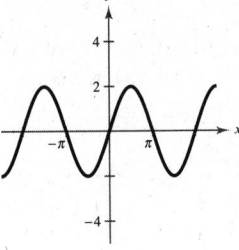

63. $f(x) = \sin^2 x$

$g(x) = \dfrac{1}{2}(1 - \cos 2x)$

$f(x) = g(x)$

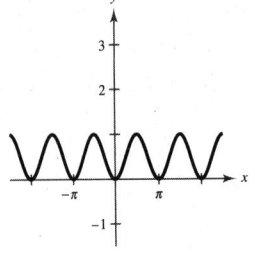

64. $f(x) = \cos^2 \dfrac{\pi x}{2}$

$g(x) = \dfrac{1}{2}(1 + \cos \pi x)$

It appears that $f(x) = g(x)$. That is,

$\cos^2 \dfrac{\pi x}{2} = \dfrac{1}{2}(1 + \cos \pi x)$.

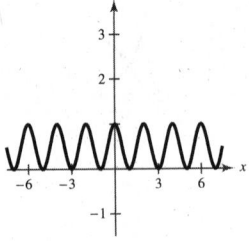

65. $g(x) = e^{-x^2/2} \sin x$

$-e^{-x^2/2} \leq g(x) \leq e^{-x^2/2}$

The damping factor is $y = e^{-x^2/2}$.

As $x \to \infty$, $g(x) \to 0$.

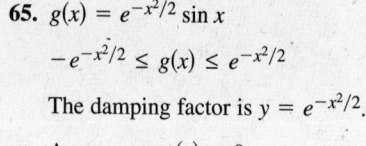

66. $f(x) = e^{-x} \cos x$

Damping factor: e^{-x}

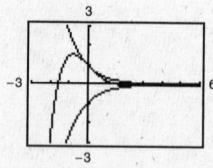

As $x \to \infty$, $f(x) \to 0$.

67. $f(x) = 2^{-x/4} \cos \pi x$

$-2^{-x/4} \leq f(x) \leq 2^{-x/4}$

Damping factor: $y = 2^{-x/4}$.

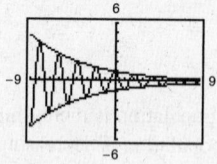

As $x \to \infty$, $f(x) \to 0$.

68. $h(x) = 2^{-x^2/4} \sin x$

Damping factor: $2^{-x^2/4}$

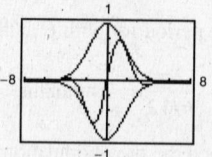

As $x \to \infty$, $h(x) \to 0$.

69. $y = \dfrac{6}{x} + \cos x$, $x > 0$

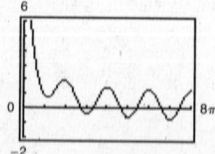

As $x \to 0$, $y \to \infty$.

70. $y = \dfrac{4}{x} + \sin 2x$, $x > 0$

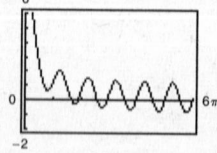

As $x \to 0$, $y \to \infty$.

71. $g(x) = \dfrac{\sin x}{x}$

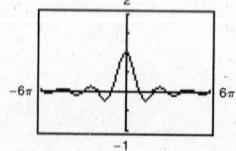

As $x \to 0$, $g(x) \to 1$.

72. $f(x) = \dfrac{1 - \cos x}{x}$

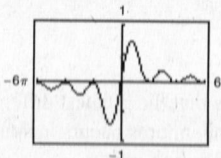

As $x \to 0$, $f(x) \to 0$.

73. $f(x) = \sin \dfrac{1}{x}$

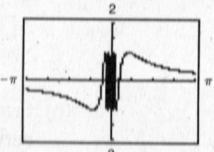

As $x \to 0$, $f(x)$ oscillates
between -1 and 1.

74. $h(x) = x \sin \dfrac{1}{x}$

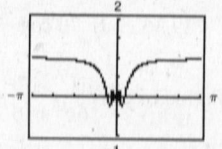

As $x \to 0$, $h(x)$ oscillates.

75. $\tan x = \dfrac{7}{d}$

$d = \dfrac{7}{\tan x} = 7 \cot x$

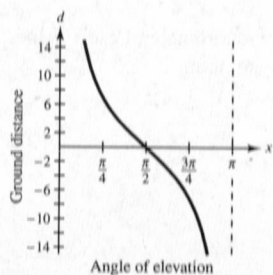

76. $\cos x = \dfrac{27}{d}$

$d = \dfrac{27}{\cos x} = 27 \sec x$, $-\dfrac{\pi}{2} < x < \dfrac{\pi}{2}$

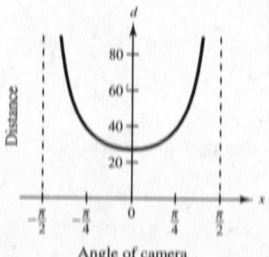

77. $C = 5000 + 2000 \sin \dfrac{\pi t}{12}$, $R = 25{,}000 + 15{,}000 \cos \dfrac{\pi t}{12}$

(a)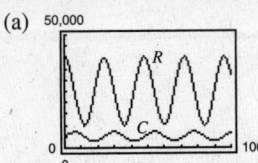

(b) As the predator population increases, the number of prey decreases. When the number of prey is small, the number of predators decreases.

(c) The period for both C and R is:

$$p = \frac{2\pi}{\pi/12} = 24 \text{ months}$$

When the prey population is highest, the predator population is increasing most rapidly.
When the prey population is lowest, the predator population is decreasing most rapidly.
When the predator population is lowest, the prey population is increasing most rapidly.
When the predator population is highest, the prey population is decreasing most rapidly.

In addition, weather, food sources for the prey, hunting, all affect the populations of both the predator and the prey.

78. $S = 74 + 3t - 40 \cos \dfrac{\pi t}{6}$

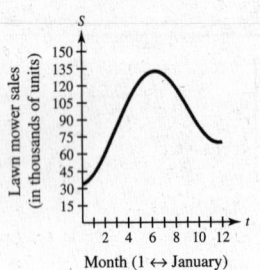

79. $H(t) = 54.33 - 20.38 \cos \dfrac{\pi t}{6} - 15.69 \sin \dfrac{\pi t}{6}$

$L(t) = 39.36 - 15.70 \cos \dfrac{\pi t}{6} - 14.16 \sin \dfrac{\pi t}{6}$

(a) Period of $\cos \dfrac{\pi t}{6}$: $\dfrac{2\pi}{\pi/6} = 12$

Period of $\sin \dfrac{\pi t}{6}$: $\dfrac{2\pi}{\pi/6} = 12$

Period of $H(t)$: 12 months

Period of $L(t)$: 12 months

(b) From the graph, it appears that the greatest difference between high and low temperatures occurs in summer. The smallest difference occurs in winter.

(c) The highest high and low temperatures appear to occur around the middle of July, roughly one month after the time when the sun is northernmost in the sky.

80. (a) 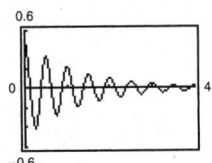 $y = \dfrac{1}{2} e^{-t/4} \cos 4t$

(b) The displacement is a damped sine wave.
$y \to 0$ as t increases.

81. True. Since

$$y = \csc x = \frac{1}{\sin x},$$

for a given value of x, the y-coordinate of $\csc x$ is the reciprocal of the y-coordinate of $\sin x$.

82. True.

$$y = \sec x = \frac{1}{\cos x}$$

If the reciprocal of $y = \sin x$ is translated $\pi/2$ units to the left, we have

$$y = \frac{1}{\sin\left(x + \dfrac{\pi}{2}\right)} = \frac{1}{\cos x} = \sec x.$$

83. As $x \to \dfrac{\pi}{2}$ from the left, $f(x) = \tan x \to \infty$.

As $x \to \dfrac{\pi}{2}$ from the right, $f(x) = \tan x \to -\infty$.

84. As $x \to \pi$ from the left, $f(x) = \csc x \to \infty$.

As $x \to \pi$ from the right, $f(x) = \csc x \to -\infty$.

85. $f(x) = x - \cos x$

(a)

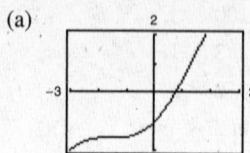

The zero between 0 and 1 occurs at $x \approx 0.7391$.

(b) $x_n = \cos(x_{n-1})$

$x_0 = 1$

$x_1 = \cos 1 \approx 0.5403$

$x_2 = \cos 0.5403 \approx 0.8576$

$x_3 = \cos 0.8576 \approx 0.6543$

$x_4 = \cos 0.6543 \approx 0.7935$

$x_5 = \cos 0.7935 \approx 0.7014$

$x_6 = \cos 0.7014 \approx 0.7640$

$x_7 = \cos 0.7640 \approx 0.7221$

$x_8 = \cos 0.7221 \approx 0.7504$

$x_9 = \cos 0.7504 \approx 0.7314$

$\vdots$

This sequence appears to be approaching the zero of f: $x \approx 0.7391$.

86. $y = \tan x$

$$y = x + \frac{2x^3}{3!} + \frac{16x^5}{5!}$$

The graphs are nearly the same for $-1.1 < x < 1.1$.

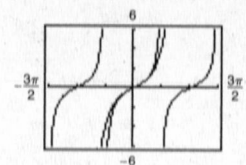

87. $y_1 = \sec x$

$$y_2 = 1 + \frac{x^2}{2!} + \frac{5x^4}{4!}$$

The graph appears to coincide on the interval $-1.1 \le x \le 1.1$.

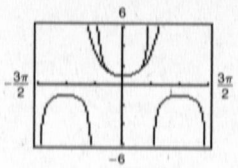

88. (a) $y_1 = \dfrac{4}{\pi}\left(\sin \pi x + \dfrac{1}{3}\sin 3\pi x\right)$ $y_2 = \dfrac{4}{\pi}\left(\sin \pi x + \dfrac{1}{3}\sin 3\pi x + \dfrac{1}{5}\sin 5\pi x\right)$

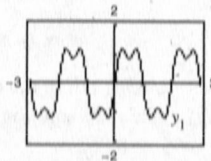
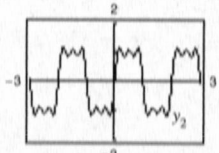

—**CONTINUED**—

88. —CONTINUED—

(b) $y_3 = \dfrac{4}{\pi}\left(\sin \pi x + \dfrac{1}{3}\sin 3\pi x + \dfrac{1}{5}\sin 5\pi x + \dfrac{1}{7}\sin 7\pi x\right)$

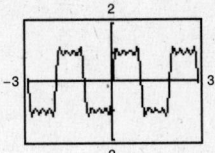

(c) $y_4 = \dfrac{4}{\pi}\left(\sin \pi x + \dfrac{1}{3}\sin 3\pi x + \dfrac{1}{5}\sin 5\pi x + \dfrac{1}{7}\sin 7\pi x + \dfrac{1}{9}\sin 9\pi x\right)$

89. $e^{2x} = 54$

$2x = \ln 54$

$x = \dfrac{\ln 54}{2} \approx 1.994$

90. $8^{3x} = 98$

$3x = \log_8 98$

$x = \dfrac{\ln 98}{3 \ln 8} \approx 0.735$

91. $\dfrac{300}{1 + e^{-x}} = 100$

$\dfrac{300}{100} = 1 + e^{-x}$

$3 = 1 + e^{-x}$

$2 = e^{-x}$

$\ln 2 = -x$

$x = -\ln 2 \approx -0.693$

92. $\left(1 + \dfrac{0.15}{365}\right)^{365t} = 5$

$1 + \dfrac{0.15}{365} \approx 1.00041096$

$1.00041096^{365t} = 5$

$365t = \log_{1.00041096} 5$

$t = \dfrac{1}{365}\left(\dfrac{\log_{10} 5}{\log_{10} 1.00041096}\right) \approx 10.732$

93. $\ln(3x - 2) = 73$

$3x - 2 = e^{73}$

$3x = 2 + e^{73}$

$x = \dfrac{2 + e^{73}}{3}$

$\approx 1.684 \times 10^{31}$

94. $\ln(14 - 2x) = 68$

$14 - 2x = e^{68}$

$14 - e^{68} = 2x$

$x = \dfrac{14 - e^{68}}{2} \approx -1.702 \times 10^{29}$

95. $\ln(x^2 + 1) = 3.2$

$x^2 + 1 = e^{3.2}$

$x^2 = e^{3.2} - 1$

$x = \pm\sqrt{e^{3.2} - 1} \approx \pm 4.851$

96. $\ln\sqrt{x + 4} = 5$

$\dfrac{1}{2}\ln(x + 4) = 5$

$\ln(x + 4) = 10$

$x + 4 = e^{10}$

$x = e^{10} - 4$

$\approx 22{,}022.466$

97. $\log_8 x + \log_8(x - 1) = \dfrac{1}{3}$

$\log_8[x(x - 1)] = \dfrac{1}{3}$

$x(x - 1) = 8^{1/3}$

$x^2 - x = 2$

$x^2 - x - 2 = 0$

$(x - 2)(x + 1) = 0$

$x = 2, -1$

$x = -1$ is extraneous (not in the domain of $\log_8 x$) so only $x = 2$ is a solution.

98. $\log_6 x + \log_6(x^2 - 1) = \log_6(64x)$

$\log_6(x(x^2 - 1)) = \log_6(64x)$

$x(x^2 - 1) = 64x$

$x^2 - 1 = 64$

$x = \pm\sqrt{65}$

Since $-\sqrt{65}$ is not in the domain of $\log_6 x$, the only solution is $x = \sqrt{65} \approx 8.062$.

Section 4.7   Inverse Trigonometric Functions

■ You should know the definitions, domains, and ranges of $y = \arcsin x$, $y = \arccos x$, and $y = \arctan x$.

Function	Domain	Range
$y = \arcsin x \implies x = \sin y$	$-1 \leq x \leq 1$	$-\dfrac{\pi}{2} \leq y \leq \dfrac{\pi}{2}$
$y = \arccos x \implies x = \cos y$	$-1 \leq x \leq 1$	$0 \leq y \leq \pi$
$y = \arctan x \implies x = \tan y$	$-\infty < x < \infty$	$-\dfrac{\pi}{2} < x < \dfrac{\pi}{2}$

■ You should know the inverse properties of the inverse trigonometric functions.

$$\sin(\arcsin x) = x \quad \text{and} \quad \arcsin(\sin y) = y, \; -\frac{\pi}{2} \leq y \leq \frac{\pi}{2}$$

$$\cos(\arccos x) = x \quad \text{and} \quad \arccos(\cos y) = y, \; 0 \leq y \leq \pi$$

$$\tan(\arctan x) = x \quad \text{and} \quad \arctan(\tan y) = y, \; -\frac{\pi}{2} < y < \frac{\pi}{2}$$

■ You should be able to use the triangle technique to convert trigonometric functions of inverse trigonometric functions into algebraic expressions.

Vocabulary Check

Function	Alternative Notation	Domain	Range
1. $y = \arcsin x$	$y = \sin^{-1} x$	$-1 \leq x \leq 1$	$-\dfrac{\pi}{2} \leq y \leq \dfrac{\pi}{2}$
2. $y = \arccos x$	$y = \cos^{-1} x$	$-1 \leq x \leq 1$	$0 \leq y \leq \pi$
3. $y = \arctan x$	$y = \tan^{-1} x$	$-\infty < x < \infty$	$-\dfrac{\pi}{2} < y < \dfrac{\pi}{2}$

---

1. $y = \arcsin \dfrac{1}{2} \implies \sin y = \dfrac{1}{2}$ for $-\dfrac{\pi}{2} \leq y \leq \dfrac{\pi}{2} \implies y = \dfrac{\pi}{6}$

2. $y = \arcsin 0 \implies \sin y = 0$ for $-\dfrac{\pi}{2} \leq y \leq \dfrac{\pi}{2} \implies y = 0$

3. $y = \arccos \dfrac{1}{2} \implies \cos y = \dfrac{1}{2}$ for $0 \leq y \leq \pi \implies y = \dfrac{\pi}{3}$

4. $y = \arccos 0 \implies \cos y = 0$ for $0 \leq y \leq \pi \implies y = \dfrac{\pi}{2}$

5. $y = \arctan \dfrac{\sqrt{3}}{3} \implies \tan y = \dfrac{\sqrt{3}}{3}$ for

$-\dfrac{\pi}{2} < y < \dfrac{\pi}{2} \implies y = \dfrac{\pi}{6}$

6. $y = \arctan(-1) \implies \tan y = -1$ for

$-\dfrac{\pi}{2} < y < \dfrac{\pi}{2} \implies y = -\dfrac{\pi}{4}$

7. $y = \arccos\left(-\frac{\sqrt{3}}{2}\right) \Rightarrow \cos y = -\frac{\sqrt{3}}{2}$ for

$0 \le y \le \pi \Rightarrow y = \frac{5\pi}{6}$

8. $y = \arcsin\left(-\frac{\sqrt{2}}{2}\right) \Rightarrow \sin y = -\frac{\sqrt{2}}{2}$ for

$-\frac{\pi}{2} \le y \le \frac{\pi}{2} \Rightarrow y = -\frac{\pi}{4}$

9. $y = \arctan\left(-\sqrt{3}\right) \Rightarrow \tan y = -\sqrt{3}$ for

$-\frac{\pi}{2} < y < \frac{\pi}{2} \Rightarrow y = -\frac{\pi}{3}$

10. $y = \arctan\left(\sqrt{3}\right) \Rightarrow \tan y = \sqrt{3}$ for

$-\frac{\pi}{2} < y < \frac{\pi}{2} \Rightarrow y = \frac{\pi}{3}$

11. $y = \arccos\left(-\frac{1}{2}\right) \Rightarrow \cos y = -\frac{1}{2}$ for

$0 \le y \le \pi \Rightarrow y = \frac{2\pi}{3}$

12. $y = \arcsin\frac{\sqrt{2}}{2} \Rightarrow \sin y = \frac{\sqrt{2}}{2}$ for

$-\frac{\pi}{2} \le y \le \frac{\pi}{2} \Rightarrow y = \frac{\pi}{4}$

13. $y = \arcsin\frac{\sqrt{3}}{2} \Rightarrow \sin y = \frac{\sqrt{3}}{2}$ for

$-\frac{\pi}{2} \le y \le \frac{\pi}{2} \Rightarrow y = \frac{\pi}{3}$

14. $y = \arctan\left(-\frac{\sqrt{3}}{3}\right) \Rightarrow \tan y = -\frac{\sqrt{3}}{3}$ for

$-\frac{\pi}{2} < y < \frac{\pi}{2} \Rightarrow y = -\frac{\pi}{6}$

15. $y = \arctan 0 \Rightarrow \tan y = 0$ for $-\frac{\pi}{2} < y < \frac{\pi}{2} \Rightarrow y = 0$

16. $y = \arccos 1 \Rightarrow \cos y = 1$ for $0 \le y \le \pi \Rightarrow y = 0$

17. $f(x) = \sin x$

$g(x) = \arcsin x$

$y = x$

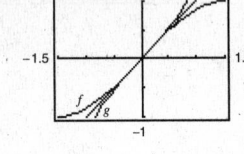

18. $f(x) = \tan x$ and $g(x) = \arctan x$

Graph $y_1 = \tan x$.

Graph $y_2 = \tan^{-1} x$.

Graph $y_3 = x$.

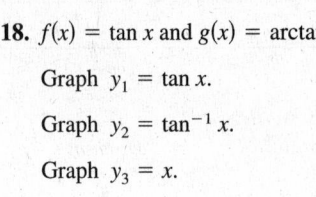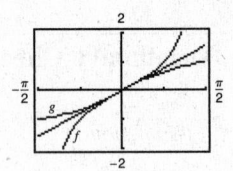

19. $\arccos 0.28 = \cos^{-1} 0.28 \approx 1.29$

20. $\arcsin 0.45 \approx 0.47$

21. $\arcsin(-0.75) = \sin^{-1}(-0.75) \approx -0.85$

22. $\arccos(-0.7) \approx 2.35$

23. $\arctan(-3) = \tan^{-1}(-3) \approx -1.25$

24. $\arctan 15 \approx 1.50$

25. $\arcsin 0.31 = \sin^{-1} 0.31 \approx 0.32$

26. $\arccos 0.26 \approx 1.31$

27. $\arccos(-0.41) = \cos^{-1}(-0.41) \approx 1.99$

28. $\arcsin(-0.125) \approx -0.13$

29. $\arctan 0.92 = \tan^{-1} 0.92 \approx 0.74$

30. $\arctan 2.8 \approx 1.23$

31. $\arcsin\left(\frac{3}{4}\right) = \sin^{-1}(0.75) \approx 0.85$

32. $\arccos\left(-\frac{1}{3}\right) \approx 1.91$

33. $\arctan\left(\frac{7}{2}\right) = \tan^{-1}(3.5) \approx 1.29$

34. $\arctan\left(-\frac{95}{7}\right) \approx -1.50$

35. This is the graph of $y = \arctan x$. The coordinates are

$\left(-\sqrt{3}, -\frac{\pi}{3}\right), \left(-\frac{\sqrt{3}}{3}, -\frac{\pi}{6}\right)$, and $\left(1, \frac{\pi}{4}\right)$.

36. $\arccos(-1) = \pi$

$\arccos\left(-\dfrac{1}{2}\right) = \dfrac{2\pi}{3}$

$\cos\left(\dfrac{\pi}{6}\right) = \dfrac{\sqrt{3}}{2}$

37. $\tan\theta = \dfrac{x}{4}$

$\theta = \arctan\dfrac{x}{4}$

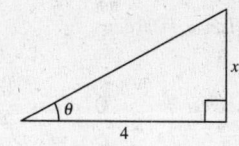

38. $\cos\theta = \dfrac{4}{x}$

$\theta = \arccos\dfrac{4}{x}$

39. $\sin\theta = \dfrac{x+2}{5}$

$\theta = \arcsin\left(\dfrac{x+2}{5}\right)$

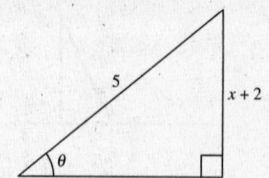

40. $\tan\theta = \dfrac{x+1}{10}$

$\theta = \arctan\left(\dfrac{x+1}{10}\right)$

41. $\cos\theta = \dfrac{x+3}{2x}$

$\theta = \arccos\left(\dfrac{x+3}{2x}\right)$

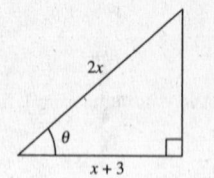

42. $\tan\theta = \dfrac{x-1}{x^2-1} = \dfrac{1}{x+1}$

$\theta = \arctan\dfrac{1}{x+1}$

$x \neq 1$

43. $\sin(\arcsin 0.3) = 0.3$

44. $\tan(\arctan 25) = 25$

45. $\cos[\arccos(-0.1)] = -0.1$

46. $\sin[\arcsin(-0.2)] = -0.2$

47. $\arcsin(\sin 3\pi) = \arcsin(0) = 0$

Note: 3π is not in the range of the arcsine function.

48. $\arccos\left(\cos\dfrac{7\pi}{2}\right) = \arccos 0 = \dfrac{\pi}{2}$

Note: $\dfrac{7\pi}{2}$ is not in the range of the arccosine function.

49. Let $y = \arctan\dfrac{3}{4}$,

$\tan y = \dfrac{3}{4},\ 0 < y < \dfrac{\pi}{2}$,

and $\sin y = \dfrac{3}{5}$.

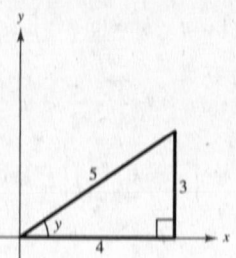

50. Let $u = \arcsin\dfrac{4}{5}$,

$\sin u = \dfrac{4}{5},\ 0 < u < \dfrac{\pi}{2}$,

$\sec\left(\arcsin\dfrac{4}{5}\right) = \sec u = \dfrac{5}{3}$.

51. Let $y = \arctan 2$,

$\tan y = 2 = \dfrac{2}{1},\ 0 < y < \dfrac{\pi}{2}$,

and $\cos y = \dfrac{1}{\sqrt{5}} = \dfrac{\sqrt{5}}{5}$.

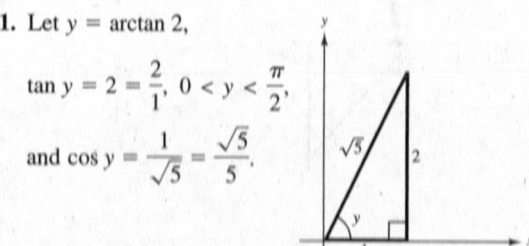

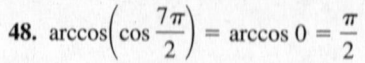

52. Let $u = \arccos \dfrac{\sqrt{5}}{5}$,

$\cos u = \dfrac{\sqrt{5}}{5}, 0 < u < \dfrac{\pi}{2}$,

$\sin\left(\arccos \dfrac{\sqrt{5}}{5}\right) = \sin u = \dfrac{2}{\sqrt{5}} = \dfrac{2\sqrt{5}}{5}$.

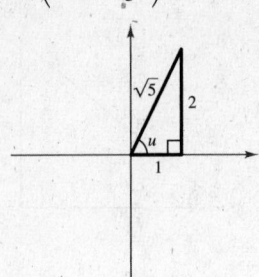

53. Let $y = \arcsin \dfrac{5}{13}$,

$\sin y = \dfrac{5}{13}, \ 0 < y < \dfrac{\pi}{2}$, and $\cos y = \dfrac{12}{13}$.

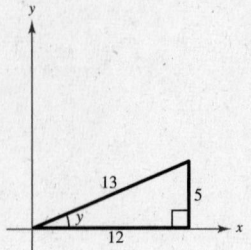

54. Let $u = \arctan\left(-\dfrac{5}{12}\right)$,

$\tan u = -\dfrac{5}{12}, -\dfrac{\pi}{2} < u < 0$,

$\csc\left[\arctan\left(-\dfrac{5}{12}\right)\right] = \csc u = -\dfrac{13}{5}$.

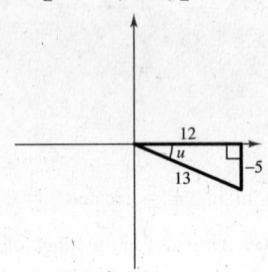

55. Let $y = \arctan\left(-\dfrac{3}{5}\right)$,

$\tan y = -\dfrac{3}{5}, \ -\dfrac{\pi}{2} < y < 0$, and $\sec y = \dfrac{\sqrt{34}}{5}$.

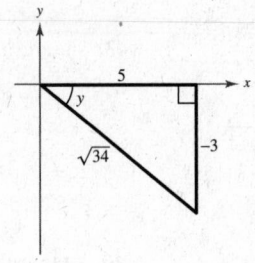

56. Let $u = \arcsin\left(-\dfrac{3}{4}\right)$,

$\sin u = -\dfrac{3}{4}, -\dfrac{\pi}{2} < u < 0$,

$\tan\left[\arcsin\left(-\dfrac{3}{4}\right)\right] = \tan u = -\dfrac{3}{\sqrt{7}} = -\dfrac{3\sqrt{7}}{7}$.

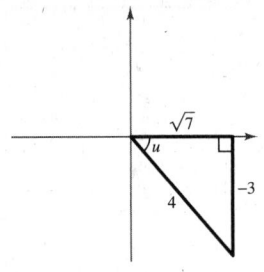

57. Let $y = \arccos\left(-\dfrac{2}{3}\right)$,

$\cos y = -\dfrac{2}{3}, \ \dfrac{\pi}{2} < y < \pi$, and $\sin y = \dfrac{\sqrt{5}}{3}$.

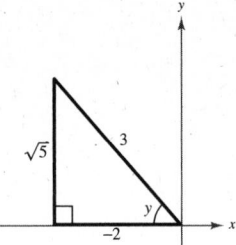

58. Let $u = \arctan \dfrac{5}{8}$,

$\tan u = \dfrac{5}{8}, 0 < u < \dfrac{\pi}{2}$,

$\cot\left(\arctan \dfrac{5}{8}\right) = \cot u = \dfrac{8}{5}$.

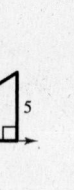

59. Let $y = \arctan x$,

$\tan y = x = \dfrac{x}{1}$,

and $\cot y = \dfrac{1}{x}$.

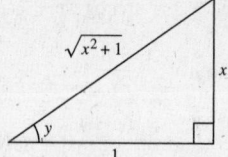

60. Let $u = \arctan x$, $\tan u = x = \dfrac{x}{1}$,

$\sin(\arctan x) = \sin u = \dfrac{x}{\sqrt{x^2 + 1}}$.

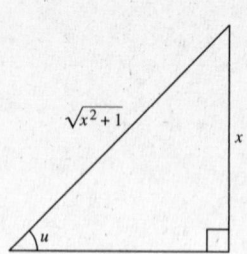

61. Let $y = \arcsin(2x)$,

$\sin y = 2x = \dfrac{2x}{1}$,

and $\cos y = \sqrt{1 - 4x^2}$.

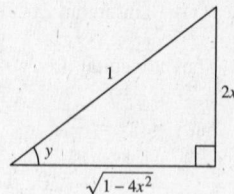

62. Let $u = \arctan 3x$,

$\tan u = 3x = \dfrac{3x}{1}$,

$\sec(\arctan 3x) = \sec u = \sqrt{9x^2 + 1}$.

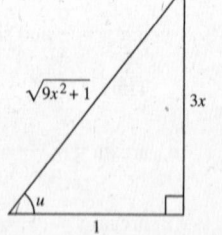

63. Let $y = \arccos x$,

$\cos y = x = \dfrac{x}{1}$,

and $\sin y = \sqrt{1 - x^2}$.

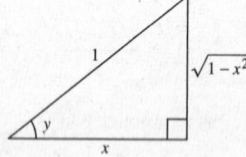

64. Let $u = \arcsin(x - 1)$,

$\sin u = x - 1 = \dfrac{x - 1}{1}$,

$\sec[\arcsin(x - 1)] = \sec u = \dfrac{1}{\sqrt{2x - x^2}}$.

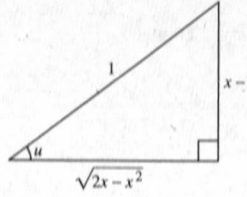

65. Let $y = \arccos\left(\dfrac{x}{3}\right)$,

$\cos y = \dfrac{x}{3}$,

and $\tan y = \dfrac{\sqrt{9 - x^2}}{x}$.

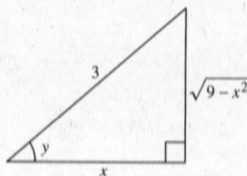

66. Let $u = \arctan \dfrac{1}{x}$,

$\tan u = \dfrac{1}{x}$,

$\cot\left(\arctan \dfrac{1}{x}\right) = \cot u = x$.

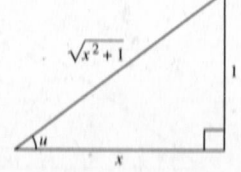

67. Let $y = \arctan \dfrac{x}{\sqrt{2}}$,

$\tan y = \dfrac{x}{\sqrt{2}}$,

and $\csc y = \dfrac{\sqrt{x^2 + 2}}{x}$.

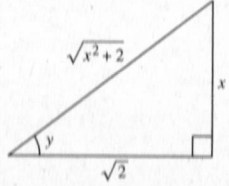

68. Let $u = \arcsin \dfrac{x-h}{r}$,

$\sin u = \dfrac{x-h}{r}$,

$\cos\left(\arcsin \dfrac{x-h}{r}\right) = \cos u = \dfrac{\sqrt{r^2 - (x-h)^2}}{r}$.

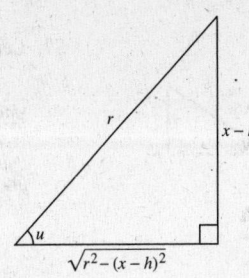

69. $f(x) = \sin(\arctan 2x)$, $g(x) = \dfrac{2x}{\sqrt{1 + 4x^2}}$

They are equal. Let $y = \arctan 2x$,

$\tan y = 2x = \dfrac{2x}{1}$,

and $\sin y = \dfrac{2x}{\sqrt{1 + 4x^2}}$.

$g(x) = \dfrac{2x}{\sqrt{1 + 4x^2}} = f(x)$

The graph has horizontal asymptotes at $y = \pm 1$.

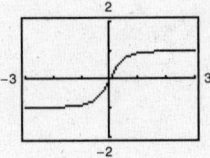

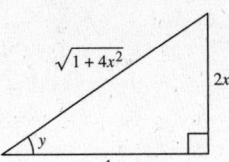

70. $f(x) = \tan\left(\arccos \dfrac{x}{2}\right)$

$g(x) = \dfrac{\sqrt{4 - x^2}}{x}$

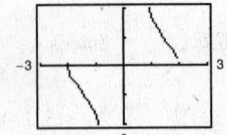

Asymptote: $x = 0$

These are equal because:

Let $u = \arccos \dfrac{x}{2}$.

$f(x) = \tan\left(\arccos \dfrac{x}{2}\right) = \tan u$

$= \dfrac{\sqrt{4 - x^2}}{x} = g(x)$

Thus, $f(x) = g(x)$.

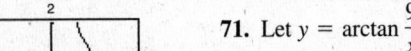

71. Let $y = \arctan \dfrac{9}{x}$.

$\tan y = \dfrac{9}{x}$ and $\sin y = \dfrac{9}{\sqrt{x^2 + 81}}, x > 0; \dfrac{-9}{\sqrt{x^2 + 81}}, x < 0.$

Thus,

$\arcsin y = \dfrac{9}{\sqrt{x^2 + 81}}, x > 0;$

$\arcsin y = \dfrac{-9}{\sqrt{x^2 + 81}}, x < 0.$

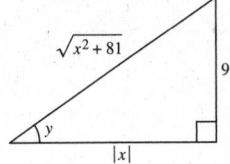

72. If $\arcsin \dfrac{\sqrt{36 - x^2}}{6} = u$,

then $\sin u = \dfrac{\sqrt{36 - x^2}}{6}$,

$\arcsin \dfrac{\sqrt{36 - x^2}}{6} = \arccos \dfrac{x}{6}$.

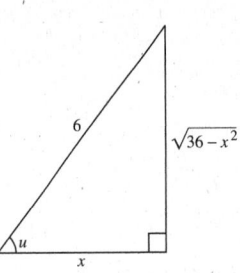

73. Let $y = \arccos \dfrac{3}{\sqrt{x^2 - 2x + 10}}$. Then,

$\cos y = \dfrac{3}{\sqrt{x^2 - 2x + 10}} = \dfrac{3}{\sqrt{(x - 1)^2 + 9}}$

and $\sin y = \dfrac{|x - 1|}{\sqrt{(x - 1)^2 + 9}}$.

Thus, $y = \arcsin \dfrac{|x - 1|}{\sqrt{x^2 - 2x + 10}}$.

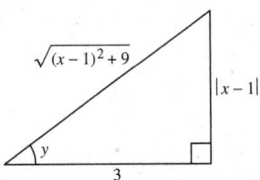

74. If $\arccos \dfrac{x-2}{2} = u$,

then $\cos u = \dfrac{x-2}{2}$,

$\arccos \dfrac{x-2}{2} = \arctan \dfrac{\sqrt{4x-x^2}}{x-2}$.

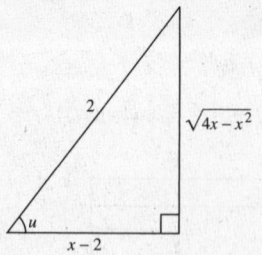

75. $y = 2\arccos x$

Domain: $-1 \le x \le 1$

Range: $0 \le y \le 2\pi$

This is the graph of $f(x) = \arccos x$ with a factor of 2.

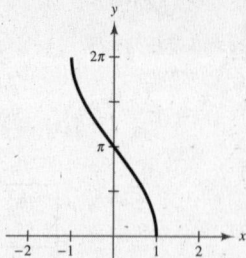

76. $y = \arcsin \dfrac{x}{2}$

Domain: $-2 \le x \le 2$

Range: $-\dfrac{\pi}{2} \le y \le \dfrac{\pi}{2}$

This is the graph of $f(x) = \arcsin x$ with a horizontal stretch of a factor of 2.

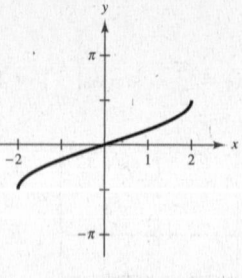

77. $f(x) = \arcsin(x-1)$

Domain: $0 \le x \le 2$

Range: $-\dfrac{\pi}{2} \le y \le \dfrac{\pi}{2}$

This is the graph of $g(x) = \arcsin(x)$ shifted one unit to the right.

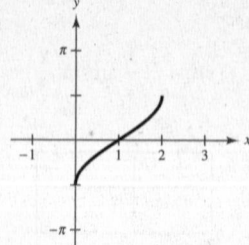

78. $g(t) = \arccos(t+2)$

Domain: $-3 \le t \le -1$

Range: $0 \le y \le \pi$

This is the graph of $y = \arccos t$ shifted two units to the left.

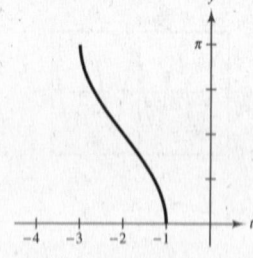

79. $f(x) = \arctan 2x$

Domain: all real numbers

Range: $-\dfrac{\pi}{2} < y < \dfrac{\pi}{2}$

This is the graph of $g(x) = \arctan(x)$ with a horizontal stretch of a factor of 2.

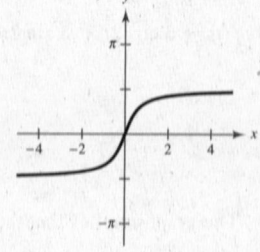

80. $f(x) = \dfrac{\pi}{2} + \arctan x$

Domain: all real numbers

Range: $0 < y \le \pi$

This is the graph of $y = \arctan x$ shifted upward $\pi/2$ units.

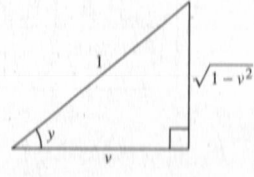

81. $h(v) = \tan(\arccos v) = \dfrac{\sqrt{1-v^2}}{v}$

Domain: $-1 \le v \le 1, v \ne 0$

Range: all real numbers

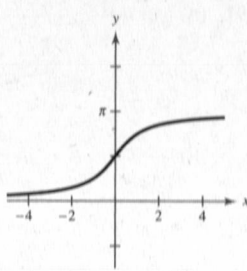

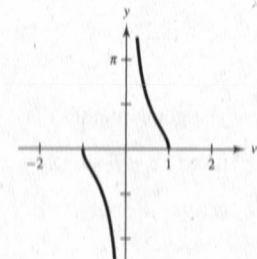

82. $f(x) = \arccos \dfrac{x}{4}$

Domain: $-4 \le x \le 4$

Range: $0 \le y \le \pi$

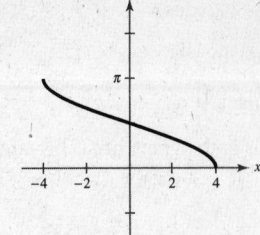

83. $f(x) = 2 \arccos(2x)$

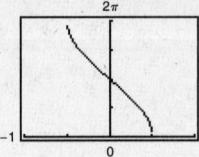

84. $f(x) = \pi \arcsin(4x)$

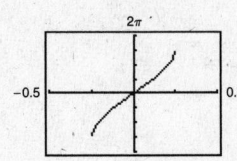

85. $f(x) = \arctan(2x - 3)$

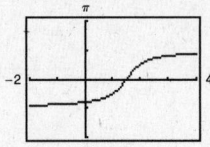

86. $f(x) = -3 + \arctan(\pi x)$

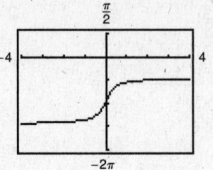

87. $f(x) = \pi - \arcsin\!\left(\dfrac{2}{3}\right) \approx 2.412$

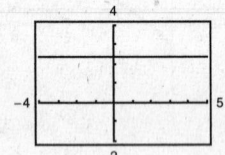

88. $f(x) = \dfrac{\pi}{2} + \arccos\!\left(\dfrac{1}{\pi}\right) \approx 2.82$

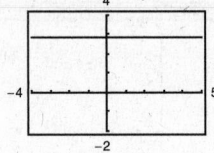

89. $f(t) = 3\cos 2t + 3\sin 2t = \sqrt{3^2 + 3^2}\,\sin\!\left(2t + \arctan \dfrac{3}{3}\right)$

$\qquad\qquad = 3\sqrt{2}\,\sin(2t + \arctan 1)$

$\qquad\qquad = 3\sqrt{2}\,\sin\!\left(2t + \dfrac{\pi}{4}\right)$

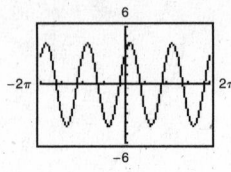

The graph implies that the identity is true.

90. $f(t) = 4\cos \pi t + 3\sin \pi t$

$\qquad = \sqrt{4^2 + 3^2}\,\sin\!\left(\pi t + \arctan \dfrac{4}{3}\right)$

$\qquad = 5\sin\!\left(\pi t + \arctan \dfrac{4}{3}\right)$

The graph implies that

$A\cos \omega t + B\sin \omega t = \sqrt{A^2 + B^2}\,\sin\!\left(\omega t + \arctan \dfrac{A}{B}\right)$

is true.

91. (a) $\sin \theta = \dfrac{5}{s}$

$\qquad \theta = \arcsin \dfrac{5}{s}$

(b) $s = 40$: $\theta = \arcsin \dfrac{5}{40} \approx 0.13$

$\qquad s = 20$: $\theta = \arcsin \dfrac{5}{20} \approx 0.25$

92. (a) $\tan \theta = \dfrac{s}{750}$

$\theta = \arctan \dfrac{s}{750}$

(b) When $s = 300$,

$\theta = \arctan \dfrac{300}{750} \approx 0.38 \approx 21.8°$.

When $s = 1200$,

$\theta = \arctan \dfrac{1200}{750} \approx 1.01 \approx 58.0°$.

93. $\beta = \arctan \dfrac{3x}{x^2 + 4}$

(a)

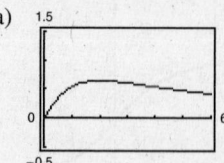

(b) β is maximum when $x = 2$ feet.

(c) The graph has a horizontal asymptote at $\beta = 0$.
As x increases, β decreases.

94. (a) $\tan \theta = \dfrac{11}{17}$

$\theta = \arctan \dfrac{11}{17} \approx 0.5743 \approx 32.9°$

(b) $r = \dfrac{1}{2}(40) = 20$

$\tan \theta = \dfrac{h}{r} = \dfrac{h}{20}$

$h = 20 \tan \theta = 20 \cdot \dfrac{11}{17} \approx 12.94$ feet

95.

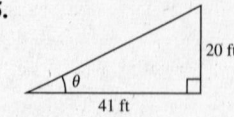

(a) $\tan \theta = \dfrac{20}{41}$

$\theta = \arctan\left(\dfrac{20}{41}\right) \approx 26.0°$

(b) $\tan 26° = \dfrac{h}{50}$

$h = 50 \tan 26° \approx 24.39$ feet

96. (a) $\tan \theta = \dfrac{6}{x}$

$\theta = \arctan \dfrac{6}{x}$

(b) $x = 7$ miles

$\theta = \arctan \dfrac{6}{7} \approx 0.71 \approx 40.6°$

$x = 1$ mile

$\theta = \arctan \dfrac{6}{1} \approx 1.41 \approx 80.5°$

97. (a) $\tan \theta = \dfrac{x}{20}$

$\theta = \arctan \dfrac{x}{20}$

(b) $x = 5$: $\theta = \arctan \dfrac{5}{20} \approx 14.0°$

$x = 12$: $\theta = \arctan \dfrac{12}{20} \approx 31.0°$

98. False.

$\dfrac{5\pi}{6}$ is not in the range of $\arcsin(x)$.

$\arcsin \dfrac{1}{2} = \dfrac{\pi}{6}$

99. False.

$\dfrac{5\pi}{4}$ is not in the range of the arctangent function.

$\arctan 1 = \dfrac{\pi}{4}$

100. False.

$\arctan x$ is defined for all real x, but $\arcsin x$ and $\arccos x$ require $-1 \le x \le 1$.

Also, for example, $\arctan 1 \ne \dfrac{\arcsin 1}{\arccos 1}$.

Since $\arctan 1 = \dfrac{\pi}{4}$, but $\dfrac{\arcsin 1}{\arccos 1} = \dfrac{\pi/2}{0} =$ undefined.

101. $y = \text{arccot } x$ if and only if $\cot y = x$.

Domain: $-\infty < x < \infty$

Range: $0 < x < \pi$

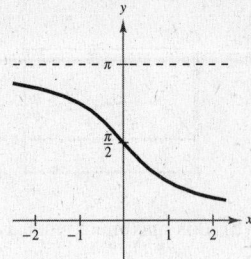

102. $y = \text{arcsec } x$ if and only if $\sec y = x$ where

$x \le -1 \cup x \ge 1$ and $0 \le y < \dfrac{\pi}{2}$ and $\dfrac{\pi}{2} < y \le \pi$.

The domain of $y = \text{arcsec } x$ is $(-\infty, -1] \cup [1, \infty)$

and the range is $\left[0, \dfrac{\pi}{2}\right) \cup \left(\dfrac{\pi}{2}, \pi\right]$.

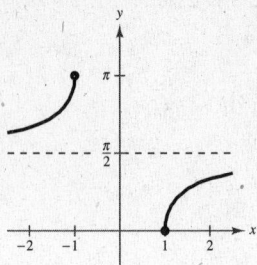

103. $y = \text{arccsc } x$ if and only if $\csc y = x$.

Domain: $(-\infty, -1] \cup [1, \infty)$

Range: $\left[-\dfrac{\pi}{2}, 0\right) \cup \left(0, \dfrac{\pi}{2}\right]$

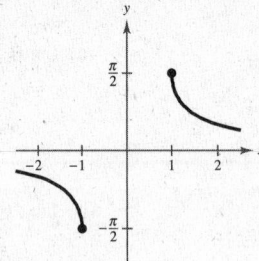

104. (a) $y = \text{arcsec } \sqrt{2} \implies \sec y = \sqrt{2}$ and $0 \le y < \dfrac{\pi}{2} \cup \dfrac{\pi}{2} < y \le \pi \implies y = \dfrac{\pi}{4}$

(b) $y = \text{arcsec } 1 \implies \sec y = 1$ and $0 \le y < \dfrac{\pi}{2} \cup \dfrac{\pi}{2} < y \le \pi \implies y = 0$

(c) $y = \text{arccot}\left(-\sqrt{3}\right) \implies \cot y = -\sqrt{3}$ and $0 < y < \pi \implies y = \dfrac{5\pi}{6}$

(d) $y = \text{arccsc } 2 \implies \csc y = 2$ and $-\dfrac{\pi}{2} \le y < 0 \cup 0 < y \le \dfrac{\pi}{2} \implies y = \dfrac{\pi}{6}$

105. Area $= \arctan b - \arctan a$

(a) $a = 0, b = 1$

Area $= \arctan 1 - \arctan 0 = \dfrac{\pi}{4} - 0 = \dfrac{\pi}{4}$

(b) $a = -1, b = 1$

Area $= \arctan 1 - \arctan(-1)$

$= \dfrac{\pi}{4} - \left(-\dfrac{\pi}{4}\right) = \dfrac{\pi}{2}$

(c) $a = 0, b = 3$

Area $= \arctan 3 - \arctan 0$

$\approx 1.25 - 0 = 1.25$

(d) $a = -1, b = 3$

Area $= \arctan 3 - \arctan(-1)$

$\approx 1.25 - \left(-\dfrac{\pi}{4}\right) \approx 2.03$

106. $f(x) = \sqrt{x}$

$g(x) = 6 \arctan x$

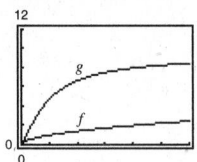

As x increases to infinity, g approaches 3π, but f has no maximum. Using the solve feature of the graphing utility, you find $a \approx 87.54$.

107. $f(x) = \sin(x), f^{-1}(x) = \arcsin(x)$

  (a) $f \cdot f^{-1} = \sin(\arcsin x)$ $f^{-1} \cdot f = \arcsin(\sin x)$

  (b) The graphs coincide with the graph of $y = x$ only for certain values of x.

    $f \cdot f^{-1} = x$ over its entire domain, $-1 \le x \le 1$.

    $f^{-1} \cdot f = x$ over the region $-\dfrac{\pi}{2} \le x \le \dfrac{\pi}{2}$, corresponding to the region where $\sin x$ is one-to-one and thus has an inverse.

108. (a) Let $y = \arcsin(-x)$. Then,

$$\sin y = -x$$
$$-\sin y = x$$
$$\sin(-y) = x$$
$$-y = \arcsin x$$
$$y = -\arcsin x.$$

    Therefore, $\arcsin(-x) = -\arcsin x$.

  (c) Let $y_2 = \dfrac{\pi}{2} - y_1$.

$$\arctan x + \arctan \frac{1}{x} = y_1 + y_2$$
$$= y_1 + \left(\frac{\pi}{2} - y_1\right) = \frac{\pi}{2}$$

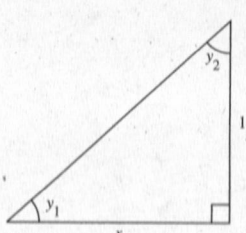

  (e) $\arcsin x = \arcsin \dfrac{x}{1} = \arctan \dfrac{x}{\sqrt{1 - x^2}}$

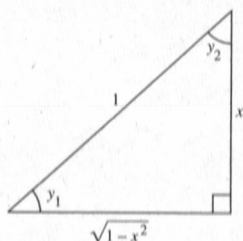

(b) Let $y = \arctan(-x)$. Then,

$$\tan y = -x, \quad -\frac{\pi}{2} < y < \frac{\pi}{2}$$
$$-\tan y = x$$
$$\tan(-y) = x, \quad -\frac{\pi}{2} < -y < \frac{\pi}{2}$$
$$\arctan(\tan(-y)) = \arctan x$$
$$-y = \arctan x$$
$$y = -\arctan x$$

Thus, $\arctan(-x) = -\arctan(x)$.

(d) Let $\alpha = \arcsin x$ and $\beta = \arccos x$, then $\sin \alpha = x$ and $\cos \beta = x$. Thus, $\sin \alpha = \cos \beta$ which implies that α and β are complementary angles and we have

$$\alpha + \beta = \frac{\pi}{2}$$

$$\arcsin x + \arccos x = \frac{\pi}{2}.$$

109. $(8.2)^{3.4} \approx 1279.284$

110. $10(14)^{-2} = \dfrac{10}{14^2} = \dfrac{10}{196} \approx 0.051$

111. $(1.1)^{50} \approx 117.391$

112. $16^{-2\pi} = \dfrac{1}{16^{2\pi}} \approx 2.718 \times 10^{-8}$

113.
$$\sin \theta = \frac{3}{4} = \frac{\text{opp}}{\text{hyp}}$$
$$(\text{adj})^2 + (3)^2 = (4)^2$$
$$(\text{adj})^2 + 9 = 16$$
$$(\text{adj})^2 = 7$$
$$\text{adj} = \sqrt{7}$$
$$\cos \theta = \frac{\sqrt{7}}{4}$$
$$\tan \theta = \frac{3}{\sqrt{7}} = \frac{3\sqrt{7}}{7}$$
$$\cot \theta = \frac{\sqrt{7}}{3}$$
$$\sec \theta = \frac{4}{\sqrt{7}} = \frac{4\sqrt{7}}{7}$$
$$\csc \theta = \frac{4}{3}$$

114. $\tan \theta = 2$
$$\text{hyp} = \sqrt{1^2 + 2^2} = \sqrt{5}$$
$$\cos \theta = \frac{1}{\sqrt{5}}$$
$$\sin \theta = \frac{2}{\sqrt{5}}$$
$$\cot \theta = \frac{1}{2}$$
$$\sec \theta = \sqrt{5}$$
$$\csc \theta = \frac{1}{2}\sqrt{5}$$

115.
$$\cos \theta = \frac{5}{6} = \frac{\text{adj}}{\text{hyp}}$$
$$(\text{opp})^2 + (5)^2 = (6)^2$$
$$(\text{opp})^2 + 25 = 36$$
$$(\text{opp})^2 = 11$$
$$\text{opp} = \sqrt{11}$$
$$\sin \theta = \frac{\sqrt{11}}{6}$$
$$\tan \theta = \frac{\sqrt{11}}{5}$$
$$\cot \theta = \frac{5}{\sqrt{11}} = \frac{5\sqrt{11}}{11}$$
$$\sec \theta = \frac{6}{5}$$
$$\csc \theta = \frac{6}{\sqrt{11}} = \frac{6\sqrt{11}}{11}$$

116. $\sec \theta = 3$
$$\text{opp} = \sqrt{3^2 - 1^2}$$
$$= \sqrt{8}$$
$$= 2\sqrt{2}$$
$$\cos \theta = \frac{1}{3}$$
$$\sin \theta = \frac{2\sqrt{2}}{3}$$
$$\tan \theta = 2\sqrt{2}$$
$$\sec \theta = 3$$
$$\csc \theta = \frac{3}{2\sqrt{2}} = \frac{3\sqrt{2}}{4}$$
$$\cot \theta = \frac{1}{2\sqrt{2}} = \frac{\sqrt{2}}{4}$$

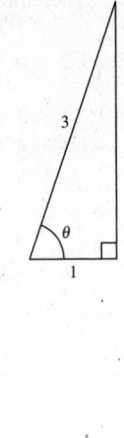

117. Let $x = $ the number of people presently in the group. Each person's share is now $250,000/x$.
If two more join the group, each person's share would then be $250,000/(x + 2)$.

$$\begin{array}{c}\text{Share per person with} \\ \text{two more people}\end{array} = \begin{array}{c}\text{Original share} \\ \text{per person}\end{array} - 6250$$

$$\frac{250,000}{x + 2} = \frac{250,000}{x} - 6250$$

$$250,000x = 250,000(x + 2) - 6250x(x + 2)$$

$$250,000x = 250,000x + 500,000 - 6250x^2 - 12500x$$

$$6250x^2 + 12500x - 500,000 = 0$$

$$6250(x^2 + 2x - 80) = 0$$

$$6250(x + 10)(x - 8) = 0$$

$$x = -10 \quad \text{or} \quad x = 8$$

$x = -10$ is not possible.

There were 8 people in the original group.

118. Rate downstream: $18 + x$

Rate upstream: $18 - x$

$$\text{rate} \times \text{time} = \text{distance} \implies t = \frac{d}{r}$$

(Time to go upstream) + (Time to go downstream) = 4

$$\frac{35}{18 - x} + \frac{35}{18 + x} = 4$$

$$35(18 + x) + 35(18 - x) = 4(18 - x)(18 + x)$$

$$630 + 35x + 630 - 35x = 4(324 - x^2)$$

$$1260 = 4(324 - x^2)$$

$$315 = 324 - x^2$$

$$x^2 = 9$$

$$x = \pm 3$$

The speed of the current is 3 miles per hour.

119. (a) $A = 15,000\left(1 + \dfrac{0.035}{4}\right)^{(4)(10)} \approx \$21,253.63$

(b) $A = 15,000\left(1 + \dfrac{0.035}{12}\right)^{(12)(10)} \approx \$21,275.17$

(c) $A = 15,000\left(1 + \dfrac{0.035}{365}\right)^{(365)(10)} \approx \$21,285.66$

(d) $A = 15,000e^{(0.035)(10)} \approx \$21,286.01$

120. Data: $(2, 742,000), (4, 632,000)$

To find: $(8, y)$

Assume: $\qquad P = P_0 \cdot e^{-rt}$

$$742,000 = P_0 e^{-r \cdot 2}$$

$$632,000 = P_0 e^{-r \cdot 4}$$

Then: $e^{-r \cdot 2} = \dfrac{P_0 e^{-r \cdot 4}}{P_0 e^{-r \cdot 2}} = \dfrac{632}{742}$

$$y = P_0 e^{-r \cdot 8} = P_0 e^{-r \cdot 4} \cdot e^{-r \cdot 4}$$

$$= 632,000 \cdot (e^{-r \cdot 2})^2$$

$$= 632,000 \cdot \left(\frac{632}{742}\right)^2$$

$$= 458,504.31$$

Section 4.8 Applications and Models

- You should be able to solve right triangles.
- You should be able to solve right triangle applications.
- You should be able to solve applications of simple harmonic motion.

Vocabulary Check

1. elevation; depression

2. bearing

3. harmonic motion

---

1. Given: $A = 20°$, $b = 10$

$$\tan A = \frac{a}{b} \implies a = b \tan A = 10 \tan 20° \approx 3.64$$

$$\cos A = \frac{b}{c} \implies c = \frac{b}{\cos A} = \frac{10}{\cos 20°} \approx 10.64$$

$$B = 90° - 20° = 70°$$

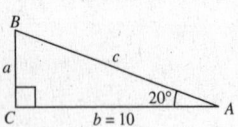

2. Given: $B = 54°$, $c = 15$

$$A = 90° - B$$

$$= 90° - 54° = 36°$$

$$\sin B = \frac{b}{c} \implies b = c \sin B$$

$$= 15 \sin 54° \approx 12.14$$

$$\cos B = \frac{a}{c} \implies a = c \cos B = 15 \cos 54° \approx 8.82$$

3. Given: $B = 71°$, $b = 24$

$$\tan B = \frac{b}{a} \implies a = \frac{b}{\tan B} = \frac{24}{\tan 71°} \approx 8.26$$

$$\sin B = \frac{b}{c} \implies c = \frac{b}{\sin B} = \frac{24}{\sin 71°} \approx 25.38$$

$$A = 90° - 71° = 19°$$

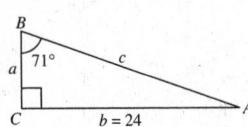

4. Given: $A = 8.4°$, $a = 40.5$

$$B = 90° - A$$

$$= 90° - 8.4° = 81.6°$$

$$\tan A = \frac{a}{b} \implies b = \frac{a}{\tan A}$$

$$= \frac{40.5}{\tan 8.4°} \approx 274.27$$

$$\sin A = \frac{a}{c} \implies c = \frac{a}{\sin A} = \frac{40.5}{\sin 8.4°} \approx 277.24$$

5. Given: $a = 6$, $b = 10$

$$c^2 = a^2 + b^2 \implies c = \sqrt{36 + 100}$$

$$= 2\sqrt{34} \approx 11.66$$

$$\tan A = \frac{a}{b} = \frac{6}{10} \implies A = \arctan \frac{3}{5} \approx 30.96°$$

$$B = 90° - 30.96° = 59.04°$$

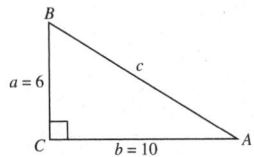

6. Given: $a = 25$, $c = 35$

$$b = \sqrt{c^2 - a^2}$$

$$= \sqrt{35^2 - 25^2}$$

$$= \sqrt{600} \approx 24.49$$

$$\sin A = \frac{a}{c} \implies A = \arcsin \frac{a}{c}$$

$$= \arcsin \frac{25}{35} \approx 45.58°$$

$$\cos B = \frac{a}{c} \implies B = \arccos \frac{a}{c} = \arccos \frac{25}{35} \approx 44.42°$$

7. Given: $b = 16$, $c = 52$

$a = \sqrt{52^2 - 16^2}$

$\quad = \sqrt{2448} = 12\sqrt{17} \approx 49.48$

$\cos A = \dfrac{16}{52}$

$\quad\quad A = \arccos \dfrac{16}{52} \approx 72.08°$

$\quad\quad B = 90° - 72.08° \approx 17.92°$

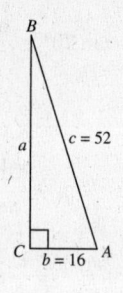

8. Given: $b = 1.32$, $c = 9.45$

$a = \sqrt{c^2 - b^2} = \sqrt{87.5601} \approx 9.36$

$\cos A = \dfrac{b}{c} \Rightarrow A = \arccos \dfrac{b}{c} = \arccos \dfrac{1.32}{9.45} \approx 81.97°$

$\sin B = \dfrac{b}{c} \Rightarrow B = \arcsin \dfrac{b}{c}$

$\quad\quad\quad\quad\quad = \arcsin \dfrac{1.32}{9.45}$

$\quad\quad\quad\quad\quad \approx 8.03°$

9. Given: $A = 12° \, 15'$, $c = 430.5$

$B = 90° - 12° \, 15' = 77° \, 45'$

$\sin 12° \, 15' = \dfrac{a}{430.5}$

$\quad\quad a = 430.5 \sin 12° \, 15' \approx 91.34$

$\cos 12° \, 15' = \dfrac{b}{430.5}$

$\quad\quad b = 430.5 \cos 12° \, 15' \approx 420.70$

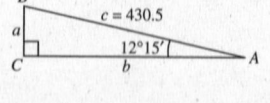

10. Given: $B = 65° \, 12'$, $a = 14.2$

$A = 90° - B = 90° - 65° \, 12' = 24° \, 48'$

$\cos B = \dfrac{a}{c} \Rightarrow c = \dfrac{a}{\cos B} = \dfrac{14.2}{\cos 65° \, 12'} \approx 33.85$

$\tan B = \dfrac{b}{a} \Rightarrow b = a \tan B = 14.2 \tan 65° \, 12' \approx 30.73$

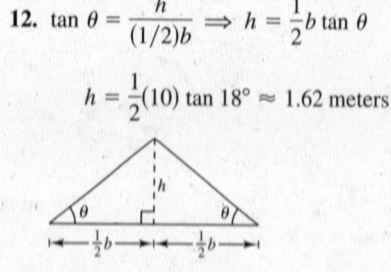

11. $\tan \theta = \dfrac{h}{(1/2)b} \Rightarrow h = \dfrac{1}{2} b \tan \theta$

$h = \dfrac{1}{2}(4) \tan 52° \approx 2.56$ inches

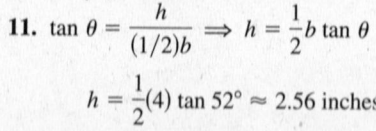

12. $\tan \theta = \dfrac{h}{(1/2)b} \Rightarrow h = \dfrac{1}{2} b \tan \theta$

$h = \dfrac{1}{2}(10) \tan 18° \approx 1.62$ meters

13. $\tan \theta = \dfrac{h}{(1/2)b} \Rightarrow h = \dfrac{1}{2} b \tan \theta$

$h = \dfrac{1}{2}(46) \tan 41° \approx 19.99$ inches

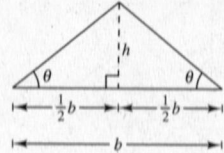

14. $\tan \theta = \dfrac{h}{(1/2)b} \Rightarrow h = \dfrac{1}{2} b \tan \theta$

$h = \dfrac{1}{2}(11) \tan 27° \approx 2.80$ feet

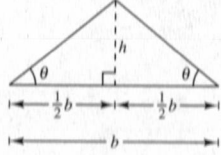

15. $\tan 25° = \dfrac{50}{x}$

$x = \dfrac{50}{\tan 25°}$

≈ 107.2 feet

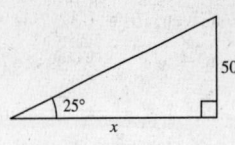

16. $\tan 20° = \dfrac{600}{x}$

$x = \dfrac{600}{\tan 20}$

≈ 1648.5 feet

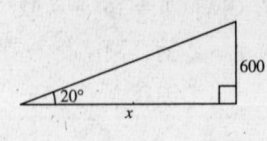

17. $\sin 80° = \dfrac{h}{20}$

$20 \sin 80° = h$

$h \approx 19.7$ feet

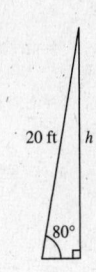

18. $\tan 33° = \dfrac{h}{125}$

$h = 125 \tan 33$

≈ 81.2 feet

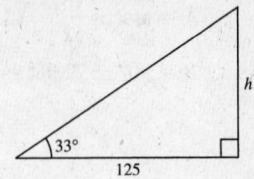

19. (a)

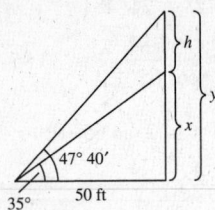

(b) Let the height of the church $= x$ and the height of the church and steeple $= y$. Then,

$\tan 35° = \dfrac{x}{50}$ and $\tan 47°40' = \dfrac{y}{50}$

$x = 50 \tan 35°$ and $y = 50 \tan 47°40'$

$h = y - x = 50(\tan 47°40' - \tan 35°).$

(c) $h \approx 19.9$ feet

20. $\tan 51° = \dfrac{h}{100}$

$h = 100 \tan 51°$

≈ 123.5 feet

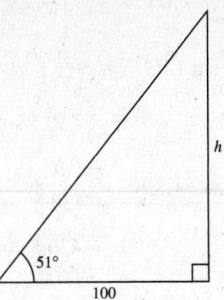

21. $\sin 34° = \dfrac{x}{4000}$

$x = 4000 \sin 34°$

≈ 2236.8 feet

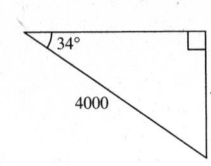

22. $\tan \theta = \dfrac{75}{50}$

$\theta = \arctan \dfrac{3}{2} \approx 56.3°$

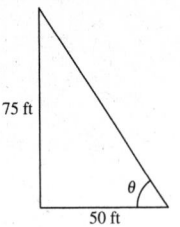

23. (a)

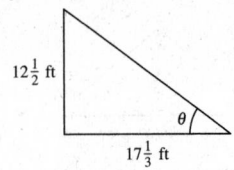

(b) $\tan \theta = \dfrac{12\frac{1}{2}}{17\frac{1}{3}}$

(c) $\theta = \arctan \dfrac{12\frac{1}{2}}{17\frac{1}{3}} \approx 35.8°$

The angle of elevation of the sum is 35.8°.

24. $12{,}500 + 4000 = 16{,}500$

$\sin \theta = \dfrac{4000}{16{,}500}$

$\theta = \arcsin\left(\dfrac{4000}{16{,}500}\right)$

$\theta \approx 14.03°$

Angle of depression $= \alpha \approx 90° - 14.03° = 75.97°$

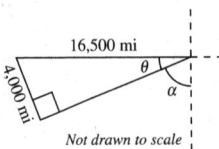

25. 1200 feet + 150 feet − 400 feet = 950 feet

$$5 \text{ miles} = 5 \text{ miles}\left(\frac{5280 \text{ feet}}{1 \text{ mile}}\right) = 26,400 \text{ feet}$$

$$\tan \theta = \frac{950}{26,400}$$

$$\theta = \arctan\left(\frac{950}{26,400}\right) \approx 2.06°$$

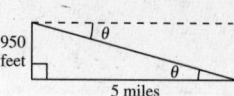

Not drawn to scale

26. (a) Since the airplane speed is

$$\left(275\frac{\text{ft}}{\text{sec}}\right)\left(60\frac{\text{sec}}{\text{min}}\right) = 16,500\frac{\text{ft}}{\text{min}},$$

after one minute its distance travelled is 16,500 feet.

$$\sin 18° = \frac{a}{16,500}$$

$$a = 16,500 \sin 18° \approx 5099 \text{ ft}$$

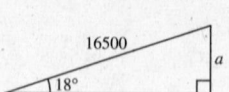

(b) $$\sin 18° = \frac{10,000}{275s}$$

$$s = \frac{10,000}{275(\sin 18°)}$$

$$\approx 117.7 \text{ seconds}$$

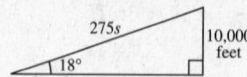

27. $$\sin 10.5° = \frac{x}{4}$$

$$x = 4 \sin 10.5° \approx 0.73 \text{ mile}$$

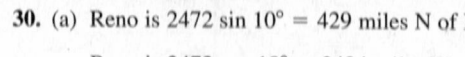

28.

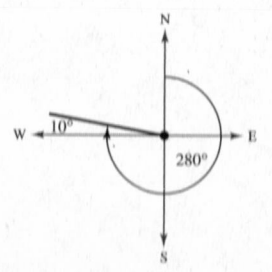

Angle of grade: $$\tan \theta = \frac{12x}{100x}$$

$$\theta = \arctan 0.12 \approx 6.8°$$

Change in elevation:

$$\sin \theta = \frac{y}{21,120}$$

$$y = 21,120 \sin \theta$$

$$= 21,120 \sin(\arctan 0.12)$$

$$\approx 2516.3 \text{ feet}$$

29. The plane has traveled 1.5(600) = 900 miles.

$$\sin 38° = \frac{a}{900} \implies a \approx 554 \text{ miles north}$$

$$\cos 38° = \frac{b}{900} \implies b \approx 709 \text{ miles east}$$

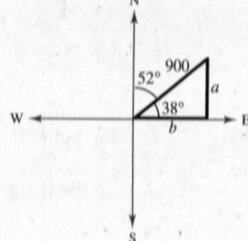

30. (a) Reno is 2472 sin 10° = 429 miles N of Miami.

Reno is 2472 cos 10° = 2434 miles W of Miami.

(b) The return heading is 280°.

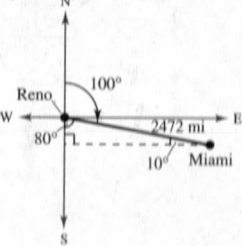

31.

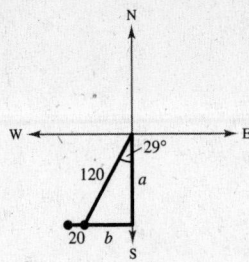

(a) $\cos 29° = \dfrac{a}{120} \implies a \approx 104.95$ nautical miles south

$\sin 29° = \dfrac{b}{120} \implies b \approx 58.18$ nautical miles west

(b) $\tan \theta = \dfrac{20 + b}{a} \approx \dfrac{78.18}{104.95} \implies \theta \approx 36.7°$

Bearing: S 36.7° W

Distance: $d \approx \sqrt{104.95^2 + 78.18^2}$

≈ 130.9 nautical miles from port

32.

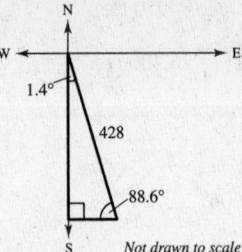

Not drawn to scale

(a) $t = \dfrac{428}{20} = 21.4$ hours

(b) After 12 hours, the yacht will have traveled 240 nautical miles.

$240 \sin 1.4° \approx 5.9$ miles E

$240 \cos 1.4° \approx 239.9$ miles S

(c) Bearing from N is 178.6°.

33. $\theta = 32°$, $\phi = 68°$

(a) $\alpha = 90° - 32° = 58°$

Bearing from A to C: N 58° E

(b) $\beta = \theta = 32°$

$\gamma = 90° - \phi = 22°$

$C = \beta + \gamma = 54°$

$\tan C = \dfrac{d}{50} \implies \tan 54°$

$= \dfrac{d}{50} \implies d \approx 68.82$ meters

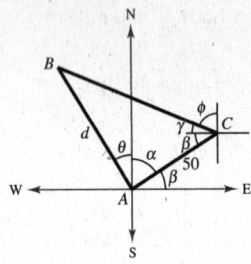

34. $\tan 14° = \dfrac{d}{x} \implies x = d \cot 14°$

$\tan 34° = \dfrac{d}{y} = \dfrac{d}{30 - x}$

$= \dfrac{d}{30 - d \cot 14°}$

$\cot 34° = \dfrac{30 - d \cot 14°}{d}$

$d \cot 34° = 30 - d \cot 14°$

$d = \dfrac{30}{\cot 34° + \cot 14°}$

≈ 5.46 kilometers

35. $\tan \theta = \dfrac{45}{30} \implies \theta \approx 56.3°$

Bearing: N 56.3° W

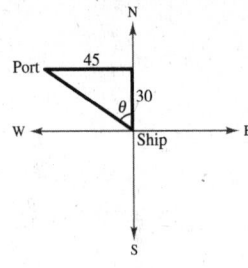

36. Bearing $= 180° + \arctan\left(\dfrac{85}{160}\right)$

$= 208.0°$ or $528°$ W

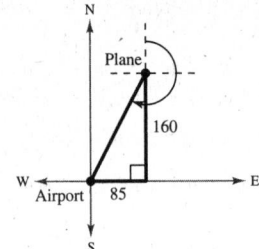

37. $\tan 6.5° = \dfrac{350}{d} \implies d \approx 3071.91$ ft

$\tan 4° = \dfrac{350}{D} \implies D \approx 5005.23$ ft

Distance between ships: $D - d \approx 1933.3$ ft

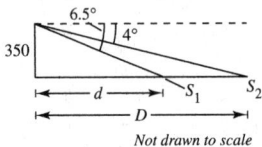

Not drawn to scale

38. $\cot 55 = \dfrac{d}{10} \implies d \approx 7$ kilometers

$\cot 28° = \dfrac{D}{10} \implies D \approx 18.8$ kilometers

Distance between towns:

$D - d = 18.8 - 7 = 11.8$ kilometers

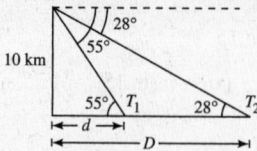

40. $\tan 2.5° = \dfrac{h}{x}$

$x = \dfrac{h}{\tan 2.5°}$

$\tan 9° = \dfrac{h}{x - 17}$

$x = \dfrac{h}{\tan 9°} + 17$

$\dfrac{h}{\tan 2.5°} = \dfrac{h}{\tan 9°} + 17$

$h = \dfrac{17}{\left(\dfrac{1}{\tan 2.5°} - \dfrac{1}{\tan 9°}\right)} \approx 1.025$ miles

≈ 5410 feet

42. $L_1 = 2x - y = 8 \implies m_1 = 2$

$L_2 = x - 5y = -4 \implies m_2 = \dfrac{1}{5}$

$\tan \alpha = \left| \dfrac{m_2 - m_1}{1 + m_2 m_1} \right|$

$\alpha = \arctan \left| \dfrac{m_2 - m_1}{1 + m_2 m_1} \right| = \arctan \left| \dfrac{(1/5) - 2}{1 + (1/5)(2)} \right|$

$= \arctan \left(\dfrac{9}{7} \right) \approx 52.1°$

44. $\tan \theta = \dfrac{a\sqrt{2}}{a} = \sqrt{2}$

$\theta = \arctan \sqrt{2} \approx 54.7°$

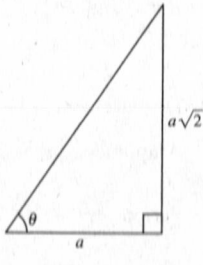

39. $\tan 57° = \dfrac{a}{x} \implies x = a \cot 57°$

$\tan 16° = \dfrac{a}{x + (55/6)}$

$\tan 16° = \dfrac{a}{a \cot 57° + (55/6)}$

$\cot 16° = \dfrac{a \cot 57° + (55/6)}{a}$

$a \cot 16° - a \cot 57° = \dfrac{55}{6} \implies a \approx 3.23$ miles

$\approx 17{,}054$ ft

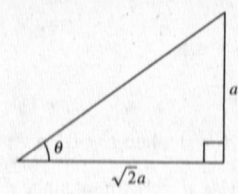

41. L_1: $3x - 2y = 5 \implies y = \dfrac{3}{2}x - \dfrac{5}{2} \implies m_1 = \dfrac{3}{2}$

L_2: $x + y = 1 \implies y = -x + 1 \implies m_2 = -1$

$\tan \alpha = \left| \dfrac{-1 - (3/2)}{1 + (-1)(3/2)} \right| = \left| \dfrac{-5/2}{-1/2} \right| = 5$

$\alpha = \arctan 5 \approx 78.7°$

43. The diagonal of the base has a length of $\sqrt{a^2 + a^2} = \sqrt{2}a$. Now, we have

$\tan \theta = \dfrac{a}{\sqrt{2}a} = \dfrac{1}{\sqrt{2}}$

$\theta = \arctan \dfrac{1}{\sqrt{2}}$

$\theta \approx 35.3°$.

45. $\sin 36° = \dfrac{d}{25} \implies d \approx 14.69$

Length of side: $2d \approx 29.4$ inches

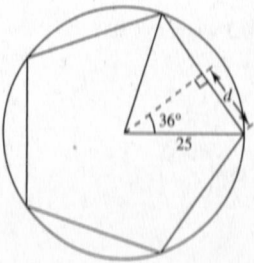

46.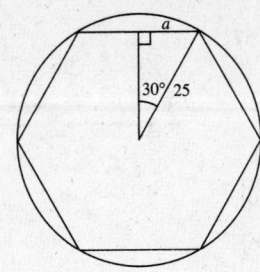

$$\sin 30° = \frac{a}{25}$$

$$a = 25 \sin 30° = 12.5$$

$$\text{Length of side} = 2a = 2(12.5)$$

$$= 25 \text{ inches}$$

47.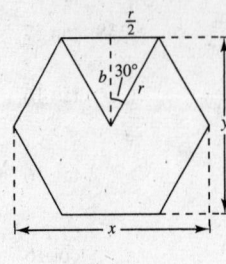

$$\cos 30° = \frac{b}{r}$$

$$b = r \cos 30°$$

$$b = \frac{\sqrt{3}r}{2}$$

$$y = 2b = 2\left(\frac{\sqrt{3}r}{2}\right) = \sqrt{3}r$$

48.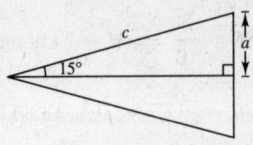

$$c = \frac{35}{2} = 17.5$$

$$\sin 15° = \frac{a}{c}$$

$$a = c \sin 15° = 17.5 \sin 15°$$

$$\approx 4.53$$

$$\text{Distance} = 2a \approx 9.06 \text{ centimeters}$$

49.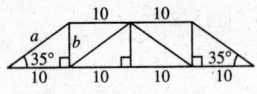

$$\tan 35° = \frac{b}{10}$$

$$b = 10 \tan 35° \approx 7$$

$$\cos 35° = \frac{10}{a}$$

$$a = \frac{10}{\cos 35°} \approx 12.2$$

50. $\tan \theta = \frac{12}{18}$

$$\theta = \arctan \frac{2}{3} = 0.588 \text{ rad} \approx 33.7°$$

$$\cos \theta = \frac{18}{a}$$

$$a = \frac{18}{\cos \theta} \approx 21.6 \text{ feet}$$

$$f \approx \frac{21.6}{2} = 10.8 \text{ feet}$$

$$\phi \approx 90 - 33.7 = 56.3°$$

$$\sin \phi = \frac{6}{b}$$

$$b = \frac{6}{\sin \phi} \approx 7.2 \text{ feet}$$

$$c = \sqrt{10.8^2 + 7.2^2} \approx 13 \text{ feet}$$

51. $d = 0$ when $t = 0$, $a = 4$, period $= 2$

Use $d = a \sin \omega t$ since $d = 0$ when $t = 0$.

$$\frac{2\pi}{\omega} = 2 \implies \omega = \pi$$

Thus, $d = 4 \sin(\pi t)$.

52. Displacement at $t = 0$ is $0 \implies d = a \sin \omega t$.

Amplitude: $|a| = 3$

Period: $\frac{2\pi}{\omega} = 6 \implies \omega = \frac{\pi}{3}$

$$d = 3 \sin\left(\frac{\pi t}{3}\right)$$

53. $d = 3$ when $t = 0$, $a = 3$, period $= 1.5$

Use $d = a \cos \omega t$ since $d = 3$ when $t = 0$.

$$\frac{2\pi}{\omega} = 1.5 \implies \omega = \frac{4\pi}{3}$$

Thus, $d = 3 \cos\left(\frac{4\pi}{3}t\right) = 3 \cos\left(\frac{4\pi t}{3}\right)$.

54. Displacement at $t = 0$ is $2 \implies d = a \cos \omega t$.

Amplitude: $|a| = 2$

Period: $\frac{2\pi}{\omega} = 10 \implies \omega = \frac{\pi}{5}$

$$d = 2 \cos\left(\frac{\pi t}{5}\right)$$

55. $d = 4 \cos 8\pi t$

(a) Maximum displacement = amplitude = 4

(b) Frequency = $\dfrac{\omega}{2\pi} = \dfrac{8\pi}{2\pi}$

$\qquad\qquad\;\; = 4$ cycles per unit of time

(c) $d = 4 \cos 40\pi = 4$

(d) $8\pi t = \dfrac{\pi}{2} \Longrightarrow t = \dfrac{1}{16}$

56. $d = \dfrac{1}{2} \cos 20\pi t$

(a) Maximum displacement: $|a| = \left|\dfrac{1}{2}\right| = \dfrac{1}{2}$

(b) Frequency: $\dfrac{\omega}{2\pi} = \dfrac{20\pi}{2\pi} = 10$ cycles per unit of time

(c) $t = 5 \Longrightarrow d = \dfrac{1}{2} \cos 100\pi \approx \dfrac{1}{2}$

(d) Least positive value for t for which $d = 0$

$\dfrac{1}{2} \cos 20\pi t = 0$

$\cos 20\pi t = 0$

$20\pi t = \arccos 0$

$20\pi t = \dfrac{\pi}{2}$

$t = \dfrac{\pi}{2} \cdot \dfrac{1}{20\pi} = \dfrac{1}{40}$

57. $d = \dfrac{1}{16} \sin 120\pi t$

(a) Maximum displacement = amplitude = $\dfrac{1}{16}$

(b) Frequency = $\dfrac{\omega}{2\pi} = \dfrac{120\pi}{2\pi}$

$\qquad\qquad\;\; = 60$ cycles per unit of time

(c) $d = \dfrac{1}{16} \sin 600\pi = 0$

(d) $120\pi t = \pi \Longrightarrow t = \dfrac{1}{120}$

58. $d = \dfrac{1}{64} \sin 792\pi t$

(a) Maximum displacement: $|a| = \left|\dfrac{1}{64}\right| = \dfrac{1}{64}$

(b) Frequency: $\dfrac{\omega}{2\pi} = \dfrac{792\pi}{2\pi} = 396$ cycles per unit of time

(c) $t = 5 \Longrightarrow d = \dfrac{1}{64} \sin(3960\pi) = 0$

(d) Least positive value for t for which $d = 0$

$\dfrac{1}{64} \sin 792\pi t = 0$

$\sin 792\pi t = 0$

$792\pi t = \arcsin 0$

$792\pi t = \pi$

$t = \dfrac{\pi}{792\pi} = \dfrac{1}{792}$

59. $\qquad\quad d = a \sin \omega t$

Frequency = $\dfrac{\omega}{2\pi}$

$264 = \dfrac{\omega}{2\pi}$

$\omega = 2\pi(264) = 528\pi$

60. At $t = 0$, buoy is at its high point $\Longrightarrow d = a \cos \omega t$.

Distance from high to low = $2|a| = 3.5$

$|a| = \dfrac{7}{4}$

Returns to high point every 10 seconds:

Period: $\dfrac{2\pi}{\omega} = 10 \Longrightarrow \omega = \dfrac{\pi}{5}$

$d = \dfrac{7}{4} \cos \dfrac{\pi t}{5}$

61. $y = \dfrac{1}{4} \cos 16t, \ t > 0$

(a)

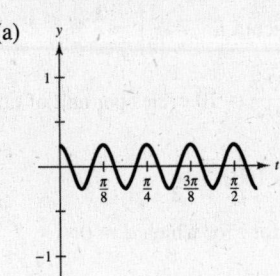

(b) Period: $\dfrac{2\pi}{16} = \dfrac{\pi}{8}$

(c) $\dfrac{1}{4} \cos 16t = 0$ when $16t = \dfrac{\pi}{2} \implies t = \dfrac{\pi}{32}$

62. (a)

θ	L_1	L_2	$L_1 + L_2$
0.1	$\dfrac{2}{\sin 0.1}$	$\dfrac{3}{\cos 0.1}$	23.0
0.2	$\dfrac{2}{\sin 0.2}$	$\dfrac{3}{\cos 0.2}$	13.1
0.3	$\dfrac{2}{\sin 0.3}$	$\dfrac{3}{\cos 0.3}$	9.9
0.4	$\dfrac{2}{\sin 0.4}$	$\dfrac{3}{\cos 0.4}$	8.4

(b)

θ	L_1	L_2	$L_1 + L_2$
0.5	$\dfrac{2}{\sin 0.5}$	$\dfrac{3}{\cos 0.5}$	7.6
0.6	$\dfrac{2}{\sin 0.6}$	$\dfrac{3}{\cos 0.6}$	7.2
0.7	$\dfrac{2}{\sin 0.7}$	$\dfrac{3}{\cos 0.7}$	7.0
0.8	$\dfrac{2}{\sin 0.8}$	$\dfrac{3}{\cos 0.8}$	7.1

The minimum length of the elevator is 7.0 meters.

(c) $L = L_1 + L_2 = \dfrac{2}{\sin \theta} + \dfrac{3}{\cos \theta}$

(d)

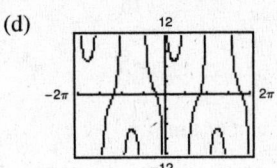

From the graph, it appears that the minimum length is 7.0 meters, which agrees with the estimate of part (b).

63. (a) and (b)

Base 1	Base 2	Altitude	Area
8	$8 + 16 \cos 10°$	$8 \sin 10°$	22.1
8	$8 + 16 \cos 20°$	$8 \sin 20°$	42.5
8	$8 + 16 \cos 30°$	$8 \sin 30°$	59.7
8	$8 + 16 \cos 40°$	$8 \sin 40°$	72.7
8	$8 + 16 \cos 50°$	$8 \sin 50°$	80.5
8	$8 + 16 \cos 60°$	$8 \sin 60°$	83.1
8	$8 + 16 \cos 70°$	$8 \sin 70°$	80.7

The maximum occurs when $\theta = 60°$ and is approximately 83.1 square feet.

(c) $A(\theta) = [8 + (8 + 16 \cos \theta)] \left[\dfrac{8 \sin \theta}{2} \right]$

$= (16 + 16 \cos \theta)(4 \sin \theta)$

$= 64(1 + \cos \theta)(\sin \theta)$

(d)

The maximum of 83.1 square feet occurs when $\theta = \dfrac{\pi}{3} = 60°.$

64. (a)

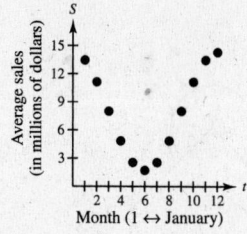

(b) $a = \dfrac{1}{2}(14.3 - 1.7) = 6.3$

$\dfrac{2\pi}{b} = 12 \implies b = \dfrac{\pi}{6}$

Shift: $d = 14.3 - 6.3 = 8$

$S = d + a \cos bt$

$S = 8 + 6.3 \cos\left(\dfrac{\pi t}{6}\right)$

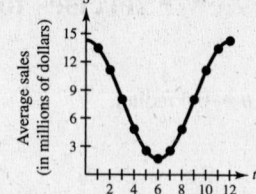

Note: Another model is $S = 8 + 6.3 \sin\left(\dfrac{\pi t}{6} + \dfrac{\pi}{2}\right)$.

The model is a good fit.

(c) Period: $\dfrac{2\pi}{\pi/6} = 12$

This corresponds to the 12 months in a year. Since the sales of outerwear is seasonal, this is reasonable.

(d) The amplitude represents the maximum displacement from average sales of 8 million dollars. Sales are greatest in December (cold weather + Christmas) and least in June.

65. False. Since the tower is not exactly vertical, a right triangle with sides 191 feet and d is not formed.

66. False. One period is the time for one complete cycle of the motion.

67. No. N 24° E means 24° east of north.

68. Aeronautical bearings are always taken clockwise from North (rather than the acute angle from a north-south line).

69. $m = 4$, passes through $(-1, 2)$

$y - 2 = 4(x - (-1))$

$y - 2 = 4x + 4$

$\qquad y = 4x + 6$

70. Linear equation $m = -\dfrac{1}{2}$ through $\left(\dfrac{1}{3}, 0\right)$

$y = -\dfrac{1}{2}x + b$

$0 = -\dfrac{1}{2}\left(\dfrac{1}{3}\right) + b$

$0 = -\dfrac{1}{6} + b$

$b = \dfrac{1}{6}$

$y = -\dfrac{1}{2}x + \dfrac{1}{6}$

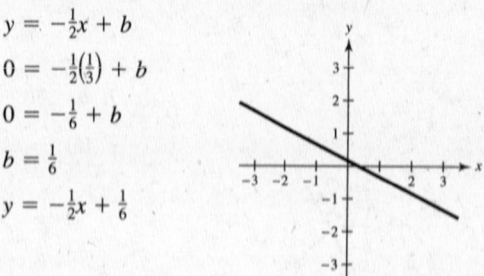

71. Passes through $(-2, 6)$ and $(3, 2)$

$m = \dfrac{2 - 6}{3 - (-2)} = -\dfrac{4}{5}$

$y - 6 = -\dfrac{4}{5}[x - (-2)]$

$y - 6 = -\dfrac{4}{5}x - \dfrac{8}{5}$

$y = -\dfrac{4}{5}x + \dfrac{22}{5}$

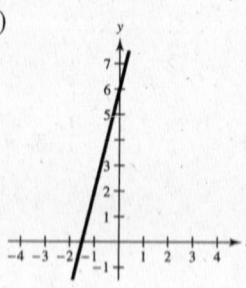

72. Linear equation through $\left(\dfrac{1}{4}, -\dfrac{2}{3}\right)$ and $\left(-\dfrac{1}{2}, \dfrac{1}{3}\right)$

$m = \dfrac{(1/3) - (-2/3)}{(-1/2) - (1/4)}$

$\quad = \dfrac{1}{-3/4}$

$\quad = -\dfrac{4}{3}$

$y + \dfrac{2}{3} = -\dfrac{4}{3}\left(x - \dfrac{1}{4}\right)$

$\qquad y = -\dfrac{4}{3}x - \dfrac{1}{3}$

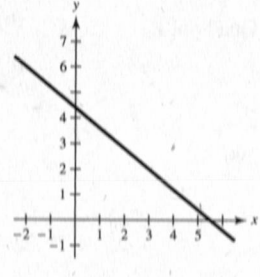

Review Exercises for Chapter 4

1. $\theta \approx 0.5$ radian

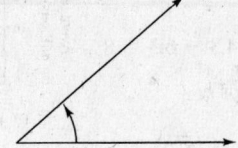

2. $\theta \approx 4.5$ radians

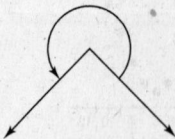

3. $\theta = \dfrac{11\pi}{4}$

(a)

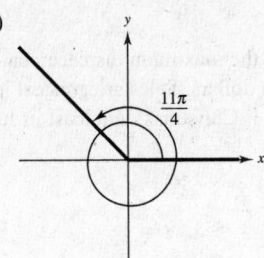

(b) The angle lies in Quadrant II.

(c) Coterminal angles:

$$\frac{11\pi}{4} - 2\pi = \frac{3\pi}{4}$$

$$\frac{3\pi}{4} - 2\pi = -\frac{5\pi}{4}$$

4. $\theta = \dfrac{2\pi}{9}$

(a)

(b) Quadrant I

(c) $\dfrac{2\pi}{9} + 2\pi = \dfrac{20\pi}{9}$

$$\frac{2\pi}{9} - 2\pi = -\frac{16\pi}{9}$$

5. $\theta = -\dfrac{4\pi}{3}$

(a)

(b) The angle lies in Quadrant II.

(c) Coterminal angles:

$$-\frac{4\pi}{3} + 2\pi = \frac{2\pi}{3}$$

$$-\frac{4\pi}{3} - 2\pi = -\frac{10\pi}{3}$$

6. $\theta = -\dfrac{23\pi}{3}$

(a)

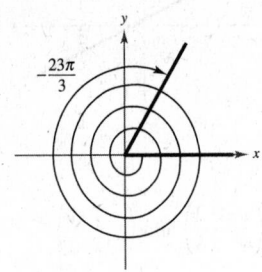

(b) Quadrant I

(c) $-\dfrac{23\pi}{3} + 8\pi = \dfrac{\pi}{3}$

$$-\frac{23\pi}{3} + 2\pi = -\frac{17\pi}{3}$$

7. $\theta = 70°$

(a)

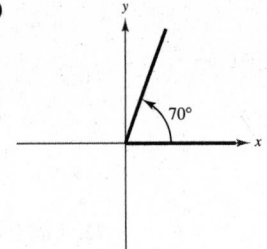

(b) The angle lies in Quadrant I.

(c) Coterminal angles:

$70° + 360° = 430°$

$70° - 360° = -290°$

8. $\theta = 280°$

(a)

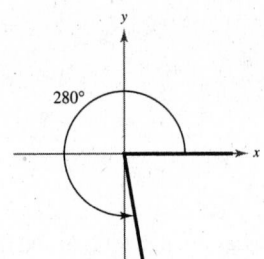

(b) Quadrant IV

(c) $280° + 360° = 640°$

$280° - 360° = -80°$

9. $\theta = -110°$

(a)

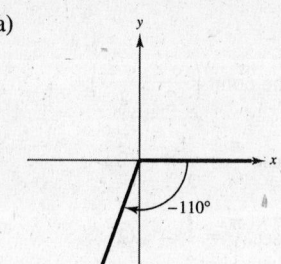

(b) The angle lies in Quadrant III.

(c) Coterminal angles:

$-110° + 360° = 250°$

$-110° - 360° = -470°$

10. $\theta = -405°$

(a)

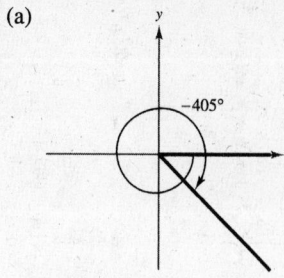

(b) Quadrant IV

(c) $-405° + 720° = 315°$

$-405° + 360° = -45°$

11. $480° = 480° \cdot \dfrac{\pi \text{ rad}}{180°}$

$= \dfrac{8\pi}{3}$ radians

≈ 8.378 radians

12. $-127.5° \cdot \dfrac{\pi}{180°} \approx -2.225$

13. $-33° \, 45' = -33.75° = -33.75° \cdot \dfrac{\pi \text{ rad}}{180°}$

$= -\dfrac{3\pi}{16}$ radian ≈ -0.589 radian

14. $196° \, 77' = \left(196 + \dfrac{77}{60}\right)° \cdot \dfrac{\pi}{180°} \approx 3.443$

15. $\dfrac{5\pi \text{ rad}}{7} = \dfrac{5\pi \text{ rad}}{7} \cdot \dfrac{180°}{\pi \text{ rad}} \approx 128.571°$

16. $-\dfrac{11\pi}{6} \cdot \dfrac{180°}{\pi} = -330.000°$

17. $-3.5 \text{ rad} = -3.5 \text{ rad} \cdot \dfrac{180°}{\pi \text{ rad}} \approx -200.535°$

18. $5.7 \cdot \dfrac{180°}{\pi} \approx 326.586°$

19. $138° = \dfrac{138\pi}{180} = \dfrac{23\pi}{30}$ radians

$s = r\theta = 20\left(\dfrac{23\pi}{30}\right) \approx 48.17$ inches

20. $60° = \dfrac{60\pi}{180}$ radians

$s = r\theta = 11 \cdot \left(\dfrac{60}{180}\right)\pi$

$= \dfrac{11}{3}\pi$ meters

$s \approx 11.52$ meters

21. (a) Angular speed $= \dfrac{\left(33\frac{1}{3}\right)(2\pi) \text{ radians}}{1 \text{ minute}}$

$= 66\frac{2}{3}\pi$ radians per minute

(b) Linear speed $= \dfrac{6\left(66\frac{2}{3}\pi\right) \text{ inches}}{1 \text{ minute}}$

$= 400\pi$ inches per minute

22. (linear speed) $=$ (angular speed) $\cdot$ (radius)

$= (5\pi \text{ rad/s}) \cdot (13.5 \text{ inches})$

$= 67.5\pi$ inches per second

≈ 212.1 inches per second

≈ 12.05 miles per hour

23. $120° = \dfrac{120\pi}{180} = \dfrac{2\pi}{3}$ radians

$A = \dfrac{1}{2}r^2\theta = \dfrac{1}{2}(18)^2\left(\dfrac{2\pi}{3}\right) \approx 339.29$ square inches

24. $A = \dfrac{1}{2}\theta r^2 = \dfrac{1}{2}\left(\dfrac{5\pi}{6}\right)6.5^2$

$A \approx 55.31$ square millimeters

25. $t = \dfrac{2\pi}{3}$ corresponds to the point $\left(-\dfrac{1}{2}, \dfrac{\sqrt{3}}{2}\right)$.

26. $t = \dfrac{3\pi}{4}, \; (x, y) = \left(-\dfrac{\sqrt{2}}{2}, \dfrac{\sqrt{2}}{2}\right)$

27. $t = \dfrac{5\pi}{6}$ corresponds to the point $\left(-\dfrac{\sqrt{3}}{2}, \dfrac{1}{2}\right)$.

28. $t = -\dfrac{4\pi}{3}$, $(x, y) = \left(-\dfrac{1}{2}, \dfrac{\sqrt{3}}{2}\right)$

29. $t = \dfrac{7\pi}{6}$ corresponds to the point $\left(-\dfrac{\sqrt{3}}{2}, -\dfrac{1}{2}\right)$.

$$\sin \frac{7\pi}{6} = y = -\frac{1}{2} \qquad \csc \frac{7\pi}{6} = \frac{1}{y} = -2$$

$$\cos \frac{7\pi}{6} = x = -\frac{\sqrt{3}}{2} \qquad \sec \frac{7\pi}{6} = \frac{1}{x} = -\frac{2\sqrt{3}}{3}$$

$$\tan \frac{7\pi}{6} = \frac{y}{x} = \frac{1}{\sqrt{3}} = \frac{\sqrt{3}}{3} \qquad \cot \frac{7\pi}{6} = \frac{x}{y} = \sqrt{3}$$

30. $t = \dfrac{\pi}{4}$ corresponds to the point $\left(\dfrac{\sqrt{2}}{2}, \dfrac{\sqrt{2}}{2}\right)$.

$$\sin \frac{\pi}{4} = y = \frac{\sqrt{2}}{2} \qquad \csc \frac{\pi}{4} = \frac{1}{y} = \sqrt{2}$$

$$\cos \frac{\pi}{4} = x = \frac{\sqrt{2}}{2} \qquad \sec \frac{\pi}{4} = \frac{1}{x} = \sqrt{2}$$

$$\tan \frac{\pi}{4} = \frac{y}{x} = 1 \qquad \cot \frac{\pi}{4} = \frac{x}{y} = 1$$

31. $t = -\dfrac{2\pi}{3}$ corresponds to the point $\left(-\dfrac{1}{2}, -\dfrac{\sqrt{3}}{2}\right)$.

$$\sin\left(-\frac{2\pi}{3}\right) = y = -\frac{\sqrt{3}}{2} \qquad \csc\left(-\frac{2\pi}{3}\right) = \frac{1}{y} = -\frac{2\sqrt{3}}{3}$$

$$\cos\left(-\frac{2\pi}{3}\right) = x = -\frac{1}{2} \qquad \sec\left(-\frac{2\pi}{3}\right) = \frac{1}{x} = -2$$

$$\tan\left(-\frac{2\pi}{3}\right) = \frac{y}{x} = \sqrt{3} \qquad \cot\left(-\frac{2\pi}{3}\right) = \frac{x}{y} = \frac{\sqrt{3}}{3}$$

32. $t = 2\pi$ corresponds to the point $(1, 0)$.

$$\sin 2\pi = y = 0 \qquad \csc 2\pi = \frac{1}{y} \text{ is undefined.}$$

$$\cos 2\pi = x = 1 \qquad \sec 2\pi = \frac{1}{x} = 1$$

$$\tan 2\pi = \frac{y}{x} = 0 \qquad \cot 2\pi = \frac{x}{y} \text{ is undefined.}$$

33. $\sin \dfrac{11\pi}{4} = \sin \dfrac{3\pi}{4} = \dfrac{\sqrt{2}}{2}$

34. $\cos 4\pi = \cos 0 = 1$

35. $\sin\left(-\dfrac{17\pi}{6}\right) = \sin\left(-\dfrac{5\pi}{6}\right) = -\dfrac{1}{2}$

36. $\cos\left(-\dfrac{13\pi}{3}\right) = \cos\left(\dfrac{5\pi}{3}\right) = \dfrac{1}{2}$

37. $\tan 33 \approx -75.3130$

38. $\csc 10.5 = \dfrac{1}{\sin 10.5} \approx -1.1368$

39. $\sec\left(\dfrac{12\pi}{5}\right) = \dfrac{1}{\cos\left(\dfrac{12\pi}{5}\right)} \approx 3.2361$

40. $\sin\left(-\dfrac{\pi}{9}\right) \approx -0.3420$

41. $\text{opp} = 4$, $\text{adj} = 5$, $\text{hyp} = \sqrt{4^2 + 5^2} = \sqrt{41}$

$$\sin \theta = \frac{\text{opp}}{\text{hyp}} = \frac{4}{\sqrt{41}} = \frac{4\sqrt{41}}{41} \qquad \csc \theta = \frac{\text{hyp}}{\text{opp}} = \frac{\sqrt{41}}{4}$$

$$\cos \theta = \frac{\text{adj}}{\text{hyp}} = \frac{5}{\sqrt{41}} = \frac{5\sqrt{41}}{41} \qquad \sec \theta = \frac{\text{hyp}}{\text{adj}} = \frac{\sqrt{41}}{5}$$

$$\tan \theta = \frac{\text{opp}}{\text{adj}} = \frac{4}{5} \qquad \cot \theta = \frac{\text{adj}}{\text{opp}} = \frac{5}{4}$$

42. $\text{adj} = 6$, $\text{opp} = 6$

$$\text{hyp} = \sqrt{6^2 + 6^2} = 6\sqrt{2}$$

$$\sin \theta = \frac{\text{opp}}{\text{hyp}} = \frac{6}{6\sqrt{2}} = \frac{\sqrt{2}}{2} \qquad \csc \theta = \frac{\text{hyp}}{\text{opp}} = \frac{6\sqrt{2}}{6} = \sqrt{2}$$

$$\cos \theta = \frac{\text{adj}}{\text{hyp}} = \frac{6}{6\sqrt{2}} = \frac{\sqrt{2}}{2} \qquad \sec \theta = \frac{\text{hyp}}{\text{adj}} = \frac{6\sqrt{2}}{6} = \sqrt{2}$$

$$\tan \theta = \frac{\text{opp}}{\text{adj}} = \frac{6}{6} = 1 \qquad \cot \theta = \frac{\text{adj}}{\text{opp}} = \frac{6}{6} = 1$$

43. $\text{adj} = 4$, $\text{hyp} = 8$, $\text{opp} = \sqrt{8^2 - 4^2} = \sqrt{48} = 4\sqrt{3}$

$$\sin \theta = \frac{\text{opp}}{\text{hyp}} = \frac{4\sqrt{3}}{8} = \frac{\sqrt{3}}{2} \qquad \csc \theta = \frac{\text{hyp}}{\text{opp}} = \frac{8}{4\sqrt{3}} = \frac{2\sqrt{3}}{3}$$

$$\cos \theta = \frac{\text{adj}}{\text{hyp}} = \frac{4}{8} = \frac{1}{2} \qquad \sec \theta = \frac{\text{hyp}}{\text{adj}} = \frac{8}{4} = 2$$

$$\tan \theta = \frac{\text{opp}}{\text{adj}} = \frac{4\sqrt{3}}{4} = \sqrt{3} \qquad \cot \theta = \frac{\text{adj}}{\text{opp}} = \frac{4}{4\sqrt{3}} = \frac{\sqrt{3}}{3}$$

44. opp = 5, hyp = 9

$$\text{adj} = \sqrt{9^2 - 5^2} = 2\sqrt{14}$$

$$\sin \theta = \frac{\text{opp}}{\text{hyp}} = \frac{5}{9}$$

$$\cos \theta = \frac{\text{adj}}{\text{hyp}} = \frac{2\sqrt{14}}{9}$$

$$\tan \theta = \frac{\text{opp}}{\text{adj}} = \frac{5}{2\sqrt{14}} = \frac{5\sqrt{14}}{28}$$

$$\csc \theta = \frac{\text{hyp}}{\text{opp}} = \frac{9}{5}$$

$$\sec \theta = \frac{\text{hyp}}{\text{adj}} = \frac{9}{2\sqrt{14}} = \frac{9\sqrt{14}}{28}$$

$$\cot \theta = \frac{\text{adj}}{\text{opp}} = \frac{2\sqrt{14}}{5}$$

45. $\sin \theta = \dfrac{1}{3}$

(a) $\csc \theta = \dfrac{1}{\sin \theta} = 3$

(b) $\sin^2 \theta + \cos^2 \theta = 1$

$$\left(\frac{1}{3}\right)^2 + \cos^2 \theta = 1$$

$$\cos^2 \theta = 1 - \frac{1}{9}$$

$$\cos^2 \theta = \frac{8}{9}$$

$$\cos \theta = \sqrt{\frac{8}{9}}$$

$$\cos \theta = \frac{2\sqrt{2}}{3}$$

(c) $\sec \theta = \dfrac{1}{\cos \theta} = \dfrac{3}{2\sqrt{2}} = \dfrac{3\sqrt{2}}{4}$

(d) $\tan \theta = \dfrac{\sin \theta}{\cos \theta} = \dfrac{1/3}{(2\sqrt{2})/3} = \dfrac{1}{2\sqrt{2}} = \dfrac{\sqrt{2}}{4}$

46. $\tan \theta = 4$

(a) $\cot \theta = \dfrac{1}{\tan \theta} = \dfrac{1}{4}$

(b) $\sec \theta = \sqrt{1 + \tan^2 \theta} = \sqrt{1 + 16} = \sqrt{17}$

(c) $\cos \theta = \dfrac{1}{\sec \theta} = \dfrac{1}{\sqrt{17}} = \dfrac{\sqrt{17}}{17}$

(d) $\csc \theta = \sqrt{1 + \cot^2 \theta} = \sqrt{1 + \dfrac{1}{16}} = \dfrac{\sqrt{17}}{4}$

47. $\csc \theta = 4$

(a) $\sin \theta = \dfrac{1}{\csc \theta} = \dfrac{1}{4}$

(b) $\sin^2 \theta + \cos^2 \theta = 1$

$$\left(\frac{1}{4}\right)^2 + \cos^2 \theta = 1$$

$$\cos^2 \theta = 1 - \frac{1}{16}$$

$$\cos^2 \theta = \frac{15}{16}$$

$$\cos \theta = \sqrt{\frac{15}{16}}$$

$$\cos \theta = \frac{\sqrt{15}}{4}$$

(c) $\sec \theta = \dfrac{1}{\cos \theta} = \dfrac{4}{\sqrt{15}} = \dfrac{4\sqrt{15}}{15}$

(d) $\tan \theta = \dfrac{\sin \theta}{\cos \theta} = \dfrac{1/4}{\sqrt{15}/4} = \dfrac{1}{\sqrt{15}} = \dfrac{\sqrt{15}}{15}$

48. $\csc \theta = 5$

(a) $\sin \theta = \dfrac{1}{\csc \theta} = \dfrac{1}{5}$

(b) $\cot \theta = \sqrt{\csc^2 \theta - 1} = \sqrt{25 - 1} = 2\sqrt{6}$

(c) $\tan \theta = \dfrac{1}{\cot \theta} = \dfrac{1}{2\sqrt{6}} = \dfrac{\sqrt{6}}{12}$

(d) $\sec(90° - \theta) = \csc \theta = 5$

49. $\tan 33° \approx 0.6494$

50. $\csc 11° = \dfrac{1}{\sin 11°} \approx 5.2408$

51. $\sin 34.2° \approx 0.5621$

52. $\sec 79.3° = \dfrac{1}{\cos 79.3°} \approx 5.3860$

53. $\cot 15° \, 14' = \dfrac{1}{\tan\left(15 + \frac{14}{60}\right)}$
≈ 3.6722

54. $\cos 78° \, 11' \, 58'' = \cos\left(78 + \dfrac{11}{60} + \dfrac{58}{3600}\right)°$
≈ 0.2045

55. $\sin 1° \, 10' = \dfrac{x}{3.5}$

$x = 3.5 \sin 1° \, 10' \approx 0.07$ kilometer or 71.3 meters

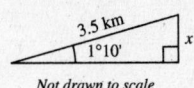

Not drawn to scale

56.

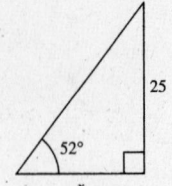

$\tan 52° = \dfrac{25}{x}$

$x = \dfrac{25}{\tan 52°} \approx 19.5$ feet

57. $x = 12, y = 16, r = \sqrt{144 + 256} = \sqrt{400} = 20$

$\sin \theta = \dfrac{y}{r} = \dfrac{4}{5} \qquad \csc \theta = \dfrac{r}{y} = \dfrac{5}{4}$

$\cos \theta = \dfrac{x}{r} = \dfrac{3}{5} \qquad \sec \theta = \dfrac{r}{x} = \dfrac{5}{3}$

$\tan \theta = \dfrac{y}{x} = \dfrac{4}{3} \qquad \cot \theta = \dfrac{x}{y} = \dfrac{3}{4}$

58. $(x, y) = (3, -4)$

$r = \sqrt{3^2 + (-4)^2} = 5$

$\sin \theta = \dfrac{y}{r} = -\dfrac{4}{5} \qquad \csc \theta = \dfrac{r}{y} = -\dfrac{5}{4}$

$\cos \theta = \dfrac{x}{r} = \dfrac{3}{5} \qquad \sec \theta = \dfrac{r}{x} = \dfrac{5}{3}$

$\tan \theta = \dfrac{y}{x} = -\dfrac{4}{3} \qquad \cot \theta = \dfrac{x}{y} = -\dfrac{3}{4}$

59. $x = \dfrac{2}{3}, y = \dfrac{5}{2}$

$r = \sqrt{\left(\dfrac{2}{3}\right)^2 + \left(\dfrac{5}{2}\right)^2} = \dfrac{\sqrt{241}}{6}$

$\sin \theta = \dfrac{y}{r} = \dfrac{5/2}{\sqrt{241}/6} = \dfrac{15}{\sqrt{241}} = \dfrac{15\sqrt{241}}{241} \qquad \csc \theta = \dfrac{r}{y} = \dfrac{\sqrt{241}/6}{5/2} = \dfrac{2\sqrt{241}}{30} = \dfrac{\sqrt{241}}{15}$

$\cos \theta = \dfrac{x}{r} = \dfrac{2/3}{\sqrt{241}/6} = \dfrac{4}{\sqrt{241}} = \dfrac{4\sqrt{241}}{241} \qquad \sec \theta = \dfrac{r}{x} = \dfrac{\sqrt{241}/6}{2/3} = \dfrac{\sqrt{241}}{4}$

$\tan \theta = \dfrac{y}{x} = \dfrac{5/2}{2/3} = \dfrac{15}{4} \qquad \cot \theta = \dfrac{x}{y} = \dfrac{2/3}{5/2} = \dfrac{4}{15}$

60. $(x, y) = \left(-\dfrac{10}{3}, -\dfrac{2}{3}\right)$

$r = \sqrt{\left(-\dfrac{10}{3}\right)^2 + \left(-\dfrac{2}{3}\right)^2} = \dfrac{2\sqrt{26}}{3}$

$\sin \theta = \dfrac{y}{r} = \dfrac{-2/3}{(2\sqrt{26})/3} = -\dfrac{\sqrt{26}}{26} \qquad \csc \theta = \dfrac{r}{y} = \dfrac{(2\sqrt{26})/3}{-2/3} = -\sqrt{26}$

$\cos \theta = \dfrac{x}{r} = \dfrac{-10/3}{(2\sqrt{26})/3} = -\dfrac{5\sqrt{26}}{26} \qquad \sec \theta = \dfrac{r}{x} = \dfrac{(2\sqrt{26})/3}{-10/3} = -\dfrac{\sqrt{26}}{5}$

$\tan \theta = \dfrac{y}{x} = \dfrac{-2/3}{-10/3} = \dfrac{1}{5} \qquad \cot \theta = \dfrac{x}{y} = \dfrac{-10/3}{-2/3} = 5$

61. $x = -0.5, y = 4.5$

$$r = \sqrt{(-0.5)^2 + (4.5)^2} = \sqrt{20.5} = \frac{\sqrt{82}}{2}$$

$$\sin \theta = \frac{y}{r} = \frac{4.5}{\sqrt{82}/2} = \frac{9\sqrt{82}}{82} \qquad \csc \theta = \frac{r}{y} = \frac{\sqrt{82}/2}{4.5} = \frac{\sqrt{82}}{9}$$

$$\cos \theta = \frac{x}{r} = \frac{-0.5}{\sqrt{82}/2} = \frac{-\sqrt{82}}{82} \qquad \sec \theta = \frac{r}{x} = \frac{\sqrt{82}/2}{-0.5} = -\sqrt{82}$$

$$\tan \theta = \frac{y}{x} = \frac{4.5}{-0.5} = -9 \qquad \cot \theta = \frac{x}{y} = \frac{-0.5}{4.5} = -\frac{1}{9}$$

62. $(x, y) = (0.3, 0.4)$

$$r = \sqrt{(0.3)^2 + (0.4)^2} = 0.5$$

$$\sin \theta = \frac{y}{r} = \frac{0.4}{0.5} = \frac{4}{5} = 0.8 \qquad \csc \theta = \frac{r}{y} = \frac{0.5}{0.4} = \frac{5}{4} = 1.25$$

$$\cos \theta = \frac{x}{r} = \frac{0.3}{0.5} = \frac{3}{5} = 0.6 \qquad \sec \theta = \frac{r}{x} = \frac{0.5}{0.3} = \frac{5}{3} \approx 1.67$$

$$\tan \theta = \frac{y}{x} = \frac{0.4}{0.3} = \frac{4}{3} \approx 1.33 \qquad \cot \theta = \frac{x}{y} = \frac{0.3}{0.4} = \frac{3}{4} = 0.75$$

63. $(x, 4x), \; x > 0$

$$x' = x, y' = 4x$$

$$r = \sqrt{x^2 + (4x)^2} = \sqrt{17}x$$

$$\sin \theta = \frac{y'}{r} = \frac{4x}{\sqrt{17}x} = \frac{4\sqrt{17}}{17} \qquad \csc \theta = \frac{r}{y'} = \frac{\sqrt{17}x}{4x} = \frac{\sqrt{17}}{4}$$

$$\cos \theta = \frac{x'}{r} = \frac{x}{\sqrt{17}x} = \frac{\sqrt{17}}{17} \qquad \sec \theta = \frac{r}{x'} = \frac{\sqrt{17}x}{x} = \sqrt{17}$$

$$\tan \theta = \frac{y'}{x'} = \frac{4x}{x} = 4 \qquad \cot \theta = \frac{x'}{y'} = \frac{x}{4x} = \frac{1}{4}$$

64. $(x', y') = (-2x, -3x), x > 0$

$$r = \sqrt{(-2x)^2 + (-3x)^2} = \sqrt{13}x$$

$$\sin \theta = \frac{y'}{r} = \frac{-3x}{\sqrt{13}x} = -\frac{3\sqrt{13}}{13}$$

$$\cos \theta = \frac{x'}{r} = \frac{-2x}{\sqrt{13}x} = -\frac{2\sqrt{13}}{13}$$

$$\tan \theta = \frac{y'}{x'} = \frac{-3x}{-2x} = \frac{3}{2}$$

$$\csc \theta = \frac{r}{y'} = \frac{\sqrt{13}x}{-3x} = -\frac{\sqrt{13}}{3}$$

$$\sec \theta = \frac{r}{x'} = \frac{\sqrt{13}x}{-2x} = -\frac{\sqrt{13}}{2}$$

$$\cot \theta = \frac{x'}{y'} = \frac{-2x}{-3x} = \frac{2}{3}$$

65. $\sec \theta = \frac{6}{5}, \; \tan \theta < 0 \implies \theta$ is in Quadrant IV.

$$r = 6, x = 5, y = -\sqrt{36 - 25} = -\sqrt{11}$$

$$\sin \theta = \frac{y}{r} = -\frac{\sqrt{11}}{6}$$

$$\cos \theta = \frac{x}{r} = \frac{5}{6}$$

$$\tan \theta = \frac{y}{x} = -\frac{\sqrt{11}}{5}$$

$$\csc \theta = \frac{r}{y} = -\frac{6\sqrt{11}}{11}$$

$$\sec \theta = \frac{6}{5}$$

$$\cot \theta = -\frac{5\sqrt{11}}{11}$$

66. $\csc \theta = \dfrac{3}{2}$, $\cos \theta < 0$

θ is in Quadrant II.

$\sin \theta = \dfrac{1}{\csc \theta} = \dfrac{2}{3}$

$\cos \theta = -\sqrt{1 - \sin^2 \theta} = -\dfrac{\sqrt{5}}{3}$

$\tan \theta = \dfrac{\sin \theta}{\cos \theta} = -\dfrac{2\sqrt{5}}{5}$

$\sec \theta = \dfrac{1}{\cos \theta} = -\dfrac{3\sqrt{5}}{5}$

$\cot \theta = \dfrac{1}{\tan \theta} = -\dfrac{\sqrt{5}}{2}$

67. $\sin \theta = \dfrac{3}{8}$, $\cos \theta < 0 \implies \theta$ is in Quadrant II.

$y = 3, r = 8, x = -\sqrt{55}$

$\sin \theta = \dfrac{y}{r} = \dfrac{3}{8}$

$\cos \theta = \dfrac{x}{r} = -\dfrac{\sqrt{55}}{8}$

$\tan \theta = \dfrac{y}{x} = -\dfrac{3}{\sqrt{55}} = -\dfrac{3\sqrt{55}}{55}$

$\csc \theta = \dfrac{8}{3}$

$\sec \theta = -\dfrac{8}{\sqrt{55}} = -\dfrac{8\sqrt{55}}{55}$

$\cot \theta = -\dfrac{\sqrt{55}}{3}$

68. $\tan \theta = \dfrac{5}{4}$, $\cos \theta < 0$

θ is in Quadrant III.

$\sec \theta = -\sqrt{1 + \tan^2 \theta} = -\sqrt{1 + \left(\dfrac{25}{16}\right)} = -\dfrac{\sqrt{41}}{4}$

$\cos \theta = \dfrac{1}{\sec \theta} = -\dfrac{4\sqrt{41}}{41}$

$\sin \theta = -\sqrt{1 - \cos^2 \theta} = -\sqrt{1 - \dfrac{16}{41}} = -\dfrac{5\sqrt{41}}{41}$

$\csc \theta = \dfrac{1}{\sin \theta} = -\dfrac{\sqrt{41}}{5}$

$\cot \theta = \dfrac{1}{\tan \theta} = \dfrac{4}{5}$

69. $\cos \theta = \dfrac{x}{r} = \dfrac{-2}{5} \implies y^2 = 21$

$\sin \theta > 0 \implies \theta$ is in Quadrant II $\implies y = \sqrt{21}$

$\sin \theta = \dfrac{y}{r} = \dfrac{\sqrt{21}}{5}$

$\tan \theta = \dfrac{y}{x} = -\dfrac{\sqrt{21}}{2}$

$\csc \theta = \dfrac{r}{y} = \dfrac{5}{\sqrt{21}} = \dfrac{5\sqrt{21}}{21}$

$\sec \theta = \dfrac{r}{x} = \dfrac{5}{-2} = -\dfrac{5}{2}$

$\cot \theta = \dfrac{x}{y} = \dfrac{-2}{\sqrt{21}} = -\dfrac{2\sqrt{21}}{21}$

70. $\sin \theta = -\dfrac{2}{4} = -\dfrac{1}{2}$, $\cos \theta > 0$

θ is in Quadrant IV.

$\csc \theta = \dfrac{1}{\sin \theta} = -2$

$\cos \theta = \sqrt{1 - \sin^2 \theta} = \sqrt{1 - \left(\dfrac{1}{4}\right)} = \dfrac{\sqrt{3}}{2}$

$\sec \theta = \dfrac{1}{\cos \theta} = \dfrac{2\sqrt{3}}{3}$

$\tan \theta = \dfrac{\sin \theta}{\cos \theta} = -\dfrac{\sqrt{3}}{3}$

$\cot \theta = \dfrac{1}{\tan \theta} = -\sqrt{3}$

71. $\theta = 264°$

$\theta' = 264° - 180° = 84°$

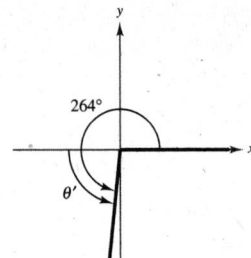

72. $\theta = 635° = 720° - 85°$

$\theta' = 85°$

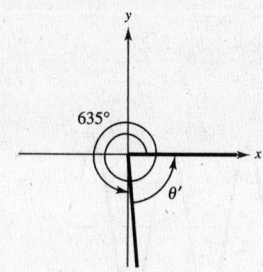

73. $\theta = -\dfrac{6\pi}{5}$

$-\dfrac{6\pi}{5} + 2\pi = \dfrac{4\pi}{5}$

$\theta' = \pi - \dfrac{4\pi}{5} = \dfrac{\pi}{5}$

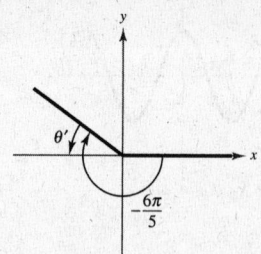

74. $\theta = \dfrac{17\pi}{3} = \dfrac{18\pi}{3} - \dfrac{\pi}{3}$

$= 6\pi - \dfrac{\pi}{3}$

$\theta' = \dfrac{\pi}{3}$

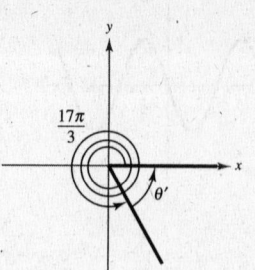

75. $\sin\dfrac{\pi}{3} = \dfrac{\sqrt{3}}{2}$

$\cos\dfrac{\pi}{3} = \dfrac{1}{2}$

$\tan\dfrac{\pi}{3} = \sqrt{3}$

76. $\sin\dfrac{\pi}{4} = \dfrac{\sqrt{2}}{2}$

$\cos\dfrac{\pi}{4} = \dfrac{\sqrt{2}}{2}$

$\tan\dfrac{\pi}{4} = \dfrac{2}{\sqrt{2}/2} = 1$

77. $\sin\left(-\dfrac{7\pi}{3}\right) = -\sin\dfrac{\pi}{3} = -\dfrac{\sqrt{3}}{2}$

$\cos\left(-\dfrac{7\pi}{3}\right) = \cos\dfrac{\pi}{3} = \dfrac{1}{2}$

$\tan\left(-\dfrac{7\pi}{3}\right) = -\tan\dfrac{\pi}{3} = -\sqrt{3}$

78. $\sin\left(-\dfrac{5\pi}{4}\right) = \sin\dfrac{\pi}{4} = \dfrac{\sqrt{2}}{2}$

$\cos\left(-\dfrac{5\pi}{4}\right) = -\cos\dfrac{\pi}{4} = -\dfrac{\sqrt{2}}{2}$

$\tan\left(-\dfrac{5\pi}{4}\right) = -\tan\dfrac{\pi}{4}$

$= \dfrac{2}{-\sqrt{2}/2} = -1$

79. $\sin 495° = \sin 45° = \dfrac{\sqrt{2}}{2}$

$\cos 495° = -\cos 45° = -\dfrac{\sqrt{2}}{2}$

$\tan 495° = -\tan 45° = -1$

80. $\sin(-150°) = -\dfrac{1}{2}$

$\cos(-150°) = -\dfrac{\sqrt{3}}{2}$

$\tan(-150°) = \dfrac{-1/2}{-\sqrt{3}/2} = \dfrac{\sqrt{3}}{3}$

81. $\sin(-240°) = \sin 60° = \dfrac{\sqrt{3}}{2}$

$\cos(-240°) = -\cos 60° = -\dfrac{1}{2}$

$\tan(-240°) = -\tan 60° = -\sqrt{3}$

82. $\sin(315°) = -\dfrac{\sqrt{2}}{2}$

$\cos(315°) = \dfrac{\sqrt{2}}{2}$

$\tan(315°) = \dfrac{-\sqrt{2}/2}{\sqrt{2}/2} = -1$

83. $\sin 4 \approx -0.7568$

84. $\tan 3 \approx -0.1425$

85. $\sin(-3.2) \approx 0.0584$

86. $\cot(-4.8) = \dfrac{1}{\tan(-4.8)} \approx 0.0878$

87. $\sec\left(\dfrac{12\pi}{5}\right) = \dfrac{1}{\cos\left(\dfrac{12\pi}{5}\right)} \approx 3.2361$

88. $\tan\left(\dfrac{-25\pi}{7}\right) \approx 4.3813$

89. $y = \sin x$

Amplitude: 1

Period: 2π

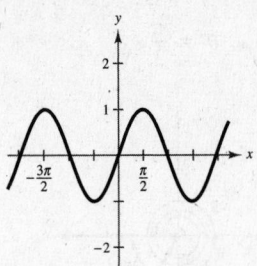

90. $y = \cos x$

Amplitude: 1

Period: 2π

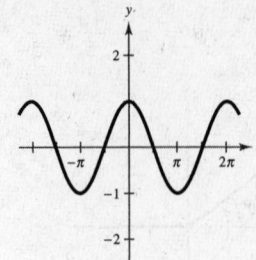

91. $f(x) = 5 \sin \dfrac{2x}{5}$

Amplitude: 5

Period: $\dfrac{2\pi}{2/5} = 5\pi$

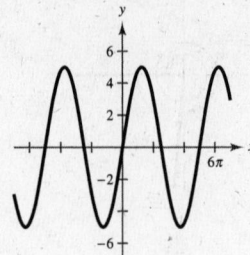

92. $f(x) = 8 \cos\left(-\dfrac{x}{4}\right)$

Amplitude: 8

Period: $\dfrac{2\pi}{1/4} = 8\pi$

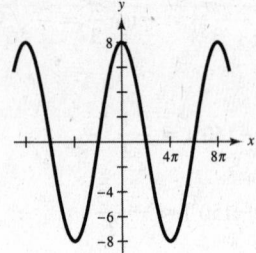

93. $y = 2 + \sin x$

Shift the graph of $y = \sin x$ two units upward.

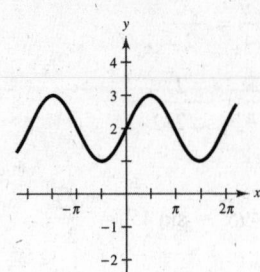

94. $y = -4 - \cos \pi x$

Amplitude: 1

Period: $\dfrac{2\pi}{\pi} = 2$

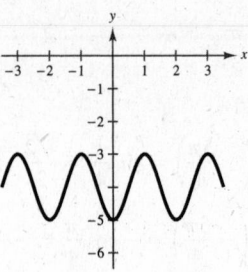

95. $g(t) = \dfrac{5}{2} \sin(t - \pi)$

Amplitude: $\dfrac{5}{2}$

Period: 2π

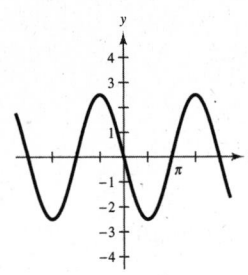

96. $g(t) = 3 \cos(t + \pi)$

Amplitude: 3

Period: 2π

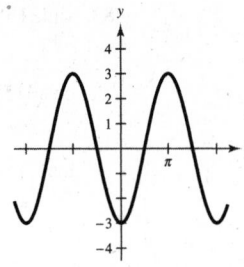

97. $y = a \sin bx$

(a) $a = 2$,

$\dfrac{2\pi}{b} = \dfrac{1}{264} \implies b = 528\pi$

$y = 2 \sin(528\pi x)$

(b) $f = \dfrac{1}{1/264}$

$= 264$ cycles per second.

98. (a) $S(t) = 18.09 + 1.41 \sin\left(\dfrac{\pi t}{6} + 4.60\right)$

(b) Period $= \dfrac{2\pi}{\pi/6} = (2)(6) = 12$

12 months = 1 year, so this is expected.

(c) Amplitude: 1.41

The amplitude represents the maximum change in the time of sunset from the average time ($d = 18.09$).

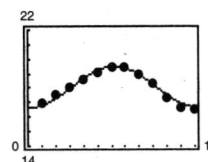

99. $f(x) = \tan x$

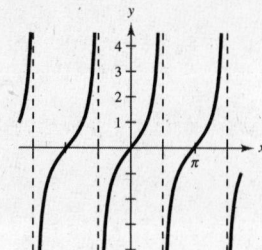

100. $f(t) = \tan\left(t - \dfrac{\pi}{4}\right)$

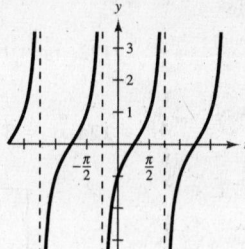

101. $f(x) = \cot x$

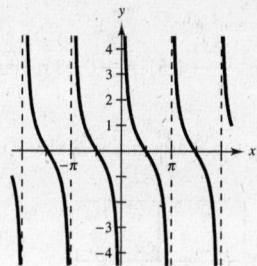

102. $g(t) = 2 \cot 2t$

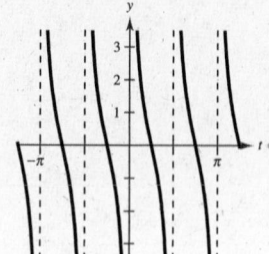

103. $f(x) = \sec x$

Graph $y = \cos x$ first.

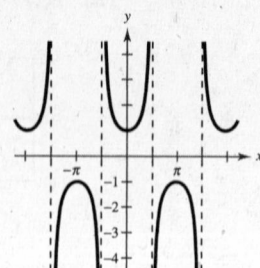

104. $h(t) = \sec\left(t - \dfrac{\pi}{4}\right)$

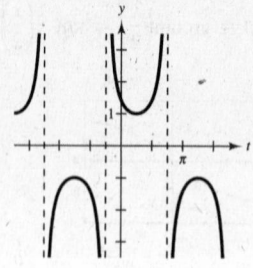

105. $f(x) = \csc x$

Graph $y = \sin x$ first.

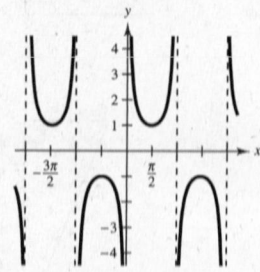

106. $f(t) = 3 \csc\left(2t + \dfrac{\pi}{4}\right)$

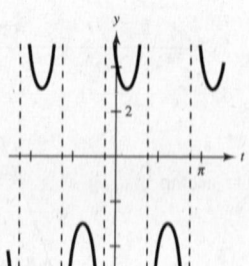

107. $f(x) = x \cos x$

Graph $y = x$ and $y = -x$ first.

As $x \to \infty$, $f(x) \to \infty$.

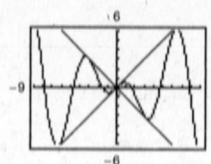

108. $g(x) = x^4 \cos x$

Damping factor: x^4

As $x \to \infty$, $f(x) \to \infty$.

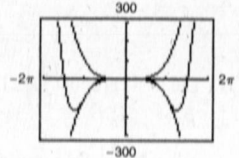

109. $\arcsin\left(-\dfrac{1}{2}\right) = -\arcsin\dfrac{1}{2} = -\dfrac{\pi}{6}$

110. $\arcsin(-1) = -\dfrac{\pi}{2}$

111. $\arcsin 0.4 \approx 0.41$ radian

112. $\arcsin(0.213) \approx 0.21$ radian

113. $\sin^{-1}(-0.44) \approx -0.46$ radian

114. $\sin^{-1}(0.89) \approx 1.10$ radians

115. $\arccos\dfrac{\sqrt{3}}{2} = \dfrac{\pi}{6}$

116. $\arccos\left(\dfrac{\sqrt{2}}{2}\right) = \dfrac{\pi}{4}$

117. $\cos^{-1}(-1) = \pi$

118. $\cos^{-1}\left(\dfrac{\sqrt{3}}{2}\right) = \dfrac{\pi}{6}$

119. arccos $0.324 \approx 1.24$ radians

120. arccos$(-0.888) \approx 2.66$ radians

121. $\tan^{-1}(-1.5) \approx -0.98$ radian

122. $\tan^{-1}(8.2) \approx 1.45$ radians

123. $f(x) = 2 \arcsin x = 2 \sin^{-1}(x)$

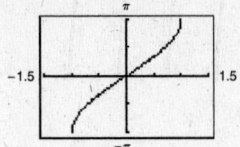

124. $y = 3 \arccos x$

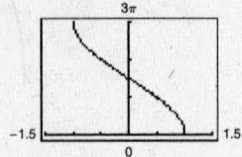

125. $f(x) = \arctan\left(\dfrac{x}{2}\right) = \tan^{-1}\left(\dfrac{x}{2}\right)$

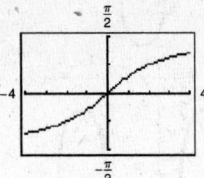

126. $f(x) = -\arcsin 2x$

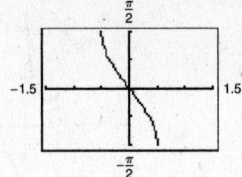

127. $\cos\left(\arctan \dfrac{3}{4}\right) = \dfrac{4}{5}$

Use a right triangle. Let
$\theta = \arctan \dfrac{3}{4}$ then $\tan \theta = \dfrac{3}{4}$
and $\cos \theta = \dfrac{4}{5}$.

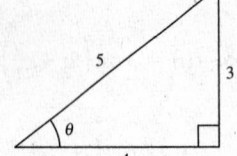

128. Let $u = \arccos \dfrac{3}{5}$.

$\tan\left(\arccos \dfrac{3}{5}\right) = \tan u = \dfrac{4}{3}$

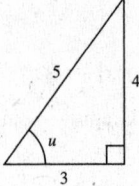

129. $\sec\left(\arctan \dfrac{12}{5}\right) = \dfrac{13}{5}$

Use a right triangle. Let $\theta = \arctan \dfrac{12}{5}$
then $\tan \theta = \dfrac{12}{5}$ and $\sec \theta = \dfrac{13}{5}$.

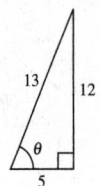

130. Let $u = \arcsin\left(-\dfrac{12}{13}\right)$.

$\cot\left[\arcsin\left(-\dfrac{12}{13}\right)\right] = \cot u = -\dfrac{5}{12}$

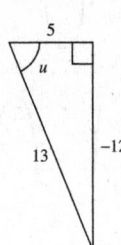

131. Let $y = \arccos\left(\dfrac{x}{2}\right)$. Then

$\cos y = \dfrac{x}{2}$ and $\tan y = \tan\left(\arccos\left(\dfrac{x}{2}\right)\right) = \dfrac{\sqrt{4 - x^2}}{x}$.

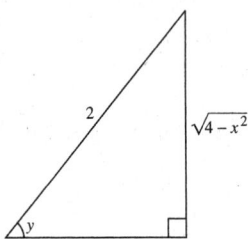

132. $\sec(\arcsin(x - 1))$

$\theta = \arcsin(x - 1) \implies -\dfrac{\pi}{2} \le \theta \le \dfrac{\pi}{2}$

$\sin \theta = x - 1$

$\cos \theta = \sqrt{1^2 - (x - 1)^2} = \sqrt{x(2 - x)}$

$\sec \theta = \dfrac{1}{\sqrt{x(2 - x)}}$

133. $\tan \theta = \dfrac{70}{30}$

$\theta = \arctan\left(\dfrac{70}{30}\right) \approx 66.8°$

134. $\tan 21° = \dfrac{h}{25}$

$h = 25 \tan 21° \approx 9.6$ feet

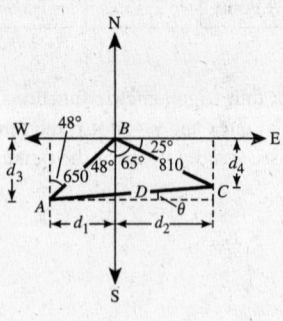

135. $\sin 48° = \dfrac{d_1}{650} \Rightarrow d_1 \approx 483$
$\cos 25° = \dfrac{d_2}{810} \Rightarrow d_2 \approx 734$ $\Big\}$ $d_1 + d_2 \approx 1217$

$\cos 48° = \dfrac{d_3}{650} \Rightarrow d_3 \approx 435$
$\sin 25° = \dfrac{d_4}{810} \Rightarrow d_4 \approx 342$ $\Big\}$ $d_3 - d_4 \approx 93$

$\tan \theta \approx \dfrac{93}{1217} \Rightarrow \theta \approx 4.4°$

$\sec 4.4° \approx \dfrac{D}{1217} \Rightarrow D \approx 1217 \sec 4.4° \approx 1221$

The distance is 1221 miles and the bearing is 85.6°.

136. Amplitude: $\dfrac{1.5}{2} = 0.75$ inches

Period: 3 seconds

$d = a \cos bt$

$a = 0.75$

$b = \dfrac{2\pi}{3}$

$d = 0.75 \cos\left(\dfrac{2\pi t}{3}\right)$

137. False. The sine or cosine functions are often useful for modeling simple harmonic motion.

138. True. The inverse sine, $y = \arcsin x$, is defined where $-1 \le x \le 1$ and $-\dfrac{\pi}{2} \le y \le \dfrac{\pi}{2}$.

139. False. For each θ there corresponds exactly one value of y.

140. False. The range of arctan is $\left(-\dfrac{\pi}{2}, \dfrac{\pi}{2}\right)$, so $\arctan(-1) = -\dfrac{\pi}{4}$.

141. $y = 3 \sin x$

Amplitude: 3

Period: 2π

Matches graph (d)

142. $y = -3 \sin x$ matches graph (a).

Period: 2π

Amplitude: 3

143. $y = 2 \sin \pi x$

Amplitude: 2

Period: 2

Matches graph (b)

144. $y = 2 \sin \dfrac{x}{2}$ matches graph (c).

Period: 4π

Amplitude: 2

145. $f(\theta) = \sec \theta$ is undefined at the zeros of $g(\theta) = \cos \theta$ since $\sec \theta = \dfrac{1}{\cos \theta}$.

146. (a)

θ	0.1	0.4	0.7	1.0	1.3
$\tan\left(\theta - \dfrac{\pi}{2}\right)$	-9.9666	-2.3652	-1.1872	-0.6421	-0.2776
$-\cot\theta$	-9.9666	-2.3652	-1.1872	-0.6421	-0.2776

(b) $\tan\left(\theta - \dfrac{\pi}{2}\right) = -\cot\theta$

147. The ranges for the other four trigonometric functions are not bounded. For $y = \tan x$ and $y = \cot x$, the range is $(-\infty, \infty)$. For $y = \sec x$ and $y = \csc x$, the range is $(-\infty, -1] \cup [1, \infty)$.

148. $y = Ae^{-kt} \cos bt = \frac{1}{5} e^{-t/10} \cos 6t$

(a) A is changed from $\frac{1}{5}$ to $\frac{1}{3}$: The displacement is increased.

(b) k is changed from $\frac{1}{10}$ to $\frac{1}{3}$: The friction damps the oscillations more rapidly.

(c) b is changed from 6 to 9: The frequency of oscillation is increased.

149. $A = \frac{1}{2} r^2 \theta, \ s = r\theta$

(a) $A = \frac{1}{2} r^2 (0.8) = 0.4 r^2, \ r > 0$

$s = r(0.8) = 0.8r, \ r > 0$

As r increases, the area function increases more rapidly.

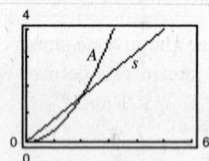

(b) $A = \frac{1}{2}(10)^2 \theta = 50\theta, \ \theta > 0$

$s = 10\theta, \ \theta > 0$

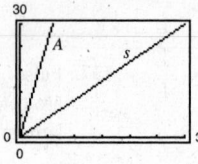

150. Answers will vary.

Problem Solving for Chapter 4

1. (a) $8:57 - 6:45 = 2$ hours 12 minutes $= 132$ minutes

$\dfrac{132}{48} = \dfrac{11}{4}$ revolutions

$\theta = \left(\dfrac{11}{4}\right)(2\pi) = \dfrac{11\pi}{2}$ radians or $990°$

(b) $s = r\theta = 47.25(5.5\pi) \approx 816.42$ feet

2. Gear 1: $\dfrac{24}{32}(360°) = 270° = \dfrac{3\pi}{2}$ radians

Gear 2: $\dfrac{24}{26}(360°) \approx 332.308° \approx 5.80$ radians

Gear 3: $\dfrac{24}{22}(360°) \approx 392.727° \approx 6.85$ radians

Gear 4: $\dfrac{40}{32}(360°) = 450° = \dfrac{5\pi}{2}$ radians

Gear 5: $\dfrac{24}{19}(360°) \approx 454.737° \approx 7.94$ radians

3. (a) $\sin 39° = \dfrac{3000}{d}$

$d = \dfrac{3000}{\sin 39°} \approx 4767$ feet

(b) $\tan 39° = \dfrac{3000}{x}$

$x = \dfrac{3000}{\tan 39°} \approx 3705$ feet

(c) $\tan 63° = \dfrac{w + 3705}{3000}$

$3000 \tan 63° = w + 3705$

$w = 3000 \tan 63° - 3705 \approx 2183$ feet

4. (a) $\triangle ABC$, $\triangle ADE$, and $\triangle AFG$ are all similar triangles since they all have the same angles. $\angle A$ is part of all three triangles and $\angle C = \angle E = \angle G = 90°$. Thus, $\angle B = \angle D = \angle F$.

(b) Since the triangles are similar, the ratios of corresponding sides are equal.

$$\frac{BC}{AB} = \frac{DE}{AD} = \frac{FG}{AF}$$

(c) Since the ratios: $\dfrac{\text{opp}}{\text{hyp}} = \dfrac{BC}{AB} = \dfrac{DE}{AD} = \dfrac{FG}{AF} = \sin A$ it does not matter which triangle is used to calculate $\sin A$.

Any triangle similar to these three triangles could be used to find $\sin A$. The value of $\sin A$ would not change.

(d) Since the values of all six trigonometric functions can be found by taking the ratios of the sides of a right triangle, similar triangles would yield the same values.

5. (a) $h(x) = \cos^2 x$

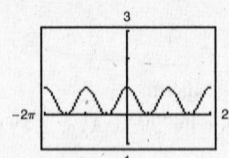

h is even.

(b) $h(x) = \sin^2 x$

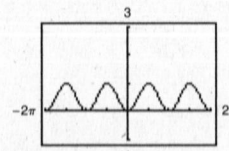

h is even.

6. Given: f is an even function and g is an odd function.

(a) $h(x) = [f(x)]^2$

$h(-x) = [f(-x)]^2$

$\qquad = [f(x)]^2$ since f is even

$\qquad = h(x)$

Thus, h is an even function.

(b) $h(x) = [g(x)]^2$

$h(-x) = [g(-x)]^2$

$\qquad = [-g(x)]^2$ since g is odd

$\qquad = [g(x)]^2$

$\qquad = h(x)$

Thus, h is an even function.

Conjecture: The square of either an even function or an odd function is an even function.

7. If we alter the model so that $h = 1$ when $t = 0$, we can use either a sine or a cosine model.

$a = \frac{1}{2}[\text{max} - \text{min}] = \frac{1}{2}[101 - 1] = 50$

$d = \frac{1}{2}[\text{max} + \text{min}] = \frac{1}{2}[101 + 1] = 51$

$b = 8\pi$

For the cosine model we have: $h = 51 - 50\cos(8\pi t)$

For the sine model we have: $h = 51 - 50\sin\left(8\pi t + \dfrac{\pi}{2}\right)$

Notice that we needed the horizontal shift so that the sine value was one when $t = 0$.

Another model would be: $h = 51 + 50\sin\left(8\pi t + \dfrac{3\pi}{2}\right)$

Here we wanted the sine value to be 1 when $t = 0$.

8. $P = 100 - 20\cos\left(\dfrac{8\pi}{3}t\right)$

(a)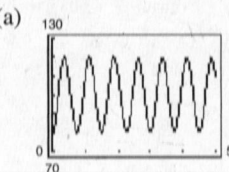

(b) Period $= \dfrac{2\pi}{8\pi/3} = \dfrac{6}{8} = \dfrac{3}{4}$ sec

This is the time between heartbeats.

(c) Amplitude: 20

The blood pressure ranges between $100 - 20 = 80$ and $100 + 20 = 120$.

(d) Pulse rate $= \dfrac{60 \text{ sec/min}}{\frac{3}{4} \text{ sec/beat}} = 80$ beats/min

(e) Period $= \dfrac{60}{64} = \dfrac{15}{16}$ sec

$64 = \dfrac{60}{2\pi/b} \implies b = \dfrac{64}{60} \cdot 2\pi = \dfrac{32}{15}\pi$

9. Physical (23 days): $P = \sin \dfrac{2\pi t}{23}, t \geq 0$

Emotional (28 days): $E = \sin \dfrac{2\pi t}{28}, t \geq 0$

Intellectual (33 days): $I = \sin \dfrac{2\pi t}{33}, t \geq 0$

(a)

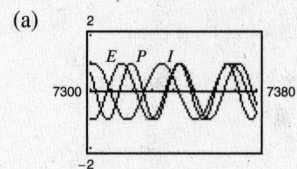

(b) Number of days since birth until September 1, 2006:

$$t = \underbrace{365 \times 20}_{20 \text{ years}} + \underbrace{5}_{\text{leap years}} + \underbrace{11}_{\substack{\text{remaining} \\ \text{July days}}} + \underbrace{31}_{\text{August days}} + \underbrace{1}_{\substack{\text{day in} \\ \text{September}}}$$

$t = 7348$

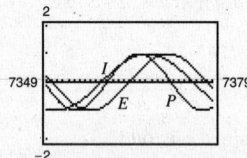

All three drop early in the month, then peak toward the middle of the month, and drop again toward the latter part of the month.

(c) For September 22, 2006, use $t = 7369$.

$P \approx 0.631$

$E \approx 0.901$

$I \approx 0.945$

10. $f(x) = 2 \cos 2x + 3 \sin 3x$

$g(x) = 2 \cos 2x + 3 \sin 4x$

(a)

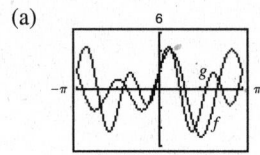

(b) The period of $f(x)$ is 2π.

The period of $g(x)$ is π.

(c) $h(x) = A \cos \alpha x + B \sin \beta x$ is periodic since the sine and cosine functions are periodic.

11. (a) Both graphs have a period of 2 and intersect when $x = 5.35$. They should also intersect when $x = 5.35 - 2 = 3.35$ and $x = 5.35 + 2 = 7.35$.

(b) The graphs intersect when $x = 5.35 - 3(2) = -0.65$.

(c) Since $13.35 = 5.35 + 4(2)$ and $-4.65 = 5.35 - 5(2)$ the graphs will intersect again at these values. Therefore $f(13.35) = g(-4.65)$.

12. (a) $f(t - 2c) = f(t)$ is true since this is a two period horizontal shift.

(b) $f\left(t + \frac{1}{2}c\right) = f\left(\frac{1}{2}t\right)$ is not true.

$f\left(t + \frac{1}{2}c\right)$ is a horizontal translation of $f(t)$.

$f\left(\frac{1}{2}t\right)$ is a doubling of the period of $f(t)$.

(c) $f\left(\frac{1}{2}(t + c)\right) = f\left(\frac{1}{2}t\right)$ is not true.

$f\left(\frac{1}{2}(t + c)\right) = f\left(\frac{1}{2}t + \frac{1}{2}c\right)$ is a horizontal

translation of $f\left(\frac{1}{2}t\right)$ by half a period.

For example, $\sin\left[\frac{1}{2}(\pi + 2\pi)\right] \neq \sin\left(\frac{1}{2}\pi\right)$.

13.

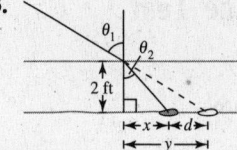

(a) $\dfrac{\sin \theta_1}{\sin \theta_2} = 1.333$

$\sin \theta_2 = \dfrac{\sin \theta_1}{1.333} = \dfrac{\sin 60°}{1.333} \approx 0.6497$

$\theta_2 \approx 40.52°$

(b) $\tan \theta_2 = \dfrac{x}{2} \implies x = 2 \tan 40.52° \approx 1.71$ feet

$\tan \theta_1 = \dfrac{y}{2} \implies y = 2 \tan 60° \approx 3.46$ feet

(c) $d = y - x = 3.46 - 1.71 = 1.75$ feet

(d) As you more closer to the rock, θ_1 decreases, which causes y to decrease, which in turn causes d to decrease.

14. $\arctan x \approx x - \dfrac{x^3}{3} + \dfrac{x^5}{5} - \dfrac{x^7}{7}$

(a)

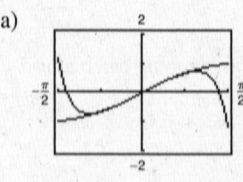

The graphs are nearly the same for $-1 < x < 1$.

(b)

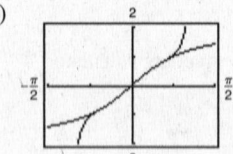

The accuracy of the approximation improved slightly by adding the next term $(x^9/9)$.

Chapter 4   Practice Test

1. Express 350° in radian measure.

2. Express $(5\pi)/9$ in degree measure.

3. Convert 135° 14′ 12″ to decimal form.

4. Convert −22.569° to D° M′ S″ form.

5. If $\cos \theta = \frac{2}{3}$, use the trigonometric identities to find $\tan \theta$.

6. Find θ given $\sin \theta = 0.9063$.

7. Solve for x in the figure below.

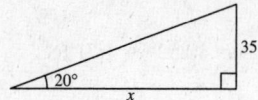

8. Find the reference angle θ' for $\theta = (6\pi)/5$.

9. Evaluate csc 3.92.

10. Find $\sec \theta$ given that θ lies in Quadrant III and $\tan \theta = 6$.

11. Graph $y = 3 \sin \dfrac{x}{2}$.

12. Graph $y = -2 \cos(x - \pi)$.

13. Graph $y = \tan 2x$.

14. Graph $y = -\csc\left(x + \dfrac{\pi}{4}\right)$.

15. Graph $y = 2x + \sin x$, using a graphing calculator.

16. Graph $y = 3x \cos x$, using a graphing calculator.

17. Evaluate arcsin 1.

18. Evaluate $\arctan(-3)$.

19. Evaluate $\sin\left(\arccos \dfrac{4}{\sqrt{35}}\right)$.

20. Write an algebraic expression for $\cos\left(\arcsin \dfrac{x}{4}\right)$.

For Exercises 21–23, solve the right triangle.

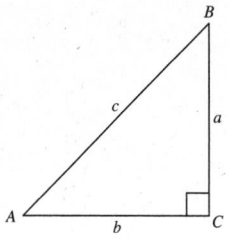

21. $A = 40°$, $c = 12$

22. $B = 6.84°$, $a = 21.3$

23. $a = 5$, $b = 9$

24. A 20-foot ladder leans against the side of a barn. Find the height of the top of the ladder if the angle of elevation of the ladder is 67°.

25. An observer in a lighthouse 250 feet above sea level spots a ship off the shore. If the angle of depression to the ship is 5°, how far out is the ship?

CHAPTER 5
Analytic Trigonometry

CHAPTER 5
Analytic Trigonometry

Section 5.1   Using Fundamental Identities

■ You should know the fundamental trigonometric identities.

(a) Reciprocal Identities

$$\sin u = \frac{1}{\csc u} \qquad\qquad \csc u = \frac{1}{\sin u}$$

$$\cos u = \frac{1}{\sec u} \qquad\qquad \sec u = \frac{1}{\cos u}$$

$$\tan u = \frac{1}{\cot u} = \frac{\sin u}{\cos u} \qquad\qquad \cot u = \frac{1}{\tan u} = \frac{\cos u}{\sin u}$$

(b) Pythagorean Identities

$$\sin^2 u + \cos^2 u = 1$$
$$1 + \tan^2 u = \sec^2 u$$
$$1 + \cot^2 u = \csc^2 u$$

(c) Cofunction Identities

$$\sin\left(\frac{\pi}{2} - u\right) = \cos u \qquad\qquad \cos\left(\frac{\pi}{2} - u\right) = \sin u$$

$$\tan\left(\frac{\pi}{2} - u\right) = \cot u \qquad\qquad \cot\left(\frac{\pi}{2} - u\right) = \tan u$$

$$\sec\left(\frac{\pi}{2} - u\right) = \csc u \qquad\qquad \csc\left(\frac{\pi}{2} - u\right) = \sec u$$

(d) Even/Odd Identities

$$\sin(-x) = -\sin x \qquad\qquad \csc(-x) = -\csc x$$
$$\cos(-x) = \cos x \qquad\qquad \sec(-x) = \sec x$$
$$\tan(-x) = -\tan x \qquad\qquad \cot(-x) = -\cot x$$

■ You should be able to use these fundamental identities to find function values.

■ You should be able to convert trigonometric expressions to equivalent forms by using the fundamental identities.

Vocabulary Check

1. $\tan u$ 2. $\cos u$ 3. $\cot u$

4. $\csc u$ 5. $\cot^2 u$ 6. $\sec^2 u$

7. $\cos u$ 8. $\csc u$ 9. $\cos u$

10. $-\tan u$

---

1. $\sin x = \dfrac{\sqrt{3}}{2}$, $\cos x = -\dfrac{1}{2}$ $\Longrightarrow$ x is in Quadrant II.

$\tan x = \dfrac{\sin x}{\cos x} = \dfrac{\sqrt{3}/2}{-1/2} = -\sqrt{3}$

$\cot x = \dfrac{1}{\tan x} = -\dfrac{1}{\sqrt{3}} = -\dfrac{\sqrt{3}}{3}$

$\sec x = \dfrac{1}{\cos x} = \dfrac{1}{-1/2} = -2$

$\csc x = \dfrac{1}{\sin x} = \dfrac{1}{\sqrt{3}/2} = \dfrac{2}{\sqrt{3}} = \dfrac{2\sqrt{3}}{3}$

2. $\tan x = \dfrac{\sqrt{3}}{3}$, $\cos x = -\dfrac{\sqrt{3}}{2}$

x is in Quadrant III.

$\sin x = -\sqrt{1 - \left(-\dfrac{\sqrt{3}}{2}\right)^2} = -\sqrt{\dfrac{1}{4}} = -\dfrac{1}{2}$

$\csc x = \dfrac{1}{\sin x} = -2$

$\sec x = \dfrac{1}{\cos x} = -\dfrac{2}{\sqrt{3}} = -\dfrac{2\sqrt{3}}{3}$

$\cot x = \dfrac{1}{\tan x} = \dfrac{3}{\sqrt{3}} = \sqrt{3}$

3. $\sec \theta = \sqrt{2}$, $\sin \theta = -\dfrac{\sqrt{2}}{2}$ $\Longrightarrow$ θ is in Quadrant IV.

$\cos \theta = \dfrac{1}{\sec \theta} = \dfrac{1}{\sqrt{2}} = \dfrac{\sqrt{2}}{2}$

$\tan \theta = \dfrac{\sin \theta}{\cos \theta} = \dfrac{-\sqrt{2}/2}{\sqrt{2}/2} = -1$

$\cot \theta = \dfrac{1}{\tan \theta} = -1$

$\csc \theta = \dfrac{1}{\sin \theta} = -\sqrt{2}$

4. $\csc \theta = \dfrac{5}{3}$, $\tan \theta = \dfrac{3}{4}$

θ is in Quadrant I.

$\sin \theta = \dfrac{1}{\csc \theta} = \dfrac{3}{5}$

$\cos \theta = \dfrac{\sin \theta}{\tan \theta} = \dfrac{3}{5} \cdot \dfrac{4}{3} = \dfrac{4}{5}$

$\sec \theta = \dfrac{1}{\cos \theta} = \dfrac{5}{4}$

$\cot \theta = \dfrac{1}{\tan \theta} = \dfrac{4}{3}$

5. $\tan x = \dfrac{5}{12}$, $\sec x = -\dfrac{13}{12}$ $\Longrightarrow$ x is in

Quadrant III.

$\cos x = \dfrac{1}{\sec x} = -\dfrac{12}{13}$

$\sin x = -\sqrt{1 - \cos^2 x} = -\sqrt{1 - \dfrac{144}{169}} = -\dfrac{5}{13}$

$\cot x = \dfrac{1}{\tan x} = \dfrac{12}{5}$

$\csc x = \dfrac{1}{\sin x} = -\dfrac{13}{5}$

6. $\cot \phi = -3$, $\sin \phi = \dfrac{\sqrt{10}}{10}$

ϕ is in Quadrant II.

$\cos \phi = \cot \phi \sin \phi = -\dfrac{3\sqrt{10}}{10}$

$\tan \phi = \dfrac{1}{\cot \phi} = -\dfrac{1}{3}$

$\csc \phi = \dfrac{1}{\sin \phi} = \sqrt{10}$

$\sec \phi = \dfrac{1}{\cos \phi} = -\dfrac{10}{3\sqrt{10}} = -\dfrac{\sqrt{10}}{3}$

7. $\sec \phi = \dfrac{3}{2}$, $\csc \phi = -\dfrac{3\sqrt{5}}{5}$ $\Longrightarrow$ ϕ is in Quadrant IV.

$\sin \phi = \dfrac{1}{\csc \phi} = \dfrac{1}{-3\sqrt{5}/5} = -\dfrac{\sqrt{5}}{3}$

$\cos \phi = \dfrac{1}{\sec \phi} = \dfrac{1}{3/2} = \dfrac{2}{3}$

$\tan \phi = \dfrac{\sin \phi}{\cos \phi} = \dfrac{-\sqrt{5}/3}{2/3} = -\dfrac{\sqrt{5}}{2}$

$\cot \phi = \dfrac{1}{\tan \phi} = \dfrac{1}{-\sqrt{5}/2} = -\dfrac{2}{\sqrt{5}} = -\dfrac{2\sqrt{5}}{5}$

8. $\cos\left(\dfrac{\pi}{2} - x\right) = \dfrac{3}{5}$, $\cos x = \dfrac{4}{5}$, x is in Quadrant I.

$\sin x = \sqrt{1 - \left(\dfrac{4}{5}\right)^2} = \dfrac{3}{5}$

$\tan x = \dfrac{\sin x}{\cos x} = \dfrac{3}{5} \cdot \dfrac{5}{4} = \dfrac{3}{4}$

$\csc x = \dfrac{1}{\sin x} = \dfrac{5}{3}$

$\sec x = \dfrac{1}{\cos x} = \dfrac{5}{4}$

$\cot x = \dfrac{1}{\tan x} = \dfrac{4}{3}$

9. $\sin(-x) = -\frac{1}{3} \Rightarrow \sin x = \frac{1}{3}$, $\tan x = -\frac{\sqrt{2}}{4} \Rightarrow x$ is

in Quadrant II.

$\cos x = -\sqrt{1 - \sin^2 x} = -\sqrt{1 - \frac{1}{9}} = -\frac{2\sqrt{2}}{3}$

$\cot x = \frac{1}{\tan x} = \frac{1}{-\sqrt{2}/4} = -2\sqrt{2}$

$\sec x = \frac{1}{\cos x} = \frac{1}{-2\sqrt{2}/3} = -\frac{3\sqrt{2}}{4}$

$\csc x = \frac{1}{\sin x} = \frac{1}{1/3} = 3$

10. $\sec x = 4$, $\sin x > 0$

x is in Quadrant I.

$\cos x = \frac{1}{\sec x} = \frac{1}{4}$

$\sin x = \sqrt{1 - \left(\frac{1}{4}\right)^2} = \frac{\sqrt{15}}{4}$

$\tan x = \frac{\sin x}{\cos x} = \frac{\sqrt{15}}{4} \cdot \frac{4}{1} = \sqrt{15}$

$\csc x = \frac{1}{\sin x} = \frac{4}{\sqrt{15}} = \frac{4\sqrt{15}}{15}$

$\cot x = \frac{1}{\tan x} = \frac{1}{\sqrt{15}} = \frac{\sqrt{15}}{15}$

11. $\tan \theta = 2$, $\sin \theta < 0 \Rightarrow \theta$ is in Quadrant III.

$\sec \theta = -\sqrt{\tan^2 \theta + 1} = -\sqrt{4 + 1} = -\sqrt{5}$

$\cos \theta = \frac{1}{\sec \theta} = -\frac{1}{\sqrt{5}} = -\frac{\sqrt{5}}{5}$

$\sin \theta = -\sqrt{1 - \cos^2 \theta}$

$\quad = -\sqrt{1 - \frac{1}{5}} = -\frac{2}{\sqrt{5}} = -\frac{2\sqrt{5}}{5}$

$\csc \theta = \frac{1}{\sin \theta} = -\frac{\sqrt{5}}{2}$

$\cot \theta = \frac{1}{\tan \theta} = \frac{1}{2}$

12. $\csc \theta = -5$, $\cos \theta < 0$

θ is in Quadrant III.

$\sin \theta = \frac{1}{\csc \theta} = -\frac{1}{5}$

$\cos \theta = -\sqrt{1 - \left(-\frac{1}{5}\right)^2} = \frac{-2\sqrt{6}}{5}$

$\tan \theta = \frac{\sin \theta}{\cos \theta} = -\frac{1}{5} \cdot -\frac{5}{2\sqrt{6}} = \frac{\sqrt{6}}{12}$

$\sec \theta = \frac{1}{\cos \theta} = -\frac{5}{2\sqrt{6}} = -\frac{5\sqrt{6}}{12}$

$\cot \theta = \frac{1}{\tan \theta} = \frac{12}{\sqrt{6}} = 2\sqrt{6}$

13. $\sin \theta = -1$, $\cot \theta = 0 \Rightarrow \theta = \frac{3\pi}{2}$

$\cos \theta = \sqrt{1 - \sin^2 \theta} = 0$

$\sec \theta$ is undefined.

$\tan \theta$ is undefined.

$\csc \theta = -1$

14. $\tan \theta$ is undefined, $\sin \theta > 0$.

$\theta = \frac{\pi}{2}$

$\tan \theta = \frac{\sin \theta}{\cos \theta}$ is undefined $\Rightarrow \cos \theta = 0$

$\sin \theta = \sqrt{1 - 0^2} = 1$

$\csc \theta = \frac{1}{\sin \theta} = 1$

$\sec \theta = \frac{1}{\cos \theta}$ is undefined.

$\cot \theta = \frac{\cos \theta}{\sin \theta} = \frac{0}{1} = 0$

15. $\sec x \cos x = \sec x \cdot \frac{1}{\sec x} = 1$

The expression is matched with (d).

16. $\tan x \csc x = \frac{\sin x}{\cos x} \cdot \frac{1}{\sin x} = \frac{1}{\cos x} = \sec x$

Matches (a).

17. $\cot^2 x - \csc^2 x = \cot^2 x - (1 + \cot^2 x) = -1$

The expression is matched with (b).

18. $(1 - \cos^2 x)(\csc x) = (\sin^2 x)\frac{1}{\sin x} = \sin x$

Matches (f).

19. $\dfrac{\sin(-x)}{\cos(-x)} = \dfrac{-\sin x}{\cos x} = -\tan x$

The expression is matched with (e).

20. $\dfrac{\sin[(\pi/2) - x]}{\cos[(\pi/2) - x]} = \dfrac{\cos x}{\sin x} = \cot x$

Matches (c).

21. $\sin x \sec x = \sin x \cdot \dfrac{1}{\cos x} = \tan x$

The expression is matched with (b).

22. $\cos^2 x(\sec^2 x - 1) = \cos^2 x(\tan^2 x)$

$$= \cos^2 x\left(\dfrac{\sin^2 x}{\cos^2 x}\right)$$

$$= \sin^2 x$$

Matches (c).

23. $\sec^4 x - \tan^4 x = (\sec^2 x + \tan^2 x)(\sec^2 x - \tan^2 x)$

$$= (\sec^2 x + \tan^2 x)(1) = \sec^2 x + \tan^2 x$$

The expression is matched with (f).

24. $\cot x \sec x = \dfrac{\cos x}{\sin x} \cdot \dfrac{1}{\cos x} = \dfrac{1}{\sin x} = \csc x$

Matches (a).

25. $\dfrac{\sec^2 x - 1}{\sin^2 x} = \dfrac{\tan^2 x}{\sin^2 x} = \dfrac{\sin^2 x}{\cos^2 x} \cdot \dfrac{1}{\sin^2 x} = \sec^2 x$

The expression is matched with (e).

26. $\dfrac{\cos^2[(\pi/2) - x]}{\cos x} = \dfrac{\sin^2 x}{\cos x} = \dfrac{\sin x}{\cos x}\sin x = \tan x \sin x$

Matches (d).

27. $\cot \theta \sec \theta = \dfrac{\cos \theta}{\sin \theta} \cdot \dfrac{1}{\cos \theta} = \dfrac{1}{\sin \theta} = \csc \theta$

28. $\cos \beta \tan \beta = \cos \beta \dfrac{\sin \beta}{\cos \beta} = \sin \beta$

29. $\sin \phi(\csc \phi - \sin \phi) = (\sin \phi)\dfrac{1}{\sin \phi} - \sin^2 \phi$

$$= 1 - \sin^2 \phi = \cos^2 \phi$$

30. $\sec^2 x(1 - \sin^2 x) = \sec^2 x - \sec^2 x \sin^2 x$

$$= \sec^2 x - \dfrac{1}{\cos^2 x} \cdot \sin^2 x$$

$$= \sec^2 x - \dfrac{\sin^2 x}{\cos^2 x}$$

$$= \sec^2 x - \tan^2 x$$

$$= 1$$

31. $\dfrac{\cot x}{\csc x} = \dfrac{\cos x/\sin x}{1/\sin x}$

$$= \dfrac{\cos x}{\sin x} \cdot \dfrac{\sin x}{1}$$

$$= \cos x$$

32. $\dfrac{\csc \theta}{\sec \theta} = \dfrac{1/(\sin \theta)}{1/(\cos \theta)} = \dfrac{\cos \theta}{\sin \theta} = \cot \theta$

33. $\dfrac{1 - \sin^2 x}{\csc^2 x - 1} = \dfrac{\cos^2 x}{\cot^2 x} = \cos^2 x \tan^2 x = (\cos^2 x)\dfrac{\sin^2 x}{\cos^2 x}$

$$= \sin^2 x$$

34. $\dfrac{1}{\tan^2 x + 1} = \dfrac{1}{\sec^2 x} = \dfrac{1}{1/(\cos^2 x)} = \cos^2 x$

35. $\sec \alpha \dfrac{\sin \alpha}{\tan \alpha} = \dfrac{1}{\cos \alpha}(\sin \alpha) \cot \alpha$

$$= \dfrac{1}{\cos \alpha}(\sin \alpha)\left(\dfrac{\cos \alpha}{\sin \alpha}\right) = 1$$

36. $\dfrac{\tan^2 \theta}{\sec^2 \theta} = \dfrac{\sin^2 \theta}{\cos^2 \theta} \cdot \dfrac{1}{\sec^2 \theta}$

$$= \dfrac{\sin^2 \theta}{\cos^2 \theta} \cdot \dfrac{1}{1/(\cos^2 \theta)} = \dfrac{\sin^2 \theta \cos^2 \theta}{\cos^2 \theta} = \sin^2 \theta$$

37. $\cos\left(\dfrac{\pi}{2} - x\right)\sec x = (\sin x)(\sec x)$

$$= (\sin x)\left(\dfrac{1}{\cos x}\right) = \dfrac{\sin x}{\cos x} = \tan x$$

38. $\cot\left(\dfrac{\pi}{2} - x\right)\cos x = \tan x \cos x = \dfrac{\sin x}{\cos x} \cdot \cos x = \sin x$

39. $\dfrac{\cos^2 y}{1 - \sin y} = \dfrac{1 - \sin^2 y}{1 - \sin y}$

$$= \dfrac{(1 + \sin y)(1 - \sin y)}{1 - \sin y} = 1 + \sin y$$

40. $(\cos t)(1 + \tan^2 t) = (\cos t)(\sec^2 t) = \dfrac{\cos t}{\cos^2 t} = \dfrac{1}{\cos t} = \sec t$

41. $\sin \beta \tan \beta + \cos \beta = (\sin \beta)\dfrac{\sin \beta}{\cos \beta} + \cos \beta$

$$= \dfrac{\sin^2 \beta}{\cos \beta} + \dfrac{\cos^2 \beta}{\cos \beta}$$

$$= \dfrac{\sin^2 \beta + \cos^2 \beta}{\cos \beta}$$

$$= \dfrac{1}{\cos \beta}$$

$$= \sec \beta$$

42. $\csc \phi \tan \phi + \sec \phi = \dfrac{1}{\sin \phi} \cdot \dfrac{\sin \phi}{\cos \phi} + \sec \phi$

$$= \dfrac{1}{\cos \phi} + \sec \phi$$

$$= 2 \sec \phi$$

43. $\cot u \sin u + \tan u \cos u = \dfrac{\cos u}{\sin u}(\sin u) + \dfrac{\sin u}{\cos u}(\cos u)$

$$= \cos u + \sin u$$

44. $\sin \theta \sec \theta + \cos \theta \csc \theta = \dfrac{\sin \theta}{\cos \theta} + \dfrac{\cos \theta}{\sin \theta}$

$$= \dfrac{\sin^2 \theta + \cos^2 \theta}{\cos \theta \sin \theta}$$

$$= \dfrac{1}{\cos \theta \sin \theta}$$

$$= \sec \theta \csc \theta$$

45. $\tan^2 x - \tan^2 x \sin^2 x = \tan^2 x(1 - \sin^2 x)$

$$= \tan^2 x \cos^2 x$$

$$= \dfrac{\sin^2 x}{\cos^2 x} \cdot \cos^2 x$$

$$= \sin^2 x$$

46. $\sin^2 x \csc^2 x - \sin^2 x = \sin^2 x(\csc^2 x - 1)$

$$= \sin^2 x \cot^2 x$$

$$= \sin^2 x \cdot \dfrac{\cos^2 x}{\sin^2 x}$$

$$= \cos^2 x$$

47. $\sin^2 x \sec^2 x - \sin^2 x = \sin^2 x(\sec^2 x - 1)$

$$= \sin^2 x \tan^2 x$$

48. $\cos^2 x + \cos^2 x \tan^2 x = \cos^2 x(1 + \tan^2 x)$

$$= \cos^2 x(\sec^2 x)$$

$$= \cos^2 x\left(\dfrac{1}{\cos^2 x}\right)$$

$$= 1$$

49. $\dfrac{\sec^2 x - 1}{\sec x - 1} = \dfrac{(\sec x + 1)(\sec x - 1)}{\sec x - 1}$

$$= \sec x + 1$$

50. $\dfrac{\cos^2 x - 4}{\cos x - 2} = \dfrac{(\cos x + 2)(\cos x - 2)}{\cos x - 2}$

$$= \cos x + 2$$

51. $\tan^4 x + 2 \tan^2 x + 1 = (\tan^2 x + 1)^2$

$$= (\sec^2 x)^2$$

$$= \sec^4 x$$

52. $1 - 2\cos^2 x + \cos^4 x = (1 - \cos^2 x)^2$

$$= (\sin^2 x)^2$$

$$= \sin^4 x$$

53. $\sin^4 x - \cos^4 x = (\sin^2 x + \cos^2 x)(\sin^2 x - \cos^2 x)$

$\qquad\qquad\quad = (1)(\sin^2 x - \cos^2 x)$

$\qquad\qquad\quad = \sin^2 x - \cos^2 x$

54. $\sec^4 x - \tan^4 x = (\sec^2 x + \tan^2 x)(\sec^2 x - \tan^2 x)$

$\qquad\qquad\quad = (\sec^2 x + \tan^2 x)(1)$

$\qquad\qquad\quad = \sec^2 x + \tan^2 x$

55. $\csc^3 x - \csc^2 x - \csc x + 1 = \csc^2 x(\csc x - 1) - 1(\csc x - 1)$

$\qquad\qquad\qquad\qquad\qquad\quad = (\csc^2 x - 1)(\csc x - 1)$

$\qquad\qquad\qquad\qquad\qquad\quad = \cot^2 x(\csc x - 1)$

56. $\sec^3 x - \sec^2 x - \sec x + 1 = \sec^2 x(\sec x - 1) - (\sec x - 1)$

$\qquad\qquad\qquad\qquad\qquad\quad = (\sec^2 x - 1)(\sec x - 1)$

$\qquad\qquad\qquad\qquad\qquad\quad = \tan^2 x(\sec x - 1)$

57. $(\sin x + \cos x)^2 = \sin^2 x + 2 \sin x \cos x + \cos^2 x$

$\qquad\qquad\qquad\quad = (\sin^2 x + \cos^2 x) + 2 \sin x \cos x$

$\qquad\qquad\qquad\quad = 1 + 2 \sin x \cos x$

58. $(\cot x + \csc x)(\cot x - \csc x) = \cot^2 x - \csc^2 x$

$\qquad\qquad\qquad\qquad\qquad\quad = -1$

59. $(2 \csc x + 2)(2 \csc x - 2) = 4 \csc^2 x - 4$

$\qquad\qquad\qquad\qquad\quad = 4(\csc^2 x - 1)$

$\qquad\qquad\qquad\qquad\quad = 4 \cot^2 x$

60. $(3 - 3 \sin x)(3 + 3 \sin x) = 9 - 9 \sin^2 x$

$\qquad\qquad\qquad\qquad\quad = 9(1 - \sin^2 x)$

$\qquad\qquad\qquad\qquad\quad = 9 \cos^2 x$

61. $\dfrac{1}{1 + \cos x} + \dfrac{1}{1 - \cos x} = \dfrac{1 - \cos x + 1 + \cos x}{(1 + \cos x)(1 - \cos x)}$

$\qquad\qquad\qquad\qquad\quad = \dfrac{2}{1 - \cos^2 x}$

$\qquad\qquad\qquad\qquad\quad = \dfrac{2}{\sin^2 x}$

$\qquad\qquad\qquad\qquad\quad = 2 \csc^2 x$

62. $\dfrac{1}{\sec x + 1} - \dfrac{1}{\sec x - 1} = \dfrac{\sec x - 1 - (\sec x + 1)}{(\sec x + 1)(\sec x - 1)}$

$\qquad\qquad\qquad\qquad\quad = \dfrac{\sec x - 1 - \sec x - 1}{\sec^2 x - 1}$

$\qquad\qquad\qquad\qquad\quad = \dfrac{-2}{\tan^2 x}$

$\qquad\qquad\qquad\qquad\quad = -2\left(\dfrac{1}{\tan^2 x}\right)$

$\qquad\qquad\qquad\qquad\quad = -2 \cot^2 x$

63. $\dfrac{\cos x}{1 + \sin x} + \dfrac{1 + \sin x}{\cos x} = \dfrac{\cos^2 x + (1 + \sin x)^2}{\cos x(1 + \sin x)} = \dfrac{\cos^2 x + 1 + 2 \sin x + \sin^2 x}{\cos x(1 + \sin x)}$

$\qquad\qquad\qquad\qquad\quad = \dfrac{2 + 2 \sin x}{\cos x(1 + \sin x)}$

$\qquad\qquad\qquad\qquad\quad = \dfrac{2(1 + \sin x)}{\cos x(1 + \sin x)}$

$\qquad\qquad\qquad\qquad\quad = \dfrac{2}{\cos x}$

$\qquad\qquad\qquad\qquad\quad = 2 \sec x$

64. $\tan x - \dfrac{\sec^2 x}{\tan x} = \dfrac{\tan^2 x - \sec^2 x}{\tan x}$

$\qquad\qquad\qquad = \dfrac{-1}{\tan x} = -\cot x$

65. $\dfrac{\sin^2 y}{1 - \cos y} = \dfrac{1 - \cos^2 y}{1 - \cos y}$

$\qquad\quad = \dfrac{(1 + \cos y)(1 - \cos y)}{1 - \cos y} = 1 + \cos y$

66. $\dfrac{5}{\tan x + \sec x} \cdot \dfrac{\tan x - \sec x}{\tan x - \sec x} = \dfrac{5(\tan x - \sec x)}{\tan^2 x - \sec^2 x}$

$\qquad\qquad\qquad\qquad = \dfrac{5(\tan x - \sec x)}{-1}$

$\qquad\qquad\qquad\qquad = 5(\sec x - \tan x)$

67. $\dfrac{3}{\sec x - \tan x} \cdot \dfrac{\sec x + \tan x}{\sec x + \tan x} = \dfrac{3(\sec x + \tan x)}{\sec^2 x - \tan^2 x}$

$\qquad\qquad\qquad\qquad = \dfrac{3(\sec x + \tan x)}{1}$

$\qquad\qquad\qquad\qquad = 3(\sec x + \tan x)$

68. $\dfrac{\tan^2 x}{\csc x + 1} \cdot \dfrac{\csc x - 1}{\csc x - 1} = \dfrac{\tan^2 x(\csc x - 1)}{\csc^2 x - 1} = \dfrac{\tan^2 x(\csc x - 1)}{\cot^2 x} = \tan^2 x(\csc x - 1)\tan^2 x = \tan^4 x(\csc x - 1)$

69. $y_1 = \cos\left(\dfrac{\pi}{2} - x\right), \ y_2 = \sin x$

x	0.2	0.4	0.6	0.8	1.0	1.2	1.4
y_1	0.1987	0.3894	0.5646	0.7174	0.8415	0.9320	0.9854
y_2	0.1987	0.3894	0.5646	0.7174	0.8415	0.9320	0.9854

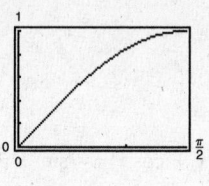

Conclusion: $y_1 = y_2$

70. $y_1 = \sec x - \cos x, \ y_2 = \sin x \tan x$

x	0.2	0.4	0.6	0.8	1.0	1.2	1.4
y_1	0.0403	0.1646	0.3863	0.7386	1.3105	2.3973	5.7135
y_2	0.0403	0.1646	0.3863	0.7386	1.3105	2.3973	5.7135

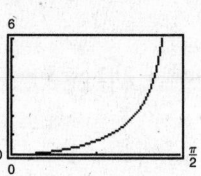

It appears that $y_1 = y_2$.

71. $y_1 = \dfrac{\cos x}{1 - \sin x}, \ y_2 = \dfrac{1 + \sin x}{\cos x}$

x	0.2	0.4	0.6	0.8	1.0	1.2	1.4
y_1	1.2230	1.5085	1.8958	2.4650	3.4082	5.3319	11.6814
y_2	1.2230	1.5085	1.8958	2.4650	3.4082	5.3319	11.6814

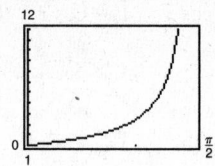

Conclusion: $y_1 = y_2$

72. $y_1 = \sec^4 x - \sec^2 x, \ y_2 = \tan^2 x + \tan^4 x$

x	0.2	0.4	0.6	0.8	1.0	1.2	1.4
y_1	0.0428	0.2107	0.6871	2.1841	8.3087	50.3869	1163.6143
y_2	0.0428	0.2107	0.6871	2.1841	8.3087	50.3869	1163.6143

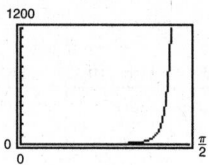

It appears that $y_1 = y_2$.

73. $y_1 = \cos x \cot x + \sin x = \csc x$

$\cos x \cot x + \sin x = \cos x\left(\dfrac{\cos x}{\sin x}\right) + \sin x$

$\qquad\qquad\qquad\quad = \dfrac{\cos^2 x}{\sin x} + \dfrac{\sin^2 x}{\sin x}$

$\qquad\qquad\qquad\quad = \dfrac{\cos^2 x + \sin^2 x}{\sin x} = \dfrac{1}{\sin x} = \csc x$

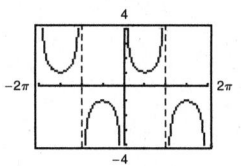

74. $y_1 = \sec x \csc x - \tan x = \cot x$

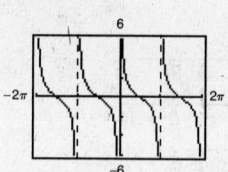

$$\sec x \csc x - \tan x = \frac{1}{\cos x} \cdot \frac{1}{\sin x} - \frac{\sin x}{\cos x}$$

$$= \frac{1}{\cos x \sin x} - \frac{\sin^2 x}{\cos x \sin x}$$

$$= \frac{1 - \sin^2 x}{\cos x \sin x}$$

$$= \frac{\cos^2 x}{\cos x \sin x} = \frac{\cos x}{\sin x} = \cot x$$

75. $y_1 = \frac{1}{\sin x}\left(\frac{1}{\cos x} - \cos x\right) = \tan x$

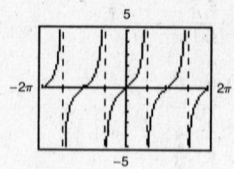

$$\frac{1}{\sin x}\left(\frac{1}{\cos x} - \cos x\right) = \frac{1}{\sin x \cos x} - \frac{\cos x}{\sin x}$$

$$= \frac{1 - \cos^2 x}{\sin x \cos x} = \frac{\sin^2 x}{\sin x \cos x} = \frac{\sin x}{\cos x} = \tan x$$

76. $y_1 = \frac{1}{2}\left(\frac{1 + \sin \theta}{\cos \theta} + \frac{\cos \theta}{1 + \sin \theta}\right)$

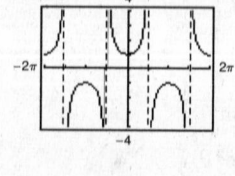

$$\frac{1}{2}\left(\frac{1 + \sin \theta}{\cos \theta} + \frac{\cos \theta}{1 + \sin \theta}\right) = \frac{1}{2}\left(\frac{(1 + \sin \theta)(1 + \sin \theta)}{(\cos \theta)(1 + \sin \theta)} + \frac{(\cos \theta)(\cos \theta)}{(\cos \theta)(1 + \sin \theta)}\right)$$

$$= \frac{1}{2}\left(\frac{1 + 2\sin \theta + \sin^2 \theta + \cos^2 \theta}{(\cos \theta)(1 + \sin \theta)}\right)$$

$$= \frac{1}{2}\left(\frac{1 + 2\sin \theta + 1}{(\cos \theta)(1 + \sin \theta)}\right)$$

$$= \frac{1}{2}\left(\frac{2 + 2\sin \theta}{(\cos \theta)(1 + \sin \theta)}\right)$$

$$= \frac{1 + \sin \theta}{(\cos \theta)(1 + \sin \theta)} = \frac{1}{\cos \theta} = \sec \theta$$

77. Let $x = 3 \cos \theta$, then

$$\sqrt{9 - x^2} = \sqrt{9 - (3 \cos \theta)^2} = \sqrt{9 - 9 \cos^2 \theta} = \sqrt{9(1 - \cos^2 \theta)}$$

$$= \sqrt{9 \sin^2 \theta} = 3 \sin \theta.$$

78. Let $x = 2 \cos \theta$.

$$\sqrt{64 - 16x^2} = \sqrt{64 - 16(2 \cos \theta)^2}$$

$$= \sqrt{64(1 - \cos^2 \theta)}$$

$$= \sqrt{64 \sin^2 \theta}$$

$$= 8 \sin \theta$$

79. Let $x = 3 \sec \theta$, then

$$\sqrt{x^2 - 9} = \sqrt{(3 \sec \theta)^2 - 9}$$

$$= \sqrt{9 \sec^2 \theta - 9}$$

$$= \sqrt{9(\sec^2 \theta - 1)}$$

$$= \sqrt{9 \tan^2 \theta}$$

$$= 3 \tan \theta.$$

80. Let $x = 2 \sec \theta$.

$$\begin{aligned}
\sqrt{x^2 - 4} &= \sqrt{(2 \sec \theta)^2 - 4} \\
&= \sqrt{4(\sec^2 \theta - 1)} \\
&= \sqrt{4 \tan^2 \theta} \\
&= 2 \tan \theta
\end{aligned}$$

81. Let $x = 5 \tan \theta$, then

$$\begin{aligned}
\sqrt{x^2 + 25} &= \sqrt{(5 \tan \theta)^2 + 25} \\
&= \sqrt{25 \tan^2 \theta + 25} \\
&= \sqrt{25(\tan^2 \theta + 1)} \\
&= \sqrt{25 \sec^2 \theta} \\
&= 5 \sec \theta.
\end{aligned}$$

82. Let $x = 10 \tan \theta$.

$$\begin{aligned}
\sqrt{x^2 + 100} &= \sqrt{(10 \tan \theta)^2 + 100} \\
&= \sqrt{100(\tan^2 \theta + 1)} \\
&= \sqrt{100 \sec^2 \theta} \\
&= 10 \sec \theta
\end{aligned}$$

83. Let $x = 3 \sin \theta$, then $\sqrt{9 - x^2} = 3$ becomes

$$\begin{aligned}
\sqrt{9 - (3 \sin \theta)^2} &= 3 \\
\sqrt{9 - 9 \sin^2 \theta} &= 3 \\
\sqrt{9(1 - \sin^2 \theta)} &= 3 \\
\sqrt{9 \cos^2 \theta} &= 3 \\
3 \cos \theta &= 3 \\
\cos \theta &= 1 \\
\sin \theta = \sqrt{1 - \cos^2 \theta} &= \sqrt{1 - (1)^2} = 0.
\end{aligned}$$

84. $x = 6 \sin \theta$

$$\begin{aligned}
3 &= \sqrt{36 - x^2} \\
&= \sqrt{36 - (6 \sin \theta)^2} \\
&= \sqrt{36(1 - \sin^2 \theta)} \\
&= \sqrt{36 \cos^2 \theta} \\
&= 6 \cos \theta
\end{aligned}$$

$$\cos \theta = \frac{3}{6} = \frac{1}{2}$$

$$\begin{aligned}
\sin \theta &= \pm \sqrt{1 - \cos^2 \theta} \\
&= \pm \sqrt{1 - \left(\frac{1}{2}\right)^2} \\
&= \pm \sqrt{\frac{3}{4}} \\
&= \pm \frac{\sqrt{3}}{2}
\end{aligned}$$

85. Let $x = 2 \cos \theta$, then $\sqrt{16 - 4x^2} = 2\sqrt{2}$ becomes

$$\begin{aligned}
\sqrt{16 - 4(2 \cos \theta)^2} &= 2\sqrt{2} \\
\sqrt{16 - 16 \cos^2 \theta} &= 2\sqrt{2} \\
\sqrt{16(1 - \cos^2 \theta)} &= 2\sqrt{2} \\
\sqrt{16 \sin^2 \theta} &= 2\sqrt{2} \\
4 \sin \theta &= 2\sqrt{2}
\end{aligned}$$

$$\sin \theta = \frac{\sqrt{2}}{2}$$

$$\begin{aligned}
\cos \theta &= \sqrt{1 - \sin^2 \theta} \\
&= \sqrt{1 - \frac{1}{2}} \\
&= \sqrt{\frac{1}{2}} \\
&= \frac{\sqrt{2}}{2}.
\end{aligned}$$

86. $\qquad x = 10 \cos \theta$

$$\begin{aligned}
-5\sqrt{3} &= \sqrt{100 - x^2} \\
&= \sqrt{100 - (10 \cos \theta)^2} \\
&= \sqrt{100(1 - \cos^2 \theta)} \\
&= \sqrt{100 \sin^2 \theta} \\
&= 10 \sin \theta
\end{aligned}$$

$$\sin \theta = -\frac{5\sqrt{3}}{10} = -\frac{\sqrt{3}}{2}$$

$$\cos \theta = \sqrt{1 - \sin^2 \theta} = \sqrt{1 - \left(-\frac{\sqrt{3}}{2}\right)^2} = \frac{1}{2}$$

87. $\sin \theta = \sqrt{1 - \cos^2 \theta}$

Let $y_1 = \sin x$ and $y_2 = \sqrt{1 - \cos^2 x}$, $0 \le x \le 2\pi$.

$y_1 = y_2$ for $0 \le x \le \pi$, so we have

$\sin \theta = \sqrt{1 - \cos^2 \theta}$ for $0 \le \theta \le \pi$.

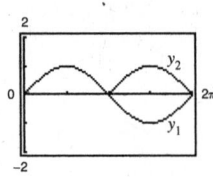

88. $\cos \theta = -\sqrt{1 - \sin^2 \theta}$

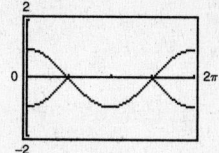

$\dfrac{\pi}{2} \le \theta \le \dfrac{3\pi}{2}$

89. $\sec \theta = \sqrt{1 + \tan^2 \theta}$

Let $y_1 = \dfrac{1}{\cos x}$ and $y_2 = \sqrt{1 + \tan^2 x}$, $0 \le x \le 2\pi$.

$y_1 = y_2$ for $0 \le x < \dfrac{\pi}{2}$ and $\dfrac{3\pi}{2} < x \le 2\pi$, so we have

$\sec \theta = \sqrt{1 + \tan^2 \theta}$ for $0 \le \theta < \dfrac{\pi}{2}$ and $\dfrac{3\pi}{2} < \theta < 2\pi$.

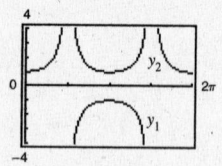

90. $\csc \theta = \sqrt{1 + \cot^2 \theta}$

$0 < \theta < \pi$

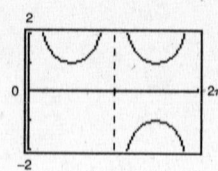

91. $\ln|\cos x| - \ln|\sin x| = \ln \dfrac{|\cos x|}{|\sin x|} = \ln|\cot x|$

92. $\ln|\sec x| + \ln|\sin x| = \ln|\sec x \sin x|$

$\qquad = \ln\left|\dfrac{1}{\cos x} \cdot \sin x\right|$

$\qquad = \ln|\tan x|$

93. $\ln|\cot t| + \ln(1 + \tan^2 t) = \ln\big[|\cot t|(1 + \tan^2 t)\big]$

$\qquad = \ln|\cot t \sec^2 t|$

$\qquad = \ln\left|\dfrac{\cos t}{\sin t} \cdot \dfrac{1}{\cos^2 t}\right|$

$\qquad = \ln\left|\dfrac{1}{\sin t \cos t}\right| = \ln|\csc t \sec t|$

94. $\ln(\cos^2 t) + \ln(1 + \tan^2 t) = \ln[\cos^2 t(1 + \tan^2 t)]$

$\qquad = \ln[\cos^2 t \sec^2 t]$

$\qquad = \ln\left(\cos^2 t \cdot \dfrac{1}{\cos^2 t}\right)$

$\qquad = \ln(1) = 0$

95. (a) $\csc^2 132° - \cot^2 132° \approx 1.8107 - 0.8107 = 1$

(b) $\csc^2 \dfrac{2\pi}{7} - \cot^2 \dfrac{2\pi}{7} \approx 1.6360 - 0.6360 = 1$

96. $\tan^2 \theta + 1 = \sec^2 \theta$

(a) $\qquad\qquad \theta = 346°$

$(\tan 346°)^2 + 1 \approx 1.0622$

$(\sec 346°)^2 = \left(\dfrac{1}{\cos 346°}\right)^2 \approx 1.0622$

(b) $\qquad\qquad \theta = 3.1$

$(\tan 3.1)^2 + 1 \approx 1.00173$

$(\sec 3.1)^2 = \left(\dfrac{1}{\cos 3.1}\right)^2 \approx 1.00173$

97. $\cos\left(\dfrac{\pi}{2} - \theta\right) = \sin \theta$

(a) $\qquad\qquad \theta = 80°$

$\cos(90° - 80°) = \sin 80°$

$0.9848 = 0.9848$

(b) $\qquad\qquad \theta = 0.8$

$\cos\left(\dfrac{\pi}{2} - 0.8\right) = \sin 0.8$

$0.7174 = 0.7174$

98. $\sin(-\theta) = -\sin \theta$

   (a) $\theta = 250°$

      $\sin(-250°) \approx 0.9397$

      $-(\sin 250°) \approx 0.9397$

   (b) $\theta = \dfrac{1}{2}$

      $\sin\left(-\dfrac{1}{2}\right) \approx -0.4794$

      $-\left(\sin \dfrac{1}{2}\right) \approx -0.4794$

99. $\mu W \cos \theta = W \sin \theta$

$$\mu = \frac{W \sin \theta}{W \cos \theta} = \tan \theta$$

100. $\csc x \cot x - \cos x = \dfrac{1}{\sin x}\left(\dfrac{\cos x}{\sin x}\right) - \cos x$

$$= \frac{\cos x}{\sin^2 x} - \cos x$$

$$= \frac{\cos x - \sin^2 x \, \cos x}{\sin^2 x}$$

$$= \frac{\cos x(1 - \sin^2 x)}{\sin^2 x}$$

$$= \frac{\cos x \cos^2 x}{\sin^2 x} = \cos x \cot^2 x$$

101. True. For example, $\sin(-x) = -\sin x$ means that the graph of $\sin x$ is symmetric about the origin.

102. False. A cofunction identity can be used to transform a tangent function so that it can be represented by a cotangent function.

103. As $x \to \dfrac{\pi^-}{2}$, $\sin x \to 1$ and $\csc x \to 1$.

104. As $x \to 0^+$, $\cos x \to 1$ and $\sec x = \dfrac{1}{\cos x} \to 1$.

105. As $x \to \dfrac{\pi^-}{2}$, $\tan x \to \infty$ and $\cot x \to 0$.

106. As $x \to \pi^+$, $\sin x \to 0$ and $\csc x = \dfrac{1}{\sin x} \to -\infty$.

107. $\cos \theta = \sqrt{1 - \sin^2 \theta}$ *is not* an identity.

$$\cos^2 \theta + \sin^2 \theta = 1 \implies \cos \theta = \pm\sqrt{1 - \sin^2 \theta}$$

108. The equation *is not* an identity.

$$\cot \theta = \pm\sqrt{\csc^2 \theta - 1}$$

109. $\dfrac{\sin k\theta}{\cos k\theta} = \tan \theta$ *is not* an identity.

$$\frac{\sin k\theta}{\cos k\theta} = \tan k\theta$$

110. The equation *is not* an identity.

$$\frac{1}{5 \cos \theta} = \frac{1}{5}\left(\frac{1}{\cos \theta}\right) = \frac{1}{5} \sec \theta \neq 5 \sec \theta$$

111. $\sin \theta \csc \theta = 1$ *is* an identity.

$$\sin \theta \cdot \frac{1}{\sin \theta} = 1, \text{ provided } \sin \theta \neq 0.$$

112. The equation *is not* an identity. The angles are not the same.

$$\sin \theta \csc \phi = \sin \theta \cdot \frac{1}{\sin \phi} = \frac{\sin \theta}{\sin \phi} \neq 1, \text{ in general}$$

113. Let (x, y) be any point on the terminal side of θ.

Then, $r = \sqrt{x^2 + y^2}$ and

$$\sin^2 \theta + \cos^2 \theta = \left(\frac{y}{r}\right)^2 + \left(\frac{x}{r}\right)^2$$

$$= \frac{y^2 + x^2}{r^2}$$

$$= \frac{r^2}{r^2}$$

$$= 1.$$

114. Divide both sides of $\sin^2 \theta + \cos^2 \theta = 1$ by $\cos^2 \theta$:

$$\frac{\sin^2 \theta}{\cos^2 \theta} + \frac{\cos^2 \theta}{\cos^2 \theta} = \frac{1}{\cos^2 \theta}$$

$$\tan^2 \theta + 1 = \sec^2 \theta$$

Divide both sides of $\sin^2 \theta + \cos^2 \theta = 1$ by $\sin^2 \theta$:

$$\frac{\sin^2 \theta}{\sin^2 \theta} + \frac{\cos^2 \theta}{\sin^2 \theta} = \frac{1}{\sin^2 \theta}$$

$$1 + \cot^2 \theta = \csc^2 \theta$$

Discussion for remembering identities will vary, but one key is first to learn the identities that concern the sine and cosine functions thoroughly, and then to use these as a basis to establish the other identities when necessary.

115. $\left(\sqrt{x} + 5\right)\left(\sqrt{x} - 5\right) = \left(\sqrt{x}\right)^2 - (5)^2 = x - 25$

116. $\left(2\sqrt{z} + 3\right)^2 = \left(2\sqrt{z}\right)^2 + 2\left(2\sqrt{z}\right)(3) + (3)^2$

$$= 4z + 12\sqrt{z} + 9$$

117. $\dfrac{1}{x + 5} + \dfrac{x}{x - 8} = \dfrac{(x - 8) + x(x + 5)}{(x + 5)(x - 8)}$

$$= \frac{x^2 + 6x - 8}{(x + 5)(x - 8)}$$

118. $\dfrac{6x}{x - 4} - \dfrac{3}{4 - x} = \dfrac{6x}{x - 4} + \dfrac{3}{x - 4}$

$$= \frac{6x + 3}{x - 4}$$

$$= \frac{3(2x + 1)}{x - 4}$$

119. $\dfrac{2x}{x^2 - 4} - \dfrac{7}{x + 4} = \dfrac{2x(x + 4) - 7(x^2 - 4)}{(x^2 - 4)(x + 4)}$

$$= \frac{2x^2 + 8x - 7x^2 + 28}{(x^2 - 4)(x + 4)}$$

$$= \frac{-5x^2 + 8x + 28}{(x^2 - 4)(x + 4)}$$

120. $\dfrac{x}{x^2 - 25} + \dfrac{x^2}{x - 5} = \dfrac{x}{(x - 5)(x + 5)} + \dfrac{x^2(x + 5)}{(x - 5)(x + 5)}$

$$= \frac{x + x^3 + 5x^2}{(x - 5)(x + 5)}$$

$$= \frac{x(1 + x^2 + 5x)}{(x - 5)(x + 5)}$$

$$= \frac{x(x^2 + 5x + 1)}{x^2 - 25}$$

121. $f(x) = \dfrac{1}{2} \sin(\pi x)$

Amplitude: $\dfrac{1}{2}$

Period: $\dfrac{2\pi}{\pi} = 2$

Key points:

$(0, 0), \left(\dfrac{1}{2}, \dfrac{1}{2}\right), (1, 0), \left(\dfrac{3}{2}, -\dfrac{1}{2}\right), (2, 0)$

122. $f(x) = -2 \tan\left(\dfrac{\pi x}{2}\right)$

Amplitude: 2

Period: $\dfrac{\pi}{\pi/2} = 2$

Two consecutive vertical asymptotes: $x = -1, x = 1$

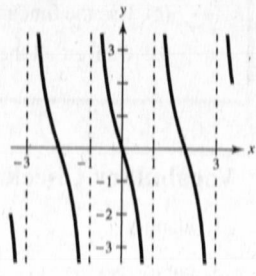

Key points: $\left(-\dfrac{1}{2}, 2\right), (0, 0), \left(\dfrac{1}{2}, 2\right)$

123. $f(x) = \frac{1}{2} \sec\left(x + \frac{\pi}{4}\right)$

Sketch the graph of $y = \frac{1}{2} \cos\left(x + \frac{\pi}{4}\right)$ first.

Amplitude: $\frac{1}{2}$

Period: 2π

One cycle: $x + \frac{\pi}{4} = 0 \implies x = -\frac{\pi}{4}$

$$x + \frac{\pi}{4} = 2\pi \implies x = \frac{7\pi}{4}$$

The x-intercepts of $y = \frac{1}{2} \cos\left(x + \frac{\pi}{4}\right)$ correspond to the vertical asymptotes of $f(x) = \frac{1}{2} \sec\left(x + \frac{\pi}{4}\right)$.

$$x = \frac{\pi}{4}, x = \frac{5\pi}{4}, \ldots$$

124. $f(x) = \frac{3}{2} \cos(x - \pi) + 3$

Using $y = a \cos bx$, $a = \frac{3}{2}$ so the amplitude is $\frac{3}{2}$.

$b = 1$ so the period is $\frac{2\pi}{1} = 2\pi$.

$(x - \pi)$ shifts the graph right by π and $+3$ shifts the graph upward by 3.

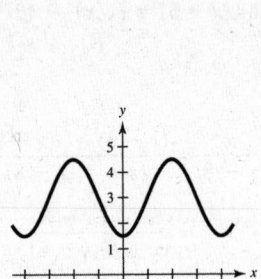

Section 5.2   Verifying Trigonometric Identities

- You should know the difference between an expression, a conditional equation, and an identity.

- You should be able to solve trigonometric identities, using the following techniques.

  (a) Work with *one* side at a time. Do not "cross" the equal sign.

  (b) Use algebraic techniques such as combining fractions, factoring expressions, rationalizing denominators, and squaring binomials.

  (c) Use the fundamental identities.

  (d) Convert all the terms into sines and cosines.

Vocabulary Check

1. identity

2. conditional equation

3. $\tan u$

4. $\cot u$

5. $\cos^2 u$

6. $\sin u$

7. $-\csc u$

8. $\sec u$

1. $\sin t \csc t = \sin t \left(\dfrac{1}{\sin t}\right) = 1$

2. $\sec y \cos y = \left(\dfrac{1}{\cos y}\right) \cos y = 1$

3. $(1 + \sin \alpha)(1 - \sin \alpha) = 1 - \sin^2 \alpha = \cos^2 \alpha$

4. $\cot^2 y(\sec^2 y - 1) = \cot^2 y \tan^2 y = 1$

5. $\cos^2 \beta - \sin^2 \beta = (1 - \sin^2 \beta) - \sin^2 \beta$
$$= 1 - 2 \sin^2 \beta$$

6. $\cos^2 \beta - \sin^2 \beta = \cos^2 \beta - (1 - \cos^2 \beta)$
$$= 2 \cos^2 \beta - 1$$

7. $\sin^2 \alpha - \sin^4 \alpha = \sin^2 \alpha (1 - \sin^2 \alpha)$
$$= (1 - \cos^2 \alpha)(\cos^2 \alpha)$$
$$= \cos^2 \alpha - \cos^4 \alpha$$

8. $\cos x + \sin x \tan x = \cos x + \sin x \left(\dfrac{\sin x}{\cos x} \right)$
$$= \frac{\cos^2 x + \sin^2 x}{\cos x}$$
$$= \frac{1}{\cos x}$$
$$= \sec x$$

9. $\dfrac{\csc^2 \theta}{\cot \theta} = \csc^2 \theta \left(\dfrac{1}{\cot \theta} \right)$
$$= \csc^2 \theta \tan \theta$$
$$= \left(\frac{1}{\sin^2 \theta} \right) \left(\frac{\sin \theta}{\cos \theta} \right)$$
$$= \left(\frac{1}{\sin \theta} \right) \left(\frac{1}{\cos \theta} \right)$$
$$= \csc \theta \sec \theta$$

10. $\dfrac{\cot^3 t}{\csc t} = \dfrac{\cot t \cot^2 t}{\csc t}$
$$= \frac{\cot t (\csc^2 t - 1)}{\csc t}$$
$$= \frac{\dfrac{\cos t}{\sin t}(\csc^2 t - 1)}{\dfrac{1}{\sin t}}$$
$$= \frac{\cos t \sin t}{\sin t}(\csc^2 t - 1)$$
$$= \cos t (\csc^2 t - 1)$$

11. $\dfrac{\cot^2 t}{\csc t} = \dfrac{\cos^2 t}{\sin^2 t} \cdot \sin t$
$$= \frac{\cos^2 t}{\sin t}$$
$$= \frac{1 - \sin^2 t}{\sin t} = \frac{1}{\sin t} - \frac{\sin^2 t}{\sin t}$$
$$= \csc t - \sin t$$

12. $\dfrac{1}{\tan \beta} + \tan \beta = \dfrac{1 + \tan^2 \beta}{\tan \beta}$
$$= \frac{\sec^2 \beta}{\tan \beta}$$

13. $\sin^{1/2} x \cos x - \sin^{5/2} x \cos x = \sin^{1/2} x \cos x (1 - \sin^2 x) = \sin^{1/2} x \cos x \cdot \cos^2 x = \cos^3 x \sqrt{\sin x}$

14. $\sec^6 x (\sec x \tan x) - \sec^4 x (\sec x \tan x) = \sec^4 x (\sec x \tan x)(\sec^2 x - 1)$
$$= \sec^4 x (\sec x \tan x) \tan^2 x$$
$$= \sec^5 x \tan^3 x$$

15. $\dfrac{1}{\sec x \tan x} = \cos x \cot x = \cos x \cdot \dfrac{\cos x}{\sin x}$
$$= \frac{\cos^2 x}{\sin x}$$
$$= \frac{1 - \sin^2 x}{\sin x}$$
$$= \frac{1}{\sin x} - \sin x$$
$$= \csc x - \sin x$$

16. $\dfrac{\sec \theta - 1}{1 - \cos \theta} = \dfrac{\sec \theta - 1}{1 - (1/\sec \theta)} \cdot \dfrac{\sec \theta}{\sec \theta}$
$$= \frac{\sec \theta (\sec \theta - 1)}{\sec \theta - 1}$$
$$= \sec \theta$$

17. $\csc x - \sin x = \dfrac{1}{\sin x} - \dfrac{\sin^2 x}{\sin x}$

$$= \dfrac{1 - \sin^2 x}{\sin x}$$

$$= \dfrac{\cos^2 x}{\sin x}$$

$$= \dfrac{\cos x}{1} \cdot \dfrac{\cos x}{\sin x}$$

$$= \cos x \cot x$$

18. $\sec x - \cos x = \dfrac{1}{\cos x} - \cos x$

$$= \dfrac{1 - \cos^2 x}{\cos x}$$

$$= \dfrac{\sin^2 x}{\cos x}$$

$$= \sin x \cdot \dfrac{\sin x}{\cos x}$$

$$= \sin x \tan x$$

19. $\dfrac{1}{\tan x} + \dfrac{1}{\cot x} = \dfrac{\cot x + \tan x}{\tan x \cot x}$

$$= \dfrac{\cot x + \tan x}{1}$$

$$= \tan x + \cot x$$

20. $\dfrac{1}{\sin x} - \dfrac{1}{\csc x} = \dfrac{\csc x - \sin x}{\sin x \csc x}$

$$= \dfrac{\csc x - \sin x}{1}$$

$$= \csc x - \sin x$$

21. $\dfrac{\cos \theta \cot \theta}{1 - \sin \theta} - 1 = \dfrac{\cos \theta \cot \theta - (1 - \sin \theta)}{1 - \sin \theta}$

$$= \dfrac{\cos \theta \left(\dfrac{\cos \theta}{\sin \theta}\right) - 1 + \sin \theta}{1 - \sin \theta} \cdot \dfrac{\sin \theta}{\sin \theta}$$

$$= \dfrac{\cos^2 \theta - \sin \theta + \sin^2 \theta}{\sin \theta (1 - \sin \theta)}$$

$$= \dfrac{1 - \sin \theta}{\sin \theta (1 - \sin \theta)}$$

$$= \dfrac{1}{\sin \theta}$$

$$= \csc \theta$$

22. $\dfrac{1 + \sin \theta}{\cos \theta} + \dfrac{\cos \theta}{1 + \sin \theta} = \dfrac{(1 + \sin \theta)^2 + \cos^2 \theta}{\cos \theta (1 + \sin \theta)}$

$$= \dfrac{1 + 2 \sin \theta + \sin^2 \theta + \cos^2 \theta}{\cos \theta (1 + \sin \theta)}$$

$$= \dfrac{2 + 2 \sin \theta}{\cos \theta (1 + \sin \theta)}$$

$$= \dfrac{2(1 + \sin \theta)}{\cos \theta (1 + \sin \theta)}$$

$$= \dfrac{2}{\cos \theta}$$

$$= 2 \sec \theta$$

23. $\dfrac{1}{\sin x + 1} + \dfrac{1}{\csc x + 1} = \dfrac{\csc x + 1 + \sin x + 1}{(\sin x + 1)(\csc x + 1)}$

$$= \dfrac{\sin x + \csc x + 2}{\sin x \csc x + \sin x + \csc x + 1}$$

$$= \dfrac{\sin x + \csc x + 2}{1 + \sin x + \csc x + 1}$$

$$= \dfrac{\sin x + \csc x + 2}{\sin x + \csc x + 2}$$

$$= 1$$

24. $\cos x - \dfrac{\cos x}{1 - \tan x} = \dfrac{\cos x (1 - \tan x) - \cos x}{1 - \tan x}$

$$= \dfrac{-\cos x \tan x}{1 - \tan x}$$

$$= \dfrac{-\cos x (\sin x / \cos x)}{1 - (\sin x / \cos x)} \cdot \dfrac{\cos x}{\cos x}$$

$$= \dfrac{-\sin x \cos x}{\cos x - \sin x}$$

$$= \dfrac{\sin x \cos x}{\sin x - \cos x}$$

25. $\tan\left(\dfrac{\pi}{2} - \theta\right) \tan \theta = \cot \theta \tan \theta$

$$= \left(\dfrac{1}{\tan \theta}\right) \tan \theta$$

$$= 1$$

26. $\dfrac{\cos[(\pi/2) - x]}{\sin[(\pi/2) - x]} = \dfrac{\sin x}{\cos x} = \tan x$

27. $\dfrac{\csc(-x)}{\sec(-x)} = \dfrac{1/\sin(-x)}{1/\cos(-x)}$

$$= \dfrac{\cos(-x)}{\sin(-x)}$$

$$= \dfrac{\cos x}{-\sin x}$$

$$= -\cot x$$

28. $(1 + \sin y)[1 + \sin(-y)] = (1 + \sin y)(1 - \sin y)$

$$= 1 - \sin^2 y$$

$$= \cos^2 y$$

29. $\dfrac{\tan x \cot x}{\cos x} = \dfrac{1}{\cos x} = \sec x$

30. $\dfrac{\tan x + \tan y}{1 - \tan x \tan y} = \dfrac{\dfrac{1}{\cot x} + \dfrac{1}{\cot y}}{1 - \dfrac{1}{\cot x} \cdot \dfrac{1}{\cot y}} \cdot \dfrac{\cot x \cot y}{\cot x \cot y}$

$$= \dfrac{\cot y + \cot x}{\cot x \cot y - 1}$$

31. $\dfrac{\tan x + \cot y}{\tan x \cot y} = \dfrac{\dfrac{1}{\cot x} + \dfrac{1}{\tan y}}{\dfrac{1}{\cot x} \cdot \dfrac{1}{\tan y}} \cdot \dfrac{\cot x \tan y}{\cot x \tan y}$

$$= \tan y + \cot x$$

32. $\dfrac{\cos x - \cos y}{\sin x + \sin y} + \dfrac{\sin x - \sin y}{\cos x + \cos y} = \dfrac{(\cos x - \cos y)(\cos x + \cos y) + (\sin x - \sin y)(\sin x + \sin y)}{(\sin x + \sin y)(\cos x + \cos y)}$

$$= \dfrac{\cos^2 x - \cos^2 y + \sin^2 x - \sin^2 y}{(\sin x + \sin y)(\cos x + \cos y)}$$

$$= \dfrac{(\cos^2 x + \sin^2 x) - (\cos^2 y + \sin^2 y)}{(\sin x + \sin y)(\cos x + \cos y)}$$

$$= 0$$

33. $\sqrt{\dfrac{1 + \sin \theta}{1 - \sin \theta}} = \sqrt{\dfrac{1 + \sin \theta}{1 - \sin \theta} \cdot \dfrac{1 + \sin \theta}{1 + \sin \theta}}$

$$= \sqrt{\dfrac{(1 + \sin \theta)^2}{1 - \sin^2 \theta}}$$

$$= \sqrt{\dfrac{(1 + \sin \theta)^2}{\cos^2 \theta}}$$

$$= \dfrac{1 + \sin \theta}{|\cos \theta|}$$

34. $\sqrt{\dfrac{1 - \cos \theta}{1 + \cos \theta}} = \sqrt{\dfrac{1 - \cos \theta}{1 + \cos \theta} \cdot \dfrac{1 - \cos \theta}{1 - \cos \theta}}$

$$= \sqrt{\dfrac{(1 - \cos \theta)^2}{1 - \cos^2 \theta}}$$

$$= \sqrt{\dfrac{(1 - \cos \theta)^2}{\sin^2 \theta}}$$

$$= \dfrac{1 - \cos \theta}{|\sin \theta|}$$

35. $\cos^2 \beta + \cos^2\left(\dfrac{\pi}{2} - \beta\right) = \cos^2 \beta + \sin^2 \beta = 1$

36. $\sec^2 y - \cot^2\left(\dfrac{\pi}{2} - y\right) = \sec^2 y - \tan^2 y = 1$

37. $\sin t \csc\left(\dfrac{\pi}{2} - t\right) = \sin t \sec t = \sin t\left(\dfrac{1}{\cos t}\right)$

$$= \dfrac{\sin t}{\cos t} = \tan t$$

38. $\sec^2\left(\dfrac{\pi}{2} - x\right) - 1 = \csc^2 x - 1 = \cot^2 x$

39. (a)

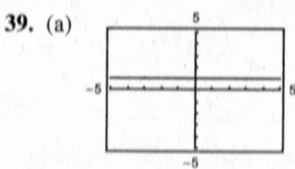

Let $y_1 = \dfrac{2}{(\cos x)^2} - \dfrac{2(\sin x)^2}{(\cos x)^2} - (\sin x)^2 - (\cos x)^2$
and $y_2 = 1$.

Identity

(b)

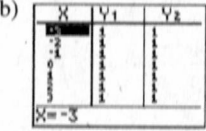

Identity

—CONTINUED—

39. —CONTINUED—

(c) $2\sec^2 x - 2\sec^2 x\sin^2 x - \sin^2 x - \cos^2 x = 2\sec^2 x(1 - \sin^2 x) - (\sin^2 x + \cos^2 x)$

$$= 2\sec^2 x(\cos^2 x) - 1$$

$$= 2 \cdot \frac{1}{\cos^2 x} \cdot \cos^2 x - 1$$

$$= 2 - 1$$

$$= 1$$

40. (a)

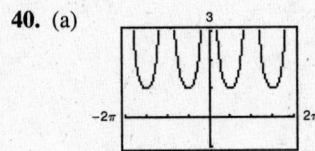

Identity

(b)

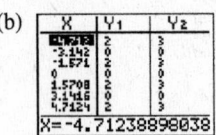

Identity

(c) $\csc x(\csc x - \sin x) + \dfrac{\sin x - \cos x}{\sin x} + \cot x = \csc^2 x - \csc x\sin x + 1 - \dfrac{\cos x}{\sin x} + \cot x$

$$= \csc^2 x - 1 + 1 - \cot x + \cot x$$

$$= \csc^2 x$$

41. (a)

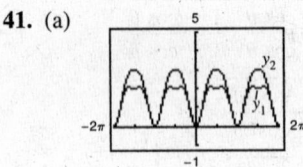

Let $y_1 = 2 + (\cos x)^2 - 3(\cos x)^4$ and

$y_2 = (\sin x)^2(3 + 2(\cos x)^2)$.

Not an identity

(b)

Not an identity

(c) $2 + \cos^2 x - 3\cos^4 x = (1 - \cos^2 x)(2 + 3\cos^2 x)$

$$= \sin^2 x(2 + 3\cos^2 x)$$

$$\neq \sin^2 x(3 + 2\cos^2 x)$$

42. (a)

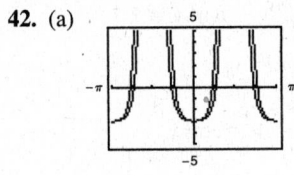

Not an identity

(b)

Not an identity

(c) $\tan^4 x + \tan^2 x - 3 = \dfrac{\sin^4 x}{\cos^4 x} + \dfrac{\sin^2 x}{\cos^2 x} - 3$

$$= \frac{1}{\cos^2 x}\left(\frac{\sin^4 x}{\cos^2 x} + \sin^2 x\right) - 3$$

$$= \frac{1}{\cos^2 x}\left(\frac{\sin^4 x + \sin^2 x\cos^2 x}{\cos^2 x}\right) - 3$$

$$= \frac{1}{\cos^2 x}\left(\frac{\sin^2 x}{\cos^2 x}\right)(\sin^2 x + \cos^2 x) - 3$$

$$= \frac{1}{\cos^2 x}\left(\frac{\sin^2 x}{\cos^2 x} \cdot 1\right) - 3$$

$$= \sec^2 x\tan^2 x - 3$$

$$\neq \sec^2 x(4\tan^2 x - 3)$$

43. (a)

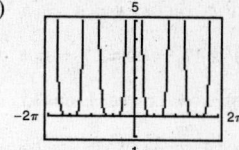

Let $y_1 = \dfrac{1}{(\sin x)^4} - \dfrac{2}{(\sin x)^2} + 1$ and $y_2 = \dfrac{1}{(\tan x)^4}$.

Identity

(b)

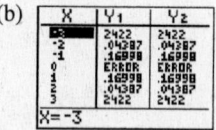

Identity

(c) $\csc^4 x - 2\csc^2 x + 1 = (\csc^2 x - 1)^2$
$$= (\cot^2 x)^2 = \cot^4 x$$

44. (a)

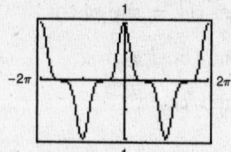

Identity

(b)

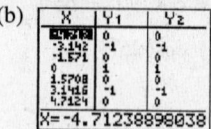

Identity

(c) $(\sin^4 \beta - 2\sin^2 \beta + 1)\cos \beta = (\sin^2 \beta - 1)^2 \cos \beta$
$$= (-\cos^2 \beta)^2 \cos \beta$$
$$= \cos^5 \beta$$

45. (a)

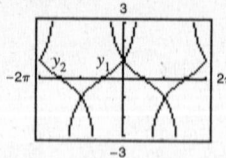

Let $y_1 = \dfrac{\cos x}{(1 - \sin x)}$ and $y_2 = \dfrac{(1 - \sin x)}{\cos x}$.

Not an identity

(b)

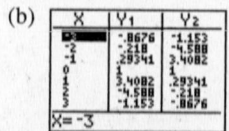

Not an identity

(c) $\dfrac{\cos x}{1 - \sin x} = \dfrac{\cos x}{1 - \sin x} \cdot \dfrac{1 + \sin x}{1 + \sin x}$

$$= \dfrac{\cos x(1 + \sin x)}{1 - \sin^2 x}$$

$$= \dfrac{\cos x(1 + \sin x)}{\cos^2 x} = \dfrac{1 + \sin x}{\cos x}$$

46. (a)

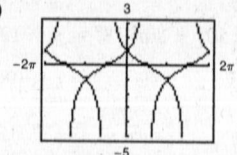

Not an identity

(b)

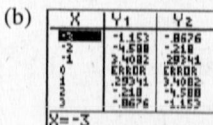

Not an identity

(c) $\dfrac{\cot \alpha}{\csc \alpha + 1}$ is the reciprocal of $\dfrac{\csc \alpha + 1}{\cot \alpha}$.

They will only be equivalent at isolated points in their respective domains. Hence, not an identity.

47. $\tan^3 x \sec^2 x - \tan^3 x = \tan^3 x(\sec^2 x - 1)$
$$= \tan^3 x \tan^2 x$$
$$= \tan^5 x$$

48. $(\tan^2 x + \tan^4 x)\sec^2 x = \left(\dfrac{\sin^2 x}{\cos^2 x} + \dfrac{\sin^4 x}{\cos^4 x}\right)\dfrac{1}{\cos^2 x}$

$$= \dfrac{1}{\cos^4 x}\left(\sin^2 x + \dfrac{\sin^4 x}{\cos^2 x}\right)$$

$$= \dfrac{1}{\cos^4 x}\left(\dfrac{\sin^2 x \cos^2 x + \sin^4 x}{\cos^2 x}\right)$$

$$= \dfrac{1}{\cos^4 x}\left(\dfrac{\sin^2 x(\cos^2 x + \sin^2 x)}{\cos^2 x}\right)$$

$$= \dfrac{1}{\cos^4 x}\left(\dfrac{\sin^2 x}{\cos^2 x} \cdot 1\right) = \sec^4 x \cdot \tan^2 x$$

49. $(\sin^2 x - \sin^4 x) \cos x = \sin^2 x (1 - \sin^2 x) \cos x$

$$= \sin^2 x \cos^2 x \cos x$$

$$= \sin^2 x \cos^3 x$$

50. $\sin^4 x + \cos^4 x = \sin^2 x \sin^2 x + \cos^4 x$

$$= (1 - \cos^2 x)(1 - \cos^2 x) + \cos^4 x$$

$$= 1 - 2 \cos^2 x + \cos^4 x + \cos^4 x$$

$$= 1 - 2 \cos^2 x + 2 \cos^4 x$$

51. $\sin^2 25° + \sin^2 65° = \sin^2 25° + \cos^2(90° - 65°)$

$$= \sin^2 25° + \cos^2 25°$$

$$= 1$$

52. $\cos^2 55° + \cos^2 35° = \cos^2 55° + \sin^2(90° - 35°)$

$$= \cos^2 55° + \sin^2 55°$$

$$= 1$$

53. $\cos^2 20° + \cos^2 52° + \cos^2 38° + \cos^2 70° = \cos^2 20° + \cos^2 52° + \sin^2(90° - 38°) + \sin^2(90° - 70°)$

$$= \cos^2 20° + \cos^2 52° + \sin^2 52° + \sin^2 20°$$

$$= (\cos^2 20° + \sin^2 20°) + (\cos^2 52° + \sin^2 52°)$$

$$= 1 + 1$$

$$= 2$$

54. $\sin^2 12° + \sin^2 40° + \sin^2 50° + \sin^2 78° = \sin^2 12° + \sin^2 78° + \sin^2 40° + \sin^2 50°$

$$= \cos^2(90° - 12°) + \sin^2 78° + \cos^2(90° - 40°) + \sin^2 50°$$

$$= \cos^2 78° + \sin^2 78° + \cos^2 50° + \sin^2 50°$$

$$= 1 + 1 = 2$$

55. $\cos x - \csc x \cot x = \cos x - \dfrac{1}{\sin x} \dfrac{\cos x}{\sin x}$

$$= \cos x \left(1 - \dfrac{1}{\sin^2 x} \right)$$

$$= \cos x (1 - \csc^2 x)$$

$$= -\cos x (\csc^2 x - 1)$$

$$= -\cos x \cot^2 x$$

56. (a) $\dfrac{h \sin(90° - \theta)}{\sin \theta} = \dfrac{h \cos \theta}{\sin \theta} = h \cot \theta$

(b)

θ	10°	20°	30°	40°	50°	60°	70°	80°	90°
s	28.36	13.74	8.66	5.96	4.20	2.89	1.82	0.88	0

(c) Greatest: 10°, Least: 90°

(d) Noon

57. False. For the equation to be an identity, it must be true for all values of θ in the domain.

58. True. An identity is an equation that is true for all real values in the domain of the variable.

59. Since $\sin^2 \theta = 1 - \cos^2 \theta$, then $\sin \theta = \pm \sqrt{1 - \cos^2 \theta}$;

$\sin \theta \neq \sqrt{1 - \cos^2 \theta}$ if θ lies in Quadrant III or IV. One such angle is $\theta = \dfrac{7\pi}{4}$.

60. $\tan \theta = \sqrt{\sec^2 \theta - 1}$

True identity: $\tan \theta = \pm \sqrt{\sec^2 \theta - 1}$

$\tan \theta = \sqrt{\sec^2 \theta - 1}$ is not true for $\pi/2 < \theta < \pi$ or $3\pi/2 < \theta < 2\pi$. Thus, the equation is not true for $\theta = 3\pi/4$.

61. $(2 + 3i) - \sqrt{-26} = 2 + 3i - \sqrt{26}i$

$\qquad\qquad\qquad = 2 + (3 - \sqrt{26})i$

62. $(2 - 5i)^2 = (2 - 5i)(2 - 5i)$

$\qquad\qquad = 4 - 20i + 25i^2$

$\qquad\qquad = 4 - 20i - 25$

$\qquad\qquad = -21 - 20i$

63. $\sqrt{-16}(1 + \sqrt{-4}) = 4i(1 + 2i)$

$\qquad\qquad\qquad = 4i + 8i^2$

$\qquad\qquad\qquad = 4i - 8$

$\qquad\qquad\qquad = -8 + 4i$

64. $(3 + 2i)^3 = (3 + 2i)(3 + 2i)(3 + 2i)$

$\qquad\qquad = (9 + 12i + 4i^2)(3 + 2i)$

$\qquad\qquad = (5 + 12i)(3 + 2i)$

$\qquad\qquad = 15 + 10i + 36i + 24i^2$

$\qquad\qquad = -9 + 46i$

65. $x^2 + 6x - 12 = 0$

$\qquad a = 1, b = 6, c = -12$

$\qquad x = \dfrac{-6 \pm \sqrt{6^2 - 4(1)(-12)}}{2(1)}$

$\qquad = \dfrac{-6 \pm \sqrt{36 + 48}}{2}$

$\qquad = \dfrac{-6 \pm \sqrt{84}}{2}$

$\qquad = \dfrac{-6 \pm 2\sqrt{21}}{2}$

$\qquad = -3 \pm \sqrt{21}$

66. $x^2 + 5x - 7 = 0$

$\qquad a = 1, b = 5, c = -7$

$\qquad x = \dfrac{-5 \pm \sqrt{5^2 + (4)(1)(7)}}{2(1)}$

$\qquad x = \dfrac{-5 \pm \sqrt{53}}{2}$

67. $3x^2 - 6x - 12 = 0$

$\qquad 3(x^2 - 2x - 4) = 0$

$\qquad x^2 - 2x - 4 = 0$

$\qquad a = 1, b = -2, c = -4$

$\qquad x = \dfrac{-(-2) \pm \sqrt{(-2)^2 - 4(1)(-4)}}{2(1)}$

$\qquad = \dfrac{2 \pm \sqrt{4 + 16}}{2}$

$\qquad = \dfrac{2 \pm \sqrt{20}}{2}$

$\qquad = \dfrac{2 \pm 2\sqrt{5}}{2}$

$\qquad = 1 \pm \sqrt{5}$

68. $8x^2 - 4x - 3 = 0$

$\qquad a = 8, b = -4, c = -3$

$\qquad x = \dfrac{-(-4) \pm \sqrt{(-4)^2 + 4(8)(3)}}{2(8)}$

$\qquad = \dfrac{4 \pm \sqrt{112}}{16}$

$\qquad x = \dfrac{4 \pm 4\sqrt{7}}{16}$

$\qquad x = \dfrac{1}{4}(1 \pm \sqrt{7})$

Section 5.3 Solving Trigonometric Equations

- You should be able to identify and solve trigonometric equations.
- A trigonometric equation is a conditional equation. It is true for a specific set of values.
- To solve trigonometric equations, use algebraic techniques such as collecting like terms, extracting square roots, factoring, squaring, converting to quadratic type, using formulas, and using inverse functions. Study the examples in this section.

Vocabulary Check

1. general   2. quadratic   3. extraneous

1. $2 \cos x - 1 = 0$

   (a) $2 \cos \dfrac{\pi}{3} - 1 = 2\left(\dfrac{1}{2}\right) - 1 = 0$

   (b) $2 \cos \dfrac{5\pi}{3} - 1 = 2\left(\dfrac{1}{2}\right) - 1 = 0$

2. $\sec x - 2 = 0$

   (a) $x = \dfrac{\pi}{3}$

   $$\sec \dfrac{\pi}{3} - 2 = \dfrac{1}{\cos(\pi/3)} - 2$$

   $$= \dfrac{1}{1/2} - 2 = 2 - 2 = 0$$

   (b) $x = \dfrac{5\pi}{3}$

   $$\sec \dfrac{5\pi}{3} - 2 = \dfrac{1}{\cos(5\pi/3)} - 2$$

   $$= \dfrac{1}{1/2} - 2 = 2 - 2 = 0$$

3. $3 \tan^2 2x - 1 = 0$

   (a) $3\left[\tan 2\left(\dfrac{\pi}{12}\right)\right]^2 - 1 = 3 \tan^2 \dfrac{\pi}{6} - 1$

   $$= 3\left(\dfrac{1}{\sqrt{3}}\right)^2 - 1$$

   $$= 0$$

   (b) $3\left[\tan 2\left(\dfrac{5\pi}{12}\right)\right]^2 - 1 = 3 \tan^2 \dfrac{5\pi}{6} - 1$

   $$= 3\left(-\dfrac{1}{\sqrt{3}}\right)^2 - 1$$

   $$= 0$$

4. $2 \cos^2 4x - 1 = 0$

   (a) $x = \dfrac{\pi}{16}$

   $$2 \cos^2\left[4\left(\dfrac{\pi}{16}\right)\right] - 1 = 2 \cos^2 \dfrac{\pi}{4} - 1$$

   $$= 2\left(\dfrac{\sqrt{2}}{2}\right)^2 - 1$$

   $$= 2\left(\dfrac{1}{2}\right) - 1 = 1 - 1 = 0$$

   (b) $x = \dfrac{3\pi}{16}$

   $$2 \cos^2\left[4\left(\dfrac{3\pi}{16}\right)\right] - 1 = 2 \cos^2 \dfrac{3\pi}{4} - 1$$

   $$= 2\left(-\dfrac{\sqrt{2}}{2}\right)^2 - 1$$

   $$= 2\left(\dfrac{1}{2}\right) - 1 = 0$$

5. $2 \sin^2 x - \sin x - 1 = 0$

(a) $2 \sin^2 \dfrac{\pi}{2} - \sin \dfrac{\pi}{2} - 1 = 2(1)^2 - 1 - 1$

$$= 0$$

(b) $2 \sin^2 \dfrac{7\pi}{6} - \sin \dfrac{7\pi}{6} - 1 = 2\left(-\dfrac{1}{2}\right)^2 - \left(-\dfrac{1}{2}\right) - 1$

$$= \dfrac{1}{2} + \dfrac{1}{2} - 1$$

$$= 0$$

6. $\csc^4 x - 4 \csc^2 x = 0$

(a) $x = \dfrac{\pi}{6}$

$$\csc^4 \dfrac{\pi}{6} - 4 \csc^2 \dfrac{\pi}{6} = \dfrac{1}{\sin^4(\pi/6)} - \dfrac{4}{\sin^2(\pi/6)}$$

$$= \dfrac{1}{(1/2)^4} - \dfrac{4}{(1/2)^2}$$

$$= 16 - 16 = 0$$

(b) $x = \dfrac{5\pi}{6}$

$$\csc^4 \dfrac{5\pi}{6} - 4 \csc \dfrac{5\pi}{6} = \dfrac{1}{\sin^4(5\pi/6)} - \dfrac{4}{\sin^2(5\pi/6)}$$

$$= \dfrac{1}{(1/2)^4} - \dfrac{4}{(1/2)^2}$$

$$= 16 - 16 = 0$$

7. $2 \cos x + 1 = 0$

$$2 \cos x = -1$$

$$\cos x = -\dfrac{1}{2}$$

$$x = \dfrac{2\pi}{3} + 2n\pi$$

$$\text{or } x = \dfrac{4\pi}{3} + 2n\pi$$

8. $2 \sin x + 1 = 0$

$$\sin x = -\dfrac{1}{2}$$

$$x = \dfrac{7\pi}{6} + 2n\pi$$

$$\text{or } x = \dfrac{11\pi}{6} + 2n\pi$$

9. $\sqrt{3} \csc x - 2 = 0$

$$\sqrt{3} \csc x = 2$$

$$\csc x = \dfrac{2}{\sqrt{3}}$$

$$x = \dfrac{\pi}{3} + 2n\pi$$

$$\text{or } x = \dfrac{2\pi}{3} + 2n\pi$$

10. $\tan x + \sqrt{3} = 0$

$$\tan x = -\sqrt{3}$$

$$x = \dfrac{2\pi}{3} + n\pi$$

11. $3 \sec^2 x - 4 = 0$

$$\sec^2 x = \dfrac{4}{3}$$

$$\sec x = \pm \dfrac{2}{\sqrt{3}}$$

$$x = \dfrac{\pi}{6} + n\pi$$

$$\text{or } x = \dfrac{5\pi}{6} + n\pi$$

12. $3 \cot^2 x - 1 = 0$

$$\cot^2 x = \dfrac{1}{3}$$

$$\cot x = \pm \dfrac{1}{\sqrt{3}}$$

$$x = \dfrac{\pi}{3} + n\pi$$

$$\text{or } x = \dfrac{2\pi}{3} + n\pi$$

13. $\sin x(\sin x + 1) = 0$

$$\sin x = 0 \quad \text{or} \quad \sin x = -1$$

$$x = n\pi \qquad x = \dfrac{3\pi}{2} + 2n\pi$$

14. $(3 \tan^2 x - 1)(\tan^2 x - 3) = 0$

$$3 \tan^2 x - 1 = 0 \qquad \text{or} \quad \tan^2 x - 3 = 0$$

$$\tan x = \pm \dfrac{1}{\sqrt{3}} \qquad\qquad \tan x = \pm \sqrt{3}$$

$$x = \dfrac{\pi}{6} + n\pi \qquad\qquad x = \dfrac{\pi}{3} + n\pi$$

$$\text{or } x = \dfrac{5\pi}{6} + n\pi \qquad \text{or } x = \dfrac{2\pi}{3} + n\pi$$

15. $4\cos^2 x - 1 = 0$

$$\cos^2 x = \frac{1}{4}$$

$$\cos^2 x = \pm\frac{1}{2}$$

$$x = \frac{\pi}{3} + n\pi \quad \text{or} \quad x = \frac{2\pi}{3} + n\pi$$

16.
$$\sin^2 x = 3\cos^2 x$$

$$\sin^2 x - 3(1 - \sin^2 x) = 0$$

$$4\sin^2 x = 3$$

$$\sin x = \pm\frac{\sqrt{3}}{2}$$

$$x = \frac{\pi}{3} + n\pi \quad \text{or} \quad x = \frac{2\pi}{3} + n\pi$$

17. $2\sin^2 2x = 1$

$$\sin 2x = \pm\frac{1}{\sqrt{2}} = \pm\frac{\sqrt{2}}{2}$$

$$2x = \frac{\pi}{4} + 2n\pi, \; 2x = \frac{3\pi}{4} + 2n\pi,$$

$$2x = \frac{5\pi}{4} + 2n\pi, \; 2x = \frac{7\pi}{4} + 2n\pi$$

Thus, $x = \dfrac{\pi}{8} + n\pi, \dfrac{3\pi}{8} + n\pi, \dfrac{5\pi}{8} + n\pi, \dfrac{7\pi}{8} + n\pi.$

We can combine these as follows:

$$x = \frac{\pi}{8} + \frac{n\pi}{2}, x = \frac{3\pi}{8} + \frac{n\pi}{2}$$

18. $\tan^2 3x = 3$

$$\tan 3x = \pm\sqrt{3}$$

$$3x = \frac{\pi}{3} + n\pi \Rightarrow x = \frac{\pi}{9} + \frac{n\pi}{3}$$

or

$$3x = \frac{2\pi}{3} + n\pi \Rightarrow x = \frac{2\pi}{9} + \frac{n\pi}{3}$$

19. $\tan 3x(\tan x - 1) = 0$

$$\tan 3x = 0 \quad \text{or} \quad \tan x - 1 = 0$$

$$3x = n\pi \qquad\qquad \tan x = 1$$

$$x = \frac{n\pi}{3} \qquad\qquad\quad x = \frac{\pi}{4} + n\pi$$

20. $\cos 2x(2\cos x + 1) = 0$

$$\cos 2x = 0 \qquad \text{or} \qquad 2\cos x + 1 = 0$$

$$2x = \frac{\pi}{2} + n\pi \qquad\qquad \cos x = -\frac{1}{2}$$

$$x = \frac{\pi}{4} + \frac{n\pi}{2} \qquad\qquad x = \frac{2\pi}{3} + 2n\pi$$

$$\text{or} \qquad\qquad x = \frac{4\pi}{3} + 2n\pi$$

21.
$$\cos^3 x = \cos x$$

$$\cos^3 x - \cos x = 0$$

$$\cos x(\cos^2 x - 1) = 0$$

$$\cos x = 0 \quad \text{or} \quad \cos^2 x - 1 = 0$$

$$x = \frac{\pi}{2}, \frac{3\pi}{2} \qquad\qquad \cos x = \pm 1$$

$$x = 0, \pi$$

22. $\sec^2 x - 1 = 0$

$$\sec^2 x = 1$$

$$\sec x = \pm 1$$

$$x = 0 \text{ or } x = \pi$$

23.
$$3\tan^3 x - \tan x = 0$$

$$\tan x(3\tan^2 x - 1) = 0$$

$$\tan x = 0 \quad \text{or} \quad 3\tan^2 x - 1 = 0$$

$$x = 0, \pi \qquad\qquad \tan x = \pm\frac{\sqrt{3}}{3}$$

$$x = \frac{\pi}{6}, \frac{5\pi}{6}, \frac{7\pi}{6}, \frac{11\pi}{6}$$

24.
$$2\sin^2 x = 2 + \cos x$$
$$2 - 2\cos^2 x = 2 + \cos x$$
$$2\cos^2 x + \cos x = 0$$
$$\cos x(2\cos x + 1) = 0$$

$\cos x = 0$ or $2\cos x + 1 = 0$

$x = \dfrac{\pi}{2}, \dfrac{3\pi}{2}$ $2\cos x = -1$

$$\cos x = -\frac{1}{2}$$

$$x = \frac{2\pi}{3}, \frac{4\pi}{3}$$

25.
$$\sec^2 x - \sec x - 2 = 0$$
$$(\sec x - 2)(\sec x + 1) = 0$$

$\sec x - 2 = 0$ or $\sec x + 1 = 0$

$\sec x = 2$ $\sec x = -1$

$x = \dfrac{\pi}{3}, \dfrac{5\pi}{3}$ $x = \pi$

26.
$$\sec x \csc x = 2\csc x$$
$$\sec x \csc x - 2\csc x = 0$$
$$\csc x(\sec x - 2) = 0$$

$\csc x = 0$ or $\sec x - 2 = 0$

No solution $\sec x = 2$

$$x = \frac{\pi}{3}, \frac{5\pi}{3}$$

27. $2\sin x + \csc x = 0$

$$2\sin x + \frac{1}{\sin x} = 0$$

$$2\sin^2 x + 1 = 0$$

$$\sin^2 x = -\frac{1}{2} \implies \text{No solution}$$

28.
$$\sec x + \tan x = 1$$

$$\frac{1}{\cos x} + \frac{\sin x}{\cos x} = 1$$

$$1 + \sin x = \cos x$$

$$(1 + \sin x)^2 = \cos^2 x$$

$$1 + 2\sin x + \sin^2 x = \cos^2 x$$

$$1 + 2\sin x + \sin^2 x = 1 - \sin^2 x$$

$$2\sin^2 x + 2\sin x = 0$$

$$2\sin x(\sin x + 1) = 0$$

$\sin x = 0$ or $\sin x + 1 = 0$

$x = 0, \pi$ $\sin x = -1$

(π is extraneous.) $x = \dfrac{3\pi}{2}$

$$\left(\frac{3\pi}{2} \text{ is extraneous.}\right)$$

$x = 0$ is the only solution.

29.
$$2\cos^2 x + \cos x - 1 = 0$$
$$(2\cos x - 1)(\cos x + 1) = 0$$

$2\cos x - 1 = 0$ or $\cos x + 1 = 0$

$\cos x = \dfrac{1}{2}$ $\cos x = -1$

$x = \dfrac{\pi}{3}, \dfrac{5\pi}{3}$. $x = \pi$

30. $2\sin^2 x + 3\sin x + 1 = 0$

$(2\sin x + 1)(\sin x + 1) = 0$

$2\sin x + 1 = 0$ or $\sin x + 1 = 0$

$\sin x = -\dfrac{1}{2}$ $\sin x = -1$

$x = \dfrac{3\pi}{2}$

$x = \dfrac{7\pi}{6}, \dfrac{11\pi}{6}$

31.
$$2\sec^2 x + \tan^2 x - 3 = 0$$
$$2(\tan^2 x + 1) + \tan^2 x - 3 = 0$$
$$3\tan^2 x - 1 = 0$$

$$\tan x = \pm\frac{\sqrt{3}}{3}$$

$$x = \frac{\pi}{6}, \frac{5\pi}{6}, \frac{7\pi}{6}, \frac{11\pi}{6}$$

32. $\cos x + \sin x \tan x = 2$

$$\cos x + \sin x\left(\frac{\sin x}{\cos x}\right) = 2$$

$$\frac{\cos^2 x + \sin^2 x}{\cos x} = 2$$

$$\frac{1}{\cos x} = 2$$

$$\cos x = \frac{1}{2}$$

$$x = \frac{\pi}{3}, \frac{5\pi}{3}$$

33.

$$\csc x + \cot x = 1$$

$$(\csc x + \cot x)^2 = 1^2$$

$$\csc^2 x + 2 \csc x \cot x + \cot^2 x = 1$$

$$\cot^2 x + 1 + 2 \csc x \cot x + \cot^2 x = 1$$

$$2 \cot^2 x + 2 \csc x \cot x = 0$$

$$2 \cot x(\cot x + \csc x) = 0$$

$$2 \cot x = 0 \qquad \text{or} \quad \cot x + \csc x = 0$$

$$x = \frac{\pi}{2}, \frac{3\pi}{2} \qquad\qquad \frac{\cos x}{\sin x} = -\frac{1}{\sin x}$$

$$\cos x = -1$$

$$x = \pi$$

By checking in the original equation, we find that $x = \pi$ and $x = 3\pi/2$ are extraneous. The only solution to the equation in the interval $[0, 2\pi)$ is $x = \pi/2$.

34. $\sin x - 2 = \cos x - 2$

$$\sin x = \cos x$$

$$\frac{\sin x}{\cos x} = 1$$

$$\tan x = 1$$

$$x = \tan^{-1} 1$$

$$x = \frac{\pi}{4}, \frac{5\pi}{4}$$

35. $\cos 2x = \frac{1}{2}$

$$2x = \frac{\pi}{3} + 2n\pi \quad \text{or} \quad 2x = \frac{5\pi}{3} + 2n\pi$$

$$x = \frac{\pi}{6} + n\pi \qquad\qquad x = \frac{5\pi}{6} + n\pi$$

36. $\sin 2x = -\dfrac{\sqrt{3}}{2}$

$$2x = \frac{4\pi}{3} + 2n\pi \qquad \text{or} \qquad 2x = \frac{5\pi}{3} + 2n\pi$$

$$x = \frac{2\pi}{3} + n\pi \qquad\qquad x = \frac{5\pi}{6} + n\pi$$

37. $\tan 3x = 1$

$$3x = \frac{\pi}{4} + 2n\pi \qquad \text{or} \qquad 3x = \frac{5\pi}{4} + 2n\pi$$

$$x = \frac{\pi}{12} + \frac{2n\pi}{3} \qquad\qquad x = \frac{5\pi}{12} + \frac{2n\pi}{3}$$

These can be combined as $x = \dfrac{\pi}{12} + \dfrac{n\pi}{3}$.

38. $\sec 4x = 2$

$$4x = \frac{\pi}{3} + 2n\pi \qquad \text{or} \qquad 4x = \frac{5\pi}{3} + 2n\pi$$

$$x = \frac{\pi}{12} + \frac{n\pi}{2} \qquad\qquad x = \frac{5\pi}{12} + \frac{n\pi}{2}$$

39. $\cos\left(\dfrac{x}{2}\right) = \dfrac{\sqrt{2}}{2}$

$$\frac{x}{2} = \frac{\pi}{4} + 2n\pi \quad \text{or} \quad \frac{x}{2} = \frac{7\pi}{4} + 2n\pi$$

$$x = \frac{\pi}{2} + 4n\pi \qquad\qquad x = \frac{7\pi}{2} + 4n\pi$$

40. $\sin \dfrac{x}{2} = -\dfrac{\sqrt{3}}{2}$

$$\dfrac{x}{2} = \dfrac{4\pi}{3} + 2n\pi \qquad \text{or} \qquad \dfrac{x}{2} = \dfrac{5\pi}{3} + 2n\pi$$

$$x = \dfrac{8\pi}{3} + 4n\pi \qquad\qquad x = \dfrac{10\pi}{3} + 4n\pi$$

41. $y = \sin \dfrac{\pi x}{2} + 1$

From the graph in the textbook we see that the curve has x-intercepts at $x = -1$ and at $x = 3$.

In general, we have: $\sin\left(\dfrac{\pi x}{2}\right) = -1$

$$\dfrac{\pi x}{2} = \dfrac{3\pi}{2} + 2n\pi$$

$$x = 3 + 4n$$

42. $\qquad\qquad y = \sin \pi x + \cos \pi x$

$$\sin \pi x + \cos \pi x = 0$$

$$\sin \pi x = -\cos \pi x$$

$$\pi x = -\dfrac{\pi}{4} + n\pi$$

$$x = -\dfrac{1}{4} + n$$

For $-1 < x < 3$ the intercepts are $-\dfrac{1}{4}, \dfrac{3}{4}, \dfrac{7}{4}, \dfrac{11}{4}$.

43. $y = \tan^2\left(\dfrac{\pi x}{6}\right) - 3$

From the graph in the textbook we see that the curve has x-intercepts at $x = \pm 2$.

In general, we have: $\tan^2\left(\dfrac{\pi x}{6}\right) = 3$

$$\tan\left(\dfrac{\pi x}{6}\right) = \pm\sqrt{3}$$

$$\dfrac{\pi x}{6} = \pm\dfrac{\pi}{3} + n\pi$$

$$x = \pm 2 + 6n$$

44. $\qquad\qquad y = \sec^4\left(\dfrac{\pi x}{8}\right) - 4$

$$\sec^4\left(\dfrac{\pi x}{8}\right) - 4 = 0$$

$$\sec^4\left(\dfrac{\pi x}{8}\right) = 4$$

$$\sec\left(\dfrac{\pi x}{8}\right) = \pm\sqrt{2}$$

$$\dfrac{\pi x}{8} = \dfrac{\pi}{4} + \dfrac{\pi}{2}n$$

$$x = 2 + 4n$$

For $-3 < x < 3$ the intercepts are -2 and 2.

45. Graph $y_1 = 2 \sin x + \cos x$.

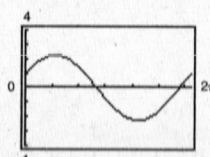

The x-intercepts occur at
$x \approx 2.678$ and $x \approx 5.820$.

46. $4 \sin^3 x + 2 \sin^2 x - 2 \sin x - 1 = 0$

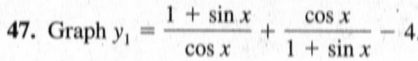

$x \approx 0.785, 2.356, 3.665, 3.927, 5.498, 5.760$

47. Graph $y_1 = \dfrac{1 + \sin x}{\cos x} + \dfrac{\cos x}{1 + \sin x} - 4$.

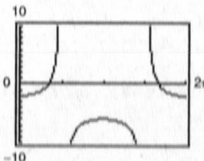

The x-intercepts occur at
$x = \dfrac{\pi}{3} \approx 1.047$ and $x = \dfrac{5\pi}{3} \approx 5.236$.

48. $\dfrac{\cos x \cot x}{1 - \sin x} = 3$

$$y_1 = \left(\dfrac{\dfrac{\cos x}{\tan x}}{1 - \sin x}\right) - 3$$

$x \approx 0.524, 2.618$

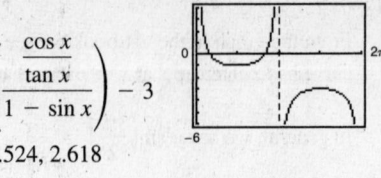

49. $x \tan x - 1 = 0$

Graph $y_1 = x \tan x - 1$.

The x-intercepts occur at $x \approx 0.860$ and $x \approx 3.426$.

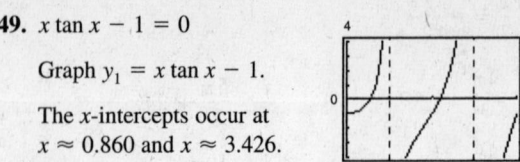

50. $x \cos x - 1 = 0$

$x \approx 4.917$

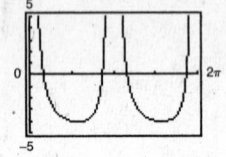

51. $\sec^2 x + 0.5 \tan x - 1 = 0$

Graph $y_1 = \dfrac{1}{(\cos x)^2} + 0.5 \tan x - 1$.

The x-intercepts occur at

$x = 0$, $x \approx 2.678$,

$x = \pi \approx 3.142$, and

$x \approx 5.820$.

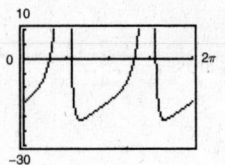

52. $\csc^2 x + 0.5 \cot x - 5 = 0$

$$y_1 = \left(\dfrac{1}{\sin x}\right)^2 + \dfrac{1}{2 \tan x} - 5$$

$x \approx 0.515, 2.726, 3.657, 5.868$

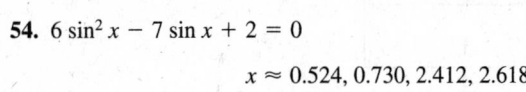

53. Graph $y_1 = 2 \tan^2 x + 7 \tan x - 15$.

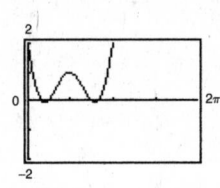

The x-intercepts occur at $x \approx 0.983$, $x \approx 1.768$, $x \approx 4.124$ and $x \approx 4.910$.

54. $6 \sin^2 x - 7 \sin x + 2 = 0$

$x \approx 0.524, 0.730, 2.412, 2.618$

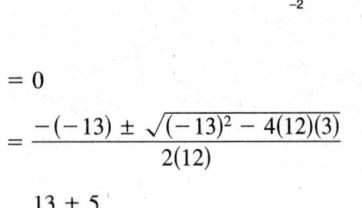

55. $12 \sin^2 x - 13 \sin x + 3 = 0$

$$\sin x = \dfrac{-(-13) \pm \sqrt{(-13)^2 - 4(12)(3)}}{2(12)}$$

$$= \dfrac{13 \pm 5}{24}$$

$\sin x = \dfrac{1}{3}$ or $\sin x = \dfrac{3}{4}$

$x \approx 0.3398, 2.8018$ $\qquad x \approx 0.8481, 2.2935$

Graph $y_1 = 12 \sin^2 x - 13 \sin x + 3$.

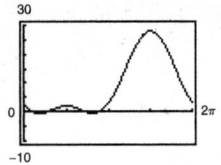

The x-intercepts occur at

$x \approx 0.3398$, $x \approx 0.8481$, $x \approx 2.2935$, and $x \approx 2.8018$.

56. $3\tan^2 x + 4\tan x - 4 = 0$

$$\tan x = \frac{-4 \pm \sqrt{4^2 - 4(3)(-4)}}{2(3)} = \frac{-4 \pm \sqrt{64}}{6} = -2, \frac{2}{3}$$

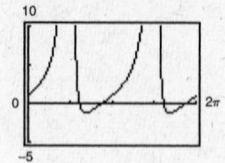

$$\tan x = -2 \qquad\qquad \tan x = \frac{2}{3}$$

$$x = \arctan(-2) + n\pi \qquad x = \arctan\left(\frac{2}{3}\right) + n\pi$$

$$\approx -1.1071 + n\pi \qquad\qquad \approx 0.5880 + n\pi$$

The values of x in $[0, 2\pi)$ are 0.5880, 3.7296, 2.0344, 5.1760.

57. $\tan^2 x + 3\tan x + 1 = 0$

$$\tan x = \frac{-3 \pm \sqrt{3^2 - 4(1)(1)}}{2(1)} = \frac{-3 \pm \sqrt{5}}{2}$$

$$\tan x = \frac{-3 - \sqrt{5}}{2} \quad \text{or} \quad \tan x = \frac{-3 + \sqrt{5}}{2}$$

$$x \approx 1.9357,\ 5.0773 \qquad x \approx 2.7767,\ 5.9183$$

Graph $y_1 = \tan^2 x + 3\tan x + 1$.

The x-intercepts occur at $x \approx 1.9357$, $x \approx 2.7767$, $x \approx 5.0773$, and $x \approx 5.9183$.

58. $4\cos^2 x - 4\cos x - 1 = 0$

$$\cos x = \frac{4 \pm \sqrt{(-4)^2 - 4(4)(-1)}}{2(4)}$$

$$= \frac{4 \pm \sqrt{32}}{8} = \frac{1 \pm \sqrt{2}}{2}$$

$$\cos x = \frac{1 - \sqrt{2}}{2} \qquad\qquad \cos x = \frac{1 + \sqrt{2}}{2}$$

$$x = \arccos\left(\frac{1 - \sqrt{2}}{2}\right) \qquad \text{No solution}$$

$$\approx 1.7794 \qquad\qquad \left(\frac{1 + \sqrt{2}}{2} > 1\right)$$

Solutions in $[0, 2\pi)$ are $\arccos\left(\dfrac{1 - \sqrt{2}}{2}\right)$ and

$2\pi - \arccos\left(\dfrac{1 - \sqrt{2}}{2}\right)$: 1.7794, 4.5038.

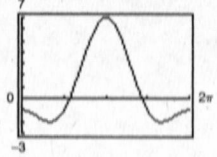

59. $\tan^2 x - 6\tan x + 5 = 0$

$$(\tan x - 1)(\tan x - 5) = 0$$

$$\tan x - 1 = 0 \quad \text{or} \quad \tan x - 5 = 0$$

$$\tan x = 1 \qquad\qquad \tan x = 5$$

$$x = \frac{\pi}{4}, \frac{5\pi}{4} \qquad\qquad x = \arctan 5,\ \arctan 5 + \pi$$

60. $\sec^2 x + \tan x - 3 = 0$

$1 + \tan^2 x + \tan x - 3 = 0$

$\tan^2 x + \tan x - 2 = 0$

$(\tan x + 2)(\tan x - 1) = 0$

$\tan x + 2 = 0$	$\tan x - 1 = 0$
$\tan x = -2$	$\tan x = 1$
$x = \arctan(-2) + n\pi$	$x = \arctan(1) + n\pi$
$\approx -1.1071 + n\pi$	$= \dfrac{\pi}{4} + n\pi$

Solutions in $[0, 2\pi)$ are $\arctan(-2) + \pi$, $\arctan(-2) + 2\pi$, $\dfrac{\pi}{4}, \dfrac{5\pi}{4}$.

61. $2\cos^2 x - 5\cos x + 2 = 0$

$(2\cos x - 1)(\cos x - 2) = 0$

| $2\cos x - 1 = 0$ or $\cos x - 2 = 0$ |

$\cos x = \dfrac{1}{2}$ $\cos x = 2$

$x = \dfrac{\pi}{3}, \dfrac{5\pi}{3}$ No solution

62. $2\sin^2 x - 7\sin x + 3 = 0$

$(\sin x - 3)(2\sin x - 1) = 0$

$\sin x - 3 = 0$ $2\sin x - 1 = 0$

No solution $\sin x = \dfrac{1}{2}$

$x = \dfrac{\pi}{6}, \dfrac{5\pi}{6}$

Solutions in $[0, 2\pi)$ are $\dfrac{\pi}{6}, \dfrac{5\pi}{6}$.

63. (a) $f(x) = \sin x + \cos x$

Maximum: $\left(\dfrac{\pi}{4}, \sqrt{2}\right)$

Minimum: $\left(\dfrac{5\pi}{4}, -\sqrt{2}\right)$

(b) $\cos x - \sin x = 0$

$\cos x = \sin x$

$1 = \dfrac{\sin x}{\cos x}$

$\tan x = 1$

$x = \dfrac{\pi}{4}, \dfrac{5\pi}{4}$

$f\left(\dfrac{\pi}{4}\right) = \sin\dfrac{\pi}{4} + \cos\dfrac{\pi}{4} = \dfrac{\sqrt{2}}{2} + \dfrac{\sqrt{2}}{2} = \sqrt{2}$

$f\left(\dfrac{5\pi}{4}\right) = \sin\dfrac{5\pi}{4} + \cos\dfrac{5\pi}{4} = -\sin\dfrac{\pi}{4} + \left(-\cos\dfrac{\pi}{4}\right) = -\dfrac{\sqrt{2}}{2} - \dfrac{\sqrt{2}}{2} = -\sqrt{2}$

Therefore, the maximum point in the interval $[0, 2\pi)$ is $\left(\pi/4, \sqrt{2}\right)$ and the minimum point is $\left(5\pi/4, -\sqrt{2}\right)$.

64. (a) $f(x) = 2 \sin x + \cos 2x$

   max: (0.5240, 1.5)   min: (1.5708, 1.0)

   max: (2.6180, 1.5)   min: (4.7124, −3.0)

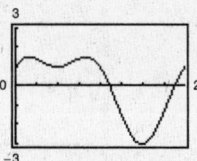

(b) $2 \cos x - 4 \sin x \cos x = 0$

   $2 \cos x (1 - 2 \sin x) = 0$

   $2 \cos x = 0$ $1 - 2 \sin x = 0$

   $x = \dfrac{\pi}{2}, \dfrac{3\pi}{2}$ $\sin x = \dfrac{1}{2}$

   $\approx 1.5708, 4.7124$ $x = \dfrac{\pi}{6}, \dfrac{5\pi}{6}$

                                 $\approx 0.5240, 2.6180$

65. $f(x) = \tan \dfrac{\pi x}{4}$

Since $\tan \pi/4 = 1$, $x = 1$ is the smallest nonnegative fixed point.

66. Graph $y = \cos x$ and $y = x$ on the same set of axes. Their point of intersection gives the value of c such that $f(c) = c \implies \cos c = c$.

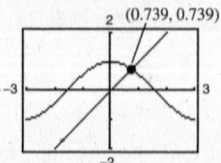

$c \approx 0.739$

67. $f(x) = \cos \dfrac{1}{x}$

(a) The domain of $f(x)$ is all real numbers x except $x = 0$.

(b) The graph has y-axis symmetry and a horizontal asymptote at $y = 1$.

(c) As $x \to 0$, $f(x)$ oscillates between -1 and 1.

(d) There are infinitely many solutions in the interval $[-1, 1]$. They occur at $x = \dfrac{2}{(2n + 1)\pi}$ where n is any integer.

(e) The greatest solution appears to occur at $x \approx 0.6366$.

68. $f(x) = \dfrac{\sin x}{x}$

(a) Domain: all real numbers except $x = 0$.

(b) The graph has y-axis symmetry.

(c) As $x \to 0, f(x) \to 1$.

(d) $\dfrac{\sin x}{x} = 0$ has four solutions in the interval $[-8, 8]$.

   $\sin x \left(\dfrac{1}{x} \right) = 0$

   $\sin x = 0$

   $x = -2\pi, -\pi, \pi, 2\pi$

69. $y = \dfrac{1}{12}(\cos 8t - 3 \sin 8t)$

   $\dfrac{1}{12}(\cos 8t - 3 \sin 8t) = 0$

   $\cos 8t = 3 \sin 8t$

   $\dfrac{1}{3} = \tan 8t$

   $8t \approx 0.32175 + n\pi$

   $t \approx 0.04 + \dfrac{n\pi}{8}$

In the interval $0 \le t \le 1, t \approx 0.04, 0.43,$ and 0.83.

70. $y_1 = 1.56e^{-0.22t} \cos 4.9t$

Right-most point of intersection: $(1.96, -1)$

The displacement does not exceed one foot from equilibrium after $t = 1.96$ seconds.

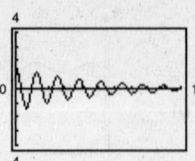

71. $S = 74.50 + 43.75 \sin \dfrac{\pi t}{6}$

t	1	2	3	4	5	6	7	8	9	10	11	12
S	96.4	112.4	118.3	112.4	96.4	74.5	52.6	36.6	30.8	36.6	52.6	74.5

Sales exceed 100,000 units during February, March, and April.

72. Graph $y_1 = 58.3 + 32 \cos\left(\dfrac{\pi t}{6}\right)$

$\qquad y_2 = 75.$

Left point of intersection: $(1.95, 75)$

Right point of intersection: $(10.05, 75)$

So, sales exceed 7500 in January, November, and December.

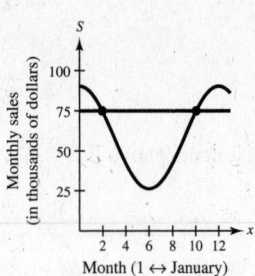

73.   Range = 300 feet

$\qquad\qquad v_0 = 100$ feet per second

$\qquad\qquad r = \tfrac{1}{32}v_0^{\,2} \sin 2\theta$

$\tfrac{1}{32}(100)^2 \sin 2\theta = 300$

$\qquad\qquad \sin 2\theta = 0.96$

$\qquad\qquad 2\theta = \arcsin(0.96) \approx 73.74°$

$\qquad\qquad \theta \approx 36.9°$

or

$2\theta = 180° - \arcsin(0.96) \approx 106.26°$

$\theta \approx 53.1°$

74. Range = 1000 yards = 3000 feet

$\qquad\qquad v_0 = 1200$ feet per second

$\qquad\qquad f = \tfrac{1}{32}v_0^{\,2} \sin 2\theta$

$3000 = \tfrac{1}{32}(1200)^2 \sin 2\theta$

$\sin 2\theta \approx 0.066667$

$\qquad 2\theta \approx 3.8°$

$\qquad\quad \theta \approx 1.9°$

75. $h(t) = 53 + 50 \sin\left(\dfrac{\pi}{16}t - \dfrac{\pi}{2}\right)$

(a) $h(t) = 53$ when $50 \sin\left(\dfrac{\pi}{16}t - \dfrac{\pi}{2}\right) = 0.$

$\dfrac{\pi}{16}t - \dfrac{\pi}{2} = 0 \qquad \text{or} \qquad \dfrac{\pi}{16}t - \dfrac{\pi}{2} = \pi$

$\qquad \dfrac{\pi}{16}t = \dfrac{\pi}{2} \qquad\qquad\qquad \dfrac{\pi}{16}t = \dfrac{3\pi}{2}$

$\qquad\qquad t = 8 \qquad\qquad\qquad\qquad t = 24$

The Ferris wheel will be 53 feet above ground at 8 seconds and at 24 seconds.

—CONTINUED—

75. —CONTINUED—

(b) The person will be at the top of the Ferris wheel when

$$\sin\left(\frac{\pi}{16}t - \frac{\pi}{2}\right) = 1$$

$$\frac{\pi}{16}t - \frac{\pi}{2} = \frac{\pi}{2}$$

$$\frac{\pi}{16}t = \pi$$

$$t = 16.$$

The first time this occurs is after 16 seconds. The period of this function is $\dfrac{2\pi}{\pi/16} = 32$. During

160 seconds, 5 cycles will take place and the person will be at the top of the ride 5 times, spaced 32 seconds apart. The times are: 16 seconds, 48 seconds, 80 seconds, 112 seconds, and 144 seconds.

76. (a)

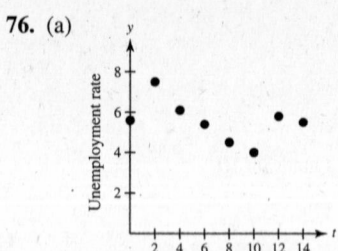

Year (0 ↔ 1990)

(b) By graphing the curves, we see that
   (1) $r = 1.24 \sin(0.47t + 0.40) + 5.45$ best fits the data.

(c) The constant term gives the average unemployment rate of 5.45%.

(d) Period: $\dfrac{2\pi}{0.47} = 13.4$ years

(e) $r = 5$ when $t = 20$ which corresponds to the year 2010.

77. $A = 2x \cos x$, $0 < x < \dfrac{\pi}{2}$

(a)

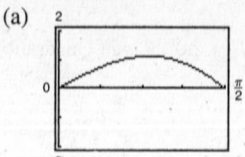

The maximum area of $A \approx 1.12$ occurs when $x \approx 0.86$.

(b) $A \geq 1$ for $0.6 < x < 1.1$

78. $f(x) = 3\sin(0.6x - 2)$

(a) Zero: $\sin(0.6x - 2) = 0$

$$0.6x - 2 = 0$$

$$0.6x = 2$$

$$x = \frac{2}{0.6} = \frac{10}{3}$$

(c) $-0.45x^2 + 5.52x - 13.70 = 0$

$$x = \frac{-5.52 \pm \sqrt{(5.52)^2 - 4(-0.45)(-13.70)}}{2(-0.45)}$$

$$x \approx 3.46, 8.81$$

The zero of g on $[0, 6]$ is 3.46. The zero is close to the zero $\frac{10}{3} \approx 3.33$ of f.

(b) $g(x) = -0.45x^2 + 5.52x - 13.70$

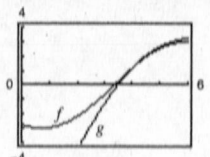

For $3.5 \leq x \leq 6$ the approximation appears to be good.

79. True. The period of $2\sin 4t - 1$ is $\dfrac{\pi}{2}$ and the period of $2\sin t - 1$ is 2π.

In the interval $[0, 2\pi)$ the first equation has four cycles whereas the second equation has only one cycle, thus the first equation has four times the x-intercepts (solutions) as the second equation.

80. False.

sin $x = 3.4$ has no solution since
3.4 is outside the range of sin.
Also, 3.4 is outside the domain of
arcsin, so $x = \arcsin(3.4)$ is an
invalid equation.

81. $y_1 = 2 \sin x$

$y_2 = 3x + 1$

From the graph we see that there is
only one point of intersection.

82. By inspecting the graphs of y_1 and
y_2, it appears they intersect at three
points.

83.
$$C = 90° - 66° = 24°$$

$$\cos 66° = \frac{22.3}{a}$$

$$a \cos 66° = 22.3$$

$$a = \frac{22.3}{\cos 66°} \approx 54.8$$

$$\tan 66° = \frac{b}{22.3}$$

$$b = 22.3 \tan 66° \approx 50.1$$

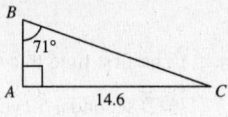

84. Given: $A = 90°, B = 71°, b = 14.6$

$$C = 90° - 71° = 19°$$

$$\sin 71° = \frac{14.6}{a}$$

$$a = \frac{14.6}{\sin 71} \approx 15.4$$

$$\tan 71° = \frac{14.6}{c}$$

$$c = \frac{14.6}{\tan 71°} \approx 5.0$$

85. $\theta = 390°$, $\theta' = 390° - 360° = 30°$, θ is in Quadrant I.

$$\sin 390° = \sin 30° = \frac{1}{2}$$

$$\cos 390° = \cos 30° = \frac{\sqrt{3}}{2}$$

$$\tan 390° = \tan 30° = \frac{1}{\sqrt{3}} = \frac{\sqrt{3}}{3}$$

86. $600°$

$600° - 360° = 240°$, Quadrant III

Reference angle: $60°$

$$\sin 600° = -\sin 60° = -\frac{\sqrt{3}}{2}$$

$$\cos 600° = -\cos 60° = -\frac{1}{2}$$

$$\tan 600° = \tan 60° = \sqrt{3}$$

87. $\theta = -1845°$, $\theta' = 45°$, θ is in Quadrant IV.

$$\sin(-1845°) = -\sin 45° = -\frac{\sqrt{2}}{2}$$

$$\cos(-1845°) = \cos 45° = \frac{\sqrt{2}}{2}$$

$$\tan(-1845°) = -\tan 45° = -1$$

88. $-1410°$

$$-1410° + 4(360°) = 30°, \quad \text{Quadrant I}$$

$$\sin(-1410°) = \sin 30° = \frac{1}{2}$$

$$\cos(-1410°) = \cos 30° = \frac{\sqrt{3}}{2}$$

$$\tan(-1410°) = \tan 30° = \frac{\sqrt{3}}{3}$$

89. $\tan \theta = \dfrac{250 \text{ feet}}{2 \text{ miles}} \times \dfrac{1 \text{ mile}}{5280 \text{ feet}} \approx 0.02367$

$$\theta \approx 1.36°$$

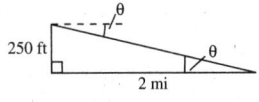

Not drawn to scale

90.
$$h = y - x$$

$$\tan 39.75° = \frac{y}{100}$$

$$100 \tan 39.75° = y$$

$$\tan 28° = \frac{x}{100}$$

$$100 \tan 28° = x$$

$$h = 100 \tan 39.75° - 100 \tan 28°$$

$$h \approx 30 \text{ feet}$$

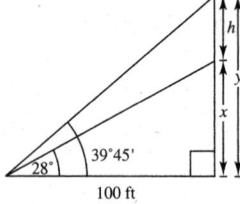

91. Answers will vary.

Section 5.4   Sum and Difference Formulas

■ You should know the sum and difference formulas.

$\sin(u \pm v) = \sin u \cos v \pm \cos u \sin v$

$\cos(u \pm v) = \cos u \cos v \mp \sin u \sin v$

$\tan(u \pm v) = \dfrac{\tan u \pm \tan v}{1 \mp \tan u \tan v}$

■ You should be able to use these formulas to find the values of the trigonometric functions of angles whose sums or differences are special angles.

■ You should be able to use these formulas to solve trigonometric equations.

Vocabulary Check

1. $\sin u \cos v - \cos u \sin v$

2. $\cos u \cos v - \sin u \sin v$

3. $\dfrac{\tan u + \tan v}{1 - \tan u \tan v}$

4. $\sin u \cos v + \cos u \sin v$

5. $\cos u \cos v + \sin u \sin v$

6. $\dfrac{\tan u - \tan v}{1 + \tan u \tan v}$

---

1. (a) $\cos(120° + 45°) = \cos 120° \cos 45° - \sin 120° \sin 45°$

$$= \left(-\frac{1}{2}\right)\left(\frac{\sqrt{2}}{2}\right) - \left(\frac{\sqrt{3}}{2}\right)\left(\frac{\sqrt{2}}{2}\right)$$

$$= \frac{-\sqrt{2} - \sqrt{6}}{4}$$

(b) $\cos 120° + \cos 45° = -\dfrac{1}{2} + \dfrac{\sqrt{2}}{2} = \dfrac{-1 + \sqrt{2}}{2}$

2. (a) $\sin(135° - 30°) = \sin 135° \cos 30° - \cos 135° \sin 30°$

$$= \left(\frac{\sqrt{2}}{2}\right)\left(\frac{\sqrt{3}}{2}\right) - \left(-\frac{\sqrt{2}}{2}\right)\left(\frac{1}{2}\right)$$

$$= \frac{\sqrt{6} + \sqrt{2}}{4}$$

(b) $\sin 135° - \cos 30° = \dfrac{\sqrt{2}}{2} - \dfrac{\sqrt{3}}{2} = \dfrac{\sqrt{2} - \sqrt{3}}{2}$

3. (a) $\cos\left(\dfrac{\pi}{4} + \dfrac{\pi}{3}\right) = \cos\dfrac{\pi}{4}\cos\dfrac{\pi}{3} - \sin\dfrac{\pi}{4}\sin\dfrac{\pi}{3}$

$$= \frac{\sqrt{2}}{2} \cdot \frac{1}{2} - \frac{\sqrt{2}}{2} \cdot \frac{\sqrt{3}}{2}$$

$$= \frac{\sqrt{2} - \sqrt{6}}{4}$$

(b) $\cos\dfrac{\pi}{4} + \cos\dfrac{\pi}{3} = \dfrac{\sqrt{2}}{2} + \dfrac{1}{2} = \dfrac{\sqrt{2} + 1}{2}$

4. (a) $\sin\left(\dfrac{3\pi}{4} + \dfrac{5\pi}{6}\right) = \sin\dfrac{3\pi}{4}\cos\dfrac{5\pi}{6} + \cos\dfrac{3\pi}{4}\sin\dfrac{5\pi}{6}$

$$= \left(\frac{\sqrt{2}}{2}\right)\left(-\frac{\sqrt{3}}{2}\right) + \left(-\frac{\sqrt{2}}{2}\right)\left(\frac{1}{2}\right)$$

$$= -\frac{\sqrt{6} + \sqrt{2}}{4}$$

(b) $\sin\dfrac{3\pi}{4} + \sin\dfrac{5\pi}{6} = \dfrac{\sqrt{2}}{2} + \dfrac{1}{2} = \dfrac{\sqrt{2} + 1}{2}$

5. (a) $\sin\left(\dfrac{7\pi}{6} - \dfrac{\pi}{3}\right) = \sin\dfrac{5\pi}{6} = \sin\dfrac{\pi}{6} = \dfrac{1}{2}$

(b) $\sin\dfrac{7\pi}{6} - \sin\dfrac{\pi}{3} = -\dfrac{1}{2} - \dfrac{\sqrt{3}}{2} = \dfrac{-1 - \sqrt{3}}{2}$

6. (a) $\sin(315° - 60°) = \sin 315° \cos 60° - \cos 315° \sin 60°$

$$= \frac{-\sqrt{2}}{2} \cdot \frac{1}{2} - \frac{\sqrt{2}}{2} \cdot \frac{\sqrt{3}}{2}$$

$$= -\frac{\sqrt{2}}{4} - \frac{\sqrt{6}}{4}$$

(b) $\sin 315° - \sin 60° = \dfrac{-\sqrt{2}}{2} - \dfrac{\sqrt{3}}{2} = \dfrac{-\sqrt{2} - \sqrt{3}}{2}$

7. $\sin 105° = \sin(60° + 45°)$

$$= \sin 60° \cos 45° + \cos 60° \sin 45°$$

$$= \frac{\sqrt{3}}{2} \cdot \frac{\sqrt{2}}{2} + \frac{1}{2} \cdot \frac{\sqrt{2}}{2}$$

$$= \frac{\sqrt{2}}{4}(\sqrt{3} + 1)$$

$\cos 105° = \cos(60° + 45°)$

$$= \cos 60° \cos 45° - \sin 60° \sin 45°$$

$$= \frac{1}{2} \cdot \frac{\sqrt{2}}{2} - \frac{\sqrt{3}}{2} \cdot \frac{\sqrt{2}}{2}$$

$$= \frac{\sqrt{2}}{4}(1 - \sqrt{3})$$

$\tan 105° = \tan(60° + 45°)$

$$= \frac{\tan 60° + \tan 45°}{1 - \tan 60° \tan 45°}$$

$$= \frac{\sqrt{3} + 1}{1 - \sqrt{3}} = \frac{\sqrt{3} + 1}{1 - \sqrt{3}} \cdot \frac{1 + \sqrt{3}}{1 + \sqrt{3}}$$

$$= \frac{4 + 2\sqrt{3}}{-2} = -2 - \sqrt{3}$$

8. $165° = 135° + 30°$

$\sin 165° = \sin(135° + 30°)$

$$= \sin 135° \cos 30° + \sin 30° \cos 135°$$

$$= \sin 45° \cos 30° - \sin 30° \cos 45°$$

$$= \frac{\sqrt{2}}{2} \cdot \frac{\sqrt{3}}{2} - \frac{1}{2} \cdot \frac{\sqrt{2}}{2}$$

$$= \frac{\sqrt{2}}{4}(\sqrt{3} - 1)$$

$\cos 165° = \cos(135° + 30°)$

$$= \cos 135° \cos 30° - \sin 135° \sin 30°$$

$$= -\cos 45° \cos 30° - \sin 45° \sin 30°$$

$$= -\frac{\sqrt{2}}{2} \cdot \frac{\sqrt{3}}{2} - \frac{\sqrt{2}}{2} \cdot \frac{1}{2}$$

$$= -\frac{\sqrt{2}}{4}(\sqrt{3} + 1)$$

$\tan 165° = \tan(135° + 30°)$

$$= \frac{\tan 135° + \tan 30°}{1 - \tan 135° \tan 30°}$$

$$= \frac{-\tan 45° + \tan 30°}{1 + \tan 45° \tan 30°}$$

$$= \frac{-1 + \frac{\sqrt{3}}{3}}{1 + \frac{\sqrt{3}}{3}}$$

$$= -2 + \sqrt{3}$$

9. $\sin 195° = \sin(225° - 30°)$

$$= \sin 225° \cos 30° - \cos 225° \sin 30°$$

$$= -\sin 45° \cos 30° + \cos 45° \sin 30°$$

$$= -\frac{\sqrt{2}}{2} \cdot \frac{\sqrt{3}}{2} + \frac{\sqrt{2}}{2} \cdot \frac{1}{2}$$

$$= \frac{\sqrt{2}}{4}(1 - \sqrt{3})$$

$\cos 195° = \cos(225° - 30°)$

$$= \cos 225° \cos 30° + \sin 225° \sin 30°$$

$$= -\cos 45° \cos 30° - \sin 45° \sin 30°$$

$$= -\frac{\sqrt{2}}{2} \cdot \frac{\sqrt{3}}{2} - \frac{\sqrt{2}}{2} \cdot \frac{1}{2}$$

$$= -\frac{\sqrt{2}}{4}(\sqrt{3} + 1)$$

$\tan 195° = \tan(225° - 30°)$

$$= \frac{\tan 225° - \tan 30°}{1 + \tan 225° \tan 30°}$$

$$= \frac{\tan 45° - \tan 30°}{1 + \tan 45° \tan 30°}$$

$$= \frac{1 - \left(\frac{\sqrt{3}}{3}\right)}{1 + \left(\frac{\sqrt{3}}{3}\right)} = \frac{3 - \sqrt{3}}{3 + \sqrt{3}} \cdot \frac{3 - \sqrt{3}}{3 - \sqrt{3}}$$

$$= \frac{12 - 6\sqrt{3}}{6} = 2 - \sqrt{3}$$

10. $255° = 300° - 45°$

$\sin 255° = \sin(300° - 45°)$

$\qquad = \sin 300° \cos 45° - \sin 45° \cos 300°$

$\qquad = -\sin 60° \cos 45° - \sin 45° \cos 60°$

$\qquad = -\dfrac{\sqrt{3}}{2} \cdot \dfrac{\sqrt{2}}{2} - \dfrac{\sqrt{2}}{2} \cdot \dfrac{1}{2}$

$\qquad = -\dfrac{\sqrt{2}}{4}\left(\sqrt{3} + 1\right)$

$\cos 255° = \cos(300° - 45°)$

$\qquad = \cos 300° \cos 45° + \sin 300° \sin 45°$

$\qquad = \cos 60° \cos 45° - \sin 60° \sin 45°$

$\qquad = \dfrac{1}{2} \cdot \dfrac{\sqrt{2}}{2} - \dfrac{\sqrt{3}}{2} \cdot \dfrac{\sqrt{2}}{2}$

$\qquad = \dfrac{\sqrt{2}}{4}\left(1 - \sqrt{3}\right)$

$\tan 255° = \tan(300° - 45°)$

$\qquad = \dfrac{\tan 300° - \tan 45°}{1 + \tan 300° \tan 45°}$

$\qquad = \dfrac{-\tan 60° - \tan 45°}{1 - \tan 60° \tan 45°}$

$\qquad = \dfrac{-\sqrt{3} - 1}{1 - \sqrt{3}} = 2 + \sqrt{3}$

11. $\sin \dfrac{11\pi}{12} = \sin\left(\dfrac{3\pi}{4} + \dfrac{\pi}{6}\right)$

$\qquad = \sin \dfrac{3\pi}{4} \cos \dfrac{\pi}{6} + \cos \dfrac{3\pi}{4} \sin \dfrac{\pi}{6}$

$\qquad = \dfrac{\sqrt{2}}{2} \cdot \dfrac{\sqrt{3}}{2} + \left(-\dfrac{\sqrt{2}}{2}\right)\dfrac{1}{2}$

$\qquad = \dfrac{\sqrt{2}}{4}\left(\sqrt{3} - 1\right)$

$\cos \dfrac{11\pi}{12} = \cos\left(\dfrac{3\pi}{4} + \dfrac{\pi}{6}\right)$

$\qquad = \cos \dfrac{3\pi}{4} \cos \dfrac{\pi}{6} - \sin \dfrac{3\pi}{4} \sin \dfrac{\pi}{6}$

$\qquad = -\dfrac{\sqrt{2}}{2} \cdot \dfrac{\sqrt{3}}{2} - \dfrac{\sqrt{2}}{2} \cdot \dfrac{1}{2} = -\dfrac{\sqrt{2}}{4}\left(\sqrt{3} + 1\right)$

$\tan \dfrac{11\pi}{4} = \tan\left(\dfrac{3\pi}{4} + \dfrac{\pi}{6}\right)$

$\qquad = \dfrac{\tan \dfrac{3\pi}{4} + \tan \dfrac{\pi}{6}}{1 - \tan \dfrac{3\pi}{4} \tan \dfrac{\pi}{6}}$

$\qquad = \dfrac{-1 + \dfrac{\sqrt{3}}{3}}{1 - (-1)\dfrac{\sqrt{3}}{3}}$

$\qquad = \dfrac{-3 + \sqrt{3}}{3 + \sqrt{3}} \cdot \dfrac{3 - \sqrt{3}}{3 - \sqrt{3}}$

$\qquad = \dfrac{-12 + 6\sqrt{3}}{6} = -2 + \sqrt{3}$

12. $\dfrac{7\pi}{12} = \dfrac{\pi}{3} + \dfrac{\pi}{4}$

$\sin \dfrac{7\pi}{12} = \sin\left(\dfrac{\pi}{3} + \dfrac{\pi}{4}\right)$

$\qquad = \sin \dfrac{\pi}{3} \cos \dfrac{\pi}{4} + \sin \dfrac{\pi}{4} \cos \dfrac{\pi}{3}$

$\qquad = \dfrac{\sqrt{3}}{2} \cdot \dfrac{\sqrt{2}}{2} + \dfrac{\sqrt{2}}{2} \cdot \dfrac{1}{2}$

$\qquad = \dfrac{\sqrt{2}}{4}\left(\sqrt{3} + 1\right)$

$\cos \dfrac{7\pi}{12} = \cos\left(\dfrac{\pi}{3} + \dfrac{\pi}{4}\right)$

$\qquad = \cos \dfrac{\pi}{3} \cos \dfrac{\pi}{4} - \sin \dfrac{\pi}{3} \sin \dfrac{\pi}{4}$

$\qquad = \dfrac{1}{2} \cdot \dfrac{\sqrt{2}}{2} - \dfrac{\sqrt{3}}{2} \cdot \dfrac{\sqrt{2}}{2}$

$\qquad = \dfrac{\sqrt{2}}{4}\left(1 - \sqrt{3}\right)$

$\tan \dfrac{7\pi}{12} = \tan\left(\dfrac{\pi}{3} + \dfrac{\pi}{4}\right)$

$\qquad = \dfrac{\tan \dfrac{\pi}{3} + \tan \dfrac{\pi}{4}}{1 - \tan \dfrac{\pi}{3} \tan \dfrac{\pi}{4}}$

$\qquad = \dfrac{\sqrt{3} + 1}{1 - \sqrt{3}}$

$\qquad = -2 - \sqrt{3}$

13. $\sin \dfrac{17\pi}{12} = \sin\left(\dfrac{9\pi}{4} - \dfrac{5\pi}{6}\right)$

$\qquad = \sin \dfrac{9\pi}{4} \cos \dfrac{5\pi}{6} - \cos \dfrac{9\pi}{4} \sin \dfrac{5\pi}{6}$

$\qquad = \dfrac{\sqrt{2}}{2}\left(-\dfrac{\sqrt{3}}{2}\right) - \left(\dfrac{\sqrt{2}}{2}\right)\left(\dfrac{1}{2}\right)$

$\qquad = -\dfrac{\sqrt{2}}{4}\left(\sqrt{3} + 1\right)$

$\cos \dfrac{17\pi}{12} = \cos\left(\dfrac{9\pi}{4} - \dfrac{5\pi}{6}\right)$

$\qquad = \cos \dfrac{9\pi}{4} \cos \dfrac{5\pi}{6} + \sin \dfrac{9\pi}{4} \sin \dfrac{5\pi}{6}$

$\qquad = \dfrac{\sqrt{2}}{2}\left(-\dfrac{\sqrt{3}}{2}\right) + \dfrac{\sqrt{2}}{2}\left(\dfrac{1}{2}\right)$

$\qquad = \dfrac{\sqrt{2}}{4}\left(1 - \sqrt{3}\right)$

$\tan \dfrac{17\pi}{12} = \tan\left(\dfrac{9\pi}{4} - \dfrac{5\pi}{6}\right)$

$\qquad = \dfrac{\tan(9\pi/4) - \tan(5\pi/6)}{1 + \tan(9\pi/4)\tan(5\pi/6)}$

$\qquad = \dfrac{1 - \left(-\sqrt{3}/3\right)}{1 + \left(-\sqrt{3}/3\right)}$

$\qquad = \dfrac{3 + \sqrt{3}}{3 - \sqrt{3}} \cdot \dfrac{3 + \sqrt{3}}{3 + \sqrt{3}}$

$\qquad = \dfrac{12 + 6\sqrt{3}}{6} = 2 + \sqrt{3}$

14. $-\dfrac{\pi}{12} = \dfrac{\pi}{6} - \dfrac{\pi}{4}$

$\sin\left(-\dfrac{\pi}{12}\right) = \sin\left(\dfrac{\pi}{6} - \dfrac{\pi}{4}\right)$

$\qquad = \sin \dfrac{\pi}{6} \cos \dfrac{\pi}{4} - \sin \dfrac{\pi}{4} \cos \dfrac{\pi}{6}$

$\qquad = \dfrac{1}{2} \cdot \dfrac{\sqrt{2}}{2} - \dfrac{\sqrt{2}}{2} \cdot \dfrac{\sqrt{3}}{2}$

$\qquad = \dfrac{\sqrt{2}}{4}\left(1 - \sqrt{3}\right)$

$\cos\left(-\dfrac{\pi}{12}\right) = \cos\left(\dfrac{\pi}{6} - \dfrac{\pi}{4}\right)$

$\qquad = \cos \dfrac{\pi}{6} \cos \dfrac{\pi}{4} + \sin \dfrac{\pi}{6} \sin \dfrac{\pi}{4}$

$\qquad = \dfrac{\sqrt{3}}{2} \cdot \dfrac{\sqrt{2}}{2} + \dfrac{1}{2} \cdot \dfrac{\sqrt{2}}{2}$

$\qquad = \dfrac{\sqrt{2}}{4}\left(\sqrt{3} + 1\right)$

$\tan\left(-\dfrac{\pi}{12}\right) = \tan\left(\dfrac{\pi}{6} - \dfrac{\pi}{4}\right)$

$\qquad = \dfrac{\tan \dfrac{\pi}{6} - \tan \dfrac{\pi}{4}}{1 + \tan \dfrac{\pi}{6} \tan \dfrac{\pi}{4}}$

$\qquad = \dfrac{\dfrac{\sqrt{3}}{3} - 1}{1 + \dfrac{\sqrt{3}}{3}}$

$\qquad = -2 + \sqrt{3}$

15. $285 = 225 + 60$

$\sin 285° = \sin(225° + 60°)$

$\qquad = \sin 225° \cos 60° + \cos 225° \sin 60°$

$\qquad = -\dfrac{\sqrt{2}}{2}\left(\dfrac{1}{2}\right) - \dfrac{\sqrt{2}}{2}\left(\dfrac{\sqrt{3}}{2}\right) = -\dfrac{\sqrt{2}}{4}\left(\sqrt{3} + 1\right)$

$\cos 285° = \cos(225° + 60°)$

$\qquad = \cos 225° \cos 60° - \sin 225° \sin 60°$

$\qquad = -\dfrac{\sqrt{2}}{2}\left(\dfrac{1}{2}\right) - \left(-\dfrac{\sqrt{2}}{2}\right)\left(\dfrac{\sqrt{3}}{2}\right) = \dfrac{\sqrt{2}}{4}\left(\sqrt{3} - 1\right)$

$\tan 285° = \tan(225° + 60°)$

$\qquad = \dfrac{\tan 225° + \tan 60°}{1 - \tan 225° \tan 60°} = \dfrac{1 + \sqrt{3}}{1 - \sqrt{3}} \cdot \dfrac{1 + \sqrt{3}}{1 + \sqrt{3}}$

$\qquad = \dfrac{4 + 2\sqrt{3}}{-2} = -2 - \sqrt{3} = -\left(2 + \sqrt{3}\right)$

16. $-105 = 30° - 135°$

$$\sin(30° - 135°) = \sin 30° \cos 135° - \cos 30° \sin 135°$$
$$= \sin 30°(-\cos 45°) - \cos 30° \sin 45°$$
$$= \left(\frac{1}{2}\right)\left(-\frac{\sqrt{2}}{2}\right) - \left(\frac{\sqrt{3}}{2}\right)\left(\frac{\sqrt{2}}{2}\right)$$
$$= -\frac{\sqrt{2}}{4}\left(1 + \sqrt{3}\right)$$

$$\cos(30° - 135°) = \cos 30° \cos 135° + \sin 30° \sin 135°$$
$$= \cos 30°(-\cos 45°) + \sin 30° \sin 45°$$
$$= \left(\frac{\sqrt{3}}{2}\right)\left(-\frac{\sqrt{2}}{2}\right) + \left(\frac{1}{2}\right)\left(\frac{\sqrt{2}}{2}\right)$$
$$= \frac{\sqrt{2}}{4}\left(1 - \sqrt{3}\right)$$

$$\tan(30° - 135°) = \frac{\tan 30° - \tan 135°}{1 + \tan 30° \tan 135°}$$
$$= \frac{\tan 30° - (-\tan 45°)}{1 + \tan 30°(-\tan 45°)}$$
$$= \frac{\frac{\sqrt{3}}{3} - (-1)}{1 + \left(\frac{\sqrt{3}}{3}\right)(-1)} = 2 + \sqrt{3}$$

17. $-165° = -(120° + 45°)$

$$\sin(-165°) = \sin[-(120° + 45°)]$$
$$= -\sin(120° + 45°)$$
$$= -[\sin 120° \cos 45° + \cos 120° \sin 45°]$$
$$= -\left[\frac{\sqrt{3}}{2} \cdot \frac{\sqrt{2}}{2} - \frac{1}{2} \cdot \frac{\sqrt{2}}{2}\right]$$
$$= -\frac{\sqrt{2}}{4}\left(\sqrt{3} - 1\right)$$

$$\cos(-165°) = \cos[-(120° + 45°)]$$
$$= \cos(120° + 45°)$$
$$= \cos 120° \cos 45° - \sin 120° \sin 45°$$
$$= -\frac{1}{2} \cdot \frac{\sqrt{2}}{2} - \frac{\sqrt{3}}{2} \cdot \frac{\sqrt{2}}{2}$$
$$= -\frac{\sqrt{2}}{4}\left(1 + \sqrt{3}\right)$$

$$\tan(-165°) = \tan[-(120° + 45°)]$$
$$= -\tan(120° + 45°)$$
$$= -\frac{\tan 120° + \tan 45°}{1 - \tan 120° \tan 45°}$$
$$= -\frac{-\sqrt{3} + 1}{1 - (-\sqrt{3})(1)}$$
$$= -\frac{1 - \sqrt{3}}{1 + \sqrt{3}} \cdot \frac{1 - \sqrt{3}}{1 - \sqrt{3}}$$
$$= -\frac{4 - 2\sqrt{3}}{-2}$$
$$= 2 - \sqrt{3}$$

18. $15° = 45° - 30°$

$$\sin 15° = \sin(45° - 30°) = \sin 45° \cos 30° - \cos 45° \sin 30°$$
$$= \left(\frac{\sqrt{2}}{2}\right)\left(\frac{\sqrt{3}}{2}\right) - \left(\frac{\sqrt{2}}{2}\right)\left(\frac{1}{2}\right) = \frac{\sqrt{2}(\sqrt{3} - 1)}{4} = \frac{\sqrt{2}}{4}\left(\sqrt{3} - 1\right)$$

$$\cos 15° = \cos(45° - 30°) = \cos 45° \cos 30° + \sin 45° \sin 30°$$
$$= \left(\frac{\sqrt{2}}{2}\right)\left(\frac{\sqrt{3}}{2}\right) + \left(\frac{\sqrt{2}}{2}\right)\left(\frac{1}{2}\right) = \frac{\sqrt{2}(\sqrt{3} + 1)}{4} = \frac{\sqrt{2}}{4}\left(\sqrt{3} + 1\right)$$

$$\tan 15° = \tan(45° - 30°) = \frac{\tan 45° - \tan 30°}{1 + \tan 45° \tan 30°}$$
$$= \frac{1 - \frac{\sqrt{3}}{3}}{1 + (1)\left(\frac{\sqrt{3}}{3}\right)} = \frac{\frac{3 - \sqrt{3}}{3}}{\frac{3 + \sqrt{3}}{3}} = \frac{3 - \sqrt{3}}{3 + \sqrt{3}} \cdot \frac{3 - \sqrt{3}}{3 - \sqrt{3}} = \frac{12 - 6\sqrt{3}}{6} = 2 - \sqrt{3}$$

19. $\dfrac{13\pi}{12} = \dfrac{3\pi}{4} + \dfrac{\pi}{3}$

$\sin\dfrac{13\pi}{12} = \sin\left(\dfrac{3\pi}{4} + \dfrac{\pi}{3}\right)$

$\qquad = \sin\dfrac{3\pi}{4}\cos\dfrac{\pi}{3} + \cos\dfrac{3\pi}{4}\sin\dfrac{\pi}{3}$

$\qquad = \dfrac{\sqrt{2}}{2}\cdot\dfrac{1}{2} + \left(-\dfrac{\sqrt{2}}{2}\right)\left(\dfrac{\sqrt{3}}{2}\right)$

$\qquad = \dfrac{\sqrt{2}}{4}\left(1 - \sqrt{3}\right)$

$\cos\dfrac{13\pi}{12} = \cos\left(\dfrac{3\pi}{4} + \dfrac{\pi}{3}\right)$

$\qquad = \cos\dfrac{3\pi}{4}\cos\dfrac{\pi}{3} - \sin\dfrac{3\pi}{4}\sin\dfrac{\pi}{3}$

$\qquad = -\dfrac{\sqrt{2}}{2}\cdot\dfrac{1}{2} - \dfrac{\sqrt{2}}{2}\cdot\dfrac{\sqrt{3}}{2} = -\dfrac{\sqrt{2}}{4}\left(1 + \sqrt{3}\right)$

$\tan\dfrac{13\pi}{12} = \tan\left(\dfrac{3\pi}{4} + \dfrac{\pi}{3}\right)$

$\qquad = \dfrac{\tan\left(\dfrac{3\pi}{4}\right) + \tan\left(\dfrac{\pi}{3}\right)}{1 - \tan\left(\dfrac{3\pi}{4}\right)\tan\left(\dfrac{\pi}{3}\right)}$

$\qquad = \dfrac{-1 + \sqrt{3}}{1 - (-1)(\sqrt{3})}$

$\qquad = -\dfrac{1 - \sqrt{3}}{1 + \sqrt{3}}\cdot\dfrac{1 - \sqrt{3}}{1 - \sqrt{3}}$

$\qquad = -\dfrac{4 - 2\sqrt{3}}{-2}$

$\qquad = 2 - \sqrt{3}$

20. $-\dfrac{7\pi}{12} = -\dfrac{\pi}{3} - \dfrac{\pi}{4}$

$\sin\left(-\dfrac{7\pi}{12}\right) = \sin\left(-\dfrac{\pi}{3} - \dfrac{\pi}{4}\right) = \sin\left(-\dfrac{\pi}{3}\right)\cos\left(\dfrac{\pi}{4}\right) - \cos\left(-\dfrac{\pi}{3}\right)\sin\left(\dfrac{\pi}{4}\right)$

$\qquad = \left(-\dfrac{\sqrt{3}}{2}\right)\left(\dfrac{\sqrt{2}}{2}\right) - \left(\dfrac{1}{2}\right)\left(\dfrac{\sqrt{2}}{2}\right) = -\dfrac{\sqrt{2}}{4}\left(\sqrt{3} + 1\right)$

$\cos\left(-\dfrac{7\pi}{12}\right) = \cos\left(-\dfrac{\pi}{3} - \dfrac{\pi}{4}\right) = \cos\left(-\dfrac{\pi}{3}\right)\cos\left(\dfrac{\pi}{4}\right) + \sin\left(-\dfrac{\pi}{3}\right)\sin\left(\dfrac{\pi}{4}\right)$

$\qquad = \left(\dfrac{1}{2}\right)\left(\dfrac{\sqrt{2}}{2}\right) + \left(-\dfrac{\sqrt{3}}{2}\right)\left(\dfrac{\sqrt{2}}{2}\right) = \dfrac{\sqrt{2}}{4}\left(1 - \sqrt{3}\right)$

$\tan\left(-\dfrac{7\pi}{12}\right) = \tan\left(-\dfrac{\pi}{3} - \dfrac{\pi}{4}\right) = \dfrac{\tan\left(-\dfrac{\pi}{3}\right) - \tan\left(\dfrac{\pi}{4}\right)}{1 + \tan\left(-\dfrac{\pi}{3}\right)\tan\left(\dfrac{\pi}{4}\right)}$

$\qquad = \dfrac{-\sqrt{3} - 1}{1 + (-\sqrt{3})(1)} = 2 + \sqrt{3}$

21. $-\dfrac{13\pi}{12} = -\left(\dfrac{3\pi}{4} + \dfrac{\pi}{3}\right)$

$$\sin\left[-\left(\dfrac{3\pi}{4} + \dfrac{\pi}{3}\right)\right] = -\sin\left(\dfrac{3\pi}{4} + \dfrac{\pi}{3}\right)$$

$$= -\left[\sin\dfrac{3\pi}{4}\cos\dfrac{\pi}{3} + \cos\dfrac{3\pi}{4}\sin\dfrac{\pi}{3}\right]$$

$$= -\left[\dfrac{\sqrt{2}}{2}\left(\dfrac{1}{2}\right) + \left(-\dfrac{\sqrt{2}}{2}\right)\left(\dfrac{\sqrt{3}}{2}\right)\right]$$

$$= -\dfrac{\sqrt{2}}{4}\left(1 - \sqrt{3}\right) = \dfrac{\sqrt{2}}{4}\left(\sqrt{3} - 1\right)$$

$$\cos\left[-\left(\dfrac{3\pi}{4} + \dfrac{\pi}{3}\right)\right] = \cos\left(\dfrac{3\pi}{4} + \dfrac{\pi}{3}\right)$$

$$= \cos\dfrac{3\pi}{4}\cos\dfrac{\pi}{3} - \sin\dfrac{3\pi}{4}\sin\dfrac{\pi}{3}$$

$$= -\dfrac{\sqrt{2}}{2}\left(\dfrac{1}{2}\right) - \dfrac{\sqrt{2}}{2}\left(\dfrac{\sqrt{3}}{2}\right) = -\dfrac{\sqrt{2}}{4}(\sqrt{3} + 1)$$

$$\tan\left[-\left(\dfrac{3\pi}{4} + \dfrac{\pi}{3}\right)\right] = -\tan\left(\dfrac{3\pi}{4} + \dfrac{\pi}{3}\right)$$

$$= -\dfrac{\tan\dfrac{3\pi}{4} + \tan\dfrac{\pi}{3}}{1 - \tan\dfrac{3\pi}{4}\tan\dfrac{\pi}{3}} = -\dfrac{-1 + \sqrt{3}}{1 - \left(-\sqrt{3}\right)}$$

$$= \dfrac{1 - \sqrt{3}}{1 + \sqrt{3}} \cdot \dfrac{1 - \sqrt{3}}{1 - \sqrt{3}} = \dfrac{4 - 2\sqrt{3}}{-2} = -2 + \sqrt{3}$$

22. $\dfrac{5\pi}{12} = \dfrac{\pi}{4} + \dfrac{\pi}{6}$

$$\sin\left(\dfrac{\pi}{4} + \dfrac{\pi}{6}\right) = \sin\dfrac{\pi}{4}\cos\dfrac{\pi}{6} + \cos\dfrac{\pi}{4}\sin\dfrac{\pi}{6}$$

$$= \left(\dfrac{\sqrt{2}}{2}\right)\left(\dfrac{\sqrt{3}}{2}\right) + \left(\dfrac{\sqrt{2}}{2}\right)\left(\dfrac{1}{2}\right) = \dfrac{\sqrt{2}}{4}(\sqrt{3} + 1)$$

$$\cos\left(\dfrac{\pi}{4} + \dfrac{\pi}{6}\right) = \cos\dfrac{\pi}{4}\cos\dfrac{\pi}{6} - \sin\dfrac{\pi}{4}\sin\dfrac{\pi}{6}$$

$$= \left(\dfrac{\sqrt{2}}{2}\right)\left(\dfrac{\sqrt{3}}{2}\right) - \left(\dfrac{\sqrt{2}}{2}\right)\left(\dfrac{1}{2}\right) = \dfrac{\sqrt{2}}{4}(\sqrt{3} - 1)$$

$$\tan\left(\dfrac{\pi}{4} + \dfrac{\pi}{6}\right) = \dfrac{\tan\dfrac{\pi}{4} + \tan\dfrac{\pi}{6}}{1 - \tan\dfrac{\pi}{4}\tan\dfrac{\pi}{6}} = \dfrac{1 + \dfrac{\sqrt{3}}{3}}{1 - (1)\left(\dfrac{\sqrt{3}}{3}\right)} = \sqrt{3} + 2$$

23. $\cos 25° \cos 15° - \sin 25° \sin 15° = \cos(25° + 15°) = \cos 40°$

24. $\sin 140° \cos 50° + \cos 140° \sin 50° = \sin(140° + 50°) = \sin 190°$

25. $\dfrac{\tan 325° - \tan 86°}{1 + \tan 325° \tan 86°} = \tan(325° - 86°) = \tan 239°$

26. $\dfrac{\tan 140° - \tan 60°}{1 + \tan 140° \tan 60°} = \tan(140° - 60°) = \tan 80°$

27. $\sin 3 \cos 1.2 - \cos 3 \sin 1.2 = \sin(3 - 1.2) = \sin 1.8$

28. $\cos \dfrac{\pi}{7} \cos \dfrac{\pi}{5} - \sin \dfrac{\pi}{7} \sin \dfrac{\pi}{5} = \cos\left(\dfrac{\pi}{7} + \dfrac{\pi}{5}\right)$

$$= \cos \dfrac{12\pi}{35}$$

29. $\dfrac{\tan 2x + \tan x}{1 - \tan 2x \tan x} = \tan(2x + x) = \tan 3x$

30. $\cos 3x \cos 2y + \sin 3x \sin 2y = \cos(3x - 2y)$

31. $\sin 330° \cos 30° - \cos 330° \sin 30° = \sin(330° - 30°)$

$$= \sin 300°$$

$$= -\dfrac{\sqrt{3}}{2}$$

32. $\cos 15° \cos 60° + \sin 15° \sin 60° = \cos(15° - 60°)$

$$= \cos(-45°) = \dfrac{\sqrt{2}}{2}$$

33. $\sin \dfrac{\pi}{12} \cos \dfrac{\pi}{4} + \cos \dfrac{\pi}{12} \sin \dfrac{\pi}{4} = \sin\left(\dfrac{\pi}{12} + \dfrac{\pi}{4}\right)$

$$= \sin \dfrac{\pi}{3}$$

$$= \dfrac{\sqrt{3}}{2}$$

34. $\cos \dfrac{\pi}{16} \cos \dfrac{3\pi}{16} - \sin \dfrac{\pi}{16} \sin \dfrac{3\pi}{16} = \cos\left(\dfrac{\pi}{16} + \dfrac{3\pi}{16}\right)$

$$= \cos \dfrac{\pi}{4} = \dfrac{\sqrt{2}}{2}$$

35. $\dfrac{\tan 25° + \tan 110°}{1 - \tan 25° \tan 110°} = \tan(25° + 110°)$

$$= \tan 135°$$

$$= -1$$

36. $\dfrac{\tan\left(\dfrac{5\pi}{4}\right) - \tan\left(\dfrac{\pi}{12}\right)}{1 + \tan\left(\dfrac{5\pi}{4}\right) \tan\left(\dfrac{\pi}{12}\right)} = \tan\left(\dfrac{5\pi}{4} - \dfrac{\pi}{12}\right)$

$$= \tan\left(\dfrac{7\pi}{6}\right)$$

$$= \tan\left(\dfrac{\pi}{6}\right) = \dfrac{\sqrt{3}}{3}$$

For Exercises 37–44, we have:

$\sin u = \frac{5}{13}$, u in Quadrant II $\Rightarrow$ $\cos u = -\frac{12}{13}$, $\tan u = -\frac{5}{12}$

$\cos v = -\frac{3}{5}$, v in Quadrant II $\Rightarrow$ $\sin v = \frac{4}{5}$, $\tan v = -\frac{4}{3}$,

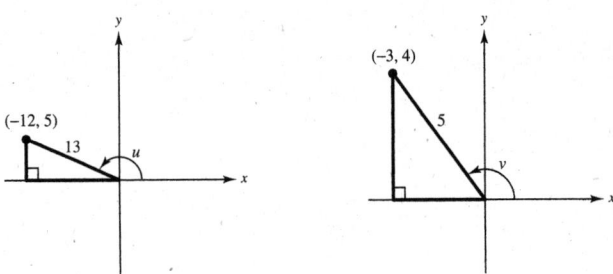

Figures for Exercises 37–44

37. $\sin(u + v) = \sin u \cos v + \cos u \sin v$

$$= \left(\frac{5}{13}\right)\left(-\frac{3}{5}\right) + \left(-\frac{12}{13}\right)\left(\frac{4}{5}\right)$$

$$= -\frac{63}{65}$$

38. $\cos(u - v) = \cos u \cos v + \sin u \sin v$

$$= \left(-\frac{12}{13}\right)\left(-\frac{3}{5}\right) + \left(\frac{5}{13}\right)\left(\frac{4}{5}\right)$$

$$= \frac{36}{65} + \frac{20}{65} = \frac{56}{65}$$

39. $\cos(u + v) = \cos u \cos v - \sin u \sin v$

$$= \left(-\frac{12}{13}\right)\left(-\frac{3}{5}\right) - \left(\frac{5}{13}\right)\left(\frac{4}{5}\right)$$

$$= \frac{16}{65}$$

40. $\sin(v - u) = \sin v \cos u - \cos v \sin u$

$$= \left(\frac{4}{5}\right)\left(-\frac{12}{13}\right) - \left(-\frac{3}{5}\right)\left(\frac{5}{13}\right)$$

$$= -\frac{48}{65} + \frac{15}{65} = -\frac{33}{65}$$

41. $\tan(u + v) = \dfrac{\tan u + \tan v}{1 - \tan u \tan v} = \dfrac{-\frac{5}{12} + \left(-\frac{4}{3}\right)}{1 - \left(-\frac{5}{12}\right)\left(-\frac{4}{3}\right)} = \dfrac{-\frac{21}{12}}{1 - \frac{5}{9}}$

$$= \left(-\frac{7}{4}\right)\left(\frac{9}{4}\right) = -\frac{63}{16}$$

42. $\csc(u - v) = \dfrac{1}{\sin(u - v)} = \dfrac{1}{-\sin(v - u)}$

$$= \dfrac{1}{-\left(-\frac{33}{65}\right)} = \frac{65}{33}$$

43. $\sec(v - u) = \dfrac{1}{\cos(v - u)} = \dfrac{1}{\cos v \cos u + \sin v \sin u}$

$$= \dfrac{1}{\left(-\frac{3}{5}\right)\left(-\frac{12}{13}\right) + \left(\frac{4}{5}\right)\left(\frac{5}{13}\right)} = \dfrac{1}{\left(\frac{36}{65}\right) + \left(\frac{20}{65}\right)} = \dfrac{1}{\frac{56}{65}}$$

$$= \frac{65}{56}$$

44. $\tan(u + v) = \dfrac{\tan u + \tan v}{1 - \tan u \tan v} = \dfrac{\left(-\frac{5}{12}\right) + \left(-\frac{4}{3}\right)}{1 - \left(-\frac{5}{12}\right)\left(-\frac{4}{3}\right)}$

$$= \dfrac{-\frac{7}{4}}{\frac{4}{9}} = -\frac{63}{16}$$

$$\cot(u + v) = \dfrac{1}{\tan(u + v)} = \dfrac{1}{-\frac{63}{16}} = -\frac{16}{63}$$

For Exercises 45–50, we have:

$\sin u = -\frac{7}{25}$, u in Quadrant III $\implies \cos u = -\frac{24}{25}$, $\tan u = \frac{7}{24}$

$\cos v = -\frac{4}{5}$, v in Quadrant III $\implies \sin v = -\frac{3}{5}$, $\tan v = \frac{3}{4}$

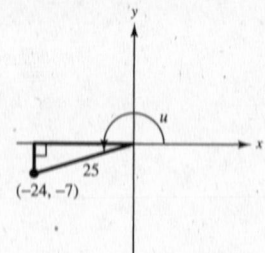 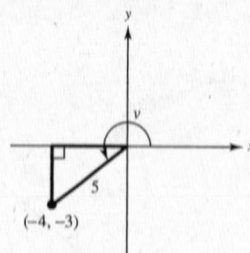

Figures for Exercises 45–50

45. $\cos(u + v) = \cos u \cos v - \sin u \sin v$

$$= \left(-\frac{24}{25}\right)\left(-\frac{4}{5}\right) - \left(-\frac{7}{25}\right)\left(-\frac{3}{5}\right)$$

$$= \frac{3}{5}$$

46. $\sin(u + v) = \sin u \cos v + \cos u \sin v$

$$= \left(-\frac{7}{25}\right)\left(-\frac{4}{5}\right) + \left(-\frac{24}{25}\right)\left(-\frac{3}{5}\right)$$

$$= \frac{28}{125} + \frac{72}{125} = \frac{100}{125} = \frac{4}{5}$$

47. $\tan(u - v) = \dfrac{\tan u - \tan v}{1 + \tan u \tan v}$

$= \dfrac{\frac{7}{24} - \frac{3}{4}}{1 + \left(\frac{7}{24}\right)\left(\frac{3}{4}\right)} = \dfrac{-\frac{11}{24}}{\frac{39}{32}} = -\dfrac{44}{117}$

48. $\tan(v - u) = \dfrac{\tan v - \tan u}{1 + \tan v \tan u} = \dfrac{\left(\frac{3}{4}\right) - \left(\frac{7}{24}\right)}{1 + \left(\frac{3}{4}\right)\left(\frac{7}{24}\right)}$

$= \dfrac{\frac{11}{24}}{\frac{39}{32}} = \dfrac{44}{117}$

$\cot(v - u) = \dfrac{1}{\tan(v - u)} = \dfrac{1}{\frac{44}{117}} = \dfrac{117}{44}$

49. $\sec(u + v) = \dfrac{1}{\cos(u + v)} = \dfrac{1}{\frac{3}{5}} = \dfrac{5}{3}$

Use Exercise 45 for $\cos(u + v)$.

50. $\cos(u - v) = \cos u \cos v + \sin u \sin v = \left(-\dfrac{24}{25}\right)\left(-\dfrac{4}{5}\right) + \left(-\dfrac{7}{25}\right)\left(-\dfrac{3}{5}\right)$

$= \dfrac{96}{125} + \dfrac{21}{125} = \dfrac{117}{125}$

51. $\sin(\arcsin x + \arccos x) = \sin(\arcsin x)\cos(\arccos x) + \sin(\arccos x)\cos(\arcsin x)$

$= x \cdot x + \sqrt{1 - x^2} \cdot \sqrt{1 - x^2}$

$= x^2 + 1 - x^2$

$= 1$

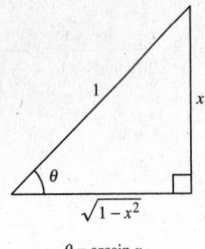

$\theta = \arcsin x$

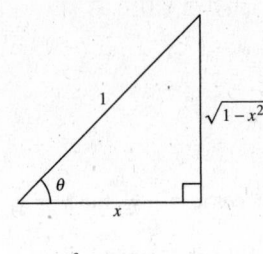

$\theta = \arccos x$

52. Let

$u = \arctan 2x$ and $v = \arccos x$

$\tan u = 2x \qquad \cos v = x.$

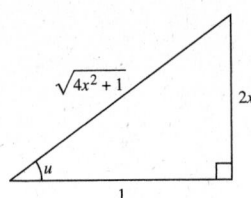

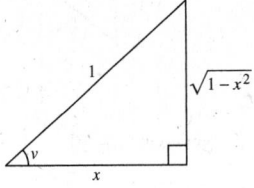

$\sin(\arctan 2x - \arccos x) = \sin(u - v)$

$= \sin u \cos v - \cos u \sin v$

$= \dfrac{2x}{\sqrt{4x^2 + 1}}(x) - \dfrac{1}{\sqrt{4x^2 + 1}}\left(\sqrt{1 - x^2}\right)$

$= \dfrac{2x^2 - \sqrt{1 - x^2}}{\sqrt{4x^2 + 1}}$

53. $\cos(\arccos x + \arcsin x) = \cos(\arccos x)\cos(\arcsin x) - \sin(\arccos x)\sin(\arcsin x)$

$$= x \cdot \sqrt{1 - x^2} - \sqrt{1 - x^2} \cdot x$$

$$= 0$$

(Use the triangles in Exercise 51.)

54. Let

$$u = \arccos x \quad \text{and} \quad v = \arctan x$$

$$\cos u = x \qquad\qquad \tan v = x.$$

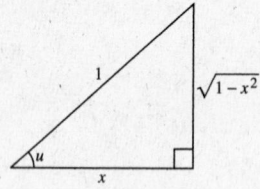 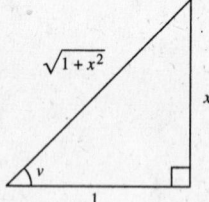

$$\cos(\arccos x - \arctan x) = \cos(u - v)$$

$$= \cos u \cos v + \sin u \sin v$$

$$= (x)\left(\frac{1}{\sqrt{1 + x^2}}\right) + \left(\sqrt{1 - x^2}\right)\left(\frac{x}{\sqrt{1 + x^2}}\right)$$

$$= \frac{x + x\sqrt{1 - x^2}}{\sqrt{1 + x^2}}$$

55. $\sin(3\pi - x) = \sin 3\pi \cos x - \sin x \cos 3\pi$

$$= (0)(\cos x) - (-1)(\sin x)$$

$$= \sin x$$

56. $\sin\left(\dfrac{\pi}{2} + x\right) = \sin\dfrac{\pi}{2}\cos x + \sin x \cos\dfrac{\pi}{2}$

$$= (1)(\cos x) + (\sin x)(0)$$

$$= \cos x$$

57. $\sin\left(\dfrac{\pi}{6} + x\right) = \sin\dfrac{\pi}{6}\cos x + \cos\dfrac{\pi}{6}\sin x$

$$= \frac{1}{2}\left(\cos x + \sqrt{3}\sin x\right)$$

58. $\cos\left(\dfrac{5\pi}{4} - x\right) = \cos\dfrac{5\pi}{4}\cos x + \sin\dfrac{5\pi}{4}\sin x$

$$= -\frac{\sqrt{2}}{2}(\cos x + \sin x)$$

59. $\cos(\pi - \theta) + \sin\left(\dfrac{\pi}{2} + \theta\right) = \cos\pi\cos\theta + \sin\pi\sin\theta + \sin\dfrac{\pi}{2}\cos\theta + \cos\dfrac{\pi}{2}\sin\theta$

$$= (-1)(\cos\theta) + (0)(\sin\theta) + (1)(\cos\theta) + (\sin\theta)(0)$$

$$= -\cos\theta + \cos\theta$$

$$= 0$$

60. $\tan\left(\dfrac{\pi}{4} - \theta\right) = \dfrac{\tan\dfrac{\pi}{4} - \tan\theta}{1 + \tan\dfrac{\pi}{4}\tan\theta} = \dfrac{1 - \tan\theta}{1 + \tan\theta}$

61. $\cos(x + y) \cos(x - y) = (\cos x \cos y - \sin x \sin y)(\cos x \cos y + \sin x \sin y)$

$$= \cos^2 x \cos^2 y - \sin^2 x \sin^2 y$$
$$= \cos^2 x(1 - \sin^2 y) - \sin^2 x \sin^2 y$$
$$= \cos^2 x - \cos^2 x \sin^2 y - \sin^2 x \sin^2 y$$
$$= \cos^2 x - \sin^2 y(\cos^2 x + \sin^2 x)$$
$$= \cos^2 x - \sin^2 y$$

62. $\sin(x + y) \sin(x - y) = (\sin x \cos y + \sin y \cos x)(\sin x \cos y - \sin y \cos x)$

$$= \sin^2 x \cos^2 y - \sin^2 y \cos^2 x$$
$$= \sin^2 x(1 - \sin^2 y) - \sin^2 y \cos^2 x$$
$$= \sin^2 x - \sin^2 x \sin^2 y - \sin^2 y \cos^2 x$$
$$= \sin^2 x - \sin^2 y(\sin^2 x + \cos^2 x)$$
$$= \sin^2 x - \sin^2 y$$

63. $\sin(x + y) + \sin(x - y) = \sin x \cos y + \cos x \sin y + \sin x \cos y - \cos x \sin y$

$$= 2 \sin x \cos y$$

64. $\cos(x + y) + \cos(x - y) = \cos x \cos y - \sin x \sin y + \cos x \cos y + \sin x \sin y$

$$= 2 \cos x \cos y$$

65. $\cos\left(\dfrac{3\pi}{2} - x\right) = \cos\dfrac{3\pi}{2} \cos x + \sin\dfrac{3\pi}{2} \sin x$

$$= (0)(\cos x) + (-1)(\sin x)$$
$$= -\sin x$$

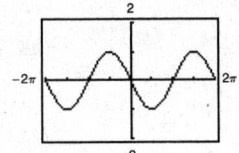

66. $\cos(\pi + x) = \cos \pi \cos x - \sin \pi \sin x$

$$= (-1)\cos x - (0)\sin x$$
$$= -\cos x$$

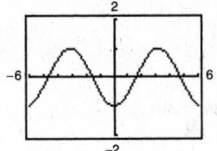

67. $\sin\left(\dfrac{3\pi}{2} + \theta\right) = \sin\dfrac{3\pi}{2} \cos \theta + \cos\dfrac{3\pi}{2} \sin \theta$

$$= (-1)(\cos \theta) + (0)(\sin \theta)$$
$$= -\cos \theta$$

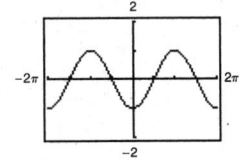

68. $\tan(\pi + \theta) = \dfrac{\tan \pi + \tan \theta}{1 - \tan \pi \tan \theta}$

$$= \dfrac{0 + \tan \theta}{1 - (0) \tan \theta}$$
$$= \tan \theta$$

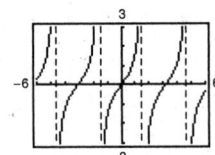

69.
$$\sin\left(x + \frac{\pi}{3}\right) + \sin\left(x - \frac{\pi}{3}\right) = 1$$

$$\sin x \cos \frac{\pi}{3} + \cos x \sin \frac{\pi}{3} + \sin x \cos \frac{\pi}{3} - \cos x \sin \frac{\pi}{3} = 1$$

$$2 \sin x(0.5) = 1$$

$$\sin x = 1$$

$$x = \frac{\pi}{2}$$

70.
$$\sin\left(x + \frac{\pi}{6}\right) - \sin\left(x - \frac{\pi}{6}\right) = \frac{1}{2}$$

$$\sin x \cos \frac{\pi}{6} + \cos x \sin \frac{\pi}{6} - \left(\sin x \cos \frac{\pi}{6} - \cos x \sin \frac{\pi}{6}\right) = \frac{1}{2}$$

$$2 \cos x(0.5) = \frac{1}{2}$$

$$\cos x = \frac{1}{2}$$

$$x = \frac{\pi}{3}, \frac{5\pi}{3}$$

71.
$$\cos\left(x + \frac{\pi}{4}\right) - \cos\left(x - \frac{\pi}{4}\right) = 1$$

$$\cos x \cos \frac{\pi}{4} - \sin x \sin \frac{\pi}{4} - \left(\cos x \cos \frac{\pi}{4} + \sin x \sin \frac{\pi}{4}\right) = 1$$

$$-2 \sin x\left(\frac{\sqrt{2}}{2}\right) = 1$$

$$-\sqrt{2} \sin x = 1$$

$$\sin x = -\frac{1}{\sqrt{2}}$$

$$\sin x = -\frac{\sqrt{2}}{2}$$

$$x = \frac{5\pi}{4}, \frac{7\pi}{4}$$

72.
$$\tan(x + \pi) + 2\sin(x + \pi) = 0$$

$$\frac{\tan x + \tan \pi}{1 - \tan x \tan \pi} + 2(\sin x \cos \pi + \cos x \sin \pi) = 0$$

$$\frac{\tan x + 0}{1 - \tan x(0)} + 2[\sin x(-1) + \cos x(0)] = 0$$

$$\frac{\tan x}{1} - 2\sin x = 0$$

$$\frac{\sin x}{\cos x} = 2\sin x$$

$$\sin x = 2\sin x \cos x$$

$$\sin x(1 - 2\cos x) = 0$$

$$\sin x = 0 \quad \text{or} \quad \cos x = \frac{1}{2}$$

$$x = 0, \pi \qquad x = \frac{\pi}{3}, \frac{5\pi}{3}$$

73. Analytically: $\cos\left(x + \frac{\pi}{4}\right) + \cos\left(x - \frac{\pi}{4}\right) = 1$

$$\cos x \cos \frac{\pi}{4} - \sin x \sin \frac{\pi}{4} + \cos x \cos \frac{\pi}{4} + \sin x \sin \frac{\pi}{4} = 1$$

$$2\cos x\left(\frac{\sqrt{2}}{2}\right) = 1$$

$$\sqrt{2}\cos x = 1$$

$$\cos x = \frac{1}{\sqrt{2}}$$

$$\cos x = \frac{\sqrt{2}}{2}$$

$$x = \frac{\pi}{4}, \frac{7\pi}{4}$$

Graphically: Graph $y_1 = \cos\left(x + \frac{\pi}{4}\right) + \cos\left(x - \frac{\pi}{4}\right)$ and $y_2 = 1$.

The points of intersection occur at $x = \frac{\pi}{4}$ and $x = \frac{7\pi}{4}$.

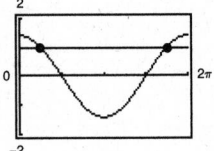

74. $\tan(x + \pi) - \cos\left(x + \frac{\pi}{2}\right) = 0$

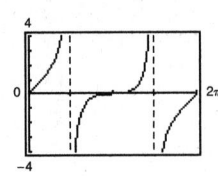

Answers: $(0, 0), (3.14, 0) \implies x = 0, \pi$

75. $y = \dfrac{1}{3} \sin 2t + \dfrac{1}{4} \cos 2t$

(a) $a = \dfrac{1}{3}, \; b = \dfrac{1}{4}, \; B = 2$

$C = \arctan \dfrac{b}{a} = \arctan \dfrac{3}{4} \approx 0.6435$

$y \approx \sqrt{\left(\dfrac{1}{3}\right)^2 + \left(\dfrac{1}{4}\right)^2} \, \sin(2t + 0.6435)$

$ = \dfrac{5}{12} \sin(2t + 0.6435)$

(b) Amplitude: $\dfrac{5}{12}$ feet

(c) Frequency: $\dfrac{1}{\text{period}} = \dfrac{B}{2\pi} = \dfrac{2}{2\pi} = \dfrac{1}{\pi}$ cycle per second

76.

$y_1 = A \cos 2\pi\left(\dfrac{t}{T} - \dfrac{x}{\lambda}\right)$

$y_2 = A \cos 2\pi\left(\dfrac{t}{T} + \dfrac{x}{\lambda}\right)$

$y_1 + y_2 = A \cos 2\pi\left(\dfrac{t}{T} - \dfrac{x}{\lambda}\right) + A \cos 2\pi\left(\dfrac{t}{T} + \dfrac{x}{\lambda}\right)$

$y_1 + y_2 = A\left[\cos 2\pi\dfrac{t}{T}\cos 2\pi\dfrac{x}{\lambda} + \sin 2\pi\dfrac{t}{T}\sin 2\pi\dfrac{x}{\lambda}\right] + A\left[\cos 2\pi\dfrac{t}{T}\cos 2\pi\dfrac{x}{\lambda} - \sin 2\pi\dfrac{t}{T}\sin 2\pi\dfrac{x}{\lambda}\right]$

$ = 2A \cos 2\pi\dfrac{t}{T}\cos 2\pi\dfrac{x}{\lambda}$

77. False.

$\sin(u \pm v) = \sin u \cos v \pm \cos u \sin v$

78. False.

$\cos(u \pm v) = \cos u \cos v \mp \sin u \sin v$

79. False. $\cos\left(x - \dfrac{\pi}{2}\right) = \cos x \cos \dfrac{\pi}{2} + \sin x \sin \dfrac{\pi}{2}$

$\phantom{\text{False. }\cos\left(x - \dfrac{\pi}{2}\right)} = (\cos x)(0) + (\sin x)(1)$

$\phantom{\text{False. }\cos\left(x - \dfrac{\pi}{2}\right)} = \sin x$

80. True.

$\sin\left(x - \dfrac{\pi}{2}\right) = -\sin\left(\dfrac{\pi}{2} - x\right) = -\cos x$

81. $\cos(n\pi + \theta) = \cos n\pi \cos \theta - \sin n\pi \sin \theta$

$ = (-1)^n(\cos \theta) - (0)(\sin \theta)$

$ = (-1)^n(\cos \theta), \text{ where } n \text{ is an integer.}$

82. $\sin(n\pi + \theta) = \sin n\pi \cos \theta + \sin \theta \cos n\pi$

$ = (0)(\cos \theta) + (\sin \theta)(-1)^n$

$ = (-1)^n(\sin \theta), \text{ where } n \text{ is an integer.}$

83. $C = \arctan \dfrac{b}{a} \implies \sin C = \dfrac{b}{\sqrt{a^2 + b^2}}, \; \cos C = \dfrac{a}{\sqrt{a^2 + b^2}}$

$\sqrt{a^2 + b^2} \, \sin(B\theta + C) = \sqrt{a^2 + b^2}\left(\sin B\theta \cdot \dfrac{a}{\sqrt{a^2 + b^2}} + \dfrac{b}{\sqrt{a^2 + b^2}} \cdot \cos B\theta\right) = a \sin B\theta + b \cos B\theta$

84. $C = \arctan \dfrac{a}{b} \implies \sin C = \dfrac{a}{\sqrt{a^2 + b^2}}, \; \cos C = \dfrac{b}{\sqrt{a^2 + b^2}}$

$\sqrt{a^2 + b^2} \, \cos(B\theta - C) = \sqrt{a^2 + b^2}\left(\cos B\theta \cdot \dfrac{b}{\sqrt{a^2 + b^2}} + \sin B\theta \cdot \dfrac{a}{\sqrt{a^2 + b^2}}\right)$

$\phantom{\sqrt{a^2 + b^2} \, \cos(B\theta - C)} = b \cos B\theta + a \sin B\theta$

$\phantom{\sqrt{a^2 + b^2} \, \cos(B\theta - C)} = a \sin B\theta + b \cos B\theta$

85. $\sin \theta + \cos \theta$

$a = 1, \ b = 1, \ B = 1$

(a) $C = \arctan \dfrac{b}{a} = \arctan 1 = \dfrac{\pi}{4}$

$\quad \sin \theta + \cos \theta = \sqrt{a^2 + b^2} \sin(B\theta + C)$

$\qquad\qquad\qquad = \sqrt{2} \sin\left(\theta + \dfrac{\pi}{4}\right)$

(b) $C = \arctan \dfrac{a}{b} = \arctan 1 = \dfrac{\pi}{4}$

$\quad \sin \theta + \cos \theta = \sqrt{a^2 + b^2} \cos(B\theta - C)$

$\qquad\qquad\qquad = \sqrt{2} \cos\left(\theta - \dfrac{\pi}{4}\right)$

86. $3 \sin 2\theta + 4 \cos 2\theta$

$a = 3, b = 4, B = 2$

(a) $C = \arctan \dfrac{b}{a} = \arctan \dfrac{4}{3} \approx 0.9273$

$\quad 3 \sin 2\theta + 4 \cos 2\theta = \sqrt{a^2 + b^2} \sin(B\theta + C)$

$\qquad\qquad\qquad\qquad \approx 5 \sin(2\theta + 0.9273)$

(b) $C = \arctan \dfrac{a}{b} = \arctan \dfrac{3}{4} \approx 0.6435$

$\quad 3 \sin 2\theta + 4 \cos 2\theta = \sqrt{a^2 + b^2} \cos(B\theta - C)$

$\qquad\qquad\qquad\qquad \approx 5 \cos(2\theta - 0.6435)$

87. $12 \sin 3\theta + 5 \cos 3\theta$

$a = 12, \ b = 5, \ B = 3$

(a) $C = \arctan \dfrac{b}{a} = \arctan \dfrac{5}{12} \approx 0.3948$

$\quad 12 \sin 3\theta + 5 \cos 3\theta = \sqrt{a^2 + b^2} \sin(B\theta + C)$

$\qquad\qquad\qquad\qquad \approx 13 \sin(3\theta + 0.3948)$

(b) $C = \arctan \dfrac{a}{b} = \arctan \dfrac{12}{5} \approx 1.1760$

$\quad 12 \sin 3\theta + 5 \cos 3\theta = \sqrt{a^2 + b^2} \cos(B\theta - C)$

$\qquad\qquad\qquad\qquad \approx 13 \cos(3\theta - 1.1760)$

88. $\sin 2\theta - \cos 2\theta$

$a = 1, b = -1, B = 2$

(a) $C = \arctan \dfrac{b}{a} = \arctan(-1) = -\dfrac{\pi}{4}$

$\quad \sin 2\theta - \cos 2\theta = \sqrt{a^2 + b^2} \sin(B\theta + C)$

$\qquad\qquad\qquad\quad = \sqrt{2} \sin\left(2\theta - \dfrac{\pi}{4}\right)$

(b) $C = \arctan \dfrac{a}{b} = \arctan(-1) = -\dfrac{\pi}{4}$

$\quad \sin 2\theta - \cos 2\theta = \sqrt{a^2 + b^2} \cos(B\theta - C)$

$\qquad\qquad\qquad\quad = \sqrt{2} \cos\left(2\theta + \dfrac{\pi}{4}\right)$

89. $C = \arctan \dfrac{b}{a} = \dfrac{\pi}{2} \implies a = 0$

$\sqrt{a^2 + b^2} = 2 \implies b = 2$

$B = 1$

$2 \sin\left(\theta + \dfrac{\pi}{2}\right) = (0)(\sin\theta) + (2)(\cos\theta) = 2 \cos \theta$

90. $C = \arctan \dfrac{a}{b} = -\dfrac{3\pi}{4} \implies a = b, a < 0, b < 0$

$\sqrt{a^2 + b^2} = 5 \implies a = b = \dfrac{-5\sqrt{2}}{2}$

$B = 1$

$5 \cos\left(\theta + \dfrac{3\pi}{4}\right) = -\dfrac{5\sqrt{2}}{2} \sin \theta - \dfrac{5\sqrt{2}}{2} \cos \theta$

91. $\dfrac{\cos(x + h) - \cos x}{h} = \dfrac{\cos x \cos h - \sin x \sin h - \cos x}{h}$

$\qquad\qquad\qquad\qquad = \dfrac{\cos x \cos h - \cos x - \sin x \sin h}{h}$

$\qquad\qquad\qquad\qquad = \dfrac{\cos x(\cos h - 1) - \sin x \sin h}{h}$

$\qquad\qquad\qquad\qquad = \dfrac{\cos x(\cos h - 1)}{h} - \dfrac{\sin x \sin h}{h}$

92. (a) The domains of f and g are the sets of real numbers, $h \neq 0$.

(c) The graphs are the same.

(b)

h	0.01	0.02	0.05	0.1	0.2	0.5
$f(h)$	-0.504	-0.509	-0.521	-0.542	-0.583	-0.691
$g(h)$	-0.504	-0.509	-0.521	-0.542	-0.583	-0.691

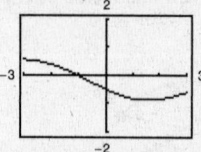

(d) As $h \to 0$, $f(h)$ approaches -0.5.

As $h \to 0$, $g(h)$ approaches -0.5.

93.

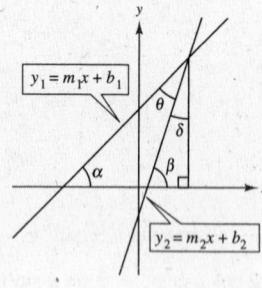

$m_1 = \tan \alpha$ and $m_2 = \tan \beta$

$\beta + \delta = 90° \implies \delta = 90° - \beta$

$\alpha + \theta + \delta = 90° \implies \alpha + \theta + (90° - \beta) = 90° \implies \theta = \beta - \alpha$

Therefore, $\theta = \arctan m_2 - \arctan m_1$.

For $y = x$ and $y = \sqrt{3}x$ we have $m_1 = 1$ and $m_2 = \sqrt{3}$.

$\theta = \arctan \sqrt{3} - \arctan 1$

$\quad = 60° - 45°$

$\quad = 15°$

94. For $m_2 > m_1 > 0$, the angle θ between the lines is:

$\theta = \arctan\left(\dfrac{m_2 - m_1}{1 + m_1 m_2}\right)$

$m_2 = 1$

$m_1 = \dfrac{1}{\sqrt{3}}$

$\theta = \arctan\left(\dfrac{1 - \dfrac{1}{\sqrt{3}}}{1 + \dfrac{1}{\sqrt{3}}}\right) = \arctan\left(2 - \sqrt{3}\right) = 15°$

95.

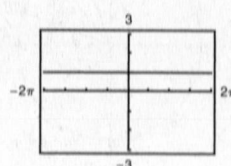

Conjecture: $\sin^2\left(\theta + \dfrac{\pi}{4}\right) + \sin^2\left(\theta - \dfrac{\pi}{4}\right) = 1$

$\sin^2\left(\theta + \dfrac{\pi}{4}\right) + \sin^2\left(\theta - \dfrac{\pi}{4}\right) = \left[\sin\theta\cos\dfrac{\pi}{4} + \cos\theta\sin\dfrac{\pi}{4}\right]^2 + \left[\sin\theta\cos\dfrac{\pi}{4} - \cos\theta\sin\dfrac{\pi}{4}\right]^2$

$\qquad = \left[\dfrac{\sin\theta}{\sqrt{2}} + \dfrac{\cos\theta}{\sqrt{2}}\right]^2 + \left[\dfrac{\sin\theta}{\sqrt{2}} - \dfrac{\cos\theta}{\sqrt{2}}\right]^2$

$\qquad = \dfrac{\sin^2\theta}{2} + \sin\theta\cos\theta + \dfrac{\cos^2\theta}{2} + \dfrac{\sin^2\theta}{2} - \sin\theta\cos\theta + \dfrac{\cos^2\theta}{2}$

$\qquad = \sin^2\theta + \cos^2\theta$

$\qquad = 1$

96. (a) To prove the identity for $\sin(u + v)$ we first need to prove the identity for $\cos(u - v)$. Assume $0 < v < u < 2\pi$ and locate u, v, and $u - v$ on the unit circle.

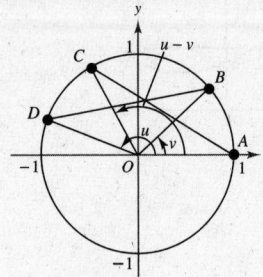

The coordinates of the points on the circle are:

$A = (1, 0)$, $B = (\cos v, \sin v)$, $C = (\cos(u - v), \sin(u - v))$, and $D = (\cos u, \sin u)$.

Since $\angle DOB = \angle COA$, chords AC and BD are equal. By the distance formula we have:

$$\sqrt{[\cos(u - v) - 1]^2 + [\sin(u - v) - 0]^2} = \sqrt{(\cos u - \cos v)^2 + (\sin u - \sin v)^2}$$

$$\cos^2(u - v) - 2\cos(u - v) + 1 + \sin^2(u - v) = \cos^2 u - 2\cos u \cos v + \cos^2 v + \sin^2 u - 2\sin u \sin v + \sin^2 v$$

$$[\cos^2(u - v) + \sin^2(u - v)] + 1 - 2\cos(u - v) = (\cos^2 u + \sin^2 u) + (\cos^2 v + \sin^2 v) - 2\cos u \cos v - 2\sin u \sin v$$

$$2 - 2\cos(u - v) = 2 - 2\cos u \cos v - 2\sin u \sin v$$

$$-2\cos(u - v) = -2(\cos u \cos v + \sin u \sin v)$$

$$\cos(u - v) = \cos u \cos v + \sin u \sin v$$

Now, to prove the identity for $\sin(u + v)$, use cofunction identities.

$$\sin(u + v) = \cos\left[\frac{\pi}{2} - (u + v)\right] = \cos\left[\left(\frac{\pi}{2} - u\right) - v\right]$$

$$= \cos\left(\frac{\pi}{2} - u\right)\cos v + \sin\left(\frac{\pi}{2} - u\right)\sin v$$

$$= \sin u \cos v + \cos u \sin v$$

(b) First, prove $\cos(u - v) = \cos u \cos v + \sin u \sin v$ using the figure containing points

$A(1, 0)$

$B(\cos(u - v), \sin(u - v))$

$C(\cos v, \sin v)$

$D(\cos u, \sin u)$

on the unit circle.

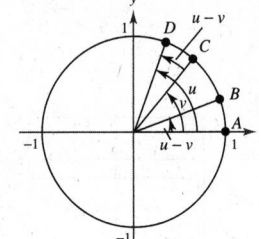

Since chords AB and CD are each subtended by angle $u - v$, their lengths are equal. Equating $[d(A, B)]^2 = [d(C, D)]^2$ we have $(\cos(u - v) - 1)^2 + \sin^2(u - v) = (\cos u - \cos v)^2 + (\sin u - \sin v)^2$. Simplifying and solving for $\cos(u - v)$, we have $\cos(u - v) = \cos u \cos v + \sin u \sin v$.

Using $\sin\theta = \cos\left(\frac{\pi}{2} - \theta\right)$ we have

$$\sin(u - v) = \cos\left[\frac{\pi}{2} - (u - v)\right] = \cos\left[\left(\frac{\pi}{2} - u\right) - (-v)\right]$$

$$= \cos\left(\frac{\pi}{2} - u\right)\cos(-v) + \sin\left(\frac{\pi}{2} - u\right)\sin(-v)$$

$$= \sin u \cos v - \cos u \sin v.$$

97. $f(x) = 5(x - 3)$

$$y = 5(x - 3)$$

$$\frac{y}{5} = x - 3$$

$$\frac{y}{5} + 3 = x$$

$$\frac{x}{5} + 3 = y$$

$$f^{-1}(x) = \frac{x + 15}{5}$$

$$f(f^{-1}(x)) = f\left(\frac{x + 15}{5}\right) = 5\left[\frac{x + 15}{5} - 3\right]$$

$$= 5\left(\frac{x + 15}{5}\right) - 5(3)$$

$$= x + 15 - 15$$

$$= x$$

$$f^{-1}(f(x)) = f^{-1}(5(x - 3)) = \frac{5(x - 3) + 15}{5}$$

$$= \frac{5x - 15 + 15}{5}$$

$$= \frac{5x}{5}$$

$$= x$$

98. $f(x) = \dfrac{7 - x}{8}$

$$y = \frac{7 - x}{8}$$

$$8y = 7 - x$$

$$x = 7 - 8y \implies f^{-1}(x) = -8x + 7$$

$$f(f^{-1}(x)) = \frac{7 - f^{-1}(x)}{8}$$

$$= \frac{7 - (-8x + 7)}{8}$$

$$= x$$

$$f^{-1}(f(x)) = -8\left(\frac{7 - x}{8}\right) + 7$$

$$= x$$

99. $f(x) = x^2 - 8$

f is not one-to-one so f^{-1} does not exist.

100. $f(x) = \sqrt{x - 16}, \, x \geq 16$

$$y = \sqrt{x - 16}$$

$$y^2 = x - 16$$

$$x = y^2 + 16 \implies f^{-1}(x) = x^2 + 16, \, x \geq 0$$

$$f(f^{-1}(x)) = \sqrt{(x^2 + 16) - 16} = x$$

$$f^{-1}(f(x)) = \left(\sqrt{x - 16}\right)^2 + 16 = x$$

101. $\log_3 3^{4x-3} = 4x - 3$

102. $\log_8 8^{3x^2} = 3x^2$

103. $e^{\ln(6x-3)} = 6x - 3$

104. $12x + e^{\ln x(x-2)} = 12x + x(x - 2)$

$$= 12x + x^2 - 2x$$

$$= x^2 + 10x$$

Section 5.5 Multiple-Angle and Product-to-Sum Formulas

■ You should know the following double-angle formulas.

(a) $\sin 2u = 2 \sin u \cos u$

(b) $\cos 2u = \cos^2 u - \sin^2 u$

$\quad = 2 \cos^2 u - 1$

$\quad = 1 - 2 \sin^2 u$

(c) $\tan 2u = \dfrac{2 \tan u}{1 - \tan^2 u}$

■ You should be able to reduce the power of a trigonometric function.

(a) $\sin^2 u = \dfrac{1 - \cos 2u}{2}$

(b) $\cos^2 u = \dfrac{1 + \cos 2u}{2}$

(c) $\tan^2 u = \dfrac{1 - \cos 2u}{1 + \cos 2u}$

■ You should be able to use the half-angle formulas. The signs of $\sin \dfrac{u}{2}$ and $\cos \dfrac{u}{2}$ depend on the quadrant in which $\dfrac{u}{2}$ lies.

(a) $\sin \dfrac{u}{2} = \pm \sqrt{\dfrac{1 - \cos u}{2}}$

(b) $\cos \dfrac{u}{2} = \pm \sqrt{\dfrac{1 + \cos u}{2}}$

(c) $\tan \dfrac{u}{2} = \dfrac{1 - \cos u}{\sin u} = \dfrac{\sin u}{1 + \cos u}$

■ You should be able to use the product-sum formulas.

(a) $\sin u \sin v = \dfrac{1}{2}[\cos(u - v) - \cos(u + v)]$

(b) $\cos u \cos v = \dfrac{1}{2}[\cos(u - v) + \cos(u + v)]$

(c) $\sin u \cos v = \dfrac{1}{2}[\sin(u + v) + \sin(u - v)]$

(d) $\cos u \sin v = \dfrac{1}{2}[\sin(u + v) - \sin(u - v)]$

■ You should be able to use the sum-product formulas.

(a) $\sin x + \sin y = 2 \sin\left(\dfrac{x + y}{2}\right) \cos\left(\dfrac{x - y}{2}\right)$

(b) $\sin x - \sin y = 2 \cos\left(\dfrac{x + y}{2}\right) \sin\left(\dfrac{x - y}{2}\right)$

(c) $\cos x + \cos y = 2 \cos\left(\dfrac{x + y}{2}\right) \cos\left(\dfrac{x - y}{2}\right)$

(d) $\cos x - \cos y = -2 \sin\left(\dfrac{x + y}{2}\right) \sin\left(\dfrac{x - y}{2}\right)$

Vocabulary Check

1. $2 \sin u \cos u$

2. $\cos^2 u$

3. $\cos^2 u - \sin^2 u = 2 \cos^2 u - 1 = 1 - 2 \sin^2 u$

4. $\tan^2 u$

5. $\pm \sqrt{\dfrac{1 - \cos u}{2}}$

6. $\dfrac{1 - \cos u}{\sin u} = \dfrac{\sin u}{1 + \cos u}$

7. $\dfrac{1}{2}[\cos(u - v) + \cos(u + v)]$

8. $\dfrac{1}{2}[\sin(u + v) + \sin(u - v)]$

9. $2 \sin\left(\dfrac{u + v}{2}\right) \cos\left(\dfrac{u - v}{2}\right)$

10. $-2 \sin\left(\dfrac{u + v}{2}\right) \sin\left(\dfrac{u - v}{2}\right)$

Figure for Exercises 1–8

$$\sin \theta = \frac{\sqrt{17}}{17}$$

$$\cos \theta = \frac{4\sqrt{17}}{17}$$

$$\tan \theta = \frac{1}{4}$$

1. $\sin \theta = \dfrac{\sqrt{17}}{17}$

2. $\tan \theta = \dfrac{1}{4}$

3. $\cos 2\theta = 2 \cos^2 \theta - 1$

$$= 2\left(\frac{4\sqrt{17}}{17}\right)^2 - 1$$

$$= \frac{32}{17} - 1$$

$$= \frac{15}{17}$$

4. $\sin 2\theta = 2 \sin \theta \cos \theta$

$$= 2\left(\frac{1}{\sqrt{17}}\right)\left(\frac{4}{\sqrt{17}}\right)$$

$$= \frac{8}{17}$$

5. $\tan 2\theta = \dfrac{2 \tan \theta}{1 - \tan^2 \theta}$

$$= \frac{2\left(\frac{1}{4}\right)}{1 - \left(\frac{1}{4}\right)^2}$$

$$= \frac{\frac{1}{2}}{1 - \frac{1}{16}}$$

$$= \frac{1}{2} \cdot \frac{16}{15}$$

$$= \frac{8}{15}$$

6. $\sec 2\theta = \dfrac{1}{\cos 2\theta}$

$$= \frac{1}{\cos^2 \theta - \sin^2 \theta}$$

$$= \frac{1}{\left(\frac{4}{\sqrt{17}}\right)^2 - \left(\frac{1}{\sqrt{17}}\right)^2}$$

$$= \frac{1}{\frac{16}{17} - \frac{1}{17}}$$

$$= \frac{17}{15}$$

7. $\csc 2\theta = \dfrac{1}{\sin 2\theta} = \dfrac{1}{2 \sin \theta \cos \theta} = \dfrac{1}{2\left(\frac{\sqrt{17}}{17}\right)\left(\frac{4\sqrt{17}}{17}\right)}$

$$= \frac{17}{8}$$

8. $\cot 2\theta = \dfrac{1}{\tan 2\theta} = \dfrac{1 - \tan^2 \theta}{2 \tan \theta} = \dfrac{1 - \left(\frac{1}{4}\right)^2}{2\left(\frac{1}{4}\right)}$

$$= \frac{15}{8}$$

9.

$$\sin 2x - \sin x = 0$$
$$2 \sin x \cos x - \sin x = 0$$
$$\sin x(2 \cos x - 1) = 0$$
$$\sin x = 0 \quad \text{or} \quad 2 \cos x - 1 = 0$$
$$x = 0, \pi \qquad \cos x = \frac{1}{2}$$
$$x = \frac{\pi}{3}, \frac{5\pi}{3}$$
$$x = 0, \frac{\pi}{3}, \pi, \frac{5\pi}{3}$$

10.

$$\sin 2x + \cos x = 0$$
$$2 \sin x \cos x + \cos x = 0$$
$$\cos x(2 \sin x + 1) = 0$$
$$\cos x = 0 \quad \text{or} \quad 2 \sin x + 1 = 0$$
$$x = \frac{\pi}{2}, \frac{3\pi}{2} \qquad \sin x = -\frac{1}{2}$$
$$x = \frac{7\pi}{6}, \frac{11\pi}{6}$$

11. $4 \sin x \cos x = 1$

$$2 \sin 2x = 1$$
$$\sin 2x = \frac{1}{2}$$
$$2x = \frac{\pi}{6} + 2n\pi \quad \text{or} \quad 2x = \frac{5\pi}{6} + 2n\pi$$
$$x = \frac{\pi}{12} + n\pi \qquad x = \frac{5\pi}{12} + n\pi$$
$$x = \frac{\pi}{12}, \frac{13\pi}{12} \qquad x = \frac{5\pi}{12}, \frac{17\pi}{12}$$

12.

$$\sin 2x \sin x = \cos x$$
$$2 \sin x \cos x \sin x - \cos x = 0$$
$$\cos x(2 \sin^2 x - 1) = 0$$
$$\cos x = 0 \quad \text{or} \quad 2 \sin^2 x - 1 = 0$$
$$x = \frac{\pi}{2}, \frac{3\pi}{2} \qquad \sin^2 x = \frac{1}{2}$$
$$\sin x = \pm\frac{\sqrt{2}}{2}$$
$$x = \frac{\pi}{4}, \frac{3\pi}{4}, \frac{5\pi}{4}, \frac{7\pi}{4}$$

13.

$$\cos 2x - \cos x = 0$$
$$\cos 2x = \cos x$$
$$\cos^2 x - \sin^2 x = \cos x$$
$$\cos^2 x - (1 - \cos^2 x) - \cos x = 0$$
$$2 \cos^2 x - \cos x - 1 = 0$$
$$(2 \cos x + 1)(\cos x - 1) = 0$$
$$2 \cos x + 1 = 0 \qquad \text{or} \quad \cos x - 1 = 0$$
$$\cos x = -\frac{1}{2} \qquad \cos x = 1$$
$$x = \frac{2\pi}{3}, \frac{4\pi}{3} \qquad x = 0$$

14.

$$\cos 2x + \sin x = 0$$
$$1 - 2 \sin^2 x + \sin x = 0$$
$$2 \sin^2 x - \sin x - 1 = 0$$
$$(2 \sin x + 1)(\sin x - 1) = 0$$
$$2 \sin x + 1 = 0 \qquad \text{or} \quad \sin x - 1 = 0$$
$$\sin x = -\frac{1}{2} \qquad \sin x = 1$$
$$x = \frac{7\pi}{6}, \frac{11\pi}{6} \qquad x = \frac{\pi}{2}$$

15. $\tan 2x - \cot x = 0$

$$\frac{2 \tan x}{1 - \tan^2 x} = \cot x$$

$$2 \tan x = \cot x(1 - \tan^2 x)$$

$$2 \tan x = \cot x - \cot x \tan^2 x$$

$$2 \tan x = \cot x - \tan x$$

$$3 \tan x = \cot x$$

$$3 \tan x - \cot x = 0$$

$$3 \tan x - \frac{1}{\tan x} = 0$$

$$\frac{3 \tan^2 x - 1}{\tan x} = 0$$

$$\frac{1}{\tan x}(3 \tan^2 x - 1) = 0$$

$$\cot x(3 \tan^2 x - 1) = 0$$

$\cot x = 0$ or $3 \tan^2 x - 1 = 0$

$x = \dfrac{\pi}{2}, \dfrac{3\pi}{2}$ $\tan^2 x = \dfrac{1}{3}$

$$\tan x = \pm\frac{\sqrt{3}}{3}$$

$$x = \frac{\pi}{6}, \frac{5\pi}{6}, \frac{7\pi}{6}, \frac{11\pi}{6}$$

$x = \dfrac{\pi}{6}, \dfrac{\pi}{2}, \dfrac{5\pi}{6}, \dfrac{7\pi}{6}, \dfrac{3\pi}{2}, \dfrac{11\pi}{6}$

16. $\tan 2x - 2 \cos x = 0$

$$\frac{2 \tan x}{1 - \tan^2 x} = 2 \cos x$$

$$2 \tan x = 2 \cos x(1 - \tan^2 x)$$

$$2 \tan x = 2 \cos x - 2 \cos x \tan^2 x$$

$$2 \tan x = 2 \cos x - 2 \cos x \frac{\sin^2 x}{\cos^2 x}$$

$$2 \tan x = 2 \cos x - 2 \frac{\sin^2 x}{\cos x}$$

$$\tan x = \cos x - \frac{\sin^2 x}{\cos x}$$

$$\frac{\sin x}{\cos x} = \cos x - \frac{\sin^2 x}{\cos x}$$

$$\frac{\sin x}{\cos x} + \frac{\sin^2 x}{\cos x} - \cos x = 0$$

$$\frac{\sin x + \sin^2 x - \cos^2 x}{\cos x} = 0$$

$$\frac{1}{\cos x}[\sin x + \sin^2 x - (1 - \sin^2 x)] = 0$$

$$\sec x[2 \sin^2 x + \sin x - 1] = 0$$

$$\sec x(2 \sin x - 1)(\sin x + 1) = 0$$

$\sec x = 0$ or $2 \sin x - 1 = 0$ or $\sin x + 1 = 0$

No solution $\sin x = \dfrac{1}{2}$ $\sin x = -1$

$x = \dfrac{\pi}{6}, \dfrac{5\pi}{6}$ $x = \dfrac{3\pi}{2}$

Also, values for which $\cos x = 0$ need to be checked.

$\dfrac{\pi}{2}, \dfrac{3\pi}{2}$ are solutions.

$x = \dfrac{\pi}{6}, \dfrac{\pi}{2}, \dfrac{5\pi}{6}, \dfrac{3\pi}{2}$

17. $\sin 4x = -2 \sin 2x$

$$\sin 4x + 2 \sin 2x = 0$$

$$2 \sin 2x \cos 2x + 2 \sin 2x = 0$$

$$2 \sin 2x(\cos 2x + 1) = 0$$

$2 \sin 2x = 0$ or $\cos 2x + 1 = 0$

$\sin 2x = 0$ $\cos 2x = -1$

$2x = n\pi$ $2x = \pi + 2n\pi$

$x = \dfrac{n}{2}\pi$ $x = \dfrac{\pi}{2} + n\pi$

$x = 0, \dfrac{\pi}{2}, \pi, \dfrac{3\pi}{2}$ $x = \dfrac{\pi}{2}, \dfrac{3\pi}{2}$

18.
$$(\sin 2x + \cos 2x)^2 = 1$$
$$\sin^2 2x + 2 \sin 2x \cos 2x + \cos^2 2x = 1$$
$$2 \sin 2x \cos 2x = 0$$
$$\sin 4x = 0$$
$$4x = n\pi$$
$$x = \frac{n\pi}{4}$$
$$x = 0, \frac{\pi}{4}, \frac{\pi}{2}, \frac{3\pi}{4}, \pi, \frac{5\pi}{4}, \frac{3\pi}{2}, \frac{7\pi}{4}$$

19. $6 \sin x \cos x = 3(2 \sin x \cos x)$
$$= 3 \sin 2x$$

20. $6 \cos^2 x - 3 = 3(2 \cos^2 x - 1)$
$$= 3 \cos 2x$$

21. $4 - 8 \sin^2 x = 4(1 - 2 \sin^2 x)$
$$= 4 \cos 2x$$

22. $(\cos x + \sin x)(\cos x - \sin x) = \cos^2 x - \sin^2 x$
$$= \cos 2x$$

23. $\sin u = -\dfrac{4}{5}, \pi < u < \dfrac{3\pi}{2} \implies \cos u = -\dfrac{3}{5}$

$$\sin 2u = 2 \sin u \cos u = 2\left(-\frac{4}{5}\right)\left(-\frac{3}{5}\right) = \frac{24}{25}$$

$$\cos 2u = \cos^2 u - \sin^2 u = \frac{9}{25} - \frac{16}{25} = -\frac{7}{25}$$

$$\tan 2u = \frac{2 \tan u}{1 - \tan^2 u} = \frac{2\left(\frac{4}{3}\right)}{1 - \frac{16}{9}} = \frac{8}{3}\left(-\frac{9}{7}\right) = -\frac{24}{7}$$

24. $\cos u = -\dfrac{2}{3}, \dfrac{\pi}{2} < u < \pi$

$$\sin 2u = 2 \sin u \cos u = 2 \cdot \frac{\sqrt{5}}{3}\left(-\frac{2}{3}\right) = -\frac{4\sqrt{5}}{9}$$

$$\cos 2u = \cos^2 u - \sin^2 u = \frac{4}{9} - \frac{5}{9} = -\frac{1}{9}$$

$$\tan 2u = \frac{2 \tan u}{1 - \tan^2 u} = \frac{2\left(-\frac{\sqrt{5}}{2}\right)}{1 - \frac{5}{4}} = 4\sqrt{5}$$

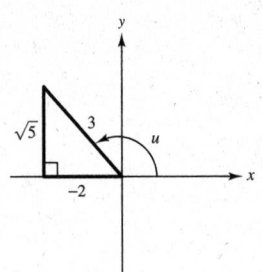

$$\csc u = 3, \frac{\pi}{2} < u < \pi$$

25. $\tan u = \dfrac{3}{4}, 0 < u < \dfrac{\pi}{2} \implies \sin u = \dfrac{3}{5} \text{ and } \cos u = \dfrac{4}{5}$

$$\sin 2u = 2 \sin u \cos u = 2\left(\frac{3}{5}\right)\left(\frac{4}{5}\right) = \frac{24}{25}$$

$$\cos 2u = \cos^2 u - \sin^2 u = \frac{16}{25} - \frac{9}{25} = \frac{7}{25}$$

$$\tan 2u = \frac{2 \tan u}{1 - \tan^2 u} = \frac{2\left(\frac{3}{4}\right)}{1 - \frac{9}{16}} = \frac{3}{2}\left(\frac{16}{7}\right) = \frac{24}{7}$$

26. $\cot u = -4, \dfrac{3\pi}{2} < u < 2\pi$

$\sin 2u = 2 \sin u \cos u = 2\left(-\dfrac{1}{\sqrt{17}}\right)\left(\dfrac{4}{\sqrt{17}}\right) = -\dfrac{8}{17}$

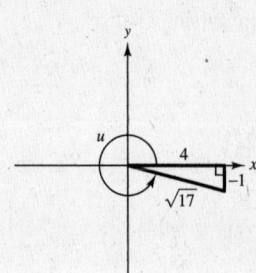

$\cos 2u = \cos^2 u - \sin^2 u$

$= \left(\dfrac{4}{\sqrt{17}}\right)^2 - \left(-\dfrac{1}{\sqrt{17}}\right)^2 = \dfrac{15}{17}$

$\tan 2u = \dfrac{2 \tan u}{1 - \tan^2 u} = \dfrac{2\left(-\dfrac{1}{4}\right)}{1 - \left(-\dfrac{1}{4}\right)^2} = -\dfrac{8}{15}$

27. $\sec u = -\dfrac{5}{2}, \dfrac{\pi}{2} < u < \pi \implies \sin u = \dfrac{\sqrt{21}}{5}$ and $\cos u = -\dfrac{2}{5}$

$\sin 2u = 2 \sin u \cos u = 2\left(\dfrac{\sqrt{21}}{5}\right)\left(-\dfrac{2}{5}\right) = -\dfrac{4\sqrt{21}}{25}$

$\cos 2u = \cos^2 u - \sin^2 u = \left(-\dfrac{2}{5}\right)^2 - \left(\dfrac{\sqrt{21}}{5}\right)^2 = -\dfrac{17}{25}$

$\tan 2u = \dfrac{2 \tan u}{1 - \tan^2 u} = \dfrac{2\left(-\dfrac{\sqrt{21}}{2}\right)}{1 - \left(-\dfrac{\sqrt{21}}{2}\right)^2}$

$= \dfrac{-\sqrt{21}}{1 - \dfrac{21}{4}} = \dfrac{4\sqrt{21}}{17}$

28. $\sin 2u = 2 \sin u \cos u = 2 \cdot \dfrac{1}{3}\left(-\dfrac{2\sqrt{2}}{3}\right) = -\dfrac{4\sqrt{2}}{9}$

$\cos 2u = \cos^2 u - \sin^2 u = \left(-\dfrac{2\sqrt{2}}{3}\right)^2 - \left(\dfrac{1}{3}\right)^2 = \dfrac{7}{9}$

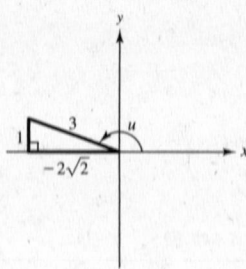

$\tan 2u = \dfrac{2 \tan u}{1 - \tan^2 u} = \dfrac{2\left(-\dfrac{\sqrt{2}}{4}\right)}{1 - \left(-\dfrac{\sqrt{2}}{4}\right)^2} = -\dfrac{4\sqrt{2}}{7}$

29. $\cos^4 x = (\cos^2 x)(\cos^2 x) = \left(\dfrac{1 + \cos 2x}{2}\right)\left(\dfrac{1 + \cos 2x}{2}\right) = \dfrac{1 + 2\cos 2x + \cos^2 2x}{4}$

$= \dfrac{1 + 2\cos 2x + \dfrac{1 + \cos 4x}{2}}{4}$

$= \dfrac{2 + 4\cos 2x + 1 + \cos 4x}{8}$

$= \dfrac{3 + 4\cos 2x + \cos 4x}{8}$

$= \dfrac{1}{8}(3 + 4\cos 2x + \cos 4x)$

30. $\sin^8 x = \sin^4 x \sin^4 x = (\sin^2 x \sin^2 x)(\sin^2 x \sin^2 x)$

$$\sin^2 x = \frac{1 - \cos 2x}{2}$$

$$\sin^4 x = \left(\frac{1 - \cos 2x}{2}\right)\left(\frac{1 - \cos 2x}{2}\right)$$

$$= \frac{1}{4}(1 - 2\cos 2x + \cos^2 2x)$$

$$= \frac{1}{4}\left(1 - 2\cos 2x + \left(\frac{1 + \cos 4x}{2}\right)\right)$$

$$= \frac{1}{8}(3 - 4\cos 2x + \cos 4x)$$

$\sin^8 x = \sin^4 x \sin^4 x$

$$= \frac{1}{64}(3 - 4\cos 2x + \cos 4x)(3 - 4\cos 2x + \cos 4x)$$

$$= \frac{1}{64}(9 - 24\cos 2x + 16\cos^2 2x + 6\cos 4x - 8\cos 2x \cos 4x + \cos^2 4x)$$

$$= \frac{1}{64}\left[9 - 24\cos 2x + 16\left(\frac{1 + \cos 4x}{2}\right) + 6\cos 4x - (8)\left(\frac{1}{2}\right)(\cos 6x + \cos 2x) + \left(\frac{1 + \cos 8x}{2}\right)\right]$$

$$= \frac{1}{64}\left[\frac{35}{2} - 28\cos 2x + 14\cos 4x - 4\cos 6x + \frac{1}{2}\cos 8x\right]$$

$$= \frac{1}{128}[35 - 56\cos 2x + 28\cos 4x - 8\cos 6x + \cos 8x]$$

In the above, we used $\cos 2x \cos 4x = \frac{1}{2}(\cos 6x + \cos 2x)$.

31. $(\sin^2 x)(\cos^2 x) = \left(\frac{1 - \cos 2x}{2}\right)\left(\frac{1 + \cos 2x}{2}\right)$

$$= \frac{1 - \cos^2 2x}{4}$$

$$= \frac{1}{4}\left(1 - \frac{1 + \cos 4x}{2}\right)$$

$$= \frac{1}{8}(2 - 1 - \cos 4x)$$

$$= \frac{1}{8}(1 - \cos 4x)$$

32. $\sin^4 x \cos^4 x = \sin^2 x \sin^2 x \cos^2 x \cos^2 x$

$$= (\sin^2 x \cos^2 x)(\sin^2 x \cos^2 x)$$

$$= \left(\frac{1}{4}\sin^2 2x\right)\left(\frac{1}{4}\sin^2 2x\right)$$

$$= \left[\frac{1}{4}\left(\frac{1 - \cos 4x}{2}\right)\right]\left[\frac{1}{4}\left(\frac{1 - \cos 4x}{2}\right)\right]$$

$$= \frac{1}{64}[1 - 2\cos 4x + \cos^2 4x]$$

$$= \frac{1}{64}\left[1 - 2\cos 4x + \left(\frac{1 + \cos 8x}{2}\right)\right]$$

$$= \frac{1}{64}\left[\frac{3}{2} - 2\cos 4x + \frac{1}{2}\cos 8x\right]$$

$$= \frac{1}{128}(3 - 4\cos 4x + \cos 8x)$$

33. $\sin^2 x \cos^4 x = \sin^2 x \cos^2 x \cos^2 x = \left(\dfrac{1 - \cos 2x}{2}\right)\left(\dfrac{1 + \cos 2x}{2}\right)\left(\dfrac{1 + \cos 2x}{2}\right)$

$= \dfrac{1}{8}(1 - \cos 2x)(1 + \cos 2x)(1 + \cos 2x)$

$= \dfrac{1}{8}(1 - \cos^2 2x)(1 + \cos 2x)$

$= \dfrac{1}{8}(1 + \cos 2x - \cos^2 2x - \cos^3 2x)$

$= \dfrac{1}{8}\left[1 + \cos 2x - \left(\dfrac{1 + \cos 4x}{2}\right) - \cos 2x\left(\dfrac{1 + \cos 4x}{2}\right)\right]$

$= \dfrac{1}{16}[2 + 2\cos 2x - 1 - \cos 4x - \cos 2x - \cos 2x \cos 4x]$

$= \dfrac{1}{16}(1 + \cos 2x - \cos 4x - \cos 2x \cos 4x)$

34. $\sin^4 x \cos^2 x = \sin^2 x \sin^2 x \cos^2 x$

$= \left(\dfrac{1 - \cos 2x}{2}\right)\left(\dfrac{1 - \cos 2x}{2}\right)\left(\dfrac{1 + \cos 2x}{2}\right)$

$= \dfrac{1}{8}(1 - \cos 2x)(1 - \cos^2 2x)$

$= \dfrac{1}{8}(1 - \cos 2x - \cos^2 2x + \cos^3 2x)$

$= \dfrac{1}{8}\left[1 - \cos 2x - \left(\dfrac{1 + \cos 4x}{2}\right) + \cos 2x\left(\dfrac{1 + \cos 4x}{2}\right)\right]$

$= \dfrac{1}{16}[2 - 2\cos 2x - 1 - \cos 4x + \cos 2x + \cos 2x \cos 4x]$

$= \dfrac{1}{16}\left[1 - \cos 2x - \cos 4x + \dfrac{1}{2}\cos 2x + \dfrac{1}{2}\cos 6x\right]$

$= \dfrac{1}{32}[2 - 2\cos 2x - 2\cos 4x + \cos 2x + \cos 6x]$

$= \dfrac{1}{32}[2 - \cos 2x - 2\cos 4x + \cos 6x]$

Figure for Exercises 35–40

$\sin \theta = \frac{8}{17}$

$\cos \theta = \frac{15}{17}$

35. $\cos \dfrac{\theta}{2} = \sqrt{\dfrac{1 + \cos \theta}{2}} = \sqrt{\dfrac{1 + \frac{15}{17}}{2}} = \sqrt{\dfrac{32}{34}} = \sqrt{\dfrac{16}{17}} = \dfrac{4\sqrt{17}}{17}$

36. $\sin \dfrac{\theta}{2} = \sqrt{\dfrac{1 - \cos \theta}{2}} = \sqrt{\dfrac{1 - \frac{15}{17}}{2}} = \sqrt{\dfrac{\frac{2}{17}}{2}} = \dfrac{1}{\sqrt{17}} = \dfrac{\sqrt{17}}{17}$

37. $\tan\dfrac{\theta}{2} = \dfrac{\sin\theta}{1 + \cos\theta} \doteq \dfrac{8/17}{1 + (15/17)} = \dfrac{8}{17} \cdot \dfrac{17}{32} = \dfrac{1}{4}$

38. $\sec\dfrac{\theta}{2} = \dfrac{1}{\cos(\theta/2)} = \dfrac{1}{\sqrt{(1 + \cos\theta)/2}}$

$\qquad\qquad = \dfrac{1}{\sqrt{[1 + (15/17)]/2}} = \dfrac{1}{\sqrt{16/17}}$

$\qquad\qquad = \dfrac{\sqrt{17}}{4}$

39. $\csc\dfrac{\theta}{2} = \dfrac{1}{\sin(\theta/2)} = \dfrac{1}{\sqrt{(1 - \cos\theta)/2}}$

$\qquad\quad = \dfrac{1}{\sqrt{[1 - (15/17)]/2}} = \dfrac{1}{\sqrt{1/17}} = \sqrt{17}$

40. $\cot\dfrac{\theta}{2} = \dfrac{1}{\tan(\theta/2)} = \dfrac{\sin\theta}{1 - \cos\theta} = \dfrac{8/17}{1 - (15/17)}$

$\qquad\quad = \dfrac{8/17}{2/17} = 4$

41. $\sin 75° = \sin\left(\dfrac{1}{2} \cdot 150°\right) = \sqrt{\dfrac{1 - \cos 150°}{2}} \doteq \sqrt{\dfrac{1 + (\sqrt{3}/2)}{2}}$

$\qquad\quad = \dfrac{1}{2}\sqrt{2 + \sqrt{3}}$

$\quad\cos 75° = \cos\left(\dfrac{1}{2} \cdot 150°\right) = \sqrt{\dfrac{1 + \cos 150°}{2}} = \sqrt{\dfrac{1 - (\sqrt{3}/2)}{2}}$

$\qquad\quad = \dfrac{1}{2}\sqrt{2 - \sqrt{3}}$

$\quad\tan 75° = \tan\left(\dfrac{1}{2} \cdot 150°\right) = \dfrac{\sin 150°}{1 + \cos 150°} = \dfrac{1/2}{1 - (\sqrt{3}/2)}$

$\qquad\quad = \dfrac{1}{2 - \sqrt{3}} \cdot \dfrac{2 + \sqrt{3}}{2 + \sqrt{3}} = \dfrac{2 + \sqrt{3}}{4 - 3} = 2 + \sqrt{3}$

42. $\sin 165 = \sin\left(\dfrac{1}{2} \cdot 330\right) = \sqrt{\dfrac{1 - \cos 330}{2}} = \sqrt{\dfrac{1 - (\sqrt{3}/2)}{2}} = \dfrac{1}{2}\sqrt{2 - \sqrt{3}}$

$\quad\cos 165 = \cos\left(\dfrac{1}{2} \cdot 330\right) = -\sqrt{\dfrac{1 + \cos 330}{2}} = -\sqrt{\dfrac{1 + (\sqrt{3}/2)}{2}} = -\dfrac{1}{2}\sqrt{2 + \sqrt{3}}$

$\quad\tan 165 = \tan\left(\dfrac{1}{2} \cdot 330\right) = \dfrac{\sin 330}{1 + \cos 330} = \dfrac{-1/2}{1 + (\sqrt{3}/2)} = \dfrac{-1}{2 + \sqrt{3}} = \sqrt{3} - 2$

43. $\sin 112° 30' = \sin\left(\dfrac{1}{2} \cdot 225°\right) = \sqrt{\dfrac{1 - \cos 225°}{2}} = \sqrt{\dfrac{1 + (\sqrt{2}/2)}{2}} = \dfrac{1}{2}\sqrt{2 + \sqrt{2}}$

$\quad\cos 112° 30' = \cos\left(\dfrac{1}{2} \cdot 225°\right) = -\sqrt{\dfrac{1 + \cos 225°}{2}} = -\sqrt{\dfrac{1 - (\sqrt{2}/2)}{2}} = -\dfrac{1}{2}\sqrt{2 - \sqrt{2}}$

$\quad\tan 112° 30' = \tan\left(\dfrac{1}{2} \cdot 225°\right) = \dfrac{\sin 225°}{1 + \cos 225°} = \dfrac{-\sqrt{2}/2}{1 - (\sqrt{2}/2)} = -1 - \sqrt{2}$

44. $\sin 67° 30' = \sin\left(\dfrac{1}{2} \cdot 135°\right) = \sqrt{\dfrac{1 - \cos 135°}{2}} = \sqrt{\dfrac{1 + (\sqrt{2}/2)}{2}} = \dfrac{1}{2}\sqrt{2 + \sqrt{2}}$

$\quad\cos 67° 30' = \cos\left(\dfrac{1}{2} \cdot 135°\right) = \sqrt{\dfrac{1 + \cos 135°}{2}} = \sqrt{\dfrac{1 - (\sqrt{2}/2)}{2}} = \dfrac{1}{2}\sqrt{2 - \sqrt{2}}$

$\quad\tan 67° 30' = \tan\left(\dfrac{1}{2} \cdot 135°\right) = \dfrac{\sin 135°}{1 + \cos 135°} = \dfrac{\sqrt{2}/2}{1 - (\sqrt{2}/2)} = 1 + \sqrt{2}$

45. $\sin\dfrac{\pi}{8} = \sin\left[\dfrac{1}{2}\left(\dfrac{\pi}{4}\right)\right] = \sqrt{\dfrac{1 - \cos\dfrac{\pi}{4}}{2}} = \dfrac{1}{2}\sqrt{2 - \sqrt{2}}$

$\cos\dfrac{\pi}{8} = \cos\left[\dfrac{1}{2}\left(\dfrac{\pi}{4}\right)\right] = \sqrt{\dfrac{1 + \cos\dfrac{\pi}{4}}{2}} = \dfrac{1}{2}\sqrt{2 + \sqrt{2}}$

$\tan\dfrac{\pi}{8} = \tan\left[\dfrac{1}{2}\left(\dfrac{\pi}{4}\right)\right] = \dfrac{\sin\dfrac{\pi}{4}}{1 + \cos\dfrac{\pi}{4}} = \dfrac{\dfrac{\sqrt{2}}{2}}{1 + \dfrac{\sqrt{2}}{2}} = \sqrt{2} - 1$

46. $\sin\dfrac{\pi}{12} = \sin\left[\dfrac{1}{2}\left(\dfrac{\pi}{6}\right)\right] = \sqrt{\dfrac{1 - \cos\dfrac{\pi}{6}}{2}} = \sqrt{\dfrac{1 - \dfrac{\sqrt{3}}{2}}{2}}$

$= \dfrac{1}{2}\sqrt{2 - \sqrt{3}}$

$\cos\dfrac{\pi}{12} = \cos\left[\dfrac{1}{2}\left(\dfrac{\pi}{6}\right)\right] = \sqrt{\dfrac{1 + \cos\dfrac{\pi}{6}}{2}} = \dfrac{1}{2}\sqrt{2 + \sqrt{3}}$

$\tan\dfrac{\pi}{12} = \tan\left[\dfrac{1}{2}\left(\dfrac{\pi}{6}\right)\right]\dfrac{\sin\dfrac{\pi}{6}}{1 + \cos\dfrac{\pi}{6}} = \dfrac{\dfrac{1}{2}}{1 + \dfrac{\sqrt{3}}{2}} = 2 - \sqrt{3}$

47. $\sin\dfrac{3\pi}{8} = \sin\left(\dfrac{1}{2}\cdot\dfrac{3\pi}{4}\right) = \sqrt{\dfrac{1 - \cos\dfrac{3\pi}{4}}{2}} = \sqrt{\dfrac{1 + \dfrac{\sqrt{2}}{2}}{2}} = \dfrac{1}{2}\sqrt{2 + \sqrt{2}}$

$\cos\dfrac{3\pi}{8} = \cos\left(\dfrac{1}{2}\cdot\dfrac{3\pi}{4}\right) = \sqrt{\dfrac{1 + \cos\dfrac{3\pi}{4}}{2}} = \sqrt{\dfrac{1 - \dfrac{\sqrt{2}}{2}}{2}} = \dfrac{1}{2}\sqrt{2 - \sqrt{2}}$

$\tan\dfrac{3\pi}{8} = \tan\left(\dfrac{1}{2}\cdot\dfrac{3\pi}{4}\right) = \dfrac{\sin\dfrac{3\pi}{4}}{1 + \cos\dfrac{3\pi}{4}} = \dfrac{\dfrac{\sqrt{2}}{2}}{1 - \dfrac{\sqrt{2}}{2}} = \dfrac{\dfrac{\sqrt{2}}{2}}{\dfrac{(2 - \sqrt{2})}{2}} = \dfrac{\sqrt{2}}{2 - \sqrt{2}} = \sqrt{2} + 1$

48. $\sin\dfrac{7\pi}{12} = \sin\left[\dfrac{1}{2}\left(\dfrac{7\pi}{6}\right)\right] = \sqrt{\dfrac{1 - \cos(7\pi(6))}{2}} = \sqrt{\dfrac{1 + \dfrac{\sqrt{3}}{2}}{2}} = \dfrac{1}{2}\sqrt{2 + \sqrt{3}}$

$\cos\dfrac{7\pi}{12} = \cos\left[\dfrac{1}{2}\left(\dfrac{7\pi}{6}\right)\right] = -\sqrt{\dfrac{1 + \cos\dfrac{7\pi}{6}}{2}} = -\sqrt{\dfrac{1 - \dfrac{\sqrt{3}}{2}}{2}} = -\dfrac{1}{2}\sqrt{2 - \sqrt{3}}$

$\tan\dfrac{7\pi}{12} = \tan\left[\dfrac{1}{2}\left(\dfrac{7\pi}{6}\right)\right] = \dfrac{\sin\dfrac{7\pi}{6}}{1 + \cos\dfrac{7\pi}{6}} = \dfrac{-\dfrac{1}{2}}{1 - \dfrac{\sqrt{3}}{2}} = -2 - \sqrt{3}$

49. $\sin u = \dfrac{5}{13},\ \dfrac{\pi}{2} < u < \pi \Rightarrow \cos u = -\dfrac{12}{13}$

$\sin\left(\dfrac{u}{2}\right) = \sqrt{\dfrac{1 - \cos u}{2}} = \sqrt{\dfrac{1 + \dfrac{12}{13}}{2}} = \dfrac{5\sqrt{26}}{26}$

$\cos\left(\dfrac{u}{2}\right) = \sqrt{\dfrac{1 + \cos u}{2}} = \sqrt{\dfrac{1 - \dfrac{12}{13}}{2}} = \dfrac{\sqrt{26}}{26}$

$\tan\left(\dfrac{u}{2}\right) = \dfrac{\sin u}{1 + \cos u} = \dfrac{\dfrac{5}{13}}{1 - \dfrac{12}{13}} = 5$

50. $\cos u = \dfrac{3}{5},\ 0 < u < \dfrac{\pi}{2}$

$\sin\left(\dfrac{u}{2}\right) = \sqrt{\dfrac{1 - \cos u}{2}} = \sqrt{\dfrac{1 - \dfrac{3}{5}}{2}} = \dfrac{\sqrt{5}}{5}$

$\cos\left(\dfrac{u}{2}\right) = \sqrt{\dfrac{1 + \cos u}{2}} = \sqrt{\dfrac{1 + \dfrac{3}{5}}{2}} = \dfrac{2\sqrt{5}}{5}$

51. $\tan u = -\dfrac{5}{8}$, $\dfrac{3\pi}{2} < u < 2\pi \implies \sin u = -\dfrac{5}{\sqrt{89}}$ and $\cos u = \dfrac{8}{\sqrt{89}}$

$$\sin\left(\frac{u}{2}\right) = \sqrt{\frac{1 - \cos u}{2}} = \sqrt{\frac{1 - \dfrac{8}{\sqrt{89}}}{2}} \cdot \sqrt{\frac{\sqrt{89} - 8}{2\sqrt{89}}} = \sqrt{\frac{89 - 8\sqrt{89}}{178}}$$

$$\cos\left(\frac{u}{2}\right) = -\sqrt{\frac{1 + \cos u}{2}} = -\sqrt{\frac{1 + \dfrac{8}{\sqrt{89}}}{2}} = -\sqrt{\frac{\sqrt{89} + 8}{2\sqrt{89}}} = -\sqrt{\frac{89 + 8\sqrt{89}}{178}}$$

$$\tan\left(\frac{u}{2}\right) = \frac{1 - \cos u}{\sin u} = \frac{1 - \dfrac{8}{\sqrt{89}}}{-\dfrac{5}{\sqrt{89}}} = \frac{8 - \sqrt{89}}{5}$$

52. $\cot u = 3$, $\pi < u < \dfrac{3\pi}{2}$

$$\sin\left(\frac{u}{2}\right) = \sqrt{\frac{1 - \cos u}{2}} = \sqrt{\frac{1 + \dfrac{3}{\sqrt{10}}}{2}} = \sqrt{\frac{10 + 3\sqrt{10}}{20}} = \frac{1}{2}\sqrt{\frac{10 + 3\sqrt{10}}{5}}$$

$$\cos\left(\frac{u}{2}\right) = -\sqrt{\frac{1 + \cos u}{2}} = -\sqrt{\frac{1 - \dfrac{3}{\sqrt{10}}}{2}} = -\sqrt{\frac{10 - 3\sqrt{10}}{20}} = -\frac{1}{2}\sqrt{\frac{10 - 3\sqrt{10}}{5}}$$

$$\tan\left(\frac{u}{2}\right) = \frac{1 - \cos u}{\sin u} = \frac{1 + \dfrac{3}{\sqrt{10}}}{-\dfrac{1}{\sqrt{10}}} = -\sqrt{10} - 3$$

53. $\csc u = -\dfrac{5}{3}$, $\pi < u < \dfrac{3\pi}{2} \implies \sin u = -\dfrac{3}{5}$ and $\cos u = -\dfrac{4}{5}$

$$\sin\left(\frac{u}{2}\right) = \sqrt{\frac{1 - \cos u}{2}} = \sqrt{\frac{1 + \frac{4}{5}}{2}} = \frac{3\sqrt{10}}{10}$$

$$\cos\left(\frac{u}{2}\right) = -\sqrt{\frac{1 + \cos u}{2}} = -\sqrt{\frac{1 - \frac{4}{5}}{2}} = -\frac{\sqrt{10}}{10}$$

$$\tan\left(\frac{u}{2}\right) = \frac{1 - \cos u}{\sin u} = \frac{1 + \frac{4}{5}}{-\frac{3}{5}} = -3$$

54. $\sec u = -\dfrac{7}{2}$, $\dfrac{\pi}{2} < u < \pi$

$$\sin\left(\frac{u}{2}\right) = \sqrt{\frac{1 - \cos u}{2}} = \sqrt{\frac{1 + \frac{2}{7}}{2}} = \frac{3\sqrt{14}}{14}$$

$$\cos\left(\frac{u}{2}\right) = \sqrt{\frac{1 + \cos u}{2}} = \sqrt{\frac{1 - \frac{2}{7}}{2}} = \frac{\sqrt{70}}{14}$$

$$\tan\left(\frac{u}{2}\right) = \frac{1 - \cos u}{\sin u} = \frac{1 + \frac{2}{7}}{3\dfrac{\sqrt{5}}{7}} = \frac{3\sqrt{5}}{5}$$

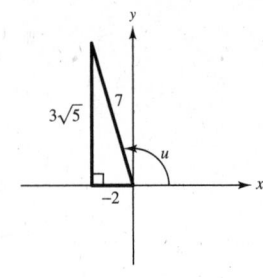

55. $\sqrt{\dfrac{1-\cos 6x}{2}} = |\sin 3x|$

56. $\sqrt{\dfrac{1+\cos 4x}{2}} = \left|\cos \dfrac{4x}{2}\right| = |\cos 2x|$

57. $-\sqrt{\dfrac{1-\cos 8x}{1+\cos 8x}} = -\dfrac{\sqrt{\dfrac{1-\cos 8x}{2}}}{\sqrt{\dfrac{1+\cos 8x}{2}}}$

$$= -\left|\dfrac{\sin 4x}{\cos 4x}\right|$$

$$= -|\tan 4x|$$

58. $-\sqrt{\dfrac{1-\cos(x-1)}{2}} = -\left|\sin\left(\dfrac{x-1}{2}\right)\right|$

59. $\sin \dfrac{x}{2} + \cos x = 0$

$$\pm\sqrt{\dfrac{1-\cos x}{2}} = -\cos x$$

$$\dfrac{1-\cos x}{2} = \cos^2 x$$

$$0 = 2\cos^2 x + \cos x - 1$$

$$= (2\cos x - 1)(\cos x + 1)$$

$$\cos x = \dfrac{1}{2} \quad \text{or} \quad \cos x = -1$$

$$x = \dfrac{\pi}{3}, \dfrac{5\pi}{3} \qquad x = \pi$$

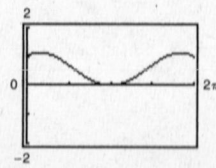

By checking these values in the original equation, we see that $x = \pi/3$ and $x = 5\pi/3$ are extraneous, and $x = \pi$ is the only solution.

60. $h(x) = \sin \dfrac{x}{2} + \cos x - 1$

$$\sin \dfrac{x}{2} + \cos x - 1 = 0$$

$$\pm\sqrt{\dfrac{1-\cos x}{2}} = 1 - \cos x$$

$$\dfrac{1-\cos x}{2} = 1 - 2\cos x + \cos^2 x$$

$$1 - \cos x = 2 - 4\cos x + 2\cos^2 x$$

$$2\cos^2 x - 3\cos x + 1 = 0$$

$$(2\cos x - 1)(\cos x - 1) = 0$$

$$2\cos x - 1 = 0 \quad \text{or} \quad \cos x - 1 = 0$$

$$\cos x = \dfrac{1}{2} \qquad\qquad \cos x = 1$$

$$x = \dfrac{\pi}{3}, \dfrac{5\pi}{3} \qquad\qquad x = 0$$

$0, \dfrac{\pi}{3}$, and $\dfrac{5\pi}{3}$ are all solutions to the equation.

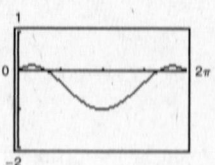

61.
$$\cos \frac{x}{2} - \sin x = 0$$

$$\pm\sqrt{\frac{1 + \cos x}{2}} = \sin x$$

$$\frac{1 + \cos x}{2} = \sin^2 x$$

$$1 + \cos x = 2 \sin^2 x$$

$$1 + \cos x = 2 - 2 \cos^2 x$$

$$2 \cos^2 x + \cos x - 1 = 0$$

$$(2 \cos x - 1)(\cos x + 1) = 0$$

$$2 \cos x - 1 = 0 \quad \text{or} \quad \cos x + 1 = 0$$

$$\cos x = \frac{1}{2} \qquad \cos x = -1$$

$$x = \frac{\pi}{3}, \frac{5\pi}{3} \qquad x = \pi$$

$$x = \frac{\pi}{3}, \pi, \frac{5\pi}{3}$$

$\pi/3$, π, and $5\pi/3$ are all solutions to the equation.

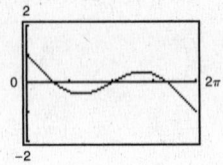

62.
$$g(x) = \tan \frac{x}{2} - \sin x$$

$$\tan \frac{x}{2} - \sin x = 0$$

$$\frac{1 - \cos x}{\sin x} = \sin x$$

$$1 - \cos x = \sin^2 x$$

$$1 - \cos x = 1 - \cos^2 x$$

$$\cos^2 x - \cos x = 0$$

$$\cos x(\cos x - 1) = 0$$

$$\cos x = 0 \quad \text{or} \quad \cos x - 1 = 0$$

$$x = \frac{\pi}{2}, \frac{3\pi}{2} \qquad \cos x = 1$$

$$x = 0$$

$0, \dfrac{\pi}{2},$ and $\dfrac{3\pi}{2}$ are all solutions to the equation.

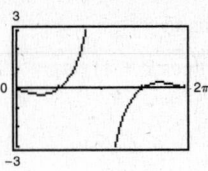

63. $6 \sin \dfrac{\pi}{4} \cos \dfrac{\pi}{4} = 6 \cdot \dfrac{1}{2}\left[\sin\left(\dfrac{\pi}{4} + \dfrac{\pi}{4}\right) + \sin\left(\dfrac{\pi}{4} - \dfrac{\pi}{4}\right)\right] = 3\left(\sin \dfrac{\pi}{2} + \sin 0\right)$

64. $4 \cos \dfrac{\pi}{3} \sin \dfrac{5\pi}{6} = 4 \cdot \dfrac{1}{2}\left[\sin\left(\dfrac{\pi}{3} + \dfrac{5\pi}{6}\right) - \sin\left(\dfrac{\pi}{3} - \dfrac{5\pi}{6}\right)\right] = 2\left[\sin\left(\dfrac{7\pi}{6}\right) - \sin\left(-\dfrac{\pi}{2}\right)\right] = 2\left[\sin\left(\dfrac{7\pi}{6}\right) + \sin\left(\dfrac{\pi}{2}\right)\right]$

65. $10 \cos 75° \cos 15° = 10\left(\dfrac{1}{2}\right)[\cos(75° - 15°) + \cos(75° + 15°)] = 5[\cos 60° + \cos 90°]$

66. $6 \sin 45° \cos 15° = 6\left(\dfrac{1}{2}\right)(\sin 60° + \sin 30°) = 3(\sin 60° + \sin 30°)$

67. $\cos 4\theta \sin 6\theta = \dfrac{1}{2}[\sin(4\theta + 6\theta) - \sin(4\theta - 6\theta)] = \dfrac{1}{2}[\sin 10\theta - \sin(-2\theta)] = \dfrac{1}{2}(\sin 10\theta + \sin 2\theta)$

68. $3 \sin 2\alpha \sin 3\alpha = 3 \cdot \dfrac{1}{2}[\cos(2\alpha - 3\alpha) - \cos(2\alpha + 3\alpha)] = \dfrac{3}{2}[\cos(-\alpha) - \cos 5\alpha] = \dfrac{3}{2}(\cos \alpha - \cos 5\alpha)$

69. $5 \cos(-5\beta) \cos 3\beta = 5 \cdot \dfrac{1}{2}[\cos(-5\beta - 3\beta) + \cos(-5\beta + 3\beta)] = \dfrac{5}{2}[\cos(-8\beta) + \cos(-2\beta)]$

$$= \dfrac{5}{2}(\cos 8\beta + \cos 2\beta)$$

70. $\cos 2\theta \cos 4\theta = \dfrac{1}{2}[\cos(2\theta - 4\theta) + \cos(2\theta + 4\theta)] = \dfrac{1}{2}[\cos(-2\theta) + \cos 6\theta] = \dfrac{1}{2}(\cos 2\theta + \cos 6\theta)$

71. $\sin(x + y) \sin(x - y) = \dfrac{1}{2}(\cos 2y - \cos 2x)$

72. $\sin(x + y) \cos(x - y) = \dfrac{1}{2}(\sin 2x + \sin 2y)$

73. $\cos(\theta - \pi)\sin(\theta + \pi) = \frac{1}{2}[\sin 2\theta - \sin(-2\pi)]$

$\qquad = \frac{1}{2}(\sin 2\theta + \sin 2\pi)$

74. $\sin(\theta + \pi)\sin(\theta - \pi) = \frac{1}{2}(\cos 2\pi - \cos 2\theta)$

75. $\sin 5\theta - \sin 3\theta = 2\cos\left(\dfrac{5\theta + 3\theta}{2}\right)\sin\left(\dfrac{5\theta - 3\theta}{2}\right)$

$\qquad = 2\cos 4\theta \sin\theta$

76. $\sin 3\theta + \sin\theta = 2\sin\left(\dfrac{3\theta + \theta}{2}\right)\cos\left(\dfrac{3\theta - \theta}{2}\right)$

$\qquad = 2\sin 2\theta \cos\theta$

77. $\cos 6x + \cos 2x = 2\cos\left(\dfrac{6x + 2x}{2}\right)\cos\left(\dfrac{6x - 2x}{2}\right) = 2\cos 4x \cos 2x$

78. $\sin x + \sin 5x = 2\sin\left(\dfrac{x + 5x}{2}\right)\cos\left(\dfrac{x - 5x}{2}\right) = 2\sin 3x \cos(-2x) = 2\sin 3x \cos 2x$

79. $\sin(\alpha + \beta) - \sin(\alpha - \beta) = 2\cos\left(\dfrac{\alpha + \beta + \alpha - \beta}{2}\right)\sin\left(\dfrac{\alpha + \beta - \alpha + \beta}{2}\right) = 2\cos\alpha\sin\beta$

80. $\cos(\phi + 2\pi) + \cos\phi = 2\cos\left(\dfrac{\phi + 2\pi + \phi}{2}\right)\cos\left(\dfrac{\phi + 2\pi - \phi}{2}\right) = 2\cos(\phi + \pi)\cos(\pi)$

81. $\cos\left(\theta + \dfrac{\pi}{2}\right) - \cos\left(\theta - \dfrac{\pi}{2}\right) = -2\sin\left[\dfrac{\left(\theta + \dfrac{\pi}{2}\right) + \left(\theta - \dfrac{\pi}{2}\right)}{2}\right]\sin\left[\dfrac{\left(\theta + \dfrac{\pi}{2}\right) - \left(\theta - \dfrac{\pi}{2}\right)}{2}\right] = -2\sin\theta\sin\dfrac{\pi}{2}$

82. $\sin\left(x + \dfrac{\pi}{2}\right) + \sin\left(x - \dfrac{\pi}{2}\right) = 2\sin\left(\dfrac{x + \dfrac{\pi}{2} + x - \dfrac{\pi}{2}}{2}\right)\cos\left(\dfrac{x + \dfrac{\pi}{2} - \left(x - \dfrac{\pi}{2}\right)}{2}\right) = 2\sin x\cos\dfrac{\pi}{2}$

83. $\sin 60° + \sin 30° = 2\sin\left(\dfrac{60° + 30°}{2}\right)\cos\left(\dfrac{60° - 30°}{2}\right) = 2\sin 45°\cos 15°$

$\sin 60° + \sin 30° = \dfrac{\sqrt{3}}{2} + \dfrac{1}{2} = \dfrac{\sqrt{3} + 1}{2}$

84. $\cos 120° + \cos 30° = 2\cos\left(\dfrac{120° + 30°}{2}\right)\cos\left(\dfrac{120° - 30°}{2}\right) = 2\cos 75°\cos 45°$

$\cos 120° + \cos 30° = -\dfrac{1}{2} + \dfrac{\sqrt{3}}{2} = \dfrac{\sqrt{3} - 1}{2}$

85. $\cos\dfrac{3\pi}{4} - \cos\dfrac{\pi}{4} = -2\sin\left(\dfrac{\dfrac{3\pi}{4} + \dfrac{\pi}{4}}{2}\right)\sin\left(\dfrac{\dfrac{3\pi}{4} - \dfrac{\pi}{4}}{2}\right) = -2\sin\dfrac{\pi}{2}\sin\dfrac{\pi}{4}$

$\cos\dfrac{3\pi}{4} - \cos\dfrac{\pi}{4} = -\dfrac{\sqrt{2}}{2} - \dfrac{\sqrt{2}}{2} = -\sqrt{2}$

86. $\sin\dfrac{5\pi}{4} - \sin\dfrac{3\pi}{4} = 2\cos\left(\dfrac{\dfrac{5\pi}{4} + \dfrac{3\pi}{4}}{2}\right)\sin\left(\dfrac{\dfrac{5\pi}{4} - \dfrac{3\pi}{4}}{2}\right) = 2\cos\pi\sin\dfrac{\pi}{4}$

$\sin\dfrac{5\pi}{4} - \sin\dfrac{3\pi}{4} = -\dfrac{\sqrt{2}}{2} - \dfrac{\sqrt{2}}{2} = -\sqrt{2}$

87.
$$\sin 6x + \sin 2x = 0$$

$$2 \sin\left(\frac{6x + 2x}{2}\right) \cos\left(\frac{6x - 2x}{2}\right) = 0$$

$$2(\sin 4x) \cos 2x = 0$$

$$\sin 4x = 0 \quad \text{or} \quad \cos 2x = 0$$

$$4x = n\pi \qquad 2x = \frac{\pi}{2} + n\pi$$

$$x = \frac{n\pi}{4} \qquad x = \frac{\pi}{4} + \frac{n\pi}{2}$$

In the interval $[0, 2\pi)$ we have

$$x = 0, \frac{\pi}{4}, \frac{\pi}{2}, \frac{3\pi}{4}, \pi, \frac{5\pi}{4}, \frac{3\pi}{2}, \frac{7\pi}{4}.$$

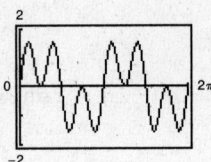

88.
$$h(x) = \cos 2x - \cos 6x$$

$$\cos 2x - \cos 6x = 0$$

$$-2 \sin 4x \sin(-2x) = 0$$

$$2 \sin 4x \sin 2x = 0$$

$\sin 4x = 0$	or $\quad \sin 2x = 0$
$4x = n\pi$	$2x = n\pi$
$x = \dfrac{n\pi}{4}$	$x = \dfrac{n\pi}{2}$
$x = 0, \dfrac{\pi}{4}, \dfrac{\pi}{2}, \dfrac{3\pi}{4}, \pi, \dfrac{5\pi}{4}, \dfrac{3\pi}{2}, \dfrac{7\pi}{4}$	$x = 0, \dfrac{\pi}{2}, \pi, \dfrac{3\pi}{2}$

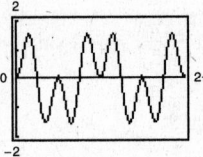

89.
$$\frac{\cos 2x}{\sin 3x - \sin x} - 1 = 0$$

$$\frac{\cos 2x}{\sin 3x - \sin x} = 1$$

$$\frac{\cos 2x}{2 \cos 2x \sin x} = 1$$

$$2 \sin x = 1$$

$$\sin x = \frac{1}{2}$$

$$x = \frac{\pi}{6}, \frac{5\pi}{6}$$

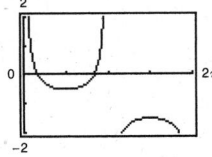

90.
$$f(x) = \sin^2 3x - \sin^2 x$$

$$\sin^2 3x - \sin^2 x = 0$$

$$(\sin 3x + \sin x)(\sin 3x - \sin x) = 0$$

$$(2 \sin 2x \cos x)(2 \cos 2x \sin x) = 0$$

$$\sin 2x = 0 \implies x = 0, \frac{\pi}{2}, \pi, \frac{3\pi}{2} \quad \text{or}$$

$$\cos x = 0 \implies x = \frac{\pi}{2}, \frac{3\pi}{2} \quad \text{or}$$

$$\cos 2x = 0 \implies x = \frac{\pi}{4}, \frac{3\pi}{4}, \frac{5\pi}{4}, \frac{7\pi}{4} \quad \text{or}$$

$$\sin x = 0 \implies x = 0, \pi$$

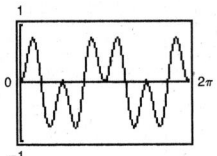

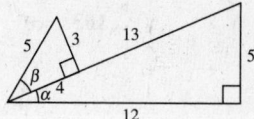

Figure for Exercises 91–94

91. $\sin^2 \alpha = \left(\dfrac{5}{13}\right)^2 = \dfrac{25}{169}$

$\sin^2 \alpha = 1 - \cos^2 \alpha = 1 - \left(\dfrac{12}{13}\right)^2$

$\qquad = 1 - \dfrac{144}{169} = \dfrac{25}{169}$

92. $\cos^2 \alpha = (\cos \alpha)^2 = \left(\dfrac{12}{13}\right)^2 = \dfrac{144}{169}$

$\cos^2 \alpha = 1 - \sin^2 \alpha$

$\qquad = 1 - \left(\dfrac{5}{13}\right)^2$

$\qquad = 1 - \dfrac{25}{169} = \dfrac{144}{169}$

93. $\sin \alpha \cos \beta = \left(\dfrac{5}{13}\right)\left(\dfrac{4}{5}\right) = \dfrac{4}{13}$

$\sin \alpha \cos \beta = \cos\left(\dfrac{\pi}{2} - \alpha\right)\sin\left(\dfrac{\pi}{2} - \beta\right)$

$\qquad = \left(\dfrac{5}{13}\right)\left(\dfrac{4}{5}\right) = \dfrac{4}{13}$

94. $\cos \alpha \sin \beta = \left(\dfrac{12}{13}\right)\left(\dfrac{3}{5}\right) = \dfrac{36}{65}$

$\cos \alpha \sin \beta = \sin\left(\dfrac{\pi}{2} - \alpha\right)\cos\left(\dfrac{\pi}{2} - \beta\right)$

$\qquad = \left(\dfrac{12}{13}\right)\left(\dfrac{3}{5}\right) = \dfrac{36}{65}$

95. $\csc 2\theta = \dfrac{1}{\sin 2\theta}$

$\qquad = \dfrac{1}{2\sin\theta\cos\theta}$

$\qquad = \dfrac{1}{\sin\theta} \cdot \dfrac{1}{2\cos\theta}$

$\qquad = \dfrac{\csc\theta}{2\cos\theta}$

96. $\sec 2\theta = \dfrac{1}{\cos 2\theta} = \dfrac{1}{\cos^2\theta - \sin^2\theta}$

$\qquad = \dfrac{\dfrac{1}{\cos^2\theta}}{1 - \dfrac{\sin^2\theta}{\cos^2\theta}}$

$\qquad = \dfrac{\sec^2\theta}{1 - \tan^2\theta}$

$\qquad = \dfrac{\sec^2\theta}{1 - (\sec^2\theta - 1)}$

$\qquad = \dfrac{\sec^2\theta}{2 - \sec^2\theta}$

97. $\cos^2 2\alpha - \sin^2 2\alpha = \cos[2(2\alpha)]$

$\qquad = \cos 4\alpha$

98. $\cos^4 x - \sin^4 x = (\cos^2 x - \sin^2 x)(\cos^2 x + \sin^2 x)$

$\qquad = (\cos 2x)(1)$

$\qquad = \cos 2x$

99. $(\sin x + \cos x)^2 = \sin^2 x + 2\sin x \cos x + \cos^2 x$

$\qquad = (\sin^2 x + \cos^2 x) + 2\sin x \cos x$

$\qquad = 1 + \sin 2x$

100. $\sin\left(\dfrac{\alpha}{3}\right)\cos\left(\dfrac{\alpha}{3}\right) = \dfrac{1}{2}\left[2\left(\sin\left(\dfrac{\alpha}{3}\right)\cos\left(\dfrac{\alpha}{3}\right)\right)\right]$

$\qquad = \dfrac{1}{2}\sin\dfrac{2\alpha}{3}$

101. $1 + \cos 10y = 1 + \cos^2 5y - \sin^2 5y$

$\qquad = 1 + \cos^2 5y - (1 - \cos^2 5y)$

$\qquad = 2\cos^2 5y$

102. $\dfrac{\cos 3\beta}{\cos \beta} = \dfrac{\cos(2\beta + \beta)}{\cos \beta}$

$\qquad = \dfrac{\cos 2\beta \cos \beta - \sin 2\beta \sin \beta}{\cos \beta}$

$\qquad = \dfrac{(1 - 2\sin^2 \beta)\cos \beta - (2\cos \beta \sin \beta)\sin \beta}{\cos \beta}$

$\qquad = 1 - 2\sin^2 \beta - 2\sin^2 \beta$

$\qquad = 1 - 4\sin^2 \beta$

103. $\sec \dfrac{u}{2} = \dfrac{1}{\cos \dfrac{u}{2}}$

$\qquad = \pm\sqrt{\dfrac{2}{1 + \cos u}}$

$\qquad = \pm\sqrt{\dfrac{2\sin u}{\sin u(1 + \cos u)}}$

$\qquad = \pm\sqrt{\dfrac{2\sin u}{\sin u + \sin u \cos u}}$

$\qquad = \pm\sqrt{\dfrac{\dfrac{2\sin u}{\cos u}}{\dfrac{\sin u}{\cos u} + \dfrac{\sin u \cos u}{\cos u}}}$

$\qquad = \pm\sqrt{\dfrac{2\tan u}{\tan u + \sin u}}$

104. $\tan \dfrac{u}{2} = \dfrac{1 - \cos u}{\sin u}$

$\qquad = \dfrac{1}{\sin u} - \dfrac{\cos u}{\sin u}$

$\qquad = \csc u - \cot u$

105. $\dfrac{\sin x \pm \sin y}{\cos x + \cos y} = \dfrac{2\sin\left(\dfrac{x \pm y}{2}\right)\cos\left(\dfrac{x \mp y}{2}\right)}{2\cos\left(\dfrac{x + y}{2}\right)\cos\left(\dfrac{x - y}{2}\right)}$

$\qquad = \tan\left(\dfrac{x \pm y}{2}\right)$

106. $\dfrac{\sin x + \sin y}{\cos x - \cos y} = \dfrac{2\sin\left(\dfrac{x + y}{2}\right)\cos\left(\dfrac{x - y}{2}\right)}{-2\sin\left(\dfrac{x + y}{2}\right)\sin\left(\dfrac{x - y}{2}\right)}$

$\qquad = -\cot\left(\dfrac{x - y}{2}\right)$

107. $\dfrac{\cos 4x + \cos 2x}{\sin 4x + \sin 2x} = \dfrac{2\cos\left(\dfrac{4x + 2x}{2}\right)\cos\left(\dfrac{4x - 2x}{2}\right)}{2\sin\left(\dfrac{4x + 2x}{2}\right)\cos\left(\dfrac{4x - 2x}{2}\right)}$

$\qquad = \dfrac{2\cos 3x \cos x}{2\sin 3x \cos x} = \cot 3x$

108. $\dfrac{\cos t + \cos 3t}{\sin 3t - \sin t} = \dfrac{2\cos\left(\dfrac{4t}{2}\right)\cos\left(-\dfrac{2t}{2}\right)}{2\cos\left(\dfrac{4t}{2}\right)\sin\left(\dfrac{2t}{2}\right)}$

$\qquad = \dfrac{\cos(-t)}{\sin(t)}$

$\qquad = \dfrac{\cos(t)}{\sin(t)} = \cot t$

109. $\sin\left(\dfrac{\pi}{6} + x\right) + \sin\left(\dfrac{\pi}{6} - x\right) = 2\sin\dfrac{\pi}{6}\cos x$

$\qquad = 2 \cdot \dfrac{1}{2}\cos x$

$\qquad = \cos x$

110. $\cos\left(\dfrac{\pi}{3} + x\right) + \cos\left(\dfrac{\pi}{3} - x\right) = 2\cos\left(\dfrac{\dfrac{\pi}{3} + x + \dfrac{\pi}{3} - x}{2}\right)\cos\left(\dfrac{\dfrac{\pi}{3} + x - \left(\dfrac{\pi}{3} - x\right)}{2}\right)$

$\qquad = 2\cos\left(\dfrac{\pi}{3}\right)\cos(x)$

$\qquad = 2\left(\dfrac{1}{2}\right)\cos x = \cos x$

111.

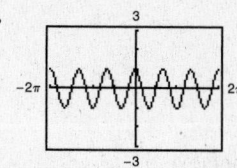

Let $y_1 = \cos(3x)$ and

$y_2 = (\cos x)^3 - 3(\sin x)^2 \cos x$.

$\cos 3\beta = \cos(2\beta + \beta)$

$= \cos 2\beta \cos\beta - \sin 2\beta \sin \beta$

$= (\cos^2 \beta - \sin^2 \beta) \cos \beta - 2 \sin \beta \cos \beta \sin \beta$

$= \cos^3 \beta - \sin^2 \beta \cos \beta - 2 \sin^2 \beta \cos \beta$

$= \cos^3 \beta - 3 \sin^2 \beta \cos \beta$

112. $\sin 4\beta = 2 \sin 2\beta \cos 2\beta$

$= 2(2 \sin \beta \cos \beta)(1 - 2 \sin^2 \beta)$

$= 4 \sin \beta \cos \beta (1 - 2 \sin^2 \beta)$

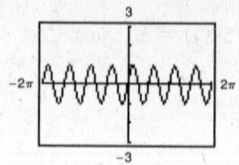

113.

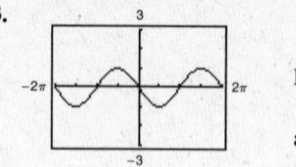

Let $y_1 = \dfrac{(\cos 4x - \cos 2x)}{(2 \sin 3x)}$

and $y_2 = -\sin x$.

$\dfrac{\cos 4x - \cos 2x}{2 \sin 3x} = \dfrac{-2 \sin\left(\dfrac{4x + 2x}{2}\right) \sin\left(\dfrac{4x - 2x}{2}\right)}{2 \sin 3x}$

$= \dfrac{-2 \sin 3x \sin x}{2 \sin 3x} = -\sin x$

114. $\dfrac{\cos 3x - \cos x}{\sin 3x - \sin x} = \dfrac{-2 \sin\left(\dfrac{3x + x}{2}\right) \sin\left(\dfrac{3x - x}{2}\right)}{2 \cos\left(\dfrac{3x + x}{2}\right) \sin\left(\dfrac{3x - x}{2}\right)}$

$= \dfrac{-2 \sin 2x \sin x}{2 \cos 2x \sin x}$

$= -\tan 2x$

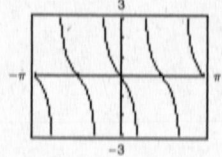

115. $\sin^2 x = \dfrac{1 - \cos 2x}{2} = \dfrac{1}{2} - \dfrac{\cos 2x}{2}$

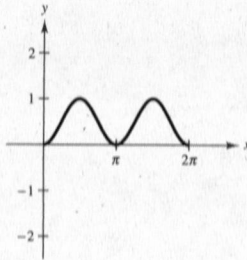

116. $f(x) = \cos^2 x = \dfrac{1 + \cos 2x}{2} = \dfrac{1}{2} + \dfrac{\cos 2x}{2}$

Shifted upward by $\dfrac{1}{2}$ unit.

Amplitude: $|a| = \dfrac{1}{2}$

Period: $\dfrac{2\pi}{2} = \pi$

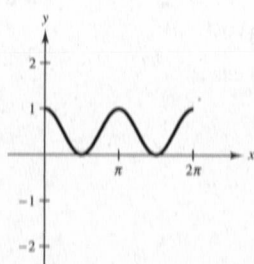

117. $\sin(2 \arcsin x) = 2 \sin(\arcsin x) \cos(\arcsin x)$

$= 2x \sqrt{1 - x^2}$

118. $\cos(2\arccos x) = \cos^2(\arccos x) - \sin^2(\arccos x)$
$$= x^2 - (1 - x^2) = 2x^2 - 1$$

119. $\dfrac{1}{32}(75)^2 \sin 2\theta = 130$

$$\sin 2\theta = \frac{130(32)}{75^2}$$

$$\theta = \frac{1}{2}\sin^{-1}\left(\frac{130(32)}{75^2}\right)$$

$$\theta \approx 23.85°$$

120. (a) $A = \dfrac{1}{2}bh$

$\cos\dfrac{\theta}{2} = \dfrac{h}{10} \implies h = 10\cos\dfrac{\theta}{2}$

$\sin\dfrac{\theta}{2} = \dfrac{(1/2)b}{10} \implies \dfrac{1}{2}b = 10\sin\dfrac{\theta}{2}$

$A = 10\sin\dfrac{\theta}{2}10\cos\dfrac{\theta}{2} \implies A = 100\sin\dfrac{\theta}{2}\cos\dfrac{\theta}{2}$

(b) $A = 100\sin\dfrac{\theta}{2}\cos\dfrac{\theta}{2}$

$A = 50\left(2\sin\dfrac{\theta}{2}\cos\dfrac{\theta}{2}\right)$

$A = 50\sin\theta$

When $\theta = \pi/2$, $\sin\theta = 1 \implies$ the area is a maximum.

$A = 50\sin\dfrac{\pi}{2} = 50(1) = 50$ square feet

121. $\sin\dfrac{\theta}{2} = \dfrac{1}{M}$

(a) $\sin\dfrac{\theta}{2} = 1$

$\dfrac{\theta}{2} = \arcsin 1$

$\dfrac{\theta}{2} = \dfrac{\pi}{2}$

$\theta = \pi$

(c) $\dfrac{S}{760} = 1$

$S = 760$ miles per hour

$\dfrac{S}{760} = 4.5$

$S = 3420$ miles per hour

(b) $\sin\dfrac{\theta}{2} = \dfrac{1}{4.5}$

$\dfrac{\theta}{2} = \arcsin\left(\dfrac{1}{4.5}\right)$

$\theta = 2\arcsin\left(\dfrac{1}{4.5}\right)$

$\theta \approx 0.4482$

(d) $\sin\dfrac{\theta}{2} = \dfrac{1}{M}$

$\dfrac{\theta}{2} = \arcsin\left(\dfrac{1}{M}\right)$

$\theta = 2\arcsin\left(\dfrac{1}{M}\right)$

122. $\dfrac{x}{2} = 2r\sin^2\dfrac{\theta}{2} = 2r\left(\dfrac{1 - \cos\theta}{2}\right)$

$= r(1 - \cos\theta)$

So, $x = 2r(1 - \cos\theta)$.

123. False. For $u < 0$,

$\sin 2u = -\sin(-2u)$

$= -2\sin(-u)\cos(-u)$

$= -2(-\sin u)\cos u$

$= 2\sin u\cos u.$

124. False. If $90° < u < 180°$,

$\dfrac{u}{2}$ is in the first quadrant and

$\sin\dfrac{u}{2} = \sqrt{\dfrac{1 - \cos u}{2}}.$

125. (a) $y = 4 \sin \dfrac{x}{2} + \cos x$

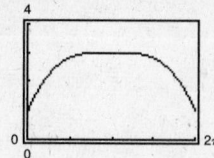

Maximum: $(\pi, 3)$

(b) $\qquad 2 \cos \dfrac{x}{2} - \sin x = 0$

$$2\left(\pm\sqrt{\dfrac{1 + \cos x}{2}}\right) = \sin x$$

$$4\left(\dfrac{1 + \cos x}{2}\right) = \sin^2 x$$

$$2(1 + \cos x) = 1 - \cos^2 x$$

$$\cos^2 x + 2 \cos x + 1 = 0$$

$$(\cos x + 1)^2 = 0$$

$$\cos x = -1$$

$$x = \pi$$

126. $f(x) = \cos 2x - 2 \sin x$

(a)

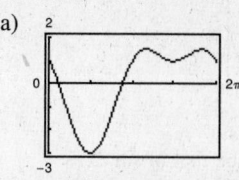

Maximum points: $(3.6652, 1.5)$, $(5.7596, 1.5)$

Minimum points: $(1.5708, -3)$, $(4.7124, 1)$

(b) $-2 \cos x(2 \sin x + 1) = 0$

$-2 \cos x = 0 \qquad$ or $\quad 2 \sin x + 1 = 0$

$\cos x = 0 \qquad\qquad\qquad \sin x = -\dfrac{1}{2}$

$x = \dfrac{\pi}{2}, \dfrac{3\pi}{2} \qquad\qquad x = \dfrac{7\pi}{6}, \dfrac{11\pi}{6}$

$\dfrac{\pi}{2} \approx 1.5708 \qquad\qquad \dfrac{7\pi}{6} \approx 3.6652$

$\dfrac{3\pi}{2} \approx 4.7124 \qquad\qquad \dfrac{11\pi}{2} \approx 5.7596$

127. $f(x) = \sin^4 x + \cos^4 x$

(a) $\sin^4 x + \cos^4 x = (\sin^2 x)^2 + (\cos^2 x)^2$

$$= \left(\dfrac{1 - \cos 2x}{2}\right)^2 + \left(\dfrac{1 + \cos 2x}{2}\right)^2$$

$$= \dfrac{1}{4}[(1 - \cos 2x)^2 + (1 + \cos 2x)^2]$$

$$= \dfrac{1}{4}(1 - 2 \cos 2x + \cos^2 2x + 1 + 2 \cos 2x + \cos^2 2x)$$

$$= \dfrac{1}{4}(2 + 2 \cos^2 2x)$$

$$= \dfrac{1}{4}\left[2 + 2\left(\dfrac{1 + \cos 2(2x)}{2}\right)\right]$$

$$= \dfrac{1}{4}(3 + \cos 4x)$$

(b) $\sin^4 x + \cos^4 x = (\sin^2 x)^2 + \cos^4 x$

$$= (1 - \cos^2 x)^2 + \cos^4 x$$

$$= 1 - 2 \cos^2 x + \cos^4 x + \cos^4 x$$

$$= 2 \cos^4 x - 2 \cos^2 x + 1$$

(c) $\sin^4 x + \cos^4 x = \sin^4 x + 2 \sin^2 x \cos^2 x + \cos^4 x - 2 \sin^2 x \cos^2 x$

$$= (\sin^2 x + \cos^2 x)^2 - 2 \sin^2 x \cos^2 x$$

$$= 1 - 2 \sin^2 x \cos^2 x$$

(d) $1 - 2 \sin^2 x \cos^2 x = 1 - (2 \sin x \cos x)(\sin x \cos x)$

$$= 1 - (\sin 2x)\left(\dfrac{1}{2} \sin 2x\right)$$

$$= 1 - \dfrac{1}{2} \sin^2 2x$$

(e) No, it does not mean that one of you is wrong. There is often more than one way to rewrite a trigonometric expression.

128. (a)

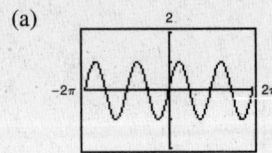

(b) The graph appears to be that of sin 2x.

(c) $2 \sin x \left[2 \cos^2\left(\frac{x}{2}\right) - 1 \right] = 2 \sin x \cos x$

$$= \sin 2x$$

129. (a)

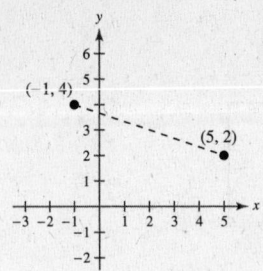

(b) $d = \sqrt{(-1-5)^2 + (4-2)^2} = \sqrt{(-6)^2 + (2)^2}$

$$= \sqrt{40} = 2\sqrt{10}$$

(c) Midpoint: $\left(\frac{5 + (-1)}{2}, \frac{2+4}{2} \right) = (2, 3)$

130. (a)

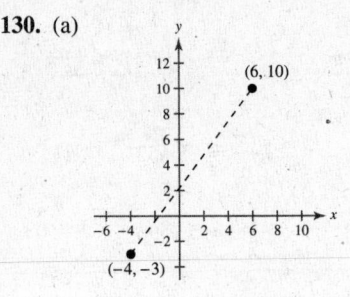

(b) $d = \sqrt{(-4-6)^2 + (-3-10)^2}$

$$= \sqrt{(-10)^2 + (-13)^2}$$

$$= \sqrt{100 + 169} = \sqrt{269}$$

(c) Midpoint: $\left(\frac{-4+6}{2}, \frac{-3+10}{2} \right) = \left(1, \frac{7}{2} \right)$

131. (a)

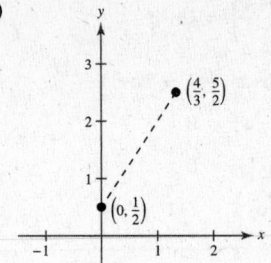

(b) $d = \sqrt{\left(\frac{4}{3} - 0\right)^2 + \left(\frac{5}{2} - \frac{1}{2}\right)^2} = \sqrt{\frac{16}{9} + 4}$

$$= \sqrt{\frac{52}{9}} = \frac{2\sqrt{13}}{3}$$

(c) Midpoint: $\left(\frac{0 + \frac{4}{3}}{2}, \frac{\frac{1}{2} + \frac{5}{2}}{2} \right) = \left(\frac{2}{3}, \frac{3}{2} \right)$

132. (a)

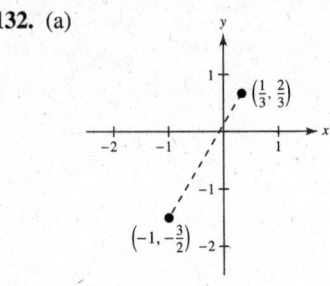

(b) $d = \sqrt{\left(\frac{1}{3} + 1\right)^2 + \left(\frac{2}{3} + \frac{3}{2}\right)^2} = \sqrt{\left(\frac{4}{3}\right)^2 + \left(\frac{13}{6}\right)^2}$

$$= \sqrt{\frac{16}{9} + \frac{169}{36}} = \sqrt{\frac{233}{36}} = \frac{1}{6}\sqrt{233}$$

(c) Midpoint:

$$\left(\frac{\frac{1}{3} + (-1)}{2}, \frac{\frac{2}{3} + \left(-\frac{3}{2}\right)}{2} \right) = \left(\frac{-\frac{2}{3}}{2}, \frac{-\frac{5}{6}}{2} \right) = \left(-\frac{1}{3}, \frac{-5}{12} \right)$$

133. (a) Complement: $90° - 55° = 35°$

Supplement: $180° - 55° = 125°$

(b) Complement: Not possible. $162° > 90°$

Supplement: $180° - 162° = 18°$

134. (a) The supplement is $180° - 109° = 71°$.

There is no complement.

(b) The supplement is $180° - 78° = 102°$.

The complement is $90° - 78° = 12°$.

135. (a) Complement: $\dfrac{\pi}{2} - \dfrac{\pi}{18} = \dfrac{4\pi}{9}$

Supplement: $\pi - \dfrac{\pi}{18} = \dfrac{17\pi}{18}$

(b) Complement: $\dfrac{\pi}{2} - \dfrac{9\pi}{20} = \dfrac{\pi}{20}$

Supplement: $\pi - \dfrac{9\pi}{20} = \dfrac{11\pi}{20}$

136. (a) The supplement is $\pi - 0.95 = 2.19$.

The complement is $\dfrac{\pi}{2} - 0.95 = 0.62$.

(b) The supplement is $\pi - 2.76 = 0.38$.

There is no complement.

137. Let x = profit for September,
then $x + 0.16x$ = profit for October.

$$x + (x + 0.16x) = 507{,}600$$
$$2.16x = 507{,}600$$
$$x = 235{,}000$$
$$x + 0.16x = 272{,}600$$

Profit for September: \$235,000

Profit for October: \$272,600

138. Let x = number of gallons of 100% concentrate.

$$0.30(55 - x) + 1.00x = 0.50(55)$$
$$16.50 - 0.30x + x = 27.50$$
$$0.70x = 11$$
$$x \approx 15.7 \text{ gallons}$$

139.

$$d^2 = 90^2 + 90^2$$
$$= 16{,}200$$
$$d = \sqrt{16{,}200}$$
$$= 90\sqrt{2}$$
$$\approx 127 \text{ feet}$$

Review Exercises for Chapter 5

1. $\dfrac{1}{\cos x} = \sec x$

2. $\dfrac{1}{\sin x} = \csc x$

3. $\dfrac{1}{\sec x} = \cos x$

4. $\dfrac{1}{\tan x} = \cot x$

5. $\dfrac{\cos x}{\sin x} = \cot x$

6. $\sqrt{1 + \tan^2 x} = \sqrt{\sec^2 x} = |\sec x|$

7. $\sin x = \dfrac{3}{5}, \; \cos x = \dfrac{4}{5}$

$\tan x = \dfrac{\sin x}{\cos x} = \dfrac{\frac{3}{5}}{\frac{4}{5}} = \dfrac{3}{4}$

$\cot x = \dfrac{1}{\tan x} = \dfrac{4}{3}$

$\sec x = \dfrac{1}{\cos x} = \dfrac{5}{4}$

$\csc x = \dfrac{1}{\sin x} = \dfrac{5}{3}$

8. $\tan \theta = \dfrac{2}{3}, \; \sec \theta = \dfrac{\sqrt{13}}{3}$

θ is in Quadrant I.

$\cos \theta = \dfrac{1}{\sec \theta} = \dfrac{3}{\sqrt{13}} = \dfrac{3\sqrt{13}}{13}$

$\sin \theta = \sqrt{1 - \cos^2 \theta} = \sqrt{1 - \dfrac{9}{13}} = \sqrt{\dfrac{4}{13}} = \dfrac{2\sqrt{13}}{13}$

$\csc \theta = \dfrac{1}{\sin \theta} = \dfrac{\sqrt{13}}{2}$

$\cot \theta = \dfrac{1}{\tan \theta} = \dfrac{3}{2}$

9. $\sin\left(\dfrac{\pi}{2} - x\right) = \dfrac{\sqrt{2}}{2} \Rightarrow \cos x = \dfrac{1}{\sqrt{2}} = \dfrac{\sqrt{2}}{2}$

$\sin x = -\dfrac{\sqrt{2}}{2}$

$\tan x = \dfrac{\sin x}{\cos x} = \dfrac{-\dfrac{1}{\sqrt{2}}}{\dfrac{1}{\sqrt{2}}} = -1$

$\cot x = \dfrac{1}{\tan x} = -1$

$\sec x = \dfrac{1}{\cos x} = \sqrt{2}$

$\csc x = \dfrac{1}{\sin x} = -\sqrt{2}$

10. $\csc\left(\dfrac{\pi}{2} - \theta\right) = \sec\theta = 9,\ \sin\theta = \dfrac{4\sqrt{5}}{9}$

θ is in Quadrant I.

$\cos\theta = \dfrac{1}{\sec\theta} = \dfrac{1}{9}$

$\tan\theta = \dfrac{\sin\theta}{\cos\theta} = \dfrac{\dfrac{4\sqrt{5}}{9}}{\dfrac{1}{9}} = 4\sqrt{5}$

$\csc\theta = \dfrac{1}{\sin\theta} = \dfrac{9}{4\sqrt{5}} = \dfrac{9\sqrt{5}}{20}$

$\cot\theta = \dfrac{1}{\tan\theta} = \dfrac{1}{4\sqrt{5}} = \dfrac{\sqrt{5}}{20}$

11. $\dfrac{1}{\cot^2 x + 1} = \dfrac{1}{\csc^2 x} = \sin^2 x$

12. $\dfrac{\tan\theta}{1 - \cos^2\theta} = \dfrac{\dfrac{\sin\theta}{\cos\theta}}{\sin^2\theta} = \dfrac{1}{\sin\theta\cos\theta}$

$= \csc\theta\sec\theta$

13. $\tan^2 x(\csc^2 x - 1) = \tan^2 x(\cot^2 x) = \tan^2 x\left(\dfrac{1}{\tan^2 x}\right) = 1$

14. $\cot^2 x(\sin^2 x) = \dfrac{\cos^2 x}{\sin^2 x}\sin^2 x = \cos^2 x$

15. $\dfrac{\sin\left(\dfrac{\pi}{2} - \theta\right)}{\sin\theta} = \dfrac{\cos\theta}{\sin\theta} = \cot\theta$

16. $\dfrac{\cot\left(\dfrac{\pi}{2} - u\right)}{\cos u} = \dfrac{\tan u}{\cos u} = \tan u\sec u$

17. $\cos^2 x + \cos^2 x\cot^2 x = \cos^2 x(1 + \cot^2 x) = \cos^2 x(\csc^2 x)$

$= \cos^2 x\left(\dfrac{1}{\sin^2 x}\right) = \dfrac{\cos^2 x}{\sin^2 x} = \cot^2 x$

18. $\tan^2\theta\csc^2\theta - \tan^2\theta = \tan^2\theta(\csc^2\theta - 1)$

$= \tan^2\theta\cot^2\theta = 1$

19. $(\tan x + 1)^2\cos x = (\tan^2 x + 2\tan x + 1)\cos x$

$= (\sec^2 x + 2\tan x)\cos x$

$= \sec^2 x\cos x + 2\left(\dfrac{\sin x}{\cos x}\right)\cos x = \sec x + 2\sin x$

20. $(\sec x - \tan x)^2 = \sec^2 x - 2\sec x\tan x + \tan^2 x$

$= 1 + \tan^2 x - 2\sec x\tan x + \tan^2 x$

$= 1 - 2\sec x\tan x + 2\tan^2 x$

21. $\dfrac{1}{\csc\theta + 1} - \dfrac{1}{\csc\theta - 1} = \dfrac{(\csc\theta - 1) - (\csc\theta + 1)}{(\csc\theta + 1)(\csc\theta - 1)}$

$= \dfrac{-2}{\csc^2\theta - 1}$

$= \dfrac{-2}{\cot^2\theta}$

$= -2\tan^2\theta$

22. $\dfrac{\cos^2 x}{1 - \sin x} = \dfrac{\cos^2 x}{(1 - \sin x)}\cdot\dfrac{(1 + \sin x)}{(1 + \sin x)}$

$= \dfrac{\cos^2 x(1 + \sin x)}{1 - \sin^2 x}$

$= 1 + \sin x$

23. $\csc^2 x - \csc x\cot x = \dfrac{1}{\sin^2 x} - \left(\dfrac{1}{\sin x}\right)\left(\dfrac{\cos x}{\sin x}\right)$

$= \dfrac{1 - \cos x}{\sin^2 x}$

24. $\sin^{-1/2} x \cos x = \dfrac{1}{\sqrt{\sin x}}(\cos x) = \dfrac{\sqrt{\sin x}}{\sin x}(\cos x)$

$\qquad\qquad = \sqrt{\sin x}\left(\dfrac{\cos x}{\sin x}\right) = \sqrt{\sin x}\,\cot x$

25. $\cos x(\tan^2 x + 1) = \cos x \sec^2 x$

$\qquad\qquad = \dfrac{1}{\sec x} \sec^2 x$

$\qquad\qquad = \sec x$

26. $\sec^2 x \cot x - \cot x = \cot x(\sec^2 x - 1) = \cot x \tan^2 x$

$\qquad\qquad = \left(\dfrac{1}{\tan x}\right) \tan^2 x = \tan x$

27. $\cos\left(x + \dfrac{\pi}{2}\right) = \cos x \cos \dfrac{\pi}{2} - \sin x \sin \dfrac{\pi}{2}$

$\qquad\qquad = (\cos x)(0) - (\sin x)(1)$

$\qquad\qquad = -\sin x$

28. $\cot\left(\dfrac{\pi}{2} - x\right) = \tan x$ by the Cofunction Identity

29. $\dfrac{1}{\tan \theta \csc \theta} = \dfrac{1}{\dfrac{\sin \theta}{\cos \theta} \cdot \dfrac{1}{\sin \theta}} = \cos \theta$

30. $\dfrac{1}{\tan x \csc x \sin x} = \dfrac{1}{(\tan x)\left(\dfrac{1}{\sin x}\right)(\sin x)} = \dfrac{1}{\tan x}$

$\qquad\qquad = \cot x$

31. $\sin^5 x \cos^2 x = \sin^4 x \cos^2 x \sin x$

$\qquad\qquad = (1 - \cos^2 x)^2 \cos^2 x \sin x$

$\qquad\qquad = (1 - 2\cos^2 x + \cos^4 x) \cos^2 x \sin x$

$\qquad\qquad = (\cos^2 x - 2\cos^4 x + \cos^6 x) \sin x$

32. $\cos^3 x \sin^2 x = \cos x \cos^2 x \sin^2 x$

$\qquad\qquad = \cos x(1 - \sin^2 x) \sin^2 x$

$\qquad\qquad = \cos x(\sin^2 x - \sin^4 x)$

$\qquad\qquad = (\sin^2 x - \sin^4 x) \cos x$

33. $\sin x = \sqrt{3} - \sin x$

$\qquad \sin x = \dfrac{\sqrt{3}}{2}$

$\qquad x = \dfrac{\pi}{3} + 2\pi n, \dfrac{2\pi}{3} + 2\pi n$

34. $4 \cos \theta = 1 + 2 \cos \theta$

$\qquad 2 \cos \theta = 1$

$\qquad \cos \theta = \dfrac{1}{2}$

$\qquad \theta = \dfrac{\pi}{3} + 2n\pi \quad \text{or} \quad \dfrac{5\pi}{3} + 2n\pi$

35. $3\sqrt{3} \tan u = 3$

$\qquad \tan u = \dfrac{1}{\sqrt{3}}$

$\qquad u = \dfrac{\pi}{6} + n\pi$

36. $\dfrac{1}{2} \sec x - 1 = 0$

$\qquad \dfrac{1}{2} \sec x = 1$

$\qquad \sec x = 2$

$\qquad \cos x = \dfrac{1}{2}$

$\qquad x = \dfrac{\pi}{3} + 2n\pi \quad \text{or} \quad \dfrac{5\pi}{3} + 2n\pi$

37. $3 \csc^2 x = 4$

$\qquad \csc^2 x = \dfrac{4}{3}$

$\qquad \sin x = \pm\dfrac{\sqrt{3}}{2}$

$\qquad x = \dfrac{\pi}{3} + 2\pi n, \dfrac{2\pi}{3} + 2\pi n, \dfrac{4\pi}{3} + 2\pi n, \dfrac{5\pi}{3} + 2\pi n$

These can be combined as:

$\qquad x = \dfrac{\pi}{3} + n\pi \quad \text{or} \quad x = \dfrac{2\pi}{3} + n\pi$

38. $4 \tan^2 u - 1 = \tan^2 u$

$3 \tan^2 u - 1 = 0$

$\tan^2 u = \dfrac{1}{3}$

$\tan u = \pm \dfrac{1}{\sqrt{3}} = \pm \dfrac{\sqrt{3}}{3}$

$u = \dfrac{\pi}{6} + n\pi \quad \text{or} \quad \dfrac{5\pi}{6} + n\pi$

39. $2 \cos^2 x - \cos x = 1$

$2 \cos^2 x - \cos x - 1 = 0$

$(2 \cos x + 1)(\cos x - 1) = 0$

$2 \cos x + 1 = 0 \qquad\qquad \cos x - 1 = 0$

$\cos x = -\dfrac{1}{2} \qquad\qquad \cos x = 1$

$x = \dfrac{2\pi}{3}, \dfrac{4\pi}{3} \qquad\qquad x = 0$

40. $2 \sin^2 x - 3 \sin x = -1$

$2 \sin^2 x - 3 \sin x + 1 = 0$

$(2 \sin x - 1)(\sin x - 1) = 0$

$2 \sin x - 1 = 0 \quad \text{or} \quad \sin x - 1 = 0$

$\sin x = \dfrac{1}{2} \qquad\qquad \sin x = 1$

$x = \dfrac{\pi}{6}, \dfrac{5\pi}{6} \qquad\qquad x = \dfrac{\pi}{2}$

41. $\cos^2 x + \sin x = 1$

$1 - \sin^2 x + \sin x - 1 = 0$

$-\sin x (\sin x - 1) = 0$

$\sin x = 0 \qquad \sin x - 1 = 0$

$x = 0, \pi \qquad\quad \sin x = 1$

$x = \dfrac{\pi}{2}$

42. $\sin^2 x + 2 \cos x = 2$

$1 - \cos^2 x + 2 \cos x = 2$

$0 = \cos^2 x - 2 \cos x + 1$

$0 = (\cos x - 1)^2$

$\cos x - 1 = 0$

$\cos x = 1$

$x = 0$

43. $2 \sin 2x - \sqrt{2} = 0$

$\sin 2x = \dfrac{\sqrt{2}}{2}$

$2x = \dfrac{\pi}{4} + 2\pi n, \dfrac{3\pi}{4} + 2\pi n$

$x = \dfrac{\pi}{8} + \pi n, \dfrac{3\pi}{8} + \pi n$

$x = \dfrac{\pi}{8}, \dfrac{3\pi}{8}, \dfrac{9\pi}{8}, \dfrac{11\pi}{8}$

44. $\sqrt{3} \tan 3x = 0$

$\tan 3x = 0$

$3x = 0, \pi, 2\pi, 3\pi, 4\pi, 5\pi$

$x = 0, \dfrac{\pi}{3}, \dfrac{2\pi}{3}, \pi, \dfrac{4\pi}{3}, \dfrac{5\pi}{3}$

45. $\cos 4x (\cos x - 1) = 0$

$\cos 4x = 0 \qquad\qquad\qquad \cos x - 1 = 0$

$4x = \dfrac{\pi}{2} + 2\pi n, \dfrac{3\pi}{2} + 2\pi n \qquad \cos x = 1$

$x = \dfrac{\pi}{8} + \dfrac{\pi}{2}n, \dfrac{3\pi}{8} + \dfrac{\pi}{2}n \qquad\qquad x = 0$

$x = 0, \dfrac{\pi}{8}, \dfrac{3\pi}{8}, \dfrac{5\pi}{8}, \dfrac{7\pi}{8}, \dfrac{9\pi}{8}, \dfrac{11\pi}{8}, \dfrac{13\pi}{8}, \dfrac{15\pi}{8}$

46. $3 \csc^2 5x = -4$

$\csc^2 5x = -\dfrac{4}{3}$

$\csc 5x = \pm \sqrt{-\dfrac{4}{3}}$

No real solution

47. $\sin^2 x - 2 \sin x = 0$

$\sin x (\sin x - 2) = 0$

$\sin x = 0 \qquad \sin x - 2 = 0$

$x = 0, \pi \qquad \text{No solution}$

48. $2 \cos^2 x + 3 \cos x = 0$

$\cos x (2 \cos x + 3) = 0$

$\cos x = 0 \quad \text{or} \quad 2 \cos x + 3 = 0$

$x = \dfrac{\pi}{2}, \dfrac{3\pi}{2} \qquad\qquad 2 \cos x = -3$

$\cos x = -\dfrac{3}{2}$

No solution

49. $\tan^2 \theta + \tan \theta - 12 = 0$

$(\tan \theta + 4)(\tan \theta - 3) = 0$

$\tan \theta + 4 = 0 \qquad\qquad\qquad \tan \theta - 3 = 0$

$\theta = \arctan(-4) + n\pi \qquad\qquad \theta = \arctan 3 + n\pi$

$\theta = \arctan(-4) + \pi, \arctan(-4) + 2\pi, \arctan 3, \arctan 3 + \pi$

50. $\sec^2 x + 6 \tan x + 4 = 0$

$1 + \tan^2 x + 6 \tan x + 4 = 0$

$\tan^2 x + 6 \tan x + 5 = 0$

$(\tan x + 5)(\tan x + 1) = 0$

$\tan x + 5 = 0 \quad \text{or} \qquad \tan x + 1 = 0$

$\tan x = -5 \qquad\qquad\quad \tan x = -1$

$x = \arctan(-5) + \pi \qquad x = \dfrac{3\pi}{4}, \dfrac{7\pi}{4}$

$x = \arctan(-5) + 2\pi$

51. $\sin 285° = \sin(315° - 30°)$

$= \sin 315° \cos 30° - \cos 315° \sin 30°$

$= \left(-\dfrac{\sqrt{2}}{2}\right)\left(\dfrac{\sqrt{3}}{2}\right) - \left(\dfrac{\sqrt{2}}{2}\right)\left(\dfrac{1}{2}\right)$

$= -\dfrac{\sqrt{2}}{4}\left(\sqrt{3} + 1\right)$

$\cos 285° = \cos(315° - 30°)$

$= \cos 315° \cos 30° + \sin 315° \sin 30°$

$= \left(\dfrac{\sqrt{2}}{2}\right)\left(\dfrac{\sqrt{3}}{2}\right) + \left(-\dfrac{\sqrt{2}}{2}\right)\left(\dfrac{1}{2}\right)$

$= \dfrac{\sqrt{2}}{4}\left(\sqrt{3} - 1\right)$

$\tan 285° = \tan(315° - 30°) = \dfrac{\tan 315° - \tan 30°}{1 + \tan 315° \tan 30°}$

$= \dfrac{(-1) - \left(\dfrac{\sqrt{3}}{3}\right)}{1 + (-1)\left(\dfrac{\sqrt{3}}{3}\right)} = -2 - \sqrt{3}$

52. $\sin(345°) = \sin(300° + 45°)$

$= \sin 300° \cos 45° + \cos 300° \sin 45°$

$= -\dfrac{\sqrt{3}}{2} \cdot \dfrac{\sqrt{2}}{2} + \dfrac{1}{2} \cdot \dfrac{\sqrt{2}}{2}$

$= \dfrac{\sqrt{2}}{4}\left(-\sqrt{3} + 1\right) = \dfrac{\sqrt{2}}{4}\left(1 - \sqrt{3}\right)$

$\cos(345°) = \cos(300° + 45°)$

$= \cos 300° \cos 45° - \sin 300° \sin 45°$

$= \dfrac{1}{2} \cdot \dfrac{\sqrt{2}}{2} - \left(-\dfrac{\sqrt{3}}{2}\right)\dfrac{\sqrt{2}}{2}$

$= \dfrac{\sqrt{2}}{4}\left(1 + \sqrt{3}\right)$

$\tan(345°) = \tan(300° + 45°)$

$= \dfrac{\tan 300° + \tan 45°}{1 - \tan 300° \tan 45°} = \dfrac{-\sqrt{3} + 1}{1 + \sqrt{3}(1)} \cdot \dfrac{1 - \sqrt{3}}{1 - \sqrt{3}}$

$= \dfrac{4 - 2\sqrt{3}}{-2} = -2 + \sqrt{3}$

53. $\sin \dfrac{25\pi}{12} = \sin\left(\dfrac{11\pi}{6} + \dfrac{\pi}{4}\right) = \sin \dfrac{11\pi}{6} \cos \dfrac{\pi}{4} + \cos \dfrac{11\pi}{6} \sin \dfrac{\pi}{4}$

$= \left(-\dfrac{1}{2}\right)\left(\dfrac{\sqrt{2}}{2}\right) + \left(\dfrac{\sqrt{3}}{2}\right)\left(\dfrac{\sqrt{2}}{2}\right) = \dfrac{\sqrt{2}}{4}\left(\sqrt{3} - 1\right)$

$\cos \dfrac{25\pi}{12} = \cos\left(\dfrac{11\pi}{6} + \dfrac{\pi}{4}\right) = \cos \dfrac{11\pi}{6} \cos \dfrac{\pi}{4} - \sin \dfrac{11\pi}{6} \sin \dfrac{\pi}{4}$

$= \left(\dfrac{\sqrt{3}}{2}\right)\left(\dfrac{\sqrt{2}}{2}\right) - \left(-\dfrac{1}{2}\right)\left(\dfrac{\sqrt{2}}{2}\right) = \dfrac{\sqrt{2}}{4}\left(\sqrt{3} + 1\right)$

$\tan \dfrac{25\pi}{12} = \tan\left(\dfrac{11\pi}{6} + \dfrac{\pi}{4}\right) = \dfrac{\tan \dfrac{11\pi}{6} + \tan \dfrac{\pi}{4}}{1 - \tan \dfrac{11\pi}{6} \tan \dfrac{\pi}{4}}$

$= \dfrac{\left(-\dfrac{\sqrt{3}}{3}\right) + 1}{1 - \left(-\dfrac{\sqrt{3}}{3}\right)(1)} = 2 - \sqrt{3}$

54. $\sin\left(\dfrac{19\pi}{12}\right) = \sin\left(\dfrac{11\pi}{6} - \dfrac{\pi}{4}\right)$

$\qquad = \sin\dfrac{11\pi}{6}\cos\dfrac{\pi}{4} - \cos\dfrac{11\pi}{6}\sin\dfrac{\pi}{4}$

$\qquad = -\dfrac{1}{2}\cdot\dfrac{\sqrt{2}}{2} - \dfrac{\sqrt{3}}{2}\cdot\dfrac{\sqrt{2}}{2}$

$\qquad = -\dfrac{\sqrt{2}}{4}(1 + \sqrt{3}) = -\dfrac{\sqrt{2}}{4}(\sqrt{3} + 1)$

$\cos\left(\dfrac{19\pi}{12}\right) = \cos\left(\dfrac{11\pi}{6} - \dfrac{\pi}{4}\right)$

$\qquad = \cos\dfrac{11\pi}{6}\cos\dfrac{\pi}{4} + \sin\dfrac{11\pi}{6}\sin\dfrac{\pi}{4}$

$\qquad = \dfrac{\sqrt{3}}{2}\cdot\dfrac{\sqrt{2}}{2} + \left(-\dfrac{1}{2}\right)\dfrac{\sqrt{2}}{2}$

$\qquad = \dfrac{\sqrt{2}}{4}(\sqrt{3} - 1)$

$\tan\left(\dfrac{19\pi}{12}\right) = \tan\left(\dfrac{11\pi}{6} - \dfrac{\pi}{4}\right)$

$\qquad = \dfrac{\tan\dfrac{11\pi}{6} - \tan\dfrac{\pi}{4}}{1 + \tan\dfrac{11\pi}{6}\tan\dfrac{\pi}{4}}$

$\qquad = \dfrac{-\dfrac{\sqrt{3}}{3} - 1}{1 + \left(-\dfrac{\sqrt{3}}{3}\right)(1)} = \dfrac{-\sqrt{3} - 3}{3 - \sqrt{3}}\cdot\dfrac{3 + \sqrt{3}}{3 + \sqrt{3}}$

$\qquad = \dfrac{-(12 + 6\sqrt{3})}{6} = -2 - \sqrt{3}$

55. $\sin 60° \cos 45° - \cos 60° \sin 45° = \sin(60° - 45°) = \sin 15°$

56. $\cos 45° \cos 120° - \sin 45° \sin 120° = \cos(45° + 120°) = \cos 165°$

57. $\dfrac{\tan 25° + \tan 10°}{1 - \tan 25° \tan 10°} = \tan(25° + 10°)$

$\qquad = \tan 35°$

58. $\dfrac{\tan 68° - \tan 115°}{1 + \tan 68° \tan 115°} = \tan(68° - 115°) = \tan(-47°)$

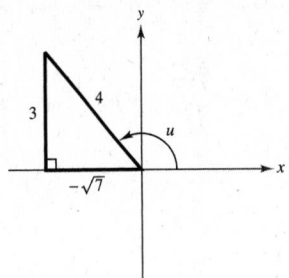
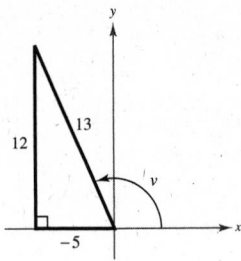

Figures for Exercises 59–64

59. $\sin(u + v) = \sin u \cos v + \cos u \sin v$

$\qquad = \left(\dfrac{3}{4}\right)\left(-\dfrac{5}{13}\right) + \left(-\dfrac{\sqrt{7}}{4}\right)\left(\dfrac{12}{13}\right)$

$\qquad = -\dfrac{3}{52}(5 + 4\sqrt{7})$

60. $\tan(u + v) = \dfrac{\tan u + \tan v}{1 - \tan u \tan v} = \dfrac{\left(-\dfrac{3}{\sqrt{7}}\right) + \left(-\dfrac{12}{5}\right)}{1 - \left(-\dfrac{3}{\sqrt{7}}\right)\left(-\dfrac{12}{5}\right)}$

$\qquad = \dfrac{15 + 12\sqrt{7}}{36 - 5\sqrt{7}}\cdot\dfrac{36 + 5\sqrt{7}}{36 + 5\sqrt{7}} = \dfrac{960 + 507\sqrt{7}}{1121}$

61. $\cos(u - v) = \cos u \cos v + \sin u \sin v$

$$= \left(-\frac{\sqrt{7}}{4}\right)\left(-\frac{5}{13}\right) + \left(\frac{3}{4}\right)\left(\frac{12}{13}\right)$$

$$= \frac{1}{52}(5\sqrt{7} + 36)$$

62. $\sin(u - v) = \sin u \cos v - \cos u \sin v$

$$= \left(\frac{3}{4}\right)\left(-\frac{5}{13}\right) - \left(-\frac{\sqrt{7}}{4}\right)\left(\frac{12}{13}\right)$$

$$= \frac{-15 + 12\sqrt{7}}{52} = \frac{12\sqrt{7} - 15}{52}$$

63. $\cos(u + v) = \cos u \cos v - \sin u \sin v$

$$= \left(-\frac{\sqrt{7}}{4}\right)\left(-\frac{5}{13}\right) - \left(\frac{3}{4}\right)\left(\frac{12}{13}\right)$$

$$= \frac{1}{52}(5\sqrt{7} - 36)$$

64. $\tan(u - v) = \dfrac{\tan u - \tan v}{1 + \tan u \tan v} = \dfrac{\left(-\dfrac{3}{\sqrt{7}}\right) - \left(-\dfrac{12}{5}\right)}{1 + \left(-\dfrac{3}{\sqrt{7}}\right)\left(-\dfrac{12}{5}\right)}$

$$= \frac{-15 + 12\sqrt{7}}{36 + 5\sqrt{7}} \cdot \frac{36 - 5\sqrt{7}}{36 - 5\sqrt{7}}$$

$$= \frac{-960 + 507\sqrt{7}}{1121}$$

65. $\cos\left(x + \dfrac{\pi}{2}\right) = \cos x \cos \dfrac{\pi}{2} - \sin x \sin \dfrac{\pi}{2}$

$$= \cos x(0) - \sin x(1)$$

$$= -\sin x$$

66. $\sin\left(x - \dfrac{3\pi}{2}\right) = \sin x \cos \dfrac{3\pi}{2} - \cos x \sin \dfrac{3\pi}{2}$

$$= \sin x(0) - \cos x(-1)$$

$$= \cos x$$

67. $\cot\left(\dfrac{\pi}{2} - x\right) = \tan x$ by the cofunction identity.

68. $\sin(\pi - x) = \sin \pi \cos x - \cos \pi \sin x$

$$= 0 \cdot \cos x - (-1)\sin x$$

$$= \sin x$$

69. $\cos 3x = \cos(2x + x)$

$$= \cos 2x \cos x - \sin 2x \sin x$$

$$= (\cos^2 x - \sin^2 x) \cos x - (2 \sin x \cos x) \sin x$$

$$= \cos^3 x - \sin^2 x \cos x - 2 \sin^2 x \cos x$$

$$= \cos^3 x - 3 \sin^2 x \cos x$$

$$= \cos^3 x - 3(1 - \cos^2 x) \cos x$$

$$= \cos^3 x - 3 \cos x + 3 \cos^3 x$$

$$= 4 \cos^3 x - 3 \cos x$$

70. $\dfrac{\sin(\alpha + \beta)}{\cos \alpha \cos \beta} = \dfrac{\sin \alpha \cos \beta + \cos \alpha \sin \beta}{\cos \alpha \cos \beta}$

$$= \frac{\sin \alpha \cos \beta}{\cos \alpha \cos \beta} + \frac{\cos \alpha \sin \beta}{\cos \alpha \cos \beta}$$

$$= \frac{\sin \alpha}{\cos \alpha} + \frac{\sin \beta}{\cos \beta}$$

$$= \tan \alpha + \tan \beta$$

71. $\sin\left(x + \dfrac{\pi}{4}\right) - \sin\left(x - \dfrac{\pi}{4}\right) = 1$

$$2 \cos x \sin \frac{\pi}{4} = 1$$

$$\cos x = \frac{\sqrt{2}}{2}$$

$$x = \frac{\pi}{4}, \frac{7\pi}{4}$$

72.
$$\cos\left(x + \frac{\pi}{6}\right) - \cos\left(x - \frac{\pi}{6}\right) = 1$$

$$\left(\cos x \cos\frac{\pi}{6} - \sin x \sin\frac{\pi}{6}\right) - \left(\cos x \cos\frac{\pi}{6} + \sin x \sin\frac{\pi}{6}\right) = 1$$

$$-2\sin x \sin\frac{\pi}{6} = 1$$

$$-2\sin x \left(\frac{1}{2}\right) = 1$$

$$\sin x = -1$$

$$x = \frac{3\pi}{2}$$

73. $\sin\left(x + \frac{\pi}{2}\right) - \sin\left(x - \frac{\pi}{2}\right) = \sqrt{3}$

$$2\cos x \sin\frac{\pi}{2} = \sqrt{3}$$

$$\cos x = \frac{\sqrt{3}}{2}$$

$$x = \frac{\pi}{6}, \frac{11\pi}{6}$$

74.
$$\cos\left(x + \frac{3\pi}{4}\right) - \cos\left(x - \frac{3\pi}{4}\right) = 0$$

$$\left(\cos x \cos\frac{3\pi}{4} - \sin x \sin\frac{3\pi}{4}\right) - \left(\cos x \cos\frac{3\pi}{4} + \sin x \sin\frac{3\pi}{4}\right) = 0$$

$$-2\sin x \sin\frac{3\pi}{4} = 0$$

$$-2\sin x \left(\frac{\sqrt{2}}{2}\right) = 0$$

$$-\sqrt{2}\sin x = 0$$

$$\sin x = 0$$

$$x = 0, \pi$$

75. $\sin u = -\frac{4}{5}, \ \pi < u < \frac{3\pi}{2}$

$$\cos u = -\sqrt{1 - \sin^2 u} = \frac{-3}{5}$$

$$\tan u = \frac{\sin u}{\cos u} = \frac{4}{3}$$

$$\sin 2u = 2\sin u \cos u = 2\left(-\frac{4}{5}\right)\left(-\frac{3}{5}\right) = \frac{24}{25}$$

$$\cos 2u = \cos^2 u - \sin^2 u = \left(-\frac{3}{5}\right)^2 - \left(-\frac{4}{5}\right)^2 = -\frac{7}{25}$$

$$\tan 2u = \frac{2\tan u}{1 - \tan^2 u} = \frac{2\left(\frac{4}{3}\right)}{1 - \left(\frac{4}{3}\right)^2} = -\frac{24}{7}$$

76. $\cos u = -\frac{2}{\sqrt{5}}, \ \frac{\pi}{2} < u < \pi \implies \sin u = \frac{1}{\sqrt{5}}$ and

$$\tan u = -\frac{1}{2}$$

$$\sin 2u = 2\sin u \cos u = 2\left(\frac{1}{\sqrt{5}}\right)\left(-\frac{2}{\sqrt{5}}\right) = -\frac{4}{5}$$

$$\cos 2u = \cos^2 u - \sin^2 u = \left(-\frac{2}{\sqrt{5}}\right)^2 - \left(\frac{1}{\sqrt{5}}\right)^2 = \frac{3}{5}$$

$$\tan 2u = \frac{2\tan u}{1 - \tan^2 u} = \frac{2\left(-\frac{1}{2}\right)}{1 - \left(-\frac{1}{2}\right)^2} = \frac{-1}{\frac{3}{4}} = -\frac{4}{3}$$

77. $\sin 4x = 2 \sin 2x \cos 2x$

$\qquad = 2[2 \sin x \cos x(\cos^2 x - \sin^2 x)]$

$\qquad = 4 \sin x \cos x(2 \cos^2 x - 1)$

$\qquad = 8 \cos^3 x \sin x - 4 \cos x \sin x$

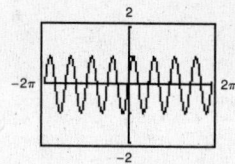

78. $\dfrac{1 - \cos 2x}{1 + \cos 2x} = \dfrac{1 - (1 - 2 \sin^2 x)}{1 + (2 \cos^2 x - 1)}$

$\qquad = \dfrac{2 \sin^2 x}{2 \cos^2 x}$

$\qquad = \tan^2 x$

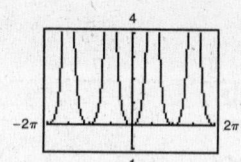

79. $\tan^2 2x = \dfrac{\sin^2 2x}{\cos^2 2x} = \dfrac{\dfrac{1 - \cos 4x}{2}}{\dfrac{1 + \cos 4x}{2}} = \dfrac{1 - \cos 4x}{1 + \cos 4x}$

80. $\cos^2 3x = \dfrac{1 + \cos 6x}{2}$

81. $\sin^2 x \tan^2 x = \sin^2 x \left(\dfrac{\sin^2 x}{\cos^2 x}\right) = \dfrac{\sin^4 x}{\cos^2 x}$

$\qquad = \dfrac{\left(\dfrac{1 - \cos 2x}{2}\right)^2}{\dfrac{1 + \cos 2x}{2}} = \dfrac{\dfrac{1 - 2 \cos 2x + \cos^2 2x}{4}}{\dfrac{1 + \cos 2x}{2}}$

$\qquad = \dfrac{1 - 2 \cos 2x + \dfrac{1 + \cos 4x}{2}}{2(1 + \cos 2x)}$

$\qquad = \dfrac{2 - 4 \cos 2x + 1 + \cos 4x}{4(1 + \cos 2x)}$

$\qquad = \dfrac{3 - 4 \cos 2x + \cos 4x}{4(1 + \cos 2x)}$

82. $\cos^2 x \tan^2 x = \sin^2 x = \dfrac{1 - \cos 2x}{2}$

83. $\sin(-75°) = -\sqrt{\dfrac{1 - \cos 150°}{2}} = -\sqrt{\dfrac{1 - \left(-\dfrac{\sqrt{3}}{2}\right)}{2}} = -\dfrac{\sqrt{2 + \sqrt{3}}}{2}$

$\qquad = -\dfrac{1}{2}\sqrt{2 + \sqrt{3}}$

$\cos(-75°) = \sqrt{\dfrac{1 + \cos 150°}{2}} = \sqrt{\dfrac{1 + \left(-\dfrac{\sqrt{3}}{2}\right)}{2}} = \dfrac{\sqrt{2 - \sqrt{3}}}{2}$

$\qquad = \dfrac{1}{2}\sqrt{2 - \sqrt{3}}$

$\tan(-75°) = -\left(\dfrac{1 - \cos 150°}{\sin 150°}\right) = -\left(\dfrac{1 - \left(-\dfrac{\sqrt{3}}{2}\right)}{\dfrac{1}{2}}\right) = -\left(2 + \sqrt{3}\right)$

$\qquad = -2 - \sqrt{3}$

84. $\sin 15° = \sin\left(\dfrac{30°}{2}\right) = \sqrt{\dfrac{1 - \cos 30°}{2}} = \sqrt{\dfrac{1 - \dfrac{\sqrt{3}}{2}}{2}} = \dfrac{1}{2}\sqrt{2 - \sqrt{3}}$

$\cos 15° = \cos\left(\dfrac{30°}{2}\right) = \sqrt{\dfrac{1 + \cos 30°}{2}} = \sqrt{\dfrac{1 + \dfrac{\sqrt{3}}{2}}{2}} = \dfrac{1}{2}\sqrt{2 + \sqrt{3}}$

$\tan 15° = \tan\left(\dfrac{30°}{2}\right) = \dfrac{1 - \cos 30°}{\sin 30°} = \dfrac{1 - \dfrac{\sqrt{3}}{2}}{\dfrac{1}{2}} = 2 - \sqrt{3}$

85. $\sin\left(\dfrac{19\pi}{12}\right) = -\sqrt{\dfrac{1 - \cos\dfrac{19\pi}{6}}{2}} = -\sqrt{\dfrac{1 - \left(-\dfrac{\sqrt{3}}{2}\right)}{2}} = -\dfrac{\sqrt{2 + \sqrt{3}}}{2} = -\dfrac{1}{2}\sqrt{2 + \sqrt{3}}$

$\cos\left(\dfrac{19\pi}{12}\right) = \sqrt{\dfrac{1 + \cos\dfrac{19\pi}{6}}{2}} = \sqrt{\dfrac{1 + \left(-\dfrac{\sqrt{3}}{2}\right)}{2}} = \dfrac{\sqrt{2 - \sqrt{3}}}{2} = \dfrac{1}{2}\sqrt{2 - \sqrt{3}}$

$\tan\left(\dfrac{19\pi}{12}\right) = \dfrac{1 - \cos\dfrac{19\pi}{6}}{\sin\dfrac{19\pi}{6}} = \dfrac{1 - \left(-\dfrac{\sqrt{3}}{2}\right)}{-\dfrac{1}{2}} = -2 - \sqrt{3}$

86. $\sin\left(-\dfrac{17\pi}{12}\right) = \sin\left(\dfrac{-\dfrac{17\pi}{6}}{2}\right) = \sqrt{\dfrac{1 - \cos\left(-\dfrac{17\pi}{6}\right)}{2}} = \sqrt{\dfrac{1 - \left(-\dfrac{\sqrt{3}}{2}\right)}{2}} = \dfrac{1}{2}\sqrt{2 + \sqrt{3}}$

$\cos\left(-\dfrac{17\pi}{12}\right) = \cos\left(\dfrac{-\dfrac{17\pi}{6}}{2}\right) = -\sqrt{\dfrac{1 + \cos\left(-\dfrac{17\pi}{6}\right)}{2}} = -\sqrt{\dfrac{1 + \left(-\dfrac{\sqrt{3}}{2}\right)}{2}} = -\dfrac{1}{2}\sqrt{2 - \sqrt{3}}$

$\tan\left(-\dfrac{17\pi}{12}\right) = \tan\left(\dfrac{-\left(\dfrac{17\pi}{6}\right)}{2}\right) = \dfrac{1 - \cos\left(-\dfrac{17\pi}{6}\right)}{\sin\left(-\dfrac{17\pi}{6}\right)} = \dfrac{1 - \left(-\dfrac{\sqrt{3}}{2}\right)}{-\dfrac{1}{2}} = -2 - \sqrt{3}$

87. Given $\sin u = \dfrac{3}{5}, 0 < u < \dfrac{\pi}{2} \Rightarrow \cos u = \dfrac{4}{5}$ and $\dfrac{u}{2}$ is in Quadrant I.

$\sin\left(\dfrac{u}{2}\right) = \sqrt{\dfrac{1 - \cos u}{2}} = \sqrt{\dfrac{1 - 4/5}{2}} = \sqrt{\dfrac{1}{10}} = \dfrac{\sqrt{10}}{10}$

$\cos\left(\dfrac{u}{2}\right) = \sqrt{\dfrac{1 + \cos u}{2}} = \sqrt{\dfrac{1 + 4/5}{2}} = \sqrt{\dfrac{9}{10}} = \dfrac{3\sqrt{10}}{10}$

$\tan\left(\dfrac{u}{2}\right) = \dfrac{1 - \cos u}{\sin u} = \dfrac{1 - 4/5}{3/5} = \dfrac{1}{3}$

88. $\tan u = \dfrac{5}{8}, \pi < u < \dfrac{3\pi}{2}$

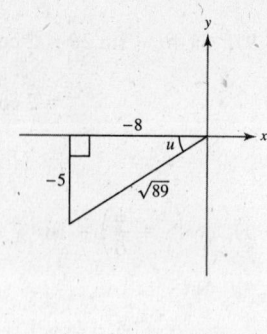

$$\sin u = \frac{-5}{\sqrt{89}}$$

$$\cos u = \frac{-8}{\sqrt{89}}$$

$$\sin \frac{u}{2} = \sqrt{\frac{1 - \cos u}{2}} = \sqrt{\frac{1 - \left(-8/\sqrt{89}\right)}{2}} = \sqrt{\frac{\sqrt{89} + 8}{2\sqrt{89}}} = \sqrt{\frac{89 + 8\sqrt{89}}{178}}$$

$$\cos \frac{u}{2} = -\sqrt{\frac{1 + \cos u}{2}} = -\sqrt{\frac{1 + \left(-8/\sqrt{89}\right)}{2}} = -\sqrt{\frac{\sqrt{89} - 8}{2\sqrt{89}}} = -\sqrt{\frac{89 - 8\sqrt{89}}{178}}$$

$$\tan \frac{u}{2} = \frac{1 - \cos u}{\sin u} = \frac{1 - \dfrac{-8}{\sqrt{89}}}{\dfrac{-5}{\sqrt{89}}} = \frac{\sqrt{89} + 8}{-5} = \frac{-8 - \sqrt{89}}{5}$$

89. Given $\cos u = -\dfrac{2}{7}, \dfrac{\pi}{2} < u < \pi \Rightarrow \sin u = \dfrac{3\sqrt{5}}{7}$ and $\dfrac{u}{2}$ is in Quadrant I.

$$\sin\left(\frac{u}{2}\right) = \sqrt{\frac{1 - \cos u}{2}} = \sqrt{\frac{1 - (-2/7)}{2}} = \sqrt{\frac{9}{14}} = \frac{3}{\sqrt{14}} = \frac{3\sqrt{14}}{14}$$

$$\cos\left(\frac{u}{2}\right) = \sqrt{\frac{1 + \cos u}{2}} = \sqrt{\frac{1 + (-2/7)}{2}} = \sqrt{\frac{5}{14}} = \frac{\sqrt{70}}{14}$$

$$\tan\left(\frac{u}{2}\right) = \frac{1 - \cos u}{\sin u} = \frac{1 - (-2/7)}{3\sqrt{5}/7} = \frac{9/7}{3\sqrt{5}/7} = \frac{3}{\sqrt{5}} = \frac{3\sqrt{5}}{5}$$

90. $\sec u = -6, \dfrac{\pi}{2} < u < \pi, \qquad \sin u = \sqrt{1 - \dfrac{1}{36}} = \dfrac{\sqrt{35}}{6}, \qquad \cos u = -\dfrac{1}{6}$

$$\sin \frac{u}{2} = \sqrt{\frac{1 - \cos u}{2}} = \sqrt{\frac{1 + (1/6)}{2}} = \sqrt{\frac{7}{12}} = \frac{\sqrt{21}}{6}$$

$$\cos \frac{u}{2} = \sqrt{\frac{1 + \cos u}{2}} = \sqrt{\frac{1 - (1/6)}{2}} = \sqrt{\frac{5}{12}} = \frac{\sqrt{15}}{6}$$

$$\tan \frac{u}{2} = \frac{1 - \cos u}{\sin u} = \frac{1 + (1/6)}{\sqrt{35}/6} = \frac{7}{6} \cdot \frac{6}{\sqrt{35}} = \frac{7}{\sqrt{35}} = \frac{\sqrt{35}}{5} \quad \text{or}$$

$$\tan \frac{u}{2} = \frac{\sin(u/2)}{\cos(u/2)} = \frac{\sqrt{21}/6}{\sqrt{15}/6} = \frac{\sqrt{21}}{\sqrt{15}} = \frac{\sqrt{35}}{5}$$

91. $-\sqrt{\dfrac{1 + \cos 10x}{2}} = -\left|\cos \dfrac{10x}{2}\right| = -|\cos 5x|$

92. $\dfrac{\sin 6x}{1 + \cos 6x} = \tan 3x$

93. $\cos \dfrac{\pi}{6} \sin \dfrac{\pi}{6} = \dfrac{1}{2}\left[\sin \dfrac{\pi}{3} - \sin 0\right] = \dfrac{1}{2} \sin \dfrac{\pi}{3}$

94. $6 \sin 15° \sin 45° = 6\left(\dfrac{1}{2}\right)[\cos(15° - 45°) - \cos(15° + 45°)]$

$$= 3[\cos(-30°) - \cos 60°]$$

$$= 3(\cos 30° - \cos 60°)$$

95. $\cos 5\theta \cos 3\theta = \dfrac{1}{2}[\cos 2\theta + \cos 8\theta]$

96. $4 \sin 3\alpha \cos 2\alpha = 4\left(\dfrac{1}{2}\right)[\sin(3\alpha + 2\alpha) + \sin(3\alpha - 2\alpha)]$

$$= 2(\sin 5\alpha + \sin \alpha)$$

97. $\sin 4\theta - \sin 2\theta = 2\cos\left(\dfrac{4\theta + 2\theta}{2}\right)\sin\left(\dfrac{4\theta - 2\theta}{2}\right)$

$\qquad\qquad\qquad = 2\cos 3\theta \sin\theta$

98. $\cos 3\theta + \cos 2\theta = 2\cos\left(\dfrac{3\theta + 2\theta}{2}\right)\cos\left(\dfrac{3\theta - 2\theta}{2}\right)$

$\qquad\qquad\qquad = 2\cos\dfrac{5\theta}{2}\cos\dfrac{\theta}{2}$

99. $\cos\left(x + \dfrac{\pi}{6}\right) - \cos\left(x - \dfrac{\pi}{6}\right) = -2\sin x\ \sin\dfrac{\pi}{6}$

100. $\sin\left(x + \dfrac{\pi}{4}\right) - \sin\left(x - \dfrac{\pi}{4}\right) = 2\cos\left[\dfrac{\left(x + \dfrac{\pi}{4}\right) + \left(x - \dfrac{\pi}{4}\right)}{2}\right]\sin\left[\dfrac{\left(x + \dfrac{\pi}{4}\right) - \left(x - \dfrac{\pi}{4}\right)}{2}\right] = 2\cos x \sin\dfrac{\pi}{4}$

101. $\qquad r = \dfrac{1}{32}v_0{}^2 \sin 2\theta$

range $= 100$ feet

$\qquad v_0 = 80$ feet per second

$\qquad r = \dfrac{1}{32}(80)^2 \sin 2\theta = 100$

$\sin 2\theta = 0.5$

$\qquad 2\theta = 30°$

$\qquad\qquad \theta = 15°$ or $\dfrac{\pi}{12}$

102. Volume V of the trough will be the area A of the isosceles triangle times the length l of the trough.

$\qquad V = A \cdot l$

(a) $\qquad A = \dfrac{1}{2}bh$

$\qquad \cos\dfrac{\theta}{2} = \dfrac{h}{0.5} \implies h = 0.5\cos\dfrac{\theta}{2}$

$\qquad \sin\dfrac{\theta}{2} = \dfrac{\dfrac{b}{2}}{0.5} \implies \dfrac{b}{2} = 0.5\sin\dfrac{\theta}{2}$

$\qquad A = 0.5\sin\dfrac{\theta}{2}\,0.5\cos\dfrac{\theta}{2}$

$\qquad\quad = (0.5)^2\sin\dfrac{\theta}{2}\cos\dfrac{\theta}{2}$

$\qquad\quad = 0.25\sin\dfrac{\theta}{2}\cos\dfrac{\theta}{2}$ square meters

$\qquad V = (0.25)(4)\sin\dfrac{\theta}{2}\cos\dfrac{\theta}{2}$ cubic meters

$\qquad\quad = \sin\dfrac{\theta}{2}\cos\dfrac{\theta}{2}$ cubic meters

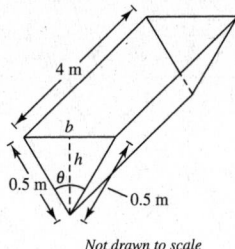

4 m

b

h

0.5 m $\quad \theta$

0.5 m

Not drawn to scale

(b) $V = \sin\dfrac{\theta}{2}\cos\dfrac{\theta}{2} = \dfrac{1}{2}\left(2\sin\dfrac{\theta}{2}\cos\dfrac{\theta}{2}\right) = \dfrac{1}{2}\sin\theta$ cubic meters

$\qquad$ Volume is maximum when $\theta = \dfrac{\pi}{2}$.

103. $y = 1.5 \sin 8t - 0.5 \cos 8t$

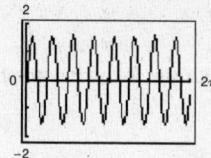

104. $y = 1.5 \sin 8t - 0.5 \cos 8t = \frac{1}{2}(3 \sin 8t - 1 \cos 8t)$

Using the identity

$a \sin B\theta + b \cos B\theta = \sqrt{a^2 + b^2} \sin(B\theta + C),$

$C = \arctan \dfrac{b}{a}, a > 0$

(Exercise 83, Section 5.4), we have

$y = \frac{1}{2}\sqrt{(3)^2 + (-1)^2} \sin\left(8t + \arctan\left(-\frac{1}{3}\right)\right)$

$= \dfrac{\sqrt{10}}{2} \sin\left(8t - \arctan\left(\frac{1}{3}\right)\right).$

105. Amplitude $= \dfrac{\sqrt{10}}{2}$ feet

106. Frequency $= \dfrac{1}{\frac{2\pi}{8}} = \dfrac{4}{\pi}$ cycles per second

107. False. If $\dfrac{\pi}{2} < \theta < \pi$, then $\dfrac{\pi}{4} < \dfrac{\theta}{2} < \dfrac{\pi}{2}$ and $\dfrac{\theta}{2}$ is in Quadrant I.

$\cos \dfrac{\theta}{2} > 0$

108. False. The correct identity is

$\sin(x + y) = \sin x \cos y + \cos x \sin y.$

109. True. $4 \sin(-x)\cos(-x) = 4(-\sin x)\cos x$

$= -4 \sin x \cos x = -2(2 \sin x \cos x)$

$= -2 \sin 2x$

110. True. It can be verified using a product-to-sum identity.

$4 \sin 45° \cos 15° = 4 \cdot \frac{1}{2}[\sin 60° + \sin 30°]$

$= 2\left[\dfrac{\sqrt{3}}{2} + \dfrac{1}{2}\right] = \sqrt{3} + 1$

111. Reciprocal Identities: $\sin \theta = \dfrac{1}{\csc \theta}$ $\quad \csc \theta = \dfrac{1}{\sin \theta}$

$\cos \theta = \dfrac{1}{\sec \theta}$ $\quad \sec \theta = \dfrac{1}{\cos \theta}$

$\tan \theta = \dfrac{1}{\cot \theta}$ $\quad \cot \theta = \dfrac{1}{\tan \theta}$

Quotient Identities: $\quad \tan \theta = \dfrac{\sin \theta}{\cos \theta}$ $\quad \cot \theta = \dfrac{\cos \theta}{\sin \theta}$

Pythagorean Identities: $\sin^2 \theta + \cos^2 \theta = 1$

$1 + \tan^2 \theta = \sec^2 \theta$

$1 + \cot^2 \theta = \csc^2 \theta$

112. No. For an equation to be an identity, the equation must be true for all real numbers. $\sin \theta = \frac{1}{2}$ has an infinite number of solutions but is not an identity.

113. $a \sin x - b = 0$

$\sin x = \dfrac{b}{a}$

If $|b| > |a|$, then $\left|\dfrac{b}{a}\right| > 1$ and there is no solution since $|\sin x| \le 1$ for all x.

114. $S = 6hs + \dfrac{3}{2}s^2\left(\dfrac{\sqrt{3} - \cos\theta}{\sin\theta}\right)$, $0° < \theta \le 90°$

     where $h = 2.4$ inches, $s = 0.75$ inch, and θ is the given angle.

(a) For a surface area of 12 square inches,

$$S = 6(2.4)(0.75) + \dfrac{3}{2}(0.75)^2\left(\dfrac{\sqrt{3} - \cos\theta}{\sin\theta}\right) = 12$$

$$= 10.8 + 0.84375\left(\dfrac{\sqrt{3} - \cos\theta}{\sin\theta}\right) = 12$$

$$0.84375\left(\dfrac{\sqrt{3} - \cos\theta}{\sin\theta}\right) = 1.2.$$

     Using the solve function of a graphing calculator gives

     $\theta = 49.91479°$ or $\theta = 59.86118°$.

(b) Using a graphing calculator yields the following graph:

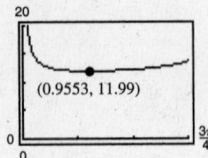

     Using the minimum function yields

     $\theta = 0.9553$ radians   or $\theta = 54.73466°$.

115. The graph of y_1 is a vertical shift of the graph of y_2 one unit upward so $y_1 = y_2 + 1$.

116. $y_1 = \dfrac{\cos 3x}{\cos x}$, $y_2 = (2\sin x)^2$

     If the graph of y_2 is reflected in the x-axis and then shifted upward by one unit, it coincides with the graph of y_1. Therefore,

$$\dfrac{\cos 3x}{\cos x} = -(2\sin x)^2 + 1.$$

     So, $y_1 = 1 - y_2$.

117. $y = \sqrt{x + 3} + 4\cos x$

     Zeros: $x \approx -1.8431, 2.1758,$
     $3.9903, 8.8935, 9.8820$

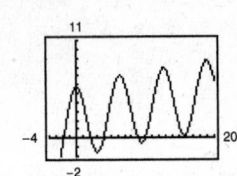

118. $y = 2 - \dfrac{1}{2}x^2 + 3\sin\dfrac{\pi x}{2}$

     Approximate roots:
     $-3.1395, -2.0000,$
     $-0.4378, 2.0000$

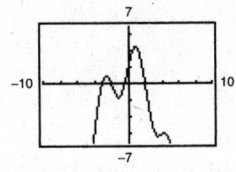

$$y = 2 - \dfrac{1}{2}x^2 + 3\sin\dfrac{\pi x}{2}$$

Problem Solving for Chapter 5

1. (a) Since $\sin^2\theta + \cos^2\theta = 1$ and $\cos^2\theta = 1 - \sin^2\theta$:

$$\cos\theta = \pm\sqrt{1 - \sin^2\theta}$$

$$\tan\theta = \dfrac{\sin\theta}{\cos\theta} = \pm\dfrac{\sin\theta}{\sqrt{1 - \sin^2\theta}}$$

$$\cot\theta = \dfrac{1}{\tan\theta} = \pm\dfrac{\sqrt{1 - \sin^2\theta}}{\sin\theta}$$

$$\sec\theta = \dfrac{1}{\cos\theta} = \pm\dfrac{1}{\sqrt{1 - \sin^2\theta}}$$

$$\cos\theta = \dfrac{1}{\sin\theta}$$

We also have the following relationships:

$$\cos\theta = \sin\left(\dfrac{\pi}{2} - \theta\right)$$

$$\tan\theta = \dfrac{\sin\theta}{\sin\left(\dfrac{\pi}{2} - \theta\right)}$$

$$\cot\theta = \dfrac{\sin\left(\dfrac{\pi}{2} - \theta\right)}{\sin\theta}$$

$$\sec\theta = \dfrac{1}{\sin\left(\dfrac{\pi}{2} - \theta\right)}$$

$$\csc\theta = \dfrac{1}{\sin\theta}$$

—CONTINUED—

1. **—CONTINUED—**

(b) $\sin \theta = \pm\sqrt{1 - \cos^2 \theta}$

$$\tan \theta = \frac{\sin \theta}{\cos \theta} = \pm\frac{\sqrt{1 - \cos^2 \theta}}{\cos \theta}$$

$$\csc \theta = \frac{1}{\sin \theta} = \pm\frac{1}{\sqrt{1 - \cos^2 \theta}}$$

$$\sec \theta = \frac{1}{\cos \theta}$$

$$\cot \theta = \frac{1}{\tan \theta} = \pm\frac{\cos \theta}{\sqrt{1 - \cos^2 \theta}}$$

We also have the following relationships:

$$\sin \theta = \cos\left(\frac{\pi}{2} - \theta\right)$$

$$\tan \theta = \frac{\cos[(\pi/2) - \theta]}{\cos \theta}$$

$$\csc \theta = \frac{1}{\cos[(\pi/2) - \theta]}$$

$$\sec \theta = \frac{1}{\cos \theta}$$

$$\cot \theta = \frac{\cos \theta}{\cos[(\pi/2) - \theta]}$$

2. $\cos\left[\dfrac{(2n + 1)\pi}{2}\right] = \cos\left(\dfrac{2n\pi + \pi}{2}\right)$

$$= \cos\left(n\pi + \frac{\pi}{2}\right)$$

$$= \cos n\pi \cos\frac{\pi}{2} - \sin n\pi \sin\frac{\pi}{2}$$

$$= (\pm 1)(0) - (0)(1)$$

$$= 0$$

Thus, $\cos\left[\dfrac{(2n + 1)\pi}{2}\right] = 0$ for all integers n.

3. $\sin\left[\dfrac{(12n + 1)\pi}{6}\right] = \sin\left[\dfrac{1}{6}(12n\pi + \pi)\right]$

$$= \sin\left(2n\pi + \frac{\pi}{6}\right)$$

$$= \sin\frac{\pi}{6} = \frac{1}{2}$$

Thus, $\sin\left[\dfrac{(12n + 1)\pi}{6}\right] = \dfrac{1}{2}$ for all integers n.

4. $p(t) = \dfrac{1}{4\pi}[p_1(t) + 30p_2(t) + p_3(t) + p_5(t) + 30p_6(t)]$

(a) $p_1(t) = \sin(524\pi t)$

$$p_2(t) = \frac{1}{2}\sin(1048\pi t)$$

$$p_3(t) = \frac{1}{3}\sin(1572\pi t)$$

$$p_5(t) = \frac{1}{5}\sin(2620\pi t)$$

$$p_6(t) = \frac{1}{6}\sin(3144\pi t)$$

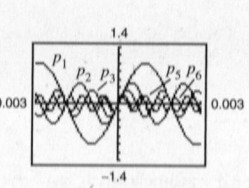

The graph of

$$p(t) = \frac{1}{4\pi}\left[\sin(524\pi t) + 15\sin(1048\pi t) + \frac{1}{3}\sin(1572\pi t) + \frac{1}{5}\sin(2620\pi t) + 5\sin(3144\pi t)\right]$$

yields the graph shown in the text and to the right.

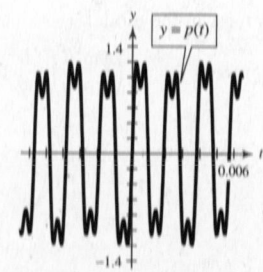

—CONTINUED—

4. —CONTINUED—

(b) Function   Period

$$p_1(t) \quad \frac{2\pi}{524\pi} = \frac{1}{262} \approx 0.0038$$

$$p_2(t) \quad \frac{2\pi}{1048\pi} = \frac{1}{524} \approx 0.0019$$

$$p_3(t) \quad \frac{2\pi}{1572\pi} = \frac{1}{786} \approx 0.0013$$

$$p_5(t) \quad \frac{2\pi}{2620\pi} = \frac{1}{1310} \approx 0.0008$$

$$p_6(t) \quad \frac{2\pi}{3144\pi} = \frac{1}{1572} \approx 0.0006$$

The graph of p appears to be periodic with a period of $\frac{1}{262} \approx 0.0038$.

(c)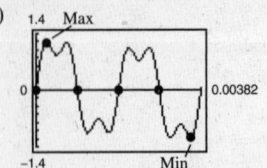

Over one cycle, $0 \leq t < \frac{1}{262}$, we have four t-intercepts:

$t = 0$, $t \approx 0.00096$, $t \approx 0.00191$, and $t \approx 0.00285$

(d) The absolute maximum value of p over one cycle is $p \approx 1.1952$, and the absolute minimum value of p over one cycle is $p \approx -1.1952$.

---

5. From the figure, it appears that $u + v = w$. Assume that u, v, and w are all in Quadrant I. From the figure:

$$\tan u = \frac{s}{3s} = \frac{1}{3}$$

$$\tan v = \frac{s}{2s} = \frac{1}{2}$$

$$\tan w = \frac{s}{s} = 1$$

$$\tan(u + v) = \frac{\tan u + \tan v}{1 - \tan u \tan v}$$

$$= \frac{1/3 + 1/2}{1 - (1/3)(1/2)}$$

$$= \frac{5/6}{1 - (1/6)}$$

$$= 1 = \tan w.$$

Thus, $\tan(u + v) = \tan w$. Because u, v, and w are all in Quadrant I, we have

$$\arctan[\tan(u + v)] = \arctan[\tan w] u + v = w.$$

---

6. $y = -\dfrac{16}{v_0^2 \cos^2 \theta} x^2 + (\tan \theta)x + h_0$

Let $h_0 = 0$ and take half of the horizontal distance:

$$\frac{1}{2}\left(\frac{1}{32} v_0^2 \sin 2\theta\right) = \frac{1}{64} v_0^2 (2 \sin \theta \cos \theta) = \frac{1}{32} v_0^2 \sin \theta \cos \theta$$

Substitute this expression for x in the model.

$$y = -\frac{16}{v_0^2 \cos^2 \theta}\left(\frac{1}{32}v_0^2 \sin \theta \cos \theta\right)^2 + \left(\frac{\sin \theta}{\cos \theta}\right)\left(\frac{1}{32} v_0^2 \sin \theta \cos \theta\right)$$

$$= -\frac{1}{64} v_0^2 \sin^2 \theta + \frac{1}{32} v_0^2 \sin^2 \theta$$

$$= \frac{1}{64} v_0^2 \sin^2 \theta$$

7.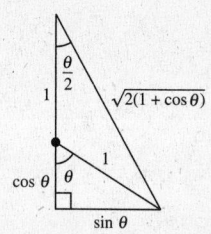

The hypotenuse of the larger right triangle is:

$$\sqrt{\sin^2\theta + (1 + \cos\theta)^2} = \sqrt{\sin^2\theta + 1 + 2\cos\theta + \cos^2\theta}$$

$$= \sqrt{2 + 2\cos\theta}$$

$$= \sqrt{2(1 + \cos\theta)}$$

$$\sin\left(\frac{\theta}{2}\right) = \frac{\sin\theta}{\sqrt{2(1+\cos\theta)}} = \frac{\sin\theta}{\sqrt{2(1+\cos\theta)}} \cdot \frac{\sqrt{1-\cos\theta}}{\sqrt{1-\cos\theta}}$$

$$= \frac{\sin\theta\sqrt{1-\cos\theta}}{\sqrt{2(1-\cos^2\theta)}} = \frac{\sin\theta\sqrt{1-\cos\theta}}{\sqrt{2}\sin\theta} = \sqrt{\frac{1-\cos\theta}{2}}$$

$$\cos\left(\frac{\theta}{2}\right) = \frac{1+\cos\theta}{\sqrt{2(1+\cos\theta)}} = \frac{\sqrt{(1+\cos\theta)^2}}{\sqrt{2(1+\cos\theta)}} = \sqrt{\frac{1+\cos\theta}{2}}$$

$$\tan\left(\frac{\theta}{2}\right) = \frac{\sin\theta}{1+\cos\theta}$$

8. $F = \dfrac{0.6W\sin(\theta + 90°)}{\sin 12°}$

(a) $F = \dfrac{0.6W(\sin\theta\cos 90° + \cos\theta\sin 90°)}{\sin 12°}$

$= \dfrac{0.6W[(\sin\theta)(0) + (\cos\theta)(1)]}{\sin 12°}$

$= \dfrac{0.6W\cos\theta}{\sin 12°}$

(b) Let $y_1 = \dfrac{0.6(185)\cos x}{\sin 12°}$.

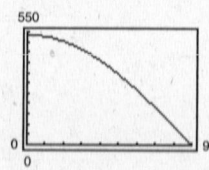

(c) The force is maximum (533.88 pounds) when $\theta = 0°$. The force is minimum (0 pounds) when $\theta = 90°$.

9. Seward: $D = 12.2 - 6.4\cos\left[\dfrac{\pi(t + 0.2)}{182.6}\right]$

New Orleans: $D = 12.2 - 1.9\cos\left[\dfrac{\pi(t + 0.2)}{182.6}\right]$

(a)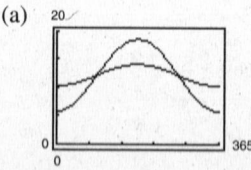

(b) The graphs intersect when $t \approx 91$ and when $t \approx 274$. These values correspond to April 1 and October 1, the spring equinox and the fall equinox.

(c) Seward has the greater variation in the number of daylight hours. This is determined by the amplitudes, 6.4 and 1.9.

(d) Period: $\dfrac{2\pi}{\pi/182.6} = 365.2$ days

10. $d = 35 - 28\cos\dfrac{\pi}{6.2}t$ when $t = 0$ corresponds to 12:00 A.M.

(a) The high tides occur when $\cos\dfrac{\pi}{6.2}t = -1$. Solving yields $t = 6.2$ or $t = 18.6$.

These t-values correspond to 6:12 A.M. and 6:36 P.M.

The low tide occurs when $\cos\dfrac{\pi}{6.2}t = 1$. Solving yields $t = 0$ and $t = 12.4$ which corresponds to 12:00 A.M. and 12:24 P.M.

(b) The water depth is never 3.5 feet. At low tide the depth is $d = 35 - 28 = 7$ feet.

(c)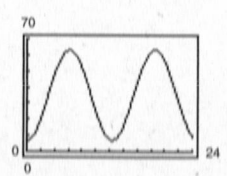

11. (a) Let $y_1 = \sin x$ and $y_2 = 0.5$.

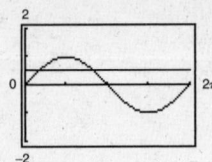

$\sin x \geq 0.5$ on the interval $\left[\dfrac{\pi}{6}, \dfrac{5\pi}{6}\right]$.

(b) Let $y_1 = \cos x$ and $y_2 = -0.5$.

$\cos x \leq -0.5$ on the interval $\left[\dfrac{2\pi}{3}, \dfrac{4\pi}{3}\right]$.

(c) Let $y_1 = \tan x$ and $y_2 = \sin x$.

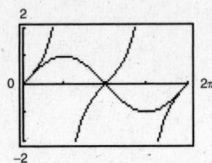

$\tan x < \sin x$ on the intervals $\left(\dfrac{\pi}{2}, \pi\right)$ and $\left(\dfrac{3\pi}{2}, 2\pi\right)$.

(d) Let $y_1 = \cos x$ and $y_2 = \sin x$.

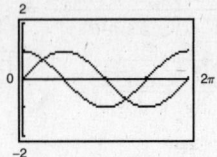

$\cos x \geq \sin x$ on the intervals $\left[0, \dfrac{\pi}{4}\right]$ and $\left[\dfrac{5\pi}{4}, 2\pi\right]$.

12. (a) $n = \dfrac{\sin\left(\dfrac{\theta}{2} + \dfrac{\alpha}{2}\right)}{\sin\dfrac{\theta}{2}}$

$= \dfrac{\sin\left(\dfrac{\theta}{2}\right)\cos\left(\dfrac{\alpha}{2}\right) + \cos\left(\dfrac{\theta}{2}\right)\sin\left(\dfrac{\alpha}{2}\right)}{\sin\left(\dfrac{\theta}{2}\right)}$

$= \cos\left(\dfrac{\alpha}{2}\right) + \cot\left(\dfrac{\theta}{2}\right)\sin\left(\dfrac{\alpha}{2}\right)$

For $\alpha = 60°$, $n = \cos 30° + \cot\left(\dfrac{\theta}{2}\right)\sin 30°$

$n = \dfrac{\sqrt{3}}{2} + \dfrac{1}{2}\cot\left(\dfrac{\theta}{2}\right)$.

(b) For glass, $n = 1.50$.

$$1.50 = \dfrac{\sqrt{3}}{2} + \dfrac{1}{2}\cot\left(\dfrac{\theta}{2}\right)$$

$$2\left(1.50 - \dfrac{\sqrt{3}}{2}\right) = \cot\left(\dfrac{\theta}{2}\right)$$

$$\dfrac{1}{3 - \sqrt{3}} = \tan\left(\dfrac{\theta}{2}\right)$$

$$\theta = 2\tan^{-1}\left(\dfrac{1}{3 - \sqrt{3}}\right)$$

$$\theta \approx 76.52°$$

13. (a) $\sin(u + v + w) = \sin[(u + v) + w]$

$= \sin(u + v)\cos w + \cos(u + v)\sin w$

$= [\sin u \cos v + \cos u \sin v]\cos w + [\cos u \cos v - \sin u \sin v]\sin w$

$= \sin u \cos v \cos w + \cos u \sin v \cos w + \cos u \cos v \sin w - \sin u \sin v \sin w$

(b) $\tan(u + v + w) = \tan[(u + v) + w]$

$= \dfrac{\tan(u + v) + \tan w}{1 - \tan(u + v)\tan w}$

$= \dfrac{\left[\dfrac{\tan u + \tan v}{1 - \tan u \tan v}\right] + \tan w}{1 - \left[\dfrac{\tan u + \tan v}{1 - \tan u \tan v}\right]\tan w} \cdot \dfrac{(1 - \tan u \tan v)}{(1 - \tan u \tan v)}$

$= \dfrac{\tan u + \tan v + (1 - \tan u \tan v)\tan w}{(1 - \tan u \tan v) - (\tan u + \tan v)\tan w}$

$= \dfrac{\tan u + \tan v + \tan w - \tan u \tan v \tan w}{1 - \tan u \tan v - \tan u \tan w - \tan v \tan w}$

14. (a) $\cos(3\theta) = \cos(2\theta + \theta)$

$\qquad = \cos 2\theta \cos\theta - \sin 2\theta \sin\theta$

$\qquad = (1 - 2\sin^2\theta)\cos\theta - (2\sin\theta\cos\theta)\sin\theta$

$\qquad = \cos\theta - 4\sin^2\theta\cos\theta$

$\qquad = \cos\theta(1 - 4\sin^2\theta)$

(b) $\cos(4\theta) = \cos(2\theta + 2\theta)$

$\qquad = \cos 2\theta \cos 2\theta - \sin 2\theta \sin 2\theta$

$\qquad = \cos^2 2\theta - \sin^2 2\theta$

$\qquad = (1 - \sin^2 2\theta) - \sin^2 2\theta$

$\qquad = 1 - 2\sin^2 2\theta$

$\qquad = 1 - 2(2\sin\theta\cos\theta)^2$

$\qquad = 1 - 8\sin^2\theta\cos^2\theta$

15. $h_1 = 3.75\sin 733t + 7.5$

$\quad h_2 = 3.75\sin 733\left(t + \dfrac{4\pi}{3}\right) + 7.5$

(a)

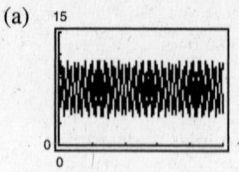

(b) The period for h_1 and h_2 is $\dfrac{2\pi}{733} \approx 0.0086$.

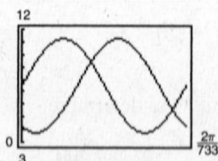

The graphs intersect twice per cycle.

There are $\dfrac{1}{2\pi/733} \approx 116.66$ cycles in the interval $[0, 1]$, so the graphs intersect approximately 233.3 times.

Chapter 5   Practice Test

1. Find the value of the other five trigonometric functions, given $\tan x = \frac{4}{11}$, $\sec x < 0$.

2. Simplify $\dfrac{\sec^2 x + \csc^2 x}{\csc^2 x(1 + \tan^2 x)}$.

3. Rewrite as a single logarithm and simplify $\ln|\tan \theta| - \ln|\cot \theta|$.

4. True or false:
$$\cos\left(\frac{\pi}{2} - x\right) = \frac{1}{\csc x}$$

5. Factor and simplify: $\sin^4 x + (\sin^2 x)\cos^2 x$

6. Multiply and simplify: $(\csc x + 1)(\csc x - 1)$

7. Rationalize the denominator and simplify:
$$\frac{\cos^2 x}{1 - \sin x}$$

8. Verify:
$$\frac{1 + \cos \theta}{\sin \theta} + \frac{\sin \theta}{1 + \cos \theta} = 2 \csc \theta$$

9. Verify:
$$\tan^4 x + 2 \tan^2 x + 1 = \sec^4 x$$

10. Use the sum or difference formulas to determine:

   (a) $\sin 105°$ (b) $\tan 15°$

11. Simplify: $(\sin 42°)\cos 38° - (\cos 42°)\sin 38°$

12. Verify $\tan\left(\theta + \dfrac{\pi}{4}\right) = \dfrac{1 + \tan \theta}{1 - \tan \theta}$.

13. Write $\sin(\arcsin x - \arccos x)$ as an algebraic expression in x.

14. Use the double-angle formulas to determine:

   (a) $\cos 120°$ (b) $\tan 300°$

15. Use the half-angle formulas to determine:

   (a) $\sin 22.5°$ (b) $\tan \dfrac{\pi}{12}$

16. Given $\sin = 4/5$, θ lies in Quadrant II, find $\cos(\theta/2)$.

17. Use the power-reducing identities to write $(\sin^2 x)\cos^2 x$ in terms of the first power of cosine.

18. Rewrite as a sum: $6(\sin 5\theta)\cos 2\theta$.

19. Rewrite as a product: $\sin(x + \pi) + \sin(x - \pi)$.

20. Verify $\dfrac{\sin 9x + \sin 5x}{\cos 9x - \cos 5x} = -\cot 2x$.

21. Verify:
$$(\cos u) \sin v = \tfrac{1}{2}[\sin(u + v) - \sin(u - v)].$$

22. Find all solutions in the interval $[0, 2\pi)$:

   $4 \sin^2 x = 1$

23. Find all solutions in the interval $[0, 2\pi)$:

   $\tan^2 \theta + \left(\sqrt{3} - 1\right)\tan \theta - \sqrt{3} = 0$

24. Find all solutions in the interval $[0, 2\pi)$:

   $\sin 2x = \cos x$

25. Use the quadratic formula to find all solutions in the interval $[0, 2\pi)$:

   $\tan^2 x - 6 \tan x + 4 = 0$

CHAPTER 6
Additional Topics in Trigonometry

C H A P T E R   6
Additional Topics in Trigonometry

Section 6.1   Law of Sines

■ If ABC is any oblique triangle with sides a, b, and c, then

$$\frac{a}{\sin A} = \frac{b}{\sin B} = \frac{c}{\sin C}.$$

■ You should be able to use the Law of Sines to solve an oblique triangle for the remaining three parts, given:

(a) Two angles and any side (AAS or ASA)

(b) Two sides and an angle opposite one of them (SSA)

    1. If A is acute and $h = b \sin A$:

      (a) $a < h$, no triangle is possible.

      (b) $a = h$ or $a > b$, one triangle is possible.

      (c) $h < a < b$, two triangles are possible.

    2. If A is obtuse and $h = b \sin A$:

      (a) $a \leq b$, no triangle is possible.

      (b) $a > b$, one triangle is possible.

■ The area of any triangle equals one-half the product of the lengths of two sides and the sine of their included angle.

$$A = \tfrac{1}{2}ab \sin C = \tfrac{1}{2}ac \sin B = \tfrac{1}{2}bc \sin A$$

Vocabulary Check

1. oblique

2. $\dfrac{b}{\sin B}$

3. $\dfrac{1}{2} ac \sin B$

1.

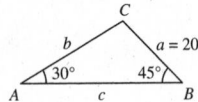

Given: $A = 30°$, $B = 45°$, $a = 20$

$C = 180° - A - B = 105°$

$b = \dfrac{a}{\sin A}(\sin B) = \dfrac{20 \sin 45°}{\sin 30°} = 20\sqrt{2} \approx 28.28$

$c = \dfrac{a}{\sin A}(\sin C) = \dfrac{20 \sin 105°}{\sin 30°} \approx 38.64$

2.

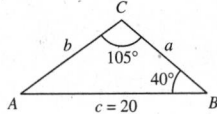

Given: $B = 40°$, $C = 105°$, $c = 20$

$A = 180° - B - C = 35°$

$a = \dfrac{c}{\sin C}(\sin A) = \dfrac{20 \sin 35°}{\sin 105°} \approx 11.88$

$b = \dfrac{c}{\sin C}(\sin B) = \dfrac{20 \sin 40°}{\sin 105°} \approx 13.31$

3.

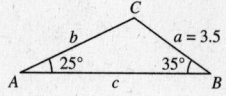

Given: $A = 25°, B = 35°, a = 3.5$

$C = 180° - A - B = 120°$

$b = \dfrac{a}{\sin A}(\sin B) = \dfrac{3.5}{\sin 25°}(\sin 35°) \approx 4.75$

$c = \dfrac{a}{\sin A}(\sin C) = \dfrac{3.5}{\sin 25°}(\sin 120°) \approx 7.17$

4.

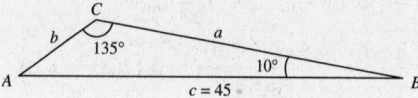

Given: $B = 10°, C = 135°, c = 45$

$A = 180° - B - C = 35°$

$a = \dfrac{c}{\sin C}(\sin A) = \dfrac{45 \sin 35°}{\sin 135°} \approx 36.50$

$b = \dfrac{c}{\sin C}(\sin B) = \dfrac{45 \sin 10°}{\sin 135°} \approx 11.05$

5. Given: $A = 36°, a = 8, b = 5$

$\sin B = \dfrac{b \sin A}{a} = \dfrac{5 \sin 36°}{8} \approx 0.36737 \implies B \approx 21.55°$

$C = 180° - A - B \approx 180° - 36° - 21.55 = 122.45°$

$c = \dfrac{a}{\sin A}(\sin C) = \dfrac{8}{\sin 36°}(\sin 122.45°) \approx 11.49$

6. Given: $A = 60°, a = 9, c = 10$

$\sin C = \dfrac{c \sin A}{a} = \dfrac{10 \sin 60°}{9} \approx 0.9623 \implies C \approx 74.21°$ or $C \approx 105.79°$

Case 1

$C \approx 74.21°$

$B = 180° - A - C \approx 45.79°$

$b = \dfrac{a}{\sin A}(\sin B) \approx \dfrac{9 \sin 45.79°}{\sin 60°} \approx 7.45$

Case 2

$C \approx 105.79°$

$B = 180° - A - C \approx 14.21°$

$b = \dfrac{a}{\sin A}(\sin B) \approx \dfrac{9 \sin 14.21°}{\sin 60°} \approx 2.55$

7. Given: $A = 102.4°, C = 16.7°, a = 21.6$

$B = 180° - A - C = 60.9°$

$b = \dfrac{a}{\sin A}(\sin B) = \dfrac{21.6}{\sin 102.4°}(\sin 60.9°) \approx 19.32$

$c = \dfrac{a}{\sin A}(\sin C) = \dfrac{21.6}{\sin 102.4°}(\sin 16.7°) \approx 6.36$

8. Given: $A = 24.3°, C = 54.6°, c = 2.68$

$B = 180° - A - C = 101.1°$

$a = \dfrac{c}{\sin C}(\sin A) = \dfrac{2.68 \sin 24.3°}{\sin 54.6°} \approx 1.35$

$b = \dfrac{c}{\sin C}(\sin B) = \dfrac{2.68 \sin 101.1°}{\sin 54.6°} \approx 3.23$

9. Given: $A = 83° 20', C = 54.6°, c = 18.1$

$B = 180° - A - C = 180° - 83° 20' - 54° 36' = 42° 4'$

$a = \dfrac{c}{\sin C}(\sin A) = \dfrac{18.1}{\sin 54.6°}(\sin 83° 20') \approx 22.05$

$b = \dfrac{c}{\sin C}(\sin B) = \dfrac{18.1}{\sin 54.6°}(\sin 42° 4') \approx 14.88$

10. Given: $A = 5° 40', B = 8° 15', b = 4.8$

$C = 180° - A - B = 166° 5'$

$a = \dfrac{b}{\sin B}(\sin A) = \dfrac{4.8 \sin 5° 40'}{\sin 8° 15'} \approx 3.30$

$c = \dfrac{b}{\sin B}(\sin C) = \dfrac{4.8 \sin 166° 5'}{\sin 8° 15'} \approx 8.05$

11. Given: $B = 15° \, 30'$, $a = 4.5$, $b = 6.8$

$$\sin A = \frac{a \sin B}{b} = \frac{4.5 \sin 15° \, 30'}{6.8} \approx 0.17685 \implies A \approx 10° \, 11'$$

$$C = 180° - A - B \approx 180° - 10° \, 11' - 15° \, 30' = 154° \, 19'$$

$$c = \frac{b}{\sin B} (\sin C) = \frac{6.8}{\sin 15° \, 30'} (\sin 154° \, 19') \approx 11.03$$

12. Given: $B = 2° \, 45'$, $b = 6.2$, $c = 5.8$

$$\sin C = \frac{c \sin B}{b} = \frac{5.8 \sin 2° \, 45'}{6.2} \approx 0.04488 \implies C \approx 2.57° \text{ or } 2° \, 34'$$

$$A = 180 - B - C \approx 174.68°, \text{ or } 174° \, 41'$$

$$a = \frac{b}{\sin B} (\sin A) \approx \frac{6.2 \sin 174.68°}{\sin 2° \, 45'} \approx 11.99$$

13. Given: $C = 145°$, $b = 4$, $c = 14$

$$\sin B = \frac{b \sin C}{c} = \frac{4 \sin 145°}{14} \approx 0.16388 \implies B \approx 9.43°$$

$$A = 180° - B - C \approx 180° - 9.43° - 145° = 25.57°$$

$$a = \frac{c}{\sin C} (\sin A) \approx \frac{14}{\sin 145°} (\sin 25.57°) \approx 10.53$$

14. Given: $A = 100°$, $a = 125$, $c = 10$

$$\sin C = \frac{c \sin A}{a} = \frac{10 \sin 100°}{125} \approx 0.07878 \implies C \approx 4.52°$$

$$B = 180° - A - C \approx 75.48°$$

$$b = \frac{a}{\sin A} (\sin B) \approx \frac{125 \sin 75.48°}{\sin 100°} \approx 122.87$$

15. Given: $A = 110° \, 15'$, $a = 48$, $b = 16$

$$\sin B = \frac{b \sin A}{a} = \frac{16 \sin 110° \, 15'}{48} \approx 0.31273 \implies B \approx 18° \, 13'$$

$$C = 180° - A - B \approx 180° - 110° \, 15' - 18° \, 13' = 51° \, 32'$$

$$c = \frac{a}{\sin A} (\sin C) = \frac{48}{\sin 110° \, 15'} (\sin 51° \, 32') \approx 40.06$$

16. Given: $C = 85° \, 20'$, $a = 35$, $c = 50$

$$\sin A = \frac{a \sin C}{c} = \frac{35 \sin 85° \, 20'}{50} \approx 0.6977 \implies A \approx 44.24°, \text{ or } 44° \, 14'$$

$$B = 180° - A - C \approx 50.43°, \text{ or } 50° \, 26'$$

$$b = \frac{C \sin B}{\sin C} \approx \frac{50 \sin 50.43°}{\sin 85° \, 20'} \approx 38.67$$

17. Given: $A = 55°$, $B = 42°$, $c = \frac{3}{4}$

$$C = 180° - A - B = 83°$$

$$a = \frac{c}{\sin C} (\sin A) = \frac{0.75}{\sin 83°} (\sin 55°) \approx 0.62$$

$$b = \frac{c}{\sin C} (\sin B) = \frac{0.75}{\sin 83°} (\sin 42°) \approx 0.51$$

18. Given: $B = 28°$, $C = 104°$, $a = 3\frac{5}{8}$

$$A = 180° - B - C = 48°$$

$$b = \frac{a \sin B}{\sin A} = \frac{3\frac{5}{8} \sin 28°}{\sin 48°} \approx 2.29$$

$$c = \frac{a \sin C}{\sin A} = \frac{3\frac{5}{8} \sin 104°}{\sin 48°} \approx 4.73$$

19. Given: $A = 110°$, $a = 125$, $b = 100$

$$\sin B = \frac{b \sin A}{a} = \frac{100 \sin 110°}{125} \approx 0.75175 \implies B \approx 48.74°$$

$$C = 180° - A - B \approx 21.26°$$

$$c = \frac{a \sin C}{\sin A} \approx \frac{125 \sin 21.26°}{\sin 110°} \approx 48.23$$

20. Given: $a = 125$, $b = 200$, $A = 110°$

No triangle is formed because A is obtuse and $a < b$.

21. Given: $a = 18$, $b = 20$, $A = 76°$

$h = 20 \sin 76° \approx 19.41$

Since $a < h$, no triangle is formed.

22. Given: $A = 76°$, $a = 34$, $b = 21$

$$\sin B = \frac{b \sin A}{a} = \frac{21 \sin 76°}{34} \approx 0.5993 \implies B \approx 36.82°$$

$$C = 180° - A - B \approx 67.18°$$

$$c = \frac{a \sin C}{\sin A} \approx \frac{34 \sin 67.18°}{\sin 76°} \approx 32.30$$

23. Given: $A = 58°$, $a = 11.4$, $c = 12.8$

$$\sin B = \frac{b \sin A}{a} = \frac{12.8 \sin 58°}{11.4} \approx 0.9522 \implies B \approx 72.21° \text{ or } B \approx 107.79°$$

Case 1

$B \approx 72.21°$

$C = 180° - A - B \approx 49.79°$

$$c = \frac{a}{\sin A}(\sin C) \approx \frac{11.4 \sin 49.79°}{\sin 58°} \approx 10.27$$

Case 2

$B \approx 107.79°$

$C = 180° - A - B \approx 14.21°$

$$c = \frac{a}{\sin A}(\sin C) \approx \frac{11.4 \sin 14.21°}{\sin 58°} \approx 3.30$$

24. Given: $a = 4.5$, $b = 12.8$, $A = 58°$

$h = 12.8 \sin 58° \approx 10.86$

Since $a < h$, no triangle is formed.

25. Given: $A = 36°$, $a = 5$

(a) One solution if $b \le 5$ or $b = \dfrac{5}{\sin 36°}$

(b) Two solutions if $5 < b < \dfrac{5}{\sin 36°}$

(c) No solution if $b > \dfrac{5}{\sin 36°}$

26. Given: $A = 60°$, $a = 10$

(a) One solution if $b \le 10$ or $b = \dfrac{10}{\sin 60°}$

(b) Two solutions if $10 < b < \dfrac{10}{\sin 60°}$

(c) No solutions if $b > \dfrac{10}{\sin 60°}$

27. Given: $A = 10°$, $a = 10.8$

(a) One solution if $b \le 10.8$ or $b = \dfrac{10.8}{\sin 10°}$

(b) Two solutions if $10.8 < b < \dfrac{10.8}{\sin 10°}$

(c) No solution if $b > \dfrac{10.8}{\sin 10°}$

28. Given: $A = 88°$, $a = 315.6$

(a) One solution if $b \le 315.6$ or $b = \dfrac{315.6}{\sin 88°}$

(b) Two solutions if $315.6 < b < \dfrac{315.6}{\sin 88°}$

(c) No solutions if $b > \dfrac{315.6}{\sin 88°}$

29. Area $= \dfrac{1}{2}ab \sin C = \dfrac{1}{2}(4)(6) \sin 120° \approx 10.4$

30. Area $= \frac{1}{2}ac \sin B = \frac{1}{2}(62)(20) \sin 130° \approx 474.9$

31. Area $= \frac{1}{2}bc \sin A = \frac{1}{2}(57)(85) \sin 43° \, 45' \approx 1675.2$

32. $A = 5° \, 15'$, $b = 4.5$, $c = 22$

Area $= \frac{1}{2}bc \sin A$

$= \left(\frac{1}{2}\right)(4.5)(22) \sin 5.25° \approx 4.5$

33. Area $= \frac{1}{2}ac \sin B = \frac{1}{2}(105)(64)\sin(72°30') \approx 3204.5$

34. $C = 84° \, 30'$, $a = 16$, $b = 20$

Area $= \frac{1}{2}ab \sin C$

$= \left(\frac{1}{2}\right)(16)(20) \sin 84.5° \approx 159.3$

35. $C = 180° - 23° - 94° = 63°$

$h = \dfrac{35}{\sin 63°}(\sin 23°) \approx 15.3$ meters

36. (a)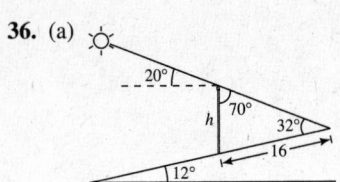

(b) $\dfrac{h}{\sin 32°} = \dfrac{16}{\sin 70°}$

(c) $h = \dfrac{16 \sin 32°}{\sin 70°} \approx 9.0$ meters

37. $\dfrac{\sin(42° - \theta)}{10} = \dfrac{\sin 48°}{17}$

$\sin(42° - \theta) \approx 0.43714$

$42° - \theta \approx 25.9°$

$\theta \approx 16.1°$

38.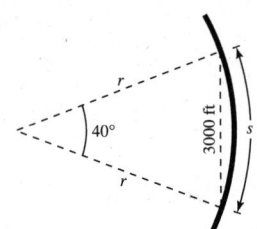

Given: $A = 46°$, $a = 720$, $b = 500$

$\sin B = \dfrac{b \sin A}{a} = \dfrac{500 \sin 46°}{720} \approx 0.50 \implies B \approx 30°$

The bearing from C to B is 240°.

39. Given: $c = 100$

$A = 74° - 28° = 46°$,

$B = 180° - 41° - 74° = 65°$,

$C = 180° - 46° - 65° = 69°$

$a = \dfrac{c}{\sin C}(\sin A) = \dfrac{100}{\sin 69°}(\sin 46°) \approx 77$ meters

40.

(b) $r = \dfrac{3000 \sin\left[\frac{1}{2}(180° - 40°)\right]}{\sin 40°} \approx 4385.71$ feet

(c) $s \approx 40\left(\dfrac{\pi}{180}\right)4385.71 \approx 3061.80$ feet

41. (a)

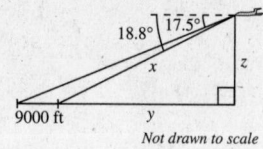

Not drawn to scale

(b) $\dfrac{x}{\sin 17.5^\circ} = \dfrac{9000}{\sin 1.3^\circ}$

$x \approx 119{,}289.1261 \text{ feet} \approx 22.6 \text{ miles}$

(c) $\dfrac{y}{\sin 71.2^\circ} = \dfrac{x}{\sin 90^\circ}$

$y = x \sin 71.2^\circ \approx 119{,}289.1261 \sin 71.2^\circ$

$\approx 112{,}924.963 \text{ feet} \approx 21.4 \text{ miles}$

(d) $z = x \sin 18.8^\circ \approx 119{,}289.1261 \sin 18.8^\circ$

$\approx 38{,}443 \text{ feet} \approx 7.3 \text{ miles}$

42. Given: $A = 15^\circ, B = 135^\circ, c = 30$

$C = 180^\circ - A - B = 30^\circ$

From Pine Knob:

$b = \dfrac{c \sin B}{\sin C} = \dfrac{30 \sin 135^\circ}{\sin 30^\circ} \approx 42.4 \text{ kilometers}$

From Colt Station:

$a = \dfrac{c \sin A}{\sin C} = \dfrac{30 \sin 15^\circ}{\sin 30^\circ} \approx 15.5 \text{ kilometers}$

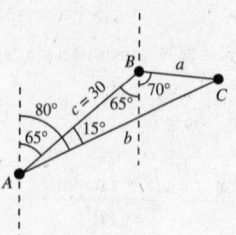

43.

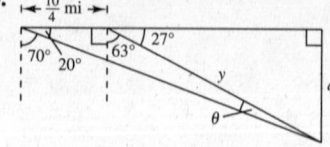

In 15 minutes the boat has traveled

$(10 \text{ mph})\left(\tfrac{1}{4} \text{ hr}\right) = \tfrac{10}{4} \text{ miles}.$

$\theta = 180^\circ - 20^\circ - (90^\circ + 63^\circ)$

$\theta = 7^\circ$

$\dfrac{10/4}{\sin 7^\circ} = \dfrac{y}{\sin 20^\circ}$

$y \approx 7.0161$

$\sin 27^\circ = \dfrac{d}{7.0161}$

$d \approx 3.2 \text{ miles}$

44. (a) $\sin \alpha = \dfrac{5.45}{58.36} \approx 0.0934 \implies \alpha \approx 5.36^\circ$

(c) $\dfrac{d}{\sin(84.64^\circ - \theta)} = \dfrac{58.36}{\sin \theta}$ or

$d = \dfrac{58.36 \sin(84.64^\circ - \theta)}{\sin \theta}$

(b) $\dfrac{\sin \beta}{d} = \dfrac{\sin \theta}{58.36}$

$\sin \beta = \dfrac{d \sin \theta}{58.36}$

$\beta = \sin^{-1}\left(\dfrac{d \sin \theta}{58.36}\right)$

(d)

θ	10°	20°	30°	40°	50°	60°
d	324.1	154.2	95.19	63.80	43.30	28.10

45. True. If one angle of a triangle is obtuse, then there is less than 90° left for the other two angles, so it cannot contain a right angle. It must be oblique.

46. False. Two sides and one opposite angle do not necessarily determine a unique triangle.

47. (a) $\dfrac{\sin \alpha}{9} = \dfrac{\sin \beta}{18}$

$\sin \alpha = 0.5 \sin \beta$

$\alpha = \arcsin(0.5 \sin \beta)$

(b)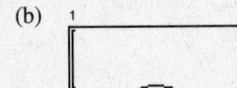

Domain: $0 < \beta < \pi$

Range: $0 < \alpha \le \dfrac{\pi}{6}$

(c) $\gamma = \pi - \alpha - \beta = \pi - \beta - \arcsin(0.5 \sin \beta)$

$\dfrac{c}{\sin \gamma} = \dfrac{18}{\sin \beta}$

$c = \dfrac{18 \sin \gamma}{\sin \beta} = \dfrac{18 \sin[\pi - \beta - \arcsin(0.5 \sin \beta)]}{\sin \beta}$

(d)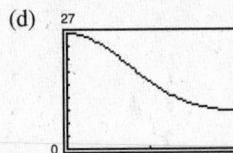

Domain: $0 < \beta < \pi$

Range: $9 < c < 27$

(e)

β	0.4	0.8	1.2	1.6	2.0	2.4	2.8
α	0.1960	0.3669	0.4848	0.5234	0.4720	0.3445	0.1683
c	25.95	23.07	19.19	15.33	12.29	10.31	9.27

As $\beta \to 0$, $c \to 27$

As $\beta \to \pi$, $c \to 9$

48. (a) $A = \dfrac{1}{2}(30)(20) \sin\left(\theta + \dfrac{\theta}{2}\right) - \dfrac{1}{2}(8)(20) \sin \dfrac{\theta}{2} - \dfrac{1}{2}(8)(30) \sin \theta$

$= 300 \sin \dfrac{3\theta}{2} - 80 \sin \dfrac{\theta}{2} - 120 \sin \theta$

$= 20\left[15 \sin \dfrac{3\theta}{2} - 4 \sin \dfrac{\theta}{2} - 6 \sin \theta \right]$

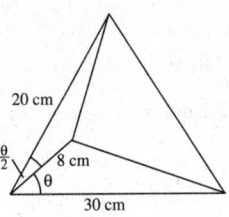

(b)

(c) Domain: $0 \le \theta \le 1.6690$

The domain would increase in length and the area would have a greater maximum value if the 8-centimeter line segment were decreased.

49. $\sin x \cot x = \sin x \dfrac{\cos x}{\sin x} = \cos x$

50. $\tan x \cos x \sec x = \tan x \cos x \dfrac{1}{\cos x} = \tan x$

51. $1 - \sin^2\left(\dfrac{\pi}{2} - x\right) = 1 - \cos^2 x = \sin^2 x$

52. $1 + \cot^2\left(\dfrac{\pi}{2} - x\right) = 1 + \tan^2 x = \sec^2 x$

53. $6 \sin 8\theta \cos 3\theta = (6)\left(\dfrac{1}{2}\right)[\sin(8\theta + 3\theta) + \sin(8\theta - 3\theta)]$

$= 3(\sin 11\theta + \sin 5\theta)$

54. $2 \cos 5\theta \sin 2\theta = 2 \cdot \dfrac{1}{2}[\sin(5\theta + 2\theta) - \sin(5\theta - 2\theta)]$

$= \sin 7\theta - \sin 3\theta$

Section 6.2   Law of Cosines

- If ABC is any oblique triangle with sides a, b, and c, the following equations are valid.

  (a) $a^2 = b^2 + c^2 - 2bc \cos A$ or $\cos A = \dfrac{b^2 + c^2 - a^2}{2bc}$

  (b) $b^2 = a^2 + c^2 - 2ac \cos B$ or $\cos B = \dfrac{a^2 + c^2 - b^2}{2ac}$

  (c) $c^2 = a^2 + b^2 - 2ab \cos C$ or $\cos C = \dfrac{a^2 + b^2 - c^2}{2ab}$

- You should be able to use the Law of Cosines to solve an oblique triangle for the remaining three parts, given:

  (a) Three sides (SSS)

  (b) Two sides and their included angle (SAS)

- Given any triangle with sides of length a, b, and c, the area of the triangle is

  $$\text{Area} = \sqrt{s(s-a)(s-b)(s-c)}, \text{ where } s = \frac{a+b+c}{2}. \qquad \text{(Heron's Formula)}$$

Vocabulary Check

1. Cosines

2. $b^2 = a^2 + c^2 - 2ac \cos B$

3. Heron's Area

---

1. Given: $a = 7, b = 10, c = 15$

$$\cos C = \frac{a^2 + b^2 - c^2}{2ab} = \frac{49 + 100 - 225}{2(7)(10)} \approx -0.5429 \implies C \approx 122.88°$$

$$\sin B = \frac{b \sin C}{c} = \frac{10 \sin 122.88°}{15} \approx 0.5599 \implies B \approx 34.05°$$

$$A \approx 180° - 34.05° - 122.88° \approx 23.07°$$

2. Given: $a = 8, b = 3, c = 9$

$$\cos C = \frac{a^2 + b^2 - c^2}{2ab} = \frac{8^2 + 3^2 - 9^2}{2(8)(3)} \approx -0.16667 \implies C = 99.59°$$

$$\sin A = \frac{a \sin C}{c} = \frac{8 \sin 99.59°}{9} = 0.8765 \implies A = 61.22°$$

$$B \approx 180° - 61.22° - 99.59° \approx 19.19°$$

3. Given: $A = 30°, \ b = 15, \ c = 30$

$$a^2 = b^2 + c^2 - 2bc \cos A$$

$$= 225 + 900 - 2(15)(30) \cos 30° \approx 345.5771$$

$$a \approx 18.59$$

$$\cos C = \frac{a^2 + b^2 - c^2}{2ab} \approx \frac{(18.59)^2 + 15^2 - 30^2}{2(18.59)(15)} \approx -0.5907 \implies C \approx 126.21°$$

$$B \approx 180° - 30° - 126.21° = 13.79°$$

4. Given: $C = 105°$, $a = 10$, $b = 4.5$

$c^2 = a^2 + b^2 - 2ab \cos C = 10^2 + 4.5^2 - 2(10)(4.5) \cos 105° \approx 143.5437 \implies c \approx 11.98$

$\cos B = \dfrac{a^2 + c^2 - b^2}{2ac} \approx \dfrac{10^2 + (12.0)^2 - (4.5)^2}{2(10)(12.0)} \approx 0.93187 \implies B \approx 21.27°$

$A = 180° - 105° - 21.27° \approx 53.73°$

5. $a = 11$, $b = 14$, $c = 20$

$\cos C = \dfrac{a^2 + b^2 - c^2}{2ab} = \dfrac{121 + 196 - 400}{2(11)(14)} \approx -0.2695 \implies C \approx 105.63°$

$\sin B = \dfrac{b \sin C}{c} = \dfrac{14 \sin 105.63°}{20} \approx 0.6741 \implies B \approx 42.38°$

$A \approx 180° - 42.38° - 105.63° \approx 31.99°$

6. Given: $a = 55$, $b = 25$, $c = 72$

$\cos C = \dfrac{a^2 + b^2 - c^2}{2ab} = \dfrac{55^2 + 25^2 - 72^2}{2(55)(25)} \approx -0.5578 \implies C \approx 123.91°$

$\cos A = \dfrac{b^2 + c^2 - a^2}{2bc} = \dfrac{25^2 + 72^2 - 55^2}{2(25)(72)} \approx 0.7733 \implies A \approx 39.35°$

$B = 180° - 123.91° - 39.35° \approx 16.74°$

7. Given: $a = 75.4$, $b = 52$, $c = 52$

$\cos A = \dfrac{b^2 + c^2 - a^2}{2bc} = \dfrac{52^2 + 52^2 - 75.4^2}{2(52)(52)} = -0.05125 \implies A \approx 92.94°$

$\sin B = \dfrac{b \sin A}{a} \approx \dfrac{52(0.9987)}{75.4} \approx 0.68876 \implies B \approx 43.53°$

$C = B \approx 43.53°$

8. Given: $a = 1.42$, $b = 0.75$, $c = 1.25$

$\cos A = \dfrac{b^2 + c^2 - a^2}{2bc} = \dfrac{(0.75)^2 + (1.25)^2 - (1.42)^2}{2(0.75)(1.25)} = 0.05792 \implies A \approx 86.68°$

$\cos B = \dfrac{a^2 + c^2 - b^2}{2ac} = \dfrac{(1.42)^2 + (1.25)^2 - (0.75)^2}{2(1.42)(1.25)} \approx 0.84969 \implies B \approx 31.82°$

$C = 180° - 86.68° - 31.82° \approx 61.50°$

9. Given: $A = 135°$, $b = 4$, $c = 9$

$a^2 = b^2 + c^2 - 2bc \cos A = 16 + 81 - 2(4)(9)\cos 135° \approx 147.9117 \implies a \approx 12.16$

$\sin B = \dfrac{b \sin A}{a} = \dfrac{4 \sin 135°}{12.16} \approx 0.2326 \implies B \approx 13.45°$

$C \approx 180° - 135° - 13.45° \approx 31.55°$

10. Given: $A = 55°$, $b = 3$, $c = 10$

$a^2 = b^2 + c^2 - 2bc \cos A = 3^2 + 10^2 - 2(3)(10) \cos 55° \approx 74.585 \implies a \approx 8.64$

$\sin B = \dfrac{b \sin A}{a} \approx \dfrac{3 \sin 55°}{8.636} \approx 0.2846 \implies A \approx 16.53°$

$C \approx 180° - 16.53° - 55° \approx 108.47°$

11. Given: $B = 10° 35'$, $a = 40$, $c = 30$

$b^2 = a^2 + c^2 - 2ac \cos B = 1600 + 900 - 2(40)(30)\cos 10° 35' \approx 140.8268 \implies b \approx 11.87$

$\sin C = \dfrac{c \sin B}{b} = \dfrac{30 \sin 10° 35'}{11.87} \approx 0.4642 \implies C \approx 27.66° \approx 27° 40'$

$A \approx 180° - 10° 35' - 27° 40' = 141° 45'$

12. Given: $B = 75° 20'$, $a = 6.2$, $c = 9.5$

$b^2 = a^2 + c^2 - 2ac \cos B = (6.2)^2 + (9.5)^2 - 2(6.2)(9.5) \cos 75° 20' \approx 98.8636 \implies b \approx 9.94$

$\sin A = \dfrac{a \sin B}{b} \approx \dfrac{6.2 \sin 75° 20'}{9.94} \approx 0.6034 \implies A \approx 37.1°, \text{ or } 37° 6'$

$C \approx 180° - 75° 20' - 37° 6' \approx 67° 34'$

13. Given: $B = 125° 40'$, $a = 32$, $c = 32$

$b^2 = a^2 + c^2 - 2ac \cos B \approx 32^2 + 32^2 - 2(32)(32) \cos 125° 40' \approx 3242.1888 \implies b \approx 56.94$

$A = C \implies 2A = 180° - 125° 40' = 54° 20' \implies A = C = 27° 10'$

14. Given: $C = 15° 15'$, $a = 6.25$, $b = 2.15$

$c^2 = a^2 + b^2 - 2ab \cos C = (6.25)^2 + (2.15)^2 - 2(6.25)(2.15) \cos 15° 15' \approx 17.7563 \implies c \approx 4.21$

$\cos A = \dfrac{b^2 + c^2 - a^2}{2bc} \approx \dfrac{(2.15)^2 + (4.2138)^2 - (6.25)^2}{2(2.15)(4.2138)} \approx -0.9208 \implies A \approx 157.04° \text{ or } 157° 2'$

$B \approx 180° - 15° 15' - 157.04° \approx 7.7° \text{ or } 7° 43'$

15. $C = 43°$, $a = \dfrac{4}{9}$, $b = \dfrac{7}{9}$

$c^2 = a^2 + b^2 - 2ab \cos C = \left(\dfrac{4}{9}\right)^2 + \left(\dfrac{7}{9}\right)^2 - 2\left(\dfrac{4}{9}\right)\left(\dfrac{7}{9}\right)\cos 43° \approx 0.2968 \implies c \approx 0.54$

$\sin A = \dfrac{a \sin C}{c} = \dfrac{(4/9) \sin 43°}{0.5448} \approx 0.5564 \implies A \approx 33.80°$

$B \approx 180° - 43° - 33.8° \approx 103.20°$

16. Given: $C = 103°$, $a = \dfrac{3}{8}$, $b = \dfrac{3}{4}$

$c^2 = a^2 + b^2 - 2ab \cos C = \left(\dfrac{3}{8}\right)^2 + \left(\dfrac{3}{4}\right)^2 - 2\left(\dfrac{3}{8}\right)\left(\dfrac{3}{4}\right)\cos 103° \approx 0.8297 \implies c \approx 0.91$

$\cos A = \dfrac{b^2 + c^2 - a^2}{2bc} \approx \dfrac{\left(\dfrac{3}{4}\right)^2 + (0.91)^2 - \left(\dfrac{3}{8}\right)^2}{2\left(\dfrac{3}{4}\right)(0.91)} \approx 0.9160 \implies A \approx 23.65°$

$B \approx 180° - 23.65° - 103° \approx 53.35°$

17. $d^2 = 5^2 + 8^2 - 2(5)(8)\cos 45° \approx 32.4315 \implies d \approx 5.69$

$2\phi = 360° - 2(45°) = 270° \implies \phi = 135°$

$c^2 = 5^2 + 8^2 - 2(5)(8)\cos 135° \approx 145.5685 \implies c \approx 12.07$

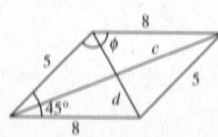

18.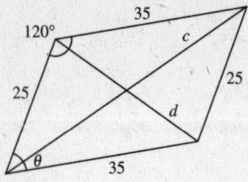

$$c^2 = 25^2 + 35^2 - 2(25)(35) \cos 120°$$

$$= 2725 \implies c \approx 52.20$$

$$\theta = \frac{1}{2}[360° - 2(120°)] = 60°$$

$$d^2 = 25^2 + 35^2 - 2(25)(35) \cos 60°$$

$$= 975 \implies d \approx 31.22$$

19.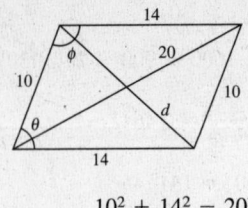

$$\cos \phi = \frac{10^2 + 14^2 - 20^2}{2(10)(14)}$$

$$\phi \approx 111.8°$$

$$2\theta \approx 360° - 2(111.8°)$$

$$\theta = 68.2°$$

$$d^2 = 10^2 + 14^2 - 2(10)(14) \cos 68.2°$$

$$d \approx 13.86$$

20. $\cos \theta = \dfrac{40^2 + 60^2 - 80^2}{2(40)(60)} = -\dfrac{1}{4} \implies \theta \approx 104.5°$

$$\phi \approx \frac{1}{2}[360° - 2(104.5°)] \approx 75.5°$$

$$c^2 \approx 40^2 + 60^2 - 2(40)(60) \cos 75.5° = 3998$$

$$c \approx 63.23$$

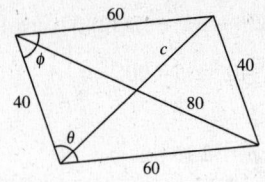

21. $\cos \alpha = \dfrac{(12.5)^2 + (15)^2 - 10^2}{2(12.5)(15)} = 0.75 \implies \alpha \approx 41.41°$

$$\cos \beta = \frac{10^2 + 15^2 - (12.5)^2}{2(10)(15)} = 0.5625 \implies \beta \approx 55.77°$$

$$z = 180° - \alpha - \beta = 82.82°$$

$$u = 180° - z = 97.18°$$

$$b^2 = 12.5^2 + 10^2 - 2(12.5)(10)\cos 97.18° \approx 287.4967 \implies b \approx 16.96$$

$$\cos \delta = \frac{12.5^2 + 16.96^2 - 10^2}{2(12.5)(16.96)} \approx 0.8111 \implies \delta \approx 35.80°$$

$$\theta = \alpha + \delta = 41.41° + 35.80° \approx 77.2°$$

$$2\phi = 360° - 2\theta \implies \phi = \frac{360° - 2(77.21°)}{2} = 102.8°$$

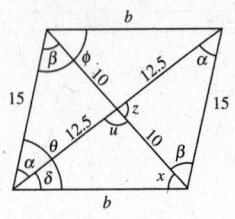

22. $\cos \alpha = \dfrac{25^2 + 17.5^2 - 25^2}{2(25)(17.5)}$

$$\alpha \approx 69.512°$$

$$\beta \approx 180 - \alpha \approx 110.488°$$

$$a^2 = 17.5^2 + 25^2 - 2(17.5)(25) \cos 110.488°$$

$$a \approx 35.18$$

$$z = 180 - 2\alpha \approx 40.975°$$

$$\cos \mu = \frac{25^2 + 35.18^2 - 17.5^2}{2(25)(35.18)}$$

$$\mu \approx 27.775°$$

$$\theta = \mu + z \approx 68.7°$$

$$\omega = 180° - \mu - \beta \approx 41.738°$$

$$\phi = \omega + \alpha \approx 111.3°$$

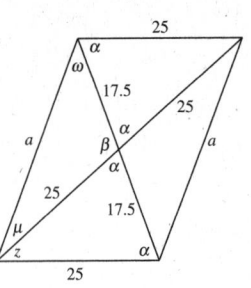

23. $a = 5, \ b = 7, \ c = 10 \implies s = \dfrac{a + b + c}{2} = 11$

Area $= \sqrt{s(s - a)(s - b)(s - c)} = \sqrt{11(6)(4)(1)} \approx 16.25$

24. $a = 12, \ b = 15, \ c = 9 \implies s = \dfrac{12 + 15 + 9}{2} = 18$

Area $= \sqrt{18(6)(3)(9)} = 54$

25. $a = 2.5, b = 10.2, c = 9 \implies s = \dfrac{a + b + c}{2} = 10.85$

Area $= \sqrt{s(s - a)(s - b)(s - c)} = \sqrt{10.85(8.35)(0.65)(1.85)} \approx 10.4$

26. Given: $a = 75.4, b = 52, c = 52$

$s = \dfrac{75.4 + 52 + 52}{2} = 89.7$

Area $= \sqrt{s(s - a)(s - b)(s - c)} = \sqrt{89.7(14.3)(37.7)(37.7)} \approx 1350.2$

27. $a = 12.32, b = 8.46, c = 15.05 \implies s = \dfrac{a + b + c}{2} = 17.915$

Area $= \sqrt{s(s - a)(s - b)(s - c)} = \sqrt{17.915(5.595)(9.455)(2.865)} \approx 52.11$

28. Given: $a = 3.05, b = 0.75, c = 2.45$

$s = \dfrac{3.05 + 0.75 + 2.45}{2} = 3.125$

Area $= \sqrt{s(s - a)(s - b)(s - c)} = \sqrt{3.125(0.075)(2.375)(0.675)} \approx 0.61$

29. $\cos B = \dfrac{1700^2 + 3700^2 - 3000^2}{2(1700)(3700)} \implies B \approx 52.9°$

Bearing: $90° - 52.9° = $ N $37.1°$ E

$\cos C = \dfrac{1700^2 + 3000^2 - 3700^2}{2(1700)(3000)} \implies C \approx 100.2°$

Bearing: $90° - 26.9° = $ S $63.1°$ E

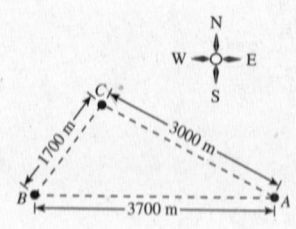

30. Distance from Franklin to Rosemount:

$d = \sqrt{810^2 + 648^2 - 2(810)(648)\cos(137°)}$

≈ 1357.8 miles

Bearing from Franklin to Rosemount:

N $(75° - \theta)$ E

$\cos \theta \approx \dfrac{(1357.8)^2 + 810^2 - 648^2}{2(1357.8)(810)}$

≈ 0.9456

$\theta \approx 19.0°$

Bearing from Franklin to Rosemont: N $56.0°$ E

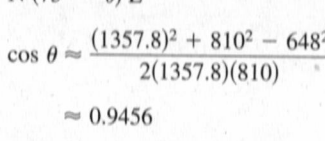

31. $b^2 = 220^2 + 250^2 - 2(220)(250)\cos 105° \implies b \approx 373.3$ meters

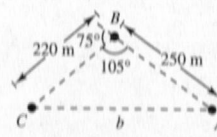

32.

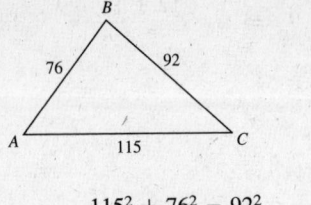

$$\cos A = \frac{115^2 + 76^2 - 92^2}{2(115)(76)} \approx 0.6028 \implies A \approx 52.9°$$

$$\cos C = \frac{115^2 + 92^2 - 76^2}{2(115)(92)} \approx 0.75203 \implies c \approx 41.2°$$

33.

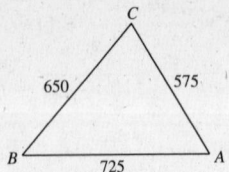

The largest angle is across from the largest side.

$$\cos C = \frac{650^2 + 575^2 - 725^2}{2(650)(575)}$$

$$C \approx 72.3°$$

34. $\cos \theta = \dfrac{2^2 + 3^2 - (4.5)^2}{2(2)(3)} \approx -0.60417$

$\theta \approx 127.2°$

35. $C = 180° - 53° - 67° = 60°$

$c^2 = a^2 + b^2 - 2ab \cos C$

$\quad = 36^2 + 48^2 - 2(36)(48)(0.5)$

$\quad = 1872$

$c \approx 43.3$ mi

36. The angles at the base of the tower are 96° and 84°. The longer guy wire g_1 is given by:

$g_1{}^2 = 75^2 + 100^2 - 2(75)(100) \cos 96°$

$\quad \approx 17{,}192.9 \implies g_1 \approx 131.1$ feet

The shorter guy wire g_2 is given by:

$g_2{}^2 = 75^2 + 100^2 - 2(75)(100) \cos 84$

$\quad \approx 14{,}057.1 \implies g_2 \approx 118.6$ feet

37. (a) $\cos \theta = \dfrac{273^2 + 178^2 - 235^2}{2(273)(178)}$

$\quad \theta \approx 58.4°$

Bearing: N 58.4° W

(b) $\cos \phi = \dfrac{235^2 + 178^2 - 273^2}{2(235)(178)}$

$\quad \phi \approx 81.5°$

Bearing: S 81.5° W

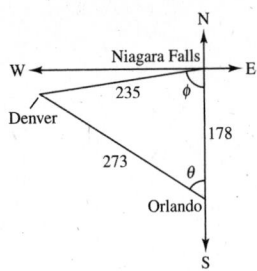

38.

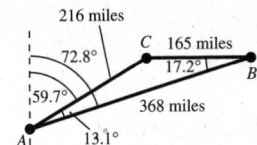

$a = 165, b = 216, c = 368$

$\cos B = \dfrac{165^2 + 368^2 - 216^2}{2(165)(368)} \approx 0.9551$

$B \approx 17.2°$

$\cos A = \dfrac{216^2 + 368^2 - 165^2}{2(216)(368)} \approx 0.9741$

$A \approx 13.1°$

(a) Bearing of Minneapolis (C) from Phoenix (A)

$\dot{N} (90° - 17.2° - 13.1°) E$

N 59.7 E

(b) Bearing of Albany (B) from Phoenix (A)

N $(90° - 17.2°)$ E

N 72.8° E

39. $d^2 = 60.5^2 + 90^2 - 2(60.5)(90) \cos 45° \approx 4059.8572 \implies d \approx 63.7$ ft

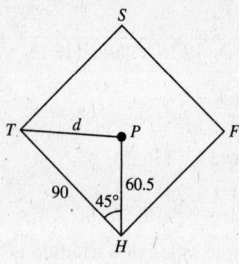

40. $d = \sqrt{330^2 + 420^2 - 2(330)(420) \cos 8°}$

≈ 103.9 feet

41. $a^2 = 35^2 + 20^2 - 2(35)(20)\cos 42° \implies a \approx 24.2$ miles

42. $a = \sqrt{20^2 + 20^2 - 2(20)(20) \cos 11°}$

≈ 3.8 miles

43. $\overline{RS} = \sqrt{8^2 + 10^2} = \sqrt{164} = 2\sqrt{41} \approx 12.8$ ft

$\overline{PQ} = \frac{1}{2}\sqrt{16^2 + 10^2} = \frac{1}{2}\sqrt{356} = \sqrt{89} \approx 9.4$ ft

$\tan P = \dfrac{10}{16} = \dfrac{\overline{QS}}{\overline{PS}} = \dfrac{\overline{QS}}{8} \implies \overline{QS} = 5$

44. (a)
$$7^2 = 1.5^2 + x^2 - 2(1.5)\,x \cos \theta$$
$$49 = 2.25 + x^2 - 3x \cos \theta$$
$$x^2 - 3x \cos \theta - 46.75 = 0$$

(b) $x = \dfrac{3 \cos \theta \pm \sqrt{(-3 \cos \theta)^2 - 4(1)(-46.75)}}{2(1)}$

$x = \dfrac{1}{2}\left(3 \cos \theta + \sqrt{9 \cos^2 \theta + 187}\right)$

(c)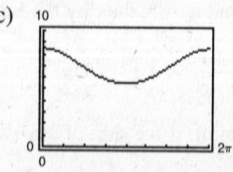

(d) Maximum: 8.5 inches

45. $d^2 = 10^2 + 7^2 - 2(10)(7) \cos \theta$

$\theta = \arccos\left[\dfrac{10^2 + 7^2 - d^2}{2(10)(7)}\right]$

$s = \dfrac{360° - \theta}{360°}(2\pi r) = \dfrac{(360° - \theta)\pi}{45°}$

d (inches)	9	10	12	13	14	15	16
θ (degrees)	60.9°	69.5°	88.0°	98.2°	109.6°	122.9°	139.8°
s (inches)	20.88	20.28	18.99	18.28	17.48	16.55	15.37

46.

$\dfrac{x}{\sin 20°} = \dfrac{10}{\sin 120°}$

$x = \dfrac{10 \sin 20°}{\sin 120°} \approx 3.95$ feet

47. $a = 200$

$b = 500$

$c = 600 \implies s = \dfrac{200 + 500 + 600}{2} = 650$

Area $= \sqrt{650(450)(150)(50)} \approx 46{,}837.5$ square feet

48. area $= 2\left[\dfrac{1}{2}(70)(100)\sin 70°\right]$

≈ 6577.8 square meters

(The area of the parallelogram is the sum of the areas of two triangles.)

49. $s = \dfrac{510 + 840 + 1120}{2} = 1235$

Area $= \sqrt{1235(1235 - 510)(1235 - 840)(1235 - 1120)}$

$\approx 201{,}674$ square yards

Cost $\approx \left(\dfrac{201{,}674}{4840}\right)(2000) \approx \$83{,}336.36$

50. area $= \sqrt{s(s - a)(s - b)(s - c)}$

$s = \dfrac{(a + b + c)}{2} = \dfrac{(2490 + 1860 + 1350)}{2} = 2850$

area $= \sqrt{(2850(360)(990)(1500)} = 1234346.0 \text{ ft}^2$

$\dfrac{1234346.0 \text{ ft}^2}{(43560 \text{ ft}^2/\text{acre})} = 28.33669 \text{ acre}$

$(28.33669 \text{ acre})(\$2200/\text{acre}) = \$62{,}340.71$

51. False. The average of the three sides of a triangle is

$\dfrac{a + b + c}{3}$, not $\dfrac{a + b + c}{2} = s$.

52. False. To solve an SSA triangle, the Law of Sines is needed.

53. False. If $a = 10$, $b = 16$, and $c = 5$, then by the Law of Cosines, we would have:

$\cos A = \dfrac{16^2 + 5^2 - 10^2}{2(16)(5)} = 1.13125 > 1$

This is not possible. In general, if the sum of any two sides is less than the third side, then they cannot form a triangle. Here $10 + 5$ is less than 16.

54. (a) Working with $\triangle ODC$, we have $\cos \alpha = \dfrac{a/2}{R}$.

This implies that $2R = \dfrac{a}{\cos \alpha}$.

Since we know that

$\dfrac{a}{\sin A} = \dfrac{b}{\sin B} = \dfrac{c}{\sin C}$,

we can complete the proof by showing that $\cos \alpha = \sin A$. The solution of the system

$A + B + C = 180°$

$\alpha - C + A = \beta$

$\alpha + \beta = B$

is $\alpha = 90° - A$. Therefore:

$2R = \dfrac{a}{\cos \alpha} = \dfrac{a}{\cos(90° - A)} = \dfrac{a}{\sin A}$.

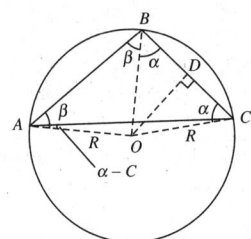

(b) By Heron's Formula, the area of the triangle is

Area $= \sqrt{s(s - a)(s - b)(s - c)}$.

We can also find the area by dividing the area into six triangles and using the fact that the area is $\frac{1}{2}$ the base times the height. Using the figure as given, we have

Area $= \dfrac{1}{2}xr + \dfrac{1}{2}xr + \dfrac{1}{2}yr + \dfrac{1}{2}yr + \dfrac{1}{2}zr + \dfrac{1}{2}zr$

$= r(x + y + z)$

$= rs.$

Therefore: $rs = \sqrt{s(s - a)(s - b)(s - c)} \Longrightarrow$

$r = \sqrt{\dfrac{(s - a)(s - b)(s - c)}{s}}.$

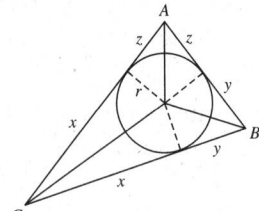

55. $a = 25, b = 55, c = 72$

(a) Area of triangle: $s = \dfrac{1}{2}(25 + 55 + 72) = 76$

Area $= \sqrt{76(51)(21)(4)} \approx 570.60$

(c) Area of inscribed circle:

$$r = \sqrt{\dfrac{(s-a)(s-b)(s-c)}{s}}$$

$$= \sqrt{\dfrac{(51)(21)(4)}{76}} \approx 7.51 \quad \text{(see \#54)}$$

Area $= \pi r^2 \approx 177.09$

(b) Area of circumscribed circle:

$$\cos C = \dfrac{25^2 + 55^2 - 72^2}{2(25)(55)} \approx -0.5578 \Rightarrow C \approx 123.9°$$

$$R = \dfrac{1}{2}\left(\dfrac{c}{\sin C}\right) \approx 43.37 \quad \text{(see \#54)}$$

Area $= \pi R^2 \approx 5909.2$

56. Given: $a = 200$ ft, $b = 250$ ft, $c = 325$ ft

$$s = \dfrac{200 + 250 + 325}{2} \approx 387.5$$

Radius of the inscribed circle: $r = \sqrt{\dfrac{(s-a)(s-b)(s-c)}{s}} = \sqrt{\dfrac{(187.5)(137.5)(62.5)}{387.5}} \approx 64.5$ ft (see \#54)

Circumference of an inscribed circle: $C = 2\pi r \approx 2\pi(64.5) \approx 405.2$ ft

57. $\dfrac{1}{2}bc(1 + \cos A) = \dfrac{1}{2}bc\left[1 + \dfrac{b^2 + c^2 - a^2}{2bc}\right]$

$\qquad = \dfrac{1}{2}bc\left[\dfrac{2bc + b^2 + c^2 - a^2}{2bc}\right]$

$\qquad = \dfrac{1}{4}\left[(b+c)^2 - a^2\right]$

$\qquad = \dfrac{1}{4}\left[(b+c) + a\right]\left[(b+c) - a\right]$

$\qquad = \dfrac{b+c+a}{2} \cdot \dfrac{b+c-a}{2}$

$\qquad = \dfrac{a+b+c}{2} \cdot \dfrac{-a+b+c}{2}$

58. $\dfrac{1}{2}bc(1 - \cos A) = \dfrac{1}{2}bc\left[1 + \dfrac{a^2 - (b^2 + c^2)}{2bc}\right]$

$\qquad = \dfrac{1}{2}bc\left[\dfrac{2bc + a^2 - b^2 - c^2}{2bc}\right]$

$\qquad = \dfrac{a^2 - (b^2 - 2bc + c^2)}{4}$

$\qquad = \dfrac{a^2 - (b-c)^2}{4}$

$\qquad = \left(\dfrac{a - (b-c)}{2}\right)\left(\dfrac{a + (b-c)}{2}\right)$

$\qquad = \dfrac{a - b + c}{2} \cdot \dfrac{a + b - c}{2}$

59. $\arcsin(-1) = -\dfrac{\pi}{2}$

60. $\arccos 0 = \dfrac{\pi}{2}$

61. $\arctan\sqrt{3} = \dfrac{\pi}{3}$

62. $\arctan(-\sqrt{3}) = -\arctan\sqrt{3}$

$\qquad = -\dfrac{\pi}{3}$

63. $\arcsin\left(-\dfrac{\sqrt{3}}{2}\right) = -\dfrac{\pi}{3}$

64. $\arccos\left(-\dfrac{\sqrt{3}}{2}\right) = \pi - \arccos\dfrac{\sqrt{3}}{2}$

$\qquad = \pi - \dfrac{\pi}{6} = \dfrac{5\pi}{6}$

65. Let $\theta = \arcsin 2x$, then

$\sin\theta = 2x = \dfrac{2x}{1}$ and

$\sec\theta = \dfrac{1}{\sqrt{1 - 4x^2}}$.

66. Let $u = \arccos 3x$

$\cos u = 3x = \dfrac{3x}{1}$.

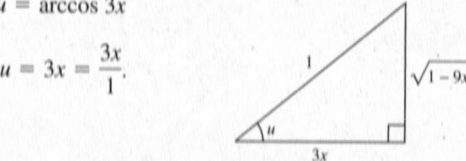

$$\tan(\arccos 3x) = \tan u = \dfrac{\sqrt{1 - 9x^2}}{3x}$$

67. Let $\theta = \arctan(x - 2)$, then

$$\tan \theta = x - 2 = \frac{x - 2}{1} \text{ and}$$

$$\cot \theta = \frac{1}{x - 2}.$$

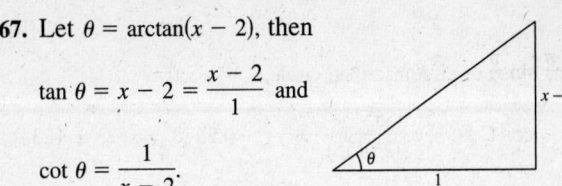

68. Let $u = \arcsin \dfrac{x - 1}{2}$,

$$\sin u = \frac{x - 1}{2}.$$

$$\cos\left(\arcsin \frac{x - 1}{2}\right) = \cos u$$

$$= \frac{\sqrt{4 - (x - 1)^2}}{2}$$

69. $5 = \sqrt{25 - x^2}, x = 5 \sin \theta$

$$5 = \sqrt{25 - (5 \sin \theta)^2}$$

$$5 = \sqrt{25(1 - \sin^2 \theta)}$$

$$5 = 5 \cos \theta$$

$$\cos \theta = 1$$

$$\sec \theta = \frac{1}{\cos \theta} = 1$$

$\csc \theta$ is undefined.

70. $x = 2 \cos \theta, -\dfrac{\pi}{2} < \theta < \dfrac{\pi}{2}$

$$-\sqrt{2} = \sqrt{4 - x^2}$$

$$-\sqrt{2} = \sqrt{4 - (2 \cos \theta)^2}$$

$$-\sqrt{2} = \sqrt{4 - 4 \cos^2 \theta}$$

$$-\sqrt{2} = \sqrt{4(1 - \cos^2 \theta)}$$

$$-\sqrt{2} = \sqrt{4 \sin^2 \theta}$$

$$-\sqrt{2} = 2 \sin \theta$$

$$-\frac{\sqrt{2}}{2} = \sin \theta \Longrightarrow \cos \theta = \frac{\sqrt{2}}{2}$$

$$\sec \theta = \frac{1}{\cos \theta} = \frac{1}{\sqrt{2}/2} = \sqrt{2}$$

$$\csc \theta = \frac{1}{\sin \theta} = \frac{1}{-\sqrt{2}/2} = -\sqrt{2}$$

71. $-\sqrt{3} = \sqrt{x^2 - 9}, x = 3 \sec \theta$

$$-\sqrt{3} = \sqrt{(3 \sec \theta)^2 - 9}$$

$$-\sqrt{3} = \sqrt{9(\sec^2 \theta - 1)}$$

$$-\sqrt{3} = 3 \tan \theta$$

$$\tan \theta = -\frac{\sqrt{3}}{3}$$

$$\sec \theta = \sqrt{1 + \tan^2 \theta} = \sqrt{1 + \left(-\frac{\sqrt{3}}{3}\right)^2} = \frac{2\sqrt{3}}{3}$$

$$\cot \theta = \frac{1}{\tan \theta} = -\sqrt{3}$$

$$\csc \theta = -\sqrt{1 + \cot^2 \theta} = -\sqrt{1 + \left(-\sqrt{3}\right)^2} = -2$$

72. $x = 6 \tan \theta, -\dfrac{\pi}{2} < \theta < \dfrac{\pi}{2}$

$$12 = \sqrt{36 + x^2}$$

$$12 = \sqrt{36 + (6 \tan \theta)^2}$$

$$12 = \sqrt{36 + 36 \tan^2 \theta}$$

$$12 = \sqrt{36(1 + \tan^2 \theta)}$$

$$12 = \sqrt{36 \sec^2 \theta}$$

$$12 = 6 \sec \theta$$

$$2 = \sec \theta$$

$$\cos \theta = \frac{1}{2}$$

$$\sin^2 \theta + \left(\frac{1}{2}\right)^2 = 1$$

$$\sin^2 \theta = 1 - \frac{1}{4} = \frac{3}{4}$$

$$\sin \theta = \pm \sqrt{\frac{3}{4}} = \pm \frac{\sqrt{3}}{2}$$

$$\csc \theta = \frac{1}{\sin \theta} = \frac{1}{\pm \sqrt{3}/2} = \pm \frac{2}{\sqrt{3}} = \pm \frac{2\sqrt{3}}{3}$$

73. $\cos\dfrac{5\pi}{6} - \cos\dfrac{\pi}{3} = -2\sin\left(\dfrac{\dfrac{5\pi}{6} + \dfrac{\pi}{3}}{2}\right)\sin\left(\dfrac{\dfrac{5\pi}{6} - \dfrac{\pi}{3}}{2}\right) = -2\sin\dfrac{7\pi}{12}\sin\dfrac{\pi}{4}$

74. $\sin\left(x - \dfrac{\pi}{2}\right) - \sin\left(x + \dfrac{\pi}{2}\right) = 2\cos\left(\dfrac{x - \dfrac{\pi}{2} + x + \dfrac{\pi}{2}}{2}\right)\sin\left(\dfrac{x - \dfrac{\pi}{2} - \left(x + \dfrac{\pi}{2}\right)}{2}\right)$

$$= 2\cos\left(\dfrac{2x}{2}\right)\sin\left(\dfrac{-\pi}{2}\right)$$

$$= 2\cos x \sin\left(-\dfrac{\pi}{2}\right)$$

Section 6.3   Vectors in the Plane

- A vector **v** is the collection of all directed line segments that are equivalent to a given directed line segment $\overrightarrow{PQ}$.
- You should be able to *geometrically* perform the operations of vector addition and scalar multiplication.
- The component form of the vector with initial point $P = (p_1, p_2)$ and terminal point $Q = (q_1, q_2)$ is
  $$\overrightarrow{PQ} = \langle q_1 - p_1, q_2 - p_2 \rangle = \langle v_1, v_2 \rangle = \mathbf{v}.$$
- The magnitude of $\mathbf{v} = \langle v_1, v_2 \rangle$ is given by $\|\mathbf{v}\| = \sqrt{v_1{}^2 + v_2{}^2}$.
- If $\|\mathbf{v}\| = 1$, **v** is a unit vector.
- You should be able to perform the operations of scalar multiplication and vector addition in component form.
  - (a) $\mathbf{u} + \mathbf{v} = \langle u_1 + v_1, u_2 + v_2 \rangle$ (b) $k\mathbf{u} = \langle ku_1, ku_2 \rangle$
- You should know the following properties of vector addition and scalar multiplication.

  (a) $\mathbf{u} + \mathbf{v} = \mathbf{v} + \mathbf{u}$ (b) $(\mathbf{u} + \mathbf{v}) + \mathbf{w} = \mathbf{u} + (\mathbf{v} + \mathbf{w})$

  (c) $\mathbf{u} + \mathbf{0} = \mathbf{u}$ (d) $\mathbf{u} + (-\mathbf{u}) = \mathbf{0}$

  (e) $c(d\mathbf{u}) = (cd)\mathbf{u}$ (f) $(c + d)\mathbf{u} = c\mathbf{u} + d\mathbf{u}$

  (g) $c(\mathbf{u} + \mathbf{v}) = c\mathbf{u} + c\mathbf{v}$ (h) $1(\mathbf{u}) = \mathbf{u}, \, 0\mathbf{u} = \mathbf{0}$

  (i) $\|c\mathbf{v}\| = |c| \, \|\mathbf{v}\|$

- A unit vector in the direction of **v** is $\mathbf{u} = \dfrac{\mathbf{v}}{\|\mathbf{v}\|}$.
- The standard unit vectors are $\mathbf{i} = \langle 1, 0 \rangle$ and $\mathbf{j} = \langle 0, 1 \rangle$. $\mathbf{v} = \langle v_1, v_2 \rangle$ can be written as $\mathbf{v} = v_1\mathbf{i} + v_2\mathbf{j}$.
- A vector **v** with magnitude $\|\mathbf{v}\|$ and direction θ can be written as $\mathbf{v} = a\mathbf{i} + b\mathbf{j} = \|\mathbf{v}\|(\cos \theta)\mathbf{i} + \|\mathbf{v}\|(\sin \theta)\mathbf{j}$, where $\tan \theta = b/a$.

Vocabulary Check

1. directed line segment

2. initial; terminal

3. magnitude

4. vector

5. standard position

6. unit vector

7. multiplication; addition

8. resultant

9. linear combination; horizontal; vertical

1. $\mathbf{v} = \langle 4 - 0, 1 - 0 \rangle = \langle 4, 1 \rangle$

   $\mathbf{u} = \mathbf{v}$

2. $\mathbf{u} = \langle -3 - 0, -4 - 4 \rangle = \langle -3, -8 \rangle$

   $\mathbf{v} = \langle 0 - 3, -5 - 3 \rangle = \langle -3, -8 \rangle$

   $\mathbf{u} = \mathbf{v}$

3. Initial point: $(0, 0)$

   Terminal point: $(3, 2)$

   $\mathbf{v} = \langle 3 - 0, 2 - 0 \rangle = \langle 3, 2 \rangle$

   $\|\mathbf{v}\| = \sqrt{3^2 + 2^2} = \sqrt{13}$

4. Initial point: $(0, 0)$

   Terminal point: $(-4, -2)$

   $\mathbf{v} = \langle -4 - 0, -2 - 0 \rangle = \langle -4, -2 \rangle$

   $\|\mathbf{v}\| = \sqrt{(-4)^2 + (-2)^2} = \sqrt{20} = 2\sqrt{5}$

5. Initial point: $(2, 2)$

   Terminal point: $(-1, 4)$

   $\mathbf{v} = \langle -1 - 2, 4 - 2 \rangle = \langle -3, 2 \rangle$

   $\|\mathbf{v}\| = \sqrt{(-3)^2 + 2^2} = \sqrt{13}$

6. Initial point: $(-1, -1)$

   Terminal point: $(3, 5)$

   $\mathbf{v} = \langle 3 - (-1), 5 - (-1) \rangle = \langle 4, 6 \rangle$

   $\|\mathbf{v}\| = \sqrt{4^2 + 6^2} = \sqrt{52} = 2\sqrt{13}$

7. Initial point: $(3, -2)$

   Terminal point: $(3, 3)$

   $\mathbf{v} = \langle 3 - 3, 3 - (-2) \rangle = \langle 0, 5 \rangle$

   $\|\mathbf{v}\| = \sqrt{0^2 + 5^2} = \sqrt{25} = 5$

8. Initial point: $(-4, -1)$

   Terminal point: $(3, -1)$

   $\mathbf{v} = \langle 3 - (-4), -1 - (-1) \rangle = \langle 7, 0 \rangle$

   $\|\mathbf{v}\| = \sqrt{7^2 + 0^2} = 7$

9. Initial point: $(-1, 5)$

   Terminal point: $(15, 12)$

   $\mathbf{v} = \langle 15 - (-1), 12 - 5 \rangle = \langle 16, 7 \rangle$

   $\|\mathbf{v}\| = \sqrt{16^2 + 7^2} = \sqrt{305}$

10. Initial point: $(1, 11)$

   Terminal point: $(9, 3)$

   $\mathbf{v} = \langle 9 - 1, 3 - 11 \rangle = \langle 8, -8 \rangle$

   $\|\mathbf{v}\| = \sqrt{8^2 + (-8)^2} = 8\sqrt{2}$

11. Initial point: $(-3, -5)$

   Terminal point: $(5, 1)$

   $\mathbf{v} = \langle 5 - (-3), 1 - (-5) \rangle = \langle 8, 6 \rangle$

   $\|\mathbf{v}\| = \sqrt{8^2 + 6^2} = \sqrt{100} = 10$

12. Initial point: $(-3, 11)$

   Terminal point: $(9, 40)$

   $\mathbf{v} = \langle 9 - (-3), 40 - 11 \rangle = \langle 12, 29 \rangle$

   $\|\mathbf{v}\| = \sqrt{12^2 + 29^2} = \sqrt{985}$

13. Initial point: $(1, 3)$

   Terminal point: $(-8, -9)$

   $\mathbf{v} = \langle -8 - 1, -9 - 3 \rangle = \langle -9, -12 \rangle$

   $\|\mathbf{v}\| = \sqrt{(-9)^2 + (-12)^2} = \sqrt{225} = 15$

14. Initial point: $(-2, 7)$

   Terminal point: $(5, -17)$

   $\mathbf{v} = \langle 5 - (-2), -17 - 7 \rangle = \langle 7, -24 \rangle$

   $\|\mathbf{v}\| = \sqrt{7^2 + (-24)^2} = 25$

15.

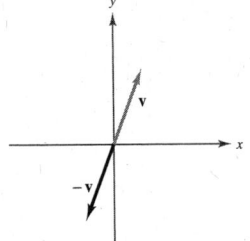

16. $5\mathbf{v}$

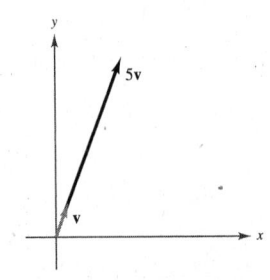

17.

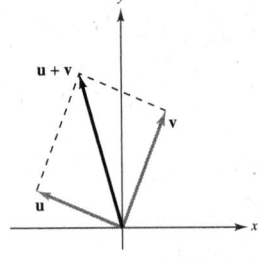

18. **u − v**

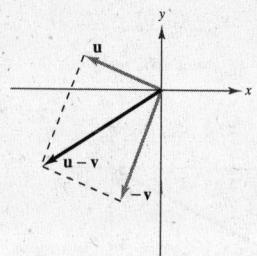

19. **u + 2v**

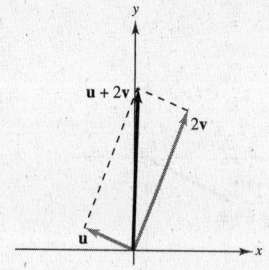

20. **v − ½u**

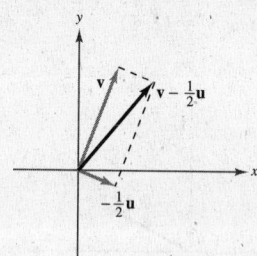

21. **u** = ⟨2, 1⟩, **v** = ⟨1, 3⟩

(a) **u** + **v** = ⟨3, 4⟩

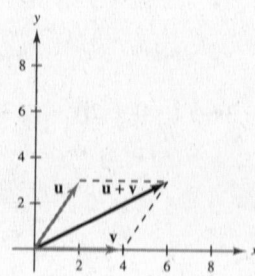

(b) **u** − **v** = ⟨1, −2⟩

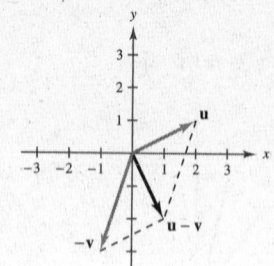

(c) 2**u** − 3**v** = ⟨4, 2⟩ − ⟨3, 9⟩ = ⟨1, −7⟩

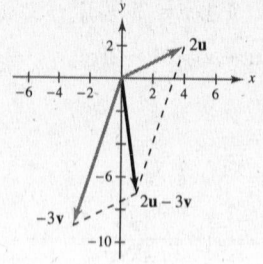

22. **u** = ⟨2, 3⟩, **v** = ⟨4, 0⟩

(a) **u** + **v** = ⟨6, 3⟩

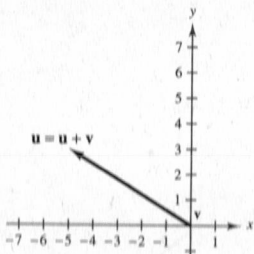

(b) **u** − **v** = ⟨−2, 3⟩

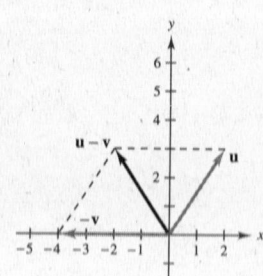

(c) 2**u** − 3**v** = ⟨4, 6⟩ − ⟨12, 0⟩

$$= ⟨−8, 6⟩$$

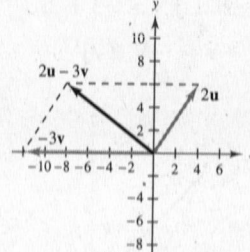

23. **u** = ⟨−5, 3⟩, **v** = ⟨0, 0⟩

(a) **u** + **v** = ⟨−5, 3⟩ = **u**

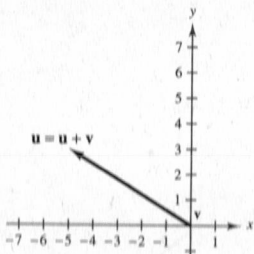

(b) **u** − **v** = ⟨−5, 3⟩ = **u**

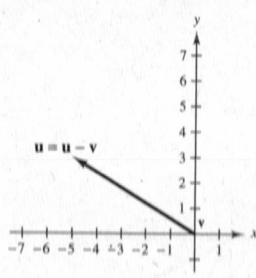

(c) 2**u** − 3**v** = 2**u** = ⟨−10, 6⟩

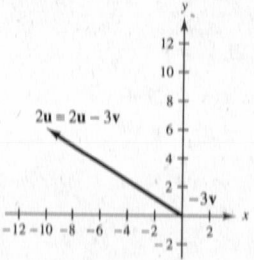

24. $\mathbf{u} = \langle 0, 0 \rangle, \mathbf{v} = \langle 2, 1 \rangle$

   (a) $\mathbf{u} + \mathbf{v} = \langle 2, 1 \rangle$

   (b) $\mathbf{u} - \mathbf{v} = \langle -2, -1 \rangle$

   (c) $2\mathbf{u} - 3\mathbf{v} = \langle 0, 0 \rangle - \langle 6, 3 \rangle$
   $= \langle -6, -3 \rangle$

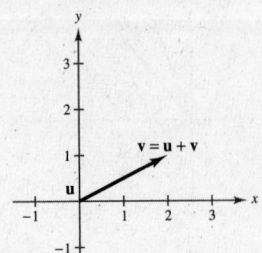
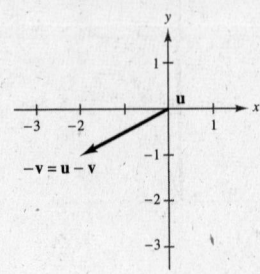
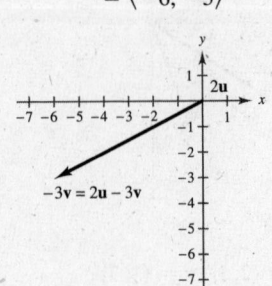

25. $\mathbf{u} = \mathbf{i} + \mathbf{j}, \mathbf{v} = 2\mathbf{i} - 3\mathbf{j}$

   (a) $\mathbf{u} + \mathbf{v} = 3\mathbf{i} - 2\mathbf{j}$

   (b) $\mathbf{u} - \mathbf{v} = -\mathbf{i} + 4\mathbf{j}$

   (c) $2\mathbf{u} - 3\mathbf{v} = (2\mathbf{i} + 2\mathbf{j}) - (6\mathbf{i} - 9\mathbf{j})$
   $= -4\mathbf{i} + 11\mathbf{j}$

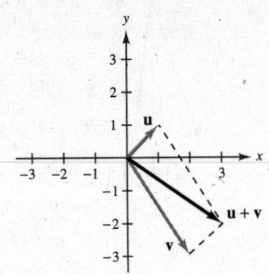
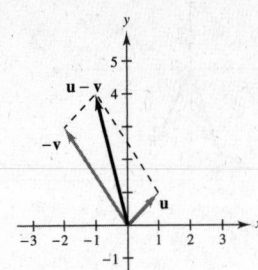
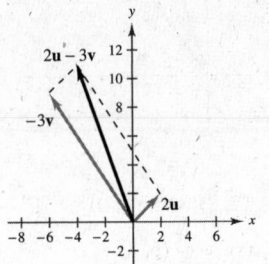

26. $\mathbf{u} = -2\mathbf{i} + \mathbf{j}, \mathbf{v} = -\mathbf{i} + 2\mathbf{j}$

   (a) $\mathbf{u} + \mathbf{v} = -3\mathbf{i} + 3\mathbf{j}$

   (b) $\mathbf{u} - \mathbf{v} = -\mathbf{i} - \mathbf{j}$

   (c) $2\mathbf{u} - 3\mathbf{v} = (-4\mathbf{i} + 2\mathbf{j}) - (-3\mathbf{i} + 6\mathbf{j})$
   $= -\mathbf{i} - 4\mathbf{j}$

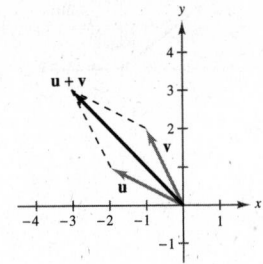
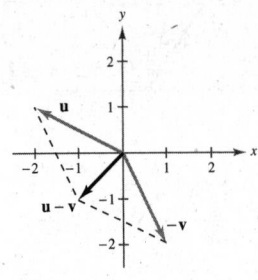
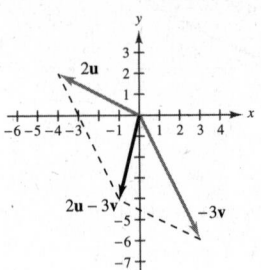

27. $\mathbf{u} = 2\mathbf{i}, \mathbf{v} = \mathbf{j}$

   (a) $\mathbf{u} + \mathbf{v} = 2\mathbf{i} + \mathbf{j}$

   (b) $\mathbf{u} - \mathbf{v} = 2\mathbf{i} - \mathbf{j}$

   (c) $2\mathbf{u} - 3\mathbf{v} = 4\mathbf{i} - 3\mathbf{j}$

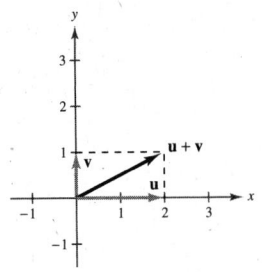
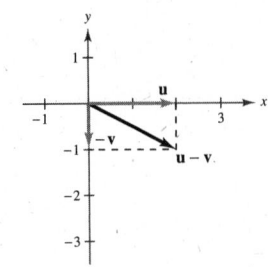
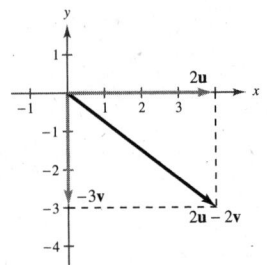

28. $\mathbf{u} = 3\mathbf{j}, \mathbf{v} = 2\mathbf{i}$

(a) $\mathbf{u} + \mathbf{v} = 2\mathbf{i} + 3\mathbf{j}$

(b) $\mathbf{u} - \mathbf{v} = -2\mathbf{i} + 3\mathbf{j}$

(c) $2\mathbf{u} - 3\mathbf{v} = 6\mathbf{j} - 6\mathbf{i}$

$= -6\mathbf{i} + 6\mathbf{j}$

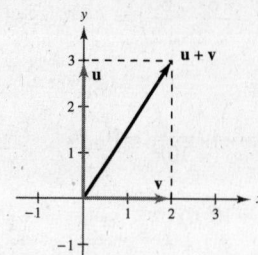

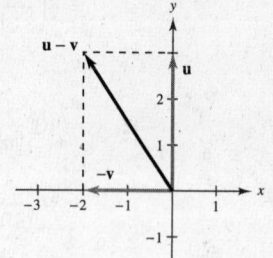

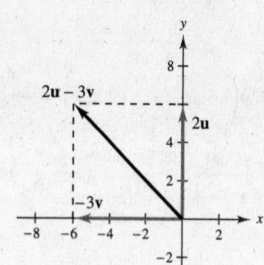

29. $\mathbf{v} = \dfrac{1}{\|\mathbf{u}\|}\mathbf{u} = \dfrac{1}{\sqrt{3^2 + 0^2}}\langle 3, 0\rangle = \dfrac{1}{3}\langle 3, 0\rangle = \langle 1, 0\rangle$

30. $\mathbf{u} = \langle 0, -2\rangle$

$\mathbf{v} = \dfrac{1}{\|\mathbf{u}\|}\mathbf{u} = \dfrac{1}{\sqrt{0^2 + (-2)^2}}\langle 0, -2\rangle$

$= \dfrac{1}{2}\langle 0, -2\rangle = \langle 0, -1\rangle$

31. $\mathbf{u} = \dfrac{1}{\|\mathbf{v}\|}\mathbf{v} = \dfrac{1}{\sqrt{(-2)^2 + 2^2}}\langle -2, 2\rangle = \dfrac{1}{2\sqrt{2}}\langle -2, 2\rangle$

$= \left\langle -\dfrac{1}{\sqrt{2}}, \dfrac{1}{\sqrt{2}}\right\rangle$

$= \left\langle -\dfrac{\sqrt{2}}{2}, \dfrac{\sqrt{2}}{2}\right\rangle$

32. $\mathbf{v} = \langle 5, -12\rangle$

$\mathbf{u} = \dfrac{1}{\|\mathbf{v}\|}\mathbf{v} = \dfrac{1}{\sqrt{5^2 + (-12)^2}}\langle 5, -12\rangle$

$= \dfrac{1}{13}\langle 5, -12\rangle$

$= \left\langle \dfrac{5}{13}, -\dfrac{12}{13}\right\rangle$

33. $\mathbf{u} = \dfrac{1}{\|\mathbf{v}\|}\mathbf{v} = \dfrac{1}{\sqrt{6^2 + (-2)^2}}(6\mathbf{i} - 2\mathbf{j}) = \dfrac{1}{\sqrt{40}}(6\mathbf{i} - 2\mathbf{j})$

$= \dfrac{1}{2\sqrt{10}}(6\mathbf{i} - 2\mathbf{j}) = \dfrac{3}{\sqrt{10}}\mathbf{i} - \dfrac{1}{\sqrt{10}}\mathbf{j}$

$= \dfrac{3\sqrt{10}}{10}\mathbf{i} - \dfrac{\sqrt{10}}{10}\mathbf{j}$

34. $\mathbf{v} = \mathbf{i} + \mathbf{j}$

$\mathbf{u} = \dfrac{1}{\|\mathbf{v}\|}\mathbf{v}$

$= \dfrac{1}{\sqrt{1^2 + 1^2}}(\mathbf{i} + \mathbf{j}) = \dfrac{1}{\sqrt{2}}(\mathbf{i} + \mathbf{j}) = \dfrac{\sqrt{2}}{2}\mathbf{i} + \dfrac{\sqrt{2}}{2}\mathbf{j}$

35. $\mathbf{u} = \dfrac{1}{\|\mathbf{w}\|}\mathbf{w} = \dfrac{1}{4}(4\mathbf{j}) = \mathbf{j}$

36. $\mathbf{w} = -6\mathbf{i}$

$\mathbf{v} = \dfrac{1}{\|\mathbf{w}\|}\mathbf{w} = \dfrac{1}{\sqrt{(-6)^2 + 0^2}}(-6\mathbf{i})$

$= \dfrac{1}{6}(-6\mathbf{i}) = -\mathbf{i}$

37. $\mathbf{u} = \dfrac{1}{\|\mathbf{w}\|}\mathbf{w} = \dfrac{1}{\sqrt{1^2 + (-2)^2}}(\mathbf{i} - 2\mathbf{j}) = \dfrac{1}{\sqrt{5}}(\mathbf{i} - 2\mathbf{j})$

$= \dfrac{1}{\sqrt{5}}\mathbf{i} - \dfrac{2}{\sqrt{5}}\mathbf{j} = \dfrac{\sqrt{5}}{5}\mathbf{i} - \dfrac{2\sqrt{5}}{5}\mathbf{j}$

38. $\mathbf{w} = 7\mathbf{j} - 3\mathbf{i}$

$\mathbf{v} = \dfrac{1}{\|\mathbf{w}\|}\mathbf{w} = \dfrac{1}{\sqrt{(-3)^2 + 7^2}}(-3\mathbf{i} + 7\mathbf{j})$

$= -\dfrac{3}{\sqrt{58}}\mathbf{i} + \dfrac{7}{\sqrt{58}}\mathbf{j} = -\dfrac{3\sqrt{58}}{58}\mathbf{i} + \dfrac{7\sqrt{58}}{58}\mathbf{j}$

39. $5\left(\dfrac{1}{\|\mathbf{u}\|}\mathbf{u}\right) = 5\left(\dfrac{1}{\sqrt{3^2 + 3^2}}\langle 3, 3 \rangle\right) = \dfrac{5}{3\sqrt{2}}\langle 3, 3 \rangle$

$\qquad = \left\langle \dfrac{5}{\sqrt{2}}, \dfrac{5}{\sqrt{2}} \right\rangle = \left\langle \dfrac{5\sqrt{2}}{2}, \dfrac{5\sqrt{2}}{2} \right\rangle$

40. $\mathbf{v} = 6\left(\dfrac{1}{\|\mathbf{u}\|}\mathbf{u}\right) = 6\left(\dfrac{1}{\sqrt{(-3)^2 + 3^2}}\langle -3, 3 \rangle\right)$

$\qquad = 6\left(\dfrac{1}{3\sqrt{2}}\langle -3, 3 \rangle\right) = \left\langle -\dfrac{6}{\sqrt{2}}, \dfrac{6}{\sqrt{2}} \right\rangle = \left\langle -3\sqrt{2}, 3\sqrt{2} \right\rangle$

41. $9\left(\dfrac{1}{\|\mathbf{u}\|}\mathbf{u}\right) = 9\left(\dfrac{1}{\sqrt{2^2 + 5^2}}\langle 2, 5 \rangle\right) = \dfrac{9}{\sqrt{29}}\langle 2, 5 \rangle$

$\qquad = \left\langle \dfrac{18}{\sqrt{29}}, \dfrac{45}{\sqrt{29}} \right\rangle = \left\langle \dfrac{18\sqrt{29}}{29}, \dfrac{45\sqrt{29}}{29} \right\rangle$

42. $\mathbf{v} = 10\left(\dfrac{1}{\|\mathbf{u}\|}\mathbf{u}\right) = 10\left(\dfrac{1}{\sqrt{0^2 + (-10)^2}}\langle -10, 0 \rangle\right)$

$\qquad = 10\left(\dfrac{1}{10}\langle -10, 0 \rangle\right) = \langle -10, 0 \rangle$

43. $\mathbf{u} = \langle 4 - (-3), 5 - 1 \rangle$

$\qquad = \langle 7, 4 \rangle$

$\qquad = 7\mathbf{i} + 4\mathbf{j}$

44. $\mathbf{u} = \langle 3 - 0, 6 - (-2) \rangle$

$\qquad \mathbf{u} = \langle 3, 8 \rangle$

$\qquad \mathbf{u} = 3\mathbf{i} + 8\mathbf{j}$

45. $\mathbf{u} = \langle 2 - (-1), 3 - (-5) \rangle$

$\qquad = \langle 3, 8 \rangle$

$\qquad = 3\mathbf{i} + 8\mathbf{j}$

46. $\mathbf{u} = \langle 0 - (-6), 1 - 4 \rangle$

$\qquad \mathbf{u} = \langle 6, -3 \rangle$

$\qquad \mathbf{u} = 6\mathbf{i} - 3\mathbf{j}$

47. $\mathbf{v} = \dfrac{3}{2}\mathbf{u}$

$\qquad = \dfrac{3}{2}(2\mathbf{i} - \mathbf{j})$

$\qquad = 3\mathbf{i} - \dfrac{3}{2}\mathbf{j} = \left\langle 3, -\dfrac{3}{2} \right\rangle$

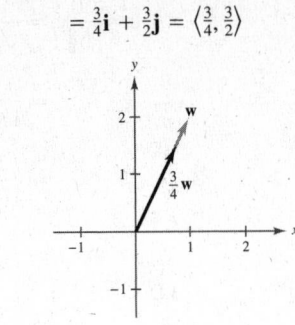

48. $\mathbf{v} = \dfrac{3}{4}\mathbf{w} = \dfrac{3}{4}(\mathbf{i} + 2\mathbf{j})$

$\qquad = \dfrac{3}{4}\mathbf{i} + \dfrac{3}{2}\mathbf{j} = \left\langle \dfrac{3}{4}, \dfrac{3}{2} \right\rangle$

49. $\mathbf{v} = \mathbf{u} + 2\mathbf{w}$

$\qquad = (2\mathbf{i} - \mathbf{j}) + 2(\mathbf{i} + 2\mathbf{j})$

$\qquad = 4\mathbf{i} + 3\mathbf{j} = \langle 4, 3 \rangle$

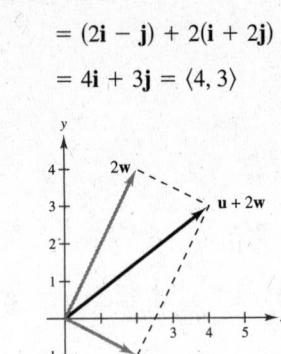

50. $\mathbf{v} = -\mathbf{u} + \mathbf{w}$

$\qquad = -(2\mathbf{i} - \mathbf{j}) + (\mathbf{i} + 2\mathbf{j})$

$\qquad = -\mathbf{i} + 3\mathbf{j} = \langle -1, 3 \rangle$

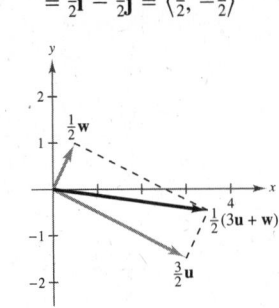

51. $\mathbf{v} = \dfrac{1}{2}(3\mathbf{u} + \mathbf{w})$

$\qquad = \dfrac{1}{2}(6\mathbf{i} - 3\mathbf{j} + \mathbf{i} + 2\mathbf{j})$

$\qquad = \dfrac{7}{2}\mathbf{i} - \dfrac{1}{2}\mathbf{j} = \left\langle \dfrac{7}{2}, -\dfrac{1}{2} \right\rangle$

52. $\mathbf{v} = \mathbf{u} - 2\mathbf{w}$

$\qquad = (2\mathbf{i} - \mathbf{j}) - 2(\mathbf{i} + 2\mathbf{j})$

$\qquad = -5\mathbf{j} = \langle 0, -5 \rangle$

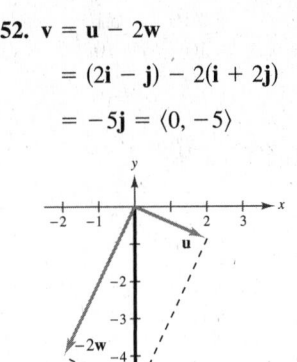

53. $\mathbf{v} = 3(\cos 60°\mathbf{i} + \sin 60°\mathbf{j})$

$\|\mathbf{v}\| = 3,\ \theta = 60°$

54. $\mathbf{v} = 8(\cos 135°\,\mathbf{i} + \sin 135°\,\mathbf{j})$

$\|\mathbf{v}\| = 8,\ \theta = 135°$

55. $\mathbf{v} = 6\mathbf{i} - 6\mathbf{j}$

$\|\mathbf{v}\| = \sqrt{6^2 + (-6)^2} = \sqrt{72}$

$\quad = 6\sqrt{2}$

$\tan \theta = \dfrac{-6}{6} = -1$

Since $\mathbf{v}$ lies in Quadrant IV,
$\theta = 315°$.

56. $\mathbf{v} = -5\mathbf{i} + 4\mathbf{j}$

$\|\mathbf{v}\| = \sqrt{(-5)^2 + 4^2} = \sqrt{41}$

$\tan \theta = -\dfrac{4}{5}$

Since $\mathbf{v}$ lies in Quadrant II,
$\theta = 141.3°$.

57. $\mathbf{v} = \langle 3 \cos 0°, 3 \sin 0° \rangle$

$\quad = \langle 3, 0 \rangle$

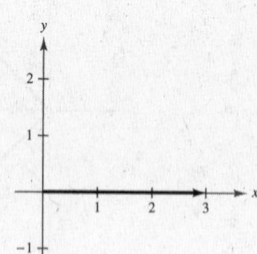

58. $\mathbf{v} = \langle \cos 45°, \sin 45° \rangle$

$\quad = \left\langle \dfrac{\sqrt{2}}{2}, \dfrac{\sqrt{2}}{2} \right\rangle$

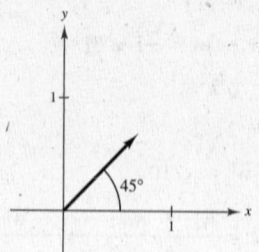

59. $\mathbf{v} = \left\langle \dfrac{7}{2} \cos 150°, \dfrac{7}{2} \sin 150° \right\rangle$

$\quad = \left\langle -\dfrac{7\sqrt{3}}{4}, \dfrac{7}{4} \right\rangle$

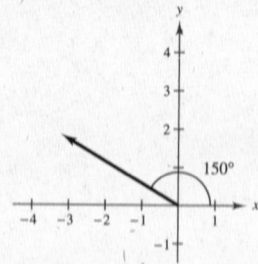

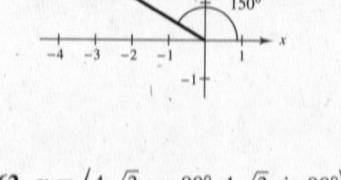

60. $\mathbf{v} = \left\langle \dfrac{5}{2} \cos 45°, \dfrac{5}{2} \sin 45° \right\rangle$

$\quad = \left\langle \dfrac{5\sqrt{2}}{4}, \dfrac{5\sqrt{2}}{4} \right\rangle$

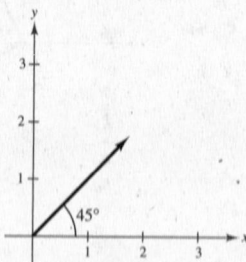

61. $\mathbf{v} = \langle 3\sqrt{2} \cos 150°,\ 3\sqrt{2} \sin 150° \rangle$

$\quad = \left\langle -\dfrac{3\sqrt{6}}{2}, \dfrac{3\sqrt{2}}{2} \right\rangle$

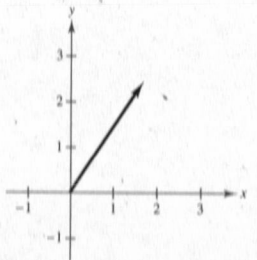

62. $\mathbf{v} = \langle 4\sqrt{3} \cos 90°,\ 4\sqrt{3} \sin 90° \rangle$

$\quad = \langle 0, 4\sqrt{3} \rangle$

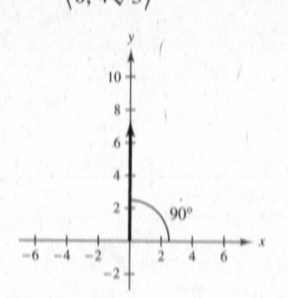

63. $\mathbf{v} = 2\left(\dfrac{1}{\sqrt{1^2 + 3^2}} \right)(\mathbf{i} + 3\mathbf{j})$

$\quad = \dfrac{2}{\sqrt{10}}(\mathbf{i} + 3\mathbf{j})$

$\quad = \dfrac{\sqrt{10}}{5}\mathbf{i} + \dfrac{3\sqrt{10}}{5}\mathbf{j} = \left\langle \dfrac{\sqrt{10}}{5}, \dfrac{3\sqrt{10}}{5} \right\rangle$

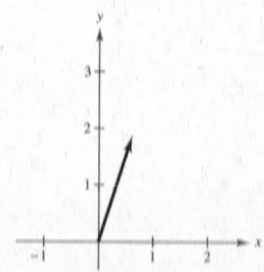

64. $\mathbf{v} = 3\left(\dfrac{1}{\sqrt{3^2 + 4^2}} \right)(3\mathbf{i} + 4\mathbf{j})$

$\quad = \dfrac{3}{5}(3\mathbf{i} + 4\mathbf{j})$

$\quad = \dfrac{9}{5}\mathbf{i} + \dfrac{12}{5}\mathbf{j} = \left\langle \dfrac{9}{5}, \dfrac{12}{5} \right\rangle$

65. $\mathbf{u} = \langle 5 \cos 0°, 5 \sin 0° \rangle = \langle 5, 0 \rangle$

$\mathbf{v} = (5 \cos 90°, 5 \sin 90°) = \langle 0, 5 \rangle$

$\mathbf{u} + \mathbf{v} = \langle 5, 5 \rangle$

66. $\mathbf{u} = \langle 4 \cos 60°, 4 \sin 60° \rangle = \langle 2, 2\sqrt{3} \rangle$

$\mathbf{v} = (4 \cos 90°, 4 \sin 90°) = \langle 0, 4 \rangle$

$\mathbf{u} + \mathbf{v} = \langle 2, 4 + 2\sqrt{3} \rangle$

67. $\mathbf{u} = \langle 20 \cos 45°, 20 \sin 45° \rangle = \langle 10\sqrt{2}, 10\sqrt{2} \rangle$

$\mathbf{v} = \langle 50 \cos 180°, 50 \sin 180° \rangle = \langle -50, 0 \rangle$

$\mathbf{u} + \mathbf{v} = \langle 10\sqrt{2} - 50, 10\sqrt{2} \rangle$

68. $\mathbf{u} = \langle 50 \cos 30°, 50 \sin 30° \rangle = \langle 25\sqrt{3}, 25 \rangle$

$\approx \langle 43.301, 25 \rangle$

$\mathbf{v} = \langle 30 \cos 110°, 30 \sin 110° \rangle \approx \langle -10.261, 28.191 \rangle$

$\mathbf{u} + \mathbf{v} \approx \langle 33.04, 53.19 \rangle$

69. $\mathbf{v} = \mathbf{i} + \mathbf{j}$

$\mathbf{w} = 2\mathbf{i} - 2\mathbf{j}$

$\mathbf{u} = \mathbf{v} - \mathbf{w} = -\mathbf{i} + 3\mathbf{j}$

$\|\mathbf{v}\| = \sqrt{2}$

$\|\mathbf{w}\| = 2\sqrt{2}$

$\|\mathbf{v} - \mathbf{w}\| = \sqrt{10}$

$\cos \alpha = \dfrac{\|\mathbf{v}\|^2 + \|\mathbf{w}\|^2 - \|\mathbf{v} - \mathbf{w}\|^2}{2\|\mathbf{v}\|\,\|\mathbf{w}\|} = \dfrac{2 + 8 - 10}{2\sqrt{2} \cdot 2\sqrt{2}} = 0$

$\alpha = 90°$

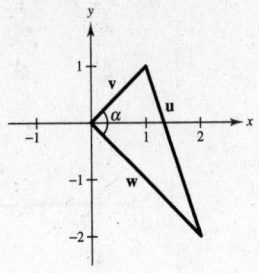

70. $\mathbf{v} = \mathbf{i} + 2\mathbf{j}$

$\mathbf{w} = 2\mathbf{i} - \mathbf{j}$

$\mathbf{u} = \mathbf{v} - \mathbf{w} = -\mathbf{i} + 3\mathbf{j}$

$\cos \theta = \dfrac{\|\mathbf{v}\|^2 + \|\mathbf{w}\|^2 - \|\mathbf{v} - \mathbf{w}\|^2}{2\|\mathbf{v}\|\,\|\mathbf{w}\|} = \dfrac{5 + 5 - 10}{2\sqrt{5}\sqrt{5}} = 0$

$\theta = 90°$

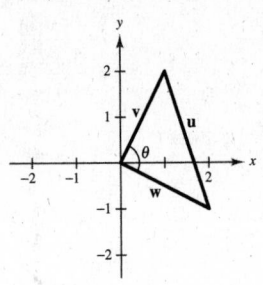

71. Force One: $\mathbf{u} = 45\mathbf{i}$

Force Two: $\mathbf{v} = 60 \cos \theta \mathbf{i} + 60 \sin \theta \mathbf{j}$

Resultant Force: $\mathbf{u} + \mathbf{v} = (45 + 60 \cos \theta)\mathbf{i} + 60 \sin \theta \mathbf{j}$

$\|\mathbf{u} + \mathbf{v}\| = \sqrt{(45 + 60 \cos \theta)^2 + (60 \sin \theta)^2} = 90$

$2025 + 5400 \cos \theta + 3600 = 8100$

$5400 \cos \theta = 2475$

$\cos \theta = \dfrac{2475}{5400} \approx 0.4583$

$\theta \approx 62.7°$

72. Force One: **u** = 3000**i**
Force Two: **v** = 1000 cos θ**i** + 1000 sin θ**j**
Resultant Force: **u** + **v** = (3000 + 1000 cos θ)**i** + 1000 sin θ**j**

$$\|\mathbf{u} + \mathbf{v}\| = \sqrt{(3000 + 1000 \cos \theta)^2 + (1000 \sin \theta)^2} = 3750$$

$$9{,}000{,}000 + 6{,}000{,}000 \cos \theta + 1{,}000{,}000 = 14{,}062{,}500$$

$$6{,}000{,}000 \cos \theta = 4{,}062{,}500$$

$$\cos \theta = \frac{4{,}062{,}500}{6{,}000{,}000} \approx 0.6771$$

$$\theta \approx 47.4°$$

73.   **u** = 300**i**

$$\mathbf{v} = (125 \cos 45°)\mathbf{i} + (125 \sin 45°)\mathbf{j} = \frac{125}{\sqrt{2}}\mathbf{i} + \frac{125}{\sqrt{2}}\mathbf{j}$$

$$\mathbf{R} = \mathbf{u} + \mathbf{v} = \left(300 + \frac{125}{\sqrt{2}}\right)\mathbf{i} + \frac{125}{\sqrt{2}}\mathbf{j}$$

$$\|\mathbf{R}\| = \sqrt{\left(300 + \frac{125}{\sqrt{2}}\right)^2 + \left(\frac{125}{\sqrt{2}}\right)^2} \approx 398.32 \text{ newtons}$$

$$\tan \theta = \frac{\dfrac{125}{\sqrt{2}}}{300 + \left(\dfrac{125}{\sqrt{2}}\right)} \implies \theta \approx 12.8°$$

74.   **u** = (2000 cos 30°)**i** + (2000 sin 30°**j**)

$$\approx 1732.05\mathbf{i} + 1000\mathbf{j}$$

$$\mathbf{v} = (900 \cos(-45°))\mathbf{i} + (900 \sin(-45°))\mathbf{j}$$

$$\approx 636.4\mathbf{i} + -636.4\mathbf{j}$$

u + **v** ≈ 2368.4**i** + 363.6**j**

$$\|\mathbf{u} + \mathbf{v}\| \approx \sqrt{(2368.4)^2 + (363.6)^2} \approx 2396.19$$

$$\tan \theta = \frac{363.6}{2368.4} \approx 0.1535 \implies \theta \approx 8.7°$$

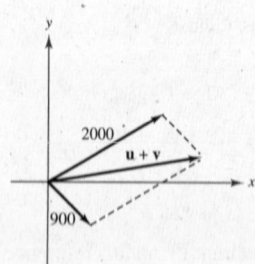

75. **u** = (75 cos 30°)**i** + (75 sin 30°)**j** ≈ 64.95**i** + 37.5**j**

v = (100 cos 45°)**i** + (100 sin 45°)**j** ≈ 70.71**i** + 70.71**j**

w = (125 cos 120°)**i** + (125 sin 120°)**j** ≈ -62.5**i** + 108.3**j**

u + **v** + **w** ≈ 73.16**i** + 216.5**j**

‖**u** + **v** + **w**‖ ≈ 228.5 pounds

$$\tan \theta \approx \frac{216.5}{73.16} \approx 2.9593$$

$$\theta \approx 71.3°$$

76.
$$\mathbf{u} = (70 \cos 30°)\mathbf{i} - (70 \sin 30°)\mathbf{j} \approx 60.62\mathbf{i} - 35\mathbf{j}$$

$$\mathbf{v} = (40 \cos 45°)\mathbf{i} + (40 \sin 45°)\mathbf{j} \approx 28.28\mathbf{i} + 28.28\mathbf{j}$$

$$\mathbf{w} = (60 \cos 135°)\mathbf{i} + (60 \sin 135°)\mathbf{j} \approx -42.43\mathbf{i} + 42.43\mathbf{j}$$

$$\mathbf{u} + \mathbf{v} + \mathbf{w} = 46.48\mathbf{i} + 35.71\mathbf{j}$$

$$\|\mathbf{u} + \mathbf{v} + \mathbf{w}\| \approx 58.61 \text{ pounds}$$

$$\tan \theta \approx \frac{35.71}{46.47} \approx 0.7683$$

$$\theta \approx 37.5°$$

77. Horizontal component of velocity: $70 \cos 35° \approx 57.34$ feet per second

Vertical component of velocity: $70 \sin 35° \approx 40.15$ feet per second

78. Horizontal component of velocity: $1200 \cos 6° \approx 1193.4$ ft/sec

Vertical component of velocity: $1200 \sin 6° \approx 125.4$ ft/sec

79. Cable $\overrightarrow{AC}$: $\mathbf{u} = \|\mathbf{u}\|(\cos 50°\mathbf{i} - \sin 50°\mathbf{j})$

Cable $\overrightarrow{BC}$: $\mathbf{v} = \|\mathbf{v}\|(-\cos 30°\mathbf{i} - \sin 30°\mathbf{j})$

Resultant: $\mathbf{u} + \mathbf{v} = -2000\mathbf{j}$

$$\|\mathbf{u}\| \cos 50° - \|\mathbf{v}\| \cos 30° = 0$$

$$-\|\mathbf{u}\| \sin 50° - \|\mathbf{v}\|\sin 30° = -2000$$

Solving this system of equations yields:

$T_{AC} = \|\mathbf{u}\| \approx 1758.8$ pounds

$T_{BC} = \|\mathbf{v}\| \approx 1305.4$ pounds

80. Rope $\overrightarrow{AC}$: $\mathbf{u} = 10\mathbf{i} - 24\mathbf{j}$

The vector lies in Quadrant IV and its reference angle is $\arctan\left(\frac{12}{5}\right)$.

$$\mathbf{u} = \|\mathbf{u}\|\left[\cos\left(\arctan \tfrac{12}{5}\right)\mathbf{i} - \sin\left(\arctan \tfrac{12}{5}\right)\mathbf{j}\right]$$

Rope $\overrightarrow{BC}$: $\mathbf{v} = -20\mathbf{i} - 24\mathbf{j}$

The vector lies in Quadrant III and its reference angle is $\arctan\left(\frac{6}{5}\right)$.

$$\mathbf{v} = \|\mathbf{v}\|\left[-\cos\left(\arctan \tfrac{6}{5}\right)\mathbf{i} - \sin\left(\arctan \tfrac{6}{5}\right)\mathbf{j}\right]$$

Resultant: $\mathbf{u} + \mathbf{v} = -5000\mathbf{j}$

$$\|\mathbf{u}\| \cos\left(\arctan \tfrac{12}{5}\right) - \|\mathbf{v}\| \cos\left(\arctan \tfrac{6}{5}\right) = 0$$

$$-\|\mathbf{u}\| \sin\left(\arctan \tfrac{12}{5}\right) - \|\mathbf{v}\| \sin\left(\arctan \tfrac{6}{5}\right) = -5000$$

Solving this system of equations yields: $T_{AC} = \|\mathbf{u}\| \approx 3611.1$ pounds

$$T_{BC} = \|\mathbf{v}\| \approx 2169.5 \text{ pounds}$$

81. Towline 1: $\mathbf{u} = \|\mathbf{u}\|(\cos 18°\mathbf{i} + \sin 18°\mathbf{j})$

Towline 2: $\mathbf{v} = \|\mathbf{u}\|(\cos 18°\mathbf{i} - \sin 18°\mathbf{j})$

Resultant: $\mathbf{u} + \mathbf{v} = 6000\mathbf{i}$

$$\|\mathbf{u}\| \cos 18° + \|\mathbf{u}\| \cos 18° = 6000$$

$$\|\mathbf{u}\| \approx 3154.4$$

Therefore, the tension on each towline is $\|\mathbf{u}\| \approx 3154.4$ pounds.

82. Rope 1: $\mathbf{u} = \|\mathbf{u}\| (\cos 70°\mathbf{i} - \sin 70°\mathbf{j})$

Rope 2: $\mathbf{v} = \|\mathbf{u}\| (-\cos 70°\mathbf{i} - \sin 70°\mathbf{j})$

Resultant: $\mathbf{u} + \mathbf{v} = -100\mathbf{j}$

$$-\|\mathbf{u}\| \sin 70° - \|\mathbf{u}\| \sin 70° = -100$$

$$\|\mathbf{u}\| \approx 53.2$$

Therefore, the tension of each rope is $\|\mathbf{u}\| \approx 53.2$ pounds.

83. Airspeed: $\mathbf{u} = (875 \cos 58°)\mathbf{i} - (875 \sin 58°)\mathbf{j}$

Groundspeed: $\mathbf{v} = (800 \cos 50°)\mathbf{i} - (800 \sin 50°)\mathbf{j}$

Wind: $\mathbf{w} = \mathbf{v} - \mathbf{u} = (800 \cos 50° - 875 \cos 58°)\mathbf{i} + (-800 \sin 50° + 875 \sin 58°)\mathbf{j}$

$$\approx 50.5507\mathbf{i} + 129.2065\mathbf{j}$$

Wind speed: $\|\mathbf{w}\| \approx \sqrt{(50.5507)^2 + (129.2065)^2}$

$$\approx 138.7 \text{ kilometers per hour}$$

Wind direction: $\tan \theta \approx \dfrac{129.2065}{50.5507}$

$$\theta \approx 68.6°; \quad 90° - \theta = 21.4°$$

Bearing: N 21.4° E

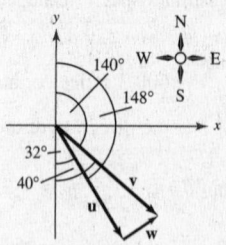

84. (a)

(b) The velocity vector $\mathbf{v}_w$ of the wind has a magnitude of 60 and a direction angle of 45°.

$$\mathbf{v}_w = \|\mathbf{v}_w\|(\cos \theta)\mathbf{i} + \|\mathbf{v}_w\|(\sin \theta)\mathbf{j}$$

$$= 60(\cos 45°)\mathbf{i} + 60(\sin 45°)\mathbf{j}$$

$$= 60[(\cos 45°)\mathbf{i} + (\sin 45°)\mathbf{j}]$$

$$= 60\langle \cos 45°, \sin 45° \rangle, \text{ or } \langle 30\sqrt{2}, 30\sqrt{2} \rangle$$

(c) The velocity vector $\mathbf{v}_j$ of the jet has a magnitude of 580 and a direction angle of 118°.

$$\mathbf{v}_j = \|\mathbf{v}_j\|(\cos \theta)\mathbf{i} + \|\mathbf{v}_j\|(\sin \theta)\mathbf{j}$$

$$= 580(\cos 118°)\mathbf{i} + 580(\sin 118°)\mathbf{j}$$

$$= 580[(\cos 118°)\mathbf{i} + (\sin 118°)\mathbf{j}]$$

$$= 580\langle \cos 118°, \sin 118° \rangle$$

—CONTINUED—

84. —CONTINUED—

(d) The velocity of the jet (in the wind) is

$$\mathbf{v} = \mathbf{v}_w + \mathbf{v}_j$$

$$= 60\langle\cos 45°, \sin 45°\rangle + 580\langle\cos 118°, \sin 118°\rangle$$

$$= \langle 60 \cos 45° + 580 \cos 118°, 60 \sin 45° + 580 \sin 118°\rangle$$

$$\approx \langle -229.87, 554.54 \rangle$$

The resultant speed of the jet is

$$\|\mathbf{v}\| = \sqrt{(-229.87)^2 + (554.54)^2}$$

$$\approx 600.3 \text{ miles per hour}$$

(e) If θ is the direction of the flight path, then

$$\tan\theta = \frac{554.54}{-229.87} \approx -2.4124$$

Because θ lies in the Quadrant II, $\theta = 180° + \arctan(-2.4124) \approx 180° - 67.5° = 112.5°$. The true bearing of the jet is $112.5° - 90° = 22.5°$ west of north, or $360° - 22.5° = 337.5°$.

85. $W = FD = (100 \cos 50°)(30) = 1928.4$ foot–pounds

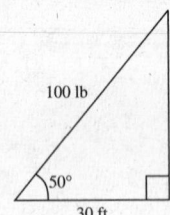

100 lb

50°

30 ft

86. Horizontal force: $\mathbf{u} = \|\mathbf{u}\|\mathbf{i}$

Weight: $\mathbf{w} = -\mathbf{j}$

Rope: $\mathbf{t} = \|\mathbf{t}\| (\cos 135°\mathbf{i} + \sin 135°\mathbf{j})$

$$\mathbf{u} + \mathbf{w} + \mathbf{t} = \mathbf{0} \implies \|\mathbf{u}\| + \|\mathbf{t}\| \cos 135° = 0$$

$$-1 + \|\mathbf{t}\| \sin 135° = 0$$

$\|\mathbf{t}\| \approx \sqrt{2}$ pounds

$\|\mathbf{u}\| \approx 1$ pound

87. True. See Example 1.

88. True.

$$\|\mathbf{u}\| = \sqrt{a^2 + b^2} = 1 \implies a^2 + b^2 = 1$$

89. (a) The angle between them is $0°$.

(b) The angle between them is $180°$.

(c) No. At most it can be equal to the sum when the angle between them is $0°$.

90. $\mathbf{F}_1 = \langle 10, 0 \rangle$, $\mathbf{F}_2 = 5\langle \cos\theta, \sin\theta \rangle$

(a) $\mathbf{F}_1 + \mathbf{F}_2 = \langle 10 + 5 \cos\theta, 5 \sin\theta \rangle$

$$\|\mathbf{F}_1 + \mathbf{F}_2\| = \sqrt{(10 + 5 \cos\theta)^2 + (5 \sin\theta)^2}$$

$$= \sqrt{100 + 100 \cos\theta + 25 \cos^2\theta + 25 \sin^2\theta}$$

$$= 5\sqrt{4 + 4 \cos\theta + \cos^2\theta + \sin^2\theta}$$

$$= 5\sqrt{4 + 4 \cos\theta + 1}$$

$$= 5\sqrt{5 + 4 \cos\theta}$$

(b)

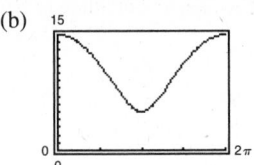

15

0

0

2π

(c) Range: $[5, 15]$

Maximum is 15 when $\theta = 0$.

Minimum is 5 when $\theta = \pi$.

(d) The magnitude of the resultant is never 0 because the magnitudes of $\mathbf{F}_1$ and $\mathbf{F}_2$ are not the same.

91. Let $\mathbf{v} = (\cos\theta)\mathbf{i} + (\sin\theta)\mathbf{j}$.

$\|\mathbf{v}\| = \sqrt{\cos^2\theta + \sin^2\theta} = \sqrt{1} = 1$

Therefore, $\mathbf{v}$ is a unit vector for any value of θ.

92. The following program is written for a *TI-82* or *TI-83* or *TI-83 Plus* graphing calculator. The program sketches two vectors $\mathbf{u} = a\mathbf{i} + b\mathbf{j}$ and $\mathbf{v} = c\mathbf{i} + d\mathbf{j}$ in standard position, and then sketches the vector difference $\mathbf{u} - \mathbf{v}$ using the parallelogram law.

```
PROGRAM:  SUBVECT
:Input "ENTER A", A
:Input "ENTER B", B
:Input "ENTER C", C
:Input "ENTER D", D
:Line (0, 0, A, B)
:Line (0, 0, C, D)
:Pause
:A – C→E
:B – D→F
:Line (A, B, C, D)
:Line (A, B, E, F)
:Line (0, 0, E, F)
:Pause
:ClrDraw
:Stop
```

93. $\mathbf{u} = \langle 5 - 1, 2 - 6 \rangle = \langle 4, -4 \rangle$

$\mathbf{v} = \langle 9 - 4, 4 - 5 \rangle = \langle 5, -1 \rangle$

$\mathbf{u} - \mathbf{v} = \langle -1, -3 \rangle$ or $\mathbf{v} - \mathbf{u} = \langle 1, 3 \rangle$

94. $\mathbf{u} = \langle 80 - 10, 80 - 60 \rangle = \langle 70, 20 \rangle$

$\mathbf{v} = \langle -20 - (-100), 70 - 0 \rangle = \langle 80, 70 \rangle$

$\mathbf{u} - \mathbf{v} = \langle 70 - 80, 20 - 70 \rangle = \langle -10, -50 \rangle$

$\mathbf{v} - \mathbf{u} = \langle 80 - 70, 70 - 20 \rangle = \langle 10, 50 \rangle$

95. $\sqrt{x^2 - 64} = \sqrt{(8\sec\theta)^2 - 64}$

$\qquad = \sqrt{64(\sec^2\theta - 1)}$

$\qquad = 8\sqrt{\tan^2\theta}$

$\qquad = 8\tan\theta$ for $0 < \theta < \dfrac{\pi}{2}$

96. $x = 8\sin\theta$

$\sqrt{64 - x^2} = \sqrt{64 - (8\sin^2\theta)}$

$\qquad = \sqrt{64 - 64\sin^2\theta}$

$\qquad = 8\sqrt{1 - \sin^2\theta}$

$\qquad = 8\sqrt{\cos^2\theta}$

$\qquad = 8\cos\theta$ for $0 < \theta < \dfrac{\pi}{2}$

97. $\sqrt{x^2 + 36} = \sqrt{(6\tan\theta)^2 + 36}$

$\qquad = \sqrt{36(\tan^2\theta + 1)}$

$\qquad = 6\sqrt{\sec^2\theta}$

$\qquad = 6\sec\theta$ for $0 < \theta < \dfrac{\pi}{2}$

98. $x = 5\sec\theta$

$\sqrt{(x^2 - 25)^3} = \sqrt{[(5\sec\theta)^2 - 25]^3}$

$\qquad = \sqrt{(25\sec^2\theta - 25)^3}$

$\qquad = \sqrt{[25(\sec^2\theta - 1)]^3}$

$\qquad = \sqrt{(25\tan^2\theta)^3}$

$\qquad = \sqrt{15{,}625\tan^6\theta}$

$\qquad = 125\tan^3\theta$ for $0 < \theta < \dfrac{\pi}{2}$

99. $\cos x(\cos x + 1) = 0$

$\cos x = 0$ or $\cos x + 1 = 0$

$x = \dfrac{\pi}{2} + n\pi \qquad \cos x = -1$

$\qquad\qquad\qquad x = \pi + 2n\pi$

100. $\sin x(2 \sin x + \sqrt{2}) = 0$

$\sin x = 0 \qquad\qquad 2 \sin x + \sqrt{2} = 0$

$x = 0 + n\pi \qquad\qquad\qquad \sin x = -\dfrac{\sqrt{2}}{2}$

$$x = \dfrac{5\pi}{4} + 2\pi n, \dfrac{7\pi}{4} + 2\pi n$$

$x = n\pi, \dfrac{5\pi}{4} + 2\pi n, \dfrac{7\pi}{4} + 2\pi n$

101. $3 \sec x \sin x - 2\sqrt{3} \sin x = 0$

$\sin x(3 \sec x - 2\sqrt{3}) = 0$

$\sin x = 0 \quad \text{or} \quad 3 \sec x - 2\sqrt{3} = 0$

$x = n\pi \qquad\qquad \sec x = \dfrac{2\sqrt{3}}{3}$

$$\cos x = \dfrac{3}{2\sqrt{3}} = \dfrac{\sqrt{3}}{2}$$

$$x = \dfrac{\pi}{6} + 2n\pi$$

$$x = \dfrac{11\pi}{6} + 2n\pi$$

102. $\cos x \csc x + \cos x \sqrt{2} = 0$

$\cos x(\csc x + \sqrt{2}) = 0$

$\cos x = 0 \qquad\qquad \csc x + \sqrt{2} = 0$

$x = \dfrac{\pi}{2} + n\pi \qquad\qquad \csc x = -\sqrt{2}$

$$x = \dfrac{5\pi}{4} + 2\pi n, \dfrac{7\pi}{4} + 2\pi n$$

$x = \dfrac{\pi}{2} + n\pi, \dfrac{5\pi}{4} + 2\pi n, \dfrac{7\pi}{4} + 2\pi n$

Section 6.4   Vectors and Dot Products

■ Know the definition of the dot product of $\mathbf{u} = \langle u_1, u_2 \rangle$ and $\mathbf{v} = \langle v_1, v_2 \rangle$.

$\mathbf{u} \cdot \mathbf{v} = u_1 v_1 + u_2 v_2$

■ Know the following properties of the dot product:

1. $\mathbf{u} \cdot \mathbf{v} = \mathbf{v} \cdot \mathbf{u}$

2. $\mathbf{0} \cdot \mathbf{v} = 0$

3. $\mathbf{u} \cdot (\mathbf{v} + \mathbf{w}) = \mathbf{u} \cdot \mathbf{v} + \mathbf{u} \cdot \mathbf{w}$

4. $\mathbf{v} \cdot \mathbf{v} = \|\mathbf{v}\|^2$

5. $c(\mathbf{u} \cdot \mathbf{v}) = c\mathbf{u} \cdot \mathbf{v} = \mathbf{u} \cdot c\mathbf{v}$

■ If θ is the angle between two nonzero vectors $\mathbf{u}$ and $\mathbf{v}$, then

$$\cos \theta = \dfrac{\mathbf{u} \cdot \mathbf{v}}{\|\mathbf{u}\| \|\mathbf{v}\|}.$$

■ The vectors $\mathbf{u}$ and $\mathbf{v}$ are orthogonal if $\mathbf{u} \cdot \mathbf{v} = 0$.

■ Know the definition of vector components.

$\mathbf{u} = \mathbf{w}_1 + \mathbf{w}_2$ where $\mathbf{w}_1$ and $\mathbf{w}_2$ are orthogonal, and $\mathbf{w}_1$ is parallel to $\mathbf{v}$. $\mathbf{w}_1$ is called the projection of $\mathbf{u}$ onto $\mathbf{v}$

and is denoted by $\mathbf{w}_1 = \text{proj}_{\mathbf{v}}\mathbf{u} = \left(\dfrac{\mathbf{u} \cdot \mathbf{v}}{\|\mathbf{v}\|^2}\right)\mathbf{v}$. Then we have $\mathbf{w}_2 = \mathbf{u} - \mathbf{w}_1$.

■ Know the definition of work.

1. Projection form: $w = \|\text{proj}_{\overrightarrow{PQ}} \mathbf{F}\| \|\overrightarrow{PQ}\|$

2. Dot product form: $w = \mathbf{F} \cdot \overrightarrow{PQ}$

Vocabulary Check

1. dot product

2. $\dfrac{\mathbf{u} \cdot \mathbf{v}}{\|\mathbf{u}\| \, \|\mathbf{v}\|}$

3. orthogonal

4. $\left(\dfrac{\mathbf{u} \cdot \mathbf{v}}{\|\mathbf{v}\|^2}\right)\mathbf{v}$

5. $\|\text{proj}_{\overrightarrow{PQ}} \, \mathbf{F}\| \, \|\overrightarrow{PQ}\|; \; \mathbf{F} \cdot \overrightarrow{PQ}$

---

1. $\mathbf{u} = \langle 6, 1 \rangle, \; \mathbf{v} = \langle -2, 3 \rangle$

    $\mathbf{u} \cdot \mathbf{v} = 6(-2) + 1(3) = -9$

2. $\mathbf{u} = \langle 5, 12 \rangle, \; \mathbf{v} = \langle -3, 2 \rangle$

    $\mathbf{u} \cdot \mathbf{v} = 5(-3) + 12(2) = 9$

3. $\mathbf{u} = \langle -4, 1 \rangle, \; \mathbf{v} = \langle 2, -3 \rangle$

    $\mathbf{u} \cdot \mathbf{v} = -4(2) + 1(-3) = -11$

4. $\mathbf{u} = \langle -2, 5 \rangle, \; \mathbf{v} = \langle -1, -2 \rangle$

    $\mathbf{u} \cdot \mathbf{v} = (-2)(-1) + 5(-2)$

          $= 2 - 10 = -8$

5. $\mathbf{u} = 4\mathbf{i} - 2\mathbf{j}, \; \mathbf{v} = \mathbf{i} - \mathbf{j}$

    $\mathbf{u} \cdot \mathbf{v} = 4(1) + (-2)(-1) = 6$

6. $\mathbf{u} = 3\mathbf{i} + 4\mathbf{j}, \; \mathbf{v} = 7\mathbf{i} - 2\mathbf{j}$

    $\mathbf{u} \cdot \mathbf{v} = 3(7) + 4(-2) = 13$

7. $\mathbf{u} = 3\mathbf{i} + 2\mathbf{j}, \; \mathbf{v} = -2\mathbf{i} - 3\mathbf{j}$

    $\mathbf{u} \cdot \mathbf{v} = 3(-2) + 2(-3) = -12$

8. $\mathbf{u} = \mathbf{i} - 2\mathbf{j}, \; \mathbf{v} = -2\mathbf{i} + \mathbf{j}$

    $\mathbf{u} \cdot \mathbf{v} = 1(-2) + (-2)1 = -4$

9. $\mathbf{u} = \langle 2, 2 \rangle$

    $\mathbf{u} \cdot \mathbf{u} = 2(2) + 2(2) = 8$

    The result is a scalar.

10. $\mathbf{u} = \langle 2, 2 \rangle, \; \mathbf{v} = \langle -3, 4 \rangle$

    $3\mathbf{u} \cdot \mathbf{v} = 3[(2)(-3) + (2)(4)] = 3(2) = 6$

    The result is a scalar.

11. $\mathbf{u} = \langle 2, 2 \rangle, \; \mathbf{v} = \langle -3, 4 \rangle$

    $(\mathbf{u} \cdot \mathbf{v})\mathbf{v} = [(2)(-3) + 2(4)]\langle -3, 4 \rangle$

           $= 2\langle -3, 4 \rangle = \langle -6, 8 \rangle$

    The result is a vector.

12. $\mathbf{u} = \langle 2, 2 \rangle, \; \mathbf{v} = \langle -3, 4 \rangle, \; \mathbf{w} = \langle 1, -2 \rangle$

    $(\mathbf{v} \cdot \mathbf{u})\mathbf{w} = [(-3)(2) + (4)(2)]\langle 1, -2 \rangle$

           $= 2\langle 1, -2 \rangle$

           $= \langle 2, -4 \rangle$ vector

13. $\mathbf{u} = \langle 2, 2 \rangle, \; \mathbf{v} = \langle -3, 4 \rangle, \; \mathbf{w} = \langle 1, -2 \rangle$

    $(3\mathbf{w} \cdot \mathbf{v})\mathbf{u} = [3(1)(-3) + 3(-2)(4)]\langle 2, 2 \rangle$

           $= -33\langle 2, 2 \rangle$

           $= \langle -66, -66 \rangle$

    The result is a vector.

14. $\mathbf{u} = \langle 2, 2 \rangle, \; \mathbf{v} = \langle -3, 4 \rangle, \; \mathbf{w} = \langle 1, -2 \rangle$

    $2\mathbf{v} = \langle -6, 8 \rangle$

    $(\mathbf{u} \cdot 2\mathbf{v})\mathbf{w} = [(2)(-6) + (2)(8)]\langle 1, -2 \rangle$

             $= 4\langle 1, -2 \rangle$

             $= \langle 4, -8 \rangle$ vector

15. $\mathbf{w} = \langle 1, -2 \rangle$

    $\|\mathbf{w}\| - 1 = \sqrt{(1)^2 + (-2)^2} - 1 = \sqrt{5} - 1$

    The result is a scalar.

16. $\mathbf{u} = \langle 2, 2 \rangle$

    $2 - \|\mathbf{u}\| = 2 - \sqrt{\mathbf{u} \cdot \mathbf{u}}$

              $= 2 - \sqrt{(2)(2) + (2)(2)}$

              $= 2 - \sqrt{8}$

              $= 2 - 2\sqrt{2}$ scalar

17. $\mathbf{u} = \langle 2, 2 \rangle, \; \mathbf{v} = \langle -3, 4 \rangle, \; \mathbf{w} = \langle 1, -2 \rangle$

    $(\mathbf{u} \cdot \mathbf{v}) - (\mathbf{u} \cdot \mathbf{w}) = [2(-3) + 2(4)] - [2(1) + 2(-2)]$

                     $= 2 - (-2)$

                     $= 4$

    The result is a scalar.

18. $\mathbf{u} = \langle 2, 2 \rangle$, $\mathbf{v} = \langle -3, 4 \rangle$, $\mathbf{w} = \langle 1, -2 \rangle$

$(\mathbf{v} \cdot \mathbf{u}) - (\mathbf{w} \cdot \mathbf{v}) = [(-3)(2) + (4)(2)] - [(1)(-3) + (-2)(4)]$

$\qquad = 2 - (-11)$

$\qquad = 13$ scalar

19. $\mathbf{u} = \langle -5, 12 \rangle$

$\|\mathbf{u}\| = \sqrt{\mathbf{u} \cdot \mathbf{u}} = \sqrt{(-5)^2 + 12^2} = 13$

20. $\mathbf{u} = \langle 2, -4 \rangle$

$\|\mathbf{u}\| = \sqrt{\mathbf{u} \cdot \mathbf{u}}$

$\qquad = \sqrt{2(2) + (-4)(-4)}$

$\qquad = \sqrt{20} = 2\sqrt{5}$

21. $\mathbf{u} = 20\mathbf{i} + 25\mathbf{j}$

$\|\mathbf{u}\| = \sqrt{\mathbf{u} \cdot \mathbf{u}} = \sqrt{(20)^2 + (25)^2} = \sqrt{1025} = 5\sqrt{41}$

22. $\mathbf{u} = 12\mathbf{i} - 16\mathbf{j}$

$\|\mathbf{u}\| = \sqrt{\mathbf{u} \cdot \mathbf{u}} = \sqrt{12(12) + (-16)(-16)}$

$\qquad = \sqrt{400} = 20$

23. $\mathbf{u} = 6\mathbf{j}$

$\|\mathbf{u}\| = \sqrt{\mathbf{u} \cdot \mathbf{u}} = \sqrt{(0)^2 + (6)^2} = \sqrt{36} = 6$

24. $\mathbf{u} = -21\mathbf{i}$

$\|\mathbf{u}\| = \sqrt{\mathbf{u} \cdot \mathbf{u}} = \sqrt{(-21)(-21) + 0(0)}$

$\qquad = \sqrt{21^2} = 21$

25. $\mathbf{u} = \langle 1, 0 \rangle$, $\mathbf{v} = \langle 0, -2 \rangle$

$\cos \theta = \dfrac{\mathbf{u} \cdot \mathbf{v}}{\|\mathbf{u}\| \, \|\mathbf{v}\|} = \dfrac{0}{(1)(2)} = 0$

$\theta = 90°$

26. $\mathbf{u} = \langle 3, 2 \rangle$, $\mathbf{v} = \langle 4, 0 \rangle$

$\cos \theta = \dfrac{\mathbf{u} \cdot \mathbf{v}}{\|\mathbf{u}\| \, \|\mathbf{v}\|} = \dfrac{3(4) + 2(0)}{\sqrt{13}\,(4)}$

$\qquad = \dfrac{3}{\sqrt{13}} \approx 0.83205$

$\qquad \theta \approx 33.69°$

27. $\mathbf{u} = 3\mathbf{i} + 4\mathbf{j}$, $\mathbf{v} = -2\mathbf{j}$

$\cos \theta = \dfrac{\mathbf{u} \cdot \mathbf{v}}{\|\mathbf{u}\| \, \|\mathbf{v}\|} = -\dfrac{8}{(5)(2)}$

$\qquad \theta = \arccos\left(-\dfrac{4}{5}\right)$

$\qquad \theta \approx 143.13°$

28. $\mathbf{u} = 2\mathbf{i} - 3\mathbf{j}$, $\mathbf{v} = \mathbf{i} - 2\mathbf{j}$

$\cos \theta = \dfrac{\mathbf{u} \cdot \mathbf{v}}{\|\mathbf{u}\| \, \|\mathbf{v}\|}$

$\qquad = \dfrac{2(1) + (-3)(-2)}{\sqrt{2^2 + 3^2}\sqrt{1^2 + 2^2}}$

$\qquad = \dfrac{8}{\sqrt{65}} \approx 0.992278$

$\theta \approx 7.13°$

29. $\mathbf{u} = 2\mathbf{i} - \mathbf{j}$, $\mathbf{v} = 6\mathbf{i} + 4\mathbf{j}$

$\cos \theta = \dfrac{\mathbf{u} \cdot \mathbf{v}}{\|\mathbf{u}\| \, \|\mathbf{v}\|} = \dfrac{8}{\sqrt{5}\sqrt{52}} = 0.4961$

$\qquad \theta = 60.26°$

30. $\mathbf{u} = -6\mathbf{i} - 3\mathbf{j}$, $\mathbf{v} = -8\mathbf{i} + 4\mathbf{j}$

$\cos \mathbf{u} = \dfrac{\mathbf{u} \cdot \mathbf{v}}{\|\mathbf{u}\| \, \|\mathbf{v}\|} = \dfrac{-6(-8) + (-3)(4)}{\sqrt{45}\sqrt{80}} = \dfrac{36}{60} = 0.6$

$\qquad \theta \approx 53.13°$

31. $\mathbf{u} = 5\mathbf{i} + 5\mathbf{j}$, $\mathbf{v} = -6\mathbf{i} + 6\mathbf{j}$

$\cos \theta = \dfrac{\mathbf{u} \cdot \mathbf{v}}{\|\mathbf{u}\| \, \|\mathbf{v}\|} = 0$

$\qquad \theta = 90°$

32. $\mathbf{u} = 2\mathbf{i} - 3\mathbf{j}$, $\mathbf{v} = 4\mathbf{i} + 3\mathbf{j}$

$\cos \theta = \dfrac{\mathbf{u} \cdot \mathbf{v}}{\|\mathbf{u}\| \, \|\mathbf{v}\|} = \dfrac{2(4) + (-3)(3)}{\sqrt{13}\sqrt{25}} \approx -0.0555$

$\qquad \theta \approx 93.18°$

33. $\mathbf{u} = \left(\cos\dfrac{\pi}{3}\right)\mathbf{i} + \left(\sin\dfrac{\pi}{3}\right)\mathbf{j} = \dfrac{1}{2}\mathbf{i} + \dfrac{\sqrt{3}}{2}\mathbf{j}$

$\mathbf{v} = \left(\cos\dfrac{3\pi}{4}\right)\mathbf{i} + \left(\sin\dfrac{3\pi}{4}\right)\mathbf{j} = -\dfrac{\sqrt{2}}{2}\mathbf{i} + \dfrac{\sqrt{2}}{2}\mathbf{j}$

$\|\mathbf{u}\| = \|\mathbf{v}\| = 1$

$\cos\theta = \dfrac{\mathbf{u}\cdot\mathbf{v}}{\|\mathbf{u}\|\,\|\mathbf{v}\|} = \mathbf{u}\cdot\mathbf{v}$

$\qquad = \left(\dfrac{1}{2}\right)\left(-\dfrac{\sqrt{2}}{2}\right) + \left(\dfrac{\sqrt{3}}{2}\right)\left(\dfrac{\sqrt{2}}{2}\right) = \dfrac{-\sqrt{2}+\sqrt{6}}{4}$

$\theta = \arccos\left(\dfrac{-\sqrt{2}+\sqrt{6}}{4}\right) = 75° = \dfrac{5\pi}{12}$

34. $\mathbf{u} = \cos\left(\dfrac{\pi}{4}\right)\mathbf{i} + \sin\left(\dfrac{\pi}{4}\right)\mathbf{j} = \dfrac{\sqrt{2}}{2}\mathbf{i} + \dfrac{\sqrt{2}}{2}\mathbf{j}$

$\mathbf{v} = \cos\left(\dfrac{\pi}{2}\right)\mathbf{i} + \sin\left(\dfrac{\pi}{2}\right)\mathbf{j} = \mathbf{j}$

$\cos\theta = \dfrac{\mathbf{u}\cdot\mathbf{v}}{\|\mathbf{u}\|\,\|\mathbf{v}\|} = \dfrac{\dfrac{\sqrt{2}}{2}(0) + \dfrac{\sqrt{2}}{2}(1)}{1\cdot 1} = \dfrac{\sqrt{2}}{2}$

$\theta = \dfrac{\pi}{4}$

35. $\mathbf{u} = 3\mathbf{i} + 4\mathbf{j}$

$\mathbf{v} = -7\mathbf{i} + 5\mathbf{j}$

$\cos\theta = \dfrac{\mathbf{u}\cdot\mathbf{v}}{\|\mathbf{u}\|\,\|\mathbf{v}\|}$

$\qquad = \dfrac{3(-7) + 4(5)}{3\sqrt{74}}$

$\qquad = \dfrac{-1}{5\sqrt{74}} \approx -0.0232$

$\theta \approx 91.3°$

36. $\mathbf{u} = 6\mathbf{i} + 3\mathbf{j},\ \mathbf{v} = -4\mathbf{i} + 4\mathbf{j}$

$\mathbf{u}\cdot\mathbf{v} = \|\mathbf{u}\|\,\|\mathbf{v}\|\cos\theta$

$(6)(-4) + (3)(4) = \sqrt{6^2 + 3^2}\cdot\sqrt{(-4)^2 + (4)^2}\cdot\cos\theta$

$\qquad\qquad -12 = \sqrt{45}\cdot\sqrt{32}\cos\theta$

$\qquad\qquad -12 = 12\sqrt{10}\cos\theta$

$\qquad\qquad \dfrac{-1}{\sqrt{10}} = \cos\theta$

$\cos^{-1}\left(\dfrac{-1}{\sqrt{10}}\right) = \theta \Rightarrow \theta \approx 108.4°$

37. $\mathbf{u} = 5\mathbf{i} + 5\mathbf{j}$

$\mathbf{v} = -8\mathbf{i} + 8\mathbf{j}$

$\cos\theta = \dfrac{\mathbf{u}\cdot\mathbf{v}}{\|\mathbf{u}\|\,\|\mathbf{v}\|}$

$\qquad = \dfrac{5(-8) + 5(8)}{\sqrt{50}\sqrt{128}}$

$\qquad = 0$

$\theta = 90°$

38. $\mathbf{u} = 2\mathbf{i} - 3\mathbf{j},\ \mathbf{v} = 8\mathbf{i} + 3\mathbf{j}$

$\mathbf{u}\cdot\mathbf{v} = \|\mathbf{u}\|\,\|\mathbf{v}\|\cos\theta$

$(2)(8) + (-3)(3) = \sqrt{2^2 + (-3)^2}\cdot\sqrt{8^2 + 3^2}\cdot\cos\theta$

$\qquad\qquad 7 = \sqrt{13}\cdot\sqrt{73}\cos\theta$

$\qquad\qquad \dfrac{7}{\sqrt{13}\cdot\sqrt{73}} = \cos\theta$

$\cos^{-1}\left[\dfrac{7}{\sqrt{13}\cdot\sqrt{73}}\right] = \theta \Rightarrow \theta \approx 76.9°$

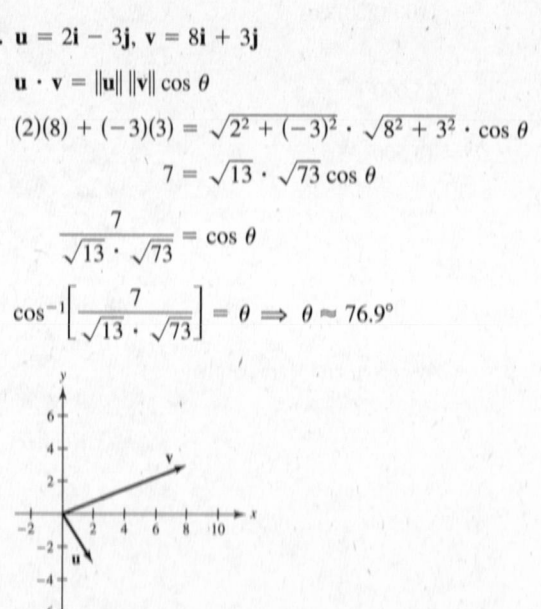

39. $P = (1, 2), \quad Q = (3, 4), \quad R = (2, 5)$

$\overrightarrow{PQ} = \langle 2, 2 \rangle, \quad \overrightarrow{PR} = \langle 1, 3 \rangle, \quad \overrightarrow{QR} = \langle -1, 1 \rangle$

$\cos \alpha = \dfrac{\overrightarrow{PQ} \cdot \overrightarrow{PR}}{\|\overrightarrow{PQ}\| \, \|\overrightarrow{PR}\|} = \dfrac{8}{(2\sqrt{2})(\sqrt{10})} \implies \alpha = \arccos \dfrac{2}{\sqrt{5}} \approx 26.57°$

$\cos \beta = \dfrac{\overrightarrow{PQ} \cdot \overrightarrow{QR}}{\|\overrightarrow{PQ}\| \, \|\overrightarrow{QR}\|} = 0 \implies \beta = 90°.$ Thus, $\gamma = 180° - 26.57° - 90° = 63.43°.$

40. $P = (-3, -4), Q = (1, 7), R = (8, 2)$

$\overrightarrow{PQ} = \langle 4, 11 \rangle, \quad \overrightarrow{QR} = \langle 7, -5 \rangle,$

$\overrightarrow{PR} = \langle 11, 6 \rangle, \quad \overrightarrow{QP} = \langle -4, -11 \rangle$

$\cos \alpha = \dfrac{\overrightarrow{PQ} \cdot \overrightarrow{PR}}{\|\overrightarrow{PQ}\| \, \|\overrightarrow{PR}\|} = \dfrac{110}{(\sqrt{137})(\sqrt{157})} \implies \alpha \approx 41.41°$

$\cos \beta = \dfrac{\overrightarrow{QR} \cdot \overrightarrow{QP}}{\|\overrightarrow{QR}\| \, \|\overrightarrow{QP}\|} = \dfrac{27}{(\sqrt{74})(\sqrt{137})} \implies \beta \approx 74.45°$

$\gamma \approx 180° - 41.41° - 74.45° = 64.14°$

41. $P = (-3, 0), Q = (2, 2), R = (0, 6)$

$\overrightarrow{QP} = \langle -5, -2 \rangle, \overrightarrow{PR} = \langle 3, 6 \rangle, \overrightarrow{QR} = \langle -2, 4 \rangle, \overrightarrow{PQ} = \langle 5, 2 \rangle$

$\cos \alpha = \dfrac{\overrightarrow{PQ} \cdot \overrightarrow{PR}}{\|\overrightarrow{PQ}\| \, \|\overrightarrow{PR}\|} = \dfrac{27}{\sqrt{29}\sqrt{45}} \implies \alpha \approx 41.63°$

$\cos \beta = \dfrac{\overrightarrow{QP} \cdot \overrightarrow{QR}}{\|\overrightarrow{QP}\| \, \|\overrightarrow{PR}\|} = \dfrac{2}{\sqrt{29}\sqrt{20}} \implies \beta \approx 85.24°$

$\delta = 180° - 41.63° - 85.24° = 53.13°$

42. $P = (-3, 5), Q = (-1, 9), R = (7, 9)$

$\overrightarrow{PQ} = \langle 2, 4 \rangle, \quad \overrightarrow{QR} = \langle 8, 0 \rangle,$

$\overrightarrow{PR} = \langle 10, 4 \rangle, \quad \overrightarrow{QP} = \langle -2, -4 \rangle$

$\cos \alpha = \dfrac{\overrightarrow{PQ} \cdot \overrightarrow{PR}}{\|\overrightarrow{PQ}\| \, \|\overrightarrow{PR}\|} = \dfrac{36}{(\sqrt{20})(\sqrt{116})} \implies \alpha \approx 41.6°$

$\cos \beta = \dfrac{\overrightarrow{QR} \cdot \overrightarrow{QP}}{\|\overrightarrow{QR}\| \, \|\overrightarrow{QP}\|} = \dfrac{-16}{8(\sqrt{20})} \implies \beta \approx 116.6°$

$\gamma \approx 180° - 41.6° - 116.6° = 21.8°$

43. $\mathbf{u} \cdot \mathbf{v} = \|\mathbf{u}\| \, \|\mathbf{v}\| \cos \theta$

$= (4)(10) \cos \dfrac{2\pi}{3}$

$= 40 \left(-\dfrac{1}{2} \right)$

$= -20$

44. $\|\mathbf{u}\| = 100, \|\mathbf{v}\| = 250, \theta = \dfrac{\pi}{6}$

$\mathbf{u} \cdot \mathbf{v} = \|\mathbf{u}\| \, \|\mathbf{v}\| \cos \theta$

$= (100)(250) \cos \dfrac{\pi}{6}$

$= 25{,}000 \cdot \dfrac{\sqrt{3}}{2}$

$= 12{,}500\sqrt{3}$

45. $\mathbf{u} \cdot \mathbf{v} = \|\mathbf{u}\| \, \|\mathbf{v}\| \cos \theta$

$= (9)(36) \cos \dfrac{3\pi}{4}$

$= 324 \left(-\dfrac{\sqrt{2}}{2} \right)$

$= -162\sqrt{2} \approx -229.1$

46. $\|\mathbf{u}\| = 4$

$\|\mathbf{v}\| = 12$

$\theta = \dfrac{\pi}{3}$

$\mathbf{u} \cdot \mathbf{v} = \|\mathbf{u}\| \, \|\mathbf{v}\| \cos \theta$

$= (4)(12) \cos \dfrac{\pi}{3}$

$= (4)(12) \left(\dfrac{1}{2} \right) = 24$

47. $\mathbf{u} = \langle -12, 30 \rangle, \quad \mathbf{v} = \left\langle \dfrac{1}{2}, -\dfrac{5}{4} \right\rangle$

$\mathbf{u} = -24\mathbf{v} \implies \mathbf{u}$ and $\mathbf{v}$ are parallel.

48. $\mathbf{u} = \langle 3, 15 \rangle, \mathbf{v} = \langle -1, 5 \rangle$

$\mathbf{u} \neq k\mathbf{v} \implies$ Not parallel

$\mathbf{u} \cdot \mathbf{v} \neq 0 \implies$ Not orthogonal

Neither

49. $\mathbf{u} = \dfrac{1}{4}(3\mathbf{i} - \mathbf{j}), \mathbf{v} = 5\mathbf{i} + 6\mathbf{j}$

$\mathbf{u} \neq k\mathbf{v} \implies$ Not parallel

$\mathbf{u} \cdot \mathbf{v} \neq 0 \implies$ Not orthogonal

Neither

50. $\mathbf{u} = 1, \mathbf{v} = -2\mathbf{i} + 2\mathbf{j}$

     $\mathbf{u} \neq k\mathbf{v} \implies$ Not parallel

     $\mathbf{u} \cdot \mathbf{v} \neq 0 \implies$ Not orthogonal

     Neither

51. $\mathbf{u} = 2\mathbf{i} - 2\mathbf{j}, \mathbf{v} = -\mathbf{i} - \mathbf{j}$

     $\mathbf{u} \cdot \mathbf{v} = 0 \implies \mathbf{u}$ and $\mathbf{v}$ are orthogonal.

52. $\mathbf{u} = \langle \cos\theta, \sin\theta \rangle$

     $\mathbf{v} = \langle \sin\theta, -\cos\theta \rangle$

     $\mathbf{u} \cdot \mathbf{v} = 0 \implies \mathbf{u}$ and $\mathbf{v}$ are orthogonal.

53. $\mathbf{u} = \langle 2, 2 \rangle, \mathbf{v} = \langle 6, 1 \rangle$

$$\mathbf{w}_1 = \text{proj}_{\mathbf{v}}\mathbf{u} = \left(\frac{\mathbf{u} \cdot \mathbf{v}}{\|\mathbf{v}\|^2} \right)\mathbf{v} = \frac{14}{37}\langle 6, 1 \rangle = \frac{1}{37}\langle 84, 14 \rangle$$

$$\mathbf{w}_2 = \mathbf{u} - \mathbf{w}_1 = \langle 2, 2 \rangle - \frac{14}{37}\langle 6, 1 \rangle = \left\langle -\frac{10}{37}, \frac{60}{37} \right\rangle = \frac{10}{37}\langle -1, 6 \rangle = \frac{1}{37}\langle -10, 60 \rangle$$

$$\mathbf{u} = \frac{1}{37}\langle 84, 14 \rangle + \frac{1}{37}\langle -10, 60 \rangle = \langle 2, 2 \rangle$$

54. $\mathbf{u} = \langle 4, 2 \rangle, \mathbf{v} = \langle 1, -2 \rangle$

$$\mathbf{w}_1 = \text{proj}_{\mathbf{v}}\mathbf{u} = \left(\frac{\mathbf{u} \cdot \mathbf{v}}{\|\mathbf{v}\|^2} \right)\mathbf{v} = 0\langle 1, -2 \rangle = \langle 0, 0 \rangle$$

$$\mathbf{w}_2 = \mathbf{u} - \mathbf{w}_1 = \langle 4, 2 \rangle - \langle 0, 0 \rangle = \langle 4, 2 \rangle$$

$$\mathbf{u} = \langle 4, 2 \rangle + \langle 0, 0 \rangle = \langle 4, 2 \rangle$$

55. $\mathbf{u} = \langle 0, 3 \rangle, \mathbf{v} = \langle 2, 15 \rangle$

$$\mathbf{w}_1 = \text{proj}_{\mathbf{v}}\mathbf{u} = \left(\frac{\mathbf{u} \cdot \mathbf{v}}{\|\mathbf{v}\|^2} \right)\mathbf{v} = \frac{45}{229}\langle 2, 15 \rangle$$

$$\mathbf{w}_2 = \mathbf{u} - \mathbf{w}_1 = \langle 0, 3 \rangle - \frac{45}{229}\langle 2, 15 \rangle = \left\langle -\frac{90}{229}, \frac{12}{229} \right\rangle$$

$$= \frac{6}{229}\langle -15, 2 \rangle$$

$$\mathbf{u} = \frac{45}{229}\langle 2, 15 \rangle + \frac{6}{229}\langle -15, 2 \rangle = \langle 0, 3 \rangle$$

56. $\mathbf{u} = \langle -3, -2 \rangle, \mathbf{v} = \langle -4, -1 \rangle$

$$\mathbf{w}_1 = \text{proj}_{\mathbf{v}}\mathbf{u} = \left(\frac{\mathbf{u} \cdot \mathbf{v}}{\|\mathbf{v}\|^2} \right)\mathbf{v} = \left(\frac{14}{17} \right)\langle -4, -1 \rangle$$

$$\mathbf{w}_2 = \mathbf{u} - \mathbf{w}_1 = \langle -3, -2 \rangle - \frac{14}{17}\langle -4, -1 \rangle = \frac{5}{17}\langle 1, -4 \rangle$$

$$\mathbf{u} = \frac{14}{17}\langle -4, -1 \rangle + \frac{5}{17}\langle 1, -4 \rangle = \langle -3, -2 \rangle$$

57. $\text{proj}_{\mathbf{v}}\mathbf{u} = \mathbf{0}$ since they are perpendicular.

     Since $\mathbf{u}$ and $\mathbf{v}$ are orthogonal, $\mathbf{u} \cdot \mathbf{v} = \mathbf{0}$ and $\text{proj}_{\mathbf{v}}\mathbf{u} = \mathbf{0}$.

     $\text{proj}_{\mathbf{v}}\mathbf{u} = \dfrac{\mathbf{u} \cdot \mathbf{v}}{\|\mathbf{v}\|^2}\mathbf{v} = \mathbf{0}$, since $\mathbf{u} \cdot \mathbf{v} = \mathbf{0}$.

58. Because $\mathbf{u}$ and $\mathbf{v}$ are orthogonal, the projection of $\mathbf{u}$ onto $\mathbf{v}$ is $\mathbf{0}$.

     $\text{proj}_{\mathbf{w}}\,\mathbf{u} = \dfrac{\mathbf{u} \cdot \mathbf{v}}{\|\mathbf{v}\|^2}\mathbf{v} = \mathbf{0}$ since $\mathbf{u} \cdot \mathbf{v} = \mathbf{0}$.

59. $\mathbf{u} = \langle 3, 5 \rangle$

     For $\mathbf{v}$ to be orthogonal to $\mathbf{u}$, $\mathbf{u} \cdot \mathbf{v}$ must equal 0.

     Two possibilities: $\langle -5, 3 \rangle$ and $\langle 5, -3 \rangle$

60. $\mathbf{u} = \langle -8, 3 \rangle$

     For $\mathbf{v}$ to be orthogonal to $\mathbf{u}$, $\mathbf{u} \cdot \mathbf{v}$ must be equal to 0.

     Two possibilities: $\langle 3, 8 \rangle, \langle -3, -8 \rangle$

61. $\mathbf{u} = \frac{1}{2}\mathbf{i} - \frac{2}{3}\mathbf{j}$

     For $\mathbf{u}$ and $\mathbf{v}$ to be orthogonal, $\mathbf{u} \cdot \mathbf{v}$ must equal 0.

     Two possibilities: $\mathbf{v} = \frac{2}{3}\mathbf{i} + \frac{1}{2}\mathbf{j}$ and $\mathbf{v} = -\frac{2}{3}\mathbf{i} - \frac{1}{2}\mathbf{j}$

62. $\mathbf{u} = -\frac{5}{2}\mathbf{i} - 3\mathbf{j}$

     For $\mathbf{v}$ to be orthogonal to $\mathbf{u}$, $\mathbf{u} \cdot \mathbf{v}$ must be equal to 0.

     Two possibilities: $\mathbf{v} = 3\mathbf{i} - \frac{5}{2}\mathbf{j}$ and $\mathbf{v} = -3\mathbf{i} + \frac{5}{2}\mathbf{j}$

63. $\mathbf{w} = \|\operatorname{proj}_{\overrightarrow{PQ}}\mathbf{v}\|\|\overrightarrow{PQ}\|$ where $\overrightarrow{PQ} = \langle 4, 7 \rangle$ and $\mathbf{v} = \langle 1, 4 \rangle$.

$$\operatorname{proj}_{\overrightarrow{PQ}}\mathbf{v} = \left(\frac{\mathbf{v} \cdot \overrightarrow{PQ}}{\|\overrightarrow{PQ}\|^2}\right)\overrightarrow{PQ} = \left(\frac{32}{65}\right)\langle 4, 7 \rangle$$

$$\mathbf{w} = \|\operatorname{proj}_{\overrightarrow{PQ}}\mathbf{v}\|\|\overrightarrow{PQ}\| = \left(\frac{32\sqrt{65}}{65}\right)\left(\sqrt{65}\right) = 32$$

64. $P = (1, 3),\ Q = (-3, 5),\ \mathbf{v} = -2\mathbf{i} + 3\mathbf{j}$

$$\text{work} = \mathbf{v} \cdot \overrightarrow{PQ}$$

$$= (-2\mathbf{i} + 3\mathbf{j}) \cdot (-4\mathbf{i} + 2\mathbf{j})$$

$$= (-2)(-4) + 3(2) = 14$$

65. (a) $\mathbf{u} = \langle 1650, 3200 \rangle,\ \mathbf{v} = \langle 15.25, 10.50 \rangle$

$\mathbf{u} \cdot \mathbf{v} = 1650(15.25) + 3200(10.50) = \$58{,}762.50$

This gives the total revenue that can be earned by selling all of the pans.

(b) Increase prices by 5%: $1.05\mathbf{v}$ The operation is scalar multiplication.

$\mathbf{u} \cdot 1.05\mathbf{v} = 1.05\mathbf{u} \cdot \mathbf{v}$

$= 1.05[1650(15.25) + 3200(10.50)]$

$= 1.05(58{,}762.50)$

$= 61{,}700.63$

66. (a) $\mathbf{u} = \langle 3240, 2450 \rangle,\ \mathbf{v} = \langle 1.75, 1.25 \rangle$

$\mathbf{u} \cdot \mathbf{v} = (3240)(1.75) + (2450)(1.25) = 8732.5$

The fast food stand sold $8732.50 of hamburgers and hot dogs in one month.

(b) Increase prices by 2.5%:

$1.025\mathbf{v}$ scalar multiplication

67. (a) Force due to gravity:

$\mathbf{F} = -30{,}000\mathbf{j}$

Unit vector along hill:

$\mathbf{v} = (\cos d)\mathbf{i} + (\sin d)\mathbf{j}$

Projection of $\mathbf{F}$ onto $\mathbf{v}$:

$$\mathbf{w}_1 = \operatorname{proj}_{\mathbf{v}}\mathbf{F} = \left(\frac{\mathbf{F} \cdot \mathbf{v}}{\|\mathbf{v}\|^2}\right)\mathbf{v} = (\mathbf{F} \cdot \mathbf{v})\mathbf{v} = -30{,}000 \sin d\,\mathbf{v}$$

The magnitude of the force is $30{,}000 \sin d$.

(b)

d	0°	1°	2°	3°	4°	5°	6°	7°	8°	9°	10°
Force	0	523.6	1047.0	1570.1	2092.7	2614.7	3135.9	3656.1	4175.2	4693.0	5209.4

(c) Force perpendicular to the hill when $d = 5°$:

$\text{Force} = \sqrt{(30{,}000)^2 - (2614.7)^2} \approx 29{,}885.8$ pounds

68. Force due to gravity: $\mathbf{F} = -5400\mathbf{j}$

Unit vector along hill: $\mathbf{v} = (\cos 10°)\mathbf{i} + (\sin 10°)\mathbf{j}$

Projection of $\mathbf{F}$ onto $\mathbf{v}$: $\quad \mathbf{w}_1 = \operatorname{proj}_{\mathbf{v}}\mathbf{F}$

$$= \left(\frac{\mathbf{F} \cdot \mathbf{v}}{\|\mathbf{v}\|^2}\right)\mathbf{v}$$

$= (\mathbf{F} \cdot \mathbf{v})\mathbf{v}$ because $\mathbf{v}$ is a unit vector, $\|\mathbf{v}\| = 1$

$= [(0)(\cos 10°) + (-5400)(\sin 10°)]\mathbf{v}$

$= -5400(\sin 10°)\mathbf{v} = -937.7\mathbf{v}$

The magnitude of the force is 937.7, so a force of 937.7 pounds is required to keep the vehicle from rolling down the hill.

Force perpendicular to the hill: $\text{Force} = \sqrt{(5400)^2 - (937)^2} \approx 5318.0$ pounds

69. $\mathbf{w} = (245)(3) = 735$ newton-meters

70. work $= (2400)(5) = 12{,}000$ foot-pounds

71. $\mathbf{w} = (\cos 30°)(45)(20) \approx 779.4$ foot-pounds

72. work $= (\cos 35°)(15{,}691)(800)$

$\approx 10{,}282{,}652$ newton-meters

73. $\mathbf{w} = (\cos 30°)(250)(100) \approx 21{,}650.64$ foot-pounds

74. work $= (\cos \theta)\|\mathbf{F}\|\,\|\overrightarrow{PQ}\|$

$= (\cos 20°)(25 \text{ pounds})(50 \text{ feet})$

$= 1174.62$ foot-pounds

75. False. Work is represented by a scalar.

76. True.

$W = \mathbf{F} \cdot \overrightarrow{PQ} = 0$ if $\mathbf{F}$ and $\overrightarrow{PQ}$ are orthogonal.

77. (a) $\mathbf{u} \cdot \mathbf{v} = 0 \implies \mathbf{u}$ and $\mathbf{v}$ are orthogonal and $\theta = \dfrac{\pi}{2}$.

(b) $\mathbf{u} \cdot \mathbf{v} > 0 \implies \cos \theta > 0 \implies 0 \le \theta < \dfrac{\pi}{2}$

(c) $\mathbf{u} \cdot \mathbf{v} < 0 \implies \cos \theta < 0 \implies \dfrac{\pi}{2} < \theta \le \pi$

78. (a) $\text{proj}_\mathbf{v}\mathbf{u} = \mathbf{u} \implies \mathbf{u}$ and $\mathbf{v}$ are parallel.

(b) $\text{proj}_\mathbf{v}\mathbf{u} = 0 \implies \mathbf{u}$ and $\mathbf{v}$ are orthogonal.

79. In a rhombus, $\|\mathbf{u}\| = \|\mathbf{v}\|$. The diagonals are $\mathbf{u} + \mathbf{v}$ and $\mathbf{u} - \mathbf{v}$.

$(\mathbf{u} + \mathbf{v}) \cdot (\mathbf{u} - \mathbf{v}) = (\mathbf{u} + \mathbf{v}) \cdot \mathbf{u} - (\mathbf{u} + \mathbf{v}) \cdot \mathbf{v}$

$= \mathbf{u} \cdot \mathbf{u} + \mathbf{v} \cdot \mathbf{u} - \mathbf{u} \cdot \mathbf{v} - \mathbf{v} \cdot \mathbf{v}$

$= \|\mathbf{u}\|^2 - \|\mathbf{v}\|^2 = 0$

Therefore, the diagonals are orthogonal.

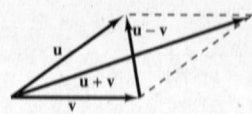

80. Let $\mathbf{u} = \langle u_1, u_2 \rangle$ and $\mathbf{v} = \langle v_1, v_2 \rangle$.

$\mathbf{u} - \mathbf{v} = \langle u_1 - v_1, u_2 - v_2 \rangle$

$\|\mathbf{u} - \mathbf{v}\|^2 = (u_1 - v_1)^2 + (u_2 - v_2)^2$

$= u_1^2 - 2u_1v_1 + v_1^2 + u_2^2 - 2u_2v_2 + v_2^2$

$= u_1^2 + u_2^2 + v_1^2 + v_2^2 - 2u_1v_1 - 2u_2v_2$

$= \|\mathbf{u}\|^2 + \|\mathbf{v}\|^2 - 2(u_1v_1 + u_2v_2)$

$= \|\mathbf{u}\|^2 + \|\mathbf{v}\|^2 - 2\mathbf{u} \cdot \mathbf{v}$

81. $\sqrt{42} \cdot \sqrt{24} = \sqrt{1008}$

$= \sqrt{144 \cdot 7}$

$= 12\sqrt{7}$

82. $\sqrt{18} \cdot \sqrt{112} = \sqrt{18 \cdot 112}$

$= \sqrt{2 \cdot 3^2 \cdot 2^4 \cdot 7}$

$= (3 \cdot 2^2)\sqrt{2 \cdot 7}$

$= 12\sqrt{14}$

83. $\sqrt{-3}\,\sqrt{-8} = \left(\sqrt{3}i\right)\left(2\sqrt{2}i\right)$

$= 2\sqrt{6}i^2$

$= -2\sqrt{6}$

84. $\sqrt{-12} \cdot \sqrt{-96} = i\sqrt{12}\,\cdot i\sqrt{96}$

$= i^2\sqrt{12 \cdot 96}$

$= \sqrt{2^2 \cdot 3 \cdot 2^5 \cdot 3}$

$= \sqrt{2 \cdot 2^6 \cdot 3^2}$

$= (2^3 \cdot 3)\sqrt{2}$

$= -24\sqrt{2}$

85. $\sin 2x - \sqrt{3}\sin x = 0$

$2\sin x \cos x - \sqrt{3}\sin x = 0$

$\sin x\left(2\cos x - \sqrt{3}\right) = 0$

$\sin x = 0 \quad \text{or} \quad 2\cos x - \sqrt{3} = 0$

$x = 0, \pi \qquad\qquad \cos x = \dfrac{\sqrt{3}}{2}$

$x = \dfrac{\pi}{6}, \dfrac{11\pi}{6}$

86. $\sin 2x + \sqrt{2} \cos x = 0$

$2 \sin x \cos x + \sqrt{2} \cos x = 0$

$\cos x(2 \sin x + \sqrt{2}) = 0$

$\cos x = 0 \qquad\qquad 2 \sin x + \sqrt{2} = 0$

$x = \dfrac{\pi}{2}, \dfrac{3\pi}{2} \qquad\qquad \sin x = -\dfrac{\sqrt{2}}{2}$

$x = \dfrac{5\pi}{4}, \dfrac{7\pi}{4}$

$x = \dfrac{\pi}{2}, \dfrac{5\pi}{4}, \dfrac{3\pi}{2}, \dfrac{7\pi}{4}$

87. $2 \tan x = \tan 2x$

$2 \tan x = \dfrac{2 \tan x}{1 - \tan^2 x}$

$2 \tan x(1 - \tan^2 x) = 2 \tan x$

$2 \tan x(1 - \tan^2 x) - 2 \tan x = 0$

$2 \tan x[(1 - \tan^2 x) - 1] = 0$

$2 \tan x(-\tan^2 x) = 0$

$-2 \tan^3 x = 0$

$\tan x = 0$

$x = 0, \pi$

88. $\cos 2x - 3 \sin x = 2$

$1 - 2 \sin^2 x - 3 \sin x - 2 = 0$

$2 \sin^2 x + 3 \sin x + 1 = 0$

$(2 \sin x + 1)(\sin x + 1) = 0$

$2 \sin x + 1 = 0 \qquad\qquad \sin x + 1 = 0$

$\sin x = -\dfrac{1}{2} \qquad\qquad \sin x = -1$

$x = \dfrac{7\pi}{6}, \dfrac{11\pi}{6} \qquad\qquad x = \dfrac{3\pi}{2}$

$x = \dfrac{7\pi}{6}, \dfrac{3\pi}{2}, \dfrac{11\pi}{6}$

For Exercises 89–92:

$\sin u = -\dfrac{12}{13}, u$ in **Quadrant IV** $\Rightarrow \cos u = \dfrac{5}{13}$ $\qquad \cos v = \dfrac{24}{25}, v$ in **Quadrant IV** $\Rightarrow \sin v = -\dfrac{7}{25}$

89. $\sin(u - v) = \sin u \cos v - \cos u \sin v$

$= \left(-\dfrac{12}{13}\right)\left(\dfrac{24}{25}\right) - \left(\dfrac{5}{13}\right)\left(-\dfrac{7}{25}\right)$

$= -\dfrac{253}{325}$

90. $\sin u = -\dfrac{12}{13}, \cos u = \sqrt{1 - \left(-\dfrac{12}{13}\right)^2} = \dfrac{5}{13}$

$\cos v = \dfrac{24}{25}, \sin v = -\sqrt{1 - \left(\dfrac{24}{25}\right)^2} = -\dfrac{7}{25}$

$\sin(u + v) = \sin u \cos v + \cos u \sin v$

$= \left(-\dfrac{12}{13}\right)\left(\dfrac{24}{25}\right) + \left(\dfrac{5}{13}\right)\left(-\dfrac{7}{25}\right)$

$= -\dfrac{323}{325}$

91. $\cos(v - u) = \cos v \cos u + \sin v \sin u$

$= \left(\dfrac{24}{25}\right)\left(\dfrac{5}{13}\right) + \left(-\dfrac{7}{25}\right)\left(-\dfrac{12}{13}\right)$

$= \dfrac{204}{325}$

92. $\sin u = -\dfrac{12}{13}, \cos u = \dfrac{5}{13}, \tan u = -\dfrac{12}{5}$

$\cos v = \dfrac{24}{25}, \sin v = -\dfrac{7}{25}, \tan v = -\dfrac{7}{24}$

$\tan(u - v) = \dfrac{\tan u - \tan v}{1 + \tan u \tan v}$

$= \dfrac{\left(-\dfrac{12}{5}\right) - \left(-\dfrac{7}{24}\right)}{1 + \left(-\dfrac{12}{5}\right)\left(-\dfrac{7}{24}\right)} = \dfrac{-\dfrac{253}{120}}{\dfrac{17}{10}}$

$= -\dfrac{253}{204}$

Section 6.5 Trigonometric Form of a Complex Number

- You should be able to graphically represent complex numbers and know the following facts about them.
- The absolute value of the complex number $z = a + bi$ is $|z| = \sqrt{a^2 + b^2}$.
- The trigonometric form of the complex number $z = a + bi$ is $z = r(\cos \theta + i \sin \theta)$ where
  - (a) $a = r \cos \theta$
  - (b) $b = r \sin \theta$
  - (c) $r = \sqrt{a^2 + b^2}$; r is called the modulus of z.
  - (d) $\tan \theta = \dfrac{b}{a}$; θ is called the argument of z.
- Given $z_1 = r_1(\cos \theta_1 + i \sin \theta_1)$ and $z_2 = r_2(\cos \theta_2 + i \sin \theta_2)$:
  - (a) $z_1 z_2 = r_1 r_2 [\cos(\theta_1 + \theta_2) + i \sin(\theta_1 + \theta_2)]$
  - (b) $\dfrac{z_1}{z_2} = \dfrac{r_1}{r_2}[\cos(\theta_1 - \theta_2) + i \sin(\theta_1 - \theta_2)]$, $z_2 \neq 0$
- You should know DeMoivre's Theorem: If $z = r(\cos \theta + i \sin \theta)$, then for any positive integer n,
  $$z^n = r^n (\cos n\theta + i \sin n\theta).$$
- You should know that for any positive integer n, $z = r(\cos \theta + i \sin \theta)$ has n distinct nth roots given by
  $$\sqrt[n]{r}\left[\cos\left(\frac{\theta + 2\pi k}{n}\right) + i \sin\left(\frac{\theta + 2\pi k}{n}\right)\right]$$
  where $k = 0, 1, 2, \ldots, n - 1$.

Vocabulary Check

1. absolute value

2. trigonometric form; modulus; argument

3. DeMoivre's

4. n^{th} root

1. $|-7i| = \sqrt{0^2 + (-7)^2}$
   $= \sqrt{49} = 7$

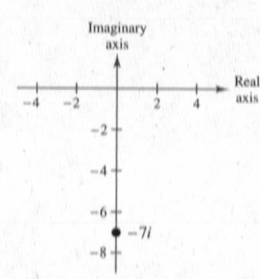

2. $|-7| = \sqrt{(-7)^2 + 0^2} = \sqrt{49} = 7$

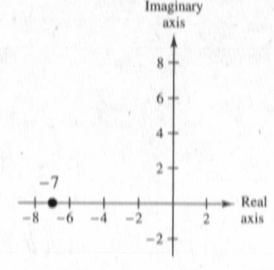

3. $|-4 + 4i| = \sqrt{(-4)^2 + (4)^2}$
   $= \sqrt{32} = 4\sqrt{2}$

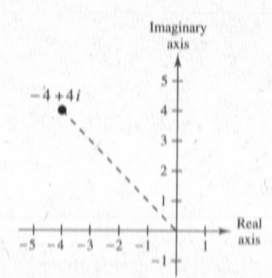

4. $|5 - 12i| = \sqrt{5^2 + (-12)^2}$

$\qquad\quad = \sqrt{169} = 13$

5. $|6 - 7i| = \sqrt{6^2 + (-7)^2}$

$\qquad\quad = \sqrt{85}$

6. $|-8 + 3i| = \sqrt{(-8)^2 + (3)^2}$

$\qquad\qquad = \sqrt{73}$

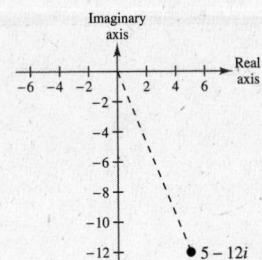

7. $z = 3i$

$r = \sqrt{0^2 + 3^2} = \sqrt{9} = 3$

$\tan \theta = \dfrac{3}{0}$, undefined $\implies \theta = \dfrac{\pi}{2}$

$z = 3\left(\cos \dfrac{\pi}{2} + i \sin \dfrac{\pi}{2}\right)$

8. $z = -2$

$r = \sqrt{(-2)^2 + 0^2} = \sqrt{4} = 2$

$\tan \theta = \dfrac{0}{-2} \implies \theta = \pi$

$z = 2(\cos \pi + i \sin \pi)$

9. $z = 3 - i$

$r = \sqrt{(3)^2 + (-1)^2} = \sqrt{10}$

$\tan \theta = -\dfrac{1}{3}$, θ is in Quadrant IV.

$\theta \approx 5.96$ radians

$z \approx \sqrt{10}(\cos 5.96 + i \sin 5.96)$

10. $z = -1 + \sqrt{3}i$

$r = \sqrt{(-1)^2 + \left(\sqrt{3}\right)^2} = \sqrt{4} = 2$

$\tan \theta = \dfrac{\sqrt{3}}{-1} = -\sqrt{3} \implies \theta = \dfrac{2\pi}{3}$

$z = 2\left(\cos \dfrac{2\pi}{3} + i \sin \dfrac{2\pi}{3}\right)$

11. $z = 3 - 3i$

$r = \sqrt{3^2 + (-3)^2} = \sqrt{18} = 3\sqrt{2}$

$\tan \theta = \dfrac{-3}{3} = -1$, θ is in Quadrant IV $\implies \theta = \dfrac{7\pi}{4}$.

$z = 3\sqrt{2}\left(\cos \dfrac{7\pi}{4} + i \sin \dfrac{7\pi}{4}\right)$

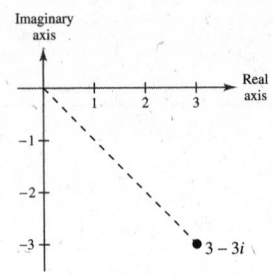

12. $z = 2 + 2i$

$r = \sqrt{2^2 + 2^2} = \sqrt{8} = 2\sqrt{2}$

$\tan \theta = \dfrac{2}{2} = 1 \implies \theta = \dfrac{\pi}{4}$

$z = 2\sqrt{2}\left(\cos \dfrac{\pi}{4} + i \sin \dfrac{\pi}{4}\right)$

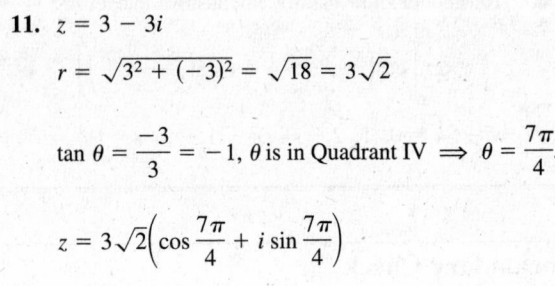

13. $z = \sqrt{3} + i$

$r = \sqrt{\left(\sqrt{3}\right)^2 + 1^2} = \sqrt{4} = 2$

$\tan \theta = \dfrac{1}{\sqrt{3}} = \dfrac{\sqrt{3}}{3} \implies \theta = \dfrac{\pi}{6}$

$z = 2\left(\cos \dfrac{\pi}{6} + i \sin \dfrac{\pi}{6}\right)$

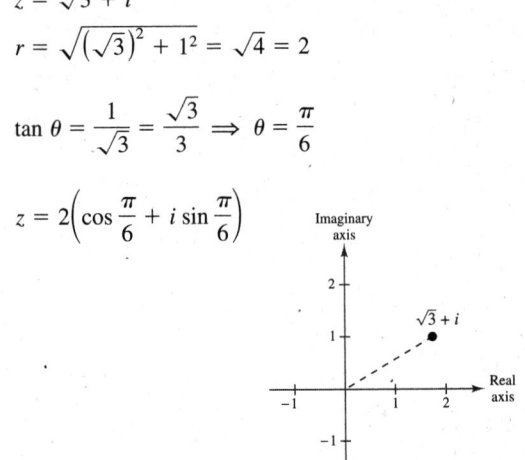

14. $z = 4 - 4\sqrt{3}i$

$r = \sqrt{4^2 + (-4\sqrt{3})^2} = 8$

$\tan \theta = \dfrac{-4\sqrt{3}}{4} = -\sqrt{3} \implies \theta = \dfrac{5\pi}{3}$

$z = 8\left(\cos \dfrac{5\pi}{3} + i \sin \dfrac{5\pi}{3}\right)$

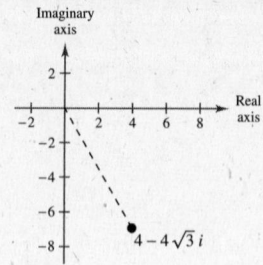

15. $z = -2(1 + \sqrt{3}i)$

$r = \sqrt{(-2)^2 + (-2\sqrt{3})^2} = \sqrt{16} = 4$

$\tan \theta = \dfrac{\sqrt{3}}{1} = \sqrt{3}, \theta \text{ is in Quadrant III} \implies \theta = \dfrac{4\pi}{3}.$

$z = 4\left(\cos \dfrac{4\pi}{3} + i \sin \dfrac{4\pi}{3}\right)$

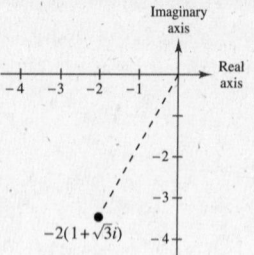

16. $z = \dfrac{5}{2}(\sqrt{3} - i)$

$r = \sqrt{\left(\dfrac{5}{2}\sqrt{3}\right)^2 + \left(\dfrac{5}{2}(-1)\right)^2} = \sqrt{\dfrac{100}{4}} = \sqrt{25} = 5$

$\tan \theta = \dfrac{-1}{\sqrt{3}} = \dfrac{-\sqrt{3}}{3} \implies \theta = \dfrac{11\pi}{6}$

$z = 5\left(\cos \dfrac{11\pi}{6} + i \sin \dfrac{11\pi}{6}\right)$

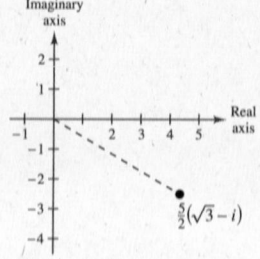

17. $z = -5i$

$r = \sqrt{0^2 + (-5)^2} = \sqrt{25} = 5$

$\tan \theta = \dfrac{-5}{0}, \text{ undefined} \implies \theta = \dfrac{3\pi}{2}$

$z = 5\left(\cos \dfrac{3\pi}{2} + i \sin \dfrac{3\pi}{2}\right)$

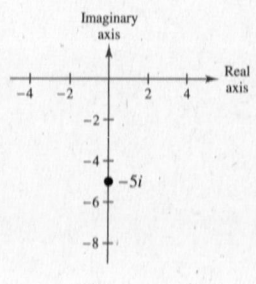

18. $z = 4i$

$r = \sqrt{0^2 + 4^2} = \sqrt{16^2} = 4$

$\tan \theta = \dfrac{4}{0}, \text{ undefined} \implies \theta = \dfrac{\pi}{2}$

$z = 4\left(\cos \dfrac{\pi}{2} + i \sin \dfrac{\pi}{2}\right)$

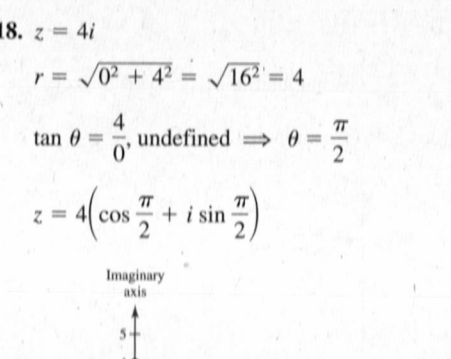

19. $z = -7 + 4i$

$r = \sqrt{(-7)^2 + (4)^2} = \sqrt{65}$

$\tan \theta = \dfrac{4}{-7}, \theta \text{ is in Quadrant II} \implies \theta \approx 2.62.$

$z \approx \sqrt{65}(\cos 2.62 + i \sin 2.62)$

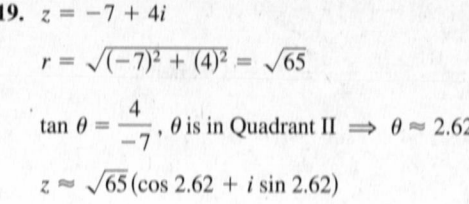
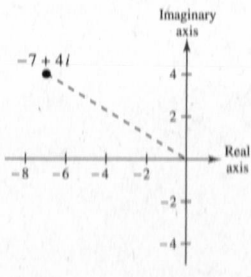

20. $z = 3 - i$

$r = \sqrt{(3)^2 + (-1)^2} = \sqrt{10}$

$\tan\theta = \dfrac{-1}{3} = \theta \approx 5.96 \text{ radians}$

$z = \sqrt{10}\,(\cos 5.96 + i\sin 5.96)$

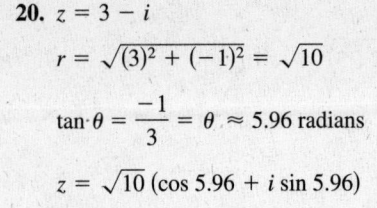

21. $z = 7 + 0i$

$r = \sqrt{(7)^2 + (0)^2} = \sqrt{49} = 7$

$\tan\theta = \dfrac{0}{7} = 0 \implies \theta = 0$

$z = 7(\cos 0 + i\sin 0)$

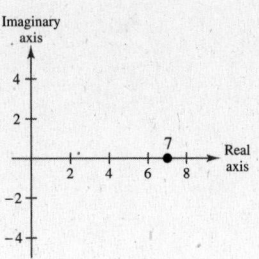

22. $z = 4$

$r = \sqrt{4^2 + 0^2} = \sqrt{16^2} = 4$

$\tan\theta = \dfrac{0}{4} = 0 \implies \theta = 0$

$z = 4(\cos 0 + i\sin 0)$

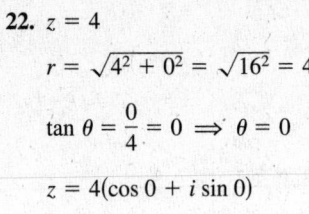

23. $z = 3 + \sqrt{3}i$

$r = \sqrt{(3)^2 + \left(\sqrt{3}\right)^2} = \sqrt{12}$

$\qquad = 2\sqrt{3}$

$\tan\theta = \dfrac{\sqrt{3}}{3} \implies \theta = \dfrac{\pi}{6}$

$z = 2\sqrt{3}\left(\cos\dfrac{\pi}{6} + i\sin\dfrac{\pi}{6}\right)$

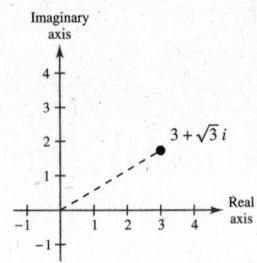

24. $z = 2\sqrt{2} - i$

$r = \sqrt{(2\sqrt{2})^2 + (-1)^2} = \sqrt{9} = 3$

$\tan\theta = \dfrac{-1}{2\sqrt{2}} = -\dfrac{\sqrt{2}}{4} \implies \theta \approx 5.94 \text{ radians}$

$z = 3(\cos 5.94 + i\sin 5.94)$

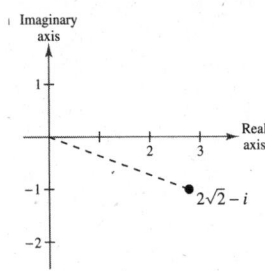

25. $z = -3 - i$

$r = \sqrt{(-3)^2 + (-1)^2} = \sqrt{10}$

$\tan\theta = \dfrac{-1}{-3} = \dfrac{1}{3},\ \theta \text{ is in Quadrant III} \implies \theta \approx 3.46.$

$z \approx \sqrt{10}\,(\cos 3.46 + i\sin 3.46)$

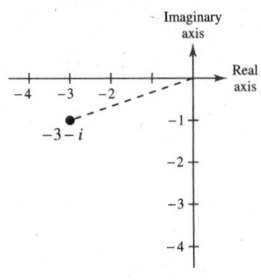

26. $z = 1 + 3i$

$r = \sqrt{1^2 + 3^2} = \sqrt{10}$

$\tan \theta = \frac{3}{1} = 3 \implies \theta \approx 1.25$ radians

$z \approx \sqrt{10}\,(\cos 1.25 + i \sin 1.25)$

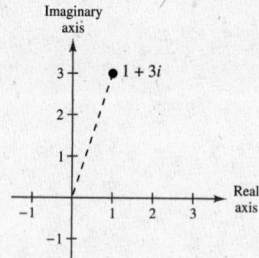

27. $z = 5 + 2i$

$r = \sqrt{5^2 + 2^2} = \sqrt{29}$

$\tan \theta = \frac{2}{5}$

$\theta \approx 0.38$

$z \approx \sqrt{29}(\cos 0.38 + i \sin 0.38)$

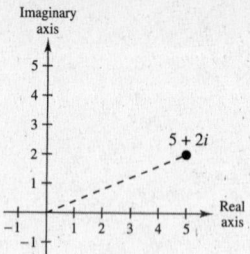

28. $z = 8 + 3i$

$r = \sqrt{8^2 + 3^2} = \sqrt{73}$

$\tan \theta = \frac{3}{8}$

$\theta = 0.36$

$z \approx \sqrt{73}(\cos 0.36 + i \sin 0.36)$

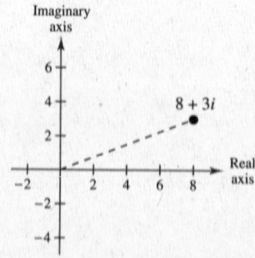

29. $z = -8 - 5\sqrt{3}i$

$r = \sqrt{(-8)^2 + \left(-5\sqrt{3}\right)^2} = \sqrt{139}$

$\tan \theta = \frac{5\sqrt{3}}{8}$

$\theta \approx 3.97$

$z \approx \sqrt{139}(\cos 3.97 + i \sin 3.97)$

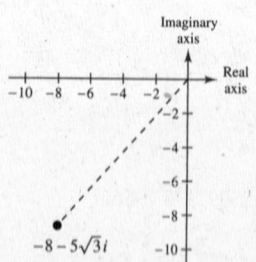

30. $z = -9 - 2\sqrt{10}i$

$r = \sqrt{(9)^2 + \left(-2\sqrt{10}\right)^2} = \sqrt{121}$

$r = 11$

$\tan \theta = \frac{-2\sqrt{10}}{9}$

$\theta = 3.75$

$z \approx 11(\cos 3.75 + i \sin 3.75)$

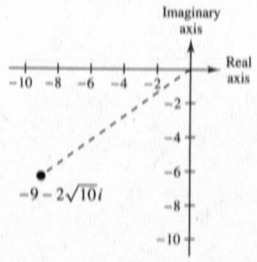

31. $3(\cos 120° + i \sin 120°) = 3\left(-\dfrac{1}{2} + \dfrac{\sqrt{3}}{2}i\right)$

$$= -\frac{3}{2} + \frac{3\sqrt{3}}{2}i$$

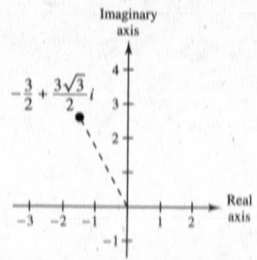

32. $5(\cos 135° + i \sin 135°) = 5\left[-\dfrac{\sqrt{2}}{2} + i\left(\dfrac{\sqrt{2}}{2}\right)\right]$

$= -\dfrac{5\sqrt{2}}{2} + \dfrac{5\sqrt{2}}{2}i$

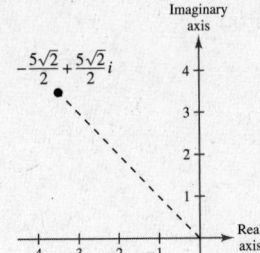

33. $\dfrac{3}{2}(\cos 300° + i \sin 300°) = \dfrac{3}{2}\left[\dfrac{1}{2} + i\left(-\dfrac{\sqrt{3}}{2}\right)\right]$

$= \dfrac{3}{4} - \dfrac{3\sqrt{3}}{4}i$

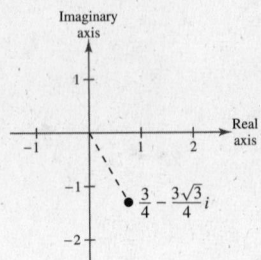

34. $\dfrac{1}{4}(\cos 225° + i \sin 225°) = \dfrac{1}{4}\left(-\dfrac{\sqrt{2}}{2} + i\left(-\dfrac{\sqrt{2}}{2}\right)\right)$

$= -\dfrac{\sqrt{2}}{8} - i\dfrac{\sqrt{2}}{8}$

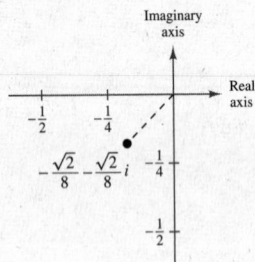

35. $3.75\left(\cos \dfrac{3\pi}{4} + i \sin \dfrac{3\pi}{4}\right) = -\dfrac{15\sqrt{2}}{8} + \dfrac{15\sqrt{2}}{8}i$

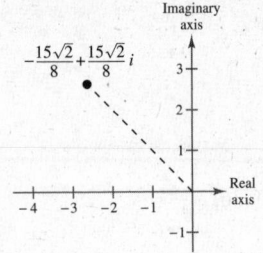

36. $6\left(\cos \dfrac{5\pi}{12} + i \sin \dfrac{5\pi}{12}\right) \approx 1.5529 + 5.7956i$

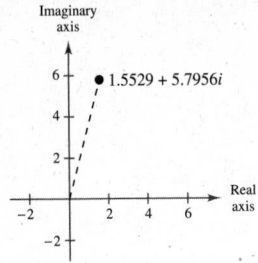

37. $8\left(\cos \dfrac{\pi}{2} + i \sin \dfrac{\pi}{2}\right) = 8(0 + i) = 8i$

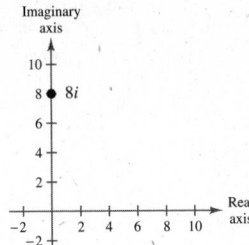

38. $7(\cos 0° + i \sin 0°) = 7$

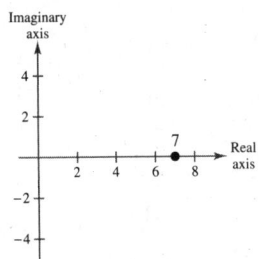

39. $3[\cos(18°45') + i \sin(18°45')] \approx 2.8408 + 0.9643i$

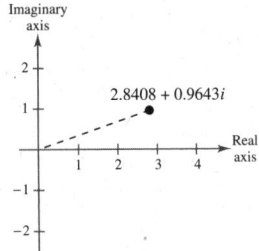

40. $6[\cos(230° 30') + i \sin(230° 30')] \approx -3.8165 - 4.6297i$

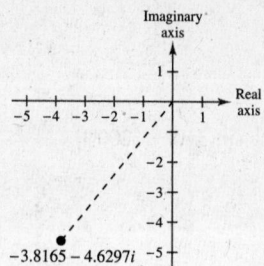

41. $5\left(\cos \dfrac{\pi}{9} + i \sin \dfrac{\pi}{9}\right) \approx 4.6985 + 1.7101i$

42. $10\left(\cos \dfrac{2\pi}{5} + i \sin \dfrac{2\pi}{5}\right) \approx 3.0902 + 9.5106i$

43. $3(\cos 165.5° + i \sin 165.5°) \approx -2.9044 + 0.7511i$

44. $9(\cos 58° + i \sin 58°) \approx 4.7693 + 7.6324i$

45. $z = \dfrac{\sqrt{2}}{2}(1 + i) = \cos 45° + i \sin 45°$

$z^2 = \cos 90° + i \sin 90° = i$

$z^3 = \cos 135° + i \sin 135° = \dfrac{\sqrt{2}}{2}(-1 + i)$

$z^4 = \cos 180° + i \sin 180° = -1$

The absolute value of each is 1, and consecutive powers of z are each 45° apart.

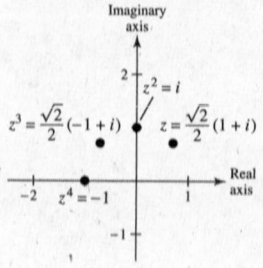

46. $z = \dfrac{1}{2}\left(1 + \sqrt{3}i\right)$

$z^n = r^n(\cos n\theta + i \sin n\theta)$

$r = \sqrt{\left(\dfrac{1}{2}\right)^2 + \left(\dfrac{\sqrt{3}}{2}\right)^2} = 1$

$\tan \theta = \sqrt{3}$

$\theta = \dfrac{\pi}{3}$

$z = 1\left(\cos\dfrac{\pi}{3} + i \sin\dfrac{\pi}{3}\right) = \dfrac{1}{2} + \dfrac{\sqrt{3}}{2}i$

$z^2 = 1^2\left(\cos\dfrac{2\pi}{3} + i \sin\dfrac{2\pi}{3}\right) = -\dfrac{1}{2} + \dfrac{\sqrt{3}}{2}i$

$z^3 = 1^3(\cos \pi + i \sin \pi) = -1$

$z^4 = 1^4\left(\cos\dfrac{4\pi}{3} + i \sin\dfrac{4\pi}{3}\right) = -\dfrac{1}{2} - \dfrac{\sqrt{3}}{2}i$

The absolute value of each is 1 and consecutive powers of z are each $\pi/3$ radians apart.

47. $\left[2\left(\cos\dfrac{\pi}{4} + i \sin\dfrac{\pi}{4}\right)\right]\left[6\left(\cos\dfrac{\pi}{12} + i \sin\dfrac{\pi}{12}\right)\right] = (2)(6)\left[\cos\left(\dfrac{\pi}{4} + \dfrac{\pi}{12}\right) + i \sin\left(\dfrac{\pi}{4} + \dfrac{\pi}{12}\right)\right]$

$$= 12\left(\cos\dfrac{\pi}{3} + i \sin\dfrac{\pi}{3}\right)$$

48. $\left[\dfrac{3}{4}\left(\cos\dfrac{\pi}{3} + i \sin\dfrac{\pi}{3}\right)\right]\left[4\left(\cos\dfrac{3\pi}{4} + i \sin\dfrac{3\pi}{4}\right)\right] = \left(\dfrac{3}{4}\right)(4)\left[\cos\left(\dfrac{\pi}{3} + \dfrac{3\pi}{4}\right) + i \sin\left(\dfrac{\pi}{3} + \dfrac{3\pi}{4}\right)\right]$

$$= 3\left(\cos\dfrac{13\pi}{12} + i \sin\dfrac{13\pi}{12}\right)$$

49. $\left[\dfrac{5}{3}(\cos 140° + i \sin 140°)\right]\left[\dfrac{2}{3}(\cos 60° + i \sin 60°)\right] = \left(\dfrac{5}{3}\right)\left(\dfrac{2}{3}\right)[\cos(140° + 60°) + i \sin(140° + 60°)]$

$$= \dfrac{10}{9}(\cos 200° + i \sin 200°)$$

50. $[0.5(\cos 100° + i \sin 100°)][0.8(\cos 300° + i \sin 300°)] = (0.5)(0.8)[\cos(100° + 300°) + i \sin(100° + 300°)]$

$$= 0.4(\cos 400° + i \sin 400°)$$

$$= 0.4(\cos 40° + i \sin 40°)$$

51. $[0.45(\cos 310° + i \sin 310°)][0.60(\cos 200° + i \sin 200°)] = (0.45)(0.60)[\cos(310° + 200°) + i \sin(310° + 200°)]$

$$= 0.27(\cos 510° + i \sin 510°)$$

$$= 0.27(\cos 150° + i \sin 150°)$$

52. $(\cos 5° + i \sin 5°)(\cos 20° + i \sin 20°) = \cos(5° + 20°) + i \sin(5° + 20°) = \cos 25° + i \sin 25°$

53. $\dfrac{\cos 50° + i \sin 50°}{\cos 20° + i \sin 20°} = \cos(50° - 20°) + i \sin(50° - 20°) = \cos 30° + i \sin 30°$

54. $\dfrac{2(\cos 120° + i \sin 120°)}{4(\cos 40° + i \sin 40°)} = \dfrac{2}{4}[\cos(120° - 40°) + i \sin(120° - 40°)]$

$$= \dfrac{1}{2}(\cos 80° + i \sin 80°)$$

55. $\dfrac{\cos\dfrac{5\pi}{3} + i \sin\dfrac{5\pi}{3}}{\cos \pi + i \sin \pi} = \cos\left(\dfrac{5\pi}{3} - \pi\right) + i \sin\left(\dfrac{5\pi}{3} - \pi\right) = \cos\left(\dfrac{2\pi}{3}\right) + i \sin\left(\dfrac{2\pi}{3}\right)$

56. $\dfrac{5[\cos (4.3) + i \sin(4.3)]}{4[\cos (2.1) + i \sin(2.1)]} = \dfrac{5}{4}[\cos(4.3 - 2.1) + i \sin(4.3 - 2.1)]$

$$= \dfrac{5}{4}[\cos(2.2) + i \sin(2.2)]$$

57. $\dfrac{12(\cos 52° + i \sin 52°)}{3(\cos 110° + i \sin 110°)} = 4[\cos(52° - 110°) + i \sin(52° - 110°)]$

$$= 4[\cos(-58°) + i \sin(-58°)]$$

$$= 4(\cos 302° + i \sin 302°)$$

58. $\dfrac{6[\cos 40° + i \sin 40°]}{7[\cos 100° + i \sin 100°]} = \dfrac{6}{7}[\cos(40° - 100°) + i \sin(40° - 100°)]$

$$= \dfrac{6}{7}[\cos 300° + i \sin 300°]$$

59. (a) $2 + 2i = 2\sqrt{2}\left(\cos\dfrac{\pi}{4} + i \sin\dfrac{\pi}{4}\right)$

$$1 - i = \sqrt{2}\left[\cos\left(-\dfrac{\pi}{4}\right) + i \sin\left(-\dfrac{\pi}{4}\right)\right] = \sqrt{2}\left(\cos\dfrac{7\pi}{4} + i \sin\dfrac{7\pi}{4}\right)$$

(b) $(2 + 2i)(1 - i) = \left[2\sqrt{2}\left(\cos\dfrac{\pi}{4} + i \sin\dfrac{\pi}{4}\right)\right]\left[\sqrt{2}\left(\cos\left(\dfrac{7\pi}{4}\right) + i \sin\left(\dfrac{7\pi}{4}\right)\right)\right] = 4(\cos 2\pi + i \sin 2\pi)$

$$= 4(\cos 0 + i \sin 0) = 4$$

(c) $(2 + 2i)(1 - i) = 2 - 2i + 2i - 2i^2 = 2 + 2 = 4$

60. (a) $\sqrt{3} + i = 2(\cos 30° + i \sin 30°)$

$1 + i = \sqrt{2}(\cos 45° + i \sin 45°)$

(b) $(\sqrt{3} + i)(1 + i) = [2(\cos 30° + i \sin 30°)][\sqrt{2}(\cos 45° + i \sin 45°)]$

$\qquad = 2\sqrt{2}(\cos 75° + i \sin 75°)$

$\qquad = 2\sqrt{2}\left[\left(\dfrac{\sqrt{6} - \sqrt{2}}{4}\right) + \left(\dfrac{\sqrt{6} + \sqrt{2}}{4}\right)i\right]$

$\qquad = (\sqrt{3} - 1) + (\sqrt{3} + 1)i \approx 0.732 + 2.732i$

(c) $(\sqrt{3} + i)(1 + i) = \sqrt{3} + (\sqrt{3} + 1)i + i^2 = (\sqrt{3} - 1) + (\sqrt{3} + 1)i \approx 0.732 + 2.732i$

61. (a) $-2i = 2\left[\cos\left(-\dfrac{\pi}{2}\right) + i \sin\left(-\dfrac{\pi}{2}\right)\right] = 2\left(\cos\dfrac{3\pi}{2} + i \sin\dfrac{3\pi}{2}\right)$

$1 + i = \sqrt{2}\left(\cos\dfrac{\pi}{4} + i \sin\dfrac{\pi}{4}\right)$

(b) $-2i(1 + i) = 2\left[\cos\left(\dfrac{3\pi}{2}\right) + i \sin\left(\dfrac{3\pi}{2}\right)\right]\left[\sqrt{2}\left(\cos\dfrac{\pi}{4} + i \sin\dfrac{\pi}{4}\right)\right]$

$\qquad = 2\sqrt{2}\left[\cos\left(\dfrac{7\pi}{4}\right) + i \sin\left(\dfrac{7\pi}{4}\right)\right]$

$\qquad = 2\sqrt{2}\left[\dfrac{1}{\sqrt{2}} - \dfrac{1}{\sqrt{2}}i\right] = 2 - 2i$

(c) $-2i(1 + i) = -2i - 2i^2 = -2i + 2 = 2 - 2i$

62. (a) $\qquad 4 = 4(\cos 0 + i \sin 0)$

$1 - \sqrt{3}i = 2\left(\cos\left(-\dfrac{\pi}{3}\right) + i \sin\left(\dfrac{-\pi}{3}\right)\right)$

(c) $4(1 - \sqrt{3}i) = 4 - 4\sqrt{3}i$

(b) $4(1 - \sqrt{3}i) = 8\left(\cos\left(-\dfrac{\pi}{3}\right) + i \sin\left(-\dfrac{\pi}{3}\right)\right)$

$\qquad = 8\left(\dfrac{1}{2} + i\left(-\dfrac{\sqrt{3}}{2}\right)\right)$

$\qquad = 4 - 4\sqrt{3}i$

63. (a) $3 + 4i \approx 5(\cos 0.93 + i \sin 0.93)$

$1 - \sqrt{3}i = 2\left(\cos\dfrac{5\pi}{3} + i \sin\dfrac{5\pi}{3}\right)$

(c) $\dfrac{3 + 4i}{1 - \sqrt{3}i} = \dfrac{3 + 4i}{1 - \sqrt{3}i} \cdot \dfrac{1 + \sqrt{3}i}{1 + \sqrt{3}i}$

$\qquad = \dfrac{3 + (4 + 3\sqrt{3})i + 4\sqrt{3}i^2}{1 + 3}$

$\qquad = \dfrac{3 - 4\sqrt{3}}{4} + \dfrac{4 + 3\sqrt{3}}{4}i$

$\qquad \approx -0.982 + 2.299i$

(b) $\dfrac{3 + 4i}{1 - \sqrt{3}i} \approx \dfrac{5(\cos 0.93 + i \sin 0.93)}{2\left(\cos\dfrac{5\pi}{3} + i \sin\dfrac{5\pi}{3}\right)}$

$\qquad \approx 2.5[\cos(-4.31) + i \sin(-4.31)]$

$\qquad = \dfrac{5}{2}(\cos 1.97 + i \sin 1.97)$

$\qquad \approx -0.982 + 2.299i$

64. (a) $1 + \sqrt{3}i = 2\left(\cos\dfrac{\pi}{3} + i \sin\dfrac{\pi}{3}\right)$

$6 - 3i \approx 3\sqrt{5}[\cos(-0.464) + i \sin(-0.464)]$

(b) $\dfrac{1 + \sqrt{3}i}{6 - 3i} \approx \dfrac{2}{3\sqrt{5}}\left[\cos\left(\dfrac{\pi}{3} + 0.464\right) + i \sin\left(\dfrac{\pi}{3} + 0.464\right)\right] \approx \dfrac{2\sqrt{5}}{15}[\cos 1.51 + i \sin 1.51] \approx 0.018 + 0.298i$

(c) $\dfrac{1 + \sqrt{3}i}{6 - 3i} \cdot \dfrac{6 + 3i}{6 + 3i} = \dfrac{(6 - 3\sqrt{3}) + i(3 + 6\sqrt{3})}{45} = \dfrac{2 - \sqrt{3}}{15} + i\dfrac{1 + 2\sqrt{3}}{15} \approx 0.018 + 0.298i$

65. (a) $5 = 5(\cos 0 + i \sin 0)$

$2 + 3i \approx \sqrt{13}(\cos 0.98 + i \sin 0.98)$

(b) $\dfrac{5}{2 + 3i} \approx \dfrac{5(\cos 0 + i \sin 0)}{\sqrt{13}(\cos 0.98 + i \sin 0.98)} = \dfrac{5}{\sqrt{13}}[\cos(-0.98) + i \sin(-0.98)] = \dfrac{5}{\sqrt{13}}(\cos 5.30 + i \sin 5.30) \approx 0.769 - 1.154i$

(c) $\dfrac{5}{2 + 3i} = \dfrac{5}{2 + 3i} \cdot \dfrac{2 - 3i}{2 - 3i} = \dfrac{10 - 15i}{13} = \dfrac{10}{13} - \dfrac{15}{13}i \approx 0.769 - 1.154i$

66. (a) $4i = 4(\cos 90° + i \sin 90°)$

$-4 + 2i = 2\sqrt{5}(\cos 153.4° + i \sin 153.4°)$

(b) $\dfrac{4i}{-4 + 2i} = \dfrac{4(\cos 90° + i \sin 90°)}{2\sqrt{5}(\cos 153.4° + i \sin 153.4°)}$

$= \dfrac{2\sqrt{5}}{5}(\cos 296.6° + i \sin 296.6°)$

$\approx 0.400 - 0.800i$

(c) $\dfrac{4i}{-4 + 2i} = \dfrac{4i}{-4 + 2i} \cdot \dfrac{-4 - i}{-4 - 2i}$

$= \dfrac{8 - 16i}{20} = \dfrac{2}{5} - \dfrac{4}{5}i = 0.400 - 0.800i$

67. Let $z = x + iy$ such that:

$|z| = 2 \implies 2 = \sqrt{x^2 + y^2}$

$\implies 4 = x^2 + y^2$:

circle with radius of 2

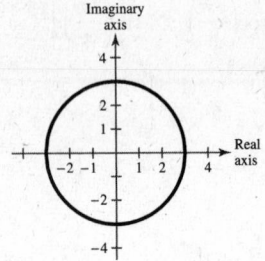

68. $|z| = 3$

69. Let $\theta = \dfrac{\pi}{6}$.

Since $r \geq 0$, we have the portion of the line $\theta = \pi/6$ in Quadrant I.

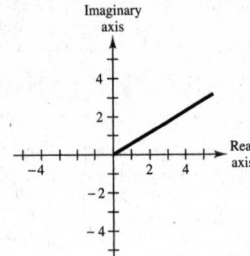

70. $\theta = \dfrac{5\pi}{4}$

Since $r \geq 0$, we have the portion of the line $\theta = 5\pi/4$ in Quadrant III.

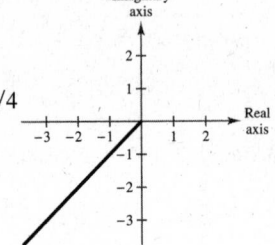

71. $(1 + i)^5 = \left[\sqrt{2}\left(\cos \dfrac{\pi}{4} + i \sin \dfrac{\pi}{4} \right) \right]^5$

$= (\sqrt{2})^5 \left(\cos \dfrac{5\pi}{4} + i \sin \dfrac{5\pi}{4} \right)$

$= 4\sqrt{2}\left(-\dfrac{\sqrt{2}}{2} - \dfrac{\sqrt{2}}{2}i \right)$

$= -4 - 4i$

72. $(2 + 2i)^6 = \left[2\sqrt{2}\left(\cos \dfrac{\pi}{4} + i \sin \dfrac{\pi}{4} \right) \right]^6$

$= (2\sqrt{2})^6 \left(\cos \dfrac{6\pi}{4} + i \sin \dfrac{6\pi}{4} \right)$

$= 512\left(\cos \dfrac{3\pi}{2} + i \sin \dfrac{3\pi}{2} \right)$

$= -512i$

73. $(-1 + i)^{10} = \left[\sqrt{2} \left(\cos \frac{3\pi}{4} + i \sin \frac{3\pi}{4} \right) \right]^{10} = (\sqrt{2})^{10} \left(\cos \frac{30\pi}{4} + i \sin \frac{30\pi}{4} \right)$

$$= 32 \left[\cos \left(\frac{3\pi}{2} + 6\pi \right) + i \sin \left(\frac{3\pi}{2} + 6\pi \right) \right] = 32 \left(\cos \frac{3\pi}{2} + i \sin \frac{3\pi}{2} \right)$$

$$= 32[0 + i(-1)] = -32i$$

74. $(3 - 2i)^8 = \left[\sqrt{13} \left(\cos \left(-\arctan \left(\tfrac{2}{3} \right) \right) + i \sin \left(-\arctan \left(\tfrac{2}{3} \right) \right) \right) \right]^8$

$$= (\sqrt{13})^8 \left[\cos \left(-8 \arctan \left(\tfrac{2}{3} \right) \right) + i \sin \left(-8 \arctan \left(\tfrac{2}{3} \right) \right) \right]$$

$$= -239 + 28{,}560i$$

75. $2(\sqrt{3} + i)^7 = 2 \left[2 \left(\cos \frac{\pi}{6} + i \sin \frac{\pi}{6} \right) \right]^7$

$$= 2 \left[2^7 \left(\cos \frac{7\pi}{6} + i \sin \frac{7\pi}{6} \right) \right]$$

$$= 256 \left(-\frac{\sqrt{3}}{2} - \frac{1}{2}i \right)$$

$$= -128\sqrt{3} - 128i$$

76. $4(1 - \sqrt{3}i)^3 = 4 \left[2 \left(\cos \frac{5\pi}{3} + i \sin \frac{5\pi}{3} \right) \right]^3$

$$= 4[2^3(\cos 5\pi + i \sin 5\pi)]$$

$$= 32(-1)$$

$$= -32$$

77. $[5(\cos 20° + i \sin 20°)]^3 = 5^3(\cos 60° + i \sin 60°) = \frac{125}{2} + \frac{125\sqrt{3}}{2}i$

78. $[3(\cos 150° + i \sin 150°)]^4 = 3^4(\cos 600° + i \sin 600°)$

$$= 81(\cos 240° + i \sin 240°)$$

$$= 81(-\cos 60° - i \sin 60°)$$

$$= -\frac{81}{2} - \frac{81\sqrt{3}}{2}i$$

79. $\left(\cos \frac{\pi}{4} + i \sin \frac{\pi}{4} \right)^{12} = \cos \frac{12\pi}{4} + i \sin \frac{12\pi}{4}$

$$= \cos 3\pi + i \sin 3\pi$$

$$= -1$$

80. $\left[2 \left(\cos \frac{\pi}{2} + i \sin \frac{\pi}{2} \right) \right]^8 = 2^8(\cos 4\pi + i \sin 4\pi)$

$$= 256(\cos 0 + i \sin 0)$$

$$= 256$$

81. $[5(\cos 3.2 + i \sin 3.2)]^4 = 5^4(\cos 12.8 + i \sin 12.8)$

$$\approx 608.02 + 144.69i$$

82. $(\cos 0 + i \sin 0)^{20} = \cos 0 + i \sin 0$

$$= 1$$

83. $(3 - 2i)^5 \approx [3.6056[\cos(-0.588) + i \sin(-0.588)]]^5$

$$\approx (3.6056)^5[\cos(-2.94) + i \sin(-2.94)]$$

$$\approx -597 - 122i$$

84. $(\sqrt{5} - 4i)^3 \approx \left[\sqrt{21}(\cos(-1.06106) + i \sin(-1.06106) \right]^3$

$$\approx (\sqrt{21})^3(\cos[(3)(-1.06106)] + i \sin[(3)(-1.06106)])$$

$$\approx -96.15 + 4.00i$$

85. $[3(\cos 15° + i \sin 15°)]^4 = 81(\cos 60° + i \sin 60°)$

$$= \frac{81}{2} + \frac{81\sqrt{3}}{2}i$$

86. $[2(\cos 10° + i \sin 10°)]^8 = 256(\cos 80° + i \sin 80°)$

$$\approx 44.45 + 252.11i$$

87. $\left[2\left(\cos\dfrac{\pi}{10} + i\sin\dfrac{\pi}{10}\right)\right]^5 = 2^5\left(\cos\dfrac{\pi}{2} + i\sin\dfrac{\pi}{2}\right)$

$\qquad\qquad\qquad\qquad\qquad = 32i$

88. $\left[2\left(\cos\dfrac{\pi}{8} + i\sin\dfrac{\pi}{8}\right)\right]^6 = 64\left(\cos\dfrac{3\pi}{4} + i\sin\dfrac{3\pi}{4}\right)$

$\qquad\qquad\qquad\qquad\qquad = -32\sqrt{2} + 32\sqrt{2}i$

89. (a) Square roots of $5(\cos 120° + i\sin 120°)$:

$\qquad \sqrt{5}\left[\cos\left(\dfrac{120° + 360°k}{2}\right) + i\sin\left(\dfrac{120° + 360°k}{2}\right)\right]$, $k = 0, 1$

$\qquad k = 0$: $\sqrt{5}(\cos 60° + i\sin 60°)$

$\qquad k = 1$: $\sqrt{5}(\cos 240° + i\sin 240°)$

(c) $\dfrac{\sqrt{5}}{2} + \dfrac{\sqrt{15}}{2}i, \ -\dfrac{\sqrt{5}}{2} - \dfrac{\sqrt{15}}{2}i$

(b)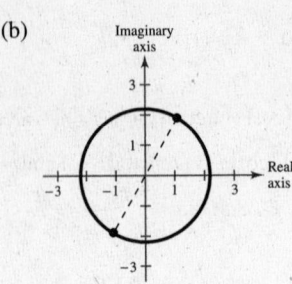

90. (a) Square roots of $16(\cos 60° + i\sin 60°)$:

$\qquad \sqrt{16}\left[\cos\left(\dfrac{60° + k\,360°}{2}\right) + i\sin\left(\dfrac{60° + k\,360°}{2}\right)\right]$, $k = 0, 1$

$\qquad k = 0$: $4(\cos 30° + i\sin 30°)$

$\qquad k = 1$: $4(\cos 210° + i\sin 210°)$

(c) $2\sqrt{3} + 2i, -2\sqrt{3} - 2i$

(b)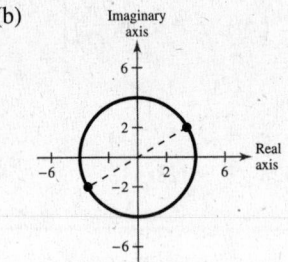

91. (a) Cube roots of $8\left(\cos\dfrac{2\pi}{3} + i\sin\dfrac{2\pi}{3}\right)$:

$\qquad \sqrt[3]{8}\left[\cos\left(\dfrac{(2\pi/3) + 2\pi k}{3}\right) + i\sin\left(\dfrac{(2\pi/3) + 2\pi k}{3}\right)\right]$, $k = 0, 1, 2$

$\qquad k = 0$: $2\left(\cos\dfrac{2\pi}{9} + i\sin\dfrac{2\pi}{9}\right)$

$\qquad k = 1$: $2\left(\cos\dfrac{8\pi}{9} + i\sin\dfrac{8\pi}{9}\right)$

$\qquad k = 2$: $2\left(\cos\dfrac{14\pi}{9} + i\sin\dfrac{14\pi}{9}\right)$

(c) $1.5321 + 1.2856i, -1.8794 + 0.6840i, 0.3473 - 1.9696i$

(b)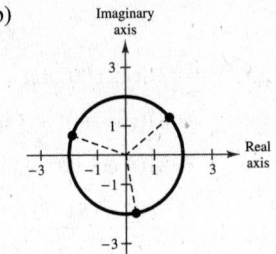

92. (a) Fifth roots of $32\left(\cos\dfrac{5\pi}{6} + i\sin\dfrac{5\pi}{6}\right)$:

$\qquad \sqrt[5]{32}\left[\cos\left(\dfrac{(5\pi/6) + 2k\pi}{5}\right) + i\sin\left(\dfrac{(5\pi/6) + 2k\pi}{5}\right)\right]$, $k = 0, 1, 2, 3, 4$

$\qquad k = 0$: $2\left(\cos\dfrac{\pi}{6} + i\sin\dfrac{\pi}{6}\right)$

$\qquad k = 1$: $2\left(\cos\dfrac{17\pi}{30} + i\sin\dfrac{17\pi}{30}\right)$

$\qquad k = 2$: $2\left(\cos\dfrac{29\pi}{30} + i\sin\dfrac{29\pi}{30}\right)$

$\qquad k = 3$: $2\left(\cos\dfrac{41\pi}{30} + i\sin\dfrac{41\pi}{30}\right)$

$\qquad k = 4$: $2\left(\cos\dfrac{53\pi}{30} + i\sin\dfrac{53\pi}{30}\right)$

(b)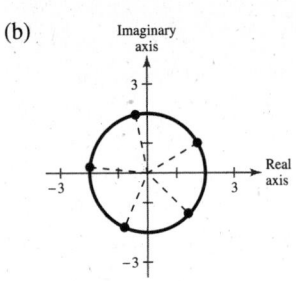

(c) $\sqrt{3} + i, -0.4158 + 1.9563i,$
$\qquad -1.9890 + 0.2091i, -0.8135 - 1.8271i,$
$\qquad 1.4863 - 1.3383i$

93. (a) Square roots of $-25i = 25\left(\cos\dfrac{3\pi}{2} + i\sin\dfrac{3\pi}{2}\right)$:

(b)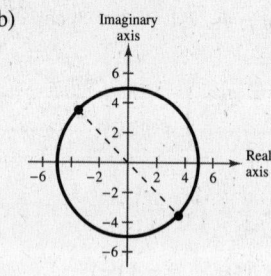

$$\sqrt{25}\left[\cos\left(\dfrac{\dfrac{3\pi}{2}+2k\pi}{2}\right) + i\sin\left(\dfrac{\dfrac{3\pi}{2}+2k\pi}{2}\right)\right],\ k = 0, 1$$

$k = 0:\ 5\left(\cos\dfrac{3\pi}{4} + i\sin\dfrac{3\pi}{4}\right)$

$k = 1:\ 5\left(\cos\dfrac{7\pi}{4} + i\sin\dfrac{7\pi}{4}\right)$

(c) $-\dfrac{5\sqrt{2}}{2} + \dfrac{5\sqrt{2}}{2}i,\ \dfrac{5\sqrt{2}}{2} - \dfrac{5\sqrt{2}}{2}i$

94. (a) Fourth roots of $625i = 625\left(\cos\dfrac{\pi}{2} + i\sin\dfrac{\pi}{2}\right)$:

(b)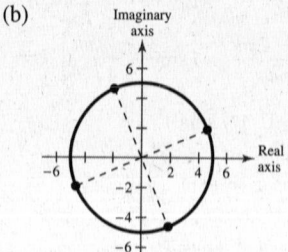

$$\sqrt[4]{625}\left[\cos\left(\dfrac{\dfrac{\pi}{2}+2k\pi}{4}\right) + i\sin\left(\dfrac{\dfrac{\pi}{2}+2k\pi}{2}\right)\right]$$

$k = 0, 1, 2, 3$

$k = 0:\ 5\left(\cos\dfrac{\pi}{8} + i\sin\dfrac{\pi}{8}\right)$

$k = 1:\ 5\left(\cos\dfrac{5\pi}{8} + i\sin\dfrac{5\pi}{8}\right)$

$k = 2:\ 5\left(\cos\dfrac{9\pi}{8} + i\sin\dfrac{9\pi}{8}\right)$

$k = 3:\ 5\left(\cos\dfrac{13\pi}{8} + i\sin\dfrac{13\pi}{8}\right)$

(c) $4.6194 + 1.9134i,\ -1.9134 + 4.6194i,$
$-4.6194 - 1.9134i,\ 1.9134 - 4.6194i$

95. (a) Cube roots of $-\dfrac{125}{2}(1 + \sqrt{3}i) = 125\left(\cos\dfrac{4\pi}{3} + i\sin\dfrac{4\pi}{3}\right)$:

(b)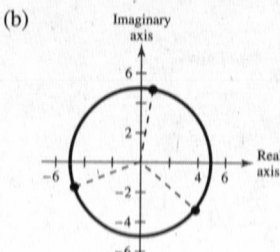

$$\sqrt[3]{125}\left[\cos\left(\dfrac{\dfrac{4\pi}{3}+2k\pi}{3}\right) + i\sin\left(\dfrac{\dfrac{4\pi}{3}+2k\pi}{3}\right)\right],\ k = 0, 1, 2$$

$k = 0:\ 5\left(\cos\dfrac{4\pi}{9} + i\sin\dfrac{4\pi}{9}\right)$

$k = 1:\ 5\left(\cos\dfrac{10\pi}{9} + i\sin\dfrac{10\pi}{9}\right)$

$k = 2:\ 5\left(\cos\dfrac{16\pi}{9} + i\sin\dfrac{16\pi}{9}\right)$

(c) $0.8682 + 4.9240i,\ -4.6985 - 1.7101i,\ 3.8302 - 3.2140i$

96. (a) Cube roots of $-4\sqrt{2}(1 - i) = 8\left(\cos\dfrac{3\pi}{4} + i\sin\dfrac{3\pi}{4}\right)$:

(b)

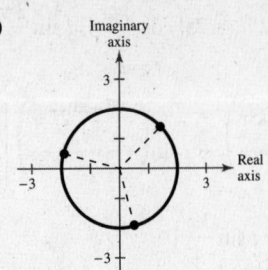

$$\sqrt[3]{8}\left[\cos\left(\dfrac{\dfrac{3\pi}{4} + 2k\pi}{3}\right) + i\sin\left(\dfrac{\dfrac{3\pi}{4} + 2k\pi}{3}\right)\right],\ k = 0, 1, 2$$

$k = 0$: $2\left(\cos\dfrac{\pi}{4} + i\sin\dfrac{\pi}{4}\right)$

$k = 1$: $2\left(\cos\dfrac{11\pi}{12} + i\sin\dfrac{11\pi}{12}\right)$

$k = 2$: $2\left(\cos\dfrac{19\pi}{12} + i\sin\dfrac{19\pi}{12}\right)$

(c) $\sqrt{2} + \sqrt{2}i,\ -1.9319 + 0.5176i,\ 0.5176 - 1.9319i$

97. (a) Fourth roots of $16 = 16(\cos 0 + i\sin 0)$:

(b)

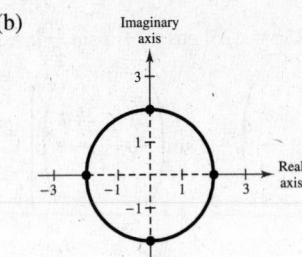

$$\sqrt[4]{16}\left[\cos\dfrac{0 + 2\pi k}{4} + i\sin\dfrac{0 + 2\pi k}{4}\right],\ k = 0, 1, 2, 3$$

$k = 0$: $2(\cos 0 + i\sin 0)$

$k = 1$: $2\left(\cos\dfrac{\pi}{2} + i\sin\dfrac{\pi}{2}\right)$

$k = 2$: $2(\cos\pi + i\sin\pi)$

$k = 3$: $2\left(\cos\dfrac{3\pi}{2} + i\sin\dfrac{3\pi}{2}\right)$

(c) $2, 2i, -2, -2i$

98. (a) Fourth roots of $i = \cos\dfrac{\pi}{2} + i\sin\dfrac{\pi}{2}$:

(b)

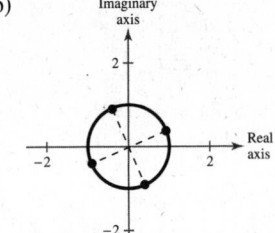

$$\sqrt[4]{1}\left[\cos\left(\dfrac{\dfrac{\pi}{2} + 2k\pi}{4}\right) + i\sin\left(\dfrac{\dfrac{\pi}{2} + 2k\pi}{4}\right)\right],\ k = 0, 1, 2, 3$$

$k = 0$: $\cos\dfrac{\pi}{8} + i\sin\dfrac{\pi}{8}$

$k = 1$: $\cos\dfrac{5\pi}{8} + i\sin\dfrac{5\pi}{8}$

$k = 2$: $\cos\dfrac{9\pi}{8} + i\sin\dfrac{9\pi}{8}$

$k = 3$: $\cos\dfrac{13\pi}{8} + i\sin\dfrac{13\pi}{8}$

(c) $0.9239 + 0.3827i,\ -0.3827 + 0.9239i,$
$-0.9239 - 0.3827i,\ 0.3827 - 0.9239i$

99. (a) Fifth roots of $1 = \cos 0 + i \sin 0$:

$$\cos\left(\frac{2k\pi}{5}\right) + i \sin\left(\frac{2k\pi}{5}\right), k = 0, 1, 2, 3, 4$$

$k = 0$: $\cos 0 + i \sin 0$

$k = 1$: $\cos \dfrac{2\pi}{5} + i \sin \dfrac{2\pi}{5}$

$k = 2$: $\cos \dfrac{4\pi}{5} + i \sin \dfrac{4\pi}{5}$

$k = 3$: $\cos \dfrac{6\pi}{5} + i \sin \dfrac{6\pi}{5}$

$k = 4$: $\cos \dfrac{8\pi}{5} + i \sin \dfrac{8\pi}{5}$

(b)

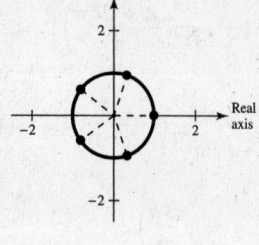

(c) $1, 0.3090 + 0.9511i, -0.8090 + 0.5878i, -0.8090 - 0.5878i, 0.3090 - 0.9511i$

100. (a) Cube roots of $1000 = 1000(\cos 0 + i \sin 0)$:

$$\sqrt[3]{1000}\left(\cos \frac{2k\pi}{3} + i \sin \frac{2k\pi}{3}\right)$$

$k = 0, 1, 2$

$k = 0$: $10(\cos 0 + i \sin 0)$

$k = 1$: $10\left(\cos \dfrac{2\pi}{3} + i \sin \dfrac{2\pi}{3}\right)$

$k = 2$: $10\left(\cos \dfrac{4\pi}{3} + i \sin \dfrac{4\pi}{3}\right)$

(b)

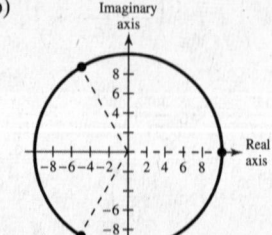

(c) $10, -5 + 5\sqrt{3}i, -5 - 5\sqrt{3}i$

101. (a) Cube roots of $-125 = 125(\cos \pi + i \sin \pi)$:

$$\sqrt[3]{125}\left[\cos\left(\frac{\pi + 2\pi k}{3}\right) + i \sin\left(\frac{\pi + 2\pi k}{3}\right)\right], k = 0, 1, 2$$

$k = 0$: $5\left(\cos \dfrac{\pi}{3} + i \sin \dfrac{\pi}{3}\right)$

$k = 1$: $5(\cos \pi + i \sin \pi)$

$k = 2$: $5\left(\cos \dfrac{5\pi}{3} + i \sin \dfrac{5\pi}{3}\right)$

(b)

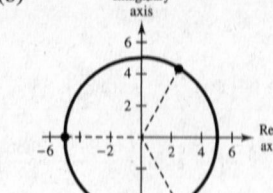

(c) $\dfrac{5}{2} + \dfrac{5\sqrt{3}}{2}i, -5, \dfrac{5}{2} - \dfrac{5\sqrt{3}}{2}i$

102. (a) Fourth roots of $-4 = 4(\cos \pi + i \sin \pi)$:

$$\sqrt[4]{4}\left[\cos\left(\frac{\pi + 2k\pi}{4}\right) + i \sin\left(\frac{\pi + 2k\pi}{4}\right)\right]$$

$k = 0, 1, 2, 3$

$k = 0$: $\sqrt{2}\left(\cos\frac{\pi}{4} + i \sin\frac{\pi}{4}\right)$

$k = 1$: $\sqrt{2}\left(\cos\frac{3\pi}{4} + i \sin\frac{3\pi}{4}\right)$

$k = 2$: $\sqrt{2}\left(\cos\frac{5\pi}{4} + i \sin\frac{5\pi}{4}\right)$

$k = 3$: $\sqrt{2}\left(\cos\frac{7\pi}{4} + i \sin\frac{7\pi}{4}\right)$

(b)

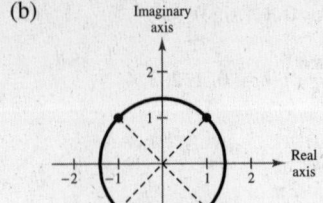

(c) $1 + i, -1 + i, -1 - i, 1 - i$

103. (a) Fifth roots of $128(-1 + i) = 128\sqrt{2}\left(\cos\frac{3\pi}{4} + i \sin\frac{3\pi}{4}\right) = 2^{15/2}\left(\cos\frac{3\pi}{4} + i \sin\frac{3\pi}{4}\right)$

$$2^{3/2}\left[\cos\left(\frac{\frac{3\pi}{4} + 2\pi k}{5}\right) + i \sin\left(\frac{\frac{3\pi}{4} + 2\pi k}{5}\right)\right], \; k = 0, 1, 2, 3, 4$$

$k = 0$: $2\sqrt{2}\left(\cos\frac{3\pi}{20} + i \sin\frac{3\pi}{20}\right)$

$k = 1$: $2\sqrt{2}\left(\cos\frac{11\pi}{20} + i \sin\frac{11\pi}{20}\right)$

$k = 2$: $2\sqrt{2}\left(\cos\frac{19\pi}{20} + i \sin\frac{19\pi}{20}\right)$

$k = 3$: $2\sqrt{2}\left(\cos\frac{27\pi}{20} + i \sin\frac{27\pi}{20}\right)$

$k = 4$: $2\sqrt{2}\left(\cos\frac{7\pi}{4} + i \sin\frac{7\pi}{4}\right)$

(b)

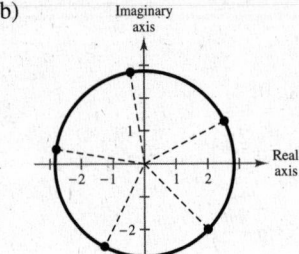

(c) $2.5201 + 1.2841i, -0.4425 + 2.7936i,$
$-2.7936 + 0.4425i, -1.2841 - 2.5201i, 2 - 2i$

104. (a) Sixth roots of $64i = 64\left(\cos\frac{\pi}{2} + i \sin\frac{\pi}{2}\right)$:

$$\sqrt[6]{64}\left[\cos\left(\frac{(\pi/2) + 2k\pi}{6}\right) + i \sin\left(\frac{(\pi/2) + 2k\pi}{6}\right)\right]$$

$k = 0, 1, 2, 3, 4, 5$

$k = 0$: $2\left(\cos\frac{\pi}{12} + i \sin\frac{\pi}{12}\right)$

$k = 1$: $2\left(\cos\frac{5\pi}{12} + i \sin\frac{5\pi}{12}\right)$

$k = 2$: $2\left(\cos\frac{3\pi}{4} + i \sin\frac{3\pi}{4}\right)$

$k = 3$: $2\left(\cos\frac{13\pi}{12} + i \sin\frac{13\pi}{12}\right)$

$k = 4$: $2\left(\cos\frac{17\pi}{12} + i \sin\frac{17\pi}{12}\right)$

$k = 5$: $2\left(\cos\frac{7\pi}{4} + i \sin\frac{7\pi}{4}\right)$

(b)

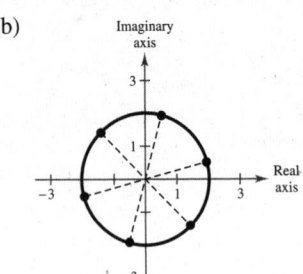

(c) $1.9319 + 0.5176i, 0.5176 + 1.9319i, -\sqrt{2} + \sqrt{2}i,$
$-1.9319 - 0.5176i, -0.5176 - 1.9319i, \sqrt{2} - \sqrt{2}i$

105. $x^4 + i = 0$

$\qquad x^4 = -i$

The solutions are the fourth roots of $i = \cos\dfrac{3\pi}{2} + i\sin\dfrac{3\pi}{2}$:

$$\sqrt[4]{1}\left[\cos\left(\dfrac{\dfrac{3\pi}{2} + 2k\pi}{4}\right) + i\sin\left(\dfrac{\dfrac{3\pi}{2} + 2k\pi}{4}\right)\right],\ k = 0, 1, 2, 3$$

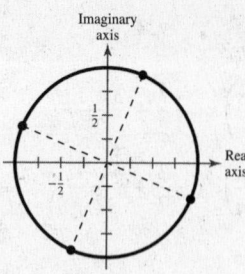

$k = 0$: $\cos\dfrac{3\pi}{8} + i\sin\dfrac{3\pi}{8} \approx 0.3827 + 0.9239i$

$k = 1$: $\cos\dfrac{7\pi}{8} + i\sin\dfrac{7\pi}{8} \approx -0.9239 + 0.3827i$

$k = 2$: $\cos\dfrac{11\pi}{8} + i\sin\dfrac{11\pi}{8} \approx -0.3827 - 0.9239i$

$k = 3$: $\cos\dfrac{15\pi}{8} + i\sin\dfrac{15\pi}{8} \approx 0.9239 - 0.3827i$

106. $x^3 + 1 = 0$

$\qquad x^3 = -1$

The solutions are the cube roots of
$-1 = \cos\pi + i\sin\pi$:

$$\cos\left(\dfrac{\pi + 2k\pi}{3}\right) + i\sin\left(\dfrac{\pi + 2k\pi}{3}\right)$$

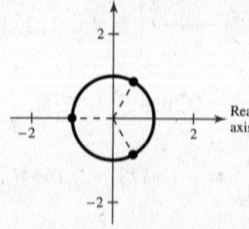

$k = 0, 1, 2$

$k = 0$: $\cos\dfrac{\pi}{3} + i\sin\dfrac{\pi}{3} = \dfrac{1}{2} + \dfrac{\sqrt{3}}{2}i$

$k = 1$: $\cos\pi + i\sin\pi = -1$

$k = 2$: $\cos\dfrac{5\pi}{3} + i\sin\dfrac{5\pi}{3} = \dfrac{1}{2} - \dfrac{\sqrt{3}}{2}i$

107. $x^5 + 243 = 0$

$\qquad x^5 = -243$

The solutions are the fifth roots of $-243 = 243(\cos\pi + i\sin\pi)$:

$$\sqrt[5]{243}\left[\cos\left(\dfrac{\pi + 2k\pi}{5}\right) + i\sin\left(\dfrac{\pi + 2k\pi}{5}\right)\right],\ k = 0, 1, 2, 3, 4$$

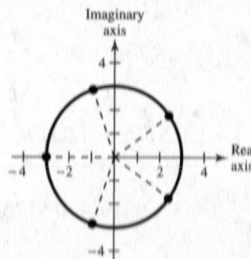

$k = 0$: $3\left(\cos\dfrac{\pi}{5} + i\sin\dfrac{\pi}{5}\right) \approx 2.4271 + 1.7634i$

$k = 1$: $3\left(\cos\dfrac{3\pi}{5} + i\sin\dfrac{3\pi}{5}\right) \approx -0.9271 + 2.8532i$

$k = 2$: $3(\cos\pi + i\sin\pi) = -3$

$k = 3$: $3\left(\cos\dfrac{7\pi}{5} + i\sin\dfrac{7\pi}{5}\right) \approx -0.9271 - 2.8532i$

$k = 4$: $3\left(\cos\dfrac{9\pi}{5} + i\sin\dfrac{9\pi}{5}\right) \approx 2.4271 - 1.7634i$

108. $x^3 - 27 = 0$

$$x^3 = 27$$

The solutions are the cube roots of $27 = 27(\cos 0 + i \sin 0)$:

$$\sqrt[3]{27}\left[\cos\left(\frac{2k\pi}{3}\right) + i \sin\left(\frac{2k\pi}{3}\right)\right]$$

$k = 0, 1, 2$

$k = 0$: $3(\cos 0 + i \sin 0) = 3$

$k = 1$: $3\left(\cos\dfrac{2\pi}{3} + i \sin\dfrac{2\pi}{3}\right) = -\dfrac{3}{2} + \dfrac{3\sqrt{3}}{2}i$

$k = 2$: $3\left(\cos\dfrac{4\pi}{3} + i \sin\dfrac{4\pi}{3}\right) = -\dfrac{3}{2} - \dfrac{3\sqrt{3}}{2}i$

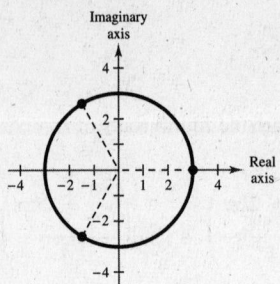

109. $x^4 + 16i = 0$

$$x^4 = -16i$$

The solutions are the fourth roots of $-16i = 16\left(\cos\dfrac{3\pi}{2} + i \sin\dfrac{3\pi}{2}\right)$:

$$\sqrt[4]{16}\left[\cos\frac{\frac{3\pi}{2} + 2\pi k}{4} + i \sin\frac{\frac{3\pi}{2} + 2\pi k}{4}\right], k = 0, 1, 2, 3$$

$k = 0$: $2\left(\cos\dfrac{3\pi}{8} + i \sin\dfrac{3\pi}{8}\right) \approx 0.7654 + 1.8478i$

$k = 1$: $2\left(\cos\dfrac{7\pi}{8} + i \sin\dfrac{7\pi}{8}\right) \approx -1.8478 + 0.7654i$

$k = 2$: $2\left(\cos\dfrac{11\pi}{8} + i \sin\dfrac{11\pi}{8}\right) \approx -0.7654 - 1.8478i$

$k = 3$: $2\left(\cos\dfrac{15\pi}{8} + i \sin\dfrac{15\pi}{8}\right) \approx 1.8478 - 0.7654i$

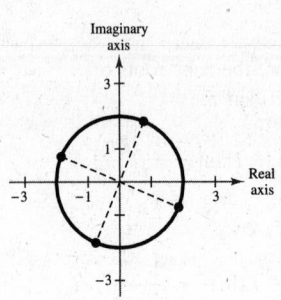

110. $x^6 + 64i = 0$

$$x^6 = -64i$$

The solutions are the sixth roots of $-64i = 64\left[\cos\dfrac{3\pi}{2} + i \sin\dfrac{3\pi}{2}\right]$:

$$\sqrt[6]{64}\left[\cos\left(\frac{(3\pi/2) + 2\pi k}{6}\right) + i \sin\left(\frac{(3\pi/2) + 2\pi k}{6}\right)\right]$$

$k = 0, 1, 2, 3, 4, 5$

$k = 0$: $2\left(\cos\dfrac{\pi}{4} + i \sin\dfrac{\pi}{4}\right) = \sqrt{2} + \sqrt{2}\,i$

$k = 1$: $2\left(\cos\dfrac{7\pi}{12} + i \sin\dfrac{7\pi}{12}\right) \approx -0.5176 + 1.9319i$

$k = 2$: $2\left(\cos\dfrac{11\pi}{12} + i \sin\dfrac{11\pi}{12}\right) \approx -1.9319 + 0.5176i$

$k = 3$: $2\left(\cos\dfrac{5\pi}{4} + i \sin\dfrac{5\pi}{4}\right) = -\sqrt{2} - \sqrt{2}\,i$

$k = 4$: $2\left(\cos\dfrac{19\pi}{12} + i \sin\dfrac{19\pi}{12}\right) = 0.5176 - 1.9319i$

$k = 5$: $2\left(\cos\dfrac{23\pi}{12} + i \sin\dfrac{23\pi}{12}\right) = 1.9319 - 0.5176i$

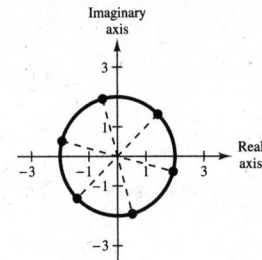

111. $x^3 - (1 - i) = 0$

$$x^3 = 1 - i = \sqrt{2}\left(\cos\frac{7\pi}{4} + i\sin\frac{7\pi}{4}\right)$$

The solutions are the cube roots of $1 - i$:

$$\sqrt[3]{\sqrt{2}}\left[\cos\left(\frac{(7\pi/4) + 2\pi k}{3}\right) + i\sin\left(\frac{(7\pi/4) + 2\pi k}{3}\right)\right], \ k = 0, 1, 2$$

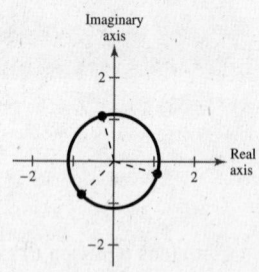

$k = 0$: $\sqrt[6]{2}\left(\cos\frac{7\pi}{12} + i\sin\frac{7\pi}{12}\right) \approx -0.2905 + 1.0842i$

$k = 1$: $\sqrt[6]{2}\left(\cos\frac{5\pi}{4} + i\sin\frac{5\pi}{4}\right) \approx -0.7937 - 0.7937i$

$k = 2$: $\sqrt[6]{2}\left(\cos\frac{23\pi}{12} + i\sin\frac{23\pi}{12}\right) \approx 1.0842 - 0.2905i$

112. $x^4 + (1 + i) = 0$

$$x^4 = -1 - i = \sqrt{2}(\cos 225° + i\sin 225°)$$

The solutions are the fourth roots of $-1 - i$:

$$\sqrt[4]{\sqrt{2}}\left[\cos\left(\frac{225° + 360°k}{4}\right) + i\sin\left(\frac{225° + 360°k}{4}\right)\right]$$

$k = 0, 1, 2, 3$

$k = 0$: $\sqrt[8]{2}(\cos 56.25° + i\sin 56.25°) \approx 0.6059 + 0.9067i$

$k = 1$: $\sqrt[8]{2}(\cos 146.25° + i\sin 146.25°) \approx -0.9067 + 0.6059i$

$k = 2$: $\sqrt[8]{2}(\cos 236.25° + i\sin 236.25°) \approx -0.6059 - 0.9067i$

$k = 3$: $\sqrt[8]{2}(\cos 326.25° + i\sin 326.25°) \approx 0.9067 - 0.6059i$

113. True, by the definition of the absolute value of a complex number.

114. False. They are equally spaced along the circle centered at the origin with radius $\sqrt[n]{r}$.

115. True. $z_1 z_2 = r_1 r_2[\cos(\theta_1 + \theta_2) + i\sin(\theta_1 + \theta_2)]$ and $z_1 z_2 = 0$ if and only if $r_1 = 0$ and/or $r_2 = 0$.

116. False. The complex number must be converted to trigonometric form before applying DeMoivre's Theorem.

$$\left(4 + \sqrt{6}i\right)^8 = \left[\sqrt{22}\left(\cos\left(\arctan\frac{\sqrt{6}}{4}\right) + i\sin\left(\arctan\frac{\sqrt{6}}{4}\right)\right)\right]^8$$

117.
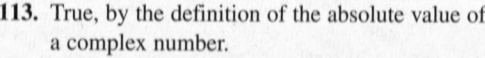
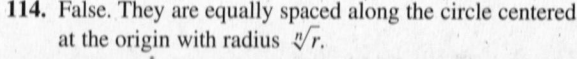

$$\frac{z_1}{z_2} = \frac{r_1(\cos\theta_1 + i\sin\theta_1)}{r_2(\cos\theta_2 + i\sin\theta_2)} \cdot \frac{\cos\theta_2 - i\sin\theta_2}{\cos\theta_2 - i\sin\theta_2}$$

$$= \frac{r_1}{r_2(\cos^2\theta_2 + \sin^2\theta_2)}[\cos\theta_1\cos\theta_2 + \sin\theta_1\sin\theta_2 + i(\sin\theta_1\cos\theta_2 - \sin\theta_2\cos\theta_1)]$$

$$= \frac{r_1}{r_2}[\cos(\theta_1 - \theta_2) + i\sin(\theta_1 - \theta_2)]$$

118. $\bar{z} = r[\cos(-\theta) + i\sin(-\theta)]$

$\qquad = r[\cos\theta + -i\sin\theta]$

$\qquad = r\cos\theta - ir\sin\theta$

which is the complex conjugate of
$r(\cos\theta + i\sin\theta) = r\cos\theta + ir\sin\theta$.

119. (a) $z\bar{z} = [r(\cos\theta + i\sin\theta)][r(\cos(-\theta) + i\sin(-\theta))]$

$\qquad = r^2[\cos(\theta - \theta) + i\sin(\theta - \theta)]$

$\qquad = r^2[\cos 0 + i\sin 0]$

$\qquad = r^2$

(b) $\dfrac{z}{\bar{z}} = \dfrac{r(\cos\theta + i\sin\theta)}{r[\cos(-\theta) + i\sin(-\theta)]}$

$\qquad = \dfrac{r}{r}[\cos(\theta - (-\theta)) + i\sin(\theta - (-\theta))]$

$\qquad = \cos 2\theta + i\sin 2\theta$

120. $z = r(\cos\theta + i\sin\theta)$

$-z = -r(\cos\theta + i\sin\theta)$

$\qquad = r(-\cos\theta + -i\sin\theta)$

$\qquad = r(\cos(\theta + \pi) + i\sin(\theta + \pi))$

121. $-\dfrac{1}{2}(1 + \sqrt{3}i) = -\left(\cos\dfrac{4\pi}{3} + i\sin\dfrac{4\pi}{3}\right)$

$\left[-\dfrac{1}{2}(1 + \sqrt{3}i)\right]^6 = \left[-\left(\cos\dfrac{4\pi}{3} + i\sin\dfrac{4\pi}{3}\right)\right]^6$

$\qquad = \cos 8\pi + i\sin 8\pi$

$\qquad = 1$

122.
$$2^{-1/4}(1 - i) = 2^{-1/4}\left[\sqrt{2}\left(\cos\dfrac{7\pi}{4} + i\sin\dfrac{7\pi}{4}\right)\right]$$

$$= 2^{1/4}\left(\cos\dfrac{7\pi}{4} + i\sin\dfrac{7\pi}{4}\right)$$

$$\left[2^{1/4}\left(\cos\dfrac{7\pi}{4} + i\sin\dfrac{7\pi}{4}\right)\right]^4 = (2^{1/4})^4(\cos 7\pi + i\sin 7\pi)$$

$$= 2(\cos\pi + i\sin\pi)$$

$$= -2$$

123. (a) $2(\cos 30° + i\sin 30°)$

$\qquad 2(\cos 150° + i\sin 150°)$

$\qquad 2(\cos 270° + i\sin 270°)$

(b) These are the cube roots of $8i$.

124. (a) $3(\cos 45° + i\sin 45°)$

$\qquad 3(\cos 135° + i\sin 135°)$

$\qquad 3(\cos 225° + i\sin 225°)$

$\qquad 3(\cos 315° + i\sin 315°)$

(b) These are the fourth roots of -81.

(c) The fourth roots of -81:

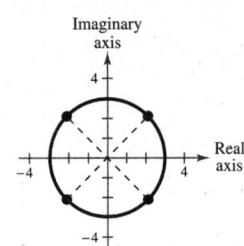

125. $A = 22°, a = 8$

$B = 90° - A = 68°$

$\tan 22° = \dfrac{8}{b} \implies b = \dfrac{8}{\tan 22°} \approx 19.80$

$\sin 22° = \dfrac{8}{c} \implies c = \dfrac{8}{\sin 22°} \approx 21.36$

126. $B = 66°, a = 33.5$

$A = 90° - 66° = 24°$

$b = \dfrac{a \sin B}{\sin A} = \dfrac{(33.5) \sin 66°}{\sin 24°} \approx 75.24$

$c = \dfrac{a \sin C}{\sin A} = \dfrac{(33.5) \sin 90°}{\sin 24°} \approx 82.36$

127. $A = 30°, b = 112.6$

$B = 90° - A = 60°$

$\tan 30° = \dfrac{a}{112.6} \implies a = 112.6 \tan 30° \approx 65.01$

$\cos 30° = \dfrac{112.6}{c} \implies c = \dfrac{112.6}{\cos 30°} \approx 130.02$

128. $B = 6°, b = 211.2$

$A = 90° - 6° = 84°$

$a = \dfrac{b \sin A}{\sin B} = \dfrac{(211.2) \sin 84°}{\sin 6°} \approx 2009.43$

$c = \dfrac{b \sin C}{\sin B} = \dfrac{(211.2) \sin 90°}{\sin 6°} \approx 2020.50$

129. $A = 42°15' = 42.25°, c = 11.2$

$B = 90° - A = 47°45'$

$\sin 42.25° = \dfrac{a}{11.2} \implies a = 11.2 \sin 42.25° \approx 7.53$

$\cos 42.25° = \dfrac{b}{11.2} \implies b = 11.2 \cos 42.25° \approx 8.29$

130. $B = 81° 30', c = 6.8$

$A = 90° - 81° 30' = 8° 30'$

$a = \dfrac{c \sin A}{\sin C} = \dfrac{(6.8) \sin 8° 30'}{1} \approx 1.01$

$b = \dfrac{c \sin B}{\sin C} = \dfrac{(6.8) \sin 81° 30'}{1} \approx 6.73$

131. $d = 16 \cos \dfrac{\pi}{4} t$

Maximum displacement: $|16| = 16$

$16 \cos \dfrac{\pi}{4} t = 0 \implies \dfrac{\pi}{4} t = \dfrac{\pi}{2} \implies t = 2$

132. $d = \dfrac{1}{8} \cos 12\pi t$

Maximum displacement: $\dfrac{1}{8}$

$d = 0$ when $12\pi t = \dfrac{\pi}{2}$, or $t = \dfrac{1}{24}$

133. $d = \dfrac{1}{16} \sin\left(\dfrac{5}{4}\pi t\right)$

Maximum displacement: $\left|\dfrac{1}{16}\right| = \dfrac{1}{16}$

$\dfrac{1}{16} \sin\left(\dfrac{5}{4}\pi t\right) = 0$

$\dfrac{5}{4}\pi t = \pi$

$t = \dfrac{4}{5}$

134. $d = \dfrac{1}{12} \sin 60\pi t$

Maximum displacement: $\dfrac{1}{12}$

$d = 0$ when $60\pi t = \pi$, or $t = \dfrac{1}{60}$

135. $6 \sin 8\theta \cos 3\theta = (6)\left(\dfrac{1}{2}\right)[\sin(8\theta + 3\theta) + \sin(8\theta - 3\theta)]$

$\qquad = 3(\sin 11\theta + \sin 5\theta)$

136. $2 \cos 5\theta \sin 2\theta = 2 \cdot \dfrac{1}{2}[\sin(5\theta + 2\theta) - \sin(5\theta - 2\theta)]$

$\qquad = \sin 7\theta - \sin 3\theta$

Review Exercises for Chapter 6

1. Given: $A = 35°, B = 71°, a = 8$

$C = 180° - 35° - 71° = 74°$

$b = \dfrac{a \sin B}{\sin A} = \dfrac{8 \sin 71°}{\sin 35°} \approx 13.19$

$c = \dfrac{a \sin C}{\sin A} = \dfrac{8 \sin 74°}{\sin 35°} \approx 13.41$

2. Given: $A = 22°, B = 121°, a = 17$

$C = 180 - A - B = 37°$

$b = \dfrac{a \sin B}{\sin A} = \dfrac{17 \sin 121°}{\sin 22°} \approx 38.90$

$c = \dfrac{a \sin C}{\sin A} = \dfrac{17 \sin 37°}{\sin 22°} \approx 27.31$

3. Given: $B = 72°, C = 82°, b = 54$

$A = 180° - 72° - 82° = 26°$

$a = \dfrac{b \sin A}{\sin B} = \dfrac{54 \sin 26°}{\sin 72°} \approx 24.89$

$c = \dfrac{b \sin C}{\sin B} = \dfrac{54 \sin 82°}{\sin 72°} \approx 56.23$

4. Given: $B = 10°, C = 20°, c = 33$

$A = 180° - B - C = 150°$

$a = \dfrac{c \sin A}{\sin C} = \dfrac{33 \sin 150°}{\sin 20°} \approx 48.24$

$b = \dfrac{c \sin B}{\sin C} = \dfrac{33 \sin 10°}{\sin 20°} \approx 16.75$

5. Given: $A = 16°, B = 98°, c = 8.4$

$C = 180° - 16° - 98° = 66°$

$a = \dfrac{c \sin A}{\sin C} = \dfrac{8.4 \sin 16°}{\sin 66°} \approx 2.53$

$b = \dfrac{c \sin B}{\sin C} = \dfrac{8.4 \sin 98°}{\sin 66°} \approx 9.11$

6. Given: $A = 95°, B = 45°, c = 104.8$

$C = 180° - A - B = 40°$

$a = \dfrac{c \sin A}{\sin C} = \dfrac{104.8 \sin 95°}{\sin 40°} \approx 162.42$

$b = \dfrac{c \sin B}{\sin C} = \dfrac{104.8 \sin 45°}{\sin 40°} \approx 115.29$

7. Given: $A = 24°, C = 48°, b = 27.5$

$B = 180° - 24° - 48° = 108°$

$a = \dfrac{b \sin A}{\sin B} = \dfrac{27.5 \sin 24°}{\sin 108°} \approx 11.76$

$c = \dfrac{b \sin C}{\sin B} = \dfrac{27.5 \sin 48°}{\sin 108°} \approx 21.49$

8. Given: $B = 64°, C = 36°, a = 367$

$A = 180° - B - C = 80°$

$b = \dfrac{a \sin B}{\sin A} = \dfrac{367 \sin 64°}{\sin 80°} \approx 334.95$

$c = \dfrac{a \sin C}{\sin A} = \dfrac{367 \sin 36°}{\sin 80°} \approx 219.04$

9. Given: $B = 150°, b = 30, c = 10$

$\sin C = \dfrac{c \sin B}{b} = \dfrac{10 \sin 150°}{30} \approx 0.1667 \implies C \approx 9.59°$

$A \approx 180° - 150° - 9.59° = 20.41°$

$a = \dfrac{b \sin A}{\sin B} = \dfrac{30 \sin 20.41°}{\sin 150°} \approx 20.92$

10. Given: $B = 150°, a = 10, b = 3$

$\sin A = \dfrac{a \sin B}{b} = \dfrac{10 \sin 150°}{3} \approx 1.67 > 1$

No solution

11. $A = 75°, a = 51.2, b = 33.7$

$\sin B = \dfrac{b \sin A}{a} = \dfrac{33.7 \sin 75°}{51.2} \approx 0.6358 \implies B \approx 39.48°$

$C \approx 180° - 75° - 39.48° = 65.52°$

$c = \dfrac{a \sin C}{\sin A} = \dfrac{51.2 \sin 65.52°}{\sin 75°} \approx 48.24$

12. Given: $B = 25°$, $a = 6.2$, $b = 4$

$\sin A = \dfrac{a \sin B}{b} \approx 0.65506 \implies A \approx 40.92°$ or $139.08°$

Case 1: $A \approx 40.92°$ 　　　　　　Case 2: $A \approx 139.08°$

$C \approx 180° - 25° - 40.92° = 114.08°$　$C \approx 180° - 25° - 139.08° = 15.92°$

$c \approx 8.64$ 　　　　　　　　　　　$c \approx 2.60$

13. Area $= \frac{1}{2}bc \sin A = \frac{1}{2}(5)(7)\sin 27° \approx 7.9$

14. $B = 80°$, $a = 4$, $c = 8$

Area $= \frac{1}{2}ac \sin B = \frac{1}{2}(4)(8)(0.9848) \approx 15.8$

15. Area $= \frac{1}{2}ab \sin C = \frac{1}{2}(16)(5)\sin 123° \approx 33.5$

16. $A = 11°$, $b = 22$, $c = 21$

Area $= \frac{1}{2}bc \sin A \approx \frac{1}{2}(22)(21)(0.1908) \approx 44.1$

17. $\tan 17° = \dfrac{h}{x + 50} \implies h = (x + 50)\tan 17°$

$h = x \tan 17° + 50 \tan 17°$

$\tan 31° = \dfrac{h}{x} \implies h = x \tan 31°$

$x \tan 17° + 50 \tan 17° = x \tan 31°$

$50 \tan 17° = x(\tan 31° - \tan 17°)$

$\dfrac{50 \tan 17°}{\tan 31° - \tan 17°} = x$

$x \approx 51.7959$

$h = x \tan 31° \approx 51.7959 \tan 31° \approx 31.1$ meters

The height of the building is approximately 31.1 meters.

18. $16^2 = w^2 + 12^2 - 2w(12) \cos 140°$

$w^2 - (24 \cos 140°)w - 112 = 0 \implies w \approx 4.83$

19. $\dfrac{h}{\sin 17°} = \dfrac{75}{\sin 45°}$

$h = \dfrac{75 \sin 17°}{\sin 45°}$

$h \approx 31.01$ feet

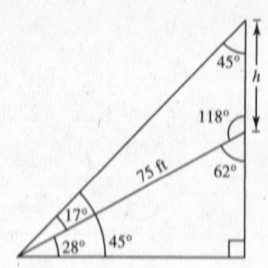

20. The triangle of base 400 feet formed by the two angles of sight to the tree has base angles of $90° - 22° 30' = 67° 30'$, or $67.5°$, and $90° - 15° = 75°$. The angle at the tree measures $180° - 67.5° - 75° = 37.5°$.

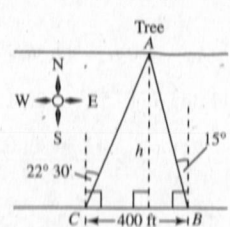

$b = \dfrac{400 \sin 75°}{\sin 37.5°} \approx 634.683$

$h = 634.683 \sin 67.5°$

$h \approx 586.4$

The width of the river is about 586.4 feet.

21. Given: $a = 5, b = 8, c = 10$

$$\cos C = \frac{a^2 + b^2 - c^2}{2ab} = -0.1375 \implies C \approx 97.90°$$

$$\cos B = \frac{a^2 + c^2 - b^2}{2ac} = 0.61 \implies B \approx 52.41°$$

$$A = 180° - B - C \approx 29.69°$$

22. Given: $a = 80, b = 60, c = 100$

$$\cos C = \frac{a^2 + b^2 - c^2}{2ab} = \frac{6400 + 3600 - 10{,}000}{2(80)(60)}$$

$$= 0 \implies C = 90°$$

$$\sin A = \frac{80}{100} = 0.8 \implies A \approx 53.13°$$

$$\sin B = \frac{60}{100} = 0.6 \implies B \approx 36.87°$$

23. Given: $a = 2.5, b = 5.0, c = 4.5$

$$\cos B = \frac{a^2 + c^2 - b^2}{2ac} = 0.0667 \implies B \approx 86.18°$$

$$\cos C = \frac{a^2 + b^2 - c^2}{2ab} = 0.44 \implies C \approx 63.90°$$

$$A = 180° - B - C \approx 29.92°$$

24. Given: $a = 16.4, b = 8.8, c = 12.2$

$$\cos A = \frac{b^2 + c^2 - a^2}{2bc} = \frac{8.8^2 + 12.2^2 - 16.4^2}{2(8.8)(12.2)} \approx -0.1988 \implies A \approx 101.47°$$

$$\sin B = \frac{b \sin A}{a} \approx \frac{8.8 \sin 101.47°}{16.4} \approx 0.5259 \implies B \approx 31.73°$$

$$C \approx 180° - 101.47° - 31.73° = 46.80°$$

25. Given: $B = 110°, a = 4, c = 4$

$$b = \sqrt{a^2 + c^2 - 2ac \cos B} \approx 6.55$$

$$A = C = \tfrac{1}{2}(180° - 110°) = 35°$$

26. Given: $B = 150°, a = 10, c = 20$

$$b^2 = 10^2 + 20^2 - 2(10)(20)\cos 150° \implies b \approx 29.09$$

$$\sin A = \frac{a \sin B}{b} \approx \frac{10 \sin 150°}{29.09} \implies A \approx 9.90°$$

$$C \approx 180° - 150° - 9.90° = 20.10°$$

27. Given: $C = 43°, a = 22.5, b = 31.4$

$$c = \sqrt{a^2 + b^2 - 2ab \cos C} \approx 21.42$$

$$\cos B = \frac{a^2 + c^2 - b^2}{2ac} \approx -0.02169 \implies B \approx 91.24°$$

$$A = 180° - B - C \approx 45.76°$$

28. Given: $A = 62°, b = 11.34, c = 19.52$

$$a^2 = 11.34^2 + 19.52^2 - 2(11.34)(19.52) \cos 62° \implies a \approx 17.37$$

$$\sin B = \frac{b \sin A}{a} \approx \frac{11.34 \sin 62°}{17.37} \implies B \approx 35.20°$$

$$C \approx 180° - 62° - 35.20° = 82.80°$$

29.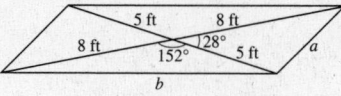

$a^2 = 5^2 + 8^2 - 2(5)(8)\cos 28° \approx 18.364$

$a \approx 4.3$ feet

$b^2 = 8^2 + 5^2 - 2(8)(5)\cos 152° \approx 159.636$

$b \approx 12.6$ feet

30.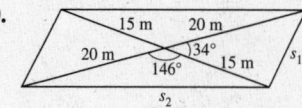

$s_1^2 = 15^2 + 20^2 + 2 \cdot 15 \cdot 20 \cos 34° \approx 127.58$

$s_1 \approx 11.3$ meters

$s_2^2 = 15^2 + 20^2 + 2 \cdot 15 \cdot 20 \cos 146° \approx 1122.42$

$s_2 \approx 33.5$ meters

31. Length of AC $= \sqrt{300^2 + 425^2 - 2(300)(425) \cos 115°}$

≈ 615.1 meters

32. $d^2 = 850^2 + 1060^2 - 2(850)(1060) \cos 72°$

$\approx 1{,}289{,}251$

$d \approx 1135$ miles

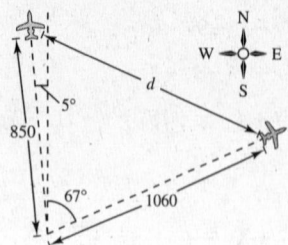

33. $a = 4, \ b = 5, \ c = 7$

$s = \dfrac{a + b + c}{2} = \dfrac{4 + 5 + 7}{2} = 8$

Area $= \sqrt{s(s - a)(s - b)(s - c)}$

$= \sqrt{8(4)(3)(1)} \approx 9.80$

34. $a = 15, b = 8, c = 10$

$s = \dfrac{15 + 8 + 10}{2} = 16.5$

Area $= \sqrt{16.5(1.5)(8.5)(6.5)} \approx 36.979$

35. $a = 12.3, b = 15.8, c = 3.7$

$s = \dfrac{a + b + c}{2} = \dfrac{12.3 + 15.8 + 3.7}{2} = 15.9$

Area $= \sqrt{s(s - a)(s - b)(s - c)}$

$= \sqrt{15.9(3.6)(0.1)(12.2)} = 8.36$

36. $a = 38.1, b = 26.7, c = 19.4$

$s = \dfrac{38.1 + 26.7 + 19.4}{2} = 42.1$

Area $= \sqrt{42.1(4)(15.4)(22.7)} \approx 242.630$

37. $\|\mathbf{u}\| = \sqrt{(4 - (-2))^2 + (6 - 1)^2} = \sqrt{61}$

$\|\mathbf{v}\| = \sqrt{(6 - 0)^2 + (3 - (-2))^2} = \sqrt{61}$

$\mathbf{u}$ is directed along a line with a slope of $\dfrac{6 - 1}{4 - (-2)} = \dfrac{5}{6}$.

$\mathbf{v}$ is directed along a line with a slope of $\dfrac{3 - (-2)}{6 - 0} = \dfrac{5}{6}$.

Since $\mathbf{u}$ and $\mathbf{v}$ have identical magnitudes and directions, $\mathbf{u} = \mathbf{v}$.

38. $\|\mathbf{u}\| = \sqrt{(3 - 1)^2 + (-2 - 4)^2} = 2\sqrt{10}$

$\|\mathbf{v}\| = \sqrt{(-1 - (-3))^2 + (-4 - 2)^2} = 2\sqrt{10}$

$\mathbf{u}$ is directed along a line with a slope of $\dfrac{-2 - 4}{3 - 1} = -3$.

$\mathbf{v}$ is directed along a line with a slope of $\dfrac{-4 - 2}{-1 - (-3)} = -3$.

Since $\mathbf{u}$ and $\mathbf{v}$ have identical magnitudes and directions, $\mathbf{u} = \mathbf{v}$.

39. Initial point: $(-5, 4)$

Terminal point: $(2, -1)$

$\mathbf{v} = \langle 2 - (-5), -1 - 4 \rangle = \langle 7, -5 \rangle$

40. Initial point: $(0, 1)$

Terminal point: $\left(6, \frac{7}{2}\right)$

$\mathbf{v} = \left\langle 6 - 0, \frac{7}{2} - 1 \right\rangle = \left\langle 6, \frac{5}{2} \right\rangle$

41. Initial point: $(0, 10)$

Terminal point: $(7, 3)$

$\mathbf{v} = \langle 7 - 0, 3 - 10 \rangle = \langle 7, -7 \rangle$

42. Initial point: $(1, 5)$

Terminal point: $(15, 9)$

$\mathbf{v} = \langle 15 - 1, 9 - 5 \rangle = \langle 14, 4 \rangle$

43. $\|\mathbf{v}\| = 8$, $\theta = 120°$

$\langle 8 \cos 120°, 8 \sin 120° \rangle = \langle -4, 4\sqrt{3} \rangle$

44. $\|\mathbf{v}\| = \frac{1}{2}$, $\theta = 225°$

$\left\langle \frac{1}{2} \cos 225°, \frac{1}{2} \sin 225° \right\rangle = \left\langle -\frac{\sqrt{2}}{4}, -\frac{\sqrt{2}}{4} \right\rangle$

45. $\mathbf{u} = \langle -1, -3 \rangle$, $\mathbf{v} = \langle -3, 6 \rangle$

(a) $\mathbf{u} + \mathbf{v} = \langle -1, -3 \rangle + \langle -3, 6 \rangle = \langle -4, 3 \rangle$

(b) $\mathbf{u} - \mathbf{v} = \langle -1, -3 \rangle - \langle -3, 6 \rangle = \langle 2, -9 \rangle$

(c) $3\mathbf{u} = 3\langle -1, -3 \rangle = \langle -3, -9 \rangle$

(d) $2\mathbf{v} + 5\mathbf{u} = 2\langle -3, 6 \rangle + 5\langle -1, -3 \rangle$

$\qquad = \langle -6, 12 \rangle + \langle -5, -15 \rangle = \langle -11, -3 \rangle$

46. $\mathbf{u} = \langle 4, 5 \rangle$, $\mathbf{v} = \langle 0, -1 \rangle$

(a) $\mathbf{u} + \mathbf{v} = \langle 4 + 0, 5 + (-1) \rangle = \langle 4, 4 \rangle$

(b) $\mathbf{u} - \mathbf{v} = \langle 4 - 0, 5 - (-1) \rangle = \langle 4, 6 \rangle$

(c) $3\mathbf{u} = \langle 3(4), 3(5) \rangle = \langle 12, 15 \rangle$

(d) $2\mathbf{v} + 5\mathbf{u} = \langle 2(0), 2(-1) \rangle + \langle 5(4), 5(5) \rangle$

$\qquad = \langle 0 + 20, -2 + 25 \rangle = \langle 20, 23 \rangle$

47. $\mathbf{u} = \langle -5, 2 \rangle$, $\mathbf{v} = \langle 4, 4 \rangle$

(a) $\mathbf{u} + \mathbf{v} = \langle -5, 2 \rangle + \langle 4, 4 \rangle = \langle -1, 6 \rangle$

(b) $\mathbf{u} - \mathbf{v} = \langle -5, 2 \rangle - \langle 4, 4 \rangle = \langle -9, -2 \rangle$

(c) $3\mathbf{u} = 3\langle -5, 2 \rangle = \langle -15, 6 \rangle$

(d) $2\mathbf{v} + 5\mathbf{u} = 2\langle 4, 4 \rangle + 5\langle -5, 2 \rangle$

$\qquad = \langle 8, 8 \rangle + \langle -25, 10 \rangle = \langle -17, 18 \rangle$

48. $\mathbf{u} = \langle 1, -8 \rangle$, $\mathbf{v} = \langle 3, -2 \rangle$

(a) $\mathbf{u} + \mathbf{v} = \langle 1 + 3, -8 + (-2) \rangle = \langle 4, -10 \rangle$

(b) $\mathbf{u} - \mathbf{v} = \langle 1 - 3, -8 - (-2) \rangle = \langle -2, -6 \rangle$

(c) $3\mathbf{u} = \langle 3(1), 3(-8) \rangle = \langle 3, -24 \rangle$

(d) $2\mathbf{v} + 5\mathbf{u} = \langle 2(3), 2(-2) \rangle + \langle 5(1), 5(-8) \rangle$

$\qquad = \langle 6 + 5, -4 + (-40) \rangle = \langle 11, -44 \rangle$

49. $\mathbf{u} = 2\mathbf{i} - \mathbf{j}$, $\mathbf{v} = 5\mathbf{i} + 3\mathbf{j}$

(a) $\mathbf{u} + \mathbf{v} = (2\mathbf{i} - \mathbf{j}) + (5\mathbf{i} + 3\mathbf{j}) = 7\mathbf{i} + 2\mathbf{j}$

(b) $\mathbf{u} - \mathbf{v} = (2\mathbf{i} - \mathbf{j}) - (5\mathbf{i} + 3\mathbf{j}) = -3\mathbf{i} - 4\mathbf{j}$

(c) $3\mathbf{u} = 3(2\mathbf{i} - \mathbf{j}) = 6\mathbf{i} - 3\mathbf{j}$

(d) $2\mathbf{v} + 5\mathbf{u} = 2(5\mathbf{i} + 3\mathbf{j}) + 5(2\mathbf{i} - \mathbf{j})$

$\qquad = (10\mathbf{i} + 6\mathbf{j}) + (10\mathbf{i} - 5\mathbf{j}) = 20\mathbf{i} + \mathbf{j}$

50. $\mathbf{u} = -7\mathbf{i} - 3\mathbf{j}$, $\mathbf{v} = 4\mathbf{i} - \mathbf{j}$

(a) $\mathbf{u} + \mathbf{v} = -7\mathbf{i} - 3\mathbf{j} + 4\mathbf{i} - \mathbf{j} = -3\mathbf{i} - 4\mathbf{j}$

(b) $\mathbf{u} - \mathbf{v} = -7\mathbf{i} - 3\mathbf{j} - 4\mathbf{i} + \mathbf{j} = -11\mathbf{i} - 2\mathbf{j}$

(c) $3\mathbf{u} = 3(-7\mathbf{i} - 3\mathbf{j}) = -21\mathbf{i} - 9\mathbf{j}$

(d) $2\mathbf{v} + 5\mathbf{u} = 8\mathbf{i} - 2\mathbf{j} - 35\mathbf{i} - 15\mathbf{j}$

$\qquad = -27\mathbf{i} - 17\mathbf{j}$

51. $\mathbf{u} = 4\mathbf{i}$, $\mathbf{v} = -\mathbf{i} + 6\mathbf{j}$

(a) $\mathbf{u} + \mathbf{v} = 4\mathbf{i} + (-\mathbf{i} + 6\mathbf{j}) = 3\mathbf{i} + 6\mathbf{j}$

(b) $\mathbf{u} - \mathbf{v} = 4\mathbf{i} - (-\mathbf{i} + 6\mathbf{j}) = 5\mathbf{i} - 6\mathbf{j}$

(c) $3\mathbf{u} = 3(4\mathbf{i}) = 12\mathbf{i}$

(d) $2\mathbf{v} + 5\mathbf{u} = 2(-\mathbf{i} + 6\mathbf{j}) + 5(4\mathbf{i})$

$\qquad = (-2\mathbf{i} + 12\mathbf{j}) + 20\mathbf{i} = 18\mathbf{i} + 12\mathbf{j}$

52. $\mathbf{u} = -6\mathbf{j}$, $\mathbf{v} = \mathbf{i} + \mathbf{j}$

(a) $\mathbf{u} + \mathbf{v} = -6\mathbf{j} + \mathbf{i} + \mathbf{j} = \mathbf{i} - 5\mathbf{j}$

(b) $\mathbf{u} - \mathbf{v} = -6\mathbf{j} - \mathbf{i} - \mathbf{j} = -\mathbf{i} - 7\mathbf{j}$

(c) $3\mathbf{u} = -18\mathbf{j}$

(d) $2\mathbf{v} + 5\mathbf{u} = 2\mathbf{i} + 2\mathbf{j} - 30\mathbf{j}$

$\qquad = 2\mathbf{i} - 28\mathbf{j}$

53. $\mathbf{u} = 6\mathbf{i} - 5\mathbf{j}, \mathbf{v} = 10\mathbf{i} + 3\mathbf{j}$

$2\mathbf{u} + \mathbf{v} = 2(6\mathbf{i} - 5\mathbf{j}) + (10\mathbf{i} + 3\mathbf{j})$

$\quad = 22\mathbf{i} - 7\mathbf{j}$

$\quad = \langle 22, -7 \rangle$

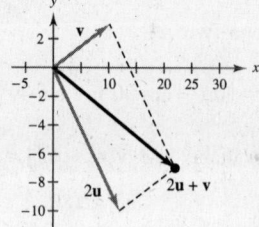

54. $\mathbf{u} = 6\mathbf{i} - 5\mathbf{j}, \ \mathbf{v} = 10\mathbf{i} + 3\mathbf{j}$

$4\mathbf{u} - 5\mathbf{v} = (24\mathbf{i} - 20\mathbf{j}) - (50\mathbf{i} + 15\mathbf{j})$

$\quad = -26\mathbf{i} - 35\mathbf{j}$

$\quad = \langle -26, -35 \rangle$

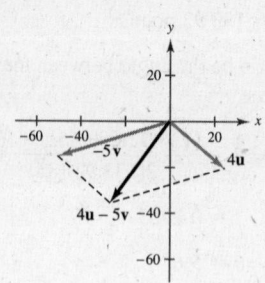

55. $\mathbf{v} = 10\mathbf{i} + 3\mathbf{j}$

$3\mathbf{v} = 3(10\mathbf{i} + 3\mathbf{j})$

$\quad = 30\mathbf{i} + 9\mathbf{j}$

$\quad = \langle 30, 9 \rangle$

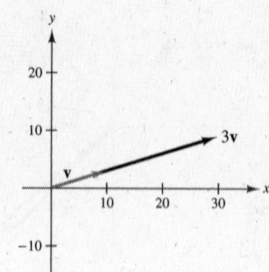

56. $\mathbf{v} = 10\mathbf{i} + 3\mathbf{j}$

$\frac{1}{2}\mathbf{v} = 5\mathbf{i} + \frac{3}{2}\mathbf{j} = \left\langle 5, \frac{3}{2} \right\rangle$

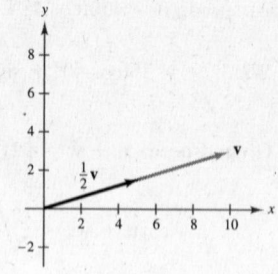

57. $\mathbf{u} = \langle -3, 4 \rangle = -3\mathbf{i} + 4\mathbf{j}$

58. $\mathbf{u} = \langle -6, -8 \rangle = -6\mathbf{i} - 8\mathbf{j}$

59. Initial point: $(3, 4)$

Terminal point: $(9, 8)$

$\mathbf{u} = (9 - 3)\mathbf{i} + (8 - 4)\mathbf{j} = 6\mathbf{i} + 4\mathbf{j}$

60. Initial point: $(-2, 7)$

Terminal point: $(5, -9)$

$\mathbf{u} = \langle 5 - (-2), -9 - 7 \rangle = \langle 7, -16 \rangle = 7\mathbf{i} - 16\mathbf{j}$

61. $\mathbf{v} = -10\mathbf{i} + 10\mathbf{j}$

$\|\mathbf{v}\| = \sqrt{(-10)^2 + (10)^2} = \sqrt{200} = 10\sqrt{2}$

$\tan \theta = \dfrac{10}{-10} = -1 \implies \theta = 135°$ since

$\mathbf{v}$ is in Quadrant II,

$\mathbf{v} = 10\sqrt{2}(\mathbf{i} \cos 135° + \mathbf{j} \sin 135°)$

62. $\mathbf{v} = 4\mathbf{i} - \mathbf{j}$

$\|\mathbf{v}\| = \sqrt{4^2 + (-1)^2} = \sqrt{17}$

$\tan \theta = \dfrac{-1}{4}, \theta$ in Quadrant IV $\implies \theta \approx 346°$

$\mathbf{v} \approx \sqrt{17}(\cos 346° \, \mathbf{i} + \sin 346° \, \mathbf{j})$

63. $\mathbf{v} = 7(\cos 60° \, \mathbf{i} + \sin 60° \, \mathbf{j})$

$\|\mathbf{v}\| = 7$

$\theta = 60°$

64. $\mathbf{v} = 3(\cos 150° \mathbf{i} + \sin 150° \, \mathbf{j})$

$\|\mathbf{v}\| = 3, \theta = 150°$

65. $\mathbf{v} = 5\mathbf{i} + 4\mathbf{j}$

$\|\mathbf{v}\| = \sqrt{5^2 + 4^2} = \sqrt{41}$

$\tan \theta = \dfrac{4}{5} \implies \theta \approx 38.7°$

66. $\mathbf{v} = -4\mathbf{i} + 7\mathbf{j}$

$\|\mathbf{v}\| = \sqrt{(-4)^2 + 7^2} = \sqrt{65}$

$\tan \theta = \dfrac{7}{-4}, \theta$ in Quadrant II $\implies \theta \approx 119.7°$

67. $\mathbf{v} = -3\mathbf{i} - 3\mathbf{j}$

$\|\mathbf{v}\| = \sqrt{(-3)^2 + (-3)^2} = 3\sqrt{2}$

$\tan \theta = \dfrac{-3}{-3} = 1 \implies \theta = 225°$

68. $\mathbf{v} = 8\mathbf{i} - \mathbf{j}$

$\|\mathbf{v}\| = \sqrt{8^2 + (-1)^2} = \sqrt{65}$

$\tan \theta = \dfrac{-1}{8}, \theta$ in Quadrant IV $\implies \theta \approx 352.9°$

69. Magnitude of resultant:

$$c = \sqrt{85^2 + 50^2 - 2(85)(50)\cos 165°}$$

$$\approx 133.92 \text{ pounds}$$

Let θ be the angle between the resultant and the 85-pound force.

$$\cos\theta \approx \frac{(133.92)^2 + 85^2 - 50^2}{2(133.92)(85)}$$

$$\approx 0.9953$$

$$\Rightarrow \theta \approx 5.6°$$

70. Rope One:

$$\mathbf{u} = \|\mathbf{u}\|(\cos 30°\mathbf{i} - \sin 30°\mathbf{j}) = \|\mathbf{u}\|\left(\frac{\sqrt{3}}{2}\mathbf{i} - \frac{1}{2}\mathbf{j}\right)$$

Rope Two:

$$\mathbf{v} = \|\mathbf{u}\|(-\cos 30°\mathbf{i} - \sin 30°\mathbf{j}) = \|\mathbf{u}\|\left(-\frac{\sqrt{3}}{2}\mathbf{i} - \frac{1}{2}\mathbf{j}\right)$$

Resultant: $\mathbf{u} + \mathbf{v} = -\|\mathbf{u}\|\mathbf{j} = -180\mathbf{j}$

$$\|\mathbf{u}\| = 180$$

Therefore, the tension on each rope is $\|\mathbf{u}\| = 180$ lb.

71. Airspeed: $\mathbf{u} = 430(\cos 45°\mathbf{i} - \sin 45°\mathbf{j}) = 215\sqrt{2}(\mathbf{i} - \mathbf{j})$

Wind: $\mathbf{w} = 35(\cos 60° + \sin 60°\mathbf{j}) = \dfrac{35}{2}(\mathbf{i} + \sqrt{3}\mathbf{j})$

Groundspeed: $\mathbf{u} + \mathbf{w} = \left(215\sqrt{2} + \dfrac{35}{2}\right)\mathbf{i} + \left(\dfrac{35\sqrt{3}}{2} - 215\sqrt{2}\right)\mathbf{j}$

$$\|\mathbf{u} + \mathbf{w}\| = \sqrt{\left(215\sqrt{2} + \frac{35}{2}\right)^2 + \left(\frac{35\sqrt{3}}{2} - 215\sqrt{2}\right)^2}$$

$$\approx 422.30 \text{ miles per hour}$$

Bearing: $\tan\theta' = \dfrac{17.5\sqrt{3} - 215\sqrt{2}}{215\sqrt{2} + 17.5}$

$$\theta' \approx -40.4°$$

$$\theta = 90° + |\theta'| = 130.4°$$

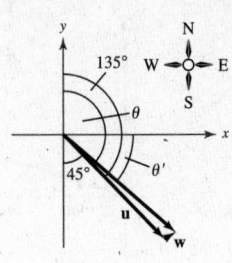

72. Airspeed: $\mathbf{u} = 724(\cos 60°\mathbf{i} + \sin 60°\mathbf{j})$

$$= 362(\mathbf{i} + \sqrt{3}\mathbf{j})$$

Wind: $\mathbf{w} = 32\mathbf{i}$

Groundspeed $= \mathbf{u} + \mathbf{w} = (394\mathbf{i} + 362\sqrt{3}\mathbf{j})$

$$\|\mathbf{u} + \mathbf{w}\| = \sqrt{(394)^2 + (362\sqrt{3})^2} \approx 740.5 \text{ km/hr}$$

$$\tan\theta = \frac{362\sqrt{3}}{394} \Rightarrow \theta \approx 57.9°$$

Bearing: N 32.1° E

73. $\mathbf{u} = \langle 6, 7\rangle, \mathbf{v} = \langle -3, 9\rangle$

$$\mathbf{u} \cdot \mathbf{v} = 6(-3) + 7(9) = 45$$

74. $\mathbf{u} = \langle -7, 12\rangle, \mathbf{v} = \langle -4, -14\rangle$

$$\mathbf{u} \cdot \mathbf{v} = -7(-4) + 12(-14) = -140$$

75. $\mathbf{u} = 3\mathbf{i} + 7\mathbf{j}, \mathbf{v} = 11\mathbf{i} - 5\mathbf{j}$

$$\mathbf{u} \cdot \mathbf{v} = 3(11) + 7(-5) = -2$$

76. $\mathbf{u} = -7\mathbf{i} + 2\mathbf{j}, \mathbf{v} = 16\mathbf{i} - 12\mathbf{j}$

$$\mathbf{u} \cdot \mathbf{v} = -7(16) + 2(-12) = -136$$

77. $\mathbf{u} = \langle -3, 4\rangle$

$$2\mathbf{u} = \langle -6, 8\rangle$$

$$2\mathbf{u} \cdot \mathbf{u} = (-6)(-3) + 8(4) = 50$$

The result is a scalar.

78. $\mathbf{v} = \langle 2, 1\rangle$

$$\|\mathbf{v}\|^2 = \mathbf{v} \cdot \mathbf{v} = 2^2 + 1^2 = 5; \text{ scalar}$$

79. $\mathbf{u} = \langle -3, 4 \rangle$, $\mathbf{v} = \langle 2, 1 \rangle$

$\mathbf{u} \cdot \mathbf{v} = (-3)(2) + 4(1) = -2$

$\mathbf{u}(\mathbf{u} \cdot \mathbf{v}) = \mathbf{u}(-2) = -2\mathbf{u} = \langle 6, -8 \rangle$

The result is a vector.

80. $\mathbf{u} = \langle -3, 4 \rangle$, $\mathbf{v} = \langle 2, 1 \rangle$

$3\mathbf{u} \cdot \mathbf{v} = 3(-3(2) + 4(1)) = 3(-2) = -6$; scalar

81. $\mathbf{u} = \cos\dfrac{7\pi}{4}\mathbf{i} + \sin\dfrac{7\pi}{4}\mathbf{j} = \left\langle \dfrac{1}{\sqrt{2}}, -\dfrac{1}{\sqrt{2}} \right\rangle$

$\mathbf{v} = \cos\dfrac{5\pi}{6}\mathbf{i} + \sin\dfrac{5\pi}{6}\mathbf{j} = \left\langle -\dfrac{\sqrt{3}}{2}, \dfrac{1}{2} \right\rangle$

$\cos\theta = \dfrac{\mathbf{u} \cdot \mathbf{v}}{\|\mathbf{u}\|\,\|\mathbf{v}\|} = \dfrac{-\sqrt{3} - 1}{2\sqrt{2}} \implies \theta = \dfrac{11\pi}{12}$

82. $\mathbf{u} = \cos 45° \, \mathbf{i} + \sin 45° \, \mathbf{j}$

$\mathbf{v} = \cos 300° \, \mathbf{i} + \sin 300° \, \mathbf{j}$

Angle between $\mathbf{u}$ and $\mathbf{v}$: $60° + 45° = 105°$

83. $\mathbf{u} = \langle 2\sqrt{2}, -4 \rangle$, $\mathbf{v} = \langle -\sqrt{2}, 1 \rangle$

$\cos\theta = \dfrac{\mathbf{u} \cdot \mathbf{v}}{\|\mathbf{u}\|\,\|\mathbf{v}\|} = \dfrac{-8}{\left(\sqrt{24}\right)\left(\sqrt{3}\right)} \implies \theta \approx 160.5°$

84. $\mathbf{u} = \langle 3, \sqrt{3} \rangle$, $\mathbf{v} = \langle 4, 3\sqrt{3} \rangle$

$\cos\theta = \dfrac{\mathbf{u} \cdot \mathbf{v}}{\|\mathbf{u}\|\,\|\mathbf{v}\|} = \dfrac{21}{\sqrt{12}\sqrt{43}} \implies \theta \approx 22.4°$

85. $\mathbf{u} = \langle -3, 8 \rangle$

$\mathbf{v} = \langle 8, 3 \rangle$

$\mathbf{u} \cdot \mathbf{v} = -3(8) + 8(3) = 0$

$\mathbf{u}$ and $\mathbf{v}$ are orthogonal.

86. $\mathbf{u} = \langle \frac{1}{4}, -\frac{1}{2} \rangle$, $\mathbf{v} = \langle -2, 4 \rangle$

$\mathbf{v} = -8\mathbf{u} \implies$ Parallel

87. $\mathbf{u} = -\mathbf{i}$

$\mathbf{v} = \mathbf{i} + 2\mathbf{j}$

$\mathbf{u} \cdot \mathbf{v} \neq 0 \implies$ Not orthogonal

$\mathbf{v} \neq k\mathbf{u} \implies$ Not parallel

Neither

88. $\mathbf{u} = -2\mathbf{i} + \mathbf{j}$, $\mathbf{v} = 3\mathbf{i} + 6\mathbf{j}$

$\mathbf{u} \cdot \mathbf{v} = 0 \implies$ Orthogonal

89. $\mathbf{u} = \langle -4, 3 \rangle$, $\mathbf{v} = \langle -8, -2 \rangle$

$\mathbf{w}_1 = \text{proj}_\mathbf{v}\mathbf{u} = \left(\dfrac{\mathbf{u} \cdot \mathbf{v}}{\|\mathbf{v}\|^2}\right)\mathbf{v} = \left(\dfrac{26}{68}\right)\langle -8, -2 \rangle = -\dfrac{13}{17}\langle 4, 1 \rangle$

$\mathbf{w}_2 = \mathbf{u} - \mathbf{w}_1 = \langle -4, 3 \rangle - \left(-\dfrac{13}{17}\right)\langle 4, 1 \rangle = \dfrac{16}{17}\langle -1, 4 \rangle$

$\mathbf{u} = \mathbf{w}_1 + \mathbf{w}_2 = -\dfrac{13}{17}\langle 4, 1 \rangle + \dfrac{16}{17}\langle -1, 4 \rangle$

90. $\mathbf{u} = \langle 5, 6 \rangle$, $\mathbf{v} = \langle 10, 0 \rangle$

$\mathbf{w}_1 = \text{proj}_\mathbf{v}\mathbf{u} = \left(\dfrac{\mathbf{u} \cdot \mathbf{v}}{\|\mathbf{v}\|^2}\right)\mathbf{v} = \dfrac{50}{100}\langle 10, 0 \rangle = \langle 5, 0 \rangle$

$\mathbf{w}_2 = \mathbf{u} - \mathbf{w}_1 = \langle 5, 6 \rangle - \langle 5, 0 \rangle = \langle 0, 6 \rangle$

$\mathbf{u} = \mathbf{w}_1 + \mathbf{w}_2 = \langle 5, 0 \rangle + \langle 0, 6 \rangle$

91. $\mathbf{u} = \langle 2, 7 \rangle$, $\mathbf{v} = \langle 1, -1 \rangle$

$\mathbf{w}_1 = \text{proj}_\mathbf{v}\mathbf{u} = \left(\dfrac{\mathbf{u} \cdot \mathbf{v}}{\|\mathbf{v}\|^2}\right)\mathbf{v} = -\dfrac{5}{2}\langle 1, -1 \rangle$

$\qquad = \dfrac{5}{2}\langle -1, 1 \rangle$

$\mathbf{w}_2 = \mathbf{u} - \mathbf{w}_1 = \langle 2, 7 \rangle - \left(\dfrac{5}{2}\right)\langle -1, 1 \rangle$

$\qquad = \dfrac{9}{2}\langle 1, 1 \rangle$

$\mathbf{u} = \mathbf{w}_1 + \mathbf{w}_2 = \dfrac{5}{2}\langle -1, 1 \rangle + \dfrac{9}{2}\langle 1, 1 \rangle$

92. $\mathbf{u} = \langle -3, 5 \rangle$, $\mathbf{v} = \langle -5, 2 \rangle$

$\mathbf{w}_1 = \text{proj}_{\mathbf{v}}\mathbf{u} = \left(\dfrac{\mathbf{u} \cdot \mathbf{v}}{\|\mathbf{v}\|^2} \right)\mathbf{v} = \dfrac{25}{29}\langle -5, 2 \rangle$

$\mathbf{w}_2 = \mathbf{u} - \mathbf{w}_1 = \langle -3, 5 \rangle - \dfrac{25}{29}\langle -5, 2 \rangle = \dfrac{19}{29}\langle 2, 5 \rangle$

$\mathbf{u} = \mathbf{w}_1 + \mathbf{w}_2 = \dfrac{25}{29}\langle -5, 2 \rangle + \dfrac{19}{25}\langle 2, 5 \rangle$

93. $P = (5, 3)$, $Q = (8, 9) \Rightarrow \overrightarrow{PQ} = \langle 3, 6 \rangle$

$W = \mathbf{v} \cdot \overrightarrow{PQ} = \langle 2, 7 \rangle \cdot \langle 3, 6 \rangle = 48$

94. work $= \mathbf{v} \cdot \overrightarrow{PQ}$

$= (3\mathbf{i} - 6\mathbf{j}) \cdot (-10\mathbf{i} + 17\mathbf{j})$

$= -30 - 102$

$= -132$

95. $w = (18,000)\left(\dfrac{48}{12}\right) = 72,000$ foot-pounds

96. $W = \cos \theta \, \|\mathbf{F}\| \, \|\overrightarrow{PQ}\|$

$= (\cos 20°)(25 \text{ pounds})(12 \text{ ft})$

$= 281.9$ foot-pounds

97. $|7i| = \sqrt{0^2 + 7^2} = 7$

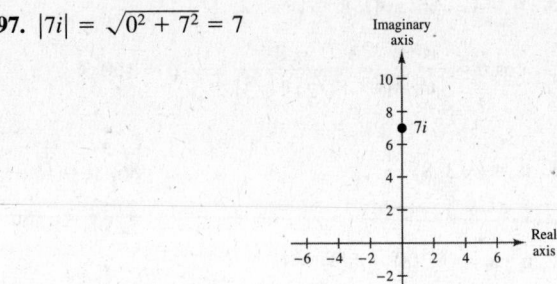

98. $|-6i| = 6$

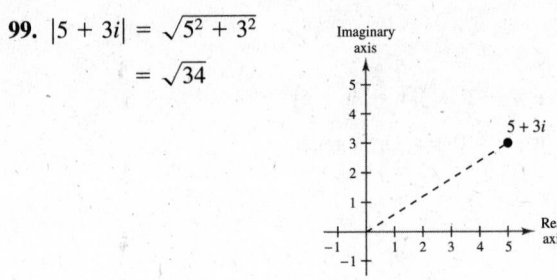

99. $|5 + 3i| = \sqrt{5^2 + 3^2}$

$= \sqrt{34}$

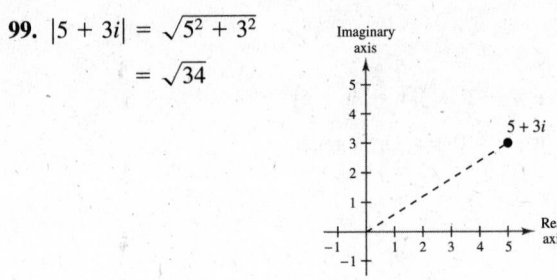

100. $|-10 - 4i| = \sqrt{(-10)^2 + (-4)^2}$

$= 2\sqrt{29}$

101. $5 - 5i$

$r = \sqrt{5^2 + (-5)^2} = \sqrt{50} = 5\sqrt{2}$

$\tan \theta = \dfrac{-5}{5} = -1 \Rightarrow \theta = \dfrac{7\pi}{4}$ since the

complex number is in Quadrant IV.

$5 - 5i = 5\sqrt{2}\left(\cos \dfrac{7\pi}{4} + i \sin \dfrac{7\pi}{4} \right)$

102. $z = 5 + 12i$

$|z| = \sqrt{5^2 + 12^2} = 13$

$\tan \theta = \dfrac{12}{5} \Rightarrow \theta \approx 1.176$

$z \approx 13(\cos 1.176 + i \sin 1.176)$

103. $-3\sqrt{3} + 3i$

$r = \sqrt{(-3\sqrt{3})^2 + 3^2} = \sqrt{36} = 6$

$\tan \theta = \dfrac{3}{-3\sqrt{3}} = -\dfrac{1}{\sqrt{3}} \Rightarrow \theta = \dfrac{5\pi}{6}$

since the complex number is in Quadrant II.

$-3\sqrt{3} + 3i = 6\left(\cos \dfrac{5\pi}{6} + i \sin \dfrac{5\pi}{6} \right)$

104. $z = -7$

$|z| = 7$

$\tan \theta = \dfrac{0}{-7} = 0 \implies \theta = \pi$

$z = 7(\cos \pi + i \sin \pi)$

105. (a) $z_1 = 2\sqrt{3} - 2i = 4\left(\cos \dfrac{11\pi}{6} + i \sin \dfrac{11\pi}{6}\right)$

$z_2 = -10i = 10\left(\cos \dfrac{3\pi}{2} + i \sin \dfrac{3\pi}{2}\right)$

(b) $z_1 z_2 = \left[4\left(\cos \dfrac{11\pi}{6} + i \sin \dfrac{11\pi}{6}\right)\right]\left[10\left(\cos \dfrac{3\pi}{2} + i \sin \dfrac{3\pi}{2}\right)\right]$

$= 40\left(\cos \dfrac{10\pi}{3} + i \sin \dfrac{10\pi}{3}\right)$

$\dfrac{z_1}{z_2} = \dfrac{4\left(\cos \dfrac{11\pi}{6} + i \sin \dfrac{11\pi}{6}\right)}{10\left(\cos \dfrac{3\pi}{2} + i \sin \dfrac{3\pi}{2}\right)} = \dfrac{2}{5}\left(\cos \dfrac{\pi}{3} + i \sin \dfrac{\pi}{3}\right)$

106. (a) $z_1 = -3(1 + i) = 3\sqrt{2}\left(\cos \dfrac{5\pi}{4} + i \sin \dfrac{5\pi}{4}\right)$

$z_2 = 2(\sqrt{3} + i) = 4\left(\cos \dfrac{\pi}{6} + i \sin \dfrac{\pi}{6}\right)$

(b) $z_1 z_2 = \left[3\sqrt{2}\left(\cos \dfrac{5\pi}{4} + i \sin \dfrac{5\pi}{4}\right)\right]\left[4\left(\cos \dfrac{\pi}{6} + i \sin \dfrac{\pi}{6}\right)\right]$

$= 12\sqrt{2}\left(\cos \dfrac{17\pi}{12} + i \sin \dfrac{17\pi}{12}\right)$

$\dfrac{z_1}{z_2} = \dfrac{3\sqrt{2}\left[\cos \dfrac{5\pi}{4} + i \sin \dfrac{5\pi}{4}\right]}{4\left[\cos \dfrac{\pi}{6} + i \sin \dfrac{\pi}{6}\right]} = \dfrac{3\sqrt{2}}{4}\left(\cos \dfrac{13\pi}{12} + i \sin \dfrac{13\pi}{12}\right)$

107. $\left[5\left(\cos \dfrac{\pi}{12} + i \sin \dfrac{\pi}{12}\right)\right]^4 = 5^4\left(\cos \dfrac{4\pi}{12} + i \sin \dfrac{4\pi}{12}\right)$

$= 625\left(\cos \dfrac{\pi}{3} + i \sin \dfrac{\pi}{3}\right)$

$= 625\left(\dfrac{1}{2} + \dfrac{\sqrt{3}}{2}i\right)$

$= \dfrac{625}{2} + \dfrac{625\sqrt{3}}{2}i$

108. $\left[2\left(\cos \dfrac{4\pi}{15} + i \sin \dfrac{4\pi}{15}\right)\right]^5 = 2^5\left(\cos \dfrac{4\pi}{3} + i \sin \dfrac{4\pi}{3}\right)$

$= 32\left(-\dfrac{1}{2} - \dfrac{\sqrt{3}}{2}i\right)$

$= -16 - 16\sqrt{3}i$

109. $(2 + 3i)^6 \approx \left[\sqrt{13}(\cos 56.3° + i \sin 56.3°)\right]^6$

$= 13^3(\cos 337.9° + i \sin 337.9°)$

$\approx 13^3(0.9263 - 0.3769i)$

$\approx 2035 - 828i$

110. $(1 - i)^8 = \left[\sqrt{2}(\cos 315° + i \sin 315°)\right]^8$

$= 16(\cos 2520° + i \sin 2520°)$

$= 16(\cos 0° + i \sin 0°)$

$= 16$

111. Sixth roots of $-729i = 729\left(\cos\dfrac{3\pi}{2} + i \sin\dfrac{3\pi}{2}\right)$:

(a) and (c)

$$\sqrt[6]{729}\left[\cos\left(\dfrac{\dfrac{3\pi}{2} + 2k\pi}{6}\right) + i \sin\left(\dfrac{\dfrac{3\pi}{2} + 2k\pi}{6}\right)\right], k = 0, 1, 2, 3, 4, 5$$

$k = 0: \ 3\left(\cos\dfrac{\pi}{4} + i \sin\dfrac{\pi}{4}\right) = \dfrac{3\sqrt{2}}{2} + \dfrac{3\sqrt{2}}{2}i$

$k = 1: \ 3\left(\cos\dfrac{7\pi}{12} + i \sin\dfrac{7\pi}{12}\right) \approx -0.776 + 2.898i$

$k = 2: \ 3\left(\cos\dfrac{11\pi}{12} + i \sin\dfrac{11\pi}{12}\right) \approx -2.898 + 0.776i$

$k = 3: \ 3\left(\cos\dfrac{5\pi}{4} + i \sin\dfrac{5\pi}{4}\right) = -\dfrac{3\sqrt{2}}{2} - \dfrac{3\sqrt{2}}{2}i$

$k = 4: \ 3\left(\cos\dfrac{19\pi}{12} + i \sin\dfrac{19\pi}{12}\right) \approx 0.776 - 2.898i$

$k = 5: \ 3\left(\cos\dfrac{23\pi}{12} + i \sin\dfrac{23\pi}{12}\right) \approx 2.898 - 0.776i$

(b)

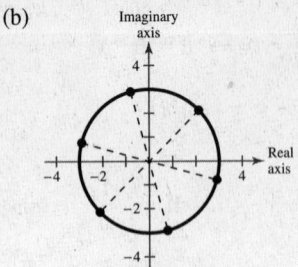

112. (a) $256i = 256\left(\cos\dfrac{\pi}{2} + i \sin\dfrac{\pi}{2}\right)$

Fourth roots of $256i$:

$$\sqrt[4]{256}\left(\cos\dfrac{\dfrac{\pi}{2} + 2\pi k}{4} + i \sin\dfrac{\dfrac{\pi}{2} + 2\pi k}{4}\right), k = 0, 1, 2, 3$$

$k = 0: \ 4\left(\cos\dfrac{\pi}{8} + i \sin\dfrac{\pi}{8}\right)$

$k = 1: \ 4\left(\cos\dfrac{5\pi}{8} + i \sin\dfrac{5\pi}{8}\right)$

$k = 2: \ 4\left(\cos\dfrac{9\pi}{8} + i \sin\dfrac{9\pi}{8}\right)$

$k = 3: \ 4\left(\cos\dfrac{13\pi}{8} + i \sin\dfrac{13\pi}{8}\right)$

(b)

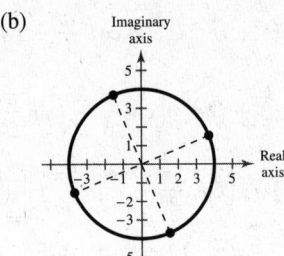

(c) $3.696 + 1.531i$

$-1.531 + 3.696i$

$-3.696 - 1.531i$

$1.531 - 3.696i$

113. Cube roots of $8 = 8(\cos 0 + i \sin 0), k = 0, 1, 2$

(a) and (c)

$$\sqrt[3]{8}\left[\cos\left(\dfrac{0 + 2\pi k}{3}\right) + i \sin\left(\dfrac{0 + 2\pi k}{3}\right)\right]$$

$k = 0: \ 2(\cos 0 + i \sin 0) = 2$

$k = 1: \ 2\left(\cos\dfrac{2\pi}{3} + i \sin\dfrac{2\pi}{3}\right) = -1 + \sqrt{3}i$

$k = 2: \ 2\left(\cos\dfrac{4\pi}{3} + i \sin\dfrac{4\pi}{3}\right) = -1 - \sqrt{3}i$

(b)

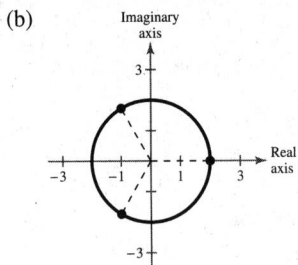

114. (a) $-1024 = 1024(\cos \pi + i \sin \pi)$

Fifth roots of -1024:

$$\sqrt[5]{1024}\left(\cos \frac{\pi + 2\pi k}{5} + i \sin\left(\frac{\pi + 2\pi k}{5}\right)\right), k = 0, 1, 2, 3, 4$$

$k = 0$: $4\left(\cos \dfrac{\pi}{5} + i \sin \dfrac{\pi}{5}\right)$

$k = 1$: $4\left(\cos \dfrac{3\pi}{5} + i \sin \dfrac{3\pi}{5}\right)$

$k = 2$: $4(\cos \pi + i \sin \pi)$

$k = 3$: $4\left(\cos \dfrac{7\pi}{5} + i \sin \dfrac{7\pi}{5}\right)$

$k = 4$: $4\left(\cos \dfrac{9\pi}{5} + i \sin \dfrac{9\pi}{5}\right)$

(b)

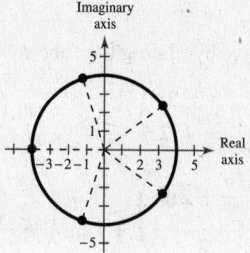

(c) $3.236 + 2.351i$

$-1.236 + 3.804i$

-4

$-1.236 - 3.804i$

$3.236 - 2.351i$

115. $x^4 + 81 = 0$

$x^4 = -81$ Solve by finding the fourth roots of -81.

$-81 = 81(\cos \pi + i \sin \pi)$

$$\sqrt[4]{-81} = \sqrt[4]{81}\left[\cos\left(\frac{\pi + 2\pi k}{4}\right) + i \sin\left(\frac{\pi + 2\pi k}{4}\right)\right], k = 0, 1, 2, 3$$

$k = 0$: $3\left(\cos \dfrac{\pi}{4} + i \sin \dfrac{\pi}{4}\right) = \dfrac{3\sqrt{2}}{2} + \dfrac{3\sqrt{2}}{2}i$

$k = 1$: $3\left(\cos \dfrac{3\pi}{4} + i \sin \dfrac{3\pi}{4}\right) = -\dfrac{3\sqrt{2}}{2} + \dfrac{3\sqrt{2}}{2}i$

$k = 2$: $3\left(\cos \dfrac{5\pi}{4} + i \sin \dfrac{5\pi}{4}\right) = -\dfrac{3\sqrt{2}}{2} - \dfrac{3\sqrt{2}}{2}i$

$k = 3$: $3\left(\cos \dfrac{7\pi}{4} + i \sin \dfrac{7\pi}{4}\right) = \dfrac{3\sqrt{2}}{2} - \dfrac{3\sqrt{2}}{2}i$

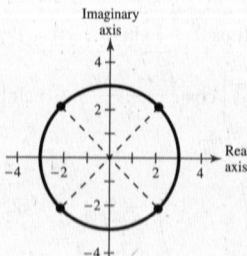

116. $x^5 - 32 = 0$

$x^5 = 32$

$32 = 32(\cos 0 + i \sin 0)$

$$\sqrt[5]{32} = \sqrt[5]{32}\left[\cos\left(0 + \frac{2\pi k}{5}\right) + i \sin\left(0 + \frac{2\pi k}{5}\right)\right]$$

$k = 0, 1, 2, 3, 4$

$k = 0$: $2(\cos 0 + i \sin 0) = 2$

$k = 1$: $2\left(\cos \dfrac{2\pi}{5} + i \sin \dfrac{2\pi}{5}\right) = 0.6180 + 1.9021i$

$k = 2$: $2\left(\cos \dfrac{4\pi}{5} + i \sin \dfrac{4\pi}{5}\right) = -1.6180 + 1.1756i$

$k = 3$: $2\left(\cos \dfrac{6\pi}{5} + i \sin \dfrac{6\pi}{5}\right) = -1.6180 - 1.1756i$

$k = 4$: $2\left(\cos \dfrac{8\pi}{5} + i \sin \dfrac{8\pi}{5}\right) = 0.6180 - 1.9021i$

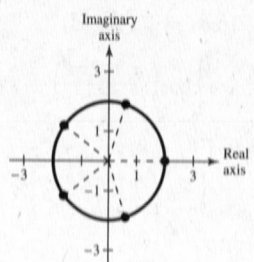

117. $x^3 + 8i = 0$

$x^3 = -8i$ Solve by finding the cube roots of $-8i$.

$$-8i = 8\left(\cos\frac{3\pi}{2} + i\sin\frac{3\pi}{2}\right)$$

$$\sqrt[3]{-8i} = \sqrt[3]{8}\left[\cos\left(\frac{\frac{3\pi}{2}+2\pi k}{3}\right) + i\sin\left(\frac{\frac{3\pi}{2}+2\pi k}{3}\right)\right], \; k = 0, 1, 2$$

$k = 0: 2\left(\cos\dfrac{\pi}{2} + i\sin\dfrac{\pi}{2}\right) = 2i$

$k = 1: 2\left(\cos\dfrac{7\pi}{6} + i\sin\dfrac{7\pi}{6}\right) = -\sqrt{3} - i$

$k = 2: 2\left(\cos\dfrac{11\pi}{6} + i\sin\dfrac{11\pi}{6}\right) = \sqrt{3} - i$

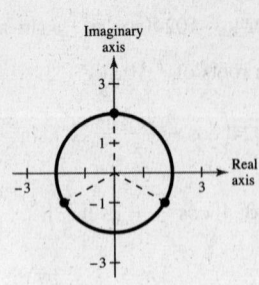

118. $(x^3 - 1)(x^2 + 1) = 0$

$x^3 - 1 = 0$

$x^2 + 1 = 0$

$x^3 = 1$

$1 = 1(\cos 0 + i\sin 0)$

$$\sqrt[3]{1} = \sqrt[3]{1}\left[\cos\left(\frac{0+2\pi k}{3}\right) + i\sin\left(\frac{0+2\pi k}{3}\right)\right], \; k = 0, 1, 2$$

$1(\cos 0 + i\sin 0) = 1$

$1\left(\cos\dfrac{2\pi}{3} + i\sin\dfrac{2\pi}{3}\right) = -\dfrac{1}{2} + \dfrac{\sqrt{3}}{2}i$

$1\left(\cos\dfrac{4\pi}{3} + i\sin\dfrac{4\pi}{3}\right) = -\dfrac{1}{2} - \dfrac{\sqrt{3}}{2}i$

$x^2 + 1 = 0$

$x^2 = -1$

$-1 = 1(\cos\pi + i\sin\pi)$

$$\sqrt{-1} = \sqrt{1}\left[\cos\left(\frac{\pi+2\pi k}{2}\right) + i\sin\left(\frac{\pi+2\pi k}{2}\right)\right], \; k = 0, 1$$

$k = 0, 1$

$1\left(\cos\dfrac{\pi}{2} + i\sin\dfrac{\pi}{2}\right) = i$

$1\left(\cos\dfrac{3\pi}{2} + i\sin\dfrac{3\pi}{2}\right) = -i$

119. True. Sin 90° is defined in the Law of Sines.

120. False. There may be no solution, one solution, or two solutions.

121. True, by the definition of a unit vector.

$\mathbf{u} = \dfrac{\mathbf{v}}{\|\mathbf{v}\|}$ so $\mathbf{v} = \|\mathbf{v}\|\mathbf{u}$

122. False, $a = b = 0$.

123. False. $x = \sqrt{3} + i$ is a solution to $x^3 - 8i = 0$, not $x^2 - 8i = 0$.

Also, $(\sqrt{3} + i)^2 - 8i = 2 + (2\sqrt{3} - 8)i \neq 0$.

124. $\dfrac{a}{\sin A} = \dfrac{b}{\sin B} = \dfrac{c}{\sin C}$ or $\dfrac{\sin A}{a} = \dfrac{\sin B}{b} = \dfrac{\sin C}{c}$

125. $a^2 = b^2 + c^2 - 2bc \cos A$

$b^2 = a^2 + c^2 - 2ac \cos B$

$c^2 = a^2 + b^2 - 2ab \cos C$

126. A vector in the plane has both a magnitude and a direction.

127. A and C appear to have the same magnitude and direction.

128. $\|\mathbf{u} + \mathbf{v}\|$ is larger in figure (a) since the angle between $\mathbf{u}$ and $\mathbf{v}$ is acute rather than obtuse.

129. If $k > 0$, the direction of $k\mathbf{u}$ is the same, and the magnitude is $k\|\mathbf{u}\|$.

If $k < 0$, the direction of $k\mathbf{u}$ is the opposite direction of $\mathbf{u}$, and the magnitude is $|k|\,\|\mathbf{u}\|$.

130. The sum of $\mathbf{u}$ and $\mathbf{v}$ lies on the diagonal of the parallelogram with $\mathbf{u}$ and $\mathbf{v}$ as its adjacent sides.

131. (a) The trigonometric form of the three roots shown is:

$4(\cos 60° + i \sin 60°)$

$4(\cos 180° + i \sin 180°)$

$4(\cos 300° + i \sin 300°)$

(b) Since there are three evenly spaced roots on the circle of radius 4, they are cube roots of a complex number of modulus $4^3 = 64$.

Cubing them yields -64.

$[4(\cos 60° + i \sin 60°)]^3 = -64$

$[4(\cos 180° + i \sin 180°)]^3 = -64$

$[4(\cos 300° + i \sin 300°)]^3 = -64$

132. (a) The trigonometric forms of the four roots shown are:

$4(\cos 60° + i \sin 60°)$

$4(\cos 150° + i \sin 150°)$

$4(\cos 240° + i \sin 240°)$

$4(\cos 330° + i \sin 330°)$

(b) Since there are four evenly spaced roots on the circle of radius 4, they are fourth roots of a complex number of modulus 4^4. In this case, raising them to the fourth power yields $-128 - 128\sqrt{3}i$.

133. $z_1 = 2(\cos \theta + i \sin \theta)$

$z_2 = 2(\cos(\pi - \theta) + i \sin(\pi - \theta))$

$z_1 z_2 = (2)(2)[\cos(\theta + (\pi - \theta)) + i \sin(\theta + (\pi - \theta))]$

$= 4(\cos \pi + i \sin \pi)$

$= -4$

$\dfrac{z_1}{z_2} = \dfrac{2(\cos \theta + i \sin \theta)}{2(\cos(\pi - \theta) + i \sin(\pi - \theta))}$

$= 1[\cos(\theta - (\pi - \theta)) + i \sin(\theta - (\pi - \theta))]$

$= \cos(2\theta - \pi) + i \sin(2\theta - \pi)$

$= \cos 2\theta \cos \pi + \sin 2\theta \sin \pi + i(\sin 2\theta \cos \pi - \cos 2\theta \sin \pi)$

$= -\cos 2\theta - i \sin 2\theta$

134. (a) z has 4 fourth roots. Three are not shown.

(b) The roots are located on the circle at $\theta = 30° + 90°k$, $k = 0, 1, 2, 3$.
The three roots not shown are located at $120°, 210°, 300°$.

Problem Solving for Chapter 6

1. $\overrightarrow{PQ}^2 = 4.7^2 + 6^2 - 2(4.7)(6) \cos 25°$

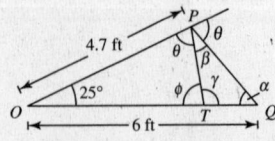

$\overrightarrow{PQ} \approx 2.6409$ feet

$\dfrac{\sin \alpha}{4.7} = \dfrac{\sin 25°}{2.6409} \Rightarrow \alpha \approx 48.78°$

$\theta + \beta = 180° - 25° - 48.78° = 106.22°$

$(\theta + \beta) + \theta = 180° \Rightarrow \theta = 180° - 106.22° = 73.78°$

$\beta = 106.22° - 73.78° = 32.44°$

$\gamma = 180° - \alpha - \beta = 180° - 48.78° - 32.44° = 98.78°$

$\phi = 180° - \gamma = 180° - 98.78° = 81.22°$

$\dfrac{\overrightarrow{PT}}{\sin 25°} = \dfrac{4.7}{\sin 81.22°}$

$\overrightarrow{PT} \approx 2.01$ feet

2.

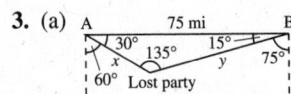

$\dfrac{3}{4}$ mile = 1320 yards

$x^2 = 1320^2 + 300^2 - 2(1320)(300)\cos 10°$

$x \approx 1025.881$ yards ≈ 0.58 mile

$\dfrac{\sin \theta}{1320} = \dfrac{\sin 10°}{1025.881}$

$\sin \theta \approx 0.2234$

$\theta = 180° - \sin^{-1}(0.2234)$

$\theta \approx 167.09°$

Bearing: $\theta - 55° - 90° \approx 22.09°$

$\qquad$ S 22.09° E

3. (a)

(b) $\dfrac{x}{\sin 15°} = \dfrac{75}{\sin 135°}$ and $\dfrac{y}{\sin 30°} = \dfrac{75}{\sin 135°}$

$\quad x \approx 27.45$ miles $\qquad y \approx 53.03$ miles

(c)

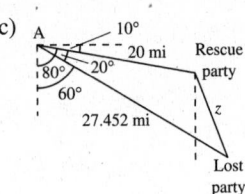

$z^2 = (27.45)^2 + (20)^2 - 2(27.45)(20) \cos 20°$

$z \approx 11.03$ miles

$\dfrac{\sin \theta}{27.45} = \dfrac{\sin 20°}{11.03}$

$\sin \theta \approx 0.8511$

$\theta = 180° - \sin^{-1}(0.8511)$

$\theta \approx 121.7°$

To find the bearing, we have $\theta - 10° - 90° \approx 21.7°$.

Bearing: S 21.7° E

4. (a)

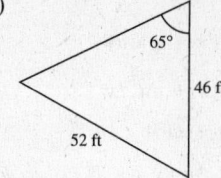

(b) $\dfrac{\sin C}{46} = \dfrac{\sin 65°}{52}$

$\sin C = \dfrac{46 \sin 65°}{52} \approx 0.801734$

$C \approx 53.296°$

$A = 180° - B - C = 61.704°$

$\dfrac{a}{\sin 61.704°} = \dfrac{52}{\sin 65°}$

$a = \dfrac{52 \sin 61.704°}{\sin 65°}$

$a \approx 50.52 \text{ feet}$

(c) Area $= \dfrac{1}{2}(46)(52)\sin 61.704° \approx 1053.09$ square feet

Number of bags: $\dfrac{1053.09}{50} \approx 21.06$

To entirely cover the courtyard, you would need to buy 22 bags.

5. If $\mathbf{u} \neq 0$, $\mathbf{v} \neq 0$, and $\mathbf{u} + \mathbf{v} \neq 0$, then $\left\| \dfrac{\mathbf{u}}{\|\mathbf{u}\|} \right\| = \left\| \dfrac{\mathbf{v}}{\|\mathbf{v}\|} \right\| = \left\| \dfrac{\mathbf{u} + \mathbf{v}}{\|\mathbf{u} + \mathbf{v}\|} \right\| = 1$ since all of these are magnitudes of **unit** vectors.

(a) $\mathbf{u} = \langle 1, -1 \rangle$, $\quad \mathbf{v} = \langle -1, 2 \rangle$, $\quad \mathbf{u} + \mathbf{v} = \langle 0, 1 \rangle$

$\|\mathbf{u}\| = \sqrt{2}$, $\quad \|\mathbf{v}\| = \sqrt{5}$, $\quad \|\mathbf{u} + \mathbf{v}\| = 1$

(b) $\mathbf{u} = \langle 0, 1 \rangle$, $\quad \mathbf{v} = \langle 3, -3 \rangle$, $\quad \mathbf{u} + \mathbf{v} = \langle 3, -2 \rangle$

$\|\mathbf{u}\| = 1$, $\quad \|\mathbf{v}\| = \sqrt{18} = 3\sqrt{2}$, $\|\mathbf{u} + \mathbf{v}\| = \sqrt{13}$

(c) $\mathbf{u} = \left\langle 1, \dfrac{1}{2} \right\rangle$, $\quad \mathbf{v} = \langle 2, 3 \rangle$, $\mathbf{u} + \mathbf{v} = \left\langle 3, \dfrac{7}{2} \right\rangle$

$\|\mathbf{u}\| = \dfrac{\sqrt{5}}{2}$, $\quad \|\mathbf{v}\| = \sqrt{13}$, $\|\mathbf{u} + \mathbf{v}\| = \sqrt{9 + \dfrac{49}{4}} = \dfrac{\sqrt{85}}{2}$

(d) $\mathbf{u} = \langle 2, -4 \rangle$, $\quad \mathbf{v} = \langle 5, 5 \rangle$, $\quad \mathbf{u} + \mathbf{v} = \langle 7, 1 \rangle$

$\|\mathbf{u}\| = \sqrt{20} = 2\sqrt{5}$, $\|\mathbf{v}\| = \sqrt{50} = 5\sqrt{2}$, $\|\mathbf{u} + \mathbf{v}\| = \sqrt{50} = 5\sqrt{2}$

6. (a) $\mathbf{u} = -120\mathbf{j}$

$\mathbf{v} = 40\mathbf{i}$

(b) $\mathbf{s} = \mathbf{u} + \mathbf{v} = 40\mathbf{i} - 120\mathbf{j}$

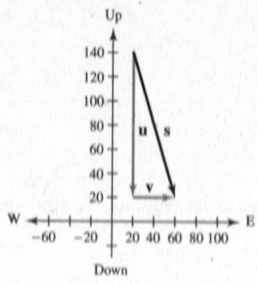

(c) $\|\mathbf{s}\| = \sqrt{40^2 + (-120)^2} = \sqrt{16000} = 40\sqrt{10}$

≈ 126.49 miles per hour

This represents the actual rate of the skydiver's fall.

(d) $\tan\theta = \dfrac{120}{40} \Rightarrow \theta = \tan^{-1} 3 \Rightarrow \theta \approx 71.565°$

(e)

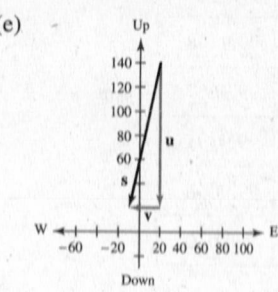

$\mathbf{s} = 30\mathbf{i} - 120\mathbf{j}$

$\|\mathbf{s}\| = \sqrt{30^2 + (-120)^2}$

$= \sqrt{15300}$

≈ 123.69 miles per hour

7. Initial point: $(0, 0)$

Terminal point: $\left(\dfrac{u_1 + v_1}{2}, \dfrac{u_2 + v_2}{2}\right)$

$\mathbf{w} = \left\langle \dfrac{u_1 + v_1}{2}, \dfrac{u_2 + v_2}{2} \right\rangle = \dfrac{1}{2}(\mathbf{u} + \mathbf{v})$

Initial point: $(\mathbf{u}_1, \mathbf{u}_2)$

Terminal point: $\dfrac{1}{2}(\mathbf{u}_1 + \mathbf{v}_1, \mathbf{u}_2 + \mathbf{v}_2)$

$\mathbf{w} = \left\langle \dfrac{u_1 + v_1}{2} - u_1, \dfrac{u_2 + v_2}{2} - u_2 \right\rangle$

$= \left\langle \dfrac{v_1 - u_1}{2}, \dfrac{v_2 - u_2}{2} \right\rangle = \dfrac{1}{2}(\mathbf{v} - \mathbf{u})$

8. Let $\mathbf{u} \cdot \mathbf{v} = 0$ and $\mathbf{u} \cdot \mathbf{w} = 0$.

Then, $\mathbf{u} \cdot (c\mathbf{v} + d\mathbf{w}) = \mathbf{u} \cdot c\mathbf{v} + \mathbf{u} \cdot d\mathbf{w}$

$= c\mathbf{u} \cdot \mathbf{v} + d\mathbf{u} \cdot \mathbf{w}$

$= c(0) + d(0)$

$= 0.$

Thus for all scalars c and d, $\mathbf{u}$ is orthogonal to $c\mathbf{v} + d\mathbf{w}$.

9. $W = (\cos \theta)\|F\| \|\overrightarrow{PQ}\|$ and $\|F_1\| = \|F_2\|$

(a)

If $\theta_1 = -\theta_2$ then the work is the same since $\cos(-\theta) = \cos \theta$.

(b)

If $\theta_1 = 60°$ then $W_1 = \dfrac{1}{2}\|F_1\| \|\overrightarrow{PQ}\|$

If $\theta_2 = 30°$ then $W_2 = \dfrac{\sqrt{3}}{2}\|F_2\| \|\overrightarrow{PQ}\|$

$W_2 = \sqrt{3}\, W_1$

The amount of work done by F_2 is $\sqrt{3}$ times as great as the amount of work done by F_1.

10. (a)

θ	$100 \sin \theta$	$100 \cos \theta$
$0.5°$	0.8727	99.9962
$1.0°$	1.7452	99.9848
$1.5°$	2.6177	99.9657
$2.0°$	3.4899	99.9391
$2.5°$	4.3619	99.9048
$3.0°$	5.2336	99.8630

(b) *No,* the airplane's speed does *not* equal the sum of the vertical and horizontal components of its velocity. To find speed:

speed $= \sqrt{(\|\mathbf{v}\| \sin\theta)^2 + (\|\mathbf{v}\| \cos\theta)^2}$

(c) (i) speed $= \sqrt{5.235^2 + 149.909^2} \approx 150$ miles per hour

(ii) speed $= \sqrt{10.463^2 + 149.634^2} \approx 150$ miles per hour

Chapter 6   Practice Test

For Exercises 1 and 2, use the Law of Sines to find the remaining sides and angles of the triangle.

1. $A = 40°$, $B = 12°$, $b = 100$

2. $C = 150°$, $a = 5$, $c = 20$

3. Find the area of the triangle: $a = 3$, $b = 6$, $C = 130°$.

4. Determine the number of solutions to the triangle: $a = 10$, $b = 35$, $A = 22.5°$.

For Exercises 5 and 6, use the Law of Cosines to find the remaining sides and angles of the triangle.

5. $a = 49$, $b = 53$, $c = 38$

6. $C = 29°$, $a = 100$, $b = 300$

7. Use Heron's Formula to find the area of the triangle: $a = 4.1$, $b = 6.8$, $c = 5.5$.

8. A ship travels 40 miles due east, then adjusts its course 12° southward. After traveling 70 miles in that direction, how far is the ship from its point of departure?

9. $\mathbf{w} = 4\mathbf{u} - 7\mathbf{v}$ where $\mathbf{u} = 3\mathbf{i} + \mathbf{j}$ and $\mathbf{v} = -\mathbf{i} + 2\mathbf{j}$. Find $\mathbf{w}$.

10. Find a unit vector in the direction of $\mathbf{v} = 5\mathbf{i} - 3\mathbf{j}$.

11. Find the dot product and the angle between $\mathbf{u} = 6\mathbf{i} + 5\mathbf{j}$ and $\mathbf{v} = 2\mathbf{i} - 3\mathbf{j}$.

12. $\mathbf{v}$ is a vector of magnitude 4 making an angle of 30° with the positive x-axis. Find $\mathbf{v}$ in component form.

13. Find the projection of $\mathbf{u}$ onto $\mathbf{v}$ given $\mathbf{u} = \langle 3, -1 \rangle$ and $\mathbf{v} = \langle -2, 4 \rangle$.

14. Give the trigonometric form of $z = 5 - 5i$.

15. Give the standard form of $z = 6(\cos 225° + i \sin 225°)$.

16. Multiply $[7(\cos 23° + i \sin 23°)][4(\cos 7° + i \sin 7°)]$.

17. Divide $\dfrac{9\left(\cos \dfrac{5\pi}{4} + i \sin \dfrac{5\pi}{4}\right)}{3(\cos \pi + i \sin \pi)}$.

18. Find $(2 + 2i)^8$.

19. Find the cube roots of $8\left(\cos \dfrac{\pi}{3} + i \sin \dfrac{\pi}{3}\right)$.

20. Find all the solutions to $x^4 + i = 0$.

CHAPTER 7
Systems of Equations and Inequalities

C H A P T E R 7
Systems of Equations and Inequalities

Section 7.1   Linear and Nonlinear Systems of Equations

■ You should be able to solve systems of equations by the method of substitution.

    1. Solve one of the equations for one of the variables.

    2. Substitute this expression into the other equation and solve.

    3. Back-substitute into the first equation to find the value of the other variable.

    4. Check your answer in each of the original equations.

■ You should be able to find solutions graphically. (See Example 5 in textbook.)

Vocabulary Check

1. system of equations

2. solution

3. solving

4. substitution

5. point of intersection

6. break-even

---

1. $\begin{cases} 4x - y = 1 \\ 6x + y = -6 \end{cases}$

  (a) $4(0) - (-3) \neq 1$

    $(0, -3)$ *is not* a solution.

  (b) $4(-1) - (-4) \neq 1$

    $(-1, -4)$ *is not* a solution.

  (c) $4\left(-\frac{3}{2}\right) - (-2) \neq 1$

    $\left(-\frac{3}{2}, -2\right)$ *is not* a solution.

  (d) $4\left(-\frac{1}{2}\right) - (-3) = 1$

    $6\left(-\frac{1}{2}\right) + (-3) = -6$

    $\left(-\frac{1}{2}, -3\right)$ *is* a solution.

2. $\begin{cases} 4x^2 + y = 3 \\ -x - y = 11 \end{cases}$

  (a) $4(2)^2 + (-13) \overset{?}{=} 3$

    $16 - 13 = 3$

    $-2 - (-13) \overset{?}{=} 11$

    $-2 + 13 = 11$

    $(2, -13)$ *is* a solution.

  (b) $4(2)^2 + (-9) \overset{?}{=} 3$

    $16 - 9 \neq 3$

    $(2, -9)$ *is not* a solution.

  (c) $4\left(-\frac{3}{2}\right)^2 + \left(-\frac{31}{3}\right) \overset{?}{=} 3$

    $\frac{36}{4} - \frac{31}{3} \neq 3$

    $\left(-\frac{3}{2}, -\frac{31}{3}\right)$ *is not* a solution.

  (d) $4\left(-\frac{7}{4}\right)^2 + \left(-\frac{37}{4}\right) \overset{?}{=} 3$

    $\frac{49}{4} - \frac{37}{4} = 3$

    $-\left(-\frac{7}{4}\right) - \left(-\frac{37}{4}\right) \overset{?}{=} 11$

    $\frac{7}{4} + \frac{37}{4} = 11$

    $\left(-\frac{7}{4}, -\frac{37}{4}\right)$ *is* a solution.

3. $\begin{cases} y = -2e^x \\ 3x - y = 2 \end{cases}$

(a) $0 \neq -2e^{-2}$

$(-2, 0)$ *is not* a solution.

(b) $-2 = -2e^0$

$3(0) - (-2) = 2$

$(0, -2)$ *is* a solution.

(c) $-3 \neq -2e^0$

$(0, -3)$ *is not* a solution.

(d) $2 \neq -2e^{-1}$

$(-1, 2)$ *is not* a solution.

4. $\begin{cases} -\log x + 3 = y \\ \frac{1}{9}x + y = \frac{28}{9} \end{cases}$

(a) $-\log 9 + 3 \neq \frac{37}{9}$

$\left(9, \frac{37}{9}\right)$ *is not* a solution.

(b) $-\log 10 + 3 = 2$

$\frac{1}{9}(10) + 2 = \frac{28}{9}$

$(10, 2)$ *is* a solution.

(c) $-\log(1) + 3 = 3$

$\frac{1}{9}(1) + 3 = \frac{28}{9}$

$(1, 3)$ *is* a solution.

(d) $-\log 2 + 3 \neq 4$

$(2, 4)$ *is not* a solution.

5. $\begin{cases} 2x + y = 6 & \text{Equation 1} \\ -x + y = 0 & \text{Equation 2} \end{cases}$

Solve for y in Equation 1: $y = 6 - 2x$

Substitute for y in Equation 2: $-x + (6 - 2x) = 0$

Solve for x: $-3x + 6 = 0 \implies x = 2$

Back-substitute $x = 2$: $y = 6 - 2(2) = 2$

Solution: $(2, 2)$

6. $\begin{cases} x - y = -4 & \text{Equation 1} \\ x + 2y = 5 & \text{Equation 2} \end{cases}$

Solve for x in Equation 1: $x = y - 4$

Substitute for x in Equation 2: $(y - 4) + 2y = 5$

Solve for y: $3y - 4 = 5 \implies y = 3$

Back-substitute $y = 3$: $x = 3 - 4 = -1$

Solution: $(-1, 3)$

7. $\begin{cases} x - y = -4 & \text{Equation 1} \\ x^2 - y = -2 & \text{Equation 2} \end{cases}$

Solve for y in Equation 1: $y = x + 4$

Substitute for y in Equation 2: $x^2 - (x + 4) = -2$

Solve for x: $x^2 - x - 2 = 0 \implies (x + 1)(x - 2) = 0 \implies x = -1, 2$

Back-substitute $x = -1$: $y = -1 + 4 = 3$

Back-substitute $x = 2$: $y = 2 + 4 = 6$

Solutions: $(-1, 3), (2, 6)$

8. $\begin{cases} 3x + y = 2 & \text{Equation 1} \\ x^3 - 2 + y = 0 & \text{Equation 2} \end{cases}$

Solve for y in Equation 1: $y = 2 - 3x$

Substitute for y in Equation 2: $x^3 - 2 + (2 - 3x) = 0$

$$x^3 - 3x = 0$$

Solve for x: $x^3 - 3x = 0 \implies x(x^2 - 3) = 0 \implies x = 0, \pm\sqrt{3}$

Back-substitute $x = 0$: $y = 2 - 3(0) = 2$

Back-substitute $x = \sqrt{3}$: $y = 2 - 3\sqrt{3}$

Back-substitute $x = -\sqrt{3}$: $y = 2 - 3(-\sqrt{3}) = 2 + 3\sqrt{3}$

Solutions: $(0, 2), \left(\sqrt{3}, 2 - 3\sqrt{3}\right), \left(-\sqrt{3}, 2 + 3\sqrt{3}\right)$

9. $\begin{cases} -2x + y = -5 & \text{Equation 1} \\ x^2 + y^2 = 25 & \text{Equation 2} \end{cases}$

Solve for y in Equation 1: $y = 2x - 5$

Substitute for y in Equation 2: $x^2 + (2x - 5)^2 = 25$

Solve for x:

$$5x^2 - 20x = 0 \implies 5x(x - 4) = 0 \implies x = 0, 4$$

Back-substitute $x = 0$: $y = 2(0) - 5 = -5$

Back-substitute $x = 4$: $y = 2(4) - 5 = 3$

Solutions: $(0, -5), (4, 3)$

10. $\begin{cases} x + y = 0 & \text{Equation 1} \\ x^3 - 5x - y = 0 & \text{Equation 2} \end{cases}$

Solve for y in Equation 1: $y = -x$

Substitute for y in Equation 2: $x^3 - 5x - (-x) = 0$

Solve for x:

$$x^3 - 4x = 0 \implies x(x^2 - 4) = 0 \implies x = 0, \pm 2$$

Back-substitute $x = 0$: $y = -0 = 0$

Back-substitute $x = 2$: $y = -2$

Back-substitute $x = -2$: $y = -(-2) = 2$

Solutions: $(0, 0), (2, -2), (-2, 2)$

11. $\begin{cases} x^2 + y = 0 & \text{Equation 1} \\ x^2 - 4x - y = 0 & \text{Equation 2} \end{cases}$

Solve for y in Equation 1: $y = -x^2$

Substitute for y in Equation 2: $x^2 - 4x - (-x^2) = 0$

Solve for x: $2x^2 - 4x = 0 \implies 2x(x - 2) = 0 \implies x = 0, 2$

Back-substitute $x = 0$: $y = -0^2 = 0$

Back-substitute $x = 2$: $y = -2^2 = -4$

Solutions: $(0, 0), (2, -4)$

12. $\begin{cases} y = -2x^2 + 2 & \text{Equation 1} \\ y = 2(x^4 - 2x^2 + 1) & \text{Equation 2} \end{cases}$

Substitute for y in Equation 1:

$$2(x^4 - 2x^2 + 1) = -2x^2 + 2$$

Solve for x: $x^4 - 2x^2 + 1 + x^2 - 1 = 0$

$$x^4 - x^2 = 0$$

$$x^2(x^2 - 1) = 0 \implies x = 0, \pm 1$$

Back-substitute $x = 0$: $y = -2(0)^2 + 2 = 2$

Back-substitute $x = 1$: $y = -2(1)^2 + 2 = 0$

Back-substitute $x = -1$: $y = -2(-1)^2 + 2 = 0$

Solutions: $(0, 2), (1, 0), (-1, 0)$

13. $\begin{cases} y = x^3 - 3x^2 + 1 & \text{Equation 1} \\ y = x^2 - 3x + 1 & \text{Equation 2} \end{cases}$

Substitute for y in Equation 2:

$$x^3 - 3x^2 + 1 = x^2 - 3x + 1$$

$$x^3 - 4x^2 + 3x = 0$$

$$x(x - 1)(x - 3) = 0 \implies x = 0, 1, 3$$

Back-substitute $x = 0$: $y = 0^3 - 3(0)^2 + 1 = 1$

Back-substitute $x = 1$: $y = 1^3 - 3(1)^2 + 1 = -1$

Back-substitute $x = 3$: $y = 3^3 - 3(3)^2 + 1 = 1$

Solutions: $(0, 1), (1, -1), (3, 1)$

14. $\begin{cases} y = x^3 - 3x^2 + 4 & \text{Equation 1} \\ y = -2x + 4 & \text{Equation 2} \end{cases}$

Substitute for y in Equation 1: $-2x + 4 = x^3 - 3x^2 + 4$

Solve for x: $0 = x^3 - 3x^2 + 2x$

$$0 = x(x^2 - 3x + 2)$$

$$0 = x(x - 2)(x - 1) \implies x = 0, 1, 2$$

Back-substitute $x = 0$: $y = -2(0) + 4 = 4$

Back-substitute $x = 1$: $y = -2(1) + 4 = 2$

Back-substitute $x = 2$: $y = -2(2) + 4 = 0$

Solutions: $(0, 4), (1, 2), (2, 0)$

15. $\begin{cases} x - y = 0 & \text{Equation 1} \\ 5x - 3y = 10 & \text{Equation 2} \end{cases}$

Solve for y in Equation 1: $y = x$

Substitute for y in Equation 2: $5x - 3x = 10$

Solve for x: $2x = 10 \implies x = 5$

Back-substitute in Equation 1: $y = x = 5$

Solution: $(5, 5)$

16. $\begin{cases} x + 2y = 1 & \text{Equation 1} \\ 5x - 4y = -23 & \text{Equation 2} \end{cases}$

Solve for x in Equation 1: $x = 1 - 2y$

Substitute for x in Equation 2: $5(1 - 2y) - 4y = -23$

Solve for y: $-14y = -28 \implies y = 2$

Back-substitute $y = 2$: $x = 1 - 2y = 1 - 2(2) = -3$

Solution: $(-3, 2)$

17. $\begin{cases} 2x - y + 2 = 0 & \text{Equation 1} \\ 4x + y - 5 = 0 & \text{Equation 2} \end{cases}$

Solve for y in Equation 1: $y = 2x + 2$

Substitute for y in Equation 2: $4x + (2x + 2) - 5 = 0$

Solve for x: $6x - 3 = 0 \implies x = \frac{1}{2}$

Back-substitute $x = \frac{1}{2}$: $y = 2x + 2 = 2\left(\frac{1}{2}\right) + 2 = 3$

Solution: $\left(\frac{1}{2}, 3\right)$

18. $\begin{cases} 6x - 3y - 4 = 0 & \text{Equation 1} \\ x + 2y - 4 = 0 & \text{Equation 2} \end{cases}$

Solve for x in Equation 2: $x = 4 - 2y$

Substitute for x in Equation 1: $6(4 - 2y) - 3y - 4 = 0$

Solve for y: $24 - 12y - 3y - 4 = 0 \implies -15y = -20 \implies y = \frac{4}{3}$

Back-substitute $y = \frac{4}{3}$: $x = 4 - 2y = 4 - 2\left(\frac{4}{3}\right) = \frac{4}{3}$

Solution: $\left(\frac{4}{3}, \frac{4}{3}\right)$

19. $\begin{cases} 1.5x + 0.8y = 2.3 & \text{Equation 1} \\ 0.3x - 0.2y = 0.1 & \text{Equation 2} \end{cases}$

Multiply the equations by 10.

$15x + 8y = 23$ Revised Equation 1

$3x - 2y = 1$ Revised Equation 2

Solve for y in revised Equation 2: $y = \frac{3}{2}x - \frac{1}{2}$

Substitute for y in revised Equation 1: $15x + 8\left(\frac{3}{2}x - \frac{1}{2}\right) = 23$

Solve for x: $15x + 12x - 4 = 23 \implies 27x = 27 \implies x = 1$

Back-substitute $x = 1$: $y = \frac{3}{2}(1) - \frac{1}{2} = 1$

Solution: $(1, 1)$

20. $\begin{cases} 0.5x + 3.2y = 9.0 & \text{Equation 1} \\ 0.2x - 1.6y = -3.6 & \text{Equation 2} \end{cases}$

Multiply the equations by 10.

$5x + 32y = 90$ Revised Equation 1

$2x - 16y = -36$ Revised Equation 2

Solve for x in revised Equation 2: $x = 8y - 18$

Substitute for x in revised Equation 1: $5(8y - 18) + 32y = 90$

Solve for y: $40y - 90 + 32y = 90 \implies 72y = 180$

$$\implies y = \frac{5}{2}$$

Back-substitute $y = \frac{5}{2}$: $x = 8\left(\frac{5}{2}\right) - 18 = 2$

Solution: $\left(2, \frac{5}{2}\right)$

21. $\begin{cases} \frac{1}{5}x + \frac{1}{2}y = 8 & \text{Equation 1} \\ x + y = 20 & \text{Equation 2} \end{cases}$

Solve for x in Equation 2: $x = 20 - y$

Substitute for x in Equation 1: $\frac{1}{5}(20 - y) + \frac{1}{2}y = 8$

Solve for y: $4 + \frac{3}{10}y = 8 \implies y = \frac{40}{3}$

Back-substitute $y = \frac{40}{3}$: $x = 20 - y = 20 - \frac{40}{3} = \frac{20}{3}$

Solution: $\left(\frac{20}{3}, \frac{40}{3}\right)$

22. $\begin{cases} \frac{1}{2}x + \frac{3}{4}y = 10 & \text{Equation 1} \\ \frac{3}{4}x - y = 4 & \text{Equation 2} \end{cases}$

Solve for y in Equation 2: $y = \frac{3}{4}x - 4$

Substitute for y in Equation 1: $\frac{1}{2}x + \frac{3}{4}\left(\frac{3}{4}x - 4\right) = 10$

Solve for x: $\frac{1}{2}x + \frac{9}{16}x - 3 = 10 \implies \frac{17}{16}x = 13 \implies x = \frac{208}{17}$

Back-substitute $x = \frac{208}{17}$: $y = \frac{3}{4}\left(\frac{208}{17}\right) - 4 = \frac{88}{17}$

Solution: $\left(\frac{208}{17}, \frac{88}{17}\right)$

23. $\begin{cases} 6x + 5y = -3 & \text{Equation 1} \\ -x - \frac{5}{6}y = -7 & \text{Equation 2} \end{cases}$

Solve for x in Equation 2: $x = 7 - \frac{5}{6}y$

Substitute for x in Equation 1: $6\left(7 - \frac{5}{6}y\right) + 5y = -3$

Solve for y: $42 - 5y + 5y = -3 \implies 42 = -3$ (False)

No solution

24. $\begin{cases} -\frac{2}{3}x + y = 2 & \text{Equation 1} \\ 2x - 3y = 6 & \text{Equation 2} \end{cases}$

Solve for y in Equation 1: $y = \frac{2}{3}x + 2$

Substitute for y in Equation 2: $2x - 3\left(\frac{2}{3}x + 2\right) = 6$

Solve for x: $2x - 2x - 6 = 6 \implies 0 = 12$ Inconsistent

No solution

25. $\begin{cases} x^2 - y = 0 & \text{Equation 1} \\ 2x + y = 0 & \text{Equation 2} \end{cases}$

Solve for y in Equation 2: $y = -2x$

Substitute for y in Equation 1: $x^2 - (-2x) = 0$

Solve for x: $x^2 + 2x = 0 \implies x(x + 2) = 0 \implies x = 0, -2$

Back-substitute $x = 0$: $y = -2(0) = 0$

Back-substitute $x = -2$: $y = -2(-2) = 4$

Solutions: $(0, 0), (-2, 4)$

26. $\begin{cases} x - 2y = 0 & \text{Equation 1} \\ 3x - y^2 = 0 & \text{Equation 2} \end{cases}$

Solve for x in Equation 1: $x = 2y$

Substitute for x in Equation 2: $3(2y) - y^2 = 0$

Solve for y: $6y - y^2 = 0 \implies y(6 - y) = 0 \implies y = 0, 6$

Back-substitute $y = 0$: $x = 2(0) = 0$

Back-substitute $y = 6$: $x = 2(6) = 12$

Solutions: $(0, 0), (12, 6)$

27. $\begin{cases} x - y = -1 & \text{Equation 1} \\ x^2 - y = -4 & \text{Equation 2} \end{cases}$

Solve for y in Equation 1: $y = x + 1$

Substitute for y in Equation 2: $x^2 - (x + 1) = -4$

Solve for x: $x^2 - x - 1 = -4 \implies x^2 - x + 3 = 0$

The Quadratic Formula yields no real solutions.

28. $\begin{cases} y = -x & \text{Equation 1} \\ y = x^3 + 3x^2 + 2x & \text{Equation 2} \end{cases}$

Substitute for y in Equation 2: $-x = x^3 + 3x^2 + 2x$

Solve for x: $x^3 + 3x^2 + 3x = 0 \implies x(x^2 + 3x + 3) = 0$

$$\implies x = 0, \frac{-3 \pm i\sqrt{3}}{2}$$

Back-substitute $x = 0$: $y = 0$

The only real solution is $(0, 0)$.

29. $\begin{cases} -x + 2y = 2 \implies y = \dfrac{x + 2}{2} \\ 3x + y = 15 \implies y = -3x + 15 \end{cases}$

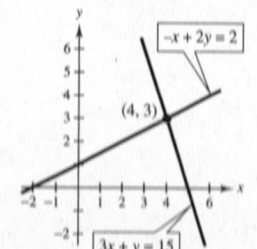

Point of intersection: $(4, 3)$

30. $\begin{cases} x + y = 0 \\ 3x - 2y = 10 \end{cases}$

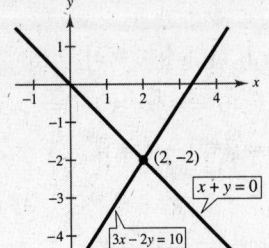

Point of intersection: $(2, -2)$

31. $\begin{cases} x - 3y = -2 \implies y = \frac{1}{3}(x + 2) \\ 5x + 3y = 17 \implies y = \frac{1}{3}(-5x + 17) \end{cases}$

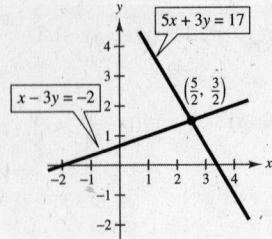

Point of intersection: $\left(\frac{5}{2}, \frac{3}{2}\right)$

32. $\begin{cases} -x + 2y = 1 \\ x - y = 2 \end{cases}$

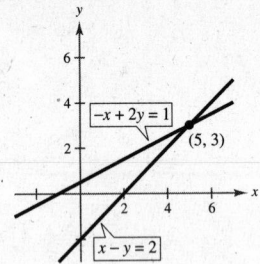

Point of intersection: $(5, 3)$

33. $\begin{cases} x + y = 4 \implies y = -x + 4 \\ x^2 + y^2 - 4x = 0 \implies (x - 2)^2 + y^2 = 4 \end{cases}$

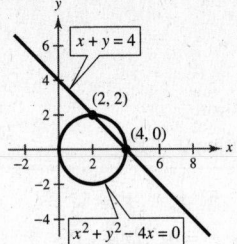

Points of intersection: $(2, 2), (4, 0)$

34. $\begin{cases} -x + y = 3 \\ x^2 - 6x - 27 + y^2 = 0 \end{cases}$

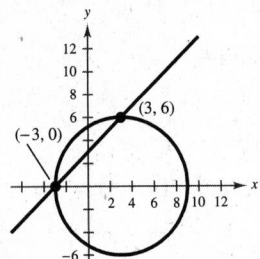

Points of intersection: $(-3, 0), (3, 6)$

35. $\begin{cases} x - y + 3 = 0 \implies y = x + 3 \\ y = x^2 - 4x + 7 \implies y = (x - 2)^2 + 3 \end{cases}$

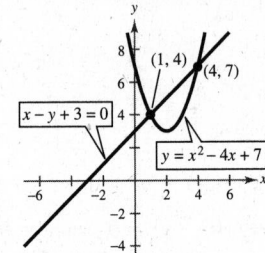

Points of intersection: $(1, 4), (4, 7)$

36. $\begin{cases} y^2 - 4x + 11 = 0 \\ -\frac{1}{2}x + y = -\frac{1}{2} \end{cases}$

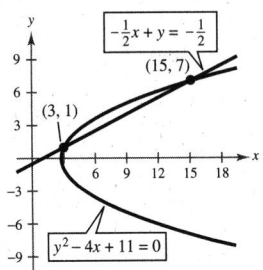

Points of intersection: $(3, 1), (15, 7)$

37. $\begin{cases} 7x + 8y = 24 \implies y = -\frac{7}{8}x + 3 \\ x - 8y = 8 \implies y = \frac{1}{8}x - 1 \end{cases}$

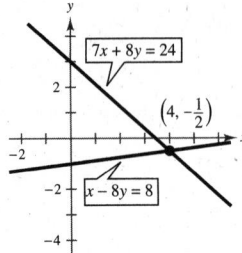

Point of intersection: $\left(4, -\frac{1}{2}\right)$

38. $\begin{cases} x - y = 0 \\ 5x - 2y = 6 \end{cases}$

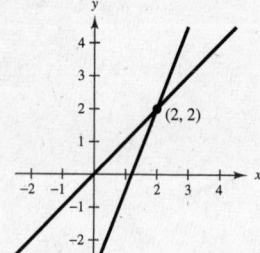

Point of intersection: $(2, 2)$

39. $\begin{cases} 3x - 2y = 0 \implies y = \dfrac{3}{2}x \\ x^2 - y^2 = 4 \implies \dfrac{x^2}{4} - \dfrac{y^2}{4} = 1 \end{cases}$

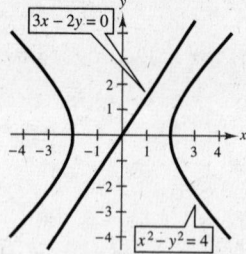

No points of intersection $\implies$ No solution

40. $\begin{cases} 2x - y + 3 = 0 \\ x^2 + y^2 - 4x = 0 \end{cases}$

No points of intersection
so, no solution

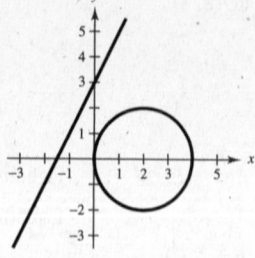

41. $\begin{cases} x^2 + y^2 = 25 \\ 3x^2 - 16y = 0 \implies y = \dfrac{3}{16}x^2 \end{cases}$

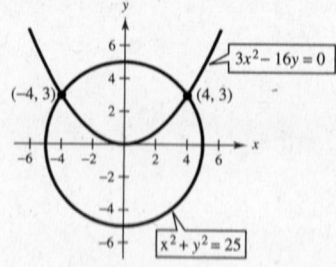

Points of intersection: $(-4, 3)$ and $(4, 3)$

Algebraically we have:

$$x^2 = 25 - y^2$$
$$\tfrac{16}{3}y = 25 - y^2$$
$$16y = 75 - 3y^2$$
$$3y^2 + 16y - 75 = 0$$
$$(3y + 25)(y - 3) = 0$$
$$y = -\tfrac{25}{3} \implies x^2 = -\tfrac{400}{9}, \quad \text{No real solution}$$
$$y = 3 \implies x^2 = 16$$

Solutions: $(\pm 4, 3)$

42. $\begin{cases} x^2 + y^2 = 25 \\ (x - 8)^2 + y^2 = 41 \end{cases}$

Points of intersection:
$(3, 4), (3, -4)$

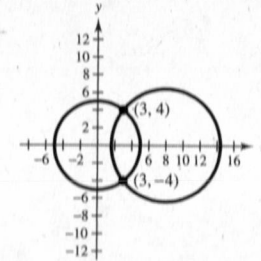

43. $\begin{cases} y = e^x \\ x - y + 1 = 0 \implies y = x + 1 \end{cases}$

Point of intersection: $(0, 1)$

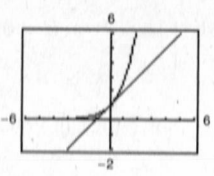

44. $\begin{cases} y = -4e^{-x} \\ y + 3x + 8 = 0 \end{cases}$

Point of intersection:
$(-0.49, -6.53)$

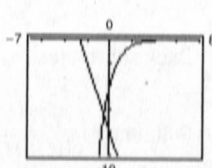

45. $\begin{cases} x + 2y = 8 & \Rightarrow y = -\dfrac{1}{2}x + 4 \\ \\ y = \log_2 x \Rightarrow y = \dfrac{\ln x}{\ln 2} \end{cases}$

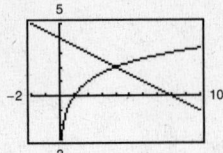

Point of intersection: $(4, 2)$

46. $\begin{cases} y = -2 + \ln(x - 1) \\ 3y + 2x = 9 \end{cases}$

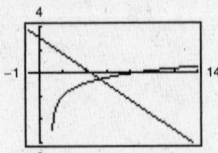

Point of intersection: $(5.31, -0.54)$

47. $\begin{cases} x^2 + y^2 = 169 \Rightarrow y_1 = \sqrt{169 - x^2} \text{ and } y_2 = -\sqrt{169 - x^2} \\ x^2 - 8y = 104 \Rightarrow y_3 = \dfrac{1}{8}x^2 - 13 \end{cases}$

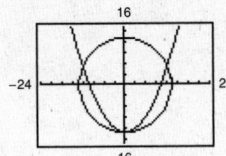

Points of intersection: $(0, -13), (\pm 12, 5)$

48. $\begin{cases} x^2 + y^2 = 4 \Rightarrow y_1 = \sqrt{4 - x^2}, y_2 = -\sqrt{4 - x^2} \\ 2x^2 - y = 2 \Rightarrow y_3 = 2x^2 - 2 \end{cases}$

Points of intersection:

$(0, -2), (1.32, 1.5), (-1.32, 1.5)$

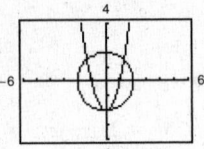

49. $\begin{cases} y = 2x & \text{Equation 1} \\ y = x^2 + 1 & \text{Equation 2} \end{cases}$

Substitute for y in Equation 2: $2x = x^2 + 1$

Solve for x: $x^2 - 2x + 1 = (x - 1)^2 = 0 \Rightarrow x = 1$

Back-substitute $x = 1$ in Equation 1: $y = 2x = 2$

Solution: $(1, 2)$

50. $\begin{cases} x + y = 4 & \text{Equation 1} \\ x^2 + y = 2 & \text{Equation 2} \end{cases}$

Solve for y in Equation 1: $y = 4 - x$

Substitute for y in Equation 2: $x^2 + (4 - x) = 2$

Solve for x: $x^2 - x + 2 = 0$

No real solutions because the discriminant in the Quadratic Formula is negative.

Inconsistent; no solution

51. $\begin{cases} 3x - 7y + 6 = 0 & \text{Equation 1} \\ x^2 - y^2 = 4 & \text{Equation 2} \end{cases}$

Solve for y in Equation 1: $y = \dfrac{3x + 6}{7}$

Substitute for y in Equation 2: $x^2 - \left(\dfrac{3x + 6}{7}\right)^2 = 4$

Solve for x: $x^2 - \left(\dfrac{9x^2 + 36x + 36}{49}\right) = 4$

$$49x^2 - (9x^2 + 36x + 36) = 196$$

$$40x^2 - 36x - 232 = 0$$

$$4(10x - 29)(x + 2) = 0 \Rightarrow x = \dfrac{29}{10}, -2$$

Back-substitute $x = \dfrac{29}{10}$: $y = \dfrac{3x + 6}{7} = \dfrac{3(29/10) + 6}{7} = \dfrac{21}{10}$

Back-substitute $x = -2$: $y = \dfrac{3x + 6}{7} = \dfrac{3(-2) + 6}{7} = 0$

Solutions: $\left(\dfrac{29}{10}, \dfrac{21}{10}\right), (-2, 0)$

52. $\begin{cases} x^2 + y^2 = 25 & \text{Equation 1} \\ 2x + y = 10 & \text{Equation 2} \end{cases}$

Solve for y in Equation 2: $y = 10 - 2x$

Substitute for y in Equation 1: $x^2 + (10 - 2x)^2 = 25$

Solve for x: $x^2 + 100 - 40x + 4x^2 = 25 \implies x^2 - 8x + 15 = 0$

$$\implies (x - 5)(x - 3) = 0 \implies x = 3, 5$$

Back-substitute $x = 3$: $y = 10 - 2(3) = 4$

Back-substitute $x = 5$: $y = 10 - 2(5) = 0$

Solutions: $(3, 4), (5, 0)$

53. $\begin{cases} x - 2y = 4 & \text{Equation 1} \\ x^2 - y = 0 & \text{Equation 2} \end{cases}$

Solve for y in Equation 2: $y = x^2$

Substitute for y in Equation 1: $x - 2x^2 = 4$

Solve for x: $0 = 2x^2 - x + 4 \implies x = \dfrac{1 \pm \sqrt{1 - 4(2)(4)}}{2(2)} \implies x = \dfrac{1 \pm \sqrt{-31}}{4}$

The discriminant in the Quadratic Formula is negative.

No real solution

54. $\begin{cases} y = (x + 1)^3 \\ y = \sqrt{x - 1} \end{cases}$

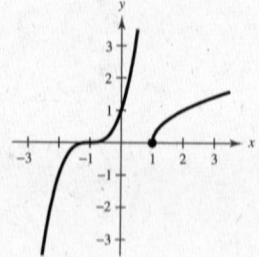

No points of intersection, so no solution

55. $\begin{cases} y - e^{-x} = 1 \implies y = e^{-x} + 1 \\ y - \ln x = 3 \implies y = \ln x + 3 \end{cases}$

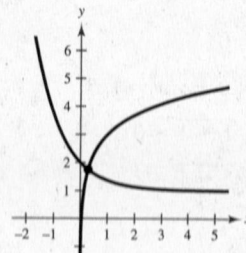

Point of intersection: approximately $(0.287, 1.751)$

56. $\begin{cases} x^2 + y = 4 \implies y = 4 - x^2 \\ e^x - y = 0 \implies y = e^x \end{cases}$

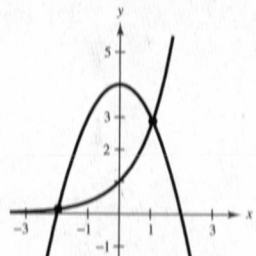

Points of intersection (solutions):

approximately $(-1.96, 0.14), (1.06, 2.88)$

57. $\begin{cases} y = x^4 - 2x^2 + 1 & \text{Equation 1} \\ y = 1 - x^2 & \text{Equation 2} \end{cases}$

Substitute for y in Equation 1: $1 - x^2 = x^4 - 2x^2 + 1$

Solve for x: $x^4 - x^2 = 0 \implies x^2(x^2 - 1) = 0$

$$\implies x = 0, \pm 1$$

Back-substitute $x = 0$: $1 - x^2 = 1 - 0^2 = 1$

Back-substitute $x = 1$: $1 - x^2 = 1 - 1^2 = 0$

Back-substitute $x = -1$: $1 - x^2 = 1 - (-1)^2 = 0$

Solutions: $(0, 1), (\pm 1, 0)$

58. $\begin{cases} y = x^3 - 2x^2 + x - 1 & \text{Equation 1} \\ y = -x^2 + 3x - 1 & \text{Equation 2} \end{cases}$

Substitute for y in Equation 1:

$$-x^2 + 3x - 1 = x^3 - 2x^2 + x - 1$$

Solve for x: $0 = x^3 - x^2 - 2x$

$$0 = x(x^2 - x - 2)$$

$$0 = x(x - 2)(x + 1) \implies x = 0, 2, -1$$

Back-substitute $x = 0$ in Equation 2:

$$y = -0^2 + 3(0) - 1 = -1$$

Back-substitute $x = 2$ in Equation 2:

$$y = -2^2 + 3(2) - 1 = 1$$

Back-substitute $x = -1$ in Equation 2:

$$y = -(-1)^2 + 3(-1) - 1 = -5$$

Solutions: $(0, -1), (2, 1), (-1, -5)$

60. $\begin{cases} x - 2y = 1 & \text{Equation 1} \\ y = \sqrt{x - 1} & \text{Equation 2} \end{cases}$

Substitute for y in Equation 1: $x - 2\sqrt{x - 1} = 1$

Solve for x: $x - 1 = 2\sqrt{x - 1}$

$$(x - 1)^2 = 4(x - 1)$$

$$x^2 - 2x + 1 = 4x - 4$$

$$x^2 - 6x + 5 = 0$$

$$(x - 1)(x - 5) = 0 \implies x = 1, 5$$

Back-substitute $x = 1$: $y = \sqrt{1 - 1} = 0$

Back-substitute $x = 5$: $y = \sqrt{5 - 1} = 2$

Solutions: $(1, 0), (5, 2)$

59. $\begin{cases} xy - 1 = 0 & \text{Equation 1} \\ 2x - 4y + 7 = 0 & \text{Equation 2} \end{cases}$

Solve for y in Equation 1: $y = \dfrac{1}{x}$

Substitute for y in Equation 2: $2x - 4\left(\dfrac{1}{x}\right) + 7 = 0$

Solve for x:

$$2x^2 - 4 + 7x = 0 \implies (2x - 1)(x + 4) = 0$$

$$\implies x = \frac{1}{2}, -4$$

Back-substitute $x = \dfrac{1}{2}$: $y = \dfrac{1}{1/2} = 2$

Back-substitute $x = -4$: $y = \dfrac{1}{-4} = -\dfrac{1}{4}$

Solutions: $\left(\dfrac{1}{2}, 2\right), \left(-4, -\dfrac{1}{4}\right)$

61. $C = 8650x + 250,000, \ R = 9950x$

$$R = C$$

$$9950x = 8650x + 250,000$$

$$1300x = 250,000$$

$$x \approx 192 \text{ units}$$

62. $C = 5.5\sqrt{x} + 10,000, \ R = 3.29x$

$$R = C$$

$$3.29x = 5.5\sqrt{x} + 10,000$$

$$3.29x - 5.5\sqrt{x} - 10,000 = 0$$

Let $u = \sqrt{x}$.

$$3.29u^2 - 5.5u - 10,000 = 0$$

$$u = \frac{5.5 \pm \sqrt{(-5.5)^2 - 4(3.29)(-10,000)}}{2(3.29)}$$

$$u = \frac{5.5 \pm \sqrt{131,630.25}}{6.58}$$

$$u \approx 55.974, -54.302$$

Choosing the positive value for u, we have

$$x = u^2 \implies x = (55.974)^2 \approx 3133 \text{ units.}$$

63. $C = 35.45x + 16,000,\ R = 55.95x$

  (a) $R = C$

    $55.95x = 35.45x + 16,000$

    $20.50x = 16,000$

       $x \approx 781$ units

  (b) $P = R - C$

    $60,000 = 55.95x - (35.45x + 16,000)$

    $60,000 = 20.50x - 16,000$

    $76,000 = 20.50x$

       $x \approx 3708$ units

64. $C = 2.16x + 5000,\ R = 3.49x$

  (a) $R = C$

    $2.16x + 5000 = 3.49x$

    $5000 = 1.33x$

       $x \approx 3760$

  3760 items must be sold to break even.

  (b) $P = R - C$

    $8500 = 3.49x - (2.16x + 5000)$

    $8500 = 1.33x - 5000$

    $13,500 = 1.33x$

    $10,151 \approx x$

  10,151 items must be sold to make a profit of $8500.

65. $\begin{cases} R = 360 - 24x & \text{Equation 1} \\ R = 24 + 18x & \text{Equation 2} \end{cases}$

  (a) Substitute for R in Equation 2: $360 - 24x = 24 + 18x$

    Solve for x: $336 = 42x \implies x = 8$ weeks

  (b)

Weeks	1	2	3	4	5	6	7	8	9	10
$R = 360 - 24x$	336	312	288	264	240	216	192	168	144	120
$R = 24 + 18x$	42	60	78	96	114	132	150	168	186	204

The rentals are equal when $x = 8$ weeks.

66. (a) $\begin{cases} S = 25x + 100 & \text{Rock CD} \\ S = -50x + 475 & \text{Rap CD} \end{cases}$

    $25x + 100 = -50x + 475$

    $75x + 100 = 475$

        $75x = 375$

         $x = 5$

  Conclusion: It takes 5 weeks for the sales of the two CDs to become equal.

  (b)

Number of weeks, x	0	1	2	3	4	5	6
Sales, S (rock)	100	125	150	175	200	225	250
Sales, S (rap)	475	425	375	325	275	225	175

By inspecting the table, we can see that the two sales figures are equal when $x = 5$.

67. $0.06x = 0.03x + 350$

  $0.03x = 350$

    $x \approx \$11,666.67$

To make the straight commission offer the better offer, you would have to sell more than $11,666.67 per week.

68. $p = 1.45 + 0.00014x^2$

  $p = (2.388 - 0.007x)^2$

The market equilibrium (point of intersection) is approximately $(99.99, 2.85)$.

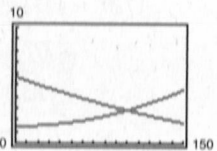

69. (a) $\begin{cases} x + y = 25{,}000 \\ 0.06x + 0.085y = 2000 \end{cases}$

(b) $y_1 = 25{,}000 - x$

$y_2 = \dfrac{2000 - 0.06x}{0.085}$

As the amount at 6% increases, the amount at 8.5% decreases. The amount of interest is fixed at $2000.

(c) The point of intersection occurs when $x = 5000$, so the most that can be invested at 6% and still earn $2000 per year in interest is $5000.

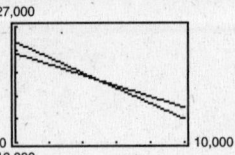

70. $\begin{cases} V = (D - 4)^2, & 5 \le D \le 40 & \text{Doyle Log Rule} \\ V = 0.79D^2 - 2D - 4, & 5 \le D \le 40 & \text{Scribner Log Rule} \end{cases}$

(a)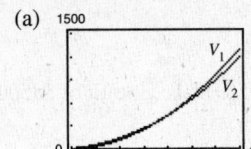

(b) The graphs intersect when $D \approx 24.7$ inches.

(c) For large logs, the Doyle Log Rule gives a greater volume for a given diameter.

71.

t	8	9	10	11	12	13
Solar	70	69	66	65	64	63
Wind	31	46	57	68	105	108

(a) Solar: $C \approx 0.1429t^2 - 4.46t + 96.8$

Wind: $C \approx 16.371t - 102.7$

(b)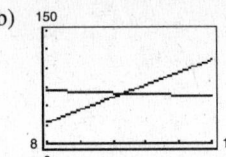

(c) Point of intersection: $(10.3, 66.01)$

During the year 2000, the consumption of solar energy will equal the consumption of wind energy.

(d) $0.1429t^2 - 4.46t + 96.8 = 16.371t - 102.7$

$0.1429t^2 - 20.831t + 199.5 = 0$

By the Quadratic Formula we obtain $t \approx 10.3$ and $t \approx 135.47$.

(e) The results are the same for $t \approx 10.3$. The other "solution", $t \approx 135.47$, is too large to consider as a reasonable answer.

(f) Answers will vary.

72. (a) For Alabama, $P = 17.4t + 4273.2$.

For Colorado, $P = 84.9t + 3467.9$.

(b) The lines appear to intersect at $(11.93, 4480.79)$.

Colorado's population exceeded Alabama's just after this point.

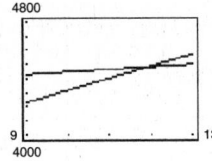

(c) Using the equations from part (a),

$17.4t + 4273.2 = 84.9t + 3467.9$

$4273.2 = 67.5t + 3467.9$

$805.3 = 67.5t$

$11.93 = t.$

73. $2l + 2w = 30 \implies l + w = 15$

$\qquad l = w + 3 \implies (w + 3) + w = 15$

$\qquad\qquad\qquad\qquad 2w = 12$

$\qquad\qquad\qquad\qquad w = 6$

$l = w + 3 = 9$

Dimensions: 6×9 meters

74. $2l + 2w = 280 \implies l + w = 140$

$\qquad w = l - 20 \implies l + (l - 20) = 140$

$\qquad\qquad\qquad\qquad 2l = 160$

$\qquad\qquad\qquad\qquad l = 80$

$w = l - 20 = 80 - 20 = 60$

Dimensions: 60×80 centimeters

75. $2l + 2w = 42 \implies l + w = 21$

$\qquad w = \frac{3}{4}l \implies l + \frac{3}{4}l = 21$

$\qquad\qquad\qquad\quad \frac{7}{4}l = 21$

$\qquad\qquad\qquad\quad l = 12$

$w = \frac{3}{4}l = 9$

Dimensions: 9×12 inches

76. $2l + 2w = 210 \implies l + w = 105$

$\qquad l = \frac{3}{2}w \implies \frac{3}{2}w + w = 105$

$\qquad\qquad\qquad\qquad \frac{5}{2}w = 105$

$\qquad\qquad\qquad\qquad w = 42$

$l = \frac{3}{2}(42) = 63$

Dimensions: 42×63 feet

77. $2l + 2w = 40 \implies l + w = 20 \implies w = 20 - l$

$\qquad lw = 96 \implies l(20 - l) = 96$

$\qquad\qquad\qquad 20l - l^2 = 96$

$\qquad\qquad\qquad 0 = l^2 - 20l + 96$

$\qquad\qquad\qquad 0 = (l - 8)(l - 12)$

$\qquad\qquad\qquad l = 8 \text{ or } l = 12$

If $l = 8$, then $w = 12$.

If $l = 12$, then $w = 8$.

Since the length is supposed to be greater than the width, we have $l = 12$ kilometers and $w = 8$ kilometers.
Dimensions: 8×12 kilometers

78. $A = \frac{1}{2}bh$

$\quad 1 = \frac{1}{2}a^2$

$\quad a^2 = 2$

$\quad a = \sqrt{2}$

The dimensions are $\sqrt{2} \times \sqrt{2} \times 2$ inches.

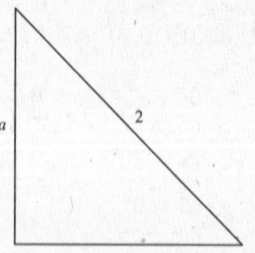

79. False. To solve a system of equations by substitution, you can solve for either variable in one of the two equations and then back-substitute.

80. False. The system can have at most four solutions because a parabola and a circle can intersect at most four times.

81. To solve a system of equations by substitution, use the following steps.

1. Solve one of the equations for one variable in terms of the other.

2. Substitute this expression into the other equation to obtain an equation in one variable.

3. Solve this equation.

4. Back-substitute the value(s) found in Step 3 into the expression found in Step 1 to find the value(s) of the other variable.

5. Check your solution(s) in each of the original equations.

82. For a linear system the result will be a contradictory equation such as $0 = N$, where N is a nonzero real number.
For a nonlinear system there may be an equation with imaginary solutions.

83. $y = x^2$

(a) Line with two points of intersection

$\qquad y = 2x$

$\qquad (0, 0)$ and $(2, 4)$

(b) Line with one point of intersection

$\qquad y = 0$

$\qquad (0, 0)$

(c) Line with no points of intersection

$\qquad y = x - 2$

84. (a) $b = 1$ 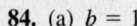 $b = 2$ $b = 3$ $b = 4$

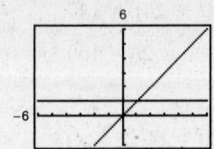   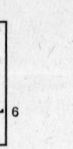   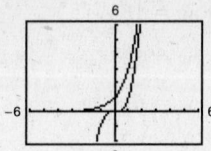   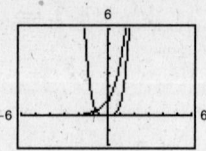

(b) Three

85. $(-2, 7), (5, 5)$

$$m = \frac{5 - 7}{5 - (-2)} = -\frac{2}{7}$$

$$y - 7 = -\frac{2}{7}(x - (-2))$$

$$7y - 49 = -2x - 4$$

$$2x + 7y - 45 = 0$$

86. $(3.5, 4), (10, 6)$

$$m = \frac{6 - 4}{10 - 3.5} = \frac{2}{6.5}$$

$$y - 6 = \frac{2}{6.5}(x - 10)$$

$$6.5y - 39 = 2x - 20$$

$$2x - 6.5y + 19 = 0$$

87. $(6, 3), (10, 3)$

$$m = \frac{3 - 3}{10 - 6} = 0 \implies \text{The line is horizontal.}$$

$$y = 3$$

$$y - 3 = 0$$

88. $(4, -2), (4, 5)$

$$x = 4$$

$$x - 4 = 0$$

89. $\left(\frac{3}{5}, 0\right), (4, 6)$

$$m = \frac{6 - 0}{4 - (3/5)} = \frac{6}{17/5} = \frac{30}{17}$$

$$y - 6 = \frac{30}{17}(x - 4)$$

$$17y - 102 = 30x - 120$$

$$0 = 30x - 17y - 18$$

$$30x - 17y - 18 = 0$$

90. $\left(-\frac{7}{3}, 8\right), \left(\frac{5}{2}, \frac{1}{2}\right)$

$$m = \frac{8 - (1/2)}{-(7/3) - (5/2)} = \frac{15/2}{-29/6} = -\frac{45}{29}$$

$$y - \frac{1}{2} = -\frac{45}{29}\left(x - \frac{5}{2}\right)$$

$$29y - \frac{29}{2} = -45x + \frac{225}{2}$$

$$45x + 29y - 127 = 0$$

91. $f(x) = \dfrac{5}{x - 6}$

Domain: All real numbers except $x = 6$

Horizontal asymptote: $y = 0$

Vertical asymptote: $x = 6$

92. $f(x) = \dfrac{2x - 7}{3x + 2}$

Domain: All real numbers except $x = -\dfrac{2}{3}$

Horizontal asymptote: $y = \dfrac{2}{3}$

Vertical asymptote: $x = -\dfrac{2}{3}$

93. $f(x) = \dfrac{x^2 + 2}{x^2 - 16}$

Domain: All real numbers except $x = \pm 4$.

Horizontal asymptote: $y = 1$

Vertical asymptotes: $x = \pm 4$

94. $f(x) = 3 - \dfrac{2}{x^2}$

Domain: All real numbers except $x = 0$

Horizontal asymptote: $y = 3$

Vertical asymptote: $x = 0$

Section 7.2   Two-Variable Linear Systems

■   You should be able to solve a linear system by the method of elimination.

    1. Obtain coefficients for either x or y that differ only in sign. This is done by multiplying all the terms of one or both equations by appropriate constants.

    2. Add the equations to eliminate one of the variables and then solve for the remaining variable.

    3. Use back-substitution into either original equation and solve for the other variable.

    4. Check your answer.

■   You should know that for a system of two linear equations, one of the following is true.

    1. There are infinitely many solutions; the lines are identical. The system is consistent. The slopes are equal.

    2. There is no solution; the lines are parallel. The system is inconsistent. The slopes are equal.

    3. There is one solution; the lines intersect at one point. The system is consistent. The slopes are not equal.

Vocabulary Check

1. elimination

2. equivalent

3. consistent; inconsistent

4. equilibrium price

---

1. $\begin{cases} 2x + y = 5 & \text{Equation 1} \\ x - y = 1 & \text{Equation 2} \end{cases}$

Add to eliminate y: $3x = 6 \implies x = 2$

Substitute $x = 2$ in Equation 2: $2 - y = 1 \implies y = 1$

Solution: $(2, 1)$

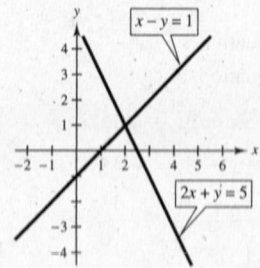

2. $\begin{cases} x + 3y = 1 & \text{Equation 1} \\ -x + 2y = 4 & \text{Equation 2} \end{cases}$

Add to eliminate x: $\quad\begin{aligned} x + 3y &= 1 \\ -x + 2y &= 4 \\ \hline 5y &= 5 \implies y = 1 \end{aligned}$

Substitute $y = 1$ in Equation 1: $x + 3(1) = 1 \implies x = -2$

Solution: $(-2, 1)$

3. $\begin{cases} x + y = 0 & \text{Equation 1} \\ 3x + 2y = 1 & \text{Equation 2} \end{cases}$

Multiply Equation 1 by -2: $-2x - 2y = 0$

Add this to Equation 2 to eliminate y: $x = 1$

Substitute $x = 1$ in Equation 1: $1 + y = 0 \implies y = -1$

Solution: $(1, -1)$

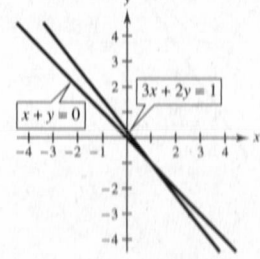

4. $\begin{cases} 2x - y = 3 & \text{Equation 1} \\ 4x + 3y = 21 & \text{Equation 2} \end{cases}$

Multiply Equation 1 by 3: $6x - 3y = 9$

Add this to Equation 2 to eliminate y:
$$\begin{aligned} 6x - 3y &= 9 \\ 4x + 3y &= 21 \\ \hline 10x &= 30 \\ \Rightarrow x &= 3 \end{aligned}$$

Substitute $x = 3$ in Equation 1: $2(3) - y = 3 \Rightarrow y = 3$

Solution: $(3, 3)$

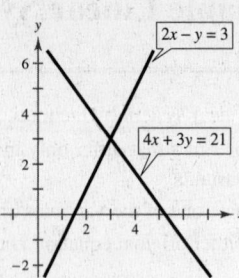

5. $\begin{cases} x - y = 2 & \text{Equation 1} \\ -2x + 2y = 5 & \text{Equation 2} \end{cases}$

Multiply Equation 1 by 2: $2x - 2y = 4$

Add this to Equation 2: $0 = 9$

There are no solutions.

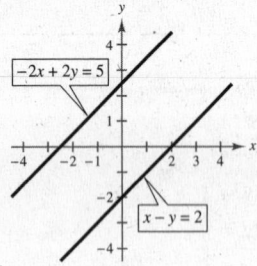

6. $\begin{cases} 3x + 2y = 3 & \text{Equation 1} \\ 6x + 4y = 14 & \text{Equation 2} \end{cases}$

Multiply Equation 1 by -2: $-6x - 4y = -6$

Add this to Equation 2:
$$\begin{aligned} -6x - 4y &= -6 \\ 6x + 4y &= 14 \\ \hline 0 &= 8 \end{aligned}$$

There are no solutions.

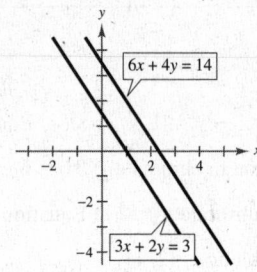

7. $\begin{cases} 3x - 2y = 5 & \text{Equation 1} \\ -6x + 4y = -10 & \text{Equation 2} \end{cases}$

Multiply Equation 1 by 2 and add to Equation 2: $0 = 0$

The equations are dependent. There are infinitely many solutions.

Let $x = a$, then $y = \dfrac{3a - 5}{2} = \dfrac{3}{2}a - \dfrac{5}{2}$.

Solution: $\left(a, \dfrac{3}{2}a - \dfrac{5}{2}\right)$ where a is any real number.

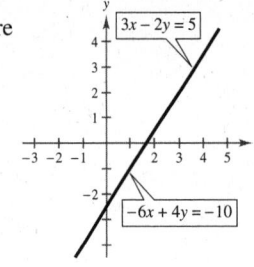

8. $\begin{cases} 9x - 3y = -15 & \text{Equation 1} \\ -3x + y = 5 & \text{Equation 2} \end{cases}$

Multiply Equation 2 by 3: $-9x + 3y = 15$

Add this to Equation 1:
$$\begin{aligned} 9x - 3y &= -15 \\ -9x + 3y &= 15 \\ \hline 0 &= 0 \end{aligned}$$

There are infinitely many solutions. Let $x = a$.

$-3a + y = 5 \Rightarrow y = 3a + 5$

Solution: $(a, 3a + 5)$, where a is any real number.

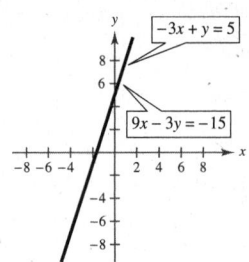

9. $\begin{cases} 9x + 3y = 1 & \text{Equation 1} \\ 3x - 6y = 5 & \text{Equation 2} \end{cases}$

Multiply Equation 2 by (-3): $9x + 3y = 1$
$$-9x + 18y = -15$$

Add to eliminate x: $21y = -14 \implies y = -\frac{2}{3}$

Substitute $y = -\frac{2}{3}$ in Equation 1: $9x + 3\left(-\frac{2}{3}\right) = 1$
$$x = \frac{1}{3}$$

Solution: $\left(\frac{1}{3}, -\frac{2}{3}\right)$

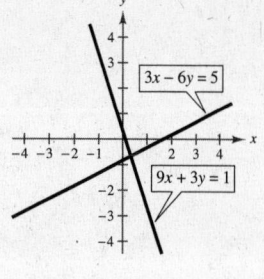

10. $\begin{cases} 5x + 3y = -18 & \text{Equation 1} \\ 2x - 6y = 1 & \text{Equation 2} \end{cases}$

Multiply Equation 1 by 2: $10x + 6y = -36$

Add this to Equation 2 to eliminate y:

$$10x + 6y = -36$$
$$\underline{2x - 6y = 1}$$
$$12x = -35 \implies x = -\frac{35}{12}$$

Substitute $x = -\frac{35}{12}$ in Equation 2:

$$2\left(-\frac{35}{12}\right) - 6y = 1 \implies y = -\frac{41}{36}$$

Solution: $\left(-\frac{35}{12}, -\frac{41}{36}\right)$

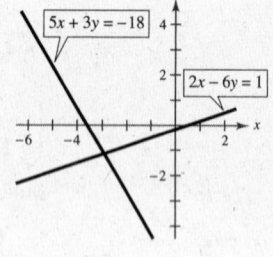

11. $\begin{cases} x + 2y = 4 & \text{Equation 1} \\ x - 2y = 1 & \text{Equation 2} \end{cases}$

Add to eliminate y:

$$2x = 5$$
$$x = \frac{5}{2}$$

Substitute $x = \frac{5}{2}$ in Equation 1:

$$\frac{5}{2} + 2y = 4 \implies y = \frac{3}{4}$$

Solution: $\left(\frac{5}{2}, \frac{3}{4}\right)$

12. $\begin{cases} 3x - 5y = 2 & \text{Equation 1} \\ 2x + 5y = 13 & \text{Equation 2} \end{cases}$

Add to eliminate y: $3x - 5y = 2$

$$\underline{2x + 5y = 13}$$
$$5x = 15 \implies x = 3$$

Substitute $x = 3$ in Equation 1: $3(3) - 5y = 2 \implies y = \frac{7}{5}$

Solution: $\left(3, \frac{7}{5}\right)$

13. $\begin{cases} 2x + 3y = 18 & \text{Equation 1} \\ 5x - y = 11 & \text{Equation 2} \end{cases}$

Multiply Equation 2 by 3: $15x - 3y = 33$

Add this to Equation 1 to eliminate y:

$$17x = 51 \implies x = 3$$

Substitute $x = 3$ in Equation 1:

$$6 + 3y = 18 \implies y = 4$$

Solution: $(3, 4)$

14. $\begin{cases} x + 7y = 12 & \text{Equation 1} \\ 3x - 5y = 10 & \text{Equation 2} \end{cases}$

Multiply Equation 1 by -3: $-3x - 21y = -36$

Add this to Equation 2 to eliminate x:

$$-3x - 21y = -36$$
$$\underline{3x - 5y = 10}$$
$$-26y = -26$$
$$\implies y = 1$$

Substitute $y = 1$ in Equation 1: $x + 7 = 12 \implies x = 5$

Solution: $(5, 1)$

15. $\begin{cases} 3x + 2y = 10 & \text{Equation 1} \\ 2x + 5y = 3 & \text{Equation 2} \end{cases}$

Multiply Equation 1 by 2 and Equation 2 by (-3):

$\begin{cases} 6x + 4y = 20 \\ -6x - 15y = -9 \end{cases}$

Add to eliminate x: $-11y = 11 \implies y = -1$

Substitute $y = -1$ in Equation 1:

$3x - 2 = 10 \implies x = 4$

Solution: $(4, -1)$

16. $\begin{cases} 2r + 4s = 5 & \text{Equation 1} \\ 16r + 50s = 55 & \text{Equation 2} \end{cases}$

Multiply Equation 1 by (-8): $-16r - 32s = -40$

Add this to Equation 2 to eliminate r:

$\begin{array}{r} -16r - 32s = -40 \\ \underline{16r + 50s = 55} \\ 18s = 15 \\ \implies s = \frac{5}{6} \end{array}$

Substitute $s = \frac{5}{6}$ in Equation 1:

$2r + 4\left(\frac{5}{6}\right) = 5 \implies r = \frac{5}{6}$

Solution: $\left(\frac{5}{6}, \frac{5}{6}\right)$

17. $\begin{cases} 5u + 6v = 24 & \text{Equation 1} \\ 3u + 5v = 18 & \text{Equation 2} \end{cases}$

Multiply Equation 1 by 5 and Equation 2 by -6:

$\begin{cases} 25u + 30v = 120 \\ -18u - 30v = -108 \end{cases}$

Add to eliminate v: $7u = 12 \implies u = \frac{12}{7}$

Substitute $u = \frac{12}{7}$ in Equation 1:

$5\left(\frac{12}{7}\right) + 6v = 24 \implies 6v = \frac{108}{7} \implies v = \frac{18}{7}$

Solution: $\left(\frac{12}{7}, \frac{18}{7}\right)$

18. $\begin{cases} 3x + 11y = 4 & \text{Equation 1} \\ -2x - 5y = 9 & \text{Equation 2} \end{cases}$

Multiply Equation 1 by 2 and Equation 2 by 3:

$\begin{cases} 6x + 22y = 8 \\ -6x - 15y = 27 \end{cases}$

Add to eliminate x: $\begin{array}{r} 6x + 22y = 8 \\ \underline{-6x - 15y = 27} \\ 7y = 35 \implies y = 5 \end{array}$

Substitute $y = 5$ in Equation 1: $3x + 11(5) = 4$

$\implies x = -17$

Solution: $(-17, 5)$

19. $\begin{cases} \frac{9}{5}x + \frac{6}{5}y = 4 & \text{Equation 1} \\ 9x + 6y = 3 & \text{Equation 2} \end{cases}$

Multiply Equation 1 by 10 and Equation 2 by -2:

$\begin{cases} 18x + 12y = 40 \\ -18x - 12y = -6 \end{cases}$

Add to eliminate x and y: $0 = 34$

Inconsistent

No solution

20. $\begin{cases} \frac{3}{4}x + y = \frac{1}{8} & \text{Equation 1} \\ \frac{9}{4}x + 3y = \frac{3}{8} & \text{Equation 2} \end{cases}$

Multiply Equation 1 by -3:

$\begin{cases} -\frac{9}{4}x - 3y = -\frac{3}{8} \\ \frac{9}{4}x + 3y = \frac{3}{8} \end{cases}$

Add these two together to obtain $0 = 0$.

The original equations are dependent. They have infinitely many solutions.

Set $x = a$ in $\frac{3}{4}x + y = \frac{1}{8}$ and solve for y.

The points on the line have the form $\left(a, \frac{1}{8} - \frac{3}{4}a\right)$.

21. $\begin{cases} \frac{x}{4} + \frac{y}{6} = 1 & \text{Equation 1} \\ x - y = 3 & \text{Equation 2} \end{cases}$

Multiply Equation 1 by 6: $\frac{3}{2}x + y = 6$

Add this to Equation 2 to eliminate y:

$\frac{5}{2}x = 9 \implies x = \frac{18}{5}$

Substitute $x = \frac{18}{5}$ in Equation 2:

$\frac{18}{5} - y = 3$

$y = \frac{3}{5}$

Solution: $\left(\frac{18}{5}, \frac{3}{5}\right)$

22. $\begin{cases} \frac{2}{3}x + \frac{1}{6}y = \frac{2}{3} & \text{Equation 1} \\ 4x + y = 4 & \text{Equation 2} \end{cases}$

Multiply Equation 1 by (-6): $-4x - y = -4$

Add this to Equation 2: $-4x - y = -4$

$$\underline{\ 4x + y = \ \ 4}$$
$$0 = \ \ 0$$

There are infinitely many solutions. Let $x = a$.

$4a + y = 4 \implies y = 4 - 4a$

Solution: $(a, 4 - 4a)$ where a is any real number

24. $\begin{cases} 7x + 8y = 6 & \text{Equation 1} \\ -14x - 16y = -12 & \text{Equation 2} \end{cases}$

Multiply Equation 1 by 2:

$$\begin{cases} 14x + 16y = 12 \\ -14x - 16y = -12 \end{cases}$$

Add these two together to obtain $0 = 0$.

The original equations are dependent. They have infinitely many solutions.

Set $x = a$ in $7x + 8y = 6$ and solve for y.

The points on the line have the form $\left(a, \frac{3}{4} - \frac{7}{8}a\right)$.

26. $\begin{cases} 0.2x - 0.5y = -27.8 & \text{Equation 1} \\ 0.3x + 0.4y = 68.7 & \text{Equation 2} \end{cases}$

Multiply Equation 1 by 4 and Equation 2 by 5:

$$\begin{cases} 0.8x - 2y = -111.2 \\ 1.5x + 2y = 343.5 \end{cases}$$

Add these to eliminate y: $0.8x - 2y = -111.2$

$$\underline{ 1.5x + 2y = \ \ \ 343.5}$$
$$2.3x = \ \ \ 232.3$$
$$\implies x = \ \ \ 101$$

Substitute $x = 101$ in Equation 1:

$0.2(101) - 0.5y = -27.8 \implies y = 96$

Solution: $(101, 96)$

23. $\begin{cases} -5x + 6y = -3 & \text{Equation 1} \\ 20x - 24y = 12 & \text{Equation 2} \end{cases}$

Multiply Equation 1 by 4:

$$\begin{cases} -20x + 24y = -12 \\ 20x - 24y = 12 \end{cases}$$

Add to eliminate x and y: $0 = 0$

The equations are dependent. There are infinitely many solutions.

Let $x = a$, then

$$-5a + 6y = -3 \implies y = \frac{5a - 3}{6} = \frac{5}{6}a - \frac{1}{2}.$$

Solution: $\left(a, \frac{5}{6}a - \frac{1}{2}\right)$ where a is any real number

25. $\begin{cases} 0.05x - 0.03y = 0.21 & \text{Equation 1} \\ 0.07x + 0.02y = 0.16 & \text{Equation 2} \end{cases}$

Multiply Equation 1 by 200 and Equation 2 by 300:

$$\begin{cases} 10x - 6y = 42 \\ 21x + 6y = 48 \end{cases}$$

Add to eliminate y: $31x = 90$

$$x = \frac{90}{31}$$

Substitute $x = \frac{90}{31}$ in Equation 2:

$$0.07\left(\frac{90}{31}\right) + 0.02y = 0.16$$
$$y = -\frac{67}{31}$$

Solution: $\left(\frac{90}{31}, -\frac{67}{31}\right)$

27. $\begin{cases} 4b + 3m = 3 & \text{Equation 1} \\ 3b + 11m = 13 & \text{Equation 2} \end{cases}$

Multiply Equation 1 by 3 and Equation 2 by (-4):

$$\begin{cases} 12b + 9m = 9 \\ -12b - 44m = -52 \end{cases}$$

Add to eliminate b: $-35m = -43$

$$m = \frac{43}{35}$$

Substitute $m = \frac{43}{35}$ in Equation 1:

$4b + 3\left(\frac{43}{35}\right) = 3 \implies b = -\frac{6}{35}$

Solution: $\left(-\frac{6}{35}, \frac{43}{35}\right)$

28. $\begin{cases} 2x + 5y = 8 & \text{Equation 1} \\ 5x + 8y = 10 & \text{Equation 2} \end{cases}$

Multiply Equation 1 by 5 and Equation 2 by (-2):

$$\begin{cases} 10x + 25y = 40 \\ -10x - 16y = -20 \end{cases}$$

Add to eliminate x:

$$\begin{array}{r} 10x + 25y = 40 \\ -10x - 16y = -20 \\ \hline 9y = 20 \Rightarrow y = \dfrac{20}{9} \end{array}$$

Substitute $y = \dfrac{20}{9}$ in Equation 1: $2x + 5\left(\dfrac{20}{9}\right) = 8$

$$\Rightarrow x = -\dfrac{14}{9}$$

Solution: $\left(-\dfrac{14}{9}, \dfrac{20}{9}\right)$

30. $\begin{cases} \dfrac{x-1}{2} + \dfrac{y+2}{3} = 4 & \text{Equation 1} \\ x - 2y = 5 & \text{Equation 2} \end{cases}$

Multiply Equation 1 by 6:

$$3(x-1) + 2(y+2) = 24 \Rightarrow 3x + 2y = 23$$

Add this to Equation 2 to eliminate y:

$$\begin{array}{r} 3x + 2y = 23 \\ x - 2y = 5 \\ \hline 4x = 28 \\ \Rightarrow x = 7 \end{array}$$

Substitute $x = 7$ in Equation 2: $7 - 2y = 5 \Rightarrow y = 1$

Solution: $(7, 1)$

32. $\begin{cases} -7x + 6y = -4 \\ 14x - 12y = 8 \end{cases}$

$-7x + 6y = -4 \Rightarrow 6y = 7x - 4 \Rightarrow y = \frac{7}{6}x - \frac{2}{3}$;

The graph contains $\left(0, -\frac{2}{3}\right)$ and $(4, 4)$.

$14x - 12y = 8 \Rightarrow -12y = -14x + 8 \Rightarrow y = \frac{7}{6}x - \frac{2}{3}$;

The graph is the same as the previous graph.

The graph of the system matches (a).

Number of solutions: Infinite

Consistent

29. $\begin{cases} \dfrac{x+3}{4} + \dfrac{y-1}{3} = 1 & \text{Equation 1} \\ 2x - y = 12 & \text{Equation 2} \end{cases}$

Multiply Equation 1 by 12 and Equation 2 by 4:

$$\begin{cases} 3x + 4y = 7 \\ 8x - 4y = 48 \end{cases}$$

Add to eliminate y: $11x = 55 \Rightarrow x = 5$

Substitute $x = 5$ into Equation 2:

$$2(5) - y = 12 \Rightarrow y = -2$$

Solution: $(5, -2)$

31. $\begin{cases} 2x - 5y = 0 \\ x - y = 3 \end{cases}$

Multiply Equation 2 by -5:

$$\begin{cases} 2x - 5y = 0 \\ -5x + 5y = -15 \end{cases}$$

Add to eliminate y: $-3x = -15 \Rightarrow x = 5$

Matches graph (b).

Number of solutions: One

Consistent

33. $\begin{cases} 2x - 5y = 0 \\ 2x - 3y = -4 \end{cases}$

Multiply Equation 1 by -1:

$$\begin{cases} -2x + 5y = 0 \\ 2x - 3y = -4 \end{cases}$$

Add to eliminate x: $2y = -4 \Rightarrow y = -2$

Matches graph (c).

Number of solutions: One

Consistent

34. $\begin{cases} 7x - 6y = -6 \\ -7x + 6y = -4 \end{cases}$

$7x - 6y = -6 \implies -6y = -7x - 6 \implies y = \frac{7}{6}x + 1;$

The graph contains $(0, 1)$ and $\left(3, \frac{9}{2}\right)$.

$-7x + 6y = -4 \implies 6y = 7x - 4 \implies y = \frac{7}{6}x - \frac{2}{3};$

The graph contains $\left(0, -\frac{2}{3}\right)$ and is parallel to the previous graph.

The graph of the system matches (d).

Number of solutions: None

Inconsistent

35. $\begin{cases} 3x - 5y = 7 & \text{Equation 1} \\ 2x + y = 9 & \text{Equation 2} \end{cases}$

Multiply Equation 2 by 5:

$\qquad 10x + 5y = 45$

Add this to Equation 1:

$\qquad 13x = 52 \implies x = 4$

Back-substitute $x = 4$ into Equation 2:

$\qquad 2(4) + y = 9 \implies y = 1$

Solution: $(4, 1)$

36. $\begin{cases} -x + 3y = 17 & \text{Equation 1} \\ 4x + 3y = 7 & \text{Equation 2} \end{cases}$

Subtract Equation 2 from Equation 1 to eliminate y:

$$\begin{array}{r} -x + 3y = 17 \\ \underline{-4x - 3y = -7} \\ -5x = 10 \implies x = -2 \end{array}$$

Substitute $x = -2$ in Equation 1:

$-(-2) + 3y = 17 \implies y = 5$

Solution: $(-2, 5)$

37. $\begin{cases} y = 2x - 5 & \text{Equation 1} \\ y = 5x - 11 & \text{Equation 2} \end{cases}$

Since both equations are solved for y, set them equal to one another and solve for x.

$\qquad 2x - 5 = 5x - 11$

$\qquad 6 = 3x$

$\qquad 2 = x$

Back-substitute $x = 2$ into Equation 1:

$\qquad y = 2(2) - 5 = -1$

Solution: $(2, -1)$

38. $\begin{cases} 7x + 3y = 16 & \text{Equation 1} \\ y = x + 2 & \text{Equation 2} \end{cases}$

Substitute for y in Equation 1:

$\qquad 7x + 3(x + 2) = 16$

$\qquad 7x + 3x + 6 = 16$

$\qquad 10x = 10 \implies x = 1$

Substitute $x = 1$ in Equation 2: $y = 1 + 2 = 3$

Solution: $(1, 3)$

39. $\begin{cases} x - 5y = 21 & \text{Equation 1} \\ 6x + 5y = 21 & \text{Equation 2} \end{cases}$

Add the equations: $7x = 42 \implies x = 6$

Back-substitute $x = 6$ into Equation 1:

$\qquad 6 - 5y = 21 \implies -5y = 15 \implies y = -3$

Solution: $(6, -3)$

40. $\begin{cases} y = -3x - 8 & \text{Equation 1} \\ y = 15 - 2x & \text{Equation 2} \end{cases}$

Since both equations are solved for y, set them equal to one another and solve for x:

$\qquad -3x - 8 = 15 - 2x$

$\qquad -x = 23$

$\qquad x = -23$

Back-substitute $x = -23$ into Equation 1:

$\qquad y = -3(-23) - 8 = 61$

Solution: $(-23, 61)$

41. $\begin{cases} -2x + 8y = 19 & \text{Equation 1} \\ y = x - 3 & \text{Equation 2} \end{cases}$

Substitute the expression for y from Equation 2 into Equation 1.

$\qquad -2x + 8(x - 3) = 19 \implies -2x + 8x - 24 = 19$

$\qquad\qquad\qquad\qquad\qquad\qquad 6x = 43$

$\qquad\qquad\qquad\qquad\qquad\qquad x = \frac{43}{6}$

Back-substitute $x = \frac{43}{6}$ into Equation 2:

$\qquad y = \frac{43}{6} - 3 \implies y = \frac{25}{6}$

Solution: $\left(\frac{43}{6}, \frac{25}{6}\right)$

42. $\begin{cases} 4x - 3y = 6 & \text{Equation 1} \\ -5x + 7y = -1 & \text{Equation 2} \end{cases}$

Multiply Equation 1 by 5 and Equation 2 by 4:

$\begin{cases} 20x - 15y = 30 \\ -20x + 28y = -4 \end{cases}$

Add to eliminate x:

$\begin{array}{r} 20x - 15y = 30 \\ \underline{-20x + 28y = -4} \\ 13y = 26 \implies y = 2 \end{array}$

Back-substitute $y = 2$ into Equation 1:

$4x - 3(2) = 6 \implies x = 3$

Solution: $(3, 2)$

43. Let $r_1 =$ the air speed of the plane
and $r_2 =$ the wind air speed.

$\begin{array}{lll} 3.6(r_1 - r_2) = 1800 & \text{Equation 1} \implies & r_1 - r_2 = 500 \\ 3(r_1 + r_2) = 1800 & \text{Equation 2} \implies & \underline{r_1 + r_2 = 600} \\ & & 2r_1 \quad = 1100 \quad \text{Add the equations.} \\ & & r_1 \quad = 550 \\ & & 550 + r_2 = 600 \\ & & r_2 = 50 \end{array}$

The air speed of the plane is 550 mph and the speed of the wind is 50 mph.

44. Let $x =$ the speed of the plane that leaves first and $y =$ the speed of the plane that leaves second.

$\begin{cases} y - x = 80 & \text{Equation 1} \\ 2x + \frac{3}{2}y = 3200 & \text{Equation 2} \end{cases}$

$\begin{array}{r} -2x + 2y = 160 \\ \underline{2x + \frac{3}{2}y = 3200} \\ \frac{7}{2}y = 3360 \\ y = 960 \end{array}$

$960 - x = 80$

$x = 880$

Solution: First plane: 880 kilometers per hour; Second plane: 960 kilometers per hour

45. $\begin{aligned} 50 - 0.5x &= 0.125x \\ 50 &= 0.625x \\ x &= 80 \text{ units} \\ p &= \$10 \end{aligned}$

Solution: $(80, 10)$

46. Supply = Demand

$\begin{aligned} 25 + 0.1x &= 100 - 0.05x \\ 0.15x &= 75 \\ x &= 500 \\ p &= 75 \end{aligned}$

Equilibrium point: $(500, 75)$

47. $140 - 0.00002x = 80 + 0.00001x$

$\qquad 60 = 0.00003x$

$\qquad x = 2{,}000{,}000 \text{ units}$

$\qquad p = \$100.00$

Solution: $(2{,}000{,}000, 100)$

48.   Supply = Demand

$225 + 0.0005x = 400 - 0.0002x$

$\qquad 0.0007x = 175$

$\qquad x = 250{,}000$

$\qquad p = 350$

Equilibrium point: $(250{,}000, 350)$

49. Let x = number of calories in a cheeseburger

$\qquad y$ = number of calories in a small order of french fries

$$\begin{cases} 2x + y = 850 & \text{Equation 1} \\ 3x + 2y = 1390 & \text{Equation 2} \end{cases}$$

Multiply Equation 1 by -2:

$$\begin{cases} -4x - 2y = -1700 \\ \underline{3x + 2y = 1390} \end{cases}$$

$\quad -x \qquad = -310 \qquad$ Add the equations.

$\qquad x = \quad 310$

$\qquad y = \quad 230$

Solution: The cheeseburger contains 310 calories and the
fries contain 230 calories.

50. Let x = Vitamin C in a glass of apple juice

$\qquad y$ = Vitamin C in a glass of orange juice.

$$\begin{cases} x + y = 185 & \text{Equation 1} \\ 2x + 3y = 452 & \text{Equation 2} \end{cases}$$

Multiply Equation 1 by -2; then add the equations:

$$\begin{cases} -2x - 2y = -370 \\ \underline{2x + 3y = 452} \end{cases}$$

$\qquad\qquad y = \quad 82$

Then $x = 185 - 82 = 103$.

The point $(103, 82)$ is the solution of the system.

Apple juice has 103mg of Vitamin C, while orange juice
has 82 mg.

51. Let x = the number of liters at 20%

Let y = the number of liters at 50%.

(a) $\begin{cases} x + y = 10 \\ 0.2x + 0.5y = 0.3(10) \end{cases}$

$-2 \cdot$ Equation 1: $\quad -2x - 2y = -20$

$10 \cdot$ Equation 2: $\quad \underline{2x + 5y = 30}$

$\qquad\qquad\qquad 3y = \quad 10$

$\qquad\qquad\qquad y = \quad \frac{10}{3}$

$\qquad\qquad x + \frac{10}{3} = \quad 10$

$\qquad\qquad\qquad x = \quad \frac{20}{3}$

(b)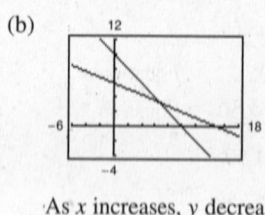

As x increases, y decreases.

(c) In order to obtain the specified concentration of
the final mixture, $6\frac{2}{3}$ liters of the 20% solution and
$3\frac{1}{3}$ liters of the 50% solution are required.

52. Let x = the number of gallons of 87 octane gasoline; y = the number of gallons of 92 octane gasoline.

(a) $\begin{cases} x + y = 500 & \text{Equation 1} \\ 87x + 92y = 44{,}500 & \text{Equation 2} \end{cases}$

(b)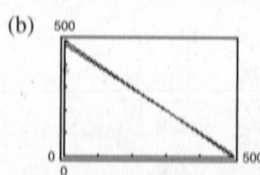

As the amount of 87 octane gasoline increases,
the amount of 92 octane gasoline decreases.

(c) (-87)Equation 1: $\qquad -87x - 87y = -43{,}500$

$\qquad$ Equation 2: $\qquad \underline{87x + 92y = 44{,}500}$

$\qquad\qquad\qquad\qquad\qquad 5y = \quad 1000$

$\qquad\qquad\qquad\qquad\qquad y = \quad 200$

$\qquad\qquad\qquad\qquad x + 200 = \quad 500$

$\qquad\qquad\qquad\qquad\qquad x = \quad 300$

Solution: 87 octane: 300 gallons;

$\qquad\qquad\quad$ 92 octane: 200 gallons

53. Let x = amount invested at 7.5%

y = amount invested at 9%.

$$\begin{cases} x + y = 12{,}000 & \text{Equation 1} \\ 0.075x + 0.09y = 990 & \text{Equation 2} \end{cases}$$

Multiply Equation 1 by 9 and Equation 2 by -100.

$$\begin{cases} 9x + 9y = 108{,}000 \\ -7.5x - 9y = -99{,}000 \end{cases}$$

$$\begin{aligned} 1.5x &= 9{,}000 \qquad \text{Add the equations.} \\ x &= \$6000 \\ y &= \$6000 \end{aligned}$$

The most that can be invested at 7.5% is $6000.

54. Let x = the amount invested at 5.75%; y = the amount invested at 6.25%.

$$\begin{cases} x + y = 32{,}000 & \text{Equation 1} \Rightarrow (-5.75)\text{Equation 1}: & -5.75x - 5.75y = -184{,}000 \\ 0.0575x + 0.0625y = 1900 & \text{Equation 2} \Rightarrow (100)\text{Equation 2}: & 5.75x + 6.25y = 190{,}000 \end{cases}$$

$$\begin{aligned} 0.5y &= 6000 \\ y &= 12{,}000 \\ x + 12{,}000 &= 32{,}000 \\ x &= 20{,}000 \end{aligned}$$

The amount that should be invested in the bond that pays 5.75% interest is $20,000.

55. Let x = number of student tickets

y = number of adult tickets.

$$\begin{cases} x + y = 1435 & \text{Equation 1} \\ 1.50x + 5.00y = 3552.50 & \text{Equation 2} \end{cases}$$

Multiply Equation 1 by -1.50.

$$\begin{cases} -1.50x - 1.50y = -2152.50 \\ 1.50x + 5.00y = 3552.50 \end{cases}$$

$$\begin{aligned} 3.50y &= 1400.00 \qquad \text{Add the equations.} \\ y &= 400 \\ x &= 1035 \end{aligned}$$

Solution: 1035 student tickets and 400 adult tickets were sold.

56. Let x = the number of jackets sold before noon; y = the number of jackets sold after noon.

$$\begin{cases} x + y = 214 & \text{Equation 1} \Rightarrow (-31.95)\text{Equation 1}: & -31.95x - 31.95y = -6837.30 \\ 31.95x + 18.95y = 5108.30 & \text{Equation 2} \Rightarrow & \text{Equation 2}: & 31.95x + 18.95y = 5108.30 \end{cases}$$

$$\begin{aligned} -13y &= -1729 \\ y &= 133 \\ x + 133 &= 214 \\ x &= 81 \end{aligned}$$

So, 81 jackets were sold before noon and 133 jackets were sold after noon.

57. $\begin{cases} 5b + 10a = 20.2 \implies -10b - 20a = -40.4 \\ 10b + 30a = 50.1 \implies \underline{10b + 30a = 50.1} \end{cases}$

$$10a = 9.7$$
$$a = 0.97$$
$$b = 2.10$$

Least squares regression line: $y = 0.97x + 2.10$

58. $\begin{cases} 5b + 10a = 11.7 \implies -10b - 20a = -23.4 \\ 10b + 30a = 25.6 \implies \underline{10b + 30a = 25.6} \end{cases}$

$$10a = 2.2$$
$$a = 0.22$$
$$5b + 10(0.22) = 11.7$$
$$b = 1.9$$

Least squares regression line: $y = 0.22x + 1.9$

59. $\begin{cases} 7b + 21a = 35.1 \implies -21b - 63a = -105.3 \\ 21b + 91a = 114.2 \implies \underline{21b + 91a = 114.2} \end{cases}$

$$28a = 8.9$$
$$a = \frac{89}{280}$$
$$b = \frac{1137}{280}$$

Least squares regression line: $y = \frac{1}{280}(89x + 1137)$

$$y \approx 0.32x + 4.1$$

60. $\begin{cases} 6b + 15a = 23.6 \implies -15b - 37.5a = -59 \\ 15b + 55a = 48.8 \implies \underline{15b + 55a = 48.8} \end{cases}$

$$17.5a = -10.2$$
$$a \approx -0.583$$
$$b \approx 5.390$$

Least squares regression line: $y = -0.583x + 5.390$

61. $(0, 4), (1, 3), (1, 1), (2, 0)$

$n = 4, \displaystyle\sum_{i=1}^{4} x_i = 4, \sum_{i=1}^{4} y_i = 8, \sum_{i=1}^{4} x_i^2 = 6, \sum_{i=1}^{4} x_i y_i = 4$

$\begin{cases} 4b + 4a = 8 \implies 4b + 4a = 8 \\ 4b + 6a = 4 \implies \underline{-4b - 6a = -4} \end{cases}$

$$-2a = 4$$
$$a = -2$$
$$b = 4$$

Least squares regression line: $y = -2x + 4$

62. $\begin{cases} 8b + 28a = 8 \implies -224b - 784a = -224 \\ 28b + 116a = 37 \implies \underline{224b + 928a = 296} \end{cases}$

$$144a = 72$$
$$a = \tfrac{1}{2}$$
$$8b + 28\left(\tfrac{1}{2}\right) = 8$$
$$b = -\tfrac{3}{4}$$

Least squares regression line: $y = \tfrac{1}{2}x - \tfrac{3}{4}$

63. $(5, 66.65), (6, 70.93), (7, 75.31), (8, 78.62), (9, 81.33), (10, 85.89), (11, 88.27)$

(a) $n = 7, \displaystyle\sum_{i=1}^{7} x_i = 56, \sum_{i=1}^{7} x_i^2 = 476, \sum_{i=1}^{7} y_i = 547, \sum_{i=1}^{7} x_i y_i = 4476.8$

$\begin{cases} 7b + 56a = 547 \\ 56b + 476a = 4476.8 \end{cases}$

Multiply Equation 1 by -8.

$\begin{cases} -56b - 448a = -4376 \\ \underline{56b + 476a = 4476.8} \end{cases}$

$$28a = 100.8$$
$$a = 3.6$$
$$b \approx 49.343$$

Least squares regression line: $y \approx 3.6t + 49.343$

(b) $y \approx 3.6t + 49.343$, This agrees with part (a).

(c)

t	Actual room rate	Model approximation
5	\$66.65	\$67.34
6	\$70.93	\$70.94
7	\$75.31	\$74.54
8	\$78.62	\$78.14
9	\$81.33	\$81.74
10	\$85.89	\$85.34
11	\$88.27	\$88.94

The model is a good fit to the data.

(d) When $t = 12$: $y \approx \$92.54$

This is a little off from the actual rate.

(e) $3.6t + 49.343 = 100$

$$3.6t = 50.657$$
$$t \approx 14.1$$

According to the model, room rates will average \$100.00 during the year 2004.

64. (a) $(1.0, 32), (1.5, 41), (2.0, 48), (2.5, 53)$

$$\begin{cases} 4b + 7a = 174 \implies -7b - 12.25a = -304.5 \\ 7b + 13.5a = 322 \implies \quad 7b + 13.5a = \quad 322 \end{cases}$$

$$\begin{aligned} 1.25a &= \quad 17.5 \\ a &= \quad 14 \\ 4b + 98 &= \quad 174 \\ b &= \quad 19 \end{aligned}$$

Least squares regression line: $y = 14x + 19$

(b) When $x = 1.6$: $y = 14(1.6) + 19 = 41.4$ bushels per acre.

65. False. Two lines that coincide have infinitely many points of intersection.

66. False. Solving a system of equations algebraically will always give an exact solution.

67. No, it is not possible for a consistent system of linear equations to have exactly two solutions. Either the lines will intersect once or they will coincide and then the system would have infinite solutions.

68. Answers will vary.

(a) No solution

$$\begin{cases} x + y = 10 \\ x + y = 20 \end{cases}$$

(b) Infinite number of solutions

$$\begin{cases} x + y = 3 \\ 2x + 2y = 6 \end{cases}$$

69. $\begin{cases} 100y - x = \quad 200 & \text{Equation 1} \\ 99y - x = -198 & \text{Equation 2} \end{cases}$

Subtract Equation 2 from Equation 1 to eliminate x:

$$\begin{aligned} 100y - x &= 200 \\ \underline{-99y + x} &= \underline{198} \\ y &= 398 \end{aligned}$$

Substitute $y = 398$ into Equation 1:

$100(398) - x = 200 \implies x = 39{,}600$

Solution: $(39{,}600, \; 398)$

The lines are not parallel. The scale on the axes must be changed to see the point of intersection.

70. $\begin{cases} 21x - 20y = \quad 0 & \text{Equation 1} \\ 13x - 12y = 120 & \text{Equation 2} \end{cases}$

Multiply Equation 2 by $\left(-\frac{5}{3}\right)$: $-\frac{65}{3}x + 20y = -200$

Add this to Equation 1 to eliminate y:

$$-\frac{2}{3}x = -200 \implies x = 300$$

Back-substitute $x = 300$ in Equation 1:

$$21(300) - 20y = 0 \implies y = 315$$

Solution: $(300, 315)$

The lines are not parallel. It is necessary to change the scale on the axes to see the point of intersection.

71. $\begin{cases} 4x - 8y = -3 & \text{Equation 1} \\ 2x + ky = \quad 16 & \text{Equation 2} \end{cases}$

Multiply Equation 2 by -2: $-4x - 2ky = -32$

Add this to Equation 1:

$$\begin{aligned} 4x - 8y &= \quad -3 \\ \underline{-4x - 2ky} &= \underline{-32} \\ -8y - 2ky &= -35 \end{aligned}$$

The system is inconsistent if $-8y - 2ky = 0$. This occurs when $k = -4$.

72. $\begin{cases} 15x + 3y = 6 \implies \quad 30x + 6y = 12 \\ -10x + ky = 9 \implies \underline{-30x + 3ky = 27} \\ \qquad\qquad\qquad\qquad (6 + 3k)y = 39 \end{cases}$

If $k = -2$, then we would have $0 = 39$ and the system would be inconsistent.

73. $-11 - 6x \geq 33$

$-6x \geq 44$

$x \leq -\frac{22}{3}$

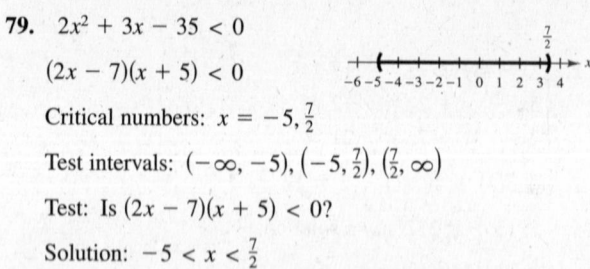

74. $2(x - 3) > -5x + 1$

$2x - 6 > -5x + 1$

$7x > 7$

$x > 1$

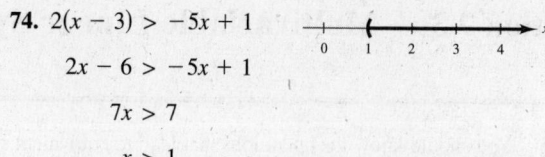

75. $8x - 15 \leq -4(2x - 1)$

$8x - 15 \leq -8x + 4$

$16x \leq 19$

$x \leq \frac{19}{16}$

76. $-6 \leq 3x - 10 < 6$

$4 \leq 3x < 16$

$\frac{4}{3} \leq x < \frac{16}{3}$

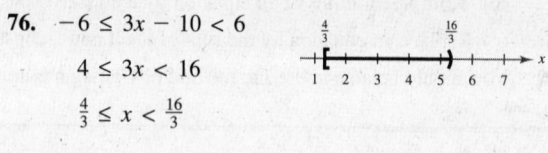

77. $|x - 8| < 10$

$-10 < x - 8 < 10$

$-2 < x < 18$

78. $|x + 10| \geq -3$

All real numbers x

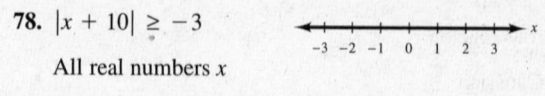

79. $2x^2 + 3x - 35 < 0$

$(2x - 7)(x + 5) < 0$

Critical numbers: $x = -5, \frac{7}{2}$

Test intervals: $(-\infty, -5), \left(-5, \frac{7}{2}\right), \left(\frac{7}{2}, \infty\right)$

Test: Is $(2x - 7)(x + 5) < 0$?

Solution: $-5 < x < \frac{7}{2}$

80. $3x^2 + 12x > 0$

$3x(x + 4) > 0$

Critical numbers: $x = 0, -4$

Test Intervals: $(-\infty, -4), (-4, 0), (0, \infty)$

Test: Is $3x(x + 4) > 0$?

Solution: $x < -4, \ x > 0$

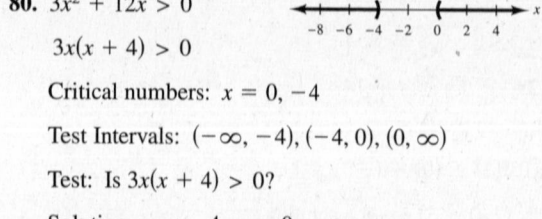

81. $\ln x + \ln 6 = \ln(6x)$

82. $\ln x - 5 \ln(x + 3) = \ln x - \ln(x + 3)^5$

$$= \ln \frac{x}{(x + 3)^5}$$

83. $\log_9 12 - \log_9 x = \log_9\left(\dfrac{12}{x}\right)$

84. $\dfrac{1}{4} \log_6 3x = \log_6 \sqrt[4]{3x}$

85. $\begin{cases} 2x - y = 4 \implies y = 2x - 4 \\ -4x + 2y = -12 \end{cases}$

$-4x + 2(2x - 4) = -12$

$-4x + 4x - 8 = -12$

$-8 = -12$

There are no solutions.

86. $30x - 40y - 33 = 0$

$10x + 20y - 21 = 0 \implies y = -\frac{1}{2}x + \frac{21}{20}$

$30x - 40\left(-\frac{1}{2}x + \frac{21}{20}\right) - 33 = 0$

$30x + 20x - 42 - 33 = 0$

$50x = 75$

$x = \frac{3}{2}$

$y = -\frac{1}{2}\left(\frac{3}{2}\right) + \frac{21}{20} = \frac{6}{20} = \frac{3}{10}$

Solution: $\left(\frac{3}{2}, \frac{3}{10}\right)$

87. Answers will vary.

Section 7.3   Multivariable Linear Systems

- You should know the operations that lead to equivalent systems of equations:

  (a) Interchange any two equations.

  (b) Multiply all terms of an equation by a nonzero constant.

  (c) Replace an equation by the sum of itself and a constant multiple of any other equation in the system.

- You should be able to use the method of Gaussian elimination with back-substitution.

Vocabulary Check

1. row-echelon

2. ordered triple

3. Gaussian

4. row operation

5. nonsquare

6. position

---

1. $\begin{cases} 3x - y + z = 1 \\ 2x \quad - 3z = -14 \\ \quad 5y + 2z = 8 \end{cases}$

(a) $3(2) - (0) + (-3) \neq 1$

   $(2, 0, -3)$ *is not* a solution.

(b) $3(-2) - (0) + 8 \neq 1$

   $(-2, 0, 8)$ *is not* a solution.

(c) $3(0) - (-1) + 3 \neq 1$

   $(0, -1, 3)$ *is not* a solution.

(d) $3(-1) - (0) + 4 = 1$

   $2(-1) \quad - 3(4) = -14$

   $\quad 5(0) + 2(4) = 8$

   $(-1, 0, 4)$ *is a* solution.

2. $\begin{cases} 3x + 4y - z = 17 \\ 5x - y + 2z = -2 \\ 2x - 3y + 7z = -21 \end{cases}$

(a) $3(3) + 4(-1) - 2 \neq 17$

   $(3, -1, 2)$ *is not* a solution.

(b) $3(1) + 4(3) - (-2) = 17$

   $5(1) - 3 + 2(-2) = -2$

   $2(1) - 3(3) + 7(-2) = -21$

   $(1, 3, -2)$ *is a* solution.

(c) $3(4) + 4(1) - (-3) \neq 17$

   $(4, 1, -3)$ *is not* a solution.

(d) $3(1) + 4(-2) - 2 \neq 17$

   $(1, -2, 2)$ *is not* a solution.

3. $\begin{cases} 4x + y - z = 0 \\ -8x - 6y + z = -\frac{7}{4} \\ 3x - y = -\frac{9}{4} \end{cases}$

(a) $4\left(\frac{1}{2}\right) + \left(-\frac{3}{4}\right) - \left(-\frac{7}{4}\right) \neq 0$

   $\left(\frac{1}{2}, -\frac{3}{4}, -\frac{7}{4}\right)$ *is not* a solution.

(b) $4\left(-\frac{3}{2}\right) + \left(\frac{5}{4}\right) - \left(-\frac{5}{4}\right) \neq 0$

   $\left(-\frac{3}{2}, \frac{5}{4}, -\frac{5}{4}\right)$ *is not* a solution.

(c) $4\left(-\frac{1}{2}\right) + \left(\frac{3}{4}\right) - \left(-\frac{5}{4}\right) = 0$

   $-8\left(-\frac{1}{2}\right) - 6\left(\frac{3}{4}\right) + \left(-\frac{5}{4}\right) = -\frac{7}{4}$

   $3\left(-\frac{1}{2}\right) - \left(\frac{3}{4}\right) = -\frac{9}{4}$

   $\left(-\frac{1}{2}, \frac{3}{4}, -\frac{5}{4}\right)$ *is a* solution.

(d) $4\left(-\frac{1}{2}\right) + \left(\frac{1}{6}\right) - \left(-\frac{3}{4}\right) \neq 0$

   $\left(-\frac{1}{2}, \frac{1}{6}, -\frac{3}{4}\right)$ *is not* a solution.

4. $\begin{cases} -4x - y - 8z = -6 \\ \qquad\quad y + z = 0 \\ \quad 4x - 7y \qquad = 6 \end{cases}$

(a) $-4(-2) - (-2) - 8(2) = -6$

$$-2 + 2 = 0$$

$$4(-2) - 7(-2) = 6$$

$(-2, -2, 2)$ *is* a solution.

(b) $-4\left(-\frac{33}{2}\right) - (-10) - 8(10) \neq -6$

$\left(-\frac{33}{2}, -10, 10\right)$ *is not* a solution.

(c) $-4\left(\frac{1}{8}\right) - \left(-\frac{1}{2}\right) - 8\left(\frac{1}{2}\right) \neq -6$

$\left(\frac{1}{8}, -\frac{1}{2}, \frac{1}{2}\right)$ *is not* a solution.

(d) $-4\left(-\frac{11}{2}\right) - (-4) - 8(4) = -6$

$$-4 + 4 = 0$$

$$4\left(-\frac{11}{2}\right) - 7(-4) = 6$$

$\left(-\frac{11}{2}, -4, 4\right)$ *is* a solution.

5. $\begin{cases} 2x - y + 5z = 24 & \text{Equation 1} \\ \qquad\quad y + 2z = 6 & \text{Equation 2} \\ \qquad\qquad\quad z = 4 & \text{Equation 3} \end{cases}$

Back-substitute $z = 4$ into Equation 2:

$$y + 2(4) = 6$$

$$y = -2$$

Back-substitute $y = -2$ and $z = 4$ into Equation 1:

$$2x - (-2) + 5(4) = 24$$

$$2x + 22 = 24$$

$$x = 1$$

Solution: $(1, -2, 4)$

6. $\begin{cases} 4x - 3y - 2z = 21 & \text{Equation 1} \\ \qquad\quad 6y - 5z = -8 & \text{Equation 2} \\ \qquad\qquad\quad z = -2 & \text{Equation 3} \end{cases}$

Back-substitute $z = -2$ in Equation 2:

$$6y - 5(-2) = -8$$

$$y = -3$$

Back-substitute $z = -2$ and $y = -3$ in Equation 1:

$$4x - 3(-3) - 2(-2) = 21$$

$$4x + 13 = 21$$

$$x = 2$$

Solution: $(2, -3, -2)$

7. $\begin{cases} 2x + y - 3z = 10 & \text{Equation 1} \\ \qquad\quad y + z = 12 & \text{Equation 2} \\ \qquad\qquad z = 2 & \text{Equation 3} \end{cases}$

Substitute $z = 2$ into Equation 2: $y + (2) = 12 \implies y = 10$

Substitute $y = 10$ and $z = 2$ into Equation 1:

$$2x + (10) - 3(2) = 10$$

$$2x + 4 = 10$$

$$2x = 6$$

$$x = 3$$

Solution: $(3, 10, 2)$

8. $\begin{cases} x - y + 2z = 22 & \text{Equation 1} \\ \quad 3y - 8z = -9 & \text{Equation 2} \\ \qquad\quad z = -3 & \text{Equation 3} \end{cases}$

Back-substitute $z = -3$ in Equation 2:

$$3y - 8(-3) = -9$$

$$3y + 24 = -9$$

$$y = -11$$

Back-substitute $z = -3$ and $y = -11$ in Equation 1:

$$x - (-11) + 2(-3) = 22$$

$$x + 5 = 22$$

$$x = 17$$

Solution: $(17, -11, -3)$

9. $\begin{cases} 4x - 2y + z = 8 & \text{Equation 1} \\ \qquad -y + z = 4 & \text{Equation 2} \\ \qquad\qquad z = 2 & \text{Equation 3} \end{cases}$

Substitute $z = 2$ into Equation 2:

$$-y + (2) = 4 \implies y = -2$$

Substitute $y = -2$ and $z = 2$ into Equation 1:

$$4x - 2(-2) + (2) = 8$$

$$4x + 6 = 8$$

$$4x = 2$$

$$x = \frac{1}{2}$$

Solution: $\left(\frac{1}{2}, -2, 2\right)$

10. $\begin{cases} 5x - 8z = 22 \\ 3y - 5z = 10 \\ z = -4 \end{cases}$

Back-substitute $z = -4$ in Equation 2:

$3y - 5(-4) = 10 \implies y = -\frac{10}{3}$

Back-substitute $z = -4$ in Equation 1:

$5x - 8(-4) = 22 \implies x = -2$

Solution: $\left(-2, -\frac{10}{3}, -4\right)$

11. $\begin{cases} x - 2y + 3z = 5 & \text{Equation 1} \\ -x + 3y - 5z = 4 & \text{Equation 2} \\ 2x - 3z = 0 & \text{Equation 3} \end{cases}$

Add Equation 1 to Equation 2:

$\begin{cases} x - 2y + 3z = 5 \\ y - 2z = 9 \\ 2x - 3z = 0 \end{cases}$

This is the first step in putting the system in row-echelon form.

12. $\begin{cases} x - 2y + 3z = 5 & \text{Equation 1} \\ -x + 3y - 5z = 4 & \text{Equation 2} \\ 2x - 3z = 0 & \text{Equation 3} \end{cases}$

Add -2 times Equation 1 to Equation 3:

$\begin{cases} x - 2y + 3z = 5 \\ -x + 3y - 5z = 4 \\ 4y - 9z = -10 \end{cases}$

This is the first step in putting the system in row-echelon form.

13. $\begin{cases} x + y + z = 6 & \text{Equation 1} \\ 2x - y + z = 3 & \text{Equation 2} \\ 3x - z = 0 & \text{Equation 3} \end{cases}$

$\begin{cases} x + y + z = 6 \\ -3y - z = -9 & -2\text{Eq.1} + \text{Eq.2} \\ -3y - 4z = -18 & -3\text{Eq.1} + \text{Eq.3} \end{cases}$

$\begin{cases} x + y + z = 6 \\ -3y - z = -9 \\ -3z = -9 & -\text{Eq.2} + \text{Eq.3} \end{cases}$

$\begin{cases} x + y + z = 6 \\ -3y - z = -9 \\ z = 3 & -\frac{1}{3}\text{Eq.3} \end{cases}$

$-3y - 3 = -9 \implies y = 2$

$x + 2 + 3 = 6 \implies x = 1$

Solution: $(1, 2, 3)$

14. $\begin{cases} x + y + z = 3 & \text{Equation 1} \\ x - 2y + 4z = 5 & \text{Equation 2} \\ 3y + 4z = 5 & \text{Equation 3} \end{cases}$

$\begin{cases} x + y + z = 3 \\ -3y + 3z = 2 & (-1)\text{Eq.1} + \text{Eq.2} \\ 3y + 4z = 5 \end{cases}$

$\begin{cases} x + y + z = 3 \\ -3y + 3z = 2 \\ 7z = 7 & \text{Eq.2} + \text{Eq.3} \end{cases}$

$\begin{cases} x + y + z = 3 \\ y - z = -\frac{2}{3} & (-\frac{1}{3})\text{Eq.2} \\ z = 1 & (\frac{1}{7})\text{Eq.3} \end{cases}$

$y - 1 = -\frac{2}{3} \implies y = \frac{1}{3}$

$x + \frac{1}{3} + 1 = 3 \implies x = \frac{5}{3}$

Solution: $\left(\frac{5}{3}, \frac{1}{3}, 1\right)$

15. $\begin{cases} 2x + 2z = 2 \\ 5x + 3y = 4 \\ 3y - 4z = 4 \end{cases}$

$\begin{cases} x + z = 1 & \frac{1}{2}\text{Eq.1} \\ 5x + 3y = 4 \\ 3y - 4z = 4 \end{cases}$

$\begin{cases} x + z = 1 \\ 3y - 5z = -1 & -5\text{Eq.1} + \text{Eq.2} \\ 3y - 4z = 4 \end{cases}$

$\begin{cases} x + z = 1 \\ 3y - 5z = -1 \\ z = 5 & -\text{Eq.2} + \text{Eq.3} \end{cases}$

$3y - 5(5) = -1 \implies y = 8$

$x + 5 = 1 \implies x = -4$

Solution: $(-4, 8, 5)$

16. $\begin{cases} x + y - z = -1 \\ 2x + 4y + z = 1 \\ x - 2y - 3z = 2 \end{cases}$ Interchange equations.

$\begin{cases} x + y - z = -1 \\ 2y + 3z = 3 \\ -3y - 2z = 3 \end{cases}$ (-2)Eq.1 + Eq.2
(-1)Eq.1 + Eq.3

$\begin{cases} x + y - z = -1 \\ 2y + 3z = 3 \\ -6y - 4z = 6 \end{cases}$ 2 Eq.3

$\begin{cases} x + y - z = -1 \\ 2y + 3z = 3 \\ 5z = 15 \end{cases}$ 3Eq.2 + Eq.3

$\begin{cases} x + y - z = -1 \\ y + \frac{3}{2}z = \frac{3}{2} \\ z = 3 \end{cases}$ $(\frac{1}{2})$Eq.2
$(\frac{1}{5})$Eq.3

$y + \frac{3}{2}(3) = \frac{3}{2} \implies y = -3$
$x - 3 - 3 = -1 \implies x = 5$

Solution: $(5, -3, 3)$

17. $\begin{cases} 3x + 3y \phantom{{}+ 3z} = 9 \\ 2x \phantom{{}+ 3y} - 3z = 10 \\ 6y + 4z = -12 \end{cases}$ Interchange equations.

$\begin{cases} x + y \phantom{{}- 3z} = 3 \\ 2x \phantom{{}+ y} - 3z = 10 \\ 6y + 4z = -12 \end{cases}$ $\frac{1}{3}$Eq.1

$\begin{cases} x + y \phantom{{}- 3z} = 3 \\ -2y - 3z = 4 \\ 6y + 4z = -12 \end{cases}$ -2Eq.1 + Eq.2

$\begin{cases} x + y \phantom{{}- 3z} = 3 \\ -2y - 3z = 4 \\ -5z = 0 \end{cases}$ 3Eq.2 + Eq.3

$\begin{cases} x + y \phantom{{}- 3z} = 3 \\ -2y - 3z = 4 \\ z = 0 \end{cases}$ $-\frac{1}{5}$Eq.3

$-2y - 3(0) = 4 \implies y = -2$

$x - 2 = 3 \implies x = 5$

Solution: $(5, -2, 0)$

18. $\begin{cases} x + 4y + z = 0 \\ 2x + 4y - z = 7 \\ 2x - 4y + 2z = -6 \end{cases}$ Interchange equations.

$\begin{cases} x + 4y + z = 0 \\ -4y - 3z = 7 \\ -12y = -6 \end{cases}$ (-2)Eq.1 + Eq.2
(-2)Eq.1 + Eq.3

$\begin{cases} x + 4y + z = 0 \\ -4y - 3z = 7 \\ 9z = -27 \end{cases}$ (-3)Eq.2 + Eq.3

$\begin{cases} x + 4y + z = 0 \\ y + \frac{3}{4}z = -\frac{7}{4} \\ z = -3 \end{cases}$ $(-\frac{1}{4})$Eq.2
$(\frac{1}{9})$Eq.3

$y + \frac{3}{4}(-3) = -\frac{7}{4} \implies y = \frac{1}{2}$

$x + 4\left(\frac{1}{2}\right) + (-3) = 0 \implies x = 1$

Solution: $\left(1, \frac{1}{2}, -3\right)$

19. $\begin{cases} x - 2y + 2z = -9 \\ 2x + y - z = 7 \\ 3x - y + z = 5 \end{cases}$ Interchange equations.

$\begin{cases} x - 2y + 2z = -9 \\ 5y - 5z = 25 \\ 5y - 5z = 32 \end{cases}$ -2Eq.1 + Eq.2
-3Eq.1 + Eq.3

$\begin{cases} x - 2y + 2z = -9 \\ 5y - 5z = 25 \\ 0 = 7 \end{cases}$ $-$Eq.2 + Eq.3

Inconsistent, no solution

20. $\begin{cases} x - 11y + 4z = 3 \\ 5x - 3y + 2z = 3 \\ 2x + 4y - z = 7 \end{cases}$ Interchange equations.

$\begin{cases} x - 11y + 4z = 3 \\ 52y - 18z = -12 \\ 26y - 9z = 1 \end{cases}$ (-5)Eq.1 + Eq.2
(-2)Eq.1 + Eq.3

$\begin{cases} x - 11y + 4z = 3 \\ 52y - 18z = -12 \\ 0 = 7 \end{cases}$ $(-\frac{1}{2})$Eq.2 + Eq.3

Inconsistent, no solution

21. $\begin{cases} 3x - 5y + 5z = 1 \\ 5x - 2y + 3z = 0 \\ 7x - y + 3z = 0 \end{cases}$

$\begin{cases} 6x - 10y + 10z = 2 \qquad 2\text{Eq.1} \\ 5x - 2y + 3z = 0 \\ 7x - y + 3z = 0 \end{cases}$

$\begin{cases} x - 8y + 7z = 2 \qquad -\text{Eq.2} + \text{Eq.1} \\ 5x - 2y + 3z = 0 \\ 7x - y + 3z = 0 \end{cases}$

$\begin{cases} x - 8y + 7z = 2 \\ 38y - 32z = -10 \qquad -5\text{Eq.1} + \text{Eq.2} \\ 55y - 46z = -14 \qquad -7\text{Eq.1} + \text{Eq.3} \end{cases}$

$\begin{cases} x - 8y + 7z = 2 \\ 2090y - 1760z = -550 \qquad 55\text{Eq.2} \\ -2090y + 1748z = 532 \qquad -38\text{Eq.3} \end{cases}$

$\begin{cases} x - 8y + 7z = 2 \\ 2090y - 1760z = -550 \\ -12z = -18 \qquad \text{Eq.2} + \text{Eq.3} \end{cases}$

$-12z = -18 \implies z = \frac{3}{2}$

$38y - 32\left(\frac{3}{2}\right) = -10 \implies y = 1$

$x - 8(1) + 7\left(\frac{3}{2}\right) = 2 \implies x = -\frac{1}{2}$

Solution: $\left(-\frac{1}{2}, 1, \frac{3}{2}\right)$

22. $\begin{cases} 2x + y + 3z = 1 \qquad \text{Equation 1} \\ 2x + 6y + 8z = 3 \qquad \text{Equation 2} \\ 6x + 8y + 18z = 5 \qquad \text{Equation 3} \end{cases}$

$\begin{cases} 2x + y + 3z = 1 \\ 5y + 5z = 2 \qquad (-1)\text{Eq.1} + \text{Eq.2} \\ 5y + 9z = 2 \qquad (-3)\text{Eq.1} + \text{Eq.3} \end{cases}$

$\begin{cases} 2x + y + 3z = 1 \\ 5y + 5z = 2 \\ 4z = 0 \qquad (-1)\text{Eq.2} + \text{Eq.3} \end{cases}$

$\begin{cases} x + \frac{1}{2}y + \frac{3}{2}z = \frac{1}{2} \qquad \left(\frac{1}{2}\right)\text{Eq.1} \\ y + z = \frac{2}{5} \qquad \left(\frac{1}{5}\right)\text{Eq.2} \\ z = 0 \qquad \left(\frac{1}{4}\right)\text{Eq.3} \end{cases}$

$y + 0 = \frac{2}{5} \implies y = \frac{2}{5}$

$x + \frac{1}{2}\left(\frac{2}{5}\right) + \frac{3}{2}(0) = \frac{1}{2} \implies x = \frac{3}{10}$

Solution: $\left(\frac{3}{10}, \frac{2}{5}, 0\right)$

23. $\begin{cases} x + 2y - 7z = -4 \\ 2x + y + z = 13 \\ 3x + 9y - 36z = -33 \end{cases}$

$\begin{cases} x + 2y - 7z = -4 \\ -3y + 15z = 21 \qquad -2\text{Eq.1} + \text{Eq.2} \\ 3y - 15z = -21 \qquad -3\text{Eq.1} + \text{Eq.3} \end{cases}$

$\begin{cases} x + 2y - 7z = -4 \\ -3y + 15z = 21 \\ 0 = 0 \qquad \text{Eq.2} + \text{Eq.3} \end{cases}$

$\begin{cases} x + 2y - 7z = -4 \\ y - 5z = -7 \qquad -\frac{1}{3}\text{Eq.2} \end{cases}$

$\begin{cases} x + 3z = 10 \qquad -2\text{Eq.2} + \text{Eq.1} \\ y - 5z = -7 \end{cases}$

Let $z = a$, then:

$y = 5a - 7$

$x = -3a + 10$

Solution: $(-3a + 10, 5a - 7, a)$

24. $\begin{cases} 2x + y - 3z = 4 \qquad \text{Equation 1} \\ 4x + 2z = 10 \qquad \text{Equation 2} \\ -2x + 3y - 13z = -8 \qquad \text{Equation 3} \end{cases}$

$\begin{cases} 2x + y - 3z = 4 \qquad (-2)\text{Eq.1} + \text{Eq.2} \\ -2y + 8z = 2 \qquad \text{Eq.1} + \text{Eq.3} \\ 4y - 16z = -4 \end{cases}$

$\begin{cases} 2x + y - 3z = 4 \\ -2y + 8z = 2 \\ 0 = 0 \qquad 2\text{Eq.2} + \text{Eq.3} \end{cases}$

$\begin{cases} 2x + z = 5 \qquad \left(\frac{1}{2}\right)\text{Eq.2} + \text{Eq.1} \\ -2y + 8z = 2 \end{cases}$

$\begin{cases} x + z/2 = \frac{5}{2} \qquad \left(\frac{1}{2}\right)\text{Eq.1} \\ y - 4z = -1 \qquad \left(-\frac{1}{2}\right)\text{Eq. 2} \end{cases}$

$z = a$

$y - 4a = -1 \implies y = 4a - 1$

$x + \frac{1}{2}a = \frac{5}{2} \implies x = -\frac{1}{2}a + \frac{5}{2}$

Solution: $\left(-\frac{1}{2}a + \frac{5}{2}, 4a - 1, a\right)$

25. $\begin{cases} 3x - 3y + 6z = 6 \\ x + 2y - z = 5 \\ 5x - 8y + 13z = 7 \end{cases}$

$\begin{cases} x - y + 2z = 2 \qquad \frac{1}{3}\text{Eq.1} \\ x + 2y - z = 5 \\ 5x - 8y + 13z = 7 \end{cases}$

$\begin{cases} x - y + 2z = 2 \\ 3y - 3z = 3 \qquad -\text{Eq.1} + \text{Eq.2} \\ -3y + 3z = -3 \qquad -5\text{Eq.1} + \text{Eq.3} \end{cases}$

$\begin{cases} x - y + 2z = 2 \\ y - z = 1 \qquad \frac{1}{3}\text{Eq.2} \\ 0 = 0 \qquad \text{Eq.2} + \text{Eq.3} \end{cases}$

$\begin{cases} x + z = 3 \qquad \text{Eq.2} + \text{Eq.1} \\ y - z = 1 \end{cases}$

Let $z = a$, then:

$y = a + 1$

$x = -a + 3$

Solution: $(-a + 3, a + 1, a)$

26. $\begin{cases} x \quad\;\; + \;2z = 5 & \text{Equation 1} \\ 3x - y - \;\; z = 1 & \text{Equation 2} \\ 6x - y + 5z = 16 & \text{Equation 3} \end{cases}$

$\begin{cases} x \quad\;\; + 2z = \;\;\;5 \\ \;\;\; -y - 7z = -14 & (-3)\text{Eq.1} + \text{Eq.2} \\ \;\;\; -y - 7z = -14 & (-6)\text{Eq.1} + \text{Eq.3} \end{cases}$

$\begin{cases} x \quad\;\; + 2z = \;\;\;5 \\ \;\;\; -y - 7z = -14 \\ \quad\quad\quad 0 = \;\;\;0 & (-1)\text{Eq.2} + \text{Eq.3} \end{cases}$

$\begin{cases} x \quad\;\; + 2z = \;\;\;5 \\ \quad\; y + 7z = \;\;14 & (-1)\text{Eq.2} \end{cases}$

$z = a$

$y + 7a = 14 \implies y = -7a + 14$

$x + 2a = \;\;\;5 \implies x = -2a + 5$

Solution: $\left(-2a + 5, -7a + 14, a \right)$

27. $\begin{cases} x - 2y + 5z = 2 \\ 4x \quad\quad\; - \;\; z = 0 \end{cases}$

Let $z = a$, then $x = \frac{1}{4}a$.

$\frac{1}{4}a - 2y + \;\;5a = 2$

$a - 8y + 20a = 8$

$\quad\quad -8y = -21a + 8$

$\quad\quad\quad\; y = \frac{21}{8}a - 1$

Answer: $\left(\frac{1}{4}a, \frac{21}{8}a - 1, a \right)$

To avoid fractions, we could go back and let $z = 8a$, then $4x - 8a = 0 \implies x = 2a$.

$2a - 2y + 5(8a) = 2$

$\quad\quad -2y + 42a = 2$

$\quad\quad\quad\quad\;\; y = 21a - 1$

Solution: $(2a, 21a - 1, 8a)$

28. $\begin{cases} x - \;\;3y + \;\;2z = \;\;18 & \text{Equation 1} \\ 5x - 13y + 12z = \;\;80 & \text{Equation 2} \end{cases}$

$\begin{cases} x - \;3y + 2z = \;\;18 \\ \quad\;\; 2y + 2z = -10 & (-5)\text{Eq.1} + \text{Eq.2} \end{cases}$

$\begin{cases} x - 3y + 2z = \;\;18 \\ \quad\;\; y + \;\; z = -5 & \left(\frac{1}{2}\right)\text{Eq.2} \end{cases}$

$\begin{cases} x \quad\quad\;\; + 5z = \;\;\;3 & 3\text{Eq.2} + \text{Eq.1} \\ \quad\; y + \;\; z = -5 \end{cases}$

Let $z = a$, then: $\quad y + \;\; a = -5 \implies y = -a - 5$

$\quad\quad\quad\quad\quad\quad\quad\;\; x + 5a = \;\;\;3 \implies x = -5a + 3$

Solution: $(-5a + 3, -a - 5, a)$

29. $\begin{cases} 2x - 3y + \;\; z = -2 \\ -4x + 9y \quad\quad = \;\;7 \end{cases}$

$\begin{cases} 2x - 3y + \;\; z = -2 \\ \quad\quad 3y + 2z = \;\;3 & 2\text{Eq.1} + \text{Eq.2} \end{cases}$

$\begin{cases} 2x \quad\quad\; + 3z = 1 & \text{Eq.2} + \text{Eq.1} \\ \quad\; 3y + 2z = 3 \end{cases}$

Let $z = a$, then:

$y = -\frac{2}{3}a + 1$

$x = -\frac{3}{2}a + \frac{1}{2}$

Solution: $\left(-\frac{3}{2}a + \frac{1}{2}, -\frac{2}{3}a + 1, a \right)$

30. $\begin{cases} 2x + \;\;3y + \;\;3z = \;\;7 & \text{Equation 1} \\ 4x + 18y + 15z = 44 & \text{Equation 2} \end{cases}$

$\begin{cases} 2x + 3y + 3z = \;\;7 \\ \quad\;\; 12y + 9z = 30 & (-2)\text{Eq.1} + \text{Eq.2} \end{cases}$

$\begin{cases} 2x \quad\quad\; + \frac{3}{4}z = -\frac{1}{2} & \left(-\frac{1}{4}\right)\text{Eq.2} + \text{Eq.1} \\ \quad\; 12y + 9z = 30 \end{cases}$

$\begin{cases} x + \quad\quad \frac{3}{8}z = -\frac{1}{4} & \left(\frac{1}{2}\right)\text{Eq.1} \\ \quad\; y + \frac{3}{4}z = \;\;\frac{5}{2} & \left(\frac{1}{12}\right)\text{Eq.2} \end{cases}$

Let $z = a$, then:

$y + \frac{3}{4}a = \;\;\frac{5}{2} \implies y = -\frac{3}{4}a + \frac{5}{2}$

$x + \frac{3}{8}a = -\frac{1}{4} \implies x = -\frac{3}{8}a - \frac{1}{4}$

Solution: $\left(-\frac{3}{8}a - \frac{1}{4}, -\frac{3}{4}a + \frac{5}{2}, a \right)$

31.
$$\begin{cases} x & + 3w = 4 \\ 2y - z - w = 0 \\ 3y & - 2w = 1 \\ 2x - y + 4z & = 5 \end{cases}$$

$$\begin{cases} x & + 3w = 4 \\ 2y - z - w = 0 \\ 3y & - 2w = 1 \\ -y + 4z - 6w = -3 \quad -2\text{Eq.1} + \text{Eq.4} \end{cases}$$

$$\begin{cases} x & + 3w = 4 \\ y - 4z + 6w = 3 \quad -\text{Eq.4 and interchange} \\ 2y - z - w = 0 \quad \text{the equations.} \\ 3y & - 2w = 1 \end{cases}$$

$$\begin{cases} x & + 3w = 4 \\ y - 4z + 6w = 3 \\ 7z - 13w = -6 \quad -\text{Eq.2} + \text{Eq.3} \\ 12z - 20w = -8 \quad -3\text{Eq.2} + \text{Eq.4} \end{cases}$$

$$\begin{cases} x & + 3w = 4 \\ y - 4z + 6w = 3 \\ z - 3w = -2 \quad -\frac{1}{2}\text{Eq.4} + \text{Eq.3} \\ 12z - 20w = -8 \end{cases}$$

$$\begin{cases} x & + 3w = 4 \\ y - 4z + 6w = 3 \\ z - 3w = -2 \\ 16w = 16 \quad -12\text{Eq.3} + \text{Eq.4} \end{cases}$$

$$16w = 16 \implies w = 1$$
$$z - 3(1) = -2 \implies z = 1$$
$$y - 4(1) + 6(1) = 3 \implies y = 1$$
$$x + 3(1) = 4 \implies x = 1$$

Solution: $(1, 1, 1, 1)$

32.
$$\begin{cases} x + y + z + w = 6 & \text{Equation 1} \\ 2x + 3y - w = 0 & \text{Equation 2} \\ -3x + 4y + z + 2w = 4 & \text{Equation 3} \\ x + 2y - z + w = 0 & \text{Equation 4} \end{cases}$$

$$\begin{cases} x + y + z + w = 6 \\ y - 2z - 3w = -12 \quad (-2)\text{Eq.1} + \text{Eq.2} \\ 7y + 4z + 5w = 22 \quad 3\text{Eq.1} + \text{Eq.3} \\ y - 2z = -6 \quad (-1)\text{Eq.1} + \text{Eq.4} \end{cases}$$

$$\begin{cases} x + y + z + w = 6 \\ y - 2z - 3w = -12 \\ 18z + 26w = 106 \quad (-7)\text{Eq.2} + \text{Eq.3} \\ 3w = 6 \quad (-1)\text{Eq.2} + \text{Eq.4} \end{cases}$$

$$\begin{cases} x + y + z + w = 6 \\ y - 2z - 3w = -12 \\ z + \frac{13}{9}w = \frac{53}{9} \quad \left(\frac{1}{18}\right)\text{Eq.3} \\ w = 2 \quad \left(\frac{1}{3}\right)\text{Eq.4} \end{cases}$$

$$z + \frac{13}{9}(2) = \frac{53}{9} \implies z = 3$$
$$y - 2(3) - 3(2) = -12 \implies y = 0$$
$$x + 0 + 3 + 2 = 6 \implies x = 1$$

Solution: $(1, 0, 3, 2)$

33.
$$\begin{cases} x + 4z = 1 \\ x + y + 10z = 10 \\ 2x - y + 2z = -5 \end{cases}$$

$$\begin{cases} x + 4z = 1 \\ y + 6z = 9 \quad -\text{Eq.1} + \text{Eq.2} \\ -y - 6z = -7 \quad -2\text{Eq.1} + \text{Eq.3} \end{cases}$$

$$\begin{cases} x + 4z = 1 \\ y + 6z = 9 \\ 0 = 2 \quad \text{Eq.2} + \text{Eq.3} \end{cases}$$

No solution, inconsistent

34.
$$\begin{cases} 2x - 2y - 6z = -4 & \text{Equation 1} \\ -3x + 2y + 6z = 1 & \text{Equation 2} \\ x - y - 5z = -3 & \text{Equation 3} \end{cases}$$

$$\begin{cases} x - y - 5z = -3 \\ -3x + 2y + 6z = 1 \quad \text{Interchange equations.} \\ 2x - 2y - 6z = -4 \end{cases}$$

$$\begin{cases} x - y - 5z = -3 \\ -y - 9z = -8 \quad 3\text{Eq.1} + \text{Eq.2} \\ 4z = 2 \quad (-2)\text{Eq.1} + \text{Eq.3} \end{cases}$$

$$\begin{cases} x - y - 5z = -3 \\ y + 9z = 8 \quad (-1)\text{Eq.2} \\ z = \frac{1}{2} \quad \left(\frac{1}{4}\right)\text{Eq.3} \end{cases}$$

$$y + 9\left(\tfrac{1}{2}\right) = 8 \implies y = \tfrac{7}{2}$$
$$x - \tfrac{7}{2} - 5\left(\tfrac{1}{2}\right) = -3 \implies x = 3$$

Solution: $\left(3, \tfrac{7}{2}, \tfrac{1}{2}\right)$

35. $\begin{cases} 2x + 3y = 0 \\ 4x + 3y - z = 0 \\ 8x + 3y + 3z = 0 \end{cases}$

$\begin{cases} 2x + 3y = 0 \\ -3y - z = 0 & -2\text{Eq.1} + \text{Eq.2} \\ -9y + 3z = 0 & -4\text{Eq.1} + \text{Eq.3} \end{cases}$

$\begin{cases} 2x + 3y = 0 \\ -3y - z = 0 \\ 6z = 0 & -3\text{Eq.2} + \text{Eq.3} \end{cases}$

$6z = 0 \implies z = 0$

$-3y - 0 = 0 \implies y = 0$

$2x + 3(0) = 0 \implies x = 0$

Solution: $(0, 0, 0)$

36. $\begin{cases} 4x + 3y + 17z = 0 \\ 5x + 4y + 22z = 0 \\ 4x + 2y + 19z = 0 \end{cases}$

$\begin{cases} 5x + 4y + 22z = 0 \\ 4x + 3y + 17z = 0 \\ 4x + 2y + 19z = 0 \end{cases}$ Interchange equations.

$\begin{cases} x + y + 5z = 0 & (-1)\text{Eq.2} + \text{Eq.1} \\ 4x + 3y + 17z = 0 \\ 4x + 2y + 19z = 0 \end{cases}$

$\begin{cases} x + y + 5z = 0 \\ -y - 3z = 0 & (-4)\text{Eq.1} + \text{Eq.2} \\ -2y - z = 0 & (-4)\text{Eq.1} + \text{Eq.3} \end{cases}$

$\begin{cases} x + y + 5z = 0 \\ y + 3z = 0 & (-1)\text{Eq.2} \\ 5z = 0 & (-2)\text{Eq.2} + \text{Eq.3} \end{cases}$

$\begin{cases} x + y + 5z = 0 \\ y + 3z = 0 \\ z = 0 & \left(\frac{1}{5}\right)\text{Eq.3} \end{cases}$

$y + 3(0) = 0 \implies y = 0$

$x + 0 + 5(0) = 0 \implies x = 0$

Solution: $(0, 0, 0)$

37. $\begin{cases} 12x + 5y + z = 0 \\ 23x + 4y - z = 0 \end{cases}$

$\begin{cases} 24x + 10y + 2z = 0 & 2\text{Eq.1} \\ 23x + 4y - z = 0 \end{cases}$

$\begin{cases} x + 6y + 3z = 0 & -\text{Eq.2} + \text{Eq.1} \\ 23x + 4y - z = 0 \end{cases}$

$\begin{cases} x + 6y + 3z = 0 \\ -134y - 70z = 0 & -23\text{Eq.1} + \text{Eq.2} \end{cases}$

$\begin{cases} x + 6y + 3z = 0 \\ -67y - 35z = 0 & \frac{1}{2}\text{Eq.2} \end{cases}$

To avoid fractions, let $z = 67a$, then:

$-67y - 35(67a) = 0$

$y = -35a$

$x + 6(-35a) + 3(67a) = 0$

$x = 9a$

Solution: $(9a, -35a, 67a)$

38. $\begin{cases} 2x - y - z = 0 & \text{Equation 1} \\ -2x + 6y + 4z = 2 & \text{Equation 2} \end{cases}$

$\begin{cases} 2x - y - z = 0 \\ 5y + 3z = 2 & \text{Eq.1} + \text{Eq.2} \end{cases}$

$\begin{cases} x - \frac{1}{2}y - \frac{1}{2}z = 0 & \left(\frac{1}{2}\right)\text{Eq.1} \\ y + \frac{3}{5}z = \frac{2}{5} & \left(\frac{1}{5}\right)\text{Eq.2} \end{cases}$

Let $z = a$, then:

$y + \frac{3}{5}a = \frac{2}{5} \implies y = -\frac{3}{5}a + \frac{2}{5}$

$x - \frac{1}{2}\left(-\frac{3}{5}a + \frac{2}{5}\right) - \frac{1}{2}a = 0 \implies x = \frac{1}{5}a + \frac{1}{5}$

Solution: $\left(\frac{1}{5}a + \frac{1}{5}, -\frac{3}{5}a + \frac{2}{5}, a\right)$

39. $s = \frac{1}{2}at^2 + v_0 t + s_0$

$(1, 128), (2, 80), (3, 0)$

$128 = \frac{1}{2}a + v_0 + s_0 \implies a + 2v_0 + 2s_0 = 256$

$80 = 2a + 2v_0 + s_0 \implies 2a + 2v_0 + s_0 = 80$

$0 = \frac{9}{2}a + 3v_0 + s_0 \implies 9a + 6v_0 + 2s_0 = 0$

Solving this system yields $a = -32$, $v_0 = 0$, $s_0 = 144$.

Thus, $s = \frac{1}{2}(-32)t^2 + (0)t + 144 = -16t^2 + 144$.

40. $s = \frac{1}{2}at^2 + v_0t + s_0$

(1, 48), (2, 64), (3, 48)

$$\begin{cases} 48 = \frac{1}{2}a + v_0 + s_0 \\ 64 = 2a + 2v_0 + s_0 \\ 48 = \frac{9}{2}a + 3v_0 + s_0 \end{cases}$$

$$\begin{cases} a + 2v_0 + 2s_0 = 96 \quad \text{2Eq. 1} \\ 2a + 2v_0 + s_0 = 64 \\ 9a + 6v_0 + 2s_0 = 96 \quad \text{2Eq. 3} \end{cases}$$

$$\begin{cases} a + 2v_0 + 2s_0 = 96 \\ -2v_0 - 3s_0 = -128 \quad (-2)\text{Eq.1} + \text{Eq.2} \\ -12v_0 - 16s_0 = -768 \quad (-9)\text{Eq.1} + \text{Eq.3} \end{cases}$$

$$\begin{cases} a + 2v_0 + 2s_0 = 96 \\ -2v_0 - 3s_0 = -128 \\ 2s_0 = 0 \quad (-6)\text{Eq.2} + \text{Eq.3} \end{cases}$$

$$\begin{cases} a + 2v_0 + 2s_0 = 96 \\ v_0 + 1.5s_0 = 64 \quad (-0.5)\text{Eq.2} \\ s_0 = 0 \quad (0.5)\text{Eq.3} \end{cases}$$

$$v_0 + 1.5(0) = 64 \implies v_0 = 64$$
$$a + 2(64) + 2(0) = 96 \implies a = -32$$

Thus, $s = \frac{1}{2}(-32)t^2 + 64t + 0$

$$= -16t^2 + 64t.$$

41. $s = \frac{1}{2}at^2 + v_0t + s_0$

(1, 452), (2, 372), (3, 260)

$452 = \frac{1}{2}a + v_0 + s_0 \implies a + 2v_0 + 2s_0 = 904$

$372 = 2a + 2v_0 + s_0 \implies 2a + 2v_0 + s_0 = 372$

$260 = \frac{9}{2}a + 3v_0 + s_0 \implies 9a + 6v_0 + 2s_0 = 520$

Solving this system yields $a = -32$, $v_0 = -32$, $s_0 = 500$.

Thus, $s = \frac{1}{2}(-32)t^2 + (-32)t + 500$

$$= -16t^2 - 32t + 500.$$

42. $s = \frac{1}{2}at^2 + v_0t + s_0$

(1, 132), (2, 100), (3, 36)

$$\begin{cases} 132 = \frac{1}{2}a + v_0 + s_0 \\ 100 = 2a + 2v_0 + s_0 \\ 36 = \frac{9}{2}a + 3v_0 + s_0 \end{cases}$$

$$\begin{cases} a + 2v_0 + 2s_0 = 264 \quad \text{2Eq. 1} \\ 2a + 2v_0 + s_0 = 100 \\ 9a + 6v_0 + 2s_0 = 72 \quad \text{2Eq. 3} \end{cases}$$

$$\begin{cases} a + 2v_0 + 2s_0 = 264 \\ -2v_0 - 3s_0 = -428 \quad (-2)\text{Eq.1} + \text{Eq.2} \\ -12v_0 - 16s_0 = -2304 \quad (-9)\text{Eq.1} + \text{Eq.3} \end{cases}$$

$$\begin{cases} a + 2v_0 + 2s_0 = 264 \\ -2v_0 - 3s_0 = -428 \\ 2s_0 = 264 \quad (-6)\text{Eq.2} + \text{Eq.3} \end{cases}$$

$$\begin{cases} a + 2v_0 + 2s_0 = 264 \\ v_0 + 1.5s_0 = 214 \quad (-0.5)\text{Eq.2} \\ s_0 = 132 \quad (0.5)\text{Eq.3} \end{cases}$$

$$v_0 + 1.5(132) = 214 \implies v_0 = 16$$
$$a + 2(16) + 2(132) = 264 \implies a = -32$$

Thus, $s = \frac{1}{2}(-32)t^2 + 16t + 132$

$$= -16t^2 + 16t + 132.$$

43. $y = ax^2 + bx + c$ passing through (0, 0), (2, −2), (4, 0)

(0, 0): $0 = c$

(2, −2): $-2 = 4a + 2b + c \implies -1 = 2a + b$

(4, 0): $0 = 16a + 4b + c \implies 0 = 4a + b$

Solution: $a = \frac{1}{2}$, $b = -2$, $c = 0$

The equation of the parabola is $y = \frac{1}{2}x^2 - 2x$.

44. $y = ax^2 + bx + c$ passing through $(0, 3), (1, 4), (2, 3)$

$(0, 3)$: $3 = c$

$(1, 4)$: $4 = a + b + c \implies 1 = a + b$

$(2, 3)$: $3 = 4a + 2b + c \implies 0 = 2a + b$

Solution: $a = -1, b = 2, c = 3$

The equation of the parabola is $y = -x^2 + 2x + 3$.

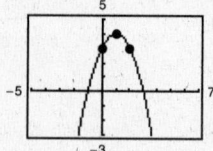

45. $y = ax^2 + bx + c$ passing through $(2, 0), (3, -1), (4, 0)$

$(2, 0)$: $0 = 4a + 2b + c$

$(3, -1)$: $-1 = 9a + 3b + c$

$(4, 0)$: $0 = 16a + 4b + c$

$$\begin{cases} 0 = 4a + 2b + c \\ -1 = 5a + b \qquad -\text{Eq.1} + \text{Eq.2} \\ 0 = 12a + 2b \qquad -\text{Eq.1} + \text{Eq.3} \end{cases}$$

$$\begin{cases} 0 = 4a + 2b + c \\ -1 = 5a + b \\ 2 = 2a \qquad -2\text{Eq.2} + \text{Eq.3} \end{cases}$$

Solution: $a = 1, b = -6, c = 8$

The equation of the parabola is $y = x^2 - 6x + 8$.

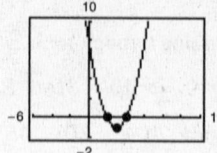

46. $y = ax^2 + bx + c$ passing through $(1, 3), (2, 2), (3, -3)$

$(1, 3)$: $3 = a + b + c$

$(2, 2)$: $2 = 4a + 2b + c$

$(3, -3)$: $-3 = 9a + 3b + c$

$$\begin{cases} a + b + c = 3 \\ 3a + b = -1 \qquad (-1)\text{Eq.1} + \text{Eq.2} \\ 8a + 2b = -6 \qquad (-1)\text{Eq.1} + \text{Eq.3} \end{cases}$$

$$\begin{cases} a + b + c = 3 \\ 3a + b = -1 \\ 2a = -4 \qquad (-2)\text{Eq.2} + \text{Eq.3} \end{cases}$$

Solution: $a = -2, b = 5, c = 0$

The equation of the parabola is $y = -2x^2 + 5x$.

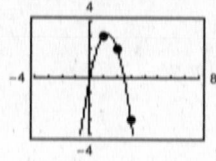

47. $x^2 + y^2 + Dx + Ey + F = 0$ passing through $(0, 0), (2, 2), (4, 0)$

$(0, 0)$: $F = 0$

$(2, 2)$: $8 + 2D + 2E + F = 0 \implies D + E = -4$

$(4, 0)$: $16 + 4D + F = 0 \implies D = -4$ and $E = 0$

The equation of the circle is $x^2 + y^2 - 4x = 0$.

To graph, let $y_1 = \sqrt{4x - x^2}$ and $y_2 = -\sqrt{4x - x^2}$.

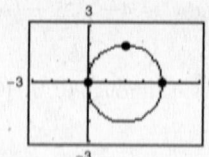

48. $x^2 + y^2 + Dx + Ey + F = 0$ passing through $(0, 0), (0, 6), (3, 3)$

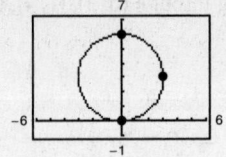

$(0, 0)$: $F = 0$

$(0, 6)$: $36 + 6E + F = 0 \implies E = -6$

$(3, 3)$: $18 + 3D + 3E + F = 0 \implies D = 0$

The equation of the circle is $x^2 + y^2 - 6y = 0$. To graph, complete the square first, then solve for y.

$$x^2 + (y^2 - 6y + 9) = 9$$
$$x^2 + (y - 3)^2 = 9$$
$$(y - 3)^2 = 9 - x^2$$
$$y - 3 = \pm\sqrt{9 - x^2}$$
$$y = 3 \pm \sqrt{9 - x^2}$$

Let $y_1 = 3 + \sqrt{9 - x^2}$ and $y_2 = 3 - \sqrt{9 - x^2}$.

49. $x^2 + y^2 + Dx + Ey + F = 0$ passing through $(-3, -1), (2, 4), (-6, 8)$

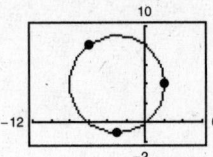

$(-3, -1)$: $10 - 3D - E + F = 0 \implies 10 = 3D + E - F$

$(2, 4)$: $20 + 2D + 4E + F = 0 \implies 20 = -2D - 4E - F$

$(-6, 8)$: $100 - 6D + 8E + F = 0 \implies 100 = 6D - 8E - F$

Solution: $D = 6, E = -8, F = 0$

The equation of the circle is $x^2 + y^2 + 6x - 8y = 0$. To graph, complete the squares first, then solve for y.

$$(x^2 + 6x + 9) + (y^2 - 8y + 16) = 0 + 9 + 16$$
$$(x + 3)^2 + (y - 4)^2 = 25$$
$$(y - 4)^2 = 25 - (x + 3)^2$$
$$y - 4 = \pm\sqrt{25 - (x + 3)^2}$$
$$y = 4 \pm \sqrt{25 - (x + 3)^2}$$

Let $y_1 = 4 + \sqrt{25 - (x + 3)^2}$ and $y_2 = 4 - \sqrt{25 - (x + 3)^2}$.

50. $x^2 + y^2 + Dx + Ey + F = 0$ passing through $(0, 0), (0, -2), (3, 0)$

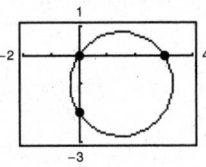

$(0, 0)$: $F = 0$

$(0, -2)$: $4 - 2E + F = 0 \implies E = 2$

$(3, 0)$: $9 + 3D + F = 0 \implies D = -3$

The equation of the circle is $x^2 + y^2 - 3x + 2y = 0$. To graph, complete the squares first, then solve for y.

$$\left(x^2 - 3x + \tfrac{9}{4}\right) + (y^2 + 2y + 1) = \tfrac{9}{4} + 1$$
$$\left(x - \tfrac{3}{2}\right)^2 + (y + 1)^2 = \tfrac{13}{4}$$
$$(y + 1)^2 = \tfrac{13}{4} - \left(x - \tfrac{3}{2}\right)^2$$
$$y + 1 = \pm\sqrt{\tfrac{13}{4} - \left(x - \tfrac{3}{2}\right)^2}$$
$$y = -1 \pm \sqrt{\tfrac{13}{4} - \left(x - \tfrac{3}{2}\right)^2}$$

Let $y_1 = -1 + \sqrt{\tfrac{13}{4} - \left(x - \tfrac{3}{2}\right)^2}$ and $y_2 = -1 - \sqrt{\tfrac{13}{4} - \left(x - \tfrac{3}{2}\right)^2}$.

51. Let x = number of touchdowns.

Let y = number of extra-point kicks.

Let z = number of field goals.

$$\begin{cases} x + y + z = 13 \\ 6x + y + 3z = 45 \\ x - y \quad\;\; = 0 \\ x \quad\;\; - 6z = 0 \end{cases}$$

$$\begin{cases} x + y + z = 13 \\ \quad -5y - 3z = -33 \quad -6\text{Eq.1} + \text{Eq.2} \\ \quad -2y - z = -13 \quad -\text{Eq.1} + \text{Eq.3} \\ \quad - y - 7z = -13 \quad -\text{Eq.1} + \text{Eq.4} \end{cases}$$

$$\begin{cases} x + y + z = 13 \\ \quad - y - 7z = -13 \quad \text{Interchange Eq.2 and Eq.4.} \\ \quad -2y - z = -13 \\ \quad -5y - 3z = -33 \end{cases}$$

$$\begin{cases} x + y + z = 13 \\ \quad y + 7z = 13 \quad -\text{Eq.2} \\ \quad -2y - z = -13 \\ \quad -5y - 3z = -33 \end{cases}$$

$$\begin{cases} x + y + z = 13 \\ \quad y + 7z = 13 \\ \quad\quad 13z = 13 \quad 2\text{Eq.2} + \text{Eq.3} \\ \quad\quad 32z = 32 \quad 5\text{Eq.2} + \text{Eq.4} \end{cases}$$

$z = 1$

$y + 7(1) = 13 \implies y = 6$

$x + 6 + 1 = 13 \implies x = 6$

Thus, 6 touchdowns, 6 extra-point kicks, and 1 field goal were scored.

52. Let x = number of 2-point baskets.

Let y = number of 3-point baskets.

Let z = number of free throws.

$$\begin{cases} 2x + 3y + z = 70 \\ x \quad\;\; - z = 2 \\ \quad -2y + z = 1 \end{cases}$$

Add Equation 2 to Equation 3, and then add Equation 1 to Equation 2:

$$\begin{cases} 2x + 3y + z = 70 \\ 3x + 3y \quad\;\; = 72 \\ x - 2y \quad\;\; = 3 \end{cases}$$

Divide Equation 2 by 3:

$$\begin{cases} 2x + 3y + z = 70 \\ x + y \quad\;\; = 24 \\ x - 2y \quad\;\; = 3 \end{cases}$$

Subtract Equation 3 from Equation 2: $3y = 21 \implies y = 7$

Back-substitute into Equation 2: $x = 24 - 7 = 17$

Back-substitute into Equation 1: $z = 70 - 2(17) - 3(7) = 15$

There were 17 two-point baskets, 7 three-pointers, and 15 free-throws.

53. Let x = amount at 8%.

Let y = amount at 9%.

Let z = amount at 10%.

$$\begin{cases} x + y + z = 775{,}000 \\ 0.08x + 0.09y + 0.10z = 67{,}500 \\ x = 4z \end{cases}$$

$y + 5z = 775{,}000$

$0.09y + 0.42z = 67{,}500$

$z = 75{,}000$

$y = 775{,}000 - 5z = 400{,}000$

$x = 4z = 300{,}000$

$300{,}000 was borrowed at 8%.

$400{,}000 was borrowed at 9%.

$75{,}000 was borrowed at 10%.

54. Let x = amount at 8%.

Let y = amount at 9%.

Let z = amount at 10%.

$$\begin{cases} x + y + z = 800{,}000 \\ 0.08x + 0.09y + 0.10z = 67{,}000 \\ x = 5z \end{cases}$$

$$\begin{cases} y + 6z = 800{,}000 \\ 0.09y + 0.5z = 67{,}000 \end{cases}$$

$z = 125{,}000$

$y = 800{,}000 - 6(125{,}000) = 50{,}000$

$x = 5(125{,}000) = 625{,}000$

Solution: $x = \$625{,}000$ at 8%

$y = \$50{,}000$ at 9%

$z = \$125{,}000$ at 10%

55. Let C = amount in certificates of deposit.

Let M = amount in municipal bonds.

Let B = amount in blue-chip stocks.

Let G = amount in growth or speculative stocks.

$$\begin{cases} C + M + B + G = 500{,}000 \\ 0.10C + 0.08M + 0.12B + 0.13G = 0.10(500{,}000) \\ B + G = \frac{1}{4}(500{,}000) \end{cases}$$

This system has infinitely many solutions.

Let $G = s$, then $B = 125{,}000 - s$

$M = 125{,}000 + \frac{1}{2}s$

$C = 250{,}000 - \frac{1}{2}s$

One possible solution is to let $s = 50{,}000$.

Certificates of deposit: $225{,}000

Municipal bonds: $150{,}000

Blue-chip stocks: $75{,}000

Growth or speculative stocks: $50{,}000

56. Let C = amount in certificates of deposit.

Let M = amount in municipal bonds.

Let B = amount in blue-chip stocks.

Let G = amount in growth or speculative stocks.

$$\begin{cases} C + M + B + G = 500{,}000 \\ 0.09C + 0.05M + 0.12B + 0.14G = 0.10(500{,}000) \\ B + G = \frac{1}{4}(500{,}000) \end{cases}$$

This system has infinitely many solutions.

Let $G = s$, then $B = 125{,}000 - s$

$M = \frac{1}{2}s - 31{,}250$

$C = 406{,}250 - \frac{1}{2}s.$

Solution:

$406{,}250 - \frac{1}{2}s$ in certificates of deposit,

$-31{,}250 + \frac{1}{2}s$ in municipal bonds,

$125{,}000 - s$ in blue-chip stocks,

s in growth stocks

One possible solution is to let $s = \$100{,}000$.

Certificates of deposit: $356{,}250

Municipal bonds: $18{,}750

Blue-chip stocks: $25{,}000

Growth or speculative stocks: $100{,}000

57. Let x = pounds of brand X.

Let y = pounds of brand Y.

Let z = pounds of brand Z.

Fertilizer A: $\frac{1}{3}y + \frac{2}{9}z = 5$

Fertilizer B: $\frac{1}{2}x + \frac{2}{3}y + \frac{5}{9}z = 13$

Fertilizer C: $\frac{1}{2}x \quad\quad + \frac{2}{9}z = 4$

$\begin{cases} \frac{1}{2}x + \frac{2}{3}y + \frac{5}{9}z = 13 & \text{Interchange Eq.1 and Eq.2.} \\ \quad\quad \frac{1}{3}y + \frac{2}{9}z = 5 \\ \frac{1}{2}x \quad\quad + \frac{2}{9}z = 4 \end{cases}$

$\begin{cases} \frac{1}{2}x + \frac{2}{3}y + \frac{5}{9}z = 13 \\ \quad\quad \frac{1}{3}y + \frac{2}{9}z = 5 \\ \quad\quad -\frac{2}{3}y - \frac{1}{3}z = -9 & -\text{Eq.1} + \text{Eq.3} \end{cases}$

$\begin{cases} \frac{1}{2}x + \frac{2}{3}y + \frac{5}{9}z = 13 \\ \quad\quad \frac{1}{3}y + \frac{2}{9}z = 5 \\ \quad\quad\quad\quad \frac{1}{9}z = 1 & 2\text{Eq.2} + \text{Eq.3} \end{cases}$

$z = 9$

$\frac{1}{3}y + \frac{2}{9}(9) = 5 \implies y = 9$

$\frac{1}{2}x + \frac{2}{3}(9) + \frac{5}{9}(9) = 13 \implies x = 4$

4 pounds of brand X, 9 pounds of brand Y, and 9 pounds of brand Z are needed to obtain the desired mixture.

58. Let x = liters of spray X.

Let y = liters of spray Y.

Let z = liters of spray Z.

Chemical A: $\left.\begin{array}{l} \frac{1}{5}x + \frac{1}{2}z = 12 \\ \frac{2}{5}x + \frac{1}{2}z = 16 \end{array}\right\} \implies x = 20, z = 16$

Chemical B:

Chemical C: $\frac{2}{5}x + y = 26 \implies y = 18$

20 liters of spray X, 18 liters of spray Y, and 16 liters of spray Z are needed to get the desired mixture.

59. Let x = pounds of Vanilla coffee.

Let y = pounds of Hazelnut coffee.

Let z = pounds of French Roast coffee.

$\begin{cases} x + \quad y + z = 10 \\ 2x + 2.50y + 3z = 26 \\ \quad\quad y - z = 0 \end{cases}$

$\begin{cases} x + \quad y + z = 10 \\ \quad\quad 0.5y + z = 6 & -2\text{Eq.1} + \text{Eq.2} \\ \quad\quad y - z = 0 \end{cases}$

$\begin{cases} x + \quad y + z = 10 \\ \quad\quad 0.5y + z = 6 \\ \quad\quad\quad\quad -3z = -12 & -2\text{Eq.2} + \text{Eq.3} \end{cases}$

$z = 4$

$0.5y + 4 = 6 \implies y = 4$

$x + 4 + 4 = 10 \implies x = 2$

2 pounds of Vanilla coffee, 4 pounds of Hazelnut coffee, and 4 pounds of French Roast coffee are needed to obtain the desired mixture.

60. Each centerpiece costs \$30.

Let x = number of roses in a centerpiece.

Let y = number of lilies.

Let z = number of irises.

$\begin{cases} x + \quad y + z = 12 \\ 2.5x + 4y + 2z = 30 \\ x - 2y - 2z = 0 \end{cases}$

$\begin{cases} x + \quad y + z = 12 \\ 3.5x + 2y \quad\quad = 30 & \text{Eq.3} + \text{Eq.2} \\ 3x \quad\quad\quad = 24 & 2\text{Eq.1} + \text{Eq.3} \end{cases}$

$3x = 24 \implies x = 8$

$3.5x + 2y = 30 \implies y = \frac{1}{2}(30 - 3.5(8))$

$\quad\quad\quad\quad\quad\quad = \frac{1}{2}(30 - 28) = \frac{1}{2}(2) = 1$

$x + y + z = 12 \implies z = 12 - 8 - 1 = 3$

The point $(8, 1, 3)$ is the solution of the system of equations.

Each centerpiece should contain 8 roses, 1 lily, and 3 irises.

61. Let x = number of television ads.

Let y = number of radio ads.

Let z = number of local newspaper ads.

$$\begin{cases} x + y + z = 60 \\ 1000x + 200y + 500z = 42{,}000 \\ x - y - z = 0 \end{cases}$$

$$\begin{cases} x + y + z = 60 \\ -800y - 500z = -18{,}000 \quad -1000\text{Eq.1} + \text{Eq.2} \\ -2y - 2z = -60 \quad -\text{Eq.1} + \text{Eq.3} \end{cases}$$

$$\begin{cases} x + y + z = 60 \\ -2y - 2z = -60 \quad \text{Interchange} \\ -800y - 500z = -18{,}000 \quad \text{Eq.2 and Eq.3.} \end{cases}$$

$$\begin{cases} x + y + z = 60 \\ -2y - 2z = -60 \\ 300z = 6000 \quad -400\text{Eq.2} + \text{Eq.3} \end{cases}$$

$z = 20$

$-2y - 2(20) = -60 \implies y = 10$

$x + 10 + 20 = 60 \implies x = 30$

30 television ads, 10 radio ads, and 20 newspaper ads can be run each month.

62. Let x = number of rock songs.

Let y = number of dance songs.

Let z = number of pop songs.

$$\begin{cases} x + y + z = 32 \\ x - 2z = 0 \\ y - z = -4 \end{cases}$$

$$\begin{cases} x + y + z = 32 \\ -y - 3z = -32 \quad (-1)\text{Eq.1} + \text{Eq.2} \\ y - z = -4 \end{cases}$$

$$\begin{cases} x + y + z = 32 \\ -y - 3z = -32 \\ -4z = -36 \quad \text{Eq.2} + \text{Eq.3} \end{cases}$$

$-4z = -36 \implies z = 9$

$-y - 3(9) = -32 \implies y = 5$

$x + 5 + 9 = 32 \implies x = 18$

Play 18 rock songs, 5 dance songs, and 9 pop songs.

63. (a) To use 2 liters of the 50% solution:

Let x = amount of 10% solution.

Let y = amount of 20% solution.

$x + y = 8 \implies y = 8 - x$

$x(0.10) + y(0.20) + 2(0.50) = 10(0.25)$

$0.10x + 0.20(8 - x) + 1 = 2.5$

$0.10x + 1.6 - 0.20x + 1 = 2.5$

$-0.10x = -0.1$

$x = 1$ liter of 10% solution

$y = 7$ liters of 20% solution

Given: 2 liters of 50% solution

(b) To use as little of the 50% solution as possible, the chemist should use no 10% solution.

Let x = amount of 20% solution.

Let y = amount of 50% solution.

$x + y = 10 \implies y = 10 - x$

$x(0.20) + y(0.50) = 10(0.25)$

$x(0.20) + (10 - x)(0.50) = 10(0.25)$

$x(0.20) + 5 - 0.50x = 2.5$

$-0.30x = -2.5$

$x = 8\frac{1}{3}$ liters of 20% solution

$y = 1\frac{2}{3}$ liters of 50% solution

(c) To use as much of the 50% solution as possible, the chemist should use no 20% solution.

Let x = amount of 10% solution.

Let y = amount of 50% solution.

$x + y = 10 \implies y = 10 - x$

$x(0.10) + y(0.50) = 10(0.25)$

$0.10x + 0.50(10 - x) = 2.5$

$0.10x + 5 - 0.50x = 2.5$

$-0.40x = -2.5$

$x = 6\frac{1}{4}$ liters of 10% solution

$y = 3\frac{3}{4}$ liters of 50% solution

64. Let x = amount of 10% solution.

Let y = amount of 15% solution.

Let z = amount of 25% solution.

$$\begin{cases} x + y + z = 12 \\ 0.10x + 0.15y + 0.25z = 0.20 \cdot 12 \end{cases}$$

$$\begin{cases} x + y + z = 12 \\ 2x + 3y + 5z = 48 \end{cases} \quad 20\text{Eq.2}$$

(a) If $z = 4$,

$$\begin{cases} x + y + 4 = 12 \\ 2x + 3y + 20 = 48 \end{cases}$$

$$\begin{cases} x + y = 8 \\ 2x + 3y = 28 \end{cases}$$

$$\begin{cases} x + y = 8 \\ y = 12 \end{cases} \quad \text{Eq.2} + (-2)\text{Eq.1}$$

$y = 12 \implies x = 8 - 12 = -4$, but $x \geq 0$.

There is no solution; 4 gallons of the 25% solution is not enough.

(c) $$\begin{cases} x + y + z = 12 \\ 2x + 3y + 5z = 48 \end{cases}$$

$$\begin{cases} x + y + z = 12 \\ y + 3z = 24 \end{cases} \quad (-2)\text{Eq.1} + \text{Eq.2}$$

$y + 3z = 24 \implies z = 8 - \frac{1}{3}y \implies z$ is largest when $y = 0$.

$y = 0$ and $z = 8 \implies x = 12 - 0 - 8 = 4$.

The 12-gallon mixture made with the largest portion of the 25% solution contains 4 gallons of the 10% solution, none of the 15% solution, and 8 gallons of the 25% solution.

(b) $$\begin{cases} x + y + z = 12 \\ 2x + 3y + 5z = 48 \end{cases}$$

Minimize z while $x \geq 0$, $y \geq 0$, and $z \geq 0$.

$$\begin{cases} x + y + z = 12 \\ -x + 2z = 12 \end{cases} \quad (-3)\text{Eq.1} + \text{Eq.2}$$

$-x + 2z = 12 \implies z = 6 + \frac{1}{2}x \implies z$ is smallest when $x = 0$.

$x = 0$ and $z = 6 \implies y = 6$

The 12-gallon mixture using the least amount of the 25% solution is made using none of the 10% solution and 6 gallons each of the 15% and 25% solution.

65. $$\begin{cases} I_1 - I_2 + I_3 = 0 & \text{Equation 1} \\ 3I_1 + 2I_2 = 7 & \text{Equation 2} \\ 2I_2 + 4I_3 = 8 & \text{Equation 3} \end{cases}$$

$$\begin{cases} I_1 - I_2 + I_3 = 0 \\ 5I_2 - 3I_3 = 7 \\ 2I_2 + 4I_3 = 8 \end{cases} \quad (-3)\text{Eq.1} + \text{Eq.2}$$

$$\begin{cases} I_1 - I_2 + I_3 = 0 \\ 10I_2 - 6I_3 = 14 \\ 10I_2 + 20I_3 = 40 \end{cases} \quad \begin{matrix} \\ 2\text{Eq.2} \\ 5\text{Eq.3} \end{matrix}$$

$$\begin{cases} I_1 - I_2 + I_3 = 0 \\ 10I_2 - 6I_3 = 14 \\ 26I_3 = 26 \end{cases} \quad (-1)\text{Eq.2} + \text{Eq.3}$$

$26I_3 = 26 \implies I_3 = 1$

$10I_2 - 6(1) = 14 \implies I_2 = 2$

$I_1 - 2 + 1 = 0 \implies I_1 = 1$

Solution: $I_1 = 1$, $I_2 = 2$, $I_3 = 1$

66. (a) $\begin{cases} t_1 - 2t_2 = 0 \\ t_1 \quad - 2a = 128 \\ t_2 + a = 32 \end{cases} \Rightarrow \begin{array}{r} 2t_2 - 2a = 128 \\ -2t_2 - 2a = -64 \end{array}$

$$-4a = 64$$
$$a = -16$$
$$t_2 = 48$$
$$t_1 = 96$$

So, $t_1 = 96$ pounds

$t_2 = 48$ pounds

$a = -16$ feet per second squared.

(b) $\begin{cases} t_1 - 2t_2 = 0 & \text{Equation 1} \\ t_1 \quad - 2a = 128 & \text{Equation 2} \\ t_2 + 2a = 64 & \text{Equation 3} \end{cases}$

$\begin{cases} t_1 - 2t_2 = 0 \\ \quad 2t_2 - 2a = 128 \quad (-1)\text{Eq.1} + \text{Eq.2} \\ \quad t_2 + 2a = 64 \end{cases}$

$\begin{cases} t_1 - 2t_2 = 0 \\ \quad 2t_2 - 2a = 128 \\ \quad 3a = 0 \quad (-\frac{1}{2})\text{Eq.2} + \text{Eq.3} \end{cases}$

$3a = 0 \Rightarrow a = 0$

$2t_2 - 2(0) = 128 \Rightarrow t_2 = 64$

$t_1 - 2(64) = 0 \Rightarrow t_1 = 128$

Solution: $a = 0$ ft/sec^2

$t_1 = 128$ lb

$t_2 = 64$ lb

The system is stable.

67. $(-4, 5), (-2, 6), (2, 6), (4, 2)$

$$n = 4, \sum_{i=1}^{4} x_i = 0, \sum_{i=1}^{4} x_i^2 = 40, \sum_{i=1}^{4} x_i^3 = 0, \sum_{i=1}^{4} x_i^4 = 544, \sum_{i=1}^{4} y_i = 19, \sum_{i=1}^{4} x_i y_i = -12, \sum_{i=1}^{4} x_i^2 y_i = 160$$

$\begin{cases} 4c \quad + 40a = 19 \\ \quad 40b \quad = -12 \\ 40c \quad + 544a = 160 \end{cases}$

$\begin{cases} 4c \quad + 40a = 19 \\ \quad 40b \quad = -12 \\ \quad 144a = -30 \quad -10\text{Eq.1} + \text{Eq.3} \end{cases}$

$144a = -30 \Rightarrow a = -\frac{5}{24}$

$40b = -12 \Rightarrow b = -\frac{3}{10}$

$4c + 40\left(-\frac{5}{24}\right) = 19 \Rightarrow c = \frac{41}{6}$

Least squares regression parabola: $y = -\frac{5}{24}x^2 - \frac{3}{10}x + \frac{41}{6}$

68. $\begin{cases} 5c \quad + 10a = 8 \\ \quad 10b \quad = 12 \\ 10c \quad + 34a = 22 \end{cases}$

$\begin{cases} 5c \quad + 10a = 8 \\ \quad 10b \quad = 12 \\ \quad 14a = 6 \quad (-2)\text{Eq.1} + \text{Eq.3} \end{cases}$

$14a = 6 \Rightarrow a = \frac{3}{7}$

$10b = 12 \Rightarrow b = \frac{6}{5}$

$5c + 10\left(\frac{3}{7}\right) = 8 \Rightarrow c = \frac{26}{35}$

Least squares regression parabola: $y = \frac{3}{7}x^2 + \frac{6}{5}x + \frac{26}{35}$

69. $(0, 0), (2, 2), (3, 6), (4, 12)$

$$n = 4, \sum_{i=1}^{4} x_i = 9, \sum_{i=1}^{4} x_i^2 = 29, \sum_{i=1}^{4} x_i^3 = 99, \sum_{i=1}^{4} x_i^4 = 353, \sum_{i=1}^{4} y_i = 20, \sum_{i=1}^{4} x_i y_i = 70, \sum_{i=1}^{4} x_i^2 y_i = 254$$

$$\begin{cases} 4c + 9b + 29a = 20 \\ 9c + 29b + 99a = 70 \\ 29c + 99b + 353a = 254 \end{cases}$$

$$\begin{cases} 9c + 29b + 99a = 70 \\ 4c + 9b + 29a = 20 \\ 29c + 99b + 353a = 254 \end{cases} \quad \text{Interchange equations.}$$

$$\begin{cases} c + 11b + 41a = 30 & -2\text{Eq.2} + \text{Eq.1} \\ -35b - 135a = -100 & -4\text{Eq.1} + \text{Eq.2} \\ -220b - 836a = -616 & -29\text{Eq.1} + \text{Eq.3} \end{cases}$$

$$\begin{cases} c + 11b + 41a = 30 \\ 1540b + 5940a = 4400 & -44\text{Eq.2} \\ -1540b - 5852a = -4312 & 7\text{Eq.3} \end{cases}$$

$$\begin{cases} c + 11b + 41a = 30 \\ 1540b + 5940a = 4400 \\ 88a = 88 & \text{Eq.2} + \text{Eq.3} \end{cases}$$

$$88a = 88 \implies a = 1$$

$$1540b + 5940(1) = 4400 \implies b = -1$$

$$c + 11(-1) + 41(1) = 30 \implies c = 0$$

Least squares regression parabola: $y = x^2 - x$

70. $\begin{cases} 4c + 6b + 14a = 25 \\ 6c + 14b + 36a = 21 \\ 14c + 36b + 98a = 33 \end{cases}$

$$\begin{cases} 4c + 6b + 14a = 25 \\ -10b - 30a = 33 & 3\text{Eq.1} - 2\text{Eq.2} \\ -60b - 196a = 218 & 14\text{Eq.1} - 4\text{Eq.3} \end{cases}$$

$$\begin{cases} 4c + 6b + 14a = 25 \\ -10b - 30a = 33 \\ -16a = 20 & (-6)\text{Eq.2} + \text{Eq.3} \end{cases}$$

$$-16a = 20 \implies a = -\frac{5}{4}$$

$$-10b - 30\left(-\frac{5}{4}\right) = 33 \implies b = \frac{9}{20}$$

$$4c + 6\left(\frac{9}{20}\right) + 14\left(-\frac{5}{4}\right) = 25 \implies c = \frac{199}{20}$$

Least squares regression parabola: $y = -\frac{5}{4}x^2 + \frac{9}{20}x + \frac{199}{20}$

71. (a) (100, 75), (120, 68), (140, 55)

$$n = 3, \sum_{i=1}^{3} x_i = 360, \sum_{i=1}^{3} x_i^2 = 44{,}000, \sum_{i=1}^{3} x_i^3 = 5{,}472{,}000$$

$$\sum_{i=1}^{3} x_i^4 = 691{,}520{,}000, \sum_{i=1}^{3} y_i = 198, \sum_{i=1}^{3} x_i y_i = 23{,}360,$$

$$\sum_{i=1}^{3} x_i^2 y_i = 2{,}807{,}200$$

$$
\begin{aligned}
3c + \quad 360b + \quad\quad 44{,}000a &= \quad\quad 198 \\
360c + \quad 44{,}000b + \quad 5{,}472{,}000a &= \quad 23{,}360 \\
44{,}000c + 5{,}472{,}000b + 691{,}520{,}000a &= 2{,}807{,}200
\end{aligned}
$$

Solving this system yields $a = -0.0075$, $b = 1.3$ and $c = 20$.

Least squares regression parabola:
$y = -0.0075x^2 + 1.3x + 20$

(b)

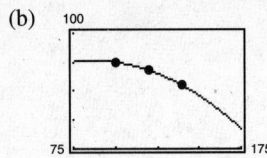

(c)

x	Actual Percent y	Model Approximation y
100	75	75
120	68	68
140	55	55

The model is a good fit to the actual data.
The values are the same.

(d) For $x = 170$:

$$y = -0.0075(170)^2 + 1.3(170) + 20$$

$$= 24.25\%$$

(e) For $y = 40$:

$$40 = -0.0075x^2 + 1.3x + 20$$

$$0.0075x^2 - 1.3x + 20 = 0$$

By the Quadratic Formula we have $x \approx 17$ or $x \approx 156$.

Choosing the value that fits with our data, we have 156 females.

72. (30, 55), (40, 105), (50, 188)

(a)
$$
\begin{cases}
3c + \quad 120b + \quad\quad 5000a = \quad\quad 348 \\
120c + \quad 5000b + \quad 216{,}000a = \quad 15{,}250 \\
5000c + 216{,}000b + 9{,}620{,}000a = 687{,}500
\end{cases}
$$

$$
\begin{cases}
3c + \quad 120b + \quad\quad 5000a = \quad\quad 348 \\
\quad\quad 200b + \quad\quad 16{,}000a = \quad\quad 1330 \quad (-40)\text{Eq.1} + \text{Eq.2} \\
\quad\quad 48{,}000b + 3{,}860{,}000a = 322{,}500 \quad (-5000)\text{Eq.1} + (3)\text{Eq.3}
\end{cases}
$$

$$
\begin{cases}
3c + 120b + \quad 5000a = \quad 348 \\
\quad\quad 200b + 16{,}000a = 1330 \\
\quad\quad\quad\quad 20{,}000a = 3300 \quad (-240)\text{Eq.2} + \text{Eq.3}
\end{cases}
$$

$$20{,}000a = 3300 \implies a = 0.165$$

$$200b + 16{,}000(0.165) = 1330 \implies b = -6.55$$

$$3c + 120(-6.55) + 5000(0.165) = 348 \implies c = 103$$

Least-squares regression parabola: $y = 0.165x^2 - 6.55x + 103$

(b)

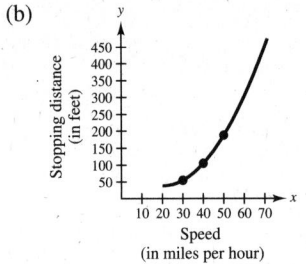

(c) When $x = 70$, $y = 453$ feet.

73. Let x = number of touchdowns.

Let y = number of extra-point kicks.

Let z = number of two-point conversions.

Let w = number of field goals.

$$\begin{cases} x + y + z + w = 16 \\ 6x + y + 2z + 3w = 32 + 29 \\ x \qquad\quad - 4w = 0 \qquad \Longrightarrow x = 4w \\ \qquad\quad 2z - w = 0 \qquad \Longrightarrow z = \tfrac{1}{2}w \end{cases}$$

$$\begin{cases} 4w + y + \tfrac{1}{2}w + w = 16 \Longrightarrow 5.5w + y = 16 \\ 6(4w) + y + 2(\tfrac{1}{2})w + 3w = 61 \Longrightarrow 28w + y = 61 \end{cases}$$

$$28w + y = 61$$
$$\underline{-5.5w - y = -16}$$
$$22.5w \quad = 45$$
$$w = 2$$
$$y = 5$$
$$x = 4w = 8$$
$$z = \tfrac{1}{2}w = 1$$

Thus, 8 touchdowns, 5 extra-point kicks, 1 two-point conversion, and 2 field goals were scored.

74. Let t = number of touchdowns.

Let x = number of extra-points.

Let f = number of field goals.

Let s = number of safeties.

$$\begin{cases} t + x + f + s = 22 \\ 6t + x + 3f + 2s = 74 \\ t - x \qquad\qquad = 0 \\ \qquad\quad f - 3s = 0 \end{cases}$$

$$\begin{cases} 2t \qquad + f + s = 22 & \text{Eq.1 + Eq.3} \\ 7t \qquad + 3f + 2s = 74 & \text{Eq.2 + Eq.3} \\ t - x \qquad\qquad = 0 \\ \qquad\quad f - 3s = 0 \end{cases}$$

$$\begin{cases} 2t \qquad\quad + 4s = 22 & \text{Eq.1 − Eq.4} \\ 7t \qquad + 3f + 2s = 74 \\ t - x \qquad\qquad = 0 \\ \qquad\quad f - 3s = 0 \end{cases}$$

$$\begin{cases} 2t \qquad\quad + 4s = 22 & \text{Eq.2 − (3)Eq.4} \\ 7t \qquad\quad + 11s = 74 \\ t - x \qquad\qquad = 0 \\ \qquad\quad f - 3s = 0 \end{cases}$$

$$\begin{cases} t \qquad\quad + 2s = 11 & (\tfrac{1}{2})\text{Eq.1} \\ 7t \qquad\quad + 11s = 74 \\ t - x \qquad\qquad = 0 \\ \qquad\quad f - 3s = 0 \end{cases}$$

$$\begin{cases} t \qquad\quad + 2s = 11 \\ \qquad\qquad - 3s = -3 & (-7)\text{Eq.1 + Eq.2} \\ t - x \qquad\qquad = 0 \\ \qquad\quad f - 3s = 0 \end{cases}$$

$$-3s = -3 \Longrightarrow s = 1$$

$$t + 2(1) = 11 \Longrightarrow t = 9$$

$$9 - x = 0 \Longrightarrow x = 9$$

$$f - 3(1) = 0 \Longrightarrow f = 3$$

There were 9 touchdowns, each with an extra point; and there were 3 field goals and 1 safety.

75.
$$\begin{cases} \quad y + \lambda = 0 \\ x \quad + \lambda = 0 \end{cases} \Longrightarrow x = y = -\lambda$$
$$\begin{cases} x + y - 10 = 0 \Longrightarrow 2x - 10 = 0 \end{cases}$$
$$x = 5$$
$$y = 5$$
$$\lambda = -5$$

76.
$$\begin{cases} 2x + \lambda = 0 \\ 2y + \lambda = 0 \end{cases} \quad x = y = -\frac{\lambda}{2}$$
$$\begin{cases} x + y - 4 = 0 \Longrightarrow 2x - 4 = 0 \end{cases}$$
$$2x = 4$$
$$x = 2$$
$$y = 2$$
$$\lambda = -4$$

77. $\begin{cases} 2x - 2x\lambda = 0 \implies 2x(1 - \lambda) = 0 \implies \lambda = 1 \text{ or } x = 0 \\ -2y + \quad \lambda = 0 \\ \quad y - \quad x^2 = 0 \end{cases}$

If $\lambda = 1$:

$2y = \lambda \implies y = \dfrac{1}{2}$

$x^2 = y \implies x = \pm\sqrt{\dfrac{1}{2}} = \pm\dfrac{\sqrt{2}}{2}$

If $x = 0$:

$x^2 = y \implies y = 0$

$2y = \lambda \implies \lambda = 0$

Solution: $x = \pm\dfrac{\sqrt{2}}{2}$ or $x = 0$

$\qquad\qquad y = \dfrac{1}{2} \qquad\qquad y = 0$

$\qquad\qquad \lambda = 1 \qquad\qquad \lambda = 0$

78. $\begin{cases} 2 + 2y + 2\lambda = 0 \\ 2x + 1 + \quad \lambda = 0 \implies \lambda = -2x - 1 \\ 2x + y - 100 = 0 \end{cases}$

$2 + 2y + 2(-2x - 1) = 0 \implies -4x + 2y = 0 \implies -4x + 2y = \quad 0$

$2x + y - 100 = 0 \implies 2x + y = 100 \implies \qquad\qquad \underline{4x + 2y = 200}$

$\qquad\qquad\qquad\qquad\qquad\qquad\qquad\qquad\qquad\qquad 4y = 200$

$\qquad\qquad\qquad\qquad\qquad\qquad\qquad\qquad\qquad\quad y = \quad 50$

$\qquad\qquad\qquad\qquad\qquad\qquad\qquad\qquad\qquad\quad x = \quad 25$

$\lambda = -2(25) - 1 = -51$

79. False. Equation 2 does not have a leading coefficient of 1.

80. True. If a system of three linear equations is inconsistent, then it has no points in common to all three equations.

81. No, they are not equivalent. There are two arithmetic errors. The constant in the second equation should be -11 and the coefficient of z in the third equation should be 2.

82. When using Gaussian elimination to solve a system of linear equations, a system has no solution when there is a row representing a contradictory equation such as $0 = N$, where N is a nonzero real number.

For instance: $\begin{array}{ll} x + y = 3 & \text{Equation 1} \\ -x - y = 3 & \text{Equation 2} \end{array}$

$\begin{array}{ll} x + y = 0 & \\ \quad\ 0 = 6 & \text{Eq.1 + Eq.2} \end{array}$

No solution

83. There are an infinite number of linear systems that have $(4, -1, 2)$ as their solution. Two such systems are as follows:

$\begin{cases} 3x + \ y - \ z = 9 \\ \ x + 2y - \ z = 0 \\ -x + \ y + 3z = 1 \end{cases}$ $\begin{cases} x + \ y + \ z = 5 \\ x \qquad\quad - 2z = 0 \\ \quad\ 2y + \ z = 0 \end{cases}$

84. There are an infinite number of linear systems that have $(-5, -2, 1)$ as their solution. Two systems are:

$\begin{cases} \ x + \ y + \ z = -6 \\ -2x - \ y + 3z = \ 15 \\ \ x + 4y - \ z = -14 \end{cases}$ $\begin{cases} 2x - \ y - \ z = -9 \\ -x + 2y + 2z = \ 3 \\ -3x + \ y - 2z = \ 11 \end{cases}$

85. There are an infinite number of linear systems that have $\left(3, -\frac{1}{2}, \frac{7}{4}\right)$ as their solution. Two such systems are as follows:

$$\begin{cases} x + 2y - 4z = -5 \\ -x - 4y + 8z = 13 \\ x + 6y + 4z = 7 \end{cases} \qquad \begin{cases} x + 2y + 4z = 9 \\ y + 2z = 3 \\ x \qquad - 4z = -4 \end{cases}$$

86. There are an infinite number of linear systems that have $\left(-\frac{3}{2}, 4, -7\right)$ as their solution. Two systems are:

$$\begin{cases} 2x - y + 3z = -28 \\ -6x + 4y + z = 18 \\ -4x - 2y - 3z = 19 \end{cases} \qquad \begin{cases} 4x + y - 2z = 12 \\ 4y + 2z = 2 \\ -2x + y + z = 0 \end{cases}$$

87. $(0.075)(85) = 6.375$

88. $225 = \dfrac{x}{100}(150)$

$225 = 1.5x$

$150\% = x$

89. $(0.005)n = 400$

$n = 80{,}000$

90. $(0.48)n = 132$

$n = 275$

91. $(7 - i) + (4 + 2i) = (7 + 4) + (-i + 2i) = 11 + i$

92. $(-6 + 3i) - (1 + 6i) = (-6 - 1) + (3 - 6)i$

$= -7 - 3i$

93. $(4 - i)(5 + 2i) = 20 + 8i - 5i - 2i^2 = 20 + 3i + 2$

$= 22 + 3i$

94. $(1 + 2i)(3 - 4i) = 3 - 4i + 6i - 8i^2$

$= 3 + 2i - 8(-1) = 11 + 2i$

95. $\dfrac{i}{1 + i} + \dfrac{6}{1 - i} = \dfrac{i(1 - i) + 6(1 + i)}{(1 + i)(1 - i)}$

$= \dfrac{i - i^2 + 6 + 6i}{1 - i^2}$

$= \dfrac{7 + 7i}{2}$

$= \dfrac{7}{2} + \dfrac{7}{2}i$

96. $\dfrac{i}{4 + i} - \dfrac{2i}{8 - 3i} = \dfrac{i}{4 + i}\left(\dfrac{4 - i}{4 - i}\right) - \dfrac{2i}{8 - 3i}\left(\dfrac{8 + 3i}{8 + 3i}\right)$

$= \dfrac{1 + 4i}{17} - \dfrac{-6 + 16i}{73}$

$= \dfrac{73(1 + 4i) - 17(-6 + 16i)}{17(73)}$

$= \dfrac{73 + 292i + 102 - 272i}{1241}$

$= \dfrac{175}{1241} + \dfrac{20}{1241}i$

97. $f(x) = x^3 + x^2 - 12x$

(a) $x^3 + x^2 - 12x = 0$ (b)

$x(x^2 + x - 12) = 0$

$x(x + 4)(x - 3) = 0$

Zeros: $x = -4, 0, 3$

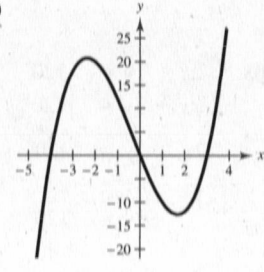

98. $f(x) = -8x^4 + 32x^2$

(a) $-8x^4 + 32x^2 = 0$ (b)

$-8x^2(x^2 - 4) = 0$

Zeros: $x = 0, \pm 2$

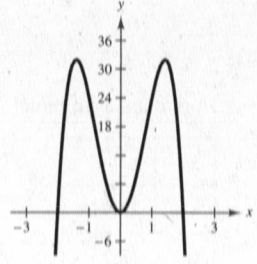

99. $f(x) = 2x^3 + 5x^2 - 21x - 36$

(a) $2x^3 + 5x^2 - 21x - 36 = 0$

$\begin{array}{r|rrrr} 3 & 2 & 5 & -21 & -36 \\ & & 6 & 33 & 36 \\ \hline & 2 & 11 & 12 & 0 \end{array}$

$f(x) = (x - 3)(2x^2 + 11x + 12)$

$= (x - 3)(x + 4)(2x + 3)$

Zeros: $x = -4, -\frac{3}{2}, 3$

(b)

100. $f(x) = 6x^3 - 29x^2 - 6x + 5$

(a) $6x^3 - 29x^2 - 6x + 5 = 0$

$$
\begin{array}{r|rrrr}
5 & 6 & -29 & -6 & 5 \\
  &   & 30 & 5 & -5 \\
\hline
  & 6 & 1   & -1 & 0
\end{array}
$$

$f(x) = (x - 5)(6x^2 + x - 1)$

$\quad\quad = (x - 5)(3x - 1)(2x + 1)$

Zeros: $x = 5, \frac{1}{3}, -\frac{1}{2}$

(b)
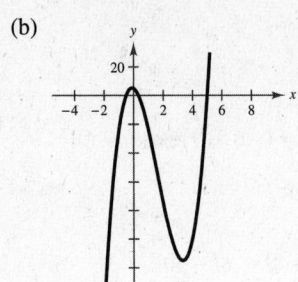

101. $y = 4^{x-4} - 5$

x	-2	0	2	4	5
y	-4.9998	-4.996	-4.938	-4	-1

Horizontal asymptote: $y = -5$

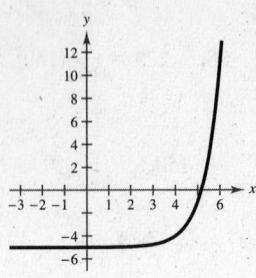

102. $y = \left(\frac{5}{2}\right)^{-x+1} - 4$

Horizontal asymptote: $y = -4$

x	y
-2	11.625
-1	2.25
0	-1.5
1	-3
2	-3.6

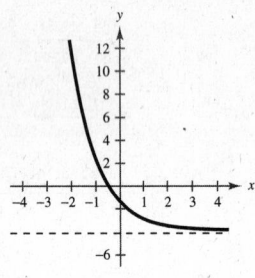

103. $y = 1.9^{-0.8x} + 3$

x	-2	-1	0	1	2
y	5.793	4.671	4	3.598	3.358

Horizontal asymptote: $y = 3$

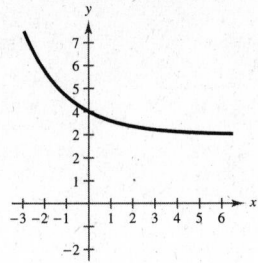

104. $y = 3.5^{-x+2} + 6$

Horizontal asymptote: $y = 6$

x	y
$-\frac{1}{2}$	28.918
0	18.25
$\frac{1}{2}$	12.548
1	9.5
2	7

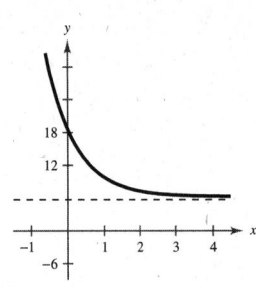

105. $\begin{cases} 2x + y = 120 & \text{Equation 1} \\ x + 2y = 120 & \text{Equation 2} \end{cases}$

$$
\begin{array}{r}
2x + y = 120 \\
-2x - 4y = -240 \quad (-2)\text{Eq.2} \\
\hline
-3y = -120 \\
y = 40
\end{array}
$$

$x + 2(40) = 120 \implies x = 40$

Solution: $(40, 40)$

106. $\begin{cases} 6x - 5y = 3 & \text{Equation 1} \\ 10x - 12y = 5 & \text{Equation 2} \end{cases}$

$\begin{cases} 72x - 60y = 36 & 12\text{Eq.1} \\ -50x + 60y = -25 & (-5)\text{Eq.2} \end{cases}$

$22x = 11$

$x = \tfrac{1}{2}$

$6\left(\tfrac{1}{2}\right) - 5y = 3 \implies y = 0$

Solution: $\left(\tfrac{1}{2}, 0\right)$

107. Answers will vary.

Section 7.4   Partial Fractions

■ You should know how to decompose a rational function $\dfrac{N(x)}{D(x)}$ into partial fractions.

(a) If the fraction is improper, divide to obtain

$$\frac{N(x)}{D(x)} = p(x) + \frac{N_1(x)}{D(x)}$$

where $p(x)$ is a polynomial.

(b) Factor the denominator completely into linear and irreducible quadratic factors.

(c) For each factor of the form $(px + q)^m$, the partial fraction decomposition includes the terms

$$\frac{A_1}{(px + q)} + \frac{A_2}{(px + q)^2} + \cdots + \frac{A_m}{(px + q)^m}.$$

(d) For each factor of the form $(ax^2 + bx + c)^n$, the partial fraction decomposition includes the terms

$$\frac{B_1x + C_1}{ax^2 + bx + c} + \frac{B_2x + C_2}{(ax^2 + bx + c)^2} + \cdots + \frac{B_nx + C_n}{(ax^2 + bx + c)^n}.$$

■ You should know how to determine the values of the constants in the numerators.

(a) Set $\dfrac{N_1(x)}{D(x)}$ = partial fraction decomposition.

(b) Multiply both sides by $D(x)$ to obtain the basic equation.

(c) For distinct linear factors, substitute the zeros of the distinct linear factors into the basic equation.

(d) For repeated linear factors, use the coefficients found in part (c) to rewrite the basic equation. Then use other values of x to solve for the remaining coefficients.

(e) For quadratic factors, expand the basic equation, collect like terms, and then equate the coefficients of like terms.

Vocabulary Check

1. partial fraction decomposition

2. improper

3. m; n; irreducible

4. basic equation

1. $\dfrac{3x - 1}{x(x - 4)} = \dfrac{A}{x} + \dfrac{B}{x - 4}$

Matches (b).

2. $\dfrac{3x - 1}{x^2(x - 4)} = \dfrac{A}{x} + \dfrac{B}{x^2} + \dfrac{C}{x - 4}$

Matches (c).

3. $\dfrac{3x - 1}{x(x^2 + 4)} = \dfrac{A}{x} + \dfrac{Bx + C}{x^2 + 4}$

Matches (d).

4. $\dfrac{3x-1}{x(x^2-4)} = \dfrac{3x-1}{x(x-2)(x+2)} = \dfrac{A}{x} + \dfrac{B}{x-2} + \dfrac{C}{x+2}$

Matches (a).

5. $\dfrac{7}{x^2-14x} = \dfrac{7}{x(x-14)} = \dfrac{A}{x} + \dfrac{B}{x-14}$

6. $\dfrac{x-2}{x^2+4x+3} = \dfrac{x-2}{(x+3)(x+1)} = \dfrac{A}{x+3} + \dfrac{B}{x+1}$

7. $\dfrac{12}{x^3-10x^2} = \dfrac{12}{x^2(x-10)} = \dfrac{A}{x} + \dfrac{B}{x^2} + \dfrac{C}{x-10}$

8. $\dfrac{x^2-3x+2}{4x^3+11x^2} = \dfrac{x^2-3x+2}{x^2(4x+11)} = \dfrac{A}{x} + \dfrac{B}{x^2} + \dfrac{C}{4x+11}$

9. $\dfrac{4x^2+3}{(x-5)^3} = \dfrac{A}{x-5} + \dfrac{B}{(x-5)^2} + \dfrac{C}{(x-5)^3}$

10. $\dfrac{6x+5}{(x+2)^4} = \dfrac{6x+5}{(x+2)(x+2)(x+2)(x+2)}$

$\qquad = \dfrac{A}{x+2} + \dfrac{B}{(x+2)^2} + \dfrac{C}{(x+2)^3} + \dfrac{D}{(x+2)^4}$

11. $\dfrac{2x-3}{x^3+10x} = \dfrac{2x-3}{x(x^2+10)} = \dfrac{A}{x} + \dfrac{Bx+C}{x^2+10}$

12. $\dfrac{x-6}{2x^3+8x} = \dfrac{x-6}{2x(x^2+4)} = \dfrac{A}{2x} + \dfrac{Bx+C}{x^2+4}$

13. $\dfrac{x-1}{x(x^2+1)^2} = \dfrac{A}{x} + \dfrac{Bx+C}{x^2+1} + \dfrac{Dx+E}{(x^2+1)^2}$

14. $\dfrac{x+4}{x^2(3x-1)^2} = \dfrac{A}{x} + \dfrac{B}{x^2} + \dfrac{C}{3x-1} + \dfrac{D}{(3x-1)^2}$

15. $\dfrac{1}{x^2-1} = \dfrac{A}{x+1} + \dfrac{B}{x-1}$

$1 = A(x-1) + B(x+1)$

Let $x = -1$: $1 = -2A \implies A = -\dfrac{1}{2}$

Let $x = 1$: $1 = 2B \implies B = \dfrac{1}{2}$

$\dfrac{1}{x^2-1} = \dfrac{1/2}{x-1} - \dfrac{1/2}{x+1} = \dfrac{1}{2}\left(\dfrac{1}{x-1} - \dfrac{1}{x+1}\right)$

16. $\dfrac{1}{4x^2-9} = \dfrac{A}{2x+3} + \dfrac{B}{2x-3}$

$1 = A(2x-3) + B(2x+3)$

Let $x = -\dfrac{3}{2}$: $1 = -6A \implies A = -\dfrac{1}{6}$

Let $x = \dfrac{3}{2}$: $1 = 6B \implies B = \dfrac{1}{6}$

$\dfrac{1}{4x^2-9} = \dfrac{1}{6}\left(\dfrac{1}{2x-3} - \dfrac{1}{2x+3}\right)$

17. $\dfrac{1}{x^2+x} = \dfrac{A}{x} + \dfrac{B}{x+1}$

$1 = A(x+1) + Bx$

Let $x = 0$: $1 = A$

Let $x = -1$: $1 = -B \implies B = -1$

$\dfrac{1}{x^2+x} = \dfrac{1}{x} - \dfrac{1}{x+1}$

18. $\dfrac{3}{x^2-3x} = \dfrac{A}{x-3} + \dfrac{B}{x}$

$3 = Ax + B(x-3)$

Let $x = 3$: $3 = 3A \implies A = 1$

Let $x = 0$: $3 = -3B \implies B = -1$

$\dfrac{3}{x^2-3x} = \dfrac{1}{x-3} - \dfrac{1}{x}$

19. $\dfrac{1}{2x^2+x} = \dfrac{A}{2x+1} + \dfrac{B}{x}$

$1 = Ax + B(2x+1)$

Let $x = -\dfrac{1}{2}$: $1 = -\dfrac{1}{2}A \implies A = -2$

Let $x = 0$: $1 = B$

$\dfrac{1}{2x^2+x} = \dfrac{1}{x} - \dfrac{2}{2x+1}$

20. $\dfrac{5}{x^2 + x - 6} = \dfrac{A}{x + 3} + \dfrac{B}{x - 2}$

$\qquad 5 = A(x - 2) + B(x + 3)$

Let $x = -3: 5 = -5A \implies A = -1$

Let $x = 2: 5 = 5B \implies B = 1$

$\dfrac{5}{x^2 + x - 6} = \dfrac{1}{x - 2} - \dfrac{1}{x + 3}$

21. $\dfrac{3}{x^2 + x - 2} = \dfrac{A}{x - 1} + \dfrac{B}{x + 2}$

$\qquad 3 = A(x + 2) + B(x - 1)$

Let $x = 1: 3 = 3A \implies A = 1$

Let $x = -2: 3 = -3B \implies B = -1$

$\dfrac{3}{x^2 + x - 2} = \dfrac{1}{x - 1} - \dfrac{1}{x + 2}$

22. $\dfrac{x + 1}{x^2 + 4x + 3} = \dfrac{x + 1}{(x + 3)(x + 1)} = \dfrac{1}{x + 3}, x \neq -1$

23. $\dfrac{x^2 + 12x + 12}{x^3 - 4x} = \dfrac{A}{x} + \dfrac{B}{x + 2} + \dfrac{C}{x - 2}$

$x^2 + 12x + 12 = A(x + 2)(x - 2) + Bx(x - 2) + Cx(x + 2)$

Let $x = 0: 12 = -4A \implies A = -3$

Let $x = -2: -8 = 8B \implies B = -1$

Let $x = 2: 40 = 8C \implies C = 5$

$\dfrac{x^2 + 12x + 12}{x^3 - 4x} = -\dfrac{3}{x} - \dfrac{1}{x + 2} + \dfrac{5}{x - 2}$

24. $\dfrac{x + 2}{x(x - 4)} = \dfrac{A}{x} + \dfrac{B}{x - 4}$

$\qquad x + 2 = A(x - 4) + Bx$

Let $x = 0: 2 = -4A \implies A = -\dfrac{1}{2}$

Let $x = 4: 6 = 4B \implies B = \dfrac{3}{2}$

$\dfrac{x + 2}{x(x - 4)} = \dfrac{1}{2}\left(\dfrac{3}{x - 4} - \dfrac{1}{x}\right)$

25. $\dfrac{4x^2 + 2x - 1}{x^2(x + 1)} = \dfrac{A}{x} + \dfrac{B}{x^2} + \dfrac{C}{x + 1}$

$4x^2 + 2x - 1 = Ax(x + 1) + B(x + 1) + Cx^2$

Let $x = 0: -1 = B$

Let $x = -1: 1 = C$

Let $x = 1: 5 = 2A + 2B + C$

$\qquad\qquad 5 = 2A - 2 + 1$

$\qquad\qquad 6 = 2A$

$\qquad\qquad 3 = A$

$\dfrac{4x^2 + 2x - 1}{x^2(x + 1)} = \dfrac{3}{x} - \dfrac{1}{x^2} + \dfrac{1}{x + 1}$

26. $\dfrac{2x - 3}{(x - 1)^2} = \dfrac{A}{x - 1} + \dfrac{B}{(x - 1)^2}$

$\qquad 2x - 3 = A(x - 1) + B$

Let $x = 1: -1 = B$

Let $x = 0: -3 = -A + B$

$\qquad\qquad -3 = -A - 1$

$\qquad\qquad 2 = A$

$\dfrac{2x - 3}{(x - 1)^2} = \dfrac{2}{x - 1} - \dfrac{1}{(x - 1)^2}$

27. $\dfrac{3x}{(x - 3)^2} = \dfrac{A}{x - 3} + \dfrac{B}{(x - 3)^2}$

$\qquad 3x = A(x - 3) + B$

Let $x = 3: 9 = B$

Let $x = 0: 0 = -3A + B$

$\qquad\qquad 0 = -3A + 9$

$\qquad\qquad 3 = A$

$\dfrac{3x}{(x - 3)^2} = \dfrac{3}{x - 3} + \dfrac{9}{(x - 3)^2}$

28. $\dfrac{6x^2 + 1}{x^2(x-1)^2} = \dfrac{A}{x} + \dfrac{B}{x^2} + \dfrac{C}{x-1} + \dfrac{D}{(x-1)^2}$

$6x^2 + 1 = Ax(x-1)^2 + B(x-1)^2 + Cx^2(x-1) + Dx^2$

Let $x = 0 : 1 = B$

Let $x = 1 : 7 = D$

Substitute B and D into the equation, expand the binomials, collect like terms, and equate the coefficients of like terms.

$-2x^2 + 2x = (A + C)x^3 + (-2A - C)x^2 + Ax$

$\qquad A = 2$

$-2A - C = -2 \Rightarrow C = -2$ or

$\qquad A + C = 0 \Rightarrow C = -2$

$\dfrac{6x^2 + 1}{x^2(x-1)^2} = \dfrac{2}{x} + \dfrac{1}{x^2} - \dfrac{2}{x-1} + \dfrac{7}{(x-1)^2}$

29. $\dfrac{x^2 - 1}{x(x^2 + 1)} = \dfrac{A}{x} + \dfrac{Bx + C}{x^2 + 1}$

$x^2 - 1 = A(x^2 + 1) + (Bx + C)x$

$\qquad = Ax^2 + A + Bx^2 + Cx$

$\qquad = (A + B)x^2 + Cx + A$

Equating coefficients of like terms gives

$1 = A + B, 0 = C,$ and $-1 = A.$

Therefore, $A = -1, B = 2,$ and $C = 0.$

$\dfrac{x^2 - 1}{x(x^2 + 1)} = -\dfrac{1}{x} + \dfrac{2x}{x^2 + 1}$

30. $\dfrac{x}{(x-1)(x^2 + x + 1)} = \dfrac{A}{x-1} + \dfrac{Bx + C}{x^2 + x + 1}$

$x = A(x^2 + x + 1) + (Bx + C)(x - 1)$

$\quad = Ax^2 + Ax + A + Bx^2 - Bx + Cx - C$

$\quad = (A + B)x^2 + (A - B + C)x + (A - C)$

Equating coefficients of like powers gives $0 = A + B$, $1 = A - B + C$, and $0 = A - C$. Substituting $-A$ for B and A for C in the second equation gives $1 = 3A$, so $A = \frac{1}{3}$, $B = -\frac{1}{3}$, and $C = \frac{1}{3}$.

$\dfrac{x}{(x-1)(x^2 + x + 1)} = \dfrac{1}{3}\left(\dfrac{1}{x-1} - \dfrac{x-1}{x^2 + x + 1}\right)$

31. $\dfrac{x}{x^3 - x^2 - 2x + 2} = \dfrac{x}{(x-1)(x^2 - 2)} = \dfrac{A}{x-1} + \dfrac{Bx + C}{x^2 - 2}$

$x = A(x^2 - 2) + (Bx + C)(x - 1)$

$\quad = Ax^2 - 2A + Bx^2 - Bx + Cx - C$

$\quad = (A + B)x^2 + (C - B)x - (2A + C)$

Equating coefficients of like terms gives $0 = A + B$, $1 = C - B$, and $0 = 2A + C$. Therefore, $A = -1$, $B = 1$, and $C = 2$.

$\dfrac{x}{x^3 - x^2 - 2x + 2} = \dfrac{-1}{x-1} + \dfrac{x+2}{x^2 - 2}$

32. $\dfrac{x + 6}{x^3 - 3x^2 - 4x + 12} = \dfrac{x + 6}{(x+2)(x-2)(x-3)} = \dfrac{A}{x+2} + \dfrac{B}{x-2} + \dfrac{C}{x-3}$

$x + 6 = A(x-2)(x-3) + B(x+2)(x-3) + C(x+2)(x-2)$

Let $x = 3: \quad 9 = 5C \quad \Rightarrow \quad \frac{9}{5} = C$

Let $x = -2: 4 = 20A \quad \Rightarrow \quad \frac{1}{5} = A$

Let $x = 2: \quad 8 = -4B \quad \Rightarrow \quad -2 = B$

$\dfrac{x + 6}{x^3 - 3x^2 - 4x + 12} = \dfrac{\frac{1}{5}}{x+2} + \dfrac{-2}{x-2} + \dfrac{\frac{9}{5}}{x-3} = \dfrac{1}{5}\left(\dfrac{1}{x+2} - \dfrac{10}{x-2} + \dfrac{9}{x-3}\right)$

33. $\dfrac{x^2}{x^4 - 2x^2 - 8} = \dfrac{x^2}{(x^2 - 4)(x^2 + 2)} = \dfrac{x^2}{(x + 2)(x - 2)(x^2 + 2)}$

$$= \frac{A}{x + 2} + \frac{B}{x - 2} + \frac{Cx + D}{x^2 + 2}$$

$$x^2 = A(x - 2)(x^2 + 2) + B(x + 2)(x^2 + 2) + (Cx + D)(x + 2)(x - 2)$$

$$= A(x^3 - 2x^2 + 2x - 4) + B(x^3 + 2x^2 + 2x + 4) + (Cx + D)(x^2 - 4)$$

$$= Ax^3 - 2Ax^2 + 2Ax - 4A + Bx^3 + 2Bx^2 + 2Bx + 4B + Cx^3 + Dx^2 - 4Cx - 4D$$

$$= (A + B + C)x^3 + (-2A + 2B + D)x^2 + (2A + 2B - 4C)x + (-4A + 4B - 4D)$$

Equating coefficients of like terms gives

$0 = A + B + C, 1 = -2A + 2B + D, 0 = 2A + 2B - 4C$, and $0 = -4A + 4B - 4D$.

Using the first and third equation, we have $A + B + C = 0$ and $A + B - 2C = 0$;
by subtraction, $C = 0$. Using the second and fourth equation, we have $-2A + 2B + D = 1$
and $-2A + 2B - 2D = 0$; by subtraction, $3D = 1$, so $D = \frac{1}{3}$. Substituting 0 for C and $\frac{1}{3}$ for
D in the first and second equations, we have

$A + B = 0$ and $-2A + 2B = \frac{2}{3}$, so $A = -\frac{1}{6}$ and $B = \frac{1}{6}$.

$$\frac{x^2}{x^4 - 2x^2 - 8} = \frac{-\frac{1}{6}}{x + 2} + \frac{\frac{1}{6}}{x - 2} + \frac{\frac{1}{3}}{x^2 + 2}$$

$$= \frac{1}{3(x^2 + 2)} - \frac{1}{6(x + 2)} + \frac{1}{6(x - 2)}$$

$$= \frac{1}{6}\left(\frac{2}{x^2 + 2} - \frac{1}{x + 2} + \frac{1}{x - 2}\right)$$

34. $\dfrac{2x^2 + x + 8}{(x^2 + 4)^2} = \dfrac{Ax + B}{x^2 + 4} + \dfrac{Cx + D}{(x^2 + 4)^2}$

$2x^2 + x + 8 = (Ax + B)(x^2 + 4) + Cx + D$

$2x^2 + x + 8 = Ax^3 + Bx^2 + (4A + C)x + (4B + D)$

Equating coefficients of like powers:

$0 = A$

$2 = B$

$1 = 4A + C \Rightarrow C = 1$

$8 = 4B + D \Rightarrow D = 0$

$$\frac{2x^2 + x + 8}{(x^2 + 4)^2} = \frac{2}{x^2 + 4} + \frac{x}{(x^2 + 4)^2}$$

35. $\dfrac{x}{16x^4 - 1} = \dfrac{x}{(4x^2 - 1)(4x^2 + 1)} = \dfrac{x}{(2x + 1)(2x - 1)(4x^2 + 1)}$

$$= \frac{A}{2x + 1} + \frac{B}{2x - 1} + \frac{Cx + D}{4x^2 + 1}$$

$$x = A(2x - 1)(4x^2 + 1) + B(2x + 1)(4x^2 + 1) + (Cx + D)(2x + 1)(2x - 1)$$

$$= A(8x^3 - 4x^2 + 2x - 1) + B(8x^3 + 4x^2 + 2x + 1) + (Cx + D)(4x^2 - 1)$$

$$= 8Ax^3 - 4Ax^2 + 2Ax - A + 8Bx^3 + 4Bx^2 + 2Bx + B + 4Cx^3 + 4Dx^2 - Cx - D$$

$$= (8A + 8B + 4C)x^3 + (-4A + 4B + 4D)x^2 + (2A + 2B - C)x + (-A + B - D)$$

—CONTINUED—

35. **—CONTINUED—**

Equating coefficients of like terms gives $0 = 8A + 8B + 4C$, $0 = -4A + 4B + 4D$, $1 = 2A + 2B - C$, and $0 = -A + B - D$.

Using the first and third equations, we have $2A + 2B + C = 0$ and $2A + 2B - C = 1$; by subtraction, $2C = -1$, so $C = -\frac{1}{2}$.

Using the second and fourth equations, we have $-A + B + D = 0$ and $-A + B - D = 0$; by subtraction $2D = 0$, so $D = 0$.

Substituting $-\frac{1}{2}$ for C and 0 for D in the first and second equations, we have $8A + 8B = 2$ and $-4A + 4B = 0$, so $A = \frac{1}{8}$ and $B = \frac{1}{8}$.

$$\frac{x}{16x^4 - 1} = \frac{\frac{1}{8}}{2x + 1} + \frac{\frac{1}{8}}{2x - 1} + \frac{\left(-\frac{1}{2}\right)x}{4x^2 + 1}$$

$$= \frac{1}{8(2x + 1)} + \frac{1}{8(2x - 1)} - \frac{x}{2(4x^2 + 1)}$$

$$= \frac{1}{8}\left(\frac{1}{2x + 1} + \frac{1}{2x - 1} - \frac{4x}{4x^2 + 1}\right)$$

36. $\dfrac{x + 1}{x^3 + x} = \dfrac{A}{x} + \dfrac{Bx + C}{x^2 + 1}$

$$= (A + B)x^2 + Cx + A$$

Equating coefficients of like powers gives $0 = A + B$, $1 = C$, and $1 = A$.
Therefore, $A = 1$, $B = -1$, and $C = 1$.

$$\frac{x + 1}{x^3 + x} = \frac{1}{x} - \frac{x - 1}{x^2 + 1}$$

37. $\dfrac{x^2 + 5}{(x + 1)(x^2 - 2x + 3)} = \dfrac{A}{x + 1} + \dfrac{Bx + C}{x^2 - 2x + 3}$

$$x^2 + 5 = A(x^2 - 2x + 3) + (Bx + C)(x + 1)$$

$$= Ax^2 - 2Ax + 3A + Bx^2 + Bx + Cx + C$$

$$= (A + B)x^2 + (-2A + B + C)x + (3A + C)$$

Equating coefficients of like terms gives $1 = A + B$, $0 = -2A + B + C$, and $5 = 3A + C$.

Subtracting both sides of the second equation from the first gives $1 = 3A - C$; combining this with the third equation gives $A = 1$ and $C = 2$. Since $A + B = 1$, we also have $B = 0$.

$$\frac{x^2 + 5}{(x + 1)(x^2 - 2x + 3)} = \frac{1}{x + 1} + \frac{2}{x^2 - 2x + 3}$$

38. $\dfrac{x^2 - 4x + 7}{(x + 1)(x^2 - 2x + 3)} = \dfrac{A}{x + 1} + \dfrac{Bx + C}{x^2 - 2x + 3}$

$$x^2 - 4x + 7 = A(x^2 - 2x + 3) + (Bx + C)(x + 1)$$

$$= Ax^2 - 2Ax + 3A + Bx^2 + Bx + Cx + C$$

$$= (A + B)x^2 + (-2A + B + C)x + (3A + C)$$

Equating coefficients of like terms gives $1 = A + B$, $-4 = -2A + B + C$, and $7 = 3A + C$.
Adding the second and third equations, and subtracting the first, gives $2 = 2C$, so $C = 1$. Therefore, $A = 2$, $B = -1$, and $C = 1$.

$$\frac{x^2 - 4x + 7}{(x + 1)(x^2 - 2x + 3)} = \frac{2}{x + 1} - \frac{x - 1}{x^2 - 2x + 3}$$

39. $\dfrac{x^2 - x}{x^2 + x + 1} = 1 + \dfrac{-2x - 1}{x^2 + x + 1} = 1 - \dfrac{2x + 1}{x^2 + x + 1}$

40. $\dfrac{x^2 - 4x}{x^2 + x + 6}$

Using long division gives $\dfrac{x^2 - 4x}{x^2 + x + 6} = 1 - \dfrac{5x + 6}{x^2 + x + 6}$.

41. $\dfrac{2x^3 - x^2 + x + 5}{x^2 + 3x + 2} = 2x - 7 + \dfrac{18x + 19}{(x + 1)(x + 2)}$

$\dfrac{18x + 19}{(x + 1)(x + 2)} = \dfrac{A}{x + 1} + \dfrac{B}{x + 2}$

$18x + 19 = A(x + 2) + B(x + 1)$

Let $x = -1$: $1 = A$

Let $x = -2$: $-17 = -B \implies B = 17$

$\dfrac{2x^3 - x^2 + x + 5}{x^2 + 3x + 2} = 2x - 7 + \dfrac{1}{x + 1} + \dfrac{17}{x + 2}$

42. $\dfrac{x^3 + 2x^2 - x + 1}{x^2 + 3x - 4}$

Using long division gives:

$\dfrac{x^3 + 2x^2 - x + 1}{x^2 + 3x - 4} = x - 1 + \dfrac{6x - 3}{x^2 + 3x - 4}$

$\dfrac{x^3 + 2x^2 - x + 1}{x^2 + 3x - 4} - x + 1 = \dfrac{6x - 3}{x^2 + 3x - 4} = \dfrac{6x - 3}{(x + 4)(x - 1)} = \left(\dfrac{A}{x + 4} + \dfrac{B}{x - 1} \right)$

$\dfrac{6x - 3}{(x + 4)(x - 1)} = \left(\dfrac{A}{x + 4} + \dfrac{B}{x - 1} \right)$

$6x - 3 = A(x - 1) + B(x + 4)$

$6x - 3 = (A + B)x + (4B - A)$

$A + B = 6 \implies A = 6 - B$

$4B - A = -3 \implies 4B - 6 + B = -3$

$5B - 6 = -3$

$5B = 3$

$B = \dfrac{3}{5}$

$A = 6 - \dfrac{3}{5} = \dfrac{30 - 3}{5} = \dfrac{27}{5}$

$\dfrac{x^3 + 2x^2 - x + 1}{x^2 + 3x - 4} = x - 1 + \left(\dfrac{\frac{27}{5}}{x + 4} + \dfrac{\frac{3}{5}}{x - 1} \right) = x - 1 + \dfrac{1}{5}\left(\dfrac{27}{x + 4} + \dfrac{3}{x - 1} \right)$

43.
$$\frac{x^4}{(x-1)^3} = \frac{x^4}{x^3 - 3x^2 + 3x - 1} = x + 3 + \frac{6x^2 - 8x + 3}{(x-1)^3}$$

$$\frac{6x^2 - 8x + 3}{(x-1)^3} = \frac{A}{x-1} + \frac{B}{(x-1)^2} + \frac{C}{(x-1)^3}$$

$$6x^2 - 8x + 3 = A(x-1)^2 + B(x-1) + C$$

Let $x = 1$: $1 = C$

Let $x = 0$: $3 = A - B + 1$ } $A - B = 2$

Let $x = 2$: $11 = A + B + 1$ } $A + B = 10$

So, $A = 6$ and $B = 4$.

$$\frac{x^4}{(x-1)^3} = x + 3 + \frac{6}{x-1} + \frac{4}{(x-1)^2} + \frac{1}{(x-1)^3}$$

44.
$$\frac{16x^4}{(2x-1)^3} = \frac{16x^4}{8x^3 - 12x^2 + 6x - 1} = 2x + 3 + \frac{24x^2 - 16x + 3}{(2x-1)^3}$$

$$\frac{24x^2 - 16x + 3}{(2x-1)^2} = \frac{A}{2x-1} + \frac{B}{(2x-1)^2} + \frac{C}{(2x-1)^3}$$

$$24x^2 - 16x + 3 = A(2x-1)^2 + B(2x-1) + C$$

Let $x = \frac{1}{2}$: $1 = C$

$$24x^2 - 16x + 3 = 4Ax^2 - 4Ax + A + 2Bx - B + 1$$

$$24x^2 - 16x + 3 = 4Ax^2 + (-4A + 2B)x + (A - B + 1)$$

Equating coefficients of like powers:

$6 = A, 3 = A - B + 1$

$\qquad 3 = 6 - B + 1$

$\qquad 4 = B$

$$\frac{16x^4}{(2x-1)^3} = 2x + 3 + \frac{6}{2x-1} + \frac{4}{(2x-1)^2} + \frac{1}{(2x-1)^3}$$

45. $\dfrac{5-x}{2x^2 + x - 1} = \dfrac{A}{2x-1} + \dfrac{B}{x+1}$

$\qquad -x + 5 = A(x + 1) + B(2x - 1)$

Let $x = \frac{1}{2}$: $\frac{9}{2} = \frac{3}{2}A \implies A = 3$

Let $x = -1$: $6 = -3B \implies B = -2$

$$\frac{5-x}{2x^2 + x - 1} = \frac{3}{2x-1} - \frac{2}{x+1}$$

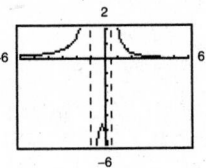

46. $\dfrac{3x^2 - 7x - 2}{x^3 - x} = \dfrac{A}{x} + \dfrac{B}{x+1} + \dfrac{C}{x-1}$

$3x^2 - 7x - 2 = A(x^2 - 1) + Bx(x - 1) + Cx(x + 1)$

Let $x = 0$: $-2 = -A \implies A = 2$

Let $x = -1$: $8 = 2B \implies B = 4$

Let $x = 1$: $-6 = 2C \implies C = -3$

$$\frac{3x^2 - 7x - 2}{x^3 - x} = \frac{2}{x} + \frac{4}{x+1} - \frac{3}{x-1}$$

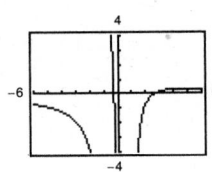

47. $\dfrac{x-1}{x^3+x^2} = \dfrac{A}{x} + \dfrac{B}{x^2} + \dfrac{C}{x+1}$

$\qquad x - 1 = Ax(x+1) + B(x+1) + Cx^2$

Let $x = -1$: $-2 = C$

Let $x = 0$: $-1 = B$

Let $x = 1$: $0 = 2A + 2B + C$

$\qquad\qquad 0 = 2A - 2 - 2$

$\qquad\qquad 2 = A$

$\dfrac{x-1}{x^3+x^2} = \dfrac{2}{x} - \dfrac{1}{x^2} - \dfrac{2}{x+1}$

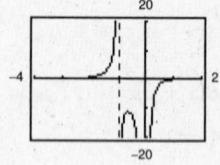

48. $\dfrac{4x^2-1}{2x(x+1)^2} = \dfrac{A}{2x} + \dfrac{B}{x+1} + \dfrac{C}{(x+1)^2}$

$\qquad 4x^2 - 1 = A(x+1)^2 + 2Bx(x+1) + 2Cx$

Let $x = 0$: $-1 = A$

Let $x = -1$: $3 = -2C \Rightarrow C = -\dfrac{3}{2}$

Let $x = 1$: $3 = 4A + 4B + 2C$

$\qquad\qquad 3 = -4 + 4B - 3$

$\qquad\qquad \dfrac{5}{2} = B$

$\dfrac{4x^2-1}{2x(x+1)^2} = \dfrac{1}{2}\left[-\dfrac{1}{x} + \dfrac{5}{x+1} - \dfrac{3}{(x+1)^2} \right]$

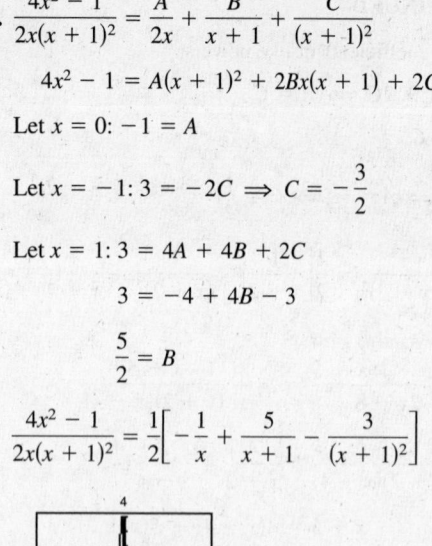

49. $\dfrac{x^2+x+2}{(x^2+2)^2} = \dfrac{Ax+B}{x^2+2} + \dfrac{Cx+D}{(x^2+2)^2}$

$x^2 + x + 2 = (Ax+B)(x^2+2) + Cx + D$

$x^2 + x + 2 = Ax^3 + Bx^2 + (2A+C)x + (2B+D)$

Equating coefficients of like powers:

$\quad 0 = A$

$\quad 1 = B$

$\quad 1 = 2A + C \Rightarrow C = 1$

$\quad 2 = 2B + D \Rightarrow D = 0$

$\dfrac{x^2+x+2}{(x^2+2)^2} = \dfrac{1}{x^2+2} + \dfrac{x}{(x^2+2)^2}$

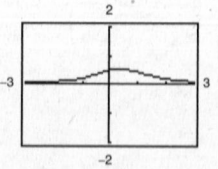

50. $\dfrac{x^3}{(x+2)^2(x-2)^2} = \dfrac{A}{x+2} + \dfrac{B}{(x+2)^2} + \dfrac{C}{x-2} + \dfrac{D}{(x-2)^2}$

$\qquad x^3 = A(x+2)(x-2)^2 + B(x-2)^2 + C(x+2)^2(x-2) + D(x+2)^2$

Let $x = -2$: $-8 = 16B \Rightarrow B = -\dfrac{1}{2}$

Let $x = 2$: $8 = 16D \Rightarrow D = \dfrac{1}{2}$

$\qquad x^3 = A(x+2)(x-2)^2 - \dfrac{1}{2}(x-2)^2 + C(x+2)^2(x-2) + \dfrac{1}{2}(x+2)^2$

$x^3 - 4x = (A+C)x^3 + (-2A+2C)x^2 + (-4A-4C)x + (8A-8C)$

—CONTINUED—

50. —CONTINUED—

Equating coefficients of like powers:

$0 = -2A + 2C \implies A = C$

$1 = A + C$

$1 = 2A \implies A = \dfrac{1}{2} \implies C = \dfrac{1}{2}$

$\dfrac{x^3}{(x+2)^2(x-2)^2} = \dfrac{1}{2}\left[\dfrac{1}{x+2} - \dfrac{1}{(x+2)^2} + \dfrac{1}{x-2} + \dfrac{1}{(x-2)^2}\right]$

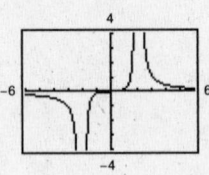

51. $\dfrac{2x^3 - 4x^2 - 15x + 5}{x^2 - 2x - 8} = 2x + \dfrac{x+5}{(x+2)(x-4)}$

$\dfrac{x+5}{(x+2)(x-4)} = \dfrac{A}{x+2} + \dfrac{B}{x-4}$

$x + 5 = A(x-4) + B(x+2)$

Let $x = -2$: $3 = -6A \implies A = -\dfrac{1}{2}$

Let $x = 4$: $9 = 6B \implies B = \dfrac{3}{2}$

$\dfrac{2x^3 - 4x^2 - 15x + 5}{x^2 - 2x - 8} = 2x + \dfrac{1}{2}\left(\dfrac{3}{x-4} - \dfrac{1}{x+2}\right)$

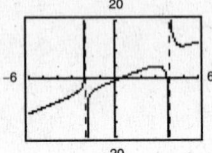

52. $\dfrac{x^3 - x + 3}{x^2 + x - 2} = x - 1 + \dfrac{2x+1}{(x+2)(x-1)}$

$\dfrac{2x+1}{(x+2)(x-1)} = \dfrac{A}{x+2} + \dfrac{B}{x-1}$

$2x + 1 = A(x-1) + B(x+2)$

Let $x = -2$: $-3 = -3A \implies A = 1$

Let $x = 1$: $3 = 3B \implies B = 1$

$\dfrac{x^3 - x + 3}{x^2 + x - 2} = x - 1 + \dfrac{1}{x+2} + \dfrac{1}{x-1}$

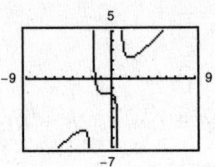

53. (a) $\dfrac{x-12}{x(x-4)} = \dfrac{A}{x} + \dfrac{B}{x-4}$

$x - 12 = A(x-4) + Bx$

Let $x = 0$: $-12 = -4A \implies A = 3$

Let $x = 4$: $-8 = 4B \implies B = -2$

$\dfrac{x-12}{x(x-4)} = \dfrac{3}{x} - \dfrac{2}{x-4}$

(b) $y = \dfrac{x-12}{x(x-4)}$ $\qquad\qquad\qquad y = \dfrac{3}{x}$ $\qquad\qquad\qquad y = -\dfrac{2}{x-4}$

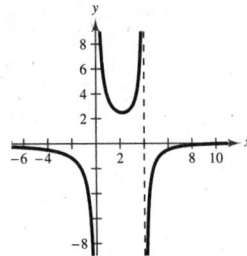 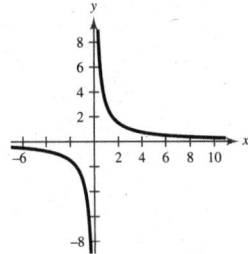 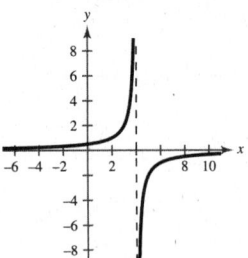

Vertical asymptotes: $x = 0$ $\qquad$ Vertical asymptote: $x = 0$ $\qquad$ Vertical asymptote: $x = 4$
and $x = 4$

(c) The combination of the vertical asymptotes of the terms of the decomposition are the same as the vertical asymptotes of the rational function.

54. (a) $y = \dfrac{2(x+1)^2}{x(x^2+1)} = \dfrac{A}{x} + \dfrac{Bx+C}{x^2+1}$

$2(x+1)^2 = A(x^2+1) + Bx^2 + Cx$

$2x^2 + 4x + 2 = (A+B)x^2 + Cx + A$

Equating coefficients of like powers gives
$2 = A + B, 4 = C,$ and $2 = A.$

Therefore, $A = 2, B = 0,$ and $C = 4.$

$\dfrac{2(x+1)^2}{x(x^2+1)} = \dfrac{2}{x} + \dfrac{4}{x^2+1}$

(b) $\dfrac{2(x+1)^2}{x(x^2+1)}$

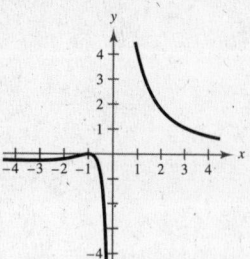

Vertical asymptote at $x = 0$

$y = \dfrac{2}{x}$ and $y = \dfrac{4}{x^2+1}$

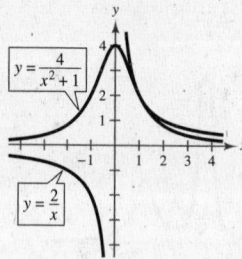

$y = \dfrac{2}{x}$ has vertical asymptote $x = 0.$

(c) The vertical asymptote of $y = \dfrac{2}{x}$ is the same as the vertical asymptote of the rational function.

55. (a) $\dfrac{2(4x-3)}{x^2-9} = \dfrac{A}{x-3} + \dfrac{B}{x+3}$

$2(4x-3) = A(x+3) + B(x-3)$

Let $x = 3$: $18 = 6A \implies A = 3$

Let $x = -3$: $-30 = -6B \implies B = 5$

$\dfrac{2(4x-3)}{x^2-9} = \dfrac{3}{x-3} + \dfrac{5}{x+3}$

(b) $y = \dfrac{2(4x-3)}{x^2-9}$ $y = \dfrac{3}{x-3}$ $y = \dfrac{5}{x+3}$

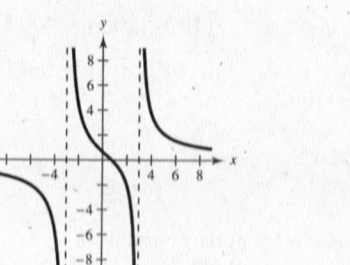   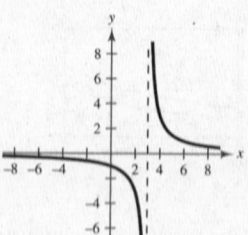 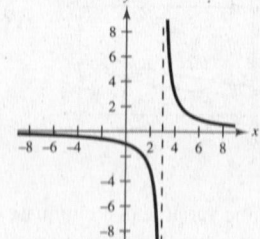

Vertical asymptotes: $x = \pm 3$ Vertical asymptote: $x = 3$ Vertical asymptote: $x = -3$

(c) The combination of the vertical asymptotes of the terms of the decomposition are the same as the vertical asymptotes of the rational function.

56. (a) $y = \dfrac{2(4x^2-15x+39)}{x^2(x^2-10x+26)} = \dfrac{A}{x} + \dfrac{B}{x^2} + \dfrac{Cx+D}{x^2-10x+26}$

$2(4x^2-15x+39) = Ax(x^2-10x+26) + B(x^2-10x+26) + Cx^3 + Dx^2$

$8x^2 - 30x + 78 = Ax^3 - 10Ax^2 + 26Ax + Bx^2 - 10Bx + 26B + Cx^3 + Dx^2$

$= (A+C)x^3 + (-10A+B+D)x^2 + (26A-10B)x + 26B$

Equating coefficients of like powers gives $0 = A + C, 8 = -10A + B + D, -30 = 26A - 10B,$ and $78 = 26B.$ Since $78 = 26B, B = 3.$ Therefore, $A = 0, B = 3, C = 0,$ and $D = 5.$

$\dfrac{2(4x^2-15x+39)}{x^2(x^2-10x+26)} = \dfrac{3}{x^2} + \dfrac{5}{x^2-10x+26}$

—CONTINUED—

56. —CONTINUED—

(b) $\dfrac{2(4x^2 - 15x + 39)}{x^2(x^2 - 10x + 26)}$

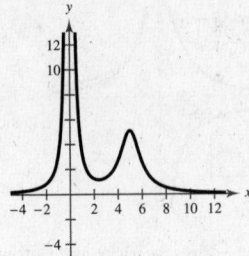

Vertical asymptote is $x = 0$.

$\dfrac{3}{x^2}$ and $\dfrac{5}{x^2 - 10x + 26}$

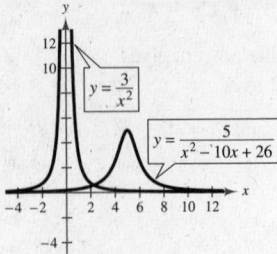

$y = \dfrac{3}{x^2}$ has vertical asymptote $x = 0$.

(c) The vertical asymptote of $y = \dfrac{3}{x^2}$ is the same as the vertical asymptote of the rational function.

57. (a) $\dfrac{2000(4 - 3x)}{(11 - 7x)(7 - 4x)} = \dfrac{A}{11 - 7x} + \dfrac{B}{7 - 4x}, \ 0 < x \le 1$

$2000(4 - 3x) = A(7 - 4x) + B(11 - 7x)$

Let $x = \dfrac{11}{7}$: $-\dfrac{10{,}000}{7} = \dfrac{5}{7}A \implies A = -2000$

Let $x = \dfrac{7}{4}$: $-2500 = -\dfrac{5}{4}B \implies B = 2000$

$\dfrac{2000(4 - 3x)}{(11 - 7x)(7 - 4x)} = \dfrac{-2000}{11 - 7x} + \dfrac{2000}{7 - 4x} = \dfrac{2000}{7 - 4x} - \dfrac{2000}{11 - 7x}, 0 < x \le 1$

(b) $Y_{\max} = \left| \dfrac{2000}{7 - 4x} \right|$

$Y_{\min} = \left| \dfrac{2000}{11 - 7x} \right|$

(c)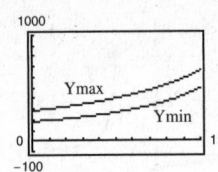

(d) $Y_{\max}(0.5) = 400°\text{F}$

$Y_{\min}(0.5) \approx 266.7°\text{F}$

58. One way to find the constants is to choose values of the variable that eliminate one or more of the constants in the basic equation so that you can solve for another constant. If necessary, you can then use these constants with other chosen values of the variable to solve for any remaining constants. Another way is to expand the basic equation and collect like terms. Then you can equate coefficients of the like terms on each side of the equation to obtain simple equations involving the constants. If necessary, you can solve these equations using substitution.

59. False. The partial fraction decomposition is

$$\dfrac{A}{x + 10} + \dfrac{B}{x - 10} + \dfrac{C}{(x - 10)^2}.$$

60. False. The expression is an improper rational expression, so you must first divide before applying partial fraction decomposition.

61. $\dfrac{1}{a^2 - x^2} = \dfrac{A}{a + x} + \dfrac{B}{a - x}$, a is a constant.

$1 = A(a - x) + B(a + x)$

Let $x = -a$: $1 = 2aA \implies A = \dfrac{1}{2a}$

Let $x = a$: $1 = 2aB \implies B = \dfrac{1}{2a}$

$\dfrac{1}{a^2 - x^2} = \dfrac{1}{2a}\left(\dfrac{1}{a + x} + \dfrac{1}{a - x} \right)$

62. $\dfrac{1}{x(x + a)} = \dfrac{A}{x} + \dfrac{B}{x + a}$, a is a constant.

$1 = A(x + a) + Bx$

Let $x = 0$: $1 = aA \implies A = \dfrac{1}{a}$

Let $x = -a$: $1 = -aB \implies B = -\dfrac{1}{a}$

$\dfrac{1}{x(x + a)} = \dfrac{1}{a}\left(\dfrac{1}{x} - \dfrac{1}{x + a} \right)$

63. $\dfrac{1}{y(a - y)} = \dfrac{A}{y} + \dfrac{B}{a - y}$

$$1 = A(a - y) + By$$

Let $y = 0$: $1 = aA \Rightarrow A = \dfrac{1}{a}$

Let $y = a$: $1 = aB \Rightarrow B = \dfrac{1}{a}$

$$\dfrac{1}{y(a - y)} = \dfrac{1}{a}\left(\dfrac{1}{y} + \dfrac{1}{a - y}\right)$$

64. $\dfrac{1}{(x + 1)(a - x)} = \dfrac{A}{x + 1} + \dfrac{B}{a - x}$, a is a positive integer.

$$1 = A(a - x) + B(x + 1)$$

Let $x = -1$: $1 = A(a + 1) \Rightarrow A = \dfrac{1}{a + 1}$

Let $x = a$: $1 = B(a + 1) \Rightarrow B = \dfrac{1}{a + 1}$

$$\dfrac{1}{(x + 1)(a - x)} = \dfrac{1}{a + 1}\left(\dfrac{1}{x + 1} + \dfrac{1}{a - x}\right)$$

65. $f(x) = x^2 - 9x + 18 = (x - 6)(x - 3)$

Intercepts:
$(0, 18), (3, 0), (6, 0)$

Graph rises to the left and
rises to the right.

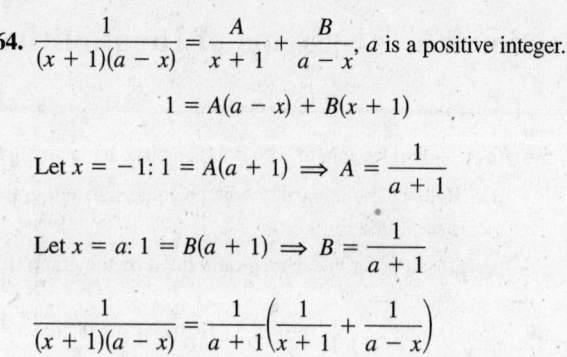

66. $f(x) = 2x^2 - 9x - 5 = (2x + 1)(x - 5)$

$$= 2\left(x - \tfrac{9}{4}\right)^2 - \tfrac{121}{8}$$

Vertex: $\left(\tfrac{9}{4}, -\tfrac{121}{8}\right)$

x-intercepts:
$\left(-\tfrac{1}{2}, 0\right), (5, 0)$

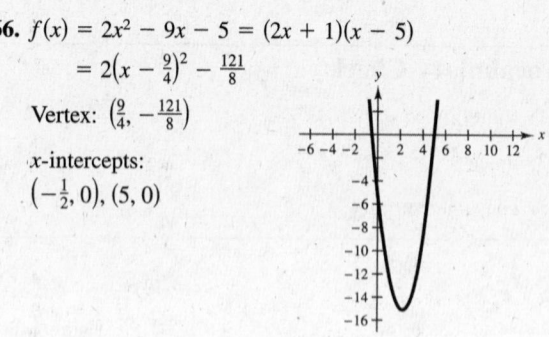

67. $f(x) = -x^2(x - 3)$

Intercepts: $(0, 0), (3, 0)$

Graph rises to the left and
falls to the right.

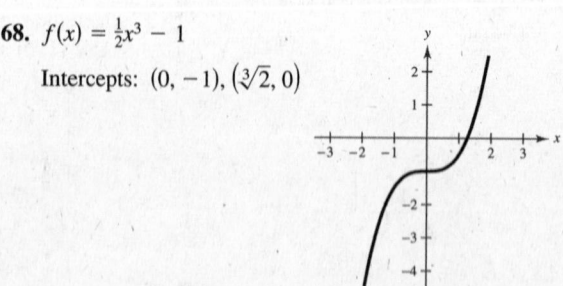

68. $f(x) = \tfrac{1}{2}x^3 - 1$

Intercepts: $(0, -1), \left(\sqrt[3]{2}, 0\right)$

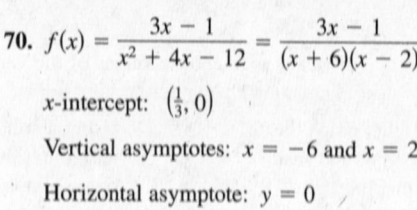

69. $f(x) = \dfrac{x^2 + x - 6}{x + 5}$

x-intercepts: $(-3, 0), (2, 0)$

y-intercept: $\left(0, -\tfrac{6}{5}\right)$

Vertical asymptote: $x = -5$

Slant asymptote: $y = x - 4$

No horizontal asymptote.

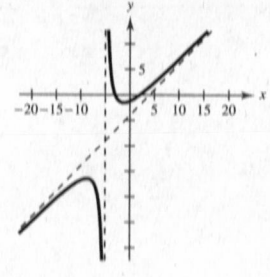

70. $f(x) = \dfrac{3x - 1}{x^2 + 4x - 12} = \dfrac{3x - 1}{(x + 6)(x - 2)}$

x-intercept: $\left(\tfrac{1}{3}, 0\right)$

Vertical asymptotes: $x = -6$ and $x = 2$

Horizontal asymptote: $y = 0$

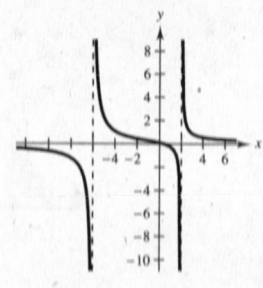

Section 7.5   Systems of Inequalities

■   You should be able to sketch the graph of an inequality in two variables.

(a) Replace the inequality with an equal sign and graph the equation. Use a dashed line for < or >, a solid line for ≤ or ≥.

(b) Test a point in each region formed by the graph. If the point satisfies the inequality, shade the whole region.

■   You should be able to sketch systems of inequalities.

Vocabulary Check

1. solution

2. graph

3. linear

4. solution

5. consumer surplus

---

1. $y < 2 - x^2$

Using a dashed line, graph $y = 2 - x^2$ and shade inside the parabola.

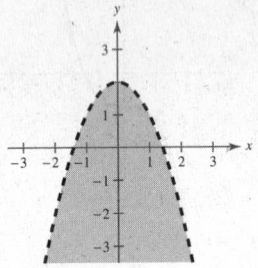

2. $y^2 - x < 0$

Using a dashed line, graph the parabola $y^2 - x = 0$, and shade the region inside this parabola. (Use $(1, 0)$ as a test point.)

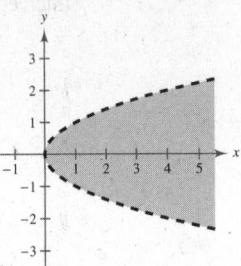

3. $x \geq 2$

Using a solid line, graph the vertical line $x = 2$ and shade to the right of this line.

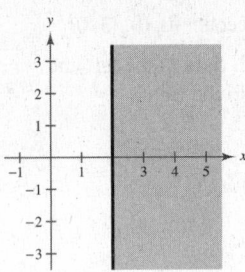

4. $x \leq 4$

Using a solid line, graph the vertical line $x = 4$, and shade to the left of this line.

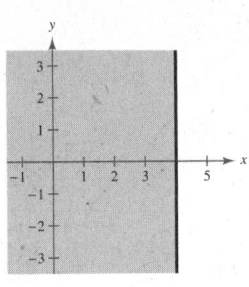

5. $y \geq -1$

Using a solid line, graph the horizontal line $y = -1$ and shade above this line.

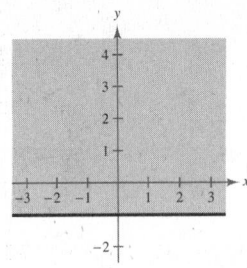

6. $y \leq 3$

Using a solid line, graph the horizontal line $y = 3$, and shade below this line.

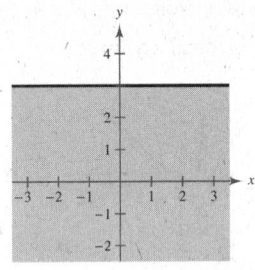

7. $y < 2 - x$

Using a dashed line, graph $y = 2 - x$, and then shade below the line. (Use $(0, 0)$ as a test point.)

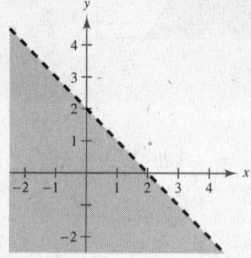

8. $y > 2x - 4$

Using a dashed line, graph $y = 2x - 4$, and shade above the line. (Use $(0, 0)$ as a test point.)

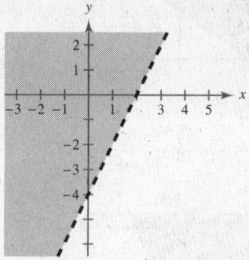

9. $2y - x \geq 4$

Using a solid line, graph $2y - x = 4$, and then shade above the line. (Use $(0, 0)$ as a test point.)

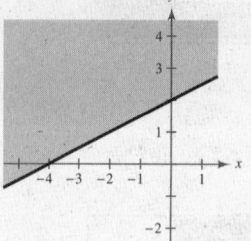

10. $5x + 3y \geq -15$

Using a solid line, graph $5x + 3y = -15$, and shade above the line. (Use $(0, 0)$ as a test point.)

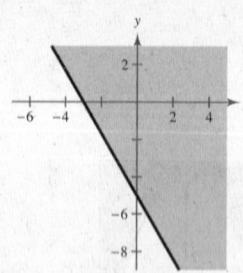

11. $(x + 1)^2 + (y - 2)^2 < 9$

Using a dashed line, sketch the circle $(x + 1)^2 + (y - 2)^2 = 9$.

Center: $(-1, 2)$

Radius: 3

Test point: $(0, 0)$

Shade the inside of the circle.

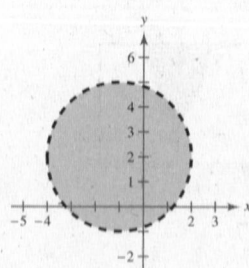

12. $(x - 1)^2 + (y - 4)^2 > 9$

Using a dashed line, graph the circle $(x - 1)^2 + (y - 4)^2 = 9$ and shade the exterior. The circle has center $(1, 4)$ and radius 3, so the origin could serve as a test point.

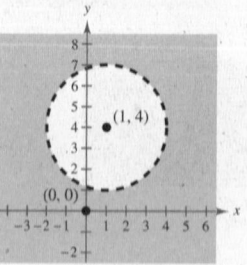

13. $y \leq \dfrac{1}{1 + x^2}$

Using a solid line, graph $y = \dfrac{1}{1 + x^2}$, and then shade below the curve. (Use $(0, 0)$ as a test point.)

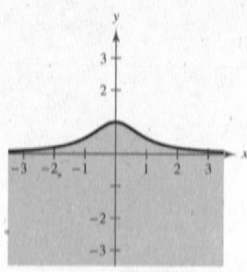

14. $y > \dfrac{-15}{x^2 + x + 4}$

Using a dashed line, graph $y = \dfrac{-15}{x^2 + x + 4}$ and then shade above the curve. (Use $(0, 0)$ as a test point.)

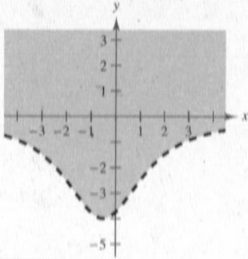

15. $y < \ln x$

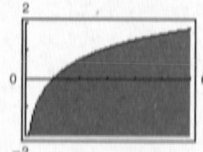

16. $y \geq 6 - \ln(x + 5)$

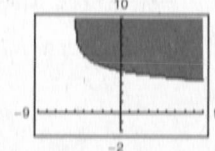

17. $y < 3^{-x-4}$

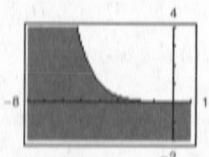

18. $y \leq 2^{2x-0.5} - 7$

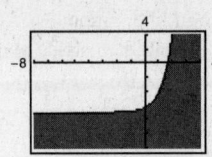

19. $y \geq \frac{2}{3}x - 1$

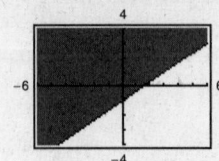

20. $y \leq 6 - \frac{3}{2}x$

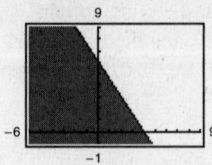

21. $y < -3.8x + 1.1$

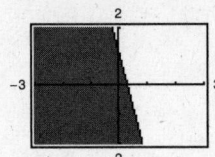

22. $y \geq -20.74 + 2.66x$

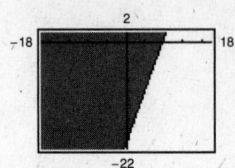

23. $x^2 + 5y - 10 \leq 0$

$$y \leq 2 - \frac{x^2}{5}$$

24. $2x^2 - y - 3 > 0$

$$y < 2x^2 - 3$$

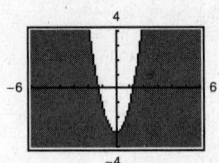

25. $\frac{5}{2}y - 3x^2 - 6 \geq 0$

$$y \geq \frac{2}{5}(3x^2 + 6)$$

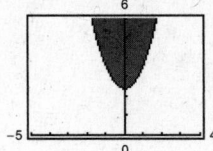

26. $-\frac{1}{10}x^2 - \frac{3}{8}y < -\frac{1}{4}$

$$y > \frac{2}{3} - \frac{4}{15}x^2$$

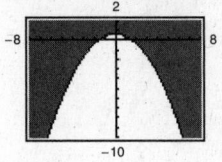

27. The line through $(-4, 0)$ and $(0, 2)$ is $y = \frac{1}{2}x + 2$. For the shaded region below the line, we have $y \leq \frac{1}{2}x + 2$.

28. The parabola through $(-2, 0)$, $(0, -4)$, $(2, 0)$ is $y = x^2 - 4$. For the shaded region inside the parabola, we have $y \geq x^2 - 4$.

29. The line through $(0, 2)$ and $(3, 0)$ is $y = -\frac{2}{3}x + 2$. For the shaded region above the line, we have

$$y \geq -\frac{2}{3}x + 2.$$

30. The circle shown is $x^2 + y^2 = 9$. For the shaded region inside the circle, we have $x^2 + y^2 \leq 9$.

31. $\begin{cases} x \geq -4 \\ y > -3 \\ y \leq -8x - 3 \end{cases}$

(a) $0 \leq -8(0) - 3$, False

$(0, 0)$ *is not* a solution.

(b) $-3 > -3$, False

$(-1, -3)$ *is not* a solution.

(c) $-4 \geq -4$, True

$0 > -3$, True

$0 \leq -8(-4) - 3$, True

$(-4, 0)$ *is* a solution.

(d) $-3 \geq -4$, True

$11 > -3$, True

$11 < -8(-3) - 3$, True

$(-3, 11)$ *is* a solution.

32. $\begin{cases} -2x + 5y \geq 3 \\ \qquad y < 4 \\ -4x + 2y < 7 \end{cases}$

(a) $-2(0) + 5(2) \geq 3$, True

$\qquad\qquad 2 < 4$, True

$\quad -4(0) + 2(2) < 7$, True

$\quad (0, 2)$ *is* a solution

(b) $-2(-6) + 5(4) \geq 3$, True

$\qquad\qquad 4 < 4$, False

$\quad (-6, 4)$ *is not* a solution.

(c) $-2(-8) + 5(-2) \geq 3$, True

$\qquad\qquad -2 < 4$, True

$\quad -4(-8) + 2(-2) < 7$, False

$\quad (-8, -2)$ *is not* a solution.

(d) $-2(-3) + 5(2) \geq 3$, True

$\qquad\qquad 2 < 4$, True

$\quad -4(-3) + 2(2) < 7$, False

$\quad (-3, 2)$ *is not* a solution.

33. $\begin{cases} 3x + y > 1 \\ -y - \frac{1}{2}x^2 \leq -4 \\ -15x + 4y > 0 \end{cases}$

(a) $\quad 3(0) + (10) > 1$, True

$\quad -10 - \frac{1}{2}(0)^2 \leq -4$, True

$\quad -15(0) + 4(10) > 0$, True

$\quad (0, 10)$ *is* a solution.

(b) $3(0) + (-1) > 1$, False $\Rightarrow (0, -1)$ *is not* a solution.

(c) $\quad 3(2) + (9) > 1$, True

$\quad -9 - \frac{1}{2}(2)^2 \leq -4$, True

$\quad -15(2) + 4(9) > 0$, True

$\quad (2, 9)$ *is* a solution.

(d) $\quad 3(-1) + 6 > 1$, True

$\quad -6 - \frac{1}{2}(-1)^2 \leq -4$, True

$\quad -15(-1) + 4(6) > 0$, True

$\quad (-1, 6)$ *is* a solution.

34. $\begin{cases} x^2 + y^2 \geq 36 \\ -3x + y \leq 10 \\ \frac{2}{3}x - y \geq 5 \end{cases}$

(a) $\quad (-1)^2 + 7^2 \geq 36$, True

$\quad -3(-1) + 7 \leq 10$, True

$\quad \frac{2}{3}(-1) - 7 \geq 5$, False

$\quad (-1, 7)$ *is not* a solution.

(c) $\quad 6^2 + 0^2 \geq 36$, True

$\quad -3(6) + 0 \leq 0$, True

$\quad \frac{2}{3}(6) - 0 \geq 5$, False

$\quad (6, 0)$ *is not* a solution.

(b) $(-5)^2 + 1^2 \geq 36$, False

$\quad (-5, 1)$ *is not* a solution.

(d) $4^2 + (-8)^2 \geq 36$, True

$\quad -3(4) - 8 \leq 10$, True

$\quad \frac{2}{3}(4) - (-8) \geq 5$, True

$\quad (4, -8)$ *is* a solution.

35. $\begin{cases} x + y \leq 1 \\ -x + y \leq 1 \\ \qquad y \geq 0 \end{cases}$

First, find the points of intersection of each pair of equations.

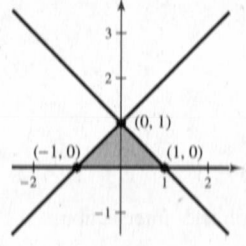

Vertex A	Vertex B	Vertex C
$x + y = 1$	$x + y = 1$	$-x + y = 1$
$-x + y = 1$	$y = 0$	$y = 0$
$(0, 1)$	$(1, 0)$	$(-1, 0)$

36. $\begin{cases} 3x + 2y < 6 \\ \quad x \qquad > 0 \\ \qquad y > 0 \end{cases}$

First, find the points of intersection of each pair of equations.

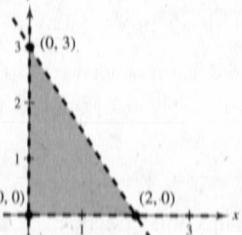

Vertex A	Vertex B	Vertex C
$3x + 2y = 6$	$x = 0$	$3x + 2y = 6$
$x = 0$	$y = 0$	$y = 0$
$(0, 3)$	$(0, 0)$	$(2, 0)$

37. $\begin{cases} x^2 + y \le 5 \\ x \quad\ \ge -1 \\ \quad\ \ y \ge 0 \end{cases}$

First, find the points
of intersection of each
pair of equations.

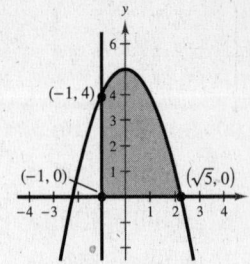

Vertex A	Vertex B	Vertex C
$x^2 + y = 5$	$x^2 + y = 5$	$x = -1$
$x = -1$	$y = 0$	$y = 0$
$(-1, 4)$	$(\pm\sqrt{5}, 0)$	$(-1, 0)$

38. $\begin{cases} 2x^2 + y \ge 2 \\ x \quad\ \ \le 2 \\ \quad\ \ y \le 1 \end{cases}$

First, find the points
of intersection of each
pair of equations.

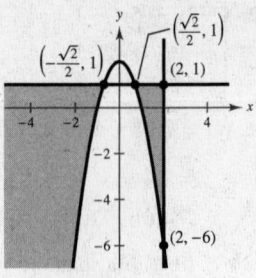

Vertex A	Vertex B	Vertex C
$2x + y = 2$	$x = 2$	$2x^2 + y = 2$
$x = 2$	$y = 1$	$y = 1$
$(2, -6)$	$(2, 1)$	

39. $\begin{cases} 2x + y > 2 \\ 6x + 3y < 2 \end{cases}$

The graphs of $2x + y = 2$
and $6x + 3y = 2$ are
parallel lines. The first
inequality has the region
above the line shaded. The
second inequality has the
region below the line shaded.
There are no points that
satisfy both inequalities.

No solution

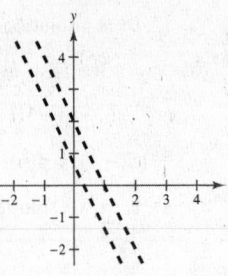

40. $\begin{cases} x - 7y > -36 \\ 5x + 2y > 5 \\ 6x - 5y > 6 \end{cases}$

First, find the points
of intersection of each
pair of equations.

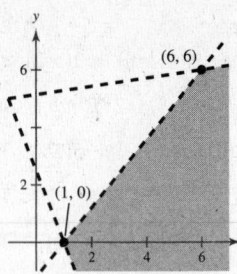

Vertex A	Vertex B	Vertex C
$x - 7y = -36$	$5x + 2y = 5$	$x - 7y = -36$
$5x + 2y = 5$	$6x - 5y = 6$	$6x - 5y = 6$
$(-1, 5)$	$(1, 0)$	$(6, 6)$

41. $\begin{cases} -3x + 2y < 6 \\ x - 4y > -2 \\ 2x + y < 3 \end{cases}$

First, find the points
of intersection of each
pair of equations.

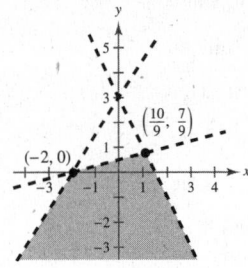

Vertex A	Point B	Vertex C
$-3x + 2y = 6$	$-3x + 2y = 6$	$x - 4y = -2$
$x - 4y = -2$	$2x + y = 3$	$2x + y = 3$
$(-2, 0)$	$(0, 3)$	$\left(\frac{10}{9}, \frac{7}{9}\right)$

Note that B is not a vertex of the solution region.

42. $\begin{cases} x - 2y < -6 \\ 5x - 3y > 9 \end{cases}$

Point of intersection:
$(0, 3)$

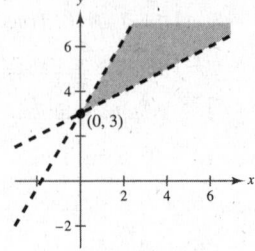

43. $\begin{cases} x > y^2 \\ x < y + 2 \end{cases}$

Points of intersection:

$$y^2 = y + 2$$
$$y^2 - y - 2 = 0$$
$$(y + 1)(y - 2) = 0$$
$$y = -1, 2$$

$(1, -1), (4, 2)$

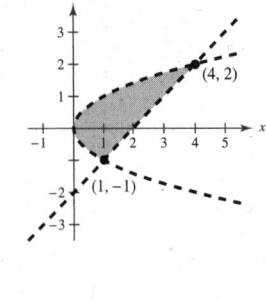

44. $\begin{cases} x - y^2 > 0 \\ x - y > 2 \end{cases}$

Points of intersection:

$$y^2 = y + 2$$
$$y^2 - y - 2 = 0$$
$$(y + 1)(y - 2) = 0$$
$$y = -1, 2$$

$(1, -1), (4, 2)$

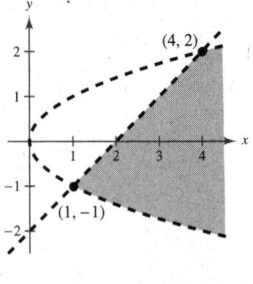

45. $\begin{cases} x^2 + y^2 \le 9 \\ x^2 + y^2 \ge 1 \end{cases}$

There are no points of intersection. The region common to both inequalities is the region between the circles.

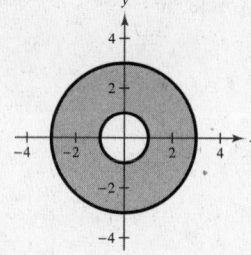

46. $\begin{cases} x^2 + y^2 \le 25 \\ 4x - 3y \le 0 \end{cases}$

Points of intersection:

$x^2 + \left(\frac{4}{3}x\right)^2 = 25$

$\frac{25}{9}x^2 = 25$

$x = \pm 3$

$(-3, -4), (3, 4)$

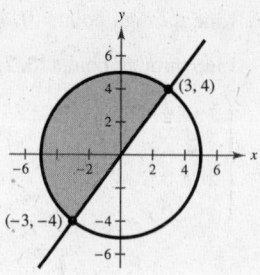

47. $3x + 4 \ge y^2$

$x - y < 0$

Points of intersection:

$x - y = 0 \Rightarrow y = x$

$3y + 4 = y^2$

$0 = y^2 - 3y - 4$

$0 = (y - 4)(y + 1)$

$y = 4$ or $y = -1$

$x = 4 \qquad x = -1$

$(4, 4)$ and $(-1, -1)$

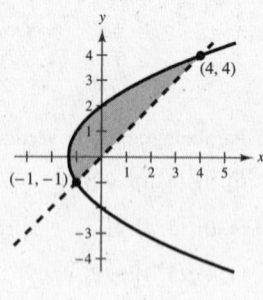

48. $\begin{cases} x < 2y - y^2 \\ 0 < x + y \end{cases}$

Points of intersection:

$-y = 2y - y^2$

$y^2 - 3y = 0$

$y(y - 3) = 0$

$y = 0, 3$

$(0, 0), (-3, 3)$

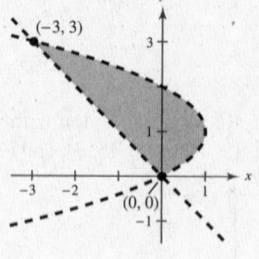

49. $\begin{cases} y \le \sqrt{3x} + 1 \\ y \ge x^2 + 1 \end{cases}$

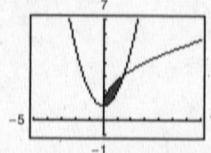

50. $\begin{cases} y < -x^2 + 2x + 3 \\ y > x^2 - 4x + 3 \end{cases}$

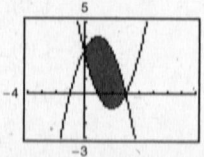

51. $\begin{cases} y < x^3 - 2x + 1 \\ y > -2x \\ x \le 1 \end{cases}$

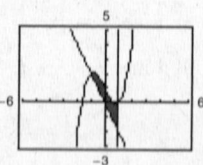

52. $\begin{cases} y \ge x^4 - 2x^2 + 1 \\ y \le 1 - x^2 \end{cases}$

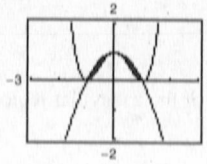

53. $\begin{cases} x^2 y \ge 1 \Rightarrow y \ge \dfrac{1}{x^2} \\ 0 < x \le 4 \\ y \le 4 \end{cases}$

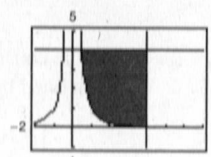

54. $\begin{cases} y \le e^{-x^2/2} \\ y \ge 0 \\ -2 \le x \le 2 \end{cases}$

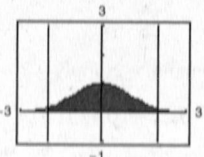

55. $\begin{cases} y \le 4 - x \\ x \ge 0 \\ y \ge 0 \end{cases}$

56. $(0, 6), (3, 0), (0, -3)$

$\begin{cases} y < 6 - 2x \\ y \ge x - 3 \\ x \ge 1 \end{cases}$

57. Line through points $(0, 4)$ and $(4, 0)$: $y = 4 - x$

Line through points $(0, 2)$ and $(8, 0)$: $y = 2 - \frac{1}{4}x$

$$\begin{cases} y \geq 4 - x \\ y \geq 2 - \frac{1}{4}x \\ x \geq 0 \\ y \geq 0 \end{cases}$$

58. Circle: $x^2 + y^2 > 4$

59. $\begin{cases} x^2 + y^2 \leq 16 \\ \quad x \geq 0 \\ \quad\ y \geq 0 \end{cases}$

60. $(0, 0), (0, 4), \left(\sqrt{8}, \sqrt{8}\right)$

$$\begin{cases} x^2 + y^2 \leq 16 \\ \quad x \leq y \\ \quad x \geq 0 \end{cases}$$

61. Rectangular region with vertices at $(2, 1), (5, 1), (5, 7),$ and $(2, 7)$

$$\begin{cases} x \geq 2 \\ x \leq 5 \\ y \geq 1 \\ y \leq 7 \end{cases}$$

This system may be written as:

$$\begin{cases} 2 \leq x \leq 5 \\ 1 \leq y \leq 7 \end{cases}$$

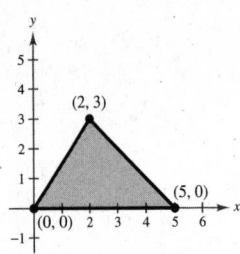

62. Parallelogram with vertices at $(0, 0), (4, 0), (1, 4), (5, 4)$

$(0, 0), (4, 0)$: $y \geq 0$

$(4, 0), (5, 4)$: $4x - y \leq 16$

$(1, 4), (5, 4)$: $y \leq 4$

$(0, 0), (1, 4)$: $4x - y \geq 0$

$$\begin{cases} 4x - y \geq 0 \\ 4x - y \leq 16 \\ 0 \leq y \leq 4 \end{cases}$$

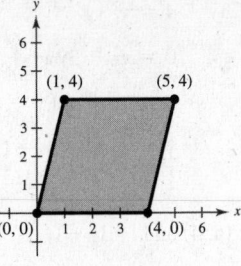

63. Triangle with vertices at $(0, 0), (5, 0), (2, 3)$

$(0, 0), (5, 0)$ Line: $y = 0$

$(0, 0), (2, 3)$ Line: $y = \frac{3}{2}x$

$(2, 3), (5, 0)$ Line: $y = -x + 5$

$$\begin{cases} y \leq \frac{3}{2}x \\ y \leq -x + 5 \\ y \geq 0 \end{cases}$$

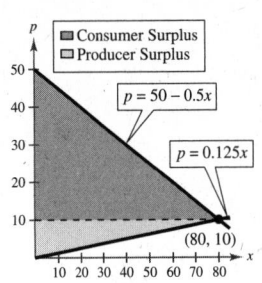

64. Triangle with vertices at $(-1, 0), (1, 0), (0, 1)$

$(-1, 0), (1, 0)$: $y \geq 0$

$(-1, 0), (0, 1)$: $y \leq x + 1$

$(0, 1), (1, 0)$: $y \leq -x + 1$

$$\begin{cases} y \leq x + 1 \\ y \leq -x + 1 \\ y \geq 0 \end{cases}$$

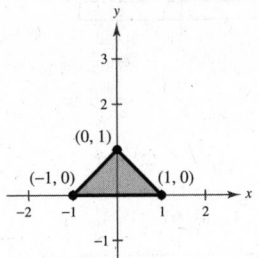

65. (a) Demand = Supply

$50 - 0.5x = 0.125x$

$50 = 0.625x$

$80 = x$

$10 = p$

Point of equilibrium: $(80, 10)$

(b) The consumer surplus is the area of the triangular region defined by

$$\begin{cases} p \leq 50 - 0.5x \\ p \geq 10 \\ x \geq 0. \end{cases}$$

Consumer surplus $= \frac{1}{2}$(base)(height) $= \frac{1}{2}(80)(40) = \$1600$

The producer surplus is the area of the triangular region defined by

$$\begin{cases} p \geq 0.125x \\ p \leq 10 \\ x \geq 0. \end{cases}$$

Producer surplus $= \frac{1}{2}$(base)(height) $= \frac{1}{2}(80)(10) = \$400$

66. (a)   Demand = Supply

$$100 - 0.05x = 25 + 0.1x$$

$$75 = 0.15x$$

$$500 = x$$

$$75 = p$$

Point of equilibrium: (500, 75)

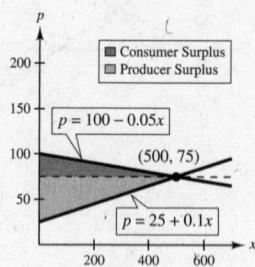

(b) The consumer surplus is the area of the triangular region defined by

$$\begin{cases} p \le 100 - 0.05x \\ p \ge 75 \\ x \ge 0. \end{cases}$$

Consumer surplus $= \frac{1}{2}$(base)(height) $= \frac{1}{2}(500)(25) = 6250$

The producer surplus is the area of the triangular region defined by

$$\begin{cases} p \le 25 + 0.1x \\ p \le 75 \\ x \le 0. \end{cases}$$

Producer surplus $= \frac{1}{2}$(base)(height) $= \frac{1}{2}(500)(50) = 12,500$

67. (a)   Demand = Supply

$$140 - 0.00002x = 80 + 0.00001x$$

$$60 = 0.00003x$$

$$2,000,000 = x$$

$$100 = p$$

Point of equilibrium: (2,000,000, 100)

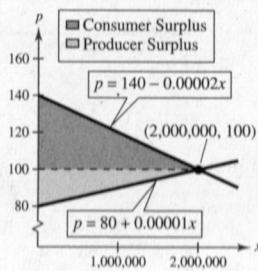

(b) The consumer surplus is the area of the triangular region defined by

$$\begin{cases} p \le 140 - 0.00002x \\ p \ge 100 \\ x \ge \quad 0. \end{cases}$$

Consumer surplus $= \frac{1}{2}$(base)(height)

$$= \frac{1}{2}(2,000,000)(40)$$

$$= \$40,000,000 \text{ or } \$40 \text{ million}$$

The producer surplus is the area of the triangular region defined by

$$\begin{cases} p \ge \quad 80 + 0.00001x \\ p \le 100 \\ x \ge \quad 0. \end{cases}$$

Producer surplus $= \frac{1}{2}$(base)(height)

$$= \frac{1}{2}(2,000,000)(20)$$

$$= \$20,000,000 \text{ or } \$20 \text{ million}$$

68. (a)   Demand = Supply

$$400 - 0.0002x = 225 + 0.0005x$$

$$175 = 0.0007x$$

$$250,000 = x$$

$$350 = p$$

Point of equilibrium: (250,000, 350)

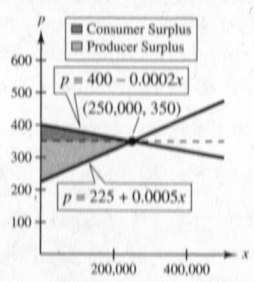

(b) The consumer surplus is the area of the triangular region defined by

$$\begin{cases} p \le 400 - 0.0002x \\ p \ge 350 \\ x \ge 0. \end{cases}$$

Consumer surplus
$= \frac{1}{2}$(base)(height) $= \frac{1}{2}(250,000)(50) = 6,250,000$

The producer surplus is the area of the triangular region defined by

$$\begin{cases} p \ge 225 + 0.0005x \\ p \le 350 \\ x \ge 0. \end{cases}$$

Producer surplus
$= \frac{1}{2}$(base)(height) $= \frac{1}{2}(250,000)(125) = 15,625,000$

69. x = number of tables

y = number of chairs

$$\begin{cases} x + \frac{3}{2}y \le 12 & \text{Assembly center} \\ \frac{4}{3}x + \frac{3}{2}y \le 15 & \text{Finishing center} \\ x \ge 0 \\ y \ge 0 \end{cases}$$

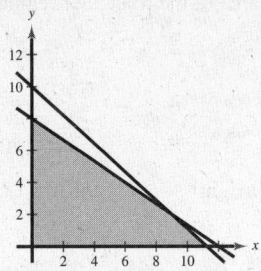

70. x = number of model A

y = number of model B

$$\begin{cases} x \ge 2y \\ 8x + 12y \le 200 \\ x \ge 4 \\ y \ge 2 \end{cases}$$

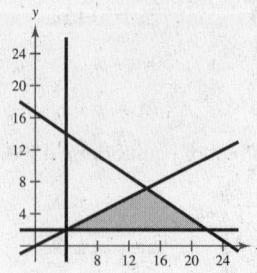

71. x = amount in smaller account

y = amount in larger account

Account constraints:

$$\begin{cases} x + y \le 20,000 \\ y \ge 2x \\ x \ge 5,000 \\ y \ge 5,000 \end{cases}$$

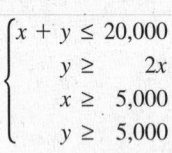 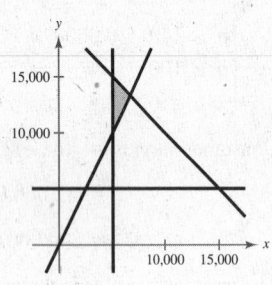

72. x = number of \$30 tickets

y = number of \$20 tickets

$$\begin{cases} x + y \le 3000 \\ 30x + 20y \ge 75,000 \\ x \le 2000 \\ x \ge 0 \\ y \ge 0 \end{cases}$$

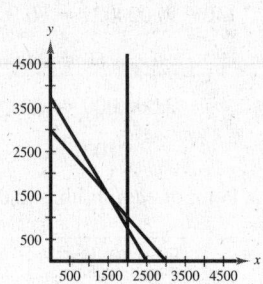

73. x = number of packages of gravel

y = number of bags of stone

$$\begin{cases} 55x + 70y \le 7500 & \text{Weight} \\ x \ge 50 \\ y \ge 40 \end{cases}$$

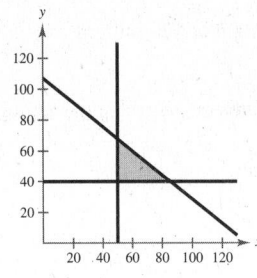

74. Let x = number of large trucks.

Let y = number of medium trucks.

The delivery requirements are:

$$\begin{cases} 6x + 4y \ge 15 \\ 3x + 6y \ge 16 \\ x \ge 0 \\ y \ge 0 \end{cases}$$

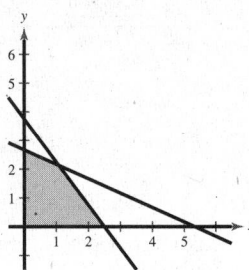

75. (a) x = number of ounces of food X

y = number of ounces of food Y

$$\begin{cases} 20x + 10y \ge 300 & \text{(calcium)} \\ 15x + 10y \ge 150 & \text{(iron)} \\ 10x + 20y \ge 200 & \text{(vitamin B)} \\ x \ge 0 \\ y \ge 0 \end{cases}$$

(b)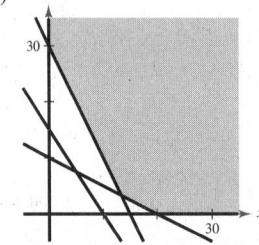

(c) Answers will vary. Some possible solutions which would satisfy the minimum daily requirements for calcium, iron, and vitamin B:

$(0, 30) \implies 30$ ounces of food Y

$(20, 0) \implies 20$ ounces of food X

$\left(13\frac{1}{3}, 3\frac{1}{3}\right) \implies 13\frac{1}{3}$ ounces of food X and $3\frac{1}{3}$ ounces of food Y

76. (a) Let y = heart rate.

$y \geq 0.5(220 - x)$

$y \leq 0.75(220 - x)$

$x \geq 20$

$x \leq 70$

(b)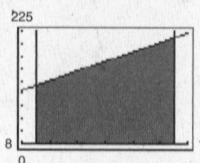

(c) Answers will vary. For example, the points $(24, 98)$ and $(24, 147)$ are on the boundary of the solution set; a person aged 24 should have a heart rate between 98 and 147.

77. (a) $(9, 125.8), (10, 145.6), (11, 164.1),$ $(12, 182.7), (13, 203.1)$

Linear model:
$y = 19.17t - 46.61$

(b) $\begin{cases} y \leq 19.17t - 46.61 \\ t \geq 8.5 \\ t \leq 13.5 \\ y \geq 0 \end{cases}$

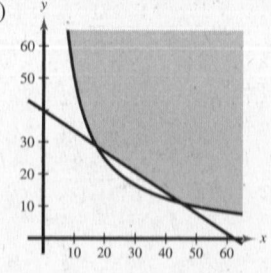

(c) Area of a trapezoid: $A = \dfrac{h}{2}(a + b)$

$h = 13.5 - 8.5 = 5$

$a = 19.17(8.5) - 46.61 = 116.335$

$b = 19.17(13.5) - 46.61 = 212.185$

$A = \dfrac{5}{2}(116.335 + 212.185)$

$= \$821.3$ billion

78. (a) $\begin{cases} xy \geq 500 \\ 2x + \pi y \geq 125 \\ x \geq 0 \\ y \geq 0 \end{cases}$ Body-building space
Track (Two semi–circles and two lengths)

(b)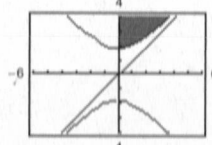

79. True. The figure is a rectangle with length of 9 units and width of 11 units.

80. False. The graph shows the solution of the system
$\begin{cases} y < 6 \\ -4x - 9y < 6 \\ 3x + y^2 \geq 2. \end{cases}$

81. The graph is a half-line on the real number line; on the rectangular coordinate system, the graph is a half-plane.

82. Test a point on either side of the boundary.

83. x = radius of smaller circle

y = radius of larger circle

(a) Constraints on circles:

$\pi y^2 - \pi x^2 \geq 10$

$y > x$

$x > 0$

(b)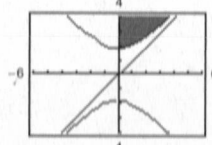

(c) The line is an asymptote to the boundary. The larger the circles, the closer the radii can be and the constraint still be satisfied.

84. (a) The boundary would be included in the solution.

(b) The solution would be the half-plane on the opposite side of the boundary.

85. $\begin{cases} x^2 + y^2 \leq 16 \Rightarrow \text{region inside the circle} \\ x + y \geq 4 \Rightarrow \text{region above the line} \end{cases}$

Matches graph (d).

86. $x^2 + y^2 \leq 16 \Rightarrow$ region inside the circle

$x + y \leq 4 \Rightarrow$ region below the line

Matches graph (b).

87. $\begin{cases} x^2 + y^2 \geq 16 \Rightarrow \text{region outside the circle} \\ x + y \geq 4 \Rightarrow \text{region above the line} \end{cases}$

Matches graph (c).

88. $x^2 + y^2 \geq 16 \Rightarrow$ region outside the circle

$x + y \leq 4 \Rightarrow$ region below the line

Matches graph (a).

89. $(-2, 6), (4, -4)$

$$m = \frac{-4 - 6}{4 - (-2)} = \frac{-10}{6} = -\frac{5}{3}$$

$$y - (-4) = -\frac{5}{3}(x - 4)$$

$$3y + 12 = -5x + 20$$

$$5x + 3y - 8 = 0$$

90. $(-8, 0), (3, -1)$

$$m = \frac{-1 - 0}{3 - (-8)} = -\frac{1}{11}$$

$$y - 0 = -\frac{1}{11}(x - (-8))$$

$$y = -\frac{1}{11}x - \frac{8}{11}$$

$$11y = -x - 8$$

$$x + 11y + 8 = 0$$

91. $\left(\frac{3}{4}, -2\right), \left(-\frac{7}{2}, 5\right)$

$$m = \frac{5 - (-2)}{-\frac{7}{2} - \frac{3}{4}} = \frac{7}{-\frac{17}{4}} = -\frac{28}{17}$$

$$y - (-2) = -\frac{28}{17}\left(x - \frac{3}{4}\right)$$

$$17y + 34 = -28x + 21$$

$$28x + 17y + 13 = 0$$

92. $\left(-\frac{1}{2}, 0\right), \left(\frac{11}{2}, 12\right)$

$$m = \frac{12 - 0}{\frac{11}{2} - \left(\frac{1}{2}\right)} = \frac{12}{6} = 2$$

$$y - 0 = 2\left(x - \left(-\frac{1}{2}\right)\right)$$

$$y = 2x + 1$$

$$2x - y + 1 = 0$$

93. $(3.4, -5.2), (-2.6, 0.8)$

$$m = \frac{0.8 - (-5.2)}{-2.6 - 3.4} = \frac{6}{-6} = -1$$

$$y - 0.8 = -1(x - (-2.6))$$

$$y - 0.8 = -x - 2.6$$

$$x + y + 1.8 = 0$$

94. $(-4.1, -3.8), (2.9, 8.2)$

$$m = \frac{8.2 - (-3.8)}{2.9 - (-4.1)} = \frac{12}{7}$$

$$y + 3.8 = \frac{12}{7}(x + 4.1)$$

$$y + 3.8 = \frac{12}{7}x + \frac{246}{35}$$

$$y = \frac{12}{7}x + \frac{113}{35}$$

$$35y = 60x + 113$$

$$60x - 35y + 113 = 0$$

95. (a) (8, 39.43), (9, 41.24), (10, 45.27), (11, 47.37), (12, 48.40), (13, 49.91)

Linear model: $y \approx 2.17t + 22.5$

Quadratic model: $y \approx -0.241t^2 + 7.23t - 3.4$

Exponential model: $y \approx 27(1.05^t)$

(b)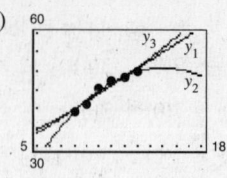

(c) The quadratic model is the best fit for the actual data.

(d) For 2008, use $t = 18$: $y \approx -0.241(18)^2 + 7.23(18) - 3.4 \approx \$48.66$

96. $A = P\left(1 + \dfrac{r}{t}\right)^{nt}$

$A = 4000\left(1 + \dfrac{0.06}{12}\right)^{5 \cdot 12}$

$\quad = 4000(1.005)^{60}$

$\quad = 5395.40061$

The amount after 5 years is $5395.40.

Section 7.6   Linear Programming

- To solve a linear programming problem:
  1. Sketch the solution set for the system of constraints.
  2. Find the vertices of the region.
  3. Test the objective function at each of the vertices.

Vocabulary Check

1. optimization

2. linear programming

3. objective

4. constraints; feasible solutions

5. vertex

---

1. $z = 4x + 3y$

At $(0, 5)$: $z = 4(0) + 3(5) = 15$

At $(0, 0)$: $z = 4(0) + 3(0) = 0$

At $(5, 0)$: $z = 4(5) + 3(0) = 20$

The minimum value is 0 at $(0, 0)$.

The maximum value is 20 at $(5, 0)$.

2. $z = 2x + 8y$

At $(0, 4)$: $z = 2(0) + 8(4) = 32$

At $(0, 0)$: $z = 2(0) + 8(0) = 0$

At $(2, 0)$: $z = 2(2) + 8(0) = 4$

The minimum value is 0 at $(0, 0)$.

The maximum value is 32 at $(0, 4)$.

3. $z = 3x + 8y$

At $(0, 5)$: $z = 3(0) + 8(5) = 40$

At $(0, 0)$: $z = 3(0) + 8(0) = 0$

At $(5, 0)$: $z = 3(5) + 8(0) = 15$

The minimum value is 0 at $(0, 0)$.

The maximum value is 40 at $(0, 5)$.

4. $z = 7x + 3y$

At $(0, 4)$: $z = 7(0) + 3(4) = 12$

At $(0, 0)$: $z = 7(0) + 3(0) = 0$

At $(2, 0)$: $z = 7(2) + 3(0) = 14$

The minimum value is 0 at $(0, 0)$.

The maximum value is 14 at $(2, 0)$.

5. $z = 3x + 2y$

At $(0, 5)$: $z = 3(0) + 2(5) = 10$

At $(4, 0)$: $z = 3(4) + 2(0) = 12$

At $(3, 4)$: $z = 3(3) + 2(4) = 17$

At $(0, 0)$: $z = 3(0) + 2(0) = 0$

The minimum value is 0 at $(0, 0)$.

The maximum value is 17 at $(3, 4)$.

6. $z = 4x + 5y$

At $(0, 2)$: $z = 4(0) + 5(2) = 10$

At $(0, 4)$: $z = 4(0) + 5(4) = 20$

At $(3, 0)$: $z = 4(3) + 5(0) = 12$

At $(4, 3)$: $z = 4(4) + 5(3) = 31$

The minimum value is 10 at $(0, 2)$.

The maximum value is 31 at $(4, 3)$.

7. $z = 5x + 0.5y$

At $(0, 5)$: $z = 5(0) + \frac{5}{2} = \frac{5}{2}$

At $(4, 0)$: $z = 5(4) + \frac{0}{2} = 20$

At $(3, 4)$: $z = 5(3) + \frac{4}{2} = 17$

At $(0, 0)$: $z = 5(0) + \frac{0}{2} = 0$

The minimum value is 0 at $(0, 0)$.

The maximum value is 20 at $(4, 0)$.

8. $z = 2x + y$

At $(0, 2)$: $z = 2(0) + 2 = 2$

At $(0, 4)$: $z = 2(0) + 4 = 4$

At $(3, 0)$: $z = 2(3) + 0 = 6$

At $(4, 3)$: $z = 2(4) + 3 = 11$

The minimum value is 2 at $(0, 2)$.

The maximum value is 11 at $(4, 3)$.

9. $z = 10x + 7y$

At $(0, 45)$: $z = 10(0) + 7(45) = 315$

At $(30, 45)$: $z = 10(30) + 7(45) = 615$

At $(60, 20)$: $z = 10(60) + 7(20) = 740$

At $(60, 0)$: $z = 10(60) + 7(0) = 600$

At $(0, 0)$: $z = 10(0) + 7(0) = 0$

The minimum value is 0 at $(0, 0)$.

The maximum value is 740 at $(60, 20)$.

10. $z = 25x + 35y$

At $(0, 400)$: $z = 25(0) + 35(400) = 14{,}000$

At $(0, 800)$: $z = 25(0) + 35(800) = 28{,}000$

At $(450, 0)$: $z = 25(450) + 35(0) = 11{,}250$

At $(900, 0)$: $z = 25(900) + 35(0) = 22{,}500$

The minimum value is 11,250 at $(450, 0)$.

The maximum value is 28,000 at $(0, 800)$.

11. $z = 25x + 30y$

At $(0, 45)$: $z = 25(0) + 30(45) = 1350$

At $(30, 45)$: $z = 25(30) + 30(45) = 2100$

At $(60, 20)$: $z = 25(60) + 30(20) = 2100$

At $(60, 0)$: $z = 25(60) + 30(0) = 1500$

At $(0, 0)$: $z = 25(0) + 30(0) = 0$

The minimum value is 0 at $(0, 0)$.

The maximum value is 2100 at any point along the line segment connecting $(30, 45)$ and $(60, 20)$.

12. $z = 15x + 20y$

At $(0, 400)$: $z = 15(0) + 20(400) = 8000$

At $(0, 800)$: $z = 15(0) + 20(800) = 16{,}000$

At $(450, 0)$: $z = 15(450) + 20(0) = 6750$

At $(900, 0)$: $z = 15(900) + 20(0) = 13{,}500$

The minimum value is 6750 at $(450, 0)$.

The maximum value is 16,000 at $(0, 800)$.

13. $z = 6x + 10y$

At $(0, 2)$: $z = 6(0) + 10(2) = 20$

At $(5, 0)$: $z = 6(5) + 10(0) = 30$

At $(0, 0)$: $z = 6(0) + 10(0) = 0$

The minimum value is 0 at $(0, 0)$.

The maximum value is 30 at $(5, 0)$.

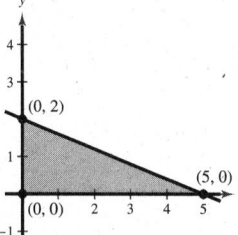

14. $z = 7x + 8y$

At $(0, 8)$: $z = 7(0) + 8(8) = 64$

At $(4, 0)$: $z = 7(4) + 8(0) = 28$

At $(0, 0)$: $z = 7(0) + 8(0) = 0$

The minimum value is 0 at $(0, 0)$.

The maximum value is 64 at $(0, 8)$.

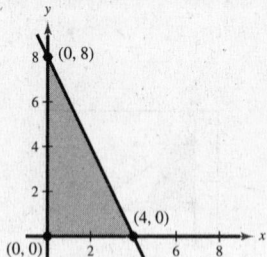

16. $z = 7x + 2y$

At $(0, 8)$: $z = 7(0) + 2(8) = 16$

At $(4, 0)$: $z = 7(4) + 2(0) = 28$

At $(0, 0)$: $z = 7(0) + 2(0) = 0$

The minimum value is 0 at $(0, 0)$.

The maximum value is 28 at $(4, 0)$.

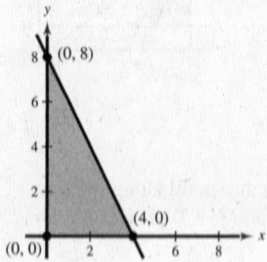

18. $z = 4x + 5y$

At $(0, 0)$: $z = 4(0) + 5(0) = 0$

At $(5, 0)$: $z = 4(5) + 5(0) = 20$

At $(4, 1)$: $z = 4(4) + 5(1) = 21$

At $(0, 3)$: $z = 4(0) + 5(3) = 15$

The minimum value is 0 at $(0, 0)$.

The maximum value is 21 at $(4, 1)$.

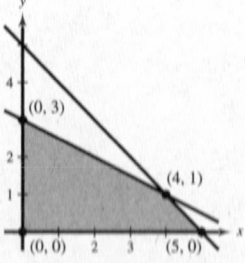

15. $z = 9x + 24y$

At $(0, 2)$: $z = 9(0) + 24(2) = 48$

At $(5, 0)$: $z = 9(5) + 24(0) = 45$

At $(0, 0)$: $z = 9(0) + 24(0) = 0$

The minimum value is 0 at $(0, 0)$.

The maximum value is 48 at $(0, 2)$.

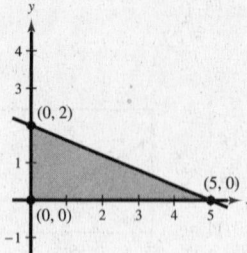

17. $z = 4x + 5y$

At $(10, 0)$: $z = 4(10) + 5(0) = 40$

At $(5, 3)$: $z = 4(5) + 5(3) = 35$

At $(0, 8)$: $z = 4(0) + 5(8) = 40$

The minimum value is 35 at $(5, 3)$.

The region is unbounded. There is no maximum.

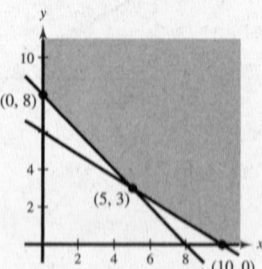

19. $z = 2x + 7y$

At $(10, 0)$: $z = 2(10) + 7(0) = 20$

At $(5, 3)$: $z = 2(5) + 7(3) = 31$

At $(0, 8)$: $z = 2(0) + 7(8) = 56$

The minimum value is 20 at $(10, 0)$.

The region is unbounded. There is no maximum.

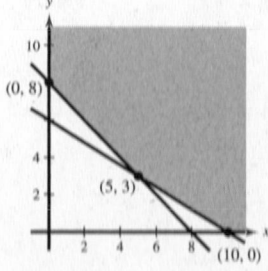

20. $z = 2x - y$

At $(0, 0)$: $z = 2(0) - 0 = 0$

At $(5, 0)$: $z = 2(5) - 0 = 10$

At $(4, 1)$: $z = 2(4) - 1 = 7$

At $(0, 3)$: $z = 2(0) - 3 = -3$

The minimum value is -3 at $(0, 3)$.

The maximum value is 10 at $(5, 0)$.

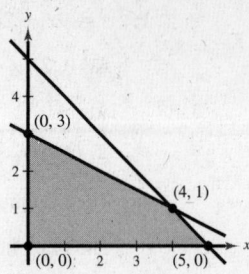

21. $z = 4x + y$

At $(36, 0)$: $z = 4(36) + 0 = 144$

At $(40, 0)$: $z = 4(40) + 0 = 160$

At $(24, 8)$: $z = 4(24) + 8 = 104$

The minimum value is 104 at $(24, 8)$.

The maximum value is 160 at $(40, 0)$.

22. $z = x$

At $(0, 0)$: $z = 0$

At $(12, 0)$: $z = 12$

At $(10, 8)$: $z = 10$

At $(6, 16)$: $z = 6$

At $(0, 20)$: $z = 0$

The minimum value is 0 at any point along the line segment connecting $(0, 0)$ and $(0, 20)$. The maximum value is 12 at $(12, 0)$.

23. $z = x + 4y$

At $(36, 0)$: $z = 36 + 4(0) = 36$

At $(40, 0)$: $z = 40 + 4(0) = 40$

At $(24, 8)$: $z = 24 + 4(8) = 56$

The minimum value is 36 at $(36, 0)$.

The maximum value is 56 at $(24, 8)$.

24. $z = y$

At $(0, 0)$: $z = 0$

At $(12, 0)$: $z = 0$

At $(10, 8)$: $z = 8$

At $(6, 16)$: $z = 16$

At $(0, 20)$: $z = 20$

The minimum value is 0 at any point along the line segment connecting $(0, 0)$ and $(12, 0)$. The maximum value is 20 at $(0, 20)$.

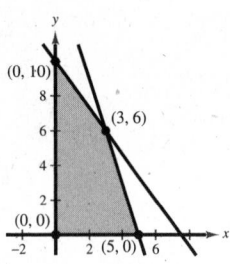

Figure for Exercises 25–28

25. $z = 2x + y$

At $(0, 10)$: $z = 2(0) + (10) = 10$

At $(3, 6)$: $z = 2(3) + (6) = 12$

At $(5, 0)$: $z = 2(5) + (0) = 10$

At $(0, 0)$: $z = 2(0) + (0) = 0$

The maximum value is 12 at $(3, 6)$.

26. $z = 5x + y$

At $(0, 10)$: $z = 5(0) + 10 = 10$

At $(3, 6)$: $z = 5(3) + 6 = 21$

At $(5, 0)$: $z = 5(5) + 0 = 25$

At $(0, 0)$: $z = 5(0) + 0 = 0$

The maximum value is 25 at $(5, 0)$.

27. $z = x + y$

At $(0, 10)$: $z = (0) + (10) = 10$

At $(3, 6)$: $z = (3) + (6) = 9$

At $(5, 0)$: $z = (5) + (0) = 5$

At $(0, 0)$: $z = (0) + (0) = 0$

The maximum value is 10 at $(0, 10)$.

28. $z = 3x + y$

At $(0, 10)$: $z = 3(0) + 10 = 10$

At $(3, 6)$: $z = 3(3) + 6 = 15$

At $(5, 0)$: $z = 3(5) + 0 = 15$

At $(0, 0)$: $z = 3(0) + 0 = 0$

The maximum value is 15 at any point along the line segment connecting $(3, 6)$ and $(5, 0)$.

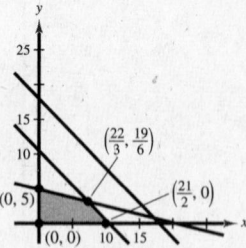

Figure for Exercises 29–32

29. $z = x + 5y$

At $(0, 5)$: $z = 0 + 5(5) = 25$

At $\left(\frac{22}{3}, \frac{19}{6}\right)$: $z = \frac{22}{3} + 5\left(\frac{19}{6}\right) = \frac{139}{6}$

At $\left(\frac{21}{2}, 0\right)$: $z = \frac{21}{2} + 5(0) = \frac{21}{2}$

At $(0, 0)$: $z = 0 + 5(0) = 0$

The maximum value is 25 at $(0, 5)$.

30. $z = 2x + 4y$

At $(0, 5)$: $z = 2(0) + 4(5) = 20$

At $\left(\frac{22}{3}, \frac{19}{6}\right)$: $z = 2\left(\frac{22}{3}\right) + 4\left(\frac{19}{6}\right) = \frac{82}{3}$

At $\left(\frac{21}{2}, 0\right)$: $z = 2\left(\frac{21}{2}\right) + 4(0) = 21$

At $(0, 0)$: $z = 2(0) + 4(0) = 0$

The maximum value is $\frac{82}{3}$ at $\left(\frac{22}{3}, \frac{19}{6}\right)$.

31. $z = 4x + 5y$

At $(0, 5)$: $z = 4(0) + 5(5) = 25$

At $\left(\frac{22}{3}, \frac{19}{6}\right)$: $z = 4\left(\frac{22}{3}\right) + 5\left(\frac{19}{6}\right) = \frac{271}{6}$

At $\left(\frac{21}{2}, 0\right)$: $z = 4\left(\frac{21}{2}\right) + 5(0) = 42$

At $(0, 0)$: $z = 4(0) + 5(0) = 0$

The maximum value is $\frac{271}{6}$ at $\left(\frac{22}{3}, \frac{19}{6}\right)$.

32. $z = 4x + y$

At $(0, 5)$: $z = 4(0) + 5 = 5$

At $\left(\frac{22}{3}, \frac{19}{6}\right)$: $z = 4\left(\frac{22}{3}\right) + \frac{19}{6} = \frac{65}{2}$

At $\left(\frac{21}{2}, 0\right)$: $z = 4\left(\frac{21}{2}\right) + 0 = 42$

At $(0, 0)$: $z = 4(0) + 0 = 0$

The maximum value is 42 at $\left(\frac{21}{2}, 0\right)$.

33. Objective function: $z = 2.5x + y$

Constraints: $x \geq 0, y \geq 0, 3x + 5y \leq 15, 5x + 2y \leq 10$

At $(0, 0)$: $z = 0$

At $(2, 0)$: $z = 5$

At $\left(\frac{20}{19}, \frac{45}{19}\right)$: $z = \frac{95}{19} = 5$

At $(0, 3)$: $z = 3$

The maximum value of 5 occurs at any point on the line segment connecting $(2, 0)$ and $\left(\frac{20}{19}, \frac{45}{19}\right)$.

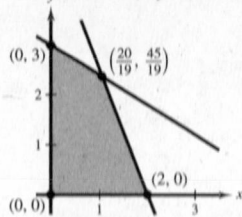

34. Objective function: $z = x + y$

Constraints: $x \geq 0$, $y \geq 0$, $-x + y \leq 1$, $-x + 2y \leq 4$

At $(0, 0)$: $z = 0 + 0 = 0$

At $(0, 1)$: $z = 0 + 1 = 1$

At $(2, 3)$: $z = 2 + 3 = 5$

The constraints do not form a closed set of points. Therefore, $z = x + y$ is unbounded.

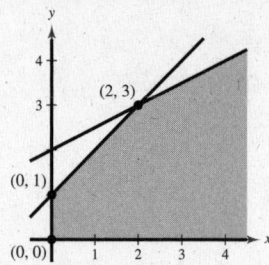

35. Objective function: $z = -x + 2y$

Constraints: $x \geq 0$, $y \geq 0$, $x \leq 10$, $x + y \leq 7$

At $(0, 0)$: $z = -0 + 2(0) = 0$

At $(0, 7)$: $z = -0 + 2(7) = 14$

At $(7, 0)$: $z = -7 + 2(0) = -7$

The constraint $x \leq 10$ is extraneous.

The maximum value of 14 occurs at $(0, 7)$.

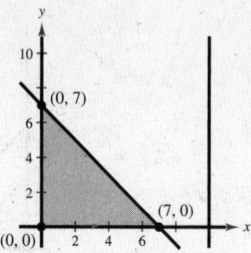

36. Objective function: $z = x + y$

Constraints: $x \geq 0$, $y \geq 0$, $-x + y \leq 0$, $-3x + y \geq 3$

The feasible set is empty.

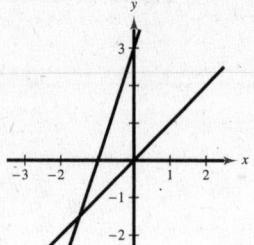

37. Objective function: $z = 3x + 4y$

Constraints: $x \geq 0$, $y \geq 0$, $x + y \leq 1$, $2x + y \leq 4$

At $(0, 0)$: $z = 3(0) + 4(0) = 0$

At $(0, 1)$: $z = 3(0) + 4(1) = 4$

At $(1, 0)$: $z = 3(1) + 4(0) = 3$

The constraint $2x + y \leq 4$ is extraneous.

The maximum value of 4 occurs at $(0, 1)$.

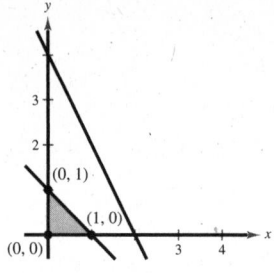

38. Objective function: $z = x + 2y$

Constraints: $x \geq 0$, $y \geq 0$, $x + 2y \leq 4$, $2x + y \leq 4$

At $(0, 0)$: $z = 0 + 2(0) = 0$

At $(0, 2)$: $z = 0 + 2(2) = 4$

At $\left(\frac{4}{3}, \frac{4}{3}\right)$: $z = \frac{4}{3} + 2\left(\frac{4}{3}\right) = 4$

At $(2, 0)$: $z = 2 + 2(0) = 2$

The maximum value is 4 at any point along the line segment connecting $(0, 2)$ and $\left(\frac{4}{3}, \frac{4}{3}\right)$.

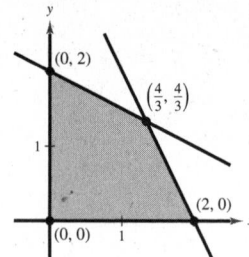

39. x = number of Model A

y = number of Model B

Constraints: $2x + 2.5y \leq 4000$

$\qquad\qquad 4x + \quad y \leq 4800$

$\qquad\qquad x + 0.75y \leq 1500$

$\qquad\qquad\qquad\quad x \geq 0$

$\qquad\qquad\qquad\quad y \geq 0$

Objective function: $P = 45x + 50y$

Vertices:

$(0, 0), (0, 1600), (750, 1000), (1050, 600), (1200, 0)$

At $(0, 0)$: $P = 45(0) + 50(0) = 0$

At $(0, 1600)$: $P = 45(0) + 50(1600) = 80,000$

At $(750, 1000)$: $P = 45(750) + 50(1000) = 83,750$

At $(1050, 600)$: $P = 45(1050) + 50(600) = 77,250$

At $(1200, 0)$: $P = 45(1200) + 50(0) = 54,000$

The optimal profit of \$83,750 occurs when 750 units of Model A and 1000 units of Model B are produced.

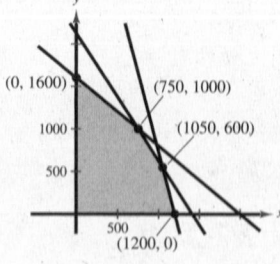

40. x = number of Model A

y = number of Model B

Constraints: $2.5x + \quad 3y \leq 4000$

$\qquad\qquad\quad 2x + \quad y \leq 2500$

$\qquad\quad 0.75x + 1.25y \leq 1500$

$\qquad\qquad\qquad\quad x \geq \quad 0$

$\qquad\qquad\qquad\quad y \geq \quad 0$

Objective function: $P = 50x + 52y$

Vertices:

$(0, 0), (0, 1200), \left(\frac{4000}{7}, \frac{6000}{7}\right), (1000, 500), (1250, 0)$

At $(0, 0)$: $P = 50(0) + 52(0) = 0$

At $(0, 1200)$: $P = 50(0) + 52(1200) = 62,400$

At $\left(\frac{4000}{7}, \frac{6000}{7}\right)$: $P = 50\left(\frac{4000}{7}\right) + 52\left(\frac{6000}{7}\right) \approx 73,142.86$

At $(1000, 500)$: $P = 50(1000) + 52(500) = 76,000$

At $(1250, 0)$: $P = 50(1250) + 52(0) = 62,500$

The optimal profit of \$76,000 occurs when 1000 units of Model A and 500 units of Model B are produced.

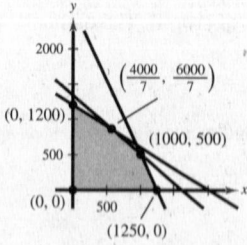

41. x = number of \$250 models

y = number of \$300 models

Constraints: $250x + 300y \leq 65,000$

$\qquad\qquad\quad x + \quad y \leq 250$

$\qquad\qquad\qquad\quad x \geq 0$

$\qquad\qquad\qquad\quad y \geq 0$

Objective function: $P = 25x + 40y$

Vertices: $(0, 0), (250, 0), (200, 50), \left(0, 216\frac{2}{3}\right)$

At $(0, 0)$: $P = 25(0) + 40(0) = 0$

At $(250, 0)$: $P = 25(250) + 40(0) = 6250$

At $(200, 50)$: $P = 25(200) + 40(50) = 7000$

At $\left(0, 216\frac{2}{3}\right)$: $P = 25(0) + 40\left(216\frac{2}{3}\right) \approx 8666.67$

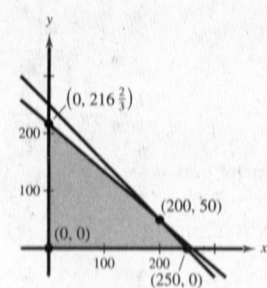

An optimal profit of \$8640 occurs when 0 units of the \$250 model and 216 units of the \$300 model are stocked in inventory. (**Note:** A merchant cannot sell $\frac{2}{3}$ of a unit.)

42. x = number of acres for crop A

y = number of acres for crop B

Constraints:

$$x + y \leq 150$$
$$x + 2y \leq 240$$
$$0.3x + 0.1y \leq 30$$
$$x \geq 0$$
$$y \geq 0$$

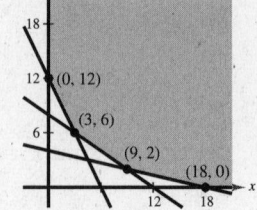

Objective function:
$P = 140x + 235y$

Vertices: $(0, 0)$, $(100, 0)$, $(0, 120)$, $(60, 90)$, $(75, 75)$

At $(0, 0)$: $P = 140(0) + 235(0) = 0$

At $(100, 0)$: $P = 140(100) + 235(0) = 14,000$

At $(0, 120)$: $P = 140(0) + 235(120) = 28,200$

At $(60, 90)$: $P = 140(60) + 235(90) = 29,550$

At $(75, 75)$: $P = 140(75) + 235(75) = 28,125$

To optimize the profit, the fruit grower should plant 60 acres of crop A and 90 acres of crop B. The optimal profit is $29,550.

43. x = number of bags of Brand X

y = number of bags of Brand Y

Constraints:

$$2x + y \geq 12$$
$$2x + 9y \geq 36$$
$$2x + 3y \geq 24$$
$$x \geq 0$$
$$y \geq 0$$

Objective function:
$C = 25x + 20y$

Vertices: $(0, 12)$, $(3, 6)$, $(9, 2)$, $(18, 0)$

At $(0, 12)$: $C = 25(0) + 20(12) = 240$

At $(3, 6)$: $C = 25(3) + 20(6) = 195$

At $(9, 2)$: $C = 25(9) + 20(2) = 265$

At $(18, 0)$: $C = 25(18) + 20(0) = 450$

To optimize cost, use three bags of Brand X and six bags of Brand Y for an optimal cost of $195.

44. (a) x = the proportion of regular gasoline

y = the proportion of premium

$C = 1.84x + 2.03y$

(b) The constraints are:

$$x + y = 1$$
$$87x + 93y \geq 89$$
$$x \geq 0$$
$$y \geq 0$$

(d) Actually the only points of the plane which satisfy all the constraints are the points of the line segment connecting $(0, 1)$ and $\left(\frac{2}{3}, \frac{1}{3}\right)$. Evaluate $C = 1.84x + 2.03y$ at the two endpoints to find that the lower cost occurs at $\left(\frac{2}{3}, \frac{1}{3}\right)$.

(c)

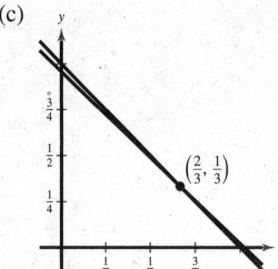

(e) The optimal cost is $C = 1.84\left(\frac{2}{3}\right) + 2.03\left(\frac{1}{3}\right) = \$1.90$.

(f) This is lower than the national average of $1.96.

45. x = number of audits

y = number of tax returns

Constraints:

$$75x + 12.5y \leq 900$$
$$10x + 2.5y \leq 155$$
$$x \geq 0$$
$$y \geq 0$$

Objective function:
$R = 2500x + 350y$

Vertices: $(0, 0)$, $(12, 0)$, $(5, 42)$, $(0, 62)$

At $(0, 0)$: $R = 2500(0) + 350(0) = 0$

At $(12, 0)$: $R = 2500(12) + 350(0) = 30,000$

At $(5, 42)$: $R = 2500(5) + 350(42) = 27,200$

At $(0, 62)$: $R = 2500(0) + 350(62) = 21,700$

The revenue will be optimal if 12 audits and 0 tax returns are done each week. The optimal revenue is $30,000.

46. The modified objective function is $R = 2000x + 350y$.

The vertices of the region are at $(0, 0)$, $(0, 62)$, $(5, 42)$, and $(12, 0)$.

At $(0, 0)$: $R = 2000(0) + 350(0) = 0$

At $(0, 62)$: $R = 2000(0) + 350(62) = 21,700$

At $(5, 42)$: $R = 2000(5) + 350(42) = 24,700$

At $(12, 0)$: $R = 2000(12) + 350(0) = 24,000$

The optimal revenue of $24,700 occurs with 5 audits and 42 tax returns.

47. $x =$ amount of Type A

$y =$ amount of Type B

Constraints: $x + y \le \quad 250,000$

$\qquad\quad x \qquad \ge \frac{1}{4}(250,000)$

$\qquad\qquad\quad y \ge \frac{1}{4}(250,000)$

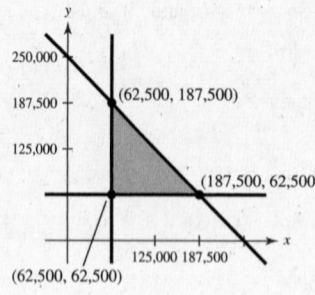

Objective Function: $P = 0.08x + 0.10y$

Vertices: $(62,500, 62,500)$, $(62,500, 187,500)$, $(187,500, 62,500)$

At $(62,500, 62,500)$: $P = 0.08(62,500) + 0.10(62,500) = \$11,250$

At $(62,500, 187,500)$: $P = 0.08(62,500) + 0.10(187,500) = \$23,750$

At $(187,500, 62,500)$: $P = 0.08(187,500) + 0.10(62,500) = \$21,250$

To obtain an optimal return the investor should allocate \$62,500 to Type A and \$187,500 to Type B. The optimal return is \$23,750.

48. $x =$ amount in investment of Type A; $y =$ amount in investment of Type B

Constraints: $x + y \le 450,000$

$\qquad\qquad\quad x \ge 225,000$

$\qquad\qquad\quad y \ge 112,500$

Objective function: $R = 0.06x + 0.1y$

Vertices: $(225,000, 112,500)$, $(337,500, 112,500)$, $(225,000, 225,000)$

At $(225,000, 112,500)$: $R = 0.06(225,000) + 0.1(112,500) = 24,750$

At $(337,500, 112,500)$: $R = 0.06(337,500) + 0.1(112,500) = 31,500$

At $(225,000, 225,000)$: $R = 0.06(225,000) + 0.1(225,000) = 36,000$

The optimal return of \$36,000 occurs for an investment of \$225,000 to Type A and \$225,000 to Type B.

49. True. The objective function has a maximum value at any point on the line segment connecting the two vertices. Both of these points are on the line $y = -x + 11$ and lie between $(4, 7)$ and $(8, 3)$.

50. True. If an objective function has a maximum value at more than one vertex, then any point on the line segment connecting the points will produce the maximum value.

51. Constraints: $x \geq 0, y \geq 0, x + 3y \leq 15, 4x + y \leq 16$

Vertex	Value of $z = 3x + ty$
$(0, 0)$	$z = 0$
$(0, 5)$	$z = 5t$
$(3, 4)$	$z = 9 + 4t$
$(4, 0)$	$z = 12$

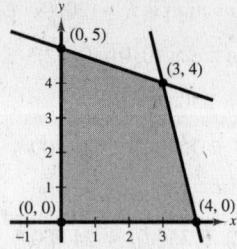

(a) For the maximum value to be at $(0, 5)$, $z = 5t$ must be greater than or equal to $z = 9 + 4t$ and $z = 12$.

$$5t \geq 9 + 4t \quad \text{and} \quad 5t \geq 12$$

$$t \geq 9 \qquad\qquad t \geq \tfrac{12}{5}$$

Thus, $t \geq 9$.

(b) For the maximum value to be at $(3, 4)$, $z = 9 + 4t$ must be greater than or equal to $z = 5t$ and $z = 12$.

$$9 + 4t \geq 5t \quad \text{and} \quad 9 + 4t \geq 12$$

$$9 \geq t \qquad\qquad 4t \geq 3$$

$$t \geq \tfrac{3}{4}$$

Thus, $\tfrac{3}{4} \leq t \leq 9$.

52. Constraints: $x \geq 0, y \geq 0, x + 2y \leq 4, x - y \leq 1$

$z = 3x + ty$

At $(0, 0)$: $z = 3(0) + t(0) = 0$

At $(1, 0)$: $z = 3(1) + t(0) = 3$

At $(2, 1)$: $z = 3(2) + t(1) = 6 + t$

At $(0, 2)$: $z = 3(0) + t(2) = 2t$

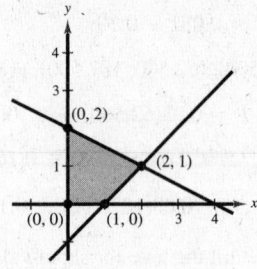

(a) For the maximum value to be at $(2, 1)$, $z = 6 + t$ must be greater than or equal to $z = 2t$ and $z = 3$.

$$6 + t \geq 2t \quad \text{and} \quad 6 + t \geq 3$$

$$6 \geq t \qquad\qquad t \geq -3$$

Thus, $-3 \leq t \leq 6$.

(b) For maximum value to be at $(0, 2)$, $z = 2t$ must be greater than or equal to $z = 6 + t$ and $z = 3$.

$$2t \geq 6 + t \quad \text{and} \quad 2t \geq 3$$

$$t \geq 6 \qquad\qquad t \geq \tfrac{3}{2}$$

Thus, $t \geq 6$.

53. There are an infinite number of objective functions that would have a maximum at $(0, 4)$. One such objective function is $z = x + 5y$.

54. There are an infinite number of objective functions that would have a maximum at $(4, 3)$. One such objective function is $z = x + y$.

55. There are an infinite number of objective functions that would have a maximum at $(5, 0)$. One such objective function is $z = 4x + y$.

56. There are an infinite number of objective functions that would have a minimum at $(5, 0)$. One such objective function is $z = -10x + y$.

57. $\dfrac{\dfrac{9}{x}}{\left(\dfrac{6}{x} + 2\right)} = \dfrac{\dfrac{9}{x}}{\dfrac{6 + 2x}{x}} = \dfrac{9}{x} \cdot \dfrac{x}{2(3 + x)} = \dfrac{9}{2(3 + x)} = \dfrac{9}{2(x + 3)}, \quad x \neq 0$

58. $\dfrac{\left(1 + \dfrac{2}{x}\right)}{x - \dfrac{4}{x}} = \dfrac{\dfrac{x + 2}{x}}{\dfrac{x^2 - 4}{x}} = \dfrac{x + 2}{x} \cdot \dfrac{x}{x^2 - 4} = \dfrac{x + 2}{x} \cdot \dfrac{x}{(x + 2)(x - 2)} = \dfrac{1}{x - 2}, \quad x \neq 0, -2$

59. $\dfrac{\left(\dfrac{4}{x^2 - 9} + \dfrac{2}{x - 2}\right)}{\left(\dfrac{1}{x + 3} + \dfrac{1}{x - 3}\right)} = \dfrac{\dfrac{4(x - 2) + 2(x^2 - 9)}{(x - 2)(x^2 - 9)}}{\dfrac{(x - 3) + (x + 3)}{x^2 - 9}}$

$= \dfrac{2x^2 + 4x - 26}{(x - 2)(x^2 - 9)} \cdot \dfrac{x^2 - 9}{2x}$

$= \dfrac{2(x^2 + 2x - 13)}{(x - 2)(2x)}$

$= \dfrac{x^2 + 2x - 13}{x(x - 2)}, \quad x \neq \pm 3$

60. $\dfrac{\left(\dfrac{1}{x + 1} + \dfrac{1}{2}\right)}{\left(\dfrac{3}{2x^2 + 4x + 2}\right)} = \dfrac{\dfrac{x + 3}{2(x + 1)}}{\dfrac{3}{2(x + 1)^2}}$

$= \dfrac{x + 3}{2(x + 1)} \cdot \dfrac{2(x + 1)^2}{3}$

$= \dfrac{(x + 3)(x + 1)}{3}, \quad x \neq -1$

61. $e^{2x} + 2e^x - 15 = 0$

$(e^x + 5)(e^x - 3) = 0$

$e^x = -5 \text{ or } e^x = 3$

No real $x = \ln 3$
solution. $x \approx 1.099$

62. $e^{2x} - 10e^x + 24 = 0$

$(e^x - 4)(e^x - 6) = 0$

$e^x = 4 \quad \text{ or } e^x = 6$

$x = \ln 4 \qquad x = \ln 6$

$x \approx 1.386 \qquad x \approx 1.792$

63. $8(62 - e^{x/4}) = 192$

$62 - e^{x/4} = 24$

$-e^{x/4} = -38$

$e^{x/4} = 38$

$\dfrac{x}{4} = \ln 38$

$x = 4 \ln 38$

$x \approx 14.550$

64. $\dfrac{150}{e^{-x} - 4} = 75$

$150 = 75e^{-x} - 300$

$75e^{-x} = 450$

$e^{-x} = 6$

$-x = \ln 6$

$x = -\ln 6$

$x \approx -1.792$

65. $7 \ln 3x = 12$

$\ln 3x = \dfrac{12}{7}$

$3x = e^{12/7}$

$x = \dfrac{e^{12/7}}{3}$

$x \approx 1.851$

66. $\ln(x + 9)^2 = 2$

$2 \ln(x + 9) = 2$

$\ln(x + 9) = 1$

$x + 9 = e$

$x = e - 9$

$x \approx -6.282$

67. $\begin{cases} -x - 2y + 3z = -23 \\ 2x + 6y - z = 17 \\ 5y + z = 8 \end{cases}$

$\begin{cases} -x - 2y + 3z = -23 \\ 2y + 5z = -29 \qquad 2\text{Eq.1} + \text{Eq.2} \\ 5y + z = 8 \end{cases}$

$\begin{cases} -x - 2y + 3z = -23 \\ 2y + 5z = -29 \\ -\frac{23}{2}z = \frac{161}{2} \qquad -\frac{5}{2}\text{Eq.2} + \text{Eq.3} \end{cases}$

$-\frac{23}{2}z = \frac{161}{2} \implies z = -7$

$2y + 5(-7) = -29 \implies y = 3$

$-x - 2(3) + 3(-7) = -23 \implies -x - 27 = -23$

$\implies x = -4$

Solution: $(-4, 3, -7)$

68. $\begin{cases} 7x - 3y + 5z = -28 \\ 4x + 4z = -16 \\ 7x + 2y - z = 0 \end{cases}$

$\begin{cases} 7x - 3y + 5z = -28 \\ 12y + 8z = 0 \qquad (-4)\text{Eq.1} + 7\text{Eq.2} \\ 5y - 6z = 28 \qquad (-1)\text{Eq.1} + \text{Eq.3} \end{cases}$

$\begin{cases} 7x - 3y + 5z = -28 \\ 12y + 8z = 0 \\ -112z = 336 \qquad (-5)\text{Eq.2} + 12\text{Eq.3} \end{cases}$

$\begin{cases} 7x - 3y + 5z = -28 \\ 3y + 2z = 0 \qquad (\frac{1}{4})\text{Eq.2} \\ z = -3 \qquad (-\frac{1}{112})\text{Eq.3} \end{cases}$

$3y + 2(-3) = 0 \implies y = 2$

$7x - 3(2) + 5(-3) = -28 \implies x = -1$

Solution: $(-1, 2, -3)$

Review Exercises for Chapter 7

1. $\begin{cases} x + y = 2 \\ x - y = 0 \implies x = y \end{cases}$

$x + x = 2$

$2x = 2$

$x = 1$

$y = 1$

Solution: $(1, 1)$

2. $\begin{cases} 2x - 3y = 3 \\ x - y = 0 \implies x = y \end{cases}$

$2y - 3y = 3$

$-y = 3$

$y = -3$

$x = -3$

Solution: $(-3, -3)$

3. $\begin{cases} 0.5x + y = 0.75 \implies y = 0.75 - 0.5x \\ 1.25x - 4.5y = -2.5 \end{cases}$

$1.25x - 4.5(0.75 - 0.5x) = -2.5$

$1.25x - 3.375 + 2.25x = -2.5$

$3.50x = 0.875$

$x = 0.25$

$y = 0.625$

Solution: $(0.25, 0.625)$

4. $\begin{cases} -x + \frac{2}{5}y = \frac{3}{5} \\ -x + \frac{1}{5}y = -\frac{4}{5} \end{cases}$

Multiply both equations by 5 to clear the denominators.

$\begin{cases} -5x + 2y = 3 \\ -5x + y = -4 \implies -5x = -4 - y \end{cases}$

$(-4 - y) + 2y = 3$

$-4 + y = 3$

$y = 7$

$-5x = -4 - 7$

$-5x = -11$

$x = \frac{11}{5}$

Solution: $\left(\frac{11}{5}, 7\right)$

5. $\begin{cases} x^2 - y^2 = 9 \\ x - y = 1 \implies x = y + 1 \end{cases}$

$(y + 1)^2 - y^2 = 9$

$2y + 1 = 9$

$y = 4$

$x = 5$

Solution: $(5, 4)$

6. $\begin{cases} x^2 + y^2 = 169 \\ 3x + 2y = 39 \implies x = \frac{1}{3}(39 - 2y) \end{cases}$

$\left[\frac{1}{3}(39 - 2y)\right]^2 + y^2 = 169$

$\frac{1}{9}(1521 - 156y + 4y^2) + y^2 = 169$

$1521 - 156y + 4y^2 + 9y^2 = 1521$

$13y^2 - 156y = 0$

$13y(y - 12) = 0 \implies y = 0, 12$

$y = 0: \ x = \frac{1}{3}(39 - 2(0)) = 13$

$y = 12: \ x = \frac{1}{3}(39 - 2(12)) = 5$

Solution: $(13, 0), (5, 12)$

7. $\begin{cases} y = 2x^2 \\ y = x^4 - 2x^2 \implies 2x^2 = x^4 - 2x^2 \end{cases}$

$0 = x^4 - 4x^2$

$0 = x^2(x^2 - 4)$

$0 = x^2(x + 2)(x - 2)$

$x = 0, x = -2, x = 2$

$y = 0, y = 8, y = 8$

Solutions: $(0, 0), (-2, 8), (2, 8)$

8. $\begin{cases} x = y + 3 \\ x = y^2 + 1 \end{cases}$

$y + 3 = y^2 + 1$

$0 = y^2 - y - 2$

$0 = (y - 2)(y + 1) \implies y = 2, -1$

$y = 2: \ x = 2 + 3 = 5$

$y = -1: \ x = -1 + 3 = 2$

Solution: $(5, 2), (2, -1)$

9. $\begin{cases} 2x - y = 10 \\ x + 5y = -6 \end{cases}$

Point of intersection: $(4, -2)$

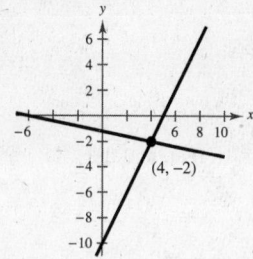

10. $\begin{cases} 8x - 3y = -3 \\ 2x + 5y = 28 \end{cases}$

The point of intersection appears to be at $(1.5, 5)$.

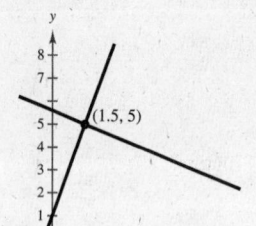

11. $\begin{cases} y = 2x^2 - 4x + 1 \\ y = x^2 - 4x + 3 \end{cases}$

Point of intersection: $(1.41, -0.66), (-1.41, 10.66)$

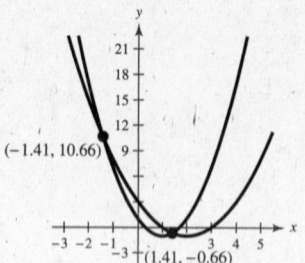

12. $y^2 - 2y + x = 0 \implies (y - 1)^2 = 1 - x \implies y = 1 \pm \sqrt{1 - x}$

$x + y = 0 \implies y = -x$

Points of intersection: $(0, 0)$ and $(-3, 3)$

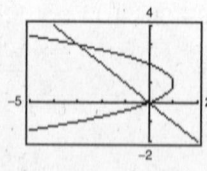

13. $\begin{cases} y = -2e^{-x} \\ 2e^x + y = 0 \implies y = -2e^x \end{cases}$

Point of intersection: $(0, -2)$

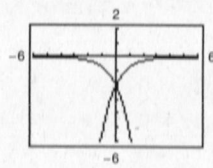

14. $y = \ln(x - 1) - 3$

$y = 4 - \frac{1}{2}x$

Point of intersection: $(9.68, -0.84)$

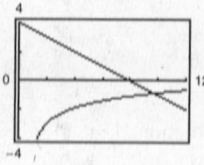

15. Let x = number of kits.

$C = 12x + 50,000$

$R = 25x$

Break-even: $R = C$

$25x = 12x + 50,000$

$13x = 50,000$

$x \approx 3846.15$

You would need to sell 3847 kits to cover your costs.

16. $\begin{cases} y = 35,000 + 0.015x \\ y = 32,500 + 0.02x \end{cases}$

$35,000 + 0.015x = 32,500 + 0.02x$

$2500 = 0.005x$

$\$500,000 = x$

For the second offer to be better, you would have to sell more than \$500,000 per year.

17.
$$2l + 2w = 480$$
$$l = 1.50w$$
$$2(1.50w) + 2w = 480$$
$$5w = 480$$
$$w = 96$$
$$l = 144$$

The dimensions are 96×144 meters.

18. $\begin{cases} 2l + 2w = 68 \\ \quad w = \frac{8}{9}l \end{cases}$

$$2l + 2\left(\frac{8}{9}\right)l = 68$$
$$\frac{34}{9}l = 68$$
$$l = 18$$
$$w = \frac{8}{9}l = 16$$

The width of the rectangle is 16 feet, and the length is 18 feet.

19. $\begin{cases} 2x - y = 2 \Rightarrow 16x - 8y = 16 \\ 6x + 8y = 39 \Rightarrow \underline{6x + 8y = 39} \end{cases}$
$$22x = 55$$
$$x = \frac{55}{22} = \frac{5}{2}$$

Back-substitute $x = \frac{5}{2}$ into Equation 1.
$$2\left(\frac{5}{2}\right) - y = 2$$
$$y = 3$$
Solution: $\left(\frac{5}{2}, 3\right)$

20. $\begin{cases} 40x + 30y = 24 \Rightarrow 40x + 30y = 24 \\ 20x - 50y = -14 \Rightarrow \underline{-40x + 100y = 28} \end{cases}$
$$130y = 52$$
$$y = \frac{2}{5}$$

Back-substitute $y = \frac{2}{5}$ in Equation 1.
$$40x + 30\left(\frac{2}{5}\right) = 24$$
$$40x = 12$$
$$x = \frac{3}{10}$$
Solution: $\left(\frac{3}{10}, \frac{2}{5}\right)$

21. $\begin{cases} 0.2x + 0.3y = 0.14 \Rightarrow 20x + 30y = 14 \Rightarrow 20x + 30y = 14 \\ 0.4x + 0.5y = 0.20 \Rightarrow 4x + 5y = 2 \Rightarrow \underline{-20x - 25y = -10} \end{cases}$
$$5y = 4$$
$$y = \frac{4}{5}$$

Back-substitute $y = \frac{4}{5}$ into Equation 2.
$$4x + 5\left(\frac{4}{5}\right) = 2$$
$$4x = -2$$
$$x = -\frac{1}{2}$$
Solution: $\left(-\frac{1}{2}, \frac{4}{5}\right) = (-0.5, 0.8)$

22. $\begin{cases} 12x + 42y = -17 \Rightarrow 36x + 126y = -51 \\ 30x - 18y = 19 \Rightarrow \underline{210x - 126y = 133} \end{cases}$
$$246x = 82$$
$$x = \frac{1}{3}$$

Back-substitute $x = \frac{1}{3}$ in Equation 1.
$$12\left(\frac{1}{3}\right) + 42y = -17$$
$$42y = -21$$
$$y = -\frac{1}{2}$$
Solution: $\left(\frac{1}{3}, -\frac{1}{2}\right)$

23. $\begin{cases} 3x - 2y = 0 \Rightarrow 3x - 2y = 0 \\ 3x + 2(y + 5) = 10 \Rightarrow \underline{3x + 2y = 0} \end{cases}$
$$6x = 0$$
$$x = 0$$

Back-substitute $x = 0$ into Equation 1.
$$3(0) - 2y = 0$$
$$2y = 0$$
$$y = 0$$
Solution: $(0, 0)$

24. $\begin{cases} 7x + 12y = 63 \\ 2x + 3(y + 2) = 21 \end{cases}$

$\begin{cases} 7x + 12y = 63 \Rightarrow -7x - 12y = -63 \\ 2x + 3y = 15 \Rightarrow \underline{8x + 12y = 60} \end{cases}$
$$x = -3$$

Back-substitute $x = -3$ in Equation 1.
$$7(-3) + 12y = 63$$
$$12y = 84$$
$$y = 7$$
Solution: $(-3, 7)$

25. $\begin{cases} 1.25x - 2y = 3.5 \Rightarrow & 5x - 8y = 14 \\ 5x - 8y = 14 \Rightarrow & \underline{-5x + 8y = -14} \\ & \qquad\qquad 0 = \quad 0 \end{cases}$

There are infinitely many solutions.

Let $y = a$, then $5x - 8a = 14 \Rightarrow x = \frac{8}{5}a + \frac{14}{5}$:

Solution: $\left(\frac{8}{5}a + \frac{14}{5}, a\right)$ where a is any real number.

26. $\begin{cases} 1.5x + 2.5y = 8.5 \Rightarrow & 3x + 5y = 17 \\ 6x + 10y = 24 \Rightarrow & \underline{-3x - 5y = -12} \\ & \qquad\qquad 0 = \quad 5 \end{cases}$

The system is inconsistent. There is no solution.

27. $\begin{cases} x + 5y = 4 \Rightarrow & x + 5y = 4 \\ x - 3y = 6 \Rightarrow & \underline{-x + 3y = -6} \\ & \qquad 8y = -2 \Rightarrow y = -\frac{1}{4} \end{cases}$

Matches graph (d). The system has one solution and is consistent.

28. $\begin{cases} -3x + y = -7 \\ 9x - 3y = 21 \end{cases}$

$-3x + y = -7 \Rightarrow y = 3x - 7$;
The graph contains $(0, -7)$ and $(2, -1)$.

$9x - 3y = 21 \Rightarrow -3y = -9x + 21 \Rightarrow y = 3x - 7$;
The graph is the same as the previous graph.

The graph of the system matches (c). The system has infinitely many solutions and is consistent.

29. $\begin{cases} 3x - y = 7 \Rightarrow & 6x - 2y = 14 \\ -6x + 2y = 8 \Rightarrow & \underline{-6x + 2y = 8} \\ & \qquad\qquad 0 \neq 22 \end{cases}$

Matches graph (b). The system has no solution and is inconsistent.

30. $\begin{cases} 2x - y = -3 \\ x + 5y = 4 \end{cases}$

$2x - y = -3 \Rightarrow -y = -2x - 3 \Rightarrow y = 2x + 3$;
The graph contains $(0, 3)$ and $(-2, -1)$.

$x + 5y = 4 \Rightarrow 5y = -x + 4 \Rightarrow y = -\frac{1}{5}x + \frac{4}{5}$;

The graph contains $\left(0, \frac{4}{5}\right)$ and $(4, 0)$.

The graph of the system matches (a). The system has one solution and is consistent.

31. $37 - 0.0002x = 22 + 0.00001x$

$\qquad\qquad 15 = 0.00021x$

$\qquad\qquad x = \dfrac{500,000}{7}, p = \dfrac{159}{7}$

Point of equilibrium: $\left(\dfrac{500,000}{7}, \dfrac{159}{7}\right)$

32. $45 + 0.0002x = 120 - 0.0001x$

$\qquad\qquad 0.0003x = 75$

$\qquad\qquad x = 250,000$ units

$\qquad\qquad p = \$95.00$

Point of equilibrium: $(250,000, 95)$

33. $\begin{cases} x - 4y + 3z = 3 \\ \quad\;\; -y + z = -1 \\ \qquad\qquad z = -5 \end{cases}$

$-y + (-5) = -1 \Rightarrow y = -4$

$x - 4(-4) + 3(-5) = 3 \Rightarrow x = 2$

Solution: $(2, -4, -5)$

34. $\begin{cases} x - 7y + 8z = 85 \\ \quad\;\; y - 9z = -35 \\ \qquad\qquad z = 3 \end{cases}$

$y - 9(3) = -35 \Rightarrow y = -8$

$x - 7(-8) + 8(3) = 85 \Rightarrow x = 5$

Solution: $(5, -8, 3)$

35. $\begin{cases} x + 2y + 6z = 4 \\ -3x + 2y - z = -4 \\ 4x + 2z = 16 \end{cases}$

$\begin{cases} x + 2y + 6z = 4 \\ 8y + 17z = 8 \\ -8y - 22z = 0 \end{cases}$ $3\text{Eq.1} + \text{Eq.2}$
   $-4\text{Eq.1} + \text{Eq.3}$

$\begin{cases} x + 2y + 6z = 4 \\ 8y + 17z = 8 \\ -5z = 8 \end{cases}$ $\text{Eq.2} + \text{Eq.3}$

$\begin{cases} x + 2y + 6z = 4 \\ 8y + 17z = 8 \\ z = -\frac{8}{5} \end{cases}$ $-\frac{1}{5}\text{Eq.3}$

$8y + 17\left(-\frac{8}{5}\right) = 8 \implies y = \frac{22}{5}$

$x + 2\left(\frac{22}{5}\right) + 6\left(-\frac{8}{5}\right) = 4 \implies x = \frac{24}{5}$

Solution: $\left(\frac{24}{5}, \frac{22}{5}, -\frac{8}{5}\right)$

36. $\begin{cases} x + 3y - z = 13 \\ 2x - 5z = 23 \\ 4x - y - 2z = 14 \end{cases}$ Equation 1
   Equation 2
   Equation 3

$\begin{cases} x + 3y - z = 13 \\ -6y - 3z = -3 \\ -13y + 2z = -38 \end{cases}$ $(-2)\text{Eq.1} + \text{Eq.2}$
   $(-4)\text{Eq.1} + \text{Eq.3}$

$\begin{cases} x + 3y - z = 13 \\ -6y - 3z = -3 \\ \frac{17}{2}z = -\frac{63}{2} \end{cases}$ $\left(-\frac{13}{6}\right)\text{Eq.2} + \text{Eq.3}$

$\begin{cases} x + 3y - z = 13 \\ y + \frac{1}{2}z = \frac{1}{2} \\ z = -\frac{63}{17} \end{cases}$ $\left(-\frac{1}{6}\right)\text{Eq.2}$
   $\left(\frac{2}{17}\right)\text{Eq.3}$

$y + \frac{1}{2}\left(-\frac{63}{17}\right) = \frac{1}{2} \implies y = \frac{40}{17}$

$x + 3\left(\frac{40}{17}\right) - \left(-\frac{63}{17}\right) = 13 \implies x = \frac{38}{17}$

Solution: $\left(\frac{38}{17}, \frac{40}{17}, -\frac{63}{17}\right)$

37. $\begin{cases} x - 2y + z = -6 \\ 2x - 3y = -7 \\ -x + 3y - 3z = 11 \end{cases}$

$\begin{cases} x - 2y + z = -6 \\ y - 2z = 5 \\ y - 2z = 5 \end{cases}$ $-2\text{Eq.1} + \text{Eq.2}$
   $\text{Eq.1} + \text{Eq.3}$

$\begin{cases} x - 2y + z = -6 \\ y - 2z = 5 \\ 0 = 0 \end{cases}$ $-\text{Eq.2} + \text{Eq.3}$

Let $z = a$, then:

$$y = 2a + 5$$
$$x - 2(2a + 5) + a = -6$$
$$x - 3a - 10 = -6$$
$$x = 3a + 4$$

Solution: $(3a + 4, 2a + 5, a)$ where a is any real number.

38. $\begin{cases} 2x + 6z = -9 \\ 3x - 2y + 11z = -16 \\ 3x - y + 7z = -11 \end{cases}$ Equation 1
   Equation 2
   Equation 3

$\begin{cases} -x + 2y - 5z = 7 \\ 3x - 2y + 11z = -16 \\ 3x - y + 7z = -11 \end{cases}$ $(-1)\text{Eq.2} + \text{Eq.1}$

$\begin{cases} -x + 2y - 5z = 7 \\ 4y - 4z = 5 \\ 5y - 8z = 10 \end{cases}$ $3\text{Eq.1} + \text{Eq.2}$
   $3\text{Eq.1} + \text{Eq.3}$

$\begin{cases} -x + 2y - 5z = 7 \\ 4y - 4z = 5 \\ -3y = 0 \end{cases}$ $(-2)\text{Eq.2} + \text{Eq.3}$

$\begin{cases} -x + 2y - 5z = 7 \\ y - z = \frac{5}{4} \\ y = 0 \end{cases}$ $\left(\frac{1}{4}\right)\text{Eq.2}$
   $\left(-\frac{1}{3}\right)\text{Eq.3}$

$(0) - z = \frac{5}{4} \implies z = -\frac{5}{4}$

$-x + 2(0) - 5\left(-\frac{5}{4}\right) = 7 \implies x = -\frac{3}{4}$

Solution: $\left(-\frac{3}{4}, 0, -\frac{5}{4}\right)$

39. $\begin{cases} 5x - 12y + 7z = 16 \implies \\ 3x - 7y + 4z = 9 \implies \end{cases}$ $\begin{cases} 15x - 36y + 21z = 48 \\ -15x + 35y - 20z = -45 \\ \hline -y + z = 3 \end{cases}$

Let $z = a$. Then $y = a - 3$ and

$$5x - 12(a - 3) + 7a = 16 \implies x = a - 4.$$

Solution: $(a - 4, \ a - 3, a)$ where a is any real number.

40. $\begin{cases} 2x + 5y - 19z = 34 \implies \\ 3x + 8y - 31z = 54 \implies \end{cases}$ $\begin{cases} 6x + 15y - 57z = 102 \\ -6x - 16y + 62z = -108 \\ \hline -y + 5z = -6 \end{cases}$

Let $z = a$. Then:

$$-y + 5a = -6 \implies y = 5a + 6$$

$2x + 5(5a + 6) - 19a = 34 \implies x = -3a + 2$

Solution: $(-3a + 2, 5a + 6, a)$ where a is any real number.

41. $y = ax^2 + bx + c$ through $(0, -5)$, $(1, -2)$, and $(2, 5)$.

$(0, -5)$: $-5 = \qquad + c \Rightarrow \qquad c = -5$

$(1, -2)$: $-2 = a + b + c \Rightarrow \begin{cases} a + b = 3 \\ 2a + b = 5 \end{cases}$

$(2, 5)$: $5 = 4a + 2b + c \Rightarrow$

$$\begin{cases} 2a + b = 5 \\ -a - b = -3 \end{cases}$$

$$a = 2$$

$$b = 1$$

The equation of the parabola is $y = 2x^2 + x - 5$.

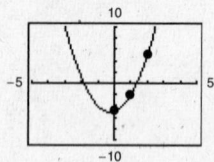

42. $y = ax^2 + bx + c$ through $(-5, 6)$, $(1, 0)$, $(2, 20)$.

$(-5, 6)$: $6 = 25a - 5b + c$

$(1, 0)$: $0 = a + b + c \Rightarrow c = -a - b$

$(2, 20)$: $20 = 4a + 2b + c$

$$\begin{cases} 24a - 6b = 6 \implies 24a - 6b = 6 \\ 3a + b = 20 \implies -24a - 8b = -160 \end{cases}$$

$$-14b = -154$$

$$b = 11$$

$$3a + 11 = 20 \Rightarrow a = 3$$

$$c = -3 - 11 \Rightarrow c = -14$$

The equation of the parabola is $y = 3x^2 + 11x - 14$.

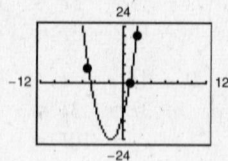

43. $x^2 + y^2 + Dx + Ey + F = 0$ through $(-1, -2)$, $(5, -2)$ and $(2, 1)$.

$(-1, -2)$: $5 - D - 2E + F = 0 \Rightarrow \begin{cases} D + 2E - F = 5 \\ 5D - 2E + F = -29 \\ 2D + E + F = -5 \end{cases}$

$(5, -2)$: $29 + 5D - 2E + F = 0 \Rightarrow$

$(2, 1)$: $5 + 2D + E + F = 0 \Rightarrow$

From the first two equations we have

$$6D = -24$$

$$D = -4.$$

Substituting $D = -4$ into the second and third equations yields:

$$\begin{aligned} -20 - 2E + F = -29 &\Rightarrow \begin{cases} -2E + F = -9 \\ -E - F = -3 \end{cases} \\ -8 + E + F = -5 &\Rightarrow \end{aligned}$$

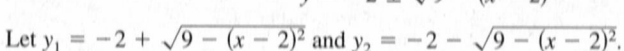

$$-3E = -12$$

$$E = 4$$

$$F = -1$$

The equation of the circle is $x^2 + y^2 - 4x + 4y - 1 = 0$.

To verify the result using a graphing utility, solve the equation for y.

$$(x^2 - 4x + 4) + (y^2 + 4y + 4) = 1 + 4 + 4$$

$$(x - 2)^2 + (y + 2)^2 = 9$$

$$(y + 2)^2 = 9 - (x - 2)^2$$

$$y = -2 \pm \sqrt{9 - (x - 2)^2}$$

Let $y_1 = -2 + \sqrt{9 - (x - 2)^2}$ and $y_2 = -2 - \sqrt{9 - (x - 2)^2}$.

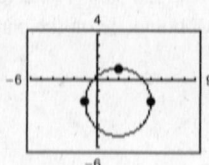

44. $x^2 + y^2 + Dx + Ey + F = 0$ through $(1, 4), (4, 3), (-2, -5)$.

$(1, 4)$: $17 + D + 4E + F = 0$

$(4, 3)$: $25 + 4D + 3E + F = 0$

$(-2, -5)$: $29 - 2D - 5E + F = 0$

$$\begin{cases} D + 4E + F = -17 & \text{Equation 1} \\ 4D + 3E + F = -25 & \text{Equation 2} \\ 2D + 5E - F = 29 & \text{Equation 3} \end{cases}$$

$$\begin{cases} D + 4E + F = -17 \\ \quad -13E - 3F = 43 & (-4)\text{Eq.1} + \text{Eq.2} \\ \quad -3E - 3F = 63 & (-2)\text{Eq.1} + \text{Eq.3} \end{cases}$$

$$\begin{cases} D + 4E + F = -17 \\ \quad -3E - 3F = 63 & \text{Interchange equations.} \\ \quad -13E - 3F = 43 \end{cases}$$

$$\begin{cases} D + 4E + F = -17 \\ \quad -3E - 3F = 63 \\ \quad\quad\quad 10F = -230 & \left(-\tfrac{13}{3}\right)\text{Eq.2} + \text{Eq.3} \end{cases}$$

$F = -23, E = 2, D = -2$

The equation of the circle is $x^2 + y^2 - 2x + 2y - 23 = 0$.

To verify the result using a graphing utility, solve the equation for y.

$(x^2 - 2x + 1) + (y^2 + 2y + 1) = 23 + 1 + 1$

$$(x - 1)^2 + (y + 1)^2 = 25$$

$$(y + 1)^2 = 25 - (x - 1)^2$$

$$y = -1 \pm \sqrt{25 - (x - 1)^2}$$

Let $y_1 = -1 + \sqrt{25 - (x - 1)^2}$ and $y_2 = -1 - \sqrt{25 - (x - 1)^2}$.

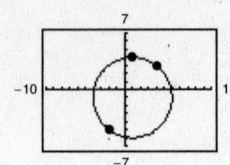

45. $(3, 101.7), (4, 108.4), (5, 121.1)$

(a) $n = 3, \sum_{i=1}^{3} x_i = 12, \sum_{i=1}^{3} x_i^2 = 50, \sum_{i=1}^{3} x_i^3 = 216, \sum_{i=1}^{3} x_i^4 = 962, \sum_{i=1}^{3} y_i = 331.2, \sum_{i=1}^{3} x_i y_i = 1344.2, \sum_{i=1}^{3} x_i^2 y_i = 5677.2$

$$\begin{cases} 3c + 12b + 50a = 331.2 \\ 12c + 50b + 216a = 1344.2 \\ 50c + 216b + 962a = 5677.2 \end{cases}$$

Solving this system yields $c = 117.6, b = -14.3, a = 3$.

Quadratic model: $y = 3x^2 - 14.3x + 117.6$

(b)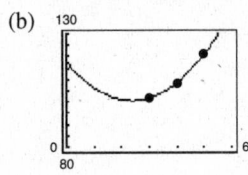

The model is a good fit to the data. The actual points lie on the parabola.

(c) For 2008, use $x = 8$:

$y = 3(8)^2 - 14.3(8) + 117.6$

$= 195.2$ million online shoppers

This answer seems reasonable.

46. From the following chart we obtain our system of equations.

	A	B	C
Mixture X	$\frac{1}{5}$	$\frac{2}{5}$	$\frac{2}{5}$
Mixture Y	0	0	1
Mixture Z	$\frac{1}{3}$	$\frac{1}{3}$	$\frac{1}{3}$
Desired Mixture	$\frac{6}{27}$	$\frac{8}{27}$	$\frac{13}{27}$

$\left.\begin{array}{l} \frac{1}{5}x + \frac{1}{3}z = \frac{6}{27} \\ \frac{2}{5}x + \frac{1}{3}z = \frac{8}{27} \end{array}\right\} x = \frac{10}{27}, z = \frac{12}{27}$

$\frac{2}{5}x + y + \frac{1}{3}z = \frac{13}{27} \implies y = \frac{5}{27}$

To obtain the desired mixture, use 10 gallons of spray X, 5 gallons of spray Y, and 12 gallons of spray Z.

47. Let $x = $ amount invested at 7%

$y = $ amount invested at 9%

$z = $ amount invested at 11%.

$y = x - 3000$ and

$z = x - 5000 \implies y + z = 2x - 8000$

$$\begin{cases} x + y + z = 40,000 \\ 0.07x + 0.09y + 0.11z = 3500 \\ y + z = 2x - 8000 \end{cases}$$

$x + (2x - 8000) = 40,000 \implies x = 16,000$

$y = 16,000 - 3000 \implies y = 13,000$

$z = 16,000 - 5000 \implies z = 11,000$

Thus, \$16,000 was invested at 7%, \$13,000 at 9% and \$11,000 at 11%.

48. $s = \frac{1}{2}at^2 + v_0 t + s_0$

(a) When $t = 1$: $s = 134$: $\frac{1}{2}a(1)^2 + v_0(1) + s_0 = 134 \implies a + 2v_0 + 2s_0 = 268$

When $t = 2$: $s = 86$: $\frac{1}{2}a(2)^2 + v_0(2) + s_0 = 86 \implies 2a + 2v_0 + s_0 = 86$

When $t = 3$: $s = 6$: $\frac{1}{2}a(3)^2 + v_0(3) + s_0 = 6 \implies 9a + 6v_0 + 2s_0 = 12$

$$\begin{cases} a + 2v_0 + 2s_0 = 268 \\ 2a + 2v_0 + s_0 = 86 \\ 9a + 6v_0 + 2s_0 = 12 \end{cases}$$

$$\begin{cases} a + 2v_0 + 2s_0 = 268 \\ \quad - 2v_0 - 3s_0 = -450 \qquad (-2)\text{Eq.1} + \text{Eq.2} \\ \quad -12v_0 - 16s_0 = -2400 \qquad (-9)\text{Eq.1} + \text{Eq.3} \end{cases}$$

$$\begin{cases} a + 2v_0 + 2s_0 = 268 \\ \quad - 2v_0 - 3s_0 = -450 \\ \quad 3v_0 + 4s_0 = 600 \qquad (-\frac{1}{4})\text{Eq.3} \end{cases}$$

$$\begin{cases} a + 2v_0 + 2s_0 = 268 \\ \quad - 2v_0 - 3s_0 = -450 \\ \quad\quad - s_0 = -150 \qquad 3\text{Eq.2} + 2\text{Eq.3} \end{cases}$$

$-s_0 = -150 \implies s_0 = 150$

$-2v_0 - 3(150) = -450 \implies v_0 = 0$

$a + 2(0) + 2(150) = 268 \implies a = -32$

The position equation is $s = \frac{1}{2}(-32)t^2 + (0)t + 150$, or $s = -16t^2 + 150$.

(b) When $t = 1$: $s = 184$: $\frac{1}{2}a(1)^2 + v_0(1) + s_0 = 184 \implies a + 2v_0 + 2s_0 = 368$

When $t = 2$: $s = 116$: $\frac{1}{2}a(2)^2 + v_0(2) + s_0 = 116 \implies 2a + 2v_0 + s_0 = 116$

When $t = 3$: $s = 16$: $\frac{1}{2}a(3)^2 + v_0(3) + s_0 = 16 \implies 9a + 6v_0 + 2s_0 = 32$

$$\begin{cases} a + 2v_0 + 2s_0 = 368 \\ 2a + 2v_0 + s_0 = 116 \\ 9a + 6v_0 + 2s_0 = 32 \end{cases}$$

$$\begin{cases} a + 2v_0 + 2s_0 = 368 \\ \quad - 2v_0 - 3s_0 = -620 \qquad (-2)\text{Eq.1} + \text{Eq.2} \\ \quad -12v_0 - 16s_0 = -3280 \qquad (-9)\text{Eq.1} + \text{Eq.3} \end{cases}$$

$$\begin{cases} a + 2v_0 + 2s_0 = 368 \\ \quad - 2v_0 - 3s_0 = -620 \\ \quad 3v_0 + 4s_0 = 820 \qquad (-\frac{1}{4})\text{Eq.3} \end{cases}$$

$$\begin{cases} a + 2v_0 + 2s_0 = 368 \\ \quad - 2v_0 - 3s_0 = -620 \\ \quad\quad - s_0 = -220 \qquad 3\text{Eq.2} + 2\text{Eq.3} \end{cases}$$

$-s_0 = -220 \implies s_0 = 220$

$-2v_0 - 3(220) = -620 \implies v_0 = -20$

$a + 2(-20) + 2(220) = 368 \implies a = -32$

The position equation is $s = \frac{1}{2}(-32)t^2 + (-20)t + 220$, or $s = -16t^2 - 20t + 220$.

49. $\dfrac{3}{x^2 + 20x} = \dfrac{3}{x(x + 20)} = \dfrac{A}{x} + \dfrac{B}{x + 20}$

50. $\dfrac{x - 8}{x^2 - 3x - 28} = \dfrac{x - 8}{(x - 7)(x + 4)} = \dfrac{A}{x - 7} + \dfrac{B}{x + 4}$

51. $\dfrac{3x - 4}{x^3 - 5x^2} = \dfrac{3x - 4}{x^2(x - 5)} = \dfrac{A}{x} + \dfrac{B}{x^2} + \dfrac{C}{x - 5}$

52. $\dfrac{x - 2}{x(x^2 + 2)^2} = \dfrac{A}{x} + \dfrac{Bx + C}{x^2 + 2} + \dfrac{Dx + E}{(x^2 + 2)^2}$

53. $\dfrac{4-x}{x^2+6x+8} = \dfrac{A}{x+2} + \dfrac{B}{x+4}$

$$4 - x = A(x+4) + B(x+2)$$

Let $x = -2$: $6 = 2A \implies A = 3$

Let $x = -4$: $8 = -2B \implies B = -4$

$$\dfrac{4-x}{x^2+6x+8} = \dfrac{3}{x+2} - \dfrac{4}{x+4}$$

54. $\dfrac{-x}{x^2+3x+2} = \dfrac{A}{x+1} + \dfrac{B}{x+2}$

$$-x = A(x+2) + B(x+1)$$

Let $x = -1$: $1 = A$

Let $x = -2$: $2 = -B \implies B = -2$

$$\dfrac{-x}{x^2+3x+2} = \dfrac{1}{x+1} - \dfrac{2}{x+2}$$

55. $\dfrac{x^2}{x^2+2x-15} = 1 - \dfrac{2x-15}{x^2+2x-15}$

$$\dfrac{-2x+15}{(x+5)(x-3)} = \dfrac{A}{x+5} + \dfrac{B}{x-3}$$

$$-2x + 15 = A(x-3) + B(x+5)$$

Let $x = -5$: $25 = -8A \implies A = -\dfrac{25}{8}$

Let $x = 3$: $9 = 8B \implies B = \dfrac{9}{8}$

$$\dfrac{x^2}{x^2+2x-15} = 1 - \dfrac{25}{8(x+5)} + \dfrac{9}{8(x-3)}$$

56. $\dfrac{9}{x^2-9} = \dfrac{A}{x-3} + \dfrac{B}{x+3}$

$$9 = A(x+3) + B(x-3)$$

Let $x = 3$: $9 = 6A \implies A = \dfrac{3}{2}$

Let $x = -3$: $9 = -6B \implies B = -\dfrac{3}{2}$

$$\dfrac{9}{x^2-9} = \dfrac{1}{2}\left(\dfrac{3}{x-3} - \dfrac{3}{x+3}\right)$$

57. $\dfrac{x^2+2x}{x^3-x^2+x-1} = \dfrac{x^2+2x}{(x-1)(x^2+1)} = \dfrac{A}{x-1} + \dfrac{Bx+C}{x^2+1}$

$$x^2 + 2x = A(x^2+1) + (Bx+C)(x-1)$$
$$= Ax^2 + A + Bx^2 - Bx + Cx - C$$
$$= (A+B)x^2 + (-B+C)x + (A-C)$$

Equating coefficients of like terms gives $1 = A + B$, $2 = -B + C$, and $0 = A - C$. Adding both sides of all three equations gives $3 = 2A$. Therefore, $A = \frac{3}{2}$, $B = -\frac{1}{2}$, and $C = \frac{3}{2}$.

$$\dfrac{x^2+2x}{x^3-x^2+x-1} = \dfrac{\frac{3}{2}}{x-1} + \dfrac{-\frac{1}{2}x+\frac{3}{2}}{x^2+1}$$

$$= \dfrac{1}{2}\left(\dfrac{3}{x-1} - \dfrac{x-3}{x^2+1}\right)$$

58. $\dfrac{4x}{3(x-1)^2} = \dfrac{A}{x-1} + \dfrac{B}{(x-1)^2}$

$$\dfrac{4}{3}x = A(x-1) + B$$

Let $x = 1$: $\dfrac{4}{3} = B$

Let $x = 2$: $\dfrac{8}{3} = A + \dfrac{4}{3} \implies A = \dfrac{4}{3}$

$$\dfrac{4x}{3(x-1)^2} = \dfrac{4}{3(x-1)} + \dfrac{4}{3(x-1)^2}$$

59. $\dfrac{3x^3+4x}{(x^2+1)^2} = \dfrac{Ax+B}{x^2+1} + \dfrac{Cx+D}{(x^2+1)^2}$

$$3x^3 + 4x = (Ax+B)(x^2+1) + Cx + D$$
$$= Ax^3 + Bx^2 + (A+C)x + (B+D)$$

Equating coefficients of like powers:

$$3 = A$$
$$0 = B$$
$$4 = 3 + C \implies C = 1$$
$$0 = B + D \implies D = 0$$

$$\dfrac{3x^3+4x}{(x^2+1)^2} = \dfrac{3x}{x^2+1} + \dfrac{x}{(x^2+1)^2}$$

60. $\dfrac{4x^2}{(x-1)(x^2+1)} = \dfrac{A}{x-1} + \dfrac{Bx+C}{x^2+1}$

$$4x^2 = A(x^2+1) + (Bx+C)(x-1)$$
$$= Ax^2 + A + Bx^2 - Bx + Cx - C$$
$$= (A+B)x^2 + (-B+C)x + (A-C)$$

Equating coefficients of like terms gives $4 = A + B$, $0 = -B + C$, and $0 = A - C$. Adding both sides of all three equations gives $4 = 2A$, so $A = 2$. Then $B = 2$ and $C = 2$.

$$\dfrac{4x^2}{(x-1)(x^2+1)} = \dfrac{2}{x-1} + \dfrac{2x+2}{x^2+1}$$

$$= 2\left(\dfrac{1}{x-1} + \dfrac{x+1}{x^2+1}\right)$$

61. $y \le 5 - \frac{1}{2}x$

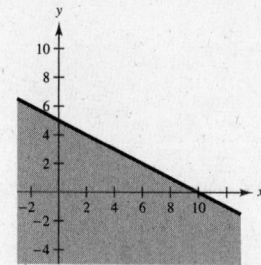

62. $3y - x \ge 7$

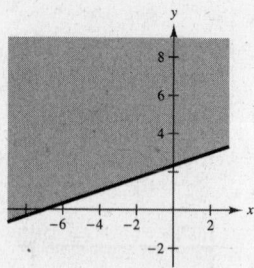

63. $y - 4x^2 > -1$

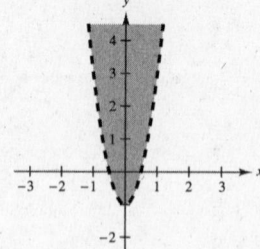

64. $y \le \dfrac{3}{x^2 + 2}$

Using a solid line, graph

$y = \dfrac{3}{x^2 + 2}$, and shade

below the curve. Use $(0, 0)$
as a test point.

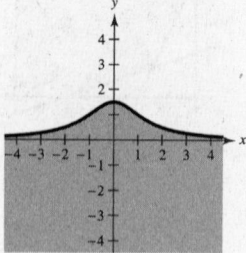

65. $\begin{cases} x + 2y \le 160 \\ 3x + y \le 180 \\ x \ge 0 \\ y \ge 0 \end{cases}$

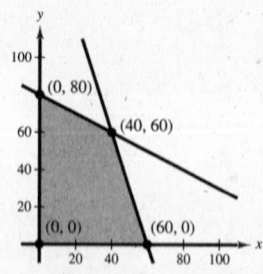

Vertex A	Vertex B	Vertex C	Vertex D
$x + 2y = 160$	$x + 2y = 160$	$3x + y = 180$	$x = 0$
$3x + y = 180$	$x = 0$	$y = 0$	$y = 0$
$(40, 60)$	$(0, 80)$	$(60, 0)$	$(0, 0)$

66. $\begin{cases} 2x + 3y \le 24 \\ 2x + y \le 16 \\ x \ge 0 \\ y \ge 0 \end{cases}$

Vertices: $(0, 0)$, $(0, 8)$, $(6, 4)$, $(8, 0)$

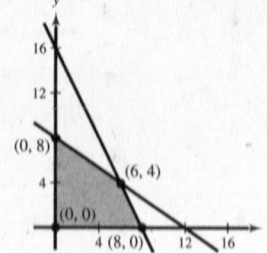

67. $\begin{cases} 3x + 2y \ge 24 \\ x + 2y \ge 12 \\ 2 \le x \le 15 \\ y \le 15 \end{cases}$

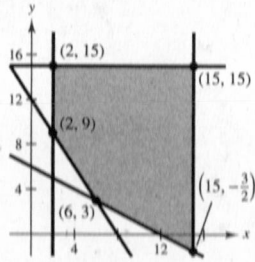

Vertex A	Vertex B	Vertex C	Vertex D	Vertex E
$3x + 2y = 24$	$3x + 2y = 24$	$x = 2$	$x = 15$	$x + 2y = 12$
$x + 2y = 12$	$x = 2$	$y = 15$	$y = 15$	$x = 15$
$(6, 3)$	$(2, 9)$	$(2, 15)$	$(15, 15)$	$\left(15, -\frac{3}{2}\right)$

68. $\begin{cases} 2x + y \geq 16 \\ x + 3y \geq 18 \\ 0 \leq x \leq 25 \\ 0 \leq y \leq 25 \end{cases}$

Vertices: $(6, 4), (0, 16), (0, 25), (25, 25), (25, 0), (18, 0)$

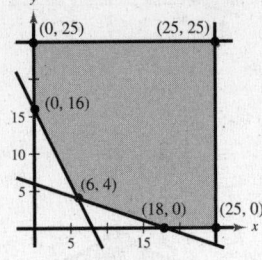

69. $\begin{cases} y < x + 1 \\ y > x^2 - 1 \end{cases}$

Vertices:

$x + 1 = x^2 - 1$

$ 0 = x^2 - x - 2 = (x + 1)(x - 2)$

$x = -1 \text{ or } x = 2$

$y = 0 \qquad y = 3$

$(-1, 0) \qquad (2, 3)$

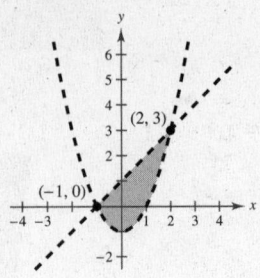

70. $\begin{cases} y \leq 6 - 2x - x^2 \\ y \geq x + 6 \end{cases}$

Vertices: $\quad x + 6 = 6 - 2x - x^2$

$x^2 + 3x = 0$

$x(x + 3) = 0 \implies x = 0, -3$

$(0, 6), (-3, 3)$

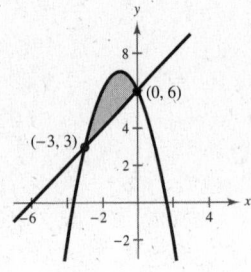

71. $\begin{cases} 2x - 3y \geq 0 \\ 2x - y \leq 8 \\ y \geq 0 \end{cases}$

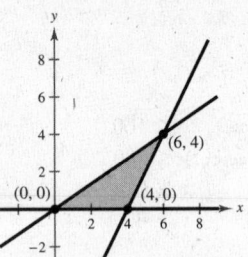

Vertex A	Vertex B	Vertex C
$2x - 3y = 0$	$2x - 3y = 0$	$2x - y = 8$
$2x - y = 8$	$y = 0$	$y = 0$
$(6, 4)$	$(0, 0)$	$(4, 0)$

72. $\begin{cases} x^2 + y^2 \leq 9 \implies y^2 \leq 9 - x^2 \\ (x - 3)^2 + y^2 \leq 9 \implies y^2 \leq 9 - (x - 3)^2 \end{cases}$

Vertices: $\qquad 9 - x^2 = 9 - (x - 3)^2$

$(x - 3)^2 - x^2 = 0$

$x^2 - 6x + 9 - x^2 = 0$

$x = \dfrac{3}{2}$

$\left(\dfrac{3}{2}, \pm\dfrac{3\sqrt{3}}{2} \right)$

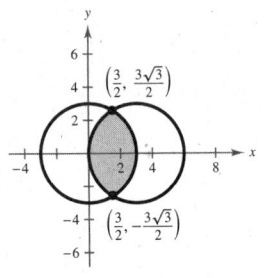

73. $x = $ number of units of Product I

$y = $ number of units of Product II

$\begin{cases} 20x + 30y \leq 24{,}000 \\ 12x + 8y \leq 12{,}400 \\ x \geq 0 \\ y \geq 0 \end{cases}$

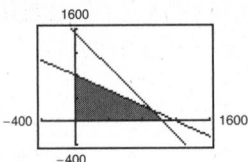

74. (a) Let x = amount of Food X.,

Let y = amount of Food Y.

$$\begin{cases} 12x + 15y \geq 300 \\ 10x + 20y \geq 280 \\ 20x + 12y \geq 300 \\ x \geq 0 \\ y \geq 0 \end{cases}$$

(b)

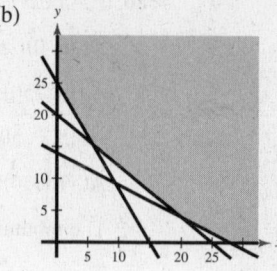

(c) Answers may vary. For example, (15, 8) or (16, 9) represent acceptable quantities (x, y) for Foods X and Y.

75. (a)

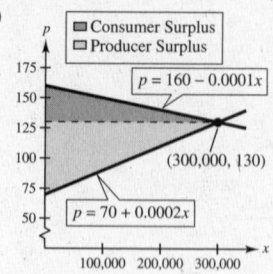

$$160 - 0.0001x = 70 + 0.0002x$$

$$90 = 0.0003x$$

$$x = 300,000 \text{ units}$$

$$p = \$130$$

Point of equilibrium: (300,000, 130)

(b) Consumer surplus: $\frac{1}{2}(300,000)(30) = \$4,500,000$

Producer surplus: $\frac{1}{2}(300,000)(60) = \$9,000,000$

76. (a)

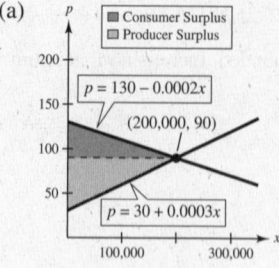

$$130 - 0.0002x = 30 + 0.0003x$$

$$100 = 0.0005x$$

$$x = 200,000 \text{ units}$$

$$p = \$90$$

Point of equilibrium: (200,000, 90)

(b) Consumer surplus: $\frac{1}{2}(200,000)(40) = \$4,000,000$

Producer surplus: $\frac{1}{2}(200,000)(60) = \$6,000,000$

77. Objective function: $z = 3x + 4y$

Constraints: $\begin{cases} x \geq 0 \\ y \geq 0 \\ 2x + 5y \leq 50 \\ 4x + y \leq 28 \end{cases}$

At $(0, 0)$: $z = 0$

At $(0, 10)$: $z = 40$

At $(5, 8)$: $z = 47$

At $(7, 0)$: $z = 21$

The minimum value is 0 at $(0, 0)$.

The maximum value is 47 at $(5, 8)$.

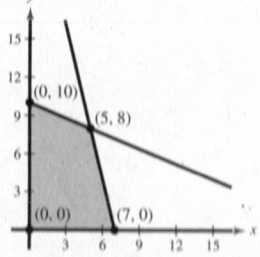

78. $z = 10x + 7y$

At $(0, 100)$: $z = 10(0) + 7(100) = 700$

At $(25, 50)$: $z = 10(25) + 7(50) = 600$

At $(75, 0)$: $z = 10(75) + 7(0) = 750$

The minimum value is 600 at $(25, 50)$.

There is no maximum value.

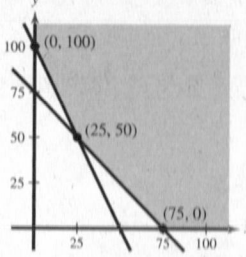

79. Objective function: $z = 1.75x + 2.25y$

Constraints: $\begin{cases} x \geq 0 \\ y \geq 0 \\ 2x + y \geq 25 \\ 3x + 2y \geq 45 \end{cases}$

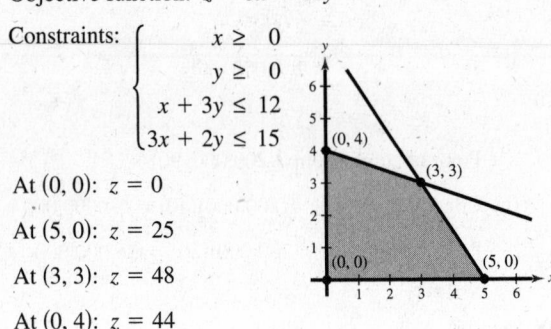

At $(0, 25)$: $z = 56.25$

At $(5, 15)$: $z = 42.5$

At $(15, 0)$: $z = 26.25$

The minimum value is 26.25 at $(15, 0)$.

Since the region in unbounded, there is no maximum value.

80. $z = 50x + 70y$

At $(0, 0)$: $z = 50(0) + 70(0) = 0$

At $(0, 750)$: $z = 50(0) + 70(750) = 52,500$

At $(500, 500)$: $z = 50(500) + 70(500) = 60,000$

At $(700, 0)$: $z = 50(700) + 70(0) = 35,000$

The minimum value is 0 at $(0, 0)$.

The maximum value is 60,000 at $(500, 500)$.

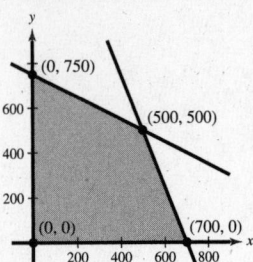

81. Objective function: $z = 5x + 11y$

Constraints: $\begin{cases} x \geq 0 \\ y \geq 0 \\ x + 3y \leq 12 \\ 3x + 2y \leq 15 \end{cases}$

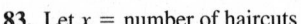

At $(0, 0)$: $z = 0$

At $(5, 0)$: $z = 25$

At $(3, 3)$: $z = 48$

At $(0, 4)$: $z = 44$

The minimum value is 0 at $(0, 0)$.

The maximum value is 48 at $(3, 3)$.

82. $z = -2x + y$

At $(0, 10)$: $z = -2(0) + 10 = 10$

At $(2, 5)$: $z = -2(2) + 5 = 1$

At $(7, 0)$: $z = -2(7) + 0 = -14$

The minimum value is -14 at $(7, 0)$.

There is no maximum value.

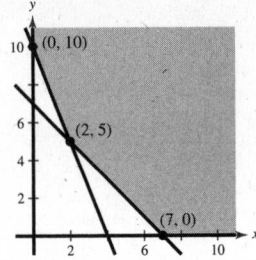

83. Let x = number of haircuts

y = number of permanents.

Objective function: Optimize $R = 25x + 70y$ subject to the following constraints:

$\begin{cases} x \geq 0 \\ y \geq 0 \\ \left(\frac{20}{60}\right)x + \left(\frac{70}{60}\right)y \leq 24 \implies 2x + 7y \leq 144 \end{cases}$

At $(0, 0)$: $R = 0$

At $(72, 0)$: $R = 1800$

At $\left(0, \frac{144}{7}\right)$: $R = 1440$

The revenue is optimal if the student does 72 haircuts and no permanents. The maximum revenue is \$1800.

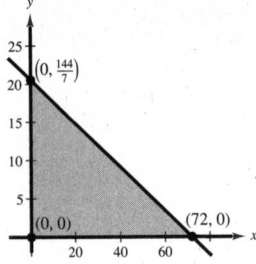

2. $2.35\left(\dfrac{180°}{\pi}\right) \approx 134.6°$

3. $\tan\theta = \dfrac{y}{x} = -\dfrac{4}{3} \implies r = 5$

Since $\sin\theta < 0$ θ is in Quadrant IV, $\implies x = 3$.

$\cos\theta = \dfrac{x}{r} = \dfrac{3}{5}$

4. $f(x) = 3 - 2\sin\pi x$

Period: $\dfrac{2\pi}{\pi} = 2$

Amplitude: $|a| = |-2| = 2$

Upward shift of 3 units (reflected in x-axis prior to shift)

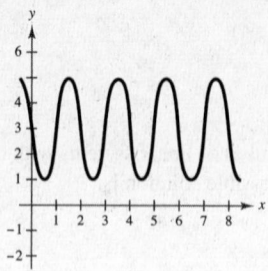

5. $g(x) = \dfrac{1}{2}\tan\left(x - \dfrac{\pi}{2}\right)$

Period: π

Asymptotes: $x = 0, x = \pi$

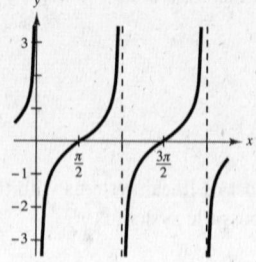

6. $h(x) = -\sec(x + \pi)$

Graph $y = -\cos(x + \pi)$ first.

Period: 2π

Amplitude: 1

Set $x + \pi = 0$ and $x + \pi = 2\pi$ for one cycle

$\qquad x = -\pi \qquad\qquad x = \pi$

The asymptotes of $h(x)$ corresponds to the x-intercepts of

$y = -\cos(x + \pi)$

$x + \pi = \dfrac{(2n + 1)\pi}{2}$

$x = \dfrac{(2n - 1)\pi}{2}$ where n is any integer

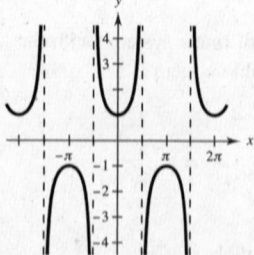

7. $h(x) = a\cos(bx + c)$

Graph is reflected in x-axis.

Amplitude: $a = -3$

Period: $2 = \dfrac{2\pi}{\pi} \implies b = \pi$

No phase shift: $c = 0$

$h(x) = -3\cos(\pi x)$

8. $f(x) = \dfrac{x}{2}\sin x, \ -3\pi \le x \le 3\pi$

$-\dfrac{x}{2} \le f(x) \le \dfrac{x}{2}$

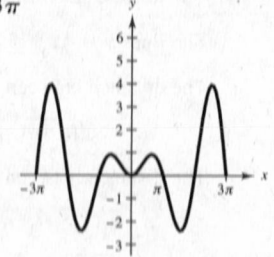

88. False. A linear programming problem either has one optional solution or infinitely many optimal solutions. (However, in real-life situations where the variables must have integer values, it is possible to have exactly ten integer-valued solutions.)

89. There are an infinite number of linear systems with the solution $(-6, 8)$. One possible solution is:

$$\begin{cases} x + y = 2 \\ x - y = -14 \end{cases}$$

90. There are an infinite number of linear systems with the solution $(5, -4)$. One possible system is:

$$\begin{cases} x - y = 9 \\ 3x + y = 11 \end{cases}$$

91. There are infinite linear systems with the solution $\left(\frac{4}{3}, 3\right)$. One possible solution is:

$$\begin{cases} 3x + y = 7 \\ -6x + 3y = 1 \end{cases}$$

92. There are an infinite number of linear systems with the solution $\left(-1, \frac{9}{4}\right)$. One possible system is:

$$\begin{cases} -x + 4y = 10 \\ 3x - 8y = -21 \end{cases}$$

93. There are an infinite number of linear systems with the solution $(4, -1, 3)$. One possible system is as follows:

$$\begin{cases} x + y + z = 6 \\ x + y - z = 0 \\ x - y - z = 2 \end{cases}$$

94. There are an infinite number of linear systems with the solution $(-3, 5, 6)$. One possible system is:

$$\begin{cases} x - 2y + z = -7 \\ 2x + y - 4z = -25 \\ -x + 3y - z = 12 \end{cases}$$

95. There are an infinite number of linear systems with the solution $\left(5, \frac{3}{2}, 2\right)$. One possible solution is:

$$\begin{cases} 2x + 2y - 3z = 7 \\ x - 2y + z = 4 \\ -x + 4y - z = -1 \end{cases}$$

96. There are an infinite number of linear systems with the solution $\left(\frac{3}{4}, -2, 8\right)$. One possible system is:

$$4x + y - z = -7$$
$$8x + 3y + 2z = 16$$
$$4x - 2y + 3z = 31$$

97. A system of linear equations is inconsistent if it has no solution.

98. The lines are distinct and parallel.

$$\begin{cases} x + 2y = 3 \\ 2x + 4y = 9 \end{cases}$$

99. If the solution to a system of equations is at fractional or irrational values, then the substitution method may yield an exact answer. The graphical method works well when the solution is at integer values, otherwise we can usually only approximate the solution.

Problem Solving for Chapter 7

1. The longest side of the triangle is a diameter of the circle and has a length of 20.

The lines $y = \frac{1}{2}x + 5$ and $y = -2x + 20$ intersect at the point $(6, 8)$.

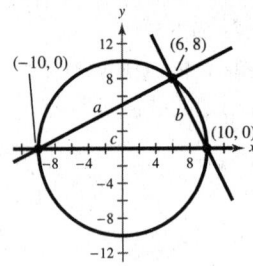

The distance between $(-10, 0)$ and $(6, 8)$ is:

$$d_1 = \sqrt{(6 - (-10))^2 + (8 - 0)^2} = \sqrt{320} = 8\sqrt{5}$$

The distance between $(6, 8)$ and $(10, 0)$ is:

$$d_2 = \sqrt{(10 - 6)^2 + (0 - 8)^2} = \sqrt{80} = 4\sqrt{5}$$

Since $\left(\sqrt{320}\right)^2 + \left(\sqrt{80}\right)^2 = (20)^2$

$$400 = 400$$

the sides of the triangle satisfy the Pythagorean Theorem. Thus, the triangle is a right triangle.

2. The system will have infinite solutions when the lines coincide, or are identical.

$$\begin{cases} 3x - 5y = 8 \implies 6x - 10y = 16 \\ 2x + k_1 y = k_2 \implies 6x + 3k_1 y = 3k_2 \end{cases}$$

$$3k_1 = -10 \implies k_1 = -\tfrac{10}{3}$$

$$3k_2 = 16 \implies k_2 = \tfrac{16}{3}$$

3. The system will have exactly one solution when the slopes of the line are *not* equal.

$$\begin{cases} ax + by = e \implies y = -\dfrac{a}{b}x + \dfrac{e}{b} \\ cx + dy = f \implies y = -\dfrac{c}{d}x + \dfrac{f}{d} \end{cases}$$

$$-\dfrac{a}{b} \neq -\dfrac{c}{d}$$

$$\dfrac{a}{b} \neq \dfrac{c}{d}$$

$$ad \neq bc$$

4. (a) $\begin{cases} x - 4y = -3 \qquad \text{Eq. 1} \\ 5x - 6y = 13 \qquad \text{Eq. 2} \end{cases}$

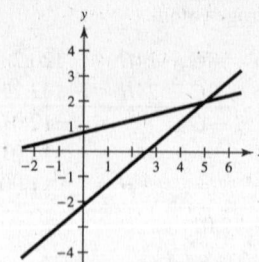

$$\begin{cases} x - 4y = -3 \\ 14y = 28 \qquad -5\text{Eq.1} + \text{Eq.2} \end{cases}$$

$$y = 2$$

$$x - 4(2) = -3 \implies x = 5$$

Solution: $(5, 2)$

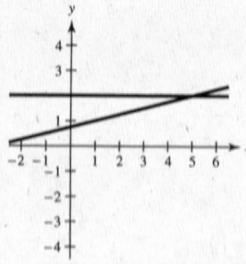

(b) $\begin{cases} 2x - 3y = 7 \qquad \text{Eq.1} \\ -4x + 6y = -14 \qquad \text{Eq.2} \end{cases}$

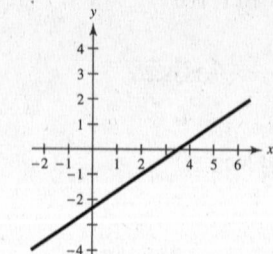

$$\begin{cases} 2x - 3y = 7 \\ 0 = 0 \qquad 2\text{Eq.1} + \text{Eq.2} \end{cases}$$

The lines coincide. Infinite solutions.

Let $y = a$, then $2x - 3a = 7 \implies x = \dfrac{3a + 7}{2}$

Solution: $\left(\dfrac{3a + 7}{2}, a \right)$

The solution(s) remain the same at each step of the process.

5. There are a finite number of solutions.

(a) If both equations are linear, then the maximum number of solutions to a finite system is *one*.

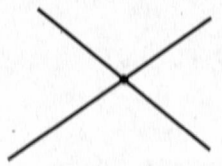

(b) If one equation is linear and the other is quadratic, then the maximum number of solutions is *two*.

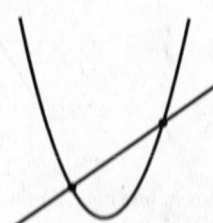

(c) If both equations are quadratic, then the maximum number of solutions to a finite system is *four*.

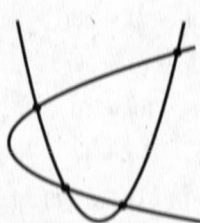

6. B = total votes cast for Bush

K = total votes cast for Kerry

N = total votes cast for Nader

$$\begin{cases} B + K + N = 118,304,000 \\ B - K \qquad = 3,320,000 \\ \qquad\quad N = 0.003(118,304,000) \end{cases}$$

$N = 354,912$

$$\begin{cases} B + K + 354,912 = 118,304,000 \\ B - K \qquad\qquad\;\; = 3,320,000 \end{cases}$$

$$\begin{cases} B + K = 117,949,088 \\ B - K = 3,320,000 \end{cases}$$

$2B = 121,269,088$

$B = 60,634,544$

$K = 57,314,544$

Bush: 60,634,544 votes

Kerry: 57,314,544 votes

Nader: 354,912 votes

7. The point where the two sections meet is at a depth of 10.1 feet. The distance between $(0, -10.1)$ and $(252.5, 0)$ is:

$$d = \sqrt{(252.5 - 0)^2 + (0 - (-10.1))^2} = \sqrt{63858.26}$$

$d \approx 252.7$

Each section is approximately 252.7 feet long.

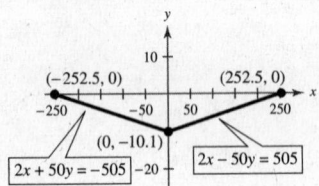

8. Let C = weight of a carbon atom.

Let H = weight of a hydrogen atom.

$$\begin{cases} 2C + 6H = 30.07 \implies \quad 8C + 24H = \quad 120.28 \\ 3C + 8H = 44.097 \implies -9C - 24H = -132.291 \end{cases}$$

$$\underline{\qquad\qquad\qquad\qquad\qquad\qquad}$$

$$-C \qquad\quad = -12.011$$

$$C = \quad 12.011$$

$$H = \quad\; 1.008$$

Each carbon atom weighs 12.011 u.

Each hydrogen atom weighs 1.008 u.

9. Let x = cost of the cable, per foot.

Let y = cost of a connector.

$$\begin{cases} 6x + 2y = 15.50 \implies \quad 6x + 2y = \quad 15.50 \\ 3x + 2y = 10.25 \implies -3x - 2y = -10.25 \end{cases}$$

$$\underline{\qquad\qquad\qquad\qquad\qquad\qquad}$$

$$3x \qquad = \quad 5.25$$

$$x = \quad 1.75$$

$$y = \quad 2.50$$

For a four-foot cable with a connector on each end the cost should be $4(1.75) + 2(2.50) = \$12.00$

10. Let t = time that the 9:00 A.M. bus is on the road.

Then $t - \frac{1}{4}$ = time that the 9:15 A.M. bus is on the road.

$$\begin{cases} d = 30t \\ d = 40\left(t - \frac{1}{4}\right) \end{cases}$$

$$40\left(t - \tfrac{1}{4}\right) = 30t$$

$$40t - 10 = 30t$$

$$10t = 10$$

$$t = 1$$

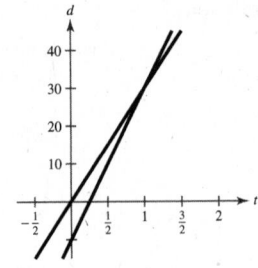

The 9:15 A.M. bus will catch up with the 9:00 A.M. bus in **one** hour. At that point both buses have traveled 30 miles and are 5 **miles** from the airport.

11. Let $X = \dfrac{1}{x}$, $Y = \dfrac{1}{y}$, and $Z = \dfrac{1}{z}$.

(a) $\begin{cases} \dfrac{12}{x} - \dfrac{12}{y} = 7 \implies 12X - 12Y = 7 \implies 12X - 12Y = 7 \\[3mm] \dfrac{3}{x} + \dfrac{4}{y} = 0 \implies 3X + 4Y = 0 \implies \underline{9X + 12Y = 0} \end{cases}$

$$21X \qquad = 7$$

$$X = \frac{1}{3}$$

$$Y = -\frac{1}{4}$$

Thus, $\dfrac{1}{x} = \dfrac{1}{3} \implies x = 3$ and $\dfrac{1}{y} = -\dfrac{1}{4} \implies y = -4$.

Solution: $(3, -4)$

(b) $\begin{cases} \dfrac{2}{x} + \dfrac{1}{y} - \dfrac{3}{z} = 4 \implies 2X + Y - 3Z = 4 \qquad \text{Eq.1} \\[3mm] \dfrac{4}{x} + \dfrac{2}{z} = 10 \implies 4X + 2Z = 10 \qquad \text{Eq.2} \\[3mm] -\dfrac{2}{x} + \dfrac{3}{y} - \dfrac{13}{z} = -8 \implies -2X + 3Y - 13Z = -8 \qquad \text{Eq.3} \end{cases}$

$$\begin{cases} 2X + Y - 3Z = 4 \\ {-2Y} + 8Z = 2 \\ 4Y - 16Z = -4 \end{cases} \qquad \begin{array}{l} -2\text{Eq.1} + \text{Eq.2} \\ \text{Eq.1} + \text{Eq.3} \end{array}$$

$$\begin{cases} 2X + Y - 3Z = 4 \\ {-2Y} + 8Z = 2 \\ 0 = 0 \end{cases} \qquad 2\text{Eq.2} + \text{Eq.3}$$

The system has infinite solutions.

Let $Z = a$, then $Y = 4a - 1$ and $X = \dfrac{-a + 5}{2}$.

Then $\dfrac{1}{z} = a \implies z = \dfrac{1}{a}$, $\dfrac{1}{y} = 4a - 1 \implies y = \dfrac{1}{4a - 1}$

$x = \dfrac{-a + 5}{2} \implies x = \dfrac{2}{-a + 5}$.

Solution: $\left(\dfrac{2}{-a + 5}, \dfrac{1}{4a - 1}, \dfrac{1}{a} \right)$, $a \neq 5, \dfrac{1}{4}, 0$

12. Solution: $(-1, 2, -3)$

$x + 2y - 3z = a \implies (-1) + 2(2) - 3(-3) = 12 = a$

$-x - y + z = b \implies -(-1) - 2 + (-3) = -4 = b$

$2x + 3y - 2z = c \implies 2(-1) + 3(2) - 2(-3) = 10 = c$

Thus, $a = 12$, $b = -4$, and $c = 10$.

13. Solution: $(1, -1, 2)$

$$\begin{cases} 4x - 2y + 5z = 16 & \text{Equation 1} \\ x + y \quad\;\; = 0 & \text{Equation 2} \\ -x - 3y + 2z = 6 & \text{Equation 3} \end{cases}$$

(a) $\begin{cases} 4x - 2y + 5z = 16 \\ x + y \quad\;\; = 0 \end{cases}$

$\begin{cases} x + y \quad\;\; = 0 \\ 4x - 2y + 5z = 16 \end{cases}$ Interchange the equations.

$\begin{cases} x + y \quad\;\; = 0 \\ \quad -6y + 5z = 16 \quad -4\text{Eq.1} + \text{Eq.2} \end{cases}$

Let $z = a$, then $y = \dfrac{5a - 16}{6}$ and $x = \dfrac{-5a + 16}{6}$.

Solution: $\left(\dfrac{-5a + 16}{6}, \dfrac{5a - 16}{6}, a \right)$

When $a = 2$ we have the original solution.

(c) $\begin{cases} x + y \quad\;\; = 0 \\ -x - 3y + 2z = 6 \end{cases}$

$\begin{cases} x + y \quad\;\; = 0 \\ \quad -2y + 2z = 6 \quad\quad \text{Eq.1} + \text{Eq.2} \end{cases}$

Let $z = c$, then $y = c - 3$ and $x = -c + 3$

Solution: $(-c + 3, c - 3, c)$

When $c = 2$ we have the original solution.

(b) $\begin{cases} 4x - 2y + 5z = 16 \\ -x - 3y + 2z = 6 \end{cases}$

$\begin{cases} -x - 3y + 2z = 6 \\ 4x - 2y + 5z = 16 \end{cases}$ Interchange the equations.

$\begin{cases} -x - \quad 3y + 2z = 6 \\ \quad\quad -14y + 13z = 40 \quad 4\text{Eq.1} + \text{Eq.2} \end{cases}$

Let $z = b$, then $y = \dfrac{13b - 40}{14}$ and $x = \dfrac{-11b + 36}{14}$

Solution: $\left(\dfrac{-11b + 36}{14}, \dfrac{13b - 40}{14}, b \right)$

When $b = 2$ we have the original solution.

(d) Each of these systems has infinite solutions.

14. $\begin{cases} x_1 - x_2 + 2x_3 + 2x_4 + 6x_5 = 6 \\ 3x_1 - 2x_2 + 4x_3 + 4x_4 + 12x_5 = 14 \\ \quad\quad x_2 - x_3 - x_4 - 3x_5 = -3 \\ 2x_1 - 2x_2 + 4x_3 + 5x_4 + 15x_5 = 10 \\ 2x_1 - 2x_2 + 4x_3 + 4x_4 + 13x_5 = 13 \end{cases}$

$\begin{cases} x_1 - x_2 + 2x_3 + 2x_4 + 6x_5 = 6 \\ x_1 \quad\quad\quad\quad\quad\quad\quad = 2 \quad\quad -2\text{Eq.1} + \text{Eq.2} \\ \quad\quad x_2 - x_3 - x_4 - 3x_5 = -3 \\ 2x_1 - 2x_2 + 4x_3 + 5x_4 + 15x_5 = 10 \\ 2x_1 - 2x_2 + 4x_3 + 4x_4 + 13x_5 = 13 \end{cases}$

$\begin{cases} x_1 + x_2 \quad\quad\quad\quad\quad = 0 \quad\quad \text{Eq.1} + 2\text{Eq.3} \\ x_1 \quad\quad\quad\quad\quad\quad\quad = 2 \\ \quad\quad x_2 - x_3 - x_4 - 3x_5 = -3 \\ 2x_1 - 2x_2 + 4x_3 + 5x_4 + 15x_5 = 10 \\ 2x_1 - 2x_2 + 4x_3 + 4x_4 + 13x_5 = 13 \end{cases}$

$\begin{cases} \quad x_2 \quad\quad\quad\quad\quad = -2 \quad\quad \text{Eq.1} - \text{Eq.2} \\ x_1 \quad\quad\quad\quad\quad\quad\quad = 2 \\ \quad\quad x_2 - x_3 - x_4 - 3x_5 = -3 \\ 2x_1 - 2x_2 + 4x_3 + 5x_4 + 15x_5 = 10 \\ 2x_1 - 2x_2 + 4x_3 + 4x_4 + 13x_5 = 13 \end{cases}$

—CONTINUED—

14. —CONTINUED—

Substitute into the subsequent equations and simplify:

$$\begin{cases} x_1 & = & 2 \\ x_2 & = & -2 \\ (-2) - x_3 - x_4 - 3x_5 & = & -3 \\ 2(2) - 2(-2) + 4x_3 + 5x_4 + 15x_5 & = & 10 \\ 2(2) - 2(-2) + 4x_3 + 4x_4 + 13x_5 & = & 13 \end{cases}$$

$$\begin{cases} x_1 & = & 2 \\ x_2 & = & -2 \\ -x_3 - x_4 - 3x_5 & = & -1 \\ 4x_3 + 5x_4 + 15x_5 & = & 2 \\ 4x_3 + 4x_4 + 13x_5 & = & 5 \end{cases}$$

$$\begin{cases} x_1 & = & 2 \\ x_2 & = & -2 \\ x_3 + x_4 + 3x_5 & = & 1 & -\text{Eq.3} \\ x_4 + 3x_5 & = & -2 & \text{Eq.4} + (4)\text{Eq.3} \\ x_5 & = & 1 & \text{Eq.5} + (4)\text{Eq.3} \end{cases}$$

$$\begin{cases} x_1 & = & 2 \\ x_2 & = & -2 \\ x_3 & = & 3 & \text{Eq.3} - \text{Eq.4} \\ x_4 & = & -5 & \text{Eq.4} - (3)\text{Eq.5} \\ x_5 & = & 1 \end{cases}$$

15. t = amount of terrestrial vegetation in kilograms

a = amount of aquatic vegetation in kilograms

$$a + t \le 32$$
$$0.15a \ge 1.9$$
$$193a + 4(193)t \ge 11,000$$

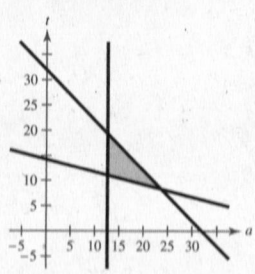

16. x = number of inches by which a person's height exceeds 4 feet 10 inches

y = person's weight in pounds

(a) $\begin{cases} y \ge 91 + 3.7x \\ y \le 119 + 4.8x \\ x \ge 0, \ y \ge 0 \end{cases}$ (b)

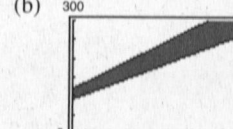

(c) For someone 6 feet tall, $x = 14$ inches.

Minimum weight: $91 + 3.7(14) = 142.8$ pounds

Maximum weight: $119 + 4.8(14) = 186.2$ pounds

17. (a) x = HDL cholesterol (good)

y = LDL cholesterol (bad)

$$\begin{cases} 0 < y < 130 \\ x \ge 35 \\ x + y \le 200 \end{cases}$$

(b)

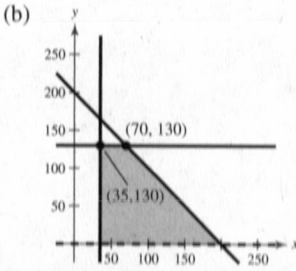

(c) $y = 120$ *is* in the region since $0 < y < 130$.

$x = 90$ *is* in the region since $35 < x < 200$.

$x + y = 210$ *is not* in the region since $x + y < 200$.

(d) If the LDL reading is 150 and the HDL reading is 40, then $x \ge 35$ and $x + y \le 200$ but $y \not< 130$.

(e) $\dfrac{x + y}{x} < 4$

$x + y < 4x$

$y < 3x$

The point $(50, 120)$ is in the region and $120 < 3(50)$.

Chapter 7   Practice Test

For Exercises 1–3, solve the given system by the method of substitution.

1. $\begin{cases} x + y = 1 \\ 3x - y = 15 \end{cases}$

2. $\begin{cases} x - 3y = -3 \\ x^2 + 6y = 5 \end{cases}$

3. $\begin{cases} x + y + z = 6 \\ 2x - y + 3z = 0 \\ 5x + 2y - z = -3 \end{cases}$

4. Find the two numbers whose sum is 110 and product is 2800.

5. Find the dimensions of a rectangle if its perimeter is 170 feet and its area is 1500 square feet.

For Exercises 6–8, solve the linear system by elimination.

6. $\begin{cases} 2x + 15y = 4 \\ x - 3y = 23 \end{cases}$

7. $\begin{cases} x + y = 2 \\ 38x - 19y = 7 \end{cases}$

8. $\begin{cases} 0.4x + 0.5y = 0.112 \\ 0.3x - 0.7y = -0.131 \end{cases}$

9. Herbert invests \$17,000 in two funds that pay 11% and 13% simple interest, respectively. If he receives \$2080 in yearly interest, how much is invested in each fund?

10. Find the least squares regression line for the points $(4, 3)$, $(1, 1)$, $(-1, -2)$, and $(-2, -1)$.

For Exercises 11–12, solve the system of equations.

11. $\begin{cases} x + y = -2 \\ 2x - y + z = 11 \\ 4y - 3z = -20 \end{cases}$

12. $\begin{cases} 3x + 2y - z = 5 \\ 6x - y + 5z = 2 \end{cases}$

13. Find the equation of the parabola $y = ax^2 + bx + c$ passing through the points $(0, -1)$, $(1, 4)$ and $(2, 13)$.

For Exercises 14–15, write the partial fraction decomposition of the rational functions.

14. $\dfrac{10x - 17}{x^2 - 7x - 8}$

15. $\dfrac{x^2 + 4}{x^4 + x^2}$

16. Graph $x^2 + y^2 \geq 9$.

17. Graph the solution of the system.

$$\begin{cases} x + y \leq 6 \\ \quad x \geq 2 \\ \quad y \geq 0 \end{cases}$$

18. Derive a set of inequalities to describe the triangle with vertices $(0, 0)$, $(0, 7)$, and $(2, 3)$.

19. Find the maximum value of the objective function, $z = 30x + 26y$, subject to the following constraints.

$$\begin{cases} \quad x \geq 0 \\ \quad y \geq 0 \\ 2x + 3y \leq 21 \\ 5x + 3y \leq 30 \end{cases}$$

20. Graph the system of inequalities.

$$\begin{cases} \quad x^2 + y^2 \leq 4 \\ (x - 2)^2 + y^2 \geq 4 \end{cases}$$

For Exercises 21–22, write the partial fraction decomposition for the rational expression.

21. $\dfrac{1 - 2x}{x^2 + x}$

22. $\dfrac{6x - 17}{(x - 3)^2}$

C H A P T E R 8
Matrices and Determinants

C H A P T E R   8
Matrices and Determinants

Section 8.1   Matrices and Systems of Equations

■ You should be able to use elementary row operations to produce a row-echelon form (or reduced row-echelon form) of a matrix.

1. Interchange two rows.

2. Multiply a row by a nonzero constant.

3. Add a multiple of one row to another row.

■ You should be able to use either Gaussian elimination with back-substitution or Gauss-Jordan elimination to solve a system of linear equations.

Vocabulary Check

1. matrix

2. square

3. main diagonal

4. row; column

5. augmented

6. coefficient

7. row-equivalent

8. reduced row-echelon form

9. Gauss-Jordan elimination

---

1. Since the matrix has one row and two columns, its order is 1×2.

2. Since the matrix has one row and four columns, its order is 1×4.

3. Since the matrix has three rows and one column, its order is 3×1.

4. Since the matrix has three rows and four columns, its order is 3×4.

5. Since the matrix has two rows and two columns, its order is 2×2.

6. Since the matrix has two rows and three columns, its order is 2×3.

7. $\begin{cases} 4x - 3y = -5 \\ -x + 3y = 12 \end{cases}$

$$\begin{bmatrix} 4 & -3 & \vdots & -5 \\ -1 & 3 & \vdots & 12 \end{bmatrix}$$

8. $\begin{cases} 7x + 4y = 22 \\ 5x - 9y = 15 \end{cases}$

$$\begin{bmatrix} 7 & 4 & \vdots & 22 \\ 5 & -9 & \vdots & 15 \end{bmatrix}$$

9. $\begin{cases} x + 10y - 2z = 2 \\ 5x - 3y + 4z = 0 \\ 2x + y = 6 \end{cases}$

$$\begin{bmatrix} 1 & 10 & -2 & \vdots & 2 \\ 5 & -3 & 4 & \vdots & 0 \\ 2 & 1 & 0 & \vdots & 6 \end{bmatrix}$$

10. $\begin{cases} -x - 8y + 5z = 8 \\ -7x - 15z = -38 \\ 3x - y + 8z = 20 \end{cases}$

$$\begin{bmatrix} -1 & -8 & 5 & \vdots & 8 \\ -7 & 0 & -15 & \vdots & -38 \\ 3 & -1 & 8 & \vdots & 20 \end{bmatrix}$$

11. $\begin{cases} 7x - 5y + z = 13 \\ 19x - 8z = 10 \end{cases}$

$$\begin{bmatrix} 7 & -5 & 1 & \vdots & 13 \\ 19 & 0 & -8 & \vdots & 10 \end{bmatrix}$$

12. $\begin{cases} 9x + 2y - 3z = 20 \\ -25y + 11z = -5 \end{cases}$

$$\begin{bmatrix} 9 & 2 & -3 & \vdots & 20 \\ 0 & -25 & 11 & \vdots & -5 \end{bmatrix}$$

13. $\begin{bmatrix} 1 & 2 & \vdots & 7 \\ 2 & -3 & \vdots & 4 \end{bmatrix}$

$\begin{cases} x + 2y = 7 \\ 2x - 3y = 4 \end{cases}$

14. $\begin{bmatrix} 7 & -5 & \vdots & 0 \\ 8 & 3 & \vdots & -2 \end{bmatrix}$

$\begin{cases} 7x - 5y = 0 \\ 8x + 3y = -2 \end{cases}$

15. $\begin{bmatrix} 2 & 0 & 5 & \vdots & -12 \\ 0 & 1 & -2 & \vdots & 7 \\ 6 & 3 & 0 & \vdots & 2 \end{bmatrix}$

$\begin{cases} 2x + 5z = -12 \\ y - 2z = 7 \\ 6x + 3y = 2 \end{cases}$

16. $\begin{bmatrix} 4 & -5 & -1 & \vdots & 18 \\ -11 & 0 & 6 & \vdots & 25 \\ 3 & 8 & 0 & \vdots & -29 \end{bmatrix}$

$\begin{cases} 4x - 5y - z = 18 \\ -11x \quad\;\; + 6z = 25 \\ 3x + 8y \qquad = -29 \end{cases}$

17. $\begin{bmatrix} 9 & 12 & 3 & 0 & \vdots & 0 \\ -2 & 18 & 5 & 2 & \vdots & 10 \\ 1 & 7 & -8 & 0 & \vdots & -4 \\ 3 & 0 & 2 & 0 & \vdots & -10 \end{bmatrix}$

$\begin{cases} 9x + 12y + 3z \qquad = 0 \\ -2x + 18y + 5z + 2w = 10 \\ x + 7y - 8z \qquad = -4 \\ 3x \quad\;\; + 2z \qquad = -10 \end{cases}$

18. $\begin{bmatrix} 6 & 2 & -1 & -5 & \vdots & -25 \\ -1 & 0 & 7 & 3 & \vdots & 7 \\ 4 & -1 & -10 & 6 & \vdots & 23 \\ 0 & 8 & 1 & -11 & \vdots & -21 \end{bmatrix}$

$\begin{cases} 6x + 2y - z - 5w = -25 \\ -x \quad\;\; + 7z + 3w = 7 \\ 4x - y - 10z + 6w = 23 \\ 8y + z - 11w = -21 \end{cases}$

19. $\begin{bmatrix} 1 & 4 & 3 \\ 2 & 10 & 5 \end{bmatrix}$

$-2R_1 + R_2 \rightarrow \begin{bmatrix} 1 & 4 & 3 \\ 0 & \boxed{2} & -1 \end{bmatrix}$

20. $\begin{bmatrix} 3 & 6 & 8 \\ 4 & -3 & 6 \end{bmatrix}$

$\tfrac{1}{3}R_1 \rightarrow \begin{bmatrix} 1 & \boxed{2} & \tfrac{8}{3} \\ 4 & -3 & 6 \end{bmatrix}$

21. $\begin{bmatrix} 1 & 1 & 4 & -1 \\ 3 & 8 & 10 & 3 \\ -2 & 1 & 12 & 6 \end{bmatrix}$

$\begin{matrix} -3R_1 + R_2 \rightarrow \\ 2R_1 + R_3 \rightarrow \end{matrix} \begin{bmatrix} 1 & 1 & 4 & -1 \\ 0 & 5 & \boxed{-2} & \boxed{6} \\ 0 & 3 & \boxed{20} & \boxed{4} \end{bmatrix}$

$\tfrac{1}{5}R_2 \rightarrow \begin{bmatrix} 1 & 1 & 4 & -1 \\ 0 & 1 & -\tfrac{2}{5} & \tfrac{6}{5} \\ 0 & 3 & \boxed{20} & \boxed{4} \end{bmatrix}$

22. $\begin{bmatrix} 2 & 4 & 8 & 3 \\ 1 & -1 & -3 & 2 \\ 2 & 6 & 4 & 9 \end{bmatrix}$

$\tfrac{1}{2}R_1 \rightarrow \begin{bmatrix} 1 & \boxed{2} & \boxed{4} & \boxed{\tfrac{3}{2}} \\ 1 & -1 & -3 & 2 \\ 2 & 6 & 4 & 9 \end{bmatrix}$

$\begin{matrix} -R_1 + R_2 \rightarrow \\ -2R_1 + R_3 \rightarrow \end{matrix} \begin{bmatrix} 1 & 2 & 4 & \tfrac{3}{2} \\ 0 & \boxed{-3} & -7 & \tfrac{1}{2} \\ 0 & 2 & \boxed{-4} & \boxed{6} \end{bmatrix}$

23. $\begin{bmatrix} -2 & 5 & 1 \\ 3 & -1 & -8 \end{bmatrix} \rightarrow \begin{bmatrix} 13 & 0 & -39 \\ 3 & -1 & -8 \end{bmatrix}$

Add 5 times Row 2 to Row 1.

24. $\begin{bmatrix} 3 & -1 & -4 \\ -4 & 3 & 7 \end{bmatrix} \rightarrow \begin{bmatrix} 3 & -1 & -4 \\ 5 & 0 & -5 \end{bmatrix}$

Add 3 times Row 1 to Row 2.

25. $\begin{bmatrix} 0 & -1 & -5 & 5 \\ -1 & 3 & -7 & 6 \\ 4 & -5 & 1 & 3 \end{bmatrix} \rightarrow \begin{bmatrix} -1 & 3 & -7 & 6 \\ 0 & -1 & -5 & 5 \\ 0 & 7 & -27 & 27 \end{bmatrix}$

Interchange Row 1 and Row 2. Then add 4 times the new Row 1 to Row 3.

26. $\begin{bmatrix} -1 & -2 & 3 & -2 \\ 2 & -5 & 1 & -7 \\ 5 & 4 & -7 & 6 \end{bmatrix} \rightarrow \begin{bmatrix} -1 & -2 & 3 & -2 \\ 0 & -9 & 7 & -11 \\ 0 & -6 & 8 & -4 \end{bmatrix}$

Add 2 times Row 1 to Row 2.
Add 5 times Row 1 to Row 3.

27. $\begin{bmatrix} 1 & 2 & 3 \\ 2 & -1 & -4 \\ 3 & 1 & -1 \end{bmatrix}$

(a) $\begin{bmatrix} 1 & 2 & 3 \\ 0 & -5 & -10 \\ 3 & 1 & -1 \end{bmatrix}$

(b) $\begin{bmatrix} 1 & 2 & 3 \\ 0 & -5 & -10 \\ 0 & -5 & -10 \end{bmatrix}$

(c) $\begin{bmatrix} 1 & 2 & 3 \\ 0 & -5 & -10 \\ 0 & 0 & 0 \end{bmatrix}$

(d) $\begin{bmatrix} 1 & 2 & 3 \\ 0 & 1 & 2 \\ 0 & 0 & 0 \end{bmatrix}$

(e) $\begin{bmatrix} 1 & 0 & -1 \\ 0 & 1 & 2 \\ 0 & 0 & 0 \end{bmatrix}$ This matrix is in reduced row-echelon form.

28. $\begin{bmatrix} 7 & 1 \\ 0 & 2 \\ -3 & 4 \\ 4 & 1 \end{bmatrix}$

(a) $\begin{bmatrix} 7 & 1 \\ 0 & 2 \\ -3 & 4 \\ 1 & 5 \end{bmatrix}$

(b) $\begin{bmatrix} 1 & 5 \\ 0 & 2 \\ -3 & 4 \\ 7 & 1 \end{bmatrix}$

(c) $\begin{bmatrix} 1 & 5 \\ 0 & 2 \\ 0 & 19 \\ 7 & 1 \end{bmatrix}$

(d) $\begin{bmatrix} 1 & 5 \\ 0 & 2 \\ 0 & 19 \\ 0 & -34 \end{bmatrix}$

(e) $\begin{bmatrix} 1 & 5 \\ 0 & 1 \\ 0 & 19 \\ 0 & -34 \end{bmatrix}$

(f) $\begin{bmatrix} 1 & 0 \\ 0 & 1 \\ 0 & 0 \\ 0 & 0 \end{bmatrix}$ This matrix is in reduced row-echelon form.

29. $\begin{bmatrix} 1 & 0 & 0 & 0 \\ 0 & 1 & 1 & 5 \\ 0 & 0 & 0 & 0 \end{bmatrix}$

This matrix is in reduced row-echelon form.

30. $\begin{bmatrix} 1 & 3 & 0 & 0 \\ 0 & 0 & 1 & 8 \\ 0 & 0 & 0 & 0 \end{bmatrix}$

This matrix is in reduced row-echelon form.

31. $\begin{bmatrix} 2 & 0 & 4 & 0 \\ 0 & -1 & 3 & 6 \\ 0 & 0 & 1 & 5 \end{bmatrix}$

The first nonzero entries in Rows 1 and 2 are not 1. The matrix is not in row-echelon form.

32. $\begin{bmatrix} 1 & 0 & 2 & 1 \\ 0 & 1 & -3 & 10 \\ 0 & 0 & 1 & 0 \end{bmatrix}$

This matrix is in row-echelon form.

33. $\begin{bmatrix} 1 & 1 & 0 & 5 \\ -2 & -1 & 2 & -10 \\ 3 & 6 & 7 & 14 \end{bmatrix}$

$\begin{matrix} 2R_1 + R_2 \to \\ -3R_1 + R_3 \to \end{matrix} \begin{bmatrix} 1 & 1 & 0 & 5 \\ 0 & 1 & 2 & 0 \\ 0 & 3 & 7 & -1 \end{bmatrix}$

$-3R_2 + R_3 \to \begin{bmatrix} 1 & 1 & 0 & 5 \\ 0 & 1 & 2 & 0 \\ 0 & 0 & 1 & -1 \end{bmatrix}$

34. $\begin{bmatrix} 1 & 2 & -1 & 3 \\ 3 & 7 & -5 & 14 \\ -2 & -1 & -3 & 8 \end{bmatrix}$

$\begin{matrix} -3R_1 + R_2 \to \\ 2R_1 + R_3 \to \end{matrix} \begin{bmatrix} 1 & 2 & -1 & 3 \\ 0 & 1 & -2 & 5 \\ 0 & 3 & -5 & 14 \end{bmatrix}$

$-3R_2 + R_3 \to \begin{bmatrix} 1 & 2 & -1 & 3 \\ 0 & 1 & -2 & 5 \\ 0 & 0 & 1 & -1 \end{bmatrix}$

35. $\begin{bmatrix} 1 & -1 & -1 & 1 \\ 5 & -4 & 1 & 8 \\ -6 & 8 & 18 & 0 \end{bmatrix}$

$\begin{matrix} -5R_1 + R_2 \to \\ 6R_1 + R_3 \to \end{matrix} \begin{bmatrix} 1 & -1 & -1 & 1 \\ 0 & 1 & 6 & 3 \\ 0 & 2 & 12 & 6 \end{bmatrix}$

$-2R_2 + R_3 \to \begin{bmatrix} 1 & -1 & -1 & 1 \\ 0 & 1 & 6 & 3 \\ 0 & 0 & 0 & 0 \end{bmatrix}$

36. $\begin{bmatrix} 1 & -3 & 0 & -7 \\ -3 & 10 & 1 & 23 \\ 4 & -10 & 2 & -24 \end{bmatrix}$

$\begin{matrix} 3R_1 + R_2 \to \\ -4R_1 + R_3 \to \end{matrix} \begin{bmatrix} 1 & -3 & 0 & -7 \\ 0 & 1 & 1 & 2 \\ 0 & 2 & 2 & 4 \end{bmatrix}$

$-2R_2 + R_3 \to \begin{bmatrix} 1 & -3 & 0 & -7 \\ 0 & 1 & 1 & 2 \\ 0 & 0 & 0 & 0 \end{bmatrix}$

37. Use the reduced row-echelon form feature of a graphing utility.

$\begin{bmatrix} 3 & 3 & 3 \\ -1 & 0 & -4 \\ 2 & 4 & -2 \end{bmatrix} \Rightarrow \begin{bmatrix} 1 & 0 & 0 \\ 0 & 1 & 0 \\ 0 & 0 & 1 \end{bmatrix}$

38. Use the reduced row-echelon form feature of a graphing utility.

$\begin{bmatrix} 1 & 3 & 2 \\ 5 & 15 & 9 \\ 2 & 6 & 10 \end{bmatrix} \Rightarrow \begin{bmatrix} 1 & 3 & 0 \\ 0 & 0 & 1 \\ 0 & 0 & 0 \end{bmatrix}$

39. Use the reduced row-echelon form feature of a graphing utility.

$$\begin{bmatrix} 1 & 2 & 3 & -5 \\ 1 & 2 & 4 & -9 \\ -2 & -4 & -4 & 3 \\ 4 & 8 & 11 & -14 \end{bmatrix} \Rightarrow \begin{bmatrix} 1 & 2 & 0 & 0 \\ 0 & 0 & 1 & 0 \\ 0 & 0 & 0 & 1 \\ 0 & 0 & 0 & 0 \end{bmatrix}$$

40. Use the reduced row-echelon form feature of a graphing utility.

$$\begin{bmatrix} -2 & 3 & -1 & -2 \\ 4 & -2 & 5 & 8 \\ 1 & 5 & -2 & 0 \\ 3 & 8 & -10 & -30 \end{bmatrix} \Rightarrow \begin{bmatrix} 1 & 0 & 0 & 0 \\ 0 & 1 & 0 & 0 \\ 0 & 0 & 1 & 0 \\ 0 & 0 & 0 & 1 \end{bmatrix}$$

41. Use the reduced row-echelon form feature of a graphing utility.

$$\begin{bmatrix} -3 & 5 & 1 & 12 \\ 1 & -1 & 1 & 4 \end{bmatrix} \Rightarrow \begin{bmatrix} 1 & 0 & 3 & 16 \\ 0 & 1 & 2 & 12 \end{bmatrix}$$

42. Use the reduced row-echelon form feature of a graphing utility.

$$\begin{bmatrix} 5 & 1 & 2 & 4 \\ -1 & 5 & 10 & -32 \end{bmatrix} \Rightarrow \begin{bmatrix} 1 & 0 & 0 & 2 \\ 0 & 1 & 2 & -6 \end{bmatrix}$$

43. $\begin{cases} x - 2y = 4 \\ \quad\quad y = -3 \end{cases}$

$x - 2(-3) = 4$

$\quad\quad x = -2$

Solution: $(-2, -3)$

44. $\begin{cases} x + 5y = 0 \\ \quad\quad y = -1 \end{cases}$

$x + 5(-1) = 0$

$\quad\quad x = 5$

Solution: $(5, -1)$

45. $\begin{cases} x - y + 2z = 4 \\ \quad y - z = 2 \\ \quad\quad z = -2 \end{cases}$

$y - (-2) = 2$

$\quad\quad y = 0$

$x - 0 + 2(-2) = 4$

$\quad\quad x = 8$

Solution: $(8, 0, -2)$

46. $\begin{cases} x + 2y - 2z = -1 \\ \quad y + z = 9 \\ \quad\quad z = -3 \end{cases}$

$y + (-3) = 9$

$\quad\quad y = 12$

$x + 2(12) - 2(-3) = -1$

$\quad\quad x = -31$

Solution: $(-31, 12, -3)$

47. $\begin{bmatrix} 1 & 0 & \vdots & 3 \\ 0 & 1 & \vdots & -4 \end{bmatrix}$

$x = 3$

$y = -4$

Solution: $(3, -4)$

48. $\begin{bmatrix} 1 & 0 & \vdots & -6 \\ 0 & 1 & \vdots & 10 \end{bmatrix}$

$x = -6$

$y = 10$

Solution: $(-6, 10)$

49. $\begin{bmatrix} 1 & 0 & 0 & \vdots & -4 \\ 0 & 1 & 0 & \vdots & -10 \\ 0 & 0 & 1 & \vdots & 4 \end{bmatrix}$

$x = -4$

$y = -10$

$z = 4$

Solution: $(-4, -10, 4)$

50. $\begin{bmatrix} 1 & 0 & 0 & \vdots & 5 \\ 0 & 1 & 0 & \vdots & -3 \\ 0 & 0 & 1 & \vdots & 0 \end{bmatrix}$

$x = 5$

$y = -3$

$z = 0$

Solution: $(5, -3, 0)$

51. $\begin{cases} x + 2y = 7 \\ 2x + y = 8 \end{cases}$

$$\begin{bmatrix} 1 & 2 & \vdots & 7 \\ 2 & 1 & \vdots & 8 \end{bmatrix}$$

$-2R_1 + R_2 \rightarrow \begin{bmatrix} 1 & 2 & \vdots & 7 \\ 0 & -3 & \vdots & -6 \end{bmatrix}$

$-\frac{1}{3}R_2 \rightarrow \begin{bmatrix} 1 & 2 & \vdots & 7 \\ 0 & 1 & \vdots & 2 \end{bmatrix}$

$\begin{cases} x + 2y = 7 \\ \quad\quad y = 2 \end{cases}$

$y = 2$

$x + 2(2) = 7 \Rightarrow x = 3$

Solution: $(3, 2)$

52. $\begin{cases} 2x + 6y = 16 \\ 2x + 3y = 7 \end{cases}$

$$\begin{bmatrix} 2 & 6 & \vdots & 16 \\ 2 & 3 & \vdots & 7 \end{bmatrix}$$

$$-R_1 + R_2 \rightarrow \begin{bmatrix} 2 & 6 & \vdots & 16 \\ 0 & -3 & \vdots & -9 \end{bmatrix}$$

$$\begin{matrix} \frac{1}{2}R_1 \rightarrow \\ -\frac{1}{3}R_2 \rightarrow \end{matrix} \begin{bmatrix} 1 & 3 & \vdots & 8 \\ 0 & 1 & \vdots & 3 \end{bmatrix}$$

$\begin{cases} x + 3y = 8 \\ \quad\; y = 3 \end{cases}$

$y = 3$

$x + 3(3) = 8 \implies x = -1$

Solution: $(-1, 3)$

53. $\begin{cases} 3x - 2y = -27 \\ x + 3y = 13 \end{cases}$

$$\begin{bmatrix} 3 & -2 & \vdots & -27 \\ 1 & 3 & \vdots & 13 \end{bmatrix}$$

$$\begin{matrix} R_1 \\ R_2 \end{matrix} \begin{bmatrix} 1 & 3 & \vdots & 13 \\ 3 & -2 & \vdots & -27 \end{bmatrix}$$

$$-3R_1 + R_2 \rightarrow \begin{bmatrix} 1 & 3 & \vdots & 13 \\ 0 & -11 & \vdots & -66 \end{bmatrix}$$

$$-\frac{1}{11}R_2 \rightarrow \begin{bmatrix} 1 & 3 & \vdots & 13 \\ 0 & 1 & \vdots & 6 \end{bmatrix}$$

$\begin{cases} x + 3y = 13 \\ \quad\; y = 6 \end{cases}$

$y = 6$

$x + 3(6) = 13 \implies x = -5$

Solution: $(-5, 6)$

54. $\begin{cases} -x + y = 4 \\ 2x - 4y = -34 \end{cases}$

$$\begin{bmatrix} -1 & 1 & \vdots & 4 \\ 2 & -4 & \vdots & -34 \end{bmatrix}$$

$$\begin{matrix} (-1)R_1 \rightarrow \\ (\frac{1}{2})R_2 \rightarrow \end{matrix} \begin{bmatrix} 1 & -1 & \vdots & -4 \\ 1 & -2 & \vdots & -17 \end{bmatrix}$$

$$-R_1 + R_2 \rightarrow \begin{bmatrix} 1 & -1 & \vdots & -4 \\ 0 & -1 & \vdots & -13 \end{bmatrix}$$

$$(-1)R_2 \rightarrow \begin{bmatrix} 1 & -1 & \vdots & -4 \\ 0 & 1 & \vdots & 13 \end{bmatrix}$$

$\begin{cases} x - y = -4 \\ \quad\; y = 13 \end{cases}$

$y = 13$

$x - 13 = -4 \implies x = 9$

Solution: $(9, 13)$

55. $\begin{cases} -2x + 6y = -22 \\ x + 2y = -9 \end{cases}$

$$\begin{bmatrix} -2 & 6 & \vdots & -22 \\ 1 & 2 & \vdots & -9 \end{bmatrix}$$

$$\begin{matrix} R_1 \\ R_2 \end{matrix} \begin{bmatrix} 1 & 2 & \vdots & -9 \\ -2 & 6 & \vdots & -22 \end{bmatrix}$$

$$2R_1 + R_2 \rightarrow \begin{bmatrix} 1 & 2 & \vdots & -9 \\ 0 & 10 & \vdots & -40 \end{bmatrix}$$

$$\frac{1}{10}R_2 \rightarrow \begin{bmatrix} 1 & 2 & \vdots & -9 \\ 0 & 1 & \vdots & -4 \end{bmatrix}$$

$\begin{cases} x + 2y = -9 \\ \quad\; y = -4 \end{cases}$

$y = -4$

$x + 2(-4) = -9 \implies x = -1$

Solution: $(-1, -4)$

56. $\begin{cases} 5x - 5y = -5 \\ -2x - 3y = 7 \end{cases}$

$$\begin{bmatrix} 5 & -5 & \vdots & -5 \\ -2 & -3 & \vdots & 7 \end{bmatrix}$$

$$\frac{1}{5}R_1 \rightarrow \begin{bmatrix} 1 & -1 & \vdots & -1 \\ -2 & -3 & \vdots & 7 \end{bmatrix}$$

$$2R_1 + R_2 \rightarrow \begin{bmatrix} 1 & -1 & \vdots & -1 \\ 0 & -5 & \vdots & 5 \end{bmatrix}$$

$$-\frac{1}{5}R_2 \rightarrow \begin{bmatrix} 1 & -1 & \vdots & -1 \\ 0 & 1 & \vdots & -1 \end{bmatrix}$$

$\begin{cases} x - y = -1 \\ \quad\; y = -1 \end{cases}$

$y = -1$

$x - (-1) = -1 \implies x = -2$

Solution: $(-2, -1)$

57. $\begin{cases} -x + 2y = 1.5 \\ 2x - 4y = 3.0 \end{cases}$

$$\begin{bmatrix} -1 & 2 & \vdots & 1.5 \\ 2 & -4 & \vdots & 3.0 \end{bmatrix}$$

$$2R_1 + R_2 \rightarrow \begin{bmatrix} -1 & 2 & \vdots & 1.5 \\ 0 & 0 & \vdots & 6.0 \end{bmatrix}$$

The system is inconsistent and there is no solution.

58. $\begin{cases} x - 3y = 5 \\ -2x + 6y = -10 \end{cases}$

$$\begin{bmatrix} 1 & -3 & \vdots & 5 \\ -2 & 6 & \vdots & -10 \end{bmatrix}$$

$2R_1 + R_2 \rightarrow \begin{bmatrix} 1 & -3 & \vdots & 5 \\ 0 & 0 & \vdots & 0 \end{bmatrix}$

$x - 3y = 5$

$\quad y = a$

$\quad x = 3a + 5$

Solution: $(3a + 5, a)$ where a is a real number

59. $\begin{cases} x \quad\;\; - 3z = -2 \\ 3x + y - 2z = 5 \\ 2x + 2y + z = 4 \end{cases}$

$$\begin{bmatrix} 1 & 0 & -3 & \vdots & -2 \\ 3 & 1 & -2 & \vdots & 5 \\ 2 & 2 & 1 & \vdots & 4 \end{bmatrix}$$

$\begin{matrix} -3R_1 + R_2 \rightarrow \\ -2R_1 + R_3 \rightarrow \end{matrix} \begin{bmatrix} 1 & 0 & -3 & \vdots & -2 \\ 0 & 1 & 7 & \vdots & 11 \\ 0 & 2 & 7 & \vdots & 8 \end{bmatrix}$

$-2R_2 + R_3 \rightarrow \begin{bmatrix} 1 & 0 & -3 & \vdots & -2 \\ 0 & 1 & 7 & \vdots & 11 \\ 0 & 0 & -7 & \vdots & -14 \end{bmatrix}$

$-\frac{1}{7}R_3 \rightarrow \begin{bmatrix} 1 & 0 & -3 & \vdots & -2 \\ 0 & 1 & 7 & \vdots & 11 \\ 0 & 0 & 1 & \vdots & 2 \end{bmatrix}$

$\begin{cases} x \quad\;\; - 3z = -2 \\ y + 7z = 11 \\ \quad\;\; z = 2 \end{cases}$

$z = 2$

$y + 7(2) = 11 \implies y = -3$

$x - 3(2) = -2 \implies x = 4$

Solution: $(4, -3, 2)$

60. $\begin{cases} 2x - y + 3z = 24 \\ \quad\;\; 2y - z = 14 \\ 7x - 5y \quad\;\; = 6 \end{cases}$

$$\begin{bmatrix} 2 & -1 & 3 & \vdots & 24 \\ 0 & 2 & -1 & \vdots & 14 \\ 7 & -5 & 0 & \vdots & 6 \end{bmatrix}$$

$R_3 + (-3)R_1 \rightarrow \begin{bmatrix} 1 & -2 & -9 & \vdots & -66 \\ 0 & 2 & -1 & \vdots & 14 \\ 7 & -5 & 0 & \vdots & 6 \end{bmatrix}$

$-7R_1 + R_3 \rightarrow \begin{bmatrix} 1 & -2 & -9 & \vdots & -66 \\ 0 & 2 & -1 & \vdots & 14 \\ 0 & 9 & 63 & \vdots & 468 \end{bmatrix}$

$4R_2 \rightarrow \begin{bmatrix} 1 & -2 & -9 & \vdots & -66 \\ 0 & 8 & -4 & \vdots & 56 \\ 0 & 9 & 63 & \vdots & 468 \end{bmatrix}$

$-R_3 + R_2 \rightarrow \begin{bmatrix} 1 & -2 & -9 & \vdots & -66 \\ 0 & -1 & -67 & \vdots & -412 \\ 0 & 9 & 63 & \vdots & 468 \end{bmatrix}$

$9R_2 + R_3 \rightarrow \begin{bmatrix} 1 & -2 & -9 & \vdots & -66 \\ 0 & -1 & -67 & \vdots & -412 \\ 0 & 0 & -540 & \vdots & -3240 \end{bmatrix}$

$\begin{matrix} -R_2 \rightarrow \\ -\frac{1}{540}R_3 \rightarrow \end{matrix} \begin{bmatrix} 1 & -2 & -9 & \vdots & -66 \\ 0 & 1 & 67 & \vdots & 412 \\ 0 & 0 & 1 & \vdots & 6 \end{bmatrix}$

$\begin{cases} x - 2y - 9z = -66 \\ y + 67z = 412 \\ \quad\;\; z = 6 \end{cases}$

$z = 6$

$y + 67(6) = 412 \implies y = 10$

$x - 2(10) - 9(6) = -66 \implies x = 8$

Solution: $(8, 10, 6)$

61. $\begin{cases} -x + y - z = -14 \\ 2x - y + z = 21 \\ 3x + 2y + z = 19 \end{cases}$

$$\begin{bmatrix} -1 & 1 & -1 & \vdots & -14 \\ 2 & -1 & 1 & \vdots & 21 \\ 3 & 2 & 1 & \vdots & 19 \end{bmatrix}$$

$-R_1 \rightarrow \begin{bmatrix} 1 & -1 & 1 & \vdots & 14 \\ 2 & -1 & 1 & \vdots & 21 \\ 3 & 2 & 1 & \vdots & 19 \end{bmatrix}$

$\begin{matrix} \\ -2R_1 + R_2 \rightarrow \\ -3R_1 + R_3 \rightarrow \end{matrix} \begin{bmatrix} 1 & -1 & 1 & \vdots & 14 \\ 0 & 1 & -1 & \vdots & -7 \\ 0 & 5 & -2 & \vdots & -23 \end{bmatrix}$

$\begin{matrix} \\ \\ -5R_2 + R_3 \rightarrow \end{matrix} \begin{bmatrix} 1 & -1 & 1 & \vdots & 14 \\ 0 & 1 & -1 & \vdots & -7 \\ 0 & 0 & 3 & \vdots & 12 \end{bmatrix}$

$\begin{matrix} \\ \\ \frac{1}{3}R_3 \rightarrow \end{matrix} \begin{bmatrix} 1 & -1 & 1 & \vdots & 14 \\ 0 & 1 & -1 & \vdots & -7 \\ 0 & 0 & 1 & \vdots & 4 \end{bmatrix}$

$\begin{cases} x - y + z = 14 \\ y - z = -7 \\ z = 4 \end{cases}$

$$z = 4$$

$$y - 4 = -7 \implies y = -3$$

$$x - (-3) + 4 = 14 \implies x = 7$$

Solution: $(7, -3, 4)$

62. $\begin{cases} 2x + 2y - z = 2 \\ x - 3y + z = -28 \\ -x + y = 14 \end{cases}$

$$\begin{bmatrix} 2 & 2 & -1 & \vdots & 2 \\ 1 & -3 & 1 & \vdots & -28 \\ -1 & 1 & 0 & \vdots & 14 \end{bmatrix}$$

$\begin{matrix} R_2 \\ R_1 \end{matrix} \begin{bmatrix} 1 & -3 & 1 & \vdots & -28 \\ 2 & 2 & -1 & \vdots & 2 \\ -1 & 1 & 0 & \vdots & 14 \end{bmatrix}$

$\begin{matrix} \\ R_3 \\ R_2 \end{matrix} \begin{bmatrix} 1 & -3 & 1 & \vdots & -28 \\ -1 & 1 & 0 & \vdots & 14 \\ 2 & 2 & -1 & \vdots & 2 \end{bmatrix}$

$\begin{matrix} R_1 + R_2 \rightarrow \\ -2R_1 + R_3 \rightarrow \end{matrix} \begin{bmatrix} 1 & -3 & 1 & \vdots & -28 \\ 0 & -2 & 1 & \vdots & -14 \\ 0 & 8 & -3 & \vdots & 58 \end{bmatrix}$

$\begin{matrix} \\ \\ 4R_2 + R_3 \rightarrow \end{matrix} \begin{bmatrix} 1 & -3 & 1 & \vdots & -28 \\ 0 & -2 & 1 & \vdots & -14 \\ 0 & 0 & 1 & \vdots & 2 \end{bmatrix}$

$\begin{matrix} \\ -\frac{1}{2}R_2 \rightarrow \\ \\ \end{matrix} \begin{bmatrix} 1 & -3 & 1 & \vdots & -28 \\ 0 & 1 & -\frac{1}{2} & \vdots & 7 \\ 0 & 0 & 1 & \vdots & 2 \end{bmatrix}$

$\begin{cases} x - 3y + z = -28 \\ y - \frac{1}{2}z = 7 \\ z = 2 \end{cases}$

$$z = 2$$

$$y - \frac{1}{2}(2) = 7 \implies y = 8$$

$$x - 3(8) + 2 = -28 \implies x = -6$$

Solution: $(-6, 8, 2)$

63. $\begin{cases} x + 2y - 3z = -28 \\ 4y + 2z = 0 \\ -x + y - z = -5 \end{cases}$

$$\begin{bmatrix} 1 & 2 & -3 & \vdots & -28 \\ 0 & 4 & 2 & \vdots & 0 \\ -1 & 1 & -1 & \vdots & -5 \end{bmatrix}$$

$\begin{matrix} \\ \frac{1}{4}R_2 \rightarrow \\ R_1 + R_3 \rightarrow \end{matrix} \begin{bmatrix} 1 & 2 & -3 & \vdots & -28 \\ 0 & 1 & \frac{1}{2} & \vdots & 0 \\ 0 & 3 & -4 & \vdots & -33 \end{bmatrix}$

$\begin{matrix} \\ \\ -3R_2 + R_3 \rightarrow \end{matrix} \begin{bmatrix} 1 & 2 & -3 & \vdots & -28 \\ 0 & 1 & \frac{1}{2} & \vdots & 0 \\ 0 & 0 & -\frac{11}{2} & \vdots & -33 \end{bmatrix}$

$\begin{matrix} \\ \\ -\frac{2}{11}R_3 \rightarrow \end{matrix} \begin{bmatrix} 1 & 2 & -3 & \vdots & -28 \\ 0 & 1 & \frac{1}{2} & \vdots & 0 \\ 0 & 0 & 1 & \vdots & 6 \end{bmatrix}$

$\begin{cases} x + 2y - 3z = -28 \\ y + \frac{1}{2}z = 0 \\ z = 6 \end{cases}$

$$z = 6$$

$$y + \frac{1}{2}(6) = 0 \implies y = -3$$

$$x + 2(-3) - 3(6) = -28 \implies x = -4$$

Solution: $(-4, -3, 6)$

64. $\begin{cases} 3x - 2y + z = 15 \\ -x + y + 2z = -10 \\ x - y - 4z = 14 \end{cases}$

$$\begin{bmatrix} 3 & -2 & 1 & \vdots & 15 \\ -1 & 1 & 2 & \vdots & -10 \\ 1 & -1 & -4 & \vdots & 14 \end{bmatrix}$$

$\begin{matrix} R_3 \\ \\ R_1 \end{matrix}$ $\begin{bmatrix} 1 & -1 & -4 & \vdots & 14 \\ -1 & 1 & 2 & \vdots & -10 \\ 3 & -2 & 1 & \vdots & 15 \end{bmatrix}$

$\begin{matrix} R_1 + R_2 \rightarrow \\ -3R_1 + R_3 \rightarrow \end{matrix}$ $\begin{bmatrix} 1 & -1 & -4 & \vdots & 14 \\ 0 & 0 & -2 & \vdots & 4 \\ 0 & 1 & 13 & \vdots & -27 \end{bmatrix}$

$\begin{matrix} R_3 \\ R_2 \end{matrix}$ $\begin{bmatrix} 1 & -1 & -4 & \vdots & 14 \\ 0 & 1 & 13 & \vdots & -27 \\ 0 & 0 & -2 & \vdots & 4 \end{bmatrix}$

$-\frac{1}{2}R_3 \rightarrow$ $\begin{bmatrix} 1 & -1 & -4 & \vdots & 14 \\ 0 & 1 & 13 & \vdots & -27 \\ 0 & 0 & 1 & \vdots & -2 \end{bmatrix}$

$\begin{cases} x - y - 4z = 14 \\ y + 13z = -27 \\ z = -2 \end{cases}$

$z = -2$

$y + 13(-2) = -27 \implies y = -1$

$x - (-1) - 4(-2) = 14 \implies x = 5$

Solution: $(5, -1, -2)$

65. $\begin{cases} x + y - 5z = 3 \\ x - 2z = 1 \\ 2x - y - z = 0 \end{cases}$

$$\begin{bmatrix} 1 & 1 & -5 & \vdots & 3 \\ 1 & 0 & -2 & \vdots & 1 \\ 2 & -1 & -1 & \vdots & 0 \end{bmatrix}$$

$\begin{matrix} -R_1 + R_2 \rightarrow \\ -2R_1 + R_3 \rightarrow \end{matrix}$ $\begin{bmatrix} 1 & 1 & -5 & \vdots & 3 \\ 0 & -1 & 3 & \vdots & -2 \\ 0 & -3 & 9 & \vdots & -6 \end{bmatrix}$

$-R_2 \rightarrow$ $\begin{bmatrix} 1 & 1 & -5 & \vdots & 3 \\ 0 & 1 & -3 & \vdots & 2 \\ 0 & -3 & 9 & \vdots & -6 \end{bmatrix}$

$\begin{matrix} -R_2 + R_1 \rightarrow \\ \\ 3R_2 + R_3 \rightarrow \end{matrix}$ $\begin{bmatrix} 1 & 0 & -2 & \vdots & 1 \\ 0 & 1 & -3 & \vdots & 2 \\ 0 & 0 & 0 & \vdots & 0 \end{bmatrix}$

$\begin{cases} x - 2z = 1 \\ y - 3z = 2 \end{cases}$

Let $z = a$.

$y - 3a = 2 \implies y = 3a + 2$

$x - 2a = 1 \implies x = 2a + 1$

Solution: $(2a + 1, 3a + 2, a)$ where a is any real number.

66. $\begin{cases} 2x + 3z = 3 \\ 4x - 3y + 7z = 5 \\ 8x - 9y + 15z = 9 \end{cases}$

$$\begin{bmatrix} 2 & 0 & 3 & \vdots & 3 \\ 4 & -3 & 7 & \vdots & 5 \\ 8 & -9 & 15 & \vdots & 9 \end{bmatrix}$$

$\begin{matrix} -2R_1 + R_2 \rightarrow \\ -4R_1 + R_3 \rightarrow \end{matrix}$ $\begin{bmatrix} 2 & 0 & 3 & \vdots & 3 \\ 0 & -3 & 1 & \vdots & -1 \\ 0 & -9 & 3 & \vdots & -3 \end{bmatrix}$

$-3R_2 + R_3 \rightarrow$ $\begin{bmatrix} 2 & 0 & 3 & \vdots & 3 \\ 0 & -3 & 1 & \vdots & -1 \\ 0 & 0 & 0 & \vdots & 0 \end{bmatrix}$

$\begin{matrix} \frac{1}{2}R_1 \rightarrow \\ -\frac{1}{3}R_2 \rightarrow \end{matrix}$ $\begin{bmatrix} 1 & 0 & \frac{3}{2} & \vdots & \frac{3}{2} \\ 0 & 1 & -\frac{1}{3} & \vdots & \frac{1}{3} \\ 0 & 0 & 0 & \vdots & 0 \end{bmatrix}$

$z = a$

$y = \frac{1}{3}a + \frac{1}{3}$

$x = -\frac{3}{2}a + \frac{3}{2}$

Solution: $\left(-\frac{3}{2}a + \frac{3}{2}, \frac{1}{3}a + \frac{1}{3}, a\right)$ where a is a real number

67. $\begin{cases} x + 2y + z + 2w = 8 \\ 3x + 7y + 6z + 9w = 26 \end{cases}$

$$\begin{bmatrix} 1 & 2 & 1 & 2 & \vdots & 8 \\ 3 & 7 & 6 & 9 & \vdots & 26 \end{bmatrix}$$

$-3R_1 + R_2 \rightarrow$ $\begin{bmatrix} 1 & 2 & 1 & 2 & \vdots & 8 \\ 0 & 1 & 3 & 3 & \vdots & 2 \end{bmatrix}$

$-2R_2 + R_1 \rightarrow$ $\begin{bmatrix} 1 & 0 & -5 & -4 & \vdots & 4 \\ 0 & 1 & 3 & 3 & \vdots & 2 \end{bmatrix}$

$\begin{cases} x - 5z - 4w = 4 \\ y + 3z + 3w = 2 \end{cases}$

Let $w = a$ and $z = b$.

$y + 3b + 3a = 2 \implies y = 2 - 3b - 3a$

$x - 5b - 4a = 4 \implies x = 4 + 5b + 4a$

Solution: $(4 + 5b + 4a, 2 - 3b - 3a, b, a)$
where a and b are real numbers

68. $\begin{cases} 4x + 12y - 7z - 20w = 22 \\ 3x + 9y - 5z - 28w = 30 \end{cases}$

$$\begin{bmatrix} 4 & 12 & -7 & -20 & : & 22 \\ 3 & 9 & -5 & -28 & : & 30 \end{bmatrix}$$

$-R_2 + R_1 \rightarrow \begin{bmatrix} 1 & 3 & -2 & 8 & : & -8 \\ 3 & 9 & -5 & -28 & : & 30 \end{bmatrix}$

$-3R_1 + R_2 \rightarrow \begin{bmatrix} 1 & 3 & -2 & 8 & : & -8 \\ 0 & 0 & 1 & -52 & : & 54 \end{bmatrix}$

$2R_2 + R_1 \rightarrow \begin{bmatrix} 1 & 3 & 0 & -96 & : & 100 \\ 0 & 0 & 1 & -52 & : & 54 \end{bmatrix}$

$w = a$

$z = 52a + 54$

$y = b$

$x = -3b + 96a + 100$

Solution: $(-3b + 96a + 100, b, 52a + 54, a)$
where a and b are real numbers

69. $\begin{cases} -x + y = -22 \\ 3x + 4y = 4 \\ 4x - 8y = 32 \end{cases}$

$$\begin{bmatrix} -1 & 1 & : & -22 \\ 3 & 4 & : & 4 \\ 4 & -8 & : & 32 \end{bmatrix}$$

$-R_1 \rightarrow \begin{bmatrix} 1 & -1 & : & 22 \\ 3 & 4 & : & 4 \\ 4 & -8 & : & 32 \end{bmatrix}$

$\begin{matrix} -3R_1 + R_2 \rightarrow \\ -4R_1 + R_3 \rightarrow \end{matrix} \begin{bmatrix} 1 & -1 & : & 22 \\ 0 & 7 & : & -62 \\ 0 & -4 & : & -56 \end{bmatrix}$

$\begin{matrix} \frac{1}{7}R_2 \rightarrow \\ -\frac{1}{4}R_3 \rightarrow \end{matrix} \begin{bmatrix} 1 & -1 & : & 22 \\ 0 & 1 & : & -\frac{62}{7} \\ 0 & 1 & : & 14 \end{bmatrix}$

$-R_2 + R_3 \rightarrow \begin{bmatrix} 1 & -1 & : & 22 \\ 0 & 1 & : & -\frac{62}{7} \\ 0 & 0 & : & \frac{160}{7} \end{bmatrix}$

The system is inconsistent and there is no solution.

70. $\begin{cases} x + 2y = 0 \\ x + y = 6 \\ 3x - 2y = 8 \end{cases}$

$$\begin{bmatrix} 1 & 2 & : & 0 \\ 1 & 1 & : & 6 \\ 3 & -2 & : & 8 \end{bmatrix}$$

$\begin{matrix} -R_1 + R_2 \rightarrow \\ -3R_1 + R_3 \rightarrow \end{matrix} \begin{bmatrix} 1 & 2 & 0 \\ 0 & -1 & 6 \\ 0 & -8 & 8 \end{bmatrix}$

$-8R_2 + R_3 \rightarrow \begin{bmatrix} 1 & 2 & 0 \\ 0 & -1 & 6 \\ 0 & 0 & -40 \end{bmatrix}$

The system in inconsistent and there is no solution.

71. Use the reduced row-echelon form feature of a graphing utility.

$\begin{cases} 3x + 3y + 12z = 6 \\ x + y + 4z = 2 \\ 2x + 5y + 20z = 10 \\ -x + 2y + 8z = 4 \end{cases}$ $\begin{bmatrix} 3 & 3 & 12 & : & 6 \\ 1 & 1 & 4 & : & 2 \\ 2 & 5 & 20 & : & 10 \\ -1 & 2 & 8 & : & 4 \end{bmatrix} \Rightarrow \begin{bmatrix} 1 & 0 & 0 & : & 0 \\ 0 & 1 & 4 & : & 2 \\ 0 & 0 & 0 & : & 0 \\ 0 & 0 & 0 & : & 0 \end{bmatrix} \Rightarrow \begin{cases} x = 0 \\ y + 4z = 2 \end{cases}$

Let $z = a$.

$y = 2 - 4a$

$x = 0$

Solution: $(0, 2 - 4a, a)$ where a is any real number

72. Use the reduced row-echelon form feature of a graphing utility.

$$\begin{cases} 2x + 10y + 2z = 6 \\ x + 5y + 2z = 6 \\ x + 5y + z = 3 \\ -3x - 15y - 3z = -9 \end{cases}$$

$$\begin{bmatrix} 2 & 10 & 2 & \vdots & 6 \\ 1 & 5 & 2 & \vdots & 6 \\ 1 & 5 & 1 & \vdots & 3 \\ -3 & -15 & -3 & \vdots & -9 \end{bmatrix} \Rightarrow \begin{bmatrix} 1 & 5 & 0 & \vdots & 0 \\ 0 & 0 & 1 & \vdots & 3 \\ 0 & 0 & 0 & \vdots & 0 \\ 0 & 0 & 0 & \vdots & 0 \end{bmatrix}$$

$$\begin{cases} z = 3 \\ x + 5y = 0 \end{cases}$$

$$z = 3$$

$$y = a$$

$$x + 5a = 0 \implies x = -5a$$

Solution: $(-5a, a, 3)$ where a is a real number

73. Use the reduced row-echelon form feature of a graphing utility.

$$\begin{cases} 2x + y - z + 2w = -6 \\ 3x + 4y + w = 1 \\ x + 5y + 2z + 6w = -3 \\ 5x + 2y - z - w = 3 \end{cases} \begin{bmatrix} 2 & 1 & -1 & 2 & \vdots & -6 \\ 3 & 4 & 0 & 1 & \vdots & 1 \\ 1 & 5 & 2 & 6 & \vdots & -3 \\ 5 & 2 & -1 & -1 & \vdots & 3 \end{bmatrix} \Rightarrow \begin{bmatrix} 1 & 0 & 0 & 0 & \vdots & 1 \\ 0 & 1 & 0 & 0 & \vdots & 0 \\ 0 & 0 & 1 & 0 & \vdots & 4 \\ 0 & 0 & 0 & 1 & \vdots & -2 \end{bmatrix}$$

$$x = 1$$

$$y = 0$$

$$z = 4$$

$$w = -2$$

Solution: $(1, 0, 4, -2)$

74. Use the reduced row-echelon form feature of a graphing utility.

$$\begin{cases} x + 2y + 2z + 4w = 11 \\ 3x + 6y + 5z + 12w = 30 \\ x + 3y - 3z + 2w = -5 \\ 6x - y - z + w = -9 \end{cases}$$

$$\begin{bmatrix} 1 & 2 & 2 & 4 & \vdots & 11 \\ 3 & 6 & 5 & 12 & \vdots & 30 \\ 1 & 3 & -3 & 2 & \vdots & -5 \\ 6 & -1 & -1 & 1 & \vdots & -9 \end{bmatrix} \Rightarrow \begin{bmatrix} 1 & 0 & 0 & 0 & \vdots & -1 \\ 0 & 1 & 0 & 0 & \vdots & 1 \\ 0 & 0 & 1 & 0 & \vdots & 3 \\ 0 & 0 & 0 & 1 & \vdots & 1 \end{bmatrix}$$

$$\begin{cases} x = -1 \\ y = 1 \\ z = 3 \\ w = 1 \end{cases}$$

$$w = 1$$

$$z = 3$$

$$y = 1$$

$$x = -1$$

Solution: $(-1, 1, 3, 1)$

75. Use the reduced row-echelon form feature of a graphing utility.

$$\begin{cases} x + y + z + w = 0 \\ 2x + 3y + z - 2w = 0 \\ 3x + 5y + z = 0 \end{cases}$$
$$\begin{bmatrix} 1 & 1 & 1 & 1 & \vdots & 0 \\ 2 & 3 & 1 & -2 & \vdots & 0 \\ 3 & 5 & 1 & 0 & \vdots & 0 \end{bmatrix} \Rightarrow \begin{bmatrix} 1 & 0 & 2 & 0 & \vdots & 0 \\ 0 & 1 & -1 & 0 & \vdots & 0 \\ 0 & 0 & 0 & 1 & \vdots & 0 \end{bmatrix}$$

$$\begin{cases} x + 2z = 0 \\ y - z = 0 \\ w = 0 \end{cases}$$

Let $z = a$. Then $x = -2a$ and $y = a$.

Solution: $(-2a, a, a, 0)$ where a is a real number

76. $\begin{cases} x + 2y + z + 3w = 0 \\ x - y + w = 0 \\ y - z + 2w = 0 \end{cases}$

$$\begin{bmatrix} 1 & 2 & 1 & 3 & \vdots & 0 \\ 1 & -1 & 0 & 1 & \vdots & 0 \\ 0 & 1 & -1 & 2 & \vdots & 0 \end{bmatrix} \Rightarrow \begin{bmatrix} 1 & 0 & 0 & 2 & \vdots & 0 \\ 0 & 1 & 0 & 1 & \vdots & 0 \\ 0 & 0 & 1 & -1 & \vdots & 0 \end{bmatrix}$$

$$\begin{cases} x + 2w = 0 \\ y + w = 0 \\ z - w = 0 \end{cases}$$

$w = a,\ z = a,\ y = -a,\ x = -2a$

Solution: $(-2a, -a, a, a)$ where a is a real number

77. (a) $\begin{cases} x - 2y + z = -6 \\ y - 5z = 16 \\ z = -3 \end{cases}$

$y - 5(-3) = 16$

$y = 1$

$x - 2(1) + (-3) = -6$

$x = -1$

Solution: $(-1, 1, -3)$

(b) $\begin{cases} x + y - 2z = 6 \\ y + 3z = -8 \\ z = -3 \end{cases}$

$y + 3(-3) = -8$

$y = 1$

$x + (1) - 2(-3) = 6$

$x = -1$

Solution: $(-1, 1, -3)$

Both systems yield the same solution, namely $(-1, 1, -3)$.

78. (a) $\begin{cases} x - 3y + 4z = -11 \\ y - z = -4 \\ z = 2 \end{cases}$

$y - 2 = -4$

$y = -2$

$x - 3(-2) + 4(2) = -11$

$x = -25$

(b) $\begin{cases} x + 4y = -11 \\ y + 3z = 4 \\ z = 2 \end{cases}$

$y + 3(2) = 4$

$y = -2$

$x + 4(-2) = -11$

$x = -3$

The systems do *not* yield the same solution.

79. (a) $\begin{cases} x - 4y + 5z = 27 \\ y - 7z = -54 \\ z = 8 \end{cases}$

$$y - 7(8) = -54$$

$$y = 2$$

$$x - 4(2) + 5(8) = 27$$

$$x = -5$$

Solution: $(-5, 2, 8)$

The systems do *not* yield the same solution.

(b) $\begin{cases} x - 6y + z = 15 \\ y + 5z = 42 \\ z = 8 \end{cases}$

$$y + 5(8) = 42$$

$$y = 2$$

$$x - 6(2) + (8) = 15$$

$$x = 19$$

Solution: $(19, 2, 8)$

80. (a) $\begin{cases} x + 3y - z = 19 \\ y + 6z = -18 \\ z = -4 \end{cases}$

$$y + 6(-4) = -18$$

$$y = 6$$

$$x + 3(6) - (-4) = 19$$

$$x = -3$$

The systems do *not* yield the same solution.

(b) $\begin{cases} x - y + 3z = -15 \\ y - 2z = 14 \\ z = -4 \end{cases}$

$$y - 2(-4) = 14$$

$$y = 6$$

$$x - 6 + 3(-4) = -15$$

$$x = 3$$

81. $\begin{cases} x + 3y + z = 3 \\ x + 5y + 5z = 1 \\ 2x + 6y + 3z = 8 \end{cases}$

$$\begin{bmatrix} 1 & 3 & 1 & \vdots & 3 \\ 1 & 5 & 5 & \vdots & 1 \\ 2 & 6 & 3 & \vdots & 8 \end{bmatrix}$$

$\begin{matrix} \\ -R_1 + R_2 \rightarrow \\ -2R_1 + R_3 \rightarrow \end{matrix} \begin{bmatrix} 1 & 3 & 1 & \vdots & 3 \\ 0 & 2 & 4 & \vdots & -2 \\ 0 & 0 & 1 & \vdots & 2 \end{bmatrix}$

$\frac{1}{2}R_2 \rightarrow \begin{bmatrix} 1 & 3 & 1 & \vdots & 3 \\ 0 & 1 & 2 & \vdots & -1 \\ 0 & 0 & 1 & \vdots & 2 \end{bmatrix}$ This is a matrix in row-echelon form.

$\begin{bmatrix} 1 & 3 & \frac{3}{2} & \vdots & 4 \\ 0 & 1 & \frac{7}{4} & \vdots & -\frac{3}{2} \\ 0 & 0 & 1 & \vdots & 2 \end{bmatrix}$ The row-echelon form feature of a graphing utility yields this form.

There are infinitely many matrices in row-echelon form that correspond to the original system of equations. All such matrices will yield the same solution, namely $(16, -5, 2)$.

82. $\begin{cases} I_1 - I_2 + I_3 = 0 \\ 3I_1 + 4I_2 = 18 \\ I_2 + 3I_3 = 6 \end{cases}$

$$\begin{bmatrix} 1 & -1 & 1 & \vdots & 0 \\ 3 & 4 & 0 & \vdots & 18 \\ 0 & 1 & 3 & \vdots & 6 \end{bmatrix}$$

$-3R_1 + R_2 \rightarrow \begin{bmatrix} 1 & -1 & 1 & \vdots & 0 \\ 0 & 7 & -3 & \vdots & 18 \\ 0 & 1 & 3 & \vdots & 6 \end{bmatrix}$

$\begin{matrix} R_3 \\ R_2 \end{matrix} \begin{bmatrix} 1 & -1 & 1 & \vdots & 0 \\ 0 & 1 & 3 & \vdots & 6 \\ 0 & 7 & -3 & \vdots & 18 \end{bmatrix}$

$-7R_2 + R_3 \rightarrow \begin{bmatrix} 1 & -1 & 1 & \vdots & 0 \\ 0 & 1 & 3 & \vdots & 6 \\ 0 & 0 & -24 & \vdots & -24 \end{bmatrix}$

$-\frac{1}{24}R_3 \rightarrow \begin{bmatrix} 1 & -1 & 1 & \vdots & 0 \\ 0 & 1 & 3 & \vdots & 6 \\ 0 & 0 & 1 & \vdots & 1 \end{bmatrix}$

$\begin{cases} I_1 - I_2 + I_3 = 0 \\ I_2 + 3I_3 = 6 \\ I_3 = 1 \end{cases}$

$$I_3 = 1$$

$$I_2 + 3(1) = 6 \implies I_2 = 3$$

$$I_1 - 3 + 1 = 0 \implies I_1 = 2$$

83. $\dfrac{4x^2}{(x+1)^2(x-1)} = \dfrac{A}{x-1} + \dfrac{B}{x+1} + \dfrac{C}{(x+1)^2}$

$4x^2 = A(x+1)^2 + B(x+1)(x-1) + C(x-1)$

$4x^2 = A(x^2 + 2x + 1) + B(x^2 - 1) + C(x-1)$

$4x^2 = (A+B)x^2 + (2A+C)x + (A-B-C)$

System of equations:
$$A + B = 4$$
$$2A + C = 0$$
$$A - B - C = 0$$

$$\begin{bmatrix} 1 & 1 & 0 & \vdots & 4 \\ 2 & 0 & 1 & \vdots & 0 \\ 1 & -1 & -1 & \vdots & 0 \end{bmatrix} \xrightarrow{\text{rref}} \begin{bmatrix} 1 & 0 & 0 & \vdots & 1 \\ 0 & 1 & 0 & \vdots & 3 \\ 0 & 0 & 1 & \vdots & -2 \end{bmatrix}$$

Thus, $A = 1, B = 3, C = -2$.

So, $\dfrac{4x^2}{(x+1)^2(x-1)} = \dfrac{1}{x-1} + \dfrac{3}{x+1} - \dfrac{2}{(x+1)^2}$.

84. $\dfrac{8x^2}{(x-1)^2(x+1)} = \dfrac{A}{x+1} + \dfrac{B}{x-1} + \dfrac{C}{(x-1)^2}$

$8x^2 = A(x-1)^2 + B(x-1)(x+1) + C(x+1)$

$8x^2 = A(x^2 - 2x + 1) + B(x^2 - 1) + C(x+1)$

$8x^2 = (A+B)x^2 + (-2A+C)x + (A-B+C)$

System of equations:
$$A + B = 8$$
$$-2A + C = 0$$
$$A - B + C = 0$$

$$\begin{bmatrix} 1 & 1 & 0 & \vdots & 8 \\ -2 & 0 & 1 & \vdots & 0 \\ 1 & -1 & 1 & \vdots & 0 \end{bmatrix} \xrightarrow{\text{rref}} \begin{bmatrix} 1 & 0 & 0 & \vdots & 2 \\ 0 & 1 & 0 & \vdots & 6 \\ 0 & 0 & 1 & \vdots & 4 \end{bmatrix}$$

$A = 2, B = 6, C = 4$

$\dfrac{8x^2}{(x-1)^2(x+1)} = \dfrac{2}{x+1} + \dfrac{6}{x-1} + \dfrac{4}{(x-1)^2}$

85. x = amount at 7%

y = amount at 8%,

z = amount at 10%

$z = 4x \implies -4x + z = 0$

$$\begin{cases} x + y + z = 1,500,000 \\ 0.07x + 0.08y + 0.10z = 130,500 \\ -4x + + z = 0 \end{cases}$$

$$\begin{bmatrix} 1 & 1 & 1 & 1,500,000 \\ 0.07 & 0.08 & 0.10 & 130,500 \\ -4 & 0 & 1 & 0 \end{bmatrix}$$

$$\begin{matrix} -0.07R_1 + R_2 \to \\ 4R_1 + R_3 \to \end{matrix} \begin{bmatrix} 1 & 1 & 1 & \vdots & 1,500,000 \\ 0 & 0.01 & 0.03 & \vdots & 25,500 \\ 0 & 4 & 5 & \vdots & 6,000,000 \end{bmatrix}$$

$$100R_2 \to \begin{bmatrix} 1 & 1 & 1 & \vdots & 1,500,000 \\ 0 & 1 & 3 & \vdots & 2,550,000 \\ 0 & 4 & 5 & \vdots & 6,000,000 \end{bmatrix}$$

$$-4R_2 + R_3 \to \begin{bmatrix} 1 & 1 & 1 & \vdots & 1,500,000 \\ 0 & 1 & 3 & \vdots & 2,550,000 \\ 0 & 0 & -7 & \vdots & -4,200,000 \end{bmatrix}$$

$$-\tfrac{1}{7}R_3 \to \begin{bmatrix} 1 & 1 & 1 & \vdots & 1,500,000 \\ 0 & 1 & 3 & \vdots & 2,550,000 \\ 0 & 0 & 1 & \vdots & 600,000 \end{bmatrix}$$

$$\begin{cases} x + y + z = 1,500,000 \\ y + 3z = 2,550,000 \\ z = 600,000 \end{cases}$$

$y + 3(600,000) = 2,550,000 \implies y = 750,000$

$x + 750,000 + 600,000 = 1,500,000 \implies x = 150,000$

Solution: \$150,000 at 7%, \$750,000 at 8%, and \$600,000 at 10%

86. x = amount at 9%, y = amount at 10%,
z = amount at 12%

$$x + \quad y \quad\quad z = 500{,}000$$

$$0.09x + 0.10y + 0.12z = 52{,}000$$

$$2.5x - \quad y \quad\quad = \quad 0$$

$$\begin{bmatrix} 1 & 1 & 1 & \vdots & 500{,}000 \\ 0.09 & 0.10 & 0.12 & \vdots & 52{,}000 \\ 2.5 & -1 & 0 & \vdots & 0 \end{bmatrix}$$

$$\begin{matrix} \\ -0.09R_1 + R_2 \rightarrow \\ -2.5R_1 + R_3 \rightarrow \end{matrix} \begin{bmatrix} 1 & 1 & 1 & \vdots & 500{,}000 \\ 0 & 0.01 & 0.03 & \vdots & 7{,}000 \\ 0 & -3.5 & -2.5 & \vdots & -1{,}250{,}000 \end{bmatrix}$$

$$\begin{matrix} \\ 100R_2 \rightarrow \\ 2R_3 \rightarrow \end{matrix} \begin{bmatrix} 1 & 1 & 1 & \vdots & 500{,}000 \\ 0 & 1 & 3 & \vdots & 700{,}000 \\ 0 & -7 & -5 & \vdots & -2{,}500{,}000 \end{bmatrix}$$

$$\begin{matrix} -R_2 + R_1 \rightarrow \\ \\ 7R_2 + R_3 \rightarrow \end{matrix} \begin{bmatrix} 1 & 0 & -2 & \vdots & -200{,}000 \\ 0 & 1 & 3 & \vdots & 700{,}000 \\ 0 & 0 & 16 & \vdots & 2{,}400{,}000 \end{bmatrix}$$

$$\begin{matrix} \\ \\ \tfrac{1}{16}R_3 \rightarrow \end{matrix} \begin{bmatrix} 1 & 0 & -2 & \vdots & -200{,}000 \\ 0 & 1 & 3 & \vdots & 700{,}000 \\ 0 & 0 & 1 & \vdots & 150{,}000 \end{bmatrix}$$

$$\begin{cases} x \quad - 2z = -200{,}000 \\ y + 3z = \quad 700{,}000 \\ \quad z = \quad 150{,}000 \end{cases}$$

$$y + 3(150{,}000) = 700{,}000 \implies y = 250{,}000$$

$$x - 2(150{,}000) = -200{,}000 \implies x = 100{,}000$$

Solution: $(100{,}000, 250{,}000, 150{,}000)$

Answer: \$100,000 at 9%, \$250,000 at 10%, \$150,000 at 12%

88. $f(x) = ax^2 + bx + c$

$f(1) = a + b + c = 9$

$f(2) = 4a + 2b + c = 8$

$f(3) = 9a + 3b + c = 5$

$$\begin{bmatrix} 1 & 1 & 1 & \vdots & 9 \\ 4 & 2 & 1 & \vdots & 8 \\ 9 & 3 & 1 & \vdots & 5 \end{bmatrix}$$

$$\begin{matrix} -4R_1 + R_2 \rightarrow \\ -9R_1 + R_3 \rightarrow \end{matrix} \begin{bmatrix} 1 & 1 & 1 & \vdots & 9 \\ 0 & -2 & -3 & \vdots & -28 \\ 0 & -6 & -8 & \vdots & -76 \end{bmatrix}$$

$$\begin{matrix} -\tfrac{1}{2}R_2 \rightarrow \\ \\ \end{matrix} \begin{bmatrix} 1 & 1 & 1 & \vdots & 9 \\ 0 & 1 & \tfrac{3}{2} & \vdots & 14 \\ 0 & -6 & -8 & \vdots & -76 \end{bmatrix}$$

$$\begin{matrix} \\ \\ 6R_2 + R_3 \rightarrow \end{matrix} \begin{bmatrix} 1 & 1 & 1 & \vdots & 9 \\ 0 & 1 & \tfrac{3}{2} & \vdots & 14 \\ 0 & 0 & 1 & \vdots & 8 \end{bmatrix}$$

87. $y = ax^2 + bx + c$

$$\begin{cases} a + \quad b + c = 8 \\ 4a + 2b + c = 13 \\ 9a + 3b + c = 20 \end{cases}$$

$$\begin{bmatrix} 1 & 1 & 1 & \vdots & 8 \\ 4 & 2 & 1 & \vdots & 13 \\ 9 & 3 & 1 & \vdots & 20 \end{bmatrix}$$

$$\begin{matrix} -4R_1 + R_2 \rightarrow \\ -9R_1 + R_3 \rightarrow \end{matrix} \begin{bmatrix} 1 & 1 & 1 & \vdots & 8 \\ 0 & -2 & -3 & \vdots & -19 \\ 0 & -6 & -8 & \vdots & -52 \end{bmatrix}$$

$$\begin{matrix} -\tfrac{1}{2}R_2 \rightarrow \\ -3R_2 + R_3 \rightarrow \end{matrix} \begin{bmatrix} 1 & 1 & 1 & \vdots & 8 \\ 0 & 1 & \tfrac{3}{2} & \vdots & \tfrac{19}{2} \\ 0 & 0 & 1 & \vdots & 5 \end{bmatrix}$$

$$\begin{cases} a + b + \quad c = 8 \\ \quad b + \tfrac{3}{2}c = \tfrac{19}{2} \\ \quad\quad c = 5 \end{cases}$$

$$c = 5$$

$$b + \tfrac{3}{2}(5) = \tfrac{19}{2} \implies b = 2$$

$$a + 2 + 5 = 8 \implies a = 1$$

Equation of parabola: $y = x^2 + 2x + 5$

$$\begin{cases} a + b + \quad c = 9 \\ \quad b + \tfrac{3}{2}c = 14 \\ \quad\quad c = 8 \end{cases}$$

$$c = 8$$

$$b + \tfrac{3}{2}(8) = 14 \implies b = 2$$

$$a + (2) + (8) = 9 \implies a = -1$$

Equation of parabola: $y = -x^2 + 2x + 8$

89. (a) (0, 5.0), (15, 9.6), (30, 12.4)

$y = ax^2 + bx + c$

$$\begin{cases} \quad\quad\quad\quad c = 5 \\ 225a + 15b + c = 9.6 \implies 225a + 15b = 4.6 \\ 900a + 30b + c = 12.4 \implies 900a + 30b = 7.4 \end{cases}$$

$$\begin{bmatrix} 225 & 15 & \vdots & 4.6 \\ 900 & 30 & \vdots & 7.4 \end{bmatrix}$$

$-4R_1 + R_2 \rightarrow \begin{bmatrix} 225 & 15 & \vdots & 4.6 \\ 0 & -30 & \vdots & -11 \end{bmatrix}$

$\begin{matrix} \frac{1}{225}R_1 \rightarrow \\ \left(-\frac{1}{30}\right)R_2 \rightarrow \end{matrix} \begin{bmatrix} 1 & \frac{1}{15} & \vdots & \frac{23}{1125} \\ 0 & 1 & \vdots & \frac{11}{30} \end{bmatrix}$

$$\begin{cases} a + \frac{1}{15}b = \frac{23}{1125} \\ \quad\quad b = \frac{11}{30} \end{cases}$$

$a + \frac{1}{15}\left(\frac{11}{30}\right) = \frac{23}{1125} \implies a = -\frac{1}{250} = -0.004$

Equation of parabola: $y = -0.004x^2 + 0.367x + 5$.

(b)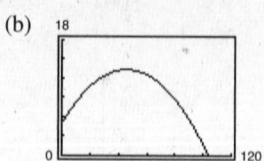

(c) The maximum height is approximately 13 feet and the ball strikes the ground at approximately 104 feet.

(d) The maximum occurs at the vertex.

$x = -\dfrac{b}{2a} = \dfrac{-0.367}{2(-0.004)} = 45.875$

$y = -0.004(45.875)^2 + 0.367(45.875) + 5$

$\quad = 13.418$ feet

The ball strikes the ground when $y = 0$.

$-0.004x^2 + 0.367x + 5 = 0$

By the Quadratic Formula and using the positive value for x we have $x \approx 103.793$ feet.

(e) The values found in part (d) are more accurate, but still very close to the estimates found in part (c).

90. (a) $f(x) = at^2 + bt + c$

$f(7) = 49a + 7b + c = 2.8$

$f(9) = 81a + 9b + c = 3.3$

$f(11) = 121a + 11b + c = 5.3$

$$\begin{cases} 49a + 7b + c = 2.8 \\ 81a + 9b + c = 3.3 \\ 121a + 11b + c = 5.3 \end{cases}$$

$$\begin{bmatrix} 49 & 7 & 1 & 2.8 \\ 81 & 9 & 1 & 3.3 \\ 121 & 11 & 1 & 5.3 \end{bmatrix}$$

$\frac{1}{49}R_1 \rightarrow \begin{bmatrix} 1 & \frac{1}{7} & \frac{1}{49} & \frac{2}{35} \\ 81 & 9 & 1 & 3.3 \\ 121 & 11 & 1 & 5.3 \end{bmatrix}$

$\begin{matrix} -81R_1 + R_2 \rightarrow \\ -121R_1 + R_3 \rightarrow \end{matrix} \begin{bmatrix} 1 & \frac{1}{7} & \frac{1}{49} & \frac{2}{35} \\ 0 & -\frac{18}{7} & -\frac{32}{49} & -\frac{93}{70} \\ 0 & -\frac{44}{7} & -\frac{72}{49} & -\frac{113}{70} \end{bmatrix}$

$-\frac{7}{18}R_2 \rightarrow \begin{bmatrix} 1 & \frac{1}{7} & \frac{1}{49} & \frac{2}{35} \\ 0 & 1 & \frac{16}{63} & \frac{31}{60} \\ 0 & -\frac{44}{7} & -\frac{72}{49} & -\frac{113}{70} \end{bmatrix}$

$\frac{44}{7}R_2 + R_3 \rightarrow \begin{bmatrix} 1 & \frac{1}{7} & \frac{1}{49} & \frac{2}{35} \\ 0 & 1 & \frac{16}{63} & \frac{31}{60} \\ 0 & 0 & \frac{8}{63} & \frac{49}{30} \end{bmatrix}$

$a + \frac{1}{7}b + \frac{1}{49}c = \frac{2}{35}$

$\quad\quad b + \frac{16}{63}c = \frac{31}{60}$

$\quad\quad\quad\quad \frac{8}{63}c = \frac{49}{30}$

$c = \frac{49}{30} \cdot \frac{63}{8} = \frac{1029}{80} = 12.86$

$b + \frac{16}{63}(12.86) = \frac{31}{60} \implies b = -2.75$

$a + \frac{1}{7}(-2.75) + \frac{1}{49}(12.86) = \frac{2}{35} \implies a = 0.1875$

Equation of parabola: $y = 0.1875t^2 - 2.75t + 12.86$

(b)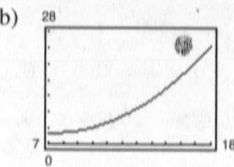

(c) For 2003, $t = 13$.

$y = 0.1875(13^2) - 2.75(13) + 12.86 = 8.8$

When compared to the actual value of 6.3, this is not very accurate.

(d) For 2008, $t = 18$.

$y = 0.1875(18^2) - 2.75(18) + 12.86 = 24.11$

The model estimates that in 2008, 24.11 million people will participate in snowboarding. This indicates that the number of participants will almost triple in 5 years which is probably not a reasonable estimate.

91. (a)

$x_1 + x_3 = 600$

$x_1 = x_2 + x_4 \implies x_1 - x_2 - x_4 = 0$

$x_2 + x_5 = 500$

$x_3 + x_6 = 600$

$x_4 + x_7 = x_6 \implies x_4 - x_6 + x_7 = 0$

$x_5 + x_7 = 500$

$$\left[\begin{array}{ccccccc:c}
1 & 0 & 1 & 0 & 0 & 0 & 0 & 600 \\
1 & -1 & 0 & -1 & 0 & 0 & 0 & 0 \\
0 & 1 & 0 & 0 & 1 & 0 & 0 & 500 \\
0 & 0 & 1 & 0 & 0 & 1 & 0 & 600 \\
0 & 0 & 0 & 1 & 0 & -1 & 1 & 0 \\
0 & 0 & 0 & 0 & 1 & 0 & 1 & 500
\end{array}\right]$$

$$\begin{array}{l}
\\
-R_1 + R_2 \rightarrow \\
R_2 + R_3 \rightarrow \\
R_3 + R_4 \rightarrow \\
R_4 + R_5 \rightarrow \\
-R_5 + R_6 \rightarrow
\end{array}
\left[\begin{array}{ccccccc:c}
1 & 0 & 1 & 0 & 0 & 0 & 0 & 600 \\
0 & -1 & -1 & -1 & 0 & 0 & 0 & -600 \\
0 & 0 & -1 & -1 & 1 & 0 & 0 & -100 \\
0 & 0 & 0 & -1 & 1 & 1 & 0 & 500 \\
0 & 0 & 0 & 0 & 1 & 0 & 1 & 500 \\
0 & 0 & 0 & 0 & 0 & 0 & 0 & 0
\end{array}\right]$$

$$\begin{array}{l}
\\
-R_3 + R_2 \rightarrow \\
-R_4 + R_3 \rightarrow \\
-R_4 \rightarrow \\
\\
\end{array}
\left[\begin{array}{ccccccc:c}
1 & 0 & 1 & 0 & 0 & 0 & 0 & 600 \\
0 & -1 & 0 & 0 & -1 & 0 & 0 & -500 \\
0 & 0 & -1 & 0 & 0 & -1 & 0 & -600 \\
0 & 0 & 0 & 1 & -1 & -1 & 0 & -500 \\
0 & 0 & 0 & 0 & 1 & 0 & 1 & 500 \\
0 & 0 & 0 & 0 & 0 & 0 & 0 & 0
\end{array}\right]$$

$$\begin{array}{l}
\\
-R_2 \rightarrow \\
-R_3 \rightarrow \\
\\
\\
\end{array}
\left[\begin{array}{ccccccc:c}
1 & 0 & 1 & 0 & 0 & 0 & 0 & 600 \\
0 & 1 & 0 & 0 & 1 & 0 & 0 & 500 \\
0 & 0 & 1 & 0 & 0 & 1 & 0 & 600 \\
0 & 0 & 0 & 1 & -1 & -1 & 0 & -500 \\
0 & 0 & 0 & 0 & 1 & 0 & 1 & 500 \\
0 & 0 & 0 & 0 & 0 & 0 & 0 & 0
\end{array}\right]$$

$$\begin{cases}
x_1 + x_3 = 600 \\
\quad x_2 + x_5 = 500 \\
\quad\quad x_3 + x_6 = 600 \\
x_4 - x_5 - x_6 = -500 \\
\quad\quad x_5 + x_7 = 500
\end{cases}$$

Let $x_7 = t$ and $x_6 = s$, then $x_5 = 500 - t$,

$x_4 = -500 + s + (500 - t) = s - t$,

$x_3 = 600 - s$, $x_2 = 500 - (500 - t) = t$,

$x_1 = 600 - (600 - s) = s$.

Solution: $(s, t, 600 - s, s - t, 500 - t, s, t)$

(b) $s = 0, t = 0$: $x_1 = 0, x_2 = 0, x_3 = 600, x_4 = 0, x_5 = 500, x_6 = 0, x_7 = 0$

(c) $s = 0, t = -500$: $x_1 = 0, x_2 = -500, x_3 = 600, x_4 = 500, x_5 = 1000, x_6 = 0, x_7 = -500$

92. (a) $x_1 + x_2 = 300$

$x_1 + x_3 = 150 + x_4 \Rightarrow x_1 + x_3 - x_4 = 150$

$x_2 + 200 = x_3 + x_5 \Rightarrow x_2 - x_3 - x_5 = -200$

$x_4 + x_5 = 350$

$$\begin{bmatrix} 1 & 1 & 0 & 0 & 0 & : & 300 \\ 1 & 0 & 1 & -1 & 0 & : & 150 \\ 0 & 1 & -1 & 0 & -1 & : & -200 \\ 0 & 0 & 0 & 1 & 1 & : & 350 \end{bmatrix}$$

$-R_1 + R_2 \rightarrow \begin{bmatrix} 1 & 1 & 0 & 0 & 0 & : & 300 \\ 0 & -1 & 1 & -1 & 0 & : & -150 \\ 0 & 1 & -1 & 0 & -1 & : & -200 \\ 0 & 0 & 0 & 1 & 1 & : & 350 \end{bmatrix}$

$R_2 + R_3 \rightarrow \begin{bmatrix} 1 & 1 & 0 & 0 & 0 & : & 300 \\ 0 & -1 & 1 & -1 & 0 & : & -150 \\ 0 & 0 & 0 & -1 & -1 & : & -350 \\ 0 & 0 & 0 & 1 & 1 & : & 350 \end{bmatrix}$

$\begin{matrix} \\ -R_2 \rightarrow \\ -R_3 \rightarrow \\ R_3 + R_4 \rightarrow \end{matrix} \begin{bmatrix} 1 & 1 & 0 & 0 & 0 & : & 300 \\ 0 & 1 & -1 & 1 & 0 & : & 150 \\ 0 & 0 & 0 & 1 & 1 & : & 350 \\ 0 & 0 & 0 & 0 & 0 & : & 0 \end{bmatrix}$

$\begin{cases} x_1 + x_2 = 300 \\ x_2 - x_3 + x_4 = 150 \\ x_4 + x_5 = 350 \end{cases}$

Let $x_5 = t$.

$x_4 + t = 350 \Rightarrow x_4 = 350 - t$

Let $x_3 = s$.

$x_2 - s + (350 - t) = 150 \Rightarrow x_2 = -200 + s + t$

$x_1 + (-200 + s + t) = 300 \Rightarrow x_1 = 500 - s - t$

Solution: $x_1 = 500 - s - t, x_2 = -200 + s + t, x_3 = s, x_4 = 350 - t, x_5 = t$,
where s and t are real numbers.

(b) When $x_2 = 200$ and $x_3 = 50$,

$x_2 = -200 + s + t$

$200 = -200 + 50 + t \Rightarrow t = 350.$

$x_1 = 100, x_2 = 200, x_3 = 50, x_4 = 0, x_5 = 350$

(c) When $x_2 = 150$ and $x_3 = 0$,

$x_2 = -200 + s + t$

$150 = -200 + 0 + t \Rightarrow t = 350.$

$x_1 = 150, x_2 = 150, x_3 = 0, x_4 = 0, x_5 = 350$

93. False. It is a 2×4 matrix.

94. False. The rows are in the wrong order. To change this matrix to reduced row-echelon form, interchange Row 1 and Row 4, and interchange Row 2 and Row 3.

95. False. Gaussian elimination reduces a matrix until a row-echelon form is obtained and Gauss-Jordan elimination reduces a matrix until a reduced row-echelon form is obtained.

96. $z = a$

$y = -4a + 1$

$x = -3a - 2$

One possible system is:

$\begin{cases} x + y + 7z = (-3a - 2) + (-4a + 1) + 7a = -1 \\ x + 2y + 11z = (-3a - 2) + 2(-4a + 1) + 11a = 0 \\ 2x + y + 10z = 2(-3a - 2) + (-4a + 1) + 10a = -3 \end{cases}$ or $\begin{cases} x + y + 7z = -1 \\ x + 2y + 11z = 0 \\ 2x + y + 10z = -3 \end{cases}$

(Note that the coefficients of x, y, and z have been chosen so that the a-terms cancel.)

97. (a) In the row-echelon form of an augmented matrix that corresponds to an inconsistent system of linear equations, there exists a row consisting of all zeros except for the entry in the last column.

(b) In the row-echelon form of an augmented matrix that corresponds to a system with an infinite number of solutions, there are fewer rows with nonzero entries than there are variables and no row has the first non-zero value in the last column.

98. 1. Interchange two rows.

2. Multiply a row by a nonzero constant.

3. Add a multiple of one row to another row.

99. They are the same.

100. A matrix in row-echelon form is in reduced row-echelon form if every column that has a leading 1 has zeros in every position above and below its leading 1.

101. $f(x) = \dfrac{2x^2 - 4x}{3x - x^2} = \dfrac{2x - 4}{3 - x}, x \neq 0$

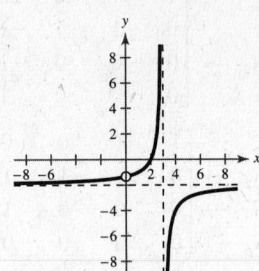

x	-2	-1	0	1	2	3	4	5
$f(x)$	-1.6	-1.5	undef.	-1	0	undef.	-4	-3

Vertical asymptote: $x = 3$

Horizontal asymptote: $y = -2$

Intercept: $(2, 0)$

102. $f(x) = \dfrac{x^2 - 2x + 1}{x^2 - 1} = \dfrac{(x - 1)(x - 1)}{(x - 1)(x + 1)} = \dfrac{x - 1}{x + 1}$

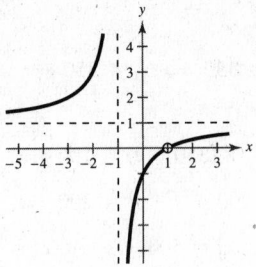

The graph has a vertical asymptote at $x = -1$ and a discontinuity at $x = 1$.

Since the degrees of the numerator and the denominator are the same, there is a horizontal asymptote at $y = 1$.

103. $f(x) = 2^{x-1}$

x	-1	0	1	2	3
$f(x)$	$\frac{1}{4}$	$\frac{1}{2}$	1	2	4

Horizontal asymptote: $y = 0$

Intercept: $\left(0, \frac{1}{2}\right)$

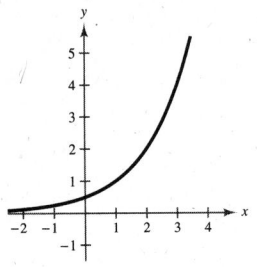

104. $g(x) = 3^{-x+2}$

x	-1	0	1	2	3	4
y	27	9	3	1	$\frac{1}{3}$	$\frac{1}{9}$

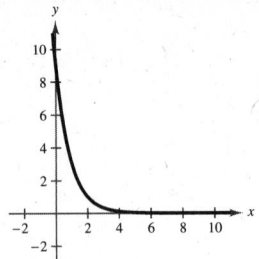

105. $h(x) = \ln(x - 1)$

x	1.5	2	3	4	5
$h(x)$	-0.693	0	0.693	1.099	1.386

Vertical asymptote: $x = 1$

Intercept: $(2, 0)$

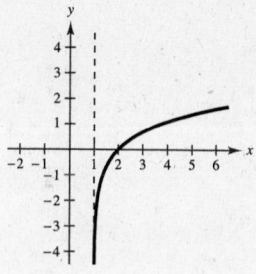

106. $f(x) = 3 + \ln x \Rightarrow y - 3 = \ln x \Rightarrow e^{y-3} = x$

x	0.05	0.14	0.37	1	2.72
y	0	1	2	3	4

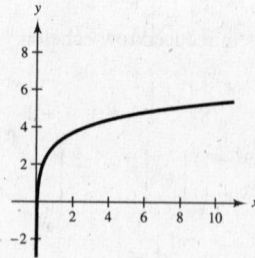

Section 8.2   Operations with Matrices

■ $A = B$ if and only if they have the same order and $a_{ij} = b_{ij}$.

■   You should be able to perform the operations of matrix addition, scalar multiplication, and matrix multiplication.

■   Some properties of matrix addition and scalar multiplication are:

    (a) $A + B = B + A$

    (b) $A + (B + C) = (A + B) + C$

    (c) $(cd)A = c(dA)$

    (d) $1A = A$

    (e) $c(A + B) = cA + cB$

    (f) $(c + d)A = cA + dA$

■   You should remember that $AB \neq BA$ in general.

■   Some properties of matrix multiplication are:

    (a) $A(BC) = (AB)C$

    (b) $A(B + C) = AB + AC$

    (c) $(A + B)C = AC + BC$

    (d) $c(AB) = (cA)B = A(cB)$

■   You should know that I_n, the identity matrix of order n, is an $n \times n$ matrix consisting of ones on its main diagonal and zeros elsewhere. If A is an $n \times n$ matrix, then $AI_n = I_nA = A$.

Vocabulary Check

1. equal

2. scalars

3. zero; O

4. identity

5. (a) (iii)   (b) (iv)   (c) (i)
    (d) (v)   (e) (ii)

6. (a) (ii)   (b) (iv)   (c) (i)   (d) (iii)

1. $x = -4$, $y = 22$

2. $x = 13$, $y = 12$

3. $2x + 1 = 5$, $3x = 6$, $3y - 5 = 4$

     $x = 2$, $y = 3$

4. $x + 2 = 2x + 6$ $2y = 18$

        $-4 = x$ $y = 9$

        $2x = -8$ $y + 2 = 11$

         $x = -4$ $y = 9$

5. (a) $A + B = \begin{bmatrix} 1 & -1 \\ 2 & -1 \end{bmatrix} + \begin{bmatrix} 2 & -1 \\ -1 & 8 \end{bmatrix} = \begin{bmatrix} 1+2 & -1-1 \\ 2-1 & -1+8 \end{bmatrix} = \begin{bmatrix} 3 & -2 \\ 1 & 7 \end{bmatrix}$

   (b) $A - B = \begin{bmatrix} 1 & -1 \\ 2 & -1 \end{bmatrix} - \begin{bmatrix} 2 & -1 \\ -1 & 8 \end{bmatrix} = \begin{bmatrix} 1-2 & -1+1 \\ 2+1 & -1-8 \end{bmatrix} = \begin{bmatrix} -1 & 0 \\ 3 & -9 \end{bmatrix}$

   (c) $3A = 3\begin{bmatrix} 1 & -1 \\ 2 & -1 \end{bmatrix} = \begin{bmatrix} 3(1) & 3(-1) \\ 3(2) & 3(-1) \end{bmatrix} = \begin{bmatrix} 3 & -3 \\ 6 & -3 \end{bmatrix}$

   (d) $3A - 2B = \begin{bmatrix} 3 & -3 \\ 6 & -3 \end{bmatrix} - 2\begin{bmatrix} 2 & -1 \\ -1 & 8 \end{bmatrix} = \begin{bmatrix} 3 & -3 \\ 6 & -3 \end{bmatrix} + \begin{bmatrix} -4 & 2 \\ 2 & -16 \end{bmatrix} = \begin{bmatrix} -1 & -1 \\ 8 & -19 \end{bmatrix}$

6. (a) $A + B = \begin{bmatrix} 1 & 2 \\ 2 & 1 \end{bmatrix} + \begin{bmatrix} -3 & -2 \\ 4 & 2 \end{bmatrix} = \begin{bmatrix} 1-3 & 2-2 \\ 2+4 & 1+2 \end{bmatrix} = \begin{bmatrix} -2 & 0 \\ 6 & 3 \end{bmatrix}$

   (b) $A - B = \begin{bmatrix} 1 & 2 \\ 2 & 1 \end{bmatrix} - \begin{bmatrix} -3 & -2 \\ 4 & 2 \end{bmatrix} = \begin{bmatrix} 1+3 & 2+2 \\ 2-4 & 1-2 \end{bmatrix} = \begin{bmatrix} 4 & 4 \\ -2 & -1 \end{bmatrix}$

   (c) $3A = 3\begin{bmatrix} 1 & 2 \\ 2 & 1 \end{bmatrix} = \begin{bmatrix} 3(1) & 3(2) \\ 3(2) & 3(1) \end{bmatrix} = \begin{bmatrix} 3 & 6 \\ 6 & 3 \end{bmatrix}$

   (d) $3A - 2B = \begin{bmatrix} 3 & 6 \\ 6 & 3 \end{bmatrix} - 2\begin{bmatrix} -3 & -2 \\ 4 & 2 \end{bmatrix} = \begin{bmatrix} 3+6 & 6+4 \\ 6-8 & 3-4 \end{bmatrix} = \begin{bmatrix} 9 & 10 \\ -2 & -1 \end{bmatrix}$

7. $A = \begin{bmatrix} 6 & -1 \\ 2 & 4 \\ -3 & 5 \end{bmatrix}$, $B = \begin{bmatrix} 1 & 4 \\ -1 & 5 \\ 1 & 10 \end{bmatrix}$

   (a) $A + B = \begin{bmatrix} 7 & 3 \\ 1 & 9 \\ -2 & 15 \end{bmatrix}$ (b) $A - B = \begin{bmatrix} 5 & -5 \\ 3 & -1 \\ -4 & -5 \end{bmatrix}$ (c) $3A = \begin{bmatrix} 18 & -3 \\ 6 & 12 \\ -9 & 15 \end{bmatrix}$

   (d) $3A - 2B = \begin{bmatrix} 18 & -3 \\ 6 & 12 \\ -9 & 15 \end{bmatrix} - \begin{bmatrix} 2 & 8 \\ -2 & 10 \\ 2 & 20 \end{bmatrix} = \begin{bmatrix} 16 & -11 \\ 8 & 2 \\ -11 & -5 \end{bmatrix}$

8. (a) $A + B = \begin{bmatrix} 2 & 1 & 1 \\ -1 & -1 & 4 \end{bmatrix} + \begin{bmatrix} 2 & -3 & 4 \\ -3 & 1 & -2 \end{bmatrix} = \begin{bmatrix} 2+2 & 1-3 & 1+4 \\ -1-3 & -1+1 & 4-2 \end{bmatrix} = \begin{bmatrix} 4 & -2 & 5 \\ -4 & 0 & 2 \end{bmatrix}$

   (b) $A - B = \begin{bmatrix} 2 & 1 & 1 \\ -1 & -1 & 4 \end{bmatrix} - \begin{bmatrix} 2 & -3 & 4 \\ -3 & 1 & -2 \end{bmatrix} = \begin{bmatrix} 2-2 & 1-(-3) & 1-4 \\ -1-(-3) & -1-1 & 4-(-2) \end{bmatrix} = \begin{bmatrix} 0 & 4 & -3 \\ 2 & -2 & 6 \end{bmatrix}$

   (c) $3A = 3\begin{bmatrix} 2 & 1 & 1 \\ -1 & -1 & 4 \end{bmatrix} = \begin{bmatrix} 3(2) & 3(1) & 3(1) \\ 3(-1) & 3(-1) & 3(4) \end{bmatrix} = \begin{bmatrix} 6 & 3 & 3 \\ -3 & -3 & 12 \end{bmatrix}$

   (d) $3A - 2B = \begin{bmatrix} 6 & 3 & 3 \\ -3 & -3 & 12 \end{bmatrix} - 2\begin{bmatrix} 2 & -3 & 4 \\ -3 & 1 & -2 \end{bmatrix} = \begin{bmatrix} 6 & 3 & 3 \\ -3 & -3 & 12 \end{bmatrix} + \begin{bmatrix} -4 & 6 & -8 \\ 6 & -2 & 4 \end{bmatrix} = \begin{bmatrix} 2 & 9 & -5 \\ 3 & -5 & 16 \end{bmatrix}$

9. $A = \begin{bmatrix} 2 & 2 & -1 & 0 & 1 \\ 1 & 1 & -2 & 0 & -1 \end{bmatrix}$, $B = \begin{bmatrix} 1 & 1 & -1 & 1 & 0 \\ -3 & 4 & 9 & -6 & -7 \end{bmatrix}$

(a) $A + B = \begin{bmatrix} 3 & 3 & -2 & 1 & 1 \\ -2 & 5 & 7 & -6 & -8 \end{bmatrix}$

(b) $A - B = \begin{bmatrix} 1 & 1 & 0 & -1 & 1 \\ 4 & -3 & -11 & 6 & 6 \end{bmatrix}$

(c) $3A = \begin{bmatrix} 6 & 6 & -3 & 0 & 3 \\ 3 & 3 & -6 & 0 & -3 \end{bmatrix}$

(d) $3A - 2B = \begin{bmatrix} 6 & 6 & -3 & 0 & 3 \\ 3 & 3 & -6 & 0 & -3 \end{bmatrix} - \begin{bmatrix} 2 & 2 & -2 & 2 & 0 \\ -6 & 8 & 18 & -12 & -14 \end{bmatrix} = \begin{bmatrix} 4 & 4 & -1 & -2 & 3 \\ 9 & -5 & -24 & 12 & 11 \end{bmatrix}$

10. (a) $A + B = \begin{bmatrix} -1 & 4 & 0 \\ 3 & -2 & 2 \\ 5 & 4 & -1 \\ 0 & 8 & -6 \\ -4 & -1 & 0 \end{bmatrix} + \begin{bmatrix} -3 & 5 & 1 \\ 2 & -4 & -7 \\ 10 & -9 & -1 \\ 3 & 2 & -4 \\ 0 & 1 & -2 \end{bmatrix}$

$= \begin{bmatrix} -1-3 & 4+5 & 0+1 \\ 3+2 & -2-4 & 2-7 \\ 5+10 & 4-9 & -1-1 \\ 0+3 & 8+2 & -6-4 \\ -4+0 & -1+1 & 0-2 \end{bmatrix} = \begin{bmatrix} -4 & 9 & 1 \\ 5 & -6 & -5 \\ 15 & -5 & -2 \\ 3 & 10 & -10 \\ -4 & 0 & -2 \end{bmatrix}$

(b) $A - B = \begin{bmatrix} -1 & 4 & 0 \\ 3 & -2 & 2 \\ 5 & 4 & -1 \\ 0 & 8 & -6 \\ -4 & -1 & 0 \end{bmatrix} - \begin{bmatrix} -3 & 5 & 1 \\ 2 & -4 & -7 \\ 10 & -9 & -1 \\ 3 & 2 & -4 \\ 0 & 1 & -2 \end{bmatrix}$

$= \begin{bmatrix} -1+3 & 4-5 & 0-1 \\ 3-2 & -2+4 & 2+7 \\ 5-10 & 4+9 & -1+1 \\ 0-3 & 8-2 & -6+4 \\ -4-0 & -1-1 & 0+2 \end{bmatrix} = \begin{bmatrix} 2 & -1 & -1 \\ 1 & 2 & 9 \\ -5 & 13 & 0 \\ -3 & 6 & -2 \\ -4 & -2 & 2 \end{bmatrix}$

(c) $3A = 3\begin{bmatrix} -1 & 4 & 0 \\ 3 & -2 & 2 \\ 5 & 4 & -1 \\ 0 & 8 & -6 \\ -4 & -1 & 0 \end{bmatrix} = \begin{bmatrix} -3 & 12 & 0 \\ 9 & -6 & 6 \\ 15 & 12 & -3 \\ 0 & 24 & -18 \\ -12 & -3 & 0 \end{bmatrix}$

(d) $3A - 2B = \begin{bmatrix} -3 & 12 & 0 \\ 9 & -6 & 6 \\ 15 & 12 & -3 \\ 0 & 24 & -18 \\ -12 & -3 & 0 \end{bmatrix} - 2\begin{bmatrix} -3 & 5 & 1 \\ 2 & -4 & -7 \\ 10 & -9 & -1 \\ 3 & 2 & -4 \\ 0 & 1 & -2 \end{bmatrix}$

$= \begin{bmatrix} -3 & 12 & 0 \\ 9 & -6 & 6 \\ 15 & 12 & -3 \\ 0 & 24 & -18 \\ -12 & -3 & 0 \end{bmatrix} + \begin{bmatrix} 6 & -10 & -2 \\ -4 & 8 & 14 \\ -20 & 18 & 2 \\ -6 & -4 & 8 \\ 0 & -2 & 4 \end{bmatrix} = \begin{bmatrix} 3 & 2 & -2 \\ 5 & 2 & 20 \\ -5 & 30 & -1 \\ -6 & 20 & -10 \\ -12 & -5 & 4 \end{bmatrix}$

11. $A = \begin{bmatrix} 6 & 0 & 3 \\ -1 & -4 & 0 \end{bmatrix}$, $B = \begin{bmatrix} 8 & -1 \\ 4 & -3 \end{bmatrix}$

(a) $A + B$ is not possible. A and B do not have the same order.

(b) $A - B$ is not possible. A and B do not have the same order.

(c) $3A = \begin{bmatrix} 18 & 0 & 9 \\ -3 & -12 & 0 \end{bmatrix}$

(d) $3A - 2B$ is not possible. A and B do not have the same order.

12. (a) $A + B$ is not possible. A and B do not have the same order.

(b) $A - B$ is not possible. A and B do not have the same order.

(c) $3A = 3\begin{bmatrix} 3 \\ 2 \\ -1 \end{bmatrix} = \begin{bmatrix} 9 \\ 6 \\ -3 \end{bmatrix}$

(d) $3A - 2B$ is not possible. A and B do not have the same order.

13. $\begin{bmatrix} -5 & 0 \\ 3 & -6 \end{bmatrix} + \begin{bmatrix} 7 & 1 \\ -2 & -1 \end{bmatrix} + \begin{bmatrix} -10 & -8 \\ 14 & 6 \end{bmatrix} = \begin{bmatrix} -5 + 7 + (-10) & 0 + 1 + (-8) \\ 3 + (-2) + 14 & -6 + (-1) + 6 \end{bmatrix} = \begin{bmatrix} -8 & -7 \\ 15 & -1 \end{bmatrix}$

14. $\begin{bmatrix} 6 & 8 \\ -1 & 0 \end{bmatrix} + \begin{bmatrix} 0 & 5 \\ -3 & -1 \end{bmatrix} + \begin{bmatrix} -11 & -7 \\ 2 & -1 \end{bmatrix} = \begin{bmatrix} 6 + 0 + (-11) & 8 + 5 + (-7) \\ -1 + (-3) + 2 & 0 + (-1) + (-1) \end{bmatrix} = \begin{bmatrix} -5 & 6 \\ -2 & -2 \end{bmatrix}$

15. $4\left(\begin{bmatrix} -4 & 0 & 1 \\ 0 & 2 & 3 \end{bmatrix} - \begin{bmatrix} 2 & 1 & -2 \\ 3 & -6 & 0 \end{bmatrix}\right) = 4\begin{bmatrix} -6 & -1 & 3 \\ -3 & 8 & 3 \end{bmatrix} = \begin{bmatrix} -24 & -4 & 12 \\ -12 & 32 & 12 \end{bmatrix}$

16. $\frac{1}{2}([5 \quad -2 \quad 4 \quad 0] + [14 \quad 6 \quad -18 \quad 9]) = \frac{1}{2}[5 + 14 \quad -2 + 6 \quad 4 + (-18) \quad 0 + 9]$

$$= \frac{1}{2}[19 \quad 4 \quad -14 \quad 9]$$

$$= \left[\frac{19}{2} \quad 2 \quad -7 \quad \frac{9}{2}\right]$$

17. $-3\left(\begin{bmatrix} 0 & -3 \\ 7 & 2 \end{bmatrix} + \begin{bmatrix} -6 & 3 \\ 8 & 1 \end{bmatrix}\right) - 2\begin{bmatrix} 4 & -4 \\ 7 & -9 \end{bmatrix} = -3\begin{bmatrix} -6 & 0 \\ 15 & 3 \end{bmatrix} - \begin{bmatrix} 8 & -8 \\ 14 & -18 \end{bmatrix} = \begin{bmatrix} 18 & 0 \\ -45 & -9 \end{bmatrix} - \begin{bmatrix} 8 & -8 \\ 14 & -18 \end{bmatrix} = \begin{bmatrix} 10 & 8 \\ -59 & 9 \end{bmatrix}$

18. $-1\begin{bmatrix} 4 & 11 \\ -2 & -1 \\ 9 & 3 \end{bmatrix} + \frac{1}{6}\left(\begin{bmatrix} -5 & -1 \\ 3 & 4 \\ 0 & 13 \end{bmatrix} + \begin{bmatrix} 7 & 5 \\ -9 & -1 \\ 6 & -1 \end{bmatrix}\right) = \begin{bmatrix} -4 & -11 \\ 2 & 1 \\ -9 & -3 \end{bmatrix} + \frac{1}{6}\begin{bmatrix} -5 + 7 & -1 + 5 \\ 3 + (-9) & 4 + (-1) \\ 0 + 6 & 13 + (-1) \end{bmatrix}$

$$= \begin{bmatrix} -4 & -11 \\ 2 & 1 \\ -9 & -3 \end{bmatrix} + \frac{1}{6}\begin{bmatrix} 2 & 4 \\ -6 & 3 \\ 6 & 12 \end{bmatrix}$$

$$= \begin{bmatrix} -4 & -11 \\ 2 & 1 \\ -9 & -3 \end{bmatrix} + \begin{bmatrix} \frac{1}{3} & \frac{2}{3} \\ -1 & \frac{1}{2} \\ 1 & 2 \end{bmatrix}$$

$$= \begin{bmatrix} -4 + \frac{1}{3} & -11 + \frac{2}{3} \\ 2 + (-1) & 1 + \frac{1}{2} \\ -9 + 1 & -3 + 2 \end{bmatrix} = \begin{bmatrix} -\frac{11}{3} & -\frac{31}{3} \\ 1 & \frac{3}{2} \\ -8 & -1 \end{bmatrix}$$

19. $\frac{3}{7}\begin{bmatrix} 2 & 5 \\ -1 & -4 \end{bmatrix} + 6\begin{bmatrix} -3 & 0 \\ 2 & 2 \end{bmatrix} \approx \begin{bmatrix} -17.143 & 2.143 \\ 11.571 & 10.286 \end{bmatrix}$

20. $55\left(\begin{bmatrix} 14 & -11 \\ -22 & 19 \end{bmatrix} + \begin{bmatrix} -22 & 20 \\ 13 & 6 \end{bmatrix}\right) = \begin{bmatrix} -440 & 495 \\ -495 & 1375 \end{bmatrix}$

21. $-\begin{bmatrix} 3.211 & 6.829 \\ -1.004 & 4.914 \\ 0.055 & -3.889 \end{bmatrix} - \begin{bmatrix} -1.630 & -3.090 \\ 5.256 & 8.335 \\ -9.768 & 4.251 \end{bmatrix} = \begin{bmatrix} -1.581 & -3.739 \\ -4.252 & -13.249 \\ 9.713 & -0.362 \end{bmatrix}$

22. $-12\left(\begin{bmatrix} 6 & 20 \\ 1 & -9 \\ -2 & 5 \end{bmatrix} + \begin{bmatrix} 14 & -15 \\ -8 & -6 \\ 7 & 0 \end{bmatrix} + \begin{bmatrix} -31 & -19 \\ 16 & 10 \\ 24 & -10 \end{bmatrix}\right) = \begin{bmatrix} 132 & 168 \\ -108 & 60 \\ -348 & 60 \end{bmatrix}$

23. $X = 3\begin{bmatrix} -2 & -1 \\ 1 & 0 \\ 3 & -4 \end{bmatrix} - 2\begin{bmatrix} 0 & 3 \\ 2 & 0 \\ -4 & -1 \end{bmatrix} = \begin{bmatrix} -6 & -3 \\ 3 & 0 \\ 9 & -12 \end{bmatrix} - \begin{bmatrix} 0 & 6 \\ 4 & 0 \\ -8 & -2 \end{bmatrix} = \begin{bmatrix} -6 & -9 \\ -1 & 0 \\ 17 & -10 \end{bmatrix}$

24. $2X = 2A - B$

$X = A - \frac{1}{2}B = \begin{bmatrix} -2 & -1 \\ 1 & 0 \\ 3 & -4 \end{bmatrix} - \frac{1}{2}\begin{bmatrix} 0 & 3 \\ 2 & 0 \\ -4 & -1 \end{bmatrix} = \begin{bmatrix} -2 & -1 \\ 1 & 0 \\ 3 & -4 \end{bmatrix} - \begin{bmatrix} 0 & \frac{3}{2} \\ 1 & 0 \\ -2 & -\frac{1}{2} \end{bmatrix} = \begin{bmatrix} -2 & -\frac{5}{2} \\ 0 & 0 \\ 5 & -\frac{7}{2} \end{bmatrix}$

25. $X = -\frac{3}{2}A + \frac{1}{2}B = -\frac{3}{2}\begin{bmatrix} -2 & -1 \\ 1 & 0 \\ 3 & -4 \end{bmatrix} + \frac{1}{2}\begin{bmatrix} 0 & 3 \\ 2 & 0 \\ -4 & -1 \end{bmatrix} = \begin{bmatrix} 3 & \frac{3}{2} \\ -\frac{3}{2} & 0 \\ -\frac{9}{2} & 6 \end{bmatrix} + \begin{bmatrix} 0 & \frac{3}{2} \\ 1 & 0 \\ -2 & -\frac{1}{2} \end{bmatrix} = \begin{bmatrix} 3 & 3 \\ -\frac{1}{2} & 0 \\ -\frac{13}{2} & \frac{11}{2} \end{bmatrix}$

26. $2A + 4B = -2X$

$X = -A - 2B = -1\begin{bmatrix} -2 & -1 \\ 1 & 0 \\ 3 & -4 \end{bmatrix} - 2\begin{bmatrix} 0 & 3 \\ 2 & 0 \\ -4 & -1 \end{bmatrix} = \begin{bmatrix} 2 & 1 \\ -1 & 0 \\ -3 & 4 \end{bmatrix} + \begin{bmatrix} 0 & -6 \\ -4 & 0 \\ 8 & 2 \end{bmatrix} = \begin{bmatrix} 2 & -5 \\ -5 & 0 \\ 5 & 6 \end{bmatrix}$

27. A is 3×2 and B is 3×3. AB is not possible.

28. A is 2×4, B is 2×2. AB is not possible.

29. A is 3×3, B is $3 \times 2 \implies AB$ is 3×2.

$\begin{bmatrix} 0 & -1 & 0 \\ 4 & 0 & 2 \\ 8 & -1 & 7 \end{bmatrix}\begin{bmatrix} 2 & 1 \\ -3 & 4 \\ 1 & 6 \end{bmatrix} = \begin{bmatrix} (0)(2) + (-1)(-3) + (0)(1) & (0)(1) + (-1)(4) + (0)(6) \\ (4)(2) + (0)(-3) + (2)(1) & (4)(1) + (0)(4) + (2)(6) \\ (8)(2) + (-1)(-3) + (7)(1) & (8)(1) + (-1)(4) + (7)(6) \end{bmatrix} = \begin{bmatrix} 3 & -4 \\ 10 & 16 \\ 26 & 46 \end{bmatrix}$

30. A is 3×2, B is $2 \times 2 \implies AB$ is 3×2.

$AB = \begin{bmatrix} -1 & 3 \\ 4 & -5 \\ 0 & 2 \end{bmatrix}\begin{bmatrix} 1 & 2 \\ 0 & 7 \end{bmatrix} = \begin{bmatrix} -1 & 19 \\ 4 & -27 \\ 0 & 14 \end{bmatrix}$

31. A is 3×3, B is 3×3 $\implies$ AB is 3×3.

$$\begin{bmatrix} 1 & 0 & 0 \\ 0 & 4 & 0 \\ 0 & 0 & -2 \end{bmatrix}\begin{bmatrix} 3 & 0 & 0 \\ 0 & -1 & 0 \\ 0 & 0 & 5 \end{bmatrix} = \begin{bmatrix} (1)(3) + (0)(0) + (0)(0) & (1)(0) + (0)(-1) + (0)(0) & (1)(0) + (0)(0) + (0)(5) \\ (0)(3) + (4)(0) + (0)(0) & (0)(0) + (4)(-1) + (0)(0) & (0)(0) + (4)(0) + (0)(5) \\ (0)(3) + (0)(0) + (-2)(0) & (0)(0) + (0)(-1) + (-2)(0) & (0)(0) + (0)(0) + (-2)(5) \end{bmatrix}$$

$$= \begin{bmatrix} 3 & 0 & 0 \\ 0 & -4 & 0 \\ 0 & 0 & -10 \end{bmatrix}$$

32. A is 3×3, B is 3×3 $\implies$ AB is 3×3.

$$AB = \begin{bmatrix} 5 & 0 & 0 \\ 0 & -8 & 0 \\ 0 & 0 & 7 \end{bmatrix}\begin{bmatrix} \frac{1}{5} & 0 & 0 \\ 0 & -\frac{1}{8} & 0 \\ 0 & 0 & \frac{1}{2} \end{bmatrix} = \begin{bmatrix} 1 & 0 & 0 \\ 0 & 1 & 0 \\ 0 & 0 & \frac{7}{2} \end{bmatrix}$$

33. A is 3×3, B is 3×3 $\implies$ AB is 3×3.

$$\begin{bmatrix} 0 & 0 & 5 \\ 0 & 0 & -3 \\ 0 & 0 & 4 \end{bmatrix}\begin{bmatrix} 6 & -11 & 4 \\ 8 & 16 & 4 \\ 0 & 0 & 0 \end{bmatrix} = \begin{bmatrix} (0)(6) + (0)(8) + (5)(0) & (0)(-11) + (0)(16) + (5)(0) & (0)(4) + (0)(4) + (5)(0) \\ (0)(6) + (0)(8) + (-3)(0) & (0)(-11) + (0)(16) + (-3)(0) & (0)(4) + (0)(4) + (-3)(0) \\ (0)(6) + (0)(8) + (4)(0) & (0)(-11) + (0)(16) + (4)(0) & (0)(4) + (0)(4) + (4)(0) \end{bmatrix}$$

$$= \begin{bmatrix} 0 & 0 & 0 \\ 0 & 0 & 0 \\ 0 & 0 & 0 \end{bmatrix}$$

34. A is 2×1, B is 1×4 $\implies$ AB is 2×4.

$$\begin{bmatrix} 10 \\ 12 \end{bmatrix}\begin{bmatrix} 6 & -2 & 1 & 6 \end{bmatrix} = \begin{bmatrix} 60 & -20 & 10 & 60 \\ 72 & -24 & 12 & 72 \end{bmatrix}$$

35. $\begin{bmatrix} 5 & 6 & -3 \\ -2 & 5 & 1 \\ 10 & -5 & 5 \end{bmatrix}\begin{bmatrix} 1 & -1 & 2 \\ 8 & 1 & 4 \\ 4 & -2 & 9 \end{bmatrix} = \begin{bmatrix} 41 & 7 & 7 \\ 42 & 5 & 25 \\ -10 & -25 & 45 \end{bmatrix}$

36. $\begin{bmatrix} 11 & -12 & 4 \\ 14 & 10 & 12 \\ 6 & -2 & 9 \end{bmatrix}\begin{bmatrix} 12 & 10 \\ -5 & 12 \\ 15 & 16 \end{bmatrix} = \begin{bmatrix} 252 & 30 \\ 298 & 452 \\ 217 & 180 \end{bmatrix}$

37. $\begin{bmatrix} -3 & 8 & -6 & 8 \\ -12 & 15 & 9 & 6 \\ 5 & -1 & 1 & 5 \end{bmatrix}\begin{bmatrix} 3 & 1 & 6 \\ 24 & 15 & 14 \\ 16 & 10 & 21 \\ 8 & -4 & 10 \end{bmatrix} = \begin{bmatrix} 151 & 25 & 48 \\ 516 & 279 & 387 \\ 47 & -20 & 87 \end{bmatrix}$

38. A is 3×3, B is 4×2. AB is not possible.

39. A is 2×4 and B is 2×4 $\implies$ AB is not possible.

40. $\begin{bmatrix} 15 & -18 \\ -4 & 12 \\ -8 & 22 \end{bmatrix}\begin{bmatrix} -7 & 22 & 1 \\ 8 & 16 & 24 \end{bmatrix} = \begin{bmatrix} -249 & 42 & -417 \\ 124 & 104 & 284 \\ 232 & 176 & 520 \end{bmatrix}$

41. (a) $AB = \begin{bmatrix} 1 & 2 \\ 4 & 2 \end{bmatrix}\begin{bmatrix} 2 & -1 \\ -1 & 8 \end{bmatrix} = \begin{bmatrix} (1)(2) + (2)(-1) & (1)(-1) + (2)(8) \\ (4)(2) + (2)(-1) & (4)(-1) + (2)(8) \end{bmatrix} = \begin{bmatrix} 0 & 15 \\ 6 & 12 \end{bmatrix}$

(b) $BA = \begin{bmatrix} 2 & -1 \\ -1 & 8 \end{bmatrix}\begin{bmatrix} 1 & 2 \\ 4 & 2 \end{bmatrix} = \begin{bmatrix} (2)(1) + (-1)(4) & (2)(2) + (-1)(2) \\ (-1)(1) + (8)(4) & (-1)(2) + (8)(2) \end{bmatrix} = \begin{bmatrix} -2 & 2 \\ 31 & 14 \end{bmatrix}$

(c) $A^2 = \begin{bmatrix} 1 & 2 \\ 4 & 2 \end{bmatrix}\begin{bmatrix} 1 & 2 \\ 4 & 2 \end{bmatrix} = \begin{bmatrix} (1)(1) + (2)(4) & (1)(2) + (2)(2) \\ (4)(1) + (2)(4) & (4)(2) + (2)(2) \end{bmatrix} = \begin{bmatrix} 9 & 6 \\ 12 & 12 \end{bmatrix}$

42. (a) $AB = \begin{bmatrix} 2 & -1 \\ 1 & 4 \end{bmatrix}\begin{bmatrix} 0 & 0 \\ 3 & -3 \end{bmatrix} = \begin{bmatrix} 2(0) + (-1)3 & 2(0) + (-1)(-3) \\ 1(0) + 4(3) & 1(0) + 4(-3) \end{bmatrix} = \begin{bmatrix} -3 & 3 \\ 12 & -12 \end{bmatrix}$

   (b) $BA = \begin{bmatrix} 0 & 0 \\ 3 & -3 \end{bmatrix}\begin{bmatrix} 2 & -1 \\ 1 & 4 \end{bmatrix} = \begin{bmatrix} 0(2) + (0)1 & 0(-1) + (0)(4) \\ 3(2) + (-3)(1) & 3(-1) + (-3)4 \end{bmatrix} = \begin{bmatrix} 0 & 0 \\ 3 & -15 \end{bmatrix}$

   (c) $A^2 = \begin{bmatrix} 2 & -1 \\ 1 & 4 \end{bmatrix}\begin{bmatrix} 2 & -1 \\ 1 & 4 \end{bmatrix} = \begin{bmatrix} 2(2) + (-1)(1) & 2(-1) + (-1)4 \\ 1(2) + 4(1) & 1(-1) + 4(4) \end{bmatrix} = \begin{bmatrix} 3 & -6 \\ 6 & 15 \end{bmatrix}$

43. (a) $AB = \begin{bmatrix} 3 & -1 \\ 1 & 3 \end{bmatrix}\begin{bmatrix} 1 & -3 \\ 3 & 1 \end{bmatrix} = \begin{bmatrix} (3)(1) + (-1)(3) & (3)(-3) + (-1)(1) \\ (1)(1) + (3)(3) & (1)(-3) + (3)(1) \end{bmatrix} = \begin{bmatrix} 0 & -10 \\ 10 & 0 \end{bmatrix}$

   (b) $BA = \begin{bmatrix} 1 & -3 \\ 3 & 1 \end{bmatrix}\begin{bmatrix} 3 & -1 \\ 1 & 3 \end{bmatrix} = \begin{bmatrix} (1)(3) + (-3)(1) & (1)(-1) + (-3)(3) \\ (3)(3) + (1)(1) & (3)(-1) + (1)(3) \end{bmatrix} = \begin{bmatrix} 0 & -10 \\ 10 & 0 \end{bmatrix}$

   (c) $A^2 = \begin{bmatrix} 3 & -1 \\ 1 & 3 \end{bmatrix}\begin{bmatrix} 3 & -1 \\ 1 & 3 \end{bmatrix} = \begin{bmatrix} (3)(3) + (-1)(1) & (3)(-1) + (-1)(3) \\ (1)(3) + (3)(1) & (1)(-1) + (3)(3) \end{bmatrix} = \begin{bmatrix} 8 & -6 \\ 6 & 8 \end{bmatrix}$

44. (a) $AB = \begin{bmatrix} 1 & -1 \\ 1 & 1 \end{bmatrix}\begin{bmatrix} 1 & 3 \\ -3 & 1 \end{bmatrix} = \begin{bmatrix} 1(1) + (-1)(-3) & 1(3) + (-1)(1) \\ 1(1) + 1(-3) & 1(3) + 1(1) \end{bmatrix} = \begin{bmatrix} 4 & 2 \\ -2 & 4 \end{bmatrix}$

   (b) $BA = \begin{bmatrix} 1 & 3 \\ -3 & 1 \end{bmatrix}\begin{bmatrix} 1 & -1 \\ 1 & 1 \end{bmatrix} = \begin{bmatrix} 1(1) + (3)1 & 1(-1) + 3(1) \\ -3(1) + (1)(1) & -3(-1) + 1(1) \end{bmatrix} = \begin{bmatrix} 4 & 2 \\ -2 & 4 \end{bmatrix}$

   (c) $A^2 = \begin{bmatrix} 1 & -1 \\ 1 & 1 \end{bmatrix}\begin{bmatrix} 1 & -1 \\ 1 & 1 \end{bmatrix} = \begin{bmatrix} 1(1) + (-1)(1) & 1(-1) + (-1)(1) \\ 1(1) + (1)(1) & 1(-1) + 1(1) \end{bmatrix} = \begin{bmatrix} 0 & -2 \\ 2 & 0 \end{bmatrix}$

45. (a) $AB = \begin{bmatrix} 7 \\ 8 \\ -1 \end{bmatrix}\begin{bmatrix} 1 & 1 & 2 \end{bmatrix} = \begin{bmatrix} 7(1) & 7(1) & 7(2) \\ 8(1) & 8(1) & 8(2) \\ -1(1) & -1(1) & -1(2) \end{bmatrix} = \begin{bmatrix} 7 & 7 & 14 \\ 8 & 8 & 16 \\ -1 & -1 & -2 \end{bmatrix}$

   (b) $BA = \begin{bmatrix} 1 & 1 & 2 \end{bmatrix}\begin{bmatrix} 7 \\ 8 \\ -1 \end{bmatrix} = [(1)(7) + (1)(8) + (2)(-1)] = [13]$

   (c) A^2 is not possible.

46. (a) $AB = \begin{bmatrix} 3 & 2 & 1 \end{bmatrix}\begin{bmatrix} 2 \\ 3 \\ 0 \end{bmatrix} = [3(2) + 2(3) + 1(0)] = [12]$

   (b) $BA = \begin{bmatrix} 2 \\ 3 \\ 0 \end{bmatrix}\begin{bmatrix} 3 & 2 & 1 \end{bmatrix} = \begin{bmatrix} 2(3) & 2(2) & 2(1) \\ 3(3) & 3(2) & 3(1) \\ 0(3) & 0(2) & 0(1) \end{bmatrix} = \begin{bmatrix} 6 & 4 & 2 \\ 9 & 6 & 3 \\ 0 & 0 & 0 \end{bmatrix}$

   (c) The number of columns of A does not equal the number of rows of A; the multiplication is not possible.

47. $\begin{bmatrix} 3 & 1 \\ 0 & -2 \end{bmatrix}\begin{bmatrix} 1 & 0 \\ -2 & 2 \end{bmatrix}\begin{bmatrix} 1 & 0 \\ 2 & 4 \end{bmatrix} = \begin{bmatrix} 1 & 2 \\ 4 & -4 \end{bmatrix}\begin{bmatrix} 1 & 0 \\ 2 & 4 \end{bmatrix} = \begin{bmatrix} 5 & 8 \\ -4 & -16 \end{bmatrix}$

48. $3\left(\begin{bmatrix} 6 & 5 & -1 \\ 1 & -2 & 0 \end{bmatrix}\begin{bmatrix} 0 & 3 \\ -1 & -3 \\ 4 & 1 \end{bmatrix}\right) = -3\left(\begin{bmatrix} 6(0) + 5(-1) + (-1)(4) & 6(3) + 5(-3) + (-1)(1) \\ 1(0) + (-2)(-1) + (0)(4) & 1(3) + (-2)(-3) + (0)(1) \end{bmatrix}\right)$

$= -3\begin{bmatrix} -9 & 2 \\ 2 & 9 \end{bmatrix}$

$= \begin{bmatrix} 27 & -6 \\ -6 & -27 \end{bmatrix}$

49. $\begin{bmatrix} 0 & 2 & -2 \\ 4 & 1 & 2 \end{bmatrix} \left(\begin{bmatrix} 4 & 0 \\ 0 & -1 \\ -1 & 2 \end{bmatrix} + \begin{bmatrix} -2 & 3 \\ -3 & 5 \\ 0 & -3 \end{bmatrix} \right) = \begin{bmatrix} 0 & 2 & -2 \\ 4 & 1 & 2 \end{bmatrix} \begin{bmatrix} 2 & 3 \\ -3 & 4 \\ -1 & -1 \end{bmatrix} = \begin{bmatrix} -4 & 10 \\ 3 & 14 \end{bmatrix}$

50. $\begin{bmatrix} 3 \\ -1 \\ 5 \\ 7 \end{bmatrix} ([5 \quad -6] + [7 \quad -1] + [-8 \quad 9]) = \begin{bmatrix} 3 \\ -1 \\ 5 \\ 7 \end{bmatrix} [4 \quad 2]$

$$= \begin{bmatrix} 3(4) & 3(2) \\ (-1)(4) & (-1)(2) \\ 5(4) & 5(2) \\ 7(4) & 7(2) \end{bmatrix} = \begin{bmatrix} 12 & 6 \\ -4 & -2 \\ 20 & 10 \\ 28 & 14 \end{bmatrix}$$

51. (a) $\begin{bmatrix} -1 & 1 \\ -2 & 1 \end{bmatrix} \begin{bmatrix} x_1 \\ x_2 \end{bmatrix} = \begin{bmatrix} 4 \\ 0 \end{bmatrix}$

(b) $\begin{bmatrix} -1 & 1 & \vdots & 4 \\ -2 & 1 & \vdots & 0 \end{bmatrix}$

$-R_2 + R_1 \rightarrow \begin{bmatrix} 1 & 0 & \vdots & 4 \\ -2 & 1 & \vdots & 0 \end{bmatrix}$

$2R_1 + R_2 \rightarrow \begin{bmatrix} 1 & 0 & \vdots & 4 \\ 0 & 1 & \vdots & 8 \end{bmatrix}$

$X = \begin{bmatrix} 4 \\ 8 \end{bmatrix}$

52. (a) $\begin{bmatrix} 2 & 3 \\ 1 & 4 \end{bmatrix} \begin{bmatrix} x_1 \\ x_2 \end{bmatrix} = \begin{bmatrix} 5 \\ 10 \end{bmatrix}$

(b) $\begin{matrix} R_2 \\ R_1 \end{matrix} \begin{bmatrix} 1 & 4 & \vdots & 10 \\ 2 & 3 & \vdots & 5 \end{bmatrix}$

$-2R_1 + R_2 \rightarrow \begin{bmatrix} 1 & 4 & \vdots & 10 \\ 0 & -5 & \vdots & -15 \end{bmatrix}$

$-\frac{1}{5}R_2 \rightarrow \begin{bmatrix} 1 & 4 & \vdots & 10 \\ 0 & 1 & \vdots & 3 \end{bmatrix}$

$\begin{matrix} -4R_2 + R_1 \rightarrow \\ -\frac{1}{5}R_2 \rightarrow \end{matrix} \begin{bmatrix} 1 & 0 & \vdots & -2 \\ 0 & 1 & \vdots & 3 \end{bmatrix}$

$X = \begin{bmatrix} -2 \\ 3 \end{bmatrix}$

53. (a) $\begin{bmatrix} -2 & -3 \\ 6 & 1 \end{bmatrix} \begin{bmatrix} x_1 \\ x_2 \end{bmatrix} = \begin{bmatrix} -4 \\ -36 \end{bmatrix}$

(b) $\begin{bmatrix} -2 & -3 & \vdots & -4 \\ 6 & 1 & \vdots & -36 \end{bmatrix}$

$3R_1 + R_2 \rightarrow \begin{bmatrix} -2 & -3 & \vdots & -4 \\ 0 & -8 & \vdots & -48 \end{bmatrix}$

$\begin{matrix} -\frac{1}{2}R_1 \rightarrow \\ -\frac{1}{8}R_2 \rightarrow \end{matrix} \begin{bmatrix} 1 & \frac{3}{2} & \vdots & 2 \\ 0 & 1 & \vdots & 6 \end{bmatrix}$

$-\frac{3}{2}R_2 + R_1 \rightarrow \begin{bmatrix} 1 & 0 & \vdots & -7 \\ 0 & 1 & \vdots & 6 \end{bmatrix}$

$X = \begin{bmatrix} -7 \\ 6 \end{bmatrix}$

54. (a) $\begin{bmatrix} -4 & 9 \\ 1 & -3 \end{bmatrix} \begin{bmatrix} x_1 \\ x_2 \end{bmatrix} = \begin{bmatrix} -13 \\ 12 \end{bmatrix}$

(b) $\begin{matrix} R_1 \\ R_2 \end{matrix} \begin{bmatrix} 1 & -3 & \vdots & 12 \\ -4 & 9 & \vdots & -13 \end{bmatrix}$

$4R_1 + R_2 \rightarrow \begin{bmatrix} 1 & -3 & \vdots & 12 \\ 0 & -3 & \vdots & 35 \end{bmatrix}$

$-\frac{1}{3}R_2 \rightarrow \begin{bmatrix} 1 & -3 & \vdots & 12 \\ 0 & 1 & \vdots & -\frac{35}{3} \end{bmatrix}$

$3R_2 + R_1 \rightarrow \begin{bmatrix} 1 & 0 & \vdots & -23 \\ 0 & 1 & \vdots & -\frac{35}{3} \end{bmatrix}$

$X = \begin{bmatrix} -23 \\ -\frac{35}{3} \end{bmatrix}$

55. (a) $A = \begin{bmatrix} 1 & -2 & 3 \\ -1 & 3 & -1 \\ 2 & -5 & 5 \end{bmatrix} \begin{bmatrix} x_1 \\ x_2 \\ x_3 \end{bmatrix} = \begin{bmatrix} 9 \\ -6 \\ 17 \end{bmatrix}$

(b)
$$\begin{bmatrix} 1 & -2 & 3 & \vdots & 9 \\ -1 & 3 & -1 & \vdots & -6 \\ 2 & -5 & 5 & \vdots & 17 \end{bmatrix}$$

$$\begin{matrix} R_1 + R_2 \to \\ -2R_2 + R_3 \to \end{matrix} \begin{bmatrix} 1 & -2 & 3 & \vdots & 9 \\ 0 & 1 & 2 & \vdots & 3 \\ 0 & -1 & -1 & \vdots & -1 \end{bmatrix}$$

$$\begin{matrix} 2R_2 + R_1 \to \\ \\ R_2 + R_3 \to \end{matrix} \begin{bmatrix} 1 & 0 & 7 & \vdots & 15 \\ 0 & 1 & 2 & \vdots & 3 \\ 0 & 0 & 1 & \vdots & 2 \end{bmatrix}$$

$$\begin{matrix} -7R_3 + R_1 \to \\ -2R_3 + R_2 \to \end{matrix} \begin{bmatrix} 1 & 0 & 0 & \vdots & 1 \\ 0 & 1 & 0 & \vdots & -1 \\ 0 & 0 & 1 & \vdots & 2 \end{bmatrix}$$

$$X = \begin{bmatrix} 1 \\ -1 \\ 2 \end{bmatrix}$$

56. (a) $\begin{bmatrix} 1 & 1 & -3 \\ -1 & 2 & 0 \\ 1 & -1 & 1 \end{bmatrix} \begin{bmatrix} x_1 \\ x_2 \\ x_3 \end{bmatrix} = \begin{bmatrix} 9 \\ 6 \\ -5 \end{bmatrix}$

(b)
$$\begin{bmatrix} 1 & 1 & -3 & \vdots & 9 \\ -1 & 2 & 0 & \vdots & 6 \\ 1 & -1 & 1 & \vdots & -5 \end{bmatrix}$$

$$\begin{matrix} R_1 + R_2 \to \\ -R_1 + R_3 \to \end{matrix} \begin{bmatrix} 1 & 1 & -3 & \vdots & 9 \\ 0 & 3 & -3 & \vdots & 15 \\ 0 & -2 & 4 & \vdots & -14 \end{bmatrix}$$

$$\begin{matrix} \frac{1}{3}R_2 \to \\ -\frac{1}{2}R_3 \to \end{matrix} \begin{bmatrix} 1 & 1 & -3 & \vdots & 9 \\ 0 & 1 & -1 & \vdots & 5 \\ 0 & 1 & -2 & \vdots & 7 \end{bmatrix}$$

$$-R_2 + R_3 \to \begin{bmatrix} 1 & 1 & -3 & \vdots & 9 \\ 0 & 1 & -1 & \vdots & 5 \\ 0 & 0 & -1 & \vdots & 2 \end{bmatrix}$$

$$\begin{matrix} -R_2 + R_1 \to \\ \\ -R_3 \to \end{matrix} \begin{bmatrix} 1 & 0 & -2 & \vdots & 4 \\ 0 & 1 & -1 & \vdots & 5 \\ 0 & 0 & 1 & \vdots & -2 \end{bmatrix}$$

$$\begin{matrix} 2R_3 + R_1 \to \\ R_3 + R_2 \to \end{matrix} \begin{bmatrix} 1 & 0 & 0 & \vdots & 0 \\ 0 & 1 & 0 & \vdots & 3 \\ 0 & 0 & 1 & \vdots & -2 \end{bmatrix}$$

$$X = \begin{bmatrix} 0 \\ 3 \\ -2 \end{bmatrix}$$

57. (a) $\begin{bmatrix} 1 & -5 & 2 \\ -3 & 1 & -1 \\ 0 & -2 & 5 \end{bmatrix} \begin{bmatrix} x_1 \\ x_2 \\ x_3 \end{bmatrix} = \begin{bmatrix} -20 \\ 8 \\ -16 \end{bmatrix}$

(b)
$$\begin{bmatrix} 1 & -5 & 2 & \vdots & -20 \\ -3 & 1 & -1 & \vdots & 8 \\ 0 & -2 & 5 & \vdots & -16 \end{bmatrix}$$

$$3R_1 + R_2 \to \begin{bmatrix} 1 & -5 & 2 & \vdots & -20 \\ 0 & -14 & 5 & \vdots & -52 \\ 0 & -2 & 5 & \vdots & -16 \end{bmatrix}$$

$$-R_3 + R_2 \to \begin{bmatrix} 1 & -5 & 2 & \vdots & -20 \\ 0 & -12 & 0 & \vdots & -36 \\ 0 & -2 & 5 & \vdots & -16 \end{bmatrix}$$

$$-\frac{1}{12}R_2 \to \begin{bmatrix} 1 & -5 & 2 & \vdots & -20 \\ 0 & 1 & 0 & \vdots & 3 \\ 0 & -2 & 5 & \vdots & -16 \end{bmatrix}$$

$$\begin{matrix} 5R_2 + R_1 \to \\ \\ 2R_2 + R_3 \to \end{matrix} \begin{bmatrix} 1 & 0 & 2 & \vdots & -5 \\ 0 & 1 & 0 & \vdots & 3 \\ 0 & 0 & 5 & \vdots & -10 \end{bmatrix}$$

$$\frac{1}{5}R_3 \to \begin{bmatrix} 1 & 0 & 2 & \vdots & -5 \\ 0 & 1 & 0 & \vdots & 3 \\ 0 & 0 & 1 & \vdots & -2 \end{bmatrix}$$

$$-2R_3 + R_1 \to \begin{bmatrix} 1 & 0 & 0 & \vdots & -1 \\ 0 & 1 & 0 & \vdots & 3 \\ 0 & 0 & 1 & \vdots & -2 \end{bmatrix}$$

$$X = \begin{bmatrix} -1 \\ 3 \\ -2 \end{bmatrix}$$

58. (a) $\begin{bmatrix} 1 & -1 & 4 \\ 1 & 3 & 0 \\ 0 & -6 & 5 \end{bmatrix} \begin{bmatrix} x_1 \\ x_2 \\ x_3 \end{bmatrix} = \begin{bmatrix} 17 \\ -11 \\ 40 \end{bmatrix}$

(b)
$$\begin{bmatrix} 1 & -1 & 4 & \vdots & 17 \\ 1 & 3 & 0 & \vdots & -11 \\ 0 & -6 & 5 & \vdots & 40 \end{bmatrix}$$

$$-R_1 + R_2 \to \begin{bmatrix} 1 & -1 & 4 & \vdots & 17 \\ 0 & 4 & -4 & \vdots & -28 \\ 0 & -6 & 5 & \vdots & 40 \end{bmatrix}$$

$$\frac{1}{4}R_2 \to \begin{bmatrix} 1 & -1 & 4 & \vdots & 17 \\ 0 & 1 & -1 & \vdots & -7 \\ 0 & -6 & 5 & \vdots & 40 \end{bmatrix}$$

$$6R_2 + R_3 \to \begin{bmatrix} 1 & -1 & 4 & \vdots & 17 \\ 0 & 1 & -1 & \vdots & -7 \\ 0 & 0 & -1 & \vdots & -2 \end{bmatrix}$$

$$\begin{matrix} R_2 + R_1 \to \\ \\ -R_3 \to \end{matrix} \begin{bmatrix} 1 & 0 & 3 & \vdots & 10 \\ 0 & 1 & -1 & \vdots & -7 \\ 0 & 0 & 1 & \vdots & 2 \end{bmatrix}$$

$$\begin{matrix} -3R_3 + R_1 \to \\ R_3 + R_2 \to \end{matrix} \begin{bmatrix} 1 & 0 & 0 & \vdots & 4 \\ 0 & 1 & 0 & \vdots & -5 \\ 0 & 0 & 1 & \vdots & 2 \end{bmatrix}$$

$$X = \begin{bmatrix} 4 \\ -5 \\ 2 \end{bmatrix}$$

59. $1.2\begin{bmatrix} 70 & 50 & 25 \\ 35 & 100 & 70 \end{bmatrix} = \begin{bmatrix} 84 & 60 & 30 \\ 42 & 120 & 84 \end{bmatrix}$

60. $1.10\begin{bmatrix} 100 & 90 & 70 & 30 \\ 40 & 20 & 60 & 60 \end{bmatrix} = \begin{bmatrix} 110 & 99 & 77 & 33 \\ 44 & 22 & 66 & 66 \end{bmatrix}$

61. (a)

	Farmer's Market	Fruit Stand	Fruit Farm	

$A = \begin{bmatrix} 125 & 100 & 75 \\ 100 & 175 & 125 \end{bmatrix} \begin{matrix} \text{Apples} \\ \text{Peaches} \end{matrix}$

Each entry represents the number of bushels of each type of crop that are shipped to each outlet.

(b) $B = \begin{bmatrix} 3.50 & 6.00 \end{bmatrix}$

Each entry represents the profit per bushel for each type of crop.

(c) $BA = \begin{bmatrix} 3.50 & 6.00 \end{bmatrix}\begin{bmatrix} 125 & 100 & 75 \\ 100 & 175 & 125 \end{bmatrix}$

$\qquad = \begin{bmatrix} \$1037.50 & \$1400.00 & \$1012.50 \end{bmatrix}$

The entries in the matrix represent the profits for both crops at each of the three outlets.

62. $BA = \begin{bmatrix} \$39.50 & \$44.50 & \$56.50 \end{bmatrix}\begin{bmatrix} 5,000 & 4,000 \\ 6,000 & 10,000 \\ 8,000 & 5,000 \end{bmatrix} = \begin{bmatrix} \$916,500 & \$885,500 \end{bmatrix}$

The entries represent the costs of the three models of the product at the two warehouses.

63. $ST = \begin{bmatrix} 3 & 2 & 2 & 3 & 0 \\ 0 & 2 & 3 & 4 & 3 \\ 4 & 2 & 1 & 3 & 2 \end{bmatrix}\begin{bmatrix} 840 & 1100 \\ 1200 & 1350 \\ 1450 & 1650 \\ 2650 & 3000 \\ 3050 & 3200 \end{bmatrix} = \begin{bmatrix} \$15,770 & \$18,300 \\ \$26,500 & \$29,250 \\ \$21,260 & \$24,150 \end{bmatrix}$

The entries represent the wholesale and retail inventory values of the inventories at the three outlets.

64. $P^2 = \begin{bmatrix} 0.6 & 0.1 & 0.1 \\ 0.2 & 0.7 & 0.1 \\ 0.2 & 0.2 & 0.8 \end{bmatrix}\begin{bmatrix} 0.6 & 0.1 & 0.1 \\ 0.2 & 0.7 & 0.1 \\ 0.2 & 0.2 & 0.8 \end{bmatrix} = \begin{bmatrix} 0.40 & 0.15 & 0.15 \\ 0.28 & 0.53 & 0.17 \\ 0.32 & 0.32 & 0.68 \end{bmatrix}$

The P^2 matrix gives the proportion of the voting population that changed parties or remained loyal to their party from the first election to the third.

65. $P^3 = P^2P = \begin{bmatrix} 0.40 & 0.15 & 0.15 \\ 0.28 & 0.53 & 0.17 \\ 0.32 & 0.32 & 0.68 \end{bmatrix}\begin{bmatrix} 0.6 & 0.1 & 0.1 \\ 0.2 & 0.7 & 0.1 \\ 0.2 & 0.2 & 0.8 \end{bmatrix} = \begin{bmatrix} 0.300 & 0.175 & 0.175 \\ 0.308 & 0.433 & 0.217 \\ 0.392 & 0.392 & 0.608 \end{bmatrix}$

$P^4 = P^3P = \begin{bmatrix} 0.300 & 0.175 & 0.175 \\ 0.308 & 0.433 & 0.217 \\ 0.392 & 0.392 & 0.608 \end{bmatrix}\begin{bmatrix} 0.6 & 0.1 & 0.1 \\ 0.2 & 0.7 & 0.1 \\ 0.2 & 0.2 & 0.8 \end{bmatrix} = \begin{bmatrix} 0.250 & 0.188 & 0.188 \\ 0.315 & 0.377 & 0.248 \\ 0.435 & 0.435 & 0.565 \end{bmatrix}$

$P^5 = P^4P = \begin{bmatrix} 0.250 & 0.188 & 0.188 \\ 0.315 & 0.377 & 0.248 \\ 0.435 & 0.435 & 0.565 \end{bmatrix}\begin{bmatrix} 0.6 & 0.1 & 0.1 \\ 0.2 & 0.7 & 0.1 \\ 0.2 & 0.2 & 0.8 \end{bmatrix} = \begin{bmatrix} 0.225 & 0.194 & 0.194 \\ 0.314 & 0.345 & 0.267 \\ 0.461 & 0.461 & 0.539 \end{bmatrix}$

$P^6 = \begin{bmatrix} 0.213 & 0.197 & 0.197 \\ 0.311 & 0.326 & 0.280 \\ 0.477 & 0.477 & 0.523 \end{bmatrix}$

—CONTINUED—

65. —CONTINUED—

$$P^7 = \begin{bmatrix} 0.206 & 0.198 & 0.198 \\ 0.308 & 0.316 & 0.288 \\ 0.486 & 0.486 & 0.514 \end{bmatrix}$$

$$P^8 = \begin{bmatrix} 0.203 & 0.199 & 0.199 \\ 0.305 & 0.309 & 0.292 \\ 0.492 & 0.492 & 0.508 \end{bmatrix}$$

As P is raised to higher and higher powers, the resulting matrices appear to be approaching the matrix

$$\begin{bmatrix} 0.2 & 0.2 & 0.2 \\ 0.3 & 0.3 & 0.3 \\ 0.5 & 0.5 & 0.5 \end{bmatrix}.$$

66. $ST = \begin{bmatrix} 1 & 0.5 & 0.2 \\ 1.6 & 1.0 & 0.2 \\ 2.5 & 2.0 & 1.4 \end{bmatrix} \begin{bmatrix} 12 & 10 \\ 9 & 8 \\ 8 & 7 \end{bmatrix} = \begin{bmatrix} \$18.10 & \$15.40 \\ \$29.80 & \$25.40 \\ \$59.20 & \$50.80 \end{bmatrix}$

This represents the labor cost for each boat size at each plant.

67. (a) $AB = \begin{bmatrix} 40 & 64 & 52 \\ 60 & 82 & 76 \\ 76 & 96 & 84 \end{bmatrix} \begin{bmatrix} 2.65 & 0.65 \\ 2.85 & 0.70 \\ 3.05 & 0.85 \end{bmatrix} = \begin{matrix} & \text{Sales} & \text{Profit} \\ & \begin{bmatrix} 447 & 115 \\ 624.50 & 161 \\ 731.20 & 188 \end{bmatrix} & \begin{matrix} \text{Friday} \\ \text{Saturday} \\ \text{Sunday} \end{matrix} \end{matrix}$

The entries in Column 1 represent the total sales of the three kinds of milk for Friday, Saturday, and Sunday. The entries in Column 2 represent each days' total profit.

(b) Total profit for the weekend: $115 + 161 + 188 = \$464$

68. (a) $AB = \begin{bmatrix} 580 & 840 & 320 \\ 560 & 420 & 160 \\ 860 & 1020 & 540 \end{bmatrix} \begin{bmatrix} 1.95 & 0.32 \\ 2.05 & 0.36 \\ 2.15 & 0.40 \end{bmatrix} = \begin{matrix} & \text{Sales (\$)} & \text{Profit} \\ & \begin{bmatrix} 3541 & 616 \\ 2297 & 394.4 \\ 4929 & 858.4 \end{bmatrix} & \begin{matrix} 87 \\ 89 \\ 93 \end{matrix} \end{matrix}$

The first column of AB gives the amount of sales for each octane. The second column gives the profit made by each octane.

(b) The store's profit for the weekend is $\$616 + \$394.40 + \$858.40 = \$1868.80$.

69. (a)
$$\begin{matrix} & \text{Bicycled} & \text{Jogged} & \text{Walked} \\ B = & [2 & 0.5 & 3] \end{matrix}$$
20-minute time periods

(b) $BA = \begin{bmatrix} 2 & 0.5 & 3 \end{bmatrix} \begin{bmatrix} 109 & 136 \\ 127 & 159 \\ 64 & 79 \end{bmatrix} = \begin{matrix} \begin{matrix} \text{120-pound} \\ \text{person} \end{matrix} & \begin{matrix} \text{150-pound} \\ \text{person} \end{matrix} \\ [473.5 & 588.5] \end{matrix}$ Calories burned

The first entry represents the total calories burned by the 120-pound person and the second entry represents the total calories burned by the 150-pound person.

70. (a)
$$A = \begin{matrix} & \begin{matrix} \text{Individual} \\ \text{costs} \end{matrix} & \begin{matrix} \text{Family} \\ \text{costs} \end{matrix} \\ & \begin{bmatrix} 694.32 & 1725.36 \\ 451.8 & 1187.76 \\ 489.48 & 1248.12 \end{bmatrix} & \begin{matrix} \text{Comprehensive plan} \\ \text{HMO standard plan} \\ \text{HMO plus plan} \end{matrix} \end{matrix}$$

$$B = \begin{bmatrix} 683.91 & 1699.48 \\ 463.1 & 1217.45 \\ 499.27 & 1273.08 \end{bmatrix} \begin{matrix} \text{Comprehensive plan} \\ \text{HMO standard plan} \\ \text{HMO plus plan} \end{matrix}$$

—CONTINUED—

70. —CONTINUED—

(b)

		Change in individual costs	Change in family cost	

$$A - B = \begin{bmatrix} 694.32 & 1725.36 \\ 451.8 & 1187.76 \\ 489.48 & 1248.12 \end{bmatrix} - \begin{bmatrix} 683.91 & 1699.48 \\ 463.1 & 1217.45 \\ 499.27 & 1273.08 \end{bmatrix} = \begin{bmatrix} 10.41 & 25.88 \\ -11.3 & -29.69 \\ -9.79 & -24.96 \end{bmatrix} \begin{matrix} \text{Comprehensive plan} \\ \text{HMO standard plan} \\ \text{HMO plus plan} \end{matrix}$$

Employees choosing the comprehensive plan have a decrease in cost while those choosing the other two have an increased cost.

(c) Dividing each entry of matrix A by 12 yields

$$\tfrac{1}{12}A = \begin{bmatrix} 57.86 & 143.78 \\ 37.65 & 98.98 \\ 40.79 & 104.01 \end{bmatrix}, \qquad \tfrac{1}{12}B = \begin{bmatrix} 56.99 & 141.62 \\ 38.59 & 101.45 \\ 41.61 & 106.09 \end{bmatrix}.$$

(d) If the costs increase by 4% next year, then the new cost matrix would be:

$$A + 0.04A = \begin{bmatrix} 722.09 & 1794.37 \\ 469.87 & 1235.27 \\ 509.06 & 1298.05 \end{bmatrix}$$

$$\tfrac{1}{12}(A + 0.04A) = \begin{bmatrix} \text{Monthly individual cost} & \text{Monthly family cost} \\ 60.17 & 149.53 \\ 39.16 & 102.94 \\ 42.42 & 108.17 \end{bmatrix} \begin{matrix} \text{Comprehensive plan} \\ \text{HMO standard plan} \\ \text{HMO plus plan} \end{matrix}$$

71. True.
The sum of two matrices of different orders is undefined.

72. False. For most matrices, $AB \neq BA$.

For 73–80, A is of order 2×3, B is of order 2×3, C is of order 3×2 and D is of order 2×2.

73. $A + 2C$ is not possible. A and C are not of the same order.

74. $B - 3C$ is not possible. B and C are not of the same order.

75. AB is not possible. The number of columns of A does not equal the number of rows of B.

76. BC is possible. The resulting order is 2×2.

77. $BC - D$ is possible. The resulting order is 2×2.

78. $CB - D$ is not possible. The order of CB is 3×3, but the order of D is 2×2.

79. $D(A - 3B)$ is possible. The resulting order is 2×3.

80. $(BC - D)A$ is possible. The resulting order is 2×3.

81. $AC = \begin{bmatrix} 0 & 1 \\ 0 & 1 \end{bmatrix}\begin{bmatrix} 2 & 3 \\ 2 & 3 \end{bmatrix} = \begin{bmatrix} 2 & 3 \\ 2 & 3 \end{bmatrix}$

$BC = \begin{bmatrix} 1 & 0 \\ 1 & 0 \end{bmatrix}\begin{bmatrix} 2 & 3 \\ 2 & 3 \end{bmatrix} = \begin{bmatrix} 2 & 3 \\ 2 & 3 \end{bmatrix}$

Thus, $AC = BC$ even though $A \neq B$.

82. $AB = \begin{bmatrix} 3 & 3 \\ 4 & 4 \end{bmatrix}\begin{bmatrix} 1 & -1 \\ -1 & 1 \end{bmatrix} = \begin{bmatrix} 0 & 0 \\ 0 & 0 \end{bmatrix}$

$AB = O$ and neither A nor B is O.

83. The product of two diagonal matrices of the same order is a diagonal matrix whose entries are the products of the corresponding diagonal entries of A and B.

84. (a) $A^2 = \begin{bmatrix} i & 0 \\ 0 & i \end{bmatrix}\begin{bmatrix} i & 0 \\ 0 & i \end{bmatrix} = \begin{bmatrix} (i)(i) + (0)(0) & (i)(0) + (0)(i) \\ (0)(i) + (i)(0) & (0)(0) + (i)(i) \end{bmatrix} = \begin{bmatrix} -1 & 0 \\ 0 & -1 \end{bmatrix}$ and $i^2 = -1$

$A^3 = A^2A = \begin{bmatrix} -1 & 0 \\ 0 & -1 \end{bmatrix}\begin{bmatrix} i & 0 \\ 0 & i \end{bmatrix} = \begin{bmatrix} (-1)(i) + (0)(0) & (-1)(0) + (0)(i) \\ (0)(i) + (-1)(0) & (0)(0) + (-1)(i) \end{bmatrix} = \begin{bmatrix} -i & 0 \\ 0 & -i \end{bmatrix}$ and $i^3 = -i$

$A^4 = A^3A = \begin{bmatrix} -i & 0 \\ 0 & -i \end{bmatrix}\begin{bmatrix} i & 0 \\ 0 & i \end{bmatrix} = \begin{bmatrix} (-i)(i) + (0)(0) & (i)(0) + (0)(i) \\ (0)(i) + (-i)(0) & (0)(0) + (-i)(i) \end{bmatrix} = \begin{bmatrix} 1 & 0 \\ 0 & 1 \end{bmatrix}$ and $i^4 = 1$

(b) $B = \begin{bmatrix} 0 & -i \\ i & 0 \end{bmatrix}$

$B^2 = \begin{bmatrix} 0 & -i \\ i & 0 \end{bmatrix}\begin{bmatrix} 0 & -i \\ i & 0 \end{bmatrix} = \begin{bmatrix} (0)(0) + (-i)(i) & (0)(-i) + (-i)(0) \\ (i)(0) + (0)(i) & (i)(-i) + (0)(0) \end{bmatrix} = \begin{bmatrix} 1 & 0 \\ 0 & 1 \end{bmatrix} = I$, the identity matrix

85. $3x^2 + 20x - 32 = 0$

$(3x - 4)(x + 8) = 0$

$3x - 4 = 0$ or $x + 8 = 0$

$x = \frac{4}{3}$ or $\quad x = -8$

Solutions: $\frac{4}{3}, -8$

86. $8x^2 - 10x - 3 = 0$

$(2x - 3)(4x + 1) = 0$

$2x - 3 = 0 \implies x = \frac{3}{2}$

$4x + 1 = 0 \implies x = -\frac{1}{4}$

Solutions: $-\frac{1}{4}, \frac{3}{2}$

87. $4x^3 + 10x^2 - 3x = 0$

$x(4x^2 + 10x - 3) = 0$

$x = 0$ or $4x^2 + 10x - 3 = 0$

$x = \dfrac{-10 \pm \sqrt{10^2 - 4(4)(-3)}}{2(4)} = \dfrac{-10 \pm \sqrt{148}}{8}$

$= \dfrac{-5 \pm \sqrt{37}}{4}$ by the Quadratic Formula

Solutions: $0, \dfrac{-5 \pm \sqrt{37}}{4}$

88. $3x^3 + 22x^2 - 45x = 0$

$x(3x^2 + 22x - 45) = 0$

$x(x + 9)(3x - 5) = 0$

$x = 0$

$x + 9 = 0 \implies x = -9$

$3x - 5 = 0 \implies x = \frac{5}{3}$

Solutions: $0, -9, \frac{5}{3}$

89. $3x^3 - 12x^2 + 5x - 20 = 0$

$3x^2(x - 4) + 5(x - 4) = 0$

$(x - 4)(3x^2 + 5) = 0$

$x - 4 = 0$ or $3x^2 + 5 = 0$

$x = 4 \qquad\qquad x^2 = -\dfrac{5}{3}$

$x = \pm\sqrt{-\dfrac{5}{3}} = \pm\dfrac{\sqrt{15}}{3}i$

Solutions: $4, \pm\dfrac{\sqrt{15}}{3}i$

90. $2x^3 - 5x^2 - 12x + 30 = 0$

$x^2(2x - 5) - 6(2x - 5) = 0$

$(2x - 5)(x^2 - 6) = 0$

$2x - 5 = 0 \implies x = \dfrac{5}{2}$

$x^2 - 6 = 0 \implies x^2 = 6 \implies x = \pm\sqrt{6}$

$x = \pm\sqrt{6}$

Solutions: $\dfrac{5}{2}, \pm\sqrt{6}$

91. $\begin{cases} -x + 4y = -9 & \text{Eq.1} \\ 5x - 8y = 39 & \text{Eq.2} \end{cases}$

$\begin{aligned} -5x + 20y &= -45 \qquad (5)\text{Eq.1} \\ 5x - 8y &= 39 \\ 12y &= -6 \qquad\quad \text{Add equations.} \\ y &= -\tfrac{1}{2} \end{aligned}$

$-x + 4\left(-\tfrac{1}{2}\right) = -9 \implies x = 7$

Solution: $\left(7, -\tfrac{1}{2}\right)$

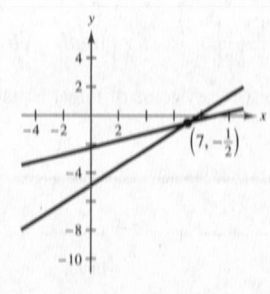

92. $\begin{cases} 8x - 3y = -17 & \text{Equation 1} \\ -6x + 7y = 27 & \text{Equation 2} \end{cases}$

$\begin{array}{ll} 48x - 18y = -102 & \text{(6)Eq.1} \\ -48x + 56y = 216 & \text{(8)Eq.2} \\ \hline 38y = 114 & \text{Add equations.} \\ y = 3 \end{array}$

$8x - 3(3) = -17 \implies x = -1$

Solution: $(-1, 3)$

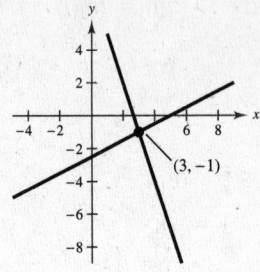

93. $\begin{cases} -x + 2y = -5 & \text{Equation 1} \\ -3x - y = -8 & \text{Equation 2} \end{cases}$

$\begin{array}{ll} -x + 2y = -5 & \\ -6x - 2y = -16 & \text{(2)Eq.2} \\ \hline -7x = -21 & \text{Add equations.} \\ x = 3 \end{array}$

$-3 + 2y = -5 \implies y = -1$

Solution: $(3, -1)$

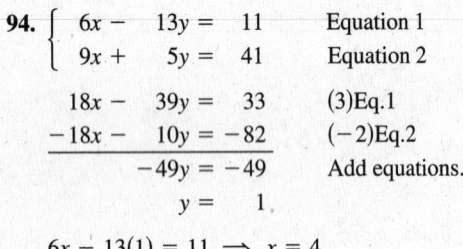

94. $\begin{cases} 6x - 13y = 11 & \text{Equation 1} \\ 9x + 5y = 41 & \text{Equation 2} \end{cases}$

$\begin{array}{ll} 18x - 39y = 33 & \text{(3)Eq.1} \\ -18x - 10y = -82 & \text{(-2)Eq.2} \\ \hline -49y = -49 & \text{Add equations.} \\ y = 1 \end{array}$

$6x - 13(1) = 11 \implies x = 4$

Solution: $(4, 1)$

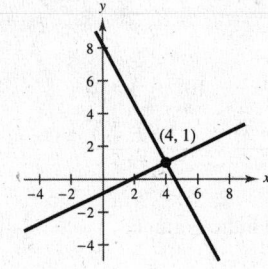

Section 8.3   The Inverse of a Square Matrix

- ■   You should know that the inverse of an $n \times n$ matrix A is the $n \times n$ matrix A^{-1}, if is exists, such that $AA^{-1} = A^{-1}A = I$, where I is the $n \times n$ identity matrix.

- ■   You should be able to find the inverse, if it exists, of a square matrix.

   (a) Write the $n \times 2n$ matrix that consists of the given matrix A on the left and the $n \times n$ identity matrix I on the right to obtain $[A \ \vdots \ I]$. Note that we separate the matrices A and I by a dotted line. We call this process **adjoining** the matrices A and I.

   (b) If possible, row reduce A to I using elementary row operations on the *entire* matrix $[A \ \vdots \ I]$. The result will be the matrix $[I \ \vdots \ A^{-1}]$. If this is not possible, then A is not invertible.

   (c) Check your work by multiplying to see that $AA^{-1} = I = A^{-1}A$.

- ■   The inverse of $A = \begin{bmatrix} a & b \\ c & d \end{bmatrix}$ is $A^{-1} = \dfrac{1}{ad - bc}\begin{bmatrix} d & -b \\ -c & a \end{bmatrix}$ if $ad - cb \neq 0$.

- ■   You should be able to use inverse matrices to solve systems of linear equations if the coefficient matrix is square and invertible.

Vocabulary Check

1. square   **2.** inverse   **3.** nonsingular; singular   **4.** $A^{-1}B$

1. $AB = \begin{bmatrix} 2 & 1 \\ 5 & 3 \end{bmatrix}\begin{bmatrix} 3 & -1 \\ -5 & 2 \end{bmatrix} = \begin{bmatrix} 6-5 & -2+2 \\ 15-15 & -5+6 \end{bmatrix} = \begin{bmatrix} 1 & 0 \\ 0 & 1 \end{bmatrix}$

$BA = \begin{bmatrix} 3 & -1 \\ -5 & 2 \end{bmatrix}\begin{bmatrix} 2 & 1 \\ 5 & 3 \end{bmatrix} = \begin{bmatrix} 6-5 & 3-3 \\ -10+10 & -5+6 \end{bmatrix} = \begin{bmatrix} 1 & 0 \\ 0 & 1 \end{bmatrix}$

2. $AB = \begin{bmatrix} 1 & -1 \\ -1 & 2 \end{bmatrix}\begin{bmatrix} 2 & 1 \\ 1 & 1 \end{bmatrix} = \begin{bmatrix} 2-1 & 1-1 \\ -2+2 & -1+2 \end{bmatrix} = \begin{bmatrix} 1 & 0 \\ 0 & 1 \end{bmatrix}$

$BA = \begin{bmatrix} 2 & 1 \\ 1 & 1 \end{bmatrix}\begin{bmatrix} 1 & -1 \\ -1 & 2 \end{bmatrix} = \begin{bmatrix} 2-1 & -2+2 \\ 1-1 & -1+2 \end{bmatrix} = \begin{bmatrix} 1 & 0 \\ 0 & 1 \end{bmatrix}$

3. $AB = \begin{bmatrix} 1 & 2 \\ 3 & 4 \end{bmatrix}\begin{bmatrix} -2 & 1 \\ \frac{3}{2} & -\frac{1}{2} \end{bmatrix} = \begin{bmatrix} -2+3 & 1-1 \\ -6+6 & 3-2 \end{bmatrix} = \begin{bmatrix} 1 & 0 \\ 0 & 1 \end{bmatrix}$

$BA = \begin{bmatrix} -2 & 1 \\ \frac{3}{2} & -\frac{1}{2} \end{bmatrix}\begin{bmatrix} 1 & 2 \\ 3 & 4 \end{bmatrix} = \begin{bmatrix} -2+3 & -4+4 \\ \frac{3}{2}-\frac{3}{2} & 3-2 \end{bmatrix} = \begin{bmatrix} 1 & 0 \\ 0 & 1 \end{bmatrix}$

4. $AB = \begin{bmatrix} 1 & -1 \\ 2 & 3 \end{bmatrix}\begin{bmatrix} \frac{3}{5} & \frac{1}{5} \\ -\frac{2}{5} & \frac{1}{5} \end{bmatrix} = \begin{bmatrix} \frac{3}{5}+\frac{2}{5} & \frac{1}{5}-\frac{1}{5} \\ \frac{6}{5}-\frac{6}{5} & \frac{2}{5}+\frac{3}{5} \end{bmatrix} = \begin{bmatrix} 1 & 0 \\ 0 & 1 \end{bmatrix}$

$BA = \begin{bmatrix} \frac{3}{5} & \frac{1}{5} \\ -\frac{2}{5} & \frac{1}{5} \end{bmatrix}\begin{bmatrix} 1 & -1 \\ 2 & 3 \end{bmatrix} = \begin{bmatrix} \frac{3}{5}+\frac{2}{5} & -\frac{3}{5}+\frac{3}{5} \\ -\frac{2}{5}+\frac{2}{5} & \frac{2}{5}+\frac{3}{5} \end{bmatrix} = \begin{bmatrix} 1 & 0 \\ 0 & 1 \end{bmatrix}$

5. $AB = \begin{bmatrix} 2 & -17 & 11 \\ -1 & 11 & -7 \\ 0 & 3 & -2 \end{bmatrix}\begin{bmatrix} 1 & 1 & 2 \\ 2 & 4 & -3 \\ 3 & 6 & -5 \end{bmatrix} = \begin{bmatrix} 2-34+33 & 2-68+66 & 4+51-55 \\ -1+22-21 & -1+44-42 & -2-33+35 \\ 6-6 & 12-12 & -9+10 \end{bmatrix} = \begin{bmatrix} 1 & 0 & 0 \\ 0 & 1 & 0 \\ 0 & 0 & 1 \end{bmatrix}$

$BA = \begin{bmatrix} 1 & 1 & 2 \\ 2 & 4 & -3 \\ 3 & 6 & -5 \end{bmatrix}\begin{bmatrix} 2 & -17 & 11 \\ -1 & 11 & -7 \\ 0 & 3 & -2 \end{bmatrix} = \begin{bmatrix} 2-1 & -17+11+6 & 11-7-4 \\ 4-4 & -34+44-9 & 22-28+6 \\ 6-6 & -51+66-15 & 33-42+10 \end{bmatrix} = \begin{bmatrix} 1 & 0 & 0 \\ 0 & 1 & 0 \\ 0 & 0 & 1 \end{bmatrix}$

6. $AB = \begin{bmatrix} -4 & 1 & 5 \\ -1 & 2 & 4 \\ 0 & -1 & -1 \end{bmatrix}\begin{bmatrix} -\frac{1}{2} & 1 & \frac{3}{2} \\ \frac{1}{4} & -1 & -\frac{11}{4} \\ -\frac{1}{4} & 1 & \frac{7}{4} \end{bmatrix} = \begin{bmatrix} 2+\frac{1}{4}-\frac{5}{4} & -4-1+5 & -6-\frac{11}{4}+\frac{35}{4} \\ \frac{1}{2}+\frac{1}{2}-1 & -1-2+4 & -\frac{3}{2}-\frac{11}{2}+7 \\ -\frac{1}{4}+\frac{1}{4} & 1-1 & \frac{11}{4}-\frac{7}{4} \end{bmatrix} = \begin{bmatrix} 1 & 0 & 0 \\ 0 & 1 & 0 \\ 0 & 0 & 1 \end{bmatrix}$

$BA = \begin{bmatrix} -\frac{1}{2} & 1 & \frac{3}{2} \\ \frac{1}{4} & -1 & -\frac{11}{4} \\ -\frac{1}{4} & 1 & \frac{7}{4} \end{bmatrix}\begin{bmatrix} -4 & 1 & 5 \\ -1 & 2 & 4 \\ 0 & -1 & -1 \end{bmatrix} = \begin{bmatrix} 2-1 & -\frac{1}{2}+2-\frac{3}{2} & -\frac{5}{2}+4-\frac{3}{2} \\ -1+1 & \frac{1}{4}-2+\frac{11}{4} & \frac{5}{4}-4+\frac{11}{4} \\ 1-1 & -\frac{1}{4}+2-\frac{7}{4} & -\frac{5}{4}+4-\frac{7}{4} \end{bmatrix} = \begin{bmatrix} 1 & 0 & 0 \\ 0 & 1 & 0 \\ 0 & 0 & 1 \end{bmatrix}$

7. $AB = \begin{bmatrix} 2 & 0 & 1 & 1 \\ 3 & 0 & 0 & 1 \\ -1 & 1 & -2 & 1 \\ 4 & -1 & 1 & 0 \end{bmatrix}\begin{bmatrix} -1 & 2 & -1 & -1 \\ -4 & 9 & -5 & -6 \\ 0 & 1 & -1 & -1 \\ 3 & -5 & 3 & 3 \end{bmatrix}$

$= \begin{bmatrix} -2+3 & 4+1-5 & -2-1+3 & -2-1+3 \\ 0 & 6-5 & 0 & 0 \\ 1-4+3 & -2+9-2-5 & 1-5+2+3 & 1-6+2+3 \\ 0 & 8-9+1 & -4+5-1 & -4+6-1 \end{bmatrix} = \begin{bmatrix} 1 & 0 & 0 & 0 \\ 0 & 1 & 0 & 0 \\ 0 & 0 & 1 & 0 \\ 0 & 0 & 0 & 1 \end{bmatrix}$

—CONTINUED—

7. —CONTINUED—

$$BA = \begin{bmatrix} -1 & 2 & -1 & -1 \\ -4 & 9 & -5 & -6 \\ 0 & 1 & -1 & -1 \\ 3 & -5 & 3 & 3 \end{bmatrix} \begin{bmatrix} 2 & 0 & 1 & 1 \\ 3 & 0 & 0 & 1 \\ -1 & 1 & -2 & 1 \\ 4 & -1 & 1 & 0 \end{bmatrix}$$

$$= \begin{bmatrix} -2+6+1-4 & 0 & -1+2-1 & -1+2-1 \\ -8+27+5-24 & -5+6 & -4+10-6 & -4+9-5 \\ 3+1-4 & 0 & 2-1 & 0 \\ 6-15-3+12 & 0 & 3-6+3 & 3-5+3 \end{bmatrix} = \begin{bmatrix} 1 & 0 & 0 & 0 \\ 0 & 1 & 0 & 0 \\ 0 & 0 & 1 & 0 \\ 0 & 0 & 0 & 1 \end{bmatrix}$$

8. $AB = \begin{bmatrix} -2 & 0 & 1 & 0 \\ 1 & -1 & -3 & 0 \\ -2 & -1 & 0 & -2 \\ 0 & 1 & 3 & -1 \end{bmatrix} \begin{bmatrix} -3 & -3 & 1 & -2 \\ 12 & 14 & -5 & 10 \\ -5 & -6 & 2 & -4 \\ -3 & -4 & 1 & -3 \end{bmatrix}$

$$= \begin{bmatrix} 6-5 & 6-6 & -2+2 & 4-4 \\ -3-12+15 & -3-14+18 & 1+5-6 & -2-10+12 \\ 6-12+6 & 6-14+8 & -2+5-2 & 4-10+6 \\ 12-15+3 & 14-18+4 & -5+6-1 & 10-12+3 \end{bmatrix}$$

$$= \begin{bmatrix} 1 & 0 & 0 & 0 \\ 0 & 1 & 0 & 0 \\ 0 & 0 & 1 & 0 \\ 0 & 0 & 0 & 1 \end{bmatrix}$$

$$BA = \begin{bmatrix} -3 & -3 & 1 & -2 \\ 12 & 14 & -5 & 10 \\ -5 & -6 & 2 & -4 \\ -3 & -4 & 1 & -3 \end{bmatrix} \begin{bmatrix} -2 & 0 & 1 & 0 \\ 1 & -1 & -3 & 0 \\ -2 & -1 & 0 & -2 \\ 0 & 1 & 3 & -1 \end{bmatrix}$$

$$= \begin{bmatrix} 6-3-2 & 3-1-2 & -3+9-6 & -2+2 \\ -24+14+10 & -14+5+10 & 12-42+30 & 10-10 \\ 10-6-4 & 6-2-4 & -5+18-12 & -4+4 \\ 6-4-2 & 4-1-3 & -3+12-9 & -2+3 \end{bmatrix}$$

$$= \begin{bmatrix} 1 & 0 & 0 & 0 \\ 0 & 1 & 0 & 0 \\ 0 & 0 & 1 & 0 \\ 0 & 0 & 0 & 1 \end{bmatrix}$$

9. $AB = \frac{1}{3}\begin{bmatrix} -2 & 2 & 3 \\ 1 & -1 & 0 \\ 0 & 1 & 4 \end{bmatrix} \begin{bmatrix} -4 & -5 & 3 \\ -4 & -8 & 3 \\ 1 & 2 & 0 \end{bmatrix} = \frac{1}{3}\begin{bmatrix} 8-8+3 & 10-16+6 & -6+6 \\ -4+4 & -5+8 & 3-3 \\ -4+4 & -8+8 & 3 \end{bmatrix}$

$$= \frac{1}{3}\begin{bmatrix} 3 & 0 & 0 \\ 0 & 3 & 0 \\ 0 & 0 & 3 \end{bmatrix} = \begin{bmatrix} 1 & 0 & 0 \\ 0 & 1 & 0 \\ 0 & 0 & 1 \end{bmatrix}$$

$$BA = \frac{1}{3}\begin{bmatrix} -4 & -5 & 3 \\ -4 & -8 & 3 \\ 1 & 2 & 0 \end{bmatrix} \begin{bmatrix} -2 & 2 & 3 \\ 1 & -1 & 0 \\ 0 & 1 & 4 \end{bmatrix} = \frac{1}{3}\begin{bmatrix} 8-5 & -8+5+3 & -12+12 \\ 8-8 & -8+8+3 & -12+12 \\ -2+2 & 2-2 & 3 \end{bmatrix} = \begin{bmatrix} 1 & 0 & 0 \\ 0 & 1 & 0 \\ 0 & 0 & 1 \end{bmatrix}$$

10. $AB = \frac{1}{3}\begin{bmatrix} -1 & 1 & 0 & -1 \\ 1 & -1 & 1 & 0 \\ -1 & 1 & 2 & 0 \\ 0 & -1 & 1 & 1 \end{bmatrix}\begin{bmatrix} -3 & 1 & 1 & -3 \\ -3 & -1 & 2 & -3 \\ 0 & 1 & 1 & 0 \\ -3 & -2 & 1 & 0 \end{bmatrix}$

$= \frac{1}{3}\begin{bmatrix} 3-3+3 & -1-1+2 & -1+2-1 & 3-3 \\ -3+3 & 1+1+1 & 1-2+1 & -3+3 \\ 3-3 & -1-1+2 & -1+2+2 & 3-3 \\ 3-3 & 1+1-2 & -2+1+1 & 3 \end{bmatrix}$

$= \begin{bmatrix} 1 & 0 & 0 & 0 \\ 0 & 1 & 0 & 0 \\ 0 & 0 & 1 & 0 \\ 0 & 0 & 0 & 1 \end{bmatrix}$

$BA = \frac{1}{3}\begin{bmatrix} -3 & 1 & 1 & -3 \\ -3 & -1 & 2 & -3 \\ 0 & 1 & 1 & 0 \\ -3 & -2 & 1 & 0 \end{bmatrix}\begin{bmatrix} -1 & 1 & 0 & -1 \\ 1 & -1 & 1 & 0 \\ -1 & 1 & 2 & 0 \\ 0 & -1 & 1 & 1 \end{bmatrix}$

$= \frac{1}{3}\begin{bmatrix} 3+1-1 & -3-1+1+3 & 1+2-3 & 3-3 \\ 3-1-2 & -3+1+2+3 & -1+4-3 & 3-3 \\ 1-1 & -1+1 & 1+2 & 0 \\ 3-2-1 & -3+2+1 & -2+2 & 3 \end{bmatrix}$

$= \begin{bmatrix} 1 & 0 & 0 & 0 \\ 0 & 1 & 0 & 0 \\ 0 & 0 & 1 & 0 \\ 0 & 0 & 0 & 1 \end{bmatrix}$

11. $[A \;\vdots\; I] = \begin{bmatrix} 2 & 0 & \vdots & 1 & 0 \\ 0 & 3 & \vdots & 0 & 1 \end{bmatrix}$

$\begin{array}{c} \frac{1}{2}R_1 \rightarrow \\ \frac{1}{3}R_2 \rightarrow \end{array} \begin{bmatrix} 1 & 0 & \vdots & \frac{1}{2} & 0 \\ 0 & 1 & \vdots & 0 & \frac{1}{3} \end{bmatrix} = [I \;\vdots\; A^{-1}]$

$A^{-1} = \begin{bmatrix} \frac{1}{2} & 0 \\ 0 & \frac{1}{3} \end{bmatrix}$

12. $[A \;\vdots\; I] = \begin{bmatrix} 1 & 2 & \vdots & 1 & 0 \\ 3 & 7 & \vdots & 0 & 1 \end{bmatrix}$

$-3R_1 + R_2 \rightarrow \begin{bmatrix} 1 & 2 & \vdots & 1 & 0 \\ 0 & 1 & \vdots & -3 & 1 \end{bmatrix}$

$-2R_2 + R_1 \rightarrow \begin{bmatrix} 1 & 0 & \vdots & 7 & -2 \\ 0 & 1 & \vdots & -3 & 1 \end{bmatrix} = [I \;\vdots\; A^{-1}]$

$A^{-1} = \begin{bmatrix} 7 & -2 \\ -3 & 1 \end{bmatrix}$

13. $[A \;\vdots\; I] = \begin{bmatrix} 1 & -2 & \vdots & 1 & 0 \\ 2 & -3 & \vdots & 0 & 1 \end{bmatrix}$

$-2R_1 + R_2 \rightarrow \begin{bmatrix} 1 & -2 & \vdots & 1 & 0 \\ 0 & 1 & \vdots & -2 & 1 \end{bmatrix}$

$2R_2 + R_1 \rightarrow \begin{bmatrix} 1 & 0 & \vdots & -3 & 2 \\ 0 & 1 & \vdots & -2 & 1 \end{bmatrix} = [I \;\vdots\; A^{-1}]$

$A^{-1} = \begin{bmatrix} -3 & 2 \\ -2 & 1 \end{bmatrix}$

14. $[A \;\vdots\; I] = \begin{bmatrix} -7 & 33 & \vdots & 1 & 0 \\ 4 & -19 & \vdots & 0 & 1 \end{bmatrix}$

$2R_2 + R_1 \rightarrow \begin{bmatrix} 1 & -5 & \vdots & 1 & 2 \\ 4 & -19 & \vdots & 0 & 1 \end{bmatrix}$

$-4R_1 + R_2 \rightarrow \begin{bmatrix} 1 & -5 & \vdots & 1 & 2 \\ 0 & 1 & \vdots & -4 & -7 \end{bmatrix}$

$5R_2 + R_1 \rightarrow \begin{bmatrix} 1 & 0 & \vdots & -19 & -33 \\ 0 & 1 & \vdots & -4 & -7 \end{bmatrix} = [I \;\vdots\; A^{-1}]$

$A^{-1} = \begin{bmatrix} -19 & -33 \\ -4 & -7 \end{bmatrix}$

15. $[A \; \vdots \; I] = \begin{bmatrix} -1 & 1 & \vdots & 1 & 0 \\ -2 & 1 & \vdots & 0 & 1 \end{bmatrix}$

$-R_2 + R_1 \rightarrow \begin{bmatrix} 1 & 0 & \vdots & 1 & -1 \\ -2 & 1 & \vdots & 0 & 1 \end{bmatrix}$

$2R_1 + R_2 \rightarrow \begin{bmatrix} 1 & 0 & \vdots & 1 & -1 \\ 0 & 1 & \vdots & 2 & -1 \end{bmatrix} = [I \; \vdots \; A^{-1}]$

$A^{-1} = \begin{bmatrix} 1 & -1 \\ 2 & -1 \end{bmatrix}$

16. $[A \; \vdots \; I] = \begin{bmatrix} 11 & 1 & \vdots & 1 & 0 \\ -1 & 0 & \vdots & 0 & 1 \end{bmatrix}$

$10R_2 + R_1 \rightarrow \begin{bmatrix} 1 & 1 & \vdots & 1 & 10 \\ -1 & 0 & \vdots & 0 & 1 \end{bmatrix}$

$R_1 + R_2 \rightarrow \begin{bmatrix} 1 & 1 & \vdots & 1 & 10 \\ 0 & 1 & \vdots & 1 & 11 \end{bmatrix}$

$-R_2 + R_1 \rightarrow \begin{bmatrix} 1 & 0 & \vdots & 0 & -1 \\ 0 & 1 & \vdots & 1 & 11 \end{bmatrix} = [I \; \vdots \; A^{-1}]$

$A^{-1} = \begin{bmatrix} 0 & -1 \\ 1 & 11 \end{bmatrix}$

17. $[A \; \vdots \; I] = \begin{bmatrix} 2 & 4 & \vdots & 1 & 0 \\ 4 & 8 & \vdots & 0 & 1 \end{bmatrix}$

$-2R_1 + R_2 \rightarrow \begin{bmatrix} 2 & 4 & \vdots & 1 & 0 \\ 0 & 0 & \vdots & -2 & 1 \end{bmatrix}$

The two zeros in the second row imply that the inverse does not exist.

18. $[A \; \vdots \; I] = \begin{bmatrix} 2 & 3 & \vdots & 1 & 0 \\ 1 & 4 & \vdots & 0 & 1 \end{bmatrix}$

$\begin{matrix} R_2 \\ R_1 \end{matrix} \begin{bmatrix} 1 & 4 & \vdots & 0 & 1 \\ 2 & 3 & \vdots & 1 & 0 \end{bmatrix}$

$-2R_1 + R_2 \rightarrow \begin{bmatrix} 1 & 4 & \vdots & 0 & 1 \\ 0 & -5 & \vdots & 1 & -2 \end{bmatrix}$

$-\frac{1}{5}R_2 \rightarrow \begin{bmatrix} 1 & 4 & \vdots & 0 & 1 \\ 0 & 1 & \vdots & -\frac{1}{5} & \frac{2}{5} \end{bmatrix}$

$-4R_2 + R_1 \rightarrow \begin{bmatrix} 1 & 0 & \vdots & \frac{4}{5} & -\frac{3}{5} \\ 0 & 1 & \vdots & -\frac{1}{5} & \frac{2}{5} \end{bmatrix} = [I \; \vdots \; A^{-1}]$

$A^{-1} = \frac{1}{5}\begin{bmatrix} 4 & -3 \\ -1 & 2 \end{bmatrix}$

19. $A = \begin{bmatrix} 2 & 7 & 1 \\ -3 & -9 & 2 \end{bmatrix}$ *A* has no inverse because it is not square.

20. $A = \begin{bmatrix} -2 & 5 \\ 6 & -15 \\ 0 & 1 \end{bmatrix}$ *A* has no inverse because it is not square.

21. $[A \; \vdots \; I] = \begin{bmatrix} 1 & 1 & 1 & \vdots & 1 & 0 & 0 \\ 3 & 5 & 4 & \vdots & 0 & 1 & 0 \\ 3 & 6 & 5 & \vdots & 0 & 0 & 1 \end{bmatrix}$

$\begin{matrix} -3R_1 + R_2 \rightarrow \\ -3R_1 + R_3 \rightarrow \end{matrix} \begin{bmatrix} 1 & 1 & 1 & \vdots & 1 & 0 & 0 \\ 0 & 2 & 1 & \vdots & -3 & 1 & 0 \\ 0 & 3 & 2 & \vdots & -3 & 0 & 1 \end{bmatrix}$

$\frac{1}{2}R_2 \rightarrow \begin{bmatrix} 1 & 1 & 1 & \vdots & 1 & 0 & 0 \\ 0 & 1 & \frac{1}{2} & \vdots & -\frac{3}{2} & \frac{1}{2} & 0 \\ 0 & 3 & 2 & \vdots & -3 & 0 & 1 \end{bmatrix}$

$-R_2 + R_1 \rightarrow \begin{bmatrix} 1 & 0 & \frac{1}{2} & \vdots & \frac{5}{2} & -\frac{1}{2} & 0 \\ 0 & 1 & \frac{1}{2} & \vdots & -\frac{3}{2} & \frac{1}{2} & 0 \\ 0 & 0 & \frac{1}{2} & \vdots & \frac{3}{2} & -\frac{3}{2} & 1 \end{bmatrix}$

$-3R_2 + R_3 \rightarrow$

$\begin{matrix} -R_3 + R_1 \rightarrow \\ -R_3 + R_2 \rightarrow \end{matrix} \begin{bmatrix} 1 & 0 & 0 & \vdots & 1 & 1 & -1 \\ 0 & 1 & 0 & \vdots & -3 & 2 & -1 \\ 0 & 0 & \frac{1}{2} & \vdots & \frac{3}{2} & -\frac{3}{2} & 1 \end{bmatrix}$

$2R_3 \rightarrow \begin{bmatrix} 1 & 0 & 0 & \vdots & 1 & 1 & -1 \\ 0 & 1 & 0 & \vdots & -3 & 2 & -1 \\ 0 & 0 & 1 & \vdots & 3 & -3 & 2 \end{bmatrix} = [I \; \vdots \; A^{-1}]$

$A^{-1} = \begin{bmatrix} 1 & 1 & -1 \\ -3 & 2 & -1 \\ 3 & -3 & 2 \end{bmatrix}$

22. $\quad [A \,\vdots\, I] = \begin{bmatrix} 1 & 2 & 2 & \vdots & 1 & 0 & 0 \\ 3 & 7 & 9 & \vdots & 0 & 1 & 0 \\ -1 & -4 & -7 & \vdots & 0 & 0 & 1 \end{bmatrix}$

$\begin{matrix} -3R_1 + R_2 \to \\ R_1 + R_3 \to \end{matrix} \begin{bmatrix} 1 & 2 & 2 & \vdots & 1 & 0 & 0 \\ 0 & 1 & 3 & \vdots & -3 & 1 & 0 \\ 0 & -2 & -5 & \vdots & 1 & 0 & 1 \end{bmatrix}$

$\begin{matrix} -2R_2 + R_1 \to \\ {} \\ 2R_2 + R_3 \to \end{matrix} \begin{bmatrix} 1 & 0 & -4 & \vdots & 7 & -2 & 0 \\ 0 & 1 & 3 & \vdots & -3 & 1 & 0 \\ 0 & 0 & 1 & \vdots & -5 & 2 & 1 \end{bmatrix}$

$\begin{matrix} 4R_3 + R_1 \to \\ -3R_3 + R_2 \to \\ {} \end{matrix} \begin{bmatrix} 1 & 0 & 0 & \vdots & -13 & 6 & 4 \\ 0 & 1 & 0 & \vdots & 12 & -5 & -3 \\ 0 & 0 & 1 & \vdots & -5 & 2 & 1 \end{bmatrix} = [I \,\vdots\, A^{-1}]$

$A^{-1} = \begin{bmatrix} -13 & 6 & 4 \\ 12 & -5 & -3 \\ -5 & 2 & 1 \end{bmatrix}$

23. $\quad [A \,\vdots\, I] = \begin{bmatrix} 1 & 0 & 0 & \vdots & 1 & 0 & 0 \\ 3 & 4 & 0 & \vdots & 0 & 1 & 0 \\ 2 & 5 & 5 & \vdots & 0 & 0 & 1 \end{bmatrix}$

$\begin{matrix} -3R_1 + R_2 \to \\ -2R_1 + R_3 \to \end{matrix} \begin{bmatrix} 1 & 0 & 0 & \vdots & 1 & 0 & 0 \\ 0 & 4 & 0 & \vdots & -3 & 1 & 0 \\ 0 & 5 & 5 & \vdots & -2 & 0 & 1 \end{bmatrix}$

$\begin{matrix} {} \\ {} \\ -\frac{5}{4}R_2 + R_3 \to \end{matrix} \begin{bmatrix} 1 & 0 & 0 & \vdots & 1 & 0 & 0 \\ 0 & 4 & 0 & \vdots & -3 & 1 & 0 \\ 0 & 0 & 5 & \vdots & \frac{7}{4} & -\frac{5}{4} & 1 \end{bmatrix}$

$\begin{matrix} {} \\ \frac{1}{4}R_2 \to \\ \frac{1}{5}R_3 \to \end{matrix} \begin{bmatrix} 1 & 0 & 0 & \vdots & 1 & 0 & 0 \\ 0 & 1 & 0 & \vdots & -\frac{3}{4} & \frac{1}{4} & 0 \\ 0 & 0 & 1 & \vdots & \frac{7}{20} & -\frac{1}{4} & \frac{1}{5} \end{bmatrix} = [I \,\vdots\, A^{-1}]$

$A^{-1} = \begin{bmatrix} 1 & 0 & 0 \\ -\frac{3}{4} & \frac{1}{4} & 0 \\ \frac{7}{20} & -\frac{1}{4} & \frac{1}{5} \end{bmatrix}$

24. $[A \,\vdots\, I] = \begin{bmatrix} 1 & 0 & 0 & \vdots & 1 & 0 & 0 \\ 3 & 0 & 0 & \vdots & 0 & 1 & 0 \\ 2 & 5 & 5 & \vdots & 0 & 0 & 1 \end{bmatrix} \begin{matrix} {} \\ -3R_1 + R_2 \to \\ -2R_1 + R_3 \to \end{matrix} \begin{bmatrix} 1 & 0 & 0 & \vdots & 1 & 0 & 0 \\ 0 & 0 & 0 & \vdots & -3 & 1 & 0 \\ 0 & 5 & 5 & \vdots & -2 & 0 & 1 \end{bmatrix}$

Since the first three entries of row 2 are all zeros, the inverse of A does not exist.

25. $[A \,\vdots\, I] = \begin{bmatrix} -8 & 0 & 0 & 0 & \vdots & 1 & 0 & 0 & 0 \\ 0 & 1 & 0 & 0 & \vdots & 0 & 1 & 0 & 0 \\ 0 & 0 & 4 & 0 & \vdots & 0 & 0 & 1 & 0 \\ 0 & 0 & 0 & -5 & \vdots & 0 & 0 & 0 & 1 \end{bmatrix}$

$\begin{matrix} -\frac{1}{8}R_1 \to \\ {} \\ \frac{1}{4}R_3 \to \\ -\frac{1}{5}R_4 \to \end{matrix} \begin{bmatrix} 1 & 0 & 0 & 0 & \vdots & -\frac{1}{8} & 0 & 0 & 0 \\ 0 & 1 & 0 & 0 & \vdots & 0 & 1 & 0 & 0 \\ 0 & 0 & 1 & 0 & \vdots & 0 & 0 & \frac{1}{4} & 0 \\ 0 & 0 & 0 & 1 & \vdots & 0 & 0 & 0 & -\frac{1}{5} \end{bmatrix} = [I \,\vdots\, A^{-1}]$

$A^{-1} = \begin{bmatrix} -\frac{1}{8} & 0 & 0 & 0 \\ 0 & 1 & 0 & 0 \\ 0 & 0 & \frac{1}{4} & 0 \\ 0 & 0 & 0 & -\frac{1}{5} \end{bmatrix}$

26. $[A \; \vdots \; I] = \begin{bmatrix} 1 & 3 & -2 & 0 & \vdots & 1 & 0 & 0 & 0 \\ 0 & 2 & 4 & 6 & \vdots & 0 & 1 & 0 & 0 \\ 0 & 0 & -2 & 1 & \vdots & 0 & 0 & 1 & 0 \\ 0 & 0 & 0 & 5 & \vdots & 0 & 0 & 0 & 1 \end{bmatrix}$

$\begin{array}{l} \\ \frac{1}{2}R_2 \rightarrow \\ \\ \frac{1}{5}R_4 \rightarrow \end{array} \begin{bmatrix} 1 & 3 & -2 & 0 & \vdots & 1 & 0 & 0 & 0 \\ 0 & 1 & 2 & 3 & \vdots & 0 & \frac{1}{2} & 0 & 0 \\ 0 & 0 & -2 & 1 & \vdots & 0 & 0 & 1 & 0 \\ 0 & 0 & 0 & 1 & \vdots & 0 & 0 & 0 & \frac{1}{5} \end{bmatrix}$

$\begin{array}{l} -3R_2 + R_1 \rightarrow \\ R_3 + R_2 \rightarrow \\ -R_4 + R_3 \rightarrow \\ \\ \end{array} \begin{bmatrix} 1 & 0 & -8 & -9 & \vdots & 1 & -\frac{3}{2} & 0 & 0 \\ 0 & 1 & 0 & 4 & \vdots & 0 & \frac{1}{2} & 1 & 0 \\ 0 & 0 & -2 & 0 & \vdots & 0 & 0 & 1 & -\frac{1}{5} \\ 0 & 0 & 0 & 1 & \vdots & 0 & 0 & 0 & \frac{1}{5} \end{bmatrix}$

$\begin{array}{l} -4R_3 + R_1 \rightarrow \\ -4R_4 + R_2 \rightarrow \\ -\frac{1}{2}R_3 \rightarrow \\ \\ \end{array} \begin{bmatrix} 1 & 0 & 0 & -9 & \vdots & 1 & -\frac{3}{2} & -4 & \frac{4}{5} \\ 0 & 1 & 0 & 0 & \vdots & 0 & \frac{1}{2} & 1 & -\frac{4}{5} \\ 0 & 0 & 1 & 0 & \vdots & 0 & 0 & -\frac{1}{2} & \frac{1}{10} \\ 0 & 0 & 0 & 1 & \vdots & 0 & 0 & 0 & \frac{1}{5} \end{bmatrix}$

$\begin{array}{l} 9R_4 + R_1 \rightarrow \\ \\ \\ \\ \end{array} \begin{bmatrix} 1 & 0 & 0 & 0 & \vdots & 1 & -\frac{3}{2} & -4 & \frac{13}{5} \\ 0 & 1 & 0 & 0 & \vdots & 0 & \frac{1}{2} & 1 & -\frac{4}{5} \\ 0 & 0 & 1 & 0 & \vdots & 0 & 0 & -\frac{1}{2} & \frac{1}{10} \\ 0 & 0 & 0 & 1 & \vdots & 0 & 0 & 0 & \frac{1}{5} \end{bmatrix} = [I \; \vdots \; A^{-1}]$

$A^{-1} = \frac{1}{10} \begin{bmatrix} 10 & -15 & -40 & 26 \\ 0 & 5 & 10 & -8 \\ 0 & 0 & -5 & 1 \\ 0 & 0 & 0 & 2 \end{bmatrix}$

27. $A = \begin{bmatrix} 1 & 2 & -1 \\ 3 & 7 & -10 \\ -5 & -7 & -15 \end{bmatrix}$

$A^{-1} = \begin{bmatrix} -175 & 37 & -13 \\ 95 & -20 & 7 \\ 14 & -3 & 1 \end{bmatrix}$

28. $A = \begin{bmatrix} 10 & 5 & -7 \\ -5 & 1 & 4 \\ 3 & 2 & -2 \end{bmatrix}$

$A^{-1} = \begin{bmatrix} -10 & -4 & 27 \\ 2 & 1 & -5 \\ -13 & -5 & 35 \end{bmatrix}$

29. $A = \begin{bmatrix} 1 & 1 & 2 \\ 3 & 1 & 0 \\ -2 & 0 & 3 \end{bmatrix}$

$A^{-1} = \frac{1}{2} \begin{bmatrix} -3 & 3 & 2 \\ 9 & -7 & -6 \\ -2 & 2 & 2 \end{bmatrix} = \begin{bmatrix} -1.5 & 1.5 & 1 \\ 4.5 & -3.5 & -3 \\ -1 & 1 & 1 \end{bmatrix}$

30. $A = \begin{bmatrix} 3 & 2 & 2 \\ 2 & 2 & 2 \\ -4 & 4 & 3 \end{bmatrix}$

$A^{-1} = \begin{bmatrix} 1 & -1 & 0 \\ 7 & -8.5 & 1 \\ -8 & 10 & -1 \end{bmatrix}$

31. $A = \begin{bmatrix} -\frac{1}{2} & \frac{3}{4} & \frac{1}{4} \\ 1 & 0 & -\frac{3}{2} \\ 0 & -1 & \frac{1}{2} \end{bmatrix}$

$A^{-1} = \begin{bmatrix} -12 & -5 & -9 \\ -4 & -2 & -4 \\ -8 & -4 & -6 \end{bmatrix}$

32. $\begin{bmatrix} -\frac{5}{6} & \frac{1}{3} & \frac{11}{6} \\ 0 & \frac{2}{3} & 2 \\ 1 & -\frac{1}{2} & -\frac{5}{2} \end{bmatrix}$

A^{-1} does not exist.

33. $A = \begin{bmatrix} 0.1 & 0.2 & 0.3 \\ -0.3 & 0.2 & 0.2 \\ 0.5 & 0.4 & 0.4 \end{bmatrix}$

$A^{-1} = \frac{5}{11}\begin{bmatrix} 0 & -4 & 2 \\ -22 & 11 & 11 \\ 22 & -6 & -8 \end{bmatrix} = \begin{bmatrix} 0 & -1.\overline{81} & 0.\overline{90} \\ -10 & 5 & 5 \\ 10 & -2.\overline{72} & -3.\overline{63} \end{bmatrix}$

34. $A = \begin{bmatrix} 0.6 & 0 & -0.3 \\ 0.7 & -1 & 0.2 \\ 1 & 0 & -0.9 \end{bmatrix}$

$A^{-1} = \begin{bmatrix} 3.75 & 0 & -1.25 \\ 3.458\overline{3} & -1 & -1.375 \\ 4.1\overline{6} & 0 & -2.5 \end{bmatrix}$

35. $A = \begin{bmatrix} 1 & 0 & 3 & 0 \\ 0 & 2 & 0 & 4 \\ 1 & 0 & 3 & 0 \\ 0 & 2 & 0 & 4 \end{bmatrix}$

A^{-1} does not exist.

36. $A = \begin{bmatrix} 4 & 8 & -7 & 14 \\ 2 & 5 & -4 & 6 \\ 0 & 2 & 1 & -7 \\ 3 & 6 & -5 & 10 \end{bmatrix}$

$A^{-1} = \begin{bmatrix} 27 & -10 & 4 & -29 \\ -16 & 5 & -2 & 18 \\ -17 & 4 & -2 & 20 \\ -7 & 2 & -1 & 8 \end{bmatrix}$

37. $A = \begin{bmatrix} -1 & 0 & 1 & 0 \\ 0 & 2 & 0 & -1 \\ 2 & 0 & -1 & 0 \\ 0 & -1 & 0 & 1 \end{bmatrix}$

$A^{-1} = \begin{bmatrix} 1 & 0 & 1 & 0 \\ 0 & 1 & 0 & 1 \\ 2 & 0 & 1 & 0 \\ 0 & 1 & 0 & 2 \end{bmatrix}$

38. $A = \begin{bmatrix} 1 & -2 & -1 & -2 \\ 3 & -5 & -2 & -3 \\ 2 & -5 & -2 & -5 \\ -1 & 4 & 4 & 11 \end{bmatrix}$

$A^{-1} = \begin{bmatrix} -24 & 7 & 1 & -2 \\ -10 & 3 & 0 & -1 \\ -29 & 7 & 3 & -2 \\ 12 & -3 & -1 & 1 \end{bmatrix}$

39. $A = \begin{bmatrix} a & b \\ c & d \end{bmatrix}, A^{-1} = \frac{1}{ad-bc}\begin{bmatrix} d & -b \\ -c & a \end{bmatrix}$

$A = \begin{bmatrix} 5 & -2 \\ 2 & 3 \end{bmatrix}$

$ad - bc = (5)(3) - (-2)(2) = 19$

$A^{-1} = \frac{1}{19}\begin{bmatrix} 3 & 2 \\ -2 & 5 \end{bmatrix} = \begin{bmatrix} \frac{3}{19} & \frac{2}{19} \\ -\frac{2}{19} & \frac{5}{19} \end{bmatrix}$

40. $A = \begin{bmatrix} 7 & 12 \\ -8 & -5 \end{bmatrix}$

$ad - bc = 7(-5) - 12(-8) = -35 + 96 = 61$

$A^{-1} = \frac{1}{61}\begin{bmatrix} -5 & -12 \\ 8 & 7 \end{bmatrix} = \begin{bmatrix} -\frac{5}{61} & -\frac{12}{61} \\ \frac{8}{61} & \frac{7}{61} \end{bmatrix}$

41. $A = \begin{bmatrix} -4 & -6 \\ 2 & 3 \end{bmatrix}$

$ad - bc = (-4)(3) - (-2)(-6) = 0$

Since $ad - bc = 0$, A^{-1} does not exist.

42. $A = \begin{bmatrix} -12 & 3 \\ 5 & -2 \end{bmatrix}$

$ad - bc = (-12)(-2) - 3(5) = 24 - 15 = 9$

$A^{-1} = \frac{1}{9}\begin{bmatrix} -2 & -3 \\ -5 & -12 \end{bmatrix} = \begin{bmatrix} -\frac{2}{9} & -\frac{1}{3} \\ -\frac{5}{9} & -\frac{4}{3} \end{bmatrix}$

43. $A = \begin{bmatrix} \frac{7}{2} & -\frac{3}{4} \\ \frac{1}{5} & \frac{4}{5} \end{bmatrix}$

$ad - bc = \left(\frac{7}{2}\right)\left(\frac{4}{5}\right) - \left(-\frac{3}{4}\right)\left(\frac{1}{5}\right) = \frac{28}{10} + \frac{3}{20} = \frac{59}{20}$

$A^{-1} = \frac{1}{59/20}\begin{bmatrix} \frac{4}{5} & \frac{3}{4} \\ -\frac{1}{5} & \frac{7}{2} \end{bmatrix} = \frac{20}{59}\begin{bmatrix} \frac{4}{5} & \frac{3}{4} \\ -\frac{1}{5} & \frac{7}{2} \end{bmatrix} = \begin{bmatrix} \frac{16}{59} & \frac{15}{59} \\ -\frac{4}{59} & \frac{70}{59} \end{bmatrix}$

44. $A = \begin{bmatrix} -\frac{1}{4} & \frac{9}{4} \\ \frac{5}{3} & \frac{8}{9} \end{bmatrix}$

$ad - bc = \left(-\frac{1}{4}\right)\left(\frac{8}{9}\right) - \left(\frac{9}{4}\right)\left(\frac{5}{3}\right) = -\frac{143}{36}$

$A^{-1} = -\frac{36}{143}\begin{bmatrix} \frac{8}{9} & -\frac{9}{4} \\ -\frac{5}{3} & -\frac{1}{4} \end{bmatrix} = \begin{bmatrix} -\frac{32}{143} & \frac{81}{143} \\ \frac{60}{143} & \frac{9}{143} \end{bmatrix}$

45. $\begin{bmatrix} x \\ y \end{bmatrix} = \begin{bmatrix} -3 & 2 \\ -2 & 1 \end{bmatrix} \begin{bmatrix} 5 \\ 10 \end{bmatrix} = \begin{bmatrix} 5 \\ 0 \end{bmatrix}$

Solution: $(5, 0)$

46. $\begin{bmatrix} x \\ y \end{bmatrix} = \begin{bmatrix} -3 & 2 \\ -2 & 1 \end{bmatrix} \begin{bmatrix} 0 \\ 3 \end{bmatrix} = \begin{bmatrix} 6 \\ 3 \end{bmatrix}$

Solution: $(6, 3)$

47. $\begin{bmatrix} x \\ y \end{bmatrix} = \begin{bmatrix} -3 & 2 \\ -2 & 1 \end{bmatrix} \begin{bmatrix} 4 \\ 2 \end{bmatrix} = \begin{bmatrix} -8 \\ -6 \end{bmatrix}$

Solution: $(-8, -6)$

48. $\begin{bmatrix} x \\ y \end{bmatrix} = \begin{bmatrix} -3 & 2 \\ -2 & 1 \end{bmatrix} \begin{bmatrix} 1 \\ -2 \end{bmatrix} = \begin{bmatrix} -7 \\ -4 \end{bmatrix}$

Solution: $(-7, -4)$

49. $\begin{bmatrix} x \\ y \\ z \end{bmatrix} = \begin{bmatrix} 1 & 1 & -1 \\ -3 & 2 & -1 \\ 3 & -3 & 2 \end{bmatrix} \begin{bmatrix} 0 \\ 5 \\ 2 \end{bmatrix} = \begin{bmatrix} 3 \\ 8 \\ -11 \end{bmatrix}$

Solution: $(3, 8, -11)$

50. $\begin{bmatrix} x \\ y \\ z \end{bmatrix} = \begin{bmatrix} 1 & 1 & -1 \\ -3 & 2 & -1 \\ 3 & -3 & 2 \end{bmatrix} \begin{bmatrix} -1 \\ 2 \\ 0 \end{bmatrix} = \begin{bmatrix} 1 \\ 7 \\ -9 \end{bmatrix}$

Solution: $(1, 7, -9)$

51. $\begin{bmatrix} x_1 \\ x_2 \\ x_3 \\ x_4 \end{bmatrix} = \begin{bmatrix} -24 & 7 & 1 & -2 \\ -10 & 3 & 0 & -1 \\ -29 & 7 & 3 & -2 \\ 12 & -3 & -1 & 1 \end{bmatrix} \begin{bmatrix} 0 \\ 1 \\ -1 \\ 2 \end{bmatrix} = \begin{bmatrix} 2 \\ 1 \\ 0 \\ 0 \end{bmatrix}$

Solution: $(2, 1, 0, 0)$

52. $\begin{bmatrix} x \\ y \\ z \\ w \end{bmatrix} = \begin{bmatrix} -24 & 7 & 1 & -2 \\ -10 & 3 & 0 & -1 \\ -29 & 7 & 3 & -2 \\ 12 & -3 & -1 & 1 \end{bmatrix} \begin{bmatrix} 1 \\ -2 \\ 0 \\ -3 \end{bmatrix} = \begin{bmatrix} -32 \\ -13 \\ -37 \\ 15 \end{bmatrix}$

Solution: $(-32, -13, -37, 15)$

53. $A = \begin{bmatrix} 3 & 4 \\ 5 & 3 \end{bmatrix}$

$A^{-1} = \dfrac{1}{9 - 20} \begin{bmatrix} 3 & -4 \\ -5 & 3 \end{bmatrix}$

$\begin{bmatrix} x \\ y \end{bmatrix} = -\dfrac{1}{11} \begin{bmatrix} 3 & -4 \\ -5 & 3 \end{bmatrix} \begin{bmatrix} -2 \\ 4 \end{bmatrix} = -\dfrac{1}{11} \begin{bmatrix} -22 \\ 22 \end{bmatrix} = \begin{bmatrix} 2 \\ -2 \end{bmatrix}$

Solution: $(2, -2)$

54. $A = \begin{bmatrix} 18 & 12 \\ 30 & 24 \end{bmatrix}$

$A^{-1} = \dfrac{1}{432 - 360} \begin{bmatrix} 24 & -12 \\ -30 & 18 \end{bmatrix}$

$\begin{bmatrix} x \\ y \end{bmatrix} = \dfrac{1}{72} \begin{bmatrix} 24 & -12 \\ -30 & 18 \end{bmatrix} \begin{bmatrix} 13 \\ 23 \end{bmatrix} = \dfrac{1}{72} \begin{bmatrix} 36 \\ 24 \end{bmatrix} = \begin{bmatrix} \frac{1}{2} \\ \frac{1}{3} \end{bmatrix}$

Solution: $\left(\frac{1}{2}, \frac{1}{3}\right)$

55. $A = \begin{bmatrix} -0.4 & 0.8 \\ 2 & -4 \end{bmatrix}$

$A^{-1} = \dfrac{1}{1.6 - 1.6} \begin{bmatrix} -4 & -0.8 \\ -2 & -0.4 \end{bmatrix}$

A^{-1} does not exist.

This implies that there is no unique solution; that is, either the system is inconsistent *or* there are infinitely many solutions.

Find the reduced row-echelon form of the matrix corresponding to the system.

$$\begin{bmatrix} -0.4 & 0.8 & \vdots & 1.6 \\ 2 & -4 & \vdots & 5 \end{bmatrix}$$

$$-2.5R_1 \rightarrow \begin{bmatrix} 1 & -2 & \vdots & -4 \\ 2 & -4 & \vdots & 5 \end{bmatrix}$$

$$-2R_1 + R_2 \rightarrow \begin{bmatrix} 1 & -2 & \vdots & -4 \\ 0 & 0 & \vdots & 13 \end{bmatrix}$$

The given system is inconsistent and there is no solution.

56. $A = \begin{bmatrix} 0.2 & -0.6 \\ -1 & 1.4 \end{bmatrix}$

$A^{-1} = \dfrac{1}{0.28 - 0.6} \begin{bmatrix} 1.4 & 0.6 \\ 1 & 0.2 \end{bmatrix}$

$\begin{bmatrix} x \\ y \end{bmatrix} = -\dfrac{1}{0.32} \begin{bmatrix} 1.4 & 0.6 \\ 1 & 0.2 \end{bmatrix} \begin{bmatrix} 2.4 \\ -8.8 \end{bmatrix}$

$= -\dfrac{1}{0.32} \begin{bmatrix} -1.92 \\ 0.64 \end{bmatrix} = \begin{bmatrix} 6 \\ -2 \end{bmatrix}$

Solution: $(6, -2)$

57. $A = \begin{bmatrix} -\frac{1}{4} & \frac{3}{8} \\ \frac{3}{2} & \frac{3}{4} \end{bmatrix}$

$A^{-1} = \dfrac{1}{-\frac{3}{16} - \frac{9}{16}} \begin{bmatrix} \frac{3}{4} & -\frac{3}{8} \\ -\frac{3}{2} & -\frac{1}{4} \end{bmatrix} = -\dfrac{4}{3}\begin{bmatrix} \frac{3}{4} & -\frac{3}{8} \\ -\frac{3}{2} & -\frac{1}{4} \end{bmatrix} = \begin{bmatrix} -1 & \frac{1}{2} \\ 2 & \frac{1}{3} \end{bmatrix}$

$\begin{bmatrix} x \\ y \end{bmatrix} = \begin{bmatrix} -1 & \frac{1}{2} \\ 2 & \frac{1}{3} \end{bmatrix}\begin{bmatrix} -2 \\ -12 \end{bmatrix} = \begin{bmatrix} -4 \\ -8 \end{bmatrix}$

Solution: $(-4, -8)$

58. $A = \begin{bmatrix} \frac{5}{6} & -1 \\ \frac{4}{3} & -\frac{7}{2} \end{bmatrix}$

$A^{-1} = \dfrac{1}{-\frac{35}{12} + \frac{4}{3}} \begin{bmatrix} -\frac{7}{2} & 1 \\ -\frac{4}{3} & \frac{5}{6} \end{bmatrix}$

$\begin{bmatrix} x \\ y \end{bmatrix} = -\dfrac{12}{19}\begin{bmatrix} -\frac{7}{2} & 1 \\ -\frac{4}{3} & \frac{5}{6} \end{bmatrix}\begin{bmatrix} -20 \\ -51 \end{bmatrix} = -\dfrac{12}{19}\begin{bmatrix} 19 \\ -\frac{95}{6} \end{bmatrix} = \begin{bmatrix} -12 \\ 10 \end{bmatrix}$

Solution: $(-12, 10)$

59. $A = \begin{bmatrix} 4 & -1 & 1 \\ 2 & 2 & 3 \\ 5 & -2 & 6 \end{bmatrix}$

Find A^{-1}.

$[A \ \vdots \ I] = \begin{bmatrix} 4 & -1 & 1 & \vdots & 1 & 0 & 0 \\ 2 & 2 & 3 & \vdots & 0 & 1 & 0 \\ 5 & -2 & 6 & \vdots & 0 & 0 & 1 \end{bmatrix}$

$\begin{matrix} R_1 \\ \\ R_3 \end{matrix}\begin{bmatrix} 5 & -2 & 6 & \vdots & 0 & 0 & 1 \\ 2 & 2 & 3 & \vdots & 0 & 1 & 0 \\ 4 & -1 & 1 & \vdots & 1 & 0 & 0 \end{bmatrix}$

$-R_3 + R_1 \rightarrow \begin{bmatrix} 1 & -1 & 5 & \vdots & -1 & 0 & 1 \\ 2 & 2 & 3 & \vdots & 0 & 1 & 0 \\ 4 & -1 & 1 & \vdots & 1 & 0 & 0 \end{bmatrix}$

$\begin{matrix} \\ -2R_1 + R_2 \rightarrow \\ -4R_1 + R_3 \rightarrow \end{matrix}\begin{bmatrix} 1 & -1 & 5 & \vdots & -1 & 0 & 1 \\ 0 & 4 & -7 & \vdots & 2 & 1 & -2 \\ 0 & 3 & -19 & \vdots & 5 & 0 & -4 \end{bmatrix}$

$-R_3 + R_2 \rightarrow \begin{bmatrix} 1 & -1 & 5 & \vdots & -1 & 0 & 1 \\ 0 & 1 & 12 & \vdots & -3 & 1 & 2 \\ 0 & 3 & -19 & \vdots & 5 & 0 & -4 \end{bmatrix}$

$\begin{matrix} R_2 + R_1 \rightarrow \\ \\ -3R_2 + R_3 \rightarrow \end{matrix}\begin{bmatrix} 1 & 0 & 17 & \vdots & -4 & 1 & 3 \\ 0 & 1 & 12 & \vdots & -3 & 1 & 2 \\ 0 & 0 & -55 & \vdots & 14 & -3 & -10 \end{bmatrix}$

$-\frac{1}{55}R_3 \rightarrow \begin{bmatrix} 1 & 0 & 17 & \vdots & -4 & 1 & 3 \\ 0 & 1 & 12 & \vdots & -3 & 1 & 2 \\ 0 & 0 & 1 & \vdots & -\frac{14}{55} & \frac{3}{55} & \frac{2}{11} \end{bmatrix}$

$\begin{matrix} -17R_3 + R_1 \rightarrow \\ -12R_3 + R_2 \rightarrow \\ \\ \end{matrix}\begin{bmatrix} 1 & 0 & 0 & \vdots & \frac{18}{55} & \frac{4}{55} & -\frac{1}{11} \\ 0 & 1 & 0 & \vdots & \frac{3}{55} & \frac{19}{55} & -\frac{2}{11} \\ 0 & 0 & 1 & \vdots & -\frac{14}{55} & \frac{3}{55} & \frac{2}{11} \end{bmatrix} = [I \ \vdots \ A^{-1}]$

$A^{-1} = \dfrac{1}{55}\begin{bmatrix} 18 & 4 & -5 \\ 3 & 19 & -10 \\ -14 & 3 & 10 \end{bmatrix}$

$\begin{bmatrix} x \\ y \\ z \end{bmatrix} = \dfrac{1}{55}\begin{bmatrix} 18 & 4 & -5 \\ 3 & 19 & -10 \\ -14 & 3 & 10 \end{bmatrix}\begin{bmatrix} -5 \\ 10 \\ 1 \end{bmatrix} = \dfrac{1}{55}\begin{bmatrix} -55 \\ 165 \\ 110 \end{bmatrix} = \begin{bmatrix} -1 \\ 3 \\ 2 \end{bmatrix}$

Solution: $(-1, 3, 2)$

60. $A = \begin{bmatrix} 4 & -2 & 3 \\ 2 & 2 & 5 \\ 8 & -5 & -2 \end{bmatrix}$

$A^{-1} = \dfrac{1}{82} \begin{bmatrix} -21 & 19 & 16 \\ -44 & 32 & 14 \\ 26 & -4 & -12 \end{bmatrix}$

$\begin{bmatrix} x \\ y \\ z \end{bmatrix} = \dfrac{1}{82} \begin{bmatrix} -21 & 19 & 16 \\ -44 & 32 & 14 \\ 26 & -4 & -12 \end{bmatrix} \begin{bmatrix} -2 \\ 16 \\ 4 \end{bmatrix} = \dfrac{1}{82} \begin{bmatrix} 410 \\ 656 \\ -164 \end{bmatrix} = \begin{bmatrix} 5 \\ 8 \\ -2 \end{bmatrix}$

Solution: $(5, 8, -2)$.

61. $A = \begin{bmatrix} 5 & -3 & 2 \\ 2 & 2 & -3 \\ 1 & -7 & 8 \end{bmatrix}$

A^{-1} does not exist. This implies that there is no unique solution; that is, either the system is inconsistent *or* the system has infinitely many solutions. Use a graphing utility to find the reduced row-echelon form of the matrix corresponding to the system.

$\begin{bmatrix} 5 & -3 & 2 & \vdots & 2 \\ 2 & 2 & -3 & \vdots & 3 \\ 1 & -7 & 8 & \vdots & -4 \end{bmatrix}$

$\begin{bmatrix} 1 & 0 & -\frac{5}{16} & \vdots & \frac{13}{16} \\ 0 & 1 & -\frac{19}{16} & \vdots & \frac{11}{16} \\ 0 & 0 & 0 & \vdots & 0 \end{bmatrix}$

$\begin{cases} x - \frac{5}{16}z = \frac{13}{16} \\ y - \frac{19}{16}z = \frac{11}{16} \end{cases}$

Let $z = a$. Then $x = \frac{5}{16}a + \frac{13}{16}$ and $y = \frac{19}{16}a + \frac{11}{16}$.

Solution: $\left(\frac{5}{16}a + \frac{13}{16}, \frac{19}{16}a + \frac{11}{16}, a \right)$ where a is a real number

62. $A = \begin{bmatrix} 2 & 3 & 5 \\ 3 & 5 & 9 \\ 5 & 9 & 17 \end{bmatrix}$

A^{-1} does not exist. This implies that there is no unique solution; that is, either the system is inconsistent *or* the system has infinitely many solutions. Use a graphing utility to find the reduced row-echelon form of the matrix corresponding to the system.

$\begin{bmatrix} 2 & 3 & 5 & \vdots & 4 \\ 3 & 5 & 9 & \vdots & 7 \\ 5 & 9 & 17 & \vdots & 13 \end{bmatrix}$

$\begin{bmatrix} 1 & 0 & -2 & \vdots & -1 \\ 0 & 1 & 3 & \vdots & 2 \\ 0 & 0 & 0 & \vdots & 0 \end{bmatrix}$

$\begin{cases} x - 2z = -1 \\ y + 3z = 2 \end{cases}$

Let $z = a$. Then $x = 2a - 1$ and $y = -3a + 2$.

Solution: $(2a - 1, -3a + 2, a)$ where a is a real number

63. $A = \begin{bmatrix} 3 & -2 & 1 \\ -4 & 1 & -3 \\ 1 & -5 & 1 \end{bmatrix}$

$A^{-1} = \begin{bmatrix} 0.56 & 0.12 & -0.2 \\ -0.04 & -0.08 & -0.2 \\ -0.76 & -0.52 & 0.2 \end{bmatrix}$

$\begin{bmatrix} x \\ y \\ z \end{bmatrix} = \begin{bmatrix} 0.56 & 0.12 & -0.2 \\ -0.04 & -0.08 & -0.2 \\ -0.76 & -0.52 & 0.2 \end{bmatrix} \begin{bmatrix} -29 \\ 37 \\ -24 \end{bmatrix} = \begin{bmatrix} -7 \\ 3 \\ -2 \end{bmatrix}$

Solution: $(-7, 3, -2)$

64. $A = \begin{bmatrix} -8 & 7 & -10 \\ 12 & 3 & -5 \\ 15 & -9 & 2 \end{bmatrix}$

$A^{-1} \approx \begin{bmatrix} -0.034 & 0.066 & -0.004 \\ -0.086 & 0.117 & -0.139 \\ -0.133 & 0.029 & -0.094 \end{bmatrix}$

$\begin{bmatrix} x \\ y \\ z \end{bmatrix} \approx \begin{bmatrix} -0.034 & 0.066 & -0.004 \\ -0.086 & 0.117 & -0.139 \\ -0.133 & 0.029 & -0.094 \end{bmatrix} \begin{bmatrix} -151 \\ 86 \\ 187 \end{bmatrix} \approx \begin{bmatrix} 10 \\ -3 \\ 5 \end{bmatrix}$

Solution: $(10, -3, 5)$

65. $A = \begin{bmatrix} 7 & -3 & 0 & 2 \\ -2 & 1 & 0 & -1 \\ 4 & 0 & 1 & -2 \\ -1 & 1 & 0 & -1 \end{bmatrix}$

$A^{-1} = \begin{bmatrix} 0 & -1 & 0 & 1 \\ -1 & -5 & 0 & 3 \\ -2 & -4 & 1 & -2 \\ -1 & -4 & 0 & 1 \end{bmatrix}$

$\begin{bmatrix} x \\ y \\ z \\ w \end{bmatrix} = \begin{bmatrix} 0 & -1 & 0 & 1 \\ -1 & -5 & 0 & 3 \\ -2 & -4 & 1 & -2 \\ -1 & -4 & 0 & 1 \end{bmatrix} \begin{bmatrix} 41 \\ -13 \\ 12 \\ -8 \end{bmatrix} = \begin{bmatrix} 5 \\ 0 \\ -2 \\ 3 \end{bmatrix}$

Solution: $(5, 0, -2, 3)$

66. $A = \begin{bmatrix} 2 & 5 & 0 & 1 \\ 1 & 4 & 2 & -2 \\ 2 & -2 & 5 & 1 \\ 1 & 0 & 0 & -3 \end{bmatrix}$

$A^{-1} \approx \begin{bmatrix} 0.338 & -0.352 & 0.141 & 0.394 \\ 0.042 & 0.164 & -0.066 & -0.117 \\ -0.141 & 0.230 & 0.108 & -0.164 \\ 0.113 & -0.117 & 0.047 & -0.202 \end{bmatrix}$

$\begin{bmatrix} x \\ y \\ z \\ w \end{bmatrix} \approx \begin{bmatrix} 0.338 & -0.352 & 0.141 & 0.394 \\ 0.042 & 0.164 & -0.066 & -0.117 \\ -0.141 & 0.230 & 0.108 & -0.164 \\ 0.113 & -0.117 & 0.047 & -0.202 \end{bmatrix} \begin{bmatrix} 11 \\ -7 \\ 3 \\ -1 \end{bmatrix} \approx \begin{bmatrix} 6.21 \\ -0.77 \\ -2.67 \\ 2.40 \end{bmatrix}$

Solution: $(6.21, -0.77, -2.67, 2.40)$

67. $A = \begin{bmatrix} 1 & 1 & 1 \\ 0.065 & 0.07 & 0.09 \\ 0 & 2 & -1 \end{bmatrix}$

$[A \;\vdots\; I] = \begin{bmatrix} 1 & 1 & 1 & \vdots & 1 & 0 & 0 \\ 0.065 & 0.07 & 0.09 & \vdots & 0 & 1 & 0 \\ 0 & 2 & -1 & \vdots & 0 & 0 & 1 \end{bmatrix}$

$200R_2 \rightarrow \begin{bmatrix} 1 & 1 & 1 & \vdots & 1 & 0 & 0 \\ 13 & 14 & 18 & \vdots & 0 & 200 & 0 \\ 0 & 2 & -1 & \vdots & 0 & 0 & 1 \end{bmatrix}$

$-13R_1 + R_2 \rightarrow \begin{bmatrix} 1 & 1 & 1 & \vdots & 1 & 0 & 0 \\ 0 & 1 & 5 & \vdots & -13 & 200 & 0 \\ 0 & 2 & -1 & \vdots & 0 & 0 & 1 \end{bmatrix}$

$\begin{matrix} -R_2 + R_1 \rightarrow \\ \\ -2R_2 + R_3 \rightarrow \end{matrix} \begin{bmatrix} 1 & 0 & -4 & \vdots & 14 & -200 & 0 \\ 0 & 1 & 5 & \vdots & -13 & 200 & 0 \\ 0 & 0 & -11 & \vdots & 26 & -400 & 1 \end{bmatrix}$

$-\tfrac{1}{11}R_3 \rightarrow \begin{bmatrix} 1 & 0 & -4 & \vdots & 14 & -200 & 0 \\ 0 & 1 & 5 & \vdots & -13 & 200 & 0 \\ 0 & 0 & 1 & \vdots & -\tfrac{26}{11} & \tfrac{400}{11} & -\tfrac{1}{11} \end{bmatrix}$

$\begin{matrix} 4R_3 + R_1 \rightarrow \\ -5R_3 + R_2 \rightarrow \\ \\ \end{matrix} \begin{bmatrix} 1 & 0 & 0 & \vdots & \tfrac{50}{11} & -\tfrac{600}{11} & -\tfrac{4}{11} \\ 0 & 1 & 0 & \vdots & -\tfrac{13}{11} & \tfrac{200}{11} & \tfrac{5}{11} \\ 0 & 0 & 1 & \vdots & -\tfrac{26}{11} & \tfrac{400}{11} & -\tfrac{1}{11} \end{bmatrix} = [I \;\vdots\; A^{-1}]$

$X = A^{-1}B = \tfrac{1}{11}\begin{bmatrix} 50 & -600 & -4 \\ -13 & 200 & 5 \\ -26 & 400 & -1 \end{bmatrix}\begin{bmatrix} 10,000 \\ 705 \\ 0 \end{bmatrix} = \begin{bmatrix} 7000 \\ 1000 \\ 2000 \end{bmatrix}$

Solution: $7000 in AAA-rated bonds, $1000 in A-rated bonds, $2000 in B-rated bonds

68. $A = \begin{bmatrix} 1 & 1 & 1 \\ 0.065 & 0.07 & 0.09 \\ 0 & 2 & -1 \end{bmatrix}$

$[A \vdots I] = \begin{bmatrix} 1 & 1 & 1 & \vdots & 1 & 0 & 0 \\ 0.065 & 0.07 & 0.09 & \vdots & 0 & 1 & 0 \\ 0 & 2 & -1 & \vdots & 0 & 0 & 1 \end{bmatrix}$

$200R_2 \to \begin{bmatrix} 1 & 1 & 1 & \vdots & 1 & 0 & 0 \\ 13 & 14 & 18 & \vdots & 0 & 200 & 0 \\ 0 & 2 & -1 & \vdots & 0 & 0 & 1 \end{bmatrix}$

$-13R_1 + R_2 \to \begin{bmatrix} 1 & 1 & 1 & \vdots & 1 & 0 & 0 \\ 0 & 1 & 5 & \vdots & -13 & 200 & 0 \\ 0 & 2 & -1 & \vdots & 0 & 0 & 1 \end{bmatrix}$

$\begin{matrix} -R_2 + R_1 \to \\ \\ -2R_2 + R_3 \to \end{matrix} \begin{bmatrix} 1 & 0 & -4 & \vdots & 14 & -200 & 0 \\ 0 & 1 & 5 & \vdots & -13 & 200 & 0 \\ 0 & 0 & -11 & \vdots & 26 & -400 & 1 \end{bmatrix}$

$-\frac{1}{11}R_3 \to \begin{bmatrix} 1 & 0 & -4 & \vdots & 14 & -200 & 0 \\ 0 & 1 & 5 & \vdots & -13 & 200 & 0 \\ 0 & 0 & 1 & \vdots & -\frac{26}{11} & \frac{400}{11} & -\frac{1}{11} \end{bmatrix}$

$\begin{matrix} 4R_3 + R_1 \to \\ -5R_3 + R_2 \to \\ \\ \end{matrix} \begin{bmatrix} 1 & 0 & 0 & \vdots & \frac{50}{11} & -\frac{600}{11} & -\frac{4}{11} \\ 0 & 1 & 0 & \vdots & -\frac{13}{11} & \frac{200}{11} & \frac{5}{11} \\ 0 & 0 & 1 & \vdots & -\frac{26}{11} & \frac{400}{11} & -\frac{1}{11} \end{bmatrix} = [I \vdots A^{-1}]$

$X = A^{-1}B = \frac{1}{11}\begin{bmatrix} 50 & -600 & -4 \\ -13 & 200 & 5 \\ -26 & 400 & -1 \end{bmatrix}\begin{bmatrix} 10{,}000 \\ 760 \\ 0 \end{bmatrix} = \begin{bmatrix} 4000 \\ 2000 \\ 4000 \end{bmatrix}$

Solution: $4000 in AAA-rated bonds, $2000 in A-rated bonds, $4000 in B-rated bonds.

69. Use the inverse matrix A^{-1} from Exercise 67.

$X = A^{-1}B = \frac{1}{11}\begin{bmatrix} 50 & -600 & -4 \\ -13 & 200 & 5 \\ -26 & 400 & -1 \end{bmatrix}\begin{bmatrix} 12{,}000 \\ 835 \\ 0 \end{bmatrix} = \begin{bmatrix} 9000 \\ 1000 \\ 2000 \end{bmatrix}$

Solution: $9000 in AAA-rated bonds, $1000 in A-rated bonds, $2000 in B-rated bonds

70. Use the inverse matrix A^{-1} from Exercise 69.

$X = A^{-1}B = \frac{1}{11}\begin{bmatrix} 50 & -600 & -4 \\ -13 & 200 & 5 \\ -26 & 400 & -1 \end{bmatrix}\begin{bmatrix} 500{,}000 \\ 38{,}000 \\ 0 \end{bmatrix} = \begin{bmatrix} 200{,}000 \\ 100{,}000 \\ 200{,}000 \end{bmatrix}$

Solution: $200,000 in AAA-rated bonds, $100,000 in A-rated bonds, and $200,000 in B-rated bonds.

71. (a)
$$A = \begin{bmatrix} 2 & 0 & 4 \\ 0 & 1 & 4 \\ 1 & 1 & -1 \end{bmatrix}$$

$$[A \;\vdots\; I] = \begin{bmatrix} 2 & 0 & 4 & \vdots & 1 & 0 & 0 \\ 0 & 1 & 4 & \vdots & 0 & 1 & 0 \\ 1 & 1 & -1 & \vdots & 0 & 0 & 1 \end{bmatrix}$$

$$\begin{matrix} R_1 \\ \\ R_3 \end{matrix} \begin{bmatrix} 1 & 1 & -1 & \vdots & 0 & 0 & 1 \\ 0 & 1 & 4 & \vdots & 0 & 1 & 0 \\ 2 & 0 & 4 & \vdots & 1 & 0 & 0 \end{bmatrix}$$

$$-2R_1 + R_3 \rightarrow \begin{bmatrix} 1 & 1 & -1 & \vdots & 0 & 0 & 1 \\ 0 & 1 & 4 & \vdots & 0 & 1 & 0 \\ 0 & -2 & 6 & \vdots & 1 & 0 & -2 \end{bmatrix}$$

$$\begin{matrix} -R_2 + R_1 \rightarrow \\ \\ 2R_2 + R_3 \rightarrow \end{matrix} \begin{bmatrix} 1 & 0 & -5 & \vdots & 0 & -1 & 1 \\ 0 & 1 & 4 & \vdots & 0 & 1 & 0 \\ 0 & 0 & 14 & \vdots & 1 & 2 & -2 \end{bmatrix}$$

$$\tfrac{1}{14}R_3 \rightarrow \begin{bmatrix} 1 & 0 & -5 & \vdots & 0 & -1 & 1 \\ 0 & 1 & 4 & \vdots & 0 & 1 & 0 \\ 0 & 0 & 1 & \vdots & \tfrac{1}{14} & \tfrac{1}{7} & -\tfrac{1}{7} \end{bmatrix}$$

$$\begin{matrix} 5R_3 + R_1 \rightarrow \\ -4R_3 + R_2 \rightarrow \\ \\ \end{matrix} \begin{bmatrix} 1 & 0 & 0 & \vdots & \tfrac{5}{14} & -\tfrac{2}{7} & \tfrac{2}{7} \\ 0 & 1 & 0 & \vdots & -\tfrac{2}{7} & \tfrac{3}{7} & \tfrac{4}{7} \\ 0 & 0 & 1 & \vdots & \tfrac{1}{14} & \tfrac{1}{7} & -\tfrac{1}{7} \end{bmatrix} = [I \;\vdots\; A^{-1}]$$

$$A^{-1} = \tfrac{1}{14} \begin{bmatrix} 5 & -4 & 4 \\ -4 & 6 & 8 \\ 1 & 2 & -2 \end{bmatrix}$$

$$\begin{bmatrix} I_1 \\ I_2 \\ I_3 \end{bmatrix} = \tfrac{1}{14} \begin{bmatrix} 5 & -4 & 4 \\ -4 & 6 & 8 \\ 1 & 2 & -2 \end{bmatrix} \begin{bmatrix} 14 \\ 28 \\ 0 \end{bmatrix} = \begin{bmatrix} -3 \\ 8 \\ 5 \end{bmatrix}$$

Solution: $I_1 = -3$ amperes, $I_2 = 8$ amperes, $I_3 = 5$ amperes

(b)
$$\begin{bmatrix} I_1 \\ I_2 \\ I_3 \end{bmatrix} = \tfrac{1}{14} \begin{bmatrix} 5 & -4 & 4 \\ -4 & 6 & 8 \\ 1 & 2 & -2 \end{bmatrix} \begin{bmatrix} 24 \\ 23 \\ 0 \end{bmatrix} = \begin{bmatrix} 2 \\ 3 \\ 5 \end{bmatrix}$$

Solution:

$I_1 = 2$ amperes, $I_2 = 3$ amperes, $I_3 = 5$ amperes

72. (a) $n = 3; \displaystyle\sum_{i=1}^{n} x_i = 7 + 9 + 11 = 27;$

$$\sum_{i=1}^{n} y_i = 182.7 + 187.2 + 191.3 = 561.2;$$

$$\sum_{i=1}^{n} x_i^2 = 49 + 81 + 121 = 251$$

$$\sum_{i=1}^{n} x_i y_i = 7(182.7) + 9(187.2) + 11(191.3) = 5068$$

System: $\begin{cases} 3b + 27a = 561.2 \\ 27b + 251a = 5068 \end{cases}$

(e) $2.15t + 167.7 = 208$

$$2.15t = 40.3$$

$$t \approx 18.7$$

Since $t = 18$ represents 2008, the model projects that the number of licensed drivers will reach 208 million during 2008.

(b) $\begin{bmatrix} 3 & 27 \\ 27 & 251 \end{bmatrix}^{-1} = \begin{bmatrix} \tfrac{251}{24} & -\tfrac{9}{8} \\ -\tfrac{9}{8} & \tfrac{1}{8} \end{bmatrix}; \quad \begin{bmatrix} \tfrac{251}{24} & -\tfrac{9}{8} \\ -\tfrac{9}{8} & \tfrac{1}{8} \end{bmatrix} \begin{bmatrix} 561.2 \\ 5068 \end{bmatrix}$

$$= \begin{bmatrix} \left(\tfrac{251}{24}\right)(561.2) + \left(-\tfrac{9}{8}\right)(5068) \\ \left(-\tfrac{9}{8}\right)(561.2) + \left(\tfrac{1}{8}\right)(5068) \end{bmatrix} = \begin{bmatrix} 167.7 \\ 2.15 \end{bmatrix}$$

$b = 167.7, a = 2.15$

The least squares regression line is $y = 2.15t + 167.7$.

(c) For 2003, $t = 13; y = 2.15(13) + 167.7 = 195.65.$

This projects about 196 million licensed drivers in 2003.

(d) The projected value is very close to the actual value.

73. True. If B is the inverse of A, then $AB = I = BA$.

74. True. If A and B are both square matrices and $AB = I_n$, it can be shown that $BA = I_n$.

75. $AA^{-1} = \begin{bmatrix} a & b \\ c & d \end{bmatrix} \left(\dfrac{1}{ad-bc} \right) \begin{bmatrix} d & -b \\ -c & a \end{bmatrix} = \dfrac{1}{ad-bc} \begin{bmatrix} a & b \\ c & d \end{bmatrix} \begin{bmatrix} d & -b \\ -c & a \end{bmatrix}$

$= \dfrac{1}{ad-bc} \begin{bmatrix} ad-bc & 0 \\ 0 & ad-bc \end{bmatrix} = \begin{bmatrix} 1 & 0 \\ 0 & 1 \end{bmatrix}$

$A^{-1}A = \dfrac{1}{ad-bc} \begin{bmatrix} d & -b \\ -c & a \end{bmatrix} \begin{bmatrix} a & b \\ c & d \end{bmatrix} = \dfrac{1}{ad-bc} \begin{bmatrix} ad-bc & 0 \\ 0 & ad-bc \end{bmatrix} = \begin{bmatrix} 1 & 0 \\ 0 & 1 \end{bmatrix}$

76. (a) Given $A = \begin{bmatrix} a_{11} & 0 \\ 0 & a_{22} \end{bmatrix}$, $A^{-1} = \begin{bmatrix} \dfrac{1}{a_{11}} & 0 \\ 0 & \dfrac{1}{a_{22}} \end{bmatrix}$.

Given $A = \begin{bmatrix} a_{11} & 0 & 0 \\ 0 & a_{22} & 0 \\ 0 & 0 & a_{33} \end{bmatrix}$, $A^{-1} = \begin{bmatrix} \dfrac{1}{a_{11}} & 0 & 0 \\ 0 & \dfrac{1}{a_{22}} & 0 \\ 0 & 0 & \dfrac{1}{a_{33}} \end{bmatrix}$.

(b) In general, the inverse of a matrix in the form of A is

$$\begin{bmatrix} \dfrac{1}{a_{11}} & 0 & 0 & \cdots & 0 \\ 0 & \dfrac{1}{a_{22}} & 0 & \cdots & 0 \\ 0 & 0 & \dfrac{1}{a_{33}} & \cdots & 0 \\ \vdots & \vdots & \vdots & \cdots & \vdots \\ 0 & 0 & 0 & \cdots & \dfrac{1}{a_{nn}} \end{bmatrix}$$

77. $|x + 7| \geq 2$

$x + 7 \leq -2$ or $x + 7 \geq 2$

$x \leq -9$ or $x \geq -5$

78. $|2x - 1| < 3$

$-3 < 2x - 1 < 3$

$-2 < 2x < 4$

$-1 < x < 2$

79. $3^{x/2} = 315$

$\ln 3^{x/2} = \ln 315$

$\dfrac{x}{2} \ln 3 = \ln 315$

$x = \dfrac{2 \ln 315}{\ln 3} \approx 10.472$

80. $2000e^{-x/5} = 400$

$e^{-x/5} = \dfrac{1}{5}$

$\ln e^{-x/5} = \ln \dfrac{1}{5}$

$-\dfrac{x}{5} = \ln \dfrac{1}{5}$

$x = -5 \ln \dfrac{1}{5} \approx -8.047$

81. $\log_2 x - 2 = 4.5$

$\log_2 x = 6.5$

$x = 2^{6.5} \approx 90.510$

82. $\ln x + \ln(x - 1) = 0$

$\ln[x(x - 1)] = 0$

$e^{\ln[x(x-1)]} = e^0$

$x(x - 1) = 1$

$x^2 - x - 1 = 0$

$x = \dfrac{1 \pm \sqrt{1 - 4(-1)}}{2}$

$x = \dfrac{1 \pm \sqrt{5}}{2}$

Choose the positive value only:

$x = \dfrac{1 + \sqrt{5}}{2} \approx 1.618$

83. Answers will vary.

Section 8.4   The Determinant of a Square Matrix

■ You should be able to determine the determinant of a matrix of order 2×2 by using the difference of the products of the diagonals.

■ You should be able to use expansion by cofactors to find the determinant of a matrix of order 3×3 or greater.

■ The determinant of a triangular matrix equals the product of the entries on the main diagonal.

Vocabulary Check

1. determinant   **2.** minor   **3.** cofactor   **4.** expanding by cofactors

---

1. 5

2. -8

3. $\begin{vmatrix} 2 & 1 \\ 3 & 4 \end{vmatrix} = 2(4) - 1(3) = 8 - 3 = 5$

4. $\begin{vmatrix} -3 & 1 \\ 5 & 2 \end{vmatrix} = (-3)(2) - (5)(1) = -11$

5. $\begin{vmatrix} 5 & 2 \\ -6 & 3 \end{vmatrix} = 5(3) - 2(-6) = 15 + 12 = 27$

6. $\begin{vmatrix} 2 & -2 \\ 4 & 3 \end{vmatrix} = (2)(3) - (4)(-2) = 14$

7. $\begin{vmatrix} -7 & 0 \\ 3 & 0 \end{vmatrix} = -7(0) - 0(3) = 0$

8. $\begin{vmatrix} 4 & -3 \\ 0 & 0 \end{vmatrix} = (4)(0) - (0)(-3) = 0$

9. $\begin{vmatrix} 2 & 6 \\ 0 & 3 \end{vmatrix} = 2(3) - 6(0) = 6$

10. $\begin{vmatrix} 2 & -3 \\ -6 & 9 \end{vmatrix} = (2)(9) - (-6)(-3) = 0$

11. $\begin{vmatrix} -3 & -2 \\ -6 & -1 \end{vmatrix} = (-3)(-1) - (-2)(-6) = 3 - 12 = -9$

12. $\begin{vmatrix} 4 & 7 \\ -2 & 5 \end{vmatrix} = (4)(5) - (-2)(7) = 34$

13. $\begin{vmatrix} 9 & 0 \\ 7 & 8 \end{vmatrix} = 9(8) - 0(7) = 72 - 0 = 72$

14. $\begin{vmatrix} 0 & 6 \\ -3 & 2 \end{vmatrix} = (0)(2) - (-3)(6) = 18$

15. $\begin{vmatrix} -\frac{1}{2} & \frac{1}{3} \\ -6 & \frac{1}{3} \end{vmatrix} = -\frac{1}{2}\left(\frac{1}{3}\right) - \frac{1}{3}(-6) = -\frac{1}{6} + 2 = \frac{11}{6}$

16. $\begin{vmatrix} \frac{2}{3} & \frac{4}{3} \\ -1 & -\frac{1}{3} \end{vmatrix} = \left(\frac{2}{3}\right)\left(-\frac{1}{3}\right) - (-1)\left(\frac{4}{3}\right) = \frac{10}{9}$

17. $\begin{vmatrix} 0.3 & 0.2 & 0.2 \\ 0.2 & 0.2 & 0.2 \\ -0.4 & 0.4 & 0.3 \end{vmatrix} = -0.002$

18. $\begin{vmatrix} 0.1 & 0.2 & 0.3 \\ -0.3 & 0.2 & 0.2 \\ 0.5 & 0.4 & 0.4 \end{vmatrix} = -0.022$

19. $\begin{vmatrix} 0.9 & 0.7 & 0 \\ -0.1 & 0.3 & 1.3 \\ -2.2 & 4.2 & 6.1 \end{vmatrix} = -4.842$

20. $\begin{vmatrix} 0.1 & 0.1 & -4.3 \\ 7.5 & 6.2 & 0.7 \\ 0.3 & 0.6 & -1.2 \end{vmatrix} = -11.217$

21. $\begin{vmatrix} 1 & 4 & -2 \\ 3 & 6 & -6 \\ -2 & 1 & 4 \end{vmatrix} = 0$

22. $\begin{vmatrix} 2 & 3 & 1 \\ 0 & 5 & -2 \\ 0 & 0 & -2 \end{vmatrix} = -20$

23. $\begin{bmatrix} 3 & 4 \\ 2 & -5 \end{bmatrix}$

(a) $M_{11} = -5$ (b) $C_{11} = M_{11} = -5$

     $M_{12} = 2$ $C_{12} = -M_{12} = -2$

     $M_{21} = 4$ $C_{21} = -M_{21} = -4$

     $M_{22} = 3$ $C_{22} = M_{22} = 3$

24. $\begin{bmatrix} 11 & 0 \\ -3 & 2 \end{bmatrix}$

(a) $M_{11} = 2$ (b) $C_{11} = M_{11} = 2$

     $M_{12} = -3$ $C_{12} = -M_{12} = 3$

     $M_{21} = 0$ $C_{21} = M_{21} = 0$

     $M_{22} = 11$ $C_{22} = M_{22} = 11$

25. $\begin{bmatrix} 3 & 1 \\ -2 & -4 \end{bmatrix}$

(a) $M_{11} = -4$ (b) $C_{11} = M_{11} = -4$

     $M_{12} = -2$ $C_{12} = -M_{12} = 2$

     $M_{21} = 1$ $C_{21} = -M_{21} = -1$

     $M_{22} = 3$ $C_{22} = M_{22} = 3$

26. $\begin{bmatrix} -6 & 5 \\ 7 & -2 \end{bmatrix}$

(a) $M_{11} = -2$ (b) $C_{11} = M_{11} = -2$

     $M_{12} = 7$ $C_{12} = -M_{12} = -7$

     $M_{21} = 5$ $C_{21} = -M_{21} = -5$

     $M_{22} = -6$ $C_{22} = M_{22} = -6$

27. $\begin{bmatrix} 4 & 0 & 2 \\ -3 & 2 & 1 \\ 1 & -1 & 1 \end{bmatrix}$

(a) $M_{11} = \begin{vmatrix} 2 & 1 \\ -1 & 1 \end{vmatrix} = 2 - (-1) = 3$

   $M_{12} = \begin{vmatrix} -3 & 1 \\ 1 & 1 \end{vmatrix} = -3 - 1 = -4$

   $M_{13} = \begin{vmatrix} -3 & 2 \\ 1 & -1 \end{vmatrix} = 3 - 2 = 1$

   $M_{21} = \begin{vmatrix} 0 & 2 \\ -1 & 1 \end{vmatrix} = 0 - (-2) = 2$

   $M_{22} = \begin{vmatrix} 4 & 2 \\ 1 & 1 \end{vmatrix} = 4 - 2 = 2$

   $M_{23} = \begin{vmatrix} 4 & 0 \\ 1 & -1 \end{vmatrix} = -4 - 0 = -4$

   $M_{31} = \begin{vmatrix} 0 & 2 \\ 2 & 1 \end{vmatrix} = 0 - 4 = -4$

   $M_{32} = \begin{vmatrix} 4 & 2 \\ -3 & 1 \end{vmatrix} = 4 - (-6) = 10$

   $M_{33} = \begin{vmatrix} 4 & 0 \\ -3 & 2 \end{vmatrix} = 8 - 0 = 8$

(b) $C_{11} = (-1)^2 M_{11} = 3$

   $C_{12} = (-1)^3 M_{12} = 4$

   $C_{13} = (-1)^4 M_{13} = 1$

   $C_{21} = (-1)^3 M_{21} = -2$

   $C_{22} = (-1)^4 M_{22} = 2$

   $C_{23} = (-1)^5 M_{23} = 4$

   $C_{31} = (-1)^4 M_{31} = -4$

   $C_{32} = (-1)^5 M_{32} = -10$

   $C_{33} = (-1)^6 M_{33} = 8$

28. $\begin{bmatrix} 1 & -1 & 0 \\ 3 & 2 & 5 \\ 4 & -6 & 4 \end{bmatrix}$

(a) $M_{11} = \begin{vmatrix} 2 & 5 \\ -6 & 4 \end{vmatrix} = 8 - (-30) = 38$

   $M_{12} = \begin{vmatrix} 3 & 5 \\ 4 & 4 \end{vmatrix} = 12 - 20 = -8$

   $M_{13} = \begin{vmatrix} 3 & 2 \\ 4 & -6 \end{vmatrix} = -18 - 8 = -26$

   $M_{21} = \begin{vmatrix} -1 & 0 \\ -6 & 4 \end{vmatrix} = -4 - 0 = -4$

   $M_{22} = \begin{vmatrix} 1 & 0 \\ 4 & 4 \end{vmatrix} = 4 - 0 = 4$

   $M_{23} = \begin{vmatrix} 1 & -1 \\ 4 & -6 \end{vmatrix} = -6 - (-4) = -2$

   $M_{31} = \begin{vmatrix} -1 & 0 \\ 2 & 5 \end{vmatrix} = -5 - 0 = -5$

   $M_{32} = \begin{vmatrix} 1 & 0 \\ 3 & 5 \end{vmatrix} = 5 - 0 = 5$

   $M_{33} = \begin{vmatrix} 1 & -1 \\ 3 & 2 \end{vmatrix} = 2 - (-3) = 5$

(b) $C_{11} = (-1)^2 M_{11} = 38$

   $C_{12} = (-1)^3 M_{12} = 8$

   $C_{13} = (-1)^4 M_{13} = -26$

   $C_{21} = (-1)^3 M_{21} = 4$

   $C_{22} = (-1)^4 M_{22} = 4$

   $C_{23} = (-1)^5 M_{23} = 2$

   $C_{31} = (-1)^4 M_{31} = -5$

   $C_{32} = (-1)^5 M_{32} = -5$

   $C_{33} = (-1)^6 M_{33} = 5$

29. $\begin{bmatrix} 3 & -2 & 8 \\ 3 & 2 & -6 \\ -1 & 3 & 6 \end{bmatrix}$

(a) $M_{11} = \begin{vmatrix} 2 & -6 \\ 3 & 6 \end{vmatrix} = 12 + 18 = 30$

$M_{12} = \begin{vmatrix} 3 & -6 \\ -1 & 6 \end{vmatrix} = 18 - 6 = 12$

$M_{13} = \begin{vmatrix} 3 & 2 \\ -1 & 3 \end{vmatrix} = 9 + 2 = 11$

$M_{21} = \begin{vmatrix} -2 & 8 \\ 3 & 6 \end{vmatrix} = -12 - 24 = -36$

$M_{22} = \begin{vmatrix} 3 & 8 \\ -1 & 6 \end{vmatrix} = 18 + 8 = 26$

$M_{23} = \begin{vmatrix} 3 & -2 \\ -1 & 3 \end{vmatrix} = 9 - 2 = 7$

$M_{31} = \begin{vmatrix} -2 & 8 \\ 2 & -6 \end{vmatrix} = 12 - 16 = -4$

$M_{32} = \begin{vmatrix} 3 & 8 \\ 3 & -6 \end{vmatrix} = -18 - 24 = -42$

$M_{33} = \begin{vmatrix} 3 & -2 \\ 3 & 2 \end{vmatrix} = 6 + 6 = 12$

(b) $C_{11} = (-1)^2 M_{11} = 30$

$C_{12} = (-1)^3 M_{12} = -12$

$C_{13} = (-1)^4 M_{13} = 11$

$C_{21} = (-1)^3 M_{21} = 36$

$C_{22} = (-1)^4 M_{22} = 26$

$C_{23} = (-1)^5 M_{23} = -7$

$C_{31} = (-1)^4 M_{31} = -4$

$C_{32} = (-1)^5 M_{32} = 42$

$C_{33} = (-1)^6 M_{33} = 12$

30. $\begin{bmatrix} -2 & 9 & 4 \\ 7 & -6 & 0 \\ 6 & 7 & -6 \end{bmatrix}$

(a) $M_{11} = \begin{vmatrix} -6 & 0 \\ 7 & -6 \end{vmatrix} = 36$

$M_{12} = \begin{vmatrix} 7 & 0 \\ 6 & -6 \end{vmatrix} = -42$

$M_{13} = \begin{vmatrix} 7 & -6 \\ 6 & 7 \end{vmatrix} = 85$

$M_{21} = \begin{vmatrix} 9 & 4 \\ 7 & -6 \end{vmatrix} = -82$

$M_{22} = \begin{vmatrix} -2 & 4 \\ 6 & -6 \end{vmatrix} = -12$

$M_{23} = \begin{vmatrix} -2 & 9 \\ 6 & 7 \end{vmatrix} = -68$

$M_{31} = \begin{vmatrix} 9 & 4 \\ -6 & 0 \end{vmatrix} = 24$

$M_{32} = \begin{vmatrix} -2 & 4 \\ 7 & 0 \end{vmatrix} = -28$

$M_{33} = \begin{vmatrix} -2 & 9 \\ 7 & -6 \end{vmatrix} = -51$

(b) $C_{11} = (-1)^2 M_{11} = 36$

$C_{12} = (-1)^3 M_{12} = 42$

$C_{13} = (-1)^4 M_{13} = 85$

$C_{21} = (-1)^3 M_{21} = 82$

$C_{22} = (-1)^4 M_{22} = -12$

$C_{23} = (-1)^5 M_{23} = 68$

$C_{31} = (-1)^4 M_{31} = 24$

$C_{32} = (-1)^5 M_{32} = 28$

$C_{33} = (-1)^6 M_{33} = -51$

31. (a) $\begin{vmatrix} -3 & 2 & 1 \\ 4 & 5 & 6 \\ 2 & -3 & 1 \end{vmatrix} = -3 \begin{vmatrix} 5 & 6 \\ -3 & 1 \end{vmatrix} - 2 \begin{vmatrix} 4 & 6 \\ 2 & 1 \end{vmatrix} + \begin{vmatrix} 4 & 5 \\ 2 & -3 \end{vmatrix} = -3(23) - 2(-8) - 22 = -75$

(b) $\begin{vmatrix} -3 & 2 & 1 \\ 4 & 5 & 6 \\ 2 & -3 & 1 \end{vmatrix} = -2 \begin{vmatrix} 4 & 6 \\ 2 & 1 \end{vmatrix} + 5 \begin{vmatrix} -3 & 1 \\ 2 & 1 \end{vmatrix} + 3 \begin{vmatrix} -3 & 1 \\ 4 & 6 \end{vmatrix} = -2(-8) + 5(-5) + 3(-22) = -75$

32. (a) $\begin{vmatrix} -3 & 4 & 2 \\ 6 & 3 & 1 \\ 4 & -7 & -8 \end{vmatrix} = -6 \begin{vmatrix} 4 & 2 \\ -7 & -8 \end{vmatrix} + 3 \begin{vmatrix} -3 & 2 \\ 4 & -8 \end{vmatrix} - 1 \begin{vmatrix} -3 & 4 \\ 4 & -7 \end{vmatrix} = -6(-18) + 3(16) - (5) = 151$

(b) $\begin{vmatrix} -3 & 4 & 2 \\ 6 & 3 & 1 \\ 4 & -7 & -8 \end{vmatrix} = 2 \begin{vmatrix} 6 & 3 \\ 4 & -7 \end{vmatrix} - \begin{vmatrix} -3 & 4 \\ 4 & -7 \end{vmatrix} - 8 \begin{vmatrix} -3 & 4 \\ 6 & 3 \end{vmatrix} = 2(-54) - (5) - 8(-33) = 151$

33. (a) $\begin{vmatrix} 5 & 0 & -3 \\ 0 & 12 & 4 \\ 1 & 6 & 3 \end{vmatrix} = 0\begin{vmatrix} 0 & -3 \\ 6 & 3 \end{vmatrix} + 12\begin{vmatrix} 5 & -3 \\ 1 & 3 \end{vmatrix} - 4\begin{vmatrix} 5 & 0 \\ 1 & 6 \end{vmatrix} = 0(18) + 12(18) - 4(30) = 96$

(b) $\begin{vmatrix} 5 & 0 & -3 \\ 0 & 12 & 4 \\ 1 & 6 & 3 \end{vmatrix} = 0\begin{vmatrix} 0 & 4 \\ 1 & 3 \end{vmatrix} + 12\begin{vmatrix} 5 & -3 \\ 1 & 3 \end{vmatrix} - 6\begin{vmatrix} 5 & -3 \\ 0 & 4 \end{vmatrix} = 0(-4) + 12(18) - 6(20) = 96$

34. (a) $\begin{vmatrix} 10 & -5 & 5 \\ 30 & 0 & 10 \\ 0 & 10 & 1 \end{vmatrix} = 0\begin{vmatrix} -5 & 5 \\ 0 & 10 \end{vmatrix} - 10\begin{vmatrix} 10 & 5 \\ 30 & 10 \end{vmatrix} + \begin{vmatrix} 10 & -5 \\ 30 & 0 \end{vmatrix} = 0(-50) - 10(-50) + 150 = 650$

(b) $\begin{vmatrix} 10 & -5 & 5 \\ 30 & 0 & 10 \\ 0 & 10 & 1 \end{vmatrix} = 10\begin{vmatrix} 0 & 10 \\ 10 & 1 \end{vmatrix} - 30\begin{vmatrix} -5 & 5 \\ 10 & 1 \end{vmatrix} + 0\begin{vmatrix} -5 & 5 \\ 0 & 10 \end{vmatrix} = 10(-100) - 30(-55) + 0(-50) = 650$

35. (a) $\begin{vmatrix} 6 & 0 & -3 & 5 \\ 4 & 13 & 6 & -8 \\ -1 & 0 & 7 & 4 \\ 8 & 6 & 0 & 2 \end{vmatrix} = -4\begin{vmatrix} 0 & -3 & 5 \\ 0 & 7 & 4 \\ 6 & 0 & 2 \end{vmatrix} + 13\begin{vmatrix} 6 & -3 & 5 \\ -1 & 7 & 4 \\ 8 & 0 & 2 \end{vmatrix} - 6\begin{vmatrix} 6 & 0 & 5 \\ -1 & 0 & 4 \\ 8 & 6 & 2 \end{vmatrix} - 8\begin{vmatrix} 6 & 0 & -3 \\ -1 & 0 & 7 \\ 8 & 6 & 0 \end{vmatrix}$

$$= -4(-282) + 13(-298) - 6(-174) - 8(-234) = 170$$

(b) $\begin{vmatrix} 6 & 0 & -3 & 5 \\ 4 & 13 & 6 & -8 \\ -1 & 0 & 7 & 4 \\ 8 & 6 & 0 & 2 \end{vmatrix} = 0\begin{vmatrix} 4 & 6 & -8 \\ -1 & 7 & 4 \\ 8 & 0 & 2 \end{vmatrix} + 13\begin{vmatrix} 6 & -3 & 5 \\ -1 & 7 & 4 \\ 8 & 0 & 2 \end{vmatrix} + 0\begin{vmatrix} 6 & -3 & 5 \\ 4 & 6 & -8 \\ 8 & 0 & 2 \end{vmatrix} + 6\begin{vmatrix} 6 & -3 & 5 \\ 4 & 6 & -8 \\ -1 & 7 & 4 \end{vmatrix}$

$$= 0 + 13(-298) + 0 + 6(674) = 170$$

36. (a) $\begin{vmatrix} 10 & 8 & 3 & -7 \\ 4 & 0 & 5 & -6 \\ 0 & 3 & 2 & 7 \\ 1 & 0 & -3 & 2 \end{vmatrix} = 0\begin{vmatrix} 8 & 3 & -7 \\ 0 & 5 & -6 \\ 0 & -3 & 2 \end{vmatrix} - 3\begin{vmatrix} 10 & 3 & -7 \\ 4 & 5 & -6 \\ 1 & -3 & 2 \end{vmatrix} + 2\begin{vmatrix} 10 & 8 & -7 \\ 4 & 0 & -6 \\ 1 & 0 & 2 \end{vmatrix} - 7\begin{vmatrix} 10 & 8 & 3 \\ 4 & 0 & 5 \\ 1 & 0 & -3 \end{vmatrix}$

$$= 0(-64) - 3(-3) + 2(-112) - 7(136) = -1167$$

(b) $\begin{vmatrix} 10 & 8 & 3 & -7 \\ 4 & 0 & 5 & -6 \\ 0 & 3 & 2 & 7 \\ 1 & 0 & -3 & 2 \end{vmatrix} = 10\begin{vmatrix} 0 & 5 & -6 \\ 3 & 2 & 7 \\ 0 & -3 & 2 \end{vmatrix} - 4\begin{vmatrix} 8 & 3 & -7 \\ 3 & 2 & 7 \\ 0 & -3 & 2 \end{vmatrix} + 0\begin{vmatrix} 8 & 3 & -7 \\ 0 & 5 & -6 \\ 0 & -3 & 2 \end{vmatrix} - 1\begin{vmatrix} 8 & 3 & -7 \\ 0 & 5 & -6 \\ 3 & 2 & 7 \end{vmatrix}$

$$= 10(24) - 4(245) + 0(-64) - 1(427) = -1167$$

37. Expand along Column 1.

$\begin{vmatrix} 2 & -1 & 0 \\ 4 & 2 & 1 \\ 4 & 2 & 1 \end{vmatrix} = 2\begin{vmatrix} 2 & 1 \\ 2 & 1 \end{vmatrix} - 4\begin{vmatrix} -1 & 0 \\ 2 & 1 \end{vmatrix} + 4\begin{vmatrix} -1 & 0 \\ 2 & 1 \end{vmatrix} = 2(0) - 4(-1) + 4(-1) = 0$

38. Expand along Row 3.

$\begin{vmatrix} -2 & 2 & 3 \\ 1 & -1 & 0 \\ 0 & 1 & 4 \end{vmatrix} = 0\begin{vmatrix} 2 & 3 \\ -1 & 0 \end{vmatrix} - 1\begin{vmatrix} -2 & 3 \\ 1 & 0 \end{vmatrix} + 4\begin{vmatrix} -2 & 2 \\ 1 & -1 \end{vmatrix}$

$$= 0(3) - 1(-3) + 4(0) = 3$$

39. Expand along Row 2.

$$\begin{vmatrix} 6 & 3 & -7 \\ 0 & 0 & 0 \\ 4 & -6 & 3 \end{vmatrix} = 0\begin{vmatrix} 3 & -7 \\ -6 & 3 \end{vmatrix} - 0\begin{vmatrix} 6 & -7 \\ 4 & 3 \end{vmatrix} + 0\begin{vmatrix} 6 & 3 \\ 4 & -6 \end{vmatrix} = 0$$

40. Expand along Column 3.

$$\begin{vmatrix} 1 & 1 & 2 \\ 3 & 1 & 0 \\ -2 & 0 & 3 \end{vmatrix} = 2\begin{vmatrix} 3 & 1 \\ -2 & 0 \end{vmatrix} - 0\begin{vmatrix} 1 & 1 \\ -2 & 0 \end{vmatrix} + 3\begin{vmatrix} 1 & 1 \\ 3 & 1 \end{vmatrix}$$

$$= 2(2) - 0(2) + 3(-2) = -2$$

41. $\begin{vmatrix} -1 & 2 & -5 \\ 0 & 3 & 4 \\ 0 & 0 & 3 \end{vmatrix} = (-1)(3)(3) = -9$ (Upper triangular)

42. Expand along Row 1.

$$\begin{vmatrix} 1 & 0 & 0 \\ -4 & -1 & 0 \\ 5 & 1 & 5 \end{vmatrix} = 1\begin{vmatrix} -1 & 0 \\ 1 & 5 \end{vmatrix} - 0\begin{vmatrix} -4 & 0 \\ 5 & 5 \end{vmatrix} + 0\begin{vmatrix} -4 & -1 \\ 5 & 1 \end{vmatrix}$$

$$= 1(-5) - 0(-20) + 0(1) = -5$$

43. Expand along Column 3.

$$\begin{vmatrix} 1 & 4 & -2 \\ 3 & 2 & 0 \\ -1 & 4 & 3 \end{vmatrix} = -2\begin{vmatrix} 3 & 2 \\ -1 & 4 \end{vmatrix} + 3\begin{vmatrix} 1 & 4 \\ 3 & 2 \end{vmatrix}$$

$$= -2(14) + 3(-10) = -58$$

44. Expand along Row 3.

$$\begin{vmatrix} 2 & -1 & 3 \\ 1 & 4 & 4 \\ 1 & 0 & 2 \end{vmatrix} = 1\begin{vmatrix} -1 & 3 \\ 4 & 4 \end{vmatrix} - 0\begin{vmatrix} 2 & 3 \\ 1 & 4 \end{vmatrix} + 2\begin{vmatrix} 2 & -1 \\ 1 & 4 \end{vmatrix} = 1(-16) - 0(5) + 2(9) = 2$$

45. $\begin{vmatrix} 2 & 4 & 6 \\ 0 & 3 & 1 \\ 0 & 0 & -5 \end{vmatrix} = (2)(3)(-5) = -30$ (Upper triangular)

46. Expand along Row 1.

$$\begin{vmatrix} -3 & 0 & 0 \\ 7 & 11 & 0 \\ 1 & 2 & 2 \end{vmatrix} = -3\begin{vmatrix} 11 & 0 \\ 2 & 2 \end{vmatrix} - 0\begin{vmatrix} 7 & 0 \\ 1 & 2 \end{vmatrix} + 0\begin{vmatrix} 7 & 11 \\ 1 & 2 \end{vmatrix}$$

$$= -3(22) - 0(14) + 0(3) = -66$$

47. Expand along Column 3.

$$\begin{vmatrix} 2 & 6 & 6 & 2 \\ 2 & 7 & 3 & 6 \\ 1 & 5 & 0 & 1 \\ 3 & 7 & 0 & 7 \end{vmatrix} = 6\begin{vmatrix} 2 & 7 & 6 \\ 1 & 5 & 1 \\ 3 & 7 & 7 \end{vmatrix} - 3\begin{vmatrix} 2 & 6 & 2 \\ 1 & 5 & 1 \\ 3 & 7 & 7 \end{vmatrix} = 6(-20) - 3(16) = -168$$

48. Expand along Row 2.

$$\begin{vmatrix} 3 & 6 & -5 & 4 \\ -2 & 0 & 6 & 0 \\ 1 & 1 & 2 & 2 \\ 0 & 3 & -1 & -1 \end{vmatrix} = -(-2)\begin{vmatrix} 6 & -5 & 4 \\ 1 & 2 & 2 \\ 3 & -1 & -1 \end{vmatrix} - 6\begin{vmatrix} 3 & 6 & 4 \\ 1 & 1 & 2 \\ 0 & 3 & -1 \end{vmatrix} = 2(-63) - 6(-3) = -108$$

49. Expand along Column 1.

$$\begin{vmatrix} 5 & 3 & 0 & 6 \\ 4 & 6 & 4 & 12 \\ 0 & 2 & -3 & 4 \\ 0 & 1 & -2 & 2 \end{vmatrix} = 5\begin{vmatrix} 6 & 4 & 12 \\ 2 & -3 & 4 \\ 1 & -2 & 2 \end{vmatrix} - 4\begin{vmatrix} 3 & 0 & 6 \\ 2 & -3 & 4 \\ 1 & -2 & 2 \end{vmatrix} = 5(0) - 4(0) = 0$$

50. Expand along Row 3.

$$\begin{vmatrix} 1 & 4 & 3 & 2 \\ -5 & 6 & 2 & 1 \\ 0 & 0 & 0 & 0 \\ 3 & -2 & 1 & 5 \end{vmatrix} = 0$$

51. Expand along Column 2, then along Column 4.

$$\begin{vmatrix} 3 & 2 & 4 & -1 & 5 \\ -2 & 0 & 1 & 3 & 2 \\ 1 & 0 & 0 & 4 & 0 \\ 6 & 0 & 2 & -1 & 0 \\ 3 & 0 & 5 & 1 & 0 \end{vmatrix} = -2\begin{vmatrix} -2 & 1 & 3 & 2 \\ 1 & 0 & 4 & 0 \\ 6 & 2 & -1 & 0 \\ 3 & 5 & 1 & 0 \end{vmatrix} = (-2)(-2)\begin{vmatrix} 1 & 0 & 4 \\ 6 & 2 & -1 \\ 3 & 5 & 1 \end{vmatrix} = 4(103) = 412$$

52. Expand along Column 1.

$$\begin{vmatrix} 5 & 2 & 0 & 0 & -2 \\ 0 & 1 & 4 & 3 & 2 \\ 0 & 0 & 2 & 6 & 3 \\ 0 & 0 & 3 & 4 & 1 \\ 0 & 0 & 0 & 0 & 2 \end{vmatrix} = 5\begin{vmatrix} 1 & 4 & 3 & 2 \\ 0 & 2 & 6 & 3 \\ 0 & 3 & 4 & 1 \\ 0 & 0 & 0 & 2 \end{vmatrix} = 5 \cdot 1\begin{vmatrix} 2 & 6 & 3 \\ 3 & 4 & 1 \\ 0 & 0 & 2 \end{vmatrix} = 5(-20) = -100$$

53. $\begin{vmatrix} 3 & 8 & -7 \\ 0 & -5 & 4 \\ 8 & 1 & 6 \end{vmatrix} = -126$

54. $\begin{vmatrix} 5 & -8 & 0 \\ 9 & 7 & 4 \\ -8 & 7 & 1 \end{vmatrix} = 223$

55. $\begin{vmatrix} 7 & 0 & -14 \\ -2 & 5 & 4 \\ -6 & 2 & 12 \end{vmatrix} = 0$

56. $\begin{vmatrix} 3 & 0 & 0 \\ -2 & 5 & 0 \\ 12 & 5 & 7 \end{vmatrix} = 105$

57. $\begin{vmatrix} 1 & -1 & 8 & 4 \\ 2 & 6 & 0 & -4 \\ 2 & 0 & 2 & 6 \\ 0 & 2 & 8 & 0 \end{vmatrix} = -336$

58. $\begin{vmatrix} 0 & -3 & 8 & 2 \\ 8 & 1 & -1 & 6 \\ -4 & 6 & 0 & 9 \\ -7 & 0 & 0 & 14 \end{vmatrix} = 7441$

59. $\begin{vmatrix} 3 & -2 & 4 & 3 & 1 \\ -1 & 0 & 2 & 1 & 0 \\ 5 & -1 & 0 & 3 & 2 \\ 4 & 7 & -8 & 0 & 0 \\ 1 & 2 & 3 & 0 & 2 \end{vmatrix} = 410$

60. $\begin{vmatrix} -2 & 0 & 0 & 0 & 0 \\ 0 & 3 & 0 & 0 & 0 \\ 0 & 0 & -1 & 0 & 0 \\ 0 & 0 & 0 & 2 & 0 \\ 0 & 0 & 0 & 0 & -4 \end{vmatrix} = -48$

61. (a) $\begin{vmatrix} -1 & 0 \\ 0 & 3 \end{vmatrix} = -3$

(b) $\begin{vmatrix} 2 & 0 \\ 0 & -1 \end{vmatrix} = -2$

(c) $\begin{bmatrix} -1 & 0 \\ 0 & 3 \end{bmatrix}\begin{bmatrix} 2 & 0 \\ 0 & -1 \end{bmatrix} = \begin{bmatrix} -2 & 0 \\ 0 & -3 \end{bmatrix}$

(d) $\begin{vmatrix} -2 & 0 \\ 0 & -3 \end{vmatrix} = 6$

62. (a) $|A| = \begin{vmatrix} -2 & 1 \\ 4 & -2 \end{vmatrix} = 0$

(b) $|B| = \begin{vmatrix} 1 & 2 \\ 0 & -1 \end{vmatrix} = -1$

(c) $AB = \begin{bmatrix} -2 & 1 \\ 4 & -2 \end{bmatrix}\begin{bmatrix} 1 & 2 \\ 0 & -1 \end{bmatrix} = \begin{bmatrix} -2 & -5 \\ 4 & 10 \end{bmatrix}$

(d) $|AB| = \begin{vmatrix} -2 & -5 \\ 4 & 10 \end{vmatrix} = 0$

63. (a) $\begin{vmatrix} 4 & 0 \\ 3 & -2 \end{vmatrix} = -8$

(b) $\begin{vmatrix} -1 & 1 \\ -2 & 2 \end{vmatrix} = 0$

(c) $\begin{bmatrix} 4 & 0 \\ 3 & -2 \end{bmatrix}\begin{bmatrix} -1 & 1 \\ -2 & 2 \end{bmatrix} = \begin{bmatrix} -4 & 4 \\ 1 & -1 \end{bmatrix}$

(d) $\begin{vmatrix} -4 & 4 \\ 1 & -1 \end{vmatrix} = 0$

64. (a) $|A| = \begin{vmatrix} 5 & 4 \\ 3 & -1 \end{vmatrix} = -17$

(b) $|B| = \begin{vmatrix} 0 & 6 \\ 1 & -2 \end{vmatrix} = -6$

(c) $AB = \begin{bmatrix} 5 & 4 \\ 3 & -1 \end{bmatrix}\begin{bmatrix} 0 & 6 \\ 1 & -2 \end{bmatrix} = \begin{bmatrix} 4 & 22 \\ -1 & 20 \end{bmatrix}$

(d) $|AB| = \begin{vmatrix} 4 & 22 \\ -1 & 20 \end{vmatrix} = 102$

65. (a) $\begin{vmatrix} 0 & 1 & 2 \\ -3 & -2 & 1 \\ 0 & 4 & 1 \end{vmatrix} = -21$

(b) $\begin{vmatrix} 3 & -2 & 0 \\ 1 & -1 & 2 \\ 3 & 1 & 1 \end{vmatrix} = -19$

(c) $\begin{bmatrix} 0 & 1 & 2 \\ -3 & -2 & 1 \\ 0 & 4 & 1 \end{bmatrix}\begin{bmatrix} 3 & -2 & 0 \\ 1 & -1 & 2 \\ 3 & 1 & 1 \end{bmatrix} = \begin{bmatrix} 7 & 1 & 4 \\ -8 & 9 & -3 \\ 7 & -3 & 9 \end{bmatrix}$

(d) $\begin{vmatrix} 7 & 1 & 4 \\ -8 & 9 & -3 \\ 7 & -3 & 9 \end{vmatrix} = 399$

66. (a) $|A| = \begin{vmatrix} 3 & 2 & 0 \\ -1 & -3 & 4 \\ -2 & 0 & 1 \end{vmatrix} = -23$

(b) $|B| = \begin{vmatrix} -3 & 0 & 1 \\ 0 & 2 & -1 \\ -2 & -1 & 1 \end{vmatrix} = 1$

(c) $AB = \begin{bmatrix} 3 & 2 & 0 \\ -1 & -3 & 4 \\ -2 & 0 & 1 \end{bmatrix}\begin{bmatrix} -3 & 0 & 1 \\ 0 & 2 & -1 \\ -2 & -1 & 1 \end{bmatrix}$

$= \begin{bmatrix} -9 & 4 & 1 \\ -5 & -10 & 6 \\ 4 & -1 & -1 \end{bmatrix}$

(d) $|AB| = \begin{vmatrix} -9 & 4 & 1 \\ -5 & -10 & 6 \\ 4 & -1 & -1 \end{vmatrix} = -23$

67. (a) $\begin{vmatrix} -1 & 2 & 1 \\ 1 & 0 & 1 \\ 0 & 1 & 0 \end{vmatrix} = 2$

(b) $\begin{vmatrix} -1 & 0 & 0 \\ 0 & 2 & 0 \\ 0 & 0 & 3 \end{vmatrix} = -6$

(c) $\begin{bmatrix} -1 & 2 & 1 \\ 1 & 0 & 1 \\ 0 & 1 & 0 \end{bmatrix}\begin{bmatrix} -1 & 0 & 0 \\ 0 & 2 & 0 \\ 0 & 0 & 3 \end{bmatrix} = \begin{bmatrix} 1 & 4 & 3 \\ -1 & 0 & 3 \\ 0 & 2 & 0 \end{bmatrix}$

(d) $\begin{vmatrix} 1 & 4 & 3 \\ -1 & 0 & 3 \\ 0 & 2 & 0 \end{vmatrix} = -12$

68. (a) $|A| = \begin{vmatrix} 2 & 0 & 1 \\ 1 & -1 & 2 \\ 3 & 1 & 0 \end{vmatrix} = 0$

(b) $|B| = \begin{vmatrix} 2 & -1 & 4 \\ 0 & 1 & 3 \\ 3 & -2 & 1 \end{vmatrix} = -7$

(c) $AB = \begin{bmatrix} 2 & 0 & 1 \\ 1 & -1 & 2 \\ 3 & 1 & 0 \end{bmatrix}\begin{bmatrix} 2 & -1 & 4 \\ 0 & 1 & 3 \\ 3 & -2 & 1 \end{bmatrix} = \begin{bmatrix} 7 & -4 & 9 \\ 8 & -6 & 3 \\ 6 & -2 & 15 \end{bmatrix}$

(d) $|AB| = \begin{vmatrix} 7 & -4 & 9 \\ 8 & -6 & 3 \\ 6 & -2 & 15 \end{vmatrix} = 0$

69. $\begin{vmatrix} w & x \\ y & z \end{vmatrix} = wz - xy$

$-\begin{vmatrix} y & z \\ w & x \end{vmatrix} = -(xy - wz) = wz - xy$

Thus, $\begin{vmatrix} w & x \\ y & z \end{vmatrix} = -\begin{vmatrix} y & z \\ w & x \end{vmatrix}$.

70. $\begin{vmatrix} w & cx \\ y & cz \end{vmatrix} = cwz - cxy = c(wz - xy)$

$c\begin{vmatrix} w & x \\ y & z \end{vmatrix} = c(wz - xy)$

So, $\begin{vmatrix} w & cx \\ y & cz \end{vmatrix} = c\begin{vmatrix} w & x \\ y & z \end{vmatrix}$.

71. $\begin{vmatrix} w & x \\ y & z \end{vmatrix} = wz - xy$

$\begin{vmatrix} w & x + cw \\ y & z + cy \end{vmatrix} = w(z + cy) - y(x + cw) = wz - xy$

Thus, $\begin{vmatrix} w & x \\ y & z \end{vmatrix} = \begin{vmatrix} w & x + cw \\ y & z + cy \end{vmatrix}$.

72. $\begin{vmatrix} w & x \\ cw & cx \end{vmatrix} = cxw - cxw = 0$

So, $\begin{vmatrix} w & x \\ cw & cx \end{vmatrix} = 0$.

73. $\begin{vmatrix} 1 & x & x^2 \\ 1 & y & y^2 \\ 1 & z & z^2 \end{vmatrix} = \begin{vmatrix} y & y^2 \\ z & z^2 \end{vmatrix} - \begin{vmatrix} x & x^2 \\ z & z^2 \end{vmatrix} + \begin{vmatrix} x & x^2 \\ y & y^2 \end{vmatrix}$

$= (yz^2 - y^2z) - (xz^2 - x^2z) + (xy^2 - x^2y)$

$= yz^2 - xz^2 - y^2z + x^2z + xy(y - x)$

$= z^2(y - x) - z(y^2 - x^2) + xy(y - x)$

$= z^2(y - x) - z(y - x)(y + x) + xy(y - x)$

$= (y - x)[z^2 - z(y + x) + xy]$

$= (y - x)[z^2 - zy - zx + xy]$

$= (y - x)[z^2 - zx - zy + xy]$

$= (y - x)[z(z - x) - y(z - x)]$

$= (y - x)(z - x)(z - y)$

74. $\begin{vmatrix} a + b & a & a \\ a & a + b & a \\ a & a & a + b \end{vmatrix} = (a + b)\begin{vmatrix} a + b & a \\ a & a + b \end{vmatrix} - a\begin{vmatrix} a & a \\ a & a + b \end{vmatrix} + a\begin{vmatrix} a & a \\ a + b & a \end{vmatrix}$

$= (a + b)[(a + b)^2 - a^2] - a[a(a + b) - a^2] + a[a^2 - a(a + b)]$

$= (a + b)^3 - a^2(a + b) - a^2(a + b) + a^3 + a^3 - a^2(a + b)$

$= (a + b)^3 - 3a^2(a + b) + 2a^3$

$= a^3 + 3a^2b + 3ab^2 + b^3 - 3a^3 - 3a^2b + 2a^3$

$= 3ab^2 + b^3 = b^2(3a + b)$

75. $\begin{vmatrix} x - 1 & 2 \\ 3 & x - 2 \end{vmatrix} = 0$

$(x - 1)(x - 2) - 6 = 0$

$x^2 - 3x - 4 = 0$

$(x + 1)(x - 4) = 0$

$x = -1 \text{ or } x = 4$

76. $\begin{vmatrix} x - 2 & -1 \\ -3 & x \end{vmatrix} = 0$

$x(x - 2) - (-3)(-1) = 0$

$x^2 - 2x - 3 = 0$

$(x + 1)(x - 3) = 0$

$x = -1 \text{ or } x = 3$

77. $\begin{vmatrix} x + 3 & 2 \\ 1 & x + 2 \end{vmatrix} = 0$

$(x + 3)(x + 2) - 2 = 0$

$x^2 + 5x + 4 = 0$

$(x + 1)(x + 4) = 0$

$x = -1 \text{ or } x = -4$

78. $\begin{vmatrix} x + 4 & -2 \\ 7 & x - 5 \end{vmatrix} = 0$

$(x + 4)(x - 5) - 7(-2) = 0$

$x^2 - x - 6 = 0$

$(x + 2)(x - 3) = 0$

$x = -2 \text{ or } x = 3$

79. $\begin{vmatrix} 4u & -1 \\ -1 & 2v \end{vmatrix} = 8uv - 1$

80. $\begin{vmatrix} 3x^2 & -3y^2 \\ 1 & 1 \end{vmatrix} = 3x^2 - (-3y^2) = 3x^2 + 3y^2$

81. $\begin{vmatrix} e^{2x} & e^{3x} \\ 2e^{2x} & 3e^{3x} \end{vmatrix} = 3e^{5x} - 2e^{5x} = e^{5x}$

82. $\begin{vmatrix} e^{-x} & xe^{-x} \\ -e^{-x} & (1-x)e^{-x} \end{vmatrix} = (1-x)e^{-2x} - (-xe^{-2x}) = e^{-2x} - xe^{-2x} + xe^{-2x} = e^{-2x}$

83. $\begin{vmatrix} x & \ln x \\ 1 & \dfrac{1}{x} \end{vmatrix} = 1 - \ln x$

84. $\begin{vmatrix} x & x\ln x \\ 1 & 1 + \ln x \end{vmatrix} = x(1 + \ln x) - x\ln x$

$$= x + x\ln x - x\ln x = x$$

85. True. If an entire row is zero, then each cofactor in the expansion is multiplied by zero.

86. True. If a square matrix has two columns that are equal, then elementary column operations can be used to create a column with all zeros.

87. Let $A = \begin{bmatrix} 1 & 3 \\ -2 & 4 \end{bmatrix}$ and $B = \begin{bmatrix} -4 & 0 \\ 3 & 5 \end{bmatrix}$.

$|A| = \begin{vmatrix} 1 & 3 \\ -2 & 4 \end{vmatrix} = 10,\ |B| = \begin{vmatrix} -4 & 0 \\ 3 & 5 \end{vmatrix} = -20,\ |A| + |B| = -10$

$A + B = \begin{bmatrix} -3 & 3 \\ 1 & 9 \end{bmatrix},\ |A + B| = \begin{vmatrix} -3 & 3 \\ 1 & 9 \end{vmatrix} = -30$

Thus, $|A + B| \neq |A| + |B|$. Your answer may differ, depending on how you choose A and B.

88. (a) $\begin{vmatrix} 4 & 5 & 6 \\ 7 & 8 & 9 \\ 10 & 11 & 12 \end{vmatrix} = 0$

$\begin{vmatrix} 33 & 34 & 35 \\ 36 & 37 & 38 \\ 39 & 40 & 41 \end{vmatrix} = 0 \qquad \begin{vmatrix} -5 & -4 & -3 \\ -2 & -1 & 0 \\ 1 & 2 & 3 \end{vmatrix} = 0$

$\begin{vmatrix} 19 & 20 & 21 & 22 \\ 23 & 24 & 25 & 26 \\ 27 & 28 & 29 & 30 \\ 31 & 32 & 33 & 34 \end{vmatrix} = 0 \qquad \begin{vmatrix} 57 & 58 & 59 & 60 \\ 61 & 62 & 63 & 64 \\ 65 & 66 & 67 & 68 \\ 69 & 70 & 71 & 72 \end{vmatrix} = 0$

For an $n \times n$ matrix $(n > 2)$ with consecutive integer entries, the determinant appears to be 0.

(b) $\begin{vmatrix} x & x+1 & x+2 \\ x+3 & x+4 & x+5 \\ x+6 & x+7 & x+8 \end{vmatrix} = x\begin{vmatrix} x+4 & x+5 \\ x+7 & x+8 \end{vmatrix} - (x+1)\begin{vmatrix} x+3 & x+5 \\ x+6 & x+8 \end{vmatrix} + (x+2)\begin{vmatrix} x+3 & x+4 \\ x+6 & x+7 \end{vmatrix}$

$$= x[(x+4)(x+8) - (x+7)(x+5)] - (x+1)[(x+3)(x+8)$$
$$\qquad - (x+6)(x+5)] + (x+2)[(x+3)(x+7) - (x+6)(x+4)]$$

$$= x[(x^2 + 12x + 32) - (x^2 + 12x + 35)] - (x+1)[(x^2 + 11x + 24)$$
$$\qquad - (x^2 + 11x + 30)] + (x+2)[(x^2 + 10x + 21) - (x^2 + 10x + 24)]$$

$$= -3x - (x+1)(-6) + (x+2)(-3)$$

$$= -3x + 6x + 6 - 3x - 6 = 0$$

89. A square matrix is a square array of numbers. The determinant of a square matrix is a real number.

90. Let $A = \begin{bmatrix} x_{11} & x_{12} & x_{13} \\ x_{21} & x_{22} & x_{23} \\ x_{31} & x_{32} & x_{33} \end{bmatrix}$ and $|A| = 5$.

$$2A = \begin{bmatrix} 2x_{11} & 2x_{12} & 2x_{13} \\ 2x_{21} & 2x_{22} & 2x_{23} \\ 2x_{31} & 2x_{32} & 2x_{33} \end{bmatrix}$$

$$|2A| = 2x_{11}\begin{vmatrix} 2x_{22} & 2x_{23} \\ 2x_{32} & 2x_{33} \end{vmatrix} - 2x_{12}\begin{vmatrix} 2x_{21} & 2x_{23} \\ 2x_{31} & 2x_{33} \end{vmatrix} + 2x_{13}\begin{vmatrix} 2x_{21} & 2x_{22} \\ 2x_{31} & 2x_{32} \end{vmatrix}$$

$$= 2[x_{11}(4x_{22}x_{33} - 4x_{32}x_{23}) - x_{12}(4x_{21}x_{33} - 4x_{31}x_{23}) + x_{13}(4x_{21}x_{32} - 4x_{31}x_{22})]$$

$$= 8[x_{11}(x_{22}x_{33} - x_{32}x_{23}) - x_{12}(x_{21}x_{33} - x_{31}x_{23}) + x_{13}(x_{21}x_{32} - x_{31}x_{22})]$$

$$= 8|A|$$

So, $|2A| = 8|A| = 8(5) = 40$.

91. (a) $\begin{vmatrix} 1 & 3 & 4 \\ -7 & 2 & -5 \\ 6 & 1 & 2 \end{vmatrix} = -115$ (b) $\begin{vmatrix} 1 & 3 & 4 \\ -2 & 2 & 0 \\ 1 & 6 & 2 \end{vmatrix} = -40$

$-\begin{vmatrix} 1 & 4 & 3 \\ -7 & -5 & 2 \\ 6 & 2 & 1 \end{vmatrix} = -115$ $-\begin{vmatrix} 1 & 6 & 2 \\ -2 & 2 & 0 \\ 1 & 3 & 4 \end{vmatrix} = -40$

Column 2 and Column 3 were interchanged.   Row 1 and Row 3 were interchanged.

92. (a) Multiplying Row 1 of the matrix $\begin{bmatrix} 1 & -3 \\ 5 & 2 \end{bmatrix}$ by -5 and adding it to Row 2 gives the matrix $\begin{bmatrix} 1 & -3 \\ 0 & 17 \end{bmatrix}$.

$$\begin{vmatrix} 1 & -3 \\ 5 & 2 \end{vmatrix} = 17 = \begin{vmatrix} 1 & -3 \\ 0 & 17 \end{vmatrix}$$

(b) Multiplying Row 2 of the matrix $\begin{bmatrix} 5 & 4 & 2 \\ 2 & -3 & 4 \\ 7 & 6 & 3 \end{bmatrix}$ by -2 and adding it to Row 1 gives the matrix $\begin{bmatrix} 1 & 10 & -6 \\ 2 & -3 & 4 \\ 7 & 6 & 3 \end{bmatrix}$.

$$\begin{vmatrix} 5 & 4 & 2 \\ 2 & -3 & 4 \\ 7 & 6 & 3 \end{vmatrix} = -11 = \begin{vmatrix} 1 & 10 & -6 \\ 2 & -3 & 4 \\ 7 & 6 & 3 \end{vmatrix}$$

93. (a) $A = \begin{bmatrix} 1 & 2 \\ 2 & -3 \end{bmatrix}, B = \begin{bmatrix} 5 & 10 \\ 2 & -3 \end{bmatrix}$

$|B| = \begin{vmatrix} 5 & 10 \\ 2 & -3 \end{vmatrix} = -35$

$5|A| = 5\begin{vmatrix} 1 & 2 \\ 2 & -3 \end{vmatrix} = -35$

Row 1 was multiplied by 5.

$|B| = 5|A|$

(b) $A = \begin{bmatrix} 1 & 2 & -1 \\ 3 & -3 & 2 \\ 7 & 1 & 3 \end{bmatrix}, B = \begin{bmatrix} 1 & 8 & -3 \\ 3 & -12 & 6 \\ 7 & 4 & 9 \end{bmatrix}$

$|B| = \begin{vmatrix} 1 & 8 & -3 \\ 3 & -12 & 6 \\ 7 & 4 & 9 \end{vmatrix} = -300$

$12|A| = 12\begin{vmatrix} 1 & 2 & -1 \\ 3 & -3 & 2 \\ 7 & 1 & 3 \end{vmatrix} = -300$

Column 2 was multiplied by 4 and Column 3 was multiplied by 3.

$|B| = (4)(3)|A| = 12|A|$

94. (a) $A = \begin{vmatrix} 7 & 0 \\ 0 & 4 \end{vmatrix}$, $|A| = 7(4) - 0 = 28$

(b) $A = \begin{vmatrix} -1 & 0 & 0 \\ 0 & 5 & 0 \\ 0 & 0 & 2 \end{vmatrix}$, $|A| = (-1)(5)(2) = -10$

(c) $A = \begin{vmatrix} 2 & 0 & 0 & 0 \\ 0 & -2 & 0 & 0 \\ 0 & 0 & 1 & 0 \\ 0 & 0 & 0 & 3 \end{vmatrix}$

Using cofactors and a_{11}, $|A| = 2 \cdot C_{11} + 0 \cdot C_{12} + 0 \cdot C_{13} + 0 \cdot C_{14}$.

$$C_{11} = \begin{vmatrix} -2 & 0 & 0 \\ 0 & 1 & 0 \\ 0 & 0 & 3 \end{vmatrix}$$

$|A| = 2C_{11} = 2(-2 \cdot 1 \cdot 3) = 2 \cdot (-6) = -12$

In each case, the determinant of the matrix is the product of the diagonal entries. From this, one would conjecture that the determinant of a diagonal matrix is the product of the diagonal entries.

95. $f(x) = x^3 - 2x$

Since f is a polynomial, the domain is all real numbers x.

96. $g(x) = \sqrt[3]{x}$

An odd root of a number is defined for all real numbers.
Domain: all real numbers x

97. $h(x) = \sqrt{16 - x^2}$

$16 - x^2 \geq 0$

$(4 + x)(4 - x) \geq 0$

Critical numbers: $x = \pm 4$

Test intervals: $(-\infty, -4), (-4, 4), (4, \infty)$

Test: Is $16 - x^2 \geq 0$?

Solution: $[-4, 4]$

Domain of h: $-4 \leq x \leq 4$

98. $A(x) = \dfrac{3}{36 - x^2}$

$36 - x^2 \neq 0 \implies x^2 \neq 36 \implies x \neq \pm 6$

Domain: all real numbers $x \neq \pm 6$

99. $g(t) = \ln(t - 1)$

$t - 1 > 0$

$t > 1$

Domain: all real numbers $t > 1$

100. $f(s) = 625e^{-0.5S}$

The exponential function $y = Ae^x$ is defined for all real numbers.

Domain: all real numbers

101.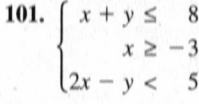

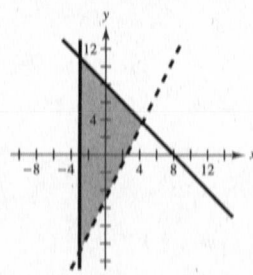

102.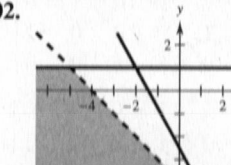

103. $[A : I] = \begin{bmatrix} -4 & 1 & \vdots & 1 & 0 \\ 8 & -1 & \vdots & 0 & 1 \end{bmatrix}$

$2R_1 + R_2 \rightarrow \begin{bmatrix} -4 & 1 & \vdots & 1 & 0 \\ 0 & 1 & \vdots & 2 & 1 \end{bmatrix}$

$-R_2 + R_1 \rightarrow \begin{bmatrix} -4 & 0 & \vdots & -1 & -1 \\ 0 & 1 & \vdots & 2 & 1 \end{bmatrix}$

$-\frac{1}{4}R_1 \rightarrow \begin{bmatrix} 1 & 0 & \vdots & \frac{1}{4} & \frac{1}{4} \\ 0 & 1 & \vdots & 2 & 1 \end{bmatrix} = [I : A^{-1}]$

$A^{-1} = \begin{bmatrix} \frac{1}{4} & \frac{1}{4} \\ 2 & 1 \end{bmatrix}$

104. $[A : I] = \begin{bmatrix} -5 & -8 & \vdots & 1 & 0 \\ 3 & 6 & \vdots & 0 & 1 \end{bmatrix}$

$\begin{matrix} R_2 \\ R_1 \end{matrix} \begin{bmatrix} 3 & 6 & \vdots & 0 & 1 \\ -5 & -8 & \vdots & 1 & 0 \end{bmatrix}$

$\frac{1}{3}R_1 \rightarrow \begin{bmatrix} 1 & 2 & \vdots & 0 & \frac{1}{3} \\ -5 & -8 & \vdots & 1 & 0 \end{bmatrix}$

$5R_1 + R_2 \rightarrow \begin{bmatrix} 1 & 2 & \vdots & 0 & \frac{1}{3} \\ 0 & 2 & \vdots & 1 & \frac{5}{3} \end{bmatrix}$

$\frac{1}{2}R_2 \rightarrow \begin{bmatrix} 1 & 2 & \vdots & 0 & \frac{1}{3} \\ 0 & 1 & \vdots & \frac{1}{2} & \frac{5}{6} \end{bmatrix}$

$-2R_2 + R_1 \rightarrow \begin{bmatrix} 1 & 0 & \vdots & -1 & -\frac{4}{3} \\ 0 & 1 & \vdots & \frac{1}{2} & \frac{5}{6} \end{bmatrix} = [I : A^{-1}]$

$A^{-1} = \begin{bmatrix} -1 & -\frac{4}{3} \\ \frac{1}{2} & \frac{5}{6} \end{bmatrix}$

105. $[A : I] = \begin{bmatrix} -7 & 2 & 9 & \vdots & 1 & 0 & 0 \\ 2 & -4 & -6 & \vdots & 0 & 1 & 0 \\ 3 & 5 & 2 & \vdots & 0 & 0 & 1 \end{bmatrix}$

$4R_2 + R_1 \rightarrow \begin{bmatrix} 1 & -14 & -15 & \vdots & 1 & 4 & 0 \\ 2 & -4 & -6 & \vdots & 0 & 1 & 0 \\ 3 & 5 & 2 & \vdots & 0 & 0 & 1 \end{bmatrix}$

$\begin{matrix} -2R_1 + R_2 \rightarrow \\ -3R_1 + R_3 \rightarrow \end{matrix} \begin{bmatrix} 1 & -14 & -15 & \vdots & 1 & 4 & 0 \\ 0 & 24 & 24 & \vdots & -2 & -7 & 0 \\ 0 & 47 & 47 & \vdots & -3 & -12 & 1 \end{bmatrix}$

$-\frac{47}{24}R_2 + R_3 \rightarrow \begin{bmatrix} 1 & -14 & -15 & \vdots & 1 & 4 & 0 \\ 0 & 24 & 24 & \vdots & -2 & -7 & 0 \\ 0 & 0 & 0 & \vdots & \frac{11}{12} & \frac{41}{24} & 1 \end{bmatrix}$

The zeros in Row 3 imply that the inverse does not exist.

106. $[A : I] = \begin{bmatrix} -6 & 2 & 0 & \vdots & 1 & 0 & 0 \\ 1 & 3 & -2 & \vdots & 0 & 1 & 0 \\ -2 & 0 & 1 & \vdots & 0 & 0 & 1 \end{bmatrix}$

$\begin{matrix} R_2 \\ R_1 \end{matrix} \begin{bmatrix} 1 & 3 & -2 & \vdots & 0 & 1 & 0 \\ -6 & 2 & 0 & \vdots & 1 & 0 & 0 \\ -2 & 0 & 1 & \vdots & 0 & 0 & 1 \end{bmatrix}$

$\begin{matrix} R_3 \\ R_2 \end{matrix} \begin{bmatrix} 1 & 3 & -2 & \vdots & 0 & 1 & 0 \\ -2 & 0 & 1 & \vdots & 0 & 0 & 1 \\ -6 & 2 & 0 & \vdots & 1 & 0 & 0 \end{bmatrix}$

$\begin{matrix} 2R_1 + R_2 \rightarrow \\ 6R_1 + R_3 \rightarrow \end{matrix} \begin{bmatrix} 1 & 3 & -2 & \vdots & 0 & 1 & 0 \\ 0 & 6 & -3 & \vdots & 0 & 2 & 1 \\ 0 & 20 & -12 & \vdots & 1 & 6 & 0 \end{bmatrix}$

$\frac{1}{6}R_2 \rightarrow \begin{bmatrix} 1 & 3 & -2 & \vdots & 0 & 1 & 0 \\ 0 & 1 & -\frac{1}{2} & \vdots & 0 & \frac{1}{3} & \frac{1}{6} \\ 0 & 20 & -12 & \vdots & 1 & 6 & 0 \end{bmatrix}$

—CONTINUED—

106. —CONTINUED—

$$\begin{bmatrix} 1 & 3 & -2 & \vdots & 0 & 1 & 0 \\ 0 & 1 & -\frac{1}{2} & \vdots & 0 & \frac{1}{3} & \frac{1}{6} \\ 0 & 0 & -2 & \vdots & 1 & -\frac{2}{3} & -\frac{10}{3} \end{bmatrix}$$

$-20R_2 + R_3 \rightarrow$ (row 3 above)

$$\begin{bmatrix} 1 & 3 & -2 & \vdots & 0 & 1 & 0 \\ 0 & 1 & -\frac{1}{2} & \vdots & 0 & \frac{1}{3} & \frac{1}{6} \\ 0 & 0 & 1 & \vdots & -\frac{1}{2} & \frac{1}{3} & \frac{5}{3} \end{bmatrix}$$

$-\frac{1}{2}R_3 \rightarrow$ (row 3 above)

$-3R_2 + R_1 \rightarrow \begin{bmatrix} 1 & 0 & -\frac{1}{2} & \vdots & 0 & 0 & -\frac{1}{2} \\ 0 & 1 & -\frac{1}{2} & \vdots & 0 & \frac{1}{3} & \frac{1}{6} \\ 0 & 0 & 1 & \vdots & -\frac{1}{2} & \frac{1}{3} & \frac{5}{3} \end{bmatrix}$

$\frac{1}{2}R_3 + R_1 \rightarrow$
$\frac{1}{2}R_3 + R_2 \rightarrow$
$\begin{bmatrix} 1 & 0 & 0 & \vdots & -\frac{1}{4} & \frac{1}{6} & \frac{1}{3} \\ 0 & 1 & 0 & \vdots & -\frac{1}{4} & \frac{1}{2} & 1 \\ 0 & 0 & 1 & \vdots & -\frac{1}{2} & \frac{1}{3} & \frac{5}{3} \end{bmatrix} = [I \vdots A^{-1}]$

$$A^{-1} = \begin{bmatrix} -\frac{1}{4} & \frac{1}{6} & \frac{1}{3} \\ -\frac{1}{4} & \frac{1}{2} & 1 \\ -\frac{1}{2} & \frac{1}{3} & \frac{5}{3} \end{bmatrix}$$

Section 8.5   Applications of Matrices and Determinants

- You should be able to use Cramer's Rule to solve a system of linear equations.
- Now you should be able to solve a system of linear equations by graphing, substitution, elimination, elementary row operations on an augmented matrix, using the inverse matrix, or Cramer's Rule.
- You should be able to find the area of a triangle with vertices (x_1, y_1), (x_2, y_2), and (x_3, y_3).

$$\text{Area} = \pm\frac{1}{2} \begin{vmatrix} x_1 & y_1 & 1 \\ x_2 & y_2 & 1 \\ x_3 & y_3 & 1 \end{vmatrix}$$

  The $\pm$ symbol indicates that the appropriate sign should be chosen so that the area is positive.

- You should be able to test to see if three points, (x_1, y_1), (x_2, y_2), and (x_3, y_3), are collinear.

$$\begin{vmatrix} x_1 & y_1 & 1 \\ x_2 & y_2 & 1 \\ x_3 & y_3 & 1 \end{vmatrix} = 0, \text{ if and only if they are collinear.}$$

- You should be able to find the equation of the line through (x_1, y_1) and (x_2, y_2) by evaluating.

$$\begin{vmatrix} x & y & 1 \\ x_1 & y_1 & 1 \\ x_2 & y_2 & 1 \end{vmatrix} = 0$$

- You should be able to encode and decode messages by using an invertible $n \times n$ matrix.

Vocabulary Check

1. Cramer's Rule   **2.** colinear   **3.** $A = \pm\frac{1}{2} \begin{vmatrix} x_1 & y_1 & 1 \\ x_2 & y_2 & 1 \\ x_3 & y_3 & 1 \end{vmatrix}$ **4.** cryptogram   **5.** uncoded; coded

1. $\begin{cases} 3x + 4y = -2 \\ 5x + 3y = 4 \end{cases}$

$$x = \frac{\begin{vmatrix} -2 & 4 \\ 4 & 3 \end{vmatrix}}{\begin{vmatrix} 3 & 4 \\ 5 & 3 \end{vmatrix}} = \frac{-22}{-11} = 2$$

$$y = \frac{\begin{vmatrix} 3 & -2 \\ 5 & 4 \end{vmatrix}}{\begin{vmatrix} 3 & 4 \\ 5 & 3 \end{vmatrix}} = \frac{22}{-11} = -2$$

Solution: $(2, -2)$

2. $\begin{cases} -4x - 7y = 47 \\ -x + 6y = -27 \end{cases}$

$$x = \frac{\begin{vmatrix} 47 & -7 \\ -27 & 6 \end{vmatrix}}{\begin{vmatrix} -4 & -7 \\ -1 & 6 \end{vmatrix}} = \frac{93}{-31} = -3$$

$$y = \frac{\begin{vmatrix} -4 & 47 \\ -1 & -27 \end{vmatrix}}{\begin{vmatrix} -4 & -7 \\ -1 & 6 \end{vmatrix}} = \frac{155}{-31} = -5$$

Solution: $(-3, -5)$

3. $\begin{cases} 3x + 2y = -2 \\ 6x + 4y = 4 \end{cases}$

Since $\begin{vmatrix} 3 & 2 \\ 6 & 4 \end{vmatrix} = 0$, Cramer's Rule does not apply.

The system is inconsistent in this case and has no solution.

4. $\begin{cases} 6x - 5y = 17 \\ -13x + 3y = -76 \end{cases}$

$$x = \frac{\begin{vmatrix} 17 & -5 \\ -76 & 3 \end{vmatrix}}{\begin{vmatrix} 6 & -5 \\ -13 & 3 \end{vmatrix}} = \frac{-329}{-47} = 7$$

$$y = \frac{\begin{vmatrix} 6 & 17 \\ -13 & -76 \end{vmatrix}}{\begin{vmatrix} 6 & -5 \\ -13 & 3 \end{vmatrix}} = \frac{-235}{-47} = 5$$

Solution: $(7, 5)$

5. $\begin{cases} -0.4x + 0.8y = 1.6 \\ 0.2x + 0.3y = 2.2 \end{cases}$

$$x = \frac{\begin{vmatrix} 1.6 & 0.8 \\ 2.2 & 0.3 \end{vmatrix}}{\begin{vmatrix} -0.4 & 0.8 \\ 0.2 & 0.3 \end{vmatrix}} = \frac{-1.28}{-0.28} = \frac{32}{7}$$

$$y = \frac{\begin{vmatrix} -0.4 & 1.6 \\ 0.2 & 2.2 \end{vmatrix}}{\begin{vmatrix} -0.4 & 0.8 \\ 0.2 & 0.3 \end{vmatrix}} = \frac{-1.20}{-0.28} = \frac{30}{7}$$

Solution: $\left(\dfrac{32}{7}, \dfrac{30}{7} \right)$

6. $\begin{cases} 2.4x - 1.3y = 14.63 \\ -4.6x + 0.5y = -11.51 \end{cases}$

$$x = \frac{\begin{vmatrix} 14.63 & -1.3 \\ -11.51 & 0.5 \end{vmatrix}}{\begin{vmatrix} 2.4 & -1.3 \\ -4.6 & 0.5 \end{vmatrix}} = \frac{-7.648}{-4.78} = \frac{8}{5}$$

$$y = \frac{\begin{vmatrix} 2.4 & 14.63 \\ -4.6 & -11.51 \end{vmatrix}}{\begin{vmatrix} 2.4 & -1.3 \\ -4.6 & 0.5 \end{vmatrix}} = \frac{39.674}{-4.78} = \frac{-83}{10}$$

Solution: $\left(\dfrac{8}{5}, -\dfrac{83}{10} \right)$

7. $\begin{cases} 4x - y + z = -5 \\ 2x + 2y + 3z = 10, \\ 5x - 2y + 6z = 1 \end{cases}$ $\quad D = \begin{vmatrix} 4 & -1 & 1 \\ 2 & 2 & 3 \\ 5 & -2 & 6 \end{vmatrix} = 55$

$$x = \frac{\begin{vmatrix} -5 & -1 & 1 \\ 10 & 2 & 3 \\ 1 & -2 & 6 \end{vmatrix}}{55} = \frac{-55}{55} = -1, \quad y = \frac{\begin{vmatrix} 4 & -5 & 1 \\ 2 & 10 & 3 \\ 5 & 1 & 6 \end{vmatrix}}{55} = \frac{165}{55} = 3, \quad z = \frac{\begin{vmatrix} 4 & -1 & -5 \\ 2 & 2 & 10 \\ 5 & -2 & 1 \end{vmatrix}}{55} = \frac{110}{55} = 2$$

Solution: $(-1, 3, 2)$

8. $\begin{cases} 4x - 2y + 3z = -2 \\ 2x + 2y + 5z = 16 \\ 8x - 5y - 2z = 4 \end{cases}$

$$D = \begin{vmatrix} 4 & -2 & 3 \\ 2 & 2 & 5 \\ 8 & -5 & -2 \end{vmatrix} = -82$$

$$x = \frac{\begin{vmatrix} -2 & -2 & 3 \\ 16 & 2 & 5 \\ 4 & -5 & -2 \end{vmatrix}}{-82} = \frac{-401}{-82} = 5$$

$$y = \frac{\begin{vmatrix} 4 & -2 & 3 \\ 2 & 16 & 5 \\ 8 & 4 & -2 \end{vmatrix}}{-82} = \frac{-656}{-82} = 8$$

$$z = \frac{\begin{vmatrix} 4 & -2 & -2 \\ 2 & 2 & 16 \\ 8 & -5 & 4 \end{vmatrix}}{-82} = \frac{164}{-82} = -2$$

Solution: $(5, 8, -2)$

9. $\begin{cases} x + 2y + 3z = -3 \\ -2x + y - z = 6, \\ 3x - 3y + 2z = -11 \end{cases}$ $D = \begin{vmatrix} 1 & 2 & 3 \\ -2 & 1 & -1 \\ 3 & -3 & 2 \end{vmatrix} = 10$

$$x = \frac{\begin{vmatrix} -3 & 2 & 3 \\ 6 & 1 & -1 \\ -11 & -3 & 2 \end{vmatrix}}{10} = \frac{-20}{10} = -2$$

$$y = \frac{\begin{vmatrix} 1 & -3 & 3 \\ -2 & 6 & -1 \\ 3 & -11 & 2 \end{vmatrix}}{10} = \frac{10}{10} = 1$$

$$z = \frac{\begin{vmatrix} 1 & 2 & -3 \\ -2 & 1 & 6 \\ 3 & -3 & -11 \end{vmatrix}}{10} = \frac{-10}{10} = -1$$

Solution: $(-2, 1, -1)$

10. $\begin{cases} 5x - 4y + z = -14 \\ -x + 2y - 2z = 10 \\ 3x + y + z = 1 \end{cases}$

$$D = \begin{vmatrix} 5 & -4 & 1 \\ -1 & 2 & -2 \\ 3 & 1 & 1 \end{vmatrix} = 33$$

$$x = \frac{\begin{vmatrix} -14 & -4 & 1 \\ 10 & 2 & -2 \\ 1 & 1 & 1 \end{vmatrix}}{33} = \frac{0}{33} = 0$$

$$y = \frac{\begin{vmatrix} 5 & -14 & 1 \\ -1 & 10 & -2 \\ 3 & 1 & 1 \end{vmatrix}}{33} = \frac{99}{33} = 3$$

$$z = \frac{\begin{vmatrix} 5 & -4 & -14 \\ -1 & 2 & 10 \\ 3 & 1 & 1 \end{vmatrix}}{33} = -\frac{66}{33} = -2$$

Solution: $(0, 3, -2)$

11. $\begin{cases} 3x + 3y + 5z = 1 \\ 3x + 5y + 9z = 2, \\ 5x + 9y + 17z = 4 \end{cases}$ $D = \begin{vmatrix} 3 & 3 & 5 \\ 3 & 5 & 9 \\ 5 & 9 & 17 \end{vmatrix} = 4$

$$x = \frac{\begin{vmatrix} 1 & 3 & 5 \\ 2 & 5 & 9 \\ 4 & 9 & 17 \end{vmatrix}}{4} = 0$$

$$y = \frac{\begin{vmatrix} 3 & 1 & 5 \\ 3 & 2 & 9 \\ 5 & 4 & 17 \end{vmatrix}}{4} = -\frac{1}{2}$$

$$z = \frac{\begin{vmatrix} 3 & 3 & 1 \\ 3 & 5 & 2 \\ 5 & 9 & 4 \end{vmatrix}}{4} = \frac{1}{2}$$

Solution: $\left(0, -\frac{1}{2}, \frac{1}{2}\right)$

12. $\begin{cases} x + 2y - z = -7 \\ 2x - 2y - 2z = -8, \\ -x + 3y + 4z = 8 \end{cases}$ $D = \begin{vmatrix} 1 & 2 & -1 \\ 2 & -2 & -2 \\ -1 & 3 & 4 \end{vmatrix} = -18$

$$x = \frac{\begin{vmatrix} -7 & 2 & -1 \\ -8 & -2 & -2 \\ 8 & 3 & 4 \end{vmatrix}}{-18} = -3, \quad y = \frac{\begin{vmatrix} 1 & -7 & -1 \\ 2 & -8 & -2 \\ -1 & 8 & 4 \end{vmatrix}}{-18} = -1, \quad z = \frac{\begin{vmatrix} 1 & 2 & -7 \\ 2 & -2 & -8 \\ -1 & 3 & 8 \end{vmatrix}}{-18} = 2$$

Solution: $(-3, -1, 2)$

13. $\begin{cases} 2x + y + 2z = 6 \\ -x + 2y - 3z = 0 \\ 3x + 2y - z = 6 \end{cases}$ $D = \begin{vmatrix} 2 & 1 & 2 \\ -1 & 2 & -3 \\ 3 & 2 & -1 \end{vmatrix} = -18$

$x = \dfrac{\begin{vmatrix} 6 & 1 & 2 \\ 0 & 2 & -3 \\ 6 & 2 & -1 \end{vmatrix}}{-18} = 1,\ y = \dfrac{\begin{vmatrix} 2 & 6 & 2 \\ -1 & 0 & -3 \\ 3 & 6 & -1 \end{vmatrix}}{-18} = 2,\ z = \dfrac{\begin{vmatrix} 2 & 1 & 6 \\ -1 & 2 & 0 \\ 3 & 2 & 6 \end{vmatrix}}{-18} = 1$

Solution: $(1, 2, 1)$

14. $\begin{cases} 2x + 3y + 5z = 4 \\ 3x + 5y + 9z = 7 \\ 5x + 9y + 17z = 13 \end{cases}$

$D = \begin{vmatrix} 2 & 3 & 5 \\ 3 & 5 & 9 \\ 5 & 9 & 17 \end{vmatrix} = 0$

Cramer's Rule does not apply.

15. Vertices: $(0, 0), (3, 1), (1, 5)$

$\text{Area} = \dfrac{1}{2}\begin{vmatrix} 0 & 0 & 1 \\ 3 & 1 & 1 \\ 1 & 5 & 1 \end{vmatrix} = \dfrac{1}{2}\begin{vmatrix} 3 & 1 \\ 1 & 5 \end{vmatrix} = 7$ square units

16. Vertices: $(0, 0), (4, 5), (5, -2)$

$\text{Area} = -\dfrac{1}{2}\begin{vmatrix} 0 & 0 & 1 \\ 4 & 5 & 1 \\ 5 & -2 & 1 \end{vmatrix} = -\dfrac{1}{2}\begin{vmatrix} 4 & 5 \\ 5 & -2 \end{vmatrix} = \dfrac{33}{2}$ square units

17. Vertices: $(-2, -3), (2, -3), (0, 4)$

$\text{Area} = \dfrac{1}{2}\begin{vmatrix} -2 & -3 & 1 \\ 2 & -3 & 1 \\ 0 & 4 & 1 \end{vmatrix} = \dfrac{1}{2}\left(-2\begin{vmatrix} -3 & 1 \\ 4 & 1 \end{vmatrix} - 2\begin{vmatrix} -3 & 1 \\ 4 & 1 \end{vmatrix}\right) = \dfrac{1}{2}(14 + 14) = 14$ square units

18. Vertices: $(-2, 1), (1, 6), (3, -1)$

$\text{Area} = -\dfrac{1}{2}\begin{vmatrix} -2 & 1 & 1 \\ 1 & 6 & 1 \\ 3 & -1 & 1 \end{vmatrix} = -\dfrac{1}{2}\left(-2\begin{vmatrix} 6 & 1 \\ -1 & 1 \end{vmatrix} - \begin{vmatrix} 1 & 1 \\ 3 & 1 \end{vmatrix} + \begin{vmatrix} 1 & 6 \\ 3 & -1 \end{vmatrix}\right) = -\dfrac{1}{2}(-14 + 2 - 19) = \dfrac{31}{2}$ square units

19. Vertices: $\left(0, \dfrac{1}{2}\right), \left(\dfrac{5}{2}, 0\right), (4, 3)$

$\text{Area} = \dfrac{1}{2}\begin{vmatrix} 0 & \frac{1}{2} & 1 \\ \frac{5}{2} & 0 & 1 \\ 4 & 3 & 1 \end{vmatrix} = \dfrac{1}{2}\left(-\dfrac{1}{2}\begin{vmatrix} \frac{5}{2} & 1 \\ 4 & 1 \end{vmatrix} + 1\begin{vmatrix} \frac{5}{2} & 0 \\ 4 & 3 \end{vmatrix}\right) = \dfrac{1}{2}\left(\dfrac{3}{4} + \dfrac{15}{2}\right) = \dfrac{33}{8}$ square units

20. Vertices: $(-4, -5), (6, 10), (6, -1)$

$\text{Area} = -\dfrac{1}{2}\begin{vmatrix} -4 & -5 & 1 \\ 6 & 10 & 1 \\ 6 & -1 & 1 \end{vmatrix} = -\dfrac{1}{2}\left(6\begin{vmatrix} -5 & 1 \\ 10 & 1 \end{vmatrix} - (-1)\begin{vmatrix} -4 & 1 \\ 6 & 1 \end{vmatrix} + \begin{vmatrix} -4 & -5 \\ 6 & 10 \end{vmatrix}\right) = 55$ square units

21. Vertices: $(-2, 4), (2, 3), (-1, 5)$

$\text{Area} = \dfrac{1}{2}\begin{vmatrix} -2 & 4 & 1 \\ 2 & 3 & 1 \\ -1 & 5 & 1 \end{vmatrix} = \dfrac{1}{2}\left[\begin{vmatrix} 2 & 3 \\ -1 & 5 \end{vmatrix} - \begin{vmatrix} -2 & 4 \\ -1 & 5 \end{vmatrix} + \begin{vmatrix} -2 & 4 \\ 2 & 3 \end{vmatrix}\right] = \dfrac{1}{2}(13 + 6 - 14) = \dfrac{5}{2}$ square units

22. Vertices: $(0, -2), (-1, 4), (3, 5)$

$$\text{Area} = -\frac{1}{2}\begin{vmatrix} 0 & -2 & 1 \\ -1 & 4 & 1 \\ 3 & 5 & 1 \end{vmatrix} = -\frac{1}{2}\left(2\begin{vmatrix} -1 & 1 \\ 3 & 1 \end{vmatrix} + \begin{vmatrix} -1 & 4 \\ 3 & 5 \end{vmatrix}\right) = -\frac{1}{2}(-8 - 17) = \frac{25}{2} \text{ square units}$$

23. Vertices: $(-3, 5), (2, 6), (3, -5)$

$$\text{Area} = -\frac{1}{2}\begin{vmatrix} -3 & 5 & 1 \\ 2 & 6 & 1 \\ 3 & -5 & 1 \end{vmatrix} = -\frac{1}{2}\left[\begin{vmatrix} 2 & 6 \\ 3 & -5 \end{vmatrix} - \begin{vmatrix} -3 & 5 \\ 3 & -5 \end{vmatrix} + \begin{vmatrix} -3 & 5 \\ 2 & 6 \end{vmatrix}\right] = -\frac{1}{2}(-28 + 0 - 28) = 28 \text{ square units}$$

24. Vertices: $(-2, 4), (1, 5), (3, -2)$

$$\text{Area} = -\frac{1}{2}\begin{vmatrix} -2 & 4 & 1 \\ 1 & 5 & 1 \\ 3 & -2 & 1 \end{vmatrix} = -\frac{1}{2}\left(-2\begin{vmatrix} 5 & 1 \\ -2 & 1 \end{vmatrix} - \begin{vmatrix} 4 & 1 \\ -2 & 1 \end{vmatrix} + 3\begin{vmatrix} 4 & 1 \\ 5 & 1 \end{vmatrix}\right) = -\frac{1}{2}(-14 - 6 - 3) = \frac{23}{2} \text{ square units}$$

25. $4 = \pm\frac{1}{2}\begin{vmatrix} -5 & 1 & 1 \\ 0 & 2 & 1 \\ -2 & y & 1 \end{vmatrix}$

$\pm 8 = -5\begin{vmatrix} 2 & 1 \\ y & 1 \end{vmatrix} - 2\begin{vmatrix} 1 & 1 \\ 2 & 1 \end{vmatrix}$

$\pm 8 = -5(2 - y) - 2(-1)$

$\pm 8 = 5y - 8$

$y = \frac{8 \pm 8}{5}$

$y = \frac{16}{5}$ or $y = 0$

26. $4 = \pm\frac{1}{2}\begin{vmatrix} -4 & 2 & 1 \\ -3 & 5 & 1 \\ -1 & y & 1 \end{vmatrix}$

$\pm 8 = \begin{vmatrix} -3 & 5 \\ -1 & y \end{vmatrix} - \begin{vmatrix} -4 & 2 \\ -1 & y \end{vmatrix} + \begin{vmatrix} -4 & 2 \\ -3 & 5 \end{vmatrix}$

$\pm 8 = -3y + 5 - (-4y + 2) - 20 + 6$

$\pm 8 = -3y + 5 + 4y - 2 - 20 + 6$

$\pm 8 = y - 11$

$y = 11 \pm 8$

$y = 19$ or $y = 3$

27. $6 = \pm\frac{1}{2}\begin{vmatrix} -2 & -3 & 1 \\ 1 & -1 & 1 \\ -8 & y & 1 \end{vmatrix}$

$\pm 12 = \begin{vmatrix} 1 & -1 \\ -8 & y \end{vmatrix} - \begin{vmatrix} -2 & -3 \\ -8 & y \end{vmatrix} + \begin{vmatrix} -2 & -3 \\ 1 & -1 \end{vmatrix}$

$\pm 12 = (y - 8) - (-2y - 24) + 5$

$\pm 12 = 3y + 21$

$y = \frac{-21 \pm 12}{3} = -7 \pm 4$

$y = -3$ or $y = -11$

28. $6 = \pm\frac{1}{2}\begin{vmatrix} 1 & 0 & 1 \\ 5 & -3 & 1 \\ -3 & y & 1 \end{vmatrix}$

$\pm 12 = \begin{vmatrix} -3 & 1 \\ y & 1 \end{vmatrix} + \begin{vmatrix} 5 & -3 \\ -3 & y \end{vmatrix}$

$\pm 12 = -3 - y + 5y - 9$

$\pm 12 = 4y - 12$

$y = \frac{12 \pm 12}{4} = 3 \pm 3$

$y = 6$ or $y = 0$

29. Vertices: $(0, 25), (10, 0), (28, 5)$

$$\text{Area} = \frac{1}{2}\begin{vmatrix} 0 & 25 & 1 \\ 10 & 0 & 1 \\ 28 & 5 & 1 \end{vmatrix} = 250 \text{ square miles}$$

30. Vertices: $(0, 30), (85, 0), (20, -50)$

$$\text{Area} = -\frac{1}{2}\begin{vmatrix} 0 & 30 & 1 \\ 85 & 0 & 1 \\ 20 & -50 & 1 \end{vmatrix} = 3100 \text{ square feet}$$

31. Points: $(3, -1), (0, -3), (12, 5)$

$$\begin{vmatrix} 3 & -1 & 1 \\ 0 & -3 & 1 \\ 12 & 5 & 1 \end{vmatrix} = 3 \begin{vmatrix} -3 & 1 \\ 5 & 1 \end{vmatrix} + 12 \begin{vmatrix} -1 & 1 \\ -3 & 1 \end{vmatrix} = 3(-8) + 12(2) = 0$$

The points are collinear.

32. Points: $(-3, -5), (6, 1), (10, 2)$

$$\begin{vmatrix} -3 & -5 & 1 \\ 6 & 1 & 1 \\ 10 & 2 & 1 \end{vmatrix} = \begin{vmatrix} 6 & 1 \\ 10 & 2 \end{vmatrix} - \begin{vmatrix} -3 & -5 \\ 10 & 2 \end{vmatrix} + \begin{vmatrix} -3 & -5 \\ 6 & 1 \end{vmatrix} = 2 - 44 + 27 = -15 \neq 0$$

The points are not collinear.

33. Points: $\left(2, -\frac{1}{2}\right), (-4, 4), (6, -3)$

$$\begin{vmatrix} 2 & -\frac{1}{2} & 1 \\ -4 & 4 & 1 \\ 6 & -3 & 1 \end{vmatrix} = \begin{vmatrix} -4 & 4 \\ 6 & -3 \end{vmatrix} - \begin{vmatrix} 2 & -\frac{1}{2} \\ 6 & -3 \end{vmatrix} + \begin{vmatrix} 2 & -\frac{1}{2} \\ -4 & 4 \end{vmatrix} = -12 + 3 + 6 = -3 \neq 0$$

The points are not collinear.

34. Points: $(0, 1), (4, -2), \left(-2, \frac{5}{2}\right)$

$$\begin{vmatrix} 0 & 1 & 1 \\ 4 & -2 & 1 \\ -2 & \frac{5}{2} & 1 \end{vmatrix} = -\begin{vmatrix} 4 & 1 \\ -2 & 1 \end{vmatrix} + \begin{vmatrix} 4 & -2 \\ -2 & \frac{5}{2} \end{vmatrix} = -6 + 6 = 0$$

The points are collinear.

35. Points: $(0, 2), (1, 2.4), (-1, 1.6)$

$$\begin{vmatrix} 0 & 2 & 1 \\ 1 & 2.4 & 1 \\ -1 & 1.6 & 1 \end{vmatrix} = -2 \begin{vmatrix} 1 & 1 \\ -1 & 1 \end{vmatrix} + \begin{vmatrix} 1 & 2.4 \\ -1 & 1.6 \end{vmatrix} = -2(2) + 4 = 0$$

The points are collinear.

36. Points: $(2, 3), (3, 3.5), (-1, 2)$

$$\begin{vmatrix} 2 & 3 & 1 \\ 3 & 3.5 & 1 \\ -1 & 2 & 1 \end{vmatrix} = \begin{vmatrix} 3 & 3.5 \\ -1 & 2 \end{vmatrix} - \begin{vmatrix} 2 & 3 \\ -1 & 2 \end{vmatrix} + \begin{vmatrix} 2 & 3 \\ 3 & 3.5 \end{vmatrix} = 9.5 - 7 + (-2) = \frac{1}{2} \neq 0$$

The points are not collinear.

37.

$$\begin{vmatrix} 2 & -5 & 1 \\ 4 & y & 1 \\ 5 & -2 & 1 \end{vmatrix} = 0$$

$$2 \begin{vmatrix} y & 1 \\ -2 & 1 \end{vmatrix} + 5 \begin{vmatrix} 4 & 1 \\ 5 & 1 \end{vmatrix} + \begin{vmatrix} 4 & y \\ 5 & -2 \end{vmatrix} = 0$$

$$2(y + 2) + 5(-1) + (-8 - 5y) = 0$$

$$-3y - 9 = 0$$

$$y = -3$$

38.

$$\begin{vmatrix} -6 & 2 & 1 \\ -5 & y & 1 \\ -3 & 5 & 1 \end{vmatrix} = 0$$

$$\begin{vmatrix} -5 & y \\ -3 & 5 \end{vmatrix} - \begin{vmatrix} -6 & 2 \\ -3 & 5 \end{vmatrix} + \begin{vmatrix} -6 & 2 \\ -5 & y \end{vmatrix} = 0$$

$$-25 + 3y + 24 - 6y + 10 = 0$$

$$-3y = -9$$

$$y = 3$$

39. Points: $(0, 0), (5, 3)$

Equation: $\begin{vmatrix} x & y & 1 \\ 0 & 0 & 1 \\ 5 & 3 & 1 \end{vmatrix} = -\begin{vmatrix} x & y \\ 5 & 3 \end{vmatrix} = 5y - 3x = 0 \implies 3x - 5y = 0$

40. Points: $(0, 0), (-2, 2)$

Equation: $\begin{vmatrix} x & y & 1 \\ 0 & 0 & 1 \\ -2 & 2 & 1 \end{vmatrix} = -\begin{vmatrix} x & y \\ -2 & 2 \end{vmatrix} = -(2x + 2y) = 0$ or $x + y = 0$

41. Points: $(-4, 3), (2, 1)$

Equation: $\begin{vmatrix} x & y & 1 \\ -4 & 3 & 1 \\ 2 & 1 & 1 \end{vmatrix} = x\begin{vmatrix} 3 & 1 \\ 1 & 1 \end{vmatrix} - y\begin{vmatrix} -4 & 1 \\ 2 & 1 \end{vmatrix} + \begin{vmatrix} -4 & 3 \\ 2 & 1 \end{vmatrix} = 2x + 6y - 10 = 0 \implies x + 3y - 5 = 0$

42. Points: $(10, 7), (-2, -7)$

Equation:

$\begin{vmatrix} x & y & 1 \\ 10 & 7 & 1 \\ -2 & -7 & 1 \end{vmatrix} = \begin{vmatrix} 10 & 7 \\ -2 & -7 \end{vmatrix} - \begin{vmatrix} x & y \\ -2 & -7 \end{vmatrix} + \begin{vmatrix} x & y \\ 10 & 7 \end{vmatrix} = -70 + 14 - (-7x + 2y) + 7x - 10y = 0$ or $7x - 6y - 28 = 0$

43. Points: $\left(-\frac{1}{2}, 3\right), \left(\frac{5}{2}, 1\right)$

Equation: $\begin{vmatrix} x & y & 1 \\ -\frac{1}{2} & 3 & 1 \\ \frac{5}{2} & 1 & 1 \end{vmatrix} = x\begin{vmatrix} 3 & 1 \\ 1 & 1 \end{vmatrix} - y\begin{vmatrix} -\frac{1}{2} & 1 \\ \frac{5}{2} & 1 \end{vmatrix} + \begin{vmatrix} -\frac{1}{2} & 3 \\ \frac{5}{2} & 1 \end{vmatrix} = 2x + 3y - 8 = 0$

44. Points: $\left(\frac{2}{3}, 4\right), (6, 12)$

Equation: $\begin{vmatrix} x & y & 1 \\ \frac{2}{3} & 4 & 1 \\ 6 & 12 & 1 \end{vmatrix} = \begin{vmatrix} \frac{2}{3} & 4 \\ 6 & 12 \end{vmatrix} - \begin{vmatrix} x & y \\ 6 & 12 \end{vmatrix} + \begin{vmatrix} x & y \\ \frac{2}{3} & 4 \end{vmatrix} = -16 - (12x - 6y) + 4x - \frac{2}{3}y = 0$ or $3x - 2y + 6 = 0$

45. The uncoded row matrices are the rows of the 7×3 matrix on the left.

$$
\begin{array}{ccc}
\text{T} & \text{R} & \text{O} \\
\text{U} & \text{B} & \text{L} \\
\text{E} & & \text{I} \\
\text{N} & & \text{R} \\
\text{I} & \text{V} & \text{E} \\
\text{R} & & \text{C} \\
\text{I} & \text{T} & \text{Y}
\end{array}
\begin{bmatrix}
20 & 18 & 15 \\
21 & 2 & 12 \\
5 & 0 & 9 \\
14 & 0 & 18 \\
9 & 22 & 5 \\
18 & 0 & 3 \\
9 & 20 & 25
\end{bmatrix}
\begin{bmatrix}
1 & -1 & 0 \\
1 & 0 & -1 \\
-6 & 2 & 3
\end{bmatrix}
=
\begin{bmatrix}
-52 & 10 & 27 \\
-49 & 3 & 34 \\
-49 & 13 & 27 \\
-94 & 22 & 54 \\
1 & 1 & -7 \\
0 & -12 & 9 \\
-121 & 41 & 55
\end{bmatrix}
$$

Solution: $-52\ 10\ 27\ -49\ 3\ 34\ -49\ 13\ 27\ -94\ 22\ 54\ 1\ 1\ -7\ 0\ -12\ 9\ -121\ 41\ 55$

46. $[16 \quad 12 \quad 5]\begin{bmatrix} 4 & 2 & 1 \\ -3 & -3 & -1 \\ 3 & 2 & 1 \end{bmatrix} = [43 \quad 6 \quad 9]$

$[1 \quad 19 \quad 5]\begin{bmatrix} 4 & 2 & 1 \\ -3 & -3 & -1 \\ 3 & 2 & 1 \end{bmatrix} = [-38 \quad -45 \quad -13]$

$[0 \quad 19 \quad 5]\begin{bmatrix} 4 & 2 & 1 \\ -3 & -3 & -1 \\ 3 & 3 & 1 \end{bmatrix} = [-42 \quad -47 \quad -14]$

$[14 \quad 4 \quad 0]\begin{bmatrix} 4 & 2 & 1 \\ -3 & -3 & -1 \\ 3 & 1 & 1 \end{bmatrix} = [44 \quad 16 \quad 10]$

$[13 \quad 15 \quad 14]\begin{bmatrix} 4 & 2 & 1 \\ -3 & -3 & -1 \\ 3 & 2 & 1 \end{bmatrix} = [49 \quad 9 \quad 12]$

$[5 \quad 25 \quad 0]\begin{bmatrix} 4 & 2 & 1 \\ -3 & -3 & -1 \\ 3 & 2 & 1 \end{bmatrix} = [-55 \quad -65 \quad -20]$

Solution: Uncoded 1×3 matrices: $[16 \quad 12 \quad 5], [1 \quad 19 \quad 5], [0 \quad 19 \quad 5], [14 \quad 4 \quad 0], [13 \quad 15 \quad 14], [5 \quad 25 \quad 0]$
Encoded 1×3 matrices: $[43 \quad 6 \quad 9], [-38 \quad -45 \quad -13], [-42 \quad -47 \quad -14],$
$[44 \quad 16 \quad 10], [49 \quad 9 \quad 12], [-55 \quad -65 \quad -20]$

Encoded message: 43 6 9 −38 −45 −13 −42 −47 −14
44 16 10 49 9 12 −55 −65 −20

In Exercises 47–50, use the matrix $A = \begin{bmatrix} 1 & 2 & 2 \\ 3 & 7 & 9 \\ -1 & -4 & -7 \end{bmatrix}$.

47. C A L L __ A T __ N O O N
$[3 \quad 1 \quad 12] [12 \quad 0 \quad 1] [20 \quad 0 \quad 14] [15 \quad 15 \quad 14]$

$[3 \quad 1 \quad 12]A = [-6 \; -35 \; -69]$

$[12 \quad 0 \quad 1]A = [11 \quad 20 \quad 17]$

$[20 \quad 0 \quad 14]A = [6 \; -16 \; -58]$

$[15 \quad 15 \quad 14]A = [46 \quad 79 \quad 67]$

Cryptogram: $-6 \quad -35 \quad -69 \quad 11 \quad 20 \quad 17 \quad 6 \quad -16 \quad -58 \quad 46 \quad 79 \quad 67$

48. I C E B E R G __ D E A D __ A H E A D
$[9 \quad 3 \quad 5] \quad [2 \quad 5 \quad 18] \quad [7 \quad 0 \quad 4] \quad [5 \quad 1 \quad 4] \quad [0 \quad 1 \quad 8] \quad [5 \quad 1 \quad 4]$

$[9 \quad 3 \quad 5]\begin{bmatrix} 1 & 2 & 2 \\ 3 & 7 & 9 \\ -1 & -4 & -7 \end{bmatrix} = [13 \quad 19 \quad 10]$

$[2 \quad 5 \quad 18]\begin{bmatrix} 1 & 2 & 2 \\ 3 & 7 & 9 \\ -1 & -4 & -7 \end{bmatrix} = [-1 \quad -33 \quad -77]$

$[7 \quad 0 \quad 4]\begin{bmatrix} 1 & 2 & 2 \\ 3 & 7 & 9 \\ -1 & -4 & -7 \end{bmatrix} = [3 \quad -2 \quad -14]$

—CONTINUED—

48. —CONTINUED—

$$[5 \quad 1 \quad 4] \begin{bmatrix} 1 & 2 & 2 \\ 3 & 7 & 9 \\ -1 & -4 & -7 \end{bmatrix} = [4 \quad 1 \quad -9]$$

$$[0 \quad 1 \quad 8] \begin{bmatrix} 1 & 2 & 2 \\ 3 & 7 & 9 \\ -1 & -4 & -7 \end{bmatrix} = [-5 \quad -25 \quad -47]$$

$$[5 \quad 1 \quad 4] \begin{bmatrix} 1 & 2 & 2 \\ 3 & 7 & 9 \\ -1 & -4 & -7 \end{bmatrix} = [4 \quad 1 \quad -9]$$

Cryptogram: 13   19   10   −1   −33   −77   3   −2   −14
4   1   −9   −5   −25   −47   4   1   −9

49. H   A   P   P   Y   _ B   I   R   T   H   D A   Y   _
[8   1   16] [16   25   0] [2   9   18] [20   8   4] [1   25   0]

$$[8 \quad 1 \quad 16] \, A = [-5 \quad -41 \quad -87]$$

$$[16 \quad 25 \quad 0] \, A = [91 \quad 207 \quad 257]$$

$$[2 \quad 9 \quad 18] \, A = [11 \quad -5 \quad -41]$$

$$[20 \quad 8 \quad 4] \, A = [40 \quad 80 \quad 84]$$

$$[1 \quad 25 \quad 0] \, A = [76 \quad 177 \quad 227]$$

Cryptogram: −5   −41   −87   91   207   257   11   −5   −41   40   80   84   76   177   227

50. O   P   E   R   A   T   I   O   N   _   O   V   E   R   L   O   A   D
[15   16   5] [18   1   20] [9   15   14] [0   15   22] [5   18   12] [15   1   4]

$$[15 \quad 16 \quad 5] \begin{bmatrix} 1 & 2 & 2 \\ 3 & 7 & 9 \\ -1 & -4 & -7 \end{bmatrix} = [58 \quad 122 \quad 139]$$

$$[18 \quad 1 \quad 20] \begin{bmatrix} 1 & 2 & 2 \\ 3 & 7 & 9 \\ -1 & -4 & -7 \end{bmatrix} = [1 \quad -37 \quad -95]$$

$$[9 \quad 15 \quad 14] \begin{bmatrix} 1 & 2 & 2 \\ 3 & 7 & 9 \\ -1 & -4 & -7 \end{bmatrix} = [40 \quad 67 \quad 55]$$

$$[0 \quad 15 \quad 22] \begin{bmatrix} 1 & 2 & 2 \\ 3 & 7 & 9 \\ -1 & -4 & -7 \end{bmatrix} = [23 \quad 17 \quad -19]$$

$$[5 \quad 18 \quad 12] \begin{bmatrix} 1 & 2 & 2 \\ 3 & 7 & 9 \\ -1 & -4 & -7 \end{bmatrix} = [47 \quad 88 \quad 88]$$

$$[15 \quad 1 \quad 4] \begin{bmatrix} 1 & 2 & 2 \\ 3 & 7 & 9 \\ -1 & -4 & -7 \end{bmatrix} = [14 \quad 21 \quad 11]$$

Cryptogram: 58   122   139   1   −37   −95   40   67   55   23   17   −19   47   88   88   14   21   11

51. $A^{-1} = \begin{bmatrix} 1 & 2 \\ 3 & 5 \end{bmatrix}^{-1} = \begin{bmatrix} -5 & 2 \\ 3 & -1 \end{bmatrix}$

$$\begin{bmatrix} 11 & 21 \\ 64 & 112 \\ 25 & 50 \\ 29 & 53 \\ 23 & 46 \\ 40 & 75 \\ 55 & 92 \end{bmatrix} \begin{bmatrix} -5 & 2 \\ 3 & -1 \end{bmatrix} = \begin{bmatrix} 8 & 1 \\ 16 & 16 \\ 25 & 0 \\ 14 & 5 \\ 23 & 0 \\ 25 & 5 \\ 1 & 18 \end{bmatrix} \begin{matrix} H & A \\ P & P \\ Y & \_ \\ N & E \\ W & \_ \\ Y & E \\ A & R \end{matrix}$$

Message: HAPPY NEW YEAR

52.

$$A^{-1} = \begin{bmatrix} -3 & 2 \\ -7 & 5 \end{bmatrix}$$

$\begin{bmatrix} -136 & 58 \end{bmatrix} \begin{bmatrix} -3 & 2 \\ -7 & 5 \end{bmatrix} = \begin{bmatrix} 2 & 18 \end{bmatrix}$ B   R

$\begin{bmatrix} -173 & 72 \end{bmatrix} \begin{bmatrix} -3 & 2 \\ -7 & 5 \end{bmatrix} = \begin{bmatrix} 15 & 14 \end{bmatrix}$ O   N

$\begin{bmatrix} -120 & 51 \end{bmatrix} \begin{bmatrix} -3 & 2 \\ -7 & 5 \end{bmatrix} = \begin{bmatrix} 3 & 15 \end{bmatrix}$ C   O

$\begin{bmatrix} -95 & 38 \end{bmatrix} \begin{bmatrix} -3 & 2 \\ -7 & 5 \end{bmatrix} = \begin{bmatrix} 19 & 0 \end{bmatrix}$ S   _

$\begin{bmatrix} -178 & 73 \end{bmatrix} \begin{bmatrix} -3 & 2 \\ -7 & 5 \end{bmatrix} = \begin{bmatrix} 23 & 9 \end{bmatrix}$ W   I

$\begin{bmatrix} -70 & 28 \end{bmatrix} \begin{bmatrix} -3 & 2 \\ -7 & 5 \end{bmatrix} = \begin{bmatrix} 14 & 0 \end{bmatrix}$ N   _

$\begin{bmatrix} -242 & 101 \end{bmatrix} \begin{bmatrix} -3 & 2 \\ -7 & 5 \end{bmatrix} = \begin{bmatrix} 19 & 21 \end{bmatrix}$ S   U

$\begin{bmatrix} -115 & 47 \end{bmatrix} \begin{bmatrix} -3 & 2 \\ -7 & 5 \end{bmatrix} = \begin{bmatrix} 16 & 5 \end{bmatrix}$ P   E

$\begin{bmatrix} -90 & 36 \end{bmatrix} \begin{bmatrix} -3 & 2 \\ -7 & 5 \end{bmatrix} = \begin{bmatrix} 18 & 0 \end{bmatrix}$ R   _

$\begin{bmatrix} -115 & 49 \end{bmatrix} \begin{bmatrix} -3 & 2 \\ -7 & 5 \end{bmatrix} = \begin{bmatrix} 2 & 15 \end{bmatrix}$ B   O

$\begin{bmatrix} -199 & 82 \end{bmatrix} \begin{bmatrix} -3 & 2 \\ -7 & 5 \end{bmatrix} = \begin{bmatrix} 23 & 12 \end{bmatrix}$ W   L

Message: BRONCOS WIN SUPER BOWL

53. $A^{-1} = \begin{bmatrix} 1 & -1 & 0 \\ 1 & 0 & -1 \\ -6 & 2 & 3 \end{bmatrix}^{-1} = \begin{bmatrix} -2 & -3 & -1 \\ -3 & -3 & -1 \\ -2 & -4 & -1 \end{bmatrix}$

$$\begin{bmatrix} 9 & -1 & -9 \\ 38 & -19 & -19 \\ 28 & -9 & -19 \\ -80 & 25 & 41 \\ -64 & 21 & 31 \\ 9 & -5 & -4 \end{bmatrix} \begin{bmatrix} -2 & -3 & -1 \\ -3 & -3 & -1 \\ -2 & -4 & -1 \end{bmatrix} = \begin{bmatrix} 3 & 12 & 1 \\ 19 & 19 & 0 \\ 9 & 19 & 0 \\ 3 & 1 & 14 \\ 3 & 5 & 12 \\ 5 & 4 & 0 \end{bmatrix} \begin{matrix} C & L & A \\ S & S & \_ \\ I & S & \_ \\ C & A & N \\ C & E & L \\ E & D & \_ \end{matrix}$$

Message: CLASS IS CANCELED

54.
$$A^{-1} = \begin{bmatrix} 11 & 2 & -8 \\ 4 & 1 & -3 \\ -8 & -1 & 6 \end{bmatrix}$$

$$[112 \quad -140 \quad 83] \begin{bmatrix} 11 & 2 & -8 \\ 4 & 1 & -3 \\ -8 & -1 & 6 \end{bmatrix} = [8 \quad 1 \quad 22] \quad \text{H} \quad \text{A} \quad \text{V}$$

$$[19 \quad -25 \quad 13] \begin{bmatrix} 11 & 2 & -8 \\ 4 & 1 & -3 \\ -8 & -1 & 6 \end{bmatrix} = [5 \quad 0 \quad 1] \quad \text{E} \quad \_ \quad \text{A}$$

$$[72 \quad -76 \quad 61] \begin{bmatrix} 11 & 2 & -8 \\ 4 & 1 & -3 \\ -8 & -1 & 6 \end{bmatrix} = [0 \quad 7 \quad 18] \quad \_ \quad \text{G} \quad \text{R} \qquad \text{Message: HAVE A GREAT WEEKEND}$$

$$[95 \quad -118 \quad 71] \begin{bmatrix} 11 & 2 & -8 \\ 4 & 1 & -3 \\ -8 & -1 & 6 \end{bmatrix} = [5 \quad 1 \quad 20] \quad \text{E} \quad \text{A} \quad \text{T}$$

$$[20 \quad 21 \quad 38] \begin{bmatrix} 11 & 2 & -8 \\ 4 & 1 & -3 \\ -8 & -1 & 6 \end{bmatrix} = [0 \quad 23 \quad 5] \quad \_ \quad \text{W} \quad \text{E}$$

$$[35 \quad -23 \quad 36] \begin{bmatrix} 11 & 2 & -8 \\ 4 & 1 & -3 \\ -8 & -1 & 6 \end{bmatrix} = [5 \quad 11 \quad 5] \quad \text{E} \quad \text{K} \quad \text{E}$$

$$[42 \quad -48 \quad 32] \begin{bmatrix} 11 & 2 & -8 \\ 4 & 1 & -3 \\ -8 & -1 & 6 \end{bmatrix} = [14 \quad 4 \quad 0] \quad \text{N} \quad \text{D} \quad \_$$

55. $A^{-1} = \begin{bmatrix} 1 & 2 & 2 \\ 3 & 7 & 9 \\ -1 & -4 & -7 \end{bmatrix}^{-1} = \begin{bmatrix} -13 & 6 & 4 \\ 12 & -5 & -3 \\ -5 & 2 & 1 \end{bmatrix}$

$$\begin{bmatrix} 20 & 17 & -15 \\ -12 & -56 & -104 \\ 1 & -25 & -65 \\ 62 & 143 & 181 \end{bmatrix} \begin{bmatrix} -13 & 6 & 4 \\ 12 & -5 & -3 \\ -5 & 2 & 1 \end{bmatrix} = \begin{bmatrix} 19 & 5 & 14 \\ 4 & 0 & 16 \\ 12 & 1 & 14 \\ 5 & 19 & 0 \end{bmatrix} \begin{matrix} \text{S} & \text{E} & \text{N} \\ \text{D} & \_ & \text{P} \\ \text{L} & \text{A} & \text{N} \\ \text{E} & \text{S} & \_ \end{matrix} \qquad \text{Message: SEND PLANES}$$

56.
$$[13 \quad -9 \quad -59] \begin{bmatrix} -13 & 6 & 4 \\ 12 & -5 & -3 \\ -5 & 2 & 1 \end{bmatrix} = [18 \quad 5 \quad 20] \quad \text{R} \quad \text{E} \quad \text{T}$$

$$[61 \quad 112 \quad 106] \begin{bmatrix} -13 & 6 & 4 \\ 12 & -5 & -3 \\ -5 & 2 & 1 \end{bmatrix} = [21 \quad 18 \quad 14] \quad \text{U} \quad \text{R} \quad \text{N}$$

$$[-17 \quad -73 \quad -131] \begin{bmatrix} -13 & 6 & 4 \\ 12 & -5 & -3 \\ -5 & 2 & 1 \end{bmatrix} = [0 \quad 1 \quad 20] \quad \_ \quad \text{A} \quad \text{T} \qquad \text{Message: RETURN AT DAWN}$$

$$[11 \quad 24 \quad 29] \begin{bmatrix} -13 & 6 & 4 \\ 12 & -5 & -3 \\ -5 & 2 & 1 \end{bmatrix} = [0 \quad 4 \quad 1] \quad \_ \quad \text{D} \quad \text{A}$$

$$[65 \quad 144 \quad 172] \begin{bmatrix} -13 & 6 & 4 \\ 12 & -5 & -3 \\ -5 & 2 & 1 \end{bmatrix} = [23 \quad 14 \quad 0] \quad \text{W} \quad \text{N} \quad \_$$

57. Let A be the 2×2 matrix needed to decode the message.

$$\begin{bmatrix} -18 & -18 \\ 1 & 16 \end{bmatrix} A = \begin{bmatrix} 0 & 18 \\ 15 & 14 \end{bmatrix} \begin{matrix} R \\ O & N \end{matrix}$$

$$A = \begin{bmatrix} -18 & -18 \\ 1 & 16 \end{bmatrix}^{-1} \begin{bmatrix} 0 & 18 \\ 15 & 14 \end{bmatrix} = \begin{bmatrix} -\frac{8}{135} & -\frac{1}{15} \\ \frac{1}{270} & \frac{1}{15} \end{bmatrix} \begin{bmatrix} 0 & 18 \\ 15 & 14 \end{bmatrix} = \begin{bmatrix} -1 & -2 \\ 1 & 1 \end{bmatrix}$$

$$\begin{bmatrix} 8 & 21 \\ -15 & -10 \\ -13 & -13 \\ 5 & 10 \\ 5 & 25 \\ 5 & 19 \\ -1 & 6 \\ 20 & 40 \\ -18 & -18 \\ 1 & 16 \end{bmatrix} \begin{bmatrix} -1 & -2 \\ 1 & 1 \end{bmatrix} = \begin{bmatrix} 13 & 5 \\ 5 & 20 \\ 0 & 13 \\ 5 & 0 \\ 20 & 15 \\ 14 & 9 \\ 7 & 8 \\ 20 & 0 \\ 0 & 18 \\ 15 & 14 \end{bmatrix} \begin{matrix} M & E \\ E & T \\ \_ & M \\ E & \_ \\ T & O \\ N & I \\ G & H \\ T & \_ \\ \_ & R \\ O & N \end{matrix}$$

Message: MEET ME TONIGHT RON

58. (a) $n = 3$; $\sum\limits_{i=1}^{n} x_i = 0 + 1 + 2 = 3$; $\sum\limits_{i=1}^{n} x_i^2 = 0^2 + 1^2 + 2^2 = 5$; $\sum\limits_{i=1}^{n} x_i^3 = 0^3 + 1^3 + 2^3 = 9$;

$$\sum\limits_{i=1}^{n} x_i^4 = 0^4 + 1^4 + 2^4 = 17; \quad \sum\limits_{i=1}^{n} y_i = 8965 + 9176 + 9406 = 27,547$$

$$\sum\limits_{i=1}^{n} x_i y_i = 0(8965) + 1(9176) + 2(9406) = 27,988$$

$$\sum\limits_{i=1}^{n} x_i^2 y_i = 0^2(8965) + 1^2(9176) + 2^2(9406) = 46,800$$

System: $\begin{cases} 3c + 3b + 5a = 27,547 \\ 3c + 5b + 9a = 27,988 \\ 5c + 9b + 17a = 46,800 \end{cases}$

(b) $D = \begin{vmatrix} 3 & 3 & 5 \\ 3 & 5 & 9 \\ 5 & 9 & 17 \end{vmatrix} = 4$

$$c = \frac{\begin{vmatrix} 27,547 & 3 & 5 \\ 27,988 & 5 & 9 \\ 46,800 & 9 & 17 \end{vmatrix}}{4} = \frac{35,860}{4} = 8965$$

$$b = \frac{\begin{vmatrix} 3 & 27,547 & 5 \\ 3 & 27,988 & 9 \\ 5 & 46,800 & 17 \end{vmatrix}}{4} = \frac{806}{4} = 201.5$$

$$a = \frac{\begin{vmatrix} 3 & 3 & 27,547 \\ 3 & 5 & 27,988 \\ 5 & 9 & 46,800 \end{vmatrix}}{4} = \frac{38}{4} = 9.5$$

The least squares regression parabola is $y = 9.5t^2 + 201.5t + 8965$.

(c)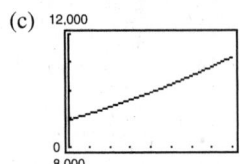

(d) The intersection of the regression parabola and the line $y = 10,000$ is about $t = 4.3$, so the number of cases waiting to be tried will reach 10,000 in about 2004.

59. False. In Cramer's Rule, the **denominator** is the determinant of the coefficient matrix.

60. True. If the determinant of the coefficient matrix is zero, the solution of the system would result in division by zero which is undefined.

61. False. If the determinant of the coefficient matrix is zero, the system has either no solution or infinitely many solutions.

62. Answers will vary. To solve a system of linear equations you can use graphing, substitution, elimination, elementary row operations on an augmented matrix (Gaussian elimination with back–substitution or Gauss-Jordan elimination), the inverse of a matrix, or Cramer's Rule.

63. $\begin{cases} -x - 7y = -22 & \text{Equation 1} \\ 5x + y = -26 & \text{Equation 2} \end{cases}$

$\begin{array}{ll} -5x - 35y = -110 & \text{(5)Eq.1} \\ \underline{5x + y = -26} & \\ \quad\quad -34y = -136 & \text{Add equations.} \\ \quad\quad\quad y = 4 \\ -x - 7(4) = -22 \\ \quad\quad\quad x = -6 \end{array}$

Solution: $(-6, 4)$

64. $\begin{cases} 3x + 8y = 11 & \text{Equation 1} \\ -2x + 12y = -16 & \text{Equation 2} \end{cases}$

$\begin{array}{ll} 9x + 24y = 33 & \text{(3)Eq.1} \\ \underline{4x - 24y = 32} & \text{(−2)Eq.2} \\ 13x \quad\quad = 65 & \text{Add equations.} \\ \quad x = \frac{65}{13} = 5 \end{array}$

$3(5) + 8y = 11 \implies 8y = -4 \implies y = -\frac{1}{2}$

Solution: $\left(5, -\frac{1}{2}\right)$

65. $\begin{cases} -x - 3y + 5z = -14 \\ 4x + 2y - z = -1 \\ 5x - 3y + 2z = -11 \end{cases}$

$A^{-1} = \begin{bmatrix} -1 & -3 & 5 \\ 4 & 2 & -1 \\ 5 & -3 & 2 \end{bmatrix}^{-1}$

$= \frac{1}{72}\begin{bmatrix} -1 & 9 & 7 \\ 13 & 27 & -19 \\ 22 & 18 & -10 \end{bmatrix}$

$\begin{bmatrix} x \\ y \\ z \end{bmatrix} = A^{-1}\begin{bmatrix} -14 \\ -1 \\ -11 \end{bmatrix} = \begin{bmatrix} -1 \\ 0 \\ -3 \end{bmatrix}$

Solution: $(-1, 0, -3)$

66. $\begin{cases} 5x - y - z = 7 \\ -2x + 3y + z = -5 \\ 4x + 10y - 5z = -37 \end{cases}$

$A^{-1} = \begin{bmatrix} 5 & -1 & -1 \\ -2 & 3 & 1 \\ 4 & 10 & -5 \end{bmatrix}^{-1} = \begin{bmatrix} \frac{25}{87} & \frac{5}{29} & -\frac{2}{87} \\ \frac{2}{29} & \frac{7}{29} & \frac{1}{29} \\ \frac{32}{87} & \frac{18}{29} & -\frac{13}{87} \end{bmatrix}$

$\begin{bmatrix} x \\ y \\ z \end{bmatrix} = A^{-1}\begin{bmatrix} 7 \\ -5 \\ -37 \end{bmatrix} = \begin{bmatrix} 2 \\ -2 \\ 5 \end{bmatrix}$

Solution: $(2, -2, 5)$

67. Objective function: $z = 6x + 4y$

Constraints:
$\quad x \geq 0$
$\quad y \geq 0$
$\quad x + 6y \leq 30$
$\quad 6x + y \leq 40$

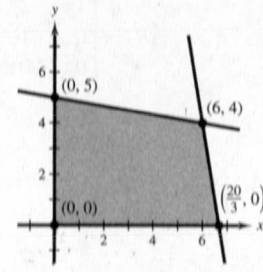

At $(0, 0)$: $z = 6(0) + 4(0) = 0$

At $(0, 5)$: $z = 6(0) + 4(5) = 20$

At $(6, 4)$: $z = 6(6) + 4(4) = 52$

At $\left(\frac{20}{3}, 0\right)$: $z = 6\left(\frac{20}{3}\right) + 4(0) = 40$

The minimum value of 0 occurs at $(0, 0)$.

The maximum value of 52 occurs at $(6, 4)$.

68. Objective function: $z = 6x + 7y$

Constraints:
$\quad x \geq 0$
$\quad y \geq 0$
$\quad 4x + 3y \geq 24$
$\quad x + 3y \geq 15$

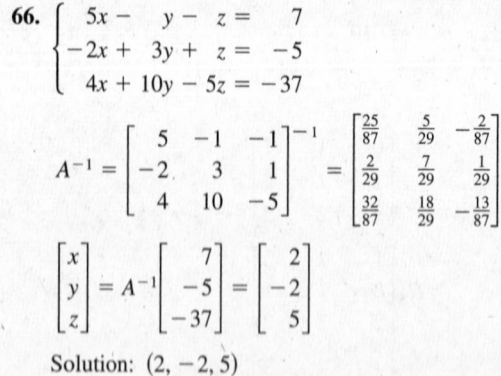

Since the region is unbounded, there is no maximum value of the objective function. To find the minimum value, check the vertices.

At $(0, 8)$: $z = 6(0) + 7(8) = 56$

At $(3, 4)$: $z = 6(3) + 7(4) = 46$

At $(15, 0)$: $z = 6(15) + 7(0) = 90$

The minimum value of 46 occurs at $(3, 4)$.

Review Exercises for Chapter 8

1. $\begin{bmatrix} -4 \\ 0 \\ 5 \end{bmatrix}$

Order: 3×1

2. $\begin{bmatrix} 3 & -1 & 0 & 6 \\ -2 & 7 & 1 & 4 \end{bmatrix}$

Since the matrix has two rows and four columns, its order is 2×4.

3. $[3]$

Order: 1×1

4. $\begin{bmatrix} 6 & 2 & -5 & 8 & 0 \end{bmatrix}$

Since the matrix has one row and five columns, its order is 1×5.

5. $\begin{cases} 3x - 10y = 15 \\ 5x + 4y = 22 \end{cases}$

$\begin{bmatrix} 3 & -10 & \vdots & 15 \\ 5 & 4 & \vdots & 22 \end{bmatrix}$

6. $\begin{cases} 8x - 7y + 4z = 12 \\ 3x - 5y + 2z = 20 \\ 5x + 3y - 3z = 26 \end{cases}$

$\begin{bmatrix} 8 & -7 & 4 & \vdots & 12 \\ 3 & -5 & 2 & \vdots & 20 \\ 5 & 3 & -3 & \vdots & 26 \end{bmatrix}$

7. $\begin{bmatrix} 5 & 1 & 7 & \vdots & -9 \\ 4 & 2 & 0 & \vdots & 10 \\ 9 & 4 & 2 & \vdots & 3 \end{bmatrix}$

$\begin{cases} 5x + y + 7z = -9 \\ 4x + 2y = 10 \\ 9x + 4y + 2z = 3 \end{cases}$

8. $\begin{bmatrix} 13 & 16 & 7 & 3 & \vdots & 2 \\ 1 & 21 & 8 & 5 & \vdots & 12 \\ 4 & 10 & -4 & 3 & \vdots & -1 \end{bmatrix}$

$\begin{cases} 13x + 16y + 7z + 3w = 2 \\ x + 21y + 8z + 5w = 12 \\ 4x + 10y - 4z + 3w = -1 \end{cases}$

9. $\begin{bmatrix} 0 & 1 & 1 \\ 1 & 2 & 3 \\ 2 & 2 & 2 \end{bmatrix}$

$\begin{matrix} R_1 \\ R_2 \end{matrix} \begin{bmatrix} 1 & 2 & 3 \\ 0 & 1 & 1 \\ 2 & 2 & 2 \end{bmatrix}$

$-2R_1 + R_3 \rightarrow \begin{bmatrix} 1 & 2 & 3 \\ 0 & 1 & 1 \\ 0 & -2 & -4 \end{bmatrix}$

$2R_2 + R_3 \rightarrow \begin{bmatrix} 1 & 2 & 3 \\ 0 & 1 & 1 \\ 0 & 0 & -2 \end{bmatrix}$

$-\frac{1}{2}R_3 \rightarrow \begin{bmatrix} 1 & 2 & 3 \\ 0 & 1 & 1 \\ 0 & 0 & 1 \end{bmatrix}$

10. $\begin{bmatrix} 4 & 8 & 16 \\ 3 & -1 & 2 \\ -2 & 10 & 12 \end{bmatrix}$

$\begin{matrix} \frac{1}{4}R_1 \rightarrow \\ \\ -\frac{1}{2}R_3 \rightarrow \end{matrix} \begin{bmatrix} 1 & 2 & 4 \\ 3 & -1 & 2 \\ 1 & -5 & -6 \end{bmatrix}$

$\begin{matrix} -3R_1 + R_2 \rightarrow \\ -R_1 + R_3 \rightarrow \end{matrix} \begin{bmatrix} 1 & 2 & 4 \\ 0 & -7 & -10 \\ 0 & -7 & -10 \end{bmatrix}$

$-R_2 + R_3 \rightarrow \begin{bmatrix} 1 & 2 & 4 \\ 0 & -7 & -10 \\ 0 & 0 & 0 \end{bmatrix}$

$-\frac{1}{7}R_2 \rightarrow \begin{bmatrix} 1 & 2 & 4 \\ 0 & 1 & \frac{10}{7} \\ 0 & 0 & 0 \end{bmatrix}$

11. $\begin{bmatrix} 1 & 2 & 3 & \vdots & 9 \\ 0 & 1 & -2 & \vdots & 2 \\ 0 & 0 & 1 & \vdots & 0 \end{bmatrix} \Rightarrow \begin{cases} x + 2y + 3z = 9 \\ y - 2z = 2 \\ z = 0 \end{cases}$

$y - 2(0) = 2 \Rightarrow y = 2$

$x + 2(2) + 3(0) = 9 \Rightarrow x = 5$

Solution: $(5, 2, 0)$

12. $\begin{cases} x + 3y - 9z = 4 \\ y - z = 10 \\ z = -2 \end{cases}$

$y - (-2) = 10$

$y = 8$

$x + 3(8) - 9(-2) = 4$

$x = -38$

Solution: $(-38, 8, -2)$

13. $\begin{bmatrix} 1 & -5 & 4 & \vdots & 1 \\ 0 & 1 & 2 & \vdots & 3 \\ 0 & 0 & 1 & \vdots & 4 \end{bmatrix} \Rightarrow \begin{cases} x - 5y + 4z = 1 \\ \quad\quad y + 2z = 3 \\ \quad\quad\quad\quad z = 4 \end{cases}$

$$y + 2(4) = 3 \implies y = -5$$

$$x - 5(-5) + 4(4) = 1 \implies x = -40$$

Solution: $(-40, -5, 4)$

14. $\begin{cases} x - 8y \quad\quad = -2 \\ \quad\quad y - z = -7 \\ \quad\quad\quad\quad z = 1 \end{cases}$

$$y - 1 = -7$$

$$y = -6$$

$$x - 8(-6) = -2$$

$$x = -50$$

Solution: $(-50, -6, 1)$

15. $\begin{bmatrix} 5 & 4 & \vdots & 2 \\ -1 & 1 & \vdots & -22 \end{bmatrix}$

$4R_2 + R_1 \rightarrow \begin{bmatrix} 1 & 8 & \vdots & -86 \\ -1 & 1 & \vdots & -22 \end{bmatrix}$

$R_1 + R_2 \rightarrow \begin{bmatrix} 1 & 8 & \vdots & -86 \\ 0 & 9 & \vdots & -108 \end{bmatrix}$

$\frac{1}{9}R_2 \rightarrow \begin{bmatrix} 1 & 8 & \vdots & -86 \\ 0 & 1 & \vdots & -12 \end{bmatrix}$

$\begin{cases} x + 8y = -86 \\ \quad\quad y = -12 \end{cases}$

$$y = -12$$

$$x + 8(-12) = -86 \implies x = 10$$

Solution: $(10, -12)$

16. $\begin{bmatrix} 2 & -5 & \vdots & 2 \\ 3 & -7 & \vdots & 1 \end{bmatrix}$

$\frac{1}{2}R_1 \rightarrow \begin{bmatrix} 1 & -\frac{5}{2} & \vdots & 1 \\ 3 & -7 & \vdots & 1 \end{bmatrix}$

$-3R_1 + R_2 \rightarrow \begin{bmatrix} 1 & -\frac{5}{2} & \vdots & 1 \\ 0 & \frac{1}{2} & \vdots & -2 \end{bmatrix}$

$2R_3 \rightarrow \begin{bmatrix} 1 & -\frac{5}{2} & \vdots & 1 \\ 0 & 1 & \vdots & -4 \end{bmatrix}$

$\begin{cases} x - \frac{5}{2}y = 1 \\ \quad\quad y = -4 \end{cases}$

$$y = -4$$

$$x - \frac{5}{2}(-4) = 1 \implies x = -9$$

Solution: $(-9, -4)$

17. $\begin{bmatrix} 0.3 & -0.1 & \vdots & -0.13 \\ 0.2 & -0.3 & \vdots & -0.25 \end{bmatrix}$

$\begin{matrix} 10R_1 \rightarrow \\ 10R_2 \rightarrow \end{matrix} \begin{bmatrix} 3 & -1 & \vdots & -1.3 \\ 2 & -3 & \vdots & -2.5 \end{bmatrix}$

$-R_2 + R_1 \rightarrow \begin{bmatrix} 1 & 2 & \vdots & 1.2 \\ 2 & -3 & \vdots & -2.5 \end{bmatrix}$

$-2R_1 + R_2 \rightarrow \begin{bmatrix} 1 & 2 & \vdots & 1.2 \\ 0 & -7 & \vdots & -4.9 \end{bmatrix}$

$-\frac{1}{7}R_2 \rightarrow \begin{bmatrix} 1 & 2 & \vdots & 1.2 \\ 0 & 1 & \vdots & 0.7 \end{bmatrix}$

$\begin{cases} x + 2y = 1.2 \\ \quad\quad y = 0.7 \end{cases}$

$$y = 0.7$$

$$x + 2(0.7) = 1.2 \implies x = -0.2$$

Solution: $(-0.2, 0.7) = \left(-\frac{1}{5}, \frac{7}{10}\right)$

18. $\begin{bmatrix} 0.2 & -0.1 & \vdots & 0.07 \\ 0.4 & -0.5 & \vdots & -0.01 \end{bmatrix}$

$\begin{matrix} 5R_1 \rightarrow \\ -2R_1 + R_2 \rightarrow \end{matrix} \begin{bmatrix} 1 & -0.5 & \vdots & 0.35 \\ 0 & -0.3 & \vdots & -0.15 \end{bmatrix}$

$-\frac{1}{0.3}R_2 \rightarrow \begin{bmatrix} 1 & -0.5 & \vdots & 0.35 \\ 0 & 1 & \vdots & 0.5 \end{bmatrix}$

$\begin{cases} x - 0.5y = 0.35 \\ \quad\quad y = 0.5 \end{cases}$

$$y = 0.5$$

$$x - 0.5(0.5) = 0.35 \implies x = 0.6$$

Solution: $(0.6, 0.5) = \left(\frac{3}{5}, \frac{1}{2}\right)$

19.
$$\begin{bmatrix} 2 & 3 & 1 & \vdots & 10 \\ 2 & -3 & -3 & \vdots & 22 \\ 4 & -2 & 3 & \vdots & -2 \end{bmatrix}$$

$$\begin{matrix} \\ -R_1 + R_2 \to \\ -2R_1 + R_3 \to \end{matrix} \begin{bmatrix} 2 & 3 & 1 & \vdots & 10 \\ 0 & -6 & -4 & \vdots & 12 \\ 0 & -8 & 1 & \vdots & -22 \end{bmatrix}$$

$$\begin{matrix} \frac{1}{2}R_1 \to \\ -\frac{1}{6}R_2 \to \\ \\ \end{matrix} \begin{bmatrix} 1 & \frac{3}{2} & \frac{1}{2} & \vdots & 5 \\ 0 & 1 & \frac{2}{3} & \vdots & -2 \\ 0 & -8 & 1 & \vdots & -22 \end{bmatrix}$$

$$\begin{matrix} \\ \\ 8R_2 + R_3 \to \end{matrix} \begin{bmatrix} 1 & \frac{3}{2} & \frac{1}{2} & \vdots & 5 \\ 0 & 1 & \frac{2}{3} & \vdots & -2 \\ 0 & 0 & \frac{19}{3} & \vdots & -38 \end{bmatrix}$$

$$\begin{matrix} \\ \\ \frac{3}{19}R_3 \to \end{matrix} \begin{bmatrix} 1 & \frac{3}{2} & \frac{1}{2} & \vdots & 5 \\ 0 & 1 & \frac{2}{3} & \vdots & -2 \\ 0 & 0 & 1 & \vdots & -6 \end{bmatrix}$$

$$z = -6$$
$$y + \tfrac{2}{3}(-6) = -2 \implies y = 2$$
$$x + \tfrac{3}{2}(2) + \tfrac{1}{2}(-6) = 5 \implies x = 5$$

Solution: $(5, 2, -6)$

20.
$$\begin{bmatrix} 2 & 3 & 3 & \vdots & 3 \\ 6 & 6 & 12 & \vdots & 13 \\ 12 & 9 & -1 & \vdots & 2 \end{bmatrix}$$

$$\begin{matrix} \\ -3R_1 + R_2 \to \\ -6R_1 + R_3 \to \end{matrix} \begin{bmatrix} 2 & 3 & 3 & \vdots & 3 \\ 0 & -3 & 3 & \vdots & 4 \\ 0 & -9 & -19 & \vdots & -16 \end{bmatrix}$$

$$\begin{matrix} R_2 + R_1 \to \\ \\ -3R_2 + R_3 \to \end{matrix} \begin{bmatrix} 2 & 0 & 6 & \vdots & 7 \\ 0 & -3 & 3 & \vdots & 4 \\ 0 & 0 & -28 & \vdots & -28 \end{bmatrix}$$

$$\begin{matrix} \frac{1}{2}R_1 \to \\ -\frac{1}{3}R_2 \to \\ -\frac{1}{28}R_3 \to \end{matrix} \begin{bmatrix} 1 & 0 & 3 & \vdots & \frac{7}{2} \\ 0 & 1 & -1 & \vdots & -\frac{4}{3} \\ 0 & 0 & 1 & \vdots & 1 \end{bmatrix}$$

$$\begin{cases} x & + 3z = \frac{7}{2} \\ y & - z = -\frac{4}{3} \\ & z = 1 \end{cases}$$

$$z = 1$$
$$y - 1 = -\tfrac{4}{3} \implies y = -\tfrac{1}{3}$$
$$x + 3(1) = \tfrac{7}{2} \implies x = \tfrac{1}{2}$$

Solution: $\left(\tfrac{1}{2}, -\tfrac{1}{3}, 1\right)$

21.
$$\begin{bmatrix} 2 & 1 & 2 & \vdots & 4 \\ 2 & 2 & 0 & \vdots & 5 \\ 2 & -1 & 6 & \vdots & 2 \end{bmatrix}$$

$$\begin{matrix} \\ -R_1 + R_2 \to \\ -R_1 + R_3 \to \end{matrix} \begin{bmatrix} 2 & 1 & 2 & \vdots & 4 \\ 0 & 1 & -2 & \vdots & 1 \\ 0 & -2 & 4 & \vdots & -2 \end{bmatrix}$$

$$\begin{matrix} -R_2 + R_1 \to \\ \\ 2R_2 + R_3 \to \end{matrix} \begin{bmatrix} 2 & 0 & 4 & \vdots & 3 \\ 0 & 1 & -2 & \vdots & 1 \\ 0 & 0 & 0 & \vdots & 0 \end{bmatrix}$$

$$\begin{matrix} \frac{1}{2}R_1 \to \\ \\ \\ \end{matrix} \begin{bmatrix} 1 & 0 & 2 & \vdots & \frac{3}{2} \\ 0 & 1 & -2 & \vdots & 1 \\ 0 & 0 & 0 & \vdots & 0 \end{bmatrix}$$

Let $z = a$, then:

$$y - 2a = 1 \implies y = 2a + 1$$
$$x + 2a = \tfrac{3}{2} \implies x = -2a + \tfrac{3}{2}$$

Solution: $\left(-2a + \tfrac{3}{2}, 2a + 1, a\right)$ where a is any real number

22.
$$\begin{bmatrix} 1 & 2 & 6 & \vdots & 1 \\ 2 & 5 & 15 & \vdots & 4 \\ 3 & 1 & 3 & \vdots & -6 \end{bmatrix}$$

$$\begin{matrix} \\ -2R_1 + R_2 \to \\ -3R_1 + R_3 \to \end{matrix} \begin{bmatrix} 1 & 2 & 6 & \vdots & 1 \\ 0 & 1 & 3 & \vdots & 2 \\ 0 & -5 & -15 & \vdots & -9 \end{bmatrix}$$

$$\begin{matrix} \\ \\ 5R_2 + R_3 \to \end{matrix} \begin{bmatrix} 1 & 2 & 6 & \vdots & 1 \\ 0 & 1 & 3 & \vdots & 2 \\ 0 & 0 & 0 & \vdots & 1 \end{bmatrix}$$

Because the last row consists of all zeros except for the last entry, the system is inconsistent and there is no solution.

23.

$$\begin{bmatrix} 2 & 1 & 1 & 0 & \vdots & 6 \\ 0 & -2 & 3 & -1 & \vdots & 9 \\ 3 & 3 & -2 & -2 & \vdots & -11 \\ 1 & 0 & 1 & 3 & \vdots & 14 \end{bmatrix}$$

$$-R_4 + R_1 \begin{bmatrix} 1 & 1 & 0 & -3 & \vdots & -8 \\ 0 & -2 & 3 & -1 & \vdots & 9 \\ 3 & 3 & -2 & -2 & \vdots & -11 \\ 1 & 0 & 1 & 3 & \vdots & 14 \end{bmatrix}$$

$$\begin{matrix} \\ \\ -3R_1 + R_3 \rightarrow \\ -R_1 + R_4 \rightarrow \end{matrix} \begin{bmatrix} 1 & 1 & 0 & -3 & \vdots & -8 \\ 0 & -2 & 3 & -1 & \vdots & 9 \\ 0 & 0 & -2 & 7 & \vdots & 13 \\ 0 & -1 & 1 & 6 & \vdots & 22 \end{bmatrix}$$

$$-3R_4 + R_2 \rightarrow \begin{bmatrix} 1 & 1 & 0 & -3 & \vdots & -8 \\ 0 & 1 & 0 & -19 & \vdots & -57 \\ 0 & 0 & -2 & 7 & \vdots & 13 \\ 0 & -1 & 1 & 6 & \vdots & 22 \end{bmatrix}$$

$$\begin{matrix} \\ \\ \\ R_2 + R_4 \rightarrow \end{matrix} \begin{bmatrix} 1 & 1 & 0 & -3 & \vdots & -8 \\ 0 & 1 & 0 & -19 & \vdots & -57 \\ 0 & 0 & -2 & 7 & \vdots & 13 \\ 0 & 0 & 1 & -13 & \vdots & -35 \end{bmatrix}$$

$$\begin{matrix} \\ \\ \curvearrowright R_4 \\ \curvearrowleft R_3 \end{matrix} \begin{bmatrix} 1 & 1 & 0 & -3 & \vdots & -8 \\ 0 & 1 & 0 & -19 & \vdots & -57 \\ 0 & 0 & 1 & -13 & \vdots & -35 \\ 0 & 0 & -2 & 7 & \vdots & 13 \end{bmatrix}$$

$$\begin{matrix} \\ \\ \\ 2R_3 + R_4 \rightarrow \end{matrix} \begin{bmatrix} 1 & 1 & 0 & -3 & \vdots & -8 \\ 0 & 1 & 0 & -19 & \vdots & -57 \\ 0 & 0 & 1 & -13 & \vdots & -35 \\ 0 & 0 & 0 & -19 & \vdots & -57 \end{bmatrix}$$

$$\begin{matrix} \\ \\ \\ \frac{1}{19}R_4 \rightarrow \end{matrix} \begin{bmatrix} 1 & 1 & 0 & -3 & \vdots & -8 \\ 0 & 1 & 0 & -19 & \vdots & -57 \\ 0 & 0 & 1 & -13 & \vdots & -35 \\ 0 & 0 & 0 & 1 & \vdots & 3 \end{bmatrix}$$

$$w = 3$$

$$z - 13(3) = -35 \implies z = 4$$

$$y - 19(3) = -57 \implies y = 0$$

$$x + 0 - 3(3) = -8 \implies x = 1$$

Solution: $(1, 0, 4, 3)$

24.

$$\begin{bmatrix} 1 & 2 & 0 & 1 & \vdots & 3 \\ 0 & -3 & 3 & 0 & \vdots & 0 \\ 4 & 4 & 1 & 2 & \vdots & 0 \\ 2 & 0 & 1 & 0 & \vdots & 3 \end{bmatrix}$$

$$\begin{matrix} -\frac{1}{3}R_2 \rightarrow \\ -4R_1 + R_3 \rightarrow \\ -2R_1 + R_4 \rightarrow \end{matrix} \begin{bmatrix} 1 & 2 & 0 & 1 & \vdots & 3 \\ 0 & 1 & -1 & 0 & \vdots & 0 \\ 0 & -4 & 1 & -2 & \vdots & -12 \\ 0 & -4 & 1 & -2 & \vdots & -3 \end{bmatrix}$$

$$\begin{matrix} \\ \\ \\ -R_3 + R_4 \rightarrow \end{matrix} \begin{bmatrix} 1 & 2 & 0 & 1 & \vdots & 3 \\ 0 & 1 & -1 & 0 & \vdots & 0 \\ 0 & -4 & 1 & -2 & \vdots & -12 \\ 0 & 0 & 0 & 0 & \vdots & 9 \end{bmatrix}$$

Because the last row consists of all zeros except for the last entry, the system is inconsistent and there is no solution.

25.

$$\begin{bmatrix} -1 & 1 & 2 & \vdots & 1 \\ 2 & 3 & 1 & \vdots & -2 \\ 5 & 4 & 2 & \vdots & 4 \end{bmatrix}$$

$$-R_1 \rightarrow \begin{bmatrix} 1 & -1 & -2 & \vdots & -1 \\ 2 & 3 & 1 & \vdots & -2 \\ 5 & 4 & 2 & \vdots & 4 \end{bmatrix}$$

$$\begin{matrix} -2R_1 + R_2 \rightarrow \\ -5R_1 + R_3 \rightarrow \end{matrix} \begin{bmatrix} 1 & -1 & -2 & \vdots & -1 \\ 0 & 5 & 5 & \vdots & 0 \\ 0 & 9 & 12 & \vdots & 9 \end{bmatrix}$$

$$\frac{1}{5}R_2 \rightarrow \begin{bmatrix} 1 & -1 & -2 & \vdots & -1 \\ 0 & 1 & 1 & \vdots & 0 \\ 0 & 9 & 12 & \vdots & 9 \end{bmatrix}$$

$$\begin{matrix} R_2 + R_1 \rightarrow \\ \\ -9R_2 + R_3 \rightarrow \end{matrix} \begin{bmatrix} 1 & 0 & -1 & \vdots & -1 \\ 0 & 1 & 1 & \vdots & 0 \\ 0 & 0 & 3 & \vdots & 9 \end{bmatrix}$$

$$\frac{1}{3}R_3 \rightarrow \begin{bmatrix} 1 & 0 & -1 & \vdots & -1 \\ 0 & 1 & 1 & \vdots & 0 \\ 0 & 0 & 1 & \vdots & 3 \end{bmatrix}$$

$$\begin{matrix} R_3 + R_1 \rightarrow \\ -R_3 + R_2 \rightarrow \\ \end{matrix} \begin{bmatrix} 1 & 0 & 0 & \vdots & 2 \\ 0 & 1 & 0 & \vdots & -3 \\ 0 & 0 & 1 & \vdots & 3 \end{bmatrix}$$

$$x = 2, y = -3, z = 3$$

Solution: $(2, -3, 3)$

26. $\begin{cases} 4x + 4y + 4z = 5 \\ 4x - 2y - 8z = 1 \\ 5x + 3y + 8z = 6 \end{cases}$

$$\begin{bmatrix} 4 & 4 & 4 & \vdots & 5 \\ 4 & -2 & -8 & \vdots & 1 \\ 5 & 3 & 8 & \vdots & 6 \end{bmatrix}$$

$\frac{1}{4}R_1 \rightarrow \begin{bmatrix} 1 & 1 & 1 & \vdots & \frac{5}{4} \\ 4 & -2 & -8 & \vdots & 1 \\ 5 & 3 & 8 & \vdots & 6 \end{bmatrix}$

$\begin{matrix} \\ -4R_1 + R_2 \rightarrow \\ -5R_1 + R_3 \rightarrow \end{matrix} \begin{bmatrix} 1 & 1 & 1 & \vdots & \frac{5}{4} \\ 0 & -6 & -12 & \vdots & -4 \\ 0 & -2 & 3 & \vdots & -\frac{1}{4} \end{bmatrix}$

$-\frac{1}{6}R_2 \rightarrow \begin{bmatrix} 1 & 1 & 1 & \vdots & \frac{5}{4} \\ 0 & 1 & 2 & \vdots & \frac{2}{3} \\ 0 & -2 & 3 & \vdots & -\frac{1}{4} \end{bmatrix}$

$\begin{matrix} -R_2 + R_1 \rightarrow \\ \\ 2R_2 + R_3 \rightarrow \end{matrix} \begin{bmatrix} 1 & 0 & -1 & \vdots & \frac{7}{12} \\ 0 & 1 & 2 & \vdots & \frac{2}{3} \\ 0 & 0 & 7 & \vdots & \frac{13}{12} \end{bmatrix}$

$\begin{matrix} \\ \\ \frac{1}{7}R_3 \rightarrow \end{matrix} \begin{bmatrix} 1 & 0 & -1 & \vdots & \frac{7}{12} \\ 0 & 1 & 2 & \vdots & \frac{2}{3} \\ 0 & 0 & 1 & \vdots & \frac{13}{84} \end{bmatrix}$

$\begin{matrix} R_3 + R_1 \rightarrow \\ -2R_3 + R_2 \rightarrow \\ \end{matrix} \begin{bmatrix} 1 & 0 & 0 & \vdots & \frac{31}{42} \\ 0 & 1 & 0 & \vdots & \frac{5}{14} \\ 0 & 0 & 1 & \vdots & \frac{13}{84} \end{bmatrix}$

$x = \frac{31}{42}$

$y = \frac{5}{14}$

$z = \frac{13}{84}$

Solution: $\left(\frac{31}{42}, \frac{5}{14}, \frac{13}{84} \right)$

28. $\begin{cases} -3x + y + 7z = -20 \\ 5x - 2y - z = 34 \\ -x + y + 4z = -8 \end{cases}$

$$\begin{bmatrix} -3 & 1 & 7 & \vdots & -20 \\ 5 & -2 & -1 & \vdots & 34 \\ -1 & 1 & 4 & \vdots & -8 \end{bmatrix}$$

$\begin{matrix} R_3 \\ \\ R_1 \end{matrix} \begin{bmatrix} -1 & 1 & 4 & \vdots & -8 \\ 5 & -2 & -1 & \vdots & 34 \\ -3 & 1 & 7 & \vdots & -20 \end{bmatrix}$

$-1R_1 \rightarrow \begin{bmatrix} 1 & -1 & -4 & \vdots & 8 \\ 5 & -2 & -1 & \vdots & 34 \\ -3 & 1 & 7 & \vdots & -20 \end{bmatrix}$

$\begin{matrix} -5R_1 + R_2 \rightarrow \\ 3R_1 + R_3 \rightarrow \end{matrix} \begin{bmatrix} 1 & -1 & -4 & \vdots & 8 \\ 0 & 3 & 19 & \vdots & -6 \\ 0 & -2 & -5 & \vdots & 4 \end{bmatrix}$

27.

$$\begin{bmatrix} 2 & -1 & 9 & \vdots & -8 \\ -1 & -3 & 4 & \vdots & -15 \\ 5 & 2 & -1 & \vdots & 17 \end{bmatrix}$$

$R_2 + R_1 \rightarrow \begin{bmatrix} 1 & -4 & 13 & \vdots & -23 \\ -1 & -3 & 4 & \vdots & -15 \\ 5 & 2 & -1 & \vdots & 17 \end{bmatrix}$

$\begin{matrix} R_1 + R_2 \rightarrow \\ -5R_1 + R_3 \rightarrow \end{matrix} \begin{bmatrix} 1 & -4 & 13 & \vdots & -23 \\ 0 & -7 & 17 & \vdots & -38 \\ 0 & 22 & -66 & \vdots & 132 \end{bmatrix}$

$\begin{matrix} R_3 \\ R_2 \end{matrix} \begin{bmatrix} 1 & -4 & 13 & \vdots & -23 \\ 0 & 22 & -66 & \vdots & 132 \\ 0 & -7 & 17 & \vdots & -38 \end{bmatrix}$

$\frac{1}{22}R_2 \rightarrow \begin{bmatrix} 1 & -4 & 13 & \vdots & -23 \\ 0 & 1 & -3 & \vdots & 6 \\ 0 & -7 & 17 & \vdots & -38 \end{bmatrix}$

$7R_2 + R_3 \rightarrow \begin{bmatrix} 1 & -4 & 13 & \vdots & -23 \\ 0 & 1 & -3 & \vdots & 6 \\ 0 & 0 & -4 & \vdots & 4 \end{bmatrix}$

$-\frac{1}{4}R_3 \rightarrow \begin{bmatrix} 1 & -4 & 13 & \vdots & -23 \\ 0 & 1 & -3 & \vdots & 6 \\ 0 & 0 & 1 & \vdots & -1 \end{bmatrix}$

$4R_2 + R_1 \rightarrow \begin{bmatrix} 1 & 0 & 1 & \vdots & 1 \\ 0 & 1 & -3 & \vdots & 6 \\ 0 & 0 & 1 & \vdots & -1 \end{bmatrix}$

$\begin{matrix} -R_3 + R_1 \rightarrow \\ 3R_3 + R_2 \rightarrow \end{matrix} \begin{bmatrix} 1 & 0 & 0 & \vdots & 2 \\ 0 & 1 & 0 & \vdots & 3 \\ 0 & 0 & 1 & \vdots & -1 \end{bmatrix}$

$x = 2, y = 3, z = -1$

Solution: $(2, 3, -1)$

$\frac{1}{3}R_2 \rightarrow \begin{bmatrix} 1 & -1 & -4 & \vdots & 8 \\ 0 & 1 & \frac{19}{3} & \vdots & -2 \\ 0 & -2 & -5 & \vdots & 4 \end{bmatrix}$

$\begin{matrix} R_2 + R_1 \rightarrow \\ \\ 2R_2 + R_3 \rightarrow \end{matrix} \begin{bmatrix} 1 & 0 & \frac{7}{3} & \vdots & 6 \\ 0 & 1 & \frac{19}{3} & \vdots & -2 \\ 0 & 0 & \frac{23}{3} & \vdots & 0 \end{bmatrix}$

$\frac{3}{23}R_3 \rightarrow \begin{bmatrix} 1 & 0 & \frac{7}{3} & \vdots & 6 \\ 0 & 1 & \frac{19}{3} & \vdots & -2 \\ 0 & 0 & 1 & \vdots & 0 \end{bmatrix}$

$\begin{matrix} -\frac{7}{3}R_3 + R_1 \rightarrow \\ \\ -\frac{19}{3}R_3 + R_2 \rightarrow \end{matrix} \begin{bmatrix} 1 & 0 & 0 & \vdots & 6 \\ 0 & 1 & 0 & \vdots & -2 \\ 0 & 0 & 1 & \vdots & 0 \end{bmatrix}$

$x = 6, y = -2, z = 0$

Solution: $(6, -2, 0)$

29. Use the reduced row-echelon form feature of a graphing utility.

$$\begin{bmatrix} 3 & -1 & 5 & -2 & : & -44 \\ 1 & 6 & 4 & -1 & : & 1 \\ 5 & -1 & 1 & 3 & : & -15 \\ 0 & 4 & -1 & -8 & : & 58 \end{bmatrix} \Rightarrow \begin{bmatrix} 1 & 0 & 0 & 0 & : & 2 \\ 0 & 1 & 0 & 0 & : & 6 \\ 0 & 0 & 1 & 0 & : & -10 \\ 0 & 0 & 0 & 1 & : & -3 \end{bmatrix}$$

$x = 2, y = 6, z = -10, w = -3$

Solution: $(2, 6, -10, -3)$

30. Use the reduced row-echelon form feature of the graphing utility.

$$\begin{bmatrix} 4 & 12 & 2 & : & 20 \\ 1 & 6 & 4 & : & 12 \\ 1 & 6 & 1 & : & 8 \\ -2 & -10 & -2 & : & -10 \end{bmatrix} \Rightarrow \begin{bmatrix} 1 & 0 & 0 & : & 0 \\ 0 & 1 & 0 & : & 0 \\ 0 & 0 & 1 & : & 0 \\ 0 & 0 & 0 & : & 1 \end{bmatrix}$$

The system is inconsistent and there is no solution.

31. $\begin{bmatrix} -1 & x \\ y & 9 \end{bmatrix} = \begin{bmatrix} -1 & 12 \\ -7 & 9 \end{bmatrix} \Rightarrow x = 12$ and $y = -7$

32. $\begin{bmatrix} -1 & 0 \\ x & 5 \\ -4 & y \end{bmatrix} = \begin{bmatrix} -1 & 0 \\ 8 & 5 \\ -4 & 0 \end{bmatrix} \Rightarrow x = 8, y = 0$

33. $\begin{bmatrix} x+3 & -4 & 4y \\ 0 & -3 & 2 \\ -2 & y+5 & 6x \end{bmatrix} = \begin{bmatrix} 5x-1 & -4 & 44 \\ 0 & -3 & 2 \\ -2 & 16 & 6 \end{bmatrix}$

$\left.\begin{array}{r} x + 3 = 5x - 1 \\ 4y = 44 \\ y + 5 = 16 \\ 6x = 6 \end{array}\right\}$ $x = 1$ and $y = 11$

34. $\begin{bmatrix} -9 & 4 & 2 & -5 \\ 0 & -3 & 7 & -4 \\ 6 & -1 & 1 & 0 \end{bmatrix} = \begin{bmatrix} -9 & 4 & x-10 & -5 \\ 0 & -3 & 7 & 2y \\ \frac{1}{2}x & -1 & 1 & 0 \end{bmatrix}$

$\left.\begin{array}{r} 6 = \frac{1}{2}x \\ 2 = x - 10 \\ -4 = 2y \end{array}\right\}$ $x = 12, y = -2$

35. (a) $A + B = \begin{bmatrix} 2 & -2 \\ 3 & 5 \end{bmatrix} + \begin{bmatrix} -3 & 10 \\ 12 & 8 \end{bmatrix} = \begin{bmatrix} -1 & 8 \\ 15 & 13 \end{bmatrix}$

(b) $A - B = \begin{bmatrix} 2 & -2 \\ 3 & 5 \end{bmatrix} - \begin{bmatrix} -3 & 10 \\ 12 & 8 \end{bmatrix} = \begin{bmatrix} 5 & -12 \\ -9 & -3 \end{bmatrix}$

(c) $4A = 4\begin{bmatrix} 2 & -2 \\ 3 & 5 \end{bmatrix} = \begin{bmatrix} 8 & -8 \\ 12 & 20 \end{bmatrix}$

(d) $A + 3B = \begin{bmatrix} 2 & -2 \\ 3 & 5 \end{bmatrix} + 3\begin{bmatrix} -3 & 10 \\ 12 & 8 \end{bmatrix} = \begin{bmatrix} 2 & -2 \\ 3 & 5 \end{bmatrix} + \begin{bmatrix} -9 & 30 \\ 36 & 24 \end{bmatrix} = \begin{bmatrix} -7 & 28 \\ 39 & 29 \end{bmatrix}$

36. (a) $A + B = \begin{bmatrix} 5 & 4 \\ -7 & 2 \\ 11 & 2 \end{bmatrix} + \begin{bmatrix} 4 & 12 \\ 20 & 40 \\ 15 & 30 \end{bmatrix} = \begin{bmatrix} 5+4 & 4+12 \\ -7+20 & 2+40 \\ 11+15 & 2+30 \end{bmatrix} = \begin{bmatrix} 9 & 16 \\ 13 & 42 \\ 26 & 32 \end{bmatrix}$

(b) $A - B = \begin{bmatrix} 5 & 4 \\ -7 & 2 \\ 11 & 2 \end{bmatrix} - \begin{bmatrix} 4 & 12 \\ 20 & 40 \\ 15 & 30 \end{bmatrix} = \begin{bmatrix} 5-4 & 4-12 \\ -7-20 & 2-40 \\ 11-15 & 2-30 \end{bmatrix} = \begin{bmatrix} 1 & -8 \\ -27 & -38 \\ -4 & -28 \end{bmatrix}$

(c) $4A = 4\begin{bmatrix} 5 & 4 \\ -7 & 2 \\ 11 & 2 \end{bmatrix} = \begin{bmatrix} 20 & 16 \\ -28 & 8 \\ 44 & 8 \end{bmatrix}$

(d) $A + 3B = \begin{bmatrix} 5 & 4 \\ -7 & 2 \\ 11 & 2 \end{bmatrix} + 3\begin{bmatrix} 4 & 12 \\ 20 & 40 \\ 15 & 30 \end{bmatrix} = \begin{bmatrix} 5 & 4 \\ -7 & 2 \\ 11 & 2 \end{bmatrix} + \begin{bmatrix} 12 & 36 \\ 60 & 120 \\ 45 & 90 \end{bmatrix} = \begin{bmatrix} 17 & 40 \\ 53 & 122 \\ 56 & 92 \end{bmatrix}$

37. (a) $A + B = \begin{bmatrix} 5 & 4 \\ -7 & 2 \\ 11 & 2 \end{bmatrix} + \begin{bmatrix} 0 & 3 \\ 4 & 12 \\ 20 & 40 \end{bmatrix} = \begin{bmatrix} 5 & 7 \\ -3 & 14 \\ 31 & 42 \end{bmatrix}$

(b) $A - B = \begin{bmatrix} 5 & 4 \\ -7 & 2 \\ 11 & 2 \end{bmatrix} - \begin{bmatrix} 0 & 3 \\ 4 & 12 \\ 20 & 40 \end{bmatrix} = \begin{bmatrix} 5 & 1 \\ -11 & -10 \\ -9 & -38 \end{bmatrix}$

(c) $4A = 4\begin{bmatrix} 5 & 4 \\ -7 & 2 \\ 11 & 2 \end{bmatrix} = \begin{bmatrix} 20 & 16 \\ -28 & 8 \\ 44 & 8 \end{bmatrix}$

(d) $A + 3B = \begin{bmatrix} 5 & 4 \\ -7 & 2 \\ 11 & 2 \end{bmatrix} + 3\begin{bmatrix} 0 & 3 \\ 4 & 12 \\ 20 & 40 \end{bmatrix} = \begin{bmatrix} 5 & 4 \\ -7 & 2 \\ 11 & 2 \end{bmatrix} + \begin{bmatrix} 0 & 9 \\ 12 & 36 \\ 60 & 120 \end{bmatrix} = \begin{bmatrix} 5 & 13 \\ 5 & 38 \\ 71 & 122 \end{bmatrix}$

38. (a) $A + B$ is not possible. A and B do not have the same order.

(b) $A - B$ is not possible. A and B do not have the same order.

(c) $4A = 4[6 \ -5 \ 7] = [24 \ -20 \ 28]$

(d) $A + 3B$ is not possible. A and B do not have the same order.

39. $\begin{bmatrix} 7 & 3 \\ -1 & 5 \end{bmatrix} + \begin{bmatrix} 10 & -20 \\ 14 & -3 \end{bmatrix} = \begin{bmatrix} 7 + 10 & 3 - 20 \\ -1 + 14 & 5 - 3 \end{bmatrix} = \begin{bmatrix} 17 & -17 \\ 13 & 2 \end{bmatrix}$

40. Since the matrices are not of the same order, the operation cannot be performed.

41. $-2\begin{bmatrix} 1 & 2 \\ 5 & -4 \\ 6 & 0 \end{bmatrix} + 8\begin{bmatrix} 7 & 1 \\ 1 & 2 \\ 1 & 4 \end{bmatrix} = \begin{bmatrix} -2 & -4 \\ -10 & 8 \\ -12 & 0 \end{bmatrix} + \begin{bmatrix} 56 & 8 \\ 8 & 16 \\ 8 & 32 \end{bmatrix} = \begin{bmatrix} 54 & 4 \\ -2 & 24 \\ -4 & 32 \end{bmatrix}$

42. $-\begin{bmatrix} 8 & -1 & 8 \\ -2 & 4 & 12 \\ 0 & -6 & 0 \end{bmatrix} - 5\begin{bmatrix} -2 & 0 & -4 \\ 3 & -1 & 1 \\ 6 & 12 & -8 \end{bmatrix} = \begin{bmatrix} -8 & 1 & -8 \\ 2 & -4 & -12 \\ 0 & 6 & 0 \end{bmatrix} + \begin{bmatrix} 10 & 0 & 20 \\ -15 & 5 & -5 \\ -30 & -60 & 40 \end{bmatrix}$

$= \begin{bmatrix} -8 + 10 & 1 + 0 & -8 + 20 \\ 2 - 15 & -4 + 5 & -12 - 5 \\ 0 - 30 & 6 - 60 & 0 + 40 \end{bmatrix} = \begin{bmatrix} 2 & 1 & 12 \\ -13 & 1 & -17 \\ -30 & -54 & 40 \end{bmatrix}$

43. $3\begin{bmatrix} 8 & -2 & 5 \\ 1 & 3 & -1 \end{bmatrix} + 6\begin{bmatrix} 4 & -2 & -3 \\ 2 & 7 & 6 \end{bmatrix} = \begin{bmatrix} 24 & -6 & 15 \\ 3 & 9 & -3 \end{bmatrix} + \begin{bmatrix} 24 & -12 & -18 \\ 12 & 42 & 36 \end{bmatrix} = \begin{bmatrix} 48 & -18 & -3 \\ 15 & 51 & 33 \end{bmatrix}$

44. $-5\begin{bmatrix} 2 & 0 \\ 7 & -2 \\ 8 & 2 \end{bmatrix} + 4\begin{bmatrix} 4 & -2 \\ 6 & 11 \\ -1 & 3 \end{bmatrix} = \begin{bmatrix} 6 & -8 \\ -11 & 54 \\ -44 & 2 \end{bmatrix}$

45. $X = 3A - 2B = 3\begin{bmatrix} -4 & 0 \\ 1 & -5 \\ -3 & 2 \end{bmatrix} - 2\begin{bmatrix} 1 & 2 \\ -2 & 1 \\ 4 & 4 \end{bmatrix}$

$= \begin{bmatrix} -14 & -4 \\ 7 & -17 \\ -17 & -2 \end{bmatrix}$

46. $X = \frac{1}{6}(4A + 3B) = \frac{1}{6}\left(4\begin{bmatrix} -4 & 0 \\ 1 & -5 \\ -3 & 2 \end{bmatrix} + 3\begin{bmatrix} 1 & 2 \\ -2 & 1 \\ 4 & 4 \end{bmatrix}\right) = \frac{1}{6}\left(\begin{bmatrix} -16 & 0 \\ 4 & -20 \\ -12 & 8 \end{bmatrix} + \begin{bmatrix} 3 & 6 \\ -6 & 3 \\ 12 & 12 \end{bmatrix}\right) = \frac{1}{6}\begin{bmatrix} -16+3 & 0+6 \\ 4-6 & -20+3 \\ -12+12 & 8+12 \end{bmatrix}$

$= \frac{1}{6}\begin{bmatrix} -13 & 6 \\ -2 & -17 \\ 0 & 20 \end{bmatrix} = \begin{bmatrix} -\frac{13}{6} & 1 \\ -\frac{1}{3} & -\frac{17}{6} \\ 0 & \frac{10}{3} \end{bmatrix}$

47. $X = \frac{1}{3}[B - 2A] = \frac{1}{3}\left(\begin{bmatrix} 1 & 2 \\ -2 & 1 \\ 4 & 4 \end{bmatrix} - 2\begin{bmatrix} -4 & 0 \\ 1 & -5 \\ -3 & 2 \end{bmatrix}\right)$

$= \frac{1}{3}\begin{bmatrix} 9 & 2 \\ -4 & 11 \\ 10 & 0 \end{bmatrix} = \begin{bmatrix} 3 & \frac{2}{3} \\ -\frac{4}{3} & \frac{11}{3} \\ \frac{10}{3} & 0 \end{bmatrix}$

48. $X = \frac{1}{3}(2A - 5B) = \frac{1}{3}\left(2\begin{bmatrix} -4 & 0 \\ 1 & -5 \\ -3 & 2 \end{bmatrix} - 5\begin{bmatrix} 1 & 2 \\ -2 & 1 \\ 4 & 4 \end{bmatrix}\right) = \frac{1}{3}\left(\begin{bmatrix} -8 & 0 \\ 2 & -10 \\ -6 & 4 \end{bmatrix} + \begin{bmatrix} -5 & -10 \\ 10 & -5 \\ -20 & -20 \end{bmatrix}\right) = \frac{1}{3}\begin{bmatrix} -8-5 & 0-10 \\ 2+10 & -10-5 \\ -6-20 & 4-20 \end{bmatrix}$

$= \frac{1}{3}\begin{bmatrix} -13 & -10 \\ 12 & -15 \\ -26 & -16 \end{bmatrix} = \begin{bmatrix} -\frac{13}{3} & -\frac{10}{3} \\ 4 & -5 \\ -\frac{26}{3} & -\frac{16}{3} \end{bmatrix}$

49. A and B are both 2×2 so AB exists.

$AB = \begin{bmatrix} 2 & -2 \\ 3 & 5 \end{bmatrix}\begin{bmatrix} -3 & 10 \\ 12 & 8 \end{bmatrix} = \begin{bmatrix} 2(-3) + (-2)(12) & 2(10) + (-2)(8) \\ 3(-3) + 5(12) & 3(10) + 5(8) \end{bmatrix} = \begin{bmatrix} -30 & 4 \\ 51 & 70 \end{bmatrix}$

50. Not possible because the number of columns of A does not equal the number of rows of B.

51. Since A is 3×2 and B is 2×2, AB exists.

$AB = \begin{bmatrix} 5 & 4 \\ -7 & 2 \\ 11 & 2 \end{bmatrix}\begin{bmatrix} 4 & 12 \\ 20 & 40 \end{bmatrix} = \begin{bmatrix} 5(4)+4(20) & 5(12)+4(40) \\ -7(4)+2(20) & -7(12)+2(40) \\ 11(4)+2(20) & 11(12)+2(40) \end{bmatrix} = \begin{bmatrix} 100 & 220 \\ 12 & -4 \\ 84 & 212 \end{bmatrix}$

52. $AB = \begin{bmatrix} 6 & -5 & 7 \end{bmatrix}\begin{bmatrix} -1 \\ 4 \\ 8 \end{bmatrix} = \begin{bmatrix} 6(-1) - 5(4) + 7(8) \end{bmatrix} = \begin{bmatrix} 30 \end{bmatrix}$

53. $\begin{bmatrix} 1 & 2 \\ 5 & -4 \\ 6 & 0 \end{bmatrix}\begin{bmatrix} 6 & -2 & 8 \\ 4 & 0 & 0 \end{bmatrix} = \begin{bmatrix} 1(6)+2(4) & 1(-2)+2(0) & 1(8)+2(0) \\ 5(6)+(-4)(4) & 5(-2)+(-4)(0) & 5(8)+(-4)(0) \\ 6(6)+(0)(4) & 6(-2)+(0)(0) & 6(8)+(0)(0) \end{bmatrix}$

$= \begin{bmatrix} 14 & -2 & 8 \\ 14 & -10 & 40 \\ 36 & -12 & 48 \end{bmatrix}$

54. Not possible because the number of columns of the first matrix does not equal the number of rows of the second matrix.

55. $\begin{bmatrix} 1 & 5 & 6 \\ 2 & -4 & 0 \end{bmatrix} \begin{bmatrix} 6 & 4 \\ -2 & 0 \\ 8 & 0 \end{bmatrix} = \begin{bmatrix} 1(6) + 5(-2) + 6(8) & 1(4) + 5(0) + 6(0) \\ 2(6) - 4(-2) + 0(8) & 2(4) - 4(0) + 0(0) \end{bmatrix}$

$$= \begin{bmatrix} 44 & 4 \\ 20 & 8 \end{bmatrix}$$

56. $\begin{bmatrix} 1 & 3 & 2 \\ 0 & 2 & -4 \\ 0 & 0 & 3 \end{bmatrix} \begin{bmatrix} 4 & -3 & 2 \\ 0 & 3 & -1 \\ 0 & 0 & 2 \end{bmatrix} = \begin{bmatrix} 1(4) & 1(-3) + 3(3) & 1(2) + 3(-1) + 2(2) \\ 0 & 2(3) & 2(-1) + (-4)(2) \\ 0 & 0 & 3(2) \end{bmatrix}$

$$= \begin{bmatrix} 4 & 6 & 3 \\ 0 & 6 & -10 \\ 0 & 0 & 6 \end{bmatrix}$$

57. $\begin{bmatrix} 4 \\ 6 \end{bmatrix} \begin{bmatrix} 6 & -2 \end{bmatrix} = \begin{bmatrix} 4(6) & 4(-2) \\ 6(6) & 6(-2) \end{bmatrix} = \begin{bmatrix} 24 & -8 \\ 36 & -12 \end{bmatrix}$

58. $\begin{bmatrix} 4 & -2 & 6 \end{bmatrix} \begin{bmatrix} -2 & 1 \\ 0 & -3 \\ 2 & 0 \end{bmatrix} = \begin{bmatrix} 4(-2) - 2(0) + 6(2) & 4(1) - 2(-3) + 6(0) \end{bmatrix}$

$$= \begin{bmatrix} 4 & 10 \end{bmatrix}$$

59. $\begin{bmatrix} 2 & 1 \\ 6 & 0 \end{bmatrix} \left(\begin{bmatrix} 4 & 2 \\ -3 & 1 \end{bmatrix} + \begin{bmatrix} -2 & 4 \\ 0 & 4 \end{bmatrix} \right) = \begin{bmatrix} 2 & 1 \\ 6 & 0 \end{bmatrix} \begin{bmatrix} 2 & 6 \\ -3 & 5 \end{bmatrix}$

$$= \begin{bmatrix} 2(2) + 1(-3) & 2(6) + 1(5) \\ 6(2) + 0 & 6(6) + 0 \end{bmatrix}$$

$$= \begin{bmatrix} 1 & 17 \\ 12 & 36 \end{bmatrix}$$

60. $-3 \begin{bmatrix} 1 & -1 \\ 4 & 2 \end{bmatrix} \left(\begin{bmatrix} 0 & 3 \\ 1 & 2 \end{bmatrix} \begin{bmatrix} 1 & 0 \\ 5 & -3 \end{bmatrix} \right) = \begin{bmatrix} -3 & 3 \\ -12 & -6 \end{bmatrix} \begin{bmatrix} 0(1) + 3(5) & 0(0) + 3(-3) \\ 1(1) + 2(5) & 1(0) + 2(-3) \end{bmatrix}$

$$= \begin{bmatrix} -3 & 3 \\ -12 & -6 \end{bmatrix} \begin{bmatrix} 15 & -9 \\ 11 & -6 \end{bmatrix}$$

$$= \begin{bmatrix} -3(15) + 3(11) & -3(-9) + 3(-6) \\ -12(15) - 6(11) & -12(-9) - 6(-6) \end{bmatrix}$$

$$= \begin{bmatrix} -12 & 9 \\ -246 & 144 \end{bmatrix}$$

61. $\begin{bmatrix} 4 & 1 \\ 11 & -7 \\ 12 & 3 \end{bmatrix} \begin{bmatrix} 3 & -5 & 6 \\ 2 & -2 & -2 \end{bmatrix} = \begin{bmatrix} 14 & -22 & 22 \\ 19 & -41 & 80 \\ 42 & -66 & 66 \end{bmatrix}$

62. $\begin{bmatrix} -2 & 3 & 10 \\ 4 & -2 & 2 \end{bmatrix} \begin{bmatrix} 1 & 1 \\ -5 & 2 \\ 3 & 2 \end{bmatrix} = \begin{bmatrix} 13 & 24 \\ 20 & 4 \end{bmatrix}$

63. $0.95A = 0.95 \begin{bmatrix} 80 & 120 & 140 \\ 40 & 100 & 80 \end{bmatrix} = \begin{bmatrix} 76 & 114 & 133 \\ 38 & 95 & 76 \end{bmatrix}$

64. $1.2A = 1.2 \begin{bmatrix} 80 & 70 & 90 & 40 \\ 50 & 30 & 80 & 20 \\ 90 & 60 & 100 & 50 \end{bmatrix} = \begin{bmatrix} 96 & 84 & 108 & 48 \\ 60 & 36 & 96 & 24 \\ 108 & 72 & 120 & 60 \end{bmatrix}$

65. $BA = \begin{bmatrix} 10.25 & 14.50 & 17.75 \end{bmatrix} \begin{bmatrix} 8200 & 7400 \\ 6500 & 9800 \\ 5400 & 4800 \end{bmatrix} = \begin{bmatrix} \$274,150 & \$303,150 \end{bmatrix}$

The merchandise shipped to warehouse 1 is worth \$274,150, and the merchandise shipped to warehouse 2 is worth \$303,150.

66. (a) $T = \begin{bmatrix} 120 & 80 & 20 \end{bmatrix}$

(b) $TC = \begin{bmatrix} 120 & 80 & 20 \end{bmatrix} \begin{bmatrix} 0.07 & 0.095 \\ 0.10 & 0.08 \\ 0.28 & 0.25 \end{bmatrix} = \begin{bmatrix} 22 & 22.8 \end{bmatrix}$

Your cost with company A is \$22.00. Your cost with company B is \$22.80.

67. $AB = \begin{bmatrix} -4 & -1 \\ 7 & 2 \end{bmatrix} \begin{bmatrix} -2 & -1 \\ 7 & 4 \end{bmatrix} = \begin{bmatrix} -4(-2) + (-1)(7) & -4(-1) + (-1)(4) \\ 7(-2) + 2(7) & 7(-1) + 2(4) \end{bmatrix}$

$= \begin{bmatrix} 1 & 0 \\ 0 & 1 \end{bmatrix} = I$

$BA = \begin{bmatrix} -2 & -1 \\ 7 & 4 \end{bmatrix} \begin{bmatrix} -4 & -1 \\ 7 & 2 \end{bmatrix} = \begin{bmatrix} -2(-4) + (-1)(7) & -2(-1) + (-1)(2) \\ 7(-4) + 4(7) & 7(-1) + 4(2) \end{bmatrix}$

$= \begin{bmatrix} 1 & 0 \\ 0 & 1 \end{bmatrix} = I$

68. $AB = \begin{bmatrix} 5 & -1 \\ 11 & -2 \end{bmatrix} \begin{bmatrix} -2 & 1 \\ -11 & 5 \end{bmatrix} = \begin{bmatrix} 1 & 0 \\ 0 & 1 \end{bmatrix} = I$

$BA = \begin{bmatrix} -2 & 1 \\ -11 & 5 \end{bmatrix} \begin{bmatrix} 5 & -1 \\ 11 & -2 \end{bmatrix} = \begin{bmatrix} 1 & 0 \\ 0 & 1 \end{bmatrix} = I$

69. $AB = \begin{bmatrix} 1 & 1 & 0 \\ 1 & 0 & 1 \\ 6 & 2 & 3 \end{bmatrix} \begin{bmatrix} -2 & -3 & 1 \\ 3 & 3 & -1 \\ 2 & 4 & -1 \end{bmatrix}$

$= \begin{bmatrix} 1(-2) + 1(3) + 0(2) & 1(-3) + 1(3) + 0(4) & 1(1) + 1(-1) + 0(-1) \\ 1(-2) + 0(3) + 1(2) & 1(-3) + 0(3) + 1(4) & 1(1) + 0(-1) + 1(-1) \\ 6(-2) + 2(3) + 3(2) & 6(-3) + 2(3) + 3(4) & 6(1) + 2(-1) + 3(-1) \end{bmatrix}$

$= \begin{bmatrix} 1 & 0 & 0 \\ 0 & 1 & 0 \\ 0 & 0 & 1 \end{bmatrix} = I$

$BA = \begin{bmatrix} -2 & -3 & 1 \\ 3 & 3 & -1 \\ 2 & 4 & -1 \end{bmatrix} \begin{bmatrix} 1 & 1 & 0 \\ 1 & 0 & 1 \\ 6 & 2 & 3 \end{bmatrix}$

$= \begin{bmatrix} -2(1) + (-3)(1) + 1(6) & -2(1) + (-3)(0) + 1(2) & -2(0) + (-3)(1) + 1(3) \\ 3(1) + 3(1) + (-1)(6) & 3(1) + 3(0) + (-1)(2) & 3(0) + 3(1) + (-1)(3) \\ 2(1) + 4(1) + (-1)(6) & 2(1) + 4(0) + (-1)(2) & 2(0) + 4(1) + (-1)(3) \end{bmatrix}$

$= \begin{bmatrix} 1 & 0 & 0 \\ 0 & 1 & 0 \\ 0 & 0 & 1 \end{bmatrix} = I$

70. $AB = \begin{bmatrix} 1 & -1 & 0 \\ -1 & 0 & -1 \\ 8 & -4 & 2 \end{bmatrix} \begin{bmatrix} -2 & 1 & \frac{1}{2} \\ -3 & 1 & \frac{1}{2} \\ 2 & -2 & -\frac{1}{2} \end{bmatrix} = \begin{bmatrix} 1 & 0 & 0 \\ 0 & 1 & 0 \\ 0 & 0 & 1 \end{bmatrix} = I$

$BA = \begin{bmatrix} -2 & 1 & \frac{1}{2} \\ -3 & 1 & \frac{1}{2} \\ 2 & -2 & -\frac{1}{2} \end{bmatrix} \begin{bmatrix} 1 & -1 & 0 \\ -1 & 0 & -1 \\ 8 & -4 & 2 \end{bmatrix} = \begin{bmatrix} 1 & 0 & 0 \\ 0 & 1 & 0 \\ 0 & 0 & 1 \end{bmatrix} = I$

71. $[A \;\vdots\; I] = \begin{bmatrix} -6 & 5 & \vdots & 1 & 0 \\ -5 & 4 & \vdots & 0 & 1 \end{bmatrix}$

$-\frac{1}{6}R_1 \rightarrow \begin{bmatrix} 1 & -\frac{5}{6} & \vdots & -\frac{1}{6} & 0 \\ -5 & 4 & \vdots & 0 & 1 \end{bmatrix}$

$5R_1 + R_2 \rightarrow \begin{bmatrix} 1 & -\frac{5}{6} & \vdots & -\frac{1}{6} & 0 \\ 0 & -\frac{1}{6} & \vdots & -\frac{5}{6} & 1 \end{bmatrix}$

$-6R_2 \rightarrow \begin{bmatrix} 1 & -\frac{5}{6} & \vdots & -\frac{1}{6} & 0 \\ 0 & 1 & \vdots & 5 & -6 \end{bmatrix}$

$\frac{5}{6}R_2 + R_1 \rightarrow \begin{bmatrix} 1 & 0 & \vdots & 4 & -5 \\ 0 & 1 & \vdots & 5 & -6 \end{bmatrix} = [I \;\vdots\; A^{-1}]$

$A^{-1} = \begin{bmatrix} 4 & -5 \\ 5 & -6 \end{bmatrix}$

72. $[A \;\vdots\; I] = \begin{bmatrix} -3 & -5 & \vdots & 1 & 0 \\ 2 & 3 & \vdots & 0 & 1 \end{bmatrix}$

$2R_2 + R_1 \rightarrow \begin{bmatrix} 1 & 1 & \vdots & 1 & 2 \\ 2 & 3 & \vdots & 0 & 1 \end{bmatrix}$

$-2R_1 + R_2 \rightarrow \begin{bmatrix} 1 & 1 & \vdots & 1 & 2 \\ 0 & 1 & \vdots & -2 & -3 \end{bmatrix}$

$-R_2 + R_1 \rightarrow \begin{bmatrix} 1 & 0 & \vdots & 3 & 5 \\ 0 & 1 & \vdots & -2 & -3 \end{bmatrix} = [I \;\vdots\; A^{-1}]$

$A^{-1} = \begin{bmatrix} 3 & 5 \\ -2 & -3 \end{bmatrix}$

73. $[A \;\vdots\; I] = \begin{bmatrix} -1 & -2 & -2 & \vdots & 1 & 0 & 0 \\ 3 & 7 & 9 & \vdots & 0 & 1 & 0 \\ 1 & 4 & 7 & \vdots & 0 & 0 & 1 \end{bmatrix}$

$-R_1 \rightarrow \begin{bmatrix} 1 & 2 & 2 & \vdots & -1 & 0 & 0 \\ 3 & 7 & 9 & \vdots & 0 & 1 & 0 \\ 1 & 4 & 7 & \vdots & 0 & 0 & 1 \end{bmatrix}$

$\begin{matrix} \\ -3R_1 + R_2 \rightarrow \\ -R_1 + R_3 \rightarrow \end{matrix} \begin{bmatrix} 1 & 2 & 2 & \vdots & -1 & 0 & 0 \\ 0 & 1 & 3 & \vdots & 3 & 1 & 0 \\ 0 & 2 & 5 & \vdots & 1 & 0 & 1 \end{bmatrix}$

$\begin{matrix} -2R_2 + R_1 \rightarrow \\ \\ -2R_2 + R_3 \rightarrow \end{matrix} \begin{bmatrix} 1 & 0 & -4 & \vdots & -7 & -2 & 0 \\ 0 & 1 & 3 & \vdots & 3 & 1 & 0 \\ 0 & 0 & -1 & \vdots & -5 & -2 & 1 \end{bmatrix}$

$\begin{matrix} -4R_3 + R_1 \rightarrow \\ 3R_3 + R_2 \rightarrow \\ -R_3 \rightarrow \end{matrix} \begin{bmatrix} 1 & 0 & 0 & \vdots & 13 & 6 & -4 \\ 0 & 1 & 0 & \vdots & -12 & -5 & 3 \\ 0 & 0 & 1 & \vdots & 5 & 2 & -1 \end{bmatrix} = [I \;\vdots\; A^{-1}]$

$A^{-1} = \begin{bmatrix} 13 & 6 & -4 \\ -12 & -5 & 3 \\ 5 & 2 & -1 \end{bmatrix}$

74. $[A \; \vdots \; I] = \begin{bmatrix} 0 & -2 & 1 & \vdots & 1 & 0 & 0 \\ -5 & -2 & -3 & \vdots & 0 & 1 & 0 \\ 7 & 3 & 4 & \vdots & 0 & 0 & 1 \end{bmatrix}$

$\begin{matrix} R_3 \\ \\ R_1 \end{matrix} \begin{bmatrix} 7 & 3 & 4 & \vdots & 0 & 0 & 1 \\ -5 & -2 & -3 & \vdots & 0 & 1 & 0 \\ 0 & -2 & 1 & \vdots & 1 & 0 & 0 \end{bmatrix}$

$R_2 + R_1 \rightarrow \begin{bmatrix} 2 & 1 & 1 & \vdots & 0 & 1 & 1 \\ -5 & -2 & -3 & \vdots & 0 & 1 & 0 \\ 0 & -2 & 1 & \vdots & 1 & 0 & 0 \end{bmatrix}$

$5R_1 + 2R_2 \rightarrow \begin{bmatrix} 2 & 1 & 1 & \vdots & 0 & 1 & 1 \\ 0 & 1 & -1 & \vdots & 0 & 7 & 5 \\ 0 & -2 & 1 & \vdots & 1 & 0 & 0 \end{bmatrix}$

$\begin{matrix} -R_2 + R_1 \rightarrow \\ \\ 2R_2 + R_3 \rightarrow \end{matrix} \begin{bmatrix} 2 & 0 & 2 & \vdots & 0 & -6 & -4 \\ 0 & 1 & -1 & \vdots & 0 & 7 & 5 \\ 0 & 0 & -1 & \vdots & 1 & 14 & 10 \end{bmatrix}$

$\begin{matrix} \frac{1}{2}R_1 \rightarrow \\ \\ -R_3 \rightarrow \end{matrix} \begin{bmatrix} 1 & 0 & 1 & \vdots & 0 & -3 & -2 \\ 0 & 1 & -1 & \vdots & 0 & 7 & 5 \\ 0 & 0 & 1 & \vdots & -1 & -14 & -10 \end{bmatrix}$

$\begin{matrix} -R_3 + R_1 \rightarrow \\ R_3 + R_2 \rightarrow \end{matrix} \begin{bmatrix} 1 & 0 & 0 & \vdots & 1 & 11 & 8 \\ 0 & 1 & 0 & \vdots & -1 & -7 & -5 \\ 0 & 0 & 1 & \vdots & -1 & -14 & -10 \end{bmatrix} = [I \; \vdots \; A^{-1}]$

$A^{-1} = \begin{bmatrix} 1 & 11 & 8 \\ -1 & -7 & -5 \\ -1 & -14 & -10 \end{bmatrix}$

75. $\begin{bmatrix} 2 & 0 & 3 \\ -1 & 1 & 1 \\ 2 & -2 & 1 \end{bmatrix}^{-1} = \begin{bmatrix} \frac{1}{2} & -1 & -\frac{1}{2} \\ \frac{1}{2} & -\frac{2}{3} & -\frac{5}{6} \\ 0 & \frac{2}{3} & \frac{1}{3} \end{bmatrix}$

76. $A = \begin{bmatrix} 1 & 4 & 6 \\ 2 & -3 & 1 \\ -1 & 18 & 16 \end{bmatrix}$

A^{-1} does not exist.

77. $\begin{bmatrix} 1 & 3 & 1 & 6 \\ 4 & 4 & 2 & 6 \\ 3 & 4 & 1 & 2 \\ -1 & 2 & -1 & -2 \end{bmatrix}^{-1} = \begin{bmatrix} -3 & 6 & -\frac{11}{2} & \frac{7}{2} \\ 1 & -2 & 2 & -1 \\ 7 & -15 & \frac{29}{2} & -\frac{19}{2} \\ -1 & \frac{5}{2} & -\frac{5}{2} & \frac{3}{2} \end{bmatrix}$

$= \begin{bmatrix} -3 & 6 & -5.5 & 3.5 \\ 1 & -2 & 2 & -1 \\ 7 & -15 & 14.5 & -9.5 \\ -1 & 2.5 & -2.5 & 1.5 \end{bmatrix}$

78. $A = \begin{bmatrix} 8 & 0 & 2 & 8 \\ 4 & -2 & 0 & -2 \\ 1 & 2 & 1 & 4 \\ -1 & 4 & 1 & 1 \end{bmatrix}$

$A^{-1} = \begin{bmatrix} -2.5 & 3 & 7 & -2 \\ -4 & 4.5 & 11 & -3 \\ 14.5 & -16 & -40 & 12 \\ -1 & 1 & 3 & -1 \end{bmatrix}$

79. $A = \begin{bmatrix} -7 & 2 \\ -8 & 2 \end{bmatrix}$

$A^{-1} = \dfrac{1}{-7(2) - 2(-8)} \begin{bmatrix} 2 & -2 \\ 8 & -7 \end{bmatrix} = \dfrac{1}{2} \begin{bmatrix} 2 & -2 \\ 8 & -7 \end{bmatrix} = \begin{bmatrix} 1 & -1 \\ 4 & -\frac{7}{2} \end{bmatrix}$

80. $A = \begin{bmatrix} 10 & 4 \\ 7 & 3 \end{bmatrix}$

$ad - bc = (10)(3) - (4)(7) = 2$

$A^{-1} = \dfrac{1}{10(3) - 4(7)} \begin{bmatrix} 3 & -4 \\ -7 & 10 \end{bmatrix} = \dfrac{1}{2} \begin{bmatrix} 3 & -4 \\ -7 & 10 \end{bmatrix} = \begin{bmatrix} \frac{3}{2} & -2 \\ -\frac{7}{2} & 5 \end{bmatrix}$

81. $A = \begin{bmatrix} -\frac{1}{2} & 20 \\ \frac{3}{10} & -6 \end{bmatrix}$

$A^{-1} = \dfrac{1}{-\frac{1}{2}(-6) - 20\left(\frac{3}{10}\right)} \begin{bmatrix} -6 & -20 \\ -\frac{3}{10} & -\frac{1}{2} \end{bmatrix} = -\dfrac{1}{3} \begin{bmatrix} -6 & -20 \\ -\frac{3}{10} & -\frac{1}{2} \end{bmatrix}$

$\quad = \begin{bmatrix} 2 & \frac{20}{3} \\ \frac{1}{10} & \frac{1}{6} \end{bmatrix}$

82. $A = \begin{bmatrix} -\frac{3}{4} & \frac{5}{2} \\ -\frac{4}{5} & -\frac{8}{3} \end{bmatrix}$

$ad - bc = \left(-\frac{3}{4}\right)\left(-\frac{8}{3}\right) - \left(\frac{5}{2}\right)\left(-\frac{4}{5}\right) = 2 + 2 = 4$

$A^{-1} = \dfrac{1}{4} \begin{bmatrix} -\frac{8}{3} & -\frac{5}{2} \\ \frac{4}{5} & -\frac{3}{4} \end{bmatrix} = \begin{bmatrix} -\frac{2}{3} & -\frac{5}{8} \\ \frac{1}{5} & -\frac{3}{16} \end{bmatrix}$

83. $\begin{cases} -x + 4y = 8 \\ 2x - 7y = -5 \end{cases}$

$\begin{bmatrix} x \\ y \end{bmatrix} = \begin{bmatrix} -1 & 4 \\ 2 & -7 \end{bmatrix}^{-1} \begin{bmatrix} 8 \\ -5 \end{bmatrix} = \begin{bmatrix} 7 & 4 \\ 2 & 1 \end{bmatrix} \begin{bmatrix} 8 \\ -5 \end{bmatrix}$

$\quad = \begin{bmatrix} 7(8) + 4(-5) \\ 2(8) + 1(-5) \end{bmatrix} = \begin{bmatrix} 36 \\ 11 \end{bmatrix}$

Solution: $(36, 11)$

84. $\begin{cases} 5x - y = 13 \\ -9x + 2y = -24 \end{cases}$

$\begin{bmatrix} x \\ y \end{bmatrix} = \begin{bmatrix} 5 & -1 \\ -9 & 2 \end{bmatrix}^{-1} \begin{bmatrix} 13 \\ -24 \end{bmatrix} = \begin{bmatrix} 2 & 1 \\ 9 & 5 \end{bmatrix} \begin{bmatrix} 13 \\ -24 \end{bmatrix} = \begin{bmatrix} 2 \\ -3 \end{bmatrix}$

Solution: $(2, -3)$

85. $\begin{cases} -3x + 10y = 8 \\ 5x - 17y = -13 \end{cases}$

$\begin{bmatrix} x \\ y \end{bmatrix} = \begin{bmatrix} -3 & 10 \\ 5 & -17 \end{bmatrix}^{-1} \begin{bmatrix} 8 \\ -13 \end{bmatrix} = \begin{bmatrix} -17 & -10 \\ -5 & -3 \end{bmatrix} \begin{bmatrix} 8 \\ -13 \end{bmatrix}$

$\quad = \begin{bmatrix} -17(8) + (-10)(-13) \\ -5(8) + (-3)(-13) \end{bmatrix} = \begin{bmatrix} -6 \\ -1 \end{bmatrix}$

Solution: $(-6, -1)$

86. $\begin{cases} 4x - 2y = -10 \\ -19x + 9y = 47 \end{cases}$

$\begin{bmatrix} x \\ y \end{bmatrix} = \begin{bmatrix} 4 & -2 \\ -19 & 9 \end{bmatrix}^{-1} \begin{bmatrix} -10 \\ 47 \end{bmatrix}$

$\quad = \begin{bmatrix} -\frac{9}{2} & -1 \\ -\frac{19}{2} & -2 \end{bmatrix} \begin{bmatrix} -10 \\ 47 \end{bmatrix} = \begin{bmatrix} -2 \\ 1 \end{bmatrix}$

Solution: $(-2, 1)$

87. $\begin{cases} 3x + 2y - z = 6 \\ x - y + 2z = -1 \\ 5x + y + z = 7 \end{cases}$

$\begin{bmatrix} x \\ y \\ z \end{bmatrix} = \begin{bmatrix} 3 & 2 & -1 \\ 1 & -1 & 2 \\ 5 & 1 & 1 \end{bmatrix}^{-1} \begin{bmatrix} 6 \\ -1 \\ 7 \end{bmatrix} = \begin{bmatrix} -1 & -1 & 1 \\ 3 & \frac{8}{3} & -\frac{7}{3} \\ 2 & \frac{7}{3} & -\frac{5}{3} \end{bmatrix} \begin{bmatrix} 6 \\ -1 \\ 7 \end{bmatrix}$

$\quad = \begin{bmatrix} -1(6) - 1(-1) + 1(7) \\ 3(6) + \frac{8}{3}(-1) - \frac{7}{3}(7) \\ 2(6) + \frac{7}{3}(-1) - \frac{5}{3}(7) \end{bmatrix} = \begin{bmatrix} 2 \\ -1 \\ -2 \end{bmatrix}$

Solution: $(2, -1, -2)$

88. $\begin{cases} -x + 4y - 2z = 12 \\ 2x - 9y + 5z = -25 \\ -x + 5y - 4z = 10 \end{cases}$

$\begin{bmatrix} x \\ y \\ z \end{bmatrix} = \begin{bmatrix} -1 & 4 & -2 \\ 2 & -9 & 5 \\ -1 & 5 & -4 \end{bmatrix}^{-1} \begin{bmatrix} 12 \\ -25 \\ 10 \end{bmatrix}$

$\quad = \begin{bmatrix} -11 & -6 & -2 \\ -3 & -2 & -1 \\ -1 & -1 & -1 \end{bmatrix} \begin{bmatrix} 12 \\ -25 \\ 10 \end{bmatrix} = \begin{bmatrix} -2 \\ 4 \\ 3 \end{bmatrix}$

Solution: $(-2, 4, 3)$

89. $\begin{cases} -2x + y + 2z = -13 \\ -x - 4y + z = -11 \\ -y - z = 0 \end{cases}$

$\begin{bmatrix} x \\ y \\ z \end{bmatrix} = \begin{bmatrix} -2 & 1 & 2 \\ -1 & -4 & 1 \\ 0 & -1 & -1 \end{bmatrix}^{-1} \begin{bmatrix} -13 \\ -11 \\ 0 \end{bmatrix} = \begin{bmatrix} -\frac{5}{9} & \frac{1}{9} & -1 \\ \frac{1}{9} & -\frac{2}{9} & 0 \\ -\frac{1}{9} & \frac{2}{9} & -1 \end{bmatrix} \begin{bmatrix} -13 \\ -11 \\ 0 \end{bmatrix}$

$\quad = \begin{bmatrix} -\frac{5}{9}(-13) + \frac{1}{9}(-11) - 1(0) \\ \frac{1}{9}(-13) - \frac{2}{9}(-11) + 0(0) \\ -\frac{1}{9}(-13) + \frac{2}{9}(-11) - 1(0) \end{bmatrix} = \begin{bmatrix} 6 \\ 1 \\ -1 \end{bmatrix}$

Solution: $(6, 1, -1)$

90. $\begin{cases} 3x - y + 5z = -14 \\ -x + y + 6z = 8 \\ -8x + 4y - z = 44 \end{cases}$

$$\begin{bmatrix} x \\ y \\ z \end{bmatrix} = \begin{bmatrix} 3 & -1 & 5 \\ -1 & 1 & 6 \\ -8 & 4 & -1 \end{bmatrix}^{-1} \begin{bmatrix} -14 \\ 8 \\ 44 \end{bmatrix} = \begin{bmatrix} \frac{25}{6} & -\frac{19}{6} & \frac{11}{6} \\ \frac{49}{6} & -\frac{37}{6} & \frac{23}{6} \\ -\frac{2}{3} & \frac{2}{3} & -\frac{1}{3} \end{bmatrix} \begin{bmatrix} -14 \\ 8 \\ 44 \end{bmatrix} = \begin{bmatrix} -3 \\ 5 \\ 0 \end{bmatrix}$$

Solution: $(-3, 5, 0)$

91. $\begin{cases} x + 2y = -1 \\ 3x + 4y = -5 \end{cases}$

$$\begin{bmatrix} x \\ y \end{bmatrix} = \begin{bmatrix} 1 & 2 \\ 3 & 4 \end{bmatrix}^{-1} \begin{bmatrix} -1 \\ -5 \end{bmatrix} = \begin{bmatrix} -2 & 1 \\ \frac{3}{2} & -\frac{1}{2} \end{bmatrix} \begin{bmatrix} -1 \\ -5 \end{bmatrix} = \begin{bmatrix} -3 \\ 1 \end{bmatrix}$$

Solution: $(-3, 1)$

92. $\begin{cases} x + 3y = 23 \\ -6x + 2y = -18 \end{cases}$

$$\begin{bmatrix} x \\ y \end{bmatrix} = \begin{bmatrix} 1 & 3 \\ -6 & 2 \end{bmatrix}^{-1} \begin{bmatrix} 23 \\ -18 \end{bmatrix} = \begin{bmatrix} 0.1 & -0.15 \\ 0.3 & 0.05 \end{bmatrix} \begin{bmatrix} 23 \\ -18 \end{bmatrix} = \begin{bmatrix} 5 \\ 6 \end{bmatrix}$$

$x = 5, y = 6$

Solution: $(5, 6)$

93. $\begin{cases} -3x - 3y - 4z = 2 \\ y + z = -1 \\ 4x + 3y + 4z = -1 \end{cases}$

$$\begin{bmatrix} x \\ y \\ z \end{bmatrix} = \begin{bmatrix} -3 & -3 & -4 \\ 0 & 1 & 1 \\ 4 & 3 & 4 \end{bmatrix}^{-1} \begin{bmatrix} 2 \\ -1 \\ -1 \end{bmatrix} = \begin{bmatrix} 1 & 0 & 1 \\ 4 & 4 & 3 \\ -4 & -3 & -3 \end{bmatrix} \begin{bmatrix} 2 \\ -1 \\ -1 \end{bmatrix} = \begin{bmatrix} 1 \\ 1 \\ -2 \end{bmatrix}$$

Solution: $(1, 1, -2)$

94. $\begin{cases} x - 3y - 2z = 8 \\ -2x + 7y + 3z = -19 \\ x - y - 3z = 3 \end{cases}$

$$\begin{bmatrix} x \\ y \\ z \end{bmatrix} = \begin{bmatrix} 1 & -3 & -2 \\ -2 & 7 & 3 \\ 1 & -1 & -3 \end{bmatrix}^{-1} \begin{bmatrix} 8 \\ -19 \\ 3 \end{bmatrix} = \begin{bmatrix} -18 & -7 & 5 \\ -3 & -1 & 1 \\ -5 & -2 & 1 \end{bmatrix} \begin{bmatrix} 8 \\ -19 \\ 3 \end{bmatrix} = \begin{bmatrix} 4 \\ -2 \\ 1 \end{bmatrix}$$

$x = 4, y = -2, z = 1$

Solution: $(4, -2, 1)$

95. $\begin{vmatrix} 8 & 5 \\ 2 & -4 \end{vmatrix} = 8(-4) - 5(2) = -42$

96. $\begin{vmatrix} -9 & 11 \\ 7 & -4 \end{vmatrix} = (-9)(-4) - (11)(7) = -41$

97. $\begin{vmatrix} 50 & -30 \\ 10 & 5 \end{vmatrix} = 50(5) - (-30)(10) = 550$

98. $\begin{vmatrix} 14 & -24 \\ 12 & -15 \end{vmatrix} = (14)(-15) - (-24)(12) = 78$

99. $\begin{bmatrix} 2 & -1 \\ 7 & 4 \end{bmatrix}$

    (a) $M_{11} = 4$ (b) $C_{11} = M_{11} = 4$

        $M_{12} = 7$ $C_{12} = -M_{12} = -7$

        $M_{21} = -1$ $C_{21} = -M_{21} = 1$

        $M_{22} = 2$ $C_{22} = M_{22} = 2$

100. $\begin{bmatrix} 3 & 6 \\ 5 & -4 \end{bmatrix}$

    (a) $M_{11} = -4$ (b) $C_{11} = M_{11} = -4$

        $M_{12} = 5$ $C_{12} = -M_{12} = -5$

        $M_{21} = 6$ $C_{21} = -M_{21} = -6$

        $M_{22} = 3$ $C_{22} = M_{22} = 3$

101. $\begin{bmatrix} 3 & 2 & -1 \\ -2 & 5 & 0 \\ 1 & 8 & 6 \end{bmatrix}$

    (a) $M_{11} = \begin{vmatrix} 5 & 0 \\ 8 & 6 \end{vmatrix} = 30$

        $M_{12} = \begin{vmatrix} -2 & 0 \\ 1 & 6 \end{vmatrix} = -12$

        $M_{13} = \begin{vmatrix} -2 & 5 \\ 1 & 8 \end{vmatrix} = -21$

        $M_{21} = \begin{vmatrix} 2 & -1 \\ 8 & 6 \end{vmatrix} = 20$

        $M_{22} = \begin{vmatrix} 3 & -1 \\ 1 & 6 \end{vmatrix} = 19$

        $M_{23} = \begin{vmatrix} 3 & 2 \\ 1 & 8 \end{vmatrix} = 22$

        $M_{31} = \begin{vmatrix} 2 & -1 \\ 5 & 0 \end{vmatrix} = 5$

        $M_{32} = \begin{vmatrix} 3 & -1 \\ -2 & 0 \end{vmatrix} = -2$

        $M_{33} = \begin{vmatrix} 3 & 2 \\ -2 & 5 \end{vmatrix} = 19$

    (b) $C_{11} = M_{11} = 30$

        $C_{12} = -M_{12} = 12$

        $C_{13} = M_{13} = -21$

        $C_{21} = -M_{21} = -20$

        $C_{22} = M_{22} = 19$

        $C_{23} = -M_{23} = -22$

        $C_{31} = M_{31} = 5$

        $C_{32} = -M_{32} = 2$

        $C_{33} = M_{33} = 19$

102. $\begin{bmatrix} 8 & 3 & 4 \\ 6 & 5 & -9 \\ -4 & 1 & 2 \end{bmatrix}$

    (a) $M_{11} = \begin{vmatrix} 5 & -9 \\ 1 & 2 \end{vmatrix} = 19$

        $M_{12} = \begin{vmatrix} 6 & -9 \\ -4 & 2 \end{vmatrix} = -24$

        $M_{13} = \begin{vmatrix} 6 & 5 \\ -4 & 1 \end{vmatrix} = 26$

        $M_{21} = \begin{vmatrix} 3 & 4 \\ 1 & 2 \end{vmatrix} = 2$

        $M_{22} = \begin{vmatrix} 8 & 4 \\ -4 & 2 \end{vmatrix} = 32$

        $M_{23} = \begin{vmatrix} 8 & 3 \\ -4 & 1 \end{vmatrix} = 20$

        $M_{31} = \begin{vmatrix} 3 & 4 \\ 5 & -9 \end{vmatrix} = -47$

        $M_{32} = \begin{vmatrix} 8 & 4 \\ 6 & -9 \end{vmatrix} = -96$

        $M_{33} = \begin{vmatrix} 8 & 3 \\ 6 & 5 \end{vmatrix} = 22$

    (b) $C_{11} = M_{11} = 19$

        $C_{12} = -M_{12} = 24$

        $C_{13} = M_{13} = 26$

        $C_{21} = -M_{21} = -2$

        $C_{22} = M_{22} = 32$

        $C_{23} = -M_{23} = -20$

        $C_{31} = M_{31} = -47$

        $C_{32} = -M_{32} = 96$

        $C_{33} = M_{33} = 22$

103. Expand using Column 2.

$$\begin{vmatrix} -2 & 4 & 1 \\ -6 & 0 & 2 \\ 5 & 3 & 4 \end{vmatrix} = -4\begin{vmatrix} -6 & 2 \\ 5 & 4 \end{vmatrix} - 3\begin{vmatrix} -2 & 1 \\ -6 & 2 \end{vmatrix}$$

$$= -4(-34) - 3(2) = 130$$

104. Expand using Row 3.

$$\begin{vmatrix} 4 & 7 & -1 \\ 2 & -3 & 4 \\ -5 & 1 & -1 \end{vmatrix} = -5\begin{vmatrix} 7 & -1 \\ -3 & 4 \end{vmatrix} - 1\begin{vmatrix} 4 & -1 \\ 2 & 4 \end{vmatrix} - 1\begin{vmatrix} 4 & 7 \\ 2 & -3 \end{vmatrix}$$

$$= -5(25) - (18) - (-26) = -117$$

105. Expand along Row 1.

$$\begin{vmatrix} 3 & 0 & -4 & 0 \\ 0 & 8 & 1 & 2 \\ 6 & 1 & 8 & 2 \\ 0 & 3 & -4 & 1 \end{vmatrix} = 3\begin{vmatrix} 8 & 1 & 2 \\ 1 & 8 & 2 \\ 3 & -4 & 1 \end{vmatrix} + (-4)\begin{vmatrix} 0 & 8 & 2 \\ 6 & 1 & 2 \\ 0 & 3 & 1 \end{vmatrix}$$

$$= 3[8(8 - (-8)) - 1(1 - 6) + 2(-4 - 24)] - 4[0 - 6(8 - 6) + 0]$$

$$= 3[128 + 5 - 56] - 4[-12]$$

$$= 279$$

106. Expand using Row 1, then use Row 3 of each 3×3 matrix.

$$\begin{vmatrix} -5 & 6 & 0 & 0 \\ 0 & 1 & -1 & 2 \\ -3 & 4 & -5 & 1 \\ 1 & 6 & 0 & 3 \end{vmatrix} = -5\begin{vmatrix} 1 & -1 & 2 \\ 4 & -5 & 1 \\ 6 & 0 & 3 \end{vmatrix} - 6\begin{vmatrix} 0 & -1 & 2 \\ -3 & -5 & 1 \\ 1 & 0 & 3 \end{vmatrix}$$

$$= -5[6(-1 + 10) + 3(-5 + 4)] - 6[(-1 + 10) + 3(0 - 3)]$$

$$= -5(54 - 3) - 6(9 - 9)$$

$$= -255$$

107. $\begin{cases} 5x - 2y = 6 \\ -11x + 3y = -23 \end{cases}$

$$x = \frac{\begin{vmatrix} 6 & -2 \\ -23 & 3 \end{vmatrix}}{\begin{vmatrix} 5 & -2 \\ -11 & 3 \end{vmatrix}} = \frac{-28}{-7} = 4, \qquad y = \frac{\begin{vmatrix} 5 & 6 \\ -11 & -23 \end{vmatrix}}{\begin{vmatrix} 5 & -2 \\ -11 & 3 \end{vmatrix}} = \frac{-49}{-7} = 7$$

Solution: $(4, 7)$

108. $\begin{cases} 3x + 8y = -7 \\ 9x - 5y = 37 \end{cases}$

$$x = \frac{\begin{vmatrix} -7 & 8 \\ 37 & -5 \end{vmatrix}}{\begin{vmatrix} 3 & 8 \\ 9 & -5 \end{vmatrix}} = \frac{-261}{-87} = 3, \qquad y = \frac{\begin{vmatrix} 3 & -7 \\ 9 & 37 \end{vmatrix}}{\begin{vmatrix} 3 & 8 \\ 9 & -5 \end{vmatrix}} = \frac{174}{-87} = -2$$

Solution: $(3, -2)$

109. $\begin{cases} -2x + 3y - 5z = -11 \\ 4x - y + z = -3 \\ -x - 4y + 6z = 15 \end{cases}$

$$D = \begin{vmatrix} -2 & 3 & -5 \\ 4 & -1 & 1 \\ -1 & -4 & 6 \end{vmatrix} = -2(-1)^2\begin{vmatrix} -1 & 1 \\ -4 & 6 \end{vmatrix} + 4(-1)^3\begin{vmatrix} 3 & -5 \\ -4 & 6 \end{vmatrix} - 1(-1)^4\begin{vmatrix} 3 & -5 \\ -1 & 1 \end{vmatrix}$$

$$= -2(-2) - 4(-2) - (-2) = 14$$

$$x = \frac{\begin{vmatrix} -11 & 3 & -5 \\ -3 & -1 & 1 \\ 15 & -4 & 6 \end{vmatrix}}{14} = \frac{-11(-1)^2\begin{vmatrix} -1 & 1 \\ -4 & 6 \end{vmatrix} - 3(-1)^3\begin{vmatrix} 3 & -5 \\ -4 & 6 \end{vmatrix} + 15(-1)^4\begin{vmatrix} 3 & -5 \\ -1 & 1 \end{vmatrix}}{14}$$

$$= \frac{-11(-2) + 3(-2) + 15(-2)}{14} = \frac{-14}{14} = -1$$

$$y = \frac{\begin{vmatrix} -2 & -11 & -5 \\ 4 & -3 & 1 \\ -1 & 15 & 6 \end{vmatrix}}{14} = \frac{-2(-1)^2\begin{vmatrix} -3 & 1 \\ 15 & 6 \end{vmatrix} + 4(-1)^3\begin{vmatrix} -11 & -5 \\ 15 & 6 \end{vmatrix} - 1(-1)^4\begin{vmatrix} -11 & -5 \\ -3 & 1 \end{vmatrix}}{14}$$

$$= \frac{-2(-33) - 4(9) - 1(-26)}{14} = \frac{56}{14} = 4$$

$$z = \frac{\begin{vmatrix} -2 & 3 & -11 \\ 4 & -1 & -3 \\ -1 & -4 & 15 \end{vmatrix}}{14} = \frac{-2(-1)^2\begin{vmatrix} -1 & -3 \\ -4 & 15 \end{vmatrix} + 4(-1)^3\begin{vmatrix} 3 & -11 \\ -4 & 15 \end{vmatrix} - 1(-1)^4\begin{vmatrix} 3 & -11 \\ -1 & -3 \end{vmatrix}}{14}$$

$$= \frac{-2(-27) - 4(1) - 1(-20)}{14} = \frac{70}{14} = 5$$

Solution: $(-1, 4, 5)$

110. $\begin{cases} 5x - 2y + z = 15 \\ 3x - 3y - z = -7, \\ 2x - y - 7z = -3 \end{cases}$ $D = \begin{vmatrix} 5 & -2 & 1 \\ 3 & -3 & -1 \\ 2 & -1 & -7 \end{vmatrix} = 65$

$$x = \frac{\begin{vmatrix} 15 & -2 & 1 \\ -7 & -3 & -1 \\ -3 & -1 & -7 \end{vmatrix}}{65} = \frac{390}{65} = 6, \quad y = \frac{\begin{vmatrix} 5 & 15 & 1 \\ 3 & -7 & -1 \\ 2 & -3 & -7 \end{vmatrix}}{65} = \frac{520}{65} = 8, \quad z = \frac{\begin{vmatrix} 5 & -2 & 15 \\ 3 & -3 & -7 \\ 2 & -1 & -3 \end{vmatrix}}{65} = \frac{65}{65} = 1$$

Solution: $(6, 8, 1)$

111. $(1, 0), (5, 0), (5, 8)$

$$\text{Area} = \frac{1}{2}\begin{vmatrix} 1 & 0 & 1 \\ 5 & 0 & 1 \\ 5 & 8 & 1 \end{vmatrix} = \frac{1}{2}\left(1\begin{vmatrix} 0 & 1 \\ 8 & 1 \end{vmatrix} + 1\begin{vmatrix} 5 & 0 \\ 5 & 8 \end{vmatrix}\right) = \frac{1}{2}(-8 + 40) = \frac{1}{2}(32) = 16 \text{ square units}$$

112. $(-4, 0), (4, 0), (0, 6)$

$$\text{Area} = \frac{1}{2}\begin{vmatrix} -4 & 0 & 1 \\ 4 & 0 & 1 \\ 0 & 6 & 1 \end{vmatrix} = \frac{1}{2}(48) = 24 \text{ square units}$$

113. $(1, -4), (-2, 3), (0, 5)$

$$\text{Area} = -\frac{1}{2}\begin{vmatrix} 1 & -4 & 1 \\ -2 & 3 & 1 \\ 0 & 5 & 1 \end{vmatrix}$$

$$= -\frac{1}{2}\left(-5\begin{vmatrix} 1 & 1 \\ -2 & 1 \end{vmatrix} + \begin{vmatrix} 1 & -4 \\ -2 & 3 \end{vmatrix}\right)$$

$$= -\frac{1}{2}(-5(3) + (-5)) = 10 \text{ square units}$$

114. $\left(\frac{3}{2}, 1\right), \left(4, -\frac{1}{2}\right), (4, 2)$

$$\text{Area} = \frac{1}{2}\begin{vmatrix} \frac{3}{2} & 1 & 1 \\ 4 & -\frac{1}{2} & 1 \\ 4 & 2 & 1 \end{vmatrix} = \frac{1}{2}\left(\frac{25}{4}\right) = \frac{25}{8} \text{ square units}$$

115. $(-1, 7), (3, -9), (-3, 15)$

$$\begin{vmatrix} -1 & 7 & 1 \\ 3 & -9 & 1 \\ -3 & 15 & 1 \end{vmatrix} = 0$$

The points are collinear.

116. Points: $(0, -5), (-2, -6), (8, -1)$

$$\begin{vmatrix} 0 & -5 & 1 \\ -2 & -6 & 1 \\ 8 & -1 & 1 \end{vmatrix} = \begin{vmatrix} -2 & -6 \\ 8 & -1 \end{vmatrix} - \begin{vmatrix} 0 & -5 \\ 8 & -1 \end{vmatrix} + \begin{vmatrix} 0 & -5 \\ -2 & -6 \end{vmatrix}$$

$$= 50 - 40 - 10 = 0$$

The points are collinear.

117. $(-4, 0), (4, 4)$

$$\begin{vmatrix} x & y & 1 \\ -4 & 0 & 1 \\ 4 & 4 & 1 \end{vmatrix} = 0$$

$$1\begin{vmatrix} -4 & 0 \\ 4 & 4 \end{vmatrix} - 1\begin{vmatrix} x & y \\ 4 & 4 \end{vmatrix} + 1\begin{vmatrix} x & y \\ -4 & 0 \end{vmatrix} = 0$$

$$-16 - (4x - 4y) + 4y = 0$$

$$-4x + 8y - 16 = 0$$

$$x - 2y + 4 = 0$$

118. $(2, 5), (6, -1)$

$$\begin{vmatrix} x & y & 1 \\ 2 & 5 & 1 \\ 6 & -1 & 1 \end{vmatrix} = 0$$

$$6x + 4y - 32 = 0$$

$$3x + 2y - 16 = 0$$

119. $\left(-\frac{5}{2}, 3\right), \left(\frac{7}{2}, 1\right)$

$$\begin{vmatrix} x & y & 1 \\ -\frac{5}{2} & 3 & 1 \\ \frac{7}{2} & 1 & 1 \end{vmatrix} = 0$$

$$1\begin{vmatrix} -\frac{5}{2} & 3 \\ \frac{7}{2} & 1 \end{vmatrix} - 1\begin{vmatrix} x & y \\ \frac{7}{2} & 1 \end{vmatrix} + 1\begin{vmatrix} x & y \\ -\frac{5}{2} & 3 \end{vmatrix} = 0$$

$$-13 - \left(x - \frac{7}{2}y\right) + \left(3x + \frac{5}{2}y\right) = 0$$

$$2x + 6y - 13 = 0$$

120. $(-0.8, 0.2), (0.7, 3.2)$

$$\begin{vmatrix} x & y & 1 \\ -0.8 & 0.2 & 1 \\ 0.7 & 3.2 & 1 \end{vmatrix} = 0$$

$-3x + 1.5y - 2.7 = 0$ Multiply both sides by $-\frac{10}{3}$.

$10x - 5y + 9 = 0$

121. L   O   O   K   _   O   U   T   _   B   E   L   O   W   _
[12 15 15] [11 0 15] [21 20 0] [2 5 12] [15 23 0]

$$A = \begin{bmatrix} 2 & -2 & 0 \\ 3 & 0 & -3 \\ -6 & 2 & 3 \end{bmatrix}$$

$$[12 \quad 15 \quad 15] \begin{bmatrix} 2 & -2 & 0 \\ 3 & 0 & -3 \\ -6 & 2 & 3 \end{bmatrix} = [-21 \quad 6 \quad 0]$$

$$[11 \quad 0 \quad 15] \begin{bmatrix} 2 & -2 & 0 \\ 3 & 0 & -3 \\ -6 & 2 & 3 \end{bmatrix} = [-68 \quad 8 \quad 45]$$

$$[21 \quad 20 \quad 0] \begin{bmatrix} 2 & -2 & 0 \\ 3 & 0 & -3 \\ -6 & 2 & 3 \end{bmatrix} = [102 \quad -42 \quad -60]$$

$$[2 \quad 5 \quad 12] \begin{bmatrix} 2 & -2 & 0 \\ 3 & 0 & -3 \\ -6 & 2 & 3 \end{bmatrix} = [-53 \quad 20 \quad 21]$$

$$[15 \quad 23 \quad 0] \begin{bmatrix} 2 & -2 & 0 \\ 3 & 0 & -3 \\ -6 & 2 & 3 \end{bmatrix} = [99 \quad -30 \quad -69]$$

Cryptogram: −21 6 0 −68 8 45 102
−42 −60 −53 20 21 99 −30 −69

122. R   E   T   U   R   N   _   T   O   _   B   A   S   E   _
[18 5 20] [21 18 14] [0 20 15] [0 2 1] [19 5 0]

$$A = \begin{bmatrix} 2 & 1 & 0 \\ -6 & -6 & -2 \\ 3 & 2 & 1 \end{bmatrix}$$

$$[18 \quad 5 \quad 20]A = [66 \quad 28 \quad 10]$$

$$[21 \quad 18 \quad 14]A = [-24 \ -59 \ -22]$$

$$[0 \quad 20 \quad 15]A = [-75 \ -90 \ -25]$$

$$[0 \quad 2 \quad 1]A = [-9 \ -10 \ -3]$$

$$[19 \quad 5 \quad 0]A = [8 \ -11 \ -10]$$

Cryptogram: 66 28 10 −24 −59 −22 −75 −90 −25 −9 −10 −3 8 −11 −10

123. $A^{-1} = \begin{bmatrix} -1 & 2 & -3 \\ 2 & 1 & 0 \\ 4 & -2 & 5 \end{bmatrix}$

$\begin{bmatrix} -5 & 11 & -2 \end{bmatrix} \begin{bmatrix} -1 & 2 & -3 \\ 2 & 1 & 0 \\ 4 & -2 & 5 \end{bmatrix} = \begin{bmatrix} 19 & 5 & 5 \end{bmatrix}$ S   E   E

$\begin{bmatrix} 370 & -265 & 225 \end{bmatrix} \begin{bmatrix} -1 & 2 & -3 \\ 2 & 1 & 0 \\ 4 & -2 & 5 \end{bmatrix} = \begin{bmatrix} 0 & 25 & 15 \end{bmatrix}$ __   Y   O

$\begin{bmatrix} -57 & 48 & -33 \end{bmatrix} \begin{bmatrix} -1 & 2 & -3 \\ 2 & 1 & 0 \\ 4 & -2 & 5 \end{bmatrix} = \begin{bmatrix} 21 & 0 & 6 \end{bmatrix}$ U   __   F

$\begin{bmatrix} 32 & -15 & 20 \end{bmatrix} \begin{bmatrix} -1 & 2 & -3 \\ 2 & 1 & 0 \\ 4 & -2 & 5 \end{bmatrix} = \begin{bmatrix} 18 & 9 & 4 \end{bmatrix}$ R   I   D

$\begin{bmatrix} 245 & -171 & 147 \end{bmatrix} \begin{bmatrix} -1 & 2 & -3 \\ 2 & 1 & 0 \\ 4 & -2 & 5 \end{bmatrix} = \begin{bmatrix} 1 & 25 & 0 \end{bmatrix}$ A   Y   __

Message: SEE YOU FRIDAY

124. $A^{-1} = \begin{bmatrix} -1 & 2 & -3 \\ 2 & 1 & 0 \\ 4 & -2 & 5 \end{bmatrix}$

$\begin{bmatrix} 145 & -105 & 92 \\ 264 & -188 & 160 \\ 23 & -16 & 15 \\ 129 & -84 & 78 \\ -9 & 8 & -5 \\ 159 & -118 & 100 \\ 219 & -152 & 133 \\ 370 & -265 & 225 \\ -105 & 84 & -63 \end{bmatrix} \begin{bmatrix} -1 & 2 & -3 \\ 2 & 1 & 0 \\ 4 & -2 & 5 \end{bmatrix} = \begin{bmatrix} 13 & 1 & 25 \\ 0 & 20 & 8 \\ 5 & 0 & 6 \\ 15 & 18 & 3 \\ 5 & 0 & 2 \\ 5 & 0 & 23 \\ 9 & 20 & 8 \\ 0 & 25 & 15 \\ 21 & 0 & 0 \end{bmatrix}$

M	A	Y
_	T	H
E	_	F
O	R	C
E	_	B
E	_	W
I	T	H
_	Y	O
U	_	_

Message: MAY THE FORCE BE WITH YOU

125. False. The matrix must be square.

126. True. Expand along Row 3.

$\begin{vmatrix} a_{11} & a_{12} & a_{13} \\ a_{21} & a_{22} & a_{23} \\ a_{31} + c_1 & a_{32} + c_2 & a_{33} + c_3 \end{vmatrix} = (a_{31} + c_1) \begin{vmatrix} a_{12} & a_{13} \\ a_{22} & a_{23} \end{vmatrix} - (a_{32} + c_2) \begin{vmatrix} a_{11} & a_{13} \\ a_{21} & a_{23} \end{vmatrix} + (a_{33} + c_3) \begin{vmatrix} a_{11} & a_{12} \\ a_{21} & a_{22} \end{vmatrix}$

$= a_{31} \begin{vmatrix} a_{12} & a_{13} \\ a_{22} & a_{23} \end{vmatrix} - a_{32} \begin{vmatrix} a_{11} & a_{13} \\ a_{21} & a_{23} \end{vmatrix} + a_{33} \begin{vmatrix} a_{11} & a_{12} \\ a_{21} & a_{22} \end{vmatrix}$

$\quad + c_1 \begin{vmatrix} a_{12} & a_{13} \\ a_{22} & a_{23} \end{vmatrix} - c_2 \begin{vmatrix} a_{11} & a_{13} \\ a_{21} & a_{23} \end{vmatrix} + c_3 \begin{vmatrix} a_{11} & a_{12} \\ a_{21} & a_{22} \end{vmatrix}$

$= \begin{vmatrix} a_{11} & a_{12} & a_{13} \\ a_{21} & a_{22} & a_{23} \\ a_{31} & a_{32} & a_{33} \end{vmatrix} + \begin{vmatrix} a_{11} & a_{12} & a_{13} \\ a_{21} & a_{22} & a_{23} \\ c_1 & c_2 & c_3 \end{vmatrix}$

Note: Expand each of these matrices along Row 3 to see the previous step.

127. The matrix must be square and its determinant nonzero to have an inverse.

128. If A is a square matrix, the cofactor C_{ij} of the entry a_{ij} is $(-1)^{i+j}M_{ij}$, where M_{ij} is the determinant obtained by deleting the ith row and jth column of A. The determinant of A is the sum of the entries of any row or column of A multiplied by their respective cofactors.

129. No. Each matrix is in row-echelon form, but the third matrix cannot be achieved from the first or second matrix with elementary row operations. Also, the first two matrices describe a system of equations with one solution. The third matrix describes a system with infinitely many solutions.

130. The part of the matrix corresponding to the coefficients of the system reduces to a matrix in which the number of rows with nonzero entries is the same as the number of variables.

131.

$$\begin{vmatrix} 2 - \lambda & 5 \\ 3 & -8 - \lambda \end{vmatrix} = 0$$

$$(2 - \lambda)(-8 - \lambda) - 15 = 0$$

$$-16 + 6\lambda + \lambda^2 - 15 = 0$$

$$\lambda^2 + 6\lambda - 31 = 0$$

$$\lambda = \frac{-6 \pm \sqrt{36 - 4(-31)}}{2}$$

$$\lambda = -3 \pm 2\sqrt{10}$$

Problem Solving for Chapter 8

1. $A = \begin{bmatrix} 0 & -1 \\ 1 & 0 \end{bmatrix}$ $T = \begin{bmatrix} 1 & 2 & 3 \\ 1 & 4 & 2 \end{bmatrix}$

(a) $AT = \begin{bmatrix} -1 & -4 & -2 \\ 1 & 2 & 3 \end{bmatrix}$ $AAT = \begin{bmatrix} -1 & -2 & -3 \\ -1 & -4 & -2 \end{bmatrix}$

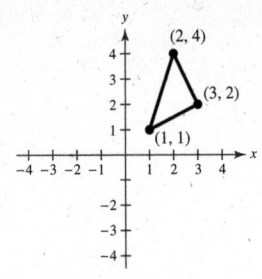

Original Triangle

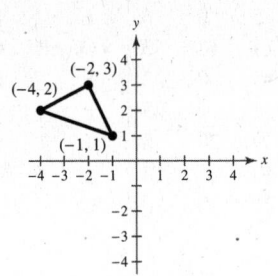

AT Triangle

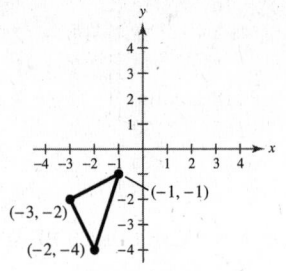

AAT Triangle

The transformation A interchanges the x and y coordinates and then takes the negative of the x coordinate. A represents a counterclockwise rotation by $90°$.

(b) $A^{-1}(AAT) = (A^{-1}A)(AT) = (I)(AT) = AT$

$A^{-1}(AT) = (A^{-1}A)T = IT = T$

$A^{-1} = \begin{bmatrix} 0 & 1 \\ -1 & 0 \end{bmatrix}$

A^{-1} represents a clockwise rotation by $90°$.

2. (a)

$$
\begin{array}{c}
\quad\quad\quad\quad\quad 2000 \\
\begin{array}{ccc}
0\text{--}17 & 18\text{--}64 & 65+
\end{array} \\
\begin{bmatrix}
4.64\% & 11.79\% & 2.62\% \\
5.91\% & 14.03\% & 2.94\% \\
9.09\% & 22.11\% & 4.42\% \\
1.75\% & 3.98\% & 0.72\% \\
4.30\% & 9.96\% & 1.74\%
\end{bmatrix}
\begin{array}{l}
\text{Northeast} \\
\text{Midwest} \\
\text{South} \\
\text{Mountain} \\
\text{Pacific}
\end{array}
\end{array}
$$

$$
\begin{array}{c}
\quad\quad\quad\quad\quad 2015 \\
\begin{array}{ccc}
0\text{--}17 & 18\text{--}64 & 65+
\end{array} \\
\begin{bmatrix}
4.06\% & 10.99\% & 2.63\% \\
5.12\% & 13.23\% & 3.26\% \\
8.36\% & 22.25\% & 5.63\% \\
1.69\% & 4.07\% & 1.05\% \\
4.81\% & 10.74\% & 2.12\%
\end{bmatrix}
\begin{array}{l}
\text{Northeast} \\
\text{Midwest} \\
\text{South} \\
\text{Mountain} \\
\text{Pacific}
\end{array}
\end{array}
$$

(b) Change in Percent of Population from 2000 to 2015

$$
\begin{array}{c}
\begin{array}{ccc}
0\text{--}17 & 18\text{--}64 & 65+
\end{array} \\
\begin{bmatrix}
-0.58\% & -0.80\% & 0.01\% \\
-0.79\% & -0.80\% & 0.32\% \\
-0.73\% & 0.14\% & 1.21\% \\
-0.06\% & 0.09\% & 0.33\% \\
0.51\% & 0.78\% & 0.38\%
\end{bmatrix}
\begin{array}{l}
\text{Northeast} \\
\text{Midwest} \\
\text{South} \\
\text{Mountain} \\
\text{Pacific}
\end{array}
\end{array}
$$

(c) All regions show growth in the 65+ age bracket, especially the South. The South, Mountain and Pacific regions show growth in the 18–64 age bracket. Only the Pacific region shows growth in the 0–17 age bracket.

3. (a) $A^2 = \begin{bmatrix} 1 & 0 \\ 0 & 0 \end{bmatrix}\begin{bmatrix} 1 & 0 \\ 0 & 0 \end{bmatrix} = \begin{bmatrix} 1 & 0 \\ 0 & 0 \end{bmatrix} = A$

A *is* idempotent.

(b) $A^2 = \begin{bmatrix} 0 & 1 \\ 1 & 0 \end{bmatrix}\begin{bmatrix} 0 & 1 \\ 1 & 0 \end{bmatrix} = \begin{bmatrix} 1 & 0 \\ 0 & 1 \end{bmatrix} \neq A$

A is *not* idempotent.

(c) $A^2 = \begin{bmatrix} 2 & 3 \\ -1 & -2 \end{bmatrix}\begin{bmatrix} 2 & 3 \\ -1 & -2 \end{bmatrix} = \begin{bmatrix} 1 & 0 \\ 0 & 1 \end{bmatrix} \neq A$

A is *not* idempotent.

(d) $A^2 = \begin{bmatrix} 2 & 3 \\ 1 & 2 \end{bmatrix}\begin{bmatrix} 2 & 3 \\ 1 & 2 \end{bmatrix} = \begin{bmatrix} 7 & 12 \\ 4 & 7 \end{bmatrix} \neq A$

A is *not* idempotent.

4. $A = \begin{bmatrix} 1 & 2 \\ -2 & 1 \end{bmatrix}$

(a) $A^2 - 2A + 5I = \begin{bmatrix} 1 & 2 \\ -2 & 1 \end{bmatrix}\begin{bmatrix} 1 & 2 \\ -2 & 1 \end{bmatrix} - 2\begin{bmatrix} 1 & 2 \\ -2 & 1 \end{bmatrix} + 5\begin{bmatrix} 1 & 0 \\ 0 & 1 \end{bmatrix}$

$$= \begin{bmatrix} -3 & 4 \\ -4 & -3 \end{bmatrix} + \begin{bmatrix} -2 & -4 \\ 4 & -2 \end{bmatrix} + \begin{bmatrix} 5 & 0 \\ 0 & 5 \end{bmatrix}$$

$$= \begin{bmatrix} 0 & 0 \\ 0 & 0 \end{bmatrix} = 0$$

(b) $A^{-1} = \dfrac{1}{(1) - (-4)}\begin{bmatrix} 1 & -2 \\ 2 & 1 \end{bmatrix} = \dfrac{1}{5}\begin{bmatrix} 1 & -2 \\ 2 & 1 \end{bmatrix}$

$\dfrac{1}{5}(2I - A) = \dfrac{1}{5}\left[\begin{bmatrix} 2 & 0 \\ 0 & 2 \end{bmatrix} - \begin{bmatrix} 1 & 2 \\ -2 & 1 \end{bmatrix}\right] = \dfrac{1}{5}\begin{bmatrix} 1 & -2 \\ 2 & 1 \end{bmatrix}$

Thus, $A^{-1} = \dfrac{1}{5}(2I - A)$.

(c) $A^2 - 2A + 5I = 0$

$A^2 - 2A = -5I$

$(A - 2I)A = -5I$

$-\dfrac{1}{5}(A - 2I)A = I$

$\dfrac{1}{5}(2I - A)A = I$

Thus, $A^{-1} = \dfrac{1}{5}(2I - A)$.

5. (a) $\begin{bmatrix} 0.70 & 0.15 & 0.15 \\ 0.20 & 0.80 & 0.15 \\ 0.10 & 0.05 & 0.70 \end{bmatrix} \begin{bmatrix} 25{,}000 \\ 30{,}000 \\ 45{,}000 \end{bmatrix} = \begin{bmatrix} 28{,}750 \\ 35{,}750 \\ 35{,}500 \end{bmatrix}$

Gold Cable Company: 28,750 households

Galaxy Cable Company: 35,750 households

Nonsubscribers: 35,500 households

(c) $\begin{bmatrix} 0.70 & 0.15 & 0.15 \\ 0.20 & 0.80 & 0.15 \\ 0.10 & 0.05 & 0.70 \end{bmatrix} \begin{bmatrix} 30{,}812.5 \\ 39{,}675 \\ 29{,}512.5 \end{bmatrix} \approx \begin{bmatrix} 31{,}947 \\ 42{,}329 \\ 25{,}724 \end{bmatrix}$

Gold Cable Company: 31,947 households

Galaxy Cable Company: 42,329 households

Nonsubscribers: 25,724 households

(b) $\begin{bmatrix} 0.70 & 0.15 & 0.15 \\ 0.20 & 0.80 & 0.15 \\ 0.10 & 0.05 & 0.70 \end{bmatrix} \begin{bmatrix} 28{,}750 \\ 35{,}750 \\ 35{,}500 \end{bmatrix} \approx \begin{bmatrix} 30{,}813 \\ 39{,}675 \\ 29{,}513 \end{bmatrix}$

Gold Cable Company: 30,813 households

Galaxy Cable Company: 39,675 households

Nonsubscribers: 29,513 households

(d) Both cable companies are increasing the number of subscribers, while the number of nonsubscribers is decreasing each year.

6. $A = \begin{bmatrix} 3 & x \\ -2 & -3 \end{bmatrix} \Rightarrow A^{-1} = \dfrac{1}{-9+2x} \begin{bmatrix} -3 & -x \\ 2 & 3 \end{bmatrix}$

If $A = A^{-1}$, then $\begin{bmatrix} \dfrac{-3}{-9+2x} & \dfrac{-x}{-9+2x} \\[2mm] \dfrac{2}{-9+2x} & \dfrac{3}{-9+2x} \end{bmatrix} = \begin{bmatrix} 3 & x \\ -2 & -3 \end{bmatrix}$.

Equating the first entry in Row 1 yields $\dfrac{-3}{-9+2x} = 3 \implies -3 = -27 + 6x \implies x = 4$.

Now check $x = 4$ in the other entries:

$\dfrac{-4}{-9+2(4)} = 4$ ✓

$\dfrac{2}{-9+2(4)} = -2$ ✓

$\dfrac{3}{-9+2(4)} = -3$ ✓

Thus, $x = 4$.

7. If $A = \begin{bmatrix} 4 & x \\ -2 & -3 \end{bmatrix}$ is singular then

$ad - bc = -12 + 2x = 0$.

Thus, $x = 6$.

8. From Exercise 3 we have the singular matrix

$A = \begin{bmatrix} 1 & 0 \\ 0 & 0 \end{bmatrix}$ where $A^2 = A$.

Also, $A = \begin{bmatrix} 1 & 1 \\ 0 & 0 \end{bmatrix}$ has this property.

9. $(a - b)(b - c)(c - a) = -a^2b + a^2c + ab^2 - ac^2 - b^2c + bc^2$

$\begin{vmatrix} 1 & 1 & 1 \\ a & b & c \\ a^2 & b^2 & c^2 \end{vmatrix} = \begin{vmatrix} b & c \\ b^2 & c^2 \end{vmatrix} - \begin{vmatrix} a & c \\ a^2 & c^2 \end{vmatrix} + \begin{vmatrix} a & b \\ a^2 & b^2 \end{vmatrix} = bc^2 - b^2c - ac^2 + a^2c + ab^2 - a^2b$

Thus, $\begin{vmatrix} 1 & 1 & 1 \\ a & b & c \\ a^2 & b^2 & c^2 \end{vmatrix} = (a - b)(b - c)(c - a)$.

10. $(a - b)(b - c)(c - a)(a + b + c) = -a^3b + a^3c + ab^3 - ac^3 - b^3c + bc^3$

$$\begin{vmatrix} 1 & 1 & 1 \\ a & b & c \\ a^3 & b^3 & c^3 \end{vmatrix} = \begin{vmatrix} b & c \\ b^3 & c^3 \end{vmatrix} - \begin{vmatrix} a & c \\ a^3 & c^3 \end{vmatrix} + \begin{vmatrix} a & b \\ a^3 & b^3 \end{vmatrix} = bc^3 - b^3c - ac^3 + a^3c + ab^3 - a^3b$$

Thus, $\begin{vmatrix} 1 & 1 & 1 \\ a & b & c \\ a^3 & b^3 & c^3 \end{vmatrix} = (a - b)(b - c)(c - a)(a + b + c).$

11. $\begin{vmatrix} x & 0 & c \\ -1 & x & b \\ 0 & -1 & a \end{vmatrix} = x\begin{vmatrix} x & b \\ -1 & a \end{vmatrix} + c\begin{vmatrix} -1 & x \\ 0 & -1 \end{vmatrix} = x(ax + b) + c(1 - 0) = ax^2 + bx + c$

12. $\begin{vmatrix} x & 0 & 0 & d \\ -1 & x & 0 & c \\ 0 & -1 & x & b \\ 0 & 0 & -1 & a \end{vmatrix} = x\begin{vmatrix} x & 0 & c \\ -1 & x & b \\ 0 & -1 & a \end{vmatrix} - d\begin{vmatrix} -1 & x & 0 \\ 0 & -1 & x \\ 0 & 0 & -1 \end{vmatrix} = \underbrace{x(ax^2 + bx + c)}_{\text{From Exercise 11}} - d\left(-\begin{vmatrix} -1 & x \\ 0 & -1 \end{vmatrix}\right)$

$$= ax^3 + bx^2 + cx + d$$

13. $4S + 4N \qquad = 184$

$\quad S \qquad + 6F = 146$

$\qquad 2N + 4F = 104$

$D = \begin{vmatrix} 4 & 4 & 0 \\ 1 & 0 & 6 \\ 0 & 2 & 4 \end{vmatrix} = -64$

$N = \dfrac{\begin{vmatrix} 4 & 184 & 0 \\ 1 & 146 & 6 \\ 0 & 104 & 4 \end{vmatrix}}{-64} = \dfrac{-896}{-64} = 14$

$S = \dfrac{\begin{vmatrix} 184 & 4 & 0 \\ 146 & 0 & 6 \\ 104 & 2 & 4 \end{vmatrix}}{-64} = \dfrac{-2048}{-64} = 32$

$F = \dfrac{\begin{vmatrix} 4 & 4 & 184 \\ 1 & 0 & 146 \\ 0 & 2 & 104 \end{vmatrix}}{-64} = \dfrac{-1216}{-64} = 19$

Element	Atomic mass
Sulfur	32
Nitrogen	14
Fluoride	19

14. Let x = cost of a transformer, y = cost per foot of wire, z = cost of a light.

$x + 25y + 5z = 20$

$x + 50y + 15z = 35$

$x + 100y + 20z = 50$

$$\begin{bmatrix} 1 & 25 & 5 & \vdots & 20 \\ 1 & 50 & 15 & \vdots & 35 \\ 1 & 100 & 20 & \vdots & 50 \end{bmatrix} \xrightarrow{\text{rref}} \begin{bmatrix} 1 & 0 & 0 & \vdots & 10 \\ 0 & 1 & 0 & \vdots & 0.2 \\ 0 & 0 & 1 & \vdots & 1 \end{bmatrix}$$

By using the matrix capabilities of a graphing calculator to reduce the augmented matrix to reduced row-echelon form, we have the following costs:

Transformer $10.00

Foot of wire $ 0.20

Light $ 1.00

15. $A = \begin{bmatrix} -1 & 1 & -2 \\ 2 & 0 & 1 \end{bmatrix}$, $\qquad B = \begin{bmatrix} -3 & 0 \\ 1 & 2 \\ 1 & -1 \end{bmatrix}$

$A^T = \begin{bmatrix} -1 & 2 \\ 1 & 0 \\ -2 & 1 \end{bmatrix}$, $\qquad B^T = \begin{bmatrix} -3 & 1 & 1 \\ 0 & 2 & -1 \end{bmatrix}$

$AB = \begin{bmatrix} 2 & 4 \\ -5 & -1 \end{bmatrix}$, $\qquad (AB)^T = \begin{bmatrix} 2 & -5 \\ 4 & -1 \end{bmatrix}$

$B^T A^T = \begin{bmatrix} -3 & 1 & 1 \\ 0 & 2 & -1 \end{bmatrix} \begin{bmatrix} -1 & 2 \\ 1 & 0 \\ -2 & 1 \end{bmatrix} = \begin{bmatrix} 2 & -5 \\ 4 & -1 \end{bmatrix}$

Thus, $(AB)^T = B^T A^T$.

16. $A = \begin{bmatrix} 1 & -2 & 2 \\ 1 & 1 & -3 \\ 1 & -1 & 4 \end{bmatrix} \Rightarrow A^{-1} = \begin{bmatrix} \frac{1}{11} & \frac{6}{11} & \frac{4}{11} \\ -\frac{7}{11} & \frac{2}{11} & \frac{5}{11} \\ -\frac{2}{11} & -\frac{1}{11} & \frac{3}{11} \end{bmatrix}$

$$\begin{bmatrix} 23 & 13 & -34 \\ 31 & -34 & 63 \\ 25 & -17 & 61 \\ 24 & 14 & -37 \\ 41 & -17 & -8 \\ 20 & -29 & 40 \\ 38 & -56 & 116 \\ 13 & -11 & 1 \\ 22 & -3 & -6 \\ 41 & -53 & 85 \\ 28 & -32 & 16 \end{bmatrix} \begin{bmatrix} \frac{1}{11} & \frac{6}{11} & \frac{4}{11} \\ -\frac{7}{11} & \frac{2}{11} & \frac{5}{11} \\ -\frac{2}{11} & -\frac{1}{11} & \frac{3}{11} \end{bmatrix} = \begin{bmatrix} 0 & 18 & 5 \\ 13 & 5 & 13 \\ 2 & 5 & 18 \\ 0 & 19 & 5 \\ 16 & 20 & 5 \\ 13 & 2 & 5 \\ 18 & 0 & 20 \\ 8 & 5 & 0 \\ 5 & 12 & 5 \\ 22 & 5 & 14 \\ 20 & 8 & 0 \end{bmatrix}$$

0	18	5	13	5	13	2	5	18	0
_	R	E	M	E	M	B	E	R	_

19	5	16	20	5	13	2	5	18	0
S	E	P	T	E	M	B	E	R	_

20	8	5	0	5	12	5	22	5	14	20	8	0
T	H	E	_	E	L	E	V	E	N	T	H	_

REMEMBER SEPTEMBER THE ELEVENTH

17. (a) $[45 \quad -35]\begin{bmatrix} w & x \\ y & z \end{bmatrix} = [10 \quad 15]$

$[38 \quad -30]\begin{bmatrix} w & x \\ y & z \end{bmatrix} = [8 \quad 14]$

$45w - 35y = 10$

$45x - 35z = 15$

$38w - 30y = 8$

$38x - 30z = 14$

$\left.\begin{matrix} 45w - 35y = 10 \\ 38w - 30y = 8 \end{matrix}\right\} \Rightarrow w = 1, y = 1$

$\left.\begin{matrix} 45x - 35z = 15 \\ 38x - 30z = 14 \end{matrix}\right\} \Rightarrow x = -2, z = -3$

$A^{-1} = \begin{bmatrix} 1 & -2 \\ 1 & -3 \end{bmatrix}$

(b) $\begin{bmatrix} 45 & -35 \\ 38 & -30 \\ 18 & -18 \\ 35 & -30 \\ 81 & -60 \\ 42 & -28 \\ 75 & -55 \\ 2 & -2 \\ 22 & -21 \\ 15 & -10 \end{bmatrix} \begin{bmatrix} 1 & -2 \\ 1 & -3 \end{bmatrix} = \begin{bmatrix} 10 & 15 \\ 8 & 14 \\ 0 & 18 \\ 5 & 20 \\ 21 & 18 \\ 14 & 0 \\ 20 & 15 \\ 0 & 2 \\ 1 & 19 \\ 5 & 0 \end{bmatrix} \begin{matrix} J & O \\ H & N \\ — & R \\ E & T \\ U & R \\ N & — \\ T & O \\ — & B \\ A & S \\ E & — \end{matrix}$

JOHN RETURN TO BASE

18. $A = \begin{bmatrix} 6 & 4 & 1 \\ 0 & 2 & 3 \\ 1 & 1 & 2 \end{bmatrix}$

$A^{-1} = \begin{bmatrix} \frac{1}{16} & -\frac{7}{16} & \frac{5}{8} \\ \frac{3}{16} & \frac{11}{16} & -\frac{9}{8} \\ -\frac{1}{8} & -\frac{1}{8} & \frac{3}{4} \end{bmatrix}$

$|A| = 16$ and $|A^{-1}| = \dfrac{1}{16}$

Conjecture: $|A^{-1}| = \dfrac{1}{|A|}$

19. Let $A = \begin{bmatrix} 3 & -3 \\ 5 & -5 \end{bmatrix}$, then $|A| = 0$.

Let $A = \begin{bmatrix} 2 & 4 & -6 \\ -3 & 1 & 2 \\ 5 & -8 & 3 \end{bmatrix}$, then $|A| = 0$.

Let $A = \begin{bmatrix} 3 & -7 & 5 & -1 \\ -6 & 4 & 0 & 2 \\ 5 & 8 & -6 & -7 \\ 9 & 11 & -4 & -16 \end{bmatrix}$, then $|A| = 0$.

Conjecture: If A is an $n \times n$ matrix, each of whose rows add up to zero, then $|A| = 0$.

20. (a) Answers will vary.

$A = \begin{bmatrix} 0 & 3 \\ 0 & 0 \end{bmatrix}, \qquad B = \begin{bmatrix} 0 & 4 & -1 \\ 0 & 0 & 7 \\ 0 & 0 & 0 \end{bmatrix}$

(b) $A^2 = 0$ so $A^n = 0$ for n an integer ≥ 2.

$B^2 = \begin{bmatrix} 0 & 0 & 28 \\ 0 & 0 & 0 \\ 0 & 0 & 0 \end{bmatrix}$

$B^3 = 0$ so $B^n = 0$ for n an integer ≥ 3.

(c) $A^4 = 0$ if A is 4×4.

(d) Conjecture: If A is $n \times n$, then $A^n = 0$.

Chapter 8   Practice Test

1. Put the matrix in reduced row-echelon form.

$$\begin{bmatrix} 1 & -2 & 4 \\ 3 & -5 & 9 \end{bmatrix}$$

For Exercises 2–4, use matrices to solve the system of equations.

2. $\begin{cases} 3x + 5y = 3 \\ 2x - y = -11 \end{cases}$

3. $\begin{cases} 2x + 3y = -3 \\ 3x + 2y = 8 \\ x + y = 1 \end{cases}$

4. $\begin{cases} x + 3z = -5 \\ 2x + y = 0 \\ 3x + y - z = 3 \end{cases}$

5. Multiply $\begin{bmatrix} 1 & 4 & 5 \\ 2 & 0 & -3 \end{bmatrix} \begin{bmatrix} 1 & 6 \\ 0 & -7 \\ -1 & 2 \end{bmatrix}$.

6. Given $A = \begin{bmatrix} 9 & 1 \\ -4 & 8 \end{bmatrix}$ and $B = \begin{bmatrix} 6 & -2 \\ 3 & 5 \end{bmatrix}$, find $3A - 5B$.

7. Find $f(A)$.

$$f(x) = x^2 - 7x + 8, \quad A = \begin{bmatrix} 3 & 0 \\ 7 & 1 \end{bmatrix}$$

8. True or false:

$(A + B)(A + 3B) = A^2 + 4AB + 3B^2$ where A and B are matrices.

(Assume that A^2, AB, and B^2 exist.)

For Exercises 9–10, find the inverse of the matrix, if it exists.

9. $\begin{bmatrix} 1 & 2 \\ 3 & 5 \end{bmatrix}$

10. $\begin{bmatrix} 1 & 1 & 1 \\ 3 & 6 & 5 \\ 6 & 10 & 8 \end{bmatrix}$

11. Use an inverse matrix to solve the systems.

(a) $x + 2y = 4$
    $3x + 5y = 1$

(b) $x + 2y = 3$
    $3x + 5y = -2$

For Exercises 12–14, find the determinant of the matrix.

12. $\begin{bmatrix} 6 & -1 \\ 3 & 4 \end{bmatrix}$

13. $\begin{bmatrix} 1 & 3 & -1 \\ 5 & 9 & 0 \\ 6 & 2 & -5 \end{bmatrix}$

14. $\begin{bmatrix} 1 & 4 & 2 & 3 \\ 0 & 1 & -2 & 0 \\ 3 & 5 & -1 & 1 \\ 2 & 0 & 6 & 1 \end{bmatrix}$

15. Evaluate $\begin{vmatrix} 6 & 4 & 3 & 0 & 6 \\ 0 & 5 & 1 & 4 & 8 \\ 0 & 0 & 2 & 7 & 3 \\ 0 & 0 & 0 & 9 & 2 \\ 0 & 0 & 0 & 0 & 1 \end{vmatrix}$.

16. Use a determinant to find the area of the triangle with vertices $(0, 7)$, $(5, 0)$, and $(3, 9)$.

17. Find the equation of the line through $(2, 7)$ and $(-1, 4)$.

For Exercises 18–20, use Cramer's Rule to find the indicated value.

18. Find x.

$$\begin{cases} 6x - 7y = 4 \\ 2x + 5y = 11 \end{cases}$$

19. Find z.

$$\begin{cases} 3x \quad\;\; + z = 1 \\ \quad\;\; y + 4z = 3 \\ x - y \quad\;\;\; = 2 \end{cases}$$

20. Find y.

$$\begin{cases} 721.4x - 29.1y = 33.77 \\ 45.9x + 105.6y = 19.85 \end{cases}$$

CHAPTER 9
Sequences, Series, and Probability

CHAPTER 9
Sequences, Series, and Probability

Section 9.1　Sequences and Series

- Given the general nth term in a sequence, you should be able to find, or list, some of the terms.
- You should be able to find an expression for the apparent nth term of a sequence.
- You should be able to use and evaluate factorials.
- You should be able to use summation notation for a sum.
- You should know that the sum of the terms of a sequence is a series.

Vocabulary Check

1. infinite sequence
2. terms
3. finite
4. recursively
5. factorial
6. summation notation
7. index; upper; lower
8. series
9. nth partial sum

---

1. $a_n = 3n + 1$
   $a_1 = 3(1) + 1 = 4$
   $a_2 = 3(2) + 1 = 7$
   $a_3 = 3(3) + 1 = 10$
   $a_4 = 3(4) + 1 = 13$
   $a_5 = 3(5) + 1 = 16$

2. $a_n = 5n - 3$
   $a_1 = 5(1) - 3 = 2$
   $a_2 = 5(2) - 3 = 7$
   $a_3 = 5(3) - 3 = 12$
   $a_4 = 5(4) - 3 = 17$
   $a_5 = 5(5) - 3 = 22$

3. $a_n = 2^n$
   $a_1 = 2^1 = 2$
   $a_2 = 2^2 = 4$
   $a_3 = 2^3 = 8$
   $a_4 = 2^4 = 16$
   $a_5 = 2^5 = 32$

4. $a_n = \left(\frac{1}{2}\right)^n$
   $a_1 = \left(\frac{1}{2}\right)^1 = \frac{1}{2}$
   $a_2 = \left(\frac{1}{2}\right)^2 = \frac{1}{4}$
   $a_3 = \left(\frac{1}{2}\right)^3 = \frac{1}{8}$
   $a_4 = \left(\frac{1}{2}\right)^4 = \frac{1}{16}$
   $a_5 = \left(\frac{1}{2}\right)^5 = \frac{1}{32}$

5. $a_n = (-2)^n$
   $a_1 = (-2)^1 = -2$
   $a_2 = (-2)^2 = 4$
   $a_3 = (-2)^3 = -8$
   $a_4 = (-2)^4 = 16$
   $a_5 = (-2)^5 = -32$

6. $a_n = \left(-\frac{1}{2}\right)^n$
   $a_1 = \left(-\frac{1}{2}\right)^1 = -\frac{1}{2}$
   $a_2 = \left(-\frac{1}{2}\right)^2 = \frac{1}{4}$
   $a_3 = \left(-\frac{1}{2}\right)^3 = -\frac{1}{8}$
   $a_4 = \left(-\frac{1}{2}\right)^4 = \frac{1}{16}$
   $a_5 = \left(-\frac{1}{2}\right)^5 = -\frac{1}{32}$

7. $a_n = \dfrac{n+2}{n}$

$a_1 = \dfrac{1+2}{1} = 3$

$a_2 = \dfrac{4}{2} = 2$

$a_3 = \dfrac{5}{3}$

$a_4 = \dfrac{6}{4} = \dfrac{3}{2}$

$a_5 = \dfrac{7}{5}$

8. $a_n = \dfrac{n}{n+2}$

$a_1 = \dfrac{1}{1+2} = \dfrac{1}{3}$

$a_2 = \dfrac{2}{2+2} = \dfrac{1}{2}$

$a_3 = \dfrac{3}{3+2} = \dfrac{3}{5}$

$a_4 = \dfrac{4}{4+2} = \dfrac{2}{3}$

$a_5 = \dfrac{5}{5+2} = \dfrac{5}{7}$

9. $a_n = \dfrac{6n}{3n^2 - 1}$

$a_1 = \dfrac{6(1)}{3(1)^2 - 1} = 3$

$a_2 = \dfrac{6(2)}{3(2)^2 - 1} = \dfrac{12}{11}$

$a_3 = \dfrac{6(3)}{3(3)^2 - 1} = \dfrac{9}{13}$

$a_4 = \dfrac{6(4)}{3(4)^2 - 1} = \dfrac{24}{47}$

$a_5 = \dfrac{6(5)}{3(5)^2 - 1} = \dfrac{15}{37}$

10. $a_n = \dfrac{3n^2 - n + 4}{2n^2 + 1}$

$a_1 = \dfrac{3(1)^2 - 1 + 4}{2(1)^2 + 1} = 2$

$a_2 = \dfrac{3(2)^2 - 2 + 4}{2(2)^2 + 1} = \dfrac{14}{9}$

$a_3 = \dfrac{3(3)^2 - 3 + 4}{2(3)^2 + 1} = \dfrac{28}{19}$

$a_4 = \dfrac{3(4)^2 - 4 + 4}{2(4)^2 + 1} = \dfrac{16}{11}$

$a_5 = \dfrac{3(5)^2 - 5 + 4}{2(5)^2 + 1} = \dfrac{74}{51}$

11. $a_n = \dfrac{1 + (-1)^n}{n}$

$a_1 = 0$

$a_2 = \dfrac{2}{2} = 1$

$a_3 = 0$

$a_4 = \dfrac{2}{4} = \dfrac{1}{2}$

$a_5 = 0$

12. $a_n = 1 + (-1)^n$

$a_1 = 1 + (-1)^1 = 0$

$a_2 = 1 + (-1)^2 = 2$

$a_3 = 1 + (-1)^3 = 0$

$a_4 = 1 + (-1)^4 = 2$

$a_5 = 1 + (-1)^5 = 0$

13. $a_n = 2 - \dfrac{1}{3^n}$

$a_1 = 2 - \dfrac{1}{3} = \dfrac{5}{3}$

$a_2 = 2 - \dfrac{1}{9} = \dfrac{17}{9}$

$a_3 = 2 - \dfrac{1}{27} = \dfrac{53}{27}$

$a_4 = 2 - \dfrac{1}{81} = \dfrac{161}{81}$

$a_5 = 2 - \dfrac{1}{243} = \dfrac{485}{243}$

14. $a_n = \dfrac{2^n}{3^n}$

$a_1 = \dfrac{2^1}{3^1} = \dfrac{2}{3}$

$a_2 = \dfrac{2^2}{3^2} = \dfrac{4}{9}$

$a_3 = \dfrac{2^3}{3^3} = \dfrac{8}{27}$

$a_4 = \dfrac{2^4}{3^4} = \dfrac{16}{81}$

$a_5 = \dfrac{2^5}{3^5} = \dfrac{32}{243}$

15. $a_n = \dfrac{1}{n^{3/2}}$

$a_1 = \dfrac{1}{1} = 1$

$a_2 = \dfrac{1}{2^{3/2}}$

$a_3 = \dfrac{1}{3^{3/2}}$

$a_4 = \dfrac{1}{4^{3/2}} = \dfrac{1}{8}$

$a_5 = \dfrac{1}{5^{3/2}}$

16. $a_n = \dfrac{10}{n^{2/3}} = \dfrac{10}{\sqrt[3]{n^2}}$

$a_1 = \dfrac{10}{1} = 10$

$a_2 = \dfrac{10}{\sqrt[3]{2^2}} = \dfrac{10}{\sqrt[3]{4}}$

$a_3 = \dfrac{10}{\sqrt[3]{3^2}} = \dfrac{10}{\sqrt[3]{9}}$

$a_4 = \dfrac{10}{\sqrt[3]{4^2}} = \dfrac{10}{\sqrt[3]{16}}$

$a_5 = \dfrac{10}{\sqrt[3]{5^2}} = \dfrac{10}{\sqrt[3]{25}}$

17. $a_n = \dfrac{(-1)^n}{n^2}$

$a_1 = -\dfrac{1}{1} = -1$

$a_2 = \dfrac{1}{4}$

$a_3 = -\dfrac{1}{9}$

$a_4 = \dfrac{1}{16}$

$a_5 = -\dfrac{1}{25}$

18. $a_n = (-1)^n\left(\dfrac{n}{n+1}\right)$

$a_1 = (-1)^1 \dfrac{1}{1+1} = -\dfrac{1}{2}$

$a_2 = (-1)^2 \dfrac{2}{2+1} = \dfrac{2}{3}$

$a_3 = (-1)^3 \dfrac{3}{3+1} = -\dfrac{3}{4}$

$a_4 = (-1)^4 \dfrac{4}{4+1} = \dfrac{4}{5}$

$a_5 = (-1)^5 \dfrac{5}{5+1} = -\dfrac{5}{6}$

19. $a_n = \tfrac{2}{3}$

$a_1 = \tfrac{2}{3}$

$a_2 = \tfrac{2}{3}$

$a_3 = \tfrac{2}{3}$

$a_4 = \tfrac{2}{3}$

$a_5 = \tfrac{2}{3}$

20. $a_n = 0.3$

$a_1 = 0.3$

$a_2 = 0.3$

$a_3 = 0.3$

$a_4 = 0.3$

$a_5 = 0.3$

21. $a_n = n(n-1)(n-2)$

$a_1 = (1)(0)(-1) = 0$

$a_2 = (2)(1)(0) = 0$

$a_3 = (3)(2)(1) = 6$

$a_4 = (4)(3)(2) = 24$

$a_5 = (5)(4)(3) = 60$

22. $a_n = n(n^2 - 6)$

$a_1 = 1(1^2 - 6) = -5$

$a_2 = 2(2^2 - 6) = -4$

$a_3 = 3(3^2 - 6) = 9$

$a_4 = 4(4^2 - 6) = 40$

$a_5 = 5(5^2 - 6) = 95$

23. $a_{25} = (-1)^{25}(3(25) - 2) = -73$

24. $a_n = (-1)^{n-1}[n(n-1)]$

$a_{16} = (-1)^{16-1}[16(16-1)] = -240$

25. $a_{11} = \dfrac{4(11)}{2(11)^2 - 3} = \dfrac{44}{239}$

26. $a_n = \dfrac{4n^2 - n + 3}{n(n-1)(n+2)}$

$a_{13} = \dfrac{4(13)^2 - 13 + 3}{13(13-1)(13+2)} = \dfrac{37}{130}$

27. $a_n = \dfrac{3}{4}n$

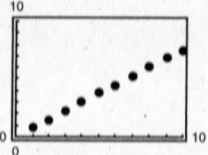

28. $a_n = 2 - \dfrac{4}{n}$

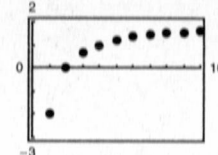

29. $a_n = 16(-0.5)^{n-1}$

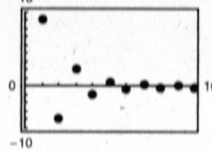

30. $a_n = 8(0.75)^{n-1}$

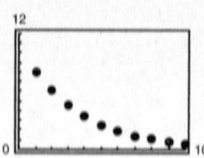

31. $a_n = \dfrac{2n}{n+1}$

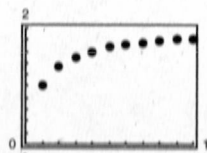

32. $a_n = \dfrac{n^2}{n^2 + 2}$

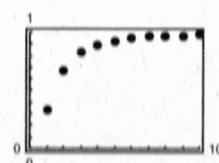

33. $a_n = \dfrac{8}{n+1}$

$a_1 = 4,\ a_{10} = \dfrac{8}{11}$

The sequence decreases.

Matches graph (c).

34. $a_n = \dfrac{8n}{n+1}$

$a_n \to 8$ as $n \to \infty$

$a_1 = 4, \quad a_3 = \dfrac{24}{4} = 6$

Matches graph (b).

35. $a_n = 4(0.5)^{n-1}$

$a_1 = 4, a_{10} = \dfrac{1}{128}$

The sequence decreases.

Matches graph (d).

36. $a_n = \dfrac{4^n}{n!}$

$a_n \to 0$ as $n \to \infty$

$a_1 = 4, a_4 = \dfrac{4^4}{4!} = \dfrac{256}{24} = 10\dfrac{2}{3}$

Matches graph (a).

37. $1, 4, 7, 10, 13, \ldots$

$a_n = 1 + (n-1)3 = 3n - 2$

38. $3, 7, 11, 15, 19, \ldots$

n: 1 2   3   4   5 . . . n

Terms: 3   7   11   15   19 . . . a_n

Apparent pattern:

Each term is one less than four times n, which implies that $a_n = 4n - 1$.

39. $0, 3, 8, 15, 24, \ldots$

$a_n = n^2 - 1$

40. $2, -4, 6, -8, 10, \ldots$

n: 1   2   3   4   5 . . . n

Terms: 2 -4 6 -8 10 . . . a_n

Apparent pattern:

Each term is the product of $(-1)^{n+1}$ and twice n, which implies that $a_n = (-1)^{n+1}(2n)$.

41. $-\dfrac{2}{3}, \dfrac{3}{4}, -\dfrac{4}{5}, \dfrac{5}{6}, -\dfrac{6}{7}, \ldots$

$a_n = (-1)^n \left(\dfrac{n+1}{n+2} \right)$

42. $\dfrac{1}{2}, \dfrac{-1}{4}, \dfrac{1}{8}, \dfrac{-1}{16}, \ldots$

n: 1   2   3   4 . . . n

Terms: $\dfrac{1}{2}$ $\dfrac{-1}{4}$ $\dfrac{1}{8}$ $\dfrac{-1}{16}$ . . . a_n

Apparent pattern:

Each term is $(-1)^{n+1}$ divided by 2 raised to the n, which implies that

$a_n = \dfrac{(-1)^{n+1}}{2^n}$.

43. $\dfrac{2}{1}, \dfrac{3}{3}, \dfrac{4}{5}, \dfrac{5}{7}, \dfrac{6}{9}, \ldots$

$a_n = \dfrac{n+1}{2n-1}$

44. $\dfrac{1}{3}, \dfrac{2}{9}, \dfrac{4}{27}, \dfrac{8}{81}, \ldots$

n: 1 2 3   4 . . . n

Terms: $\dfrac{1}{3}$ $\dfrac{2}{9}$ $\dfrac{4}{27}$ $\dfrac{8}{81}$ . . . a_n

Apparent pattern:

Each term is 2^{n-1} divided by 3 raised to the n, which implies that

$a_n = \dfrac{2^{n-1}}{3^n}$.

45. $1, \dfrac{1}{4}, \dfrac{1}{9}, \dfrac{1}{16}, \dfrac{1}{25}, \ldots$

$a_n = \dfrac{1}{n^2}$

46. $1, \dfrac{1}{2}, \dfrac{1}{6}, \dfrac{1}{24}, \dfrac{1}{120}, \ldots$

n: 1   2   3   4   5 . . . n

Terms: 1 $\dfrac{1}{2}$ $\dfrac{1}{6}$ $\dfrac{1}{24}$ $\dfrac{1}{120}$ . . . a_n

Apparent pattern:

Each term is the reciprocal of $n!$, which implies that

$a_n = \dfrac{1}{n!}$.

47. $1, -1, 1, -1, 1, \ldots$

$a_n = (-1)^{n+1}$

48. $1, 2, \dfrac{2^2}{2}, \dfrac{2^3}{6}, \dfrac{2^4}{24}, \dfrac{2^5}{120}, \ldots$

     n: 1   2   3   4   5   6 $\ldots$ n

Terms: 1   2 $\dfrac{2^2}{2}$ $\dfrac{2^3}{6}$ $\dfrac{2^4}{24}$ $\dfrac{2^5}{120}$ $\ldots$ a_n

Apparent pattern:

Each term is 2^{n-1} divided by $(n-1)!$, which implies that

$$a_n = \frac{2^{n-1}}{(n-1)!}.$$

49. $1 + \dfrac{1}{1}, 1 + \dfrac{1}{2}, 1 + \dfrac{1}{3}, 1 + \dfrac{1}{4}, 1 + \dfrac{1}{5}, \ldots$

$$a_n = 1 + \frac{1}{n}$$

50. $1 + \dfrac{1}{2}, 1 + \dfrac{3}{4}, 1 + \dfrac{7}{8}, 1 + \dfrac{15}{16}, 1 + \dfrac{31}{32}, \ldots$

     n: 1   2   3   4   5 $\ldots$ n

Terms: $1 + \dfrac{1}{2}$ $1 + \dfrac{3}{4}$ $1 + \dfrac{7}{8}$ $1 + \dfrac{15}{16}$ $1 + \dfrac{31}{32}$ $\ldots$ a_n

Apparent pattern: Each term is the sum of 1 and the quantity 1 less than 2^n divided by 2^n, which implies that

$$a_n = 1 + \frac{2^n - 1}{2^n}.$$

51. $a_1 = 28$ and $a_{k+1} = a_k - 4$

$a_1 = 28$

$a_2 = a_1 - 4 = 28 - 4 = 24$

$a_3 = a_2 - 4 = 24 - 4 = 20$

$a_4 = a_3 - 4 = 20 - 4 = 16$

$a_5 = a_4 - 4 = 16 - 4 = 12$

52. $a_1 = 15,\quad a_{k+1} = a_k + 3$

$a_1 = 15$

$a_2 = a_1 + 3 = 15 + 3 = 18$

$a_3 = a_2 + 3 = 18 + 3 = 21$

$a_4 = a_3 + 3 = 21 + 3 = 24$

$a_5 = a_4 + 3 = 24 + 3 = 27$

53. $a_1 = 3$ and $a_{k+1} = 2(a_k - 1)$

$a_1 = 3$

$a_2 = 2(a_1 - 1) = 2(3 - 1) = 4$

$a_3 = 2(a_2 - 1) = 2(4 - 1) = 6$

$a_4 = 2(a_3 - 1) = 2(6 - 1) = 10$

$a_5 = 2(a_4 - 1) = 2(10 - 1) = 18$

54. $a_1 = 32,\quad a_{k+1} = \tfrac{1}{2}a_k$

$a_1 = 32$

$a_2 = \tfrac{1}{2}a_1 = \tfrac{1}{2}(32) = 16$

$a_3 = \tfrac{1}{2}a_2 = \tfrac{1}{2}(16) = 8$

$a_4 = \tfrac{1}{2}a_3 = \tfrac{1}{2}(8) = 4$

$a_5 = \tfrac{1}{2}a_4 = \tfrac{1}{2}(4) = 2$

55. $a_1 = 6$ and $a_{k+1} = a_k + 2$

$a_1 = 6$

$a_2 = a_1 + 2 = 6 + 2 = 8$

$a_3 = a_2 + 2 = 8 + 2 = 10$

$a_4 = a_3 + 2 = 10 + 2 = 12$

$a_5 = a_4 + 2 = 12 + 2 = 14$

In general, $a_n = 2n + 4$.

56. $a_1 = 25,\quad a_{k+1} = a_k - 5$

$a_1 = 25$

$a_2 = a_1 - 5 = 25 - 5 = 20$

$a_3 = a_2 - 5 = 20 - 5 = 15$

$a_4 = a_3 - 5 = 15 - 5 = 10$

$a_5 = a_4 - 5 = 10 - 5 = 5$

In general, $a_n = 30 - 5n$.

57. $a_1 = 81$ and $a_{k+1} = \dfrac{1}{3}a_k$

$a_1 = 81$

$a_2 = \dfrac{1}{3}a_1 = \dfrac{1}{3}(81) = 27$

$a_3 = \dfrac{1}{3}a_2 = \dfrac{1}{3}(27) = 9$

$a_4 = \dfrac{1}{3}a_3 = \dfrac{1}{3}(9) = 3$

$a_5 = \dfrac{1}{3}a_4 = \dfrac{1}{3}(3) = 1$

In general,

$$a_n = 81\left(\frac{1}{3}\right)^{n-1} = 81(3)\left(\frac{1}{3}\right)^n = \frac{243}{3^n}.$$

58. $a_1 = 14,\quad a_{k+1} = (-2)a_k$

$a_1 = 14$

$a_2 = (-2)a_1 = (-2)(14) = -28$

$a_3 = (-2)a_2 = (-2)(-28) = 56$

$a_4 = (-2)a_3 = (-2)(56) = -112$

$a_5 = (-2)(a_4) = (-2)(-112) = 224$

In general, $a_n = 14(-2)^{n-1}$.

59. $a_n = \dfrac{3^n}{n!}$

$a_0 = \dfrac{3^0}{0!} = 1$

$a_1 = \dfrac{3^1}{1!} = 3$

$a_2 = \dfrac{3^2}{2!} = \dfrac{9}{2}$

$a_3 = \dfrac{3^3}{3!} = \dfrac{27}{6} = \dfrac{9}{2}$

$a_4 = \dfrac{3^4}{4!} = \dfrac{81}{24} = \dfrac{27}{8}$

60. $a_n = \dfrac{n!}{n}$

$a_0 = \dfrac{0!}{0} =$ undefined

$a_1 = \dfrac{1!}{1} = \dfrac{1}{1} = 1$

$a_2 = \dfrac{2!}{2} = \dfrac{2 \cdot 1}{2} = 1$

$a_3 = \dfrac{3!}{3} = \dfrac{3 \cdot 2 \cdot 1}{3} = 2$

$a_4 = \dfrac{4!}{4} = \dfrac{4 \cdot 3 \cdot 2 \cdot 1}{4} = 6$

61. $a_n = \dfrac{1}{(n + 1)!}$

$a_0 = \dfrac{1}{1!} = 1$

$a_1 = \dfrac{1}{2!} = \dfrac{1}{2}$

$a_2 = \dfrac{1}{3!} = \dfrac{1}{6}$

$a_3 = \dfrac{1}{4!} = \dfrac{1}{24}$

$a_4 = \dfrac{1}{5!} = \dfrac{1}{120}$

62. $a_n = \dfrac{n^2}{(n + 1)!}$

$a_0 = \dfrac{0^2}{(0 + 1)!} = \dfrac{0}{1} = 0$

$a_1 = \dfrac{1^2}{(1 + 1)!} = \dfrac{1}{2 \cdot 1} = \dfrac{1}{2}$

$a_2 = \dfrac{2^2}{(2 + 1)!} = \dfrac{4}{3 \cdot 2 \cdot 1} = \dfrac{2}{3}$

$a_3 = \dfrac{3^2}{(3 + 1)!} = \dfrac{9}{4 \cdot 3 \cdot 2 \cdot 1} = \dfrac{3}{8}$

$a_4 = \dfrac{4^2}{(4 + 1)!} = \dfrac{16}{5 \cdot 4 \cdot 3 \cdot 2 \cdot 1} = \dfrac{2}{15}$

63. $a_n = \dfrac{(-1)^{2n}}{(2n)!} = \dfrac{1}{(2n)!}$

$a_0 = \dfrac{1}{0!} = 1$

$a_1 = \dfrac{1}{2!} = \dfrac{1}{2}$

$a_2 = \dfrac{1}{4!} = \dfrac{1}{24}$

$a_3 = \dfrac{1}{6!} = \dfrac{1}{720}$

$a_4 = \dfrac{1}{8!} = \dfrac{1}{40,320}$

64. $a_n = \dfrac{(-1)^{2n+1}}{(2n + 1)!}$

$a_0 = \dfrac{(-1)^{2(0)+1}}{(2 \cdot 0 + 1)!} = \dfrac{(-1)^1}{1!} = \dfrac{-1}{1} = -1$

$a_1 = \dfrac{(-1)^{2 \cdot 1+1}}{(2 \cdot 1 + 1)!} = \dfrac{(-1)^3}{3!} = \dfrac{-1}{6} = -\dfrac{1}{6}$

$a_2 = \dfrac{(-1)^{2 \cdot 2+1}}{(2 \cdot 2 + 1)!} = \dfrac{(-1)^5}{5!} = \dfrac{-1}{120} = -\dfrac{1}{120}$

$a_3 = \dfrac{(-1)^{2 \cdot 3+1}}{(2 \cdot 3 + 1)!} = \dfrac{(-1)^7}{7!} = \dfrac{-1}{5040} = -\dfrac{1}{5040}$

$a_4 = \dfrac{(-1)^{2 \cdot 4+1}}{(2 \cdot 4 + 1)!} = \dfrac{(-1)^9}{9!} = \dfrac{-1}{362,880} = -\dfrac{1}{362,880}$

65. $\dfrac{4!}{6!} = \dfrac{1 \cdot 2 \cdot 3 \cdot 4}{1 \cdot 2 \cdot 3 \cdot 4 \cdot 5 \cdot 6} = \dfrac{1}{5 \cdot 6} = \dfrac{1}{30}$

66. $\dfrac{5!}{8!} = \dfrac{1 \cdot 2 \cdot 3 \cdot 4 \cdot 5}{1 \cdot 2 \cdot 3 \cdot 4 \cdot 5 \cdot 6 \cdot 7 \cdot 8} = \dfrac{1}{6 \cdot 7 \cdot 8} = \dfrac{1}{336}$

67. $\dfrac{10!}{8!} = \dfrac{1 \cdot 2 \cdot 3 \cdot 4 \cdot 5 \cdot 6 \cdot 7 \cdot 8 \cdot 9 \cdot 10}{1 \cdot 2 \cdot 3 \cdot 4 \cdot 5 \cdot 6 \cdot 7 \cdot 8} = \dfrac{9 \cdot 10}{1} = 90$

68. $\dfrac{25!}{23!} = \dfrac{1 \cdot 2 \cdot 3 \cdots 23 \cdot 24 \cdot 25}{1 \cdot 2 \cdot 3 \cdots 23} = \dfrac{24 \cdot 25}{1} = 600$

69. $\dfrac{(n + 1)!}{n!} = \dfrac{1 \cdot 2 \cdot 3 \cdots n \cdot (n + 1)}{1 \cdot 2 \cdot 3 \cdots n} = \dfrac{n + 1}{1}$

$\quad\quad = n + 1$

70. $\dfrac{(n + 2)!}{n!} = \dfrac{1 \cdot 2 \cdot 3 \cdots n \cdot (n + 1) \cdot (n + 2)}{1 \cdot 2 \cdot 3 \cdots n}$

$\quad\quad = (n + 1)(n + 2)$

71. $\dfrac{(2n - 1)!}{(2n + 1)!} = \dfrac{1 \cdot 2 \cdot 3 \cdots (2n - 1)}{1 \cdot 2 \cdot 3 \cdots (2n - 1) \cdot (2n) \cdot (2n + 1)}$

$\quad\quad = \dfrac{1}{2n(2n + 1)}$

72. $\dfrac{(3n + 1)!}{(3n)!} = \dfrac{1 \cdot 2 \cdot 3 \cdots (3n) \cdot (3n + 1)}{1 \cdot 2 \cdot 3 \cdots (3n)}$

$\quad\quad = \dfrac{3n + 1}{1} = 3n + 1$

73. $\displaystyle\sum_{i=1}^{5} (2i + 1) = (2 + 1) + (4 + 1) + (6 + 1) + (8 + 1) + (10 + 1) = 35$

74. $\displaystyle\sum_{i=1}^{6} (3i - 1) = (3 \cdot 1 - 1) + (3 \cdot 2 - 1) + (3 \cdot 3 - 1) + (3 \cdot 4 - 1) + (3 \cdot 5 - 1) + (3 \cdot 6 - 1) = 57$

75. $\displaystyle\sum_{k=1}^{4} 10 = 10 + 10 + 10 + 10 = 40$

76. $\displaystyle\sum_{k=1}^{5} 5 = 5 + 5 + 5 + 5 + 5 = 25$

77. $\displaystyle\sum_{i=0}^{4} i^2 = 0^2 + 1^2 + 2^2 + 3^2 + 4^2 = 30$

78. $\displaystyle\sum_{i=0}^{5} 2i^2 = 2(0^2) + 2(1^2) + 2(2^2) + 2(3^2) + 2(4^2) + 2(5^2)$
$= 110$

79. $\displaystyle\sum_{k=0}^{3} \frac{1}{k^2 + 1} = \frac{1}{1} + \frac{1}{1+1} + \frac{1}{4+1} + \frac{1}{9+1} = \frac{9}{5}$

80. $\displaystyle\sum_{j=3}^{5} \frac{1}{j^2 - 3} = \frac{1}{3^2 - 3} + \frac{1}{4^2 - 3} + \frac{1}{5^2 - 3} = \frac{124}{429}$

81. $\displaystyle\sum_{k=2}^{5} (k+1)^2(k-3) = (3)^2(-1) + (4)^2(0) + (5)^2(1) + (6)^2(2) = 88$

82. $\displaystyle\sum_{i=1}^{4} [(i-1)^2 + (i+1)^3] = [(0)^2 + (2)^3] + [(1)^2 + (3)^3] + [(2)^2 + (4)^3] + [(3)^2 + (5)^3] = 238$

83. $\displaystyle\sum_{i=1}^{4} 2^i = 2^1 + 2^2 + 2^3 + 2^4 = 30$

84. $\displaystyle\sum_{j=0}^{4} (-2)^j = (-2)^0 + (-2)^1 + (-2)^2 + (-2)^3 + (-2)^4$
$= 11$

85. $\displaystyle\sum_{j=1}^{6} (24 - 3j) = 81$

86. $\displaystyle\sum_{j=1}^{10} \frac{3}{j+1} \approx 6.06$

87. $\displaystyle\sum_{k=0}^{4} \frac{(-1)^k}{k+1} = \frac{47}{60}$

88. $\displaystyle\sum_{k=0}^{4} \frac{(-1)^k}{k!} = \frac{3}{8}$

89. $\dfrac{1}{3(1)} + \dfrac{1}{3(2)} + \dfrac{1}{3(3)} + \cdots + \dfrac{1}{3(9)} = \displaystyle\sum_{i=1}^{9} \frac{1}{3i}$

90. $\dfrac{5}{1+1} + \dfrac{5}{1+2} + \dfrac{5}{1+3} + \cdots + \dfrac{5}{1+15} = \displaystyle\sum_{i=1}^{15} \frac{5}{1+i}$

91. $\left[2\left(\dfrac{1}{8}\right) + 3\right] + \left[2\left(\dfrac{2}{8}\right) + 3\right] + \left[2\left(\dfrac{3}{8}\right) + 3\right] + \cdots + \left[2\left(\dfrac{8}{8}\right) + 3\right] = \displaystyle\sum_{i=1}^{8} \left[2\left(\dfrac{i}{8}\right) + 3\right]$

92. $\left[1 - \left(\dfrac{1}{6}\right)^2\right] + \left[1 - \left(\dfrac{2}{6}\right)^2\right] + \cdots + \left[1 - \left(\dfrac{6}{6}\right)^2\right] = \displaystyle\sum_{k=1}^{6} \left[1 - \left(\dfrac{k}{6}\right)^2\right]$

93. $3 - 9 + 27 - 81 + 243 - 729 = \displaystyle\sum_{i=1}^{6} (-1)^{i+1} 3^i$

94. $1 - \dfrac{1}{2} + \dfrac{1}{4} - \dfrac{1}{8} + \cdots - \dfrac{1}{128} = \dfrac{1}{2^0} - \dfrac{1}{2^1} + \dfrac{1}{2^2} - \dfrac{1}{2^3} + \cdots - \dfrac{1}{2^7} = \displaystyle\sum_{n=0}^{7} \left(-\dfrac{1}{2}\right)^n$

95. $\dfrac{1}{1^2} - \dfrac{1}{2^2} + \dfrac{1}{3^2} - \dfrac{1}{4^2} + \cdots - \dfrac{1}{20^2} = \displaystyle\sum_{i=1}^{20} \frac{(-1)^{i+1}}{i^2}$

96. $\dfrac{1}{1 \cdot 3} + \dfrac{1}{2 \cdot 4} + \dfrac{1}{3 \cdot 5} + \cdots + \dfrac{1}{10 \cdot 12} = \displaystyle\sum_{k=1}^{10} \frac{1}{k(k+2)}$

97. $\dfrac{1}{4} + \dfrac{3}{8} + \dfrac{7}{16} + \dfrac{15}{32} + \dfrac{31}{64} = \displaystyle\sum_{i=1}^{5} \frac{2^i - 1}{2^{i+1}}$

98. $\dfrac{1}{2} + \dfrac{2}{4} + \dfrac{6}{8} + \dfrac{24}{16} + \dfrac{120}{32} + \dfrac{720}{64} = \displaystyle\sum_{k=1}^{6} \frac{k!}{2^k}$

99. $\displaystyle\sum_{i=1}^{4} 5\left(\dfrac{1}{2}\right)^i = 5\left(\dfrac{1}{2}\right) + 5\left(\dfrac{1}{2}\right)^2 + 5\left(\dfrac{1}{2}\right)^3 + 5\left(\dfrac{1}{2}\right)^4 = \dfrac{75}{16}$

100. $\displaystyle\sum_{i=1}^{5} 2\left(\dfrac{1}{3}\right)^i = 2\left(\dfrac{1}{3}\right)^1 + 2\left(\dfrac{1}{3}\right)^2 + 2\left(\dfrac{1}{3}\right)^3 + 2\left(\dfrac{1}{3}\right)^4 + 2\left(\dfrac{1}{3}\right)^5$
$= \dfrac{242}{243}$

101. $\displaystyle\sum_{n=1}^{3} 4\left(-\dfrac{1}{2}\right)^n = 4\left(-\dfrac{1}{2}\right) + 4\left(-\dfrac{1}{2}\right)^2 + 4\left(-\dfrac{1}{2}\right)^3 = -\dfrac{3}{2}$

102. $\displaystyle\sum_{n=1}^{4} 8\left(-\dfrac{1}{4}\right)^n = 8\left(-\dfrac{1}{4}\right)^1 + 8\left(-\dfrac{1}{4}\right)^2 + 8\left(-\dfrac{1}{4}\right)^3 + 8\left(-\dfrac{1}{4}\right)^4$
$= -\dfrac{51}{32}$

103. $\displaystyle\sum_{i=1}^{\infty} 6\left(\frac{1}{10}\right)^i = 0.6 + 0.06 + 0.006 + 0.0006 + \cdots = \frac{2}{3}$

104. $\displaystyle\sum_{k=1}^{\infty} \left(\frac{1}{10}\right)^k = \frac{1}{10} + \frac{1}{10^2} + \frac{1}{10^3} + \frac{1}{10^4} + \frac{1}{10^5} + \cdots$

$$= 0.1 + 0.01 + 0.001 + 0.0001 + 0.00001 + \cdots$$

$$= 0.11111\ldots$$

$$= \frac{1}{9}$$

105. By using a calculator, we have

$$\sum_{k=1}^{10} 7\left(\frac{1}{10}\right)^k \approx 0.7777777777$$

$$\sum_{k=1}^{50} 7\left(\frac{1}{10}\right)^k \approx 0.7777777778$$

$$\sum_{k=1}^{100} 7\left(\frac{1}{10}\right)^k \approx \frac{7}{9}.$$

The terms approach zero as $n \rightarrow \infty$.

Thus, we conclude that $\displaystyle\sum_{k=1}^{\infty} 7\left(\frac{1}{10}\right)^k = \frac{7}{9}.$

106. $\displaystyle\sum_{i=1}^{\infty} 2\left(\frac{1}{10}\right)^i = 2\left(\frac{1}{10} + \frac{1}{10^2} + \frac{1}{10^3} + \frac{1}{10^4} + \cdots\right)$

$$= 2(0.1 + 0.01 + 0.001 + 0.0001 + \cdots)$$

$$= 2(0.111\ldots)$$

$$= 0.222\ldots$$

$$= \frac{2}{9}$$

107. $A_n = 5000\left(1 + \dfrac{0.08}{4}\right)^n$, $n = 1, 2, 3, \ldots$

(a) $A_1 = \$5100.00$

$A_2 = \$5202.00$

$A_3 = \$5306.04$

$A_4 = \$5412.16$

$A_5 = \$5520.40$

$A_6 = \$5630.81$

$A_7 = \$5743.43$

$A_8 = \$5858.30$

(b) $A_{40} = \$11,040.20$

108. (a) $A_1 = 100(101)[(1.01)^1 - 1] = \$101.00$

$A_2 = 100(101)[(1.01)^2 - 1] = \$203.01$

$A_3 = 100(101)[(1.01)^3 - 1] \approx \$306.04$

$A_4 = 100(101)[(1.01)^4 - 1] \approx \$410.10$

$A_5 = 100(101)[(1.01)^5 - 1] \approx \$515.20$

$A_6 = 100(101)[(1.01)^6 - 1] \approx \$621.35$

(b) $A_{60} = 100(101)[(1.01)^{60} - 1] \approx \$8248.64$

(c) $A_{240} = 100(101)[(1.01)^{240} - 1] \approx \$99,914.79$

109. (a) Linear model: $a_n \approx 60.57n - 182$

(b) Quadratic model: $a_n \approx 1.61n^2 + 26.8n - 9.5$

(c)

Year	n	Actual Data	Linear Model	Quadratic Model
1998	8	311	303	308
1999	9	357	363	362
2000	10	419	424	420
2001	11	481	484	480
2002	12	548	545	544
2003	13	608	605	611

The quadratic model is a better fit.

(d) For the year 2008 we have the following predictions:

Linear model: 908 stores

Quadratic model: 995 stores

Since the quadratic model is a better fit, the predicted number of stores in 2008 is 995.

110. (a) $a_n = 0.0457n^3 - 0.3498n^2 - 9.04n + 121.3$, $n = 5, \ldots, 13$.

$a_5 = 73.1$ $a_{10} = 41.6$

$a_6 = 64.3$ $a_{11} = 40.4$

$a_7 = 56.6$ $a_{12} = 41.4$

$a_8 = 50.0$ $a_{13} = 45.1$

$a_9 = 44.9$

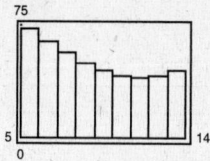

(b) The number of cases reported fluctuates.

111. (a) $a_n = 2.7698n^3 - 61.372n^2 + 600.00n + 3102.9$

$a_0 = \$3102.9$ billion $a_7 \approx \$5245.7$ billion

$a_1 \approx \$3644.3$ billion $a_8 \approx \$5393.2$ billion

$a_2 \approx \$4079.6$ billion $a_9 \approx \$5551.0$ billion

$a_3 \approx \$4425.3$ billion $a_{10} = \$5735.5$ billion

$a_4 \approx \$4698.2$ billion $a_{11} \approx \$5963.5$ billion

$a_5 \approx \$4914.8$ billion $a_{12} \approx \$6251.5$ billion

$a_6 \approx \$5091.8$ billion $a_{13} \approx \$6616.3$ billion

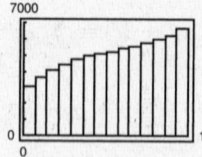

(b) The federal debt is increasing.

112. $\sum_{n=6}^{13} (46.609n^2 - 119.84n - 1125.8) = \$17,495$ million

The results from the model and the figure (which are approximations) are very similar.

113. True, $\sum_{i=1}^{4} (i^2 + 2i) = \sum_{i=1}^{4} i^2 + 2\sum_{i=1}^{4} i$ by the Properties of Sums.

114. $\sum_{j=1}^{4} 2^j = \sum_{j=3}^{6} 2^{j-2}$

True, because $2^1 + 2^2 + 2^3 + 2^4 = 2^{3-2} + 2^{4-2} + 2^{5-2} + 2^{6-2}$.

115. $a_1 = 1, a_2 = 1, a_{k+2} = a_{k+1} + a_k, k \geq 1$

$a_1 = 1$

$a_2 = 1$

$a_3 = 1 + 1 = 2$

$a_4 = 2 + 1 = 3$

$a_5 = 3 + 2 = 5$

$a_6 = 5 + 3 = 8$

$a_7 = 8 + 5 = 13$

$a_8 = 13 + 8 = 21$

$a_9 = 21 + 13 = 34$

$a_{10} = 34 + 21 = 55$

$a_{11} = 55 + 34 = 89$

$a_{12} = 89 + 55 = 144$

$b_1 = \dfrac{1}{1} = 1$

$b_2 = \dfrac{2}{1} = 2$

$b_3 = \dfrac{3}{2}$

$b_4 = \dfrac{5}{3}$

$b_5 = \dfrac{8}{5}$

$b_6 = \dfrac{13}{8}$

$b_7 = \dfrac{21}{13}$

$b_8 = \dfrac{34}{21}$

$b_9 = \dfrac{55}{34}$

$b_{10} = \dfrac{89}{55}$

116. $b_n = \dfrac{a_{n+1}}{a_n}$; $b_1 = 1, b_2 = 2, b_3 = \dfrac{3}{2}, b_4 = \dfrac{5}{3}, \ldots$

$b_2 = 1 + \dfrac{1}{b_1} = 1 + \dfrac{1}{1} = 2$

$b_3 = 1 + \dfrac{1}{b_2} = 1 + \dfrac{1}{2} = \dfrac{3}{2}$

$b_4 = 1 + \dfrac{1}{b_3} = 1 + \dfrac{2}{3} = \dfrac{5}{3}$

$b_5 = 1 + \dfrac{1}{b_4} = 1 + \dfrac{3}{5} = \dfrac{8}{5}$

$b_n = 1 + \dfrac{1}{b_{n-1}}$

117. $\dfrac{327.15 + 785.69 + 433.04 + 265.38 + 604.12 + 590.30}{6} \approx \$500.95$

118. $\bar{x} = \dfrac{1}{n}\displaystyle\sum_{i=1}^{n} x_i = \dfrac{1.899 + 1.959 + 1.919 + 1.939 + 1.999}{5}$

$\qquad = \$1.943$

119. $\displaystyle\sum_{i=1}^{n}(x_i - \bar{x}) = \sum_{i=1}^{n} x_i - \sum_{i=1}^{n} \bar{x}$

$\qquad\qquad\qquad = \left(\displaystyle\sum_{i=1}^{n} x_i\right) - n\bar{x}$

$\qquad\qquad\qquad = \left(\displaystyle\sum_{i=1}^{n} x_i\right) - n\left(\dfrac{1}{n}\displaystyle\sum_{i=1}^{n} x_i\right)$

$\qquad\qquad\qquad = 0$

120. $\displaystyle\sum_{i=1}^{n}(x_i - \bar{x})^2 = \sum_{i=1}^{n}\left(x_i^2 - 2x_i\bar{x} + \bar{x}^2\right) = \sum_{i=1}^{n} x_i^2 - 2\bar{x}\sum_{i=1}^{n} x_i + n\bar{x}^2$

$\qquad\qquad\qquad\qquad = \displaystyle\sum_{i=1}^{n} x_i^2 - 2 \cdot \dfrac{1}{n}\sum_{i=1}^{n} x_i \sum_{i=1}^{n} x_i + n \cdot \dfrac{1}{n}\sum_{i=1}^{n} x_i \cdot \dfrac{1}{n}\sum_{i=1}^{n} x_i$

$\qquad\qquad\qquad\qquad = \displaystyle\sum_{i=1}^{n} x_i^2 + \sum_{i=1}^{n} x_i \sum_{i=1}^{n} x_i\left(-\dfrac{2}{n} + \dfrac{1}{n}\right) = \sum_{i=1}^{n} x_i^2 - \dfrac{1}{n}\left(\sum_{i=1}^{n} x_i\right)^2$

121. $a_n = \dfrac{x^n}{n!}$

$a_1 = \dfrac{x^1}{1!} = x$

$a_2 = \dfrac{x^2}{2!} = \dfrac{x^2}{2}$

$a_3 = \dfrac{x^3}{3!} = \dfrac{x^3}{6}$

$a_4 = \dfrac{x^4}{4!} = \dfrac{x^4}{24}$

$a_5 = \dfrac{x^5}{5!} = \dfrac{x^5}{120}$

122. $a_n = \dfrac{(-1)^n x^{2n+1}}{2n + 1}$

$a_1 = \dfrac{(-1)^1 x^{2(1)+1}}{2(1) + 1} = -\dfrac{x^3}{3}$

$a_2 = \dfrac{(-1)^2 x^{2(2)+1}}{2(2) + 1} = \dfrac{x^5}{5}$

$a_3 = \dfrac{(-1)^3 x^{2(3)+1}}{2(3) + 1} = -\dfrac{x^7}{7}$

$a_4 = \dfrac{(-1)^4 x^{2(4)+1}}{2(4) + 1} = \dfrac{x^9}{9}$

$a_5 = \dfrac{(-1)^5 x^{2(5)+1}}{2(5) + 1} = -\dfrac{x^{11}}{11}$

123. $a_n = \dfrac{(-1)^n x^{2n}}{(2n)!}$

$a_1 = \dfrac{-x^2}{2!} = -\dfrac{x^2}{2}$

$a_2 = \dfrac{x^4}{4!} = \dfrac{x^4}{24}$

$a_3 = \dfrac{-x^6}{6!} = -\dfrac{x^6}{720}$

$a_4 = \dfrac{x^8}{8!} = \dfrac{x^8}{40,320}$

$a_5 = \dfrac{-x^{10}}{10!} = -\dfrac{x^{10}}{3,628,800}$

124. $a_n = \dfrac{(-1)^n x^{2n+1}}{(2n + 1)!}$

$a_1 = \dfrac{(-1)^1 x^{2(1)+1}}{(2(1) + 1)!} = -\dfrac{x^3}{3!} = -\dfrac{x^3}{6}$

$a_2 = \dfrac{(-1)^2 x^{2(2)+1}}{(2(2) + 1)!} = \dfrac{x^5}{5!} = \dfrac{x^5}{120}$

$a_3 = \dfrac{(-1)^3 x^{2(3)+1}}{(2(3) + 1)!} = -\dfrac{x^7}{7!} = -\dfrac{x^7}{5040}$

$a_4 = \dfrac{(-1)^4 x^{2(4)+1}}{(2(4) + 1)!} = \dfrac{x^9}{9!} = \dfrac{x^9}{362,880}$

$a_5 = \dfrac{(-1)^5 x^{2(5)+1}}{(2(5) + 1)!} = -\dfrac{x^{11}}{11!} = -\dfrac{x^{11}}{39,916,800}$

125. $f(x) = 4x - 3$ is one-to-one, so it has an inverse.

$\qquad y = 4x - 3$

$\qquad x = 4y - 3$

$\qquad \dfrac{x + 3}{4} = y$

$\qquad f^{-1}(x) = \dfrac{x + 3}{4}$

126. $g(x) = \dfrac{3}{x}$

$$y = \dfrac{3}{x}$$

$$x = \dfrac{3}{y}$$

$$xy = 3$$

$$y = \dfrac{3}{x}$$

This is a function of x, so f has in inverse.

$$f^{-1}(x) = \dfrac{3}{x}, x \neq 0$$

127. $h(x) = \sqrt{5x + 1}$ is one-to-one, so it has an inverse.

Domain: $x \geq -\dfrac{1}{5}$

Range: $y \geq 0$

$$y = \sqrt{5x + 1}, x \geq -\dfrac{1}{5}, y \geq 0$$

$$x = \sqrt{5y + 1}, x \geq 0, y \geq -\dfrac{1}{5}$$

$$x^2 = 5y + 1, x \geq 0$$

$$\dfrac{x^2 - 1}{5} = y, x \geq 0$$

$$h^{-1}(x) = \dfrac{x^2 - 1}{5} = \dfrac{1}{5}(x^2 - 1), x \geq 0$$

128. $f(x) = (x - 1)^2$

$$y = (x - 1)^2$$

$$x = (y - 1)^2$$

$$\pm\sqrt{x} = y - 1$$

$$1 \pm \sqrt{x} = y$$

This does not represent y as a function of x, so f does not have an inverse.

129. (a) $A - B = \begin{bmatrix} 6 & 5 \\ 3 & 4 \end{bmatrix} - \begin{bmatrix} -2 & 4 \\ 6 & -3 \end{bmatrix} = \begin{bmatrix} 6 - (-2) & 5 - 4 \\ 3 - 6 & 4 - (-3) \end{bmatrix} = \begin{bmatrix} 8 & 1 \\ -3 & 7 \end{bmatrix}$

(b) $4B - 3A = 4\begin{bmatrix} -2 & 4 \\ 6 & -3 \end{bmatrix} - 3\begin{bmatrix} 6 & 5 \\ 3 & 4 \end{bmatrix} = \begin{bmatrix} -8 - 18 & 16 - 15 \\ 24 - 9 & -12 - 12 \end{bmatrix} = \begin{bmatrix} -26 & 1 \\ 15 & -24 \end{bmatrix}$

(c) $AB = \begin{bmatrix} 6 & 5 \\ 3 & 4 \end{bmatrix}\begin{bmatrix} -2 & 4 \\ 6 & -3 \end{bmatrix} = \begin{bmatrix} -12 + 30 & 24 - 15 \\ -6 + 24 & 12 - 12 \end{bmatrix} = \begin{bmatrix} 18 & 9 \\ 18 & 0 \end{bmatrix}$

(d) $BA = \begin{bmatrix} -2 & 4 \\ 6 & -3 \end{bmatrix}\begin{bmatrix} 6 & 5 \\ 3 & 4 \end{bmatrix} = \begin{bmatrix} -12 + 16 & -10 + 12 \\ 36 - 9 & 30 - 12 \end{bmatrix} = \begin{bmatrix} 0 & 6 \\ 27 & 18 \end{bmatrix}$

130. (a) $A - B = \begin{bmatrix} 10 & 7 \\ -4 & 6 \end{bmatrix} - \begin{bmatrix} 0 & -12 \\ 8 & 11 \end{bmatrix} = \begin{bmatrix} 10 - 0 & 7 - (-12) \\ -4 - 8 & 6 - 11 \end{bmatrix} = \begin{bmatrix} 10 & 19 \\ -12 & -5 \end{bmatrix}$

(b) $4B - 3A = 4\begin{bmatrix} 0 & -12 \\ 8 & 11 \end{bmatrix} - 3\begin{bmatrix} 10 & 7 \\ -4 & 6 \end{bmatrix} = \begin{bmatrix} 0 - 30 & -48 - 21 \\ 32 + 12 & 44 - 18 \end{bmatrix} = \begin{bmatrix} -30 & -69 \\ 44 & 26 \end{bmatrix}$

(c) $AB = \begin{bmatrix} 10 & 7 \\ -4 & 6 \end{bmatrix}\begin{bmatrix} 0 & -12 \\ 8 & 11 \end{bmatrix} = \begin{bmatrix} 0 + 56 & -120 + 77 \\ 0 + 48 & 48 + 66 \end{bmatrix} = \begin{bmatrix} 56 & -43 \\ 48 & 114 \end{bmatrix}$

(d) $BA = \begin{bmatrix} 0 & -12 \\ 8 & 11 \end{bmatrix}\begin{bmatrix} 10 & 7 \\ -4 & 6 \end{bmatrix} = \begin{bmatrix} 0 + 48 & 0 - 72 \\ 80 - 44 & 56 + 66 \end{bmatrix} = \begin{bmatrix} 48 & -72 \\ 36 & 122 \end{bmatrix}$

131. (a) $A - B = \begin{bmatrix} -2 & -3 & 6 \\ 4 & 5 & 7 \\ 1 & 7 & 4 \end{bmatrix} - \begin{bmatrix} 1 & 4 & 2 \\ 0 & 1 & 6 \\ 0 & 3 & 1 \end{bmatrix} = \begin{bmatrix} -2-1 & -3-4 & 6-2 \\ 4-0 & 5-1 & 7-6 \\ 1-0 & 7-3 & 4-1 \end{bmatrix} = \begin{bmatrix} -3 & -7 & 4 \\ 4 & 4 & 1 \\ 1 & 4 & 3 \end{bmatrix}$

(b) $4B - 3A = 4\begin{bmatrix} 1 & 4 & 2 \\ 0 & 1 & 6 \\ 0 & 3 & 1 \end{bmatrix} - 3\begin{bmatrix} -2 & -3 & 6 \\ 4 & 5 & 7 \\ 1 & 7 & 4 \end{bmatrix} = \begin{bmatrix} 4-(-6) & 16-(-9) & 8-18 \\ 0-12 & 4-15 & 24-21 \\ 0-3 & 12-21 & 4-12 \end{bmatrix} = \begin{bmatrix} 10 & 25 & -10 \\ -12 & -11 & 3 \\ -3 & -9 & -8 \end{bmatrix}$

(c) $AB = \begin{bmatrix} -2 & -3 & 6 \\ 4 & 5 & 7 \\ 1 & 7 & 4 \end{bmatrix}\begin{bmatrix} 1 & 4 & 2 \\ 0 & 1 & 6 \\ 0 & 3 & 1 \end{bmatrix} = \begin{bmatrix} -2+0+0 & -8-3+18 & -4-18+6 \\ 4+0+0 & 16+5+21 & 8+30+7 \\ 1+0+0 & 4+7+12 & 2+42+4 \end{bmatrix} = \begin{bmatrix} -2 & 7 & -16 \\ 4 & 42 & 45 \\ 1 & 23 & 48 \end{bmatrix}$

(d) $BA = \begin{bmatrix} 1 & 4 & 2 \\ 0 & 1 & 6 \\ 0 & 3 & 1 \end{bmatrix}\begin{bmatrix} -2 & -3 & 6 \\ 4 & 5 & 7 \\ 1 & 7 & 4 \end{bmatrix} = \begin{bmatrix} -2+16+2 & -3+20+14 & 6+28+8 \\ 0+4+6 & 0+5+42 & 0+7+24 \\ 0+12+1 & 0+15+7 & 0+21+4 \end{bmatrix} = \begin{bmatrix} 16 & 31 & 42 \\ 10 & 47 & 31 \\ 13 & 22 & 25 \end{bmatrix}$

132. (a) $A - B = \begin{bmatrix} -1 & 4 & 0 \\ 5 & 1 & 2 \\ 0 & -1 & 3 \end{bmatrix} - \begin{bmatrix} 0 & 4 & 0 \\ 3 & 1 & -2 \\ -1 & 0 & 2 \end{bmatrix} = \begin{bmatrix} -1-0 & 4-4 & 0-0 \\ 5-3 & 1-1 & 2-(-2) \\ 0-(-1) & -1-0 & 3-2 \end{bmatrix} = \begin{bmatrix} -1 & 0 & 0 \\ 2 & 0 & 4 \\ 1 & -1 & 1 \end{bmatrix}$

(b) $4B - 3A = 4\begin{bmatrix} 0 & 4 & 0 \\ 3 & 1 & -2 \\ -1 & 0 & 2 \end{bmatrix} - 3\begin{bmatrix} -1 & 4 & 0 \\ 5 & 1 & 2 \\ 0 & -1 & 3 \end{bmatrix} = \begin{bmatrix} 0-(-3) & 16-12 & 0-0 \\ 12-15 & 4-3 & -8-6 \\ -4-0 & 0-(-3) & 8-9 \end{bmatrix} = \begin{bmatrix} 3 & 4 & 0 \\ -3 & 1 & -14 \\ -4 & 3 & -1 \end{bmatrix}$

(c) $AB = \begin{bmatrix} -1 & 4 & 0 \\ 5 & 1 & 2 \\ 0 & -1 & 3 \end{bmatrix}\begin{bmatrix} 0 & 4 & 0 \\ 3 & 1 & -2 \\ -1 & 0 & 2 \end{bmatrix} = \begin{bmatrix} 0+12+0 & -4+4+0 & 0-8+0 \\ 0+3-2 & 20+1+0 & 0-2+4 \\ 0-3-3 & 0-1+0 & 0+2+6 \end{bmatrix} = \begin{bmatrix} 12 & 0 & -8 \\ 1 & 21 & 2 \\ -6 & -1 & 8 \end{bmatrix}$

(d) $BA = \begin{bmatrix} 0 & 4 & 0 \\ 3 & 1 & -2 \\ -1 & 0 & 2 \end{bmatrix}\begin{bmatrix} -1 & 4 & 0 \\ 5 & 1 & 2 \\ 0 & -1 & 3 \end{bmatrix} = \begin{bmatrix} 0+20+0 & 0+4+0 & 0+8+0 \\ -3+5+0 & 12+1+2 & 0+2-6 \\ 1+0+0 & -4+0-2 & 0+0+6 \end{bmatrix} = \begin{bmatrix} 20 & 4 & 8 \\ 2 & 15 & -4 \\ 1 & -6 & 6 \end{bmatrix}$

133. $|A| = \begin{vmatrix} 3 & 5 \\ -1 & 7 \end{vmatrix} = 3(7) - 5(-1) = 26$

134. $\begin{vmatrix} -2 & 8 \\ 12 & 15 \end{vmatrix} = -2(15) - 8(12) = -126$

135. $|A| = \begin{vmatrix} 3 & 4 & 5 \\ 0 & 7 & 3 \\ 4 & 9 & -1 \end{vmatrix} = 3\begin{vmatrix} 7 & 3 \\ 9 & -1 \end{vmatrix} + 4\begin{vmatrix} 4 & 5 \\ 7 & 3 \end{vmatrix}$

$= 3[7(-1) - 3(9)] + 4[4(3) - 5(7)] = -194$

136. $|A| = 16(C_{11}) + 9(C_{21}) - 2(C_{31}) - 4(C_{41})$

$C_{11} = (-1)^{1+1}\begin{vmatrix} 8 & 3 & 7 \\ -1 & 12 & 3 \\ 6 & 2 & 1 \end{vmatrix} = \begin{vmatrix} 8 & 3 & 7 \\ -1 & 12 & 3 \\ 6 & 2 & 1 \end{vmatrix}$

$= 8\begin{vmatrix} 12 & 3 \\ 2 & 1 \end{vmatrix} - 3\begin{vmatrix} -1 & 3 \\ 6 & 1 \end{vmatrix} + 7\begin{vmatrix} -1 & 12 \\ 6 & 2 \end{vmatrix}$

$= 8(12 - 6) - 3(-1 - 18) + 7(-2 - 72) = -413$

$C_{21} = (-1)^{2+1}\begin{vmatrix} 11 & 10 & 2 \\ -1 & 12 & 3 \\ 6 & 2 & 1 \end{vmatrix} = \begin{vmatrix} -11 & -10 & -2 \\ 1 & -12 & -3 \\ -6 & -2 & -1 \end{vmatrix}$

$= -11\begin{vmatrix} -12 & -3 \\ -2 & -1 \end{vmatrix} - 1\begin{vmatrix} -10 & -2 \\ -2 & -1 \end{vmatrix} - 6\begin{vmatrix} -10 & -2 \\ -12 & -3 \end{vmatrix}$

$= -11(12 - 6) - 1(10 - 4) - 6(30 - 24) = -108$

—CONTINUED—

136. —CONTINUED—

$$C_{31} = (-1)^{3+1} \begin{vmatrix} 11 & 10 & 2 \\ 8 & 3 & 7 \\ 6 & 2 & 1 \end{vmatrix} = \begin{vmatrix} 11 & 10 & 2 \\ 8 & 3 & 7 \\ 6 & 2 & 1 \end{vmatrix}$$

$$= 11 \begin{vmatrix} 3 & 7 \\ 2 & 1 \end{vmatrix} - 8 \begin{vmatrix} 10 & 2 \\ 2 & 1 \end{vmatrix} + 6 \begin{vmatrix} 10 & 2 \\ 3 & 7 \end{vmatrix}$$

$$= 11(3 - 14) - 8(10 - 4) + 6(70 - 6) = 215$$

$$C_{41} = (-1)^{4+1} \begin{vmatrix} 11 & 10 & 2 \\ 8 & 3 & 7 \\ -1 & 12 & 3 \end{vmatrix} = \begin{vmatrix} -11 & -10 & -2 \\ -8 & -3 & -7 \\ 1 & -12 & -3 \end{vmatrix}$$

$$= -11 \begin{vmatrix} -3 & -7 \\ -12 & -3 \end{vmatrix} - (-8) \begin{vmatrix} -10 & -2 \\ -12 & -3 \end{vmatrix} + 1 \begin{vmatrix} -10 & -2 \\ -3 & -7 \end{vmatrix}$$

$$= -11(9 - 84) + 8(30 - 24) + 1(70 - 6) = 937$$

So, $|A| = 16(-413) + 9(-108) - 2(215) - 4(937)$

$$= -11,758.$$

Section 9.2   Arithmetic Sequences and Partial Sums

- ■ You should be able to recognize an arithmetic sequence, find its common difference, and find its nth term.
- ■ You should be able to find the nth partial sum of an arithmetic sequence by using the formula

$$S_n = \frac{n}{2}(a_1 + a_n).$$

Vocabulary Check

1. arithmetic; common

2. $a_n = dn + c$

3. sum of a finite arithmetic sequence

1. 10, 8, 6, 4, 2, . . .

Arithmetic sequence, $d = -2$

2. 4, 7, 10, 13, 16, . . .

Arithmetic sequence, $d = 3$

3. 1, 2, 4, 8, 16, . . .

Not an arithmetic sequence

4. 80, 40, 20, 10, 5, . . .

Not an arithmetic sequence

5. $\frac{9}{4}$, 2, $\frac{7}{4}$, $\frac{3}{2}$, $\frac{5}{4}$, . . .

Arithmetic sequence, $d = -\frac{1}{4}$

6. 3, $\frac{5}{2}$, 2, $\frac{3}{2}$, 1, . . .

Arithmetic sequence, $d = -\frac{1}{2}$

7. $\frac{1}{3}$, $\frac{2}{3}$, 1, $\frac{4}{3}$, $\frac{5}{6}$, . . .

Not an arithmetic sequence

8. 5.3, 5.7, 6.1, 6.5, 6.9, . . .

Arithmetic sequence, $d = 0.4$

9. ln 1, ln 2, ln 3, ln 4, ln 5, . . .

Not an arithmetic sequence

10. 1^2, 2^2, 3^2, 4^2, 5^2, . . .

Not an arithmetic sequence

11. $a_n = 5 + 3n$

8, 11, 14, 17, 20

Arithmetic sequence, $d = 3$

12. $a_n = 100 - 3n$

97, 94, 91, 88, 85

Arithmetic sequence, $d = -3$

13. $a_n = 3 - 4(n - 2)$

$7, 3, -1, -5, -9$

Arithmetic sequence, $d = -4$

14. $a_n = 1 + (n - 1)4$

$1, 5, 9, 13, 17$

Arithmetic sequence, $d = 4$

15. $a_n = (-1)^n$

$-1, 1, -1, 1, -1$

Not an arithmetic sequence

16. $a_n = 2^{n-1}$

$1, 2, 4, 8, 16$

Not an arithmetic sequence

17. $a_n = \dfrac{(-1)^n 3}{n}$

$-3, \dfrac{3}{2}, -1, \dfrac{3}{4}, -\dfrac{3}{5}$

Not an arithmetic sequence

18. $a_n = (2^n)n$

$2, 8, 24, 64, 160$

Not an arithmetic sequence

19. $a_1 = 1$, $d = 3$

$a_n = a_1 + (n - 1)d = 1 + (n - 1)(3) = 3n - 2$

20. $a_1 = 15$, $d = 4$

$a_n = a_1 + (n - 1)d = 15 + (n - 1)4$

$\qquad = 4n + 11$

21. $a_1 = 100$, $d = -8$

$a_n = a_1 + (n - 1)d = 100 + (n - 1)(-8)$

$\qquad = -8n + 108$

22. $a_1 = 0$, $d = -\frac{2}{3}$

$a_n = a_1 + (n - 1)d = (n - 1)\left(-\frac{2}{3}\right)$

$\qquad = -\frac{2}{3}n + \frac{2}{3}$

23. $a_1 = x$, $d = 2x$

$a_n = a_1 + (n - 1)d = x + (n - 1)(2x) = 2xn - x$

24. $a_1 = -y$, $d = 5y$

$a_n = a_1 + (n - 1)d = -y + (n - 1)(5y)$

$\qquad = 5yn - 6y$

25. $4, \frac{3}{2}, -1, -\frac{7}{2}, \ldots$

$d = -\frac{5}{2}$

$a_n = a_1 + (n - 1)d = 4 + (n - 1)\left(-\frac{5}{2}\right) = -\frac{5}{2}n + \frac{13}{2}$

26. $10, 5, 0, -5, -10, \ldots$

$d = -5$

$a_n = a_1 + (n - 1)d = 10 + (n - 1)(-5) = -5n + 15$

27. $a_1 = 5$, $a_4 = 15$

$a_4 = a_1 + 3d \implies 15 = 5 + 3d \implies d = \frac{10}{3}$

$a_n = a_1 + (n - 1)d = 5 + (n - 1)\left(\frac{10}{3}\right) = \frac{10}{3}n + \frac{5}{3}$

28. $a_1 = -4$, $a_5 = 16$

$a_n = a_1 + (n - 1)d$

$16 = -4 + 4d$

$d = 5$

$a_n = a_1 + (n - 1)d = -4 + (n - 1)5$

$\qquad = 5n - 9$

29. $a_3 = 94$, $a_6 = 85$

$a_6 = a_3 + 3d \implies 85 = 94 + 3d \implies d = -3$

$a_1 = a_3 - 2d \implies a_1 = 94 - 2(-3) = 100$

$a_n = a_1 + (n - 1)d = 100 + (n - 1)(-3)$

$\qquad\qquad = -3n + 103$

30. $a_5 = 190$, $a_{10} = 115$

$a_{10} = a_5 + 5d \implies 115 = 190 + 5d \implies d = -15$

$a_1 = a_5 - 4d \implies a_1 = 190 - 4(-15) = 250$

$a_n = a_1 + (n - 1)d = 250 + (n - 1)(-15)$

$\qquad\qquad = -15n + 265$

31. $a_1 = 5, d = 6$

$a_1 = 5$

$a_2 = 5 + 6 = 11$

$a_3 = 11 + 6 = 17$

$a_4 = 17 + 6 = 23$

$a_5 = 23 + 6 = 29$

32. $a_1 = 5, d = -\frac{3}{4}$

$a_1 = 5$

$a_2 = 5 - \frac{3}{4} = \frac{17}{4}$

$a_3 = \frac{17}{4} - \frac{3}{4} = \frac{14}{4} = \frac{7}{2}$

$a_4 = \frac{7}{2} - \frac{3}{4} = \frac{11}{4}$

$a_5 = \frac{11}{4} - \frac{3}{4} = \frac{8}{4} = 2$

33. $a_1 = -2.6, d = -0.4$

$a_1 = -2.6$

$a_2 = -2.6 + (-0.4) = -3.0$

$a_3 = -3.0 + (-0.4) = -3.4$

$a_4 = -3.4 + (-0.4) = -3.8$

$a_5 = -3.8 + (-0.4) = -4.2$

34. $a_1 = 16.5, d = 0.25$

$a_1 = 16.5$

$a_2 = 16.5 + 0.25 = 16.75$

$a_3 = 16.75 + 0.25 = 17$

$a_4 = 17 + 0.25 = 17.25$

$a_5 = 17.25 + 0.25 = 17.5$

35. $a_1 = 2, a_{12} = 46$

$46 = 2 + (12 - 1)d$

$44 = 11d$

$4 = d$

$a_1 = 2$

$a_2 = 2 + 4 = 6$

$a_3 = 6 + 4 = 10$

$a_4 = 10 + 4 = 14$

$a_5 = 14 + 4 = 18$

36. $a_4 = 16, a_{10} = 46$

$16 = a_4 = a_1 + (n - 1)d = a_1 + 3d$

$46 = a_{10} = a_1 + (n - 1)d = a_1 + 9d$

Answer: $a_1 = 1, d = 5$

$a_1 = 1$

$a_2 = 1 + 5 = 6$

$a_3 = 6 + 5 = 11$

$a_4 = 11 + 5 = 16$

$a_5 = 16 + 5 = 21$

37. $a_8 = 26, a_{12} = 42$

$a_{12} = a_8 + 4d$

$42 = 26 + 4d \Longrightarrow d = 4$

$a_8 = a_1 + 7d$

$26 = a_1 + 28 \Longrightarrow a_1 = -2$

$a_1 = -2$

$a_2 = -2 + 4 = 2$

$a_3 = 2 + 4 = 6$

$a_4 = 6 + 4 = 10$

$a_5 = 10 + 4 = 14$

38. $a_3 = 19, a_{15} = -1.7$

$19 = a_3 = a_1 + (n - 1)d = a_1 + 2d$

$-1.7 = a_{15} = a_1 + (n - 1)d = a_1 + 14d$

Answer: $a_1 = 22.45, d = -1.725$

$a_1 = 22.45$

$a_2 = 22.45 - 1.725 = 20.725$

$a_3 = 20.725 - 1.725 = 19$

$a_4 = 19 - 1.725 = 17.275$

$a_5 = 17.275 - 1.725 = 15.55$

39. $a_1 = 15, a_{k+1} = a_k + 4$

$a_2 = 15 + 4 = 19$

$a_3 = 19 + 4 = 23$

$a_4 = 23 + 4 = 27$

$a_5 = 27 + 4 = 31$

$d = 4$

$c = a_1 - d = 15 - 4 = 11$

$a_n = 4n + 11$

40. $a_1 = 6, a_{k+1} = a_k + 5$

$a_2 = 6 + 5 = 11$

$a_3 = 11 + 5 = 16$

$a_4 = 16 + 5 = 21$

$a_5 = 21 + 5 = 26$

$d = 5$

$a_n = dn + c$

$a_n = 5n + c$

$c = a_1 - d$

$= 6 - 5$

$= 1$

So, $a_n = 5n + 1$.

41. $a_1 = 200, a_{k+1} = a_k - 10$

$a_2 = 200 - 10 = 190$

$a_3 = 190 - 10 = 180$

$a_4 = 180 - 10 = 170$

$a_5 = 170 - 10 = 160$

$d = -10$

$c = a_1 - d = 200 - (-10) = 210$

$a_n = -10n + 210$

42. $a_1 = 72, a_{k+1} = a_k - 6$

$a_2 = 72 - 6 = 66$

$a_3 = 66 - 6 = 60$

$a_4 = 60 - 6 = 54$

$a_5 = 54 - 6 = 48$

$d = -6$

$a_n = dn + c$

$a_n = -6n + c$

$c = a_1 - d$

$\quad = 72 - (-6)$

$\quad = 78$

So, $a_n = -6n + 78$.

43. $a_1 = \frac{5}{8}, a_{k+1} = a_k - \frac{1}{8}$

$a_1 = \frac{5}{8}$

$a_2 = \frac{5}{8} - \frac{1}{8} = \frac{1}{2}$

$a_3 = \frac{1}{2} - \frac{1}{8} = \frac{3}{8}$

$a_4 = \frac{3}{8} - \frac{1}{8} = \frac{1}{4}$

$a_5 = \frac{1}{4} - \frac{1}{8} = \frac{1}{8}$

$d = -\frac{1}{8}$

$c = a_1 - d = \frac{5}{8} - \left(-\frac{1}{8}\right) = \frac{3}{4}$

$a_n = -\frac{1}{8}n + \frac{3}{4}$

44. $a_1 = 0.375, a_{k+1} = a_k + 0.25$

$a_2 = 0.375 + 0.25 = 0.625$

$a_3 = 0.625 + 0.25 = 0.875$

$a_4 = 0.875 + 0.25 = 1.125$

$a_5 = 1.125 + 0.25 = 1.375$

$d = 0.25$

$a_n = dn + c$

$a_n = 0.25n + c$

$c = a_1 - d$

$\quad = 0.375 - 0.25$

$\quad = 0.125$

So, $a_n = 0.25n + 0.125$.

45. $a_1 = 5, a_2 = 11 \Rightarrow d = 11 - 5 = 6$

$a_n = a_1 + (n - 1)d \Rightarrow a_{10} = 5 + 9(6) = 59$

46. $a_1 = 3, a_2 = 13$

$d = a_2 - a_1 = 13 - 3 = 10$

$a_n = dn + c, a_n = 10n + c$

$c = a_1 - d = 3 - 10 = -7$

$a_n = 10n - 7, a_9 = 10(9) - 7 = 83$

47. $a_1 = 4.2, a_2 = 6.6 \Rightarrow d = 6.6 - 4.2 = 2.4$

$a_n = a_1 + (n - 1)d \Rightarrow a_7 = 4.2 + 6(2.4) = 18.6$

48. $a_1 = -0.7, a_2 = -13.8$

$d = a_2 - a_1 = -13.8 + 0.7 = -13.1$

$a_n = dn + c, a_n = -13.1n + c$

$c = a_1 - d = -0.7 + 13.1 = 12.4$

$a_n = -13.1n + 12.4, a_8 = -92.4$

49. $a_n = -\frac{3}{4}n + 8$

$d = -\frac{3}{4}$ so the sequence is decreasing and $a_1 = 7\frac{1}{4}$.

Matches (b).

50. $a_n = 3n - 5$

$d = 3$ so the sequence is increasing and $a_1 = -2$.

Matches (d).

51. $a_n = 2 + \frac{3}{4}n$

$d = \frac{3}{4}$ so the sequence is increasing and $a_1 = 2\frac{3}{4}$.

Matches (c).

52. $a_n = 25 - 3n$

$d = -3$ so the sequence is decreasing and $a_1 = 22$.

Matches (a).

53. $a_n = 15 - \frac{3}{2}n$

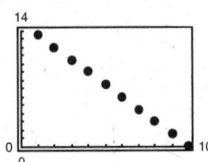

54. $a_n = -5 + 2n$

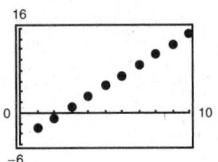

55. $a_n = 0.2n + 3$

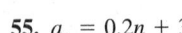
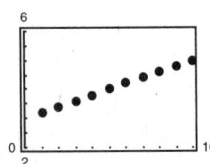

56. $a_n = -0.3n + 8$

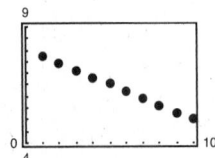

57. $8, 20, 32, 44, \ldots$

$a_1 = 8, \ d = 12, \ n = 10$

$a_{10} = 8 + 9(12) = 116$

$S_{10} = \frac{10}{2}(8 + 116) = 620$

58. 2, 8, 14, 20, . . . , $n = 25$

$d = 6, c = 2 - 6 = -4$

$a_n = 6n - 4$

$a_1 = 2$ and $a_{25} = 146$

$S_{25} = \frac{25}{2}(2 + 146) = 1850$

59. 4.2, 3.7, 3.2, 2.7, . . .

$a_1 = 4.2, d = -0.5, n = 12$

$a_{12} = 4.2 + 11(-0.5) = -1.3$

$S_{12} = \frac{12}{2}[4.2 + (-1.3)] = 17.4$

60. 0.5, 0.9, 1.3, 1.7, . . . , $n = 10$

$d = 0.4, c = 0.1$

$a_n = 0.4n + 0.1$

$a_1 = 0.5$ and $a_{10} = 4.1$

$S_{10} = \frac{10}{2}(0.5 + 4.1) = 23$

61. 40, 37, 34, 31, . . .

$a_1 = 40, d = -3, n = 10$

$a_{10} = 40 + 9(-3) = 13$

$S_{10} = \frac{10}{2}(40 + 13) = 265$

62. 75, 70, 65, 60, . . . , $n = 25$

$d = -5, c = 80$

$a_n = -5n + 80$

$a_1 = 75$ and $a_{25} = -45$

$S_{25} = \frac{25}{2}(75 - 45) = 375$

63. $a_1 = 100, \ a_{25} = 220, \ n = 25$

$S_n = \frac{n}{2}[a_1 + a_n]$

$S_{25} = \frac{25}{2}(100 + 220) = 4000$

64. $a_1 = 15, a_{100} = 307, n = 100$

$S_{100} = \frac{100}{2}(15 + 307) = 16{,}100$

65. $a_n = 2n - 1$

$a_1 = 1, a_{100} = 199$

$\displaystyle\sum_{n=1}^{100} (2n - 1) = \frac{100}{2}(1 + 199) = 10{,}000$

66. $a_0 = -10, a_{60} = 50, n = 60$

$\displaystyle\sum_{i=0}^{60} (i - 10) = \frac{60}{2}(-10 + 50)$

$= 1200$

67. $a_1 = 1, \ a_{50} = 50, \ n = 50$

$\displaystyle\sum_{n=1}^{50} n = \frac{50}{2}(1 + 50) = 1275$

68. $a_n' = 2n$

$a_1 = 2, a_{100} = 200, n = 100$

$\displaystyle\sum_{n=1}^{100} 2n = \frac{100}{2}(2 + 200) = 10{,}100$

69. $a_{10} = 60, a_{100} = 600, n = 91$

$\displaystyle\sum_{n=10}^{100} 6n = \frac{91}{2}(60 + 600) = 30{,}030$

70. $a_n = 7n$

$a_{51} = 357, a_{100} = 700$

$\displaystyle\sum_{n=51}^{100} 7n = \frac{50}{2}(357 + 700) = 26{,}425$

71. $\displaystyle\sum_{n=11}^{30} n - \sum_{n=1}^{10} n = \frac{20}{2}(11 + 30) - \frac{10}{2}(1 + 10) = 355$

72. $\displaystyle\sum_{n=51}^{100} n - \sum_{n=1}^{50} n = \frac{50}{2}(51 + 100) - \frac{50}{2}(1 + 50)$

$= 3775 - 1275 = 2500$

73. $a_1 = 1, a_{400} = 799, n = 400$

$\displaystyle\sum_{n=1}^{400} (2n - 1) = \frac{400}{2}(1 + 799) = 160{,}000$

74. $a_n = 1000 - n$

$a_1 = 999, a_{250} = 750, n = 250$

$\displaystyle\sum_{n=1}^{250} (1000 - n) = \frac{250}{2}(999 + 750) = 218{,}625$

75. $\displaystyle\sum_{n=1}^{20} (2n + 5) = 520$

76. $a_0 = 1000, \ a_{50} = 750, \ n = 51$

$\displaystyle\sum_{n=0}^{50} (1000 - 5n) = \frac{51}{2}(1000 + 750) = 44{,}625$

77. $\displaystyle\sum_{n=1}^{100} \frac{n + 4}{2} = 2725$

78. $a_0 = \dfrac{1}{2}, a_{100} = \dfrac{-73}{4}, n = 101$

$\displaystyle\sum_{n=0}^{100} \frac{8 - 3n}{16} = \frac{101}{2}\left(\frac{1}{2} - \frac{73}{4}\right) = -896.375$

79. $\displaystyle\sum_{i=1}^{60}\left(250-\tfrac{8}{3}i\right)=10{,}120$

80. $a_1=4.525,\ a_{200}=9.5,\ n=200$

$$\sum_{j=1}^{200}(4.5+0.025j)=\tfrac{200}{2}(4.525+9.5)=1402.5$$

81. (a) $a_1=32{,}500,\ d=1500$

$a_6=a_1+5d=32{,}500+5(1500)=\$40{,}000$

(b) $S_6=\tfrac{6}{2}[32{,}500+40{,}000]=\$217{,}500$

82. (a) $a_1=36{,}800,\ d=1750$

$a_6=a_1+5d=36{,}800+5(1750)=\$45{,}550$

(b) $S_6=\tfrac{6}{2}[36{,}800+45{,}550]=\$247{,}050$

83. $a_1=20,\ d=4,\ n=30$

$a_{30}=20+29(4)=136$

$S_{30}=\tfrac{30}{2}(20+136)=2340$ seats

84. $a_1=15,\ d=3,\ n=36$

$a_{36}=15+35(3)=120$

$S_{36}=\tfrac{36}{2}(15+120)=2430$ seats

85. $a_1=14,\ a_{18}=31$

$S_{18}=\tfrac{18}{2}(14+31)=405$ bricks

86. $a_1=14,\ a_{28}=0.5,\ n=28$

$S_{28}=\tfrac{28}{2}(14+0.5)=203$ bricks

87. $4.9,\ 14.7,\ 24.5,\ 34.3,\ \ldots$

$d=9.8$

$a_{10}=4.9+9(9.8)=93.1$ meters

$S_{10}=\tfrac{10}{2}(4.9+93.1)=490$ meters

88. $a_1=16,\ a_2=48,\ a_3=80,\ a_4=112$

$d=32$

$a_n=dn+c=32n+c$

$c=a_1-d=16-32=-16$

$a_n=32n-16$

Distance $=\displaystyle\sum_{n=1}^{7}(32n-16)=784$ ft

89. (a) $a_1=200,\ a_2=175\ \Rightarrow\ d=-25$

$c=200-(-25)=225$

$a_n=-25n+225$

(b) $a_8=-25(8)+225=25$

$S_8=\tfrac{8}{2}(200+25)=\$900$

90. (a) $a_1=1200,\ a_2=1100,\ a_3=1000$

$d=-100$

$a_n=dn+c$

$a_n=-100n+c$

$c=a_1-d=1200+100=1300$

$a_n=-100n+1300$

(b) Total prize money $=\displaystyle\sum_{n=1}^{12}(-100n+1300)$

$=\tfrac{12}{2}(1200+100)$

$=\$7800$

91. $a_n=1500n+6500$

$a_1=8000,\ a_6=15{,}500$

$S_6=\tfrac{6}{2}(8000+15{,}500)=\$70{,}500$

The cost of gasoline, labor, equipment, insurance, and maintenance are a few economic factors that could prevent the company from meeting its goals, but the biggest unknown variable is the amount of annual snowfall.

92. $a_1 = 15,000$

$d = 5,000$

$n = 1, \ldots, 10$

$a_n = dn + c = 5000n + c$

$c = a_1 - d = 15,000 - 5000 = 10,000$

$a_n = 5000n + 10,000$

Total sales $= \displaystyle\sum_{n=1}^{10} (5000n + 10,000) = \frac{10}{2}(15,000 + 60,000) = \$375,000$

93. (a)

Monthly Payment	Unpaid Balance
$a_1 = 200 + 0.01(2000) = \$220$	$1800
$a_2 = 200 + 0.01(1800) = \$218$	$1600
$a_3 = 200 + 0.01(1600) = \$216$	$1400
$a_4 = 200 + 0.01(1400) = \$214$	$1200
$a_5 = 200 + 0.01(1200) = \$212$	$1000
$a_6 = 200 + 0.01(1000) = \$210$	$800

(b) $a_n = -2n + 222 \implies a_{10} = 202$

$S_{10} = \frac{10}{2}(220 + 202) = \$2110$

Interest paid: $110

94. (a) Borrowed Amount $= a_0 = \$5,000$

Amount of Balance Paid Per Month $= \$250$

Unpaid Balance $= a_n = 5000 - 250n$

Interest $= I =$ Balance Before Payment $\cdot 1\% = a_{n-1} \cdot 0.01$

Total Payment $= \$250 + I$

Month (n)	1	2	3	4	5	6
Interest (I)	$50	$47.50	$45.00	$42.50	$40.00	$37.50
Total Payment ($250 + I$)	$300	$297.50	$295.00	$292.50	$290.00	$287.50
Unpaid Balance (a_n)	$4750	$4500	$4250	$4000	$3750	$3500

Month (n)	7	8	9	10	11	12
Interest (I)	$35.00	$32.50	$30.00	$27.50	$25.00	$22.50
Total Payment ($250 + I$)	$285.00	$282.50	$280.00	$277.50	$275.00	$272.50
Unpaid Balance (a_n)	$3250	$3000	$2750	$2500	$2250	$2000

(b) Total Interest Paid $= \displaystyle\sum_{n=1}^{20} [5000 - 250(n-1)] \cdot 0.01 = \frac{20}{2}[(5000)(0.01) + (250)(0.01)] = \$525$

95. (a) Using (5, 23,078) and (6, 24,176) we have $d = 1098$ and $c = 23,078 - 5(1098) = 17,588$.

$$a_n \approx 1098n + 17,588$$

(b) $a_n \approx 1114.95n + 17,795.07$
The models are similar.

(c)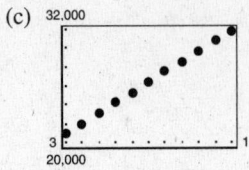

(d) For 2004 use $n = 14$: \$32,960
For 2005 use $n = 15$: \$34,058

(e) Answers will vary.

96. (a) $n = 7$ is 1997.

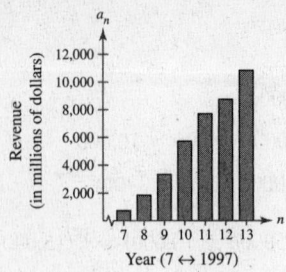

Year (7 $\leftrightarrow$ 1997)

(b) $a_n = \text{Revenue} = 1726.93n - 11,718.43$

(c) Total revenue $= \sum_{n=7}^{13} (1726.93n - 11,718.43)$

$$= \frac{7}{2}(370.08 + 10,731.66)$$

$$= \$38,856 \text{ million}$$

(d) $a_{18} = 1726.93(18) - 11,718.43 = \$19,366.31 \text{ million}$

97. True; given a_1 and a_2 then $d = a_2 - a_1$ and $a_n = a_1 + (n - 1)d$.

98. True, by the formula for the sum of a finite arithmetic sequence,

$$S_n = \frac{n}{2}(a_1 + a_n).$$

99. A sequence is arithmetic if the differences between consecutive terms are the same.

$$a_{n+1} - a_n = d \text{ for } n \geq 1$$

100. First term plus $(n - 1)$ times the common difference

101. (a) $a_n = 2 + 3n$

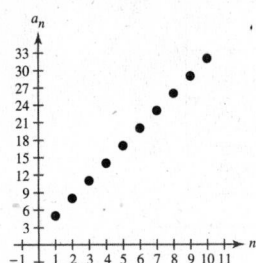

(c) The graph of $a_n = 2 + 3n$ contains only points at the positive integers. The graph of $y = 3x + 2$ is a solid line which contains these points.

(b) $y = 3x + 2$

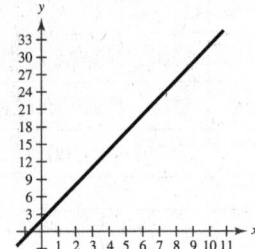

(d) The slope $m = 3$ is equal to the common difference $d = 3$. In general, these should be equal.

102. (a) $1 + 3 = 4$

$1 + 3 + 5 = 9$

$1 + 3 + 5 + 7 = 16$

$1 + 3 + 5 + 7 + 9 = 25$

$1 + 3 + 5 + 7 + 9 + 11 = 36$

(b) $S_n = n^2$

$S_7 = 1 + 3 + 5 + 7 + 9 + 11 + 13 = 49 = 7^2$

(c) $S_n = \frac{n}{2}[1 + (2n - 1)] = \frac{n}{2}(2n) = n^2$

103.
$$S_{20} = \frac{20}{2}\{a_1 + [a_1 + (20 - 1)(3)]\} = 650$$

$$10(2a_1 + 57) = 650$$

$$2a_1 + 57 = 65$$

$$2a_1 = 8$$

$$a_1 = 4$$

104. Let $S_n = \dfrac{n}{2}(a_1 + a_n)$ be the sum of the first n terms of the original sequence.

$$S_n' = \frac{n}{2}(a_1 + 5 + a_n + 5)$$

$$= \frac{n}{2}(a_1 + a_n + 10)$$

$$= \frac{n}{2}(a_1 + a_n) + \frac{n}{2}(10)$$

$$= \frac{n}{2}(a_1 + a_n) + 5n$$

$$= S_n + 5n$$

105. $2x - 4y = 3$

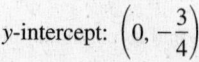

Slope: $m = \dfrac{1}{2}$

y-intercept: $\left(0, -\dfrac{3}{4}\right)$

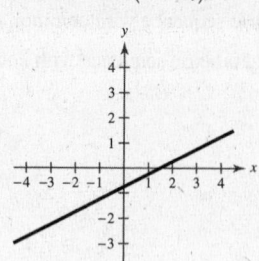

106. $9x + y = -8$

$y = -9x - 8$

Slope: -9

y-intercept: $(0, -8)$

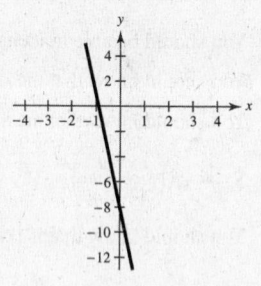

107. $x - 7 = 0$

$x = 7$

Vertical line

No slope

No y-intercept

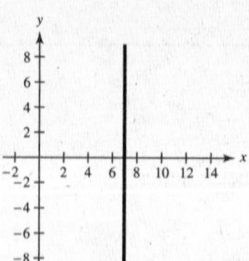

108. $y + 11 = 0$

$y = -11$

Slope: 0

y-intercept: $(0, -11)$

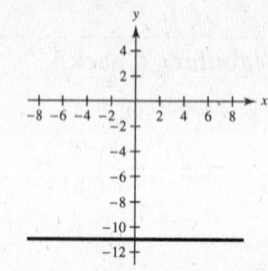

109. $\begin{cases} 2x - y + 7z = -10 & \text{Equation 1} \\ 3x + 2y - 4z = 17 & \text{Equation 2} \\ 6x - 5y + z = -20 & \text{Equation 3} \end{cases}$

$\begin{cases} x - \frac{1}{2}y + \frac{7}{2}z = -5 & \frac{1}{2}\text{Eq.1} \\ 3x + 2y - 4z = 17 \\ 6x - 5y + z = -20 \end{cases}$

$\begin{cases} x - \frac{1}{2}y + \frac{7}{2}z = -5 \\ \frac{7}{2}y - \frac{29}{2}z = 32 & (-3)\text{Eq.1} + \text{Eq.2} \\ -2y - 20z = 10 & (-6)\text{Eq.1} + \text{Eq.3} \end{cases}$

$\begin{cases} x - \frac{1}{2}y + \frac{7}{2}z = -5 \\ -2y - 20z = 10 \\ \frac{7}{2}y - \frac{29}{2}z = 32 \end{cases}$

$\begin{cases} x - \frac{1}{2}y + \frac{7}{2}z = -5 \\ y + 10z = -5 & (-\frac{1}{2})\text{Eq.2} \\ 7y - 29z = 64 & 2\,\text{Eq.3} \end{cases}$

$\begin{cases} x \quad + \frac{17}{2}z = -\frac{15}{2} & (\frac{1}{2})\text{Eq.2} + \text{Eq.1} \\ y + 10z = -5 \\ -99z = 99 & (-7)\text{Eq.2} + \text{Eq.3} \end{cases}$

$\begin{cases} x \quad + \frac{17}{2}z = -\frac{15}{2} \\ y + 10z = -5 \\ z = -1 & (-\frac{1}{99})\text{Eq.3} \end{cases}$

$\begin{cases} x = 1 & (-\frac{17}{2})\text{Eq.3} + \text{Eq.1} \\ y = 5 & (-10)\text{Eq.3} + \text{Eq.2} \\ z = -1 \end{cases}$

Answer: $x = 1, y = 5, z = -1$

110. $\begin{bmatrix} -1 & 4 & 10 & : & 4 \\ 5 & -3 & 1 & : & 31 \\ 8 & 2 & -3 & : & -5 \end{bmatrix}$

$5R_1 + R_2 \rightarrow \begin{bmatrix} -1 & 4 & 10 & : & 4 \\ 0 & 17 & 51 & : & 51 \\ 8 & 2 & -3 & : & -5 \end{bmatrix}$

$8R_1 + R_3 \rightarrow \begin{bmatrix} -1 & 4 & 10 & : & 4 \\ 0 & 17 & 51 & : & 51 \\ 0 & 34 & 77 & : & 27 \end{bmatrix}$

$2R_2 - R_3 \rightarrow \begin{bmatrix} -1 & 4 & 10 & : & 4 \\ 0 & 17 & 51 & : & 51 \\ 0 & 0 & 25 & : & 75 \end{bmatrix}$

$-R_1 \rightarrow \begin{bmatrix} 1 & -4 & -10 & : & -4 \\ 0 & 17 & 51 & : & 51 \\ 0 & 0 & 25 & : & 75 \end{bmatrix}$

$\frac{1}{17}R_2 \rightarrow \begin{bmatrix} 1 & -4 & -10 & : & -4 \\ 0 & 1 & 3 & : & 3 \\ 0 & 0 & 25 & : & 75 \end{bmatrix}$

$\frac{1}{25}R_3 \rightarrow \begin{bmatrix} 1 & -4 & -10 & : & -4 \\ 0 & 1 & 3 & : & 3 \\ 0 & 0 & 1 & : & 3 \end{bmatrix}$

$R_2 - 3R_3 \rightarrow \begin{bmatrix} 1 & -4 & -10 & : & -4 \\ 0 & 1 & 0 & : & -6 \\ 0 & 0 & 1 & : & 3 \end{bmatrix}$

$R_1 + 4R_2 + 10R_3 \rightarrow \begin{bmatrix} 1 & 0 & 0 & : & 2 \\ 0 & 1 & 0 & : & -6 \\ 0 & 0 & 1 & : & 3 \end{bmatrix}$

$x = 2, y = -6, z = 3$

111. Answers will vary.

Section 9.3   Geometric Sequences and Series

- You should be able to identify a geometric sequence, find its common ratio, and find the *n*th term.

- You should know that the *n*th term of a geometric sequence with common ratio *r* is given by $a_n = a_1 r^{n-1}$.

- You should know that the *n*th partial sum of a geometric sequence with common ratio $r \neq 1$ is given by

$$S_n = a_1 \left(\frac{1 - r^n}{1 - r} \right).$$

- You should know that if $|r| < 1$, then

$$\sum_{n=1}^{\infty} a_1 r^{n-1} = \sum_{n=0}^{\infty} a_1 r^n = \frac{a_1}{1 - r}.$$

Vocabulary Check

1. geometric; common

2. $a_n = a_1 r^{n-1}$

3. $S_n = a_1 \left(\dfrac{1 - r^n}{1 - r} \right)$

4. geometric series

5. $S = \dfrac{a_1}{1 - r}$

---

1. 5, 15, 45, 135, . . .

Geometric sequence, $r = 3$

2. 3, 12, 48, 192, . . .

Geometric sequence, $r = 4$

3. 3, 12, 21, 30, . . .

Not a geometric sequence

Note: It is an arithmetic sequence with $d = 9$.

4. 36, 27, 18, 9, . . .

Not a geometric sequence

5. $1, -\frac{1}{2}, \frac{1}{4}, -\frac{1}{8}, \ldots$

Geometric sequence, $r = -\frac{1}{2}$

6. 5, 1, 0.2, 0.04, ...

Geometric sequence, $r = \frac{1}{5} = 0.2$

7. $\frac{1}{8}, \frac{1}{4}, \frac{1}{2}, 1, \ldots$

Geometric sequence, $r = 2$

8. $9, -6, 4, -\frac{8}{3}, \ldots$

Geometric sequence, $r = -\frac{2}{3}$

9. $1, \frac{1}{2}, \frac{1}{3}, \frac{1}{4}, \ldots$

Not a geometric sequence

10. $\frac{1}{5}, \frac{2}{7}, \frac{3}{9}, \frac{4}{11}, \ldots$

Not a geometric sequence

11. $a_1 = 2, \ r = 3$

$a_1 = 2$

$a_2 = 2(3) = 6$

$a_3 = 6(3) = 18$

$a_4 = 18(3) = 54$

$a_5 = 54(3) = 162$

12. $a_1 = 6, r = 2$

$a_1 = 6$

$a_2 = 6(2)^1 = 12$

$a_3 = 6(2)^2 = 24$

$a_4 = 6(2)^3 = 48$

$a_5 = 6(2)^4 = 96$

13. $a_1 = 1$, $r = \frac{1}{2}$

$a_1 = 1$

$a_2 = 1\left(\frac{1}{2}\right) = \frac{1}{2}$

$a_3 = \frac{1}{2}\left(\frac{1}{2}\right) = \frac{1}{4}$

$a_4 = \frac{1}{4}\left(\frac{1}{2}\right) = \frac{1}{8}$

$a_5 = \frac{1}{8}\left(\frac{1}{2}\right) = \frac{1}{16}$

14. $a_1 = 1$, $r = \frac{1}{3}$

$a_1 = 1$

$a_2 = 1\left(\frac{1}{3}\right)^1 = \frac{1}{3}$

$a_3 = 1\left(\frac{1}{3}\right)^2 = \frac{1}{9}$

$a_4 = 1\left(\frac{1}{3}\right)^3 = \frac{1}{27}$

$a_5 = 1\left(\frac{1}{3}\right)^4 = \frac{1}{81}$

15. $a_1 = 5$, $r = -\frac{1}{10}$

$a_1 = 5$

$a_2 = 5\left(-\frac{1}{10}\right) = -\frac{1}{2}$

$a_3 = \left(-\frac{1}{2}\right)\left(-\frac{1}{10}\right) = \frac{1}{20}$

$a_4 = \frac{1}{20}\left(-\frac{1}{10}\right) = -\frac{1}{200}$

$a_5 = \left(-\frac{1}{200}\right)\left(-\frac{1}{10}\right) = \frac{1}{2000}$

16. $a_1 = 6$, $r = -\frac{1}{4}$

$a_1 = 6$

$a_2 = 6\left(-\frac{1}{4}\right)^1 = -\frac{3}{2}$

$a_3 = 6\left(-\frac{1}{4}\right)^2 = \frac{3}{8}$

$a_4 = 6\left(-\frac{1}{4}\right)^3 = -\frac{3}{32}$

$a_5 = 6\left(-\frac{1}{4}\right)^4 = \frac{3}{128}$

17. $a_1 = 1$, $r = e$

$a_1 = 1$

$a_2 = 1(e) = e$

$a_3 = (e)(e) = e^2$

$a_4 = (e^2)(e) = e^3$

$a_5 = (e^3)(e) = e^4$

18. $a_1 = 3$, $r = \sqrt{5}$

$a_1 = 3$

$a_2 = 3\left(\sqrt{5}\right)^1 = 3\sqrt{5}$

$a_3 = 3\left(\sqrt{5}\right)^2 = 15$

$a_4 = 3\left(\sqrt{5}\right)^3 = 15\sqrt{5}$

$a_5 = 3\left(\sqrt{5}\right)^4 = 75$

19. $a_1 = 2$, $r = \frac{x}{4}$

$a_1 = 2$

$a_2 = 2\left(\frac{x}{4}\right) = \frac{x}{2}$

$a_3 = \left(\frac{x}{2}\right)\left(\frac{x}{4}\right) = \frac{x^2}{8}$

$a_4 = \left(\frac{x^2}{8}\right)\left(\frac{x}{4}\right) = \frac{x^3}{32}$

$a_5 = \left(\frac{x^3}{32}\right)\left(\frac{x}{4}\right) = \frac{x^4}{128}$

20. $a_1 = 5$, $r = 2x$

$a_1 = 5$

$a_2 = 5(2x)^1 = 10x$

$a_3 = 5(2x)^2 = 20x^2$

$a_4 = 5(2x)^3 = 40x^3$

$a_5 = 5(2x)^4 = 80x^4$

21. $a_1 = 64$, $a_{k+1} = \frac{1}{2}a_k$

$a_1 = 64$

$a_2 = \frac{1}{2}(64) = 32$

$a_3 = \frac{1}{2}(32) = 16$

$a_4 = \frac{1}{2}(16) = 8$

$a_5 = \frac{1}{2}(8) = 4$

$r = \frac{1}{2}$

$a_n = 64\left(\frac{1}{2}\right)^{n-1} = 128\left(\frac{1}{2}\right)^n$

22. $a_1 = 81$, $a_{k+1} = \frac{1}{3}a_k$

$a_1 = 81$

$a_2 = \frac{1}{3}(81) = 27$

$a_3 = \frac{1}{3}(27) = 9$

$a_4 = \frac{1}{3}(9) = 3$

$a_5 = \frac{1}{3}(3) = 1$

$a_n = 81\left(\frac{1}{3}\right)^{n-1} = 243\left(\frac{1}{3}\right)^n$

23. $a_1 = 7$, $a_{k+1} = 2a_k$

$a_1 = 7$

$a_2 = 2(7) = 14$

$a_3 = 2(14) = 28$

$a_4 = 2(28) = 56$

$a_5 = 2(56) = 112$

$r = 2$

$a_n = 7(2)^{n-1} = \frac{7}{2}(2)^n$

24. $a_1 = 5$, $a_{k+1} = -2a_k$

$a_1 = 5$

$a_2 = -2(5) = -10$

$a_3 = -2(-10) = 20$

$a_4 = -2(20) = -40$

$a_5 = -2(-40) = 80$

$a_n = 5(-2)^{n-1} = -\frac{5}{2}(-2)^n$

25. $a_1 = 6$, $a_{k+1} = -\frac{3}{2}a_k$

$a_1 = 6$

$a_2 = -\frac{3}{2}(6) = -9$

$a_3 = -\frac{3}{2}(-9) = \frac{27}{2}$

$a_4 = -\frac{3}{2}\left(\frac{27}{2}\right) = -\frac{81}{4}$

$a_5 = -\frac{3}{2}\left(-\frac{81}{4}\right) = \frac{243}{8}$

$r = -\frac{3}{2}$

$a_n = 6\left(-\frac{3}{2}\right)^{n-1}$ or $a_n = -4\left(-\frac{3}{2}\right)^n$

26. $a_1 = 48$, $a_{k+1} = -\frac{1}{2}a_k$

$a_1 = 48$

$a_2 = -\frac{1}{2}(48) = -24$

$a_3 = -\frac{1}{2}(-24) = 12$

$a_4 = -\frac{1}{2}(12) = -6$

$a_5 = -\frac{1}{2}(-6) = 3$

$a_n = 48\left(-\frac{1}{2}\right)^{n-1} = -96\left(-\frac{1}{2}\right)^n$

27. $a_1 = 4$, $r = \frac{1}{2}$, $n = 10$

$a_n = a_1 r^{n-1} = 4\left(\frac{1}{2}\right)^{n-1}$

$a_{10} = 4\left(\frac{1}{2}\right)^9 = \left(\frac{1}{2}\right)^7 = \frac{1}{128}$

28. $a_1 = 5$, $r = \frac{3}{2}$, $n = 8$

$a_n = a_1 r^{n-1} = 5\left(\frac{3}{2}\right)^{n-1}$

$a_8 = 5\left(\frac{3}{2}\right)^7 = \frac{10,935}{128}$

29. $a_1 = 6$, $r = -\frac{1}{3}$, $n = 12$

$a_n = a_1 r^{n-1} = 6\left(-\frac{1}{3}\right)^{n-1}$

$a_{12} = 6\left(-\frac{1}{3}\right)^{11} = -\frac{2}{3^{10}}$

30. $a_1 = 64$, $r = -\frac{1}{4}$, $n = 10$

$a_n = a_1 r^{n-1} = 64\left(-\frac{1}{4}\right)^{n-1}$

$a_{10} = 64\left(-\frac{1}{4}\right)^9 = -\frac{64}{262,144}$

31. $a_1 = 100$, $r = e^x$, $n = 9$

$a_n = a_1 r^{n-1} = 100(e^x)^{n-1}$

$a_9 = 100(e^x)^8 = 100e^{8x}$

32. $a_1 = 1$, $r = \sqrt{3}$, $n = 8$

$a_n = a_1 r^{n-1} = 1\left(\sqrt{3}\right)^{n-1}$

$a_8 = 1\left(\sqrt{3}\right)^7 = 27\sqrt{3}$

33. $a_1 = 500$, $r = 1.02$, $n = 40$

$a_n = a_1 r^{n-1} = 500(1.02)^{n-1}$

$a_{40} = 500(1.02)^{39} \approx 1082.372$

34. $a_1 = 1000$, $r = 1.005$, $n = 60$

$a_n = a_1 r^{n-1} = 1000(1.005)^{n-1}$

$a_{60} = 1000(1.005)^{59} \approx 1342.139$

35. $7, 21, 63, \ldots \Longrightarrow r = 3$

$a_n = 7(3)^{n-1}$

$a_9 = 7(3)^8 = 45,927$

36. $a_1 = 3$, $a_2 = 36$, $a_3 = 432$

$r = \dfrac{a_2}{a_1} = \dfrac{36}{3} = 12$

$a_n = a_1 r^{(n-1)}$

$a_7 = (3)(12)^6 = 8,957,952$

37. $5, 30, 180, \ldots \Longrightarrow r = 6$

$a_n = 5(6)^{n-1}$

$a_{10} = 5(6)^9 = 50,388,480$

38. $a_1 = 4$, $a_2 = 8$, $a_3 = 16$

$r = \dfrac{a_2}{a_1} = \dfrac{8}{4} = 2$

$a_n = a_1 r^{n-1}$

$a_{22} = (4)(2)^{21} = 8,388,608$

39. $a_1 = 16$, $a_4 = \dfrac{27}{4}$

$a_4 = a_1 r^3$

$\dfrac{27}{4} = 16r^3$

$\dfrac{27}{64} = r^3$

$\dfrac{3}{4} = r$

$a_n = 16\left(\dfrac{3}{4}\right)^{n-1}$

$a_3 = 16\left(\dfrac{3}{4}\right)^2 = 9$

40. $a_2 = 3, a_5 = \dfrac{3}{64}$

$a_5 = a_2 r^{(5-2)}$

$a_5 = a_2 r^3$

$\dfrac{3}{64} = 3r^3$

$\dfrac{1}{64} = r^3$

$r = \dfrac{1}{4}$

$a_2 = a_1 r^1$

$3 = a_1 \left(\dfrac{1}{4}\right)$

$a_1 = 12$

41. $a_4 = -18, a_7 = \dfrac{2}{3}$

$a_7 = a_4 r^3$

$\dfrac{2}{3} = -18r^3$

$-\dfrac{1}{27} = r^3$

$-\dfrac{1}{3} = r$

$a_6 = \dfrac{a_7}{r} = \dfrac{2/3}{-1/3} = -2$

42. $a_3 = \dfrac{16}{3}, a_5 = \dfrac{64}{27}$

$a_5 = a_3 r^{(5-3)}$

$a_5 = a_3 r^2$

$\dfrac{64}{27} = \dfrac{16}{3}r^2$

$r^2 = \dfrac{4}{9}$

$r = \dfrac{2}{3}$

$a_7 = a_5 r^{(7-5)}$

$a_7 = a_5 r^2$

$a_7 = \left(\dfrac{64}{27}\right)\left(\dfrac{2}{3}\right)^2 = \dfrac{256}{243}$

43. $a_n = 18\left(\dfrac{2}{3}\right)^{n-1}$

$a_1 = 18$ and $r = \dfrac{2}{3}$

Since $0 < r < 1$, the sequence is decreasing.

Matches (a).

44. $a_n = 18\left(-\dfrac{2}{3}\right)^{n-1}$

$r = \left(-\dfrac{2}{3}\right) > -1$, so the sequence alternates as it approaches 0.

Matches (c).

45. $a_n = 18\left(\dfrac{3}{2}\right)^{n-1}$

$a_1 = 18$ and $r = \dfrac{3}{2} > 1$, so the sequence is increasing.

Matches (b).

46. $a_n = 18\left(-\dfrac{3}{2}\right)^{n-1}$

$r = \left(-\dfrac{3}{2}\right) < -1$, so the sequence alternates as it approaches ∞.

Matches (d).

47. $a_n = 12(-0.75)^{n-1}$

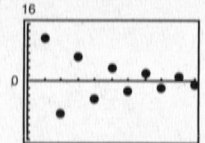

48. $a_n = 10(1.5)^{n-1}$

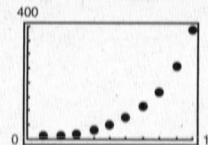

49. $a_n = 12(-0.4)^{n-1}$

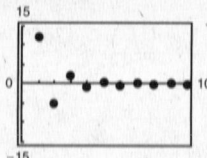

50. $a_n = 20(-1.25)^{n-1}$

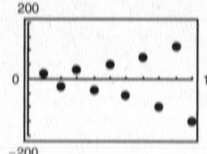

51. $a_n = 2(1.3)^{n-1}$

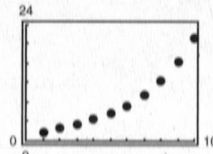

52. $a_n = 10(1.2)^{n-1}$

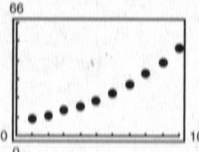

53. $\displaystyle\sum_{n=1}^{9} 2^{n-1} = 1 + 2^1 + 2^2 + \cdots + 2^8 \implies a_1 = 1, \ r = 2$

$S_9 = \dfrac{1(1 - 2^9)}{1 - 2} = 511$

54. $\sum_{n=1}^{10}\left(\frac{5}{2}\right)^{n-1} = 1 + \left(\frac{5}{2}\right)^1 + \left(\frac{5}{2}\right)^2 + \cdots + \left(\frac{5}{2}\right)^9 \Rightarrow a_1 = 1, r = \frac{5}{2}$

$$S_{10} = 1\left[\frac{1 - \left(\frac{5}{2}\right)^{10}}{1 - \left(\frac{5}{2}\right)}\right] = -\frac{2}{3}\left[1 - \left(\frac{5}{2}\right)^{10}\right] = \frac{3,254,867}{512} \approx 6357.162$$

55. $\sum_{n=1}^{9}(-2)^{n-1} \Rightarrow a_1 = 1, r = -2, n = 9$

$$S_9 = 1\left(\frac{1 - (-2)^9}{1 - (-2)}\right) = 171$$

56. $\sum_{n=1}^{8}5\left(-\frac{3}{2}\right)^{n-1} = 5 + 5\left(-\frac{3}{2}\right)^1 + 5\left(-\frac{3}{2}\right)^2 + \cdots + 5\left(-\frac{3}{2}\right)^7 \Rightarrow a_1 = 5, r = -\frac{3}{2}$

$$S_8 = 5\left[\frac{1 - \left(-\frac{3}{2}\right)^8}{1 - \left(-\frac{3}{2}\right)}\right] = 2\left[1 - \left(-\frac{3}{2}\right)^8\right] = -\frac{6305}{128} \approx -49.258$$

57. $\sum_{i=1}^{7}64\left(-\frac{1}{2}\right)^{i-1} = 64 + 64\left(-\frac{1}{2}\right)^1 + 64\left(-\frac{1}{2}\right)^2 + \cdots + 64\left(-\frac{1}{2}\right)^6 \Rightarrow a_1 = 64, r = -\frac{1}{2}$

$$S_7 = 64\left[\frac{1 - \left(-\frac{1}{2}\right)^7}{1 - \left(-\frac{1}{2}\right)}\right] = \frac{128}{3}\left[1 - \left(-\frac{1}{2}\right)^7\right] = 43$$

58. $\sum_{i=1}^{10}2\left(\frac{1}{4}\right)^{i-1} = 2 + 2\left(\frac{1}{4}\right)^1 + 2\left(\frac{1}{4}\right)^2 + \cdots + 2\left(\frac{1}{4}\right)^9 \Rightarrow a_1 = 2, r = \frac{1}{4}$

$$S_{10} = 2\left[\frac{1 - \left(\frac{1}{4}\right)^{10}}{1 - \left(\frac{1}{4}\right)}\right] = \frac{8}{3}\left[1 - \left(\frac{1}{4}\right)^{10}\right] \approx 2.667$$

59. $\sum_{i=1}^{6}32\left(\frac{1}{4}\right)^{i-1} = 32 + 32\left(\frac{1}{4}\right)^1 + 32\left(\frac{1}{4}\right)^2 + 32\left(\frac{1}{4}\right)^3 + 32\left(\frac{1}{4}\right)^4 + 32\left(\frac{1}{4}\right)^5 \Rightarrow a_1 = 32, r = \frac{1}{4}, n = 6$

$$S_6 = 32\left(\frac{1 - \left(\frac{1}{4}\right)^6}{1 - \frac{1}{4}}\right) = \frac{1365}{32}$$

60. $\sum_{i=1}^{12}16\left(\frac{1}{2}\right)^{i-1} = 16 + 16\left(\frac{1}{2}\right)^1 + 16\left(\frac{1}{2}\right)^2 + \cdots + 16\left(\frac{1}{2}\right)^{11} \Rightarrow a_1 = 16, r = \frac{1}{2}$

$$S_{12} = 16\left[\frac{1 - \left(\frac{1}{2}\right)^{12}}{1 - \left(\frac{1}{2}\right)}\right] = 32\left[1 - \left(\frac{1}{2}\right)^{12}\right] = \frac{4095}{128} \approx 31.992$$

61. $\sum_{n=0}^{20}3\left(\frac{3}{2}\right)^n = \sum_{n=1}^{21}3\left(\frac{3}{2}\right)^{n-1} = 3 + 3\left(\frac{3}{2}\right)^1 + 3\left(\frac{3}{2}\right)^2 + \cdots + 3\left(\frac{3}{2}\right)^{20} \Rightarrow a_1 = 3, r = \frac{3}{2}$

$$S_{21} = 3\left[\frac{1 - \left(\frac{3}{2}\right)^{21}}{1 - \frac{3}{2}}\right] = -6\left[1 - \left(\frac{3}{2}\right)^{21}\right] \approx 29,921.311$$

62. $\sum_{n=0}^{40}5\left(\frac{3}{5}\right)^n = 5 + \sum_{n=1}^{40}5\left(\frac{3}{5}\right)^n = 5 + \left[5\left(\frac{3}{5}\right)^1 + 5\left(\frac{3}{5}\right)^2 + 5\left(\frac{3}{5}\right)^3 + \cdots + 5\left(\frac{3}{5}\right)^{40}\right] \Rightarrow a_1 = 3, r = \frac{3}{5}$

$$S_{41} = 5 + 3\left[\frac{1 - \left(\frac{3}{5}\right)^{40}}{1 - \left(\frac{3}{5}\right)}\right] = 5 + \frac{15}{2}\left[1 - \left(\frac{3}{5}\right)^{40}\right] \approx 12.500$$

63. $\displaystyle\sum_{n=0}^{15} 2\left(\frac{4}{3}\right)^n = \sum_{n=1}^{16} 2\left(\frac{4}{3}\right)^{n-1} = 2 + 2\left(\frac{4}{3}\right)^1 + 2\left(\frac{4}{3}\right)^2 + \cdots + 2\left(\frac{4}{3}\right)^{15} \Rightarrow a_1 = 2, r = \frac{4}{3}, n = 16$

$S_{16} = 2\left(\dfrac{1 - \left(\frac{4}{3}\right)^{16}}{1 - \frac{4}{3}}\right) \approx 592.647$

64. $\displaystyle\sum_{n=0}^{20} 10\left(\frac{1}{5}\right)^n = 10 + \sum_{n=1}^{20} 10\left(\frac{1}{5}\right)^n = 10 + \left[10\left(\frac{1}{5}\right)^1 + 10\left(\frac{1}{5}\right)^2 + 10\left(\frac{1}{5}\right)^3 + \cdots + 10\left(\frac{1}{5}\right)^{20}\right] \Rightarrow a_1 = 2, r = \frac{1}{5}$

$S_{21} = 10 + 2\left[\dfrac{1 - \left(\frac{1}{5}\right)^{20}}{1 - \left(\frac{1}{5}\right)}\right] = 10 + \frac{5}{2}\left[1 - \left(\frac{1}{5}\right)^{20}\right] \approx 12.500$

65. $\displaystyle\sum_{n=0}^{5} 300(1.06)^n = \sum_{n=1}^{6} 300(1.06)^{n-1}$

$= 300 + 300(1.06)^1 + 300(1.06)^2 + 300(1.06)^3 + 300(1.06)^4 + 300(1.06)^5 \Rightarrow a_1 = 300, \ r = 1.06$

$S_6 = 300\left[\dfrac{1 - (1.06)^6}{1 - 1.06}\right] \approx 2092.596$

66. $\displaystyle\sum_{n=0}^{6} 500(1.04)^n = 500 + \sum_{n=1}^{6} 500(1.04)^n = 500 + \left[500(1.04)^1 + 500(1.04)^2 + \cdots + 500(1.04)^6\right]$

$a_1 = 520, r = 1.04$

$S_7 = 500 + 520\left[\dfrac{1 - (1.04)^6}{1 - (1.04)}\right] = 500 - 13{,}000[1 - (1.04)^6] \approx 3949.147$

67. $\displaystyle\sum_{n=0}^{40} 2\left(-\frac{1}{4}\right)^n = 2 + 2\left(-\frac{1}{4}\right) + 2\left(-\frac{1}{4}\right)^2 + \cdots + 2\left(-\frac{1}{4}\right)^{40} \Rightarrow a_1 = 2, \ r = -\frac{1}{4}, n = 41$

$S_{41} = 2\left[\dfrac{1 - \left(-\frac{1}{4}\right)^{41}}{1 - \left(-\frac{1}{4}\right)}\right] = \frac{8}{5}\left[1 - \left(-\frac{1}{4}\right)^{41}\right] \approx 1.6 = \frac{8}{5}$

68. $\displaystyle\sum_{n=0}^{50} 10\left(\frac{2}{3}\right)^{n-1} = 15 + \sum_{n=1}^{50} 10\left(\frac{2}{3}\right)^{n-1} = 15 + \left[10 + 10\left(\frac{2}{3}\right)^1 + 10\left(\frac{2}{3}\right)^2 + \cdots + 10\left(\frac{2}{3}\right)^{49}\right] \Rightarrow a_1 = 10, r = \frac{2}{3}$

$S_{51} = 15 + 10\left[\dfrac{1 - \left(\frac{2}{3}\right)^{50}}{1 - \left(\frac{2}{3}\right)}\right] = 15 + 30\left[1 - \left(\frac{2}{3}\right)^{50}\right] \approx 45.000$

69. $\displaystyle\sum_{i=1}^{10} 8\left(-\frac{1}{4}\right)^{i-1} = 8 + 8\left(-\frac{1}{4}\right)^1 + 8\left(-\frac{1}{4}\right)^2 + \cdots + 8\left(-\frac{1}{4}\right)^9 \Rightarrow a_1 = 8, \ r = -\frac{1}{4}$

$S_{10} = 8\left[\dfrac{1 - \left(-\frac{1}{4}\right)^{10}}{1 - \left(-\frac{1}{4}\right)}\right] = \frac{32}{5}\left[1 - \left(-\frac{1}{4}\right)^{10}\right] \approx 6.400$

70. $\displaystyle\sum_{i=0}^{25} 8\left(-\frac{1}{2}\right)^i = 8 + \sum_{i=1}^{25} 8\left(-\frac{1}{2}\right)^i = 8 + \left[-4 + 8\left(-\frac{1}{2}\right)^2 + 8\left(-\frac{1}{2}\right)^3 + \cdots + 8\left(-\frac{1}{2}\right)^{25}\right] \Rightarrow a_1 = -4, r = -\frac{1}{2}$

$S_{26} = 8 - 4\left[\dfrac{1 - \left(-\frac{1}{2}\right)^{25}}{1 - \left(-\frac{1}{2}\right)}\right] = 8 - \frac{8}{3}\left[1 - \left(-\frac{1}{2}\right)^{25}\right] \approx 5.333$

71. $\displaystyle\sum_{i=1}^{10} 5\left(-\frac{1}{3}\right)^{i-1} = 5 + 5\left(-\frac{1}{3}\right)^1 + 5\left(-\frac{1}{3}\right)^2 + \cdots + 5\left(-\frac{1}{3}\right)^9 \Rightarrow a_1 = 5, r = -\frac{1}{3}, n = 10$

$S_{10} = 5\left(\dfrac{1 - \left(-\frac{1}{3}\right)^{10}}{1 - \left(-\frac{1}{3}\right)}\right) \approx 3.750$

72. $\displaystyle\sum_{i=1}^{100} 15\left(\frac{2}{3}\right)^{i-1} = 15 + 15\left(\frac{2}{3}\right)^{1} + 15\left(\frac{2}{3}\right)^{2} + \cdots + 15\left(\frac{2}{3}\right)^{99} \Rightarrow a_1 = 15, r = \frac{2}{3}$

$$S_{100} = 15\left[\frac{1 - \left(\frac{2}{3}\right)^{100}}{1 - \left(\frac{2}{3}\right)}\right] = 45\left[1 - \left(\frac{2}{3}\right)^{100}\right] \approx 45.000$$

73. $5 + 15 + 45 + \cdots + 3645$

$r = 3$ and $3645 = 5(3)^{n-1}$

$729 = 3^{n-1} \Rightarrow 6 = n - 1 \Rightarrow n = 7$

Thus, the sum can be written as $\displaystyle\sum_{n=1}^{7} 5(3)^{n-1}$.

74. $7 + 14 + 28 + \cdots + 896$

$a_1 = 7, r = 2$

$7(2)^{n-1} = 896$

$2^{n-1} = 128$

$2^{n-1} = 2^7$

$n - 1 = 7$

$n = 8$

Thus, the sum can be written as $\displaystyle\sum_{n=1}^{8} 7(2)^{n-1}$.

75. $2 - \dfrac{1}{2} + \dfrac{1}{8} - \cdots + \dfrac{1}{2048}$

$r = -\dfrac{1}{4}$ and $\dfrac{1}{2048} = 2\left(-\dfrac{1}{4}\right)^{n-1}$

By trial and error, we find that $n = 7$.

Thus, the sum can be written as $\displaystyle\sum_{n=1}^{7} 2\left(-\frac{1}{4}\right)^{n-1}$.

76. $15 - 3 + \dfrac{3}{5} - \cdots - \dfrac{3}{625}$

$a_1 = 15, r = -\dfrac{1}{5}$

$15\left(-\dfrac{1}{5}\right)^{n-1} = -\dfrac{3}{625}$

$\left(-\dfrac{1}{5}\right)^{n-1} = -\dfrac{1}{3125}$

$\left(-\dfrac{1}{5}\right)^{n} = \dfrac{1}{15,625}$

By trial and error, we find that $n = 6$.
Thus, the sum can be written as

$$\sum_{n=1}^{6} 15\left(-\frac{1}{5}\right)^{n-1}.$$

77. $0.1 + 0.4 + 1.6 + \cdots + 102.4$

$r = 4$ and $102.4 = 0.1(4)^{n-1}$

$1024 = 4^{n-1} \Rightarrow 5 = n - 1 \Rightarrow n = 6$

Thus, the sum can be written as $\displaystyle\sum_{n=1}^{6} 0.1(4)^{n-1}$.

78. $32 + 24 + 18 + \cdots + 10.125$

$a_1 = 32, r = \dfrac{3}{4}$

$32\left(\dfrac{3}{4}\right)^{n-1} = 10.125 = \dfrac{81}{8}$

$\left(\dfrac{3}{4}\right)^{n-1} = \dfrac{81}{256}$

$\left(\dfrac{3}{4}\right)^{n-1} = \left(\dfrac{3}{4}\right)^{4}$

$n - 1 = 4$

$n = 5$

Thus, the sum can be written as

$$\sum_{n=1}^{5} 32\left(\frac{3}{4}\right)^{n-1}.$$

79. $\displaystyle\sum_{n=0}^{\infty} \left(\frac{1}{2}\right)^n = 1 + \left(\frac{1}{2}\right)^1 + \left(\frac{1}{2}\right)^2 + \cdots$

$a_1 = 1,\; r = \dfrac{1}{2}$

$\displaystyle\sum_{n=0}^{\infty} \left(\frac{1}{2}\right)^n = \frac{a_1}{1-r} = \frac{1}{1-\left(\frac{1}{2}\right)} = 2$

80. $\displaystyle\sum_{n=0}^{\infty} 2\left(\frac{2}{3}\right)^n = 2 + 2\left(\frac{2}{3}\right)^1 + 2\left(\frac{2}{3}\right)^2 + \cdots$

$a_1 = 2,\; r = \dfrac{2}{3}$

$\displaystyle\sum_{n=0}^{\infty} 2\left(\frac{2}{3}\right)^n = \frac{a_1}{1-r} = \frac{2}{1-\frac{2}{3}} = 6$

81. $\displaystyle\sum_{n=0}^{\infty} \left(-\frac{1}{2}\right)^n = 1 + \left(-\frac{1}{2}\right)^1 + \left(-\frac{1}{2}\right)^2 + \cdots$

$a_1 = 1,\; r = -\dfrac{1}{2}$

$\displaystyle\sum_{n=0}^{\infty} \left(-\frac{1}{2}\right)^n = \frac{a_1}{1-r} = \frac{1}{1-\left(-\frac{1}{2}\right)} = \frac{2}{3}$

82. $\displaystyle\sum_{n=0}^{\infty} 2\left(-\frac{2}{3}\right)^n = 2 + 2\left(-\frac{2}{3}\right)^1 + 2\left(-\frac{2}{3}\right)^2 + \cdots$

$a_1 = 2,\; r = -\dfrac{2}{3}$

$\displaystyle\sum_{n=0}^{\infty} 2\left(-\frac{2}{3}\right)^n = \frac{a_1}{1-r} = \frac{2}{1-\left(-\frac{2}{3}\right)} = \frac{6}{5}$

83. $\displaystyle\sum_{n=0}^{\infty} 4\left(\frac{1}{4}\right)^n = 4 + 4\left(\frac{1}{4}\right)^1 + 4\left(\frac{1}{4}\right)^2 + \cdots$

$a_1 = 4,\; r = \dfrac{1}{4}$

$\displaystyle\sum_{n=0}^{\infty} 4\left(\frac{1}{4}\right)^n = \frac{a_1}{1-r} = \frac{4}{1-\left(\frac{1}{4}\right)} = \frac{16}{3}$

84. $\displaystyle\sum_{n=0}^{\infty} \left(\frac{1}{10}\right)^n = 1 + \left(\frac{1}{10}\right)^1 + \left(\frac{1}{10}\right)^2 + \cdots$

$a_1 = 1,\; r = \dfrac{1}{10}$

$\displaystyle\sum_{n=0}^{\infty} \left(\frac{1}{10}\right)^n = \frac{a_1}{1-r} = \frac{1}{1-\frac{1}{10}} = \frac{10}{9}$

85. $\displaystyle\sum_{n=0}^{\infty} (0.4)^n = 1 + (0.4)^1 + (0.4)^2 + \cdots$

$a_1 = 1,\; r = 0.4$

$\displaystyle\sum_{n=0}^{\infty} (0.4)^n = \frac{1}{1-0.4} = \frac{5}{3}$

86. $\displaystyle\sum_{n=0}^{\infty} 4(0.2)^n = 4 + 4(0.2)^1 + 4(0.2)^2 + \cdots$

$a_1 = 4,\; r = 0.2$

$\displaystyle\sum_{n=0}^{\infty} 4(0.2)^n = \frac{4}{1-0.2} = 5$

87. $\displaystyle\sum_{n=0}^{\infty} -3(0.9)^n = -3 - 3(0.9)^1 - 3(0.9)^2 - \cdots$

$a_1 = -3,\; r = 0.9$

$\displaystyle\sum_{n=0}^{\infty} -3(0.9)^n = \frac{-3}{1-0.9} = -30$

88. $\displaystyle\sum_{n=0}^{\infty} \left[-10(0.2)^n\right] = -10 - 10(0.2)^1 - 10(0.2)^2 - \cdots$

$a_1 = -10,\; r = 0.2$

$\displaystyle\sum_{n=0}^{\infty} -10(0.2)^n = \frac{-10}{1-0.2} = -12.5$

89. $8 + 6 + \dfrac{9}{2} + \dfrac{27}{8} + \cdots = \displaystyle\sum_{n=0}^{\infty} 8\left(\frac{3}{4}\right)^n = \frac{8}{1-\frac{3}{4}} = 32$

90. $9 + 6 + 4 + \dfrac{8}{3} + \cdots$

$a_1 = 9,\; r = \dfrac{2}{3}$

$\displaystyle\sum_{n=0}^{\infty} 9\left(\frac{2}{3}\right)^n = \frac{9}{1-\frac{2}{3}} = 27$

91. $\dfrac{1}{9} - \dfrac{1}{3} + 1 - 3 + \cdots = \displaystyle\sum_{n=0}^{\infty} \frac{1}{9}(-3)^n$

The sum is undefined because

$|r| = |-3| = 3 > 1.$

92. $\dfrac{-125}{36} + \dfrac{25}{6} - 5 + 6 - \cdots = \displaystyle\sum_{n=0}^{\infty} -\dfrac{125}{36}\left(-\dfrac{6}{5}\right)^n$

The sum is undefined because

$$|r| = \left|-\dfrac{6}{5}\right| = \dfrac{6}{5} > 1.$$

93. $0.\overline{36} = \displaystyle\sum_{n=0}^{\infty} 0.36(0.01)^n = \dfrac{0.36}{1 - 0.01} = \dfrac{0.36}{0.99} = \dfrac{36}{99} = \dfrac{4}{11}$

94. $0.\overline{297} = \displaystyle\sum_{n=0}^{\infty} 0.297(0.001)^n = \dfrac{0.297}{1 - 0.001} = \dfrac{0.297}{0.999}$

$\qquad = \dfrac{297}{999} = \dfrac{11}{37}$

95. $0.3\overline{18} = 0.3 + \displaystyle\sum_{n=0}^{\infty} 0.018(0.01)^n = \dfrac{3}{10} + \dfrac{0.018}{1 - 0.01}$

$\qquad = \dfrac{3}{10} + \dfrac{0.018}{0.99} = \dfrac{3}{10} + \dfrac{18}{990} = \dfrac{3}{10} + \dfrac{2}{110}$

$\qquad = \dfrac{35}{110} = \dfrac{7}{22}$

96. $1.3\overline{8} = 1.3 + \displaystyle\sum_{n=0}^{\infty} 0.08(0.1)^n = 1.3 + \dfrac{0.08}{1 - 0.1} = 1.3 + \dfrac{0.08}{0.9} = 1\dfrac{3}{10} + \dfrac{4}{45} = 1\dfrac{7}{18} = \dfrac{25}{18}$

97. $f(x) = 6\left[\dfrac{1 - (0.5)^x}{1 - (0.5)}\right]$, $\displaystyle\sum_{n=0}^{\infty} 6\left(\dfrac{1}{2}\right)^n = \dfrac{6}{1 - \frac{1}{2}} = 12$

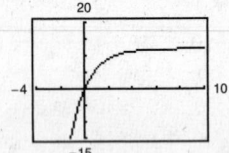

The horizontal asymptote of $f(x)$ is $y = 12$.
This corresponds to the sum of the series.

98. $f(x) = 2\left[\dfrac{1 - (0.8)^x}{1 - (0.8)}\right]$, $\displaystyle\sum_{n=0}^{\infty} 2\left(\dfrac{4}{5}\right)^n = \dfrac{2}{1 - \frac{4}{5}} = 10$

The horizontal asymptote of $f(x)$ is $y = 10$.
This corresponds to the sum of the series.

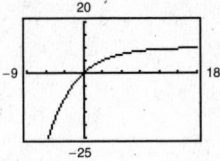

99. (a) $a_n \approx 1190.88(1.006)^n$

(b) The population is growing at a rate of 0.6% per year.

(c) For 2010, let $n = 20$: $a_n = 1190.88(1.006)^{20}$

$\qquad\qquad\qquad\qquad\qquad \approx 1342.2$ million

(d) $1190.88(1.006)^n = 1320$

$\qquad 1.006^n = \dfrac{1320}{1190.88}$

$\qquad \ln 1.006^n = \ln\left(\dfrac{1320}{1190.88}\right)$

$\qquad n \ln 1.006 = \ln\left(\dfrac{1320}{1190.88}\right)$

$\qquad n = \dfrac{\ln\left(\dfrac{1320}{1190.88}\right)}{\ln 1.006} \approx 17.21$

This corresponds with the year 2008.

100. $A = P\left(1 + \dfrac{r}{n}\right)^{nt} = 1000\left(1 + \dfrac{0.06}{n}\right)^{n(10)}$

(a) $n = 1$, $A = 1000(1 + 0.06)^{10} \approx \$1790.85$

(b) $n = 2$, $A = 1000\left(1 + \dfrac{0.06}{2}\right)^{2(10)} \approx \$1806.11$

(c) $n = 4$, $A = 1000\left(1 + \dfrac{0.06}{4}\right)^{4(10)} \approx \$1814.02$

(d) $n = 12$, $A = 1000\left(1 + \dfrac{0.06}{12}\right)^{12(10)} \approx \$1819.40$

(e) $n = 365$, $A = 1000\left(1 + \dfrac{0.06}{365}\right)^{365(10)} \approx \$1822.03$

101. $A = P\left(1 + \dfrac{r}{n}\right)^{nt} = 2500\left(1 + \dfrac{0.02}{n}\right)^{n(20)}$

(a) $n = 1$: $A = 2500\left(1 + \dfrac{0.02}{1}\right)^{(1)(20)} \approx \$3714.87$

(b) $n = 2$: $A = 2500\left(1 + \dfrac{0.02}{2}\right)^{(2)(20)} \approx \$3722.16$

(c) $n = 4$: $A = 2500\left(1 + \dfrac{0.02}{4}\right)^{(4)(20)} \approx \$3725.85$

(d) $n = 12$: $A = 2500\left(1 + \dfrac{0.02}{12}\right)^{(12)(20)} \approx \$3728.32$

(e) $n = 365$: $A = 2500\left(1 + \dfrac{0.02}{365}\right)^{(365)(20)} \approx \$3729.52$

102. $V_5 = 135{,}000(0.70)^5 = \$22{,}689.45$

103. $A = \displaystyle\sum_{n=1}^{60} 100\left(1 + \dfrac{0.06}{12}\right)^n = \sum_{n=1}^{60} 100(1.005)^n = 100(1.005) \cdot \dfrac{[1 - 1.005^{60}]}{[1 - 1.005]} \approx \$7011.89$

104. $A = \displaystyle\sum_{n=1}^{60} 50\left(1 + \dfrac{0.08}{12}\right)^n$

$= 50(1.006666667)\left(\dfrac{1 - (1.006666667)^{60}}{1 - 1.006666667}\right)$

$\approx \$3698.34$

105. Let $N = 12t$ be the total number of deposits.

$A = P\left(1 + \dfrac{r}{12}\right) + P\left(1 + \dfrac{r}{12}\right)^2 + \cdots + P\left(1 + \dfrac{r}{12}\right)^N$

$= \left(1 + \dfrac{r}{12}\right)\left[P + P\left(1 + \dfrac{r}{12}\right) + \cdots + P\left(1 + \dfrac{r}{12}\right)^{N-1}\right]$

$= P\left(1 + \dfrac{r}{12}\right)\displaystyle\sum_{n=1}^{N}\left(1 + \dfrac{r}{12}\right)^{n-1}$

$= P\left(1 + \dfrac{r}{12}\right)\left[\dfrac{1 - \left(1 + \dfrac{r}{12}\right)^N}{1 - \left(1 + \dfrac{r}{12}\right)}\right]$

$= P\left(1 + \dfrac{r}{12}\right)\left(-\dfrac{12}{r}\right)\left[1 - \left(1 + \dfrac{r}{12}\right)^N\right]$

$= P\left(\dfrac{12}{r} + 1\right)\left[-1 + \left(1 + \dfrac{r}{12}\right)^N\right]$

$= P\left[\left(1 + \dfrac{r}{12}\right)^N - 1\right]\left(1 + \dfrac{12}{r}\right)$

$= P\left[\left(1 + \dfrac{r}{12}\right)^{12t} - 1\right]\left(1 + \dfrac{12}{r}\right)$

106. Let $N = 12t$ be the total number of deposits.

$A = Pe^{r/12} + Pe^{2r/12} + \cdots + Pe^{Nr/12}$

$= \displaystyle\sum_{n=1}^{N} Pe^{r/12 \cdot n}$

$= Pe^{r/12}\dfrac{(1 - (e^{r/12})^N)}{(1 - e^{r/12})}$

$= Pe^{r/12}\dfrac{(1 - (e^{r/12})^{12t})}{1 - e^{r/12}}$

$= \dfrac{Pe^{r/12}(e^{rt} - 1)}{(e^{r/12} - 1)}$

107. $P = \$50$, $r = 7\%$, $t = 20$ years

(a) Compounded monthly:

$A = 50\left[\left(1 + \dfrac{0.07}{12}\right)^{12(20)} - 1\right]\left(1 + \dfrac{12}{0.07}\right)$

$\approx \$26{,}198.27$

(b) Compounded continuously:

$A = \dfrac{50e^{0.07/12}(e^{0.07(20)} - 1)}{e^{0.07/12} - 1} \approx \$26{,}263.88$

108. $P = \$75, r = 3\%, t = 25$ years

    (a) Compounded monthly: $A = 75\left[\left(1 + \dfrac{0.03}{12}\right)^{12(25)} - 1\right]\left(1 + \dfrac{12}{0.03}\right) \approx \$33,534.21$

    (b) Compounded continuously: $A = \dfrac{75e^{0.03/12}(e^{0.03(25)} - 1)}{e^{0.03/12} - 1} \approx \$33,551.91$

109. $P = \$100, r = 10\%, t = 40$ years

    (a) Compounded monthly: $A = 100\left[\left(1 + \dfrac{0.10}{12}\right)^{12(40)} - 1\right]\left(1 + \dfrac{12}{0.10}\right) \approx \$637,678.02$

    (b) Compounded continuously: $A = \dfrac{100e^{0.10/12}(e^{(0.10)(40)} - 1)}{e^{0.10/12} - 1} \approx \$645,861.43$

110. $P = \$20, r = 6\%, t = 50$ years

    (a) Compounded monthly: $A = 20\left[\left(1 + \dfrac{0.06}{12}\right)^{12(50)} - 1\right]\left(1 + \dfrac{12}{0.06}\right) \approx \$76,122.54$

    (b) Compounded continuously: $A = \dfrac{20e^{0.06/12}(e^{0.06(50)} - 1)}{e^{0.06/12} - 1} \approx \$76,533.16$

111. $P = W\displaystyle\sum_{n=1}^{12t}\left[\left(1 + \dfrac{r}{12}\right)^{-1}\right]^{n}$

$= W\left(1 + \dfrac{r}{12}\right)^{-1}\left[\dfrac{1 - \left(1 + \dfrac{r}{12}\right)^{-12t}}{1 - \left(1 + \dfrac{r}{12}\right)^{-1}}\right]$

$= W\left(\dfrac{1}{1 + \dfrac{r}{12}}\right)\dfrac{\left[1 - \left(1 + \dfrac{r}{12}\right)^{-12t}\right]}{1 - \dfrac{1}{\left(1 + \dfrac{r}{12}\right)}}$

$= W\dfrac{\left[1 - \left(1 + \dfrac{r}{12}\right)^{-12t}\right]}{\left(1 + \dfrac{r}{12}\right) - 1}$

$= W\left(\dfrac{12}{r}\right)\left[1 - \left(1 + \dfrac{r}{12}\right)^{-12t}\right]$

112. $W = \$2000, t = 20, r = 9\%$

$P = W\left(\dfrac{12}{r}\right)\left[1 - \left(1 + \dfrac{r}{12}\right)^{-12t}\right]$

$P = 2000\left(\dfrac{12}{0.09}\right)\left[1 - \left(1 + \dfrac{0.09}{12}\right)^{-12(20)}\right] \approx \$222,289.91$

113. $\displaystyle\sum_{n=1}^{\infty} 400(0.75)^{n} = \dfrac{300}{1 - 0.75} = \$1200$

114. $a_1 = 250(0.80) = 200$

$r = 80\% = 0.80$

Amount put back into economy $= \displaystyle\sum_{n=1}^{\infty} 250(0.80)^{n}$

$= \dfrac{200}{1 - 0.80}$

$= \dfrac{200}{0.20}$

$= \$1000$

115. $\displaystyle\sum_{n=1}^{\infty} 600(0.725)^n = \frac{435}{1 - 0.725} \approx \$1581.82$

116. $a_1 = 450(0.775) = 348.75$

$r = 77.5\% = 0.775$

Amount put back into economy $= \displaystyle\sum_{n=1}^{\infty} 450(0.775)^n$

$= \dfrac{348.75}{1 - 0.775}$

$= \dfrac{348.75}{0.225}$

$= \$1550$

117. $64 + 32 + 16 + 8 + 4 + 2 = 126$

Total area of shaded region is approximately 126 square inches.

118. $a_n = 54.6e^{0.172n}, n = 4, 5, \ldots, 13$

$r = e^{0.172n}$

$a_1 = 54.6e^{0.172} = 64.84721$

$S_n = \displaystyle\sum_{i=1}^{n} a_1 r^{i-1} = a_1\left(\frac{1 - r^n}{1 - r}\right)$

$S = S_{13} - S_3$

$= 64.84721\left(\dfrac{1 - e^{(0.172)(13)}}{1 - e^{0.172}}\right) - 64.84721\left(\dfrac{1 - e^{(0.172)(3)}}{1 - e^{0.172}}\right)$

$= 2887.141484 - 233.336893 = 2653.80$

The total sales over the 10-year period is \$2653.80 million.

119. $a_n = 30,000(1.05)^{n-1}$

$T = \displaystyle\sum_{n=1}^{40} 30,000(1.05)^{n-1} = 30,000\dfrac{(1 - 1.05^{40})}{(1 - 1.05)} \approx \$3,623,993.23$

120. (a) Total distance $= \left[\displaystyle\sum_{n=0}^{\infty} 32(0.81)^n\right] - 16 = \dfrac{32}{1 - 0.81} - 16 \approx 152.42$ feet

(b) $t = 1 + 2\displaystyle\sum_{n=1}^{\infty} (0.9)^n = 1 + 2\left[\dfrac{0.9}{1 - 0.9}\right] = 19$ seconds

121. False. A sequence is geometric if the ratios of consecutive terms are the same.

122. False. $a_n = a_1 r^{n-1}$, NOT ra_1^{n-1}

The nth-term of a geometric sequence can be found by multiplying its first term by its common ratio raised to the $(n - 1)$th power.

123. Given a real number r between -1 and 1, as the exponent n increases, r^n approaches zero.

124. *Sample answer:*

$\displaystyle\sum_{n=1}^{199} 4(-1)^{n-1}$ and $\displaystyle\sum_{n=1}^{8} -\frac{4}{85}(-2)^{n-1}$

125. $g(x) = x^2 - 1$

$g(x + 1) = (x + 1)^2 - 1$

$= x^2 + 2x + 1 - 1 = x^2 + 2x$

126. $f(x) = 3x + 1$

$f(x + 1) = 3(x + 1) + 1 = 3x + 4$

127. $f(x) = 3x + 1, g(x) = x^2 - 1$

$f(g(x + 1)) = f(x^2 + 2x)$

$= 3(x^2 + 2x) + 1$

$= 3x^2 + 6x + 1$

128. $g(x) = x^2 - 1$

$g(f(x + 1)) = g(3x + 4)$ From Exercise 126

$= (3x + 4)^2 - 1$

$= 9x^2 + 24x + 15$

129. $9x^3 - 64x = x(9x^2 - 64) = x(3x + 8)(3x - 8)$

130. $x^2 + 4x - 63$ Does not factor

131. $6x^2 - 13x - 5 = (3x + 1)(2x - 5)$

132. $16x^2 - 4x^4 = 4x^2(4 - x^2)$

$= 4x^2(2 + x)(2 - x)$

133. $\dfrac{3}{x + 3} \cdot \dfrac{x(x + 3)}{x - 3} = \dfrac{3x}{x - 3}, x \neq -3$

134. $\dfrac{x - 2}{x + 7} \cdot \dfrac{\overset{1}{2x(x + 7)}}{\underset{3}{6x(x - 2)}} = \dfrac{1}{3}, x \neq -7, 2$

135. $\dfrac{x}{3} \div \dfrac{3x}{6x + 3} = \dfrac{x}{3} \cdot \dfrac{3(2x + 1)}{3x} = \dfrac{2x + 1}{3}, x \neq 0, -\dfrac{1}{2}$

136. $\dfrac{x - 5}{x - 3} \div \dfrac{10 - 2x}{2(3 - x)} = \dfrac{x - 5}{x - 3} \cdot \dfrac{2(x - 3)}{2(x - 5)} = 1, x \neq 3, 5$

137. $5 + \dfrac{7}{x + 2} + \dfrac{2}{x - 2} = \dfrac{5(x + 2)(x - 2) + 7(x - 2) + 2(x + 2)}{(x + 2)(x - 2)}$

$= \dfrac{5(x^2 - 4) + 7(x - 2) + 2(x + 2)}{(x + 2)(x - 2)}$

$= \dfrac{5x^2 - 20 + 7x - 14 + 2x + 4}{(x + 2)(x - 2)} = \dfrac{5x^2 + 9x - 30}{(x + 2)(x - 2)}$

138. $8 - \dfrac{x - 1}{x + 4} - \dfrac{4}{x - 1} - \dfrac{x + 4}{(x - 1)(x + 4)} = \dfrac{8(x - 1)(x + 4) - (x - 1)^2 - 4(x + 4) - (x + 4)}{(x - 1)(x + 4)}$

$= \dfrac{8(x^2 + 3x - 4) - (x^2 - 2x + 1) - 4x - 16 - x - 4}{(x - 1)(x + 4)}$

$= \dfrac{8x^2 + 24x - 32 - x^2 + 2x - 1 - 4x - 16 - x - 4}{(x - 1)(x + 4)}$

$= \dfrac{7x^2 + 21x - 53}{(x - 1)(x + 4)}$

139. Answers will vary.

Section 9.4   Mathematical Induction

■   You should be sure that you understand the principle of mathematical induction. If P_n is a statement involving the positive integer n, where P_1 is true and the truth of P_k implies the truth of P_{k+1} for every positive k, then P_n is true for all positive integers n.

■   You should be able to verify (by induction) the formulas for the sums of powers of integers and be able to use these formulas.

■   You should be able to calculate the first and second differences of a sequence.

■   You should be able to find the quadratic model for a sequence, when it exists.

Vocabulary Check

1. mathematical induction   **2.** first

3. arithmetic   **4.** second

---

1. $P_k = \dfrac{5}{k(k + 1)}$

$$P_{k+1} = \dfrac{5}{(k + 1)[(k + 1) + 1]} = \dfrac{5}{(k + 1)(k + 2)}$$

2. $P_k = \dfrac{1}{2(k + 2)}$

$$P_{k+1} = \dfrac{1}{2(k + 1 + 2)} = \dfrac{1}{2(k + 3)}$$

3. $P_k = \dfrac{k^2(k + 1)^2}{4}$

$$P_{k+1} = \dfrac{(k + 1)^2[(k + 1) + 1]^2}{4} = \dfrac{(k + 1)^2(k + 2)^2}{4}$$

4. $P_k = \dfrac{k}{3}(2k + 1)$

$$P_{k+1} = \dfrac{k + 1}{3}[2(k + 1) + 1] = \dfrac{k + 1}{3}(2k + 3)$$

5. 1. When $n = 1$, $S_1 = 2 = 1(1 + 1)$.

     2. Assume that

$$S_k = 2 + 4 + 6 + 8 + \cdots + 2k = k(k + 1).$$

Then,

$$S_{k+1} = 2 + 4 + 6 + 8 + \cdots + 2k + 2(k + 1)$$
$$= S_k + 2(k + 1) = k(k + 1) + 2(k + 1) = (k + 1)(k + 2).$$

Therefore, we conclude that the formula is valid for all positive integer values of n.

6. 1. When $n = 1$, $S_1 = 3 = 1(2 \cdot 1 + 1)$.

     2. Assume that

$$S_k = 3 + 7 + 11 + 15 + \cdots + (4k - 1) = k(2k + 1).$$

Then,

$$S_{k+1} = S_k + a_{k+1} = (3 + 7 + 11 + 15 + \cdots + (4k - 1)) + [4(k + 1) - 1]$$
$$= k(2k + 1) + (4k + 3)$$
$$= 2k^2 + 5k + 3$$
$$= (k + 1)(2k + 3)$$
$$= (k + 1)[2(k + 1) + 1].$$

Therefore, we conclude that this formula is valid.

7. 1. When $n = 1$, $S_1 = 2 = \frac{1}{2}(5(1) - 1)$.

2. Assume that

$$S_k = 2 + 7 + 12 + 17 + \cdots + (5k - 3) = \frac{k}{2}(5k - 1).$$

Then,

$$S_{k+1} = 2 + 7 + 12 + 17 + \cdots + (5k - 3) + [5(k + 1) - 3]$$

$$= S_k + (5k + 5 - 3) = \frac{k}{2}(5k - 1) + 5k + 2$$

$$= \frac{5k^2 - k + 10k + 4}{2} = \frac{5k^2 + 9k + 4}{2}$$

$$= \frac{(k + 1)(5k + 4)}{2} = \frac{(k + 1)}{2}[5(k + 1) - 1].$$

Therefore, we conclude that this formula is valid for all positive integer values of n.

8. 1. When $n = 1$,

$$S_1 = 1 = \frac{1}{2}(3 \cdot 1 - 1).$$

2. Assume that

$$S_k = 1 + 4 + 7 + 10 + \cdots + (3k - 2) = \frac{k}{2}(3k - 1).$$

Then,

$$S_{k+1} = S_k + a_{k+1} = (1 + 4 + 7 + 10 + \cdots + (3k - 2)) + (3(k + 1) - 2)$$

$$= \frac{k}{2}(3k - 1) + (3k + 1)$$

$$= \frac{3k^2 - k + 6k + 2}{2}$$

$$= \frac{3k^2 + 5k + 2}{2}$$

$$= \frac{(k + 1)(3k + 2)}{2}$$

$$= \frac{k + 1}{2}[3(k + 1) - 1].$$

Therefore, we conclude that this formula is valid for all positive integer values of n.

9. 1. When $n = 1$, $S_1 = 1 = 2^1 - 1$.

2. Assume that

$$S_k = 1 + 2 + 2^2 + 2^3 + \cdots + 2^{k-1} = 2^k - 1.$$

Then,

$$S_{k+1} = 1 + 2 + 2^2 + 2^3 + \cdots + 2^{k-1} + 2^k$$

$$= S_k + 2^k = 2^k - 1 + 2^k = 2(2^k) - 1 = 2^{k+1} - 1.$$

Therefore, we conclude that this formula is valid for all positive integer values of n.

10. 1. When $n = 1$, $S_1 = 2 = 3^1 - 1$.

2. Assume that

$$S_k = 2(1 + 3 + 3^2 + 3^3 + \cdots + 3^{k-1}) = 3^k - 1.$$

Then,

$$
\begin{aligned}
S_{k+1} &= S_k + a_{k+1} \\
&= [2(1 + 3 + 3^2 + 3^3 + \cdots + 3^{k-1})] + 2 \cdot 3^{k+1-1} \\
&= 3^k - 1 + 2 \cdot 3^k \\
&= 3 \cdot 3^k - 1 \\
&= 3^{k+1} - 1.
\end{aligned}
$$

Therefore, we conclude that this formula is valid for all positive integer values of n.

11. 1. When $n = 1$, $S_1 = 1 = \dfrac{1(1 + 1)}{2}$.

2. Assume that

$$S_k = 1 + 2 + 3 + 4 + \cdots + k = \frac{k(k + 1)}{2}.$$

Then,

$$
\begin{aligned}
S_{k+1} &= 1 + 2 + 3 + 4 + \cdots + k + (k + 1) \\
&= S_k + (k + 1) = \frac{k(k + 1)}{2} + \frac{2(k + 1)}{2} = \frac{(k + 1)(k + 2)}{2}.
\end{aligned}
$$

Therefore, we conclude that this formula is valid for all positive integer values of n.

12. $S_n = 1^3 + 2^3 + 3^3 + 4^3 + \cdots + n^3 = \dfrac{n^2(n + 1)^2}{4}$.

1. When $n = 1$, $S_n = 1^3 = 1 = \dfrac{1^2(1 + 1)^2}{4}$.

2. Assume that

$$S_k = 1^3 + 2^3 + 3^3 + 4^3 + \cdots + k^3 = \frac{k^2(k + 1)^2}{4}.$$

Then,

$$
\begin{aligned}
S_{k+1} &= 1^3 + 2^3 + 3^3 + 4^3 + \cdots + k^3 + (k + 1)^3 \\
&= S_k + (k + 1)^3 = \frac{k^2(k + 1)^2}{4} + (k + 1)^3 = \frac{k^2(k + 1)^2 + 4(k + 1)^3}{4} \\
&= \frac{(k + 1)^2(k^2 + 4k + 4)}{4} = \frac{(k + 1)^2(k + 2)^2}{4} = \frac{(k + 1)^2[(k + 1) + 1]^2}{4}.
\end{aligned}
$$

Therefore, we conclude that this formula is valid for all positive integer values of n.

13. 1. When $n = 1$, $S_1 = 1 = \dfrac{(1)^2(1 + 1)^2(2(1)^2 + 2(1) - 1)}{12}$.

2. Assume that

$$S_k = \sum_{i=1}^{k} i^5 = \frac{k^2(k + 1)^2(2k^2 + 2k - 1)}{12}.$$

Then,

$$S_{k+1} = \sum_{i=1}^{k+1} i^5 = \left(\sum_{i=1}^{k} i^5 \right) + (k + 1)^5$$

$$= \frac{k^2(k + 1)^2(2k^2 + 2k - 1)}{12} + \frac{12(k + 1)^5}{12}$$

$$= \frac{(k + 1)^2[k^2(2k^2 + 2k - 1) + 12(k + 1)^3]}{12}$$

$$= \frac{(k + 1)^2[2k^4 + 2k^3 - k^2 + 12(k^3 + 3k^2 + 3k + 1)]}{12}$$

$$= \frac{(k + 1)^2[2k^4 + 14k^3 + 35k^2 + 36k + 12]}{12}$$

$$= \frac{(k + 1)^2(k^2 + 4k + 4)(2k^2 + 6k + 3)}{12}$$

$$= \frac{(k + 1)^2(k + 2)^2[2(k + 1)^2 + 2(k + 1) - 1]}{12}.$$

Therefore, we conclude that this formula is valid for all positive integer values of n.

Note: The easiest way to complete the last two steps is to "work backwards." Start with the desired expression for S_{k+1} and multiply out to show that it is equal to the expression you found for $S_k + (k + 1)^5$.

14. 1. When $n = 1$,

$$S_1 = 1^4 = \frac{1(1 + 1)(2 \cdot 1 + 1)(3 \cdot 1^2 + 3 \cdot 1 - 1)}{30}.$$

2. Assume that

$$S_k = \sum_{i=1}^{k} i^4 = \frac{k(k + 1)(2k + 1)(3k^2 + 3k - 1)}{30}.$$

Then,

$$S_{k+1} = S_k + a_{k+1} = S_k + (k + 1)^4$$

$$= \frac{k(k + 1)(2k + 1)(3k^2 + 3k - 1)}{30} + (k + 1)^4$$

$$= \frac{k(k + 1)(2k + 1)(3k^2 + 3k - 1) + 30(k + 1)^4}{30}$$

$$= \frac{(k + 1)[k(2k + 1)(3k^2 + 3k - 1) + 30(k + 1)^3]}{30}$$

$$= \frac{(k + 1)(6k^4 + 39k^3 + 91k^2 + 89k + 30)}{30}$$

$$= \frac{(k + 1)(k + 2)(2k + 3)(3k^2 + 9k + 5)}{30}$$

$$= \frac{(k + 1)(k + 2)(2(k + 1) + 1)(3(k + 1)^2 + 3(k + 1) - 1)}{30}.$$

Therefore, we conclude that this formula is valid for all positive integer values of n.

15. 1. When $n = 1$, $S_1 = 2 = \dfrac{1(2)(3)}{3}$.

2. Assume that

$$S_k = 1(2) + 2(3) + 3(4) + \cdots + k(k+1) = \frac{k(k+1)(k+2)}{3}.$$

Then,

$$S_{k+1} = 1(2) + 2(3) + 3(4) + \cdots + k(k+1) + (k+1)(k+2)$$

$$= S_k + (k+1)(k+2) = \frac{k(k+1)(k+2)}{3} + \frac{3(k+1)(k+2)}{3}$$

$$= \frac{(k+1)(k+2)(k+3)}{3}.$$

Therefore, we conclude that this formula is valid for all positive integer values of n.

16. 1. When $n = 1$,

$$S_1 = \frac{1}{3} = \frac{1}{2 \cdot 1 + 1}.$$

2. Assume that

$$S_k = \sum_{i=1}^{k} \frac{1}{(2i-1)(2i+1)} = \frac{k}{2k+1}.$$

Then,

$$S_{k+1} = S_k + a_{k+1} = S_k + \frac{1}{(2(k+1)-1)(2(k+1)+1)}$$

$$= \frac{k}{2k+1} + \frac{1}{(2k+1)(2k+3)}$$

$$= \frac{k(2k+3)+1}{(2k+1)(2k+3)}$$

$$= \frac{2k^2 + 3k + 1}{(2k+1)(2k+3)}$$

$$= \frac{(2k+1)(k+1)}{(2k+1)(2k+3)}$$

$$= \frac{k+1}{2(k+1)+1}.$$

Therefore, we conclude that this formula is valid for all positive integer values of n.

17. 1. When $n = 4$, $4! = 24$ and $2^4 = 16$, thus $4! > 2^4$.

2. Assume

$k! > 2^k$, $k > 4$.

Then,

$(k+1)! = k!(k+1) > 2^k(2)$ since $k! > 2^k$ and $k+1 > 2$.

Thus, $(k+1)! > 2^{k+1}$.

Therefore, by extended mathematical induction, the inequality is valid for all integers n such that $n \geq 4$.

18. 1. When $n = 7$, $\left(\dfrac{4}{3}\right)^7 \approx 7.4915 > 7$.

2. Assume that $\left(\dfrac{4}{3}\right)^k > k, k > 7$.

Then, $\left(\dfrac{4}{3}\right)^{k+1} = \left(\dfrac{4}{3}\right)^k\left(\dfrac{4}{3}\right) > k\left(\dfrac{4}{3}\right) = k + \dfrac{k}{3} > k + 1$ for $k > 7$.

Thus, $\left(\dfrac{4}{3}\right)^{k+1} > k + 1$.

Therefore, the inequality $\left(\dfrac{4}{3}\right)^n > n$ is valid for all integers n such that $n \geq 7$.

19. 1. When $n = 2$, $\dfrac{1}{\sqrt{1}} + \dfrac{1}{\sqrt{2}} \approx 1.707$ and $\sqrt{2} \approx 1.414$, thus $\dfrac{1}{\sqrt{1}} + \dfrac{1}{\sqrt{2}} > \sqrt{2}$.

2. Assume that

$$\dfrac{1}{\sqrt{1}} + \dfrac{1}{\sqrt{2}} + \dfrac{1}{\sqrt{3}} + \cdots + \dfrac{1}{\sqrt{k}} > \sqrt{k}, k > 2.$$

Then,

$$\dfrac{1}{\sqrt{1}} + \dfrac{1}{\sqrt{2}} + \dfrac{1}{\sqrt{3}} + \cdots + \dfrac{1}{\sqrt{k}} + \dfrac{1}{\sqrt{k+1}} > \sqrt{k} + \dfrac{1}{\sqrt{k+1}}.$$

Now it is sufficient to show that

$$\sqrt{k} + \dfrac{1}{\sqrt{k+1}} > \sqrt{k+1}, k > 2,$$

or equivalently $\left(\text{multiplying by } \sqrt{k+1}\right)$,

$$\sqrt{k}\sqrt{k+1} + 1 > k + 1.$$

This is true because

$$\sqrt{k}\sqrt{k+1} + 1 > \sqrt{k}\sqrt{k} + 1 = k + 1.$$

Therefore,

$$\dfrac{1}{\sqrt{1}} + \dfrac{1}{\sqrt{2}} + \dfrac{1}{\sqrt{3}} + \cdots + \dfrac{1}{\sqrt{k}} + \dfrac{1}{\sqrt{k+1}} > \sqrt{k+1}.$$

Therefore, by extended mathematical induction, the inequality is valid for all integers n such that $n \geq 2$.

20. 1. When $n = 1$, $\left(\dfrac{x}{y}\right)^2 < \left(\dfrac{x}{y}\right)$ and $(0 < x < y)$.

2. Assume that

$$\left(\dfrac{x}{y}\right)^{k+1} < \left(\dfrac{x}{y}\right)^k$$

$$\left(\dfrac{x}{y}\right)^{k+1} < \left(\dfrac{x}{y}\right)^k \Rightarrow \left(\dfrac{x}{y}\right)\left(\dfrac{x}{y}\right)^{k+1} < \left(\dfrac{x}{y}\right)\left(\dfrac{x}{y}\right)^k \Rightarrow \left(\dfrac{x}{y}\right)^{k+2} < \left(\dfrac{x}{y}\right)^{k+1}.$$

Therefore, $\left(\dfrac{x}{y}\right)^{n+1} < \left(\dfrac{x}{y}\right)^n$ for all integers $n \geq 1$.

21. $(1 + a)^n \geq na, n \geq 1$ and $a > 0$

Since a is positive, then all of the terms in the binomial expansion are positive.

$(1 + a)^n = 1 + na + \cdots + na^{n-1} + a^n > na$

22. $2n^2 > (n + 1)^2, n \geq 3$

1. For $n = 3$, the statement is true, because
   $2(3)^2 = 18 > (3 + 1)^2 = 16.$

2. Assuming that $2k^2 > (k + 1)^2$ you need to show that
   $2(k + 1)^2 > (k + 2)^2$. For $n = k$, you have

   $(k + 2)^2 = k^2 + 4k + 4$

   $\qquad = k^2 + 2k + 1 + 2k + 3$

   $\qquad = (k + 1)^2 + 2k + 3.$

   By the assumption $(k + 1)^2 < 2k^2$, you have

   $(k + 1)^2 + 2k + 3 < 2k^2 + 2k + 3.$

   Because $2k + 3 < 4k + 2$, or $1 < 2k$ for all $k > 3$,
   you can say that

   $2k^2 + 2k + 3 < 2k^2 + 4k + 2 = 2(k + 1)^2.$

   It follows that $(k + 2)^2 < 2k^2 + 2k + 3 < 2(k + 1)^2$

   or $2(k + 1)^2 > (k + 2)^2.$

Therefore, $2n^2(n + 1)^2$ for all $n \geq 3$.

23. 1. When $n = 1$, $(ab)^1 = a^1b^1 = ab.$

2. Assume that $(ab)^k = a^kb^k$.

   Then, $(ab)^{k+1} = (ab)^k(ab)$

   $\qquad\qquad = a^kb^kab$

   $\qquad\qquad = a^{k+1}b^{k+1}.$

Thus, $(ab)^n = a^nb^n.$

24. 1. When $n = 1$, $\left(\dfrac{a}{b}\right)^1 = \dfrac{a^1}{b^1}.$

2. Assume that $\left(\dfrac{a}{b}\right)^k = \dfrac{a^k}{b^k}.$

   Then, $\left(\dfrac{a}{b}\right)^{k+1} = \left(\dfrac{a}{b}\right)^k\left(\dfrac{a}{b}\right) = \dfrac{a^k}{b^k} \cdot \dfrac{a}{b} = \dfrac{a^{k+1}}{b^{k+1}}.$

Thus, $\left(\dfrac{a}{b}\right)^n = \dfrac{a^n}{b^n}.$

25. 1. When $n = 2$, $(x_1x_2)^{-1} = \dfrac{1}{x_1x_2} = \dfrac{1}{x_1} \cdot \dfrac{1}{x_2} = x_1^{-1}x_2^{-1}.$

2. Assume that

   $(x_1x_2x_3 \cdots x_k)^{-1} = x_1^{-1}x_2^{-1}x_3^{-1} \cdots x_k^{-1}.$

   Then,

   $(x_1x_2x_3 \cdots x_kx_{k+1})^{-1} = [(x_1x_2x_3 \cdots x_k)x_{k+1}]^{-1}$

   $\qquad\qquad\qquad = (x_1x_2x_3 \ldots x_k)^{-1}x_{k+1}^{-1}$

   $\qquad\qquad\qquad = x_1^{-1}x_2^{-1}x_3^{-1} \cdots x_k^{-1}x_{k+1}^{-1}.$

Thus, the formula is valid.

26. 1. When $n = 1$, $\ln x_1 = \ln x_1.$

2. Assume that $\ln(x_1x_2x_3 \ldots x_k) = \ln x_1 + \ln x_2 + \ln x_3 + \cdots + \ln x_k.$

   Then, $\ln(x_1x_2x_3 \ldots x_k x_{k+1}) = \ln[(x_1x_2x_3 \ldots x_k)x_{k+1}]$

   $\qquad\qquad\qquad = \ln(x_1x_2x_3 \ldots x_k) + \ln x_{k+1}$

   $\qquad\qquad\qquad = \ln x_1 + \ln x_2 + \ln x_3 + \cdots + \ln x_k + \ln x_{k+1}.$

Thus, $\ln(x_1x_2x_3 \ldots x_n) = \ln x_1 + \ln x_2 + \ln x_3 + \cdots \ln x_n.$

27. 1. When $n = 1$, $x(y_1) = xy_1.$

2. Assume that

   $x(y_1 + y_2 + \cdots + y_k) = xy_1 + xy_2 + \cdots + xy_k.$

   Then,

   $xy_1 + xy_2 + \cdots + xy_k + xy_{k+1} = x(y_1 + y_2 + \cdots + y_k) + xy_{k+1}$

   $\qquad\qquad\qquad = x[(y_1 + y_2 + \cdots + y_k) + y_{k+1}]$

   $\qquad\qquad\qquad = x(y_1 + y_2 + \cdots + y_k + y_{k+1}).$

Hence, the formula holds.

28. 1. When $n = 1$, $a + bi$ and $a - bi$ are complex conjugates by definition.

2. Assume that $(a + bi)^k$ and $(a - bi)^k$ are complex conjugates. That is, if $(a + bi)^k = c + di$, then $(a - bi)^k = c - di$.

   Then,

   $$(a + bi)^{k+1} = (a + bi)^k(a + bi) = (c + di)(a + bi)$$
   $$= (ac - bd) + i(bc + ad)$$

   and $(a - bi)^{k+1} = (a - bi)^k (a - bi) = (c - di)(a - bi)$
   $$= (ac - bd) - i(bc + ad).$$

   This implies that $(a + bi)^{k+1}$ and $(a - bi)^{k+1}$ are complex conjugates.

   Therefore, $(a + bi)^n$ and $(a - bi)^n$ are complex conjugates for $n \geq 1$.

29. 1. When $n = 1$, $[1^3 + 3(1)^2 + 2(1)] = 6$ and 3 is a factor.

2. Assume that 3 is a factor of $k^3 + 3k^2 + 2k$.

   Then,

   $$(k + 1)^3 + 3(k + 1)^2 + 2(k + 1) = k^3 + 3k^2 + 3k + 1 + 3k^2 + 6k + 3 + 2k + 2$$
   $$= (k^3 + 3k^2 + 2k) + (3k^2 + 9k + 6)$$
   $$= (k^3 + 3k^2 + 2k) + 3(k^2 + 3k + 2).$$

   Since 3 is a factor of $(k^3 + 3k^2 + 2k)$, our assumption, and 3 is a factor of $3(k^2 + 3k + 2)$, we conclude that 3 is a factor of the whole sum.

   Thus, 3 is a factor of $(n^3 + 3n^2 + 2n)$ for every positive integer n.

30. Prove 3 is a factor of $n^3 - n + 3$ for all positive integers n.

1. When $n = 1, 1^3 - 1 + 3 = 3$ and 3 is a factor.

2. Assume that 3 is a factor of $k^3 - k + 3$.

   Then,

   $$(k + 1)^3 - (k + 1) + 3 = k^3 + 3k^2 + 3k + 1 - k - 1 + 3$$
   $$= k^3 + 3k^2 + 2k + 3$$
   $$= (k^3 - k + 3) + 3k^2 + 3k$$
   $$= (k^3 - k + 3) + 3k(k + 1).$$

   Since 3 is a factor of each term, 3 is a factor of the sum.

   Thus, 3 is a factor of $n^3 - n + 3$ for all positive integers n.

31. A factor of $n^4 - n + 4$ is 2.

1. When $n = 1$, $1^4 - 1 + 4 = 4$ and 2 is a factor.

2. Assume that 2 is a factor of $k^4 - k + 4$.

   Then,

   $$(k + 1)^4 - (k + 1) + 4 = k^4 + 4k^3 + 6k^2 + 4k + 1 - k - 1 + 4$$
   $$= (k^4 - k + 4) + (4k^3 + 6k^2 + 4k)$$
   $$= (k^4 - k + 4) + 2(2k^3 + 3k^2 + 2k).$$

   Since 2 is a factor of $k^4 - k + 4$, our assumption, and 2 is a factor of $2(2k^3 + 3k^2 + 2k)$, we conclude that 2 is a factor of the entire expression.

   Thus, 2 is a factor of $n^4 - n + 4$ for every positive integer n.

32. Prove 3 is a factor of $2^{2n+1} + 1$ for all positive integers n.

1. When $n = 1, 2^{2 \cdot 1 + 1} + 1 = 2^3 + 1 = 8 + 1 = 9$ and 3 is a factor.

2. Assume 3 is a factor of $2^{2k+1} + 1$.

Then,

$$2^{2(k+1)+1} + 1 = 2^{2k+2+1} + 1$$
$$= 2^{(2k+1)+2} + 1$$
$$= 2^{2k+1} \cdot 2^2 + 1$$
$$= 4 \cdot 2^{2k+1} + 1$$
$$= 4(2^{2k+1} + 1) - 3$$

Since 3 is a factor of each term, 3 is a factor of the sum.

Thus, 3 is a factor of $2^{2n+1} + 1$ for all positive integers n.

33. A factor of $2^{4n-2} + 1$ is 5.

1. When $n = 1$,

$2^{4(1)-2} + 1 = 5$ and 5 is a factor.

2. Assume that 5 is a factor of $2^{4k-2} + 1$.

Then,

$$2^{4(k+1)-2} + 1 = 2^{4k+4-2} + 1$$
$$= 2^{4k-2} \cdot 2^4 + 1$$
$$= 2^{4k-2} \cdot 16 + 1$$
$$= (2^{4k-2} + 1) + 15 \cdot 2^{4k-2}.$$

Since 5 is a factor of $2^{4k-2} + 1$, our assumption, and 5 is a factor of $15 \cdot 2^{4k-2}$, we conclude that 5 is a factor of the entire expression.

Thus, 5 is a factor of $2^{4n-2} + 1$ for every positive integer n.

34. 1. When $n = 1, (2^{2(1)-1} + 3^{2(1)-1}) = 2 + 3 = 5$ and 5 is a factor.

2. Assume that 5 is a factor of $(2^{2k-1} + 3^{2k-1})$.

Then, $2^{2(k+1)-1} + 3^{2(k+1)-1} = 2^{2k+2-1} + 3^{2k+2-1}$
$$= 2^{2k-1}2^2 + 3^{2k-1}3^2$$
$$= 4 \cdot 2^{2k-1} + 9 \cdot 3^{2k-1}$$
$$= (2^{2k-1} + 3^{2k-1}) + (2^{2k-1} + 3^{2k-1})$$
$$+ (2^{2k-1} + 3^{2k-1}) + (2^{2k-1} + 3^{2k-1}) + 5 \cdot 3^{2k-1}.$$

Since 5 is a factor of each set in parentheses and 5 is a factor of $5 \cdot 3^{2k-1}$, then 5 is a factor of the whole sum.

Thus, 5 is a factor of $(2^{2n-1} + 3^{2n-1})$ for every positive integer n.

35. $S_n = 1 + 5 + 9 + 13 + \cdots + (4n - 3)$

$S_1 = 1 = 1 \cdot 1$

$S_2 = 1 + 5 = 6 = 2 \cdot 3$

$S_3 = 1 + 5 + 9 = 15 = 3 \cdot 5$

$S_4 = 1 + 5 + 9 + 13 = 28 = 4 \cdot 7$

From this sequence, it appears that $S_n = n(2n - 1)$. This can be verified by mathematical induction. The formula has already been verified for $n = 1$. Assume that the formula is valid for $n = k$. Then,

$S_{k+1} = [1 + 5 + 9 + 13 + \cdots + (4k - 3)] + [4(k + 1) - 3]$

$\qquad = k(2k - 1) + (4k + 1)$

$\qquad = 2k^2 + 3k + 1$

$\qquad = (k + 1)(2k + 1)$

$\qquad = (k + 1)[2(k + 1) - 1].$

Thus, the formula is valid.

36. $S_n = 25 + 22 + 19 + 16 + \cdots + (-3n + 28)$

$S_1 = 25 = \dfrac{1}{2}(50)$

$S_2 = 25 + 22 = 47 = \dfrac{2}{2}(47)$

$S_3 = 25 + 22 + 19 = 66 = \dfrac{3}{2}(44)$

$S_4 = 25 + 22 + 19 + 16 = 82 = \dfrac{4}{2}(41)$

From the sequence, it appears that

$S_n = \dfrac{n}{2}(-3n + 53).$

This can be verified by mathematical induction. The formula has already been verified for $n = 1$. Assume that the formula is valid for $n = k$. Then,

$S_{k+1} = [25 + 22 + 19 + 16 + \cdots + (-3k + 28)] + [-3(k + 1) + 28]$

$\qquad = \dfrac{k}{2}(-3k + 53) + (-3k + 25)$

$\qquad = \dfrac{1}{2}(-3k^2 + 47k + 50)$

$\qquad = -\dfrac{1}{2}(3k^2 - 47k - 50)$

$\qquad = -\dfrac{1}{2}(k + 1)(3k - 50)$

$\qquad = \dfrac{k + 1}{2}[-3(k + 1) + 53].$

Thus, the formula is valid.

37. $S_n = 1 + \dfrac{9}{10} + \dfrac{81}{100} + \dfrac{729}{1000} + \cdots + \left(\dfrac{9}{10}\right)^{n-1}$

Since this series is geometric, we have

$$S_n = \sum_{i=1}^{n} \left(\frac{9}{10}\right)^{i-1} = \frac{1 - \left(\dfrac{9}{10}\right)^n}{1 - \dfrac{9}{10}} = 10\left[1 - \left(\frac{9}{10}\right)^n\right]$$

$$= 10 - 10\left(\frac{9}{10}\right)^n.$$

38. $S_n = 3 - \dfrac{9}{2} + \dfrac{27}{4} - \dfrac{81}{8} + \cdots + 3\left(-\dfrac{3}{2}\right)^{n-1}$

Since the series is geometric, we have

$$S_n = \sum_{i=1}^{n} 3\left(-\frac{3}{2}\right)^{i-1} = 3\left[\frac{1 - \left(-\dfrac{3}{2}\right)^n}{1 - \left(-\dfrac{3}{2}\right)}\right] = \frac{6}{5}\left[1 - \left(-\frac{3}{2}\right)^n\right].$$

39. $S_n = \dfrac{1}{4} + \dfrac{1}{12} + \dfrac{1}{24} + \dfrac{1}{40} + \cdots + \dfrac{1}{2n(n+1)}$

$S_1 = \dfrac{1}{4} = \dfrac{1}{2(2)}$

$S_2 = \dfrac{1}{4} + \dfrac{1}{12} = \dfrac{4}{12} = \dfrac{2}{6} = \dfrac{2}{2(3)}$

$S_3 = \dfrac{1}{4} + \dfrac{1}{12} + \dfrac{1}{24} = \dfrac{9}{24} = \dfrac{3}{8} = \dfrac{3}{2(4)}$

$S_4 = \dfrac{1}{4} + \dfrac{1}{12} + \dfrac{1}{24} + \dfrac{1}{40} = \dfrac{16}{40} = \dfrac{4}{10} = \dfrac{4}{2(5)}$

From this sequence, it appears that

$S_n = \dfrac{n}{2(n+1)}.$

This can be verified by mathematical induction. The formula has already been verified for $n = 1$.
Assume that the formula is valid for $n = k$. Then,

$$S_{k+1} = \left[\frac{1}{4} + \frac{1}{12} + \frac{1}{40} + \cdots + \frac{1}{2k(k+1)}\right] + \frac{1}{2(k+1)(k+2)}$$

$$= \frac{k}{2(k+1)} + \frac{1}{2(k+1)(k+2)}$$

$$= \frac{k(k+2) + 1}{2(k+1)(k+2)}$$

$$= \frac{k^2 + 2k + 1}{2(k+1)(k+2)}$$

$$= \frac{(k+1)^2}{2(k+1)(k+2)}$$

$$= \frac{k+1}{2(k+2)}.$$

Thus, the formula is valid.

40. $S_n = \dfrac{1}{2 \cdot 3} + \dfrac{1}{3 \cdot 4} + \dfrac{1}{4 \cdot 5} + \dfrac{1}{5 \cdot 6} + \cdots + \dfrac{1}{(n+1)(n+2)}$

$S_1 = \dfrac{1}{6} = \dfrac{1}{2 \cdot 3}$

$S_2 = \dfrac{1}{6} + \dfrac{1}{12} = \dfrac{1}{4} = \dfrac{2}{2 \cdot 4}$

$S_3 = \dfrac{1}{6} + \dfrac{1}{12} + \dfrac{1}{20} = \dfrac{3}{10} = \dfrac{3}{2 \cdot 5}$

$S_4 = \dfrac{1}{6} + \dfrac{1}{12} + \dfrac{1}{20} + \dfrac{1}{30} = \dfrac{1}{3} = \dfrac{4}{2 \cdot 6}$

From this sequence, it appears that

$S_n = \dfrac{n}{2(n+2)}.$

This can be verified by mathematical induction. The formula has already been verified for $n = 1$.
Assume that the formula is valid for $n = k$. Then,

$S_{k+1} = \left[\dfrac{1}{6} + \dfrac{1}{12} + \dfrac{1}{20} + \dfrac{1}{30} + \cdots + \dfrac{1}{(k+1)(k+2)} \right] + \dfrac{1}{(k+2)(k+3)}$

$= \dfrac{k}{2(k+2)} + \dfrac{1}{(k+2)(k+3)}$

$= \dfrac{k(k+3) + 2}{2(k+2)(k+3)}$

$= \dfrac{k^2 + 3k + 2}{2(k+2)(k+3)}$

$= \dfrac{(k+1)(k+2)}{2(k+2)(k+3)}$

$= \dfrac{k+1}{2[(k+1)+2]}.$

Thus, the formula is valid.

41. $\displaystyle\sum_{n=1}^{15} n = \dfrac{15(15+1)}{2} = 120$

42. $\displaystyle\sum_{n=1}^{30} n = \dfrac{30(30+1)}{2} = 465$

43. $\displaystyle\sum_{n=1}^{6} n^2 = \dfrac{6(6+1)[2(6)+1]}{6} = 91$

44. $\displaystyle\sum_{n=1}^{10} n^3 = \dfrac{10^2(10+1)^2}{4} = 3025$

45. $\displaystyle\sum_{n=1}^{5} n^4 = \dfrac{5(5+1)[2(5)+1][3(5)^2+3(5)-1]}{30} = 979$

46. $\displaystyle\sum_{n=1}^{8} n^5 = \dfrac{8^2(8+1)^2(2(8)^2+2(8)-1)}{12} = 61{,}776$

47. $\displaystyle\sum_{n=1}^{6} (n^2 - n) = \sum_{n=1}^{6} n^2 - \sum_{n=1}^{6} n$

$= \dfrac{6(6+1)[2(6)+1]}{6} - \dfrac{6(6+1)}{2}$

$= 91 - 21 = 70$

48. $\displaystyle\sum_{n=1}^{20} (n^3 - n) = \sum_{n=1}^{20} n^3 - \sum_{n=1}^{20} n$

$= \dfrac{(20)^2(20+1)^2}{4} - \dfrac{20(20+1)}{2}$

$= \dfrac{(20)^2(21)^2 - 2(20)(21)}{4} = 43{,}890$

49. $\displaystyle\sum_{i=1}^{6} (6i - 8i^3) = 6\sum_{i=1}^{6} i - 8\sum_{i=1}^{6} i^3 = 6\left[\dfrac{6(6+1)}{2} \right] - 8\left[\dfrac{(6)^2(6+1)^2}{4} \right] = 6(21) - 8(441) = -3402$

50. $\sum_{j=1}^{10}\left(3 - \frac{1}{2}j + \frac{1}{2}j^2\right) = \sum_{j=1}^{10} 3 - \frac{1}{2}\sum_{j=1}^{10} j + \frac{1}{2}\sum_{j=1}^{10} j^2$

$$= 3(10) - \frac{1}{2} \cdot \frac{10(10+1)}{2} + \frac{1}{2} \cdot \frac{10(10+1)(2\cdot 10+1)}{6}$$

$$= \frac{3(10)(12) - 3(10)(11) + 10(11)(21)}{12} = 195$$

51. $a_1 = 0, a_n = a_{n-1} + 3$

$a_1 = a_1 = 0$

$a_2 = a_1 + 3 = 0 + 3 = 3$

$a_3 = a_2 + 3 = 3 + 3 = 6$

$a_4 = a_3 + 3 = 6 + 3 = 9$

$a_5 = a_4 + 3 = 9 + 3 = 12$

$a_6 = a_5 + 3 = 12 + 3 = 15$

a_n: 0   3   6   9   12   15

First differences: 3   3   3   3   3

Second differences: 0   0   0   0

Since the first differences are equal, the sequence has a linear model.

52. $a_1 = 2, a_n = a_{n-1} + 2$

$a_1 = a_1 = 2$

$a_2 = a_1 + 2 = 2 + 2 = 4$

$a_3 = a_2 + 2 = 4 + 2 = 6$

$a_4 = a_3 + 2 = 6 + 2 = 8$

$a_5 = a_4 + 2 = 8 + 2 = 10$

$a_6 = a_5 + 2 = 10 + 2 = 12$

a_n: 2   4   6   8   10   12

First differences: 2   2   2   2   2

Second differences: 0   0   0   0

Since the first differences are equal, the sequence has a linear model.

53. $a_1 = 3, a_n = a_{n-1} - n$

$a_1 = a_1 = 3$

$a_2 = a_1 - 2 = 3 - 2 = 1$

$a_3 = a_2 - 3 = 1 - 3 = -2$

$a_4 = a_3 - 4 = -2 - 4 = -6$

$a_5 = a_4 - 5 = -6 - 5 = -11$

$a_6 = a_5 - 6 = -11 - 6 = -17$

a_n: 3   1   −2   −6   −11   −17

First differences: −2   −3   −4   −5   −6

Second differences: −1   −1   −1   −1

Since the second differences are all the same, the sequence has a quadratic model.

54. $a_2 = -3, a_n = -2a_{n-1}$

$a_2 = -3 \Rightarrow -3 = -2a_1$

$a_1 = \frac{3}{2}$

$a_2 = -3$

$a_3 = -2a_2 = -2(-3) = 6$

$a_4 = -2a_3 = -2(6) = -12$

$a_5 = -2a_4 = -2(-12) = 24$

$a_6 = -2a_5 = -2(24) = -48$

$a_7 = -2a_6 = -2(-48) = 96$

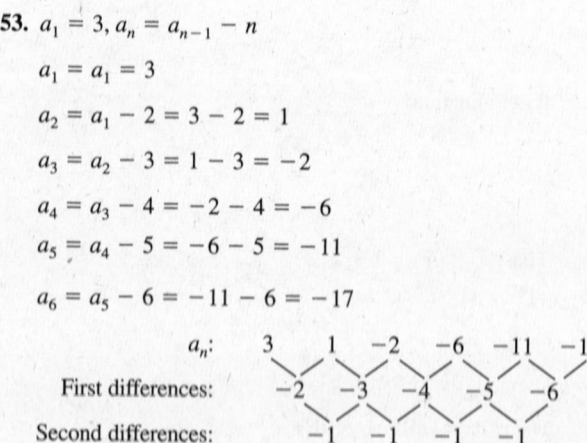

a_n: $\frac{3}{2}$ −3   6   −12   24   −48   96

First differences: $-\frac{9}{2}$ 9   −18   36   −72   144

Second differences: $\frac{27}{2}$ −27   54   −108   216

Since neither the first differences nor the second differences are equal, the sequence does not have a linear or quadratic model.

55. $a_0 = 2, a_n = (a_{n-1})^2$

$a_0 = 2$

$a_1 = a_0^2 = 2^2 = 4$

$a_2 = a_1^2 = 4^2 = 16$

$a_3 = a_2^2 = 16^2 = 256$

$a_4 = a_3^2 = 256^2 = 65{,}536$

$a_5 = a_4^2 = 65{,}536^2 = 4{,}294{,}967{,}296$

a_n: 2 4 16 256 65,536 4,294,967,296

First differences: 2 12 240 65,280 4,294,901,760

Second differences: 10 228 65,040 4,294,836,480

Since neither the first differences nor the second differences are equal, the sequence does not have a linear or quadratic model.

56. $a_0 = 0, a_n = a_{n-1} + n$

$a_0 = 0$

$a_1 = a_0 + 1 = 0 + 1 = 1$

$a_2 = a_1 + 2 = 1 + 2 = 3$

$a_3 = a_2 + 3 = 3 + 3 = 6$

$a_4 = a_3 + 4 = 6 + 4 = 10$

$a_5 = a_4 + 5 = 10 + 5 = 15$

a_n: 0 1 3 6 10 15

First differences: 1 2 3 4 5

Second differences: 1 1 1 1

Since the second differences are equal, the sequence has a quadratic model.

57. $a_0 = 3, a_1 = 3, a_4 = 15$

Let $a_n = an^2 + bn + c$.

Thus: $a_0 = a(0)^2 + b(0) + c = 3 \implies c = 3$

$a_1 = a(1)^2 + b(1) + c = 3 \implies a + b + c = 3$

$a + b = 0$

$a_4 = a(4)^2 + b(4) + c = 15 \implies 16a + 4b + c = 15$

$16a + 4b = 12$

$4a + b = 3$

By elimination: $-a - b = 0$

$\underline{\quad 4a + b = 3 \quad}$

$3a \quad\;\; = 3$

$a = 1 \implies b = -1$

Thus, $a_n = n^2 - n + 3$.

58. $a_0 = 7, a_1 = 6, a_3 = 10$

Let $a_n = an^2 + bn + c$. Then:

$a_0 = a(0)^2 + b(0) + c = 7 \implies c = 7$

$a_1 = a(1)^2 + b(1) + c = 6 \implies a + b + c = 6$

$a + b = -1$

$a_3 = a(3)^2 + b(3) + c = 10 \implies 9a + 3b + c = 10$

$9a + 3b = 3$

$3a + b = 1$

By elimination: $-a - b = 1$

$\underline{\quad 3a + b = 1 \quad}$

$2a = 2$

$a = 1 \implies b = -2$

Thus, $a_n = n^2 - 2n + 7$.

59. $a_0 = -3, a_2 = 1, a_4 = 9$

Let $a_n = an^2 + bn + c$.

Then: $a_0 = a(0)^2 + b(0) + c = -3 \implies c = -3$

$a_2 = a(2)^2 + b(2) + c = 1 \implies 4a + 2b + c = 1$

$4a + 2b = 4$

$2a + b = 2$

$a_4 = a(4)^2 + b(4) + c = 9 \implies 16a + 4b + c = 9$

$16a + 4b = 12$

$4a + b = 3$

By elimination: $-2a - b = -2$

$\underline{\quad 4a + b = \;\; 3 \quad}$

$2a \quad\;\; = 1$

$a = \frac{1}{2} \implies b = 1$

Thus, $a_n = \frac{1}{2}n^2 + n - 3$.

60. $a_0 = 3, a_2 = 0, a_6 = 36$

Let $a_n = an^2 + bn + c$. Then:

$a_0 = a(0)^2 + b(0) + c = 3 \implies \qquad\qquad c = 3$
$a_2 = a(2)^2 + b(2) + c = 0 \implies 4a + 2b + c = 0$
$\qquad\qquad\qquad\qquad\qquad\qquad 4a + 2b \qquad = -3$

$a_6 = a(6)^2 + b(6) + c = 36 \implies 36a + 6b + c = 36$
$\qquad\qquad\qquad\qquad\qquad\qquad 36a + 6b \qquad = 33$
$\qquad\qquad\qquad\qquad\qquad\qquad 12a + 2b \qquad = 11$

By elimination: $-4a - 2b = 3$
$\qquad\qquad\qquad\dfrac{12a + 2b = 11}{8a \qquad = 14}$
$\qquad\qquad\qquad\qquad a = \tfrac{7}{4} \implies b = -5$

Thus, $a_n = \tfrac{7}{4}n^2 - 5n + 3$.

61. (a)
$$120.3 \quad 122.5 \quad 124.9 \quad 127.1 \quad 129.4 \quad 130.3$$
First differences: $\quad 2.2 \qquad 2.4 \qquad 2.2 \qquad 2.3 \qquad 0.9$

(b) The first differences are not equal, but are fairly close to each other, so a linear model can be used. If we let $m = 2.2$, then $b = 120.3 - 2.2(8) = 102.7$

$\quad a_n \approx 2.2n + 102.7$

(c) $a_n \approx 2.08n + 103.9$ is obtained by using the regression feature of a graphing utility.

(d) For 2008, let $n = 18$.

$\quad a_n \approx 2.2(18) + 102.7 = 142.3$

$\quad a_n = 2.08(18) + 103.9 = 141.34$

These are very similar.

62. Answers will vary. See page 626.   **63.** True. P_7 may be false.   **64.** False. P_1 must be proven to be true.

65. True. If the second differences are all zero, then the first differences are all the same, so the sequence is arithmetic.   **66.** False. It has $n - 2$ second differences.

67. $(2x^2 - 1)^2 = (2x^2 - 1)(2x^2 - 1) = 4x^4 - 4x^2 + 1$ **68.** $(2x - y)^2 = 4x^2 - 4xy + y^2$

69. $(5 - 4x)^3 = -64x^3 + 240x^2 - 300x + 125$ **70.** $(2x - 4y)^3 = 8x^3 - 48x^2y + 96xy^2 - 64y^3$

71. $f(x) = \dfrac{x}{x + 3}$

(a) Domain: All real numbers x except $x = -3$

(b) Intercept: $(0, 0)$

(c) Vertical asymptote: $x = -3$

Horizontal asymptote: $y = 1$

(d)

x	-5	-4	-2	-1	1
$f(x)$	$\tfrac{5}{2}$	4	-2	$-\tfrac{1}{2}$	$\tfrac{1}{4}$

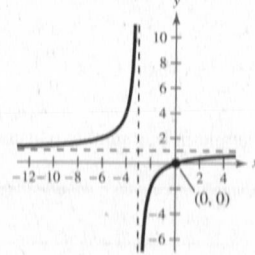

72. $g(x) = \dfrac{x^2}{x^2 - 4}$

(a) Domain: All real numbers x except $x = \pm 2$

(b) Intercept: $(0, 0)$

(c) Vertical asymptotes: $x = -2, x = 2$

Horizontal asymptote: $y = 1$

(d)

x	-4	-3	-1.5	0	1.5	3	4
$g(x)$	$\frac{4}{3}$	$\frac{9}{5}$	$-\frac{9}{7}$	0	$-\frac{9}{7}$	$\frac{9}{5}$	$\frac{4}{3}$

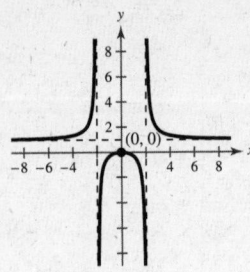

73. $h(t) = \dfrac{t - 7}{t}$

(a) Domain: All real numbers t except $t = 0$

(b) Intercept: $(7, 0)$

(c) Vertical asymptote: $t = 0$

Horizontal asymptote: $y = 1$

(d)

t	-2	-1	1	2	3
$h(t)$	$\frac{9}{2}$	8	-6	$-\frac{5}{2}$	$-\frac{4}{3}$

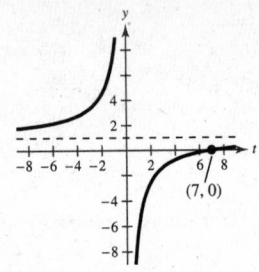

74. $f(x) = \dfrac{5 + x}{1 - x}$

(a) Domain: All real numbers x except $x = 1$

(b) x-intercept: $(-5, 0)$

y-intercept: $(0, 5)$

(c) Vertical asymptote: $x = 1$

Horizontal asymptote: $y = -1$

(d)

x	-8	-5	-2	0	2	3	5	7
$f(x)$	$-\frac{1}{3}$	0	1	5	-7	-4	$-\frac{5}{2}$	-2

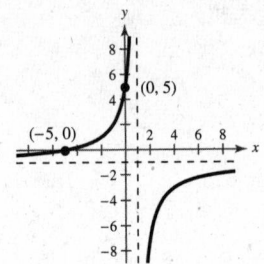

Section 9.5 The Binomial Theorem

■ You should be able to use the formula

$$(x + y)^n = x^n + nx^{n-1}y + \frac{n(n-1)}{2!}x^{n-2}y^2 + \cdots + {}_nC_r\, x^{n-r}y^r + \cdots + y^n$$

where ${}_nC_r = \dfrac{n!}{(n-r)!\,r!}$, to expand $(x + y)^n$. Also, ${}_nC_r = \dbinom{n}{r}$.

■ You should be able to use Pascal's Triangle in binomial expansion.

Vocabulary Check

1. binomial coefficients

2. Binomial Theorem/Pascal's Triangle

3. $\dbinom{n}{r}$ or ${}_nC_r$

4. expanding a binomial

1. $_5C_3 = \dfrac{5!}{3!2!} = \dfrac{5 \cdot 4}{2 \cdot 1} = 10$

2. $_8C_6 = \dfrac{8!}{6! \cdot 2!} = \dfrac{8 \cdot 7}{2 \cdot 1} = 28$

3. $_{12}C_0 = \dfrac{12!}{0!12!} = 1$

4. $_{20}C_{20} = \dfrac{20!}{20! \cdot 0!} = 1$

5. $_{20}C_{15} = \dfrac{20!}{15!5!} = \dfrac{20 \cdot 19 \cdot 18 \cdot 17 \cdot 16}{5 \cdot 4 \cdot 3 \cdot 2 \cdot 1} = 15{,}504$

6. $_{12}C_5 = \dfrac{12!}{5! \cdot 7!} = \dfrac{(12 \cdot 11 \cdot 10 \cdot 9 \cdot 8) \cdot 7!}{5!7!} = \dfrac{12 \cdot 11 \cdot 10 \cdot 9 \cdot 8}{5 \cdot 4 \cdot 3 \cdot 2 \cdot 1} = 792$

7. $\dbinom{10}{4} = \dfrac{10!}{6!4!} = \dfrac{10 \cdot 9 \cdot 8 \cdot 7 \cdot 6!}{6!(24)} = 210$

8. $\dbinom{10}{6} = \dfrac{10!}{6! \cdot 4!} = \dfrac{(10 \cdot 9 \cdot 8 \cdot 7) \cdot 6!}{6! \cdot 4!} = \dfrac{10 \cdot 9 \cdot 8 \cdot 7}{4 \cdot 3 \cdot 2 \cdot 1} = 210$

9. $\dbinom{100}{98} = \dfrac{100!}{2!98!} = \dfrac{100 \cdot 99}{2 \cdot 1} = 4950$

10. $\dbinom{100}{2} = \dfrac{100!}{98! \cdot 2!} = \dfrac{(100 \cdot 99) \cdot 98!}{98! \cdot 2!} = \dfrac{100 \cdot 99}{2 \cdot 1}$

$= 4950$

11.
```
            1
          1   1
        1   2   1
      1   3   3   1
    1   4   6   4   1
  1   5  10  10   5   1
1   6  15  20  15   6   1
1  7  21  35  35  21  7  1
1  8  28  56  70 (56) 28  8  1
```

$\dbinom{8}{5} = 56$, the 6th entry in the 8th row.

12.
```
            1
          1   1
        1   2   1
      1   3   3   1
    1   4   6   4   1
  1   5  10  10   5   1
1   6  15  20  15   6   1
1  7  21  35  35  21  7  1
1  8  28  56  70  56  28 (8)  1
```

$\dbinom{8}{7} = 8$, the 8th entry in the 8th row.

13.
```
          1
        1   1
      1   2   1
    1   3   3   1
  1   4   6   4   1
1   5  10  10   5   1
1  6  15  20  15   6   1
1  7  21  35 (35) 21  7  1
```

$_7C_4 = 35$, the 5th entry in the 7th row.

14.
```
          1
        1   1
      1   2   1
    1   3   3   1
  1   4   6   4   1
1   5  10  10   5   1
1  6  15 (20) 15   6   1
```

$_6C_3 = 20$, the 4th entry in the 6th row.

15. $(x + 1)^4 = {_4C_0}x^4 + {_4C_1}x^3(1) + {_4C_2}x^2(1)^2 + {_4C_3}x(1)^3 + {_4C_4}(1)^4$

$= x^4 + 4x^3 + 6x^2 + 4x + 1$

16. $(x + 1)^6 = {_6C_0}x^6 + {_6C_1}x^5(1) + {_6C_2}x^4(1)^2 + {_6C_3}x^3(1)^3 + {_6C_4}x^2(1)^4 + {_6C_5}x(1)^5 + {_6C_6}(1)^6$

$= x^6 + 6x^5 + 15x^4 + 20x^3 + 15x^2 + 6x + 1$

17. $(a + 6)^4 = {_4C_0}a^4 + {_4C_1}a^3(6) + {_4C_2}a^2(6)^2 + {_4C_3}a(6)^3 + {_4C_4}(6)^4$

$= 1a^4 + 4a^3(6) + 6a^2(6)^2 + 4a(6)^3 + 1(6)^4$

$= a^4 + 24a^3 + 216a^2 + 864a + 1296$

18. $(a + 5)^5 = {}_5C_0a^5 + {}_5C_1a^4(5) + {}_5C_2a^3(5)^2 + {}_5C_3a^2(5)^3 + {}_5C_4a(5)^4 + {}_5C_5(5)^5$

$\qquad = a^5 + 25a^4 + 250a^3 + 1250a^2 + 3125a + 3125$

19. $(y - 4)^3 = {}_3C_0y^3 - {}_3C_1y^2(4) + {}_3C_2y(4)^2 - {}_3C_3(4)^3$

$\qquad = 1y^3 - 3y^2(4) + 3y(4)^2 - 1(4)^3$

$\qquad = y^3 - 12y^2 + 48y - 64$

20. $(y - 2)^5 = {}_5C_0y^5 - {}_5C_1y^4(2) + {}_5C_2y^3(2)^2 - {}_5C_3y^2(2)^3 + {}_5C_4y(2)^4 - {}_5C_5(2)^5$

$\qquad = y^5 - 10y^4 + 40y^3 - 80y^2 + 80y - 32$

21. $(x + y)^5 = {}_5C_0x^5 + {}_5C_1x^4y + {}_5C_2x^3y^2 + {}_5C_3x^2y^3 + {}_5C_4xy^4 + {}_5C_5y^5$

$\qquad = x^5 + 5x^4y + 10x^3y^2 + 10x^2y^3 + 5xy^4 + y^5$

22. $(c + d)^3 = {}_3C_0c^3 + {}_3C_1c^2d + {}_3C_2cd^2 + {}_3C_3d^3$

$\qquad = c^3 + 3c^2d + 3cd^2 + d^3$

23. $(r + 3s)^6 = {}_6C_0r^6 + {}_6C_1r^5(3s) + {}_6C_2r^4(3s)^2 + {}_6C_3r^3(3s)^3 + {}_6C_4r^2(3s)^4 + {}_6C_5r(3s)^5 + {}_6C_6(3s)^6$

$\qquad = 1r^6 + 6r^5(3s) + 15r^4(3s)^2 + 20r^3(3s)^3 + 15r^2(3s)^4 + 6r(3s)^5 + 1(3s)^6$

$\qquad = r^6 + 18r^5s + 135r^4s^2 + 540r^3s^3 + 1215r^2s^4 + 1458rs^5 + 729s^6$

24. $(x + 2y)^4 = {}_4C_0 x^4 + {}_4C_1x^3(2y) + {}_4C_2x^2(2y)^2 + {}_4C_3x(2y)^3 + {}_4C_4(2y)^4$

$\qquad = x^4 + 4x^3(2y) + 6x^2(4y^2) + 4x(8y^3) + 16y^4$

$\qquad = x^4 + 8x^3y + 24x^2y^2 + 32xy^3 + 16y^4$

25. $(3a - 4b)^5 = {}_5C_0(3a)^5 - {}_5C_1(3a)^4(4b) + {}_5C_2(3a)^3(4b)^2 - {}_5C_3(3a)^2(4b)^3 + {}_5C_4(3a)(4b)^4 - {}_5C_5(4b)^5$

$\qquad = (1)(243a^5) - 5(81a^4)(4b) + 10(27a^3)(16b^2) - 10(9a^2)(64b^3) + 5(3a)(256b^4) - (1)(1024b^5)$

$\qquad = 243a^5 - 1620a^4b + 4320a^3b^2 - 5760a^2b^3 + 3840ab^4 - 1024b^5$

26. $(2x - 5y)^5 = {}_5C_0(2x)^5 + {}_5C_1(2x)^4(-5y) + {}_5C_2(2x)^3(-5y)^2 + {}_5C_3(2x)^2(-5y)^3 + {}_5C_4(2x)(-5y)^4 + {}_5C_5(-5y)^5$

$\qquad = (2x)^5 + 5(2x)^4(-5y) + 10(2x)^3(-5y)^2 + 10(2x)^2(-5y)^3 + 5(2x)(-5y)^4 + (-5y)^5$

$\qquad = 32x^5 - 400x^4y + 2000x^3y^2 - 5000x^2y^3 + 6250xy^4 - 3125y^5$

27. $(2x + y)^3 = {}_3C_0(2x)^3 + {}_3C_1(2x)^2(y) + {}_3C_2(2x)(y^2) + {}_3C_3(y^3)$

$\qquad = (1)(8x^3) + (3)(4x^2)(y) + (3)(2x)(y^2) + (1)(y^3)$

$\qquad = 8x^3 + 12x^2y + 6xy^2 + y^3$

28. $(7a + b)^3 = {}_3C_0(7a)^3 + {}_3C_1(7a)^2(b) + {}_3C_2(7a)(b)^2 + {}_3C_3(b)^3$

$\qquad = (7a)^3 + 3(7a)^2(b) + 3(7a)(b)^2 + (b)^3$

$\qquad = 343a^3 + 147a^2b + 21ab^2 + b^3$

29. $(x^2 + y^2)^4 = {}_4C_0(x^2)^4 + {}_4C_1(x^2)^3(y^2) + {}_4C_2(x^2)^2(y^2)^2 + {}_4C_3(x^2)(y^2)^3 + {}_4C_4(y^2)^4$

$$= (1)(x^8) + (4)(x^6y^2) + (6)(x^4y^4) + (4)(x^2y^6) + (1)(y^8)$$

$$= x^8 + 4x^6y^2 + 6x^4y^4 + 4x^2y^6 + y^8$$

30. $(x^2 + y^2)^6 = {}_6C_0(x^2)^6 + {}_6C_1(x^2)^5(y^2) + {}_6C_2(x^2)^4(y^2)^2 + {}_6C_3(x^2)^3(y^2)^3 + {}_6C_4(x^2)^2(y^2)^4 + {}_6C_5(x^2)(y^2)^5 + {}_6C_6(y^2)^6$

$$= x^{12} + 6x^{10}y^2 + 15x^8y^4 + 20x^6y^6 + 15x^4y^8 + 6x^2y^{10} + y^{12}$$

31. $\left(\dfrac{1}{x} + y\right)^5 = {}_5C_0\left(\dfrac{1}{x}\right)^5 + {}_5C_1\left(\dfrac{1}{x}\right)^4 y + {}_5C_2\left(\dfrac{1}{x}\right)^3 y^2 + {}_5C_3\left(\dfrac{1}{x}\right)^2 y^3 + {}_5C_4\left(\dfrac{1}{x}\right)y^4 + {}_5C_5 y^5$

$$= \dfrac{1}{x^5} + \dfrac{5y}{x^4} + \dfrac{10y^2}{x^3} + \dfrac{10y^3}{x^2} + \dfrac{5y^4}{x} + y^5$$

32. $\left(\dfrac{1}{x} + 2y\right)^6 = {}_6C_0\left(\dfrac{1}{x}\right)^6 + {}_6C_1\left(\dfrac{1}{x}\right)^5(2y) + {}_6C_2\left(\dfrac{1}{x}\right)^4(2y)^2 + {}_6C_3\left(\dfrac{1}{x}\right)^3(2y)^3 + {}_6C_4\left(\dfrac{1}{x}\right)^2(2y)^4 + {}_6C_5\left(\dfrac{1}{x}\right)(2y)^5 + {}_6C_6(2y)^6$

$$= 1\left(\dfrac{1}{x}\right)^6 + 6(2)\left(\dfrac{1}{x}\right)^5 y + 15(4)\left(\dfrac{1}{x}\right)^4 y^2 + 20(8)\left(\dfrac{1}{x}\right)^3 y^3 + 15(16)\left(\dfrac{1}{x}\right)^2 y^4 + 6(32)\left(\dfrac{1}{x}\right)y^5 + 1(64)y^6$$

$$= \dfrac{1}{x^6} + \dfrac{12y}{x^5} + \dfrac{60y^2}{x^4} + \dfrac{160y^3}{x^3} + \dfrac{240y^4}{x^2} + \dfrac{192y^5}{x} + 64y^6$$

33. $2(x - 3)^4 + 5(x - 3)^2 = 2[x^4 - 4(x^3)(3) + 6(x^2)(3^2) - 4(x)(3^3) + 3^4] + 5[x^2 - 2(x)(3) + 3^2]$

$$= 2(x^4 - 12x^3 + 54x^2 - 108x + 81) + 5(x^2 - 6x + 9)$$

$$= 2x^4 - 24x^3 + 113x^2 - 246x + 207$$

34. $3(x + 1)^5 - 4(x + 1)^3 = 3[{}_5C_0 x^5 + {}_5C_1 x^4(1) + {}_5C_2 x^3(1)^2 + {}_5C_3 x^2(1)^3 + {}_5C_4 x(1)^4 + {}_5C_5(1)^5]$

$$- 4[{}_3C_0 x^3 + {}_3C_1 x^2(1) + {}_3C_2 x(1)^2 + {}_3C_3(1)^3]$$

$$= 3[(1)x^5 + 5x^4 + 10x^3 + 10x^2 + 5x + 1] - 4[(1)x^3 + 3x^2 + 3x + 1]$$

$$= 3x^5 + 15x^4 + 26x^3 + 18x^2 + 3x - 1$$

35. 5th Row of Pascal's Triangle: 1   5   10   10   5   1

$(2t - s)^5 = 1(2t)^5 - 5(2t)^4(s) + 10(2t)^3(s)^2 - 10(2t)^2(s)^3 + 5(2t)(s)^4 - 1(s)^5$

$$= 32t^5 - 80t^4s + 80t^3s^2 - 40t^2s^3 + 10ts^4 - s^5$$

36. 4th Row of Pascal's Triangle: 1   4   6   4   1

$(3 - 2z)^4 = 3^4 - 4(3)^3(2z) + 6(3)^2(2z)^2 - 4(3)(2z)^3 + (2z)^4$

$$= 81 - 216z + 216z^2 - 96z^3 + 16z^4$$

37. 5th Row of Pascal's Triangle: 1   5   10   10   5   1

$(x + 2y)^5 = 1x^5 + 5x^4(2y) + 10x^3(2y)^2 + 10x^2(2y)^3 + 5x(2y)^4 + 1(2y)^5$

$$= x^5 + 10x^4y + 40x^3y^2 + 80x^2y^3 + 80xy^4 + 32y^5$$

38. 6th Row of Pascal's Triangle: 1   6   15   20   15   6   1

$(2v + 3)^6 = (2v)^6 + 6(2v)^5(3) + 15(2v)^4(3)^2 + 20(2v)^3(3)^3 + 15(2v)^2(3)^4 + 6(2v)(3)^5 + (3)^6$

$$= 64v^6 + 576v^5 + 2160v^4 + 4320v^3 + 4860v^2 + 2916v + 729$$

39. The 4th term in the expansion of $(x + y)^{10}$ is

$$_{10}C_3 x^{10-3} y^3 = 120 x^7 y^3.$$

40. The 7th term in the expansion of $(x - y)^6$ is

$$_6C_6 x^{6-6}(-y)^6 = 1 \cdot x^0 y^6 = y^6.$$

41. The 3rd term in the expansion of $(x - 6y)^5$ is

$$_5C_2 x^{5-2}(-6y)^2 = 10 x^3 (36y^2) = 360 x^3 y^2.$$

42. The 4th term in the expansion of $(x - 10z)^7$ is

$$_7C_3 x^{7-3}(-10z)^3 = 35 \cdot x^4(-1000z^3) = -35,000 x^4 z^3.$$

43. The 8th term in the expansion of $(4x + 3y)^9$ is

$$_9C_7(4x)^{9-7}(3y)^7 = 36(16x^2)(2187y^7)$$
$$= 1,259,712 x^2 y^7.$$

44. The 5th term in the expansion of $(5a + 6b)^5$ is

$$_5C_4(5a)^{5-4}(6b)^4 = 5 \cdot (5a)(1296b^4) = 32,400 ab^4.$$

45. The 9th term in the expansion of $(10x - 3y)^{12}$ is

$$_{12}C_8(10x)^{12-8}(-3y)^8 = 495(10,000x^4)(6561y^8)$$
$$= 32,476,950,000 x^4 y^8.$$

46. The 7th term in the expansion of $(7x + 2y)^{15}$ is

$$_{15}C_6(7x)^{15-6}(2y)^6 = 5005 \cdot (40,353,607x^9)(64y^6)$$
$$\approx 1.293 \times 10^{13} x^9 y^6.$$

47. The term involving x^5 in the expansion of $(x + 3)^{12}$ is

$$_{12}C_7 x^5(3)^7 = \frac{12!}{7!5!} \cdot 3^7 x^5 = 1,732,104 x^5.$$

The coefficient is 1,732,104.

48. The term involving x^8 in the expansion of $(x^2 + 3)^{12}$ is

$$_{12}C_8(x^2)^4(3)^8 = \frac{12!}{(12 - 8)!8!} \cdot 3^8 x^8 = 3,247,695 x^8.$$

The coefficient is 3,247,695.

49. The term involving $x^8 y^2$ in the expansion of $(x - 2y)^{10}$ is

$$_{10}C_2 x^8(-2y)^2 = \frac{10!}{2!8!} \cdot 4 x^8 y^2 = 180 x^8 y^2.$$

The coefficient is 180.

50. The term involving $x^2 y^8$ in the expansion of $(4x - y)^{10}$ is

$$_{10}C_8(4x)^2(-y)^8 = \frac{10!}{(10 - 8)!8!} \cdot 16 x^2 y^8 = 720 x^2 y^8.$$

The coefficient is 720.

51. The term involving $x^4 y^5$ in the expansion of $(3x - 2y)^9$ is

$$_9C_5(3x)^4(-2y)^5 = \frac{9!}{5!4!}(81x^4)(-32y^5) = -326,592 x^4 y^5.$$

The coefficient is $-326,592$.

52. The term involving $x^6 y^2$ in the expansion of $(2x - 3y)^8$ is

$$_8C_2(2x)^6(-3y)^2 = \frac{8!}{(8 - 2)!2!}(64x^6)(9y^2) = 16,128 x^6 y^2.$$

The coefficient is 16,128.

53. The term involving $x^8 y^6 = (x^2)^4 y^6$ in the expansion of

$(x^2 + y)^{10}$ is $_{10}C_6(x^2)^4 y^6 = \frac{10!}{4!6!}(x^2)^4 y^6 = 210 x^8 y^6.$

The coefficient is 210.

54. The term involving $z^4 t^8$ in the expansion of $(z^2 - t)^{10}$ is

$$_{10}C_8(z^2)^2(-t)^8 = \frac{10!}{(10 - 8)!8!} z^4 t^8 = 45 z^4 t^8.$$

The coefficient is 45.

55. $\left(\sqrt{x} + 3\right)^4 = \left(\sqrt{x}\right)^4 + 4\left(\sqrt{x}\right)^3(3) + 6\left(\sqrt{x}\right)^2(3)^2 + 4\left(\sqrt{x}\right)(3)^3 + (3)^4$

$$= x^2 + 12x\sqrt{x} + 54x + 108\sqrt{x} + 81$$
$$= x^2 + 12x^{3/2} + 54x + 108x^{1/2} + 81$$

56. $\left(2\sqrt{t} - 1\right)^3 = \left(2\sqrt{t}\right)^3 + 3\left(2\sqrt{t}\right)^2(-1) + 3\left(2\sqrt{t}\right)(-1)^2 + (-1)^3$

$$= 8t^{3/2} - 12t + 6t^{1/2} - 1$$

57. $(x^{2/3} - y^{1/3})^3 = (x^{2/3})^3 - 3(x^{2/3})^2(y^{1/3}) + 3(x^{2/3})(y^{1/3})^2 - (y^{1/3})^3$

$$= x^2 - 3x^{4/3}y^{1/3} + 3x^{2/3}y^{2/3} - y$$

58. $(u^{3/5} + 2)^5 = (u^{3/5})^5 + 5(u^{3/5})^4(2) + 10(u^{3/5})^3(2)^2 + 10(u^{3/5})^2(2)^3 + 5(u^{3/5})(2)^4 + 2^5$

$\qquad = u^3 + 10u^{12/5} + 40u^{9/5} + 80u^{6/5} + 80u^{3/5} + 32$

59. $\dfrac{f(x + h) - f(x)}{h} = \dfrac{(x + h)^3 - x^3}{h}$

$\qquad = \dfrac{x^3 + 3x^2h + 3xh^2 + h^3 - x^3}{h}$

$\qquad = \dfrac{h(3x^2 + 3xh + h^2)}{h}$

$\qquad = 3x^2 + 3xh + h^2, h \neq 0$

60. $\dfrac{f(x + h) - f(x)}{h} = \dfrac{(x + h)^4 - x^4}{h}$

$\qquad = \dfrac{x^4 + 4x^3h + 6x^2h^2 + 4xh^3 + h^4 - x^4}{h}$

$\qquad = \dfrac{h(4x^3 + 6x^2h + 4xh^2 + h^3)}{h}$

$\qquad = 4x^3 + 6x^2h + 4xh^2 + h^3, h \neq 0$

61. $\dfrac{f(x + h) - f(x)}{h} = \dfrac{\sqrt{x + h} - \sqrt{x}}{h}$

$\qquad = \dfrac{\sqrt{x + h} - \sqrt{x}}{h} \cdot \dfrac{\sqrt{x + h} + \sqrt{x}}{\sqrt{x + h} + \sqrt{x}}$

$\qquad = \dfrac{(x + h) - x}{h(\sqrt{x + h} + \sqrt{x})}$

$\qquad = \dfrac{1}{\sqrt{x + h} + \sqrt{x}}, h \neq 0$

62. $\dfrac{f(x + h) - f(x)}{h} = \dfrac{\dfrac{1}{x + h} - \dfrac{1}{x}}{h}$

$\qquad = \dfrac{\dfrac{x - (x + h)}{x(x + h)}}{h}$

$\qquad = \dfrac{\dfrac{-h}{x(x + h)}}{h}$

$\qquad = -\dfrac{1}{x(x + h)}, h \neq 0$

63. $(1 + i)^4 = {}_4C_0(1)^4 + {}_4C_1(1)^3i + {}_4C_2(1)^2i^2 + {}_4C_3(1)i^3 + {}_4C_4i^4$

$\qquad = 1 + 4i - 6 - 4i + 1$

$\qquad = -4$

64. $(2 - i)^5 = {}_5C_0 2^5 - {}_5C_1 2^4i + {}_5C_2 2^3i^2 - {}_5C_3 2^2i^3 + {}_5C_4 2i^4 - {}_5C_5 i^5$

$\qquad = 32 - 80i - 80 + 40i + 10 - i$

$\qquad = -38 - 41i$

65. $(2 - 3i)^6 = {}_6C_0 2^6 - {}_6C_1 2^5(3i) + {}_6C_2 2^4(3i)^2 - {}_6C_3 2^3(3i)^3 + {}_6C_4 2^2(3i)^4 - {}_6C_5 2(3i)^5 + {}_6C_6(3i)^6$

$\qquad = (1)(64) - (6)(32)(3i) + 15(16)(-9) - 20(8)(-27i) + 15(4)(81) - 6(2)(243i) + (1)(-729)$

$\qquad = 64 - 576i - 2160 + 4320i + 4860 - 2916i - 729$

$\qquad = 2035 + 828i$

66. $\left(5 + \sqrt{-9}\right)^3 = (5 + 3i)^3$

$\qquad = 5^3 + 3 \cdot 5^2(3i) + 3 \cdot 5(3i)^2 + (3i)^3$

$\qquad = 125 + 225i - 135 - 27i$

$\qquad = -10 + 198i$

67. $\left(-\dfrac{1}{2} + \dfrac{\sqrt{3}}{2}i\right)^3 = \dfrac{1}{8}\left[(-1)^3 + 3(-1)^2(\sqrt{3}i) + 3(-1)(\sqrt{3}i)^2 + (\sqrt{3}i)^3\right]$

$\qquad\qquad\qquad = \dfrac{1}{8}\left[-1 + 3\sqrt{3}i + 9 - 3\sqrt{3}i\right]$

$\qquad\qquad\qquad = 1$

68. $\left(5 - \sqrt{3}i\right)^4 = 5^4 - 4 \cdot 5^3\left(\sqrt{3}i\right) + 6 \cdot 5^2\left(\sqrt{3}i\right)^2 - 4 \cdot 5\left(\sqrt{3}i\right)^3 + \left(\sqrt{3}i\right)^4$

$\qquad\qquad\quad = 625 - 500\sqrt{3}i - 450 + 60\sqrt{3}i + 9$

$\qquad\qquad\quad = 184 - 440\sqrt{3}i$

69. $(1.02)^8 = (1 + 0.02)^8$

$\qquad = 1 + 8(0.02) + 28(0.02)^2 + 56(0.02)^3 + 70(0.02)^4 + 56(0.02)^5 + 28(0.02)^6 + 8(0.02)^7 + (0.02)^8$

$\qquad = 1 + 0.16 + 0.0112 + 0.000448 + \cdots \approx 1.172$

70. $(2.005)^{10} = (2 + 0.005)^{10} = 2^{10} + 10(2)^9(0.005) + 45(2)^8(0.005)^2 + 120(2)^7(0.005)^3 + 210(2)^6(0.005)^4$

$\qquad\qquad\qquad\qquad\quad + 252(2)^5(0.005)^5 + 210(2)^4(0.005)^6 + 120(2)^3(0.005)^7 + 45(2)^2(0.005)^8$

$\qquad\qquad\qquad\qquad\quad + 10(2)(0.005)^9 + (0.005)^{10}$

$\qquad\qquad\qquad\qquad = 1024 + 25.6 + 0.288 + 0.00192 + 0.0000084 + \cdots$

$\qquad\qquad\qquad\qquad \approx 1049.890$

71. $(2.99)^{12} = (3 - 0.01)^{12}$

$\qquad = 3^{12} - 12(3)^{11}(0.01) + 66(3)^{10}(0.01)^2 - 220(3)^9(0.01)^3 + 495(3)^8(0.01)^4$

$\qquad\quad - 792(3)^7(0.01)^5 + 924(3)^6(0.01)^6 - 792(3)^5(0.01)^7 + 495(3)^4(0.01)^8$

$\qquad\quad - 220(3)^3(0.01)^9 + 66(3)^2(0.01)^{10} - 12(3)(0.01)^{11} + (0.01)^{12}$

$\qquad \approx 531{,}441 - 21{,}257.64 + 389.7234 - 4.3303 + 0.0325 - 0.0002 + \cdots \approx 510{,}568.785$

72. $(1.98)^9 = (2 - 0.02)^9 = 2^9 - 9(2)^8(0.02) + 36(2)^7(0.02)^2 - 84(2)^6(0.02)^3 + 126(2)^5(0.02)^4$

$\qquad\qquad\qquad\qquad\quad - 126(2)^4(0.02)^5 + 84(2)^3(0.02)^6 - 36(2)^2(0.02)^7 + 9(2)(0.02)^8 - (0.02)^9$

$\qquad\qquad\qquad\quad = 512 - 46.08 + 1.8432 - 0.043008 + 0.00064512 + 0.0000064512 + \cdots$

$\qquad\qquad\qquad\quad \approx 467.721$

73. $f(x) = x^3 - 4x$

$\quad g(x) = f(x + 4)$

$\qquad\quad = (x + 4)^3 - 4(x + 4)$

$\qquad\quad = x^3 + 3x^2(4) + 3x(4)^2 + (4)^3 - 4x - 16$

$\qquad\quad = x^3 + 12x^2 + 48x + 64 - 4x - 16$

$\qquad\quad = x^3 + 12x^2 + 44x + 48$

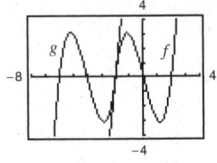

The graph of g is the same as the graph of f shifted four units to the left.

74. $f(x) = -x^4 + 4x^2 - 1$, $g(x) = f(x - 3)$

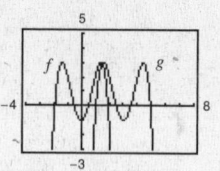

$$g(x) = f(x - 3)$$
$$= -(x - 3)^4 + 4(x - 3)^2 - 1$$
$$= -(x^4 + 4x^3(-3) + 6x^2(-3)^2 + 4x(-3)^3 + (-3)^4) + 4(x^2 - 6x + 9) - 1$$
$$= -x^4 + 12x^3 - 54x^2 + 108x - 81 + 4x^2 - 24x + 36 - 1$$
$$= -x^4 + 12x^3 - 50x^2 + 84x - 46$$

The graph of g is the same as the graph of f shifted three units to the right.

75. $_7C_4\left(\frac{1}{2}\right)^4\left(\frac{1}{2}\right)^3 = \frac{7!}{3!4!}\left(\frac{1}{16}\right)\left(\frac{1}{8}\right) = 35\left(\frac{1}{16}\right)\left(\frac{1}{8}\right) \approx 0.273$

76. $_{10}C_3\left(\frac{1}{4}\right)^3\left(\frac{3}{4}\right)^7 = 120\left(\frac{1}{64}\right)\left(\frac{2187}{16,384}\right) \approx 0.2503$

77. $_8C_4\left(\frac{1}{3}\right)^4\left(\frac{2}{3}\right)^4 = \frac{8!}{4!4!}\left(\frac{1}{81}\right)\left(\frac{16}{81}\right) = 70\left(\frac{1}{81}\right)\left(\frac{16}{81}\right) \approx 0.171$

78. $_8C_4\left(\frac{1}{2}\right)^4\left(\frac{1}{2}\right)^4 = 70\left(\frac{1}{16}\right)\left(\frac{1}{16}\right) \approx 0.273$

79. (a) $f(t) \approx 0.0025t^3 - 0.015t^2 + 0.88t + 7.7$

(b)
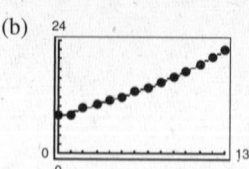

(c) $g(t) = f(t + 10) = 0.0025(t + 10)^3 - 0.015(t + 10)^2$
$$+ 0.88(t + 10) + 7.7$$
$$= 0.0025(t^3 + 30t^2 + 300t + 1000)$$
$$- 0.015(t^2 + 20t + 100) + 0.88(t + 10) + 7.7$$
$$= 0.0025t^3 + 0.06t^2 + 1.33t + 17.5$$

(d)
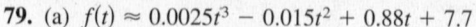

(e) For 2008 use $t = 18$ in $f(t)$ and $t = 8$ in $g(t)$.

$f(18) = 33.26$ gallons

$g(8) = 33.26$ gallons

Both models yield the same answer.

(f) The trend is for the per capita consumption of bottled water to increase. This may be due to the increasing concern with contaminants in tap water.

80. $f(t) = 0.031t^2 + 0.82t + 6.1$

(a) $g(t) = f(t + 10)$
$$= 0.031(t + 10)^2 + 0.82(t + 10) + 6.1$$
$$= 0.031(t^2 + 20t + 100) + 0.82(t + 10) + 6.1$$
$$= 0.031t^2 + 1.44t + 17.4$$

(b)
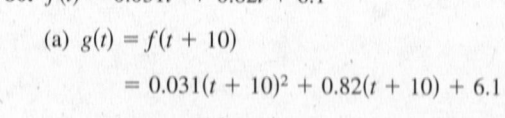

(c) $f(t)$: $f(17) = 2007$

$g(t)$: $g(7) = 2007$

81. True. The coefficients from the Binomial Theorem can be used to find the numbers in Pascal's Triangle.

82. False. Expanding binomials that represent differences is just as accurate as expanding binomials that represent sums, but for differences the coefficient signs are alternating.

83. False.

The coefficient of the x^{10}-term is $_{12}C_7(3)^7 = 1,732,104$.

The coefficient of the x^{14}-term is $_{12}C_5(3)^5 = 192,456$.

84. The first and last numbers in each row are 1. Every other number in each row is formed by adding the two numbers immediately above the number.

85.

```
                        1
                     1     1
                  1     2     1
               1     3     3     1
            1     4     6     4     1
         1     5    10    10     5     1
      1     6    15    20    15     6     1
   1     7    21    35    35    21     7     1
 1     8    28    56    70    56    28     8     1
1     9    36    84   126   126    84    36     9     1
1    10    45   120   210   252   210   120    45    10     1
```

86. $(n + 1)$ terms

87. The signs of the terms in the expansion of $(x - y)^n$ alternate from positive to negative.

88. The functions $f(x) = (1 - x)^3$ and $k(x) = 1 - 3x + 3x^2 + x^3$ (choices (a) and (d)) have identical graphs, because $k(x)$ is the expansion of $f(x)$.

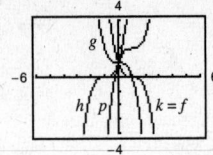

89.
$$_nC_{n-r} = \frac{n!}{(n - (n - r))!(n - r)!}$$
$$= \frac{n!}{r!(n - r)!}$$
$$= \frac{n!}{(n - r)!r!}$$
$$= {_nC_r}$$

90. $0 = (1 - 1)^n = {_nC_0} - {_nC_1} + {_nC_2} - {_nC_3} + \cdots \pm {_nC_n}$

91.
$$_nC_r + {_nC_{r-1}} = \frac{n!}{(n - r)!r!} + \frac{n!}{(n - r + 1)!(r - 1)!}$$
$$= \frac{n!(n - r + 1)!(r - 1)! + n!(n - r)!r!}{(n - r)!r!(n - r + 1)!(r - 1)!}$$
$$= \frac{n![(n - r + 1)!(r - 1)! + r!(n - r)!]}{(n - r)!r!(n - r + 1)!(r - 1)!}$$
$$= \frac{n!(r - 1)![(n - r + 1)! + r(n - r)!]}{(n - r)!r!(n - r + 1)!(r - 1)!}$$
$$= \frac{n!(n - r)![(n - r + 1) + r]}{(n - r)!r!(n - r + 1)!}$$
$$= \frac{n![n + 1]}{r!(n - r + 1)!}$$
$$= \frac{(n + 1)!}{[(n + 1) - r]!r!}$$
$$= {_{n+1}C_r}$$

92. ${_nC_0} + {_nC_1} + {_nC_2} + {_nC_3} + \cdots + {_nC_n} = (1 + 1)^n = 2^n$

93. The graph of $f(x) = x^2$ is shifted three units to the right. Thus, $g(x) = (x - 3)^2$.

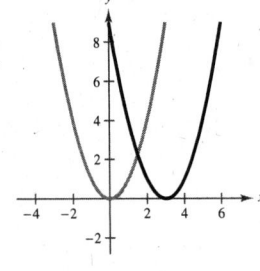

94. The graph of $f(x) = x^2$ has been reflected in the x–axis, shifted two units to the left, and shifted three units upward. Thus, $g(x) = -(x + 2)^2 + 3$.

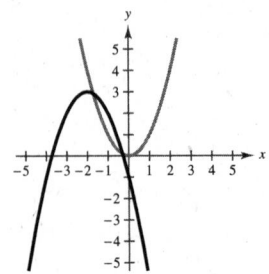

95. The graph of $f(x) = \sqrt{x}$ is shifted two units to the left and shifted one unit upward. Thus, $g(x) = \sqrt{x+2} + 1$.

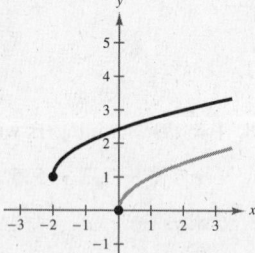

96. The graph of $f(x) = \sqrt{x}$ has been reflected in the *x*–axis, shifted one unit to the left, and shifted two units downward. Thus, $g(x) = -\sqrt{x+1} - 2$.

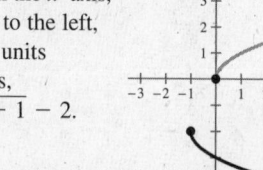

97. $A^{-1} = \dfrac{1}{(-6)(4) - (5)(-5)} \begin{bmatrix} 4 & -5 \\ 5 & -6 \end{bmatrix} = \begin{bmatrix} 4 & -5 \\ 5 & -6 \end{bmatrix}$

98. $[A \ \vdots \ I] = \begin{bmatrix} 1.2 & -2.3 & \vdots & 1 & 0 \\ -2 & 4 & \vdots & 0 & 1 \end{bmatrix}$

$0.1R_2 + R_1 \rightarrow \begin{bmatrix} 1 & -1.9 & \vdots & 1 & 0.1 \\ -2 & 4 & \vdots & 0 & 1 \end{bmatrix}$

$2R_1 + R_2 \rightarrow \begin{bmatrix} 1 & -1.9 & \vdots & 1 & 0.1 \\ 0 & 0.2 & \vdots & 2 & 1.2 \end{bmatrix}$

$5R_2 \rightarrow \begin{bmatrix} 1 & -1.9 & \vdots & 1 & 0.1 \\ 0 & 1 & \vdots & 10 & 6 \end{bmatrix}$

$1.9R_2 + R_1 \rightarrow \begin{bmatrix} 1 & 0 & \vdots & 20 & 11.5 \\ 0 & 1 & \vdots & 10 & 6 \end{bmatrix} = [I \ \vdots \ A^{-1}]$

$A^{-1} = \begin{bmatrix} 20 & 11.5 \\ 10 & 6 \end{bmatrix}$

Section 9.6   Counting Principles

- ■ You should know The Fundamental Counting Principle.

- ■ $_nP_r = \dfrac{n!}{(n-r)!}$ is the number of permutations of *n* elements taken *r* at a time.

- ■ Given a set of *n* objects that has n_1 of one kind, n_2 of a second kind, and so on, the number of distinguishable permutations is

$$\frac{n!}{n_1! n_2! \ldots n_k!}.$$

- ■ $_nC_r = \dfrac{n!}{(n-r)!r!}$ is the number of combinations of *n* elements taken *r* at a time.

Vocabulary Check

1. Fundamental Counting Principle

2. permutation

3. $_nP_r = \dfrac{n!}{(n-r)!}$

4. distinguishable permutations

5. combinations

1. Odd integers: 1, 3, 5, 7, 9, 11

6 ways

2. Even integers: 2, 4, 6, 8, 10, 12

6 ways

3. Prime integers: 2, 3, 5, 7, 11

5 ways

4. Greater than 9: 10, 11, 12

3 ways

5. Divisible by 4: 4, 8, 12

3 ways

6. Divisible by 3: 3, 6, 9, 12

4 ways

7. Sum is 9: $1 + 8, 2 + 7, 3 + 6, 4 + 5, 5 + 4,$

$6 + 3, 7 + 2, 8 + 1$

8 ways

8. Two *distinct* integers whose sum is 8:

$1 + 7, 2 + 6, 3 + 5, 5 + 3, 6 + 2, 7 + 1$

6 ways

9. Amplifiers: 3 choices

Compact disc players: 2 choices

Speakers: 5 choices

Total: $3 \cdot 2 \cdot 5 = 30$ ways

10. Chemist: 5 choices

Statistician: 3 choices

Total: $5 \cdot 3 = 15$ ways

11. Math courses: 2

Science courses: 3

Social sciences and
humanities courses: 5

Total: $2 \cdot 3 \cdot 5 = 30$ schedules

12. 1st position: 2

2nd position: 1

3rd position: 6

4th position: 5

5th position: 4

6th position: 3

7th position: 2

8th position: 1

Total: $2!6! = 1440$ ways

13. $2^6 = 64$

14. $2^{12} = 4096$ ways

15. $26 \cdot 26 \cdot 26 \cdot 10 \cdot 10 \cdot 10 \cdot 10 = 175,760,000$
distinct license plate numbers

16. $24 \cdot 24 \cdot 10 \cdot 10 \cdot 10 \cdot 10 = 5,760,000$
distinct license plates

17. (a) $9 \cdot 10 \cdot 10 = 900$

(b) $9 \cdot 9 \cdot 8 = 648$

(c) $9 \cdot 10 \cdot 2 = 180$

(d) $6 \cdot 10 \cdot 10 = 600$

18. (a) $9 \cdot 10 \cdot 10 \cdot 10 = 9000$ numbers

(b) $9 \cdot 9 \cdot 8 \cdot 7 = 4536$ numbers

(c) $4 \cdot 10 \cdot 10 \cdot 10 = 4000$ numbers

(d) $9 \cdot 10 \cdot 10 \cdot 5 = 4500$ numbers

19. $40^3 = 64,000$

20. $50^3 = 125,000$ combinations

21. (a) $8 \cdot 7 \cdot 6 \cdot 5 \cdot 4 \cdot 3 \cdot 2 \cdot 1 = 40,320$

(b) $8 \cdot 1 \cdot 6 \cdot 1 \cdot 4 \cdot 1 \cdot 2 \cdot 1 = 384$

22. (a) $8! = 40,320$ orders

(b) $4!4! = 576$ orders

23. $_nP_r = \dfrac{n!}{(n-r)!}$

So, $_4P_4 = \dfrac{4!}{0!} = 4! = 24.$

24. $_nP_r = \dfrac{n!}{(n-r)!}$

$_5P_5 = \dfrac{5!}{(5-5)!} = \dfrac{5!}{0!} = 120$

25. $_8P_3 = \dfrac{8!}{5!} = 8 \cdot 7 \cdot 6 = 336$

26. $_{20}P_2 = \dfrac{20!}{18!} = 20 \cdot 19 = 380$

27. $_5P_4 = \dfrac{5!}{1!} = 120$

28. $_7P_4 = \dfrac{7!}{3!} = 7 \cdot 6 \cdot 5 \cdot 4 = 840$

29. $14 \cdot {}_nP_3 = {}_{n+2}P_4$ **Note:** $n \geq 3$ for this to be defined.

$$14\left(\frac{n!}{(n-3)!}\right) = \frac{(n+2)!}{(n-2)!}$$

$14n(n-1)(n-2) = (n+2)(n+1)n(n-1)$ (We can divide here by $n(n-1)$ since $n \neq 0, n \neq 1$.)

$$14(n-2) = (n+2)(n+1)$$
$$14n - 28 = n^2 + 3n + 2$$
$$0 = n^2 - 11n + 30$$
$$0 = (n-5)(n-6)$$
$$n = 5 \text{ or } n = 6$$

30. ${}_nP_5 = 18 \cdot {}_{n-2}P_4$ **Note:** $n \geq 6$ for this to be defined.

$$\frac{n!}{(n-5)!} = 18\left(\frac{(n-2)!}{(n-6)!}\right)$$

$n(n-1)(n-2)(n-3)(n-4) = 18(n-2)(n-3)(n-4)(n-5)$ $\left(\begin{array}{l}\text{We can divide by } (n-2), (n-3), \\ (n-4) \text{ since } n \neq 2, n \neq 3, \text{ and } n \neq 4.\end{array}\right)$

$$n^2 - n = 18n - 90$$
$$n^2 - 19n + 90 = 0$$
$$(n-9)(n-10) = 0$$
$$n = 9 \text{ or } n = 10$$

31. ${}_{20}P_5 = 1,860,480$

32. ${}_{100}P_5 = 9,034,502,400$

33. ${}_{100}P_3 = 970,200$

34. ${}_{10}P_8 = 1,814,400$

35. ${}_{20}C_5 = 15,504$

36. ${}_{10}C_7 = 120$

37. $5! = 120$ ways

38. $6! = 720$ ways

39. ${}_{12}P_4 = \dfrac{12!}{8!} = 12 \cdot 11 \cdot 10 \cdot 9 = 11,880$ ways

40. $4! = 24$ orders

41. $\dfrac{7!}{2!1!3!1!} = \dfrac{7!}{2!3!} = 420$

42. $\dfrac{8!}{3!5!} = 56$

43. $\dfrac{7!}{2!1!1!1!1!1!} = \dfrac{7!}{2!} = 7 \cdot 6 \cdot 5 \cdot 4 \cdot 3 = 2520$

44. $\dfrac{11!}{1!4!4!2!} = \dfrac{11!}{4!4!2!} = 34,650$

45.

ABCD	BACD	CABD	DABC
ABDC	BADC	CADB	DACB
ACBD	BCAD	CBAD	DBAC
ACDB	BCDA	CBDA	DBCA
ADBC	BDAC	CDAB	DCAB
ADCB	BDCA	CDBA	DCBA

46.

ABCD

ACBD

DBCA

DCBA

47. ${}_{15}P_9 = \dfrac{15!}{6!} = 1,816,214,400$

different batting orders

48. ${}_6P_3 = \dfrac{6!}{3!} = 120$

49. ${}_{40}C_{12} = \dfrac{40!}{28!12!} = 5,586,853,480$ ways

50. $_{100}C_{14} = \dfrac{100!}{(100 - 14)!14!}$

$\qquad\quad = \dfrac{100!}{86!14!}$

$\qquad\quad = 4.42 \times 10^{16}$

51. $_6C_2 = 15$

The 15 ways are listed below.

AB, AC, AD, AE, AF, BC, BD, BE,

BF, CD, CE, CF, DE, DF, EF

52. $_{20}C_5 = 15{,}504$ groups

53. $_{35}C_5 = \dfrac{35!}{30!5!} = 324{,}632$ ways

54. $_{40}C_6 = 3{,}838{,}380$ ways

55. There are 7 good units and 3 defective units.

(a) $_7C_4 = \dfrac{7!}{3!4!} = 35$ ways

(b) $_7C_2 \cdot {}_3C_2 = \dfrac{7!}{5!2!} \cdot \dfrac{3!}{1!2!} = 21 \cdot 3 = 63$ ways

(c) $_7C_4 + {}_7C_3 \cdot {}_3C_1 + {}_7C_2 \cdot {}_3C_2 = \dfrac{7!}{3!4!} + \dfrac{7!}{4!3!} \cdot \dfrac{3!}{2!1!} + \dfrac{7!}{5!2!} \cdot \dfrac{3!}{1!2!}$

$\qquad\qquad\qquad\qquad\qquad\quad = 35 + 35 \cdot 3 + 21 \cdot 3$

$\qquad\qquad\qquad\qquad\qquad\quad = 203$ ways

56. (a) $_3C_2 = \dfrac{3!}{2!1!} = 3$ relationships

(b) $_8C_2 = \dfrac{8!}{2!6!} = \dfrac{8 \cdot 7}{2} = 28$ relationships

(c) $_{12}C_2 = \dfrac{12!}{2!10!} = \dfrac{12 \cdot 11}{2} = 66$ relationships

(d) $_{20}C_2 = \dfrac{20!}{2!18!} = \dfrac{20 \cdot 19}{2} = 190$ relationships

57. (a) Select type of card for three of a kind: $_{13}C_1$

Select three of four cards for three of a kind: $_4C_3$

Select type of card for pair: $_{12}C_1$

Select two of four cards for pair: $_4C_2$

$\quad _{13}C_1 \cdot {}_4C_3 \cdot {}_{12}C_1 \cdot {}_4C_2 = \dfrac{13!}{(13 - 1)!1!} \cdot \dfrac{4!}{(4 - 3)!3!} \cdot \dfrac{12!}{(12 - 1)!1!} \cdot \dfrac{4!}{(4 - 2)!2!} = 3744$

(b) Select two jacks: $_4C_2$

Select three aces: $_4C_3$

$\quad _4C_2 \cdot {}_4C_3 = \dfrac{4!}{(4 - 2)!2!} \cdot \dfrac{4!}{(4 - 3)!3!} = 24$

58. (a) $_8C_4 = \dfrac{8!}{(8 - 4)!4!} = \dfrac{8!}{4!4!} = \dfrac{8 \cdot 7 \cdot 6 \cdot 5}{4 \cdot 3 \cdot 2} = 70$ ways

(b) $_3C_2 \cdot {}_5C_2 = \dfrac{3!}{(3 - 2)!2!} \cdot \dfrac{5!}{(5 - 2)!2!} = 3 \cdot 10 = 30$ ways

59. $_7C_1 \cdot {}_{12}C_3 \cdot {}_{20}C_2 = \dfrac{7!}{(7 - 1)!1!} \cdot \dfrac{12!}{(12 - 3)!3!} \cdot \dfrac{20!}{(20 - 2)!2!} = 292{,}600$

60. (a) $(195)(99)(89)(105)(74) \approx 1.335 \times 10^{10}$ different faces

(b) $(89)(105)(74) = 691{,}530$ different faces

61. $_5C_2 - 5 = 10 - 5 = 5$ diagonals

62. $_6C_2 - 6 = 15 - 6 = 9$ diagonals

63. $_8C_2 - 8 = 28 - 8 = 20$ diagonals

64. $_{10}C_2 - 10 = 45 - 10 = 35$ diagonals

65. (a) $_{53}C_5 \cdot (42) = 120{,}526{,}770$

    (b) 1. If the jackpot is won, then there is only one winning number.

    (c) There are $22{,}957{,}480$ possible winning numbers in the state lottery, which is less than the possible number of winning Powerball numbers.

66. (a) Permutation because order matters

    (b) Combination because order does not matter

    (c) Permutation because order matters

    (d) Combination because order does not matter

67. False.
It is an example of a combination.

68. True by the definition of the Fundamental Counting Principle

69. $_nC_r = {_nC_{n-r}}$ They are the same.

70. $_{10}P_6 > {_{10}C_6}$
Changing the order of any of the six elements selected results in a different permutation but the same combination.

71. $_nP_{n-1} = \dfrac{n!}{(n-(n-1))!} = \dfrac{n!}{1!} = \dfrac{n!}{0!} = {_nP_n}$

72. $_nC_n = \dfrac{n!}{(n-n)!n!} = \dfrac{n!}{0!n!} = \dfrac{n!}{n!0!} = \dfrac{n!}{(n-0)!0!} = {_nC_0}$

73. $_nC_{n-1} = \dfrac{n!}{(n-(n-1))!(n-1)!} = \dfrac{n!}{(1)!(n-1)!}$

$\qquad = \dfrac{n!}{(n-1)!1!} = {_nC_1}$

74. $_nC_r = \dfrac{n!}{(n-r)!r!}$

$\qquad = \dfrac{n(n-1)(n-2)\cdots(n-r+1)(n-r)!}{(n-r)!r!}$

$\qquad = \dfrac{n(n-1)(n-2)\cdots(n-r+1)}{r!}$

$\qquad = \dfrac{_nP_r}{r!}$

75. $_{100}P_{80} \approx 3.836 \times 10^{139}$

This number is too large for some calculators to evaluate.

76. The symbol $_nP_r$ denotes the number of ways to choose and order r elements out of a collection of n elements.

77. $f(x) = 3x^2 + 8$

    (a) $f(3) = 3(3)^2 + 8 = 35$

    (b) $f(0) = 3(0)^2 + 8 = 8$

    (c) $f(-5) = 3(-5)^2 + 8 = 83$

78. $g(x) = \sqrt{x-3} + 2$

    (a) $g(3) = \sqrt{3-3} + 2 = 2$

    (b) $g(7) = \sqrt{7-3} + 2 = 4$

    (c) $g(x+1) = \sqrt{x+1-3} + 2 = \sqrt{x-2} + 2$

79. $f(x) = -|x-5| + 6$

    (a) $f(-5) = -|-5-5| + 6 = -10 + 6 = -4$

    (b) $f(-1) = -|-1-5| + 6 = -6 + 6 = 0$

    (c) $f(11) = -|11-5| + 6 = -6 + 6 = 0$

80. $f(x) = \begin{cases} x^2 - 2x + 5, & x \le -4 \\ -x^2 - 2, & x > -4 \end{cases}$

    (a) $f(-4) = (-4)^2 - 2(-4) + 5 = 29$

    (b) $f(-1) = -(-1)^2 - 2 = -3$

    (c) $f(-20) = (-20)^2 - 2(-20) + 5 = 445$

81. $\sqrt{x - 3} = x - 6$

$\left(\sqrt{x - 3}\right)^2 = (x - 6)^2$

$x - 3 = x^2 - 12x + 36$

$0 = x^2 - 13x + 39$

By the Quadratic Formula we have: $x = \dfrac{13 \pm \sqrt{13}}{2}$

$x = \dfrac{13 - \sqrt{13}}{2}$ is extraneous.

The only valid solution is $x = \dfrac{13 + \sqrt{13}}{2} \approx 8.30$.

82. $\dfrac{4}{t} + \dfrac{3}{2t} = 1$

$\dfrac{4}{t}(2t) + \dfrac{3}{2t}(2t) = 1(2t)$

$8 + 3 = 2t$

$5.5 = t$

83. $\log_2(x - 3) = 5$

$x - 3 = 2^5$

$x - 3 = 32$

$x = 35$

84. $e^{x/3} = 16$

$\dfrac{x}{3} = \ln 16$

$x = 3 \ln 16 \approx 8.32$

Section 9.7   Probability

You should know the following basic principles of probability.

■ If an event E has $n(E)$ equally likely outcomes and its sample space has $n(S)$ equally likely outcomes, then the probability of event E is

$$P(E) = \frac{n(E)}{n(S)}, \text{ where } 0 \le P(E) \le 1.$$

■ If A and B are mutually exclusive events, then $P(A \cup B) = P(A) + P(B)$.

If A and B are not mutually exclusive events, then $P(A \cup B) = P(A) + P(B) - P(A \cap B)$.

■ If A and B are independent events, then the probability that both A and B will occur is $P(A)P(B)$.

■ The complement of an event A is denoted by A' and its probability is $P(A') = 1 - P(A)$.

Vocabulary Check

1. experiment; outcomes

2. sample space

3. probability

4. impossible; certain

5. mutually exclusive

6. independent

7. complement

8. (a) iii   (b) i   (c) iv   (d) ii

---

1. $\{(H, 1), (H, 2), (H, 3), (H, 4), (H, 5), (H, 6),$
$(T, 1), (T, 2), (T, 3), (T, 4), (T, 5), (T, 6)\}$

2. $\{2, 3, 4, 5, 6, 7, 8, 9, 10, 11, 12\}$

3. $\{ABC, ACB, BAC, BCA, CAB, CBA\}$

4. $\{(\text{red, red}), (\text{red, blue}), (\text{red, yellow}), (\text{blue, blue}), (\text{blue, yellow})\}$

5. {AB, AC, AD, AE, BC, BD, BE, CD, CE, DE}

6. {SSS, SSF, SFS, FSS, SFF, FFS, FSF, FFF}

7. $E = \{HHT, HTH, THH\}$

$$P(E) = \frac{n(E)}{n(S)} = \frac{3}{8}$$

8. $E = \{HHH, HHT, HTH, HTT\}$

$$P(E) = \frac{n(E)}{n(S)} = \frac{4}{8} = \frac{1}{2}$$

9. $E = \{HHH, HHT, HTH, HTT, THH, THT, TTH\}$

$$P(E) = \frac{n(E)}{n(S)} = \frac{7}{8}$$

10. $E = \{HHH, HHT, HTH, THH\}$

$$P(E) = \frac{n(E)}{n(S)} = \frac{4}{8} = \frac{1}{2}$$

11. $E = \{K\clubsuit, K\blacklozenge, K\blacktriangledown, K\spadesuit, Q\clubsuit, Q\blacklozenge, Q\blacktriangledown, Q\spadesuit, J\clubsuit, J\blacklozenge, J\blacktriangledown, J\spadesuit\}$

$$P(E) = \frac{n(E)}{n(S)} = \frac{12}{52} = \frac{3}{13}$$

12. The probability that the card is *not* a face card is the complement of getting a face card. (See Exercise 11.)

$$P(E') = 1 - P(E) = 1 - \frac{3}{13} = \frac{10}{13}$$

13. $E = \{K\blacklozenge, K\blacktriangledown, Q\blacklozenge, Q\blacktriangledown, J\blacklozenge, J\blacktriangledown\}$

$$P(E) = \frac{n(E)}{n(S)} = \frac{6}{52} = \frac{3}{26}$$

14. There are six possible cards in each of 4 suits: $6 \cdot 4 = 24$

$$P(E) = \frac{n(E)}{n(S)} = \frac{24}{52} = \frac{6}{13}$$

15. $E = \{(1, 3), (2, 2), (3, 1)\}$

$$P(E) = \frac{n(E)}{n(S)} = \frac{3}{36} = \frac{1}{12}$$

16. $E = \{(1, 6), (2, 5), (2, 6), (3, 4), (3, 5), (3, 6), (4, 3),$
$(4, 4), (4, 5), (4, 6), (5, 2), (5, 3), (5, 4), (5, 6),$
$(6, 1), (6, 2), (6, 3), (6, 4), (6, 5), (6, 6)\}$

$$P(E) = \frac{n(E)}{n(S)} = \frac{21}{36} = \frac{7}{12}$$

17. Use the complement.

$$E' = \{(5, 6), (6, 5), (6, 6)\}$$

$$P(E') = \frac{n(E')}{n(S)} = \frac{3}{36} = \frac{1}{12}$$

$$P(E) = 1 - P(E') = 1 - \frac{1}{12} = \frac{11}{12}$$

18. $E = \{(1, 1), (1, 2), (2, 1), (6, 6)\}$

$$P(E) = \frac{n(E)}{n(S)} = \frac{4}{36} = \frac{1}{9}$$

19. $E_3 = \{(1, 2), (2, 1)\}, \; n(E_3) = 2$
$E_5 = \{(1, 4), (2, 3), (3, 2), (4, 1)\}, \; n(E_5) = 4$
$E_7 = \{(1, 6), (2, 5), (3, 4), (4, 3), (5, 2), (6, 1)\}, \; n(E_7) = 6$
$E = E_3 \cup E_5 \cup E_7$
$n(E) = 2 + 4 + 6 = 12$

$$P(E) = \frac{n(E)}{n(S)} = \frac{12}{36} = \frac{1}{3}$$

20. $E = \{(1, 1), (1, 2), (1, 4), (1, 6), (2, 1), (2, 3), (2, 5),$
$(3, 2), (3, 4), (3, 6), (4, 1), (4, 3), (4, 5), (5, 2), (5, 4),$
$(5, 6), (6, 1), (6, 3), (6, 5)\}$

$$P(E) = \frac{n(E)}{n(S)} = \frac{19}{36}$$

21. $P(E) = \frac{{}_3C_2}{{}_6C_2} = \frac{3}{15} = \frac{1}{5}$

22. $P(E) = \frac{{}_2C_2}{{}_6C_2} = \frac{1}{15}$

23. $P(E) = \frac{{}_4C_2}{{}_6C_2} = \frac{6}{15} = \frac{2}{5}$

24. $P(E) = \frac{{}_1C_1 \cdot {}_2C_1 + {}_1C_1 \cdot {}_3C_1 + {}_2C_1 \cdot {}_3C_1}{{}_6C_2}$

$$= \frac{2 + 3 + 6}{15} = \frac{11}{15}$$

25. $P(E') = 1 - P(E) = 1 - 0.7 = 0.3$

26. $P(E') = 1 - P(E) = 1 - 0.36 = 0.64$ **27.** $P(E') = 1 - P(E) = 1 - \frac{1}{4} = \frac{3}{4}$ **28.** $1 - P(E) = 1 - \frac{2}{3} = \frac{1}{3}$

29. $P(E) = 1 - P(E')$

$\quad\quad = 1 - 0.14 = 0.86$

30. $1 - P(E') = 1 - 0.92 = 0.08$ **31.** $P(E) = 1 - P(E') = 1 - \frac{17}{35} = \frac{18}{35}$

32. $1 - P(E') = 1 - \frac{61}{100} = \frac{39}{100}$

33. (a) $\frac{290}{500} = 0.58 = 58\%$

(b) $\frac{478}{500} = 0.956 = 95.6\%$

(c) $\frac{2}{500} = 0.004 = 0.4\%$

34. (a) $\frac{34}{100} = 0.34 = 34\%$

(b) $\frac{45}{100} = 0.45 = 45\%$

(c) $\frac{23}{100} = 0.23 = 23\%$

35. (a) $0.24(1011) \approx 243$ adults

(b) $2\% = \frac{1}{50}$

(c) $52\% + 12\% = 64\% = \frac{16}{25}$

36. (a) $59\% = \frac{59}{100}$

(b) $6\% + 11\% = 17\% = \frac{17}{100}$

(c) $1 - \frac{13}{100} = \frac{87}{100}$

37. (a) $\frac{672}{1254} = \frac{112}{209}$

(b) $\frac{582}{1254} = \frac{97}{209}$

(c) $\frac{672 - 124}{1254} = \frac{548}{1254} = \frac{274}{627}$

38. (a) $\frac{71 + 53}{202} = \frac{124}{202} = \frac{62}{101}$

(b) $1 - \frac{62}{101} = \frac{39}{101}$

(c) $\frac{24}{202} = \frac{12}{101}$

39. $p + p + 2p = 1$

$\quad\quad p = 0.25$

Taylor: $0.50 = \frac{1}{2}$

Moore: $0.25 = \frac{1}{4}$

Jenkins: $0.25 = \frac{1}{4}$

40. $1 - 0.37 - 0.44 = 0.19 = 19\%$

41. (a) $\frac{_{15}C_{10}}{_{20}C_{10}} = \frac{3003}{184,756} = \frac{21}{1292} \approx 0.016$

(b) $\frac{_{15}C_8 \cdot {_5}C_2}{_{20}C_{10}} = \frac{64,350}{184,756} = \frac{225}{646} \approx 0.348$

(c) $\frac{_{15}C_9 \cdot {_5}C_1}{_{20}C_{10}} + \frac{_{15}C_{10}}{_{20}C_{10}} = \frac{25,025 + 3003}{184,756} = \frac{28,028}{184,756} = \frac{49}{323} \approx 0.152$

42. Total ways to insert paychecks: $5! = 120$ ways

5 correct: 1 way

4 correct: not possible

3 correct: $_5C_3 = 10$ ways (because once you choose the three envelopes that will contain the correct paychecks, there is only one way to insert the paychecks so that the other two are wrong)

2 correct: $_5C_3 \cdot 2 = 20$ ways (because once you choose the two envelopes that will contain the correct paychecks, there are two ways to fill the next envelope incorrectly, then only one incorrect way to insert the remaining paychecks)

1 correct: $5 \cdot 3 \cdot 3 = 45$ ways (five ways to choose which envelope is paired with the correct paycheck, three ways to fill the next envelope incorrectly, then three ways to fill the envelope whose correct paycheck was placed in the second envelope, and only one way to fill the remaining two envelopes such that both are incorrect)

0 correct: $120 - 1 - 10 - 20 - 45 = 44$ ways

(a) $\frac{45}{120} = \frac{3}{8}$ (b) $\frac{45 + 20 + 10 + 1}{120} = \frac{19}{30}$

43. (a) $\dfrac{1}{{}_5P_5} = \dfrac{1}{120}$

(b) $\dfrac{1}{{}_4P_4} = \dfrac{1}{24}$

44. (a) $\dfrac{{}_8C_2 \cdot {}_{100}C_5}{{}_{108}C_7} = \dfrac{\dfrac{8!}{6!2!} \cdot \dfrac{100!}{95!5!}}{\dfrac{108!}{101!7!}} = 0.076$

(b) $\dfrac{{}_8C_2 \cdot {}_{25}C_2 \cdot {}_{25}C_3}{{}_{108}C_7} = \dfrac{\dfrac{8!}{6!2!} \cdot \dfrac{25!}{23!2!} \cdot \dfrac{25!}{22!3!}}{\dfrac{108!}{101!7!}} = 0.00069$

45. (a) $\dfrac{20}{52} = \dfrac{5}{13}$

(b) $\dfrac{26}{52} = \dfrac{1}{2}$

(c) $\dfrac{16}{52} = \dfrac{4}{13}$

46. $\dfrac{{}_{13}C_1 \cdot {}_4C_3 \cdot {}_{12}C_1 \cdot {}_4C_2}{{}_{52}C_5} = \dfrac{13 \cdot 4 \cdot 12 \cdot 6}{2{,}598{,}960}$

$= \dfrac{3744}{2{,}598{,}960}$

$= \dfrac{6}{4165}$

47. (a) $\dfrac{{}_9C_4}{{}_{12}C_4} = \dfrac{126}{495} = \dfrac{14}{55}$ (4 good units)

(b) $\dfrac{{}_9C_2 \cdot {}_3C_2}{{}_{12}C_4} = \dfrac{108}{495} = \dfrac{12}{55}$ (2 good units)

(c) $\dfrac{{}_9C_3 \cdot {}_3C_1}{{}_{12}C_4} = \dfrac{252}{495} = \dfrac{28}{55}$ (3 good units)

At least 2 good units: $\dfrac{12}{55} + \dfrac{28}{55} + \dfrac{14}{55} = \dfrac{54}{55}$

48. (a) $P(EE) = \dfrac{20}{40} \cdot \dfrac{20}{40} = \dfrac{1}{4}$

(b) $P(EO \text{ or } OE) = 2\left(\dfrac{20}{40}\right)\left(\dfrac{20}{40}\right) = \dfrac{1}{2}$

(c) $P(N_1 < 30, N_2 < 30) = \dfrac{29}{40} \cdot \dfrac{29}{40} = \dfrac{841}{1600}$

(d) $P(N_1 N_1) = \dfrac{40}{40} \cdot \dfrac{1}{40} = \dfrac{1}{40}$

49. $(0.78)^3 \approx 0.4746$

50. $(0.32)^2 = 0.1024$

51. (a) $P(SS) = (0.985)^2 \approx 0.9702$

(b) $P(S) = 1 - P(FF) = 1 - (0.015)^2 \approx 0.9998$

(c) $P(FF) = (0.015)^2 \approx 0.0002$

52. (a) $P(AA) = (0.90)^2 = 0.81$

(b) $P(NN) = (0.10)^2 = 0.01$

(c) $P(A) = 1 - P(NN) = 1 - 0.01 = 0.99$

53. (a) $P(BBBB) = \left(\frac{1}{2}\right)^4 = \frac{1}{16}$

(b) $P(BBBB) + P(GGGG) = \left(\frac{1}{2}\right)^4 + \left(\frac{1}{2}\right)^4 = \frac{1}{8}$

(c) $P(\text{at least one boy}) = 1 - P(\text{no boys})$

$= 1 - P(GGGG) = 1 - \frac{1}{16} = \frac{15}{16}$

54. (a) $\frac{1}{38}$

(b) $\frac{18}{38} = \frac{9}{19}$

(c) $\frac{2}{38} + \frac{18}{38} = \frac{20}{38} = \frac{10}{19}$

(d) $\frac{1}{38} \cdot \frac{1}{38} = \frac{1}{1444}$

(e) $\frac{18}{38} \cdot \frac{18}{38} \cdot \frac{18}{38} = \frac{5832}{54{,}872} = \frac{729}{6859}$

(f) a. $\frac{1}{37}$

b. $\frac{18}{37}$

c. $\frac{1}{37} + \frac{18}{37} = \frac{19}{37}$

d. $\frac{1}{37} \cdot \frac{1}{37} = \frac{1}{1369}$

e. $\frac{18}{37} \cdot \frac{18}{37} \cdot \frac{18}{37} = \frac{5832}{50{,}653}$

The probabilities are better for European roulette.

55. $1 - \dfrac{(45)^2}{(60)^2} = 1 - \left(\dfrac{45}{60}\right)^2 = 1 - \left(\dfrac{3}{4}\right)^2 = 1 - \dfrac{9}{16} = \dfrac{7}{16}$

56. (a) If the *center* of the coin falls within the circle of radius $d/2$ around a vertex, the coin will cover the vertex.

$$P(\text{coin covers a vertex}) = \dfrac{\substack{\text{Area in which coin may fall}\\\text{so that it covers a vertex}}}{\text{Total area}}$$

$$= \dfrac{n\left[\pi\left(\dfrac{d}{2}\right)^2\right]}{nd^2} = \dfrac{\pi}{4}$$

(b) Experimental results will vary.

57. True. Two events are independent if the occurance of one has no effect on the occurance of the other.

58. False. The complement of the event is to roll a number greater than or equal to 3 and its probability is 2/3.

59. (a) As you consider successive people with distinct birthdays, the probabilities must decrease to take into account the birth dates already used. Because the birth dates of people are independent events, multiply the respective probabilities of distinct birthdays.

(b) $\dfrac{365}{365} \cdot \dfrac{364}{365} \cdot \dfrac{363}{365} \cdot \dfrac{362}{365}$

(c) $P_1 = \dfrac{365}{365} = 1$

$P_2 = \dfrac{365}{365} \cdot \dfrac{364}{365} = \dfrac{364}{365}P_1 = \dfrac{365 - (2 - 1)}{365}P_1$

$P_3 = \dfrac{365}{365} \cdot \dfrac{364}{365} \cdot \dfrac{363}{365} = \dfrac{363}{365}P_2 = \dfrac{365 - (3 - 1)}{365}P_2$

$P_n = \dfrac{365}{365} \cdot \dfrac{364}{365} \cdot \dfrac{363}{365} \cdots \dfrac{365 - (n - 1)}{365} = \dfrac{365 - (n - 1)}{365}P_{n-1}$

(d) Q_n is the probability that the birthdays are not distinct which is equivalent to at least two people having the same birthday.

(e)

n	10	15	20	23	30	40	50
P_n	0.88	0.75	0.59	0.49	0.29	0.11	0.03
Q_n	0.12	0.25	0.41	(0.51)	0.71	0.89	0.97

(f) 23, see the chart above.

60. If a weather forecast indicates that the probability of rain is 40%, this means the meteorological records indicate that over an extended period of time with similar weather conditions it will rain 40% of the time.

61. $6x^2 + 8 = 0$

$6x^2 = -8$

$x^2 = -\dfrac{4}{3}$

No real solution

62. $4x^2 + 6x - 12 = 0$

$2x^2 + 3x - 6 = 0$

$x = \dfrac{-b \pm \sqrt{b^2 - 4ac}}{2a} = \dfrac{-3 \pm \sqrt{3^2 - 4(2)(-6)}}{2(2)}$

$= \dfrac{-3 \pm \sqrt{57}}{4}$

63. $x^3 - x^2 - 3x = 0$

$x(x^2 - x - 3) = 0$

$x = 0 \quad \text{or} \quad x^2 - x - 3 = 0$

$x = \dfrac{1 \pm \sqrt{1 - 4(1)(-3)}}{2(1)} = \dfrac{1 \pm \sqrt{13}}{2}$

64. $x^5 + x^3 - 2x = 0$

$x(x^4 + x^2 - 2) = 0$

$x(x^2 + 2)(x^2 - 1) = 0$

$x = 0$

$x^2 - 1 = 0 \Longrightarrow x = \pm 1$

$x = 0, \pm 1$

65. $\dfrac{12}{x} = -3$

$12 = -3x$

$-4 = x$

66. $\dfrac{32}{x} = 2x$

$32 = 2x^2$

$16 = x^2$

$\pm 4 = x$

67. $\dfrac{2}{x - 5} = 4$

$2 = 4(x - 5)$

$2 = 4x - 20$

$22 = 4x$

$\dfrac{11}{2} = x$

68. $\dfrac{3}{2x + 3} - 4 = \dfrac{-1}{2x + 3}$

$\dfrac{3}{2x + 3} + \dfrac{1}{2x + 3} = 4$

$\dfrac{4}{2x + 3} = 4$

$4 = 4(2x + 3)$

$4 = 8x + 12$

$8x = -8$

$x = -1$

69. $\dfrac{3}{x - 2} + \dfrac{x}{x + 2} = 1$

$3(x + 2) + x(x - 2) = 1(x - 2)(x + 2)$

$3x + 6 + x^2 - 2x = x^2 - 4$

$x^2 + x + 6 = x^2 - 4$

$x + 6 = -4$

$x = -10$

70. $\dfrac{2}{x} - \dfrac{5}{x - 2} = \dfrac{-13}{x^2 - 2x}$

$\dfrac{2(x - 2) - 5x}{x^2 - 2x} = \dfrac{-13}{x^2 - 2x}$

$2x - 4 - 5x = -13$

$-4 - 3x = -13$

$3x = 9$

$x = 3$

71. $\begin{cases} y \geq -3 \\ x \geq -1 \\ -x - y \geq -8 \end{cases}$

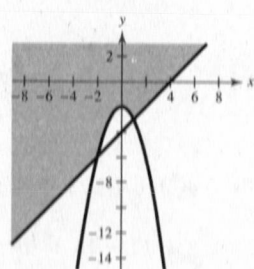

72.

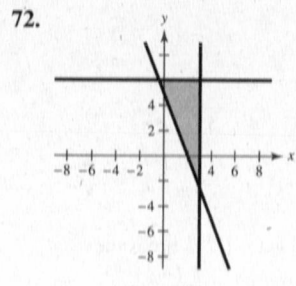

73. $\begin{cases} x^2 + y \geq -2 \\ y \geq x - 4 \end{cases}$

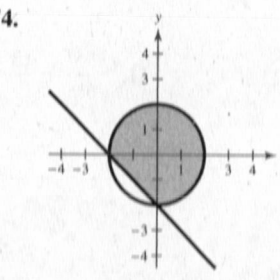

74.

Review Exercises for Chapter 9

1. $a_n = 2 + \dfrac{6}{n}$

$a_1 = 2 + \dfrac{6}{1} = 8$

$a_2 = 2 + \dfrac{6}{2} = 5$

$a_3 = 2 + \dfrac{6}{3} = 4$

$a_4 = 2 + \dfrac{6}{4} = \dfrac{7}{2}$

$a_5 = 2 + \dfrac{6}{5} = \dfrac{16}{5}$

2. $a_n = \dfrac{(-1)^n 5n}{2n - 1}$

$a_1 = \dfrac{(-1)^1 5(1)}{2(1) - 1} = -5$

$a_2 = \dfrac{(-1)^2 5(2)}{2(2) - 1} = \dfrac{10}{3}$

$a_3 = \dfrac{(-1)^3 5(3)}{2(3) - 1} = -3$

$a_4 = \dfrac{(-1)^4 5(4)}{2(4) - 1} = \dfrac{20}{7}$

$a_5 = \dfrac{(-1)^5 5(5)}{2(5) - 1} = -\dfrac{25}{9}$

3. $a_n = \dfrac{72}{n!}$

$a_1 = \dfrac{72}{1!} = 72$

$a_2 = \dfrac{72}{2!} = 36$

$a_3 = \dfrac{72}{3!} = 12$

$a_4 = \dfrac{72}{4!} = 3$

$a_5 = \dfrac{72}{5!} = \dfrac{3}{5}$

4. $a_n = n(n - 1)$

$a_1 = 1(1 - 1) = 0$

$a_2 = 2(2 - 1) = 2$

$a_3 = 3(3 - 1) = 6$

$a_4 = 4(4 - 1) = 12$

$a_5 = 5(5 - 1) = 20$

5. $-2, 2, -2, 2, -2, \ldots$

$a_n = 2(-1)^n$

6. $-1, 2, 7, 14, 23, \ldots$

n:	1	2	3	4	5 $\ldots n$
Terms:	-1	2	7	14	23 $\ldots a_n$

Apparent pattern: Each term is 2 less than the square of n, which implies that $a_n = n^2 - 2$.

7. $4, 2, \frac{4}{3}, 1, \frac{4}{5}, \ldots$

$a_n = \dfrac{4}{n}$

8. $1, -\dfrac{1}{2}, \dfrac{1}{3}, -\dfrac{1}{4}, \dfrac{1}{5}, \ldots$

n:	1	2	3	4	5 $\ldots n$
Terms:	1	$-\dfrac{1}{2}$	$\dfrac{1}{3}$	$-\dfrac{1}{4}$	$\dfrac{1}{5} \cdots a_n$

Apparent pattern: Each term is $(-1)^{n+1}$ times the reciprocal of n, which implies that $a_n = \dfrac{(-1)^{n+1}}{n}$.

9. $5! = 5 \cdot 4 \cdot 3 \cdot 2 \cdot 1 = 120$

10. $3! \cdot 2! = (3 \cdot 2 \cdot 1) \cdot (2 \cdot 1) = 12$

11. $\dfrac{3! \, 5!}{6!} = \dfrac{(3 \cdot 2 \cdot 1)5!}{6 \cdot 5!} = 1$

12. $\dfrac{7! \cdot 6!}{6! \cdot 8!} = \dfrac{7! \cdot 6!}{6!(8 \cdot 7!)} = \dfrac{1}{8}$

13. $\displaystyle\sum_{i=1}^{6} 5 = 6(5) = 30$

14. $\displaystyle\sum_{k=2}^{5} 4k = 4(2) + 4(3) + 4(4) + 4(5)$

$= 8 + 12 + 16 + 20 = 56$

15. $\displaystyle\sum_{j=1}^{4}\frac{6}{j^2}=\frac{6}{1^2}+\frac{6}{2^2}+\frac{6}{3^2}+\frac{6}{4^2}=6+\frac{3}{2}+\frac{2}{3}+\frac{3}{8}=\frac{205}{24}$

16. $\displaystyle\sum_{i=1}^{8}\frac{i}{i+1}=\frac{1}{1+1}+\frac{2}{2+1}+\frac{3}{3+1}+\frac{4}{4+1}+\frac{5}{5+1}+\frac{6}{6+1}+\frac{7}{7+1}+\frac{8}{8+1}$

$$=\frac{1}{2}+\frac{2}{3}+\frac{3}{4}+\frac{4}{5}+\frac{5}{6}+\frac{6}{7}+\frac{7}{8}+\frac{8}{9}\approx 6.17$$

17. $\displaystyle\sum_{k=1}^{10}2k^3=2(1)^3+2(2)^3+2(3)^3+\cdots+2(10)^3=6050$

18. $\displaystyle\sum_{j=0}^{4}(j^2+1)=(0^2+1)+(1^2+1)+(2^2+1)+(3^2+1)+(4^2+1)$

$$=1+2+5+10+17=35$$

19. $\displaystyle\frac{1}{2(1)}+\frac{1}{2(2)}+\frac{1}{2(3)}+\cdots+\frac{1}{2(20)}=\sum_{k=1}^{20}\frac{1}{2k}$

20. $\displaystyle\frac{1}{2}+\frac{2}{3}+\frac{3}{4}+\cdots+\frac{9}{10}=\sum_{k=1}^{9}\frac{k}{k+1}$

21. $\displaystyle\sum_{i=1}^{\infty}\frac{5}{10^i}=0.5+0.05+0.005+0.0005+\cdots=0.5555\cdots=\frac{5}{9}$

22. $\displaystyle\sum_{i=1}^{\infty}\frac{3}{10^i}=\sum_{i=1}^{\infty}3\left(\frac{1}{10^i}\right)=\frac{\frac{3}{10}}{1-\frac{1}{10}}=\frac{1}{3}$

23. $\displaystyle\sum_{k=1}^{\infty}\frac{2}{100^k}=0.02+0.0002+0.000002+\cdots=0.020202\cdots=\frac{2}{99}$

24. $\displaystyle\sum_{k=2}^{\infty}\frac{9}{10^k}=\sum_{k=2}^{\infty}9\left(\frac{1}{10^k}\right)=\frac{\frac{9}{100}}{1-\frac{1}{10}}=\frac{1}{10}$

25. $A_n=10,000\left(1+\dfrac{0.08}{12}\right)^n$

(a) $A_1\approx\$10,066.67$

$A_2\approx\$10,133.78$

$A_3\approx\$10,201.34$

$A_4\approx\$10,269.35$

$A_5\approx\$10,337.81$

$A_6\approx\$10,406.73$

$A_7\approx\$10,476.10$

$A_8\approx\$10,545.95$

$A_9\approx\$10,616.25$

$A_{10}\approx\$10,687.03$

(b) $A_{120}\approx\$22,196.40$

26. $a_4=734.52$

$a_5=750.25$

$a_6=768.12$

$a_7=788.13$

$a_8=810.28$

$a_9=834.57$

$a_{10}=861.00$

$a_{11}=889.57$

$a_{12}=920.28$

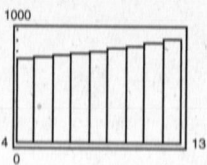

27. $5, 3, 1, -1, -3, \ldots$

Arithmetic sequence, $d = -2$

28. $0, 1, 3, 6, 10, \ldots$

Not an arithmetic sequence

29. $\frac{1}{2}, 1, \frac{3}{2}, 2, \frac{5}{2}, \ldots$

Arithmetic sequence, $d = \frac{1}{2}$

30. $\frac{9}{9}, \frac{8}{9}, \frac{7}{9}, \frac{6}{9}, \frac{5}{9}, \ldots$

Arithmetic sequence, $d = -\frac{1}{9}$

31. $a_1 = 4, \ d = 3$

$a_1 = 4$

$a_2 = 4 + 3 = 7$

$a_3 = 7 + 3 = 10$

$a_4 = 10 + 3 = 13$

$a_5 = 13 + 3 = 16$

32. $a_1 = 6, d = -2$

$a_1 = 6$

$a_2 = 6 - 2 = 4$

$a_3 = 4 - 2 = 2$

$a_4 = 2 - 2 = 0$

$a_5 = 0 - 2 = -2$

33. $a_1 = 25, \ a_{k+1} = a_k + 3$

$a_1 = 25$

$a_2 = 25 + 3 = 28$

$a_3 = 28 + 3 = 31$

$a_4 = 31 + 3 = 34$

$a_5 = 34 + 3 = 37$

34. $a_1 = 4.2, a_{k+1} = a_k + 0.4$

$a_1 = 4.2$

$a_2 = 4.2 + 0.4 = 4.6$

$a_3 = 4.6 + 0.4 = 5.0$

$a_4 = 5.0 + 0.4 = 5.4$

$a_5 = 5.4 + 0.4 = 5.8$

35. $a_1 = 7, d = 12$

$a_n = 7 + (n - 1)12$

$= 7 + 12n - 12$

$= 12n - 5$

36. $a_1 = 25, d = -3$

$a_n = dn + c$

$a_n = -3n + c$

$c = a_1 - d = 25 - (-3) = 28$

So, $a_n = -3n + 28$.

37. $a_1 = y, d = 3y$

$a_n = y + (n - 1)3y$

$= y + 3ny - 3y$

$= 3ny - 2y$

38. $a_1 = -2x, d = x$

$a_n = dn + c$

$a_n = xn + c$

$c = a_1 - d = -2x - x = -3x$

So, $a_n = xn - 3x$.

39. $a_2 = 93, a_6 = 65$

$a_6 = a_2 + 4d \implies 65 = 93 + 4d \implies -28 = 4d \implies d = -7$

$a_1 = a_2 - d \implies a_1 = 93 - (-7) = 100$

$a_n = a_1 + (n - 1)d = 100 + (n - 1)(-7) = -7n + 107$

40. $a_7 = 8, a_{13} = 6$

$a_{13} = a_7 + 6d \implies 6 = 8 + 6d \implies d = -\frac{1}{3}$

$a_1 = a_7 - 6d \implies a_1 = 8 - 6\left(-\frac{1}{3}\right) \implies a_1 = 10$

$a_n = a_1 + (n - 1)d \implies a_n = 10 + (n - 1)\left(-\frac{1}{3}\right) \implies a_n = -\frac{1}{3}n + \frac{31}{3}$

41. $\sum_{j=1}^{10} (2j - 3)$ is arithmetic. Therefore, $a_1 = -1, a_{10} = 17, S_{10} = \frac{10}{2}[-1 + 17] = 80$.

42. $\sum_{j=1}^{8} (20 - 3j) = \sum_{j=1}^{8} 20 - 3\sum_{j=1}^{8} j = 8(20) - 3\left[\frac{(8)(9)}{2}\right] = 52$

43. $\sum_{k=1}^{11} \left(\frac{2}{3}k + 4\right)$ is arithmetic. Therefore, $a_1 = \frac{14}{3}, a_{11} = \frac{34}{3}, S_{11} = \frac{11}{2}\left[\frac{14}{3} + \frac{34}{3}\right] = 88$.

44. $\displaystyle\sum_{k=1}^{25}\left(\frac{3k+1}{4}\right) = \frac{3}{4}\sum_{k=1}^{25}k + \sum_{k=1}^{25}\frac{1}{4} = \frac{3}{4}\left[\frac{(25)(26)}{2}\right] + 25\left(\frac{1}{4}\right) = 250$

45. $\displaystyle\sum_{k=1}^{100} 5k$ is arithmetic. Therefore, $a_1 = 5$, $a_{100} = 500$, $S_{500} = \frac{100}{2}(5 + 500) = 25{,}250$.

46. $\displaystyle\sum_{n=20}^{80} n = \sum_{n=1}^{80} n - \sum_{n=1}^{19} n = \frac{(80)(81)}{2} - \frac{(19)(20)}{2} = 3050$

47. $a_n = 34{,}000 + (n - 1)(2250)$

(a) $a_5 = 34{,}000 + 4(2250) = \$43{,}000$

(b) $S_5 = \frac{5}{2}(34{,}000 + 43{,}000) = \$192{,}500$

48. $a_1 = 123$, $d = 112 - 123 = -11$

$n = 8$

$a_8 = 123 + 7(-11) = 46$

$S_8 = \frac{8}{2}(123 + 46) = 676$

49. $5, 10, 20, 40, \ldots$

The sequence *is* geometric, $r = 2$.

50. $54, -18, 6, -2, \ldots$

Geometric sequence, $r = -\frac{18}{54} = -\frac{1}{3}$

51. $\frac{1}{3}, -\frac{2}{3}, \frac{4}{3}, -\frac{8}{3}, \ldots$

The sequence *is* geometric, $r = -2$

52. $\frac{1}{4}, \frac{2}{5}, \frac{3}{6}, \frac{4}{7}, \ldots$

Not a geometric sequence

53. $a_1 = 4$, $r = -\frac{1}{4}$

$a_1 = 4$

$a_2 = 4\left(-\frac{1}{4}\right) = -1$

$a_3 = -1\left(-\frac{1}{4}\right) = \frac{1}{4}$

$a_4 = \frac{1}{4}\left(-\frac{1}{4}\right) = -\frac{1}{16}$

$a_5 = -\frac{1}{16}\left(-\frac{1}{4}\right) = \frac{1}{64}$

54. $a_1 = 2$, $r = 2$

$a_1 = 2$

$a_2 = 2(2) = 4$

$a_3 = 4(2) = 8$

$a_4 = 8(2) = 16$

$a_5 = 16(2) = 32$

55. $a_1 = 9$, $a_3 = 4$

$a_3 = a_1 r^2$

$4 = 9r^2$

$\frac{4}{9} = r^2 \implies r = \pm\frac{2}{3}$

$a_1 = 9$ $\qquad\qquad$ $a_1 = 9$

$a_2 = 9\left(\frac{2}{3}\right) = 6$ $\qquad$ $a_2 = 9\left(-\frac{2}{3}\right) = -6$

$a_3 = 6\left(\frac{2}{3}\right) = 4$ or $a_3 = -6\left(-\frac{2}{3}\right) = 4$

$a_4 = 4\left(\frac{2}{3}\right) = \frac{8}{3}$ $\qquad$ $a_4 = 4\left(-\frac{2}{3}\right) = -\frac{8}{3}$

$a_5 = \frac{8}{3}\left(\frac{2}{3}\right) = \frac{16}{9}$ $\qquad$ $a_5 = -\frac{8}{3}\left(-\frac{2}{3}\right) = \frac{16}{9}$

56. $a_1 = 2$, $a_3 = 12$

$a_3 = a_1 r^2$

$12 = 2r^2$

$6 = r^2$

$\pm\sqrt{6} = r$

$a_1 = 2$ $\qquad\qquad$ $a_1 = 2$

$a_2 = 2\left(\sqrt{6}\right) = 2\sqrt{6}$ $\qquad$ $a_2 = 2\left(-\sqrt{6}\right) = -2\sqrt{6}$

$a_3 = 2\sqrt{6}\left(\sqrt{6}\right) = 12$ or $a_3 = -2\sqrt{6}\left(-\sqrt{6}\right) = 12$

$a_4 = 12\left(\sqrt{6}\right) = 12\sqrt{6}$ $\qquad$ $a_4 = 12\left(-\sqrt{6}\right) = -12\sqrt{6}$

$a_5 = 12\sqrt{6}\left(\sqrt{6}\right) = 72$ $\qquad$ $a_5 = -12\sqrt{6}\left(-\sqrt{6}\right) = 72$

57. $a_1 = 16, a_2 = -8$

$a_2 = a_1 r \implies -8 = 16r \implies r = -\frac{1}{2}$

$a_n = 16\left(-\frac{1}{2}\right)^{n-1}$

$a_{20} = 16\left(-\frac{1}{2}\right)^{19} \approx -3.052 \times 10^{-5}$

58. $a_3 = 6, a_4 = 1$

$a_3 r = a_4$

$6r = 1$

$r = \frac{1}{6}$

$a_3 = a_1 r^2$

$6 = a_1\left(\frac{1}{6}\right)^2$

$6 = a_1\left(\frac{1}{36}\right)$

$a_1 = 216$

$a_n = 216\left(\frac{1}{6}\right)^{n-1}$

$a_{20} = 216\left(\frac{1}{6}\right)^{19} = 3.545 \times 10^{-13}$

59. $a_1 = 100, r = 1.05$

$a_n = 100(1.05)^{n-1}$

$a_{20} = 100(1.05)^{19} \approx 252.695$

60. $a_1 = 5, r = 0.2$

$a_n = 5(0.2)^{n-1}$

$a_{20} = 5(0.2)^{19} \approx 2.62 \times 10^{-13}$

61. $\displaystyle\sum_{i=1}^{7} 2^{i-1} = \frac{1-2^7}{1-2} = 127$

62. $\displaystyle\sum_{i=1}^{5} 3^{i-1} = 1\left(\frac{1-3^5}{1-3}\right) = 121$

63. $\displaystyle\sum_{i=1}^{4}\left(\frac{1}{2}\right)^i = \frac{1}{2} + \frac{1}{4} + \frac{1}{8} + \frac{1}{16} = \frac{15}{16}$

64. $\displaystyle\sum_{i=1}^{6}\left(\frac{1}{3}\right)^{i-1} = \left(\frac{1-\left(\frac{1}{3}\right)^6}{1-\frac{1}{3}}\right) = \frac{1-\frac{1}{729}}{1-\frac{1}{3}} = \frac{364}{243}$

65. $\displaystyle\sum_{i=1}^{5}(2)^{i-1} = 1 + 2 + 4 + 8 + 16 = 31$

66. $\displaystyle\sum_{i=1}^{4} 6(3)^i = 6(3)\left(\frac{1-3^4}{1-3}\right) = 720$

67. $\displaystyle\sum_{i=1}^{10} 10\left(\frac{3}{5}\right)^{i-1} \approx 24.85$

68. $\displaystyle\sum_{i=1}^{15} 20(0.2)^{i-1} = 25$

69. $\displaystyle\sum_{i=1}^{25} 100(1.06)^{i-1} \approx 5486.45$

70. $\displaystyle\sum_{i=1}^{20} 8\left(\frac{6}{5}\right)^{i-1} = 1493.50$

71. $\displaystyle\sum_{i=1}^{\infty}\left(\frac{7}{8}\right)^{i-1} = \frac{1}{1-\frac{7}{8}} = 8$

72. $\displaystyle\sum_{i=1}^{\infty}\left(\frac{1}{3}\right)^{i-1} = \frac{1}{1-\frac{1}{3}} = \frac{3}{2}$

73. $\displaystyle\sum_{i=1}^{\infty}(0.1)^{i-1} = \frac{1}{1-0.1} = \frac{10}{9}$

74. $\displaystyle\sum_{i=1}^{\infty}(0.5)^{i-1} = \frac{1}{1-0.5} = 2$

75. $\displaystyle\sum_{k=1}^{\infty} 4\left(\frac{2}{3}\right)^{k-1} = \frac{4}{1-\frac{2}{3}} = 12$

76. $\displaystyle\sum_{k=1}^{\infty} 1.3\left(\frac{1}{10}\right)^{k-1} = \frac{1.3}{1-\frac{1}{10}} = \frac{13}{9}$

77. (a) $a_t = 120{,}000(0.7)^t$

(b) $a_5 = 120{,}000(0.7)^5$

$= \$20{,}168.40$

78. Monthly: $A = P\left[\left(1 + \dfrac{r}{12}\right)^{12t} - 1\right]\left(1 + \dfrac{12}{r}\right)$

$= 200\left[\left(1 + \dfrac{0.06}{12}\right)^{12 \cdot 10} - 1\right]\left(1 + \dfrac{12}{0.06}\right)$

$= \$32{,}939.75$

Continuously: $A = \dfrac{Pe^{r/12}(e^{rt} - 1)}{e^{r/12} - 1}$

$= \dfrac{200e^{0.06/12}(e^{(0.06)(10)} - 1)}{e^{0.06/12} - 1} = \$32{,}967.03$

79. 1. When $n = 1, 3 = 1(1 + 2)$.

2. Assume that $S_k = 3 + 5 + 7 + \cdots + (2k + 1) = k(k + 2)$.

Then, $S_{k+1} = 3 + 5 + 7 + \cdots + (2k + 1) + [2(k + 1) + 1] = S_k + (2k + 3)$

$$= k(k + 2) + 2k + 3$$

$$= k^2 + 4k + 3$$

$$= (k + 1)(k + 3)$$

$$= (k + 1)[(k + 1) + 2].$$

Therefore, by mathematical induction, the formula is valid for all positive integer values of n.

80. 1. When $n = 1, S_1 = 1 = \dfrac{1}{4}(1 + 3) = 1$.

2. Assume that $S_k = 1 + \dfrac{3}{2} + 2 + \dfrac{5}{2} + \cdots + \dfrac{1}{2}(k + 1) = \dfrac{k}{4}(k + 3)$. Then,

$$S_{k+1} = S_k + a_{k+1} = \left(1 + \frac{3}{2} + 2 + \frac{5}{2} + \cdots + \frac{1}{2}(k + 1)\right) + \frac{1}{2}(k + 2)$$

$$= \frac{k}{4}(k + 3) + \frac{1}{2}(k + 2)$$

$$= \frac{k(k + 3) + 2(k + 2)}{4}$$

$$= \frac{k^2 + 5k + 4}{4}$$

$$= \frac{(k + 1)(k + 4)}{4}$$

$$= \frac{k + 1}{4}[(k + 1) + 3].$$

Thus, the formula holds for all positive integers n.

81. 1. When $n = 1, a = a\left(\dfrac{1 - r}{1 - r}\right)$.

2. Assume that $S_k = \displaystyle\sum_{i=0}^{k-1} ar^i = \dfrac{a(1 - r^k)}{1 - r}$.

Then, $S_{k+1} = \displaystyle\sum_{i=0}^{k} ar^i = \left(\sum_{i=0}^{k-1} ar^i\right) + ar^k = \dfrac{a(1 - r^k)}{1 - r} + ar^k$

$$= \frac{a(1 - r^k + r^k - r^{k+1})}{1 - r} = \frac{a(1 - r^{k+1})}{1 - r}.$$

Therefore, by mathematical induction, the formula is valid for all positive integer values of n.

82. 1. When $n = 1$, $S_1 = a + 0 \cdot d = a = \frac{1}{2}[2a + (1 - 1)d] = a$.

2. Assume that $S_k = \sum_{k=0}^{i-1} (a + kd) = \frac{i}{2}[2a + (i - 1)d]$. Then,

$$S_{k+1} = S_k + a_{k+1}$$

$$\sum_{k=0}^{i+1-1} (a + kd) = \frac{i}{2}[2a + (i - 1)d] + [a + id]$$

$$= \frac{2ia + i(i - 1)d + 2a + 2id}{2} = \frac{2a(i + 1) + id(i + 1)}{2} = \left(\frac{i + 1}{2}\right)[2a + id].$$

Thus, the formula holds for all positive integers n.

83. $S_1 = 9 = 1(9) = 1[2(1) + 7]$

$S_2 = 9 + 13 = 22 = 2(11) = 2[2(2) + 7]$

$S_3 = 9 + 13 + 17 = 39 = 3(13) = 3[2(3) + 7]$

$S_4 = 9 + 13 + 17 + 21 = 60 = 4(15) = 4[2(4) + 7]$

$S_n = n(2n + 7)$

84. $S_1 = 68 = 4 \cdot 17$

$S_2 = 68 + 60 = 128 = 8 \cdot 16$

$S_3 = 68 + 60 + 52 = 180 = 12 \cdot 15$

$S_4 = 68 + 60 + 52 + 44 = 224 = 16 \cdot 14$

$S_n = 4n(18 - n)$

85. $S_1 = 1$

$S_2 = 1 + \frac{3}{5} = \frac{8}{5}$

$S_3 = 1 + \frac{3}{5} + \frac{9}{25} = \frac{49}{25}$

$S_4 = 1 + \frac{3}{5} + \frac{9}{25} + \frac{27}{125} = \frac{272}{125}$

Since the series is geometric,

$$S_n = \frac{1 - \left(\frac{3}{5}\right)^n}{1 - \frac{3}{5}} = \frac{5}{2}\left[1 - \left(\frac{3}{5}\right)^n\right].$$

86. $S_1 = 12$

$S_2 = 12 - 1 = 11$

$S_3 = 12 - 1 + \frac{1}{12} = \frac{133}{12}$

$S_4 = 12 - 1 + \frac{1}{12} - \frac{1}{144} = \frac{1595}{144}$

Since the series is geometric,

$$S_n = 12\frac{1 - \left(-\frac{1}{12}\right)^n}{\left[1 - \left(-\frac{1}{12}\right)\right]} = \frac{144}{13}\left[1 - \left(-\frac{1}{12}\right)^n\right].$$

87. $\sum_{n=1}^{30} n = \frac{30(31)}{2} = 465$

88. $\sum_{n=1}^{10} n^2 = \frac{10(10 + 1)(2 \cdot 10 + 1)}{6} = \frac{10(11)(21)}{6} = 385$

89. $\sum_{n=1}^{7} (n^4 - n) = \sum_{n=1}^{7} n^4 - \sum_{n=1}^{7} n = \frac{(7)(8)(15)[(3)(49) + 21 - 1]}{30} - \frac{(7)(8)}{2}$

$$= \frac{(7)(8)(15)(167)}{30} - \frac{(7)(8)}{2}$$

$$= 4676 - 28 = 4648$$

90. $\sum_{n=1}^{6} (n^5 - n^2) = \sum_{n=1}^{6} n^5 - \sum_{n=1}^{6} n^2$

$$= \frac{(6)^2(6 + 1)^2[2(6)^2 + 2(6) - 1]}{12} - \frac{6(6 + 1)[2(6) + 1]}{6}$$

$$= \frac{(6)^2(7)^2(83)}{12} - \frac{6(7)(13)}{6}$$

$$= \frac{(6)^2(7)^2(83) - 2(6)(7)(13)}{12} = 12,110$$

91. $a_1 = f(1) = 5, \quad a_n = a_{n-1} + 5$

$a_1 = 5$

$a_2 = 5 + 5 = 10$

$a_3 = 10 + 5 = 15$

$a_4 = 15 + 5 = 20$

$a_5 = 20 + 5 = 25$

n:	1	2	3	4	5
a_n:	5	10	15	20	25

First differences: 5   5   5   5

Second differences: 0   0   0

The sequence has a linear model.

93. $a_1 = f(1) = 16, \quad a_n = a_{n-1} - 1$

$a_1 = 16$

$a_2 = 16 - 1 = 15$

$a_3 = 15 - 1 = 14$

$a_4 = 14 - 1 = 13$

$a_5 = 13 - 1 = 12$

n:	1	2	3	4	5
a_n:	16	15	14	13	12

First differences: -1 -1 -1 -1

Second differences: 0   0   0

The sequence has a linear model.

95. $_6C_4 = \dfrac{6!}{2!4!} = 15$

97. $_8C_5 = \dfrac{8!}{3!5!} = 56$

99. $\dbinom{7}{3} = 35$

```
            1
          1   1
        1   2   1
      1   3   3   1
    1   4   6   4   1
  1   5  10  10   5   1
1   6  15  20  15   6   1
1  7  21  35 (35) 21  7  1
```

$\dbinom{7}{3} = 35$, the 5th entry in the 7th row

92. $a_1 = -3$

$a_n = a_{n-1} - 2n$

$a_1 = -3$

$a_2 = a_1 - 2(2) = -3 - 4 = -7$

$a_3 = a_2 - 2(3) = -7 - 6 = -13$

$a_4 = a_3 - 2(4) = -13 - 8 = -21$

$a_5 = a_4 - 2(5) = -21 - 10 = -31$

a_n:	-3	-7	-13	-21	-31

First differences: -4 -6 -8 -10

Second differences: -2 -2 -2

Since the second differences are all the same, the sequence has a quadratic model.

94. $a_0 = 0, \quad a_n = n - a_{n-1}$

$a_0 = 0$

$a_1 = 1 - a_0 = 1 - 0 = 1$

$a_2 = 2 - a_1 = 2 - 1 = 1$

$a_3 = 3 - a_2 = 3 - 1 = 2$

$a_4 = 4 - a_3 = 4 - 2 = 2$

a_n:	0	1	1	2	2

First differences: 1   0   1   0

Second differences: -1 1 -1

Since neither the first differences nor the second differences are equal, the sequence does not have a linear or a quadratic model.

96. $_{10}C_7 = \dfrac{10!}{7!3!} = \dfrac{10 \cdot 9 \cdot 8 \cdot 7!}{7!3!}$

$= \dfrac{10 \cdot 9 \cdot 8}{3 \cdot 2 \cdot 1} = 120$

98. $_{12}C_3 = \dfrac{12!}{3!9!} = \dfrac{12 \cdot 11 \cdot 10 \cdot 9!}{3! \cdot 9!} = \dfrac{12 \cdot 11 \cdot 10}{3 \cdot 2 \cdot 1} = 220$

100.

```
              1
            1   1
          1   2   1
        1   3   3   1
      1   4   6   4   1
    1   5  10  10   5   1
  1   6  15  20  15   6   1
1   7  21  35  35  21   7   1
1  8  28  56  70  56  28  8  1
1 9 36 84 (126)126 84 36 9 1
```

$\dbinom{9}{4} = 126$, the 5th entry in the 9th row

101. $\binom{8}{6} = 28$

$$
\begin{array}{ccccccccccccccccc}
& & & & & & & & 1 \\
& & & & & & & 1 & & 1 \\
& & & & & & 1 & & 2 & & 1 \\
& & & & & 1 & & 3 & & 3 & & 1 \\
& & & & 1 & & 4 & & 6 & & 4 & & 1 \\
& & & 1 & & 5 & & 10 & & 10 & & 5 & & 1 \\
& & 1 & & 6 & & 15 & & 20 & & 15 & & 6 & & 1 \\
& 1 & & 7 & & 21 & & 35 & & 35 & & 21 & & 7 & & 1 \\
1 & & 8 & & 28 & & 56 & & 70 & & 56 & & \circled{28} & & 8 & & 1
\end{array}
$$

$\binom{8}{6} = 28$, the 7th entry in the 8th row

102.

$$
\begin{array}{ccccccccccc}
& & & & 1 \\
& & & 1 & & 1 \\
& & 1 & & 2 & & 1 \\
& 1 & & 3 & & 3 & & 1 \\
1 & & 4 & & 6 & & 4 & & 1 \\
1 & 5 & 10 & \circled{10} & 5 & 1
\end{array}
$$

$\binom{5}{3} = 10$, the 4th entry in the 5th row

103. $(x + 4)^4 = x^4 + 4x^3(4) + 6x^2(4)^2 + 4x(4)^3 + 4^4$

$\qquad\qquad = x^4 + 16x^3 + 96x^2 + 256x + 256$

104. $(x - 3)^6 = {}_6C_0(x)^6(-3)^0 + {}_6C_1(x)^5(-3) + {}_6C_2(x)^4(-3)^2 + {}_6C_3(x)^3(-3)^3 + {}_6C_4(x)^2(-3)^4 + {}_6C_5(x)(-3)^5 + {}_6C_6(x)^0(-3)^6$

$\qquad\qquad = x^6 - 18x^5 + 135x^4 - 540x^3 + 1215x^2 - 1458x + 729$

105. $(a - 3b)^5 = a^5 - 5a^4(3b) + 10a^3(3b)^2 - 10a^2(3b)^3 + 5a(3b)^4 - (3b)^5$

$\qquad\qquad = a^5 - 15a^4b + 90a^3b^2 - 270a^2b^3 + 405ab^4 - 243b^5$

106. $(3x + y^2)^7 = {}_7C_0(3x)^7 + {}_7C_1(3x)^6(y^2) + {}_7C_2(3x)^5(y^2)^2 + {}_7C_3(3x)^4(y^2)^3 + {}_7C_4(3x)^3(y^2)^4 + {}_7C_5(3x)^2(y^2)^5$

$\qquad\qquad\quad + {}_7C_6(3x)(y^2)^6 + {}_7C_7(y^2)^7$

$\qquad\quad = (3x)^7 + 7(3x)^6y^2 + 21(3x)^5(y^2)^2 + 35(3x)^4(y^2)^3 + 35(3x)^3(y^2)^4 + 21(3x)^2(y^2)^5 + 7(3x)(y^2)^6 + (y^2)^7$

$\qquad\quad = 2187x^7 + 5103x^6y^2 + 5103x^5y^4 + 2835x^4y^6 + 945x^3y^8 + 189x^2y^{10} + 21xy^{12} + y^{14}$

107. $(5 + 2i)^4 = (5)^4 + 4(5)^3(2i) + 6(5)^2(2i)^2 + 4(5)(2i)^3 + (2i)^4$

$\qquad\qquad = 625 + 1000i + 600i^2 + 160i^3 + 16i^4$

$\qquad\qquad = 625 + 1000i - 600 - 160i + 16 = 41 + 840i$

108. $(4 - 5i)^3 = {}_3C_0(4^3) + {}_3C_1(4^2)(-5i) + {}_3C_2(4)(-5i)^2 + {}_3C_3(-5i)^3$

$\qquad\qquad = 4^3 - 3(4)^2(5i) + 3(4)(5i)^2 - (5i)^3$

$\qquad\qquad = 64 - 240i - 300 + 125i$

$\qquad\qquad = -236 - 115i$

109.

First number:	1	2	3	4	5	6	7	8	9	10	11
Second number:	11	10	9	8	7	6	5	4	3	2	1

From this list, you can see that a total of 12 occurs 11 different ways.

110. ${}_6C_1 \cdot {}_5C_1 \cdot {}_6C_1 = 6 \cdot 5 \cdot 6 = 180$

111. $(10)(10)(10)(10) = 10{,}000$ different telephone numbers

112. ${}_3C_1 \cdot {}_4C_1 \cdot {}_6C_1 = 3 \cdot 4 \cdot 6 = 72$

113. ${}_{10}P_3 = \dfrac{10!}{7!} = \dfrac{10 \cdot 9 \cdot 8 \cdot 7!}{7!}$

$\qquad\qquad = 10 \cdot 9 \cdot 8 = 720$ different ways

114. $_{32}C_{12} = \dfrac{32!}{20!12!} = 225{,}792{,}840$

115. $_8C_3 = \dfrac{8!}{5!3!} = 56$

116. Breads: $_5C_1 = 5$

Meats: $_7C_0 + _7C_1 + _7C_2 + _7C_3 + _7C_4 + _7C_5 + _7C_6 + _7C_7 = 1 + 7 + 21 + 35 + 35 + 21 + 7 + 1 = 128$

Cheese: $_3C_0 + _3C_1 + _3C_2 + _3C_3 = 1 + 3 + 3 + 1 = 8$

Vegetables: $_6C_0 + _6C_1 + _6C_2 + _6C_3 + _6C_4 + _6C_5 + _6C_6 = 1 + 6 + 15 + 20 + 15 + 6 + 1 = 64$

$5 \cdot 128 \cdot 8 \cdot 64 = 327{,}680$

117. $(1)\left(\dfrac{1}{9}\right) = \dfrac{1}{9}$

118. $P(E) = \dfrac{n(E)}{n(S)} = \dfrac{1}{5!} = \dfrac{1}{120}$

119. (a) $25\% + 18\% = 43\%$

(b) $100\% - 18\% = 82\%$

120. (a) $\dfrac{208}{500} = 0.416$ or 41.6%

(b) $\dfrac{400}{500} = 0.8$ or 80%

(c) $\dfrac{37}{500} = 0.074$ or 7.4%

121. $\left(\dfrac{1}{6}\right)\left(\dfrac{1}{6}\right)\left(\dfrac{1}{6}\right) = \dfrac{1}{216}$

122. $\left(\dfrac{6}{6}\right)\left(\dfrac{5}{6}\right)\left(\dfrac{4}{6}\right)\left(\dfrac{3}{6}\right)\left(\dfrac{2}{6}\right)\left(\dfrac{1}{6}\right) = \dfrac{6!}{6^6} = \dfrac{720}{46{,}656} = \dfrac{5}{324}$

123. $1 - \dfrac{13}{52} = 1 - \dfrac{1}{4} = \dfrac{3}{4}$

124. $1 - P(HHHHH) = 1 - \left(\dfrac{1}{2}\right)^5 = \dfrac{31}{32}$

125. True. $\dfrac{(n+2)!}{n!} = \dfrac{(n+2)(n+1)n!}{n!} = (n+2)(n+1)$

126. True by Properties of Sums

127. True. $\displaystyle\sum_{k=1}^{8} 3k = 3\sum_{k=1}^{8} k$ by the Properties of Sums.

128. True because $2^1 + 2^2 + 2^3 + 2^4 + 2^5 + 2^6 = 2^{3-2} + 2^{4-2} + 2^{5-2} + 2^{6-2} + 2^{7-2} + 2^{8-2}$

129. False. If $r = 0$ or $r = 1$, then $_nP_r = {}_nC_r$.

130. The domain of an infinite sequence is the set of natural numbers.

131. (a) Odd-numbered terms are negative.

(b) Even-numbered terms are negative.

132. (a) Arithmetic. There is a constant difference between consecutive terms.

(b) Geometric. Each term is a constant multiple of the previous term. In this case the common ratio is greater than 1.

133. Each term of the sequence is defined in terms of preceding terms.

134. Increased powers of real numbers between 0 and 1 approach zero.

135. $a_n = 4\left(\dfrac{1}{2}\right)^{n-1}$

$a_1 = 4$, $a_2 = 2$, $a_{10} = \dfrac{1}{128}$

The sequence is geometric and is decreasing.

Matches graph (d).

136. $a_n = 4\left(-\dfrac{1}{2}\right)^{n-1}$

$a_1 = 4$ and a_n fluctuates from positive to negative.

Matches graph (a).

137. $a_n = \displaystyle\sum_{k=1}^{n} 4\left(\dfrac{1}{2}\right)^{k-1}$

$a_1 = 4$ and $a_n \to 8$ as $n \to \infty$

Matches graph (b).

138 . $a_n = \sum_{k=1}^{n} 4\left(-\frac{1}{2}\right)^{k-1}$

$a_1 = 4$ and $a_n \to \frac{8}{3}$ as $n \to \infty$.

Matches graph (c).

139. $S_6 = S_5 + S_4 + S_3 = 130 + 70 + 40 = 240$

$S_7 = S_6 + S_5 + S_4 = 240 + 130 + 70 = 440$

$S_8 = S_7 + S_6 + S_5 = 440 + 240 + 130 = 810$

$S_9 = S_8 + S_7 + S_6 = 810 + 440 + 240 = 1490$

$S_{10} = S_9 + S_8 + S_7 = 1490 + 810 + 440 = 2740$

140. $0 \le p \le 1$, closed interval

Problem Solving for Chapter 9

1. $x_0 = 1$ and $x_n = \frac{1}{2}x_{n-1} + \frac{1}{x_{n-1}}, n = 1, 2, \dots$

$x_0 = 1$

$x_1 = \frac{1}{2}(1) + \frac{1}{1} = \frac{3}{2} = 1.5$

$x_2 = \frac{1}{2}\left(\frac{3}{2}\right) + \frac{1}{3/2} = \frac{17}{12} = 1.41\overline{6}$

$x_3 = \frac{1}{2}\left(\frac{17}{12}\right) + \frac{1}{17/12} = \frac{577}{408} \approx 1.414215686$

$x_4 = \frac{1}{2}\left(\frac{577}{408}\right) + \frac{1}{577/408} \approx 1.414213562$

$x_5 = \frac{1}{2}x_4 + \frac{1}{x_4} \approx 1.414213562$

$x_6 \approx x_7 \approx x_8 \approx x_9 \approx 1.414213562$

Conjecture: $x_n \to \sqrt{2}$ as $n \to \infty$

2. $a_n = \frac{n+1}{n^2+1}$

(a)

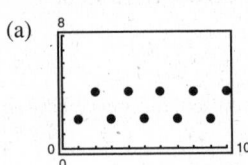

(b) $a_n \to 0$ as $n \to \infty$

(c)

n	1	10	100	1000	10,000
a_n	1	$\frac{11}{101}$	$\frac{101}{10,001}$	$\frac{1001}{1,000,001}$	$\frac{10,001}{100,000,001}$

(d) $a_n \to 0$ as $n \to \infty$

3. $a_n = 3 + (-1)^n$

(a)

(b) $a_n = \begin{cases} 2, & \text{if } n \text{ is odd} \\ 4, & \text{if } n \text{ is even} \end{cases}$

(c)

n	1	10	101	1000	10,001
a_n	2	4	2	4	2

(d) As $n \to \infty$, a_n oscillates between 2 and 4 and does not approach a fixed value.

4. Let $a_n = dn + c$, an arithmetic sequence with a common difference of d.

(a) If C is added to each term, then the resulting sequence, $b_n = a_n + C = dn + c + C$ is still arithmetic with a common difference of d.

(b) If each term is multiplied by a nonzero constant C, then the resulting sequence, $b_n = C(dn + c) = Cdn + Cc$ is still arithmetic. The common difference is Cd.

(c) If each term is squared, the resulting sequence, $b_n = a_n^2 = (dn + c)^2$ is not arithmetic.

5. (a)

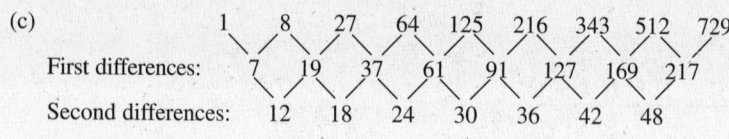

First differences: 3   5   7   9   11   13   15   17

In general, $b_n = 2n + 1$ for the first differences.

(b) Find the second differences of the perfect cubes.

(c)

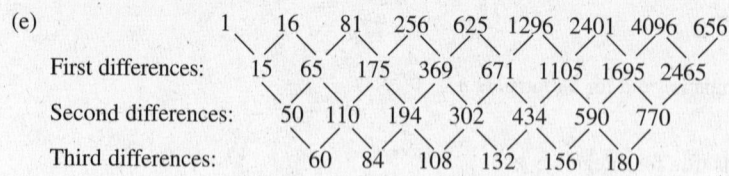

First differences: 7   19   37   61   91   127   169   217

Second differences: 12   18   24   30   36   42   48

In general, $c_n = 6(n + 1) = 6n + 6$ for the second differences.

(d) Find the third differences of the perfect fourth powers.

(e)

       1   16   81   256   625   1296   2401   4096   6561

First differences: 15   65   175   369   671   1105   1695   2465

Second differences: 50   110   194   302   434   590   770

Third differences: 60   84   108   132   156   180

In general, $d_n = 24n + 36$ for the third differences.

6. Distance: $\displaystyle\sum_{n=1}^{\infty} 20\left(\frac{1}{2}\right)^{n-1} = \frac{20}{1 - \frac{1}{2}} = 40$

Time: $\displaystyle\sum_{n=1}^{\infty} \left(\frac{1}{2}\right)^{n-1} = \frac{1}{1 - \frac{1}{2}} = 2$

In two seconds, both Achilles and the tortoise will be 40 feet away from Achilles starting point.

7. Side lengths: $1, \dfrac{1}{2}, \dfrac{1}{4}, \dfrac{1}{8}, \ldots$

$S_n = \left(\dfrac{1}{2}\right)^{n-1}$ for $n \geq 1$

Areas: $\dfrac{\sqrt{3}}{4}, \dfrac{\sqrt{3}}{4}\left(\dfrac{1}{2}\right)^2, \dfrac{\sqrt{3}}{4}\left(\dfrac{1}{4}\right)^2, \dfrac{\sqrt{3}}{4}\left(\dfrac{1}{8}\right)^2, \ldots$

$A_n = \dfrac{\sqrt{3}}{4}\left[\left(\dfrac{1}{2}\right)^{n-1}\right]^2 = \dfrac{\sqrt{3}}{4}\left(\dfrac{1}{2}\right)^{2n-2} = \dfrac{\sqrt{3}}{4}S_n^{\,2}$

8. $a_n = \begin{cases} \dfrac{a_{n-1}}{2}, & \text{if } a_{n-1} \text{ is even} \\ 3a_{n-1} + 1, & \text{if } a_{n-1} \text{ is odd} \end{cases}$

(a) $a_1 = 7$

$a_2 = 3(7) + 1 = 22$

$a_3 = \frac{22}{2} = 11$

$a_4 = 3(11) + 1 = 34$

$a_5 = \frac{34}{2} = 17$

$a_6 = 3(17) + 1 = 52$

$a_7 = \frac{52}{2} = 26$

$a_8 = \frac{26}{2} = 13$

$a_9 = 3(13) + 1 = 40$

$a_{10} = \frac{40}{2} = 20$

$a_{11} = \frac{20}{2} = 10$

$a_{12} = \frac{10}{2} = 5$

$a_{13} = 3(5) + 1 = 16$

$a_{14} = \frac{16}{2} = 8$

$a_{15} = \frac{8}{2} = 4$

$a_{16} = \frac{4}{2} = 2$

$a_{17} = \frac{2}{2} = 1$

$a_{18} = 3(1) + 1 = 4$

$a_{19} = \frac{4}{2} = 2$

$a_{20} = \frac{2}{2} = 1$

(b) $a_1 = 4$

$a_2 = 2$

$a_3 = 1$

$a_4 = 4$

$a_5 = 2$

$a_6 = 1$

$a_7 = 4$

$a_8 = 2$

$a_9 = 1$

$a_{10} = 4$

$a_1 = 5$

$a_2 = 16$

$a_3 = 8$

$a_4 = 4$

$a_5 = 2$

$a_6 = 1$

$a_7 = 4$

$a_8 = 2$

$a_9 = 1$

$a_{10} = 4$

$a_1 = -3$

$a_2 = -8$

$a_3 = -4$

$a_4 = -2$

$a_5 = -1$

$a_6 = -2$

$a_7 = -1$

$a_8 = -2$

$a_9 = -1$

$a_{10} = -2$

Eventually the terms repeat; 4, 2, 1 if a_1 is a positive integer and $-2, -1$ if a_1 is a negative integer.

9. The numbers 1, 5, 12, 22, 35, 51, . . . can be written recursively as $P_n = P_{n-1} + (3n - 2)$. Show that $P_n = n(3n - 1)/2$.

1. For $n = 1$: $1 = \dfrac{1(3 - 1)}{2}$

2. Assume $P_k = \dfrac{k(3k - 1)}{2}$.

   Then, $P_{k+1} = P_k + [3(k + 1) - 2]$

   $= \dfrac{k(3k - 1)}{2} + (3k + 1) = \dfrac{k(3k - 1) + 2(3k + 1)}{2}$

   $= \dfrac{3k^2 + 5k + 2}{2} = \dfrac{(k + 1)(3k + 2)}{2}$

   $= \dfrac{(k + 1)[3(k + 1) - 1]}{2}.$

   Therefore, by mathematical induction, the formula is valid for all integers $n \geq 1$.

10. (a) If P_3 is true and P_k implies P_{k+1}, then P_n is true for integers $n \geq 3$.

   (b) If $P_1, P_2, P_3, \ldots, P_{50}$ are all true, then you can draw *no* conclusion about P_n in general other than it is true for $1 \leq n \leq 50$.

   (c) If $P_1, P_2,$ and P_3 are all true, but the truth of P_k does not imply that P_{k+1} is true, then P_n is false for some values of $n \geq 4$. You can only conclude that it is true for $P_1, P_2,$ and P_3.

   (d) If P_2 is true and P_{2k} implies P_{2k+2}, then P_{2n} is true for all integers $n \geq 1$.

11. (a) The Fibonacci sequence is defined as follows: $f_1 = 1, f_2 = 1, f_n = f_{n-2} + f_{n-1}$ for $n \geq 3$.

   By this definition $f_3 = f_1 + f_2 = 2, f_4 = f_2 + f_3 = 3, f_5 = f_4 + f_3 = 5, f_6 = f_5 + f_4 = 8, \ldots$

   1. For $n = 2$: $f_1 + f_2 = 2$ and $f_4 - 1 = 2$

   2. Assume $f_1 + f_2 + \ldots + f_k = f_{k+2} - 1$.

      Then, $f_1 + f_2 + f_3 + \ldots + f_k + f_{k+1} = f_{k+2} - 1 + f_{k+1} = (f_{k+2} + f_{k+1}) - 1 = f_{k+3} - 1 = f_{(k+1)+2} - 1.$

   Therefore, by mathematical induction, the formula is valid for all integers $n \geq 2$.

   (b) $S_{20} = f_{22} - 1 = 17{,}711 - 1 = 17{,}710$

12. (a) Odds against choosing a red marble $= \dfrac{\text{number of non-red marbles}}{\text{number of red marbles}}$

   $$\dfrac{4}{1} = \dfrac{x}{6}$$

   $$24 = x \quad \text{(number of non-red marbles)}$$

   $$\text{Total marbles} = 6 + 24 = 30$$

   (b) Odds in favor of choosing a blue marble $= \dfrac{\text{number of blue marbles}}{\text{number of yellow marbles}} = \dfrac{3}{7}$

   Odds against choosing a blue marble $= \dfrac{\text{number of yellow marbles}}{\text{number of blue marbles}} = \dfrac{7}{3}$

   (c) $P(E) = \dfrac{n(E)}{n(S)} = \dfrac{n(E)}{n(E) + n(E')} = \dfrac{n(E)/n(E')}{n(E)/n(E') + n(E')/n(E')}$

   $$P(E) = \dfrac{\text{odds in favor of } E}{\text{odds in favor of } E + 1}$$

—CONTINUED—

12. **—CONTINUED—**

(d) $\quad P(E) = \dfrac{n(E)}{n(S)} \qquad\qquad P(E') = \dfrac{n(E')}{n(S)}$

$\quad n(S)P(E) = n(E) \qquad n(S)P(E') = n(E')$

Odds in favor of event $E = \dfrac{n(E)}{n(E')} = \dfrac{n(S)P(E)}{n(S)P(E')} = \dfrac{P(E)}{P(E')}$

13. $\dfrac{1}{3}$

14. $1 - \dfrac{\text{Area of triangle}}{\text{Area of circle}} = 1 - \dfrac{\frac{1}{2}(12)(6)}{\pi(6)^2} = 1 - \dfrac{1}{\pi}$

$\qquad\qquad\qquad\qquad\qquad\qquad \approx 0.682$

$\qquad\qquad\qquad\qquad\qquad\qquad = 68.2\%$

15. (a) $V = \left(\dfrac{1}{_{47}C_5(27)}\right)(12{,}000{,}000) + \left(1 - \dfrac{1}{_{47}C_5(27)}\right)(-1)$

$\qquad\quad \approx -\$0.71$

(b) $\quad V = \dfrac{1}{36}(1) + \dfrac{1}{36}(4) + \dfrac{1}{36}(9) + \dfrac{1}{36}(16) + \dfrac{1}{36}(25) + \dfrac{1}{36}(36) + \dfrac{30}{36}(0) \approx 2.53$

$\qquad \dfrac{60}{2.53} \approx 24 \text{ turns}$

Chapter 9 Practice Test

1. Write out the first five terms of the sequence $a_n = \dfrac{2n}{(n+2)!}$.

2. Write an expression for the nth term of the sequence $\frac{4}{3}, \frac{5}{9}, \frac{6}{27}, \frac{7}{81}, \frac{8}{243}, \ldots$

3. Find the sum $\displaystyle\sum_{i=1}^{6} (2i - 1)$.

4. Write out the first five terms of the arithmetic sequence where $a_1 = 23$ and $d = -2$.

5. Find a_n for the arithmetic sequence with $a_1 = 12$, $d = 3$, and $n = 50$.

6. Find the sum of the first 200 positive integers.

7. Write out the first five terms of the geometric sequence with $a_1 = 7$ and $r = 2$.

8. Evaluate $\displaystyle\sum_{n=1}^{10} 6\left(\frac{2}{3}\right)^{n-1}$.

9. Evaluate $\displaystyle\sum_{n=0}^{\infty} (0.03)^n$.

10. Use mathematical induction to prove that $1 + 2 + 3 + 4 + \cdots + n = \dfrac{n(n+1)}{2}$.

11. Use mathematical induction to prove that $n! > 2^n$, $n \geq 4$.

12. Evaluate $_{13}C_4$.

13. Expand $(x + 3)^5$.

14. Find the term involving x^7 in $(x - 2)^{12}$.

15. Evaluate $_{30}P_4$.

16. How many ways can six people sit at a table with six chairs?

17. Twelve cars run in a race. How many different ways can they come in first, second, and third place? (Assume that there are no ties.)

18. Two six-sided dice are tossed. Find the probability that the total of the two dice is less than 5.

19. Two cards are selected at random form a deck of 52 playing cards without replacement. Find the probability that the first card is a King and the second card is a black ten.

20. A manufacturer has determined that for every 1000 units it produces, 3 will be faulty. What is the probability that an order of 50 units will have one or more faulty units?

C H A P T E R 1 0
Topics in Analytic Geometry

CHAPTER 10
Topics in Analytic Geometry

Section 10.1   Lines

- The **inclination** of a nonhorizontal line is the positive angle θ, $(\theta < 180°)$ measured counterclockwise from the x-axis to the line. A horizontal line has an inclination of zero.
- If a nonvertical line has inclination of θ and slope m, then $m = \tan \theta$.
- If two nonperpendicular lines have slopes m_1 and m_2, then the angle between the lines is given by
$$\tan \theta = \left| \frac{m_2 - m_1}{1 + m_1 m_2} \right|.$$
- The distance between a point (x_1, y_1) and a line $Ax + By + C = 0$ is given by
$$d = \frac{|Ax_1 + By_1 + C|}{\sqrt{A^2 + B^2}}.$$

Vocabulary Check

1. inclination
2. $\tan \theta$
3. $\left| \dfrac{m_2 - m_1}{1 + m_1 m_2} \right|$
4. $\dfrac{|Ax_1 + By_1 + C|}{\sqrt{A^2 + B^2}}$

---

1. $m = \tan \dfrac{\pi}{6} = \dfrac{\sqrt{3}}{3}$

2. $m = \tan \dfrac{\pi}{4} = 1$

3. $m = \tan \dfrac{3\pi}{4} = -1$

4. $m = \tan \dfrac{2\pi}{3} = -\sqrt{3}$

5. $m = \tan \dfrac{\pi}{3} = \sqrt{3}$

6. $m = \tan \dfrac{5\pi}{6} = -\dfrac{\sqrt{3}}{3}$

7. $m = \tan 1.27 \approx 3.2236$

8. $m = \tan 2.88 \approx -0.2677$

9. $m = -1$

$-1 = \tan \theta$

$\theta = 180° + \arctan(-1)$

$= \dfrac{3\pi}{4}$ radians $= 135°$

10. $-2 = \tan \theta$

$\theta = \tan^{-1}(-2) + \pi$

≈ 2.034 radians $\approx 116.6°$

11. $m = 1$

$1 = \tan \theta$

$\theta = \dfrac{\pi}{4}$ radian $= 45°$

12. $2 = \tan \theta$

$\theta = \tan^{-1} 2$

≈ 1.107 radians $\approx 63.4°$

13. $m = \frac{3}{4}$

$\frac{3}{4} = \tan \theta$

$\theta = \arctan\left(\frac{3}{4}\right) \approx 0.6435$ radian $\approx 36.9°$

14. $-\frac{5}{2} = \tan \theta$

$\theta = \tan^{-1}\left(-\frac{5}{2}\right) + \pi \approx 1.9513$ radians $\approx 111.8°$

15. $(6, 1), (10, 8)$

$$m = \frac{8 - 1}{10 - 6} = \frac{7}{4}$$

$$\frac{7}{4} = \tan \theta$$

$$\theta = \arctan\left(\frac{7}{4}\right) \approx 1.0517 \text{ radians} \approx 60.3°$$

16. $m = \frac{8 - (-3)}{12 - (-4)} = \frac{11}{16}$

$$\frac{11}{16} = \tan \theta$$

$$\theta = \tan^{-1} \frac{11}{16} \approx 0.6023 \text{ radian} \approx 34.5°$$

17. $(-2, 20), (10, 0)$

$$m = \frac{0 - 20}{10 - (-2)} = -\frac{20}{12} = -\frac{5}{3}$$

$$-\frac{5}{3} = \tan \theta$$

$$\theta = \pi + \arctan\left(-\frac{5}{3}\right) \approx 2.1112 \text{ radians} \approx 121.0°$$

18. $m = \frac{100 - 0}{0 - 50} = -2$

$$-2 = \tan \theta$$

$$\theta = \tan^{-1}(-2) + \pi \approx 2.0344 \text{ radians} \approx 116.6°$$

19. $6x - 2y + 8 = 0$

$$y = 3x + 4 \Rightarrow m = 3$$

$$3 = \tan \theta$$

$$\theta = \arctan 3 \approx 1.2490 \text{ radians} \approx 71.6°$$

20. $4x + 5y - 9 = 0$

$$y = -\tfrac{4}{5}x + \tfrac{9}{5} \Rightarrow m = -\tfrac{4}{5}$$

$$-\tfrac{4}{5} = \tan \theta$$

$$\theta = \tan^{-1}\left(-\tfrac{4}{5}\right) + \pi$$

$$\approx 2.4669 \text{ radians} \approx 141.3°$$

21. $5x + 3y = 0$

$$y = -\frac{5}{3}x \Rightarrow m = -\frac{5}{3}$$

$$-\frac{5}{3} = \tan \theta$$

$$\theta = \pi + \arctan\left(-\frac{5}{3}\right) \approx 2.1112 \text{ radians} \approx 121.0°$$

22. $x - y - 10 = 0$

$$y = x - 10 \Rightarrow m = 1$$

$$1 = \tan \theta$$

$$\theta = \tan^{-1} 1 = 45° = \frac{\pi}{4} \text{ radian}$$

23. $3x + y = 3 \Rightarrow y = -3x + 3 \Rightarrow m_1 = -3$

$$x - y = 2 \Rightarrow y = x - 2 \Rightarrow m_2 = 1$$

$$\tan \theta = \left|\frac{1 - (-3)}{1 + (-3)(1)}\right| = 2$$

$$\theta = \arctan 2 \approx 1.1071 \text{ radians} \approx 63.4°$$

24. $x + 3y = 2 \Rightarrow y = -\frac{1}{3}x + \frac{2}{3} \Rightarrow m_1 = -\frac{1}{3}$

$$x - 2y = -3 \Rightarrow y = \frac{1}{2}x + \frac{3}{2} \Rightarrow m_2 = \frac{1}{2}$$

$$\tan \theta = \left|\frac{(1/2) - (-1/3)}{1 + (-1/3)(1/2)}\right| = 1$$

$$\theta = \tan^{-1} 1 = 45° = \frac{\pi}{4} \text{ radian}$$

25. $x - y = 0 \Rightarrow y = x \Rightarrow m_1 = 1$

$$3x - 2y = -1 \Rightarrow y = \frac{3}{2}x + \frac{1}{2} \Rightarrow m_2 = \frac{3}{2}$$

$$\tan \theta = \left|\frac{\frac{3}{2} - 1}{1 + \left(\frac{3}{2}\right)(1)}\right| = \frac{1}{5}$$

$$\theta = \arctan \frac{1}{5} \approx 0.1974 \text{ radian} \approx 11.3°$$

26. $2x - y = 2 \Rightarrow y = 2x - 2 \Rightarrow m_1 = 2$

$$4x + 3y = 24 \Rightarrow y = -\frac{4}{3}x + 8 \Rightarrow m_2 = -\frac{4}{3}$$

$$\tan \theta = \left|\frac{(-4/3) - 2}{1 + (2)(-4/3)}\right| = 2$$

$$\theta = \tan^{-1} 2 \approx 63.4° \approx 1.1071 \text{ radians}$$

27. $x - 2y = 7 \Rightarrow y = \dfrac{1}{2}x - \dfrac{7}{2} \Rightarrow m_1 = \dfrac{1}{2}$

$6x + 2y = 5 \Rightarrow y = -3x + \dfrac{5}{2} \Rightarrow m_2 = -3$

$\tan \theta = \left| \dfrac{-3 - \frac{1}{2}}{1 + \left(\frac{1}{2}\right)(-3)} \right| = 7$

$\theta = \arctan 7 \approx 1.4289 \text{ radians} \approx 81.9°$

28. $5x + 2y = 16 \Rightarrow y = -\dfrac{5}{2}x + 8 \Rightarrow m_1 = -\dfrac{5}{2}$

$3x - 5y = -1 \Rightarrow y = \dfrac{3}{5}x + \dfrac{1}{5} \Rightarrow m_2 = \dfrac{3}{5}$

$\tan \theta = \left| \dfrac{(-5/2) - (3/5)}{1 + (-5/2)(3/5)} \right| = \dfrac{31}{5}$

$\theta = \tan^{-1} \dfrac{31}{5} \approx 80.8° \approx 1.4109 \text{ radians}$

29. $x + 2y = 8 \Rightarrow y = -\dfrac{1}{2}x + 4 \Rightarrow m_1 = -\dfrac{1}{2}$

$x - 2y = 2 \Rightarrow y = \dfrac{1}{2}x - 1 \Rightarrow m_2 = \dfrac{1}{2}$

$\tan \theta = \left| \dfrac{\frac{1}{2} - \left(-\frac{1}{2}\right)}{1 + \left(-\frac{1}{2}\right)\left(\frac{1}{2}\right)} \right| = \dfrac{4}{3}$

$\theta = \arctan\left(\dfrac{4}{3}\right) \approx 0.9273 \text{ radian} \approx 53.1°$

30. $3x - 5y = 3 \Rightarrow y = \dfrac{3}{5}x - \dfrac{3}{5} \Rightarrow m_1 = \dfrac{3}{5}$

$3x + 5y = 12 \Rightarrow y = -\dfrac{3}{5}x + \dfrac{12}{5} \Rightarrow m_2 = -\dfrac{3}{5}$

$\tan \theta = \left| \dfrac{(3/5) - (-3/5)}{1 + (3/5)(-3/5)} \right| = \dfrac{15}{8}$

$\theta = \tan^{-1} \dfrac{15}{8} \approx 61.9° \approx 1.0808 \text{ radians}$

31. $0.05x - 0.03y = 0.21 \Rightarrow y = \dfrac{5}{3}x - 7 \Rightarrow m_1 = \dfrac{5}{3}$

$0.07x + 0.02y = 0.16 \Rightarrow y = -\dfrac{7}{2}x + 8 \Rightarrow m_2 = -\dfrac{7}{2}$

$\tan \theta = \left| \dfrac{\left(-\frac{7}{2}\right) - \left(\frac{5}{3}\right)}{1 + \left(\frac{5}{3}\right)\left(-\frac{7}{2}\right)} \right| = \dfrac{31}{29}$

$\theta = \arctan\left(\dfrac{31}{29}\right) \approx 0.8187 \text{ radian} \approx 46.9°$

32. $0.02x - 0.05y = -0.19 \Rightarrow y = \dfrac{2}{5}x + \dfrac{19}{5} \Rightarrow m_1 = \dfrac{2}{5}$

$0.03x + 0.04y = 0.52 \Rightarrow y = -\dfrac{3}{4}x + 13 \Rightarrow m_2 = -\dfrac{3}{4}$

$\tan \theta = \left| \dfrac{(-3/4) - (2/5)}{1 + (2/5)(-3/4)} \right| \approx \dfrac{23}{14}$

$\theta = \tan^{-1}\left(\dfrac{23}{14}\right) \approx 58.7° \approx 1.0240 \text{ radians}$

33. Let $A = (2, 1)$, $B = (4, 4)$, and $C = (6, 2)$.

Slope of AB: $m_1 = \dfrac{1 - 4}{2 - 4} = \dfrac{3}{2}$

Slope of BC: $m_2 = \dfrac{4 - 2}{4 - 6} = -1$

Slope of AC: $m_3 = \dfrac{1 - 2}{2 - 6} = \dfrac{1}{4}$

$\tan A = \left| \dfrac{\frac{1}{4} - \frac{3}{2}}{1 + \left(\frac{3}{2}\right)\left(\frac{1}{4}\right)} \right| = \dfrac{\frac{5}{4}}{\frac{11}{8}} = \dfrac{10}{11}$

$A = \arctan\left(\dfrac{10}{11}\right) \approx 42.3°$

$\tan B = \left| \dfrac{\frac{3}{2} - (-1)}{1 + (-1)\left(\frac{3}{2}\right)} \right| = \dfrac{\frac{5}{2}}{\frac{1}{2}} = 5$

$B = \arctan 5 \approx 78.7°$

$\tan C = \left| \dfrac{-1 - \frac{1}{4}}{1 + \left(\frac{1}{4}\right)(-1)} \right| = \dfrac{\frac{5}{4}}{\frac{3}{4}} = \dfrac{5}{3}$

$C = \arctan\left(\dfrac{5}{3}\right) \approx 59.0°$

34. Let $A = (-3, 2)$, $B = (1, 3)$, and $C = (2, 0)$.

Slope of AB: $m_1 = \dfrac{2 - 3}{-3 - 1} = \dfrac{1}{4}$

Slope of BC: $m_2 = \dfrac{3 - 0}{1 - 2} = -3$

Slope of AC: $m_3 = \dfrac{2 - 0}{-3 - 2} = -\dfrac{2}{5}$

$\tan A = \left| \dfrac{(1/4) - (-2/5)}{1 + (-2/5)(1/4)} \right| = \dfrac{13/20}{18/20} = \dfrac{13}{18}$

$A = \tan^{-1}\left(\dfrac{13}{18}\right) \approx 35.8°$

$\tan C = \left| \dfrac{-3 - (-2/5)}{1 + (-3)(-2/5)} \right| \approx \dfrac{13/5}{11/5} = \dfrac{13}{11}$

$C = \tan^{-1}\left(\dfrac{13}{11}\right) \approx 49.8°$

$B = 180° - A - C \approx 180° - 35.8° - 49.8°$

$= 94.4°$

35. Let $A = (-4, -1)$, $B = (3, 2)$, and $C = (1, 0)$.

Slope of AB:　$m_1 = \dfrac{-1 - 2}{-4 - 3} = \dfrac{3}{7}$

Slope of BC:　$m_2 = \dfrac{2 - 0}{3 - 1} = 1$

Slope of AC:　$m_3 = \dfrac{-1 - 0}{-4 - 1} = \dfrac{1}{5}$

$\tan A = \left| \dfrac{\frac{1}{5} - \frac{3}{7}}{1 + \left(\frac{3}{7}\right)\left(\frac{1}{5}\right)} \right| = \dfrac{\frac{8}{35}}{\frac{38}{35}} = \dfrac{4}{19}$

$A = \arctan\left(\dfrac{4}{19}\right) \approx 11.9°$

$\tan B = \left| \dfrac{1 - \frac{3}{7}}{1 + \left(\frac{3}{7}\right)(1)} \right| = \dfrac{\frac{4}{7}}{\frac{10}{7}} = \dfrac{2}{5}$

$B = \arctan\left(\dfrac{2}{5}\right) \approx 21.8°$

$C = 180° - A - B$

$\approx 180° - 11.9° - 21.8° = 146.3°$

36. Let $A = (-3, 4)$, $B = (2, 1)$, and $C = (-2, 2)$.

Slope of AB: $m_1 = \dfrac{4 - 1}{-3 - 2} = -\dfrac{3}{5}$

Slope of BC: $m_2 = \dfrac{1 - 2}{2 - (-2)} = -\dfrac{1}{4}$

Slope of AC: $m_3 = \dfrac{4 - 2}{-3 - (-2)} = -2$

$\tan A = \left| \dfrac{(-3/5) - (-2)}{1 + (-3/5)(-2)} \right| = \dfrac{7}{11}$

$A = \tan^{-1}\left(\dfrac{7}{11}\right) \approx 32.5°$

$\tan B = \left| \dfrac{(-3/5) - (-1/4)}{1 + (-3/5)(-1/4)} \right| = \dfrac{7}{23}$

$B = \tan^{-1}\left(\dfrac{7}{23}\right) \approx 16.9°$

$C = 180° - A - B \approx 180° - 32.5° - 16.9°$

$= 130.6°$

37. $(0, 0) \implies x_1 = 0$ and $y_1 = 0$

$4x + 3y = 0 \implies A = 4, B = 3,$ and $C = 0$

$d = \dfrac{|4(0) + 3(0) + 0|}{\sqrt{4^2 + 3^2}} = \dfrac{0}{5} = 0$

Note: The point is *on* the line.

38. $(0, 0) \implies x_1 = 0$　and $y_1 = 0$

$2x - y - 4 = 0 \implies A = 2, B = -1,$ and $C = -4$

$d = \dfrac{|2(0) + (-1)(0) + (-4)|}{\sqrt{2^2 + (-1)^2}}$

$= \dfrac{4}{\sqrt{5}} = \dfrac{4\sqrt{5}}{5} \approx 1.7889$

39. $(2, 3) \implies x_1 = 2$ and $y_1 = 3$

$4x + 3y - 10 = 0 \implies A = 4, B = 3,$ and $C = -10$

$d = \dfrac{|4(2) + 3(3) + (-10)|}{\sqrt{4^2 + 3^2}} = \dfrac{7}{5}$

40. $(-2, 1) \implies x_1 = -2$ and $y_1 = 1$

$x - y - 2 = 0 \implies A = 1, B = -1,$ and $C = -2$

$d = \dfrac{|1(-2) + (-1)(1) + (-2)|}{\sqrt{1^2 + (-1)^2}}$

$= \dfrac{5}{\sqrt{2}} = \dfrac{5\sqrt{2}}{2} \approx 3.5355$

41. $(6, 2) \implies x_1 = 6$ and $y_1 = 2$

$x + 1 = 0 \implies A = 1, B = 0,$ and $C = 1$

$d = \dfrac{|1(6) + 0(2) + 1|}{\sqrt{1^2 + 0^2}} = 7$

42. $(10, 8) \implies x_1 = 10$ and $y_1 = 8$

$y - 4 = 0 \implies A = 0, B = 1,$ and $C = -4$

$d = \dfrac{|0(10) + 1(8) + (-4)|}{\sqrt{0^2 + 1^2}} = \dfrac{4}{1} = 4$

43. $(0, 8) \implies x_1 = 0$ and $y_1 = 8$

$6x - y = 0 \implies A = 6, B = -1,$ and $C = 0$

$d = \dfrac{|6(0) + (-1)(8) + 0|}{\sqrt{6^2 + (-1)^2}}$

$= \dfrac{8}{\sqrt{37}} = \dfrac{8\sqrt{37}}{37} \approx 1.3152$

44. $(4, 2) \implies x_1 = 4$ and $y_1 = 2$

$x - y - 20 = 0 \implies A = 1, B = -1,$ and $C = -20$

$d = \dfrac{|1(4) + (-1)(2) + (-20)|}{\sqrt{1^2 + (-1)^2}}$

$= \dfrac{18}{\sqrt{2}} = 9\sqrt{2} \approx 12.7279$

45. $A = (0, 0)$, $B = (1, 4)$, $C = (4, 0)$

(a)

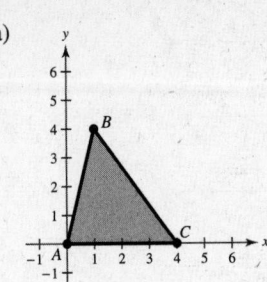

(b) The slope the line through AC is $m = \dfrac{0 - 0}{4 - 0} = 0$.

The equation of the line through AC is $y = 0$.

The distance between the line and $B = (1, 4)$ is

$$d = \frac{|0(1) + (1)(4) + 0|}{\sqrt{0^2 + 1^2}} = 4.$$

(c) The distance between A and C is 4.

$$A = \frac{1}{2}(4)(4) = 8 \text{ square units}$$

46. (a)

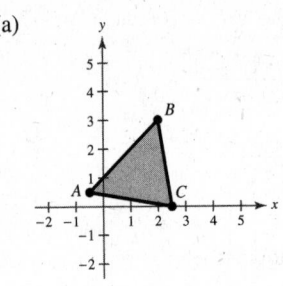

(b) The slope of the line through AC is $m = \dfrac{0 + 2}{0 - 5} = -\dfrac{2}{5}$.

The equation of the line is $y - 0 = -\dfrac{2}{5}(x - 0) \Rightarrow 2x + 5y = 0$.

The altitude from vertex B to side AC is the distance between the line through AC and

$$B = (4, 5) \Rightarrow d = \frac{|2(4) + 5(5) + 0|}{\sqrt{2^2 + 5^2}} = \frac{33}{\sqrt{29}} = \frac{33\sqrt{29}}{29}.$$

(c) The distance between A and C is $d = \sqrt{(0 - 5)^2 + (0 + 2)^2} = \sqrt{29}$, which is the length of the base of the triangle. So,

$$A = \frac{1}{2}\sqrt{29}\left(\frac{33\sqrt{29}}{29}\right) = \frac{33}{2} \text{ square units.}$$

47. $A = \left(-\dfrac{1}{2}, \dfrac{1}{2}\right)$, $B = (2, 3)$, $C = \left(\dfrac{5}{2}, 0\right)$

(a)

(b) The slope of the line through AC is $m = \dfrac{\frac{1}{2} - 0}{\left(-\frac{1}{2}\right) - \frac{5}{2}} = -\dfrac{1}{6}$.

The equation of the line through AC is $y - 0 = -\dfrac{1}{6}\left(x - \dfrac{5}{2}\right) \Rightarrow 2x + 12y - 5 = 0$.

The distance between the line and $B = (2, 3)$ is $d = \dfrac{|2(2) + 12(3) + (-5)|}{\sqrt{2^2 + 12^2}} = \dfrac{35}{\sqrt{148}} = \dfrac{35\sqrt{37}}{74}$.

(c) The distance between A and C is $d = \sqrt{\left[\left(-\dfrac{1}{2}\right) - \left(\dfrac{5}{2}\right)\right]^2 + \left[\left(\dfrac{1}{2}\right) - 0\right]^2} = \dfrac{\sqrt{37}}{2}$.

$$A = \frac{1}{2}\left(\frac{\sqrt{37}}{2}\right)\left(\frac{35\sqrt{37}}{74}\right) = \frac{35}{8} \text{ square units}$$

48. (a)

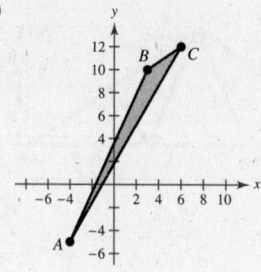

(b) The slope of the line through AC is $m = \dfrac{12 - (-5)}{6 - (-4)} = \dfrac{17}{10}$.

The equation of the line through AC is $y - 12 = \dfrac{17}{10}(x - 6) \implies 17x - 10y + 18 = 0$.

The altitude from vertex B to side AC is the distance between the line through AC and

$$B = (3, 10) \implies d = \frac{|17(3) + (-10)(10) + 18|}{\sqrt{17^2 + (-10)^2}} = \frac{31}{\sqrt{389}} = \frac{31\sqrt{389}}{389}.$$

(c) The distance between A and C is $d = \sqrt{(6 + 4)^2 + (12 + 5)^2} = \sqrt{389}$, which is the length of the base of the triangle.

$$A = \frac{1}{2}\left(\sqrt{389}\right)\left(\frac{31\sqrt{389}}{389}\right) = \frac{31}{2}$$

49. $x + y = 1 \implies (0, 1)$ is a point on the line $\implies x_1 = 0$ and $y_1 = 1$

$x + y = 5 \implies A = 1, B = 1,$ and $C = -5$

$$d = \frac{|1(0) + 1(1) + (-5)|}{\sqrt{1^2 + 1^2}} = \frac{4}{\sqrt{2}} = 2\sqrt{2}$$

50. $3x - 4y = 1$

$3x - 4y = 10$

A point on $3x - 4y = 10$ is $\left(0, -\frac{5}{2}\right)$. The distance between $\left(0, -\frac{5}{2}\right)$ and $3x - 4y = 1$ is:

$A = 3, B = -4, C = -1, x_1 = 0, y_1 = -\dfrac{5}{2}$

$$d = \frac{|3(0) + (-4)(-5/2) - 1|}{\sqrt{3^2 + (-4)^2}} = \frac{9}{5}$$

51. Slope: $m = \tan 0.1 \approx 0.1003$

Change in elevation: $\sin 0.1 = \dfrac{x}{2(5280)}$

$x \approx 1054$ feet

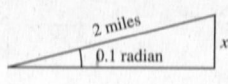

Not drawn to scale

52. Slope: $m = \tan 0.2 \approx 0.2027$

Change in elevation:

$\sin 0.20 = \dfrac{x}{5280} \implies x = 5280 \sin 0.20 \approx 1049$ feet

53. Slope $= \dfrac{3}{5}$

Inclination $= \tan^{-1}\dfrac{3}{5} \approx 31.0°$

54. (a)

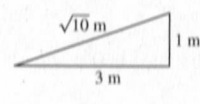

(b) $m = \dfrac{1}{3}$

$\dfrac{1}{3} = \tan \theta$

$\tan^{-1}\left(\dfrac{1}{3}\right) = \theta$

or $\theta \approx 18.4°$

(c) Use similar triangles:

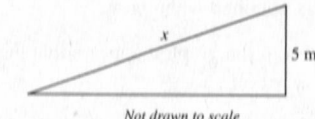

Not drawn to scale

$\dfrac{x}{5} = \dfrac{\sqrt{10}}{1}$

$x = 5\sqrt{10} \approx 15.8$ m

55. $\tan \gamma = \dfrac{6}{9}$

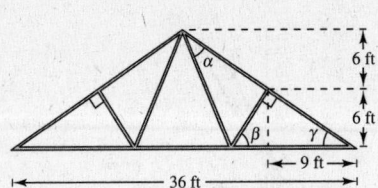
6 ft
6 ft
◄—9 ft—►
◄——————— 36 ft ———————►

$\gamma = \arctan\left(\dfrac{2}{3}\right) \approx 33.69°$

$\beta = 90 - \gamma \approx 56.31°$

Also, since the right triangles containing α and β are equal, $\alpha = \gamma \approx 33.69°$.

56. (a) $m = \tan$

$0.709 = \tan \theta$

$\tan^{-1} 0.709 = \theta$

$\theta \approx 0.6167$ radian, or $35.34°$

(b) $\sin \theta = \dfrac{\text{elev } \Delta}{896.5}$

$896.5 \sin \theta = \text{elev } \Delta$

$\text{elev } \Delta = 896.5 \sin 0.6167 \approx 518.5$ ft

(c) $m = 0.709$ and y-intercept $= (0, 0)$, so $y = 0.709x$.

(d)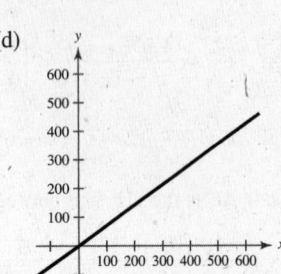

57. True. The inclination of a line is related to its slope by $m = \tan \theta$. If the angle is greater than $\pi/2$ but less than π, then the angle is in the second quadrant where the tangent function is negative.

58. False. Substitute $m_1 = \tan \theta_1$ and $m_2 = \tan \theta_2$ into the formula for the angle between two lines.

59. (a) $(0, 0) \implies x_1 = 0$ and $y_1 = 0$

$y = mx + 4 \implies 0 = mx - y + 4$

$d = \dfrac{|m(0) + (-1)(0) + 4|}{\sqrt{m^2 + (-1)^2}} = \dfrac{4}{\sqrt{m^2 + 1}}$

(b)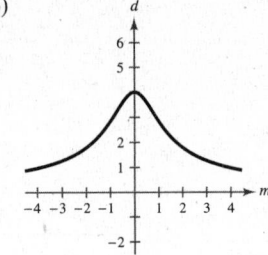

(c) The maximum distance of 4 occurs when the slope m is 0 and the line through $(0, 4)$ is horizontal.

(d) The graph has a horizontal asymptote at $d = 0$. As the slope becomes larger, the distance between the origin and the line, $y = mx + 4$, becomes smaller and approaches 0.

60. Slope m and y-intercept $(0, 4)$

(a) $(x_1, y_1) = (3, 1)$ and line: $y = mx + 4$

$A = -m, B = 1, C = -4$

$d = \dfrac{|(-m)(3) + (1)(1) + (-4)|}{\sqrt{(-m)^2 + 1^2}} = \dfrac{3|m + 1|}{\sqrt{m^2 + 1}}$

(b)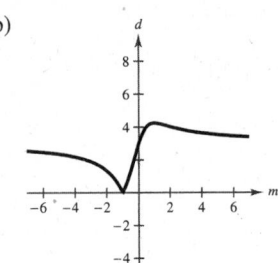

(c) From the graph it appears that the maximum distance is obtained when $m = 1$.

(d) From the graph it appears that the distance is 0 when $m = -1$.

(e) The asymptote of the graph in part (b) is $d = 3$. As the line approaches the vertical, the distance approaches 3.

61. $f(x) = (x - 7)^2$

x-intercept: $0 = (x - 7)^2 \Rightarrow x = 7$

$(7, 0)$

y-intercept: $y = (0 - 7)^2 = 49$

$(0, 49)$

62. $f(x) = (x + 9)^2$

$f(x) = (x + 9)^2 = 0 \Rightarrow x = -9$

x-intercept: $(-9, 0)$

$f(0) = (0 + 9)^2 = 81$

y-intercept: $(0, 81)$

63. $f(x) = (x - 5)^2 - 5$

x-intercepts: $0 = (x - 5)^2 - 5$

$5 = (x - 5)^2$

$\pm\sqrt{5} = x - 5$

$5 \pm \sqrt{5} = x$

$\left(5 \pm \sqrt{5}, 0\right)$

y-intercept: $y = (0 - 5)^2 - 5 = 20$

$(0, 20)$

64. $f(x) = (x + 11)^2 + 12$

$f(x) = (x + 11)^2 + 12 = 0$

$(x + 11)^2 = -12$

No solution

x-intercept: none

$f(0) = (0 + 11)^2 + 12 = 133$

y-intercept: $(0, 133)$

65. $f(x) = x^2 - 7x - 1$

x-intercepts: $0 = x^2 - 7x - 1$

$x = \dfrac{7 \pm \sqrt{53}}{2}$ by the Quadratic Formula

$\left(\dfrac{7 \pm \sqrt{53}}{2}, 0\right)$

y-intercept: $y = 0^2 - 7(0) - 1 = -1$

$(0, -1)$

66. $f(x) = x^2 + 9x - 22$

$f(x) = x^2 + 9x - 22 = 0$

$(x + 11)(x - 2) = 0$

$x = -11, 2$

x-intercepts: $(-11, 0), (2, 0)$

$f(0) = -22$

y-intercept: $(0, -22)$

67. $f(x) = 3x^2 + 2x - 16$

$= 3\left(x^2 + \tfrac{2}{3}x\right) - 16$

$= 3\left(x^2 + \tfrac{2}{3}x + \tfrac{1}{9}\right) - \tfrac{1}{3} - 16$

$= 3\left(x + \tfrac{1}{3}\right)^2 - \tfrac{49}{3}$

Vertex: $\left(-\tfrac{1}{3}, -\tfrac{49}{3}\right)$

68. $f(x) = 2x^2 - x - 21$

$= 2\left[x^2 - \tfrac{1}{2}x - \tfrac{21}{2}\right] = 2\left[x^2 - \tfrac{1}{2}x + \tfrac{1}{16} - \tfrac{1}{16} - \tfrac{21}{2}\right]$

$= 2\left[\left(x - \tfrac{1}{4}\right)^2 - \tfrac{169}{16}\right]$

$= 2\left(x - \tfrac{1}{4}\right)^2 - \tfrac{169}{8}$

Vertex: $\left(\tfrac{1}{4}, -\tfrac{169}{8}\right)$

69. $f(x) = 5x^2 + 34x - 7$

$= 5\left(x^2 + \tfrac{34}{5}x\right) - 7$

$= 5\left(x^2 + \tfrac{34}{5}x + \tfrac{289}{25}\right) - \tfrac{289}{5} - 7$

$= 5\left(x + \tfrac{17}{5}\right)^2 - \tfrac{324}{5}$

Vertex: $\left(-\tfrac{17}{5}, -\tfrac{324}{5}\right)$

70. $f(x) = -x^2 - 8x - 15$

$= -\left[x^2 + 8x + 15\right] = -\left[x^2 + 8x + 16 - 16 + 15\right]$

$= -\left[(x + 4)^2 - 1\right] = -(x + 4)^2 + 1$

Vertex: $(-4, 1)$

71. $f(x) = 6x^2 - x - 12$

$= 6\left(x^2 - \tfrac{1}{6}x\right) - 12$

$= 6\left(x^2 - \tfrac{1}{6}x + \tfrac{1}{144}\right) - \tfrac{1}{24} - 12$

$= 6\left(x - \tfrac{1}{12}\right)^2 - \tfrac{289}{24}$

Vertex: $\left(\tfrac{1}{12}, -\tfrac{289}{24}\right)$

72. $f(x) = -8x^2 - 34x - 21$

$= -8\left[x^2 + \tfrac{17}{4}x + \tfrac{21}{8}\right]$

$= -8\left[x^2 + \tfrac{17}{4}x + \tfrac{289}{64} - \tfrac{289}{64} + \tfrac{21}{8}\right]$

$= -8\left[\left(x + \tfrac{17}{8}\right)^2 - \tfrac{121}{64}\right]$

$= -8\left(x + \tfrac{17}{8}\right)^2 + \tfrac{121}{8}$

Vertex: $\left(-\tfrac{17}{8}, \tfrac{121}{8}\right)$

73. $f(x) = (x - 4)^2 + 3$

Vertex: $(4, 3)$

y-intercept: $(0, 19)$

x-intercept: None

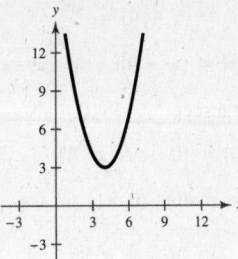

74. $f(x) = 6 - (x + 1)^2$

Vertex: $(-1, 6)$

x	-4	-3	-2	-1	0	1	2
$g(x)$	-3	2	5	6	5	2	-3

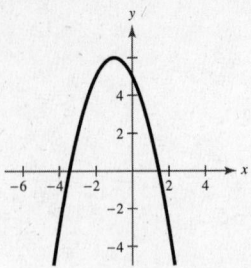

75. $g(x) = 2x^2 - 3x + 1$

$$= 2\left(x^2 - \frac{3}{2}x + \frac{9}{16}\right) - \frac{9}{8} + 1$$

$$= 2\left(x - \frac{3}{4}\right)^2 - \frac{1}{8}$$

Vertex: $\left(\frac{3}{4}, -\frac{1}{8}\right)$

y-intercept: $(0, 1)$

x-intercept: $\left(\frac{1}{2}, 0\right), (1, 0)$

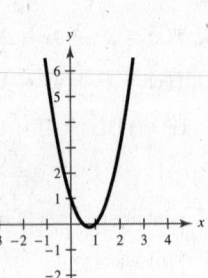

76. $g(x) = -x^2 + 6x - 8$

$$\frac{-b}{2a} = \frac{-6}{2(-1)} = 3 \implies \text{Vertex} = (3, g(3)) = (3, 1)$$

x	0	1	2	3	4	5	6
$g(x)$	-8	-3	0	1	0	-3	-8

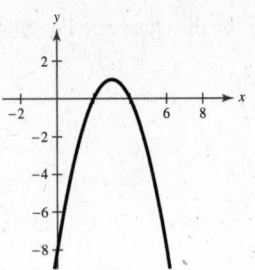

Section 10.2 Introduction to Conics: Parabolas

- A **parabola** is the set of all points (x, y) that are equidistant from a fixed line (**directrix**) and a fixed point (**focus**) not on the line.

- The standard equation of a parabola with vertex (h, k) and:
  - (a) Vertical axis $x = h$ and directrix $y = k - p$ is: $(x - h)^2 = 4p(y - k), p \neq 0$
  - (b) Horizontal axis $y = k$ and directrix $x = h - p$ is: $(y - k)^2 = 4p(x - h), p \neq 0$

- The tangent line to a parabola at a point P makes **equal angles** with:
  - (a) the line through P and the focus.
  - (b) the axis of the parabola.

Vocabulary Check

1. conic

2. locus

3. parabola; directrix; focus

4. axis

5. vertex

6. focal chord

7. tangent

1. A circle is formed when a plane intersects the top or bottom half of a double-napped cone and is perpendicular to the axis of the cone.

2. An ellipse is formed when a plane intersects only the top or bottom half of a double-napped cone but is not perpendicular to the axis of the cone, not parallel to the side of the cone, and does not intersect the vertex.

3. A parabola is formed when a plane intersects the top or bottom half of a double-napped cone, is parallel to the side of the cone, and does not intersect the vertex.

4. A hyperbola is formed when a plane intersects both halves of a double-napped cone, is parallel to the axis of the cone, and does not intersect the vertex.

5. $y^2 = -4x$

   Vertex: $(0, 0)$

   Opens to the left since p is negative; matches graph (e).

6. $x^2 = 2y$

   Vertex: $(0, 0)$

   $p = \frac{1}{2} > 0$

   Opens upward; matches graph (b).

7. $x^2 = -8y$

   Vertex: $(0, 0)$

   Opens downward since p is negative; matches graph (d).

8. $y^2 = -12x$

   Vertex: $(0, 0)$

   $p = -3 < 0$

   Opens to the left; matches graph (f).

9. $(y - 1)^2 = 4(x - 3)$

   Vertex: $(3, 1)$

   Opens to the right since p is positive; matches graph (a).

10. $(x + 3)^2 = -2(y - 1)$

   Vertex: $(-3, 1)$

   $p = -\frac{1}{2} < 0$

   Opens downward; matches graph (c).

11. $y = \frac{1}{2}x^2$

   $x^2 = 2y$

   $x^2 = 4\left(\frac{1}{2}\right)y \implies h = 0, k = 0, p = \frac{1}{2}$

   Vertex: $(0, 0)$

   Focus: $\left(0, \frac{1}{2}\right)$

   Directrix: $y = -\frac{1}{2}$

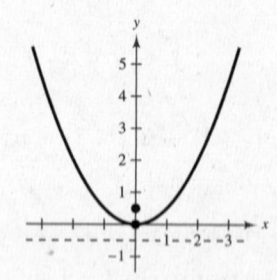

12. $y = -2x^2 \implies x^2 = 4\left(-\frac{1}{8}\right)y$

   Vertex: $(0, 0)$

   Focus: $\left(0, -\frac{1}{8}\right)$

   Directrix: $y = \frac{1}{8}$

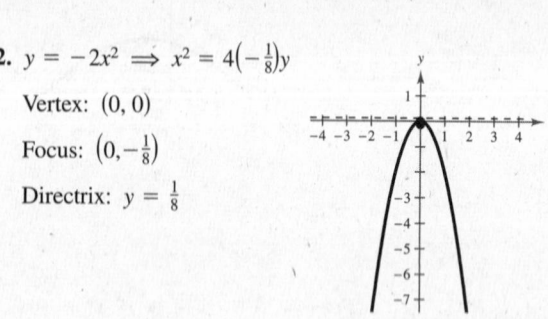

13. $y^2 = -6x$

   $y^2 = 4\left(-\frac{3}{2}\right)x \implies h = 0, k = 0, p = -\frac{3}{2}$

   Vertex: $(0, 0)$

   Focus: $\left(-\frac{3}{2}, 0\right)$

   Directrix: $x = \frac{3}{2}$

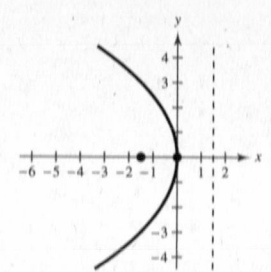

14. $y^2 = 3x \implies 4\left(\frac{3}{4}\right)x$

   Vertex: $(0, 0)$

   Focus: $\left(\frac{3}{4}, 0\right)$

   Directrix: $x = -\frac{3}{4}$

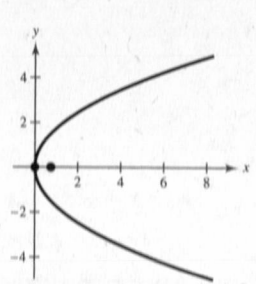

15. $x^2 + 6y = 0$

$$x^2 = -6y = 4\left(-\tfrac{3}{2}\right)y \implies h = 0, k = 0, p = -\tfrac{3}{2}$$

Vertex: $(0, 0)$

Focus: $\left(0, -\tfrac{3}{2}\right)$

Directrix: $y = \tfrac{3}{2}$

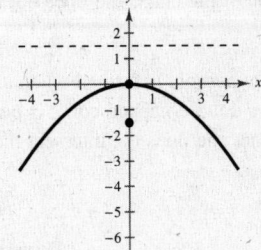

16. $x + y^2 = 0$

$$y^2 = -x = 4\left(-\tfrac{1}{4}\right)x$$

Vertex: $(0, 0)$

Focus: $\left(-\tfrac{1}{4}, 0\right)$

Directrix: $x = \tfrac{1}{4}$

17. $(x - 1)^2 + 8(y + 2) = 0$

$$(x - 1)^2 = 4(-2)(y + 2)$$

$h = 1, k = -2, p = -2$

Vertex: $(1, -2)$

Focus: $(1, -4)$

Directrix: $y = 0$

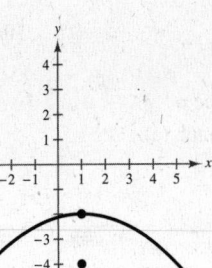

18. $(x + 5) + (y - 1)^2 = 0$

$$(y - 1)^2 = 4\left(-\tfrac{1}{4}\right)(x + 5)$$

Vertex: $(-5, 1)$

Focus: $\left(-5 + \left(-\tfrac{1}{4}\right), 1\right) \implies \left(-\tfrac{21}{4}, 1\right)$

Directrix: $x = -5 - \left(-\tfrac{1}{4}\right) = -\tfrac{19}{4}$

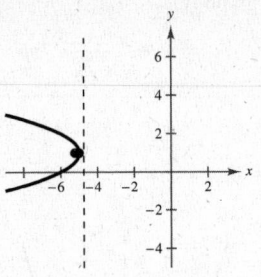

19. $\left(x + \tfrac{3}{2}\right)^2 = 4(y - 2)$

$$\left(x + \tfrac{3}{2}\right)^2 = 4(1)(y - 2)$$

$h = -\tfrac{3}{2}, k = 2, p = 1$

Vertex: $\left(-\tfrac{3}{2}, 2\right)$

Focus: $\left(-\tfrac{3}{2}, 3\right)$

Directrix: $y = 1$

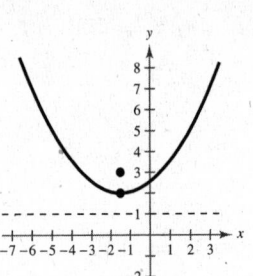

20. $\left(x + \tfrac{1}{2}\right)^2 = 4(y - 1) = 4(1)(y - 1)$

Vertex: $\left(-\tfrac{1}{2}, 1\right)$

Focus: $\left(-\tfrac{1}{2}, 1 + 1\right) \implies \left(-\tfrac{1}{2}, 2\right)$

Directrix: $y = 1 - 1 = 0$

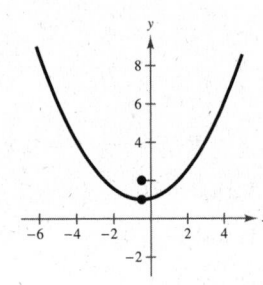

21.

$$y = \tfrac{1}{4}(x^2 - 2x + 5)$$

$$4y = x^2 - 2x + 5$$

$$4y - 5 + 1 = x^2 - 2x + 1$$

$$4y - 4 = (x - 1)^2$$

$$(x - 1)^2 = 4(1)(y - 1)$$

$h = 1, k = 1, p = 1$

Vertex: $(1, 1)$

Focus: $(1, 2)$

Directrix: $y = 0$

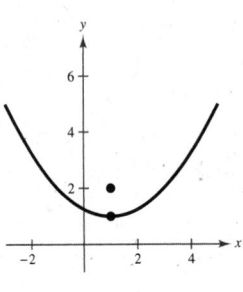

22.

$$x = \tfrac{1}{4}(y^2 + 2y + 33)$$

$$4x = y^2 + 2y + 1 - 1 + 33 = (y + 1)^2 + 32$$

$$(y + 1)^2 = 4(1)(x - 8)$$

Vertex: $(8, -1)$

Focus: $(9, -1)$

Directrix: $x = 7$

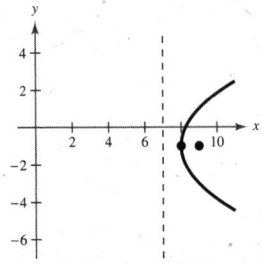

23. $y^2 + 6y + 8x + 25 = 0$

$\qquad y^2 + 6y + 9 = -8x - 25 + 9$

$\qquad (y + 3)^2 = 4(-2)(x + 2)$

$h = -2, k = -3, p = -2$

Vertex: $(-2, -3)$

Focus: $(-4, -3)$

Directrix: $x = 0$

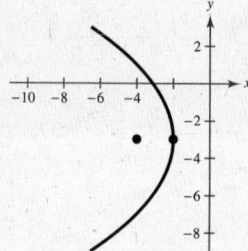

24. $y^2 - 4y - 4x = 0$

$\qquad y^2 - 4y + 4 = 4x + 4$

$\qquad (y - 2)^2 = 4(1)(x + 1)$

Vertex: $(-1, 2)$

Focus: $(0, 2)$

Directrix: $x = -2$

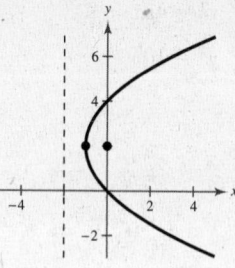

25. $x^2 + 4x + 6y - 2 = 0$

$\qquad x^2 + 4x = -6y + 2$

$\qquad x^2 + 4x + 4 = -6y + 2 + 4$

$\qquad (x + 2)^2 = -6(y - 1)$

$\qquad (x + 2)^2 = 4\left(-\frac{3}{2}\right)(y - 1)$

$h = -2, k = 1, p = -\frac{3}{2}$

Vertex: $(-2, 1)$

Focus: $\left(-2, -\frac{1}{2}\right)$

Directrix: $y = \frac{5}{2}$

On a graphing calculator, enter:

$y_1 = -\frac{1}{6}(x^2 + 4x - 2)$

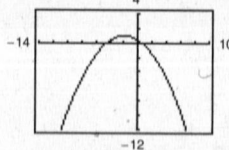

26. $x^2 - 2x + 8y + 9 = 0$

$\qquad x^2 - 2x + 1 = -8y - 9 + 1$

$\qquad (x - 1)^2 = -8(y + 1) = 4(-2)(y + 1)$

Vertex: $(1, -1)$

Focus: $(1, -3)$

Directrix: $y = 1$

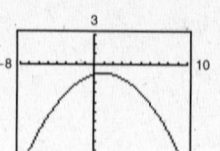

27. $y^2 + x + y = 0$

$\qquad y^2 + y + \frac{1}{4} = -x + \frac{1}{4}$

$\qquad \left(y + \frac{1}{2}\right)^2 = 4\left(-\frac{1}{4}\right)\left(x - \frac{1}{4}\right)$

$h = \frac{1}{4}, k = -\frac{1}{2}, p = -\frac{1}{4}$

Vertex: $\left(\frac{1}{4}, -\frac{1}{2}\right)$

Focus: $\left(0, -\frac{1}{2}\right)$

Directrix: $x = \frac{1}{2}$

To use a graphing calculator, enter:

$y_1 = -\frac{1}{2} + \sqrt{\frac{1}{4} - x}$

$y_2 = -\frac{1}{2} - \sqrt{\frac{1}{4} - x}$

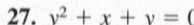

28. $y^2 - 4x - 4 = 0$

$\qquad y^2 = 4x + 4 = 4(1)(x + 1)$

Vertex: $(-1, 0)$

Focus: $(0, 0)$

Directrix: $x = -2$

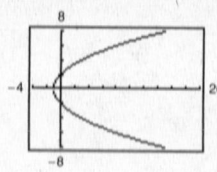

29. Vertex: $(0, 0) \implies h = 0, k = 0$

Graph opens upward.

$x^2 = 4py$

Point on graph: $(3, 6)$

$3^2 = 4p(6)$

$9 = 24p$

$\frac{3}{8} = p$

Thus, $x^2 = 4\left(\frac{3}{8}\right)y \implies x^2 = \frac{3}{2}y.$

30. Point: $(-2, 6)$

$x = ay^2$

$-2 = a(6)^2$

$-\frac{1}{18} = a$

$x = -\frac{1}{18}y^2$

$y^2 = -18x$

31. Vertex: $(0, 0) \implies h = 0, k = 0$

Focus: $\left(0, -\frac{3}{2}\right) \implies p = -\frac{3}{2}$

$x^2 = 4py$

$x^2 = 4\left(-\frac{3}{2}\right)y$

$x^2 = -6y$

32. Focus: $\left(\frac{5}{2}, 0\right) \implies p = \frac{5}{2}$

$y^2 = 4px$

$y^2 = 10x$

33. Vertex: $(0, 0) \implies h = 0, k = 0$

Focus: $(-2, 0) \implies p = -2$

$y^2 = 4px$

$y^2 = 4(-2)x$

$y^2 = -8x$

34. Focus: $(0, -2) \implies p = -2$

$x^2 = 4py$

$x^2 = -8y$

35. Vertex: $(0, 0) \implies h = 0, k = 0$

Directrix: $y = -1 \implies p = 1$

$x^2 = 4py$

$x^2 = 4(1)y$

$x^2 = 4y$

36. Directrix: $y = 3 \implies p = -3$

$x^2 = 4py$

$x^2 = -12y$

37. Vertex: $(0, 0) \implies h = 0, k = 0$

Directrix: $x = 2 \implies p = -2$

$y^2 = 4px$

$y^2 = 4(-2)x$

$y^2 = -8x$

38. Directrix: $x = -3 \implies p = 3$

$y^2 = 4px$

$y^2 = 12x$

39. Vertex: $(0, 0) \implies h = 0, k = 0$

Horizontal axis and passes through the point $(4, 6)$

$y^2 = 4px$

$6^2 = 4p(4)$

$36 = 16p \implies p = \frac{9}{4}$

$y^2 = 4\left(\frac{9}{4}\right)x$

$y^2 = 9x$

40. Vertical axis

Passes through: $(-3, -3)$

$x^2 = 4py$

$(-3)^2 = 4p(-3)$

$9 = -12p$

$p = -\frac{3}{4}$

$x^2 = -3y$

41. Vertex: $(3, 1)$ and opens downward. Passes through $(2, 0)$ and $(4, 0)$.

$y = -(x - 2)(x - 4)$

$\quad = -x^2 + 6x - 8$

$\quad = -(x - 3)^2 + 1$

$(x - 3)^2 = -(y - 1)$

42. Vertex: $(5, 3) \implies h = 5, k = 3$

Passes through: $(4.5, 4)$

$(y - k)^2 = 4p(x - h)$

$(y - 3)^2 = 4p(x - 5)$

$1 = 4p(4.5 - 5)$

$p = -\frac{1}{2}$

$(y - 3)^2 = -2(x - 5)$

43. Vertex: $(-4, 0)$ and opens to the right. Passes through $(0, 4)$.

$(y - 0)^2 = 4p(x + 4)$

$4^2 = 4p(0 + 4)$

$16 = 16p$

$1 = p$

$y^2 = 4(x + 4)$

44. Vertex: $(3, -3) \implies h = 3, k = -3$

Passes through: $(0, 0)$

$(x - h)^2 = 4p(y - k)$

$(x - 3)^2 = 4p(y + 3)$

$(0 - 3)^2 = 4p(0 + 3)$

$9 = 12p$

$p = \frac{3}{4}$

$(x - 3)^2 = 3(y + 3)$

45. Vertex: $(5, 2)$

Focus: $(3, 2)$

Horizontal axis

$p = 3 - 5 = -2$

$(y - 2)^2 = 4(-2)(x - 5)$

$(y - 2)^2 = -8(x - 5)$

46. Vertex: $(-1, 2) \implies h = -1, k = 2$

Focus: $(-1, 0) \implies p = -2$

$(x - h)^2 = 4p(y - k)$

$(x + 1)^2 = 4(-2)(y - 2)$

$(x + 1)^2 = -8(y - 2)$

47. Vertex: $(0, 4)$

Directrix: $y = 2$

Vertical axis

$$p = 4 - 2 = 2$$

$$(x - 0)^2 = 4(2)(y - 4)$$

$$x^2 = 8(y - 4)$$

48. Vertex: $(-2, 1) \Rightarrow h = -2,$
$k = 1$

Directrix: $x = 1 \Rightarrow p = -3$

$$(y - k)^2 = 4p(x - h)$$

$$(y - 1)^2 = 4(-3)(x + 2)$$

$$(y - 1)^2 = -12(x + 2)$$

49. Focus: $(2, 2)$

Directrix: $x = -2$

Horizontal axis

Vertex: $(0, 2)$

$$p = 2 - 0 = 2$$

$$(y - 2)^2 = 4(2)(x - 0)$$

$$(y - 2)^2 = 8x$$

50. Focus: $(0, 0)$

Directrix: $y = 8 \Rightarrow p = -4$

$$\Rightarrow h = 0, k = 4$$

$$(x - h)^2 = 4p(y - k)$$

$$x^2 = 4(-4)(y - 4)$$

$$x^2 = -16(y - 4)$$

51. $(y - 3)^2 = 6(x + 1)$

For the upper half of the parabola:

$$y - 3 = \sqrt{6(x + 1)}$$

$$y = \sqrt{6(x + 1)} + 3$$

52. $(y + 1)^2 = 2(x - 4)$

$$y + 1 = \pm\sqrt{2(x - 4)}$$

$$y = -1 \pm \sqrt{2(x - 4)}$$

Lower half of parabola:

$$y = -1 - \sqrt{2(x - 4)}$$

53. $y^2 - 8x = 0 \Rightarrow y = \pm\sqrt{8x}$

$x - y + 2 = 0 \Rightarrow y = x + 2$

The point of tangency is $(2, 4)$.

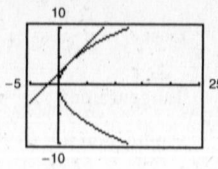

54. $x^2 + 12y = 0 \Rightarrow y_1 = -\frac{1}{12}x^2$

$x + y - 3 = 0 \Rightarrow y_2 = 3 - x$

Using the trace or intersect feature, the point of tangency is $(6, -3)$.

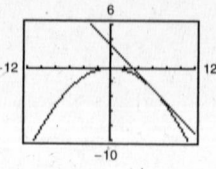

55. $x^2 = 2y \Rightarrow p = \frac{1}{2}$

Point: $(4, 8)$

Focus: $\left(0, \frac{1}{2}\right)$

$$d_1 = \frac{1}{2} - b$$

$$d_2 = \sqrt{(4 - 0)^2 + \left(8 - \frac{1}{2}\right)^2}$$

$$= \frac{17}{2}$$

$$d_1 = d_2 \Rightarrow b = -8$$

Slope: $m = \dfrac{8 - (-8)}{4 - 0} = 4$

$$y = 4x - 8 \Rightarrow 0 = 4x - y - 8$$

x-intercept: $(2, 0)$

56.

$$x^2 = 2y$$

$$x^2 = 4\left(\frac{1}{2}\right)y$$

$$4\left(\frac{1}{2}\right)y = x^2$$

$$p = \frac{1}{2}$$

Focus: $\left(0, \frac{1}{2}\right)$

$$d_1 = \frac{1}{2} - b$$

$$d_2 = \sqrt{(-3 - 0)^2 + \left(\frac{9}{2} - \frac{1}{2}\right)^2} = 5$$

$$\frac{1}{2} - b = 5$$

$$b = -\frac{9}{2}$$

$$m = \frac{-(9/2) - (9/2)}{0 + 3} = -3$$

Tangent line: $y = -3x - \dfrac{9}{2} \Rightarrow 6x + 2y + 9 = 0$

x-intercept: $\left(-\dfrac{3}{2}, 0\right)$

57. $y = -2x^2 \Rightarrow x^2 = -\frac{1}{2}y \Rightarrow p = -\frac{1}{8}$

Point: $(-1, -2)$

Focus: $\left(0, -\frac{1}{8}\right)$

$d_1 = b - \left(-\frac{1}{8}\right) = b + \frac{1}{8}$

$d_2 = \sqrt{(-1-0)^2 + \left(-2 - \left(-\frac{1}{8}\right)\right)^2}$

$\quad = \frac{17}{8}$

$d_1 = d_2 \Rightarrow b = 2$

Slope: $m = \frac{-2-2}{-1-0} = 4$

$\quad y = 4x + 2 \Rightarrow 0 = 4x - y + 2$

x-intercept: $\left(-\frac{1}{2}, 0\right)$

58. $\quad y = -2x^2$

$-\frac{1}{2}y = x^2$

$4\left(-\frac{1}{8}\right)y = x^2$

$p = -\frac{1}{8}$

Focus: $\left(0, -\frac{1}{8}\right)$

$d_1 = \frac{1}{8} + b$

$d^2 = \sqrt{(2-0)^2 + \left(-8 - \left(-\frac{1}{8}\right)\right)^2} = \frac{65}{8}$

$\frac{1}{8} + b = \frac{65}{8}$

$b = \frac{64}{8} = 8$

$m = \frac{-8-8}{2-0} = -8$

Tangent line: $y = -8x + 8 \Rightarrow 8x + y - 8 = 0$

x-intercept: $(1, 0)$

59. $\quad (x - 106)^2 = -\frac{4}{5}(R - 14{,}045)$

$x^2 - 212x + 11{,}236 = -\frac{4}{5}R + 11{,}236$

$\quad\quad R = 265x - \frac{5}{4}x^2$

The revenue is maximum when $x = 106$ units.

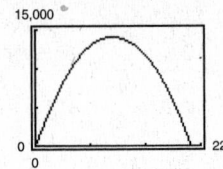

60. Maximum revenue occurs at $x = 135$.

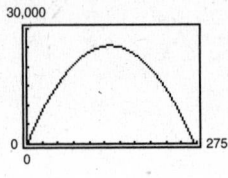

61. Vertex: $(0, 0) \Rightarrow h = 0, k = 0$

Focus: $(0, 4.5) \Rightarrow p = 4.5$

$(x - h)^2 = 4p(y - k)$

$(x - 0)^2 = 4(4.5)(y - 0)$

$\quad x^2 = 18y$ or $y = \frac{1}{18}x^2$

62. (a)

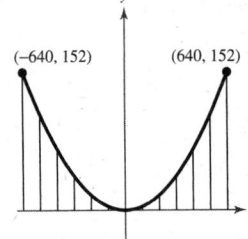

(b) Vertex: $(0, 0)$; opens upward

$y - 0 = a(x - 0)^2$

$152 = a(640)^2$

$\frac{152}{640^2} = a$

$\frac{19}{51{,}200} = a$

An equation of the cables is

$$y = \frac{19}{51{,}200}x^2.$$

(c)

Distance, x	Height, y
0	0
250	23.19
400	59.38
500	92.77
1000	371.09

63. (a) Vertex: $(0, 0) \implies h = 0, k = 0$

Points on the parabola: $(\pm 16, -0.4)$

$$x^2 = 4py$$

$$(\pm 16)^2 = 4p(-0.4)$$

$$256 = -1.6p$$

$$-160 = p$$

$$x^2 = 4(-160y)$$

$$x^2 = -640y$$

$$y = -\frac{1}{640}x^2$$

(b) When $y = -0.1$ we have

$$-0.1 = -\frac{1}{640}x^2$$

$$64 = x^2$$

$$\pm 8 = x.$$

Thus, 8 feet away from the center of the road, the road surface is 0.1 foot lower than in the middle.

64. Vertex: $(0, 0)$

$$(y - 0)^2 = 4p(x - 0)$$

$$y^2 = 4px$$

At $(1000, 800)$: $800^2 = 4p(1000) \implies p = 160$

$$y^2 = 4(160)x$$

$$y^2 = 640x$$

65. (a) $V = 17,500\sqrt{2}$ mi/hr

$$\approx 24,750 \text{ mi/hr}$$

(b) $p = -4100$, $(h, k) = (0, 4100)$

$$(x - 0)^2 = 4(-4100)(y - 4100)$$

$$x^2 = -16,400(y - 4100)$$

66. (a)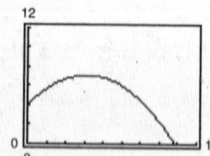

(b) Highest point: $(6.25, 7.125)$

Range: 15.69 feet

67. (a) $x^2 = -\dfrac{(32)^2}{16}(y - 75)$

$$x^2 = -64(y - 75)$$

(b) When $y = 0$, $x^2 = -64(-75) = 4800$.

Thus, $x = \sqrt{4800} = 40\sqrt{3} \approx 69.3$ feet.

68. $\dfrac{540 \text{ mi}}{1 \text{ hr}} \cdot \dfrac{5280 \text{ ft}}{1 \text{ mi}} \cdot \dfrac{1 \text{ hr}}{60 \text{ min}} \cdot \dfrac{1 \text{ min}}{60 \text{ s}} = 792 \text{ ft/s}$

$s = 30,000$

The crate hits the ground when $y = 0$.

$$x^2 = \frac{-v^2}{16}(y - s)$$

$$x^2 = -\frac{(792)^2}{16}(0 - 30,000)$$

$$x^2 = 1,176,120,000$$

$$x \approx 34,295$$

The distance is about 34,295 feet.

69. False. It is not possible for a parabola to intersect its directrix. If the graph crossed the directrix there would exist points closer to the directrix than the focus.

70. True. If the axis (line connecting the vertex and focus) is horizontal, then the directrix must be vertical.

71. (a)

As p increases, the graph becomes wider.

(b) $(0, 1), (0, 2), (0, 3), (0, 4)$

(c) 4, 8, 12, 16. The chord passing through the focus and parallel to the directrix has length $|4p|$.

(d) This provides an easy way to determine two additional points on the graph, each of which is $|2p|$ units away from the focus on the chord.

72. (a) $A = \dfrac{8}{3}(2)^{1/2}(4)^{3/2} = \dfrac{8}{3}\left(\sqrt{2}\right)(8) = \dfrac{64\sqrt{2}}{3}$ square units

(b) As p approaches zero, the parabola becomes narrower and narrower, thus the area becomes smaller and smaller.

73. $y - y_1 = \dfrac{x_1}{2p}(x - x_1)$

Slope: $m = \dfrac{x_1}{2p}$

74. *Sample answer:* Any light ray (or other electromagnetic radiation) that enters a parabolic reflector (a surface for which any cross section containing the axis is a parabola) in a direction parallel to the axis of the surface will be reflected to the focus of the surface (the focus of any of the cross-sectional parabolas). Conversely, any ray projected from the focus in a direction that intersects the parabolic surface will be reflected in a direction parallel to the axis.

75. $f(x) = x^3 - 2x^2 + 2x - 4$

Possible rational zeros: $\pm 1, \pm 2, \pm 4$

76. $f(x) = 2x^3 + 4x^2 - 3x + 10$

Rational zeros $\dfrac{p}{q}$: $p = $ factor of 10, $q = $ factor of 2

Possible rational zeros: $\pm\dfrac{1}{2}, \pm 1, \pm 2, \pm\dfrac{5}{2}, \pm 5, \pm 10$

77. $f(x) = 2x^5 + x^2 + 16$

Possible rational zeros: $\pm 1, \pm 2, \pm 4, \pm 8, \pm 16, \pm\dfrac{1}{2}$

78. $f(x) = 3x^3 - 12x + 22$

Rational zeros $\dfrac{p}{q}$: $p = $ factor of 22, $q = $ factor of 3

Possible rational zeros: $\pm\dfrac{1}{3}, \pm\dfrac{2}{3}, \pm 1, \pm 2, \pm\dfrac{11}{3}, \pm\dfrac{22}{3},$ $\pm 11, \pm 22$

79. $f(x) = (x - 3)[x - (2 + i)][x - (2 - i)]$
$= (x - 3)[(x - 2) - i][(x - 2) + i]$
$= (x - 3)(x^2 - 4x + 5)$
$= x^3 - 7x^2 + 17x - 15$

80. $f(x) = 2x^3 - 3x^2 + 50x - 75$

$$\begin{array}{r|rrrr} \frac{3}{2} & 2 & -3 & 50 & -75 \\ & & 3 & 0 & 75 \\ \hline & 2 & 0 & 50 & 0 \end{array}$$

$2x^2 + 50 = 0 \implies x^2 = -25 \implies x = \pm 5i$

Zeros: $x = \frac{3}{2}, \pm 5i$

81. $g(x) = 6x^4 + 7x^3 - 29x^2 - 28x + 20$

Possible rational roots: $\pm 1, \pm 2, \pm 4, \pm 5, \pm 10, \pm 20,$
$\pm\frac{1}{2}, \pm\frac{5}{2}, \pm\frac{1}{3}, \pm\frac{2}{3}, \pm\frac{4}{3}, \pm\frac{5}{3}, \pm\frac{10}{3}, \pm\frac{20}{3}, \pm\frac{1}{6}, \pm\frac{5}{6}$

$x = \pm 2$ are both solutions.

$$\begin{array}{r|rrrrr} 2 & 6 & 7 & -29 & -28 & 20 \\ & & 12 & 38 & 18 & -20 \\ \hline & 6 & 19 & 9 & -10 & 0 \\ -2 & 6 & 19 & 9 & -10 & \\ & & -12 & -14 & 10 & \\ \hline & 6 & 7 & -5 & 0 & \end{array}$$

$g(x) = (x - 2)(x + 2)(6x^2 + 7x - 5)$

$= (x - 2)(x + 2)(2x - 1)(3x + 5)$

The zeros of $g(x)$ are $x = \pm 2, x = \frac{1}{2}, x = -\frac{5}{3}$.

82. $h(x) = 2x^4 + x^3 - 19x^2 - 9x + 9$

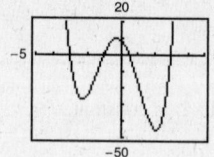

Zeros: $x = \pm 3, -1, \dfrac{1}{2}$

83. $A = 35°, a = 10, b = 7$

$\dfrac{\sin B}{7} = \dfrac{\sin 35°}{10} \implies \sin B \approx 0.4015 \implies B \approx 23.67°$

$C \approx 180° - 35° - 23.67° = 121.33°$

$\dfrac{c}{\sin 121.33°} = \dfrac{10}{\sin 35°} \implies c \approx 14.89$

84. $B = 54°, b = 18, c = 11$

Because B is acute and $18 > 11$, one triangle is possible.

$\sin C = \dfrac{c \sin B}{b} = \dfrac{11 \sin 54}{18} \approx 0.49440 \implies C \approx 29.63°$

$A = 180° - B - C \approx 180° - 54° - 29.63° = 96.37°$

$a = \dfrac{b}{\sin B}(\sin A) = \dfrac{18}{\sin 54}(\sin 96.37°) \approx 22.11$

85. $A = 40°, B = 51°, c = 3$

$C = 180° - 40° - 51° = 89°$

$\dfrac{a}{\sin 40°} = \dfrac{3}{\sin 89°} \implies a \approx 1.93$

$\dfrac{b}{\sin 51°} = \dfrac{3}{\sin 89°} \implies b \approx 2.33$

86. $B = 26°, C = 104°, a = 19$

$A = 180° - B - C \approx 180° - 26° - 104° = 50°$

$b = \dfrac{a}{\sin A}(\sin B) = \dfrac{19}{\sin 50}(\sin 26°) \approx 10.87$

$c = \dfrac{a}{\sin A}(\sin C) = \dfrac{19}{\sin 50}(\sin 104°) \approx 24.07$

87. $a = 7, b = 10, c = 16$

$\cos C = \dfrac{7^2 + 10^2 - 16^2}{2(7)(10)} \approx -0.7643 \implies C \approx 139.84°$

$\dfrac{\sin B}{10} = \dfrac{\sin 139.84°}{16} \implies \sin B \approx 0.4031 \implies B \approx 23.77°$

$A = 180° - B - C \implies A \approx 16.39°$

88. $a = 58, b = 28, c = 75$

$\cos A = \dfrac{b^2 + c^2 - a^2}{2bc} = \dfrac{784 + 5625 - 3364}{2(28)(75)} = 0.725 \implies A \approx 43.53°$

$\cos B = \dfrac{a^2 + c^2 - b^2}{2ac} = \dfrac{3364 + 5625 - 784}{2(58)(75)} = 0.943103 \implies B \approx 19.42°$

$C = 180° - A - B \approx 180° - 43.53° - 19.42° = 117.05°$

89. $A = 65°, b = 5, c = 12$

$a^2 = 5^2 + 12^2 - 2(5)(12) \cos 65° \implies a \approx 10.8759 \approx 10.88$

$\dfrac{\sin B}{5} = \dfrac{\sin 65°}{10.8759} \implies \sin B \approx 0.4167 \implies B \approx 24.62°$

$C = 180° - A - B \implies C \approx 90.38°$

90. $B = 71°, a = 21, c = 29$

$b^2 = a^2 + c^2 - 2ac \cos B$

$b^2 = 441 + 841 - 2(21)(29) \cos 71° \approx 885.458$

$b \approx 29.76$

$\cos A = \dfrac{b^2 + c^2 - a^2}{2bc} \approx \dfrac{(29.76)^2 + 841 - 441}{2(29.76)(29)} \approx 0.74484 \implies A \approx 41.85°$

$C = 180° - A - B \approx 180° - 41.85° - 71° = 67.15°$

Section 10.3   Ellipses

- An **ellipse** is the set of all points (x, y) the sum of whose distances from two distinct fixed points (**foci**) is constant.
- The standard equation of an ellipse with center (h, k) and major and minor axes of lengths $2a$ and $2b$ is:

    (a) $\dfrac{(x - h)^2}{a^2} + \dfrac{(y - k)^2}{b^2} = 1$ if the major axis is horizontal.

    (b) $\dfrac{(x - h)^2}{b^2} + \dfrac{(y - k)^2}{a^2} = 1$ is the major axis is vertical.

- $c^2 = a^2 - b^2$ where c is the distance from the center to a focus.

- The eccentricity of an ellipse is $e = \dfrac{c}{a}$.

Vocabulary Check

1. ellipse; foci

2. major axis, center

3. minor axis

4. eccentricity

---

1. $\dfrac{x^2}{4} + \dfrac{y^2}{9} = 1$

Center: $(0, 0)$

$a = 3, b = 2$

Vertical major axis

Matches graph (b).

2. $\dfrac{x^2}{9} + \dfrac{y^2}{4} = 1$

Center: $(0, 0)$

$a = 3, b = 2$

Horizontal major axis

Matches graph (c).

3. $\dfrac{x^2}{4} + \dfrac{y^2}{25} = 1$

Center: $(0, 0)$

$a = 5, b = 2$

Vertical major axis

Matches graph (d).

4. $\dfrac{y^2}{4} + \dfrac{x^2}{4} = 1$

Center: $(0, 0)$

Circle of radius: 2

Matches graph (f).

5. $\dfrac{(x - 2)^2}{16} + (y + 1)^2 = 1$

Center: $(2, -1)$

$a = 4, b = 1$

Horizontal major axis

Matches graph (a).

6. $\dfrac{(x + 2)^2}{9} + \dfrac{(y + 2)^2}{4} = 1$

Center: $(-2, -2)$

$a = 3, b = 2$

Horizontal major axis

Matches graph (e).

7. $\dfrac{x^2}{25} + \dfrac{y^2}{16} = 1$

Ellipse

Center: $(0, 0)$

$a = 5, b = 4, c = 3$

Vertices: $(\pm 5, 0)$

Foci: $(\pm 3, 0)$

$e = \dfrac{3}{5}$

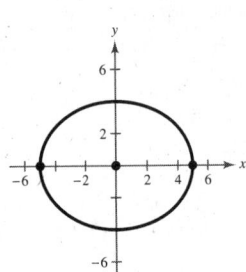

8. $\dfrac{x^2}{81} + \dfrac{y^2}{144} = 1$

$a = 12, b = 9,$

$c = \sqrt{63} = 3\sqrt{7}$

Ellipse

Center: $(0, 0)$

Vertices: $(0, \pm 12)$

Foci: $\left(0, \pm 3\sqrt{7}\right)$

Eccentricity: $e = \dfrac{\sqrt{7}}{4}$

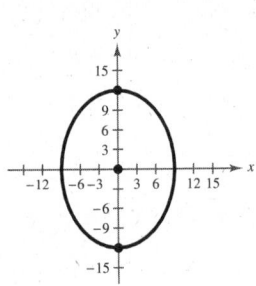

9. $\dfrac{x^2}{25} + \dfrac{y^2}{25} = 1 \implies x^2 + y^2 = 25$

Circle

Center: $(0, 0)$

Radius: 5

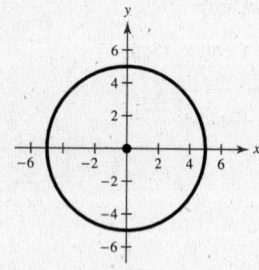

10. $\dfrac{x^2}{9} + \dfrac{y^2}{9} = 1 \implies x^2 + y^2 = 9$

Circle

Center: $(0, 0)$

Radius: 3

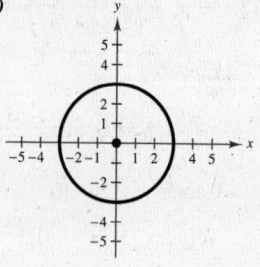

11. $\dfrac{x^2}{5} + \dfrac{y^2}{9} = 1$

Ellipse

$a = 3, b = \sqrt{5}, c = 2$

Center: $(0, 0)$

Vertices: $(0, \pm 3)$

Foci: $(0, \pm 2)$

$e = \dfrac{2}{3}$

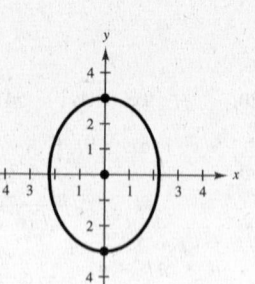

12. $\dfrac{x^2}{64} + \dfrac{y^2}{28} = 1$

$a = 8, b = \sqrt{28} = 2\sqrt{7},$

$c = 6$

Ellipse

Center: $(0, 0)$

Vertices: $(\pm 8, 0)$

Foci: $(\pm 6, 0)$

Eccentricity: $e = \dfrac{3}{4}$

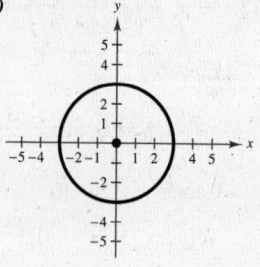

13. $\dfrac{(x + 3)^2}{16} + \dfrac{(y - 5)^2}{25} = 1$

Ellipse

$a = 5, b = 4, c = 3$

Center: $(-3, 5)$

Vertices: $(-3, 10)(-3, 0)$

Foci: $(-3, 8)(-3, 2)$

$e = \dfrac{3}{5}$

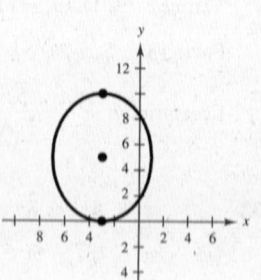

14. $\dfrac{(x - 4)^2}{12} + \dfrac{(y + 3)^2}{16} = 1$

$a = 4, b = 2\sqrt{3}, c = 2$

Ellipse

Center: $(4, -3)$

Vertices: $(4, 1), (4, -7)$

Foci: $(4, -1), (4, -5)$

$e = \dfrac{2}{4} = \dfrac{1}{2}$

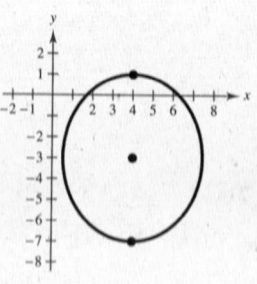

15. $\dfrac{x^2}{4/9} + \dfrac{(y + 1)^2}{4/9} = 1 \implies x^2 + (y + 1)^2 = \dfrac{4}{9}$

Circle

Center: $(0, -1)$

Radius: $\dfrac{2}{3}$

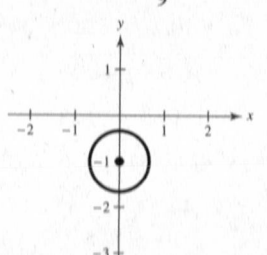

16. $\dfrac{(x + 5)^2}{9/4} + (y - 1)^2 = 1$

Ellipse

$a = \dfrac{3}{2}, b = 1, c = \dfrac{\sqrt{5}}{2}$

Center: $(-5, 1)$

Vertices: $\left(-\dfrac{7}{2}, 1\right), \left(-\dfrac{13}{2}, 1\right)$

Foci: $\left(-5 + \dfrac{\sqrt{5}}{2}, 1\right), \left(-5 - \dfrac{\sqrt{5}}{2}, 1\right)$

$e = \dfrac{\sqrt{5}}{3}$

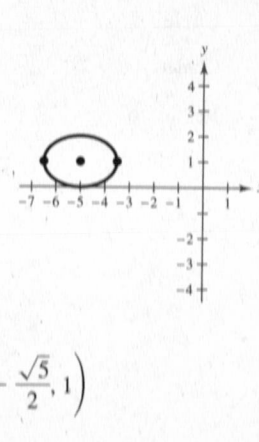

17. $\dfrac{(x + 2)^2}{1} + \dfrac{(y + 4)^2}{1/4} = 1$

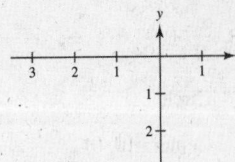

Ellipse

$a = 1, b = \dfrac{1}{2}, c = \dfrac{\sqrt{3}}{2}$

Center: $(-2, -4)$

Vertices: $(-1, -4), (-3, -4)$

Foci: $\left(-2 \pm \dfrac{\sqrt{3}}{2}, -4\right) = \left(\dfrac{-4 \pm \sqrt{3}}{2}, -4\right)$

$e = \dfrac{\sqrt{3}}{2}$

18. $\dfrac{(x - 3)^2}{25/4} + \dfrac{(y - 1)^2}{25/4} = 1$

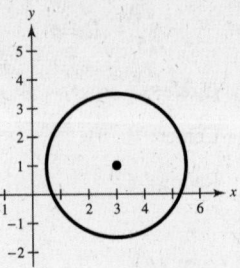

Circle

Center: $(3, 1)$

Radius: $\dfrac{5}{2}$

19. $9x^2 + 4y^2 + 36x - 24y + 36 = 0$

$9(x^2 + 4x + 4) + 4(y^2 - 6y + 9) = -36 + 36 + 36$

$9(x + 2)^2 + 4(y - 3)^2 = 36$

$\dfrac{(x + 2)^2}{4} + \dfrac{(y - 3)^2}{9} = 1$

Ellipse

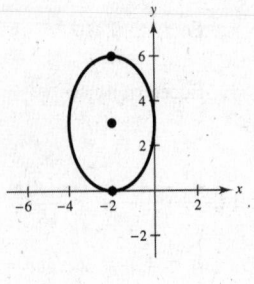

$a = 3, b = 2, c = \sqrt{5}$

Center: $(-2, 3)$

Vertices: $(-2, 6), (-2, 0)$

Foci: $\left(-2, 3 \pm \sqrt{5}\right)$

$e = \dfrac{\sqrt{5}}{3}$

20. $9x^2 + 4y^2 - 54x + 40y + 37 = 0$

$9(x^2 - 6x + 9) + 4(y^2 + 10y + 25) = -37 + 81 + 100$

$\dfrac{(x - 3)^2}{16} + \dfrac{(y + 5)^2}{36} = 1$

$a = 6, b = 4,$

$c = \sqrt{20} = 2\sqrt{5}$

Ellipse

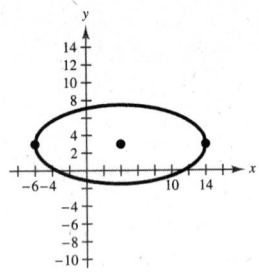

Center: $(3, -5)$

Vertices: $(3, 1), (3, -11)$

Foci: $\left(3, -5 \pm 2\sqrt{5}\right)$

Eccentricity: $e = \dfrac{\sqrt{5}}{3}$

21. $x^2 + y^2 - 2x + 4y - 31 = 0$

$(x^2 - 2x + 1) + (y^2 + 4y + 4) = 31 + 1 + 4$

$(x - 1)^2 + (y + 2)^2 = 36$

$\dfrac{(x - 1)^2}{36} + \dfrac{(y + 2)^2}{36} = 1$

Circle

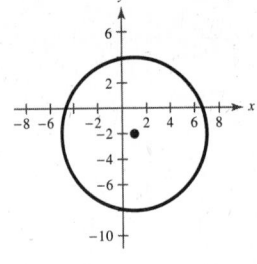

Center: $(1, -2)$

Radius: 6

22. $x^2 + 5y^2 - 8x - 30y - 39 = 0$

$(x^2 - 8x + 16) + 5(y^2 - 6y + 9) = 39 + 16 + 45$

$(x - 4)^2 + 5(y - 3)^2 = 100$

$\dfrac{(x - 4)^2}{100} + \dfrac{(y - 3)^2}{20} = 1$

Ellipse

Center: $(4, 3)$

$a = 10, b = \sqrt{20} = 2\sqrt{5},$

$c = \sqrt{80} = 4\sqrt{5}$

Foci: $\left(4 \pm 4\sqrt{5}, 3\right)$

Vertices: $(14, 3), (-6, 3)$

$e = \dfrac{4\sqrt{5}}{10} = \dfrac{2\sqrt{5}}{5}$

23. $3x^2 + y^2 + 18x - 2y - 8 = 0$

$3(x^2 + 6x + 9) + (y^2 - 2y + 1) = 8 + 27 + 1$

$3(x + 3)^2 + (y - 1)^2 = 36$

$\dfrac{(x + 3)^2}{12} + \dfrac{(y - 1)^2}{36} = 1$

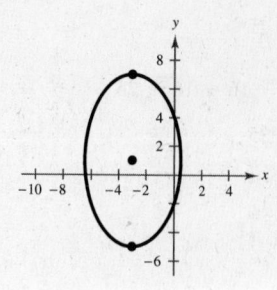

Ellipse

$a = 6, b = \sqrt{12} = 2\sqrt{3}, c = \sqrt{24} = 2\sqrt{6}$

Center: $(-3, 1)$

Vertices: $(-3, 7), (-3, -5)$

Foci: $\left(-3, 1 \pm 2\sqrt{6}\right)$

Eccentricity: $e = \dfrac{\sqrt{6}}{3}$

24. $6x^2 + 2y^2 + 18x - 10y + 2 = 0$

$6\left(x^2 + 3x + \dfrac{9}{4}\right) + 2\left(y^2 - 5y + \dfrac{25}{4}\right) = -2 + \dfrac{27}{2} + \dfrac{25}{2}$

$6\left(x + \dfrac{3}{2}\right)^2 + 2\left(y - \dfrac{5}{2}\right)^2 = 24$

$\dfrac{\left(x + \frac{3}{2}\right)^2}{4} + \dfrac{\left(y - \frac{5}{2}\right)^2}{12} = 1$

$a = \sqrt{12} = 2\sqrt{3}, b = 2, c = \sqrt{8} = 2\sqrt{2}$

Ellipse

Center: $\left(-\dfrac{3}{2}, \dfrac{5}{2}\right)$

Foci: $\left(-\dfrac{3}{2}, \dfrac{5}{2} \pm 2\sqrt{2}\right)$

Vertices: $\left(-\dfrac{3}{2}, \dfrac{5}{2} \pm 2\sqrt{3}\right)$

$e = \dfrac{2\sqrt{2}}{2\sqrt{3}} = \dfrac{\sqrt{6}}{3}$

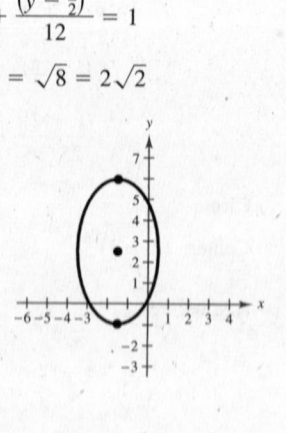

25. $x^2 + 4y^2 - 6x + 20y - 2 = 0$

$(x^2 - 6x + 9) + 4\left(y^2 + 5y + \dfrac{25}{4}\right) = 2 + 9 + 25$

$(x - 3)^2 + 4\left(y + \dfrac{5}{2}\right)^2 = 36$

$\dfrac{(x - 3)^2}{36} + \dfrac{\left(y + \frac{5}{2}\right)^2}{9} = 1$

Ellipse

$a = 6, b = 3, c = \sqrt{27} = 3\sqrt{3}$

Center: $\left(3, -\dfrac{5}{2}\right)$

Vertices: $\left(9, -\dfrac{5}{2}\right), \left(-3, -\dfrac{5}{2}\right)$

Foci: $\left(3 \pm 3\sqrt{3}, -\dfrac{5}{2}\right)$

Eccentricity: $e = \dfrac{\sqrt{3}}{2}$

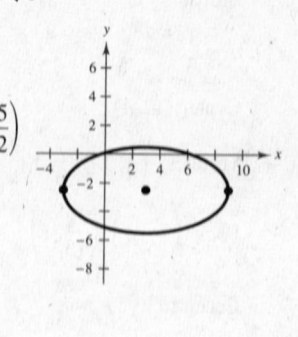

26. $x^2 + y^2 - 4x + 6y - 3 = 0$

$(x^2 - 4x + 4) + (y^2 + 6y + 9) = 3 + 4 + 9$

$\dfrac{(x - 2)^2}{16} + \dfrac{(y + 3)^2}{16} = 1$

Circle

Center: $(2, -3)$

Radius: 4

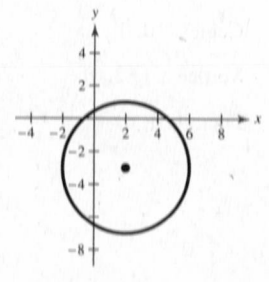

27. $9x^2 + 9y^2 + 18x - 18y + 14 = 0$

$9(x^2 + 2x + 1) + 9(y^2 - 2y + 1) = -14 + 9 + 9$

$9(x + 1)^2 + 9(y - 1)^2 = 4$

$(x + 1)^2 + (y - 1)^2 = \dfrac{4}{9}$

$\dfrac{(x + 1)^2}{4/9} + \dfrac{(y - 1)^2}{4/9} = 1$

Circle

Center: $(-1, 1)$

Radius: $\dfrac{2}{3}$

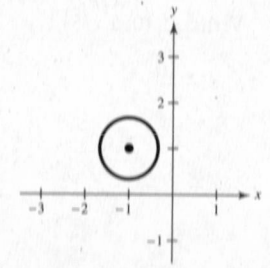

28. $16x^2 + 25y^2 - 32x + 50y + 16 = 0$

$16(x^2 - 2x + 1) + 25(y^2 + 2y + 1) = -16 + 16 + 25$

$16(x - 1)^2 + 25(y + 1)^2 = 25$

$\dfrac{(x - 1)^2}{25/16} + (y + 1)^2 = 1$

$a^2 = \dfrac{25}{16}, b^2 = 1, c^2 = \dfrac{9}{16}$

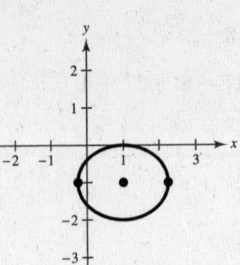

Ellipse

Center: $(1, -1)$

Foci: $\left(\dfrac{7}{4}, -1\right), \left(\dfrac{1}{4}, -1\right)$

Vertices: $\left(\dfrac{9}{4}, -1\right), \left(-\dfrac{1}{4}, -1\right)$

$e = \dfrac{3}{5}$

29. $9x^2 + 25y^2 - 36x - 50y + 60 = 0$

$9(x^2 - 4x + 4) + 25(y^2 - 2y + 1) = -60 + 36 + 25$

$9(x - 2)^2 + 25(y - 1)^2 = 1$

$\dfrac{(x - 2)^2}{1/9} + \dfrac{(y - 1)^2}{1/25} = 1$

Ellipse

$a = \dfrac{1}{3}, b = \dfrac{1}{5}, c = \dfrac{4}{15}$

Center: $(2, 1)$

Vertices: $\left(\dfrac{5}{3}, 1\right), \left(\dfrac{7}{3}, 1\right)$

Foci: $\left(\dfrac{34}{15}, 1\right), \left(\dfrac{26}{15}, 1\right)$

Eccentricity: $e = \dfrac{4}{5}$

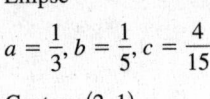

30. $16x^2 + 16y^2 - 64x + 32y + 55 = 0$

$16(x^2 - 4x + 4) + 16(y^2 + 2y + 1) = -55 + 64 + 16$

$16(x - 2)^2 + 16(y + 1)^2 = 25$

$(x - 2)^2 + (y + 1)^2 = \dfrac{25}{16}$

$\dfrac{(x - 2)^2}{25/16} + \dfrac{(y + 1)^2}{25/16} = 1$

Circle

Center: $(2, -1)$

Radius: $\dfrac{5}{4}$

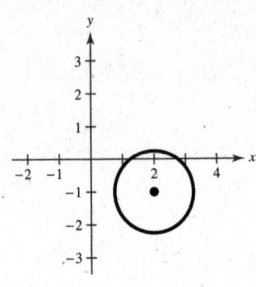

31. $5x^2 + 3y^2 = 15$

$\dfrac{x^2}{3} + \dfrac{y^2}{5} = 1$

Center: $(0, 0)$

$a = \sqrt{5}, b = \sqrt{3}, c = \sqrt{2}$

Foci: $\left(0, \pm\sqrt{2}\right)$

Vertices: $\left(0, \pm\sqrt{5}\right)$

$e = \dfrac{\sqrt{10}}{5}$

To graph, solve for y.

$y^2 = \dfrac{15 - 5x^2}{3}$

$y_1 = \sqrt{\dfrac{15 - 5x^2}{3}}$

$y_2 = -\sqrt{\dfrac{15 - 5x^2}{3}}$

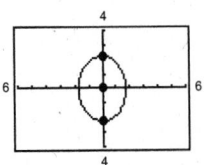

32. $3x^2 + 4y^2 = 12$

$\dfrac{x^2}{4} + \dfrac{y^2}{3} = 1$

$a^2 = 4, b^2 = 3, c^2 = 1$

Center: $(0, 0)$

Vertices: $(\pm 2, 0)$

Foci: $(\pm 1, 0)$

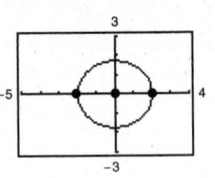

33. $12x^2 + 20y^2 - 12x + 40y - 37 = 0$

$$12\left(x^2 - x + \frac{1}{4}\right) + 20(y^2 + 2y + 1) = 37 + 3 + 20$$

$$12\left(x - \frac{1}{2}\right)^2 + 20(y + 1)^2 = 60$$

$$\frac{\left(x - \frac{1}{2}\right)^2}{5} + \frac{(y + 1)^2}{3} = 1$$

$a = \sqrt{5}, b = \sqrt{3}, c = \sqrt{2}$

Center: $\left(\frac{1}{2}, -1\right)$

Foci: $\left(\frac{1}{2} \pm \sqrt{2}, -1\right)$

Vertices: $\left(\frac{1}{2} \pm \sqrt{5}, -1\right)$

$e = \frac{\sqrt{10}}{5}$

To graph, solve for y.

$$(y + 1)^2 = 3\left[1 - \frac{(x - 0.5)^2}{5}\right]$$

$$y_1 = -1 + \sqrt{3\left[1 - \frac{(x - 0.5)^2}{5}\right]}$$

$$y_2 = -1 - \sqrt{3\left[1 - \frac{(x - 0.5)^2}{5}\right]}$$

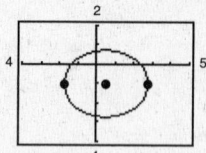

34. $36x^2 + 9y^2 + 48x - 36y - 72 = 0$

$$36\left(x^2 + \frac{4}{3}x + \frac{4}{9}\right) + 9(y^2 - 4y + 4) = 72 + 16 + 36$$

$$\frac{\left(x + \frac{2}{3}\right)^2}{\frac{31}{9}} + \frac{(y - 2)^2}{\frac{124}{9}} = 1$$

$a^2 = \frac{124}{9}, b^2 = \frac{31}{9}, c^2 = \frac{31}{3}$

Center: $\left(-\frac{2}{3}, 2\right)$

Vertices: $\left(-\frac{2}{3}, 2 \pm \frac{2\sqrt{31}}{3}\right)$

Foci: $\left(-\frac{2}{3}, 2 \pm \frac{\sqrt{93}}{3}\right)$

35. Center: $(0, 0)$

$a = 4, b = 2$

Vertical major axis

$$\frac{(x - h)^2}{b^2} + \frac{(y - k)^2}{a^2} = 1$$

$$\frac{x^2}{4} + \frac{y^2}{16} = 1$$

36. Vertices: $(\pm 2, 0) \Rightarrow a = 2$

Endpoints of minor axis:

$$\left(0, \pm\frac{3}{2}\right) \Rightarrow b = \frac{3}{2}$$

$$\frac{x^2}{a^2} + \frac{y^2}{b^2} = 1$$

$$\frac{x^2}{2^2} + \frac{y^2}{(3/2)^2} = 1$$

$$\frac{x^2}{4} + \frac{4y^2}{9} = 1$$

37. Vertices: $(\pm 6, 0)$

$a = 6, c = 2 \Rightarrow b = \sqrt{32} = 4\sqrt{2}$

Foci: $(\pm 2, 0)$

Horizontal major axis

Center: $(0, 0)$

$$\frac{(x - h)^2}{a^2} + \frac{(y - k)^2}{b^2} = 1$$

$$\frac{x^2}{36} + \frac{y^2}{32} = 1$$

38. Vertices: $(0, \pm 8) \Rightarrow a = 8$

Foci: $(0, \pm 4) \Rightarrow c = 4$

$b^2 = a^2 - c^2 = 64 - 16 = 48$

$$\frac{x^2}{b^2} + \frac{y^2}{a^2} = 1$$

$$\frac{x^2}{48} + \frac{y^2}{64} = 1$$

39. Foci: $(\pm 5, 0) \implies c = 5$

Center: $(0, 0)$

Horizontal major axis

Major axis of length 12 $\implies 2a = 12$

$$a = 6$$

$6^2 - b^2 = 5^2 \implies b^2 = 11$

$$\frac{(x - h)^2}{a^2} + \frac{(y - k)^2}{b^2} = 1$$

$$\frac{x^2}{36} + \frac{y^2}{11} = 1$$

40. Foci: $(\pm 2, 0) \implies c = 2$

Major axis length: $8 \implies a = 4$

$b^2 = a^2 - c^2 = 16 - 4 = 12$

$$\frac{x^2}{a^2} + \frac{y^2}{b^2} = 1$$

$$\frac{x^2}{16} + \frac{y^2}{12} = 1$$

41. Vertices: $(0, \pm 5) \implies a = 5$

Center: $(0, 0)$

Vertical major axis

$$\frac{(x - h)^2}{b^2} + \frac{(y - k)^2}{a^2} = 1$$

$$\frac{x^2}{b^2} + \frac{y^2}{25} = 1$$

Point: $(4, 2)$

$$\frac{4^2}{b^2} + \frac{2^2}{25} = 1$$

$$\frac{16}{b^2} = 1 - \frac{4}{25} = \frac{21}{25}$$

$$400 = 21b^2$$

$$\frac{400}{21} = b^2$$

$$\frac{x^2}{400/21} + \frac{y^2}{25} = 1$$

$$\frac{21x^2}{400} + \frac{y^2}{25} = 1$$

42. Major axis vertical

Passes through: $(0, 4)$ and $(2, 0)$

$a = 4, b = 2$

$$\frac{x^2}{b^2} + \frac{y^2}{a^2} = 1$$

$$\frac{x^2}{4} + \frac{y^2}{16} = 1$$

43. Center: $(2, 3)$

$a = 3, b = 1$

Vertical major axis

$$\frac{(x - h)^2}{b^2} + \frac{(y - k)^2}{a^2} = 1$$

$$\frac{(x - 2)^2}{1} + \frac{(y - 3)^2}{9} = 1$$

44. Vertices: $(4, \pm 4) \implies a = 4$

Center: $(4, 0) \implies h = 4, k = 0$

Endpoints of minor axis: $(1, 0), (7, 0) \implies b = 3$

$$\frac{(x - h)^2}{b^2} + \frac{(y - k)^2}{a^2} = 1$$

$$\frac{(x - 4)^2}{9} + \frac{y^2}{16} = 1$$

45. Center: $(-2, 3)$

$a = 4, b = 3$

Horizontal major axis

$$\frac{(x - h)^2}{a^2} + \frac{(y - k)^2}{b^2} = 1$$

$$\frac{(x + 2)^2}{16} + \frac{(y - 3)^2}{9} = 1$$

46. Vertices: $(0, -1), (4, -1) \Rightarrow a = 2$

Center: $(2, -1) \Rightarrow h = 2, k = -1$

Endpoints of minor axis: $(2, 0), (2, -2) \Rightarrow b = 1$

$$\frac{(x - h)^2}{a^2} + \frac{(y - k)^2}{b^2} = 1$$

$$\frac{(x - 2)^2}{4} + \frac{(y + 1)^2}{1} = 1$$

47. Vertices: $(0, 4), (4, 4) \Rightarrow a = 2$

Minor axis of length $2 \Rightarrow b = 1$

Center: $(2, 4) = (h, k)$

$$\frac{(x - h)^2}{a^2} + \frac{(y - k)^2}{b^2} = 1$$

$$\frac{(x - 2)^2}{4} + \frac{(y - 4)^2}{1} = 1$$

48. Foci: $(0, 0), (4, 0) \Rightarrow c = 2, h = 2, k = 0$

Major axis length: $8 \Rightarrow a = 4$

$b^2 = a^2 - c^2 = 16 - 4 = 12$

$$\frac{(x - h)^2}{a^2} + \frac{(y - k)^2}{b^2} = 1$$

$$\frac{(x - 2)^2}{16} + \frac{y^2}{12} = 1$$

49. Foci: $(0, 0), (0, 8) \Rightarrow c = 4$

Major axis of length $16 \Rightarrow a = 8$

$b^2 = a^2 - c^2 = 64 - 16 = 48$

Center: $(0, 4) = (h, k)$

$$\frac{(x - h)^2}{b^2} + \frac{(y - k)^2}{a^2} = 1$$

$$\frac{x^2}{48} + \frac{(y - 4)^2}{64} = 1$$

50. Center: $(2, -1) \Rightarrow h = 2, k = -1$

Vertex: $\left(2, \frac{1}{2}\right) \Rightarrow a = \frac{3}{2}$

Minor axis length: $2 \Rightarrow b = 1$

$$\frac{(x - h)}{b^2} + \frac{(y - k)^2}{a^2} = 1$$

$$\frac{(x - 2)^2}{1} + \frac{(y + 1)^2}{\left(\frac{3}{2}\right)^2} = 1$$

$$(x - 2)^2 + \frac{4(y + 1)^2}{9} = 1$$

51. Center: $(0, 4)$

Vertices: $(-4, 4), (4, 4) \Rightarrow a = 4$

$a = 2c \Rightarrow 4 = 2c \Rightarrow c = 2$

$2^2 = 4^2 - b^2 \Rightarrow b^2 = 12$

Horizontal major axis

$$\frac{(x - h)^2}{a^2} + \frac{(y - k)^2}{b^2} = 1$$

$$\frac{x^2}{16} + \frac{(y - 4)^2}{12} = 1$$

52. Center: $(3, 2) \Rightarrow h = 3, k = 2$

$a = 3c$

Foci: $(1, 2), (5, 2) \Rightarrow c = 2, a = 6$

$b^2 = a^2 - c^2 = 36 - 4 = 32$

$$\frac{(x - h)^2}{a^2} + \frac{(y - k)^2}{b^2} = 1$$

$$\frac{(x - 3)^2}{36} + \frac{(y - 2)^2}{32} = 1$$

53. Vertices: $(0, 2), (4, 2) \Rightarrow a = 2$

Center: $(2, 2)$

Endpoints of the minor axis: $(2, 3), (2, 1) \Rightarrow b = 1$

Horizontal major axis

$$\frac{(x - h)^2}{a^2} + \frac{(y - k)^2}{b^2} = 1$$

$$\frac{(x - 2)^2}{4} + \frac{(y - 2)^2}{1} = 1$$

54. Vertices: $(5, 0), (5, 12) \Rightarrow a = 6$

Endpoints of the minor axis:

$(1, 6), (9, 6) \Rightarrow b = 4$

Center: $(5, 6) \Rightarrow h = 5, k = 6$

$$\frac{(x - h)^2}{b^2} + \frac{(y - k)^2}{a^2} = 1$$

$$\frac{(x - 5)^2}{16} + \frac{(y - 6)^2}{36} = 1$$

55. Vertices: $(\pm 5, 0) \Rightarrow a = 5$

Eccentricity: $\frac{3}{5} \Rightarrow c = \frac{3}{5} a = 3$

$b^2 = a^2 - c^2 = 25 - 9 = 16$

Center: $(0, 0) = (h, k)$

$$\frac{(x - h)^2}{a^2} + \frac{(y - k)^2}{b^2} = 1$$

$$\frac{x^2}{25} + \frac{y^2}{16} = 1$$

56. Vertices: $(0, \pm 8) \Rightarrow a = 8$

$e = \frac{1}{2} \Rightarrow \frac{c}{a} = \frac{1}{2}, c = 4$

$b^2 = a^2 - c^2 = 64 - 16 = 48$

Center: $(0, 0)$

$$\frac{x^2}{b^2} + \frac{y^2}{a^2} = 1$$

$$\frac{x^2}{48} + \frac{y^2}{64} = 1$$

57. (a)

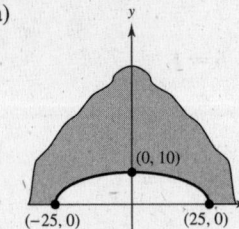

(b) $a = 25, b = 10$

$$\frac{x^2}{a^2} + \frac{y^2}{b^2} = 1$$

$$\frac{x^2}{625} + \frac{y^2}{100} = 1$$

(c) When $x = \pm 4$:

$$\frac{4^2}{625} + \frac{y^2}{100} = 1$$

$$y^2 = 100\left(1 - \frac{16}{625}\right) = \frac{2436}{25}$$

$$y = \sqrt{\frac{2436}{25}} \approx 9.87 \text{ feet} > 9 \text{ feet}$$

Yes. If the truck travels down the center of the tunnel, it will clear the opening of the arch.

58. The tacks should be placed at the foci and the length of the string is the length of the major axis, $2a$.

Center: $(0, 0)$

$a = 3, b = 2, c = \sqrt{5}$

Foci (Positions of the tacks): $\left(\pm\sqrt{5}, 0\right)$

Length of string: 6 feet

59. (a) $a = \dfrac{35.88}{2} = 17.94$

$$e = \frac{c}{a} = 0.967$$

$$c = ea \approx 17.35$$

$$b^2 = a^2 - c^2 \approx 20.82$$

$$\frac{x^2}{a^2} + \frac{y^2}{b^2} = 1$$

$$\frac{x^2}{321.84} + \frac{y^2}{20.82} = 1$$

(b)

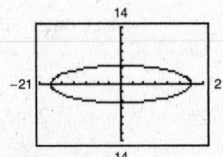

(c) The sun's center is at a focus of the orbit, 17.35 astronomical units from the center of the orbit.

Apogee $\approx 17.35 + \frac{1}{2}(35.88) = 35.29$ astronomical units

Perigee $\approx \frac{1}{2}(35.88) - 17.35 + = 0.59$ astronomical units

60. $a + c = 6378 + 947 = 7325$

$a - c = 6378 + 228 = 6606$

Solving this system for a and c yields $a = 6965.5$ and $c = 359.5$.

$$e = \frac{c}{a} = \frac{359.5}{6965.5} \approx 0.052$$

61. (a) The equation is the bottom half of the ellipse.

$$\frac{\theta^2}{(0.2)^2} + \frac{y^2}{(1.6)^2} = 1$$

$$y = -1.6\sqrt{1 - \frac{\theta^2}{0.04}}$$

$$= -8\sqrt{0.04 - \theta^2}$$

(b)

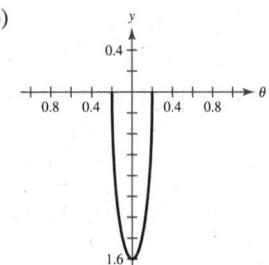

(c) The bottom half models the motion of the pendulum.

62. For $\dfrac{x^2}{a^2} + \dfrac{y^2}{b^2} = 1$, we have $c^2 = a^2 - b^2$.

When $x = c$: $\dfrac{c^2}{a^2} + \dfrac{y^2}{b^2} = 1$

$$y^2 = b^2\left(1 - \dfrac{a^2 - b^2}{a^2}\right) = \dfrac{b^4}{a^2}$$

$$y = \dfrac{b^2}{a}$$

Length of latus rectum: $2y = \dfrac{2b^2}{a}$

63. $\dfrac{x^2}{9} + \dfrac{y^2}{16} = 1$

$a = 4, b = 3, c = \sqrt{7}$

Points on the ellipse:

$(\pm 3, 0), (0, \pm 4)$

Length of latus recta:

$$\dfrac{2b^2}{a} = \dfrac{2(3)^2}{4} = \dfrac{9}{2}$$

Additional points: $\left(\pm\dfrac{9}{4}, -\sqrt{7}\right), \left(\pm\dfrac{9}{4}, \sqrt{7}\right)$

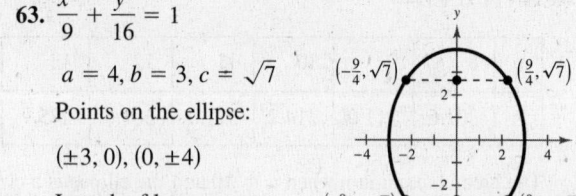

64. $\dfrac{x^2}{4} + \dfrac{y^2}{1} = 1$

$a = 2, b = 1, c = \sqrt{3}$

Points on the ellipse:

$(\pm 2, 0), (0, \pm 1)$

Length of latera recta:

$$\dfrac{2b^2}{a} = \dfrac{2(1)^2}{2} = 1$$

Additional points: $\left(-\sqrt{3}, \pm\dfrac{1}{2}\right), \left(\sqrt{3}, \pm\dfrac{1}{2}\right)$

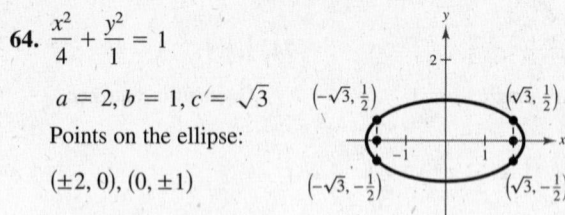

65. $5x^2 + 3y^2 = 15$

$$\dfrac{x^2}{3} + \dfrac{y^2}{5} = 1$$

$a = \sqrt{5}, b = \sqrt{3}, c = \sqrt{2}$

Points on the ellipse:

$(\pm\sqrt{3}, 0), (0, \pm\sqrt{5})$

Length of latus recta:

$$\dfrac{2b^2}{a} = \dfrac{2 \cdot 3}{\sqrt{5}} = \dfrac{6\sqrt{5}}{5}$$

Additional points: $\left(\pm\dfrac{3\sqrt{5}}{5}, \pm\sqrt{2}\right)$

66. $9x^2 + 4y^2 = 36$

$$\dfrac{x^2}{4} + \dfrac{y^2}{9} = 1$$

$a = 3, b = 2, c = \sqrt{5}$

Points on the ellipse: $(\pm 2, 0), (0, \pm 3)$

Length of latera recta: $\dfrac{2b^2}{a} = \dfrac{2 \cdot 2^2}{3} = \dfrac{8}{3}$

Additional points: $\left(\pm\dfrac{4}{3}, -\sqrt{5}\right), \left(\pm\dfrac{4}{3}, \sqrt{5}\right)$

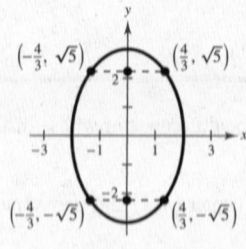

67. False. The graph of $\dfrac{x^2}{4} + y^4 = 1$ is not an ellipse. The degree on y is 4, not 2.

68. True. If e is close to 1, the ellipse is elongated and the foci are close to the vertices.

69. $\dfrac{x^2}{a^2} + \dfrac{y^2}{b^2} = 1$

(a) $a + b = 20 \Rightarrow b = 20 - a$

$A = \pi ab = \pi a(20 - a)$

(b) $264 = \pi a(20 - a)$

$0 = -\pi a^2 + 20\pi a - 264$

$0 = \pi a^2 - 20\pi a + 264$

By the Quadratic Formula: $a \approx 14$ or $a \approx 6$. Choosing the larger value of a, we have $a \approx 14$ and $b \approx 6$. The equation of an ellipse with an area of 264 is

$$\dfrac{x^2}{196} + \dfrac{y^2}{36} = 1.$$

—CONTINUED—

69. —CONTINUED—

(c)

a	8	9	10	11	12	13
A	301.6	311.0	314.2	311.0	301.6	285.9

The area is maximum when $a = 10$ and the ellipse is a circle.

(d)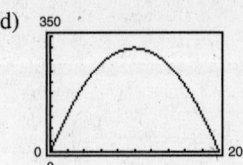

The area is maximum (314.16) when $a = b = 10$ and the ellipse is a circle.

70. (a) Length of string $= 2a$

(b) By keeping the string taut, the sum of the distances from the two fixed points is constant (equal to the length of the string).

71. 80, 40, 20, 10, 5, . . .

Geometric, $r = \frac{1}{2}$

72. 66, 55, 44, 33, 22, . . .

Arithmetic sequence

73. $-\frac{1}{2}, \frac{1}{2}, \frac{3}{2}, \frac{5}{2}, \frac{7}{2}, \ldots$

Arithmetic, $d = 1$

74. $\frac{1}{4}, \frac{1}{2}, 1, 2, 4$

Geometric sequence

75. $\displaystyle\sum_{n=0}^{6} (-3)^n = 1 - 3 + 9 - 27 + 81 - 243 + 729$

$= 547$

76. $\displaystyle\sum_{n=0}^{6} 3^n = \sum_{n=1}^{7} 3^{(n-1)} \implies a_1 = 1, r = 3$

$S_7 = \dfrac{1(1 - 3^7)}{1 - 3} = 1093$

77. $\displaystyle\sum_{n=0}^{10} 5\left(\frac{4}{3}\right)^n = 5\dfrac{\left(1 - \left(\frac{4}{3}\right)^{11}\right)}{1 - \frac{4}{3}}$

≈ 340.15

78. $\displaystyle\sum_{n=1}^{10} 4\left(\frac{3}{4}\right)^{n-1} \implies a_1 = 4, r = \frac{3}{4}$

$S_{10} = \dfrac{4\left(1 - \left(\frac{3}{4}\right)^{10}\right)}{1 - \frac{3}{4}} \approx 15.10$

Section 10.4   Hyperbolas

- A **hyperbola** is the set of all points   the difference of whose distances from two distinct fixed points (**foci**) is constant.

- The standard equation of a hyperbola with center (h, k) and transverse and conjugate axes of lengths $2a$ and $2b$ is:

  (a) $\dfrac{(x - h)^2}{a^2} - \dfrac{(y - k)^2}{b^2} = 1$ if the traverse axis is horizontal.

  (b) $\dfrac{(y - k)^2}{a^2} - \dfrac{(x - h)^2}{b^2} = 1$ if the traverse axis is vertical.

- $c^2 = a^2 + b^2$ where c is the distance from the center to a focus.

- The asymptotes of a hyperbola are:

  (a) $y = k \pm \dfrac{b}{a}(x - h)$ if the transverse axis is horizontal.

  (b) $y = k \pm \dfrac{a}{b}(x - h)$ if the transverse axis is vertical.

- The eccentricity of a hyperbola is $e = \dfrac{c}{a}$.

- To classify a nondegenerate conic from its general equation $Ax^2 + Cy^2 + Dx + Ey + F = 0$:
  (a) If $A = C$ ($A \neq 0, C \neq 0$), then it is a circle.
  (b) If $AC = 0$ ($A = 0$ or $C = 0$, but not both), then it is a parabola.
  (c) If $AC > 0$, then it is an ellipse.

Vocabulary Check

1. hyperbola

2. branches

3. transverse axis; center

4. asymptotes

5. $Ax^2 + Cy^2 + Dx + Ey + F = 0$

---

1. $\dfrac{y^2}{9} - \dfrac{x^2}{25} = 1$

Center: $(0, 0)$

$a = 3, b = 5$

Vertical transverse axis

Matches graph (b).

2. $\dfrac{y^2}{25} - \dfrac{x^2}{9} = 1$

Center: $(0, 0)$

$a = 5, b = 3$

Vertical transverse axis

Matches graph (c).

3. $\dfrac{(x-1)^2}{16} - \dfrac{y^2}{4} = 1$

Center: $(1, 0)$

$a = 4, b = 2$

Horizontal transverse axis

Matches graph (a).

4. $\dfrac{(x+1)^2}{16} - \dfrac{(y-2)^2}{9} = 1$

Center: $(-1, 2)$

$a = 4, b = 3$

Horizontal transverse axis

Matches graph (d).

5. $x^2 - y^2 = 1$

$a = 1, b = 1, c = \sqrt{2}$

Center: $(0, 0)$

Vertices: $(\pm 1, 0)$

Foci: $\left(\pm\sqrt{2}, 0\right)$

Asymptotes: $y = \pm x$

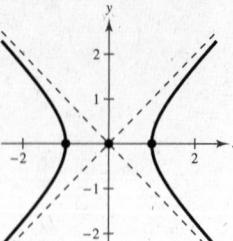

6. $\dfrac{x^2}{9} - \dfrac{y^2}{25} = 1$

$a = 3, b = 5$

$c = \sqrt{3^2 + 5^2} = \sqrt{34}$

Center: $(0, 0)$

Vertices: $(\pm 3, 0)$

Foci: $\left(\pm\sqrt{34}, 0\right)$

Asymptotes: $y = \pm\dfrac{5}{3}x$

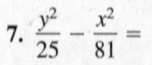

7. $\dfrac{y^2}{25} - \dfrac{x^2}{81} = 1$

$a = 5, b = 9, c = \sqrt{106}$

Center: $(0, 0)$

Vertices: $(0, \pm 5)$

Foci: $\left(0, \pm\sqrt{106}\right)$

Asymptotes: $y = \pm\dfrac{5}{9}x$

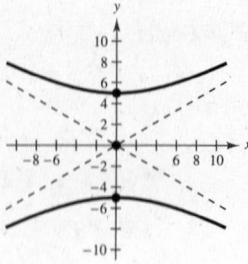

8. $\dfrac{x^2}{36} - \dfrac{y^2}{4} = 1$

$a = 6, b = 2,$

$c = \sqrt{36 + 4} = 2\sqrt{10}$

Center: $(0, 0)$

Vertices: $(\pm 6, 0)$

Foci: $\left(\pm 2\sqrt{10}, 0\right)$

Asymptotes: $y = \pm\dfrac{1}{3}x$

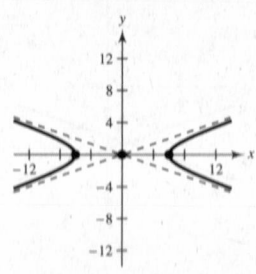

9. $\dfrac{(x-1)^2}{4} - \dfrac{(y+2)^2}{1} = 1$

$a = 2, b = 1, c = \sqrt{5}$

Center: $(1, -2)$

Vertices: $(-1, -2), (3, -2)$

Foci: $\left(1 \pm \sqrt{5}, -2\right)$

Asymptotes: $y = -2 \pm \dfrac{1}{2}(x - 1)$

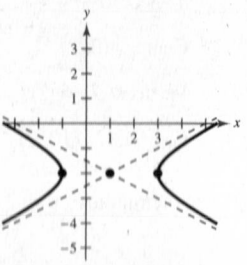

10. $\dfrac{(x + 3)^2}{144} - \dfrac{(y - 2)^2}{25} = 1$

$a = 12, b = 5$

$c = \sqrt{144 + 25} = 13$

Center: $(-3, 2)$

Vertices: $(9, 2), (-15, 2)$

Foci: $(10, 2), (-16, 2)$

Asymptotes: $y = 2 \pm \dfrac{5}{12}(x + 3)$

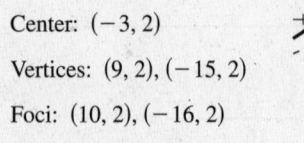

11. $\dfrac{(y + 6)^2}{1/9} - \dfrac{(x - 2)^2}{1/4} = 1$

$a = \dfrac{1}{3}, b = \dfrac{1}{2}, c = \dfrac{\sqrt{13}}{6}$

Center: $(2, -6)$

Vertices: $\left(2, -\dfrac{17}{3}\right), \left(2, -\dfrac{19}{3}\right)$

Foci: $\left(2, -6 \pm \dfrac{\sqrt{13}}{6}\right)$

Asymptotes: $y = -6 \pm \dfrac{2}{3}(x - 2)$

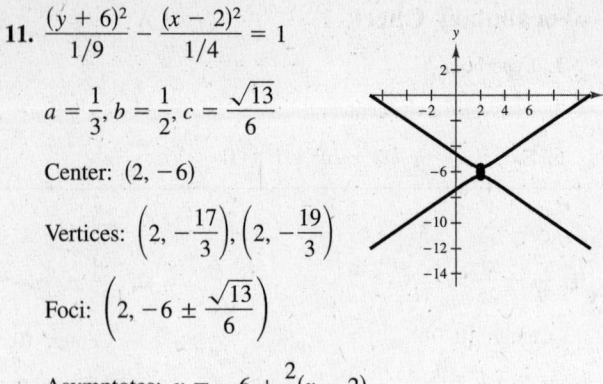

12. $\dfrac{(y - 1)^2}{1/4} - \dfrac{(x + 3)^2}{1/16} = 1$

$a = \dfrac{1}{2}, b = \dfrac{1}{4}$

$c = \sqrt{\dfrac{1}{4} + \dfrac{1}{16}} = \dfrac{\sqrt{5}}{4}$

Center: $(-3, 1)$

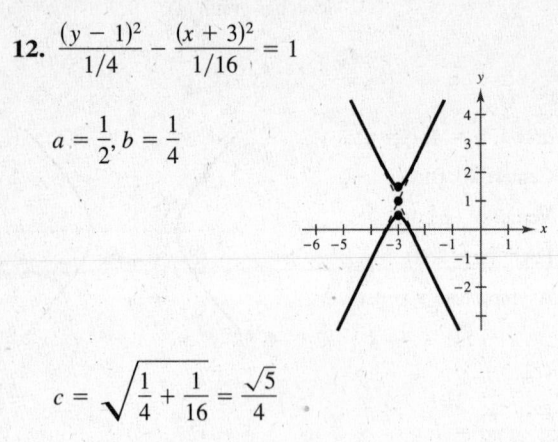

13. $9x^2 - y^2 - 36x - 6y + 18 = 0$

$9(x^2 - 4x + 4) - (y^2 + 6y + 9) = -18 + 36 - 9$

$9(x - 2)^2 - (y + 3)^2 = 9$

$\dfrac{(x - 2)^2}{1} - \dfrac{(y + 3)^2}{9} = 1$

$a = 1, b = 3, c = \sqrt{10}$

Center: $(2, -3)$

Vertices: $(1, -3), (3, -3)$

Foci: $\left(2 \pm \sqrt{10}, -3\right)$

Asymptotes:
$y = -3 \pm 3(x - 2)$

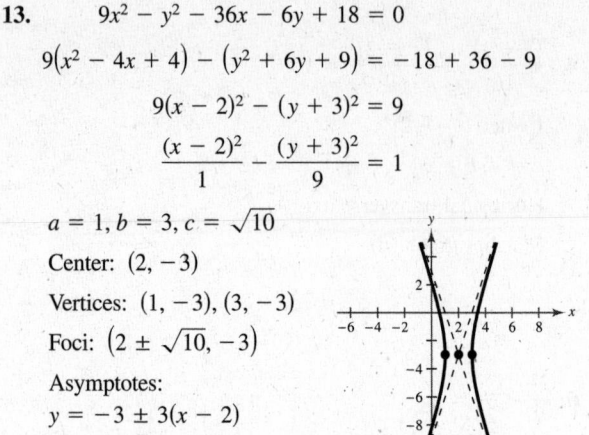

14. $x^2 - 9y^2 + 36y - 72 = 0$

$x^2 - 9(y^2 - 4y + 4) = 72 - 36$

$x^2 - 9(y - 2)^2 = 36$

$\dfrac{x^2}{36} - \dfrac{(y - 2)^2}{4} = 1$

$a = 6, b = 2,$

$c = \sqrt{36 + 4} = 2\sqrt{10}$

Center: $(0, 2)$

Vertices: $(\pm 6, 2)$

Foci: $\left(\pm 2\sqrt{10}, 2\right)$

Asymptotes: $y = 2 \pm \dfrac{1}{3}x$

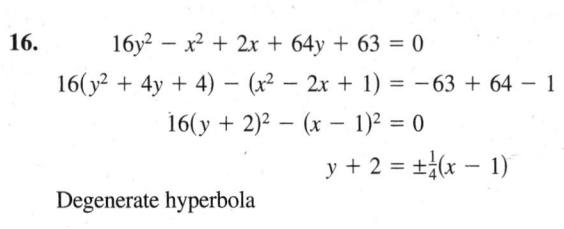

15. $x^2 - 9y^2 + 2x - 54y - 80 = 0$

$(x^2 + 2x + 1) - 9(y^2 + 6y + 9) = 80 + 1 - 81$

$(x + 1)^2 - 9(y + 3)^2 = 0$

$y + 3 = \pm\dfrac{1}{3}(x + 1)$

Degenerate hyperbola is two lines intersecting at $(-1, -3)$.

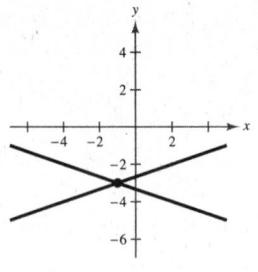

16. $16y^2 - x^2 + 2x + 64y + 63 = 0$

$16(y^2 + 4y + 4) - (x^2 - 2x + 1) = -63 + 64 - 1$

$16(y + 2)^2 - (x - 1)^2 = 0$

$y + 2 = \pm\dfrac{1}{4}(x - 1)$

Degenerate hyperbola

The graph is two lines intersecting at $(1, -2)$.

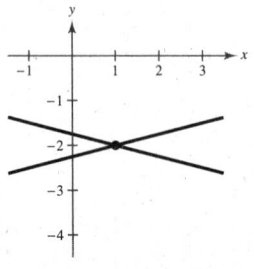

17. $2x^2 - 3y^2 = 6$

$$\frac{x^2}{3} - \frac{y^2}{2} = 1$$

$a = \sqrt{3}, b = \sqrt{2}, c = \sqrt{5}$

Center: $(0, 0)$

Vertices: $(\pm\sqrt{3}, 0)$

Foci: $(\pm\sqrt{5}, 0)$

Asymptotes: $y = \pm\sqrt{\dfrac{2}{3}}\,x = \pm\dfrac{\sqrt{6}}{3}x$

To use a graphing calculator, solve for y first.

$$y^2 = \frac{2x^2 - 6}{3}$$

$\left. \begin{aligned} y_1 &= \sqrt{\frac{2x^2 - 6}{3}} \\[2mm] y_2 &= -\sqrt{\frac{2x^2 - 6}{3}} \end{aligned} \right\}$ Hyperbola

$\left. \begin{aligned} y_3 &= \frac{\sqrt{6}}{3}x \\[2mm] y_4 &= -\frac{\sqrt{6}}{3}x \end{aligned} \right\}$ Asymptotes

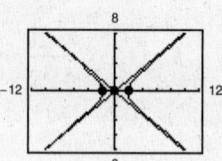

18. $6y^2 - 3x^2 = 18$

$$\frac{y^2}{3} - \frac{x^2}{6} = 1$$

$a = \sqrt{3}, b = \sqrt{6}, c = \sqrt{3 + 6} = 3$

Center: $(0, 0)$

Vertices: $(0, \pm\sqrt{3})$

Foci: $(0, \pm 3)$

Asymptotes: $y = \pm\dfrac{\sqrt{2}}{2}x$

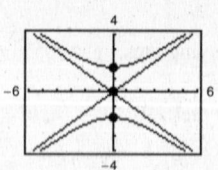

19. $9y^2 - x^2 + 2x + 54y + 62 = 0$

$9(y^2 + 6y + 9) - (x^2 - 2x + 1) = -62 - 1 + 81$

$9(y + 3)^2 - (x - 1)^2 = 18$

$$\frac{(y + 3)^2}{2} - \frac{(x - 1)^2}{18} = 1$$

$a = \sqrt{2}, b = 3\sqrt{2}, c = 2\sqrt{5}$

Center: $(1, -3)$

Vertices: $(1, -3 \pm \sqrt{2})$

Foci: $(1, -3 \pm 2\sqrt{5})$

Asymptotes: $y = -3 \pm \dfrac{1}{3}(x - 1)$

To use a graphing calculator, solve for y first.

$9(y + 3)^2 = 18 + (x - 1)^2$

$$y = -3 \pm \sqrt{\frac{18 + (x - 1)^2}{9}}$$

$\left. \begin{aligned} y_1 &= -3 + \frac{1}{3}\sqrt{18 + (x - 1)^2} \\[2mm] y_2 &= -3 - \frac{1}{3}\sqrt{18 + (x - 1)^2} \end{aligned} \right\}$ Hyperbola

$\left. \begin{aligned} y_3 &= -3 + \frac{1}{3}(x - 1) \\[2mm] y_4 &= -3 - \frac{1}{3}(x - 1) \end{aligned} \right\}$ Asymptotes

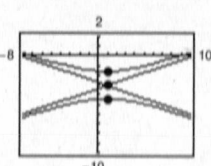

20.
$$9x^2 - y^2 + 54x + 10y + 55 = 0$$
$$9(x^2 + 6x + 9) - (y^2 - 10y + 25) = -55 + 81 - 25$$

$$\frac{(x + 3)^2}{1/9} - \frac{(y - 5)^2}{1} = 1$$

$$a = \frac{1}{3}, b = 1, c = \frac{\sqrt{10}}{3}$$

Center: $(-3, 5)$

Vertices: $\left(-3 \pm \frac{1}{3}, 5\right) \Rightarrow \left(-\frac{10}{3}, 5\right), \left(-\frac{8}{3}, 5\right)$

Foci: $\left(-3 \pm \frac{\sqrt{10}}{3}, 5\right)$

Asymptotes:
$y = 5 \pm 3(x + 3)$

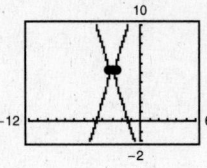

21. Vertices: $(0, \pm 2) \Rightarrow a = 2$

Foci: $(0, \pm 4) \Rightarrow c = 4$

$b^2 = c^2 - a^2 = 16 - 4 = 12$

Center: $(0, 0) = (h, k)$

$$\frac{(y - k)^2}{a^2} - \frac{(x - h)^2}{b^2} = 1$$

$$\frac{y^2}{4} - \frac{x^2}{12} = 1$$

22. Vertices: $(\pm 4, 0) \Rightarrow a = 4$

Foci: $(\pm 6, 0) \Rightarrow c = 6$

$b^2 = c^2 - a^2 = 36 - 16 = 20 \Rightarrow b = 2\sqrt{5}$

$$\frac{x^2}{a^2} - \frac{y^2}{b^2} = 1$$

$$\frac{x^2}{16} - \frac{y^2}{20} = 1$$

23. Vertices: $(\pm 1, 0) \Rightarrow a = 1$

Asymptotes: $y = \pm 5x \Rightarrow \frac{b}{a} = 5, b = 5$

Center: $(0, 0) = (h, k)$

$$\frac{(x - h)^2}{a^2} - \frac{(y - k)^2}{b^2} = 1$$

$$\frac{x^2}{1} - \frac{y^2}{25} = 1$$

24. Vertices: $(0, \pm 3) \Rightarrow a = 3$

Asymptotes: $y = \pm 3x \Rightarrow \frac{a}{b} = 3, b = 1$

$$\frac{y^2}{a^2} - \frac{x^2}{b^2} = 1$$

$$\frac{y^2}{9} - \frac{x^2}{1} = 1$$

25. Foci: $(0, \pm 8) \Rightarrow c = 8$

Asymptotes: $y = \pm 4x \Rightarrow \frac{a}{b} = 4 \Rightarrow a = 4b$

Center: $(0, 0) = (h, k)$

$c^2 = a^2 + b^2 \Rightarrow 64 = 16b^2 + b^2$

$$\frac{64}{17} = b^2 \Rightarrow a^2 = \frac{1024}{17}$$

$$\frac{(y - k)^2}{a^2} - \frac{(x - h)^2}{b^2} = 1$$

$$\frac{y^2}{1024/17} - \frac{x^2}{64/17} = 1$$

$$\frac{17y^2}{1024} - \frac{17x^2}{64} = 1$$

26. Foci: $(\pm 10, 0) \Rightarrow c = 10$ $a = 4(2) = 8$

Asymptotes: $y = \pm \frac{3}{4}x \Rightarrow \frac{b}{a} = \frac{3m}{4m}$ $b = 3(2) = 6$

$c^2 = a^2 + b^2 \Rightarrow 100 = (3m)^2 + (4m)^2$ $\frac{x^2}{a^2} - \frac{y^2}{b^2} = 1$

$100 = 25m^2$

$2 = m$ $\frac{x^2}{64} - \frac{y^2}{36} = 1$

27. Vertices: $(2, 0), (6, 0) \implies a = 2$

Foci: $(0, 0), (8, 0) \implies c = 4$

$b^2 = c^2 - a^2 = 16 - 4 = 12$

Center: $(4, 0) = (h, k)$

$\dfrac{(x - h)^2}{a^2} - \dfrac{(y - k)^2}{b^2} = 1$

$\dfrac{(x - 4)^2}{4} - \dfrac{y^2}{12} = 1$

28. Vertices: $(2, 3), (2, -3) \implies a = 3$

Center: $(2, 0)$

Foci: $(2, 6), (2, -6) \implies c = 6$

$b^2 = c^2 - a^2 = 36 - 9 = 27$

$\dfrac{(y - k)^2}{a^2} - \dfrac{(x - h)^2}{b^2} = 1$

$\dfrac{y^2}{9} - \dfrac{(x - 2)^2}{27} = 1$

29. Vertices: $(4, 1), (4, 9) \implies a = 4$

Foci: $(4, 0), (4, 10) \implies c = 5$

$b^2 = c^2 - a^2 = 25 - 16 = 9$

Center: $(4, 5) = (h, k)$

$\dfrac{(y - k)^2}{a^2} - \dfrac{(x - h)^2}{b^2} = 1$

$\dfrac{(y - 5)^2}{16} - \dfrac{(x - 4)^2}{9} = 1$

30. Vertices: $(-2, 1), (2, 1) \implies a = 2$

Center: $(0, 1)$

Foci: $(-3, 1), (3, 1) \implies c = 3$

$b^2 = c^2 - a^2 = 9 - 4 = 5$

$\dfrac{(x - h)^2}{a^2} - \dfrac{(y - k)^2}{b^2} = 1$

$\dfrac{x^2}{4} - \dfrac{(y - 1)^2}{5} = 1$

31. Vertices: $(2, 3), (2, -3) \implies a = 3$

Passes through the point: $(0, 5)$

Center: $(2, 0) = (h, k)$

$\dfrac{(y - k)^2}{a^2} - \dfrac{(x - h)^2}{b^2} = 1$

$\dfrac{y^2}{9} - \dfrac{(x - 2)^2}{b^2} = 1 \implies$

$\dfrac{(x - 2)^2}{b^2} = \dfrac{y^2}{9} - 1 = \dfrac{y^2 - 9}{9} \implies$

$b^2 = \dfrac{9(x - 2)^2}{y^2 - 9} = \dfrac{9(-2)^2}{25 - 9} = \dfrac{36}{16} = \dfrac{9}{4}$

$\dfrac{y^2}{9} - \dfrac{(x - 2)^2}{9/4} = 1$

$\dfrac{y^2}{9} - \dfrac{4(x - 2)^2}{9} = 1$

32. Vertices: $(-2, 1), (2, 1) \implies a = 2$

Center: $(0, 1)$

Point on curve: $(5, 4)$

$\dfrac{(x - h)^2}{a^2} - \dfrac{(y - k)^2}{b^2} = 1$

$\dfrac{x^2}{4} - \dfrac{(y - 1)^2}{b^2} = 1$

$\dfrac{25}{4} - \dfrac{9}{b^2} = 1$

$b^2 = \dfrac{12}{7}$

$\dfrac{x^2}{4} - \dfrac{(y - 1)^2}{12/7} = 1$

$\dfrac{x^2}{4} - \dfrac{7(y - 1)^2}{12} = 1$

33. Vertices: $(0, 4), (0, 0) \implies a = 2$

Passes through the point $\left(\sqrt{5}, -1 \right)$

Center: $(0, 2) = (h, k)$

$\dfrac{(y - k)^2}{a^2} - \dfrac{(x - h)^2}{b^2} = 1$

$\dfrac{(y - 2)^2}{4} - \dfrac{x^2}{b^2} = 1 \implies \dfrac{x^2}{b^2} = \dfrac{(y - 2)^2}{4} - 1 = \dfrac{(y - 2)^2 - 4}{4}$

$\implies b^2 = \dfrac{4x^2}{(y - 2)^2 - 4} = \dfrac{4\left(\sqrt{5} \right)^2}{(-1 - 2)^2 - 4} = \dfrac{20}{5} = 4$

$\dfrac{(y - 2)^2}{4} - \dfrac{x^2}{4} = 1$

34. Vertices: $(1, \pm 2) \implies a = 2$

Center: $(1, 0)$

Point on curve: $\left(0, \sqrt{5}\right)$

$$\frac{(y-k)^2}{a^2} - \frac{(x-h)^2}{b^2} = 1$$

$$\frac{y^2}{4} - \frac{(x-1)^2}{b^2} = 1$$

$$\frac{5}{4} - \frac{1}{b^2} = 1$$

$$b^2 = 4$$

$$\frac{y^2}{4} - \frac{(x-1)^2}{4} = 1$$

35. Vertices: $(1, 2), (3, 2) \implies a = 1$

Asymptotes: $y = x, y = 4 - x$

$$\frac{b}{a} = 1 \implies \frac{b}{1} = 1 \implies b = 1$$

Center: $(2, 2) = (h, k)$

$$\frac{(x-h)^2}{a^2} - \frac{(y-k)^2}{b^2} = 1$$

$$\frac{(x-2)^2}{1} - \frac{(y-2)^2}{1} = 1$$

36. Vertices: $(3, 0), (3, 6) \implies a = 3$

Center: $(3, 3)$

Asymptotes: $y = 6 - x, y = x$

$$\frac{a}{b} = 1 \implies b = 3$$

$$\frac{(y-k)^2}{a^2} - \frac{(x-h)^2}{b^2} = 1$$

$$\frac{(y-3)^2}{9} - \frac{(x-3)^2}{9} = 1$$

37. Vertices: $(0, 2), (6, 2) \implies a = 3$

Asymptotes: $y = \frac{2}{3}x, y = 4 - \frac{2}{3}x$

$$\frac{b}{a} = \frac{2}{3} \implies b = 2$$

Center: $(3, 2) = (h, k)$

$$\frac{(x-h)^2}{a^2} - \frac{(y-k)^2}{b^2} = 1$$

$$\frac{(x-3)^2}{9} - \frac{(y-2)^2}{4} = 1$$

38. Vertices: $(3, 0), (3, 4) \implies a = 2$

Asymptotes: $y = \frac{2}{3}x, y = 4 - \frac{2}{3}x$

$$\frac{a}{b} = \frac{2}{3} \implies b = 3$$

Center: $(3, 2) = (h, k)$

$$\frac{(y-k)^2}{a^2} - \frac{(x-h)^2}{b^2} = 1$$

$$\frac{(y-2)^2}{4} - \frac{(x-3)^2}{9} = 1$$

39. (a) Vertices: $(\pm 1, 0) \implies a = 1$

Horizontal transverse axis

Center: $(0, 0)$

$$\frac{x^2}{a^2} - \frac{y^2}{b^2} = 1$$

Point on the graph: $(2, 13)$

$$\frac{2^2}{1^2} - \frac{13^2}{b^2} = 1$$

$$4 - \frac{169}{b^2} = 1$$

$$3b^2 = 169$$

$$b^2 = \frac{169}{3} \approx 56.33$$

Thus we have $\dfrac{x^2}{1} - \dfrac{y^2}{56.33} = 1.$

(b) When $y = 5$: $x^2 = 1 + \dfrac{5^2}{56.33}$

$$x = \sqrt{1 + \frac{25}{56.33}} \approx 1.2016$$

Width: $2x \approx 2.403$ feet

40.

$$2c = 4 \text{ mi} = 21{,}120 \text{ ft}$$

$$c = 10{,}560 \text{ ft}$$

$$(1100 \text{ ft/s})(18 \text{ s}) = 19{,}800 \text{ ft}$$

The lightening occurred 19,800 feet further from B than from A:

$$d_2 - d_1 = 2a = 19{,}800 \text{ ft}$$

$$a = 9900 \text{ ft}$$

$$b^2 = c^2 - a^2 = (10{,}560)^2 - (9900)^2$$

$$b^2 = 13{,}503{,}600$$

$$\frac{x^2}{(9900)^2} - \frac{y^2}{13{,}503{,}600} = 1$$

$$\frac{x^2}{98{,}010{,}000} - \frac{y^2}{13{,}503{,}600} = 1$$

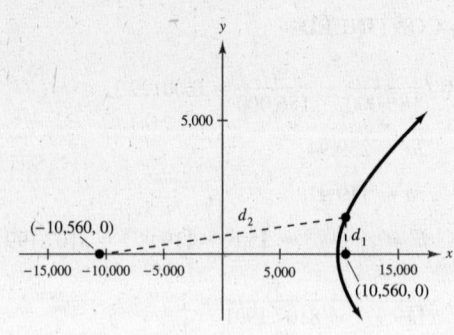

41. Since listening station C heard the explosion 4 seconds after listening station A, and since listening station B heard the explosion one second after listening station A, and sound travels 1100 feet per second, the explosion is located in Quadrant IV on the line $x = 3300$. The locus of all points 4400 feet closer to A than C is one branch of the hyperbola.

$$\frac{x^2}{a^2} - \frac{y^2}{b^2} = 1 \text{ where } c = 3300 \text{ feet and } a = \frac{4400}{2} = 2200 \text{ feet}, \ b^2 = c^2 - a^2 = 6{,}050{,}000.$$

When $x = 3300$ we have $\dfrac{3300^2}{2200^2} - \dfrac{y^2}{6{,}050{,}000} = 1$.

Solving for y: $y^2 = 6{,}050{,}000\left(\dfrac{3300^2}{2200^2} - 1\right)$

$$= 7{,}562{,}500$$

$$y = \pm 2750$$

Since the explosion is in Quadrant IV, its coordinates are $(3300, -2750)$.

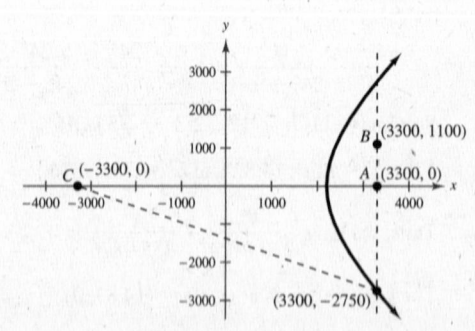

42. (a) Foci: $(\pm 150, 0) \Rightarrow c = 150$

Center: $(0, 0) = (h, k)$

$$\frac{d_2}{186{,}000} - \frac{d_1}{186{,}000} = 0.001 \Rightarrow 2a = 186, \ a = 93$$

$$b^2 = c^2 - a^2 = 150^2 - 93^2 = 13{,}851$$

$$\frac{x^2}{93^2} - \frac{y^2}{13{,}851} = 1$$

$$x^2 = 93^2\left(1 + \frac{75^2}{13{,}851}\right) \approx 12{,}161$$

$$x \approx 110.3 \text{ miles}$$

(b) $c - a = 150 - 93 = 57$ miles

(c) $\dfrac{270}{186{,}000} - \dfrac{30}{186{,}000} \approx 0.00129$ second

—CONTINUED—

42. —CONTINUED—

(d) $\dfrac{d_2}{186,000} - \dfrac{d_1}{186,000} = 0.00129$

$2a \approx 239.94$

$a \approx 119.97$

$b^2 = c^2 - a^2 = 150^2 - 119.97^2 = 8107.1991$

$\dfrac{x^2}{119.97^2} - \dfrac{y^2}{8107.1991} = 1$

$x^2 = 119.97^2\left(1 + \dfrac{60^2}{8107.1991}\right)$

$x \approx 144.2$ miles

Position: $(144.2, 60)$

43. Center: $(0, 0) = (h, k)$

Focus: $(24, 0) \implies c = 24$

Solution point: $(24, 24)$

$24^2 = a^2 + b^2 \implies b^2 = 24^2 - a^2$

$\dfrac{(x - h)^2}{a^2} - \dfrac{(y - k)^2}{b^2} = 1$

$\dfrac{x^2}{a^2} - \dfrac{y^2}{24^2 - a^2} = 1 \implies \dfrac{24^2}{a^2} - \dfrac{24^2}{24^2 - a^2} = 1$

Solving yields $a = 12\sqrt{2\left(3 - \sqrt{5}\right)}$ OR

$12\left(\sqrt{5} - 1\right) \approx 14.83$ and $b^2 \approx 355.9876$.

Thus, we have $\dfrac{x^2}{220.0124} - \dfrac{y^2}{355.9876} = 1$.

The right vertex is at $(a, 0) \approx (14.83, 0)$.

44. (a) $x^2 + y^2 - 200x - 52,500 = 0$

$Ax^2 + Cy^2 + Dx + Ey + F = 0$

$A = 1, C = 1, D = -200, E = 0, F = -52,500$

$A = C$: circle

(b) $(x^2 - 200x + 10,000) + (y^2) = 52,500 + 10,000$

$(x - 100)^2 + y^2 = 62,500$

$\dfrac{(x - 100)^2}{62,500} + \dfrac{y^2}{62,500} = 1$

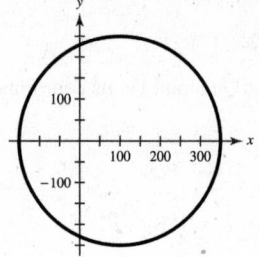

(c) $d = \sqrt{(-100 - 0)^2 + (150 - 0)^2} = 50\sqrt{13}$

$d \approx 180.28$ meters

45. $x^2 + y^2 - 6x + 4y + 9 = 0$

$A = 1, C = 1$

$A = C \implies$ Circle

46. $x^2 + 4y^2 - 6x + 16y + 21 = 0$

$A = 1, C = 4$

$AC = (1)(4) > 0$ and $A \neq C \implies$ Ellipse

47. $4x^2 - y^2 - 4x - 3 = 0$

$A = 4, C = -1$

$AC = (4)(-1) = -4 < 0 \implies$ Hyperbola

48. $y^2 - 6y - 4x + 21 = 0$

$A = 0, C = 1$

$AC = (0)(1) = 0 \implies$ Parabola

49. $y^2 - 4x^2 + 4x - 2y - 4 = 0$

$A = -4, C = 1$

$AC = (-4)(1) = -4 < 0 \implies$ Hyperbola

50. $x^2 + y^2 - 4x + 6y - 3 = 0$

$A = 1, C = 1$

$A = C \implies$ Circle

51. $x^2 - 4x - 8y + 2 = 0$

$A = 1, C = 0$

$AC = (1)(0) = 0 \implies$ Parabola

52. $4x^2 + y^2 - 8x + 3 = 0$

$A = 4, C = 1$

$AC = 4 > 0$ and $A \neq C \implies$ Ellipse

53. $4x^2 + 3y^2 + 8x - 24y + 51 = 0$

$A = 4, C = 3$

$AC = 4(3) = 12 > 0$ and $A \neq C \implies$ Ellipse

54. $4y^2 - 2x^2 - 4y - 8x - 15 = 0$

$AC = (-2)(4) < 0 \implies$ Hyperbola

55. $25x^2 - 10x - 200y - 119 = 0$

$A = 25, C = 0$

$AC = 25(0) = 0 \implies$ Parabola

56. $4y^2 + 4x^2 - 24x + 35 = 0$

$A = C = 4 \implies$ Circle

57. $4x^2 + 16y^2 - 4x - 32y + 1 = 0$

$A = 4, C = 16$

$AC = (4)(16) = 64 > 0$ and $A \neq C \implies$ Ellipse

58. $2y^2 + 2x + 2y + 1 = 0$

$A = 0, C = 2$

$AC = 0$, but $C \neq 0 \implies$ Parabola

59. $100x^2 + 100y^2 - 100x + 400y + 409 = 0$

$A = 100, C = 100$

$A = C \implies$ Circle

60. $4x^2 - y^2 + 4x + 2y - 1 = 0$

$A = 4, C = -1$

$AC = (4)(-1) = -4 < 0 \implies$ Hyperbola

61. True. For a hyperbola, $c^2 = a^2 + b^2$ or

$$e^2 = \frac{c^2}{a^2} = 1 + \frac{b^2}{a^2}.$$

The larger the ratio of b to a, the larger the eccentricity $e = c/a$ of the hyperbola.

62. False. For the trivial solution of two intersecting lines to occur, the standard form of the equation of the hyperbola would be equal to zero.

$$\frac{(x - h)^2}{a^2} - \frac{(y - k)^2}{b^2} = 0 \quad \text{or} \quad \frac{(y - k)^2}{a^2} - \frac{(x - h)^2}{b^2} = 0$$

63. Let (x, y) be such that the difference of the distances from $(c, 0)$ and $(-c, 0)$ is $2a$ (again only deriving one of the forms).

$$2a = \left| \sqrt{(x + c)^2 + y^2} - \sqrt{(x - c)^2 + y^2} \right|$$

$$2a + \sqrt{(x - c)^2 + y^2} = \sqrt{(x + c)^2 + y^2}$$

$$4a^2 + 4a\sqrt{(x - c)^2 + y^2} + (x - c)^2 + y^2 = (x + c)^2 + y^2$$

$$4a\sqrt{(x - c)^2 + y^2} = 4cx - 4a^2$$

$$a\sqrt{(x - c)^2 + y^2} = cx - a^2$$

$$a^2(x^2 - 2cx + c^2 + y^2) = c^2x^2 - 2a^2cx + a^4$$

$$a^2(c^2 - a^2) = (c^2 - a^2)x^2 - a^2y^2$$

Let $b^2 = c^2 - a^2$. Then $a^2b^2 = b^2x^2 - a^2y^2 \implies 1 = \dfrac{x^2}{a^2} - \dfrac{y^2}{b^2}.$

64. The extended diagonals of the central rectangle are the asymptotes of the hyperbola.

65.
$$9x^2 - 54x - 4y^2 + 8y + 41 = 0$$
$$9(x^2 - 6x + 9) - 4(y^2 - 2y + 1) = -41 + 81 - 4$$
$$9(x - 3)^2 - 4(y - 1)^2 = 36$$
$$\frac{(x - 3)^2}{4} - \frac{(y - 1)^2}{9} = 1$$
$$\frac{(y - 1)^2}{9} = \frac{(x - 3)^2}{4} - 1$$
$$(y - 1)^2 = 9\left[\frac{(x - 3)^2}{4} - 1\right]$$

The bottom half of the hyperbola is:
$$y - 1 = -\sqrt{9\left[\frac{(x - 3)^2}{4} - 1\right]}$$
$$y = 1 - 3\sqrt{\frac{(x - 3)^2}{4} - 1}$$

66.

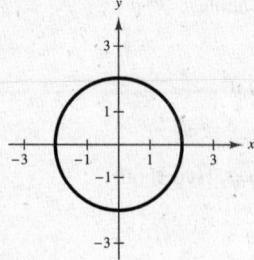

Value of C   Possible number of points of intersection

$C > 2$

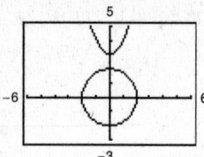

$C = 2$

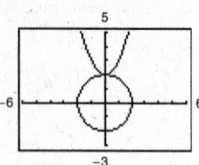

$-2 < C < 2$

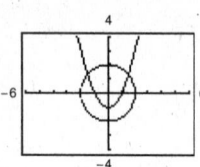

$C = -2$

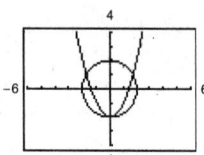

—CONTINUED—

66. **—CONTINUED—**

$C < -2$

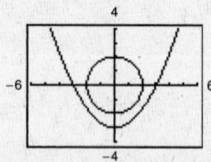   or   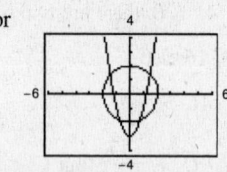   or

For $C \le -2$, we need to analyze the two curves to determine the number of points of intersection.

$C = -2$:

$$x^2 + y^2 = 4 \quad \text{and} \quad y = x^2 - 2$$

$$x^2 = y + 2$$

Substitute: $(y + 2) + y^2 = 4$

$$y^2 + y - 2 = 0$$

$$(y + 2)(y - 1) = 0$$

$$y = -2, 1$$

$x^2 = y + 2$	$x^2 = y + 2$
$x^2 = -2 + 2$	$x^2 = 1 + 2$
$x^2 = 0$	$x^2 = 3$
$x = 0$	$x = \pm\sqrt{3}$
$(0, -2)$	$(-\sqrt{3}, 1), (\sqrt{3}, 1)$

There are three points of intersection when $C = -2$.

$C < -2$:

$$x^2 + y^2 = 4 \quad \text{and} \quad y = x^2 + C$$

$$x^2 = y - C$$

Substitute: $(y - C) + y^2 = 4$

$$y^2 + y - 4 - C = 0$$

$$y = \frac{-1 \pm \sqrt{(1)^2 - (4)(1)(-C - 4)}}{2}$$

$$y = \frac{-1 \pm \sqrt{1 + 4(C + 4)}}{2}$$

If $1 + 4(C + 4) < 0$, there are no real solutions (no points of intersection):

$$1 + 4C + 16 < 0$$

$$4C < -17$$

$$C < \frac{-17}{4}, \text{ no points of intersection}$$

If $1 + 4(C + 4) = 0$, there is one real solution (two points of intersection):

$$1 + 4C + 16 = 0$$

$$4C = -17$$

$$C = \frac{-17}{4}, \text{ two points of intersection}$$

—CONTINUED—

66. **—CONTINUED—**

If $1 + 4(C + 4) > 0$, there are two real solutions (four points of intersection):

$$1 + 4C + 16 > 0$$

$$4C > -17$$

$$C > \frac{-17}{4}, \text{ (but } C < -2\text{), four points of intersection}$$

Summary:

a. no points of intersection: $C > 2$ or $C < \frac{-17}{4}$

b. one point of intersection: $C = 2$

c. two points of intersection: $-2 < C < 2$ or $C = \frac{-17}{4}$

d. three points of intersection: $C = -2$

e. four points of intersection: $\frac{-17}{4} < C < -2$

67. $x^3 - 16x = x(x^2 - 16) = x(x + 4)(x - 4)$

68. $x^2 + 14x + 49 = x^2 + 2(7)x + 7^2 = (x + 7)^2$

69. $2x^3 - 24x^2 + 72x = 2x(x^2 - 12x + 36) = 2x(x - 6)^2$

70. $6x^3 - 11x^2 - 10x = x(6x^2 - 11x - 10)$

$$= x(6x^2 - 15x + 4x - 10)$$

$$= x[3x(2x - 5) + 2(2x - 5)]$$

$$= x(3x + 2)(2x - 5)$$

71. $16x^3 + 54 = 2(8x^3 + 27)$

$$= 2[(2x)^3 + (3)^3]$$

$$= 2(2x + 3)(4x^2 - 6x + 9)$$

72. $4 - x + 4x^2 - x^3 = (4 + 4x^2) - (x + x^3)$

$$= 4(1 + x^2) - x(1 + x^2)$$

$$= (4 - x)(1 + x^2)$$

73. $y = 2 \cos x + 1$

Amplitude: 2

Period: 2π

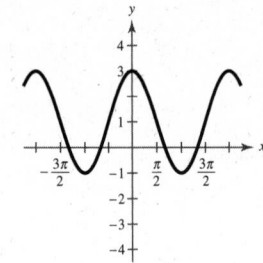

74. $y = \sin \pi x$

Period: $\frac{2\pi}{\pi} = 2$

Amplitude: 1

Key points: $(0, 0)$, $\left(\frac{1}{2}, 1\right)$,

$(1, 0)$, $\left(\frac{3}{2}, -1\right)$, $(2, 0)$

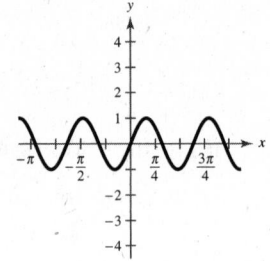

75. $y = \tan 2x$

Period: $\frac{\pi}{2}$

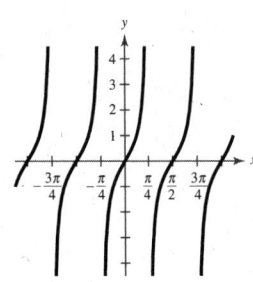

76. $y = -\frac{1}{2} \sec x$

Graph $y = -\frac{1}{2} \cos x$ first.

Period: 2π

One cycle: 0 to 2π

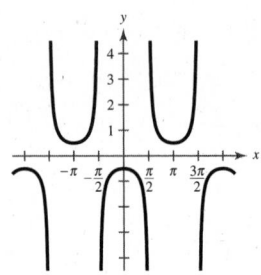

Section 10.5   Rotation of Conics

■ The general second-degree equation $Ax^2 + Bxy + Cy^2 + Dx + Ey + F = 0$ can be rewritten as $A'(x')^2 + C'(y')^2 + D'x' + E'y' + F' = 0$ by rotating the coordinate axes through the angle θ, where $\cot 2\theta = (A - C)/B$ and the following quantities are invariant under rotation:

   1. $F = F'$

   2. $A + C = A' + C'$

   3. $B^2 - 4AC = (B')^2 - 4A'C'$

■ $x = x'\cos\theta - y'\sin\theta$
   $y = x'\sin\theta + y'\cos\theta$

■ The graph of the nondegenerate equation $Ax^2 + Bxy + Cy^2 + Dx + Ey + F = 0$ is:

   (a) An ellipse or circle if $B^2 - 4AC < 0$.

   (b) A parabola if $B^2 - 4AC = 0$.

   (c) A hyperbola if $B^2 - 4AC > 0$.

Vocabulary Check

1. rotation of axes

2. $A'(x')^2 + C'(y')^2 + D'x' + E'y' + F' = 0$

3. invariant under rotation

4. discriminant

---

1. $\theta = 90°$; Point: $(0, 3)$

$x = x'\cos\theta - y'\sin\theta$ $y = x'\sin\theta + y'\cos\theta$

$0 = x'\cos 90° - y'\sin 90°$ $3 = x'\sin 90° - y'\cos 90°$

$0 = y'$ $3 = x'$

So, $(x', y') = (3, 0)$.

2. $\theta = 45°$; Point: $(3, 3)$

$\begin{aligned} x &= x'\cos\theta - y'\sin\theta \\ y &= x'\sin\theta + y'\cos\theta \end{aligned} \Rightarrow \begin{cases} 3 = x'\cos 45° - y'\sin 45° \\ 3 = x'\sin 45° + y'\cos 45° \end{cases}$

Solving the system yields $(x', y') = (3\sqrt{2}, 0)$.

3. $\theta = 30°$; Point: $(1, 3)$

$\begin{aligned} x &= x'\cos\theta - y'\sin\theta \\ y &= x'\sin\theta + y'\cos\theta \end{aligned} \Rightarrow \begin{cases} 1 = x'\cos 30° - y'\sin 30° \\ 3 = x'\sin 30° + y'\cos 30° \end{cases}$

Solving the system yields $(x', y') = \left(\dfrac{3 + \sqrt{3}}{2}, \dfrac{3\sqrt{3} - 1}{2}\right)$.

4. $\theta = 60°$; Point: $(3, 1)$

$\begin{aligned} x &= x'\cos\theta - y'\sin\theta \\ y &= x'\sin\theta + y'\cos\theta \end{aligned} \Rightarrow \begin{cases} 3 = x'\cos 60° - y'\sin 60° \\ 1 = x'\sin 60° + y'\cos 60° \end{cases}$

Solving this system yields $(x', y') = \left(\dfrac{3 + \sqrt{3}}{2}, \dfrac{1 - 3\sqrt{3}}{2}\right)$.

5. $\theta = 45°$; Point $(2, 1)$

$\begin{aligned} x &= x'\cos\theta - y'\sin\theta \\ y &= x'\sin\theta + y'\cos\theta \end{aligned} \Rightarrow \begin{cases} 2 = x'\cos 45° - y'\sin 45° \\ 1 = x'\sin 45° + y'\cos 45° \end{cases}$

Solving the system yields $(x', y') = \left(\dfrac{3\sqrt{2}}{2}, -\dfrac{\sqrt{2}}{2}\right)$.

6. $\theta = 30°$; Point: $(2, 4)$

$\begin{aligned} x &= x'\cos\theta - y'\sin\theta \\ y &= x'\sin\theta + y'\cos\theta \end{aligned} \Rightarrow \begin{cases} 2 = x'\cos 30° - y'\sin 30° \\ 4 = x'\sin 30° + y'\cos 30° \end{cases}$

Solving this system yields $(x', y') = \left(\sqrt{3} + 2, 2\sqrt{3} - 1\right)$.

7. $xy + 1 = 0$, $A = 0$, $B = 1$, $C = 0$

$$\cot 2\theta = \frac{A - C}{B} = 0 \implies 2\theta = \frac{\pi}{2} \implies \theta = \frac{\pi}{4}$$

$$x = x'\cos\frac{\pi}{4} - y'\sin\frac{\pi}{4} \qquad\qquad y = x'\sin\frac{\pi}{4} + y'\cos\frac{\pi}{4}$$

$$= x'\left(\frac{\sqrt{2}}{2}\right) - y'\left(\frac{\sqrt{2}}{2}\right) \qquad\qquad = x'\left(\frac{\sqrt{2}}{2}\right) + y'\left(\frac{\sqrt{2}}{2}\right)$$

$$= \frac{x' - y'}{\sqrt{2}} \qquad\qquad\qquad\qquad = \frac{x' + y'}{\sqrt{2}}$$

$$xy + 1 = 0$$

$$\left(\frac{x' - y'}{\sqrt{2}}\right)\left(\frac{x' + y'}{\sqrt{2}}\right) + 1 = 0$$

$$\frac{(y')^2}{2} - \frac{(x')^2}{2} = 1$$

8. $xy - 2 = 0$, $A = 0$, $B = 1$, $C = 0$

$$\cot 2\theta = \frac{A - C}{B} = 0 \implies 2\theta = \frac{\pi}{2} \implies \theta = \frac{\pi}{4}$$

$$x = x'\cos\frac{\pi}{4} - y'\sin\frac{\pi}{4} \qquad\qquad y = x'\sin\frac{\pi}{4} + y'\cos\frac{\pi}{4}$$

$$= \frac{x' - y'}{\sqrt{2}} \qquad\qquad\qquad\qquad = \frac{x' + y'}{\sqrt{2}}$$

$$xy - 2 = 0$$

$$\left(\frac{x' - y'}{\sqrt{2}}\right)\left(\frac{x' + y'}{\sqrt{2}}\right) - 2 = 0$$

$$\frac{(x')^2 - (y')^2}{2} = 2$$

$$\frac{(x')^2}{4} - \frac{(y')^2}{4} = 1$$

9. $x^2 - 2xy + y^2 - 1 = 0$, $A = 1$, $B = -2$, $C = 1$

$$\cot 2\theta = \frac{A - C}{B} = 0 \implies 2\theta = \frac{\pi}{2} \implies \theta = \frac{\pi}{4}$$

$$x = x'\cos\frac{\pi}{4} - y'\sin\frac{\pi}{4} \qquad\qquad y = x'\sin\frac{\pi}{4} + y'\cos\frac{\pi}{4}$$

$$= x'\left(\frac{\sqrt{2}}{2}\right) - y'\left(\frac{\sqrt{2}}{2}\right) \qquad\qquad = x'\left(\frac{\sqrt{2}}{2}\right) + y'\left(\frac{\sqrt{2}}{2}\right)$$

$$= \frac{x' - y'}{\sqrt{2}} \qquad\qquad\qquad\qquad = \frac{x' + y'}{\sqrt{2}}$$

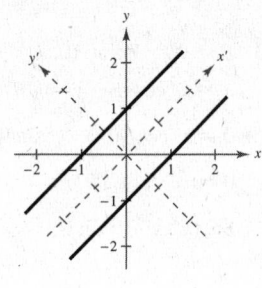

$$x^2 - 2xy + y^2 - 1 = 0$$

$$\left(\frac{x' - y'}{\sqrt{2}}\right)^2 - 2\left(\frac{x' - y'}{\sqrt{2}}\right)\left(\frac{x' + y'}{\sqrt{2}}\right) + \left(\frac{x' + y'}{\sqrt{2}}\right)^2 - 1 = 0$$

$$\frac{(x')^2 - 2(x')(y') + (y')^2}{2} - \frac{2((x')^2 - (y')^2)}{2} + \frac{(x')^2 + 2(x')(y') + (y')^2}{2} - 1 = 0$$

$$2(y')^2 - 1 = 0$$

$$(y')^2 = \frac{1}{2}$$

$$y' = \pm\sqrt{\frac{1}{2}} = \pm\frac{\sqrt{2}}{2}$$

The graph is two parallel lines.

Alternate solution:

$$x^2 - 2xy + y^2 - 1 = 0$$

$$(x - y)^2 = 1$$

$$x - y = \pm 1$$

$$y = x \pm 1$$

10. $xy + x - 2y + 3 = 0$

$A = 0, B = 1, C = 0$

$\cot 2\theta = \dfrac{A - C}{B} = 0 \implies 2\theta = \dfrac{\pi}{2} \implies \theta = \dfrac{\pi}{4}$

$x = x' \cos \dfrac{\pi}{4} - y' \sin \dfrac{\pi}{4}$ $\qquad$ $y = x' \sin \dfrac{\pi}{4} + y' \cos \dfrac{\pi}{4}$

$\quad = x'\left(\dfrac{\sqrt{2}}{2}\right) - y'\left(\dfrac{\sqrt{2}}{2}\right)$ $\qquad\quad = x'\left(\dfrac{\sqrt{2}}{2}\right) + y'\left(\dfrac{\sqrt{2}}{2}\right)$

$\quad = \dfrac{x' - y'}{\sqrt{2}}$ $\qquad\qquad\qquad\qquad = \dfrac{x' + y'}{\sqrt{2}}$

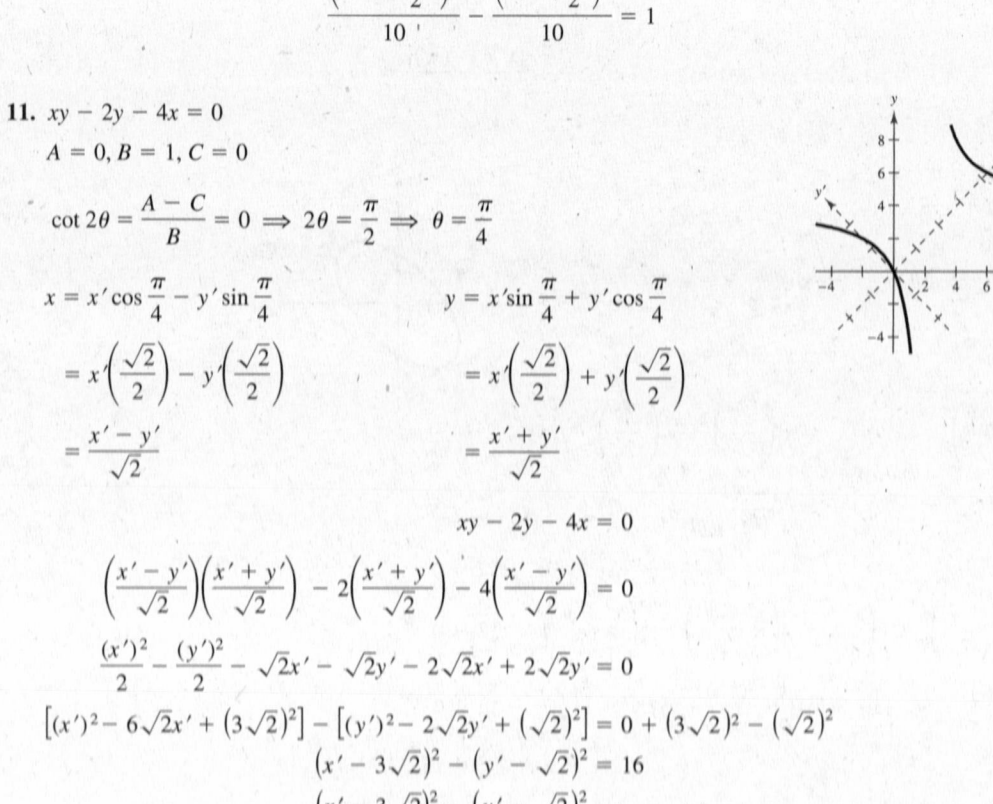

$$xy + x - 2y + 3 = 0$$

$$\left(\dfrac{x' - y'}{\sqrt{2}}\right)\left(\dfrac{x' + y'}{\sqrt{2}}\right) + \left(\dfrac{x' - y'}{\sqrt{2}}\right) - 2\left(\dfrac{x' + y'}{\sqrt{2}}\right) + 3 = 0$$

$$\dfrac{(x')^2}{2} - \dfrac{(y')^2}{2} + \dfrac{x'}{\sqrt{2}} - \dfrac{y'}{\sqrt{2}} - \dfrac{2x'}{\sqrt{2}} - \dfrac{2y'}{\sqrt{2}} + 3 = 0$$

$$\left[(x')^2 - \sqrt{2}x' + \left(\dfrac{\sqrt{2}}{2}\right)^2\right] - \left[(y')^2 + 3\sqrt{2}y' + \left(\dfrac{3\sqrt{2}}{2}\right)^2\right] = -6 + \left(\dfrac{\sqrt{2}}{2}\right)^2 - \left(\dfrac{3\sqrt{2}}{2}\right)^2$$

$$\left(x' - \dfrac{\sqrt{2}}{2}\right)^2 - \left(y' + \dfrac{3\sqrt{2}}{2}\right)^2 = -10$$

$$\dfrac{\left(y' + \dfrac{3\sqrt{2}}{2}\right)^2}{10} - \dfrac{\left(x' - \dfrac{\sqrt{2}}{2}\right)^2}{10} = 1$$

11. $xy - 2y - 4x = 0$

$A = 0, B = 1, C = 0$

$\cot 2\theta = \dfrac{A - C}{B} = 0 \implies 2\theta = \dfrac{\pi}{2} \implies \theta = \dfrac{\pi}{4}$

$x = x' \cos \dfrac{\pi}{4} - y' \sin \dfrac{\pi}{4}$ $\qquad$ $y = x' \sin \dfrac{\pi}{4} + y' \cos \dfrac{\pi}{4}$

$\quad = x'\left(\dfrac{\sqrt{2}}{2}\right) - y'\left(\dfrac{\sqrt{2}}{2}\right)$ $\qquad\quad = x'\left(\dfrac{\sqrt{2}}{2}\right) + y'\left(\dfrac{\sqrt{2}}{2}\right)$

$\quad = \dfrac{x' - y'}{\sqrt{2}}$ $\qquad\qquad\qquad\qquad = \dfrac{x' + y'}{\sqrt{2}}$

$$xy - 2y - 4x = 0$$

$$\left(\dfrac{x' - y'}{\sqrt{2}}\right)\left(\dfrac{x' + y'}{\sqrt{2}}\right) - 2\left(\dfrac{x' + y'}{\sqrt{2}}\right) - 4\left(\dfrac{x' - y'}{\sqrt{2}}\right) = 0$$

$$\dfrac{(x')^2}{2} - \dfrac{(y')^2}{2} - \sqrt{2}x' - \sqrt{2}y' - 2\sqrt{2}x' + 2\sqrt{2}y' = 0$$

$$\left[(x')^2 - 6\sqrt{2}x' + \left(3\sqrt{2}\right)^2\right] - \left[(y')^2 - 2\sqrt{2}y' + \left(\sqrt{2}\right)^2\right] = 0 + \left(3\sqrt{2}\right)^2 - \left(\sqrt{2}\right)^2$$

$$\left(x' - 3\sqrt{2}\right)^2 - \left(y' - \sqrt{2}\right)^2 = 16$$

$$\dfrac{\left(x' - 3\sqrt{2}\right)^2}{16} - \dfrac{\left(y' - \sqrt{2}\right)^2}{16} = 1$$

12. $2x^2 - 3xy - 2y^2 + 10 = 0$

$A = 2, B = -3, C = -2$

$\cot 2\theta = \dfrac{A - C}{B} = -\dfrac{4}{3} \implies \theta \approx 71.57°$

$\cos 2\theta = -\dfrac{4}{5}$

$\sin \theta = \sqrt{\dfrac{1 - \cos 2\theta}{2}} = \sqrt{\dfrac{1 - (-4/5)}{2}} = \dfrac{3}{\sqrt{10}}$

$\cos \theta = \sqrt{\dfrac{1 + \cos 2\theta}{2}} = \sqrt{\dfrac{1 + (-4/5)}{2}} = \dfrac{1}{\sqrt{10}}$

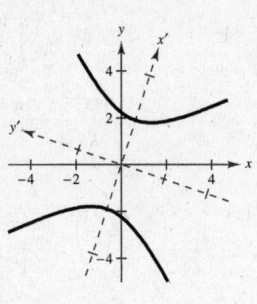

$x = x' \cos \theta - y' \sin \theta \qquad\qquad y = x' \sin \theta + y' \cos \theta$

$\quad = x'\left(\dfrac{1}{\sqrt{10}}\right) - y'\left(\dfrac{3}{\sqrt{10}}\right) \qquad = x'\left(\dfrac{3}{\sqrt{10}}\right) + y'\left(\dfrac{1}{\sqrt{10}}\right)$

$\quad = \dfrac{x' - 3y'}{\sqrt{10}} \qquad\qquad\qquad\quad = \dfrac{3x' + y'}{\sqrt{10}}$

$$2x^2 - 3xy - 2y^2 + 10 = 0$$

$$2\left(\dfrac{x' - 3y'}{\sqrt{10}}\right)^2 - 3\left(\dfrac{x' - 3y'}{\sqrt{10}}\right)\left(\dfrac{3x' + y'}{\sqrt{10}}\right) - 2\left(\dfrac{3x' + y'}{\sqrt{10}}\right)^2 + 10 = 0$$

$$\dfrac{(x')^2}{5} - \dfrac{6x'y'}{5} + \dfrac{9(y')^2}{5} - \dfrac{9(x')^2}{10} + \dfrac{24x'y'}{10} + \dfrac{9(y')^2}{10} - \dfrac{9(x')^2}{5} - \dfrac{6x'y'}{5} - \dfrac{(y')^2}{5} + 10 = 0$$

$$-\dfrac{5}{2}(x')^2 + \dfrac{5}{2}(y')^2 = -10$$

$$\dfrac{(x')^2}{4} - \dfrac{(y')^2}{4} = 1$$

13. $5x^2 - 6xy + 5y^2 - 12 = 0$

$A = 5, B = -6, C = 5$

$\cot 2\theta = \dfrac{A - C}{B} = 0 \implies 2\theta = \dfrac{\pi}{2} \implies \theta = \dfrac{\pi}{4}$

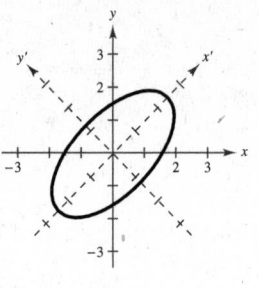

$x = x' \cos \dfrac{\pi}{4} - y' \sin \dfrac{\pi}{4} \qquad\qquad y = x' \sin \dfrac{\pi}{4} + y' \cos \dfrac{\pi}{4}$

$\quad = x'\left(\dfrac{\sqrt{2}}{2}\right) - y'\left(\dfrac{\sqrt{2}}{2}\right) \qquad\qquad = x'\left(\dfrac{\sqrt{2}}{2}\right) + y'\left(\dfrac{\sqrt{2}}{2}\right)$

$\quad = \dfrac{x' - y'}{\sqrt{2}} \qquad\qquad\qquad\qquad = \dfrac{x' + y'}{\sqrt{2}}$

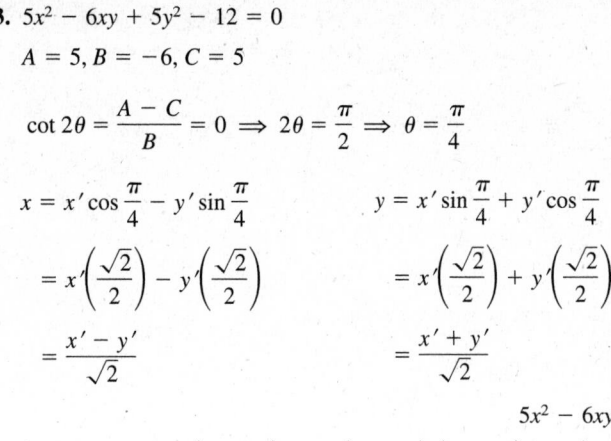

$$5x^2 - 6xy + 5y^2 - 12 = 0$$

$$5\left(\dfrac{x' - y'}{\sqrt{2}}\right)^2 - 6\left(\dfrac{x' - y'}{\sqrt{2}}\right)\left(\dfrac{x' + y'}{\sqrt{2}}\right) + 5\left(\dfrac{x' + y'}{\sqrt{2}}\right)^2 - 12 = 0$$

$$\dfrac{5(x')^2}{2} - 5x'y' + \dfrac{5(y')^2}{2} - 3(x')^2 + 3(y')^2 + \dfrac{5(x')^2}{2} + 5x'y' + \dfrac{5(y')^2}{2} - 12 = 0$$

$$2(x')^2 + 8(y')^2 = 12$$

$$\dfrac{(x')^2}{6} + \dfrac{(y')^2}{3/2} = 1$$

14. $13x^2 + 6\sqrt{3}xy + 7y^2 - 16 = 0$

$A = 13, B = 6\sqrt{3}, C = 7$

$\cot 2\theta = \dfrac{A - C}{B} = \dfrac{1}{\sqrt{3}} \implies 2\theta = \dfrac{\pi}{3} \implies \theta = \dfrac{\pi}{6}$

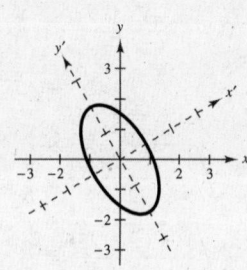

$x = x' \cos \dfrac{\pi}{6} - y' \sin \dfrac{\pi}{6}$ $\qquad$ $y = x' \sin \dfrac{\pi}{6} + y' \cos \dfrac{\pi}{6}$

$\quad = x'\left(\dfrac{\sqrt{3}}{2}\right) - y'\left(\dfrac{1}{2}\right)$ $\qquad\quad = x'\left(\dfrac{1}{2}\right) + y'\left(\dfrac{\sqrt{3}}{2}\right)$

$\quad = \dfrac{\sqrt{3}x' - y'}{2}$ $\qquad\qquad\quad = \dfrac{x' + \sqrt{3}y'}{2}$

$$13x^2 + 6\sqrt{3}xy + 7y^2 - 16 = 0$$

$$13\left(\frac{\sqrt{3}x' - y'}{2}\right)^2 + 6\sqrt{3}\left(\frac{\sqrt{3}x' - y'}{2}\right)\left(\frac{x' + \sqrt{3}y'}{2}\right) + 7\left(\frac{x' + \sqrt{3}y'}{2}\right)^2 - 16 = 0$$

$$\frac{39(x')^2}{4} - \frac{13\sqrt{3}x'y'}{2} + \frac{13(y')^2}{4} + \frac{18(x')^2}{4} + \frac{18\sqrt{3}x'y'}{4} - \frac{6\sqrt{3}x'y'}{4}$$

$$-\frac{18(y')^2}{4} + \frac{7(x')^2}{4} + \frac{7\sqrt{3}x'y'}{2} + \frac{21(y')^2}{4} - 16 = 0$$

$$16(x')^2 + 4(y')^2 = 16$$

$$\frac{(x')^2}{1} + \frac{(y')^2}{4} = 1$$

15. $3x^2 - 2\sqrt{3}xy + y^2 + 2x + 2\sqrt{3}y = 0$

$A = 3, B = -2\sqrt{3}, C = 1$

$\cot 2\theta = \dfrac{A - C}{B} = -\dfrac{1}{\sqrt{3}} \implies \theta = 60°$

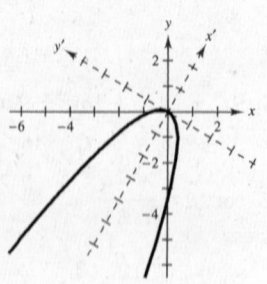

$x = x' \cos 60° - y' \sin 60°$ $\qquad\qquad y = x' \sin 60° + y' \cos 60°$

$\quad = x'\left(\dfrac{1}{2}\right) - y'\left(\dfrac{\sqrt{3}}{2}\right) = \dfrac{x' - \sqrt{3}y'}{2}$ $\qquad = x'\left(\dfrac{\sqrt{3}}{2}\right) + y'\left(\dfrac{1}{2}\right) = \dfrac{\sqrt{3}x' + y'}{2}$

$$3x^2 - 2\sqrt{3}xy + y^2 + 2x + 2\sqrt{3}y = 0$$

$$3\left(\frac{x' - \sqrt{3}y'}{2}\right)^2 - 2\sqrt{3}\left(\frac{x' - \sqrt{3}y'}{2}\right)\left(\frac{\sqrt{3}x' + y'}{2}\right) + \left(\frac{\sqrt{3}x' + y'}{2}\right)^2 + 2\left(\frac{x' - \sqrt{3}y'}{2}\right) + 2\sqrt{3}\left(\frac{\sqrt{3}x' + y'}{2}\right) = 0$$

$$\frac{3(x')^2}{4} - \frac{6\sqrt{3}x'y'}{4} + \frac{9(y')^2}{4} - \frac{6(x')^2}{4} + \frac{4\sqrt{3}x'y'}{4} + \frac{6(y')^2}{4} + \frac{3(x')^2}{4} + \frac{2\sqrt{3}x'y'}{4} + \frac{(y')^2}{4}$$

$$+ x' - \sqrt{3}y' + 3x' + \sqrt{3}y' = 0$$

$$4(y')^2 + 4x' = 0$$

$$(y')^2 = -x'$$

16. $16x^2 - 24xy + 9y^2 - 60x - 80y + 100 = 0$

$A = 16, B = -24, C = 9$

$\cot 2\theta = \dfrac{A - C}{B} = -\dfrac{7}{24} \implies \theta \approx 53.13°$

$\cos 2\theta = -\dfrac{7}{25}$

$\sin \theta = \sqrt{\dfrac{1 - \cos 2\theta}{2}} = \sqrt{\dfrac{1 - (-7/25)}{2}} = \dfrac{4}{5}$

$\cos \theta = \sqrt{\dfrac{1 + \cos 2\theta}{2}} = \sqrt{\dfrac{1 + (-7/25)}{2}} = \dfrac{3}{5}$

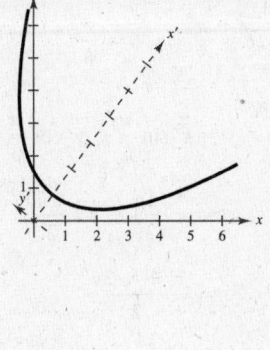

$x = x'\cos\theta - y'\sin\theta \qquad\qquad y = x'\sin\theta + y'\cos\theta$

$\quad = x'\left(\dfrac{3}{5}\right) - y'\left(\dfrac{4}{5}\right) \qquad\qquad = x'\left(\dfrac{4}{3}\right) + y'\left(\dfrac{3}{5}\right) = \dfrac{4x' + 3y'}{5}$

$16x^2 - 24xy + 9y^2 - 60x - 80y + 100 = 0$

$16\left(\dfrac{3x' - 4y'}{5}\right)^2 - 24\left(\dfrac{3x' - 4y'}{5}\right)\left(\dfrac{4x' + 3y'}{5}\right) + 9\left(\dfrac{4x' + 3y'}{5}\right)^2 - 60\left(\dfrac{3x' - 4y'}{5}\right) - 80\left(\dfrac{4x' + 3y'}{5}\right) + 100 = 0$

$\dfrac{144(x')^2}{25} - \dfrac{384x'y'}{25} + \dfrac{256(y')^2}{25} - \dfrac{288(x')^2}{25} + \dfrac{168x'y'}{25} + \dfrac{288(y')^2}{25} + \dfrac{144(x')^2}{25} + \dfrac{216x'y'}{25}$

$\qquad\qquad + \dfrac{81(y')^2}{25} - 36x' + 48y' - 64x' - 48y' + 100 = 0$

$\qquad\qquad\qquad 25(y')^2 - 100x' + 100 = 0$

$\qquad\qquad\qquad\qquad (y')^2 = 4(x' - 1)$

17. $9x^2 + 24xy + 16y^2 + 90x - 130y = 0$

$A = 9, B = 24, C = 16$

$\cot 2\theta = \dfrac{A - C}{B} = -\dfrac{7}{24} \implies \theta \approx 53.13°$

$\cos 2\theta = -\dfrac{7}{25}$

$\sin \theta = \sqrt{\dfrac{1 - \cos 2\theta}{2}} = \sqrt{\dfrac{1 - \left(-\dfrac{7}{25}\right)}{2}} = \dfrac{4}{5}$

$\cos \theta = \sqrt{\dfrac{1 + \cos 2\theta}{2}} = \sqrt{\dfrac{1 + \left(-\dfrac{7}{25}\right)}{2}} = \dfrac{3}{5}$

$x = x'\cos\theta - y'\sin\theta \qquad\qquad y = x'\sin\theta + y'\cos\theta$

$\quad = x'\left(\dfrac{3}{5}\right) - y'\left(\dfrac{4}{5}\right) = \dfrac{3x' - 4y'}{5} \qquad\qquad = x'\left(\dfrac{4}{5}\right) + y'\left(\dfrac{3}{5}\right)$

$\qquad\qquad\qquad\qquad\qquad\qquad = \dfrac{4x' + 3y'}{5}$

—CONTINUED—

17. —CONTINUED—

$$9x^2 + 24xy + 16y^2 + 90x - 130y = 0$$

$$9\left(\frac{3x' - 4y'}{5}\right)^2 + 24\left(\frac{3x' - 4y'}{5}\right)\left(\frac{4x' + 3y'}{5}\right) + 16\left(\frac{4x' + 3y'}{5}\right)^2 + 90\left(\frac{3x' - 4y'}{5}\right) - 130\left(\frac{4x' + 3y'}{5}\right) = 0$$

$$\frac{81(x')^2}{25} - \frac{216x'y'}{25} + \frac{144(y')^2}{25} + \frac{288(x')^2}{25} - \frac{168x'y'}{25} - \frac{288(y')^2}{25} + \frac{256(x')^2}{25} + \frac{384x'y'}{25} + \frac{144(y')^2}{25}$$

$$+ 54x' - 72y' - 104x' - 78y' = 0$$

$$25(x')^2 - 50x' - 150y' = 0$$

$$(x')^2 - 2x' = 6y'$$

$$(x')^2 - 2x' + 1 = 6y' + 1$$

$$(x' - 1)^2 = 6\left(y' + \frac{1}{6}\right)$$

18. $9x^2 + 24xy + 16y^2 + 80x - 60y = 0$

$A = 9, B = 24, C = 16$

$$\cot 2\theta = \frac{A - C}{B} = -\frac{7}{24} \implies \theta \approx 53.13°$$

$$\cos 2\theta = -\frac{7}{25}$$

$$\sin \theta = \sqrt{\frac{1 - \cos 2\theta}{2}} = \sqrt{\frac{1 - (-7/25)}{2}} = \frac{4}{5}$$

$$\cos \theta = \sqrt{\frac{1 + \cos 2\theta}{2}} = \sqrt{\frac{1 + (-7/25)}{2}} = \frac{3}{5}$$

$x = x' \cos \theta - y' \sin \theta \qquad y = x' \sin \theta + y' \cos \theta$

$$= x'\left(\frac{3}{5}\right) - y'\left(\frac{4}{5}\right) \qquad = x'\left(\frac{4}{5}\right) + y'\left(\frac{3}{5}\right)$$

$$= \frac{3x' - 4y'}{5} \qquad\qquad = \frac{4x' + 3y}{5}$$

$$9x^2 + 24xy + 16y^2 + 80x - 60y = 0$$

$$9\left(\frac{3x' - 4y'}{5}\right)^2 + 24\left(\frac{3x' - 4y'}{5}\right)\left(\frac{4x' + 3y'}{5}\right) + 16\left(\frac{4x' + 3y'}{5}\right)^2 + 80\left(\frac{3x' - 4y'}{5}\right) - 60\left(\frac{4x' + 3y'}{5}\right) = 0$$

$$\frac{81(x')^2}{25} - \frac{216x'y'}{25} + \frac{144(y')^2}{25} + \frac{288(x')^2}{25} - \frac{168x'y'}{25} - \frac{288(y')^2}{25} + \frac{256(x')^2}{25} + \frac{384x'y'}{25} + \frac{144(y')^2}{25}$$

$$+ 48x' - 64x' - 48x' - 36x' = 0$$

$$25(x')^2 - 100y' = 0$$

$$(x')^2 = 4y'$$

$$\frac{1}{4}(x')^2 = y'$$

19. $x^2 + 2xy + y^2 = 20$

$A = 1, B = 2, C = 1$

$\cot 2\theta = \dfrac{A - C}{B} = \dfrac{1 - 1}{2} = 0 \implies \theta = \dfrac{\pi}{4}$ or $45°$

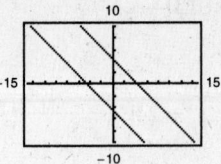

To graph the conic using a graphing calculator, we need to solve for y in terms of x.

$(x + y)^2 = 20$

$x + y = \pm\sqrt{20}$

$y = -x \pm \sqrt{20}$

Use $y_1 = -x + \sqrt{20}$ and $y_2 = -x - \sqrt{20}$.

20. $x^2 - 4xy + 2y^2 = 6$

$A = 1, B = -4, C = 2$

$\cot 2\theta = \dfrac{A - C}{B} = \dfrac{1 - 2}{-4} = \dfrac{1}{4}$

$\dfrac{1}{\tan 2\theta} = \dfrac{1}{4}$

$\tan 2\theta = 4$

$2\theta \approx 75.96$

$\theta \approx 37.98°$

To graph conic with a graphing calculator, we need to solve for y in terms of x.

$x^2 - 4xy + 2y^2 = 6$

$y^2 - 2xy + x^2 = 3 - \dfrac{x^2}{2} + x^2$

$(y - x)^2 = 3 + \dfrac{x^2}{2}$

$y - x = \pm\sqrt{3 + \dfrac{x^2}{2}}$

$y = x \pm \sqrt{3 + \dfrac{x^2}{2}}$

Enter $y_1 = x + \sqrt{3 + \dfrac{x^2}{2}}$ and $y_2 = x - \sqrt{3 + \dfrac{x^2}{2}}$.

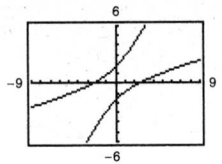

21. $17x^2 + 32xy - 7y^2 = 75$

$\cot 2\theta = \dfrac{A - C}{B} = \dfrac{17 + 7}{32} = \dfrac{24}{32} = \dfrac{3}{4} \implies \theta \approx 26.57°$

Solve for y in terms of x by completing the square.

$-7y^2 + 32xy = -17x^2 + 75$

$y^2 - \dfrac{32}{7}xy = \dfrac{17}{7}x^2 - \dfrac{75}{7}$

$y^2 - \dfrac{32}{7}xy + \dfrac{256}{49}x^2 = \dfrac{119}{49}x^2 - \dfrac{525}{49} + \dfrac{256}{49}x^2$

$\left(y - \dfrac{16}{7}x\right)^2 = \dfrac{375x^2 - 525}{49}$

$y = \dfrac{16}{7}x \pm \sqrt{\dfrac{375x^2 - 525}{49}}$

$y = \dfrac{16x \pm 5\sqrt{15x^2 - 21}}{7}$

Use $y_1 = \dfrac{16x + 5\sqrt{15x^2 - 21}}{7}$

and $y_2 = \dfrac{16x - 5\sqrt{15x^2 - 21}}{7}$.

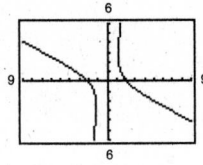

22. $40x^2 + 36xy + 25y^2 = 52$

$A = 40, B = 36, C = 25$

$\cot 2\theta = \dfrac{A - C}{B} = \dfrac{40 - 25}{36} = \dfrac{5}{12}$

$\dfrac{1}{\tan 2\theta} = \dfrac{5}{12}$

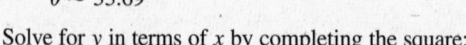

$\tan 2\theta = \dfrac{12}{5}$

$2\theta \approx 67.38°$

$\theta \approx 33.69°$

Solve for y in terms of x by completing the square:

$25y^2 + 36xy = 52 - 40x^2$

$y^2 + \dfrac{36}{25}xy = \dfrac{52}{25} - \dfrac{40}{25}x^2$

$y^2 + \dfrac{36}{25}xy + \dfrac{324}{625}x^2 = \dfrac{52}{25} - \dfrac{40}{25}x^2 + \dfrac{324}{625}x^2$

$\left(y + \dfrac{18}{25}x\right)^2 = \dfrac{1300 - 676x^2}{625}$

$y + \dfrac{18}{25}x = \pm\sqrt{\dfrac{1300 - 676x^2}{625}}$

$y = \dfrac{-18x \pm \sqrt{1300 - 676x^2}}{25}$

Enter $y_1 = \dfrac{-18x + \sqrt{1300 - 676x^2}}{25}$ and

$y_2 = \dfrac{-18x - \sqrt{1300 - 676x^2}}{25}$.

23. $32x^2 + 48xy + 8y^2 = 50$

$\cot 2\theta = \dfrac{A - C}{B} = \dfrac{24}{48} = \dfrac{1}{2} \implies \theta \approx 31.72°$

Solve for y in terms of x by completing the square.

$8y^2 + 48xy = -32x^2 + 50$

$y^2 + 6xy = -4x^2 + \dfrac{25}{4}$

$y^2 + 6xy + 9x^2 = -4x^2 + \dfrac{25}{4} + 9x^2$

$(y + 3x)^2 = 5x^2 + \dfrac{25}{4}$

$y + 3x = \pm\sqrt{5x^2 + \dfrac{25}{4}}$

$y = -3x \pm \sqrt{5x^2 + \dfrac{25}{4}}$

Use $y_1 = -3x + \sqrt{5x^2 + \dfrac{25}{4}}$ and

$y_2 = -3x - \sqrt{5x^2 + \dfrac{25}{4}}$.

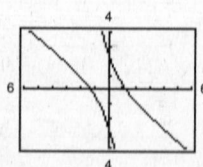

24. $24x^2 + 18xy + 12y^2 = 34$

$A = 24, B = 18, C = 12$

$\cot 2\theta = \dfrac{A - C}{B} = \dfrac{24 - 12}{18} = \dfrac{2}{3}$

$\tan 2\theta = \dfrac{3}{2}$

$2\theta \approx 56.31°$

$\theta \approx 28.15°$

Solve for y in terms of x by completing the square:

$12x^2 + 9xy + 6y^2 = 17$

$6\left(y^2 + \dfrac{3}{2}xy + \dfrac{9}{16}x^2\right) = 17 - 12x^2 + \dfrac{27}{8}x^2 = 17 - \dfrac{69}{8}x^2$

$\left(y + \dfrac{3}{4}x\right)^2 = \dfrac{136 - 69x^2}{48}$

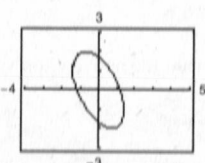

$y = -\dfrac{3}{4}x \pm \sqrt{\dfrac{136 - 69x^2}{48}} = \dfrac{-9x \pm \sqrt{3(136 - 69x^2)}}{12}$

Enter $y_1 = \dfrac{-9x + \sqrt{3(136 - 69x^2)}}{12}$ and $y_2 = \dfrac{-9x - \sqrt{3(136 - 69x^2)}}{12}$.

25. $4x^2 - 12xy + 9y^2 + (4\sqrt{13} - 12)x - (6\sqrt{13} + 8)y = 91$

$A = 4, B = -12, C = 9$

$\cot 2\theta = \dfrac{A - C}{B} = \dfrac{4 - 9}{-12} = \dfrac{5}{12}$

$\dfrac{1}{\tan 2\theta} = \dfrac{5}{12}$

$\tan 2\theta = \dfrac{12}{5}$

$2\theta \approx 67.38°$

$\theta \approx 33.69°$

Solve for y in terms of x with the quadratic formula:

$4x^2 - 12xy + 9y^2 + (4\sqrt{13} - 12)x - (6\sqrt{13} + 8)y = 91$

$9y^2 - (12x + 6\sqrt{13} + 8)y + (4x^2 + 4\sqrt{13}x - 12x - 91) = 0$

$a = 9, b = -(12x + 6\sqrt{13} + 8), c = 4x^2 + 4\sqrt{13}x - 12x - 91$

$y = \dfrac{-b \pm \sqrt{b^2 - 4ac}}{2a}$

$y = \dfrac{(12x + 6\sqrt{13} + 8) \pm \sqrt{(12x + 6\sqrt{13} + 8)^2 - 4(9)(4x^2 + 4\sqrt{13}x - 12x - 91)}}{18}$

$= \dfrac{(12x + 6\sqrt{13} + 8) \pm \sqrt{624x + 3808 + 96\sqrt{13}}}{18}$

Enter $y_1 = \dfrac{12x + 6\sqrt{13} + 8 + \sqrt{624x + 3808 + 96\sqrt{13}}}{18}$

and $y_2 = \dfrac{12x + 6\sqrt{13} + 8 - \sqrt{624x + 3808 + 96\sqrt{13}}}{18}$.

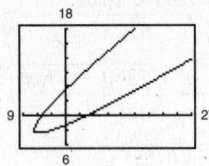

26. $6x^2 - 4xy + 8y^2 + (5\sqrt{5} - 10)x - (7\sqrt{5} + 5)y = 80$

$A = 6, B = -4, C = 8$

$\cot 2\theta = \dfrac{A - C}{B} = \dfrac{6 - 8}{-4} = \dfrac{1}{2}$

$\tan 2\theta = 2$

$2\theta \approx 63.43°$

$\theta \approx 31.72°$

Solve for y in terms of x using the quadratic formula.

$8y^2 - (4x + 7\sqrt{5} + 5)y + 6x^2 + (5\sqrt{5} - 10)x - 80 = 0$

$y = \dfrac{1}{16}\left[4x + 7\sqrt{5} + 5 \pm \sqrt{(4x + 7\sqrt{5} + 5)^2 - 32(6x^2 + (5\sqrt{5} - 10)x - 80)}\right]$

Enter y_1 and y_2 from the above expression.

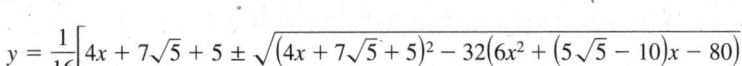

27. $xy + 2 = 0$

$B^2 - 4AC = 1 \implies$ The graph is a hyperbola.

$\cot 2\theta = \dfrac{A - C}{B} = 0 \implies \theta = 45°$

Matches graph (e).

28. $x^2 + 2xy + y^2 = 0$

$(x + y)^2 = 0$

$x + y = 0$

$y = -x$

The graph is a line. Matches graph (f).

29. $-2x^2 + 3xy + 2y^2 + 3 = 0$

$B^2 - 4AC = (3)^2 - 4(-2)(2) = 25 \implies$

The graph is a hyperbola.

$\cot 2\theta = \dfrac{A - C}{B} = -\dfrac{4}{3} \implies \theta \approx -18.43°$

Matches graph (b).

30. $x^2 - xy + 3y^2 - 5 = 0$

$A = 1, B = -1, C = 3$

$B^2 - 4AC = (-1)^2 - 4(1)(3) = -11$

The graph is an ellipse.

$\cot 2\theta = \dfrac{A - C}{B} = \dfrac{1 - 3}{-1} = 2 \implies \theta \approx 13.28°$

Matches graph (a).

31. $3x^2 + 2xy + y^2 - 10 = 0$

$B^2 - 4AC = (2)^2 - 4(3)(1) = -8 \implies$

The graph is an ellipse or circle.

$\cot 2\theta = \dfrac{A - C}{B} = 1 \implies \theta = 22.5°$

Matches graph (d).

32. $x^2 - 4xy + 4y^2 + 10x - 30 = 0$

$A = 1, B = -4, C = 4$

$B^2 - 4AC = (-4)^2 - 4(1)(4) = 0$

The graph is a parabola.

$\cot 2\theta = \dfrac{A - C}{B} = \dfrac{1 - 4}{-4} = \dfrac{3}{4} \implies \theta \approx 26.57°$

Matches graph (c).

33. (a) $16x^2 - 8xy + y^2 - 10x + 5y = 0$

$B^2 - 4AC = (-8)^2 - 4(16)(1) = 0$

The graph is a parabola.

(b) $y^2 + (-8x + 5)y + (16x^2 - 10x) = 0$

$y = \dfrac{-(-8x + 5) \pm \sqrt{(-8x + 5)^2 - 4(1)(16x^2 - 10x)}}{2(1)}$

$\quad = \dfrac{(8x - 5) \pm \sqrt{(8x - 5)^2 - 4(16x^2 - 10x)}}{2}$

(c)

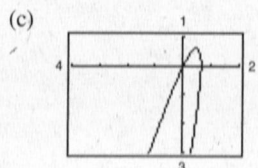

34. (a) $x^2 - 4xy - 2y^2 - 6 = 0$

$A = 1, B = -4, C = -2$

$B^2 - 4AC = (-4)^2 - 4(1)(-2) = 24 > 0$

The graph is a hyperbola.

(b) $-2y^2 - 4xy + x^2 - 6 = 0$

$y = -\dfrac{1}{4}\left[4x \pm \sqrt{16x^2 + 8(x^2 - 6)}\right]$

(c)

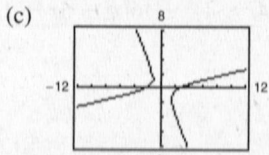

35. (a) $12x^2 - 6xy + 7y^2 - 45 = 0$

$B^2 - 4AC = (-6)^2 - 4(12)(7) = -300 < 0$

The graph is an ellipse.

(b) $7y^2 + (-6x)y + (12x^2 - 45) = 0$

$y = \dfrac{-(-6x) \pm \sqrt{(-6x)^2 - 4(7)(12x^2 - 45)}}{2(7)}$

$\quad = \dfrac{6x \pm \sqrt{36x^2 - 28(12x^2 - 45)}}{14}$

(c)

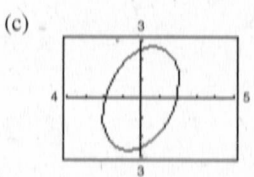

36. (a) $2x^2 + 4xy + 5y^2 + 3x - 4y - 20 = 0$

$A = 2, B = 4, C = 5$

$B^2 - 4AC = 4^2 - 4(2)(5) = 16 - 40 = -24 < 0$

The graph is an ellipse.

(b) $5y^2 + (4x - 4)y + 2x^2 + 3x - 20 = 0$

$y = \dfrac{1}{10}\left[-(4x - 4) \pm \sqrt{(4x - 4)^2 - 20(2x^2 + 3x - 20)}\right]$

(c)

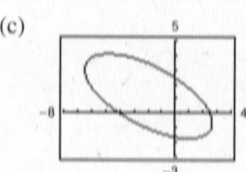

37. (a) $x^2 - 6xy - 5y^2 + 4x - 22 = 0$

$B^2 - 4AC = (-6)^2 - 4(1)(-5) = 56 > 0$

The graph is a hyperbola.

(b) $-5y^2 + (-6x)y + (x^2 + 4x - 22) = 0$

$$y = \frac{-(-6x) \pm \sqrt{(-6x)^2 - 4(-5)(x^2 + 4x - 22)}}{2(-5)}$$

$$= \frac{6x \pm \sqrt{36x^2 + 20(x^2 + 4x - 22)}}{-10}$$

$$= \frac{-6x \pm \sqrt{36x^2 + 20(x^2 + 4x - 22)}}{10}$$

(c)

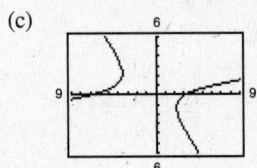

38. (a) $36x^2 - 60xy + 25y^2 + 9y = 0$

$A = 36, B = -60, C = 25$

$B^2 - 4AC = (-60)^2 - 4(36)(25) = 0$

The graph is a parabola.

(b) $25y^2 - (60x - 9)y + 36x^2 = 0$

$$y = \frac{1}{50}\left[60x - 9 \pm \sqrt{(60x - 9)^2 - 3600x^2}\right]$$

(c)

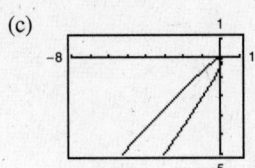

39. (a) $x^2 + 4xy + 4y^2 - 5x - y - 3 = 0$

$B^2 - 4AC = (4)^2 - 4(1)(4) = 0$

The graph is a parabola.

(b) $4y^2 + (4x - 1)y + (x^2 - 5x - 3) = 0$

$$y = \frac{-(4x - 1) \pm \sqrt{(4x - 1)^2 - 4(4)(x^2 - 5x - 3)}}{2(4)}$$

$$= \frac{-(4x - 1) \pm \sqrt{(4x - 1)^2 - 16(x^2 - 5x - 3)}}{8}$$

(c)

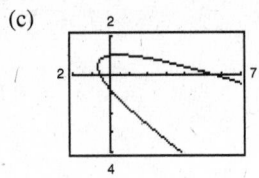

40. (a) $x^2 + xy + 4y^2 + x + y - 4 = 0$

$A = 1, B = 1, C = 4$

$B^2 - 4AC = 1^2 - 4(1)(4) = -15$

The graph is an ellipse.

(b) $4y^2 + (x + 1)y + x^2 + x - 4 = 0$

$$y = \frac{1}{8}\left[-(x + 1) \pm \sqrt{(x + 1)^2 - 16(x^2 + x - 4)}\right]$$

(c)

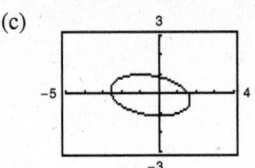

41. $y^2 - 9x^2 = 0$

$y^2 = 9x^2$

$y = \pm 3x$

Two intersecting lines

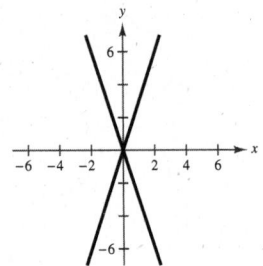

42. $x^2 + y^2 - 2x + 6y + 10 = 0$

$(x^2 - 2x + 1) + (y^2 + 6y + 9) = -10 + 1 + 9$

$(x - 1)^2 + (y + 3)^2 = 0$

Point at $(1, -3)$

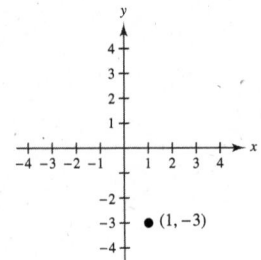

43. $x^2 + 2xy + y^2 - 1 = 0$

$\qquad (x + y)^2 - 1 = 0$

$\qquad (x + y)^2 = 1$

$\qquad x + y = \pm 1$

$\qquad y = -x \pm 1$

Two parallel lines

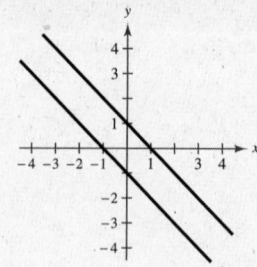

44. $x^2 - 10xy + y^2 = 0$

$\qquad y^2 - 10xy + 25x^2 = 25x^2 - x^2$

$\qquad (y - 5x)^2 = 24x^2$

$\qquad y - 5x = \pm\sqrt{24x^2}$

$\qquad y = 5x \pm 2\sqrt{6}x$

$\qquad y = \left(5 \pm 2\sqrt{6}\right)x$

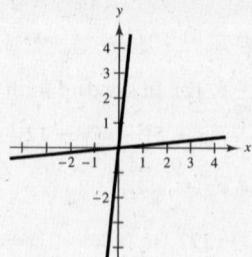

45. $\qquad -x^2 + y^2 + 4x - 6y + 4 = 0 \implies (y - 3)^2 - (x - 2)^2 = 1$

$\qquad \underline{\quad x^2 + y^2 - 4x - 6y + 12 = 0 \implies (x - 2)^2 + (y - 3)^2 = 1}$

$\qquad\qquad 2y^2 - 12y + 16 = 0$

$\qquad\qquad 2(y - 2)(y - 4) = 0$

$\qquad\qquad y = 2 \text{ or } y = 4$

For $y = 2$: $x^2 + 2^2 - 4x - 6(2) + 12 = 0$

$\qquad\qquad x^2 - 4x + 4 = 0$

$\qquad\qquad (x - 2)^2 = 0$

$\qquad\qquad x = 2$

For $y = 4$: $x^2 + 4^2 - 4x - 6(4) + 12 = 0$

$\qquad\qquad x^2 - 4x + 4 = 0$

$\qquad\qquad (x - 2)^2 = 0$

$\qquad\qquad x = 2$

The points of intersection are $(2, 2)$ and $(2, 4)$.

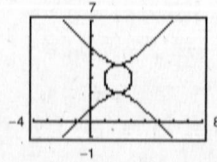

46. $-x^2 - y^2 - 8x + 20y - 7 = 0 \implies (x + 4)^2 + (y - 10)^2 = 109$

$\underline{\quad x^2 + 9y^2 + 8x + 4y + 7 = 0 \implies (x + 4)^2 + 9\left(y + \frac{2}{9}\right)^2 = \frac{85}{9}}$

$\qquad 8y^2 \qquad + 24y \qquad = 0$

$\qquad 8y(y + 3) = 0$

$\qquad\qquad y = 0 \text{ or } y = -3$

When $y = 0$: $x^2 + 9(0)^2 + 8x + 4(0) + 7 = 0$

$\qquad\qquad (x + 7)(x + 1) = 0$

$\qquad\qquad x = -7, -1$

When $y = -3$: $x^2 + 9(-3)^2 + 8x + 4(-3) + 7 = 0$

$\qquad\qquad x^2 + 8x + 76 = 0$

$\qquad\qquad$ No real solution

Points of intersection: $(-7, 0), (-1, 0)$

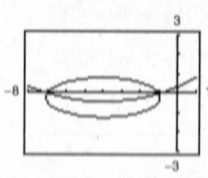

47.
$$-4x^2 - y^2 - 16x + 24y - 16 = 0$$
$$\underline{4x^2 + y^2 + 40x - 24y + 208 = 0}$$
$$24x \qquad + 192 = 0$$
$$x = -8$$

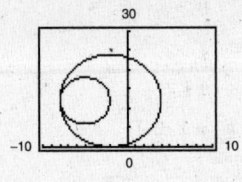

When $x = -8$: $\quad 4(-8)^2 + y^2 + 40(-8) - 24y + 208 = 0$
$$y^2 - 24y + 144 = 0$$
$$(y - 12)^2 = 0$$
$$y = 12$$

The point of intersection is $(-8, 12)$. In standard form the equations are:

$$\frac{(x + 2)^2}{36} + \frac{(y - 12)^2}{144} = 1 \text{ and } \frac{(x + 5)^2}{9} + \frac{(y - 12)^2}{36} = 1$$

48.
$$x^2 - 4y^2 - 20x - 64y - 172 = 0 \implies (x - 10)^2 - 4(y + 8)^2 = 16$$
$$\underline{16x^2 + 4y^2 - 320x + 64y + 1600 = 0 \implies 16(x - 10)^2 + 4(y + 8)^2 = 256}$$
$$17x^2 \qquad - 340x \qquad 1428 = 0$$
$$(17x - 238)(x - 6) = 0$$
$$x = 6 \text{ or } x = 14$$

When $x = 6$: $\quad 6^2 - 4y^2 - 20(6) - 64y - 172 = 0$
$$-4y^2 - 64y - 256 = 0$$
$$y^2 + 16y + 64 = 0$$
$$(y + 8)^2 = 0$$
$$y = -8$$

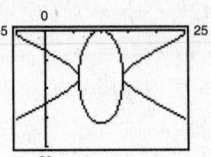

When $x = 14$: $\quad 14^2 - 4y^2 - 20(14) - 64y - 172 = 0$
$$-4y^2 - 64y - 256 = 0$$
$$y^2 + 16y + 64 = 0$$
$$(y + 8)^2 = 0$$
$$y = -8$$

Points of intersection: $(6, -8), (14, -8)$

49.
$$x^2 - y^2 - 12x + 16y - 64 = 0$$
$$\underline{x^2 + y^2 - 12x - 16y + 64 = 0}$$
$$2x^2 \qquad - 24x \qquad = 0$$
$$2x(x - 12) = 0$$
$$x = 0 \text{ or } x = 12$$

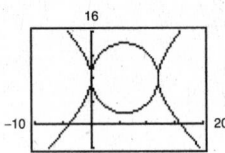

When $x = 0$: $\quad 0^2 + y^2 - 12(0) - 16y + 64 = 0$
$$y^2 - 16y + 64 = 0$$
$$(y - 8)^2 = 0$$
$$y = 8$$

When $x = 12$: $12^2 + y^2 - 12(12) - 16y + 64 = 0$
$$y^2 - 16y + 64 = 0$$
$$(y - 8)^2 = 0$$
$$y = 8$$

The points of intersection are $(0, 8)$ and $(12, 8)$. The standard forms of the equations are:

$$\frac{(x - 6)^2}{36} - \frac{(y - 8)^2}{36} = 1 \text{ and } (x - 6)^2 + (y - 8)^2 = 36$$

50. $x^2 + 4y^2 - 2x - 8y + 1 = 0 \implies (x - 1)^2 + 4(y - 1)^2 = 4$

$\underline{-x^2 \qquad\quad + 2x - 4y - 1 = 0} \implies y = -\frac{1}{4}(x - 1)^2$

$\qquad 4y^2 \qquad\ - 12y \qquad = 0$

$\qquad\qquad\quad 4y(y - 3) = 0$

$\qquad\qquad\qquad\quad y = 0 \text{ or } y = 3$

When $y = 0$: $x^2 + 4(0)^2 - 2x - 8(0) + 1 = 0$

$\qquad\qquad\qquad\quad x^2 - 2x + 1 = 0$

$\qquad\qquad\qquad\quad (x - 1)^2 = 0$

$\qquad\qquad\qquad\qquad\quad x = 1$

When $y = 3$: $-x^2 + 2x - 4(3) - 1 = 0$

$\qquad\qquad\qquad\quad x^2 - 2x + 13 = 0$

$\qquad\qquad$ No real solution

Point of intersection: $(1, 0)$

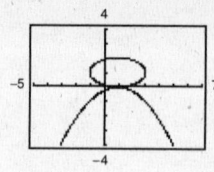

51. $-16x^2 -\quad y^2 + 24y -\ \ 80 = 0$

$\underline{\ \ 16x^2 + 25y^2 \qquad\quad - 400 = 0}$

$\qquad 24y^2 + 24y - 480 = 0$

$\qquad 24(y + 5)(y - 4) = 0$

$\qquad\qquad\quad y = -5 \text{ or } y = 4$

When $y = -5$: $16x^2 + 25(-5)^2 - 400 = 0$

$\qquad\qquad\qquad\qquad 16x^2 = -225$

$\qquad\qquad$ No real solution

When $y = 4$: $16x^2 + 25(4)^2 - 400 = 0$

$\qquad\qquad\qquad\qquad 16x^2 = 0$

$\qquad\qquad\qquad\qquad\ \ x = 0$

The point of intersection is $(0, 4)$.

In standard form the equations are:

$$\frac{x^2}{4} + \frac{(y - 12)^2}{64} = 1$$

$$\frac{x^2}{25} + \frac{y^2}{16} = 1$$

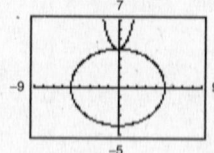

52. $16x^2 - y^2 \qquad\quad + 16y - 128 = 0 \implies 16x^2 - (y - 8)^2 = 64$

$\underline{\qquad\quad y^2 - 48x - 16y -\ \ 32 = 0} \implies (y - 8)^2 - 48x = 96$

$16x^2 \qquad - 48x \qquad\quad - 160 = 0$

$\qquad 16(x^2 - 3x - 10) = 0$

$\qquad\quad (x - 5)(x + 2) = 0$

$\qquad\qquad\qquad\quad x = 5 \text{ or } x = -2$

When $x = 5$: $y^2 - 48(5) - 16y - 32 = 0$

$\qquad\qquad\qquad\ y^2 - 16y - 272 = 0$

$\qquad\qquad\qquad\qquad\quad y = 8 \pm 4\sqrt{21}$

When $x = -2$: $y^2 - 48(-2) - 16y - 32 = 0$

$\qquad\qquad\qquad\ y^2 - 16y + 64 = 0$

$\qquad\qquad\qquad\qquad (y - 8)^2 = 0$

$\qquad\qquad\qquad\qquad\qquad y = 8$

Points of intersection: $\left(5, 8 + 4\sqrt{21}\right), \left(5, 8 - 4\sqrt{21}\right), (-2, 8)$

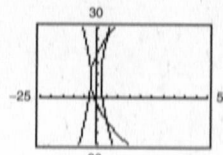

53. $x^2 \qquad + y^2 - 4 = 0$

$\underline{\qquad 3x - y^2 \qquad = 0}$

$x^2 + 3x \qquad - 4 = 0$

$(x + 4)(x - 1) = 0$

$x = -4 \quad \text{or} \quad x = 1$

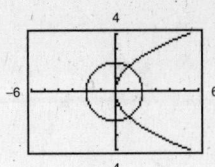

When $x = -4$: $3(-4) - y^2 = 0$

$y^2 = -12$

No real solution

When $x = 1$: $3(1) - y^2 = 0$

$y^2 = 3$

$y = \pm\sqrt{3}$

The points of intersection are $\left(1, \sqrt{3}\right)$ and $\left(1, -\sqrt{3}\right)$.

The standard forms of the equations are:

$x^2 + y^2 = 4$

$y^2 = 3x$

54. $4x^2 + 9y^2 - 36y = 0 \implies 4x^2 + 9(y - 2)^2 = 36$

$x^2 + 9y - 27 = 0 \implies y = -\dfrac{x^2}{9} + 3$

$4(27 - 9y) + 9y^2 - 36y = 0$

$9y^2 - 72y + 108 = 0$

$9(y - 6)(y - 2) = 0$

$y = 6 \text{ or } y = 2$

When $y = 6$: $x^2 = 27 - 9(6) = -27$

No real solution

When $y = 2$: $x^2 = 27 - 9(2) = 9$

$x = \pm 3$

Points of intersection: $(3, 2), (-3, 2)$

55. $x^2 + 2y^2 - 4x + 6y - 5 = 0$

$-x + y - 4 = 0 \implies y = x + 4$

$x^2 + 2(x + 4)^2 - 4x + 6(x + 4) - 5 = 0$

$x^2 + 2(x^2 + 8x + 16) - 4x + 6x + 24 - 5 = 0$

$3x^2 + 18x + 51 = 0$

$3(x^2 + 6x + 17) = 0$

$x^2 + 6x + 17 = 0$

$x^2 + 6x + 9 = -17 + 9$

$(x + 3)^2 = -8$

No real solution

No points of intersection

The standard forms of the equations are:

$\dfrac{(x - 2)^2}{\frac{27}{2}} + \dfrac{\left(y + \frac{3}{2}\right)^2}{\frac{27}{4}} = 1$

$x - y = -4$

56. $x^2 + 2y^2 - 4x + 6y - 5 = 0 \implies 2(x - 2)^2 + 4\left(y + \frac{3}{2}\right)^2 = 27$

$x^2 - 4x - y + 4 = 0 \implies y = x^2 - 4x + 4$

$y - 4 + 2y^2 + 6y - 5 = 0$

$2y^2 + 7y - 9 = 0$

$(2y + 9)(y - 1) = 0$

$y = -\dfrac{9}{2} \text{ or } y = 1$

When $y = 1$: $x^2 - 4x - 1 + 4 = 0$

$(x - 3)(x - 1) = 0$

$x = 1 \text{ or } x = 3$

When $y = -\dfrac{9}{2}$: $x^2 - 4x - \left(-\dfrac{9}{2}\right) + 4 = 0$

$x^2 - 4x + \dfrac{17}{2} = 0$

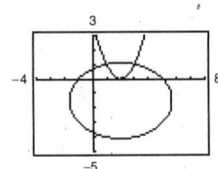

No real solution

Points of intersection: $(1, 1), (3, 1)$

57.

$$xy + x - 2y + 3 = 0 \implies y = \frac{-x - 3}{x - 2}$$

$$x^2 + 4y^2 - 9 = 0$$

$$x^2 + 4\left(\frac{-x - 3}{x - 2}\right)^2 = 9$$

$$x^2(x - 2)^2 + 4(-x - 3)^2 = 9(x - 2)^2$$

$$x^2(x^2 - 4x + 4) + 4(x^2 + 6x + 9) = 9(x^2 - 4x + 4)$$

$$x^4 - 4x^3 + 4x^2 + 4x^2 + 24x + 36 = 9x^2 - 36x + 36$$

$$x^4 - 4x^3 - x^2 + 60x = 0$$

$$x(x + 3)(x^2 - 7x + 20) = 0$$

$$x = 0 \text{ or } x = -3$$

Note: $x^2 - 7x + 20 = 0$ has no real solution.

When $x = 0$: $y = \dfrac{-0 - 3}{0 - 2} = \dfrac{3}{2}$

When $x = -3$: $y = \dfrac{-(-3) - 3}{-3 - 2} = 0$

The points of intersection are $\left(0, \dfrac{3}{2}\right), (-3, 0)$.

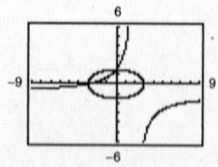

58. $5x^2 - 2xy + 5y^2 - 12 = 0$

$$x + y - 1 = 0 \implies y = 1 - x$$

$$5x^2 - 2x(1 - x) + 5(1 - x)^2 - 12 = 0$$

$$5x^2 - 2x + 2x^2 + 5(1 - 2x + x^2) - 12 = 0$$

$$5x^2 - 2x + 2x^2 + 5 - 10x + 5x^2 - 12 = 0$$

$$12x^2 - 12x - 7 = 0$$

$$x = \frac{3 \pm \sqrt{30}}{6}$$

When $x = \dfrac{3 + \sqrt{30}}{6}$: $y = 1 - \dfrac{3 + \sqrt{30}}{6} = \dfrac{3 - \sqrt{30}}{6}$

When $x = \dfrac{3 - \sqrt{30}}{6}$: $y = 1 - \dfrac{3 - \sqrt{30}}{6} = \dfrac{3 + \sqrt{30}}{6}$

Points of intersection:

$$\left(\frac{1}{6}(3 + \sqrt{30}), \frac{1}{6}(3 - \sqrt{30})\right), \left(\frac{1}{6}(3 - \sqrt{30}), \frac{1}{6}(3 + \sqrt{30})\right)$$

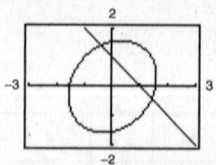

59. $x^2 + xy + ky^2 + 6x + 10 = 0$

$B^2 - 4AC = 1^2 - 4(1)(k) = 1 - 4k > 0 \implies -4k > -1 \implies k < \frac{1}{4}$

True. For the graph to be a hyperbola, the discriminant must be greater than zero.

60. False. The coefficients of the new equation, after rotation of axes, are obtained by making the substitutions.

$x = x' \cos \theta - y' \sin \theta$

$y = x' \sin \theta + y' \cos \theta$

61. $r^2 = x^2 + y^2 = (x' \cos \theta - y' \sin \theta)^2 + (y' \cos \theta + x' \sin \theta)^2$

$\qquad = (x')^2 \cos^2 \theta - 2x'y' \cos \theta \sin \theta + (y')^2 \sin^2 \theta + (y')^2 \cos^2 \theta + 2x'y' \cos \theta \sin \theta + (x')^2 \sin^2 \theta$

$\qquad = (x')^2(\cos^2 \theta + \sin^2 \theta) + (y')^2(\sin^2 \theta + \cos^2 \theta) = (x')^2 + (y')^2$

Thus, $(x')^2 + (y')^2 = r^2$.

62. In Exercise 14, the equation of the rotated ellipse is:

$$\frac{(x')^2}{1} + \frac{(y')^2}{4} = 1$$

$a^2 = 4 \implies a = 2$

$b^2 = 1 \implies b = 1$

Length of major axis is $2a = 2(2) = 4$.

Length of minor axis is $2b = 2(1) = 2$.

63. $f(x) = |x + 3|$

Shift the graph of $y = |x|$ three units to the left.

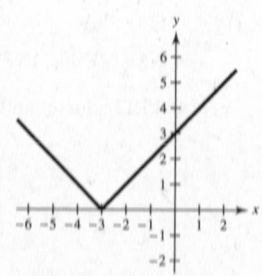

64. $f(x) = |x - 4| + 1$

The graph of the function $f(x)$ is the graph of $|x|$ shifted four units to the right and one unit upward.

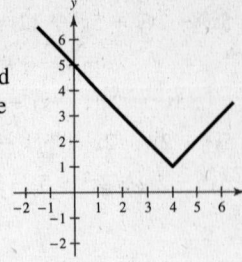

65. $g(x) = \sqrt{4 - x^2}$

$y^2 = 4 - x^2$

$x^2 + y^2 = 4$

$g(x)$ is the top half of this circle since $y \geq 0$.

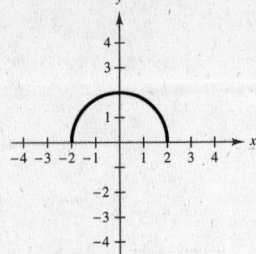

66. $g(x) = \sqrt{3x - 2}$

x-intercept: $0 = \sqrt{3x - 2}$

$0^2 = 3x - 2$

$2 = 3x$

$\frac{2}{3} = x, \ \left(\frac{2}{3}, 0\right)$

Domain: $\left[\frac{2}{3}, \infty\right)$

x	$\frac{2}{3}$	1	2	$\frac{11}{3}$
y	0	1	2	3

67. $h(t) = -(t - 2)^3 + 3$

Reflect the graph of $y = x^3$ about the x-axis, shift it to the right two units, and upward three units.

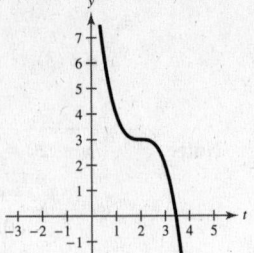

68. $h(t) = \frac{1}{2}(t + 4)^3$

y-intercept:

$h(0) = \frac{1}{2}(0 + 4)^3$

$= 32, \ (0, 32)$

x-intercept: $0 = \frac{1}{2}(t + 4)^3$

$0 = t + 4$

$t = -4, \ (-4, 0)$

t	-6	-4	-2	-1	0	1
$h(t)$	-4	0	4	$\frac{27}{2}$	32	$\frac{125}{2}$

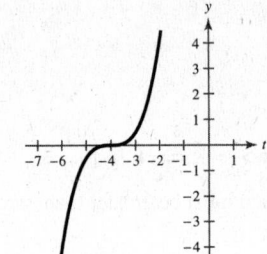

69. $f(t) = [\![t - 5]\!] + 1$

Shift the graph of $y = [\![x]\!]$ five units to the right and upward one unit.

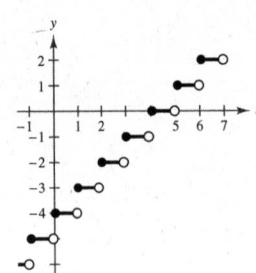

70. $f(t) = -2[\![t]\!] + 3$

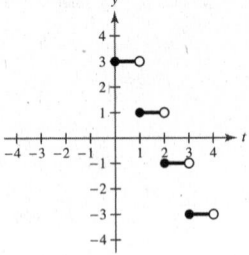

71. Area $= \frac{1}{2}ab \sin C$

$= \frac{1}{2}(8)(12) \sin 110°$

≈ 45.11 square units

72. $\sin 70° = \dfrac{h}{25}$

$h = 25 \sin 70°$

Area $= \dfrac{1}{2}(\text{base})(\text{height})$

$= \dfrac{1}{2}(16)(25 \sin 70°)$

≈ 187.9

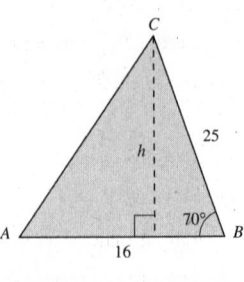

73. $s = \dfrac{a + b + c}{2} = \dfrac{11 + 18 + 10}{2} = 19.5$

Area $= \sqrt{s(s - a)(s - b)(s - c)} = \sqrt{(19.5)(8.5)(1.5)(9.5)} \approx 48.60$ square units

74. Law of Cosines:

$35^2 = 23^2 + 27^2 - (2)(23)(27) \cos \theta$

$\cos(\theta) \approx \dfrac{11}{414}$

$\theta = \cos^{-1}\left(\dfrac{11}{414}\right)$

$\dfrac{h}{23} = \sin \theta = \sin\left(\cos^{-1}\left(\dfrac{11}{414}\right)\right)$

$h = 23 \sin\left(\cos^{-1}\left(\dfrac{11}{414}\right)\right)$

Area $= \dfrac{1}{2}(\text{base})(\text{height}) = \left(\dfrac{1}{2}\right)(27)\left(23 \sin\left(\cos^{-1}\left(\dfrac{11}{414}\right)\right)\right) \approx 310.4$

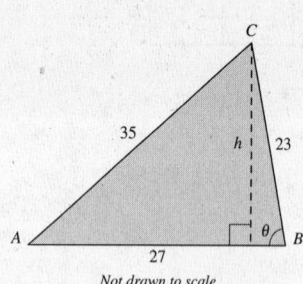

Not drawn to scale

Section 10.6   Parametric Equations

■ If f and g are continuous functions of t on an interval I, then the set of ordered pairs $(f(t), g(t))$ is a *plane curve C*. The equations $x = f(t)$ and $y = g(t)$ are *parametric equations* for C and t is the *parameter.*

■ To eliminate the parameter:
(a) Solve for t in one equation and substitute into the second equation.
(b) Use trigonometric identities.

■ You should be able to find the parametric equations for a graph.

Vocabulary Check

1. plane curve; parametric; parameter

2. orientation

3. eliminating the parameter

1. $x = \sqrt{t}, y = 3 - t$

(a)

t	0	1	2	3	4
x	0	1	$\sqrt{2}$	$\sqrt{3}$	2
y	3	2	1	0	-1

(c) $x = \sqrt{t} \quad \Rightarrow \quad x^2 = t$

$y = 3 - t \quad \Rightarrow \quad y = 3 - x^2$

The graph of the parametric equations only shows the right half of the parabola, whereas the rectangular equation yields the entire parabola.

(b)

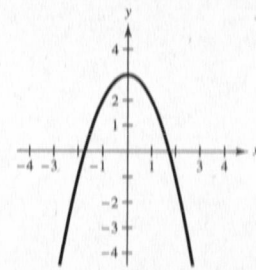

2. $x = 4\cos^2\theta$, $y = 2\sin\theta$

(a)

θ	$-\pi/2$	$-\pi/4$	0	$\pi/4$	$\pi/2$
x	0	2	4	2	0
y	-2	$-\sqrt{2}$	0	$\sqrt{2}$	2

(b)

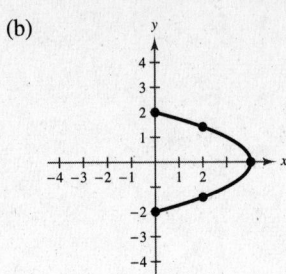

(c) $\dfrac{x}{4} = \cos^2\theta$, $\dfrac{y}{2} = \sin\theta$

$\cos^2\theta + \sin^2\theta = 1$

$\dfrac{x}{4} + \left(\dfrac{y}{2}\right)^2 = 1$

$\dfrac{x}{4} + \dfrac{y^2}{4} = 1$

$x = -y^2 + 4$

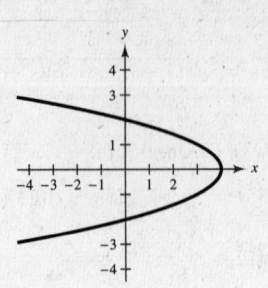

The rectangular version of the graph continues into the second and third quadrants.

3. (a) $x = 3t - 3$, $y = 2t + 1$

t	-2	-1	0	1	2
x	-9	-6	-3	0	3
y	-3	-1	1	3	5

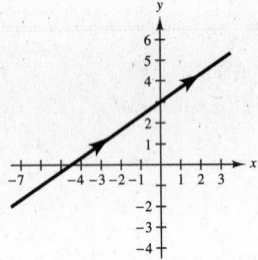

(b) $x = 3t - 3 \implies t = \dfrac{x+3}{3}$

$y = 2t + 1 \implies y = \dfrac{2}{3}(x+3) + 1 = \dfrac{2}{3}x + 3$

4. (a) $x = 3 - 2t$, $y = 2 + 3t$

t	-3	-2	-1	0	1	2	3
x	9	7	5	3	1	-1	-3
y	-7	-4	-1	2	5	8	11

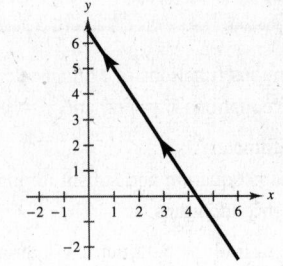

(b) $x = 3 - 2t \implies t = -\dfrac{1}{2}x + \dfrac{3}{2}$

$y = 2 + 3t$

$y = 2 + 3\left(-\dfrac{1}{2}x + \dfrac{3}{2}\right)$

$y = 2 - \dfrac{3}{2}x + \dfrac{9}{2}$

$2y = 4 - 3x + 9$

$3x + 2y - 13 = 0$

5. (a) $x = \dfrac{1}{4}t$, $y = t^2$

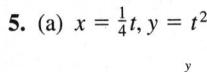

t	-2	-1	0	1	2
x	$-\dfrac{1}{2}$	$-\dfrac{1}{4}$	0	$\dfrac{1}{4}$	$\dfrac{1}{2}$
y	4	1	0	1	4

(b) $x = \dfrac{1}{4}t \implies t = 4x$

$y = t^2 \implies y = 16x^2$

6. (a) $x = t$, $y = t^3$

t	-3	-2	-1	0	1	2	3
x	-3	-2	-1	0	1	2	3
y	-27	-8	-1	0	1	8	27

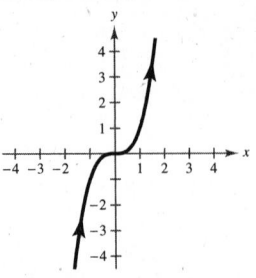

(b) $x = t$, $y = t^3$, $y = x^3$

7. (a) $x = t + 2, y = t^2$

t	-2	-1	0	1	2
x	0	1	2	3	4
y	4	1	0	1	4

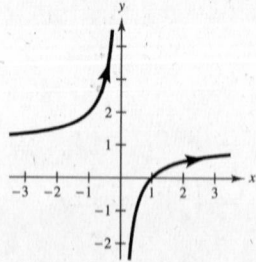

(b) $x = t + 2 \implies t = x - 2$

$y = t^2 \implies y = (x - 2)^2 = x^2 - 4x + 4$

8. (a) $x = \sqrt{t}, y = 1 - t$

t	0	1	2	3
x	0	1	$\sqrt{2}$	$\sqrt{3}$
y	1	0	-1	-2

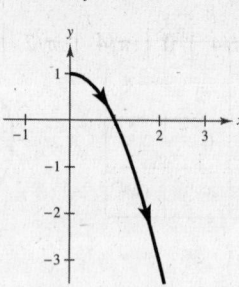

(b) $x = \sqrt{t} \implies x^2 = t, \ t \geq 0$

$y = 1 - t = 1 - x^2, \ x \geq 0$

9. (a) $x = t + 1, y = \dfrac{t}{t + 1}$

t	-3	-2	0	1	2
x	-2	-1	1	2	3
y	$\frac{3}{2}$	2	0	$\frac{1}{2}$	$\frac{2}{3}$

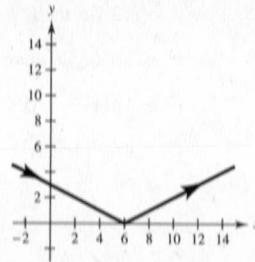

(b) $x = t + 1 \implies t = x - 1$

$y = \dfrac{t}{t + 1} \implies y = \dfrac{x - 1}{x}$

10. (a) $x = t - 1, y = \dfrac{t}{t - 1}$

t	-3	-2	-1	0	2	3
x	-4	-3	-2	-1	1	2
y	$\frac{3}{4}$	$\frac{2}{3}$	$\frac{1}{2}$	0	2	$\frac{3}{2}$

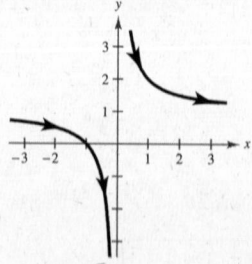

(b) $x = t - 1 \implies t = x + 1$

$y = \dfrac{t}{t - 1} = \dfrac{x + 1}{x + 1 - 1} = \dfrac{x + 1}{x}$

11. (a) $x = 2(t + 1), y = |t - 2|$

t	0	2	4	6	8	10
x	2	6	10	14	18	22
y	2	0	2	4	6	8

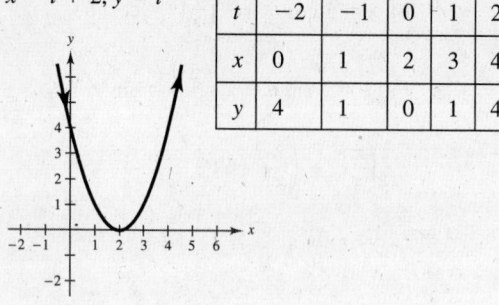

(b) $x = 2(t + 1) \implies \dfrac{x}{2} - 1 = t \ $ or $ \ t = \dfrac{x - 2}{2}$

$y = |t - 2| \implies y = \left| \dfrac{x}{2} - 1 - 2 \right| = \left| \dfrac{x}{2} - 3 \right|$

12. (a) $x = |t - 1|, y = t + 2$

t	-3	-2	-1	0	1	2	3
x	4	3	2	1	0	1	2
y	-1	0	1	2	3	4	5

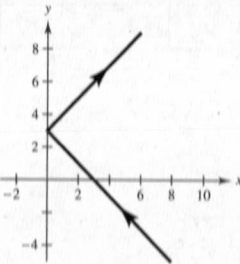

(b) $x = |t - 1|$

$y = t + 2 \implies t = y - 2 \implies x = |y - 3|$

OR $y = x + 3, x \geq 0$ and $y = -x + 3, x \geq 0$

13. (a) $x = 3 \cos \theta$, $y = 3 \sin \theta$

θ	0	$\dfrac{\pi}{2}$	π	$\dfrac{3\pi}{2}$	2π
x	3	0	-3	0	3
y	0	3	0	-3	0

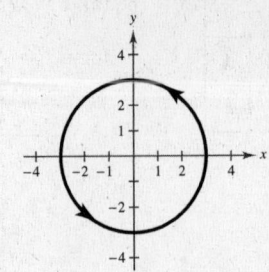

(b) $x = 3 \cos \theta \implies \left(\dfrac{x}{3}\right)^2 = \cos^2 \theta$

$y = 3 \sin \theta \implies \left(\dfrac{y}{3}\right)^2 = \sin^2 \theta$

$\left(\dfrac{x}{3}\right)^2 + \left(\dfrac{y}{3}\right)^2 = 1$

$\dfrac{x^2}{9} + \dfrac{y^2}{9} = 1$

14. (a) $x = 2 \cos \theta$, $y = 3 \sin \theta$

θ	0	$\pi/4$	$\pi/2$	$3\pi/4$	π	$5\pi/4$	$3\pi/2$	$7\pi/4$	2π
x	2	$\sqrt{2}$	0	$-\sqrt{2}$	-2	$-\sqrt{2}$	0	$\sqrt{2}$	2
y	0	$3\sqrt{2}/2$	3	$3\sqrt{2}/2$	0	$-3\sqrt{2}/2$	-3	$-3\sqrt{2}/2$	0

(b) $x = 2 \cos \theta$

$y = 3 \sin \theta$

$\cos^2 \theta + \sin^2 \theta = 1$

$\left(\dfrac{x}{2}\right)^2 + \left(\dfrac{y}{3}\right)^2 = 1$

$\dfrac{x^2}{4} + \dfrac{y^2}{9} = 1$

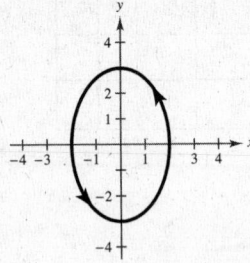

15. (a) $x = 4 \sin 2\theta$, $y = 2 \cos 2\theta$

θ	0	$\dfrac{\pi}{4}$	$\dfrac{\pi}{2}$	$\dfrac{3\pi}{4}$	π
x	0	4	0	-4	0
y	2	0	-2	0	2

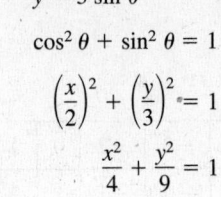

(b) $x = 4 \sin 2\theta \implies \left(\dfrac{x}{4}\right)^2 = \sin^2 2\theta$

$y = 2 \cos 2\theta \implies \left(\dfrac{y}{2}\right)^2 = \cos^2 2\theta$

$\left(\dfrac{x}{4}\right)^2 + \left(\dfrac{y}{2}\right)^2 = 1$

$\dfrac{x^2}{16} + \dfrac{y^2}{4} = 1$

16. (a) $x = \cos \theta$, $y = 2 \sin 2\theta$

θ	0	$\pi/4$	$\pi/2$	$3\pi/4$	π	$5\pi/4$	$3\pi/2$	$7\pi/4$	2π
x	1	$\sqrt{2}/2$	0	$-\sqrt{2}/2$	-1	$-\sqrt{2}/2$	0	$\sqrt{2}/2$	1
y	0	2	0	-2	0	2	0	-2	0

(b) $x = \cos \theta$

$y = 2 \sin 2\theta$

$y = 2 \sin 2\theta$

$y = 2(2 \sin \theta \cos \theta)$

$y^2 = 16 \sin^2 \theta \cos^2 \theta$

$y^2 = 16(1 - x^2)x^2$

$y^2 = 16x^2(1 - x^2)$

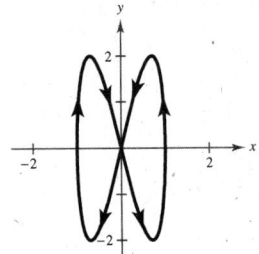

17. (a) $x = 4 + 2\cos\theta$, $y = -1 + \sin\theta$

θ	0	$\dfrac{\pi}{2}$	π	$\dfrac{3\pi}{2}$	2π
x	6	4	2	4	6
y	-1	0	-1	-2	-1

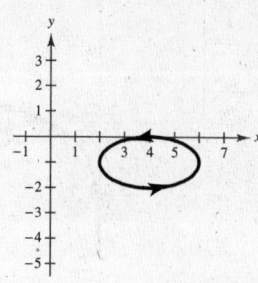

(b) $x = 4 + 2\cos\theta \Rightarrow \left(\dfrac{x-4}{2}\right)^2 = \cos^2\theta$

$y = -1 + \sin\theta \Rightarrow (y+1)^2 = \sin^2\theta$

$\dfrac{(x-4)^2}{4} + \dfrac{(y+1)^2}{1} = 1$

18. (a) $x = 4 + 2\cos\theta$, $y = 2 + 3\sin\theta$

θ	0	$\pi/4$	$\pi/2$	$3\pi/4$	π	$5\pi/4$	$3\pi/2$	$7\pi/4$	2π
x	6	$4+\sqrt{2}$	4	$4-\sqrt{2}$	2	$4-\sqrt{2}$	4	$4+\sqrt{2}$	6
y	2	$2 + \left(3\sqrt{2}/2\right)$	5	$2 + \left(3\sqrt{2}/2\right)$	2	$2 - \left(3\sqrt{2}/2\right)$	-1	$2 - \left(3\sqrt{2}/2\right)$	2

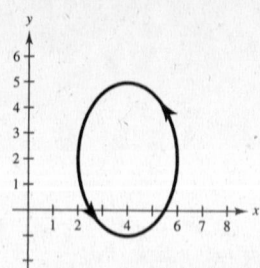

(b) $x = 4 + 2\cos\theta$

$y = 2 + 3\sin\theta$

$\cos^2\theta + \sin^2\theta = 1$

$\dfrac{(x-4)^2}{4} + \dfrac{(y-2)^2}{9} = 1$

19. (a) $x = e^{-t}$, $y = e^{3t}$

t	-2	-1	0	1	2
x	7.3891	2.7183	1	0.3679	0.1353
y	0.0025	0.0498	1	20.0855	403.4288

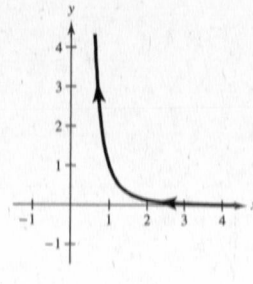

(b) $x = e^{-t} \Rightarrow \dfrac{1}{x} = e^t$

$y = e^{3t} \Rightarrow y = (e^t)^3$

$y = \left(\dfrac{1}{x}\right)^3$

$y = \dfrac{1}{x^3}$, $x > 0$, $y > 0$

20. (a) $x = e^{2t}$, $y = e^t$

t	-3	-2	-1	0	1	2
x	0.0025	0.0183	0.1353	1	7.3891	54.5982
y	0.0498	0.1353	0.3679	1	2.7183	7.3891

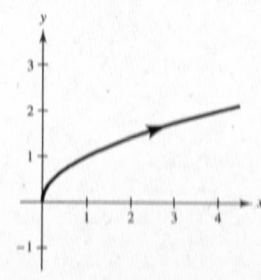

(b) $x = e^{2t}$

$y = e^t \Rightarrow y^2 = e^{2t}$

$x = e^{2t} = y^2$

$y^2 = x$, $y > 0$

21. (a) $x = t^3, y = 3 \ln t$

t	$\frac{1}{2}$	1	2	3	4
x	$\frac{1}{8}$	1	8	27	64
y	-2.0794	0	2.0794	3.2958	4.1589

(b) $x = t^3 \implies x^{1/3} = t$

$y = 3 \ln t \implies y = \ln t^3$

$y = \ln(x^{1/3})^3$

$y = \ln x$

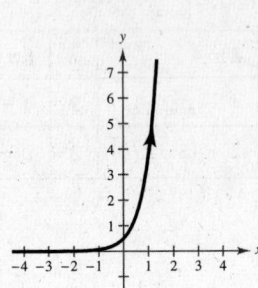

22. (a) $x = \ln 2t, y = 2t^2$

t	1	2	3	4
x	0.6931	1.3863	1.7918	2.0794
y	2	8	18	32

(b) $x = \ln 2t \implies t = \frac{1}{2}e^x$

$y = 2t^2$

$y = 2t^2 = 2\left(\frac{1}{2}e^x\right)^2 = \frac{1}{2}e^{2x}$

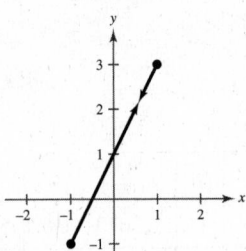

23. By eliminating the parameter, each curve becomes
$y = 2x + 1$.

(a) $x = t$

$y = 2t + 1$

There are no restrictions on x and y.

Domain: $(-\infty, \infty)$

Orientation: Left to right

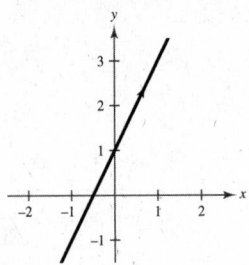

(b) $x = \cos\theta \implies -1 \le x \le 1$

$y = 2\cos\theta + 1 \implies -1 \le y \le 3$

The graph oscillates.

Domain: $[-1, 1]$

Orientation: Depends on θ

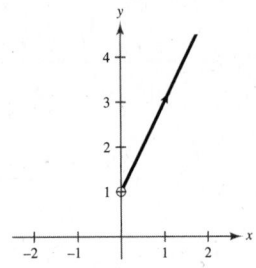

(c) $x = e^{-t} \implies x > 0$

$y = 2e^{-t} + 1 \implies y > 1$

Domain: $(0, \infty)$

Orientation: Downward or right to left

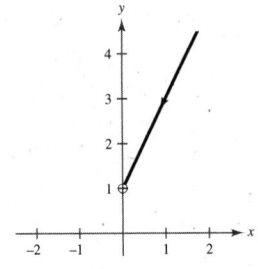

(d) $x = e^t \implies x > 0$

$y = 2e^t + 1 \implies y > 1$

Domain: $(0, \infty)$

Orientation: Upward or left to right

24. By eliminating the parameter, each curve represents a portion of $y = x^2 - 1$.

(a) $x = t$

$y = t^2 - 1$

There are no restrictions on x.

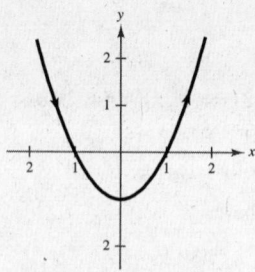

Domain: $(-\infty, \infty)$
Orientation: Left to right

(b) $x = t^2 \implies x \geq 0$

$y = t^4 - 1$

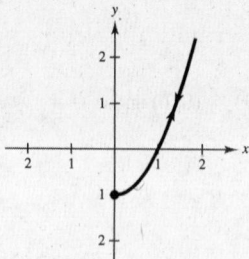

Domain: $[0, \infty)$
Orientation: Depends on t

(c) $x = \sin t \implies -1 \leq x \leq 1$

$y = \sin^2 t - 1$

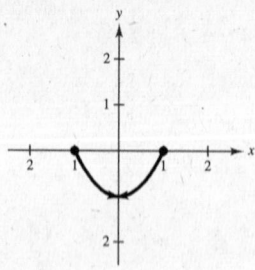

Domain: $[-1, 1]$
Orientation: Depends on t

(d) $x = e^t \implies x > 0$

$y = e^{2t} - 1$

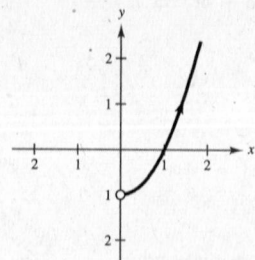

Domain: $(0, \infty)$
Orientation: Left to right

25. $x = x_1 + t(x_2 - x_1),\ y = y_1 + t(y_2 - y_1)$

$$\frac{x - x_1}{x_2 - x_1} = t$$

$$y = y_1 + \left(\frac{x - x_1}{x_2 - x_1}\right)(y_2 - y_1)$$

$$y - y_1 = \frac{y_2 - y_1}{x_2 - x_1}(x - x_1) = m(x - x_1)$$

26. $x = h + r\cos\theta,\ y = k + r\sin\theta$

$$\cos\theta = \frac{x - h}{r},\ \sin\theta = \frac{y - k}{r}$$

$$\cos^2\theta + \sin^2\theta = \frac{(x - h)^2}{r^2} + \frac{(y - k)^2}{r^2} = 1$$

$$(x - h)^2 + (y - k)^2 = r^2$$

27. $x = h + a\cos\theta,\ y = k + b\sin\theta$

$$\frac{x - h}{a} = \cos\theta,\ \frac{y - k}{b} = \sin\theta$$

$$\frac{(x - h)^2}{a^2} + \frac{(y - k)^2}{b^2} = 1$$

28. $x = h + a\sec\theta,\ y = k + b\tan\theta$

$$\frac{x - h}{a} = \sec\theta,\ \frac{y - k}{b} = \tan\theta$$

$$\frac{(x - h)^2}{a^2} - \frac{(y - k)^2}{b^2} = 1$$

29. From Exercise 25 we have:

$x = 0 + t(6 - 0) = 6t$

$y = 0 + t(-3 - 0) = -3t$

30. Line through $(2, 3)$ and $(6, -3)$

From Exercise 25 we have:

$x = x_1 + t(x_2 - x_1) = 2 + t(6 - 2) = 2 + 4t$

$y = y_1 + t(y_2 - y_1) = 3 + t(-3 - 3) = 3 - 6t$

31. From Exercise 26 we have:

$x = 3 + 4 \cos \theta$

$y = 2 + 4 \sin \theta$

32. Circle with center $(-3, 2)$; radius: 5

From Exercise 26 we have:

$x = h + r \cos \theta = -3 + 5 \cos \theta$

$y = k + r \sin \theta = 2 + 5 \sin \theta$

33. Vertices: $(\pm 4, 0) \implies (h, k) = (0, 0)$ and $a = 4$

Foci: $(\pm 3, 0) \implies c = 3$

$c^2 = a^2 - b^2 \implies 9 = 16 - b^2 \implies b = \sqrt{7}$

From Exercise 27 we have:

$x = 4 \cos \theta$

$y = \sqrt{7} \sin \theta$

34. Ellipse

Vertices: $(4, 7), (4, -3) \implies (h, k) = (4, 2), a = 5$

Foci: $(4, 5), (4, -1) \implies c = 3$

$b^2 = a^2 - c^2 = 25 - 9 = 16 \implies b = 4$

From Exercise 27 we have:

$x = h + b \cos \theta = 4 + 4 \cos \theta$

$y = k + a \sin \theta = 2 + 5 \sin \theta$

35. Vertices: $(\pm 4, 0) \implies (h, k) = (0, 0)$ and $a = 4$

Foci: $(\pm 5, 0) \implies c = 5$

$c^2 = a^2 + b^2 \implies 25 = 16 + b^2 \implies b = 3$

From Exercise 28 we have:

$x = 4 \sec \theta$

$y = 3 \tan \theta$

36. Hyperbola

Vertices: $(\pm 2, 0) \implies (h, k) = (0, 0), a = 2$

Foci: $(\pm 4, 0) \implies c = 4$

$b^2 = c^2 - a^2 = 16 - 4 \implies b = 2\sqrt{3}$

From Exercise 28 we have:

$x = h + a \sec \theta = 2 \sec \theta$

$y = k + b \tan \theta = 2\sqrt{3} \tan \theta$

37. $y = 3x - 2$

(a) $t = x \implies x = t$ and $y = 3t - 2$

(b) $t = 2 - x \implies x = -t + 2$ and

$\quad y = 3(-t + 2) - 2 = -3t + 4$

38. $x = 3y - 2$

(a) $t = x, x = t, y = \frac{1}{3}(t + 2)$

(b) $t = 2 - x, x = 2 - t, y = \frac{1}{3}(x + 2) = \frac{1}{3}(4 - t)$

39. $y = x^2$

(a) $t = x \implies x = t$ and $y = t^2$

(b) $t = 2 - x \implies x = -t + 2$ and

$\quad y = (-t + 2)^2 = t^2 - 4t + 4$

40. $y = x^3$

(a) $t = x, x = t, y = t^3$

(b) $t = 2 - x, x = 2 - t, y = (2 - t)^3$

41. $y = x^2 + 1$

(a) $t = x \implies x = t$ and $y = t^2 + 1$

(b) $t = 2 - x \implies x = -t + 2$ and

$\quad y = (-t + 2)^2 + 1 = t^2 - 4t + 5$

42. $y = 2 - x$

(a) $t = x, x = t, y = 2 - t$

(b) $t = 2 - x, x = 2 - t, y = 2 - (2 - t) = t$

43. $y = \dfrac{1}{x}$

(a) $t = x \implies x = t$ and $y = \dfrac{1}{t}$

(b) $t = 2 - x \implies x = -t + 2$ and $y = \dfrac{1}{-t + 2} = \dfrac{-1}{t - 2}$

44. $y = \dfrac{1}{2x}$

(a) $t = x, x = t, y = \dfrac{1}{2t}$

(b) $t = 2 - x, x = 2 - t, y = \dfrac{1}{2(2 - t)} = \dfrac{1}{4 - 2t}$

45. $x = 4(\theta - \sin \theta)$
$y = 4(1 - \cos \theta)$

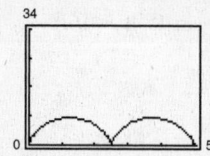

46. $x = \theta + \sin \theta$
$y = 1 - \cos \theta$

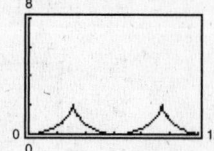

47. $x = \theta - \frac{3}{2} \sin \theta$
$y = 1 - \frac{3}{2} \cos \theta$

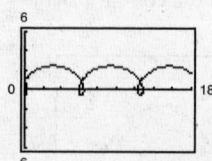

48. $x = 2\theta - 4 \sin \theta$
$y = 2 - 4 \cos \theta$

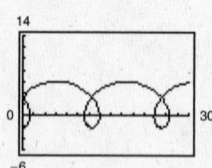

49. $x = 3 \cos^3 \theta$
$y = 3 \sin^3 \theta$

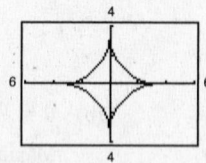

50. $x = 8\theta - 4 \sin \theta$
$y = 8 - 4 \cos \theta$

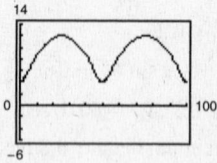

51. $x = 2 \cot \theta$
$y = 2 \sin^2 \theta$

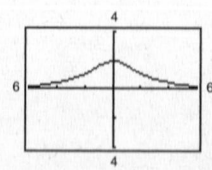

52. $x = \dfrac{3t}{1 + t^3}$

$y = \dfrac{3t^2}{1 + t^3}$

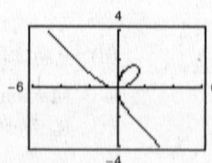

53. $x = 2 \cos \theta \implies -2 \leq x \leq 2$

$y = \sin 2\theta \implies -1 \leq y \leq 1$

Matches graph (b).

Domain: $[-2, 2]$

Range: $[-1, 1]$

54. $x = 4 \cos^3 \theta \implies -4 \leq x \leq 4$

$y = 6 \sin^3 \theta \implies -6 \leq y \leq 6$

Matches graph (c).

Domain: $[-4, 4]$

Range: $[-6, 6]$

55. $x = \frac{1}{2}(\cos \theta + \theta \sin \theta)$

$y = \frac{1}{2}(\sin \theta - \theta \cos \theta)$

Matches graph (d).

Domain: $(-\infty, \infty)$

Range: $(-\infty, \infty)$

56. $x = \frac{1}{2} \cot \theta \implies -\infty < x < \infty$

$y = 4 \sin \theta \cos \theta \implies -2 \leq y \leq 2$

Matches graph (a).

Domain: $(-\infty, \infty)$

Range: $[-2, 2]$

57. $x = (v_0 \cos \theta)t$ and $y = h + (v_0 \sin \theta)t - 16t^2$

(a) $\theta = 60°$, $v_0 = 88$ ft/sec

$x = (88 \cos 60°)t$ and $y = (88 \sin 60°)t - 16t^2$

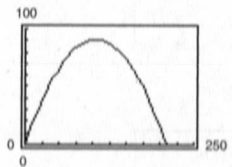

Maximum height: 90.7 feet
Range: 209.6 feet

(b) $\theta = 60°$, $v_0 = 132$ ft/sec

$x = (132 \cos 60°)t$ and $y = (132 \sin 60°)t - 16t^2$

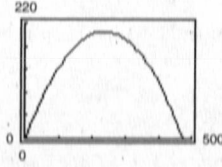

Maximum height: 204.2 feet
Range: 471.6 feet

—CONTINUED—

57. —CONTINUED—

(c) $\theta = 45°$, $v_0 = 88$ ft/sec

$x = (88 \cos 45°)t$ and $y = (88 \sin 45°)t - 16t^2$

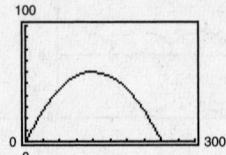

Maximum height: 60.5 ft

Range: 242.0 ft

(d) $\theta = 45°$, $v_0 = 132$ ft/sec

$x = (132 \cos 45°)t$ and $y = (132 \sin 45°)t - 16t^2$

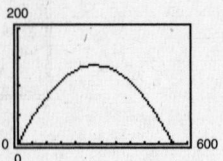

Maximum height: 136.1 ft

Range: 544.5 ft

58. $x = (v_0 \cos \theta)t$

$y = h + (v_0 \sin \theta)t - 16t^2$

(a) $\theta = 15°$, $v_0 = 60$ ft/sec

Maximum height: 3.8 feet

Range: 56.3 feet

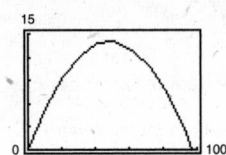

(b) $\theta = 15°$, $v_0 = 100$ ft/sec

Maximum height: 10.5 feet

Range: 156.3 feet

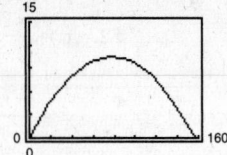

(c) $\theta = 30°$, $v_0 = 60$ ft/sec

Maximum height: 14.1 feet

Range: 97.4 feet

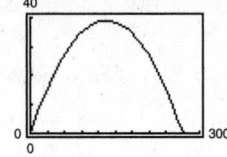

(d) $\theta = 30°$, $v_0 = 100$ ft/sec

Maximum height: 39.1 feet

Range: 270.6 feet

59. (a) 100 miles per hour $= 100\left(\frac{5280}{3600}\right)$ ft/sec $= \frac{440}{3}$ ft/sec

$x = \left(\frac{440}{3} \cos \theta\right)t \approx (146.67 \cos \theta)t$

$y = 3 + \left(\frac{440}{3} \sin \theta\right)t - 16t^2 \approx 3 + (146.67 \sin \theta)t - 16t^2$

(b) For $\theta = 15°$, we have:

$x = \left(\frac{440}{3} \cos 15°\right)t \approx 141.7t$

$y = 3 + \left(\frac{440}{3} \sin 15°\right)t - 16t^2 \approx 3 + 38.0t - 16t^2$

The ball hits the ground inside the ballpark, so it is not a home run.

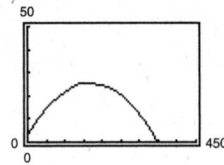

(c) For $\theta = 23°$, we have:

$x = \left(\frac{440}{3} \cos 23°\right)t \approx 135.0t$

$y = 3 + \left(\frac{440}{3} \sin 23°\right)t - 16t^2 \approx 3 + 57.3t - 16t^2$

The ball easily clears the 7-foot fence at 408 feet so it is a home run.

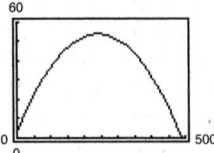

(d) Find θ so that $y = 7$ when $x = 408$ by graphing the parametric equations for θ values between 15° and 23°. This occurs when $\theta \approx 19.3°$.

60. (a) $x = (v_0 \cos \theta)t$

$y = h + (v_0 \sin \theta)t - 16t^2$

$h = 5, v_0 = 240, \theta = 10°$

$x = (240 \cos 10°)t$

$y = 5 + (240 \sin 10°)t - 16t^2$

(b) $y = 5 + (240 \sin 10°)t - 16t^2 = 0$

$$t = \frac{-240 \sin 10° \pm \sqrt{(240 \sin 10°)^2 - 4(-16)5}}{2(-16)}$$

$t \approx -0.1149, 2.7196$

Distance traveled before arrow hits ground:

$(240 \cos 10°)(2.7196) \approx 643$ feet

(c)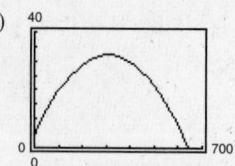

Maximum height: 32.1 feet

(d) Time arrow is in the air: approximately 2.72 seconds (see part b)

61. $x = (v_0 \cos \theta)t \implies t = \dfrac{x}{v_0 \cos \theta}$

$y = h + (v_0 \sin \theta)t - 16t^2$

$= h + (v_0 \sin \theta)\left(\dfrac{x}{v_0 \cos \theta}\right) - 16\left(\dfrac{x}{v_0 \cos \theta}\right)^2$

$= h + (\tan \theta)x - \dfrac{16x^2}{v_0^2 \cos^2 \theta}$

$= -\dfrac{16 \sec^2 \theta}{v_0^2}x^2 + (\tan \theta)x + h$

62. $y = 7 + x - 0.02x^2$

(a) Exercise 61 result: $y = -\dfrac{16 \sec^2 \theta}{v_0^2}x^2 + (\tan \theta)x + h$

$h = 7$

$\tan \theta = 1 \implies \theta = 45°$

$\dfrac{16 \sec^2 45°}{v_0^2} = 0.02 \implies v_0 = 40$

$x = (v_0 \cos \theta)t = (40 \cos 45°)t$

$y = h + (v_0 \sin \theta)t - 16t^2 \approx 7 + (40 \sin 45°)t - 16t^2$

(b)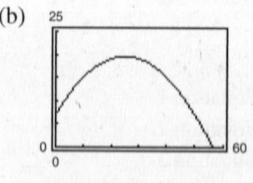

(c) Maximum height: 19.5 feet

Range: 56.2 feet

63. When the circle has rolled θ radians, the center is at $(a\theta, a)$.

$\sin \theta = \sin(180° - \theta)$

$= \dfrac{|AC|}{b} = \dfrac{|BD|}{b} \implies |BD| = b \sin \theta$

$\cos \theta = -\cos(180° - \theta)$

$= \dfrac{|AP|}{-b} \implies |AP| = -b \cos \theta$

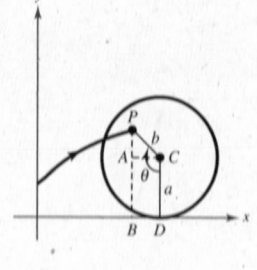

Therefore, $x = a\theta - b \sin \theta$ and $y = a - b \cos \theta$.

64. The coordinates of point (x, y) can be thought of as the sum of two vectors:

From origin to center of small circle: $\langle 3 \cos \theta, 3 \sin \theta \rangle$

From center of small circle to point (x, y): $\langle \cos \beta, \sin \beta \rangle$

Because the small circle rotates by 2θ when its center has rotated by θ, we have $\beta = \pi + 3\theta$.

$x = 3 \cos \theta + \cos(\pi + 3\theta) = 3 \cos \theta - \cos 3\theta$

$y = 3 \sin \theta + \sin(\pi + 3\theta) = 3 \sin \theta - \sin 3\theta$

65. True

$x = t$

$y = t^2 + 1 \implies y = x^2 + 1$

$x = 3t$

$y = 9t^2 + 1 \implies y = x^2 + 1$

66. False. Since $t^2 \geq 0$, the graph is that of $y = x$ for $x \geq 0$.

67. The use of parametric equations is useful when graphing two functions simultaneously on the same coordinate system. For example, this is useful when tracking the path of an object so the position and the time associated with that position can be determined.

68. For selected values of t, prepare a table of values for $x(t)$ and $y(t)$. Plot the points $(x(t), y(t))$ in the table. Sketch a curve through the points in order of increasing t (this is the *orientation* of the curve).

69.
$$\begin{array}{rl} 5x - 7y = & 11 \implies \\ -3x + y = & -13 \implies \end{array} \begin{array}{rl} 5x - 7y = & 11 \\ -21x + 7y = & -91 \\ \hline -16x \quad\quad = & -80 \\ x = & 5 \end{array}$$

$$5(5) - 7y = 11 \implies y = 2$$

Solution: $(5, 2)$

70. $\begin{cases} 3x + 5y = \ \ \ 9 \implies 6x + 10y = \ \ \ 18 \\ 4x - 2y = -14 \implies 20x - 10y = -70 \end{cases}$

$$\begin{array}{rl} 26x \quad\quad = & -52 \\ x = & -2 \end{array}$$

$$3(-2) + 5y = 9 \implies y = 3$$

Solution: $(-2, 3)$

71. $\begin{array}{rl} 3a - 2b + \ c = & 8 \implies 9a - 6b + 3c = \ \ 24 \\ 2a + \ b - 3c = & -3 \implies 2a + \ b - 3c = -3 \\ \hline & 11a - 5b \quad\quad = \ \ 21 \end{array}$

$$\begin{array}{rl} 2a + \ b - 3c = & -3 \implies 6a + 3b - 9c = \ -9 \\ a - 3b + 9c = & 16 \implies \ \ a - 3b + 9c = \ \ 16 \\ \hline & 7a \quad\quad = \ \ 7 \\ & a \quad\quad = \ \ 1 \end{array}$$

$$11(1) - 5b = 21 \implies b = -2$$

$$3(1) - 2(-2) + c = 8 \implies c = 1$$

Solution: $(1, -2, 1)$

72. $\begin{cases} 5u + 7v + 9w = \ \ 4 & \text{Equation 1} \\ u - 2v - 3w = \ \ 7 & \text{Equation 2} \\ 8u - 2v + \ w = 20 & \text{Equation 3} \end{cases}$

$\begin{cases} u - 2v - 3w = \ \ 7 & \text{Interchange Eq.1 and Eq.2} \\ 5u + 7v + 9w = \ \ 4 \\ 8u - 2v + \ w = 20 \end{cases}$

$\begin{cases} u - \ 2v - \ 3w = \ \ \ \ 7 \\ \quad 17v + 24w = -31 & (-5)\text{Eq.1} + \text{Eq.2} \\ \quad 14v + 25w = -36 & (-8)\text{Eq.1} + \text{Eq.3} \end{cases}$

$\begin{cases} u - \ 2v - \ 3w = \ \ \ \ 7 \\ \quad\quad 17v + 24w = \ -31 \\ \quad\quad\quad\quad 89w = -178 & (-14)\text{Eq.2} + (17)\text{Eq.3} \end{cases}$

$$89w = -178 \implies w = -2$$
$$17v + 24(-2) = -31 \implies v = 1$$
$$u - 2(1) - 3(-2) = 7 \implies u = 3$$

Solution: $(3, 1, -2)$

73. $\theta = 105°$

$\quad \theta' = 180° - 105° = 75°$

74. $\theta = 230°$

$\quad \theta' = \theta - 180°$

$\quad\quad = 230° - 180° = 50°$

75. $\theta = -\dfrac{2\pi}{3}$

$\quad \theta' = -\dfrac{2\pi}{3} + \pi = \dfrac{\pi}{3}$

76. $\theta = \dfrac{5\pi}{6}$

$\quad \theta' = \pi - \theta = \pi - \dfrac{5\pi}{6} = \dfrac{\pi}{6}$

Section 10.7   Polar Coordinates

- In polar coordinates you do not have unique representation of points. The point (r, θ) can be represented by $(r, \theta \pm 2n\pi)$ or by $(-r, \theta \pm (2n + 1)\pi)$ where n is any integer. The pole is represented by $(0, \theta)$ where θ is any angle.

- To convert from polar coordinates to rectangular coordinates, use the following relationships.
  $$x = r \cos \theta$$
  $$y = r \sin \theta$$

- To convert from rectangular coordinates to polar coordinates, use the following relationships.
  $$r = \pm\sqrt{x^2 + y^2}$$
  $$\tan \theta = y/x$$

  If θ is in the same quadrant as the point (x, y), then r is positive. If θ is in the opposite quadrant as the point (x, y), then r is negative.

- You should be able to convert rectangular equations to polar form and vice versa.

Vocabulary Check

1. pole

2. directed distance; directed angle

3. polar

4. $x = r \cos \theta, \quad \tan \theta = \dfrac{y}{x}$

$\quad\quad y = r \sin \theta, \quad r^2 = x^2 + y^2$

1. Polar coordinates: $\left(4, -\dfrac{\pi}{3}\right)$

Additional representations:

$\left(4, -\dfrac{\pi}{3} + 2\pi\right) = \left(4, \dfrac{5\pi}{3}\right)$

$\left(-4, -\dfrac{\pi}{3} - \pi\right) = \left(-4, -\dfrac{4\pi}{3}\right)$

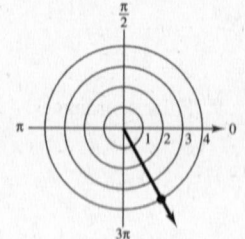

2. $\left(-1, -\dfrac{3\pi}{4}\right)$

$\left(1, \dfrac{\pi}{4}\right)$

$\left(-1, \dfrac{5\pi}{4}\right)$

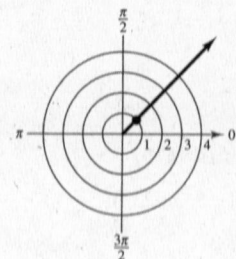

3. Polar coordinates: $\left(0, -\dfrac{7\pi}{6}\right)$

Additional representations:

$\left(0, -\dfrac{7\pi}{6} + 2\pi\right) = \left(0, \dfrac{5\pi}{6}\right)$

$\left(0, -\dfrac{7\pi}{6} + \pi\right) = \left(0, -\dfrac{\pi}{6}\right)$ or $(0, \theta)$ for any $\theta, -2\pi < \theta < 2\pi$

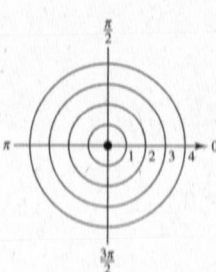

4. $\left(16, \dfrac{5\pi}{2}\right)$

$\left(16, \dfrac{\pi}{2}\right)$

$\left(-16, \dfrac{3\pi}{2}\right)$

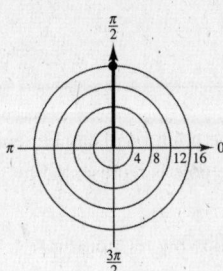

5. Polar coordinates: $\left(\sqrt{2}, 2.36\right)$

Additional representations:

$\left(\sqrt{2}, 2.36 - 2\pi\right) \approx \left(\sqrt{2}, -3.92\right)$

$\left(-\sqrt{2}, 2.36 - \pi\right) \approx \left(-\sqrt{2}, -0.78\right)$

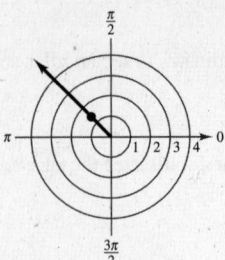

6. $(-3, -1.57)$

$(3, 1.5716)$

$(-3, 4.7132)$

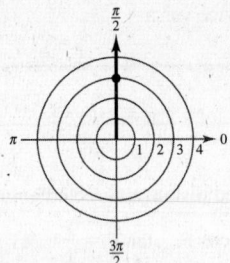

7. Polar coordinates: $\left(2\sqrt{2}, 4.71\right)$

Additional representations:

$\left(2\sqrt{2}, 4.71 - 2\pi\right) \approx \left(2\sqrt{2}, -1.57\right)$

$\left(-2\sqrt{2}, 2\pi - 4.71\right) \approx \left(-2\sqrt{2}, 1.57\right)$

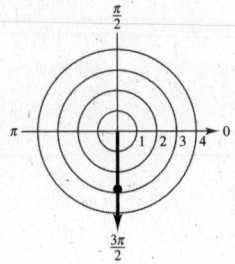

8. $(-5, -2.36)$

$(5, 0.7816)$

$(-5, 3.9232)$

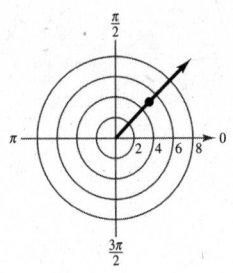

9. Polar coordinates: $\left(3, \dfrac{\pi}{2}\right)$

$x = 3 \cos \dfrac{\pi}{2} = 0$

$y = 3 \sin \dfrac{\pi}{2} = 3$

Rectangular coordinates: $(0, 3)$

10. Polar coordinates: $\left(3, \dfrac{3\pi}{2}\right) = (r, \theta)$

$x = r \cos \theta = 3 \cos \dfrac{3\pi}{2} = 0$

$y = r \sin \theta = 3 \sin \dfrac{3\pi}{2} = -3$

Rectangular coordinates: $(0, -3)$

11. Polar coordinates: $\left(-1, \dfrac{5\pi}{4}\right)$

$x = -1 \cos\left(\dfrac{5\pi}{4}\right) = \dfrac{\sqrt{2}}{2}, \; y = -1 \sin\left(\dfrac{5\pi}{4}\right) = \dfrac{\sqrt{2}}{2}$

Rectangular coordinates: $\left(\dfrac{\sqrt{2}}{2}, \dfrac{\sqrt{2}}{2}\right)$

12. Polar coordinates: $(0, -\pi) = (r, \theta)$

$x = r \cos \theta = 0$

$y = r \sin \theta = 0$

Rectangular coordinates: $(0, 0)$

13. Polar coordinates: $\left(2, \dfrac{3\pi}{4}\right)$

$x = 2 \cos \dfrac{3\pi}{4} = -\sqrt{2}$

$y = 2 \sin \dfrac{3\pi}{4} = \sqrt{2}$

Rectangular coordinates: $\left(-\sqrt{2}, \sqrt{2}\right)$

14. Polar coordinates: $\left(-2, \dfrac{7\pi}{6}\right) = (r, \theta)$

$x = r \cos \theta = -2 \cos \dfrac{7\pi}{6} = \sqrt{3}$

$y = r \sin \theta = -2 \sin \dfrac{7\pi}{6} = 1$

Rectangular coordinates: $\left(\sqrt{3}, 1\right)$

15. Polar coordinates: $(-2.5, 1.1)$

$x = -2.5 \cos 1.1 \approx -1.1340$

$y = -2.5 \sin 1.1 \approx -2.2280$

Rectangular coordinates: $(-1.1340, -2.2280)$

16. Polar coordinates: $(8.25, 3.5) = (r, \theta)$

$x = r \cos \theta = 8.25 \cos 3.5 \approx -7.7258$

$y = r \sin \theta = 8.25 \sin 3.5 \approx -2.8940$

Rectangular coordinates: $(-7.7258, -2.8940)$

17. Rectangular coordinates: $(1, 1)$

$r = \pm\sqrt{2}, \ \tan \theta = 1, \ \theta = \dfrac{\pi}{4} \text{ or } \dfrac{5\pi}{4}$

Polar coordinates: $\left(\sqrt{2}, \dfrac{\pi}{4}\right), \left(-\sqrt{2}, \dfrac{5\pi}{4}\right)$

18. Rectangular coordinates: $(-3, -3)$

$r = 3\sqrt{2}, \ \tan \theta = 1, \ \theta = \dfrac{5\pi}{4}$

Polar coordinates: $\left(3\sqrt{2}, \dfrac{5\pi}{4}\right)$

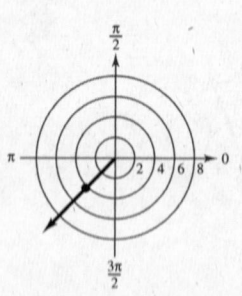

19. Rectangular coordinates: $(-6, 0)$

$r = \pm 6, \ \tan \theta = 0, \ \theta = 0 \text{ or } \pi$

Polar coordinates: $(6, \pi), (-6, 0)$

20. Rectangular coordinates: $(0, -5)$

$r = 5, \ \tan \theta \text{ undefined}, \ \theta = \dfrac{\pi}{2}$

Polar coordinates: $\left(5, \dfrac{3\pi}{2}\right)$

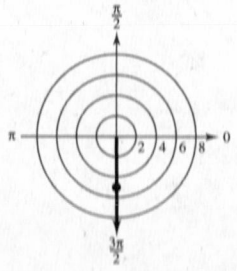

21. Rectangular coordinates: $(-3, 4)$

$r = \pm\sqrt{9 + 16} = \pm 5, \ \tan \theta = -\dfrac{4}{3}, \ \theta \approx 2.2143, \ 5.3559$

Polar coordinates: $(5, 2.2143), (-5, 5.3559)$

22. Rectangular coordinates: $(3, -1)$

$r = \sqrt{9+1} = \sqrt{10}$, $\tan\theta = -\frac{1}{3}$, $\theta \approx -0.322$ or 5.9614

Polar coordinates: $\left(\sqrt{10}, 5.961\right)$

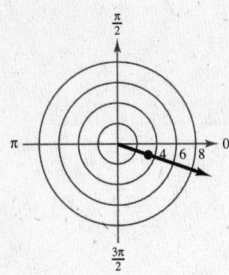

23. Rectangular coordinates: $\left(-\sqrt{3}, -\sqrt{3}\right)$

$r = \pm\sqrt{3+3} = \pm\sqrt{6}$, $\tan\theta = 1$, $\theta = \frac{\pi}{4}$ or $\frac{5\pi}{4}$

Polar coordinates: $\left(\sqrt{6}, \frac{5\pi}{4}\right), \left(-\sqrt{6}, \frac{\pi}{4}\right)$

24. Rectangular coordinates: $\left(\sqrt{3}, -1\right)$

$r = \sqrt{3+1} = 2$, $\tan\theta = -\frac{1}{\sqrt{3}}$, $\theta = \frac{11\pi}{6}$

Polar coordinates: $\left(2, \frac{11\pi}{6}\right)$

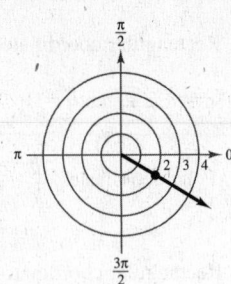

25. Rectangular coordinates: $(6, 9)$

$r = \pm\sqrt{6^2 + 9^2} = \pm\sqrt{117} = \pm 3\sqrt{13}$

$\tan\theta = \frac{9}{6}$, $\theta \approx 0.9828$, 4.1244

Polar coordinates: $\left(3\sqrt{13}, 0.9828\right), \left(-3\sqrt{13}, 4.1244\right)$

26. Rectangular coordinates: $(5, 12)$

$r = \sqrt{25 + 144} = 13$, $\tan\theta = \frac{12}{5}$, $\theta \approx 1.176$

Polar coordinates: $(13, 1.176)$

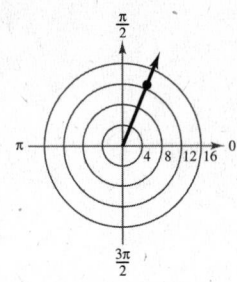

27. Rectangular: $(3, -2)$

$(3, -2) \blacktriangleright$ Pol

$\approx (3.606, -0.5880)$

or $\left(\sqrt{13}, -0.5880\right)$

or $\left(\sqrt{13}, 5.6952\right)$

28. Rectangular coordinates: $(-5, 2)$

$R \blacktriangleright Pr(-5, 2) \approx 5.385$

$R \blacktriangleright P\theta(-5, 2) \approx 2.761$

$\approx (5.385, 2.761)$

29. Rectangular: $\left(\sqrt{3}, 2\right)$

$\left(\sqrt{3}, 2\right) \blacktriangleright$ Pol

$\approx (2.646, 0.8571)$

or $\left(\sqrt{7}, 0.8571\right)$

30. Rectangular coordinates: $\left(3\sqrt{2}, 3\sqrt{2}\right)$

$R \blacktriangleright Pr\left(3\sqrt{2}, 3\sqrt{2}\right) = 6$

$R \blacktriangleright P\theta\left(3\sqrt{2}, 3\sqrt{2}\right) \approx 0.785$

$= \left(6, \frac{\pi}{4}\right)$

31. Rectangular: $\left(\frac{5}{2}, \frac{4}{3}\right)$

$\left(\frac{5}{2}, \frac{4}{3}\right) \blacktriangleright$ Pol

$\approx (2.833, 0.4900)$

or $\left(\frac{17}{6}, 0.4900\right)$

32. Rectangular coordinates: $\left(\frac{7}{4}, \frac{3}{2}\right)$

$R \blacktriangleright Pr\left(\frac{7}{4}, \frac{3}{2}\right) \approx 2.305$

$R \blacktriangleright P\theta\left(\frac{7}{4}, \frac{3}{2}\right) \approx 0.709$

$\approx (2.305, 0.709)$

33. $x^2 + y^2 = 9$

$r = 3$

34. $x^2 + y^2 = 16$

$r = 4$

35. $y = 4$

$r \sin \theta = 4$

$r = 4 \csc \theta$

36. $y = x$

$r \cos \theta = r \sin \theta$

$1 = \tan \theta$

$\theta = \dfrac{\pi}{4}$

37. $x = 10$

$r \cos \theta = 10$

$r = 10 \sec \theta$

38. $x = 4a$

$r \cos \theta = 4a$

$r = 4a \sec \theta$

39. $3x - y + 2 = 0$

$3r \cos \theta - r \sin \theta + 2 = 0$

$r(3 \cos \theta - \sin \theta) = -2$

$r = \dfrac{-2}{3 \cos \theta - \sin \theta}$

40. $3x + 5y - 2 = 0$

$3r \cos \theta + 5r \sin \theta - 2 = 0$

$r(3 \cos \theta + 5 \sin \theta) = 2$

$r = \dfrac{2}{3 \cos \theta + 5 \sin \theta}$

41. $xy = 16$

$(r \cos \theta)(r \sin \theta) = 16$

$r^2 = 16 \sec \theta \csc \theta = 32 \csc 2\theta$

42. $2xy = 1$

$2(r \cos \theta)(r \sin \theta) = 1$

$2r^2 \cos \theta \sin \theta = 1$

$r^2 = \dfrac{1}{2 \cos \theta \sin \theta}$

$r^2 = \dfrac{1}{2} \sec \theta \csc \theta$

$r^2 = \csc 2\theta$

43. $y^2 - 8x - 16 = 0$

$r^2 \sin^2 \theta - 8r \cos \theta - 16 = 0$

By the Quadratic Formula, we have:

$r = \dfrac{-(-8 \cos \theta) \pm \sqrt{(-8 \cos \theta)^2 - 4(\sin^2 \theta)(-16)}}{2 \sin^2 \theta}$

$= \dfrac{8 \cos \theta \pm \sqrt{64 \cos^2 \theta + 64 \sin^2 \theta}}{2 \sin^2 \theta}$

$= \dfrac{8 \cos \theta \pm \sqrt{64(\cos^2 \theta + \sin^2 \theta)}}{2 \sin^2 \theta}$

$= \dfrac{8 \cos \theta \pm 8}{2 \sin^2 \theta}$

$= \dfrac{4(\cos \theta \pm 1)}{1 - \cos^2 \theta}$

$r = \dfrac{4(\cos \theta + 1)}{(1 + \cos \theta)(1 - \cos \theta)} = \dfrac{4}{1 - \cos \theta}$

or

$r = \dfrac{4(\cos \theta - 1)}{(1 + \cos \theta)(1 - \cos \theta)} = \dfrac{-4}{1 + \cos \theta}$

44. $(x^2 + y^2)^2 = 9(x^2 - y^2)$

$(r^2)^2 = 9(r^2 \cos^2 \theta - r^2 \sin^2 \theta)$

$= 9r^2(\cos^2 \theta - \sin^2 \theta)$

$r^2 = 9 \cos 2\theta$

45. $x^2 + y^2 = a^2$

$r^2 = a^2$

$r = a$

46. $x^2 + y^2 = 9a^2$

$r = 3a$

47. $x^2 + y^2 - 2ax = 0$

$r^2 - 2a\, r \cos \theta = 0$

$r(r - 2a \cos \theta) = 0$

$r - 2a \cos \theta = 0$

$r = 2a \cos \theta$

48. $x^2 + y^2 - 2ay = 0$

$r^2 - 2ar \sin \theta = 0$

$r = 2a \sin \theta$

49. $r = 4 \sin \theta$

$r^2 = 4r \sin \theta$

$x^2 + y^2 = 4y$

$x^2 + y^2 - 4y = 0$

50. Because $x = r \cos \theta$ and r is given as $2 \cos \theta$:

$x = 2 \cos \theta \cos \theta = 2 \cos^2 \theta$

$r = 2 \cos \theta$

$r^2 = 4 \cos^2 \theta$

$x^2 + y^2 = 2\,(2 \cos^2 \theta)$

$x^2 + y^2 = 2x$

$x^2 + y^2 - 2x = 0$

51. $\theta = \dfrac{2\pi}{3}$

$\tan \theta = \tan \dfrac{2\pi}{3}$

$\dfrac{y}{x} = -\sqrt{3}$

$y = -\sqrt{3}x$

$\sqrt{3}x + y = 0$

52. $\theta = \dfrac{5\pi}{3}$

$\tan \theta = -\sqrt{3}$

$\dfrac{y}{x} = -\sqrt{3}$

$y = -\sqrt{3}\, x$

$\sqrt{3}\, x + y = 0$

53. $r = 4$

$r^2 = 16$

$x^2 + y^2 = 16$

54. $r = 10$

$r^2 = 100$

$x^2 + y^2 = 100$

55. $r = 4 \csc \theta$

$r \sin \theta = 4$

$y = 4$

56. $r = -3 \sec \theta$

$\dfrac{r}{\sec \theta} = -3$

$r \cos \theta = -3$

$x = -3$

57. $r^2 = \cos \theta$

$r^3 = r \cos \theta$

$\left(\pm \sqrt{x^2 + y^2} \right)^3 = x$

$\pm \left(x^2 + y^2 \right)^{3/2} = x$

$\left(x^2 + y^2 \right)^3 = x^2$

$x^2 + y^2 = x^{2/3}$

$x^2 + y^2 - x^{2/3} = 0$

58. $r^2 = \sin 2\theta = 2 \sin \theta \cos \theta$

$r^2 = 2 \left(\dfrac{y}{r} \right) \left(\dfrac{x}{r} \right) = \dfrac{2xy}{r^2}$

$r^4 = 2xy$

$\left(x^2 + y^2 \right)^2 = 2xy$

59. $r = 2 \sin 3\theta$

$r = 2 \sin(\theta + 2\theta)$

$r = 2[\sin \theta \cos 2\theta + \cos \theta \sin 2\theta]$

$r = 2[\sin \theta (1 - 2 \sin^2 \theta) + \cos \theta (2 \sin \theta \cos \theta)]$

$r = 2[\sin \theta - 2 \sin^3 \theta + 2 \sin \theta \cos^2 \theta]$

$r = 2[\sin \theta - 2 \sin^3 \theta + 2 \sin \theta (1 - \sin^2 \theta)]$

$r = 2(3 \sin \theta - 4 \sin^3 \theta)$

$r^4 = 6r^3 \sin \theta - 8r^3 \sin^3 \theta$

$\left(x^2 + y^2 \right)^2 = 6 \left(x^2 + y^2 \right) y - 8y^3$

$\left(x^2 + y^2 \right)^2 = 6x^2 y - 2y^3$

60. $r = 3 \cos 2\theta$

$$r = 3(2\cos^2\theta - 1)$$

$$r = 3\left(\frac{2x^2}{r^2} - 1\right)$$

$$r = 3\left(\frac{2x^2 - r^2}{r^2}\right)$$

$$r = 3\left(\frac{2x^2 - x^2 - y^2}{x^2 + y^2}\right)$$

$$r = 3\left(\frac{x^2 - y^2}{x^2 + y^2}\right)$$

$$r^2 = 9\left(\frac{x^2 - y^2}{x^2 + y^2}\right)^2$$

$$x^2 + y^2 = 9\frac{(x^2 - y^2)^2}{(x^2 + y^2)^2}$$

$$(x^2 + y^2)^3 = 9(x^2 - y^2)^2$$

61. $r = \dfrac{2}{1 + \sin\theta}$

$$r(1 + \sin\theta) = 2$$

$$r + r\sin\theta = 2$$

$$r = 2 - r\sin\theta$$

$$\pm\sqrt{x^2 + y^2} = 2 - y$$

$$x^2 + y^2 = (2 - y)^2$$

$$x^2 + y^2 = 4 - 4y + y^2$$

$$x^2 + 4y - 4 = 0$$

62. $r = \dfrac{1}{1 - \cos\theta}$

$$r - r\cos\theta = 1$$

$$\sqrt{x^2 + y^2} - x = 1$$

$$x^2 + y^2 = 1 + 2x + x^2$$

$$y^2 = 2x + 1$$

63. $r = \dfrac{6}{2 - 3\sin\theta}$

$$r(2 - 3\sin\theta) = 6$$

$$2r = 6 + 3r\sin\theta$$

$$2(\pm\sqrt{x^2 + y^2}) = 6 + 3y$$

$$4(x^2 + y^2) = (6 + 3y)^2$$

$$4x^2 + 4y^2 = 36 + 36y + 9y^2$$

$$4x^2 - 5y^2 - 36y - 36 = 0$$

64. $r = \dfrac{6}{2\cos\theta - 3\sin\theta}$

$$r = \frac{6}{2(x/r) - 3(y/r)}$$

$$r = \frac{6r}{2x - 3y}$$

$$1 = \frac{6}{2x - 3y}$$

$$2x - 3y = 6$$

65. The graph of the polar equation consists of all points that are six units from the pole.

$$r = 6$$

$$r^2 = 36$$

$$x^2 + y^2 = 36$$

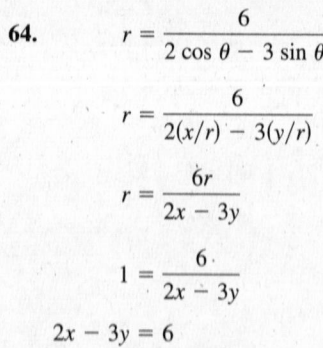

66. The graph of the polar equation consists of all points that are eight units from the pole.

$$r = 8$$

$$r^2 = 64$$

$$x^2 + y^2 = 64$$

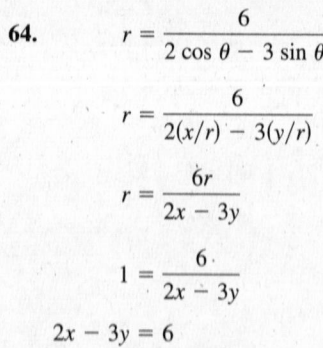

67. The graph of the polar equation consists of all points that make an angle of $\pi/6$ with the polar axis.

$$\theta = \frac{\pi}{6}$$

$$\tan\theta = \tan\frac{\pi}{6}$$

$$\frac{y}{x} = \frac{\sqrt{3}}{3}$$

$$y = \frac{\sqrt{3}}{3}x$$

$$3y = \sqrt{3}x$$

$$-\sqrt{3}x + 3y = 0$$

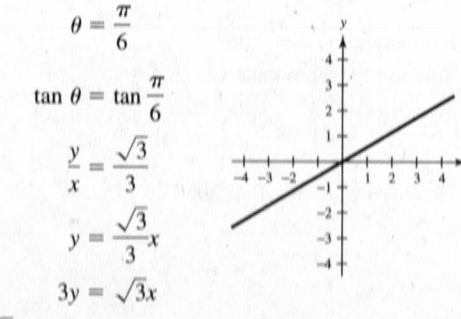

68. The graph of the polar equation consists of all points that make an angle of $3\pi/4$ with the polar axis.

$$\theta = \frac{3\pi}{4}$$

$$\tan\theta = \tan\frac{3\pi}{4}$$

$$\frac{y}{x} = -1$$

$$y = -x$$

$$x + y = 0$$

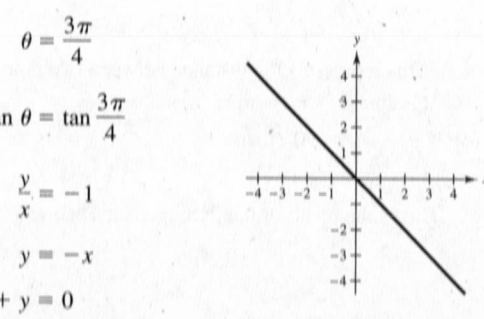

69. The graph of the polar equation is not evident by simple inspection. Convert to rectangular form first.

$$r = 3 \sec \theta$$

$$r \cos\theta = 3$$

$$x = 3$$

$$x - 3 = 0$$

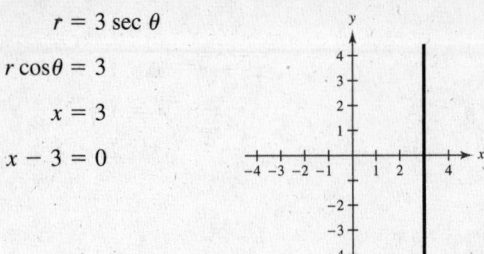

70. The graph of the polar equation is not evident by simple inspection. Convert to rectangular form first.

$$r = 2 \csc \theta$$

$$r \sin \theta = 2$$

$$y = 2$$

$$y - 2 = 0$$

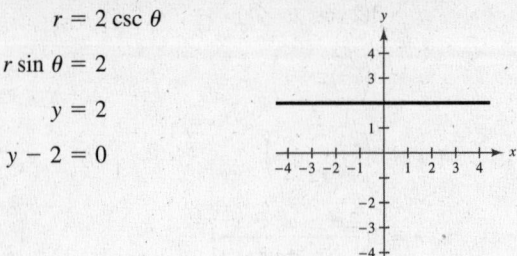

71. True. Because r is a directed distance, then the point (r, θ) can be represented as $(r, \theta \pm 2n\pi)$.

72. False. (r_1, θ) and (r_2, θ) represent the same point only if $r_1 = r_2$.

73.
$$r = 2(h \cos \theta + k \sin \theta)$$
$$r = 2\left(h\left(\frac{x}{r}\right) + k\left(\frac{y}{r}\right)\right)$$
$$r = \frac{2hx + 2ky}{r}$$
$$r^2 = 2hx + 2ky$$
$$x^2 + y^2 = 2hx + 2ky$$
$$x^2 - 2hx + y^2 - 2ky = 0$$
$$\left(x^2 - 2hx + h^2\right) + \left(y^2 - 2ky + k^2\right) = h^2 + k^2$$
$$(x - h)^2 + (y - k)^2 = h^2 + k^2$$

Center: (h, k)

Radius: $\sqrt{h^2 + k^2}$

74.
$$r = \cos \theta + 3 \sin \theta$$
$$r = \frac{x}{r} + \frac{3y}{r}$$
$$r^2 = x + 3y$$
$$x^2 + y^2 = x + 3y$$
$$x^2 - x + y^2 - 3y = 0$$
$$\left(x - \frac{1}{2}\right)^2 + \left(y - \frac{3}{2}\right)^2 = \frac{5}{2}$$

The graph is a circle.

75. (a) $(r_1, \theta_1) = (x_1, y_1)$ where $x_1 = r_1 \cos \theta_1$ and $y_1 = r_1 \sin \theta_1$.

$(r_2, \theta_2) = (x_2, y_2)$ where $x_2 = r_2 \cos \theta_2$ and $y_2 = r_2 \sin \theta_2$.

$$d = \sqrt{(x_1 - x_2)^2 + (y_1 - y_2)^2}$$
$$= \sqrt{x_1{}^2 - 2x_1x_2 + x_2{}^2 + y_1{}^2 - 2y_1y_2 + y_2{}^2}$$
$$= \sqrt{(x_1{}^2 + y_1{}^2) + (x_2{}^2 + y_2{}^2) - 2(x_1x_2 + y_1y_2)}$$
$$= \sqrt{r_1{}^2 + r_2{}^2 - 2(r_1r_2 \cos \theta_1 \cos \theta_2 + r_1r_2 \sin \theta_1 \sin \theta_2)}$$
$$= \sqrt{r_1{}^2 + r_2{}^2 - 2r_1r_2 \cos(\theta_1 - \theta_2)}$$

(b) If $\theta_1 = \theta_2$, then
$$d = \sqrt{r_1{}^2 + r_2{}^2 - 2r_1r_2}$$
$$= \sqrt{(r_1 - r_2)^2}$$
$$= |r_1 - r_2|.$$

This represents the distance between two points on the line $\theta = \theta_1 = \theta_2$.

(c) If $\theta_1 - \theta_2 = 90°$, then
$$d = \sqrt{r_1{}^2 + r_2{}^2}.$$
This is the result of the Pythagorean Theorem.

(d) The results should be the same. For example, use the points

$$\left(3, \frac{\pi}{6}\right) \text{ and } \left(4, \frac{\pi}{3}\right).$$

The distance is $d \approx 2.053$.

Now use the representations

$$\left(-3, \frac{7\pi}{6}\right) \text{ and } \left(-4, \frac{4\pi}{3}\right).$$

The distance is still $d \approx 2.053$.

76. (a) For horizontal moves, just the x-coordinate changes.
For vertical moves, just the y-coordinate changes.

(b) For horizontal moves, both r and θ change.
For vertical moves, both r and θ change.

(c) Unlike r and θ, x and y measure horizontal and vertical changes, respectively.

77. $\log_6 \dfrac{x^2 z}{3y} = \log_6 x^2 z - \log_6 3y$

$$= \log_6 x^2 + \log_6 z - (\log_6 3 + \log_6 y)$$

$$= 2\log_6 x + \log_6 z - \log_6 3 - \log_6 y$$

78. $\log_4 \dfrac{\sqrt{2x}}{y} = \log_4 \sqrt{2x} - \log_4 y$

$$= \frac{1}{2}\log_4 2x - \log_4 y$$

$$= \frac{1}{2}\log_4 2 + \frac{1}{2}\log_4 x - \log_4 y$$

$$= \frac{1}{4} + \frac{1}{2}\log_4 x - \log_4 y$$

79. $\ln x(x+4)^2 = \ln x + \ln(x+4)^2$

$$= \ln x + 2\ln(x+4)$$

80. $\ln 5x^2(x^2+1) = \ln 5 + \ln x^2 + \ln(x^2+1)$

$$= \ln 5 + 2\ln x + \ln(x^2+1)$$

81. $\log_7 x - \log_7 3y = \log_7 \dfrac{x}{3y}$

82. $\log_5 a + 8\log_5(x+1) = \log_5 a + \log_5(x+1)^8$

$$= \log_5 a(x+1)^8$$

83. $\frac{1}{2}\ln x + \ln(x-2) = \ln \sqrt{x} + \ln(x-2)$

$$= \ln \sqrt{x}(x-2)$$

84. $\ln 6 + \ln y - \ln(x-3) = \ln 6y - \ln(x-3)$

$$= \ln \dfrac{6y}{x-3}$$

85. $\begin{cases} 5x - 7y = -11 \\ -3x + y = -3 \end{cases}$

By Cramer's Rule we have:

$$x = \frac{\begin{vmatrix} -11 & -7 \\ -3 & 1 \end{vmatrix}}{\begin{vmatrix} 5 & -7 \\ -3 & 1 \end{vmatrix}} = \frac{-32}{-16} = 2$$

$$y = \frac{\begin{vmatrix} 5 & -11 \\ -3 & -3 \end{vmatrix}}{\begin{vmatrix} 5 & -7 \\ -3 & 1 \end{vmatrix}} = \frac{-48}{-16} = 3$$

Solution: $(2, 3)$

86. $\begin{cases} 3x - 5y = 10 \\ 4x - 2y = -5 \end{cases}$

$$x = \frac{\begin{vmatrix} 10 & -5 \\ -5 & -2 \end{vmatrix}}{\begin{vmatrix} 3 & -5 \\ 4 & -2 \end{vmatrix}} = -\frac{45}{14}$$

$$y = \frac{\begin{vmatrix} 3 & 10 \\ 4 & -5 \end{vmatrix}}{\begin{vmatrix} 3 & -5 \\ 4 & -2 \end{vmatrix}} = -\frac{55}{14}$$

Solution: $\left(-\dfrac{45}{14}, -\dfrac{55}{14}\right)$

87. $\begin{cases} 3a - 2b + c = 0 \\ 2a + b - 3c = 0 \\ a - 3b + 9c = 8 \end{cases}$

$\begin{vmatrix} 3 & -2 & 1 \\ 2 & 1 & -3 \\ 1 & -3 & 9 \end{vmatrix} = 35$

By Cramer's Rule we have:

$a = \dfrac{\begin{vmatrix} 0 & -2 & 1 \\ 0 & 1 & -3 \\ 8 & -3 & 9 \end{vmatrix}}{35} = \dfrac{40}{35} = \dfrac{8}{7}$

$b = \dfrac{\begin{vmatrix} 3 & 0 & 1 \\ 2 & 0 & -3 \\ 1 & 8 & 9 \end{vmatrix}}{35} = \dfrac{88}{35}$

$c = \dfrac{\begin{vmatrix} 3 & -2 & 0 \\ 2 & 1 & 0 \\ 1 & -3 & 8 \end{vmatrix}}{35} = \dfrac{56}{35} = \dfrac{8}{5}$

Solution: $\left(\dfrac{8}{7}, \dfrac{88}{35}, \dfrac{8}{5} \right)$

88. $\begin{cases} 5u + 7v + 9w = 15 \\ u - 2v - 3w = 7 \\ 8u - 2v + w = 0 \end{cases} \Rightarrow \begin{vmatrix} 5 & 7 & 9 \\ 1 & -2 & -3 \\ 8 & -2 & 1 \end{vmatrix} = -89$

$u = \dfrac{\begin{vmatrix} 15 & 7 & 9 \\ 7 & -2 & -3 \\ 0 & -2 & 1 \end{vmatrix}}{-89} = \dfrac{-295}{-89} = \dfrac{295}{89},$

$v = \dfrac{\begin{vmatrix} 5 & 15 & 9 \\ 1 & 7 & -3 \\ 8 & 0 & 1 \end{vmatrix}}{-89} = \dfrac{-844}{-89} = \dfrac{844}{89},$

$w = \dfrac{\begin{vmatrix} 5 & 7 & 15 \\ 1 & -2 & 7 \\ 8 & -2 & 0 \end{vmatrix}}{-89} = \dfrac{672}{-89} = -\dfrac{672}{89}$

Solution: $\left(\dfrac{295}{89}, \dfrac{844}{89}, -\dfrac{672}{89} \right)$

89. $\begin{cases} -x + y + 2z = 1 \\ 2x + 3y + z = -2 \\ 5x + 4y + 2z = 4 \end{cases}$

$\begin{vmatrix} -1 & 1 & 2 \\ 2 & 3 & 1 \\ 5 & 4 & 2 \end{vmatrix} = -15$

By Cramer's Rule we have:

$x = \dfrac{\begin{vmatrix} 1 & 1 & 2 \\ -2 & 3 & 1 \\ 4 & 4 & 2 \end{vmatrix}}{-15} = \dfrac{-30}{-15} = 2$

$y = \dfrac{\begin{vmatrix} -1 & 1 & 2 \\ 2 & -2 & 1 \\ 5 & 4 & 2 \end{vmatrix}}{-15} = \dfrac{45}{-15} = -3$

$z = \dfrac{\begin{vmatrix} -1 & 1 & 1 \\ 2 & 3 & -2 \\ 5 & 4 & 4 \end{vmatrix}}{-15} = \dfrac{-45}{-15} = 3$

Solution: $(2, -3, 3)$

90. $\begin{cases} 2x_1 + x_2 + 2x_3 = 4 \\ 2x_1 + 2x_2 = 5 \\ 2x_1 - x_2 + 6x_3 = 2 \end{cases}$

$D = \begin{vmatrix} 2 & 1 & 2 \\ 2 & 2 & 0 \\ 2 & -1 & 6 \end{vmatrix} = 0$

Cramer's Rule does not apply.

91. Points: $(4, -3), (6, -7), (-2, -1)$

$\begin{vmatrix} 4 & -3 & 1 \\ 6 & -7 & 1 \\ -2 & -1 & 1 \end{vmatrix} = -20 \neq 0$

The points are not collinear.

92. Points: $(-2, 4), (0, 1), (4, -5)$

$\begin{vmatrix} -2 & 4 & 1 \\ 0 & 1 & 1 \\ 4 & -5 & 1 \end{vmatrix} = 0 \Rightarrow$ collinear

93. Points: $(-6, -4), (-1, -3), (1.5, -2.5)$

$$\begin{vmatrix} -6 & -4 & 1 \\ -1 & -3 & 1 \\ 1.5 & -2.5 & 1 \end{vmatrix} = 0$$

The points are collinear.

94. Points: $(-2.3, 5), (-0.5, 0), (1.5, -3)$

$$\begin{vmatrix} -2.3 & 5 & 1 \\ -0.5 & 0 & 1 \\ 1.5 & -3 & 1 \end{vmatrix} = 4.6 \Rightarrow \text{not collinear}$$

Section 10.8   Graphs of Polar Equations

- When graphing polar equations:
  1. Test for symmetry.
     (a) $\theta = \pi/2$: Replace (r, θ) by $(r, \pi - \theta)$ or $(-r, -\theta)$.
     (b) Polar axis: Replace (r, θ) by $(r, -\theta)$ or $(-r, \pi - \theta)$.
     (c) Pole: Replace (r, θ) by $(r, \pi + \theta)$ or $(-r, \theta)$.
     (d) $r = f(\sin \theta)$ is symmetric with respect to the line $\theta = \pi/2$.
     (e) $r = f(\cos \theta)$ is symmetric with respect to the polar axis.
  2. Find the θ values for which $|r|$ is maximum.
  3. Find the θ values for which $r = 0$.
  4. Know the different types of polar graphs.
     (a) Limaçons $(0 < a, 0 < b)$
     $$r = a \pm b \cos \theta$$
     $$r = a \pm b \sin \theta$$
     (b) Rose curves, $n \geq 2$
     $$r = a \cos n\theta$$
     $$r = a \sin n\theta$$
     (c) Circles
     $$r = a \cos \theta$$
     $$r = a \sin \theta$$
     $$r = a$$
     (d) Lemniscates
     $$r^2 = a^2 \cos 2\theta$$
     $$r^2 = a^2 \sin 2\theta$$
  5. Plot additional points.

Vocabulary Check

1. $\theta = \dfrac{\pi}{2}$ **2.** polar axis   **3.** convex limaçon

4. circle   **5.** lemniscate   **6.** cardioid

---

1. $r = 3 \cos 2\theta$
Rose curve with 4 petals

2. $r = 5 - 5 \sin \theta$
Cardioid

3. $r = 3(1 - 2 \cos \theta)$
Limaçon with inner loop

4. $r^2 = 16 \cos 2\theta$
Lemniscate

5. $r = 6 \sin 2\theta$
Rose curve with 4 petals

6. $r = 3 \cos \theta$
Circle

7. $r = 5 + 4 \cos \theta$

$\theta = \dfrac{\pi}{2}$: $-r = 5 + 4 \cos(-\theta)$

      $-r = 5 + 4 \cos \theta$

      Not an equivalent equation

Polar axis: $r = 5 + 4 \cos(-\theta)$

      $r = 5 + 4 \cos \theta$

      Equivalent equation

Pole: $-r = 5 + 4 \cos \theta$

      Not an equivalent equation

Answer: Symmetric with respect to polar axis

8. $r = 16 \cos 3\theta$

$\theta = \dfrac{\pi}{2}$: $-r = 16 \cos(3(-\theta))$

      $-r = 16 \cos(-3\theta)$

      $-r = 16 \cos 3\theta$

      Not an equivalent equation

Polar axis: $r = 16 \cos(3(-\theta))$

      $r = 16 \cos(-3\theta)$

      $r = 16 \cos 3\theta$

      Equivalent equation

Pole: $-r = 16 \cos 3\theta$

      Not an equivalent equation

Answer: Symmetric with respect to polar axis

9. $r = \dfrac{2}{1 + \sin \theta}$

$\theta = \dfrac{\pi}{2}$: $r = \dfrac{2}{1 + \sin(\pi - \theta)}$

$r = \dfrac{2}{1 + \sin \pi \cos \theta - \cos \pi \sin \theta}$

$r = \dfrac{2}{1 + \sin \theta}$

Equivalent equation

Polar axis: $r = \dfrac{2}{1 + \sin(-\theta)}$

$r = \dfrac{2}{1 - \sin \theta}$

Not an equivalent equation

Pole: $-r = \dfrac{2}{1 + \sin \theta}$

Answer: Symmetric with respect to $\theta = \pi/2$

10. $r = \dfrac{3}{2 + \cos \theta}$

$\theta = \dfrac{\pi}{2}$: $-r = \dfrac{3}{2 + \cos(-\theta)}$

Not an equivalent equation

Polar axis: $r = \dfrac{3}{2 + \cos(-\theta)}$

Equivalent equation

Pole: $-r = \dfrac{3}{2 + \cos \theta}$

Not an equivalent equation

Answer: Symmetric with respect to polar axis

11. $r^2 = 16 \cos 2\theta$

$\theta = \dfrac{\pi}{2}$: $(-r)^2 = 16 \cos 2(-\theta)$

$r^2 = 16 \cos 2\theta$

Equivalent equation

Polar axis: $r^2 = 16 \cos 2(-\theta)$

$r^2 = 16 \cos 2\theta$

Equivalent equation

Pole: $(-r)^2 = 16 \cos 2\theta$

$r^2 = 16 \cos 2\theta$

Equivalent equation

Answer: Symmetric with respect to $\theta = \dfrac{\pi}{2}$, the
polar axis, and the pole

12. $r^2 = 36 \sin 2\theta$

$\theta = \dfrac{\pi}{2}$: $(-r)^2 = 36 \sin(-2\theta)$

Not an equivalent equation

Polar axis: $r^2 = 36 \sin(-2\theta)$

Not an equivalent equation

Pole: $(-r)^2 = 36 \sin 2\theta$

Equivalent equation

Answer: Symmetric with respect to pole

13. $|r| = |10(1 - \sin \theta)| = 10|1 - \sin \theta| \le 10(2) = 20$

$|1 - \sin \theta| = 2$

$1 - \sin \theta = 2$ or $1 - \sin \theta = -2$

$\sin \theta = -1$ $\sin \theta = 3$

$\theta = \dfrac{3\pi}{2}$ Not possible

Maximum: $|r| = 20$ when $\theta = \dfrac{3\pi}{2}$

$0 = 10(1 - \sin \theta)$

$\sin \theta = 1$

$\theta = \dfrac{\pi}{2}$

Zero: $r = 0$ when $\theta = \dfrac{\pi}{2}$

14. $|r| = |6 + 12 \cos \theta| \le |6| + |12 \cos \theta|$

$= 6 + 12|\cos \theta| \le 18$

$\cos \theta = 1$

$\theta = 0$

Maximum: $|r| = 18$ when $\theta = 0$

$0 = 6 + 12 \cos \theta$

$\cos \theta = -\dfrac{1}{2}$

$\theta = \dfrac{2\pi}{3}, \dfrac{4\pi}{3}$

Zero: $r = 0$ when $\theta = \dfrac{2\pi}{3}, \dfrac{4\pi}{3}$

15. $|r| = |4 \cos 3\theta| = 4|\cos 3\theta| \leq 4$

$|\cos 3\theta| = 1$

$\cos 3\theta = \pm 1$

$$\theta = 0, \frac{\pi}{3}, \frac{2\pi}{3}$$

Maximum: $|r| = 4$ when $\theta = 0, \frac{\pi}{3}, \frac{2\pi}{3}$

$0 = 4 \cos 3\theta$

$\cos 3\theta = 0$

$$\theta = \frac{\pi}{6}, \frac{\pi}{2}, \frac{5\pi}{6}$$

Zero: $r = 0$ when $\theta = \frac{\pi}{6}, \frac{\pi}{2}, \frac{5\pi}{6}$

16. $|r| = |3 \sin 2\theta| = 3|\sin 2\theta| \leq 3$

$|\sin 2\theta| = 1$

$\sin 2\theta = \pm 1$

$$\theta = \frac{\pi}{4}, \frac{3\pi}{4}, \frac{5\pi}{4}, \frac{7\pi}{4}$$

Maximum: $|r| = 3$ when $\theta = \frac{\pi}{4}, \frac{3\pi}{4}, \frac{5\pi}{4}, \frac{7\pi}{4}$

$0 = 3 \sin 2\theta$

$\sin 2\theta = 0$

$$\theta = 0, \frac{\pi}{2}, \pi, \frac{3\pi}{2}$$

Zero: $r = 0$ when $\theta = 0, \frac{\pi}{2}, \pi, \frac{3\pi}{2}$

17. Circle: $r = 5$

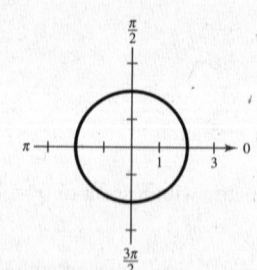

18. Circle: $r = 2$

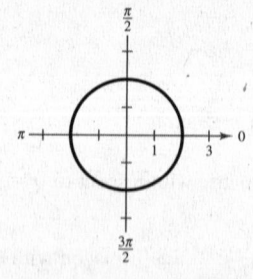

19. Circle: $r = \frac{\pi}{6}$

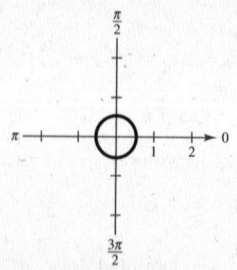

20. $r = -\frac{3\pi}{4}$

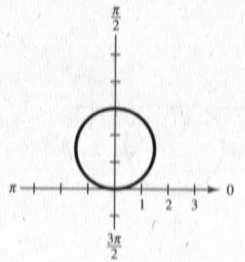

21. $r = 3 \sin \theta$

Symmetric with respect to $\theta = \pi/2$

Circle with a radius of $3/2$

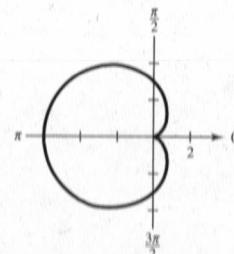

22. $r = 4 \cos \theta$

Symmetric with respect to polar axis

Circle with radius 2

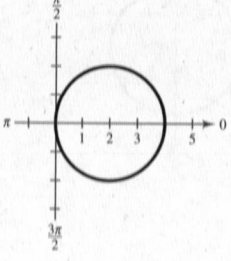

23. $r = 3(1 - \cos \theta)$

Symmetric with respect to the polar axis

$\dfrac{a}{b} = \dfrac{3}{3} = 1 \Rightarrow$ Cardioid

$|r| = 6$ when $\theta = \pi$

$r = 0$ when $\theta = 0$

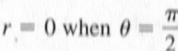

24. $r = 4(1 - \sin \theta)$

Symmetric with respect to $\pi/2$

$\dfrac{a}{b} = \dfrac{4}{4} = 1 \Rightarrow$ Cardioid

$|r| = 8$ when $\theta = \dfrac{3\pi}{2}$

$r = 0$ when $\theta = \dfrac{\pi}{2}$

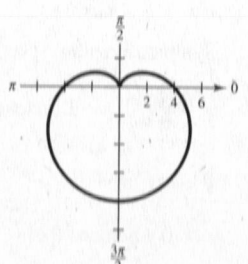

25. $r = 4(1 + \sin \theta)$

Symmetric with respect to $\theta = \dfrac{\pi}{2}$

$\dfrac{a}{b} = \dfrac{4}{4} = 1 \implies$ Cardioid

$|r| = 8$ when $\theta = \dfrac{\pi}{2}$

$r = 0$ when $\theta = \dfrac{3\pi}{2}$

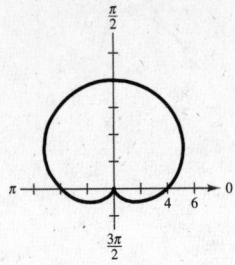

26. $r = 2(1 + \cos \theta)$

Symmetric with respect to polar axis

$\dfrac{a}{b} = \dfrac{2}{2} = 1 \implies$ Cardioid

$|r| = 4$ when $\theta = 0$

$r = 0$ when $\theta = \pi$

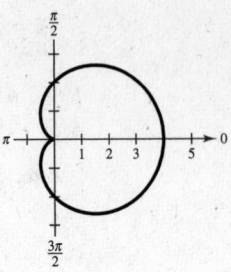

27. $r = 3 + 6 \sin \theta$

Symmetric with respect to $\theta = \dfrac{\pi}{2}$

$\dfrac{a}{b} = \dfrac{3}{6} < 1 \implies$ Limaçon with inner loop

$|r| = 9$ when $\theta = \dfrac{\pi}{2}$

$r = 0$ when $\theta = \dfrac{7\pi}{6}, \dfrac{11\pi}{6}$

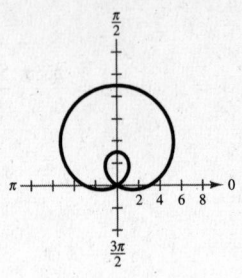

28. $r = 4 - 3 \sin \theta$

Symmetric with respect to $\pi/2$

$a = 4, b = 3$

$\dfrac{a}{b} = \dfrac{4}{3} \implies$ Dimpled limaçon

$|r| = 7$ when $\theta = \dfrac{3\pi}{2}$

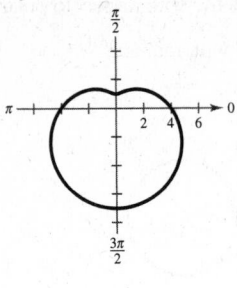

29. $r = 1 - 2 \sin \theta$

Symmetric with respect to $\theta = \dfrac{\pi}{2}$

$\dfrac{a}{b} = \dfrac{1}{2} < 1 \implies$ Limaçon with inner loop

$|r| = 3$ when $\theta = \dfrac{3\pi}{2}$

$r = 0$ when $\theta = \dfrac{\pi}{6}, \dfrac{5\pi}{6}$

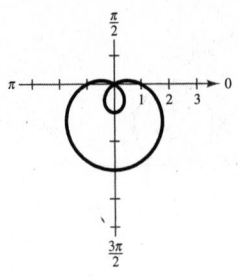

30. $r = 1 - 2 \cos \theta$

Symmetric with respect to the polar axis

$\dfrac{a}{b} = \dfrac{1}{2} \implies$ Limaçon with inner loop

$|r| = 3$ when $\theta = \pi$

$r = 0$ when $\theta = \dfrac{\pi}{3}, \dfrac{5\pi}{3}$

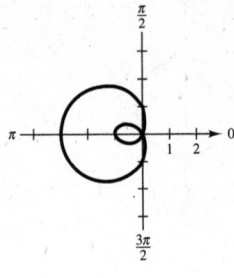

31. $r = 3 - 4 \cos \theta$

Symmetric with respect to the polar axis

$\dfrac{a}{b} = \dfrac{3}{4} < 1 \implies$ Limaçon with inner loop

$|r| = 7$ when $\theta = \pi$

$r = 0$ when $\cos \theta = \dfrac{3}{4}$ or $\theta \approx 0.723, 5.560$

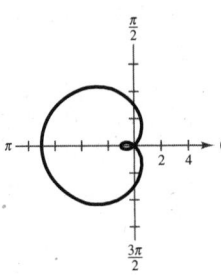

32. $r = 4 + 3\cos\theta$

Symmetric with respect to the polar axis

$\dfrac{a}{b} = \dfrac{4}{3} > 1 \implies$ Dimpled limaçon

$|r| = 7$ when $\theta = 0$

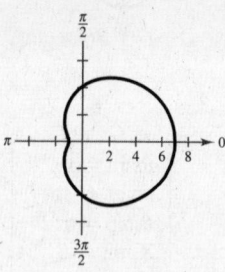

33. $r = 5\sin 2\theta$

Symmetric with respect to $\theta = \pi/2$, the polar axis, and the pole

Rose curve ($n = 2$) with 4 petals

$|r| = 5$ when $\theta = \dfrac{\pi}{4}, \dfrac{3\pi}{4}, \dfrac{5\pi}{4}, \dfrac{7\pi}{4}$

$r = 0$ when $\theta = 0, \dfrac{\pi}{2}, \pi$

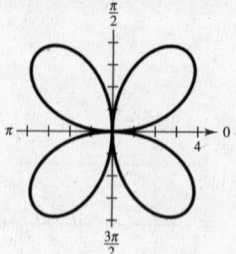

34. $r = 3\cos 2\theta$

Symmetric with respect to the polar axis

Rose curve ($n = 2$) with four petals

$|r| = 3$ when $\theta = 0, \dfrac{\pi}{2}, \pi, \dfrac{3\pi}{2}$

$r = 0$ when $\theta = \dfrac{\pi}{4}, \dfrac{3\pi}{4}, \dfrac{5\pi}{4}, \dfrac{7\pi}{4}$

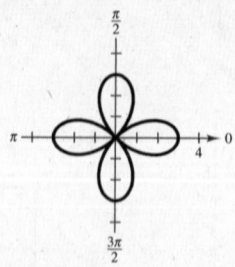

35. $r = 2\sec\theta$

$r = \dfrac{2}{\cos\theta}$

$r\cos\theta = 2$

$x = 2 \implies$ Line

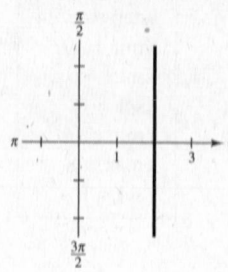

36. $r = 5\csc\theta$

$r\sin\theta = 5$

$y = 5 \implies$ Line

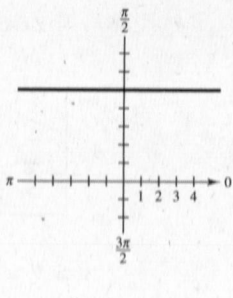

37. $r = \dfrac{3}{\sin\theta - 2\cos\theta}$

$r(\sin\theta - 2\cos\theta) = 3$

$y - 2x = 3$

$y = 2x + 3 \implies$ Line

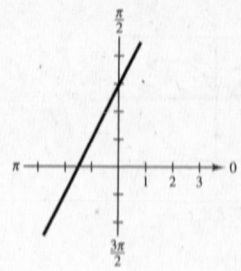

38. $r = \dfrac{6}{2\sin\theta - 3\cos\theta}$

$r(2\sin\theta - 3\cos\theta) = 6$

$2y - 3x = 6$

$y = \dfrac{3}{2}x + 3 \implies$ Line

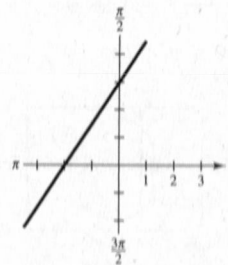

39. $r^2 = 9 \cos 2\theta$

Symmetric with respect to the polar axis, $\theta = \pi/2$, and the pole

Lemniscate

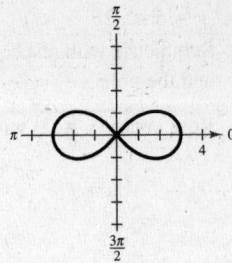

40. $r^2 = 4 \sin \theta$

$r = 2\sqrt{\sin \theta}$

$r = -2\sqrt{\sin \theta}$

$0 \le \theta \le \pi$

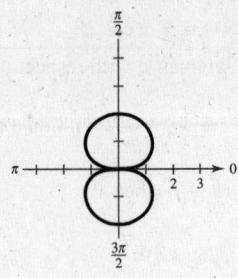

41. $r = 8 \cos \theta$

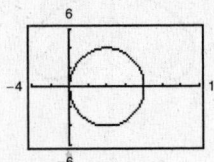

$0 \le \theta \le 2\pi$

θmin = 0
θmax = 2π
θstep = $\pi/24$
Xmin = -4
Xmax = 14
Xscl = 2
Ymin = -6
Ymax = 6
Yscl = 2

42. $r = \cos 2\theta$

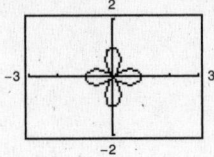

$0 \le \theta \le 2\pi$

θmin = 0
θmax = 2π
θstep = $\pi/24$
Xmin = -3
Xmax = 3
Xscl = 1
Ymin = -2
Ymax = 2
Yscl = 1

43. $r = 3(2 - \sin \theta)$

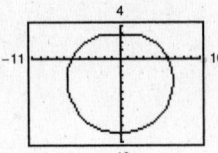

$0 \le \theta \le 2\pi$

θmin = 0
θmax = 2π
θstep = $\pi/24$
Xmin = -10
Xmax = 10
Xscl = 1
Ymin = -10
Ymax = 4
Yscl = 1

44. $r = 2 \cos(3\theta - 2)$

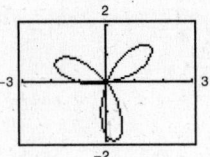

$0 \le \theta \le \pi$

θmin = 0
θmax = π
θstep = $\pi/24$
Xmin = -2
Xmax = 2
Xscl = 1
Ymin = -2
Ymax = 2
Yscl = 1

45. $r = 8 \sin \theta \cos^2 \theta$

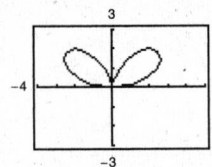

$0 \le \theta \le 2\pi$

θmin = 0
θmax = 2π
θstep = $\pi/24$
Xmin = -4
Xmax = 4
Xscl = 1
Ymin = -3
Ymax = 3
Yscl = 1

46. $r = 2 \csc \theta + 5 = \dfrac{2}{\sin \theta} + 5$

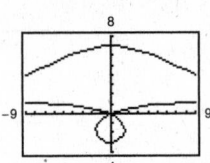

$0 \le \theta \le 2\pi$

θmin = 0
θmax = 2π
θstep = $\pi/24$
Xmin = -9
Xmax = 9
Xscl = 1
Ymin = -4
Ymax = 8
Yscl = 1

47. $r = 3 - 4 \cos \theta$

$0 \le \theta < 2\pi$

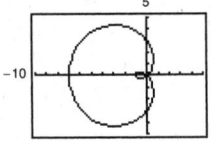

48. $r = 5 + 4 \cos \theta$

$0 \le \theta < 2\pi$

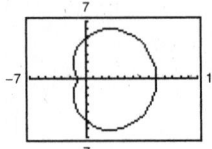

49. $r = 2 \cos\left(\dfrac{3\theta}{2}\right)$

$0 \le \theta < 4\pi$

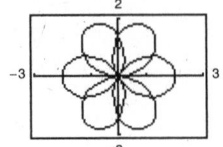

50. $r = 3 \sin\left(\dfrac{5\theta}{2}\right)$

$0 \le \theta < 4\pi$

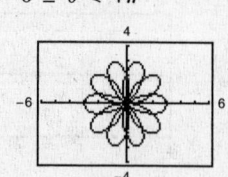

51. $r^2 = 9 \sin 2\theta$

$0 \le \theta < \pi$

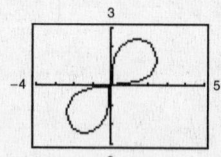

52. $r^2 = \dfrac{1}{\theta}$

$0 < \theta < \infty$

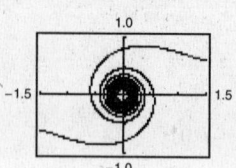

53.
$$r = 2 - \sec\theta = 2 - \frac{1}{\cos\theta}$$
$$r\cos\theta = 2\cos\theta - 1$$
$$r(r\cos\theta) = 2r\cos\theta - r$$
$$\left(\pm\sqrt{x^2 + y^2}\right)x = 2x - \left(\pm\sqrt{x^2 + y^2}\right)$$
$$\left(\pm\sqrt{x^2 + y^2}\right)(x + 1) = 2x$$
$$\left(\pm\sqrt{x^2 + y^2}\right) = \frac{2x}{x + 1}$$
$$x^2 + y^2 = \frac{4x^2}{(x + 1)^2}$$
$$y^2 = \frac{4x^2}{(x + 1)^2} - x^2$$
$$= \frac{4x^2 - x^2(x + 1)^2}{(x + 1)^2} = \frac{4x^2 - x^2(x^2 + 2x + 1)}{(x + 1)^2}$$
$$= \frac{-x^4 - 2x^3 + 3x^2}{(x + 1)^2} = \frac{-x^2(x^2 + 2x - 3)}{(x + 1)^2}$$
$$y = \pm\sqrt{\frac{x^2(3 - 2x - x^2)}{(x + 1)^2}} = \pm\left|\frac{x}{x + 1}\right|\sqrt{3 - 2x - x^2}$$

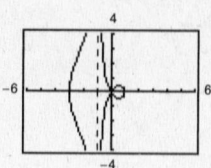

The graph has an asymptote at $x = -1$.

54.
$$r = 2 + \csc\theta = 2 + \frac{1}{\sin\theta}$$
$$r\sin\theta = 2\sin\theta + 1$$
$$r(r\sin\theta) = 2r\sin\theta + r$$
$$\left(\pm\sqrt{x^2 + y^2}\right)(y) = 2y + \left(\pm\sqrt{x^2 + y^2}\right)$$
$$\left(\pm\sqrt{x^2 + y^2}\right)(y - 1) = 2y$$
$$\left(\pm\sqrt{x^2 + y^2}\right) = \frac{2y}{y - 1}$$
$$x^2 + y^2 = \frac{4y^2}{(y - 1)^2}$$
$$x^2 = \frac{y^2(3 + 2y - y^2)}{(y - 1)^2}$$
$$x = \pm\sqrt{\frac{y^2(3 + 2y - y^2)}{(y - 1)^2}} = \pm\left|\frac{y}{y - 1}\right|\sqrt{3 + 2y - y^2}$$

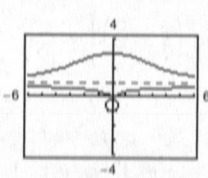

The graph has an asymptote at $y = 1$.

55. $r = \dfrac{3}{\theta}$

$\theta = \dfrac{3}{r} = \dfrac{3 \sin \theta}{r \sin \theta} = \dfrac{3 \sin \theta}{y}$

$y = \dfrac{3 \sin \theta}{\theta}$

As $\theta \to 0, y \to 3$

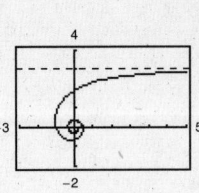

56. $r = 2 \cos 2\theta \sec \theta = \dfrac{2 \cos 2\theta}{\cos \theta}$

$r = \dfrac{2(\cos^2 \theta - \sin^2 \theta)}{\cos \theta}$

$r \cos \theta = 2(\cos^2 \theta - \sin^2 \theta)$

$x = 2(\cos^2 \theta - \sin^2 \theta)$

As $\theta \to \dfrac{\pi}{2}, x \to -2$.

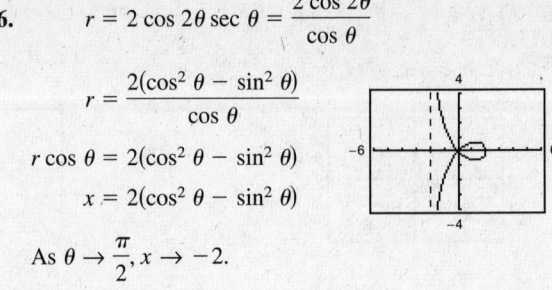

57. True. For a graph to have polar axis symmetry, replace (r, θ) by $(r, -\theta)$ or $(-r, \pi - \theta)$.

58. False. For a graph to be symmetric about the pole, one portion of the graph coincides with the other portion when rotated π radians about the pole.

59. $r = 6 \cos \theta$

(a) $0 \le \theta \le \dfrac{\pi}{2}$

Upper half of circle

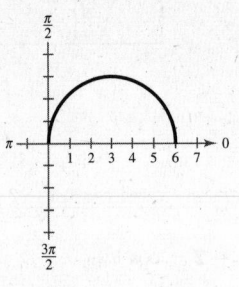

(b) $\dfrac{\pi}{2} \le \theta \le \pi$

Lower half of circle

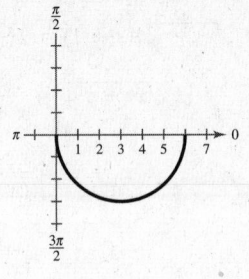

(c) $-\dfrac{\pi}{2} \le \theta \le \dfrac{\pi}{2}$

Entire circle

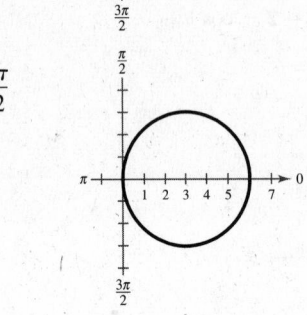

(d) $\dfrac{\pi}{4} \le \theta \le \dfrac{3\pi}{4}$

Left half of circle

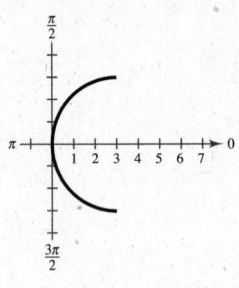

60. $r = 6[1 + \cos(\theta - \phi)]$

(a)

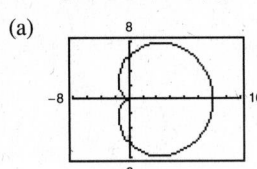

(b)

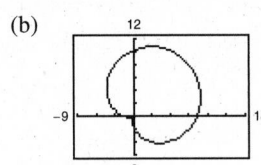

(c)

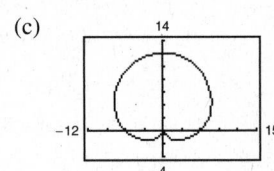

The angle ϕ has the effect of rotating the graph by the angle ϕ. For part (c),

$$r = 6\left[1 + \cos\left(\theta - \dfrac{\pi}{2}\right)\right] = 6(1 + \sin \theta).$$

61. Let the curve $r = f(\theta)$ be rotated by ϕ to form the curve $r = g(\theta)$. If (r_1, θ_1) is a point on $r = f(\theta)$, then $(r_1, \theta_1 + \phi)$ is on $r = g(\theta)$. That is, $g(\theta_1 + \phi) = r_1 = f(\theta_1)$. Letting $\theta = \theta_1 + \phi$, or $\theta_1 = \theta - \phi$, we see that $g(\theta) = g(\theta_1 + \phi) = f(\theta_1) = f(\theta - \phi)$.

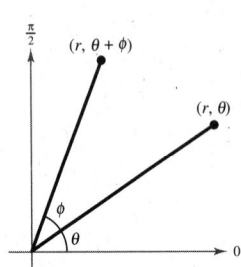

62. Use the result of Exercise 61.

(a) Rotation: $\phi = \dfrac{\pi}{2}$

Original graph: $r = f(\sin\theta)$

Rotated graph: $r = f\left(\sin\left(\theta - \dfrac{\pi}{2}\right)\right) = f(-\cos\theta)$

(b) Rotation: $\phi = \pi$

Original graph: $r = f(\sin\theta)$

Rotated graph: $r = f(\sin(\theta - \pi)) = f(-\sin\theta)$

(c) Rotation: $\phi = \dfrac{3\pi}{2}$

Original graph: $r = f(\sin\theta)$

Rotated graph: $r = f\left(\sin\left(\theta - \dfrac{3\pi}{2}\right)\right) = f(\cos\theta)$

63. (a) $r = 2 - \sin\left(\theta - \dfrac{\pi}{4}\right)$

$\quad = 2 - \left[\sin\theta\cos\dfrac{\pi}{4} - \cos\theta\sin\dfrac{\pi}{4}\right]$

$\quad = 2 - \dfrac{\sqrt{2}}{2}(\sin\theta - \cos\theta)$

(c) $r = 2 - \sin(\theta - \pi)$

$\quad = 2 - [\sin\theta\cos\pi - \cos\theta\sin\pi]$

$\quad = 2 + \sin\theta$

(b) $r = 2 - \sin\left(\theta - \dfrac{\pi}{2}\right)$

$\quad = 2 - \left[\sin\theta\cos\dfrac{\pi}{2} - \cos\theta\sin\dfrac{\pi}{2}\right]$

$\quad = 2 + \cos\theta$

(d) $r = 2 - \sin\left(\theta - \dfrac{3\pi}{2}\right)$

$\quad = 2 - \left[\sin\theta\cos\dfrac{3\pi}{2} - \cos\theta\sin\dfrac{3\pi}{2}\right]$

$\quad = 2 - \cos\theta$

64. $r = 2\sin 2\theta$

(a) $r = 2\sin\left[2\left(\theta - \dfrac{\pi}{6}\right)\right]$

$\quad = 2\left[2\sin\left(\theta - \dfrac{\pi}{6}\right)\cos\left(\theta - \dfrac{\pi}{6}\right)\right]$

$\quad = 4\sin\left(\theta - \dfrac{\pi}{6}\right)\cos\left(\theta - \dfrac{\pi}{6}\right)$

(c) $r = 2\sin\left[2\left(\theta - \dfrac{2\pi}{3}\right)\right]$

$\quad = 2\left[2\sin\left(\theta - \dfrac{2\pi}{3}\right)\cos\left(\theta - \dfrac{2\pi}{3}\right)\right]$

$\quad = 4\sin\left(\theta - \dfrac{2\pi}{3}\right)\cos\left(\theta - \dfrac{2\pi}{3}\right)$

(b) $r = 2\sin\left[2\left(\theta - \dfrac{\pi}{2}\right)\right]$

$\quad = 2\sin(2\theta - \pi)$

$\quad = -2\sin 2\theta$

$\quad = -2(2\sin\theta\cos\theta)$

$\quad = -4\sin\theta\cos\theta$

(d) $r = 2\sin[2(\theta - \pi)]$

$\quad = 2\sin(2\theta - 2\pi)$

$\quad = 2\sin 2\theta$

$\quad = 2[2\sin\theta\cos\theta]$

$\quad = 4\sin\theta\cos\theta$

65. (a) $r = 1 - \sin\theta$

(b) $r = 1 - \sin\left(\theta - \dfrac{\pi}{4}\right)$

Rotate the graph in part (a) through the angle $\dfrac{\pi}{4}$.

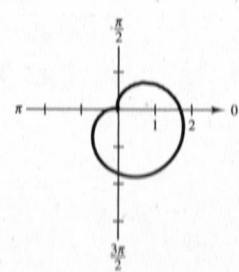

66. (a) $r = 3 \sec \theta$ **(b)** $r = 3 \sec\left(\theta - \dfrac{\pi}{4}\right)$

$$r = \frac{3}{\cos \theta}$$

$$r \cos \theta = 3 \implies x = 3$$

$$r = \frac{3}{\cos(\theta - (\pi/4))}$$

$$r = \frac{3}{\cos \theta \cos(\pi/4) + \sin \theta \sin(\pi/4)}$$

$$\frac{\sqrt{2}}{2} r \cos \theta + \frac{\sqrt{2}}{2} r \sin \theta = 3$$

$$\frac{\sqrt{2}}{2} x + \frac{\sqrt{2}}{2} y = 3$$

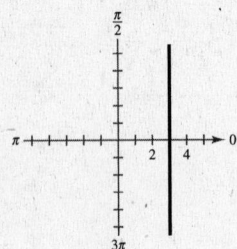

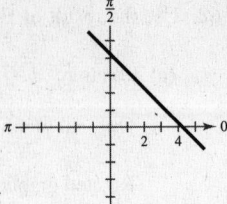

(c) $r = 3 \sec\left(\theta + \dfrac{\pi}{3}\right)$

$$r = \frac{3}{\cos(\theta + (\pi/3))}$$

$$r = \frac{3}{\cos \theta \cos(\pi/3) - \sin \theta \sin(\pi/3)}$$

$$\frac{1}{2} r \cos \theta - \frac{\sqrt{3}}{2} r \sin \theta = 3$$

$$\frac{1}{2} x - \frac{\sqrt{3}}{2} y = 3$$

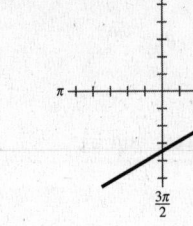

(d) $r = 3 \sec\left(\theta - \dfrac{\pi}{2}\right)$

$$r = \frac{3}{\cos(\theta - (\pi/2))}$$

$$r = \frac{3}{\cos \theta \cos(\pi/2) + \sin \theta \sin(\pi/2)}$$

$$r \sin \theta = 3 \implies y = 3$$

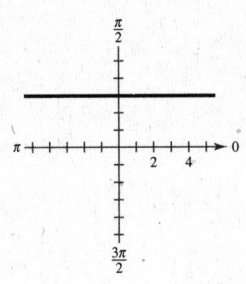

67. $r = 2 + k \sin \theta$

$k = 0$: $r = 2$

    Circle

$k = 1$: $r = 2 + \sin \theta$

    Convex limaçon

$k = 2$: $r = 2 + 2 \sin \theta$

    Cardioid

$k = 3$: $r = 2 + 3 \sin \theta$

    Limaçon with inner loop

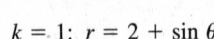

68. $r = 3 \sin k\theta$

(a) $r = 3 \sin 1.5\theta$

    $0 \leq \theta < 4\pi$

(b) $r = 3 \sin 2.5\theta$

    $0 \leq \theta < 4\pi$

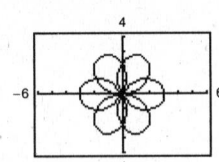

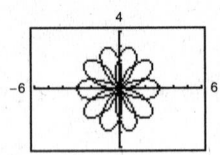

(c) Yes. $r = 3 \sin(k\theta)$.

Find the minimum value of θ, $(\theta > 0)$, that is a multiple of 2π that makes $k\theta$ a multiple of 2π.

69. $\quad y = \dfrac{x^2 - 9}{x + 1}$ **70.** $\;\; y = 6 + \dfrac{4}{x^2 + 4}$ **71.** $\qquad y = 5 - \dfrac{3}{x - 2}$ **72.** $\;\; y = \dfrac{x^3 - 27}{x^2 + 4}$

$\dfrac{x^2 - 9}{x + 1} = 0$ No zeros $5 - \dfrac{3}{x - 2} = 0$ Zero: $x = 3$

$x^2 - 9 = 0$

$\qquad x^2 = 9 \qquad\qquad\qquad\qquad\qquad\qquad\qquad 5 = \dfrac{3}{x - 2}$

$\qquad\quad x = \pm 3 \qquad\qquad\qquad\qquad\qquad\qquad 5(x - 2) = 3$

$\qquad\qquad\qquad\qquad\qquad\qquad\qquad\qquad\quad 5x - 10 = 3$

$\qquad\qquad\qquad\qquad\qquad\qquad\qquad\qquad\qquad 5x = 13$

$\qquad\qquad\qquad\qquad\qquad\qquad\qquad\qquad\qquad\;\; x = \dfrac{13}{5}$

73. Vertices: $(-4, 2), (2, 2) \Rightarrow$ Center at $(-1, 2)$ and $a = 3$ **74.** Foci: $(3, 2), (3, -4)$; Major axis of length 8

Minor axis of length 4: $2b = 4 \Rightarrow b = 2$ Center: $(h, k) = (3, -1)$

Horizontal major axis   Vertical major axis

$\dfrac{(x - h)^2}{a^2} + \dfrac{(y - k)^2}{b^2} = 1$ $a = 4, c = 3, b^2 = a^2 - c^2 = 16 - 9 \Rightarrow b = \sqrt{7}$

$\dfrac{(x + 1)^2}{9} + \dfrac{(y - 2)^2}{4} = 1$ 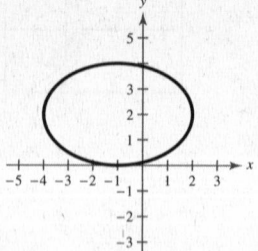 $\dfrac{(x - h)^2}{b^2} + \dfrac{(y - k)^2}{a^2} = 1$

$\qquad\qquad\qquad\qquad\qquad\qquad\qquad\qquad\qquad\qquad\qquad\qquad \dfrac{(x - 3)^2}{7} + \dfrac{(y + 1)^2}{16} = 1$ 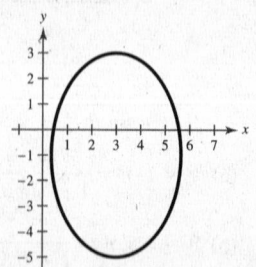

Section 10.9   Polar Equations of Conics

■   The graph of a polar equation of the form

$$r = \frac{ep}{1 \pm e \cos \theta} \quad \text{or} \quad r = \frac{ep}{1 \pm e \sin \theta}$$

is a conic, where $e > 0$ is the eccentricity and $|p|$ is the distance between the focus (pole) and the directrix.

    (a)   If $e < 1$, the graph is an ellipse.

    (b)   If $e = 1$, the graph is a parabola.

    (c)   If $e > 1$, the graph is a hyperbola.

■   Guidelines for finding polar equations of conics:

    (a)   Horizontal directrix above the pole: $r = \dfrac{ep}{1 + e \sin \theta}$

    (b)   Horizontal directrix below the pole: $r = \dfrac{ep}{1 - e \sin \theta}$

    (c)   Vertical directrix to the right of the pole: $r = \dfrac{ep}{1 + e \cos \theta}$

    (d)   Vertical directrix to the left of the pole: $r = \dfrac{ep}{1 - e \cos \theta}$

Vocabulary Check

1. conic

2. eccentricity; e

3. vertical; right

4. (a) iii (b) i (c) ii

---

1. $r = \dfrac{4e}{1 + e \cos \theta}$

    (a) $e = 1, r = \dfrac{4}{1 + \cos \theta}$, parabola

    (b) $e = 0.5, r = \dfrac{2}{1 + 0.5 \cos \theta} = \dfrac{4}{2 + \cos \theta}$, ellipse

    (c) $e = 1.5, r = \dfrac{6}{1 + 1.5 \cos \theta} = \dfrac{12}{2 + 3 \cos \theta}$, hyperbola

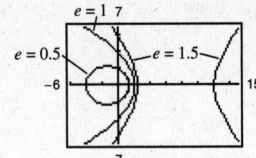

2. $r = \dfrac{4e}{1 - e \cos \theta}$

    $e = 1, r = \dfrac{4}{1 - \cos \theta}$, parabola

    $e = 0.5, r = \dfrac{2}{1 - 0.5 \cos \theta}$, ellipse

    $e = 1.5, r = \dfrac{6}{1 - 1.5 \cos \theta}$, hyperbola

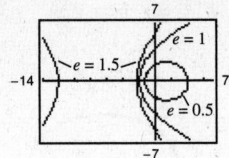

3. $r = \dfrac{4e}{1 - e \sin \theta}$

    (a) $e = 1, r = \dfrac{4}{1 - \sin \theta}$, parabola

    (b) $e = 0.5, r = \dfrac{2}{1 - 0.5 \sin \theta} = \dfrac{4}{2 - \sin \theta}$, ellipse

    (c) $e = 1.5, r = \dfrac{6}{1 - 1.5 \sin \theta} = \dfrac{12}{2 - 3 \sin \theta}$, hyperbola

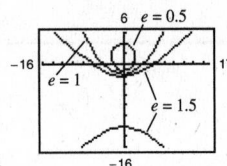

4. $r = \dfrac{4e}{1 + e \sin \theta}$

    $e = 1, r = \dfrac{4}{1 + \sin \theta}$, parabola

    $e = 0.5, r = \dfrac{2}{1 + 0.5 \sin \theta}$, ellipse

    $e = 1.5, r = \dfrac{6}{1 + 1.5 \sin \theta}$, hyperbola

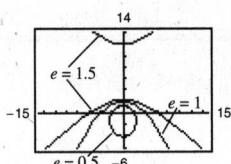

5. $r = \dfrac{2}{1 + \cos \theta}$

$e = 1 \implies$ Parabola

Vertical directrix to the right
of the pole
Matches graph (f).

6. $r = \dfrac{3}{2 - \cos \theta}$

$e = \dfrac{1}{2} \implies$ Ellipse

Vertical directrix to the left
of the pole
Matches graph (c).

7. $r = \dfrac{3}{1 + 2 \sin \theta}$

$e = 2 \implies$ Hyperbola

Matches graph (d).

8. $r = \dfrac{2}{1 - \sin \theta}$

$e = 1 \implies$ Parabola

Horizontal directrix below pole
Matches graph (e).

9. $r = \dfrac{4}{2 + \cos \theta}$

$= \dfrac{2}{1 + 0.5 \cos \theta}$

$e = 0.5 \implies$ Ellipse
Matches graph (a).

10. $r = \dfrac{4}{1 - 3 \sin \theta}$

$e = 3 \implies$ Hyperbola

Horizontal directrix below pole
Matches graph (b).

11. $r = \dfrac{2}{1 - \cos\theta}$

$e = 1$, the graph is a parabola.

Vertex: $(1, \pi)$

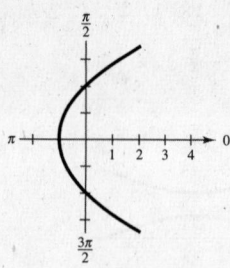

12. $r = \dfrac{3}{1 + \sin\theta}$

$e = 1 \implies$ Parabola

Vertex: $\left(\dfrac{3}{2}, \dfrac{\pi}{2}\right)$

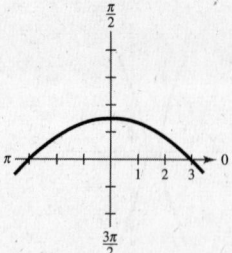

13. $r = \dfrac{5}{1 + \sin\theta}$

$e = 1$, the graph is a parabola.

Vertex: $\left(\dfrac{5}{2}, \dfrac{\pi}{2}\right)$

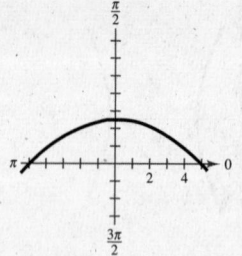

14. $r = \dfrac{6}{1 + \cos\theta}$

$e = 1 \implies$ Parabola
Vertex: $(3, 0)$

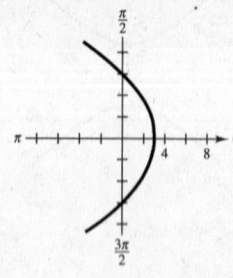

15. $r = \dfrac{2}{2 - \cos\theta} = \dfrac{1}{1 - (1/2)\cos\theta}$

$e = \dfrac{1}{2} < 1$, the graph is an ellipse.

Vertices: $(2, 0), \left(\dfrac{2}{3}, \pi\right)$

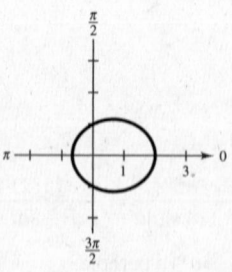

16. $r = \dfrac{3}{3 + \sin\theta} = \dfrac{1}{1 + (1/3)\sin\theta}$

$e = \dfrac{1}{3} < 1 \implies$ Ellipse

Vertices: $\left(\dfrac{3}{4}, \dfrac{\pi}{2}\right), \left(\dfrac{3}{2}, \dfrac{3\pi}{2}\right)$

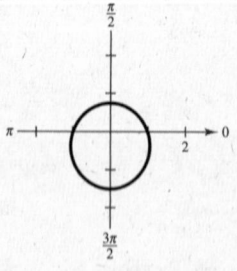

17. $r = \dfrac{6}{2 + \sin\theta} = \dfrac{3}{1 + (1/2)\sin\theta}$

$e = \dfrac{1}{2} < 1$, the graph is an ellipse.

Vertices: $\left(2, \dfrac{\pi}{2}\right), \left(6, \dfrac{3\pi}{2}\right)$

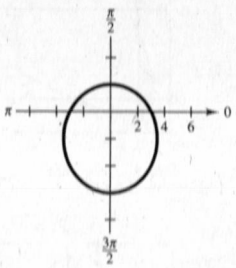

18. $r = \dfrac{9}{3 - 2\cos\theta} = \dfrac{3}{1 - (2/3)\cos\theta}$

$e = \dfrac{2}{3} < 1 \implies$ Ellipse

Vertices: $(9, 0), \left(\dfrac{9}{5}, \pi\right)$

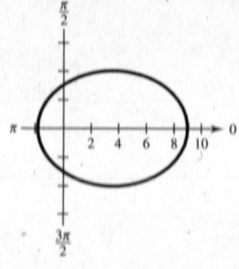

19. $r = \dfrac{3}{2 + 4\sin\theta} = \dfrac{3/2}{1 + 2\sin\theta}$

$e = 2 > 1$, the graph is a hyperbola.

Vertices: $\left(\dfrac{1}{2}, \dfrac{\pi}{2}\right), \left(-\dfrac{3}{2}, \dfrac{3\pi}{2}\right)$

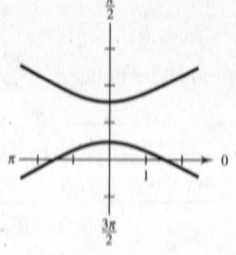

20. $r = \dfrac{5}{-1 + 2\cos\theta} = \dfrac{-5}{1 - 2\cos\theta}$

$e = 2 > 1 \implies$ Hyperbola

Vertices: $(5, 0), \left(-\dfrac{5}{3}, \pi\right)$

21. $r = \dfrac{3}{2 - 6\cos\theta} = \dfrac{3/2}{1 - 3\cos\theta}$

$e = 3 > 1$, the graph is a hyperbola.

Vertices: $\left(-\dfrac{3}{4}, 0\right), \left(\dfrac{3}{8}, \pi\right)$

22. $r = \dfrac{3}{2 + 6\sin\theta} = \dfrac{3/2}{1 + 3\sin\theta}$

$e = 3 > 1 \implies$ Hyperbola

Vertices: $\left(\dfrac{3}{8}, \dfrac{\pi}{2}\right), \left(-\dfrac{3}{4}, \dfrac{3\pi}{2}\right)$

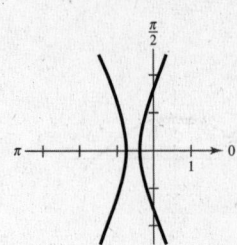

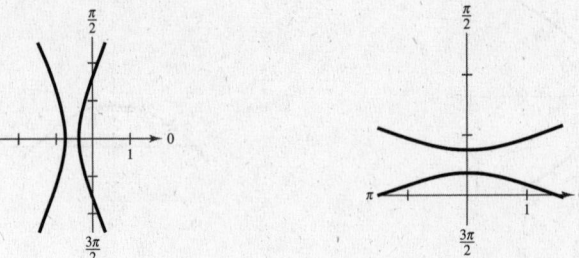

23. $r = \dfrac{4}{2 - \cos\theta} = \dfrac{2}{1 - (1/2)\cos\theta}$

$e = \dfrac{1}{2} < 1$, the graph is an ellipse.

Vertices: $(4, 0), \left(\dfrac{4}{3}, \pi\right)$

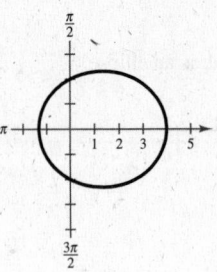

24. $r = \dfrac{2}{2 + 3\sin\theta} = \dfrac{1}{1 + (3/2)\sin\theta}$

$e = \dfrac{3}{2} > 1 \implies$ Hyperbola

Vertices: $\left(\dfrac{2}{5}, \dfrac{\pi}{2}\right), \left(-2, \dfrac{3\pi}{2}\right)$

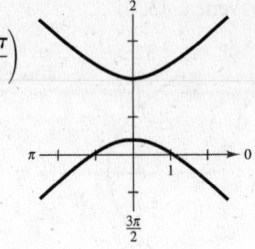

25. $r = \dfrac{-1}{1 - \sin\theta}$

$e = 1 \implies$ Parabola

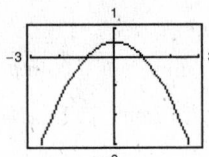

26. $r = \dfrac{-5}{2 + 4\sin\theta} = \dfrac{-(5/2)}{1 + 2\sin\theta}$

$e = 2 \implies$ Hyperbola

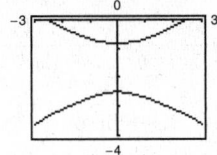

27. $r = \dfrac{3}{-4 + 2\cos\theta}$

$e = \dfrac{1}{2} \implies$ Ellipse

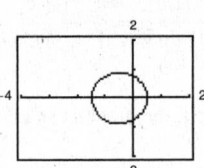

28. $r = \dfrac{4}{1 - 2\cos\theta}$

$e = 2 \implies$ Hyperbola

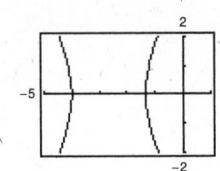

29. $r = \dfrac{2}{1 - \cos\left(\theta - \dfrac{\pi}{4}\right)}$

Rotate the graph in Exercise 11 through the angle $\pi/4$.

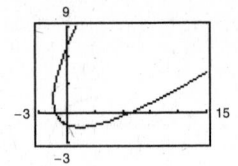

30. $r = \dfrac{3}{3 + \sin\left(\theta - \dfrac{\pi}{3}\right)}$

Rotate the graph in Exercise 16 through the angle $\pi/3$.

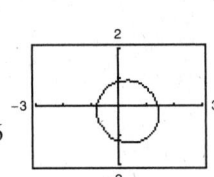

31. $r = \dfrac{6}{2 + \sin\left(\theta + \dfrac{\pi}{6}\right)}$

Rotate the graph in Exercise 17 through the angle $-\pi/6$.

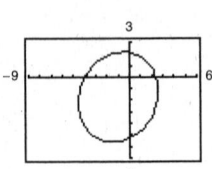

32. $r = \dfrac{5}{-1 + 2\cos\left(\theta + \dfrac{2\pi}{3}\right)}$

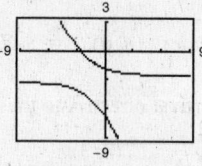

Rotate the graph in Exercise 20 through the angle $-2\pi/3$.

33. Parabola: $e = 1$

Directrix: $x = -1$

Vertical directrix to the left of the pole

$r = \dfrac{1(1)}{1 - 1\cos\theta} = \dfrac{1}{1 - \cos\theta}$

34. Parabola: $e = 1$

Directrix: $y = -2$

$p = 2$

Horizontal directrix below the pole

$r = \dfrac{1(2)}{1 - 1\sin\theta} = \dfrac{2}{1 - \sin\theta}$

35. Ellipse: $e = \dfrac{1}{2}$

Directrix: $y = 1$

$p = 1$

Horizontal directrix above the pole

$r = \dfrac{(1/2)(1)}{1 + (1/2)\sin\theta} = \dfrac{1}{2 + \sin\theta}$

36. Ellipse: $e = \dfrac{3}{4}$

Directrix: $y = -3$

$p = 3$

Horizontal directrix below the pole

$r = \dfrac{(3/4)(3)}{1 - (3/4)\sin\theta} = \dfrac{9}{4 - 3\sin\theta}$

37. Hyperbola: $e = 2$

Directrix: $x = 1$

$p = 1$

Vertical directrix to the right of the pole

$r = \dfrac{2(1)}{1 + 2\cos\theta} = \dfrac{2}{1 + 2\cos\theta}$

38. Hyperbola: $e = \dfrac{3}{2}$

Directrix: $x = -1$

$p = 1$

Vertical directrix to the left of the pole

$r = \dfrac{(3/2)(1)}{1 - (3/2)\cos\theta} = \dfrac{3}{2 - 3\cos\theta}$

39. Parabola

Vertex: $\left(1, -\dfrac{\pi}{2}\right) \Rightarrow e = 1, p = 2$

Horizontal directrix below the pole

$r = \dfrac{1(2)}{1 - 1\sin\theta} = \dfrac{2}{1 - \sin\theta}$

40. Parabola

Vertex: $(6, 0) \Rightarrow e = 1, p = 12$

Vertical directrix to the right of the pole

$r = \dfrac{1(12)}{1 + 1\cos\theta} = \dfrac{12}{1 + \cos\theta}$

41. Parabola

Vertex: $(5, \pi) \Rightarrow e = 1, p = 10$

Vertical directrix to the left of the pole

$r = \dfrac{1(10)}{1 - 1\cos\theta} = \dfrac{10}{1 - \cos\theta}$

42. Parabola

Vertex: $\left(10, \dfrac{\pi}{2}\right) \Rightarrow e = 1, p = 20$

Horizontal directrix above the pole

$r = \dfrac{1(20)}{1 + 1\sin\theta} = \dfrac{20}{1 + \sin\theta}$

43. Ellipse: Vertices $(2, 0), (10, \pi)$

Center: $(4, \pi)$; $c = 4, a = 6, e = \dfrac{2}{3}$

Vertical directrix to the right of the pole

$r = \dfrac{(2/3)p}{1 + (2/3)\cos\theta} = \dfrac{2p}{3 + 2\cos\theta}$

$2 = \dfrac{2p}{3 + 2\cos 0}$

$p = 5$

$r = \dfrac{2(5)}{3 + 2\cos\theta} = \dfrac{10}{3 + 2\cos\theta}$

44. Ellipse

Vertices: $\left(2, \dfrac{\pi}{2}\right), \left(4, \dfrac{3\pi}{2}\right)$

Center: $\left(1, \dfrac{3\pi}{2}\right)$; $c = 1$, $a = 3$, $e = \dfrac{1}{3}$

Horizontal directrix above the axis

$r = \dfrac{1/3p}{1 + (1/3)\sin\theta} = \dfrac{p}{3 + \sin\theta}$

$2 = \dfrac{p}{3 + \sin(\pi/2)}$

$p = 8$

$r = \dfrac{8}{3 + \sin\theta}$

46. Hyperbola

Vertices: $(2, 0), (8, 0)$

Center: $(5, 0)$; $c = 5$, $a = 3$, $e = \dfrac{5}{3}$

Vertical directrix to the right of the pole

$r = \dfrac{(5/3)p}{1 + (5/3)\cos\theta} = \dfrac{5p}{3 + 5\cos\theta}$

$2 = \dfrac{5p}{3 + 5\cos 0}$

$p = \dfrac{16}{5}$

$r = \dfrac{5(16/5)}{3 + 5\cos\theta} = \dfrac{16}{3 + 5\cos\theta}$

48. Hyperbola

Vertices: $\left(4, \dfrac{\pi}{2}\right), \left(1, \dfrac{\pi}{2}\right)$

Center: $\left(\dfrac{5}{2}, \dfrac{\pi}{2}\right)$; $c = \dfrac{5}{2}$, $a = \dfrac{3}{2}$, $e = \dfrac{5/2}{3/2} = \dfrac{5}{3}$

Horizontal directrix above the pole

$r = \dfrac{(5/3)p}{1 + (5/3)\sin\theta} = \dfrac{5p}{3 + 5\sin\theta}$

$1 = \dfrac{5p}{3 + 5\sin(\pi/2)}$

$p = \dfrac{8}{5}$

$r = \dfrac{5(8/5)}{3 + 5\sin\theta} = \dfrac{8}{3 + 5\sin\theta}$

45. Ellipse: Vertices $(20, 0), (4, \pi)$

Center: $(8, 0)$; $c = 8$, $a = 12$, $e = \dfrac{2}{3}$

Vertical directrix to the left of the pole

$r = \dfrac{(2/3)p}{1 - (2/3)\cos\theta} = \dfrac{2p}{3 - 2\cos\theta}$

$20 = \dfrac{2p}{3 - 2\cos 0}$

$p = 10$

$r = \dfrac{2(10)}{3 - 2\cos\theta} = \dfrac{20}{3 - 2\cos\theta}$

47. Hyperbola: Vertices $\left(1, \dfrac{3\pi}{2}\right), \left(9, \dfrac{3\pi}{2}\right)$

Center: $\left(5, \dfrac{3\pi}{2}\right)$; $c = 5$, $a = 4$, $e = \dfrac{5}{4}$

Horizontal directrix below the pole

$r = \dfrac{(5/4)p}{1 - (5/4)\sin\theta} = \dfrac{5p}{4 - 5\sin\theta}$

$1 = \dfrac{5p}{4 - 5\sin(3\pi/2)}$

$p = \dfrac{9}{5}$

$r = \dfrac{5(9/5)}{4 - 5\sin\theta} = \dfrac{9}{4 - 5\sin\theta}$

49. When $\theta = 0$, $r = c + a = ea + a = a(1 + e)$.

Therefore,

$$a(1 + e) = \dfrac{ep}{1 - e\cos 0}$$

$$a(1 + e)(1 - e) = ep$$

$$a(1 - e^2) = ep.$$

Thus, $r = \dfrac{ep}{1 - e\cos\theta} = \dfrac{(1 - e^2)a}{1 - e\cos\theta}$.

50. Minimum distance occurs when $\theta = \pi$.

$$r = \frac{(1 - e^2)a}{1 - e \cos \pi} = \frac{(1 - e)(1 + e)a}{1 + e} = a(1 - e)$$

Maximum distance occurs when $\theta = 0$.

$$r = \frac{(1 - e^2)a}{1 - e \cos 0} = \frac{(1 - e)(1 + e)a}{1 - e} = a(1 + e)$$

51. $r = \dfrac{[1 - (0.0167)^2](95.956 \times 10^6)}{1 - 0.0167 \cos \theta} \approx \dfrac{9.5929 \times 10^7}{1 - 0.0167 \cos \theta}$

Perihelion distance:
$r = 95.956 \times 10^6(1 - 0.0167) \approx 9.4354 \times 10^7$ miles

Aphelion distance:
$r = 95.956 \times 10^6(1 + 0.0167) \approx 9.7558 \times 10^7$ miles

52. $r = \dfrac{[1 - (0.0542)^2](1.427 \times 10^9)}{1 - 0.0542 \cos \theta} \approx \dfrac{1.4228 \times 10^9}{1 - 0.0542 \cos \theta}$

Perihelion distance: $r = 1.427 \times 10^9(1 - 0.0542) \approx 1.3497 \times 10^9$ kilometers

Aphelion distance: $r = 1.427 \times 10^9(1 + 0.0542) \approx 1.5043 \times 10^9$ kilometers

53. $r = \dfrac{[1 - (0.0068)^2](108.209 \times 10^6)}{1 - 0.0068 \cos \theta} \approx \dfrac{1.0820 \times 10^8}{1 - 0.0068 \cos \theta}$

Perihelion distance: $r = 108.209 \times 10^6(1 - 0.0068) \approx 1.0747 \times 10^8$ kilometers

Aphelion distance: $r = 108.209 \times 10^6(1 + 0.0068) \approx 1.0894 \times 10^8$ kilometers

54. $r = \dfrac{[1 - (0.2056)^2](35.98 \times 10^6)}{1 - 0.2056 \cos \theta} \approx \dfrac{3.4459 \times 10^7}{1 - 0.2056 \cos \theta}$

Perihelion distance: $r = 35.98 \times 10^6(1 - 0.2056) \approx 2.8583 \times 10^7$ miles

Aphelion distance: $r = 35.98 \times 10^6(1 + 0.2056) \approx 4.3377 \times 10^7$ miles

55. $r = \dfrac{[1 - (0.0934)^2](141.63 \times 10^6)}{1 - 0.0934 \cos \theta} \approx \dfrac{1.4039 \times 10^8}{1 - 0.0934 \cos \theta}$

Perihelion distance: $r = 141.63 \times 10^6(1 - 0.0934) \approx 1.2840 \times 10^8$ miles

Aphelion distance: $r = 141.63 \times 10^6(1 + 0.0934) \approx 1.5486 \times 10^8$ miles

56. $r = \dfrac{[1 - (0.0484)^2](778.41 \times 10^6)}{1 - 0.0484 \cos \theta} \approx \dfrac{7.7659 \times 10^8}{1 - 0.0484 \cos \theta}$

Perihelion distance: $r = 778.41 \times 10^6(1 - 0.0484) \approx 7.4073 \times 10^8$ kilometers

Aphelion distance: $r = 778.41 \times 10^6(1 + 0.0484) \approx 8.1609 \times 10^8$ kilometers

57. $e \approx 0.847, a \approx \dfrac{4.42}{2} = 2.21$

$2a = \dfrac{0.847p}{1 + 0.847} + \dfrac{0.847p}{1 - 0.847} \approx 5.9945p \approx 4.42$

$p \approx 0.737, ep \approx 0.624$

$r = \dfrac{0.624}{1 + 0.847 \sin \theta}$

To find the closest point to the sun, let $\theta = \dfrac{\pi}{2}$.

$r = \dfrac{0.624}{1 + 0.847 \sin(\pi/2)} \approx 0.338$ astronomical units

58. (a) $r = \dfrac{ep}{1 + e\sin\theta}$

Since the graph is a parabola, $e = 1$. The distance between the vertex and the focus (pole) is 4100, so the distance between the focus (pole) and the directrix is $p = 8200$.

$r = \dfrac{8200}{1 + \sin\theta}$

(c) When $\theta = 30°$, $r = \dfrac{8200}{1 + \sin 30°} \approx 5466.7$

Distance between surface of Earth and satellite:

$5466.7 - 4000 \approx 1467$ miles

(b)

(d) When $\theta = 60°$, $r = \dfrac{8200}{1 + \sin 60°} \approx 4394.4$

Distance between surface of Earth and satellite:

$4394.4 - 4000 \approx 394$ miles

59. True. The graphs represent the same hyperbola, although the graphs are not traced out in the same order as θ goes from 0 to 2π.

60. False. The graph has a horizontal directrix below the pole.

61. True. See Exercise 63.

$e = \frac{2}{3} < 1$

62. Answers will vary.

63.
$$\frac{x^2}{a^2} + \frac{y^2}{b^2} = 1$$
$$\frac{r^2\cos^2\theta}{a^2} + \frac{r^2\sin^2\theta}{b^2} = 1$$
$$\frac{r^2\cos^2\theta}{a^2} + \frac{r^2(1 - \cos^2\theta)}{b^2} = 1$$
$$r^2 b^2 \cos^2\theta + r^2 a^2 - r^2 a^2 \cos^2\theta = a^2 b^2$$
$$r^2(b^2 - a^2)\cos^2\theta + r^2 a^2 = a^2 b^2$$

Since $b^2 - a^2 = -c^2$, we have:

$$-r^2 c^2 \cos^2\theta + r^2 a^2 = a^2 b^2$$
$$-r^2\left(\frac{c}{a}\right)^2 \cos^2\theta + r^2 = b^2, \ e = \frac{c}{a}$$
$$-r^2 e^2 \cos^2\theta + r^2 = b^2$$
$$r^2(1 - e^2\cos^2\theta) = b^2$$
$$r^2 = \frac{b^2}{1 - e^2\cos^2\theta}$$

64.
$$\frac{x^2}{a^2} - \frac{y^2}{b^2} = 1$$
$$\frac{r^2\cos^2\theta}{a^2} - \frac{r^2\sin^2\theta}{b^2} = 1$$
$$\frac{r^2\cos^2\theta}{a^2} - \frac{r^2(1 - \cos^2\theta)}{b^2} = 1$$
$$r^2 b^2 \cos^2\theta - r^2 a^2 + r^2 a^2 \cos^2\theta = a^2 b^2$$
$$r^2(b^2 + a^2)\cos^2\theta - r^2 a^2 = a^2 b^2$$
$$a^2 + b^2 = c^2$$
$$r^2 c^2 \cos^2\theta - r^2 a^2 = a^2 b^2$$
$$r^2\left(\frac{c}{a}\right)^2 \cos^2\theta - r^2 = b^2, \ e = \frac{c}{a}$$
$$r^2 e^2 \cos^2\theta - r^2 = b^2$$
$$r^2(e^2 \cos^2\theta - 1) = b^2$$
$$r^2 = \frac{b^2}{e^2\cos^2\theta - 1}$$
$$= \frac{-b^2}{1 - e^2\cos^2\theta}$$

65. $\dfrac{x^2}{169} + \dfrac{y^2}{144} = 1$

$a = 13, b = 12, c = 5, e = \dfrac{5}{13}$

$r^2 = \dfrac{144}{1 - (25/169)\cos^2\theta} = \dfrac{24{,}336}{169 - 25\cos^2\theta}$

66. $\dfrac{x^2}{25} + \dfrac{y^2}{16} = 1$

$a = 5, b = 4, c = 3, e = \dfrac{3}{5}$

$r^2 = \dfrac{400}{25 - 9\cos^2\theta}$

67. $\dfrac{x^2}{9} - \dfrac{y^2}{16} = 1$

$a = 3, b = 4, c = 5, e = \dfrac{5}{3}$

$r^2 = \dfrac{-16}{1 - (25/9)\cos^2\theta} = \dfrac{144}{25\cos^2\theta - 9}$

68. $\dfrac{x^2}{36} - \dfrac{y^2}{4} = 1$

$a = 6, b = 2, c = 2\sqrt{10}, e = \dfrac{\sqrt{10}}{3}$

$r^2 = \dfrac{-4}{1 - (10/9)\cos^2\theta} = \dfrac{-36}{9 - 10\cos^2\theta}$

$= \dfrac{36}{10\cos^2\theta - 9}$

69. One focus: $(5, 0)$

Vertices: $(4, 0), (4, 0)$

$a = 4, c = 5 \implies b = 3$ and $e = \dfrac{5}{4}$

$$\dfrac{x^2}{16} - \dfrac{y^2}{9} = 1$$

$r^2 = \dfrac{-9}{1 - (25/16)\cos^2\theta} = \dfrac{-144}{16 - 25\cos^2\theta}$

70. Ellipse

One focus: $(4, 0)$

Vertices: $(5, 0), (5, \pi)$

$a = 5, c = 4, b = 3, e = \dfrac{4}{5}$

$r^2 = \dfrac{9}{1 - (16/25)\cos^2\theta} = \dfrac{225}{25 - 16\cos^2\theta}$

71. $r = \dfrac{4}{1 - 0.4\cos\theta}$

(a) Since $e < 1$, the conic is an ellipse.

(b) $r = \dfrac{4}{1 + 0.4\cos\theta}$ has a vertical directrix to the right

of the pole and $r = \dfrac{4}{1 - 0.4\sin\theta}$ has a horizontal

directrix below the pole. The given polar equation,

$r = \dfrac{4}{1 - 0.4\cos\theta}$, has a vertical directrix to the left

of the pole.

(c)

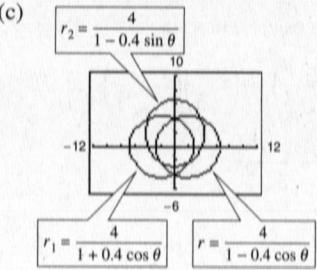

72. If e remains fixed and p changes, then the lengths of both the major axis and the minor axis change.

For example, graph $r = \dfrac{5}{1 - (2/3)\sin\theta}$, with $e = \dfrac{2}{3}$ and $p = \dfrac{15}{2}$, and

graph $r = \dfrac{6\frac{2}{3}}{1 - (2/3)\sin\theta}$, with $e = \dfrac{2}{3}$ and $p = 10$ on the same set of coordinate axes.

The first ellipse has a major axis of length 18 and a minor axis of length $6\sqrt{5}$, and the second ellipse has a major axis of length 21.6 and a minor axis of length $7.2\sqrt{5}$.

73. $4\sqrt{3}\tan\theta - 3 = 1$

$\qquad 4\sqrt{3}\tan\theta = 4$

$\qquad\qquad \tan\theta = \dfrac{1}{\sqrt{3}}$

$\qquad\qquad\quad \theta = \dfrac{\pi}{6} + n\pi$

74. $6\cos x - 2 = 1$

$\qquad \cos x = \dfrac{1}{2}$

$\qquad x = \dfrac{\pi}{3} + 2\pi n, \dfrac{5\pi}{3} + 2\pi n$

75. $12\sin^2\theta = 9$

$\qquad \sin^2\theta = \dfrac{3}{4}$

$\qquad \sin\theta = \pm\dfrac{\sqrt{3}}{2}$

$\qquad \theta = \dfrac{\pi}{3} + n\pi, \dfrac{2\pi}{3} + n\pi$

76. $9\csc^2 x - 10 = 2$

$\qquad \csc^2 x = \dfrac{4}{3}$

$\qquad \sin x = \pm\dfrac{\sqrt{3}}{2}$

$\qquad x = \dfrac{\pi}{3} + \pi n, \dfrac{2\pi}{3} + \pi n$

77. $2\cot x = 5\cos\dfrac{\pi}{2}$

$\qquad 2\cot x = 0$

$\qquad \cot x = 0$

$\qquad x = \dfrac{\pi}{2} + n\pi$

78. $\sqrt{2}\sec\theta = 2\csc\dfrac{\pi}{4}$

$\qquad \cos\theta = \dfrac{1}{2}$

$\qquad \theta = \dfrac{\pi}{3} + 2\pi n, \dfrac{5\pi}{3} + 2\pi n$

For 79–82 use the following:

u and v are in **Quadrant IV;** $\quad \sin u = -\dfrac{3}{5} \Rightarrow \cos u = \dfrac{4}{5}; \quad \cos v = \dfrac{1}{\sqrt{2}} \Rightarrow \sin v = -\dfrac{1}{\sqrt{2}}$

79. $\cos(u + v) = \cos u \cos v - \sin u \sin v$

$\qquad = \left(\dfrac{4}{5}\right)\left(\dfrac{1}{\sqrt{2}}\right) - \left(-\dfrac{3}{5}\right)\left(-\dfrac{1}{\sqrt{2}}\right)$

$\qquad = \dfrac{4}{5\sqrt{2}} - \dfrac{3}{5\sqrt{2}}$

$\qquad = \dfrac{1}{5\sqrt{2}}$

$\qquad = \dfrac{\sqrt{2}}{10}$

80. $\sin u = -\dfrac{3}{5}, \cos u = \dfrac{4}{5}$

$\qquad \cos v = \dfrac{1}{\sqrt{2}} = \dfrac{\sqrt{2}}{2}, \sin v = -\dfrac{\sqrt{2}}{2}$

$\qquad \sin(u + v) = \sin u \cos v + \cos u \sin v$

$\qquad\qquad = \left(-\dfrac{3}{5}\right)\left(\dfrac{\sqrt{2}}{2}\right) + \left(\dfrac{4}{5}\right)\left(-\dfrac{\sqrt{2}}{2}\right)$

$\qquad\qquad = \dfrac{-7\sqrt{2}}{10}$

81. $\cos(u - v) = \cos u \cos v + \sin u \sin v$

$\qquad = \left(\dfrac{4}{5}\right)\left(\dfrac{1}{\sqrt{2}}\right) + \left(-\dfrac{3}{5}\right)\left(-\dfrac{1}{\sqrt{2}}\right)$

$\qquad = \dfrac{4}{5\sqrt{2}} + \dfrac{3}{5\sqrt{2}}$

$\qquad = \dfrac{7}{5\sqrt{2}}$

$\qquad = \dfrac{7\sqrt{2}}{10}$

82. $\sin u = -\dfrac{3}{5}, \cos u = \dfrac{4}{5}$

$\qquad \cos v = \dfrac{1}{\sqrt{2}} = \dfrac{\sqrt{2}}{2}, \sin v = -\dfrac{\sqrt{2}}{2}$

$\qquad \sin(u - v) = \sin u \cos v - \cos u \sin v$

$\qquad\qquad = \left(-\dfrac{3}{5}\right)\left(\dfrac{\sqrt{2}}{2}\right) - \left(\dfrac{4}{5}\right)\left(-\dfrac{\sqrt{2}}{2}\right)$

$\qquad\qquad = \dfrac{\sqrt{2}}{10}$

83. $\sin u = \dfrac{4}{5}, \dfrac{\pi}{2} < u < \pi \Longrightarrow \cos u = -\dfrac{3}{5}$

$\sin 2u = 2 \sin u \cos u$ $\qquad$ $\cos 2u = \cos^2 u - \sin^2 u$ $\qquad$ $\tan 2u = \dfrac{\sin 2u}{\cos 2u}$

$\qquad = 2\left(\dfrac{4}{5}\right)\left(-\dfrac{3}{5}\right)$ $\qquad\qquad = \left(-\dfrac{3}{5}\right)^2 - \left(\dfrac{4}{5}\right)^2$ $\qquad\qquad = \dfrac{-24/25}{-7/25}$

$\qquad = -\dfrac{24}{25}$ $\qquad\qquad\qquad = \dfrac{9}{25} - \dfrac{16}{25} = -\dfrac{7}{25}$ $\qquad\qquad = \dfrac{24}{7}$

84. $\tan u = -\sqrt{3}, \dfrac{3\pi}{2} < u < 2\pi$

$\qquad \sin u = -\dfrac{\sqrt{3}}{2}, \cos u = \dfrac{1}{2}$

$\qquad \sin 2u = 2 \sin u \cos u = 2\left(-\dfrac{\sqrt{3}}{2}\right)\left(\dfrac{1}{2}\right) = -\dfrac{\sqrt{3}}{2}$

$\qquad \cos 2u = \cos^2 u - \sin^2 u = \left(\dfrac{1}{2}\right)^2 - \left(-\dfrac{\sqrt{3}}{2}\right)^2 = -\dfrac{1}{2}$

$\qquad \tan 2u = \dfrac{2 \tan u}{1 - \tan^2 u} = \dfrac{2(-\sqrt{3})}{1 - \left(-\sqrt{3}\right)^2} = \sqrt{3}$

85. $a_1 = 0, d = -\dfrac{1}{4}$

$\qquad a_n = a_1 + (n-1)d$

$\qquad\quad = 0 + (n-1)\left(-\dfrac{1}{4}\right)$

$\qquad\quad = -\dfrac{1}{4}n + \dfrac{1}{4}$

86. $a_n = a_1 + d(n-1)$

$\quad a_n = 13 + 3(n-1)$

$\quad a_n = 13 + 3n - 3$

$\quad a_n = 10 + 3n$

87. $a_3 = 27, a_8 = 72$

$\quad a_8 = a_3 + 5d$

$\quad 72 = 27 + 5d \Longrightarrow d = 9$

$\quad a_1 = 27 - 2(9) = 9$

$\quad a_n = a_1 + (n-1)d$

$\qquad = 9 + (n-1)(9)$

$\qquad = 9n$

88. $a_n = a_k + d(n-k)$

$\quad a_4 = a_1 + d(4-1)$

$\quad 9.5 = 5 + d(3)$

$\qquad d = 1.5$

$\quad a_n = 5 + 1.5(n-1)$

89. ${}_{12}C_9 = \dfrac{12!}{(12-9)!9!} = \dfrac{12 \cdot 11 \cdot 10}{3!} = 220$

90. ${}_{18}C_{16} = \dfrac{18!}{16!(2!)} = \dfrac{18 \cdot 17}{2} = 153$

91. ${}_{10}P_3 = \dfrac{10!}{(10-3)!} = \dfrac{10!}{7!} = 10 \cdot 9 \cdot 8 = 720$

92. ${}_{29}P_2 = 29 \cdot 28 = 812$

Review Exercises for Chapter 10

1. Points: $(-1, 2)$ and $(2, 5)$

$\quad m = \dfrac{5-2}{2-(-1)} = \dfrac{3}{3} = 1$

$\quad \tan \theta = 1 \Longrightarrow \theta = \dfrac{\pi}{4} \text{ radian} = 45°$

2. $m = \dfrac{4-7}{3-(-2)} = -\dfrac{3}{5} = \tan \theta$

$\quad \theta = \pi + \arctan\left(-\dfrac{3}{5}\right)$

$\quad \approx 2.6012, \text{ or about } 149.04°$

3. $y = 2x + 4 \Longrightarrow m = 2$

$\quad \tan \theta = 2 \Longrightarrow \theta = \arctan 2 \approx 1.1071 \text{ radians} \approx 63.43°$

4. $6x - 7y - 5 = 0$

$\quad m = \dfrac{6}{7} = \tan \theta$

$\quad \theta = \arctan \dfrac{6}{7} \approx 0.7086, \text{ or about } 40.60°$

5. $4x + y = 2 \implies y = -4x + 2 \implies m_1 = -4$

$-5x + y = -1 \implies y = 5x - 1 \implies m_2 = 5$

$\tan \theta = \left| \dfrac{5 - (-4)}{1 + (-4)(5)} \right| = \dfrac{9}{19}$

$\theta = \arctan \dfrac{9}{19} \approx 0.4424$ radian $\approx 25.35°$

6. $-5x + 3y = 3$

$-2x + 3y = 1$

$m_1 = \dfrac{5}{3}$

$m_2 = \dfrac{2}{3}$

$\tan \theta = \left| \dfrac{(5/3) - (2/3)}{1 + (5/3)(2/3)} \right| = \dfrac{9}{19}$

$\theta \approx 0.4424$, or about $25.35°$

7. $2x - 7y = 8 \implies y = \dfrac{2}{7}x - \dfrac{8}{7} \implies m_1 = \dfrac{2}{7}$

$0.4x + y = 0 \implies y = -0.4x \implies m_2 = -0.4$

$\tan \theta = \left| \dfrac{-0.4 - (2/7)}{1 + (2/7)(-0.4)} \right| = \dfrac{24}{31}$

$\theta = \arctan \left(\dfrac{24}{31} \right) \approx 0.6588$ radian $\approx 37.75°$

8. $0.02x + 0.07y = 0.18$

$0.09x - 0.04y = 0.17$

$m_1 = -\dfrac{2}{7}$

$m_2 = \dfrac{9}{4}$

$\tan \theta = \left| \dfrac{(9/4) - (-2/7)}{1 + (-2/7)(9/4)} \right| = \dfrac{71}{10}$

$\theta \approx 1.4309$, or about $81.98°$

9. $(1, 2) \implies x_1 = 1, y_1 = 2$

$x - y - 3 = 0 \implies A = 1, B = -1, C = -3$

$d = \dfrac{|1(1) + (-1)(2) + (-3)|}{\sqrt{1^2 + (-1)^2}} = \dfrac{4}{\sqrt{2}} = 2\sqrt{2}$

10. $(0, 4) \implies x_1 = 0, y_1 = 4$

$x + 2y - 2 = 0 \implies A = 1, B = 2, C = -2$

$d = \dfrac{|1(0) + (2)(4) + (-2)|}{\sqrt{1^2 + 2^2}} = \dfrac{6}{\sqrt{5}} = \dfrac{6\sqrt{5}}{5}$

11. Hyperbola

12. A parabola is formed.

13. Vertex: $(0, 0) = (h, k)$

Focus: $(4, 0) \implies p = 4$

$(y - k)^2 = 4p(x - h)$

$(y - 0)^2 = 4(4)(x - 0)$

$y^2 = 16x$

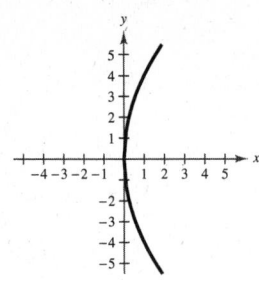

14. Vertex: $(2, 0) = (h, k)$

Focus: $(0, 0) \implies p = -2$

$(y - k)^2 = 4p(x - h)$

$y^2 = -8(x - 2)$

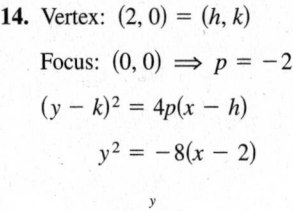

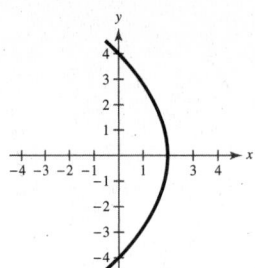

15. Vertex: $(0, 2) = (h, k)$

Directrix: $x = -3 \implies p = 3$

$(y - k)^2 = 4p(x - h)$

$(y - 2)^2 = 12x$

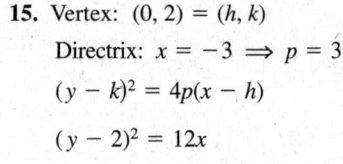

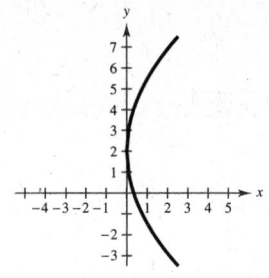

16. Vertex: $(2, 2) = (h, k)$

Directrix: $y = 0 \implies p = 2$

$(x - h)^2 = 4p(y - k)$

$(x - 2)^2 = 8(y - 2)$

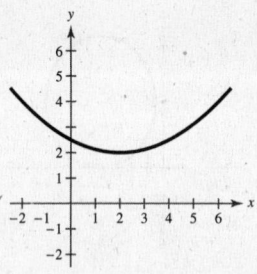

17. $x^2 = -2y \implies p = -\dfrac{1}{2}$

Focus: $\left(0, -\dfrac{1}{2}\right)$

$d_1 = b + \dfrac{1}{2}$

$d_2 = \sqrt{(2 - 0)^2 + \left(-2 + \dfrac{1}{2}\right)^2}$

$= \sqrt{4 + \dfrac{9}{4}} = \dfrac{5}{2}$

$d_1 = d_2$

$b + \dfrac{1}{2} = \dfrac{5}{2}$

$b = 2$

The slope of the line is

$m = \dfrac{-2 - 2}{2 - 0} = -2.$

Tangent line: $y = -2x + 2$

x-intercept: $(1, 0)$

18. $x^2 = -2y$

$p = -\dfrac{1}{2}$

Focus: $\left(0, -\dfrac{1}{2}\right)$

Tangent line through point $(-4, -8)$:

Slope: m

y-intercept: $(0, b)$

$d_1 = b + \dfrac{1}{2}$

$d_2 = \sqrt{(-4 - 0)^2 + \left(-8 + \dfrac{1}{2}\right)^2} = \dfrac{17}{2}$

$d_1 = d_2 \implies b = 8$

$m = \dfrac{-8 - 8}{-4 - 0} = 4$

$y = 4x + 8$

x-intercept of tangent line: $(-2, 0)$

19. Parabola

Opens downward

Vertex: $(0, 12)$

$(x - h)^2 = 4p(y - k)$

$x^2 = 4p(y - 12)$

Solution points: $(\pm 4, 10)$

$16 = 4p(10 - 12)$

$16 = -8p$

$-2 = p$

$x^2 = -8(y - 12)$

To find the x-intercepts, let $y = 0$.

$x^2 = 96$

$x = \pm\sqrt{96} = \pm 4\sqrt{6}$

At the base, the archway is $2(4\sqrt{6}) = 8\sqrt{6}$ meters wide.

20. $y^2 = 4px$

$p = 1.5$

$y^2 = 6x$

21. Vertices: $(-3, 0), (7, 0) \Rightarrow a = 5$
$(h, k) = (2, 0)$

Foci: $(0, 0), (4, 0) \Rightarrow c = 2$

$b^2 = a^2 - c^2 = 25 - 4 = 21$

$\dfrac{(x - h)^2}{a^2} + \dfrac{(y - k)^2}{b^2} = 1$

$\dfrac{(x - 2)^2}{25} + \dfrac{y^2}{21} = 1$

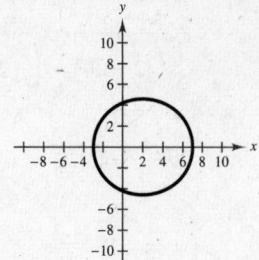

22. Vertices: $(2, 0), (2, 4) \Rightarrow a = 2, (h, k) = (2, 2)$

Foci: $(2, 1), (2, 3) \Rightarrow c = 1$

$b^2 = a^2 - c^2 = 4 - 1 = 3$

$\dfrac{(x - h)^2}{b^2} + \dfrac{(y - k)^2}{a^2} = 1$

$\dfrac{(x - 2)^2}{3} + \dfrac{(y - 2)^2}{4} = 1$

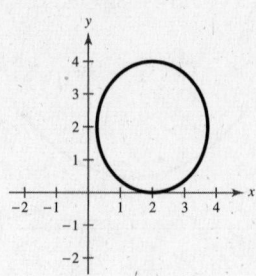

23. Vertices: $(0, 1), (4, 1) \Rightarrow a = 2, (h, k) = (2, 1)$
Endpoints of minor axis: $(2, 0), (2, 2) \Rightarrow b = 1$

$\dfrac{(x - h)^2}{a^2} + \dfrac{(y - k)^2}{b^2} = 1$

$\dfrac{(x - 2)^2}{4} + (y - 1)^2 = 1$

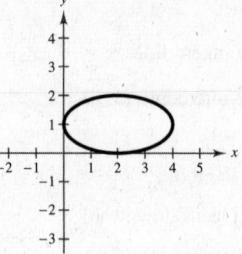

24. Vertices: $(-4, -1), (-4, 11) \Rightarrow a = 6, (h, k) = (-4, 5)$
Endpoints of the minor axis: $(-6, 5), (-2, 5) \Rightarrow b = 2$

$\dfrac{(x - h)^2}{b^2} + \dfrac{(y - k)^2}{a^2} = 1$

$\dfrac{(x + 4)^2}{4} + \dfrac{(y - 5)^2}{36} = 1$

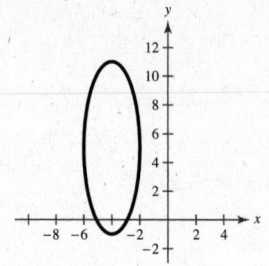

25. $2a = 10 \Rightarrow a = 5$

$b = 4$

$c^2 = a^2 - b^2 = 25 - 16 = 9 \Rightarrow c = 3$

The foci occur 3 feet from the center of the arch on a line connecting the tops of the pillars.

26. $\dfrac{x^2}{324} + \dfrac{y^2}{196} = 1$

$a = \sqrt{324} = 18, b = \sqrt{196} = 14$

$c = \sqrt{a^2 - b^2} = \sqrt{128} = 8\sqrt{2}$

Longest distance: $2a = 36$ feet

Shortest distance: $2b = 28$ feet

Distance between foci: $2c = 16\sqrt{2}$ feet

27. $\dfrac{(x + 2)^2}{81} + \dfrac{(y - 1)^2}{100} = 1$

$a = 10, b = 9, c = \sqrt{19}$

Center: $(-2, 1)$

Vertices: $(-2, 11)$ and $(-2, -9)$

Foci: $\left(-2, 1 \pm \sqrt{19}\right)$

Eccentricity: $e = \dfrac{\sqrt{19}}{10}$

28. $\dfrac{(x - 5)^2}{1} + \dfrac{(y + 3)^2}{36} = 1$

Center: $(5, -3)$

$a = 6, b = 1, c = \sqrt{a^2 - b^2} = \sqrt{35}$

Vertices: $(5, 3), (5, -9)$

Foci: $\left(5, -3 \pm \sqrt{35}\right)$

Eccentricity: $e = \dfrac{c}{a} = \dfrac{\sqrt{35}}{6}$

29. $16x^2 + 9y^2 - 32x + 72y + 16 = 0$

$16(x^2 - 2x + 1) + 9(y^2 + 8y + 16) = -16 + 16 + 144$

$16(x - 1)^2 + 9(y + 4)^2 = 144$

$$\frac{(x - 1)^2}{9} + \frac{(y + 4)^2}{16} = 1$$

$a = 4, b = 3, c = \sqrt{7}$

Center: $(1, -4)$

Vertices: $(1, 0)$ and $(1, -8)$

Foci: $\left(1, -4 \pm \sqrt{7}\right)$

Eccentricity: $e = \dfrac{\sqrt{7}}{4}$

30. $4x^2 + 25y^2 + 16x - 150y + 141 = 0$

$4(x^2 + 4x + 4) + 25(y^2 - 6y + 9) = -141 + 16 + 225$

$$\frac{(x + 2)^2}{25} + \frac{(y - 3)^2}{4} = 1$$

Center: $(-2, 3)$

$a = 5, b = 2, c = \sqrt{a^2 - b^2} = \sqrt{21}$

Vertices: $(3, 3), (-7, 3)$

Foci: $\left(-2 \pm \sqrt{21}, 3\right)$

Eccentricity: $e = \dfrac{c}{a} = \dfrac{\sqrt{21}}{5}$

31. Vertices: $(0, \pm 1) \Rightarrow a = 1, (h, k) = (0, 0)$

Foci: $(0, \pm 3) \Rightarrow c = 3$

$b^2 = c^2 - a^2 = 9 - 1 = 8$

$$\frac{(y - k)^2}{a^2} - \frac{(x - h)^2}{b^2} = 1$$

$$y^2 - \frac{x^2}{8} = 1$$

32. Vertices: $(2, 2), (-2, 2) \Rightarrow a = 2, (h, k) = (0, 2)$

Foci: $(4, 2), (-4, 2) \Rightarrow c = 4$

$b^2 = c^2 - a^2 = 16 - 4 = 12$

$$\frac{(x - h)^2}{a^2} - \frac{(y - k)^2}{b^2} = 1$$

$$\frac{x^2}{4} - \frac{(y - 2)^2}{12} = 1$$

33. Foci: $(0, 0), (8, 0) \Rightarrow c = 4, (h, k) = (4, 0)$

Asymptotes: $y = \pm 2(x - 4) \Rightarrow \dfrac{b}{a} = 2, b = 2a$

$b^2 = c^2 - a^2 \Rightarrow 4a^2 = 16 - a^2 \Rightarrow$

$a^2 = \dfrac{16}{5}, b^2 = \dfrac{64}{5}$

$$\frac{(x - h)^2}{a^2} - \frac{(y - k)^2}{b^2} = 1$$

$$\frac{(x - 4)^2}{16/5} - \frac{y^2}{64/5} = 1$$

$$\frac{5(x - 4)^2}{16} - \frac{5y^2}{64} = 1$$

34. Foci: $(3, \pm 2) \Rightarrow c = 2, (h, k) = (3, 0)$

Asymptotes: $y = \pm 2(x - 3) \Rightarrow \dfrac{a}{b} = 2, a = 2b$

$b^2 = c^2 - a^2 = 4 - 4b^2 \Rightarrow b^2 = \dfrac{4}{5}, a^2 = \dfrac{16}{5}$

$$\frac{(y - k)^2}{a^2} - \frac{(x - h)^2}{b^2} = 1$$

$$\frac{y^2}{16/5} - \frac{(x - 3)^2}{4/5} = 1 \Rightarrow \frac{5y^2}{16} - \frac{5(x - 3)^2}{4} = 1$$

35. $\dfrac{(x - 3)^2}{16} - \dfrac{(y + 5)^2}{4} = 1$

$a = 4, b = 2, c = \sqrt{20} = 2\sqrt{5}$

Center: $(3, -5)$

Vertices: $(7, -5)$ and $(-1, -5)$

Foci: $\left(3 \pm 2\sqrt{5}, -5\right)$

Asymptotes: $y = -5 \pm \dfrac{1}{2}(x - 3)$

$$y = \frac{1}{2}x - \frac{13}{2} \quad \text{or} \quad y = -\frac{1}{2}x - \frac{7}{2}$$

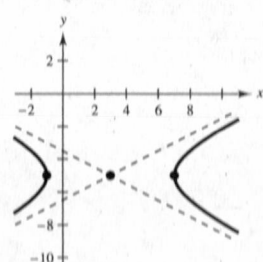

36. $\dfrac{(y-1)^2}{4} - x^2 = 1$

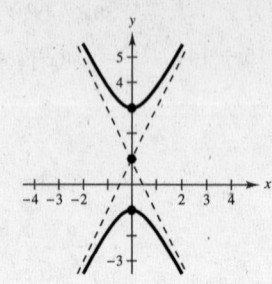

Center: $(0, 1)$

$a = 2, b = 1, c = \sqrt{a^2 + b^2} = \sqrt{5}$

Vertices: $(0, 3), (0, -1)$

Foci: $\left(0, 1 \pm \sqrt{5}\right)$

Asymptotes: $y = 1 \pm 2x$

37. $9x^2 - 16y^2 - 18x - 32y - 151 = 0$

$9(x^2 - 2x + 1) - 16(y^2 + 2y + 1) = 151 + 9 - 16$

$9(x - 1)^2 - 16(y + 1)^2 = 144$

$\dfrac{(x - 1)^2}{16} - \dfrac{(y + 1)^2}{9} = 1$

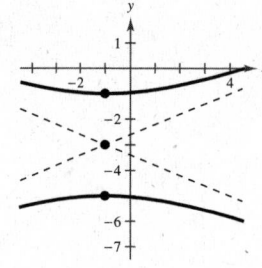

$a = 4, b = 3, c = 5$

Center: $(1, -1)$

Vertices: $(5, -1)$ and $(-3, -1)$

Foci: $(6, -1)$ and $(-4, -1)$

Asymptotes: $y = -1 \pm \dfrac{3}{4}(x - 1)$

$y = \dfrac{3}{4}x - \dfrac{7}{4} \quad \text{or} \quad y = -\dfrac{3}{4}x - \dfrac{1}{4}$

38. $-4x^2 + 25y^2 - 8x + 150y + 121 = 0$

$-4(x^2 + 2x + 1) + 25(y^2 + 6y + 9) = -121 - 4 + 225$

$\dfrac{(y + 3)^2}{4} - \dfrac{(x + 1)^2}{25} = 1$

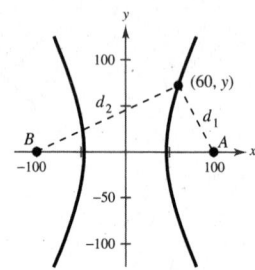

Center: $(-1, -3)$

$a = 2, b = 5, c = \sqrt{a^2 + b^2} = \sqrt{29}$

Vertices: $(-1, -1), (-1, -5)$

Foci: $\left(-1, -3 \pm \sqrt{29}\right)$

Asymptotes: $y = -3 \pm \dfrac{2}{5}(x + 1)$

39. Foci: $(\pm 100, 0) \implies c = 100$

Center: $(0, 0)$

$\dfrac{d_2}{186,000} - \dfrac{d_1}{186,000} = 0.0005 \implies d_2 - d_1 = 93 = 2a \implies a = 46.5$

$b^2 = c^2 - a^2 = 100^2 - 46.5^2 = 7837.75$

$\dfrac{x^2}{2162.25} - \dfrac{y^2}{7837.75} = 1$

$y^2 = 7837.75\left(\dfrac{60^2}{2162.25} - 1\right) \approx 5211.5736$

$y \approx 72$ miles

40. $BD = AD + 6\left(\dfrac{1100}{5280}\right)$

$CD = AD + 8\left(\dfrac{1100}{5280}\right)$

$2a = CD - BD = 2\left(\dfrac{1100}{5280}\right)$

$a = \dfrac{5}{24}, c = 2 \implies b^2 = \dfrac{2279}{576}$

Thus, we have $\dfrac{576x^2}{25} - \dfrac{576y^2}{2279} = 1$ (x and y in miles) or $\dfrac{x^2}{1,210,000} - \dfrac{y^2}{110,303,600} = 1$ (x and y in feet).

OR:

$CD = AD + 8\left(\dfrac{1100}{5280}\right)$

$BD = AD + 6\left(\dfrac{1100}{5280}\right)$

$2a = BD - AD = 6\left(\dfrac{1100}{5280}\right)$

$a = 3\left(\dfrac{5}{24}\right) = \dfrac{5}{8}, c = 1 \implies b^2 = \dfrac{39}{64}$

Center: $(1, 0)$

$\dfrac{64(x-1)^2}{25} - \dfrac{64y^2}{39} = 1$ (x and y in miles) or $\dfrac{(x-5280)^2}{10,890,000} - \dfrac{y^2}{16,988,400} = 1$ (x and y in feet).

41. $5x^2 - 2y^2 + 10x - 4y + 17 = 0$

$AC = 5(-2) = -10 < 0$

The graph is a hyperbola.

42. $-4y^2 + 5x + 3y + 7 = 0$

$AC = (0)(-4) = 0 \implies$ Parabola

43. $3x^2 + 2y^2 - 12x + 12y + 29 = 0$

$A = 3, C = 2$

$AC = 3(2) = 6 > 0$

The graph is an ellipse.

44. $4x^2 + 4y^2 - 4x + 8y - 11 = 0$

$A = 4, C = 4$

$A = C \implies$ Circle

45. $xy - 4 = 0$

$A = C = 0, B = 1$

$B^2 - 4AC = 1^2 - 4(0)(0) = 1 > 0$

The graph is a hyperbola.

$\cot 2\theta = 0 \implies 2\theta = \dfrac{\pi}{2} \implies \theta = \dfrac{\pi}{4}$

$x = x'\cos\dfrac{\pi}{4} - y'\sin\dfrac{\pi}{4} = \dfrac{x' - y'}{\sqrt{2}}$

$y = x'\sin\dfrac{\pi}{4} + y'\cos\dfrac{\pi}{4} = \dfrac{x' + y'}{\sqrt{2}}$

$\left(\dfrac{x' - y'}{\sqrt{2}}\right)\left(\dfrac{x' + y'}{\sqrt{2}}\right) - 4 = 0$

$\dfrac{(x')^2 - (y')^2}{2} = 4$

$\dfrac{(x')^2}{8} - \dfrac{(y')^2}{8} = 1$

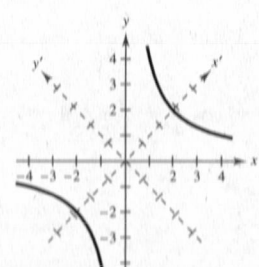

46. $x^2 - 10xy + y^2 + 1 = 0$

$B^2 - 4AC = (-10)^2 - 4(1)(1) = 96 > 0 \implies$ Hyperbola

$\cot 2\theta = \dfrac{A - C}{B} = \dfrac{1 - 1}{-10} = 0 \implies 2\theta = \dfrac{\pi}{2} \implies \theta = \dfrac{\pi}{4}$

$x = x' \cos \dfrac{\pi}{4} - y' \sin \dfrac{\pi}{4} = \dfrac{1}{\sqrt{2}}(x' - y')$

$y = x' \sin \dfrac{\pi}{4} + y' \cos \dfrac{\pi}{4} = \dfrac{1}{\sqrt{2}}(x' + y')$

$\dfrac{1}{2}(x' - y')^2 - 5(x' - y')(x' + y') + \dfrac{1}{2}(x' + y')^2 + 1 = 0$

$6(y')^2 - 4(x')^2 + 1 = 0$

$\dfrac{(x')^2}{1/4} - \dfrac{(y')^2}{1/6} = 1$

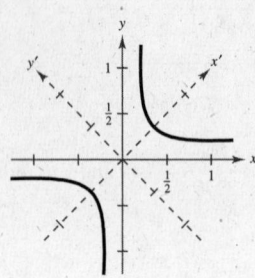

47. $5x^2 - 2xy + 5y^2 - 12 = 0$

$A = C = 5, B = -2$

$B^2 - 4AC = (-2)^2 - 4(5)(5) = -96 < 0$

The graph is an ellipse.

$\cot 2\theta = 0 \implies 2\theta = \dfrac{\pi}{2} \implies \theta = \dfrac{\pi}{4}$

$x = x' \cos \dfrac{\pi}{4} - y' \sin \dfrac{\pi}{4} = \dfrac{x' - y'}{\sqrt{2}}$

$y = x' \sin \dfrac{\pi}{4} + y' \cos \dfrac{\pi}{4} = \dfrac{x' + y'}{\sqrt{2}}$

$5\left(\dfrac{x' - y'}{\sqrt{2}}\right)^2 - 2\left(\dfrac{x' - y'}{\sqrt{2}}\right)\left(\dfrac{x' + y'}{\sqrt{2}}\right) + 5\left(\dfrac{x' + y'}{\sqrt{2}}\right)^2 - 12 = 0$

$\dfrac{5}{2}[(x')^2 - 2(x'y') + (y')^2] - [(x')^2 - (y')^2] + \dfrac{5}{2}[(x')^2 + 2(x'y') + (y')^2] = 12$

$4(x')^2 + 6(y')^2 = 12$

$\dfrac{(x')^2}{3} + \dfrac{(y')^2}{2} = 1$

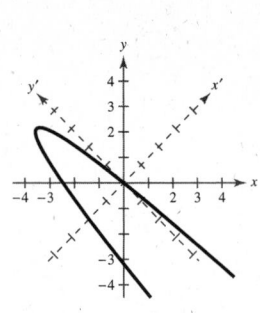

48. $4x^2 + 8xy + 4y^2 + 7\sqrt{2}x + 9\sqrt{2}y = 0$

$B^2 - 4AC = 8^2 - 4(4)(4) = 0 \implies$ Parabola

$\cot 2\theta = \dfrac{A - C}{B} = \dfrac{4 - 4}{8} = 0 \implies 2\theta = \dfrac{\pi}{2} \implies \theta = \dfrac{\pi}{4}$

$x = x' \cos \dfrac{\pi}{4} - y' \sin \dfrac{\pi}{4} = \dfrac{1}{\sqrt{2}}(x' - y')$

$y = x' \sin \dfrac{\pi}{4} + y' \cos \dfrac{\pi}{4} = \dfrac{1}{\sqrt{2}}(x' + y')$

—CONTINUED—

48. **—CONTINUED—**

$$2(x' - y')^2 + 4(x' - y')(x' + y') + 2(x' + y')^2 + 7(x' - y') + 9(x' + y') = 0$$

$$8(x')^2 + 16x' + 2y' = 0$$

$$y' = -4(x')^2 - 8x'$$

$$y' = -4((x')^2 + 2x' + 1) + 4$$

$$y' = -4(x' + 1)^2 + 4$$

$$y' - 4 = -4(x' + 1)^2$$

$$-4(x' + 1)^2 = y' - 4$$

$$y' = -4(x' + 1)^2 + 4$$

49. (a) $16x^2 - 24xy + 9y^2 - 30x - 40y = 0$

$B^2 - 4AC = (-24)^2 - 4(16)(9) = 0$

The graph is a parabola.

(b) To use a graphing utility, we need to solve for y in terms of x.

$9y^2 + (-24x - 40)y + (16x^2 - 30x) = 0$

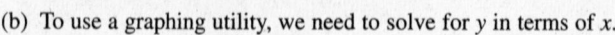

(c)

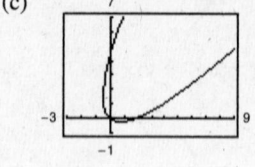

$$= \frac{(24x + 40) \pm \sqrt{(24x + 40)^2 - 36(16x^2 - 30x)}}{18}$$

50. (a) $13x^2 - 8xy + 7y^2 - 45 = 0$

$B^2 - 4AC = (-8)^2 - 4(13)(7) = -300 < 0 \implies$ Ellipse

(c)

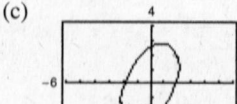

(b) Use the Quadratic Formula to solve for y in terms of x:

$7y^2 - 8xy + 13x^2 - 45 = 0$

$$y = \frac{1}{14}\left[8x \pm \sqrt{64x^2 - 28(13x^2 - 45)}\right]$$

51. (a) $x^2 + y^2 + 2xy + 2\sqrt{2}x - 2\sqrt{2}y + 2 = 0$

$B^2 - 4AC = 2^2 - 4(1)(1) = 0$

The graph is a parabola.

(c)

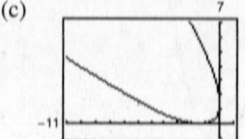

(b) To use a graphing utility, we need to solve for y in terms of x.

$y^2 + \left(2x - 2\sqrt{2}\right)y + \left(x^2 + 2\sqrt{2}x + 2\right) = 0$

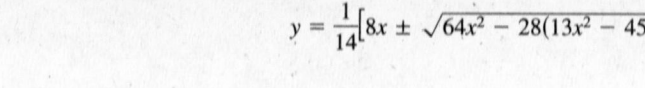

52. (a) $x^2 - 10xy + y^2 + 1 = 0$

Since $B^2 - 4AC = (-10)^2 - 4(1)(1) > 0 \implies$ Hyperbola

(c)

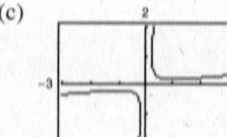

(b) Use the Quadratic Formula to solve for y in terms of x:

$y^2 - 10xy + x^2 + 1 = 0$

$$y = \tfrac{1}{2}\left[10x \pm \sqrt{100x^2 - 4(x^2 + 1)}\right]$$

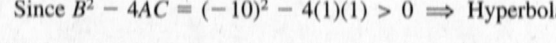

53. $x = 3t - 2$, $y = 7 - 4t$

t	-3	-2	0	1	2	3
x	-11	-8	-2	1	4	7
y	19	15	7	3	-1	-5

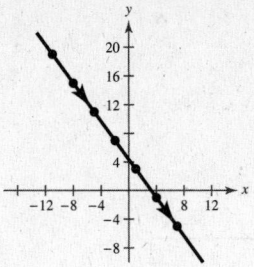

54. $x = \dfrac{1}{5}t$ and $y = \dfrac{4}{t - 1}$

t	-1	0	2	3	4	5
x	$-\frac{1}{5}$	0	$\frac{2}{5}$	$\frac{3}{5}$	$\frac{4}{5}$	1
y	-2	-4	4	2	$\frac{4}{3}$	1

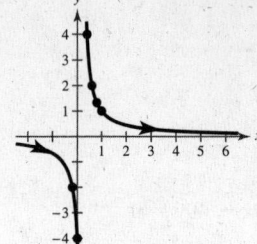

55. (a)

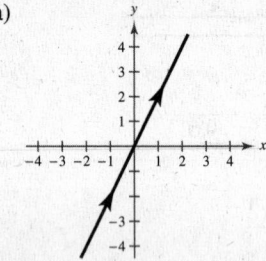

(b) $x = 2t \implies \dfrac{x}{2} = t$

$y = 4t \implies y = 4\left(\dfrac{x}{2}\right) = 2x$

56. (a)

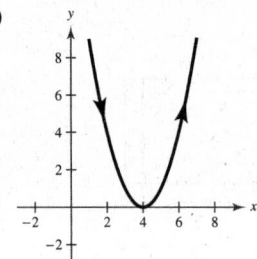

(b) $x = 1 + 4t$, $y = 2 - 3t$

$t = \dfrac{x - 1}{4}$

$y = 2 - 3\left(\dfrac{x - 1}{4}\right)$

$3x + 4y = 11$

57. (a)

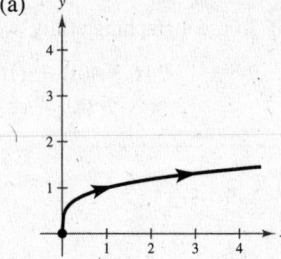

(b) $x = t^2$, $x \geq 0$

$y = \sqrt{t} \implies y^2 = t$

$x = (y^2)^2 \implies x$

$\quad = y^4 \implies y = \sqrt[4]{x}$

58. (a)

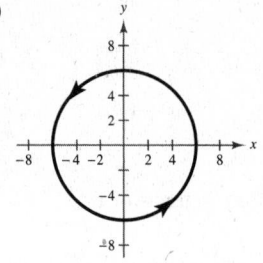

(b) $x = t + 4$, $y = t^2$

$t = x - 4$

$y = (x - 4)^2$

59. (a)

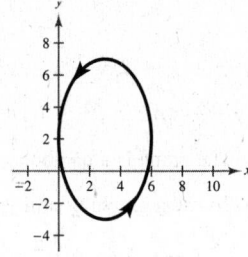

(b) $x = 6 \cos \theta$, $y = 6 \sin \theta$

$\cos \theta = \dfrac{x}{6}$, $\sin \theta = \dfrac{y}{6}$

$\dfrac{x^2}{36} + \dfrac{y^2}{36} = 1$

$x^2 + y^2 = 36$

60. (a)

(b) $x = 3 + 3 \cos \theta$, $y = 2 + 5 \sin \theta$

$\cos \theta = \dfrac{x - 3}{3}$, $\sin \theta = \dfrac{y - 2}{5}$

$\dfrac{(x - 3)^2}{9} + \dfrac{(y - 2)^2}{25} = 1$

61. Center: $(5, 4)$

Radius: 6

$x = h + r \cos \theta = 5 + 6 \cos \theta$

$y = k + r \sin \theta = 4 + 6 \sin \theta$

62. $(h, k) = (-3, 4)$

$2a = 8 \implies a = 4$

$2b = 6 \implies b = 3$

$\dfrac{(x + 3)^2}{16} + \dfrac{(y - 4)^2}{9} = 1$

$x = -3 + 4 \cos \theta$

$y = 4 + 3 \sin \theta$

This solution is not unique.

63. Hyperbola

Vertices: $(0, \pm 4)$

Foci: $(0, \pm 5)$

Center: $(0, 0)$

$a = 4, c = 5, b = \sqrt{c^2 - a^2} = 3$

$x = 3 \tan \theta, y = 4 \sec \theta$

64. $y = \overline{QB} - \overline{QA}$

$\overline{QP} = \text{arc } QC = r\theta$

$\overline{QA} = r\theta \sin(90° - \theta)$

$\quad = r\theta \cos \theta$

$\overline{QB} = r \sin \theta$

Therefore,

$y = r \sin \theta - r\theta \cos \theta = r(\sin \theta - \theta \cos \theta).$

Similarly, $x = \overline{OB} + \overline{AP}.$

Therefore, $x = r \cos \theta + r\theta \sin \theta = r(\cos \theta + \theta \sin \theta).$

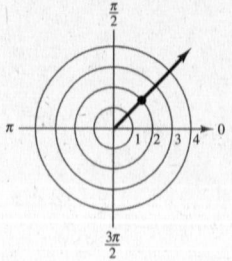

65. Polar coordinates: $\left(2, \dfrac{\pi}{4}\right)$

Additional polar representations: $\left(2, -\dfrac{7\pi}{4}\right), \left(-2, \dfrac{5\pi}{4}\right)$

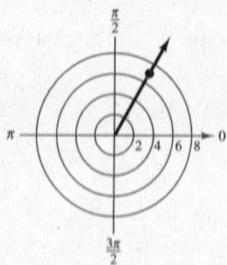

66. Polar coordinates: $\left(-5, -\dfrac{\pi}{3}\right) = \left(5, \dfrac{2\pi}{3}\right)$ or $\left(-5, \dfrac{5\pi}{3}\right)$

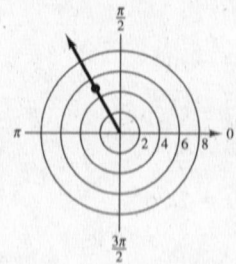

67. Polar coordinates: $(-7, 4.19)$

Additional polar representations: $(7, 1.05), (-7, -2.09)$

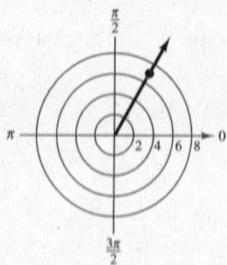

68. Polar coordinates:

$\left(\sqrt{3}, 2.62\right) \approx \left(\sqrt{3}, -3.66\right)$ or $\left(-\sqrt{3}, 5.76\right)$

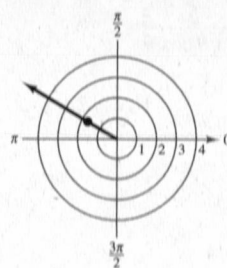

69. Polar coordinates: $\left(-1, \dfrac{\pi}{3}\right)$

$x = -1 \cos \dfrac{\pi}{3} = -\dfrac{1}{2}$

$y = -1 \sin \dfrac{\pi}{3} = -\dfrac{\sqrt{3}}{2}$

Rectangular coordinates: $\left(-\dfrac{1}{2}, -\dfrac{\sqrt{3}}{2}\right)$

70. Polar coordinates: $\left(2, \dfrac{5\pi}{4}\right) = (r, \theta)$

$x = r\cos\theta = 2\cos\dfrac{5\pi}{4} = -\sqrt{2}$

$y = r\sin\theta = 2\sin\dfrac{5\pi}{4} = -\sqrt{2}$

Rectangular coordinates: $\left(-\sqrt{2}, -\sqrt{2}\right)$

71. Polar coordinates: $\left(3, \dfrac{3\pi}{4}\right)$

$x = 3\cos\dfrac{3\pi}{4} = -\dfrac{3\sqrt{2}}{2}$

$y = 3\sin\dfrac{3\pi}{4} = \dfrac{3\sqrt{2}}{2}$

Rectangular coordinates: $\left(-\dfrac{3\sqrt{2}}{2}, \dfrac{3\sqrt{2}}{2}\right)$

72. Polar coordinates: $\left(0, \dfrac{\pi}{2}\right) = (r, \theta)$

$x = r\cos\theta = 0\cos\dfrac{\pi}{2} = 0$

$y = r\sin\theta = 0\sin\dfrac{\pi}{2} = 0$

Rectangular coordinates: $(0, 0)$

73. Rectangular coordinates: $(0, 2)$

$r = \pm\sqrt{0^2 + 2^2} = \pm 2$

$\tan\theta$ is undefined $\Rightarrow \theta = \dfrac{\pi}{2}, \dfrac{3\pi}{2}$

Polar coordinates: $\left(2, \dfrac{\pi}{2}\right)$ or $\left(-2, \dfrac{3\pi}{2}\right)$

74. Rectangular coordinates: $\left(-\sqrt{5}, \sqrt{5}\right)$

Polar coordinates:

$r = \sqrt{\left(-\sqrt{5}\right)^2 + \left(\sqrt{5}\right)^2} = \sqrt{10}$

$\tan\theta = -1, \theta = \dfrac{3\pi}{4}$

$\left(\sqrt{10}, \dfrac{3\pi}{4}\right)$

75. Rectangular coordinates: $(4, 6)$

$r = \pm\sqrt{4^2 + 6^2} = \pm\sqrt{52} = \pm 2\sqrt{13}$

$\tan\theta = \dfrac{6}{4} \Rightarrow \theta \approx 0.9828,\ 4.1244$

Polar coordinates: $\left(2\sqrt{13}, 0.9828\right)$ or $\left(-2\sqrt{13}, 4.1244\right)$

76. Rectangular coordinates: $(3, -4)$

Polar coordinates:

$r = \sqrt{3^2 + (-4)^2} = 5$

$\tan\theta = -\dfrac{4}{3}, \theta \approx -0.9273$

$(5, 5.356)$

77. $x^2 + y^2 = 49$

$r^2 = 49$

$r = 7$

78. $x^2 + y^2 = 20$

$x^2 + y^2 = r^2$

$r^2 = 20$

$r = 2\sqrt{5}$

79. $x^2 + y^2 - 6y = 0$

$r^2 - 6r\sin\theta = 0$

$r(r - 6\sin\theta) = 0$

$\qquad r = 0$ or $r = 6\sin\theta$

Since $r = 6\sin\theta$ contains $r = 0$, we just have $r = 6\sin\theta$.

80. $x^2 + y^2 - 4x = 0$

$r^2 - 4r\cos\theta = 0$

$\qquad r = 4\cos\theta$

81. $xy = 5$

$(r\cos\theta)(r\sin\theta) = 5$

$r^2 = \dfrac{5}{\sin\theta\cos\theta}$

$\quad = \dfrac{10}{\sin 2\theta} = 10\csc 2\theta$

82. $\qquad xy = -2$

$r\cos\theta\, r\sin\theta = -2$

$r^2\cos\theta\,\sin\theta = -2$

$\qquad r^2 = \dfrac{-2}{\cos\theta\sin\theta}$

$\qquad r^2 = -2\sec\theta\,\csc\theta$

$\qquad r^2 = -4\csc 2\theta$

83. $\qquad r = 5$

$\qquad r^2 = 25$

$x^2 + y^2 = 25$

84. $r = 12$

$r^2 = 144$

$\quad = x^2 + y^2$ or $x^2 + y^2 = 144$

85.
$$r = 3 \cos \theta$$
$$r^2 = 3r \cos \theta$$
$$x^2 + y^2 = 3x$$

86. Because $y = r \sin \theta$ and r is given as $8 \sin \theta$,
$$y = 8 \sin \theta \sin \theta = 8 \sin^2 \theta.$$
$$r = 8 \sin \theta$$
$$r^2 = 64 \sin^2 \theta$$
$$r^2 = 8(8 \sin^2 \theta)$$
$$x^2 + y^2 = 8y$$
$$x^2 + y^2 - 8y = 0$$

87.
$$r^2 = \sin \theta$$
$$r^3 = r \sin \theta$$
$$\left(\pm \sqrt{x^2 + y^2} \right)^3 = y$$
$$(x^2 + y^2)^3 = y^2$$
$$x^2 + y^2 = y^{2/3}$$

88.
$$r^2 = \cos 2\theta$$
$$r^2 = \left(\frac{x}{r} \right)^2 - \left(\frac{y}{r} \right)^2$$
$$(x^2 + y^2)^2 = x^2 - y^2$$

89. $r = 4$

Circle of radius 4 centered at the pole

Symmetric with respect to $\theta = \pi/2$, the polar axis and the pole

Maximum value of $|r| = 4$, for all values of θ

Zeros: None

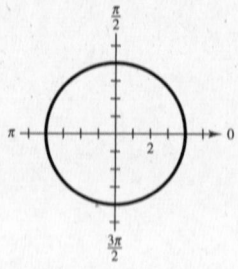

90. $r = 11$

Symmetry: $\theta = \dfrac{\pi}{2}$, polar axis, pole

Maximum value of $|r|$: 11, for all values of θ

Zeros of r: none

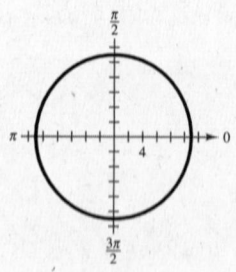

91. $r = 4 \sin 2\theta$

Rose curve ($n = 2$) with 4 petals

Symmetric with respect to $\theta = \pi/2$, the polar axis, and the pole

Maximum value of $|r| = 4$ when $\theta = \dfrac{\pi}{4}, \dfrac{3\pi}{4}, \dfrac{5\pi}{4}, \dfrac{7\pi}{4}$

Zeros: $r = 0$ when $\theta = 0, \dfrac{\pi}{2}, \pi, \dfrac{3\pi}{2}$

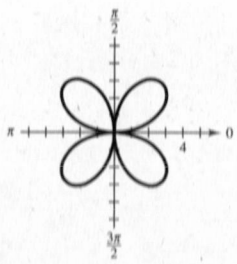

92. $r = \cos 5\theta$

Symmetry: polar axis

Maximum value of $|r|$: $|r| = 1$ when

$\theta = 0, \dfrac{2\pi}{5}, \dfrac{4\pi}{5}, \dfrac{6\pi}{5}, \dfrac{8\pi}{5}$

Zeros of r: $r = 0$ when $\theta = \dfrac{\pi}{10}, \dfrac{3\pi}{10}, \dfrac{\pi}{2}, \dfrac{7\pi}{10}, \dfrac{9\pi}{10}$

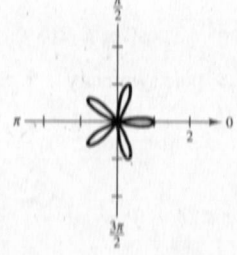

93. $r = -2(1 + \cos \theta)$

Symmetric with respect to the polar axis

Maximum value of $|r| = 4$ when $\theta = 0$

Zeros: $r = 0$ when $\theta = \pi$

$\dfrac{a}{b} = \dfrac{2}{2} = 1 \implies$ Cardioid

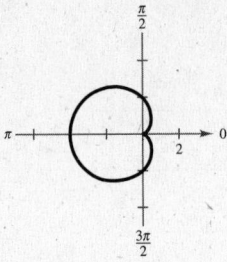

94. $r = 3 - 4 \cos \theta$

Symmetry: polar axis

Maximum value of $|r|$: $|r| = 7$ when $\theta = \pi$

Zeros of r: $r = 0$ when $\theta = \arccos \dfrac{3}{4}, 2\pi - \arccos \dfrac{3}{4}$

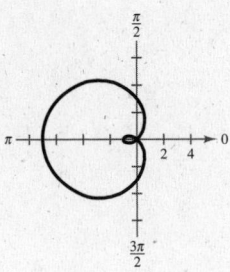

95. $r = 2 + 6 \sin \theta$

Limaçon with inner loop

$r = f(\sin \theta) \implies \theta = \dfrac{\pi}{2}$ symmetry

Maximum value: $|r| = 8$ when $\theta = \dfrac{\pi}{2}$

Zeros: $2 + 6 \sin \theta = 0 \implies \sin \theta = -\dfrac{1}{3} \implies \theta \approx 3.4814, 5.9433$

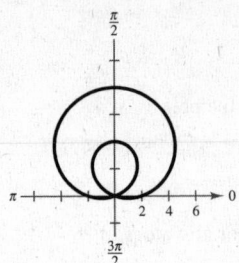

96. $r = 5 - 5 \cos \theta$

$r = 5(1 - \cos \theta)$

Symmetry: polar axis

Maximum values of $|r|$: $|r| = 10$ when $\theta = \pi$

Zeros of r: $r = 0$ when $\theta = 0, 2\pi$

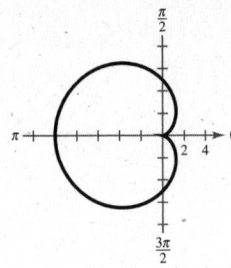

97. $r = -3 \cos 2\theta$

Rose curve with 4 petals

$r = f(\cos \theta) \implies$ polar axis symmetry

$\theta = \dfrac{\pi}{2}$: $r = -3 \cos 2(\pi - \theta) = -3 \cos(2\pi - 2\theta) = -3 \cos 2\theta$

    Equivalent equation $\implies \theta = \dfrac{\pi}{2}$ symmetry

Pole: $r = -3 \cos 2(\pi + \theta) = -3 \cos(2\pi + 2\theta) = -3 \cos 2\theta$

    Equivalent equation $\implies$ pole symmetry

Maximum value: $|r| = 3$ when $\theta = 0, \dfrac{\pi}{2}, \pi, \dfrac{3\pi}{2}$

Zeros: $-3 \cos 2\theta = 0$ when $\cos 2\theta = 0 \implies \theta = \dfrac{\pi}{4}, \dfrac{3\pi}{4}, \dfrac{5\pi}{4}, \dfrac{7\pi}{4}$

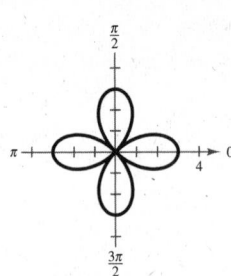

98. $r = \cos 2\theta$

Symmetry: polar axis

Maximum value of $|r|$: $|r| = 1$ when $\theta = 0, \dfrac{\pi}{2}, \pi, \dfrac{3\pi}{2}$

Zeros of r: $r = 0$ when $\theta = \dfrac{\pi}{4}, \dfrac{3\pi}{4}, \dfrac{5\pi}{4}, \dfrac{7\pi}{4}$

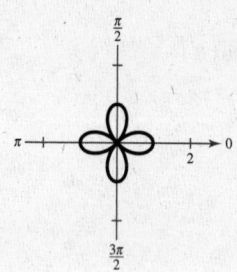

99. $r = 3(2 - \cos \theta)$

$= 6 - 3\cos \theta$

$\dfrac{a}{b} = \dfrac{6}{3} = 2$

The graph is a convex limaçon.

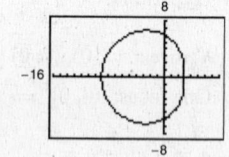

100. $r = 3(1 - 2\cos \theta)$

$r = 3 - 6\cos \theta$

Limaçon with inner loop.

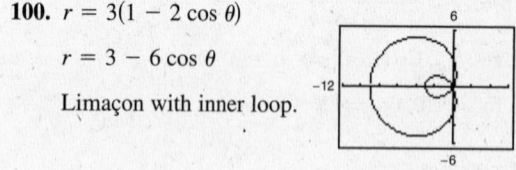

101. $r = 4\cos 3\theta$

The graph is a rose curve with 3 petals.

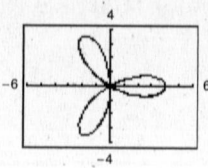

102. $r^2 = 9\cos 2\theta$

Lemniscate

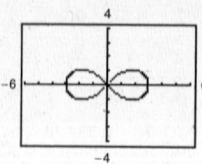

103. $r = \dfrac{1}{1 + 2\sin \theta}, e = 2$

Hyperbola symmetric with respect to $\theta = \dfrac{\pi}{2}$ and having

vertices at $\left(\dfrac{1}{3}, \dfrac{\pi}{2}\right)$ and $\left(-1, \dfrac{3\pi}{2}\right)$.

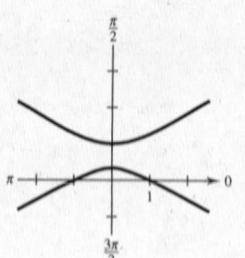

104. $r = \dfrac{2}{1 + \sin \theta}$

$e = 1 \Rightarrow$ parabola

Vertex: $\left(1, \dfrac{\pi}{2}\right)$

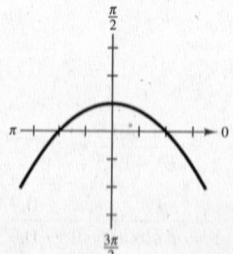

105. $r = \dfrac{4}{5 - 3\cos \theta}$

$r = \dfrac{4/5}{1 - (3/5)\cos \theta}, e = \dfrac{3}{5}$

Ellipse symmetric with respect to the polar axis and having vertices at $(2, 0)$ and $(1/2, \pi)$.

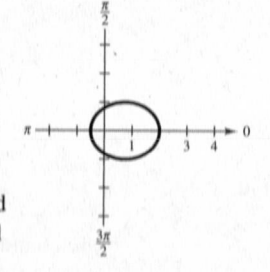

106. $r = \dfrac{16}{4 + 5\cos \theta}$

$r = \dfrac{4}{1 + (5/4)\cos \theta}$

$e = \dfrac{5}{4} > 1 \Rightarrow$ Hyperbola

Vertices: $\left(\dfrac{16}{9}, 0\right), (-16, \pi)$

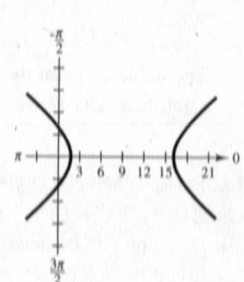

107. Parabola: $r = \dfrac{ep}{1 - e\cos\theta}$, $e = 1$

Vertex: $(2, \pi)$

Focus: $(0, 0) \implies p = 4$

$r = \dfrac{4}{1 - \cos\theta}$

108. Parabola: $r = \dfrac{ep}{1 + e\sin\theta}$, $e = 1$

Vertex: $\left(2, \dfrac{\pi}{2}\right)$

Focus: $(0, 0) \implies p = 4$

$r = \dfrac{4}{1 + \sin\theta}$

109. Ellipse: $r = \dfrac{ep}{1 - e\cos\theta}$

Vertices: $(5, 0), (1, \pi) \implies a = 3$

One focus: $(0, 0) \implies c = 2$

$e = \dfrac{c}{a} = \dfrac{2}{3}, p = \dfrac{5}{2}$

$r = \dfrac{(2/3)(5/2)}{1 - (2/3)\cos\theta} = \dfrac{5/3}{1 - (2/3)\cos\theta}$

$= \dfrac{5}{3 - 2\cos\theta}$

110. Hyperbola: $r = \dfrac{ep}{1 + e\cos\theta}$

Vertices: $(1, 0), (7, 0) \implies a = 3$

One focus: $(0, 0) \implies c = 4$

$e = \dfrac{c}{a} = \dfrac{4}{3}, p = \dfrac{7}{4}$

$r = \dfrac{(4/3)(7/4)}{1 + (4/3)\cos\theta} = \dfrac{7/3}{1 + (4/3)\cos\theta} = \dfrac{7}{3 + 4\cos\theta}$

111. $a + c = 122{,}800 + 4000 \implies a + c = 126{,}800$

$a - c = 119 + 4000 \implies a - c = 4{,}119$

$2a = 130{,}919$

$a = 65{,}459.5$

$c = 61{,}340.5$

$e = \dfrac{c}{a} = \dfrac{61{,}340.5}{65{,}459.5} \approx 0.937$

$r = \dfrac{ep}{1 - e\cos\theta} \approx \dfrac{0.937p}{1 - 0.937\cos\theta}$

$r = 126{,}800$ when $\theta = 0$

$126{,}800 = \dfrac{ep}{1 - e\cos 0}$

$ep = 126{,}800\left(1 - \dfrac{61{,}340.5}{65{,}459.5}\right) \approx 7978.81$

Thus, $r \approx \dfrac{7978.81}{1 - 0.937\cos\theta}$.

When $\theta = \dfrac{\pi}{3}$, $r \approx \dfrac{7978.81}{1 - 0.937\cos(\pi/3)} \approx 15{,}011.87$ miles.

The distance from the surface of Earth and the satellite is $15{,}011.87 - 4000 \approx 11{,}011.87$ miles.

112. Parabola: $r = \dfrac{ep}{1 + e\sin\theta}$, $e = 1$

Vertex: $\left(6{,}000{,}000, \dfrac{\pi}{2}\right)$

Focus: $(0, 0) \implies p = 12{,}000{,}000$

$r = \dfrac{12{,}000{,}000}{1 + \sin\theta}$

$\theta = -\dfrac{\pi}{3}$

$r \approx 89{,}600{,}000$ miles

113. False. When classifying equations of the form $Ax^2 + Bxy + Cy^2 + Dx + Ey + F = 0$, its graph can be determined by its discriminant. For a graph to be a parabola, its discriminant, $B^2 - 4AC$, must equal zero. So, if $B = 0$, then A **or** C equals 0, but not both.

114. False.

$\dfrac{x^2}{4} - y^4 = 1$ is a fourth-degree equation.

The equation of a hyperbola is a second degree equation.

115. False. The following are **two** sets of parametric equations for the line.

$x = t, y = 3 - 2t$

$x = 3t, y = 3 - 6t$

116. False.

$(r, \theta), (r, \theta + 2\pi), (-r, \theta + \pi)$, etc.

All represent the same point.

117. $2a = 10 \implies a = 5$

b must be less than 5; $0 < b < 5$.

As b approaches 5, the ellipse becomes more circular and approaches a circle of radius 5.

118. The orientation would be reversed.

119. $x = 4 \cos t$ and $y = 3 \sin t$

(a) $x = 4 \cos 2t$ and $y = 3 \sin 2t$

The speed would double.

(b) $x = 5 \cos t$ and $y = 3 \sin t$

The elliptical orbit would be flatter. The length of the major axis is greater.

120. (a) $\left(-4, \dfrac{\pi}{6}\right), \left(4, \dfrac{\pi}{6}\right)$: symmetric about the pole

(b) $\left(4, -\dfrac{\pi}{6}\right), \left(4, \dfrac{\pi}{6}\right)$: symmetric about the polar axis

(c) $\left(-4, -\dfrac{\pi}{6}\right), \left(4, \dfrac{\pi}{6}\right)$: symmetric about the $\theta = \dfrac{\pi}{2}$ axis.

121. (a) $x^2 + y^2 = 25$

$r = 5$

The graphs are the same. They are both circles centered at $(0, 0)$ with a radius of 5.

(b) $x - y = 0 \implies y = x$

$\theta = \dfrac{\pi}{4}$

The graphs are the same. They are both lines with slope 1 and intercept $(0, 0)$.

122. Area of the circle: $A = 100\pi$

Area of the ellipse: $A = \pi ab = \pi a(10) = 2(100\pi) \implies a = 20$

Length of major axis: $2a = 40$

Problem Solving for Chapter 10

1. (a) $\theta = \pi - 1.10 - 0.84 \approx 1.2016$ radians

(b) $\sin 0.84 = \dfrac{x}{3250} \implies x = 3250 \sin 0.84 \approx 2420$ feet

$\sin 1.10 = \dfrac{y}{6700} \implies y = 6700 \sin 1.10 \approx 5971$ feet

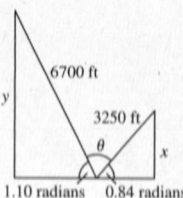

2.

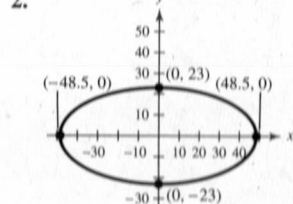

(a) Let $(0, 0)$ represent the center of the ellipse. Then

$2a = 97 \implies a = 48.5$ and $2b = 46 \implies b = 23$.

$$\dfrac{x^2}{(48.5)^2} + \dfrac{y^2}{23^2} = 1$$

$$\dfrac{x^2}{2352.25} + \dfrac{y^2}{529} = 1$$

(b) $c^2 = a^2 - b^2 = 2352.25 - 529 = 1823.25$

$c \approx 42.7$

The foci are $2c \approx 85.4$ feet apart.

(c) $A = \pi ab = \pi(48.5)(23) = 1115.5\pi \approx 3504.45$ square feet

3. Since the axis of symmetry is the x-axis, the vertex is $(h, 0)$ and $y^2 = 4p(x - h)$. Also, since the focus is $(0, 0)$,

$0 - h = p \Rightarrow h = -p$ and $y^2 = 4p(x + p)$.

4. Let (x, x) be the corner of the square in Quadrant I.

$A = 4x^2$

$\dfrac{x^2}{a^2} + \dfrac{x^2}{b^2} = 1 \Rightarrow x^2 = \dfrac{a^2 b^2}{a^2 + b^2}$

Thus, $A = \dfrac{4a^2 b^2}{a^2 + b^2}.$

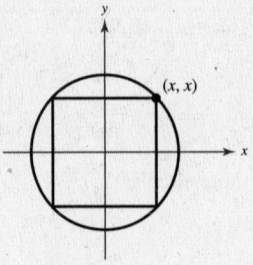

5. (a)

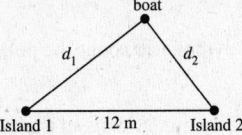

Since $d_1 + d_2 \leq 20$, by definition, the outer bound that the boat can travel is an ellipse. The islands are the foci.

(c) $d_1 + d_2 = 2a = 20 \Rightarrow a = 10$

The boat traveled 20 miles. The vertex is $(10, 0)$.

(b)

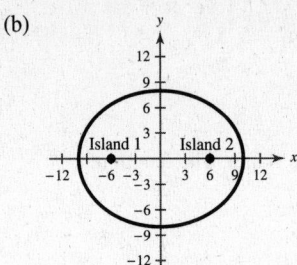

Island 1 is located at $(-6, 0)$ and Island 2 is located at $(6, 0)$.

(d) $c = 6, a = 10 \Rightarrow b^2 = a^2 - c^2 = 64$

$\dfrac{x^2}{100} + \dfrac{y^2}{64} = 1$

6. Foci: $(2, 2)$ and $(10, 2)$ $\Rightarrow$ Center is $(6, 2)$ and $c = 4$

$|d_2 - d_1| = 2a = 6 \Rightarrow a = 3$

$c^2 = a^2 + b^2 \Rightarrow 16 = 9 + b^2 \Rightarrow b^2 = 7$

Horizontal transverse axis

$\dfrac{(x - 6)^2}{9} - \dfrac{(y - 2)^2}{7} = 1$

7. $Ax^2 + Cy^2 + Dx + Ey + F = 0$

Assume that the conic is *not* degenerate.

(a) $A = C, A \neq 0$

$$Ax^2 + Ay^2 + Dx + Ey + F = 0$$

$$x^2 + y^2 + \frac{D}{A}x + \frac{E}{A}y + \frac{F}{A} = 0$$

$$\left(x^2 + \frac{D}{A}x + \frac{D^2}{4A^2}\right) + \left(y^2 + \frac{E}{A}y + \frac{E^2}{4A^2}\right) = -\frac{F}{A} + \frac{D^2}{4A^2} + \frac{E^2}{4A^2}$$

$$\left(x + \frac{D}{2A}\right)^2 + \left(y + \frac{E}{2A}\right)^2 = \frac{D^2 + E^2 - 4AF}{4A^2}$$

This is a circle with center $\left(-\dfrac{D}{2A}, -\dfrac{E}{2A}\right)$ and radius $\dfrac{\sqrt{D^2 + E^2 - 4AF}}{2|A|}.$

(b) $A = 0$ or $C = 0$ (but not both). Let $C = 0$.

$$Ax^2 + Dx + Ey + F = 0$$

$$x^2 + \frac{D}{A}x = -\frac{E}{A}y - \frac{F}{A}$$

$$x^2 + \frac{D}{A}x + \frac{D^2}{4A^2} = -\frac{E}{A}y - \frac{F}{A} + \frac{D^2}{4A^2}$$

$$\left(x + \frac{D}{2A}\right)^2 = -\frac{E}{A}\left(y + \frac{F}{E} - \frac{D^2}{4AE}\right)$$

This is a parabola with vertex $\left(-\dfrac{D}{2A}, \dfrac{D^2 - 4AF}{4AE}\right).$

$A = 0$ yields a similar result.

—CONTINUED—

7. —CONTINUED—

(c) $AC > 0 \implies A$ and C are either both positive or are both negative (if that is the case, move the terms to the other side of the equation so that they are both positive).

$$Ax^2 + Cy^2 + Dx + Ey + F = 0$$

$$A\left(x^2 + \frac{D}{A}x + \frac{D^2}{4A^2}\right) + C\left(y^2 + \frac{E}{C}y + \frac{E^2}{4C^2}\right) = -F + \frac{D^2}{4A} + \frac{E^2}{4C}$$

$$A\left(x + \frac{D}{2A}\right)^2 + C\left(y + \frac{E}{2C}\right)^2 = \frac{CD^2 + AE^2 - 4ACF}{4AC}$$

$$\frac{\left(x + \dfrac{D}{2A}\right)^2}{\dfrac{CD^2 + AE^2 - 4ACF}{4A^2C}} + \frac{\left(y + \dfrac{E}{2C}\right)^2}{\dfrac{CD^2 + AE^2 - 4ACF}{4AC^2}} = 1$$

Since A and C are both positive, $4A^2C$ and $4AC^2$ are both positive. $CD^2 + AE^2 - 4ACF$ must be positive or the conic is degenerate. Thus, we have an ellipse with center $\left(-\dfrac{D}{2A}, -\dfrac{E}{2C}\right)$.

(d) $AC < 0 \implies A$ and C have opposite signs. Let's assume that A is positive and C is negative. (If A is negative and C is positive, move the terms to the other side of the equation.) From part (c) we have

$$\frac{\left(x + \dfrac{D}{2A}\right)^2}{\dfrac{CD^2 + AE^2 - 4ACF}{4A^2C}} + \frac{\left(y + \dfrac{E}{2C}\right)^2}{\dfrac{CD^2 + AE^2 - 4ACF}{4AC^2}} = 1.$$

Since $A > 0$ and $C < 0$, the first denominator is positive if $CD^2 + AE^2 - 4ACF < 0$ and is negative if $CD^2 + AE^2 - 4ACF > 0$, since $4A^2C$ is negative. The second denominator would have the *opposite* sign since $4AC^2 > 0$. Thus, we have a hyperbola with center $\left(-\dfrac{D}{2A}, -\dfrac{E}{2C}\right)$.

8. (a) The first model describes linear motion, whereas the second model describes parabolic motion.

(b) $x = (v_0 \cos \theta)t \implies t = \dfrac{x}{v_0 \cos \theta}$

$y = (v_0 \sin \theta)t \implies t = \dfrac{y}{v_0 \sin \theta}$

$\dfrac{x}{v_0 \cos \theta} = \dfrac{y}{v_0 \sin \theta}$

$(v_0 \cos \theta)y = (v_0 \sin \theta)x$

$y = (\tan \theta)x$

$x = (v_0 \cos \theta)t \implies t = \dfrac{x}{v_0 \cos \theta}$

$y = h + (v_0 \sin \theta)t - 16t^2$

$y = h + (v_0 \sin \theta)\left(\dfrac{x}{v_0 \cos \theta}\right) - 16\left(\dfrac{x}{v_0 \cos \theta}\right)^2$

$y = h + (\tan \theta)x - \left(\dfrac{16}{v_0^2 \cos^2 \theta}\right)x^2$

(c) In the case $x = (v_0 \cos \theta)t$, $y = (v_0 \sin \theta)t$, the path of the projectile is not affected by changing the velocity v. When the parameter is eliminated, we just have $y = (\tan \theta)x$. The path is only affected by the angle θ.

9. To change the orientation, we can just replace t with $-t$.

$$x = \cos(-t) = \cos t$$
$$y = 2\sin(-t) = -2\sin t$$

10. $x = (a - b)\cos t + b\cos\left(\dfrac{a - b}{b}t\right)$

$y = (a - b)\sin t - b\sin\left(\dfrac{a - b}{b}t\right)$

(a) $a = 2, b = 1$

$x = \cos t + \cos t = 2\cos t$

$y = \sin t - \sin t = 0$

The graph oscillates between -2 and 2 on the x-axis.

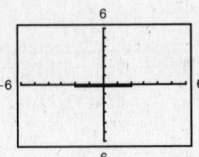

(b) $a = 3, b = 1$

$x = 2\cos t + \cos 2t$

$y = 2\sin t - \sin 2t$

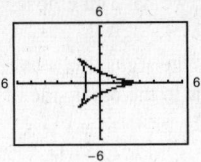

(c) $a = 4, b = 1$

$x = 3\cos t + \cos 3t$

$y = 3\sin t - \sin 3t$

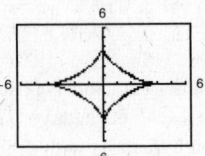

(d) $a = 10, b = 1$

$x = 9\cos t + \cos 9t$

$y = 9\sin t - \sin 9t$

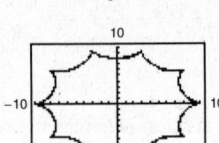

(e) $a = 3, b = 2$

$x = \cos t + 2\cos\dfrac{t}{2}$

$y = \sin t - 2\sin\dfrac{t}{2}$

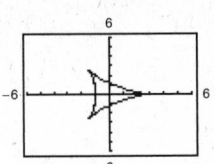

The graph looks the same as the graph in part (b), but is oriented clockwise instead of counterclockwise.

(f) $a = 4, b = 3$

$x = \cos t + 3\cos\dfrac{t}{3}$

$y = \sin t - 3\sin\dfrac{t}{3}$

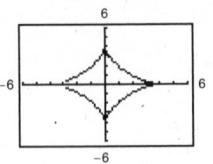

The graph is the same as the graph in part (c), but is oriented clockwise instead of counterclockwise.

11. (a) $y^2 = \dfrac{t^2(1 - t^2)^2}{(1 + t^2)^2}, \; x^2 = \dfrac{(1 - t^2)^2}{(1 + t^2)^2}$

$\dfrac{1 - x}{1 + x} = \dfrac{1 - \left(\dfrac{1 - t^2}{1 + t^2}\right)}{1 + \left(\dfrac{1 - t^2}{1 + t^2}\right)} = \dfrac{2t^2}{2} = t^2$

Thus, $y^2 = x^2\left(\dfrac{1 - x}{1 + x}\right)$.

(b)

$r^2 \sin^2 \theta = r^2 \cos^2 \theta\left(\dfrac{1 - r \cos \theta}{1 + r \cos \theta}\right)$

$\sin^2 \theta(1 + r \cos \theta) = \cos^2 \theta(1 - r \cos \theta)$

$r \cos \theta \sin^2 \theta + \sin^2 \theta = \cos^2 \theta - r \cos^3 \theta$

$r \cos \theta(\sin^2 \theta + \cos^2 \theta) = \cos^2 \theta - \sin^2 \theta$

$r \cos \theta = \cos 2\theta$

$r = \cos 2\theta \cdot \sec \theta$

(c)

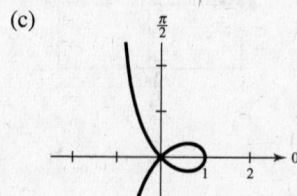

12. $r = 2 \cos\left(\dfrac{1}{2}\theta\right)$ $r = 3 \sin\left(\dfrac{5\theta}{2}\right)$ $r = -\cos(\sqrt{2}\theta)$ $r = -2 \sin\left(\dfrac{4\theta}{7}\right)$

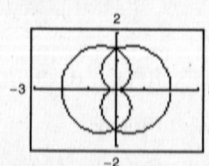 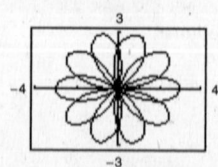 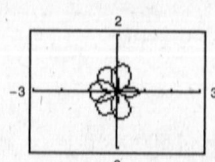 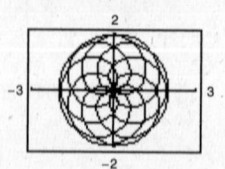

The graphs all contain overlapping loops or petals.

13. $r = a \sin \theta + b \cos \theta$

$r^2 = r(a \sin \theta + b \cos \theta)$

$r^2 = ar \sin \theta + br \cos \theta$

$x^2 + y^2 = ay + bx$

$x^2 + y^2 - bx - ay = 0$

$\left(x^2 - bx + \dfrac{b^2}{4}\right) + \left(y^2 - ay + \dfrac{a^2}{4}\right) = \dfrac{a^2}{4} + \dfrac{b^2}{4}$

$\left(x - \dfrac{b}{2}\right)^2 + \left(y - \dfrac{a}{2}\right)^2 = \dfrac{a^2 + b^2}{4}$

This represents a circle with center $\left(\dfrac{b}{2}, \dfrac{a}{2}\right)$ and radius $r = \dfrac{1}{2}\sqrt{a^2 + b^2}$.

14. $r = e^{\cos \theta} - 2 \cos 4\theta + \sin^5\left(\dfrac{\theta}{12}\right)$

(a) No, the graph appears to have a period of 2π but does not. For example, $r(\pi) \neq r(3\pi)$.

(b) By using the table feature of the calculator we have $r \approx 4.077$ when $\theta \approx 5.54$ for $0 \leq \theta \leq 2\pi$ and $r \approx 4.46$ when $\theta \approx 11.83$ for $0 \leq \theta \leq 4\pi$. The graph is not periodic. As θ increases the value of r changes.

15.

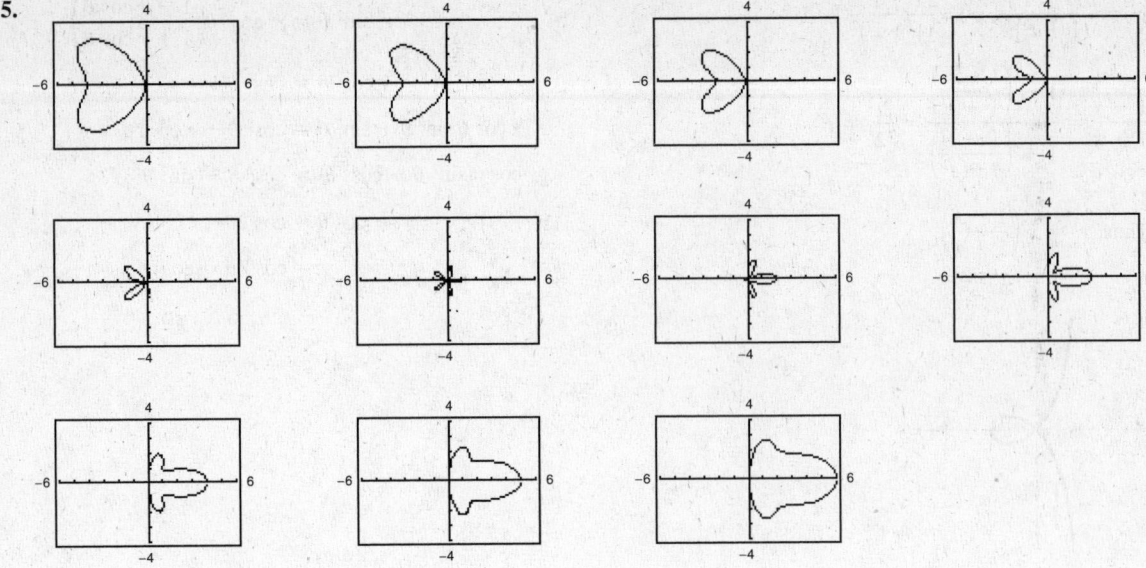

$n = 1, 2, 3, 4, 5$ produce "bells"; $n = -1, -2, -3, -4, -5$ produce "hearts".

16. (a) Neptune: $a = \dfrac{9.000 \times 10^9}{2} = 4.500 \times 10^9$

$e = 0.0086$

$r_{\text{Neptune}} = \dfrac{(1 - 0.0086^2)(4.500 \times 10^9)}{1 - (0.0086)\cos\theta}$

$r_{\text{Neptune}} = \dfrac{4.4997 \times 10^9}{1 - (0.0086)\cos\theta}$

Pluto: $a = \dfrac{10.0813 \times 10^9}{2} = 5.4065 \times 10^9$

$e = 0.2488$

$r_{\text{Pluto}} = \dfrac{(1 - 0.2488^2)(5.4065 \times 10^9)}{1 - (0.2488)\cos\theta}$

$r_{\text{Pluto}} = \dfrac{5.0718 \times 10^9}{1 - (0.2488)\cos\theta}$

(b) Neptune:

perihelion: $a(1 - e) = 4.500 \times 10^9(1 - 0.0086)$

$= 4.461 \times 10^9$ km

aphelion: $a(1 + e) = 4.500 \times 10^9(1 + 0.0086)$

$= 4.539 \times 10^9$ km

Pluto:

perihelion: $a(1 - e) = 5.4065 \times 10^9(1 - 0.2488)$

$= 4.061 \times 10^9$ km

aphelion: $a(1 + e) = 5.4065 \times 10^9(1 + 0.2488)$

$= 6.752 \times 10^9$ km

(d) If the orbits were in the same plane, then they would intersect. Furthermore, since the orbital periods differ (Neptune = 164.79 years, Pluto = 247.68 years), then the two planets would ultimately collide if the orbits intersect.

The orbital inclination of Pluto is significantly larger than that of Neptune (17.16° vs. 1.769°), so further analysis is required to determine if the orbits intersect.

(c)

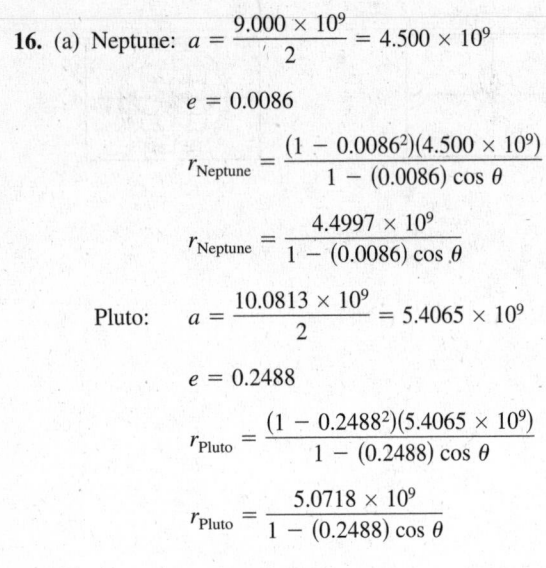

	Neptune	Pluto
(e) perihelion	4.461×10^9 km	4.061×10^9 km

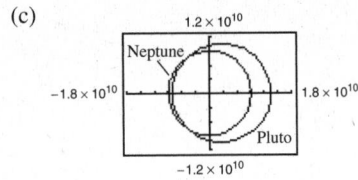

Pluto is sometimes closer to the sun than Neptune (for about 20 years of its 248-year orbit). At the time of its discovery, Pluto was more distant than Neptune. At that time, Pluto was the most distant planet (the ninth in distance) and was also the ninth planet discovered.

Chapter 10　Practice Test

1. Find the angle, θ, between the lines $3x + 4y = 12$ and $4x - 3y = 12$.

2. Find the distance between the point $(5, -9)$ and the line $3x - 7y = 21$.

3. Find the vertex, focus and directrix of the parabola $x^2 - 6x - 4y + 1 = 0$.

4. Find an equation of the parabola with its vertex at $(2, -5)$ and focus at $(2, -6)$.

5. Find the center, foci, vertices, and eccentricity of the ellipse $x^2 + 4y^2 - 2x + 32y + 61 = 0$.

6. Find an equation of the ellipse with vertices $(0, \pm 6)$ and eccentricity $e = \frac{1}{2}$.

7. Find the center, vertices, foci, and asymptotes of the hyperbola $16y^2 - x^2 - 6x - 128y + 231 = 0$.

8. Find an equation of the hyperbola with vertices at $(\pm 3, 2)$ and foci at $(\pm 5, 2)$.

9. Rotate the axes to eliminate the xy-term. Sketch the graph of the resulting equation, showing both sets of axes.

   $5x^2 + 2xy + 5y^2 - 10 = 0$

10. Use the discriminant to determine whether the graph of the equation is a parabola, ellipse, or hyperbola.

    (a) $6x^2 - 2xy + y^2 = 0$

    (b) $x^2 + 4xy + 4y^2 - x - y + 17 = 0$

11. Convert the polar point $\left(\sqrt{2}, \dfrac{3\pi}{4} \right)$ to rectangular coordinates.

12. Convert the rectangular point $\left(\sqrt{3}, -1 \right)$ to polar coordinates.

13. Convert the rectangular equation $4x - 3y = 12$ to polar form.

14. Convert the polar equation $r = 5 \cos \theta$ to rectangular form.

15. Sketch the graph of $r = 1 - \cos \theta$.

16. Sketch the graph of $r = 5 \sin 2\theta$.

17. Sketch the graph of $r = \dfrac{3}{6 - \cos \theta}$.

18. Find a polar equation of the parabola with its vertex at $\left(6, \dfrac{\pi}{2} \right)$ and focus at $(0, 0)$.

For Exercises 19 and 20, eliminate the parameter and write the corresponding rectangular equation.

19. $x = 3 - 2 \sin \theta, \ y = 1 + 5 \cos \theta$

20. $x = e^{2t}, \ y = e^{4t}$

CHAPTER 11
Analytic Geometry in Three Dimensions

CHAPTER 11
Analytic Geometry in Three Dimensions

Section 11.1   The Three-Dimensional Coordinate System

- You should be able to plot points in the three-dimensional coordinate system.
- The distance between the points (x_1, y_1, z_1) and (x_2, y_2, z_2) is
  $$d = \sqrt{(x_2 - x_1)^2 + (y_2 - y_1)^2 + (z_2 - z_1)^2}.$$
- The midpoint of the line segment joining the points (x_1, y_1, z_1) and (x_2, y_2, z_2) is
  $$\left(\frac{x_1 + x_2}{2}, \frac{y_1 + y_2}{2}, \frac{z_1 + z_2}{2}\right).$$
- The equation of the sphere with center (h, k, j) and radius r is
  $$(x - h)^2 + (y - k)^2 + (z - j)^2 = r^2.$$
- You should be able to find the trace of a surface in space.

Vocabulary Check

1. three-dimensional

2. xy-plane, xz-plane, yz-plane

3. octants

4. Distance Formula

5. $\left(\dfrac{x_1 + x_2}{2}, \dfrac{y_1 + y_2}{2}, \dfrac{z_1 + z_2}{2}\right)$

6. sphere

7. surface, space

8. trace

---

1. $A(-1, 4, 3), B(1, 3, -2), C(-3, 0, -2)$

2. $A(6, 2, -3), B(2, -1, 2)\ C(-2, 3, 0)$

3.
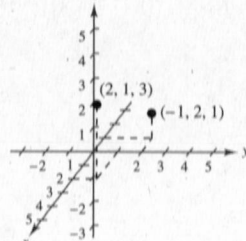

4.
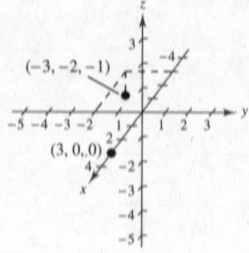

5.
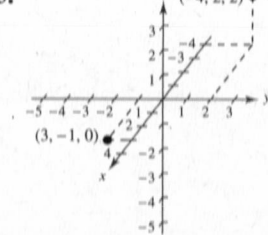

6.
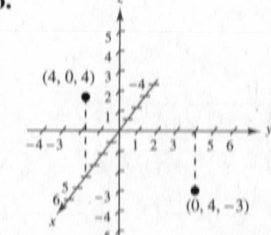

7. $x = -3, y = 3, z = 4$: $(-3, 3, 4)$

8. $x = 6, y = -1, z = -1 \Rightarrow (6, -1, -1)$

9. $y = z = 0, x = 10$: $(10, 0, 0)$

10. $x = 0, y = 2, z = 8 \Rightarrow (0, 2, 8)$

11. Octant IV

12. Octant VI

13. Octants I, II, III, IV
(above the *xy*-plane)

14. Octants III, IV, VII, or VIII

15. Octants II, IV, VI, VIII

16. Octants I, II, VII, or VIII

17. $d = \sqrt{(5 - 0)^2 + (2 - 0)^2 + (6 - 0)^2}$

$= \sqrt{25 + 4 + 36}$

$= \sqrt{65}$ units

18. $d = \sqrt{(7 - 1)^2 + (0 - 0)^2 + (4 - 0)^2}$

$= \sqrt{36 + 16}$

$= \sqrt{52}$

$= 2\sqrt{13}$

19. $d = \sqrt{(7 - 3)^2 + (4 - 2)^2 + (8 - 5)^2}$

$= \sqrt{4^2 + 2^2 + 3^2}$

$= \sqrt{16 + 4 + 9}$

$= \sqrt{29}$

≈ 5.385

20. $d = \sqrt{(4 - 2)^2 + (1 - 1)^2 + (9 - 6)^2}$

$= \sqrt{4 + 9}$

$= \sqrt{13}$

21. $d = \sqrt{[6 - (-1)]^2 + [0 - 4]^2 + [-9 - (-2)]^2}$

$= \sqrt{7^2 + 4^2 + 7^2}$

$= \sqrt{49 + 16 + 49}$

$= \sqrt{114}$

≈ 10.677

22. $d = \sqrt{(1 - (-2))^2 + (1 - (-3))^2 + (-7 - (-7))^2}$

$= \sqrt{9 + 16}$

$= \sqrt{25}$

$= 5$

23. $d = \sqrt{(1 - 0)^2 + [0 - (-3)]^2 + (-10 - 0)^2}$

$= \sqrt{1 + 9 + 100}$

$= \sqrt{110} \approx 10.488$

24. $d = \sqrt{(2 - 0)^2 + (-4 - 6)^2 + (0 - (-3))^2}$

$= \sqrt{4 + 100 + 9}$

$= \sqrt{113}$

25. $d_1 = \sqrt{(-2 - 0)^2 + (5 - 0)^2 + (2 - 2)^2} = \sqrt{4 + 25} = \sqrt{29}$

$d_2 = \sqrt{(0 - 0)^2 + (4 - 0)^2 + (0 - 2)^2} = \sqrt{16 + 4} = \sqrt{20} = 2\sqrt{5}$

$d_3 = \sqrt{(0 + 2)^2 + (4 - 5)^2 + (0 - 2)^2} = \sqrt{4 + 1 + 4} = \sqrt{9} = 3$

$d_1{}^2 = d_2{}^2 + d_3{}^2 = 29$

26. $d_1 = \sqrt{(-4 - 2)^2 + (4 + 1)^2 + (1 - 2)^2} = \sqrt{36 + 25 + 1} = \sqrt{62}$

$d_2 = \sqrt{(-4 + 2)^2 + (4 - 5)^2 + (1 - 0)^2} = \sqrt{4 + 1 + 1} = \sqrt{6}$

$d_3 = \sqrt{(2 + 2)^2 + (-1 - 5)^2 + (2 - 0)^2} = \sqrt{16 + 36 + 4} = \sqrt{56} = 2\sqrt{14}$

$d_1^2 = d_2^2 + d_3^2 = 62$

27. $d_1 = \sqrt{(2 - 0)^2 + (2 - 0)^2 + (1 - 0)^2} = \sqrt{4 + 4 + 1} = \sqrt{9} = 3$

$d_2 = \sqrt{(2 - 0)^2 + (-4 - 0)^2 + (4 - 0)^2} = \sqrt{4 + 16 + 16} = \sqrt{36} = 6$

$d_3 = \sqrt{(2 - 2)^2 + (-4 - 2)^2 + (4 - 1)^2} = \sqrt{36 + 9} = \sqrt{45} = 3\sqrt{5}$

$d_1^2 + d_2^2 = 9 + 36 = 45 = d_3^2$

28. $d_1 = \sqrt{(1 - 1)^2 + (3 - 0)^2 + (1 - 1)^2} = 3$

$d_2 = \sqrt{(1 - 1)^2 + (0 - 0)^2 + (3 - 1)^2} = 2$

$d_3 = \sqrt{(1 - 1)^2 + (0 - 3)^2 + (3 - 1)^2} = \sqrt{13}$

$d_3^2 = 13 = d_1^2 + d_2^2$

29. $d_1 = \sqrt{(5 - 1)^2 + (-1 + 3)^2 + (2 + 2)^2} = \sqrt{16 + 4 + 16} = \sqrt{36} = 6$

$d_2 = \sqrt{(5 + 1)^2 + (-1 - 1)^2 + (2 - 2)^2} = \sqrt{36 + 4} = \sqrt{40} = 2\sqrt{10}$

$d_3 = \sqrt{(-1 - 1)^2 + (1 + 3)^2 + (2 + 2)^2} = \sqrt{4 + 16 + 16} = \sqrt{36} = 6$

$d_1 = d_3$ Isosceles triangle

30. $d_1 = \sqrt{(7 - 5)^2 + (1 - 3)^2 + (3 - 4)^2} = \sqrt{4 + 4 + 1} = \sqrt{9} = 3$

$d_2 = \sqrt{(3 - 7)^2 + (5 - 1)^2 + (3 - 3)^2} = \sqrt{16 + 16} = \sqrt{32} = 4\sqrt{2}$

$d_3 = \sqrt{(3 - 5)^2 + (5 - 3)^2 + (3 - 4)^2} = \sqrt{4 + 4 + 1} = \sqrt{9} = 3$

$d_1 = d_3 = 3.$ Isosceles triangle

31. $\left(\dfrac{3 + 0}{2}, \dfrac{-2 + 0}{2}, \dfrac{4 + 0}{2}\right) = \left(\dfrac{3}{2}, -1, 2\right)$

32. Midpoint: $\left(\dfrac{1 + 2}{2}, \dfrac{5 + 2}{2}, \dfrac{-1 + 2}{2}\right) = \left(\dfrac{3}{2}, \dfrac{7}{2}, \dfrac{1}{2}\right)$

33. Midpoint: $\left(\dfrac{3 - 3}{2}, \dfrac{-6 + 4}{2}, \dfrac{10 + 4}{2}\right) = (0, -1, 7)$

34. Midpoint: $\left(\dfrac{-1 + 3}{2}, \dfrac{5 + 7}{2}, \dfrac{-3 - 1}{2}\right) = (1, 6, -2)$

35. Midpoint: $\left(\dfrac{6 - 4}{2}, \dfrac{-2 + 2}{2}, \dfrac{5 + 6}{2}\right) = \left(1, 0, \dfrac{11}{2}\right)$

36. Midpoint: $\left(\dfrac{-3 - 6}{2}, \dfrac{5 + 4}{2}, \dfrac{5 + 8}{2}\right) = \left(-\dfrac{9}{2}, \dfrac{9}{2}, \dfrac{13}{2}\right)$

37. Midpoint: $\left(\dfrac{-2 + 7}{2}, \dfrac{8 - 4}{2}, \dfrac{10 + 2}{2}\right) = \left(\dfrac{5}{2}, 2, 6\right)$

38. Midpoint: $\left(\dfrac{9 + 9}{2}, \dfrac{-5 - 2}{2}, \dfrac{1 - 4}{2}\right) = \left(9, -\dfrac{7}{2}, -\dfrac{3}{2}\right)$

39. $(x - 3)^2 + (y - 2)^2 + (z - 4)^2 = 16$

40. $(x + 3)^2 + (y - 4)^2 + (z - 3)^2 = 4$

41. $(x - 0)^2 + (y - 4)^2 + (z - 3)^2 = 3^2$

$x^2 + (y - 4)^2 + (z - 3)^2 = 9$

42. $(x - 2)^2 + (y + 1)^2 + (z - 8)^2 = 36$

43. Radius $= \dfrac{\text{Diameter}}{2} = 5$

$(x + 3)^2 + (y - 7)^2 + (z - 5)^2 = 5^2 = 25$

44. Radius $= \dfrac{\text{Diameter}}{2} = 4: (x - 0)^2 + (y - 5)^2 + (z + 9)^2 = 4^2 = 16$

45. Center: $\left(\dfrac{3+0}{2}, \dfrac{0+0}{2}, \dfrac{0+6}{2}\right) = \left(\dfrac{3}{2}, 0, 3\right)$

Radius: $\sqrt{\left(3 - \dfrac{3}{2}\right)^2 + (0-0)^2 + (0-3)^2} = \sqrt{\dfrac{9}{4} + 9} = \sqrt{\dfrac{45}{4}}$

Sphere: $\left(x - \dfrac{3}{2}\right)^2 + (y-0)^2 + (z-3)^2 = \dfrac{45}{4}$

46. Center: $\left(\dfrac{2-1}{2}, \dfrac{-2+4}{2}, \dfrac{2+6}{2}\right) = \left(\dfrac{1}{2}, 1, 4\right)$

Radius: $\sqrt{\left(2 - \dfrac{1}{2}\right)^2 + (-2-1)^2 + (2-4)^2} = \sqrt{\dfrac{9}{4} + 9 + 4} = \sqrt{\dfrac{61}{4}}$

Sphere: $\left(x - \dfrac{1}{2}\right)^2 + (y-1)^2 + (z-4)^2 = \dfrac{61}{4}$

47. $\left(x^2 - 5x + \dfrac{25}{4}\right) + y^2 + z^2 = \dfrac{25}{4}$

$\quad \left(x - \dfrac{5}{2}\right)^2 + y^2 + z^2 = \dfrac{25}{4}$

Center: $\left(\dfrac{5}{2}, 0, 0\right)$

Radius: $\dfrac{5}{2}$

48. $x^2 + y^2 - 8y + 16 + z^2 = 16$

$\quad x^2 + (y-4)^2 + z^2 = 16$

Center: $(0, 4, 0)$

Radius: 4

49. $(x^2 - 4x + 4) + (y^2 + 2y + 1) + (z^2 - 6z + 9) = -10 + 4 + 1 + 9$

$\qquad (x-2)^2 + (y+1)^2 + (z-3)^2 = 4$

Center: $(2, -1, 3)$

Radius: 2

50. $(x^2 - 6x + 9) + (y^2 + 4y + 4) + z^2 = -9 + 9 + 4$

$(x-3)^2 + (y+2)^2 + z^2 = 4$

Center: $(3, -2, 0)$

Radius: 2

51. $(x^2 + 4x + 4) + y^2 + (z^2 - 8z + 16) = -19 + 4 + 16$

$\qquad (x+2)^2 + y^2 + (z-4)^2 = 1$

Center: $(-2, 0, 4)$

Radius: 1

52. $x^2 + (y^2 - 8y + 16) + (z^2 - 6z + 9) = -13 + 16 + 9$

$x^2 + (y-4)^2 + (z-3)^2 = 12$

Center: $(0, 4, 3)$

Radius: $\sqrt{12} = 2\sqrt{3}$

53.

$$x^2 + y^2 + z^2 - 2x - \tfrac{2}{3}y - 8z = -\tfrac{73}{9}$$

$$(x^2 - 2x + 1) + \left(y^2 - \tfrac{2}{3}y + \tfrac{1}{9}\right) + (z^2 - 8z + 16) = -\tfrac{73}{9} + 1 + \tfrac{1}{9} + 16$$

$$(x - 1)^2 + \left(y - \tfrac{1}{3}\right)^2 + (z - 4)^2 = 9$$

Center: $\left(1, \tfrac{1}{3}, 4\right)$

Radius: 3

54. $x^2 + y^2 + z^2 - x - 3y - 2z = -\tfrac{5}{2}$

$$\left(x^2 - x + \tfrac{1}{4}\right) + \left(y^2 - 3y + \tfrac{9}{4}\right) + (z^2 - 2z + 1) = -\tfrac{5}{2} + \tfrac{1}{4} + \tfrac{9}{4} + 1$$

$$\left(x - \tfrac{1}{2}\right)^2 + \left(y - \tfrac{3}{2}\right)^2 + (z - 1)^2 = 1$$

Center: $\left(\tfrac{1}{2}, \tfrac{3}{2}, 1\right)$

Radius: 1

55.

$$9x^2 - 6x + 9y^2 + 18y + 9z^2 = -1$$

$$x^2 - \tfrac{2}{3}x + \tfrac{1}{9} + y^2 + 2y + 1 + z^2 = -\tfrac{1}{9} + \tfrac{1}{9} + 1$$

$$\left(x - \tfrac{1}{3}\right)^2 + (y + 1)^2 + z^2 = 1$$

Center: $\left(\tfrac{1}{3}, -1, 0\right)$

Radius: 1

56. $x^2 - x + \dfrac{1}{4} + y^2 - 8y + 16 + z^2 + 2z + 1 = \dfrac{-33}{4} + \dfrac{1}{4} + 16 + 1$

$$\left(x - \dfrac{1}{2}\right)^2 + (y - 4)^2 + (z + 1)^2 = 9$$

Center: $\left(\dfrac{1}{2}, 4, -1\right)$

Radius: 3

57.

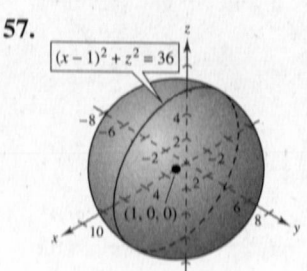

58. $yz -$ trace $(x = 0)$: $(y + 3)^2 + z^2 = 25$ Circle

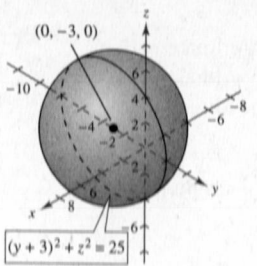

59.

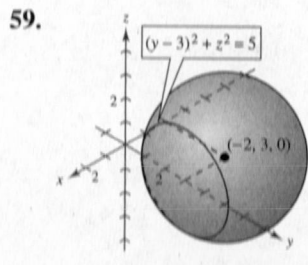

60. xy − trace $(z = 0)$: $x^2 + (y − 1)^2 = 3$ Circle

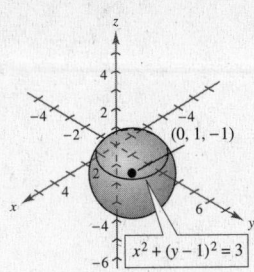

$x^2 + (y − 1)^2 = 3$

61.

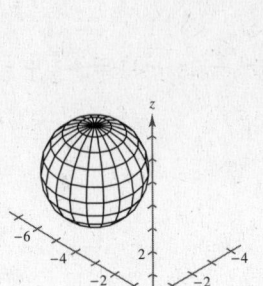

62. $x^2 + y^2 + 6y + (z^2 − 8z + 16) = −21 + 16$

$x^2 + y^2 + 6y + (z − 4)^2 = −5$

$z_1 = 4 + \sqrt{−5 − x^2 − y^2 − 6y}$

$z_2 = 4 − \sqrt{−5 − x^2 − y^2 − 6y}$

63. The length of each side is 3. Thus, $(x, y, z) = (3, 3, 3)$.

64. $x = 4, y = 4, z = 8$ $(4, 4, 8)$

65. $d = 165 \Rightarrow r = \frac{165}{2} = 82.5$

$x^2 + y^2 + z^2 = \left(\frac{165}{2}\right)^2$

66. (a) $x^2 + y^2 + z^2 = 3963^2$.

   (b) Assume the north and south poles are on the z-axis. Lines of longitude that run north-south are traces of planes containing the z-axis. These shapes are circles of radius 3963 miles.

   (c) Latitudes are traces of planes perpendicular to the z-axis. These shapes are circles.

   (d) The prime meridian is a trace of a plane containing the z-axis. It is a semi-circular arc running from pole to pole.

   (e) The equator is the trace of the plane containing the x- and y-axes.

67. False. x is the directed distance from the yz-plane to P.

68. False. The trace could be a single point, or empty.

69. In the xy-plane, the z-coordinate is 0.
In the xz-plane, the y-coordinate is 0.
In the yz-plane, the x-coordinate is 0.

70. It is a plane.

71. The trace is a circle, or a single point.

72. The trace will be a line in the xy-plane (unless the plane is the xy-plane).

73. $x_m = \dfrac{x_2 + x_1}{2} \Rightarrow x_2 = 2x_m − x_1$

   Similarly for y_2 and z_2,

   $(x_2, y_2, z_2) = (2x_m − x_1, 2y_m − y_1, 2z_m − z_1)$.

74. $x_2 = 2x_m − x_1 = 2(5) − 3 = 7$

$y_2 = 2y_m − y_1 = 2(8) − 0 = 16$

$z_2 = 2z_m − z_1 = 2(7) − 2 = 12$

$(7, 16, 12)$

75. $v^2 + 3v + \dfrac{9}{4} = 2 + \dfrac{9}{4}$

$\left(v + \dfrac{3}{2}\right)^2 = \dfrac{17}{4}$

$v + \dfrac{3}{2} = \pm\dfrac{\sqrt{17}}{2}$

$v = -\dfrac{3}{2} \pm \dfrac{\sqrt{17}}{2}$

76. $z^2 - 7z + \dfrac{49}{4} = 19 + \dfrac{49}{4}$

$\left(z - \dfrac{7}{2}\right)^2 = \dfrac{125}{4}$

$z - \dfrac{7}{2} = \pm\dfrac{5\sqrt{5}}{2}$

$z = \dfrac{7}{2} \pm \dfrac{5}{2}\sqrt{5}$

77. $x^2 - 5x + \dfrac{25}{4} = -5 + \dfrac{25}{4}$

$\left(x - \dfrac{5}{2}\right)^2 = \dfrac{5}{4}$

$x - \dfrac{5}{2} = \pm\dfrac{\sqrt{5}}{2}$

$x = \dfrac{5}{2} \pm \dfrac{\sqrt{5}}{2}$

78. $x^2 + 3x + \dfrac{9}{4} = 1 + \dfrac{9}{4}$

$\left(x + \dfrac{3}{2}\right)^2 = \dfrac{13}{4}$

$x + \dfrac{3}{2} = \pm\dfrac{\sqrt{13}}{2}$

$x = \dfrac{-3}{2} \pm \dfrac{\sqrt{13}}{2}$

79. $4y^2 + 4y = 9$

$y^2 + y + \dfrac{1}{4} = \dfrac{9}{4} + \dfrac{1}{4}$

$\left(y + \dfrac{1}{2}\right)^2 = \dfrac{10}{4}$

$y + \dfrac{1}{2} = \pm\dfrac{\sqrt{10}}{2}$

$y = -\dfrac{1}{2} \pm \dfrac{\sqrt{10}}{2}$

80. $x^2 + \dfrac{5}{2}x + \dfrac{25}{16} = 4 + \dfrac{25}{16}$

$\left(x + \dfrac{5}{4}\right)^2 = \dfrac{89}{16}$

$x + \dfrac{5}{4} = \pm\dfrac{\sqrt{89}}{4}$

$x = \dfrac{-5}{4} \pm \dfrac{\sqrt{89}}{4}$

81. $\mathbf{v} = 3\mathbf{i} - 3\mathbf{j}$, Quadrant IV

$\|\mathbf{v}\| = \sqrt{3^2 + (-3)^2}$

$= \sqrt{18}$

$= 3\sqrt{2}$

$\tan\theta = -\dfrac{3}{3} = -1 \implies$

$\theta = -45° \text{ or } 315°$

82. $\mathbf{v} = \langle -1, 2 \rangle$ Quadrant II

$\|\mathbf{v}\| = \sqrt{1^2 + 2^2} = \sqrt{5}$

$\tan\theta = \dfrac{2}{-1} \implies \theta \approx 116.6°$

83. $\mathbf{v} = 4\mathbf{i} + 5\mathbf{j}$, Quadrant I

$\|\mathbf{v}\| = \sqrt{16 + 25} = \sqrt{41}$

$\tan\theta = \dfrac{5}{4} \implies \theta \approx 51.34°$

84. $\mathbf{v} = \langle 10, -7 \rangle$ Quadrant IV

$\|\mathbf{v}\| = \sqrt{100 + 49} = \sqrt{149}$

$\tan\theta = \dfrac{-7}{10} \implies \theta \approx 325.0°$

85. $\mathbf{u} \cdot \mathbf{v} = \langle -4, 1 \rangle \cdot \langle 3, 5 \rangle$

$= -4(3) + 1(5)$

$= -7$

86. $\mathbf{u} \cdot \mathbf{v} = \langle -1, 0 \rangle \cdot \langle -2, -6 \rangle$

$= 2 + 0$

$= 2$

87. $a_0 = 1, a_n = a_{n-1} + n^2$

$a_1 = 1 + 1^2 = 2$

$a_2 = 2 + 2^2 = 6$

$a_3 = 6 + 3^2 = 15$

$a_4 = 15 + 4^2 = 31$

	1		2		6		15		31
First differences:		1		4		9		16	
Second differences:			3		5		7		

Neither model

88. $a_0 = 0, a_n = a_{n-1} - 1$

$a_1 = 0 - 1 = -1$

$a_2 = -1 - 1 = -2$

$a_3 = -3$

$a_4 = -4$

	0		−1		−2		−3		−4
First difference		−1		−1		−1		−1	
Second difference			0		0		0		

Linear model

89. $a_1 = -1, a_n = a_{n-1} + 3$

$a_2 = -1 + 3 = 2$

$a_3 = 2 + 3 = 5$

$a_4 = 5 + 3 = 8$

$a_5 = 8 + 3 = 11$

	−1		2		5		8		11
First differences:		3		3		3		3	
Second differences:			0		0		0		

Linear model

90. $a_1 = 4, a_n = a_{n-1} - 2n$

$a_2 = 4 - 2(2) = 0$

$a_3 = 0 - 2(3) = -6$

$a_4 = -6 - 2(4) = -14$

$a_5 = -14 - 2(5) = -24$

	4		0		−6		−14		−24
First difference		−4		−6		−8		−10	
Second difference			−2		−2		−2		

Quadratic model

91. $(x + 5)^2 + (y - 1)^2 = 49$

92. $(x - 3)^2 + (y + 6)^2 = 81$

93. $(y - 1)^2 = 4p(x - 4), \ p = -3$

$(y - 1)^2 = 4(-3)(x - 4)$

$(y - 1)^2 = -12(x - 4)$

94. $(x - h)^2 = 4p(y - k) \ \ p = -5, (h, k) = (-2, 5)$

$(x + 2)^2 = 4(-5)(y - 5)$

$(x + 2)^2 = -20(y - 5)$

95. $a = 3, b = 2$, center: $(3, 3)$, horizontal major axis

$$\frac{(x - 3)^2}{9} + \frac{(y - 3)^2}{4} = 1$$

96. Center: $(0, 3)$ Vertical major axis length $9 \implies a = \dfrac{9}{2}$

$c = 3 \implies b^2 = a^2 - c^2 = \dfrac{81}{4} - 9 = \dfrac{45}{4}$

$\dfrac{(x - 0)^2}{(45/4)} + \dfrac{(y - 3)^2}{(81/4)} = 1$

97. Center: $(6, 0)$, horizontal transverse axis

$a = 2, c = 6, b^2 = c^2 - a^2 = 36 - 4 = 32$

$\dfrac{(x - 6)^2}{4} - \dfrac{y^2}{32} = 1$

98. Center: $(3, 5)$ Vertical transverse axis

$a = 4, c = 5, b^2 = c^2 - a^2 = 25 - 16 = 9$

$\dfrac{(y - 5)^2}{16} - \dfrac{(x - 3)^2}{9} = 1$

Section 11.2   Vectors in Space

- Vectors in space $\mathbf{v} = \langle v_1, v_2, v_3 \rangle$ have many of the same properties as vectors in the plane.
- The dot product of two vectors $\mathbf{u} = \langle u_1, u_2, u_3 \rangle$ and $\mathbf{v} = \langle v_1, v_2, v_3 \rangle$ in space is $\mathbf{u} \cdot \mathbf{v} = u_1 v_1 + u_2 v_2 + u_3 v_3$.
- Two nonzero vectors $\mathbf{u}$ and $\mathbf{v}$ are said to be parallel if there is some scalar c such that $\mathbf{u} = c\mathbf{v}$.
- You should be able to use vectors to solve real life problems.

Vocabulary Check

1. zero

2. $\mathbf{v} = v_1 \mathbf{i} + v_2 \mathbf{j} + v_3 \mathbf{k}$

3. component form

4. orthogonal

5. parallel

---

1. $\mathbf{v} = \langle 0 - 2, 3 - 0, 2 - 1 \rangle = \langle -2, 3, 1 \rangle$

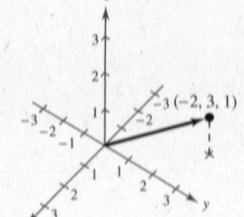

2. (a) $\mathbf{v} = \langle 1 - 1, 4 - 4, 0 - 4 \rangle = \langle 0, 0, -4 \rangle$

(b)

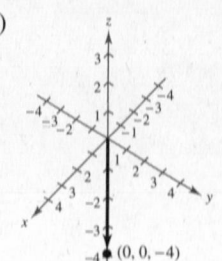

3. (a) $\mathbf{v} = \langle 1 - (-6), -1 - 4, 3 - (-2) \rangle$

$\qquad = \langle 7, -5, 5 \rangle$

(b) $\|\mathbf{v}\| = \sqrt{7^2 + (-5)^2 + 5^2}$

$\qquad = \sqrt{49 + 25 + 25}$

$\qquad = \sqrt{99}$

$\qquad = 3\sqrt{11}$

(c) $\dfrac{\mathbf{v}}{\|\mathbf{v}\|} = \dfrac{1}{3\sqrt{11}} \langle 7, -5, 5 \rangle = \dfrac{\sqrt{11}}{33} \langle 7, -5, 5 \rangle$

4. (a) $\mathbf{v} = \langle 0 + 7, 0 - 3, 2 - 5 \rangle = \langle 7, -3, -3 \rangle$

(b) $\|\mathbf{v}\| = \sqrt{49 + 9 + 9} = \sqrt{67}$

(c) Unit vector:

$\qquad \dfrac{1}{\sqrt{67}} \langle 7, -3, -3 \rangle = \dfrac{\sqrt{67}}{67} \langle 7, -3, -3 \rangle$

5. (a)

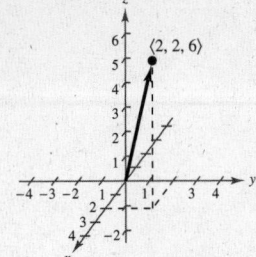

(b)

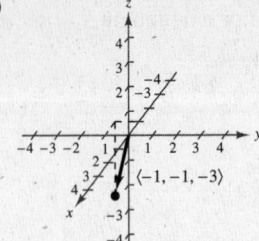

(c)

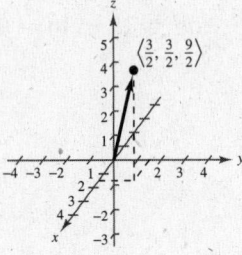

(d)

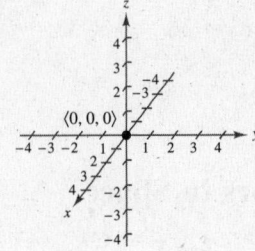

6. $\mathbf{v} = \langle -1, 2, 2 \rangle$

(a) $-\mathbf{v} = \langle 1, -2, -2 \rangle$

(b) $2\mathbf{v} = \langle -2, 4, 4 \rangle$

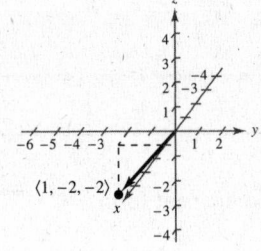

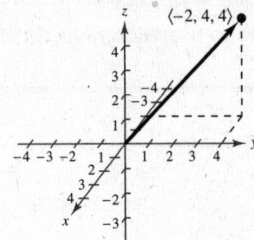

(c) $\dfrac{1}{2}\mathbf{v} = \left\langle -\dfrac{1}{2}, 1, 1 \right\rangle$

(d) $\dfrac{5}{2}\mathbf{v} = \left\langle \dfrac{-5}{2}, 5, 5 \right\rangle$

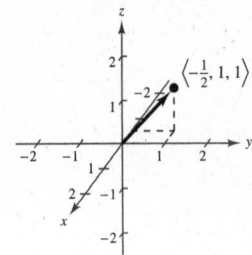

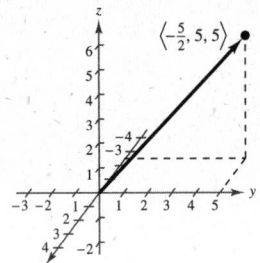

7. $\mathbf{z} = \mathbf{u} - 2\mathbf{v} = \langle -1, 3, 2 \rangle - 2\langle 1, -2, -2 \rangle = \langle -3, 7, 6 \rangle$

8. $\mathbf{z} = 7\langle -1, 3, 2 \rangle + \langle 1, -2, -2 \rangle - \dfrac{1}{5}\langle 5, 0, -5 \rangle = \langle -7, 19, 13 \rangle$

9. $2\mathbf{z} - 4\mathbf{u} = \mathbf{w} \implies \mathbf{z} = \dfrac{1}{2}(4\mathbf{u} + \mathbf{w}) = \dfrac{1}{2}(4\langle -1, 3, 2 \rangle + \langle 5, 0, -5 \rangle) = \left\langle \dfrac{1}{2}, 6, \dfrac{3}{2} \right\rangle$

10. $\mathbf{z} = -\mathbf{u} - \mathbf{v} = -\langle -1, 3, 2 \rangle - \langle 1, -2, -2 \rangle = \langle 0, -1, 0 \rangle$

11. $\|\mathbf{v}\| = \|\langle 7, 8, 7 \rangle\|$

$\quad = \sqrt{49 + 64 + 49} = \sqrt{162} = 9\sqrt{2}$

12. $\|\mathbf{v}\| = \sqrt{(-2)^2 + 0^2 + (-5)^2} = \sqrt{4 + 25} = \sqrt{29}$

13. $\|\mathbf{v}\| = \sqrt{4^2 + (-3)^2 + (-7)^2}$

$\quad = \sqrt{16 + 9 + 49} = \sqrt{74}$

14. $\|\mathbf{v}\| = \sqrt{2^2 + (-1)^2 + 6^2} = \sqrt{41}$

15. $\mathbf{v} = \langle 1 - 1, 0 - (-3), -1 - 4 \rangle = \langle 0, 3, -5 \rangle$

$\quad \|\mathbf{v}\| = \sqrt{0 + 3^2 + (-5)^2} = \sqrt{34}$

16. $\mathbf{v} = \langle 1 - 0, 2 - (-1), -2 - 0 \rangle = \langle 1, 3, -2 \rangle$

$\quad \|\mathbf{v}\| = \sqrt{1 + 9 + 4} = \sqrt{14}$

17. (a) $\dfrac{\mathbf{u}}{\|\mathbf{u}\|} = \dfrac{\langle 8, 3, -1 \rangle}{\sqrt{74}}$

$\quad\quad = \dfrac{1}{\sqrt{74}}(8\mathbf{i} + 3\mathbf{j} - \mathbf{k}) = \dfrac{\sqrt{74}}{74}\langle 8, 3, -1 \rangle$

$\quad$ (b) $-\dfrac{1}{\sqrt{74}}(8\mathbf{i} + 3\mathbf{j} - \mathbf{k}) = -\dfrac{\sqrt{74}}{74}\langle 8, 3, -1 \rangle$

18. (a) $\dfrac{\mathbf{u}}{\|\mathbf{u}\|} = \dfrac{\langle -3, 5, 10 \rangle}{\sqrt{134}} = \dfrac{1}{\sqrt{134}}(-3\mathbf{i} + 5\mathbf{j} + 10\mathbf{k})$

$\quad$ (b) $\dfrac{-1}{\sqrt{134}}(-3\mathbf{i} + 5\mathbf{j} + 10\mathbf{k})$

19. $\mathbf{u} \cdot \mathbf{v} = \langle 4, 4, -1 \rangle \cdot \langle 2, -5, -8 \rangle$

$\quad = 8 - 20 + 8 = -4$

20. $\mathbf{u} \cdot \mathbf{v} = 3(4) + (-1)(-10) + 6(1) = 28$

21. $\mathbf{u} \cdot \mathbf{v} = \langle 2, -5, 3 \rangle \cdot \langle 9, 3, -1 \rangle$

$\quad = 18 - 15 - 3 = 0$

22. $\mathbf{u} \cdot \mathbf{v} = 0(6) + 3(-4) + (-6)(-2) = 0$

23. $\cos\theta = \dfrac{\mathbf{u} \cdot \mathbf{v}}{\|\mathbf{u}\|\,\|\mathbf{v}\|} = \dfrac{-8}{\sqrt{8}\sqrt{25}} \implies \theta \approx 124.45°$

24. $\cos\theta = \dfrac{\mathbf{u} \cdot \mathbf{v}}{\|\mathbf{u}\|\,\|\mathbf{v}\|} = \dfrac{5}{\sqrt{10}\sqrt{6}} \implies \theta \approx 49.80°$

25. $\cos\theta = \dfrac{\mathbf{u} \cdot \mathbf{v}}{\|\mathbf{u}\|\,\|\mathbf{v}\|} = \dfrac{-120}{\sqrt{1700}\sqrt{73}} \implies \theta \approx 109.92°$

26. $\cos\theta = \dfrac{\mathbf{u} \cdot \mathbf{v}}{\|\mathbf{u}\|\,\|\mathbf{v}\|} = \dfrac{100}{\sqrt{464}\sqrt{125}} \implies \theta \approx 65.47°$

27. $-\frac{3}{2}\langle 8, -4, -10 \rangle = \langle -12, 6, 15 \rangle \implies$ parallel

28. $\mathbf{u} \cdot \mathbf{v} = -2 - 3 - 5 = -10 \neq 0$ and

$\quad \mathbf{u} \neq c\mathbf{v} \implies$ neither

29. $\mathbf{u} \cdot \mathbf{v} = 3 - 5 + 2 = 0 \implies$ orthogonal

30. $-8\mathbf{u} = -8\langle -1, \frac{1}{2}, -1 \rangle = \langle 8, -4, 8 \rangle = \mathbf{v} \implies$ parallel

31. $\mathbf{v} = \langle 7 - 5, 3 - 4, -1 - 1 \rangle = \langle 2, -1, -2 \rangle$

$\quad \mathbf{u} = \langle 4 - 7, 5 - 3, 3 - (-1) \rangle = \langle -3, 2, 4 \rangle$

$\quad$ Since $\mathbf{u}$ and $\mathbf{v}$ are not parallel, the points are not collinear.

32. $\mathbf{v} = \langle -4 - (-2), 8 - 7, 1 - 4 \rangle = \langle -2, 1, -3 \rangle$

$\quad \mathbf{u} = \langle 0 - (-4), 6 - 8, 7 - 1 \rangle = \langle 4, -2, 6 \rangle$

$\quad$ Since $\mathbf{u} = -2\mathbf{v}$, the points are collinear.

33. $\mathbf{v} = \langle -1 - 1, 2 - 3, 5 - 2 \rangle = \langle -2, -1, 3 \rangle$

$\quad \mathbf{u} = \langle 3 - (-1), 4 - 2, -1 - 5 \rangle = \langle 4, 2, -6 \rangle$

$\quad$ Since $\mathbf{u} = -2\mathbf{v}$, the points are collinear.

34. $\mathbf{v} = \langle -1 - 0, 5 - 4, 6 - 4 \rangle = \langle -1, 1, 2 \rangle$

$\quad \mathbf{u} = \langle -2 - (-1), 6 - 5, 7 - 6 \rangle = \langle -1, 1, 1 \rangle$

$\quad$ Since $\mathbf{u}$ and $\mathbf{v}$ are not parallel, the points are not collinear.

35. $\mathbf{v} = \langle 2, -4, 7 \rangle = \langle q_1 - 1, q_2 - 5, q_3 - 0 \rangle \Rightarrow$

$$\left.\begin{array}{l} 2 = q_1 - 1 \\ -4 = q_2 - 5 \\ 7 = q_3 \end{array}\right\} \Rightarrow \left.\begin{array}{l} q_1 = 3 \\ q_2 = 1 \\ q_3 = 7 \end{array}\right\} \Rightarrow$$

Terminal point is $(3, 1, 7)$.

36. $\langle 4, -1, -1 \rangle = \langle x - 6, y + 4, z - 3 \rangle \Rightarrow (x, y, z) = (10, -5, 2)$

37. $\mathbf{v} = \left\langle 4, \frac{3}{2}, -\frac{1}{4} \right\rangle = \left\langle q_1 - 2, q_2 - 1, q_3 + \frac{3}{2} \right\rangle$

$4 = q_1 - 2 \Rightarrow q_1 = 6$

$\frac{3}{2} = q_2 - 1 \Rightarrow q_2 = \frac{5}{2}$

$-\frac{1}{4} = q_3 + \frac{3}{2} \Rightarrow q_3 = -\frac{7}{4}$

Terminal point: $\left(6, \frac{5}{2}, -\frac{7}{4}\right)$

38. $\left\langle \frac{5}{2}, -\frac{1}{2}, 4 \right\rangle = \left\langle x - 3, y - 2, z + \frac{1}{2} \right\rangle \Rightarrow (x, y, z) = \left(\frac{11}{2}, \frac{3}{2}, \frac{7}{2}\right)$

39. $c\mathbf{u} = c\mathbf{i} + 2c\mathbf{j} + 3c\mathbf{k}$

$\|c\mathbf{u}\| = \sqrt{c^2 + 4c^2 + 9c^2} = |c|\sqrt{14} = 3 \Rightarrow$

$c = \pm\dfrac{3}{\sqrt{14}} = \pm\dfrac{3\sqrt{14}}{14}$

40. $\|c\,\mathbf{u}\| = |c|\,\|\mathbf{u}\| = |c|\sqrt{4 + 4 + 16} = |c|\sqrt{24} = 12$

$\Rightarrow |c| = \dfrac{12}{\sqrt{24}} = \dfrac{6}{\sqrt{6}} = \sqrt{6} \Rightarrow c = \pm\sqrt{6}$

41. $\mathbf{v} = \langle q_1, q_2, q_3 \rangle$

Since $\mathbf{v}$ lies in the yz-plane, $q_1 = 0$. Since $\mathbf{v}$ makes an angle of $45°$, $q_2 = q_3$. Finally, $\|\mathbf{v}\| = 4$ implies that $q_2^2 + q_3^2 = 16$. Thus, $q_2 = q_3 = 2\sqrt{2}$ and $\mathbf{v} = \langle 0, 2\sqrt{2}, 2\sqrt{2} \rangle$, or $q_2 = 2\sqrt{2}$ and $q_3 = -2\sqrt{2}$ and $\mathbf{v} = \langle 0, 2\sqrt{2}, -2\sqrt{2} \rangle$.

42. $\mathbf{v}$ lies in xz-plane $\Rightarrow y = 0$.

$\mathbf{v} = 10\langle \sin 60°, 0, \cos 60° \rangle = \langle 5\sqrt{3}, 0, 5 \rangle$, or

$\mathbf{v} = 10\langle -\sin 60°, 0, \cos 60° \rangle = \langle -5\sqrt{3}, 0, 5 \rangle$

43. $\overrightarrow{AB} = \langle 0, 70, 115 \rangle$. $F_1 = C_1\langle 0, 70, 115 \rangle$

$\overrightarrow{AC} = \langle -60, 0, 115 \rangle$. $F_2 = C_2\langle -60, 0, 115 \rangle$

$\overrightarrow{AD} = \langle 45, -65, 115 \rangle$. $F_3 = C_3\langle 45, -65, 115 \rangle$

$F_1 + F_2 + F_3 = \langle 0, 0, -500 \rangle$. Thus,

$$-60C_2 + 45C_3 = 0$$

$$70C_1 - 65C_3 = 0$$

$$115C_1 + 115C_2 + 115C_3 = -500$$

Solving this system yields $C_1 = \dfrac{-104}{69}$, $C_2 = \dfrac{-28}{23}$, $C_3 = \dfrac{-112}{69}$.

Thus,

$$\|\mathbf{F}_1\| \approx 202.919 \quad \text{N}$$

$$\|\mathbf{F}_2\| \approx 157.909 \quad \text{N}$$

$$\|\mathbf{F}_3\| \approx 226.521 \quad \text{N}$$

44. (a) $\sin \theta = \dfrac{18}{L}$

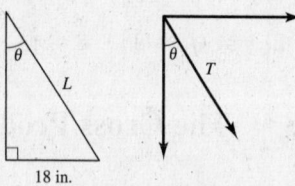

$\theta = \sin^{-1}\left(\dfrac{18}{L}\right)$

$T = \dfrac{8}{\cos \theta} = \dfrac{8}{\cos\left(\sin^{-1}\left(\dfrac{18}{L}\right)\right)} = \dfrac{8}{\dfrac{\sqrt{L^2 - 18^2}}{L}} = \dfrac{8L}{\sqrt{L^2 - 18^2}}$

Domain: $L > 18$

(b)

L	20	25	30	35	40	45	50
T	18.4	11.5	10	9.3	9.0	8.7	8.6

(c)

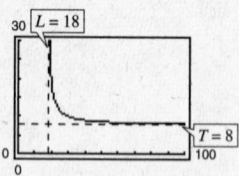

Vertical asymptote: $L = 18$

Horizontal asymptote: $T = 8$

The minimum tension in each cable is 8 pounds and the minimum cable length is 18 inches.

(d) $10 = \dfrac{8}{\cos\left(\sin^{-1}\left(\dfrac{18}{L}\right)\right)} \Rightarrow \cos\left(\sin^{-1}\left(\dfrac{18}{L}\right)\right) = \dfrac{8}{10} = \dfrac{4}{5}$

$\sin^{-1}\left(\dfrac{18}{L}\right) = \cos^{-1}\left(\dfrac{4}{5}\right)$

$\dfrac{18}{L} = \sin\left(\cos^{-1}\left(\dfrac{4}{5}\right)\right) = \dfrac{3}{5}$

$L = \dfrac{90}{3} = 30$ inches

45. True. $\cos \theta = 0 \Rightarrow \theta = 90°$ **46.** True

47. If $\mathbf{u} \cdot \mathbf{v} < 0$, then $\cos \theta < 0$ and the angle between $\mathbf{u}$ and $\mathbf{v}$ is obtuse, $180° > \theta > 90°$.

48. Let $\mathbf{v} = \langle v_1, v_2, v_3 \rangle$ and $\mathbf{u} = \langle u_1, u_2, u_3 \rangle$.

Then $t\mathbf{v} = \langle tv_1, tv_2, tv_3 \rangle$

$\mathbf{u} + t\mathbf{v} = \langle u_1 + tv_1, u_2 + tv_2, u_3 + tv_3 \rangle$

and $s\mathbf{u} + t\mathbf{v} = \langle su_1 + tv_1, su_2 + tv_2, su_3 + tv_3 \rangle$

The endpoints of these three vectors are collinear, as indicated in the figure.

So, the figure is a line.

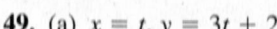

49. (a) $x = t, y = 3t + 2$

(b) $x = t - 1, y = 3(t - 1) + 2 = 3t - 1$

50. (a) $x = t, y = \dfrac{2}{t}$

(b) $x = t - 1, y = \dfrac{2}{t - 1}$

51. (a) $x = t, y = t^2 - 8$

(b) $x = t - 1, y = (t - 1)^2 - 8 = t^2 - 2t - 7$

52. (a) $x = t, y = 4t^3$

(b) $x = t - 1, y = 4(t - 1)^3$

Section 11.3   The Cross Product of Two Vectors

- The cross product of two vectors $\mathbf{u} = u_1\mathbf{i} + u_2\mathbf{j} + u_3\mathbf{k}$ and $\mathbf{v} = v_1\mathbf{i} + v_2\mathbf{j} + v_3\mathbf{k}$ is given by

$$\mathbf{u} \times \mathbf{v} = (u_2v_3 - u_3v_2)\mathbf{i} - (u_1v_3 - u_3v_1)\mathbf{j} + (u_1v_2 - u_2v_1)\mathbf{k}$$

$$= \begin{vmatrix} \mathbf{i} & \mathbf{j} & \mathbf{k} \\ u_1 & u_2 & u_3 \\ v_1 & v_2 & v_3 \end{vmatrix}.$$

- The cross product satisfies the following algebraic properties.

    (a) $\mathbf{u} \times \mathbf{v} = -(\mathbf{v} \times \mathbf{u})$

    (b) $\mathbf{u} \times (\mathbf{v} + \mathbf{w}) = (\mathbf{u} \times \mathbf{v}) + (\mathbf{u} \times \mathbf{w})$

    (c) $c(\mathbf{u} \times \mathbf{v}) = (c\mathbf{u}) \times \mathbf{v} = \mathbf{u} \times (c\mathbf{v})$

    (d) $\mathbf{u} \times \mathbf{0} = \mathbf{0} \times \mathbf{u} = \mathbf{0}$

    (e) $\mathbf{u} \times \mathbf{u} = \mathbf{0}$

    (f) $\mathbf{u} \cdot (\mathbf{v} \times \mathbf{w}) = (\mathbf{u} \times \mathbf{v}) \cdot \mathbf{w}$

- The following geometric properties of the cross product are valid, where θ is the angle between the vectors $\mathbf{u}$ and $\mathbf{v}$:

    (a) $\mathbf{u} \times \mathbf{v}$ is orthogonal to both $\mathbf{u}$ and $\mathbf{v}$.

    (b) $\|\mathbf{u} \times \mathbf{v}\| = \|\mathbf{u}\| \|\mathbf{v}\| \sin \theta$

    (c) $\mathbf{u} \times \mathbf{v} = \mathbf{0}$ if and only if $\mathbf{u}$ and $\mathbf{v}$ are scalar multiples.

    (d) $\|\mathbf{u} \times \mathbf{v}\|$ is the area of the parallelogram having $\mathbf{u}$ and $\mathbf{v}$ as sides.

- The absolute value of the triple scalar product is the volume of the parallelepiped having $\mathbf{u}$, $\mathbf{v}$, and $\mathbf{w}$ as sides.

$$\mathbf{u} \cdot (\mathbf{v} \times \mathbf{w}) = \begin{vmatrix} u_1 & u_2 & u_3 \\ v_1 & v_2 & v_3 \\ w_1 & w_2 & w_3 \end{vmatrix}$$

Vocabulary Check

1. cross product

2. 0

3. $\|\mathbf{u}\| \|\mathbf{v}\| \sin \theta$

4. triple scalar product

1. $\mathbf{j} \times \mathbf{i} = \begin{vmatrix} \mathbf{i} & \mathbf{j} & \mathbf{k} \\ 0 & 1 & 0 \\ 1 & 0 & 0 \end{vmatrix} = -\mathbf{k}$

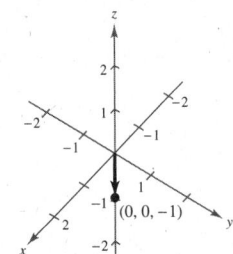

2. $\mathbf{k} \times \mathbf{j} = \begin{vmatrix} \mathbf{i} & \mathbf{j} & \mathbf{k} \\ 0 & 0 & 1 \\ 0 & 1 & 0 \end{vmatrix} = -\mathbf{i}$

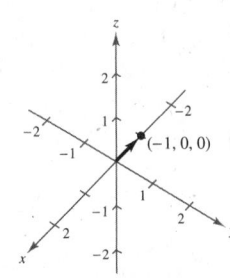

3. $\mathbf{i} \times \mathbf{k} = \begin{vmatrix} \mathbf{i} & \mathbf{j} & \mathbf{k} \\ 1 & 0 & 0 \\ 0 & 0 & 1 \end{vmatrix} = -\mathbf{j}$

4. $\mathbf{k} \times \mathbf{i} = \begin{vmatrix} \mathbf{i} & \mathbf{j} & \mathbf{k} \\ 0 & 0 & 1 \\ 1 & 0 & 0 \end{vmatrix} = \mathbf{j}$

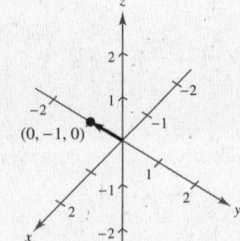

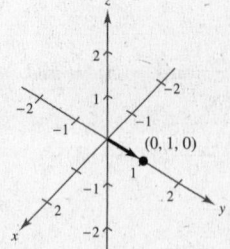

5. $\mathbf{u} \times \mathbf{v} = \begin{vmatrix} \mathbf{i} & \mathbf{j} & \mathbf{k} \\ 3 & -2 & 5 \\ 0 & -1 & 1 \end{vmatrix} = \langle 3, -3, -3 \rangle$

$(\mathbf{u} \times \mathbf{v}) \cdot \mathbf{u} = \langle 3, -3, -3 \rangle \cdot \langle 3, -2, 5 \rangle = 0$

$(\mathbf{u} \times \mathbf{v}) \cdot \mathbf{v} = \langle 3, -3, -3 \rangle \cdot \langle 0, -1, 1 \rangle = 0$

6. $\mathbf{u} \times \mathbf{v} = \begin{vmatrix} \mathbf{i} & \mathbf{j} & \mathbf{k} \\ 6 & 8 & 3 \\ 4 & -1 & -4 \end{vmatrix} = \langle -29, 36, -38 \rangle$

7. $\mathbf{u} \times \mathbf{v} = \begin{vmatrix} \mathbf{i} & \mathbf{j} & \mathbf{k} \\ -10 & 0 & 6 \\ 7 & 0 & 0 \end{vmatrix} = \langle 0, 42, 0 \rangle$

$(\mathbf{u} \times \mathbf{v}) \cdot \mathbf{u} = \langle 0, 42, 0 \rangle \cdot \langle -10, 0, 6 \rangle = 0$

$(\mathbf{u} \times \mathbf{v}) \cdot \mathbf{v} = \langle 0, 42, 0 \rangle \cdot \langle 7, 0, 0 \rangle = 0$

8. $\mathbf{u} \times \mathbf{v} = \begin{vmatrix} \mathbf{i} & \mathbf{j} & \mathbf{k} \\ -5 & 5 & 11 \\ 2 & 2 & 3 \end{vmatrix} = \langle -7, 37, -20 \rangle$

9. $\mathbf{u} \times \mathbf{v} = \begin{vmatrix} \mathbf{i} & \mathbf{j} & \mathbf{k} \\ 6 & 2 & 1 \\ 1 & 3 & -2 \end{vmatrix} = \langle -7, 13, 16 \rangle$

$= -7\mathbf{i} + 13\mathbf{j} + 16\mathbf{k}$

10. $\mathbf{u} \times \mathbf{v} = \begin{vmatrix} \mathbf{i} & \mathbf{j} & \mathbf{k} \\ 1 & \frac{3}{2} & -\frac{5}{2} \\ \frac{1}{2} & -\frac{3}{4} & \frac{1}{4} \end{vmatrix} = \left\langle -\frac{3}{2}, -\frac{3}{2}, -\frac{3}{2} \right\rangle = -\frac{3}{2}\mathbf{i} - \frac{3}{2}\mathbf{j} - \frac{3}{2}\mathbf{k}$

11. $\mathbf{u} \times \mathbf{v} = \begin{vmatrix} \mathbf{i} & \mathbf{j} & \mathbf{k} \\ 0 & 0 & 6 \\ -1 & 3 & 1 \end{vmatrix} = \langle -18, -6, 0 \rangle$

$= -18\mathbf{i} - 6\mathbf{j}$

12. $\mathbf{u} \times \mathbf{v} = \begin{vmatrix} \mathbf{i} & \mathbf{j} & \mathbf{k} \\ \frac{2}{3} & 0 & 0 \\ 0 & \frac{1}{3} & -3 \end{vmatrix} = 2\mathbf{j} + \frac{2}{9}\mathbf{k}$

13. $\mathbf{u} \times \mathbf{v} = \begin{vmatrix} \mathbf{i} & \mathbf{j} & \mathbf{k} \\ -1 & 0 & 1 \\ 0 & 1 & -2 \end{vmatrix} = \langle -1, -2, -1 \rangle$

$= -\mathbf{i} - 2\mathbf{j} - \mathbf{k}$

14. $\mathbf{u} \times \mathbf{v} = \begin{vmatrix} \mathbf{i} & \mathbf{j} & \mathbf{k} \\ 1 & 0 & -2 \\ 0 & -1 & 1 \end{vmatrix} = (0 - 2)\mathbf{i} - (1 - 0)\mathbf{j} + (-1 - 0)\mathbf{k} = -2\mathbf{i} - \mathbf{j} - \mathbf{k}$

15. $\mathbf{u} \times \mathbf{v} = \begin{vmatrix} \mathbf{i} & \mathbf{j} & \mathbf{k} \\ 3 & 1 & 0 \\ 0 & 1 & 1 \end{vmatrix} = \mathbf{i} - 3\mathbf{j} + 3\mathbf{k}$

$\|\mathbf{u} \times \mathbf{v}\| = \sqrt{19}$

Unit vector $= \dfrac{\mathbf{u} \times \mathbf{v}}{\|\mathbf{u} \times \mathbf{v}\|} = \dfrac{1}{\sqrt{19}}(\mathbf{i} - 3\mathbf{j} + 3\mathbf{k})$

$= \dfrac{\sqrt{19}}{19}\langle 1, -3, 3 \rangle$

16. $\mathbf{u} \times \mathbf{v} = \begin{vmatrix} \mathbf{i} & \mathbf{j} & \mathbf{k} \\ 1 & 2 & 0 \\ 1 & 0 & -3 \end{vmatrix} = -6\mathbf{i} + 3\mathbf{j} - 2\mathbf{k}.$

$\|\mathbf{u} \times \mathbf{v}\| = \sqrt{36 + 9 + 4} = 7$

Unit vector $= \dfrac{\mathbf{u} \times \mathbf{v}}{\|\mathbf{u} \times \mathbf{v}\|} = -\dfrac{6}{7}\mathbf{i} + \dfrac{3}{7}\mathbf{j} - \dfrac{2}{7}\mathbf{k}$

17. $\mathbf{u} \times \mathbf{v} = \begin{vmatrix} \mathbf{i} & \mathbf{j} & \mathbf{k} \\ -3 & 2 & -5 \\ \frac{1}{2} & -\frac{3}{4} & \frac{1}{10} \end{vmatrix} = \left\langle -\dfrac{71}{20}, -\dfrac{11}{5}, \dfrac{5}{4} \right\rangle$

Consider the parallel vector $\langle -71, -44, 25 \rangle = \mathbf{w}$.

$\|\mathbf{w}\| = \sqrt{71^2 + 44^2 + 25^2} = \sqrt{7602}$

Unit vector $= \dfrac{1}{\sqrt{7602}}\langle -71, -44, 25 \rangle$

$= \dfrac{\sqrt{7602}}{7602}\langle -71, -44, 25 \rangle$

18. $\mathbf{u} \times \mathbf{v} = \begin{vmatrix} \mathbf{i} & \mathbf{j} & \mathbf{k} \\ 7 & -14 & 5 \\ 14 & 28 & -15 \end{vmatrix} = 70\mathbf{i} + 175\mathbf{j} + 392\mathbf{k}$

$\|\mathbf{u} \times \mathbf{v}\| = \sqrt{70^2 + 175^2 + 392^2}$

$= \sqrt{189{,}189} = 21\sqrt{429}$

Unit vector $= \dfrac{\mathbf{u} \times \mathbf{v}}{\|\mathbf{u} \times \mathbf{v}\|} = \dfrac{1}{21\sqrt{429}}\langle 70, 175, 392 \rangle$

$= \dfrac{1}{3\sqrt{429}}\langle 10, 25, 56 \rangle$

$= \dfrac{\sqrt{429}}{1287}\langle 10, 25, 56 \rangle$

19. $\mathbf{u} \times \mathbf{v} = \begin{vmatrix} \mathbf{i} & \mathbf{j} & \mathbf{k} \\ 1 & 1 & -1 \\ 1 & 1 & 1 \end{vmatrix} = 2\mathbf{i} - 2\mathbf{j}$

$\|\mathbf{u} \times \mathbf{v}\| = 2\sqrt{2}$

Unit vector $= \dfrac{\mathbf{u} \times \mathbf{v}}{\|\mathbf{u} \times \mathbf{v}\|} = \dfrac{1}{2\sqrt{2}}(2\mathbf{i} - 2\mathbf{j})$

$= \dfrac{1}{\sqrt{2}}\mathbf{i} - \dfrac{1}{\sqrt{2}}\mathbf{j}$

$= \dfrac{\sqrt{2}}{2}\mathbf{i} - \dfrac{\sqrt{2}}{2}\mathbf{j}$

20. $\mathbf{u} \times \mathbf{v} = \begin{vmatrix} \mathbf{i} & \mathbf{j} & \mathbf{k} \\ 1 & -2 & 2 \\ 2 & -1 & -2 \end{vmatrix} = 6\mathbf{i} + 6\mathbf{j} + 3\mathbf{k}$

$\|\mathbf{u} \times \mathbf{v}\| = \sqrt{36 + 36 + 9} = 9$

Unit vector $= \dfrac{\mathbf{u} \times \mathbf{v}}{\|\mathbf{u} \times \mathbf{v}\|} = \dfrac{1}{9}(6\mathbf{i} + 6\mathbf{j} + 3\mathbf{k})$

$= \dfrac{2}{3}\mathbf{i} + \dfrac{2}{3}\mathbf{j} + \dfrac{1}{3}\mathbf{k}$

21. $\mathbf{u} \times \mathbf{v} = \begin{vmatrix} \mathbf{i} & \mathbf{j} & \mathbf{k} \\ 0 & 0 & 1 \\ 1 & 0 & 1 \end{vmatrix} = \mathbf{j}$

Area $= \|\mathbf{u} \times \mathbf{v}\| = \|\mathbf{j}\| = 1$ square unit

22. $\mathbf{u} \times \mathbf{v} = \begin{vmatrix} \mathbf{i} & \mathbf{j} & \mathbf{k} \\ 1 & 2 & 2 \\ 1 & 0 & 1 \end{vmatrix} = 2\mathbf{i} + \mathbf{j} - 2\mathbf{k}$

Area $= \|\mathbf{u} \times \mathbf{v}\| = \|2\mathbf{i} + \mathbf{j} - 2\mathbf{k}\|$

$= \sqrt{4 + 1 + 4} = 3$ square units

23. $\mathbf{u} \times \mathbf{v} = \begin{vmatrix} \mathbf{i} & \mathbf{j} & \mathbf{k} \\ 3 & 4 & 6 \\ 2 & -1 & 5 \end{vmatrix} = 26\mathbf{i} - 3\mathbf{j} - 11\mathbf{k}$

Area $= \|\mathbf{u} \times \mathbf{v}\| = \sqrt{26^2 + (-3)^2 + (-11)^2}$

$= \sqrt{806}$ square units

24. $\mathbf{u} \times \mathbf{v} = \begin{vmatrix} \mathbf{i} & \mathbf{j} & \mathbf{k} \\ -2 & 3 & 2 \\ 1 & 2 & 4 \end{vmatrix} = \langle 8, 10, -7 \rangle$

Area $= \|\mathbf{u} \times \mathbf{v}\| = \sqrt{8^2 + 10^2 + (-7)^2}$

$= \sqrt{213}$ square units

25. $\mathbf{u} \times \mathbf{v} = \begin{vmatrix} \mathbf{i} & \mathbf{j} & \mathbf{k} \\ 2 & 2 & -3 \\ 0 & 2 & 3 \end{vmatrix} = \langle 12, -6, 4 \rangle$

Area $= \|\mathbf{u} \times \mathbf{v}\| = \sqrt{12^2 + (-6)^2 + 4^2}$

$= 14$ square units

26. $\mathbf{u} \times \mathbf{v} = \begin{vmatrix} \mathbf{i} & \mathbf{j} & \mathbf{k} \\ 4 & -3 & 2 \\ 5 & 0 & 1 \end{vmatrix} = \langle -3, 6, 15 \rangle$

Area $= \|\mathbf{u} \times \mathbf{v}\| = \sqrt{(-3)^2 + 6^2 + 15^2}$

$= \sqrt{270} = 3\sqrt{30}$ square units

27. (a) $\overrightarrow{AB} = \langle 3 - 2, 1 - (-1), 2 - 4 \rangle = \langle 1, 2, -2 \rangle$ is parallel to

$\overrightarrow{DC} = \langle 0 - (-1), 5 - 3, 6 - 8 \rangle = \langle 1, 2, -2 \rangle$.

$\overrightarrow{AD} = \langle -3, 4, 4 \rangle$ is parallel to $\overrightarrow{BC} = \langle -3, 4, 4 \rangle$.

(b) $\overrightarrow{AB} \times \overrightarrow{AD} = \begin{vmatrix} \mathbf{i} & \mathbf{j} & \mathbf{k} \\ 1 & 2 & -2 \\ -3 & 4 & 4 \end{vmatrix} = \langle 16, 2, 10 \rangle$

Area $= \|\overrightarrow{AB} \times \overrightarrow{AD}\| = \sqrt{16^2 + 2^2 + 10^2} = \sqrt{360} = 6\sqrt{10}$ square units

(c) $\overrightarrow{AB} \cdot \overrightarrow{AD} = \langle 1, 2, -2 \rangle \cdot \langle -3, 4, 4 \rangle \neq 0 \implies$ not a rectangle

28. (a) $\overrightarrow{AB} = \langle 1, 2, 3 \rangle$

$\overrightarrow{CD} = \langle 1, 2, 3 \rangle$

Opposites are parallel and same length. Thus ABCD form a parallelogram.

(b) $\overrightarrow{AB} \times \overrightarrow{AC} = \begin{vmatrix} \mathbf{i} & \mathbf{j} & \mathbf{k} \\ 1 & 2 & 3 \\ 5 & 4 & 1 \end{vmatrix} = \langle -10, 14, -6 \rangle$

Area $= \|\overrightarrow{AB} \times \overrightarrow{AC}\| = \sqrt{(-10)^2 + 14^2 + (-6)^2} = 2\sqrt{83}$ square units

(c) $\overrightarrow{AB} \cdot \overrightarrow{AC} = 5 + 8 + 3 = 16 \neq 0 \implies$ not a rectangle.

29. $\mathbf{u} = \langle 1, 2, 3 \rangle, \mathbf{v} = \langle -3, 0, 0 \rangle$

$\mathbf{u} \times \mathbf{v} = \begin{vmatrix} \mathbf{i} & \mathbf{j} & \mathbf{k} \\ 1 & 2 & 3 \\ -3 & 0 & 0 \end{vmatrix} = \langle 0, -9, 6 \rangle$

Area $= \frac{1}{2}\|\mathbf{u} \times \mathbf{v}\| = \frac{1}{2}\sqrt{81 + 36} = \frac{3}{2}\sqrt{13}$ square units

30. $\mathbf{u} = \langle 2 - 1, 0 - (-4), 2 - 3 \rangle = \langle 1, 4, -1 \rangle$

$\mathbf{v} = \langle -2 - 1, 2 - (-4), 0 - 3 \rangle = \langle -3, 6, -3 \rangle$

$\mathbf{u} \times \mathbf{v} = \begin{vmatrix} \mathbf{i} & \mathbf{j} & \mathbf{k} \\ 1 & 4 & -1 \\ -3 & 6 & -3 \end{vmatrix} = \langle -6, 6, 18 \rangle$

Area $= \frac{1}{2}\|\mathbf{u} \times \mathbf{v}\| = \frac{1}{2}\sqrt{(-6)^2 + 6^2 + 18^2}$

$= \frac{1}{2}\sqrt{396} = 3\sqrt{11}$ square units

31. $\mathbf{u} = \langle -2 - 2, -2 - 3, 0 - (-5) \rangle = \langle -4, -5, 5 \rangle$

$\mathbf{v} = \langle 3 - 2, 0 - 3, 6 - (-5) \rangle = \langle 1, -3, 11 \rangle$

$\mathbf{u} \times \mathbf{v} = \begin{vmatrix} \mathbf{i} & \mathbf{j} & \mathbf{k} \\ -4 & -5 & 5 \\ 1 & -3 & 11 \end{vmatrix} = \langle -40, 49, 17 \rangle$

Area $= \frac{1}{2}\|\mathbf{u} \times \mathbf{v}\| = \frac{1}{2}\sqrt{(-40)^2 + 49^2 + 17^2}$

$= \frac{1}{2}\sqrt{4290}$ square units

32. $\mathbf{u} = \langle -2 - 2, -4 - 4, 0 - 0 \rangle = \langle -4, -8, 0 \rangle$

$\mathbf{v} = \langle 0 - 2, 0 - 4, 4 - 0 \rangle = \langle -2, -4, 4 \rangle$

$\mathbf{u} \times \mathbf{v} = \begin{vmatrix} \mathbf{i} & \mathbf{j} & \mathbf{k} \\ -4 & -8 & 0 \\ -2 & -4 & 4 \end{vmatrix} = \langle -32, 16, 0 \rangle$

Area $= \frac{1}{2}\|\mathbf{u} \times \mathbf{v}\| = \frac{1}{2}\sqrt{(-32)^2 + 16^2}$

$= \frac{1}{2}\sqrt{1280} = 8\sqrt{5}$ square units

33. $\mathbf{u} \cdot (\mathbf{v} \times \mathbf{w}) = \begin{vmatrix} 2 & 3 & 3 \\ 4 & 4 & 0 \\ 0 & 0 & 4 \end{vmatrix}$

$$= 2(16) - 3(16) + 3(0) = -16$$

34. $\mathbf{u} \cdot (\mathbf{v} \times \mathbf{w}) = \begin{vmatrix} 2 & 0 & 1 \\ 0 & 3 & 0 \\ 0 & 0 & 1 \end{vmatrix} = 6$

35. $\mathbf{u} \cdot (\mathbf{v} \times \mathbf{w}) = \begin{vmatrix} 2 & 3 & 1 \\ 1 & -1 & 0 \\ 4 & 3 & 1 \end{vmatrix}$

$$= 2(-1) - 3(1) + 1(7) = 2$$

36. $\mathbf{u} \cdot (\mathbf{v} \times \mathbf{w}) = \begin{vmatrix} 1 & 4 & -7 \\ 2 & 0 & 4 \\ 0 & -3 & 6 \end{vmatrix}$

$$= 1(0 + 12) - 4(12 - 0) - 7(-6) = 6$$

37. $\mathbf{u} \cdot (\mathbf{v} \times \mathbf{w}) = \begin{vmatrix} 1 & 1 & 0 \\ 0 & 1 & 1 \\ 1 & 0 & 1 \end{vmatrix} = 1 + 1 = 2$

Volume $= |\mathbf{u} \cdot (\mathbf{v} \times \mathbf{w})| = 2$ cubic units

38. $\mathbf{u} \cdot (\mathbf{v} \times \mathbf{w}) = \begin{vmatrix} 1 & 1 & 3 \\ 0 & 3 & 3 \\ 3 & 0 & 3 \end{vmatrix}$

$$= 1(9) - 1(-9) + 3(-9) = -9$$

Volume $= |\mathbf{u} \cdot (\mathbf{v} \times \mathbf{w})| = |-9| = 9$ cubic units

39. $\mathbf{u} \cdot (\mathbf{v} \times \mathbf{w}) = \begin{vmatrix} 0 & 2 & 2 \\ 0 & 0 & -2 \\ 3 & 0 & 2 \end{vmatrix}$

$$= 0 - 2(6) + 2(0) = -12$$

Volume $= |\mathbf{u} \cdot (\mathbf{v} \times \mathbf{w})| = 12$ cubic units

40. $\mathbf{u} \cdot (\mathbf{v} \times \mathbf{w}) = \begin{vmatrix} 1 & 2 & -1 \\ -1 & 2 & 2 \\ 2 & 0 & 1 \end{vmatrix}$

$$= 1(2) - 2(-1 - 4) - 1(0 - 4) = 16$$

Volume $= |\mathbf{u} \cdot (\mathbf{v} \times \mathbf{w})| = 16$ cubic units

41. $\mathbf{u} = \langle 4, 0, 0 \rangle$, $\mathbf{v} = \langle 0, -2, 3 \rangle$, $\mathbf{w} = \langle 0, 5, 3 \rangle$

$\mathbf{u} \cdot (\mathbf{v} \times \mathbf{w}) = \begin{vmatrix} 4 & 0 & 0 \\ 0 & -2 & 3 \\ 0 & 5 & 3 \end{vmatrix} = 4(-21) = -84$

Volume $= |-84| = 84$ cubic units

42. $\overrightarrow{AB} = \langle 1, 1, 0 \rangle$, $\overrightarrow{AC} = \langle 1, 0, 2 \rangle$, $\overrightarrow{AD} = \langle 0, 1, 1 \rangle$

$\mathbf{u} \cdot (\mathbf{v} \times \mathbf{w}) = \begin{vmatrix} 1 & 1 & 0 \\ 1 & 0 & 2 \\ 0 & 1 & 1 \end{vmatrix} = 1(-2) - 1(1) = -3$

Volume $= 3$ cubic units

43. $\mathbf{V} = \dfrac{1}{2}(-\cos 40° \, \mathbf{j} - \sin 40° \, \mathbf{k})$

$\mathbf{F} = -p\mathbf{k}$

(a) $\mathbf{V} \times \mathbf{F} = \begin{vmatrix} \mathbf{i} & \mathbf{j} & \mathbf{k} \\ 0 & -\frac{1}{2}\cos 40° & -\frac{1}{2}\sin 40° \\ 0 & 0 & -p \end{vmatrix}$

$$= \frac{1}{2}p \cos 40° \, \mathbf{i}$$

$T = \|\mathbf{V} \times \mathbf{F}\| = \dfrac{p}{2} \cos 40°$

(b)

p	T
15	5.75
20	7.66
25	9.58
30	11.49
35	13.41
40	15.32
45	17.24

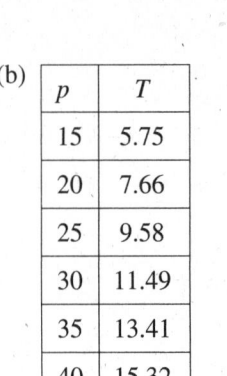

44. $\mathbf{V} = 0.16(-\cos 30° \, \mathbf{j} - \sin 30° \, \mathbf{k})$

$\mathbf{F} = -2000\mathbf{k}$

$$\mathbf{V} \times \mathbf{F} = \begin{vmatrix} \mathbf{i} & \mathbf{j} & \mathbf{k} \\ 0 & -0.16 \cos 30° & -0.16 \sin 30° \\ 0 & 0 & -2000 \end{vmatrix}$$

$= (-2000)(-0.16) \cos 30° \, \mathbf{i}$

$= 160\sqrt{3} \, \mathbf{i}$

$T = \|\mathbf{V} \times \mathbf{F}\| = 160\sqrt{3} \text{ ft-lb}$

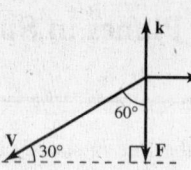

45. True. The cross product is not defined for two-dimensional vectors.

46. False. $\mathbf{u} \times \mathbf{v} = -(\mathbf{v} \times \mathbf{u})$

47. If the magnitudes of two vectors are doubled, the magnitude of the cross product will be four times as large.

48. $\mathbf{u} \times \mathbf{v} = \begin{vmatrix} \mathbf{i} & \mathbf{j} & \mathbf{k} \\ \cos \alpha & \sin \alpha & 0 \\ \cos \beta & \sin \beta & 0 \end{vmatrix} = (\cos \alpha \, \sin \beta - \sin \alpha \, \cos \beta)\mathbf{k}$

Area of triangle formed by the unit vectors $\mathbf{u}$ and $\mathbf{v}$ is

$\frac{1}{2}(\text{base})(\text{height}) = \frac{1}{2}(1) \sin(\alpha - \beta)$.

The area is also given by $\frac{1}{2}\|\mathbf{u} \times \mathbf{v}\| = \frac{1}{2}|\cos \alpha \, \sin \beta - \sin \alpha \, \cos \beta|$

Notice that $\cos \alpha \, \sin \beta - \sin \alpha \, \cos \beta$ is negative.

Thus, $\sin(\alpha - \beta) = \sin \alpha \, \cos \beta - \cos \alpha \, \sin \beta$.

49. $\cos 480° = \cos 120° = -\frac{1}{2}$

50. $\tan 300° = -\sqrt{3}$

51. $\sin 690° = \sin 330° = -\frac{1}{2}$

52. $\cos 930° = \cos 210° = -\frac{\sqrt{3}}{2}$

53. $\sin \frac{19\pi}{6} = \sin\left(\frac{7\pi}{6}\right) = -\frac{1}{2}$

54. $\cos \frac{17\pi}{6} = \cos \frac{5\pi}{6} = -\frac{\sqrt{3}}{2}$

55. $\tan \frac{15\pi}{4} = \tan \frac{7\pi}{4} = -1$

56. $\tan \frac{10\pi}{3} = \tan \frac{4\pi}{3} = \sqrt{3}$

57. $z = 6x + 4y$

At $(0, 5)$: $z = 6(0) + 4(5) = 20$

At $(0, 0)$: $z = 6(0) + 4(0) = 0$

At $\left(\frac{20}{3}, 0\right)$: $z = 6\left(\frac{20}{3}\right) + 4(0) = 40$

At $(6, 4)$: $z = 6(6) + 4(4) = 52$

The maximum value of z, $z = 52$, is found at $(6, 4)$.

The minimum value of z, $z = 0$, is found at $(0, 0)$.

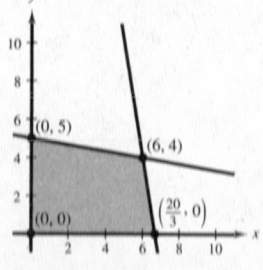

58. $(0, 8)$: $z = 6 \cdot 0 + 7 \cdot 8 = 56$

$(3, 4)$: $z = 6 \cdot 3 + 7 \cdot 4 = 46$

$(15, 0)$: $z = 6 \cdot 15 + 7 \cdot 0 = 90$

Minimum: 46 at $(3, 4)$.

Maximum: Unbounded

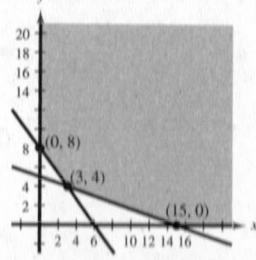

Section 11.4   Lines and Planes in Space

■ The parametric equations of the line in space parallel to the vector $\langle a, b, c \rangle$ and passing through the point (x_1, y_1, z_1) are

$$x = x_1 + at, \quad y = y_1 + bt, \quad z = z_1 + ct.$$

■ The standard equation of the plane in space containing the point (x_1, y_1, z_1) and having normal vector (a, b, c) is

$$a(x - x_1) + b(y - y_1) + c(z - z_1) = 0.$$

■ You should be able to find the angle between two planes by calculating the angle between their normal vectors.

■ You should be able to sketch a plane in space.

■ The distance between a point Q and a plane having normal $\mathbf{n}$ is

$$D = \|\text{proj}_{\mathbf{n}}\overrightarrow{PQ}\| = \frac{|\overrightarrow{PQ} \cdot \mathbf{n}|}{\|\mathbf{n}\|}$$

where P is a point in the plane.

Vocabulary Check

1. direction, $\dfrac{\overrightarrow{PQ}}{t}$

2. parametric equations

3. symmetric equations

4. normal

5. $a(x - x_1) + b(y - y_1) + c(z - z_1) = 0$

---

1. $x = x_1 + at = 0 + t$

  $y = y_1 + bt = 0 + 2t$

  $z = z_1 + ct = 0 + 3t$

  (a) Parametric equations: $x = t, y = 2t, z = 3t$

  (b) Symmetric equations: $\dfrac{x}{1} = \dfrac{y}{2} = \dfrac{z}{3}$

2. (a) $x = x_1 + at = 3 + 3t$

  $y = y_1 + bt = -5 - 7t$

  $z = z_1 + ct = 1 - 10t$

  Parametric equations:
  $x = 3 + 3t, y = -5 - 7t, z = 1 - 10t$

  (b) Symmetric equations: $\dfrac{x - 3}{3} = \dfrac{y + 5}{-7} = \dfrac{z - 1}{-10}$

3. $x = x_1 + at = -4 + \dfrac{1}{2}t, \ y = y_1 + bt = 1 + \dfrac{4}{3}t, \ z = z_1 + ct = 0 - t$

  (a) Parametric equations: $x = -4 + \dfrac{1}{2}t, y = 1 + \dfrac{4}{3}t, z = -t$

  Equivalently: $x = -4 + 3t, y = 1 + 8t, z = -6t$

  (b) Symmetric equations: $\dfrac{x + 4}{3} = \dfrac{y - 1}{8} = \dfrac{z}{-6}$

4. $x = x_1 + at = 5 + 4t$, $y = y_1 + bt = 0 + 0t$, $z = z_1 + ct = 10 + 3t$

(a) Parametric equations: $x = 5 + 4t$, $y = 0$, $z = 10 + 3t$

(b) Symmetric equations: $\dfrac{x - 5}{4} = \dfrac{z - 10}{3}$, $y = 0$

5. $x = x_1 + at = 2 + 2t$, $y = y_1 + bt = -3 - 3t$, $z = z_1 + ct = 5 + t$

(a) Parametric equations: $x = 2 + 2t$, $y = -3 - 3t$, $z = 5 + t$

(b) Symmetric equations: $\dfrac{x - 2}{2} = \dfrac{y + 3}{-3} = z - 5$

6. (a) $\mathbf{v} = \langle 3, -2, 1 \rangle$

$x = 1 + 3t$, $y = -2t$, $z = 1 + t$

(b) Symmetric equations: $\dfrac{x - 1}{3} = \dfrac{y}{-2} = \dfrac{z - 1}{1}$

7. (a) $\mathbf{v} = \langle 1 - 2, 4 - 0, -3 - 2 \rangle = \langle -1, 4, -5 \rangle$

Point: $(2, 0, 2)$

$x = 2 - t$, $y = 4t$, $z = 2 - 5t$

(b) $\dfrac{x - 2}{-1} = \dfrac{y}{4} = \dfrac{z - 2}{-5}$

8. (a) $\mathbf{v} = \langle 8, 5, 12 \rangle$

Point: $(2, 3, 0)$

Parametric equations: $x = 2 + 8t$, $y = 3 + 5t$, $z = 12t$

(b) Symmetric equations: $\dfrac{x - 2}{8} = \dfrac{y - 3}{5} = \dfrac{z}{12}$

9. (a) $\mathbf{v} = \langle 1 - (-3), -2 - 8, 16 - 15 \rangle = \langle 4, -10, 1 \rangle$

Point: $(-3, 8, 15)$

$x = -3 + 4t$, $y = 8 - 10t$, $z = 15 + t$

(b) $\dfrac{x + 3}{4} = \dfrac{y - 8}{-10} = \dfrac{z - 15}{1}$

10. $(2, 3, -1)$, $(1, -5, 3)$

(a) Let $P = (2, 3, -1)$, $Q = (1, -5, 3)$

$\vec{V} = \overrightarrow{PQ} = \langle 1 - 2, -5 - 3, 3 - (-1) \rangle = \langle -1, -8, 4 \rangle$

Direction numbers: $a = -1$, $b = -8$, $c = 4$

Choose P as the initial point:

$x = 2 - t$, $y = 3 - 8t$, $z = -1 + 4t$

(b) $\dfrac{x - 2}{-1} = \dfrac{y - 3}{-8} = \dfrac{z + 1}{4}$

11. $(3, 1, 2), (-1, 1, 5)$

(a) $\mathbf{v} = \langle -1 - 3, 1 - 1, 5 - 2 \rangle = \langle -4, 0, 3 \rangle$

Parametric: $x = 3 - 4t, y = 1, z = 2 + 3t$

(b) Since $b = 0$, there are no symmetric equations.

12. $(2, -1, 5), (2, 1, -3)$

(a) Let $P = (2, -1, 5), Q = (2, 1, -3)$

$\vec{V} = \vec{PQ} = \langle 2 - 2, 1 - (-1), -3 - 5 \rangle = \langle 0, 2, -8 \rangle$

Direction numbers: $a = 0, b = 2, c = -8$

Choose P as the initial point:

$x = 2, y = -1 + 2t, z = 5 - 8t$

(b) Since the direction number $a = 0$, no set of symmetric equations are possible.

13. $\left(-\dfrac{1}{2}, 2, \dfrac{1}{2}\right), \left(1, -\dfrac{1}{2}, 0\right)$

(a) $\mathbf{v} = \left\langle 1 - \left(-\dfrac{1}{2}\right), -\dfrac{1}{2} - 2, 0 - \dfrac{1}{2} \right\rangle = \left\langle \dfrac{3}{2}, -\dfrac{5}{2}, -\dfrac{1}{2} \right\rangle$

Direction numbers: $3, -5, -1$

Parametric: $x = -\dfrac{1}{2} + 3t, y = 2 - 5t, z = \dfrac{1}{2} - t$

(b) Symmetric: $\dfrac{2x + 1}{6} = \dfrac{y - 2}{-5} = \dfrac{2z - 1}{-2}$

14. (a) $\mathbf{v} = \left\langle 3 - \left(-\dfrac{3}{2}\right), -5 - \dfrac{3}{2}, -4 - 2 \right\rangle = \left\langle \dfrac{9}{2}, -\dfrac{13}{2}, -6 \right\rangle$, or $\langle 9, -13, -12 \rangle$

Point: $(3, -5, -4)$

Parametric equations: $x = 3 + 9t, y = -5 - 13t, z = -4 - 12t$

(b) Symmetric equations: $\dfrac{x - 3}{9} = \dfrac{y + 5}{-13} = \dfrac{z + 4}{-12}$

15.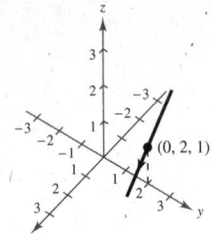

16.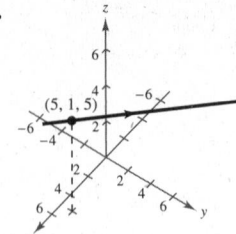

17. $a(x - x_1) + b(y - y_1) + c(z - z_1) = 0$

$1(x - 2) + 0(y - 1) + 0(z - 2) = 0$

$x - 2 = 0$

18. $a(x - x_0) + b(y - y_0) + c(z - z_0) = 0$

$0(x - 1) + 0(y - 0) + 1(z + 3) = 0$

$z + 3 = 0$

19. $-2(x - 5) + 1(y - 6) - 2(z - 3) = 0$

$-2x + y - 2z + 10 = 0$

20. $0(x - 0) - 3(y - 0) + 5(z - 0) = 0$

$-3y + 5z = 0$

21. $\mathbf{n} = \langle -1, -2, 1 \rangle \implies -1(x - 2) - 2(y - 0) + 1(z - 0) = 0$

$$-x - 2y + z + 2 = 0$$

22. $\mathbf{n} = \langle -1, 1, -2 \rangle$

$-1(x - 0) + 1(y - 0) - 2(z - 6) = 0$

$-x + y - 2z + 12 = 0$

23. $\mathbf{u} = \langle 1 - 0, 2 - 0, 3 - 0 \rangle = \langle 1, 2, 3 \rangle$

$\mathbf{v} = \langle -2 - 0, 3 - 0, 3 - 0 \rangle = \langle -2, 3, 3 \rangle$

$$\mathbf{n} = \mathbf{u} \times \mathbf{v} = \begin{vmatrix} \mathbf{i} & \mathbf{j} & \mathbf{k} \\ 1 & 2 & 3 \\ -2 & 3 & 3 \end{vmatrix} = \langle -3, -9, 7 \rangle$$

$-3(x - 0) - 9(y - 0) + 7(z - 0) = 0$

$-3x - 9y + 7z = 0$

$3x + 9y - 7z = 0$

24. $\mathbf{u} = \langle 2, -6, 2 \rangle, \mathbf{v} = \langle -3, -3, 0 \rangle$

$$\mathbf{u} \times \mathbf{v} = \begin{vmatrix} \mathbf{i} & \mathbf{j} & \mathbf{k} \\ 2 & -6 & 2 \\ -3 & -3 & 0 \end{vmatrix} = \langle 6, -6, -24 \rangle$$

$\mathbf{n} = \langle -1, 1, 4 \rangle$

Plane: $-1(x - 4) + 1(y + 1) + 4(z - 3) = 0$

$$-x + y + 4z - 7 = 0$$

25. $\mathbf{u} = \langle 3 - 2, 4 - 3, 2 + 2 \rangle = \langle 1, 1, 4 \rangle$

$\mathbf{v} = \langle 1 - 2, -1 - 3, 0 + 2 \rangle = \langle -1, -4, 2 \rangle$

$$\mathbf{n} = \mathbf{u} \times \mathbf{v} = \begin{vmatrix} \mathbf{i} & \mathbf{j} & \mathbf{k} \\ 1 & 1 & 4 \\ -1 & -4 & 2 \end{vmatrix} = \langle 18, -6, -3 \rangle$$

$18(x - 2) - 6(y - 3) - 3(z + 2) = 0$

$18x - 6y - 3z - 24 = 0$

$6x - 2y - z - 8 = 0$

26. $\mathbf{u} = \langle 4, 0, 2 \rangle, \mathbf{v} = \langle 1, 2, -5 \rangle$

$$\mathbf{u} \times \mathbf{v} = \begin{vmatrix} \mathbf{i} & \mathbf{j} & \mathbf{k} \\ 4 & 0 & 2 \\ 1 & 2 & -5 \end{vmatrix} = \langle -4, 22, 8 \rangle$$

$\mathbf{n} = \langle -2, 11, 4 \rangle$

Plane: $-2(x - 1) + 11(y + 1) + 4(z - 2) = 0$

$-2x + 11y + 4z + 5 = 0$

27. $\mathbf{n} = \mathbf{j}$: $0(x - 2) + 1(y - 5) + 0(z - 3) = 0$

$$y - 5 = 0$$

28. $\langle -1 - 2, 1 - 2, -1 - 1 \rangle = \langle -3, -1, -2 \rangle$ and $\langle 2, -3, 1 \rangle$ are parallel to plane.

$$\mathbf{n} = \begin{vmatrix} \mathbf{i} & \mathbf{j} & \mathbf{k} \\ -3 & -1 & -2 \\ 2 & -3 & 1 \end{vmatrix} = \langle -7, -1, 11 \rangle$$

$-7(x - 2) - 1(y - 2) + 11(z - 1) = 0$

$-7x - y + 11z + 5 = 0$

29. $\mathbf{n}_1 = \langle 5, -3, 1 \rangle, \mathbf{n}_2 = \langle 1, 4, 7 \rangle$

$\mathbf{n}_1 \cdot \mathbf{n}_2 = 5 - 12 + 7 = 0$; orthogonal

30. $\mathbf{n}_1 = \langle 3, 1, -4 \rangle, \mathbf{n}_2 = \langle -9, -3, 12 \rangle$

$3\mathbf{n}_1 = \langle 9, 3, -12 \rangle = -\mathbf{n}_2 \implies$ parallel planes

31. $\mathbf{n}_1 = \langle 2, 0, -1 \rangle, \mathbf{n}_2 = \langle 4, 1, 8 \rangle$

$\mathbf{n}_1 \cdot \mathbf{n}_2 = 8 - 8 = 0$; orthogonal

32. $\mathbf{n}_1 = \langle 1, -5, -1 \rangle$

$\mathbf{n}_2 = \langle 5, -25, -5 \rangle = 5\mathbf{n}_1 \implies$ parallel

33. (a) $\mathbf{n}_1 = \langle 3, -4, 5 \rangle$, $\mathbf{n}_2 = \langle 1, 1, -1 \rangle$; normal vectors to planes

$$\cos \theta = \frac{|\mathbf{n}_1 \cdot \mathbf{n}_2|}{\|\mathbf{n}_1\| \|\mathbf{n}_2\|} = \frac{|-6|}{\sqrt{50}\sqrt{3}} = \frac{6}{\sqrt{150}} \implies \theta \approx 60.67°$$

(b) $3x - 4y + 5z = 6 \qquad$ Equation 1

$\qquad x + y - z = 2 \qquad$ Equation 2

(-3) times Equation 2 added to Equation 1 gives

$-7y + 8z = 0$

$$y = \frac{8}{7}z.$$

Substituting back into Equation 2, $x = 2 - y + z = 2 - \frac{8}{7}z + z = 2 - \frac{1}{7}z$.

Letting $t = z/7$, we obtain $x = 2 - t, y = 8t, z = 7t$.

34. (a) $\mathbf{n}_1 = \langle 1, -3, 1 \rangle$, $\mathbf{n}_2 = \langle 2, 0, 5 \rangle$

$$\cos \theta = \frac{|\mathbf{n}_1 \cdot \mathbf{n}_2|}{\|\mathbf{n}_1\| \|\mathbf{n}_2\|} = \frac{|7|}{\sqrt{11}\sqrt{29}} = \frac{7}{\sqrt{319}} \implies \theta \approx 66.93°$$

(b) $2x + 5z + 3 = 0 \implies x = \frac{1}{2}(-5z - 3)$

Then $3y = x + z + 2 = \frac{1}{2}(-5z - 3) + z + 2 = -\frac{3}{2}z + \frac{1}{2} \implies y = -\frac{1}{2}z + \frac{1}{6}$

Let $z = t$. Parametric equations: $x = -\frac{5}{2}t - \frac{3}{2}, y = -\frac{1}{2}t + \frac{1}{6}, z = t$

or equivalently, let $z = 2t$ and you obtain $x = -5t - \frac{3}{2}, y = -t + \frac{1}{6}, z = 2t$.

35. (a) $\mathbf{n}_1 = \langle 1, 1, -1 \rangle$, $\mathbf{n}_2 = \langle 2, -5, -1 \rangle$; normal vectors to planes

$$\cos \theta = \frac{|\mathbf{n}_1 \cdot \mathbf{n}_2|}{\|\mathbf{n}_1\| \|\mathbf{n}_2\|} = \frac{|-2|}{\sqrt{3}\sqrt{30}} = \frac{2}{\sqrt{90}} \implies \theta \approx 77.83°$$

(b) $\qquad x + y - z = 0 \qquad$ Equation 1

$2x - 5y - z = 1 \qquad$ Equation 2

(-2) times Equation 1 added to Equation 2 gives

$-7y + z = 1$

$$y = \frac{z - 1}{7}.$$

Substituting back into Equation 1, $x = z - y = z - \frac{z-1}{7} = \frac{6z}{7} + \frac{1}{7} = \frac{1}{7}(6z + 1)$.

Letting $z = t$, $x = \frac{6t + 1}{7}, y = \frac{t - 1}{7}$.

Equivalently, let $y = t, z = 7t + 1$ and $x = 6t + 1$.

36. The planes are parallel because $\mathbf{n}_1 = \langle 2, 4, -2 \rangle$ is a multiple of $\mathbf{n}_2 = \langle -3, -6, 3 \rangle$.

The planes do not intersect.

37. $x + 2y + 3z = 6$

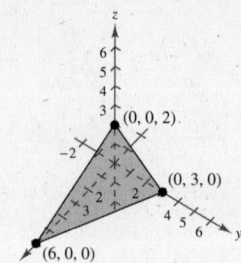

38. $2x - y + 4z = 4$

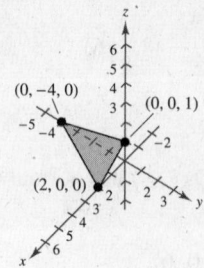

39. $x + 2y = 4$

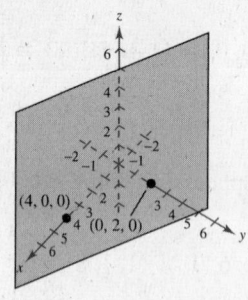

40. $y + z = 5$

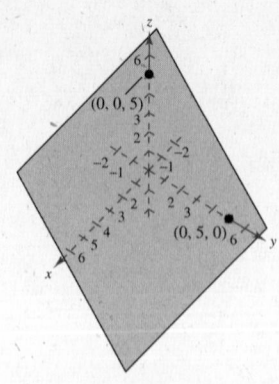

41. $3x + 2y - z = 6$

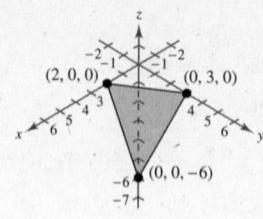

42. $x - 3z = 6$

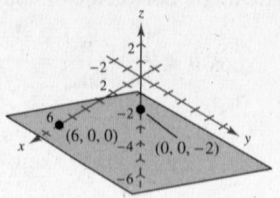

43. $D = \dfrac{\left| \overrightarrow{PQ} \cdot \mathbf{n} \right|}{\| \mathbf{n} \|}$

$P = (1, 0, 0)$ on plane, $Q = (0, 0, 0)$,

$\mathbf{n} = \langle 8, -4, 1 \rangle$, $\overrightarrow{PQ} = \langle -1, 0, 0 \rangle$

$D = \dfrac{\left| \langle -1, 0, 0 \rangle \cdot \langle 8, -4, 1 \rangle \right|}{\sqrt{64 + 16 + 1}} = \dfrac{|-8|}{\sqrt{81}} = \dfrac{8}{9}$

44. $P = (4, 0, 0)$ on plane, $Q = (3, 2, 1)$, $\mathbf{n} = \langle 1, -1, 2 \rangle$

$\overrightarrow{PQ} = \langle -1, 2, 1 \rangle$

$D = \dfrac{\left| \overrightarrow{PQ} \cdot \mathbf{n} \right|}{\| \mathbf{n} \|} = \dfrac{|-1|}{\sqrt{6}} = \dfrac{1}{\sqrt{6}} = \dfrac{\sqrt{6}}{6}$

45. $D = \dfrac{\left| \overrightarrow{PQ} \cdot \mathbf{n} \right|}{\| \mathbf{n} \|}$

$P = (2, 0, 0)$ on plane, $Q = (4, -2, -2)$,

$\mathbf{n} = \langle 2, -1, 1 \rangle$, $\overrightarrow{PQ} = \langle 2, -2, -2 \rangle$

$D = \dfrac{\left| \langle 2, -2, -2 \rangle \cdot \langle 2, -1, 1 \rangle \right|}{\sqrt{6}} = \dfrac{4}{\sqrt{6}} = \dfrac{2\sqrt{6}}{3}$

46. $P = (6, 0, 0)$ on plane, $Q = (-1, 2, 5)$,

$\overrightarrow{PQ} = \langle -7, 2, 5 \rangle$, $\mathbf{n} = \langle 2, 3, 1 \rangle$

$D = \dfrac{\left| \overrightarrow{PQ} \cdot \mathbf{n} \right|}{\| \mathbf{n} \|} = \dfrac{|-3|}{\sqrt{14}} = \dfrac{3}{\sqrt{14}} = \dfrac{3\sqrt{14}}{14}$

47. (a) $z = 0.81x + 0.36y + 0.2$

Year	x	y	z (Actual)	z (Model)
1999	6.2	7.3	7.8	7.85
2000	6.1	7.1	7.7	7.70
2001	5.9	7.0	7.4	7.50
2002	5.7	7.0	7.3	7.34
2003	5.6	6.9	7.2	7.22

(b) The approximations are very similar to the actual values of z.

(c) If the consumption of the two types of milk increases (or decreases), so does the consumption of the third type of milk.

48. The plane containing $P(6, 0, 0)$, $S(0, 0, 0)$, $T(-1, -1, 8)$ has normal vector

$$\langle 6, 0, 0 \rangle \times \langle -1, -1, 8 \rangle = \begin{vmatrix} \mathbf{i} & \mathbf{j} & \mathbf{k} \\ 6 & 0 & 0 \\ -1 & -1 & 8 \end{vmatrix} = \langle 0, -48, -6 \rangle$$

or $\mathbf{n}_1 = \langle 0, 8, 1 \rangle$.

The plane containing $P(6, 0, 0)$, $Q(6, 6, 0)$, and $R(7, 7, 8)$ has normal vector

$$\langle 0, -6, 0 \rangle \times \langle 1, 1, 8 \rangle = \begin{vmatrix} \mathbf{i} & \mathbf{j} & \mathbf{k} \\ 0 & -6 & 0 \\ 1 & 1 & 8 \end{vmatrix} = \langle -48, 0, 6 \rangle,$$

or $\mathbf{n}_2 = \langle -8, 0, 1 \rangle$.

The angle between two adjacent sides is given by

$$\cos \theta = \frac{|\mathbf{n}_1 \cdot \mathbf{n}_2|}{\|\mathbf{n}_1\| \|\mathbf{n}_2\|} = \frac{1}{\sqrt{65}\sqrt{65}} = \frac{1}{65} \implies \theta \approx 89.12°.$$

49. False. They might be skew lines, such as:

$L_1: x = t, y = 0, z = 0$ (x-axis)

and $L_2: x = 0, y = t, z = 1$

50. True

51. The lines are parallel:

$$-\tfrac{3}{2}\langle 10, -18, 20 \rangle = \langle -15, 27, -30 \rangle$$

52. (a) Sphere: $(x - 4)^2 + (y + 1)^2 + (z - 1)^2 = 4$

(b) Two planes parallel to given plane. Let $Q = (x, y, z)$ be a point on one of these planes, and pick $P = (0, 0, 10)$ on the given plane. By the distance formula,

$$2 = \frac{|\overrightarrow{PQ} \cdot \mathbf{n}|}{\|\mathbf{n}\|} = \frac{|\langle x, y, z - 10 \rangle \cdot \langle 4, -3, 1 \rangle|}{\sqrt{26}}$$

$$\pm 2\sqrt{26} = 4x - 3y + z - 10$$

$4x - 3y + z = 10 \pm 2\sqrt{26}$ (Two planes parallel to given plane)

53. $x^2 + y^2 = 10^2 = 100$

54. $\theta = \dfrac{3\pi}{4} \implies \tan \theta = -1 = \dfrac{y}{x} \implies y = -x$ (line)

55.
$$r = 3 \cos \theta$$
$$r^2 = 3r \cos \theta$$
$$x^2 + y^2 = 3x$$

56. $r = \dfrac{1}{2 - \cos \theta} \implies 2r - r \cos \theta = 1 \implies 2\sqrt{x^2 + y^2} - x = 1$

$\implies 2\sqrt{x^2 + y^2} = x + 1 \implies 4(x^2 + y^2) = x^2 + 2x + 1 \implies 3x^2 + 4y^2 = 2x + 1$

57. $r^2 = 49$

$r = 7$

58. $x^2 + y^2 - 4x = 0$

$\quad r^2 - 4r\cos\theta = 0$

$\quad\quad r - 4\cos\theta = 0 \implies r = 4\cos\theta$

59. $\quad\quad y = 5$

$\quad r\sin\theta = 5$

$\quad\quad r = 5\csc\theta$

60. $\quad\quad 2x - y + 1 = 0$

$\quad 2r\cos\theta - r\sin\theta = -1$

$\quad r(2\cos\theta - \sin\theta) = -1$

$$r = \frac{1}{\sin\theta - 2\cos\theta}$$

Review Exercises for Chapter 11

1. (a) and (b)

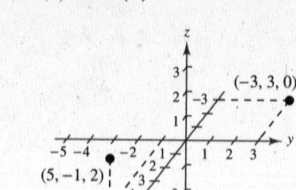

2.

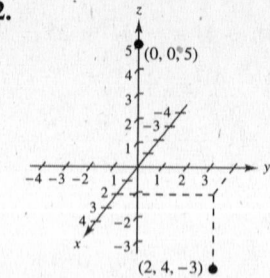

3. $(-5, 4, 0)$

4. y-axis $\implies x = z = 0$

$\quad (0, -7, 0)$

5. $d = \sqrt{(5-4)^2 + (2-0)^2 + (1-7)^2}$

$\quad = \sqrt{1 + 4 + 36}$

$\quad = \sqrt{41}$

6. $d = \sqrt{(2-(-1))^2 + (3-(-3))^2 + (-4-0)^2}$

$\quad = \sqrt{9 + 36 + 16}$

$\quad = \sqrt{61}$

7. $d_1 = \sqrt{(3-0)^2 + (-2-3)^2 + (0-2)^2} = \sqrt{9 + 25 + 4} = \sqrt{38}$

$\quad d_2 = \sqrt{(0-0)^2 + (5-3)^2 + (-3-2)^2} = \sqrt{4 + 25} = \sqrt{29}$

$\quad d_3 = \sqrt{(0-3)^2 + (5-(-2))^2 + (-3-0)^2} = \sqrt{9 + 49 + 9} = \sqrt{67}$

$\quad d_1{}^2 + d_2{}^2 = 38 + 29 = 67 = d_3{}^2$

8. $d_1 = \sqrt{(4-0)^2 + (3-0)^2 + (2-4)^2} = \sqrt{16 + 9 + 4} = \sqrt{29}$

$\quad d_2 = \sqrt{(4-4)^2 + (5-3)^2 + (5-2)^2} = \sqrt{4 + 9} = \sqrt{13}$

$\quad d_3 = \sqrt{(4-0)^2 + (5-0)^2 + (5-4)^2} = \sqrt{16 + 25 + 1} = \sqrt{42}$

$\quad d_1^2 + d_2^2 = d_3^2 = 42$

9. Midpoint: $\left(\dfrac{8+5}{2}, \dfrac{-2+6}{2}, \dfrac{3+7}{2}\right) = \left(\dfrac{13}{2}, 2, 5\right)$

10. Midpoint: $\left(\dfrac{7+1}{2}, \dfrac{1-1}{2}, \dfrac{-4+2}{2}\right) = (4, 0, -1)$

11. Midpoint: $\left(\dfrac{10-8}{2}, \dfrac{6-2}{2}, \dfrac{-12-6}{2}\right) = (1, 2, -9)$ **12.** Midpoint: $\left(\dfrac{-5-7}{2}, \dfrac{-3-9}{2}, \dfrac{1-5}{2}\right) = (-6, -6, -2)$

13. $(x-2)^2 + (y-3)^2 + (z-5)^2 = 1$ **14.** $(x-3)^2 + (y+2)^2 + (z-4)^2 = 16$

15. Radius: 6   **16.** Radius $= \dfrac{15}{2}$

$\quad (x-1)^2 + (y-5)^2 + (z-2)^2 = 36$ $\quad x^2 + (y-4)^2 + (z+1)^2 = \dfrac{225}{4}$

17. $(x^2 - 4x + 4) + (y^2 - 6y + 9) + z^2 = -4 + 4 + 9$

$\qquad\qquad (x-2)^2 + (y-3)^2 + z^2 = 9$

Center: $(2, 3, 0)$

Radius: 3

18. $(x^2 - 10x + 25) + (y^2 + 6y + 9) + (z^2 - 4z + 4) = -34 + 25 + 9 + 4$

$\qquad\qquad (x-5)^2 + (y+3)^2 + (z-2)^2 = 4$

Center: $(5, -3, 2)$

Radius: 2

19. (a) xz-trace $(y = 0)$: $x^2 + z^2 = 7$, circle   (b) yz-trace $(x = 0)$: $(y-3)^2 + z^2 = 16$, circle

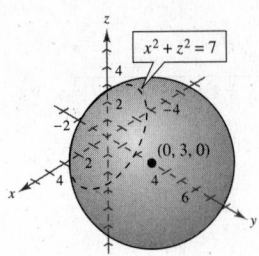   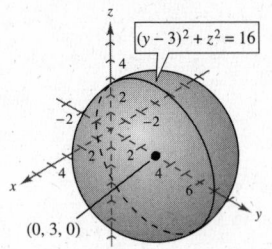

20. (a) xy-trace $(z = 0)$: $(x+2)^2 + (y-1)^2 = 9$ circle   (b) yz-trace $(x = 0)$: $4 + (y-1)^2 + z^2 = 9$

$\qquad\qquad\qquad\qquad\qquad (y-1)^2 + z^2 = 5$ circle

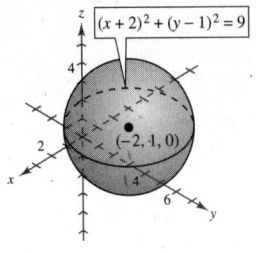   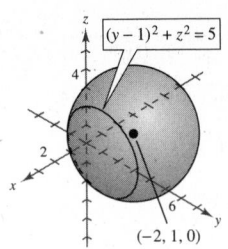

21. Initial point: $(2, -1, 4)$

Terminal point: $(3, 3, 0)$

(a) $\mathbf{v} = \langle 3-2, 3-(-1), 0-4 \rangle = \langle 1, 4, -4 \rangle$

(b) $\|\mathbf{v}\| = \sqrt{(1)^2 + (4)^2 + (-4)^2} = \sqrt{33}$

(c) $\mathbf{u} = \dfrac{\mathbf{v}}{\|\mathbf{v}\|} = \left\langle \dfrac{1}{\sqrt{33}}, \dfrac{4}{\sqrt{33}}, -\dfrac{4}{\sqrt{33}} \right\rangle$

22. (a) $\overrightarrow{PQ} = \langle -3-2, 2-(-1), 3-2 \rangle = \langle -5, 3, 1 \rangle$

(b) $\|\overrightarrow{PQ}\| = \sqrt{35}$

(c) Unit vector: $\dfrac{1}{\sqrt{35}} \langle -5, 3, 1 \rangle = \dfrac{\sqrt{35}}{35} \langle -5, 3, 1 \rangle$

23. Initial point: $(7, -4, 3)$

Terminal point: $(-3, 2, 10)$

(a) $\mathbf{v} = \langle -3 - 7, 2 - (-4), 10 - 3 \rangle = \langle -10, 6, 7 \rangle$

(b) $\|\mathbf{v}\| = \sqrt{(-10)^2 + (6)^2 + (7)^2} = \sqrt{185}$

(c) $\mathbf{u} = \dfrac{\mathbf{v}}{\|\mathbf{v}\|} = \left\langle -\dfrac{10}{\sqrt{185}}, \dfrac{6}{\sqrt{185}}, \dfrac{7}{\sqrt{185}} \right\rangle$

24. (a) $\overrightarrow{PQ} = \langle 5 - 0, -8 - 3, 6 - (-1) \rangle = \langle 5, -11, 7 \rangle$

(b) $\|\overrightarrow{PQ}\| = \sqrt{195}$

(c) Unit vector: $\dfrac{1}{\sqrt{195}} \langle 5, -11, 7 \rangle = \dfrac{\sqrt{195}}{195} \langle 5, -11, 7 \rangle$

25. $\mathbf{u} \cdot \mathbf{v} = -1(0) + 4(-6) + 3(5) = -9$

26. $\mathbf{u} \cdot \mathbf{v} = 8(2) - 4(5) + 2(2) = 0$

27. $\mathbf{u} \cdot \mathbf{v} = 2(1) - 1(0) + 1(-1) = 1$

28. $\mathbf{u} \cdot \mathbf{v} = 2(1) + 1(-3) - 2(2) = -5$

29. Since $\mathbf{u} \cdot \mathbf{v} = 0$, the angle is $90°$.

30. $\cos \theta = \dfrac{\mathbf{u} \cdot \mathbf{v}}{\|\mathbf{u}\| \, \|\mathbf{v}\|} = \dfrac{12 + 5 - 2}{\sqrt{11}\sqrt{45}}$

$\quad = \dfrac{15}{\sqrt{11}\sqrt{45}} \implies \theta \approx 47.61°$

31. Since $-\frac{2}{3} \langle 39, -12, 21 \rangle = \langle -26, 8, -14 \rangle$, the vectors are parallel.

32. $\mathbf{u} \cdot \mathbf{v} = \langle 8, 5, -8 \rangle \cdot \left\langle -2, 4, \frac{1}{2} \right\rangle$

$\quad = -16 + 20 - 4 = 0$

Orthogonal

33. First two points: $\mathbf{u} = \langle -3, 4, 1 \rangle$

Last two points: $\mathbf{v} = \langle 0, -2, 6 \rangle$

Since $\mathbf{u} \neq c\mathbf{v}$, the points are not collinear.

34. First two points: $\langle -1, 5, 4 \rangle$

Last two points: $\langle 2, -10, -8 \rangle$

Since, $\langle 2, -10, -8 \rangle = -2\langle -1, 5, 4 \rangle$, the 3 points are collinear.

35. Let $\mathbf{a}$, $\mathbf{b}$, and $\mathbf{c}$ be the three force vectors determined by $A(0, 10, 10)$, $B(-4, -6, 10)$ and $C(4, -6, 10)$.

$\mathbf{a} = \|\mathbf{a}\| \dfrac{\langle 0, 10, 10 \rangle}{10\sqrt{2}} = \|\mathbf{a}\| \left\langle 0, \dfrac{1}{\sqrt{2}}, \dfrac{1}{\sqrt{2}} \right\rangle$

$\mathbf{b} = \|\mathbf{b}\| \dfrac{\langle -4, -6, 10 \rangle}{\sqrt{152}} = \|\mathbf{b}\| \left\langle \dfrac{-2}{\sqrt{38}}, \dfrac{-3}{\sqrt{38}}, \dfrac{5}{\sqrt{38}} \right\rangle$

$\mathbf{c} = \|\mathbf{c}\| \dfrac{\langle 4, -6, 10 \rangle}{\sqrt{152}} = \|\mathbf{c}\| \left\langle \dfrac{2}{\sqrt{38}}, \dfrac{-3}{\sqrt{38}}, \dfrac{5}{\sqrt{38}} \right\rangle$

Must have $\mathbf{a} + \mathbf{b} + \mathbf{c} = 300\mathbf{k}$. Thus:

$\dfrac{-2}{\sqrt{38}} \|\mathbf{b}\| + \dfrac{2}{\sqrt{38}} \|\mathbf{c}\| = 0$

$\dfrac{1}{\sqrt{2}} \|\mathbf{a}\| - \dfrac{3}{\sqrt{38}} \|\mathbf{b}\| - \dfrac{3}{\sqrt{38}} \|\mathbf{c}\| = 0$

$\dfrac{1}{\sqrt{2}} \|\mathbf{a}\| + \dfrac{5}{\sqrt{38}} \|\mathbf{b}\| + \dfrac{5}{\sqrt{38}} \|\mathbf{c}\| = 300.$

—CONTINUED—

35. —CONTINUED—

From the first equation $\|\mathbf{b}\| = \|\mathbf{c}\|$. From the second equation, $\dfrac{1}{\sqrt{2}}\|\mathbf{a}\| = \dfrac{6}{\sqrt{38}}\|\mathbf{b}\|$.

From the third equation, $\dfrac{1}{\sqrt{2}}\|\mathbf{a}\| = 300 - \dfrac{10}{\sqrt{38}}\|\mathbf{b}\|$. Thus,

$$\frac{6}{\sqrt{38}}\|\mathbf{b}\| = 300 - \frac{10}{\sqrt{38}}\|\mathbf{b}\| \implies \frac{16}{\sqrt{38}}\|\mathbf{b}\| = 300 \text{ and } \|\mathbf{b}\| = \|\mathbf{c}\| = \frac{75\sqrt{38}}{4} \approx 115.58.$$

Finally, $\|\mathbf{a}\| = \sqrt{2}\left(\dfrac{6}{\sqrt{38}}\right)\left(\dfrac{75\sqrt{38}}{4}\right) = \dfrac{225\sqrt{2}}{2} \approx 159.10.$

36. Let $\mathbf{a}, \mathbf{b}, \mathbf{c}$ be the three force vectors determined by $A(0, 10, 10)$, $B(-4, -6, 10)$ and $C(4, -6, 10)$.

$$\mathbf{a} = \|\mathbf{a}\| \langle 0, 10, 10 \rangle / 10\sqrt{2} = \|\mathbf{a}\|\left\langle 0, \frac{1}{\sqrt{2}}, \frac{1}{\sqrt{2}}\right\rangle$$

$$\mathbf{b} = \|\mathbf{b}\| \langle -4, -6, 10 \rangle / \sqrt{152} = \|\mathbf{b}\|\left\langle \frac{-2}{\sqrt{38}}, \frac{-3}{\sqrt{38}}, \frac{5}{\sqrt{38}}\right\rangle$$

$$\mathbf{c} = \|\mathbf{c}\| \langle 4, -6, 10 \rangle / \sqrt{152} = \|\mathbf{c}\|\left\langle \frac{2}{\sqrt{38}}, \frac{-3}{\sqrt{38}}, \frac{5}{\sqrt{38}}\right\rangle$$

We must have $\mathbf{a} + \mathbf{b} + \mathbf{c} = 200\mathbf{k}$. Thus,

$$\frac{-2}{\sqrt{38}}\|\mathbf{b}\| + \frac{2}{\sqrt{38}}\|\mathbf{c}\| = 0$$

$$\frac{1}{\sqrt{2}}\|\mathbf{a}\| - \frac{3}{\sqrt{38}}\|\mathbf{b}\| - \frac{3}{\sqrt{38}}\|\mathbf{c}\| = 0$$

$$\frac{1}{\sqrt{2}}\|\mathbf{a}\| + \frac{5}{\sqrt{38}}\|\mathbf{b}\| + \frac{5}{\sqrt{38}}\|\mathbf{c}\| = 200$$

Solving this system, $\|\mathbf{a}\| \approx 106.1$, $\|\mathbf{b}\| = \|\mathbf{c}\| = 77.1$.

Thus, the tensions are 106.1, 77.1 and 77.1 pounds.

37. $\mathbf{u} \times \mathbf{v} = \begin{vmatrix} \mathbf{i} & \mathbf{j} & \mathbf{k} \\ -2 & 8 & 2 \\ 1 & 1 & -1 \end{vmatrix} = \langle -10, 0, -10 \rangle$

38. $\mathbf{u} \times \mathbf{v} = \begin{vmatrix} \mathbf{i} & \mathbf{j} & \mathbf{k} \\ 10 & 15 & 5 \\ 5 & -3 & 0 \end{vmatrix} = \langle 15, 25, -105 \rangle$

39. $\mathbf{u} \times \mathbf{v} = \begin{vmatrix} \mathbf{i} & \mathbf{j} & \mathbf{k} \\ -3 & 2 & -5 \\ 10 & -15 & 2 \end{vmatrix} = \langle -71, -44, 25 \rangle$

$\|\mathbf{u} \times \mathbf{v}\| = \sqrt{7602}$

Unit vector: $\dfrac{1}{\sqrt{7602}}\langle -71, -44, 25 \rangle$

40. $\mathbf{u} \times \mathbf{v} = \begin{vmatrix} \mathbf{i} & \mathbf{j} & \mathbf{k} \\ 0 & 0 & 4 \\ 1 & 0 & 12 \end{vmatrix} = 4\mathbf{j} \implies$ unit vector: $\mathbf{j} = \langle 0, 1, 0 \rangle$

41. First two points: $\langle 3, 2, 3 \rangle$

Last two points: $\langle 3, 2, 3 \rangle$

First and third points: $\langle -2, 2, 0 \rangle$

$$\begin{vmatrix} \mathbf{i} & \mathbf{j} & \mathbf{k} \\ 3 & 2 & 3 \\ -2 & 2 & 0 \end{vmatrix} = \langle -6, -6, 10 \rangle$$

Area $= |\langle -6, -6, 10 \rangle| = \sqrt{36 + 36 + 100}$

$\qquad = \sqrt{172} = 2\sqrt{43}$ square units

42. $\mathbf{u} = \langle 1, 0, 1 \rangle$, $\mathbf{v} = \langle 1, 0, 1 \rangle$ opposite sides parallel and equal length.

Adjacent sides: $\mathbf{u} = \langle 1, 0, 1 \rangle$, $\mathbf{w} = \langle 0, 2, 0 \rangle$

$$\mathbf{u} \times \mathbf{w} = \begin{vmatrix} \mathbf{i} & \mathbf{j} & \mathbf{k} \\ 1 & 0 & 1 \\ 0 & 2 & 0 \end{vmatrix} = \langle -2, 0, 2 \rangle$$

Area $= |\mathbf{u} \times \mathbf{w}| = \sqrt{4 + 4} = 2\sqrt{2}$ square units

43. The parallelogram is determined by the three vectors with initial point $(0, 0, 0)$.

$\mathbf{u} = \langle 3, 0, 0 \rangle$, $\mathbf{v} = \langle 2, 0, 5 \rangle$, $\mathbf{w} = \langle 0, 5, 1 \rangle$

$$\mathbf{u} \cdot (\mathbf{v} \times \mathbf{w}) = \begin{vmatrix} 3 & 0 & 0 \\ 2 & 0 & 5 \\ 0 & 5 & 1 \end{vmatrix} = -75$$

Volume $= |-75| = 75$ cubic units

44. $\mathbf{u} = \langle 2, 0. 0 \rangle$, $\mathbf{v} = \langle 0, 4, 0 \rangle$, $\mathbf{w} = \langle 0, 0, 6 \rangle$

$$\mathbf{u} \cdot (\mathbf{v} \times \mathbf{w}) = \begin{vmatrix} 2 & 0 & 0 \\ 0 & 4 & 0 \\ 0 & 0 & 6 \end{vmatrix} = 48$$

Volume $= |\mathbf{u} \cdot (\mathbf{v} \times \mathbf{w})| = 48$ cubic units

45. $\mathbf{v} = \langle 3 + 1, 6 - 3, -1 - 5 \rangle = \langle 4, 3, -6 \rangle$, point: $(-1, 3, 5)$

(a) Parametric equations: $x = -1 + 4t, y = 3 + 3t, z = 5 - 6t$

(b) Symmetric equations: $\dfrac{x + 1}{4} = \dfrac{y - 3}{3} = \dfrac{z - 5}{-6}$

46. (a) $\mathbf{v} = \langle 5, 20, -3 \rangle$

$\qquad x = 5t, \; y = -10 + 20t, \; z = 3 - 3t$

(b) $\dfrac{x}{5} = \dfrac{y + 10}{20} = \dfrac{z - 3}{-3}$

47. Use $2\mathbf{v} = \langle -4, 5, 2 \rangle$, point: $(0, 0, 0)$.

(a) Parametric equations: $x = -4t, y = 5t, z = 2t$

(b) Symmetric equations: $\dfrac{x}{-4} = \dfrac{y}{5} = \dfrac{z}{2}$

48. (a) $\mathbf{v} = \langle 1, 1, 1 \rangle$

$\qquad x = 3 + t, y = 2 + t, z = 1 + t$

(b) $\dfrac{x - 3}{1} = \dfrac{y - 2}{1} = \dfrac{z - 1}{1}$ or

$\qquad x - 3 = y - 2 = z - 1$

49. $\mathbf{u} = \langle 5, 0, 2 \rangle$, $\mathbf{v} = \langle 2, 3, 8 \rangle$

$$\mathbf{u} \times \mathbf{v} = \begin{vmatrix} \mathbf{i} & \mathbf{j} & \mathbf{k} \\ 5 & 0 & 2 \\ 2 & 3 & 8 \end{vmatrix} = \langle -6, -36, 15 \rangle$$

$\mathbf{n} = \langle 2, 12, -5 \rangle$

$a(x - x_0) + b(y - y_0) + c(z - z_0) = 0$

$2(x - 0) + 12(y - 0) - 5(z - 0) = 0$

$\qquad\qquad\qquad 2x + 12y - 5z = 0$

50. $\mathbf{u} = \langle 5, -5, -2 \rangle$, $\mathbf{v} = \langle 3, 5, 2 \rangle$

$$\mathbf{n} = \mathbf{u} \times \mathbf{v} = \begin{vmatrix} \mathbf{i} & \mathbf{j} & \mathbf{k} \\ 5 & -5 & -2 \\ 3 & 5 & 2 \end{vmatrix} = \langle 0, -16, 40 \rangle$$

Plane: $0(x + 1) - 16(y - 3) + 40(z - 4) = 0$

$\qquad\qquad -2(y - 3) + 5(z - 4) = 0$

$\qquad\qquad -2y + 5z - 14 = 0$

51. $\mathbf{n} = \mathbf{k}$, normal vector

Plane: $0(x - 5) + 0(y - 3) + 1(z - 2) = 0$

$\qquad\qquad\qquad z - 2 = 0$

52. $\mathbf{n} = \langle -1, 1, -2 \rangle$, point: $(0, 0, 6)$

$$-1(x - 0) + 1(y - 0) - 2(z - 6) = 0$$
$$-x + y - 2z + 12 = 0$$
$$x - y + 2z - 12 = 0$$

53. $3x - 2y + 3z = 6$

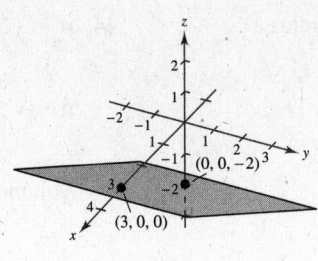

54. $5x - y - 5z = 5$

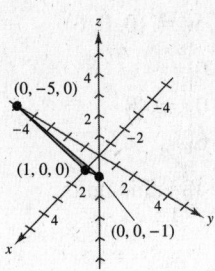

55. $2x - 3z = 6$

56. $4y - 3z = 12$

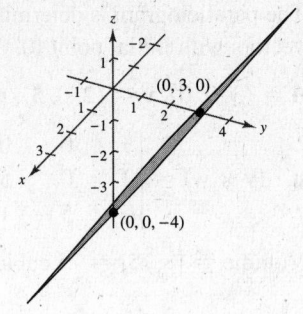

57. $\mathbf{n} = \langle 2, -20, 6 \rangle$, $P = (0, 0, 1)$ in plane, $Q = (2, 3, 10)$, $\overrightarrow{PQ} = \langle 2, 3, 9 \rangle$

$$D = \frac{|\overrightarrow{PQ} \cdot \mathbf{n}|}{\|\mathbf{n}\|} = \frac{|-2|}{\sqrt{440}} = \frac{1}{\sqrt{110}} = \frac{\sqrt{110}}{110} \approx 0.0953$$

58. $D = \dfrac{|\overrightarrow{PQ} \cdot \mathbf{n}|}{\|\mathbf{n}\|}$

$Q = (1, 2, 3)$, $P = (2, 0, 0)$ on plane. $\overrightarrow{PQ} = \langle -1, 2, 3 \rangle$, $\mathbf{n} = \langle 2, -1, 1 \rangle$

$$D = \frac{|\langle -1, 2, 3 \rangle \cdot \langle 2, -1, 1 \rangle|}{\sqrt{6}} = \frac{1}{\sqrt{6}} = \frac{\sqrt{6}}{6}$$

59. $\mathbf{n} = \langle 1, -10, 3 \rangle$, $P = (2, 0, 0)$ in plane, $Q = (0, 0, 0)$, $\overrightarrow{PQ} = \langle -2, 0, 0 \rangle$

$$D = \frac{|\overrightarrow{PQ} \cdot \mathbf{n}|}{\|\mathbf{n}\|} = \frac{|-2|}{\sqrt{1 + 100 + 9}} = \frac{2}{\sqrt{110}} = \frac{2\sqrt{110}}{110} = \frac{\sqrt{110}}{55} \approx 0.191$$

60. $D = \dfrac{|\overrightarrow{PQ} \cdot \mathbf{n}|}{\|\mathbf{n}\|}$

$Q = (0, 0, 0)$, $P = (0, 0, 12)$ on plane. $\overrightarrow{PQ} = \langle 0, 0, -12 \rangle$, $\mathbf{n} = \langle 2, 3, 1 \rangle$

$$D = \frac{|\langle 0, 0, -12 \rangle \cdot \langle 2, 3, 1 \rangle|}{\sqrt{14}} = \frac{12}{\sqrt{14}} = \frac{6\sqrt{14}}{7}$$

61. False. $\mathbf{a} \times \mathbf{b} = -(\mathbf{b} \times \mathbf{a})$

62. True. See page 831.

63. $\mathbf{u} \cdot \mathbf{u} = \langle 3, -2, 1 \rangle \cdot \langle 3, -2, 1 \rangle$

$$= 9 + 4 + 1$$
$$= 14$$
$$= \|\mathbf{u}\|^2$$

64. $\mathbf{u} \times \mathbf{v} = \begin{vmatrix} \mathbf{i} & \mathbf{j} & \mathbf{k} \\ 3 & -2 & 1 \\ 2 & -4 & -3 \end{vmatrix} = \langle 10, 11, -8 \rangle$

$\mathbf{v} \times \mathbf{u} = \begin{vmatrix} \mathbf{i} & \mathbf{j} & \mathbf{k} \\ 2 & -4 & -3 \\ 3 & -2 & 1 \end{vmatrix} = \langle -10, -11, 8 \rangle$

Thus, $\mathbf{u} \times \mathbf{v} = -(\mathbf{v} \times \mathbf{u})$.

65. $\mathbf{u} \cdot (\mathbf{v} + \mathbf{w}) = \langle 3, -2, 1 \rangle \cdot \langle 1, -2, -1 \rangle = 6$

$\mathbf{u} \cdot \mathbf{v} + \mathbf{u} \cdot \mathbf{w} = 11 + (-5) = 6$

66. $\mathbf{u} \times (\mathbf{v} + \mathbf{w}) = \mathbf{u} \times \langle 1, -2, -1 \rangle = \begin{vmatrix} \mathbf{i} & \mathbf{j} & \mathbf{k} \\ 3 & -2 & 1 \\ 1 & -2 & -1 \end{vmatrix} = \langle 4, 4, -4 \rangle$

$\mathbf{u} \times \mathbf{v} = \langle 10, 11, -8 \rangle$ (Exercise 64)

$\mathbf{u} \times \mathbf{w} = \begin{vmatrix} \mathbf{i} & \mathbf{j} & \mathbf{k} \\ 3 & -2 & 1 \\ -1 & 2 & 2 \end{vmatrix} = \langle -6, -7, 4 \rangle$

$(\mathbf{u} \times \mathbf{v}) + (\mathbf{u} \times \mathbf{w}) = \langle 10, 11, -8 \rangle + \langle -6, -7, 4 \rangle = \langle 4, 4, -4 \rangle$

$= \mathbf{u} \times (\mathbf{v} + \mathbf{w})$

Problem Solving for Chapter 11

1. (a)

(b) $\mathbf{w} = a\mathbf{u} + b\mathbf{v} = a\langle 1, 1, 0 \rangle + b\langle 0, 1, 1 \rangle$

$\mathbf{0} = \langle a, a + b, b \rangle \Longrightarrow a = b = 0$

(c) $\mathbf{w} = \langle 1, 2, 1 \rangle = a\langle 1, 1, 0 \rangle + b\langle 0, 1, 1 \rangle$

$1 = a$

$2 = a + b$

$1 = b$

Hence, $a = b = 1$.

(d) $\mathbf{w} = \langle 1, 2, 3 \rangle = a\langle 1, 1, 0 \rangle + b\langle 0, 1, 1 \rangle$

$1 = a$

$2 = a + b$

$3 = b$

Impossible

2. This set is a sphere:

$(x - x_1)^2 + (y - y_1)^2 + (z - z_1)^2 = 16$

3. Programs will vary. See online website.

4. (a) $\mathbf{u} + \mathbf{v} = \langle 4, 7.5, -2 \rangle$

(b) $\|\mathbf{u} + \mathbf{v}\| \approx 8.7321$

(c) $\|\mathbf{u}\| = \sqrt{26} \approx 5.0990$

(d) $\|\mathbf{v}\| \approx 9.0139$

5. The largest angle in a triangle is always opposite the longest side of the triangle. First, determine the lengths of the three sides. Then, once the largest angle has been identified, use the fact that $\cos \theta = \dfrac{\mathbf{u} \cdot \mathbf{v}}{\|\mathbf{u}\| \, \|\mathbf{v}\|}$, where $\mathbf{u}$ and $\mathbf{v}$ are defined to be the vectors that form θ. If $\mathbf{u} \cdot \mathbf{v} = 0$, the angle is a right angle. If $\mathbf{u} \cdot \mathbf{v} > 0$, the angle is acute. If $\mathbf{u} \cdot \mathbf{v} < 0$, the angle is obtuse.

(a) A: $(1, 2, 0)$

   B: $(0, 0, 0)$

   C: $(-2, 1, 0)$

   $d(AB) = \sqrt{5}, d(AC) = \sqrt{10}, d(BC) = \sqrt{5}$

   Angle B is largest.

   $\overrightarrow{BA} = \langle 1, 2, 0 \rangle, \overrightarrow{BC} = \langle -2, 1, 0 \rangle$

   $\overrightarrow{BA} \cdot \overrightarrow{BC} = 0 \implies$ The triangle is a right triangle.

(b) A: $(-3, 0, 0)$

   B: $(0, 0, 0)$

   C: $(1, 2, 3)$

   $d(AB) = 3, d(AC) = \sqrt{29}, d(BC) = \sqrt{14}$

   Angle B is largest.

   $\overrightarrow{BA} = \langle -3, 0, 0 \rangle, \overrightarrow{BC} = \langle 1, 2, 3 \rangle$

   $\overrightarrow{BA} \cdot \overrightarrow{BC} = -3 < 0 \implies$ The triangle is an obtuse triangle.

(c) A: $(2, -3, 4)$

   B: $(0, 1, 2)$

   C: $(-1, 2, 0)$

   $d(AB) = \sqrt{24}, d(AC) = \sqrt{50}, d(BC) = \sqrt{6}$

   Angle B is largest.

   $\overrightarrow{BA} = \langle 2, -4, 2 \rangle, \overrightarrow{BC} = \langle -1, 1, -2 \rangle$

   $\overrightarrow{BA} \cdot \overrightarrow{BC} = -10 < 0 \implies$ The triangle is an obtuse triangle.

(d) A: $(2, -7, 3)$

   B: $(-1, 5, 8)$

   C: $(4, 6, -1)$

   $d(AB) = \sqrt{178}, d(AC) = \sqrt{189}, d(BC) = \sqrt{107}$

   Angle B is largest.

   $\overrightarrow{BA} = \langle 3, -12, -5 \rangle, \overrightarrow{BC} = \langle 5, 1, -9 \rangle$

   $\overrightarrow{BA} \cdot \overrightarrow{BC} = 48 \implies$ The triangle is an acute triangle.

6. $\overrightarrow{PQ_1} = \langle 0 - 0, -1 - 0, 0 - 4 \rangle = \langle 0, -1, -4 \rangle$

$\overrightarrow{PQ_2} = \left\langle \dfrac{\sqrt{3}}{2} - 0, \dfrac{1}{2} - 0, 0 - 4 \right\rangle = \left\langle \dfrac{\sqrt{3}}{2}, \dfrac{1}{2}, -4 \right\rangle$

$\overrightarrow{PQ_3} = \left\langle -\dfrac{\sqrt{3}}{2} - 0, \dfrac{1}{2} - 0, 0 - 4 \right\rangle = \left\langle -\dfrac{\sqrt{3}}{2}, \dfrac{1}{2}, -4 \right\rangle$

Note that $\|\overrightarrow{PQ_1}\| = \|\overrightarrow{PQ_2}\| = \|\overrightarrow{PQ_3}\| = \sqrt{17}$.

Unit vectors:

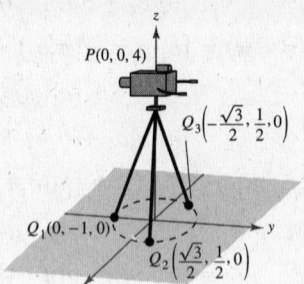

$\mathbf{u}_1 = \dfrac{1}{\sqrt{17}} \langle 0, -1, -4 \rangle$

$\mathbf{u}_2 = \dfrac{1}{\sqrt{17}} \left\langle \dfrac{\sqrt{3}}{2}, \dfrac{1}{2}, -4 \right\rangle$

$\mathbf{u}_3 = \dfrac{1}{\sqrt{17}} \left\langle -\dfrac{\sqrt{3}}{2}, \dfrac{1}{2}, -4 \right\rangle$

The unit vectors $\mathbf{u}_1, \mathbf{u}_2, \mathbf{u}_3$ give the directions of the force in each leg. Since the legs are the same length, the total weight is distributed equally among the legs. So,

$\mathbf{F}_1 = \dfrac{120}{3} \mathbf{u}_1 = \dfrac{40}{\sqrt{17}} \langle 0, -1, -4 \rangle$

$\mathbf{F}_2 = \dfrac{120}{3} \mathbf{u}_2 = \dfrac{40}{\sqrt{17}} \left\langle \dfrac{\sqrt{3}}{2}, \dfrac{1}{2}, -4 \right\rangle$

$\mathbf{F}_3 = \dfrac{120}{3} \mathbf{u}_3 = \dfrac{40}{\sqrt{17}} \left\langle -\dfrac{\sqrt{3}}{2}, \dfrac{1}{2}, -4 \right\rangle$

7. Let A lie on the y-axis and the wall on the x-axis. Then, $A = (0, 10, 0)$, $B = (8, 0, 6)$, $C = (-10, 0, 6)$ and

$\overrightarrow{AB} = \langle 8, -10, 6 \rangle, \overrightarrow{AC} = \langle -10, -10, 6 \rangle$.

$\|\overrightarrow{AB}\| = \sqrt{8^2 + (-10)^2 + 6^2} = 10\sqrt{2}$

$\|\overrightarrow{AC}\| = \sqrt{(-10)^2 + (-10)^2 + 6^2} = 2\sqrt{59}$

$\mathbf{F}_1 = 420 \dfrac{\overrightarrow{AB}}{\|\overrightarrow{AB}\|} = \dfrac{420}{10\sqrt{2}} \langle 8, -10, 6 \rangle = \dfrac{84}{\sqrt{2}} \langle 4, -5, 3 \rangle$

$\mathbf{F}_2 = 650 \dfrac{\overrightarrow{AC}}{\|\overrightarrow{AC}\|} = \dfrac{650}{2\sqrt{59}} \langle -10, -10, 6 \rangle = \dfrac{650}{\sqrt{59}} \langle -5, -5, 3 \rangle$

$\mathbf{F} = \mathbf{F}_1 + \mathbf{F}_2 = \left\langle \dfrac{(4)(84)}{\sqrt{2}} + \dfrac{(-5)(650)}{\sqrt{59}}, \dfrac{(-5)(84)}{\sqrt{2}} + \dfrac{(-5)(650)}{\sqrt{59}}, \dfrac{(3)(84)}{\sqrt{2}} + \dfrac{(3)(650)}{\sqrt{59}} \right\rangle$

$\approx \langle -185.526, -720.099, 432.059 \rangle$

$\|\mathbf{F}\| \approx 860.0 \text{ lb}$

8. Note that $\cos \theta = \dfrac{\mathbf{u} \cdot \mathbf{v}}{\|\mathbf{u}\| \|\mathbf{v}\|}$. Therefore,

$$\|\mathbf{u}\| \|\mathbf{v}\| \sin \theta = \|\mathbf{u}\| \|\mathbf{v}\| \sqrt{1 - \cos^2 \theta} = \|\mathbf{u}\| \|\mathbf{v}\| \sqrt{1 - \dfrac{(\mathbf{u} \cdot \mathbf{v})^2}{\|\mathbf{u}\|^2 \|\mathbf{v}\|^2}} = \sqrt{\|\mathbf{u}\|^2 \|\mathbf{v}\|^2 - (\mathbf{u} \cdot \mathbf{v})^2}$$

$$= \sqrt{(u_1^2 + u_2^2 + u_3^2)(v_1^2 + v_2^2 + v_3^2) - (u_1 v_1 + u_2 v_2 + u_3 v_3)^2}$$

$$= \sqrt{(u_2 v_3 - u_3 v_2)^2 + (u_1 v_3 - u_3 v_1)^2 + (u_1 v_2 - u_2 v_1)^2}$$

$$\|\mathbf{u}\| \|\mathbf{v}\| \sin \theta = \|\mathbf{u} \times \mathbf{v}\|$$

Since $\mathbf{u}$ and $\mathbf{v}$ are orthogonal, $\theta = 90°$ and $\sin \theta = 1$. So, $\|\mathbf{u}\| \|\mathbf{v}\| = \|\mathbf{u} \times \mathbf{v}\|$, $\mathbf{u}, \mathbf{v}$ orthogonal.

9. $\mathbf{u} = \langle a_1, b_1, c_1 \rangle$, $\mathbf{v} = \langle a_2, b_2, c_2 \rangle$, $\mathbf{w} = \langle a_3, b_3, c_3 \rangle$

$$\mathbf{v} \times \mathbf{w} = \begin{vmatrix} \mathbf{i} & \mathbf{j} & \mathbf{k} \\ a_2 & b_2 & c_2 \\ a_3 & b_3 & c_3 \end{vmatrix} = (b_2 c_3 - b_3 c_2)\mathbf{i} - (a_2 c_3 - a_3 c_2)\mathbf{j} + (a_2 b_3 - a_3 b_2)\mathbf{k}$$

$$\mathbf{u} \times (\mathbf{v} \times \mathbf{w}) = \begin{vmatrix} \mathbf{i} & \mathbf{j} & \mathbf{k} \\ a_1 & b_1 & c_1 \\ (b_2 c_3 - b_3 c_2) & (a_3 c_2 - a_2 c_3) & (a_2 b_3 - a_3 b_2) \end{vmatrix}$$

$$\mathbf{u} \times (\mathbf{v} \times \mathbf{w}) = [b_1(a_2 b_3 - a_3 b_2) - c_1(a_3 c_2 - a_2 c_3)]\mathbf{i} - [a_1(a_2 b_3 - a_3 b_2) - c_1(b_2 c_3 - b_3 c_2)]\mathbf{j}$$

$$+ [a_1(a_3 c_2 - a_2 c_3) - b_1(b_2 c_3 - b_3 c_2)]\mathbf{k}$$

$$= [a_2(a_1 a_3 + b_1 b_3 + c_1 c_3) - a_3(a_1 a_2 + b_1 b_2 + c_1 c_2)]\mathbf{i}$$

$$+ [b_2(a_1 a_3 + b_1 b_3 + c_1 c_3) - b_3(a_1 a_2 + b_1 b_2 + c_1 c_2)]\mathbf{j}$$

$$+ [c_2(a_1 a_3 + b_1 b_3 + c_1 c_3) - c_3(a_1 a_2 + b_1 b_2 + c_1 c_2)]\mathbf{k}$$

$$= (\mathbf{u} \cdot \mathbf{w})\mathbf{v} - (\mathbf{u} \cdot \mathbf{v})\mathbf{w}$$

10. $\begin{vmatrix} u_1 & u_2 & u_3 \\ v_1 & v_2 & v_3 \\ w_1 & w_2 & w_3 \end{vmatrix} = u_1 \begin{vmatrix} v_2 & v_3 \\ w_2 & w_3 \end{vmatrix} - u_2 \begin{vmatrix} v_1 & v_3 \\ w_1 & w_3 \end{vmatrix} + u_3 \begin{vmatrix} v_1 & v_2 \\ w_1 & w_2 \end{vmatrix}$

$$= u_1 \begin{vmatrix} v_2 & v_3 \\ w_2 & w_3 \end{vmatrix}(\mathbf{i} \cdot \mathbf{i}) - u_2 \begin{vmatrix} v_1 & v_3 \\ w_1 & w_3 \end{vmatrix}(\mathbf{j} \cdot \mathbf{j}) + u_3 \begin{vmatrix} v_1 & v_2 \\ w_1 & w_2 \end{vmatrix}(\mathbf{k} \cdot \mathbf{k})$$

$$= (u_1 \mathbf{i}) \cdot \left[\begin{vmatrix} v_2 & v_3 \\ w_2 & w_3 \end{vmatrix}\mathbf{i} \right] + (u_2 \mathbf{j}) \cdot \left[-\begin{vmatrix} v_1 & v_3 \\ w_1 & w_3 \end{vmatrix}\mathbf{j} \right] + (u_3 \mathbf{k}) \cdot \left[\begin{vmatrix} v_1 & v_2 \\ w_1 & w_2 \end{vmatrix}\mathbf{k} \right]$$

$$= (u_1 \mathbf{i} + u_2 \mathbf{j} + u_3 \mathbf{k}) \cdot \left[\begin{vmatrix} v_2 & v_3 \\ w_2 & w_3 \end{vmatrix}\mathbf{i} - \begin{vmatrix} v_1 & v_3 \\ w_1 & w_3 \end{vmatrix}\mathbf{j} + \begin{vmatrix} v_1 & v_2 \\ w_1 & w_2 \end{vmatrix}\mathbf{k} \right]$$

$$= \mathbf{u} \cdot \begin{vmatrix} \mathbf{i} & \mathbf{j} & \mathbf{k} \\ v_1 & v_2 & v_3 \\ w_1 & w_2 & w_3 \end{vmatrix} = \mathbf{u} \cdot (\mathbf{v} \times \mathbf{w})$$

11.

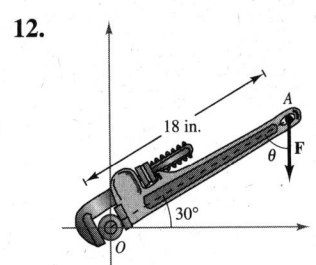

$\|\mathbf{v} \times \mathbf{w}\|$ = area of base and $\|\text{proj}_{\mathbf{v} \times \mathbf{w}} \mathbf{u}\|$ = height of parallelepiped

Therefore, the volume is

$$V = (\text{height})(\text{area of base}) = \|\text{proj}_{\mathbf{v} \times \mathbf{w}} \mathbf{u}\| \, \|\mathbf{v} \times \mathbf{w}\|$$

$$= \left| \frac{\mathbf{u} \cdot (\mathbf{v} \times \mathbf{w})}{\|\mathbf{v} \times \mathbf{w}\|} \right| \|\mathbf{v} \times \mathbf{w}\|$$

$$= |\mathbf{u} \cdot (\mathbf{v} \times \mathbf{w})|.$$

12.

(a) $O = (0, 0, 0)$

$A = (18 \cos 30°, 18 \sin 30°, 0) = (9\sqrt{3}, 9, 0)$

$\overrightarrow{OA} = \langle 9\sqrt{3}, 9, 0 \rangle$

$\|\mathbf{M}\| = \|\overrightarrow{OA} \times \mathbf{F}\| = \|\overrightarrow{OA}\| \, \|\mathbf{F}\| \sin \theta$

$\qquad = \left(\sqrt{(9\sqrt{3})^2 + 9^2 + 0^2} \right)(60) \sin \theta$

$\qquad = 1080 \sin \theta$

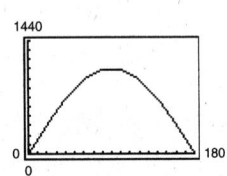

—CONTINUED—

12. —CONTINUED—

(b) $\|\vec{M}\| = 1080 \sin(45°) = (1080)\left(\dfrac{\sqrt{2}}{2}\right) = 540\sqrt{2} \approx 763.7$ in-lb

(c) $\|\vec{M}\| = 1080 \sin \theta$ has its maximum value at $\theta = 90°$. In order to generate the maximum torque, the force should be applied in a direction perpendicular to the wrench handle.

13. (a) In inches: $\vec{AB} = -15\mathbf{j} + 12\mathbf{k}$

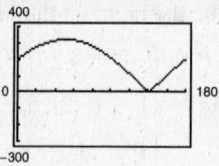

In feet: $\vec{AB} = -\dfrac{5}{4}\mathbf{j} + \mathbf{k}$

$\mathbf{F} = -200(\cos \theta \mathbf{j} + \sin \theta \mathbf{k})$

(b) $\vec{AB} \times \mathbf{F} = \begin{vmatrix} \mathbf{i} & \mathbf{j} & \mathbf{k} \\ 0 & -\dfrac{5}{4} & 1 \\ 0 & -200 \cos \theta & -200 \sin \theta \end{vmatrix} = (250 \sin \theta + 200 \cos \theta)\mathbf{i}$

$\|\vec{AB} \times \mathbf{F}\| = |250 \sin \theta + 200 \cos \theta| = 25|10 \sin \theta + 8 \cos \theta|$

(c) When $\theta = 30°$: $\|\vec{AB} \times \mathbf{F}\| = 25\left[10\left(\dfrac{1}{2}\right) + 8\left(\dfrac{\sqrt{3}}{2}\right)\right] = 25\left(5 + 4\sqrt{3}\right) \approx 298.2$

(d) From the graph we see that the maximum value occurs when $\theta \approx 51.34°$.

(e) From the graph we see that the zero occurs when $\theta \approx 141.34°$, the angle making $\vec{AB}$ parallel to $\mathbf{F}$.

14. The area of the triangle is one-half of the area of any of the 3 parallelograms having the following adjacent sides:

$\mathbf{b}$ and $\mathbf{c}$, $-\mathbf{b}$ and $\mathbf{a}$, $-\mathbf{c}$ and $-\mathbf{a}$

So,

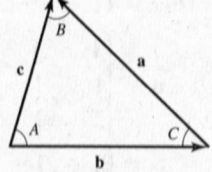

$$\text{Area} = \frac{\|\mathbf{b} \times \mathbf{c}\|}{2} = \frac{\|(-\mathbf{a}) \times (-\mathbf{c})\|}{2} = \frac{\|\mathbf{a} \times (-\mathbf{b})\|}{2}$$

$\|\mathbf{b} \times \mathbf{c}\| = \|(-\mathbf{a}) \times (-\mathbf{c})\| = \|\mathbf{a} \times (-\mathbf{b})\|$

$\|\mathbf{b}\| \|\mathbf{c}\| \sin A = \|\mathbf{a}\| \|\mathbf{c}\| \sin B = \|\mathbf{a}\| \|\mathbf{b}\| \sin C$

Divide by $\|\mathbf{a}\| \|\mathbf{b}\| \|\mathbf{c}\|$:

$$\frac{\sin A}{\|\mathbf{a}\|} = \frac{\sin B}{\|\mathbf{b}\|} = \frac{\sin C}{\|\mathbf{c}\|}$$

15. First insect: $x = 6 + t, y = 8 - t, z = 3 + t$

Second insect: $x = 1 + t, y = 2 + t, z = 2t$

(a) When $t = 0$ the first insect is located at $(6, 8, 3)$ and the second insect is located at $(1, 2, 0)$.

$d = \sqrt{(1 - 6)^2 + (2 - 8)^2 + (0 - 3)^2} = \sqrt{70}$ inches

(b) $d = \sqrt{[(1 + t) - (6 + t)]^2 + [(2 + t) - (8 - t)]^2 + [2t - (3 + t)]^2}$

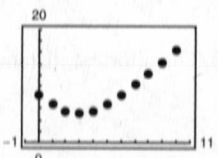

$= \sqrt{(-5)^2 + (2t - 6)^2 + (t - 3)^2}$

$= \sqrt{5t^2 - 30t + 70}$

t	0	1	2	3	4	5	6	7	8	9	10
d	$\sqrt{70}$	$\sqrt{45}$	$\sqrt{30}$	5	$\sqrt{30}$	$\sqrt{45}$	$\sqrt{70}$	$\sqrt{105}$	$\sqrt{150}$	$\sqrt{205}$	$\sqrt{270}$

—CONTINUED—

15. —CONTINUED—

(c) The distance between the two insects appears to lessen in the first 3 seconds, but then begins to increase with time.

(d) When $t = 3$, the insects get within 5 inches of each other.

16. (a) $x = -2 + 4t, y = 3, z = 1 - t \implies$ direction vector $\mathbf{u} = \langle 4, 0, -1 \rangle$

Let P be the point on the line with $t = 0$:

$$P = (-2 + 4 \cdot 0, 3, 1 - 0) = (-2, 3, 1), Q = (1, 5, -2)$$

$$\overrightarrow{PQ} = \langle 1 - (-2), 5 - 3, -2 - 1 \rangle = \langle 3, 2, -3 \rangle$$

$$\overrightarrow{PQ} \times \mathbf{u} = \begin{vmatrix} \mathbf{i} & \mathbf{j} & \mathbf{k} \\ 3 & 2 & -3 \\ 4 & 0 & -1 \end{vmatrix} = \begin{vmatrix} 2 & -3 \\ 0 & -1 \end{vmatrix} \mathbf{i} - \begin{vmatrix} 3 & -3 \\ 4 & -1 \end{vmatrix} \mathbf{j} + \begin{vmatrix} 3 & 2 \\ 4 & 0 \end{vmatrix} \mathbf{k}$$

$$= -2\mathbf{i} - 9\mathbf{j} - 8\mathbf{k} = \langle -2, -9, -8 \rangle$$

$$\|\overrightarrow{PQ} \times \mathbf{u}\| = \sqrt{(-2)^2 + (-9)^2 + (-8)^2} = \sqrt{149}$$

$$\|\mathbf{u}\| = \sqrt{4^2 + 0^2 + (-1)^2} = \sqrt{17}$$

$$D = \frac{\|\overrightarrow{PQ} \times \mathbf{u}\|}{\|\mathbf{u}\|} = \frac{\sqrt{149}}{\sqrt{17}} = \frac{\sqrt{2533}}{17} \approx 2.9605$$

(b) $x = 2t, y = -3 + t, z = 2 + 2t \implies$ direction vector $\mathbf{u} = \langle 2, 1, 2 \rangle$

Let P be the point on the line with $t = 0$:

$$P = (2 \cdot 0, -3 + 0, 2 + 2 \cdot 0) = (0, -3, 2), Q = (1, -2, 4)$$

$$\overrightarrow{PQ} = \langle 1 - 0, -2 - (-3), 4 - 2 \rangle = \langle 1, 1, 2 \rangle$$

$$\overrightarrow{PQ} \times \mathbf{u} = \begin{vmatrix} \mathbf{i} & \mathbf{j} & \mathbf{k} \\ 1 & 1 & 2 \\ 2 & 1 & 2 \end{vmatrix} = \begin{vmatrix} 1 & 2 \\ 1 & 2 \end{vmatrix} \mathbf{i} - \begin{vmatrix} 1 & 2 \\ 2 & 2 \end{vmatrix} \mathbf{j} + \begin{vmatrix} 1 & 1 \\ 2 & 1 \end{vmatrix} \mathbf{k}$$

$$= 0\mathbf{i} + 2\mathbf{j} - \mathbf{k} = \langle 0, 2, -1 \rangle$$

$$\|\overrightarrow{PQ} \times \mathbf{u}\| = \sqrt{0^2 + 2^2 + (-1)^2} = \sqrt{5}$$

$$\|\mathbf{u}\| = \sqrt{2^2 + 1^2 + 2^2} = \sqrt{9} = 3$$

$$D = \frac{\|\overrightarrow{PQ} \times \mathbf{u}\|}{\|\mathbf{u}\|} = \frac{\sqrt{5}}{3} \approx 0.7454$$

17. (a) $\mathbf{u} = \langle 0, 1, 1 \rangle$ direction vector of line determined by P_1 and P_2

$$D = \frac{\|\overrightarrow{P_1Q} \times \mathbf{u}\|}{\|\mathbf{u}\|} = \frac{\|\langle 2, 0, -1 \rangle \times \langle 0, 1, 1 \rangle\|}{\sqrt{2}}$$

$$= \frac{\|\langle 1, -2, 2 \rangle\|}{\sqrt{2}} = \frac{3}{\sqrt{2}} = \frac{3\sqrt{2}}{2}$$

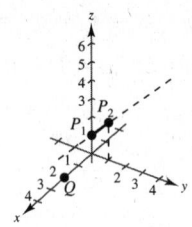

(b) The shortest distance to the line **segment** is $\|P_1Q\| = \|\langle 2, 0, -1 \rangle\| = \sqrt{5}$.

18. (a) $x = -t + 3, y = \dfrac{1}{2}t + 1, z = 2t - 1 \Rightarrow$ direction vector $\mathbf{u} = \left\langle -1, \dfrac{1}{2}, 2 \right\rangle$

Let $Q = (4, 3, s)$.

Let P be the point on the line corresponding to $t = 0$: $P = (3, 1, -1)$

$\overrightarrow{PQ} = \langle 4 - 3, 3 - 1, s - (-1) \rangle = \langle 1, 2, s + 1 \rangle$

$$\overrightarrow{PQ} \times \mathbf{u} = \begin{vmatrix} \mathbf{i} & \mathbf{j} & \mathbf{k} \\ 1 & 2 & s + 1 \\ -1 & \dfrac{1}{2} & 2 \end{vmatrix} = \begin{vmatrix} 2 & s + 1 \\ \dfrac{1}{2} & 2 \end{vmatrix}\mathbf{i} - \begin{vmatrix} 1 & s + 1 \\ -1 & 2 \end{vmatrix}\mathbf{j} + \begin{vmatrix} 1 & 2 \\ -1 & \dfrac{1}{2} \end{vmatrix}\mathbf{k}$$

$$= \dfrac{7 - s}{2}\mathbf{i} - (s + 3)\mathbf{j} + \dfrac{5}{2}\mathbf{k} = \left\langle \dfrac{7 - s}{2}, -s - 3, \dfrac{5}{2} \right\rangle$$

$$\|\overrightarrow{PQ} \times \mathbf{u}\| = \sqrt{\left(\dfrac{7 - s}{2}\right)^2 + (-s - 3)^2 + \left(\dfrac{5}{2}\right)^2} = \dfrac{\sqrt{5s^2 + 10s + 110}}{2}$$

$$\|\mathbf{u}\| = \sqrt{(-1)^2 + \left(\dfrac{1}{2}\right)^2 + 2^2} = \dfrac{\sqrt{21}}{2}$$

$$D = \dfrac{\|\overrightarrow{PQ} \times \mathbf{u}\|}{\|\mathbf{u}\|} = \dfrac{\sqrt{5s^2 + 10s + 110}}{\sqrt{21}} = \dfrac{\sqrt{105s^2 + 210s + 2310}}{21}$$

(b)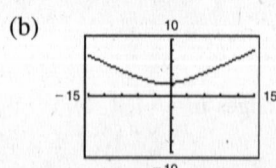

Minimum distance is $D = 2.23607$ at $s = -1$.

(c) $D = \dfrac{\sqrt{105s^2 + 210s + 2310}}{21}$

As s approaches very large (very positive) or very small (very negative) values, the expression under the radical is dominated by the term $105s^2$:

$105s^2 + 210s + 2310 \approx 105s^2, |s| =$ "large"

So, at large values of $|s|$,

$$D \approx \dfrac{\sqrt{105s^2}}{21} = \pm\dfrac{\sqrt{105}}{21}s.$$

Asymptotes: $D = \pm\dfrac{\sqrt{105}}{21}(s + 1)$

Chapter 11   Practice Test

1. Find the lengths of the sides of the triangle with vertices $(0, 0, 0)$, $(1, 2, -4)$, and $(0, -2, -1)$. Show that the triangle is a right triangle.

2. Find the standard form of the equation of a sphere having center $(0, 4, 1)$ and radius 5.

3. Find the center and radius of the sphere $x^2 + y^2 + z^2 + 2x - 4z - 11 = 0$.

4. Find the vector $\mathbf{u} - 3\mathbf{v}$ given $\mathbf{u} = \langle 1, 0, -1 \rangle$ and $\mathbf{v} = \langle 4, 3, -6 \rangle$.

5. Find the length of $\frac{1}{2}\mathbf{v}$ if $\mathbf{v} = \langle 2, 4, -6 \rangle$.

6. Find the dot product of $\mathbf{u} = \langle 2, 1, -3 \rangle$ and $\mathbf{v} = \langle 1, 1, -2 \rangle$.

7. Determine whether $\mathbf{u} = \langle 1, 1, -1 \rangle$ and $\mathbf{v} = \langle -3, -3, 3 \rangle$ are orthogonal, parallel, or neither.

8. Find the cross product of $\mathbf{u} = \langle -1, 0, 2 \rangle$ and $\mathbf{v} = \langle 1, -1, 3 \rangle$. What is $\mathbf{v} \times \mathbf{u}$?

9. Use the triple scalar product to find the volume of the parallelepiped having adjacent edges $\mathbf{u} = \langle 1, 1, 1 \rangle$, $\mathbf{v} = \langle 0, -1, 1 \rangle$, and $\mathbf{w} = \langle 1, 0, 4 \rangle$.

10. Find a set of parametric equations for the line through the points $(0, -3, 3)$ and $(2, -3, 4)$.

11. Find an equation of the plane passing through $(1, 2, 3)$ and perpendicular to the vector $\mathbf{n} = \langle 1, -1, 0 \rangle$.

12. Find an equation of the plane passing through the three points $A = (0, 0, 0)$, $B = (1, 1, 1)$, and $C = (1, 2, 3)$.

13. Determine whether the planes $x + y - z = 12$ and $3x - 4y - z = 9$ are parallel, orthogonal or neither.

14. Find the distance between the point $(1, 1, 1)$ and the plane $x + 2y + z = 6$.

CHAPTER 12
Limits and an Introduction to Calculus

CHAPTER 12
Limits and an Introduction to Calculus

Section 12.1   Introduction to Limits

- ■ If $f(x)$ becomes arbitrarily close to a unique number L as x approaches c from either side, then the limit of $f(x)$ as x approaches c is L:
$$\lim_{x \to c} f(x) = L.$$

- ■ You should be able to use a calculator to find a limit.
- ■ You should be able to use a graph to find a limit.
- ■ You should understand how limits can fail to exist:
  - (a) $f(x)$ approaches a different number from the right of c than it approaches from the left of c.
  - (b) $f(x)$ increases or decreases without bound as x approaches c.
  - (c) $f(x)$ oscillates between two fixed values as x approaches c.
- ■ You should know and be able to use the elementary properties of limits.

Vocabulary Check

1. limit   **2.** oscillates   **3.** direct substitution

---

1. (a)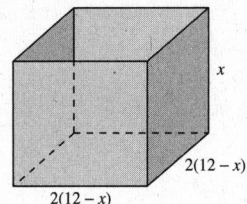

(b) $V = (\text{base})\text{height} = (24 - 2x)^2 x = 4x(12 - x)^2$

(d)

Maximum at $x = 4$

(c) $\lim_{x \to 4} V = 1024$

x	3	3.5	3.9	4	4.1	4.5	5
V	972.0	1011.5	1023.5	1024.0	1023.5	1012.5	980.0

2. (a)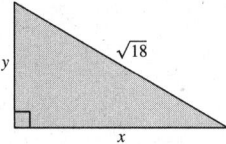

(b) $x^2 + y^2 = 18 \implies y = \sqrt{18 - x^2}$

Area $= \frac{1}{2}bh = \frac{1}{2}x\sqrt{18 - x^2}$

(d)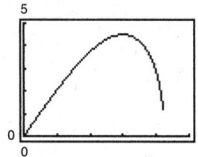

(c)

x	2	2.5	2.9	3	3.1	3.5	4
A	3.74	4.28	4.49	4.5	4.49	4.20	2.83

$\lim_{x \to 3} A(x) = 4.5$

3. $\lim\limits_{x\to 2} (5x + 4) = 14$

x	1.9	1.99	1.999	2	2.001	2.01	2.1
$f(x)$	13.5	13.95	13.995	14	14.005	14.05	14.5

The limit is reached.

4. $\lim\limits_{x\to 2}(x^2 - 3x + 1) = -1$,   The limit is reached.

x	1.9	1.99	1.999	2	2.001	2.01	2.1
$f(x)$	-1.09	-1.0099	-1.000999	-1	-0.998999	-0.9899	-0.89

5. $\lim\limits_{x\to 3} \dfrac{x - 3}{x^2 - 9} = \dfrac{1}{6}$

x	2.9	2.99	2.999	3	3.001	3.01	3.1
$f(x)$	0.1695	0.1669	0.16669	?	0.16664	0.1664	0.1639

The limit is not reached.

6. $\lim\limits_{x\to -1} \dfrac{x + 1}{x^2 - x - 2} = -\dfrac{1}{3}$,   The limit is not reached.

x	-1.1	-1.01	-1.001	-1.0	-0.999	-0.99	-0.9
$f(x)$	-0.3226	-0.3322	-0.3332	?	-0.3334	-0.3344	-0.3348

7. $f(x) = \dfrac{x - 1}{x^2 + 2x - 3}$

x	0.9	0.99	0.999	1	1.001	1.01	1.1
$f(x)$	0.2564	0.2506	0.2501	?	0.2499	0.2494	0.2439

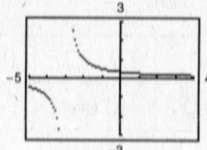

$$\lim\limits_{x\to 1} \dfrac{x - 1}{x^2 + 2x - 3} = \dfrac{1}{4}$$

8. $\lim\limits_{x\to -2} \dfrac{x + 2}{x^2 + 5x + 6} = 1$

x	-2.1	-2.01	-2.001	-2.0	-1.999	-1.99	-1.9
$f(x)$	1.1111	1.0101	1.0010	?	0.9990	0.9901	0.9091

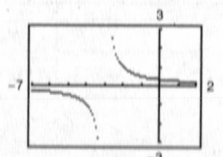

9. $f(x) = \dfrac{\sqrt{x+5} - \sqrt{5}}{x}$

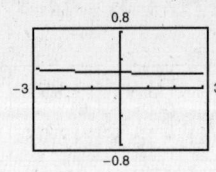

x	-0.1	-0.01	-0.001	0	0.001	0.01	0.1
$f(x)$	0.2247	0.2237	0.2236	?	0.2236	0.2235	0.2225

$$\lim_{x \to 0} \frac{\sqrt{x+5} - \sqrt{5}}{x} \approx 0.2236 \left(\text{Actual limit: } \frac{1}{2\sqrt{5}} \right)$$

10. $\displaystyle\lim_{x \to -3} \dfrac{\sqrt{1-x} - 2}{x+3} = -\dfrac{1}{4}$

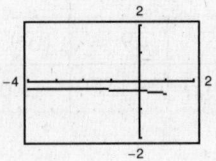

x	-3.1	-3.01	-3.001	-3.0	-2.999	-2.99	-2.9
$f(x)$	-0.2485	-0.2498	-0.25	?	-0.25	-0.2502	-0.2516

11. $f(x) = \dfrac{\dfrac{x}{x+2} - 2}{x+4}$

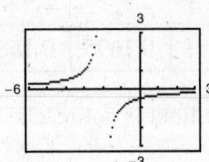

x	-4.1	-4.01	-4.001	-4	-3.999	-3.99	-3.9
$f(x)$	0.4762	0.4975	0.4998	?	0.5003	0.5025	0.5263

$$\lim_{x \to -4} \frac{\dfrac{x}{x+2} - 2}{x+4} = \frac{1}{2}$$

12. $\displaystyle\lim_{x \to 2} \dfrac{\dfrac{1}{x+2} - \dfrac{1}{4}}{x-2} = -\dfrac{1}{16}$

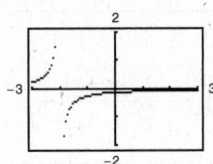

x	1.9	1.99	1.999	2.0	2.001	2.01	2.1
$f(x)$	-0.0641	-0.0627	-0.0625	?	-0.0625	-0.0623	-0.0610

13. $f(x) = \dfrac{\sin x}{x}$

Make sure your calculator is set in radian mode.

x	-0.1	-0.01	-0.001	0	0.001	0.01	0.1
$f(x)$	0.9983	0.99998	0.9999998	?	0.9999998	0.99998	0.9983

$$\lim_{x \to 0} \frac{\sin x}{x} = 1$$

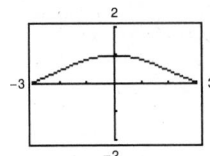

14. $\lim\limits_{x \to 0} \dfrac{\cos x - 1}{x} = 0$

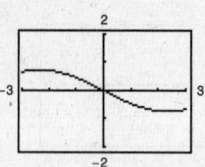

x	-0.1	-0.01	-0.001	0	0.001	0.01	0.1
$f(x)$	0.050	0.005	0.0005	?	-0.0005	-0.005	-0.050

15. $f(x) = \begin{cases} 2x + 1, & x < 2 \\ x + 3, & x \geq 2 \end{cases}$

The limit exists as x
approaches 2:

$\lim\limits_{x \to 2} f(x) = 5$

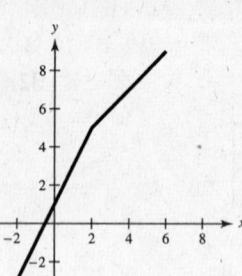

16.

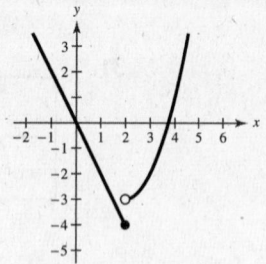

$\lim\limits_{x \to 2} f(x)$ does not exist.

17. $\lim\limits_{x \to -4} (x^2 - 3) = 13$

18. $\lim\limits_{x \to 2} \dfrac{3x^2 - 12}{x - 2} = 12$

19. $\lim\limits_{x \to -2} \dfrac{|x + 2|}{x + 2}$ does not exist. $f(x) = \dfrac{|x + 2|}{x + 2}$ equals -1 to the left of -2, and equals 1 to the right of -2.

20. The limit does not exist because $f(x)$ does not approach a real number as x approaches 1.

21. The limit does not exist because $f(x)$ oscillates between 2 and -2.

22. $\lim\limits_{x \to -1} \sin\left(\dfrac{\pi x}{2}\right) = -1$

23.

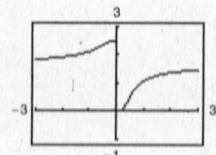

$\lim\limits_{x \to 0} \dfrac{5}{2 + e^{1/x}}$ does not exist.

24.

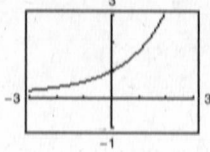

$\lim\limits_{x \to 0} \dfrac{e^x - 1}{x} = 1$

25.

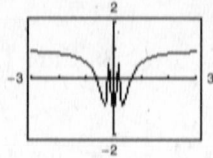

$\lim\limits_{x \to 0} \cos \dfrac{1}{x}$ does not exist.

The graph oscillates between -1 and 1.

26.

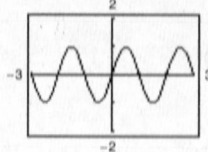

$\lim\limits_{x \to 1} \sin \pi x = 0$

27.

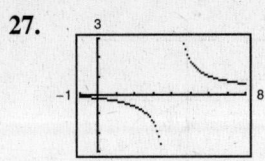

$$\lim_{x \to 4} \frac{\sqrt{x + 3} - 1}{x - 4} \text{ does not exist.}$$

28.

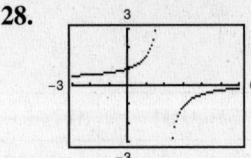

$$\lim_{x \to 2} \frac{\sqrt{x + 5} - 4}{x - 2} \text{ does not exist.}$$

29.

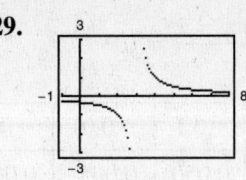

$$\lim_{x \to 1} \frac{x - 1}{x^2 - 4x + 3} = -\frac{1}{2}$$

30.

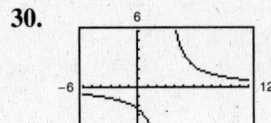

$$\lim_{x \to 3} \frac{7}{x - 3} \text{ does not exist.}$$

31.

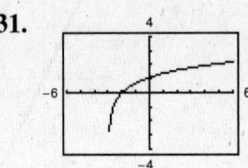

$$\lim_{x \to 4} \ln(x + 3) \approx 1.946$$

(Exact limit is ln 7.)

32.

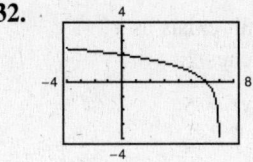

$$\lim_{x \to -1} \ln(7 - x) = \ln(7 - (-1))$$
$$= \ln 8$$

33. (a) $\lim_{x \to c} [-2g(x)] = -2(6) = -12$

(b) $\lim_{x \to c} [f(x) + g(x)] = 3 + 6 = 9$

(c) $\lim_{x \to c} \dfrac{f(x)}{g(x)} = \dfrac{3}{6} = \dfrac{1}{2}$

(d) $\lim_{x \to c} \sqrt{f(x)} = \sqrt{3}$

34. (a) $\lim_{x \to c} [f(x) + g(x)]^2 = (5 - 2)^2 = 9$

(b) $\lim_{x \to c} [6f(x)g(x)] = 6(5)(-2) = -60$

(c) $\lim_{x \to c} \dfrac{5g(x)}{4f(x)} = \dfrac{5(-2)}{4(5)} = -\dfrac{1}{2}$

(d) $\lim_{x \to c} \dfrac{1}{\sqrt{f(x)}} = \dfrac{1}{\sqrt{5}} = \dfrac{\sqrt{5}}{5}$

35. (a) $\lim_{x \to 2} f(x) = 2^3 = 8$

(b) $\lim_{x \to 2} g(x) = \dfrac{\sqrt{2^2 + 5}}{2(2^2)} = \dfrac{3}{8}$

(c) $\lim_{x \to 2} [f(x)g(x)] = 8\left(\dfrac{3}{8}\right) = 3$

(d) $\lim_{x \to 2} [g(x) - f(x)] = \dfrac{3}{8} - 8 = -\dfrac{61}{8}$

36. (a) $\lim_{x \to 2} f(x) = \dfrac{2}{3 - 2} = 2$

(b) $\lim_{x \to 2} g(x) = \sin(\pi 2) = 0$

(c) $\lim_{x \to 2} [f(x)g(x)] = 2(0) = 0$

(d) $\lim_{x \to 2} [g(x) - f(x)] = 0 - 2 = -2$

37. $\lim_{x \to 5} (10 - x^2) = 10 - 5^2 = -15$

38. $\lim_{x \to -2} \left(\dfrac{1}{2}x^3 - 5x\right) = \dfrac{1}{2}(-2)^3 - 5(-2) = 6$

39. $\lim_{x \to -3} (2x^2 + 4x + 1) = 2(-3)^2 + 4(-3) + 1 = 7$

40. $\lim_{x \to -2} (x^3 - 6x + 5) = (-2)^3 - 6(-2) + 5 = 9$

41. $\lim_{x \to 3} \left(-\dfrac{9}{x}\right) = -\dfrac{9}{3} = -3$

42. $\lim_{x \to -5} \dfrac{6}{x + 2} = \dfrac{6}{-5 + 2} = \dfrac{6}{-3} = -2$

43. $\lim_{x \to -3} \dfrac{3x}{x^2 + 1} = -\dfrac{9}{10}$

44. $\lim_{x \to 4} \dfrac{x - 1}{x^2 + 2x + 3} = \dfrac{4 - 1}{16 + 8 + 3} = \dfrac{3}{27} = \dfrac{1}{9}$

45. $\lim_{x \to -2} \dfrac{5x + 3}{2x - 9} = \dfrac{5(-2) + 3}{2(-2) - 9} = \dfrac{-7}{-13} = \dfrac{7}{13}$

46. $\lim_{x \to 3} \dfrac{x^2 + 1}{x} = \dfrac{9 + 1}{3} = \dfrac{10}{3}$

47. $\lim_{x \to -1} \sqrt{x + 2} = \sqrt{-1 + 2} = 1$

48. $\lim_{x \to 3} \sqrt[3]{x^2 - 1} = \sqrt[3]{9 - 1} = 2$

49. $\lim_{x \to 7} \dfrac{5x}{\sqrt{x + 2}} = \dfrac{5(7)}{\sqrt{7 + 2}} = \dfrac{35}{3}$

50. $\lim_{x \to 8} \dfrac{\sqrt{x + 1}}{x - 4} = \dfrac{\sqrt{8 + 1}}{x - 4} = \dfrac{3}{4}$

51. $\lim_{x \to 3} e^x = e^3 \approx 20.0855$

52. $\lim_{x \to e} \ln x = \ln e = 1$

53. $\lim_{x \to \pi} \sin 2x = \sin 2\pi = 0$

54. $\lim_{x \to \pi} \tan x = \tan \pi = 0$

55. $\lim_{x \to 1/2} \arcsin x = \arcsin \dfrac{1}{2} = \dfrac{\pi}{6} \approx 0.5236$

56. $\lim_{x \to 1} \arccos \dfrac{x}{2} = \arccos \dfrac{1}{2} = \dfrac{\pi}{3} \approx 1.0472$

57. True

58. True (assuming the limits exist).

59. Answers will vary.

60. In general you cannot use a graphing utility to determine whether a limit can be reached. It is important to analyze a function analytically.

61. (a) No. The limit may or may not exist. And if it does exist, it may not equal 4.

(b) No. $f(2)$ may or may not exist. And if $f(2)$ exists, it may not equal 4.

62. $\lim_{x \to 5} f(x) = 12$ means that the values of f approach 12 as x approaches 5.

63. $f(x) = \dfrac{x - 9}{\sqrt{x} - 3}$

(a)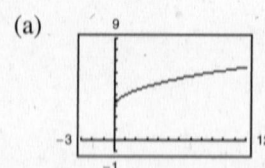

$\lim_{x \to 9} = \dfrac{x - 9}{\sqrt{x} - 3} = 6$

(b) Domain: $x \geq 0$, $x \neq 9$

(c) $\sqrt{x} - 3 \neq 0 \Rightarrow x \neq 9$; $\sqrt{x} \Rightarrow x \geq 0$

Domain: $x \geq 0$, $x \neq 9$

(d) It may not be clear from a graph that a function is not defined at a single point. Examining a function graphically and algebraically ensures that you will find all points at which the function is not defined.

64. (a)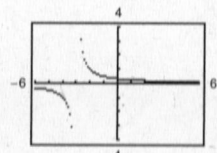

$\lim_{x \to 3} f(x) \approx 0.1667$, $\left(\dfrac{1}{6} \right)$

(b) Domain: all $x \neq \pm 3$

(c) $\dfrac{x - 3}{x^2 - 9} = \dfrac{x - 3}{(x - 3)(x + 3)}$, Domain: all $x \neq \pm 3$

(d) From the graph, it is not apparent that $x = 3$ is excluded from the domain of $f(x)$.

65. $\dfrac{5 - x}{3x - 15} = \dfrac{5 - x}{-3(5 - x)} = -\dfrac{1}{3}$, $x \neq 5$

66. $\dfrac{x^2 - 81}{9 - x} = \dfrac{(x - 9)(x + 9)}{9 - x} = -x - 9$, $x \neq 9$

67. $\dfrac{15x^2 + 7x - 4}{15x^2 + x - 2} = \dfrac{(3x - 1)(5x + 4)}{(3x - 1)(5x + 2)}$

$= \dfrac{5x + 4}{5x + 2}, \; x \neq \dfrac{1}{3}$

68. $\dfrac{x^2 - 12x + 36}{x^2 - 7x + 6} = \dfrac{(x - 6)(x - 6)}{(x - 6)(x - 1)} = \dfrac{x - 6}{x - 1}, \; x \neq 6$

69. $\dfrac{x^3 + 27}{x^2 + x - 6} = \dfrac{(x + 3)(x^2 - 3x + 9)}{(x + 3)(x - 2)}$

$= \dfrac{x^2 - 3x + 9}{x - 2}, \; x \neq -3$

70. $\dfrac{x^3 - 8}{x^2 - 4} = \dfrac{(x - 2)(x^2 + 2x + 4)}{(x - 2)(x + 2)}$

$= \dfrac{x^2 + 2x + 4}{x + 2}, \; x \neq 2$

71. (a)

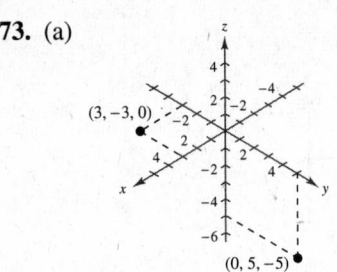

(b) $d = \sqrt{(3 - 3)^2 + (2 - 2)^2 + (8 - 7)^2} = 1$

(c) Midpoint: $\left(\dfrac{3 + 3}{2}, \dfrac{2 + 2}{2}, \dfrac{7 + 8}{2}\right) = \left(3, 2, \dfrac{15}{2}\right)$

72. (a)

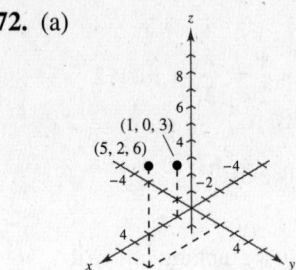

(b) $d = \sqrt{(5 - 1)^2 + (2 - 0)^2 + (6 - 3)^2}$

$= \sqrt{16 + 4 + 9} = \sqrt{29}$

(c) $M = \left(\dfrac{1 + 5}{2}, \dfrac{0 + 2}{2}, \dfrac{3 + 6}{2}\right) = \left(3, 1, \dfrac{9}{2}\right)$

73. (a)

(b) $d = \sqrt{(0 - 3)^2 + [5 - (-3)]^2 + (-5 - 0)^2}$

$= \sqrt{98} = 7\sqrt{2}$

(c) Midpoint:

$\left(\dfrac{3 + 0}{2}, \dfrac{-3 + 5}{2}, \dfrac{0 + (-5)}{2}\right) = \left(\dfrac{3}{2}, 1, -\dfrac{5}{2}\right)$

74. (a)

(b) $d = \sqrt{(2 - 0)^2 + [0 - (-4)]^2 + (-9 - 0)^2}$

$= \sqrt{4 + 16 + 81} = \sqrt{101}$

(c) $M = \left(\dfrac{0 + 2}{2}, \dfrac{-4 + 0}{2}, \dfrac{0 + (-9)}{2}\right)$

$= \left(1, -2, -\dfrac{9}{2}\right)$

Section 12.2   Techniques for Evaluating Limits

- ■ You can use direct substitution to find the limit of a polynomial function $p(x)$:
$$\lim_{x \to c} p(x) = p(c).$$

- ■ You can use direct substitution to find the limit of a rational function $r(x) = \dfrac{p(x)}{q(x)}$, as long as $q(c) \neq 0$:
$$\lim_{x \to c} r(x) = r(c) = \frac{p(c)}{q(c)}, q(c) \neq 0.$$

- ■ You should be able to use cancellation techniques to find a limit.
- ■ You should know how to use rationalization techniques to find a limit.
- ■ You should know how to use technology to find a limit.
- ■ You should be able to calculate one-sided limits.

Vocabulary Check

1. dividing out technique

2. indeterminate form

3. one-sided limit

4. difference quotient

1. $g(x) = \dfrac{-2x^2 + x}{x}$, $g_2(x) = -2x + 1$

   (a) $\lim\limits_{x \to 0} g(x) = 1$

   (b) $\lim\limits_{x \to -1} g(x) = 3$

   (c) $\lim\limits_{x \to -2} g(x) = 5$

2. $h(x) = \dfrac{x^2 - 3x}{x}$

   $h_2(x) = x - 3$

   (a) $\lim\limits_{x \to -2} h(x) = -5$

   (b) $\lim\limits_{x \to 0} h(x) = -3$

   (c) $\lim\limits_{x \to 3} h(x) = 0$

3. $g(x) = \dfrac{x^3 - x}{x - 1}$, $g_2(x) = x^2 + x = x(x + 1)$

   (a) $\lim\limits_{x \to 1} g(x) = 2$

   (b) $\lim\limits_{x \to -1} g(x) = 0$

   (c) $\lim\limits_{x \to 0} g(x) = 0$

4. $f(x) = \dfrac{x^2 - 1}{x + 1}$

   $f_2(x) = x - 1$

   (a) $\lim\limits_{x \to 1} f(x) = 0$

   (b) $\lim\limits_{x \to 2} f(x) = 1$

   (c) $\lim\limits_{x \to -1} f(x) = -2$

5. $\lim\limits_{x \to 6} \dfrac{x - 6}{x^2 - 36} = \lim\limits_{x \to 6} \dfrac{x - 6}{(x - 6)(x + 6)}$

$$= \lim\limits_{x \to 6} \frac{1}{x + 6} = \frac{1}{12}$$

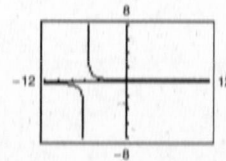

6. $\lim\limits_{x \to 5} \dfrac{5 - x}{x^2 - 25} = \lim\limits_{x \to 5} \dfrac{5 - x}{(x - 5)(x + 5)}$

$$= \lim\limits_{x \to 5} \frac{-1}{x + 5} = \frac{-1}{10}$$

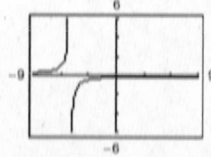

7. $\lim\limits_{x \to -1} \dfrac{1 - 2x - 3x^2}{1 + x} = \lim\limits_{x \to -1} \dfrac{(1 + x)(1 - 3x)}{1 + x}$

$\qquad\qquad = \lim\limits_{x \to -1} (1 - 3x) = 4$

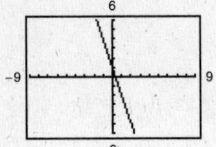

8. $\lim\limits_{x \to -2} \dfrac{2x^2 + 3x - 2}{(x + 2)} = \lim\limits_{x \to -2} \dfrac{(x + 2)(2x - 1)}{(x + 2)}$

$\qquad\qquad = \lim\limits_{x \to -2} (2x - 1) = -5$

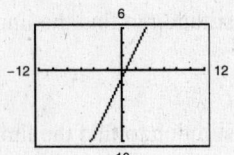

9. $\lim\limits_{t \to 2} \dfrac{t^3 - 8}{t - 2} = \lim\limits_{t \to 2} \dfrac{(t - 2)(t^2 + 2t + 4)}{t - 2}$

$\qquad\qquad = \lim\limits_{t \to 2} (t^2 + 2t + 4)$

$\qquad\qquad = 4 + 4 + 4 = 12$

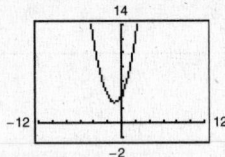

10. $\lim\limits_{t \to -3} \dfrac{t^3 + 27}{t + 3} = \lim\limits_{t \to -3} \dfrac{(t + 3)(t^2 - 3t + 9)}{t + 3}$

$\qquad\qquad = \lim\limits_{t \to -3} (t^2 - 3t + 9) = 27$

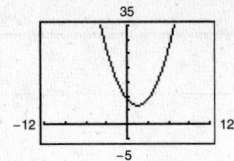

11. $\lim\limits_{x \to 2} \dfrac{x^5 - 32}{x - 2} = 80$

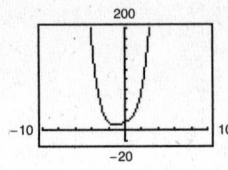

12. $\lim\limits_{x \to 1} \dfrac{x^4 - 1}{x - 1} = \lim\limits_{x \to 1} \dfrac{(x^2 - 1)(x^2 + 1)}{x - 1}$

$\qquad\qquad = \lim\limits_{x \to 1} \dfrac{(x - 1)(x + 1)(x^2 + 1)}{(x - 1)}$

$\qquad\qquad = \lim\limits_{x \to 1} (x + 1)(x^2 + 1) = 4$

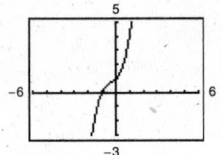

13. $\lim\limits_{y \to 0} \dfrac{\sqrt{5 + y} - \sqrt{5}}{y} = \lim\limits_{y \to 0} \dfrac{\sqrt{5 + y} - \sqrt{5}}{y} \cdot \dfrac{\sqrt{5 + y} + \sqrt{5}}{\sqrt{5 + y} + \sqrt{5}}$

$\qquad\qquad = \lim\limits_{y \to 0} \dfrac{(5 + y) - 5}{y\left(\sqrt{5 + y} + \sqrt{5}\right)}$

$\qquad\qquad = \lim\limits_{y \to 0} \dfrac{1}{\sqrt{5 + y} + \sqrt{5}}$

$\qquad\qquad = \dfrac{1}{2\sqrt{5}}$

$\qquad\qquad = \dfrac{\sqrt{5}}{10}$

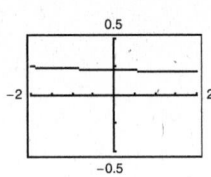

14. $\lim\limits_{z \to 0} \dfrac{\sqrt{7-z} - \sqrt{7}}{z}\left(\dfrac{\sqrt{7-z} + \sqrt{7}}{\sqrt{7-z} + \sqrt{7}}\right) = \lim\limits_{z \to 0} \dfrac{(7-z) - 7}{z\left(\sqrt{7-z} + \sqrt{7}\right)}$

$$= \lim\limits_{z \to 0} \dfrac{-1}{\sqrt{7-z} + \sqrt{7}}$$

$$= \dfrac{-1}{2\sqrt{7}} = -\dfrac{\sqrt{7}}{14} \approx -0.1890$$

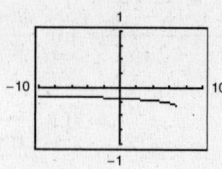

15. $\lim\limits_{x \to 0} = \dfrac{\sqrt{x+3} - \sqrt{3}}{x}$

$$= \lim\limits_{x \to 0} \dfrac{\sqrt{x+3} - \sqrt{3}}{x} \cdot \dfrac{\sqrt{x+3} + \sqrt{3}}{\sqrt{x+3} + \sqrt{3}}$$

$$= \lim\limits_{x \to 0} \dfrac{1}{\sqrt{x+3} + \sqrt{3}} = \dfrac{1}{2\sqrt{3}} = \dfrac{\sqrt{3}}{6}$$

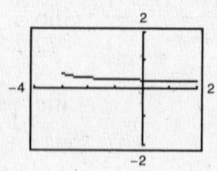

16. $\lim\limits_{x \to 0} \dfrac{\sqrt{x+4} - 2}{x} = \lim\limits_{x \to 0} \dfrac{\sqrt{x+4} - 2}{x} \cdot \dfrac{\sqrt{x+4} + 2}{\sqrt{x+4} + 2}$

$$= \lim\limits_{x \to 0} \dfrac{x+4-4}{x\left(\sqrt{x+4} + 2\right)}$$

$$= \lim\limits_{x \to 0} \dfrac{x}{x\left(\sqrt{x+4} + 2\right)}$$

$$= \lim\limits_{x \to 0} \dfrac{1}{\sqrt{x+4} + 2} = \dfrac{1}{\sqrt{4} + 2} = \dfrac{1}{4}$$

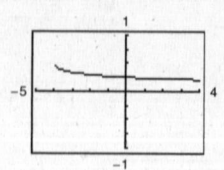

17. $\lim\limits_{x \to 0} \dfrac{\sqrt{2x+1} - 1}{x} = \lim\limits_{x \to 0} \dfrac{\sqrt{2x+1} - 1}{x} \cdot \dfrac{\sqrt{2x+1} + 1}{\sqrt{2x+1} + 1}$

$$= \lim\limits_{x \to 0} \dfrac{2}{\sqrt{2x+1} + 1} = \dfrac{2}{2} = 1$$

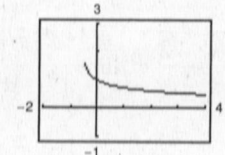

18. $\lim\limits_{x \to 9} \dfrac{3 - \sqrt{x}}{x - 9} = \lim\limits_{x \to 9} \dfrac{3 - \sqrt{x}}{x - 9} \cdot \dfrac{3 + \sqrt{x}}{3 + \sqrt{x}}$

$$= \lim\limits_{x \to 9} \dfrac{9 - x}{(x - 9)\left(3 + \sqrt{x}\right)}$$

$$= \lim\limits_{x \to 9} \dfrac{-1}{3 + \sqrt{x}} = \dfrac{-1}{3 + \sqrt{9}} = -\dfrac{1}{6}$$

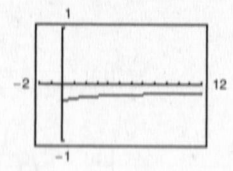

19. $\lim\limits_{x \to -3} \dfrac{\sqrt{x+7} - 2}{x + 3} = \lim\limits_{x \to -3} \dfrac{\sqrt{x+7} - 2}{x + 3} \cdot \dfrac{\sqrt{x+7} + 2}{\sqrt{x+7} + 2}$

$$= \lim\limits_{x \to -3} \dfrac{(x+7) - 4}{(x + 3)\left(\sqrt{x+7} + 2\right)}$$

$$= \lim\limits_{x \to -3} \dfrac{1}{\sqrt{x+7} + 2}$$

$$= \dfrac{1}{4}$$

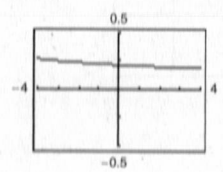

20. $\lim\limits_{x \to 2} \dfrac{4 - \sqrt{18 - x}}{x - 2} = \lim\limits_{x \to 2} \dfrac{4 - \sqrt{18 - x}}{x - 2} \cdot \dfrac{4 + \sqrt{18 - x}}{4 + \sqrt{18 - x}}$

$\qquad = \lim\limits_{x \to 2} \dfrac{16 - (18 - x)}{(x - 2)(4 + \sqrt{18 - x})}$

$\qquad = \lim\limits_{x \to 2} \dfrac{1}{4 + \sqrt{18 - x}} = \dfrac{1}{8} = 0.125$

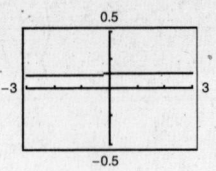

21. $\lim\limits_{x \to 0} \dfrac{1/(1 + x) - 1}{x} = \lim\limits_{x \to 0} \dfrac{1 - (1 + x)}{(1 + x)x}$

$\qquad = \lim\limits_{x \to 0} \dfrac{-1}{1 + x} = -1$

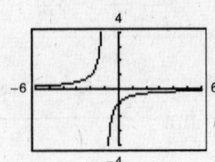

22. $\lim\limits_{x \to 0} \dfrac{\dfrac{1}{4 - x} - \dfrac{1}{4}}{x} = \lim\limits_{x \to 0} \dfrac{4 - (4 - x)}{(4 - x)4x}$

$\qquad = \lim\limits_{x \to 0} \dfrac{1}{(4 - x)4} = \dfrac{1}{16}$

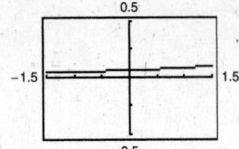

23. $f(x) = \dfrac{\dfrac{1}{x + 4} - \dfrac{1}{4}}{x}$

$\qquad \lim\limits_{x \to 0} f(x) = -\dfrac{1}{16}, \quad (-0.0625)$

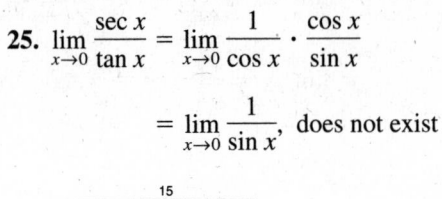

24. $\lim\limits_{x \to 0} \dfrac{\dfrac{1}{2 + x} - \dfrac{1}{2}}{x} = \lim\limits_{x \to 0} \dfrac{(2 + x)\left[\dfrac{1}{2 + x} - \dfrac{1}{2}\right]}{x(2 + x)}$

$\qquad = \lim\limits_{x \to 0} \dfrac{1 - \dfrac{2 + x}{2}}{x(2 + x)}$

$\qquad = \lim\limits_{x \to 0} \dfrac{-\dfrac{x}{2}}{x(2 + x)}$

$\qquad = \lim\limits_{x \to 0} -\dfrac{1}{2(2 + x)} = -\dfrac{1}{4}$

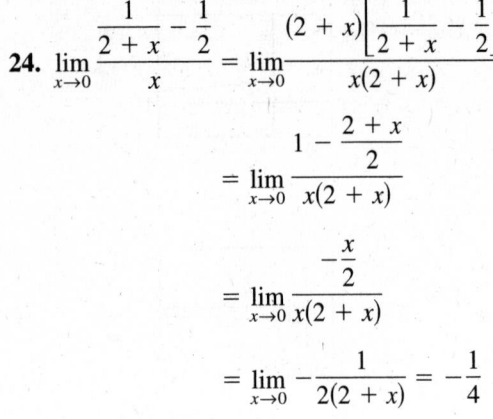

25. $\lim\limits_{x \to 0} \dfrac{\sec x}{\tan x} = \lim\limits_{x \to 0} \dfrac{1}{\cos x} \cdot \dfrac{\cos x}{\sin x}$

$\qquad = \lim\limits_{x \to 0} \dfrac{1}{\sin x}$, does not exist

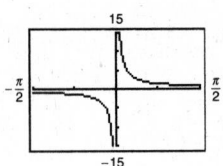

26. $\lim\limits_{x \to \pi/2} \dfrac{1 - \sin x}{\cos x} = \lim\limits_{x \to \pi/2} \dfrac{1 - \sin x}{\cos x} \cdot \dfrac{1 + \sin x}{1 + \sin x}$

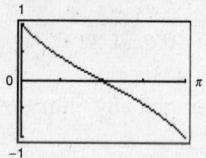

$\qquad = \lim\limits_{x \to \pi/2} \dfrac{1 - \sin^2 x}{\cos x(1 + \sin x)} = \lim\limits_{x \to \pi/2} \dfrac{\cos^2 x}{\cos x(1 + \sin x)}$

$\qquad = \lim\limits_{x \to \pi/2} \dfrac{\cos x}{1 + \sin x} = 0$

27. $f(x) = \dfrac{e^{2x} - 1}{x}$

$\quad \lim\limits_{x \to 0} f(x) = 2$

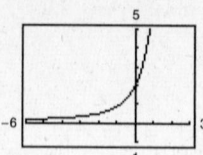

28. $\lim\limits_{x \to 0} \dfrac{1 - e^{-x}}{x} = 1$

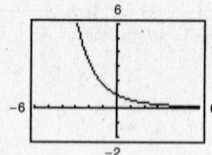

29. $\lim\limits_{x \to 0^+} x \ln x = 0$

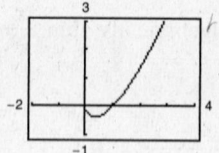

30. $\lim\limits_{x \to 0^+} x^2 \ln x = 0$

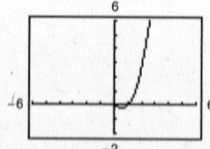

31. $\lim\limits_{x \to 0} \dfrac{\sin 2x}{x} = 2$

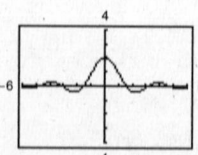

32. $\lim\limits_{x \to 0} \dfrac{\sin 3x}{x} = 3$

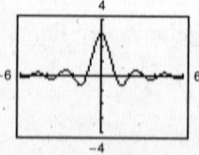

33. $\lim\limits_{x \to 0} \dfrac{\tan x}{x} = 1$

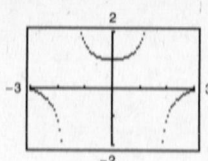

34. $\lim\limits_{x \to 0} \dfrac{1 - \cos 2x}{x} = 0$

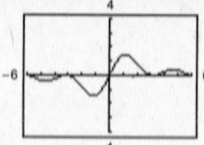

35. $\lim\limits_{x \to 1} \dfrac{1 - \sqrt[3]{x}}{1 - x} = \dfrac{1}{3} \approx 0.333$

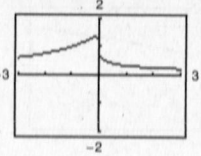

36. $\lim\limits_{x \to 1} \dfrac{\sqrt[3]{x} - x}{x - 1} \approx -0.667, \left(-\dfrac{2}{3}\right)$

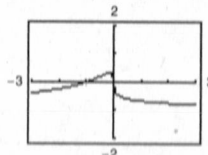

37. $f(x) = (1 - x)^{2/x}$

$\quad \lim\limits_{x \to 0} f(x) \approx 0.135$

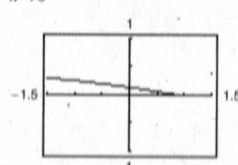

38. $\lim\limits_{x \to 0} (1 + 2x)^{1/x} \approx 7.389$

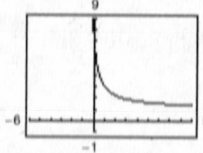

39. $f(x) = \dfrac{x-1}{x^2-1}$

(a) Graphically, $\displaystyle\lim_{x\to 1^-} \dfrac{x-1}{x^2-1} = \dfrac{1}{2}$.

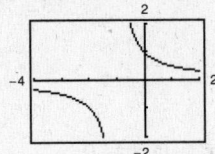

(b)

x	0.5	0.9	0.99	0.999	1
$f(x)$	0.6667	0.5263	0.5025	0.5003	0.5

Numerically, $\displaystyle\lim_{x\to 1^-} \dfrac{x-1}{x^2-1} = \dfrac{1}{2}$.

(c) Algebraically, $\displaystyle\lim_{x\to 1^-} \dfrac{x-1}{x^2-1} = \lim_{x\to 1^-} \dfrac{x-1}{(x-1)(x+1)} = \lim_{x\to 1^-} \dfrac{1}{x+1} = \dfrac{1}{2}$.

40. $\displaystyle\lim_{x\to 5^+} \dfrac{5-x}{25-x^2} = 0.1$

(a)

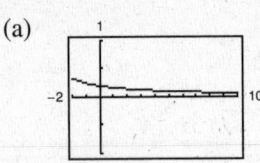

(b)

x	5.1	5.01	5.001	5
$f(x)$	0.099	0.0999	0.09999	?

(c) Algebraically, $\displaystyle\lim_{x\to 5^+} \dfrac{5-x}{25-x^2} = \lim_{x\to 5^+} \dfrac{(5-x)}{(5-x)(5+x)} = \lim_{x\to 5^+} \dfrac{1}{5+x} = \dfrac{1}{10}$.

41. $f(x) = \dfrac{4-\sqrt{x}}{x-16}$

(a) Graphically, $\displaystyle\lim_{x\to 16^+} \dfrac{4-\sqrt{x}}{x-16} = -\dfrac{1}{8}$.

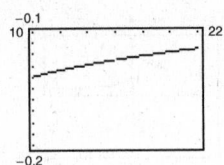

(b)

x	16	16.001	16.01	16.1	16.5
$f(x)$	?	-0.1250	-0.1250	-0.1248	-0.1240

(c) Algebraically, $\displaystyle\lim_{x\to 16^+} \dfrac{4-\sqrt{x}}{x-16} = \lim_{x\to 16^+} \dfrac{4-\sqrt{x}}{(\sqrt{x}-4)(\sqrt{x}+4)}$

$$= \lim_{x\to 16^+} \dfrac{-1}{\sqrt{x}+4}$$

$$= \dfrac{-1}{4+4} = -\dfrac{1}{8}.$$

42. $\lim\limits_{x \to 0^-} \dfrac{\sqrt{x+2} - \sqrt{2}}{x} \approx 0.3536$

(a)

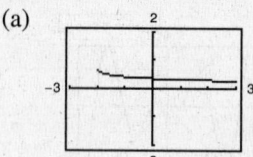

(b)

x	-1.0	-0.1	-0.01	-0.001	0
$f(x)$	0.4142	0.3581	0.3540	0.3536	?

(c) Algebraically, $\lim\limits_{x \to 0^-} \dfrac{\sqrt{x+2} - \sqrt{2}}{x} \cdot \dfrac{\sqrt{x+2} + \sqrt{2}}{\sqrt{x+2} + \sqrt{2}}$

$= \lim\limits_{x \to 0^-} \dfrac{(x+2) - 2}{x(\sqrt{x+2} + \sqrt{2})}$

$= \lim\limits_{x \to 0^-} \dfrac{1}{\sqrt{x+2} + \sqrt{2}} = \dfrac{1}{2\sqrt{2}} = \dfrac{\sqrt{2}}{4} \approx 0.3536.$

43. $f(x) = \dfrac{|x-6|}{x-6}$

$\lim\limits_{x \to 6^+} f(x) = 1$

$\lim\limits_{x \to 6^-} f(x) = -1$

Limit does not exist.

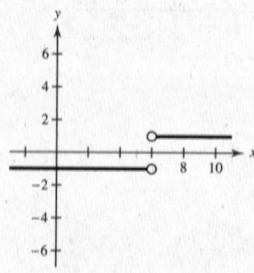

44. $\lim\limits_{x \to 2^-} \dfrac{|x-2|}{x-2} = -1$

$\lim\limits_{x \to 2^+} \dfrac{|x-2|}{x-2} = 1$

$\lim\limits_{x \to 2} \dfrac{|x-2|}{x-2} = 1$ does not exist.

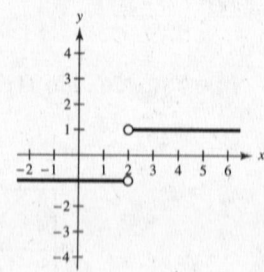

45. $f(x) = \dfrac{1}{x^2 + 1}$

$\lim\limits_{x \to 1^-} \dfrac{1}{x^2+1} = \lim\limits_{x \to 1^+} \dfrac{1}{x^2+1}$

$= \lim\limits_{x \to 1} \dfrac{1}{x^2+1} = \dfrac{1}{2}$

46. $\lim\limits_{x \to 1^-} \dfrac{1}{x^2-1}$ does not exist.

$\lim\limits_{x \to 1^+} \dfrac{1}{x^2-1}$ does not exist.

$\lim\limits_{x \to 1} \dfrac{1}{x^2-1}$ does not exist.

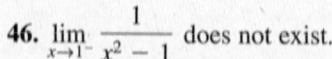

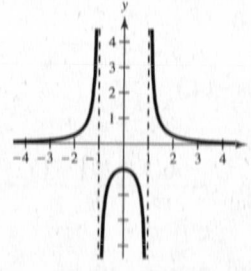

47. $\lim\limits_{x \to 2^-} f(x) = 2 - 1 = 1$

$\lim\limits_{x \to 2^+} f(x) = 2(2) - 3 = 1$

$\lim\limits_{x \to 2} f(x) = 1$

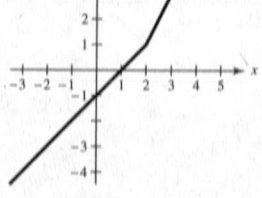

48. $\lim\limits_{x \to 1^-} f(x) = 2(1) + 1 = 3$

$\lim\limits_{x \to 1^+} f(x) = 4 - 1 = 3$

$\lim\limits_{x \to 1} f(x) = 3$

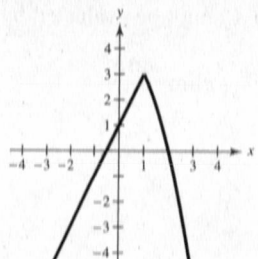

49. $f(x) = \begin{cases} 4 - x^2, & x \le 1 \\ 3 - x, & x > 1 \end{cases}$

$\lim\limits_{x \to 1^-} f(x) = 4 - 1 = 3$

$\lim\limits_{x \to 1^+} f(x) = 3 - 1 = 2$

$\lim\limits_{x \to 1} f(x)$ does not exist.

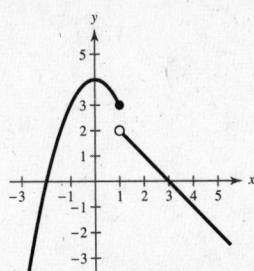

50. $\lim\limits_{x \to 0^-} f(x) = 4 - 0 = 4$

$\lim\limits_{x \to 0^+} f(x) = 0 + 4 = 4$

$\lim\limits_{x \to 0} f(x) = 4$

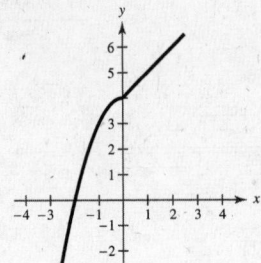

51. $\lim\limits_{x \to 0} f(x) = 0$

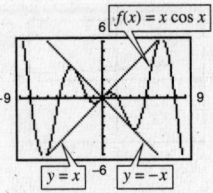

52. $\lim\limits_{x \to 0} f(x) = 0$

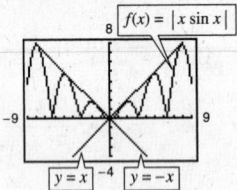

53. $\lim\limits_{x \to 0} f(x) = 0$

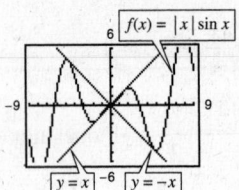

54. $\lim\limits_{x \to 0} f(x) = 0$

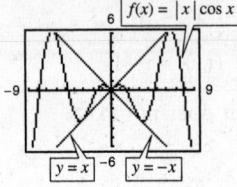

55. $\lim\limits_{x \to 0} f(x) = 0$

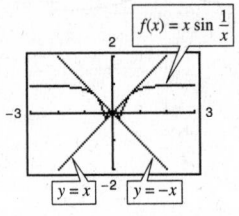

56. $\lim\limits_{x \to 0} f(x) = 0$

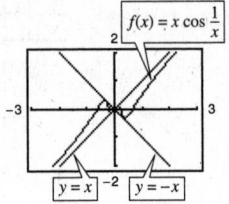

57. (a) Can be evaluated by direct substitution:

$$\lim\limits_{x \to 0} x^2 \sin x^2 = 0^2 \sin 0^2 = 0$$

(b) Cannot be evaluated by direct substitution:

$$\lim\limits_{x \to 0} \frac{\sin x^2}{x^2} = 1$$

58. (a) Can be evaluated by direct substitution.

$$\lim\limits_{x \to 0} \frac{x}{\cos x} = \frac{0}{\cos 0} = 0$$

(b) Cannot be evaluated by direct substitution.

$$\lim\limits_{x \to 0} \frac{1 - \cos x}{x} = 0$$

(See Section 12.1, Exercise 14.)

59. $\lim\limits_{h \to 0} \dfrac{f(x + h) - f(x)}{h} = \lim\limits_{h \to 0} \dfrac{3(x + h) - 1 - (3x - 1)}{h}$

$= \lim\limits_{h \to 0} \dfrac{3x + 3h - 1 - 3x + 1}{h}$

$= \lim\limits_{h \to 0} \dfrac{3h}{h} = 3$

60. $\lim\limits_{h \to 0} \dfrac{f(x + h) - f(x)}{h} = \lim\limits_{h \to 0} \dfrac{[5 - 6(x + h)] - (5 - 6x)}{h}$

$= \lim\limits_{h \to 0} -\dfrac{6h}{h} = -6$

61. $\lim\limits_{h \to 0} \dfrac{f(x + h) - f(x)}{h} = \lim\limits_{h \to 0} \dfrac{\sqrt{x + h} - \sqrt{x}}{h} \cdot \left(\dfrac{\sqrt{x + h} + \sqrt{x}}{\sqrt{x + h} + \sqrt{x}} \right)$

$\qquad\qquad = \lim\limits_{h \to 0} \dfrac{(x + h) - x}{h\left(\sqrt{x + h} + \sqrt{x} \right)}$

$\qquad\qquad = \lim\limits_{h \to 0} \dfrac{1}{\sqrt{x + h} + \sqrt{x}} = \dfrac{1}{2\sqrt{x}}$

62. $\lim\limits_{h \to 0} \dfrac{f(x + h) - f(x)}{h} = \lim\limits_{h \to 0} \dfrac{\sqrt{x + h - 2} - \sqrt{x - 2}}{h} \cdot \dfrac{\sqrt{x + h - 2} + \sqrt{x - 2}}{\sqrt{x + h - 2} + \sqrt{x - 2}}$

$\qquad\qquad = \lim\limits_{h \to 0} \dfrac{(x + h - 2) - (x - 2)}{h\left[\sqrt{x + h - 2} + \sqrt{x - 2} \right]}$

$\qquad\qquad = \lim\limits_{h \to 0} \dfrac{1}{\sqrt{x + h - 2} + \sqrt{x - 2}}$

$\qquad\qquad = \dfrac{1}{2\sqrt{x - 2}}$

63. $\lim\limits_{h \to 0} \dfrac{f(x + h) - f(x)}{h} = \lim\limits_{h \to 0} \dfrac{((x + h)^2 - 3(x + h)) - (x^2 - 3x)}{h}$

$\qquad\qquad = \lim\limits_{h \to 0} \dfrac{x^2 + 2xh + h^2 - 3x - 3h - x^2 + 3x}{h}$

$\qquad\qquad = \lim\limits_{h \to 0} \dfrac{2xh + h^2 - 3h}{h}$

$\qquad\qquad = \lim\limits_{h \to 0} (2x + h - 3) = 2x - 3$

64. $\lim\limits_{h \to 0} \dfrac{f(x + h) - f(x)}{h} = \lim\limits_{h \to 0} \dfrac{[4 - 2(x + h) - (x + h)^2] - [4 - 2x - x^2]}{h}$

$\qquad\qquad = \lim\limits_{h \to 0} \dfrac{4 - 2x - 2h - x^2 - 2xh - h^2 - 4 + 2x + x^2}{h}$

$\qquad\qquad = \lim\limits_{h \to 0} \dfrac{-2h - 2xh - h^2}{h} = \lim\limits_{h \to 0} (-2 - 2x - h) = -2 - 2x$

65. $\lim\limits_{h \to 0} \dfrac{f(x + h) - f(x)}{h} = \lim\limits_{h \to 0} \dfrac{1/(x + h + 2) - 1/(x + 2)}{h}$

$\qquad\qquad = \lim\limits_{h \to 0} \dfrac{(x + 2) - (x + h + 2)}{h(x + h + 2)(x + 2)}$

$\qquad\qquad = \lim\limits_{h \to 0} \dfrac{-h}{h(x + h + 2)(x + 2)}$

$\qquad\qquad = \lim\limits_{h \to 0} \dfrac{-1}{(x + h + 2)(x + 2)}$

$\qquad\qquad = \dfrac{-1}{(x + 2)^2}$

66. $\lim\limits_{h \to 0} \dfrac{f(x+h) - f(x)}{h} = \lim\limits_{h \to 0} \dfrac{\dfrac{1}{x+h-1} - \dfrac{1}{x-1}}{h}$

$= \lim\limits_{h \to 0} \dfrac{x - 1 - (x + h - 1)}{h(x + h - 1)(x - 1)}$

$= \lim\limits_{h \to 0} \dfrac{-1}{(x + h - 1)(x - 1)}$

$= \dfrac{-1}{(x-1)^2}$

67. $\lim\limits_{t \to 1} \dfrac{(-16(1) + 128) - (-16t^2 + 128)}{1 - t} = \lim\limits_{t \to 1} \dfrac{16t^2 - 16}{1 - t}$

$= \lim\limits_{t \to 1} \dfrac{16(t - 1)(t + 1)}{1 - t}$

$= \lim\limits_{t \to 1} -16(t + 1)$

$= -32 \dfrac{\text{ft}}{\text{sec}}$

68. $v(2) = \lim\limits_{t \to 2} \dfrac{s(2) - s(t)}{2 - t} = \lim\limits_{t \to 2} \dfrac{(-64 + 128) - (-16t^2 + 128)}{2 - t}$

$= \lim\limits_{t \to 2} \dfrac{16t^2 - 64}{2 - t} = \lim\limits_{t \to 2} \dfrac{16(t + 2)(t - 2)}{2 - t}$

$= \lim\limits_{t \to 2} -16(t + 2) = -64 \text{ feet per second}$

69. $\lim\limits_{t \to 2^-} f(t) = 30.80$, $\lim\limits_{t \to 2^+} f(t) = 33.88$

Thus, the limit of f as $t \to 2$ does not exist.

70. $\lim\limits_{x \to 1^-} f(x) = 10.75$

$\lim\limits_{x \to 1^+} f(x) = 14.70$

Thus, $\lim\limits_{x \to 1} f(x)$ does not exist.

71. $C(t) = 0.75 - 0.50[\![-(t - 1)]\!]$

(a)

(b)

t	3	3.3	3.4	3.5	3.6	3.7	4
C	1.75	2.25	2.25	2.25	2.25	2.25	2.25

$\lim\limits_{t \to 3.5} C(t) = 2.25$

(c)

t	2	2.5	2.9	3	3.1	3.5	4
C	1.25	1.75	1.75	1.75	2.25	2.25	2.25

No, $\lim\limits_{t \to 3} C(t)$ does not exist.

$\lim\limits_{t \to 3^-} C(t) = 1.75$, $\lim\limits_{t \to 3^+} C(t) = 2.25$

72. (a)

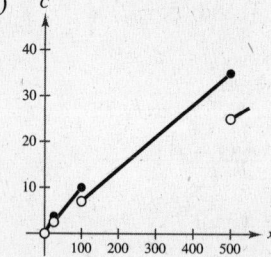

(b) i. $\lim\limits_{x\to 15^-} C(x) = (0.15)(15) = 2.25$

$\lim\limits_{x\to 15^+} C(x) = (0.15)(15) = 2.25$ $\Rightarrow \lim\limits_{x\to 15} C(x) = 2.25$

ii. $\lim\limits_{x\to 99^-} C(x) = (0.10)(99) = 9.9$

$\lim\limits_{x\to 99^+} C(x) = (0.10)(99) = 9.9$ $\Rightarrow \lim\limits_{x\to 99} C(x) = 9.9$

iii. $\lim\limits_{x\to 305^-} C(x) = (0.07)(305) = 21.35$

$\lim\limits_{x\to 305^+} C(x) = (0.07)(305) = 21.35$ $\Rightarrow \lim\limits_{x\to 305} C(x) = 21.35$

(c) i.

x	24.9	24.99	24.999	25.001	25.01	25.1
$C(x)$	3.735	3.7485	3.74985	2.5001	2.501	2.51

ii.

x	99.9	99.99	99.999	100.001	100.01	100.1
$C(x)$	9.99	9.999	9.9999	7.00007	7.0007	7.007

iii.

x	499.9	499.99	499.999	500.001	500.01	500.1
$C(x)$	34.993	34.9993	34.99993	25.00005	25.0005	25.005

(d) According to the graph, $C(x)$ is discontinuous at $x = 25$, $x = 100$, and $x = 500$. Therefore, the limits $\lim\limits_{x\to a} C(x)$, $a \in \{25, 100, 500\}$, do not exist.

73. True

74. False. The value of f at c has no bearing on the limit. See Exercise 49.

75. Many answers possible

(a)

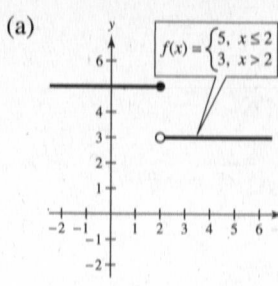

(b)

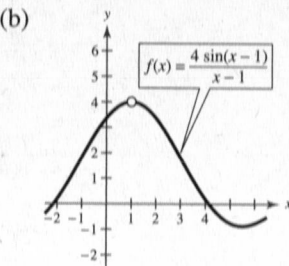

76. Answers will vary.

77. Slope of line through $(4, -6)$ and $(3, -4)$:

$$\frac{-6 + 4}{4 - 3} = -2$$

Slope of perpendicular line: $\frac{1}{2}$

Equation: $\quad y + 10 = \frac{1}{2}(x - 6)$

$$2y - x + 26 = 0$$

78. Slope between $(3, -3)$ and $(5, -2)$ is

$$\frac{-2 - (-3)}{5 - 3} = \frac{1}{2}.$$

Line: $\quad y + 1 = \frac{1}{2}(x - 1)$

$$2y + 2 = x - 1$$
$$2y - x + 3 = 0$$

79. $s = r\theta$

$$= (8.5)\left[45\left(\frac{\pi}{180}\right)\right]$$

$$= 2.125\pi \text{ inches}$$

$$\approx 6.676 \text{ inches}$$

80. $s = r\theta$, θ in radians

$$s = (26)(101°)\left(\frac{\pi}{180°}\right) \approx 45.832 \text{ mm}$$

81. $A = \dfrac{\theta r^2}{2}$

$$= \frac{1}{2}\left[135\left(\frac{\pi}{180}\right)\right](9)^2$$

$$= \frac{1}{2}\left(\frac{3\pi}{4}\right)(81)$$

$$= 30.375\pi \text{ square centimeters}$$

$$\approx 95.426 \text{ square centimeters}$$

82. $A = \dfrac{1}{2}r^2\theta$, θ in radians

$$A = \left(\frac{1}{2}\right)(3.4)^2(81°)\left(\frac{\pi}{180°}\right) \approx 8.171 \text{ ft}^2$$

83. $r = \dfrac{3}{1 + \cos\theta}$, $e = 1$, Parabola

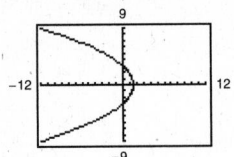

84. $r = \dfrac{12}{3 + 2\sin\theta} = \dfrac{4}{1 + (2/3)\sin\theta}$, Ellipse $\left(e = \dfrac{2}{3}\right)$

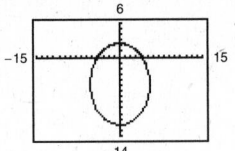

85. $r = \dfrac{9}{2 + 3\cos\theta} = \dfrac{9/2}{1 + (3/2)\cos\theta}$, $e = \dfrac{3}{2}$,

Hyperbola

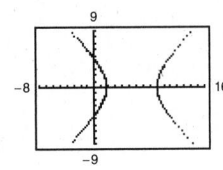

86. $r = \dfrac{4}{4 + \cos\theta} = \dfrac{1}{1 + (1/4)\cos\theta}$, Ellipse $\left(e = \dfrac{1}{4}\right)$

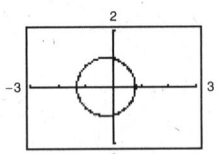

87. $r = \dfrac{5}{1 - \sin\theta}, \; e = 1,$

Parabola

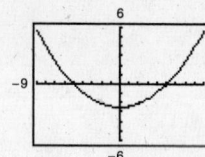

88. $r = \dfrac{6}{3 - 4\sin\theta} = \dfrac{2}{1 - (4/3)\sin\theta}$

Hyperbola $\left(e = \dfrac{4}{3}\right)$

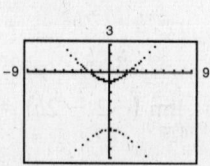

89. $\langle 7, -2, 3 \rangle \cdot \langle -1, 4, 5 \rangle = -7 - 8 + 15$

$\qquad\qquad\qquad = 0 \implies$ orthogonal

90. $\langle 5, 5, 0 \rangle \cdot \langle 0, 5, 1 \rangle = 25 \neq 0.$

Not multiples of each other; neither parallel nor orthogonal

91. $-3\langle -4, 3, -6 \rangle = \langle 12, -9, 18 \rangle \implies$ parallel

92. $\langle 2, -3, 1 \rangle \cdot \langle -2, 2, 2 \rangle = -6 \neq 0.$

Not multiples of each other; neither parallel nor orthogonal

Section 12.3   The Tangent Line Problem

- ■   You should be able to visually approximate the slope of a graph.

- ■   The slope m of the graph of f at the point $(x, f(x))$ is given by

$$m = \lim_{h \to 0} \frac{f(x + h) - f(x)}{h}$$

   provided this limit exists.

- ■   You should be able to use the limit definition to find the slope of a graph.

- ■   The derivative of f at x is given by

$$f'(x) = \lim_{h \to 0} \frac{f(x + h) - f(x)}{h}$$

   provided this limit exists. Notice that this is the same limit as that for the tangent line slope.

- ■   You should be able to use the limit definition to find the derivative of a function.

Vocabulary Check

   1. calculus   **2.** tangent line   **3.** secant line

   4. difference quotient   **5.** derivative

1. Slope is 0 at (x, y).   **2.** Slope is -1 at (x, y).   **3.** Slope is $\frac{1}{2}$ at (x, y).   **4.** Slope is -2 at (x, y).

5. $m_{\text{sec}} = \dfrac{g(3 + h) - g(3)}{h} = \dfrac{(3 + h)^2 - 4(3 + h) - (-3)}{h} = \dfrac{h^2 + 2h}{h}$

$\qquad m = \lim_{h \to 0} \dfrac{h^2 + 2h}{h} = \lim_{h \to 0} \dfrac{h(h + 2)}{h} = \lim_{h \to 0} (h + 2) = 2$

6. $m_{\text{sec}} = \dfrac{f(3 + h) - f(3)}{h} = \dfrac{10(3 + h) - 2(3 + h)^2 - 12}{h}$

$\quad = \dfrac{-2h - 2h^2}{h} = -2 - 2h, h \neq 0$

$\quad m = \lim\limits_{h \to 0} (-2 - 2h) = -2$

7. $m_{\text{sec}} = \dfrac{g(1 + h) - g(1)}{h} = \dfrac{5 - 2(1 + h) - 3}{h} = \dfrac{-2h}{h}$

$\quad m = \lim\limits_{h \to 0} \dfrac{-2h}{h} = -2$

8. $m_{\text{sec}} = \dfrac{h(-1 + k) - h(-1)}{k} = \dfrac{2(-1 + k) + 5 - 3}{k} = \dfrac{2k}{k}$

$\quad m = \lim\limits_{k \to 0} \dfrac{2k}{k} = 2$

9. $m_{\text{sec}} = \dfrac{g(2 + h) - g(2)}{h} = \dfrac{[4/(2 + h)] - 2}{h} = \dfrac{4 - 2(2 + h)}{(2 + h)h} = \dfrac{-2}{2 + h}, h \neq 0$

$\quad m = \lim\limits_{h \to 0} \left(\dfrac{-2}{2 + h} \right) = -1$

10. $m_{\text{sec}} = \dfrac{g(4 + h) - g(4)}{h} = \dfrac{\dfrac{1}{4 + h - 2} - \dfrac{1}{2}}{h} = \dfrac{\dfrac{1}{2 + h} - \dfrac{1}{2}}{h}$

$\quad = \dfrac{-h}{(2 + h)2h} = \dfrac{-1}{2(2 + h)}, h \neq 0$

$\quad m = \lim\limits_{h \to 0} \left(\dfrac{-1}{2(2 + h)} \right) = -\dfrac{1}{4}$

11. $m_{\text{sec}} = \dfrac{h(9 + k) - h(9)}{k} = \dfrac{\sqrt{9 + k} - 3}{k} \cdot \dfrac{\sqrt{9 + k} + 3}{\sqrt{9 + k} + 3} = \dfrac{(9 + k) - 9}{k[\sqrt{9 + k} + 3]} = \dfrac{1}{\sqrt{9 + k} + 3}, k \neq 0$

$\quad m = \lim\limits_{k \to 0} \dfrac{1}{\sqrt{9 + k} + 3} = \dfrac{1}{6}$

12. $m_{\text{sec}} = \dfrac{h(-1 + k) - h(-1)}{k} = \dfrac{\sqrt{-1 + k + 10} - 3}{k} \cdot \dfrac{\sqrt{k + 9} + 3}{\sqrt{k + 9} + 3} = \dfrac{(k + 9) - 9}{k[\sqrt{k + 9} + 3]} = \dfrac{1}{\sqrt{k + 9} + 3}, k \neq 0$

$\quad m = \lim\limits_{k \to 0} \dfrac{1}{\sqrt{k + 9} + 3} = \dfrac{1}{6}$

13. $f(x) = 4 - x^2$

$\quad m_{\text{sec}} = \dfrac{f(x + h) - f(x)}{h} = \dfrac{4 - (x + h)^2 - (4 - x^2)}{h} = \dfrac{-2xh - h^2}{h} = -2x - h, h \neq 0$

$\quad m = \lim\limits_{h \to 0} (-2x - h) = -2x$

(a) At $(0, 4)$, $m = -2(0) = 0$.   (b) At $(-1, 3)$, $m = -2(-1) = 2$.

14. $f(x) = x^3$

$$m_{\text{sec}} = \frac{f(x + h) - f(x)}{h} = \frac{(x + h)^3 - x^3}{h} = \frac{3x^2h + 3xh^2 + h^3}{h}$$

$$= 3x^2 + 3xh + h^2, h \neq 0$$

$$m = \lim_{h \to 0}(3x^2 + 3xh + h^2) = 3x^2$$

(a) At $(1, 1)$, $m = 3(1)^2 = 3$.   (b) At $(-2, -8)$, $m = 3(-2)^2 = 12$.

15. $f(x) = \dfrac{1}{x + 4}$

$$m_{\text{sec}} = \frac{f(x + h) - f(x)}{h} = \frac{\dfrac{1}{x + h + 4} - \dfrac{1}{x + 4}}{h} = \frac{(x + 4) - (x + 4 + h)}{(x + h + 4)(x + 4)(h)}$$

$$= \frac{-h}{(x + h + 4)(x + 4)h} = \frac{-1}{(x + h + 4)(x + 4)}, h \neq 0$$

$$m = \lim_{h \to 0} \frac{-1}{(x + h + 4)(x + 4)} = \frac{-1}{(x + 4)^2}$$

(a) At $\left(0, \dfrac{1}{4}\right)$, $m = \dfrac{-1}{(0 + 4)^2} = \dfrac{-1}{16}$.   (b) At $\left(-2, \dfrac{1}{2}\right)$, $m = \dfrac{-1}{(-2 + 4)^2} = \dfrac{-1}{4}$.

16. $f(x) = \dfrac{1}{x + 2}$

$$m_{\text{sec}} = \frac{f(x + h) - f(x)}{h} = \frac{\dfrac{1}{x + h + 2} - \dfrac{1}{x + 2}}{h}$$

$$= \frac{(x + 2) - (x + h + 2)}{h(x + h + 2)(x + 2)}$$

$$= \frac{-h}{h(x + h + 2)(x + 2)}$$

$$m = \lim_{h \to 0} \frac{-h}{h(x + h + 2)(x + 2)} = \frac{-1}{(x + 2)^2}$$

(a) At $\left(0, \dfrac{1}{2}\right)$, $m = \dfrac{1}{(0 + 2)^2} = \dfrac{-1}{4}$.   (b) At $(-1, 1)$, $m = \dfrac{1}{(-1 + 2)^2} = -1$.

17. $f(x) = \sqrt{x - 1}$

$$m_{\text{sec}} = \frac{f(x + h) - f(x)}{h} = \frac{\sqrt{x + h - 1} - \sqrt{x - 1}}{h} \cdot \frac{\sqrt{x + h - 1} + \sqrt{x - 1}}{\sqrt{x + h - 1} + \sqrt{x - 1}}$$

$$= \frac{(x + h - 1) - (x - 1)}{h(\sqrt{x + h - 1} + \sqrt{x - 1})} = \frac{1}{\sqrt{x + h - 1} + \sqrt{x - 1}}, h \neq 0$$

$$m = \lim_{h \to 0}\left(\frac{1}{\sqrt{x + h - 1} + \sqrt{x - 1}}\right) = \frac{1}{2\sqrt{x - 1}}$$

(a) At $(5, 2)$, $m = \dfrac{1}{2\sqrt{5 - 1}} = \dfrac{1}{4}$.   (b) At $(10, 3)$, $m = \dfrac{1}{2\sqrt{10 - 1}} = \dfrac{1}{6}$.

18. $f(x) = \sqrt{x - 4}$

$$m_{\text{sec}} = \frac{f(x + h) - f(x)}{h} = \frac{\sqrt{x + h - 4} - \sqrt{x - 4}}{h} \cdot \frac{\sqrt{x + h - 4} + \sqrt{x - 4}}{\sqrt{x + h - 4} + \sqrt{x - 4}}$$

$$= \frac{(x + h - 4) - (x - 4)}{h\left[\sqrt{x + h - 4} + \sqrt{x - 4}\right]}$$

$$m = \lim_{h \to 0} \frac{h}{h\left[\sqrt{x + h - 4} + \sqrt{x - 4}\right]} = \frac{1}{2\sqrt{x - 4}}$$

(a) At $(5, 1)$, $m = \frac{1}{2}$. (b) At $(8, 2)$, $m = \frac{1}{4}$.

19. $f(x) = x^2 - 2$

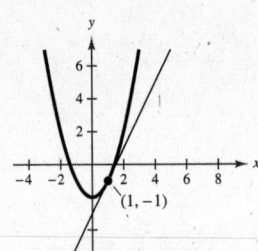

Slope at $(1, -1)$ is 2.

20. $f(x) = x^2 - 2x + 1$

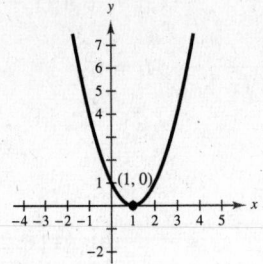

Slope at $(1, 0) \approx 0$.

21. $f(x) = \sqrt{2 - x}$

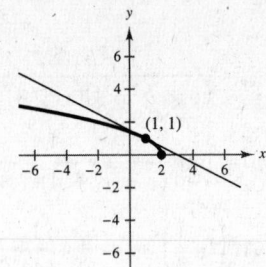

Slope at $(1, 1)$ is $-\frac{1}{2}$.

22. $f(x) = \sqrt{x + 3}$

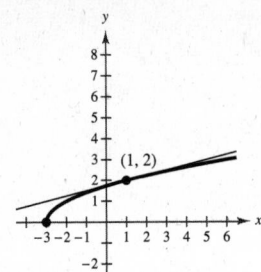

Slope at $(1, 2) \approx \frac{1}{4}$.

23. $f(x) = \dfrac{4}{x + 1}$

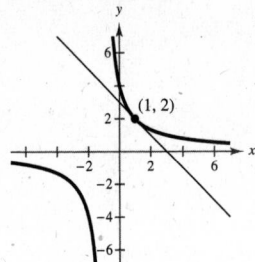

Slope at $(1, 2)$ is -1.

24. $f(x) = \dfrac{3}{2 - x}$

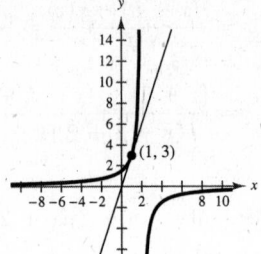

Slope at $(1, 3) \approx 3$.

25. $f'(x) = \lim_{h \to 0} \dfrac{f(x + h) - f(x)}{h} = \lim_{h \to 0} \dfrac{5 - 5}{h} = 0$

26. $f'(x) = \lim_{h0} \dfrac{f(x + h) - f(x)}{h} = \lim_{h0} \dfrac{(-1) - (-1)}{h} = 0$

27. $g'(x) = \lim_{h \to 0} \dfrac{g(x + h) - g(x)}{h} = \lim_{h \to 0} \dfrac{\left[9 - \frac{1}{3}(x + h)\right] - \left[9 - \frac{1}{3}x\right]}{h} = \lim_{h \to 0} \dfrac{-\frac{1}{3}h}{h} = -\dfrac{1}{3}$

28. $f'(x) = \lim_{h0} \dfrac{f(x+h) - f(x)}{h} = \lim_{h0} \dfrac{[-5(x+h) + 2] - (-5x + 2)}{h}$

$$= \lim_{h \to 0} \frac{-5h}{h} = -5$$

29. $f'(x) = \lim_{h \to 0} \dfrac{f(x+h) - f(x)}{h} = \lim_{h \to 0} \dfrac{[4 - 3(x+h)^2] - (4 - 3x^2)}{h}$

$$= \lim_{h \to 0} \frac{-3(x^2 + 2xh + h^2) + 3x^2}{h} = \lim_{h \to 0} \frac{-6xh - 3h^2}{h} = \lim_{h \to 0}(-6x - 3h) = -6x$$

30. $f'(x) = \lim_{h \to 0} \dfrac{f(x+h) - f(x)}{h} = \lim_{h \to 0} \dfrac{[(x+h)^2 - 3(x+h) + 4] - (x^2 - 3x + 4)}{h}$

$$= \lim_{h \to 0} \frac{x^2 + 2xh + h^2 - 3h - x^2}{h}$$

$$= \lim_{h \to 0} (2x + h - 3) = 2x - 3$$

31. $f'(x) = \lim_{h \to 0} \dfrac{f(x+h) - f(x)}{h} = \lim_{h \to 0} \dfrac{\dfrac{1}{(x+h)^2} - \dfrac{1}{x^2}}{h}$

$$= \lim_{h \to 0} \frac{x^2 - (x^2 + 2xh + h^2)}{(x+h)^2 x^2 h} = \lim_{h \to 0} \frac{-2x - h}{(x+h)^2 x^2} = -\frac{2x}{x^4} = -\frac{2}{x^3}$$

32. $f'(x) = \lim_{h \to 0} \dfrac{f(x+h) - f(x)}{h} = \lim_{h \to 0} \dfrac{\dfrac{1}{(x+h)^3} - \dfrac{1}{x^3}}{h}$

$$= \lim_{h \to 0} \frac{x^3 - (x^3 + 3x^2h + 3xh^2 + h^3)}{h(x+h)^3 x^3}$$

$$= \lim_{h \to 0} \frac{-3x^2h - 3xh^2 - h^3}{h(x+h)^3 x^3}$$

$$= \lim_{h \to 0} \frac{-3x^2 - 3xh - h^2}{(x+h)^3 x^3}$$

$$= \frac{-3x^2}{x^6} = \frac{-3}{x^4}$$

33. $f'(x) = \lim_{h \to 0} \dfrac{f(x+h) - f(x)}{h} = \lim_{h \to 0} \dfrac{\dfrac{1}{\sqrt{x+h-9}} - \dfrac{1}{\sqrt{x-9}}}{h} \cdot \dfrac{\dfrac{1}{\sqrt{x+h-9}} + \dfrac{1}{\sqrt{x-9}}}{\dfrac{1}{\sqrt{x+h-9}} + \dfrac{1}{\sqrt{x-9}}}$

$$= \lim_{h \to 0} \frac{\dfrac{1}{(x+h-9)} - \dfrac{1}{(x-9)}}{h\left[\dfrac{1}{\sqrt{x+h-9}} + \dfrac{1}{\sqrt{x-9}}\right]} = \lim_{h \to 0} \frac{(x-9) - (x+h-9)}{h(x+h-9)(x-9)\left[\dfrac{1}{\sqrt{x+h-9}} + \dfrac{1}{\sqrt{x-9}}\right]}$$

$$= \lim_{h \to 0} \frac{-1}{(x+h-9)(x-9)\left[\dfrac{1}{\sqrt{x+h-9}} + \dfrac{1}{\sqrt{x-9}}\right]} = \frac{-1}{(x-9)^2\left[\dfrac{2}{\sqrt{x-9}}\right]} = \frac{-1}{2(x-9)^{3/2}}$$

34. $h'(s) = \lim\limits_{k \to 0} \dfrac{h(s+k) - h(s)}{k} = \lim\limits_{k \to 0} \dfrac{\dfrac{1}{\sqrt{s+k+1}} - \dfrac{1}{\sqrt{s+1}}}{k}$

$= \lim\limits_{k \to 0} \dfrac{\sqrt{s+1} - \sqrt{s+k+1}}{k\sqrt{s+k+1}\sqrt{s+1}} \cdot \dfrac{\sqrt{s+1} + \sqrt{s+k+1}}{\sqrt{s+1} + \sqrt{s+k+1}}$

$= \lim\limits_{k \to 0} \dfrac{(s+1) - (s+k+1)}{k\sqrt{s+k+1}\sqrt{s+1}\left[\sqrt{s+1} + \sqrt{s+k+1}\right]}$

$= \lim\limits_{k \to 0} \dfrac{-1}{\sqrt{s+k+1}\sqrt{s+1}\left[\sqrt{s+1} + \sqrt{s+k+1}\right]}$

$= \dfrac{-1}{(s+1)2\sqrt{s+1}} = \dfrac{-1}{2(s+1)^{3/2}}$

35. $f(x) = x^2 - 1,\ (2, 3)$

(a) $m_{\text{sec}} = \dfrac{f(2+h) - f(2)}{h} = \dfrac{(2+h)^2 - 1 - 3}{h} = \dfrac{4h + h^2}{h} = 4 + h,\ h \neq 0$

(c)

$m = \lim\limits_{h \to 0} (4 + h) = 4$

(b) Tangent line: $y - 3 = 4(x - 2)$

$y = 4x - 5$

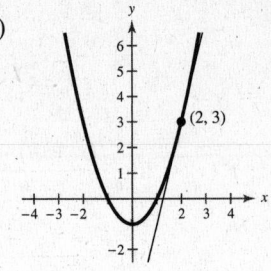

36. $f(x) = x^3 - x,\ (2, 6)$

(a) $m_{\text{sec}} = \dfrac{f(2+h) - f(2)}{h} = \dfrac{(2+h)^3 - (2+h) - 6}{h}$

$= \dfrac{h^3 + 6h^2 + 11h}{h} = h^2 + 6h + 11,\ h \neq 0$

(c)

$m = \lim\limits_{h \to 0} (h^2 + 6h + 11) = 11$

(b) Tangent line: $y - 6 = 11(x - 2)$

$y = 11x - 16$

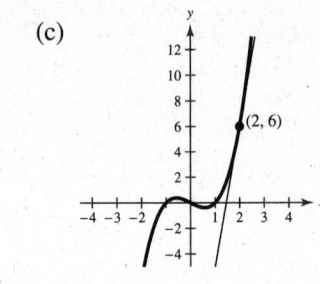

37. $f(x) = \sqrt{x + 1},\ (3, 2)$

(a) $m_{\text{sec}} = \dfrac{f(3+h) - f(3)}{h} = \dfrac{\sqrt{3+h+1} - 2}{h} \cdot \dfrac{\sqrt{4+h} + 2}{\sqrt{4+h} + 2} = \dfrac{(4+h) - 4}{h\left[\sqrt{4+h} + 2\right]} = \dfrac{1}{\sqrt{4+h} + 2}$

$m = \lim\limits_{h \to 0} \dfrac{1}{\sqrt{4+h} + 2} = \dfrac{1}{4}$

(c)

(b) Tangent line: $y - 2 = \dfrac{1}{4}(x - 3)$

$4y = x + 5$

38. $f(x) = \sqrt{x-2}, (3, 1)$

(a) $m_{sec} = \dfrac{f(3+h) - f(3)}{h} = \dfrac{\sqrt{1+h} - 1}{h} \cdot \dfrac{\sqrt{1+h} + 1}{\sqrt{1+h} + 1}$

$\qquad = \dfrac{(1+h) - 1}{h\left[\sqrt{1+h} + 1\right]} = \dfrac{h}{h\left[\sqrt{1+h} + 1\right]}$

$\quad m = \lim\limits_{h \to 0} \dfrac{h}{h\left[\sqrt{1+h} + 1\right]} = \dfrac{1}{2}$

(b) Tangent line: $y - 1 = \dfrac{1}{2}(x - 3)$

$$y = \dfrac{x}{2} - \dfrac{1}{2}$$

(c)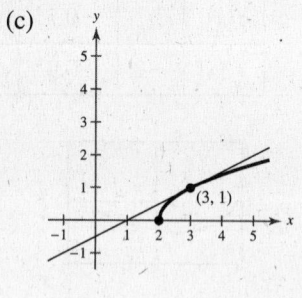

39.

x	-2	-1.5	-1	-0.5	0	0.5	1	1.5	2
$f(x)$	2	1.125	0.5	0.125	0	0.125	0.5	0.125	2
$f'(x)$	-2	-1.5	-1	-0.5	0	0.5	1	1.5	2

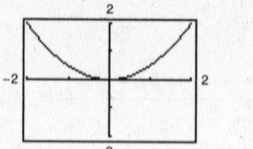

$f(x) = \frac{1}{2}x^2$

$f'(x) = x$

They appear to be the same.

40.

x	-2	-1.5	-1	-0.5	0	0.5	1	1.5	2
$f(x)$	-2	-0.844	-0.25	-0.031	0	0.031	0.25	0.844	2
$f'(x)$	3	1.688	0.75	0.188	0	0.188	0.75	1.688	3

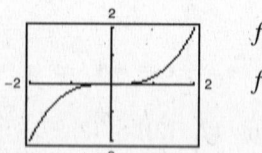

$f(x) = \frac{1}{4}x^3$

$f'(x) = \frac{3}{4}x^2$

They appear to be the same.

41.

x	-2	-1.5	-1	-0.5	0	0.5	1	1.5	2
$f(x)$	1	1.225	1.414	1.581	1.732	1.871	2	2.121	2.236
$f'(x)$	0.5	0.408	0.354	0.316	0.289	0.267	0.25	0.236	0.224

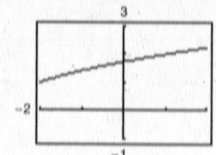

$f(x) = \sqrt{x+3}$

$f'(x) = \dfrac{1}{2\sqrt{x+3}}$

They appear to be the same.

42.

x	-2	-1.5	-1	-0.5	0	0.5	1	1.5	2
$f(x)$	0	-0.7	-1	-1.071	-1	-0.833	-0.6	-0.318	0
$f'(x)$	-2	-0.92	-0.333	0.020	0.25	0.407	0.52	0.603	0.667

$$f(x) = \frac{x^2 - 4}{x + 4}$$

$$f'(x) = \frac{x^2 + 8x + 4}{(x + 4)^2}$$

They appear to be the same.

43. Given line: $x + y = 0 \Rightarrow y = -x \Rightarrow m = -1 \Rightarrow m_{\text{tan}} = -1$ since the lines are parallel.

$$f(x) = -\frac{1}{4}x^2$$

$$m_{\text{tan}} = \lim_{h \to 0} \frac{f(x + h) - f(x)}{h} = \lim_{h \to 0} \frac{-\frac{1}{4}(x + h)^2 - (-\frac{1}{4}x^2)}{h}$$

$$= \lim_{h \to 0} \frac{-\frac{1}{4}(x^2 + 2xh + h^2) + \frac{1}{4}x^2}{h} = \lim_{h \to 0} \frac{-\frac{1}{4}h(2x + h)}{h}$$

$$= \lim_{h \to 0} -\frac{1}{4}(2x + h) = -\frac{1}{2}x$$

$$m_{\text{tan}} = -\frac{1}{2}x = -1 \Rightarrow x = 2$$

Point: $(2, f(2)) = (2, -1)$

Tangent line: $y - (-1) = -1(x - 2)$

$$y + 1 = -x + 2$$

$$y = -x + 1$$

44. Since the tangent line is parallel to $2x + y = 0$, the tangent line has a slope of $m = \dfrac{-2}{1} = -2$.

$$f'(x) = \lim_{h \to 0} \frac{f(x + h) - f(x)}{h}$$

$$= \lim_{h \to 0} \frac{(x + h)^2 + 1 - [x^2 + 1]}{h}$$

$$= \lim_{h \to 0} \frac{x^2 + 2hx + h^2 + 1 - x^2 - 1}{h}$$

$$= \lim_{h \to 0} \frac{2hx + h^2}{h} = \lim_{h \to 0}(2x + h) = 2x$$

$$f'(x) = 2x = -2 \Rightarrow x = -1$$

$$f(-1) = (-1)^2 + 1 = 2$$

The tangent line has slope $m = -2$ and passes through the point $(-1, 2)$.

$$y - 2 = -2(x - (-1))$$

$$y = -2x - 2 + 2$$

$$y = -2x, \text{ tangent line at } (-1, 2)$$

45. Given line: $6x + y + 4 = 0 \implies y = -6x - 4 \implies m = -6 \implies m_{\text{tan}} = -6$ since the lines are parallel.

$$f(x) = -\frac{1}{2}x^3$$

$$m_{\text{tan}} = \lim_{h \to 0} \frac{f(x + h) - f(x)}{h} = \lim_{h \to 0} \frac{-\frac{1}{2}(x + h)^3 - \left(-\frac{1}{2}x^3\right)}{h}$$

$$= \lim_{h \to 0} \frac{-\frac{1}{2}(x^3 + 3x^2h + 3xh^2 + h^3) + \frac{1}{2}x^3}{h} = \lim_{h \to 0} \frac{-\frac{1}{2}h(3x^2 + 3xh + h^2)}{h}$$

$$= \lim_{h \to 0} -\frac{1}{2}(3x^2 + 3xh + h^2) = -\frac{3x^2}{2}$$

$$m_{\text{tan}} = -\frac{3x^2}{2} = -6 \implies x^2 = 4 \implies x = \pm 2$$

Points: $(2, f(2)) = (2, -4)$ and $(-2, f(-2)) = (-2, 4)$

Tangent lines: $y - (-4) = -6(x - 2)$ and $y - 4 = -6[x - (-2)]$

$$y + 4 = -6x + 12 \qquad y - 4 = -6x - 12$$

$$y = -6x + 8 \qquad y = -6x - 8$$

46. Since the tangent line is parallel to $x + 2y - 6 = 0$, the tangent line has a slope of $m = -\frac{1}{2}$.

$$f'(x) = \lim_{h \to 0} \frac{(x + h)^2 - (x + h) - [x^2 - x]}{h}$$

$$= \lim_{h \to 0} \frac{x^2 + 2hx + h^2 - x - h - x^2 + x}{h}$$

$$= \lim_{h \to 0} \frac{2hx + h^2 - h}{h} = \lim_{h \to 0}(2x - 1 + h) = 2x - 1$$

$$f'(x) = 2x - 1 = -\frac{1}{2} \implies 2x = -\frac{1}{2} + 1$$

$$2x = \frac{1}{2}$$

$$x = \frac{1}{4}$$

$$f\left(\frac{1}{4}\right) = -\frac{3}{16}$$

The tangent line has slope $m = -\frac{1}{2}$ and passes through the point $\left(\frac{1}{4}, -\frac{3}{16}\right)$.

$$y - \left(-\frac{3}{16}\right) = -\frac{1}{2}\left(x - \frac{1}{4}\right)$$

$$y = -\frac{1}{2}x + \frac{1}{8} - \frac{3}{16}$$

$$y = -\frac{1}{2}x - \frac{1}{16}, \text{ tangent line at } \left(\frac{1}{4}, -\frac{3}{16}\right)$$

47. $f'(x) = \lim\limits_{h \to 0} \dfrac{f(x + h) - f(x)}{h} = \lim\limits_{h \to 0} \dfrac{[(x + h)^2 - 4(x + h) + 3] - [x^2 - 4x + 3]}{h}$

$= \lim\limits_{h \to 0} \dfrac{(x^2 + 2xh + h^2 - 4x - 4h + 3) - (x^2 - 4x + 3)}{h}$

$= \lim\limits_{h \to 0} \dfrac{2xh + h^2 - 4h}{h} = \lim\limits_{h \to 0} (2x + h - 4) = 2x - 4$

$f'(x) = 0 = 2x - 4 \implies x = 2$

f has a horizontal tangent at $(2, -1)$.

48. $f'(x) = \lim\limits_{h \to 0} \dfrac{f(x + h) - f(x)}{h} = \lim\limits_{h \to 0} \dfrac{(x + h)^3 + 3(x + h) - (x^3 + 3x)}{h}$

$= \lim\limits_{h \to 0} \dfrac{x^3 + 3x^2h + 3xh^2 + h^3 + 3x + 3h - x^3 - 3x}{h}$

$= \lim\limits_{h \to 0} \dfrac{3x^2h + 3xh^2 + h^3 + 3h}{h}$

$= \lim\limits_{h \to 0} (3x^2 + 3xh + h^2 + 3) = 3x^2 + 3$

$f'(x) = 3x^2 + 3 = 0$

Impossible; no horizontal tangents

49. $f'(x) = \lim\limits_{h \to 0} \dfrac{f(x + h) - f(x)}{h} = \lim\limits_{h \to 0} \dfrac{3(x + h)^3 - 9(x + h) - (3x^3 - 9x)}{h}$

$= \lim\limits_{h \to 0} \dfrac{9x^2h + 9xh^2 + 3h^3 - 9h}{h} = 9x^2 - 9$

$f'(x) = 0 = 9x^2 - 9 \implies x = \pm 1$

f has horizontal tangents at $(1, -6)$ and $(-1, 6)$.

50. $f'(x) = \lim\limits_{h \to 0} \dfrac{f(x + h) - f(x)}{h} = \lim\limits_{h \to 0} \dfrac{3(x + h)^4 + 4(x + h)^3 - (3x^4 + 4x^3)}{h}$

$= \lim\limits_{h \to 0} \dfrac{(12x^3h + 18x^2h^2 + 12xh^3 + 3h^4) + (12x^2h + 12xh^2 + 4h^3)}{h}$

$= 12x^3 + 12x^2$

$f'(x) = 0 = 12x^3 + 12x^2 = 12x^2(x + 1) \implies 0, -1$

f has horizontal tangents at $(0, 0)$ and $(-1, -1)$.

51.

Year	x	Revenue, y
1999	9	4463.5
2000	10	4960.1
2001	11	5156.7
2002	12	5565.9
2003	13	5911.7
2004	14	6053.2

(a) Quadratic model:

$$y \approx -21.048x^2 + 804.47x - 1054.5$$

(b)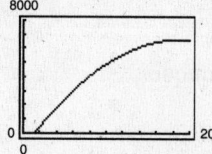

When $x = 12$ the slope is approximately 299.31. This represents a $299.31 million rate of change of revenue in 2002.

(c) Tangent line: When $x = 12$, $y \approx 5568.16$ (Model value).

$$y - 5568.16 = 299.31(x - 12)$$

$$y \approx 299.31x + 1976.44$$

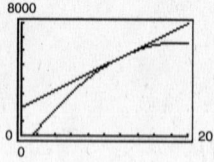

The slopes are the same.

52. (a) $N = 1.04p^2 - 81.50p + 1613.31$

(b)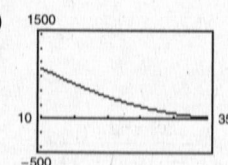

Slope $= -50.3$ for $p = 15$

Slope $= -19.1$ for $p = 30$

(c)

(d) The rate of decrease in sales decreases as the price increases.

53. $f(x) = -x^2 + 5x + 2$

Using the definition of slope, you obtain $f'(x) = -2x + 5$.

For $0 \le x \le 2, f'(x) > 0 \implies$ height increasing.

For $4 \le x \le 6, f'(x) < 0 \implies$ height decreasing.

54. $P(x) = 200 + 30x - 0.5x^2$

Using the definition of derivative,

$P'(x) = 30 - x.$

For $0 \le x \le 20, P'(x) > 0$ (profit increasing).

For $40 \le x \le 60, P'(x) < 0$ (profit decreasing).

55. True. The slope is $2x$, which is different for all x.

56. False. For example, the tangent line to $y = x^3$ at $(1, 1)$ intersects the curve at $(-2, -8)$.

57. Matches (b). (Derivative is always positive, but decreasing.)

58. Matches (a).
(Derivative approaches $-\infty$ when x approaches 0.)

59. Matches (d). (Derivative is -1 for $x < 0$, 1 for $x > 0$.)

60. Matches (c).
(Derivative decreases until origin, then increases.)

61. Answers will vary.

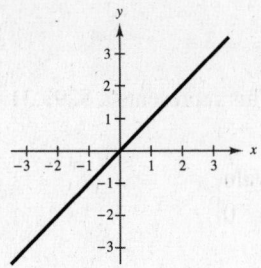

62. Answers not unique

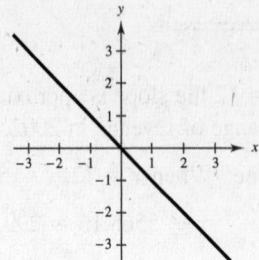

63. Answers will vary.

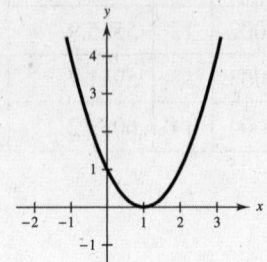

64. $f(x) = x^2$ and $g(x) = x^3$

(a) $f'(x) = 2x$

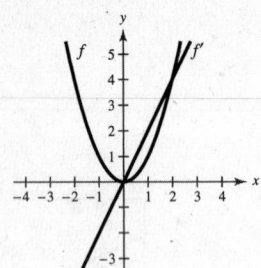

(b) $g'(x) = 3x^2$

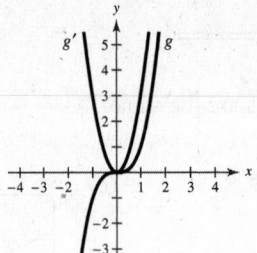

(c) Answers will vary. If
$h(x) = x^n$, $h'(x) = nx^{n-1}$.

65. $f(x) = \dfrac{1}{x^2 - x - 2} = \dfrac{1}{(x - 2)(x + 1)}$

Vertical asymptotes: $x = 2, -1$

Horizontal asymptote: $y = 0$

Intercept: $\left(0, -\dfrac{1}{2}\right)$

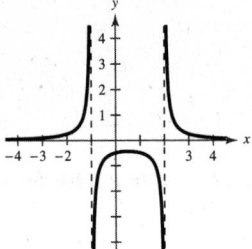

66. $f(x) = \dfrac{x - 2}{x^2 - 4x + 3} = \dfrac{x - 2}{(x - 3)(x - 1)}$

Vertical asymptotes: $x = 1, 3$

Horizontal asymptote: $y = 0$

Intercepts: $(2, 0)$, $\left(0, -\dfrac{2}{3}\right)$

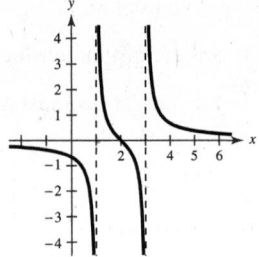

67. $f(x) = \dfrac{x^2 - x - 2}{x - 2}$

$= \dfrac{(x - 2)(x + 1)}{x - 2} = x + 1$, $x \neq 2$

Line with hole at $(2, 3)$

Intercepts: $(0, 1)$, $(-1, 0)$

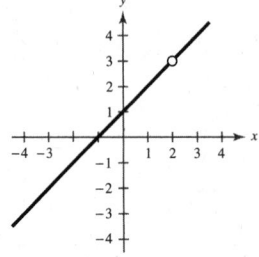

68. $f(x) = \dfrac{x^2 - 16}{x + 4} = \dfrac{(x - 4)(x + 4)}{x + 4} = x - 4, x \neq -4$

Line with hole at $(-4, -8)$

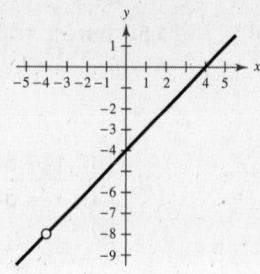

69. $\langle 1, 1, 1 \rangle \times \langle 2, 1, -1 \rangle = \begin{vmatrix} \mathbf{i} & \mathbf{j} & \mathbf{k} \\ 1 & 1 & 1 \\ 2 & 1 & -1 \end{vmatrix}$

$= \langle -2, 3, -1 \rangle$

70. $\mathbf{u} \times \mathbf{v} = \begin{vmatrix} \mathbf{i} & \mathbf{j} & \mathbf{k} \\ -10 & 0 & 6 \\ 7 & 0 & 0 \end{vmatrix} = \langle 0, 42, 0 \rangle$

71. $\langle -4, 10, 0 \rangle \times \langle 4, -1, 0 \rangle = \begin{vmatrix} \mathbf{i} & \mathbf{j} & \mathbf{k} \\ -4 & 10 & 0 \\ 4 & -1 & 0 \end{vmatrix}$

$= \langle 0, 0, -36 \rangle$

72. $\mathbf{u} \times \mathbf{v} = \begin{vmatrix} \mathbf{i} & \mathbf{j} & \mathbf{k} \\ 8 & -7 & 14 \\ -1 & 8 & 4 \end{vmatrix} = \langle -140, -46, 57 \rangle$

73. Answers will vary.

Section 12.4   Limits at Infinity and Limits of Sequences

- The limit at infinity

  $\lim\limits_{x \to \infty} f(x) = L$

  means that $f(x)$ gets arbitrarily close to L as x increases without bound.
- Similarly, the limit at infinity

  $\lim\limits_{x \to -\infty} f(x) = L$

  means that $F(x)$ gets arbitrarily close to L as x decreases without bound.
- You should be able to calculate limits at infinity, especially those arising from rational functions.
- Limits of functions can be used to evaluate limits of sequences. If f is a function such that $\lim\limits_{x \to \infty} f(x) = L$ and if a_n is a sequence such that $f(n) = a_n$, then $\lim\limits_{n \to \infty} a_n = L$.

Vocabulary Check

1. limit, infinity

2. converge

3. diverge

1. Intercept: $(0, 0)$

Horizontal asymptote: $y = 4$

Matches (c).

2. Horizontal asymptote: $y = 1$

Matches (a).

3. Horizontal asymptote: $y = 4$

Vertical asymptote: $x = 0$

Matches (d).

4. $f(x) = x + \dfrac{1}{x}$. No horizontal asymptote. Matches (b).

5. $\lim\limits_{x \to \infty} \dfrac{3}{x^2} = 0$

6. $\lim\limits_{x \to \infty} \dfrac{5}{2x} = 0$

7. $\lim\limits_{x \to \infty} \dfrac{3 + x}{3 - x} = -1$

8. $\lim\limits_{x \to \infty} \dfrac{1 - 6x}{1 + 5x} = -\dfrac{6}{5}$

9. $\lim\limits_{x \to -\infty} \dfrac{4x - 3}{2x + 1} = 2$

10. $\lim\limits_{x \to \infty} \dfrac{1 - 2x}{x + 2} = -2$

11. $\lim\limits_{x \to -\infty} \dfrac{3x^2 - 4}{1 - x^2} = -3$

12. $\lim\limits_{x \to -\infty} \dfrac{3x^2 + 1}{4x^2 - 5} = \dfrac{3}{4}$

13. $\lim\limits_{t \to \infty} \dfrac{t^2}{t + 3}$ does not exist.

14. $\lim\limits_{y \to \infty} \dfrac{4y^4}{y^2 + 3}$ does not exist.

15. $\lim\limits_{t \to \infty} \dfrac{1 - 2t + 6t^2}{5 + 3t - 4t^2} = \dfrac{6}{-4} = -\dfrac{3}{2}$

16. $\lim\limits_{x \to -\infty} \dfrac{2x^2 - 5x - 12}{1 - 6x - 8x^2} = \dfrac{2}{-8} = \dfrac{-1}{4}$

17. $\lim\limits_{x \to -\infty} \dfrac{-(x^2 + 3)}{(2 - x)^2} = \lim\limits_{x \to -\infty} \dfrac{-x^2 - 3}{x^2 - 4x + 4} = -1$

18. $\lim\limits_{x \to \infty} \dfrac{2x^2 - 6}{(x - 1)^2} = \lim\limits_{x \to \infty} \dfrac{2x^2 - 6}{x^2 - 2x + 1} = 2$

19. $\lim\limits_{x \to -\infty} \left[\dfrac{x}{(x + 1)^2} - 4 \right] = 0 - 4 = -4$

20. $\lim\limits_{x \to \infty} \left[7 + \dfrac{2x^2}{(x + 3)^2} \right] = 7 + 2 = 9$

21. $\lim\limits_{t \to \infty} \left(\dfrac{1}{3t^2} - \dfrac{5t}{t + 2} \right) = 0 - 5$
$$= -5$$

22. $\lim\limits_{x \to \infty} \left[\dfrac{x}{2x + 1} + \dfrac{3x^2}{(x - 3)^2} \right] = \dfrac{1}{2} + 3 = \dfrac{7}{2}$

23. $y = \dfrac{3x}{1 - x}$

Horizontal asymptote:
$y = -3$

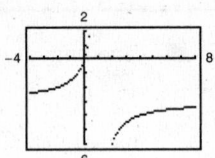

24. $y = \dfrac{x^2}{x^2 + 4}$

Horizontal asymptote:
$y = 1$

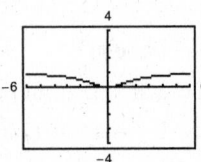

25. Horizontal asymptote:
$y = 0$

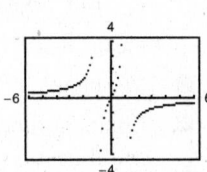

26. $y = \dfrac{2x + 1}{x^2 - 1}$

Horizontal asymptote: $y = 0$

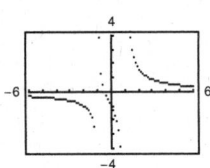

27. $y = 1 - \dfrac{3}{x^2}$

Horizontal asymptote: $y = 1$

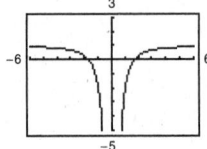

28. $y = 2 + \dfrac{1}{x}$

Horizontal asymptote: $y = 2$

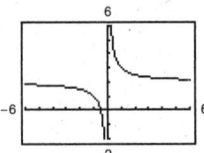

29. (a)

x	10^0	10^1	10^2	10^3	10^4	10^5	10^6
$f(x)$	-0.7321	-0.0995	-0.00999	-0.001	-1×10^{-4}	-1×10^{-5}	-1×10^{-6}

$$\lim_{x \to \infty}\left(x - \sqrt{x^2 + 2}\right) = 0$$

(b)

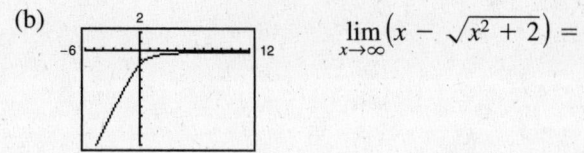

$$\lim_{x \to \infty}\left(x - \sqrt{x^2 + 2}\right) = 0$$

30. (a)

x	10^0	10^1	10^2	10^3	10^4	10^5	10^6
$f(x)$	-0.162	-0.0167	-0.00167	-1.67×10^{-4}	-1.7×10^{-5}	-1.7×10^{-6}	-2×10^{-7}

$$\lim_{x \to \infty}\left(3x - \sqrt{9x^2 + 1}\right) = 0$$

(b)

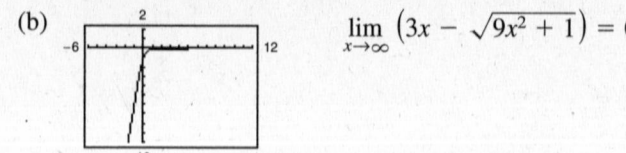

$$\lim_{x \to \infty}\left(3x - \sqrt{9x^2 + 1}\right) = 0$$

31. (a)

x	10^0	10^1	10^2	10^3	10^4	10^5	10^6
$f(x)$	-0.7082	-0.7454	-0.7495	-0.74995	-0.749995	-0.75	-0.75

$$\lim_{x \to \infty} 3\left(2x - \sqrt{4x^2 + x}\right) = -\tfrac{3}{4}$$

(b)

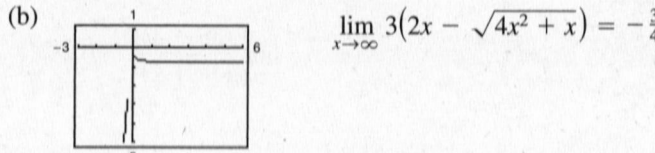

$$\lim_{x \to \infty} 3\left(2x - \sqrt{4x^2 + x}\right) = -\tfrac{3}{4}$$

32. (a)

x	10^0	10^1	10^2	10^3	10^4	10^5	10^6
$f(x)$	0.508	0.5008	0.50008	0.5	0.5	0.5	0.5

$$\lim_{x \to \infty} 4\left(4x - \sqrt{16x^2 - x}\right) = \tfrac{1}{2}$$

(b)

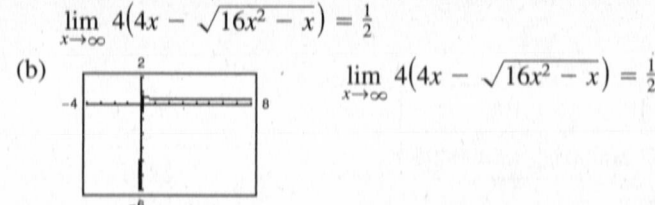

$$\lim_{x \to \infty} 4\left(4x - \sqrt{16x^2 - x}\right) = \tfrac{1}{2}$$

33. $a_n = \dfrac{n+1}{n^2+1}$

$a_1 = \dfrac{1+1}{1^2+1} = 1 \qquad a_4 = \dfrac{5}{17}$

$a_2 = \dfrac{2+1}{2^2+1} = \dfrac{3}{5} \qquad a_5 = \dfrac{6}{26} = \dfrac{3}{13}$

$a_3 = \dfrac{4}{10} = \dfrac{2}{5}$

$\lim\limits_{n\to\infty} a_n = 0$

34. $\dfrac{1}{2}, \dfrac{2}{5}, \dfrac{3}{10}, \dfrac{4}{17}, \dfrac{5}{26}$

$\lim\limits_{n\to\infty} \dfrac{n}{n^2+1} = 0$

35. $a_n = \dfrac{n}{2n+1}$

$a_1 = \dfrac{1}{3} \qquad a_4 = \dfrac{4}{9}$

$a_2 = \dfrac{2}{5} \qquad a_5 = \dfrac{5}{11}$

$a_3 = \dfrac{3}{7}$

$\lim\limits_{n\to\infty} a_n = \dfrac{1}{2}$

36. $\dfrac{3}{4}, \dfrac{7}{5}, \dfrac{11}{6}, \dfrac{15}{7}, \dfrac{19}{8}$

$\lim\limits_{n\to\infty} \dfrac{4n-1}{n+3} = 4$

37. $\dfrac{1}{5}, \dfrac{1}{2}, \dfrac{9}{11}, \dfrac{8}{7}, \dfrac{25}{17}$

$\lim\limits_{n\to\infty} \dfrac{n^2}{3n+2}$ does not exist.

38. $\dfrac{5}{2}, \dfrac{17}{4}, \dfrac{37}{6}, \dfrac{65}{8}, \dfrac{101}{10}$

$\lim\limits_{n\to\infty} \dfrac{4n^2+1}{2n}$ does not exist.

39. $2, 3, 4, 5, 6$

$\lim\limits_{n\to\infty} \dfrac{(n+1)!}{n!} = \lim\limits_{n\to\infty} (n+1)$
does not exist.

40. $a_n = \dfrac{(3n-1)!}{(3n+1)!} = \dfrac{1}{(3n+1)(3n)}$

$\dfrac{1}{12}, \dfrac{1}{42}, \dfrac{1}{90}, \dfrac{1}{156}, \dfrac{1}{240}$

$\lim\limits_{n\to\infty} \dfrac{(3n-1)!}{(3n+1)!} = 0$

41. $-1, \dfrac{1}{2}, -\dfrac{1}{3}, \dfrac{1}{4}, -\dfrac{1}{5}$

$\lim\limits_{n\to\infty} \dfrac{(-1)^n}{n} = 0$

42. $1, -\dfrac{1}{4}, \dfrac{1}{9}, -\dfrac{1}{16}, \dfrac{1}{25}$

$\lim\limits_{n\to\infty} \dfrac{(-1)^{n+1}}{n^2} = 0$

43. $\lim\limits_{n\to\infty} a_n = \lim\limits_{n\to\infty} \left[1 + \dfrac{n(n+1)}{2n^2}\right] = 1 + \dfrac{1}{2} = \dfrac{3}{2}$

n	10^0	10^1	10^2	10^3	10^4	10^5	10^6
a_n	2	1.55	1.505	1.5005	1.50005	1.500005	1.5000005

44. $\lim\limits_{n\to\infty} a_n = 12$

x	10^0	10^1	10^2	10^3	10^4	10^5	10^6
$f(x)$	20	12.8	12.08	12.008	12.0008	12.00008	12.000008

45. $\lim\limits_{n\to\infty} a_n = \frac{16}{1}\left[\frac{2}{6}\right] = \frac{16}{3}$

n	10^0	10^1	10^2	10^3	10^4	10^5	10^6
a_n	16	6.16	5.4136	5.341336	5.3341	5.33341	5.333341

46. $\lim\limits_{n\to\infty} a_n = \frac{3}{4}$

x	10^0	10^1	10^2	10^3	10^4	10^5	10^6
$f(x)$	1	0.7975	0.754975	0.75049975	0.75005	0.750005	0.7500005

47. $f(t) = \dfrac{t^2 - t + 1}{t^2 + 1}$

(a) $\lim\limits_{t\to\infty} \dfrac{t^2 - t + 1}{t^2 + 1} = 1$

(b)

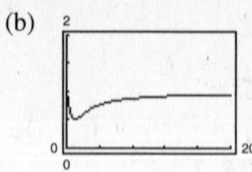

(c) Over a long period of time, the level of oxygen in the pond returns to the normal level.

48. (a) $\lim\limits_{t\to\infty} \dfrac{100t^2}{65 + t^2} = \lim\limits_{t\to\infty} \dfrac{100}{\dfrac{65}{t^2} + 1} = \dfrac{100}{0 + 1} = 100$

(b)

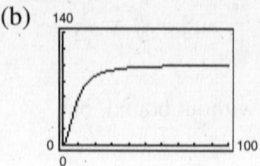

(c) The average typing speed approaches 100 words per minute as t approaches ∞.

49. (a) Average cost $= \overline{C} = \dfrac{C}{x} = 13.50 + \dfrac{45{,}750}{x}$

(b) $\overline{C}(100) = \$471$

$\overline{C}(1000) = \$59.25$

(c) $\lim\limits_{x\to\infty} \overline{C}(x) = 13.50$

As more units are produced, the fixed costs (45,750) become less dominant.

50. $C = 1.25x + 10{,}500$

(a) $\overline{C} =$ average cost $= \dfrac{C}{x} = 1.25 + \dfrac{10{,}500}{x}$

(b) $\overline{C}(100) = \$106.25$

$\overline{C}(1000) = \$11.75$

(c) As $x \to \infty$, $\overline{C} \to \$1.25$.

As more tons are recycled, the average cost per ton approaches \$1.25.

51.

Year	t	Benefit, B
1997	7	765
1998	8	780
1999	9	804
2000	10	844
2001	11	874
2002	12	896
2003	13	922

(a)

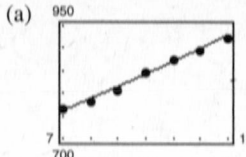

The model is a good fit to the data.

(b) For 2006: $B(16) \approx \$1032$

(c) The graph has a vertical asymptote and approaches infinity when t is slightly greater than 45.

52. $N = \dfrac{632.8 - 283.17t}{1.0 - 0.27t},\ 7 \le t \le 13$

(a)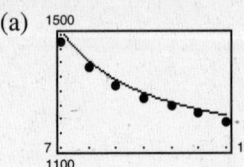

The model is a good fit to the data.

(b) $N(16) = \dfrac{632.8 - 283.17(16)}{1.0 - 0.27(16)} \approx 1174$ thousand

(c) $\lim\limits_{t\to\infty} N = \dfrac{-283.17}{-0.27} \approx 1048.78$

The number of military reserve personnel will not go below 1048.78 thousand.

53. False. $f(x) = \dfrac{x^2 + 1}{1}$ does not have a horizontal asymptote.

54. False. The limit does not exist.   **55.** True   **56.** False

57. For example, let $f(x) = \dfrac{1}{x^2}$ and $g(x) = \dfrac{1}{x^2}$.

Then, $\lim\limits_{x\to 0} \dfrac{1}{x^2}$ increases without bound, but

$\lim\limits_{x\to 0} [f(x) - g(x)] = 0.$

58.

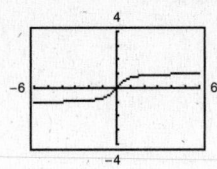

Two horizontal asymptotes: $y = \pm 1$

59. $a_n = 4\left(\dfrac{2}{3}\right)^n$

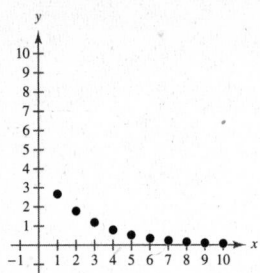

Converges to 0

60. $a_n = 3\left(\dfrac{3}{2}\right)^n$

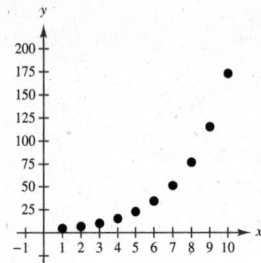

Diverges

61. $a_n = \dfrac{3[1 - (1.5)^n]}{1 - 1.5}$

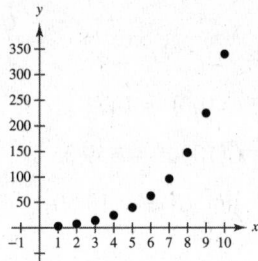

Diverges

62. $a_n = \dfrac{3[1 - (0.5)^n]}{1 - 0.5}$

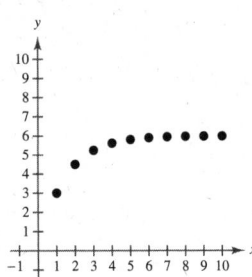

Converges to 6

63. $y = x^4$

(a) $f(x) = (x + 3)^4$

(b) $f(x) = x^4 - 1$

(c) $f(x) = -2 + x^4$

(d) $f(x) = \frac{1}{2}(x - 4)^4$

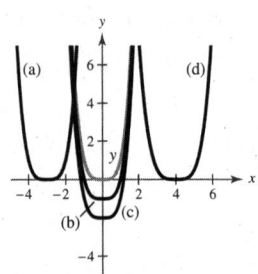

64. $y = x^3$

(a) $f(x) = (x + 2)^3$

(b) $f(x) = 3 + x^3$

(c) $f(x) = 2 - \frac{1}{4}x^3$

(d) $f(x) = 3(x + 1)^3$

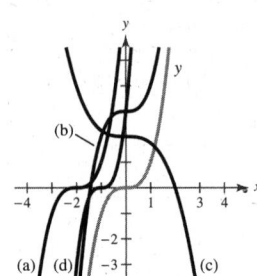

65.

$$
\begin{array}{r}
x^2 + 2x + 1 \\
x^2 - 4\overline{)x^4 + 2x^3 - 3x^2 - 8x - 4} \\
\underline{x^4 \qquad\quad - 4x^2} \\
2x^3 + \ x^2 \\
\underline{2x^3 \qquad\quad - 8x} \\
x^2 \qquad\quad - 4
\end{array}
$$

$$x^4 + 2x^3 - 3x^2 - 8x - 4 = (x^2 - 4)(x^2 + 2x + 1)$$

66.

$$
\begin{array}{r}
2x^3 + \ 4x^2 - \ 2x - 8 \\
x^2 - 2x + 1\overline{)2x^5 \qquad\ \ - \ 8x^3 \qquad\quad + \ 4x - 1} \\
\underline{2x^5 - 4x^4 + \ 2x^3} \\
4x^4 - 10x^3 \\
\underline{4x^4 - \ 8x^3 + \ 4x^2} \\
-2x^3 - \ 4x^2 + \ \ 4x \\
\underline{-2x^3 + \ 4x^2 - \ \ 2x} \\
-8x^2 + \ \ 6x - 1 \\
\underline{-8x^2 + 16x - 8} \\
-10x + 7
\end{array}
$$

$$\frac{2x^5 - 8x^3 + 4x - 1}{x^2 - 2x + 1} = 2x^3 + 4x^2 - 2x - 8 + \frac{-10x + 7}{x^2 - 2x + 1}$$

67.

$$
\begin{array}{r}
x^3 + \ 5x^2 \qquad\quad - 3 \\
3x + 2\overline{)3x^4 + 17x^3 + 10x^2 - 9x - 8} \\
\underline{3x^4 + \ 2x^3} \\
15x^3 + 10x^2 \\
\underline{15x^3 + 10x^2} \\
-9x - 8 \\
\underline{-9x - 6} \\
-2
\end{array}
$$

$$\frac{3x^4 + 17x^3 + 10x^2 - 9x - 8}{3x + 2} = x^3 + 5x^2 - 3 + \frac{-2}{3x + 2}$$

68.

$$
\begin{array}{r}
2x^2 + 11x + 14 \\
5x - 2\overline{)10x^3 + 51x^2 + 48x - 28} \\
\underline{10x^3 - \ 4x^2} \\
55x^2 + 48x \\
\underline{55x^2 - 22x} \\
70x - 28 \\
\underline{70x - 28}
\end{array}
$$

$$\frac{10x^3 + 51x^2 + 48x - 28}{5x - 2} = 2x^2 + 11x + 14, \ x \neq \frac{2}{5}$$

69. $f(x) = x^4 - x^3 - 20x^2$

$$= x^2(x^2 - x - 20)$$

$$= x^2(x - 5)(x + 4)$$

Real zeros: $0, 0, 5, -4$

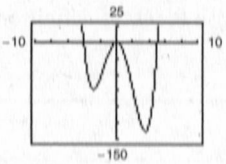

70. $x^5 + x^3 - 6x = x(x^4 + x^2 - 6) = x(x^2 + 3)(x^2 - 2)$

Real zeros: $0, \pm\sqrt{2}$

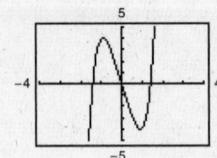

71. $f(x) = x^3 - 3x^2 + 2x - 6$

$= x^2(x - 3) + 2(x - 3)$

$= (x - 3)(x^2 + 2)$

Real zero: 3

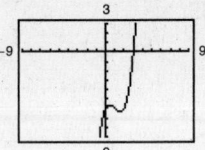

72. $x^3 - 4x^2 - 25x + 100 = x^2(x - 4) - 25(x - 4)$

$= (x^2 - 25)(x - 4)$

Real zeros: $\pm 5, 4$

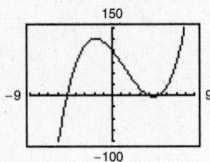

73. $\displaystyle\sum_{i=1}^{6} (2i + 3) = 5 + 7 + 9 + 11 + 13 + 15 = 60$

74. $\displaystyle\sum_{i=0}^{4} 5i^2 = 0 + 5 + 20 + 45 + 80 = 150$

75. $\displaystyle\sum_{k=1}^{10} 15 = 10(15) = 150$

76. $\displaystyle\sum_{k=0}^{8} \frac{3}{k^2 + 1} \approx 5.8791$

Section 12.5   The Area Problem

- ■   You should know the following summation formulas and properties.

    (a) $\displaystyle\sum_{i=1}^{n} c = cn$ (b) $\displaystyle\sum_{i=1}^{n} i = \frac{n(n + 1)}{2}$ (c) $\displaystyle\sum_{i=1}^{n} i^2 = \frac{n(n + 1)(2n + 1)}{6}$

    (d) $\displaystyle\sum_{i=1}^{n} i^3 = \frac{n^2(n + 1)^2}{4}$ (e) $\displaystyle\sum_{i=1}^{n} (a_i \pm b_i) = \sum_{i=1}^{n} a_i \pm \sum_{i=1}^{n} b_i$ (f) $\displaystyle\sum_{i=1}^{n} ka_i = k\sum_{i=1}^{n} a_i$

- ■   You should be able to evaluate a limit of a summation, $\displaystyle\lim_{n\to\infty} S(n)$.

- ■   You should be able to approximate the area of a region using rectangles. By increasing the number of rectangles, the approximation improves.

- ■   The area of a plane region above the x-axis bounded by f between $x = a$ and $x = b$ is the limit of the sum of the approximating rectangles:

$$A = \lim_{n\to\infty} \sum_{i=1}^{n} f\left(a + \frac{(b - a)i}{n}\right)\left(\frac{b - a}{n}\right)$$

- ■   You should be able to use the limit definition of area to find the area bounded by simple functions in the plane.

Vocabulary Check

1. $\dfrac{n(n + 1)}{2}$ **2.** $\dfrac{n^2(n + 1)^2}{4}$ **3.** area

1. $\displaystyle\sum_{i=1}^{60} 7 = 7(60) = 420$

2. $\displaystyle\sum_{i=1}^{30} i^2 = \dfrac{n(n+1)(2n+1)}{6} = \dfrac{30(31)(61)}{6} = 9455$

3. $\displaystyle\sum_{k=1}^{20} (k^3 + 2) = \dfrac{20^2(21)^2}{4} + 2(20)$

$\qquad\qquad = 44{,}100 + 40 = 44{,}140$

4. $\displaystyle\sum_{k=1}^{50} (2k+1) = 2\sum_{k=1}^{50} k + \sum_{k=1}^{50} 1 = 2\dfrac{50(51)}{2} + 50$

$\qquad\qquad\qquad\qquad = 2600$

5. $\displaystyle\sum_{j=1}^{25} (j^2 + j) = \dfrac{25(26)(51)}{6} + \dfrac{25(26)}{2} = 5850$

6. $\displaystyle\sum_{j=1}^{10} (j^3 - 3j^2) = \dfrac{10^2(11)^2}{4} - 3\left(\dfrac{10(11)(21)}{6}\right) = 1870$

7. (a) $S(n) = \displaystyle\sum_{i=1}^{n} \dfrac{i^3}{n^4} = \dfrac{1}{n^4}\left[\dfrac{n^2(n+1)^2}{4}\right] = \dfrac{n^2 + 2n + 1}{4n^2}$

(b)

n	10^0	10^1	10^2	10^3	10^4
$S(n)$	1	0.3025	0.255025	0.25050025	0.25005

(c) $\displaystyle\lim_{n\to\infty} S(n) = \dfrac{1}{4}$

8. (a) $S(n) = \displaystyle\sum_{i=1}^{n} \dfrac{i}{n^2} = \dfrac{1}{n^2}\cdot\dfrac{n(n+1)}{2} = \dfrac{n+1}{2n}$

(b)

n	10^0	10^1	10^2	10^3	10^4
$S(n)$	1	0.55	0.505	0.5005	0.50005

(c) $\displaystyle\lim_{n\to\infty} S(n) = \dfrac{1}{2}$

9. (a) $S(n) = \displaystyle\sum_{i=1}^{n} \dfrac{3}{n^3}(1 + i^2) = \dfrac{3}{n^3}\left[n + \dfrac{n(n+1)(2n+1)}{6}\right] = \dfrac{3}{n^2} + \dfrac{6n^2 + 9n + 3}{6n^2} = \dfrac{2n^2 + 3n + 7}{2n^2}$

(b)

n	10^0	10^1	10^2	10^3	10^4
$S(n)$	6	1.185	1.0154	1.0015	1.00015

(c) $\displaystyle\lim_{n\to\infty} S(n) = 1$

10. (a) $S(n) = \displaystyle\sum_{i=1}^{n} \dfrac{2i+3}{n^2} = \dfrac{1}{n^2}\left(2\left(\dfrac{n(n+1)}{2}\right) + 3n\right) = \dfrac{n+1}{n} + \dfrac{3}{n} = \dfrac{n+4}{n}$

(b)

n	10^0	10^1	10^2	10^3	10^4
$S(n)$	5	1.4	1.04	1.004	1.0004

(c) $\displaystyle\lim_{n\to\infty} S(n) = 1$

11. (a) $S(n) = \displaystyle\sum_{i=1}^{n} \left(\dfrac{i^2}{n^3} + \dfrac{2}{n}\right)\left(\dfrac{1}{n}\right) = \dfrac{1}{n}\left[\dfrac{n(n+1)(2n+1)}{6n^3} + \dfrac{2n}{n}\right] = \dfrac{1}{6n^3}(2n^2 + 3n + 1) + \dfrac{2}{n} = \dfrac{14n^2 + 3n + 1}{6n^3}$

(b)

n	10^0	10^1	10^2	10^3	10^4
$S(n)$	3	0.2385	0.02338	0.00233	0.0002333

(c) $\displaystyle\lim_{n\to\infty} S(n) = 0$

12. (a) $S(n) = \displaystyle\sum_{i=1}^{n} \left[3 - 2\left(\dfrac{i}{n}\right)\right]\dfrac{1}{n} = \dfrac{1}{n}\left[3n - \dfrac{2}{n}\dfrac{n(n+1)}{2}\right] = 3 - \dfrac{n+1}{n} = \dfrac{2n-1}{n}$

(b)

n	10^0	10^1	10^2	10^3	10^4
$S(n)$	1	1.9	1.99	1.999	1.9999

(c) $\displaystyle\lim_{n\to\infty} S(n) = 2$

13. (a) $S(n) = \sum_{i=1}^{n}\left[1 - \left(\frac{i}{n}\right)^2\right]\left(\frac{1}{n}\right) = \frac{1}{n}\left[n - \frac{1}{n^2}\left(\frac{n(n+1)(2n+1)}{6}\right)\right] = 1 - \frac{2n^2+3n+1}{6n^2} = \frac{4n^2-3n-1}{6n^2}$

(b)

n	10^0	10^1	10^2	10^3	10^4
$S(n)$	0	0.615	0.66165	0.66617	0.666617

(c) $\lim_{n\to\infty} S(n) = \frac{2}{3}$

14. (a) $S(n) = \sum_{i=1}^{n}\left(\frac{4}{n} + \frac{2i}{n^2}\right)\left(\frac{2i}{n}\right) = \frac{2}{n}\left[\frac{4}{n}\frac{n(n+1)}{2} + \frac{2}{n^2}\frac{n(n+1)(2n+1)}{6}\right]$

$= \frac{2}{n}\left[\frac{4n^2+4n}{2n} + \frac{2(2n^2+3n+1)}{6n}\right] = \frac{16n^2+18n+2}{3n^2}$

(b)

n	10^0	10^1	10^2	10^3	10^4
$S(n)$	12.0	5.94	5.3934	5.3393	5.33393

(c) $\lim_{n\to\infty} S(n) = \frac{16}{3}$

15. $f(x) = x + 4, [-1, 2], n = 6,$ width $= \frac{1}{2}$

Area $\approx \frac{1}{2}[3.5 + 4 + 4.5 + 5 + 5.5 + 6]$

$= 14.25$ square units

16. $f(x) = 2 - x^2, -1 \le x \le 1, n = 4,$ width $= \frac{1}{2}$

Area $\approx \frac{1}{2}\left[\left(2 - \left(-\frac{1}{2}\right)^2\right) + (2 - 0^2) + \left(2 - \left(\frac{1}{2}\right)^2\right) + (2 - 1^2)\right]$

$= \frac{1}{2}[1.75 + 2 + 1.75 + 1]$

$= 3.25$ square units

17. The width of each rectangle is $\frac{1}{4}$. The height is obtained by evaluating f at the right-hand endpoint of each interval.

$A \approx \sum_{i=1}^{8} f\left(\frac{i}{4}\right)\left(\frac{1}{4}\right) = \sum_{i=1}^{8} \frac{1}{4}\left(\frac{i}{4}\right)^3\left(\frac{1}{4}\right)$

$= 1.265625$ square units

18. Area $\approx \frac{1}{2}\left[\frac{1}{16} + \frac{1}{2} + \frac{27}{16} + 4\right] = 3.125$ square units

19. Width of each rectangle is $\frac{12}{n}$. The height is $f\left(\frac{12}{n}i\right) = -\frac{1}{3}\left(\frac{12}{n}i\right) + 4.$

$A = \sum_{i=1}^{n}\left[-\frac{1}{3}\left(\frac{12i}{n}\right) + 4\right]\left(\frac{12}{n}\right)$

(Note: Exact area is 24.)

n	4	8	20	50
Approximate area	18	21	22.8	23.52

20. The width of each rectangle is $3/n$. The height is

$f\left(\frac{3i}{n}\right) = 9 - \left(\frac{3i}{n}\right)^2.$

$A \approx \sum_{i=1}^{n}\left(9 - \left(\frac{3i}{n}\right)^2\right)\frac{3}{n}$

(Note: Exact area is 18.)

n	4	8	20	50
Approximate area	14.344	16.242	17.314	17.7282

21. The width of each rectangle is $\dfrac{3}{n}$. The height is $\dfrac{1}{9}\left(\dfrac{3i}{n}\right)^3$.

$$A \approx \sum_{i=1}^{n} \frac{1}{9}\left(\frac{3i}{n}\right)^3\left(\frac{3}{n}\right)$$

n	4	8	20	50
Approximate area	3.52	2.85	2.48	2.34

22. The width of each rectangle is $(2 - (-1))/n = 3/n$. The height is

$$f\left(-1 + \frac{3i}{n}\right) = 3 - \frac{1}{4}\left(-1 + \frac{3i}{n}\right)^3.$$

$$A \approx \sum_{i=1}^{n}\left[3 - \frac{1}{4}\left(-1 + \frac{3i}{n}\right)^3\right]\frac{3}{n}$$

n	4	8	20	50
Approximate area	7.113	7.614	7.8895	7.994

(**Note:** Exact area is $8\frac{1}{16} = 8.0625$.)

23. $A \approx \displaystyle\sum_{i=1}^{n} f\left(\frac{i}{n}\right)\left(\frac{1}{n}\right)$

$\quad = \displaystyle\sum_{i=1}^{n}\left[4\left(\frac{i}{n}\right) + 1\right]\left(\frac{1}{n}\right)$

$\quad = \dfrac{1}{n}\displaystyle\sum_{i=1}^{n}\left[\frac{4}{n}i + 1\right]$

$\quad = \dfrac{1}{n}\left[\dfrac{4}{n}\dfrac{n(n+1)}{2} + n\right]$

$\quad = \dfrac{1}{n}\left[2(n+1) + n\right]$

$\quad = \dfrac{3n+2}{n}$

$A = \displaystyle\lim_{n\to\infty} \dfrac{3n+2}{n} = 3$ square units

24. $A \approx \displaystyle\sum_{i=1}^{n} f\left(\frac{2i}{n}\right)\left(\frac{2}{n}\right)$

$\quad = \displaystyle\sum_{i=1}^{n}\left[3\left(\frac{2i}{n}\right) + 2\right]\frac{2}{n}$

$\quad = \dfrac{2}{n}\displaystyle\sum_{i=1}^{n}\left(\frac{6}{n}i + 2\right)$

$\quad = \dfrac{2}{n}\left[\dfrac{6}{n}\dfrac{n(n+1)}{2} + 2n\right]$

$\quad = 6\left(\dfrac{n+1}{n}\right) + 4$

$A = \displaystyle\lim_{n\to\infty}\left[6\dfrac{n+1}{n} + 4\right] = 10$ square units

25. $A \approx \displaystyle\sum_{i=1}^{n} f\left(\frac{i}{n}\right)\left(\frac{1}{n}\right)$

$\quad = \displaystyle\sum_{i=1}^{n}\left[-2\left(\frac{i}{n}\right) + 3\right]\left(\frac{1}{n}\right)$

$\quad = \dfrac{1}{n}\displaystyle\sum_{i=1}^{n}\left[-\frac{2i}{n} + 3\right]$

$\quad = \dfrac{1}{n}\left[-\dfrac{2}{n}\dfrac{n(n+1)}{2} + 3n\right]$

$\quad = \dfrac{1}{n}\left[2n - 1\right]$

$A = \displaystyle\lim_{n\to\infty} \dfrac{2n-1}{n} = 2$ square units

26. $A \approx \displaystyle\sum_{i=1}^{n} f\left(2 + \frac{3i}{n}\right)\left(\frac{3}{n}\right)$

$\quad = \displaystyle\sum_{i=1}^{n}\left[3\left(2 + \frac{3i}{n}\right) - 4\right]\frac{3}{n}$

$\quad = \dfrac{3}{n}\displaystyle\sum_{i=1}^{n}\left[2 + \frac{9}{n}i\right]$

$\quad = \dfrac{3}{n}\left[2n + \dfrac{9}{n}\dfrac{n(n+1)}{2}\right]$

$\quad = 6 + \dfrac{27}{2}\dfrac{(n+1)}{n}$

$A = \displaystyle\lim_{n\to\infty}\left[6 + \dfrac{27}{2}\dfrac{(n+1)}{n}\right] = \dfrac{39}{2}$ square units

27. $A \approx \sum_{i=1}^{n} f\left(-1 + \frac{2i}{n}\right)\left(\frac{2}{n}\right)$

$\quad = \sum_{i=1}^{n} \left[2 - \left(-1 + \frac{2i}{n}\right)^2\right]\frac{2}{n}$

$\quad = \sum_{i=1}^{n} \left[2 - 1 + \frac{4i}{n} - \frac{4i^2}{n^2}\right]\left(\frac{2}{n}\right)$

$\quad = \frac{2}{n}\sum_{i=1}^{n} 1 + \frac{8}{n^2}\sum_{i=1}^{n} i - \frac{8}{n^3}\sum_{i=1}^{n} i^2$

$\quad = \frac{2}{n}(n) + \frac{8}{n^2}\frac{n(n+1)}{2} - \frac{8}{n^3}\frac{n(n+1)(2n+1)}{6}$

$\quad A = \lim_{n\to\infty}\left[2 + 4\frac{n(n+1)}{n^2} - \frac{4}{3}\frac{n(n+1)(2n+1)}{n^3}\right] = 2 + 4 - \frac{8}{3} = \frac{10}{3}$ square units

28. $A \approx \sum_{n=1}^{\infty} f\left(\frac{i}{n}\right)\left(\frac{1}{n}\right)$

$\quad = \sum_{n=1}^{\infty} \left[\left(\frac{i}{n}\right)^2 + 2\right]\frac{1}{n}$

$\quad = \frac{1}{n}\left[\frac{1}{n^2}\frac{n(n+1)(2n+1)}{6} + 2n\right]$

$\quad = \frac{(n+1)(2n+1)}{6n^2} + 2$

$\quad A = \lim_{n\to\infty}\left[\frac{(n+1)(2n+1)}{6n^2} + 2\right] = \frac{7}{3}$ square units

29. $A \approx \sum_{i=1}^{n} g\left(1 + \frac{i}{n}\right)\left(\frac{1}{n}\right)$

$\quad = \sum_{i=1}^{n} \left[8 - \left(1 + \frac{i}{n}\right)^3\right]\frac{1}{n}$

$\quad = \sum_{i=1}^{n} \left[7 - \frac{3i}{n} - \frac{3i^2}{n^2} - \frac{i^3}{n^3}\right]\frac{1}{n}$

$\quad = \frac{7}{n}\sum_{i=1}^{n} 1 - \frac{3}{n^2}\sum_{i=1}^{n} i - \frac{3}{n^3}\sum_{i=1}^{n} i^2 - \frac{1}{n^4}\sum_{i=1}^{n} i^3$

$\quad = \frac{7}{n}(n) - \frac{3}{n^2}\frac{n(n+1)}{2} - \frac{3}{n^3}\frac{n(n+1)(2n+1)}{6} - \frac{1}{n^4}\frac{n^2(n+1)^2}{4}$

$\quad A = \lim_{n\to\infty}\left[7 - \frac{3}{2}\frac{n(n+1)}{n^2} - \frac{1}{2n^3}n(n+1)(2n+1) - \frac{1}{n^4}\frac{n^2(n+1)^2}{4}\right] = 7 - \frac{3}{2} - 1 - \frac{1}{4} = \frac{17}{4}$ square units

30. $A \approx \sum_{i=1}^{n} g\left(1 + \frac{3i}{n}\right)\left(\frac{3}{n}\right)$

$= \sum_{i=1}^{n}\left[64 - \left(1 + \frac{3i}{n}\right)^3\right]\frac{3}{n}$

$= \frac{3}{n}\sum_{i=1}^{n}\left[63 - \frac{9i}{n} - \frac{27i^2}{n^2} - \frac{27i^3}{n^3}\right]$

$= \frac{3}{n}\left[63n - \frac{9n(n+1)}{2n} - \frac{27n(n+1)(2n+1)}{6n^2} - \frac{27n^2(n+1)^2}{4n^3}\right]$

$= 189 - \frac{27(n+1)}{2n} - \frac{27(n+1)(2n+1)}{2n^2} - \frac{81(n+1)^2}{4n^2}$

$A = \lim_{n\to\infty}\left[189 - \frac{27(n+1)}{2n} - \frac{27(n+1)(2n+1)}{2n^2} - \frac{81(n+1)^2}{4n^2}\right]$

$= \frac{513}{4}$ square units

31. $A \approx \sum_{i=1}^{n} g\left(\frac{i}{n}\right)\left(\frac{1}{n}\right)$

$= \sum_{i=1}^{n}\left[2\left(\frac{i}{n}\right) - \left(\frac{i}{n}\right)^3\right]\left(\frac{1}{n}\right)$

$= \frac{1}{n}\sum_{i=1}^{n}\left[\frac{2}{n}i - \frac{1}{n^3}i^3\right]$

$= \frac{1}{n}\left[\frac{2}{n}\frac{n(n+1)}{2} - \frac{1}{n^3}\frac{n^2(n+1)^2}{4}\right]$

$= \frac{n+1}{n} - \frac{(n+1)^2}{4n^2}$

$A = \lim_{n\to\infty}\left[\frac{n+1}{n} - \frac{(n+1)^2}{4n^2}\right]$

$= 1 - \frac{1}{4} = \frac{3}{4}$ square unit

32. $A \approx \sum_{i=1}^{n} g\left(2\frac{i}{n}\right)\left(\frac{2}{n}\right)$

$= \sum_{i=1}^{n}\left[4\left(\frac{2i}{n}\right) - \left(\frac{2i}{n}\right)^3\right]\left(\frac{2}{n}\right)$

$= \frac{16}{n^2}\sum_{i=1}^{n}i - \frac{16}{n^4}\sum_{i=1}^{n}i^3$

$= \frac{16}{n^2}\frac{n(n+1)}{2} - \frac{16}{n^4}\frac{n^2(n+1)^2}{4}$

$A = \lim_{n\to\infty}\left[8\frac{n(n+1)}{n^2} - 4\frac{n^2(n+1)^2}{n^4}\right]$

$= 8 - 4 = 4$ square units

33. $A \approx \sum_{i=1}^{n} f\left(1 + \frac{3i}{n}\right)\left(\frac{3}{n}\right)$

$= \sum_{i=1}^{n}\left[\frac{1}{4}\left(1 + \frac{3i}{n}\right)^2 + \left(1 + \frac{3i}{n}\right)\right]\left(\frac{3}{n}\right)$

$= \sum_{i=1}^{n}\left(\frac{1}{4} + \frac{3}{2}\frac{i}{n} + \frac{9}{4}\frac{i^2}{n^2} + 1 + \frac{3i}{n}\right)\left(\frac{3}{n}\right)$

$= \frac{15}{4n}\sum_{i=1}^{n} 1 + \frac{27}{2n^2}\sum_{i=1}^{n}i + \frac{27}{4n^3}\sum_{i=1}^{n}i^2$

$= \frac{15}{4n}(n) + \frac{27}{2n^2}\left(\frac{n(n+1)}{2}\right) + \frac{27}{4n^3}\frac{n(n+1)(2n+1)}{6}$

$A = \lim_{n\to\infty}\left[\frac{15}{4} + \frac{27}{4}\frac{n(n+1)}{n^2} + \frac{9}{8n^3}n(n+1)(2n+1)\right]$

$= \frac{15}{4} + \frac{27}{4} + \frac{9}{4} = \frac{51}{4}$ square units

34. $A \approx \sum_{i=1}^{n} f\left(-1 + \frac{2i}{n}\right)\left(\frac{2}{n}\right)$

$= \sum_{i=1}^{n} \left[\left(-1 + \frac{2i}{n}\right)^2 - \left(-1 + \frac{2i}{n}\right)^3\right]\left(\frac{2}{n}\right)$

$= \sum_{i=1}^{n} \left[\left(1 - \frac{4i}{n} + \frac{4i^2}{n^2}\right) - \left(-1 + \frac{6i}{n} - \frac{12i^2}{n^2} + \frac{8i^3}{n^3}\right)\right]\frac{2}{n}$

$= \sum_{i=1}^{n} \left[2 - \frac{10i}{n} + \frac{16i^2}{n^2} - \frac{8i^3}{n^3}\right]\frac{2}{n}$

$= \frac{4}{n} \sum_{i=1}^{n} 1 - \frac{20}{n^2} \sum_{i=1}^{n} i + \frac{32}{n^3} \sum_{i=1}^{n} i^2 - \frac{16}{n^4} \sum_{i=1}^{n} i^3$

$= \frac{4}{n}(n) - \frac{20}{n^2} \frac{n(n + 1)}{2} + \frac{32}{n^3} \cdot \frac{n(n + 1)(2n + 1)}{6} - \frac{16}{n^4} \frac{n^2(n + 1)^2}{4}$

$A = \lim_{n \to \infty} \left[\frac{4}{n}(n) - \frac{20}{n^2} \frac{n(n + 1)}{2} + \dots\right]$

$= 4 - 10 + \frac{32}{3} - 4 = \frac{2}{3}$ square unit

35. $y = (-3.0 \cdot 10^{-6})x^3 + 0.002x^2 - 1.05x + 400$

Note that $y = 0$ when $x = 500$.

Area $\approx 105,208.33$ square feet ≈ 2.4153 acres

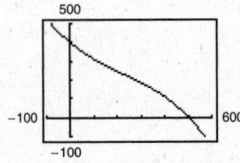

36. (a) $-4.089 \times 10^{-5}x^3 + 0.01615x^2 - 2.6716x + 452.9286$

(b)

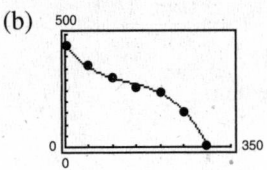

(c) Area $\approx 78,204$ square feet (Answers will vary.)

37. True. See Formula 2, page 892. **38.** False. n approaches infinity. **39.** Answers will vary.

40. Area is approximately a triangle of base 2 and height 3. Area ≈ 4. (c)

41. $\sin 2x - \sqrt{3} \sin x = 0$

$2 \sin x \cos x - \sqrt{3} \sin x = 0$

$\sin x(2 \cos x - \sqrt{3}) = 0$

$\sin x = 0 \Rightarrow x = n\pi$

$\cos x = \frac{\sqrt{3}}{2} \Rightarrow x = \frac{\pi}{6} + 2n\pi, x = \frac{11\pi}{6} + 2n\pi$

42.
$$\sin 2x + \sqrt{2}\cos x = 0$$
$$2\sin x \cos x + \sqrt{2}\cos x = 0$$
$$\cos x \left(2\sin x + \sqrt{2}\right) = 0$$

$$\cos x = 0 \implies x = \frac{\pi}{2} + n\pi$$

$$\sin x = \frac{-\sqrt{2}}{2} \implies x = \frac{5\pi}{4} + 2n\pi, \frac{7\pi}{4} + 2n\pi$$

43. $2\tan x = \tan 2x = \dfrac{2\tan x}{1 - \tan^2 x}$

$$\tan x = 0 \implies x = n\pi$$

44.
$$\cos 2x - 3\sin x = 2$$
$$1 - 2\sin^2 x - 3\sin x = 2$$
$$2\sin^2 x + 3\sin x + 1 = 0$$
$$(2\sin x + 1)(\sin x + 1) = 0$$

$$\sin x = \frac{-1}{2} \implies x = \frac{7\pi}{6} + 2n\pi, \frac{11\pi}{6} + 2n\pi$$

$$\sin x = -1 \implies x = \frac{3\pi}{2} + 2n\pi$$

45. $2\cot x = 5\cos \dfrac{\pi}{2} = 0$

$$\cot x = 0 \implies x = \frac{\pi}{2} + n\pi$$

46. $\sqrt{2}\sec x = 2\csc\left(\dfrac{\pi}{4}\right) = 2\sqrt{2}$

$$\sec x = 2$$

$$\cos x = \frac{1}{2} \implies x = \frac{\pi}{3} + 2n\pi, \frac{5\pi}{3} + 2n\pi$$

47.
$$(\mathbf{u} \cdot \mathbf{v})\mathbf{u} = (\langle 4, -5 \rangle \cdot \langle -1, -2 \rangle)\langle 4, -5 \rangle$$
$$= 6\langle 4, -5 \rangle$$
$$= \langle 24, -30 \rangle$$

48. $3\mathbf{u} \cdot \mathbf{v} = 3\langle 4, -5 \rangle \cdot \langle -1, -2 \rangle = 3(-4 + 10)$
$$= 18$$

49. $\|\mathbf{v}\| - 2 = \sqrt{5} - 2$

50.
$$\|\mathbf{u}\|^2 - \|\mathbf{v}\|^2 = (4^2 + (-5)^2) - ((-1)^2 + (-2)^2)$$
$$= (16 + 25) - (1 + 4)$$
$$= 36$$

Review Exercises for Chapter 12

1. $\displaystyle\lim_{x \to 3} (6x - 1)$

The limit (17) can be reached.

x	2.9	2.99	2.999	3	3.001	3.01	3.1
$f(x)$	16.4	16.94	16.994	17	17.006	17.06	17.6

2. $f(x) = \dfrac{x - 2}{3x^2 - 4x - 4}$

x	1.9	1.99	1.999	2	2.001	2.01	2.1
$f(x)$	0.1299	0.1255	0.1250	Undef.	0.1250	0.1245	0.1205

$\lim\limits_{x \to 2} f(x) = \dfrac{1}{8}$

The limit cannot be reached.

3. $\lim\limits_{x \to 1} (3 - x) = 2$

4. Limit does not exist.

5. $\lim\limits_{x \to 1} \dfrac{x^2 - 1}{x - 1} = 2$

6. $\lim\limits_{x \to -1} (2x^2 + 1) = 3$

7. (a) $\lim\limits_{x \to c} [f(x)]^3 = 4^3 = 64$

(b) $\lim\limits_{x \to c} [3f(x) - g(x)] = 3(4) - 5 = 7$

(c) $\lim\limits_{x \to c} [f(x)g(x)] = (4)(5) = 20$

(d) $\lim\limits_{x \to c} \dfrac{f(x)}{g(x)} = \dfrac{4}{5}$

8. (a) $\lim\limits_{x \to c} \sqrt[3]{f(x)} = \sqrt[3]{27} = 3$

(b) $\lim\limits_{x \to c} \dfrac{f(x)}{18} = \dfrac{27}{18} = \dfrac{3}{2}$

(c) $\lim\limits_{x \to c} [f(x)\, g(x)] = (27)(12) = 324$

(d) $\lim\limits_{x \to c} [f(x) - 2g(x)] = 27 - 2(12) = 3$

9. $\lim\limits_{x \to 4} \left(\dfrac{1}{2}x + 3 \right) = \dfrac{1}{2}(4) + 3 = 5$

10. $\lim\limits_{x \to -1} \sqrt{5 - x} = \sqrt{5 - (-1)}$
$= \sqrt{6}$

11. $\lim\limits_{x \to 2} \dfrac{x^2 - 1}{x^3 + 2} = \dfrac{2^2 - 1}{2^3 + 2} = \dfrac{3}{10}$

12. $\lim\limits_{x \to e} 7 = 7$

13. $\lim\limits_{x \to \pi} \sin 3x = \sin 3\pi = 0$

14. $\lim\limits_{x \to 0} \tan x = \tan 0 = 0$

15. $\lim\limits_{x \to 3}(5x - 4) = 5(3) - 4 = 11$

16. $\lim\limits_{x \to -2} (5 - 2x - x^2) = 5 - 2(-2) - (-2)^2 = 5$

17. $\lim\limits_{x \to 2}(5x - 3)(3x + 5) = (5(2) - 3)(3(2) + 5)$
$= (7)(11) = 77$

18. $\lim\limits_{x \to -3} (x^3 - 6x^2 + 3x - 1) = (-3)^3 - 6(-3)^2 + 3(-3) - 1 = -91$

19. $\lim\limits_{t \to 3} \dfrac{t^2 + 1}{t} = \dfrac{9 + 1}{3} = \dfrac{10}{3}$

20. $\lim\limits_{x \to 2} \dfrac{3x + 5}{5x - 3} = \dfrac{3(2) + 5}{5(2) - 3} = \dfrac{11}{7}$

21. $\lim\limits_{t \to -2} \dfrac{t + 2}{t^2 - 4} = \lim\limits_{t \to -2} \dfrac{t + 2}{(t + 2)(t - 2)}$
$= \lim\limits_{t \to -2} \dfrac{1}{t - 2} = -\dfrac{1}{4}$

22. $\lim\limits_{t \to 3} \dfrac{t^2 - 9}{t - 3} = \lim\limits_{t \to 3} \dfrac{(t - 3)(t + 3)}{t - 3} = \lim\limits_{t \to 3}(t + 3) = 6$

23. $\lim\limits_{x \to 5} \dfrac{x-5}{x^2 + 5x - 50} = \lim\limits_{x \to 5} \dfrac{x-5}{(x-5)(x+10)}$

$$= \lim\limits_{x \to 5} \dfrac{1}{x+10} = \dfrac{1}{15}$$

24. $\lim\limits_{x \to -1} \dfrac{x+1}{(x^2 - 5x - 6)} = \lim\limits_{x \to -1} \dfrac{(x+1)}{(x+1)(x-6)}$

$$= \lim\limits_{x \to -1} \dfrac{1}{x-6} = -\dfrac{1}{7}$$

25. $\lim\limits_{x \to -2} \dfrac{x^2 - 4}{x^3 + 8} = \lim\limits_{x \to -2} \dfrac{(x+2)(x-2)}{(x+2)(x^2 - 2x + 4)}$

$$= \lim\limits_{x \to -2} \dfrac{x-2}{x^2 - 2x + 4}$$

$$= \dfrac{-4}{12} = \dfrac{-1}{3}$$

26. $\lim\limits_{x \to 4} \dfrac{x^3 - 64}{x^2 - 16} = \lim\limits_{x \to 4} \dfrac{(x-4)(x^2 + 4x + 16)}{(x-4)(x+4)}$

$$= \lim\limits_{x \to 4} \dfrac{x^2 + 4x + 16}{x+4}$$

$$= \dfrac{16 + 16 + 16}{8} = 6$$

27. $\lim\limits_{x \to -1} \dfrac{1/(x+2) - 1}{x+1} = \lim\limits_{x \to -1} \dfrac{1 - (x+2)}{(x+2)(x+1)}$

$$= \lim\limits_{x \to -1} \dfrac{-(x+1)}{(x+2)(x+1)}$$

$$= \lim\limits_{x \to -1} \dfrac{-1}{(x+2)} = -1$$

28. $\lim\limits_{x \to 0} \dfrac{(1/(1+x) - 1)}{x} = \lim\limits_{x \to 0} \dfrac{1 - (1+x)}{x(1+x)}$

$$= \lim\limits_{x \to 0} \dfrac{-x}{x(1+x)} = -1$$

29. $\lim\limits_{u \to 0} \dfrac{\sqrt{4+u} - 2}{u} = \lim\limits_{u \to 0} \dfrac{\sqrt{4+u} - 2}{u} \cdot \dfrac{\sqrt{4+u} + 2}{\sqrt{4+u} + 2}$

$$= \lim\limits_{u \to 0} \dfrac{(4+u) - 4}{u\left(\sqrt{4+u} + 2\right)}$$

$$= \lim\limits_{u \to 0} \dfrac{1}{\sqrt{4+u} + 2} = \dfrac{1}{4}$$

30. $\lim\limits_{v \to 0} \dfrac{\sqrt{v+9} - 3}{v} = \lim\limits_{v \to 0} \dfrac{\sqrt{v+9} - 3}{v} \cdot \dfrac{\sqrt{v+9} + 3}{\sqrt{v+9} + 3}$

$$= \lim\limits_{v \to 0} \dfrac{(v+9) - 9}{v\left[\sqrt{v+9} + 3\right]}$$

$$= \lim\limits_{v \to 0} \dfrac{1}{\sqrt{v+9} + 3} = \dfrac{1}{6}$$

31. $\lim\limits_{x \to 5} \dfrac{\sqrt{x-1} - 2}{x-5} = \lim\limits_{x \to 5} \dfrac{\sqrt{x-1} - 2}{x-5} \cdot \dfrac{\sqrt{x-1} + 2}{\sqrt{x-1} + 2}$

$$= \lim\limits_{x \to 5} \dfrac{(x-1) - 4}{(x-5)\left(\sqrt{x-1} + 2\right)}$$

$$= \lim\limits_{x \to 5} \dfrac{1}{\sqrt{x-1} + 2} = \dfrac{1}{2+2} = \dfrac{1}{4}$$

32. $\lim\limits_{x \to 1} \dfrac{\sqrt{3} - \sqrt{x+2}}{1-x} = \lim\limits_{x \to 1} \dfrac{\sqrt{3} - \sqrt{x+2}}{1-x} \cdot \dfrac{\sqrt{3} + \sqrt{x+2}}{\sqrt{3} + \sqrt{x+2}}$

$$= \lim\limits_{x \to 1} \dfrac{3 - (x+2)}{(1-x)\left(\sqrt{3} + \sqrt{x+2}\right)}$$

$$= \lim\limits_{x \to 1} \dfrac{1}{\sqrt{3} + \sqrt{x+2}} = \dfrac{1}{2\sqrt{3}} = \dfrac{\sqrt{3}}{6}$$

33. (a)

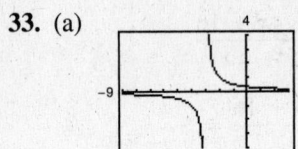

(b)

x	2.9	2.99	3	3.01	3.1
$f(x)$	0.1695	0.1669	Error	0.1664	0.1639

$\lim\limits_{x \to 3} \dfrac{x-3}{x^2-9} = \dfrac{1}{6}$

34. (a)

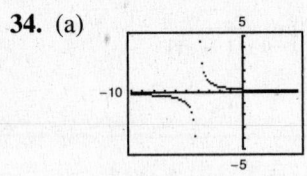

(b)

x	3.99	3.999	4	4.001	4.01
y_1	0.12516	0.12502	Error	0.12498	0.12484

$\lim\limits_{x \to 4} \dfrac{4-x}{16-x^2} = \dfrac{1}{8}$

35. (a)

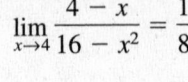

(b) (Answers will vary.)

x	-0.1	-0.01	-0.001	0	0.001	0.01	0.1
y_1	4.85 E 8	7.2 E 86	Error	Error	0	1 E -87	2.1 E -9

$\lim\limits_{x \to 0} e^{-2/x}$ does not exist.

36. (a)

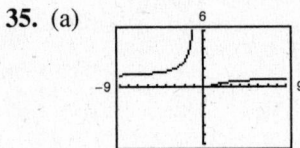

(b)

x	-0.01	-0.001	0	0.001	0.01
y_1	0	0	Error	0	0

$\lim\limits_{x \to 0} e^{-4/x^2} = 0$

37. (a)

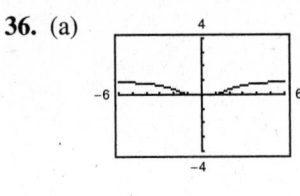

(b)

x	-0.1	-0.01	-0.001	0	0.001	0.01	0.1
y_1	1.9471	1.9995	1.999995	Error	1.999995	1.9995	1.9471

$\lim\limits_{x \to 0} \dfrac{\sin 4x}{2x} = 2$

38. (a)

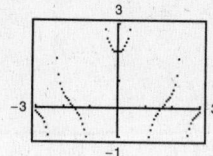

(b)

x	-0.01	-0.001	0	0.001	0.01
y_1	2.0003	2	Error	2	2.0003

$$\lim_{x \to 0} \frac{\tan 2x}{x} = 2$$

39. (a)

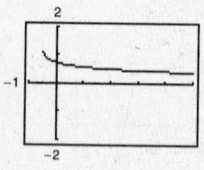

(b)

x	1.1	1.01	1.001	1.0001
$f(x)$	0.5680	0.5764	0.5773	0.5773

$$\lim_{x \to 1^+} \frac{\sqrt{2x + 1} - \sqrt{3}}{x - 1} \approx 0.577$$

$$\left(\text{Exact value: } \frac{\sqrt{3}}{3} \right)$$

40. (a)

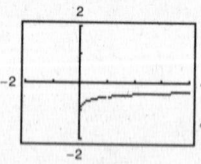

(b)

x	1.1	1.01	1.001	1.0001
$f(x)$	-0.4881	-0.4988	-0.4999	-0.5000

$$\lim_{x \to 1^+} \frac{1 - \sqrt{x}}{x - 1} = -\frac{1}{2}$$

41. $f(x) = \dfrac{|x - 3|}{x - 3}$

Limit does not exist because

$\lim_{x \to 3^+} f(x) = 1$ and

$\lim_{x \to 3^-} f(x) = -1.$

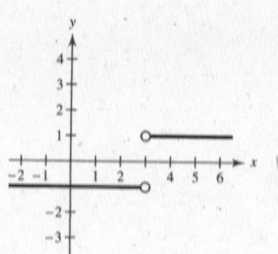

42. $\lim_{x \to 8^-} \dfrac{|8 - x|}{8 - x} = 1$

$\lim_{x \to 8^+} \dfrac{|8 - x|}{8 - x} = -1$

$\lim_{x \to 8} \dfrac{|8 - x|}{8 - x}$ does not exist.

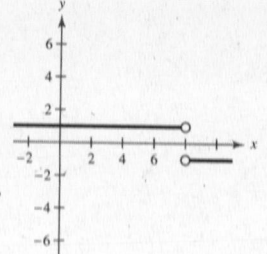

43. $f(x) = \dfrac{2}{x^2 - 4}$

Limit does not exist.

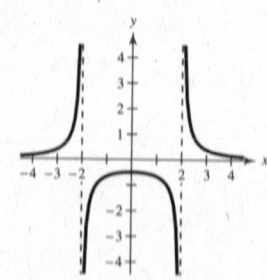

44. $\lim_{x \to -3} \dfrac{1}{x^2 + 9} = \dfrac{1}{(-3)^2 + 9}$

$$= \frac{1}{18}$$

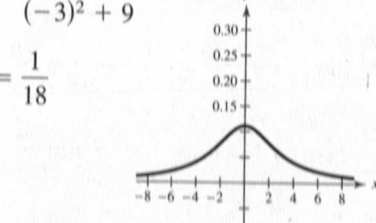

45. $\lim\limits_{x \to 5} \dfrac{|x-5|}{x-5}$ does not exist.

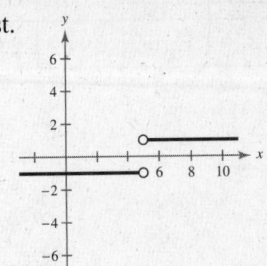

46. $\lim\limits_{x \to -2^-} \dfrac{|x+2|}{x+2} = -1$

$\lim\limits_{x \to -2^+} \dfrac{|x+2|}{x+2} = 1$

$\lim\limits_{x \to -2} \dfrac{|x+2|}{x+2}$ does not exist.

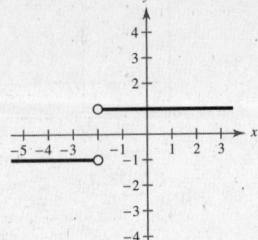

47. $\lim\limits_{x \to 2} f(x)$ does not exist.

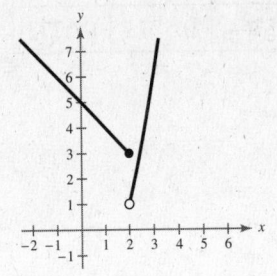

48. $\lim\limits_{x \to 0^-} f(x) = -4$

$\lim\limits_{x \to 0^+} f(x) = -6$

$\lim\limits_{x \to 0} f(x)$ does not exist.

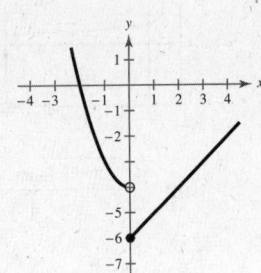

49. $f(x) = 4x + 3$

$$\lim_{h \to 0} \frac{f(x+h) - f(x)}{h} = \lim_{h \to 0} \frac{[4(x+h) + 3] - (4x + 3)}{h}$$

$$= \lim_{h \to 0} \frac{4x + 4h + 3 - 4x - 3}{h}$$

$$= \lim_{h \to 0} \frac{4h}{h}$$

$$= \lim_{h \to 0} 4 = 4$$

50. $f(x) = 11 - 2x$

$$\lim_{h \to 0} \frac{f(x+h) - f(x)}{h} = \lim_{h \to 0} \frac{11 - 2(x+h) - [11 - 2x]}{h}$$

$$= \lim_{h \to 0} \frac{11 - 2x - 2h - 11 + 2x}{h}$$

$$= \lim_{h \to 0} \frac{-2h}{h} = -2$$

51. $\lim\limits_{h \to 0} \dfrac{f(x+h) - f(x)}{h} = \lim\limits_{h \to 0} \dfrac{3(x+h) - (x+h)^2 - (3x - x^2)}{h}$

$$= \lim_{h \to 0} \frac{3x + 3h - x^2 - 2xh - h^2 - 3x + x^2}{h} = \lim_{h \to 0} \frac{3h - 2xh - h^2}{h}$$

$$= \lim_{h \to 0} (3 - 2x - h) = 3 - 2x$$

52. $\lim\limits_{h\to 0}\dfrac{f(x+h)-f(x)}{h} = \lim\limits_{h\to 0}\dfrac{[(x+h)^2 - 5(x+h) - 2] - [x^2 - 5x - 2]}{h}$

$$= \lim\limits_{h\to 0}\dfrac{2xh + h^2 - 5h}{h}$$

$$= \lim\limits_{h\to 0}\,[2x + h - 5] = 2x - 5$$

53. Slope ≈ 2

(Answers will vary.)

54. Slope $= 0$

55.

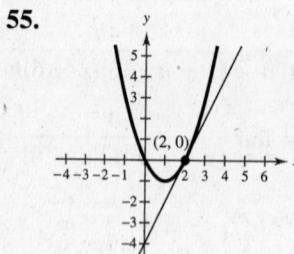

Slope at $(2, f(2))$ is approximately 2.

56.

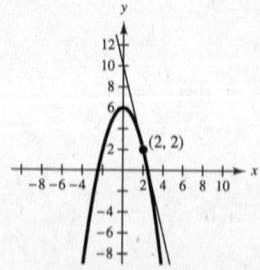

At $(2, f(2)) = (2, 2)$.

Slope $= -4$

57.

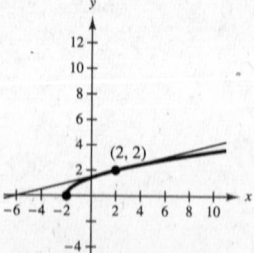

Slope is $\frac{1}{4}$ at $(2, 2)$.

58.

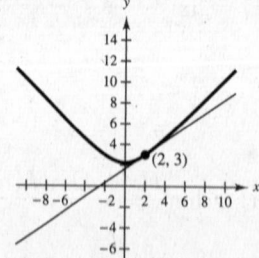

At $(2, f(2)) = (2, 3)$.

Slope $= \frac{2}{3}$

59. $f(x) = x^2 - 4x$

$m = \lim\limits_{h\to 0}\dfrac{f(x+h) - f(x)}{h}$

$= \lim\limits_{h\to 0}\dfrac{(x+h)^2 - 4(x+h) - (x^2 - 4x)}{h}$

$= \lim\limits_{h\to 0}\dfrac{x^2 + 2xh + h^2 - 4x - 4h - x^2 - 4x}{h}$

$= \lim\limits_{h\to 0}\dfrac{2xh + h^2 - 4h}{h}$

$= \lim\limits_{h\to 0}\,(2x + h - 4) = 2x - 4$

(a) At $(0, 0)$, $m = 2(0) - 4 = -4$.

(b) At $(5, 5)$, $m = 2(5) - 4 = 6$.

60. $f(x) = \dfrac{1}{4}x^4$

$m = \lim\limits_{h\to 0}\dfrac{f(x+h) - f(x)}{h}$

$= \lim\limits_{h\to 0}\dfrac{(1/4)(x+h)^4 - (1/4)x^4}{h}$

$= \lim\limits_{h\to 0}\dfrac{(1/4)[x^4 + 4x^3h + 6x^2h^2 + 4xh^3 + h^4 - x^4]}{h}$

$= \lim\limits_{h\to 0}\dfrac{1}{4}[4x^3 + 6x^2h + 4xh^2 + h^3] = x^3$

(a) At $(-2, 4)$, $m = (-2)^3 = -8$.

(b) At $\left(1, \dfrac{1}{4}\right)$, $m = (1)^3 = 1$.

61. $f(x) = \dfrac{4}{x - 6}$

$$m = \lim_{h \to 0} \frac{f(x + h) - f(x)}{h} = \lim_{h \to 0} \frac{\dfrac{4}{x + h - 6} - \dfrac{4}{x - 6}}{h}$$

$$= \lim_{h \to 0} \frac{4(x - 6) - 4(x + h - 6)}{(x + h - 6)(x - 6)h}$$

$$= \lim_{h \to 0} \frac{-4h}{(x + h - 6)(x - 6)h}$$

$$= \lim_{h \to 0} \frac{-4}{(x + h - 6)(x - 6)} = \frac{-4}{(x - 6)^2}$$

(a) At $(7, 4)$, $m = \dfrac{-4}{(7 - 6)^2} = -4$.

(b) At $(8, 2)$, $m = \dfrac{-4}{(8 - 6)^2} = -1$.

62. $f(x) = \sqrt{x}$

$$m = \lim_{h \to 0} \frac{f(x + h) - f(x)}{h}$$

$$= \lim_{h \to 0} \frac{\sqrt{x + h} - \sqrt{x}}{h} \cdot \frac{\sqrt{x + h} + \sqrt{x}}{\sqrt{x + h} + \sqrt{x}}$$

$$= \lim_{h \to 0} \frac{(x + h) - x}{h\left[\sqrt{x + h} + \sqrt{x}\right]}$$

$$= \lim_{h \to 0} \frac{1}{\sqrt{x + h} + \sqrt{x}} = \frac{1}{2\sqrt{x}}$$

(a) At $(1, 1)$, $m = \dfrac{1}{2\sqrt{1}} = \dfrac{1}{2}$.

(b) At $(4, 2)$, $m = \dfrac{1}{2\sqrt{4}} = \dfrac{1}{4}$.

63. $f'(x) = \lim\limits_{h \to 0} \dfrac{f(x + h) - f(x)}{h} = \lim\limits_{h \to 0} \dfrac{5 - 5}{h} = 0$

64. $g'(x) = \lim\limits_{h \to 0} \dfrac{g(x + h) - g(x)}{h}$

$$= \lim_{h \to 0} \frac{-3 - (-3)}{h} = \lim_{h \to 0} \frac{0}{h} = 0$$

65. $h'(x) = \lim\limits_{k \to 0} \dfrac{h(x + k) - h(x)}{k}$

$$= \lim_{k \to 0} \frac{\left[5 - \frac{1}{2}(x + k)\right] - \left[5 - \frac{1}{2}x\right]}{k}$$

$$= \lim_{k \to 0} \frac{-\frac{1}{2}k}{k} = -\frac{1}{2}$$

66. $f'(x) = \lim\limits_{h \to 0} \dfrac{f(x + h) - f(x)}{h}$

$$= \lim_{h \to 0} \frac{3(x + h) - 3x}{h} = 3$$

67. $g'(x) = \lim\limits_{h \to 0} \dfrac{g(x + h) - g(x)}{h}$

$$= \lim_{h \to 0} \frac{2(x + h)^2 - 1 - (2x^2 - 1)}{h}$$

$$= \lim_{h \to 0} \frac{2x^2 + 4xh + 2h^2 - 2x^2}{h}$$

$$= \lim_{h \to 0} (4x + 2h)$$

$$= 4x$$

68. $f'(x) = \lim\limits_{h \to 0} \dfrac{f(x + h) - f(x)}{h} = \lim\limits_{h \to 0} \dfrac{-(x + h)^3 + 4(x + h) - (-x^3 + 4x)}{h}$

$$= \lim_{h \to 0} \frac{-x^3 - 3x^2h - 3xh^2 - h^3 + 4x + 4h + x^3 - 4x}{h}$$

$$= \lim_{h \to 0} \frac{-3x^2h - 3xh^2 - h^3 + 4h}{h}$$

$$= \lim_{h \to 0} (-3x^2 - 3xh - h^2 + 4)$$

$$= -3x^2 + 4$$

69. $f'(t) = \lim_{h \to 0} \dfrac{f(t + h) - f(t)}{h}$

$= \lim_{h \to 0} \dfrac{\sqrt{t + h + 5} - \sqrt{t + 5}}{h} \cdot \dfrac{\sqrt{t + h + 5} + \sqrt{t + 5}}{\sqrt{t + h + 5} + \sqrt{t + 5}}$

$= \lim_{h \to 0} \dfrac{(t + h + 5) - (t + 5)}{h\left(\sqrt{t + h + 5} + \sqrt{t + 5}\right)}$

$= \lim_{h \to 0} \dfrac{1}{\sqrt{t + h + 5} + \sqrt{t + 5}}$

$= \dfrac{1}{2\sqrt{t + 5}}$

70. $f'(x) = \lim_{h \to 0} \dfrac{f(x + h) - f(x)}{h} = \lim_{h \to 0} \dfrac{\dfrac{1}{\sqrt{12 - x - h}} - \dfrac{1}{\sqrt{12 - x}}}{h}$

$= \lim_{h \to 0} \dfrac{\sqrt{12 - x} - \sqrt{12 - x - h}}{h\left[\sqrt{12 - x - h}\sqrt{12 - x}\right]} \cdot \dfrac{\sqrt{12 - x} + \sqrt{12 - x - h}}{\sqrt{12 - x} + \sqrt{12 - x - h}}$

$= \lim_{h \to 0} \dfrac{(12 - x) - (12 - x - h)}{h\left[\sqrt{12 - x - h}\sqrt{12 - x}\right]} \cdot \dfrac{1}{\left[\sqrt{12 - x} + \sqrt{12 - x - h}\right]}$

$= \lim_{h \to 0} \dfrac{1}{\left[\sqrt{12 - x - h}\sqrt{12 - x}\right]\left[\sqrt{12 - x} + \sqrt{12 - x - h}\right]}$

$= \dfrac{1}{(12 - x)2\sqrt{12 - x}}$

$= \dfrac{1}{2(12 - x)^{3/2}}$

71. $g'(s) = \dfrac{g(s + h) - g(s)}{h} = \lim_{h \to 0} \dfrac{\dfrac{4}{s + h + 5} - \dfrac{4}{s + 5}}{h}$

$= \lim_{h \to 0} \dfrac{4s + 20 - 4s - 4h - 20}{(s + h + 5)(s + 5)h}$

$= \lim_{h \to 0} \dfrac{-4h}{(s + h + 5)(s + 5)h}$

$= \lim_{h \to 0} \dfrac{-4}{(s + h + 5)(s + 5)} = \dfrac{-4}{(s + 5)^2}$

72. $g'(t) = \lim_{h \to 0} \dfrac{g(t + h) - g(t)}{h} = \lim_{h \to 0} \dfrac{\dfrac{6}{5 - (t + h)} - \dfrac{6}{5 - t}}{h}$

$= \lim_{h \to 0} \dfrac{30 - 6t - 30 + 6t + 6h}{h(5 - t - h)(5 - t)}$

$= \lim_{h \to 0} \dfrac{6}{(5 - t - h)(5 - t)} = \dfrac{6}{(5 - t)^2}$

73. $f(x) = 2x^2 - 1$, $(0, -1)$

(a) $m_{\tan} = \lim\limits_{h \to 0} \dfrac{f(x+h) - f(x)}{h}$

$= \lim\limits_{h \to 0} \dfrac{f(0+h) - f(0)}{h}$

$= \lim\limits_{h \to 0} \dfrac{2h^2 - 1 - (-1)}{h}$

$= \lim\limits_{h \to 0} 2h = 0$

(b) Tangent line: $\quad y - (-1) = 0(x - 0)$

$y + 1 = 0$

$y = -1$

(c)

74. $f(x) = x^2 + 10$, $(2, 14)$

(a) $f'(x) = \lim\limits_{h \to 0} \dfrac{(x+h)^2 + 10 - [x^2 + 10]}{h}$

$= \lim\limits_{h \to 0} \dfrac{x^2 + 2hx + h^2 + 10 - x^2 - 10}{h}$

$= \lim\limits_{h \to 0} \dfrac{2hx + h^2}{h}$

$= \lim\limits_{h \to 0}(2x + h) = 2x$

$f'(2) = 2 \cdot (2) = 4$

At $(2, 14)$, the slope of the tangent line is $m = f'(2) = 4$.

(b) Point: $(2, 14)$; slope: 4

$y - 14 = 4(x - 2)$

$y = 4x - 8 + 14$

$y = 4x + 6$, tangent line at $(2, 14)$

(c)

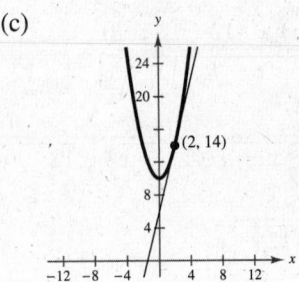

75. $\lim\limits_{x \to \infty} \dfrac{4x}{2x - 3} = \dfrac{4}{2} = 2$

76. $\lim\limits_{x \to \infty} \dfrac{7x}{14x + 2} = \dfrac{7}{14} = \dfrac{1}{2}$

77. $\lim\limits_{x \to -\infty} \dfrac{2x}{x^2 - 25} = 0$

78. $\lim\limits_{x \to -\infty} \dfrac{6x}{(x - 2)^3} = 0$

79. $\lim\limits_{x \to \infty} \dfrac{x^2}{2x + 3}$ does not exist.

80. $\lim\limits_{y \to \infty} \dfrac{3y^4}{y^2 + 1}$ does not exist.

81. $\lim\limits_{x \to \infty} \left[\dfrac{x}{(x - 2)^2} + 3\right] = 0 + 3 = 3$

82. $\lim\limits_{x \to \infty} \left[2 - \dfrac{2x^2}{(x + 1)^2}\right] = 2 - 2 = 0$

83. $\dfrac{2}{3}, \dfrac{5}{5} = 1, \dfrac{8}{7}, \dfrac{11}{9}, \dfrac{14}{11}$

$\lim\limits_{n \to \infty} a_n = \dfrac{3}{2}$

84. $\dfrac{1}{2}, \dfrac{2}{5}, \dfrac{3}{10}, \dfrac{4}{17}, \dfrac{5}{26}$

$\lim\limits_{n \to \infty} a_n = 0$

85. $a_n = \dfrac{1}{2n^2}[3 - 2n(n + 1)] = \dfrac{3}{2n^2} - \dfrac{n + 1}{n}$

$-0.5, -1.125, -1.1\overline{66}, -1.15625, -1.14$

$\lim\limits_{n \to \infty} a_n = 0 - 1 = -1$

86. $a_n = 2 + \dfrac{2}{n}(n - 1) - \dfrac{4}{n} = 4 - \dfrac{6}{n} : -2, 1, 2, \dfrac{5}{2}, \dfrac{14}{5}$

$\lim\limits_{n \to \infty} a_n = \lim\limits_{n \to \infty}\left(4 - \dfrac{6}{n}\right) = 4$

87. (a) $\displaystyle\sum_{i=1}^{n}\left(\frac{4i^2}{n^2}-\frac{i}{n}\right)\frac{1}{n}=\frac{4}{n^3}\sum_{i=1}^{n}i^2-\frac{1}{n^2}\sum_{i=1}^{n}i$

$$=\frac{4}{n^3}\frac{n(n+1)(2n+1)}{6}-\frac{1}{n^2}\frac{n(n+1)}{2}$$

$$=\frac{4n(n+1)(2n+1)-3n^2(n+1)}{6n^3}$$

$$=\frac{n(n+1)(8n+4-3n)}{6n^3}$$

$$=\frac{(n+1)(5n+4)}{6n^2}$$

(b)

n	10^0	10^1	10^2	10^3	10^4
$S(n)$	3	0.99	0.8484	0.8348	0.8335

(c) $\displaystyle\lim_{n\to\infty}S(n)=\frac{5}{6}$

88. $\displaystyle\sum_{i=1}^{n}\left[4-\left(\frac{3i}{n}\right)^2\right]\left(\frac{3i}{n^2}\right)=\frac{12}{n^2}\sum_{i=1}^{n}i-\frac{27}{n^4}\sum_{i=1}^{n}i^3=\frac{12}{n^2}\frac{n(n+1)}{2}-\frac{27}{n^4}\frac{n^2(n+1)^2}{4}$

$$=\frac{24n^2+24n-27(n^2+2n+1)}{4n^2}$$

$$=\frac{-3n^2-30n-27}{4n^2}=\frac{-3}{4n^2}(n^2+10n+9)$$

$$=\frac{-3(n+1)(n+9)}{4n^2}$$

n	10^0	10^1	10^2	10^3	10^4
$S(n)$	-15	-1.5675	-0.8257	-0.7575	-0.7508

$\displaystyle\lim_{n\to\infty}S(n)=-\frac{3}{4}$

89. Area $\approx\frac{1}{2}\left[\frac{7}{2}+3+\frac{5}{2}+2+\frac{3}{2}+1\right]=\frac{1}{2}\frac{27}{2}=\frac{27}{4}=6.75$

90. Width of rectangle: $\frac{1}{4}$; Height is f evaluated at right endpoint.

Area $\approx\frac{1}{4}\left[f\left(\frac{1}{4}\right)+f\left(\frac{1}{2}\right)+f\left(\frac{3}{4}\right)+f(1)\right]$

$=\frac{1}{4}\left[4-\left(\frac{1}{4}\right)^2+4-\left(\frac{1}{2}\right)^2+4-\left(\frac{3}{4}\right)^2+4-1\right]$

$=\frac{1}{4}\left[15-\frac{14}{16}\right]=\frac{113}{32}=3.53125$

91. $f(x)=\frac{1}{4}x^2,\ b-a=4-0=4$

$A\approx\displaystyle\sum_{i=1}^{n}f\left(\frac{4i}{n}\right)\left(\frac{4}{n}\right)$

$=\displaystyle\sum_{i=1}^{n}\frac{1}{4}\left(\frac{4i}{n}\right)^2\left(\frac{4}{n}\right)$

$=\frac{1}{n}\displaystyle\sum_{i=1}^{n}\frac{16}{n^2}i^2$

$=\frac{16}{n^3}\frac{n(n+1)(2n+1)}{6}$

$=\frac{8(n+1)(2n+1)}{3n^2}$

n	4	8	20	50
Approximate area	7.5	6.375	5.74	5.4944

(Exact area is $\frac{16}{3}\approx5.33$.)

92. $f(x) = 4x - x^2$

n	4	8	20	50
Approximate area	10	10.5	10.64	10.6624

$\left(\text{Exact area is } 10\frac{2}{3}.\right)$

93. $A = \lim\limits_{n\to\infty} \sum\limits_{i=1}^{n} \left(10 - \dfrac{10i}{n}\right)\left(\dfrac{10}{n}\right)$

$= \lim\limits_{n\to\infty} \left[\dfrac{100}{n} \sum\limits_{i=1}^{n} 1 - \dfrac{100}{n^2} \sum\limits_{i=1}^{n} i\right]$

$= \lim\limits_{n\to\infty} \left[\dfrac{100}{n}(n) - \dfrac{100}{n^2}\left(\dfrac{n(n+1)}{2}\right)\right]$

$= \lim\limits_{n\to\infty} \left[100 - 50\dfrac{n(n+1)}{n^2}\right]$

$= 100 - 50 = 50, \text{ exact area}$

94. $A = \lim\limits_{n\to\infty} \sum\limits_{i=1}^{n} \left[2\left(3 + \dfrac{3i}{n}\right) - 6\right]\left(\dfrac{3}{n}\right)$

$= \lim\limits_{n\to\infty} \sum\limits_{i=1}^{n} \dfrac{18i}{n^2} = \lim\limits_{n\to\infty} \dfrac{18}{n^2} \sum\limits_{i=1}^{n} i$

$= \lim\limits_{n\to\infty} \dfrac{18}{n^2} \dfrac{n(n+1)}{2} = 9, \text{ exact area}$

95. $A = \lim\limits_{n\to\infty} \sum\limits_{i=1}^{n} \left[\left(-1 + \dfrac{3i}{n}\right)^2 + 4\right]\left(\dfrac{3}{n}\right)$

$= \lim\limits_{n\to\infty} \sum\limits_{i=1}^{n} \left[5 - \dfrac{6i}{n} + \dfrac{9i^2}{n^2}\right]\dfrac{3}{n}$

$= \lim\limits_{n\to\infty} \left[\dfrac{15}{n} \sum\limits_{i=1}^{n} 1 - \dfrac{18}{n^2} \sum\limits_{i=1}^{n} i + \dfrac{27}{n^3} \sum\limits_{i=1}^{n} i^2\right]$

$= \lim\limits_{n\to\infty} \left[\dfrac{15}{n}(n) - \dfrac{18}{n^2} \dfrac{n(n+1)}{2} + \dfrac{27}{n^3} \dfrac{n(n+1)(2n+1)}{6}\right]$

$= 15 - 9 + 9 = 15, \text{ exact area}$

96. $A = \lim\limits_{n\to\infty} \sum\limits_{i=1}^{n} 8\left(\left(\dfrac{i}{n}\right) - \left(\dfrac{i}{n}\right)^2\right)\dfrac{1}{n}$

$= \lim\limits_{n\to\infty} \left[\dfrac{8}{n^2} \sum\limits_{i=1}^{n} i - \dfrac{8}{n^3} \sum i^2\right]$

$= \lim\limits_{n\to\infty} \left[\dfrac{8}{n^2} \dfrac{n(n+1)}{2} - \dfrac{8}{n^3} \dfrac{n(n+1)(2n+1)}{6}\right]$

$= 4 - \dfrac{8}{3} = \dfrac{4}{3}, \text{ exact area}$

97. $A = \lim\limits_{n\to\infty} \sum\limits_{i=1}^{n} 2\left[\left(-1 + \dfrac{2i}{n}\right)^2 - \left(-1 + \dfrac{2i}{n}\right)^3\right]\left(\dfrac{2}{n}\right)$

$= \lim\limits_{n\to\infty} \sum\limits_{i=1}^{n} \dfrac{4}{n}\left(1 - \dfrac{4i}{n} + \dfrac{4i^2}{n^2} - \left(-1 + \dfrac{6i}{n} - \dfrac{12i^2}{n^2} + \dfrac{8i^3}{n^3}\right)\right)$

$= \lim\limits_{n\to\infty} \sum\limits_{i=1}^{n} \dfrac{4}{n}\left(2 - \dfrac{10i}{n} + \dfrac{16i^2}{n^2} - \dfrac{8i^3}{n^3}\right)$

$= \lim\limits_{n\to\infty} \left[\dfrac{8}{n} \sum\limits_{i=1}^{n} 1 - \dfrac{40}{n^2} \sum\limits_{i=1}^{n} i + \dfrac{64}{n^3} \sum\limits_{i=1}^{n} i^2 - \dfrac{32}{n^4} \sum\limits_{i=1}^{n} i^3\right]$

$= \lim\limits_{n\to\infty} \left[\dfrac{8}{n}(n) - \dfrac{40}{n^2} \dfrac{n(n+1)}{2} + \dfrac{64}{n^3} \dfrac{n(n+1)(2n+1)}{6} - \dfrac{32}{n^4} \dfrac{n^2(n+1)^2}{4}\right] = 8 - 20 + \dfrac{64}{3} - 8 = \dfrac{4}{3}, \text{ exact area}$

98. $A = \lim\limits_{n\to\infty} \sum\limits_{i=1}^{n} \left[4 - \left(\frac{4i}{n} - 2 \right)^2 \right]\left(\frac{4}{n} \right)$

$\quad = \lim\limits_{n\to\infty} \sum\limits_{i=1}^{n} \left[\frac{16i}{n} - \frac{16i^2}{n^2} \right]\left(\frac{4}{n} \right)$

$\quad = \lim\limits_{n\to\infty} \left[\frac{16n(n+1)}{2n} - \frac{16n(n+1)(2n+1)}{6n^2} \right]\left(\frac{4}{n} \right)$

$\quad = \lim\limits_{n\to\infty} \left[8\frac{n+1}{n} - \frac{8(n+1)(2n+1)}{3n^2} \right]4 = 32 - \frac{64}{3} = 10\frac{2}{3}$

99. (a) $y = (-3.376068 \times 10^{-7})x^3 + (3.7529 \times 10^{-4})x^2 - 0.17x + 132$

    (c) Area $\approx 88{,}000$ square feet, answers will vary.

(b)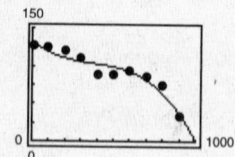

100. True (assuming all the limits exist)

101. False. The limit does not exist.

102. Answers will vary.

Problem Solving for Chapter 12

1. (a) $\lim\limits_{x\to 2} f(x) = 3; g_1, g_4$
    (b) $\lim\limits_{x\to 2^-} f(x) = 3; g_1, g_3, g_4$
    (c) $\lim\limits_{x\to 2^+} f(x) = 3; g_1, g_4$

2. $f(x) = [\![x]\!] + [\![-x]\!]$

    x not an integer: $[\![-x]\!] = -[\![x]\!] - 1$

                         $[\![x]\!] + [\![-x]\!] = -1$

    $x = n, n$ an integer: $[\![x]\!] = [\![n]\!] = n$

                         $[\![-x]\!] = [\![-n]\!] = -n$

    So, $[\![x]\!] + [\![-x]\!] = n - n = 0$

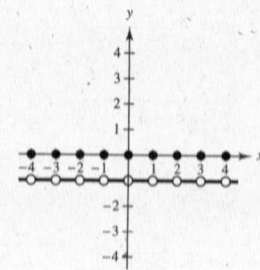

    (a) $f(x) = [\![x]\!] + [\![-x]\!] = \begin{cases} -1, & x \text{ not an integer} \\ 0, & x \text{ an integer} \end{cases}$

       $f(1) = 0, f(0) = 0, f\left(\frac{1}{2}\right) = -1, f(-2.7) = -1$

    (b) $\lim\limits_{x\to 1^-} f(x) = -1$

       $\lim\limits_{x\to 1^+} f(x) = -1 \;\Longrightarrow\; \lim\limits_{x\to 1} f(x) = -1$

       $\lim\limits_{x\to (1/2)^-} = f(x) = -1$

       $\lim\limits_{x\to (1/2)^+} = f(x) = -1 \;\Longrightarrow\; \lim\limits_{x\to (1/2)} f(x) = -1$

3.

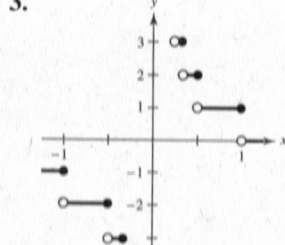

    (a) $f\left(\frac{1}{4}\right) = [\![4]\!] = 4$
    (b) $\lim\limits_{x\to 1^-} f(x) = 1$

       $f(3) = [\![\frac{1}{3}]\!] = 0$
       $\lim\limits_{x\to 1^+} f(x) = 0$

       $f(1) = [\![1]\!] = 1$
       $\lim\limits_{x\to (1/2)^-} f(x) = 2$

                     $\lim\limits_{x\to (1/2)^+} f(x) = 1$

4. (a) $P = (3, 4), 0 = (0, 0)$

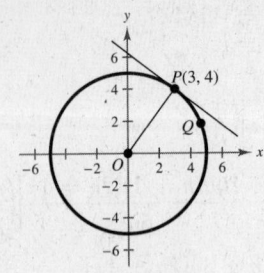

$$m = \frac{0 - 4}{0 - 3} = \frac{4}{3}$$

(b) The tangent line will be perpendicular to $\overline{PO}$.

$$y - 4 = -\frac{3}{4}(x - 3)$$

$$y = -\frac{3}{4}x + \frac{9}{4} + 4$$

$$y = -\frac{3}{4}x + \frac{25}{4}$$

(c) In the first quadrant, $y = \sqrt{25 - x^2} \implies Q = \left(x, \sqrt{25 - x^2}\right)$.

Slope of chord $\overline{PQ} = m_x = \dfrac{\sqrt{25 - x^2} - 4}{x - 3}$

(d) $\displaystyle\lim_{x \to 3} \frac{\sqrt{25 - x^2} - 4}{x - 3} = \lim_{x \to 3} \frac{\left(\sqrt{25 - x^2} - 4\right)\left(\sqrt{25 - x^2} + 4\right)}{(x - 3)\left(\sqrt{25 - x^2} + 4\right)}$

$$= \lim_{x \to 3} \frac{25 - x^2 - 16}{(x - 3)\left(\sqrt{25 + x^2} + 4\right)}$$

$$= \lim_{x \to 3} \frac{-(x^2 - 9)}{(x - 3)\left(\sqrt{25 + x^2} + 4\right)}$$

$$= \lim_{x \to 3} \frac{-(x - 3)(x + 3)}{(x - 3)\left(\sqrt{25 - x^2} + 4\right)}$$

$$= \lim_{x \to 3} \frac{-(x + 3)}{\sqrt{25 - x^2} + 4}$$

$$= \frac{-6}{\sqrt{16} + 4} = \frac{-6}{8} = -\frac{3}{4}$$

The slopes are the same.

5. Since $\displaystyle\lim_{x \to 0} \frac{\sqrt{a + bx} - \sqrt{3}}{x}$ exists $\implies \displaystyle\lim_{x \to 0}\left(\sqrt{a + bx} - \sqrt{3}\right) = 0 \implies \sqrt{a} - \sqrt{3} = 0 \implies a = 3.$

$$\lim_{x \to 0}\left(\frac{\sqrt{3 + bx} - \sqrt{3}}{x} \cdot \frac{\sqrt{3 + bx} + \sqrt{3}}{\sqrt{3 + bx} + \sqrt{3}}\right) = \lim_{x \to 0} \frac{3 + bx - 3}{x\left(\sqrt{3 + bx} + \sqrt{3}\right)}$$

$$= \lim_{x \to 0} \frac{bx}{x\left(\sqrt{3 + bx} + \sqrt{3}\right)}$$

$$= \lim_{x \to 0} \frac{b}{\sqrt{3 + bx} + \sqrt{3}} = \frac{b}{2\sqrt{3}} = \sqrt{3} \implies b = 6$$

Thus, $a = 3$ and $b = 6$.

6. (a) $f(x) = \dfrac{\sqrt{3 + x^{1/3}} - 2}{x - 1}$

(b)

Numerator: $3 + x^{1/3} \geq 0$

$$x^{1/3} \geq -3$$

$$x \geq (-3)^3$$

$$x \geq -27$$

Denominator: $x - 1 \neq 0$

$$x \neq 1$$

Domain of $f(x)$: $[-27, 1) \cup (1, \infty)$

(c) $\displaystyle\lim_{n \to -27^+} f(x) = \dfrac{\sqrt{3 + (-27)^{1/3}} - 2}{-27 - 1} = \dfrac{0 - 2}{-28} = \dfrac{1}{14}$

(d) Let $x^{1/3} = u$ or $u^3 = x$. As $x \to 1, u \to 1$.

$$f(u) = \frac{\sqrt{3 + u} - 2}{u^3 - 1}$$

$$\lim_{u \to 1} f(u) = \lim_{u \to 1} \frac{\sqrt{3 + u} - 2}{u^3 - 1}$$

$$= \lim_{u \to 1} \frac{(\sqrt{3 + u} - 2)(\sqrt{3 + u} + 2)}{(u^3 - 1)(\sqrt{3 + u} + 2)}$$

$$= \lim_{u \to 1} \frac{3 + u - 4}{(u - 1)(u^2 + u + 1)(\sqrt{3 + u} + 2)}$$

$$= \lim_{u \to 1} \frac{u - 1}{(u - 1)(u^2 + u + 1)(\sqrt{3 + u} + 2)}$$

$$= \lim_{u \to 1} \frac{1}{(u^2 + u + 1)(\sqrt{3 + u} + 2)}$$

$$= \frac{1}{(1^2 + 1 + 1)(\sqrt{3 + 1} + 2)}$$

$$= \frac{1}{(3)(4)} = \frac{1}{12}$$

7. $f(x) = \begin{cases} 0, & \text{if } x \text{ is rational} \\ 1, & \text{if } x \text{ is irrational} \end{cases}$

$\displaystyle\lim_{x \to 0} f(x)$ does not exist.

No matter how close to 0 x is, there are still an infinite number of rational and irrational numbers

$g(x) = \begin{cases} 0, & \text{if } x \text{ is rational} \\ x, & \text{if } x \text{ is irrational} \end{cases}$

$\displaystyle\lim_{x \to 0} g(x) = 0$

When x is close to 0, both parts of the function are close to zero.

8. Let (x_1, y_1) represent the coordinates of the point on the parabola $y = f(x) = x^2$.

Let (x_2, y_2) represent the coordinates of the point on the parabola $y = g(x) = -x^2 + 2x - 5$.

Then, the tangent lines for each parabola are:

$y = f'(x_1)(x - x_1) + y_1$

$y = f'(x_1)x + y_1 - f'(x_1)x_1 \Leftarrow y_1 = x_1^2$

$y = f'(x_1)x + x_1^2 - f'(x_1)x_1$

$y = g'(x_2)(x - x_2) + y_2$

$y = g'(x_2)x + y_2 - g'(x_2)x_2 \Leftarrow y_2 = -x_2^2 + 2x_2 - 5$

$y = g'(x_2)x - x_2^2 + 2x_2 - 5 - g'(x_2)x_2$

These two tangent lines are coincident (that is, they are the same lines) if they have the same slope and y-intercept:

y-intercepts: $x_1^2 - f'(x_1)x_1 = -x_2^2 + 2x_2 - 5 - g'(x_2)x_2$ (A)

Slopes: $f'(x_1) = g'(x_2)$ (B)

Find the derivatives $f'(x)$ and $g'(x)$:

$f'(x) = \lim_{h \to 0} \dfrac{(x + h)^2 - x^2}{h}$

$= \lim_{h \to 0} \dfrac{x^2 + 2hx + h^2 - x^2}{h}$

$= \lim_{h \to 0} 2x + h = 2x \Rightarrow f'(x_1) = 2x_1$ (C)

$g'(x) = \lim_{h \to 0} \dfrac{-(x + h)^2 + 2(x + h) - 5 - (-x^2 + 2x - 5)}{h}$

$= \lim_{h \to 0} \dfrac{-2hx + 2h - h^2}{h}$

$= \lim_{h \to 0} (-2x + 2 - h)$

$= -2x + 2 \Rightarrow g'(x_2) = -2x_2 + 2$ (D)

Substitute (C) and (D) into (B):

$2x_1 = -2x_2 + 2$

$x_1 = -x_2 + 1$

Substitute this result along with (C) and (D) back into (A):

$(-x_2 + 1)^2 - 2(-x_2 + 1)^2 = -x_2^2 + 2x_2 - 5 - (-2x_2 + 2)x_2$

$x_2^2 - 2x_2 + 1 - 2x_2^2 + 4x_2 - 2 = -x_2^2 + 2x_2 - 5 + 2x_2^2 - 2x_2$

$0 = 2x_2^2 - 2x_2 - 4$

$0 = x_2^2 - x_2 - 2 = (x_2 - 2)(x_2 + 1) \Rightarrow$

$x_2 = 2 \Rightarrow x_1 = -1$ or $x_2 = -1 \Rightarrow x_1 = 2$

The tangent lines are:

$y = 2(-1)x + (-1)^2 - (2)(-1)(-1)$

$y = -2x - 1$

$y = 2(2)x + (2)^2 - (2)(2)(2)$

$y = 4x - 4$

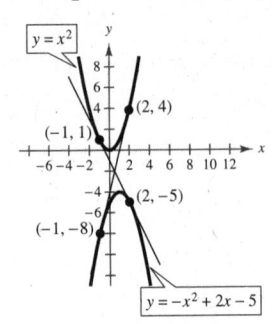

9. $f(x) = a + b\sqrt{x}$

$f(1) = 4 \implies a + b = 4$

Tangent line: $2y - 3x = 5 \implies y = \dfrac{3}{2}x + \dfrac{5}{2} \implies m_{\tan} = \dfrac{3}{2}$ at $(1, 4)$

$$m_{\tan} = \lim_{h \to 0} \frac{f(x + h) - f(x)}{h} = \lim_{h \to 0} \frac{\left(a + b\sqrt{x + h}\right) - \left(a + b\sqrt{x}\right)}{h}$$

$$= \lim_{h \to 0} \left(\frac{b\left(\sqrt{x + h} - \sqrt{x}\right)}{h} \cdot \frac{\sqrt{x + h} + \sqrt{x}}{\sqrt{x + h} + \sqrt{x}} \right)$$

$$= \lim_{h \to 0} \frac{b(x + h - x)}{h\left(\sqrt{x + h} + \sqrt{x}\right)} = \lim_{h \to 0} \frac{b}{\sqrt{x + h} + \sqrt{x}} = \frac{b}{2\sqrt{x}}$$

At $(1, 4)$, $m_{\tan} = \dfrac{b}{2} = \dfrac{3}{2} \implies b = 3$ and $a = 1$.

Thus, $f(x) = 1 + 3\sqrt{x}$.

10. **(a)** $y = f(x) = x^2$

$$f'(x) = \lim_{h \to 0} \frac{(x + h)^2 - x^2}{h} = \lim_{h \to 0} \frac{x^2 + 2xh + h^2 - x^2}{h}$$

$$= \lim_{h \to 0} (2x + h) = 2x$$

At $(2, 4)$, the slope of the tangent line is $f'(2) = 2(2) = 4$.

$y - 4 = 4(x - 2)$

$\quad y = 4x - 4$

(b) The slope of the normal line is $-\dfrac{1}{4}$.

$y - 4 = -\dfrac{1}{4}(x - 2)$

$\quad y = -\dfrac{1}{4}x + \dfrac{9}{2}$, normal line at $(2, 4)$

Substitute $y = x^2$:

$$x^2 = -\frac{1}{4}x + \frac{9}{2}$$

$4x^2 + x - 18 = 0$

$(x - 2)(4x + 9) = 0 \implies x = 2, x = -\dfrac{9}{4}$

$$y = f\left(-\frac{9}{4}\right) = \frac{81}{16}$$

The normal line intersects the parabola at $\left(-\dfrac{9}{4}, \dfrac{81}{16}\right)$.

(c) $\quad y = f(x) = x^2$

$f'(x) = 2x$

$f'(0) = 0$, slope of tangent line

The tangent line is the horizontal line through $(0, 0)$, $y = 0$. The normal line is the vertical line through $(0, 0)$, $x = 0$.

11. Slope m through $(0, 4) \implies y = mx + 4$ or $mx - y + 4 = 0$

(a) $d(m) = \dfrac{|3m + (-1)(1) + 4|}{\sqrt{m^2 + (-1)^2}} = \dfrac{|3m + 3|}{\sqrt{m^2 + 1}}$

(b)

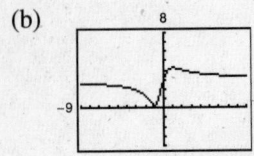

(c) $\lim\limits_{m \to \infty} d(m) = 3$ and $\lim\limits_{m \to -\infty} d(m) = 3$

This indicates that the distance between the point and the line approaches 3 as the slope approaches positive or negative infinity.

12. (a) $T_1 = -0.00303t^2 + 0.6772t + 26.5636$

(b)

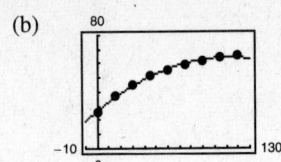

(c)

(d) $T_1(0) = -0.00303 \cdot 0^2 + 0.6772 \cdot 0 + 26.5636 = 26.5636$

$T_2(0) = \dfrac{86 \cdot 0 + 1451}{0 + 58} = 25.0172$

(e) $\lim\limits_{t \to \infty} T_2(4) = \lim\limits_{t \to \infty} \dfrac{86t + 1451}{t + 58}$

$= \lim\limits_{t \to \infty} \dfrac{86 + \dfrac{1451}{t}}{1 + \dfrac{58}{t}} = \dfrac{86 + 0}{1 + 0} = 86$

(f) The temperature approaches 86° as time goes by. The function T_1 approaches $-\infty$ as t approaches ∞ and therefore fails to model the system accurately for times greater than 120 seconds.

13. The error was probably due to the calculator being in degree mode rather than radian mode.

14. (a) Perimeter of ΔPAO:

$$AO + PA + PO = 1 + \sqrt{(x-0)^2 + (y-1)^2} + \sqrt{(x-0)^2 + (y-0)^2}$$
$$= 1 + \sqrt{x^2 + (x^2-1)^2} + \sqrt{x^2 + (x^2)^2}$$
$$= 1 + \sqrt{x^2 + (x^2-1)^2} + x\sqrt{1+x^2}$$

Perimeter of ΔPBO:

$$BO + PB + PO = 1 + \sqrt{(x-1)^2 + (y-0)^2} + \sqrt{(x-0)^2 + (y-0)^2}$$
$$= 1 + \sqrt{(x-1)^2 + (x^2)^2} + \sqrt{x^2 + (x^2)^2}$$
$$= 1 + \sqrt{x^4 + (x-1)^2} + x\sqrt{1+x^2}$$

(b) $r(x) = \dfrac{1 + \sqrt{x^2 + (x^2-1)^2} + x\sqrt{1+x^2}}{1 + \sqrt{x^4 + (x-1)^2} + x\sqrt{1+x^2}}$

x	4	2	1	0.1	0.01
Perimeter of ΔPAO:	33.01650	9.07769	3.41421	2.09554	2.00995
Perimeter of ΔPBO:	39.42413	10.47214	3.41421	2.09055	2.00990
$r(x)$	0.8375	0.8668	1	1.0024	1.0000

(c) $\displaystyle\lim_{x \to 0^+} r(x) = \dfrac{1 + \sqrt{0^2 + (0^2-1)^2} + 0\sqrt{1+0^2}}{1 + \sqrt{0^4 + (0-1)^2} + 0 \cdot \sqrt{1+0^2}} = \dfrac{1 + 1 + 0}{1 + 1 + 0} = \dfrac{2}{2} = 1$

15. (a)

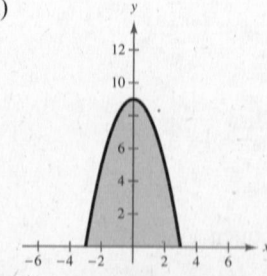

(b) If we find the area of the region in Quadrant I, bounded by $y = 9 - x^2$ and the x- and y-axes, we can find the area of the entire region by doubling this value.

Width: $\dfrac{b-a}{n} = \dfrac{3-0}{n} = \dfrac{3}{n}$

Height: $f\left(a + \dfrac{(b-a)i}{n}\right) = f\left(0 + \dfrac{3i}{n}\right) = f\left(\dfrac{3i}{n}\right) = 9 - \left(\dfrac{3i}{n}\right)^2 = 9 - \dfrac{9i^2}{n^2}$

$$\lim_{n \to \infty} \sum_{i=1}^{n} f\left(a + \dfrac{(b-a)i}{n}\right)\left(\dfrac{b-a}{n}\right) = \lim_{n \to \infty} \sum_{i=1}^{n}\left(9 - \dfrac{9i^2}{n^2}\right)\left(\dfrac{3}{n}\right)$$

$$= \lim_{n \to \infty} \sum_{i=1}^{n}\left(\dfrac{27}{n} - \dfrac{27i^2}{n^3}\right)$$

$$= \lim_{n \to \infty}\left[\left(\dfrac{27}{n}\right)(n) - \dfrac{27}{n^3}\sum_{i=1}^{n} i^2\right]$$

$$= \lim_{n \to \infty}\left[27 - \dfrac{27}{n^3}\left(\dfrac{n(n+1)(2n+1)}{6}\right)\right]$$

$$= 27 - \dfrac{27(2)}{6} = 18$$

Area $= 2(18) = 36$ square units

(c) Base $= 6$, Height $= 9$, Area $= \dfrac{2}{3}bh = \dfrac{2}{3}(6)(9) = 36$

Chapter 12   Practice Test

1. Use a graphing utility to complete the table and use the result to estimate the limit

$$\lim_{x \to 3} \frac{x - 3}{x^2 - 9}.$$

x	2.9	2.99	3	3.01	3.1
$f(x)$			?		

2. Graph the function

$$f(x) = \frac{\sqrt{x + 4} - 2}{x}$$

and estimate the limit

$$\lim_{x \to 0} \frac{\sqrt{x + 4} - 2}{x}.$$

3. Find the limit $\lim\limits_{x \to 2} e^{x-2}$ by direct substitution.

4. Find the limit $\lim\limits_{x \to 1} \dfrac{x^3 - 1}{x - 1}$ analytically.

5. Use a graphing utility to estimate the limit.

$$\lim_{x \to 0} \frac{\sin 5x}{2x}$$

6. Find the limit.

$$\lim_{x \to -2} \frac{|x + 2|}{x + 2}$$

7. Use the limit process to find the slope of the graph of $f(x) = \sqrt{x}$ at the point $(4, 2)$.

8. Find the derivative of the function $f(x) = 3x - 1$.

9. Find the limits.

(a) $\lim\limits_{x \to \infty} \dfrac{3}{x^4}$

(b) $\lim\limits_{x \to -\infty} \dfrac{x^2}{x^2 + 3}$

(c) $\lim\limits_{x \to \infty} \dfrac{|x|}{1 - x}$

10. Write the first four terms of the sequence $a_n = \dfrac{1 - n^2}{2n^2 + 1}$ and find the limit of the sequence.

11. Find the sum $\sum\limits_{i=1}^{25} (i^2 + i)$.

12. Write the sum $\sum\limits_{i=1}^{n} \dfrac{i^2}{n^3}$ as a rational function $S(n)$, and find $\lim\limits_{n \to \infty} S(n)$.

13. Find the area of the region bounded by $f(x) = 1 - x^2$ over the interval $0 \le x \le 1$.

APPENDIX A
Review of Fundamental Concepts of Algebra

APPENDIX A
Review of Fundamental Concepts of Algebra

Appendix A.1   Review of Real Numbers and Their Properties

■ You should know the following sets.

(a) The set of real numbers includes the rational numbers and the irrational numbers.

(b) The set of rational numbers includes all real numbers that can be written as the ratio p/q of two integers, where $q \neq 0$.

(c) The set of irrational numbers includes all real numbers which are not rational.

(d) The set of integers: $\{. \, . \, . \, , -3, -2, -1, 0, 1, 2, 3, . \, . \, .\}$

(e) The set of whole numbers: $\{0, 1, 2, 3, 4, . \, . \, .\}$

(f) The set of natural numbers: $\{1, 2, 3, 4, . \, . \, .\}$

■ The real number line is used to represent the real numbers.

■ Know the inequality symbols.

(a) $a < b$ means a is less than b.   (b) $a \leq b$ means a is less than or equal to b.

(c) $a > b$ means a is greater than b.   (d) $a \geq b$ means a is greater than or equal to b.

■

Interval Notation	Inequality Notation	Graph	Type
$[a, b]$	$a \leq x \leq b$		Bounded and Closed
(a, b)	$a < x < b$		Bounded and Open
$[a, b)$	$a \leq x < b$		Bounded
$(a, b]$	$a < x \leq b$		Bounded
$[a, \infty)$	$x \geq a$		Unbounded
(a, ∞)	$x > a$		Unbounded
$(-\infty, b]$	$x \leq b$		Unbounded
$(-\infty, b)$	$x < b$		Unbounded
$(-\infty, \infty)$	$-\infty < x < \infty$		Unbounded

■ You should know that $|a| = \begin{cases} a, & \text{if } a \geq 0 \\ -a, & \text{if } a < 0 \end{cases}$.

■ Know the properties of absolute value.

(a) $|a| \geq 0$ (b) $|-a| = |a|$ (c) $|ab| = |a|\,|b|$ (d) $\left|\dfrac{a}{b}\right| = \dfrac{|a|}{|b|}, \, b \neq 0$

—CONTINUED—

■ The distance between a and b on the real line is $d(a, b) = |b - a| = |a - b|$.

■ You should be able to identify the terms in an algebraic expression.

■ You should know and be able to use the basic rules of algebra.

■ Commutative Property

(a) Addition: $a + b = b + a$ (b) Multiplication: $a \cdot b = b \cdot a$

■ Associative Property

(a) Addition: $(a + b) + c = a + (b + c)$ (b) Multiplication: $(ab)c = a(bc)$

■ Identity Property

(a) Addition: 0 is the identity; $a + 0 = 0 + a = a$.   (b) Multiplication: 1 is the identity; $a \cdot 1 = 1 \cdot a = a$.

■ Inverse Property

(a) Addition: $-a$ is the additive inverse of a; $a + (-a) = -a + a = 0$.

(b) Multiplication: $1/a$ is the multiplicative inverse of a, $a \neq 0$; $a(1/a) = (1/a)a = 1$.

■ Distributive Property

(a) $a(b + c) = ab + ac$ (b) $(a + b)c = ac + bc$

■ Properties of Negation

(a) $(-1)a = -a$ (b) $-(-a) = a$ (c) $(-a)b = a(-b) = -ab$

(d) $(-a)(-b) = ab$ (e) $-(a + b) = (-a) + (-b) = -a - b$

■ Properties of Equality

(a) If $a = b$, then $a \pm c = b \pm c$.   (b) If $a = b$, then $ac = bc$.

(c) If $a \pm c = b \pm c$, then $a = b$.   (d) If $ac = bc$ and $c \neq 0$, then $a = b$.

■ Properties of Zero

(a) $a \pm 0 = a$ (b) $a \cdot 0 = 0$ (c) $0 \div a = 0/a = 0, a \neq 0$

(d) $a/0$ is undefined.   (e) If $ab = 0$, then $a = 0$ or $b = 0$.

■ Properties of Fractions $(b \neq 0, d \neq 0)$

(a) Equivalent Fractions: $a/b = c/d$ if and only if $ad = bc$.

(b) Rule of Signs: $-a/b = a/-b = -(a/b)$ and $-a/-b = a/b$

(c) Equivalent Fractions: $a/b = ac/bc, c \neq 0$

(d) Addition and Subtraction

     1. Like Denominators: $(a/b) \pm (c/b) = (a \pm c)/b$ 2. Unlike Denominators: $(a/b) \pm (c/d) = (ad \pm bc)/bd$

(e) Multiplication: $(a/b) \cdot (c/d) = (ac)/(bd)$

(f) Division: $(a/b) \div (c/d) = (a/b) \cdot (d/c) = (ad)/(bc)$ if $c \neq 0$.

Vocabulary Check

1. rational   2. irrational   3. absolute value

4. composite   5. prime   6. variables; constants

7. terms   8. coefficient   9. zero-factor property

1. $-9, -\frac{7}{2}, 5, \frac{2}{3}, \sqrt{2}, 0, 1, -4, 2, -11$

   (a) Natural numbers: $5, 1, 2$

   (b) Whole numbers: $0, 5, 1, 2$

   (c) Integers: $-9, 5, 0, 1, -4, 2, -11$

   (d) Rational numbers: $-9, -\frac{7}{2}, 5, \frac{2}{3}, 0, 1, -4, 2, -11$

   (e) Irrational numbers: $\sqrt{2}$

2. $\sqrt{5}, -7, -\frac{7}{3}, 0, 3.12, \frac{5}{4}, -3, 12, 5$

   (a) Natural numbers: $12, 5$

   (b) Whole numbers: $0, 12, 5$

   (c) Integers: $-7, 0, -3, 12, 5$

   (d) Rational numbers: $-7, -\frac{7}{3}, 0, 3.12, \frac{5}{4}, -3, 12, 5$

   (e) Irrational numbers: $\sqrt{5}$

3. $2.01, 0.666\ldots, -13, 0.010110111\ldots, 1, -6$

   (a) Natural numbers: 1

   (b) Whole numbers: 1

   (c) Integers: $-13, 1, -6$

   (d) Rational numbers: $2.01, 0.666\ldots, -13, 1, -6$

   (e) Irrational numbers: $0.010110111\ldots$

4. $2.3030030003\ldots, 0.7575, -4.63, \sqrt{10}, -75, 4$

   (a) Natural numbers: 4

   (b) Whole numbers: 4

   (c) Integers: $-75, 4$

   (d) Rational numbers: $0.7575, -4.63, -75, 4$

   (e) Irrational numbers: $2.3030030003\ldots, \sqrt{10}$

5. $-\pi, -\frac{1}{3}, \frac{6}{3}, \frac{1}{2}\sqrt{2}, -7.5, -1, 8, -22$

   (a) Natural numbers: $\frac{6}{3}$ (since it equals 2), 8

   (b) Whole numbers: $\frac{6}{3}, 8$

   (c) Integers: $\frac{6}{3}, -1, 8, -22$

   (d) Rational numbers: $-\frac{1}{3}, \frac{6}{3}, -7.5, -1, 8, -22$

   (e) Irrational numbers: $-\pi, \frac{1}{2}\sqrt{2}$

6. $25, -17, -\frac{12}{5}, \sqrt{9}, 3.12, \frac{1}{2}\pi, 7, -11.1, 13$

   (a) Natural numbers: $25, \sqrt{9}, 7, 13$

   (b) Whole numbers: $25, \sqrt{9}, 7, 13$

   (c) Integers: $25, -17, \sqrt{9}, 7, 13$

   (d) Rational numbers:

      $25, -17, -\frac{12}{5}, \sqrt{9}, 3.12, 7, -11.1, 13$

   (e) Irrational numbers: $\frac{1}{2}\pi$

7. $\frac{5}{8} = 0.625$ **8.** $\frac{1}{3} = 0.\overline{3}$ **9.** $\frac{41}{333} = 0.\overline{123}$ **10.** $\frac{6}{11} = 0.\overline{54}$

11. $-1 < 2.5$ **12.** $-6 < -2.5$ **13.** $-4 > -8$

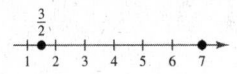

14. $-3.5 < 1$ **15.** $\frac{3}{2} < 7$ **16.** $1 < \frac{16}{3}$

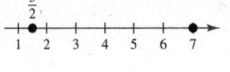

17. $\frac{5}{6} > \frac{2}{3}$ **18.** $-\frac{8}{7} < -\frac{3}{7}$

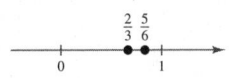

19. (a) The inequality $x \le 5$ denotes the set of all real numbers less than or equal to 5.

   (b)

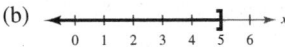

   (c) The interval is unbounded.

20. (a) The inequality $x \ge -2$ denotes the set of all real numbers greater than or equal to -2.

   (b)

   (c) The interval is unbounded.

21. (a) The inequality $x < 0$ denotes the set of all negative real numbers.

(b)

(c) The interval is unbounded.

22. (a) The inequality $x > 3$ denotes the set of all real numbers greater than 3.

(b)

(c) The interval is unbounded.

23. (a) The interval $[4, \infty)$ denotes the set of all real numbers greater than or equal to 4.

(b)

(c) The interval is unbounded.

24. (a) $(-\infty, 2)$ denotes the set of all real numbers less than 2.

(b)

(c) The interval is unbounded.

25. (a) The inequality $-2 < x < 2$ denotes the set of all real numbers greater than -2 and less than 2.

(b)

(c) The interval is bounded.

26. (a) The inequality $0 \le x \le 5$ denotes the set of all real numbers greater than or equal to zero and less than or equal to 5.

(b)

(c) The interval is bounded.

27. (a) The inequality $-1 \le x < 0$ denotes the set of all negative real numbers greater than or equal to -1.

(b)

(c) The interval is bounded.

28. (a) The inequality $0 < x \le 6$ denotes the set of all real numbers greater than zero and less than or equal to 6.

(b)

(c) The interval is bounded.

29. (a) The interval $[-2, 5)$ denotes the set of all real numbers greater than or equal to -2 and less than 5.

(b)

(c) The interval is bounded.

30. (a) The interval $(-1, 2]$ denotes the set of all real numbers greater than -1 and less than or equal to 2.

(b)

(c) The interval is bounded.

31. $-2 < x \le 4$

32. $-6 \le y < 0$

33. $y \ge 0$

34. $y \le 25$

35. $10 \le t \le 22$

36. $-3 \le k < 5$

37. $W > 65$

38. $2.5\% \le r \le 5\%$

39. $|-10| = -(-10) = 10$

40. $|0| = 0$

41. $|3 - 8| = |-5| = -(-5) = 5$

42. $|4 - 1| = |3| = 3$

43. $|-1| - |-2| = 1 - 2 = -1$

44. $-3 - |-3| = -3 - (3) = -6$

45. $\dfrac{-5}{|-5|} = \dfrac{-5}{-(-5)} = \dfrac{-5}{5} = -1$

46. $-3|-3| = -3(3) = -9$

47. If $x < -2$, then $x + 2$ is negative.

Thus $\dfrac{|x + 2|}{x + 2} = \dfrac{-(x + 2)}{x + 2} = -1$.

48. If $x > 1$, then $x - 1$ is positive.

Thus, $\dfrac{|x-1|}{x-1} = \dfrac{x-1}{x-1} = 1$.

49. $|-3| > -|-3|$ since $3 > -3$.

50. $|-4| = |4|$ since $|-4| = 4$ and $|4| = 4$.

51. $-5 = -|5|$ since $-5 = -5$.

52. $-|-6| < |-6|$ since $|-6| = 6$ and $-|-6| = -(6) = -6$.

53. $-|-2| = -|2|$ since $-2 = -2$.

54. $-(-2) > -2$ since $-(-2) = 2$.

55. $d(126, 75) = |75 - 126| = 51$

56. $d(-126, -75) = |75 - (-126)| = 51$

57. $d\left(-\dfrac{5}{2}, 0\right) = \left|0 - \left(-\dfrac{5}{2}\right)\right| = \dfrac{5}{2}$

58. $d\left(\dfrac{1}{4}, \dfrac{11}{4}\right) = \left|\dfrac{11}{4} - \dfrac{1}{4}\right| = \dfrac{5}{2}$

59. $d\left(\dfrac{16}{5}, \dfrac{112}{75}\right) = \left|\dfrac{112}{75} - \dfrac{16}{5}\right| = \dfrac{128}{75}$

60. $d(9.34, -5.65) = |-5.65 - 9.34| = 14.99$

61.

Budgeted Expense, b	Actual Expense, a	$|a - b|$	$0.05b$
\$112,700	\$113,356	\$656	$0.05(112{,}700) = \$5635$

Since \$656 < \$5635 but \$656 > \$500, the actual expense does not pass the "budget variance test."

62.

Budgeted Expense, b	Actual Expense, a	$|a - b|$	$0.05b$
\$9400	\$9772	\$372	$0.05(9400) = \$470$

Since \$372 < \$470 and \$372 < \$500, the actual expense does pass the "budget variance test."

63.

Budgeted Expense, b	Actual Expense, a	$|a - b|$	$0.05b$
\$37,640	\$37,335	\$305	$0.05(37{,}640) = \$1882$

Since \$305 < \$500 and \$305 < \$1882, the actual expense passes the "budget variance test."

64.

Budgeted Expense, b	Actual Expense, a	$|a - b|$	$0.05b$
\$2575	\$2613	\$38	$0.05(2575) = \$128.75$

Since \$38 < \$500, and \$38 < \$128.75, the actual expense passes the "budget variance test."

65. (a)

Year	Expenditures (in billions)	Surplus or Deficit (in billions)
1960	\$92.2	$|92.5 - 92.2| = \$0.3$ surplus
1970	\$195.6	$|192.8 - 195.6| = \$2.8$ deficit
1980	\$590.9	$|517.1 - 590.9| = \$73.8$ deficit
1990	\$1253.2	$|1032.0 - 1253.2| = \$221.2$ deficit
2000	\$1788.8	$|2025.2 - 1788.8| = \$236.4$ surplus

(b)

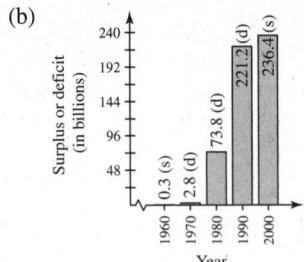

66. Total: $2213 + 3290 + 4666 + 5665 + 9784 = 25,618$

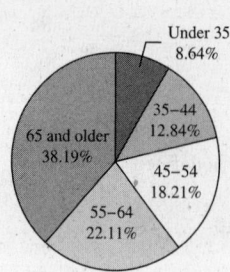

Under 35: $\dfrac{2213}{25,618} \approx 0.0863857 = 8.63857\%$

8.63857% of $360° \approx 31.1°$

35–44: $\dfrac{3290}{25,618} \approx 0.1284253 = 12.84253\%$

12.84253% of $360° \approx 46.2°$

45–54: $\dfrac{4666}{25,618} \approx 0.18213756 = 18.213756\%$

18.213756% of $360° \approx 65.6°$

55–64: $\dfrac{5665}{25,618} \approx 0.221133578 = 22.1133578\%$

22.1133578% of $360° \approx 79.6°$

65 and older: $\dfrac{9784}{25,618} \approx 0.38191896 = 38.191896\%$

38.191896% of $360° \approx 137.5°$

67. $d(x, 5) = |x - 5|$ and $d(x, 5) \le 3$, thus $|x - 5| \le 3$.

68. $d(x, -10) = |x + 10|$, and $d(x, -10) \ge 6$, thus, $|x + 10| \ge 6$.

69. $d(y, 0) = |y - 0| = |y|$ and $d(y, 0) \ge 6$, thus $|y| \ge 6$.

70. $d(y, a) = |y - a|$ and $d(y, a) \le 2$, thus $|y - a| \le 2$.

71. $d(326, 351) = |351 - 326| = 25$ miles

72. $d(48°, 82°) = |82° - 48°| = 34°$

73. $7x + 4$

Terms: $7x, 4$

Coefficient: 7

74. $6x^3 - 5x$

Terms: $6x^3, -5x$

Coefficients: $6, -5$

75. $\sqrt{3}x^2 - 8x - 11$

Terms: $\sqrt{3}x^2, -8x, -11$

Coefficients: $\sqrt{3}, -8$

76. $3\sqrt{3}x^2 + 1$

Terms: $3\sqrt{3}x^2, 1$

Coefficients: $3\sqrt{3}$

77. $4x^3 + \dfrac{x}{2} - 5$

Terms: $4x^3, \dfrac{x}{2}, -5$

Coefficients: $4, \dfrac{1}{2}$

78. $3x^4 - \dfrac{x^2}{4}$

Terms: $3x^4, -\dfrac{x^2}{4}$

Coefficients: $3, -\dfrac{1}{4}$

79. $4x - 6$

(a) $4(-1) - 6 = -4 - 6 = -10$

(b) $4(0) - 6 = 0 - 6 = -6$

80. $9 - 7x$

(a) $9 - 7(-3) = 9 + 21 = 30$

(b) $9 - 7(3) = 9 - 21 = -12$

81. $x^2 - 3x + 4$

(a) $(-2)^2 - 3(-2) + 4 = 4 + 6 + 4 = 14$

(b) $(2)^2 - 3(2) + 4 = 4 - 6 + 4 = 2$

82. $-x^2 + 5x - 4$

(a) $-(-1)^2 + 5(-1) - 4 = -1 - 5 - 4 = -10$

(b) $-(1)^2 + 5(1) - 4 = -1 + 5 - 4 = 0$

83. $\dfrac{x+1}{x-1}$

 (a) $\dfrac{1+1}{1-1} = \dfrac{2}{0}$

 Division by zero is undefined

 (b) $\dfrac{-1+1}{-1-1} = \dfrac{0}{-2} = 0$

84. $\dfrac{x}{x+2}$

 (a) $\dfrac{2}{2+2} = \dfrac{2}{4} = \dfrac{1}{2}$

 (b) $\dfrac{-2}{-2+2} = \dfrac{2}{0}$

 Division by 0 is undefined.

85. $x + 9 = 9 + x$

 Commutative Property of Addition

86. $2\left(\dfrac{1}{2}\right) = 1$

 Multiplicative Inverse Property

87. $\dfrac{1}{(h+6)}(h+6) = 1, h \neq -6$

 Multiplicative Inverse Property

88. $(x+3) - (x+3) = 0$

 Additive Inverse Property

89. $2(x+3) = 2x + 6$

 Distributive Property

90. $(z-2) + 0 = z - 2$

 Additive Identity Property

91. $1 \cdot (1 + x) = 1 + x$

 Multiplicative Identity Property

92. $(z+5)x = z \cdot x + 5 \cdot x$

 Right Distributive Property

93. $x + (y + 10) = (x + y) + 10$

 Associative Property of Addition

94. $x(3y) = (x \cdot 3)y$ Associative Property of Multiplication

 $= (3x)y$ Commutative Property of Multiplication

95. $3(t - 4) = 3 \cdot t - 3 \cdot 4$

 Distributive Property

96. $\dfrac{1}{7}(7 \cdot 12) = \left(\dfrac{1}{7} \cdot 7\right)12$ Associative Property of Multiplication

 $= 1 \cdot 12$ Multiplicative Inverse Property

 $= 12$ Multiplicative Identity Property

97. $\dfrac{3}{16} + \dfrac{5}{16} = \dfrac{8}{16} = \dfrac{1}{2}$

98. $\dfrac{6}{7} - \dfrac{4}{7} = \dfrac{6-4}{7} = \dfrac{2}{7}$

99. $\dfrac{5}{8} - \dfrac{5}{12} + \dfrac{1}{6} = \dfrac{15}{24} - \dfrac{10}{24} + \dfrac{4}{24} = \dfrac{9}{24} = \dfrac{3}{8}$

100. $\dfrac{10}{11} + \dfrac{6}{33} - \dfrac{13}{66} = \dfrac{60}{66} + \dfrac{12}{66} - \dfrac{13}{66} = \dfrac{59}{66}$

101. $12 \div \dfrac{1}{4} = 12 \cdot \dfrac{4}{1} = 12 \cdot 4 = 48$

102. $-\left(6 \cdot \dfrac{4}{8}\right) = -6 \cdot \dfrac{1}{2} = -3$

103. $\dfrac{2x}{3} - \dfrac{x}{4} = \dfrac{8x}{12} - \dfrac{3x}{12} = \dfrac{5x}{12}$

104. $\dfrac{5x}{6} \cdot \dfrac{2}{9} = \dfrac{5x}{3} \cdot \dfrac{1}{9} = \dfrac{5x}{27}$

105. (a)

n	1	0.5	0.01	0.0001	0.000001
$5/n$	5	10	500	50,000	5,000,000

 (b) The value of $5/n$ approaches infinity as n approaches 0.

106. (a)

n	1	10	100	10,000	100,000
$5/n$	5	0.5	0.05	0.0005	0.00005

 (b) The value of $5/n$ approaches 0 as n increases without bound.

107. False. If $a < b$, then $\dfrac{1}{a} > \dfrac{1}{b}$, where $a \neq b \neq 0$.

108. False. The denominators cannot be added when adding fractions.

109. (a) $|u + v| \neq |u| + |v|$ if u is positive and v is negative or vice versa.

(b) $|u + v| \leq |u| + |v|$

They are equal when u and v have the same sign. If they differ in sign, $|u + v|$ is less than $|u| + |v|$.

110. Yes. y is nonnegative if $y \geq 0$. y is positive if $y > 0$.

111. The only even prime number is 2, because its factors are itself and 1.

112.

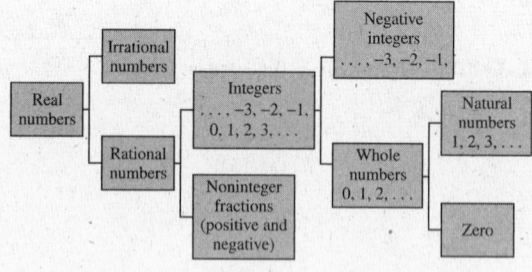

113. (a) Since $A > 0$, $-A < 0$. The expression is negative.

(b) Since $B < A$, $B - A < 0$. The expression is negative.

114. (a) Since $C < 0$, $-C > 0$. The expression is positive.

(b) Since $A > C$, $A - C > 0$. The expression is positive.

115. Yes, if a is a negative number, then $-a$ is positive. Thus, $|a| = -a$ if a is negative.

Appendix A.2   Exponents and Radicals

■ You should know the properties of exponents.

(a) $a^1 = a$

(b) $a^0 = 1, a \neq 0$

(c) $a^m a^n = a^{m+n}$

(d) $a^m / a^n = a^{m-n}, a \neq 0$

(e) $a^{-n} = 1/a^n = (1/a)^n, a \neq 0$

(f) $(a^m)^n = a^{mn}$

(g) $(ab)^n = a^n b^n$

(h) $(a/b)^n = a^n/b^n, b \neq 0$

(i) $(a/b)^{-n} = (b/a)^n, a \neq 0, b \neq 0$

(j) $|a^2| = |a|^2 = a^2$

■ You should be able to write numbers in scientific notation, $c \times 10^n$, where $1 \leq c < 10$ and n is an integer.

■ You should be able to use your calculator to evaluate expressions involving exponents.

■ You should know the properties of radicals.

(a) $\sqrt[n]{a^m} = \left(\sqrt[n]{a}\right)^m, a > 0$

(b) $\sqrt[n]{a} \cdot \sqrt[n]{b} = \sqrt[n]{ab}$

(c) $\dfrac{\sqrt[n]{a}}{\sqrt[n]{b}} = \sqrt[n]{\dfrac{a}{b}}, b \neq 0$

(d) $\sqrt[m]{\sqrt[n]{a}} = \sqrt[mn]{a}$

(e) $\left(\sqrt[n]{a}\right)^n = a$

(f) For n even, $\sqrt[n]{a^n} = |a|$.
For n odd, $\sqrt[n]{a^n} = a$.

(g) $a^{1/n} = \sqrt[n]{a}$

(h) $a^{m/n} = \left(\sqrt[n]{a}\right)^m = \sqrt[n]{a^m}, a \geq 0$

■ You should be able to simplify radicals.

(a) All possible factors have been removed from the radical sign.

(b) All fractions have radical-free denominators.

(c) The index for the radical has been reduced as far as possible.

■ You should be able to use your calculator to evaluate radicals.

Vocabulary Check

1. exponent; base

2. scientific notation

3. square root

4. principal *n*th root

5. index; radicand

6. simplest form

7. conjugates

8. rationalizing

9. power; index

1. $8^5 = 8 \times 8 \times 8 \times 8 \times 8$

2. $(-2)^7 = (-2)(-2)(-2)(-2)(-2)(-2)(-2)$

3. $(4.9)(4.9)(4.9)(4.9)(4.9)(4.9) = 4.9^6$

4. $(-10)(-10)(-10)(-10)(-10) = (-10)^5$

5. (a) $3^2 \cdot 3 = 3^3 = 27$

    (b) $3 \cdot 3^3 = 3^4 = 81$

6. (a) $\dfrac{5^5}{5^2} = 5^3 = 125$

    (b) $\dfrac{3^2}{3^4} = 3^{-2} = \dfrac{1}{3^2} = \dfrac{1}{9}$

7. (a) $(3^3)^0 = 1$

    (b) $-3^2 = -9$

8. (a) $(2^3 \cdot 3^2)^2 = 2^{3 \cdot 2} \cdot 3^{2 \cdot 2}$

    $= 2^6 \cdot 3^4 = 64 \cdot 81 = 5184$

    (b) $\left(-\dfrac{3}{5}\right)^3 \left(\dfrac{5}{3}\right)^2 = (-1)^3 \dfrac{3^3}{5^3} \cdot \dfrac{5^2}{3^2} = -1 \cdot 3^{3-2} \cdot 5^{2-3}$

    $= -3 \cdot 5^{-1} = -\dfrac{3}{5}$

9. (a) $\dfrac{3 \cdot 4^{-4}}{3^{-4} \cdot 4^{-1}} = 3^{1-(-4)} \cdot 4^{-4-(-1)} = 3^5 \cdot 4^{-3}$

    $= \dfrac{3^5}{4^3} = \dfrac{243}{64}$

    (b) $32(-2)^{-5} = \dfrac{32}{(-2)^5} = \dfrac{32}{-32} = -1$

10. (a) $\dfrac{4 \cdot 3^{-2}}{2^{-2} \cdot 3^{-1}} = 4 \cdot 2^2 \cdot 3^{-2-(-1)} = 4 \cdot 4 \cdot 3^{-1} = \dfrac{16}{3}$

    (b) $(-2)^0 = 1$

11. (a) $2^{-1} + 3^{-1} = \dfrac{1}{2} + \dfrac{1}{3} = \dfrac{3}{6} + \dfrac{2}{6} = \dfrac{5}{6}$

    (b) $(2^{-1})^{-2} = 2^{(-1)(-2)} = 2^2 = 4$

12. (a) $3^{-1} + 2^{-2} = \dfrac{1}{3} + \dfrac{1}{4} = \dfrac{4}{12} + \dfrac{3}{12} = \dfrac{7}{12}$

    (b) $(3^{-2})^2 = 3^{-4} = \dfrac{1}{3^4} = \dfrac{1}{81}$

13. $(-4)^3(5^2) = (-64)(25) = -1600$

14. $(8^{-4})(10^3) \approx 0.244$

15. $\dfrac{3^6}{7^3} = \dfrac{729}{343} \approx 2.125$

16. $\dfrac{4^3}{3^{-4}} = 4^3(3^4) = 5184$

17. When $x = 2$,

    $-3x^3 = -3(2)^3 = -24$.

18. When $x = 4$,

    $7x^{-2} = 7(4)^{-2} = \dfrac{7}{4^2} = \dfrac{7}{16}$.

19. When $x = 10$,

    $6x^0 = 6(10)^0 = 6(1) = 6$.

20. When $x = 3$,

    $5(-x)^3 = 5(-3)^3$

    $= 5(-27) = -135$.

21. When $x = -3$,

    $2x^3 = 2(-3)^3 = 2(-27) = -54$.

22. When $x = -2$,

    $-3x^4 = -3(-2)^4$

    $= -3(16) = -48$.

23. When $x = -\frac{1}{2}$,

    $4x^2 = 4\left(-\frac{1}{2}\right)^2 = 4\left(\frac{1}{4}\right) = 1$.

24. When $x = \frac{1}{3}$,

    $5(-x)^3 = 5\left(-\frac{1}{3}\right)^3 = 5\left(-\frac{1}{27}\right) = -\frac{5}{27}$.

25. (a) $(-5z)^3 = (-5)^3 z^3 = -125z^3$

    (b) $5x^4(x^2) = 5x^{4+2} = 5x^6$

26. (a) $(3x)^2 = 3^2x^2 = 9x^2$

(b) $(4x^3)^0 = 4^0 \cdot x^0 = 1,\ x \neq 0$

27. (a) $6y^2(2y^0)^2 = 6y^2(2 \cdot 1)^2 = 6y^2(4) = 24y^2$

(b) $\dfrac{3x^5}{x^3} = 3x^{5-3} = 3x^2$

28. (a) $(-z)^3(3z^4) = (-1)^3(z^3)3z^4$

$\qquad = -1 \cdot 3 \cdot z^{3+4} = -3z^7$

(b) $\dfrac{25y^8}{10y^4} = \dfrac{5}{2}y^{8-4} = \dfrac{5}{2}y^4$

29. (a) $\dfrac{7x^2}{x^3} = 7x^{2-3} = 7x^{-1} = \dfrac{7}{x}$

(b) $\dfrac{12(x+y)^3}{9(x+y)} = \dfrac{4}{3}(x+y)^{3-1} = \dfrac{4}{3}(x+y)^2$

30. (a) $\dfrac{r^4}{r^6} = r^{4-6} = r^{-2} = \dfrac{1}{r^2}$

(b) $\left(\dfrac{4}{y}\right)^3\left(\dfrac{3}{y}\right)^4 = \dfrac{4^3}{y^3} \cdot \dfrac{3^4}{y^4} = \dfrac{64 \cdot 81}{y^{3+4}} = \dfrac{5184}{y^7}$

31. (a) $(x+5)^0 = 1,\ x \neq -5$

(b) $(2x^2)^{-2} = \dfrac{1}{(2x^2)^2} = \dfrac{1}{4x^4}$

32. (a) $(2x^5)^0 = 1,\ x \neq 0$

(b) $(z+2)^{-3}(z+2)^{-1} = (z+2)^{-4} = \dfrac{1}{(z+2)^4}$

33. (a) $(-2x^2)^3(4x^3)^{-1} = \dfrac{-8x^6}{4x^3} = -2x^3$

(b) $\left(\dfrac{x}{10}\right)^{-1} = \dfrac{10}{x}$

34. (a) $(4y^{-2})(8y^4) = (4)(8)(y^{4-2}) = 32y^2$

(b) $\left(\dfrac{x^{-3}y^4}{5}\right)^{-3} = \left(\dfrac{5x^3}{y^4}\right)^3 = \dfrac{125x^9}{y^{12}}$

35. (a) $3^n \cdot 3^{2n} = 3^{n+2n} = 3^{3n}$

(b) $\left(\dfrac{a^{-2}}{b^{-2}}\right)\left(\dfrac{b}{a}\right)^3 = \left(\dfrac{b^2}{a^2}\right)\left(\dfrac{b^3}{a^3}\right) = \dfrac{b^5}{a^5}$

36. (a) $\dfrac{x^2 \cdot x^n}{x^3 \cdot x^n} = \dfrac{x^{2+n}}{x^{3+n}} = x^{2+n-3-n} = x^{-1} = \dfrac{1}{x}$

(b) $\left(\dfrac{a^{-3}}{b^{-3}}\right)\left(\dfrac{a}{b}\right)^3 = \dfrac{a^{-3}}{b^{-3}} \cdot \dfrac{a^3}{b^3} = \dfrac{a^{-3+3}}{b^{-3+3}} = \dfrac{a^0}{b^0} = 1$

37. $57{,}300{,}000 = 5.73 \times 10^7$ square miles

38. $9{,}460{,}000{,}000{,}000 = 9.460 \times 10^{12}$ kilometers

39. $0.0000899 = 8.99 \times 10^{-5}$ gram per cubic centimeter

40. $0.00003937 = 3.937 \times 10^{-5}$ inch

41. $4.568 \times 10^9 = 4{,}568{,}000{,}000$ ounces

42. $1.5 \times 10^7 = 15{,}000{,}000$ degrees Celsius

43. $1.6022 \times 10^{-19} = 0.00000000000000000016022$ coulomb

44. $9.0 \times 10^{-5} = 0.00009$ meter

45. (a) $\sqrt{25 \times 10^8} = 5 \times 10^4 = 50{,}000$

(b) $\sqrt[3]{8 \times 10^{15}} = 2 \times 10^5 = 200{,}000$

46. (a) $(1.2 \times 10^7)(5 \times 10^{-3}) = 6.0 \times 10^4 = 60{,}000$

(b) $\dfrac{(6.0 \times 10^8)}{(3.0 \times 10^{-3})} = 2.0 \times 10^{11}$

47. (a) $750\left(1 + \dfrac{0.11}{365}\right)^{800} \approx 954.448$

(b) $\dfrac{67{,}000{,}000 + 93{,}000{,}000}{0.0052} = 30{,}769{,}230{,}769.2$

$\qquad\qquad \approx 3.077 \times 10^{10}$

48. (a) $(9.3 \times 10^6)^3 (6.1 \times 10^{-4}) \approx 4.907 \times 10^{17}$

(b) $\dfrac{(2.414 \times 10^4)^6}{(1.68 \times 10^5)^5} \approx 1.479$

49. (a) $\sqrt{4.5 \times 10^9} \approx 67{,}082.039$

(b) $\sqrt[3]{6.3 \times 10^4} \approx 39.791$

50. (a) $(2.65 \times 10^{-4})^{1/3} \approx 0.064$

(b) $\sqrt{9 \times 10^{-4}} = 0.03$

51. (a) $\sqrt{9} = 3$

(b) $\sqrt[3]{\dfrac{27}{8}} = \dfrac{\sqrt[3]{27}}{\sqrt[3]{8}} = \dfrac{3}{2}$

52. (a) $27^{1/3} = \sqrt[3]{27} = 3$

(b) $36^{3/2} = 216$

53. (a) $32^{-3/5} = \dfrac{1}{32^{3/5}} = \dfrac{1}{(\sqrt[5]{32})^3} = \dfrac{1}{(2)^3} = \dfrac{1}{8}$

(b) $\left(\dfrac{16}{81}\right)^{-3/4} = \left(\dfrac{81}{16}\right)^{3/4} = \left(\sqrt[4]{\dfrac{81}{16}}\right)^3 = \left(\dfrac{3}{2}\right)^3 = \dfrac{27}{8}$

54. (a) $100^{-3/2} = \left(\sqrt{100}\right)^{-3} = 10^{-3} = \dfrac{1}{1000}$

(b) $\left(\dfrac{9}{4}\right)^{-1/2} = \left(\dfrac{4}{9}\right)^{1/2} = \dfrac{\sqrt{4}}{\sqrt{9}} = \dfrac{2}{3}$

55. (a) $\left(-\dfrac{1}{64}\right)^{-1/3} = (-64)^{1/3} = \sqrt[3]{-64} = -4$

(b) $\left(\dfrac{1}{\sqrt{32}}\right)^{-2/5} = \left(\sqrt{32}\right)^{2/5} = \sqrt[5]{\left(\sqrt{32}\right)^2} = \sqrt[5]{32} = 2$

56. (a) $\left(-\dfrac{125}{27}\right)^{-1/3} = \left(-\dfrac{27}{125}\right)^{1/3} = \dfrac{\sqrt[3]{-27}}{\sqrt[3]{125}} = \dfrac{-3}{5} = -\dfrac{3}{5}$

(b) $-\left(\dfrac{1}{125}\right)^{-4/3} = -(125)^{4/3}$

$= -(125^{1/3})^4 = -\left(\sqrt[3]{125}\right)^4$

$= -(5)^4 = -625$

57. (a) $\sqrt{57} \approx 7.550$

(b) $\sqrt[5]{-27^3} = (-27)^{3/5}$

≈ -7.225

58. (a) $\sqrt[3]{45^2} \approx 12.651$

(b) $\sqrt[6]{125} \approx 2.236$

59. (a) $(-12.4)^{-1.8} \approx -0.011$

(b) $\left(5\sqrt{3}\right)^{-2.5} \approx 0.005$

60. (a) $\dfrac{7 - (4.1)^{-3.2}}{2} \approx 3.495$

(b) $\left(\dfrac{13}{3}\right)^{-3/2} - \left(-\dfrac{3}{2}\right)^{13/3} \approx 5.906$

61. (a) $\left(\sqrt[3]{4}\right)^3 = 4^{3/3} = 4^1 = 4$

(b) $\sqrt[5]{96x^5} = \sqrt[5]{32x^5 \cdot 3}$

$= 2x\sqrt[5]{3}$

$= 2 \cdot 3^{1/5} \cdot x$

62. (a) $\sqrt{12} \cdot \sqrt{3} = \sqrt{36} = 6$

(b) $\sqrt[4]{(3x^2)^4} = \sqrt[4]{3^4 x^8}$

$= 3x^2$

63. (a) $\sqrt{8} = \sqrt{4 \cdot 2}$

$= \sqrt{4}\sqrt{2} = 2\sqrt{2}$

(b) $\sqrt[3]{24} = \sqrt[3]{8 \cdot 3}$

$= \sqrt[3]{8}\sqrt[3]{3} = 2\sqrt[3]{3}$

64. (a) $\sqrt[3]{\dfrac{16}{27}} = \dfrac{\sqrt[3]{2^3 \cdot 2}}{\sqrt[3]{3^3}} = \dfrac{2\sqrt[3]{2}}{3}$

(b) $\sqrt{\dfrac{75}{4}} = \dfrac{\sqrt{5^2 \cdot 3}}{\sqrt{2^2}} = \dfrac{5\sqrt{3}}{2}$

65. (a) $\sqrt{72x^3} = \sqrt{36x^2 \cdot 2x}$

$= 6x\sqrt{2x}$

(b) $\sqrt{\dfrac{18^2}{z^3}} = \dfrac{\sqrt{18^2}}{\sqrt{z^2 \cdot z}} = \dfrac{18}{z\sqrt{z}}$

66. (a) $\sqrt{54xy^4} = \sqrt{6 \cdot 3^2 \cdot x \cdot (y^2)^2}$

$= 3y^2\sqrt{6x}$

(b) $\sqrt{\dfrac{32a^4}{b^2}} = \dfrac{\sqrt{(2^2)^2 \cdot 2 \cdot (a^2)^2}}{\sqrt{b^2}}$

$= \dfrac{4a^2\sqrt{2}}{|b|}$

67. (a) $\sqrt[3]{16x^5} = \sqrt[3]{8x^3 \cdot 2x^2}$

$= 2x\sqrt[3]{2x^2}$

(b) $\sqrt{75x^2y^{-4}} = \sqrt{\dfrac{75x^2}{y^4}}$

$= \dfrac{\sqrt{25x^2 \cdot 3}}{\sqrt{y^4}}$

$= \dfrac{5|x|\sqrt{3}}{y^2}$

68. (a) $\sqrt[4]{3x^4y^2} = (3x^4y^2)^{1/4}$

$= 3^{1/4}|x|y^{1/2}$

$= |x|\sqrt[4]{3}\sqrt{y}$

(b) $\sqrt[5]{160x^8z^4} = (160x^8z^4)^{1/5}$

$= (2^5 \cdot 5x^5 \cdot x^3 \cdot z^4)^{1/5}$

$= 2x(5x^3z^4)^{1/5}$

$= 2x\sqrt[5]{5x^3z^4}$

69. (a) $2\sqrt{50} + 12\sqrt{8} = 2\sqrt{25 \cdot 2} + 12\sqrt{4 \cdot 2} = 2(5\sqrt{2}) + 12(2\sqrt{2}) = 10\sqrt{2} + 24\sqrt{2} = 34\sqrt{2}$

(b) $10\sqrt{32} - 6\sqrt{18} = 10\sqrt{16 \cdot 2} - 6\sqrt{9 \cdot 2} = 10(4\sqrt{2}) - 6(3\sqrt{2}) = 40\sqrt{2} - 18\sqrt{2} = 22\sqrt{2}$

70. (a) $4\sqrt{27} - \sqrt{75} = 4\sqrt{3^2 \cdot 3} - \sqrt{5^2 \cdot 3}$

$= 4 \cdot 3\sqrt{3} - 5\sqrt{3}$

$= 12\sqrt{3} - 5\sqrt{3}$

$= 7\sqrt{3}$

(b) $\sqrt[3]{16} + 3\sqrt[3]{54} = \sqrt[3]{2 \cdot 2^3} + 3\sqrt[3]{2 \cdot 3^3}$

$= 2\sqrt[3]{2} + 3 \cdot 3\sqrt[3]{2}$

$= 2\sqrt[3]{2} + 9\sqrt[3]{2}$

$= 11\sqrt[3]{2}$

71. (a) $5\sqrt{x} - 3\sqrt{x} = 2\sqrt{x}$

(b) $-2\sqrt{9y} + 10\sqrt{y} = -2(3\sqrt{y}) + 10\sqrt{y}$

$= -6\sqrt{y} + 10\sqrt{y} = 4\sqrt{y}$

72. (a) $8\sqrt{49x} - 14\sqrt{100x} = 8\sqrt{7^2 \cdot x} - 14\sqrt{10^2 \cdot x}$

$= 8 \cdot 7\sqrt{x} - 14 \cdot 10\sqrt{x} = 56\sqrt{x} - 140\sqrt{x}$

$= -84\sqrt{x}$

(b) $-3\sqrt{48x^2} + 7\sqrt{75x^2} = -3\sqrt{3 \cdot 4^2 \cdot x^2} + 7\sqrt{3 \cdot 5^2 \cdot x^2}$

$= -3 \cdot 4|x|\sqrt{3} + 7 \cdot 5|x|\sqrt{3}$

$= -12|x|\sqrt{3} + 35|x|\sqrt{3} = 23|x|\sqrt{3}$

73. (a) $3\sqrt{x+1} + 10\sqrt{x+1} = 13\sqrt{x+1}$

(b) $7\sqrt{80x} - 2\sqrt{125x} = 7\sqrt{16 \cdot 5x} - 2\sqrt{25 \cdot 5x} = 7(4\sqrt{5x}) - 2(5\sqrt{5x}) = 28\sqrt{5x} - 10\sqrt{5x} = 18\sqrt{5x}$

74. (a) $-\sqrt{x^3 - 7} + 5\sqrt{x^3 - 7} = 4\sqrt{x^3 - 7}$

(b) $11\sqrt{245x^3} - 9\sqrt{45x^3} = 11\sqrt{5 \cdot 7^2 \cdot x \cdot x^2} - 9\sqrt{5 \cdot 3^2 \cdot x \cdot x^2}$

$= 11 \cdot 7x\sqrt{5x} - 9 \cdot 3x\sqrt{5x}$

$= 77x\sqrt{5x} - 27x\sqrt{5x}$

$= 50x\sqrt{5x}$

75. $\sqrt{5} + \sqrt{3} \approx 3.968$ and

$\sqrt{5+3} = \sqrt{8} \approx 2.828$

Thus, $\sqrt{5} + \sqrt{3} > \sqrt{5+3}$.

76. $\sqrt{\dfrac{3}{11}} = \dfrac{\sqrt{3}}{\sqrt{11}}$

77. $\sqrt{3^2 + 2^2} = \sqrt{9+4}$

$= \sqrt{13} \approx 3.606$

Thus, $5 > \sqrt{3^2 + 2^2}$.

78. $\sqrt{3^2 + 4^2} = \sqrt{9 + 16}$

$= \sqrt{25} = 5$

Thus, $5 = \sqrt{3^2 + 4^2}$.

79. $\dfrac{1}{\sqrt{3}} = \dfrac{1}{\sqrt{3}} \cdot \dfrac{\sqrt{3}}{\sqrt{3}} = \dfrac{\sqrt{3}}{3}$

80. $\dfrac{5}{\sqrt{10}} = \dfrac{5}{\sqrt{10}} \cdot \dfrac{\sqrt{10}}{\sqrt{10}}$

$= \dfrac{5\sqrt{10}}{10} = \dfrac{\sqrt{10}}{2}$

81. $\dfrac{2}{5 - \sqrt{3}} = \dfrac{2}{5 - \sqrt{3}} \cdot \dfrac{5 + \sqrt{3}}{5 + \sqrt{3}} = \dfrac{2(5 + \sqrt{3})}{5^2 - (\sqrt{3})^2} = \dfrac{2(5 + \sqrt{3})}{25 - 3} = \dfrac{2(5 + \sqrt{3})}{22} = \dfrac{5 + \sqrt{3}}{11}$

82. $\dfrac{3}{\sqrt{5} + \sqrt{6}} = \dfrac{3}{\sqrt{5} + \sqrt{6}} \cdot \dfrac{\sqrt{5} - \sqrt{6}}{\sqrt{5} - \sqrt{6}} = \dfrac{3(\sqrt{5} - \sqrt{6})}{5 - 6} = \dfrac{3(\sqrt{5} - \sqrt{6})}{-1} = -3(\sqrt{5} - \sqrt{6}) = 3(\sqrt{6} - \sqrt{5})$

83. $\dfrac{\sqrt{8}}{2} = \dfrac{\sqrt{4 \cdot 2}}{2} = \dfrac{2\sqrt{2}}{2} = \dfrac{\sqrt{2}}{1} \cdot \dfrac{\sqrt{2}}{\sqrt{2}} = \dfrac{2}{\sqrt{2}}$
   84. $\dfrac{\sqrt{2}}{3} = \dfrac{\sqrt{2}}{3} \cdot \dfrac{\sqrt{2}}{\sqrt{2}} = \dfrac{2}{3\sqrt{2}}$

85. $\dfrac{\sqrt{5} + \sqrt{3}}{3} = \dfrac{\sqrt{5} + \sqrt{3}}{3} \cdot \dfrac{\sqrt{5} - \sqrt{3}}{\sqrt{5} - \sqrt{3}} = \dfrac{5 - 3}{3(\sqrt{5} - \sqrt{3})} = \dfrac{2}{3(\sqrt{5} - \sqrt{3})}$

86. $\dfrac{\sqrt{7} - 3}{4} = \dfrac{\sqrt{7} - 3}{4} \cdot \dfrac{\sqrt{7} + 3}{\sqrt{7} + 3} = \dfrac{7 - 9}{4(\sqrt{7} + 3)} = \dfrac{-2}{4(\sqrt{7} + 3)} = -\dfrac{1}{2(\sqrt{7} + 3)}$

	Radical Form	*Rational Exponent Form*
87.	$\sqrt{9} = 3$, Given	$9^{1/2} = 3$, Answer
88.	$\sqrt[3]{64} = 4$, Given	$64^{1/3} = 4$, Answer
89.	$\sqrt[5]{32} = 2$, Answer	$32^{1/5} = 2$, Given
90.	$-\sqrt{144} = -12$, Answer	$-(144^{1/2}) = -12$, Given
91.	$\sqrt[3]{-216} = -6$, Given	$(-216)^{1/3} = -6$, Answer
92.	$\sqrt[5]{-243} = -3$, Answer	$(-243)^{1/5} = -3$, Given
93.	$\sqrt[4]{81^3} = 27$, Given	$81^{3/4} = 27$, Answer
94.	$\sqrt[4]{16^5} = 32$, Answer	$16^{5/4} = 32$, Given

95. $\dfrac{(2x^2)^{3/2}}{2^{1/2}x^4} = \dfrac{2^{3/2}(x^2)^{3/2}}{2^{1/2}x^4}$
   96. $\dfrac{x^{4/3}y^{2/3}}{(xy)^{1/3}} = \dfrac{x^{4/3}y^{2/3}}{x^{1/3}y^{1/3}} = x^{3/3}y^{1/3} = xy^{1/3}$

$\qquad\quad = \dfrac{2^{3/2}x^3}{2^{1/2}x^4} = 2^{3/2 - 1/2}x^{3-4} = 2^1 x^{-1} = \dfrac{2}{x}$

97. $\dfrac{x^{-3} \cdot x^{1/2}}{x^{3/2} \cdot x^{-1}} = \dfrac{x^{1/2} \cdot x^1}{x^{3/2} \cdot x^3}$
   98. $\dfrac{5^{-1/2} \cdot 5x^{5/2}}{(5x)^{3/2}} = \dfrac{5^{-1/2} \cdot 5x^{5/2}}{5^{3/2}x^{3/2}}$

$\qquad\quad = x^{1/2 + 1 - 3/2 - 3} = x^{-3} = \dfrac{1}{x^3}, \ x > 0$
   $\qquad\quad = \dfrac{5^{1/2}x^{5/2}}{5^{3/2}x^{3/2}} = 5^{-1}x = \dfrac{x}{5}, \ x > 0$

99. (a) $\sqrt[4]{3^2} = 3^{2/4} = 3^{1/2} = \sqrt{3}$
   100. (a) $\sqrt[6]{x^3} = x^{3/6} = x^{1/2} = \sqrt{x}$

$\quad$ (b) $\sqrt[6]{(x+1)^4} = (x+1)^{4/6} = (x+1)^{2/3} = \sqrt[3]{(x+1)^2}$
   $\quad$ (b) $\sqrt[4]{(3x^2)^4} = 3x^2$

101. (a) $\sqrt{\sqrt{32}} = (32^{1/2})^{1/2}$
   102. (a) $\sqrt{\sqrt{243(x+1)}} = [(243(x+1))^{1/2}]^{1/2}$

$\qquad\qquad = 32^{1/4} = \sqrt[4]{32} = \sqrt[4]{16 \cdot 2} = 2\sqrt[4]{2}$
   $\qquad\qquad\qquad = (243(x+1))^{1/4}$

$\quad$ (b) $\sqrt{\sqrt[4]{2x}} = ((2x)^{1/4})^{1/2} = (2x)^{1/8} = \sqrt[8]{2x}$
   $\qquad\qquad\qquad = \sqrt[4]{243(x+1)}$

$\qquad\qquad\qquad = \sqrt[4]{3 \cdot 81(x+1)}$

$\qquad\qquad\qquad = 3\sqrt[4]{3(x+1)}$

$\quad$ (b) $\sqrt{\sqrt[3]{10a^7b}} = ((10a^7b)^{1/3})^{1/2}$

$\qquad\qquad\qquad = (10a^7b)^{1/6}$

$\qquad\qquad\qquad = \sqrt[6]{10a \cdot a^6 \cdot b}$

$\qquad\qquad\qquad = a\sqrt[6]{10ab}$

103. $T = 2\pi\sqrt{\dfrac{2}{32}}$

$= 2\pi\sqrt{\dfrac{1}{16}}$

$= 2\pi\left(\dfrac{1}{4}\right)$

$= \dfrac{\pi}{2} \approx 1.57$ seconds

104. Size $= 0.03\sqrt{v}$;   For $v = \dfrac{3}{4}$:

Size $= 0.03\sqrt{\dfrac{3}{4}}$

$= 0.03 \cdot \dfrac{\sqrt{3}}{\sqrt{4}}$

$= 0.03\dfrac{\sqrt{3}}{2}$

≈ 0.026 inch

105. $t = 0.03[12^{5/2} - (12 - h)^{5/2}], \ 0 \le h \le 12$

(a)

h (in centimeters)	t (in seconds)
0	0
1	2.93
2	5.48
3	7.67
4	9.53
5	11.08
6	12.32
7	13.29
8	14.00
9	14.50
10	14.80
11	14.93
12	14.96

(b) As h approaches 12, t approaches

$0.03(12^{5/2}) = 8.64\sqrt{3} \approx 14.96$ seconds.

106. Time $= \dfrac{\text{Distance}}{\text{Rate}} = \dfrac{93{,}000{,}000 \text{ miles}}{11{,}180{,}000 \text{ miles per minute}}$

≈ 8.32 minutes, or 8 minutes 19.1 seconds

107. True. When dividing variables, you subtract exponents.

108. False. When a power is raised to a power, you multiply the exponents: $(a^n)^k = a^{nk}$.

109. $1 = \dfrac{a^m}{a^m} = a^{m-m} = a^0, a \ne 0$

110. (a) 3 is also raised to the negative one power so, $(3x)^{-1} = \dfrac{1}{3x}$.

(b) When two powers have the same base, the exponents are added, $y^3 \cdot y^2 = y^5$.

(c) When a power is raised to a power, exponents are multiplied, $(a^2b^3)^4 = a^8b^{12}$.

(d) The square of a binomial contains a cross product term, $(a + b)^2 = a^2 + 2ab + b^2$.

(e) If $x < 0$, then $\sqrt{4x^2} > 0$ but $2x < 0$, $\sqrt{4x^2} = 2|x|$.

(f) Radicals can only be added together if they have the same radicand and index: $\sqrt{2} + \sqrt{2} = 2\sqrt{2}$.

111. When any positive integer is squared, the units digit is 0, 1, 4, 5, 6, or 9. Therefore, $\sqrt{5233}$ is not an integer.

112. $\left(\dfrac{2}{\sqrt{5}}\right)^2 = \dfrac{4}{5} = 0.8$

$\dfrac{2}{\sqrt{5}} \cdot \dfrac{\sqrt{5}}{\sqrt{5}} = \dfrac{2\sqrt{5}}{5} = 0.8944$

Since $0.8 \neq 0.8944$, $\left(\dfrac{2}{\sqrt{5}}\right)^2 \neq \dfrac{2}{\sqrt{5}} \cdot \dfrac{\sqrt{5}}{\sqrt{5}}$

and squaring is not equivalent to rationalizing the denominator.

Appendix A.3   Polynomials and Factoring

- Given a polynomial in x, $a_n x^n + a_{n-1} x^{n-1} + \ldots + a_1 x + a_0$, where $a_n \neq 0$, and n is a nonnegative integer, you should be able to identify the following.

  (a) Degree: n

  (b) Terms: $a_n x^n, a_{n-1} x^{n-1}, \ldots, a_1 x, a_0$

  (c) Coefficients: $a_n, a_{n-1}, \ldots, a_1, a_0$

  (d) Leading coefficient: a_n

  (e) Constant term: a_0

- You should be able to add and subtract polynomials.

- You should be able to multiply polynomials by the Distributive Properties.

- You should be able to multiply two binomials by the FOIL Method.

- You should know the special binomial products.

  (a) $(u + v)(u - v) = u^2 - v^2$

  (b) $(u \pm v)^2 = u^2 \pm 2uv + v^2$

  (c) $(u \pm v)^3 = u^3 \pm 3u^2v + 3uv^2 \pm v^3$

- You should be able to factor out all common factors, the first step in factoring.

- You should be able to factor the following special polynomial forms.

  (a) $u^2 - v^2 = (u + v)(u - v)$

  (b) $u^2 \pm 2uv + v^2 = (u \pm v)^2$

  (c) $u^3 \pm v^3 = (u \pm v)(u^2 \mp uv + v^2)$

- You should be able to factor by grouping.

- You should be able to factor some trinomials by grouping.

Vocabulary Check

1. n; a_n; a_0

2. descending

3. monomial; binomial; trinomial

4. like terms

5. First terms; Outer terms; Inner terms; Last terms

6. factoring

7. completely factored

1. (d) 12 is a polynomial of degree zero.

2. (e) $-3x^5 + 2x^3 + x$ is a polynomial of degree five.

3. (b) $1 - 2x^3 = -2x^3 + 1$ is a binomial with leading coefficient -2.

4. (a) $3x^2$ is a monomial of positive degree.

5. (f) $\frac{2}{3}x^4 + x^2 + 10$ is a trinomial with leading coefficient $\frac{2}{3}$.

6. (c) $x^3 + 3x^2 + 3x + 1$ is a third-degree polynomial with leading coefficient 1.

7. $-2x^3$; $-2x^3 + 5$;
$-2x^3 + 4x^2 - 3x + 20$, etc.
(Answers will vary.)

8. $6x^5 + 3x + 1$
(Answers will vary.)

9. $-15x^4 + 1$; $-3x^4 + 7x^2$;
$-5x^4 - 6x$, etc.
(Answers will vary.)

10. $20x^3 + 5$ (Answers will vary.)

11. (a) Standard form: $-\frac{1}{2}x^5 + 14x$

    (b) Degree: 5

    Leading coefficient: $-\frac{1}{2}$

    (c) Binomial

12. (a) Standard form: $2x^2 - x + 1$

    (b) Degree: 2

    Leading coefficient: 2

    (c) Trinomial

13. (a) Standard form: $-3x^4 + 2x^2 - 5$

    (b) Degree: 4

    Leading coefficient: -3

    (c) Trinomial

14. (a) Standard form: $7x$

    (b) Degree: 1

    Leading coefficient: 7

    (c) Monomial

15. (a) Standard form: $x^5 - 1$

    (b) Degree: 5

    Leading coefficient: 1

    (c) Binomial

16. (a) Standard form: $25y^2 - y + 1$

    (b) Degree: 2

    Leading coefficient: 25

    (c) Trinomial

17. (a) Standard form: 3

    (b) Degree: 0

    Leading coefficient: 3

    (c) Monomial

18. (a) Standard form: $t^2 + 9$

    (b) Degree: 2

    Leading coefficient: 1

    (c) Binomial

19. (a) Standard form: $-4x^5 + 6x^4 + 1$

    (b) Degree: 5

    Leading coefficient: -4

    (c) Trinomial

20. (a) Standard form: $2x + 3$

    (b) Degree: 1

    Leading coefficient: 2

    (c) Binomial

21. (a) Standard form: $4x^3y$

    (b) Degree: 4 (add the exponents
                  on x and y)

    Leading coefficient: 4

    (c) Monomial

22. (a) Standard form:
$-x^5y + 2x^2y^2 + xy^4$

    (b) Degree: 6

    Leading coefficient: -1

    (c) Trinomial

23. $2x - 3x^3 + 8$ *is* a polynomial.
Standard form: $-3x^3 + 2x + 8$

24. $2x^3 + x - 3x^{-1}$ is *not* a
polynomial because it includes a
term with a negative exponent.

25. $\dfrac{3x + 4}{x} = 3 + \dfrac{4}{x} = 3 + 4x^{-1}$ is *not*
a polynomial because it includes a
term with a negative exponent.

26. $\dfrac{x^2 + 2x - 3}{2}$ *is* a polynomial.

Standard form: $\dfrac{1}{2}x^2 + x - \dfrac{3}{2}$

27. $y^2 - y^4 + y^3$ *is* a polynomial.
Standard form: $-y^4 + y^3 + y^2$

28. $\sqrt{y^2 - y^4}$ is *not* a polynomial because of the square root.

29. $(6x + 5) - (8x + 15) = 6x + 5 - 8x - 15$

                              $= (6x - 8x) + (5 - 15)$

                              $= -2x - 10$

30. $(2x^2 + 1) - (x^2 - 2x + 1) = 2x^2 + 1 - x^2 + 2x - 1$

                                       $= (2x^2 - x^2) + 2x + (1 - 1)$

                                       $= x^2 + 2x$

31. $-(x^3 - 2) + (4x^3 - 2x) = -x^3 + 2 + 4x^3 - 2x$

                                  $= (4x^3 - x^3) - 2x + 2$

                                  $= 3x^3 - 2x + 2$

32. $-(5x^2 - 1) - (-3x^2 + 5) = -5x^2 + 1 + 3x^2 - 5$

$$= (-5x^2 + 3x^2) + (1 - 5)$$

$$= -2x^2 - 4$$

33. $(15x^2 - 6) - (-8.3x^3 - 14.7x^2 - 17) = 15x^2 - 6 + 8.3x^3 + 14.7x^2 + 17$

$$= 8.3x^3 + (15x^2 + 14.7x^2) + (-6 + 17)$$

$$= 8.3x^3 + 29.7x^2 + 11$$

34. $(15.2x^4 - 18x - 19.1) - (13.9x^4 - 9.6x + 15) = 15.2x^4 - 18x - 19.1 - 13.9x^4 + 9.6x - 15$

$$= (15.2x^4 - 13.9x^4) + (-18x + 9.6x) + (-19.1 - 15)$$

$$= 1.3x^4 - 8.4x - 34.1$$

35. $5z - [3z - (10z + 8)] = 5z - (3z - 10z - 8)$

$$= 5z - 3z + 10z + 8$$

$$= (5z - 3z + 10z) + 8$$

$$= 12z + 8$$

36. $(y^3 + 1) - [(y^2 + 1) + (3y - 7)] = y^3 + 1 - (y^2 + 1) - (3y - 7)$

$$= y^3 + 1 - y^2 - 1 - 3y + 7$$

$$= y^3 - y^2 - 3y + (1 - 1 + 7)$$

$$= y^3 - y^2 - 3y + 7$$

37. $3x(x^2 - 2x + 1) = 3x(x^2) + 3x(-2x) + 3x(1)$

$$= 3x^3 - 6x^2 + 3x$$

38. $y^2(4y^2 + 2y - 3) = y^2(4y^2) + y^2(2y) + y^2(-3)$

$$= 4y^4 + 2y^3 - 3y^2$$

39. $-5z(3z - 1) = -5z(3z) + (-5z)(-1)$

$$= -15z^2 + 5z$$

40. $(-3x)(5x + 2) = -3x(5x) + (-3x)(2)$

$$= -15x^2 - 6x$$

41. $(1 - x^3)(4x) = 1(4x) - x^3(4x)$

$$= 4x - 4x^4$$

$$= -4x^4 + 4x$$

42. $-4x(3 - x^3) = -4x(3) + (-4x)(-x^3)$

$$= -12x + 4x^4$$

$$= 4x^4 - 12x$$

43. $(2.5x^2 + 3)(3x) = (2.5x^2)(3x) + (3)(3x)$

$$= 7.5x^3 + 9x$$

44. $(2 - 3.5y)(2y^3) = 2(2y^3) + (-3.5y)(2y^3)$

$$= 4y^3 - 7y^4 = -7y^4 + 4y^3$$

45. $-4x\left(\frac{1}{8}x + 3\right) = (-4x)\left(\frac{1}{8}x\right) + (-4x)(3)$

$$= -\frac{1}{2}x^2 - 12x$$

46. $2y\left(4 - \frac{7}{8}y\right) = 2y(4) + 2y\left(-\frac{7}{8}y\right)$

$$= 8y - \frac{7}{4}y^2$$

$$= -\frac{7}{4}y^2 + 8y$$

47. $(x + 3)(x + 4) = x^2 + 4x + 3x + 12$ FOIL

$$= x^2 + 7x + 12$$

48. $(x - 5)(x + 10) = x^2 + 10x - 5x - 50$ FOIL

$$= x^2 + 5x - 50$$

49. $(3x - 5)(2x + 1) = 6x^2 + 3x - 10x - 5$ FOIL
$$= 6x^2 - 7x - 5$$

50. $(7x - 2)(4x - 3) = 28x^2 - 21x - 8x + 6$ FOIL
$$= 28x^2 - 29x + 6$$

51. Multiply:
$$
\begin{array}{r}
x^2 - x + 1 \\
x^2 + x + 1 \\
\hline
x^4 - x^3 + x^2 \\
x^3 - x^2 + x \\
x^2 - x + 1 \\
\hline
x^4 - 0x^3 + x^2 + 0x + 1 = x^4 + x^2 + 1
\end{array}
$$

52. Multiply:
$$
\begin{array}{r}
x^2 + 3x - 2 \\
x^2 - 3x - 2 \\
\hline
x^4 + 3x^3 - 2x^2 \\
-3x^3 - 9x^2 + 6x \\
- 2x^2 - 6x + 4 \\
\hline
x^4 + 0x^3 - 13x^2 + 0x + 4 = x^4 - 13x^2 + 4
\end{array}
$$

53. $(x + 10)(x - 10) = x^2 - 10^2 = x^2 - 100$

54. $(2x + 3)(2x - 3) = (2x)^2 - 3^2 = 4x^2 - 9$

55. $(x + 2y)(x - 2y) = x^2 - (2y)^2 = x^2 - 4y^2$

56. $(2x + 3y)(2x - 3y) = (2x)^2 - (3y)^2 = 4x^2 - 9y^2$

57. $(2x + 3)^2 = (2x)^2 + 2(2x)(3) + 3^2$
$$= 4x^2 + 12x + 9$$

58. $(4x + 5)^2 = (4x)^2 + 2(4x)(5) + 5^2$
$$= 16x^2 + 40x + 25$$

59. $(2x - 5y)^2 = (2x)^2 - 2(2x)(5y) + (5y)^2$
$$= 4x^2 - 20xy + 25y^2$$

60. $(5 - 8x)^2 = 5^2 + (2)(5)(-8x) + (-8x)^2$
$$= 25 - 80x + 64x^2$$

61. $(x + 1)^3 = x^3 + 3x^2(1) + 3x(1^2) + 1^3$
$$= x^3 + 3x^2 + 3x + 1$$

62. $(x - 2)^3 = x^3 - 3x^2(2) + 3x(2)^2 - 2^3$
$$= x^3 - 6x^2 + 12x - 8$$

63. $(2x - y)^3 = (2x)^3 - 3(2x)^2y + 3(2x)y^2 - y^3$
$$= 8x^3 - 12x^2y + 6xy^2 - y^3$$

64. $(3x + 2y)^3 = (3x)^3 + 3(3x)^2(2y) + 3(3x)(2y)^2 + (2y)^3$
$$= 27x^3 + 54x^2y + 36xy^2 + 8y^3$$

65. $(4x^3 - 3)^2 = (4x^3)^2 - 2(4x^3)(3) + (3)^2$
$$= 16x^6 - 24x^3 + 9$$

66. $(8x + 3)^2 = (8x)^2 + 2(8x)(3) + 3^2$
$$= 64x^2 + 48x + 9$$

67. $[(m - 3) + n][(m - 3) - n] = (m - 3)^2 - n^2$
$$= m^2 - 6m + 9 - n^2$$
$$= m^2 - n^2 - 6m + 9$$

68. $[(x + y) + 1][(x + y) - 1] = (x + y)^2 - 1^2$
$$= x^2 + 2xy + y^2 - 1$$

69. $[(x - 3) + y]^2 = (x - 3)^2 + 2y(x - 3) + y^2$
$$= x^2 - 6x + 9 + 2xy - 6y + y^2$$
$$= x^2 + 2xy + y^2 - 6x - 6y + 9$$

70. $[(x + 1) - y]^2 = (x + 1)^2 + 2(x + 1)(-y) + (-y)^2$
$$= x^2 + 2x + 1 - 2xy - 2y + y^2$$
$$= x^2 - 2xy + y^2 + 2x - 2y + 1$$

71. $(2r^2 - 5)(2r^2 + 5) = (2r^2)^2 - 5^2 = 4r^4 - 25$

72. $(3a^3 - 4b^2)(3a^3 + 4b^2) = (3a^3)^2 - (4b^2)^2$
$$= 9a^6 - 16b^4$$

73. $\left(\frac{1}{2}x - 3\right)^2 = \left(\frac{1}{2}x\right)^2 - 2\left(\frac{1}{2}x\right)(3) + 3^2$
$$= \frac{1}{4}x^2 - 3x + 9$$

74. $\left(\frac{2}{3}t + 5\right)^2 = \left(\frac{2}{3}t\right)^2 + 2\left(\frac{2}{3}t\right)(5) + (5)^2$
$$= \frac{4}{9}t^2 + \frac{20}{3}t + 25$$

75. $\left(\frac{1}{3}x - 2\right)\left(\frac{1}{3}x + 2\right) = \left(\frac{1}{3}x\right)^2 - (2)^2$
$$= \frac{1}{9}x^2 - 4$$

76. $\left(2x + \frac{1}{5}\right)\left(2x - \frac{1}{5}\right) = (2x)^2 - \left(\frac{1}{5}\right)^2$

$\qquad = 4x^2 - \frac{1}{25}$

77. $(1.2x + 3)^2 = (1.2x)^2 + 2(1.2x)(3) + 3^2$

$\qquad = 1.44x^2 + 7.2x + 9$

78. $(1.5y - 3)^2 = (1.5y)^2 + 2(1.5y)(-3) + (-3)^2$

$\qquad = 2.25y^2 - 9y + 9$

79. $(1.5x - 4)(1.5x + 4) = (1.5x)^2 - 4^2$

$\qquad = 2.25x^2 - 16$

80. $(2.5y + 3)(2.5y - 3) = (2.5y)^2 - (3)^2$

$\qquad = 6.25y^2 - 9$

81. $5x(x + 1) - 3x(x + 1) = 2x(x + 1)$

$\qquad = 2x^2 + 2x$

82. $(2x - 1)(x + 3) + 3(x + 3) = (2x + 2)(x + 3) = 2x^2 + 6x + 2x + 6$ FOIL

$\qquad = 2x^2 + 8x + 6$

83. $(u + 2)(u - 2)(u^2 + 4) = (u^2 - 4)(u^2 + 4)$

$\qquad = u^4 - 16$

84. $(x + y)(x - y)(x^2 + y^2) = (x^2 - y^2)(x^2 + y^2)$

$\qquad = (x^2)^2 - (y^2)^2 = x^4 - y^4$

85. $\left(\sqrt{x} + \sqrt{y}\right)\left(\sqrt{x} - \sqrt{y}\right) = \left(\sqrt{x}\right)^2 - \left(\sqrt{y}\right)^2$

$\qquad = x - y$

86. $\left(5 + \sqrt{x}\right)\left(5 - \sqrt{x}\right) = (5)^2 - \left(\sqrt{x}\right)^2$

$\qquad = 25 - x$

87. $\left(x - \sqrt{5}\right)^2 = x^2 - 2(x)\left(\sqrt{5}\right) + \left(\sqrt{5}\right)^2$

$\qquad = x^2 - 2\sqrt{5}x + 5$

88. $\left(x + \sqrt{3}\right)^2 = x^2 + 2x\sqrt{3} + \left(\sqrt{3}\right)^2$

$\qquad = x^2 + 2\sqrt{3}x + 3$

89. $3x + 6 = 3(x + 2)$

90. $5y - 30 = 5(y - 6)$

91. $2x^3 - 6x = 2x(x^2 - 3)$

92. $4x^3 - 6x^2 + 12x = 2x(2x^2 - 3x + 6)$

93. $x(x - 1) + 6(x - 1) = (x - 1)(x + 6)$

94. $3x(x + 2) - 4(x + 2) = (x + 2)(3x - 4)$

95. $(x + 3)^2 - 4(x + 3) = (x + 3)[(x + 3) - 4]$

$\qquad = (x + 3)(x - 1)$

96. $(3x - 1)^2 + (3x - 1) = (3x - 1 + 1)(3x - 1)$

$\qquad = 3x(3x - 1)$

97. $\frac{1}{2}x + 4 = \frac{1}{2}x + \frac{8}{2}$

$\qquad = \frac{1}{2}(x + 8)$

98. $\frac{1}{3}y + 5 = \frac{1}{3}y + \frac{15}{3} = \frac{1}{3}(y + 15)$

99. $\frac{1}{2}x^3 + 2x^2 - 5x = \frac{1}{2}x^3 + \frac{4}{2}x^2 - \frac{10}{2}x$

$\qquad = \frac{1}{2}x(x^2 + 4x - 10)$

100. $\frac{1}{3}y^4 - 5y^2 + 2y = \frac{1}{3}y^4 - \frac{15}{3}y^2 + \frac{6}{3}y$

$\qquad = \frac{1}{3}y(y^3 - 15y + 6)$

101. $\frac{2}{3}x(x - 3) - 4(x - 3) = \frac{2}{3}x(x - 3) - \frac{12}{3}(x - 3)$

$\qquad = \frac{2}{3}(x - 3)(x - 6)$

102. $\frac{4}{5}y(y + 1) - 2(y + 1) = \frac{4}{5}y(y + 1) - \frac{10}{5}(y + 1) = \frac{2}{5}(y + 1)(2y - 5)$

103. $x^2 - 81 = x^2 - 9^2$

$\qquad = (x + 9)(x - 9)$

104. $x^2 - 49 = x^2 - 7^2$

$\qquad = (x + 7)(x - 7)$

105. $32y^2 - 18 = 2(16y^2 - 9)$

$\qquad = 2[(4y)^2 - 3^2]$

$\qquad = 2(4y + 3)(4y - 3)$

106. $4 - 36y^2 = 4(1 - 9y^2)$

$\qquad = 4[1^2 - (3y)^2]$

$\qquad = 4(1 + 3y)(1 - 3y)$

107. $16x^2 - \frac{1}{9} = (4x)^2 - \left(\frac{1}{3}\right)^2$

$\qquad = \left(4x + \frac{1}{3}\right)\left(4x - \frac{1}{3}\right)$

108. $\frac{4}{25}y^2 - 64 = \left(\frac{2}{5}y\right)^2 - 8^2$

$\qquad = \left(\frac{2}{5}y + 8\right)\left(\frac{2}{5}y - 8\right)$

109. $(x - 1)^2 - 4 = (x - 1)^2 - (2)^2$

$\qquad = [(x - 1) + 2][(x - 1) - 2]$

$\qquad = (x + 1)(x - 3)$

110. $25 - (z + 5)^2 = 5^2 - (z + 5)^2$

$\qquad = (5 - (z + 5))(5 + (z + 5))$

$\qquad = (5 - z - 5)(5 + z + 5)$

$\qquad = -z(z + 10)$

111. $9u^2 - 4v^2 = (3u)^2 - (2v)^2$

$\qquad = (3u + 2v)(3u - 2v)$

112. $25x^2 - 16y^2 = (5x)^2 - (4y)^2$

$\qquad = (5x + 4y)(5x - 4y)$

113. $x^2 - 4x + 4 = x^2 - 2(2)x + 2^2$

$\qquad = (x - 2)^2$

114. $x^2 + 10x + 25 = x^2 + 2(5)(x) + 5^2 = (x + 5)^2$

115. $4t^2 + 4t + 1 = (2t)^2 + 2(2t)(1) + 1^2$

$\qquad = (2t + 1)^2$

116. $9x^2 - 12x + 4 = (3x)^2 - 2(3x)(2) + 2^2 = (3x - 2)^2$

117. $25y^2 - 10y + 1 = (5y)^2 - 2(5y)(1) + 1^2$

$\qquad = (5y - 1)^2$

118. $36y^2 - 108y + 81 = 9(4y^2 - 12y + 9)$

$\qquad = 9[(2y)^2 - 2(2y)(3) + (3)^2]$

$\qquad = 9(2y - 3)^2$

119. $9u^2 + 24uv + 16v^2 = (3u)^2 + 2(3u)(4v) + (4v)^2$

$\qquad = (3u + 4v)^2$

120. $4x^2 - 4xy + y^2 = (2x)^2 - 2(2x)y + y^2$

$\qquad = (2x - y)^2$

121. $x^2 - \frac{4}{3}x + \frac{4}{9} = x^2 - 2(x)\left(\frac{2}{3}\right) + \left(\frac{2}{3}\right)^2$

$\qquad = \left(x - \frac{2}{3}\right)^2$

122. $z^2 + z + \frac{1}{4} = z^2 + 2(z)\left(\frac{1}{2}\right) + \left(\frac{1}{2}\right)^2$

$\qquad = \left(z + \frac{1}{2}\right)^2$

123. $x^3 - 8 = x^3 - 2^3$

$\qquad = (x - 2)(x^2 + 2x + 4)$

124. $x^3 - 27 = x^3 - 3^3 = (x - 3)(x^2 + 3x + 9)$

125. $y^3 + 64 = y^3 + 4^3 = (y + 4)(y^2 - 4y + 16)$

126. $z^3 + 125 = z^3 + 5^3 = (z + 5)(z^2 - 5z + 25)$

127. $8t^3 - 1 = (2t)^3 - 1^3$

$\qquad = (2t - 1)(4t^2 + 2t + 1)$

128. $27x^3 + 8 = (3x)^3 + 2^3 = (3x + 2)(9x^2 - 6x + 4)$

129. $u^3 + 27v^3 = u^3 + (3v)^3$

$\qquad = (u + 3v)(u^2 - 3uv + 9v^2)$

130. $64x^3 - y^3 = (4x)^3 - y^3 = (4x - y)(16x^2 + 4xy + y^2)$

131. $x^2 + x - 2 = (x + 2)(x - 1)$

132. $x^2 + 5x + 6 = (x + 2)(x + 3)$

133. $s^2 - 5s + 6 = (s - 3)(s - 2)$

134. $t^2 - t - 6 = (t + 2)(t - 3)$

135. $20 - y - y^2 = -(y^2 + y - 20)$

$\qquad = -(y + 5)(y - 4)$

136. $24 + 5z - z^2 = -(z^2 - 5z - 24) = -(z - 8)(z + 3)$

137. $x^2 - 30x + 200 = (x - 20)(x - 10)$

138. $x^2 - 13x + 42 = (x - 6)(x - 7)$

139. $3x^2 - 5x + 2 = (3x - 2)(x - 1)$

140. $2x^2 - x - 1 = (2x + 1)(x - 1)$

141. $5x^2 + 26x + 5 = (5x + 1)(x + 5)$

142. $12x^2 + 7x + 1 = (3x + 1)(4x + 1)$

143. $-9z^2 + 3z + 2 = -(9z^2 - 3z - 2)$
$$= -(3z - 2)(3z + 1)$$

144. $-5u^2 - 13u + 6 = -(5u^2 + 13u - 6)$
$$= -(5u - 2)(u + 3)$$

145. $x^3 - x^2 + 2x - 2 = x^2(x - 1) + 2(x - 1)$
$$= (x - 1)(x^2 + 2)$$

146. $x^3 + 5x^2 - 5x - 25 = x^2(x + 5) - 5(x + 5)$
$$= (x + 5)(x^2 - 5)$$

147. $2x^3 - x^2 - 6x + 3 = x^2(2x - 1) - 3(2x - 1)$
$$= (2x - 1)(x^2 - 3)$$

148. $5x^3 - 10x^2 + 3x - 6 = 5x^2(x - 2) + 3(x - 2)$
$$= (x - 2)(5x^2 + 3)$$

149. $6 + 2x - 3x^3 - x^4 = 2(3 + x) - x^3(3 + x)$
$$= (3 + x)(2 - x^3)$$

150. $x^5 + 2x^3 + x^2 + 2 = x^3(x^2 + 2) + (x^2 + 2)$
$$= (x^2 + 2)(x^3 + 1)$$
$$= (x^2 + 2)(x + 1)(x^2 - x + 1)$$

151. $6x^3 - 2x + 3x^2 - 1 = 2x(3x^2 - 1) + 1(3x^2 - 1)$
$$= (3x^2 - 1)(2x + 1)$$

152. $8x^5 - 6x^2 + 12x^3 - 9 = 2x^2(4x^3 - 3) + 3(4x^3 - 3)$
$$= (4x^3 - 3)(2x^2 + 3)$$

153. $a \cdot c = (3)(8) = 24$. Rewrite the middle term, $10x = 6x + 4x$, since $(6)(4) = 24$ and $6 + 4 = 10$.
$$3x^2 + 10x + 8 = 3x^2 + 6x + 4x + 8$$
$$= 3x(x + 2) + 4(x + 2)$$
$$= (x + 2)(3x + 4)$$

154. $a \cdot c = (2)(9) = 18$. Rewrite the middle term, $9x = 6x + 3x$, since $(6)(3) = 18$ and $6 + 3 = 9$.
$$2x^2 + 9x + 9 = 2x^2 + 6x + 3x + 9$$
$$= 2x(x + 3) + 3(x + 3)$$
$$= (x + 3)(2x + 3)$$

155. $a \cdot c = (6)(-2) = -12$. Rewrite the middle term, $x = 4x - 3x$, since $4(-3) = -12$ and $4 + (-3) = 1$.
$$6x^2 + x - 2 = 6x^2 + 4x - 3x - 2$$
$$= 2x(3x + 2) - 1(3x + 2)$$
$$= (2x - 1)(3x + 2)$$

156. $a \cdot c = (6)(-15) = -90$. Rewrite the middle term, $-x = -10x + 9x$, since $(-10)(9) = -90$ and $-10 + 9 = -1$.
$$6x^2 - x - 15 = 6x^2 - 10x + 9x - 15$$
$$= 2x(3x - 5) + 3(3x - 5)$$
$$= (2x + 3)(3x - 5)$$

157. $a \cdot c = (15)(2) = 30$. Rewrite the middle term, $-11x = -6x - 5x$, since $(-6)(-5) = 30$ and $(-6) + (-5) = -11$.
$$15x^2 - 11x + 2 = 15x^2 - 6x - 5x + 2$$
$$= 3x(5x - 2) - 1(5x - 2)$$
$$= (3x - 1)(5x - 2)$$

158. $a \cdot c = (12)(1) = 12$. Rewrite the middle term, $-13x = -12x - x$, since $(-12)(-1) = 12$ and $-12 - 1 = -13$.
$$12x^2 - 13x + 1 = 12x^2 - 12x - x + 1$$
$$= 12x(x - 1) - 1(x - 1) = (x - 1)(12x - 1)$$

159. $6x^2 - 54 = 6(x^2 - 9)$

$\qquad = 6(x + 3)(x - 3)$

160. $12x^2 - 48 = 12(x^2 - 4)$

$\qquad = 12(x + 2)(x - 2)$

161. $x^3 - 4x^2 = x^2(x - 4)$

162. $x^3 - 9x = x(x^2 - 9)$

$\qquad = x(x + 3)(x - 3)$

163. $x^2 - 2x + 1 = (x - 1)^2$

164. $16 + 6x - x^2 = 16 + 8x - 2x - x^2$

$\qquad = (8 - x)(2 + x)$

165. $1 - 4x + 4x^2 = (1 - 2x)^2$

166. $-9x^2 + 6x - 1 = -9x^2 + 3x + 3x - 1$

$\qquad = (3x - 1)(-3x + 1)$

167. $2x^2 + 4x - 2x^3 = -2x(-x - 2 + x^2)$

$\qquad = -2x(x^2 - x - 2)$

$\qquad = -2x(x + 1)(x - 2)$

168. $2y^3 - 7y^2 - 15y = y(2y^2 - 7y - 15)$

$\qquad = y(2y^2 - 10y + 3y - 15)$

$\qquad = y(2y + 3)(y - 5)$

169. $9x^2 + 10x + 1 = (9x + 1)(x + 1)$

170. $13x + 6 + 5x^2 = 5x^2 + 13x + 6$

$\qquad = 5x^2 + 10x + 3x + 6$

$\qquad = (5x + 3)(x + 2)$

171. $\frac{1}{81}x^2 + \frac{2}{9}x - 8 = \frac{1}{81}x^2 + \frac{18}{81}x - \frac{648}{81}$

$\qquad = \frac{1}{81}(x^2 + 18x - 648)$

$\qquad = \frac{1}{81}(x + 36)(x - 18)$

172. $\frac{1}{8}x^2 - \frac{1}{96}x - \frac{1}{16} = \frac{1}{96}\left(12x^2 - x - 6\right)$

$\qquad = \frac{1}{96}(4x - 3)(3x + 2)$

173. $3x^3 + x^2 + 15x + 5 = x^2(3x + 1) + 5(3x + 1)$

$\qquad = (3x + 1)(x^2 + 5)$

174. $5 - x + 5x^2 - x^3 = 1(5 - x) + x^2(5 - x)$

$\qquad = (5 - x)(1 + x^2)$

175. $x^4 - 4x^3 + x^2 - 4x = x(x^3 - 4x^2 + x - 4)$

$\qquad = x[x^2(x - 4) + (x - 4)]$

$\qquad = x(x - 4)(x^2 + 1)$

176. $3u - 2u^2 + 6 - u^3 = -u^3 - 2u^2 + 3u + 6$

$\qquad = -u^2(u + 2) + 3(u + 2)$

$\qquad = (u + 2)(-u^2 + 3)$

$\qquad = (u + 2)(3 - u^2)$

177. $\frac{1}{4}x^3 + 3x^2 + \frac{3}{4}x + 9 = \frac{1}{4}x^3 + \frac{12}{4}x^2 + \frac{3}{4}x + \frac{36}{4}$

$\qquad = \frac{1}{4}(x^3 + 12x^2 + 3x + 36)$

$\qquad = \frac{1}{4}[x^2(x + 12) + 3(x + 12)]$

$\qquad = \frac{1}{4}(x + 12)(x^2 + 3)$

178. $\frac{1}{5}x^3 + x^2 - x - 5 = \frac{1}{5}(x^3 + 5x^2 - 5x - 25)$

$\qquad = \frac{1}{5}[x^2(x + 5) - 5(x + 5)]$

$\qquad = \frac{1}{5}(x^2 - 5)(x + 5)$

179. $(t - 1)^2 - 49 = (t - 1)^2 - (7)^2$

$\qquad = [(t - 1) + 7][(t - 1) - 7]$

$\qquad = (t + 6)(t - 8)$

180. $(x^2 + 1)^2 - 4x^2 = [(x^2 + 1) + 2x][(x^2 + 1) - 2x]$

$\qquad = (x^2 + 2x + 1)(x^2 - 2x + 1)$

$\qquad = (x + 1)^2(x - 1)^2$

181. $(x^2 + 8)^2 - 36x^2 = (x^2 + 8)^2 - (6x)^2$

$\qquad = [(x^2 + 8) - 6x][(x^2 + 8) + 6x]$

$\qquad = (x^2 - 6x + 8)(x^2 + 6x + 8)$

$\qquad = (x - 4)(x - 2)(x + 4)(x + 2)$

182. $2t^3 - 16 = 2(t^3 - 8) = 2(t - 2)(t^2 + 2t + 4)$

183. $5x^3 + 40 = 5(x^3 + 8)$

$\qquad = 5(x^3 + 2^3)$

$\qquad = 5(x + 2)(x^2 - 2x + 4)$

184. $4x(2x - 1) + (2x - 1)^2 = (2x - 1)[4x + (2x - 1)]$

$\qquad\qquad\qquad\qquad\qquad = (2x - 1)(6x - 1)$

185. $5(3 - 4x)^2 - 8(3 - 4x)(5x - 1) = (3 - 4x)[5(3 - 4x) - 8(5x - 1)]$

$\qquad\qquad\qquad\qquad\qquad\qquad = (3 - 4x)[15 - 20x - 40x + 8]$

$\qquad\qquad\qquad\qquad\qquad\qquad = (3 - 4x)(23 - 60x)$

186. $2(x + 1)(x - 3)^2 - 3(x + 1)^2(x - 3) = (x + 1)(x - 3)[2(x - 3) - 3(x + 1)]$

$\qquad\qquad\qquad\qquad\qquad\qquad\qquad = (x + 1)(x - 3)[2x - 6 - 3x - 3]$

$\qquad\qquad\qquad\qquad\qquad\qquad\qquad = (x + 1)(x - 3)(-x - 9)$

$\qquad\qquad\qquad\qquad\qquad\qquad\qquad = -(x + 1)(x - 3)(x + 9)$

187. $7(3x + 2)^2(1 - x)^2 + (3x + 2)(1 - x)^3 = (3x + 2)(1 - x)^2[7(3x + 2) + (1 - x)]$

$\qquad\qquad\qquad\qquad\qquad\qquad\qquad\qquad = (3x + 2)(1 - x)^2(21x + 14 + 1 - x)$

$\qquad\qquad\qquad\qquad\qquad\qquad\qquad\qquad = (3x + 2)(1 - x)^2(20x + 15)$

$\qquad\qquad\qquad\qquad\qquad\qquad\qquad\qquad = 5(3x + 2)(1 - x)^2(4x + 3)$

188. $7x(2)(x^2 + 1)(2x) - (x^2 + 1)^2(7) = 7(x^2 + 1)[4x^2 - (x^2 + 1)]$

$\qquad\qquad\qquad\qquad\qquad\qquad\quad = 7(x^2 + 1)(3x^2 - 1)$

189. $3(x - 2)^2(x + 1)^4 + (x - 2)^3(4)(x + 1)^3 = (x - 2)^2(x + 1)^3[3(x + 1) + 4(x - 2)]$

$\qquad\qquad\qquad\qquad\qquad\qquad\qquad\qquad = (x - 2)^2(x + 1)^3(3x + 3 + 4x - 8)$

$\qquad\qquad\qquad\qquad\qquad\qquad\qquad\qquad = (x - 2)^2(x + 1)^3(7x - 5)$

190. $2x(x - 5)^4 - x^2(4)(x - 5)^3 = 2x(x - 5)^3[(x - 5) - 2x]$

$\qquad\qquad\qquad\qquad\qquad\qquad = 2x(x - 5)^3(-x - 5)$

$\qquad\qquad\qquad\qquad\qquad\qquad = -2x(x - 5)^3(x + 5)$

191. $5(x^6 + 1)^4(6x^5)(3x + 2)^3 + 3(3x + 2)^2(3)(x^6 + 1)^5 = 3(x^6 + 1)^4(3x + 2)^2[10x^5(3x + 2) + 3(x^6 + 1)]$

$\qquad\qquad\qquad\qquad\qquad\qquad\qquad\qquad\qquad\qquad = 3(x^6 + 1)^4(3x + 2)^2(30x^6 + 20x^5 + 3x^6 + 3)$

$\qquad\qquad\qquad\qquad\qquad\qquad\qquad\qquad\qquad\qquad = 3(x^6 + 1)^4(3x + 2)^2(33x^6 + 20x^5 + 3)$

$\qquad\qquad\qquad\qquad\qquad\qquad\qquad\qquad\qquad\qquad = 3[(x^2)^3 + 1]^4(3x + 2)^2(33x^6 + 20x^5 + 3)$

$\qquad\qquad\qquad\qquad\qquad\qquad\qquad\qquad\qquad\qquad = 3[(x^2 + 1)(x^4 - x^2 + 1)]^4(3x + 2)^2(33x^6 + 20x^5 + 3)$

$\qquad\qquad\qquad\qquad\qquad\qquad\qquad\qquad\qquad\qquad = 3(x^2 + 1)^4(x^4 - x^2 + 1)^4(3x + 2)^2(33x^6 + 20x^5 + 3)$

192. $\dfrac{x^2}{2}(x^2 + 1)^4 - (x^2 + 1)^5 = (x^2 + 1)^4\left[\dfrac{x^2}{2} - (x^2 + 1)\right]$

$\qquad\qquad\qquad\qquad\qquad = (x^2 + 1)^4\left(-\dfrac{x^2}{2} - 1\right)$

$\qquad\qquad\qquad\qquad\qquad = -(x^2 + 1)^4\left(\dfrac{x^2}{2} + 1\right)$

193. For $x^2 + bx - 15$ to be factorable, b must equal $m + n$ where $mn = -15$.

Factors of -15	Sum of factors
$(15)(-1)$	$15 + (-1) = 14$
$(-15)(1)$	$-15 + 1 = -14$
$(3)(-5)$	$3 + (-5) = -2$
$(-3)(5)$	$-3 + 5 = 2$

The possible b-values are $14, -14, -2,$ or 2.

194. For $x^2 + bx + 50$ to be factorable, b must equal $m + n$ where $mn = 50$.

Factors of 50	Sum of factors
$(1)(50)$	$1 + 50 = 51$
$(-1)(-50)$	$-1 + (-50) = -51$
$(5)(10)$	$5 + 10 = 15$
$(-5)(-10)$	$-5 + (-10) = -15$
$(2)(25)$	$2 + 25 = 27$
$(-2)(-25)$	$-2 + (-25) = -27$

The possible b-values are $-51, 51, -15, 15, -27, 27$.

195. For $x^2 + bx - 12$ to be factorable, b must equal $m + n$ where $mn = -12$.

Factors of -12	Sum of factors
$(12)(-1)$	$12 + (-1) = 11$
$(-12)(1)$	$-12 + 1 = -11$
$(2)(-6)$	$2 + (-6) = -4$
$(-2)(6)$	$-2 + 6 = 4$
$(3)(-4)$	$3 + (-4) = -1$
$(-3)(4)$	$-3 + 4 = 1$

The possible b-values are $11, -11, -4, 4, -1, 1$.

196. For $x^2 + bx + 24$ to be factorable, b must equal $m + n$ where $mn = 24$.

Factors of 24	Sum of factors
$(1)(24)$	$1 + 24 = 25$
$(-1)(-24)$	$-1 + (-24) = -25$
$(2)(12)$	$2 + 12 = 14$
$(-2)(-12)$	$-2 + (-12) = -14$
$(3)(8)$	$3 + 8 = 11$
$(-3)(-8)$	$-3 + (-8) = -11$
$(4)(6)$	$4 + 6 = 10$
$(-4)(-6)$	$-4 + (-6) = -10$

The possible b-values are $25, -25, 14, -14$ $11, -11, 10, -10$.

197. For $2x^2 + 5x + c$ to be factorable, the factors of $2c$ must add up to 5.

Possible c-values	$2c$	Factors of $2c$ that add up to 5
2	4	$(1)(4) = 4$ and $1 + 4 = 5$
3	6	$(2)(3) = 6$ and $2 + 3 = 5$
-3	-6	$(6)(-1) = -6$ and $6 + (-1) = 5$
-7	-14	$(7)(-2) = -14$ and $7 + (-2) = 5$
-12	-24	$(8)(-3) = -24$ and $8 + (-3) = 5$

These are a few possible c-values. There are *many* correct answers.

If $c = 2$: $\quad 2x^2 + 5x + 2 = (2x + 1)(x + 2)$

If $c = 3$: $\quad 2x^2 + 5x + 3 = (2x + 3)(x + 1)$

If $c = -3$: $\quad 2x^2 + 5x - 3 = (2x - 1)(x + 3)$

If $c = -7$: $\quad 2x^2 + 5x - 7 = (2x + 7)(x - 1)$

If $c = -12$: $2x^2 + 5x - 12 = (2x - 3)(x + 4)$

198. For $3x^2 - 10x + c$ to be factorable, the factors of $3c$ must add up to -10.

Possible c-values	$3c$	Factors of $3c$ that add up to -10
3	9	$(-1)(-9) = 9$ and $-1 + (-9) = -10$
-8	-24	$(-12)(2) = -24$ and $-12 + 2 = -10$
8	24	$(-6)(-4) = 24$ and $-6 + (-4) = -10$

These are a few possible c-values. There are *many* correct answers.

If $c = 3$: $3x^2 - 10x + 3 = (3x - 1)(x - 3)$

If $c = -8$: $3x^2 - 10x - 8 = (3x + 2)(x - 4)$

If $c = 8$: $3x^2 - 10x + 8 = (3x - 4)(x - 2)$

199. For $3x^2 - x + c$ to be factorable, the factors of $3c$ must add up to -1.

Possible c-values	$3c$	Factors of $3c$ must add up to -1
-2	-6	$(2)(-3) = -6$ and $2 + (-3) = -1$
-4	-12	$(3)(-4) = -12$ and $3 + (-4) = -1$
-10	-30	$(5)(-6) = -30$ and $5 + (-6) = -1$

These are a few possible c-values. There are *many* correct answers.

If $c = -2$: $3x^2 - x - 2 = (3x + 2)(x - 1)$

If $c = -4$: $3x^2 - x - 4 = (3x - 4)(x + 1)$

If $c = -10$: $3x^2 - x - 10 = (3x + 5)(x - 2)$

200. For $2x^2 + 9x + c$ to be factorable, the factors of $2c$ must add up to 9. There are many possibilities.

Possible c-values	$2c$	Factors of $2c$ that add up to 9
4	8	$(1)(8) = 8$ and $1 + 8 = 9$
7	14	$(2)(7) = 14$ and $2 + 7 = 9$
9	18	$(3)(6) = 18$ and $3 + 6 = 9$
10	20	$(4)(5) = 20$ and $4 + 5 = 9$
-11	-22	$(-2)(11) = -22$ and $-2 + 11 = 9$
-18	-36	$(-3)(12) = -36$ and $-3 + 12 = 9$

These are a few possible c-values.

$2x^2 + 9x + 4 = (2x + 1)(x + 4)$

$2x^2 + 9x + 7 = (2x + 7)(x + 1)$

$2x^2 + 9x + 9 = (2x + 3)(x + 3)$

$2x^2 + 9x + 10 = (2x + 5)(x + 2)$

$2x^2 + 9x - 11 = (2x + 11)(x - 1)$

$2x^2 + 9x - 18 = (2x - 3)(x + 6)$

201. (a) Profit = Revenue − Cost

Profit = $95x - (73x + 25,000)$

$= 95x - 73x - 25,000 = 22x - 25,000$

(b) For $x = 5000$:

Profit = $22(5000) - 25,000$

$= 110,000 - 25,000 = \$85,000$

202. (a) Profit = Revenue − Cost

$P = 36x - (460 + 12x)$

$= 36x - 460 - 12x$

$= 24x - 460$

(b) When $x = 42$, $P = 24(42) - 460 = \$548$.

203. (a) $500(1 + r)^2 = 500(r + 1)^2 = 500(r^2 + 2r + 1)$

$= 500r^2 + 1000r + 500$

(b)

r	$2\frac{1}{2}\%$	3%	4%	$4\frac{1}{2}\%$	5%
$500(1 + r)^2$	\$525.31	\$530.45	\$540.80	\$546.01	\$551.25

(c) As r increases, the amount increases.

204. (a) $1200(1 + r)^3 = 1200(1 + 3r + 3r^2 + r^3)$

$= 1200(r^3 + 3r^2 + 3r + 1)$

$= 1200r^3 + 3600r^2 + 3600r + 1200$

(b)

r	2%	3%	$3\frac{1}{2}\%$	4%	$4\frac{1}{2}\%$
$1200(1 + r)^3$	\$1273.45	\$1311.27	\$1330.46	\$1349.84	\$1369.40

(c) Amount increases with increasing r.

205. (a) $V = l \cdot w \cdot h = (26 - 2x)(18 - 2x)(x)$

$= 2(13 - x)(2)(9 - x)(x)$

$= 4x(-1)(x - 13)(-1)(x - 9)$

$= 4x(x - 13)(x - 9)$

$= 4x^3 - 88x^2 + 468x$

(b)

x (cm)	1	2	3
V (cm³)	384	616	720

206. (a) Volume = length × width × height

$= \frac{1}{2}(45 - 3x)(15 - 2x)x$

$= \frac{1}{2}(45 - 3x)(15x - 2x^2)$

$= \frac{1}{2}[675x - 90x^2 - 45x^2 + 6x^3]$

$= \frac{1}{2}(6x^3 - 135x^2 + 675x)$

x (cm)	3	5	7
Volume (cm³)	486	375	84

(b) When $x = 3$: $V = \frac{1}{2}[6(3)^3 - 135(3)^2 + 675(3)] = \frac{1}{2}[6 \cdot 27 - 135 \cdot 9 + 2025]$

$= \frac{1}{2}[162 - 1215 + 2025] = \frac{1}{2}(972)$

$= 486$ cubic centimeters

When $x = 5$: $V = \frac{1}{2}[6(5)^3 - 135(5)^2 + 675(5)] = \frac{1}{2}[6 \cdot 125 - 135 \cdot 25 + 3375]$

$= \frac{1}{2}[750 - 3375 + 3375] = \frac{1}{2}(750)$

$= 375$ cubic centimeters

When $x = 7$: $V = \frac{1}{2}[6(7)^3 - 135(7)^2 + 675(7)] = \frac{1}{2}[6 \cdot 343 - 135 \cdot 49 + 4725]$

$= \frac{1}{2}[2058 - 6615 + 4725] = \frac{1}{2}(168)$

$= 84$ cubic centimeters

207. Area = length × width

$= (2x + 14)(22)$

$= (2x)(22) + (14)(22)$

$= 44x + 308$

208. $A = (18 + 2x)(14 + x) = 252 + 18x + 28x + 2x^2$

$= 2x^2 + 46x + 252$

209. (a) Area of shaded region = Area of outer rectangle − Area of inner rectangle

$A = 2x(2x + 6) - x(x + 4)$

$= 4x^2 + 12x - x^2 - 4x$

$= 3x^2 + 8x$

(b) Area of shaded region = Area of outer triangle − Area of inner triangle

$A = \frac{1}{2}(9x)(12x) - \frac{1}{2}(6x)(8x)$

$= 54x^2 - 24x^2$

$= 30x^2$

210. (a) $T = R + B = 1.1x + (0.0475x^2 - 0.001x + 0.23)$

$= 0.0475x^2 + 1.099x + 0.23$

(b)

x mi/hr	30	40	55
T feet	75.95	120.19	204.36

(c) Stopping distance increases at an accelerating rate as speed increases.

211. $3x^2 + 7x + 2 = (3x + 1)(x + 2)$

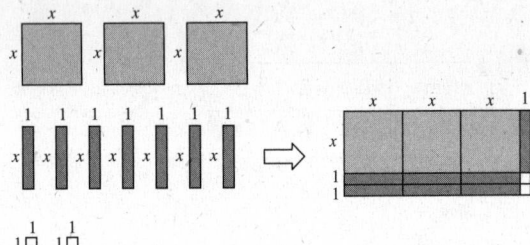

212. $x^2 + 4x + 3 = (x + 3)(x + 1)$

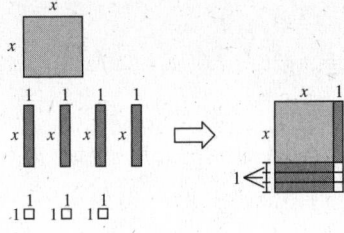

213. $2x^2 + 7x + 3 = (2x + 1)(x + 3)$

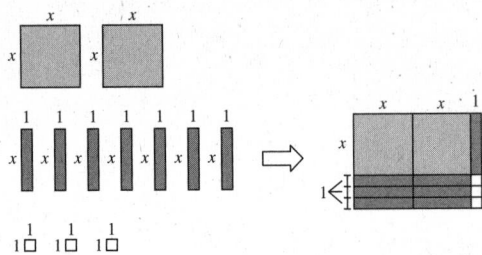

214. $x^2 + 3x + 2 = (x + 2)(x + 1)$

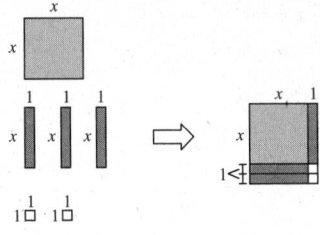

215. $A = \pi(r + 2)^2 - \pi r^2$

$= \pi[(r + 2)^2 - r^2]$

$= \pi[r^2 + 4r + 4 - r^2]$

$= \pi(4r + 4)$

$= 4\pi(r + 1)$

216. Area $= (2r)^2 - \pi r^2$

$= 4r^2 - \pi r^2$

$= r^2(4 - \pi)$

217. $A = 8(18) - 4x^2$

$= 4(36 - x^2)$

$= 4(6 - x)(6 + x)$

218. Area $= \frac{1}{2}(x + 3)\left(\frac{5}{4}\right)(x + 3) - \frac{1}{2}(5)(4)$

$\quad\quad\quad = \frac{5}{8}(x^2 + 6x + 9) - \frac{5}{8}(16)$

$\quad\quad\quad = \frac{5}{8}(x^2 + 6x + 9 - 16)$

$\quad\quad\quad = \frac{5}{8}(x^2 + 6x - 7)$

$\quad\quad\quad = \frac{5}{8}(x + 7)(x - 1)$

219. (a) $V = \pi R^2 h - \pi r^2 h$

$\quad\quad\quad = \pi h(R^2 - r^2)$

$\quad\quad\quad = \pi h(R - r)(R + r)$

    (b) The average radius is $(R + r)/2$. The thickness of the tank is $R - r$.

$$V = \pi h(R - r)(R + r) = 2\pi\left(\frac{R + r}{2}\right)(R - r)h$$

$$= 2\pi(\text{average radius})(\text{thickness})h$$

220. $kQx - kx^2 = kx(Q - x)$

221. False. $(4x^2 + 1)(3x + 1) = 12x^3 + 4x^2 + 3x + 1$

222. False.

$(4x + 3) + (-4x + 6) = 4x + 3 - 4x + 6$

$\quad\quad\quad\quad\quad\quad\quad = 3 + 6 = 9$

223. True. $a^2 - b^2 = (a + b)(a - b)$

224. False. A perfect square trinomial can be factored as the binomial sum squared.

225. Since $x^m x^n = x^{m+n}$, the degree of the product is $m + n$.

226. If the degree of one polynomial is m and the degree of the second polynomial is n (and $n > m$), the degree of the sum of the polynomials is n.

227. The unknown polynomial may be found by adding $-x^3 + 3x^2 + 2x - 1$ and $5x^2 + 8$:

$(-x^3 + 3x^2 + 2x - 1) + (5x^2 + 8) = -x^3 + (3x^2 + 5x^2) + 2x + (-1 + 8)$

$\quad\quad\quad\quad\quad\quad\quad\quad\quad\quad\quad\quad = -x^3 + 8x^2 + 2x + 7$

228. $(x + y)^2 \neq x^2 + y^2$

Let $x = 3$ and $y = 4$.

$(3 + 4)^2 = (7)^2 = 49$

$3^2 + 4^2 = 9 + 16 = 25$ $\quad\Big\rangle$ Not Equal

If either x or y is zero, then $(x + y)^2$ would equal $x^2 + y^2$.

229. $x^{2n} - y^{2n} = (x^n)^2 - (y^n)^2$

$\quad\quad\quad\quad\quad = (x^n + y^n)(x^n - y^n)$

This is not completely factored unless $n = 1$.

For $n = 2$: $(x^2 + y^2)(x^2 - y^2) = (x^2 + y^2)(x + y)(x - y)$

For $n = 3$: $(x^3 + y^3)(x^3 - y^3) = (x + y)(x^2 - xy + y^2)(x - y)(x^2 + xy + y^2)$

For $n = 4$: $(x^4 + y^4)(x^4 - y^4) = (x^4 + y^4)(x^2 + y^2)(x + y)(x - y)$

230. $x^{3n} + y^{3n} = (x^n)^3 + (y^n)^3 = (x^n + y^n)(x^{2n} - x^n y^n + y^{2n})$

Depending on the value of n, this may factor further.

231. $x^{3n} - y^{2n} = (x^n)^3 - (y^n)^2 = x^{3n} - y^{2n}$ is completely factored. For integer values of n greater than 4, the factorizations become more complicated.

232. Answers will vary. A possible answer: A polynomial is in factored form when written as a product of polynomials of lesser degree than the given polynomial.

233. Answers will vary. Some examples:

$\quad\quad x^2 - 3;\ x^2 + x + 1;\ x^2 + 16$

Appendix A.4   Rational Expressions

■ You should be able to find the domain of a rational expression.

■ You should know that a rational expression is the quotient of two polynomials.

■ You should be able to simplify rational expressions by reducing them to lowest terms. This may involve factoring both the numerator and the denominator.

■ You should be able to add, subtract, multiply, and divide rational expressions.

■ You should be able to simplify complex fractions.

■ You should be able to simplify expressions with negative or fraction exponents.

Vocabulary Check

1. domain

2. rational expression

3. complex

4. smaller

5. equivalent

6. difference quotient

---

1. The domain of the polynomial $3x^2 - 4x + 7$ is the set of all real numbers.

2. The domain of the polynomial $2x^2 + 5x - 2$ is the set of all real numbers.

3. The domain of the polynomial $4x^3 + 3, x \geq 0$ is the set of non-negative real numbers, since the polynomial is restricted to that set.

4. The domain of the polynomial $6x^2 - 9, x > 0$ is the set of all positive real numbers because the polynomial is restricted to that set.

5. The domain of $1/(x - 2)$ is the set of all real numbers x such that $x \neq 2$.

6. The domain of $(x + 1)/(2x + 1)$ is the set of all real numbers such that $x \neq -1/2$.

7. The domain of $\sqrt{x + 1}$ is the set of all real numbers x such that $x \geq -1$.

8. The domain of $\sqrt{6 - x}$ is the set of all real numbers x such that $x \leq 6$.

9. $\dfrac{5}{2x} = \dfrac{5(3x)}{(2x)(3x)} = \dfrac{5(3x)}{6x^2}, \quad x \neq 0$

   The missing factor is $3x, x \neq 0$.

10. $\dfrac{3}{4} = \dfrac{3(x + 1)}{4(x + 1)}$

    The missing factor is $(x + 1)$, where $x \neq -1$.

11. $\dfrac{15x^2}{10x} = \dfrac{5x(3x)}{5x(2)} = \dfrac{3x}{2}, \quad x \neq 0$

12. $\dfrac{18y^2}{60y^5} = \dfrac{6y^2(3)}{6y^2(10y^3)} = \dfrac{3}{10y^3}$

13. $\dfrac{3xy}{xy + x} = \dfrac{x(3y)}{x(y + 1)} = \dfrac{3y}{y + 1}, \quad x \neq 0$

14. $\dfrac{2x^2y}{xy - y} = \dfrac{2x^2y}{y(x - 1)} = \dfrac{2x^2}{x - 1}$

15. $\dfrac{4y - 8y^2}{10y - 5} = \dfrac{-4y(2y - 1)}{5(2y - 1)}$

    $= -\dfrac{4y}{5}, \quad y \neq \dfrac{1}{2}$

16. $\dfrac{9x^2 + 9x}{2x + 2} = \dfrac{9x(x + 1)}{2(x + 1)}$

    $= \dfrac{9x}{2}, \quad x \neq -1$

17. $\dfrac{x - 5}{10 - 2x} = \dfrac{x - 5}{-2(x - 5)}$

    $= -\dfrac{1}{2}, \quad x \neq 5$

18. $\dfrac{12 - 4x}{x - 3} = \dfrac{4(3 - x)}{x - 3} = -4, \quad x \neq 3$

19. $\dfrac{y^2 - 16}{y + 4} = \dfrac{(y + 4)(y - 4)}{y + 4}$

    $= y - 4, \quad y \neq -4$

20. $\dfrac{x^2 - 25}{5 - x} = \dfrac{(x + 5)(x - 5)}{-1(x - 5)}$

    $= -(x + 5), \quad x \neq 5$

21. $\dfrac{x^3 + 5x^2 + 6x}{x^2 - 4} = \dfrac{x(x + 2)(x + 3)}{(x + 2)(x - 2)} = \dfrac{x(x + 3)}{x - 2}, \quad x \neq -2$

22. $\dfrac{x^2 + 8x - 20}{x^2 + 11x + 10} = \dfrac{(x + 10)(x - 2)}{(x + 10)(x + 1)} = \dfrac{x - 2}{x - 1}, \quad x \neq -10$

23. $\dfrac{y^2 - 7y + 12}{y^2 + 3y - 18} = \dfrac{(y - 3)(y - 4)}{(y + 6)(y - 3)} = \dfrac{y - 4}{y + 6}, \quad y \neq 3$

24. $\dfrac{x^2 - 7x + 6}{x^2 + 11x + 10} = \dfrac{(x - 6)(x - 1)}{(x + 10)(x + 1)}$

25. $\dfrac{2 - x + 2x^2 - x^3}{x^2 - 4} = \dfrac{(2 - x) + x^2(2 - x)}{(x + 2)(x - 2)}$

$= \dfrac{(2 - x)(1 + x^2)}{(x + 2)(x - 2)}$

$= \dfrac{-(x - 2)(x^2 + 1)}{(x + 2)(x - 2)}$

$= -\dfrac{x^2 + 1}{x + 2}, \quad x \neq 2$

26. $\dfrac{x^2 - 9}{x^3 + x^2 - 9x - 9} = \dfrac{x^2 - 9}{x^2(x + 1) - 9(x + 1)}$

$= \dfrac{x^2 - 9}{(x^2 - 9)(x + 1)}$

$= \dfrac{1}{x + 1}, \quad x \neq \pm 3$

27. $\dfrac{z^3 - 8}{z^2 + 2z + 4} = \dfrac{(z - 2)(z^2 + 2z + 4)}{z^2 + 2z + 4} = z - 2$

28. $\dfrac{y^3 - 2y^2 - 3y}{y^3 + 1} = \dfrac{y(y - 3)(y + 1)}{(y + 1)(y^2 - y + 1)}$

$= \dfrac{y(y - 3)}{y^2 - y + 1}, \quad y \neq -1$

29.

x	0	1	2	3	4	5	6
$\dfrac{x^2 - 2x - 3}{x - 3}$	1	2	3	Undef.	5	6	7
$x + 1$	1	2	3	4	5	6	7

The expressions are equivalent except at $x = 3$.

30.

x	0	1	2	3	4	5	6
$\dfrac{x - 3}{x^2 - x - 6}$	$\dfrac{1}{2}$	$\dfrac{1}{3}$	$\dfrac{1}{4}$	Undef.	$\dfrac{1}{6}$	$\dfrac{1}{7}$	$\dfrac{1}{8}$
$\dfrac{1}{x + 2}$	$\dfrac{1}{2}$	$\dfrac{1}{3}$	$\dfrac{1}{4}$	$\dfrac{1}{5}$	$\dfrac{1}{6}$	$\dfrac{1}{7}$	$\dfrac{1}{8}$

The expressions are equivalent except at $x = 3$.

31. $\dfrac{5x^3}{2x^3 + 4} = \dfrac{5x^3}{2(x^3 + 2)}$

There are no common factors so this expression cannot be simplified. In this case factors of terms were incorrectly cancelled.

32. $\dfrac{x^3 + 25x}{x^2 - 2x - 15} = \dfrac{x(x^2 + 25)}{(x - 5)(x + 3)}$

The expression cannot be simplified.

33. $\dfrac{\pi r^2}{(2r)^2} = \dfrac{\pi r^2}{4r^2} = \dfrac{\pi}{4}, \quad r \neq 0$

34. Area of shaded portion: $\left(\dfrac{x + 5}{2}\right)^2 = \dfrac{(x + 5)^2}{4}$

Area of total figure: $(2x + 3)(x + 5)$

Ratio: $\dfrac{\dfrac{(x + 5)^2}{4}}{(2x + 3)(x + 5)} = \dfrac{\dfrac{(x + 5)}{4}}{(2x + 3)} = \dfrac{x + 5}{4(2x + 3)}$

35. $\dfrac{5}{x - 1} \cdot \dfrac{x - 1}{25(x - 2)} = \dfrac{1}{5(x - 2)}, \quad x \neq 1$

36. $\dfrac{x + 13}{x^3(3 - x)} \cdot \dfrac{x(x - 3)}{5} = \dfrac{x + 13}{x^3(x - 3)(-1)} \cdot \dfrac{x(x - 3)}{5}$

$= \dfrac{x + 13}{-5x^2} = -\dfrac{x + 13}{5x^2}, \quad x \neq 3$

37. $\dfrac{r}{r-1} \cdot \dfrac{r^2-1}{r^2} = \dfrac{r(r+1)(r-1)}{r^2(r-1)} = \dfrac{r+1}{r},\ r \neq 1, r \neq 0$

38. $\dfrac{4y-16}{5y+15} \cdot \dfrac{2y+6}{4-y} = \dfrac{4(y-4)}{5(y+3)} \cdot \dfrac{2(y+3)}{(-1)(y-4)}$

$$= \dfrac{8}{-5} = -\dfrac{8}{5},\ y \neq -3, 4$$

39. $\dfrac{t^2-t-6}{t^2+6t+9} \cdot \dfrac{t+3}{t^2-4} = \dfrac{(t-3)(t+2)(t+3)}{(t+3)^2(t+2)(t-2)} = \dfrac{t-3}{(t+3)(t-2)},\ t \neq -2$

40. $\dfrac{x^2+xy-2y^2}{x^3+x^2y} \cdot \dfrac{x}{x^2+3xy+2y^2} = \dfrac{(x+2y)(x-y)}{x^2(x+y)} \cdot \dfrac{x}{(x+2y)(x+y)} = \dfrac{x-y}{x(x+y)^2},\ x \neq -2y$

41. $\dfrac{x^2-36}{x} \div \dfrac{x^3-6x^2}{x^2+x} = \dfrac{x^2-36}{x} \cdot \dfrac{x^2+x}{x^3-6x^2}$

$$= \dfrac{(x+6)(x-6)}{x} \cdot \dfrac{x(x+1)}{x^2(x-6)}$$

$$= \dfrac{(x+6)(x+1)}{x^2},\ x \neq 6$$

42. $\dfrac{x^2-14x+49}{x^2-49} \div \dfrac{3x-21}{x+7} = \dfrac{(x-7)(x-7)}{(x+7)(x-7)} \cdot \dfrac{x+7}{3(x-7)}$

$$= \dfrac{1}{3},\ x \neq \pm 7$$

43. $\dfrac{5}{x-1} + \dfrac{x}{x-1} = \dfrac{5+x}{x-1} = \dfrac{x+5}{x-1}$

44. $\dfrac{2x-1}{x+3} + \dfrac{1-x}{x+3} = \dfrac{2x-1+1-x}{x+3} = \dfrac{x}{x+3}$

45. $6 - \dfrac{5}{x+3} = \dfrac{6(x+3)}{(x+3)} - \dfrac{5}{x+3}$

$$= \dfrac{6(x+3)-5}{x+3}$$

$$= \dfrac{6x+18-5}{x+3}$$

$$= \dfrac{6x+13}{x+3}$$

46. $\dfrac{3}{x-1} - 5 = \dfrac{3}{x-1} - \dfrac{5(x-1)}{x-1}$

$$= \dfrac{3-5(x-1)}{x-1}$$

$$= \dfrac{3-5x+5}{x-1}$$

$$= \dfrac{8-5x}{x-1}$$

47. $\dfrac{3}{x-2} + \dfrac{5}{2-x} = \dfrac{3}{x-2} - \dfrac{5}{x-2} = -\dfrac{2}{x-2}$

48. $\dfrac{2x}{x-5} - \dfrac{5}{5-x} = \dfrac{2x}{x-5} - \dfrac{5(-1)}{(-1)(5-x)}$

$$= \dfrac{2x}{x-5} - \dfrac{-5}{x-5} = \dfrac{2x+5}{x-5}$$

49. $\dfrac{1}{x^2-x-2} - \dfrac{x}{x^2-5x+6} = \dfrac{1}{(x-2)(x+1)} - \dfrac{x}{(x-2)(x-3)}$

$$= \dfrac{(x-3)-x(x+1)}{(x+1)(x-2)(x-3)} = \dfrac{x-3-x^2-x}{(x+1)(x-2)(x-3)}$$

$$= \dfrac{-x^2-3}{(x+1)(x-2)(x-3)} = -\dfrac{x^2+3}{(x+1)(x-2)(x-3)}$$

50. $\dfrac{2}{x^2-x-2} + \dfrac{10}{x^2+2x-8} = \dfrac{2}{(x-2)(x+1)} + \dfrac{10}{(x+4)(x-2)}$

$$= \dfrac{2(x+4)}{(x-2)(x+1)(x+4)} + \dfrac{10(x+1)}{(x-2)(x+1)(x+4)}$$

$$= \dfrac{2x+8+10x+10}{(x-2)(x+1)(x+4)} = \dfrac{12x+18}{(x-2)(x+1)(x+4)} = \dfrac{6(2x+3)}{(x-2)(x+1)(x+4)}$$

51. $-\dfrac{1}{x} + \dfrac{2}{x^2+1} + \dfrac{1}{x^3+x} = \dfrac{-(x^2+1)}{x(x^2+1)} + \dfrac{2x}{x(x^2+1)} + \dfrac{1}{x(x^2+1)}$

$\qquad\qquad = \dfrac{-x^2-1+2x+1}{x(x^2+1)} = \dfrac{-x^2+2x}{x(x^2+1)} = \dfrac{-x(x-2)}{x(x^2+1)}$

$\qquad\qquad = -\dfrac{x-2}{x^2+1} = \dfrac{2-x}{x^2+1}, \quad x \neq 0$

52. $\dfrac{2}{x+1} + \dfrac{2}{x-1} + \dfrac{1}{x^2-1} = \dfrac{2}{x+1} + \dfrac{2}{x-1} + \dfrac{1}{(x+1)(x-1)}$

$\qquad\qquad = \dfrac{2(x-1)}{(x+1)(x-1)} + \dfrac{2(x+1)}{(x+1)(x-1)} + \dfrac{1}{(x+1)(x-1)}$

$\qquad\qquad = \dfrac{2x-2+2x+2+1}{(x+1)(x-1)} = \dfrac{4x+1}{(x+1)(x-1)}$

53. $\dfrac{x+4}{x+2} - \dfrac{3x-8}{x+2} = \dfrac{(x+4)-(3x-8)}{x+2}$

$\qquad\qquad = \dfrac{x+4-3x+8}{x+2} = \dfrac{-2x+12}{x+2} = \dfrac{-2(x-6)}{x+2}$

The error was incorrect subtraction in the numerator.

54. $\dfrac{6-x}{x(x+2)} + \dfrac{x+2}{x^2} + \dfrac{8}{x^2(x+2)} = \dfrac{x(6-x)}{x^2(x+2)} + \dfrac{(x+2)^2}{x^2(x+2)} + \dfrac{8}{x^2(x+2)}$

$\qquad\qquad = \dfrac{6x-x^2+x^2+4x+4+8}{x^2(x+2)} = \dfrac{10x+12}{x^2(x+2)} = \dfrac{2(5x+6)}{x^2(x+2)}$

The error was an incorrect expansion of $(x+2)^2$ in the numerator.

55. $\dfrac{\left(\dfrac{x}{2}-1\right)}{(x-2)} = \dfrac{\left(\dfrac{x}{2}-\dfrac{2}{2}\right)}{\left(\dfrac{x-2}{1}\right)}$

$\qquad = \dfrac{x-2}{2} \cdot \dfrac{1}{x-2}$

$\qquad = \dfrac{1}{2}, \quad x \neq 2$

56. $\dfrac{(x-4)}{\left(\dfrac{x}{4}-\dfrac{4}{x}\right)} = \dfrac{\left(\dfrac{x-4}{1}\right)}{\left(\dfrac{x^2}{4x}-\dfrac{16}{4x}\right)} = \dfrac{\left(\dfrac{x-4}{1}\right)}{\left(\dfrac{x^2-16}{4x}\right)}$

$\qquad = \dfrac{x-4}{1} \cdot \dfrac{4x}{x^2-16}$

$\qquad = \dfrac{x-4}{1} \cdot \dfrac{4x}{(x+4)(x-4)} = \dfrac{4x}{x+4}, \quad x \neq 0, 4$

57. $\dfrac{\left[\dfrac{x^2}{(x+1)^2}\right]}{\left[\dfrac{x}{(x+1)^3}\right]} = \dfrac{x^2}{(x+1)^2} \cdot \dfrac{(x+1)^3}{x}$

$\qquad = x(x+1), \quad x \neq -1, 0$

58. $\dfrac{\left(\dfrac{x^2-1}{x}\right)}{\left[\dfrac{(x-1)^2}{x}\right]} = \dfrac{x^2-1}{x} \cdot \dfrac{x}{(x-1)^2}$

$\qquad = \dfrac{(x+1)(x-1)}{x} \cdot \dfrac{x}{(x-1)(x-1)}$

$\qquad = \dfrac{x+1}{x-1}, \quad x \neq 0$

59. $\dfrac{\left(\sqrt{x}-\dfrac{1}{2\sqrt{x}}\right)}{\sqrt{x}} = \dfrac{\left(\sqrt{x}-\dfrac{1}{2\sqrt{x}}\right)}{\sqrt{x}} \cdot \dfrac{2\sqrt{x}}{2\sqrt{x}} = \dfrac{2x-1}{2x}, \quad x > 0$

60. $\dfrac{\dfrac{t^2}{\sqrt{t^2+1}} - \sqrt{t^2+1}}{t^2} = \left[\dfrac{\dfrac{t^2}{\sqrt{t^2+1}} - \sqrt{t^2+1}}{t^2}\right] \cdot \dfrac{\sqrt{t^2+1}}{\sqrt{t^2+1}}$

$$= \dfrac{t^2 - (t^2+1)}{t^2\sqrt{t^2+1}} = -\dfrac{1}{t^2\sqrt{t^2+1}}$$

61. $x^5 - 2x^{-2} = x^{-2}(x^7 - 2) = \dfrac{x^7 - 2}{x^2}$

62. $x^5 - 5x^{-3} = x^{-3}(x^8 - 5) = \dfrac{x^8 - 5}{x^3}$

63. $x^2(x^2+1)^{-5} - (x^2+1)^{-4} = (x^2+1)^{-5}\left[x^2 - (x^2+1)\right] = -\dfrac{1}{(x^2+1)^5}$

64. $2x(x-5)^{-3} - 4x^2(x-5)^{-4} = -2x(x-5)^{-4}[-(x-5) + 2x]$

$$= -2x(x-5)^{-4}(-x + 5 + 2x)$$

$$= \dfrac{-2x(x+5)}{(x-5)^4}$$

65. $2x^2(x-1)^{1/2} - 5(x-1)^{-1/2} = (x-1)^{-1/2}\left[2x^2(x-1)^1 - 5\right] = \dfrac{2x^3 - 2x^2 - 5}{(x-1)^{1/2}}$

66. $4x^3(2x-1)^{3/2} - 2x(2x-1)^{-1/2} = (2x-1)^{-1/2}[4x^3(2x-1)^2 - 2x] = \dfrac{4x^3(2x-1)^2 - 2x}{(2x-1)^{1/2}}$

67. $\dfrac{3x^{1/3} - x^{-2/3}}{3x^{-2/3}} = \dfrac{3x^{1/3} - x^{-2/3}}{3x^{-2/3}} \cdot \dfrac{x^{2/3}}{x^{2/3}} = \dfrac{3x^1 - x^0}{3x^0} = \dfrac{3x - 1}{3}, \ x \neq 0$

68. $\dfrac{-x^3(1-x^2)^{-1/2} - 2x(1-x^2)^{1/2}}{x^4} = \dfrac{\dfrac{-x^3}{(1-x^2)^{1/2}} - 2x(1-x^2)^{1/2}}{x^4}$

$$= \dfrac{\dfrac{-x^3}{(1-x^2)^{1/2}} - \dfrac{2x(1-x^2)^{1/2}(1-x^2)^{1/2}}{(1-x^2)^{1/2}}}{x^4} = \dfrac{\dfrac{-x^3 - 2x(1-x^2)}{(1-x^2)^{1/2}}}{x^4}$$

$$= \dfrac{-x^3 - 2x + 2x^3}{(1-x^2)^{1/2}} \cdot \dfrac{1}{x^4} = \dfrac{x^3 - 2x}{(1-x^2)^{1/2}} \cdot \dfrac{1}{x^4}$$

$$= \dfrac{x(x^2 - 2)}{x^4(1-x^2)^{1/2}} = \dfrac{x^2 - 2}{x^3(1-x^2)^{1/2}}$$

69. $\dfrac{\left(\dfrac{1}{x+h} - \dfrac{1}{x}\right)}{h} = \dfrac{\left(\dfrac{1}{x+h} - \dfrac{1}{x}\right)}{h} \cdot \dfrac{x(x+h)}{x(x+h)}$

$$= \dfrac{x - (x+h)}{hx(x+h)}$$

$$= \dfrac{-h}{hx(x+h)}$$

$$= -\dfrac{1}{x(x+h)}, \ h \neq 0$$

70. $\dfrac{\left[\dfrac{1}{(x+h)^2} - \dfrac{1}{x^2}\right]}{h} = \dfrac{\left[\dfrac{1}{(x+h)^2} - \dfrac{1}{x^2}\right]}{h} \cdot \dfrac{x^2(x+h)^2}{x^2(x+h)^2}$

$$= \dfrac{x^2 - (x+h)^2}{hx^2(x+h)^2}$$

$$= \dfrac{x^2 - (x^2 + 2xh + h^2)}{hx^2(x+h)^2}$$

$$= \dfrac{-h(2x+h)}{hx^2(x+h)^2}$$

$$= -\dfrac{2x+h}{x^2(x+h)^2}, \ h \neq 0$$

71. $\dfrac{\left(\dfrac{1}{x+h-4}-\dfrac{1}{x-4}\right)}{h} = \dfrac{\left(\dfrac{1}{x+h-4}-\dfrac{1}{x-4}\right)}{h} \cdot \dfrac{(x-4)(x+h-4)}{(x-4)(x+h-4)}$

$$= \dfrac{(x-4)-(x+h-4)}{h(x-4)(x+h-4)}$$

$$= \dfrac{-h}{h(x-4)(x+h-4)}$$

$$= -\dfrac{1}{(x-4)(x+h-4)}, \quad h \neq 0$$

72. $\dfrac{\left(\dfrac{x+h}{x+h+1}-\dfrac{x}{x+1}\right)}{h} = \dfrac{\left(\dfrac{(x+h)(x+1)}{(x+h+1)(x+1)}-\dfrac{x(x+h+1)}{(x+h+1)(x+1)}\right)}{h/1}$

$$= \left(\dfrac{(x+h)(x+1)}{(x+h+1)(x+1)}-\dfrac{x(x+h+1)}{(x+h+1)(x+1)}\right) \cdot \dfrac{1}{h}$$

$$= \left(\dfrac{x^2+x+hx+h-x^2-xh-x}{(x+h+1)(x+1)}\right) \cdot \dfrac{1}{h}$$

$$= \dfrac{h}{(x+h+1)(x+1)} \cdot \dfrac{1}{h} = \dfrac{1}{(x+h+1)(x+1)}, \quad h \neq 0$$

73. $\dfrac{\sqrt{x+2}-\sqrt{x}}{2} = \dfrac{\sqrt{x+2}-\sqrt{x}}{2} \cdot \dfrac{\sqrt{x+2}+\sqrt{x}}{\sqrt{x+2}+\sqrt{x}}$

$$= \dfrac{(x+2)-x}{2\left(\sqrt{x+2}+\sqrt{x}\right)}$$

$$= \dfrac{2}{2\left(\sqrt{x+2}+\sqrt{x}\right)}$$

$$= \dfrac{1}{\sqrt{x+2}+\sqrt{x}}$$

74. $\dfrac{\sqrt{z-3}-\sqrt{z}}{3} = \dfrac{\sqrt{z-3}-\sqrt{z}}{3} \cdot \dfrac{\sqrt{z-3}+\sqrt{z}}{\sqrt{z-3}+\sqrt{z}}$

$$= \dfrac{(z-3)-z}{3\left(\sqrt{z-3}+\sqrt{z}\right)}$$

$$= \dfrac{-3}{3\left(\sqrt{z-3}+\sqrt{z}\right)}$$

$$= \dfrac{-1}{\sqrt{z-3}+\sqrt{z}}$$

75. $\dfrac{\sqrt{x+h+1}-\sqrt{x+1}}{h} = \dfrac{\sqrt{x+h+1}-\sqrt{x+1}}{h} \cdot \dfrac{\sqrt{x+h+1}+\sqrt{x+1}}{\sqrt{x+h+1}+\sqrt{x+1}}$

$$= \dfrac{(x+h+1)-(x+1)}{h\left(\sqrt{x+h+1}+\sqrt{x+1}\right)}$$

$$= \dfrac{h}{h\left(\sqrt{x+h+1}+\sqrt{x+1}\right)}$$

$$= \dfrac{1}{\sqrt{x+h+1}+\sqrt{x+1}}, \quad h \neq 0$$

76. $\dfrac{\sqrt{x+h-2}-\sqrt{x-2}}{h} \cdot \dfrac{\sqrt{x+h-2}+\sqrt{x-2}}{\left(\sqrt{x+h-2}+\sqrt{x-2}\right)} = \dfrac{x+h-2-x+2}{h\left(\sqrt{x+h-2}+\sqrt{x-2}\right)}$

$$= \dfrac{h}{h\left(\sqrt{x+h-2}+\sqrt{x-2}\right)}$$

$$= \dfrac{1}{\sqrt{x+h-2}+\sqrt{x-2}}, \quad h \neq 0$$

77. Probability $= \dfrac{\text{Shaded area}}{\text{Total area}} = \dfrac{x(x/2)}{x(2x+1)} = \dfrac{x/2}{2x+1} \cdot \dfrac{2}{2} = \dfrac{x}{2(2x+1)}$

78. Probability $= \dfrac{\text{Shaded area}}{\text{Total area}} = \dfrac{\dfrac{1}{2} \cdot \dfrac{4}{x}(x + 2)(x + x + 4)}{\dfrac{1}{2}(x + 4)\left[(x + 2) + \dfrac{4}{x}(x + 2)\right]}$

$$= \dfrac{\dfrac{4(x + 2)(2x + 4)}{x}}{(x + 4)(x + 2)\left(1 + \dfrac{4}{x}\right)} = \dfrac{\dfrac{4 \cdot 2(x + 2)^2}{x}}{(x + 4)(x + 2)\left(1 + \dfrac{4}{x}\right)}$$

$$= \dfrac{8(x + 2)^2}{x} \cdot \dfrac{1}{(x + 4)(x + 2)\left(1 + \dfrac{4}{x}\right)}$$

$$= \dfrac{8(x + 2)^2}{(x + 4)(x + 2)(x + 4)} = \dfrac{8(x + 2)}{(x + 4)^2}$$

79. (a) $\dfrac{1}{16}$ minute

(b) $x\left(\dfrac{1}{16}\right) = \dfrac{x}{16}$ minutes

(c) $\dfrac{60}{16} = \dfrac{15}{4}$ minutes

80. $\dfrac{t}{3} + \dfrac{t}{5} = \dfrac{5t + 3t}{15} = \dfrac{8t}{15}$

81. (a) $r = \dfrac{\left(\dfrac{24[48(400) - 16,000]}{48}\right)}{\left[16,000 + \dfrac{48(400)}{12}\right]} \approx 0.0909 = 9.09\%$

(b) $r = \dfrac{\left[\dfrac{24(NM - P)}{N}\right]}{\left(P + \dfrac{NM}{12}\right)} = \dfrac{24(NM - P)}{N} \cdot \dfrac{12}{12P + NM} = \dfrac{288(NM - P)}{N(12P + NM)}$

$r = \dfrac{288[48(400) - 16,000]}{48[12(16,000) + 48(400)]} \approx 0.0909 = 9.09\%$

82. (a) $r = \dfrac{\left[\dfrac{24(NM - P)}{N}\right]}{\left(P + \dfrac{NM}{12}\right)} = \dfrac{\left[\dfrac{24(60 \cdot 525 - 28,000)}{60}\right]}{\left(28,000 + \dfrac{60 \cdot 525}{12}\right)} = -\dfrac{\dfrac{2(31,500 - 28,000)}{5}}{28,000 + \dfrac{31,500}{12}}$

$= \dfrac{\dfrac{2(3500)}{5}}{28,000 + 2625} = \dfrac{2(700)}{30,625} = \dfrac{1400}{30,625} \approx 4.57\%$

(b) $r = \dfrac{\left[\dfrac{24(NM - P)}{N}\right]}{\dfrac{12P + NM}{12}} = \dfrac{24(NM - P)}{N} \cdot \dfrac{12}{12P + NM} = \dfrac{288(NM - P)}{N(12P + NM)}$

$= \dfrac{288(60 \cdot 525 - 28,000)}{60(12 \cdot 28,000 + 60 \cdot 525)} = \dfrac{288(31,500 - 28,000)}{60(336,000 + 31,500)} = \dfrac{288(3500)}{60(367,500)} = \dfrac{1,008,000}{22,050,000} \approx 4.57\%$

83. $T = 10\left(\dfrac{4t^2 + 16t + 75}{t^2 + 4t + 10}\right)$

(a)

t	0	2	4	6	8	10	12	14	16	18	20	22
T	75°	55.9°	48.3°	45°	43.3°	42.3°	41.7°	41.3°	41.1°	40.9°	40.7°	40.6°

(b) T is approaching 40°.

84. (a)

Year	2002	2003	2004	2005	2006	2007
Banking	21.9	27.0	31.4	35.6	40.2	45.6
Paying Bills	13.9	17.8	21.5	25.0	28.1	31.0

(b) The estimates and actual values are quite close for banking. For paying bills, they are close for 2002–2004, but the model generally tends to overestimate the number.

(c) $\dfrac{\dfrac{4.39t + 5.5}{0.002t^2 + 0.01t + 1.0}}{\dfrac{-0.728t^2 + 23.81t - 0.3}{-0.049t^2 + 0.61t + 1.0}} = \dfrac{4.39t + 5.5}{0.002t^2 + 0.01t + 1.0} \cdot \dfrac{-0.049t^2 + 0.61t + 1.0}{-0.728t^2 + 23.81t - 0.3}$

$= \dfrac{-0.21511t^3 + 2.6779t^2 + 4.39t - 0.2695t^2 + 3.355t + 5.5}{-0.001456t^4 + 0.04762t^3 - 0.0006t^2 - 72.8t^3 + 0.2381t^2 - 0.003t - 0.728t^2 + 23.81t - 0.3}$

$= \dfrac{-0.21511t^3 + 2.4084t^2 + 7.745t + 5.5}{-0.001456t^4 - 72.75238t^3 - 0.4905t^2 + 23.807t - 0.3}$

(d)

Year	2002	2003	2004	2005	2006	2007
Paying Bills/Banking	0.63	0.66	0.69	0.70	0.70	0.68

The ratio is approximately $\frac{2}{3}$, and it appears to peak in 2005–2006.

85. False. In order for the simplified expression to be equivalent to the original expression, the domain of the simplified expression needs to be restricted. If n is even, $x \neq \pm 1$. If n is odd, $x \neq 1$.

86. False. The two expressions are equivalent for all values of x such that $x \neq 1$.

87. Completely factor the numerator and the denominator. A rational expression is in **simplest** form if there are no common factors in the numerator and the denominator other than ± 1.

Appendix A.5   Solving Equations

> ■ You should know how to solve linear equations.
> $ax + b = 0$
>
> ■ An identity is an equation whose solution consists of every real number in its domain.
>
> ■ To solve an equation you can:
>
> (a) Add or subtract the same quantity from both sides.
>
> (b) Multiply or divide both sides by the same nonzero quantity.
>
> ■ To solve an equation that can be simplified to a linear equation:
>
> (a) Remove all symbols of grouping and all fractions.
>
> (b) Combine like terms.
>
> (c) Solve by algebra.
>
> (d) Check the answer.
>
> ■ A "solution" that does not satisfy the original equation is called an extraneous solution.
>
> ■ You should be able to solve a quadratic equation by factoring, if possible.
>
> ■ You should be able to solve a quadratic equation of the form $u^2 = d$ by extracting square roots.
>
> ■ You should be able to solve a quadratic equation by completing the square.
>
> ■ You should know and be able to use the Quadratic Formula: For $ax^2 + bx + c = 0, a \neq 0$,
>
> $$x = \frac{-b \pm \sqrt{b^2 - 4ac}}{2a}.$$
>
> ■ You should be able to solve polynomials of higher degree by factoring.
>
> ■ For equations involving radicals or fractional powers, raise both sides to the same power.
>
> ■ For equations with fractions, multiply both sides by the least common denominator to clear the fractions.
>
> ■ For equations involving absolute value, remember that the expression inside the absolute value can be positive or negative.

Vocabulary Check

1. equation

2. solve

3. identities; conditional

4. $ax + b = 0$

5. extraneous

6. quadratic equation

7. factoring; extracting square roots; completing the square; Quadratic Formula

1. $2(x - 1) = 2x - 2$ is an *identity* by the Distributive Property. It is true for all real values of x.

2. $3(x + 2) = 5x + 4$ is *conditional*. There are real values of x for which the equation is not true (for example, $x = 0$).

3. $-6(x - 3) + 5 = -2x + 10$ is *conditional*. There are real values of x for which the equation is not true.

4. $3(x + 2) - 5 = 3x + 1$ is an *identity* by simplification. It is true for all real values of x.

    $3(x + 2) - 5 = 3x + 6 - 5 = 3x + 1$

5. $4(x + 1) - 2x = 4x + 4 - 2x = 2x + 4 = 2(x + 2)$

    This is an *identity* by simplification. It is true for all real values of x.

6. $-7(x - 3) + 4x = 3(7 - x)$ is an *identity* by simplification. It is true for all real values of x.

    $-7(x - 3) + 4x = -7x + 21 + 4x$

    $= 21 - 3x = 3(7 - x)$

7. $(x - 4)^2 - 11 = x^2 - 8x + 16 - 11 = x^2 - 8x + 5$

Thus, $x^2 - 8x + 5 = (x - 4)^2 - 11$ is an *identity* by simplification. It is true for all real values of x.

8. $x^2 + 2(3x - 2) = x^2 + 6x - 4$ is an *identity* by simplification. It is true for all real values of x.

9. $3 + \dfrac{1}{x + 1} = \dfrac{4x}{x + 1}$ is *conditional*. There are real values of x for which the equation is not true.

10. $\dfrac{5}{x} + \dfrac{3}{x} = 24$ is *conditional*. There are real values of x for which the equation is not true (for example, $x = 1$).

11.
$$x + 11 = 15$$
$$x + 11 - 11 = 15 - 11$$
$$x = 4$$

12.
$$7 - x = 19$$
$$7 - x + x = 19 + x$$
$$7 = 19 + x$$
$$7 - 19 = 19 + x - 19$$
$$-12 = x$$

13.
$$7 - 2x = 25$$
$$7 - 7 - 2x = 25 - 7$$
$$-2x = 18$$
$$\frac{-2x}{-2} = \frac{18}{-2}$$
$$x = -9$$

14.
$$7x + 2 = 23$$
$$7x + 2 - 2 = 23 - 2$$
$$7x = 21$$
$$\frac{7x}{7} = \frac{21}{7}$$
$$x = 3$$

15.
$$8x - 5 = 3x + 20$$
$$8x - 3x - 5 = 3x - 3x + 20$$
$$5x - 5 = 20$$
$$5x - 5 + 5 = 20 + 5$$
$$5x = 25$$
$$\frac{5x}{5} = \frac{25}{5}$$
$$x = 5$$

16.
$$7x + 3 = 3x - 17$$
$$7x + 3 - 3 - 3x = 3x - 17 - 3 - 3x$$
$$4x = -20$$
$$x = -5$$

17.
$$2(x + 5) - 7 = 3(x - 2)$$
$$2x + 10 - 7 = 3x - 6$$
$$2x + 3 = 3x - 6$$
$$2x - 3x + 3 = 3x - 3x - 6$$
$$-x + 3 = -6$$
$$-x + 3 - 3 = -6 - 3$$
$$-x = -9$$
$$x = 9$$

18.
$$3(x + 3) = 5(1 - x) - 1$$
$$3x + 9 = 5 - 5x - 1$$
$$3x + 9 = 4 - 5x$$
$$3x + 9 + 5x - 9 = 4 - 5x + 5x - 9$$
$$8x = -5$$
$$x = -\tfrac{5}{8}$$

19.
$$x - 3(2x + 3) = 8 - 5x$$
$$x - 6x - 9 = 8 - 5x$$
$$-5x - 9 = 8 - 5x$$
$$-5x + 5x - 9 = 8 - 5x + 5x$$
$$-9 \neq 8$$

No solution

20.
$$9x - 10 = 5x + 2(2x - 5)$$
$$9x - 10 = 5x + 4x - 10$$
$$9x - 10 = 9x - 10$$

The solution is the set of all real numbers.

21.
$$\frac{5x}{4} + \frac{1}{2} = x - \frac{1}{2}$$
$$4\left(\frac{5x}{4} + \frac{1}{2}\right) = 4\left(x - \frac{1}{2}\right)$$
$$4\left(\frac{5x}{4}\right) + 4\left(\frac{1}{2}\right) = 4(x) - 4\left(\frac{1}{2}\right)$$
$$5x + 2 = 4x - 2$$
$$x = -4$$

22.
$$\frac{x}{5} - \frac{x}{2} = 3 + \frac{3x}{10}$$
$$10\left(\frac{x}{5} - \frac{x}{2}\right) = 10\left(3 + \frac{3x}{10}\right)$$
$$2x - 5x = 30 + 3x$$
$$-6x = 30$$
$$x = -5$$

23.
$$\frac{3}{2}(z + 5) - \frac{1}{4}(z + 24) = 0$$

$$4\left[\frac{3}{2}(z + 5) - \frac{1}{4}(z + 24)\right] = 4(0)$$

$$4\left(\frac{3}{2}\right)(z + 5) - 4\left(\frac{1}{4}\right)(z + 24) = 4(0)$$

$$6(z + 5) - (z + 24) = 0$$

$$6z + 30 - z - 24 = 0$$

$$5z = -6$$

$$z = -\frac{6}{5}$$

24.
$$\frac{3x}{2} + \frac{1}{4}(x - 2) = 10$$

$$4\left[\frac{3x}{2} + \frac{1}{4}(x - 2)\right] = 4(10)$$

$$4\left(\frac{3x}{2}\right) + 4\left(\frac{1}{4}\right)(x - 2) = 4(10)$$

$$6x + (x - 2) = 40$$

$$7x - 2 = 40$$

$$7x = 42$$

$$x = 6$$

25. $0.25x + 0.75(10 - x) = 3$

$$0.25x + 7.5 - 0.75x = 3$$

$$-0.50x + 7.5 = 3$$

$$-0.50x = -4.5$$

$$x = 9$$

26. $0.60x + 0.40(100 - x) = 50$

$$0.60x + 40 - 0.40x = 50$$

$$0.20x = 10$$

$$x = 50$$

27. $x + 8 = 2(x - 2) - x$

$$x + 8 = 2x - 4 - x$$

$$x + 8 = x - 4$$

$$8 \neq -4$$

Contradiction; no solution

28. $8(x + 2) - 3(2x + 1) = 2(x + 5)$

$$8x + 16 - 6x - 3 = 2x + 10$$

$$2x + 13 = 2x + 10$$

$$13 = 10$$

Contradiction; no solution

29.
$$\frac{100 - 4x}{3} = \frac{5x + 6}{4} + 6$$

$$12\left(\frac{100 - 4x}{3}\right) = 12\left(\frac{5x + 6}{4}\right) + 12(6)$$

$$4(100 - 4x) = 3(5x + 6) + 72$$

$$400 - 16x = 15x + 18 + 72$$

$$-31x = -310$$

$$x = 10$$

30.
$$\frac{17 + y}{y} + \frac{32 + y}{y} = 100$$

$$(y)\frac{17 + y}{y} + (y)\frac{32 + y}{y} = 100(y)$$

$$17 + y + 32 + y = 100y$$

$$49 + 2y = 100y$$

$$49 = 98y$$

$$\frac{1}{2} = y$$

31.
$$\frac{5x - 4}{5x + 4} = \frac{2}{3}$$

$$3(5x - 4) = 2(5x + 4)$$

$$15x - 12 = 10x + 8$$

$$5x = 20$$

$$x = 4$$

32.
$$\frac{10x + 3}{5x + 6} = \frac{1}{2}$$

$$2(10x + 3) = 1(5x + 6)$$

$$20x + 6 = 5x + 6$$

$$15x = 0$$

$$x = 0$$

33. $10 - \dfrac{13}{x} = 4 + \dfrac{5}{x}$

$\dfrac{10x - 13}{x} = \dfrac{4x + 5}{x}$

$10x - 13 = 4x + 5$

$6x = 18$

$x = 3$

34. $\dfrac{15}{x} - 4 = \dfrac{6}{x} + 3$

$\dfrac{15}{x} - \dfrac{6}{x} = 7$

$\dfrac{9}{x} = 7$

$9 = 7x$

$\dfrac{9}{7} = x$

35. $3 = 2 + \dfrac{2}{z + 2}$

$3(z + 2) = \left(2 + \dfrac{2}{z + 2}\right)(z + 2)$

$3z + 6 = 2z + 4 + 2$

$z = 0$

36. $\dfrac{1}{x} + \dfrac{2}{x - 5} = 0$ Multiply both sides by $x(x - 5)$.

$1(x - 5) + 2x = 0$

$3x - 5 = 0$

$3x = 5$

$x = \dfrac{5}{3}$

37. $\dfrac{x}{x + 4} + \dfrac{4}{x + 4} + 2 = 0$

$\dfrac{x + 4}{x + 4} + 2 = 0$

$1 + 2 = 0$

$3 \neq 0$

Contradiction; no solution

38. $\dfrac{7}{2x + 1} - \dfrac{8x}{2x - 1} = -4$ Multiply both sides by $(2x + 1)(2x - 1)$.

$7(2x - 1) - 8x(2x + 1) = -4(2x + 1)(2x - 1)$

$14x - 7 - 16x^2 - 8x = -16x^2 + 4$

$6x = 11$

$x = \dfrac{11}{6}$

39. $\dfrac{2}{(x - 4)(x - 2)} = \dfrac{1}{x - 4} + \dfrac{2}{x - 2}$ Multiply both sides by $(x - 4)(x - 2)$.

$2 = 1(x - 2) + 2(x - 4)$

$2 = x - 2 + 2x - 8$

$2 = 3x - 10$

$12 = 3x$

$4 = x$

A check reveals that $x = 4$ is an extraneous solution—it makes the denominator zero. There is no real solution.

40. $\dfrac{4}{x - 1} + \dfrac{6}{3x + 1} = \dfrac{15}{3x + 1}$ Multiply both sides by $(x - 1)(3x + 1)$.

$(x - 1)(3x + 1)\dfrac{4}{x - 1} + (x - 1)(3x + 1)\dfrac{6}{3x + 1} = (x - 1)(3x + 1)\dfrac{15}{3x + 1}$

$4(3x + 1) + 6(x - 1) = 15(x - 1)$

$12x + 4 + 6x - 6 = 15x - 15$

$18x - 2 = 15x - 15$

$3x = -13$

$x = -\dfrac{13}{3}$

41. $\dfrac{1}{x-3} + \dfrac{1}{x+3} = \dfrac{10}{x^2-9}$

$\dfrac{1}{x-3} + \dfrac{1}{x+3} = \dfrac{10}{(x+3)(x-3)}$ Multiply both sides by $(x+3)(x-3)$.

$1(x+3) + 1(x-3) = 10$

$2x = 10$

$x = 5$

42. $\dfrac{1}{x-2} + \dfrac{3}{x+3} = \dfrac{4}{x^2+x-6}$

$\dfrac{1}{x-2} + \dfrac{3}{x+3} = \dfrac{4}{(x+3)(x-2)}$ Multiply both sides by $(x+3)(x-2)$.

$(x+3) + 3(x-2) = 4$

$x + 3 + 3x - 6 = 4$

$4x - 3 = 4$

$4x = 7$

$x = \dfrac{7}{4}$

43. $\dfrac{3}{x^2-3x} + \dfrac{4}{x} = \dfrac{1}{x-3}$

$\dfrac{3}{x(x-3)} + \dfrac{4}{x} = \dfrac{1}{x-3}$ Multiply both sides by $x(x-3)$.

$3 + 4(x-3) = x$

$3 + 4x - 12 = x$

$3x = 9$

$x = 3$

A check reveals that $x = 3$ is an extraneous solution since it makes the denominator zero, so there is no solution.

44. $\dfrac{6}{x} - \dfrac{2}{x+3} = \dfrac{3(x+5)}{x(x+3)}$ Multiply both sides by $x(x+3)$.

$6(x+3) - 2x = 3(x+5)$

$6x + 18 - 2x = 3x + 15$

$4x + 18 = 3x + 15$

$x = -3$

Check: $\dfrac{6}{-3} - \dfrac{2}{-3+3} = \dfrac{3(-3+5)}{-3(-3+3)}$

$-2 - \dfrac{2}{0} = \dfrac{-6}{-3(0)}$

Division by zero is undefined. Thus, $x = -3$ is not a solution, and the original equation has no solution.

45. $(x+2)^2 + 5 = (x+3)^2$

$x^2 + 4x + 4 + 5 = x^2 + 6x + 9$

$4x + 9 = 6x + 9$

$-2x = 0$

$x = 0$

46. $(x+1)^2 + 2(x-2) = (x+1)(x-2)$

$x^2 + 2x + 1 + 2x - 4 = x^2 - x - 2$

$5x = 1$

$x = \dfrac{1}{5}$

47. $(x + 2)^2 - x^2 = 4(x + 1)$

$x^2 + 4x + 4 - x^2 = 4x + 4$

$4 = 4$

The equation is an identity; every real number is a solution.

48. $(2x + 1)^2 = 4(x^2 + x + 1)$

$4x^2 + 4x + 1 = 4x^2 + 4x + 4$

$1 = 4$

This is a contradiction. Thus, the equation has no solution.

49. $2x^2 = 3 - 8x$

General form: $2x^2 + 8x - 3 = 0$

50. $x^2 = 16x$

General form: $x^2 - 16x = 0$

51. $(x - 3)^2 = 3$

$x^2 - 6x + 9 = 3$

General form: $x^2 - 6x + 6 = 0$

52. $13 - 3(x + 7)^2 = 0$

$13 - 3(x^2 + 14x + 49) = 0$

$13 - 3x^2 - 42x - 147 = 0$

General form:

$-3x^2 - 42x - 134 = 0$

53. $\frac{1}{5}(3x^2 - 10) = 18x$

$3x^2 - 10 = 90x$

General form: $3x^2 - 90x - 10 = 0$

54. $x(x + 2) = 5x^2 + 1$

$x^2 + 2x = 5x^2 + 1$

$-4x^2 + 2x - 1 = 0$

$(-1)(-4x^2 + 2x - 1) = -1(0)$

General form: $4x^2 - 2x + 1 = 0$

55. $6x^2 + 3x = 0$

$3x(2x + 1) = 0$

$3x = 0$ or $2x + 1 = 0$

$x = 0$ or $x = -\frac{1}{2}$

56. $9x^2 - 1 = 0$

$(3x + 1)(3x - 1) = 0$

$3x + 1 = 0 \implies x = -\frac{1}{3}$

$3x - 1 = 0 \implies x = \frac{1}{3}$

57. $x^2 - 2x - 8 = 0$

$(x - 4)(x + 2) = 0$

$x - 4 = 0$ or $x + 2 = 0$

$x = 4$ or $x = -2$

58. $x^2 - 10x + 9 = 0$

$(x - 9)(x - 1) = 0$

$x - 9 = 0 \implies x = 9$

$x - 1 = 0 \implies x = 1$

59. $x^2 + 10x + 25 = 0$

$(x + 5)^2 = 0$

$x + 5 = 0$

$x = -5$

60. $4x^2 + 12x + 9 = 0$

$(2x + 3)(2x + 3) = 0$

$2x + 3 = 0 \implies x = -\frac{3}{2}$

61. $3 + 5x - 2x^2 = 0$

$(3 - x)(1 + 2x) = 0$

$3 - x = 0$ or $1 + 2x = 0$

$x = 3$ or $x = -\frac{1}{2}$

62. $2x^2 = 19x + 33$

$2x^2 - 19x - 33 = 0$

$(2x + 3)(x - 11) = 0$

$2x + 3 = 0 \implies x = -\frac{3}{2}$

$x - 11 = 0 \implies x = 11$

63. $x^2 + 4x = 12$

$x^2 + 4x - 12 = 0$

$(x + 6)(x - 2) = 0$

$x + 6 = 0$ or $x - 2 = 0$

$x = -6$ or $x = 2$

64. $-x^2 + 8x = 12$

$-x^2 + 8x - 12 = 0$

$(-1)(-x^2 + 8x - 12) = (-1)(0)$

$x^2 - 8x + 12 = 0$

$(x - 6)(x - 2) = 0$

$x - 6 = 0 \implies x = 6$

$x - 2 = 0 \implies x = 2$

65. $\frac{3}{4}x^2 + 8x + 20 = 0$

$4(\frac{3}{4}x^2 + 8x + 20) = 4(0)$

$3x^2 + 32x + 80 = 0$

$(3x + 20)(x + 4) = 0$

$3x + 20 = 0$ or $x + 4 = 0$

$x = -\frac{20}{3}$ or $x = -4$

66. $\frac{1}{8}x^2 - x - 16 = 0$

$x^2 - 8x - 128 = 0$

$(x - 16)(x + 8) = 0$

$x - 16 = 0 \implies x = 16$

$x + 8 = 0 \implies x = -8$

67. $x^2 + 2ax + a^2 = 0$

$(x + a)^2 = 0$

$x + a = 0$

$x = -a$

68. $(x + a)^2 - b^2 = 0$

$[(x + a) + b][(x + a) - b] = 0$

$[x + (a + b)][x + (a - b)] = 0$

$x + (a + b) = 0 \implies x = -a - b$

$x + (a - b) = 0 \implies x = -a + b$

69. $x^2 = 49$

$x = \pm 7$

70. $x^2 = 169$

$x = \pm\sqrt{169} = \pm 13$

71. $x^2 = 11$

$x = \pm\sqrt{11}$

72. $x^2 = 32$

$x = \pm\sqrt{32} = \pm 4\sqrt{2}$

73. $3x^2 = 81$

$x^2 = 27$

$x = \pm 3\sqrt{3}$

74. $9x^2 = 36$

$x^2 = 4$

$x = \pm\sqrt{4} = \pm 2$

75. $(x - 12)^2 = 16$

$x - 12 = \pm 4$

$x = 12 \pm 4$

$x = 16$ or $x = 8$

76. $(x + 13)^2 = 25$

$x + 13 = \pm\sqrt{25}$

$x + 13 = \pm 5$

$x = -13 \pm 5 = -8, -18$

77. $(x + 2)^2 = 14$

$x + 2 = \pm\sqrt{14}$

$x = -2 \pm \sqrt{14}$

78. $(x - 5)^2 = 30$

$x - 5 = \pm\sqrt{30}$

$x = 5 \pm \sqrt{30}$

79. $(2x - 1)^2 = 18$

$2x - 1 = \pm\sqrt{18}$

$2x = 1 \pm 3\sqrt{2}$

$x = \dfrac{1 \pm 3\sqrt{2}}{2}$

80. $(4x + 7)^2 = 44$

$4x + 7 = \pm\sqrt{44}$

$4x = -7 \pm 2\sqrt{11}$

$x = \dfrac{-7 \pm 2\sqrt{11}}{4} = -\dfrac{7}{4} \pm \dfrac{\sqrt{11}}{2}$

81. $(x - 7)^2 = (x + 3)^2$

$x - 7 = \pm(x + 3)$

$x - 7 = x + 3$ or $x - 7 = -x - 3$

$-7 \neq 3$ or $2x = 4$

$x = 2$

The only solution to the equation is $x = 2$.

82. $(x + 5)^2 = (x + 4)^2$

$x + 5 = \pm(x + 4)$

$x + 5 = +(x + 4)$ or $x + 5 = -(x + 4)$

$5 \neq 4$ or $x + 5 = -x - 4$

$2x = -9$

$x = -\dfrac{9}{2}$

The only solution to the equation is $x = -\dfrac{9}{2}$.

83. $x^2 + 4x - 32 = 0$

$x^2 + 4x = 32$

$x^2 + 4x + 2^2 = 32 + 2^2$

$(x + 2)^2 = 36$

$x + 2 = \pm 6$

$x = -2 \pm 6$

$x = 4$ or $x = -8$

84. $x^2 - 2x - 3 = 0$

$x^2 - 2x = 3$

$x^2 - 2x + (-1)^2 = 3 + (-1)^2$

$(x - 1)^2 = 4$

$x - 1 = \pm\sqrt{4}$

$x = 1 \pm 2$

$x = 3$ or $x = -1$

85. $x^2 + 12x + 25 = 0$

$x^2 + 12x = -25$

$x^2 + 12x + 6^2 = -25 + 6^2$

$(x + 6)^2 = 11$

$x + 6 = \pm\sqrt{11}$

$x = -6 \pm \sqrt{11}$

86. $x^2 + 8x + 14 = 0$

$x^2 + 8x = -14$

$x^2 + 8x + 4^2 = -14 + 16$

$(x + 4)^2 = 2$

$x + 4 = \pm\sqrt{2}$

$x = -4 \pm \sqrt{2}$

87. $9x^2 - 18x = -3$

$x^2 - 2x = -\dfrac{1}{3}$

$x^2 - 2x + 1^2 = -\dfrac{1}{3} + 1^2$

$(x - 1)^2 = \dfrac{2}{3}$

$x - 1 = \pm\sqrt{\dfrac{2}{3}}$

$x = 1 \pm \sqrt{\dfrac{2}{3}}$

$x = 1 \pm \dfrac{\sqrt{6}}{3}$

88. $9x^2 - 12x = 14$

$x^2 - \dfrac{4}{3}x = \dfrac{14}{9}$

$x^2 - \dfrac{4}{3}x + \left(-\dfrac{2}{3}\right)^2 = \dfrac{14}{9} + \dfrac{4}{9}$

$\left(x - \dfrac{2}{3}\right)^2 = \dfrac{18}{9}$

$\left(x - \dfrac{2}{3}\right)^2 = 2$

$x - \dfrac{2}{3} = \pm\sqrt{2}$

$x = \dfrac{2}{3} \pm \sqrt{2}$

89. $8 + 4x - x^2 = 0$

$-x^2 + 4x + 8 = 0$

$x^2 - 4x - 8 = 0$

$x^2 - 4x = 8$

$x^2 - 4x + 2^2 = 8 + 2^2$

$(x - 2)^2 = 12$

$x - 2 = \pm\sqrt{12}$

$x = 2 \pm 2\sqrt{3}$

90. $-x^2 + x - 1 = 0$

$x^2 - x + 1 = 0$

$x^2 - x + \dfrac{1}{4} = -1 + \dfrac{1}{4}$

$\left(x - \dfrac{1}{2}\right)^2 = -\dfrac{3}{4}$

No real solution

91. $2x^2 + 5x - 8 = 0$

$2x^2 + 5x = 8$

$x^2 + \dfrac{5}{2}x = 4$

$x^2 + \dfrac{5}{2}x + \left(\dfrac{5}{4}\right)^2 = 4 + \left(\dfrac{5}{4}\right)^2$

$\left(x + \dfrac{5}{4}\right)^2 = \dfrac{89}{16}$

$x + \dfrac{5}{4} = \pm\dfrac{\sqrt{89}}{4}$

$x = -\dfrac{5}{4} \pm \dfrac{\sqrt{89}}{4}$

$x = \dfrac{-5 \pm \sqrt{89}}{4}$

92. $4x^2 - 4x - 99 = 0$

$x^2 - x = \dfrac{99}{4}$

$x^2 - x + \left(-\dfrac{1}{2}\right)^2 = \dfrac{99}{4} + \dfrac{1}{4}$

$\left(x - \dfrac{1}{2}\right)^2 = \dfrac{100}{4}$

$\left(x - \dfrac{1}{2}\right)^2 = 25$

$x - \dfrac{1}{2} = \pm\sqrt{25}$

$x = \dfrac{1}{2} \pm 5 = \dfrac{11}{2}, -\dfrac{9}{2}$

93. $2x^2 + x - 1 = 0$

$x = \dfrac{-b \pm \sqrt{b^2 - 4ac}}{2a}$

$= \dfrac{-1 \pm \sqrt{1^2 - 4(2)(-1)}}{2(2)}$

$= \dfrac{-1 \pm 3}{4} = \dfrac{1}{2}, -1$

94. $2x^2 - x - 1 = 0$

$x = \dfrac{-b \pm \sqrt{b^2 - 4ac}}{2a}$

$= \dfrac{-(-1) \pm \sqrt{(-1)^2 - 4(2)(-1)}}{2(2)}$

$= \dfrac{1 \pm \sqrt{1 + 8}}{4}$

$= \dfrac{1 \pm 3}{4} = 1, -\dfrac{1}{2}$

95. $16x^2 + 8x - 3 = 0$

$$x = \frac{-b \pm \sqrt{b^2 - 4ac}}{2a}$$

$$= \frac{-8 \pm \sqrt{8^2 - 4(16)(-3)}}{2(16)}$$

$$= \frac{-8 \pm 16}{32} = \frac{1}{4}, -\frac{3}{4}$$

96. $25x^2 - 20x + 3 = 0$

$$x = \frac{-b \pm \sqrt{b^2 - 4ac}}{2a}$$

$$= \frac{-(-20) \pm \sqrt{(-20)^2 - 4(25)(3)}}{2(25)}$$

$$= \frac{20 \pm \sqrt{400 - 300}}{50}$$

$$= \frac{20 \pm 10}{50} = \frac{3}{5}, \frac{1}{5}$$

97. $2 + 2x - x^2 = 0$

$-x^2 + 2x + 2 = 0$

$$x = \frac{-b \pm \sqrt{b^2 - 4ac}}{2a}$$

$$= \frac{-2 \pm \sqrt{2^2 - 4(-1)(2)}}{2(-1)}$$

$$= \frac{-2 \pm 2\sqrt{3}}{-2} = 1 \pm \sqrt{3}$$

98. $x^2 - 10x + 22 = 0$

$$x = \frac{-b \pm \sqrt{b^2 - 4ac}}{2a}$$

$$= \frac{-(-10) \pm \sqrt{(-10)^2 - 4(1)(22)}}{2(1)}$$

$$= \frac{10 \pm \sqrt{100 - 88}}{2}$$

$$= \frac{10 \pm 2\sqrt{3}}{2} = 5 \pm \sqrt{3}$$

99. $x^2 + 14x + 44 = 0$

$$x = \frac{-b \pm \sqrt{b^2 - 4ac}}{2a}$$

$$= \frac{-14 \pm \sqrt{14^2 - 4(1)(44)}}{2(1)}$$

$$= \frac{-14 \pm 2\sqrt{5}}{2} = -7 \pm \sqrt{5}$$

100. $6x = 4 - x^2$

$x^2 + 6x - 4 = 0$

$$x = \frac{-b \pm \sqrt{b^2 - 4ac}}{2a}$$

$$= \frac{-6 \pm \sqrt{6^2 - 4(1)(-4)}}{2(1)}$$

$$= \frac{-6 \pm \sqrt{36 + 16}}{2}$$

$$= \frac{-6 \pm 2\sqrt{13}}{2}$$

$$= -3 \pm \sqrt{13}$$

101. $x^2 + 8x - 4 = 0$

$$x = \frac{-b \pm \sqrt{b^2 - 4ac}}{2a}$$

$$= \frac{-8 \pm \sqrt{8^2 - 4(1)(-4)}}{2(1)}$$

$$= \frac{-8 \pm 4\sqrt{5}}{2} = -4 \pm 2\sqrt{5}$$

102. $4x^2 - 4x - 4 = 0$

$x^2 - x - 1 = 0$

$$x = \frac{-b \pm \sqrt{b^2 - 4ac}}{2a}$$

$$= \frac{-(-1) \pm \sqrt{(-1)^2 - 4(1)(-1)}}{2(1)}$$

$$= \frac{1 \pm \sqrt{1 + 4}}{2}$$

$$= \frac{1}{2} \pm \frac{\sqrt{5}}{2}$$

103. $12x - 9x^2 = -3$

$-9x^2 + 12x + 3 = 0$

$$x = \frac{-b \pm \sqrt{b^2 - 4ac}}{2a}$$

$$= \frac{-12 \pm \sqrt{12^2 - 4(-9)(3)}}{2(-9)}$$

$$= \frac{-12 \pm 6\sqrt{7}}{-18} = \frac{2}{3} \pm \frac{\sqrt{7}}{3}$$

104. $16x^2 + 22 = 40x$

$8x^2 - 20x + 11 = 0$

$$x = \frac{-b \pm \sqrt{b^2 - 4ac}}{2a}$$

$$= \frac{-(-20) \pm \sqrt{(-20)^2 - 4(8)(11)}}{2(8)}$$

$$= \frac{20 \pm \sqrt{400 - 352}}{16}$$

$$= \frac{5}{4} \pm \frac{\sqrt{3}}{4}$$

105. $9x^2 + 24x + 16 = 0$

$$x = \frac{-b \pm \sqrt{b^2 - 4ac}}{2a}$$

$$= \frac{-24 \pm \sqrt{24^2 - 4(9)(16)}}{2(9)}$$

$$= \frac{-24 \pm 0}{18}$$

$$= -\frac{4}{3}$$

106. $36x^2 + 24x - 7 = 0$

$$x = \frac{-b \pm \sqrt{b^2 - 4ac}}{2a}$$

$$= \frac{-24 \pm \sqrt{24^2 - 4(36)(-7)}}{2(36)}$$

$$= \frac{-24 \pm \sqrt{576 + 1008}}{72}$$

$$= \frac{-24 \pm \sqrt{(144)(11)}}{72}$$

$$= -\frac{1}{3} \pm \frac{\sqrt{11}}{6}$$

107. $4x^2 + 4x = 7$

$4x^2 + 4x - 7 = 0$

$x = \dfrac{-b \pm \sqrt{b^2 - 4ac}}{2a}$

$\quad = \dfrac{-4 \pm \sqrt{4^2 - 4(4)(-7)}}{2(4)}$

$\quad = \dfrac{-4 \pm 8\sqrt{2}}{8} = -\dfrac{1}{2} \pm \sqrt{2}$

108. $16x^2 - 40x + 5 = 0$

$x = \dfrac{-b \pm \sqrt{b^2 - 4ac}}{2a}$

$\quad = \dfrac{-(-40) \pm \sqrt{(-40)^2 - 4(16)(5)}}{2(16)}$

$\quad = \dfrac{40 \pm \sqrt{1600 - 320}}{32}$

$\quad = \dfrac{40 \pm 16\sqrt{5}}{32}$

$\quad = \dfrac{5}{4} \pm \dfrac{\sqrt{5}}{2}$

109. $28x - 49x^2 = 4$

$-49x^2 + 28x - 4 = 0$

$x = \dfrac{-b \pm \sqrt{b^2 - 4ac}}{2a}$

$\quad = \dfrac{-28 \pm \sqrt{28^2 - 4(-49)(-4)}}{2(-49)}$

$\quad = \dfrac{-28 \pm 0}{-98} = \dfrac{2}{7}$

110. $3x + x^2 - 1 = 0$

$x^2 + 3x - 1 = 0$

$x = \dfrac{-b \pm \sqrt{b^2 - 4ac}}{2a}$

$\quad = \dfrac{-3 \pm \sqrt{3^2 - 4(1)(-1)}}{2(1)}$

$\quad = \dfrac{-3 \pm \sqrt{13}}{2} = -\dfrac{3}{2} \pm \dfrac{\sqrt{13}}{2}$

111. $8t = 5 + 2t^2$

$-2t^2 + 8t - 5 = 0$

$t = \dfrac{-b \pm \sqrt{b^2 - 4ac}}{2a}$

$\quad = \dfrac{-8 \pm \sqrt{8^2 - 4(-2)(-5)}}{2(-2)}$

$\quad = \dfrac{-8 \pm 2\sqrt{6}}{-4} = 2 \pm \dfrac{\sqrt{6}}{2}$

112. $25h^2 + 80h + 61 = 0$

$h = \dfrac{-b \pm \sqrt{b^2 - 4ac}}{2a}$

$\quad = \dfrac{-80 \pm \sqrt{80^2 - 4(25)(61)}}{2(25)}$

$\quad = \dfrac{-80 \pm \sqrt{6400 - 6100}}{50}$

$\quad = -\dfrac{8}{5} \pm \dfrac{10\sqrt{3}}{50}$

$\quad = -\dfrac{8}{5} \pm \dfrac{\sqrt{3}}{5}$

113. $(y - 5)^2 = 2y$

$y^2 - 12y + 25 = 0$

$y = \dfrac{-b \pm \sqrt{b^2 - 4ac}}{2a}$

$\quad = \dfrac{-(-12) \pm \sqrt{(-12)^2 - 4(1)(25)}}{2(1)}$

$\quad = \dfrac{12 \pm 2\sqrt{11}}{2} = 6 \pm \sqrt{11}$

114. $(z + 6)^2 = -2z$

$z^2 + 12z + 36 = -2z$

$z^2 + 14z + 36 = 0$

$z = \dfrac{-b \pm \sqrt{b^2 - 4ac}}{2a}$

$\quad = \dfrac{-14 \pm \sqrt{14^2 - 4(1)(36)}}{2(1)}$

$\quad = \dfrac{-14 \pm \sqrt{52}}{2}$

$\quad = -7 \pm \sqrt{13}$

115. $\dfrac{1}{2}x^2 + \dfrac{3}{8}x = 2$

$4x^2 + 3x = 16$

$4x^2 + 3x - 16 = 0$

$x = \dfrac{-b \pm \sqrt{b^2 - 4ac}}{2a}$

$\quad = \dfrac{-3 \pm \sqrt{3^2 - 4(4)(-16)}}{2(4)}$

$\quad = \dfrac{-3 \pm \sqrt{265}}{8}$

$\quad = -\dfrac{3}{8} \pm \dfrac{\sqrt{265}}{8}$

116. $\left(\dfrac{5}{7}x - 14\right)^2 = 8x$

$\dfrac{25}{49}x^2 - 20x + 196 = 8x$

$\dfrac{25}{49}x^2 - 28x + 196 = 0$

$25x^2 - 1372x + 9604 = 0$

$x = \dfrac{-b \pm \sqrt{b^2 - 4ac}}{2a} = \dfrac{-(-1372) \pm \sqrt{(-1372)^2 - 4(25)(9604)}}{2(25)} = \dfrac{1372 \pm \sqrt{921,984}}{50} = \dfrac{686 \pm 196\sqrt{6}}{25}$

117. $5.1x^2 - 1.7x - 3.2 = 0$

$$x = \frac{1.7 \pm \sqrt{(-1.7)^2 - 4(5.1)(-3.2)}}{2(5.1)}$$

$$x \approx 0.976, -0.643$$

118. $2x^2 - 2.50x - 0.42 = 0$

$$x = \frac{-b \pm \sqrt{b^2 - 4ac}}{2a}$$

$$= \frac{-(-2.50) \pm \sqrt{(-2.50)^2 - 4(2)(-0.42)}}{2(2)}$$

$$= \frac{2.50 \pm \sqrt{9.61}}{4}$$

$$= 1.400, -0.150$$

119. $-0.067x^2 - 0.852x + 1.277 = 0$

$$x = \frac{-(-0.852) \pm \sqrt{(-0.852)^2 - 4(-0.067)(1.277)}}{2(-0.067)}$$

$$x \approx -14.071, 1.355$$

120. $-0.005x^2 + 0.101x - 0.193 = 0$

$$x = \frac{-b \pm \sqrt{b^2 - 4ac}}{2a}$$

$$= \frac{-0.101 \pm \sqrt{(0.101)^2 - 4(-0.005)(-0.193)}}{2(-0.005)}$$

$$= \frac{-0.101 \pm \sqrt{0.006341}}{-0.01}$$

$$\approx 2.137, 18.063$$

121. $422x^2 - 506x - 347 = 0$

$$x = \frac{506 \pm \sqrt{(-506)^2 - 4(422)(-347)}}{2(422)}$$

$$x \approx 1.687, -0.488$$

122. $1100x^2 + 326x - 715 = 0$

$$x = \frac{-b \pm \sqrt{b^2 - 4ac}}{2a}$$

$$= \frac{-326 \pm \sqrt{(326)^2 - 4(1100)(-715)}}{2(1100)}$$

$$= \frac{-326 \pm \sqrt{3,252,276}}{2200} \approx 0.672, -0.968$$

123. $12.67x^2 + 31.55x + 8.09 = 0$

$$x = \frac{-31.55 \pm \sqrt{(31.55)^2 - 4(12.67)(8.09)}}{2(12.67)}$$

$$x \approx -2.200, -0.290$$

124. $-3.22x^2 - 0.08x + 28.651 = 0$

$$x = \frac{-b \pm \sqrt{b^2 - 4ac}}{2a}$$

$$= \frac{-(-0.08) \pm \sqrt{(-0.08)^2 - 4(-3.22)(28.651)}}{2(-3.22)}$$

$$= \frac{0.08 \pm \sqrt{369.031}}{-6.44} \approx -2.995, 2.971$$

125. $x^2 - 2x - 1 = 0$ Complete the square.

$$x^2 - 2x = 1$$

$$x^2 - 2x + 1^2 = 1 + 1^2$$

$$(x - 1)^2 = 2$$

$$x - 1 = \pm\sqrt{2}$$

$$x = 1 \pm \sqrt{2}$$

126. $11x^2 + 33x = 0$ Factor.

$$11(x^2 + 3x) = 0$$

$$x(x + 3) = 0$$

$$x = 0 \quad \text{or} \quad x + 3 = 0$$

$$x = -3$$

127. $(x + 3)^2 = 81$ Extract square roots.

$$x + 3 = \pm 9$$

$$x + 3 = 9 \quad \text{or} \quad x + 3 = -9$$

$$x = 6 \quad \text{or} \quad x = -12$$

128. $x^2 - 14x + 49 = 0$ Extract square roots.

$$(x - 7)^2 = 0$$

$$x - 7 = 0$$

$$x = 7$$

129. $x^2 - x - \frac{11}{4} = 0$ Complete the square.

$$x^2 - x = \frac{11}{4}$$

$$x^2 - x + \left(\frac{1}{2}\right)^2 = \frac{11}{4} + \left(\frac{1}{2}\right)^2$$

$$\left(x - \frac{1}{2}\right)^2 = \frac{12}{4}$$

$$x - \frac{1}{2} = \pm\sqrt{\frac{12}{4}}$$

$$x = \frac{1}{2} \pm \sqrt{3}$$

130. $x^2 + 3x - \frac{3}{4} = 0$ Complete the square.

$$x^2 + 3x + \left(\frac{3}{2}\right)^2 = \frac{3}{4} + \frac{9}{4}$$

$$\left(x + \frac{3}{2}\right)^2 = 3$$

$$x + \frac{3}{2} = \pm\sqrt{3}$$

$$x = -\frac{3}{2} \pm \sqrt{3}$$

131. $(x + 1)^2 = x^2$ Extract square roots.

$$x^2 = (x + 1)^2$$

$$x = \pm(x + 1)$$

For $x = +(x + 1)$:

$$0 \neq 1 \quad \text{No solution}$$

For $x = -(x + 1)$:

$$2x = -1$$

$$x = -\frac{1}{2}$$

132. $a^2x^2 - b^2 = 0$ Factor.

$$(ax + b)(ax - b) = 0$$

$$ax + b = 0 \implies x = -\frac{b}{a}$$

$$ax - b = 0 \implies x = \frac{b}{a}$$

133. $3x + 4 = 2x^2 - 7$ Quadratic Formula

$$0 = 2x^2 - 3x - 11$$

$$x = \frac{-(-3) \pm \sqrt{(-3)^2 - 4(2)(-11)}}{2(2)}$$

$$= \frac{3 \pm \sqrt{97}}{4}$$

$$= \frac{3}{4} \pm \frac{\sqrt{97}}{4}$$

134. $4x^2 + 2x + 4 = 2x + 8$ Factor.

$$4x^2 - 4 = 0$$

$$4(x^2 - 1) = 0$$

$$(x + 1)(x - 1) = 0$$

$$x + 1 = 0 \quad \text{or} \quad x - 1 = 0$$

$$x = -1 \qquad x = 1$$

135. $4x^4 - 18x^2 = 0$

$$2x^2(2x^2 - 9) = 0$$

$$2x^2 = 0 \implies x = 0$$

$$2x^2 - 9 = 0 \implies x = \pm\frac{3\sqrt{2}}{2}$$

136. $20x^3 - 125x = 0$

$$5x(4x^2 - 25) = 0$$

$$5x(2x + 5)(2x - 5) = 0$$

$$5x = 0 \implies x = 0$$

$$2x + 5 = 0 \implies x = -\frac{5}{2}$$

$$2x - 5 = 0 \implies x = \frac{5}{2}$$

137. $x^4 - 81 = 0$

$$(x^2 + 9)(x + 3)(x - 3) = 0$$

$$x^2 + 9 = 0 \implies \text{No real solution}$$

$$x + 3 = 0 \implies x = -3$$

$$x - 3 = 0 \implies x = 3$$

138.
$$x^6 - 64 = 0$$
$$(x^3 - 8)(x^3 + 8) = 0$$
$$(x - 2)(x^2 + 2x + 4)(x + 2)(x^2 - 2x + 4) = 0$$
$$x - 2 = 0 \implies x = 2$$
$$x^2 + 2x + 4 = 0 \implies \text{No real solution (by the Quadratic Formula)}$$
$$x + 2 = 0 \implies x = -2$$
$$x^2 - 2x + 4 = 0 \implies \text{No real solution (by the Quadratic Formula)}$$

139.
$$x^3 + 216 = 0$$
$$x^3 + 6^3 = 0$$
$$(x + 6)(x^2 - 6x + 36) = 0$$
$$x + 6 = 0 \implies x = -6$$
$$x^2 - 6x + 36 = 0 \implies \text{No real solution (by the Quadratic Formula)}$$

140.
$$27x^3 - 512 = 0$$
$$(3x - 8)(9x^2 + 24x + 64) = 0$$
$$3x - 8 = 0 \implies x = \tfrac{8}{3}$$
$$9x^2 + 24x + 64 = 0 \implies \text{No real solution (by the Quadratic Formula)}$$

141. $5x^3 + 30x^2 + 45x = 0$
$$5x(x^2 + 6x + 9) = 0$$
$$5x(x + 3)^2 = 0$$
$$5x = 0 \implies x = 0$$
$$x + 3 = 0 \implies x = -3$$

142. $9x^4 - 24x^3 + 16x^2 = 0$
$$x^2(9x^2 - 24x + 16) = 0$$
$$x^2(3x - 4)^2 = 0$$
$$x^2 = 0 \implies x = 0$$
$$3x - 4 = 0 \implies x = \tfrac{4}{3}$$

143.
$$x^3 - 3x^2 - x + 3 = 0$$
$$x^2(x - 3) - (x - 3) = 0$$
$$(x - 3)(x^2 - 1) = 0$$
$$(x - 3)(x + 1)(x - 1) = 0$$
$$x - 3 = 0 \implies x = 3$$
$$x + 1 = 0 \implies x = -1$$
$$x - 1 = 0 \implies x = 1$$

144.
$$x^3 + 2x^2 + 3x + 6 = 0$$
$$x^2(x + 2) + 3(x + 2) = 0$$
$$(x + 2)(x^2 + 3) = 0$$
$$x + 2 = 0 \implies x = -2$$
$$x^2 + 3 = 0 \implies \text{No real solution}$$

145.
$$x^4 - x^3 + x - 1 = 0$$
$$x^3(x - 1) + (x - 1) = 0$$
$$(x - 1)(x^3 + 1) = 0$$
$$(x - 1)(x + 1)(x^2 - x + 1) = 0$$
$$x - 1 = 0 \implies x = 1$$
$$x + 1 = 0 \implies x = -1$$
$$x^2 - x + 1 = 0 \implies \text{No real solution (by the Quadratic Formula)}$$

146.
$$x^4 + 2x^3 - 8x - 16 = 0$$
$$x^3(x + 2) - 8(x + 2) = 0$$
$$(x^3 - 8)(x + 2) = 0$$
$$(x - 2)(x^2 + 2x + 4)(x + 2) = 0$$
$$x - 2 = 0 \implies x = 2$$
$$x^2 + 2x + 4 = 0 \implies \text{No real solution (by the Quadratic Formula)}$$
$$x + 2 = 0 \implies x = -2$$

147.
$$x^4 - 4x^2 + 3 = 0$$
$$(x^2 - 3)(x^2 - 1) = 0$$
$$\left(x + \sqrt{3}\right)\left(x - \sqrt{3}\right)(x + 1)(x - 1) = 0$$
$$x + \sqrt{3} = 0 \implies x = -\sqrt{3}$$
$$x - \sqrt{3} = 0 \implies x = \sqrt{3}$$
$$x + 1 = 0 \implies x = -1$$
$$x - 1 = 0 \implies x = 1$$

148.
$$x^4 + 5x^2 - 36 = 0$$
$$(x^2 + 9)(x^2 - 4) = 0$$
$$(x^2 + 9)(x + 2)(x - 2) = 0$$
$$x^2 + 9 = 0 \implies \text{No real solution}$$
$$x + 2 = 0 \implies x = -2$$
$$x - 2 = 0 \implies x = 2$$

149.
$$4x^4 - 65x^2 + 16 = 0$$
$$(4x^2 - 1)(x^2 - 16) = 0$$
$$(2x + 1)(2x - 1)(x + 4)(x - 4) = 0$$
$$2x + 1 = 0 \implies x = -\tfrac{1}{2}$$
$$2x - 1 = 0 \implies x = \tfrac{1}{2}$$
$$x + 4 = 0 \implies x = -4$$
$$x - 4 = 0 \implies x = 4$$

150.
$$36t^4 + 29t^2 - 7 = 0$$
$$(36t^2 - 7)(t^2 + 1) = 0$$
$$\left(6t + \sqrt{7}\right)\left(6t - \sqrt{7}\right)(t^2 + 1) = 0$$
$$6t + \sqrt{7} = 0 \implies t = -\frac{\sqrt{7}}{6}$$
$$6t - \sqrt{7} = 0 \implies t = \frac{\sqrt{7}}{6}$$
$$t^2 + 1 = 0 \implies \text{No real solution}$$

151.
$$x^6 + 7x^3 - 8 = 0$$
$$(x^3 + 8)(x^3 - 1) = 0$$
$$(x + 2)(x^2 - 2x + 4)(x - 1)(x^2 + x + 1) = 0$$
$$x + 2 = 0 \implies x = -2$$
$$x^2 - 2x + 4 = 0 \implies \text{No real solution (by the Quadratic Formula)}$$
$$x - 1 = 0 \implies x = 1$$
$$x^2 + x + 1 = 0 \implies \text{No real solution (by the Quadratic Formula)}$$

152.
$$x^6 + 3x^3 + 2 = 0$$
$$(x^3 + 2)(x^3 + 1) = 0$$
$$\left(x + \sqrt[3]{2}\right)\left[x^2 - \sqrt[3]{2}x + \left(\sqrt[3]{2}\right)^2\right](x + 1)(x^2 - x + 1) = 0$$
$$x + \sqrt[3]{2} = 0 \implies x = -\sqrt[3]{2}$$
$$x^2 - \sqrt[3]{2}x + \left(\sqrt[3]{2}\right)^2 = 0 \implies \text{No real solution (by the Quadratic Formula)}$$
$$x + 1 = 0 \implies x = -1$$
$$x^2 - x + 1 = 0 \implies \text{No real solution (by the Quadratic Formula)}$$

153. $\sqrt{2x} - 10 = 0$

$\sqrt{2x} = 10$

$2x = 100$

$x = 50$

154. $4\sqrt{x} - 3 = 0$

$4\sqrt{x} = 3$

$16x = 9$

$x = \frac{9}{16}$

155. $\sqrt{x - 10} - 4 = 0$

$\sqrt{x - 10} = 4$

$x - 10 = 16$

$x = 26$

156. $\sqrt{5 - x} - 3 = 0$

$\sqrt{5 - x} = 3$

$5 - x = 9$

$x = -4$

157. $\sqrt[3]{2x + 5} + 3 = 0$

$\sqrt[3]{2x + 5} = -3$

$2x + 5 = -27$

$2x = -32$

$x = -16$

158. $\sqrt[3]{3x + 1} - 5 = 0$

$\sqrt[3]{3x + 1} = 5$

$3x + 1 = 125$

$3x = 124$

$x = \frac{124}{3}$

159. $-\sqrt{26 - 11x} + 4 = x$

$4 - x = \sqrt{26 - 11x}$

$16 - 8x + x^2 = 26 - 11x$

$x^2 + 3x - 10 = 0$

$(x + 5)(x - 2) = 0$

$x + 5 = 0 \implies x = -5$

$x - 2 = 0 \implies x = 2$

160. $x + \sqrt{31 - 9x} = 5$

$\sqrt{31 - 9x} = 5 - x$

$31 - 9x = 25 - 10x + x^2$

$0 = x^2 - x - 6$

$0 = (x - 3)(x + 2)$

$0 = x - 3 \implies x = 3$

$0 = x + 2 \implies x = -2$

161. $\sqrt{x + 1} = \sqrt{3x + 1}$

$x + 1 = 3x + 1$

$-2x = 0$

$x = 0$

162. $\sqrt{x + 5} = \sqrt{x - 5}$

$x + 5 = x - 5$

$5 = -5$

No solution

163. $(x - 5)^{3/2} = 8$

$(x - 5)^3 = 8^2$

$x - 5 = \sqrt[3]{64}$

$x = 5 + 4 = 9$

164. $(x + 3)^{3/2} = 8$

$(x + 3)^3 = 8^2$

$(x + 3)^3 = 64$

$x + 3 = \sqrt[3]{64}$

$x = -3 + 4 = 1$

165. $(x + 3)^{2/3} = 8$

$(x + 3)^2 = 8^3$

$x + 3 = \pm\sqrt{8^3}$

$x + 3 = \pm\sqrt{512}$

$x = -3 \pm 16\sqrt{2}$

166. $(x + 2)^{2/3} = 9$

$(x + 2)^2 = 9^3$

$x + 2 = \pm\sqrt{729}$

$x = -2 \pm 27$

$= -29, 25$

167. $(x^2 - 5)^{3/2} = 27$

$(x^2 - 5)^3 = 27^2$

$x^2 - 5 = \sqrt[3]{27^2}$

$x^2 = 5 + 9$

$x^2 = 14$

$x = \pm\sqrt{14}$

168. $(x^2 - x - 22)^{3/2} = 27$

$$x^2 - x - 22 = 27^{2/3}$$

$$x^2 - x - 22 = 9$$

$$x^2 - x - 31 = 0$$

$$x = \frac{-(-1) \pm \sqrt{(-1)^2 - 4(1)(-31)}}{2(1)} = \frac{1 \pm \sqrt{125}}{2} = \frac{1 \pm 5\sqrt{5}}{2}$$

169. $3x(x - 1)^{1/2} + 2(x - 1)^{3/2} = 0$

$$(x - 1)^{1/2}[3x + 2(x - 1)] = 0$$

$$(x - 1)^{1/2}(5x - 2) = 0$$

$$(x - 1)^{1/2} = 0 \implies x - 1 = 0 \implies x = 1$$

$$5x - 2 = 0 \implies x = \tfrac{2}{5}, \text{ extraneous}$$

170. $4x^2(x - 1)^{1/3} + 6x(x - 1)^{4/3} = 0$

$$2x[2x(x - 1)^{1/3} + 3(x - 1)^{4/3}] = 0$$

$$2x(x - 1)^{1/3}[2x + 3(x - 1)] = 0$$

$$2x(x - 1)^{1/3}(5x - 3) = 0$$

$$2x = 0 \implies x = 0$$

$$x - 1 = 0 \implies x = 1$$

$$5x - 3 = 0 \implies x = \tfrac{3}{5}$$

171.

$$x = \frac{3}{x} + \frac{1}{2}$$

$$(2x)(x) = (2x)\left(\frac{3}{x}\right) + (2x)\left(\frac{1}{2}\right)$$

$$2x^2 = 6 + x$$

$$2x^2 - x - 6 = 0$$

$$(2x + 3)(x - 2) = 0$$

$$2x + 3 = 0 \implies x = -\frac{3}{2}$$

$$x - 2 = 0 \implies x = 2$$

172.

$$\frac{4}{x} - \frac{5}{3} = \frac{x}{6}$$

$$(6x)\frac{4}{x} - (6x)\frac{5}{3} = (6x)\frac{x}{6}$$

$$24 - 10x = x^2$$

$$x^2 + 10x - 24 = 0$$

$$(x + 12)(x - 2) = 0$$

$$x + 12 = 0 \implies x = -12$$

$$x - 2 = 0 \implies x = 2$$

173.

$$\frac{1}{x} - \frac{1}{x + 1} = 3$$

$$x(x + 1)\frac{1}{x} - x(x + 1)\frac{1}{x + 1} = x(x + 1)(3)$$

$$x + 1 - x = 3x(x + 1)$$

$$1 = 3x^2 + 3x$$

$$0 = 3x^2 + 3x - 1$$

$$a = 3, \quad b = 3, \quad c = -1$$

$$x = \frac{-3 \pm \sqrt{(3)^2 - 4(3)(-1)}}{2(3)} = \frac{-3 \pm \sqrt{21}}{6}$$

174.

$$\frac{4}{x + 1} - \frac{3}{x + 2} = 1$$

$$4(x + 2) - 3(x + 1) = (x + 1)(x + 2)$$

$$4x + 8 - 3x - 3 = x^2 + 3x + 2$$

$$x^2 + 2x - 3 = 0$$

$$(x - 1)(x + 3) = 0$$

$$x - 1 = 0 \implies x = 1$$

$$x + 3 = 0 \implies x = -3$$

175. $\dfrac{20 - x}{x} = x$

$20 - x = x^2$

$0 = x^2 + x - 20$

$0 = (x + 5)(x - 4)$

$x + 5 = 0 \implies x = -5$

$x - 4 = 0 \implies x = 4$

176. $4x + 1 = \dfrac{3}{x}$

$(x)4x + (x)1 = (x)\dfrac{3}{x}$

$4x^2 + x = 3$

$4x^2 + x - 3 = 0$

$(4x - 3)(x + 1) = 0$

$4x - 3 = 0 \implies x = \dfrac{3}{4}$

$x + 1 = 0 \implies x = -1$

177. $\dfrac{x}{x^2 - 4} + \dfrac{1}{x + 2} = 3$

$(x + 2)(x - 2)\dfrac{x}{x^2 - 4} + (x + 2)(x - 2)\dfrac{1}{x + 2} = 3(x + 2)(x - 2)$

$x + x - 2 = 3x^2 - 12$

$3x^2 - 2x - 10 = 0$

$a = 3, \ b = -2, \ c = -10$

$x = \dfrac{-(-2) \pm \sqrt{(-2)^2 - 4(3)(-10)}}{2(3)}$

$= \dfrac{2 \pm \sqrt{124}}{6} = \dfrac{2 \pm 2\sqrt{31}}{6} = \dfrac{1 \pm \sqrt{31}}{3}$

178. $\dfrac{x + 1}{3} - \dfrac{x + 1}{x + 2} = 0$

$3(x + 2)\dfrac{x + 1}{3} - 3(x + 2)\dfrac{x + 1}{x + 2} = 0$

$(x + 2)(x + 1) - 3(x + 1) = 0$

$x^2 + 3x + 2 - 3x - 3 = 0$

$x^2 - 1 = 0$

$(x + 1)(x - 1) = 0$

$x + 1 = 0 \implies x = -1$

$x - 1 = 0 \implies x = 1$

179. $|2x - 1| = 5$

$2x - 1 = 5 \implies x = 3$

$-(2x - 1) = 5 \implies x = -2$

180. $|3x + 2| = 7$

$3x + 2 = 7 \implies x = \dfrac{5}{3}$

$-(3x + 2) = 7$

$-3x - 2 = 7 \implies x = -3$

181. $|x| = x^2 + x - 3$

First equation:

$x = x^2 + x - 3$

$x^2 - 3 = 0$

$x = \pm\sqrt{3}$

Second equation:

$-x = x^2 + x - 3$

$x^2 + 2x - 3 = 0$

$(x - 1)(x + 3) = 0$

$x - 1 = 0 \implies x = 1$

$x + 3 = 0 \implies x = -3$

Only $x = \sqrt{3}$ and $x = -3$ are solutions to the original equation. $x = -\sqrt{3}$ and $x = 1$ are extraneous.

182. $|x^2 + 6x| = 3x + 18$

First equation:

$$x^2 + 6x = 3x + 18$$
$$x^2 + 3x - 18 = 0$$
$$(x - 3)(x + 6) = 0$$
$$x - 3 = 0 \Rightarrow x = 3$$
$$x + 6 = 0 \Rightarrow x = -6$$

Second equation:

$$-(x^2 + 6x) = 3x + 18$$
$$0 = x^2 + 9x + 18$$
$$0 = (x + 3)(x + 6)$$
$$0 = x + 3 \Rightarrow x = -3$$
$$0 = x + 6 \Rightarrow x = -6$$

The solutions to the original equation are $x = \pm 3$ and $x = -6$.

183. $|x + 1| = x^2 - 5$

First equation:

$$x + 1 = x^2 - 5$$
$$x^2 - x - 6 = 0$$
$$(x - 3)(x + 2) = 0$$
$$x - 3 = 0 \Rightarrow x = 3$$
$$x + 2 = 0 \Rightarrow x = -2$$

Second equation:

$$-(x + 1) = x^2 - 5$$
$$-x - 1 = x^2 - 5$$
$$x^2 + x - 4 = 0$$
$$x = \frac{-1 \pm \sqrt{17}}{2}$$

Only $x = 3$ and $x = \dfrac{-1 - \sqrt{17}}{2}$ are solutions to the original equation. $x = -2$ and $x = \dfrac{-1 + \sqrt{17}}{2}$ are extraneous.

184. $|x - 10| = x^2 - 10x$

First equation:

$$x - 10 = x^2 - 10x$$
$$0 = x^2 - 11x + 10$$
$$0 = (x - 1)(x - 10)$$
$$0 = x - 1 \Rightarrow x = 1$$
$$0 = x - 10 \Rightarrow x = 10$$

Second equation:

$$-(x - 10) = x^2 - 10x$$
$$0 = x^2 - 9x - 10$$
$$0 = (x - 10)(x + 1)$$
$$0 = x - 10 \Rightarrow x = 10$$
$$0 = x + 1 \Rightarrow x = -1$$

The solutions to the original equation are $x = 10$ and $x = -1$. $x = 1$ is extraneous.

185. (a) Female: $y = 0.432x - 10.44$

For $y = 16$: $16 = 0.432x - 10.44$
$$26.44 = 0.432x$$
$$\frac{26.44}{0.432} = x$$
$$x \approx 61.2 \text{ inches}$$

(b) Male: $y = 0.449x - 12.15$

For $y = 19$: $19 = 0.449x - 12.15$
$$31.15 = 0.449x$$
$$69.4 \approx x$$

Yes, it is likely that both bones came from the same person because the estimated height of a male with a 19-inch thigh bone is 69.4 inches.

—CONTINUED—

185. —CONTINUED—

(c)

Height x	Female Femur Length	Male Femur Length
60	15.48	14.79
70	19.80	19.28
80	24.12	23.77
90	28.44	28.26
100	32.76	32.75
110	37.08	37.24

The lengths of the male and female femurs are approximately equal when the lengths are 100 inches.

186. $10,000 = 0.32m + 2500$

$7500 = 0.32m$

$\dfrac{7500}{0.32} = m$

$m = 23,437.5$ miles

187. $y = -0.25t + 8$

$1 = -0.25t + 8$

$0.25t = 7$

$t = 28$ hours

188. (a)

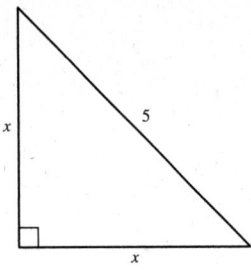

(b) $w(w + 14) = 1632$

(c) $w^2 + 14w - 1632 = 0$

$(w + 48)(w - 34) = 0$

$w = -48$ or $w = 34$

Since w must be greater than zero, we have $w = 34$ feet and the length is $w + 14 = 48$ feet.

189. $S = x^2 + 4xh$

$84 = x^2 + 4x(2)$

$0 = x^2 + 8x - 84$

$0 = (x + 14)(x - 6)$

$x = -14$ or $x = 6$

Since x must be positive, we have $x = 6$ inches. The dimensions of the box are 6 inches $\times$ 6 inches $\times$ 2 inches.

190. $x^2 + x^2 = 5^2$

$2x^2 = 25$

$x^2 = \dfrac{25}{2}$

$x = \sqrt{\dfrac{25}{2}}$

$= \dfrac{5}{\sqrt{2}}$

$= \dfrac{5\sqrt{2}}{2} \approx 3.54$ centimeters

Pythagorean Theorem

Each leg in the right triangle is approximately 3.54 centimeters.

191. *Model*:   (height)^2 + (half of side)^2 = (side)^2

Labels: height = 10 inches, side = s, half of side = $\dfrac{s}{2}$

Equation: $10^2 + \left(\dfrac{s}{2}\right)^2 = s^2$

$$100 + \frac{s^2}{4} = s^2$$

$$\frac{3}{4}s^2 = 100$$

$$s^2 = \frac{400}{3}$$

$$s = \sqrt{\frac{400}{3}} = \frac{20\sqrt{3}}{3} \approx 11.55 \text{ inches}$$

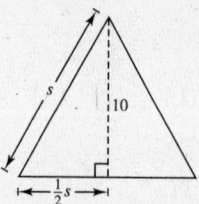

Each side of the equilateral triangle is approximately 11.55 inches long.

192.

$$d_N = (3 \text{ hours})(r + 50 \text{ mph})$$

$$d_E = (3 \text{ hours})(r \text{ mph})$$

$$d_N{}^2 + d_E{}^2 = 2440^2$$

$$9(r + 50)^2 + 9r^2 = 2440^2$$

$$18r^2 + 900r - 5,931,100 = 0$$

$$r = \frac{-900 \pm \sqrt{900^2 - 4(18)(-5,931,100)}}{2(18)} = \frac{-900 \pm 60\sqrt{118,847}}{36}$$

Using the positive value for r, we have one plane moving northbound at $r + 50 \approx 600$ miles per hour and one plane moving eastbound at $r \approx 550$ miles per hour.

193. (a) $P = 200$ million when:

$$\frac{182.45 - 3.189t}{1.00 - 0.026t} = 200$$

$$182.45 - 3.189t = 200(1.00 - 0.026t)$$

$$182.45 - 3.189t = 200 - 5.2t$$

$$2.011t = 17.55$$

$$t = 8.7$$

So the total voting-age population reached 200 million during 1998.

(b) For $P = 230$:

$$\frac{182.45 - 3.189t}{1.00 - 0.026t} = 230$$

$$182.45 - 3.189t = 230(1.00 - 0.026t)$$

$$182.45 - 3.189t = 230 - 5.98t$$

so $2.791t = 47.55$

$$t = 17$$

The model predicts that the total voting-age population will reach 230 million during 2007. This value is reasonable but the model is reaching its limit since it soon begins to rise very fast due to its asymptotic behavior.

194. When $C = 2.5$ we have:

$$2.5 = \sqrt{0.2x + 1}$$

$$6.25 = 0.2x + 1$$

$$5.25 = 0.2x$$

$$x = 26.25 = 26,250 \text{ passengers}$$

195. $37.55 = 40 - \sqrt{0.01x + 1}$

$$\sqrt{0.01x + 1} = 2.45$$

$$0.01x + 1 = 6.0025$$

$$0.01x = 5.0025$$

$$x = 500.25$$

Rounding x to the nearest whole unit yields $x \approx 500$ units.

196. When $p = \$750$, we have:

$$750 = 800 - \sqrt{0.01x + 1}$$

$$50 = \sqrt{0.01x + 1}$$

$$2500 = 0.01x + 1$$

$$0.01x = 2499$$

$$x = 249{,}900$$

So the demand is 249,900 units when the price is $750.

197. False. $x(3 - x) = 10 \implies 3x - x^2 = 10$

This is a quadratic equation. The equation cannot be written in the form $ax + b = 0$.

198. False. The product must equal zero for the Zero-Factor property to be used.

199. False—See Example 14 on page A55.

200. False. $|x| = 0$ has only one solution to check, 0.

201. Equivalent equations are derived from the substitution principle and simplification techniques. They have the same solution(s).

$2x + 3 = 8$ and $2x = 5$ are equivalent equations.

202. To transform an equation into an equivalent equation, you should first remove symbols of grouping, combine like terms, and reduce fractions. Then, as needed, you may add (or subtract) the same quantity to (from) both sides of the equation, multiply (divide) both sides of the equation by the same nonzero quantity, or interchange the two sides of the equation.

203. The student should have subtracted $15x$ from both sides so that the equation is equal to zero. By factoring out an x, there are two solutions, $x = 0$ and $x = 6$.

204. $3(x + 4)^2 + (x + 4) - 2 = 0$

(a) Let $u = x + 4$

$$3u^2 + u - 2 = 0$$

$$(3u - 2)(u + 1) = 0$$

$$3u - 2 = 0 \quad \text{or} \quad u + 1 = 0$$

$$u = \tfrac{2}{3} \qquad\qquad u = -1$$

$$x + 4 = \tfrac{2}{3} \qquad x + 4 = -1$$

$$x = -\tfrac{10}{3} \quad \text{or} \qquad x = -5$$

(b) $3(x^2 + 8x + 16) + (x + 4) - 2 = 0$

$$3x^2 + 24x + 48 + x + 4 - 2 = 0$$

$$3x^2 + 25x + 50 = 0$$

$$(3x + 10)(x + 5) = 0$$

$$3x + 10 = 0 \quad \text{or} \quad x + 5 = 0$$

$$x = -\tfrac{10}{3} \qquad\qquad x = -5$$

(c) The method of part (a) reduces the number of algebraic steps.

205. -3 and 6

One possible equation is:

$$(x - (-3))(x - 6) = 0$$

$$(x + 3)(x - 6) = 0$$

$$x^2 - 3x - 18 = 0$$

Any non-zero multiple of this equation would also have these solutions.

206. $(x - (-4))(x - (-11)) = 0$

$$(x + 4)(x + 11) = 0$$

$$x^2 + 15x + 44 = 0$$

207. 8 and 14

One possible equation is:

$(x - 8)(x - 14) = 0$

$x^2 - 22x + 112 = 0$

Any non-zero multiple of this equation would also have these solutions.

208. $x = \frac{1}{6} \implies 6x = 1 \implies 6x - 1$ is a factor.

$x = -\frac{2}{5} \implies 5x = -2 \implies 5x + 2$ is a factor.

$(6x - 1)(5x + 2) = 0$

$30x^2 + 7x - 2 = 0$

209. $1 + \sqrt{2}$ and $1 - \sqrt{2}$

One possible equation is:

$\left[x - \left(1 + \sqrt{2}\right)\right]\left[x - \left(1 - \sqrt{2}\right)\right] = 0$

$\left[(x - 1) - \sqrt{2}\right]\left[(x - 1) + \sqrt{2}\right] = 0$

$(x - 1)^2 - \left(\sqrt{2}\right)^2 = 0$

$x^2 - 2x + 1 - 2 = 0$

$x^2 - 2x - 1 = 0$

Any non-zero multiple of this equation would also have these solutions.

210. $x = -3 + \sqrt{5},\ x = -3 - \sqrt{5}$, so:

$\left(x - \left(-3 + \sqrt{5}\right)\right)\left(x - \left(-3 - \sqrt{5}\right)\right) = 0$

$\left(x + 3 - \sqrt{5}\right)\left(x + 3 + \sqrt{5}\right) = 0$

$x^2 + 6x + 4 = 0$

211. $9 + |9 - a| = b$

$|9 - a| = b - 9$

$9 - a = b - 9 \quad$ OR $\quad 9 - a = -(b - 9)$

$-a = b - 18 \qquad\qquad 9 - a = -b + 9$

$a = 18 - b \qquad\qquad\quad -a = -b$

$\qquad\qquad\qquad\qquad\qquad a = b$

Thus, $a = 18 - b$ or $a = b$. From the original equation we know that $b \geq 9$.

Some possibilities are: $b = 9,\ a = 9$

$b = 10,\ a = 8$ or $a = 10$

$b = 11,\ a = 7$ or $a = 11$

$b = 12,\ a = 6$ or $a = 12$

$b = 13,\ a = 5$ or $a = 13$

$b = 14,\ a = 4$ or $a = 14$

212. Isolate the absolute value by subtracting x from both sides of the equation. The expression inside the absolute value signs can be positive or negative, so two separate equations must be solved. Each solution must be checked since extraneous solutions may be included.

213. (a) $ax^2 + bx = 0$

$x(ax + b) = 0$

$x = 0$

$ax + b = 0 \implies x = -\dfrac{b}{a}$

(b) $ax^2 - ax = 0$

$ax(x - 1) = 0$

$ax = 0 \implies x = 0$

$x - 1 = 0 \implies x = 1$

Appendix A.6   Linear Inequalities in One Variable

■   You should know the properties of inequalities.

(a)   Transitive: $a < b$ and $b < c$ implies $a < c$.

(b)   Addition: $a < b$ and $c < d$ implies $a + c < b + d$.

(c)   Adding or Subtracting a Constant: $a \pm c < b \pm c$ if $a < b$.

(d)   Multiplying or Dividing a Constant: For $a < b$,

    1.   If $c > 0$, then $ac < bc$ and $\dfrac{a}{c} < \dfrac{b}{c}$.

    2.   If $c < 0$, then $ac > bc$ and $\dfrac{a}{c} > \dfrac{b}{c}$.

■   You should be able to solve absolute value inequalities.

(a) $|x| < a$ if and only if $-a < x < a$.

(b) $|x| > a$ if and only if $x < -a$ or $x > a$.

Vocabulary Check

1. solution set   **2.** graph   **3.** negative

4. equivalent   **5.** double   **6.** union

---

1. Interval: $[-1, 5]$

  (a) Inequality: $-1 \le x \le 5$

  (b) The interval is bounded.

2. Interval: $(2, 10]$

  (a) Inequality: $2 < x \le 10$

  (b) The interval is bounded.

3. Interval: $(11, \infty)$

  (a) Inequality: $x > 11$

  (b) The interval is unbounded.

4. Interval: $[-5, \infty)$

  (a) Inequality:
    $-5 \le x < \infty$ or $x \ge -5$

  (b) The interval is unbounded.

5. Interval: $(-\infty, -2)$

  (a) Inequality: $x < -2$

  (b) The interval is unbounded.

6. Interval: $(-\infty, 7]$

  (a) Inequality:
    $-\infty < x \le 7$ or $x \le -7$

  (b) The interval is unbounded.

7. $x < 3$

  Matches (b).

8. $x \ge 5$

  Matches (f).

9. $-3 < x \le 4$

  Matches (d).

10. $0 \le x \le \frac{9}{2}$

  Matches (c).

11. $|x| < 3 \implies -3 < x < 3$

  Matches (e).

12. $|x| > 4 \implies x > 4$ or $x < -4$

  Matches (a).

13. $5x - 12 > 0$

  (a) $x = 3$

    $5(3) - 12 \overset{?}{>} 0$

    $3 > 0$

    Yes, $x = 3$ *is* a solution.

  (b) $x = -3$

    $5(-3) - 12 \overset{?}{>} 0$

    $-27 \not> 0$

    No, $x = -3$ *is not* a solution.

  (c) $x = \frac{5}{2}$

    $5\left(\frac{5}{2}\right) - 12 \overset{?}{>} 0$

    $\frac{1}{2} > 0$

    Yes, $x = \frac{5}{2}$ *is* a solution.

  (d) $x = \frac{3}{2}$

    $5\left(\frac{3}{2}\right) - 12 \overset{?}{>} 0$

    $-\frac{9}{2} \not> 0$

    No, $x = \frac{3}{2}$ *is not* a solution.

14. $2x + 1 < -3$

(a) $x = 0$

$2(0) + 1 \overset{?}{<} -3$

$1 \not< -3$

No, $x = 0$ *is not*
a solution.

(b) $x = -\frac{1}{4}$

$2\left(-\frac{1}{4}\right) + 1 \overset{?}{<} -3$

$\frac{1}{2} \not< -3$

No, $x = -\frac{1}{4}$ *is not*
a solution.

(c) $x = -4$

$2(-4) + 1 \overset{?}{<} -3$

$-7 < -3$

Yes, $x = -4$ *is a*
solution.

(d) $x = -\frac{3}{2}$

$2\left(-\frac{3}{2}\right) + 1 \overset{?}{<} -3$

$-2 \not< -3$

No, $x = -\frac{3}{2}$ *is not*
a solution.

15. $0 < \frac{x - 2}{4} < 2$

(a) $x = 4$

$0 \overset{?}{<} \frac{4 - 2}{4} \overset{?}{<} 2$

$0 < \frac{1}{2} < 2$

Yes, $x = 4$ *is* a
solution.

(b) $x = 10$

$0 \overset{?}{<} \frac{10 - 2}{4} \overset{?}{<} 2$

$0 < 2 \not< 2$

No, $x = 10$ *is not* a
solution.

(c) $x = 0$

$0 \overset{?}{<} \frac{0 - 2}{4} \overset{?}{<} 2$

$0 \not< -\frac{1}{2} < 2$

No, $x = 0$ *is not* a
solution.

(d) $x = \frac{7}{2}$

$0 \overset{?}{<} \frac{(7/2) - 2}{4} \overset{?}{<} 2$

$0 < \frac{3}{8} < 2$

Yes, $x = \frac{7}{2}$ *is a solution.*

16. $-1 < \frac{3 - x}{2} \leq 1$

(a) $x = 0$

$-1 \overset{?}{<} \frac{3 - 0}{2} \overset{?}{\leq} 1$

$-1 < \frac{3}{2} \not\leq 1$

No, $x = 0$ *is not* a
solution.

(b) $x = -5$

$-1 \overset{?}{<} \frac{3 + 5}{2} \overset{?}{\leq} 1$

$-1 < 4 \leq 1$

No, $x = -5$ *is not* a
solution.

(c) $x = 1$

$-1 \overset{?}{<} \frac{3 - 1}{2} \overset{?}{\leq} 1$

$-1 < 1 \leq 1$

Yes, $x = 1$ *is a solution.*

(d) $x = 5$

$-1 \overset{?}{<} \frac{3 - 5}{2} \overset{?}{\leq} 1$

$-1 \overset{?}{<} -1 \leq 1$

No, $x = 5$ *is not* a
solution.

17. $|x - 10| \geq 3$

(a) $x = 13$

$|13 - 10| \overset{?}{\geq} 3$

$3 \geq 3$

Yes, $x = 13$ *is a*
solution.

(b) $x = -1$

$|-1 - 10| \overset{?}{\geq} 3$

$11 \geq 3$

Yes, $x = -1$ *is a*
solution.

(c) $x = 14$

$|14 - 10| \overset{?}{\geq} 3$

$4 \geq 3$

Yes, $x = 14$ *is a*
solution.

(d) $x = 9$

$|9 - 10| \overset{?}{\geq} 3$

$1 \not\geq 3$

No, $x = 9$ *is not* a
solution.

18. $|2x - 3| < 15$

(a) $x = -6$

$|2(-6) - 3| \overset{?}{<} 15$

$15 \not< 15$

No, $x = -6$ *is not*
a solution.

(b) $x = 0$

$|2(0) - 3| \overset{?}{<} 15$

$3 < 15$

Yes, $x = 0$ *is a*
solution.

(c) $x = 12$

$|2(12) - 3| \overset{?}{<} 15$

$21 \not< 15$

No, $x = 12$ *is not*
a solution.

(d) $x = 7$

$|2(7) - 3| \overset{?}{<} 15$

$11 < 15$

Yes, $x = 7$ *is a*
solution.

19. $4x < 12$

$\frac{1}{4}(4x) < \frac{1}{4}(12)$

$x < 3$

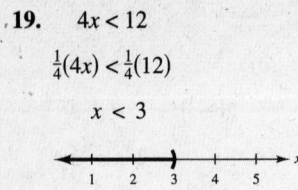

20. $10x < -40$

$x < -4$

21. $-2x > -3$

$-\frac{1}{2}(-2x) < \left(-\frac{1}{2}\right)(-3)$

$x < \frac{3}{2}$

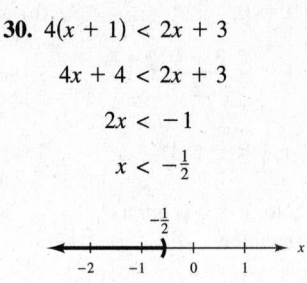

22. $-6x > 15$

$x < -\frac{15}{6}$ or $x < -\frac{5}{2}$

23. $x - 5 \geq 7$

$x \geq 12$

24. $x + 7 \leq 12$

$x \leq 5$

25. $2x + 7 < 3 + 4x$

$-2x < -4$

$x > 2$

26. $3x + 1 \geq 2 + x$

$2x \geq 1$

$x \geq \frac{1}{2}$

27. $2x - 1 \geq 1 - 5x$

$7x \geq 2$

$x \geq \frac{2}{7}$

28. $6x - 4 \leq 2 + 8x$

$-2x \leq 6$

$x \geq -3$

29. $4 - 2x < 3(3 - x)$

$4 - 2x < 9 - 3x$

$x < 5$

30. $4(x + 1) < 2x + 3$

$4x + 4 < 2x + 3$

$2x < -1$

$x < -\frac{1}{2}$

31. $\frac{3}{4}x - 6 \leq x - 7$

$-\frac{1}{4}x \leq -1$

$x \geq 4$

32. $3 + \frac{2}{7}x > x - 2$

$21 + 2x > 7x - 14$

$-5x > -35$

$x < 7$

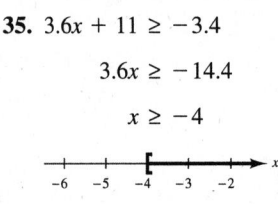

33. $\frac{1}{2}(8x + 1) \geq 3x + \frac{5}{2}$

$4x + \frac{1}{2} \geq 3x + \frac{5}{2}$

$x \geq 2$

34. $9x - 1 < \frac{3}{4}(16x - 2)$

$36x - 4 < 48x - 6$

$-12x < -2$

$x > \frac{1}{6}$

35. $3.6x + 11 \geq -3.4$

$3.6x \geq -14.4$

$x \geq -4$

36. $15.6 - 1.3x < -5.2$

$-1.3x < -20.8$

$x > 16$

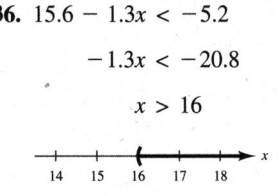

37. $1 < 2x + 3 < 9$

$-2 < 2x < 6$

$-1 < x < 3$

38. $-8 \le -(3x + 5) < 13$

$-8 \le -3x - 5 < 13$

$-3 \le -3x < 18$

$-6 < x \le 1$

39. $-4 < \dfrac{2x - 3}{3} < 4$

$-12 < 2x - 3 < 12$

$-9 < 2x < 15$

$-\dfrac{9}{2} < x < \dfrac{15}{2}$

40. $0 \le \dfrac{x + 3}{2} < 5$

$0 \le x + 3 < 10$

$-3 \le x < 7$

41. $\dfrac{3}{4} > x + 1 > \dfrac{1}{4}$

$-\dfrac{1}{4} > x > -\dfrac{3}{4}$

$-\dfrac{3}{4} < x < -\dfrac{1}{4}$

42. $-1 < 2 - \dfrac{x}{3} < 1$

$-3 < 6 - x < 3$

$-9 < -x < -3$

$3 < x < 9$

43. $3.2 \le 0.4x - 1 \le 4.4$

$4.2 \le 0.4x \le 5.4$

$10.5 \le x \le 13.5$

44. $4.5 > \dfrac{1.5x + 6}{2} > 10.5$

$9 > 1.5x + 6 > 21$

$3 > 1.5x > 15$

$2 > x > 10$

There is no solution.

45. $|x| < 6$

$-6 < x < 6$

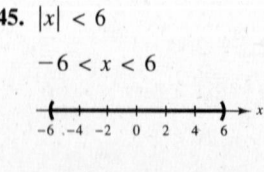

46. $|x| > 4$

$x < -4 \text{ or } x > 4$

47. $\left|\dfrac{x}{2}\right| > 1$

$\dfrac{x}{2} < -1 \text{ or } \dfrac{x}{2} > 1$

$x < -2 \qquad x > 2$

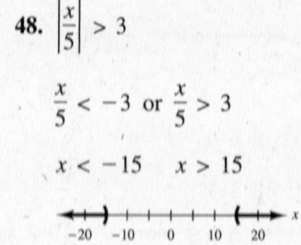

48. $\left|\dfrac{x}{5}\right| > 3$

$\dfrac{x}{5} < -3 \text{ or } \dfrac{x}{5} > 3$

$x < -15 \qquad x > 15$

49. $|x - 5| < -1$

No solution. The absolute value of a number cannot be less than a negative number.

50. There is no solution because the absolute value of a number cannot be less than a negative number.

51. $|x - 20| \le 6$

$-6 \le x - 20 \le 6$

$14 \le x \le 26$

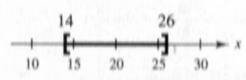

52. $|x - 8| \ge 0$

$x - 8 \ge 0 \quad \text{or} \quad -(x - 8) \ge 0$

$x \ge 8 \qquad\qquad -x + 8 \ge 0$

$-x \ge -8$

$x \le 8$

All real numbers x

53. $|3 - 4x| \geq 9$

$3 - 4x \leq -9$ or $3 - 4x \geq 9$

$-4x \leq -12$ $-4x \geq 6$

$x \geq 3$ $x \leq -\frac{3}{2}$

54. $|1 - 2x| < 5$

$-5 < 1 - 2x < 5$

$-6 < -2x < 4$

$3 > x > -2$

$-2 < x < 3$

55. $\left|\dfrac{x - 3}{2}\right| \geq 4$

$\dfrac{x - 3}{2} \leq -4$ or $\dfrac{x - 3}{2} \geq 4$

$x - 3 \leq -8$ $x - 3 \geq 8$

$x \leq -5$ $x \geq 11$

56. $\left|1 - \dfrac{2x}{3}\right| < 1$

$-1 < 1 - \dfrac{2x}{3} < 1$

$-2 < -\dfrac{2x}{3} < 0$

$3 > x > 0$

$0 < x < 3$

57. $|9 - 2x| - 2 < -1$

$|9 - 2x| < 1$

$-1 < 9 - 2x < 1$

$-10 < -2x < -8$

$5 > x > 4$

$4 < x < 5$

58. $|x + 14| + 3 > 17$

$|x + 14| > 14$

$x + 14 < -14$ or $x + 14 > 14$

$x < -28$ $x > 0$

59. $2|x + 10| \geq 9$

$|x + 10| \geq \dfrac{9}{2}$

$x + 10 \leq -\dfrac{9}{2}$ or $x + 10 \geq \dfrac{9}{2}$

$x \leq -\dfrac{29}{2}$ $x \geq -\dfrac{11}{2}$

60. $3|4 - 5x| \leq 9$

$|4 - 5x| \leq 3$

$-3 \leq 4 - 5x \leq 3$

$-7 \leq -5x \leq -1$

$\dfrac{7}{5} \geq x \geq \dfrac{1}{5}$

$\dfrac{1}{5} \leq x \leq \dfrac{7}{5}$

61. $6x > 12$

$x > 2$

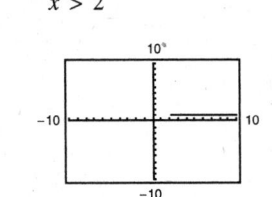

62. $3x - 1 \leq 5$

$3x \leq 6$

$x \leq 2$

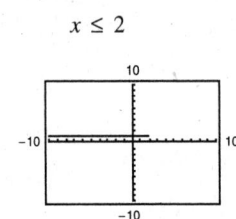

63. $5 - 2x \geq 1$

$-2x \geq -4$

$x \leq 2$

64. $3(x + 1) < x + 7$

$3x + 3 < x + 7$

$2x < 4$

$x < 2$

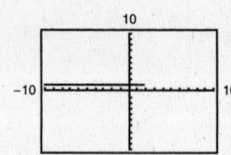

65. $|x - 8| \leq 14$

$-14 \leq x - 8 \leq 14$

$-6 \leq x \leq 22$

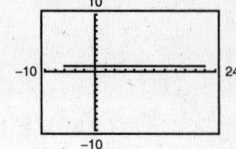

66. $|2x + 9| > 13$

$2x + 9 < -13$ or $2x + 9 > 13$

$2x < -22 \qquad 2x > 4$

$x < -11 \qquad x > 2$

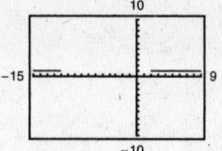

67. $2|x + 7| \geq 13$

$|x + 7| \geq \frac{13}{2}$

$x + 7 \leq -\frac{13}{2}$ or $x + 7 \geq \frac{13}{2}$

$x \leq -\frac{27}{2} \qquad x \geq -\frac{1}{2}$

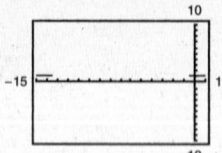

68. $\frac{1}{2}|x + 1| \leq 3$

$|x + 1| \leq 6$

$-6 \leq x + 1 \leq 6$

$-7 \leq x \leq 5$

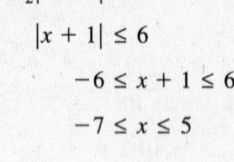

69. $y = 2x - 3$

  (a) $\qquad y \geq 1$

$\qquad 2x - 3 \geq 1$

$\qquad 2x \geq 4$

$\qquad x \geq 2$

  (b) $\qquad y \leq 0$

$\qquad 2x - 3 \leq 0$

$\qquad 2x \leq 3$

$\qquad x \leq \frac{3}{2}$

70. $y = \frac{2}{3}x + 1$

  (a) $\qquad y \leq 5$

$\qquad \frac{2}{3}x + 1 \leq 5$

$\qquad \frac{2}{3}x \leq 4$

$\qquad x \leq 6$

  (b) $\qquad y \geq 0$

$\qquad \frac{2}{3}x + 1 \geq 0$

$\qquad \frac{2}{3}x \geq -1$

$\qquad x \geq -\frac{3}{2}$

71. $y = -\frac{1}{2}x + 2$

  (a) $\quad 0 \leq y \leq 3$

$\quad 0 \leq -\frac{1}{2}x + 2 \leq 3$

$\quad -2 \leq -\frac{1}{2}x \leq 1$

$\quad 4 \geq x \geq -2$

  (b) $\qquad y \geq 0$

$\quad -\frac{1}{2}x + 2 \geq 0$

$\qquad -\frac{1}{2}x \geq -2$

$\qquad x \leq 4$

72. $y = -3x + 8$

  (a) $-1 \leq y \leq 3$

$\quad -1 \leq -3x + 8 \leq 3$

$\quad -9 \leq -3x \leq -5$

$\quad 3 \geq x \geq \frac{5}{3}$

$\quad \frac{5}{3} \leq x \leq 3$

  (b) $\qquad y \leq 0$

$\quad -3x + 8 \leq 0$

$\qquad -3x \leq -8$

$\qquad x \geq \frac{8}{3}$

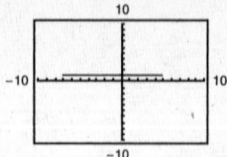

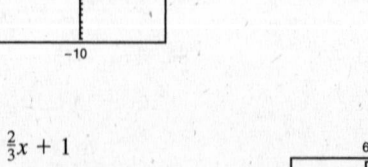

73. $y = |x - 3|$

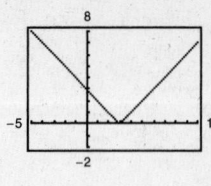

(a)　　$y \le 2$

　　$|x - 3| \le 2$

　　　$-2 \le x - 3 \le 2$

　　　　$1 \le x \le 5$

(b)　　$y \ge 4$

　　$|x - 3| \ge 4$

　　　$x - 3 \le -4$　or　$x - 3 \ge 4$

　　　　$x \le -1$　or　　　$x \ge 7$

74. $y = \left|\frac{1}{2}x + 1\right|$

(a) $y \le 4$

　$\left|\frac{1}{2}x + 1\right| \le 4$

　　$-4 \le \frac{1}{2}x + 1 \le 4$

　　　$-5 \le \frac{1}{2}x \le 3$

　　　$-10 \le x \le 6$

(b) $y \ge 1$

　$\left|\frac{1}{2}x + 1\right| \ge 1$

　　$\frac{1}{2}x + 1 \le -1$　or　$\frac{1}{2}x + 1 \ge 1$

　　　$\frac{1}{2}x \le -2$　　　　$\frac{1}{2}x \ge 0$

　　　$x \le -4$　　　　　$x \ge 0$

75. $x - 5 \ge 0$

　$x \ge 5$

　$[5, \infty)$

76. $\sqrt{x - 10}$

　$x - 10 \ge 0$

　　$x \ge 10$

　$[10, \infty)$

77. $x + 3 \ge 0$

　$x \ge -3$

　$[-3, \infty)$

78. $\sqrt{3 - x}$

　$3 - x \ge 0$

　　$3 \ge x$

　$(-\infty, 3]$

79. $7 - 2x \ge 0$

　$-2x \ge -7$

　　$x \le \frac{7}{2}$

　$\left(-\infty, \frac{7}{2}\right]$

80. $\sqrt[4]{6x + 15}$

　$6x + 15 \ge 0$

　　$6x \ge -15$

　　$x \ge -\frac{5}{2}$

　$\left[-\frac{5}{2}, \infty\right)$

81. $|x - 10| < 8$

All real numbers within 8 units of 10.

82. $|x - 8| > 4$

All real numbers more than 4 units from 8

83. The midpoint of the interval $[-3, 3]$ is 0. The interval represents all real numbers x no more than 3 units from 0.

$|x - 0| \le 3$

　$|x| \le 3$

84. The graph shows all real numbers more than 3 units from 0.

$|x - 0| > 3$

　$|x| > 3$

85. The graph shows all real numbers at least 3 units from 7.

$|x - 7| \ge 3$

86. The graph shows all real numbers no more than 4 units from -1.

$|x + 1| \le 4$

87. All real numbers within 10 units of 12

$|x - 12| < 10$

88. All real numbers at least 5 units from 8

$|x - 8| \ge 5$

89. All real numbers more than 4 units from -3

$|x - (-3)| > 4$

　$|x + 3| > 4$

90. All real numbers no more than 7 units from -6

$|x + 6| \le 7$

91. Let x = the number of checks written in a month.

Type A account charges: $6.00 + 0.25x$

Type B account charges: $4.50 + 0.50x$

$6.00 + 0.25x < 4.50 + 0.50x$

$\quad 1.50 < 0.25x$

$\quad\quad 6 < x$

If you write more than six checks a month, then the charges for the type A account are less than the charges for the type B account.

92. $3000 + 0.03x \le 0.1x$

$\quad 3000 < 0.07x$

$\quad 42{,}857 < x$

You must make more than 42,857 copies to justify buying the copier.

93. $1000(1 + r(2)) > 1062.50$

$\quad\quad 1 + 2r > 1.0625$

$\quad\quad\quad 2r > 0.0625$

$\quad\quad\quad r > 0.03125$

$\quad\quad\quad r > 3.125\%$

94. $825 < 750(1 + r(2))$

$\quad 825 < 750(1 + 2r)$

$\quad 825 < 750 + 1500r$

$\quad\ 75 < 1500r$

$\quad 0.05 < r$

The rate must be more than 5%.

95. $\quad\quad R > C$

$115.95x > 95x + 750$

$\ 20.95x > 750$

$\quad\quad x > 35.7995$

$\quad\quad x \ge 36$ units

96. $24.55x > 15.4x + 150{,}000$

$\quad\ 9.15 > 150{,}000$

$\quad\quad\ x > 16{,}393.44262$

Because the number of units x must be an integer, the product will return a profit when at least 16,394 units are sold.

97. Let x = daily sales level (in dozens) of doughnuts.

Revenue: $R = 2.95x$

Cost: $C = 150 + 1.45x$

Profit: $P = R - C$

$\quad\quad\quad = 2.95x - (150 + 1.45x)$

$\quad\quad\quad = 1.50x - 150$

$\quad 50 \le P \le 200$

$\quad 50 \le 1.50x - 150 \le 200$

$\quad 200 \le 1.50x \le 350$

$\quad 133\frac{1}{3} \le x \le 233\frac{1}{3}$

In whole dozens, $134 \le x \le 234$.

98. The goal is to lose $164 - 128 = 36$ pounds. At $1\frac{1}{2}$ pounds per week, it will take 24 weeks.

$36 \div 1\frac{1}{2} = 36 \times \frac{2}{3}$

$\quad\quad\quad = 12 \times 2$

$\quad\quad\quad = 24$

99. (a) $y = 0.067x - 5.638$

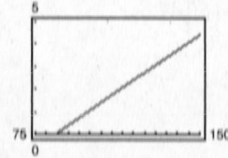

(b) From the graph we see that $y \ge 3$ when $x \ge 129$. Algebraically we have:

$\quad 3 \le 0.067x - 5.638$

$8.638 \le 0.067x$

$\quad\quad x \ge 129$

IQ scores are not a good predictor of GPAs. Other factors include study habits, class attendance, and attitude.

100. (a) and (b)

x	165	184	150	210	196	240
y	170	185	200	255	205	295
$1.3x - 36$	179	203	159	237	219	276

x	202	170	185	190	230	160
y	190	175	195	185	250	155
$1.3x - 36$	227	185	205	211	263	172

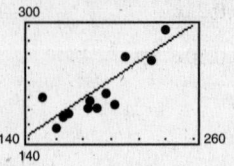

(c) One estimate is $x \geq 181$ pounds.

(d) $1.3x - 36 \geq 200$

$1.3x \geq 236$

$x \geq 181.5385 \approx 181.54$ pounds

(e) An athlete's weight is not a particularly good indicator of the athlete's maximum bench press weight. Other factors, such as muscle tone and exercise habits, influence maximum bench press weight.

101. $S = 1.05t + 31.0$, $0 \leq t \leq 12$

  (a) $32 \leq 1.05t + 31 \leq 42$

        $1 \leq 1.05t \leq 11$

      $0.95 \leq t \leq 10.48$

Rounding to the nearest year, $1 \leq t \leq 10$. The average salary was at least \$32,000 but not more than \$42,000 between 1991 and 2000.

  (b) $1.05t + 31 > 48$

      $1.05t > 17$

          $t > 16$

According to the model, the average salary will exceed \$48,000 in 2006.

102. (a) $70 \leq 1.64t + 67.2 \leq 80$

      $2.8 \leq 1.64t \leq 12.8$

      $1.7 \leq t \leq 7.8$

The number of eggs produced was between 70 and 80 billion from late 1991 until late 1997.

(b) $E = 1.64t + 67.2 \geq 95$ when $t \geq 16.95$.

So the annual egg production will exceed 95 billion in 2007.

103. $|s - 10.4| \leq \frac{1}{16}$

    $-\frac{1}{16} \leq s - 10.4 \leq \frac{1}{16}$

$-0.0625 \leq s - 10.4 \leq 0.0625$

$10.3375 \leq s \leq 10.4625$

Since $A = s^2$, we have

$(10.3375)^2 \leq$ area $\leq (10.4625)^2$

$106.864 \leq$ area ≤ 109.464.

104. $24.2 - 0.25 \leq s \leq 24.2 + 0.25$

    $23.95 \leq s \leq 24.45$

The interval containing the possible side lengths s in centimeters of the square is $[23.95, 24.45]$, so the interval containing the possible areas in square centimeters is $[23.95^2, 24.45^2]$, or $[573.6025, 597.8025]$.

105. $|x - 15| \leq \frac{1}{10}$

    $-\frac{1}{10} \leq x - 15 \leq \frac{1}{10}$

    $14.9 \leq x \leq 15.1$ gallons

$\frac{1}{10}(\$1.89) \approx \$0.19$

You might have been undercharged or overcharged by \$0.19.

106. 1 oz $= \frac{1}{16}$ lb, so $\frac{1}{2}$ oz $= \frac{1}{32}$ lb

$14.99 \cdot \frac{1}{32} = 0.4684375$, so you could have been under- or overcharged as much as \$0.47.

107.
$$\left|\frac{t - 15.6}{1.9}\right| < 1$$

$$-1 < \frac{t - 15.6}{1.9} < 1$$

$$-1.9 < t - 15.6 < 1.9$$

$$13.7 < t < 17.5$$

Two-thirds of the workers could perform the task in the time interval between 13.7 minutes and 17.5 minutes.

109.
$$|h - 50| \le 30$$

$$-30 \le h - 50 \le 30$$

$$20 \le h \le 80$$

The minimum relative humidity is 20 and the maximum is 80.

108.
$$\left|\frac{h - 68.5}{2.7}\right| \le 1$$

$$-1 \le \frac{h - 68.5}{2.7} \le 1$$

$$-2.7 \le h - 68.5 \le 2.7$$

$$65.8 \text{ inches} \le h \le 71.2 \text{ inches}$$

110. (a) Estimate from the graph: when the plate thickness is 2 millimeters, the frequency is approximately 330 vibrations per second.

(b) Estimate from the graph: when the frequency is 600, the plate thickness is approximately 3.6 millimeters.

(c) Estimate from the graph: when the frequency is between 200 and 400 vibrations per second, the plate thickness is between 1.2 and 2.4 millimeter.

(d) Estimate from the graph: when the plate thickness is less than 3 millimeters, the frequency is less than 500 vibrations per second.

111. False. If c is negative, then $ac \ge bc$.

112. False. If $-10 \le x \le 8$, then $10 \ge -x$ and $-x \ge -8$.

113. $|x - a| \ge 2$ Matches (b).

$$x - a \le -2$$

$$x \le a - 2 \text{ or}$$

$$x - a \ge 2$$

$$x \ge a + 2$$

114. $|ax - b| \le c \implies c$ must be greater than or equal to zero.

$$-c \le ax - b \le c$$

$$b - c \le ax \le b + c$$

Let $a = 1$, then $b - c = 0$ and $b + c = 10$. This is true when $b = c = 5$.
One set of values is $a = 1$, $b = 5$, $c = 5$.

(*Note:* This solution is not unique. Any positive multiple of these values will also work, such as $a = 2$, $b = c = 10$ or $a = 3$, $b = c = 15$.)

Appendix A.7 Errors and the Algebra of Calculus

■ You should be able to recognize and avoid the common algebraic errors involving parentheses, fractions, exponents, radicals, and cancellation.

■ You should be able to "unsimplify" algebraic expressions by the following methods.

(a) Unusual Factoring

(b) Rewriting with Negative Exponents

(c) Writing a Fraction as a Sum of Terms

(d) Inserting Factors or Terms

Vocabulary Check

1. numerator

2. reciprocal

1. $2x - (3y + 4) \neq 2x - 3y + 4$

Change all signs when distributing the minus sign.

$2x - (3y + 4) = 2x - 3y - 4$

2. $5z + 3(x - 2) \neq 5z + 3x - 2$

The 3 is distributed to both terms.

$5z + 3(x - 2) = 5z + 3x - 6$

3. $\dfrac{4}{16x - (2x + 1)} \neq \dfrac{4}{14x + 1}$

Change all signs when distributing the minus sign.

$\dfrac{4}{16x - (2x + 1)} = \dfrac{4}{16x - 2x - 1} = \dfrac{4}{14x - 1}$

4. $\dfrac{1 - x}{(5 - x)(-x)} \neq \dfrac{x - 1}{x(x - 5)}$

The expression on the right should be negative.

$\dfrac{1 - x}{(5 - x)(-x)} = -\dfrac{x - 1}{x(x - 5)}$

5. $(5z)(6z) \neq 30z$

z occurs twice as a factor.

$(5z)(6z) = 30z^2$

6. $x(yz) \neq (xy)(xz)$

yz is one term, not two.

$x(yz) = xyz$

7. $a\left(\dfrac{x}{y}\right) \neq \dfrac{ax}{ay}$

The fraction as a whole is multiplied by a, not the numerator and denominator separately.

$a\left(\dfrac{x}{y}\right) = \dfrac{a}{1} \cdot \dfrac{x}{y} = \dfrac{ax}{y}$

8. $(4x)^2 \neq 4x^2$

The exponent applies to the coefficient also.

$(4x)^2 = 16x^2$

9. $\sqrt{x + 9} \neq \sqrt{x} + 3$

Do not apply the radical to the terms.

$\sqrt{x + 9}$ does not simplify.

10. $\sqrt{25 - x^2} \neq 5 - x$

Do not apply radicals term-by-term.

$\sqrt{25 - x^2} = \sqrt{(5 + x)(5 - x)}$

11. $\dfrac{2x^2 + 1}{5x} \neq \dfrac{2x + 1}{5}$

Divide out common factors not common terms.

$\dfrac{2x^2 + 1}{5x}$ cannot be simplified.

12. $\dfrac{6x + y}{6x - y}$ does not simplify.

Reduce common factors of the numerator and denominator, not common factors of terms.

13. $\dfrac{1}{a^{-1} + b^{-1}} \neq \left(\dfrac{1}{a + b}\right)^{-1}$

To get rid of negative exponents:

$$\dfrac{1}{a^{-1} + b^{-1}} = \dfrac{1}{a^{-1} + b^{-1}} \cdot \dfrac{ab}{ab} = \dfrac{ab}{b + a}$$

14. $\dfrac{1}{x + y^{-1}} \neq \dfrac{y}{x + 1}$

The negative exponent is on a term of the denominator, not a factor.

$$\dfrac{1}{x + y^{-1}} = \dfrac{1}{x + (1/y)} \cdot \dfrac{y}{y} = \dfrac{y}{xy + 1}$$

15. $(x^2 + 5x)^{1/2} \neq x(x + 5)^{1/2}$

Factor within grouping symbols before applying the exponent to each factor.

$$(x^2 + 5x)^{1/2} = [x(x + 5)]^{1/2} = x^{1/2}(x + 5)^{1/2}$$

16. $x(2x - 1)^2 \neq (2x^2 - x)^2$

Factor within grouping symbols before applying the exponent to each factor.

$$x(2x - 1)^2 = x(4x^2 - 4x + 1)$$

17. $\dfrac{3}{x} + \dfrac{4}{y} = \dfrac{3}{x} \cdot \dfrac{y}{y} + \dfrac{4}{y} \cdot \dfrac{x}{x} = \dfrac{3y + 4x}{xy}$

To add fractions, they must have a common denominator.

18. $\dfrac{1}{2y} = \left(\dfrac{1}{2}\right)y$

Be careful when using a slash to denote division.

$$\left(\dfrac{1}{2}\right)y = \dfrac{1}{2} \cdot y = \dfrac{y}{2}$$

19. $\dfrac{3x + 2}{5} = \dfrac{1}{5}(3x + 2)$

The required factor is $3x + 2$.

20. $\dfrac{7x^2}{10} = \dfrac{7}{10}(x^2)$

The required factor is x^2.

21. $\dfrac{2}{3}x^2 + \dfrac{1}{3}x + 5 = \dfrac{2}{3}x^2 + \dfrac{1}{3}x + \dfrac{15}{3} = \dfrac{1}{3}(2x^2 + x + 15)$

The required factor is $2x^2 + x + 15$.

22. $\dfrac{3}{4}x + \dfrac{1}{2} = \dfrac{3}{4}x + \dfrac{2}{4} = \dfrac{1}{4}(3x + 2)$

The required factor is $3x + 2$.

23. $x^2(x^3 - 1)^4 = \dfrac{1}{3}(x^3 - 1)^4(3x^2)$

The required factor is $\dfrac{1}{3}$.

24. $x(1 - 2x^2)^3 = \dfrac{-4x}{-4}(1 - 2x^2)^3 = \left(-\dfrac{1}{4}\right)(-4x)(1 - 2x^2)^3$

$$= \left(-\dfrac{1}{4}\right)(1 - 2x^2)^3(-4x)$$

The required factor is $-\dfrac{1}{4}$.

25. $\dfrac{4x + 6}{(x^2 + 3x + 7)^3} = \dfrac{2(2x + 3)}{(x^2 + 3x + 7)^3} = \dfrac{2}{1} \cdot \dfrac{(2x + 3)}{1} \cdot \dfrac{1}{(x^2 + 3x + 7)^3} = (2)\dfrac{1}{(x^2 + 3x + 7)^3}(2x + 3)$

The required factor is 2.

26. $\dfrac{x + 1}{(x^2 + 2x - 3)^2} = \dfrac{1}{2} \cdot \dfrac{2(x + 1)}{(x^2 + 2x - 3)^2}$

$$= \left(\dfrac{1}{2}\right)\left(\dfrac{1}{(x^2 + 2x - 3)^2}\right)(2x + 2)$$

The required factor is $\dfrac{1}{2}$.

27. $\dfrac{3}{x} + \dfrac{5}{2x^2} - \dfrac{3}{2}x = \dfrac{6x}{2x^2} + \dfrac{5}{2x^2} - \dfrac{3x^3}{2x^2}$

$$= \left(\dfrac{1}{2x^2}\right)(6x + 5 - 3x^3)$$

The required factor is $\dfrac{1}{2x^2}$.

28. $\dfrac{(x-1)^2}{169} + (y+5)^2 = \dfrac{(x-1)(x-1)^2}{(x-1)(169)} + (y+5)^2$

$$= \dfrac{(x-1)^3}{169(x-1)} + (y+5)^2$$

The required factor is $(x-1)$.

29. $\dfrac{9x^2}{25} + \dfrac{16y^2}{49} = \dfrac{9}{25} \cdot \dfrac{x^2}{1} + \dfrac{16}{49} \cdot \dfrac{y^2}{1}$

$$= \dfrac{1}{25/9} \cdot \dfrac{x^2}{1} + \dfrac{1}{49/16} \cdot \dfrac{y^2}{1}$$

$$= \dfrac{x^2}{(25/9)} + \dfrac{y^2}{(49/16)}$$

The required factors are $\frac{25}{9}$ and $\frac{49}{16}$.

30. $\dfrac{3x^2}{4} - \dfrac{9y^2}{16} = \dfrac{\left(\frac{1}{3}\right)3x^2}{\left(\frac{1}{3}\right)4} - \dfrac{\left(\frac{1}{9}\right)9y^2}{\left(\frac{1}{9}\right)16} = \dfrac{x^2}{\frac{4}{3}} - \dfrac{y^2}{\frac{16}{9}}$

The required factors are $\frac{4}{3}$ and $\frac{16}{9}$.

31. $\dfrac{x^2}{1/12} - \dfrac{y^2}{2/3} = x^2\left(\dfrac{12}{1}\right) - y^2\left(\dfrac{3}{2}\right) = \dfrac{12x^2}{1} - \dfrac{3y^2}{2}$

The required factors are 1 and 2.

32. $\dfrac{x^2}{4/9} + \dfrac{y^2}{7/8} = x^2\left(\dfrac{9}{4}\right) + y^2\left(\dfrac{8}{7}\right) = \dfrac{9x^2}{4} + \dfrac{8y^2}{7}$

The required factors are 4 and 7.

33. $x^{1/3} - 5x^{4/3} = x^{1/3}(1 - 5x^{3/3}) = x^{1/3}(1 - 5x)$

The required factor is $1 - 5x$.

34. $3(2x+1)x^{1/2} + 4x^{3/2} = x^{1/2}[3(2x+1) + 4x]$

$$= x^{1/2}(6x + 3 + 4x)$$

$$= x^{1/2}(10x + 3)$$

The required factor is $10x + 3$.

35. $(1-3x)^{4/3} - 4x(1-3x)^{1/3} = (1-3x)^{1/3}[(1-3x)^1 - 4x]$

$$= (1-3x)^{1/3}(1 - 7x)$$

The required factor is $1 - 7x$.

36. $\dfrac{1}{2\sqrt{x}} + 5x^{3/2} - 10x^{5/2} = \dfrac{1}{2\sqrt{x}} + \dfrac{5x^{3/2}\left(2\sqrt{x}\right)}{2\sqrt{x}} - \dfrac{10x^{5/2}\left(2\sqrt{x}\right)}{2\sqrt{x}}$

$$= \dfrac{1}{2\sqrt{x}}\left(1 + 10x^{3/2}\sqrt{x} - 20x^{5/2}\sqrt{x}\right)$$

$$= \dfrac{1}{2\sqrt{x}}\left(1 + 10x^2 - 20x^3\right)$$

The required factor is $(1 + 10x^2 - 20x^3)$.

37. $\dfrac{1}{10}(2x+1)^{5/2} - \dfrac{1}{6}(2x+1)^{3/2} = \dfrac{3}{30}(2x+1)^{3/2}(2x+1)^1 - \dfrac{5}{30}(2x+1)^{3/2}$

$$= \dfrac{1}{30}(2x+1)^{3/2}[3(2x+1) - 5]$$

$$= \dfrac{1}{30}(2x+1)^{3/2}(6x - 2)$$

$$= \dfrac{1}{30}(2x+1)^{3/2}2(3x - 1)$$

$$= \dfrac{1}{15}(2x+1)^{3/2}(3x - 1)$$

The required factor is $3x - 1$.

38. $\dfrac{3}{7}(t+1)^{7/3} - \dfrac{3}{4}(t+1)^{4/3} = \dfrac{12}{28}(t+1)^{4/3}(t+1)^{3/3} - \dfrac{21}{28}(t+1)^{4/3}$

$$= \dfrac{3(t+1)^{4/3}}{28}[4(t+1) - 7]$$

$$= \dfrac{3(t+1)^{4/3}}{28}(4t - 3)$$

The required factor is $(4t - 3)$.

39. $\dfrac{3x^2}{(2x-1)^3} = 3x^2(2x-1)^{-3}$

40. $\dfrac{x+1}{x(6-x)^{1/2}} = (x+1)(x^{-1})(6-x)^{-1/2}$

41. $\dfrac{4}{3x} + \dfrac{4}{x^4} - \dfrac{7x}{\sqrt[3]{2x}} = 4(3x)^{-1} + 4x^{-4} - 7x(2x)^{-1/3}$

42. $\dfrac{x}{x-2} + \dfrac{1}{x^2} + \dfrac{8}{3(9x)^3} = x(x-2)^{-1} + x^{-2} + \dfrac{8}{3}(9x)^{-3}$

43. $\dfrac{16-5x-x^2}{x} = \dfrac{16}{x} - \dfrac{5x}{x} - \dfrac{x^2}{x} = \dfrac{16}{x} - 5 - x$

44. $\dfrac{x^3 - 5x^2 + 4}{x^2} = \dfrac{x^3}{x^2} - \dfrac{5x^2}{x^2} + \dfrac{4}{x^2} = x - 5 + \dfrac{4}{x^2}$

45. $\dfrac{4x^3 - 7x^2 + 1}{x^{1/3}} = \dfrac{4x^3}{x^{1/3}} - \dfrac{7x^2}{x^{1/3}} + \dfrac{1}{x^{1/3}}$

$\qquad = 4x^{3-1/3} - 7x^{2-1/3} + \dfrac{1}{x^{1/3}}$

$\qquad = 4x^{8/3} - 7x^{5/3} + \dfrac{1}{x^{1/3}}$

46. $\dfrac{2x^5 - 3x^3 + 5x - 1}{x^{3/2}} = \dfrac{2x^5}{x^{3/2}} - \dfrac{3x^3}{x^{3/2}} + \dfrac{5x}{x^{3/2}} - \dfrac{1}{x^{3/2}}$

$\qquad = 2x^{5-3/2} - 3x^{3-3/2} + 5x^{1-3/2} - x^{-3/2}$

$\qquad = 2x^{7/2} - 3x^{3/2} + \dfrac{5}{x^{1/2}} - \dfrac{1}{x^{3/2}}$

47. $\dfrac{3 - 5x^2 - x^4}{\sqrt{x}} = \dfrac{3}{\sqrt{x}} - \dfrac{5x^2}{\sqrt{x}} - \dfrac{x^4}{\sqrt{x}}$

$\qquad = \dfrac{3}{\sqrt{x}} - 5x^{2-1/2} - x^{4-1/2}$

$\qquad = \dfrac{3}{x^{1/2}} - 5x^{3/2} - x^{7/2}$

48. $\dfrac{x^3 - 5x^4}{3x^2} = \dfrac{x^3}{3x^2} - \dfrac{5x^4}{3x^2} = \dfrac{x}{3} - \dfrac{5x^2}{3}$

49. $\dfrac{-2(x^2-3)^{-3}(2x)(x+1)^3 - 3(x+1)^2(x^2-3)^{-2}}{[(x+1)^3]^2} = \dfrac{(x^2-3)^{-3}(x+1)^2[-4x(x+1) - 3(x^2-3)]}{(x+1)^6}$

$\qquad\qquad = \dfrac{-4x^2 - 4x - 3x^2 + 9}{(x^2-3)^3(x+1)^4}$

$\qquad\qquad = \dfrac{-7x^2 - 4x + 9}{(x^2-3)^3(x+1)^4}$

50. $\dfrac{x^5(-3)(x^2+1)^{-4}(2x) - (x^2+1)^{-3}(5)x^4}{(x^5)^2} = \dfrac{x^4(x^2+1)^{-4}[-6x^2 - 5(x^2+1)]}{x^{10}}$

$\qquad\qquad = \dfrac{x^4(-6x^2 - 5x^2 - 5)}{x^{10}(x^2+1)^4}$

$\qquad\qquad = \dfrac{-11x^2 - 5}{x^6(x^2+1)^4}$

51. $\dfrac{(6x+1)^3(27x^2+2) - (9x^3+2x)(3)(6x+1)^2(6)}{[(6x+1)^3]^2} = \dfrac{(6x+1)^2[(6x+1)(27x^2+2) - 18(9x^3+2x)]}{(6x+1)^6}$

$\qquad\qquad = \dfrac{162x^3 + 12x + 27x^2 + 2 - 162x^3 - 36x}{(6x+1)^4}$

$\qquad\qquad = \dfrac{27x^2 - 24x + 2}{(6x+1)^4}$

52. $\dfrac{(4x^2 + 9)^{1/2}(2) - (2x + 3)\left(\frac{1}{2}\right)(4x^2 + 9)^{-1/2}(8x)}{[(4x^2 + 9)^{1/2}]^2} = \dfrac{2(4x^2 + 9)^{-1/2}[(4x^2 + 9) - 2x(2x + 3)]}{(4x^2 + 9)}$

$$= \dfrac{2(4x^2 + 9 - 4x^2 - 6x)}{(4x^2 + 9)^{3/2}}$$

$$= \dfrac{2(9 - 6x)}{(4x^2 + 9)^{3/2}}$$

$$= \dfrac{-6(2x - 3)}{(4x^2 + 9)^{3/2}}$$

53. $\dfrac{(x + 2)^{3/4}(x + 3)^{-2/3} - (x + 3)^{1/3}(x + 2)^{-1/4}}{[(x + 2)^{3/4}]^2} = \dfrac{(x + 2)^{-1/4}(x + 3)^{-2/3}[(x + 2) - (x + 3)]}{(x + 2)^{6/4}}$

$$= \dfrac{x + 2 - x - 3}{(x + 2)^{1/4}(x + 3)^{2/3}(x + 2)^{6/4}}$$

$$= -\dfrac{1}{(x + 3)^{2/3}(x + 2)^{7/4}}$$

54. $(2x - 1)^{1/2} - (x + 2)(2x - 1)^{-1/2} = (2x - 1)^{1/2} - \dfrac{(x + 2)}{(2x - 1)^{1/2}}$

$$= \dfrac{2x - 1}{(2x - 1)^{1/2}} - \dfrac{(x + 2)}{(2x - 1)^{1/2}}$$

$$= \dfrac{2x - 1 - x - 2}{(2x - 1)^{1/2}}$$

$$= \dfrac{x - 3}{(2x - 1)^{1/2}}$$

55. $\dfrac{2(3x - 1)^{1/3} - (2x + 1)(1/3)(3x - 1)^{-2/3}(3)}{(3x - 1)^{2/3}} = \dfrac{(3x - 1)^{-2/3}[2(3x - 1) - (2x + 1)]}{(3x - 1)^{2/3}}$

$$= \dfrac{6x - 2 - 2x - 1}{(3x - 1)^{2/3}(3x - 1)^{2/3}}$$

$$= \dfrac{4x - 3}{(3x - 1)^{4/3}}$$

56. $\dfrac{(x + 1)(1/2)(2x - 3x^2)^{-1/2}(2 - 6x) - (2x - 3x^2)^{1/2}}{(x + 1)^2} = \dfrac{(x + 1)(2x - 3x^2)^{-1/2}(1 - 3x) - (2x - 3x^2)^{1/2}}{(x + 1)^2}$

$$= \dfrac{(2x - 3x^2)^{-1/2}[(x + 1)(1 - 3x) - (2x - 3x^2)]}{(x + 1)^2}$$

$$= \dfrac{x - 3x^2 + 1 - 3x - 2x + 3x^2}{(2x - 3x^2)^{1/2}(x + 1)^2}$$

$$= \dfrac{1 - 4x}{(2x - 3x^2)^{1/2}(x + 1)^2}$$

57. $\dfrac{1}{(x^2+4)^{1/2}} \cdot \dfrac{1}{2}(x^2+4)^{-1/2}(2x) = \dfrac{1}{(x^2+4)^{1/2}} \cdot \dfrac{1}{(x^2+4)^{1/2}} \cdot \dfrac{1}{2}(2x)$

$$= \dfrac{1}{(x^2+4)^1}(x)$$

$$= \dfrac{x}{x^2+4}$$

58. $\dfrac{1}{x^2-6}(2x) + \dfrac{1}{2x+5}(2) = \dfrac{2x(2x+5)+2(x^2-6)}{(x^2-6)(2x+5)}$

$$= \dfrac{4x^2+10x+2x^2-12}{(x^2-6)(2x+5)} = \dfrac{6x^2+10x-12}{(x^2-6)(2x+5)}$$

$$= \dfrac{2(3x^2+5x-6)}{(x^2-6)(2x+5)}$$

59. $(x^2+5)^{1/2}\left(\dfrac{3}{2}\right)(3x-2)^{1/2}(3) + (3x-2)^{3/2}\left(\dfrac{1}{2}\right)(x^2+5)^{-1/2}(2x) = \dfrac{9}{2}(x^2+5)^{1/2}(3x-2)^{1/2} + x(x^2+5)^{-1/2}(3x-2)^{3/2}$

$$= \dfrac{9}{2}(x^2+5)^{1/2}(3x-2)^{1/2} + \dfrac{2}{2}x(x^2+5)^{-1/2}(3x-2)^{3/2}$$

$$= \dfrac{1}{2}(x^2+5)^{-1/2}(3x-2)^{1/2}[9(x^2+5)^1 + 2x(3x-2)^1]$$

$$= \dfrac{1}{2}(x^2+5)^{-1/2}(3x-2)^{1/2}(9x^2+45+6x^2-4x)$$

$$= \dfrac{(3x-2)^{1/2}(15x^2-4x+45)}{2(x^2+5)^{1/2}}$$

60. $(3x+2)^{-1/2}(3)(x-6)^{1/2}(1) + (x-6)^3\left(-\dfrac{1}{2}\right)(3x+2)^{-3/2}(3) = 3(3x+2)^{-1/2}(x-6)^{1/2} + \left(\dfrac{-3}{2}\right)(x-6)^3(3x+2)^{-3/2}$

$$= \dfrac{3}{2}(x-6)^{1/2}(3x+2)^{-3/2}[2(3x+2)-(x-6)^{5/2}]$$

$$= \dfrac{3(x-6)^{1/2}[6x+4-(x-6)^{5/2}]}{2(3x+2)^{3/2}}$$

61. $t = \dfrac{\sqrt{x^2+4}}{2} + \dfrac{\sqrt{(4-x)^2+4}}{6}$

(a)

x	t
0.5	1.70
1.0	1.72
1.5	1.78
2.0	1.89
2.5	2.02
3.0	2.18
3.5	2.36
4.0	2.57

(b) She should swim to a point about $\frac{1}{2}$ mile down the coast to minimize the time required to reach the finish line.

(c) $\dfrac{1}{2}x(x^2+4)^{-1/2} + \dfrac{1}{6}(x-4)(x^2-8x+20)^{-1/2} = \dfrac{3}{6}x(x^2+4)^{-1/2} + \dfrac{1}{6}(x-4)(x^2-8x+20)^{-1/2}$

$$= \dfrac{1}{6}\left[3x(x^2+4)^{-1/2} + (x-4)(x^2-8x+20)^{-1/2}\right]$$

$$= \dfrac{1}{6}\left[\dfrac{3x}{(x^2+4)^{1/2}} + \dfrac{x-4}{(x^2-8x+20)^{1/2}}\right]$$

$$= \dfrac{3x\sqrt{x^2-8x+20} + (x-4)\sqrt{x^2+4}}{6\sqrt{x^2+4}\,\sqrt{x^2-8x+20}}$$

62. (a) $y_1 = x^2\left(\dfrac{1}{3}\right)(x^2 + 1)^{-2/3}(2x) + (x^2 + 1)^{1/3}(2x)$

$$= 2x(x^2 + 1)^{-2/3}\left[\dfrac{x^2}{3} + (x^2 + 1)\right]$$

$$= 2x(x^2 + 1)^{-2/3}\left[\dfrac{x^2}{3} + \dfrac{3(x^2 + 1)}{3}\right]$$

$$= \dfrac{2x}{(x^2 + 1)^{2/3}} \cdot \dfrac{4x^2 + 3}{3}$$

$$= \dfrac{2x(4x^2 + 3)}{3(x^2 + 1)^{2/3}}$$

$$= y_2$$

(b)

x	-2	-1	$-\frac{1}{2}$	0	1	2	$\frac{5}{2}$
y_1	-8.7	-2.9	-1.1	0	2.9	8.7	12.5
y_2	-8.7	-2.9	-1.1	0	2.9	8.7	12.5

63. True.

$$x^{-1} + y^{-2} = \dfrac{1}{x} + \dfrac{1}{y^2} = \dfrac{y^2 + x}{xy^2}$$

64. False. Cannot move term-by-term from denominator to numerator.

$$\dfrac{1}{x^{-2} + y^{-1}} = \dfrac{1}{\dfrac{1}{x^2} + \dfrac{1}{y}} = \dfrac{1}{\dfrac{y + x^2}{x^2 y}} = \dfrac{x^2 y}{y + x^2}$$

65. True.

$$\dfrac{1}{\sqrt{x} + 4} = \dfrac{1}{\sqrt{x} + 4} \cdot \dfrac{\sqrt{x} - 4}{\sqrt{x} - 4} = \dfrac{\sqrt{x} - 4}{x - 16}$$

66. False. $x^2 - 9$ does not factor into $(\sqrt{x} + 3)(\sqrt{x} - 3)$.

$$\dfrac{x^2 - 9}{\sqrt{x} - 3} = \dfrac{(x + 3)(x - 3)}{\sqrt{x} - 3} \cdot \dfrac{\sqrt{x} + 3}{\sqrt{x} + 3}$$

$$= \dfrac{(x + 3)(x - 3)(\sqrt{x} + 3)}{x - 9}$$

67. $x^n \cdot x^{3n} \neq x^{3n^2}$

Add exponents when multiplying powers with like bases.

$x^n \cdot x^{3n} = x^{4n}$

68. $(x^n)^{2n} + (x^{2n})^n = x^{2n^2} + x^{2n^2} = 2x^{2n^2}$

There is no error.

69. $x^{2n} + y^{2n} \neq (x^n + y^n)^2$

When squaring binomials, there is also a middle term.

$(x^n + y^n)^2 = x^{2n} + 2x^n y^n + y^{2n}$

70. $\dfrac{x^{2n} \cdot x^{3n}}{x^{3n} + x^2} = \dfrac{x^{2n+3n}}{x^{3n} + x^2} = \dfrac{x^{5n}}{x^{3n} + x^2}$

There is no error.

71. The two answers are equivalent and can be obtained by factoring.

$$\tfrac{1}{10}(2x - 1)^{5/2} + \tfrac{1}{6}(2x - 1)^{3/2} = \tfrac{1}{60}(2x - 1)^{3/2}[6(2x - 1) + 10]$$

$$= \tfrac{1}{60}(2x - 1)^{3/2}(12x + 4)$$

$$= \tfrac{4}{60}(2x - 1)^{3/2}(3x + 1)$$

$$= \tfrac{1}{15}(2x - 1)^{3/2}(3x + 1)$$

(a) $\tfrac{2}{3}x(2x - 3)^{3/2} - \tfrac{2}{15}(2x - 3)^{5/2} = \tfrac{2}{15}(2x - 3)^{3/2}[5x - (2x - 3)]$

$$= \tfrac{2}{15}(2x - 3)^{3/2}(3x + 3)$$

$$= \tfrac{2}{15}(2x - 3)^{3/2}3(x + 1)$$

$$= \tfrac{2}{5}(2x - 3)^{3/2}(x + 1)$$

(b) $\tfrac{2}{3}x(4 + x)^{3/2} - \tfrac{2}{15}(4 + x)^{5/2} = \tfrac{2}{15}(4 + x)^{3/2}[5x - (4 + x)]$

$$= \tfrac{2}{15}(4 + x)^{3/2}(4x - 4) = \tfrac{2}{15}(4 + x)^{3/2}4(x - 1) = \tfrac{8}{15}(4 + x)^{3/2}(x - 1)$$

Chapter 1   Practice Test Solutions

1. (a) Midpoint: $\left(\dfrac{-3 + 5}{2}, \dfrac{4 + (-6)}{2}\right) = (1, -1)$

 (b) Distance: $d = \sqrt{[5 - (-3)]^2 + (-6 - 4)^2}$

$= \sqrt{(8)^2 + (-10)^2}$

$= \sqrt{164} = 2\sqrt{41}$

2. $y = \sqrt{7 - x}$

Domain: $x \le 7$

x	7	6	3	-2
y	0	1	2	3

3. $[x - (-3)]^2 + (y - 5)^2 = 6^2$

$(x + 3)^2 + (y - 5)^2 = 36$

4. $m = \dfrac{-1 - 4}{3 - 2} = -5$

$y - 4 = -5(x - 2)$

$y - 4 = -5x + 10$

$y = -5x + 14$

5. $y = \dfrac{4}{3}x - 3$

6. $2x + 3y = 0$

$y = -\dfrac{2}{3}x$

$m_1 = -\dfrac{2}{3}$

$\perp m_2 = \dfrac{3}{2}$ through $(4, 1)$

$y - 1 = \dfrac{3}{2}(x - 4)$

$y - 1 = \dfrac{3}{2}x - 6$

$y = \dfrac{3}{2}x - 5$

7. $(5, 32)$ and $(9, 44)$

$m = \dfrac{44 - 32}{9 - 5} = \dfrac{12}{4} = 3$

$y - 32 = 3(x - 5)$

$y - 32 = 3x - 15$

$y = 3x + 17$

When $x = 20$, $y = 3(20) + 17$

$y = \$77.$

8. $f(x - 3) = (x - 3)^2 - 2(x - 3) + 1$

$= x^2 - 6x + 9 - 2x + 6 + 1$

$= x^2 - 8x + 16$

9. $f(3) = 12 - 11 = 1$

$\dfrac{f(x) - f(3)}{x - 3} = \dfrac{(4x - 11) - 1}{x - 3}$

$= \dfrac{4x - 12}{x - 3}$

$= \dfrac{4(x - 3)}{x - 3} = 4, x \ne 3$

10. $f(x) = \sqrt{36 - x^2} = \sqrt{(6 + x)(6 - x)}$

Domain: $[-6, 6]$, because $(6 + x)(6 - x) \ge 0$ on this interval.

Range: $[0, 6]$, because $0 \le (6 + x)(6 - x) \le 36$ on this interval.

11. (a) $6x - 5y + 4 = 0$

$$y = \frac{6x + 4}{5} \text{ is a function of } x.$$

(b) $x^2 + y^2 = 9$

$$y = \pm\sqrt{9 - x^2} \text{ is not a function of } x.$$

(c) $y^3 = x^2 + 6$

$$y = \sqrt[3]{x^2 + 6} \text{ is a function of } x.$$

12. Parabola

Vertex: $(0, -5)$

Intercepts: $(0, -5)$, $(\pm\sqrt{5}, 0)$

y-axis symmetry

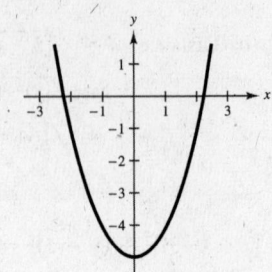

13. Intercepts: $(0, 3)$, $(-3, 0)$

x	0	1	-1	2	-2	-3	-4
y	3	4	2	5	1	0	1

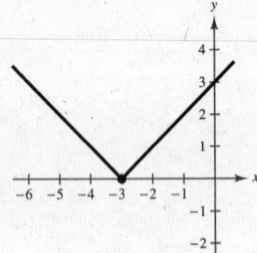

14.

x	0	1	2	3	-1	-2	-3
y	1	3	5	7	2	6	12

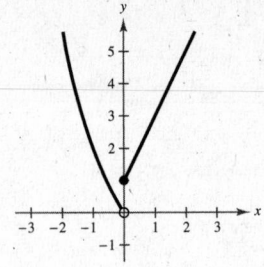

15. (a) $f(x + 2)$

Horizontal shift
two units to the left

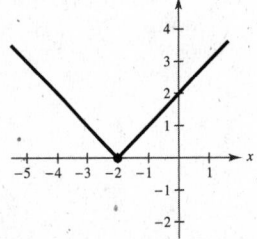

(b) $-f(x) + 2$

Reflection in the x-axis
and a vertical shift two
units upward

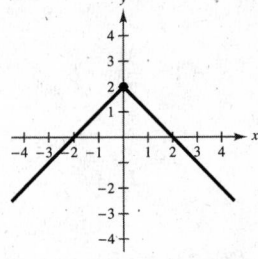

16. (a) $(g - f)(x) = g(x) - f(x)$

$$= (2x^2 - 5) - (3x + 7)$$

$$= 2x^2 - 3x - 12$$

(b) $(fg)(x) = f(x)g(x)$

$$= (3x + 7)(2x^2 - 5)$$

$$= 6x^3 + 14x^2 - 15x - 35$$

17. $f(g(x)) = f(2x + 3)$

$$= (2x + 3)^2 - 2(2x + 3) + 16$$

$$= 4x^2 + 12x + 9 - 4x - 6 + 16$$

$$= 4x^2 + 8x + 19$$

18. $f(x) = x^3 + 7$

$$y = x^3 + 7$$

$$x = y^3 + 7$$

$$x - 7 = y^3$$

$$\sqrt[3]{x - 7} = y$$

$$f^{-1}(x) = \sqrt[3]{x - 7}$$

19. (a) $f(x) = |x - 6|$ does not have an inverse.

Its graph does not pass the horizontal line test.

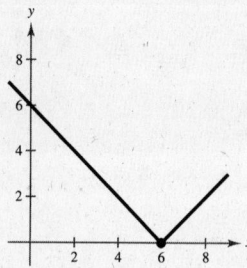

20. $f(x) = \sqrt{\dfrac{3 - x}{x}}, \ 0 < x \leq 3, \ y \geq 0$

$$y = \sqrt{\dfrac{3 - x}{x}}$$

$$x = \sqrt{\dfrac{3 - y}{y}}$$

$$x^2 = \dfrac{3 - y}{y}$$

$$x^2 y = 3 - y$$

$$x^2 y + y = 3$$

$$y(x^2 + 1) = 3$$

$$y = \dfrac{3}{x^2 + 1}$$

$$f^{-1}(x) = \dfrac{3}{x^2 + 1}, \ x \geq 0$$

(b) $f(x) = ax + b, a \neq 0$ does have an inverse.

$$y = ax + b$$

$$x = ay + b$$

$$\dfrac{x - b}{a} = y$$

$$f^{-1}(x) = \dfrac{x - b}{a}$$

(c) $f(x) = x^3 - 19$ does have an inverse.

$$y = x^3 - 19$$

$$x = y^3 - 19$$

$$x + 19 = y^3$$

$$\sqrt[3]{x + 19} = y$$

$$f^{-1}(x) = \sqrt[3]{x + 19}$$

21. False. The slopes of 3 and $\frac{1}{3}$ are not **negative** reciprocals.

22. True. Let $y = (f \circ g)(x)$. Then $x = (f \circ g)^{-1}(y)$.

Also,

$$(f \circ g)(x) = y$$

$$f(g(x)) = y$$

$$g(x) = f^{-1}(y)$$

$$x = g^{-1}(f^{-1}(y))$$

$$x = (g^{-1} \circ f^{-1})(y)$$

Since $x = x$, we have $(f \circ g)^{-1}(y) = (g^{-1} \circ f^{-1})(y)$.

23. True. It must pass the vertical line test to be a function and it must pass the horizontal line test to have an inverse.

24. $z = \dfrac{cx^3}{\sqrt{y}}$

$-1 = \dfrac{c(-1)^3}{\sqrt{25}}$

$-1 = \dfrac{-c}{5}$

$5 = c$

$z = \dfrac{5x^3}{\sqrt{y}}$

25. $y \approx 0.669x + 2.669$

Chapter 2   Practice Test Solutions

1. x-intercepts: $(1, 0), (5, 0)$

y-intercepts: $(0, 5)$

Vertex: $(3, -4)$

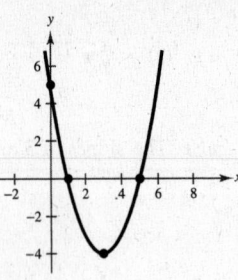

2. $a = 0.01, b = -90$

$\dfrac{-b}{2a} = \dfrac{90}{2(.01)} = 4500$ units

3. Vertex: $(1, 7)$ opening downward through $(2, 5)$

$y = a(x - 1)^2 + 7$ Standard form

$5 = a(2 - 1)^2 + 7$

$5 = a + 7$

$a = -2$

$y = -2(x - 1)^2 + 7$

$\quad = -2(x^2 - 2x + 1) + 7$

$\quad = -2x^2 + 4x + 5$

4. $y = \pm a(x - 2)(3x - 4)$ where a is any real number

$y = \pm(3x^2 - 10x + 8)$

5. Leading coefficient: -3

Degree: 5

Moves down to the right and up to the left

6. $0 = x^5 - 5x^3 + 4x$

$\quad = x(x^4 - 5x^2 + 4)$

$\quad = x(x^2 - 1)(x^2 - 4)$

$\quad = x(x + 1)(x - 1)(x + 2)(x - 2)$

$x = 0, x = \pm 1, x = \pm 2$

7. $f(x) = x(x - 3)(x + 2)$

$\quad = x(x^2 - x - 6)$

$\quad = x^3 - x^2 - 6x$

8. Intercepts: $(0, 0), \left(\pm 2\sqrt{3}, 0\right)$

Moves up to the right

Moves down to the left

Origin symmetry

x	-2	-1	0	1	2
y	16	11	0	-11	-16

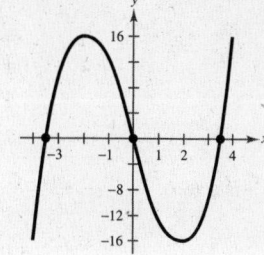

9.
$$3x^3 + 9x^2 + 20x + 62 + \frac{176}{x-3}$$

$$x - 3\ \overline{)\ 3x^4 + 0x^3 - 7x^2 + 2x - 10}$$
$$\underline{3x^4 - 9x^3}$$
$$9x^3 - 7x^2$$
$$\underline{9x^3 - 27x^2}$$
$$20x^2 + 2x$$
$$\underline{20x^2 - 60x}$$
$$62x - 10$$
$$\underline{62x - 186}$$
$$176$$

10.
$$x - 2 + \frac{5x - 13}{x^2 + 2x - 1}$$

$$x^2 + 2x - 1\ \overline{)\ x^3 + 0x^2 + 0x - 11}$$
$$\underline{x^3 + 2x^2 - x}$$
$$-2x^2 + x - 11$$
$$\underline{-2x^2 - 4x + 2}$$
$$5x - 13$$

11. -5

	3	13	0	0	12	-1
		-15	10	-50	250	-1310
	3	-2	10	-50	262	-1311

$$\frac{3x^5 + 13x^4 + 12x - 1}{x + 5} = 3x^4 - 2x^3 + 10x^2 - 50x + 262 - \frac{1311}{x + 5}$$

12. -6

	7	40	-12	15
		-42	12	0
	7	-2	0	15

$$f(-6) = 15$$

13. $0 = x^3 - 19x - 30$

Possible rational roots: $\pm 1, \pm 2, \pm 3, \pm 5, \pm 6, \pm 10, \pm 15, \pm 30$

-2

	1	0	-19	-30
		-2	4	30
	1	-2	-15	0

$x = -2$ is a zero.

$0 = (x + 2)(x^2 - 2x - 15)$

$0 = (x + 2)(x + 3)(x - 5)$

Zeros: $x = -2, x = -3, x = 5$

14. $0 = x^4 + x^3 - 8x^2 - 9x - 9$

Possible rational roots: $\pm 1, \pm 3, \pm 9$

$$
\begin{array}{r|rrrrr}
3 & 1 & 1 & -8 & -9 & -9 \\
  &   & 3 & 12 & 12 & 9 \\
\hline
  & 1 & 4 & 4 & 3 & 0
\end{array}
\qquad x = 3 \text{ is a zero.}
$$

$0 = (x - 3)(x^3 + 4x^2 + 4x + 3)$

The zeros of $x^2 + x + 1$ are $x = \dfrac{-1 \pm \sqrt{3}i}{2}$ (by the Quadratic Formula).

Zeros: $x = 3, x = -3, x = -\dfrac{1}{2} + \dfrac{\sqrt{3}}{2}i, x = -\dfrac{1}{2} - \dfrac{\sqrt{3}}{2}i$

Possible rational roots of $x^3 + 4x^2 + 4x + 3$: $\pm 1, \pm 3$

$$
\begin{array}{r|rrrr}
-3 & 1 & 4 & 4 & 3 \\
   &   & -3 & -3 & -3 \\
\hline
   & 1 & 1 & 1 & 0
\end{array}
\qquad x = -3 \text{ is a zero.}
$$

$0 = (x - 3)(x + 3)(x^2 + x + 1)$

15. $0 = 6x^3 - 5x^2 + 4x - 15$

Possible rational roots: $\pm 1, \pm 3, \pm 5, \pm 15, \pm\frac{1}{2}, \pm\frac{3}{2}, \pm\frac{5}{2}, \pm\frac{15}{2}, \pm\frac{1}{3}, \pm\frac{5}{3}, \pm\frac{1}{6}, \pm\frac{5}{6}$

16. $0 = x^3 - \frac{20}{3}x^2 + 9x - \frac{10}{3}$

$0 = 3x^3 - 20x^2 + 27x - 10$

Possible rational roots:
$\pm 1, \pm 2, \pm 5, \pm 10, \pm\frac{1}{3}, \pm\frac{2}{3}, \pm\frac{5}{3}, \pm\frac{10}{3}$

$$
\begin{array}{r|rrrr}
1 & 3 & -20 & 27 & -10 \\
  &   & 3 & -17 & 10 \\
\hline
  & 3 & -17 & 10 & 0
\end{array}
$$

$0 = (x - 1)(3x^2 - 17x + 10)$

$0 = (x - 1)(3x - 2)(x - 5)$

Zeros: $x = 1, x = \frac{2}{3}, x = 5$

17. Possible rational roots: $\pm 1, \pm 2, \pm 5, \pm 10$

$$
\begin{array}{r|rrrrr}
1 & 1 & 1 & 3 & 5 & -10 \\
  &   & 1 & 2 & 5 & 10 \\
\hline
  & 1 & 2 & 5 & 10 & 0
\end{array}
\qquad x = 1 \text{ is a zero.}
$$

$$
\begin{array}{r|rrrr}
-2 & 1 & 2 & 5 & 10 \\
   &   & -2 & 0 & -10 \\
\hline
   & 1 & 0 & 5 & 0
\end{array}
\qquad x = -2 \text{ is a zero.}
$$

$f(x) = (x - 1)(x + 2)(x^2 + 5)$

$\quad\;\;\; = (x - 1)(x + 2)(x + \sqrt{5}i)(x - \sqrt{5}i)$

18. $f(x) = (x - 2)[x - (3 + i)][x - (3 - i)]$

$\quad\;\;\; = (x - 2)[(x - 3) - i][(x - 3) + i]$

$\quad\;\;\; = (x - 2)[(x - 3)^2 - i^2]$

$\quad\;\;\; = (x - 2)[x^2 - 6x + 10]$

$\quad\;\;\; = x^3 - 8x^2 + 22x - 20$

19.
$$
\begin{array}{r|rrrr}
3i & 1 & 4 & 9 & 36 \\
   &   & 3i & 12i - 9 & -36 \\
\hline
   & 1 & 4 + 3i & 12i & 0
\end{array}
$$

20. Vertical asymptote: $x = 0$

Horizontal asymptote: $y = \frac{1}{2}$

x-intercept: $(1, 0)$

21. $y = 8$ is a horizontal asymptote since the degree on the numerator equals the degree of the denominator. There are no vertical asymptotes.

22. $x = 1$ is a vertical asymptote.

$$\frac{4x^2 - 2x + 7}{x - 1} = 4x + 2 + \frac{9}{x - 1}$$

Thus, $y = 4x + 2$ is a slant asymptote.

23. (a) $(4 - 3i) - (-2 + i) = 4 - 3i + 2 - i = 6 - 4i$

(b) $(4 - 3i)(-2 + i) = -8 + 4i + 6i - 3i^2 = -8 + 10i + 3 = -5 + 10i$

(c) $\dfrac{4 - 3i}{-2 + i} = \dfrac{4 - 3i}{-2 + i} \cdot \dfrac{-2 - i}{-2 - i} = \dfrac{-8 - 4i + 6i + 3i^2}{4 + 1}$

$\qquad = \dfrac{-11 + 2i}{5} = -\dfrac{11}{5} + \dfrac{2}{5}i$

24. $\qquad x^2 - 49 \le 0$

$(x + 7)(x - 7) \le 0$

Critical numbers: $x = -7$ and $x = 7$

Test intervals: $(-\infty, -7), (-7, 7), (7, \infty)$

Test: Is $x^2 - 49 \le 0$?

Solution set: $[-7, 7]$

25. $\dfrac{x + 3}{x - 7} \ge 0$

Critical numbers: $x = -3$ and $x = 7$

Test intervals: $(-\infty, -3), (-3, 7), (7, \infty)$

Test: Is $\dfrac{x + 3}{x - 7} \ge 0$?

Solution set: $(-\infty, -3] \cup [7, \infty)$

Chapter 3   Practice Test Solutions

1. $x^{3/5} = 8$

$x = 8^{5/3} = \left(\sqrt[3]{8}\right)^5 = 2^5 = 32$

2. $3^{x-1} = \frac{1}{81}$

$3^{x-1} = 3^{-4}$

$x - 1 = -4$

$x = -3$

3. $f(x) = 2^{-x} = \left(\frac{1}{2}\right)^x$

x	-2	-1	0	1	2
$f(x)$	4	2	1	$\frac{1}{2}$	$\frac{1}{4}$

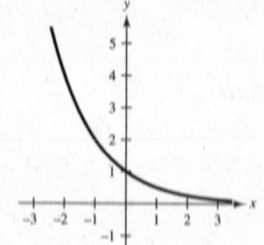

4. $g(x) = e^x + 1$

x	-2	-1	0	1	2
$g(x)$	1.14	1.37	2	3.72	8.39

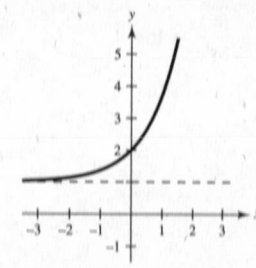

5. (a) $A = P\left(1 + \dfrac{r}{n}\right)^{nt}$

$A = 5000\left(1 + \dfrac{0.09}{12}\right)^{12(3)} \approx \$6543.23$

(b) $A = P\left(1 + \dfrac{r}{n}\right)^{nt}$

$A = 5000\left(1 + \dfrac{0.09}{4}\right)^{4(3)} \approx \$6530.25$

(c) $A = Pe^{rt}$

$A = 5000e^{(0.09)(3)} \approx \$6549.82$

6. $7^{-2} = \dfrac{1}{49}$

$\log_7 \dfrac{1}{49} = -2$

7. $x - 4 = \log_2 \dfrac{1}{64}$

$2^{x-4} = \dfrac{1}{64}$

$2^{x-4} = 2^{-6}$

$x - 4 = -6$

$x = -2$

8. $\log_b \sqrt[4]{\dfrac{8}{25}} = \dfrac{1}{4} \log_b \dfrac{8}{25}$

$= \dfrac{1}{4}[\log_b 8 - \log_b 25]$

$= \dfrac{1}{4}[\log_b 2^3 - \log_b 5^2]$

$= \dfrac{1}{4}[3 \log_b 2 - 2 \log_b 5]$

$= \dfrac{1}{4}[3(0.3562) - 2(0.8271)]$

$= -0.1464$

9. $5 \ln x - \dfrac{1}{2} \ln y + 6 \ln z = \ln x^5 - \ln \sqrt{y} + \ln z^6 = \ln\left(\dfrac{x^5 z^6}{\sqrt{y}}\right), z > 0$

10. $\log_9 28 = \dfrac{\log 28}{\log 9} \approx 1.5166$

11. $\log N = 0.6646$

$N = 10^{0.6646} \approx 4.62$

12.

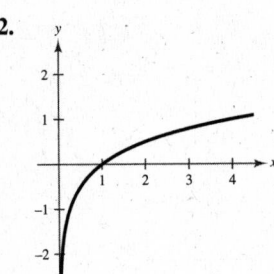

13. Domain:

$x^2 - 9 > 0$

$(x + 3)(x - 3) > 0$

$x < -3 \text{ or } x > 3$

14.

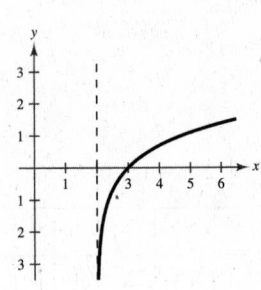

15. False. $\dfrac{\ln x}{\ln y} \neq \ln(x - y)$ since $\dfrac{\ln x}{\ln y} = \log_y x$.

16. $5^3 = 41$

$x = \log_5 41 = \dfrac{\ln 41}{\ln 5} \approx 2.3074$

17. $x - x^2 = \log_5 \frac{1}{25}$

$5^{x-x^2} = \frac{1}{25}$

$5^{x-x^2} = 5^{-2}$

$x - x^2 = -2$

$0 = x^2 - x - 2$

$0 = (x + 1)(x - 2)$

$x = -1 \text{ or } x = 2$

18. $\log_2 x + \log_2(x - 3) = 2$

$\log_2[x(x - 3)] = 2$

$x(x - 3) = 2^2$

$x^2 - 3x = 4$

$x^2 - 3x - 4 = 0$

$(x + 1)(x - 4) = 0$

$x = 4$

$x = -1 \text{ (extraneous)}$

$x = 4$ is the only solution.

19. $\dfrac{e^x + e^{-x}}{3} = 4$

$e^x(e^x + e^{-x}) = 12e^x$

$e^{2x} + 1 = 12e^x$

$e^{2x} - 12e^x + 1 = 0$

$e^x = \dfrac{12 \pm \sqrt{144 - 4}}{2}$

$e^x \approx 11.9161 \qquad \text{or} \qquad e^x \approx 0.0839$

$x = \ln 11.9161 \qquad\qquad x = \ln 0.0839$

$x \approx 2.478 \qquad\qquad\qquad x \approx -2.478$

20. $A = Pe^{et}$

$12{,}000 = 6000e^{0.13t}$

$2 = e^{0.13t}$

$0.13t = \ln 2$

$t = \dfrac{\ln 2}{0.13}$

$t \approx 5.3319 \text{ years or 5 years 4 months}$

Chapter 4   Practice Test Solutions

1. $350° = 350\left(\dfrac{\pi}{180}\right) = \dfrac{35\pi}{18}$

2. $\dfrac{5\pi}{9} = \dfrac{5\pi}{9} \cdot \dfrac{180}{\pi} = 100°$

3. $135° \, 14' \, 12'' = \left(135 + \frac{14}{60} + \frac{12}{3600}\right)^{\circ}$

$\approx 135.2367°$

4. $-22.569° = -(22° + 0.569(60)')$

$= -22° \, 34.14'$

$= -(22° \, 34' + 0.14(60)'')$

$\approx -22° \, 34' \, 8''$

5. $\cos \theta = \dfrac{2}{3}$

$x = 2, r = 3, y = \pm\sqrt{9 - 4} = \pm\sqrt{5}$

$\tan \theta = \dfrac{y}{x} = \pm\dfrac{\sqrt{5}}{2}$

6. $\sin \theta = 0.9063$

$\theta = \arcsin(0.9063)$

$\theta = 65° = \dfrac{13\pi}{36} \quad \text{or} \quad \theta = 180° - 65° = 115° = \dfrac{23\pi}{36}$

7. $\tan 20° = \dfrac{35}{x}$

$x = \dfrac{35}{\tan 20°} \approx 96.1617$

8. $\theta = \dfrac{6\pi}{5}$, θ is in Quadrant III.

Reference angle: $\dfrac{6\pi}{5} - \pi = \dfrac{\pi}{5}$ or $36°$

9. $\csc 3.92 = \dfrac{1}{\sin 3.92} \approx -1.4242$

10. $\tan \theta = 6 = \dfrac{6}{1}$, θ lies in Quandrant III.

$y = -6, x = -1, r = \sqrt{36 + 1} = \sqrt{37}$,

so $\sec \theta = \dfrac{\sqrt{37}}{-1} \approx -6.0828$.

11. Period: 4π

Amplitude: 3

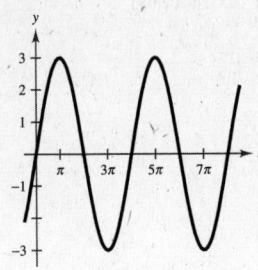

12. Period: 2π

Amplitude: 2

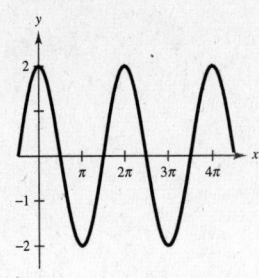

13. Period: $\dfrac{\pi}{2}$

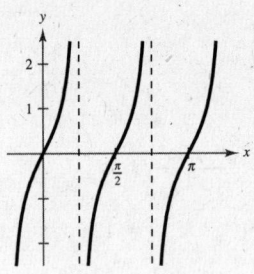

14. Period: 2π

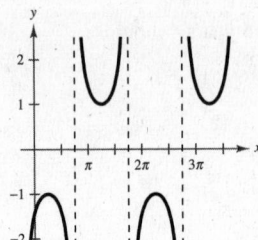

15.

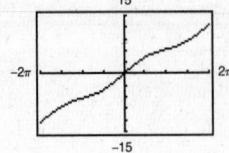

16.

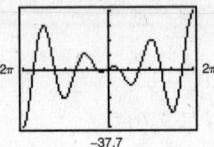

17. $\theta = \arcsin 1$

$\sin \theta = 1$

$\theta = \dfrac{\pi}{2} = 90°$

18. $\theta = \arctan(-3)$

$\tan \theta = -3$

$\theta \approx -1.249 \approx -71.565°$

19. $\sin\left(\arccos \dfrac{4}{\sqrt{35}}\right)$

$\sin \theta = \dfrac{\sqrt{19}}{\sqrt{35}} \approx 0.7368$

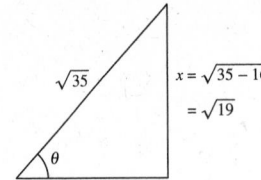

$x = \sqrt{35 - 16}$
$= \sqrt{19}$

20. $\cos\left(\arcsin \dfrac{x}{4}\right)$

$\cos \theta = \dfrac{\sqrt{16 - x^2}}{4}$

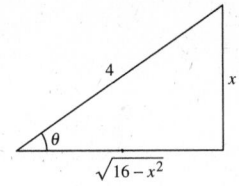

21. Given $A = 40°$, $c = 12$

$B = 90° - 40° = 50°$

$\sin 40° = \dfrac{a}{12}$

$a = 12 \sin 40° \approx 7.713$

$\cos 40° = \dfrac{b}{12}$

$b = 12 \cos 40° \approx 9.193$

22. Given $B = 6.84°$, $a = 21.3$

$A = 90° - 6.84° = 83.16°$

$\sin 83.16° = \dfrac{21.3}{c}$

$c = \dfrac{21.3}{\sin 83.16°} \approx 21.453$

$\tan 83.16° = \dfrac{21.3}{b}$

$b = \dfrac{21.3}{\tan 83.16°} \approx 2.555$

23. Given $a = 5$, $b = 9$

$c = \sqrt{25 + 81} = \sqrt{106} \approx 10.296$

$\tan A = \dfrac{5}{9}$

$A = \arctan \dfrac{5}{9} \approx 29.055°$

$B \approx 90° - 29.055° = 60.945°$

24. $\sin 67° = \dfrac{x}{20}$

$x = 20 \sin 67° \approx 18.41$ feet

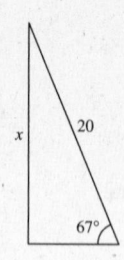

25. $\tan 5° = \dfrac{250}{x}$

$x = \dfrac{250}{\tan 5°}$

≈ 2857.513 feet

≈ 0.541 mi

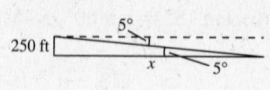

Chapter 5   Practice Test Solutions

1. $\tan x = \dfrac{4}{11}$, $\sec x < 0 \implies x$ is in Quadrant III.

$y = -4$, $x = -11$, $r = \sqrt{16 + 121} = \sqrt{137}$

$\sin x = -\dfrac{4}{\sqrt{137}} = -\dfrac{4\sqrt{137}}{137}$ $\csc x = -\dfrac{\sqrt{137}}{4}$

$\cos x = -\dfrac{11}{\sqrt{137}} = -\dfrac{11\sqrt{137}}{137}$ $\sec x = -\dfrac{\sqrt{137}}{11}$

$\tan x = \dfrac{4}{11}$ $\cot x = \dfrac{11}{4}$

2. $\dfrac{\sec^2 x + \csc^2 x}{\csc^2 x(1 + \tan^2 x)} = \dfrac{\sec^2 x + \csc^2 x}{\csc^2 x + (\csc^2 x)\tan^2 x}$

$= \dfrac{\sec^2 x + \csc^2 x}{\csc^2 x + \dfrac{1}{\sin^2 x} \cdot \dfrac{\sin^2 x}{\cos^2 x}}$

$= \dfrac{\sec^2 x + \csc^2 x}{\csc^2 x + \dfrac{1}{\cos^2 x}}$

$= \dfrac{\sec^2 x + \csc^2 x}{\csc^2 x + \sec^2 x} = 1$

3. $\ln|\tan \theta| - \ln|\cot \theta| = \ln \left| \dfrac{\tan \theta}{\cot \theta} \right| = \ln \left| \dfrac{\sin \theta/\cos \theta}{\cos \theta/\sin \theta} \right| = \ln \left| \dfrac{\sin^2 \theta}{\cos^2 \theta} \right| = \ln|\tan^2 \theta| = 2 \ln|\tan \theta|$

4. $\cos\left(\dfrac{\pi}{2} - x\right) = \dfrac{1}{\csc x}$ is true since $\cos\left(\dfrac{\pi}{2} - x\right) = \sin x = \dfrac{1}{\csc x}$.

5. $\sin^4 x + (\sin^2 x)\cos^2 x = \sin^2 x(\sin^2 x + \cos^2 x)$

$= \sin^2 x(1) = \sin^2 x$

6. $(\csc x + 1)(\csc x - 1) = \csc^2 x - 1 = \cot^2 x$

7. $\dfrac{\cos^2 x}{1 - \sin x} \cdot \dfrac{1 + \sin x}{1 + \sin x} = \dfrac{\cos^2 x(1 + \sin x)}{1 - \sin^2 x} = \dfrac{\cos^2 x(1 + \sin x)}{\cos^2 x} = 1 + \sin x$

8. $\dfrac{1 + \cos\theta}{\sin\theta} + \dfrac{\sin\theta}{1 + \cos\theta} = \dfrac{(1 + \cos\theta)^2 + \sin^2\theta}{\sin\theta(1 + \cos\theta)}$

$$= \dfrac{1 + 2\cos\theta + \cos^2\theta + \sin^2\theta}{\sin\theta(1 + \cos\theta)} = \dfrac{2 + 2\cos\theta}{\sin\theta(1 + \cos\theta)} = \dfrac{2}{\sin\theta} = 2\csc\theta$$

9. $\tan^4 x + 2\tan^2 x + 1 = (\tan^2 x + 1)^2 = (\sec^2 x)^2 = \sec^4 x$

10. (a) $\sin 105° = \sin(60° + 45°) = \sin 60° \cos 45° + \cos 60° \sin 45°$

$$= \dfrac{\sqrt{3}}{2} \cdot \dfrac{\sqrt{2}}{2} + \dfrac{1}{2} \cdot \dfrac{\sqrt{2}}{2} = \dfrac{\sqrt{2}}{4}\left(\sqrt{3} + 1\right)$$

(b) $\tan 15° = \tan(60° - 45°) = \dfrac{\tan 60° - \tan 45°}{1 + \tan 60° \tan 45°}$

$$= \dfrac{\sqrt{3} - 1}{1 + \sqrt{3}} \cdot \dfrac{1 - \sqrt{3}}{1 - \sqrt{3}} = \dfrac{2\sqrt{3} - 1 - 3}{1 - 3} = \dfrac{2\sqrt{3} - 4}{-2} = 2 - \sqrt{3}$$

11. $(\sin 42°)\cos 38° - (\cos 42°)\sin 38° = \sin(42° - 38°) = \sin 4°$

12. $\tan\left(\theta + \dfrac{\pi}{4}\right) = \dfrac{\tan\theta + \tan\left(\dfrac{\pi}{4}\right)}{1 - (\tan\theta)\tan\left(\dfrac{\pi}{4}\right)} = \dfrac{\tan\theta + 1}{1 - \tan\theta(1)} = \dfrac{1 + \tan\theta}{1 - \tan\theta}$

13. $\sin(\arcsin x - \arccos x) = \sin(\arcsin x)\cos(\arccos x) - \cos(\arcsin x)\sin(\arccos x)$

$$= (x)(x) - \left(\sqrt{1 - x^2}\right)\left(\sqrt{1 - x^2}\right) = x^2 - (1 - x^2) = 2x^2 - 1$$

14. (a) $\cos(120°) = \cos[2(60°)] = 2\cos^2 60° - 1 = 2\left(\dfrac{1}{2}\right)^2 - 1 = -\dfrac{1}{2}$

(b) $\tan(300°) = \tan[2(150°)] = \dfrac{2\tan 150°}{1 - \tan^2 150°} = \dfrac{-\dfrac{2\sqrt{3}}{3}}{1 - \left(\dfrac{1}{3}\right)} = -\sqrt{3}$

15. (a) $\sin 22.5° = \sin\dfrac{45°}{2} = \sqrt{\dfrac{1 - \cos 45°}{2}} = \sqrt{\dfrac{1 - \dfrac{\sqrt{2}}{2}}{2}} = \dfrac{\sqrt{2 - \sqrt{2}}}{2}$

(b) $\tan\dfrac{\pi}{12} = \tan\dfrac{\dfrac{\pi}{6}}{2} = \dfrac{\sin\dfrac{\pi}{6}}{1 + \cos\left(\dfrac{\pi}{6}\right)} = \dfrac{\dfrac{1}{2}}{1 + \dfrac{\sqrt{3}}{2}} = \dfrac{1}{2 + \sqrt{3}} = 2 - \sqrt{3}$

16. $\sin\theta = \dfrac{4}{5}$, θ lies in Quadrant II $\implies$ $\cos\theta = -\dfrac{3}{5}$.

$\cos\dfrac{\theta}{2} = \sqrt{\dfrac{1 + \cos\theta}{2}} = \sqrt{\dfrac{1 - \dfrac{3}{5}}{2}} = \sqrt{\dfrac{2}{10}} = \dfrac{1}{\sqrt{5}} = \dfrac{\sqrt{5}}{5}$

17. $(\sin^2 x)\cos^2 x = \dfrac{1-\cos 2x}{2} \cdot \dfrac{1+\cos 2x}{2} = \dfrac{1}{4}[1-\cos^2 2x] = \dfrac{1}{4}\left[1 - \dfrac{1+\cos 4x}{2}\right]$

$\qquad\qquad = \dfrac{1}{8}[2 - (1+\cos 4x)] = \dfrac{1}{8}[1 - \cos 4x]$

18. $6(\sin 5\theta)\cos 2\theta = 6\left\{\tfrac{1}{2}[\sin(5\theta + 2\theta) + \sin(5\theta - 2\theta)]\right\} = 3[\sin 7\theta + \sin 3\theta]$

19. $\sin(x+\pi) + \sin(x-\pi) = 2\left(\sin\dfrac{[(x+\pi)+(x-\pi)]}{2}\right)\cos\dfrac{[(x+\pi)-(x-\pi)]}{2}$

$\qquad\qquad = 2\sin x \cos \pi = -2\sin x$

20. $\dfrac{\sin 9x + \sin 5x}{\cos 9x - \cos 5x} = \dfrac{2\sin 7x \cos 2x}{-2\sin 7x \sin 2x} = -\dfrac{\cos 2x}{\sin 2x} = -\cot 2x$

21. $\tfrac{1}{2}[\sin(u+v) - \sin(u-v)] = \tfrac{1}{2}\{(\sin u)\cos v + (\cos u)\sin v - [(\sin u)\cos v - (\cos u)\sin v]\}$

$\qquad\qquad = \tfrac{1}{2}[2(\cos u)\sin v] = (\cos u)\sin v$

22. $4\sin^2 x = 1$

$\qquad \sin^2 x = \dfrac{1}{4}$

$\qquad \sin x = \pm\dfrac{1}{2}$

$\sin x = \dfrac{1}{2} \qquad$ or $\quad \sin x = -\dfrac{1}{2}$

$x = \dfrac{\pi}{6}$ or $\dfrac{5\pi}{6} \qquad x = \dfrac{7\pi}{6}$ or $\dfrac{11\pi}{6}$

23. $\tan^2 \theta + \left(\sqrt{3}-1\right)\tan\theta - \sqrt{3} = 0$

$\qquad (\tan\theta - 1)\left(\tan\theta + \sqrt{3}\right) = 0$

$\tan\theta = 1 \qquad$ or $\quad \tan\theta = -\sqrt{3}$

$\theta = \dfrac{\pi}{4}$ or $\dfrac{5\pi}{4} \qquad \theta = \dfrac{2\pi}{3}$ or $\dfrac{5\pi}{3}$

24. $\qquad\qquad \sin 2x = \cos x$

$2(\sin x)\cos x - \cos x = 0$

$\qquad \cos x(2\sin x - 1) = 0$

$\cos x = 0 \qquad$ or $\quad \sin x = \dfrac{1}{2}$

$x = \dfrac{\pi}{2}$ or $\dfrac{3\pi}{2} \qquad x = \dfrac{\pi}{6}$ or $\dfrac{5\pi}{6}$

25. $\tan^2 x - 6\tan x + 4 = 0$

$\qquad \tan x = \dfrac{-(-6) \pm \sqrt{(-6)^2 - 4(1)(4)}}{2(1)}$

$\qquad \tan x = \dfrac{6 \pm \sqrt{20}}{2} = 3 \pm \sqrt{5}$

$\tan x = 3 + \sqrt{5} \qquad$ or $\quad \tan x = 3 - \sqrt{5}$

$x \approx 1.3821$ or $4.5237 \qquad x = 0.6524$ or 3.7940

Chapter 6   Practice Test Solutions

1. $C = 180° - (40° + 12°) = 128°$

$a = \sin 40°\left(\dfrac{100}{\sin 12°}\right) \approx 309.164$

$c = \sin 128°\left(\dfrac{100}{\sin 12°}\right) \approx 379.012$

2. $\sin A = 5\left(\dfrac{\sin 150°}{20}\right) = 0.125$

$A \approx 7.181°$

$B \approx 180° - (150° + 7.181°) = 22.819°$

$b = \sin 22.819°\left(\dfrac{20}{\sin 150°}\right) \approx 15.513$

3. Area $= \frac{1}{2}ab \sin C = \frac{1}{2}(3)(6) \sin 130° \approx 6.894$ square units

4. $h = b \sin A = 35 \sin 22.5° \approx 13.394$

$a = 10$

Since $a < h$ and A is acute, the triangle has no solution.

5. $\cos A = \dfrac{(53)^2 + (38)^2 - (49)^2}{2(53)(38)} \approx 0.4598$

$A \approx 62.627°$

$\cos B = \dfrac{(49)^2 + (38)^2 - (53)^2}{2(49)(38)} \approx 0.2782$

$B \approx 73.847°$

$C \approx 180° - (62.627° + 73.847°)$

$= 43.526°$

6. $c^2 = (100)^2 + (300)^2 - 2(100)(300) \cos 29°$

≈ 47522.8176

$c \approx 218$

$\cos A = \dfrac{(300)^2 + (218)^2 - (100)^2}{2(300)(218)} \approx 0.97495$

$A \approx 12.85°$

$B \approx 180° - (12.85° + 29°) = 138.15°$

7. $s = \dfrac{a + b + c}{2} = \dfrac{4.1 + 6.8 + 5.5}{2} = 8.2$

Area $= \sqrt{s(s - a)(s - b)(s - c)}$

$= \sqrt{8.2(8.2 - 4.1)(8.2 - 6.8)(8.2 - 5.5)}$

≈ 11.273 square units

8. $x^2 = (40)^2 + (70)^2 - 2(40)(70) \cos 168°$

≈ 11977.6266

$x \approx 190.442$ miles

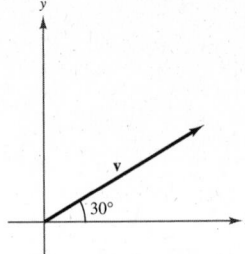

9. $\mathbf{w} = 4(3\mathbf{i} + \mathbf{j}) - 7(-\mathbf{i} + 2\mathbf{j})$

$= 19\mathbf{i} - 10\mathbf{j}$

10. $\dfrac{\mathbf{v}}{\|\mathbf{v}\|} = \dfrac{5\mathbf{i} - 3\mathbf{j}}{\sqrt{25 + 9}} = \dfrac{5}{\sqrt{34}}\mathbf{i} - \dfrac{3}{\sqrt{34}}\mathbf{j}$

$= \dfrac{5\sqrt{34}}{34}\mathbf{i} - \dfrac{3\sqrt{34}}{34}\mathbf{j}$

11. $\mathbf{u} = 6\mathbf{i} + 5\mathbf{j} \qquad \mathbf{v} = 2\mathbf{i} - 3\mathbf{j}$

$\mathbf{u} \cdot \mathbf{v} = 6(2) + 5(-3) = -3$

$\|\mathbf{u}\| = \sqrt{61}, \qquad \|\mathbf{v}\| = \sqrt{13}$

$\cos \theta = \dfrac{-3}{\sqrt{61}\sqrt{13}}$

$\theta \approx 96.116°$

12. $4(\mathbf{i} \cos 30° + \mathbf{j} \sin 30°) = 4\left(\dfrac{\sqrt{3}}{2}\mathbf{i} + \dfrac{1}{2}\mathbf{j}\right)$

$= \langle 2\sqrt{3}, 2 \rangle$

13. $\operatorname{proj}_{\mathbf{v}}\mathbf{u} = \left(\dfrac{\mathbf{u} \cdot \mathbf{v}}{\|\mathbf{v}\|^2}\right)\mathbf{v} = \dfrac{-10}{20}\langle -2, 4 \rangle = \langle 1, -2 \rangle$

14. $r = \sqrt{25 + 25} = \sqrt{50} = 5\sqrt{2}$

$\tan \theta = \dfrac{-5}{5} = -1$

Since z is in Quadrant IV, $\theta = 315°$

$z = 5\sqrt{2}(\cos 315° + i \sin 315°)$.

15. $\cos 225° = -\dfrac{\sqrt{2}}{2}, \quad \sin 225° = -\dfrac{\sqrt{2}}{2}$

$z = 6\left(-\dfrac{\sqrt{2}}{2} - i\dfrac{\sqrt{2}}{2}\right)$

$= -3\sqrt{2} - 3\sqrt{2}i$

16. $[7(\cos 23° + i \sin 23°)][4(\cos 7° + i \sin 7°)] = 7(4)[\cos(23° + 7°) + i \sin(23° + 7°)]$

$$= 28(\cos 30° + i \sin 30°)$$

17. $\dfrac{9\left(\cos \dfrac{5\pi}{4} + i \sin \dfrac{5\pi}{4}\right)}{3(\cos \pi + i \sin \pi)} = \dfrac{9}{3}\left[\cos\left(\dfrac{5\pi}{4} - \pi\right) + i \sin\left(\dfrac{5\pi}{4} - \pi\right)\right] = 3\left(\cos \dfrac{\pi}{4} + i \sin \dfrac{\pi}{4}\right)$

18. $(2 + 2i)^8 = [2\sqrt{2}(\cos 45° + i \sin 45°)]^8 = \left(2\sqrt{2}\right)^8[\cos(8)(45°) + i \sin(8)(45°)]$

$$= 4096[\cos 360° + i \sin 360°] = 4096$$

19. $z = 8\left(\cos \dfrac{\pi}{3} + i \sin \dfrac{\pi}{3}\right),\ n = 3$

The cube roots of z are: $\sqrt[3]{8}\left[\cos \dfrac{\dfrac{\pi}{3} + 2\pi k}{3} + i \sin \dfrac{\dfrac{\pi}{3} + 2\pi k}{3}\right], k = 0, 1, 2$

For $k = 0$, $\sqrt[3]{8}\left[\cos \dfrac{\dfrac{\pi}{3}}{3} + i \sin \dfrac{\dfrac{\pi}{3}}{3}\right] = 2\left(\cos \dfrac{\pi}{9} + i \sin \dfrac{\pi}{9}\right)$

For $k = 1$, $\sqrt[3]{8}\left[\cos \dfrac{\left(\dfrac{\pi}{3}\right) + 2\pi}{3} + i \sin \dfrac{\left(\dfrac{\pi}{3}\right) + 2\pi}{3}\right] = 2\left(\cos \dfrac{7\pi}{9} + i \sin \dfrac{7\pi}{9}\right)$

For $k = 2$, $\sqrt[3]{8}\left[\cos \dfrac{\dfrac{\pi}{3} + 4\pi}{3} + i \sin \dfrac{\dfrac{\pi}{3} + 4\pi}{3}\right] = 2\left(\cos \dfrac{13\pi}{9} + i \sin \dfrac{13\pi}{9}\right)$

20. $x^4 = -i = 1\left(\cos \dfrac{3\pi}{2} + i \sin \dfrac{3\pi}{2}\right)$

The fourth roots are: $\sqrt[4]{1}\left[\cos \dfrac{\left(\dfrac{3\pi}{2}\right) + 2\pi k}{4} + i \sin \dfrac{\left(\dfrac{3\pi}{2}\right) + 2\pi k}{4}\right], k = 0, 1, 2, 3$

For $k = 0$, $\cos \dfrac{\dfrac{3\pi}{2}}{4} + i \sin \dfrac{\dfrac{3\pi}{2}}{4} = \cos \dfrac{3\pi}{8} + i \sin \dfrac{3\pi}{8}$

For $k = 1$, $\cos \dfrac{\dfrac{3\pi}{2} + 2\pi}{4} + i \sin \dfrac{\dfrac{3\pi}{2} + 2\pi}{4} = \cos \dfrac{7\pi}{8} + i \sin \dfrac{7\pi}{8}$

For $k = 2$, $\cos \dfrac{\dfrac{3\pi}{2} + 4\pi}{4} + i \sin \dfrac{\dfrac{3\pi}{2} + 4\pi}{4} = \cos \dfrac{11\pi}{8} + i \sin \dfrac{11\pi}{8}$

For $k = 3$, $\cos \dfrac{\dfrac{3\pi}{2} + 6\pi}{4} + i \sin \dfrac{\dfrac{3\pi}{2} + 6\pi}{4} = \cos \dfrac{15\pi}{8} + i \sin \dfrac{15\pi}{8}$

Chapter 7   Practice Test Solutions

1. $\begin{cases} x + y = 1 \\ 3x - y = 15 \end{cases} \implies y = 3x - 15$

$x + (3x - 15) = 1$

$\qquad\qquad 4x = 16$

$\qquad\qquad\ x = 4$

$\qquad\qquad\ y = -3$

Solution: $(4, -3)$

2. $\begin{cases} x - 3y = -3 \implies x = 3y - 3 \\ x^2 + 6y = \ \ 5 \end{cases}$

$\qquad (3y - 3)^2 + 6y = 5$

$\qquad 9y^2 - 18y + 9 + 6y = 5$

$\qquad\qquad 9y^2 - 12y + 4 = 0$

$\qquad\qquad\quad (3y - 2)^2 = 0$

$\qquad\qquad\qquad\qquad y = \frac{2}{3}$

$\qquad\qquad\qquad\qquad x = -1$

Solution: $\left(-1, \frac{2}{3}\right)$

3. $\begin{cases} x + \ y + z = \ \ \ 6 \implies z = 6 - x - y \\ 2x - \ y + 3z = \ \ \ 0 \implies 2x - y + 3(6 - x - y) = \ \ \ 0 \implies -x - 4y = -18 \implies x = 18 - 4y \\ 5x + 2y - \ z = -3 \implies 5x + 2y - (6 - x - y) = -3 \implies 6x + 3y = \ \ \ 3 \end{cases}$

$6(18 - 4y) + 3y = 3$

$\qquad\qquad -21y = -105$

$\qquad\qquad\quad\ y = 5$

$\qquad\qquad\quad\ x = 18 - 4y = -2$

$\qquad\qquad\quad\ z = 6 - x - y = 3$

Solution: $(-2, 5, 3)$

4. $x + y = 110 \implies y = 110 - x$

$\qquad xy = 2800$

$\ x(110 - x) = 2800$

$\qquad\quad 0 = x^2 - 110x + 2800$

$\qquad\quad 0 = (x - 40)(x - 70)$

$x = 40 \ \ \text{or} \ \ x = 70$

$y = 70 \qquad\ y = 40$

Solution: The two numbers are 40 and 70.

5. $2x + 2y = 170 \implies y = \dfrac{170 - 2x}{2} = 85 - x$

$\qquad\qquad xy = 1500$

$\quad x(85 - x) = 1500$

$\qquad\qquad 0 = x^2 - 85x + 1500$

$\qquad\qquad 0 = (x - 25)(x - 60)$

$x = 25 \ \ \text{or} \ \ x = 60$

$y = 60 \qquad\ y = 25$

Dimensions: 60 ft $\times$ 25 ft

6. $\begin{cases} 2x + 15y = \ \ 4 \implies 2x + 15y = \ \ \ \ 4 \\ \ x - \ 3y = 23 \implies \underline{5x - 15y = \ \ 115} \end{cases}$

$\qquad\qquad\qquad\qquad\qquad 7x \quad\ \ = 119$

$\qquad\qquad\qquad\qquad\qquad\ x = \ \ 17$

$\qquad\qquad\qquad\qquad\qquad\ y = \dfrac{x - 23}{3}$

$\qquad\qquad\qquad\qquad\qquad\ \ = \ \ -2$

Solution: $(17, -2)$

7. $\begin{cases} x + \ y = 2 \implies 19x + 19y = 38 \\ 38x - 19y = 7 \implies \underline{38x - 19y = \ \ 7} \end{cases}$

$\qquad\qquad\qquad\qquad\qquad 57x \qquad = 45$

$x = \dfrac{45}{57} = \dfrac{15}{19}$

$y = 2 - x = \dfrac{38}{19} - \dfrac{15}{19} = \dfrac{23}{19}$

Solution: $\left(\dfrac{15}{19}, \dfrac{23}{19}\right)$

8. $\begin{cases} 0.4x + 0.5y = 0.112 \\ 0.3x - 0.7y = -0.131 \end{cases} \Rightarrow \begin{aligned} 0.28x + 0.35y = 0.0784 \\ 0.15x - 0.35y = -0.0655 \end{aligned}$

$$0.43x \quad\quad\quad = 0.0129$$

$$x = \frac{0.0129}{0.43} = 0.03$$

$$y = \frac{0.112 - 0.4x}{0.5} = 0.20$$

Solution: $(0.03, 0.20)$

9. Let x = amount in 11% fund and y = amount in 13% fund.

$$x + y = 17000 \implies y = 17000 - x$$

$$0.11x + 0.13y = 2080$$

$$0.11x + 0.13(17000 - x) = 2080$$

$$-0.02x = -130$$

$$x = \$6500 \quad \text{at } 11\%$$

$$y = \$10,500 \text{ at } 13\%$$

10. $(4, 3), (1, 1), (-1, -2), (-2, -1)$

Use a calculator.

$$y = ax + b = \tfrac{11}{14}x - \tfrac{1}{7}$$

11. $\begin{cases} x + y \quad\quad = -2 \\ 2x - y + z = 11 \\ \quad\quad 4y - 3z = -20 \end{cases}$

$\begin{cases} x + y \quad\quad = -2 \\ \quad -3y + z = 15 \\ \quad\quad 4y - 3z = -20 \end{cases}$ $-2\text{Eq.1} + \text{Eq.2}$

$\begin{cases} x + y \quad\quad = -2 \\ \quad y - 2z = -5 \\ \quad 4y - 3z = -20 \end{cases}$ $\text{Eq.3} + \text{Eq.2}$

$\begin{cases} x + y \quad\quad = -2 \\ \quad y - 2z = -5 \\ \quad\quad 5z = 0 \end{cases}$ $-4\text{Eq.2} + \text{Eq.3}$

$\begin{cases} x + y \quad\quad = -2 \\ \quad y - 2z = -5 \\ \quad\quad z = 0 \end{cases}$

$$y - 2(0) = -5 \implies y = -5$$
$$x + (-5) = -2 \implies x = 3$$

Solution: $(3, -5, 0)$

12. $\begin{cases} 4x - y + 5z = 4 \\ 2x + y - z = 0 \\ 2x + 4y + 8z = 0 \end{cases}$

$\begin{cases} 2x + 4y + 8z = 0 \\ 2x + y - z = 0 \\ 4x - y + 5z = 4 \end{cases}$ Interchange equations.

$\begin{cases} 2x + 4y + 8z = 0 \\ \quad -3y - 9z = 0 \\ \quad -9y - 11z = 4 \end{cases}$ $\begin{aligned} -\text{Eq.1} + \text{Eq.2} \\ -2\text{Eq.1} + \text{Eq.3} \end{aligned}$

$\begin{cases} 2x + 4y + 8z = 0 \\ \quad -3y - 9z = 0 \\ \quad\quad 16z = 4 \end{cases}$ $-3\text{Eq.2} + \text{Eq.3}$

$\begin{cases} x + 2y + 4z = 0 \\ \quad y + 3z = 0 \\ \quad\quad z = \tfrac{1}{4} \end{cases}$ $\begin{aligned} \tfrac{1}{2}\text{Eq.1} \\ -\tfrac{1}{3}\text{Eq.2} \\ \tfrac{1}{16}\text{Eq.3} \end{aligned}$

$$y + 3\left(\tfrac{1}{4}\right) = 0 \implies y = -\tfrac{3}{4}$$
$$x + 2\left(-\tfrac{3}{4}\right) + 4\left(\tfrac{1}{4}\right) = 0 \implies x = \tfrac{1}{2}$$

Solution: $\left(-\tfrac{1}{2}, -\tfrac{3}{4}, \tfrac{1}{4}\right)$

13. $\begin{cases} 3x + 2y - z = 5 \\ 6x - y + 5z = 2 \end{cases}$

$\begin{cases} 3x + 2y - z = 5 \\ \quad -5y + 7z = -8 \end{cases}$ $-2\text{Eq.1} + \text{Eq.2}$

$\begin{cases} x + \tfrac{2}{3}y - \tfrac{1}{3}z = \tfrac{5}{3} \\ \quad y - \tfrac{7}{5}z = \tfrac{8}{5} \end{cases}$ $\begin{aligned} \tfrac{1}{3}\text{Eq.1} \\ -\tfrac{1}{5}\text{Eq.2} \end{aligned}$

Let $a = z$.

Then $y = \tfrac{7}{5}a + \tfrac{8}{5}$, and

$$x + \tfrac{2}{3}\left(\tfrac{7}{5}a + \tfrac{8}{5}\right) - \tfrac{1}{3}a = \tfrac{5}{3}$$

$$x + \tfrac{3}{5}a = \tfrac{3}{5}$$

$$x = -\tfrac{3}{5}a + \tfrac{3}{5}.$$

Solution: $\left(-\tfrac{3}{5}a + \tfrac{3}{5}, \tfrac{7}{5}a + \tfrac{8}{5}, a\right)$, where a is any real number.

14. $y = ax^2 + bx + c$ passes through $(0, -1)$, $(1, 4)$, and $(2, 13)$.

At $(0, -1)$: $-1 = a(0)^2 + b(0) + c \implies c = -1$

At $(1, 4)$: $4 = a(1)^2 + b(1) - 1 \implies 5 = a + b \implies 5 = a + b$

At $(2, 13)$: $13 = a(2)^2 + b(2) - 1 \implies 14 = 4a + 2b \implies -7 = -2a - b$

$$-2 = -a$$
$$a = 2$$
$$b = 3$$

Thus, the equation of the parabola is $y = 2x^2 + 3x - 1$.

15. $s = \frac{1}{2}at^2 + v_0t + s_0$ passes through $(1, 12)$, $(2, 5)$, and $(3, 4)$.

At $(1, 12)$: $12 = \frac{1}{2}a + v_0 + s_0$

At $(2, 5)$: $5 = 2a + 2v_0 + s_0$

At $(3, 4)$: $4 = \frac{9}{2}a + 3v_0 + s_0$

$$\begin{cases} a + 2v_0 + 2s_0 = 24 \\ 2a + 2v_0 + s_0 = 5 \\ 9a + 6v_0 + 2s_0 = 8 \end{cases}$$

$$\begin{cases} a + 2v_0 + 2s_0 = 24 \\ -2v_0 - 3s_0 = -43 \quad -2\text{Eq.1} + \text{Eq.2} \\ -12v_0 - 16s_0 = -208 \quad -9\text{Eq.1} + \text{Eq.3} \end{cases}$$

$$\begin{cases} a + 2v_0 + 2s_0 = 24 \\ -2v_0 - 3s_0 = -43 \\ 2s_0 = 50 \quad -6\text{Eq.2} + \text{Eq.3} \end{cases}$$

$$\begin{cases} a + 2v_0 + 2s_0 = 24 \\ v_0 + \frac{3}{2}s_0 = \frac{43}{2} \quad -\frac{1}{2}\text{Eq.2} \\ s_0 = 25 \quad \frac{1}{2}\text{Eq.3} \end{cases}$$

$$v_0 + \frac{3}{2}(25) = \frac{43}{2} \implies v_0 = -16$$
$$a + 2(-16) + 2(25) = 24 \implies a = 6$$

Thus, $s = \frac{1}{2}(6)t^2 - 16t + 25 = 3t^2 - 16t + 25$.

16. $x^2 + y^2 \geq 9$

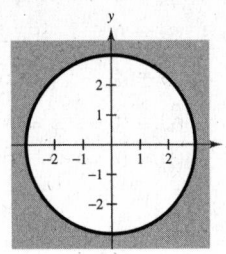

17. $\begin{cases} x + y \leq 6 \\ x \geq 2 \\ y \geq 0 \end{cases}$

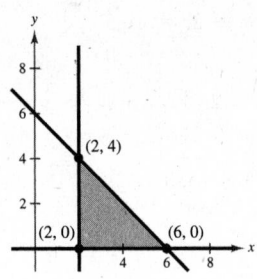

18. Line through $(0, 0)$ and $(0, 7)$:

$x = 0$

Line through $(0, 0)$ and $(2, 3)$:

$y = \frac{3}{2}x$ or $3x - 2y = 0$

Line through $(0, 7)$ and $(2, 3)$:

$y = -2x + 7$ or $2x + y = 7$

Inequalities: $\begin{cases} x \geq 0 \\ 3x - 2y \leq 0 \\ 2x + y \leq 7 \end{cases}$

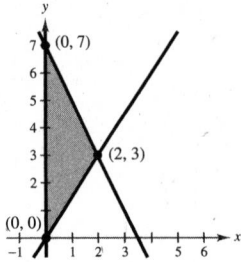

19. Vertices: $(0, 0)$, $(0, 7)$, $(6, 0)$, $(3, 5)$

$z = 30x + 26y$

At $(0, 0)$: $z = 0$

At $(0, 7)$: $z = 182$

At $(6, 0)$: $z = 180$

At $(3, 5)$: $z = 220$

The maximum value
of z occurs at $(3, 5)$
and is 220.

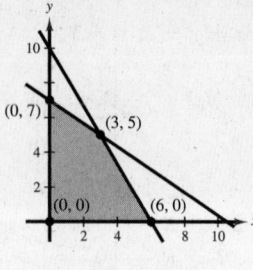

20.
$$x^2 + y^2 \leq 4$$
$$(x - 2)^2 + y^2 \geq 4$$

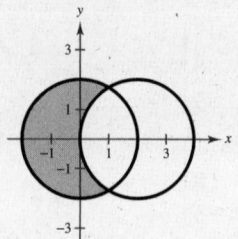

21. $\dfrac{1 - 2x}{x^2 + x} = \dfrac{1 - 2x}{x(x + 1)} = \dfrac{A}{x} + \dfrac{B}{x + 1}$

$1 - 2x = A(x + 1) + Bx$

When $x = 0$, $1 = A$.

When $x = -1$, $3 = -B \implies B = -3$.

$\dfrac{1 - 2x}{x^2 + x} = \dfrac{1}{x} - \dfrac{3}{x + 1}$

22. $\dfrac{6x - 17}{(x - 3)^2} = \dfrac{A}{x - 3} + \dfrac{B}{(x - 3)^2}$

$6x - 17 = A(x - 3) + B$

When $x = 3$, $1 = B$.

When $x = 0$, $-17 = -3A + B \implies A = 6$.

$\dfrac{6x - 17}{(x - 3)^2} = \dfrac{6}{x - 3} + \dfrac{1}{(x - 3)^2}$

Chapter 8   Practice Test Solutions

1.
$$\begin{bmatrix} 1 & -2 & 4 \\ 3 & -5 & 9 \end{bmatrix}$$

$-3R_1 + R_2 \rightarrow \begin{bmatrix} 1 & -2 & 4 \\ 0 & 1 & -3 \end{bmatrix}$

$2R_2 + R_1 \rightarrow \begin{bmatrix} 1 & 0 & -2 \\ 0 & 1 & -3 \end{bmatrix}$

2. $\begin{cases} 3x + 5y = 3 \\ 2x - y = -11 \end{cases}$

$$\begin{bmatrix} 3 & 5 & \vdots & 3 \\ 2 & -1 & \vdots & -11 \end{bmatrix}$$

$-R_2 + R_1 \rightarrow \begin{bmatrix} 1 & 6 & \vdots & 14 \\ 2 & -1 & \vdots & -11 \end{bmatrix}$

$-2R_1 + R_2 \rightarrow \begin{bmatrix} 1 & 6 & \vdots & 14 \\ 0 & -13 & \vdots & -39 \end{bmatrix}$

$-\frac{1}{13}R_2 \rightarrow \begin{bmatrix} 1 & 6 & \vdots & 14 \\ 0 & 1 & \vdots & 3 \end{bmatrix}$

$-6R_2 + R_1 \rightarrow \begin{bmatrix} 1 & 0 & \vdots & -4 \\ 0 & 1 & \vdots & 3 \end{bmatrix}$

$x = -4$, $y = 3$

Solution: $(-4, 3)$

3. $\begin{cases} 2x + 3y = -3 \\ 3x - 2y = 8 \\ x + y = 1 \end{cases}$

$$\begin{bmatrix} 2 & 3 & \vdots & -3 \\ 3 & 2 & \vdots & 8 \\ 1 & 1 & \vdots & 1 \end{bmatrix}$$

$$\begin{matrix} R_3 \to \\ \\ R_1 \to \end{matrix} \begin{bmatrix} 1 & 1 & \vdots & 1 \\ 3 & 2 & \vdots & 8 \\ 2 & 3 & \vdots & -3 \end{bmatrix}$$

$$\begin{matrix} -3R_1 + R_2 \to \\ -2R_1 + R_3 \to \end{matrix} \begin{bmatrix} 1 & 1 & \vdots & 1 \\ 0 & -1 & \vdots & 5 \\ 0 & 1 & \vdots & -5 \end{bmatrix}$$

$$-R_2 \to \begin{bmatrix} 1 & 1 & \vdots & 1 \\ 0 & 1 & \vdots & -5 \\ 0 & 1 & \vdots & -5 \end{bmatrix}$$

$$\begin{matrix} -R_2 + R_1 \to \\ \\ -R_2 + R_3 \to \end{matrix} \begin{bmatrix} 1 & 0 & \vdots & 6 \\ 0 & 1 & \vdots & -5 \\ 0 & 0 & \vdots & 0 \end{bmatrix}$$

$x = 6, y = -5$

Solution: $(6, -5)$

4. $\begin{cases} x \quad\ + 3z = -5 \\ 2x + y \quad\ = 0 \\ 3x + y - z = -3 \end{cases}$

$$\begin{bmatrix} 1 & 0 & 3 & \vdots & -5 \\ 2 & 1 & 0 & \vdots & 0 \\ 3 & 1 & -1 & \vdots & 3 \end{bmatrix}$$

$$\begin{matrix} -2R_1 + R_2 \to \\ -3R_1 + R_3 \to \end{matrix} \begin{bmatrix} 1 & 0 & 3 & \vdots & -5 \\ 0 & 1 & -6 & \vdots & 10 \\ 0 & 1 & -10 & \vdots & 18 \end{bmatrix}$$

$$-R_2 + R_3 \to \begin{bmatrix} 1 & 0 & 3 & \vdots & -5 \\ 0 & 1 & -6 & \vdots & 10 \\ 0 & 0 & -4 & \vdots & 8 \end{bmatrix}$$

$$-\tfrac{1}{4}R_3 \to \begin{bmatrix} 1 & 0 & 3 & \vdots & -5 \\ 0 & 1 & -6 & \vdots & 10 \\ 0 & 0 & 1 & \vdots & -2 \end{bmatrix}$$

$$\begin{matrix} -3R_3 + R_1 \to \\ 6R_3 + R_2 \to \end{matrix} \begin{bmatrix} 1 & 0 & 0 & \vdots & 1 \\ 0 & 1 & 0 & \vdots & -2 \\ 0 & 0 & 1 & \vdots & -2 \end{bmatrix}$$

$x = 1, y = -2, z = -2$

Solution: $(1, -2, -2)$

5. $\begin{bmatrix} 1 & 4 & 5 \\ 2 & 0 & -3 \end{bmatrix} \begin{bmatrix} 1 & 6 \\ 0 & -7 \\ -1 & 2 \end{bmatrix} = \begin{bmatrix} (1)(1) + (4)(0) + (5)(-1) & (1)(6) + (4)(-7) + (5)(2) \\ (2)(1) + (0)(0) + (-3)(-1) & (2)(6) + (0)(-7) + (-3)(2) \end{bmatrix} = \begin{bmatrix} -4 & -12 \\ 5 & 6 \end{bmatrix}$

6. $3A - 5B = 3\begin{bmatrix} 9 & 1 \\ -4 & 8 \end{bmatrix} - 5\begin{bmatrix} 6 & -2 \\ 3 & 5 \end{bmatrix}$

$= \begin{bmatrix} 27 & 3 \\ -12 & 24 \end{bmatrix} - \begin{bmatrix} 30 & -10 \\ 15 & 25 \end{bmatrix}$

$= \begin{bmatrix} -3 & 13 \\ -27 & -1 \end{bmatrix}$

7. $f(A) = \begin{bmatrix} 3 & 0 \\ 7 & 1 \end{bmatrix}^2 - 7\begin{bmatrix} 3 & 0 \\ 7 & 1 \end{bmatrix} + 8\begin{bmatrix} 1 & 0 \\ 0 & 1 \end{bmatrix}$

$= \begin{bmatrix} 3 & 0 \\ 7 & 1 \end{bmatrix}\begin{bmatrix} 3 & 0 \\ 7 & 1 \end{bmatrix} - \begin{bmatrix} 21 & 0 \\ 49 & 7 \end{bmatrix} + \begin{bmatrix} 8 & 0 \\ 0 & 8 \end{bmatrix}$

$= \begin{bmatrix} 9 & 0 \\ 28 & 1 \end{bmatrix} - \begin{bmatrix} 21 & 0 \\ 49 & 7 \end{bmatrix} + \begin{bmatrix} 8 & 0 \\ 0 & 8 \end{bmatrix}$

$= \begin{bmatrix} -4 & 0 \\ -21 & 2 \end{bmatrix}$

8. False since

$(A + B)(A + 3B) = A(A + 3B) + B(A + 3B)$

$= A^2 + 3AB + BA + 3B^2$ and, in general, $AB \neq BA$.

9.
$$\begin{bmatrix} 1 & 2 & \vdots & 1 & 0 \\ 3 & 5 & \vdots & 0 & 1 \end{bmatrix}$$

$$-3R_1 + R_2 \rightarrow \begin{bmatrix} 1 & 2 & \vdots & 1 & 0 \\ 0 & -1 & \vdots & -3 & 1 \end{bmatrix}$$

$$2R_2 + R_1 \rightarrow \begin{bmatrix} 1 & 0 & \vdots & -5 & 2 \\ 0 & -1 & \vdots & -3 & 1 \end{bmatrix}$$

$$-R_2 \rightarrow \begin{bmatrix} 1 & 0 & \vdots & -5 & 2 \\ 0 & 1 & \vdots & 3 & -1 \end{bmatrix}$$

$$A^{-1} = \begin{bmatrix} -5 & 2 \\ 3 & -1 \end{bmatrix}$$

10.
$$\begin{bmatrix} 1 & 1 & 1 & \vdots & 1 & 0 & 0 \\ 3 & 6 & 5 & \vdots & 0 & 1 & 0 \\ 6 & 10 & 8 & \vdots & 0 & 0 & 1 \end{bmatrix}$$

$$\begin{array}{l} -3R_1 + R_2 \rightarrow \\ -6R_1 + R_3 \rightarrow \end{array} \begin{bmatrix} 1 & 1 & 1 & \vdots & 1 & 0 & 0 \\ 0 & 3 & 2 & \vdots & -3 & 1 & 0 \\ 0 & 4 & 2 & \vdots & -6 & 0 & 1 \end{bmatrix}$$

$$-R_3 + R_2 \rightarrow \begin{bmatrix} 1 & 1 & 1 & \vdots & 1 & 0 & 0 \\ 0 & -1 & 0 & \vdots & 3 & 1 & -1 \\ 0 & 4 & 2 & \vdots & -6 & 0 & 1 \end{bmatrix}$$

$$\begin{array}{l} R_2 + R_1 \rightarrow \\ 4R_2 + R_3 \rightarrow \end{array} \begin{bmatrix} 1 & 0 & 1 & \vdots & 4 & 1 & -1 \\ 0 & -1 & 0 & \vdots & 3 & 1 & -1 \\ 0 & 0 & 2 & \vdots & 6 & 4 & -3 \end{bmatrix}$$

$$\begin{array}{l} -R_2 \rightarrow \\ \frac{1}{2}R_3 \rightarrow \end{array} \begin{bmatrix} 1 & 0 & 1 & \vdots & 4 & 1 & -1 \\ 0 & 1 & 0 & \vdots & -3 & -1 & 1 \\ 0 & 0 & 1 & \vdots & 3 & 2 & -\frac{3}{2} \end{bmatrix}$$

$$-R_3 + R_1 \rightarrow \begin{bmatrix} 1 & 0 & 0 & \vdots & 1 & -1 & \frac{1}{2} \\ 0 & 1 & 0 & \vdots & -3 & -1 & 1 \\ 0 & 0 & 1 & \vdots & 3 & 2 & -\frac{3}{2} \end{bmatrix}$$

$$A^{-1} = \begin{bmatrix} 1 & -1 & \frac{1}{2} \\ -3 & -1 & 1 \\ 3 & 2 & -\frac{3}{2} \end{bmatrix}$$

11. (a) $\begin{cases} x + 2y = 4 \\ 3x + 5y = 1 \end{cases}$

$$A = \begin{bmatrix} 1 & 2 \\ 3 & 5 \end{bmatrix}$$

$$A^{-1} = \frac{1}{5-6}\begin{bmatrix} 5 & -2 \\ -3 & 1 \end{bmatrix} = \begin{bmatrix} -5 & 2 \\ 3 & -1 \end{bmatrix}$$

$$\begin{bmatrix} x \\ y \end{bmatrix} = A^{-1}B = \begin{bmatrix} -5 & 2 \\ 3 & -1 \end{bmatrix}\begin{bmatrix} 4 \\ 1 \end{bmatrix} = \begin{bmatrix} -18 \\ 11 \end{bmatrix}$$

$x = -18, y = 11$

Solution: $(-18, 11)$

(b) $\begin{cases} x + 2y = 3 \\ 3x + 5y = -2 \end{cases}$

Again, $A^{-1} = \begin{bmatrix} -5 & 2 \\ 3 & -1 \end{bmatrix}$.

$$\begin{bmatrix} x \\ y \end{bmatrix} = A^{-1}B = \begin{bmatrix} -5 & 2 \\ 3 & -1 \end{bmatrix}\begin{bmatrix} 3 \\ -2 \end{bmatrix} = \begin{bmatrix} -19 \\ 11 \end{bmatrix}$$

$x = -19, y = 11$

Solution: $(-19, 11)$

12. $\begin{vmatrix} 6 & -1 \\ 3 & 4 \end{vmatrix} = 24 - (-3) = 27$

13. $\begin{vmatrix} 1 & 3 & -1 \\ 5 & 9 & 0 \\ 6 & 2 & -5 \end{vmatrix} = -1\begin{vmatrix} 5 & 9 \\ 6 & 2 \end{vmatrix} - 5\begin{vmatrix} 1 & 3 \\ 5 & 9 \end{vmatrix} = -(-44) - 5(-6) = 74$

14. Expand along Row 2.

$$\begin{vmatrix} 1 & 4 & 2 & 3 \\ 0 & 1 & -2 & 0 \\ 3 & 5 & -1 & 1 \\ 2 & 0 & 6 & 1 \end{vmatrix} = \begin{vmatrix} 1 & 2 & 3 \\ 3 & -1 & 1 \\ 2 & 6 & 1 \end{vmatrix} + 2\begin{vmatrix} 1 & 4 & 3 \\ 3 & 5 & 1 \\ 2 & 0 & 1 \end{vmatrix}$$

$$= 51 + 2(-29) = -7$$

15.
$$\begin{vmatrix} 6 & 4 & 3 & 0 & 6 \\ 0 & 5 & 1 & 4 & 8 \\ 0 & 0 & 2 & 7 & 3 \\ 0 & 0 & 0 & 9 & 2 \\ 0 & 0 & 0 & 0 & 1 \end{vmatrix} = 6 \begin{vmatrix} 5 & 1 & 4 & 8 \\ 0 & 2 & 7 & 3 \\ 0 & 0 & 9 & 2 \\ 0 & 0 & 0 & 1 \end{vmatrix} = 6(5) \begin{vmatrix} 2 & 7 & 3 \\ 0 & 9 & 2 \\ 0 & 0 & 1 \end{vmatrix} = 6(5)(2) \begin{vmatrix} 9 & 2 \\ 0 & 1 \end{vmatrix} = 6(5)(2)(9) = 540$$

16. Area $= \dfrac{1}{2} \begin{vmatrix} 0 & 7 & 1 \\ 5 & 0 & 1 \\ 3 & 9 & 1 \end{vmatrix} = \dfrac{1}{2}(31) = \dfrac{31}{2}$

17. $\begin{vmatrix} x & y & 1 \\ 2 & 7 & 1 \\ -1 & 4 & 1 \end{vmatrix} = 3x - 3y + 15 = 0$ or, equivalently, $x - y + 5 = 0$

18. $x = \dfrac{\begin{vmatrix} 4 & -7 \\ 11 & 5 \end{vmatrix}}{\begin{vmatrix} 6 & -7 \\ 2 & 5 \end{vmatrix}} = \dfrac{97}{44}$

19. $z = \dfrac{\begin{vmatrix} 3 & 0 & 1 \\ 0 & 1 & 3 \\ 1 & -1 & 2 \end{vmatrix}}{\begin{vmatrix} 3 & 0 & 1 \\ 0 & 1 & 4 \\ 1 & -1 & 0 \end{vmatrix}} = \dfrac{14}{11}$

20. $y = \dfrac{\begin{vmatrix} 721.4 & 33.77 \\ 45.9 & 19.85 \end{vmatrix}}{\begin{vmatrix} 721.4 & -29.1 \\ 45.9 & 105.6 \end{vmatrix}} = \dfrac{12,769.747}{77,515.530} \approx 0.1647$

Chapter 9   Practice Test Solutions

1. $a_n = \dfrac{2n}{(n+2)!}$

$a_1 = \dfrac{2(1)}{3!} = \dfrac{2}{6} = \dfrac{1}{3}$

$a_2 = \dfrac{2(2)}{4!} = \dfrac{4}{24} = \dfrac{1}{6}$

$a_3 = \dfrac{2(3)}{5!} = \dfrac{6}{120} = \dfrac{1}{20}$

$a_4 = \dfrac{2(4)}{6!} = \dfrac{8}{720} = \dfrac{1}{90}$

$a_5 = \dfrac{2(5)}{7!} = \dfrac{10}{5040} = \dfrac{1}{504}$

Terms: $\dfrac{1}{3}, \dfrac{1}{6}, \dfrac{1}{20}, \dfrac{1}{90}, \dfrac{1}{504}$

2. $a_n = \dfrac{n+3}{3^n}$

3. $\displaystyle\sum_{i=1}^{6} (2i - 1) = 1 + 3 + 5 + 7 + 9 + 11 = 36$

4. $a_1 = 23, d = -2$

$a_2 = 23 + (-2) = 21$

$a_3 = 21 + (-2) = 19$

$a_4 = 19 + (-2) = 17$

$a_5 = 17 + (-2) = 15$

Terms: 23, 21, 19, 17, 15

5. $a_1 = 12, d = 3, n = 50$

$a_n = a_1 + (n - 1)d$

$a_{50} = 12 + (50 - 1)3 = 159$

6. $a_1 = 1$

$a_{200} = 200$

$S_n = \dfrac{n}{2}(a_1 + a_n)$

$S_{200} = \dfrac{200}{2}(1 + 200) = 20{,}100$

7. $a_1 = 7, r = 2$

$a_2 = 7(2) = 14$

$a_3 = 7(2)^2 = 28$

$a_4 = 7(2)^3 = 56$

$a_5 = 7(2)^4 = 112$

Terms: 7, 14, 28, 56, 112

8. $\displaystyle\sum_{n=1}^{10} 6\left(\dfrac{2}{3}\right)^{n-1}, a_1 = 6, r = \dfrac{2}{3}, n = 10$

$S_n = \dfrac{a_1(1 - r^n)}{1 - r} = \dfrac{6\left[1 - \left(\frac{2}{3}\right)^{10}\right]}{1 - \frac{2}{3}} = 18\left(1 - \dfrac{1024}{59{,}049}\right) = \dfrac{116{,}050}{6561} \approx 17.6879$

9. $\displaystyle\sum_{n=0}^{\infty} (0.03)^n = \sum_{n=1}^{\infty} (0.03)^{n-1}, a_1 = 1, r = 0.03$

$S = \dfrac{a_1}{1 - r} = \dfrac{1}{1 - 0.03} = \dfrac{1}{0.97} = \dfrac{100}{97} \approx 1.0309$

10. For $n = 1, 1 = \dfrac{1(1 + 1)}{2}$.

Assume that $S_k = 1 + 2 + 3 + 4 + \cdots + k = \dfrac{k(k + 1)}{2}$.

Then $S_{k+1} = 1 + 2 + 3 + 4 + \cdots + k + (k + 1) = \dfrac{k(k + 1)}{2} + k + 1$

$= \dfrac{k(k + 1)}{2} + \dfrac{2(k + 1)}{2}$

$= \dfrac{(k + 1)(k + 2)}{2}$.

Thus, by the principle of mathematical induction, $1 + 2 + 3 + 4 + \cdots + n = \dfrac{n(n + 1)}{2}$ for all integers $n \geq 1$.

11. For $n = 4, 4! > 2^4$. Assume that $k! > 2^k$.

Then $(k + 1)! = (k + 1)(k!) > (k + 1)2^k > 2 \cdot 2^k = 2^{k+1}$.

Thus, by the extended principle of mathematical induction, $n! > 2^n$ for all integers $n \geq 4$.

12. $_{13}C_4 = \dfrac{13!}{(13-4)!4!} = 715$

13. $(x+3)^5 = x^5 + 5x^4(3) + 10x^3(3)^2 + 10x^2(3)^3 + 5x(3)^4 + (3)^5$

$\qquad\qquad = x^5 + 15x^4 + 90x^3 + 270x^2 + 405x + 243$

14. $-_{12}C_5x^7(2)^5 = -25{,}344x^7$

15. $_{30}P_4 = \dfrac{30!}{(30-4)!} = 657{,}720$

16. $6! = 720$ ways

17. $_{12}P_3 = 1320$

18. $P(2) + P(3) + P(4) = \dfrac{1}{36} + \dfrac{2}{36} + \dfrac{3}{36}$

$\qquad\qquad\qquad\qquad\quad = \dfrac{6}{36} = \dfrac{1}{6}$

19. $P(K, B10) = \dfrac{4}{52} \cdot \dfrac{2}{51} = \dfrac{2}{663}$

20. Let A = probability of no faulty units.

$\quad P(A) = \left(\dfrac{997}{1000}\right)^{50} \approx 0.8605$

$\quad P(A') = 1 - P(A) \approx 0.1395$

Chapter 10   Practice Test Solutions

1. $3x + 4y = 12 \implies y = -\dfrac{3}{4}x + 3 \implies m_1 = -\dfrac{3}{4}$

$\quad 4x - 3y = 12 \implies y = \dfrac{4}{3}x - 4 \implies m_2 = \dfrac{4}{3}$

$\quad \tan\theta = \left|\dfrac{(4/3) - (-3/4)}{1 + (4/3)(-3/4)}\right| = \left|\dfrac{25/12}{0}\right|$

Since $\tan\theta$ is undefined, the lines are perpendicular
(note that $m_2 = -1/m_1$) and $\theta = 90°$.

2. $x_1 = 5, x_2 = -9, A = 3, B = -7, C = -21$

$\quad d = \dfrac{|3(5) + (-7)(-9) + (-21)|}{\sqrt{3^2 + (-7)^2}} = \dfrac{57}{\sqrt{58}} \approx 7.484$

3. $x^2 - 6x - 4y + 1 = 0$

$\qquad x^2 - 6x + 9 = 4y - 1 + 9$

$\qquad\quad (x - 3)^2 = 4y + 8$

$\qquad\quad (x - 3)^2 = 4(1)(y + 2) \implies p = 1$

Vertex: $(3, -2)$

Focus: $(3, -1)$

Directrix: $y = -3$

4. Vertex: $(2, -5)$

Focus: $(2, -6)$

Vertical axis; opens downward with $p = -1$

$\qquad\quad (x - h)^2 = 4p(y - k)$

$\qquad\quad (x - 2)^2 = 4(-1)(y + 5)$

$\qquad\; x^2 - 4x + 4 = -4y - 20$

$\quad x^2 - 4x + 4y + 24 = 0$

5. $x^2 + 4y^2 - 2x + 32y + 61 = 0$

$(x^2 - 2x + 1) + 4(y^2 + 8y + 16) = -61 + 1 + 64$

$(x - 1)^2 + 4(y + 4)^2 = 4$

$$\frac{(x - 1)^2}{4} + \frac{(y + 4)^2}{1} = 1$$

$a = 2, b = 1, c = \sqrt{3}$

Horizontal major axis

Center: $(1, -4)$

Foci: $\left(1 \pm \sqrt{3}, -4\right)$

Vertices: $(3, -4), (-1, -4)$

Eccentricity: $e = \dfrac{\sqrt{3}}{2}$

6. Vertices: $(0, \pm 6)$

Eccentricity: $e = \dfrac{1}{2}$

Center: $(0, 0)$

Vertical major axis

$a = 6, e = \dfrac{c}{a} = \dfrac{c}{6} = \dfrac{1}{2} \Rightarrow c = 3$

$b^2 = (6)^2 - (3)^2 = 27$

$$\frac{x^2}{27} + \frac{y^2}{36} = 1$$

7. $16y^2 - x^2 - 6x - 128y + 231 = 0$

$16(y^2 - 8y + 16) - (x^2 + 6x + 9) = -231 + 256 - 9$

$16(y - 4)^2 - (x + 3)^2 = 16$

$$\frac{(y - 4)^2}{1} - \frac{(x + 3)^2}{16} = 1$$

$a = 1, b = 4, c = \sqrt{17}$

Center: $(-3, 4)$

Vertical transverse axis

Vertices: $(-3, 5), (-3, 3)$

Foci: $\left(-3, 4 \pm \sqrt{17}\right)$

Asymptotes: $y = 4 \pm \dfrac{1}{4}(x + 3)$

8. Vertices: $(\pm 3, 2)$

Foci: $(\pm 5, 2)$

Center: $(0, 2)$

Horizontal transverse axis

$a = 3, c = 5, b = 4$

$$\frac{(x - 0)^2}{9} - \frac{(y - 2)^2}{16} = 1$$

$$\frac{x^2}{9} - \frac{(y - 2)^2}{16} = 1$$

9. $5x^2 + 2xy + 5y^2 - 10 = 0$

$A = 5, B = 2, C = 5$

$\cot 2\theta = \dfrac{5 - 5}{2} = 0$

$2\theta = \dfrac{\pi}{2} \Rightarrow \theta = \dfrac{\pi}{4}$

$x = x' \cos \dfrac{\pi}{4} - y' \sin \dfrac{\pi}{4}$ $x = x' \cos \dfrac{\pi}{4} + y' \sin \dfrac{\pi}{4}$

$\quad = \dfrac{x' - y'}{\sqrt{2}}$ $\quad = \dfrac{x' + y'}{\sqrt{2}}$

$$5\left(\frac{x' - y'}{\sqrt{2}}\right)^2 + 2\left(\frac{x' - y'}{\sqrt{2}}\right)\left(\frac{x' + y'}{\sqrt{2}}\right) + 5\left(\frac{x' + y'}{\sqrt{2}}\right)^2 - 10 = 0$$

$$\frac{5(x')^2}{2} - \frac{10x'y'}{2} + \frac{5(y')^2}{2} + (x')^2 - (y')^2 + \frac{5(x')^2}{2} + \frac{10x'y'}{2} + \frac{5(y')^2}{2} - 10 = 0$$

$$6(x')^2 + 4(y')^2 - 10 = 0$$

$$\frac{3(x')^2}{5} + \frac{2(y')^2}{5} = 1$$

$$\frac{(x')^2}{5/3} + \frac{(y')^2}{5/2} = 1$$

Ellipse centered at the origin

10. (a) $6x^2 - 2xy + y^2 = 0$

$A = 6, B = -2, C = 1$

$B^2 - 4AC = (-2)^2 - 4(6)(1) = -20 < 0$

Ellipse

(b) $x^2 + 4xy + 4y^2 - x - y + 17 = 0$

$A = 1, B = 4, C = 4$

$B^2 - 4AC = (4)^2 - 4(1)(4) = 0$

Parabola

11. Polar: $\left(\sqrt{2}, \dfrac{3\pi}{4} \right)$

$x = \sqrt{2} \cos \dfrac{3\pi}{4} = \sqrt{2}\left(-\dfrac{1}{\sqrt{2}} \right) = -1$

$y = \sqrt{2} \sin \dfrac{3\pi}{4} = \sqrt{2}\left(\dfrac{1}{\sqrt{2}} \right) = 1$

Rectangular: $(-1, 1)$

12. Rectangular: $\left(\sqrt{3}, -1 \right)$

$r = \pm\sqrt{ (\sqrt{3})^2 + (-1)^2 } = \pm 2$

$\tan \theta = \dfrac{\sqrt{3}}{-1} = -\sqrt{3}$

$\theta = \dfrac{2\pi}{3}$ or $\theta = \dfrac{5\pi}{3}$

Polar: $\left(-2, \dfrac{2\pi}{3} \right)$ or $\left(2, \dfrac{5\pi}{3} \right)$

13. Rectangular: $4x - 3y = 12$

Polar: $4r \cos \theta - 3r \sin \theta = 12$

$r(4 \cos \theta - 3 \sin \theta) = 12$

$r = \dfrac{12}{4 \cos \theta - 3 \sin \theta}$

14. Polar: $r = 5 \cos \theta$

$r^2 = 5r \cos \theta$

Rectangular: $\quad x^2 + y^2 = 5x$

$x^2 + y^2 - 5x = 0$

15. $r = 1 - \cos \theta$

Cardioid

Symmetry: Polar axis

Maximum value of $|r|$: $r = 2$ when $\theta = \pi$

Zero of r: $r = 0$ when $\theta = 0$

θ	0	$\dfrac{\pi}{2}$	π	$\dfrac{3\pi}{2}$
r	0	1	2	1

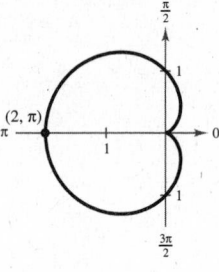

16. $r = 5 \sin 2\theta$

Rose curve with four petals

Symmetry: Polar axis, $\theta = \dfrac{\pi}{2}$, and pole

Maximum value of $|r|$: $|r| = 5$ when $\theta = \dfrac{\pi}{4}, \dfrac{3\pi}{4}, \dfrac{5\pi}{4}, \dfrac{7\pi}{4}$

Zeros of r: $r = 0$ when $\theta = 0, \dfrac{\pi}{2}, \pi, \dfrac{3\pi}{2}$

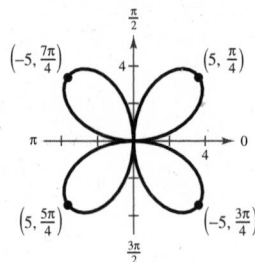

17. $r = \dfrac{3}{6 - \cos \theta}$

$r = \dfrac{1/2}{1 - (1/6) \cos \theta}$

$e = \dfrac{1}{6} < 1$, so the graph is an ellipse.

θ	0	$\dfrac{\pi}{2}$	π	$\dfrac{3\pi}{2}$
r	$\dfrac{3}{5}$	$\dfrac{1}{2}$	$\dfrac{3}{7}$	$\dfrac{1}{2}$

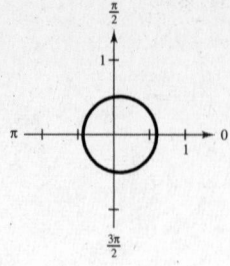

18. Parabola

Vertex: $\left(6, \dfrac{\pi}{2}\right)$

Focus: $(0, 0)$

$e = 1$

$r = \dfrac{ep}{1 + e \sin \theta}$

$r = \dfrac{p}{1 + \sin \theta}$

$6 = \dfrac{p}{1 + \sin(\pi/2)}$

$6 = \dfrac{p}{2}$

$12 = p$

$r = \dfrac{12}{1 + \sin \theta}$

19. $x = 3 - 2 \sin \theta,\ y = 1 + 5 \cos \theta$

$\dfrac{x - 3}{-2} = \sin \theta,\ \dfrac{y - 1}{5} = \cos \theta$

$\left(\dfrac{x - 3}{-2}\right)^2 + \left(\dfrac{y - 1}{5}\right)^2 = 1$

$\dfrac{(x - 3)^2}{4} + \dfrac{(y - 1)^2}{25} = 1$

20. $x = e^{2t},\ y = e^{4t}$

$x > 0,\ y > 0$

$y = (e^{2t})^2 = (x)^2 = x^2,\ x > 0, y > 0$

Chapter 11   Practice Test Solutions

1. Let $A = (0, 0, 0)$, $B = (1, 2, -4)$, $C = (0, -2, -1)$.

Side AB: $\sqrt{1^2 + 2^2 + 4^2} = \sqrt{21}$

Side AC: $\sqrt{0^2 + 2^2 + 1^2} = \sqrt{5}$

Side BC: $\sqrt{(-1)^2 + (-2 - 2)^2 + (-1 + 4)^2} = \sqrt{1 + 16 + 9} = \sqrt{26}$

$BC^2 = AB^2 + AC^2$

$26 = 21 + 5$

2. $(x - 0)^2 + (y - 4)^2 + (z - 1)^2 = 5^2$

$x^2 + (y - 4)^2 + (z - 1)^2 = 25$

3. $(x^2 + 2x + 1) + y^2 + (z^2 - 4z + 4) = 1 + 4 + 11$

$(x + 1)^2 + y^2 + (z - 2)^2 = 16$

Center: $(-1, 0, 2)$

Radius: 4

4. $\mathbf{u} - 3\mathbf{v} = \langle 1, 0, -1 \rangle - 3\langle 4, 3, -6 \rangle$

$\qquad\qquad = \langle 1, 0, -1 \rangle - \langle 12, 9, -18 \rangle$

$\qquad\qquad = \langle -11, -9, 17 \rangle$

5. $\frac{1}{2}\mathbf{v} = \frac{1}{2}\langle 2, 4, -6 \rangle = \langle 1, 2, -3 \rangle$

$\left\| \frac{1}{2}\mathbf{v} \right\| = \sqrt{1^2 + 2^2 + (-3)^2} = \sqrt{14}$

6. $\mathbf{u} \cdot \mathbf{v} = \langle 2, 1, -3 \rangle \cdot \langle 1, 1, -2 \rangle$

$\qquad\qquad = 2 + 1 + 6 = 9$

7. Because $\mathbf{v} = \langle -3, -3, 3 \rangle = -3\langle 1, 1, -1 \rangle = -3\mathbf{u}$, $\mathbf{u}$ and $\mathbf{v}$ are parallel.

8. $\mathbf{u} \times \mathbf{v} = \begin{vmatrix} \mathbf{i} & \mathbf{j} & \mathbf{k} \\ -1 & 0 & 2 \\ 1 & -1 & 3 \end{vmatrix} = \langle 2, 5, 1 \rangle$

$\mathbf{v} \times \mathbf{u} = -(\mathbf{u} \times \mathbf{v}) = \langle -2, -5, -1 \rangle$

9. $\mathbf{u} \cdot (\mathbf{v} \times \mathbf{w}) = \begin{vmatrix} 1 & 1 & 1 \\ 0 & -1 & 1 \\ 1 & 0 & 4 \end{vmatrix}$

$\qquad\qquad = 1(-4) - 1(-1) + 1(1)$

$\qquad\qquad = -4 + 1 + 1 = -2$

Volume $= |\mathbf{u} \cdot (\mathbf{v} \times \mathbf{w})| = |-2| = 2$

10. $\mathbf{v} = \langle (2 - 0), -3 - (-3), 4 - 3 \rangle = \langle 2, 0, 1 \rangle$

$x = 2 + 2t, y = -3, z = 4 + t$

11. $1(x - 1) - 1(y - 2) + 0(z - 3) = 0$

$\qquad\qquad x - 1 - y + 2 = 0$

$\qquad\qquad x - y + 1 = 0$

12. $\overrightarrow{AB} = \langle 1, 1, 1 \rangle, \overrightarrow{AC} = \langle 1, 2, 3 \rangle$

$\mathbf{n} = \overrightarrow{AB} \times \overrightarrow{AC} = \begin{vmatrix} \mathbf{i} & \mathbf{j} & \mathbf{k} \\ 1 & 1 & 1 \\ 1 & 2 & 3 \end{vmatrix} = \langle 1, -2, 1 \rangle$

Plane: $1(x - 0) - 2(y - 0) + (z - 0) = 0$

$\qquad\qquad x - 2y + z = 0$

13. $\mathbf{n}_1 = \langle 1, 1, -1 \rangle, \mathbf{n}_2 = \langle 3, -4, -1 \rangle$

$\mathbf{n}_1 \cdot \mathbf{n}_2 = 3 - 4 + 1 = 0 \implies$ Orthogonal planes

14. $\mathbf{n} = \langle 1, 2, 1 \rangle, Q = (1, 1, 1), P = (0, 0, 6)$ on plane, $\overrightarrow{PQ} = \langle 1, 1, -5 \rangle$

$D = \dfrac{|\overrightarrow{PQ} \cdot \mathbf{n}|}{\|\mathbf{n}\|} = \dfrac{|1 + 2 - 5|}{\sqrt{1 + 4 + 1}} = \dfrac{2}{\sqrt{6}} = \dfrac{\sqrt{6}}{3}$

Chapter 12   Practice Test Solutions

1.

x	2.9	2.99	3	3.01	3.1
$f(x)$	0.1695	0.1669	?	0.1664	0.1639

$\lim\limits_{x \to 3} \dfrac{x - 3}{x^2 - 9} \approx 0.1667$

2. $\lim\limits_{x \to 0} \dfrac{\sqrt{x + 4} - 2}{x} \approx \dfrac{1}{4}$

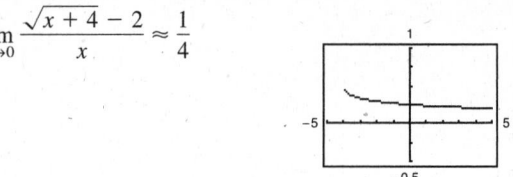

3. $\lim\limits_{x \to 2} e^{x-2} = e^{2-2} = e^0 = 1$

4. $\lim\limits_{x \to 1} \dfrac{x^3 - 1}{x - 1} = \lim\limits_{x \to 1} \dfrac{(x - 1)(x^2 + x + 1)}{x - 1}$

$\qquad\qquad = \lim\limits_{x \to 1} (x^2 + x + 1) = 3$

5. $\lim\limits_{x \to 0} \dfrac{\sin 5x}{2x} \approx 2.5$

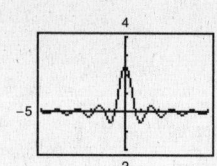

6. The limit does not exist. If

$$f(x) = \frac{|x + 2|}{x + 2},$$

then $f(x) = 1$ for $x > -2$, and $f(x) = -1$ for $x < -2$.

7.
$$\begin{aligned}
m_{\text{sec}} &= \frac{f(4 + h) - f(4)}{h} \\
&= \frac{\sqrt{4 + h} - 2}{h} \\
&= \frac{\sqrt{4 + h} - 2}{h} \cdot \frac{\sqrt{4 + h} + 2}{\sqrt{4 + h} + 2} \\
&= \frac{(4 + h) - 4}{h\left[\sqrt{4 + h} + 2\right]} \\
&= \frac{h}{h\left[\sqrt{4 + h} + 2\right]} \\
&= \frac{1}{\sqrt{4 + h} + 2}, \; h \neq 0 \\
m &= \lim_{h \to 0} \frac{1}{\sqrt{4 + h} + 2} = \frac{1}{\sqrt{4} + 2} = \frac{1}{4}
\end{aligned}$$

8.
$$\begin{aligned}
f'(x) &= \lim_{h \to 0} \frac{f(x + h) - f(x)}{h} \\
&= \lim_{h \to 0} \frac{[3(x + h) - 1] - [3x - 1]}{h} \\
&= \lim_{h \to 0} \frac{3x + 3h - 1 - 3x + 1}{h} \\
&= \lim_{h \to 0} \frac{3h}{h} = \lim_{h \to 0} 3 = 3
\end{aligned}$$

9. (a) $\lim\limits_{x \to \infty} \dfrac{3}{x^4} = 0$

(b) $\lim\limits_{x \to -\infty} \dfrac{x^2}{x^2 + 3} = 1$

(c) $\lim\limits_{x \to \infty} \dfrac{|x|}{1 - x} = -1$

10. $a_1 = 0, \; a_2 = \dfrac{1 - 4}{8 + 1} = -\dfrac{1}{3}, \; a_3 = \dfrac{1 - 9}{18 + 1} = -\dfrac{8}{19},$

$$a_4 = \frac{1 - 16}{33} = -\frac{15}{33}$$

$$\lim_{n \to \infty} a_n = \lim_{n \to \infty} \frac{1 - n^2}{2n^2 + 1} = -\frac{1}{2}$$

11. $\displaystyle\sum_{i=1}^{25} i^2 + \sum_{i=1}^{25} i = \frac{25(26)(51)}{6} + \frac{25(26)}{2} = \frac{25(26)}{6}[51 + 3] = \frac{25(26)(54)}{6} = 5850$

12. $\displaystyle\sum_{i=1}^{n} \frac{i^2}{n^3} = \frac{1}{n^3} \sum_{i=1}^{n} i^2 = \frac{1}{n^3}\left[\frac{n(n + 1)(2n + 1)}{6}\right] = \frac{2n^2 + 3n + 1}{6n^2} = S(n)$

$$\lim_{n \to \infty} S(n) = \frac{1}{3}$$

13. Width of rectangles: $\dfrac{b - a}{n} = \dfrac{1}{n}$

Height: $f\left(a + \dfrac{(b - a)i}{n}\right) = f\left(\dfrac{i}{n}\right) = 1 - \left(\dfrac{i}{n}\right)^2$

$$A \approx \sum_{i=1}^{n} \left[1 - \frac{i^2}{n^2}\right]\frac{1}{n} = \sum_{i=1}^{n} \frac{1}{n} - \sum_{i=1}^{n} \frac{i^2}{n^3} = 1 - \frac{1}{n^3}\frac{n(n + 1)(2n + 1)}{6}$$

$$A = \lim_{n \to \infty} A_n = 1 - \frac{1}{3} = \frac{2}{3}$$

PART II

Chapter 1   Chapter Test Solutions

1. Midpoint: $\left(\dfrac{-2+6}{2}, \dfrac{5+0}{2}\right) = \left(2, \dfrac{5}{2}\right)$

Distance: $d = \sqrt{(-2-6)^2 + (5-0)^2}$

$\qquad = \sqrt{64 + 25}$

$\qquad = \sqrt{89}$

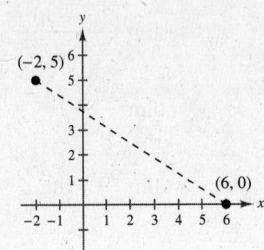

2. $\pi r^2 h = V$

$\pi(4)^2 h = 600$

$h = \dfrac{600}{16\pi}$

$h \approx 11.937$ centimeters

3. $y = 3 - 5x$

x-intercept: $\left(\frac{3}{5}, 0\right)$

y-intercept: $(0, 3)$

No axis or origin symmetry

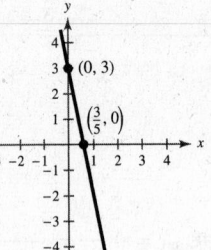

4. $y = 4 - |x|$

x-intercepts: $(\pm 4, 0)$

y-intercept: $(0, 4)$

y-axis symmetry

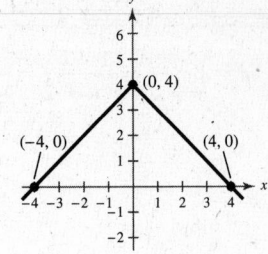

5. $y = x^2 - 1$

x-intercepts: $(\pm 1, 0)$

y-intercept: $(0, -1)$

y-axis symmetry

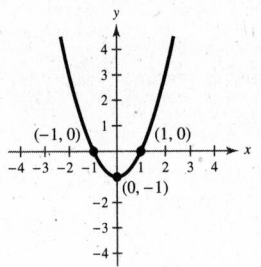

6. Center: $(1, 3)$

Radius: 4

Standard form: $(x - 1)^2 + (y - 3)^2 = 16$

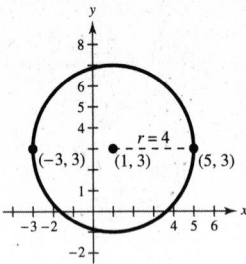

7. $(2, -3)$ and $(-4, 9)$

$m = \dfrac{9 - (-3)}{-4 - 2} = -2$

$y - (-3) = -2(x - 2)$

$y + 3 = -2x + 4$

$y = -2x + 1$

$2x + y - 1 = 0$

8. $(3, 0.8)$ and $(7, -6)$

$m = \dfrac{-6 - 0.8}{7 - 3} = -1.7$

$y - (-6) = -1.7(x - 7)$

$y + 6 = -1.7x + 11.9$

$y = -1.7x + 5.9$

$10y = -17x + 59$

$17x + 10y - 59 = 0$

9. $-4x + 7y = -5$

$$7y = 4x - 5$$

$$y = \tfrac{4}{7}x - \tfrac{5}{7} \Longrightarrow m_1 = \tfrac{4}{7}$$

(a) Parallel line: $m_2 = \tfrac{4}{7}$

$$y - 8 = \tfrac{4}{7}(x - 3)$$

$$7y - 56 = 4x - 12$$

$$-4x + 7y = 44$$

$$4x - 7y + 44 = 0$$

(b) Perpendicular line: $m_2 = -\tfrac{7}{4}$

$$y - 8 = -\tfrac{7}{4}(x - 3)$$

$$4y - 32 = -7x + 21$$

$$7x + 4y - 53 = 0$$

10. $f(x) = \dfrac{\sqrt{x + 9}}{x^2 - 81}$

(a) $f(7) = \dfrac{4}{-32} = -\dfrac{1}{8}$

(b) $f(-5) = \dfrac{2}{-56} = -\dfrac{1}{28}$

(c) $f(x - 9) = \dfrac{\sqrt{x}}{(x - 9)^2 - 81} = \dfrac{\sqrt{x}}{x^2 - 18x}$

11. $f(x) = \sqrt{100 - x^2}$

Domain: $100 - x^2 \geq 0 \Longrightarrow -10 \leq x \leq 10$ or $[-10, 10]$

12. $f(x) = 2x^6 + 5x^4 - x^2$

(a) $0, \pm 0.4314$

(b)

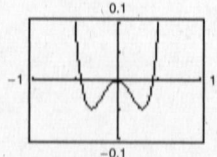

(c) Increasing on $(-0.31, 0), (0.31, \infty)$

Decreasing on $(-\infty, -0.31), (0, 0.31)$

(d) y-axis symmetry $\Longrightarrow$ The function is even.

13. $f(x) = 4x\sqrt{3 - x}$

(a) $0, 3$

(b)

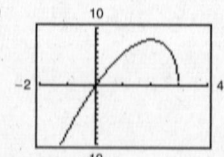

(c) Increasing on $(-\infty, 2)$

Decreasing on $(2, 3)$

(d) The function is neither odd nor even.

14. $f(x) = |x + 5|$

(a) -5

(b)

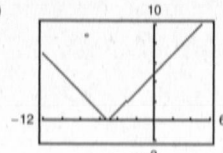

(c) Increasing on $(-5, \infty)$

Decreasing on $(-\infty, -5)$

(d) The function is neither odd nor even.

15. $f(x) = \begin{cases} 3x + 7, & x \leq -3 \\ 4x^2 - 1, & x > -3 \end{cases}$

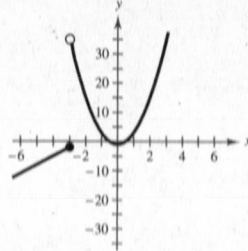

16. $h(x) = -[\![x]\!]$

Common function: $f(x) = [\![x]\!]$

Transformation: Reflection in the x-axis

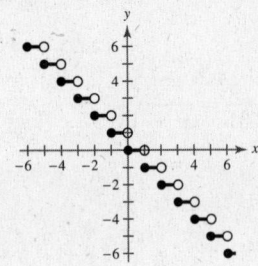

17. $h(x) = -\sqrt{x + 5} + 8$

Common function: $f(x) = \sqrt{x}$

Transformation: Reflection in the x-axis, a horizontal shift 5 units to the left, and a vertical shift 8 units upward

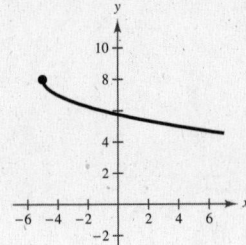

18. $f(x) = 3x^2 - 7$, $g(x) = -x^2 - 4x + 5$

(a) $(f + g)(x) = (3x^2 - 7) + (-x^2 - 4x + 5) = 2x^2 - 4x - 2$

(b) $(f - g)(x) = (3x^2 - 7) - (-x^2 - 4x + 5) = 4x^2 + 4x - 12$

(c) $(fg)(x) = (3x^2 - 7)(-x^2 - 4x + 5) = -3x^4 - 12x^3 + 22x^2 + 28x - 35$

(d) $\left(\dfrac{f}{g}\right)(x) = \dfrac{3x^2 - 7}{-x^2 - 4x + 5}, x \neq -5, 1$

(e) $(f \circ g)(x) = f(g(x)) = f(-x^2 - 4x + 5) = 3(-x^2 - 4x + 5)^2 - 7 = 3x^4 + 24x^3 + 18x^2 - 120x + 68$

(f) $(g \circ f)(x) = g(f(x)) = g(3x^2 - 7) = -(3x^2 - 7)^2 - 4(3x^2 - 7) + 5 = -9x^4 + 30x^2 - 16$

19. $f(x) = \dfrac{1}{x}$, $g(x) = 2\sqrt{x}$

(a) $(f + g)(x) = \dfrac{1}{x} + 2\sqrt{x} = \dfrac{1 + 2x^{3/2}}{x}, x > 0$

(b) $(f - g)(x) = \dfrac{1}{x} - 2\sqrt{x} = \dfrac{1 - 2x^{3/2}}{x}, x > 0$

(c) $(fg)(x) = \left(\dfrac{1}{x}\right)(2\sqrt{x}) = \dfrac{2\sqrt{x}}{x}, x > 0$

(d) $\left(\dfrac{f}{g}\right)(x) = \dfrac{\dfrac{1}{x}}{2\sqrt{x}} = \dfrac{1}{2x\sqrt{x}} = \dfrac{1}{2x^{3/2}}, x > 0$

(e) $(f \circ g)(x) = f(g(x)) = f(2\sqrt{x}) = \dfrac{1}{2\sqrt{x}} = \dfrac{\sqrt{x}}{2x}, x > 0$

(f) $(g \circ f)(x) = g(f(x)) = g\left(\dfrac{1}{x}\right) = 2\sqrt{\dfrac{1}{x}} = \dfrac{2}{\sqrt{x}} = \dfrac{2\sqrt{x}}{x}, x > 0$

20. $f(x) = x^3 + 8$

Since f is one-to-one, f has an inverse.

$$y = x^3 + 8$$
$$x = y^3 + 8$$
$$x - 8 = y^3$$
$$\sqrt[3]{x - 8} = y$$
$$f^{-1}(x) = \sqrt[3]{x - 8}$$

21. $f(x) = |x^2 - 3| + 6$

Since f is not one-to-one, f does not have an inverse.

22. $f(x) = 3x\sqrt{x} = 3x^{3/2}$

Domain: $[0, \infty)$

Range: $[0, \infty)$

The graph of $f(x)$ passes the Horizontal Line Test, so $f(x)$ is one-to-one and has an inverse.

$$f(x) = 3x^{3/2}$$
$$y = 3x^{3/2}$$
$$x = 3y^{3/2}$$
$$\frac{x}{3} = y^{3/2}$$
$$\left(\frac{x}{3}\right)^{2/3} = y$$
$$f^{-1}(x) = \left(\frac{x}{3}\right)^{2/3} = \sqrt[3]{\frac{x^2}{9}}, x \geq 0$$

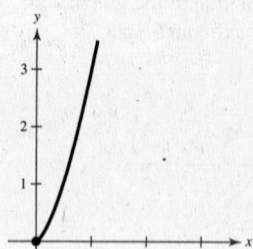

23. $v = k\sqrt{s}$

$24 = k\sqrt{16}$

$6 = k$

$v = 6\sqrt{s}$

24. $A = kxy$

$500 = k(15)(8)$

$500 = k(120)$

$\dfrac{25}{6} = k$

$A = \dfrac{25}{6}xy$

25. $b = \dfrac{k}{a}$

$32 = \dfrac{k}{1.5}$

$48 = k$

$b = \dfrac{48}{a}$

Chapter 2   Chapter Test Solutions

1. $f(x) = x^2$

(a) $g(x) = 2 - x^2$

Reflection in the x-axis followed by a vertical translation two units upward

(b) $g(x) = \left(x - \frac{3}{2}\right)^2$

Horizontal translation $\frac{3}{2}$ units to the right

2. Vertex: $(3, -6)$

$y = a(x - 3)^2 - 6$

Point on the graph: $(0, 3)$

$3 = a(0 - 3)^2 - 6$

$9 = 9a \implies a = 1$

Thus, $y = (x - 3)^2 - 6$.

3. (a) $y = -\frac{1}{20}x^2 + 3x + 5$

$= -\frac{1}{20}(x^2 - 60x + 900 - 900) + 5$

$= -\frac{1}{20}[(x - 30)^2 - 900] + 5$

$= -\frac{1}{20}(x - 30)^2 + 50$

Vertex: $(30, 50)$

The maximum height is 50 feet.

(b) The constant term, $c = 5$, determines the height at which the ball was thrown. Changing this constant results in a vertical translation of the graph, and, therefore, changes the maximum height.

4. $h(t) = -\frac{3}{4}t^5 + 2t^2$

The degree is odd and the leading coefficient is negative. The graph rises to the left and falls to the right.

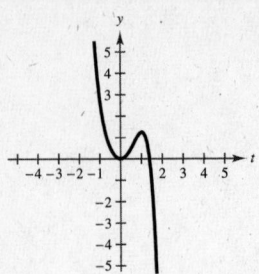

5.

$$3x + \frac{x - 1}{x^2 + 1}$$

$x^2 + 0x + 1\overline{)\,3x^3 + 0x^2 + 4x - 1}$

$\underline{3x^3 + 0x^2 + 3x}$

$\qquad\qquad\qquad x - 1$

Thus, $\dfrac{3x^3 + 4x - 1}{x^2 + 1} = 3x + \dfrac{x - 1}{x^2 + 1}$.

6. 2 | 2   0 −5   0 −3

$\qquad\quad$ 4   8   6 12

$\qquad$ 2   4   3   6   9

Thus, $\dfrac{2x^4 - 5x^2 - 3}{x - 2} = 2x^3 + 4x^2 + 3x + 6 + \dfrac{9}{x - 2}$.

7. $\sqrt{3}$ | 4   −1   −12   3

$\qquad\qquad\quad\ 4\sqrt{3}\quad 12 - \sqrt{3}\quad -3$

$\qquad\quad$ 4 $4\sqrt{3} - 1$ $-\sqrt{3}$ 0

$-\sqrt{3}$ | 4 $4\sqrt{3} - 1$ $-\sqrt{3}$

$\qquad\qquad\qquad\quad -4\sqrt{3}\qquad \sqrt{3}$

$\qquad\qquad$ 4   −1   0

$4x^3 - x^2 - 12x + 3 = (x - \sqrt{3})(x + \sqrt{3})(4x - 1)$

The real solutions are $x = \pm\sqrt{3}$ and $x = \frac{1}{4}$.

8. (a) $10i - (3 + \sqrt{-25}) = 10i - 3 - 5i$

$\qquad\qquad\qquad\qquad\qquad = -3 + 5i$

(b) $(2 + \sqrt{3}i)(2 - \sqrt{3}i) = 4 - 3i^2$

$\qquad\qquad\qquad\qquad\quad = 4 + 3$

$\qquad\qquad\qquad\qquad\quad = 7$

9. $\dfrac{5}{2 + i} = \dfrac{5}{2 + i} \cdot \dfrac{2 - i}{2 - i}$

$\qquad = \dfrac{5(2 - i)}{4 + 1}$

$\qquad = 2 - i$

10. $f(x) = x(x - 3)[x - (3 + i)][x - (3 - i)]$

$= (x^2 - 3x)[(x - 3) - i][(x - 3) + i]$

$= (x^2 - 3x)[(x - 3)^2 - i^2]$

$= (x^2 - 3x)(x^2 - 6x + 10)$

$= x^4 - 9x^3 + 28x^2 - 30x$

11. $f(x) = [x - (1 + \sqrt{3}i)][x - (1 - \sqrt{3}i)](x - 2)(x - 2)$

$= [(x - 1) - \sqrt{3}i][(x - 1) + \sqrt{3}i](x^2 - 4x + 4)$

$= [(x - 1)^2 - 3i^2](x^2 - 4x + 4)$

$= (x^2 - 2x + 4)(x^2 - 4x + 4)$

$= x^4 - 6x^3 + 16x^2 - 24x + 16$

12. $f(x) = x^3 + 2x^2 + 5x + 10$

$= x^2(x + 2) + 5(x + 2)$

$= (x + 2)(x^2 + 5)$

$= (x + 2)(x + \sqrt{5}i)(x - \sqrt{5}i)$

Zeros: $x = -2, \pm\sqrt{5}i$

13. $f(x) = x^4 - 9x^2 - 22x - 24$

Possible rational zeros: $\pm 1, \pm 2, \pm 3, \pm 4, \pm 6, \pm 8, \pm 12, \pm 24$

$$\begin{array}{r|rrrrr} -2 & 1 & 0 & -9 & -22 & -24 \\ & & -2 & 4 & 10 & 24 \\ \hline & 1 & -2 & -5 & -12 & 0 \end{array}$$

$$\begin{array}{r|rrrr} 4 & 1 & -2 & -5 & -12 \\ & & 4 & 8 & 12 \\ \hline & 1 & 2 & 3 & 0 \end{array}$$

$f(x) = (x + 2)(x - 4)(x^2 + 2x + 3)$

By the Quadratic Formula the zeros of $x^2 + 2x + 3$ are $x = -1 \pm \sqrt{2}i$. The zeros of f are: $x = -2, 4, -1 \pm \sqrt{2}i$.

14. $h(x) = \dfrac{4}{x^2} - 1$

$$= \frac{4 - x^2}{x^2}$$

$$= \frac{(2 - x)(2 + x)}{x^2}$$

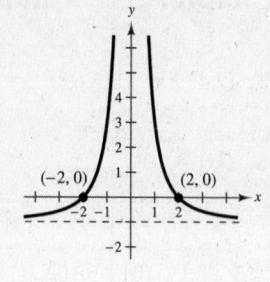

Vertical asymptote: $x = 0$

Horizontal asymptote: $y = -1$

x-intercepts: $(\pm 2, 0)$

15. $f(x) = \dfrac{2x^2 - 5x - 12}{x^2 - 16}$

$$= \frac{(2x + 3)(x - 4)}{(x + 4)(x - 4)}$$

$$= \frac{2x + 3}{x + 4}, \ x \neq 4$$

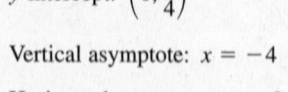

x-intercept: $\left(-\dfrac{3}{2}, 0\right)$

y-intercept: $\left(0, \dfrac{3}{4}\right)$

Vertical asymptote: $x = -4$

Horizontal asymptote: $y = 2$

16. $g(x) = \dfrac{x^2 + 2}{x - 1} = x + 1 + \dfrac{3}{x - 1}$

Vertical asymptote: $x = 1$

Slant asymptote: $y = x + 1$

y-intercept: $(0, -2)$

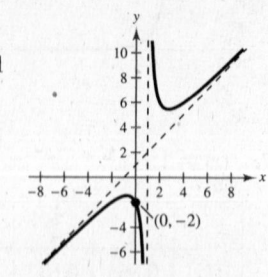

17. $2x^2 + 5x > 12$

$2x^2 + 5x - 12 > 0$

$(2x - 3)(x + 4) > 0$

Critical numbers: $x = \dfrac{3}{2}, x = -4$

Test intervals: $(-\infty, -4), \left(-4, \dfrac{3}{2}\right), \left(\dfrac{3}{2}, \infty\right)$

Test: Is $(2x - 3)(x + 4) > 0$?

Solution set: $(-\infty, -4) \cup \left(\dfrac{3}{2}, \infty\right)$

In inequality notation: $x < -4$ or $x > \dfrac{3}{2}$

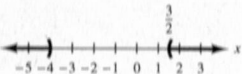

18. $\dfrac{2}{x} > \dfrac{5}{x + 6}$

$$\frac{2}{x} - \frac{5}{x + 6} > 0$$

$$\frac{2(x + 6) - 5x}{x(x + 6)} > 0$$

$$\frac{-3x + 12}{x(x + 6)} > 0$$

$$\frac{-3(x - 4)}{x(x + 6)} > 0$$

Critical numbers: $x = 4, x = 0, x = -6$

Test intervals: $(-\infty, -6), (-6, 0), (0, 4), (4, \infty)$

Test: Is $\dfrac{-3(x - 4)}{x(x + 6)} > 0$?

Solution set: $(-\infty, -6) \cup (0, 4)$

In inequality notation: $x < -6$ or $0 < x < 4$

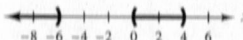

Chapter 3   Chapter Test Solutions

1. $12.4^{2.79} \approx 1123.690$ **2.** $4^{3\pi/2} \approx 687.291$ **3.** $e^{-7/10} \approx 0.497$ **4.** $e^{3.1} \approx 22.198$

5. $f(x) = 10^{-x}$

x	-1	$-\frac{1}{2}$	0	$\frac{1}{2}$	1
$f(x)$	10	3.162	1	0.316	0.1

Horizontal asymptote: $y = 0$

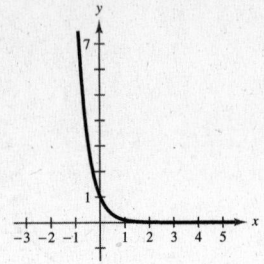

6. $f(x) = -6^{x-2}$

x	-1	0	1	2	3
$f(x)$	-0.005	-0.028	-0.167	-1	-6

Horizontal asymptote: $y = 0$

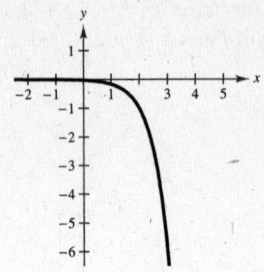

7. $f(x) = 1 - e^{2x}$

x	-1	$-\frac{1}{2}$	0	$\frac{1}{2}$	1
$f(x)$	0.865	0.632	0	-1.718	-6.389

Horizontal asymptote: $y = 1$

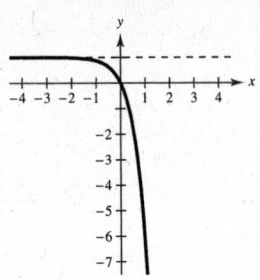

8. (a) $\log_7 7^{-0.89} = -0.89$

(b) $4.6 \ln e^2 = 4.6(2) = 9.2$

9. $f(x) = -\log x - 6$

x	$\frac{1}{2}$	1	$\frac{3}{2}$	2	4
$f(x)$	-5.699	-6	-6.176	-6.301	-6.602

Vertical asymptote: $x = 0$

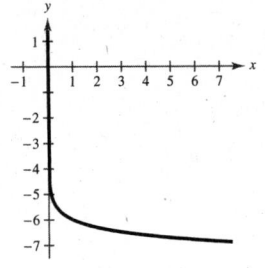

10. $f(x) = \ln(x - 4)$

x	5	7	9	11	13
$f(x)$	0	1.099	1.609	1.946	2.197

Vertical asymptote: $x = 4$

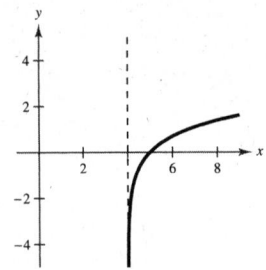

11. $f(x) = 1 + \ln(x + 6)$

x	-5	-3	-1	0	1
$f(x)$	1	2.099	2.609	2.792	2.946

Vertical asymptote: $x = -6$

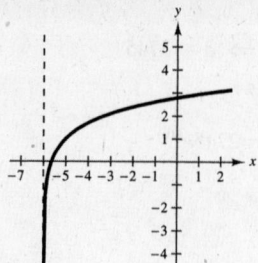

12. $\log_7 44 = \dfrac{\ln 44}{\ln 7} = \dfrac{\log 44}{\log 7} \approx 1.945$

13. $\log_{2/5} 0.9 = \dfrac{\ln 0.9}{\ln(2/5)} = \dfrac{\log 0.9}{\log(2/5)} \approx 0.115$

14. $\log_{24} 68 = \dfrac{\ln 68}{\ln 24} = \dfrac{\log 68}{\log 24} \approx 1.328$

15. $\log_2 3a^4 = \log_2 3 + \log_2 a^4 = \log_2 3 + 4 \log_2 |a|$

16. $\ln \dfrac{5\sqrt{x}}{6} = \ln\left(5\sqrt{x}\right) - \ln 6 = \ln 5 + \ln \sqrt{x} - \ln 6 = \ln 5 + \dfrac{1}{2} \ln x - \ln 6$

17. $\log\left(\dfrac{7x^2}{yz^3}\right) = \log 7x^2 - \log yz^3 = \log 7 + \log x^2 - (\log y + \log z^3) = \log 7 + 2 \log x - \log y - 3 \log z$

18. $\log_3 13 + \log_3 y = \log_3 13y$

19. $4 \ln x - 4 \ln y = \ln x^4 - \ln y^4 = \ln\left(\dfrac{x^4}{y^4}\right), \; x > 0, y > 0$

20. $2 \ln x + \ln(x - 5) - 3 \ln y = \ln x^2 + \ln(x - 5) - \ln y^3$
$$= \ln x^2(x - 5) - \ln y^3$$
$$= \ln \dfrac{x^2(x - 5)}{y^3}$$

21. $5^x = \dfrac{1}{25}$
$$5^x = 5^{-2}$$
$$x = -2$$

22. $3e^{-5x} = 132$
$$e^{-5x} = 44$$
$$-5x = \ln 44$$
$$x = \dfrac{\ln 44}{-5} \approx -0.757$$

23. $\dfrac{1025}{8 + e^{4x}} = 5$
$$1025 = 5(8 + e^{4x})$$
$$205 = 8 + e^{4x}$$
$$197 = e^{4x}$$
$$\ln 197 = 4x$$
$$\dfrac{\ln 197}{4} = x$$
$$x \approx 1.321$$

24. $\ln x = \frac{1}{2}$

$x = e^{1/2} \approx 1.649$

25. $18 + 4 \ln x = 7$

$4 \ln x = -11$

$\ln x = -\frac{11}{4}$

$x = e^{-11/4} \approx 0.064$

26. $\log x - \log(8 - 5x) = 2$

$\log \dfrac{x}{8 - 5x} = 2$

$\dfrac{x}{8 - 5x} = 10^2$

$x = 100(8 - 5x)$

$x = 800 - 500x$

$501x = 800$

$x = \dfrac{800}{501} \approx 1.597$

27. $y = ae^{bt}$

$(0, 2745):\quad 2745 = ae^{b(0)} \implies a = 2745$

$y = 2745e^{bt}$

$(9, 11{,}277):\quad 11{,}277 = 2745e^{b(9)}$

$\dfrac{11{,}277}{2745} = e^{9b}$

$\ln\!\left(\dfrac{11{,}277}{2745}\right) = 9b$

$\dfrac{1}{9}\ln\!\left(\dfrac{11{,}277}{2745}\right) = b \implies b \approx 0.1570$

Thus, $y = 2745e^{0.1570t}$.

28. $y = ae^{bt}$

$\dfrac{1}{2}a = ae^{b(21.77)}$

$\dfrac{1}{2} = e^{21.77b}$

$\ln\!\left(\dfrac{1}{2}\right) = 21.77b$

$b = \dfrac{\ln(1/2)}{21.77} \approx -0.0318$

$y = ae^{-0.0318t}$

When $t = 19$: $y = ae^{-0.0318(19)} \approx 0.55a$

Thus, 55% will remain after 19 years.

29. $H = 70.228 + 5.104x + 9.222 \ln x, \ \frac{1}{4} \le x \le 6$

(a)

x	H (cm)
$\frac{1}{4}$	58.720
$\frac{1}{2}$	66.388
1	75.332
2	86.828
3	95.671
4	103.43
5	110.59
6	117.38

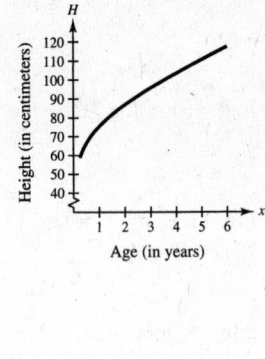

(b) When $x = 4$, $H \approx 103.43$ cm.

Chapters 1–3 Cumulative Test Solutions

1.

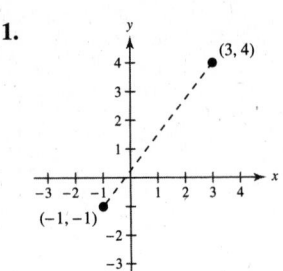

Midpoint: $\left(\dfrac{3 + (-1)}{2}, \dfrac{4 + (-1)}{2}\right) = \left(1, \dfrac{3}{2}\right)$

Distance: $d = \sqrt{(3 - (-1))^2 + (4 - (-1))^2} = \sqrt{(4)^2 + (5)^2} = \sqrt{16 + 25} = \sqrt{41}$

2. $x - 3y + 12 = 0$

Line

x-intercept: $(-12, 0)$

y-intercept: $(0, 4)$

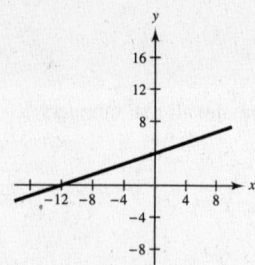

3. $y = x^2 - 9$

Parabola

x-intercepts: $(\pm 3, 0)$

y-intercept: $(0, -9)$

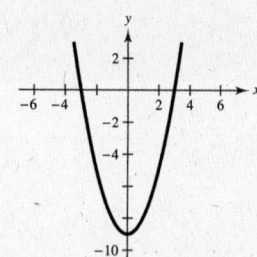

4. $y = \sqrt{4 - x}$

Domain: $x \le 4$

x-intercept: $(4, 0)$

y-intercept: $(0, 2)$

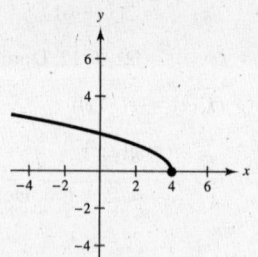

5. $\left(-\dfrac{1}{2}, 1\right)$ and $(3, 8)$

$$m = \frac{8 - 1}{3 - (-1/2)} = \frac{7}{7/2} = 2$$

$$y - 8 = 2(x - 3)$$

$$y - 8 = 2x - 6$$

$$0 = 2x - y + 2$$

6. It fails the Vertical Line Test. For some values of x there correspond two values of y.

7. $f(x) = \dfrac{x}{x - 2}$

(a) $f(6) = \dfrac{6}{4} = \dfrac{3}{2}$

(b) $f(2)$ is undefined because division by zero is undefined.

(c) $f(s + 2) = \dfrac{s + 2}{(s + 2) - 2} = \dfrac{s + 2}{s}$

8. $y = \sqrt[3]{x}$

(a) $r(x) = \dfrac{1}{2} \sqrt[3]{x}$ is a vertical shrink by a factor of $\dfrac{1}{2}$.

(b) $h(x) = \sqrt[3]{x} + 2$ is a vertical shift two units upward.

(c) $g(x) = \sqrt[3]{x + 2}$ is a horizontal shift two units to the left.

9. $f(x) = x - 3$, $g(x) = 4x + 1$

(a) $(f + g)(x) = f(x) + g(x)$

$\qquad = (x - 3) + (4x + 1)$

$\qquad = 5x - 2$

(b) $(f - g)(x) = f(x) - g(x)$

$\qquad = (x - 3) - (4x + 1)$

$\qquad = -3x - 4$

(c) $(fg)(x) = f(x)g(x)$

$\qquad = (x - 3)(4x + 1)$

$\qquad = 4x^2 - 11x - 3$

(d) $\left(\dfrac{f}{g}\right)(x) = \dfrac{f(x)}{g(x)} = \dfrac{x - 3}{4x + 1}$

Domain: all real numbers except $x = -\dfrac{1}{4}$

10. $f(x) = \sqrt{x - 1}$, $g(x) = x^2 + 1$

(a) $(f + g)(x) = f(x) + g(x)$

$\qquad = \sqrt{x - 1} + x^2 + 1$

(b) $(f - g)(x) = f(x) - g(x)$

$\qquad = \sqrt{x - 1} - x^2 - 1$

(c) $(fg)(x) = f(x)g(x)$

$\qquad = \sqrt{x - 1}(x^2 + 1) = x^2 \sqrt{x - 1} + \sqrt{x - 1}$

(d) $\left(\dfrac{f}{g}\right)(x) = \dfrac{f(x)}{g(x)} = \dfrac{\sqrt{x - 1}}{x^2 + 1}$

Domain: $x \ge 1$

11. $f(x) = 2x^2$, $g(x) = \sqrt{x + 6}$

   (a) $(f \circ g)(x) = f(g(x))$

   $\qquad = f\left(\sqrt{x + 6}\right)$

   $\qquad = 2\left(\sqrt{x + 6}\right)^2$

   $\qquad = 2(x + 6)$

   $\qquad = 2x + 12$, Domain: $x \geq -6$

   (b) $(g \circ f)(x) = g(f(x))$

   $\qquad = g(2x^2)$

   $\qquad = \sqrt{2x^2 + 6}$, Domain: all real numbers

12. $f(x) = x - 2$, $g(x) = |x|$

   (a) $(f \circ g)(x) = f(g(x))$

   $\qquad = f(|x|)$

   $\qquad = |x| - 2$, Domain: all real numbers

   (b) $(g \circ f)(x) = g(f(x))$

   $\qquad = g(x - 2)$

   $\qquad = |x - 2|$, Domain: all real numbers

13. The graph of h is a one-to-one line so h has an inverse.

   $\qquad h(x) = 5x - 2$

   $\qquad y = 5x - 2$

   $\qquad x = 5y - 2$

   $\qquad x + 2 = 5y$

   $\qquad \frac{1}{5}(x + 2) = y$

   $\qquad h^{-1}(x) = \frac{1}{5}(x + 2)$

14. $P = kS^3$

   $750 = k(27)^3 \implies k = \dfrac{750}{(27)^3} = \dfrac{250}{6561}$

   $P = \dfrac{250}{6561}S^3$

   When $S = 40$: $P = \left(\dfrac{250}{6561}\right)(40)^3 \approx 2438.65$ kilowatts

15. Vertex $(-8, 5)$

   Point $(-4, -7)$

   $\qquad y - k = a(x - h)^2$

   $\qquad y - 5 = a(x + 8)^2$

   $\qquad -7 - 5 = a(-4 + 8)^2$

   $\qquad -12 = 16a$

   $\qquad -\frac{3}{4} = a$

   $\qquad y = -\frac{3}{4}(x + 8)^2 + 5$

16. $h(x) = -(x^2 + 4x)$

   $\qquad = -(x^2 + 4x + 4 - 4)$

   $\qquad = -(x + 2)^2 + 4$

   Parabola

   Vertex: $(-2, 4)$

   Intercepts: $(-4, 0), (0, 0)$

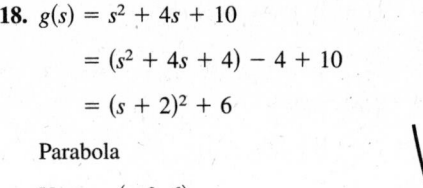

17. $f(t) = \frac{1}{4}t(t - 2)^2$

   Cubic

   Falls to the left

   Rises to the right

   Intercepts: $(0, 0), (2, 0)$

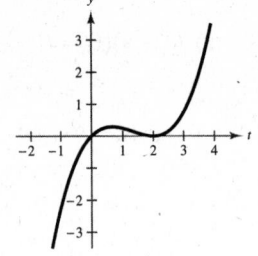

18. $g(s) = s^2 + 4s + 10$

   $\qquad = (s^2 + 4s + 4) - 4 + 10$

   $\qquad = (s + 2)^2 + 6$

   Parabola

   Vertex: $(-2, 6)$

   Intercept: $(0, 10)$

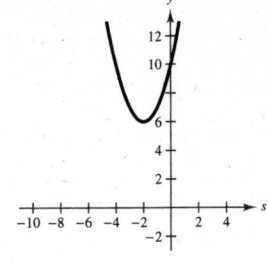

19. $f(x) = x^3 + 2x^2 + 4x + 8$

  $= x^2(x + 2) + 4(x + 2)$

  $= (x + 2)(x^2 + 4)$

$x + 2 = 0 \Longrightarrow x = -2$

$x^2 + 4 = 0 \Longrightarrow x = \pm 2i$

The zeros of $f(x)$ are -2 and $\pm 2i$.

20. $f(x) = x^4 + 4x^3 - 21x^2$

  $= x^2(x^2 + 4x - 21)$

  $= x^2(x + 7)(x - 3)$

The zeros of $f(x)$ are 0, -7, and 3.

21. $f(x) = 2x^4 - 11x^3 + 30x^2 - 62x - 40$

Possible Rational Zeros: $\pm 1, \pm 2, \pm 4, \pm 5, \pm 8, \pm 10, \pm 20, \pm 40, \pm \frac{1}{2}, \pm \frac{5}{2}$

By testing (or by looking at the graph of $f(x)$) we see that $x = 4$ and $x = -\frac{1}{2}$ are zeros.

$$
\begin{array}{r|rrrrr}
4 & 2 & -11 & 30 & -62 & -40 \\
  &   & 8 & -12 & 72 & 40 \\
\hline
  & 2 & -3 & 18 & 10 & 0
\end{array}
$$

$$
\begin{array}{r|rrrr}
-\frac{1}{2} & 2 & -3 & 18 & 10 \\
  &   & -1 & 2 & -10 \\
\hline
  & 2 & -4 & 20 & 0
\end{array}
$$

$f(x) = (x - 4)\left(x + \frac{1}{2}\right)(2x^2 - 4x + 20) = (x - 4)\left(x + \frac{1}{2}\right)(2)(x^2 - 2x + 10) = (x - 4)(2x + 1)(x^2 - 2x + 10)$

By Completing the Square (or by the Quadratic Formula) we find the zeros of $x^2 - 2x + 10$ to be $1 \pm 3i$.

$f(x) = (x - 4)(2x + 1)(x - 1 - 3i)(x - 1 + 3i)$

Zeros of $f(x)$: $4, -\frac{1}{2}, 1 + 3i, 1 - 3i$

22.

$$
\begin{array}{r}
3x - 2 + \dfrac{-3x + 2}{2x^2 + 1} \\[2mm]
2x^2 + 0x + 1 \overline{) 6x^3 - 4x^2 + 0x + 0} \\
\underline{6x^3 + 0x^2 + 3x} \\
-4x^2 - 3x + 0 \\
\underline{-4x^2 + 0x - 2} \\
-3x + 2
\end{array}
$$

Thus, $\dfrac{6x^3 - 4x^2}{2x^2 + 1} = 3x - 2 - \dfrac{3x - 2}{2x^2 + 1}$.

23.
$$
\begin{array}{r|rrrrr}
-2 & 2 & 3 & 0 & -6 & 5 \\
   &   & -4 & 2 & -4 & 20 \\
\hline
   & 2 & -1 & 2 & -10 & 25
\end{array}
$$

Thus, $\dfrac{2x^4 + 3x^3 - 6x + 5}{x + 2} = 2x^3 - x^2 + 2x - 10 + \dfrac{25}{x + 2}$.

24. $g(x) = x^3 + 3x^2 - 6$

From the graph we can see that $g(x)$ has one real zero. It is between 1 and 2 since $g(1)$ is negative and $g(2)$ is positive. The zero is $x \approx 1.20$.

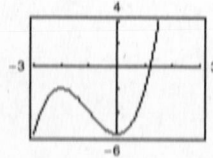

25. $f(x) = \dfrac{2x}{x^2 - 9}$

Vertical asymptotes: $x = \pm 3$

Horizontal asymptote: $y = 0$

Intercept: $(0, 0)$

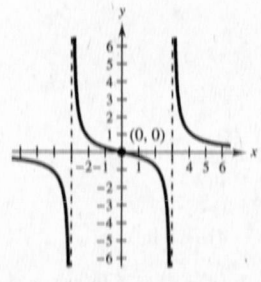

26. $f(x) = \dfrac{x^2 - 4x + 3}{x^2 - 2x - 3}$

$= \dfrac{(x-1)(x-3)}{(x+1)(x-3)}$

$= \dfrac{x-1}{x+1}, \quad x \neq 3$

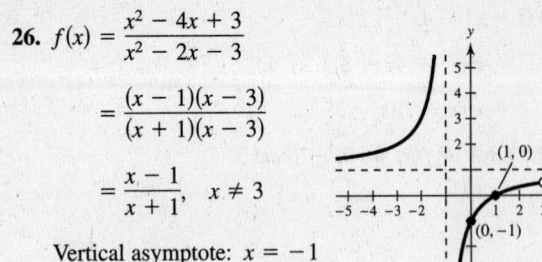

(1, 0)

(0, −1)

Vertical asymptote: $x = -1$

Horizontal asymptote: $y = 1$

x-intercept: $(1, 0)$

y-intercept: $(0, -1)$

27. $f(x) = \dfrac{x^3 + 3x^2 - 4x - 12}{x^2 - x - 2}$

$= \dfrac{(x+2)(x-2)(x+3)}{(x+1)(x-2)}$

$= \dfrac{(x+2)(x+3)}{x+1}$

$= \dfrac{x^2 + 5x + 6}{x+1}$

$= x + 4 + \dfrac{2}{x+1}, \quad x \neq 2$

(0, 6)

(−3, 0)

(−2, 0)

Vertical asymptote: $x = -1$

Slant asymptote: $y = x + 4$

x-intercepts: $(-2, 0), (-3, 0)$

y-intercept: $(0, 6)$

28. $3x^3 - 12x \leq 0$

$3x(x-2)(x+2) \leq 0$

Critical numbers: $x = 0, x = \pm 2$

Test intervals: $(-\infty, -2), (-2, 0), (0, 2), (2, \infty)$

Test: Is $3x(x-2)(x+2) \leq 0$?

By testing a value in each interval, we have the following solution set: $(-\infty, -2] \cup [0, 2]$.

In inequality form, $x \leq -2$ or $0 \leq x \leq 2$.

```
←——+——+——[——+——]——+——→ x
  −3 −2 −1  0  1  2  3
```

29. $\dfrac{1}{x+1} \geq \dfrac{1}{x+5}$

$\dfrac{1}{x+1} - \dfrac{1}{x+5} \geq 0$

$\dfrac{4}{(x+1)(x+5)} \geq 0$

Critical numbers: $x = -1, x = -5$

Test intervals: $(-\infty, -5), (-5, -1), (-1, \infty)$

Test: Is $\dfrac{4}{(x+1)(x+5)} \geq 0$?

By testing a value in each interval, we have the following solution set: $(-\infty, -5) \cup (-1, \infty)$.

In inequality form, $x < -5$ or $x > -1$.

```
←——)——+——+——+——(——+——+——→ x
 −6 −5 −4 −3 −2 −1  0  1  2
```

30. $f(x) = \left(\frac{2}{5}\right)^x$

$g(x) = -\left(\frac{2}{5}\right)^{-x+3}$

g is a reflection in the x-axis, a reflection in the y-axis, and a horizontal shift three units to the right of the graph of f.

31. $f(x) = 2.2^x$

$g(x) = -2.2^x + 4$

g is a reflection in the x-axis, and a vertical shift four units upward of the graph of f.

32. $\log 98 \approx 1.991$

33. $\log\left(\frac{6}{7}\right) \approx -0.067$

34. $\ln\sqrt{31} \approx 1.717$

35. $\ln\left(\sqrt{40} - 5\right) \approx 0.281$

36. $\ln\left(\dfrac{x^2 - 16}{x^4}\right) = \ln(x^2 - 16) - \ln x^4$

$= \ln(x+4)(x-4) - 4\ln x$

$= \ln(x+4) + \ln(x-4) - 4\ln x, \quad x > 4$

37. $2\ln x - \dfrac{1}{2}\ln(x+5) = \ln x^2 - \ln\sqrt{x+5}$

$= \ln \dfrac{x^2}{\sqrt{x+5}}, \quad x > 0$

38. $6e^{2x} = 72$

$e^{2x} = 12$

$2x = \ln 12$

$x = \dfrac{\ln 12}{2} \approx 1.242$

39. $e^{2x} - 11e^x + 24 = 0$

$(e^x - 3)(e^x - 8) = 0$

$e^x - 3 = 0 \implies e^x = 3 \implies x = \ln 3 \approx 1.099$

$e^x - 8 = 0 \implies e^x = 8 \implies x = \ln 8 \approx 2.079$

40. $\ln \sqrt{x + 2} = 3$

$\dfrac{1}{2} \ln(x + 2) = 3$

$\ln(x + 2) = 6$

$x + 2 = e^6$

$x = e^6 - 2 \approx 401.429$

41. (a)

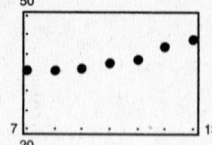

(b) $S \approx 0.274t^2 - 4.08t + 50.6$

(c)

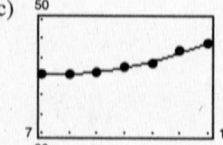

The model is a good fit to the actual data.

(d) For 2008, use $t = 18$: $S(18) \approx \$65.9$ billion

Yes, this seems reasonable, but since the model goes to infinity as t increases, it cannot be used for predictions much beyond this value.

42. $N = 175e^{kt}$

$420 = 175e^{k(8)}$

$2.4 = e^{8k}$

$\ln 2.4 = 8k$

$\dfrac{\ln 2.4}{8} = k$

$k \approx 0.1094$

$N = 175e^{0.1094t}$

$350 = 175e^{0.1094t}$

$2 = e^{0.1094t}$

$\ln 2 = 0.1094t$

$t = \dfrac{\ln 2}{0.1094} \approx 6.3$ hours to double

Chapter 4   Chapter Test Solutions

1. $\theta = \dfrac{5\pi}{4}$

(a)

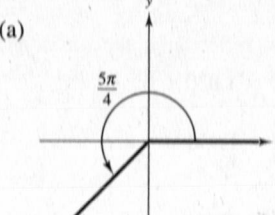

(b) $\dfrac{5\pi}{4} + 2\pi = \dfrac{13\pi}{4}$

$\dfrac{5\pi}{4} - 2\pi = -\dfrac{3\pi}{4}$

(c) $\dfrac{5\pi}{4}\left(\dfrac{180°}{\pi}\right) = 225°$

2. $90\dfrac{\text{km}}{\text{hr}} \times \dfrac{1 \text{ hr}}{60 \text{ min}} \times \dfrac{1000 \text{ m}}{1 \text{ km}} = 1500$ meters per minute

$\dfrac{\text{Revolutions}}{\text{minute}} = \dfrac{1500}{\pi}$

Circumference $= 2\pi\left(\dfrac{1}{2}\right) = \pi = \pi$ meters

Angular speed $= \left(\dfrac{1500 \text{ revolutions}}{\pi \text{ minute}}\right)\left(\dfrac{2\pi \text{ radians}}{\text{revolution}}\right)$

$= 3000$ radians per minute

3. $130° = \dfrac{130\pi}{180} = \dfrac{13\pi}{18}$ radians

$A = \dfrac{1}{2}r^2\theta = \dfrac{1}{2}(25)^2\left(\dfrac{13\pi}{18}\right) \approx 709.04$ square feet

4. $x = -2, y = 6$

$r = \sqrt{(-2)^2 + (6)^2} = 2\sqrt{10}$

$\sin\theta = \dfrac{y}{r} = \dfrac{6}{2\sqrt{10}} = \dfrac{3}{\sqrt{10}} = \dfrac{3\sqrt{10}}{10}$ $\qquad$ $\csc\theta = \dfrac{r}{y} = \dfrac{2\sqrt{10}}{6} = \dfrac{\sqrt{10}}{3}$

$\cos\theta = \dfrac{x}{r} = \dfrac{-2}{2\sqrt{10}} = -\dfrac{1}{\sqrt{10}} = -\dfrac{\sqrt{10}}{10}$ $\qquad$ $\sec\theta = \dfrac{r}{x} = \dfrac{2\sqrt{10}}{-2} = -\sqrt{10}$

$\tan\theta = \dfrac{y}{x} = \dfrac{6}{-2} = -3$ $\qquad$ $\cot\theta = \dfrac{x}{y} = \dfrac{-2}{6} = -\dfrac{1}{3}$

5.

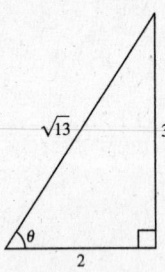

For $0 \le \theta < \dfrac{\pi}{2}$, we have:

$\sin\theta = \dfrac{\text{opp}}{\text{hyp}} = \dfrac{3}{\sqrt{13}} = \dfrac{3\sqrt{13}}{13}$

$\cos\theta = \dfrac{\text{adj}}{\text{hyp}} = \dfrac{2}{\sqrt{13}} = \dfrac{2\sqrt{13}}{13}$

$\csc\theta = \dfrac{\text{hyp}}{\text{opp}} = \dfrac{\sqrt{13}}{3}$

$\sec\theta = \dfrac{\text{hyp}}{\text{adj}} = \dfrac{\sqrt{13}}{2}$

$\cot\theta = \dfrac{\text{adj}}{\text{opp}} = \dfrac{2}{3}$

For $\pi \le \theta < \dfrac{3\pi}{2}$, we have:

$\sin\theta = -\dfrac{3\sqrt{13}}{13}$

$\cos\theta = -\dfrac{2\sqrt{13}}{13}$

$\csc\theta = -\dfrac{\sqrt{13}}{3}$

$\sec\theta = -\dfrac{\sqrt{13}}{2}$

$\cot\theta = \dfrac{2}{3}$

6. $\theta = 290°$

$\theta' = 360° - 290° = 70°$

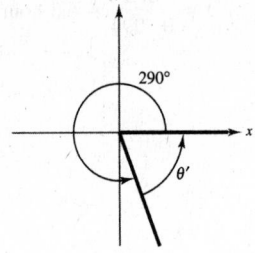

7. $\sec\theta < 0$ and $\tan\theta > 0$

$\dfrac{r}{x} < 0$ and $\dfrac{y}{x} > 0$

Quandrant III

8. $\cos\theta = -\dfrac{\sqrt{3}}{2}$

Reference angle is $30°$ and θ is in Quandrant II or III.

$\theta = 150°$ or $210°$

9. $\csc\theta = 1.030$

$\dfrac{1}{\sin\theta} = 1.030$

$\sin\theta = \dfrac{1}{1.030}$

$\theta = \arcsin\dfrac{1}{1.030}$

$\theta \approx 1.33$ and $\pi - 1.33 \approx 1.81$

10. $\cos \theta = \frac{3}{5}$, $\tan \theta < 0 \implies \theta$ lies in Quadrant IV.

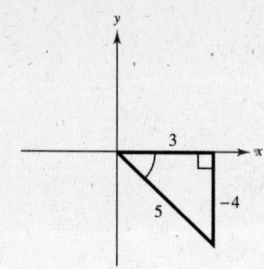

Let $x = 3$, $r = 5 \implies y = -4$.

$\sin \theta = -\frac{4}{5}$ $\csc \theta = -\frac{5}{4}$

$\cos \theta = \frac{3}{5}$ $\sec \theta = \frac{5}{3}$

$\tan \theta = -\frac{4}{3}$ $\cot \theta = -\frac{3}{4}$

11. $\sec \theta = -\frac{17}{8}$, $\sin \theta > 0 \implies \theta$ lies in Quadrant II.

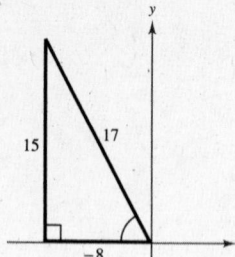

Let $r = 17$, $x = -8 \implies y = 15$.

$\sin \theta = \frac{15}{17}$ $\csc \theta = \frac{17}{15}$

$\cos \theta = -\frac{8}{17}$ $\sec \theta = -\frac{17}{8}$

$\tan \theta = -\frac{15}{8}$ $\cot \theta = -\frac{8}{15}$

12. $g(x) = -2\sin\left(x - \frac{\pi}{4}\right)$

Period: 2π

Amplitude: $|-2| = 2$

Shifted to the right by $\frac{\pi}{4}$ units and reflected in the x-axis.

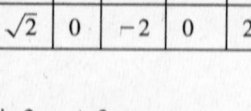

x	0	$\frac{\pi}{4}$	$\frac{3\pi}{4}$	$\frac{5\pi}{4}$	$\frac{7\pi}{4}$
y	$\sqrt{2}$	0	-2	0	2

13. $f(\alpha) = \frac{1}{2}\tan 2\alpha$

Period: $\frac{\pi}{2}$

Asymptotes:

$x = -\frac{\pi}{4}, x = \frac{\pi}{4}$

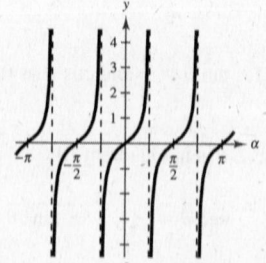

α	$-\frac{\pi}{8}$	0	$\frac{\pi}{8}$
$f(\alpha)$	$-\frac{1}{2}$	0	$\frac{1}{2}$

14. $y = \sin 2\pi x + 2\cos \pi x$

Periodic: period $= 2$

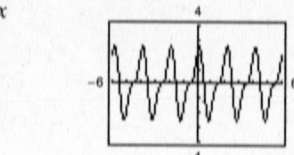

15. $y = 6e^{-0.12t}\cos(0.25t)$, $0 \le t \le 32$

Not periodic

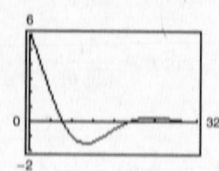

16. $f(x) = a\sin(bx + c)$

Amplitude: $2 \implies |a| = 2$

Reflected in the x-axis: $a = -2$

Period: $4\pi = \frac{2\pi}{b} \implies b = \frac{1}{2}$

Phase shift: $\frac{c}{b} = -\frac{\pi}{2} \implies c = -\frac{\pi}{4}$

$f(x) = -2\sin\left(\frac{x}{2} - \frac{\pi}{4}\right)$

17. Let $u = \arccos \frac{2}{3}$,

$\cos u = \frac{2}{3}$.

$\tan\left(\arccos \frac{2}{3}\right) = \tan u = \frac{\sqrt{5}}{2}$

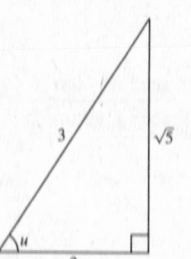

18. $f(x) = 2 \arcsin\left(\frac{1}{2}x\right)$

Domain: $[-2, 2]$

Range: $[-\pi, \pi]$

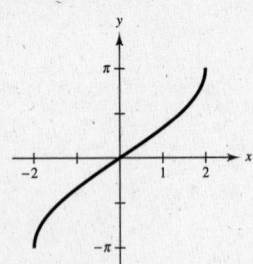

19.

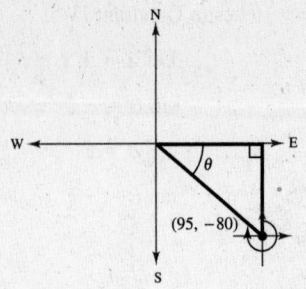

$\tan \theta = -\frac{80}{95} \implies \theta \approx -40.1°$

Bearing: $360° - (90° - 40.1°) = 310.1°$

20. $d = a \cos bt$

$a = -6$

$\frac{2\pi}{b} = 2 \implies b = \pi$

$d = -6 \cos \pi t$

Chapter 5   Chapter Test Solutions

1. $\tan \theta = \frac{3}{2}$ and $\cos \theta < 0$

θ is in Quadrant III.

$\sec \theta = -\sqrt{1 + \tan^2 \theta} = -\sqrt{1 + \left(\frac{3}{2}\right)^2} = -\frac{\sqrt{13}}{2}$

$\cos \theta = \frac{1}{\sec \theta} = -\frac{2}{\sqrt{13}} = -\frac{2\sqrt{13}}{13}$

$\sin \theta = \tan \theta \cos \theta = \left(\frac{3}{2}\right)\left(-\frac{2}{\sqrt{13}}\right) = -\frac{3}{\sqrt{13}} = -\frac{3\sqrt{13}}{13}$

$\csc \theta = \frac{1}{\sin \theta} = -\frac{\sqrt{13}}{3}$

$\cot \theta = \frac{1}{\tan \theta} = \frac{2}{3}$

2. $\csc^2 \beta(1 - \cos^2 \beta) = \frac{1}{\sin^2 \beta}(\sin^2 \beta) = 1$

3. $\frac{\sec^4 x - \tan^4 x}{\sec^2 x + \tan^2 x} = \frac{(\sec^2 x + \tan^2 x)(\sec^2 x - \tan^2 x)}{\sec^2 x + \tan^2 x}$

$= \sec^2 x - \tan^2 x = 1$

4. $\frac{\cos \theta}{\sin \theta} + \frac{\sin \theta}{\cos \theta} = \frac{\cos^2 \theta + \sin^2 \theta}{\sin \theta \cos \theta} = \frac{1}{\sin \theta \cos \theta}$

$= \csc \theta \sec \theta$

5. $y = \tan\theta,\ y = -\sqrt{\sec^2\theta - 1}$

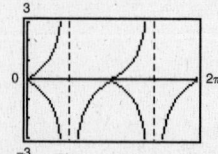

$\tan\theta = -\sqrt{\sec^2\theta - 1}$ on

$\theta = 0, \dfrac{\pi}{2} < \theta \le \pi, \dfrac{3\pi}{2} < \theta < 2\pi.$

6. $y_1 = \cos x + \sin x \tan x,\ y_2 = \sec x$

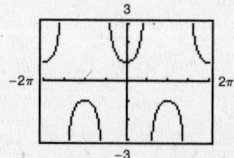

It appears that $y_1 = y_2$.

$\cos x + \sin x \tan x = \cos + \sin x \dfrac{\sin x}{\cos x}$

$= \cos + \dfrac{\sin^2 x}{\cos x}$

$= \dfrac{\cos^2 x + \sin^2 x}{\cos x}$

$= \dfrac{1}{\cos x} = \sec x$

7. $\sin\theta \sec\theta = \sin\theta \dfrac{1}{\cos\theta} = \dfrac{\sin\theta}{\cos\theta} = \tan\theta$

8. $\sec^2 x \tan^2 x + \sec^2 x = \sec^2 x(\sec^2 x - 1) + \sec^2 x$

$= \sec^4 x - \sec^2 x + \sec^2 x$

$= \sec^4 x$

9. $\dfrac{\csc\alpha + \sec\alpha}{\sin\alpha + \cos\alpha} = \dfrac{\dfrac{1}{\sin\alpha} + \dfrac{1}{\cos\alpha}}{\sin\alpha + \cos\alpha} = \dfrac{\dfrac{\cos\alpha + \sin\alpha}{\sin\alpha \cos\alpha}}{\sin\alpha + \cos\alpha} = \dfrac{1}{\sin\alpha \cos\alpha}$

$= \dfrac{\cos^2\alpha + \sin^2\alpha}{\sin\alpha \cos\alpha} = \dfrac{\cos^2\alpha}{\sin\alpha \cos\alpha} + \dfrac{\sin^2\alpha}{\sin\alpha \cos\alpha}$

$= \dfrac{\cos\alpha}{\sin\alpha} + \dfrac{\sin\alpha}{\cos\alpha} = \cot\alpha + \tan\alpha$

10. $\cos\left(x + \dfrac{\pi}{2}\right) = \cos\left(\dfrac{\pi}{2} - (-x)\right) = \sin(-x) = -\sin x$

11. $\sin(n\pi + \theta) = (-1)^n \sin\theta,\ n$ is an integer.

For n odd: $\sin(n\pi + \theta) = \sin n\pi \cos\theta + \cos n\pi \sin\theta$

$= (0)\cos\theta + (-1)\sin\theta = -\sin\theta$

For n even: $\sin(n\pi + \theta) = \sin n\pi \cos\theta + \cos n\pi \sin\theta$

$= (0)\cos\theta + (1)\sin\theta = \sin\theta$

When n is odd, $(-1)^n = -1$. When n is even $(-1)^n = 1$.

Thus, $\sin(n\pi + \theta) = (-1)^n \sin\theta$ for any integer n.

12. $(\sin x + \cos x)^2 = \sin^2 x + 2\sin x \cos x + \cos^2 x$

$= 1 + 2\sin x \cos x$

$= 1 + \sin 2x$

13. $\sin^4 x \tan^2 x = \sin^4 x \left(\dfrac{\sin^2 x}{\cos^2 x} \right) = \dfrac{\sin^6 x}{\cos^2 x} = \dfrac{(\sin^2 x)^3}{\cos^2 x}$

$= \dfrac{\left(\dfrac{1 - \cos 2x}{2} \right)^3}{\dfrac{1 + \cos 2x}{2}}$

$= \dfrac{\dfrac{1 - 3\cos 2x + 3\cos^2 2x - \cos^3 2x}{8}}{\dfrac{1 + \cos 2x}{2}}$

$= \dfrac{\dfrac{1}{4} \left[1 - 3\cos 2x + 3 \left(\dfrac{1 + \cos 4x}{2} \right) - \cos 2x \left(\dfrac{1 + \cos 4x}{2} \right) \right]}{1 + \cos 2x}$

$= \dfrac{\dfrac{1}{8} [2 - 6\cos 2x + 3 + 3\cos 4x - \cos 2x - \cos 2x \cos 4x]}{1 + \cos 2x}$

$= \dfrac{1}{8} \left[\dfrac{5 - 7\cos 2x + 3\cos 4x - \dfrac{1}{2}(\cos(-2x) + \cos(6x))}{1 + \cos 2x} \right]$

$= \dfrac{1}{16} \left[\dfrac{10 - 14\cos 2x + 6\cos 4x - \cos 2x - \cos 6x}{1 + \cos 2x} \right]$

$= \dfrac{1}{16} \left[\dfrac{10 - 15\cos 2x + 6\cos 4x - \cos 6x}{1 + \cos 2x} \right]$

14. $\dfrac{\sin 4\theta}{1 + \cos 4\theta} = \tan \dfrac{4\theta}{2} = \tan 2\theta$

15. $4\cos 2\theta \sin 4\theta = 4 \left(\dfrac{1}{2} \right) [\sin(2\theta + 4\theta) - \sin(2\theta - 4\theta)]$

$= 2[\sin 6\theta - \sin(-2\theta)]$

$= 2(\sin 6\theta + \sin 2\theta)$

16. $\sin 3\theta - \sin 4\theta = 2\cos \left(\dfrac{3\theta + 4\theta}{2} \right) \sin \left(\dfrac{3\theta - 4\theta}{2} \right)$

$= 2\cos \dfrac{7\theta}{2} \sin \left(\dfrac{-\theta}{2} \right)$

$= -2\cos \dfrac{7\theta}{2} \sin \dfrac{\theta}{2}$

17. $\tan^2 x + \tan x = 0$

$\tan x(\tan x + 1) = 0$

$\tan x = 0 \quad \text{or} \quad \tan x + 1 = 0$

$x = 0, \pi \qquad \tan x = -1$

$x = \dfrac{3\pi}{4}, \dfrac{7\pi}{4}$

18. $\sin 2\alpha - \cos \alpha = 0$

$2\sin \alpha \cos \alpha - \cos \alpha = 0$

$\cos \alpha(2\sin \alpha - 1) = 0$

$\cos \alpha = 0 \quad \text{or} \quad 2\sin \alpha - 1 = 0$

$\alpha = \dfrac{\pi}{2}, \dfrac{3\pi}{2} \qquad \sin \alpha = \dfrac{1}{2}$

$\alpha = \dfrac{\pi}{6}, \dfrac{5\pi}{6}$

19. $4\cos^2 x - 3 = 0$

$\cos^2 x = \dfrac{3}{4}$

$\cos x = \pm \sqrt{\dfrac{3}{4}} = \pm \dfrac{\sqrt{3}}{2}$

$x = \dfrac{\pi}{6}, \dfrac{5\pi}{6}, \dfrac{7\pi}{6}, \dfrac{11\pi}{6}$

20. $\csc^2 x - \csc x - 2 = 0$

$(\csc x - 2)(\csc x + 1) = 0$

$\csc x - 2 = 0 \quad \text{or} \quad \csc x + 1 = 0$

$\csc x = 2 \qquad\qquad \csc = -1$

$\dfrac{1}{\sin x} = 2 \qquad\qquad \dfrac{1}{\sin x} = -1$

$\sin x = \dfrac{1}{2} \qquad\qquad \sin x = -1$

$x = \dfrac{\pi}{6}, \dfrac{5\pi}{6} \qquad\qquad x = \dfrac{3\pi}{2}$

21. $3 \cos x - x = 0$

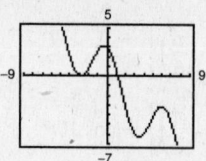

$x \approx -2.938,\ -2.663,\ 1.170$

22. $105° = 135° - 30°$

$\cos 105° = \cos(135° - 30°)$

$= \cos 135° \cos 30° + \sin 135° \sin 30°$

$= -\cos 45° \cos 30° + \sin 45° \sin 30°$

$= \left(-\dfrac{\sqrt{2}}{2}\right)\left(\dfrac{\sqrt{3}}{2}\right) + \left(\dfrac{\sqrt{2}}{2}\right)\left(\dfrac{1}{2}\right)$

$= \dfrac{-\sqrt{6} + \sqrt{2}}{4} = \dfrac{\sqrt{2} - \sqrt{6}}{4}$

23. $x = 1, y = 2, r = \sqrt{5}$

$\sin 2u = 2 \sin u \cos u$

$= 2\left(\dfrac{2}{\sqrt{5}}\right)\left(\dfrac{1}{\sqrt{5}}\right) = \dfrac{4}{5}$

$\cos 2u = \cos^2 u - \sin^2 u$

$= \left(\dfrac{1}{\sqrt{5}}\right)^2 - \left(\dfrac{2}{\sqrt{5}}\right)^2 = \dfrac{1}{5} - \dfrac{4}{5} = -\dfrac{3}{5}$

$\tan 2u = \dfrac{2 \tan u}{1 - \tan^2 u} = \dfrac{2(2)}{1 - (2)^2} = \dfrac{4}{-3} = -\dfrac{4}{3}$

24. Let $y_1 = 31 \sin\left(\dfrac{2\pi t}{365} - 1.4\right)$ and $y_2 = 20$.

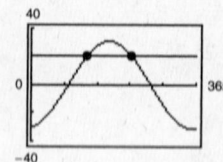

The points of intersection occur when $t \approx 123$ and $t \approx 223$. The number of days that $D > 20°$ is 100, from day 123 to day 223.

25. $28 \cos 10t + 38 = 28 \cos\left[10\left(t - \dfrac{\pi}{6}\right)\right] + 38$

$$\cos 10t = \cos\left[10\left(t - \dfrac{\pi}{6}\right)\right]$$

$$0 = \cos\left[10\left(t - \dfrac{\pi}{6}\right)\right] - \cos 10t$$

$$= -2 \sin\left(\dfrac{10(t - (\pi/6)) + 10t}{2}\right) \sin\left(\dfrac{10(t - (\pi/6)) - 10t}{2}\right)$$

$$= -2 \sin\left(10t - \dfrac{5\pi}{6}\right) \sin\left(-\dfrac{5\pi}{6}\right)$$

$$= -2 \sin\left(10t - \dfrac{5\pi}{6}\right)\left(-\dfrac{1}{2}\right)$$

$$= \sin\left(10t - \dfrac{5\pi}{6}\right)$$

$$10t - \dfrac{5\pi}{6} = n\pi \text{ where } n \text{ is any integer.}$$

$$t = \dfrac{n\pi}{10} + \dfrac{\pi}{12} \text{ where } n \text{ is any integer.}$$

The first six times the two people are at the same height are:

0.26 minutes, 0.58 minutes, 0.89 minutes, 1.20 minutes, 1.52 minutes, 1.83 minutes.

Chapter 6   Chapter Test Solutions

1. $A = 24°, B = 68°, a = 12.2$

$C = 180° - 24° - 68° = 88°$

$b = \dfrac{a \sin B}{\sin A} = \dfrac{12.2 \sin 68°}{\sin 24°} \approx 27.81$

$c = \dfrac{a \sin C}{\sin A} = \dfrac{12.2 \sin 88°}{\sin 24°} \approx 29.98$

2. $B = 104°, C = 33°, a = 18.1$

$A = 180° - 104° - 33° = 43°$

$b = \dfrac{a \sin B}{\sin A} = \dfrac{18.1 \sin 104°}{\sin 43°} \approx 25.75$

$c = \dfrac{a \sin C}{\sin A} = \dfrac{18.1 \sin 33°}{\sin 43°} \approx 14.45$

3. $A = 24°, a = 11.2, b = 13.4$

$\sin B = \dfrac{b \sin A}{a} = \dfrac{13.4 \sin 24°}{11.2} \approx 0.4866$

Two Solutions

$B \approx 29.12°$ or $B \approx 150.88°$

$C \approx 126.88°$ $C \approx 5.12°$

$c = \dfrac{a \sin C}{\sin A} = \dfrac{11.2 \sin 126.88°}{\sin 24°}$ $c = \dfrac{11.2 \sin 5.12°}{\sin 24°}$

$c \approx 22.03$ $c \approx 2.46$

4. $a = 4.0, b = 7.3, c = 12.4$

$\cos C = \dfrac{a^2 + b^2 - c^2}{2ab} = \dfrac{4^2 + 7.3^2 - 12.4^2}{2(4)(7.3)} \approx -1.4464 < -1$

No solution

5. $B = 100°, a = 15, b = 23$

$\sin A = \dfrac{a \sin B}{b} = \dfrac{15 \sin 100°}{23} \Rightarrow A \approx 39.96°$

$C \approx 180° - 100° - 39.96° = 40.04°$

$c \approx \dfrac{b \sin C}{\sin B} = \dfrac{23 \sin 40.04°}{\sin 100°} \approx 15.02$

6. $C = 123°, a = 41, b = 57$

$c^2 = 41^2 + 57^2 - 2(41)(57)\cos 123° \Rightarrow c \approx 86.46$

$\sin A = \dfrac{a \sin C}{c} = \dfrac{41 \sin 123°}{86.46} \Rightarrow A \approx 23.43°$

$B \approx 180° - 23.43° - 123° = 33.57°$

7. $a = 60, b = 70, c = 82$

$s = \dfrac{60 + 70 + 82}{2} = 106$

Area $= \sqrt{106(46)(36)(24)} \approx 2052.5$ square meters

8.

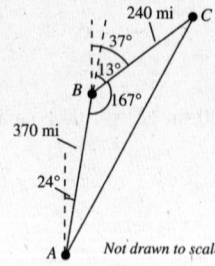

$b^2 = 370^2 + 240^2 - 2(370)(240)\cos 167°$

$b \approx 606.3$ miles

$\sin A = \dfrac{a \sin B}{b} = \dfrac{240 \sin 167°}{606.3}$

$A \approx 5.1°$

Bearing: $24° + 5.1° = 29.1°$

9. Initial point: $(-3, 7)$

Terminal point: $(11, -16)$

$\mathbf{v} = \langle 11 - (-3), -16 - 7 \rangle = \langle 14, -23 \rangle$

10. $\mathbf{v} = 12\left(\dfrac{\mathbf{u}}{\|\mathbf{u}\|}\right) = 12\left(\dfrac{\langle 3, -5 \rangle}{\sqrt{3^2 + (-5)^2}}\right) = \dfrac{12}{\sqrt{34}}\langle 3, -5 \rangle$

$= \dfrac{6\sqrt{34}}{17}\langle 3, -5 \rangle = \left\langle \dfrac{18\sqrt{34}}{17}, -\dfrac{30\sqrt{34}}{17} \right\rangle$

11. $\mathbf{u} + \mathbf{v} = \langle 3, 5 \rangle + \langle -7, 1 \rangle = \langle -4, 6 \rangle$

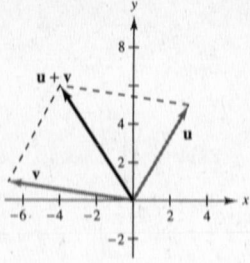

12. $\mathbf{u} - \mathbf{v} = \langle 3, 5 \rangle - \langle -7, 1 \rangle = \langle 10, 4 \rangle$

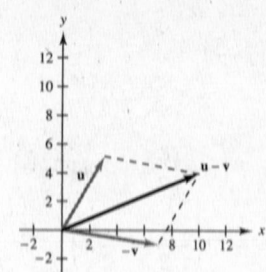

13. $5\mathbf{u} - 3\mathbf{v} = 5\langle 3, 5 \rangle - 3\langle -7, 1 \rangle = \langle 15, 25 \rangle + \langle 21, -3 \rangle$

$= \langle 36, 22 \rangle$

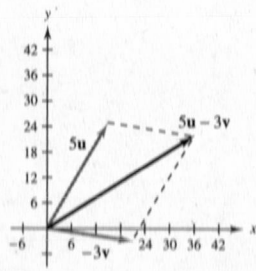

14. $\dfrac{\mathbf{u}}{\|\mathbf{u}\|} = \dfrac{\langle 4, -3 \rangle}{\sqrt{4^2 + (-3)^2}} = \dfrac{1}{5}\langle 4, -3 \rangle = \left\langle \dfrac{4}{5}, -\dfrac{3}{5} \right\rangle$

15. $\mathbf{u} = 250(\cos 45° \, \mathbf{i} + \sin 45° \, \mathbf{j})$

$\mathbf{v} = 130(\cos(-60°)\mathbf{i} + \sin(-60°)\mathbf{j})$

$\mathbf{R} = \mathbf{u} + \mathbf{v} \approx 241.7767 \, \mathbf{i} + 64.1934 \, \mathbf{j}$

$\|\mathbf{R}\| \approx \sqrt{241.7767^2 + 64.1934^2} \approx 250.15$ pounds

$\tan \theta \approx \dfrac{64.1934}{241.7767} \implies \theta \approx 14.9°$

16. $\mathbf{u} = \langle -1, 5 \rangle, \mathbf{v} = \langle 3, -2 \rangle$

$\cos \theta = \dfrac{\mathbf{u} \cdot \mathbf{v}}{\|\mathbf{u}\| \|\mathbf{v}\|} = \dfrac{-13}{\sqrt{26} \sqrt{13}} \implies \theta = 135°$

17. $\mathbf{u} = \langle 6, 10 \rangle, \mathbf{v} = \langle 2, 3 \rangle$

$\mathbf{u} \cdot \mathbf{v} = 42 \neq 0 \implies \mathbf{u}$ and $\mathbf{v}$ are not orthogonal.

18. $\mathbf{u} = \langle 6, 7 \rangle, \mathbf{v} = \langle -5, -1 \rangle$

$\mathbf{w}_1 = \text{proj}_{\mathbf{v}} \, \mathbf{u} = \left(\dfrac{\mathbf{u} \cdot \mathbf{v}}{\|\mathbf{v}\|^2} \right) \mathbf{v} = -\dfrac{37}{26} \langle -5, -1 \rangle = \dfrac{37}{26} \langle 5, 1 \rangle$

$\mathbf{w}_2 = \mathbf{u} - \mathbf{w}_1 = \langle 6, 7 \rangle - \dfrac{37}{26} \langle 5, 1 \rangle$

$= \left\langle -\dfrac{29}{26}, \dfrac{145}{26} \right\rangle$

$= \dfrac{29}{26} \langle -1, 5 \rangle$

$\mathbf{u} = \mathbf{w}_1 + \mathbf{w}_2 = \dfrac{37}{26} \langle 5, 1 \rangle + \dfrac{29}{26} \langle -1, 5 \rangle$

19. $\mathbf{F} = -500\mathbf{j}, \mathbf{v} = (\cos 12°)\mathbf{i} + (\sin 12°)\mathbf{j}$

$\mathbf{w}_1 = \text{proj}_{\mathbf{v}} \, \mathbf{F} = \left(\dfrac{\mathbf{F} \cdot \mathbf{v}}{\|\mathbf{v}\|^2} \right) \mathbf{v} = (\mathbf{F} \cdot \mathbf{v})\mathbf{v}$

$= (-500 \sin 12°)\mathbf{v}$

The magnitude of the force is $500 \sin 12° \approx 104$ pounds.

20. $z = 5 - 5i$

$|z| = \sqrt{5^2 + (-5)^2} = \sqrt{50} = 5\sqrt{2}$

$\tan \theta = \dfrac{-5}{5} = -1$ and θ is in Quadrant IV $\implies \theta = \dfrac{7\pi}{4}$

$z = 5\sqrt{2} \left(\cos \dfrac{7\pi}{4} + i \sin \dfrac{7\pi}{4} \right)$

21. $z = 6(\cos 120° + i \sin 120°)$

$= 6 \left(-\dfrac{1}{2} + \dfrac{\sqrt{3}}{2} i \right) = -3 + 3\sqrt{3} i$

22. $\left[3 \left(\cos \dfrac{7\pi}{6} + i \sin \dfrac{7\pi}{6} \right) \right]^8 = 3^8 \left(\cos \dfrac{28\pi}{3} + i \sin \dfrac{28\pi}{3} \right)$

$= 6561 \left(-\dfrac{1}{2} - \dfrac{\sqrt{3}}{2} i \right)$

$= -\dfrac{6561}{2} - \dfrac{6561\sqrt{3}}{2} i$

23. $(3 - 3i)^6 = \left[3\sqrt{2} \left(\cos \dfrac{7\pi}{4} + i \sin \dfrac{7\pi}{4} \right) \right]^6$

$= (3\sqrt{2})^6 \left(\cos \dfrac{21\pi}{2} + i \sin \dfrac{21\pi}{2} \right)$

$= 5832(0 + i)$

$= 5832i$

24. $z = 256(1 + \sqrt{3}i)$

$|z| = 256\sqrt{1^2 + (\sqrt{3})^2} = 256\sqrt{4} = 512$

$\tan \theta = \dfrac{\sqrt{3}}{1} \implies \theta = \dfrac{\pi}{3}$

$z = 512\left(\cos \dfrac{\pi}{3} + i \sin \dfrac{\pi}{3}\right)$

Fourth roots of $z = \sqrt[4]{512}\left[\cos \dfrac{\dfrac{\pi}{3} + 2\pi k}{4} + i \sin \dfrac{\dfrac{\pi}{3} + 2\pi k}{4}\right],\ k = 0, 1, 2, 3$

$k = 0:\ 4\sqrt[4]{2}\left(\cos \dfrac{\pi}{12} + i \sin \dfrac{\pi}{12}\right)$

$k = 1:\ 4\sqrt[4]{2}\left(\cos \dfrac{7\pi}{12} + i \sin \dfrac{7\pi}{12}\right)$

$k = 2:\ 4\sqrt[4]{2}\left(\cos \dfrac{13\pi}{12} + i \sin \dfrac{13\pi}{12}\right)$

$k = 3:\ 4\sqrt[4]{2}\left(\cos \dfrac{19\pi}{12} + i \sin \dfrac{19\pi}{12}\right)$

25. $x^3 - 27i = 0 \implies x = 27i$

The solutions to the equation are the cube roots of $27i = 27\left(\cos \dfrac{\pi}{2} + i \sin \dfrac{\pi}{2}\right)$.

Cube roots: $\sqrt[3]{27}\left[\cos \dfrac{\dfrac{\pi}{2} + 2\pi k}{3} + i \sin \dfrac{\dfrac{\pi}{2} + 2\pi k}{3}\right],\ k = 0, 1, 2$

$k = 0:\ 3\left(\cos \dfrac{\pi}{6} + i \sin \dfrac{\pi}{6}\right) = 3\left(\dfrac{\sqrt{3}}{2} + \dfrac{1}{2}i\right) = \dfrac{3\sqrt{3}}{2} + \dfrac{3}{2}i$

$k = 1:\ 3\left(\cos \dfrac{5\pi}{6} + i \sin \dfrac{5\pi}{6}\right) = 3\left(-\dfrac{\sqrt{3}}{2} + \dfrac{1}{2}i\right) = -\dfrac{3\sqrt{3}}{2} + \dfrac{3}{2}i$

$k = 2:\ 3\left(\cos \dfrac{3\pi}{2} + i \sin \dfrac{3\pi}{2}\right) = 3(0 - i) = -3i$

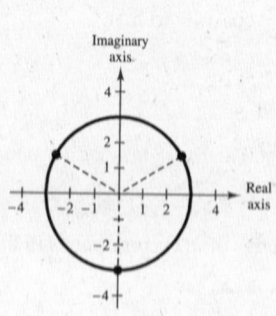

Chapters 4–6   Cumulative Test Solutions

1. (a)

(b) $-120° + 360° = 240°$

(c) $-120\left(\dfrac{\pi}{180°}\right) = -\dfrac{2\pi}{3}$

(d) $-120° + 360° = 240°$

$\theta' = 240° - 180° = 60°$

(e) $\sin(-120°) = -\sin 60° = -\dfrac{\sqrt{3}}{2}$

$\cos(-120°) = -\cos 60° = -\dfrac{1}{2}$

$\tan(-120°) = \tan 60° = \sqrt{3}$

$\csc(-120°) = \dfrac{1}{-\sin 60°} = -\dfrac{2\sqrt{3}}{3}$

$\sec(-120°) = \dfrac{1}{-\cos 60°} = -2$

$\cot(-120°) = \dfrac{1}{\tan 60°} = \dfrac{\sqrt{3}}{3}$

2. $2.35\left(\dfrac{180°}{\pi}\right) \approx 134.6°$

3. $\tan\theta = \dfrac{y}{x} = -\dfrac{4}{3} \Rightarrow r = 5$

Since $\sin\theta < 0$ θ is in Quadrant IV, $\Rightarrow x = 3$.

$\cos\theta = \dfrac{x}{r} = \dfrac{3}{5}$

4. $f(x) = 3 - 2\sin\pi x$

Period: $\dfrac{2\pi}{\pi} = 2$

Amplitude: $|a| = |-2| = 2$

Upward shift of 3 units (reflected in x-axis prior to shift)

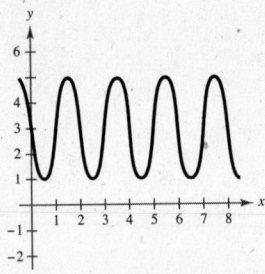

5. $g(x) = \dfrac{1}{2}\tan\left(x - \dfrac{\pi}{2}\right)$

Period: π

Asymptotes: $x = 0, x = \pi$

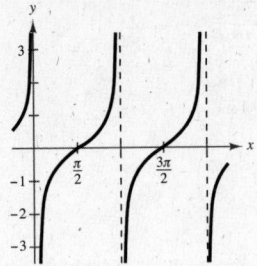

6. $h(x) = -\sec(x + \pi)$

Graph $y = -\cos(x + \pi)$ first.

Period: 2π

Amplitude: 1

Set $x + \pi = 0$ and $x + \pi = 2\pi$ for one cycle

$\qquad x = -\pi \qquad\qquad x = \pi$

The asymptotes of $h(x)$ corresponds to the x-intercepts of

$y = -\cos(x + \pi)$

$x + \pi = \dfrac{(2n + 1)\pi}{2}$

$x = \dfrac{(2n - 1)\pi}{2}$ where n is any integer

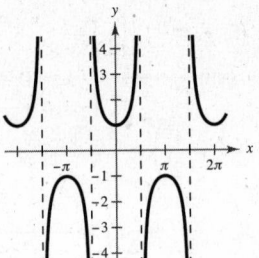

7. $h(x) = a\cos(bx + c)$

Graph is reflected in x-axis.

Amplitude: $a = -3$

Period: $2 = \dfrac{2\pi}{\pi} \Rightarrow b = \pi$

No phase shift: $c = 0$

$h(x) = -3\cos(\pi x)$

8. $f(x) = \dfrac{x}{2}\sin x, \; -3\pi \le x \le 3\pi$

$-\dfrac{x}{2} \le f(x) \le \dfrac{x}{2}$

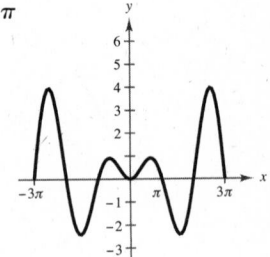

9. $\tan(\arctan 6.7) = 6.7$

10. $\tan\left(\arcsin \dfrac{3}{5}\right) = \dfrac{3}{4}$

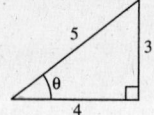

11. $y = \arccos(2x)$

$\sin y = \sin(\arccos(2x)) = \sqrt{1 - 4x^2}$

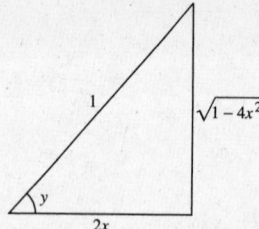

12. $\cos\left(\dfrac{\pi}{2} - x\right)\csc x = \sin x\left(\dfrac{1}{\sin x}\right) = 1$

13. $\dfrac{\sin \theta - 1}{\cos \theta} - \dfrac{\cos \theta}{\sin \theta - 1} = \dfrac{\sin \theta - 1}{\cos \theta} - \dfrac{\cos \theta(\sin \theta + 1)}{\sin^2 \theta - 1}$

$\qquad = \dfrac{\sin \theta - 1}{\cos \theta} + \dfrac{\cos \theta(\sin \theta + 1)}{\cos^2 \theta} = \dfrac{\sin \theta - 1}{\cos \theta} + \dfrac{\sin \theta + 1}{\cos \theta} = \dfrac{2 \sin \theta}{\cos \theta} = 2 \tan \theta$

14. $\cot^2 \alpha(\sec^2 \alpha - 1) = \cot^2 \alpha \tan^2 \alpha = 1$

15. $\sin(x + y) \sin(x - y) = \dfrac{1}{2}[\cos(x + y - (x - y)) - \cos(x + y + x - y)]$

$\qquad = \dfrac{1}{2}[\cos 2y - \cos 2x] = \dfrac{1}{2}[1 - 2 \sin^2 y - (1 - 2 \sin^2 x)] = \sin^2 x - \sin^2 y$

16. $\sin^2 x \cos^2 x = \left(\dfrac{1 - \cos 2x}{2}\right)\left(\dfrac{1 + \cos 2x}{2}\right)$

$\qquad = \dfrac{1}{4}(1 - \cos 2x)(1 + \cos 2x)$

$\qquad = \dfrac{1}{4}(1 - \cos^2 2x)$

$\qquad = \dfrac{1}{4}\left(1 - \dfrac{1 + \cos 4x}{2}\right)$

$\qquad = \dfrac{1}{8}(2 - (1 + \cos 4x))$

$\qquad = \dfrac{1}{8}(1 - \cos 4x)$

17. $2 \cos^2 \beta - \cos \beta = 0$

$\cos \beta(2 \cos \beta - 1) = 0$

$\cos \beta = 0 \qquad 2 \cos \beta - 1 = 0$

$\beta = \dfrac{\pi}{2}, \dfrac{3\pi}{2} \qquad \cos \beta = \dfrac{1}{2}$

$\qquad\qquad\qquad\qquad \beta = \dfrac{\pi}{3}, \dfrac{5\pi}{3}$

Answer: $\dfrac{\pi}{3}, \dfrac{\pi}{2}, \dfrac{3\pi}{2}, \dfrac{5\pi}{3}$

18. $3 \tan \theta - \cot \theta = 0$

$$3 \tan \theta - \frac{1}{\tan \theta} = 0$$

$$\frac{3 \tan^2 \theta - 1}{\tan \theta} = 0$$

$$3 \tan^2 \theta - 1 = 0$$

$$\tan^2 \theta = \frac{1}{3}$$

$$\tan \theta = \pm \frac{\sqrt{3}}{3}$$

$$\theta = \frac{\pi}{6}, \frac{5\pi}{6}, \frac{7\pi}{6}, \frac{11\pi}{6}$$

19. $\sin^2 x + 2 \sin x + 1 = 0$

$$(\sin x + 1)(\sin x + 1) = 0$$

$$\sin x + 1 = 0$$

$$\sin x = -1$$

$$x = \frac{3\pi}{2}$$

20. $\sin u = \frac{12}{13} \Rightarrow \cos u = \frac{5}{13}$ and $\tan u = \frac{12}{5}$ since u is in Quadrant I.

$\cos v = \frac{3}{5} \Rightarrow \sin v = \frac{4}{5}$ and $\tan v = \frac{4}{3}$ since v is in Quadrant I.

$$\tan(u - v) = \frac{\tan u - \tan v}{1 + \tan u \tan v} = \frac{\frac{12}{5} - \frac{4}{3}}{1 + \left(\frac{12}{5}\right)\left(\frac{4}{3}\right)} = \frac{16}{63}$$

21. $\tan \theta = \frac{1}{2}$

$$\tan 2\theta = \frac{2 \tan \theta}{1 - \tan^2 \theta} = \frac{2\left(\frac{1}{2}\right)}{1 - \left(\frac{1}{2}\right)^2} = \frac{4}{3}$$

22. $\tan \theta = \frac{4}{3} \Rightarrow \cos \theta = \pm \frac{3}{5}$

$$\sin \frac{\theta}{2} = \sqrt{\frac{1 - \cos \theta}{2}} = \sqrt{\frac{1 - \frac{3}{5}}{2}} = \frac{\sqrt{5}}{5}$$

$$\text{or} = \sqrt{\frac{1 + \frac{3}{5}}{2}} = \frac{2\sqrt{5}}{5}$$

23. $5 \sin \frac{3\pi}{4} \cos \frac{7\pi}{4} = \frac{5}{2}\left[\sin\left(\frac{3\pi}{4} + \frac{7\pi}{4}\right) + \sin\left(\frac{3\pi}{4} - \frac{7\pi}{4}\right)\right]$

$$= \frac{5}{2}\left[\sin \frac{5\pi}{2} + \sin(-\pi)\right]$$

$$= \frac{5}{2}\left(\sin \frac{5\pi}{2} - \sin \pi\right)$$

24. $\cos 8x + \cos 4x = 2 \cos\left(\frac{8x + 4x}{2}\right) \cos\left(\frac{8x - 4x}{2}\right)$

$$= 2 \cos 6x \cos 2x$$

25. Given: $A = 30°, a = 9, b = 8$

$$\frac{\sin B}{8} = \frac{\sin 30°}{9}$$

$$\sin B = \frac{8}{9}\left(\frac{1}{2}\right)$$

$$B = \arcsin\left(\frac{4}{9}\right)$$

$$B \approx 26.39°$$

$$C = 180° - A - B \approx 123.61°$$

$$\frac{c}{\sin 123.61°} = \frac{9}{\sin 30°}$$

$$c \approx 14.99$$

26. Given: $A = 30°, b = 8, c = 10$

$$a^2 = 8^2 + 10^2 - 2(8)(10)\cos 30°$$

$$a^2 \approx 25.4359$$

$$a \approx 5.04$$

$$\cos B = \frac{5.04^2 + 10^2 - 8^2}{2(5.04)(10)}$$

$$\cos B \approx 0.6091$$

$$B \approx 52.48°$$

$$C = 180° - A - B \approx 97.52°$$

27. Given: $A = 30°, C = 90°, b = 10$

$B = 180° - 30° - 90° = 60°$

$\tan 30° = \dfrac{a}{10} \Longrightarrow a = 10 \tan 30° \approx 5.77$

$\cos 30° = \dfrac{10}{c} \Longrightarrow c = \dfrac{10}{\cos 30°} \approx 11.55$

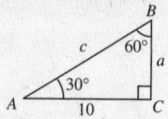

28. $a = 4, b = 8, c = 9$

$\cos C = \dfrac{4^2 + 8^2 - 9^2}{2(4)(8)} = \dfrac{-1}{64} \Longrightarrow C \approx 90.90°$

$\sin A \approx \dfrac{4 \sin 90.90°}{9} \Longrightarrow A \approx 26.38°$

$B \approx 180° - 26.38° - 90.90° = 62.72°$

29. Area $= \dfrac{1}{2}(7)(12) \sin 60° \approx 36.4$ square inches

30. $s = \dfrac{11 + 16 + 17}{2} = 22$

Area $= \sqrt{22(11)(6)(5)} \approx 85.2$ square inches

31. $\mathbf{u} = \langle 3, 5 \rangle = 3\mathbf{i} + 5\mathbf{j}$

32. $\mathbf{v} = \mathbf{i} + \mathbf{j}$

$\|\mathbf{v}\| = \sqrt{1^2 + 1^2} = \sqrt{2}$

$\mathbf{u} = \dfrac{\mathbf{v}}{\|\mathbf{v}\|} = \dfrac{1}{\sqrt{2}}(\mathbf{i} + \mathbf{j}) = \dfrac{\sqrt{2}}{2}(\mathbf{i} + \mathbf{j})$

33. $\mathbf{u} = 3\mathbf{i} + 4\mathbf{j}, \mathbf{v} = \mathbf{i} - 2\mathbf{j}$

$\mathbf{u} \cdot \mathbf{v} = 3(1) + 4(-2) = -5$

34. $\mathbf{u} = \langle 8, -2 \rangle, \mathbf{v} = \langle 1, 5 \rangle$

$\mathbf{w}_1 = \text{proj}_\mathbf{v}\,\mathbf{u} = \left(\dfrac{\mathbf{u} \cdot \mathbf{v}}{\|\mathbf{v}\|^2} \right)\mathbf{v} = \dfrac{-2}{26}\langle 1, 5 \rangle = -\dfrac{1}{13}\langle 1, 5 \rangle$

$\mathbf{w}_2 = \mathbf{u} - \mathbf{w}_1 = \langle 8, -2 \rangle - \left\langle -\dfrac{1}{13}, -\dfrac{5}{13} \right\rangle = \left\langle \dfrac{105}{13}, -\dfrac{21}{13} \right\rangle$

$= \dfrac{21}{13}\langle 5, -1 \rangle$

$\mathbf{u} = \mathbf{w}_1 + \mathbf{w}_2 = -\dfrac{1}{13}\langle 1, 5 \rangle + \dfrac{21}{13}\langle 5, -1 \rangle$

35. $r = |-2 + 2i| = \sqrt{(-2)^2 + (2)^2} = 2\sqrt{2}$

$\tan \theta = \dfrac{2}{-2} = -1$

Since $\tan \theta = -1$ and $-2 + 2i$ lies in Quadrant II,

$\theta = \dfrac{3\pi}{4}$. Thus, $-2 + 2i = 2\sqrt{2}\left(\cos\dfrac{3\pi}{4} + i \sin\dfrac{3\pi}{4} \right)$.

36. $[4(\cos 30° + i \sin 30°)][6(\cos 120° + i \sin 120°)] = (4)(6)[\cos(30° + 120°) + i \sin(30° + 120°)]$

$= 24(\cos 150° + i \sin 150°)$

$= 24\left(-\dfrac{\sqrt{3}}{2} + \dfrac{1}{2}i \right)$

$= -12\sqrt{3} + 12i$

37. $1 = 1(\cos 0 + i \sin 0)$

$$\sqrt[3]{1} = \sqrt[3]{1}\left[\cos\left(\frac{0 + 2\pi k}{3}\right) + i \sin\left(\frac{0 + 2\pi k}{3}\right)\right], k = 0, 1, 2$$

$$k = 0: \sqrt[3]{1}\left[\left(\cos\left(\frac{0 + 2\pi(0)}{3}\right) + i \sin\left(\frac{0 + 2\pi(0)}{3}\right)\right)\right] = \cos 0 + i \sin 0 = 1$$

$$k = 1: \sqrt[3]{1}\left[\left(\cos\left(\frac{0 + 2\pi(1)}{3}\right) + i \sin\left(\frac{0 + 2\pi(1)}{3}\right)\right)\right] = \cos\frac{2\pi}{3} + i \sin\frac{2\pi}{3} = -\frac{1}{2} + \frac{\sqrt{3}}{2}i$$

$$k = 2: \sqrt[3]{1}\left[\left(\cos\left(\frac{0 + 2\pi(2)}{3}\right) + i \sin\left(\frac{0 + 2\pi(2)}{3}\right)\right)\right] = \cos\frac{4\pi}{3} - i \sin\frac{4\pi}{3} = -\frac{1}{2} - \frac{\sqrt{3}}{2}i$$

38. $x^5 + 243 = 0 \implies x^5 = -243$

The solutions to the equation are the fifth roots of

$-243 = 243(\cos \pi + i \sin \pi)$, which are:

$$\sqrt[5]{243}\left[\cos\left(\frac{\pi + 2\pi k}{5} + i \sin\frac{\pi + 2\pi k}{5}\right)\right], k = 0, 1, 2, 3, 4$$

$$k = 0: 3\left(\cos\frac{\pi}{5} + i \sin\frac{\pi}{5}\right)$$

$$k = 1: 3\left(\cos\frac{3\pi}{5} + i \sin\frac{3\pi}{5}\right)$$

$$k = 2: 3(\cos \pi + i \sin \pi)$$

$$k = 3: 3\left(\cos\frac{7\pi}{5} + i \sin\frac{7\pi}{5}\right)$$

$$k = 4: 3\left(\cos\frac{9\pi}{5} + i \sin\frac{9\pi}{5}\right)$$

39. Angular speed $= \dfrac{\theta}{t} = \dfrac{2\pi(63)}{1} \approx 395.8$ radians per minute

Linear speed $= \dfrac{s}{t} = \dfrac{42\pi(63)}{1} \approx 8312.7$ inches per minute

40. Area $= \dfrac{\theta r^2}{2} = \dfrac{(114°)\left(\dfrac{\pi}{180°}\right)(8)^2}{2}$

$\approx 20.267\pi \approx 63.67$ square yards

41. Height of smaller triangle:

$\tan 16° 45' = \dfrac{h_1}{200}$

$h_1 = 200 \tan 16.75°$

≈ 60.2 feet

Height of larger triangle:

$\tan 18° = \dfrac{h_2}{200}$

$h_2 = 200 \tan 18° \approx 65.0$ feet

Height of flag: $h_2 - h_1 = 65.0 - 60.2 \approx 5$ feet

Not drawn to scale

42. $\tan \theta = \dfrac{5}{12} \implies \theta \approx 22.6°$

43. $d = a \cos bt$

$|a| = 4 \implies a = 4$

$\dfrac{2\pi}{b} = 8 \implies b = \dfrac{\pi}{4}$

$d = 4 \cos\dfrac{\pi}{4}t$

44. $\mathbf{v}_1 = 500\langle\cos 60°, \sin 60°\rangle = \langle 250, 250\sqrt{3}\rangle$

   $\mathbf{v}_2 = 50\langle\cos 30°, \sin 30°\rangle = \langle 25\sqrt{3}, 25\rangle$

   $\mathbf{v} = \mathbf{v}_1 + \mathbf{v}_2 = \langle 250 + 25\sqrt{3}, 250\sqrt{3} + 25\rangle$

   $\approx \langle 293.3, 458.0\rangle$

   $\|\mathbf{v}\| = \sqrt{(293.3)^2 + (458.0)^2} \approx 543.9$

   $\tan\theta = \dfrac{458.0}{293.3} \approx 1.56 \implies \theta \approx 57.4°$

   Bearing: $90° - 57.4° = 32.6°$

   The plane is traveling
   on a bearing of 32.6° at
   543.9 kilometers per hour.

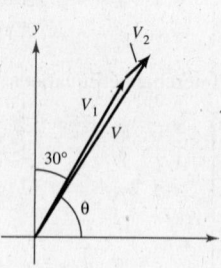

45. $\mathbf{w} = (85)(10)\cos 60° = 425$ foot-pounds

Chapter 7   Chapter Test Solutions

1. $\begin{cases} x - y = -7 \implies y = x + 7 \\ 4x + 5y = 8 \implies 4x + 5(x + 7) = 8 \end{cases}$

$9x + 35 = 8$

$9x = -27$

$x = -3 \implies y = 4$

Solution: $(-3, 4)$

2. $\begin{cases} y = x - 1 \\ y = (x - 1)^3 \end{cases}$

$x - 1 = (x - 1)^3$

$x - 1 = x^3 - 3x^2 + 3x - 1$

$0 = x^3 - 3x^2 + 2x$

$0 = x(x - 1)(x - 2)$

$x = 0, \quad x = 1, \quad x = 2$

$y = -1, \quad y = 0, \quad y = 1$

Solutions: $(0, -1), (1, 0), (2, 1)$

3. $\begin{cases} x - y = 4 \implies x = y + 4 \\ 2x - y^2 = 0 \implies 2(y + 4) - y^2 = 0 \end{cases}$

$0 = y^2 - 2y - 8$

$0 = (y + 2)(y - 4)$

$y = -2 \ \text{ or } \ y = 4$

$x = 2 \qquad\quad x = 8$

Solutions: $(2, -2), (8, 4)$

4. $\begin{cases} 2x - 3y = 0 \\ 2x + 3y = 12 \end{cases}$

Solution: $(3, 2)$

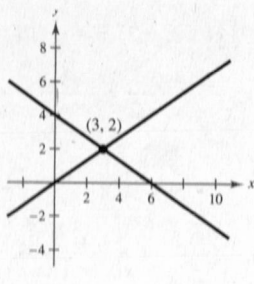

5. $\begin{cases} y = 9 - x^2 \\ y = x + 3 \end{cases}$

Solutions: $(-3, 0), (2, 5)$

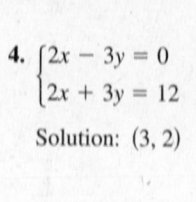

6. $\begin{cases} y - \ln x = 12 \implies y = 12 + \ln x \\ 7x - 2y + 11 = -6 \implies y = \frac{7}{2}x + \frac{17}{2} \end{cases}$

Solutions:
$(1, 12), (0.034, 8.619)$

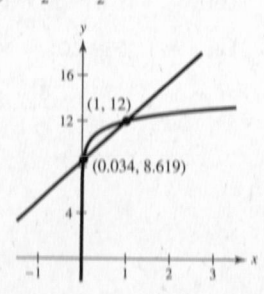

7. $\begin{cases} 2x + 3y = 17 \\ 5x - 4y = -15 \end{cases} \Rightarrow \begin{array}{r} 8x + 12y = 68 \\ 15x - 12y = -45 \\ \hline 23x = 23 \end{array}$

$$x = 1 \Rightarrow y = 5$$

Solution: $(1, 5)$

8. $\begin{cases} 2.5x - y = 6 \\ 3x + 4y = 2 \end{cases} \Rightarrow \begin{array}{r} 10x - 4y = 24 \\ 3x + 4y = 2 \\ \hline 13x = 26 \end{array}$

$$x = 2 \Rightarrow y = -1$$

Solution: $(2, -1)$

9. $\begin{cases} x - 2y + 3z = 11 \\ 2x - z = 3 \\ 3y + z = -8 \end{cases}$

$\begin{cases} x - 2y + 3z = 11 \\ 4y - 7z = -19 \\ 3y + z = -8 \end{cases}$ $\quad -2\text{Eq.1} + \text{Eq.2}$

$\begin{cases} x - 2y + 3z = 11 \\ y - 8z = -11 \\ 3y + z = -8 \end{cases}$ $\quad -\text{Eq.3} + \text{Eq.2}$

$\begin{cases} x - 2y + 3z = 11 \\ y - 8z = -11 \\ 25z = 25 \end{cases}$ $\quad -3\text{Eq.2} + \text{Eq.3}$

$\begin{cases} x - 2y + 3z = 11 \\ y - 8z = -11 \\ z = 1 \end{cases}$ $\quad \frac{1}{25}\text{Eq.3}$

$$y - 8(1) = -11 \Rightarrow y = -3$$

$$x - 2(-3) + 3(1) = 11 \Rightarrow x = 2$$

Solution: $(2, -3, 1)$

10. $\begin{cases} 3x + 2y + z = 17 & \text{Equation 1} \\ -x + y + z = 4 & \text{Equation 2} \\ x - y - z = 3 & \text{Equation 3} \end{cases}$

Interchange Equations 1 and 3

$\begin{cases} x - y - z = 3 \\ -x + y + z = 4 \\ 3x + 2y + z = 17 \end{cases}$

$\begin{cases} x - y - z = 3 \\ 0 \neq 7 & \text{Eq. 1 + Eq. 2} \\ 3x + 2y + z = 17 \end{cases}$

Inconsistent
No solution

11. $\dfrac{2x + 5}{x^2 - x - 2} = \dfrac{2x + 5}{(x - 2)(x + 1)} = \dfrac{A}{x - 2} + \dfrac{B}{x + 1}$

$$2x + 5 = A(x + 1) + B(x - 2)$$

Let $x = 2$: $9 = 3A \Rightarrow A = 3$

Let $x = -1$: $3 = -3B \Rightarrow B = -1$

$$\dfrac{2x + 5}{x^2 - x - 2} = \dfrac{3}{x - 2} - \dfrac{1}{x + 1}$$

12. $\dfrac{3x^2 - 2x + 4}{x^2(2 - x)} = \dfrac{A}{x} + \dfrac{B}{x^2} + \dfrac{C}{2 - x}$

$$3x^2 - 2x + 4 = Ax(2 - x) + B(2 - x) + Cx^2$$

Let $x = 0$: $4 = 2B \Rightarrow B = 2$

Let $x = 2$: $12 = 4C \Rightarrow C = 3$

Let $x = 1$: $5 = A + B + C = A + 2 + 3 \Rightarrow A = 0$

$$\dfrac{3x^2 - 2x + 4}{x^2(2 - x)} = \dfrac{2}{x^2} + \dfrac{3}{2 - x}$$

13. $\dfrac{x^2 + 5}{x^3 - x} = \dfrac{x^2 + 5}{x(x + 1)(x - 1)} = \dfrac{A}{x} + \dfrac{B}{x + 1} + \dfrac{C}{x - 1}$

$$x^2 + 5 = A(x + 1)(x - 1) + Bx(x - 1) + Cx(x + 1)$$

Let $x = 0$: $5 = -A \Rightarrow A = -5$

Let $x = -1$: $6 = 2B \Rightarrow B = 3$

Let $x = 1$: $6 = 2C \Rightarrow C = 3$

$$\dfrac{x^2 + 5}{x^3 - x} = -\dfrac{5}{x} + \dfrac{3}{x + 1} + \dfrac{3}{x - 1}$$

14. $\dfrac{x^2 - 4}{x^3 + 2x} = \dfrac{x^2 - 4}{x(x^2 + 2)} = \dfrac{A}{x} + \dfrac{Bx + C}{x^2 + 2}$

$$\begin{aligned} x^2 - 4 &= A(x^2 + 2) + (Bx + C)x \\ &= Ax^2 + 2A + Bx^2 + Cx \\ &= (A + B)x^2 + Cx + 2A \end{aligned}$$

Equate the coefficients of like terms:

$$1 = A + B, \quad 0 = C, \quad -4 = 2A$$

Thus, $A = -2, B = 3, C = 0$.

$$\dfrac{x^2 - 4}{x^3 + 2x} = -\dfrac{2}{x} + \dfrac{3x}{x^2 + 2}$$

15. $2x + y \le 4$

$2x - y \ge 0$

$x \ge 0$

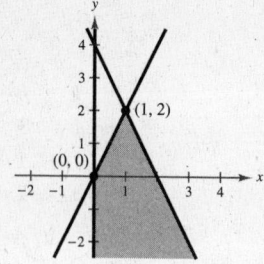

16. $y < -x^2 + x + 4$

$y > 4x$

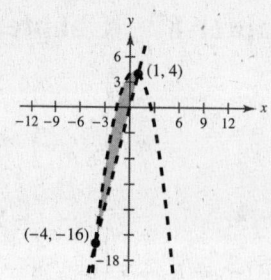

17. $x^2 + y^2 \le 16$

$x \ge 1$

$y \ge -3$

18. Maximize $z = 20x + 12y$ subject to:

$$\begin{cases} x \ge 0, y \ge 0 \\ x + 4y \le 32 \\ 3x + 2y \le 36 \end{cases}$$

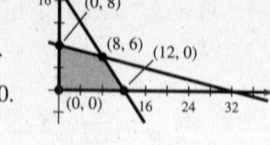

At $(0, 0)$ we have $z = 0$.

At $(0, 8)$ we have $z = 96$.

At $(8, 6)$ we have $z = 232$.

At $(12, 0)$ we have $z = 240$.

The maximum value, $z = 240$, occurs at $(12, 0)$.

The minimum value, $z = 0$ occurs at $(0, 0)$.

19. Let $x = $ amount in 8% fund.

Let $y = $ amount in 8.5% fund.

$x + y = 50{,}000 \implies y = 50{,}000 - x$

$0.08x + 0.085y = 4150$

Use substitution

$0.08x + 0.085(50{,}000 - x) = 4150$

$0.08x + 4250 - 0.085x = 4150$

$100 = 0.005x$

$20{,}000 = x \implies y = 30{,}000$

$20{,}000 is invested at 8%.

$30{,}000 is invested at 8.5%.

20. $y = ax^2 + bx + c$

$(0, 6)$: $6 = c$

$(-2, 2)$: $2 = 4a - 2b + c$

$\left(3, \frac{9}{2}\right)$: $\frac{9}{2} = 9a + 3b + c$

Solving this system yields: $a = -\frac{1}{2}$, $b = 1$, and $c = 6$.

Thus, $y = -\frac{1}{2}x^2 + x + 6$.

21. Optimize $P = 30x + 40y$ subject to:

$$\begin{cases} x \ge 0, y \ge 0 \\ 0.5x + 0.75y \le 4000 \\ 2.0x + 1.5y \le 8950 \\ 0.5x + 0.5y \le 2650 \end{cases}$$

At $(0, 0)$: $P = 0$

At $(0, 5300)$: $P = 212{,}000$

At $(2000, 3300)$: $P = 192{,}000$

At $(4475, 0)$: $P = 134{,}250$

The manufacturer should produce 5300 units of Model II and not produce any of Model I to realize an optimal profit of $212,000.

Chapter 8   Chapter Test Solutions

1.
$$\begin{bmatrix} 1 & -1 & 5 \\ 6 & 2 & 3 \\ 5 & 3 & -3 \end{bmatrix}$$

$$\begin{matrix} -6R_1 + R_2 \rightarrow \\ -5R_1 + R_3 \rightarrow \end{matrix} \begin{bmatrix} 1 & -1 & 5 \\ 0 & 8 & -27 \\ 0 & 8 & -28 \end{bmatrix}$$

$$-R_2 + R_3 \rightarrow \begin{bmatrix} 1 & -1 & 5 \\ 0 & 8 & -27 \\ 0 & 0 & -1 \end{bmatrix}$$

$$\begin{matrix} \frac{1}{8}R_2 \rightarrow \\ -R_3 \rightarrow \end{matrix} \begin{bmatrix} 1 & -1 & 5 \\ 0 & 1 & -\frac{27}{8} \\ 0 & 0 & 1 \end{bmatrix}$$

$$R_2 + R_1 \rightarrow \begin{bmatrix} 1 & 0 & \frac{13}{8} \\ 0 & 1 & -\frac{27}{8} \\ 0 & 0 & 1 \end{bmatrix}$$

$$\begin{matrix} -\frac{13}{8}R_3 + R_1 \rightarrow \\ \frac{27}{8}R_3 + R_2 \rightarrow \end{matrix} \begin{bmatrix} 1 & 0 & 0 \\ 0 & 1 & 0 \\ 0 & 0 & 1 \end{bmatrix}$$

2.
$$\begin{bmatrix} 1 & 0 & -1 & 2 \\ -1 & 1 & 1 & -3 \\ 1 & 1 & -1 & 1 \\ 3 & 2 & -3 & 4 \end{bmatrix}$$

$$\begin{matrix} R_1 + R_2 \rightarrow \\ -R_1 + R_3 \rightarrow \\ -3R_1 + R_4 \rightarrow \end{matrix} \begin{bmatrix} 1 & 0 & -1 & 2 \\ 0 & 1 & 0 & -1 \\ 0 & 1 & 0 & -1 \\ 0 & 2 & 0 & -2 \end{bmatrix}$$

$$\begin{matrix} -R_2 + R_3 \rightarrow \\ -2R_2 + R_4 \rightarrow \end{matrix} \begin{bmatrix} 1 & 0 & -1 & 2 \\ 0 & 1 & 0 & -1 \\ 0 & 0 & 0 & 0 \\ 0 & 0 & 0 & 0 \end{bmatrix}$$

3.
$$\begin{bmatrix} 4 & 3 & -2 & \vdots & 14 \\ -1 & -1 & 2 & \vdots & -5 \\ 3 & 1 & -4 & \vdots & 8 \end{bmatrix}$$

$$3R_2 + R_1 \rightarrow \begin{bmatrix} 1 & 0 & 4 & \vdots & -1 \\ -1 & -1 & 2 & \vdots & -5 \\ 3 & 1 & -4 & \vdots & 8 \end{bmatrix}$$

$$\begin{matrix} R_1 + R_2 \rightarrow \\ -3R_1 + R_3 \rightarrow \end{matrix} \begin{bmatrix} 1 & 0 & 4 & \vdots & -1 \\ 0 & -1 & 6 & \vdots & -6 \\ 0 & 1 & -16 & \vdots & 11 \end{bmatrix}$$

$$R_2 + R_3 \rightarrow \begin{bmatrix} 1 & 0 & 4 & \vdots & -1 \\ 0 & -1 & 6 & \vdots & -6 \\ 0 & 0 & -10 & \vdots & 5 \end{bmatrix}$$

$$\begin{matrix} -R_2 \rightarrow \\ -\frac{1}{10}R_3 \rightarrow \end{matrix} \begin{bmatrix} 1 & 0 & 4 & \vdots & -1 \\ 0 & 1 & -6 & \vdots & 6 \\ 0 & 0 & 1 & \vdots & -\frac{1}{2} \end{bmatrix}$$

$$\begin{matrix} -4R_3 + R_1 \rightarrow \\ 6R_3 + R_2 \rightarrow \end{matrix} \begin{bmatrix} 1 & 0 & 0 & \vdots & 1 \\ 0 & 1 & 0 & \vdots & 3 \\ 0 & 0 & 1 & \vdots & -\frac{1}{2} \end{bmatrix}$$

Solution: $\left(1, 3, -\frac{1}{2}\right)$

4. (a) $A - B = \begin{bmatrix} 5 & 4 \\ -4 & -4 \end{bmatrix} - \begin{bmatrix} 4 & -1 \\ -4 & 0 \end{bmatrix}$

$$= \begin{bmatrix} 1 & 5 \\ 0 & -4 \end{bmatrix}$$

(b) $3A = 3\begin{bmatrix} 5 & 4 \\ -4 & -4 \end{bmatrix} = \begin{bmatrix} 15 & 12 \\ -12 & -12 \end{bmatrix}$

(c) $3A - 2B = 3\begin{bmatrix} 5 & 4 \\ -4 & -4 \end{bmatrix} - 2\begin{bmatrix} 4 & -1 \\ -4 & 0 \end{bmatrix}$

$$= \begin{bmatrix} 15 & 12 \\ -12 & -12 \end{bmatrix} - \begin{bmatrix} 8 & -2 \\ -8 & 0 \end{bmatrix}$$

$$= \begin{bmatrix} 7 & 14 \\ -4 & -12 \end{bmatrix}$$

(d) $AB = \begin{bmatrix} 5 & 4 \\ -4 & -4 \end{bmatrix}\begin{bmatrix} 4 & -1 \\ -4 & 0 \end{bmatrix}$

$$= \begin{bmatrix} (5)(4) + (4)(-4) & (5)(-1) + (4)(0) \\ (-4)(4) + (-4)(-4) & (-4)(-1) + (-4)(0) \end{bmatrix}$$

$$= \begin{bmatrix} 4 & -5 \\ 0 & 4 \end{bmatrix}$$

5. $\begin{bmatrix} -6 & 4 \\ 10 & -5 \end{bmatrix}^{-1} = \dfrac{1}{(-6)(-5) - (4)(10)}\begin{bmatrix} -5 & -4 \\ -10 & -6 \end{bmatrix} = \begin{bmatrix} \frac{1}{2} & \frac{2}{5} \\ 1 & \frac{3}{5} \end{bmatrix}$

6.
$$\begin{bmatrix} -2 & 4 & -6 & \vdots & 1 & 0 & 0 \\ 2 & 1 & 0 & \vdots & 0 & 1 & 0 \\ 4 & -2 & 5 & \vdots & 0 & 0 & 1 \end{bmatrix}$$

$$\begin{matrix} R_1 + R_2 \to \\ 2R_1 + R_3 \to \end{matrix} \begin{bmatrix} -2 & 4 & -6 & \vdots & 1 & 0 & 0 \\ 0 & 5 & -6 & \vdots & 1 & 1 & 0 \\ 0 & 6 & -7 & \vdots & 2 & 0 & 1 \end{bmatrix}$$

$$\begin{matrix} -\frac{1}{2}R_1 \to \\ -R_3 + R_2 \to \end{matrix} \begin{bmatrix} 1 & -2 & 3 & \vdots & -\frac{1}{2} & 0 & 0 \\ 0 & -1 & 1 & \vdots & -1 & 1 & -1 \\ 0 & 6 & -7 & \vdots & 2 & 0 & 1 \end{bmatrix}$$

$$\begin{matrix} -2R_2 + R_1 \to \\ \\ 6R_2 + R_3 \to \end{matrix} \begin{bmatrix} 1 & 0 & 1 & \vdots & \frac{3}{2} & -2 & 2 \\ 0 & -1 & 1 & \vdots & -1 & 1 & -1 \\ 0 & 0 & -1 & \vdots & -4 & 6 & -5 \end{bmatrix}$$

$$\begin{matrix} \\ -R_2 \to \\ -R_3 \to \end{matrix} \begin{bmatrix} 1 & 0 & 1 & \vdots & \frac{3}{2} & -2 & 2 \\ 0 & 1 & -1 & \vdots & 1 & -1 & 1 \\ 0 & 0 & 1 & \vdots & 4 & -6 & 5 \end{bmatrix}$$

$$\begin{matrix} -R_3 + R_1 \to \\ R_3 + R_2 \to \end{matrix} \begin{bmatrix} 1 & 0 & 0 & \vdots & -\frac{5}{2} & 4 & -3 \\ 0 & 1 & 0 & \vdots & 5 & -7 & 6 \\ 0 & 0 & 1 & \vdots & 4 & -6 & 5 \end{bmatrix}$$

$$A^{-1} = \begin{bmatrix} -\frac{5}{2} & 4 & -3 \\ 5 & -7 & 6 \\ 4 & -6 & 5 \end{bmatrix}$$

7. $\begin{cases} -6x + 4y = 10 \\ 10x - 5y = 20 \end{cases}$

$$\begin{bmatrix} x \\ y \end{bmatrix} = \begin{bmatrix} \frac{1}{2} & \frac{2}{5} \\ 1 & \frac{3}{5} \end{bmatrix} \begin{bmatrix} 10 \\ 20 \end{bmatrix} = \begin{bmatrix} \frac{1}{2}(10) + \frac{2}{5}(20) \\ 1(10) + \frac{3}{5}(20) \end{bmatrix} = \begin{bmatrix} 13 \\ 22 \end{bmatrix}$$

Solution: $(13, 22)$

8. $\begin{vmatrix} -9 & 4 \\ 13 & 16 \end{vmatrix} = (-9)(16) - (4)(13) = -196$

9. $\begin{vmatrix} \frac{5}{2} & \frac{13}{4} \\ -8 & \frac{6}{5} \end{vmatrix} = \left(\frac{5}{2}\right)\left(\frac{6}{5}\right) - \left(\frac{13}{4}\right)(-8) = 29$

10. $\begin{vmatrix} 6 & -7 & 2 \\ 3 & -2 & 0 \\ 1 & 5 & 1 \end{vmatrix} = 2\begin{vmatrix} 3 & -2 \\ 1 & 5 \end{vmatrix} + \begin{vmatrix} 6 & -7 \\ 3 & -2 \end{vmatrix} = 2(17) + 9 = 43$

Expand along column 3.

11. $\begin{cases} 7x + 6y = 9 \\ -2x - 11y = -49 \end{cases}$ $D = \begin{vmatrix} 7 & 6 \\ -2 & -11 \end{vmatrix} = -65$

$$x = \frac{\begin{vmatrix} 9 & 6 \\ -49 & -11 \end{vmatrix}}{-65} = \frac{195}{-65} = -3$$

$$y = \frac{\begin{vmatrix} 7 & 9 \\ -2 & -49 \end{vmatrix}}{-65} = \frac{-325}{-65} = 5$$

Solution: $(-3, 5)$

12. $\begin{cases} 6x - y + 2z = -4 \\ -2x + 3y - z = 10 \\ 4x - 4y + z = -18 \end{cases}$ $D = \begin{vmatrix} 6 & -1 & 2 \\ -2 & 3 & -1 \\ 4 & -4 & 1 \end{vmatrix} = -12$

$$x = \frac{\begin{vmatrix} -4 & -1 & 2 \\ 10 & 3 & -1 \\ -18 & -4 & 1 \end{vmatrix}}{-12} = \frac{24}{-12} = -2$$

$$y = \frac{\begin{vmatrix} 6 & -4 & 2 \\ -2 & 10 & -1 \\ 4 & -18 & 1 \end{vmatrix}}{-12} = \frac{-48}{-12} = 4$$

$$z = \frac{\begin{vmatrix} 6 & -1 & -4 \\ -2 & 3 & 10 \\ 4 & -4 & -18 \end{vmatrix}}{-12} = \frac{-72}{-12} = 6$$

Solution: $(-2, 4, 6)$

13. $A = -\frac{1}{2}\begin{vmatrix} -5 & 0 & 1 \\ 4 & 4 & 1 \\ 3 & 2 & 1 \end{vmatrix} = -\frac{1}{2}(-14) = 7$

14.

$$\begin{matrix} K & N & O \\ C & K & - \\ O & N & - \\ W & O & O \\ D & - & - \end{matrix}\begin{bmatrix} 11 & 14 & 15 \\ 3 & 11 & 0 \\ 15 & 14 & 0 \\ 23 & 15 & 15 \\ 4 & 0 & 0 \end{bmatrix}\begin{bmatrix} 1 & -1 & 0 \\ 1 & 0 & -1 \\ 6 & -2 & -3 \end{bmatrix}=\begin{bmatrix} 115 & -41 & -59 \\ 14 & -3 & -11 \\ 29 & -15 & -14 \\ 128 & -53 & -60 \\ 4 & -4 & 0 \end{bmatrix}$$

Message: $[11\ 14\ 15], [3\ 11\ 0], [15\ 14\ 0], [23\ 15\ 15], [4\ 0\ 0]$

Encoded Message: $115\ -41\ -59\ 14\ -3\ -11\ 29\ -15\ -14\ 128\ -53\ -60\ 4\ -4\ 0$

15. Let $x =$ amount of 60% solution and $y =$ amount of 20% solution.

$$\begin{cases} x + y = 100 \implies y = 100 - x \\ 0.60x + 0.20y = 0.50(100) \implies 6x + 2y = 500 \end{cases}$$

By substitution, we have

$$6x + 2(100 - x) = 500$$
$$6x + 200 - 2x = 500$$
$$4x = 300$$
$$x = 75$$
$$y = 100 - x = 25$$

Answer: 75 liters of 60% solution and 25 liters of 20% solution.

Chapter 9   Chapter Test Solutions

1. $a_n = \dfrac{(-1)^n}{3n+2}$

$a_1 = -\dfrac{1}{5}$

$a_2 = \dfrac{1}{8}$

$a_3 = -\dfrac{1}{11}$

$a_4 = \dfrac{1}{14}$

$a_5 = -\dfrac{1}{17}$

2. $\dfrac{3}{1!}, \dfrac{4}{2!}, \dfrac{5}{3!}, \dfrac{6}{4!}, \dfrac{7}{5!} \cdots$

$a_n = \dfrac{n+2}{n!}$

3. $6 + 17 + 28 + 39 + \cdots$

$a_n = 11n - 5$

$a_5 = 50, a_6 = 61, a_7 = 72$

$S_5 = 6 + 17 + 28 + 39 + 50$

$= 140$

4. $a_5 = 5.4, a_{12} = 11.0$

$a_{12} = a_5 + 7d$

$11.0 = 5.4 + 7d$

$5.6 = 7d$

$0.8 = d$

$a_1 = a_5 - 4d$

$a_1 = 5.4 - 4(0.8)$

$= 2.2$

$a_n = a_1 + (n-1)d$

$= 2.2 + (n-1)(0.8)$

$= 0.8n + 1.4$

5. $a_n = 5(2)^{n-1}$

$a_1 = 5$

$a_2 = 10$

$a_3 = 20$

$a_4 = 40$

$a_5 = 80$

6. $\displaystyle\sum_{i=1}^{50}(2i^2+5)=2\sum_{i=1}^{50}i^2+\sum_{i=1}^{50}5$

$\displaystyle\qquad\qquad\qquad = 2\left[\frac{50(51)(101)}{6}\right]+50(5)$

$\displaystyle\qquad\qquad\qquad = 86{,}100$

7. $\displaystyle\sum_{n=1}^{7}(8n-5)=8\sum_{n=1}^{7}n-\sum_{n=1}^{7}5$

$\displaystyle\qquad\qquad\qquad = 8\left[\frac{(7)(8)}{2}\right]-7(5)$

$\displaystyle\qquad\qquad\qquad = 189$

8. $\displaystyle\sum_{i=1}^{\infty}4\left(\frac{1}{2}\right)^i=\frac{2}{1-\frac{1}{2}}=4$

9. $5+10+15+\cdots+5n=\dfrac{5n(n+1)}{2}$

When $n=1,\ S_1=5=\dfrac{5(1)(2)}{2}$, so the formula is valid.

Assume that $S_k=5+10+15+\cdots+5k=\dfrac{5k(k+1)}{2}$, then

$S_{k+1}=S_k+a_{k+1}$

$\qquad = \dfrac{5k(k+1)}{2}+5(k+1)$

$\qquad = \dfrac{5k(k+1)}{2}+\dfrac{10(k+1)}{2}$

$\qquad = \dfrac{5k(k+1)+10(k+1)}{2}$

$\qquad = \dfrac{5(k+1)(k+2)}{2}$

$\qquad = \dfrac{5(k+1)[(k+1)+1]}{2}.$

Thus, the formula is valid for all integers $n\ge 1$.

10. $(x+2y)^4=x^4+4x^3(2y)+6x^2(2y)^2+4x(2y)^3+(2y)^4$

$\qquad\qquad = x^4+8x^3y+24x^2y^2+32xy^3+16y^4$

11. $_nC_r\,x^{n-r}y^r={}_8C_5(2a)^3(-3b)^5$

$\qquad\qquad = 56(8a^3)(-243b^5)$

$\qquad\qquad = -108{,}864a^3b^5$

So, the coefficient of a^3b^5 is $-108{,}864$.

12. (a) $_9P_2=\dfrac{9!}{7!}=72$

(b) $_{70}P_3=\dfrac{70!}{67!}=328{,}440$

13. (a) $_{11}C_4=\dfrac{11!}{7!4!}=330$

(b) $_{66}C_4=\dfrac{66!}{62!4!}=720{,}720$

14. $(26)(10)(10)(10)=26{,}000$ distinct license plates

15. $\underbrace{(1)}_{\text{owner}}\cdot\underbrace{(3)(2)}_{\substack{\text{bow}\\\text{seats}}}\cdot\underbrace{(5)(4)(3)(2)(1)}_{\substack{\text{remaining}\\\text{seats}}}=720$ seating arrangements

16. $\dfrac{20}{300}=\dfrac{1}{15}\approx 0.0667$

17. $\dfrac{1}{_{60}C_8}\approx 3.908\times 10^{-10}$

18. $P(E')=1-P(E)$

$\qquad\qquad = 1-0.75$

$\qquad\qquad = 0.25$ or 25%

Chapters 7–9   Cumulative Test Solutions

1. $\begin{cases} y = 3 - x^2 \\ 2(y-2) = x - 1 \Rightarrow 2(3 - x^2 - 2) = x - 1 \end{cases}$

$$2(1 - x^2) = x - 1$$
$$2 - 2x^2 = x - 1$$
$$0 = 2x^2 + x - 3$$
$$0 = (2x + 3)(x - 1)$$
$$x = -\tfrac{3}{2} \text{ or } x = 1$$
$$y = \tfrac{3}{4} \qquad y = 2$$

Solutions: $\left(-\tfrac{3}{2}, \tfrac{3}{4}\right), (1, 2)$

2. $\begin{cases} x + 3y = -1 \Rightarrow \quad 4x + 12y = -4 \\ 2x + 4y = \quad 0 \Rightarrow \underline{-6x - 12y = \quad 0} \\ \qquad\qquad\qquad\qquad -2x \qquad\quad = -4 \\ \qquad\qquad\qquad\qquad x = 2 \Rightarrow y = -1 \end{cases}$

Solution: $(2, -1)$

3. $\begin{cases} -2x + 4y - z = \quad 3 \\ x - 2y + 2z = -6 \\ x - 3y - z = \quad 1 \end{cases}$

Interchange equations.

$\begin{cases} x - 2y + 2z = -6 & \text{Eq.1} \\ -2x + 4y - z = \quad 3 & \text{Eq.2} \\ x - 3y - z = \quad 1 & \text{Eq.3} \end{cases}$

$\begin{cases} x - 2y + 2z = -6 \\ \qquad\qquad 3z = -9 & 2\text{Eq.1} + \text{Eq.2} \\ \qquad -y - 3z = \quad 7 & -\text{Eq.1} + \text{Eq.3} \end{cases}$

From Equation 2 we have $z = -3$. Substituting this into Equation 3 yields $y = 2$. Using these in Equation 1 yields $x = 4$.

Solution: $(4, 2, -3)$

4. $\begin{cases} x + \quad 3y - 2z = -7 \\ -2x + \quad y - z = -5 \\ 4x + \quad y + z = \quad 3 \end{cases}$

$\begin{cases} x + \quad 3y - 2z = -7 \\ \qquad 7y - 5z = -19 & 2\text{Eq.1} + \text{Eq.2} \\ \qquad -11y + 9z = \quad 31 & -4\text{Eq.1} + \text{Eq.3} \end{cases}$

$\begin{cases} x + \quad 3y - 2z = -7 \\ \qquad y - \tfrac{5}{7}z = -\tfrac{19}{7} & \tfrac{1}{7}\text{Eq.2} \\ \qquad -11y + 9z = \quad 31 \end{cases}$

$\begin{cases} x \qquad\quad + \tfrac{1}{7}z = \quad \tfrac{8}{7} & -3\text{Eq.2} + \text{Eq.1} \\ \qquad y - \tfrac{5}{7}z = -\tfrac{19}{7} \\ \qquad\qquad \tfrac{8}{7}z = \quad \tfrac{8}{7} & 11\text{Eq.2} + \text{Eq.3} \end{cases}$

$\begin{cases} x \qquad\quad + \tfrac{1}{7}z = \quad \tfrac{8}{7} \\ \qquad y - \tfrac{5}{7}z = -\tfrac{19}{7} \\ \qquad\qquad z = \quad 1 & \tfrac{7}{8}\text{Eq.3} \end{cases}$

$\begin{cases} x \qquad\qquad = \quad 1 & -\tfrac{1}{7}\text{Eq.3} + \text{Eq.1} \\ \qquad y \qquad = -2 & \tfrac{5}{7}\text{Eq.3} + \text{Eq.2} \\ \qquad\qquad z = \quad 1 \end{cases}$

Solution: $(1, -2, 1)$

5. $\begin{cases} 2x + \quad y \geq -3 \\ x - 3y \leq \quad 2 \end{cases}$

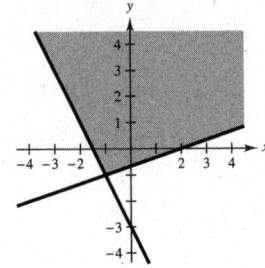

6. $\begin{cases} x - y > 6 \\ 5x + 2y < 10 \end{cases}$

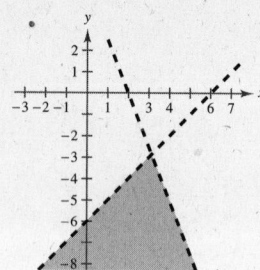

7. Objective function: $z = 3x + 2y$

Subject to: $x + 4y \le 20$

$2x + \ y \le 12$

$x \ge 0, y \ge 0$

At $(0, 0)$: $z = 0$

At $(0, 5)$: $z = 10$

At $(4, 4)$: $z = 20$

At $(6, 0)$: $z = 18$

Minimum of $z = 0$ at $(0, 0)$

Maximum of $z = 20$ at $(4, 4)$

8. $\begin{cases} x + y = 200 \implies y = 200 - x \\ 0.75x + 1.25y = 0.95(200) \end{cases}$

$0.75x + 1.25(200 - x) = 190$

$0.75x + 250 - 1.25x = 190$

$-0.50x = -60$

$x = 120$

$y = 200 - x = 80$

Answer: 120 pounds of \$0.75 seed and 80 pounds of \$1.25 seed.

9. $y = ax^2 + bx + c$

$(0, 4)$: $4 = a(0)^2 + b(0) + c \implies c = 4$

$(3, 1)$: $1 = a(3)^2 + b(3) + 4 \implies 9a + 3b = -3$

$3a + b = -1$

$(6, 4)$: $4 = a(6)^2 + b(6) + 4 \implies 36a + 6b = 0$

$6a + b = 0$

Solving the system:

$\begin{cases} 3a + b = -1 \\ 6a + b = \ \ 0 \end{cases}$ yields $a = \frac{1}{3}$ and $b = -2$.

Thus, the equation of the parabola is $y = \frac{1}{3}x^2 - 2x + 4$.

10. $\begin{cases} -x + 2y - z = \ \ 9 \\ 2x - y + 2z = -9 \\ 3x + 3y - 4z = \ \ 7 \end{cases}$ $\qquad \begin{bmatrix} -1 & 2 & -1 & \vdots & 9 \\ 2 & -1 & 2 & \vdots & -9 \\ 3 & 3 & -4 & \vdots & 7 \end{bmatrix}$

11.

$\begin{bmatrix} -1 & 2 & -1 & \vdots & 9 \\ 2 & -1 & 2 & \vdots & -9 \\ 3 & 3 & -4 & \vdots & 7 \end{bmatrix}$

$\begin{matrix} \\ 2R_1 + R_2 \to \\ 3R_1 + R_3 \to \end{matrix} \begin{bmatrix} -1 & 2 & -1 & \vdots & 9 \\ 0 & 3 & 0 & \vdots & 9 \\ 0 & 9 & -7 & \vdots & 34 \end{bmatrix}$

$\begin{matrix} -R_1 \to \\ \\ -3R_2 + R_3 \to \end{matrix} \begin{bmatrix} 1 & -2 & 1 & & -9 \\ 0 & 3 & 0 & & 3 \\ 0 & 0 & -7 & & 7 \end{bmatrix}$

$\begin{matrix} \\ \frac{1}{3}R_2 \to \\ -\frac{1}{7}R_3 \to \end{matrix} \begin{bmatrix} 1 & -2 & 1 & \vdots & -9 \\ 0 & 1 & 0 & \vdots & 3 \\ 0 & 0 & 1 & \vdots & -1 \end{bmatrix}$

$\begin{matrix} 2R_2 + R_1 \to \\ \\ \end{matrix} \begin{bmatrix} 1 & 0 & 1 & \vdots & -3 \\ 0 & 1 & 0 & \vdots & 3 \\ 0 & 0 & 1 & \vdots & -1 \end{bmatrix}$

$\begin{matrix} -R_3 + R_1 \to \\ \\ \end{matrix} \begin{bmatrix} 1 & 0 & 0 & \vdots & -2 \\ 0 & 1 & 0 & \vdots & 3 \\ 0 & 0 & 1 & \vdots & -1 \end{bmatrix}$

Solution: $(-2, 3, -1)$

12. $A + B = \begin{bmatrix} 4 & 0 \\ -1 & 2 \end{bmatrix} + \begin{bmatrix} -1 & 3 \\ 1 & 0 \end{bmatrix} = \begin{bmatrix} 3 & 3 \\ 0 & 2 \end{bmatrix}$

13. $-2B = -2\begin{bmatrix} -1 & 3 \\ 1 & 0 \end{bmatrix} = \begin{bmatrix} 2 & -6 \\ -2 & 0 \end{bmatrix}$

14. Use the result of Exercise 13.

$A - 2B = A + (-2B) = \begin{bmatrix} 4 & 0 \\ -1 & 2 \end{bmatrix} + \begin{bmatrix} 2 & -6 \\ -2 & 0 \end{bmatrix} = \begin{bmatrix} 6 & -6 \\ -3 & 2 \end{bmatrix}$

15. $AB = \begin{bmatrix} 4 & 0 \\ -1 & 2 \end{bmatrix}\begin{bmatrix} -1 & 3 \\ 1 & 0 \end{bmatrix} = \begin{bmatrix} (4)(-1) + (0)(1) & (4)(3) + (0)(0) \\ (-1)(-1) + 2(1) & (-1)(3) + (2)(0) \end{bmatrix} = \begin{bmatrix} -4 & 12 \\ 3 & -3 \end{bmatrix}$

16. $\begin{vmatrix} 8 & 0 & -5 \\ 1 & 3 & -1 \\ -2 & 6 & 4 \end{vmatrix} = 8\begin{vmatrix} 3 & -1 \\ 6 & 4 \end{vmatrix} - 5\begin{vmatrix} 1 & 3 \\ -2 & 6 \end{vmatrix}$

$$= 8(18) - 5(12)$$

$$= 84$$

Expand along Row 1.

17.

$$\begin{bmatrix} 1 & 2 & -1 & \vdots & 1 & 0 & 0 \\ 3 & 7 & -10 & \vdots & 0 & 1 & 0 \\ -5 & -7 & -15 & \vdots & 0 & 0 & 1 \end{bmatrix}$$

$$\begin{matrix} \\ -3R_1 + R_2 \to \\ 5R_1 + R_3 \to \end{matrix}\begin{bmatrix} 1 & 2 & -1 & \vdots & 1 & 0 & 0 \\ 0 & 1 & -7 & \vdots & -3 & 1 & 0 \\ 0 & 3 & -20 & \vdots & 5 & 0 & 1 \end{bmatrix}$$

$$\begin{matrix} -2R_2 + R_1 \to \\ \\ -3R_2 + R_3 \to \end{matrix}\begin{bmatrix} 1 & 0 & 13 & \vdots & 7 & -2 & 0 \\ 0 & 1 & -7 & \vdots & -3 & 1 & 0 \\ 0 & 0 & 1 & \vdots & 14 & -3 & 1 \end{bmatrix}$$

$$\begin{matrix} -13R_3 + R_1 \to \\ 7R_3 + R_2 \to \\ \end{matrix}\begin{bmatrix} 1 & 0 & 0 & \vdots & -175 & 37 & -13 \\ 0 & 1 & 0 & \vdots & 95 & -20 & 7 \\ 0 & 0 & 1 & \vdots & 14 & -3 & 1 \end{bmatrix}$$

$$\begin{bmatrix} 1 & 2 & -1 \\ 3 & 7 & -10 \\ -5 & -7 & -15 \end{bmatrix}^{-1} = \begin{bmatrix} -175 & 37 & -13 \\ 95 & -20 & 7 \\ 14 & -3 & 1 \end{bmatrix}$$

18. Let x = total sales of gym shoes (in millions),

y = total sales of jogging shoes (in millions),

z = total sales of walking shoes (in millions).

$$\begin{bmatrix} 0.09 & 0.09 & 0.03 \\ 0.06 & 0.10 & 0.05 \\ 0.12 & 0.25 & 0.12 \end{bmatrix}\begin{bmatrix} x \\ y \\ z \end{bmatrix} = \begin{bmatrix} 442.20 \\ 466.57 \\ 1088.09 \end{bmatrix}$$

$$\begin{bmatrix} x \\ y \\ z \end{bmatrix} = \begin{bmatrix} 0.09 & 0.09 & 0.03 \\ 0.06 & 0.10 & 0.05 \\ 0.12 & 0.25 & 0.12 \end{bmatrix}^{-1}\begin{bmatrix} 442.20 \\ 466.57 \\ 1088.09 \end{bmatrix}$$

$$\approx \begin{bmatrix} 2042 \\ 1733 \\ 3415 \end{bmatrix}$$

Thus, sales for each type of shoe amounted to:

Gym shoes: $2042 million

Jogging shoes: $1733 million

Walking shoes: $3415 million

19. $\begin{cases} 8x - 3y = -52 \\ 3x + 5y = 5 \end{cases}, \quad D = \begin{vmatrix} 8 & -3 \\ 3 & 5 \end{vmatrix} = 49$

$x = \dfrac{\begin{vmatrix} -52 & -3 \\ 5 & 5 \end{vmatrix}}{49} = \dfrac{-245}{49} = -5$

$y = \dfrac{\begin{vmatrix} 8 & -52 \\ 3 & 5 \end{vmatrix}}{49} = \dfrac{196}{49} = 4$

Solution: $(-5, 4)$

20.
$$\begin{cases} 5x + 4y + 3z = 7 \\ -3x - 8y + 7z = -9, \\ 7x - 5y - 6z = -53 \end{cases} \quad D = \begin{vmatrix} 5 & 4 & 3 \\ -3 & -8 & 7 \\ 7 & -5 & -6 \end{vmatrix} = 752$$

21. $A = \pm\dfrac{1}{2}\begin{vmatrix} -2 & 3 & 1 \\ 1 & 5 & 1 \\ 4 & 1 & 1 \end{vmatrix} = -\dfrac{1}{2}(-18) = 9$

$$x = \frac{\begin{vmatrix} 7 & 4 & 3 \\ -9 & -8 & 7 \\ -53 & -5 & -6 \end{vmatrix}}{752} = \frac{-2256}{752} = -3$$

$$y = \frac{\begin{vmatrix} 5 & 7 & 3 \\ -3 & -9 & 7 \\ 7 & -53 & -6 \end{vmatrix}}{752} = \frac{3008}{752} = 4$$

$$z = \frac{\begin{vmatrix} 5 & 4 & 7 \\ -3 & -8 & -9 \\ 7 & -5 & -53 \end{vmatrix}}{752} = \frac{1504}{752} = 2$$

Solution: $(-3, 4, 2)$

22. $a_n = \dfrac{(-1)^{n+1}}{2n+3}$

$a_1 = \dfrac{1}{5}$

$a_2 = -\dfrac{1}{7}$

$a_3 = \dfrac{1}{9}$

$a_4 = -\dfrac{1}{11}$

$a_5 = \dfrac{1}{13}$

23. $\dfrac{2!}{4}, \dfrac{3!}{5}, \dfrac{4!}{6}, \dfrac{5!}{7}, \dfrac{6!}{8}, \dots$

$a_n = \dfrac{(n+1)!}{n+3}$

24. $8, 12, 16, 20, \dots$

$a_n = 4n + 4$

$a_1 = 8, \; a_{20} = 84$

$S_{20} = \dfrac{20}{2}(8 + 84) = 920$

25. (a) $a_6 = 20.6$

     $a_9 = 30.2$

     $a_9 = a_6 + 3d$

     $30.2 = 20.6 + 3d$

     $9.6 = 3d$

     $3.2 = d$

     $a_{20} = a_9 + 11d = 30.2 + 11(3.2) = 65.4$

(b) $a_1 = a_6 - 5d$

     $a_1 = 20.6 - 5(3.2)$

        $= 4.6$

     $a_n = a_1 + (n-1)d$

        $= 4.6 + (n-1)(3.2)$

        $= 3.2n + 1.4$

26. $a_n = 3(2)^{n-1}$

$a_1 = 3$

$a_2 = 6$

$a_3 = 12$

$a_4 = 24$

$a_5 = 48$

27. $\displaystyle\sum_{i=6}^{\infty} 1.3\left(\dfrac{1}{10}\right)^{i-1} = \dfrac{1.3}{1 - \frac{1}{10}} = 1.3\left(\dfrac{10}{9}\right) = \dfrac{13}{9}$

28. $S_1 = 3 = 1[2(1) + 1]$

Assume that $S_k = 3 + 7 + 11 + 15 + \cdots + (4k - 1) = k(2k + 1)$.

Then, $S_{k+1} = 3 + 7 + 11 + 15 + \cdots + (4k - 1) + [4(k + 1) - 1]$

$$= S_k + (4k + 3)$$

$$= k(2k + 1) + (4k + 3)$$

$$= 2k^2 + 5k + 3$$

$$= (k + 1)(2k + 3)$$

$$= (k + 1)[2(k + 1) + 1].$$

Therefore, the formula is valid for all integers $n \geq 1$.

29. $(z - 3)^4 = z^4 - 4z^3(3) + 6z^2(3)^2 - 4z(3)^3 + (3)^4$

$$= z^4 - 12z^3 + 54z^2 - 108z + 81$$

30. $_7P_3 = \dfrac{7!}{(7 - 3)!} = \dfrac{7!}{4!} = 210$

31. $_{25}P_2 = \dfrac{25!}{(25 - 2)!} = \dfrac{25!}{23!} = 600$

32. $\dbinom{8}{4} = {_8C_4} = \dfrac{8!}{(8 - 4)!4!} = \dfrac{8!}{4!4!} = 70$

33. $_{10}C_3 = \dfrac{10!}{(10 - 3)!3!} = \dfrac{10!}{7!3!} = 120$

34. B A S K E T B A L L

$$\dfrac{10!}{2!2!2!1!1!1!1!} = 453,600 \text{ distinguishable permutations}$$

35. A N T A R C T I C A

$$\dfrac{10!}{3!2!2!1!1!1!} = 151,200 \text{ distinguishable permutations}$$

36. $_{10}P_3 = \dfrac{10!}{(10 - 3)!} = \dfrac{10!}{7!} = 720$

37. The first digit is 4 or 5, so the probability of picking it correctly is $\frac{1}{2}$.
Then there are two numbers left for the second digit so its probability is also $\frac{1}{2}$.
If these two are correct, then the third digit must be the remaining number.
The probability of winning is:

$$\left(\tfrac{1}{2}\right)\left(\tfrac{1}{2}\right)(1) = \tfrac{1}{4}$$

Chapter 10   Chapter Test Solutions

1. $2x - 7y + 3 = 0$

$$y = \frac{2}{7}x + \frac{3}{7}$$

$$\tan \theta = \frac{2}{7}$$

$$\theta \approx 0.2783 \text{ radian} \approx 15.9°$$

2. $3x + 2y - 4 = 0 \implies y = -\dfrac{3}{2}x + 2 \implies m_1 = -\dfrac{3}{2}$

$$4x - y + 6 = 0 \implies y = 4x + 6 \implies m_2 = 4$$

$$\tan \theta = \left|\frac{4 - (-3/2)}{1 + 4(-3/2)}\right| = \frac{11}{10}$$

$$\theta \approx 0.8330 \text{ radian} \approx 47.7°$$

3. $y = 5 - x \implies x + y - 5 = 0 \implies A = 1, B = 1, C = -5$

$$(x_1, y_1) = (7, 5)$$

$$d = \frac{|(1)(7) + (1)(5) + (-5)|}{\sqrt{1^2 + 1^2}} = \frac{7}{\sqrt{2}} = \frac{7\sqrt{2}}{2}$$

4. $y^2 - 4x + 4 = 0$

$$y^2 = 4(x - 1)$$

Parabola

Vertex: $(1, 0)$

Focus: $(2, 0)$

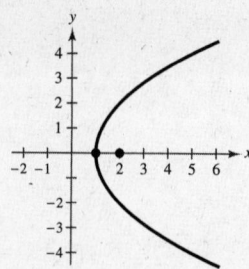

5. $x^2 - 4y^2 - 4x = 0$

$$(x - 2)^2 - 4y^2 = 4$$

$$\frac{(x - 2)^2}{4} - \frac{y^2}{1} = 1$$

Hyperbola

Center: $(2, 0)$

Horizontal transverse axis

$a = 2, b = 1,$

$c^2 = 1 + 4 = 5 \implies c = \sqrt{5}$

Vertices: $(0, 0), (4, 0)$

Foci: $\left(2 \pm \sqrt{5}, 0\right)$

Asymptotes: $y = \pm\frac{1}{2}(x - 2)$

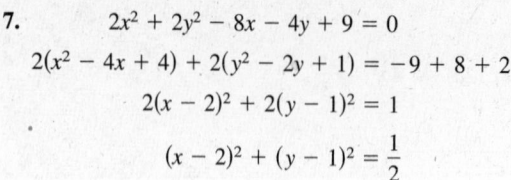

6. $9x^2 + 16y^2 + 54x - 32y - 47 = 0$

$$9(x^2 + 6x + 9) + 16(y^2 - 2y + 1) = 47 + 81 + 16$$

$$9(x + 3)^2 + 16(y - 1)^2 = 144$$

$$\frac{(x + 3)^2}{16} + \frac{(y - 1)^2}{9} = 1$$

Ellipse

Center: $(-3, 1)$

$a = 4, b = 3, c = \sqrt{7}$

Foci: $\left(-3 \pm \sqrt{7}, 1\right)$

Vertices: $(1, 1), (-7, 1)$

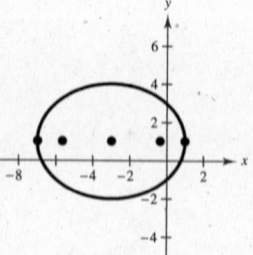

7. $2x^2 + 2y^2 - 8x - 4y + 9 = 0$

$$2(x^2 - 4x + 4) + 2(y^2 - 2y + 1) = -9 + 8 + 2$$

$$2(x - 2)^2 + 2(y - 1)^2 = 1$$

$$(x - 2)^2 + (y - 1)^2 = \frac{1}{2}$$

Circle

Center: $(2, 1)$

Radius:

$$\sqrt{\frac{1}{2}} = \frac{\sqrt{2}}{2} \approx 0.707$$

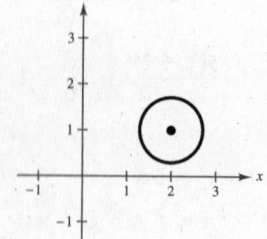

8. Parabola

Vertex: $(3, -2)$

Vertical axis

Point: $(0, 4)$

$(x - h)^2 = 4p(y - k)$

$(x - 3)^2 = 4p(y + 2)$

$(0 - 3)^2 = 4p(4 + 2)$

$9 = 24p$

$$p = \frac{9}{24} = \frac{3}{8}$$

Equation: $(x - 3)^2 = 4\left(\frac{3}{8}\right)(y + 2)$

$$(x - 3)^2 = \frac{3}{2}(y + 2)$$

9. Hyperbola

Foci: $(0, 0)$ and $(0, 4) \implies c = 2$

Asymptotes: $y = \pm\dfrac{1}{2}x + 2$

Vertical transverse axis

Center: $(0, 2) = (h, k)$

$\dfrac{a}{b} = \dfrac{1}{2} \implies 2a = b$

$c^2 = a^2 + b^2$

$4 = a^2 + (2a)^2$

$4 = 5a^2$

$\dfrac{4}{5} = a^2$

$b^2 = (2a)^2 = 4a^2 = \dfrac{16}{5}$

$\dfrac{(y - k)^2}{a^2} - \dfrac{(x - h)^2}{b^2} = 1$

$\dfrac{(y - 2)^2}{4/5} - \dfrac{x^2}{16/5} = 1$

$\dfrac{5(y - 2)^2}{4} - \dfrac{5x^2}{16} = 1$

10. (a) $x^2 + 6xy + y^2 - 6 = 0$

$A = 1, B = 6, C = 1$

$\cot 2\theta = \dfrac{1 - 1}{6} = 0$

$2\theta = 90°$

$\theta = 45°$

(b)

$x = x' \cos 45° - y' \sin 45° = \dfrac{x' - y'}{\sqrt{2}}$

$y = x' \sin 45° + y' \cos 45° = \dfrac{x' + y'}{\sqrt{2}}$

$\left(\dfrac{x' - y'}{\sqrt{2}}\right)^2 + 6\left(\dfrac{x' - y'}{\sqrt{2}}\right)\left(\dfrac{x' + y'}{\sqrt{2}}\right) + \left(\dfrac{x' + y'}{\sqrt{2}}\right)^2 - 6 = 0$

$\dfrac{1}{2}((x')^2 - 2(x')(y') + (y')^2) + 3((x')^2 - (y')^2) + \dfrac{1}{2}((x')^2 + 2(x')(y') + (y')^2) - 6 = 0$

$4(x')^2 - 2(y')^2 = 6$

$\dfrac{2(x')^2}{3} - \dfrac{(y')^2}{3} = 1$

For the graphing utility, we need to solve for y in terms of x.

$y^2 + 6xy + 9x^2 = 6 - x^2 + 9x^2$

$(y + 3x)^2 = 6 + 8x^2$

$y + 3x = \pm\sqrt{6 + 8x^2}$

$y = -3x \pm \sqrt{6 + 8x^2}$

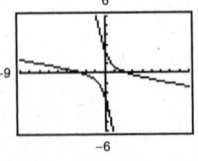

11. $x = 2 + 3 \cos \theta$

$y = 2 \sin \theta$

$x = 2 + 3 \cos \theta \implies \dfrac{x - 2}{3} = \cos \theta$

$y = 2 \sin \theta \implies \dfrac{y}{2} = \sin \theta$

$\cos^2 \theta + \sin^2 \theta = 1$

$\dfrac{(x - 2)^2}{9} + \dfrac{y^2}{4} = 1$

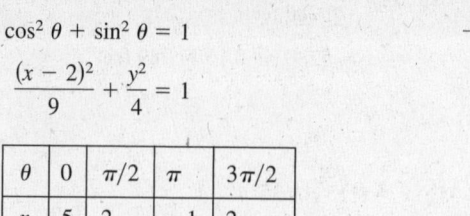

θ	0	$\pi/2$	π	$3\pi/2$
x	5	2	-1	2
y	0	2	0	-2

12. $(6, 4), (2, -3)$

$x = x_1 + t(x_2 - x_1) = 6 + t(2 - 6) = 6 - 4t$

$y = y_1 + t(y_2 - y_1) = 4 + t(-3 - 4) = 4 - 7t$

Answers are not unique. Another possible set:

$x = 6 + 4t$

$y = 4 + 7t$

13. Polar coordinates: $\left(-2, \dfrac{5\pi}{6}\right)$

$x = -2 \cos \dfrac{5\pi}{6} = -2\left(-\dfrac{\sqrt{3}}{2}\right) = \sqrt{3}$

$y = -2 \sin \dfrac{5\pi}{6} = -2\left(\dfrac{1}{2}\right) = -1$

Rectangular coordinates: $\left(\sqrt{3}, -1\right)$

14. Rectangular coordinates: $(2, -2)$

$r = \pm\sqrt{2^2 + (-2)^2} = \pm\sqrt{8} = \pm 2\sqrt{2}$

$\tan \theta = -1 \implies \theta = \dfrac{3\pi}{4}, \dfrac{7\pi}{4}$

Polar coordinates:

$\left(2\sqrt{2}, \dfrac{7\pi}{4}\right), \left(-2\sqrt{2}, \dfrac{3\pi}{4}\right), \left(2\sqrt{2}, -\dfrac{\pi}{4}\right)$

15. $x^2 + y^2 - 4y = 0$

$r^2 - 4r \sin \theta = 0$

$r^2 = 4r \sin \theta$

$r = 4 \sin \theta$

16. $r = \dfrac{4}{1 + \cos \theta}$

$e = 1 \implies$ Parabola

Vertex: $(2, 0)$

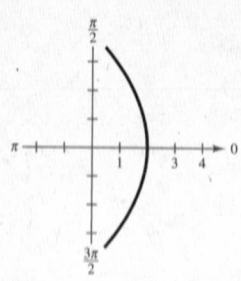

17. $r = \dfrac{4}{2 + \cos \theta} = \dfrac{2}{1 + \frac{1}{2} \cos \theta}$

$e = \dfrac{1}{2} \implies$ Ellipse

Vertex: $\left(\dfrac{4}{3}, 0\right), (4, \pi)$

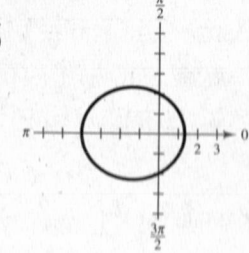

18. $r = 2 + 3 \sin \theta$

$\dfrac{a}{b} = \dfrac{2}{3} < 1$

Limaçon with inner loop

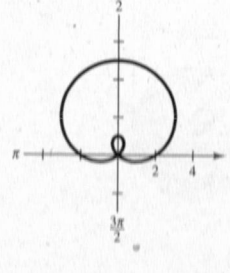

θ	0	$\dfrac{\pi}{2}$	π	$\dfrac{3\pi}{2}$
r	2	5	2	-1

19. $r = 3 \sin 2\theta$

Rose curve ($n = 2$) with four petals

$|r| = 3$ when

$\theta = \dfrac{\pi}{4}, \dfrac{3\pi}{4}, \dfrac{5\pi}{4}, \dfrac{7\pi}{4}$

$r = 0$ when $\theta = 0, \dfrac{\pi}{2}, \pi, \dfrac{3\pi}{2}$

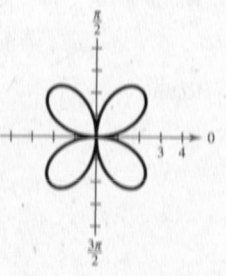

20. Ellipse, $e = \dfrac{1}{4}$, focus at the pole, directrix $y = 4$

For a horizontal directrix above the pole we have:

$$r = \frac{ep}{1 + e \sin \theta}$$

p = distance between the pole and the directrix $\Rightarrow p = 4$

Thus, $r = \dfrac{(1/4)(4)}{1 + (1/4) \sin \theta} = \dfrac{1}{1 + 0.25 \sin \theta}$.

21.

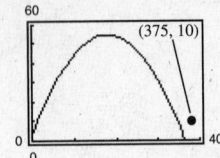

Not drawn to scale

Slope: $m = \tan 0.15 \approx 0.1511$

$\sin 0.15 = \dfrac{x}{5280 \text{ feet}}$

$x = 5280 \sin 0.15 \approx 789 \text{ feet}$

22. $x = (115 \cos \theta)t$ and $y = 3 + (115 \sin \theta)t - 16t^2$

When $\theta = 30°$: $x = (115 \cos 30°)t$

$y = 3 + (115 \sin 30°)t - 16t^2$

When $\theta = 35°$: $x = (115 \cos 35°)t$

$y = 3 + (115 \sin 35°)t - 16t^2$

The ball hits the ground inside
the ballpark, so it is not a home run.

The ball clears the 10-foot fence
at 375 feet, so it is a home run.

Chapter 11   Chapter Test Solutions

1.

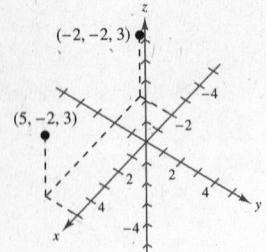

2. $AB = \sqrt{(8 - 6)^2 + (-2 - 4)^2 + (5 + 1)^2} = \sqrt{76}$

$AC = \sqrt{(8 + 4)^2 + (-2 - 3)^2 + (5 - 0)^2} = \sqrt{144 + 25 + 25} = \sqrt{194}$

$BC = \sqrt{(6 + 4)^2 + (4 - 3)^2 + (-1 - 0)^2} = \sqrt{100 + 1 + 1} = \sqrt{102}$

No. $\left(\sqrt{76}\right)^2 + \left(\sqrt{102}\right)^2 \neq \left(\sqrt{194}\right)^2$

3. Midpoint $= \left(\dfrac{8 + 6}{2}, \dfrac{-2 + 4}{2}, \dfrac{5 - 1}{2}\right) = (7, 1, 2)$

4. Diameter $= \sqrt{(8 - 6)^2 + (-2 - 4)^2 + (5 + 1)^2}$

$= \sqrt{4 + 36 + 36} = \sqrt{76}$

Radius $= \sqrt{19}$

$(x - 7)^2 + (y - 1)^2 + (z - 2)^2 = 19$

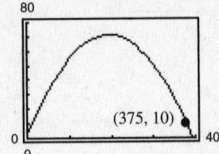

5. $\mathbf{u} = \langle 6 - 8, 4 - (-2), -1 - 5 \rangle = \langle -2, 6, -6 \rangle$

$\mathbf{v} = \langle -4 - 8, 3 - (-2), 0 - 5 \rangle = \langle -12, 5, -5 \rangle$

6. (a) $\|\mathbf{v}\| = \sqrt{(-12)^2 + 5^2 + (-5)^2} = \sqrt{194}$

(b) $\mathbf{u} \cdot \mathbf{v} = (-2)(-12) + 6(5) + (-6)(-5) = 84$

(c) $\mathbf{u} \times \mathbf{v} = \begin{vmatrix} \mathbf{i} & \mathbf{j} & \mathbf{k} \\ -2 & 6 & -6 \\ -12 & 5 & -5 \end{vmatrix} = \langle 0, 62, 62 \rangle$

7. $\cos\theta = \dfrac{\mathbf{u} \cdot \mathbf{v}}{\|\mathbf{u}\|\,\|\mathbf{v}\|} = \dfrac{84}{\sqrt{76}\sqrt{194}} \approx 0.6918 \implies \theta \approx 46.23$ or 0.8068 radians

8. (a) $x = 8 - 2t, y = -2 + 6t, z = 5 - 6t$

(b) $\dfrac{x - 8}{-2} = \dfrac{y + 2}{6} = \dfrac{z - 5}{-6}$

9. $\mathbf{u} \cdot \mathbf{v} = 0 - 2 - 6 \neq 0$ and $\mathbf{u} \neq c\mathbf{v} \implies$ neither

10. $\mathbf{u} \cdot \mathbf{v} = -2 + 3 - 1 = 0 \implies$ orthogonal

11. First two points: $\mathbf{v} = \langle 4, 8, -2 \rangle$

Last two points: $\mathbf{w} = \langle 4, 8, -2 \rangle$

Opposite sides are parallel and equal length.

Adjacent sides: $\mathbf{v}$ and $\mathbf{u} = \langle 1, -3, 3 \rangle$

Area $= \|\mathbf{u} \times \mathbf{v}\|$

$\mathbf{u} \times \mathbf{v} = \begin{vmatrix} \mathbf{i} & \mathbf{j} & \mathbf{k} \\ 1 & -3 & 3 \\ 4 & 8 & -2 \end{vmatrix} = \langle -18, 14, 20 \rangle$

$\|\mathbf{u} \times \mathbf{v}\| = \sqrt{18^2 + 14^2 + 20^2} = 2\sqrt{230} \approx 30.33$ square units

12. $\mathbf{u} = \langle 0, 8, -1 \rangle, \mathbf{v} = \langle 4, 5, -4 \rangle$

$\mathbf{n} = \mathbf{u} \times \mathbf{v} = \begin{vmatrix} \mathbf{i} & \mathbf{j} & \mathbf{k} \\ 0 & 8 & -1 \\ 4 & 5 & -4 \end{vmatrix} = \langle -27, -4, -32 \rangle$

Plane: $-27(x + 3) - 4(y + 4) - 32(z - 2) = 0$

$-27x - 4y - 32z - 33 = 0$

$27x + 4y + 32z + 33 = 0$

13. Let $A(0, 0, 5)$ be the vertex.

$\mathbf{u} = \overrightarrow{AD} = \langle 4, 0, 0 \rangle, \mathbf{v} = \overrightarrow{AB} = \langle 0, 10, 0 \rangle,$

$\mathbf{w} = \overrightarrow{AE} = \langle 0, 1, -5 \rangle$

$\mathbf{u} \cdot (\mathbf{v} \times \mathbf{w}) = \begin{vmatrix} 4 & 0 & 0 \\ 0 & 10 & 0 \\ 0 & 1 & -5 \end{vmatrix} = 4(-50) = -200$

Volume $= |-200| = 200$ cubic units

14. $2x + 3y + 4z = 12$

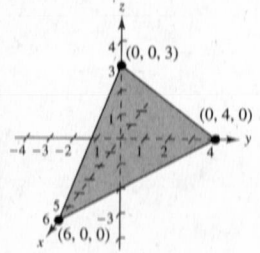

15. $5x - y - 2z = 10$

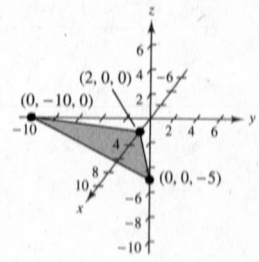

16. $\mathbf{n} = \langle 3, -2, 1 \rangle$, $Q = (2, -1, 6)$, $P = (0, 0, 6)$ in plane, $\overrightarrow{PQ} = \langle 2, -1, 0 \rangle$

$$D = \frac{|\overrightarrow{PQ} \cdot \mathbf{n}|}{\|\mathbf{n}\|} = \frac{|8|}{\sqrt{14}} = \frac{4\sqrt{14}}{7}$$

17. The normal vector to plane containing $(0, 0, 0)$, $(2, 2, 12)$ and $(10, 0, 0)$ is obtained as follows.

$\mathbf{v}_1 = \langle 2, 2, 12 \rangle$, $\mathbf{v}_2 = \langle 10, 0, 0 \rangle$

$$\mathbf{v}_1 \times \mathbf{v}_2 = \begin{vmatrix} \mathbf{i} & \mathbf{j} & \mathbf{k} \\ 2 & 2 & 12 \\ 10 & 0 & 0 \end{vmatrix} = \langle 0, 120, -20 \rangle$$

$\mathbf{n}_1 = \langle 0, 6, -1 \rangle$

The normal vector to the plane containing $(0, 0, 0)$, $(2, 2, 12)$ and $(0, 10, 0)$ is obtained as follows.

$\mathbf{u}_1 = \langle 2, 2, 12 \rangle$, $\mathbf{u}_2 = \langle 0, 10, 0 \rangle$

$$\mathbf{u}_1 \times \mathbf{u}_2 = \begin{vmatrix} \mathbf{i} & \mathbf{j} & \mathbf{k} \\ 2 & 2 & 12 \\ 0 & 10 & 0 \end{vmatrix} = \langle -120, 0, 20 \rangle$$

$\mathbf{n}_2 = \langle -6, 0, 1 \rangle$

The angle θ between two adjacent sides is given by

$$\cos \theta = \frac{|\mathbf{n}_1 \cdot \mathbf{n}_2|}{\|\mathbf{n}_1\| \|\mathbf{n}_2\|} = \frac{|-1|}{\sqrt{37}\sqrt{37}} = \frac{1}{37} \Rightarrow \theta \approx 88.45°.$$

Chapter 12   Chapter Test Solutions

1. $f(x) = \dfrac{x^2 - 1}{2x}$

$$\lim_{x \to -2} \frac{x^2 - 1}{2x} = \frac{(-2)^2 - 1}{2(-2)} = -\frac{3}{4}$$

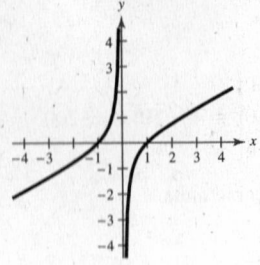

2. $f(x) = \dfrac{2x^2 - x - 3}{x + 1}$

$$= \frac{(2x - 3)(x + 1)}{x + 1} = 2x - 3, x \neq -1$$

$$\lim_{x \to -1} \frac{2x^2 - x - 3}{x + 1} = \lim_{x \to -1} \frac{(2x - 3)(x + 1)}{x + 1}$$

$$= \lim_{x \to -1} (2x - 3)$$

$$= -5$$

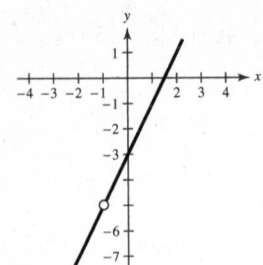

3. $f(x) = \dfrac{\sqrt{x} - 2}{x - 5}$

The graph has a vertical asymptote at $x = 5$.

$\lim\limits_{x \to 5} \dfrac{\sqrt{x} - 2}{x - 5}$ does not exist.

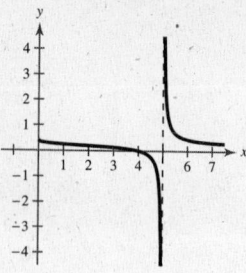

4.

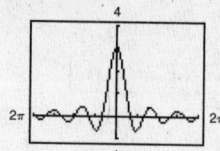

$\lim\limits_{x \to 0} \dfrac{\sin 3x}{x} = 3$

$f(x) = \dfrac{\sin 3x}{x}$

x	-0.02	-0.01	0	0.01	0.02
$f(x)$	2.9982	2.9996	?	2.9996	2.9982

5.

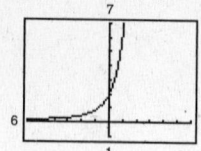

x	-0.004	-0.003	-0.002	-0.001	0	0.001	0.003	0.004
$f(x)$	1.9920	1.9940	1.9960	1.9980	?	2.0020	2.0060	2.0080

$\lim\limits_{x \to 0} \dfrac{e^{2x} - 1}{x} = 2$

$f(x) = \dfrac{e^{2x} - 1}{x}$

6. (a) $\dfrac{f(x + h) - f(x)}{h} = \dfrac{3(x + h)^2 - 5(x + h) - 2 - (3x^2 - 5x - 2)}{h}$

$\qquad = \dfrac{3x^2 + 6xh + 3h^2 - 5h - 3x^2}{h}$

$\qquad = 6x + 3h - 5$

$f'(x) = \lim\limits_{h \to 0} [6x + 3h - 5] = 6x - 5$

$f'(2) = 6(2) - 5 = 7$

(b) $\dfrac{f(x + h) - f(x)}{h} = \dfrac{[2(x + h)^3 + 6(x + h)] - [2x^3 + 6x]}{h}$

$\qquad = \dfrac{2x^3 + 6x^2h + 6xh^2 + 2h^3 + 6x + 6h - 2x^3 - 6x}{h}$

$\qquad = \dfrac{6x^2h + 6xh^2 + 2h^3 + 6h}{h}$

$\qquad = 6x^2 + 6xh + 2h^2 + 6, \ h \neq 0$

$f'(x) = \lim\limits_{h \to 0} [6x^2 + 6xh + 2h^2 + 6] = 6x^2 + 6$

$f'(-1) = 6(-1)^2 + 6 = 12$

7. $f'(x) = \lim\limits_{h \to 0} \dfrac{f(x + h) - f(x)}{h}$

$= \lim\limits_{h \to 0} \dfrac{4 - (3/4)(x + h) - [4 - (3/4)x]}{h}$

$= \lim\limits_{h \to 0} \dfrac{-(3/4)h}{h} = -\dfrac{3}{4}$

8. $f'(x) = \lim\limits_{h \to 0} \dfrac{f(x + h) - f(x)}{h}$

$= \lim\limits_{h \to 0} \dfrac{2(x + h)^2 + 4(x + h) - 1 - [2x^2 + 4x - 1]}{h}$

$= \lim\limits_{h \to 0} \dfrac{2x^2 + 4xh + 2h^2 + 4h - 2x^2}{h}$

$= \lim\limits_{h \to 0} (4x + 2h + 4) = 4x + 4$

9. $f'(x) = \lim\limits_{h \to 0} \dfrac{f(x + h) - f(x)}{h}$

$= \lim\limits_{h \to 0} \dfrac{\dfrac{1}{x + 3 + h} - \dfrac{1}{x + 3}}{h}$

$= \lim\limits_{h \to 0} \dfrac{(x + 3) - (x + 3 + h)}{h(x + 3 + h)(x + 3)}$

$= \lim\limits_{h \to 0} \dfrac{-1}{(x + 3 + h)(x + 3)}$

$= \dfrac{-1}{(x + 3)^2}$

10. $\lim\limits_{x \to \infty} \dfrac{6}{5x - 1} = 0$

11. $\lim\limits_{x \to \infty} \dfrac{1 - 3x^2}{x^2 - 5} = -3$

12. $\lim\limits_{x \to -\infty} \dfrac{3x^3}{x + 2}$ does not exist.

13. $0, \frac{3}{4}, \frac{14}{19}, \frac{12}{17}, \frac{36}{53}$

$\lim\limits_{n \to \infty} a_n = \frac{1}{2}$

14. $0, 1, 0, \frac{1}{2}, 0$

$\lim\limits_{n \to \infty} a_n = 0$

15. Width of each rectangle: $\frac{1}{2}$

Heights: $8, \frac{15}{2}, 6, \frac{7}{2}$

Area $\approx \frac{1}{2}\left[8 + \frac{15}{2} + 6 + \frac{7}{2}\right] = \frac{25}{2}$

16. Width: $\dfrac{4}{n}$, Height: $f\left(-2 + \dfrac{4i}{n}\right) = \left(-2 + \dfrac{4i}{n}\right) + 2 = \dfrac{4i}{n}$

$A \approx \sum\limits_{i=1}^{n} \left(\dfrac{4i}{n}\right)\left(\dfrac{4}{n}\right) = \dfrac{16}{n^2} \sum\limits_{i=1}^{n} i = \dfrac{16}{n^2} \dfrac{n(n + 1)}{2}$

$A = \lim\limits_{n \to \infty} \dfrac{16}{n^2} \cdot \dfrac{n(n + 1)}{2} = 8$

17. Width: $\dfrac{1}{n}$, Height: $f\left(\dfrac{i}{n}\right) = 1 - \dfrac{i^3}{n^3}$

$$A \approx \sum_{i=1}^{n} \left(1 - \frac{i^3}{n^3}\right)\left(\frac{1}{n}\right) = \sum_{i=1}^{n} \left(\frac{1}{n} - \frac{i^3}{n^4}\right)$$

$$= \frac{1}{n} \sum_{i=1}^{n} 1 - \frac{1}{n^4} \sum_{i=1}^{n} i^3$$

$$= \frac{1}{n}(n) - \frac{1}{n^4}\left(\frac{n^2(n+1)^2}{4}\right)$$

$$= 1 - \frac{(n+1)^2}{4n^2}$$

$$A = \lim_{n\to\infty}\left(1 - \frac{(n+1)^2}{4n^2}\right) = 1 - \frac{1}{4} = \frac{3}{4}$$

18. (a) $y = 8.79x^2 - 6.2x - 0.4$

(b) Velocity = Derivative = $17.58x - 6.2$

At $x = 5$, velocity ≈ 81.7 ft/sec.

Chapters 10–12   Cumulative Test Solutions

1. $\dfrac{(x-2)^2}{4} + \dfrac{(y+1)^2}{9} = 1$

Ellipse with center $(2, -1)$

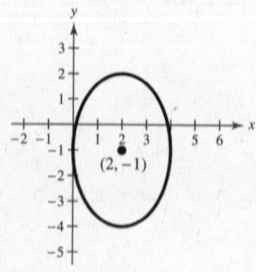

2.
$$x^2 + y^2 - 2x - 4y + 1 = 0$$
$$(x^2 - 2x + 1) + (y^2 - 4y + 4) = -1 + 1 + 4$$
$$(x-1)^2 + (y-2)^2 = 4$$

Circle

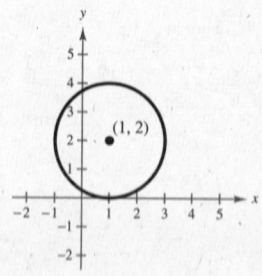

3. Ellipse

Vertices: $(0, 0)$ and $(0, 4) \implies a = 2$

Center: $(0, 2)$

Endpoint of minor axis: $(1, 2)$ and $(-1, 2) \implies b = 1$

Vertical major axis:

$$\frac{(x-0)^2}{1^2} + \frac{(y-2)^2}{2^2} = 1$$

$$\frac{x^2}{1} + \frac{(y-2)^2}{4} = 1$$

4. $x^2 - 4xy + 2y^2 = 6$

$B^2 - 4AC = 16 - 8 = 8 \implies$ Hyperbola

$\cot 2\theta = \dfrac{1-2}{-4} = \dfrac{1}{4} \implies \theta \approx 37.98°$

Graph as:

$$2y^2 - 4xy + (x^2 - 6) = 0$$

$$y = \frac{4x \pm \sqrt{16x^2 - 8(x^2 - 6)}}{4}$$

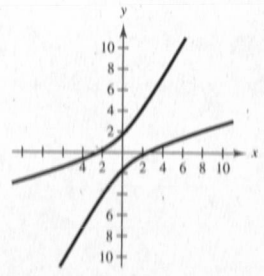

5. $x = 4 \ln t \Rightarrow t = e^{x/4}$

$y = \dfrac{1}{2}t^2$

$y = \dfrac{1}{2}(e^{x/4})^2 = \dfrac{1}{2}e^{x/2} = \dfrac{\sqrt{e^x}}{2}$

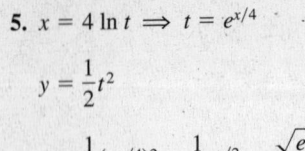

6.

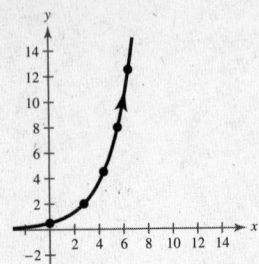

$\left(-2, \dfrac{5\pi}{4}\right), \left(2, \dfrac{\pi}{4}\right), \left(2, -\dfrac{7\pi}{4}\right)$

7. $\qquad -8x - 3y + 5 = 0$

$-8\, r \cos \theta - 3r \sin \theta + 5 = 0$

$r(8 \cos \theta + 3 \sin \theta) = 5$

$r = \dfrac{5}{8 \cos \theta + 3 \sin \theta}$

8. $\qquad r = \dfrac{2}{4 - 5 \cos \theta}$

$4r - 5r \cos \theta = 2$

$4(x^2 + y^2)^{1/2} - 5x = 2$

$16(x^2 + y^2) = (5x + 2)^2 = 25x^2 + 20x + 4$

$9x^2 + 20x - 16y^2 + 4 = 0$

9. $r = -\dfrac{\pi}{6}$, circle

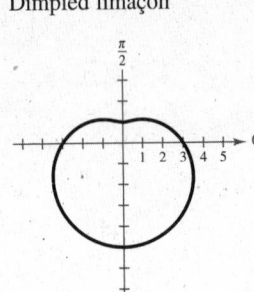

10. $r = 3 - 2 \sin \theta$

Dimpled limaçon

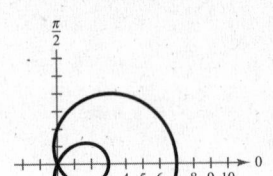

11. $r = 2 + 5 \cos \theta$

Limaçon with an inner loop

12. $(-6, 1, 3)$

13. $(0, -4, 0)$

14. $d = \sqrt{(4 - (-2))^2 + (-5 - 3)^2 + (1 - (-6))^2}$

$= \sqrt{36 + 64 + 49}$

$= \sqrt{149}$

15. $d_1 = 3, d_2 = 4, d_3 = \sqrt{4^2 + 3^2} = 5$

$d_1{}^2 + d_2{}^2 = d_3{}^2$

16. Midpoint: $\left(\dfrac{3 - 5}{2}, \dfrac{4 + 0}{2}, \dfrac{-1 + 2}{2}\right) = \left(-1, 2, \dfrac{1}{2}\right)$

17. Center $= (2, 2, 4)$

Radius $= \sqrt{2^2 + 2^2 + 4^2} = \sqrt{24}$

$(x - 2)^2 + (y - 2)^2 + (z - 4)^2 = 24$

18. *xy*-trace: $(z = 0)$

$(x - 2)^2 + (y + 1)^2 = 4$, Circle

yz-trace: $(x = 0)$

$4 + (y + 1)^2 + z^2 = 4$ or $(y + 1)^2 + z^2 = 0$, Point

$(0, -1, 0)$, Point

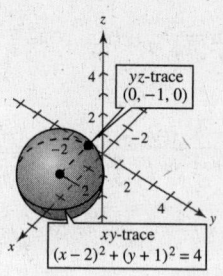

19. $\mathbf{u} \cdot \mathbf{v} = \langle 2, -6, 0 \rangle \cdot \langle -4, 5, 3 \rangle$

$= -8 - 30 = -38$

$\mathbf{u} \times \mathbf{v} = \begin{vmatrix} \mathbf{i} & \mathbf{j} & \mathbf{k} \\ 2 & -6 & 0 \\ -4 & 5 & 3 \end{vmatrix} = \langle -18, -6, -14 \rangle$

20. $\mathbf{u} \cdot \mathbf{v} \neq 0, \mathbf{u} \neq c\mathbf{v} \implies$ neither

21. $\mathbf{u} \cdot \mathbf{v} = -8 - 12 + 20 = 0 \implies$ orthogonal

22. $3\mathbf{u} = \langle -3, 18, -9 \rangle = -\mathbf{v} \implies$ parallel

23. $\overrightarrow{DA} = \langle 0, -2, 0 \rangle, \overrightarrow{DC} = \langle 2, 1, 0 \rangle, \overrightarrow{DH} = \langle 0, 0, 3 \rangle$

$\begin{vmatrix} 0 & -2 & 0 \\ 2 & 1 & 0 \\ 0 & 0 & 3 \end{vmatrix} = 12$ cubic units

24. (a) Vector is $\langle 5 + 2, 8 - 3, 25 - 0 \rangle = \langle 7, 5, 25 \rangle$.

$x = -2 + 7t, y = 3 + 5t, z = 25t$

(b) $\dfrac{x + 2}{7} = \dfrac{y - 3}{5} = \dfrac{z}{25}$

25. $\mathbf{u} = \langle -2, 3, 0 \rangle, \mathbf{v} = \langle 5, 8, 25 \rangle$

$\mathbf{u} \times \mathbf{v} = \begin{vmatrix} \mathbf{i} & \mathbf{j} & \mathbf{k} \\ -2 & 3 & 0 \\ 5 & 8 & 25 \end{vmatrix} = \langle 75, 50, -31 \rangle$

Normal to plane

Plane: $75x + 50y - 31z = 0$

26.

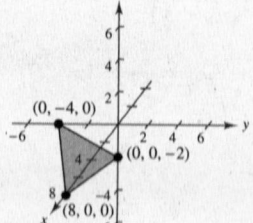

27. $\mathbf{n} = \langle 2, -5, 1 \rangle, Q = (0, 0, 25), P = (0, 0, 10)$ in plane, $\overrightarrow{PQ} = \langle 0, 0, 15 \rangle$

$D = \dfrac{|\overrightarrow{PQ} \cdot \mathbf{n}|}{\|\mathbf{n}\|} = \dfrac{15}{\sqrt{30}} = \dfrac{\sqrt{30}}{2} \approx 2.74$

28. Normal to plane containing: $(-1, -1, 3), (0, 0, 0)$ and $(2, 0, 0)$ is

$\langle -1, -1, 3 \rangle \times \langle 2, 0, 0 \rangle = \begin{vmatrix} \mathbf{i} & \mathbf{j} & \mathbf{k} \\ -1 & -1 & 3 \\ 2 & 0 & 0 \end{vmatrix} = \langle 0, 6, 2 \rangle$ or $\mathbf{n}_1 = \langle 0, 3, 1 \rangle$

Normal to front face is: $\langle 1, -1, 3 \rangle \times \langle 0, 2, 0 \rangle = \begin{vmatrix} \mathbf{i} & \mathbf{j} & \mathbf{k} \\ 1 & -1 & 3 \\ 0 & 2 & 0 \end{vmatrix} = \langle -6, 0, 2 \rangle$ or $\mathbf{n}_2 = \langle -3, 0, 1 \rangle$

Angle between sides: $\cos \theta = \dfrac{|\mathbf{n}_1 \cdot \mathbf{n}_2|}{\|\mathbf{n}_1\| \|\mathbf{n}_2\|} = \dfrac{1}{\sqrt{10}\sqrt{10}} = \dfrac{1}{10} \implies \theta \approx 84.26°$

29. $\lim\limits_{x \to 4} (5x - x^2) = 5(4) - 4^2 = 4$

30. $\lim\limits_{x \to -2^+} \dfrac{x + 2}{(x + 2)(x - 1)} = \lim\limits_{x \to -2^+} \dfrac{1}{x - 1} = -\dfrac{1}{3}$

31. $\lim\limits_{x \to 7} \dfrac{x-7}{(x-7)(x+7)} = \lim\limits_{x \to 7} \dfrac{1}{x+7} = \dfrac{1}{14}$

32. $\lim\limits_{x \to 0} \dfrac{\sqrt{x+4}-2}{x} \cdot \dfrac{\sqrt{x+4}+2}{\sqrt{x+4}+2} = \lim\limits_{x \to 0} \dfrac{(x+4)-4}{x\left(\sqrt{x+4}+2\right)} = \lim\limits_{x \to 0} \dfrac{1}{\sqrt{x+4}+2} = \dfrac{1}{2+2} = \dfrac{1}{4}$

33. $\lim\limits_{x \to 4^-} \dfrac{|x-4|}{x-4} = -1$

34. $\lim\limits_{x \to 0} \sin\left(\dfrac{\pi}{x}\right)$ does not exist.

35. $f(x) = 3 - x^2$

$m = \lim\limits_{h \to 0} \dfrac{f(x+h) - f(x)}{h}$

$\quad = \lim\limits_{h \to 0} \dfrac{3-(x+h)^2-(3-x^2)}{h} = \lim\limits_{h \to 0} \dfrac{-2xh-h^2}{h} = \lim\limits_{h \to 0}(-2x-h) = -2x$

At $(1, 2)$, $m = -2$.

36. $f(x) = \sqrt{x+3}$

$m = \lim\limits_{h \to 0} \dfrac{f(x+h) - f(x)}{h}$

$\quad = \lim\limits_{h \to 0} \dfrac{\sqrt{x+h+3}-\sqrt{x+3}}{h} \cdot \dfrac{\sqrt{x+h+3}+\sqrt{x+3}}{\sqrt{x+h+3}+\sqrt{x+3}}$

$\quad = \lim\limits_{h \to 0} \dfrac{(x+h+3)-(x+3)}{h\left[\sqrt{x+h+3}+\sqrt{x+3}\right]}$

$\quad = \lim\limits_{h \to 0} \dfrac{1}{\sqrt{x+h+3}+\sqrt{x+3}} = \dfrac{1}{2\sqrt{x+3}}$

At $(-2, 1)$, $m = \dfrac{1}{2}$.

37. $f(x) = \dfrac{1}{x+3}$

$m = \lim\limits_{h \to 0} \dfrac{f(x+h) - f(x)}{h}$

$\quad = \lim\limits_{h \to 0} \dfrac{\dfrac{1}{x+h+3} - \dfrac{1}{x+3}}{h}$

$\quad = \lim\limits_{h \to 0} \dfrac{(x+3)-(x+h+3)}{h(x+h+3)(x+3)}$

$\quad = \lim\limits_{h \to 0} \dfrac{-1}{(x+h+3)(x+3)}$

$\quad = \dfrac{-1}{(x+3)^2}$

At $\left(1, \dfrac{1}{4}\right)$, $m = \dfrac{-1}{16}$.

38. $f(x) = x^4$

$m = \lim\limits_{h \to 0} \dfrac{f(x+h) - f(x)}{h}$

$\quad = \lim\limits_{h \to 0} \dfrac{(x+h)^4 - x^4}{h}$

$\quad = \lim\limits_{h \to 0} \dfrac{4x^3h + 6x^2h^2 + 4xh^3 + h^4}{h}$

$\quad = \lim\limits_{h \to 0} \left[4x^3 + 6x^2h + 4xh^2 + h^3\right]$

$\quad = 4x^3$

At $(-1, 1)$, $m = -4$.

39. $\lim\limits_{x \to \infty} \dfrac{2x^4 - x^3 + 4}{x^2 - 9}$

Does not exist

40. $\lim\limits_{x \to \infty} \dfrac{3 - 7x}{x+4} = -7$

41. $\lim\limits_{x \to \infty} \dfrac{3x^2 + 1}{x^2 + 4} = 3$

42. $\lim\limits_{x\to\infty}\dfrac{2x}{x^2+3x-2}=0$

43. $\sum\limits_{i=1}^{50}(1-i^2)=50-\dfrac{50(51)(101)}{6}=-42{,}875$

44. $\sum\limits_{k=1}^{20}(3k^2-2k)=3\dfrac{20(21)(41)}{6}-2\dfrac{20(21)}{2}$

$$=8610-420=8190$$

45. $\sum\limits_{i=1}^{40}(12+i^3)=12(40)+\dfrac{40^2(41)^2}{4}$

$$=480+672{,}400=672{,}880$$

46. Area $\approx\frac{1}{2}[1+2+3+4+5+6]=\frac{21}{2}=10.5$ square units

47. Area $\approx\dfrac{1}{4}\left[\dfrac{1}{1+\left(-\frac{3}{4}\right)^2}+\dfrac{1}{1+\left(-\frac{1}{2}\right)^2}+\dfrac{1}{1+\left(-\frac{1}{4}\right)^2}+\dfrac{1}{1+0}+\dfrac{1}{1+\left(\frac{1}{4}\right)^2}+\dfrac{1}{1+\left(\frac{1}{2}\right)^2}+\dfrac{1}{1+\left(\frac{3}{4}\right)^2}+\dfrac{1}{1+1^2}\right]$

$$=\dfrac{1}{4}\left[2(0.64)+2(0.8)+2(0.941176)+1+\dfrac{1}{2}\right]$$

$$\approx 1.566\text{ square units}$$

48. Width: $\dfrac{1}{n}$, Height: $f\left(\dfrac{i}{n}\right)=1-\left(\dfrac{1}{n}\right)^3$

$A\approx\sum\limits_{i=1}^{n}\left(1-\left(\dfrac{i}{n}\right)^3\right)\left(\dfrac{1}{n}\right)=\dfrac{1}{n}\sum\limits_{i=1}^{n}1-\dfrac{1}{n^4}\sum\limits_{i=1}^{n}i^3=\dfrac{1}{n}(n)-\dfrac{1}{n^4}\left[\dfrac{n^2(n+1)^2}{4}\right]$

$A=\lim\limits_{n\to\infty}\left[1-\dfrac{1}{n^4}\left(\dfrac{n^2(n+1)^2}{4}\right)\right]=1-\dfrac{1}{4}=\dfrac{3}{4}$ square unit

49. Width: $\dfrac{6}{n}$, Height: $f\left(-3+\dfrac{6i}{n}\right)=\left(-3+\dfrac{6i}{n}\right)+3=\dfrac{6i}{n}$

$A\approx\sum\limits_{i=1}^{n}\left(\dfrac{6i}{n}\right)\left(\dfrac{6}{n}\right)=\dfrac{36}{n^2}\sum\limits_{i=1}^{n}i=\dfrac{36}{n^2}\dfrac{n(n+1)}{2}=\dfrac{18(n+1)}{n}$

$A=\lim\limits_{n\to\infty}\left[\dfrac{18(n+1)}{n}\right]=18$ square units

50. Width: $\dfrac{2}{n}$, Height: $f\left(-1+\dfrac{2i}{n}\right)=\left(-1+\dfrac{2i}{n}\right)^2=1-\dfrac{4i}{n}+\dfrac{4i^2}{n^2}$

$A\approx\sum\limits_{i=1}^{n}\left[1-\dfrac{4i}{n}+\dfrac{4i^2}{n^2}\right]\left(\dfrac{2}{n}\right)$

$$=\dfrac{2}{n}\sum\limits_{i=1}^{n}1-\dfrac{8}{n^2}\sum\limits_{i=1}^{n}i+\dfrac{8}{n^3}\sum\limits_{i=1}^{n}i^2$$

$$=\dfrac{2}{n}(n)-\dfrac{8}{n^2}\dfrac{n(n+1)}{2}+\dfrac{8}{n^3}\dfrac{n(n+1)(2n+1)}{6}$$

$$=2-\dfrac{4(n+1)}{n}+\dfrac{4(n+1)(2n+1)}{3n^2}$$

$A=\lim\limits_{n\to\infty}\left[2-\dfrac{4(n+1)}{n}+\dfrac{4(n+1)(2n+1)}{3n^2}\right]=2-4+\dfrac{8}{3}=\dfrac{2}{3}$ square unit